# List of Elements with Their Symbols a

| Element | Symbol | Atomic Number | Atomic Mass |
|---|---|---|---|
| Actinium | Ac | 89 | 227.03[a] |
| Aluminum | Al | 13 | 26.98 |
| Americium | Am | 95 | 243.06[a] |
| Antimony | Sb | 51 | 121.76 |
| Argon | Ar | 18 | 39.95 |
| Arsenic | As | 33 | 74.92 |
| Astatine | At | 85 | 209.99[a] |
| Barium | Ba | 56 | 137.33 |
| Berkelium | Bk | 97 | 247.07[a] |
| Beryllium | Be | 4 | 9.012 |
| Bismuth | Bi | 83 | 208.98 |
| Bohrium | Bh | 107 | 264.12[a] |
| Boron | B | 5 | 10.81 |
| Bromine | Br | 35 | 79.90 |
| Cadmium | Cd | 48 | 112.41 |
| Calcium | Ca | 20 | 40.08 |
| Californium | Cf | 98 | 251.08[a] |
| Carbon | C | 6 | 12.01 |
| Cerium | Ce | 58 | 140.12 |
| Cesium | Cs | 55 | 132.91 |
| Chlorine | Cl | 17 | 35.45 |
| Chromium | Cr | 24 | 52.00 |
| Cobalt | Co | 27 | 58.93 |
| Copernicium | Cn | 112 | 285[a] |
| Copper | Cu | 29 | 63.55 |
| Curium | Cm | 96 | 247.07[a] |
| Darmstadtium | Ds | 110 | 271[a] |
| Dubnium | Db | 105 | 262.11[a] |
| Dysprosium | Dy | 66 | 162.50 |
| Einsteinium | Es | 99 | 252.08[a] |
| Erbium | Er | 68 | 167.26 |
| Europium | Eu | 63 | 151.96 |
| Fermium | Fm | 100 | 257.10[a] |
| Flerovium | Fl | 114 | 289[a] |
| Fluorine | F | 9 | 19.00 |
| Francium | Fr | 87 | 223.02[a] |
| Gadolinium | Gd | 64 | 157.25 |
| Gallium | Ga | 31 | 69.72 |
| Germanium | Ge | 32 | 72.63 |
| Gold | Au | 79 | 196.97 |
| Hafnium | Hf | 72 | 178.49 |
| Hassium | Hs | 108 | 269.13[a] |
| Helium | He | 2 | 4.003 |
| Holmium | Ho | 67 | 164.93 |
| Hydrogen | H | 1 | 1.008 |
| Indium | In | 49 | 114.82 |
| Iodine | I | 53 | 126.90 |
| Iridium | Ir | 77 | 192.22 |
| Iron | Fe | 26 | 55.85 |
| Krypton | Kr | 36 | 83.80 |
| Lanthanum | La | 57 | 138.91 |
| Lawrencium | Lr | 103 | 262.11[a] |
| Lead | Pb | 82 | 207.2 |
| Lithium | Li | 3 | 6.94 |
| Livermorium | Lv | 116 | 292[a] |
| Lutetium | Lu | 71 | 174.97 |
| Magnesium | Mg | 12 | 24.31 |
| Manganese | Mn | 25 | 54.94 |

| Element | Symbol | Number | |
|---|---|---|---|
| Meitnerium | Mt | 109 | 268.14[a] |
| Mendelevium | Md | 101 | 258.10[a] |
| Mercury | Hg | 80 | 200.59 |
| Molybdenum | Mo | 42 | 95.96 |
| Neodymium | Nd | 60 | 144.24 |
| Neon | Ne | 10 | 20.18 |
| Neptunium | Np | 93 | 237.05[a] |
| Nickel | Ni | 28 | 58.69 |
| Niobium | Nb | 41 | 92.91 |
| Nitrogen | N | 7 | 14.01 |
| Nobelium | No | 102 | 259.10[a] |
| Osmium | Os | 76 | 190.23 |
| Oxygen | O | 8 | 16.00 |
| Palladium | Pd | 46 | 106.42 |
| Phosphorus | P | 15 | 30.97 |
| Platinum | Pt | 78 | 195.08 |
| Plutonium | Pu | 94 | 244.06[a] |
| Polonium | Po | 84 | 208.98[a] |
| Potassium | K | 19 | 39.10 |
| Praseodymium | Pr | 59 | 140.91 |
| Promethium | Pm | 61 | 145[a] |
| Protactinium | Pa | 91 | 231.04 |
| Radium | Ra | 88 | 226.03[a] |
| Radon | Rn | 86 | 222.02[a] |
| Rhenium | Re | 75 | 186.21 |
| Rhodium | Rh | 45 | 102.91 |
| Roentgenium | Rg | 111 | 272[a] |
| Rubidium | Rb | 37 | 85.47 |
| Ruthenium | Ru | 44 | 101.07 |
| Rutherfordium | Rf | 104 | 261.11[a] |
| Samarium | Sm | 62 | 150.36 |
| Scandium | Sc | 21 | 44.96 |
| Seaborgium | Sg | 106 | 266.12[a] |
| Selenium | Se | 34 | 78.96 |
| Silicon | Si | 14 | 28.09 |
| Silver | Ag | 47 | 107.87 |
| Sodium | Na | 11 | 22.99 |
| Strontium | Sr | 38 | 87.62 |
| Sulfur | S | 16 | 32.06 |
| Tantalum | Ta | 73 | 180.95 |
| Technetium | Tc | 43 | 98[a] |
| Tellurium | Te | 52 | 127.60 |
| Terbium | Tb | 65 | 158.93 |
| Thallium | Tl | 81 | 204.38 |
| Thorium | Th | 90 | 232.04 |
| Thulium | Tm | 69 | 168.93 |
| Tin | Sn | 50 | 118.71 |
| Titanium | Ti | 22 | 47.87 |
| Tungsten | W | 74 | 183.84 |
| Uranium | U | 92 | 238.03 |
| Vanadium | V | 23 | 50.94 |
| Xenon | Xe | 54 | 131.293 |
| Ytterbium | Yb | 70 | 173.05 |
| Yttrium | Y | 39 | 88.91 |
| Zinc | Zn | 30 | 65.38 |
| Zirconium | Zr | 40 | 91.22 |
| *[b] | | 113 | 284[a] |
| *[b] | | 115 | 288[a] |

[a]Mass of longest-lived or most important isotope.
[b]The names of these elements have not yet been decided.

# CHEMISTRY

## A Molecular Approach

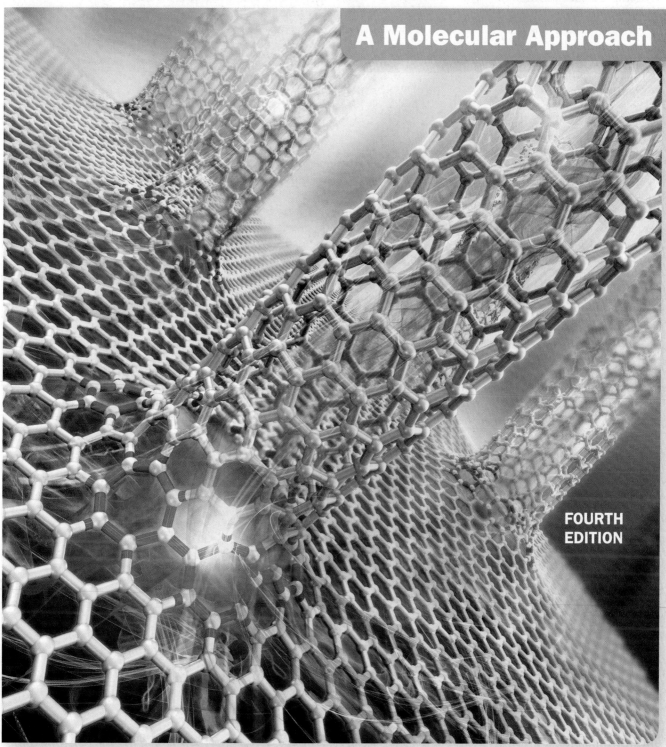

**FOURTH EDITION**

# Nivaldo J. Tro

WESTMONT COLLEGE

PEARSON

Editor in Chief: Jeanne Zalesky
Executive Editor: Terry Haugen
Director of Development: Jennifer Hart
Product Marketing Manager: Elizabeth Ellsworth
Executive Field Marketing Manager: Chris Barker
Development Editor: Erin Mulligan
Program Manager: Sarah Shefveland
Project Manager: Beth Sweeten
Editorial Assistant: Lindsey Pruett
Content Producer: Jackie Jacob
Text and Image Permissions Project Manager:
   William Opaluch

Program Management Team Lead: Kristen Flatham
Project Management Team Lead: David Zielonka
Production Management: Francesca Monaco,
   CodeMantra
Compositor: CodeMantra
Design Manager and Cover Designer: Derek Bacchus
Interior Designer: Elise Lansdon
Illustrators: Lachina, Inc.
Photo Researcher: Eric Shrader
Operations Specialist: Maura Zaldivar-Garcia
Cover and Chapter Opening Illustrations:
   Quade Paul

Tro, Nivaldo J.
  Chemistry : a molecular approach / Tro, Nivaldo J.
  Fourth edition. | Boston : Pearson, [2017] | Includes
    bibliographical references and index.
  LCCN 2015040901 | ISBN 9780134112831 (0134112830 : alk. paper)
  LCSH : Chemistry, Physical and theoretical--Textbooks.
  LCC QD453.3. T759 2017 | DDC 540--dc23 LC record available at http://lccn.loc.gov/2015040901

4  18

ISBN-10: 0-13-411283-0 / ISBN-13: 978-0-13-411283-1 (Student Edition)
ISBN-10: 0-13-412633-5 / ISBN-13: 978-0-13412633-3 (Instructor Review Copy)

www.pearsonhighered.com

# ABOUT THE AUTHOR

**NIVALDO TRO** is a professor of chemistry at Westmont College in Santa Barbara, California, where he has been a faculty member since 1990. He received his Ph.D. in chemistry from Stanford University for work on developing and using optical techniques to study the adsorption and desorption of molecules to and from surfaces in ultrahigh vacuum. He then went on to the University of California at Berkeley, where he did postdoctoral research on ultrafast reaction dynamics in solution. Since coming to Westmont, Professor Tro has been awarded grants 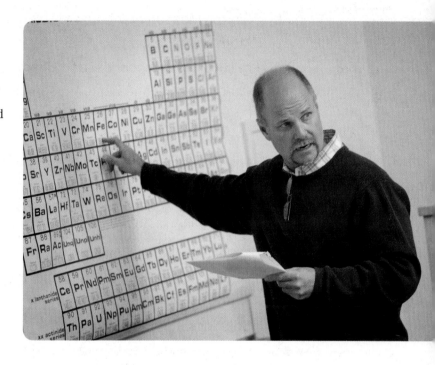 from the American Chemical Society Petroleum Research Fund, from Research Corporation, and from the National Science Foundation to study the dynamics of various processes occurring in thin adlayer films adsorbed on dielectric surfaces. He has been honored as Westmont's outstanding teacher of the year three times and has also received the college's outstanding researcher of the year award. Professor Tro lives in Santa Barbara with his wife, Ann, and their four children, Michael, Ali, Kyle, and Kaden.

TO MICHAEL, ALI, KYLE, AND KADEN

# BRIEF CONTENTS

# INTERACTIVE MEDIA CONTENTS

## Interactive Worked Examples (IWEs) https://media.pearsoncmg.com/ph/esm/esm_tro_chemistry_4/media/index.html

# Key Concept Videos (KCVs) https://media.pearsoncmg.com/ph/esm/esm_tro_chemistry_4/media/index.html

# CONTENTS

# 3 Molecules, Compounds, and Chemical Equations 86

# 4 Chemical Quantities and Aqueous Reactions 138

**5 Gases** 196

**6 Thermochemistry** 248

# 7 The Quantum-Mechanical Model of the Atom  296

# 8 Periodic Properties of the Elements  336

# 9 Chemical Bonding I: The Lewis Model 382

# 10 Chemical Bonding II: Molecular Shapes, Valence Bond Theory, and Molecular Orbital Theory 426

# 13 Solutions  570

# 14 Chemical Kinetics  622

**17 Aqueous Ionic Equilibrium** 778

**18 Free Energy and Thermodynamics** 838

# 19  Electrochemistry  888

# 20  Radioactivity and Nuclear Chemistry  938

**24 Metals and Metallurgy** 1100

# 25 Transition Metals and Coordination Compounds 1126

# PREFACE

## To the Student

As you begin this course, I invite you to think about your reasons for enrolling in it. Why are you taking general chemistry? More generally, why are you pursuing a college education? If you are like most college students taking general chemistry, part of your answer is probably that this course is required for your major and that you are pursuing a college education so you can get a good job some day. Although these are good reasons, I would like to suggest a better one. I think the primary reason for your education is to prepare you to *live a good life*. You should understand chemistry—not for what it can *get* you—but for what it can *do* to you. Understanding chemistry, I believe, is an important source of happiness and fulfillment. Let me explain.

Understanding chemistry helps you to live life to its fullest for two basic reasons. The first is *intrinsic*: through an understanding of chemistry, you gain a powerful appreciation for just how rich and extraordinary the world really is. The second reason is *extrinsic*: understanding chemistry makes you a more informed citizen—it allows you to engage with many of the issues of our day. In other words, understanding chemistry makes *you* a deeper and richer person and makes your country and the world a better place to live. These reasons have been the foundation of education from the very beginnings of civilization.

How does chemistry help prepare you for a rich life and conscientious citizenship? Let me explain with two examples. My first one comes from the very first page of Chapter 1 of this book. There, I ask the following question: What is the most important idea in all of scientific knowledge? My answer to that question is this: **the behavior of matter is determined by the properties of molecules and atoms**. That simple statement is the reason I love chemistry. We humans have been able to study the substances that compose the world around us and explain their behavior by reference to particles so small that they can hardly be imagined. If you have never realized the remarkable sensitivity of the world we *can* see to the world we *cannot*, you have missed out on a fundamental truth about our universe. To have never encountered this truth is like never having read a play by Shakespeare or seen a sculpture by Michelangelo—or, for that matter, like never having discovered that the world is round. It robs you of an amazing and unforgettable experience of the world and the human ability to understand it.

My second example demonstrates how science literacy helps you to be a better citizen. Although I am largely sympathetic to the environmental movement, a lack of science literacy within some sectors of that movement and the resulting anti-environmental backlash create confusion that impedes real progress and opens the door to what could be misinformed policies. For example, I have heard conservative pundits say that volcanoes emit more carbon dioxide—the most significant greenhouse gas—than does petroleum combustion. I have also heard a liberal environmentalist say that we have to stop using hairspray because it is causing holes in the ozone layer that will lead to global warming. Well, the claim about volcanoes emitting more carbon dioxide than petroleum combustion can be refuted by the basic tools you will learn to use in Chapter 4 of this book. We can easily show that volcanoes emit only 1/50th as much carbon dioxide as petroleum combustion. As for hairspray depleting the ozone layer and thereby leading to global warming, the chlorofluorocarbons that deplete ozone have been banned from hairspray since 1978, and ozone depletion has nothing to do with global warming anyway. People with special interests or axes to grind can conveniently distort the truth before an ill-informed public, which is why we all need to be knowledgeable.

So this is why I think you should take this course. Not just to satisfy the requirement for your major, and not just to get a good job some day, but to help you to lead a fuller life and to make the world a little better for everyone. I wish you the best as you embark on the journey to understand the world around you at the molecular level. The rewards are well worth the effort.

## To the Professor

First and foremost, thanks to all of you who adopted this book in its previous editions. You helped to make this book one of the most popular general chemistry textbooks in the world. I am grateful beyond words. Second, I have listened carefully to your feedback on the previous edition. The changes you see in this edition are the direct result of your input, as well as my own experience using the book in my general chemistry courses. If you have reviewed content or have contacted me directly, you will likely see your suggestions reflected in the changes I have made. Thank you.

Some of the most exciting changes in this edition are in the media associated with the book. I have added approximately 57 new Key Concept Videos and 61 new Interactive Worked examples to the media package. You can see a more detailed description of these videos in the following section entitled *What's New in This Edition*. This means that you now have a library of over 150 interactive videos to enhance your course. In my courses, I use these videos to implement a *before, during, after* strategy for my students. My goal is simple: *Engage students in active learning before class, during class, and after class.* To that end, I assign a key concept video *before* each class session.

The video introduces students to a key concept for that day and gets them thinking about it before they come to class. *During* class, I expand on the concept and use *Leaning Catalytics* to question my students. Instead of passively listening to a lecture, they are interacting with the concepts through questions that I pose. Some of these questions are answered individually; other times I have them pair up with a partner. This approach has changed my classroom. Students engage in the material in new ways. They have to think and process and interact. It is deeply satisfying for me to see my students so engaged. Finally, *after* class, I give them another assignment, often an interactive worked example with a follow-up question. At this point, they have to apply what they have learned to solve a problem.

The results have been fantastic. My students are enjoying the process because they are engaged before, during, and after class rather than only looking at material the night before a problem set is due. I have seen evidence of their improved learning through increases in their scores on the American Chemical Society Standard General Chemistry Exam, which I always administer as the final exam for my course.

Although we have added exciting new media elements and made other changes to the book, the book's goal remains the same: *to present a rigorous and accessible treatment of general chemistry in the context of relevance.* Teaching general chemistry would be much easier if all of our students had exactly the same level of preparation and ability. But alas, that is not the case. Even though I teach at a relatively selective institution, my courses are populated with students with a range of backgrounds and abilities in chemistry. The challenge of successful teaching, in my opinion, is figuring out how to instruct and challenge the best students while not losing those with lesser backgrounds and abilities. My strategy has always been to set the bar relatively high, while at the same time providing the motivation and support necessary to reach the high bar. That is exactly the philosophy of this book. We do not have to compromise away rigor in order to make chemistry accessible to our students. In this book, I have worked hard to combine rigor with accessibility—to create a book that does not dilute the content, yet can be used and understood by any student willing to put in the necessary effort.

*Chemistry: A Molecular Approach* is first and foremost a *student-oriented book.* My main goal is to motivate students and get them to achieve at the highest possible level. As we all know, many students take general chemistry because it is a requirement; they do not see the connection between chemistry and their lives or their intended careers. *Chemistry: A Molecular Approach* strives to make those connections consistently and effectively. Unlike other books, which often teach chemistry as something that happens only in the laboratory or in industry, this book teaches chemistry in the context of relevance. It shows students *why* chemistry is important to them, to their future careers, and to their world.

Second, *Chemistry: A Molecular Approach* is a *pedagogically driven book.* In seeking to develop problem-solving skills, a consistent approach (Sort, Strategize, Solve, and Check) is applied, usually in a two- or three-column format.

In the two-column format, the left column shows the student how to analyze the problem and devise a solution strategy. It also lists the steps of the solution, explaining the rationale for each one, while the right column shows the implementation of each step. In the three-column format, the left column outlines the general procedure for solving an important category of problems that is then applied to two side-by-side examples. This strategy allows students to see both the general pattern and the slightly different ways in which the procedure may be applied in differing contexts. The aim is to help students understand both the *concept of the problem* (through the formulation of an explicit conceptual plan for each problem) and the *solution to the problem.*

Third, *Chemistry: A Molecular Approach* is a *visual book.* Wherever possible, I use images to deepen the student's insight into chemistry. In developing chemical principles, multipart images help show the connection between everyday processes visible to the unaided eye and what atoms and molecules are actually doing. Many of these images have three parts: macroscopic, molecular, and symbolic. This combination helps students to see the relationships between the formulas they write down on paper (symbolic), the world they see around them (macroscopic), and the atoms and molecules that compose that world (molecular). In addition, most figures are designed to teach rather than just to illustrate. They are rich with annotations and labels intended to help the student grasp the most important processes and the principles that underlie them. The resulting images are rich with information but also uncommonly clear and quickly understood.

Fourth, *Chemistry: A Molecular Approach* is a "big picture" book. At the beginning of each chapter, a short paragraph helps students to see the key relationships between the different topics they are learning. Through a focused and concise narrative, I strive to make the basic ideas of every chapter clear to the student. Interim summaries are provided at selected spots in the narrative, making it easier to grasp (and review) the main points of important discussions. And to make sure that students never lose sight of the forest for the trees, each chapter includes several *Conceptual Connections,* which ask them to think about concepts and solve problems without doing any math. I want students to learn the concepts, not just plug numbers into equations to churn out the right answer. This philosophy is also integral to the *Key Concept Videos,* which concisely reinforce student appreciation of the core concepts in each chapter.

*Chemistry: A Molecular Approach* is lastly a book that delivers the depth of coverage faculty want. We do not have to cut corners and water down the material in order to get our students interested. We have to meet them where they are, challenge them to the highest level of achievement, and support them with enough pedagogy to allow them to succeed.

I hope that this book supports you in your vocation of teaching students chemistry. I am increasingly convinced of the importance of our task. Please feel free to email me with any questions or comments about the book.

Nivaldo J. Tro
**tro@westmont.edu**

# What's New in This Edition?

The book has been extensively revised and contains more small changes than can be detailed here. The most significant changes to the book and its supplements are listed below:

- NEW! Chapter 12: *Solids and Modern Materials*. This chapter contains new topics and consolidates content on materials that was found in other parts of the book in previous editions into one new chapter. All chapters beyond Chapter 12 have been renumbered.

- With the help of my colleagues, Thomas Greenbowe (University of Oregon), Kristin Ziebert (Oregon State University), and Michael Everest (Westmont College), I have added two new categories of end-of-chapter questions designed to help students build what we call "twenty-first-century skills." The first new category of questions is *Data Interpretation and Analysis*. These questions present real data in real-life situations and ask students to analyze that data. These in-depth exercises give students much needed practice in reading graphs, digesting tables, and making data-driven decisions. The second new category of questions is *Questions for Group Work*. Our group work questions give students the opportunity to work with their peers in small groups. The questions can be used in or out of the classroom, but the idea is to foster collaborative learning and to encourage students to work together as a team to solve problems.

- I have added approximately 57 new Key Concept Videos and 61 new Interactive Worked Examples to the media package that accompanies the book. (Since the previous edition had 40 Interactive worked examples, there is now a total of 158 interactive videos.) These tools are designed to help professors engage their students in active learning. Recent research has conclusively demonstrated that students learn better when they are active in the learning process, as opposed to passively listening and simply taking in content. The Key Concept Videos are brief (two to five minutes), and each introduces a key concept from a chapter. The student does not just passively listen to the video; the video stops in the middle and poses a question to the student. The student must answer the question before the video continues. Each video also includes a follow-up question that is assignable in MasteringChemistry™. The Interactive Worked Examples are similar in concept, but instead of explaining a key concept, they walk the student through one of the in-chapter worked examples from the book. Like the Key Concept Videos, Interactive Worked Examples stop in the middle and force the student to interact by completing a step in the example. The examples also have a follow-up question that is assignable in MasteringChemistry™. The power of interactivity to make connections in problem solving is immense. I did not quite realize this power until we started creating the Interactive Worked Examples and I saw how I could use the animations to make connections that are just not possible on the static page.

- There are approximately 21 new Conceptual Connection questions throughout the book.

- All the data throughout the book have been updated to reflect the most recent measurements available. These updates include *Figure 4.2 Carbon Dioxide in the Atmosphere*; *Figure 4.3 Global Temperatures*; *Figure 4.25 U.S. Energy Consumption*; the unnumbered figure in Section 6.10 of *U.S. Energy Consumption*; *Figure 6.12 Energy Consumption by Source*; *Table 6.6 Changes in National Average Pollutant Levels, 1980–2013*; *Table 14.4 Change in Pollutant Levels*; *Figure 14.19 Ozone Depletion in the Antarctic Spring*; *Figure 16.15 Sources of U.S. Energy*; *Figure 16.16 Acid Rain*; and *Figure 16.18 U.S. Sulfur Dioxide Pollutant Levels*.

- Example 4.13 *Writing Equations for Acid–Base Reactions Involving a Strong Acid* and Example 14.2 *Determining the Order and Rate Constant of a Reaction* have been expanded.

- New worked examples have been added, including Example 4.14 *Writing Equations for Acid–Base Reactions Involving a Weak Acid*; Example 18.2 *Calculating $\Delta S$ for a Change of State*; Example 12.2 *Calculating the Packing Efficiency of a Unit Cell*; and Example 12.3 *Relating Unit Cell Volume, Edge Length, and Atomic Radius*.

- Several chapter-opening sections and (or) the corresponding art, including Sections 1.1, 2.1, 5.1, 16.1, and 17.1, have been replaced or modified.

- New information about *Thermoluminescent Dosimeters*, including new Figure 20.7, has been added.

- A definition and explanation for the heat (or enthalpy) of sublimation have been added to Section 11.6.

- A new section (Section 18.4) on the *Entropy Changes Associated with State Changes* has been added. This section also includes expanded coverage on reversible and irreversible processes.

- Several sections in Chapter 20: *Radioactivity and Nuclear Chemistry* have been modified, including Sections 20.3 and 20.5 and Tables 20.1 and 20.4.

- Approximately 40 end-of-chapter problems (in addition to over 40 new problems that have been added to new Chapter 12) have been added or modified.

# Acknowledgments

The book you hold in your hands bears my name on the cover, but I am really only one member of a large team that carefully crafted this book. Most importantly, I thank my editor, Terry Haugen. Terry is a great editor and friend. He gives me the right balance of freedom and direction and always supports me in my endeavors. Thanks, Terry, for all you have done for me and for general chemistry courses throughout the world. I am also grateful to Jennifer Hart, who has now worked with me on multiple editions of several books and in several different roles, including her role overseeing development. Jennifer, your guidance, organizational skills, and wisdom are central to the success of my projects, and I am eternally grateful. I also thank my development editor Erin

Mulligan, who has worked with me on many projects over several editions. Erin is an outstanding thinker and wordsmith who also has a great eye for detail and helps me tremendously on all aspects of my projects. Thank you, Erin. Thanks also to my media editor, Jackie Jakob. Jackie has been instrumental in helping me craft and develop the Key Concept Videos and Interactive Worked Examples that accompany this text. Her expertise, vision, and attention to detail are second to none. The quality and success of these videos are largely attributable to Jackie. Thanks, Jackie.

New to this edition is Sarah Shefveland. Although we have only worked together a short while, I am already indebted to her helpfulness. I am also grateful to Lindsey Pruett, who helped with organizing reviews, as well as numerous other tasks associated with keeping the team running smoothly. I am also grateful to Jeanne Zalesky, Editor-in-Chief for Physical Sciences. She has supported me and my projects and allowed me to succeed. I am also grateful to my team of marketing managers, Chris Barker and Elizabeth Ellsworth. Chris and I go way back, and I am excited to work with him in this new role. Elizabeth and I have only worked together for a short while, but I am already impressed by her energy and ideas for marketing this book. I continue to owe a special word of thanks to Glenn and Meg Turner of Burrston House, ideal collaborators whose contributions to the first edition of the book were extremely important and much appreciated. Quade and Emiko Paul, who make my ideas come alive with their art, have been with us from the beginning, and I owe a special debt of gratitude to them. I am also grateful to Derek Bacchus and Elise Landson for their creativity and hard work in crafting the design of this text. Finally, I would like to thank Beth Sweeten and the rest of the Pearson production team. They are a first-class operation—this text has benefited immeasurably from their talents and hard work. I also thank Francesca Monaco and her coworkers at CodeMantra. I am a picky author and Francesca is endlessly patient and a true professional. I am also greatly indebted to my copy editor, Betty Pessagno, for her dedication and professionalism, and to Eric Schrader for his exemplary photo research. And of course, I am continually grateful for Paul Corey, with whom I have now worked for over 15 years and 12 projects. Paul is a man of incredible energy and vision, and it is my great privilege to work with him. Paul told me many years ago (when he first signed me on to the Pearson team) to dream big, and then he provided the resources I needed to make those dreams come true. *Thanks, Paul.* I would also like to thank my first editor at Pearson, Kent Porter-Hamann. Kent and I spent many good years together writing books, and I continue to miss her presence in my work.

I acknowledge the help of my colleagues Allan Nishimura, Kristi Lazar, David Marten, Stephen Contakes, Michael Everest, and Carrie Hill who have supported me in my department while I worked on this book. I am also grateful to Gayle Beebe and Mark Sargent, the president and provost (respectively) of Westmont College, who have allowed me the time and space to work on my books. Thank you, Gayle and Mark, for allowing me to pursue my gifts and my vision. I am also grateful to those who have supported me personally. First on that list is my wife,

Ann. Her patience and love for me are beyond description, and without her, this book would never have been written. I am also indebted to my children, Michael, Ali, Kyle, and Kaden, whose smiling faces and love of life always inspire me. I come from a large Cuban family whose closeness and support most people would envy. Thanks to my parents, Nivaldo and Sara; my siblings, Sarita, Mary, and Jorge; my siblings-in-law, Jeff, Nachy, Karen, and John; and my nephews and nieces, Germain, Danny, Lisette, Sara, and Kenny. These are the people with whom I celebrate life.

I am especially grateful to Michael Tro, who put in many hours proofreading my manuscript, working problems and quiz questions, and organizing art codes and appendices. Michael, you are an amazing kid—it is my privilege to have you work with me on this project.

I would like to thank all of the general chemistry students who have been in my classes throughout my 25 years as a professor at Westmont College. You have taught me much about teaching that is now in this book.

Lastly, I am indebted to the many reviewers, listed on the following pages, whose ideas are embedded throughout this book. They have corrected me, inspired me, and sharpened my thinking on how best to teach this subject we call chemistry. I deeply appreciate their commitment to this project. I am particularly grateful to Thomas Greenbowe, Michael Everest, Kristin Ziebart, Michael Burand, Jeff Bryan, and Ali Sezer who played particularly important roles in many of the new features of this edition. I am also grateful to the accuracy reviewers (Michael Burand, Brian Gute, Tracy Knowles, Tom McGrath, Gary Mines, and Allison Soult) who tirelessly checked page proofs for correctness.

# Reviewers of the Fourth Edition

Donald Bellew, *The University of New Mexico*
Gary Buckley, *Cameron University*
Ferman Chavez, *Oakland University*
Ted Clark, *The Ohio State University*
Guy Crundwell, *Central Connecticut State University*
Bonnie Dixon, *University of Maryland*
Jack Eichler, *University of California, Riverside*
Elda Hegmann, *Kent State University*
Clifford LeMaster, *Boise State University*
Sarah Lievens, *University of California, Davis*
Lauren McMills, *Ohio University*
Behnoush Memari, *Broward College*
Nancy Mullins, *Florida State College at Jacksonville*
George Papadantonakis, *The University of Illinois at Chicago*
David Perdian, *Broward College*
Jerry Poteat, Georgia *Perimeter College*
Raymond Sadeghi, *The University of Texas at San Antonio*
Mark Schraf, *West Virginia University*
John Selegue, *University of Kentucky*
Susan Shadle, *Boise State University*
Clarissa Sorensen, *Central New Mexico Community College*
Frank Tsung, *Boston College*

# Reviewers of Previous Editions

Patricia G. Amateis, *Virginia Tech*
Margaret R. Asirvatham, *University of Colorado*
Paul Badger, *Robert Morris University*
Monica H. Baloga, *Florida Institute of Technology*
Rebecca Barlag, *Ohio University*
Mufeed M. Basti, *North Carolina Agricultural & Technological State University*

Amy E. Beilstein, *Centre College*
Maria Benavides, *University of Houston, Downtown*
Kyle A. Beran, *University of Texas of the Permian Basin*
Thomas Bertolini, *University of Southern California*
Christine V. Bilicki, *Pasadena City College*
Silas C. Blackstock, *The University of Alabama*
Robert E. Blake, *Texas Tech University*
Angela E. Boerger, *Loyola University Chicago*
Robert S. Boikess, *Rutgers University*
Paul Brandt, *North Central College*
Michelle M. Brooks, *College of Charleston*
Joseph H. Bularzik, *Purdue University, Calumet*
Cindy M. Burkhardt, *Radford University*
Andrew E. Burns, *Kent State University at Stark*
Kim C. Calvo, *The University of Akron*
Stephen C. Carlson, *Lansing Community College*
David A. Carter, *Angelo State University*
Eric G. Chesloff, *Villanova University*
William M. Cleaver, *The University of Vermont*
Charles T. Cox, Jr., *Georgia Institute of Technology*
J. Ricky Cox, *Murray State University*
Samuel R. Cron, *Arkansas State University*
Darwin B. Dahl, *Western Kentucky University*
Robert F. Dias, *Old Dominion University*
Daniel S. Domin, *Tennessee State University*
Alan D. Earhart, *Southeast Community College*
Amina K. El-Ashmawy, *Collin College*
Joseph P. Ellison, *United States Military Academy at West Point*
Joseph M. Eridon, *Central New Mexico Community College*
Deborah B. Exton, *The University of Oregon*
William A. Faber, *Grand Rapids Community College*
Michael Ferguson, *Honolulu Community College*
Maria C. Fermin-Ennis, *Gordon College*
Oscar Navarro Fernandez, *University of Hawaii at Manoa*
Jan Florian, *Loyola University Chicago*
Andy Frazer, *University of Central Florida*
Candice E. Fulton, *Midwestern State University*
Ron Garber, *California State University at Long Beach*
Carlos D. Garcia, *The University of Texas at San Antonio*
Eric S. Goll, *Brookdale Community College*
Robert A. Gossage, *Acadia University*
Pierre Y. Goueth, *Santa Monica College*
Thomas J. Greenbowe, *Iowa State University*
Victoria Guarisco, *Middle Georgia State University*
Christin Gustafson, *Illinois Central College*
Jason A. Halfen, *University of Wisconsin-Eau Claire*
Nathan Hammer, *University of Mississippi*
Michael D. Hampton, *University of Central Florida*
Tamara Hanna, *Texas Tech University*
Lois Hansen-Polcar, *Cuyahoga Community College-Western Campus*
Tony Hascall, *Northern Arizona University*
Monte L. Helm, *Fort Lewis College*
David E. Henderson, *Trinity College*
Susan K. Henderson, *Quinnipiac University*
Peter M. Hierl, *The University of Kansas*
Paula Hjorth-Gustin, *San Diego Mesa College*
Angela Hoffman, *University of Portland*
Todd A. Hopkins, *Butler University*
Byron E. Howell, *Tyler Junior College*
Ralph Isovitsch, *Xavier University of Louisiana*
Kenneth C. Janda, *University of California, Irvine*
Milt Johnston, *University of South Florida*
Jason A. Kautz, *University of Nebraska-Lincoln*
Catherine A. Keenan, *Chaffey College*
Steven W. Keller, *University of Missouri*
Resa Kelly, *San Jose State University*
Chulsung Kim, *Georgia Gwinnett College*
Louis J. Kirschenbaum, *University of Rhode Island*
Mark Knecht, *University of Kentucky*
Bette Kreuz, *University of Michigan-Dearborn*
Sergiy Kryatov, *Tufts University*
Richard H. Langley, *Stephen F. Austin State University*
Clifford B. Lemaster, *Boise State University*
Robley Light, *Florida State University*

Adam List, *Vanderbilt University*
Christopher Lovallo, *Mount Royal University*
Eric Malina, *University of Nebraska-Lincoln*
Benjamin R. Martin, *Texas State University*
Lydia J. Martinez-Rivera, *University of Texas at San Antonio*
Marcus T. McEllistrem, *University of Wisconsin-Eau Claire*
Danny G. McGuire, *Cameron University*
Charles W. McLaughlin, *University of Nebraska, Lincoln*
Curt L. McLendon, *Saddleback College*
Robert C. McWilliams, *United States Military Academy*
David H. Metcalf, *University of Virginia*
Ray Mohseni, *East Tennessee State University*
Elisabeth A. Morlino, *University of the Science, Philadelphia*
James E. Murphy, *Santa Monica College*
Maria C. Nagan, *Truman State University*
Edward J. Neth, *University of Connecticut*
Aric Opdahl, *University of Wisconsin La Crosse*
Kenneth S. Overway, *Bates College*
Greg Owens, *University of Utah*
Naresh Pandya, *University of Hawaii*
Gerard Parkin, *Columbia University*
Jessica Parr, *University of Southern California*
Yasmin Patell, *Kansas State University*
Tom Pentecost, *Grand Valley State University*
Glenn A. Petrie, *Central Missouri State*
Norbert J. Pienta, *University of Iowa*
Louis H. Pignolet, *University of Minnesota*
Valerie Reeves, *University of New Brunswick*
Dawn J. Richardson, *Colin College*
Thomas G. Richmond, *University of Utah*
Dana L. Richter-Egger, University of Nebraska
Jason Ritchie, *University of Mississippi*
Christopher P. Roy, *Duke University*
A. Timothy Royappa, *University of West Florida*
Stephen P. Ruis, *American River College*
Alan E. Sadurski, *Ohio Northern University*
Thomas W. Schleich, *University of California, Santa Cruz*
Rod Schoonover, *CA Polytechnic State University*
Tom Selegue, *Pima Community College, West*
Anju H. Sharma, *Stevens Institute of Technology*
Sherril A. Soman, *Grand Valley State University*
Michael S. Sommer, *University of Wyoming*
Jie S. Song, *University of Michigan, Flint*
Mary Kay Sorenson, *University of Wisconsin, Milwaukee*
Stacy E. Sparks, *University of Texas, Austin*
Richard Spinney, *Ohio State University*
William H. Steel, *York College of Pennsylvania*
Vinodhkumar Subramaniam, *East Carolina University*
Jerry Suits, *University of Northern Colorado*
Tamar Y. Susskind, *Oakland Community College*
Uma Swamy, *Florida International University*
Ryan Sweeder, *Michigan State University*
Dennis Taylor, *Clemson University*
Jacquelyn Thomas, *Southwestern College*
Kathleen Thrush Shaginaw, *Villanova University*
Lydia Tien, *Monroe Community College*
David Livingstone Toppen, *California State University Northridge*
Marcy Towns, *Purdue University*
Harold Trimm, *Broome Community College*
Laura VanDorn, *University of Arizona*
Susan Varkey, *Mount Royal College*
Ramaiyer Venkatraman, *Jackson State University*
John B. Vincent, *University of Alabama, Tuscaloosa*
Kent S. Voelkner, *Lake Superior College*
Sheryl K. Wallace, *South Plains College*
Wayne E. Wesolowski, *University of Arizona*
Sarah E. West, *Notre Dame University*
John Wiginton, *University of Mississippi*
Kurt J. Winkelmann, *Florida Institute of Technology*
Troy D. Wood, *University of Buffalo*
Servet M. Yatin, *Quincy College*
Kazushige Yokoyama, *SUNY Geneseo*
Lin Zhu, *IUPUI*

**Chapter 12, Solids and Modern Materials, Advisory Board**

Michael Burand, *Oregon State University*
Christopher Collison, *Rochester Institute of Technology*
Jordan Fantini, *Denison University*
Melissa Hines, *Cornell University*
Raymond Schaak, *Penn State University*
Jennifer Shanoski, *Merritt College*
Jim Zubricky, *University of Toledo*

# Focus Group Participants

We would like to thank the following professors for contributing their valuable time to meet with the author and the publishing team in order to provide a meaningful perspective on the most important challenges they face in teaching general chemistry and give us insight into creating a general chemistry text that successfully responds to those challenges.

## Focus Group 12

Corey Beck, *Ohio University*
Jennifer Duis, *Northern Arizona University*
Alton Hassell, *Baylor University*
Tina Huang, *University of Illinois*
Amy Irwin, *Monroe Community College*
Maria Korolev, *University of Florida*
Jennifer Schwartz Poehlmann, *Stanford University*
John Selegue, *University of Kentucky*
Sarah Siegel, *Gonzaga University*
Jeff Statler, *University of Utah*

## Focus Groups 1–11

Michael R. Abraham, *University of Oklahoma*
Ramesh D. Arasasingham, *University of California, Irvine*
James A. Armstrong, *City College of San Francisco*
Silas C. Blackstock, *University of Alabama*
Roberto A. Bogomolni, *University of California, Santa Cruz*
Stacey Brydges, *University of California San Diego*
Kenneth Capps, *Central Florida Community College*
Stephen C. Carlson, *Lansing Community College*
Charles E. Carraher, *Florida Atlantic University*
Kenneth Caswell, *University of South Florida*
Robert Craig Taylor, *Oakland University*
Darwin B. Dahl, *Western Kentucky University*
Mohammed Daoudi, *University of Central Florida*
Kate Deline, *College of San Mateo*
Stephanie Dillon, *Florida State University*
Ralph C. Dougherty, *Florida State University*
William Eck, *University of Wisconsin, Marshfield/Wood County*
Robert J. Eierman, *University of Wisconsin, Eau Claire*
Amina K. El-Ashmawy, *Collin County Community College*
William A. Faber, *Grand Rapids Community College*
Richard W. Frazee, *Rowan University*
Barbara A. Gage, *Prince George's Community College*
Simon Garrett, *California State University, Northridge*
Raymond F. Glienna, *Glendale Community College*
Eric S. Goll, *Brookdale Community College*

Pierre Y. Goueth, *Santa Monica College*
W. Tandy Grubbs, *Stetson University*
Jerome E. Haky, *Florida Atlantic University*
Jason A. Halfen, *University of Wisconsin, Eau Claire*
John A. W. Harkless, *Howard University*
Paul I. Higgs, *Barry University*
Norris W. Hoffman, *University of South Alabama*
Tony Holland, *Wallace Community College*
Todd A. Hopkins, *Butler University*
Moheb Ishak, *St. Petersburg College, St. Petersburg*
Kamal Ismail, *CUNY, Bronx Community College*
Greg M. Jorgensen, *American River College*
Sharon K. Kapica, *County College of Morris*
Jason Kautz, *University of Nebraska, Lincoln*
Mark Kearley, *Florida State University*
Catherine A. Keenan, *Chaffey College*
Steven W. Keller, *University of Missouri, Columbia*
Ellen Kime-Hunt, *Riverside Community College, Riverside Campus*
Peter J. Krieger, *Palm Beach Community College, Lake Worth*
Roy A. Lacey, State *University of New York, Stony Brook*
David P. Licata, *Coastline Community College*
Michael E. Lipschutz, *Purdue University*
Patrick M. Lloyd, *CUNY, Kingsborough Community College*
Boon H. Loo, *Towson University*
James L. Mack, *Fort Valley State University*
Jeanette C. Madea, *Broward Community College, North*
Joseph L. March, *University of Alabama, Birmingham*
Jack F. McKenna, *St. Cloud State University*
Curtis L. McLendon, *Saddleback College*
Dianne Meador, *American River College*
David Metcalf, *University of Virginia*
John A. Milligan, *Los Angeles Valley College*
Alice J. Monroe, *St. Petersburg College, Clearwater*
Elisabeth A. Morlino, *University of the Science, Philadelphia*
Heino Nitsche, *University of California at Berkeley*
Pedro Patino, *University of Central Florida*
Jeremy Perotti, *Nova Southeastern University*
Norbert J. Pienta, *University of Iowa*
Jayashree Ranga, *Salem State University*
Cathrine E. Reck, *Indiana University*
Thomas Ridgway, *University of Cincinnati*
Jil Robinson, *Indiana University*
Richard Rosso, *St. John's University*
Steven Rowley, *Middlesex County College*
Benjamin E. Rusiloski, *Delaware Valley College*
Karen Sanchez, *Florida Community College, Jacksonville*
David M. Sarno, *CUNY, Queensborough Community College*
Reva A. Savkar, *Northern Virginia Community College*
Thomas W. Schleich, *University of California, Santa Cruz*
Donald L. Siegel, *Rutgers University, New Brunswick*
Mary L. Sohn, *Florida Institute of Technology*
Sherril Soman-Williams, *Grand Valley State University*
Allison Soult, *University of Kentucky*
Louise S. Sowers, *Richard Stockton College of New Jersey*
Anne Spuches, *East Carolina University*
William H. Steel, *York College of Pennsylvania*
Uma Swamy, *Florida International University*
Richard E. Sykora, *University of South Alabama*
Galina G. Talanova, *Howard University*
Claire A. Tessier, *University of Akron*
Kathleen Thrush Shaginaw, *Villanova University*
John Vincent, *University of Alabama*
Gary L. Wood, *Valdosta State University*
Servet M. Yatin, *Quincy College*
James Zubricky, *University of Toledo*

# 21st Century Skills

Tro's approach to General Chemistry addresses 21st century skills, including the ability to analyze and interpret data and the capacity to work well in groups. Concepts are covered in a thorough and approachable manner, which ensures that every explanation is relevant and helps students see the real-world importance and applications of Chemistry. Students can learn concepts and build skills through numerous resources for use before, during, and after class.

◀ **NEW!** Data Interpretation and Analysis questions present real data in real life situations and ask students to analyze that data. These in-depth exercises give students practice reading graphs, digesting tables, and making data-driven decisions.

## Data Interpretation and Analysis

### UV Radiation and Sunscreen

114. On average sunlight shines on the surface of Earth with an intensity of 910 Watts/m². Some of this energy is in the form of ultraviolet (UV) radiation. The compounds in chemical sunscreens absorb ultraviolet rays and release the energy as lower-energy infrared (IR) rays, thereby preventing the skin-damaging UV rays from reaching the skin. You may have noticed references to UV-A (320–400 nm) and UV-B (280–320 nm) protection in sunscreen packaging and advertisements. Figure a▶ lists the chemical formula and molar mass of two compounds, 2-ethylhexyl p-methoxycinnamate (2-EHMC), a UV-B sunscreen, and terephthalylidenedicamphor sulfonic acid (TDSA), a UV-A sunscreen.

Use the information provided in the figures to answer the following questions:

a. Calculate the energy absorbed by TDSA at its maximum absorption for one photon of UV radiation.

b. Calculate the energy absorbed by 2-EHMC at its maximum absorption for one photon of UV radiation.

c. Which absorbs more energy, 2-EHMC or TDSA, at their respective maximum wavelengths?

d. Why are these two compounds commonly included in sunscreen products?

e. A person does not apply sunscreen and has 0.42 m² of skin exposed to the sun for 1.00 hour. Calculate the total energy (in Joules) absorbed by the skin. Assume there are 3.066 × 10²¹ photons/meter²/sec. Assume an average wavelength of 330 nm in order to calculate the energy of a UV photon. Assume half of the photons that strike the skin are reflected.

f. In order for a sunscreen product with equal amounts of TDSA and 2-EHMC to provide adequate protection from UV-A and UV-B rays for 1.00 hour, how many molecules of TDSA and 2-EHMC must be present on the skin? Assume one molecule of TDSA and one molecule 2-EHMC can each absorb 10,000 photons per hour. Assume the photon rate is the same as in part e.

g. Calculate the mass of TDSA and 2-EHMC in grams that must be present on the exposed skin referenced in parts e and f in order to protect the skin.

| Sunscreen Compound | Chemical Formula | $\mathcal{M}$ (g/mol) |
|---|---|---|
| terephthalylidenedicamphor sulfonic acid (TDSA) | $C_{28}H_{34}O_8S_2$ | 562.70 |
| 2-ethylhexyl p-methoxycinnamate (2-EHMC) | $C_{18}H_{26}O_3$ | 290.39 |

▲ FIGURE a **Chemical Formula and Molar Mass ($\mathcal{M}$) of Chemicals Commonly Found in Sunscreens**

Organic molecules have groups of atoms bonded together that can vibrate and rotate, making more energy states available. As a result, instead of absorbing exact wavelengths of radiation as elements do, organic molecules absorb over a broad range of wavelengths. Figure b▼ plots absorbance versus wavelength for 2-EHMC and TDSA. Greater absorbance indicates that the compound absorbs more light.

▲ FIGURE b **Sunscreen Absorption**
http://mycpss.com/critical-wavelength-broad-spectrum-uv-protection/

## Data Interpretation and Analysis

### Heavy Metals in Recycled Paper Packages

142. Demand for recycled paper has increased as consumers have become more aware of the environmental issues surrounding waste disposal. Paper is a natural raw material made from renewable wood and plants. Recycled paper is made from waste paper and paperboard. Both new paper and recycled paper contain traces of heavy metals. However, some types of recycled paper contain more heavy metals than new paper due, in part, to the inks used for printing or adding color to the original paper. Metals can migrate from the paper packaging and containers used for food to the food itself. This metal migration is controlled and monitored to reduce the exposure of humans to heavy metals.

Food agencies in each country determine the limit of cadmium, zinc, nickel, and copper in packaging materials. Most countries impose a limit of heavy metals in packaging not to exceed 100.0 ppm or 100.0 mg/kg. The limit for lead in eggs, fruit-, or vegetable-packaging is lower—not to exceed—20.0 mg/kg. The table in Figure a▼ lists the results of the analysis of samples of recycled paper produced by a manufacturer during a three-month period.

The goal of the manufacturer is to reduce the amount of heavy metals—especially lead—in the recycled paper it produces and sells. From March to May, the manufacturer varied the methods of production each month to determine which method would produce paper with the lowest metal content. Each month, lab technicians cut small samples of recycled paper with an area of 1.000 dm² and a thickness of 0.0500 cm. The technicians then prepared the samples for analysis. The average density of the samples is 800.0 ± 0.0245 kg/m³.

Use the information provided in the figure to answer the following questions:

a. Did the company reduce the amount of lead in its product between March and May?

b. Which metal was not reduced?

c. Each month's sample represents a different manufacturing process. Which process would you recommend the manufacturer choose to continue to use over the next year while additional processes are tested? Why?

d. What is the total amount of the five metals for the April sample (in mg/kg)?

e. What is the total mass (in mg) of the five metals for the April sample found in the 1.000 dm² × 0.0500 cm sample?

f. Lead has four stable isotopes: ²⁰⁴Pb, ²⁰⁶Pb, ²⁰⁷Pb, ²⁰⁸Pb with % abundances of 1.40, 24.10, 22.10, and 52.40, respectively. Determine the mass (in mg) of ²⁰⁸Pb in the April sample.

g. Sketch the mass spectrum for lead.

| | Zn | Pb | Cd | Ni | Cu |
|---|---|---|---|---|---|
| March | 2.7 | 3.5 | 0.129 | 1.7 | 32.3 |
| April | 3.2 | 2.8 | 0.092 | 1.6 | 29.2 |
| May | 3.7 | 2.4 | 0.096 | 1.5 | 19.9 |

▲ FIGURE a **Heavy Metal Values in Three Samples of Recycled Paper (mg/kg ± 0.1)**

# 21st Century Skills

## Questions for Group Work

Discuss these questions with the group and record your consensus answer.

**148.** Assign one of the three simple gas laws to each member of your group. For the assigned gas law, have each member write two equations, draw a graph, and describe it in a complete sentence. Have each group member present his or her law to the group.

**149.** Review the ideal gas law. Without referring back to the text, use algebra to write the ideal gas law and solve for each of the individual variables it contains. Have each group member solve for a different variable and present answers to the group.

**150.** Hydrogen peroxide ($H_2O_2$) decomposes in the presence of a catalyst to form water and oxygen. The catalyst is added to 5.00 mL of a hydrogen peroxide solution at 25.0 °C, and 49.5 mL of gas is collected over water at a total pressure of 763.8 mmHg.

    **a.** Write and balance the chemical reaction (note: catalysts do not appear in balanced chemical equations).

    **b.** Look up the vapor pressure of water under these conditions.

    **c.** What is the partial pressure of oxygen collected over the water?

    **d.** How many moles of oxygen are collected?

    **e.** How many grams of hydrogen peroxide were in the original sample?

    **f.** What is the concentration (in mol/L) of the hydrogen peroxide solution?

    **g.** Which part of this process is conceptually most difficult for your group?

**151.** A box contains equal amounts of helium, argon, and krypton (all gases) at 25 °C. Using complete sentences, describe the temperatures, masses, average velocities, and average kinetic energy of the three kinds of gas in the mixture. What do they have in common? What are the differences? How are these properties related?

**152.** Calculate the pressure exerted by 1 mol of an ideal gas in a box that is 0.500 L and 298 K. Have each group member calculate the pressure of 1 mol of the following gases in the same box at the same temperature: He, Ne, $H_2$, $CH_4$, and $CO_2$. Compare group members' answers as well as all answers with the pressure of an ideal gas. Assuming that the van der Waals equation predictions are accurate, account for why the pressure of each gas is higher or lower than that predicted for an ideal gas.

---

## Questions for Group Work

Discuss these questions with the group and record your consensus answer.

**130.** Have each member of your group represent an atom of a metal or an atom of a nonmetal. Each group member holds a coin to represent an electron. Which group members are most reluctant to give up their electrons? Which group members are most willing to give up their electrons? Determine which kind of bond could form between each pair of group members. Tabulate your results.

**131.** Spend a few minutes reviewing the Lewis dot symbols for the atoms H through Ne. Form a circle and have each group member ask the group member on his or her right to draw the Lewis symbol for a specific atom. Keep going around until each group member can write all the Lewis dot symbols for the atoms H through Ne. Determine the formal charge for each symbol. In a complete sentence or two, describe why they are all the same.

**132.** Draw the Lewis dot symbols for the atoms Al and O. Use the Lewis model to determine the formula for the compound formed from these two atoms.

**133.** Draft a list of step-by-step instructions for writing the correct Lewis dot structure for any molecule or polyatomic ion.

**134.** Pass a piece of paper around the group and ask each group member in turn to perform the next step in the process of determining a correct Lewis structure (including formal charges on all atoms and resonance structures, if appropriate) for the following molecules and ions: $N_2H_4$, $CCl_4$, $CO_3^{2-}$, and $NH_4^+$.

▲

**NEW!** **Questions for Group Work** give students the opportunity to work with their peers in small groups. The questions can be used in or out of the classroom, and the goal is to foster collaborative learning and encourage students to work together as a team to solve problems.

# Interactive Problem-Solving Strategy

A unique and consistent step-by-step format encourages logical thinking throughout the problem-solving process, driving students to think through problems critically, rather than to simply memorize formulas.

**Example 17.6**  **Strong Acid–Strong Base Titration pH Curve**

A 50.0-mL sample of 0.200 M sodium hydroxide is titrated with 0.200 M nitric acid. Calculate pH:
(a) after adding 30.00 mL of $HNO_3$    (b) at the equivalence point

**SOLUTION**

(a) Begin by calculating the initial amount of NaOH (in moles) from the volume and molarity of the NaOH solution. Because NaOH is a strong base, it dissociates completely, so the amount of $OH^-$ is equal to the amount of NaOH.

$$moles\ NaOH = 0.0500\ L \times \frac{0.200\ mol}{1\ L}$$
$$= 0.0100\ mol$$
$$moles\ OH^- = 0.0100\ mol$$

Calculate the amount of $HNO_3$ (in moles) added at 30.0 mL from the molarity of the $HNO_3$ solution.

$$mol\ HNO_3\ added = 0.0300\ L \times \frac{0.200\ mol}{1\ L}$$
$$= 0.00600\ mol\ HNO_3$$

As $HNO_3$ is added to the solution, it neutralizes some of the $OH^-$. Calculate the number of moles of $OH^-$ remaining by setting up a table based on the neutralization reaction that shows the amount of $OH^-$ before the addition, the amount of $H_3O^+$ added, and the amounts left after the addition.

|  | $OH^-(aq)$ + | $H_3O^+(aq) \longrightarrow 2\ H_2O(l)$ |
|---|---|---|
| **Before addition** | 0.0100 mol | ≈0.00 mol |
| **Addition** | — | 0.00600 mol |
| **After addition** | 0.0040 mol | 0.00 mol |

Calculate the $OH^-$ concentration by dividing the amount of $OH^-$ remaining by the *total volume* (initial volume plus added volume).

$$[OH^-] = \frac{0.0040\ mol}{0.0500\ L + 0.0300\ L}$$
$$= 0.0500\ M$$

Calculate the pOH from $[OH^-]$.

$$pOH = -log(0.0500)$$
$$= 1.30$$

Calculate the pH from the pOH using the equation pH + pOH = 14.

$$pH = 14 - pOH$$
$$= 14 - 1.30$$
$$= 12.70$$

(b) At the equivalence point, the strong base has completely neutralized the strong acid. The $[H_3O^+]$ at 25 °C from the ionization of water is $1.00 \times 10^{-7}$ M and the pH is therefore 7.00.

$$pH = 7.00$$

**FOR PRACTICE 17.6**  Calculate the pH in the titration in Example 17.6 after the addition of 60.0 mL of 0.200 M H...

Icons appear next to examples indicating an **Interactive Worked Example** digital version is available in the eText and on mobile devices via a QR code located here, and on the back cover of your textbook.

**NEW! 61 Interactive Worked Examples** make Tro's unique problem-solving strategies interactive, bringing his award-winning teaching directly to all students using his text. In these digital, mobile versions, students are shown how to break down problems using Tro's proven "Sort, Strategize, Solve, and Check" technique, helping to learn how to consistently solve chemical problems. These examples can be utilized as an after class activity.

---

Strong Acid–Strong Base Titration pH Curve

A 50.0-mL sample of 0.200 M sodium hydroxide is titrated with 0.200 M nitric acid. Calculate: (a) the pH after adding 30.00 mL of $HNO_3$

$$moles\ NaOH = 0.0500\ L \times \frac{0.200\ mol}{1\ L}$$
$$= 0.0100\ mol$$
$$moles\ OH^- = 0.0100\ mol$$
$$moles\ HNO_3\ added = 0.0300\ L \times \frac{0.200\ mol}{1\ L}$$
$$= 0.00600\ mol\ HNO_3$$
$$[OH^-] = \frac{0.0040\ mol}{0.0500\ L + 0.0300\ L} = 0.0500\ M$$

|  | $OH^-(aq)$ + | $H_3O^+(aq) \longrightarrow 2\ H_2O(l)$ |
|---|---|---|
| Before addition | 0.0100 mol | ≈0.00 mol |
| Addition | - | 0.00600 mol |
| After addition | 0.0040 mol | ≈0.00 mol |

What is the pH of the solution at this point?
a) 0.0500
b) 1.30
c) 12.70

# A Focus on Conceptual Understanding

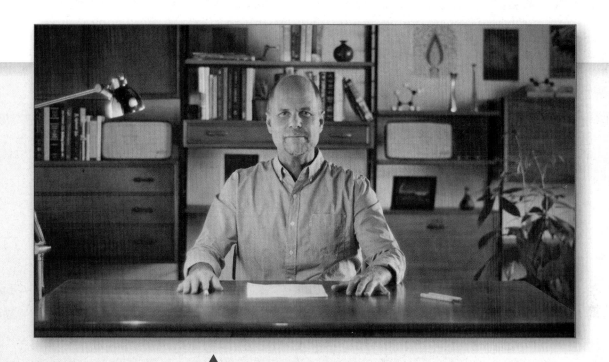

**NEW!** **57 Key Concept Videos** combine artwork from the textbook with both 2D and 3D animations to create a dynamic on-screen viewing and learning experience. These short videos include narration and brief live-action clips of author Niva Tro explaining the key concepts of each chapter. These videos can be utilized as a before class activity.

 **Electron Configurations**

## Electron Configurations

### General Energy Ordering of Orbitals for Multielectron Atoms

Energy ↑

$\overline{5s}$

$\overline{4s}$    $\overline{4p}$ $\overline{4p}$ $\overline{4p}$    $\overline{4d}$ $\overline{4d}$ $\overline{4d}$ $\overline{4d}$ $\overline{4d}$

     $\overline{3p}$ $\overline{3p}$ $\overline{3p}$    $\overline{3d}$ $\overline{3d}$ $\overline{3d}$ $\overline{3d}$ $\overline{3d}$

$\overline{3s}$

$\overline{2s}$    $\overline{2p}$ $\overline{2p}$ $\overline{2p}$

Electron configuration      Orbital diagram

$\overline{1s}$      C   $1s^2 2s^2 2p^2$

# NEW!
# Chapter 12—Solids and Modern Materials

This chapter contains new topics and consolidates content on materials that was found in other parts of the book in previous editions into one new chapter. New and consolidated topics include unit cells, carbon and silicates, ceramics, cement, glass, polymorphs, polyethylene, and the band gap in Group 4A elements.

CHAPTER **12**

## Solids and Modern Materials

Should we not suppose that in the formation of a crystal, the particles are not only established in rows and columns set in regular figures, but also by means of some polar property have turned identical sides in identical directions?

—ISAAC NEWTON (1642–1727)

12.1  Friday Night Experiments: The Discovery of Graphene 000

12.2  X-Ray Crystallography 000

12.3  Unit Cells and Basic Structures 000

12.4  The Fundamental Types of Crystalline Solids 000

12.5  The Structures of Ionic Solids 000

12.6  Network Covalent Atomic Solids: Carbon and Silicates 000

12.7  Ceramics, Cement, and Glass 000

12.8  Semiconductors and Band Theory 000

12.9  Polymers and Plastics 000

KEY LEARNING OUTCOMES 000

In this chapter, we focus on the solid state of matter. We first examine the structures of solids, keeping in mind that these structures determine the properties of solids. For example, the repeating hexagonal pattern of water molecules in crystalline ice determines the hexagonal shape of a snowflake, and the repeating cubic pattern of sodium and chloride ions in sodium chloride determines the cubic shape of salt grains. We then turn our attention to the study and development of solids with unique and useful properties, a field known as *materials science*. The ceramics that compose your coffee cups, the semiconductors in your electronic devices, and the plastics that are most likely all around you even at this moment are materials developed to have specific properties that serve specific purposes. In this chapter, we take a brief look at each of these kinds of materials.

### 12.1  Friday Night Experiments: The Discovery of graphene

In 2010, Andre Geim (1958–) and Konstantin Novoselov (1974–) were awarded the Nobel Prize in Physics for the discovery of a new material—graphene. Graphene is the thinnest known material (only one atom thick); it is also the strongest. It conducts heat and electricity, it is transparent, and it is completely impermeable to all substances, including helium. Although its many possible applications are yet to be realized, graphene may one day be used to make faster computers, foldable touchscreens, ultrathin light panels,

# Active and Adaptive

MasteringChemistry is the leading online homework, tutorial, and assessment system, designed to improve results by engaging students before, during, and after class with powerful content. Instructors can ensure students arrive ready to learn by assigning educationally effective content before class, and encourage critical thinking and retention with in-class resources. Students can further master concepts after class through traditional and adaptive homework assignments that provide hints and answer-specific feedback.

**NEW! Ready-to-Go Teaching Modules** in the Instructor Resources section help instructors efficiently make use of the available teaching tools for each chapter. Pre-built before class assignments, in-class activities, and after-class assignments are provided for ease of use. Instructors can incorporate active learning into their course with the suggested activity ideas, Learning Catalytics questions, and clicker questions.

# BEFORE CLASS

**Dynamic Study Modules** help students study effectively on their own by continuously assessing their activity and performance in real time. In this edition, 66 topics include key math skills, general chemistry skills such as nuclear chemistry, phases of matter, redox reactions, acids and bases, and organic and biochemistry skills.

**NEW! The Chemistry Primer** is a series of tutorials focused on remediating students in preparation for their first college chemistry course. The primer is offered as a pre-built assignment automatically generated as a default assignment with every newly created General Chemistry Course.

# DURING CLASS

**Learning Catalytics™** generates class discussion, guides lecture, and promotes peer-to-peer learning with real-time analytics. MasteringChemistry with eText now provides Learning Catalytics—an interactive student response tool that uses students' smartphones, tablets, or laptops to engage them in more sophisticated tasks and thinking. Instructors can:

* Pose a variety of open-ended questions that help students develop critical thinking skills
* Monitor responses to find out where students are struggling
* Use real-time data to adjust instructional strategy and try other ways of engaging students during class
* Manage student interactions by automatically grouping students for discussion, teamwork, and peer-to-peer learning

### Questions for Group Work

Tools for active learning have been enhanced throughout the book, allowing students to dive into material in an engaging and effective way.

---

*Active Classroom Learning*

## Questions for Group Work

Discuss these questions with the group and record your consensus answer.

**140.** Have each group member write a problem involving the transfer of heat from one material in Table 6.4 to another material in the table. Working as a group, solve each problem. The group member who wrote each problem may act as the group facilitator when the group is working on his or her problem. What do all of your problems have in common? How do they differ?

**141.** Classify each process as endothermic or exothermic. What is the sign of $\Delta H$ for each process? Explain your answers.
  a. gasoline burning in an engine
  b. steam condensing on a mirror
  c. water boiling in a pot

  Provide at least two additional examples of exothermic processes and two additional examples of endothermic processes. Have each member of your group provide an example.

**142.** A propane tank on a home barbeque contains $10.4 \times 10^3$ g of propane.
  a. Write the balanced chemical reaction for the combustion of gaseous propane ($C_3H_8$) to form water vapor and gaseous carbon dioxide.
  b. Use the value for $\Delta H_{rxn}$ provided in the text to calculate the total amount of heat produced when the entire contents of the tank of propane is burned.

  c. What mass of water could be warmed from 25 °C to 100 °C with this much heat?

  Solid carbon, C(s, graphite), gaseous hydrogen, $H_2(g)$, and the sugar glucose, $C_6H_{12}O_6(s)$ are all burned with oxygen in a calorimeter, and the amount of heat given off is determined for each process. How can these data be used to determine the heat of formation of glucose? Your answer should include both chemical reactions and complete sentences.

**143.** Consider the decomposition of liquid hydrogen peroxide ($H_2O_2$) to form water and oxygen.
  a. What is the heat of formation for hydrogen peroxide?
  b. What is the heat of formation for liquid water?
  c. What is the heat of formation for gaseous oxygen? Why?
  d. Write the balanced chemical equations that correspond to the $\Delta H$ values you looked up for parts a, b, and c.
  e. Write the balanced chemical equation for the decomposition of hydrogen peroxide to form water and oxygen. (Write the equation such that the coefficient on oxygen is 1.)
  f. What is the heat of reaction for the process in part e?
  g. Draw a scale diagram of this reaction (1 cm = 100 kJ) that shows the relative energies of reactants (on the left), products (on the right), and the elements in their most stable states (in the middle). Label all the energies you know.

# AFTER CLASS

**NEW!** In this edition, 200 more end of chapter problems now contain wrong answer specific feedback with links to the eText.

**Self-Assessment Quizzes** now contain wrong-answer feedback links to the eText, and an additional 10–15 multiple-choice questions authored in the ACS-exams and MCAT style in each chapter. These improvements help students optimize the use of quizzing to improve their understanding and class performance.

**Multimedia-rich Tutorials** feature specific wrong-answer feedback, hints, and a wide variety of educationally effective content to guide your students through the toughest topics in chemistry. The hallmark Hints and Feedback offer instruction similar to what students would experience in an office hours visit, allowing them to learn from their mistakes without being given the answer.

# MasteringChemistry™ **for Instructors**

## www.masteringchemistry.com

The Mastering platform was developed by scientists for science students and instructors. Mastering has been refined from data-driven insights derived from over a decade of real-world use by faculty and students.

### Gradebook
Every assignment is automatically graded. Shades of red highlight struggling students and challenging assignments.

### Gradebook Diagnostics
This screen provides you with your favorite diagnostics. With a single click, charts summarize the most difficult problems, vulnerable students, grade distribution, and even score improvement over the course.

**Adaptive Follow-Ups** are personalized assignments pairing MasteringChemistry's powerful content with Knewton's adaptive learning engine to provide personalized help to students before misconceptions take hold. These assignments address topics students struggle with on assigned homework, including core prerequisite topics.

# Instructor and Student Resources

| Resource | Available in Print | Available Online | Instructor or Student Resource | Description |
|---|---|---|---|---|
| **Selected Solutions Manual** 0134066286/ 9780134066288 | ✔ | | Student | Prepared by Kathy Shaginaw, the selected solutions manual for students contains complete, step-by-step solutions to selected odd-numbered, end-of-chapter problems. |
| **Laboratory Manual** 013406626X / 9780134066264 | ✔ | | Student | Prepared by John B. Vincent and Erica Livingston, this manual contains experiments with a focus on real-world applications. Each experiment covers one or more topics discussed within a chapter of the textbook. |
| **Study Guide** 0134066278 / 9780134066271 | ✔ | | Student | This Study Guide, prepared by Jennifer Shanoski, presents the major concepts, theories, and applications discussed in the text in a comprehensive and accessible manner. |
| **Instructor's Resource Materials** 0134074998 / 9780134074993 | | ✔ | Instructor | This resource contains the following:<br>• All illustrations, tables, and photos from the text in JPEG format.<br>• Three pre-built PowerPoint Presentations (lecture, worked examples, and images) |
| **Instructor's Resource Manual** 0134125770 / 9780134125770 | | ✔ | Instructor | Organized by chapter, this useful guide includes objectives, lecture outlines, references to figures and solved problems, as well as teaching tips. Prepared by Michael Ferguson. |
| **Test Bank** 0134126327 / 9780134126326 | | ✔ | Instructor | The Test Bank, prepared by Christine Hermann, contains more than 3,000 multiple choice, matching, and short-answer questions. |
| **Solutions Manual** 0134066251 / 9780134066257 | ✔ | | Instructor | Prepared by Kathy Shaginaw, this manual contains step-by-step solutions to all end-of-chapter exercises. With instructor permission, this manual may be made available to students. |
| **Instructor's Manual for Laboratory Manual** 0134143485 / 9780134143484 | | ✔ | Instructor | This manual provides solutions to the questions asked before each experiment in the accompanying Laboratory Manual, and provides background and suggestions for the instructor directing the lab. |

CHAPTER

# 1

# Matter, Measurement, and Problem Solving

The most incomprehensible thing about the universe is that it is comprehensible.

—ALBERT EINSTEIN (1879–1955)

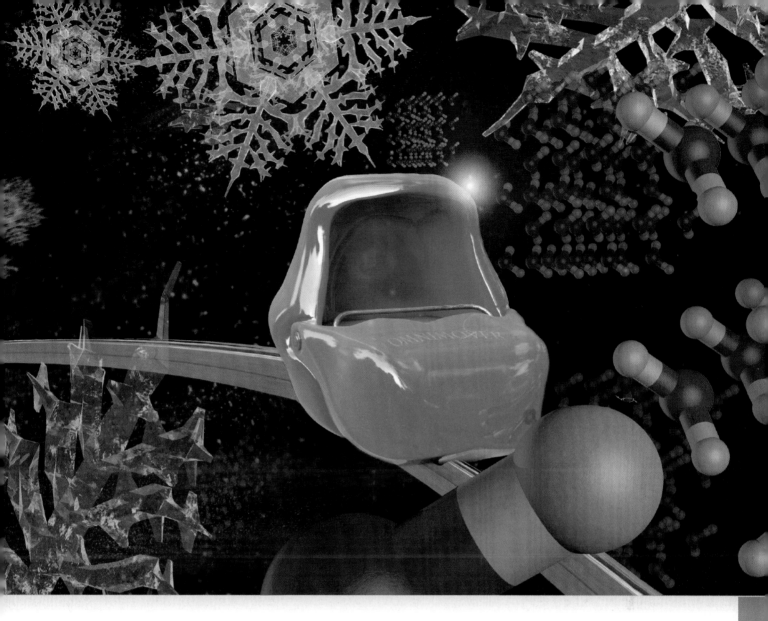

What do you think is the most important idea in all of human knowledge? There are, of course, many possible answers to this question—some practical, some philosophical, and some scientific. If we limit ourselves only to scientific answers, mine would be this: **the properties of matter are determined by the properties of atoms and molecules.** Atoms and molecules determine how matter behaves—if they were different, matter would be different. The properties of water molecules determine how water behaves, the properties of sugar molecules determine how sugar behaves, and the properties of the molecules that compose our bodies determine how our bodies behave. The understanding of matter at the molecular level gives us unprecedented control over that matter. For example, our understanding of the details of the molecules that compose living organisms has revolutionized biology over the last 50 years.

▲ *This image portrays the Disneyland ride, Adventure Thru Inner Space. The premise of the ride is that you enter a microscope and get shrunk down to the size of an atom. The red and white spheres shown here depict oxygen and hydrogen atoms bound together to form water molecules.*

## 1.1   Atoms and Molecules

**KCV**   Atoms and Molecules

As I sat in the "omnimover" and listened to the narrator's voice telling me that I was shrinking down to the size of an atom, I grew apprehensive but curious. Just minutes before, while waiting in line, I witnessed what appeared to be full-sized humans entering a microscope and emerging from the other end many times smaller. I was seven years old, and I was about to ride *Adventure Thru Inner Space*, a Disneyland ride

1

(in Tomorrowland) that simulated the process of shrinking down to the size of an atom. The ride began with darkness and shaking, but then the shaking stopped and giant snowflakes appeared. The narrator explained that you were in the process of shrinking to an ever-smaller size (which explains why the snowflakes grew larger and larger). Soon, you entered the wall of the snowflake itself and began to see water molecules all around you. These also grew larger as you continued your journey into inner space and eventually ended up within the atom itself. Although this Disneyland ride bordered on being corny, and although it has since been shut down, it was my favorite ride as a young child.

That ride sparked my interest in the world of atoms and molecules, an interest that has continued and grown to this day. I am a chemist because I am obsessed with the connection between the "stuff" around us and the atoms and molecules that compose that stuff. More specifically, I love the idea that we humans have been able to figure out the connection between the *properties of the stuff* around us and the *properties of atoms and molecules*. **Atoms** are sub-microscopic particles that are the fundamental building blocks of ordinary matter. Free atoms are rare in nature; instead they bind together in specific geometrical arrangements to form **molecules**. A good example of a molecule is the water molecule, which I remember so well from the Disneyland ride.

A water molecule is composed of one oxygen atom bound to two hydrogen atoms in the shape shown at left. The exact properties of the water molecule—the atoms that compose it, the distances between those atoms, and the geometry of how the atoms are bound together—determine the properties of water. If the molecule were different, water would be different. For example, if water contained two oxygen atoms instead of just one, it would be a molecule like this:

**Water molecule**

Oxygen atom

Hydrogen atoms

**Hydrogen peroxide molecule**

Oxygen atoms

Hydrogen atoms

The hydrogen peroxide we use as an antiseptic or bleaching agent is considerably diluted.

This molecule is hydrogen peroxide, which you may have encountered if you have ever bleached your hair. A hydrogen peroxide molecule is composed of *two* oxygen atoms and two hydrogen atoms. This seemingly small molecular difference results in a huge difference in the properties of water and hydrogen peroxide. Water is the familiar and stable liquid we all drink and bathe in. Hydrogen peroxide, in contrast, is an unstable liquid that, in its pure form, burns the skin on contact and is used in rocket fuel. When you pour water onto your hair, your hair simply becomes wet. However, if you put diluted hydrogen peroxide on your hair, a chemical reaction occurs that strips your hair of its color.

The details of how specific atoms bond to form a molecule—in a straight line, at a particular angle, in a ring, or in some other pattern—as well as the type of atoms in the molecule, determine everything about the substance that the molecule composes. If we want to understand the substances around us, we must understand the atoms and molecules that compose them—this is the central goal of chemistry. A good simple definition of **chemistry** is

The term *atoms* in this definition can be interpreted loosely to include atoms that have lost or gained electrons.

### Chemistry—the science that seeks to understand the behavior of matter by studying the behavior of atoms and molecules.

Throughout this book, we explore the connection between atoms and molecules and the matter they compose. We seek to understand how differences on the atomic or molecular level affect the properties on the macroscopic level. Before we move on, let's examine one more example that demonstrates this principle. Consider the structures of graphite and diamond shown on the next page.

Graphite is the slippery black substance (often called pencil lead) that you have probably used in a mechanical pencil. Diamond is the brilliant gemstone found in jewelry. Graphite and diamond are both composed of exactly the same atoms—carbon atoms. The striking differences between the substances are a result of how those atoms are arranged. In graphite, the atoms are arranged in sheets. The atoms within each sheet are tightly bound to each other, but the sheets are *not* tightly bound to other sheets. Therefore the sheets can slide past each other, which is why the graphite in a pencil leaves a trail as you write. In diamond, by contrast, the carbon atoms are all bound together in a three-dimensional structure where layers are strongly bound to other layers, resulting in the strong, nearly unbreakable substance. This example illustrates how even the same atoms can compose vastly different substances when they are bound together in different patterns. Such is the atomic and molecular world—small differences in atoms and molecules can result in large differences in the substances that they compose.

**Graphite structure**

**Diamond structure**

## 1.2   The Scientific Approach to Knowledge

Throughout history, humans have approached knowledge about the physical world in different ways. For example, the Greek philosopher Plato (427–347 B.C.E.) thought that the best way to learn about reality was not through the senses but through reason. He believed that the physical world was an imperfect representation of a perfect and transcendent world (a world beyond space and time). For him, true knowledge came not through observing the real physical world, but through reasoning and thinking about the ideal one.

The *scientific* approach to knowledge, however, is exactly the opposite of Plato's. Scientific knowledge is empirical—it is based on *observation* and *experiment*. Scientists observe and perform experiments on the physical world to learn about it. Some observations and experiments are qualitative (noting or describing how a process happens), but many are quantitative (measuring or quantifying something about the process). For example, Antoine Lavoisier (1743–1794), a French chemist who studied combustion (burning), made careful measurements of the mass of objects before and after burning them in closed containers. He noticed that there was no change in the total mass of material within the container during combustion. In doing so, Lavoisier made an important *observation* about the physical world.

Observations often lead scientists to formulate a **hypothesis**, a tentative interpretation or explanation of the observations. For example, Lavoisier explained his observations on combustion by hypothesizing that when a substance burns, it combines with a component of air. A good hypothesis is *falsifiable*, which means that it makes predictions that can be confirmed or refuted by further observations. Scientists test hypotheses by **experiments**, highly controlled procedures designed to generate observations that confirm or refute a hypothesis. The results of an experiment may support a hypothesis or prove it wrong—in which case the scientist must modify or discard the hypothesis.

In some cases, a series of similar observations leads to the development of a **scientific law**, a brief statement that summarizes past observations and predicts future ones. Lavoisier summarized his observations on combustion with the **law of conservation of mass**, which states, "In a chemical reaction, matter is neither created nor destroyed." This statement summarized his observations on chemical reactions and predicted the

Although some Greek philosophers, such as Aristotle, did use observation to attain knowledge, they did not emphasize experiment and measurement to the extent that modern science does.

▲ French chemist Antoine Lavoisier with his wife, Marie, who helped him in his work by illustrating his experiments and translating scientific articles from English. Lavoisier, who also made significant contributions to agriculture, industry, education, and government administration, was executed during the French Revolution. (The Metropolitan Museum of Art)

outcome of future observations on reactions. Laws, like hypotheses, are also subject to experiments, which can support them or prove them wrong.

Scientific laws are not *laws* in the same sense as civil or governmental laws. Nature does not follow laws in the way that we obey the laws against speeding or running a stop sign. Rather, scientific laws *describe* how nature behaves—they are generalizations about what nature does. For that reason, some people find it more appropriate to refer to them as *principles* rather than *laws*.

One or more well-established hypotheses may form the basis for a scientific **theory**. A scientific theory is a model for the way nature is and tries to explain not merely what nature does but why. As such, well-established theories are the pinnacle of scientific knowledge, often predicting behavior far beyond the observations or laws from which they were developed. A good example of a theory is the **atomic theory** proposed by English chemist John Dalton (1766–1844). Dalton explained the law of conservation of mass, as well as other laws and observations of the time, by proposing that matter is composed of small, indestructible particles called atoms. Since these particles are merely rearranged in chemical changes (and not created or destroyed), the total amount of mass remains the same. Dalton's theory is a model for the physical world—it gives us insight into how nature works and, therefore, *explains* our laws and observations.

Finally, the scientific approach returns to observation to test theories. For example, scientists can test the atomic theory by trying to isolate single atoms or by trying to image them (both of which, by the way, have already been accomplished). Theories are validated by experiments; however, theories can never be conclusively proven because some new observation or experiment always has the potential to reveal a flaw. Notice that the scientific approach to knowledge begins with observation and ends with observation. An experiment is in essence a highly controlled procedure for generating critical observations designed to test a theory or hypothesis. Each new set of observations has the potential to refine the original model. Figure 1.1 ▼ summarizes one way to map the scientific approach to knowledge. Scientific laws, hypotheses, and theories are all subject to continued experimentation. If a law, hypothesis, or theory is proved wrong by an experiment, it must be revised and tested with new experiments. Over time, the scientific community eliminates or corrects poor theories and laws, and valid theories and laws—those consistent with experimental results—remain.

Established theories with strong experimental support are the most powerful pieces of scientific knowledge. You may have heard the phrase "That is just a theory," as if theories are easily dismissible. Such a statement reveals a deep misunderstanding of the nature of a scientific theory. Well-established theories are as close to truth as we get in science. The idea that all matter is made of atoms is "just a theory," but it has over 200 years of experimental evidence to support it. It is a powerful piece of scientific knowledge on which many other scientific ideas are based.

One last word about the scientific approach to knowledge: some people wrongly imagine science to be a strict set of rules and procedures that automatically leads to inarguable, objective facts. This is not the case. Even our diagram of the scientific approach to knowledge is only an idealization of real science, useful to help us see the

In Dalton's time, people thought atoms were indestructible. Today, because of nuclear reactions, we know that atoms can be broken apart into their smaller components.

▼ **FIGURE 1.1 The Scientific Approach to Knowledge**

## The Scientific Approach

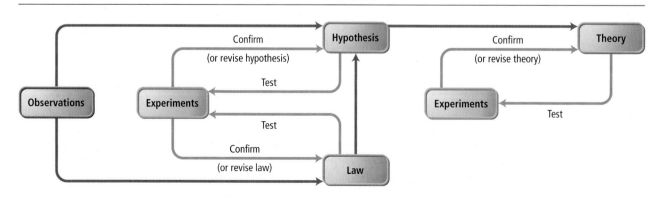

key distinctions of science. Real science requires hard work, care, creativity, and even a bit of luck. Scientific theories do not just arise out of data—men and women of genius and creativity craft theories. A great theory is not unlike a master painting, and many see a similar kind of beauty in both. (For more on this aspect of science, see the box below entitled *Thomas S. Kuhn and Scientific Revolutions*.)

**Laws and Theories** Which statement best explains the difference between a law and a theory?

CONCEPTUAL
**CONNECTION 1.1**

   **(a)** A law is truth; a theory is mere speculation.

   **(b)** A law summarizes a series of related observations; a theory gives the underlying reasons for them.

   **(c)** A theory describes *what* nature does; a law describes *why* nature does it.

# The Nature OF Science
## Thomas S. Kuhn and Scientific Revolutions

When scientists talk about science, they often talk in ways that imply that theories are "true." Further, they talk as if they arrive at theories in logical and unbiased ways. For example, a theory central to chemistry that we have discussed in this chapter is John Dalton's atomic theory—the idea that all matter is composed of atoms. Is this theory "true"? Was it reached in logical, unbiased ways? Will this theory still be around in 200 years?

The answers to these questions depend on how we view science and its development. One way to view science—let's call it the *traditional view*—is as the continual accumulation of knowledge and the building of increasingly precise theories. In this view, a scientific theory is a model of the world that reflects what is *actually in* nature. New observations and experiments result in gradual adjustments to theories. Over time, theories get better, giving us a more accurate picture of the physical world.

In the twentieth century, a different view of scientific knowledge began to develop. A book by Thomas Kuhn (1922–1996), published in 1962 and entitled *The Structure of Scientific Revolutions*, challenged the traditional view. Kuhn's ideas came from his study of the history of science, which, he argued, does not support the idea that science progresses in a smooth cumulative way. According to Kuhn, science goes through fairly quiet periods that he called *normal science*. In these periods, scientists make their data fit the reigning theory, or paradigm. Small inconsistencies are swept aside during periods of normal science. However, when too many inconsistencies and anomalies develop, a crisis

emerges. The crisis brings about a *revolution* and a new reigning theory. According to Kuhn, the new theory is usually quite different from the old one; it not only helps us to make sense of new or anomalous information, but it also enables us to see accumulated data from the past in a dramatically new way.

Kuhn further contended that theories are held for reasons that are not always logical or unbiased, and that theories are not *true* models—in the sense of a one-to-one mapping—of the physical world. Because new theories are often so different from the ones they replace, he argued, and because old theories always make good sense to those holding them, they must not be "True" with a capital *T*; otherwise "truth" would be constantly changing.

Kuhn's ideas created a controversy among scientists and science historians that continues to this day. Some, especially postmodern philosophers of science, have taken Kuhn's ideas one step further. They argue that scientific knowledge is *completely* biased and lacks any objectivity. Most scientists, including Kuhn, would disagree. Although Kuhn pointed out that scientific knowledge has *arbitrary elements*, he also said, "*Observation . . . can and must drastically restrict the range of admissible scientific belief, else there would be no science.*" In other words, saying that science contains arbitrary elements is quite different from saying that science itself is arbitrary.

#### QUESTION

*In his book, Kuhn stated, "A new theory . . . is seldom or never just an increment to what is already known." From your knowledge of the history of science, can you think of any examples that support Kuhn's statement? Do you know of any instances in which a new theory or model was drastically different from the one it replaced?*

# 1.3 The Classification of Matter

 **Classifying Matter**

**Matter** is anything that occupies space and has mass. Your desk, your chair, and even your body are all composed of matter. Less obviously, the air around you is also matter—it too occupies space and has mass. We call a specific instance of matter—such as air, water, or sand—a **substance**. We classify matter according to its **state** (its physical form) and its **composition** (the basic components that make it up).

## The States of Matter: Solid, Liquid, and Gas

The state of matter changes from solid to liquid to gas with increasing temperature.

Glass and other amorphous solids can be thought of, from one point of view, as intermediate between solids and liquids. Their atoms are fixed in position at room temperature, but they have no long-range structure and do not have distinct melting points.

Matter exists in three different states: **solid**, **liquid**, and **gas**. In *solid matter*, atoms or molecules pack closely to each other in fixed locations. Although the atoms and molecules in a solid vibrate, they do not move around or past each other. Consequently, a solid has a fixed volume and rigid shape. Ice, aluminum, and diamond are examples of solids. Solid matter may be **crystalline**, in which case its atoms or molecules are in patterns with long-range, repeating order (Figure 1.2◄), or it may be **amorphous**, in which case its atoms or molecules do not have any long-range order. Table salt and diamond are examples of *crystalline* solids; the well-ordered geometric shapes of salt and diamond crystals reflect the well-ordered geometric arrangement of their atoms (although this is not the case for *all* crystalline solids). Examples of *amorphous* solids include glass and plastic. In *liquid matter*, atoms or molecules pack about as closely as they do in solid matter, but they are free to move relative to each other, giving liquids a fixed volume but not a fixed shape. Liquids assume the shape of their containers. Water, alcohol, and gasoline are all substances that are liquids at room temperature.

**Crystalline Solid:**
Regular
three-dimensional pattern

**Diamond**
C (*s*, diamond)

▲ **FIGURE 1.2 Crystalline Solid** Diamond (first discussed in Section 1.1) is a crystalline solid composed of carbon atoms arranged in a regular, repeating pattern.

| Solid matter | Liquid matter | Gaseous matter |

▲ In a solid, the atoms or molecules are fixed in place and can only vibrate. In a liquid, although the atoms or molecules are closely packed, they can move past one another, allowing the liquid to flow and assume the shape of its container. In a gas, the atoms or molecules are widely spaced, making gases compressible as well as fluid (able to flow).

In *gaseous matter*, atoms or molecules have a lot of space between them and are free to move relative to one another, making gases *compressible* (Figure 1.3▶). When you squeeze a balloon or sit down on an air mattress, you force the atoms and molecules into a smaller space so that they are closer together. Gases always assume the shape *and* volume of their containers. Substances that are gases at room temperature include helium, nitrogen (the main component of air), and carbon dioxide.

◄ **FIGURE 1.3 The Compressibility of Gases** Gases can be compressed—squeezed into a smaller volume—because there is so much empty space between atoms or molecules in the gaseous state.

Solid–not compressible      Gas–compressible

## Classifying Matter according to Its Composition: Elements, Compounds, and Mixtures

In addition to classifying matter according to its state, we classify it according to its composition, as shown in the following chart:

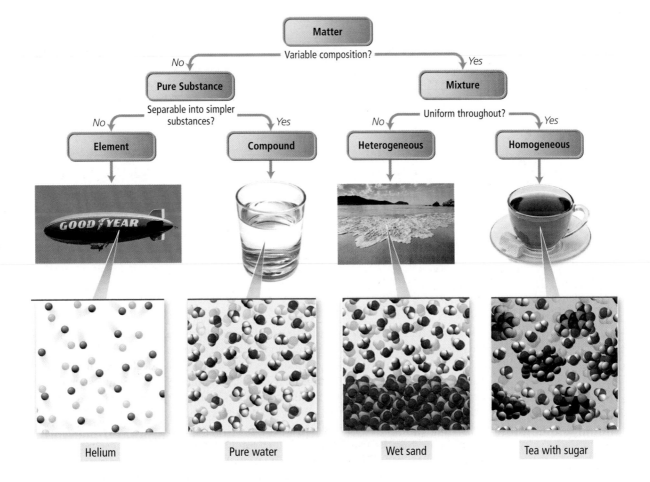

The first division in the classification of matter is between a *pure substance* and a *mixture*. A **pure substance** is made up of only one component, and its composition is invariant (it does not vary from one sample to another). The *components* of a pure substance can be individual atoms or groups of atoms joined together. For example, helium, water, and table salt (sodium chloride) are all pure substances. Each of these substances is made up of only one component: helium is made up of helium atoms, water is made up of water molecules, and sodium chloride is made up of sodium chloride units. The composition of a pure sample of any one of these substances is always exactly the same (because you can't vary the composition of a substance made up of only one component).

A **mixture**, by contrast, is composed of two or more components in proportions that can vary from one sample to another. For example, sweetened tea, composed primarily of water molecules and sugar molecules (with a few other substances mixed in), is a mixture. We can make tea slightly sweet (a small proportion of sugar to water) or very sweet (a large proportion of sugar to water) or any level of sweetness in between.

We categorize pure substances themselves into two types—*elements* and *compounds*—depending on whether or not they can be broken down (or decomposed) into simpler substances. Helium, which we just noted is a pure substance, is also a good example of an **element**, a substance that cannot be chemically broken down into simpler substances. Water, also a pure substance, is a good example of a **compound**, a substance composed of two or more elements (in this case hydrogen and oxygen) in a fixed, definite proportion. On Earth, compounds are more common than pure elements because most elements combine with other elements to form compounds.

> All known elements are listed in the periodic table in the inside front cover of this book.

We also categorize mixtures into two types—heterogeneous and homogeneous—depending on how *uniformly* the substances within them mix. Wet sand is a **heterogeneous mixture**, one in which the composition varies from one region of the mixture to another. Sweetened tea is a **homogeneous mixture**, one with the same composition throughout. Homogeneous mixtures have uniform compositions because the atoms or molecules that compose them mix uniformly. Heterogeneous mixtures are made up of distinct regions because the atoms or molecules that compose them separate. Here again we see that the properties of matter are determined by the atoms or molecules that compose it.

Classifying a substance according to its composition is not always obvious and requires that we either know the true composition of the substance or are able to test it in a laboratory. For now, we focus on relatively common substances that you are likely to have encountered. Throughout this course, you will gain the knowledge to understand the composition of a larger variety of substances.

CONCEPTUAL
**CONNECTION 1.2**

**Pure Substances and Mixtures** Let a small circle represent an atom of one type of element and a small square represent an atom of a second type of element. Make a drawing of (a) a pure substance (a compound) composed of the two elements (in a one-to-one ratio), (b) a homogeneous mixture composed of the two elements, and (c) a heterogeneous mixture composed of the two elements.

## Separating Mixtures

Chemists often want to separate a mixture into its components. Such separations can be easy or difficult, depending on the components in the mixture. In general, mixtures are separable because the different components have different physical or chemical properties. We can use various techniques that exploit these differences to achieve separation. For example, we can separate a mixture of sand and water by **decanting**—carefully pouring off—the water into another container. A homogeneous mixture of liquids can usually be separated by **distillation**, a process in which the mixture is heated to boil off the more **volatile** (easily vaporizable) liquid. The volatile liquid is then recondensed in a condenser and collected in a separate flask (Figure 1.4▶). If a mixture is composed of an insoluble solid and a liquid, we can separate the two by **filtration**, in which the mixture is poured through filter paper in a funnel (Figure 1.5▶).

**Distillation**

**Filtration**

Most volatile component boils first.

Condenser

Cooling water out

Cooling water in

Mixture of liquids with different boiling points

Vapor collected as pure liquid

▲ **FIGURE 1.4** **Separating Substances by Distillation** When a liquid mixture is heated, the component with the lowest boiling point vaporizes first, leaving behind less volatile liquids or dissolved solids. The vapor is then cooled, condensing it back to a liquid, and collected.

Stirring rod

Mixture of liquid and solid

Funnel

Filter paper traps solid.

Liquid component passes through and is collected.

▲ **FIGURE 1.5** **Separating Substances by Filtration** A solid and liquid mixture can be separated by pouring the mixture through a funnel containing filter paper designed to allow only the liquid to pass.

## 1.4 Physical and Chemical Changes and Physical and Chemical Properties

Every day we witness changes in matter: ice melts, iron rusts, gasoline burns, fruit ripens, and water evaporates. What happens to the molecules or atoms that compose these substances during such changes? The answer depends on the type of change. Changes that alter only state or appearance, but not composition, are **physical changes**. The atoms or molecules that compose a substance *do not change* their identity during a physical change. For example, when water boils, it changes its state from a liquid to a gas, but the gas remains composed of water molecules, so this is a physical change (Figure 1.6▶).

Water molecules change from liquid to gaseous state: physical change.

$H_2O(g)$

$H_2O(l)$

▶ **FIGURE 1.6** **Boiling, a Physical Change** When water boils, it turns into a gas but does not alter its chemical identity—the water molecules are the same in both the liquid and gaseous states. Boiling is a physical change, and the boiling point of water is a physical property.

A physical change results in a different form of the same substance, while a chemical change results in a completely different substance.

In Chapter 19 we will discuss *nuclear changes*, which can involve atoms of one element changing into atoms of a different element.

Iron atoms

Iron oxide (rust)

▲ **FIGURE 1.7  Rusting, a Chemical Change**  When iron rusts, the iron atoms combine with oxygen atoms to form a different chemical substance, the compound iron oxide. Rusting is a chemical change, and the tendency of iron to rust is a chemical property.

Answers to For Practice and For More Practice problems can be found in Appendix IV.

In contrast, changes that alter the composition of matter are **chemical changes**. During a chemical change, atoms rearrange, transforming the original substances into different substances. For example, the rusting of iron is a chemical change. The atoms that compose iron (iron atoms) combine with oxygen molecules from air to form iron oxide, the orange substance we call rust (Figure 1.7◀). Figure 1.8▶ illustrates other examples of physical and chemical changes.

Physical and chemical changes are manifestations of physical and chemical properties. A **physical property** is a property that a substance displays without changing its composition, whereas a **chemical property** is a property that a substance displays only by changing its composition via a chemical change. The smell of gasoline is a physical property—gasoline does not change its composition when it exhibits its odor. The flammability of gasoline, in contrast, is a chemical property—gasoline does change its composition when it burns, turning into completely new substances (primarily carbon dioxide and water). Physical properties include odor, taste, color, appearance, melting point, boiling point, and density. Chemical properties include corrosiveness, flammability, acidity, toxicity, and other such characteristics.

The differences between physical and chemical changes are not always apparent. Only chemical examination can confirm whether a particular change is physical or chemical. In many cases, however, we can identify chemical and physical changes based on what we know about the changes. Changes in the state of matter, such as melting or boiling, or changes in the physical condition of matter, such as those that result from cutting or crushing, are typically physical changes. Changes involving chemical reactions—often evidenced by temperature or color changes—are chemical changes.

---

**Example 1.1    Physical and Chemical Changes and Properties**

Determine whether each change is physical or chemical. What kind of property (chemical or physical) is demonstrated in each case?

**(a)** the evaporation of rubbing alcohol

**(b)** the burning of lamp oil

**(c)** the bleaching of hair with hydrogen peroxide

**(d)** the formation of frost on a cold night

**SOLUTION**

**(a)** When rubbing alcohol evaporates, it changes from liquid to gas, but it remains alcohol—this is a physical change. The volatility (the ability to evaporate easily) of alcohol is therefore a physical property.

**(b)** Lamp oil burns because it reacts with oxygen in air to form carbon dioxide and water—this is a chemical change. The flammability of lamp oil is therefore a chemical property.

**(c)** Applying hydrogen peroxide to hair changes pigment molecules in hair that give it color—this is a chemical change. The susceptibility of hair to bleaching is therefore a chemical property.

**(d)** Frost forms on a cold night because water vapor in air changes its state to form solid ice—this is a physical change. The temperature at which water freezes is therefore a physical property.

**FOR PRACTICE 1.1**    Determine whether each change is physical or chemical. What kind of property (chemical or physical) is demonstrated in each case?

**(a)** A copper wire is hammered flat.

**(b)** A nickel dissolves in acid to form a blue-green solution.

**(c)** Dry ice sublimes without melting.

**(d)** A match ignites when struck on a flint.

## Physical Change versus Chemical Change

**Dry ice subliming:**

$$CO_2(s) \longrightarrow CO_2(g)$$

Chemical composition unaltered
**Physical change**

$CO_2(g)$
Gaseous carbon dioxide

$CO_2(s)$
Solid carbon dioxide (dry ice)

(a)

**Sugar dissolving:**

$$C_{12}H_{22}O_{11}(s) \longrightarrow C_{12}H_{22}O_{11}(aq)$$

Chemical composition unaltered
**Physical change**

$C_{12}H_{22}O_{11}(s)$
Solid sugar

$C_{12}H_{22}O_{11}(aq)$
Dissolved sugar

(b)

**Propane gas burning:**

$$C_3H_8(g) + 5\,O_2(g) \longrightarrow$$
$$3\,CO_2(g) + 4\,H_2O(g)$$

Chemical composition altered
**Chemical change**

$CO_2(g)$, $H_2O(g)$
Carbon dioxide and water

$C_3H_8(g)$
Propane

(c)

▲ **FIGURE 1.8 Physical and Chemical Changes** **(a)** The sublimation (the state change from a solid to a gas) of dry ice (solid $CO_2$) is a physical change. **(b)** The dissolution of sugar is a physical change. **(c)** The burning of propane is a chemical change.

**Chemical and Physical Changes** The diagram on the left represents liquid water molecules in a pan. Which of the three diagrams (a, b, or c) best represents the water molecules after they have been vaporized by the boiling of liquid water?

(a)          (b)          (c)

## 1.5 Energy: A Fundamental Part of Physical and Chemical Change

The physical and chemical changes discussed in Section 1.4 are usually accompanied by energy changes. For example, when water evaporates from your skin (a physical change), the water molecules absorb energy from your body, making you feel cooler. When you burn natural gas on the stove (a chemical change), energy is released, heating the food you are cooking. Understanding the physical and chemical changes of matter—that is, understanding chemistry—requires that you understand energy changes and energy flow.

The scientific definition of **energy** is *the capacity to do work*. **Work** is defined as the action of a force through a distance. For instance, when you push a box across the floor or pedal your bicycle across the street, you have done work.

Force acts through distance; work is done.

—10 kg

High potential energy (unstable)

Kinetic energy

Low potential energy (stable)

**Heat**

▲ **FIGURE 1.9 Energy Conversions** Gravitational potential energy is converted into kinetic energy when the weight is dropped. The kinetic energy is converted mostly to thermal energy when the weight strikes the ground.

In Chapter 20 we will discuss how energy conservation is actually part of a more general law that allows for the interconvertibility of mass and energy.

The *total energy* of an object is a sum of its **kinetic energy** (the energy associated with its motion) and its **potential energy** (the energy associated with its position or composition). For example, a weight held several meters above the ground has potential energy due to its position within Earth's gravitational field (Figure 1.9 ◄). If you drop the weight, it accelerates, and its potential energy is converted to kinetic energy. When the weight hits the ground, its kinetic energy is converted primarily to **thermal energy**, the energy associated with the temperature of an object. Thermal energy is actually a type of kinetic energy because it is associated with the motion of the individual atoms or molecules that make up an object. When the weight hits the ground, its kinetic energy is essentially transferred to the atoms and molecules that compose the ground, raising the temperature of the ground ever so slightly.

The first principle to note about how energy changes as the weight falls to the ground is that *energy is neither created nor destroyed*. The potential energy of the weight becomes kinetic energy as the weight accelerates toward the ground. The kinetic energy then becomes thermal energy when the weight hits the ground. The total amount of thermal energy that is released through the process is exactly equal to the initial potential energy of the weight. The idea that energy is neither created nor destroyed is known as the **law of conservation of energy**. Although energy can change from one type into another, and although it can flow from one object to another, the *total quantity* of energy does not change—it remains constant.

The second principle to note about the raised weight and its fall is *the tendency of systems with high potential energy to change in a way that lowers their potential energy*. For this reason, objects or systems with high potential energy tend to be *unstable*. The weight lifted several meters from the ground is unstable because it contains a significant amount of potential energy. Unless restrained, the weight will naturally fall, lowering its potential energy (due to its position in Earth's gravitational field). We can harness some of the raised weight's potential energy to do work. For example, we can attach the weight to a rope that turns a paddle wheel or spins a drill as the weight falls. After the weight falls to the ground, it contains less potential energy—it has become more *stable*.

Some chemical substances are like the raised weight just described. For example, the molecules that compose gasoline have a relatively high potential energy—energy is concentrated in them just as energy is concentrated in the raised weight. The molecules in the gasoline tend to undergo chemical changes (specifically combustion) that lower the molecules' potential energy. As the energy of the molecules is released, some of it can be harnessed to do work, such as moving a car forward (Figure 1.10▲). The molecules that result from the chemical change have less potential energy than the original molecules in gasoline and are more stable.

Chemical potential energy, such as the energy contained in the molecules that compose gasoline, arises primarily from electrostatic forces between the electrically charged particles (protons and electrons) that compose atoms and molecules. We will learn more about those particles, as well as the properties of electrical charge, in Chapter 2, but for now, know that molecules contain specific, usually complex, arrangements of these charged particles. Some of these arrangements—such as the one within the molecules that compose gasoline—have a much higher potential energy than others. When gasoline undergoes combustion, the arrangement of these particles changes, creating molecules with much lower potential energy and transferring a great deal of energy (mostly in the form of heat) to the surroundings.

*Molecules in gasoline (unstable)*

*Molecules in exhaust (stable)*

*Some of released energy harnessed to do work*

*Car moves forward*

▲ **FIGURE 1.10 Using Chemical Energy to Do Work** The compounds produced when gasoline burns have less chemical potential energy than the gasoline molecules.

*Summarizing Energy:*

▶ Energy is always conserved in a physical or chemical change; it is neither created nor destroyed.

▶ Systems with high potential energy tend to change in ways that lower their potential energy, releasing energy into the surroundings.

## 1.6   The Units of Measurement

In 1999, NASA lost the $125 million *Mars Climate Orbiter*. The chairman of the commission that investigated the disaster concluded, "The root cause of the loss of the spacecraft was a failed translation of English units into metric units." As a result, the orbiter—which was supposed to monitor weather on Mars—descended too far into the Martian atmosphere and burned up. In chemistry as in space exploration, **units**—standard quantities used to specify measurements—are critical. If we get them wrong, the consequences can be disastrous.

The two most common unit systems are the **metric system**, used in most of the world, and the **English system**, used in the United States. Scientists use the **International System of Units (SI)**, which is based on the metric system.

### Standard Units

Table 1.1, on the next page, shows the standard SI base units. In this chapter, we focus on the first four of these units: the *meter*, the standard unit of length; the *kilogram*, the

▲ The $125 million *Mars Climate Orbiter* was lost in the Martian atmosphere in 1999 because two groups of engineers failed to communicate to each other the units that they used in their calculations.

The abbreviation *SI* comes from the French, *Système International d'Unités*.

▲ The Empire State Building is 443 m tall. A basketball player stands about 2 m tall.

| The velocity of light in a vacuum is $3.00 \times 10^8$ m/s.

| Scientists commonly deal with a wide range of lengths and distances. The separation between the sun and the closest star (Proxima Centauri) is about $3.8 \times 10^{16}$ m, while many chemical bonds measure about $1.5 \times 10^{-10}$ m.

▲ A nickel (5 cents) weighs about 5 g.

**TABLE 1.1 SI Base Units**

| Quantity | Unit | Symbol |
|---|---|---|
| Length | Meter | m |
| Mass | Kilogram | kg |
| Time | Second | s |
| Temperature | Kelvin | K |
| Amount of substance | Mole | mol |
| Electric current | Ampere | A |
| Luminous intensity | Candela | cd |

standard unit of mass; the *second*, the standard unit of time; and the *kelvin*, the standard unit of temperature.

## The Meter: A Measure of Length

The **meter (m)** is slightly longer than a yard (1 yard is 36 inches, while 1 meter is 39.37 inches).

Yardstick

Meterstick

Thus, a 100-yard (yd) football field measures only 91.4 m. The meter was originally defined as 1/10,000,000 of the distance from the equator to the North Pole (through Paris). The International Bureau of Weights and Measures now defines it more precisely as the distance light travels through a vacuum in a certain period of time, 1/299,792,458 second. A tall human is about 2 m tall, and the Empire State Building stands 443 m tall (including its mast).

## The Kilogram: A Measure of Mass

The **kilogram (kg)**, defined as the mass of a metal cylinder kept at the International Bureau of Weights and Measures at Sèvres, France, is a measure of *mass*, a quantity different from *weight*. The **mass** of an object is a measure of the quantity of matter within it, while the weight of an object is a measure of the *gravitational pull* on its matter. If you could weigh yourself on the moon, for example, its weaker gravity would pull on you with less force than does Earth's gravity, resulting in a lower weight. A 130-pound (lb) person on Earth would weigh only 21.5 lb on the moon. However, a person's mass—the quantity of matter in his or her body—remains the same on every planet. One kilogram of mass is the equivalent of 2.205 lb of weight on Earth, so expressed in kilograms, a 130-lb person has a mass of approximately 59 kg and this book has a mass of about 2.5 kg. A second common unit of mass is the gram (g). One gram is 1/1000 kg. A nickel (5¢) has a mass of about 5 g.

**CONCEPTUAL CONNECTION 1.4**

**The Mass of a Gas** We put a drop of water into a container and seal the container. The drop of water then vaporizes. Does the mass of the sealed container and its contents change upon vaporization?

## The Second: A Measure of Time

If you live in the United States, the **second (s)** is perhaps the SI unit most familiar to you. The International Bureau of Weights and Measures originally defined the second in terms of the day and the year, but a second is now defined more precisely as the duration of 9,192,631,770 periods of the radiation emitted from a certain transition in a cesium-133 atom. (We discuss transitions and the emission of radiation by atoms in Chapter 7.) Scientists measure time on a large range of scales. The human heart beats about once every second, the age of the universe is estimated to be about $4.32 \times 10^{17}$ s (13.7 billion years), and some molecular bonds break or form in time periods as short as $1 \times 10^{-15}$ s.

## The Kelvin: A Measure of Temperature

The **kelvin (K)** is the SI unit of **temperature**. The temperature of a sample of matter is a measure of the average kinetic energy—the energy due to motion—of the atoms or molecules that compose the matter. The molecules in a *hot* glass of water are, on average, moving faster than the molecules in a *cold* glass of water. Temperature is a measure of this molecular motion.

Temperature also determines the direction of thermal energy transfer, what we commonly call *heat*. Thermal energy transfers from hot objects to cold ones. For example, when you touch another person's warm hand (and yours is cold), thermal energy flows *from his or her hand to yours*, making your hand feel warmer. However, if you touch an ice cube, thermal energy flows *out of your hand* to the ice, cooling your hand (and possibly melting some of the ice cube).

Figure 1.11 ▼ shows the three common temperature scales. The most common in the United States is the **Fahrenheit (°F) scale**, shown on the left in Figure 1.11. On the Fahrenheit scale, water freezes at 32 °F and boils at 212 °F at sea level. Room temperature is approximately 72 °F. The Fahrenheit scale was originally determined by assigning 0 °F to the freezing point of a concentrated saltwater solution and 96 °F to normal body temperature.

Scientists and citizens of most countries other than the United States typically use the **Celsius (°C) scale**, shown in the middle in Figure 1.11. On this scale, pure water freezes at 0 °C and boils at 100 °C (at sea level). Room temperature is approximately 22 °C. The Fahrenheit scale and the Celsius scale differ both in the size of their respective degrees and the temperature each designates as "zero." Both the Fahrenheit and Celsius scales allow for negative temperatures.

Normal body temperature was later measured more accurately to be 98.6 °F.

**Temperature Scales**

◄ **FIGURE 1.11 Comparison of the Fahrenheit, Celsius, and Kelvin Temperature Scales** The Fahrenheit degree is five-ninths the size of the Celsius degree and the kelvin. The zero point of the Kelvin scale is absolute zero (the lowest possible temperature), whereas the zero point of the Celsius scale is the freezing point of water.

## The Celsius Temperature Scale

0 °C – Water freezes

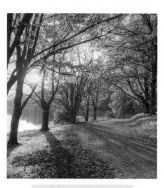

10 °C – Brisk fall day

22 °C – Room temperature

45 °C – Summer day in Death Valley

Molecular motion does not *completely* stop at absolute zero because of the uncertainty principle in quantum mechanics, which we discuss in Chapter 7.

The SI unit for temperature is the kelvin, shown in Figure 1.11. The **Kelvin scale** (sometimes also called the *absolute scale*) avoids negative temperatures by assigning 0 K to the coldest temperature possible, absolute zero. Absolute zero ($-273\,°C$ or $-459\,°F$) is the temperature at which molecular motion virtually stops. Lower temperatures do not exist. The size of the kelvin is identical to that of the Celsius degree; the only difference is the temperature that each scale designates as zero. We can convert between the temperature scales with these formulas:

Note that we give Kelvin temperatures in kelvins (*not* "degrees Kelvin") or K (*not* °K).

$$°C = \frac{(°F - 32)}{1.8}$$
$$K = °C + 273.15$$

Throughout this book we provide examples worked out in formats that are designed to help you develop problem-solving skills. The most common format uses two columns to guide you through the worked example. The left column describes the thought processes and steps used in solving the problem, while the right column shows the implementation. Example 1.2 follows this two-column format.

**Example 1.2** | **Converting between Temperature Scales**

A sick child has a temperature of 40.00 °C. What is the child's temperature in **(a)** K and **(b)** °F?

**SOLUTION**

| | |
|---|---|
| **(a)** Begin by finding the equation that relates the quantity that is given (°C) and the quantity you are trying to find (K). | $K = °C + 273.15$ |
| Since this equation gives the temperature in K directly, substitute in the correct value for the temperature in °C and calculate the answer. | $K = °C + 273.15$<br>$K = 40.00 + 273.15 = 313.15\ K$ |
| **(b)** To convert from °C to °F, first find the equation that relates these two quantities. | $°C = \frac{(°F - 32)}{1.8}$ |
| Since this equation expresses °C in terms of °F, solve the equation for °F. | $°C = \frac{(°F - 32)}{1.8}$<br>$1.8(°C) = (°F - 32)$<br>$°F = 1.8(°C) + 32$ |
| Now substitute °C into the equation and calculate the answer. *Note: The number of digits reported in this answer follows significant figure conventions, covered in Section 1.7.* | $°F = 1.8(°C) + 32$<br>$°F = 1.8(40.00\,°C) + 32 = 104.00\,°F$ |

**FOR PRACTICE 1.2** Gallium is a solid metal at room temperature but it will melt to a liquid in your hand. The melting point of gallium is 85.6 °F. What is this temperature on **(a)** the Celsius scale and **(b)** the Kelvin scale?

## Prefix Multipliers

Scientific notation (see Appendix IA) allows us to express very large or very small quantities in a compact manner by using exponents. For example, the diameter of a hydrogen atom is $1.06 \times 10^{-10}$ m. The International System of Units uses the **prefix multipliers** listed in Table 1.2 with the standard units. These multipliers change the value of the unit by powers of 10 (just like an exponent does in scientific notation). For example, the kilometer has the prefix *kilo* meaning 1000 or $10^3$. Therefore,

$$1 \text{ kilometer} = 1000 \text{ meters} = 10^3 \text{ meters}$$

Similarly, the millimeter has the prefix *milli* meaning 0.001 or $10^{-3}$.

$$1 \text{ millimeter} = 0.001 \text{ meters} = 10^{-3} \text{ meters}$$

| TABLE 1.2 SI Prefix Multipliers | | | |
|---|---|---|---|
| Prefix | Symbol | Multiplier | |
| exa | E | 1,000,000,000,000,000,000 | $(10^{18})$ |
| peta | P | 1,000,000,000,000,000 | $(10^{15})$ |
| tera | T | 1,000,000,000,000 | $(10^{12})$ |
| giga | G | 1,000,000,000 | $(10^{9})$ |
| mega | M | 1,000,000 | $(10^{6})$ |
| kilo | k | 1000 | $(10^{3})$ |
| deci | d | 0.1 | $(10^{-1})$ |
| centi | c | 0.01 | $(10^{-2})$ |
| milli | m | 0.001 | $(10^{-3})$ |
| micro | $\mu$ | 0.000001 | $(10^{-6})$ |
| nano | n | 0.000000001 | $(10^{-9})$ |
| pico | p | 0.000000000001 | $(10^{-12})$ |
| femto | f | 0.000000000000001 | $(10^{-15})$ |
| atto | a | 0.000000000000000001 | $(10^{-18})$ |

When reporting a measurement, choose a prefix multiplier close to the size of the quantity you are measuring. For example, to state the diameter of a hydrogen atom, which is $1.06 \times 10^{-10}$ m, use picometers (106 pm) or nanometers (0.106 nm) rather than micrometers or millimeters. Choose the prefix multiplier that is most convenient for a particular number.

**Prefix Multipliers**  What prefix multiplier is appropriate for reporting a measurement of $5.57 \times 10^{-5}$ m?

CONCEPTUAL
**CONNECTION 1.5**

## Derived Units: Volume and Density

A **derived unit** is a combination of other units. For example, the SI unit for speed is meters per second (m/s), a derived unit. Notice that this unit is formed from two other SI units—meters and seconds—put together. You are probably more familiar with speed measured in miles/hour or kilometers/hour—these are also examples of derived units. Two other common derived units are those for volume (SI base unit is m³) and density (SI base unit is kg/m³).

### Relationship between Length and Volume

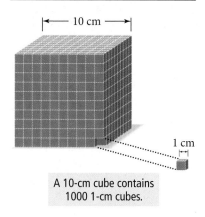

A 10-cm cube contains 1000 1-cm cubes.

▲ **FIGURE 1.12 The Relationship between Length and Volume**

## Volume

**Volume** is a measure of space. Any unit of length, when cubed (raised to the third power), becomes a unit of volume. The cubic meter ($m^3$), cubic centimeter ($cm^3$), and cubic millimeter ($mm^3$) are all units of volume. The cubic nature of volume is not always intuitive, and studies have shown that our brains are not naturally wired to process abstract concepts such as volume. For example, consider this question: how many small cubes measuring 1 cm on each side are required to construct a large cube measuring 10 cm (or 1 dm) on a side?

The answer to this question, as you can see by carefully examining the unit cube in Figure 1.12◄, is 1000 small cubes. When we go from a linear, one-dimensional distance to three-dimensional volume, we must raise both the linear dimension *and* its unit to the third power (not multiply by 3). Thus the volume of a cube is equal to the length of its edge cubed.

$$\text{volume of cube} = (\text{edge length})^3$$

A cube with a 10-cm edge has a volume of $(10\,cm)^3$ or $1000\,cm^3$, and a cube with a 100-cm edge has a volume of $(100\,cm)^3 = 1{,}000{,}000\,cm^3$.

Other common units of volume in chemistry are the **liter (L)** and the **milliliter (mL)**. One milliliter ($10^{-3}$ L) is equal to 1 $cm^3$. A gallon of gasoline contains 3.785 L. Table 1.3 lists some common units—for volume and other quantities—and their equivalents.

---

### TABLE 1.3 Some Common Units and Their Equivalents

| Length | Mass | Volume |
|---|---|---|
| 1 kilometer (km) = 0.6214 mile (mi) | 1 kilogram (kg) = 2.205 pounds (lb) | 1 liter (L) = 1000 mL = 1000 $cm^3$ |
| 1 meter (m) = 39.37 inches (in) = 1.094 yards (yd) | 1 pound (lb) = 453.59 grams (g) | 1 liter (L) = 1.057 quarts (qt) |
| 1 foot (ft) = 30.48 centimeters (cm) | 1 ounce (oz) = 28.35 grams (g) | 1 U.S. gallon (gal) = 3.785 liters (L) |
| 1 inch (in) = 2.54 centimeter (cm) (exact) | | |

---

## Density

An old riddle asks, "Which weighs more, a ton of bricks or a ton of feathers?" The answer is neither—they both weigh the same (1 ton). If you answered bricks, you confused weight with density. The **density** (*d*) of a substance is the ratio of its mass (*m*) to its volume (*V*).

> The *m* in the equation for density is in italic type, meaning that it stands for mass rather than for meters. In general, the symbols for units such as meters (m), seconds (s), or kelvins (K) appear in regular type while those for variables such as mass (*m*), volume (*V*), and time (*t*) appear in italics.

$$\text{Density} = \frac{\text{mass}}{\text{volume}} \quad \text{or} \quad d = \frac{m}{V}$$

Density is a characteristic physical property of substances (see Table 1.4 on the next page) that depends on temperature. Density is an example of an **intensive property**, one that is *independent* of the amount of the substance. The density of aluminum, for example, is the same whether you have a gram or a kilogram. Intensive properties are often used to identify substances because these properties depend only on the type of substance, not on the amount of it. For example, from Table 1.4 you can see that pure gold has a density of 19.3 $g/cm^3$. One way to determine whether a substance is pure gold is to determine its density and compare it to 19.3 $g/cm^3$. Mass, in contrast, is an **extensive property**, one that depends on the amount of the substance. If you know only the mass of a sample of gold, that information alone will not allow you to identify it as gold.

The units of density are those of mass divided by volume. Although the SI-derived unit for density is $kg/m^3$, the density of liquids and solids is most often expressed in $g/cm^3$ or g/mL. (Remember that $cm^3$ and mL are equivalent units: 1 $cm^3$ = 1 mL.) Aluminum is among the least dense metals with a density of 2.7 $g/cm^3$, while platinum is one of the densest metals with a density of 21.4 $g/cm^3$.

# Calculating Density

We can calculate the density of a substance by dividing the mass of a given amount of the substance by its volume. For example, suppose a small nugget we suspect to be gold has a mass of 22.5 g and a volume of 2.38 cm³. To find its density, we divide the mass by the volume:

$$d = \frac{m}{V} = \frac{22.5\ \text{g}}{2.38\ \text{cm}^3} = 9.45\ \text{g/cm}^3$$

In this case, the density reveals that the nugget is not pure gold because the density of gold is 19.3 g/cm³.

| TABLE 1.4 The Density of Some Common Substances at 20 °C | |
| --- | --- |
| **Substance** | **Density (g/cm³)** |
| Charcoal (from oak) | 0.57 |
| Ethanol | 0.789 |
| Ice | 0.917 (at 0 °C) |
| Water | 1.00 (at 4 °C) |
| Sugar (sucrose) | 1.58 |
| Table salt (sodium chloride) | 2.16 |
| Glass | 2.6 |
| Aluminum | 2.70 |
| Titanium | 4.51 |
| Iron | 7.86 |
| Copper | 8.96 |
| Lead | 11.4 |
| Mercury | 13.55 |
| Gold | 19.3 |
| Platinum | 21.4 |

## Example 1.3   Calculating Density

A man receives a platinum ring from his fiancée. Before the wedding, he notices that the ring feels a little light for its size, and so he decides to determine its density. He places the ring on a balance and finds that it has a mass of 3.15 g. He then finds that the ring displaces 0.233 cm³ of water. Is the ring made of platinum? (Note: The volume of irregularly shaped objects is often measured by the displacement of water. To use this method, the object is placed in water and the change in volume of the water is measured. This increase in the total volume represents the volume of water *displaced* by the object and is equal to the volume of the object.)

| | |
| --- | --- |
| Set up the problem by writing the important information that is *given* as well as the information that you are asked to *find*. In this case, we need to find the density of the ring and compare it to that of platinum. *Note: In Section 1.8, we discuss this standardized way of setting up problems.* | **GIVEN:** $m = 3.15$ g <br> $V = 0.233$ cm³ <br><br> **FIND:** Density in g/cm³ |
| Next, write down the equation that defines density. | **EQUATION** $d = \dfrac{m}{V}$ |
| Solve the problem by substituting the correct values of mass and volume into the expression for density. | **SOLUTION** <br> $d = \dfrac{m}{V} = \dfrac{3.15\ \text{g}}{0.233\ \text{cm}^3} = 13.5\ \text{g/cm}^3$ |

The density of the ring is much too low to be platinum (platinum density is 21.4 g/cm³), and the ring is therefore a fake.

**FOR PRACTICE 1.3**   The woman in Example 1.3 is shocked that the ring is fake and returns it. She buys a new ring that has a mass of 4.53 g and a volume of 0.212 cm³. Is the new ring genuine?

**FOR MORE PRACTICE 1.3**   A metal cube has an edge that is 11.4 mm long and a mass of 6.67 g. Calculate the density of the metal and use Table 1.4 to determine the likely identity of the metal.

CONCEPTUAL
**CONNECTION 1.6**

**Density**   The density of copper decreases as temperature increases (as does the density of most substances). Which change occurs in a sample of copper when it is warmed from room temperature to 95 °C?

   **(a)** The sample becomes lighter.
   **(b)** The sample becomes heavier.
   **(c)** The sample expands.
   **(d)** The sample contracts.

# Chemistry AND Medicine | Bone Density

Osteoporosis—which means *porous bone*—is a condition in which bone density becomes low. The healthy bones of a young adult have a density of about 1.0 g/cm³. Patients suffering from osteoporosis, however, can have bone densities as low as 0.22 g/cm³. These low densities indicate that the bones have deteriorated and weakened, resulting in increased susceptibility to fractures, especially hip fractures. Patients suffering from osteoporosis can also experience height loss and disfiguration such as dowager's hump, a condition in which the patient becomes hunched over due to compression of the vertebrae. Osteoporosis is most common in postmenopausal women, but it can also occur in people (including men) who have certain diseases, such as insulin-dependent diabetes, or who take certain medications, such as prednisone. Osteoporosis is usually diagnosed and monitored with hip X-rays. Low-density bones absorb less of the X-rays than do high-density bones, producing characteristic differences in the X-ray image. Treatments for osteoporosis include calcium and vitamin D supplements, drugs that prevent bone weakening, exercise and strength training, and, in extreme cases, hip-replacement surgery.

▲ Top: Severe osteoporosis can necessitate surgery to implant an artificial hip joint, seen in this X-ray image. Bottom: Views of the bone matrix in a normal bone (left) and one weakened by osteoporosis (right).

**QUESTION**

*Suppose you find a large animal bone in the woods, too large to fit in a beaker or flask. How might you approximate its density?*

## 1.7   The Reliability of a Measurement

Carbon monoxide is a colorless gas emitted by motor vehicles and found in polluted air. The table shown here lists carbon monoxide concentrations in Los Angeles County as reported by the U.S. Environmental Protection Agency (EPA) over the period 1993–2013:

| Year | Carbon Monoxide Concentration (ppm)* |
|------|:---:|
| 1993 | 6.9 |
| 1998 | 6.1 |
| 2003 | 4.1 |
| 2008 | 2.5 |
| 2013 | 1.9 |

*Second maximum, 8 hour average; ppm = *parts per million*, defined as mL pollutant per million mL of air. (Long Beach Site 060374002)

The first thing you should notice about these values is that they decrease over time. For this decrease, we can thank the Clean Air Act and its amendments, which have resulted in more efficient engines and specially blended fuels and consequently in cleaner air in all major U.S. cities over the last 30 years. The second thing you might notice is the number of digits to which the measurements are reported. The number of digits in a reported measurement indicates the certainty associated with that measurement. A less certain measurement of carbon monoxide levels might be reported as follows:

| Year | Carbon Monoxide Concentration (ppm) |
|------|-------------------------------------|
| 1993 | 7 |
| 1998 | 6 |
| 2003 | 4 |
| 2008 | 3 |
| 2013 | 2 |

Notice that the first set of data is reported to the nearest 0.1 ppm while the second set is reported to the nearest 1 ppm. Scientists report measured quantities in an agreed-upon standard way. The number of reported digits reflects the certainty in the measurement: more digits, more certainty; fewer digits, less certainty. Numbers are usually written so that the uncertainty is in the last reported digit. (We assume that uncertainty to be ±1 in the last digit unless otherwise indicated.) By reporting the 2013 carbon monoxide concentration as 1.9 ppm, the scientists mean $1.9 \pm 0.1$ ppm. The carbon monoxide concentration is between 1.8 and 2.0 ppm—it might be 2.0 ppm, for example, but it could not be 3.0 ppm. In contrast, if the reported value was 2 ppm (as in the second set of measurements), this would mean $2 \pm 1$ ppm, or between 1 and 3 ppm. In general,

**Scientific measurements are reported so that every digit is certain except the last, which is estimated.**

For example, consider the following reported number:

5.213

certain    estimated

The first three digits are certain; the last digit is estimated.

The number of digits reported in a measurement depends on the measuring device. Consider weighing a pistachio nut on two different balances (Figure 1.13▶). The balance on the top has marks every 1 g, while the balance on the bottom has marks every 0.1 g. For the balance on the top, we mentally divide the space between the 1- and 2-g marks into ten equal spaces and estimate that the pointer is at about 1.2 g. We then write the measurement as 1.2 g, indicating that we are sure of the "1" but we have estimated the ".2." The balance on the bottom, with marks every *tenth* of a gram, requires that we write the result with more digits. The pointer is between the 1.2-g mark and the 1.3-g mark. We again divide the space between the two marks into ten equal spaces and estimate the third digit. For the figure shown, we report 1.27 g.

### Example 1.4    Reporting the Correct Number of Digits

The graduated cylinder shown in the right margin has markings every 0.1 mL. Report the volume (which is read at the bottom of the meniscus) to the correct number of digits. (Note: The meniscus is the crescent-shaped surface at the top of a column of liquid.)

**SOLUTION**

Since the bottom of the meniscus is between the 4.5 and 4.6 mL markings, mentally divide the space between the markings into ten equal spaces and estimate the next digit. In this case, you should report the result as 4.57 mL.

What if you estimated a little differently and wrote 4.56 mL? In general, a one-unit difference in the last digit is acceptable because the last digit is estimated and different people might estimate it slightly differently. However, if you wrote 4.63 mL, you would have misreported the measurement.

**FOR PRACTICE 1.4**    Record the temperature on the thermometer shown in the right margin to the correct number of digits.

### Estimation in Weighing

**(a)**

Markings every 1 g
Estimated reading 1.2 g

**(b)**

Markings every 0.1 g
Estimated reading 1.27 g

▲ **FIGURE 1.13  Estimation in Weighing  (a)** This scale has markings every 1 g, so we mentally divide the space into ten equal spaces to estimate the last digit. This reading is 1.2 g. **(b)** Because this balance has markings every 0.1 g, we estimate to the hundredths place. This reading is 1.27 g.

Meniscus

## Counting Significant Figures

The precision of a measurement—which depends on the instrument used to make the measurement—must be preserved, not only when recording the measurement, but also when performing calculations that use the measurement. We can accomplish the preservation of this precision by using *significant figures*. In any reported measurement, the non–place-holding digits—those that are not simply marking the decimal place—are called **significant figures** (or **significant digits**). *The greater the number of significant figures, the greater the certainty of the measurement.* For example, the number 23.5 has three significant figures while the number 23.56 has four. To determine the number of significant figures in a number containing zeroes, we distinguish between zeroes that are significant and those that simply mark the decimal place. For example, in the number 0.0008, the leading zeroes (zeroes to the left of the first nonzero digit) mark the decimal place but *do not* add to the certainty of the measurement and are therefore not significant; this number has only one significant figure. In contrast, the trailing zeroes (zeroes at the end of a number) in the number 0.000800 *do add* to the certainty of the measurement and are therefore counted as significant; this number has three significant figures.

**To determine the number of significant figures in a number, follow these rules:**

| Significant Figure Rules | Examples | |
|---|---|---|
| **1.** All nonzero digits are significant. | 28.03 | 0.0540 |
| **2.** Interior zeroes (zeroes between two nonzero digits) are significant. | 408 | 7.0301 |
| **3.** Leading zeroes (zeroes to the left of the first nonzero digit) are not significant. They only serve to locate the decimal point. | 0.0032 | 0.00006 |
| | | not significant |
| **4.** Trailing zeroes (zeroes at the end of a number) are categorized as follows: | | |
| • Trailing zeroes after a decimal point are always significant. | 45.000 | 3.5600 |
| • Trailing zeroes before a decimal point (and after a nonzero number) are always significant. | 140.00 | 2500.55 |
| • Trailing zeroes before an *implied* decimal point are ambiguous and should be avoided by using scientific notation. | 1200 $1.2 \times 10^3$ $1.20 \times 10^3$ $1.200 \times 10^3$ | ambiguous 2 significant figures 3 significant figures 4 significant figures |
| • Some textbooks put a decimal point after one or more trailing zeroes if the zeroes are to be considered significant. We avoid that practice in this book, but you should be aware of it. | 1200. | 4 significant figures (common in some textbooks) |

## Exact Numbers

**Exact numbers** have no uncertainty and thus do not limit the number of significant figures in any calculation. We can regard an exact number as having an unlimited number of significant figures. Exact numbers originate from three sources:

• From the accurate counting of discrete objects. For example, 3 atoms means 3.00000... atoms.

• From defined quantities, such as the number of centimeters in 1 m. Because 100 cm is defined as 1 m,

$$100 \text{ cm} = 1 \text{ m means } 100.00000... \text{ cm} = 1.0000000... \text{ m}$$

• From integral numbers that are part of an equation. For example, in the equation $radius = \dfrac{diameter}{2}$, the number 2 is exact and therefore has an unlimited number of significant figures.

## Example 1.5   Determining the Number of Significant Figures in a Number

How many significant figures are in each number?

**(a)** 0.04450 m    **(b)** 5.0003 km    **(c)** 10 dm = 1 m    **(d)** $1.000 \times 10^5$ s    **(e)** 0.00002 mm    **(f)** 10,000 m

SOLUTION

| **(a)** 0.04450 m | *Four significant figures.* The two 4s and the 5 are significant (Rule 1). The trailing zero is after a decimal point and is therefore significant (Rule 4). The leading zeroes only mark the decimal place and are therefore not significant (Rule 3). |
| **(b)** 5.0003 km | *Five significant figures.* The 5 and 3 are significant (Rule 1), as are the three interior zeroes (Rule 2). |
| **(c)** 10 dm = 1 m | *Unlimited significant figures.* Defined quantities have an unlimited number of significant figures. |
| **(d)** $1.000 \times 10^5$ s | *Four significant figures.* The 1 is significant (Rule 1). The trailing zeroes are after a decimal point and therefore significant (Rule 4). |
| **(e)** 0.00002 mm | *One significant figure.* The 2 is significant (Rule 1). The leading zeroes only mark the decimal place and are therefore not significant (Rule 3). |
| **(f)** 10,000 m | *Ambiguous.* The 1 is significant (Rule 1), but the trailing zeroes occur before an implied decimal point and are therefore ambiguous (Rule 4). Without more information, you would assume one significant figure. It is better to write this as $1 \times 10^5$ to indicate one significant figure or as $1.0000 \times 10^5$ to indicate five (Rule 4). |

**FOR PRACTICE 1.5**   How many significant figures are in each number?

**(a)** 554 km    **(b)** 7 pennies    **(c)** $1.01 \times 10^5$ m    **(d)** 0.00099 s    **(e)** 1.4500 km    **(f)** 21,000 m

## Significant Figures in Calculations

When we use measured quantities in calculations, the results of the calculation must reflect the precision of the measured quantities. We should not lose or gain precision during mathematical operations. Follow these rules when carrying significant figures through calculations.

| **Rules for Calculations** | **Examples** |
|---|---|
| **1.** In multiplication or division, the result carries the same number of significant figures as the factor with the fewest significant figures. | $\underset{\text{4 (sig. figures)}}{1.052} \times \underset{\text{(5 sig. figures)}}{12.504} \times \underset{\text{(2 sig. figures)}}{0.53} = \underset{}{6.7208} = \underset{\text{(2 sig. figures)}}{6.7}$ <br><br> $\underset{\text{(5 sig. figures)}}{2.0035} \div \underset{\text{(3 sig. figures)}}{3.20} = 0.626094 = \underset{\text{(3 sig. figures)}}{0.626}$ |
| **2.** In addition or subtraction, the result carries the same number of decimal places as the quantity with the fewest decimal places. | $\begin{array}{r} 2.34\vert5 \\ 0.07\phantom{00} \\ 2.99\vert75 \\ \hline 5.41\vert25 = 5.41 \end{array}$ $\qquad \begin{array}{r} 5.9\phantom{00} \\ -0.2\vert21 \\ \hline 5.6\vert79 = 5.7 \end{array}$ <br><br> In addition and subtraction, it is helpful to draw a line next to the number with the fewest decimal places. This line determines the number of decimal places in the answer. |
| **3.** When rounding to the correct number of significant figures, round down if the last (or leftmost) digit dropped is four or less; round up if the last (or leftmost) digit dropped is five or more. | Rounding to two significant figures: <br> 5.37 rounds to 5.4 <br> 5.34 rounds to 5.3 <br> 5.35 rounds to 5.4 <br> 5.349 rounds to 5.3 <br><br> Notice in the last example (5.349) that only the *leftmost digit being dropped* (in this example, the 4) determines how you round the number. You ignore all the digits to the right of the leftmost digit you are dropping (in this example, you ignore the 9). |

A few books recommend a slightly different rounding procedure for cases in which the last digit is 5. However, the procedure presented here is consistent with electronic calculators and will be used throughout this book.

| **Rules for Calculations** | **Examples** |
|---|---|
| **4.** To avoid rounding errors in multi-step calculations, round only the final answer—do not round intermediate steps. If you write down intermediate answers, keep track of significant figures by underlining the least significant digit. | $6.78 \times 5.903 \times (5.489 - 5.01)$<br>$= 6.78 \times 5.903 \times 0.479$<br>$= 19.1707$<br>$= 19$<br><br>underline least significant digit |

Notice that for multiplication or division, the quantity with the fewest *significant figures* determines the number of *significant figures* in the answer, but for addition and subtraction, the quantity with the fewest *decimal places* determines the number of *decimal places* in the answer. In multiplication and division, we focus on significant figures, but in addition and subtraction we focus on decimal places. When a problem involves addition or subtraction, the answer may have a different number of significant figures than the initial quantities. Keep this in mind in problems that involve both addition or subtraction and multiplication or division. For example,

$$\frac{1.002 - 0.999}{3.754} = \frac{0.003}{3.754}$$

$$= 7.99 \times 10^{-4}$$

$$= 8 \times 10^{-4}$$

The answer has only one significant figure, even though the initial numbers had three or four.

## Example 1.6   Significant Figures in Calculations

Perform each calculation to the correct number of significant figures.

**(a)** $1.10 \times 0.5120 \times 4.0015 \div 3.4555$

**(b)**      $0.355$
       $+105.1$
       $\underline{-100.5820}$

**(c)** $4.562 \times 3.99870 \div (452.6755 - 452.33)$

**(d)** $(14.84 \times 0.55) - 8.02$

**SOLUTION**

| | |
|---|---|
| **(a)** Round the intermediate result (in blue) to three significant figures to reflect the three significant figures in the quantity with the fewest significant figures (1.10). | $1.10 \times 0.5120 \times 4.0015 \div 3.4555$<br>$= 0.65219$<br>$= 0.652$ |
| **(b)** Round the intermediate answer (in blue) to one decimal place to reflect the quantity with the fewest decimal places (105.1). Notice that 105.1 is *not* the quantity with the fewest significant figures, but it has the fewest decimal places and therefore determines the number of decimal places in the answer. | $\phantom{+}0.3\|55$<br>$+105.1$<br>$\underline{-100.5\|820}$<br>$\phantom{+}4.8\|730 = 4.9$ |
| **(c)** Mark the intermediate result to two decimal places to reflect the number of decimal places in the quantity within the parentheses having the fewest number of decimal places (452.33). Round the final answer to two significant figures to reflect the two significant figures in the least precisely known quantity (0.3455). | $4.562 \times 3.99870 \div (452.6755 - 452.33)$<br>$= 4.562 \times 3.99870 \div 0.3455$<br>$= 52.79904$<br>$= 53$<br><br>2 places of the decimal |

| | |
|---|---|
| **(d)** Mark the intermediate result to two significant figures to reflect the number of significant figures in the quantity within the parentheses having the fewest number of significant figures (0.55). Round the final answer to one decimal place to reflect the one decimal place in the least precisely known quantity (8.1̲62). | $(14.84 \times 0.55) - 8.02 = 8.1̲62 - 8.02$<br>$= 0.142$<br>$= 0.1$ |

**FOR PRACTICE 1.6**    Perform each calculation to the correct number of significant figures.

**(a)** $3.10007 \times 9.441 \times 0.0301 \div 2.31$

**(b)**   $\begin{array}{r} 0.881 \\ +132.1 \\ \underline{-12.02} \end{array}$

**(c)** $2.5110 \times 21.20 \div (44.11 + 1.223)$

**(d)** $(12.01 \times 0.3) + 4.811$

## Precision and Accuracy

Scientists often repeat measurements several times to increase confidence in the result. We distinguish between two different kinds of certainty—accuracy and precision—associated with such measurements. **Accuracy** refers to how close the measured value is to the actual value. **Precision** refers to how close a series of measurements are to one another or how reproducible they are. A series of measurements can be precise (close to one another in value and reproducible) but not accurate (not close to the true value). Consider the results of three students who repeatedly weighed a lead block known to have a true mass of 10.00 g (indicated by the solid horizontal blue line on the graphs in the accompanying figure).

| | Student A | Student B | Student C |
|---|---|---|---|
| **Trial 1** | 10.49 g | 9.78 g | 10.03 g |
| **Trial 2** | 9.79 g | 9.82 g | 9.99 g |
| **Trial 3** | 9.92 g | 9.75 g | 10.03 g |
| **Trial 4** | 10.31 g | 9.80 g | 9.98 g |
| **Average** | 10.13 g | 9.79 g | 10.01 g |

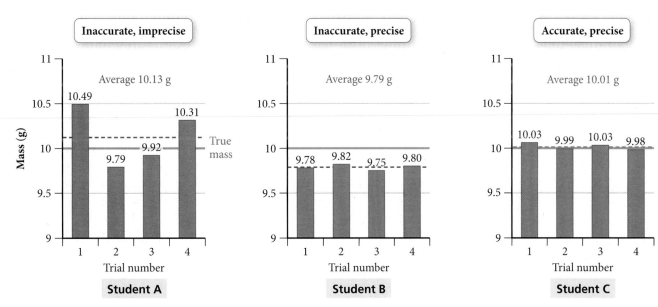

▲ Measurements are said to be precise if they are consistent with one another, but they are accurate only if they are close to the actual value.

- The results of student A are both inaccurate (not close to the true value) and imprecise (not consistent with one another). The inconsistency is the result of **random error**, error that has equal probability of being too high or too low. Almost all measurements have some degree of random error. Random error can, with enough trials, average itself out.

- The results of student B are precise (close to one another in value) but inaccurate. The inaccuracy is the result of **systematic error**, error that tends to be either too high or too low. In contrast to random error, systematic error does not average out with repeated trials. For instance, if a balance is not properly calibrated, it will systematically read too high or too low.

- The results of student C display little systematic error or random error—they are both accurate and precise.

---

## Chemistry IN YOUR DAY | Integrity in Data Gathering

Most scientists spend many hours collecting data in the laboratory. Often, the data do not turn out exactly as the scientist had expected (or hoped). A scientist may then be tempted to "fudge" the results. For example, suppose you are expecting a particular set of measurements to follow a certain pattern. After working hard over several days or weeks to make the measurements, you notice that a few of them do not quite fit the pattern that you anticipated. You might find yourself wishing that you could simply change or omit the "faulty" measurements. Altering data in this way is considered highly unethical in the scientific community and, when discovered, is punished severely.

In 2004, Dr. Hwang Woo Suk, a stem cell researcher at the Seoul National University in Korea, published a research paper in *Science* (a highly respected research journal) claiming that he and his colleagues had cloned human embryonic stem cells. As part of his evidence, he showed photographs of the cells. The paper was hailed as an incredible breakthrough, and Dr. Hwang traveled the world lecturing on his work. *Time* magazine even included him on its "people that

matter" list for 2004. Several months later, however, one of his coworkers revealed that the photographs were fraudulent. According to the coworker, the photographs came from a computer data bank of stem cell photographs, not from a cloning experiment. A university panel investigated the results and confirmed that the photographs and other data had indeed been faked. Dr. Hwang was forced to resign his prestigious post at the university.

Although not common, incidents like this do occur from time to time. They are damaging to a community that is largely built on trust. A scientist's peers (other researchers in similar fields) review all published research papers, but usually they are judging whether the data support the conclusion—they assume that the experimental measurements are authentic. The pressure to succeed sometimes leads researchers to betray that trust. However, over time, the tendency of scientists to reproduce and build upon one another's work results in the discovery of the fraudulent data. When that happens, the researchers at fault are usually banished from the community and their careers are ruined.

---

 **Solving Chemical Problems**

## 1.8  Solving Chemical Problems

Learning to solve problems is one of the most important skills you will acquire in this course. No one succeeds in chemistry—or in life, really—without the ability to solve problems. Although no simple formula applies to every chemistry problem, you can learn problem-solving strategies and begin to develop some chemical intuition. Many of the problems you will solve in this course are *unit conversion problems*, where you are given one or more quantities and asked to convert them into different units. Other problems require that you use *specific equations* to get to the information you are trying to find. In the sections that follow, you will find strategies to help you solve both of these types of problems. Of course, many problems contain both conversions and equations, requiring the combination of these strategies, and some problems may require an altogether different approach.

### Converting from One Unit to Another

In Section 1.6, we discussed the SI unit system, the prefix multipliers, and a few other units. Knowing how to work with and manipulate these units in calculations is central to solving chemical problems. In calculations, units help to determine correctness. Using units as a guide to solving problems is called **dimensional analysis**. Units should always be included in calculations; they are multiplied, divided, and canceled like any other algebraic quantity.

Consider converting 12.5 inches (in) to centimeters (cm). We know from Table 1.3 that 1 in = 2.54 cm (exact), so we can use this quantity in the calculation.

$$12.5 \text{ in} \times \frac{2.54 \text{ cm}}{1 \text{ in}} = 31.8 \text{ cm}$$

The unit, in, cancels, and we are left with cm as our final unit. The quantity $\frac{2.54 \text{ cm}}{1 \text{ in}}$ is a **conversion factor**—a fractional quantity with the units we are *converting from* on the bottom and the units we are *converting to* on the top. Conversion factors are constructed from any two equivalent quantities. In this example, 2.54 cm = 1 in, so we construct the conversion factor by dividing both sides of the equality by 1 in and canceling the units.

$$2.54 \text{ cm} = 1 \text{ in}$$

$$\frac{2.54 \text{ cm}}{1 \text{ in}} = \frac{1 \text{ in}}{1 \text{ in}}$$

$$\frac{2.54 \text{ cm}}{1 \text{ in}} = 1$$

Because the quantity $\frac{2.54 \text{ cm}}{1 \text{ in}}$ is equivalent to 1, multiplying by the conversion factor affects only the units, not the actual quantity. To convert the other way, from centimeters to inches, we must—using units as a guide—use a different form of the conversion factor. If we accidentally use the same form, we will get the wrong result, indicated by erroneous units. For example, suppose that we want to convert 31.8 cm to inches.

$$31.8 \text{ cm} \times \frac{2.54 \text{ cm}}{1 \text{ in}} = \frac{80.8 \text{ cm}^2}{\text{in}}$$

The units in the answer ($\text{cm}^2/\text{in}$), as well as the value of the answer, are obviously wrong. When we solve a problem, we always look at the final units. Are they the desired units? We must always look at the magnitude of the numerical answer as well. Does it make sense? In this case, our mistake was the form of the conversion factor. It should have been inverted so that the units cancel as follows:

$$31.8 \text{ cm} \times \frac{1 \text{ in}}{2.54 \text{ cm}} = 12.5 \text{ in}$$

We can invert conversion factors because they are equal to 1 and the inverse of 1 is 1. Therefore,

$$\frac{2.54 \text{ cm}}{1 \text{ in}} = 1 = \frac{1 \text{ in}}{2.54 \text{ cm}}$$

Most unit conversion problems take the following form:

$$\text{Information given} \times \text{conversion factor(s)} = \text{information sought}$$

$$\text{Given unit} \times \frac{\text{desired unit}}{\text{given unit}} = \text{desired unit}$$

In this book, we diagram problem solutions using a *conceptual plan*. A conceptual plan is a visual outline that helps us to see the general flow of the problem solution. For unit conversions, the conceptual plan focuses on units and the conversion from one unit to another. The conceptual plan for converting in to cm is:

The conceptual plan for converting the other way, from cm to in, is just the reverse, with the reciprocal conversion factor:

Each arrow in a conceptual plan for a unit conversion has an associated conversion factor, with the units of the previous step in the denominator and the units of the following step in the numerator. In the following section, we incorporate the idea of a conceptual plan into an overall approach to solving numerical chemical problems.

## General Problem-Solving Strategy

In this book, we introduce a standard problem-solving procedure that you can adapt to many of the problems encountered in general chemistry and beyond. One of the difficulties beginning students encounter when trying to solve problems in general chemistry is not knowing where to begin. To solve any problem, you must assess the information given in the problem and devise a way to determine the requested information. In other words, you must do the following:

- Identify the starting point (the *given* information).
- Identify the end point (what you must *find*).
- Devise a way to get from the starting point to the end point using what is given as well as what you already know or can look up. (As we just discussed, this is the *conceptual plan*.)

In graphic form, we can represent this progression as:

$$\text{Given} \longrightarrow \text{Conceptual Plan} \longrightarrow \text{Find}$$

Although no problem-solving procedure is applicable to all problems, the following four-step procedure can be helpful in working through many of the numerical problems you will encounter in this book.

1. **Sort.** Begin by sorting the information in the problem. *Given* information is the basic data provided by the problem—often one or more numbers with their associated units. *Find* indicates what information you will need for your answer.

2. **Strategize.** This is usually the most challenging part of solving a problem. In this process, you must develop a *conceptual plan*—a series of steps that take you from the given information to the information you are trying to find. You have already seen conceptual plans for simple unit conversion problems. Each arrow in a conceptual plan represents a computational step. On the left side of the arrow is the quantity you had before the step, on the right side of the arrow is the quantity you have after the step, and below the arrow is the information you need to get from one to the other—the relationship between the quantities.

   Often such relationships take the form of conversion factors or equations. These may be given in the problem, in which case you will have written them down under "Given" in step 1. Usually, however, you will need other information—which may include physical constants, formulas, or conversion factors—to help get you from what you are given to what you must find. This information comes from what you have learned or can look up in the chapter or in tables within the book.

   In some cases, you may get stuck at the strategize step. If you cannot figure out how to get from the given information to the information you are asked to find, you might try working backward. For example, you can look at the units of the quantity you are trying to find and attempt to find conversion factors to get to the units of the given quantity. You may even try a combination of strategies; work forward, backward, or some of both. If you persist, you will develop a strategy to solve the problem.

3. **Solve.** This is the easiest part of solving a problem. Once you set up the problem properly and devise a conceptual plan, you simply follow the plan to solve the problem. Carry out any mathematical operations (paying attention to the rules for significant figures in calculations) and cancel units as needed.

4. **Check.** This is the step beginning students most often overlook. Experienced problem solvers always ask, "Does this answer make sense? Are the units correct? Is the number of significant figures correct?" When solving multistep problems,

> Most problems can be solved in more than one way. The solutions in this book tend to be the most straightforward but certainly are not the only way to solve the problem.

errors can easily creep into the solution. You can catch most of these errors by simply checking the answer. For example, suppose you are calculating the number of atoms in a gold coin and end up with an answer of $1.1 \times 10^{-6}$ atoms. Could the gold coin really be composed of one-millionth of one atom?

In Examples 1.7 and 1.8, this problem-solving procedure is applied to unit conversion problems. The procedure is summarized in the left column, and two examples of applying the procedure are provided in the middle and right columns. This three-column format is used in selected examples throughout this text. It allows you to see how to apply a particular procedure to two different problems. Work through one problem first (from top to bottom) and then see how you can apply the same procedure to the other problem. Recognizing the commonalities and differences between problems is a key part of developing problem-solving skills.

## Procedure for...

| Solving Unit Conversion Problems | **Example 1.7** Unit Conversion<br>Convert 1.76 yards to centimeters. | **Example 1.8** Unit Conversion<br>Convert 1.8 quarts to cubic centimeters. |
|---|---|---|
| **SORT** Begin by sorting the information in the problem into *given* and *find*. | GIVEN: 1.76 yd<br>FIND: cm | GIVEN: 1.8 qt<br>FIND: $cm^3$ |
| **STRATEGIZE** Devise a *conceptual plan* for the problem. Begin with the *given* quantity and symbolize each conversion step with an arrow. Below each arrow, write the appropriate conversion factor for that step. Focus on the units. The conceptual plan should end at the *find* quantity and its units. In these examples, the other information you need consists of relationships between the various units. | CONCEPTUAL PLAN<br><br>yd → m → cm<br>$\dfrac{1\ m}{1.094\ yd}$  $\dfrac{1\ cm}{10^{-2}\ m}$<br><br>RELATIONSHIPS USED<br>1.094 yd = 1 m<br>$1\ cm = 10^{-2}\ m$<br>(These conversion factors are from Tables 1.2 and 1.3.) | CONCEPTUAL PLAN<br><br>qt → L → mL → $cm^3$<br>$\dfrac{1\ L}{1.057\ qt}$  $\dfrac{1\ mL}{10^{-3}\ L}$  $\dfrac{1\ cm^3}{1\ mL}$<br><br>RELATIONSHIPS USED<br>1.057 qt = 1 L<br>$1\ mL = 10^{-3}\ L$<br>$1\ mL = 1\ cm^3$<br>(These conversion factors are from Tables 1.2 and 1.3.) |
| **SOLVE** Follow the conceptual plan. Begin with the *given* quantity and its units. Multiply by the appropriate conversion factor(s), canceling units, to arrive at the *find* quantity.<br><br>Round the answer to the correct number of significant figures following the rules in Section 1.7. Remember that exact conversion factors do not limit significant figures. | SOLUTION<br>$1.76\ yd \times \dfrac{1\ m}{1.094\ yd} \times \dfrac{1\ cm}{10^{-2}\ m}$<br>$= 160.8775\ cm$<br><br>160.8775 cm = 161 cm | SOLUTION<br>$1.8\ qt \times \dfrac{1\ L}{1.057\ qt} \times \dfrac{1\ mL}{10^{-3}\ L} \times \dfrac{1\ cm^3}{1\ mL}$<br>$= 1.70293 \times 10^3\ cm^3$<br><br>$1.70293 \times 10^3\ cm^3 = 1.7 \times 10^3\ cm^3$ |
| **CHECK** Check your answer. Are the units correct? Does the answer make sense? | The units (cm) are correct. The magnitude of the answer (161) makes sense because a centimeter is a much smaller unit than a yard. | The units ($cm^3$) are correct. The magnitude of the answer (1700) makes sense because a cubic centimeter is a much smaller unit than a quart. |
| | **FOR PRACTICE 1.7**<br>Convert 288 cm to yards. | **FOR PRACTICE 1.8**<br>Convert 9255 $cm^3$ to gallons. |

## Units Raised to a Power

When building conversion factors for units raised to a power, remember to raise both the number and the unit to the power. For example, to convert from in$^2$ to cm$^2$, we construct the conversion factor as follows:

$$2.54 \text{ cm} = 1 \text{ in}$$
$$(2.54 \text{ cm})^2 = (1 \text{ in})^2$$
$$(2.54)^2 \text{ cm}^2 = 1^2 \text{ in}^2$$
$$6.45 \text{ cm}^2 = 1 \text{ in}^2$$
$$\frac{6.45 \text{ cm}^2}{1 \text{ in}^2} = 1$$

Example 1.9 demonstrates how to use conversion factors involving units raised to a power.

---

**Example** 1.9 | **Unit Conversions Involving Units Raised to a Power** ▶

Calculate the displacement (the total volume of the cylinders through which the pistons move) of a 5.70-L automobile engine in cubic inches.

| | |
|---|---|
| **SORT** Sort the information in the problem into *given* and *find*. | **GIVEN:** 5.70 L<br><br>**FIND:** in$^3$ |
| **STRATEGIZE** Write a conceptual plan. Begin with the given information and devise a path to the information that you are asked to find. For cubic units, you must cube the conversion factors. | **CONCEPTUAL PLAN**<br><br><br><br>**RELATIONSHIPS USED**<br><br>$1 \text{ mL} = 10^{-3} \text{ L}$<br>$1 \text{ mL} = 1 \text{ cm}^3$<br>$2.54 \text{ cm} = 1 \text{ in}$<br>(These conversion factors are from Tables 1.2 and 1.3.) |
| **SOLVE** Follow the conceptual plan to solve the problem. Round the answer to three significant figures to reflect the three significant figures in the least precisely known quantity (5.70 L). These conversion factors are all exact and therefore do not limit the number of significant figures. | **SOLUTION**<br><br>$5.70 \text{ L} \times \dfrac{1 \text{ mL}}{10^{-3} \text{ L}} \times \dfrac{1 \text{ cm}^3}{1 \text{ mL}} \times \dfrac{(1 \text{ in})^3}{(2.54 \text{ cm})^3}$<br><br>$= 347.835 \text{ in}^3 = 348 \text{ in}^3$ |

**CHECK** The units of the answer are correct, and the magnitude makes sense. The unit cubic inches is smaller than liters, so the volume in cubic inches should be larger than the volume in liters.

**FOR PRACTICE 1.9** How many cubic centimeters are there in 2.11 yd$^3$?

**FOR MORE PRACTICE 1.9** A vineyard has 145 acres of Chardonnay grapes. A particular soil supplement requires 5.50 g for every square meter of vineyard. How many kilograms of the soil supplement are required for the entire vineyard? (1 km$^2$ = 247 acres)

---

**Example** 1.10 | **Density as a Conversion Factor**

The mass of fuel in a jet must be calculated before each flight to ensure that the jet is not too heavy to fly. A 747 is fueled with 173,231 L of jet fuel. If the density of the fuel is 0.768 g/cm$^3$, what is the mass of the fuel in kilograms?

| | |
|---|---|
| **SORT** Begin by sorting the information in the problem into *given* and *find*. | **GIVEN:** fuel volume = 173,231 L<br><br>density of fuel = 0.768 g/cm$^3$<br><br>**FIND:** mass in kg |

**STRATEGIZE** Draw the conceptual plan by beginning with the given quantity—in this case the volume in liters (L). The overall goal of this problem is to find the mass. You can convert between volume and mass using density ($g/cm^3$). However, you must first convert the volume to $cm^3$. After you convert the volume to $cm^3$, use the density to convert to g. Finally, convert g to kg.

**CONCEPTUAL PLAN**

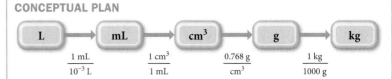

| L | mL | cm³ | g | kg |
|---|----|-----|---|----|
| | $\dfrac{1\ mL}{10^{-3}\ L}$ | $\dfrac{1\ cm^3}{1\ mL}$ | $\dfrac{0.768\ g}{cm^3}$ | $\dfrac{1\ kg}{1000\ g}$ |

**RELATIONSHIPS USED**

$1\ mL = 10^{-3}\ L$

$1\ mL = 1\ cm^3$

$d = 0.768\ g/cm^3$

$1000\ g = 1\ kg$

(These conversion factors are from Tables 1.2 and 1.3.)

**SOLVE** Follow the conceptual plan to solve the problem. Round the answer to three significant figures to reflect the three significant figures in the density.

**SOLUTION**

$$173{,}231\ \text{L} \times \frac{1\ \text{mL}}{10^{-3}\ \text{L}} \times \frac{1\ \text{cm}^3}{1\ \text{mL}} \times \frac{0.768\ \text{g}}{1\ \text{cm}^3} \times \frac{1\ \text{kg}}{1000\ \text{g}}$$

$$= 1.33 \times 10^5\ \text{kg}$$

**CHECK** The units of the answer (kg) are correct. The magnitude makes sense because the mass ($1.33 \times 10^5$ kg) is similar in magnitude to the given volume (173,231 L or $1.73231 \times 10^5$ L), as expected for a density close to 1 (0.768 g/cm$^3$).

**FOR PRACTICE 1.10**  Backpackers often use canisters of white gas to fuel a cooking stove's burner. If one canister contains 1.45 L of white gas and the density of the gas is 0.710 g/cm$^3$, what is the mass of the fuel in kilograms?

**FOR MORE PRACTICE 1.10**  A drop of gasoline has a mass of 22 mg and a density of 0.754 g/cm$^3$. What is its volume in cubic centimeters?

## Order-of-Magnitude Estimations

Calculation is an integral part of chemical problem solving. But precise numerical calculation is not always necessary, or even possible. Sometimes data are only approximate; other times we may not need a high degree of precision—a rough estimate or a simplified "back of the envelope" calculation is enough. We can also use approximate calculations to get an initial feel for a problem or to make a quick check to see whether our solution is "in the right ballpark."

One way to make such estimates is to simplify the numbers so that they can be manipulated easily. The technique known as *order-of-magnitude estimation* is based on focusing only on the exponential part of numbers written in scientific notation, according to these guidelines:

- If the decimal part of the number is less than 5, just drop it. Thus, $4.36 \times 10^5$ becomes $10^5$ and $2.7 \times 10^{-3}$ becomes $10^{-3}$.

- If the decimal part is 5 or more, round it up to 10 and rewrite the number as a power of 10. Thus, $5.982 \times 10^7$ becomes $10 \times 10^7 = 10^8$, and $6.1101 \times 10^{-3}$ becomes $10 \times 10^{-3} = 10^{-2}$

When we make these approximations, we are left with powers of 10, which are easier to multiply and divide. Of course, our answer is only as reliable as the numbers used to get it, so we should not assume that the results of an order-of-magnitude calculation are accurate to more than an order of magnitude.

Suppose, for example, that we want to estimate the number of atoms that an immortal being could have counted in the 14 billion ($1.4 \times 10^{10}$) years that the universe has been in existence, assuming a counting rate of ten atoms per second. Since a year has $3.2 \times 10^7$ seconds, we can approximate the number of atoms counted as follows:

$$\underset{\substack{\text{(number of years)}}}{10^{10}\ \text{years}} \times \underset{\substack{\text{(number of seconds} \\ \text{per year)}}}{10^7\,\frac{\text{seconds}}{\text{years}}} \times \underset{\substack{\text{(number of atoms} \\ \text{counted per second)}}}{10^1\,\frac{\text{atoms}}{\text{second}}} \approx 10^{18}\ \text{atoms}$$

A million trillion atoms ($10^{18}$) may seem like a lot, but as we discuss in Chapter 2, a speck of matter made up of a million trillion atoms is nearly impossible to see without a microscope.

In our general problem-solving procedure, the last step is to check whether the results seem reasonable. Order-of-magnitude estimations often help us catch the kinds of mistakes that can happen in a detailed calculation, such as entering an incorrect exponent or sign into a calculator, or multiplying when we should have divided.

## Problems Involving an Equation

We can solve problems involving equations in much the same way as problems involving conversions. Usually, in problems involving equations, we must find one of the variables in the equation, given the others. The *conceptual plan* concept outlined earlier can be used for problems involving equations. For example, suppose we are given the mass ($m$) and volume ($V$) of a sample and asked to calculate its density. The conceptual plan shows how the *equation* takes us from the *given* quantities to the *find* quantity:

$$d = \frac{m}{V}$$

Here, instead of a conversion factor under the arrow, this conceptual plan has an equation. The equation shows the *relationship* between the quantities on the left of the arrow and the quantities on the right. Note that at this point the equation need not be solved for the quantity on the right (although in this particular case it is). The procedure that follows, as well as the two examples, offer guidance for developing a strategy to solve problems involving equations. We again use the three-column format. Work through one problem from top to bottom and then see how you can apply the same general procedure to the second problem.

## Procedure for...

### Solving Problems Involving Equations

| | | |
|---|---|---|
| | **Example 1.11** <br> **Problems with Equations** | **Example 1.12** ▶ <br> **Problems with Equations** |
| | Find the radius ($r$) in centimeters of a spherical water droplet with a volume ($V$) of 0.058 cm³. For a sphere, $V = (4/3)\,\pi r^3$. | Find the density (in g/cm³) of a metal cylinder with a mass ($m$) of 8.3 g, a length ($l$) of 1.94 cm, and a radius ($r$) of 0.55 cm. For a cylinder, $V = \pi r^2 l$. |
| **SORT** Begin by sorting the information in the problem into *given* and *find*. | **GIVEN:** $V = 0.058 \text{ cm}^3$ <br> **FIND:** $r$ in cm | **GIVEN:** $m = 8.3 \text{ g}$ <br> $l = 1.94 \text{ cm}$ <br> $r = 0.55 \text{ cm}$ <br><br> **FIND:** $d$ in g/cm³ |
| **STRATEGIZE** Write a conceptual plan for the problem. Focus on the equation(s). The conceptual plan shows how the equation takes you from the *given* quantity (or quantities) to the *find* quantity. The conceptual plan may have several parts, involving other equations or required conversions. In these examples, you use the geometrical relationships given in the problem statements as well as the definition of density, $d = m/V$, which you learned in this chapter. | **CONCEPTUAL PLAN** <br>  <br> $$V = \frac{4}{3}\,\pi r^3$$ <br><br> **RELATIONSHIPS USED** <br> $$V = \frac{4}{3}\,\pi r^3$$ | **CONCEPTUAL PLAN** <br>  <br> $$V = \pi r^2 l$$ <br><br>  <br> $$d = m/V$$ <br><br> **RELATIONSHIPS USED** <br> $$V = \pi r^2 l$$ <br> $$d = \frac{m}{V}$$ |

| **SOLVE** Follow the conceptual plan. Solve the equation(s) for the *find* quantity (if it is not already). Gather each of the quantities that must go into the equation in the correct units. (Convert to the correct units if necessary.) Substitute the numerical values and their units into the equation(s) and calculate the answer. | **SOLUTION**<br><br>$V = \dfrac{4}{3}\pi r^3$<br><br>$r^3 = \dfrac{3}{4\pi}V$<br><br>$r = \left(\dfrac{3}{4\pi}V\right)^{1/3}$<br><br>$= \left(\dfrac{3}{4\pi}0.058\ \text{cm}^3\right)^{1/3}$<br><br>$= 0.24013\ \text{cm}$ | **SOLUTION**<br><br>$V = \pi r^2 l$<br><br>$= \pi(0.55\ \text{cm})^2(1.94\ \text{cm})$<br><br>$= 1.8436\ \text{cm}^3$<br><br>$d = \dfrac{m}{V}$<br><br>$= \dfrac{8.3\ \text{g}}{1.8436\ \text{cm}^3} = 4.50195\ \text{g/cm}^3$ |
| Round the answer to the correct number of significant figures. | $0.24013\ \text{cm} = 0.24\ \text{cm}$ | $4.50195\ \text{g/cm}^3 = 4.5\ \text{g/cm}^3$ |
| **CHECK** Check your answer. Are the units correct? Does the answer make sense? | The units (cm) are correct and the magnitude makes sense. | The units ($\text{g/cm}^3$) are correct. The magnitude of the answer seems correct for one of the lighter metals (see Table 1.4). |
| | **FOR PRACTICE 1.11** Find the radius ($r$) of an aluminum cylinder that is 2.00 cm long and has a mass of 12.4 g. For a cylinder, $V = \pi r^2 l$. | **FOR PRACTICE 1.12** Find the density, in $\text{g/cm}^3$, of a metal cube with a mass of 50.3 g and an edge length ($l$) of 2.65 cm. For a cube, $V = l^3$. |

# Self-Assessment Quiz

**Q1.** A chemist mixes sodium with water and witnesses a violent reaction between the metal and water. This is best classified as

a) an observation.     b) a law.
c) a hypothesis.     d) a theory.

**Q2.** This image represents a particulate view of a sample of matter. Classify the sample according to its composition.

a) The sample is a pure element.
b) The sample is a homogeneous mixture.
c) The sample is a compound.
d) The sample is a heterogeneous mixture.

**Q3.** Which change is a physical change?

a) wood burning
b) iron rusting
c) dynamite exploding
d) gasoline evaporating

**Q4.** Which property of rubbing alcohol is a chemical property?

a) density ($0.786\ \text{g/cm}^3$)
b) flammability
c) boiling point ($82.5\ °\text{C}$)
d) melting point ($-89\ °\text{C}$)

**Q5.** Convert $85.0\ °\text{F}$ to K.

a) $181.1\ \text{K}$    b) $358\ \text{K}$    c) $29.4\ \text{K}$    d) $302.6\ \text{K}$

**Q6.** Express the quantity $33.2 \times 10^{-4}\ \text{m}$ in mm.

a) $33.2\ \text{mm}$              b) $3.32\ \text{mm}$
c) $0.332\ \text{mm}$         d) $3.32 \times 10^{-6}\ \text{mm}$

**Q7.** What is the mass of a 1.75 L sample of a liquid that has a density of $0.921\ \text{g/mL}$?

a) $1.61 \times 10^3\ \text{g}$       b) $1.61 \times 10^{-3}\ \text{g}$
c) $1.90 \times 10^3\ \text{g}$       d) $1.90 \times 10^{-3}\ \text{g}$

**Q8.** Perform the calculation to the correct number of significant figures.

$$(43.998 \times 0.00552)/2.002$$

a) $0.121$    b) $0.12$    c) $0.12131$    d) $0.1213$

**Q9.** Perform the calculation to the correct number of significant figures.

$$(8.01 - 7.50)/3.002$$

a) $0.1698867$   b) $0.17$    c) $0.170$    d) $0.1700$

**Q10.** Convert $1285\ \text{cm}^2$ to $\text{m}^2$.

a) $1.285 \times 10^7\ \text{m}^2$       b) $12.85\ \text{m}^2$
c) $0.1285\ \text{m}^2$           d) $1.285 \times 10^5\ \text{m}^2$

*Continued—*

**Q11.** The first diagram depicts a compound in its liquid state. Which of the other diagrams best depicts the compound after it has evaporated into a gas?

(a)

(b)

(c)

(d)

**Q12.** Three samples, each of a different substance, are weighed and their volume is measured. The results are tabulated. List the substances in order of decreasing density.

|              | Mass    | Volume   |
|--------------|---------|----------|
| Substance I  | 10.0 g  | 10.0 mL  |
| Substance II | 10.0 kg | 12.0 L   |
| Substance III| 12.0 mg | 10.0 $\mu$L |

a) III > II > I
b) I > II > III
c) III > I > II
d) II > I > III

**Q13.** A solid metal sphere has a radius of 3.53 cm and a mass of 1.796 kg. What is the density of the metal in g/cm$^3$? (The volume of a sphere is $V = \frac{4}{3}\pi r^3$.)

a) 34.4 g/cm$^3$
b) 0.103 g/cm$^3$
c) 121 g/cm$^3$
d) 9.75 g/cm$^3$

**Q14.** The gas mileage of a certain German automobile is 22 km/L. Convert this quantity to miles per gallon.

a) 9.4 mi/gal
b) 1.3 × 10$^2$ mi/gal
c) 52 mi/gal
d) 3.6 mi/gal

**Q15.** A wooden block has a volume of 18.5 in$^3$. Express the volume of the cube in cm$^3$.

a) 303 cm$^3$
b) 47.0 cm$^3$
c) 1.13 cm$^3$
d) 7.28 cm$^3$

**Answers: 1.** (a) **2.** (c) **3.** (d) **4.** (b) **5.** (d) **6.** (b) **7.** (a) **8.** (a) **9.** (b) **10.** (c) **11.** (a) **12.** (c) **13.** (d) **14.** (c) **15.** (a)

# Chapter 1 IN REVIEW

## Key Terms

### Section 1.1
atoms (2)
molecules (2)
chemistry (2)

### Section 1.2
hypothesis (3)
experiment (3)
scientific law (3)
law of conservation
  of mass (3)
theory (4)
atomic theory (4)

### Section 1.3
matter (5)
substance (5)
state (5)
composition (5)
solid (6)
liquid (6)
gas (6)
crystalline (6)
amorphous (6)
pure substance (8)
mixture (8)
element (8)

compound (8)
heterogeneous mixture (8)
homogeneous mixture (8)
decanting (8)
distillation (8)
volatile (8)
filtration (8)

### Section 1.4
physical change (9)
chemical change (10)
physical property (10)
chemical property (10)

## Key Concepts

### Atoms and Molecules (1.1)

▶ All matter is composed of atoms and molecules.
▶ Chemistry is the science that investigates the properties of matter by examining the atoms and molecules that compose it.

### The Scientific Approach to Knowledge (1.2)

▶ Science begins with the observation of the physical world. A number of related observations can be summarized in a statement or generalization called a scientific law.
▶ A hypothesis is a tentative interpretation or explanation of observations. One or more well-established hypotheses may prompt the development of a scientific theory, a model for nature that explains the underlying reasons for observations and laws.
▶ Laws, hypotheses, and theories all give rise to predictions that can be tested by experiments, carefully controlled procedures designed to produce critical new observations. If scientists cannot confirm the predictions, they must modify or replace the law, hypothesis, or theory.

### The Classification of Matter (1.3)

▶ We classify matter according to its state (solid, liquid, or gas) or according to its composition (pure substance or mixture).
▶ A pure substance can either be an element, which cannot be chemically broken down into simpler substances, or a compound, which is composed of two or more elements in fixed proportions.
▶ A mixture can be either homogeneous, with the same composition throughout, or heterogeneous, with different compositions in different regions.

### The Properties of Matter (1.4)

▶ We classify the properties of matter into two types: physical and chemical. Matter displays its physical properties without changing its composition.
▶ Changes in matter in which composition does not change are physical changes. Changes in matter in which composition does change are chemical changes.

### Energy (1.5)

▶ In chemical and physical changes, matter often exchanges energy with its surroundings. In these exchanges, the total energy is always conserved; energy is neither created nor destroyed.
▶ Systems with high potential energy tend to change in the direction of lower potential energy, releasing energy into the surroundings.

### The Units of Measurement and Significant Figures (1.6, 1.7)

▶ Scientists use SI units, which are based on the metric system. The SI base units include the meter (m) for length, the kilogram (kg) for mass, the second (s) for time, and the kelvin (K) for temperature.
▶ Derived units are formed from a combination of other units. Common derived units include those for volume (cm³ or m³) and density (g/cm³).
▶ The number of digits in a reported measurement reflects the uncertainty in the measurement. Significant figures are the non–place-holding digits in a reported number.

## Key Equations and Relationships

Relationship between Kelvin (K) and Celsius (°C) Temperature Scales (1.6)

$$K = °C + 273.15$$

Relationship between Celsius (°C) and Fahrenheit (°F) Temperature Scales (1.6)

$$°C = \frac{(°F - 32)}{1.8}$$

Relationship between Density (*d*), Mass (*m*), and Volume (*V*) (1.6)

$$d = \frac{m}{V}$$

# Key Learning Outcomes

| CHAPTER OBJECTIVES | ASSESSMENT |
|---|---|
| Determining Physical and Chemical Changes and Properties (**1.4**) | Example 1.1    For Practice 1.1    Exercises 43–50 |
| Converting between the Temperature Scales: Fahrenheit, Celsius, and Kelvin (**1.6**) | Example 1.2    For Practice 1.2    Exercises 51–54 |
| Calculating the Density of a Substance (**1.6**) | Example 1.3    For Practice 1.3    For More Practice 1.3    Exercises 65–68 |
| Reporting Scientific Measurements to the Correct Digit of Uncertainty (**1.7**) | Example 1.4    For Practice 1.4    Exercises 73, 74 |
| Working with Significant Figures (**1.7**) | Examples 1.5, 1.6    For Practice 1.5, 1.6    Exercises 77, 78, 80, 83–88 |
| Using Conversion Factors (**1.8**) | Examples 1.7, 1.8, 1.9, 1.10    For Practice 1.7, 1.8, 1.9, 1.10    For More Practice 1.9, 1.10    Exercises 93, 94, 98–101, 103, 104 |
| Solving Problems Involving Equations (**1.8**) | Examples 1.11, 1.12    For Practice 1.11, 1.12    Exercises 119, 120 |

# EXERCISES

## Review Questions

1. Explain this statement in your own words and give an example. *The properties of the substances around us depend on the atoms and molecules that compose them.*

2. Explain the main goal of chemistry.

3. Describe the scientific approach to knowledge. How does it differ from other approaches?

4. Explain the differences between a hypothesis, a law, and a theory.

5. What observations did Antoine Lavoisier make? What law did he formulate?

6. What theory did John Dalton formulate?

7. What is wrong with the expression "That is just a theory," if by *theory* the speaker is referring to a scientific theory?

8. What are two different ways to classify matter?

9. How do solids, liquids, and gases differ?

10. What is the difference between a crystalline solid and an amorphous solid?

11. Explain the difference between a pure substance and a mixture.

12. Explain the difference between an element and a compound.

13. Explain the difference between a homogeneous and a heterogeneous mixture.

14. What kind of mixtures can be separated by filtration?

15. Explain how distillation is used to separate mixtures.

16. What is the difference between a physical property and a chemical property?

17. What is the difference between a physical change and a chemical change? List some examples of each.

18. Explain the significance of the law of conservation of energy.

19. What kind of energy is chemical energy? In what way is an elevated weight similar to a tank of gasoline?

20. What are the standard SI base units of length, mass, time, and temperature?

21. What are the three common temperature scales? Does the size of a degree differ among them?

22. What are prefix multipliers? List some examples.

23. What is a derived unit? List an example.

24. Explain the difference between density and mass.

25. Explain the difference between *intensive* and *extensive* properties.

26. What is the meaning of the number of digits reported in a measured quantity?

27. When multiplying or dividing measured quantities, what determines the number of significant figures in the result?

28. When adding or subtracting measured quantities, what determines the number of significant figures in the result?

29. What are the rules for rounding off the results of calculations?

30. Explain the difference between precision and accuracy.

31. Explain the difference between random error and systematic error.

32. What is dimensional analysis?

## Problems by Topic

*Note: Answers to all odd-numbered Problems, numbered in blue, can be found in Appendix III. Exercises in the Problems by Topic section are paired, with each odd-numbered problem followed by a similar even-numbered problem. Exercises in the Cumulative Problems section are also paired, but more loosely. Challenge Problems and Conceptual Problems, because of their nature, are unpaired.*

### The Scientific Approach to Knowledge

33. Classify each statement as an observation, a law, or a theory.
    a. All matter is made of tiny, indestructible particles called atoms.
    b. When iron rusts in a closed container, the mass of the container and its contents does not change.

c. In chemical reactions, matter is neither created nor destroyed.

d. When a match burns, heat is released.

**34.** Classify each statement as an observation, a law, or a theory.

a. Chlorine is a highly reactive gas.

b. If elements are listed in order of increasing mass of their atoms, their chemical reactivities follow a repeating pattern.

c. Neon is an inert (or nonreactive) gas.

d. The reactivity of elements depends on the arrangement of their electrons.

**35.** A chemist decomposes several samples of carbon monoxide into carbon and oxygen and weighs the resultant elements. The results are shown in the table.

| Sample | Mass of Carbon (g) | Mass of Oxygen (g) |
|--------|--------------------|--------------------|
| 1 | 6 | 8 |
| 2 | 12 | 16 |
| 3 | 18 | 24 |

a. Do you notice a pattern in these results?

Next, the chemist decomposes several samples of hydrogen peroxide into hydrogen and oxygen. The results are shown in the table.

| Sample | Mass of Hydrogen (g) | Mass of Oxygen (g) |
|--------|----------------------|--------------------|
| 1 | 0.5 | 8 |
| 2 | 1 | 16 |
| 3 | 1.5 | 24 |

b. Do you notice a similarity between these results and those for carbon monoxide in part a?

c. Can you formulate a law from your observations in a and b?

d. Can you formulate a hypothesis that might explain your law in c?

**36.** When astronomers observe distant galaxies, they can tell that most of them are moving away from one another. In addition, the more distant the galaxies, the more rapidly they are likely to be moving away from each other. Can you devise a hypothesis to explain these observations?

## The Classification and Properties of Matter

**37.** Classify each substance as a pure substance or a mixture. If it is a pure substance, classify it as an element or a compound. If it is a mixture, classify it as homogeneous or heterogeneous.

a. sweat

b. carbon dioxide

c. aluminum

d. vegetable soup

**38.** Classify each substance as a pure substance or a mixture. If it is a pure substance, classify it as an element or a compound. If it is a mixture, classify it as homogeneous or heterogeneous.

a. wine     b. beef stew

c. iron     d. carbon monoxide

**39.** Complete the table.

| Substance | Pure or mixture | Type |
|-----------|-----------------|------|
| aluminum | pure | element |
| apple juice | _____ | _____ |
| hydrogen peroxide | _____ | _____ |
| chicken soup | _____ | _____ |

**40.** Complete the table.

| Substance | Pure or mixture | Type |
|-----------|-----------------|------|
| water | pure | compound |
| coffee | _____ | _____ |
| ice | _____ | _____ |
| carbon | _____ | _____ |

**41.** Determine whether each molecular diagram represents a pure substance or a mixture. If it represents a pure substance, classify the substance as an element or a compound. If it represents a mixture, classify the mixture as homogeneous or heterogeneous.

(a)                    (b)

(c)                    (d)

**42.** Determine whether each molecular diagram represents a pure substance or a mixture. If it represents a pure substance, classify the substance as an element or a compound. If it represents a mixture, classify the mixture as homogeneous or heterogeneous.

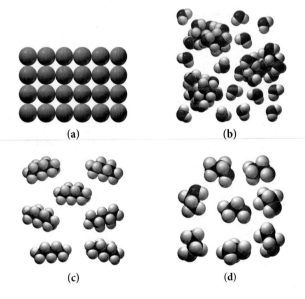

(a)                    (b)

(c)                    (d)

**43.** Classify each of the listed properties of isopropyl alcohol (also known as rubbing alcohol) as physical or chemical.

a. colorless                b. flammable

c. liquid at room temperature   d. density = 0.79 g/mL

e. mixes with water

**44.** Classify each of the listed properties of ozone (a pollutant in the lower atmosphere but part of a protective shield against UV light in the upper atmosphere) as physical or chemical.
  **a.** bluish color
  **b.** pungent odor
  **c.** very reactive
  **d.** decomposes on exposure to ultraviolet light
  **e.** gas at room temperature

**45.** Classify each property as physical or chemical.
  **a.** the tendency of ethyl alcohol to burn
  **b.** the shine on silver
  **c.** the odor of paint thinner
  **d.** the flammability of propane gas

**46.** Classify each property as physical or chemical.
  **a.** the boiling point of ethyl alcohol
  **b.** the temperature at which dry ice evaporates
  **c.** the tendency of iron to rust
  **d.** the color of gold

**47.** Classify each change as physical or chemical.
  **a.** Natural gas burns in a stove.
  **b.** The liquid propane in a gas grill evaporates because the valve was left open.
  **c.** The liquid propane in a gas grill burns in a flame.
  **d.** A bicycle frame rusts on repeated exposure to air and water.

**48.** Classify each change as physical or chemical.
  **a.** Sugar burns when heated in a skillet.
  **b.** Sugar dissolves in water.
  **c.** A platinum ring becomes dull because of continued abrasion.
  **d.** A silver surface becomes tarnished after exposure to air for a long period of time.

**49.** Based on the molecular diagram, classify each change as physical or chemical.

**(a)**

**(b)**

**(c)**

**50.** Based on the molecular diagram, classify each change as physical or chemical.

**(a)**            **(b)**

**(c)**

### Units in Measurement

**51.** Convert each temperature.
  **a.** 32 °F to °C (temperature at which water freezes)
  **b.** 77 K to °F (temperature of liquid nitrogen)
  **c.** −109 °F to °C (temperature of dry ice)
  **d.** 98.6 °F to K (body temperature)

**52.** Convert each temperature.
  **a.** 212 °F to °C (temperature of boiling water at sea level)
  **b.** 22 °C to K (approximate room temperature)
  **c.** 0.00 K to °F (coldest temperature possible, also known as absolute zero)
  **d.** 2.735 K to °C (average temperature of the universe as measured from background black body radiation)

**53.** The coldest temperature ever measured in the United States is −80 °F, recorded on January 23, 1971, in Prospect Creek, Alaska. Convert that temperature to °C and K. (Assume that −80 °F is precise to two significant figures.)

**54.** The warmest temperature ever measured in the United States is 134 °F, recorded on July 10, 1913, in Death Valley, California. Convert that temperature to °C and K.

**55.** Use the prefix multipliers to express each measurement without exponents.
  **a.** $1.2 \times 10^{-9}$ m       **b.** $22 \times 10^{-15}$ s
  **c.** $1.5 \times 10^{9}$ g        **d.** $3.5 \times 10^{6}$ L

**56.** Use prefix multipliers to express each measurement without exponents.
  **a.** $38.8 \times 10^{5}$ g      **b.** $55.2 \times 10^{-10}$ s
  **c.** $23.4 \times 10^{11}$ m     **d.** $87.9 \times 10^{-7}$ L

**57.** Use scientific notation to express each quantity with only base units (no prefix multipliers).
  **a.** 4.5 ns    **b.** 18 fs    **c.** 128 pm    **d.** 35 $\mu$m

**58.** Use scientific notation to express each quantity with only base units (no prefix multipliers).
  **a.** 35 $\mu$L    **b.** 225 Mm    **c.** 133 Tg    **d.** 1.5 cg

**59.** Complete the table.

| | | |
|---|---|---|
| **a.** 1245 kg | $1.245 \times 10^6$ g | $1.245 \times 10^9$ mg |
| **b.** 515 km | _____dm | _____cm |
| **c.** 122.355 s | _____ms | _____ks |
| **d.** 3.345 kJ | _____J | _____mJ |

**60.** Complete the table.

| | | |
|---|---|---|
| **a.** 355 km/s | _____cm/s | _____m/ms |
| **b.** 1228 g/L | _____g/mL | _____kg/mL |
| **c.** 554 mK/s | _____K/s | _____$\mu$K/ms |
| **d.** 2.554 mg/mL | _____g/L | _____$\mu$g/mL |

**61.** Express the quantity 254,998 m in each unit.
  **a.** km     **b.** Mm     **c.** mm     **d.** cm

**62.** Express the quantity $556.2 \times 10^{-12}$ s in each unit.
  **a.** ms     **b.** ns     **c.** ps     **d.** fs

**63.** How many 1-cm squares would it take to construct a square that is 1 m on each side?

**64.** How many 1-cm cubes would it take to construct a cube that is 4 cm on edge?

## Density

**65.** A new penny has a mass of 2.49 g and a volume of 0.349 $cm^3$. Is the penny made of pure copper? Explain your answer.

**66.** A titanium bicycle frame displaces 0.314 L of water and has a mass of 1.41 kg. What is the density of the titanium in $g/cm^3$?

**67.** Glycerol is a syrupy liquid often used in cosmetics and soaps. A 3.25 L sample of pure glycerol has a mass of $4.10 \times 10^3$ g. What is the density of glycerol in $g/cm^3$?

**68.** A supposedly gold nugget displaces 19.3 mL of water and has a mass of 371 g. Could the nugget be made of gold?

**69.** Ethylene glycol (antifreeze) has a density of 1.11 $g/cm^3$.
  **a.** What is the mass in g of 417 mL of ethylene glycol?
  **b.** What is the volume in L of 4.1 kg of ethylene glycol?

**70.** Acetone (nail polish remover) has a density of 0.7857 $g/cm^3$.
  **a.** What is the mass in g of 28.56 mL of acetone?
  **b.** What is the volume in mL of 6.54 g of acetone?

**71.** A small airplane takes on 245 L of fuel. If the density of the fuel is 0.821 g/mL, what mass of fuel has the airplane taken on?

**72.** Human fat has a density of 0.918 $g/cm^3$. How much volume (in $cm^3$) is gained by a person who gains 10.0 lb of pure fat?

## The Reliability of a Measurement and Significant Figures

**73.** Read each measurement to the correct number of significant figures. Laboratory glassware should always be read from the bottom of the meniscus.

(a)          (b)          (c)

**74.** Read each measurement to the correct number of significant figures. Laboratory glassware should always be read from the bottom of the meniscus. Digital balances normally display mass

to the correct number of significant figures for that particular balance.

(a)          (b)          (c)

**75.** For each number, underline the zeroes that are significant and draw an **x** through the zeroes that are not.
  **a.** 1,050,501 km          **b.** 0.0020 m
  **c.** 0.000000000000002 s     **d.** 0.001090 cm

**76.** For each number, underline the zeroes that are significant and draw an **x** through the zeroes that are not.
  **a.** 180,701 mi          **b.** 0.001040 m
  **c.** 0.005710 km          **d.** 90,201 m

**77.** How many significant figures are in each number?
  **a.** 0.000312 m   **b.** 312,000 s   **c.** $3.12 \times 10^5$ km
  **d.** 13,127 s     **e.** 2000

**78.** How many significant figures are in each number?
  **a.** 0.1111 s     **b.** 0.007 m     **c.** 108,700 km
  **d.** $1.563300 \times 10^{11}$ m     **e.** 30,800

**79.** Which numbers are exact (and therefore have an unlimited number of significant figures)?
  **a.** $\pi = 3.14$
  **b.** 12 in = 1 ft
  **c.** EPA gas mileage rating of 26 miles per gallon
  **d.** 1 gross = 144

**80.** Indicate the number of significant figures in each number. If the number is an exact number, indicate an unlimited number of significant figures.
  **a.** 318,857,056 (2014 U.S. population)     **b.** 2.54 cm = 1 in
  **c.** 11.4 $g/cm^3$ (density of lead)          **d.** 12 = 1 dozen

**81.** Round each number to four significant figures.
  **a.** 156.852     **b.** 156.842     **c.** 156.849     **d.** 156.899

**82.** Round each number to three significant figures.
  **a.** 79,845.82          **b.** $1.548937 \times 10^7$
  **c.** 2.3499999995          **d.** 0.000045389

## Significant Figures in Calculations

**83.** Calculate to the correct number of significant figures.
  **a.** $9.15 \div 4.970$          **b.** $1.54 \times 0.03060 \times 0.69$
  **c.** $27.5 \times 1.82 \div 100.04$     **d.** $(2.290 \times 10^6) \div (6.7 \times 10^4)$

**84.** Calculate to the correct number of significant figures.
  **a.** $89.3 \times 77.0 \times 0.08$     **b.** $(5.01 \times 10^5) \div (7.8 \times 10^2)$
  **c.** $4.005 \times 74 \times 0.007$     **d.** $453 \div 2.031$

**85.** Calculate to the correct number of significant figures.
  **a.** $43.7 - 2.341$          **b.** $17.6 + 2.838 + 2.3 + 110.77$
  **c.** $19.6 + 58.33 - 4.974$     **d.** $5.99 - 5.572$

**86.** Calculate to the correct number of significant figures.
  **a.** $0.004 + 0.09879$          **b.** $1239.3 + 9.73 + 3.42$
  **c.** $2.4 - 1.777$          **d.** $532 + 7.3 - 48.523$

**87.** Calculate to the correct number of significant figures.
  **a.** $(24.6681 \times 2.38) + 332.58$
  **b.** $(85.3 - 21.489) \div 0.0059$
  **c.** $(512 \div 986.7) + 5.44$
  **d.** $[(28.7 \times 10^5) \div 48.533] + 144.99$

**88.** Calculate to the correct number of significant figures.
   **a.** $[(1.7 \times 10^6) \div (2.63 \times 10^5)] + 7.33$
   **b.** $(568.99 - 232.1) \div 5.3$
   **c.** $(9443 + 45 - 9.9) \times 8.1 \times 10^6$
   **d.** $(3.14 \times 2.4367) - 2.34$

**89.** A flask containing 11.7 mL of a liquid weighs 132.8 g with the liquid in the flask and 124.1 g when empty. Calculate the density of the liquid in g/mL to the correct number of significant digits.

**90.** A flask containing 9.55 mL of a liquid weighs 157.2 g with the liquid in the flask and 148.4 g when empty. Calculate the density of the liquid in g/mL to the correct number of significant digits.

## Unit Conversions

**91.** Perform each unit conversion.
   **a.** 27.8 L to $cm^3$    **b.** 1898 mg to kg    **c.** 198 km to cm

**92.** Perform each unit conversion.
   **a.** 28.9 nm to $\mu m$    **b.** 1432 $cm^3$ to L    **c.** 1211 Tm to Gm

**93.** Perform each unit conversion.
   **a.** 154 cm to in    **b.** 3.14 kg to g
   **c.** 3.5 L to qt    **d.** 109 mm to in

**94.** Perform each unit conversion.
   **a.** 1.4 in to mm    **b.** 116 ft to cm
   **c.** 1845 kg to lb    **d.** 815 yd to km

**95.** A runner wants to run 10.0 km. Her running pace is 7.5 mi per hour. How many minutes must she run?

**96.** A cyclist rides at an average speed of 18 mi per hour. If she wants to bike 212 km, how long (in hours) must she ride?

**97.** A certain European automobile has a gas mileage of 17 km/L. What is the gas mileage in miles per gallon?

**98.** A gas can holds 5.0 gal of gasoline. Express this quantity in $cm^3$.

**99.** A house has an area of 195 $m^2$. What is its area in each unit?
   **a.** $km^2$    **b.** $dm^2$    **c.** $cm^2$

**100.** A bedroom has a volume of 115 $m^3$. What is its volume in each unit?
   **a.** $km^3$    **b.** $dm^3$    **c.** $cm^3$

**101.** The average U.S. farm occupies 435 acres. How many square miles is this? (1 acre = 43,560 $ft^2$, 1 mile = 5280 ft)

**102.** Total U.S. farmland occupies 954 million acres. How many square miles is this? (1 acre = 43,560 $ft^2$, 1 mi = 5280 ft). Total U.S. land area is 3.537 million square miles. What percentage of U.S. land is farmland?

**103.** An acetaminophen suspension for infants contains 80 mg/0.80 mL suspension. The recommended dose is 15 mg/kg body weight. How many mL of this suspension should be given to an infant weighing 14 lb? (Assume two significant figures.)

**104.** An ibuprofen suspension for infants contains 100 mg/5.0 mL suspension. The recommended dose is 10 mg/kg body weight. How many mL of this suspension should be given to an infant weighing 18 lb? (Assume two significant figures.)

# Cumulative Problems

**105.** There are exactly 60 seconds in a minute, exactly 60 minutes in an hour, exactly 24 hours in a mean solar day, and 365.24 solar days in a solar year. How many seconds are in a solar year? Give your answer with the correct number of significant figures.

**106.** Determine the number of picoseconds in 2.0 hours.

**107.** Classify each property as intensive or extensive.
   **a.** volume    **b.** boiling point
   **c.** temperature   **d.** electrical conductivity    **e.** energy

**108.** At what temperatures are the readings on the Fahrenheit and Celsius thermometers the same?

**109.** Suppose you design a new thermometer called the X thermometer. On the X scale the boiling point of water is 130 °X, and the freezing point of water is 10 °X. At what temperature are the readings on the Fahrenheit and X thermometers the same?

**110.** On a new Jekyll temperature scale, water freezes at 17 °J and boils at 97 °J. On another new temperature scale, the Hyde scale, water freezes at 0 °H and boils at 120 °H. If methyl alcohol boils at 84 °H, what is its boiling point on the Jekyll scale?

**111.** Force is defined as mass times acceleration. Starting with SI base units, derive a unit for force. Using SI prefixes, suggest a convenient unit for the force resulting from a collision with a 10-ton trailer truck moving at 55 mi per hour and for the force resulting from the collision of a molecule of mass around $10^{-20}$ kg moving almost at the speed of light ($3 \times 10^8$ m/s) with the wall of its container. (Assume a 1-second deceleration time for both collisions.)

**112.** A temperature measurement of 25 °C has three significant figures, while a temperature measurement of −196 °C has only two significant figures. Explain.

**113.** Do each calculation without your calculator and give the answers to the correct number of significant figures.

   **a.** $1.76 \times 10^{-3}/8.0 \times 10^2$
   **b.** $1.87 \times 10^{-2} + 2 \times 10^{-4} - 3.0 \times 10^{-3}$
   **c.** $[(1.36 \times 10^5)(0.000322)/0.082](129.2)$

**114.** The value of the euro was recently $1.15 U.S., and the price of 1 liter of gasoline in France is 1.42 euro. What is the price of 1 gallon of gasoline in U.S. dollars in France?

**115.** A thief uses a can of sand to replace a solid gold cylinder that sits on a weight-sensitive, alarmed pedestal. The can of sand and the gold cylinder have exactly the same dimensions (length = 22 and radius = 3.8 cm).
   **a.** Calculate the mass of each cylinder (ignore the mass of the can itself). (density of gold = 19.3 g/$cm^3$, density of sand = 3.00 g/$cm^3$)
   **b.** Does the thief set off the alarm? Explain.

**116.** The proton has a radius of approximately $1.0 \times 10^{-13}$ cm and a mass of $1.7 \times 10^{-24}$ g. Determine the density of a proton for a sphere $V = (4/3) \pi r^3$.

**117.** The density of titanium is 4.51 g/$cm^3$. What is the volume (in cubic inches) of 3.5 lb of titanium?

**118.** The density of iron is 7.86 g/$cm^3$. What is its density in pounds per cubic inch (lb/$in^3$)?

**119.** A steel cylinder has a length of 2.16 in, a radius of 0.22 in, and a mass of 41 g. What is the density of the steel in g/$cm^3$?

**120.** A solid aluminum sphere has a mass of 85 g. Use the density of aluminum to find the radius of the sphere in inches.

**121.** A backyard swimming pool holds 185 cubic yards ($yd^3$) of water. What is the mass of the water in pounds?

**122.** An iceberg has a volume of 7655 $ft^2$. What is the mass of the ice (in kg) composing the iceberg (at 0 °C)?

123. The Toyota Prius, a hybrid electric vehicle, has an EPA gas mileage rating of 52 mi/gal in the city. How many kilometers can the Prius travel on 15 L of gasoline?

124. The Honda Insight, a hybrid electric vehicle, has an EPA gas mileage rating of 41 mi/gal in the city. How many kilometers can the Insight travel on the amount of gasoline that would fit in a soda can? The volume of a soda can is 355 mL.

125. The single proton that forms the nucleus of the hydrogen atom has a radius of approximately $1.0 \times 10^{-13}$ cm. The hydrogen atom itself has a radius of approximately 52.9 pm. What fraction of the space within the atom is occupied by the nucleus?

126. A sample of gaseous neon atoms at atmospheric pressure and 0 °C contains $2.69 \times 10^{22}$ atoms per liter. The atomic radius of neon is 69 pm. What fraction of the space do the atoms themselves occupy? What does this reveal about the separation between atoms in the gaseous phase?

127. The diameter of a hydrogen atom is 212 pm. Find the length in kilometers of a row of $6.02 \times 10^{23}$ hydrogen atoms. The diameter of a ping pong ball is 4.0 cm. Find the length in kilometers of a row of $6.02 \times 10^{23}$ ping pong balls.

128. The world's record in the 100-m dash is 9.69 s, and in the 100-yd dash it is 9.21 s. Find the speed in mi/hr of the runners who set these records. (Assume three significant figures for 100 m and 100 yd.)

129. Table salt contains 39.33 g of sodium per 100 g of salt. The U.S. Food and Drug Administration (FDA) recommends that adults consume less than 2.40 g of sodium per day. A particular snack mix contains 1.25 g of salt per 100 g of the mix. What mass of the snack mix can an adult consume and still be within the FDA limit? (Assume three significant figures for 100 g.)

130. Lead metal can be extracted from a mineral called galena, which contains 86.6% lead by mass. A particular ore contains 68.5% galena by mass. If the lead can be extracted with 92.5% efficiency, what mass of ore is required to make a lead sphere with a 5.00-cm radius?

131. A length of #8 copper wire (radius = 1.63 mm) has a mass of 24.0 kg and a resistance of 2.061 ohm per km ($\Omega$/km). What is the overall resistance of the wire?

132. Rolls of aluminum foil are 304 mm wide and 0.016 mm thick. What maximum length of aluminum foil can be made from 1.10 kg of aluminum?

133. Liquid nitrogen has a density of 0.808 g/mL and boils at 77 K. Researchers often purchase liquid nitrogen in insulated 175 L tanks. The liquid vaporizes quickly to gaseous nitrogen (which has a density of 1.15 g/L at room temperature and atmospheric pressure) when the liquid is removed from the tank. Suppose that all 175 L of liquid nitrogen in a tank accidentally vaporized in a lab that measured 10.00 m × 10.00 m × 2.50 m. What maximum fraction of the air in the room could be displaced by the gaseous nitrogen?

134. Mercury is often used in thermometers. The mercury sits in a bulb on the bottom of the thermometer and rises up a thin capillary as the temperature rises. Suppose a mercury thermometer contains 3.380 g of mercury and has a capillary that is 0.200 mm in diameter. How far does the mercury rise in the capillary when the temperature changes from 0.0 °C to 25.0 °C? The density of mercury at these temperatures is 13.596 g/cm³ and 13.534 g/cm³, respectively.

# Challenge Problems

135. A force of $2.31 \times 10^4$ N is applied to a diver's face mask that has an area of 125 cm². Find the pressure in atm on the face mask.

136. The SI unit of force is the newton, derived from the base units by using the definition of force, $F = ma$. The dyne is a non-SI unit of force in which mass is measured in grams and time is measured in seconds. The relationship between the two units is 1 dyne = $10^{-5}$ N. Find the unit of length used to define the dyne.

137. Kinetic energy can be defined as $\frac{1}{2}mv^2$ or as $\frac{3}{2}PV$. Show that the derived SI units of each of these terms are those of energy. (Pressure is force/area and force is mass × acceleration.)

138. In 1999, scientists discovered a new class of black holes with masses 100 to 10,000 times the mass of our sun that occupy less space than our moon. Suppose that one of these black holes has a mass of $1 \times 10^3$ suns  and a radius equal to one-half the radius of our moon. What is the density of the black hole in g/cm³? The radius of our sun is $7.0 \times 10^5$ km and it has an average density of $1.4 \times 10^3$ kg/m³. The diameter of the moon is $2.16 \times 10^3$ mi.

139. Suppose that polluted air has carbon monoxide (CO) levels of 15.0 ppm. An average human inhales about 0.50 L of air per breath and takes about 20 breaths per minute. How many milligrams of carbon monoxide does the average person inhale in an 8-hour period at this level of carbon monoxide pollution?

Assume that the carbon monoxide has a density of 1.2 g/L. (Hint: 15.0 ppm CO means 15.0 L CO per $10^6$ L air.)

140. Nanotechnology, the field of building ultrasmall structures one atom at a time, has progressed in recent years. One potential application of nanotechnology is the construction of artificial cells. The simplest cells would probably mimic red blood cells, the body's oxygen transporters. Nanocontainers, perhaps constructed of carbon, could be pumped full of oxygen and injected into a person's bloodstream. If the person needed additional oxygen—due to a heart attack perhaps, or for the purpose of space travel—these containers could slowly release oxygen into the blood, allowing tissues that would otherwise die to remain alive. Suppose that the nanocontainers were cubic and had an edge length of 25 nm.

   a. What is the volume of one nanocontainer? (Ignore the thickness of the nanocontainer's wall.)

   b. Suppose that each nanocontainer could contain pure oxygen pressurized to a density of 85 g/L. How many grams of oxygen could each nanocontainer contain?

   c. Air typically contains about 0.28 g of oxygen per liter. An average human inhales about 0.50 L of air per breath and takes about 20 breaths per minute. How many grams of oxygen does a human inhale per hour? (Assume two significant figures.)

   d. What is the minimum number of nanocontainers that a person would need in his or her bloodstream to provide 1 hour's worth of oxygen?

   e. What is the minimum volume occupied by the number of nanocontainers calculated in part d? Is such a volume feasible, given that total blood volume in an adult is about 5 L?

**141.** Approximate the percent increase in waist size that occurs when a 155-lb person gains 40.0 lb of fat. Assume that the volume of the person can be modeled by a cylinder that is 4.0 ft tall. The average density of a human is about 1.0 g/cm³, and the density of fat is 0.918 g/cm³.

**142.** A box contains a mixture of small copper spheres and small lead spheres. The total volume of both metals is measured by the displacement of water to be 427 cm³, and the total mass is 4.36 kg. What percentage of the spheres are copper?

# Conceptual Problems

**143.** A volatile liquid (one that easily evaporates) is put into a jar, and the jar is then sealed. Does the mass of the sealed jar and its contents change upon the vaporization of the liquid?

**144.** The diagram shown first represents solid carbon dioxide, also known as dry ice. Which of the other diagrams best represents the dry ice after it has sublimed into a gas?

(a)        (b)        (c)

**145.** A cube has an edge length of 7 cm. If it is divided into 1-cm cubes, how many 1-cm cubes are there?

**146.** Substance A has density of 1.7 g/cm³. Substance B has a density of 1.7 kg/m³. Without doing any calculations, determine which substance is more dense.

**147.** For each box, examine the blocks attached to the balances. Based on their positions and sizes, determine which block is more dense (the dark block or the lighter-colored block), or if

the relative densities cannot be determined. (Think carefully about the information being shown.)

(a)        (b)        (c)

**148.** Let a triangle represent atoms of element A and a circle represent atoms of element B.
  **a.** Draw an atomic level view of a homogeneous mixture of elements A and B.
  **b.** Draw an atomic view of the compound AB in a liquid state (molecules close together).
  **c.** Draw an atomic view of the compound AB after it has undergone a physical change (such as evaporation).
  **d.** Draw an atomic view of the compound after it has undergone a chemical change (such as decomposition of AB into A and B).

**149.** Identify each statement as being most like an observation, a law, or a theory.
  **a.** All coastal areas experience two high tides and two low tides each day.
  **b.** The tides in Earth's oceans are caused mainly by the gravitational attraction of the moon.
  **c.** Yesterday, high tide in San Francisco Bay occurred at 2:43 A.M. and 3:07 P.M.
  **d.** Tides are higher at the full moon and new moon than at other times of the month.

*Active Classroom Learning*

# Questions for Group Work

Discuss these questions with the group and record your consensus answer.

**150.** Using white and black circles to represent different kinds of atoms, make a drawing that accurately represents each sample of matter: a solid element, a liquid compound, and a heterogeneous mixture. Make a drawing (clearly showing *before* and *after*) depicting your liquid compound undergoing a physical change. Make a drawing depicting your solid element undergoing a chemical change.

**151.** Look up the measurement of the approximate thickness of a human hair.
  **a.** Convert the measurement to an SI unit (if it isn't already).
  **b.** Write it in scientific notation.
  **c.** Write it without scientific notation.
  **d.** Write it with an appropriate prefix on a base unit.

Now repeat these steps using the distance from Earth to the sun.

**152.** The following statements are all true.
  **a.** Jessica's house is 5 km from the grocery store.
  **b.** Jessica's house is 4.73 km from the grocery store.
  **c.** Jessica's house is 4.73297 km from the grocery store.

How can all the statements be true? What does the number of digits in each statement communicate? What sort of device would Jessica need to make the measurement in each statement?

**153.** One inch is equal to 2.54 cm. Draw a line that is 1 in long, and mark the centimeters on the line. Draw a cube that is 1 in on each side. Draw lines on each face of the cube that are 1 cm apart. How many cubic centimeters are there in 1 in³?

**154.** Convert the height of each member in your group from feet and inches to meters. Once you have your heights in meters, calculate the sum of all the heights. Use appropriate rules for significant figures at each step.

# Data Interpretation and Analysis

## Density of Water

155. The graph in Figure a ▼ plots the density of water as a function of temperature at a pressure of 1.00 atm. The graph in Figure b ▼ plots the density of water as a function of temperature over a narrower range, between 0 °C and 10 °C.

▲ **FIGURE a** **Density of Water as a Function of Temperature at 1.00 atm.**

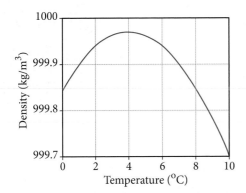

▲ **FIGURE b** **Density of Water as a Function of Temperature Between 0 °C and 10 °C.**

Use the information provided in the figures to answer the following questions:

a. Compare the density of solid ice at 0 °C to the density of liquid water at 4 °C. Use this information to explain why ice floats on water and why ice forms on the top of ponds and lakes in winter in places such as Wisconsin and Minnesota, allowing fish to survive in the water below the ice.

b. Observations by the Hubble Space Telescope estimate Pluto's mean density is between 1.8 g/cm$^3$ and 2.1 g/cm$^3$. In 2006 the International Astronomical Union reclassified Pluto from a planet to a dwarf planet in part due to Pluto's low density. By comparison, the mean density of Earth's moon is 3,346.4 kg/m$^3$. The crust of Earth's moon is composed primarily of rock containing oxygen, silicon, magnesium, iron, calcium, and aluminum. Given this information, answer the following questions:

   i. Does the mean density of Pluto support a claim that Pluto is mostly ice or does it indicate that Pluto is mostly rock?

   ii. The temperature of Pluto ranges from −233 °C to −223 °C. A cold winter day in Antarctica is −65.1 °C. How does the density of H$_2$O on Pluto compare to the density of H$_2$O in Antarctica? (Extrapolate data as needed.)

c. Antarctica contains 26.5 million cubic kilometers of ice. Assume that the average temperature of this ice during the summer months is −20 °C. If all of this ice melted, how many km$^3$ of liquid water would form at 4 °C?

d. Pluto has a mass of 1.27 × 10$^{22}$ kg and contains about 30% ice. Assume that the density of this ice is the same as it is at −175 °C. If humans traveled to Pluto on a spacecraft and were able to remove all of the ice from Pluto, and melt the ice, how many km$^3$ of liquid water would form at 4 °C? Is there more water on Antarctica or on Pluto?

# Answers to Conceptual Connections

## Laws and Theories

**1.1 (b)** A law only summarizes a series of related observations; a theory gives the underlying reasons for them.

## Pure Substances and Mixtures

**1.2**

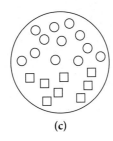

(a)        (b)        (c)

## Chemical and Physical Changes

**1.3** View **(a)** best represents the water after vaporization. Vaporization is a physical change, so the molecules must remain the same before and after the change.

## The Mass of a Gas

**1.4** No, the mass does not change. The water vaporizes and becomes a gas, but the water molecules are still present within the flask and have the same mass.

## Prefix Multipliers

**1.5** The prefix micro (10$^{-6}$) is appropriate. The measurement would be reported as 55.7 $\mu$m.

## Density

**1.6 (c)** The copper sample expands. However, because its mass remains constant while its volume increases, its density decreases.

CHAPTER **2**

# Atoms and Elements

These observations have tacitly led to the conclusion which seems universally adopted, that all bodies of sensible magnitude . . . are constituted of a vast number of extremely small particles, or atoms of matter . . . .

—JOHN DALTON (1766–1844)

If you cut a piece of graphite from the tip of a pencil into smaller and smaller pieces, how far could you go? Could you divide it forever? Would you eventually run into some basic particles that were no longer divisible, not because of their sheer smallness, but because of the nature of matter? This fundamental question about the nature of matter has been asked by thinkers for over two millennia. Their answers have varied over time. On the scale of everyday objects, matter appears continuous, or infinitely divisible. And until about 200 years ago, many scientists thought that matter was indeed continuous—but they were proven wrong. If you were to divide the graphite from your pencil tip into smaller and smaller pieces (far smaller than the eye could see), you would eventually end up with individual carbon atoms. The word *atom* comes from the Greek *atomos*, meaning "indivisible." You cannot divide a carbon atom into smaller pieces and still have carbon. Atoms compose all ordinary matter—if you want to understand matter, you must begin by understanding atoms.

▲ Scottish botanist Robert Brown observed random motion in tiny particles suspended in water. This motion later confirmed the particulate nature of matter.

## 2.1 Brownian Motion: Atoms Comfirmed

In 1827, Scottish botanist Robert Brown (1773–1858) looked through his microscope at water-suspended particles that had come from pollen grains. He noticed that the pollen particles were in continuous motion. Any one particle traveled a random, jittery path through the liquid water. Brown initially thought that the particles might be alive and were perhaps the male sexual cells of plants (similar to sperm). However, similar particles from plants long dead exhibited the same jittery motion, and so did the dust from pulverized stones. Brown concluded that the source of the motion must not come from the particles themselves. What was causing this motion?

The definitive answer to this question did not come until 1905, when Albert Einstein (1879–1955) developed a theory that quantitatively explained what was by then called *Brownian motion*. Einstein's model explained the motion Brown had observed as a result of the molecular bombardments of the particles due to the thermal energy of the surrounding water. In other words, the water molecules in liquid water—constantly in motion due to thermal energy—were continuously battering the pollen and dust particles, causing them to oscillate and move. In Einstein's model, the jittering pollen particles are like a beach ball that is thrown into a crowd at a graduation ceremony. As the eager graduates strike the ball over and over again, the ball moves through the crowd in a jittery random path. The difference is that, in the case of Brownian motion, the "crowd" is composed of molecules much too small to see.

In 1908, French physicist Jean Perrin (1870–1942) conducted experimental measurements to test Einstein's model. His measurements confirmed that Einstein's model was valid. In 1926, Perrin was awarded the Nobel Prize in Physics. During the award speech, the presenter said, "the object of the researches of Professor Jean Perrin which have gained for him the Nobel Prize in Physics for 1926 was to put a definite end to the long struggle regarding the real existence of molecules." In other words, the work of Einstein, and then Perrin, removed any lingering doubt about the particulate nature of matter.

In Einstein's day, the existence of atoms was inferred from the jittery motion first witnessed by Brown. Today, with a type of microscope called a scanning tunneling microscope (STM), we can form images of atoms themselves. In fact, STM can be used to *pick up and move* individual atoms, allowing structures and patterns to be made one atom at a time. Figure 2.1◄, for example, shows the Kanji characters for the word "atom" written with individual iron atoms on top of a copper surface. If all of the words in the books in the Library of Congress—35 million books occupying 840 miles of shelves—were written in letters the size of these Kanji characters, they would fit into an area of about five square millimeters. Scientists at IBM have also succeeded in making a short video entitled *A Boy and His Atom*, in which the main character (a boy) is animated using a few dozen atoms. In the video, which has been viewed millions of times on YouTube, the boy plays with an atom like a real boy would play with a ball.

As we discussed in Chapter 1, it was only 200 years ago that John Dalton proposed his atomic theory, and about 100 years ago that the theory was confirmed through the work of Einstein and Perrin. Yet today we can image atoms, move them, and even build tiny machines out of just a few dozen atoms (an area of research called nanotechnology). These atomic machines, and the atoms that compose them, are almost unimaginably small. To get an idea of the size of an atom, imagine picking up a grain of sand at a beach. That grain contains more atoms than you could count in a lifetime. In fact, the number of atoms in one sand grain far exceeds the number of grains on an entire beach.

In spite of their small size, atoms are the key to connecting the macroscopic and microscopic worlds. An *atom* is the smallest identifiable unit of an *element*. There are about 91 different naturally occurring elements. In addition, scientists have succeeded in making over 20 synthetic elements (elements not found in nature). In this chapter, we learn about atoms: what they are made of, how they differ from one another, and how they are structured. We also learn about the elements that are composed of these different kinds of atoms and about some of their characteristic properties. We will also discuss how the elements can be organized in a way that reveals patterns in their properties and helps us to understand what underlies those properties.

▲ **FIGURE 2.1** **Imaging Atoms** The Japanese Kanji characters for "atom" written with iron atoms (red) on a copper surface (blue). The copper atoms are not as distinct as the iron atoms, but they appear as blue ripples in the background.

The exact number of naturally occurring elements is controversial because some elements that were first discovered when they were synthesized are believed to also be present in trace amounts in nature.

## 2.2    Early Ideas about the Building Blocks of Matter

The first people to propose that matter was composed of small, indestructible particles were Leucippus (fifth century B.C., exact dates unknown) and his student Democritus (460–370 B.C.). These Greek philosophers theorized that matter is ultimately composed of small, indivisible particles they named *atomos*. Democritus wrote, "Nothing exists except atoms and empty space; everything else is opinion." Leucippus and Democritus proposed that many different kinds of atoms exist, each different in shape and size, and that they move randomly through empty space. Other influential Greek thinkers of the time, such as Plato and Aristotle, did not embrace the atomic ideas of Leucippus and Democritus. Instead, Plato and Aristotle held that matter had no smallest parts and that different substances were composed of various proportions of fire, air, earth, and water. Since there was no experimental way to test the relative merits of the competing ideas, Aristotle's view prevailed, largely because he was so influential. The idea that matter is composed of atoms took a back seat in intellectual thought for nearly 2000 years.

In the sixteenth century, modern science began to emerge. A greater emphasis on observation led Nicolaus Copernicus (1473–1543) to publish *On the Revolution of the Heavenly Orbs* in 1543. The publication of that book—which proposed that the sun, not Earth, is at the center of the universe—marks the beginning of what we now call the *scientific revolution*. The next 200 years—and the work of scientists such as Francis Bacon (1561–1626), Johannes Kepler (1571–1630), Galileo Galilei (1564–1642), Robert Boyle (1627–1691), and Isaac Newton (1642–1727)—brought rapid advancement as the scientific approach became the established way to learn about the physical world. By the early 1800s, certain observations led the English chemist John Dalton (1766–1844) to offer convincing evidence that supported the early atomic ideas of Leucippus and Democritus. However, debate continued about whether atoms actually exist until the description of Brownian motion by Einstein in 1905 and subsequent experimental verification of the description in 1908 by Perrin (see Section 2.1).

## 2.3    Modern Atomic Theory and the Laws That Led to It

KCV    **Atomic Theory**

Recall the discussion of the scientific approach to knowledge from Chapter 1. The *atomic theory* (the idea that all matter is composed of atoms) grew out of observations and laws. The three most important laws that led to the development and acceptance of the atomic theory are the law of conservation of mass, the law of definite proportions, and the law of multiple proportions.

### The Law of Conservation of Mass

In 1789, as we saw in Chapter 1, Antoine Lavoisier formulated the **law of conservation of mass**, which states:

**In a chemical reaction, matter is neither created nor destroyed.**

In other words, when a chemical reaction occurs, the total mass of the substances involved in the reaction does not change. For example, consider the reaction between sodium and chlorine to form sodium chloride. The combined mass of the sodium and chlorine that react (the *reactants*) exactly equals the mass of the sodium chloride that forms (the *product*). This law is consistent with the idea that matter is composed of small, indestructible particles. The particles rearrange during a chemical reaction, but the amount of matter is conserved because the particles themselves are indestructible (at least by chemical means).

We will see in Chapter 20 that the law of conservation of mass is a slight oversimplification. However, the changes in mass in ordinary chemical processes are so minute that they can be ignored for all practical purposes.

$$Na(s) \qquad Cl_2(g) \qquad NaCl(s)$$

7.7 g Na     11.9 g $Cl_2$     19.6 g NaCl

Total mass = 19.6 g

Mass of reactants = Mass of product

CONCEPTUAL
**CONNECTION 2.1**

**The Law of Conservation of Mass** When a log completely burns in a campfire, the mass of the ash is much less than the mass of the log. What happens to the matter that composed the log?

## The Law of Definite Proportions

In 1797, the French chemist Joseph Proust (1754–1826) made observations on the composition of compounds. He found that the elements composing a given compound always occur in fixed (or definite) proportions in all samples of the compound. In contrast, the components of a mixture can be present in any proportions whatsoever. Proust summarized his observations in the **law of definite proportions**:

The law of definite proportions is sometimes called the law of constant composition.

> **All samples of a given compound, regardless of their source or how they were prepared, have the same proportions of their constituent elements.**

For example, the decomposition of 18.0 g of water results in 16.0 g of oxygen and 2.0 g of hydrogen, or an oxygen-to-hydrogen mass ratio of:

$$\text{Mass ratio} = \frac{16.0\,\text{g O}}{2.0\,\text{g H}} = 8.0 \text{ or } 8{:}1$$

This ratio holds for any sample of pure water, regardless of its origin. The law of definite proportions applies to every compound. Consider ammonia, a compound composed of nitrogen and hydrogen. Ammonia contains 14.0 g of nitrogen for every 3.0 g of hydrogen, resulting in a nitrogen-to-hydrogen mass ratio of 4.7.

$$\text{Mass ratio} = \frac{14.0\,\text{g N}}{3.0\,\text{g H}} = 4.7 \text{ or } 4.7{:}1$$

Again, this ratio is the same for every sample of ammonia. The law of definite proportions also hints at the idea that matter is composed of atoms. Compounds have definite proportions of their constituent elements because the atoms that compose them, each with its own specific mass, occur in a definite ratio. Since the ratio of atoms is the same for all samples of a particular compound, the ratio of masses is also the same.

---

**Example 2.1**  **Law of Definite Proportions**

Two samples of carbon dioxide are decomposed into their constituent elements. One sample produces 25.6 g of oxygen and 9.60 g of carbon, and the other produces 21.6 g of oxygen and 8.10 g of carbon. Show that these results are consistent with the law of definite proportions.

**SOLUTION**

| | |
|---|---|
| To show this, for both samples calculate the mass ratio of one element to the other by dividing the mass of one element by the mass of the other. For convenience, divide the larger mass by the smaller one. | For the first sample: $$\frac{\text{Mass oxygen}}{\text{Mass carbon}} = \frac{25.6}{9.60} = 2.67 \text{ or } 2.67{:}1$$ For the second sample: $$\frac{\text{Mass oxygen}}{\text{Mass carbon}} = \frac{21.6}{8.10} = 2.67 \text{ or } 2.67{:}1$$ |

The ratios are the same for the two samples, so these results are consistent with the law of definite proportions.

**FOR PRACTICE 2.1**  Two samples of carbon monoxide are decomposed into their constituent elements. One sample produces 17.2 g of oxygen and 12.9 g of carbon, and the other sample produces 10.5 g of oxygen and 7.88 g of carbon. Show that these results are consistent with the law of definite proportions.

Answers to For Practice and For More Practice Problems can be found in Appendix IV.

---

## The Law of Multiple Proportions

In 1804, John Dalton published his **law of multiple proportions**:

> **When two elements (call them A and B) form two different compounds, the masses of element B that combine with 1 g of element A can be expressed as a ratio of small whole numbers.**

Dalton suspected that matter was composed of atoms, so that when two elements A and B combine to form more than one compound, an atom of A combines with either one, two, three, or more atoms of B ($AB_1$, $AB_2$, $AB_3$, etc.). Therefore, the masses of B that react with a fixed mass of A are always related to one another as small whole-number ratios. Consider the compounds carbon monoxide and carbon dioxide. Carbon monoxide and carbon dioxide are two compounds composed of the same two elements: carbon and oxygen. We saw in Example 2.1 that the mass ratio of oxygen to carbon in carbon dioxide is 2.67:1; therefore, 2.67 g of oxygen reacts with 1 g of carbon. In carbon monoxide, however, the mass ratio of oxygen to carbon is 1.33:1, or 1.33 g of oxygen to every 1 g of carbon.

Carbon dioxide        Mass oxygen that combines with 1 g carbon = 2.67 g

Carbon monoxide        Mass oxygen that combines with 1 g carbon = 1.33 g

The ratio of these two masses is itself a small whole number.

$$\frac{\text{Mass oxygen to 1 g carbon in carbon dioxide}}{\text{Mass oxygen to 1 g carbon in carbon monoxide}} = \frac{2.67}{1.33} = 2$$

With the help of the molecular models in the margin, we can see why the ratio is 2:1—carbon dioxide contains two oxygen atoms to every carbon atom, while carbon monoxide contains only one. Of course, neither John Dalton nor Joseph Proust had access to any kind of modern instrumentation that could detect individual atoms—Dalton supported his atomic ideas primarily by using the *masses* of samples.

## Example 2.2    Law of Multiple Proportions

Nitrogen forms several compounds with oxygen, including nitrogen dioxide and dinitrogen monoxide. Nitrogen dioxide contains 2.28 g oxygen to every 1.00 g nitrogen, while dinitrogen monoxide contains 0.570 g oxygen to every 1.00 g nitrogen. Show that these results are consistent with the law of multiple proportions.

**SOLUTION**

| Calculate the ratio of the mass of oxygen from one compound to the mass of oxygen in the other. Always divide the larger of the two masses by the smaller one. | $\dfrac{\text{Mass oxygen to 1 g nitrogen in nitrogen dioxide}}{\text{Mass oxygen to 1 g nitrogen in dinitrogen monoxide}} = \dfrac{2.28}{0.570} = 4.00$ |
|---|---|

The ratio is a small whole number (4); these results are consistent with the law of multiple proportions.

**FOR PRACTICE 2.2**    Hydrogen and oxygen form both water and hydrogen peroxide. The decomposition of a sample of water forms 0.125 g hydrogen to every 1.00 g oxygen. The decomposition of a sample of hydrogen peroxide forms 0.0625 g hydrogen to every 1.00 g oxygen. Show that these results are consistent with the law of multiple proportions.

## John Dalton and the Atomic Theory

In 1808, John Dalton explained the laws we just discussed with his **atomic theory**:

In Section 2.6, we will see that, contrary to Dalton's theory, all atoms of a given element do not have exactly the same mass.

1. Each element is composed of tiny, indestructible particles called atoms.
2. All atoms of a given element have the same mass and other properties that distinguish them from the atoms of other elements.
3. Atoms combine in simple, whole-number ratios to form compounds.
4. Atoms of one element cannot change into atoms of another element. In a chemical reaction, atoms only change the way they are *bound together* with other atoms.

Today, the evidence for the atomic theory is overwhelming. Matter is indeed composed of atoms.

## Chemistry IN YOUR DAY | Atoms and Humans

You and I are composed of atoms. We get those atoms from the food we eat. Yesterday's cheeseburger contributes to today's skin, muscle, and hair. Not only are we made of atoms, but we are made of *recycled* atoms. The carbon atoms that compose our bodies were used by other living organisms before we got them. And they will be used by still others when we are done with them. In fact, it is likely that at this moment, your body contains some (over one trillion*) carbon atoms that were at one time part of your chemistry professor.

The idea that humans are composed of atoms acting in accord with the laws of chemistry and physics has significant implications and raises important questions. If atoms compose our brains, for example, do those atoms determine our thoughts and emotions? Are our feelings caused by atoms acting according to the laws of chemistry and physics?

*This calculation assumes that all of the carbon atoms metabolized by your professor over the last 40 years have been uniformly distributed into atmospheric carbon dioxide, and subsequently incorporated into the plants you have eaten.

Richard Feynman (1918–1988), a Nobel Prize–winning physicist, said that "The most important hypothesis in all of biology is that everything that animals do, atoms do. In other words, there is nothing that living things do that cannot be understood from the point of view that they are made of atoms acting according to the laws of physics." Indeed, biology has undergone a revolution in the last 50 years, mostly through investigation of the atomic and molecular basis for life. Some people have seen the atomic view of life as a devaluation of human life. We have always wanted to distinguish ourselves from everything else, and the idea that we are made of the same basic particles as all other matter takes something away from that distinction . . . or does it?

**QUESTIONS**

*Do you find the idea that you are made of recycled atoms disturbing? Why or why not? Reductionism is the idea that complex systems can be understood by understanding their parts. Is reductionism a good way to understand humans? Is it the only way?*

**The Laws of Definite and Multiple Proportions** Explain the difference between the law of definite proportions and the law of multiple proportions.

CONCEPTUAL
**CONNECTION 2.2**

## 2.4   The Discovery of the Electron

By the end of the nineteenth century, scientists were convinced that matter is made up of atoms, the permanent, supposedly indestructible building blocks that compose everything. However, further experiments revealed that the atom itself is composed of even smaller, more fundamental particles.

### Cathode Rays

In the late 1800s, an English physicist named J. J. Thomson (1856–1940), working at Cambridge University, performed experiments to probe the properties of **cathode rays**. Thomson constructed a partially evacuated glass tube called a **cathode ray tube**, shown in Figure 2.2▼. Thomson then applied a high electrical voltage between two electrodes at either end of the tube. He found that a beam of particles, called cathode rays, traveled from the negatively charged electrode (which is called the cathode) to the positively charged one (which is called the anode).

Thomson found that the particles that compose the cathode ray have the following properties: they travel in straight lines; they are independent of the composition of the material from which they originate (the cathode); and they carry a negative **electrical charge**. Electrical charge is a fundamental property of some of the particles that compose atoms and results in attractive and repulsive forces—called *electrostatic forces*—between those particles. The area around a charged particle where these forces exist is called an *electric field*. The characteristics of electrical charge are summarized in the figure in the margin. You have probably experienced excess electrical charge when brushing your hair on a dry day. The brushing action causes the accumulation of charged particles in your hair, which repel each other, making your hair stand on end.

**Properties of Electrical Charge**

Positive (red) and negative (yellow) electrical charges attract one another.

Positive charges repel one another. Negative charges repel one another.

+1  +  (−1)  =  0

Positive and negative charges of exactly the same magnitude sum to zero when combined.

For a full explanation of electrical voltage, see Chapter 19.

High voltage

◀ **FIGURE 2.2 Cathode Ray Tube**

J. J. Thomson measured the charge-to-mass ratio of the cathode ray particles by deflecting them using electric and magnetic fields, as shown in Figure 2.3▶, on the next page. The value he measured, $-1.76 \times 10^8$ coulombs (C) per gram, implied that the cathode ray particle was about 2000 times lighter (less massive) than hydrogen, the lightest known atom. These results were revolutionary—the indestructible atom could apparently be chipped!

J. J. Thomson had discovered the **electron**, a negatively charged, low-mass particle present within all atoms. He wrote, "We have in the cathode rays matter in a new state, a state in which the subdivision of matter is carried very much further . . . a state in which all matter . . . is of one and the same kind; this matter being the substance from which all the chemical elements are built up."

The coulomb (C) is the SI unit for charge.

▶ **FIGURE 2.3 Thomson's Measurement of the Charge-to-Mass Ratio of the Electron** J. J. Thomson used electric and magnetic fields to deflect the electron beam in a cathode ray tube. By measuring the strengths at which the effects of the two fields (electric and magnetic) cancel exactly, leaving the beam undeflected, he was able to calculate the charge-to-mass ratio of the electron.

**Charge-to-Mass Ratio of the Electron**

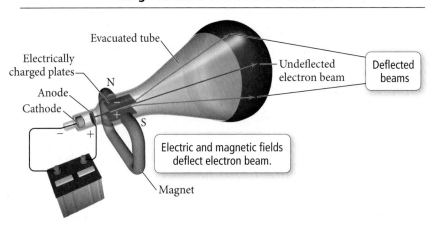

Evacuated tube

Electrically charged plates

Anode

Cathode

N

S

Undeflected electron beam

Deflected beams

Electric and magnetic fields deflect electron beam.

Magnet

# Millikan's Oil Drop Experiment: The Charge of the Electron

In 1909, American physicist Robert Millikan (1868–1953), working at the University of Chicago, performed his now famous oil drop experiment in which he deduced the charge of a single electron. The apparatus for the oil drop experiment is shown in Figure 2.4▼.

In his experiment, Millikan sprayed oil into fine droplets using an atomizer. The droplets were allowed to fall under the influence of gravity through a small hole into the lower portion of the apparatus where Millikan viewed them with the aid of a light source and a viewing microscope. During their fall, the drops acquired electrons Millikan had produced by bombarding the air in the chamber with ionizing radiation (a kind of energy described in Chapter 7). The electrons imparted a negative charge to the drops. In the lower portion of the apparatus, Millikan could create an electric field between two metal plates. Since the lower plate was negatively charged, and since Millikan could vary the strength of the electric field, he could slow or even reverse the free fall of the negatively charged drops. (Remember that like charges repel each other.)

By measuring the strength of the electric field required to halt the free fall of the drops, and by figuring out the masses of the drops themselves (determined from their radii and density), Millikan calculated the charge of each drop. He then reasoned that, since each drop must contain an integral (or whole) number of electrons, the charge of each drop must be a whole-number multiple of the electron's charge. Indeed, Millikan was correct; the measured charge on any drop is always a whole-number multiple of $-1.60 \times 10^{-19}$ C, the fundamental charge of a single electron.

▶ **FIGURE 2.4 Millikan's Measurement of the Electron's Charge** Millikan calculated the charge on oil droplets falling in an electric field. He found that it was always a whole-number multiple of $-1.60 \times 10^{-19}$ C, the charge of a single electron.

Atomizer

Positively charged plate

Ionizing radiation

Light source

Viewing microscope

Charged oil droplets are suspended in electric field.

Negatively charged plate

With this number in hand, and knowing Thomson's mass-to-charge ratio for electrons, we can deduce the mass of an electron:

$$\text{Charge} \times \frac{\text{mass}}{\text{charge}} = \text{mass}$$

$$-1.60 \times 10^{-19}\,\cancel{C} \times \frac{\text{g}}{-1.76 \times 10^{8}\,\cancel{C}} = 9.10 \times 10^{-28}\,\text{g}$$

As Thomson had correctly determined, this mass is about 2000 times lighter than hydrogen, the lightest atom.

Why did scientists work so hard to measure the charge of the electron? Since the electron is a fundamental building block of matter, scientists want to know its properties, including its charge. The magnitude of the charge of the electron is of tremendous importance because it determines how strongly an atom holds its electrons. Imagine how matter would be different if electrons had a much smaller charge, so that atoms held them more loosely. Many atoms might not even be stable. On the other hand, imagine how matter would be different if electrons had a much greater charge, so that atoms held them more tightly. Since atoms form compounds by exchanging and sharing electrons (more on this in Chapter 3), there could be fewer compounds or maybe even none. Without the abundant diversity of compounds, life would not be possible. So, the magnitude of the charge of the electron—even though it may seem like an insignificantly small number—has great importance.

---

**The Millikan Oil Drop Experiment** Suppose that one of Millikan's oil drops has a charge of $-4.8 \times 10^{-19}$ C. How many excess electrons does the drop contain?

CONCEPTUAL
**CONNECTION 2.3**

---

## 2.5 The Structure of the Atom

The discovery of negatively charged particles within atoms raised a new question. Since atoms are charge-neutral, they must contain a positive charge that neutralizes the negative charge of the electrons—but how do the positive and negative charges fit together? Are atoms just a jumble of even more fundamental particles? Are they solid spheres? Do they have some internal structure? J. J. Thomson proposed that the negatively charged electrons were small particles held within a positively charged sphere.

This model, the most popular of the time, became known as the plum-pudding model (shown at right). The model suggested by Thomson, to those of us not familiar with plum pudding (an English dessert), was like a blueberry muffin; the blueberries are the electrons, and the muffin is the positively charged sphere.

The discovery of **radioactivity**—the emission of small energetic particles from the core of certain unstable atoms—by scientists Henri Becquerel (1852–1908) and Marie Curie (1867–1934) at the end of the nineteenth century allowed researchers to experimentally probe the structure of the atom. At the time, scientists had identified three different types of radioactivity: alpha ($\alpha$) particles, beta ($\beta$) particles, and gamma ($\gamma$) rays. We will discuss these and other types of radioactivity in more detail in Chapter 20. For now, just know that $\alpha$ particles are positively charged and that they are by far the most massive of the three.

In 1909, Ernest Rutherford (1871–1937), who had worked under Thomson and subscribed to his plum-pudding model, performed an experiment in an attempt to confirm Thomson's model. Rutherford's experiment, which employed $\alpha$ particles, proved Thomson wrong instead. In the experiment, Rutherford directed the positively charged $\alpha$ particles at an ultrathin sheet of gold foil, as shown in Figure 2.5▶, on the next page.

These particles were to act as probes of the gold atoms' structure. If the gold atoms were indeed like blueberry muffins or plum pudding—with their mass and charge spread throughout the entire volume of the atom—Rutherford reasoned that these speeding probes would pass right through the gold foil with minimum deflection.

Electron

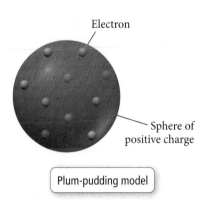
Sphere of positive charge

Plum-pudding model

Alpha particles are about 7000 times more massive than electrons.

▶ FIGURE 2.5 **Rutherford's Gold Foil Experiment** Alpha particles are directed at a thin sheet of gold foil. Most of the particles pass through the foil, but a small fraction are deflected, and a few even bounce backward.

## Rutherford's Gold Foil Experiment

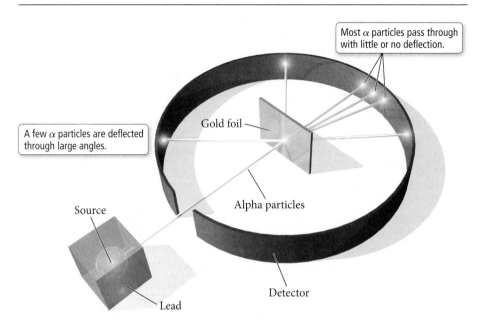

Most $\alpha$ particles pass through with little or no deflection.

A few $\alpha$ particles are deflected through large angles.

Gold foil

Alpha particles

Source

Detector

Lead

Rutherford and his coworkers performed the experiment, but the results were not what they expected. The majority of the particles did pass directly through the foil, but some particles were deflected, and some (approximately 1 in 20,000) even bounced back. The results puzzled Rutherford, who wrote that they were "about as credible as if you had fired a 15-inch shell at a piece of tissue paper and it came back and hit you." What sort of atomic structure could explain this odd behavior? Rutherford created a new model—a modern version of which is shown in Figure 2.6▼ alongside the plum-pudding model—to explain his results.

Rutherford realized that to account for the deflections he observed, the mass and positive charge of an atom must be concentrated in a space much smaller than the size of the atom itself. He concluded that, in contrast to the plum-pudding model, matter must not be as uniform as it appears. It must contain large regions of empty space dotted with small regions of very dense matter. Building on this idea, he proposed the **nuclear theory** of the atom, with three basic parts:

1. Most of the atom's mass and all of its positive charge are contained in a small core called the **nucleus**.

2. Most of the volume of the atom is empty space, throughout which tiny, negatively charged electrons are dispersed.

3. There are as many negatively charged electrons outside the nucleus as there are positively charged particles (named **protons**) within the nucleus, so the atom is electrically neutral.

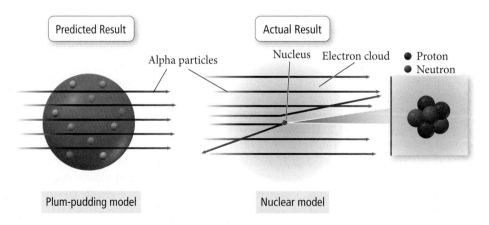

Predicted Result

Actual Result

Alpha particles

Nucleus    Electron cloud    ● Proton
● Neutron

▶ FIGURE 2.6 **The Nuclear Atom** Rutherford's results could not be explained by the plum-pudding model. Instead, they suggest that the atom has a small, dense nucleus.

Plum-pudding model

Nuclear model

Although Rutherford's model was highly successful, scientists realized that it was incomplete. For example, hydrogen atoms contain one proton, and helium atoms contain two, yet a hydrogen atom has only one-fourth the mass of a helium atom. Why? The helium atom must contain some additional mass. Later work by Rutherford and one of his students, British scientist James Chadwick (1891–1974), demonstrated that the previously unaccounted for mass was due to **neutrons**, neutral particles within the nucleus. The mass of a neutron is similar to that of a proton, but a neutron has no electrical charge. The helium atom is four times as massive as the hydrogen atom because its nucleus contains two protons *and two neutrons* (while hydrogen contains only one proton and no neutrons).

The dense nucleus contains over 99.9% of the mass of the atom but occupies very little of its volume. For now, we can think of the electrons that surround the nucleus in analogy to the water droplets that make up a cloud—although their mass is almost negligibly small, they are dispersed over a very large volume. Consequently, an atom, like a cloud, is mostly empty space.

Rutherford's nuclear theory was a success and is still valid today. The revolutionary part of this theory is the idea that matter—at its core—is much less uniform than it appears. If the nucleus of the atom were the size of the period at the end of this sentence, the average electron would be about 10 meters away. Yet the period would contain nearly all of the atom's mass. Imagine what matter would be like if atomic structure were different. What if matter were composed of atomic nuclei piled on top of each other like marbles in a box? Such matter would be incredibly dense; a single grain of sand composed of solid atomic nuclei would have a mass of 5 million kilograms (or a weight of about 11 million pounds). Astronomers believe there are some objects in the universe composed of such matter—neutron stars.

If matter really is mostly empty space, as Rutherford suggested, then why does it appear so solid? Why can we tap our knuckles on a table and feel a solid thump? Matter appears solid because the variation in its density is on such a small scale that our eyes cannot see it. Imagine a scaffolding 100 stories high and the size of a football field as shown in the margin. The volume of the scaffolding is mostly empty space. Yet if you viewed it from an airplane, it would appear as a solid mass. Matter is similar. When you tap your knuckle on the table, it is much like one giant scaffolding (your finger) crashing into another (the table). Even though they are both primarily empty space, one does not fall into the other.

## 2.6 Subatomic Particles: Protons, Neutrons, and Electrons in Atoms

**KCV** Subatomic Particles and Isotope Symbols

All atoms are composed of the same subatomic particles: protons, neutrons, and electrons. Protons and neutrons, as we discussed earlier, have nearly identical masses. In SI units, the mass of the proton is $1.67262 \times 10^{-27}$ kg, and the mass of the neutron is $1.67493 \times 10^{-27}$ kg. A more common unit to express these masses is the **atomic mass unit (amu)**, defined as 1/12 the mass of a carbon atom that contains six protons and six neutrons. The mass of a proton or neutron is approximately 1 amu. Electrons, by contrast, have an almost negligible mass of $0.00091 \times 10^{-27}$ kg or 0.00055 amu.

The proton and the electron both have electrical *charge*. We know from Millikan's oil drop experiment that the electron has a charge of $-1.60 \times 10^{-19}$ C. In atomic (or relative) units, the electron is assigned a charge of $-1$ and the proton is assigned a charge of $+1$. *The charge of the proton and the charge of the electron are equal in magnitude but opposite in sign*, so that when the two particles are paired, the charges sum to zero. The neutron has no charge.

Matter is usually charge-neutral (it has no overall charge) because protons and electrons are normally present in equal numbers. When matter does acquire charge imbalances, these imbalances usually equalize quickly, often in dramatic ways. For example, the shock you receive when touching a doorknob during dry weather is the equalization

If a proton had the mass of a baseball, an electron would have the mass of a rice grain.

Negative charge builds up on clouds.

Electrical discharge equalizes charge imbalance.

Positive charge builds up on ground.

▲ When the normal charge balance of matter is disturbed, as happens during an electrical storm, it quickly equalizes, often in dramatic ways.

of a charge imbalance that developed as you walked across the carpet. Lightning is an equalization of charge imbalances that develop during electrical storms.

A sample of matter—even a tiny sample, such as a sand grain—composed of only protons or only electrons, would have extraordinary repulsive forces inherent within it and would be unstable. Luckily, matter is not that way. Table 2.1 summarizes the properties of protons, neutrons, and electrons.

| TABLE 2.1 Subatomic Particles | | | | |
|---|---|---|---|---|
| | Mass (kg) | Mass (amu) | Charge (relative) | Charge (C) |
| Proton | $1.67262 \times 10^{-27}$ | 1.00727 | +1 | $+1.60218 \times 10^{-19}$ |
| Neutron | $1.67493 \times 10^{-27}$ | 1.00866 | 0 | 0 |
| Electron | $0.00091 \times 10^{-27}$ | 0.00055 | −1 | $-1.60218 \times 10^{-19}$ |

## Elements: Defined by Their Numbers of Protons

If all atoms are composed of the same subatomic particles, what makes the atoms of one element different from those of another? The answer is the *number* of these particles. The most important number to the *identity* of an atom is the number of protons in its nucleus. *The number of protons defines the element.* For example, an atom with two protons in its nucleus is a helium atom, an atom with six protons in its nucleus is a carbon atom (Figure 2.7▼), and an atom with 92 protons in its nucleus is a uranium atom. The number of protons in an atom's nucleus is its **atomic number** and is given the symbol $Z$. The atomic numbers of known elements range from 1 to 116 (although evidence for the synthesis of additional elements is currently being considered), as shown in the periodic table of the elements (Figure 2.8▶). In the periodic table, described in more detail in Section 2.7, the elements are arranged so that those with similar properties are in the same column.

Each element, identified by its unique atomic number, is represented with a unique **chemical symbol**, a one- or two-letter abbreviation listed directly below its atomic number on the periodic table. The chemical symbol for helium is He; for carbon, the symbol is C; and for uranium, it is U. The chemical symbol and the atomic number always go together. If the atomic number is 2, the chemical symbol *must be* He. If the atomic number is 6, the chemical symbol *must be* C. This is another way of saying that the number of protons defines the element.

### The Number of Protons Defines the Element

Helium nucleus: two protons

Carbon nucleus: six protons

▶ **FIGURE 2.7 How Elements Differ** Each element is defined by a unique atomic number (*Z*), the number of protons in the nucleus of every atom of that element.

## The Periodic Table

Atomic number (Z)

| 4 |
| :---: |
| **Be** |
| beryllium |

Chemical symbol

Name

| 1 | | | | | | | | | | | | | | | | | 2 |
| :---: | :---: | :---: | :---: | :---: | :---: | :---: | :---: | :---: | :---: | :---: | :---: | :---: | :---: | :---: | :---: | :---: | :---: |
| **H** | | | | | | | | | | | | | | | | | **He** |
| hydrogen | | | | | | | | | | | | | | | | | helium |
| 3 | 4 | | | | | | | | | | | 5 | 6 | 7 | 8 | 9 | 10 |
| **Li** | **Be** | | | | | | | | | | | **B** | **C** | **N** | **O** | **F** | **Ne** |
| lithium | beryllium | | | | | | | | | | | boron | carbon | nitrogen | oxygen | fluorine | neon |
| 11 | 12 | | | | | | | | | | | 13 | 14 | 15 | 16 | 17 | 18 |
| **Na** | **Mg** | | | | | | | | | | | **Al** | **Si** | **P** | **S** | **Cl** | **Ar** |
| sodium | magnesium | | | | | | | | | | | aluminum | silicon | phosphorus | sulfur | chlorine | argon |
| 19 | 20 | 21 | 22 | 23 | 24 | 25 | 26 | 27 | 28 | 29 | 30 | 31 | 32 | 33 | 34 | 35 | 36 |
| **K** | **Ca** | **Sc** | **Ti** | **V** | **Cr** | **Mn** | **Fe** | **Co** | **Ni** | **Cu** | **Zn** | **Ga** | **Ge** | **As** | **Se** | **Br** | **Kr** |
| potassium | calcium | scandium | titanium | vanadium | chromium | manganese | iron | cobalt | nickel | copper | zinc | gallium | germanium | arsenic | selenium | bromine | krypton |
| 37 | 38 | 39 | 40 | 41 | 42 | 43 | 44 | 45 | 46 | 47 | 48 | 49 | 50 | 51 | 52 | 53 | 54 |
| **Rb** | **Sr** | **Y** | **Zr** | **Nb** | **Mo** | **Tc** | **Ru** | **Rh** | **Pd** | **Ag** | **Cd** | **In** | **Sn** | **Sb** | **Te** | **I** | **Xe** |
| rubidium | strontium | yttrium | zirconium | niobium | molybdenum | technetium | ruthenium | rhodium | palladium | silver | cadmium | indium | tin | antimony | tellurium | iodine | xenon |
| 55 | 56 | 57 | 72 | 73 | 74 | 75 | 76 | 77 | 78 | 79 | 80 | 81 | 82 | 83 | 84 | 85 | 86 |
| **Cs** | **Ba** | **La** | **Hf** | **Ta** | **W** | **Re** | **Os** | **Ir** | **Pt** | **Au** | **Hg** | **Tl** | **Pb** | **Bi** | **Po** | **At** | **Rn** |
| cesium | barium | lanthanum | hafnium | tantalum | tungsten | rhenium | osmium | iridium | platinum | gold | mercury | thallium | lead | bismuth | polonium | astatine | radon |
| 87 | 88 | 89 | 104 | 105 | 106 | 107 | 108 | 109 | 110 | 111 | 112 | 113 | 114 | 115 | 116 | 117 | 118 |
| **Fr** | **Ra** | **Ac** | **Rf** | **Db** | **Sg** | **Bh** | **Hs** | **Mt** | **Ds** | **Rg** | **Cn** | ** | **Fl** | ** | **Lv** | ** | ** |
| francium | radium | actinium | rutherfordium | dubnium | seaborgium | bohrium | hassium | meitnerium | darmstadtium | roentgenium | copernicium | | flerovium | | livermorium | | |

| 58 | 59 | 60 | 61 | 62 | 63 | 64 | 65 | 66 | 67 | 68 | 69 | 70 | 71 |
| :---: | :---: | :---: | :---: | :---: | :---: | :---: | :---: | :---: | :---: | :---: | :---: | :---: | :---: |
| **Ce** | **Pr** | **Nd** | **Pm** | **Sm** | **Eu** | **Gd** | **Tb** | **Dy** | **Ho** | **Er** | **Tm** | **Yb** | **Lu** |
| cerium | praseodymium | neodymium | promethium | samarium | europium | gadolinium | terbium | dysprosium | holmium | erbium | thulium | ytterbium | lutetium |
| 90 | 91 | 92 | 93 | 94 | 95 | 96 | 97 | 98 | 99 | 100 | 101 | 102 | 103 |
| **Th** | **Pa** | **U** | **Np** | **Pu** | **Am** | **Cm** | **Bk** | **Cf** | **Es** | **Fm** | **Md** | **No** | **Lr** |
| thorium | protactinium | uranium | neptunium | plutonium | americium | curium | berkelium | californium | einsteinium | fermium | mendelevium | nobelium | lawrencium |

▲ **FIGURE 2.8  The Periodic Table**   Each element is represented by its symbol and atomic number. Elements in the same column have similar properties.

Most chemical symbols are based on the English name of the element. For example, the symbol for sulfur is S; for oxygen, O; and for chlorine, Cl. Several of the oldest known elements, however, have symbols based on their original Latin names. For example, the symbol for sodium is Na from the Latin *natrium*, and the symbol for tin is Sn from the Latin *stannum*. Early scientists often gave newly discovered elements names that reflect their properties. For example, argon originates from the Greek word *argos* meaning inactive, referring to argon's chemical inertness (it does not react with other elements). Chlorine originates from the Greek word *chloros* meaning pale green, referring to chlorine's pale green color. Other elements, including helium, selenium, and mercury, are named after figures from Greek or Roman mythology or astronomical bodies. Still others (such as europium, polonium, and berkelium) are named for the places where they were discovered or where their discoverers were born. More recently, elements have been named after scientists; for example, curium for Marie Curie, einsteinium for Albert Einstein, and rutherfordium for Ernest Rutherford.

| 96 |
| :---: |
| **Cm** |
| Curium |

## Isotopes: When the Number of Neutrons Varies

All atoms of a given element have the same number of protons; however, they do not necessarily have the same number of neutrons. Since neutrons have nearly the same mass as protons (1 amu), this means that—contrary to what John Dalton originally proposed in his atomic theory—all atoms of a given element *do not* have the same mass. For example, all neon atoms contain 10 protons, but they may contain 10, 11, or 12 neutrons. All three types of neon atoms exist, and each has a slightly different mass. Atoms with the same number of protons but different numbers of neutrons are

▲ Element 96 is named curium, after Marie Curie, co-discoverer of radioactivity.

**isotopes**. Some elements, such as beryllium (Be) and aluminum (Al), have only one naturally occurring isotope, while other elements, such as neon (Ne) and chlorine (Cl), have two or more.

The relative amount of each different isotope in a naturally occurring sample of a given element is roughly constant. For example, in any natural sample of neon atoms, 90.48% of them are the isotope with 10 neutrons, 0.27% are the isotope with 11 neutrons, and 9.25% are the isotope with 12 neutrons. These percentages are the **natural abundance** of the isotopes. Each element has its own characteristic natural abundance of isotopes. However, advances in mass spectrometry (see Section 2.8) have allowed accurate measurements that reveal small but significant variations in the natural abundance of isotopes for many elements.

The sum of the number of neutrons and protons in an atom is its **mass number**. We represent mass number with the symbol $A$.

$$A = \text{number of protons (p)} + \text{number of neutrons (n)}$$

For neon, with 10 protons, the mass numbers of the three different naturally occurring isotopes are 20, 21, and 22, corresponding to 10, 11, and 12 neutrons, respectively.

We symbolize isotopes using the notation:

Mass number $\longrightarrow$ $^A_Z X$ $\longleftarrow$ Chemical symbol
Atomic number $\nearrow$

where X is the chemical symbol, $A$ is the mass number, and $Z$ is the atomic number. Therefore, the symbols for the neon isotopes are:

$$^{20}_{10}\text{Ne} \quad ^{21}_{10}\text{Ne} \quad ^{22}_{10}\text{Ne}$$

Notice that the chemical symbol, Ne, and the atomic number, 10, are redundant: if the atomic number is 10, the symbol must be Ne. The mass numbers, however, are different for the different isotopes, reflecting the different number of neutrons in each one.

A second common notation for isotopes is the chemical symbol (or chemical name) followed by a dash and the mass number of the isotope.

Chemical symbol or name $\longrightarrow$ X-$A$ $\longleftarrow$ Mass number

In this notation, the neon isotopes are:

Ne-20     Ne-21     Ne-22
neon-20   neon-21   neon-22

We summarize the neon isotopes in the following table:

| Symbol | Number of Protons | Number of Neutrons | A (Mass Number) | Natural Abundance (%) |
|---|---|---|---|---|
| Ne-20 or $^{20}_{10}$Ne | 10 | 10 | 20 | 90.48 |
| Ne-21 or $^{21}_{10}$Ne | 10 | 11 | 21 | 0.27 |
| Ne-22 or $^{22}_{10}$Ne | 10 | 12 | 22 | 9.25 |

Notice that all isotopes of a given element have the same number of protons (otherwise they would be different elements). Notice also that the mass number is the *sum* of the number of protons and the number of neutrons. The number of neutrons in an isotope is therefore the difference between the mass number and the atomic number ($A$–$Z$). The different isotopes of an element generally exhibit the same chemical behavior—the three isotopes of neon, for example, all exhibit chemical inertness.

**Example** 2.3    **Atomic Numbers, Mass Numbers, and Isotope Symbols**

(a) What are the atomic number ($Z$), mass number ($A$), and symbol of the chlorine isotope with 18 neutrons?

(b) How many protons, electrons, and neutrons are present in an atom of $^{52}_{24}\text{Cr}$?

**SOLUTION**

| (a) | Look up the atomic number ($Z$) for chlorine on the periodic table. The atomic number specifies the number of protons. | $Z = 17$, so chlorine has 17 protons. |
|---|---|---|
| | The mass number ($A$) for an isotope is the sum of the number of protons and the number of neutrons. | $A =$ number of protons + number of neutrons $= 17 + 18 = 35$ |
| | The symbol for an isotope is its chemical symbol with the atomic number ($Z$) in the lower left corner and the mass number ($A$) in the upper left corner. | $^{35}_{17}\text{Cl}$ |
| (b) | For any isotope (in this case $^{52}_{24}\text{Cr}$) the atomic number located at the lower left indicates the number of protons. Since this is a neutral atom, the number of electrons equals the number of protons. | Number of protons $= Z = 24$ <br> Number of electrons $= 24$ (neutral atom) |
| | The number of neutrons is equal to the mass number (upper left) minus the atomic number (lower left). | Number of neutrons $= 52 - 24 = 28$ |

**FOR PRACTICE 2.3**

(a) What are the atomic number, mass number, and symbol for the carbon isotope with seven neutrons?

(b) How many protons and neutrons are present in an atom of $^{39}_{19}\text{K}$?

---

**Isotopes**  Carbon has two naturally occurring isotopes: C-12 (natural abundance is 98.93%) and C-13 (natural abundance is 1.07%). Using circles to represent protons and squares to represent neutrons, draw the nucleus of each isotope. How many C-13 atoms are present, on average, in a 10,000-atom sample of carbon?

CONCEPTUAL
**CONNECTION 2.4**

---

## Ions: Losing and Gaining Electrons

The number of electrons in a neutral atom is equal to the number of protons in its nucleus (designated by its atomic number $Z$). During chemical changes, however, atoms can lose or gain electrons and become charged particles called **ions**. For example, neutral lithium (Li) atoms contain three protons and three electrons; however, in many chemical reactions lithium atoms lose one electron ($e^-$) to form $\text{Li}^+$ ions.

$$\text{Li} \longrightarrow \text{Li}^+ + 1\,e^-$$

The charge of an ion is indicated in the upper right corner of the chemical symbol. Since the $\text{Li}^+$ ion contains three protons and only two electrons, its charge is 1+ (ion charges are written as the magnitude first followed by the sign of the charge; for a charge of 1+, the 1 is usually dropped and the charge is written as simply +).

Ions can also be negatively charged. For example, neutral fluorine (F) atoms contain nine protons and nine electrons; however, in many chemical reactions fluorine atoms gain one electron to form $\text{F}^-$ ions.

$$\text{F} + 1e^- \longrightarrow \text{F}^-$$

The $\text{F}^-$ ion contains nine protons and 10 electrons, resulting in a charge of 1− (written simply as –). For many elements, such as lithium and fluorine, the ion is much more common than the neutral atom. In fact, lithium and fluorine occur in nature mostly as ions.

Positively charged ions, such as $\text{Li}^+$, are **cations**, and negatively charged ions, such as $\text{F}^-$, are **anions**. Ions behave quite differently than their corresponding atoms. Neutral sodium atoms, for example, are extremely unstable, reacting violently with most things they contact. Sodium cations ($Na^+$), by contrast, are relatively inert—we eat them all the time in sodium chloride (table salt). In ordinary matter, cations and anions always occur together so that matter is charge neutral overall.

# Chemistry IN YOUR DAY | Where Did Elements Come From?

We find ourselves on a planet containing many different kinds of elements. If it were otherwise, we would not exist and would not be here to reflect on why. Where did these elements come from? The story of element formation is as old as the universe itself, and we have to go back to the very beginning to tell the story.

The birth of the universe is described by the Big Bang theory, which asserts that the universe began as a hot, dense collection of matter and energy that expanded rapidly. As it expanded, it cooled, and within the first several hours, subatomic particles formed the first atomic nuclei: hydrogen and helium. These two elements were (and continue to be) the most abundant in the universe. As the universe continued expanding, some of the hydrogen and helium clumped together under the influence of gravity to form *nebulae* (clouds of gas) that eventually gave birth to stars and galaxies. These stars and galaxies became the nurseries where all other elements formed.

Stars are fueled by nuclear fusion, which we will discuss in more detail in Chapter 20. Under the conditions within the core of a star, hydrogen nuclei can combine (or fuse) to form helium. Fusion gives off enormous quantities of energy, which is why stars emit so much heat and light. The fusion of hydrogen to helium can fuel a star for billions of years.

After it burns through large quantities of hydrogen, if a star is large enough, the helium that builds up in its core can in turn fuse to form carbon. The carbon then builds up in the core and (again, if the star is large enough) can fuse to form even heavier elements. The fusion process ends with iron, which has a highly stable nucleus. By the time iron is formed, however, the star is near the end of its existence and may enter a phase of expansion, transforming into a *supernova*. Within a supernova, which is in essence a large exploding star, a shower of neutrons allows the lighter elements (which formed during the lifetime of the star through the fusion processes just described) to capture extra neutrons. These neutrons can transform into protons (through processes that we discuss in Chapter 20), contributing ultimately to the formation of elements heavier than iron, all the way up to uranium. As the supernova continues to expand, the elements present within it are blown out into space, where they can incorporate into other nebulae and perhaps even eventually form planets that orbit stars like our own sun.

▲ Stars are born in nebulae such as the Eagle Nebula (also known as M16). This image was taken by the Hubble Space Telescope and shows a gaseous pillar in a star-forming region of the Eagle Nebula.

CONCEPTUAL CONNECTION 2.5

**The Nuclear Atom, Isotopes, and Ions** Which statement is true?
(a) For a given element, the size of an isotope with more neutrons is larger than one with fewer neutrons.
(b) For a given element, the size of an atom is the same for all of the element's isotopes.

## 2.7 Finding Patterns: The Periodic Law and the Periodic Table

The modern periodic table grew out of the work of Dmitri Mendeleev (1834–1907), a nineteenth-century Russian chemistry professor. In his time, scientists had discovered about 65 different elements, and chemists had identified many of the

Time of Discovery

| | | |
|---|---|---|
| ☐ Before 1800 | ☐ 1800–1849 | ☐ 1850–1899 |
| ☐ 1900–1949 | ☐ 1950–2012 | |

| H hydrogen | | | | | | | | | | | | | | | | | He helium |
|---|---|---|---|---|---|---|---|---|---|---|---|---|---|---|---|---|---|
| Li lithium | Be beryllium | | | | | | | | | | | B boron | C carbon | N nitrogen | O oxygen | F fluorine | Ne neon |
| Na sodium | Mg magnesium | | | | | | | | | | | Al aluminum | Si silicon | P phosphorus | S sulfur | Cl chlorine | Ar argon |
| K potassium | Ca calcium | Sc scandium | Ti titanium | V vanadium | Cr chromium | Mn manganese | Fe iron | Co cobalt | Ni nickel | Cu copper | Zn zinc | Ga gallium | Ge germanium | As arsenic | Se selenium | Br bromine | Kr krypton |
| Rb rubidium | Sr strontium | Y yttrium | Zr zirconium | Nb niobium | Mo molybdenum | Tc technetium | Ru ruthenium | Rh rhodium | Pd palladium | Ag silver | Cd cadmium | In indium | Sn tin | Sb antimony | Te tellurium | I iodine | Xe xenon |
| Cs cesium | Ba barium | La lanthanum | Hf hafnium | Ta tantalum | W tungsten | Re rhenium | Os osmium | Ir iridium | Pt platinum | Au gold | Hg mercury | Tl thallium | Pb lead | Bi bismuth | Po polonium | At astatine | Rn radon |
| Fr francium | Ra radium | Ac actinium | Rf rutherfordium | Db dubnium | Sg seaborgium | Bh bohrium | Hs hassium | Mt meitnerium | Ds darmstadtium | Rg roentgenium | Cn copernicium | | Fl flerovium | | Lv livermorium | | |

| Ce cerium | Pr praseodymium | Nd neodymium | Pm promethium | Sm samarium | Eu europium | Gd gadolinium | Tb terbium | Dy dysprosium | Ho holmium | Er erbium | Tm thulium | Yb ytterbium | Lu lutetium |
|---|---|---|---|---|---|---|---|---|---|---|---|---|---|
| Th thorium | Pa protactinium | U uranium | Np neptunium | Pu plutonium | Am americium | Cm curium | Bk berkelium | Cf californium | Es einsteinium | Fm fermium | Md mendelevium | No nobelium | Lr lawrencium |

properties of these elements—such as their relative masses, their chemical activity, and some of their physical properties. However, no one had developed any systematic way of organizing them.

In 1869, Mendeleev noticed that certain groups of elements had similar properties. He also found that when he listed elements in order of increasing mass, these similar properties recurred in a periodic pattern (Figure 2.9▼). Mendeleev summarized these observations in the **periodic law**:

▲ Many of the elements that we know today were discovered during Mendeleev's lifetime.

To be periodic means to exhibit a repeating pattern.

**When the elements are arranged in order of increasing mass, certain sets of properties recur periodically.**

### The Periodic Law

| 1 H | 2 He | 3 Li | 4 Be | 5 B | 6 C | 7 N | 8 O | 9 F | 10 Ne | 11 Na | 12 Mg | 13 Al | 14 Si | 15 P | 16 S | 17 Cl | 18 Ar | 19 K | 20 Ca |
|---|---|---|---|---|---|---|---|---|---|---|---|---|---|---|---|---|---|---|---|

Elements with similar properties recur in a regular pattern.

▲ **FIGURE 2.9 Recurring Properties** These elements are listed in order of increasing atomic number. Elements with similar properties are represented with the same color. Notice that the colors form a repeating pattern, much like musical notes form a repeating pattern on a piano keyboard.

Mendeleev organized the known elements in a table consisting of a series of rows in which mass increases from left to right. He arranged the rows so that elements with similar properties fall in the same vertical columns (Figure 2.10▶).

Since many elements had not yet been discovered, Mendeleev's table contained some gaps, which allowed him to predict the existence (and even the properties) of yet undiscovered elements. For example, Mendeleev predicted the existence of an element he called eka-silicon, which fell below silicon on the table and between gallium and arsenic (*eka* means the one beyond). In 1886, eka-silicon was discovered by German chemist Clemens Winkler (1838–1904), who named it germanium, after his home country.

### A Simple Periodic Table

| 1 H | | | | | | | 2 He |
|---|---|---|---|---|---|---|---|
| 3 Li | 4 Be | 5 B | 6 C | 7 N | 8 O | 9 F | 10 Ne |
| 11 Na | 12 Mg | 13 Al | 14 Si | 15 P | 16 S | 17 Cl | 18 Ar |
| 19 K | 20 Ca | | | | | | |

Elements with similar properties fall into columns.

▲ **FIGURE 2.10 Making a Periodic Table** We can arrange the elements in Figure 2.9 in a table in which atomic number increases from left to right and elements with similar properties (as represented by the different colors) are aligned in columns.

▲ Dmitri Mendeleev, a Russian chemistry professor who proposed the periodic law and arranged early versions of the periodic table, was honored on a Soviet postage stamp.

## Modern Periodic Table Organization

Mendeleev's original listing evolved into the modern periodic table shown in Figure 2.11▼. In the modern table, elements are listed in order of increasing atomic number rather than increasing relative mass. The modern periodic table also contains more elements than Mendeleev's original table because more have been discovered since his time. Mendeleev's periodic law was based on observation. Like all scientific laws, the periodic law summarizes many observations but does not give the underlying reason for the observations—only theories do that. For now, we accept the periodic law as it is, but in Chapters 7 and 8 we will examine a powerful theory—called quantum mechanics—that explains the law and gives the underlying reasons for it.

We can broadly classify the elements in the periodic table as metals, nonmetals, or metalloids, as shown in Figure 2.11. **Metals** lie on the lower left side and middle of the periodic table and share some common properties: they are good conductors of heat and electricity, they can be pounded into flat sheets (malleability), they can be drawn into wires (ductility), they are often shiny, and they tend to lose electrons when they undergo chemical changes. Chromium, copper, strontium, and lead are typical metals.

**Nonmetals** lie on the upper right side of the periodic table. The dividing line between metals and nonmetals is the zigzag diagonal line running from boron to astatine. Nonmetals have varied properties—some are solids at room temperature, others are liquids or gases—but typically they tend to be poor conductors of heat and electricity, and they all tend to gain electrons when they undergo chemical changes. Oxygen, carbon, sulfur, bromine, and iodine are nonmetals.

### Major Divisions of the Periodic Table

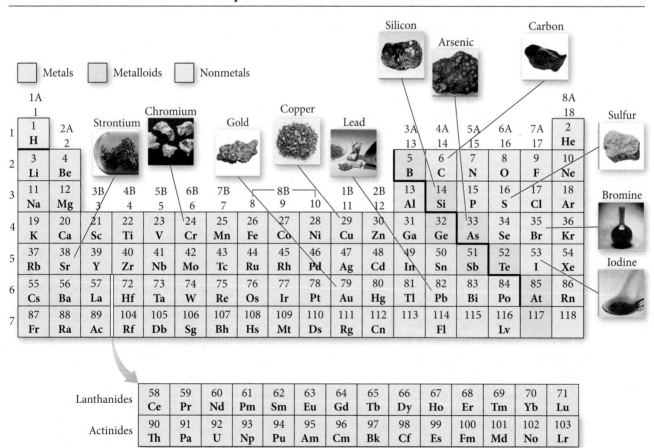

▲ **FIGURE 2.11  Metals, Nonmetals, and Metalloids**   The elements in the periodic table fall into these three broad classes.

| Main-group elements | | Transition elements | | | | | | | | | | | | Main-group elements | | | | | | |
|---|---|---|---|---|---|---|---|---|---|---|---|---|---|---|---|---|---|---|---|---|

|  |  |  |  |  |  |  |  |  |  |  |  |  |  |  |  |  |  |  |  | 8A |
|---|---|---|---|---|---|---|---|---|---|---|---|---|---|---|---|---|---|---|---|---|
| 1A 1 | *Group number* | | | | | | | | | | | | | | | | | | | 18 |

| Periods | 1A 1 | 2A 2 | | | | | | | | | | | 3A 13 | 4A 14 | 5A 15 | 6A 16 | 7A 17 | 8A 18 |
|---|---|---|---|---|---|---|---|---|---|---|---|---|---|---|---|---|---|---|
| 1 | 1 H | | | | | | | | | | | | | | | | | 2 He |
| 2 | 3 Li | 4 Be | | | | | | | | | | | 5 B | 6 C | 7 N | 8 O | 9 F | 10 Ne |
| 3 | 11 Na | 12 Mg | 3B 3 | 4B 4 | 5B 5 | 6B 6 | 7B 7 | 8B 8 | 9 | 10 | 1B 11 | 2B 12 | 13 Al | 14 Si | 15 P | 16 S | 17 Cl | 18 Ar |
| 4 | 19 K | 20 Ca | 21 Sc | 22 Ti | 23 V | 24 Cr | 25 Mn | 26 Fe | 27 Co | 28 Ni | 29 Cu | 30 Zn | 31 Ga | 32 Ge | 33 As | 34 Se | 35 Br | 36 Kr |
| 5 | 37 Rb | 38 Sr | 39 Y | 40 Zr | 41 Nb | 42 Mo | 43 Tc | 44 Ru | 45 Rh | 46 Pd | 47 Ag | 48 Cd | 49 In | 50 Sn | 51 Sb | 52 Te | 53 I | 54 Xe |
| 6 | 55 Cs | 56 Ba | 57 La | 72 Hf | 73 Ta | 74 W | 75 Re | 76 Os | 77 Ir | 78 Pt | 79 Au | 80 Hg | 81 Tl | 82 Pb | 83 Bi | 84 Po | 85 At | 86 Rn |
| 7 | 87 Fr | 88 Ra | 89 Ac | 104 Rf | 105 Db | 106 Sg | 107 Bh | 108 Hs | 109 Mt | 110 Ds | 111 Rg | 112 Cn | 113 | 114 Fl | 115 | 116 Lv | 117 | 118 |

▲ **FIGURE 2.12  The Periodic Table: Main-Group and Transition Elements**   The elements in the periodic table fall into columns. The two columns at the left and the six columns at the right comprise the main-group elements. Each of these eight columns is a group or family. The properties of main-group elements can generally be predicted from their position in the periodic table. The properties of the elements in the middle of the table, known as transition elements, are less predictable.

Many of the elements that lie along the zigzag diagonal line that divides metals and nonmetals are **metalloids** and exhibit mixed properties. Several metalloids are also classified as **semiconductors** because of their intermediate (and highly temperature-dependent) electrical conductivity. Our ability to change and control the conductivity of semiconductors makes them useful to us in the manufacture of the electronic chips and circuits central to computers, cellular telephones, and many other modern devices. Good examples of metalloids are silicon, arsenic, and antimony.

We can also divide the periodic table, as shown in Figure 2.12▲, into **main-group elements**, whose properties tend to be largely predictable based on their position in the periodic table, and **transition elements** or **transition metals**, whose properties tend to be less predictable based simply on their position in the periodic table. Main-group elements are in columns labeled with a number and the letter A. Transition elements are in columns labeled with a number and the letter B. An alternative numbering system does not use letters, but only the numbers 1–18. Both numbering systems are shown in most of the periodic tables in this book. Each column within the main-group regions of the periodic table is a **family** or **group** of elements.

The elements within a group usually have similar properties. For example, the group 8A elements, called the **noble gases**, are mostly unreactive. The most familiar noble gas is probably helium, used to fill buoyant balloons. Helium is chemically stable—it does not combine with other elements to form compounds—and is therefore safe to put into balloons. Other noble gases are neon (often used in electronic signs), argon (a small component of our atmosphere), krypton, and xenon.

The group 1A elements, the **alkali metals**, are all reactive metals. A marble-sized piece of sodium explodes violently when dropped into water. Lithium, potassium, and rubidium are also alkali metals.

The group 2A elements, the **alkaline earth metals**, are also fairly reactive, although not quite as reactive as the alkali metals. Calcium, for example, reacts fairly vigorously when dropped into water but does not explode as dramatically as sodium. Other alkaline earth metals include magnesium (a common low-density structural metal), strontium, and barium.

Metalloids are sometimes called semimetals.

**Alkali metals**

| Li |
|---|
| Na |
| K |
| Rb |
| Cs |

Halogens

The group 7A elements, the **halogens**, are very reactive nonmetals. One of the most familiar halogens is chlorine, a greenish-yellow gas with a pungent odor. Because of its reactivity, chlorine is used as a sterilizing and disinfecting agent. Other halogens include bromine, a red-brown liquid that easily evaporates into a gas; iodine, a purple solid; and fluorine, a pale-yellow gas.

## Ions and the Periodic Table

In chemical reactions, metals tend to lose electrons (forming cations) and nonmetals tend to gain them (forming anions). The number of electrons lost or gained, and therefore the charge of the resulting ion, is often predictable for a given element, especially main-group elements. Main-group elements tend to form ions that have the same number of electrons as the nearest noble gas (i.e., the noble gas that has the number of electrons closest to that of the element).

- **A main-group metal tends to lose electrons, forming a cation with the same number of electrons as the nearest noble gas.**
- **A main-group nonmetal tends to gain electrons, forming an anion with the same number of electrons as the nearest noble gas.**

For example, lithium, a metal with three electrons, tends to lose one electron, forming a 1+ cation that has the same number of electrons (two) as helium. Chlorine, a nonmetal with 17 electrons, tends to gain one electron, forming a 1− anion that has the same number of electrons (18) as argon.

In general, the alkali metals (group 1A) tend to lose one electron and form 1+ ions. The alkaline earth metals (group 2A) tend to lose two electrons and form 2+ ions. The halogens (group 7A) tend to gain one electron and form 1− ions. The oxygen family nonmetals (group 6A) tend to gain two electrons and form 2− ions. More generally, for the main-group elements that form cations with predictable charge, the charge is equal to the group number. For main-group elements that form anions with predictable charge, the charge is equal to the group number minus eight. Transition elements may form various different ions with different charges. Figure 2.13▶ shows the ions formed by the main-group elements that form ions with predictable charges. In Chapters 7 and 8, we will introduce quantum-mechanical theory, which more fully explains *why* these groups form ions as they do.

---

**Example 2.4** | **Predicting the Charge of Ions**

Predict the charges of the monoatomic (single atom) ions formed by these main-group elements.

**(a)** Al          **(b)** S

**SOLUTION**

**(a)** Aluminum is a main-group metal and tends to lose electrons to form a cation with the same number of electrons as the nearest noble gas. Aluminum atoms have 13 electrons and the nearest noble gas is neon, which has 10 electrons. Aluminum therefore loses three electrons to form a cation with a 3+ charge ($Al^{3+}$).

**(b)** Sulfur is a nonmetal and tends to gain electrons to form an anion with the same number of electrons as the nearest noble gas. Sulfur atoms have 16 electrons and the nearest noble gas is argon, which has 18 electrons. Sulfur therefore gains two electrons to form an anion with a 2− charge ($S^{2-}$).

**FOR PRACTICE 2.4** Predict the charges of the monoatomic ions formed by these main-group elements.

**(a)** N          **(b)** Rb

**Elements That Form Ions with Predictable Charges**

▲ FIGURE 2.13  **Elements That Form Ions with Predictable Charges**

---

## Chemistry AND Medicine | The Elements of Life

What kind of atoms compose us? In terms of mass, our bodies are 65% oxygen (because of the large amount of water), 18% carbon, and 10% hydrogen, with a few other elements present in smaller quantities, as shown in Figure 2.14▶ and Table 2.2. Because the atoms of different elements have different masses (more on this in Sections 2.8 and 2.9), if we consider number of atoms (instead of mass), hydrogen comes in first (because hydrogen atoms are so light), with oxygen second and carbon third.

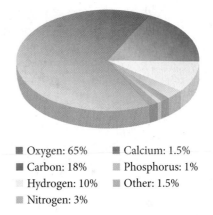

- Oxygen: 65%
- Carbon: 18%
- Hydrogen: 10%
- Nitrogen: 3%
- Calcium: 1.5%
- Phosphorus: 1%
- Other: 1.5%

▲ FIGURE 2.14  **Elemental Composition of Humans (by Mass)**

| TABLE 2.2 Approximate Percent Elemental Composition of Humans | | |
|---|---|---|
| **Element** | **% by Mass** | **% by Number of Atoms** |
| Oxygen | 65 | 26.4 |
| Carbon | 18 | 9.2 |
| Hydrogen | 10 | 62.3 |
| Nitrogen | 3 | 1.4 |
| Calcium | 1.5 | 0.2 |
| Phosphorus | 1 | 0.3 |
| Other | 1.5 | 0.2 |

Much of the *chemistry* of life revolves not around oxygen or hydrogen but around carbon, an element that forms a disproportionately large number of compounds with a few other elements such as hydrogen, oxygen, and nitrogen. We will explore the chemistry of carbon—called organic chemistry—briefly in Chapter 3 and in more detail in Chapter 21.

---

## 2.8 Atomic Mass: The Average Mass of an Element's Atoms

An important part of Dalton's atomic theory is that all atoms of a given element have the same mass. In Section 2.6, we learned that because of isotopes, the atoms of a given element often have different masses, so Dalton was not completely correct. We can, however, calculate an average mass—called the **atomic mass**—for each element.

The atomic mass of each element is listed directly beneath the element's symbol in the periodic table and represents the average mass of the isotopes that compose that element, *weighted according to the natural abundance of each isotope*. For example, the

Atomic mass is sometimes called *atomic weight* or *standard atomic weight*.

| 17 |
|:---:|
| **Cl** |
| 35.45 |
| chlorine |

In this book, we use the atomic masses recommended by IUPAC for users needing an atomic mass value for an unspecified sample. Detailed studies of the atomic masses of many samples, however, have shown that atomic masses are not constants of nature because the exact isotopic abundances in any given sample depend on the history of the sample (see the Chemistry in Your Day box later in this section).

When we use percentages in calculations, we convert them to their decimal value by dividing by 100.

periodic table lists the atomic mass of chlorine as 35.45 amu. Naturally occurring chlorine consists of 75.77% chlorine-35 atoms (mass 34.97 amu) and 24.23% chlorine-37 atoms (mass 36.97 amu). We can calculate its atomic mass accordingly:

$$\text{Atomic mass} = 0.7577(34.97 \text{ amu}) + 0.2423(36.97 \text{ amu}) = 35.45 \text{ amu}$$

Notice that the atomic mass of chlorine is closer to 35 than 37. Naturally occurring chlorine contains more chlorine-35 atoms than chlorine-37 atoms, so the weighted average mass of chlorine is closer to 35 amu than to 37 amu.

In general, we calculate the atomic mass with the equation:

$$\text{Atomic mass} = \sum_{n} (\text{fraction of isotope } n) \times (\text{mass of isotope } n)$$

$$= (\text{fraction of isotope 1} \times \text{mass of isotope 1})$$

$$+ (\text{fraction of isotope 2} \times \text{mass of isotope 2})$$

$$+ (\text{fraction of isotope 3} \times \text{mass of isotope 3}) + \ldots$$

where the fractions of each isotope are the percent natural abundances converted to their decimal values. The concept of atomic mass is useful because it allows us to assign a characteristic mass to each element, and, as we will see shortly, it allows us to quantify the number of atoms in a sample of that element.

## Example 2.5 Atomic Mass

Copper has two naturally occurring isotopes: Cu-63 with a mass of 62.9291 amu and a natural abundance of 69.17%, and Cu-65 with a mass of 64.9278 amu and a natural abundance of 30.83%. Calculate the atomic mass of copper.

**SOLUTION**

| | |
|---|---|
| Convert the percent natural abundances into decimal form by dividing by 100. | $$\text{Fraction Cu-63} = \frac{69.17}{100} = 0.6917$$ $$\text{Fraction Cu-65} = \frac{30.83}{100} = 0.3083$$ |
| Calculate the atomic mass using the equation given in the text. | $$\text{Atomic mass} = 0.6917(62.9296 \text{ amu}) + 0.3083(64.9278 \text{ amu})$$ $$= 43.5284 \text{ amu} + 20.0172 \text{ amu} = 63.5456 = 63.55 \text{ amu}$$ |

**FOR PRACTICE 2.5** Magnesium has three naturally occurring isotopes with masses of 23.99 amu, 24.99 amu, and 25.98 amu and natural abundances of 78.99%, 10.00%, and 11.01%, respectively. Calculate the atomic mass of magnesium.

**FOR MORE PRACTICE 2.5** Gallium has two naturally occurring isotopes: Ga-69 with a mass of 68.9256 amu and a natural abundance of 60.11%, and Ga-71. Use the atomic mass of gallium from the periodic table to find the mass of Ga-71.

CONCEPTUAL
**CONNECTION 2.6**

**Atomic Mass** Recall from Conceptual Connection 2.4 that carbon has two naturally occurring isotopes: C-12 (natural abundance is 98.93%; mass is 12.0000 amu) and C-13 (natural abundance is 1.07%; mass is 13.0034 amu). Without doing any calculations, determine which mass is closest to the atomic mass of carbon.

**(a)** 12.00 amu     **(b)** 12.50 amu     **(c)** 13.00 amu

## Mass Spectrometry: Measuring the Mass of Atoms and Molecules

The masses of atoms and the percent abundances of isotopes of elements are measured using **mass spectrometry**, a technique that separates particles according to their mass. In a mass spectrometer, such as the one in Figure 2.15▶, the sample (containing

**Mass Spectrometer**

▲ **FIGURE 2.15  The Mass Spectrometer**  Atoms are converted to positively charged ions, accelerated, and passed through a magnetic field that deflects their path. The heaviest ions undergo the least deflection.

the atoms whose mass is to be measured) is injected into the instrument and vaporized. The vaporized atoms are ionized by an electron beam—the electrons in the beam collide with the atoms, removing electrons and creating positively charged ions. The ions are then accelerated into a magnetic field. When ions drift through a magnetic field, they experience a force that bends their trajectory. The amount of bending depends on the mass of the ions—the trajectories of lighter ions are bent more than those of heavier ones.

In the right side of the spectrometer in Figure 2.15, you can see three different paths, each corresponding to ions of different mass. Finally, the ions strike a detector and produce an electrical signal that is recorded. The result is the separation of the ions according to their mass, producing a mass spectrum such as the one in Figure 2.16▶. The *position* of each peak on the *x*-axis indicates the *mass of the isotope* that was ionized, and the *intensity* (indicated by the height of the peak) indicates the *relative abundance of that isotope*.

The mass spectrum of an elemental sample can be used to determine the atomic mass of that sample of the element. For example, consider the mass spectrum of a naturally occurring sample of silver:

▲ **FIGURE 2.16  The Mass Spectrum of Chlorine**  The position of each peak on the *x*-axis indicates the mass of the isotope. The intensity (or height) of the peak indicates the relative abundance of the isotope. The intensity of the highest peak is usually set to 100% and the intensity of all other peaks is reported relative to the most intense one.

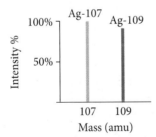

The two peaks correspond to the two isotopes of silver. We can determine the percent abundance of each isotope from the intensity of each line. However, the *total* intensity must be *normalized*—it must be made to equal 100%. We can accomplish this by dividing the intensity of each peak by the total intensity:

$$\text{Abundance of Ag-107} = \frac{100.0\%}{100.0\% + 92.90\%} \times 100\% = 51.84\%$$

$$\text{Abundance of Ag-109} = \frac{92.90\%}{100.0\% + 92.90\%} \times 100\% = 48.16\%$$

Then we can calculate the atomic mass of silver:

$$\text{Ag atomic mass} = 0.5184\,(106.905\text{ amu}) + 0.4816\,(108.904\text{ amu})$$

$$= 55.4195\text{ amu} + 52.4482\text{ amu} = 107.8677 = 107.87\text{ amu}$$

Mass spectrometry can also be used on molecules. Because molecules often fragment (break apart) during ionization, the mass spectrum of a molecule usually

# Chemistry IN YOUR DAY | Evolving Atomic Masses

A t the beginning of 2011, IUPAC (International Union of Pure and Applied Chemistry) published a new periodic table with atomic masses that looked different from previous IUPAC periodic tables. For the first time, instead of listing a single atomic mass for each element, IUPAC listed upper and lower bounds for the atomic masses of several elements (see the periodic table in this box). For example, previous IUPAC periodic tables reported the atomic mass of O (rounded to four significant figures) as 16.00. However, the new periodic table reports the atomic mass as [15.99, 16.00] denoting the upper and lower bounds for the possible atomic masses of terrestrial oxygen.

Why did this happen? The changes were necessary because developments in mass spectrometry have increasingly demonstrated that the atomic masses of several elements are not constant from one sample to another because the isotopic composition is not constant from one sample to another. In other words, the isotopic composition of a sample of a given element can vary depending on the source of the sample.

For example, the lower bound for the atomic mass of oxygen (15.99 amu) comes from measurements of oxygen from Antarctic precipitation, and the upper bound (16.00 amu) comes from measurements of oxygen in marine $N_2O$ (dinitrogen monoxide). Although we have long treated atomic masses as *constants of nature*, they are not, and the new periodic table reflects this.

So what do we do if we need an atomic mass for an element of unknown or unspecified origin? IUPAC has recommended values that apply to most samples found on Earth. The values are rounded so that atomic mass variations in samples found on Earth are plus or minus one in the last digit (just like accepted significant figure conventions). These values are adopted throughout all of the periodic tables in this book except the one shown below, which displays the upper and lower bound for those elements in which variation occurs. For further reading see IUPAC. *Pure Appl. Chem.* 2011, 83(2), 359–396.

## IUPAC Periodic Table of the Elements

| 1 | | | | | | | | | | | | | | | | | 18 |
|---|---|---|---|---|---|---|---|---|---|---|---|---|---|---|---|---|---|
| 1 **H** hydrogen [1.007; 1.009] | 2 | | | | | | | | | | | 13 | 14 | 15 | 16 | 17 | 2 **He** helium 4.003 |
| 3 **Li** lithium [6.938; 6.997] | 4 **Be** beryllium 9.012 | | Key: atomic number **Symbol** name standard atomic weight | | | | | | | | | 5 **B** boron [10.80; 10.83] | 6 **C** carbon [12.00; 12.02] | 7 **N** nitrogen [14.00; 14.01] | 8 **O** oxygen [15.99; 16.00] | 9 **F** fluorine 19.00 | 10 **Ne** neon 20.18 |
| 11 **Na** sodium 22.99 | 12 **Mg** magnesium 24.31 | 3 | 4 | 5 | 6 | 7 | 8 | 9 | 10 | 11 | 12 | 13 **Al** aluminium 26.98 | 14 **Si** silicon [28.08; 28.09] | 15 **P** phosphorus 30.97 | 16 **S** sulfur [32.05; 32.08] | 17 **Cl** chlorine [35.44; 35.46] | 18 **Ar** argon 39.95 |
| 19 **K** potassium 39.10 | 20 **Ca** calcium 40.08 | 21 **Sc** scandium 44.96 | 22 **Ti** titanium 47.87 | 23 **V** vanadium 50.94 | 24 **Cr** chromium 52.00 | 25 **Mn** manganese 54.94 | 26 **Fe** iron 55.85 | 27 **Co** cobalt 58.93 | 28 **Ni** nickel 58.69 | 29 **Cu** copper 63.55 | 30 **Zn** zinc 65.38(2) | 31 **Ga** gallium 69.72 | 32 **Ge** germanium 72.63 | 33 **As** arsenic 74.92 | 34 **Se** selenium 78.96(3) | 35 **Br** bromine 79.90 | 36 **Kr** krypton 83.80 |
| 37 **Rb** rubidium 85.47 | 38 **Sr** strontium 87.62 | 39 **Y** yttrium 88.91 | 40 **Zr** zirconium 91.22 | 41 **Nb** niobium 92.91 | 42 **Mo** molybdenum 95.96(2) | 43 **Tc** technetium | 44 **Ru** ruthenium 101.1 | 45 **Rh** rhodium 102.9 | 46 **Pd** palladium 106.4 | 47 **Ag** silver 107.9 | 48 **Cd** cadmium 112.4 | 49 **In** indium 114.8 | 50 **Sn** tin 118.7 | 51 **Sb** antimony 121.8 | 52 **Te** tellurium 127.6 | 53 **I** iodine 126.9 | 54 **Xe** xenon 131.3 |
| 55 **Cs** caesium 132.9 | 56 **Ba** barium 137.3 | 57–71 lanthanoids | 72 **Hf** hafnium 178.5 | 73 **Ta** tantalum 180.9 | 74 **W** tungsten 183.8 | 75 **Re** rhenium 186.2 | 76 **Os** osmium 190.2 | 77 **Ir** iridium 192.2 | 78 **Pt** platinum 195.1 | 79 **Au** gold 197.0 | 80 **Hg** mercury 200.6 | 81 **Tl** thallium [204.3; 204.4] | 82 **Pb** lead 207.2 | 83 **Bi** bismuth 209.0 | 84 **Po** polonium | 85 **At** astatine | 86 **Rn** radon |
| 87 **Fr** francium | 88 **Ra** radium | 89–103 actinoids | 104 **Rf** rutherfordium | 105 **Db** dubnium | 106 **Sg** seaborgium | 107 **Bh** bohrium | 108 **Hs** hassium | 109 **Mt** meitnerium | 110 **Ds** darmstadtium | 111 **Rg** roentgenium | 112 **Cn** copernicium | | 114 **Fl** flerovium | | 116 **Lv** livermorium | | |

| 57 **La** lanthanum 138.9 | 58 **Ce** cerium 140.1 | 59 **Pr** praseodymium 140.9 | 60 **Nd** neodymium 144.2 | 61 **Pm** promethium | 62 **Sm** samarium 150.4 | 63 **Eu** europium 152.0 | 64 **Gd** gadolinium 157.3 | 65 **Tb** terbium 158.9 | 66 **Dy** dysprosium 162.5 | 67 **Ho** holmium 164.9 | 68 **Er** erbium 167.3 | 69 **Tm** thulium 168.9 | 70 **Yb** ytterbium 173.1 | 71 **Lu** lutetium 175.0 |
|---|---|---|---|---|---|---|---|---|---|---|---|---|---|---|
| 89 **Ac** actinium | 90 **Th** thorium 232.0 | 91 **Pa** protactinium 231.0 | 92 **U** uranium 238.0 | 93 **Np** neptunium | 94 **Pu** plutonium | 95 **Am** americium | 96 **Cm** curium | 97 **Bk** berkelium | 98 **Cf** californium | 99 **Es** einsteinium | 100 **Fm** fermium | 101 **Md** mendelevium | 102 **No** nobelium | 103 **Lr** lawrencium |

**Notes**

- IUPAC 2009 Standard atomic weights abridged to four significant digits (Table 4 published in *Pure Appl. Chem.* 83, 359-396 (2011); doi:10.1351/PAC-REP-10-09-14). The uncertainty in the last digit of the standard atomic weight value is listed in parentheses following the value. In the absence of parentheses, the uncertainty is one in that last digit. An interval in square brackets provides the lower and upper bounds of the standard atomic weight for that element. No values are listed for elements which lack isotopes with a characteristic isotopic abundance in natural terrestrial samples. See PAC for more details.

- "Aluminum" and "cesium" are commonly used alternative spellings for "aluminium" and "caesium."

- Claims for the discovery of all the remaining elements in the last row of the Table, namely elements with atomic numbers 113, 115, 117 and 118, and for which no assignments have yet been made, are being considered by a IUPAC and IUPAP Joint Working Party.

For updates to this table, see iupac.org/reports/periodic_table/. This version is dated 1 June 2012. Copyright © 2012 IUPAC, the International Union of Pure and Applied Chemistry.

INTERNATIONAL UNION OF PURE AND APPLIED CHEMISTRY

contains many peaks representing the masses of different parts of the molecule, as well as a peak representing the mass of the molecule as a whole. The fragments that form upon ionization, and therefore the corresponding peaks that appear in the mass spectrum, are specific to the molecule, so a mass spectrum is like a molecular fingerprint.

Mass spectroscopy can be used to identify an unknown molecule and to determine how much of it is present in a particular sample. For example, mass spectrometry has been used to detect organic (carbon-containing) compounds present in meteorites, a puzzling observation that some scientists speculate may be evidence of life outside of our planet. Most scientists think that the carbon compounds in meteorites probably formed in the same way as the first organic molecules on Earth, indicating that the formation of organic molecules may be common in the universe.

Since the early 1990s, researchers have also successfully applied mass spectrometry to biological molecules, including proteins (the workhorse molecules in cells) and nucleic acids (the molecules that carry genetic information). For a long time, these molecules could not be analyzed by mass spectrometry because they were difficult to vaporize and ionize without being destroyed, but modern techniques have overcome this problem. A tumor, for example, can now be instantly analyzed by mass spectrometry to determine whether it contains specific proteins associated with cancer.

# 2.9 Molar Mass: Counting Atoms by Weighing Them

 **The Mole Concept**

Shrimp are normally sold by *count*, which indicates the number of shrimp per pound. For example, 41–50 count shrimp means that there are between 41 and 50 shrimp per pound. The smaller the count, the larger the shrimp. Big tiger prawns have counts as low as 10–15, which means that each shrimp can weigh up to 1/10 of a pound. One advantage of categorizing shrimp in this way is that we can count the shrimp by weighing them. For example, two pounds of 41–50 count shrimp contains between 82 and 100 shrimp.

A similar (but more precise) concept exists for atoms. Counting atoms is much more difficult than counting shrimp, yet as chemists we often need to know the number of atoms in a sample of a given mass. Why? *Because chemical processes happen between particles.* For elements, those particles are atoms. For example, when hydrogen and oxygen combine to form water, two hydrogen atoms combine with one oxygen atom to form one water molecule. If we want to know how much hydrogen to react with a given mass of oxygen to form water, we need to know the number of atoms in the given mass of oxygen. We also need to know the mass of hydrogen that contains exactly twice that number of atoms.

As another example, consider intravenous fluids—fluids that are delivered to patients by directly dripping them into veins. These fluids are saline (sodium chloride) solutions that must have a specific number of sodium and chloride ions per liter of fluid. The number of particles in the fluid directly influences the properties of the fluid. Administering an intravenous fluid with the wrong number of sodium and chloride ions could be fatal.

Atoms are far too small to count by any ordinary means. As we noted earlier, even if you could somehow count atoms, and counted them 24 hours a day for as long as you lived, you would barely begin to count the number of atoms in something as small as a sand grain. Therefore, if we want to know the number of atoms in anything of ordinary size, we count them by weighing.

## The Mole: A Chemist's "Dozen"

When we count large numbers of objects, we often use units such as a dozen (12 objects) or a gross (144 objects) to organize our counting and to keep our numbers more manageable. With atoms, quadrillions of which may be in a speck of dust, we need a much larger number for this purpose. The chemist's "dozen" is the **mole** (abbreviated mol). A mole is the *amount* of material containing $6.02214 \times 10^{23}$ particles.

$$1 \text{ mol} = 6.02214 \times 10^{23} \text{ particles}$$

This number is **Avogadro's number**, named after Italian physicist Amedeo Avogadro (1776–1856), and is a convenient number to use when working with atoms, molecules, and ions. In this book, we usually round Avogadro's number to four significant figures or

Twenty-two copper pennies contain approximately 1 mol of copper atoms.

Before 1982, when they became almost all zinc with only a copper coating, pennies were mostly copper.

One tablespoon of water contains approximately 1 mol of water molecules.

One tablespoon is approximately 15 mL; one mol of water occupies 18 mL.

$6.022 \times 10^{23}$. Notice that the definition of the mole is an *amount* of a substance. We often refer to the number of moles of substance as the *amount* of the substance.

The first thing to understand about the mole is that it can specify Avogadro's number of anything. For example, one mole of marbles corresponds to $6.022 \times 10^{23}$ marbles, and one mole of sand grains corresponds to $6.022 \times 10^{23}$ sand grains. *One mole of anything is $6.022 \times 10^{23}$ units of that thing.* One mole of atoms, ions, or molecules, however, makes up objects of everyday sizes. Twenty-two copper pennies, for example, contain approximately 1 mol of copper atoms, and 1 tablespoon of water contains approximately 1 mol of water molecules.

The second, and more fundamental, thing to understand about the mole is how it gets its specific value.

**The value of the mole is equal to the number of atoms in exactly 12 g of pure carbon-12 (12 g C = 1 mol C atoms = $6.022 \times 10^{23}$ C atoms).**

The definition of the mole gives us a relationship between mass (grams of carbon) and number of atoms (Avogadro's number). This relationship, as we will see shortly, allows us to count atoms by weighing them.

## Converting between Number of Moles and Number of Atoms

Converting between number of moles and number of atoms is similar to converting between dozens of eggs and number of eggs. For eggs, you use the conversion factor 1 dozen eggs = 12 eggs. For atoms, you use the conversion factor 1 mol atoms = $6.022 \times 10^{23}$ atoms. The conversion factors take the following forms:

$$\frac{1 \text{ mol atoms}}{6.022 \times 10^{23} \text{ atoms}} \quad \text{or} \quad \frac{6.022 \times 10^{23} \text{ atoms}}{1 \text{ mol atoms}}$$

Example 2.6 demonstrates how to use these conversion factors in calculations.

---

**Example 2.6**    **Converting between Number of Moles and Number of Atoms**

Calculate the number of copper atoms in 2.45 mol of copper.

| | |
|---|---|
| **SORT** You are given the amount of copper in moles and asked to find the number of copper atoms. | **GIVEN:** 2.45 mol Cu<br>**FIND:** Cu atoms |
| **STRATEGIZE** Convert between number of moles and number of atoms by using Avogadro's number as a conversion factor. | **CONCEPTUAL PLAN**<br><br>$$\boxed{\text{mol Cu}} \longrightarrow \boxed{\text{Cu atoms}}$$<br>$$\frac{6.022 \times 10^{23} \text{ Cu atoms}}{1 \text{ mol Cu}}$$<br><br>**RELATIONSHIPS USED**<br>$6.022 \times 10^{23} = 1$ mol (Avogadro's number) |
| **SOLVE** Follow the conceptual plan to solve the problem. Begin with 2.45 mol Cu and multiply by Avogadro's number to get to the number of Cu atoms. | **SOLUTION**<br>$$2.45 \text{ mol Cu} \times \frac{6.022 \times 10^{23} \text{ Cu atoms}}{1 \text{ mol Cu}} = 1.48 \times 10^{24} \text{ Cu atoms}$$ |

**CHECK** Since atoms are small, it makes sense that the answer is large. The given number of moles of copper is almost 2.5, so the number of atoms is almost 2.5 times Avogadro's number.

**FOR PRACTICE 2.6**    A pure silver ring contains $2.80 \times 10^{22}$ silver atoms. How many moles of silver atoms does it contain?

# Converting between Mass and Amount (Number of Moles)

To count atoms by weighing them, we need one other conversion factor—the mass of 1 mol of atoms. For the isotope carbon-12, we know that the mass of 1 mol of atoms is exactly 12 g, which is numerically equivalent to carbon-12's atomic mass in atomic mass units. Since the masses of all other elements are defined relative to carbon-12, the same relationship holds for all elements.

The mass of one mole of atoms of an element is its **molar mass**.

**An element's molar mass in grams per mole is numerically equal to the element's atomic mass in atomic mass units.**

For example, copper has an atomic mass of 63.55 amu and a molar mass of 63.55 g/mol. One mole of copper atoms therefore has a mass of 63.55 g. Just as the count for shrimp depends on the size of the shrimp, so the mass of 1 mol of atoms depends on the element: 1 mol of aluminum atoms (which are lighter than copper atoms) has a mass of 26.98 g, 1 mol of carbon atoms (which are even lighter than aluminum atoms) has a mass of 12.01 g, and 1 mol of helium atoms (lighter yet) has a mass of 4.003 g.

26.98 g aluminum = 1 mol aluminum = $6.022 \times 10^{23}$ Al atoms　　Al

12.01 g carbon = 1 mol carbon = $6.022 \times 10^{23}$ C atoms　　C

4.003 g helium = 1 mol helium = $6.022 \times 10^{23}$ He atoms　　He

The lighter the atom, the less mass in 1 mol of atoms.

1 dozen peas

1 dozen marbles

▲ The two dishes contain the same number of objects (12), but the masses are different because peas are less massive than marbles. Similarly, a mole of light atoms will have less mass than a mole of heavier atoms.

The molar mass of any element is the conversion factor between the mass (in grams) of that element and the amount (in moles) of that element. For carbon:

$$12.01 \text{ g C} = 1 \text{ mol C} \quad \text{or} \quad \frac{12.01 \text{ g C}}{\text{mol C}} \quad \text{or} \quad \frac{1 \text{ mol C}}{12.01 \text{ g C}}$$

Example 2.7, on the next page, demonstrates how to use these conversion factors.

**Example** 2.7 | **Converting between Mass and Amount (Number of Moles)**

Calculate the amount of carbon (in moles) contained in a 0.0265 g pencil "lead." (Assume that the pencil lead is made of pure graphite, a form of carbon.)

| | |
|---|---|
| **SORT** You are given the mass of carbon and asked to find the amount of carbon in moles. | **GIVEN:** 0.0265 g C <br> **FIND:** mol C |
| **STRATEGIZE** Convert between mass and amount (in moles) of an element by using the molar mass of the element. | **CONCEPTUAL PLAN** <br><br> g C $\longrightarrow$ mol C <br> $\dfrac{1\ \text{mol}}{12.01\ \text{g}}$ <br><br> **RELATIONSHIPS USED** <br> 12.01 g C = 1 mol C (carbon molar mass) |
| **SOLVE** Follow the conceptual plan to solve the problem. | **SOLUTION** <br><br> $0.0265\ \text{g C} \times \dfrac{1\ \text{mol C}}{12.01\ \text{g C}} = 2.21 \times 10^{-3}\ \text{mol C}$ |

**CHECK** The given mass of carbon is much less than the molar mass of carbon, so it makes sense that the answer (the amount in moles) is much less than 1 mol of carbon.

**FOR PRACTICE 2.7** Calculate the amount of copper (in moles) in a 35.8 g pure copper sheet.

**FOR MORE PRACTICE 2.7** Calculate the mass (in grams) of 0.473 mol of titanium.

We now have all the tools to count the number of atoms in a sample of an element by weighing it. First, we obtain the mass of the sample. Then we convert it to an amount in moles using the element's molar mass. Finally, we convert it to number of atoms using Avogadro's number. The conceptual plan for these kinds of calculations takes the following form:

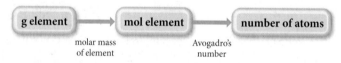

Examples 2.8 and 2.9 demonstrate these conversions.

**Example** 2.8 | **The Mole Concept—Converting between Mass and Number of Atoms**

How many copper atoms are in a copper penny with a mass of 3.10 g? (Assume that the penny is composed of pure copper.)

| | |
|---|---|
| **SORT** You are given the mass of copper and asked to find the number of copper atoms. | **GIVEN:** 3.10 g Cu <br> **FIND:** Cu atoms |
| **STRATEGIZE** Convert between the mass of an element in grams and the number of atoms of the element by first converting to moles (using the molar mass of the element) and then to number of atoms (using Avogadro's number). | **CONCEPTUAL PLAN** <br><br>  <br><br> **RELATIONSHIPS USED** <br> 63.55 g Cu = 1 mol Cu (molar mass of copper) <br> $6.022 \times 10^{23} = 1$ mol (Avogadro's number) |

| SOLVE  Follow the conceptual plan to solve the problem. Begin with 3.10 g Cu and multiply by the appropriate conversion factors to arrive at the number of Cu atoms. | SOLUTION $$3.10 \text{ g Cu} \times \frac{1 \text{ mol C}}{63.55 \text{ g Cu}} \times \frac{6.022 \times 10^{23} \text{ Cu atoms}}{1 \text{ mol Cu}} = 2.94 \times 10^{22} \text{ Cu atoms}$$ |
|---|---|

CHECK  The answer (the number of copper atoms) is less than $6.022 \times 10^{23}$ (1 mol). This is consistent with the given mass of copper atoms, which is less than the molar mass of copper.

FOR PRACTICE 2.8   How many carbon atoms are there in a 1.3-carat diamond? Diamonds are a form of pure carbon. (1 carat = 0.20 g)

FOR MORE PRACTICE 2.8   Calculate the mass of $2.25 \times 10^{22}$ tungsten atoms.

Notice that numbers with large exponents, such as $6.022 \times 10^{23}$, are almost unbelievably large. Twenty-two copper pennies contain $6.022 \times 10^{23}$ or 1 mol of copper atoms, but $6.022 \times 10^{23}$ pennies would cover Earth's entire surface to a depth of 300 m. Even objects small by everyday standards occupy a huge space when we have a mole of them. For example, a grain of sand has a mass of less than 1 mg and a diameter of less than 0.1 mm, yet 1 mol of sand grains would cover the state of Texas to a depth of several feet. For every increase of 1 in the exponent of a number, the number increases by a factor of 10, so $10^{23}$ is incredibly large. Of course 1 mol has to be a large number if it is to have practical value because atoms are so small.

## Example 2.9   The Mole Concept

An aluminum sphere contains $8.55 \times 10^{22}$ aluminum atoms. What is the sphere's radius in centimeters? The density of aluminum is $2.70 \text{ g/cm}^3$.

| SORT  You are given the number of aluminum atoms in a sphere and the density of aluminum. You are asked to find the radius of the sphere. | GIVEN:  $8.55 \times 10^{22}$ Al atoms $d = 2.70 \text{ g/cm}^3$ FIND:  radius ($r$) of sphere |
|---|---|

STRATEGIZE  The heart of this problem is density, which relates mass to volume; though you aren't given the mass directly, you are given the number of atoms, which you can use to find mass.

1. Convert from number of atoms to number of moles using Avogadro's number as a conversion factor.

2. Convert from number of moles to mass using molar mass as a conversion factor.

3. Convert from mass to volume (in $cm^3$) using density as a conversion factor.

4. Once you calculate the volume, find the radius from the volume using the formula for the volume of a sphere.

CONCEPTUAL PLAN

RELATIONSHIPS AND EQUATIONS USED

$6.022 \times 10^{23} = 1$ mol (Avogadro's number)

26.98 g Al = 1 mol Al (molar mass of aluminum)

$2.70 \text{ g/cm}^3$ (density of aluminum)

$V = \frac{4}{3}\pi r^3$ (volume of a sphere)

—*Continued on the next page*

*—Continued*

| SOLVE | SOLUTION |
|---|---|
| Finally, follow the conceptual plan to solve the problem. Begin with $8.55 \times 10^{22}$ Al atoms and multiply by the appropriate conversion factors to arrive at volume in $cm^3$. | $8.55 \times 10^{22} \text{ Al atoms} \times \dfrac{1 \text{ mol Al}}{6.022 \times 10^{23} \text{ Al atoms}}$ $\times \dfrac{26.98 \text{ g Al}}{1 \text{ mol Al}} \times \dfrac{1 \text{ cm}^3}{2.70 \text{ g Al}} = 1.4\underline{1}87 \text{ cm}^3$ |
| Then solve the equation for the volume of a sphere for $r$ and substitute the volume to calculate $r$. | $V = \dfrac{4}{3}\pi r^3$ $r = \sqrt[3]{\dfrac{3V}{4\pi}} = \sqrt[3]{\dfrac{3(1.4\underline{1}87 \text{ cm}^3)}{4\pi}} = 0.697 \text{ cm}$ |

**CHECK** The units of the answer (cm) are correct. The magnitude cannot be estimated accurately, but a radius of about one-half of a centimeter is reasonable for just over one-tenth of a mole of aluminum atoms.

**FOR PRACTICE 2.9** A titanium cube contains $2.86 \times 10^{23}$ atoms. What is the edge length of the cube? The density of titanium is $4.50 \text{ g/cm}^3$.

**FOR MORE PRACTICE 2.9** Find the number of atoms in a copper rod with a length of 9.85 cm and a radius of 1.05 cm. The density of copper is $8.96 \text{ g/cm}^3$.

CONCEPTUAL
**CONNECTION 2.7**

**Avogadro's Number** Why is Avogadro's number defined as $6.022 \times 10^{23}$ and not a simpler round number such as $1.00 \times 10^{23}$?

CONCEPTUAL
**CONNECTION 2.8**

**The Mole** Without doing any calculations, determine which sample contains the most atoms.

**(a)** a 1-g sample of copper

**(b)** a 1-g sample of carbon

**(c)** a 10-g sample of uranium

# Self-Assessment Quiz

**Q1.** Two samples of a compound containing elements A and B are decomposed. The first sample produces 15 g of A and 35 g of B. The second sample produces 25 g of A and what mass of B?

a) 11 g   b) 58 g   c) 21 g   d) 45 g

**Q2.** A compound containing only carbon and hydrogen has a carbon-to-hydrogen mass ratio of 11.89. Which carbon-to-hydrogen mass ratio is possible for another compound composed only of carbon and hydrogen?

a) 2.50   b) 3.97   c) 4.66   d) 7.89

**Q3.** Which idea came out of Rutherford's gold foil experiment?

a) Atoms contain protons and neutrons.

b) Matter is composed of atoms.

c) Elements have isotopes.

d) Atoms are mostly empty space.

**Q4.** A student re-creates the Millikan oil drop experiment and tabulates the relative charges of the oil drops in terms of a constant, $\alpha$.

| Drop #1 | $\alpha$ |
|---|---|
| Drop #2 | $\dfrac{3}{2}\alpha$ |
| Drop #3 | $\dfrac{5}{2}\alpha$ |
| Drop #4 | $3\alpha$ |

What charge for the electron (in terms of $\alpha$) is consistent with these data?

a) $\dfrac{1}{2}\alpha$   b) $\alpha$   c) $\dfrac{3}{2}\alpha$   d) $2\alpha$

**Q5.** Determine the number of protons and neutrons in the isotope Fe-58.

a) 26 protons and 58 neutrons
b) 32 protons and 26 neutrons
c) 26 protons and 32 neutrons
d) 58 protons and 58 neutrons

**Q6.** An isotope of an element contains 82 protons and 122 neutrons. What is the symbol for the isotope?

a) $^{204}_{82}Pb$  b) $^{122}_{82}Pb$  c) $^{122}_{40}Zr$  d) $^{204}_{40}Zr$

**Q7.** Determine the number of electrons in the $Cr^{3+}$ ion.

a) 24 electrons       b) 27 electrons
c) 3 electrons        d) 21 electrons

**Q8.** Which pair of elements do you expect to be most similar in their chemical properties?

a) K and Fe          b) O and Si
c) Ne and N          d) Br and I

**Q9.** Which element is *not* a main-group element?

a) Se       b) Mo       c) Sr       d) Ba

**Q10.** What is the charge of the ion most commonly formed by S?

a) 2+       b) +       c) −       d) 2−

**Q11.** A naturally occurring sample of an element contains only two isotopes. The first isotope has a mass of 68.9255 amu and a natural abundance of 60.11%. The second isotope has a mass of 70.9247 amu. Find the atomic mass of the element.

a) 70.13 amu              b) 69.72 amu
c) 84.06 amu              d) 69.93 amu

**Q12.** Which sample contains the greatest number of atoms?

a) 14 g C   b) 49 g Cr   c) 102 g Ag   d) 202 g Pb

**Q13.** Determine the number of atoms in 1.85 mL of mercury. (The density of mercury is 13.5 g/mL.)

a) $3.02 \times 10^{27}$ atoms    b) $4.11 \times 10^{20}$ atoms
c) $7.50 \times 10^{22}$ atoms    d) $1.50 \times 10^{25}$ atoms

**Q14.** A 20.0 g sample of an element contains $4.95 \times 10^{23}$ atoms. Identify the element.

a) Cr       b) O       c) Mg       d) Fe

**Q15.** Copper has two naturally occurring isotopes with masses 62.94 amu and 64.93 amu and has an atomic mass of 63.55 amu. Which mass spectrum is most likely to correspond to a naturally occurring sample of copper?

a)
(a)

b)
(b)

c)
(c)

d)
(d)

**Answers: 1.** (b)  **2.** (b)  **3.** (d)  **4.** (a)  **5.** (c)  **6.** (a)  **7.** (d)  **8.** (d)  **9.** (b)  **10.** (d)  **11.** (b)  **12.** (a)  **13.** (c)  **14.** (c)  **15.** (a)

# **Chapter 2** IN REVIEW

## Key Terms

**Section 2.3**
law of conservation of mass (47)
law of definite proportions (48)
law of multiple proportions (49)
atomic theory (50)

**Section 2.4**
cathode rays (51)
cathode ray tube (51)
electrical charge (51)
electron (51)

**Section 2.5**
radioactivity (53)
nuclear theory (54)
nucleus (54)
proton (54)
neutron (55)

**Section 2.6**
atomic mass unit (amu) (55)
atomic number (Z) (56)
chemical symbol (56)
isotope (58)
natural abundance (58)
mass number (A) (58)
ion (59)
cation (59)
anion (59)

**Section 2.7**
periodic law (61)
metal (62)
nonmetal (62)
metalloid (63)
semiconductor (63)
main-group elements (63)
transition elements
  (transition metals) (63)
family (group) (63)
noble gases (63)

alkali metals (63)
alkaline earth metals (63)
halogens (64)

**Section 2.8**
atomic mass (65)
mass spectrometry (66)

**Section 2.9**
mole (mol) (69)
Avogadro's number (69)
molar mass (71)

## Key Concepts

### Brownian Motion (2.1)

▶ Brownian motion is the erratic, jittery motion of small particles that was first observed by Robert Brown in 1827. The description of Brownian motion by Einstein in 1905 and confirmation by Perrin in 1908 removed any lingering doubt about the particulate nature of matter.

### The Atomic Theory (2.2, 2.3)

▶ Each element is composed of indestructible particles called atoms.

▶ All atoms of a given element have the same mass and other properties.

▶ Atoms combine in simple, whole-number ratios to form compounds.

▶ Atoms of one element cannot change into atoms of another element. In a chemical reaction, atoms change the way that they are bound together with other atoms to form a new substance.

### The Electron (2.4)

▶ J. J. Thomson discovered the electron in the late 1800s through experiments with cathode rays. He deduced that electrons are negatively charged, and he measured their charge-to-mass ratio.

▶ Robert Millikan measured the charge of the electron, which—in conjunction with Thomson's results—led to the calculation of the mass of an electron.

### The Nuclear Atom (2.5)

▶ In 1909, Ernest Rutherford probed the inner structure of the atom by working with a form of radioactivity called alpha radiation and developed the nuclear theory of the atom.

▶ Nuclear theory states that the atom is mainly empty space, with most of its mass concentrated in a tiny region called the nucleus and most of its volume occupied by relatively light electrons.

### Subatomic Particles (2.6)

▶ Atoms are composed of three fundamental particles: the proton (1 amu, +1 charge), the neutron (1 amu, 0 charge), and the electron (~0 amu, −1 charge).

▶ The number of protons in the nucleus of the atom is its atomic number (Z) and defines the element.

▶ The sum of the number of protons and neutrons is the mass number (A).

▶ Atoms of an element that have different numbers of neutrons (and therefore different mass numbers) are isotopes.

▶ Atoms that lose or gain electrons become charged and are ions. Cations are positively charged and anions are negatively charged.

### The Periodic Table (2.7)

▶ The periodic table tabulates all known elements in order of increasing atomic number.

▶ The periodic table is arranged so that similar elements are grouped together in columns.

▶ Elements on the left side and in the center of the periodic table are metals and tend to lose electrons in chemical changes.

▶ Elements on the upper right side of the periodic table are nonmetals and tend to gain electrons in chemical changes.

▶ Elements located on the boundary between metals and nonmetals are metalloids.

### Atomic Mass and the Mole (2.8, 2.9)

▶ The atomic mass of an element, listed directly below its symbol in the periodic table, is a weighted average of the masses of the naturally occurring isotopes of the element.

▶ One mole of an element is the amount of that element that contains Avogadro's number ($6.022 \times 10^{23}$) of atoms.

▶ Any sample of an element with a mass (in grams) that equals its atomic mass contains one mole of the element. For example, the atomic mass of carbon is 12.011 amu; therefore, 12.011 g of carbon contains 1 mol of carbon atoms.

## Key Equations and Relationships

Relationship between Mass Number (A), Number of Protons (p), and Number of Neutrons (n) (2.6)

$A$ = number of protons (p) + number of neutrons (n)

Atomic Mass (2.8)

$$\text{Atomic mass} = \sum_{n} (\text{fraction of isotope } n) \times (\text{mass of isotope } n)$$

Avogadro's Number (2.9)

$$1 \text{ mol} = 6.0221421 \times 10^{23} \text{ particles}$$

## Key Learning Outcomes

| CHAPTER OBJECTIVES | ASSESSMENT |
|---|---|
| Using the Law of Definite Proportions (**2.3**) | Example 2.1   For Practice 2.1   Exercises 31, 32 |
| Using the Law of Multiple Proportions (**2.3**) | Example 2.2   For Practice 2.2   Exercises 35–38 |
| Working with Atomic Numbers, Mass Numbers, and Isotope Symbols (**2.6**) | Example 2.3   For Practice 2.3   Exercises 51–58 |
| Predicting the Charge of Ions (**2.7**) | Example 2.4   For Practice 2.4   Exercises 59–62 |
| Calculating Atomic Mass (**2.8**) | Example 2.5   For Practice 2.5   For More Practice 2.5   Exercises 71, 72, 74–77 |
| Converting between Moles and Number of Atoms (**2.9**) | Example 2.6   For Practice 2.6   Exercises 81, 82 |
| Converting between Mass and Amount (in Moles) (**2.9**) | Example 2.7   For Practice 2.7   For More Practice 2.7   Exercises 83, 84 |
| Using the Mole Concept (**2.9**) | Examples 2.8, 2.9   For Practice 2.8, 2.9   For More Practice 2.8, 2.9   Exercises 85–94, 112, 113 |

# EXERCISES

## Review Questions

1. What is Brownian motion? How is it related to the development of the idea that matter is particulate?

2. Summarize the history of the atomic idea. How was Dalton able to convince others to accept an idea that had been controversial for 2000 years?

3. State and explain the law of conservation of mass.

4. State and explain the law of definite proportions.

5. State and explain the law of multiple proportions. How is the law of multiple proportions different from the law of definite proportions?

6. What are the main ideas in Dalton's atomic theory? How do they help explain the laws of conservation of mass, of constant composition, and of definite proportions?

7. How and by whom was the electron discovered? What basic properties of the electron were reported with its discovery?

8. Explain Millikan's oil drop experiment and how it led to the measurement of the electron's charge. Why is the magnitude of the charge of the electron so important?

9. Describe the plum-pudding model of the atom.

10. Describe Rutherford's gold foil experiment. How did the experiment prove that the plum-pudding model of the atom was wrong?

11. Describe Rutherford's nuclear model of the atom. What was revolutionary about his model?

12. If matter is mostly empty space, as suggested by Rutherford, then why does it appear so solid?

13. List the three subatomic particles that compose atoms and give the basic properties (mass and charge) of each.

14. What defines an element?

15. Explain the difference between $Z$ (the atomic number) and $A$ (the mass number).

16. Where do elements get their names?

17. What are isotopes? What is percent natural abundance of isotopes?

18. Describe the two different notations used to specify isotopes and give an example of each.

19. What is an ion? A cation? An anion?

**20.** State the periodic law. How did the periodic law lead to the periodic table?

**21.** Describe the characteristic properties of metals, nonmetals, and metalloids.

**22.** List the characteristic properties of each group.
   a. noble gases          b. alkali metals
   c. alkaline earth metals   d. halogens

**23.** How do you predict the charges of ions formed by main-group elements?

**24.** What is atomic mass? How is it calculated?

**25.** Explain how a mass spectrometer works.

**26.** What kind of information can be determined from a mass spectrum?

**27.** What is a mole? How is the mole concept useful in chemical calculations?

**28.** Why is the mass corresponding to a mole of one element different from the mass corresponding to a mole of another element?

# Problems by Topic

*Note: Answers to all odd-numbered Problems, numbered in blue, can be found in Appendix III. Exercises in the Problems by Topic section are paired, with each odd-numbered problem followed by a similar even-numbered problem. Exercises in the Cumulative Problems section are also paired, but somewhat more loosely. (Challenge Problems and Conceptual Problems, because of their nature, are unpaired.)*

## The Laws of Conservation of Mass, Definite Proportions, and Multiple Proportions

**29.** A hydrogen-filled balloon is ignited and 1.50 g of hydrogen is reacted with 12.0 g of oxygen. How many grams of water vapor form? (Assume that water vapor is the only product.)

**30.** An automobile gasoline tank holds 21 kg of gasoline. When the gasoline burns, 84 kg of oxygen is consumed, and carbon dioxide and water are produced. What is the total combined mass of carbon dioxide and water that is produced?

**31.** Two samples of carbon tetrachloride are decomposed into their constituent elements. One sample produces 38.9 g of carbon and 448 g of chlorine, and the other sample produces 14.8 g of carbon and 134 g of chlorine. Are these results consistent with the law of definite proportions? Explain your answer.

**32.** Two samples of sodium chloride are decomposed into their constituent elements. One sample produces 6.98 g of sodium and 10.7 g of chlorine, and the other sample produces 11.2 g of sodium and 17.3 g of chlorine. Are these results consistent with the law of definite proportions? Explain your answer.

**33.** The mass ratio of sodium to fluorine in sodium fluoride is 1.21:1. A sample of sodium fluoride produces 28.8 g of sodium upon decomposition. How much fluorine (in grams) forms?

**34.** Upon decomposition, one sample of magnesium fluoride produces 1.65 kg of magnesium and 2.57 kg of fluorine. A second sample produces 1.32 kg of magnesium. How much fluorine (in grams) does the second sample produce?

**35.** Two different compounds containing osmium and oxygen have the following masses of oxygen per gram of osmium: 0.168 and 0.3369 g. Show that these amounts are consistent with the law of multiple proportions.

**36.** Palladium forms three different compounds with sulfur. The mass of sulfur per gram of palladium in each compound is listed here. Show that these masses are consistent with the law of multiple proportions.

| Compound | Grams S per Gram Pd |
|---|---|
| A | 0.603 |
| B | 0.301 |
| C | 0.151 |

**37.** Sulfur and oxygen form both sulfur dioxide and sulfur trioxide. When samples of these are decomposed, the sulfur dioxide produces 3.49 g oxygen and 3.50 g sulfur, while the sulfur trioxide produces 6.75 g oxygen and 4.50 g sulfur. Calculate the mass of oxygen per gram of sulfur for each sample and show that these results are consistent with the law of multiple proportions.

**38.** Sulfur and fluorine form several different compounds including sulfur hexafluoride and sulfur tetrafluoride. Decomposition of a sample of sulfur hexafluoride produces 4.45 g of fluorine and 1.25 g of sulfur, while decomposition of a sample of sulfur tetrafluoride produces 4.43 g of fluorine and 1.87 g of sulfur. Calculate the mass of fluorine per gram of sulfur for each sample and show that these results are consistent with the law of multiple proportions.

## Atomic Theory, Nuclear Theory, and Subatomic Particles

**39.** Which statements are *consistent* with Dalton's atomic theory as it was originally stated? Why?
   a. Sulfur and oxygen atoms have the same mass.
   b. All cobalt atoms are identical.
   c. Potassium and chlorine atoms combine in a 1:1 ratio to form potassium chloride.
   d. Lead atoms can be converted into gold.

**40.** Which statements are *inconsistent* with Dalton's atomic theory as it was originally stated? Why?
   a. All carbon atoms are identical.
   b. An oxygen atom combines with 1.5 hydrogen atoms to form a water molecule.
   c. Two oxygen atoms combine with a carbon atom to form a carbon dioxide molecule.
   d. The formation of a compound often involves the destruction of one or more atoms.

**41.** Which statements are *consistent* with Rutherford's nuclear theory as it was originally stated? Why?
   a. The volume of an atom is mostly empty space.
   b. The nucleus of an atom is small compared to the size of the atom.
   c. Neutral lithium atoms contain more neutrons than protons.
   d. Neutral lithium atoms contain more protons than electrons.

**42.** Which statements are *inconsistent* with Rutherford's nuclear theory as it was originally stated? Why?
   a. Since electrons are smaller than protons, and since a hydrogen atom contains only one proton and one electron, it must follow that the volume of a hydrogen atom is mostly due to the proton.
   b. A nitrogen atom has seven protons in its nucleus and seven electrons outside of its nucleus.
   c. A phosphorus atom has 15 protons in its nucleus and 150 electrons outside of its nucleus.
   d. The majority of the mass of a fluorine atom is due to its nine electrons.

43. A chemist in an imaginary universe, where electrons have a different charge than they do in our universe, performs the Millikan oil drop experiment to measure the electron's charge. The charges of several drops are recorded here. What is the charge of the electron in this imaginary universe?

| Drop # | Charge |
|--------|--------|
| A | $-6.9 \times 10^{-19}$ C |
| B | $-9.2 \times 10^{-19}$ C |
| C | $-11.5 \times 10^{-19}$ C |
| D | $-4.6 \times 10^{-19}$ C |

44. Imagine a unit of charge called the *zorg*. A chemist performs the oil drop experiment and measures the charge of each drop in zorgs. Based on the results shown here, what is the charge of the electron in zorgs (z)? How many electrons are in each drop?

| Drop # | Charge |
|--------|--------|
| A | $-4.8 \times 10^{-9}$ z |
| B | $-9.6 \times 10^{-9}$ z |
| C | $-6.4 \times 10^{-9}$ z |
| D | $-12.8 \times 10^{-9}$ z |

45. On a dry day, your body can accumulate static charge from walking across a carpet or from brushing your hair. If your body develops a charge of $-15\ \mu C$ (microcoulombs), how many excess electrons has it acquired? What is their collective mass?

46. How many electrons are necessary to produce a charge of $-1.0$ C? What is the mass of this many electrons?

47. Which statements about subatomic particles are true?
    a. If an atom has an equal number of protons and electrons, it will be charge-neutral.
    b. Electrons are attracted to protons.
    c. Electrons are much lighter than neutrons.
    d. Protons have twice the mass of neutrons.

48. Which statements about subatomic particles are false?
    a. Protons and electrons have charges of the same magnitude but opposite signs.
    b. Protons have about the same mass as neutrons.
    c. Some atoms don't have any protons.
    d. Protons and neutrons have charges of the same magnitude but opposite signs.

49. How many electrons does it take to equal the mass of a proton?

50. A helium nucleus has two protons and two neutrons. How many electrons does it take to equal the mass of a helium nucleus?

## Isotopes and Ions

51. Write isotopic symbols in the form X-*A* (e.g., C-13) for each isotope.
    a. the silver isotope with 60 neutrons
    b. the silver isotope with 62 neutrons
    c. the uranium isotope with 146 neutrons
    d. the hydrogen isotope with one neutron

52. Write isotopic symbols in the form $^A_Z X$ for each isotope.
    a. the copper isotope with 34 neutrons
    b. the copper isotope with 36 neutrons
    c. the potassium isotope with 21 neutrons
    d. the argon isotope with 22 neutrons

53. Determine the number of protons and the number of neutrons in each isotope.
    a. $^{14}_{7}N$   b. $^{23}_{11}Na$   c. $^{222}_{86}Rn$   d. $^{208}_{82}Pb$

54. Determine the number of protons and the number of neutrons in each isotope.
    a. $^{40}_{19}K$   b. $^{226}_{88}Ra$   c. $^{99}_{43}Tc$   d. $^{33}_{15}P$

55. The amount of carbon-14 in ancient artifacts and fossils is often used to establish their age. Determine the number of protons and the number of neutrons in carbon-14 and write its symbol in the form $^A_Z X$.

56. Uranium-235 is used in nuclear fission. Determine the number of protons and the number of neutrons in uranium-235 and write its symbol in the form $^A_Z X$.

57. Determine the number of protons and the number of electrons in each ion.
    a. $Ni^{2+}$   b. $S^{2-}$   c. $Br^-$   d. $Cr^{3+}$

58. Determine the number of protons and the number of electrons in each ion.
    a. $Al^{3+}$   b. $Se^{2-}$   c. $Ga^{3+}$   d. $Sr^{2+}$

59. Predict the charge of the ion formed by each element.
    a. O   b. K   c. Al   d. Rb

60. Predict the charge of the ion formed by each element.
    a. Mg   b. N   c. F   d. Na

61. Fill in the blanks to complete the table.

| Symbol | Ion Formed | Number of Electrons in Ion | Number of Protons in Ion |
|--------|-----------|----------------------------|--------------------------|
| Ca | $Ca^{2+}$ | ____ | ____ |
| ____ | $Be^{2+}$ | 2 | ____ |
| Se | ____ | ____ | 34 |
| In | ____ | ____ | 49 |

62. Fill in the blanks to complete the table.

| Symbol | Ion Formed | Number of Electrons in Ion | Number of Protons in Ion |
|--------|-----------|----------------------------|--------------------------|
| Cl | ____ | ____ | 17 |
| Te | ____ | 54 | ____ |
| Br | $Br^-$ | ____ | ____ |
| ____ | $Sr^{2+}$ | ____ | 38 |

## The Periodic Table and Atomic Mass

63. Write the name of each element and classify it as a metal, nonmetal, or metalloid.
    a. K   b. Ba   c. I   d. O   e. Sb

64. Write the symbol for each element and classify it as a metal, nonmetal, or metalloid.
    a. gold
    b. fluorine
    c. sodium
    d. tin
    e. argon

65. Determine whether or not each element is a main-group element.
    a. tellurium
    b. potassium
    c. vanadium
    d. manganese

66. Determine whether or not each element is a transition element.
    a. Cr
    b. Br
    c. Mo
    d. Cs

67. Classify each element as an alkali metal, alkaline earth metal, halogen, or noble gas.
    a. sodium
    b. iodine
    c. calcium
    d. barium
    e. krypton

68. Classify each element as an alkali metal, alkaline earth metal, halogen, or noble gas.
    a. F
    b. Sr
    c. K
    d. Ne
    e. At

69. Which pair of elements do you expect to be most similar? Why?
    a. N and Ni
    b. Mo and Sn
    c. Na and Mg
    d. Cl and F
    e. Si and P

70. Which pair of elements do you expect to be most similar? Why?
    a. nitrogen and oxygen
    b. titanium and gallium
    c. lithium and sodium
    d. germanium and arsenic
    e. argon and bromine

71. Gallium has two naturally occurring isotopes with the following masses and natural abundances:

| Isotope | Mass (amu) | Abundance (%) |
|---------|-----------|---------------|
| Ga-69   | 68.92558  | 60.108        |
| Ga-71   | 70.92470  | 39.892        |

Sketch the mass spectrum of gallium.

72. Magnesium has three naturally occurring isotopes with the following masses and natural abundances:

| Isotope | Mass (amu) | Abundance (%) |
|---------|-----------|---------------|
| Mg-24   | 23.9850   | 78.99         |
| Mg-25   | 24.9858   | 10.00         |
| Mg-26   | 25.9826   | 11.01         |

Sketch the mass spectrum of magnesium.

73. The atomic mass of fluorine is 18.998 amu, and its mass spectrum shows a large peak at this mass. The atomic mass of chlorine is 35.45 amu, yet the mass spectrum of chlorine does not show a peak at this mass. Explain the difference.

74. The atomic mass of copper is 63.546 amu. Do any copper isotopes have a mass of 63.546 amu? Explain.

75. An element has two naturally occurring isotopes. Isotope 1 has a mass of 120.9038 amu and a relative abundance of 57.4%, and isotope 2 has a mass of 122.9042 amu. Find the atomic mass of this element and identify it.

76. An element has four naturally occurring isotopes with the masses and natural abundances given here. Find the atomic mass of the element and identify it.

| Isotope | Mass (amu) | Abundance (%) |
|---------|-----------|---------------|
| 1       | 135.90714 | 0.19          |
| 2       | 137.90599 | 0.25          |
| 3       | 139.90543 | 88.43         |
| 4       | 141.90924 | 11.13         |

77. Bromine has two naturally occurring isotopes (Br-79 and Br-81) and has an atomic mass of 79.904 amu. The mass of Br-81 is 80.9163 amu, and its natural abundance is 49.31%. Calculate the mass and natural abundance of Br-79.

78. Silicon has three naturally occurring isotopes (Si-28, Si-29, and Si-30). The mass and natural abundance of Si-28 are 27.9769 amu and 92.2%, respectively. The mass and natural abundance of Si-29 are 28.9765 amu and 4.67%, respectively. Find the mass and natural abundance of Si-30.

79. Use the mass spectrum of europium to determine the atomic mass of europium.

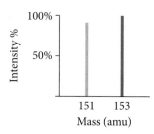

80. Use the mass spectrum of rubidium to determine the atomic mass of rubidium.

## The Mole Concept

81. How many sulfur atoms are there in 5.52 mol of sulfur?

82. How many moles of aluminum do $3.7 \times 10^{24}$ aluminum atoms represent?

83. What is the amount, in moles, of each elemental sample?
    a. 11.8 g Ar    b. 3.55 g Zn
    c. 26.1 g Ta    d. 0.211 g Li

84. What is the mass, in grams, of each elemental sample?
    a. $2.3 \times 10^{-3}$ mol Sb
    b. 0.0355 mol Ba
    c. 43.9 mol Xe
    d. 1.3 mol W

85. How many silver atoms are there in 3.78 g of silver?

86. What is the mass of $4.91 \times 10^{21}$ platinum atoms?

**87.** Calculate the number of atoms in each sample.
  **a.** 5.18 g P    **b.** 2.26 g Hg
  **c.** 1.87 g Bi    **d.** 0.082 g Sr

**88.** Calculate the number of atoms in each sample.
  **a.** 14.955 g Cr    **b.** 39.733 g S
  **c.** 12.899 g Pt    **d.** 97.552 g Sn

**89.** Calculate the mass, in grams, of each sample.
  **a.** $1.1 \times 10^{23}$ gold atoms
  **b.** $2.82 \times 10^{22}$ helium atoms
  **c.** $1.8 \times 10^{23}$ lead atoms
  **d.** $7.9 \times 10^{21}$ uranium atoms

**90.** Calculate the mass, in kg, of each sample.
  **a.** $7.55 \times 10^{26}$ cadmium atoms
  **b.** $8.15 \times 10^{27}$ nickel atoms
  **c.** $1.22 \times 10^{27}$ manganese atoms
  **d.** $5.48 \times 10^{29}$ lithium atoms

**91.** How many carbon atoms are there in a diamond (pure carbon) with a mass of 52 mg?

**92.** How many helium atoms are there in a helium blimp containing 536 kg of helium?

**93.** Calculate the average mass, in grams, of one platinum atom.

**94.** Using scanning tunneling microscopy, scientists at IBM wrote the initials of their company with 35 individual xenon atoms (as shown below). Calculate the total mass of these letters in grams.

# Cumulative Problems

**95.** A 7.83 g sample of HCN contains 0.290 g of H and 4.06 g of N. Find the mass of carbon in a sample of HCN with a mass of 3.37 g.

**96.** The ratio of sulfur to oxygen by mass in $SO_2$ is 1.0:1.0.
  **a.** Find the ratio of sulfur to oxygen by mass in $SO_3$.
  **b.** Find the ratio of sulfur to oxygen by mass in $S_2O$.

**97.** The ratio of oxygen to carbon by mass in carbon monoxide is 1.33:1.00. Find the formula of an oxide of carbon in which the ratio by mass of oxygen to carbon is 2.00:1.00.

**98.** The ratio of the mass of a nitrogen atom to the mass of an atom of $^{12}C$ is 7:6, and the ratio of the mass of nitrogen to oxygen in $N_2O$ is 7:4. Find the mass of 1 mol of oxygen atoms.

**99.** An $\alpha$ particle, $^4He^{2+}$, has a mass of 4.00151 amu. Find the value of its charge-to-mass ratio in C/kg.

**100.** Naturally occurring iodine has an atomic mass of 126.9045 amu. A 12.3849 g sample of iodine is accidentally contaminated with an additional 1.00070 g of $^{129}I$, a synthetic radioisotope of iodine used in the treatment of certain diseases of the thyroid gland. The mass of $^{129}I$ is 128.9050 amu. Find the apparent "atomic mass" of the contaminated iodine.

**101.** Use the mass spectrum of lead to estimate the atomic mass of lead. Estimate the mass and percent intensity values from the graph to three significant figures.

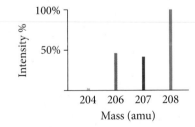

**102.** Use the mass spectrum of mercury to estimate the atomic mass of mercury. Estimate the masses and percent intensity values from the graph to three significant figures.

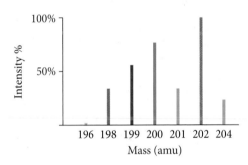

**103.** Nuclei with the same number of *neutrons* but different mass numbers are called *isotones*. Write the symbols of four isotones of $^{236}Th$.

**104.** Fill in the blanks to complete the table.

| Symbol | Z | A | Number of p | Number of $e^-$ | Number of n | Charge |
|---|---|---|---|---|---|---|
| Si | 14 | ___ | ___ | 14 | 14 | ___ |
| $S^{2-}$ | ___ | 32 | ___ | ___ | ___ | 2− |
| $Cu^{2+}$ | ___ | ___ | ___ | ___ | 34 | 2+ |
| ___ | 15 | ___ | ___ | 15 | 16 | ___ |

**105.** Fill in the blanks to complete the table.

| Symbol | Z | A | Number of p | Number of $e^-$ | Number of n | Charge |
|---|---|---|---|---|---|---|
| ___ | 8 | ___ | ___ | ___ | 8 | 2− |
| $Ca^{2+}$ | 20 | ___ | ___ | ___ | 20 | ___ |
| $Mg^{2+}$ | ___ | 25 | ___ | ___ | 13 | 2+ |
| $N^{3-}$ | ___ | 14 | ___ | 10 | ___ | ___ |

**106.** Neutron stars are composed of solid nuclear matter, primarily neutrons. Assume the radius of a neutron is approximately $1.0 \times 10^{-13}$ cm. Calculate the density of a neutron. [*Hint*: For a sphere $V = (4/3)\pi r^3$.] Assuming that a neutron star has the same density as a neutron, calculate the mass (in kg) of a small piece of a neutron star the size of a spherical pebble with a radius of 0.10 mm.

**107.** Carbon-12 contains six protons and six neutrons. The radius of the nucleus is approximately 2.7 fm (femtometers) and the radius of the atom is approximately 70 pm (picometers). Calculate the volume of the nucleus and the volume of the atom. What percentage of the carbon atom's volume is occupied by the nucleus? (Assume two significant figures.)

**108.** A penny has a thickness of approximately 1.0 mm. If you stacked Avogadro's number of pennies one on top of the other on Earth's surface, how far would the stack extend (in km)? [For comparison, the sun is about 150 million km from Earth and the nearest star (Proxima Centauri) is about 40 trillion km from Earth.]

**109.** Consider the stack of pennies in the previous problem. How much money (in dollars) would this represent? If this money were equally distributed among the world's population of 7.0 billion people, how much would each person receive? Would each person be a millionaire? A billionaire? A trillionaire?

**110.** The mass of an average blueberry is 0.75 g and the mass of an automobile is $2.0 \times 10^3$ kg. Find the number of automobiles whose total mass is the same as 1.0 mol of blueberries.

**111.** Suppose that atomic masses were based on the assignment of a mass of 12.000 g to 1 mol of carbon, rather than 1 mol of $^{12}$C. What would the atomic mass of oxygen be? (The atomic masses of carbon and oxygen based on the assignment of 12.000 g to 1 mol of $^{12}$C are 12.011 amu and 15.9994 amu, respectively.)

**112.** A pure titanium cube has an edge length of 2.78 in. How many titanium atoms does it contain? Titanium has a density of 4.50 g/cm$^3$.

**113.** A pure copper sphere has a radius of 0.935 in. How many copper atoms does it contain? [The volume of a sphere is $(4/3)\pi r^3$ and the density of copper is 8.96 g/cm$^3$.]

**114.** What is the radius (in cm) of a pure copper sphere that contains $1.14 \times 10^{24}$ copper atoms? [The volume of a sphere is $(4/3)\pi r^3$ and the density of copper is 8.96 g/cm$^3$.]

**115.** What is the edge length (in cm) of a titanium cube that contains $2.55 \times 10^{24}$ titanium atoms? The density of titanium is 4.50 g/cm$^3$.

**116.** Boron has only two naturally occurring isotopes. The mass of boron-10 is 10.01294 amu and the mass of boron-11 is 11.00931 amu. Calculate the relative abundances of the two isotopes.

**117.** Lithium has only two naturally occurring isotopes. The mass of lithium-6 is 6.01512 amu and the mass of lithium-7 is 7.01601 amu. Calculate the relative abundances of the two isotopes.

**118.** Common brass is a copper and zinc alloy containing 37.0% zinc by mass and having a density of 8.48 g/cm$^3$. A fitting composed of common brass has a total volume of 112.5 cm$^3$. How many atoms (copper and zinc) does the fitting contain?

**119.** A 67.2 g sample of a gold and palladium alloy contains $2.49 \times 10^{23}$ atoms. What is the composition (by mass) of the alloy?

**120.** Naturally occurring chlorine is composed of two isotopes: 75.76% Cl-35 (mass 34.9688 amu) and 24.24% Cl-37 (mass 36.9659 amu). Naturally occurring oxygen is composed of three isotopes: 99.757% O-16 (mass 15.9949 amu), 0.038% O-17 (mass 16.9991 amu), and 0.205% O-18 (mass 17.9991 amu). The compound dichlorine monoxide is composed of two chlorine atoms and one oxygen atom bonded together to form the Cl$_2$O molecule. How many Cl$_2$O molecules of different masses naturally exist? Give the masses of the three most abundant Cl$_2$O molecules.

**121.** Silver is composed of two naturally occurring isotopes: Ag-107 (51.839%) and Ag-109. The ratio of the masses of the two isotopes is 1.0187. What is the mass of Ag-107?

**122.** The U.S. Environmental Protection Agency (EPA) sets limits on healthful levels of air pollutants. The maximum level that the EPA considers safe for lead air pollution is 1.5 $\mu$g/m$^3$. If your lungs were filled with air containing this level of lead, how many lead atoms would be in your lungs? (Assume a total lung volume of 5.50 L.)

**123.** Pure gold is usually too soft for jewelry, so it is often alloyed with other metals. How many gold atoms are in an 0.255-ounce, 18 K gold bracelet? (18 K gold is 75% gold by mass.)

# Challenge Problems

**124.** In Section 2.9, it was stated that 1 mol of sand grains would cover the state of Texas to several feet. Estimate how many feet by assuming that the sand grains are roughly cube-shaped, each one with an edge length of 0.10 mm. Texas has a land area of 268,601 square miles.

**125.** Use the concepts in this chapter to obtain an estimate for the number of atoms in the universe. Make the following assumptions: **(a)** All of the atoms in the universe are hydrogen atoms in stars. (This is not a ridiculous assumption because over three-fourths of the atoms in the universe are in fact hydrogen. Gas and dust between the stars represent only about 15% of the visible matter of our galaxy, and planets compose a far tinier fraction.) **(b)** The sun is a typical star composed of pure hydrogen with a density of 1.4 g/cm$^3$ and a radius of $7 \times 10^8$ m.

**(c)** Each of the roughly 100 billion stars in the Milky Way galaxy contains the same number of atoms as our sun. **(d)** Each of the 10 billion galaxies in the visible universe contains the same number of atoms as our Milky Way galaxy.

**126.** On the previous page is a representation of 50 atoms of a fictitious element called westmontium (Wt). The red spheres represent Wt-296, the blue spheres Wt-297, and the green spheres Wt-298.
  **a.** Assuming that the sample is statistically representative of a naturally occurring sample, calculate the percent natural abundance of each Wt isotope.
  **b.** Draw the mass spectrum for a naturally occurring sample of Wt.
  **c.** The mass of each Wt isotope is measured relative to C-12 and tabulated. Use the mass of C-12 to convert each of the masses to amu and calculate the atomic mass of Wt.

| Isotope | Mass |
|---------|------|
| Wt-296 | $24.6630 \times \text{Mass}(^{12}\text{C})$ |
| Wt-297 | $24.7490 \times \text{Mass}(^{12}\text{C})$ |
| Wt-298 | $24.8312 \times \text{Mass}(^{12}\text{C})$ |

**127.** The ratio of oxygen to nitrogen by mass in $NO_2$ is 2.29. The ratio of fluorine to nitrogen by mass in $NF_3$ is 4.07. Find the ratio of oxygen to fluorine by mass in $OF_2$.

**128.** Naturally occurring cobalt consists of only one isotope, $^{59}\text{Co}$, whose relative atomic mass is 58.9332. A synthetic radioactive isotope of cobalt, $^{60}\text{Co}$, has a relative atomic mass of 59.9338 and is used in radiation therapy for cancer. A 1.5886 g sample of cobalt has an apparent "atomic mass" of 58.9901. Find the mass of $^{60}\text{Co}$ in this sample.

**129.** A 7.36 g sample of copper is contaminated with an additional 0.51 g of zinc. Suppose an atomic mass measurement was performed on this sample. What would be the measured atomic mass?

**130.** The ratio of the mass of O to the mass of N in $N_2O_3$ is 12:7. Another binary compound of nitrogen has a ratio of O to N of 16:7. What is its formula? What is the ratio of O to N in the next member of this series of compounds?

**131.** Naturally occurring magnesium has an atomic mass of 24.312 and consists of three isotopes. The major isotope is $^{24}\text{Mg}$, natural abundance 78.99%, relative atomic mass 23.98504. The next most abundant isotope is $^{26}\text{Mg}$, relative atomic mass 25.98259. The third most abundant isotope is $^{25}\text{Mg}$, whose natural abundance is in the ratio of 0.9083 to that of $^{26}\text{Mg}$. Find the relative atomic mass of $^{25}\text{Mg}$.

## Conceptual Problems

**132.** Which answer is an example of the law of multiple proportions? Explain.
  **a.** Two different samples of water are found to have the same ratio of hydrogen to oxygen.
  **b.** When hydrogen and oxygen react, the mass of water formed is exactly equal to the mass of hydrogen and oxygen that reacted.
  **c.** The mass ratio of oxygen to hydrogen in water is 8:1. The mass ratio of oxygen to hydrogen in hydrogen peroxide (a compound that only contains hydrogen and oxygen) is 16:1.

**133.** Lithium has two naturally occurring isotopes: Li-6 (natural abundance 7.5%) and Li-7 (natural abundance 92.5%). Using circles to represent protons and squares to represent neutrons, draw the nucleus of each isotope. How many Li-6 atoms are present, on average, in a 1000-atom sample of lithium?

**134.** As we saw in the previous problem, lithium has two naturally occurring isotopes: Li-6 (natural abundance 7.5%; mass 6.0151 amu) and Li-7 (natural abundance 92.5%; mass 7.0160 amu). Without doing any calculations, determine which mass is closest to the atomic mass of Li.
  **a.** 6.00 amu   **b.** 6.50 amu   **c.** 7.00 amu

**135.** The mole is defined as the amount of a substance containing the same number of particles as exactly 12 g of C-12. The amu is defined as 1/12 of the mass of an atom of C-12. Why is it important that both of these definitions reference the same isotope? What would be the result, for example, of defining the mole with respect to C-12, but the amu with respect to Ne-20?

**136.** Without doing any calculations, determine which of the samples contains the greatest amount of the element in moles. Which contains the greatest mass of the element?
  **a.** 55.0 g Cr   **b.** 45.0 g Ti   **c.** 60.0 g Zn

**137.** The atomic radii of the isotopes of an element are identical to one another. However, the atomic radii of the ions of an element are significantly different from the atomic radii of the neutral atom of the element. Explain.

*Active Classroom Learning*

## Questions for Group Work

**Discuss these questions with the group and record your consensus answer.**

**138.** The table shown here includes data similar to those used by Mendeleev when he created the periodic table. On a small card, write the symbol, atomic mass, and a stable compound formed by each element. Without consulting a periodic table, arrange the cards so that atomic mass increases from left to right and elements with similar properties are above and below each other. Copy the periodic table you have invented onto a piece of paper. There is one element missing. Predict its mass and a stable compound it might form.

| Element | Atomic Mass | Stable Compound |
|---------|-------------|-----------------|
| Be | 9 | $BeCl_2$ |
| S | 32 | $H_2S$ |
| F | 19 | $F_2$ |
| Ca | 40 | $CaCl_2$ |
| Li | 7 | LiCl |
| Si | 28 | $SiH_4$ |
| Cl | 35.4 | $Cl_2$ |
| B | 10.8 | $BH_3$ |

—*Continued on the next page*

*—Continued*

| Element | Atomic Mass | Stable Compound |
|---------|-------------|-----------------|
| Ge | 72.6 | $GeH_4$ |
| N | 14 | $NF_3$ |
| O | 16 | $H_2O$ |
| Ga | 69.7 | $GaH_3$ |
| As | 75 | $AsF_3$ |
| C | 12 | $CH_4$ |
| K | 39 | $KCl$ |
| Mg | 24.3 | $MgCl_2$ |
| Se | 79 | $H_2Se$ |
| Al | 27 | $AlH_3$ |
| Br | 80 | $Br_2$ |
| Na | 23 | $NaCl$ |

**139.** In a naturally occurring sample, 19.8% of boron atoms have five neutrons and 80.2% have six neutrons. What is the mass number of each boron isotope? Sketch a sample of 10 atoms that is nearly representative of a natural sample. What is the average mass of the atoms in your drawing? What is the atomic mass of boron? (Boron-10 has a mass of 10.01294 amu and boron-11 has a mass of 11.00931 amu.)

**140.** In complete sentences, describe the similarities and differences between:
**a.** different isotopes of an element
**b.** a neutral atom and an ion of the same element

**141.** Calculate the mass in grams of one mole of each of the following (the mass of a single item is given in parentheses): electrons $(9.10938 \times 10^{-28}\,g)$, protons $(1.67262 \times 10^{-24}\,g)$, neutrons $(1.67493 \times 10^{-24}\,g)$, atoms of carbon-12 $(1.992646 \times 10^{-23}\,g)$, and doughnuts (74 g). Compare the mass of one mole of carbon-12 atoms to the sum of the masses of the particles that it contains. If the doughnut mentioned in this question were made entirely of carbon, how many atoms would it contain?

# Data Interpretation and Analysis

## Heavy Metals in Recycled Paper Packages

**142.** Demand for recycled paper has increased as consumers have become more aware of the environmental issues surrounding waste disposal. Paper is a natural raw material made from renewable wood and plants. Recycled paper is made from waste paper and paperboard. Both new paper and recycled paper contain traces of heavy metals. However, some types of recycled paper contain more heavy metals than new paper due, in part, to the inks used for printing or adding color to the original paper. Metals can migrate from the paper packaging and containers used for food to the food itself. This metal migration is controlled and monitored to reduce the exposure of humans to heavy metals.

Food agencies in each country determine the limit of cadmium, zinc, nickel, and copper in packaging materials. Most countries impose a limit of heavy metals in recycled paper not to exceed 100.0 ppm or 100.0 mg/kg. The limit for lead in egg-, fruit-, or vegetable-packaging is lower—not to exceed—20.0 mg/kg. The table in Figure a▼ lists the results of the analysis of samples of recycled paper produced by a manufacturer during a three-month period.

The goal of the manufacturer is to reduce the amount of heavy metals—especially lead—in the recycled paper it produces and sells. From March to May, the manufacturer varied the methods of production each month to determine which method would produce paper with the lowest metal content. Each month, lab technicians cut small samples of recycled paper with an area of 1.000 $dm^2$ and a thickness of 0.0500 cm. The technicians then prepared the samples for analysis. The average density of the samples is 800.0 ± 0.0245 $kg/m^3$.

Use the information provided in the figure to answer the following questions:

**a.** Did the company reduce the amount of lead in its product between March and May?

**b.** Which metal was not reduced?

**c.** Each month's sample represents a different manufacturing process. Which process would you recommend the manufacturer choose to continue to use over the next year while additional processes are tested? Why?

**d.** What is the total amount of the five metals for the April sample (in mg/kg)?

**e.** What is the total mass (in mg) of the five metals for the April sample found in the 1.000 $dm^2$ × 0.0500 cm sample?

**f.** Lead has four stable isotopes: $^{204}Pb$, $^{206}Pb$, $^{207}Pb$, $^{208}Pb$ with % abundances of 1.40, 24.10, 22.10, and 52.40, respectively. Determine the mass (in mg) of $^{208}Pb$ in the April sample.

**g.** Sketch the mass spectrum for lead.

| | Zn | Pb | Cd | Ni | Cu |
|-------|-----|-----|-------|-----|------|
| March | 2.7 | 3.5 | 0.129 | 1.7 | 32.3 |
| April | 3.2 | 2.8 | 0.092 | 1.6 | 29.2 |
| May | 3.7 | 2.4 | 0.096 | 1.5 | 19.9 |

▲ FIGURE a **Heavy Metal Values in Three Samples of Recycled Paper (mg/kg ± 0.1)**

# Answers to Conceptual Connections

## The Law of Conservation of Mass

**2.1** Most of the matter that composed the log undergoes a chemical change by reacting with oxygen molecules in the air. The products of the reaction (mostly carbon dioxide and water) are released as gases into the air.

## The Laws of Definite and Multiple Proportions

**2.2** The law of definite proportions applies to two or more samples of the *same compound* and states that the ratio of one element to the other is always the same. The law of multiple proportions applies to two *different compounds* containing the same two elements (A and B) and states that the masses of B that combine with 1 g of A are always related to each other as a small whole-number ratio.

## The Millikan Oil Drop Experiment

**2.3** The drop contains three excess electrons $(3 \times (-6 \times 10^{-19}\,\text{C})$ $= -4.8 \times 10^{-19}\,\text{C})$.

## Isotopes

**2.4**

C-12 nucleus          C-13 nucleus

A 10,000-atom sample of carbon, on average, contains 107 C-13 atoms.

## The Nuclear Atom, Isotopes, and Ions

**2.5 (b)** The number of neutrons in the nucleus of an atom does not affect the atom's size because the nucleus is miniscule compared to the atom itself.

## Atomic Mass

**2.6 (a)** Since 98.93% of the atoms are C-12, we would expect the atomic mass to be very close to the mass of the C-12 isotope.

## Avogadro's Number

**2.7** Avogadro's number is defined with respect to carbon-12—it is the number equal to the number of atoms in exactly 12 g of carbon-12. If Avogadro's number was defined as $1.00 \times 10^{23}$ (a nice round number), it would correspond to 1.99 g of carbon-12 atoms (an inconvenient number). Avogadro's number is defined with respect to carbon-12 because, as you recall from Section 2.6, the amu (the basic mass unit used for all atoms) is defined relative to carbon-12. Therefore, the mass in grams of 1 mol of *any* element is equal to its atomic mass. As we have seen, these two definitions together make it possible to determine the number of atoms in a known mass of any element.

## The Mole

**2.8 (b)** The carbon sample contains more atoms than the copper sample because carbon has a lower molar mass than copper. Carbon atoms are lighter than copper atoms, so a 1-g sample of carbon contains more atoms than a 1-g sample of copper. The carbon sample also contains more atoms than the uranium sample because even though the uranium sample has 10 times the mass of the carbon sample, a uranium atom is more than 10 times as massive (238 g/mol for uranium versus 12 g/mol for carbon).

CHAPTER

# 3

# Molecules, Compounds, and Chemical Equations

Almost all aspects of life are engineered at the molecular level, and without understanding molecules we can only have a very sketchy understanding of life itself.

—FRANCIS HARRY COMPTON CRICK (1916–2004)

Periodic table (partial):

3A 13, 4A 14, 5A 15, 16, 17, 18

He 2 4.00 helium

B 5 10.81 boron | C 6 12.01 carbon | N 7 14.01 nitrogen | O 8 16.00 oxygen | F 9

Al 13 26.98 aluminum | Si 14 28.09 silicon | P 15 30.97 phosphorus | S 16

Group headers: 3B 3, 4B 4, 5B 5, 6B 6, 7B 7, —8B— 8 9 10, 1B 11, 2B 12

Ti 22 47.88 titanium | V 23 50.94 vanadium | Cr 24 52.00 chromium | Mn 25 54.94 manganese | Fe 26 55.85 iron | Co 27 58.93 cobalt | Ni 28 58.69 nickel | Cu 29 63.55 copper | Zn 30 65.39 zinc | Ga 31 69.72 gallium | Ge 32 72.61 germanium

Zr 40 91.22 zirconium | Nb 41 92.91 niobium | Mo 42 95.94 molybdenum | Tc 43 (99) technetium | Ru 44 101.07 ruthenium | Rh 45 102.91 rhodium | Pd 46 106.42 palladium | Ag 47 107.87 silver | Cd 48 112.41 cadmium | In 49 114.82 indium | Sn 50 118.71 tin

Hf 72 178.49 hafnium | Ta 73 180.95 tantalum | W 74 183.85 tungsten | Re 75 186.21 rhenium | Os 76 190.2 osmium | Ir 77 192.22 iridium | Pt 78 195.08 platinum | Au 79 196.97 gold | Hg 80 200.59 mercury | Tl 81 204.38 thallium | Pb 82 207.2 lead | Bi 208.9 bismuth

Rf 104 (261) rutherfordium | Db 105 (262) dubnium | Sg 106 (263) seaborgium | Bh 107 (262) bohrium | Hs 108 (265) hassium | Mt 109 (266) meitnerium | Ds 110 (269) | Rg 111 (272) | Cn 112 (277) | 114 (285)

Lanthanides: Ce 58 140.12 cerium | Pr 59 140.91 praseodymium | Nd 60 144.24 neodymium | Pm 61 (147) promethium | Sm 62 150.36 samarium | Eu 63 151.97 europium | Gd 64 157.25 gadolinium | Tb 65 158.93 terbium | Dy 66 162.50 dysprosium | Ho 67 164.9 | Er 68

Actinides: Th 90 (232) thorium | Pa 91 (231) protactinium | U 92 (238) uranium | Np 93 (237) neptunium | Pu 94 (244) plutonium | Am 95 (243) americium | Cm 96 (247) curium | Bk 97 (247) berkelium | Cf 98 (251) californium

H ow many different substances exist? Recall from Chapter 2 that about 91 different elements exist in nature, so there are at least 91 different substances. However, the world would be dull—not to mention lifeless—with only 91 different substances. Fortunately, elements combine with each other to form *compounds*. Just as combinations of only 26 letters in our English alphabet allow for an almost limitless number of words, each with its own specific meaning, combinations of the 91 naturally occurring elements allow for an almost limitless number of compounds, each with its own specific properties. The great diversity of substances that we find in nature is a direct result of the ability of elements to form compounds. Life, for example, could not exist with just 91 different elements. It takes compounds, in all of their diversity, to make life possible.

▲ When a balloon filled with $H_2$ and $O_2$ is ignited, the two elements react violently to form $H_2O$.

## 3.1 Hydrogen, Oxygen, and Water

Hydrogen ($H_2$) is an explosive gas used as a fuel in rocket engines. Oxygen ($O_2$), also a gas, is a natural component of the air on Earth. Oxygen itself is not flammable but must be present for combustion (burning) to occur. Hydrogen and oxygen both have extremely low boiling points, as you can see from the table on the following page. When hydrogen and oxygen combine to form the compound water ($H_2O$), however, a dramatically different substance results.

| Selected Properties | Hydrogen | Oxygen | Water |
|---|---|---|---|
| Boiling Point | −253 °C | −183 °C | 100 °C |
| State at Room Temperature | Gas | Gas | Liquid |
| Flammability | Explosive | Necessary for combustion | Used to extinguish flame |

First of all, water is a liquid rather than a gas at room temperature, and its boiling point is hundreds of degrees higher than the boiling points of hydrogen and oxygen. Second, instead of being flammable (like hydrogen gas) or supporting combustion (like oxygen gas), water actually smothers flames. Water is nothing like the hydrogen and oxygen from which it forms. The dramatic difference between the elements hydrogen and oxygen and the compound water is typical of the differences between elements and the compounds that they form. *When two or more elements combine to form a compound, an entirely new substance results.*

Consider as another example common table salt, a highly stable compound composed of sodium and chlorine. Elemental sodium, by contrast, is a highly reactive, silvery metal that can explode on contact with water. Elemental chlorine is a corrosive, greenish-yellow gas that can be fatal if inhaled. Yet the compound formed from the combination of these two elements is sodium chloride (or table salt), a flavor enhancer that tastes great on steak.

Although some of the substances that we encounter in everyday life are elements, most are compounds. As we discussed in Chapter 1, a compound is different from a mixture of elements. In a compound, elements combine in fixed, definite proportions; in a mixture, elements can mix in any proportions whatsoever. Consider the difference between a hydrogen–oxygen mixture and water in Figure 3.1▼. A hydrogen–oxygen mixture can have any proportions of hydrogen and oxygen gas. Water, by contrast, is composed of water molecules that always contain two hydrogen atoms to every one oxygen atom. Water has a definite proportion of hydrogen to oxygen.

In this chapter you will learn about compounds: how to represent them, how to name them, how to distinguish between their different types, and how to write chemical equations showing how they form and change. You will also learn how to quantify the elemental composition of a compound. This is important in determining how much of a particular element is contained within a particular compound. For example, patients with high blood pressure (hypertension) often have to reduce their sodium ion intake. The sodium ion is normally consumed in the form of sodium chloride, so a hypertension patient needs to know

## Mixtures and Compounds

> **Hydrogen and Oxygen Mixture**
> This can have any ratio of hydrogen to oxygen.

> **Water (A Compound)**
> Water molecules have a fixed ratio of hydrogen (two atoms) to oxygen (one atom).

▶ **FIGURE 3.1 Mixtures and Compounds**  The balloon in this illustration is filled with a mixture of hydrogen gas and oxygen gas. The proportions of hydrogen and oxygen are variable. The glass is filled with water, a compound of hydrogen and oxygen. The ratio of hydrogen to oxygen in water is fixed: water molecules always have two hydrogen atoms for each oxygen atom.

how much sodium is present in a given amount of sodium chloride. Similarly, an iron-mining company needs to know how much iron it can recover from a given amount of iron ore. This chapter provides the tools to understand and answer these kinds of questions.

# 3.2   Chemical Bonds

Compounds are composed of atoms held together by *chemical bonds*. Chemical bonds form because of the attractions between the charged particles (the electrons and protons) that compose atoms. We discuss these interactions in more detail in Chapter 9 (see Section 9.2). For now, remember that, as we discussed in Section 2.4, charged particles exert electrostatic forces on one another: like charges repel and opposite charges attract. These forces are responsible for chemical bonding.

We can broadly classify most chemical bonds into two types: ionic and covalent. *Ionic bonds*—which occur between metals and nonmetals—involve the *transfer* of electrons from one atom to another. *Covalent bonds*—which occur between two or more nonmetals—involve the *sharing* of electrons between two atoms.

## Ionic Bonds

Recall from Chapter 2 that metals have a tendency to lose electrons and that nonmetals have a tendency to gain them. Therefore, when a metal interacts with a nonmetal, it can transfer one or more of its electrons to the nonmetal. The metal atom then becomes a *cation* (a positively charged ion), and the nonmetal atom becomes an *anion* (a negatively charged ion) as shown in Figure 3.2▼. These oppositely charged ions attract one another

▼ **FIGURE 3.2 The Formation of an Ionic Compound** An atom of sodium (a metal) loses an electron to an atom of chlorine (a nonmetal), creating a pair of oppositely charged ions. The sodium cation is attracted to the chloride anion, and the two are held together as part of a crystalline lattice.

**The Formation of an Ionic Compound**

Sodium (a metal) loses an electron.

Chlorine (a nonmetal) gains an electron.

$e^-$

Neutral Na atom, $11e^-$

Neutral Cl atom, $17e^-$

$Na^+$ ion, $10e^-$

$Cl^-$ ion, $18e^-$

anion

cation

Sodium metal

Chlorine gas

Oppositely charged ions are held together by ionic bonds, forming a crystalline lattice.

Sodium chloride (table salt)

by electrostatic forces and form an **ionic bond**. The result is an **ionic compound**, which in the solid phase is composed of a lattice—a regular three-dimensional array—of alternating cations and anions.

## Covalent Bonds

When a nonmetal bonds with another nonmetal, neither atom transfers its electron to the other. Instead the bonding atoms *share* some of their electrons. The shared electrons have lower potential energy than they do in the isolated atoms because they interact with the nuclei of both atoms. The bond is a **covalent bond** and the covalently bound atoms compose a *molecule*. Each molecule is independent of the others—the molecules are themselves not covalently bound to one another. Therefore, we call covalently bonded compounds **molecular compounds**.

We can begin to understand the stability of a covalent bond by considering the most stable (or lowest potential energy) configuration of a negative charge interacting with two positive charges (which are separated by some small distance). Figure 3.3▼ shows that the lowest potential energy occurs when the negative charge lies *between* the two positive charges because in this arrangement the negative charge can interact with *both positive charges*. Similarly, shared electrons in a covalent chemical bond hold the bonding atoms together by attracting the positively charged nuclei of both bonding atoms.

▶ **FIGURE 3.3 The Stability of a Covalent Bond** The potential energy of a negative charge interacting with two positive charges is lowest when the negative charge is between the two positive charges.

Lowest potential energy (most stable)

CONCEPTUAL
**CONNECTION 3.1**

**Types of Chemical Bonds** What type of bond—ionic or covalent—forms between nitrogen and oxygen?

## 3.3 Representing Compounds: Chemical Formulas and Molecular Models

The quickest and easiest way to represent a compound is with its **chemical formula**, which indicates the elements present in the compound and the relative number of atoms or ions of each element. For example, $H_2O$ is the chemical formula for water—it indicates that water consists of hydrogen and oxygen atoms in a two-to-one ratio. The formula contains the symbol for each element and a subscript indicating the relative number of atoms of the element. A subscript of 1 is typically omitted. Chemical formulas generally list the more metallic (or more positively charged) elements first, followed by the less metallic (or more negatively charged) elements. Other examples of common chemical formulas include NaCl for sodium chloride, indicating sodium and chloride ions in a one-to-one ratio; $CO_2$ for carbon dioxide, indicating carbon and oxygen atoms in a one-to-two ratio; and $CCl_4$ for carbon tetrachloride, indicating carbon and chlorine in a one-to-four ratio.

## Types of Chemical Formulas

We can categorize chemical formulas into three different types: empirical, molecular, and structural. An **empirical formula** gives the *relative* number of atoms of each element in a compound. A **molecular formula** gives the *actual* number of atoms of each element in a molecule of a compound. For example, the empirical formula for hydrogen peroxide is HO, but its molecular formula is $H_2O_2$. The molecular formula is always a whole-number multiple of the empirical formula. For some compounds, the empirical formula and the molecular formula are identical. For example, the empirical and molecular formula for water is $H_2O$ because water molecules contain two hydrogen atoms and one oxygen atom, and no simpler whole-number ratio can express the relative number of hydrogen atoms to oxygen atoms.

A **structural formula** uses lines to represent covalent bonds and shows how atoms in a molecule connect or bond to each other. The structural formula for $H_2O_2$ is:

$$H—O—O—H$$

We can also write structural formulas to convey a sense of the molecule's geometry. For example, we can write the structural formula for hydrogen peroxide as:

$$
\begin{array}{c}
H \\
\backslash \\
O—O \\
\quad\ \backslash \\
\quad\ H
\end{array}
$$

This version of the formula represents the approximate angles between bonds, giving a sense of the molecule's shape.

Structural formulas can also depict the different types of bonds that occur within molecules. For example, consider the structural formula for carbon dioxide:

$$O{=}C{=}O$$

The two lines between each carbon and oxygen atom represent a double bond, which is generally stronger and shorter than a single bond (represented by a single line). A single bond corresponds to one shared electron pair, while a double bond corresponds to two shared electron pairs. We will learn more about single, double, and even triple bonds in Chapter 9.

The type of formula we use depends on how much we know about the compound and how much we want to communicate. A structural formula communicates the most information, while an empirical formula communicates the least.

**Structural Formulas**  Write the structural formula for water.

CONCEPTUAL
**CONNECTION 3.2**

---

### Example 3.1    Molecular and Empirical Formulas

Write empirical formulas for the compounds represented by the molecular formulas.

**(a)** $C_4H_8$      **(b)** $B_2H_6$      **(c)** $CCl_4$

**SOLUTION**

To determine the empirical formula from a molecular formula, divide the subscripts by the greatest common factor (the largest number that divides exactly into all of the subscripts).

**(a)** For $C_4H_8$, the greatest common factor is 4. The empirical formula is therefore $CH_2$.

**(b)** For $B_2H_6$, the greatest common factor is 2. The empirical formula is therefore $BH_3$.

**(c)** For $CCl_4$, the only common factor is 1, so the empirical formula and the molecular formula are identical.

**FOR PRACTICE 3.1**  Write the empirical formula for the compounds represented by each molecular formula.

**(a)** $C_5H_{12}$      **(b)** $Hg_2Cl_2$      **(c)** $C_2H_4O_2$

Answers to For Practice and For More Practice problems can be found in Appendix IV.

Hydrogen

Carbon

Nitrogen

Oxygen

Fluorine

Phosphorus

Sulfur

Chlorine

## Molecular Models

A *molecular model* is a more accurate and complete way to specify a compound. A **ball-and-stick molecular model** represents atoms as balls and chemical bonds as sticks; how the two connect reflects a molecule's shape. The balls are typically color-coded to specific elements. For example, carbon is customarily black, hydrogen is white, nitrogen is blue, and oxygen is red. (For a complete list of colors of elements in the molecular models used in this book, see Appendix IIA.)

In a **space-filling molecular model**, atoms fill the space between each other to more closely represent our best estimates for how a molecule might appear if scaled to visible size. Consider the following ways to represent a molecule of methane, the main component of natural gas:

$CH_4$

Molecular formula        Structural formula        Ball-and-stick model        Space-filling model

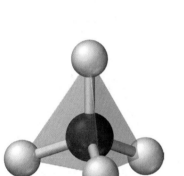

▲ A tetrahedron is a three-dimensional geometrical shape characterized by four equivalent triangular faces.

The molecular formula of methane indicates the number and type of each atom in the molecule: one carbon atom and four hydrogen atoms. The structural formula indicates how the atoms connect: the carbon atom bonds to the four hydrogen atoms. The ball-and-stick model clearly portrays the geometry of the molecule: the carbon atom sits in the center of a *tetrahedron* formed by the four hydrogen atoms. And finally, the space-filling model gives the best sense of the relative sizes of the atoms and how they merge together in bonding.

Throughout this book, you will see molecules represented in all of these ways. As you look at these representations, keep in mind what you learned in Chapter 1: the details about a molecule—the atoms that compose it, the lengths of the bonds between atoms, the angles of the bonds between atoms, and its overall shape—determine the properties of the substance that the molecule composes. Change any of these details and those properties change. Table 3.1 shows various compounds represented in the different ways we have just discussed.

**CONCEPTUAL CONNECTION 3.3**

**Representing Molecules**  Based on what you learned in Chapter 2 about atoms, what part of the atom do you think the spheres in the molecular space-filling models shown in Table 3.1 represent? If you were to superimpose a nucleus on one of these spheres, how big would you draw it?

## 3.4  An Atomic-Level View of Elements and Compounds

Recall from Chapter 1 that we categorize pure substances as either elements or compounds. We can subcategorize elements and compounds according to the basic units that compose them, as shown in Figure 3.4▶. Elements may be either atomic or molecular. Compounds may be either molecular or ionic.

**Atomic elements** exist in nature with single atoms as their basic units. Most elements fall into this category. For example, helium is composed of helium atoms, aluminum is composed of aluminum atoms, and iron is composed of iron atoms. **Molecular elements** do not normally exist in nature with single atoms as their basic units.

**TABLE 3.1 Benzene, Acetylene, Glucose, and Ammonia**

| Name of Compound | Empirical Formula | Molecular Formula | Structural Formula | Ball-and-Stick Model | Space-Filling Model |
|---|---|---|---|---|---|
| Benzene | CH | $C_6H_6$ | | | |
| Acetylene | CH | $C_2H_2$ | $H-C \equiv C-H$ | | |
| Glucose | $CH_2O$ | $C_6H_{12}O_6$ | | | |
| Ammonia | $NH_3$ | $NH_3$ | | | |

## Classification of Elements and Compounds

▲ **FIGURE 3.4  A Molecular View of Elements and Compounds**

▶ **FIGURE 3.5 Molecular Elements** The highlighted elements exist primarily as diatomic molecules (yellow) or polyatomic molecules (red).

## Molecular Elements

| | 1A 1 | 2A 2 | | | | | | | | | | | 3A 13 | 4A 14 | 5A 15 | 6A 16 | 7A 17 | 8A 18 |
|---|---|---|---|---|---|---|---|---|---|---|---|---|---|---|---|---|---|---|
| 1 | 1 H | 2 | | | | | | | | | | | | | | | | 2 He |
| 2 | 3 Li | 4 Be | | | | | | | | | | | 5 B | 6 C | 7 N | 8 O | 9 F | 10 Ne |
| 3 | 11 Na | 12 Mg | 3B 3 | 4B 4 | 5B 5 | 6B 6 | 7B 7 | 8 | 8B 9 | 10 | 1B 11 | 2B 12 | 13 Al | 14 Si | 15 P | 16 S | 17 Cl | 18 Ar |
| 4 | 19 K | 20 Ca | 21 Sc | 22 Ti | 23 V | 24 Cr | 25 Mn | 26 Fe | 27 Co | 28 Ni | 29 Cu | 30 Zn | 31 Ga | 32 Ge | 33 As | 34 Se | 35 Br | 36 Kr |
| 5 | 37 Rb | 38 Sr | 39 Y | 40 Zr | 41 Nb | 42 Mo | 43 Tc | 44 Ru | 45 Rh | 46 Pd | 47 Ag | 48 Cd | 49 In | 50 Sn | 51 Sb | 52 Te | 53 I | 54 Xe |
| 6 | 55 Cs | 56 Ba | 57 La | 72 Hf | 73 Ta | 74 W | 75 Re | 76 Os | 77 Ir | 78 Pt | 79 Au | 80 Hg | 81 Tl | 82 Pb | 83 Bi | 84 Po | 85 At | 86 Rn |
| 7 | 87 Fr | 88 Ra | 89 Ac | 104 Rf | 105 Db | 106 Sg | 107 Bh | 108 Hs | 109 Mt | 110 Ds | 111 Rg | 112 Cn | 113 | 114 Fl | 115 | 116 Lv | 117 | 118 |

☐ Elements that exist as diatomic molecules
☐ Elements that exist as polyatomic molecules

Periods

| Lanthanides | 58 Ce | 59 Pr | 60 Nd | 61 Pm | 62 Sm | 63 Eu | 64 Gd | 65 Tb | 66 Dy | 67 Ho | 68 Er | 69 Tm | 70 Yb | 71 Lu |
|---|---|---|---|---|---|---|---|---|---|---|---|---|---|---|
| Actinides | 90 Th | 91 Pa | 92 U | 93 Np | 94 Pu | 95 Am | 96 Cm | 97 Bk | 98 Cf | 99 Es | 100 Fm | 101 Md | 102 No | 103 Lr |

Diatomic chlorine molecules

▲ The basic units that compose chlorine gas are diatomic chlorine molecules.

Some ionic compounds, such as $K_2NaPO_4$, contain more than one type of metal ion.

People occasionally refer to formula units as molecules. This is incorrect because ionic compounds do not contain distinct molecules.

Instead, they exist as molecules—two or more atoms of the element bonded together. Most molecular elements exist as *diatomic* molecules. For example, hydrogen is composed of $H_2$ molecules, nitrogen is composed of $N_2$ molecules, and chlorine is composed of $Cl_2$ molecules. A few molecular elements exist as *polyatomic molecules*. Phosphorus exists as $P_4$, and sulfur exists as $S_8$. Figure 3.5▲ shows the elements that exist primarily as diatomic or polyatomic molecules.

Molecular compounds are usually composed of two or more covalently bonded nonmetals. The basic units of molecular compounds are molecules composed of the constituent atoms. For example, water is composed of $H_2O$ molecules, dry ice is composed of $CO_2$ molecules, and propane (often used as a fuel for grills) is composed of $C_3H_8$ molecules as illustrated in Figure 3.6(a)▼.

Ionic compounds are usually composed of a metal ionically bonded to one or more nonmetals. The basic unit of an ionic compound is the **formula unit**, the smallest, electrically neutral collection of ions. A formula unit is not a molecule—it does not usually exist as a discrete entity but rather as part of a larger lattice. For example, the ionic

## A Molecular Compound

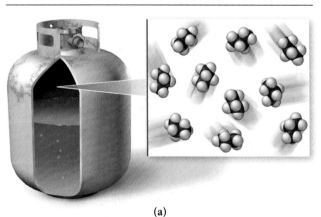

(a)

## An Ionic Compound

(b)

▲ **FIGURE 3.6 Molecular and Ionic Compounds** **(a)** Propane is a molecular compound. The basic units that compose propane gas are propane ($C_3H_8$) molecules. **(b)** Table salt (NaCl) is an ionic compound. Its formula unit is the simplest charge-neutral collection of ions: one $Na^+$ ion and one $Cl^-$ ion.

compound table salt, with the formula unit NaCl, is composed of Na$^+$ and Cl$^-$ ions in a one-to-one ratio. In table salt, Na$^+$ and Cl$^-$ ions exist in a three-dimensional alternating array. Because ionic bonds are not directional, no one Na$^+$ ion pairs with a specific Cl$^-$ ion. Rather, as illustrated in Figure 3.6(b)◀, any one Na$^+$ cation is surrounded by Cl$^-$ anions and vice versa.

Many common ionic compounds contain ions that are themselves composed of a group of covalently bonded atoms with an overall charge. For example, the active ingredient in household bleach is sodium hypochlorite, which acts to chemically alter color-causing molecules in clothes (bleaching action) and to kill bacteria (disinfection). Hypochlorite is a **polyatomic ion**—an ion composed of two or more atoms—with the formula ClO$^-$. (Note that the charge on the hypochlorite ion is a property of the whole ion, not just the oxygen atom; this is true for all polyatomic ions.) The hypochlorite ion is often found as a unit in other compounds as well (such as KClO and Mg(ClO)$_2$). Other common compounds that contain polyatomic ions include sodium bicarbonate (NaHCO$_3$), also known as baking soda; sodium nitrite (NaNO$_2$), an inhibitor of bacterial growth in packaged meats; and calcium carbonate (CaCO$_3$), the active ingredient in antacids such as TUMS®.

Active Ingredient:
Sodium Hypochlorite .....6.0%
Other Ingredients:........94.0%
Total:......................100.0%
(Yields 5.7% available chlorine)

**KEEP OUT OF REACH OF CHILDREN**
**DANGER: CORROSIVE.**
FIRST AID: IF IN EYES: Hold eye open and rinse slowly and gently with water for 15–20 minutes. Remove contact lenses, if present, after the first 5 minutes, then continue rinsing eye. IF ON SKIN OR CLOTHING: Take off contaminated clothing. Rinse skin immediately with plenty of water for 15–20 minutes. IN EITHER CASE, CALL A POISON CONTROL CENTER OR DOCTOR IMMEDIATELY FOR TREATMENT ADVICE. See back panel for additional precautionary labeling.

▲ Polyatomic ions are common in household products such as bleach, which contains sodium hypochlorite (NaClO).

## A Molecular View of Elements and Compounds
Suppose that the two elements A (represented by triangles) and B (represented by squares) form a molecular compound with the molecular formula A$_2$B and that two other elements, C (represented by circles) and D (represented by diamonds), form an ionic compound with the formula CD. Draw a molecular-level view of each compound.

CONCEPTUAL
**CONNECTION 3.4**

---

**Example 3.2** | **Classifying Substances as Atomic Elements, Molecular Elements, Molecular Compounds, or Ionic Compounds**

Classify each of the substances as an atomic element, molecular element, molecular compound, or ionic compound.

**(a)** xenon    **(b)** NiCl$_2$    **(c)** bromine    **(d)** NO$_2$    **(e)** NaNO$_3$

**SOLUTION**

**(a)** Xenon is an element. It is not a molecular element (see Figure 3.5); therefore, it is an atomic element.

**(b)** NiCl$_2$ is a compound composed of a metal (nickel is on the left side of the periodic table) and nonmetal (chlorine is on the right side of the periodic table); therefore, it is an ionic compound.

**(c)** Bromine is one of the elements that exists as a diatomic molecule (see Figure 3.5); therefore, it is a molecular element.

**(d)** NO$_2$ is a compound composed of a nonmetal and a nonmetal; therefore, it is a molecular compound.

**(e)** NaNO$_3$ is a compound composed of a metal and a polyatomic ion; therefore, it is an ionic compound.

**FOR PRACTICE 3.2**    Classify each of the substances as an atomic element, molecular element, molecular compound, or ionic compound.

**(a)** fluorine    **(b)** N$_2$O    **(c)** silver    **(d)** K$_2$O    **(e)** Fe$_2$O$_3$

CONCEPTUAL
**CONNECTION 3.5**

**Ionic and Molecular Compounds** Which statement best summarizes the difference between ionic and molecular compounds?

**(a)** Molecular compounds contain highly directional covalent bonds, which result in the formation of molecules—discrete particles that do not covalently bond to each other. Ionic compounds contain nondirectional ionic bonds, which result (in the solid phase) in the formation of ionic lattices—extended networks of alternating cations and anions.

**(b)** Molecular compounds contain covalent bonds in which one of the atoms shares an electron with the other one, resulting in a new force that holds the atoms together in a covalent molecule. Ionic compounds contain ionic bonds in which one atom donates an electron to the other, resulting in a new force that holds the ions together in pairs (in the solid phase).

**(c)** The key difference between ionic and covalent compounds is the types of elements that compose them, not the way that the atoms bond together.

**(d)** A molecular compound is composed of covalently bonded molecules. An ionic compound is composed of ionically bonded molecules (in the solid phase).

 **Naming Ionic Compounds**

▲ Ionic compounds are common in food and consumer products such as reduced-sodium salt (a mixture of NaCl and KCl) and TUMS® ($CaCO_3$).

▶ Calcite (left) is the main component of limestone, marble, and other forms of calcium carbonate ($CaCO_3$) commonly found in Earth's crust. Trona (right) is a crystalline form of hydrated sodium carbonate ($Na_3H(CO_3)_2 \cdot 2\ H_2O$).

See Figure 2.13 to review the elements that form ions with a predictable charge.

# 3.5 Ionic Compounds: Formulas and Names

Ionic compounds occur throughout Earth's crust as minerals. Examples include limestone ($CaCO_3$), a type of sedimentary rock; gibbsite [$Al(OH)_3$], a mineral; and soda ash ($Na_2CO_3$), a natural deposit. We can also find ionic compounds in the foods that we eat. Examples include sodium chloride (NaCl), which is table salt; calcium carbonate ($CaCO_3$), a source of calcium necessary for bone health; and potassium chloride (KCl), a source of potassium necessary for fluid balance and muscle function. Ionic compounds are generally very stable because the attractions between cations and anions within ionic compounds are strong and because each ion interacts with several oppositely charged ions in the crystalline lattice.

## Writing Formulas for Ionic Compounds

Since ionic compounds are charge-neutral, and since many elements form only one type of ion with a predictable charge, we can deduce the formulas for many ionic compounds from their constituent elements. For example, the formula for the ionic compound composed of sodium and chlorine must be NaCl because in compounds Na always forms 1+ cations and Cl always forms 1− anions. In order for the compound to be charge-neutral, it must contain one $Na^+$ cation for every one $Cl^-$ anion. The formula for the ionic compound composed of *calcium* and chlorine, however, is $CaCl_2$ because Ca always forms 2+ cations and Cl always forms 1− anions. In order for this compound to be charge-neutral, it must contain one $Ca^{2+}$ cation for every two $Cl^-$ anions.

*Summarizing Ionic Compound Formulas:*

▶ Ionic compounds always contain positive and negative ions.

▶ In a chemical formula, the sum of the charges of the positive ions (cations) must equal the sum of the charges of the negative ions (anions).

▶ The formula of an ionic compound reflects the smallest whole-number ratio of ions.

To write the formula for an ionic compound, follow the procedure in the left column. Two examples of how to apply the procedure are provided in the center and right columns.

## Procedure for...

| Writing Formulas for Ionic Compounds | Example 3.3 **Writing Formulas for Ionic Compounds** Write the formula for the ionic compound that forms between aluminum and oxygen. | Example 3.4 **Writing Formulas for Ionic Compounds** Write the formula for the ionic compound that forms between calcium and oxygen. |
|---|---|---|
| 1. Write the symbol for the metal cation and its charge followed by the symbol for the nonmetal anion and its charge. Determine charges from the element's group number in the periodic table (refer to Figure 2.13). | $Al^{3+}$   $O^{2-}$ | $Ca^{2+}$   $O^{2-}$ |
| 2. Adjust the subscript on each cation and anion to balance the overall charge. | $Al^{3+}$   $O^{2-}$ $\downarrow$ $Al_2O_3$ | $Ca^{2+}$   $O^{2-}$ $\downarrow$ $CaO$ |
| 3. Check that the sum of the charges of the cations equals the sum of the charges of the anions. | cations: $2(3+) = 6+$ anions: $3(2-) = 6-$ The charges cancel. | cations: $2+$ anions: $2-$ The charges cancel. |
|  | **FOR PRACTICE 3.3**  Write the formula for the compound formed between potassium and sulfur. | **FOR PRACTICE 3.4**  Write the formula for the compound formed between aluminum and nitrogen. |

## Naming Ionic Compounds

Some ionic compounds—such as NaCl (table salt) and $NaHCO_3$ (baking soda)—have **common names**, which are nicknames of sorts learned by familiarity. However, chemists have developed **systematic names** for different types of compounds including ionic ones. Even if you are not familiar with a compound, you can determine its systematic name from its chemical formula. Conversely, you can deduce the formula of a compound from its systematic name.

The first step in naming an ionic compound is to identify it as one. Remember, *ionic compounds are usually composed of metals and nonmetals*; any time you see a metal and one or more nonmetals together in a chemical formula, assume that you have an ionic compound. Ionic compounds can be categorized into two types, depending on the metal in the compound. The first type contains a metal whose charge is invariant from one compound to another. Whenever the metal in this first type of compound forms an ion, the ion always has the same charge.

Since the charge of the metal in this first type of ionic compound is always the same, it need not be specified in the name of the compound. Sodium, for instance, has a 1+ charge in all of its

**Metals Whose Charge Is Invariant from One Compound to Another**

▲ **FIGURE 3.7** **Metals with Invariant Charges** The metals highlighted in this table form cations with the same charges in all of their compounds. (Note that silver sometimes forms compounds with other charges, but these are rare.)

□ Main groups
□ Transition metals

▲ **FIGURE 3.8** **Transition Metals** Metals that can have different charges in different compounds are usually (but not always) transition metals.

compounds. Figure 3.7▲ lists some examples of these types of metals; the charges of these metals can be inferred from their group number in the periodic table.

The second type of ionic compound contains a metal with a charge that can differ in different compounds. In other words, the metal in this second type of ionic compound can form more than one kind of cation (depending on the compound), and its charge must therefore be specified for a given compound. Iron, for instance, forms a 2+ cation in some of its compounds and a 3+ cation in others. Metals of this type are often *transition metals* (Figure 3.8◀). However, some transition metals, such as Zn and Ag, form cations with the same charge in all of their compounds, and some main-group metals, such as Pb and Sn, form more than one type of cation.

## Naming Binary Ionic Compounds Containing a Metal That Forms Only One Type of Cation

**Binary compounds** contain only two different elements. The names of binary ionic compounds take the form:

| TABLE 3.2 Some Common Monoatomic Anions | | | |
| --- | --- | --- | --- |
| **Nonmetal** | **Symbol for Ion** | **Base Name** | **Anion Name** |
| Fluorine | $F^-$ | fluor | Fluoride |
| Chlorine | $Cl^-$ | chlor | Chloride |
| Bromine | $Br^-$ | brom | Bromide |
| Iodine | $I^-$ | iod | Iodide |
| Oxygen | $O^{2-}$ | ox | Oxide |
| Sulfur | $S^{2-}$ | sulf | Sulfide |
| Nitrogen | $N^{3-}$ | nitr | Nitride |
| Phosphorus | $P^{3-}$ | phosph | Phosphide |

For example, the name for KCl consists of the name of the cation, *potassium*, followed by the base name of the anion, *chlor*, with the ending *-ide*. Its full name is *potassium chloride*.

KCl     potassium chloride

The name for CaO consists of the name of the cation, *calcium*, followed by the base name of the anion, *ox*, with the ending *-ide*. Its full name is *calcium oxide*.

CaO     calcium oxide

The base names for various nonmetals, and their most common charges in ionic compounds, are shown in Table 3.2.

**Example 3.5** | **Naming Ionic Compounds Containing a Metal That Forms Only One Type of Cation**

Name the compound $CaBr_2$.

**SOLUTION**

The cation is *calcium*. The anion is from bromine, which becomes *bromide*. The correct name is *calcium bromide*.

**FOR PRACTICE 3.5**  Name the compound $Ag_3N$.

**FOR MORE PRACTICE 3.5**  Write the formula for rubidium sulfide.

## Naming Binary Ionic Compounds Containing a Metal That Forms More Than One Kind of Cation

For these types of metals, the name of the cation is followed by a Roman numeral (in parentheses) that indicates the charge of the metal in that particular compound. For example, we distinguish between $Fe^{2+}$ and $Fe^{3+}$ as follows:

$$Fe^{2+} \quad iron(II)$$

$$Fe^{3+} \quad iron(III)$$

The full names for compounds containing metals that form more than one kind of cation have the form:

| name of cation (metal) | charge of cation (metal) in Roman numerals in parentheses | base name of anion (nonmetal) + -ide |
|---|---|---|

Note that there is no space between the name of the cation and the parenthetical number indicating its charge.

You can determine the charge of the metal cation by inference from the sum of the charges of the nonmetal anions—remember that the sum of all the charges in the compound must be zero. Table 3.3 shows some of the metals that form more than one cation and their most common charges. For example, in $CrBr_3$, the charge of chromium must be 3+ in order for the compound to be charge-neutral with three $Br^-$ anions. The cation is named:

$$Cr^{3+} \quad chromium(III)$$

The full name of the compound is:

$$CrBr_3 \quad chromium(III)\ bromide$$

Similarly, in CuO the charge of copper must be 2+ in order for the compound to be charge-neutral with one $O^{2-}$ anion. The cation is therefore named:

$$Cu^{2+} \quad copper(II)$$

The full name of the compound is:

$$CuO \quad copper(II)\ oxide$$

**TABLE 3.3  Some Metals That Form Cations with Different Charges**

| Metal | Ion | Name | Older Name* |
|---|---|---|---|
| Chromium | $Cr^{2+}$ | Chromium(II) | Chromous |
|  | $Cr^{3+}$ | Chromium(III) | Chromic |
| Iron | $Fe^{2+}$ | Iron(II) | Ferrous |
|  | $Fe^{3+}$ | Iron(III) | Ferric |
| Cobalt | $Co^{2+}$ | Cobalt(II) | Cobaltous |
|  | $Co^{3+}$ | Cobalt(III) | Cobaltic |
| Copper | $Cu^+$ | Copper(I) | Cuprous |
|  | $Cu^{2+}$ | Copper(II) | Cupric |
| Tin | $Sn^{2+}$ | Tin(II) | Stannous |
|  | $Sn^{4+}$ | Tin(IV) | Stannic |
| Mercury | $Hg_2^{2+}$ | Mercury(I) | Mercurous |
|  | $Hg^{2+}$ | Mercury(II) | Mercuric |
| Lead | $Pb^{2+}$ | Lead(II) | Plumbous |
|  | $Pb^{4+}$ | Lead(IV) | Plumbic |

*An older naming system substitutes the names found in this column for the name of the metal and its charge. Under this system, chromium(II) oxide is named chromous oxide. Additionally, the suffix -ous indicates the ion with the lesser charge, and -ic indicates the ion with the greater charge. We will *not* use the older system in this text.

**Example 3.6** | **Naming Ionic Compounds Containing a Metal That Forms More Than One Kind of Cation**

Name the compound $PbCl_4$.

**SOLUTION**

The charge on Pb must be 4+ for the compound to be charge-neutral with four $Cl^-$ anions. The name for $PbCl_4$ is the name of the cation, *lead*, followed by the charge of the cation in parentheses *(IV)* and the base name of the anion, *chlor*, with the ending *-ide*. The full name is *lead(IV) chloride*.

$$PbCl_4 \quad \text{lead(IV) chloride}$$

**FOR PRACTICE 3.6**  Name the compound FeS.

**FOR MORE PRACTICE 3.6**  Write the formula for ruthenium(IV) oxide.

## Naming Ionic Compounds Containing Polyatomic Ions

We name ionic compounds that contain a polyatomic ion in the same way as other ionic compounds, except that we use the name of the polyatomic ion whenever it occurs. Table 3.4 lists common polyatomic ions and their formulas. For example, $NaNO_2$ is named according to its cation, $Na^+$ (*sodium*), and its polyatomic anion, $NO_2^-$ (*nitrite*). Its full name is *sodium nitrite*.

$$NaNO_2 \quad \text{sodium nitrite}$$

$FeSO_4$ is named according to its cation *iron*, its charge *(II)*, and its polyatomic ion *sulfate*. Its full name is *iron(II) sulfate*.

$$FeSO_4 \quad \text{iron(II) sulfate}$$

If the compound contains both a polyatomic cation and a polyatomic anion, use the names of both polyatomic ions. For example, $NH_4NO_3$ is *ammonium nitrate*.

$$NH_4NO_3 \quad \text{ammonium nitrate}$$

You should be able to recognize polyatomic ions in a chemical formula, so become familiar with the ions listed in Table 3.4. Most polyatomic ions are **oxyanions**, anions containing oxygen and another element. Notice that when a series of oxyanions contains different numbers of oxygen atoms, they are named systematically according to

**TABLE 3.4 Some Common Polyatomic Ions**

| Name | Formula | Name | Formula |
|------|---------|------|---------|
| Acetate | $C_2H_3O_2^-$ | Hypochlorite | $ClO^-$ |
| Carbonate | $CO_3^{2-}$ | Chlorite | $ClO_2^-$ |
| Hydrogen carbonate (or bicarbonate) | $HCO_3^-$ | Chlorate | $ClO_3^-$ |
| Hydroxide | $OH^-$ | Perchlorate | $ClO_4^-$ |
| Nitrite | $NO_2^-$ | Permanganate | $MnO_4^-$ |
| Nitrate | $NO_3^-$ | Sulfite | $SO_3^{2-}$ |
| Chromate | $CrO_4^{2-}$ | Hydrogen sulfite (or bisulfite) | $HSO_3^-$ |
| Dichromate | $Cr_2O_7^{2-}$ | Sulfate | $SO_4^{2-}$ |
| Phosphate | $PO_4^{3-}$ | Hydrogen sulfate (or bisulfate) | $HSO_4^-$ |
| Hydrogen phosphate | $HPO_4^{2-}$ | Cyanide | $CN^-$ |
| Dihydrogen phosphate | $H_2PO_4^-$ | Peroxide | $O_2^{2-}$ |
| Ammonium | $NH_4^+$ | | |

the number of oxygen atoms in the ion. If there are only two ions in the series, the one with more oxygen atoms has the ending *-ate* and the one with fewer has the ending *-ite*. For example, $NO_3^-$ is *nitrate* and $NO_2^-$ is *nitrite*.

$$NO_3^- \quad \text{nitr}\textit{ate}$$
$$NO_2^- \quad \text{nitr}\textit{ite}$$

If there are more than two ions in the series, then the prefixes *hypo-*, meaning *less than*, and *per-*, meaning *more than*, are used. So $ClO^-$ is hypochlorite (less oxygen than chlorite), and $ClO_4^-$ is perchlorate (more oxygen than chlorate).

$$ClO^- \quad \textit{hypo}\text{chlor}\textit{ite}$$
$$ClO_2^- \quad \text{chlor}\textit{ite}$$
$$ClO_3^- \quad \text{chlor}\textit{ate}$$
$$ClO_4^- \quad \textit{per}\text{chlor}\textit{ate}$$

> Other halides (halogen ions) form similar series with similar names. Thus, $IO_3^-$ is iodate and $BrO_3^-$ is bromate.

## Example 3.7 | Naming Ionic Compounds That Contain a Polyatomic Ion

Name the compound $Li_2Cr_2O_7$.

**SOLUTION**

The name for $Li_2Cr_2O_7$ is the name of the cation, *lithium*, followed by the name of the polyatomic ion, *dichromate*. Its full name is *lithium dichromate*.

$$Li_2Cr_2O_7 \quad \text{lithium dichromate}$$

**FOR PRACTICE 3.7**    Name the compound $Sn(ClO_3)_2$.

**FOR MORE PRACTICE 3.7**    Write the formula for cobalt(II) phosphate.

## Hydrated Ionic Compounds

The ionic compounds called **hydrates** contain a specific number of water molecules associated with each formula unit. For example, the formula for epsom salts is $MgSO_4 \cdot 7\,H_2O$, and its systematic name is magnesium sulfate heptahydrate. The seven $H_2O$ molecules associated with the formula unit are *waters of hydration*. Waters of hydration can usually be removed by heating the compound. Figure 3.9▶ shows a sample of cobalt(II) chloride hexahydrate ($CoCl_2 \cdot 6\,H_2O$) before and after heating. The hydrate is pink and the anhydrous salt (the salt without any associated water molecules) is blue. Hydrates are named just as other ionic compounds, but they are given the additional name "*prefix*hydrate," where the *prefix* indicates the number of water molecules associated with each formula unit.

Common hydrated ionic compounds and their names are as follows:

| | |
|---|---|
| $CaSO_4 \cdot \frac{1}{2}\,H_2O$ | calcium sulfate hemihydrate |
| $BaCl_2 \cdot 6\,H_2O$ | barium chloride hexahydrate |
| $CuSO_4 \cdot 5\,H_2O$ | copper(II) sulfate pentahydrate |

| Hydrate | Anhydrous |
|---|---|
| $CoCl_2 \cdot 6\,H_2O$ | $CoCl_2$ |

▲ **FIGURE 3.9** **Hydrates** Cobalt(II) chloride hexahydrate is pink. Heating the compound removes the waters of hydration, leaving the blue anhydrous cobalt(II) chloride.

> Common hydrate prefixes
> hemi = 1/2
> mono = 1
> di = 2
> tri = 3
> tetra = 4
> penta = 5
> hexa = 6
> hepta = 7
> octa = 8

## 3.6 Molecular Compounds: Formulas and Names

 **Naming Molecular Compounds**

In contrast to ionic compounds, the formula for a molecular compound *cannot* readily be determined from its constituent elements because the same combination of elements may form many different molecular compounds, each with a different formula. For example, carbon and oxygen form both $CO$ and $CO_2$, and hydrogen and oxygen form both $H_2O$ and $H_2O_2$. Nitrogen and oxygen form all of the following molecular compounds: $NO$, $NO_2$, $N_2O$, $N_2O_3$, $N_2O_4$, and $N_2O_5$. In Chapter 9, we will discuss the stability of these various combinations of the same elements. For now, we focus on naming a molecular compound based on its formula and writing its formula based on its name.

## Naming Molecular Compounds

Like ionic compounds, many molecular compounds have common names. For example, $H_2O$ and $NH_3$ have the common names *water* and *ammonia*. However, the sheer number of existing molecular compounds—numbering in the millions—necessitates a systematic approach to naming them.

The first step in naming a molecular compound is identifying it as one. Remember, *molecular compounds are composed of two or more nonmetals*. In this section, we discuss how to name binary (two-element) molecular compounds. Their names have the form:

When writing the name of a molecular compound, as when writing the formula, we first list the more metal-like element (toward the left and bottom of the periodic table). Generally, we write the name of the element with the smallest group number first. If the two elements lie in the same group, then we write the element with the greatest row number first. The prefixes given to each element indicate the number of atoms present:

|  |  |
|---|---|
| mono = 1 | hexa = 6 |
| di = 2 | hepta = 7 |
| tri = 3 | octa = 8 |
| tetra = 4 | nona = 9 |
| penta = 5 | deca = 10 |

These prefixes are the same as those used in naming hydrates.

If there is only one atom of the *first element* in the formula, the prefix *mono-* is normally omitted. For example, we name $NO_2$ according to the first element, *nitrogen*, with no prefix because *mono-* is omitted for the first element, followed by the prefix *di-*, to indicate two oxygen atoms, and the base name of the second element, *ox*, with the ending *-ide*. Its full name is *nitrogen dioxide*.

$$NO_2 \quad \text{nitrogen dioxide}$$

When a prefix ends with "o" and the base name begins with "o," the first "o" is often dropped. For example, mono-oxide becomes *monoxide*.

We name the compound $N_2O$, sometimes called laughing gas, similarly except that we use the prefix *di-* before nitrogen to indicate two nitrogen atoms and the prefix *mono-* before oxide to indicate one oxygen atom. Its full name is *dinitrogen monoxide*.

$$N_2O \quad \text{dinitrogen monoxide}$$

### Example 3.8    Naming Molecular Compounds

Name each compound.

**(a)** $NI_3$       **(b)** $PCl_5$       **(c)** $P_4S_{10}$

**SOLUTION**

**(a)** The name of the compound is the name of the first element, *nitrogen*, followed by the base name of the second element, *iod*, prefixed by *tri-* to indicate three and given the suffix *-ide*.

$$NI_3 \quad \text{nitrogen triiodide}$$

**(b)** The name of the compound is the name of the first element, *phosphorus*, followed by the base name of the second element, *chlor*, prefixed by *penta-* to indicate five and given the suffix *-ide*.

$$PCl_5 \quad \text{phosphorus pentachloride}$$

(c) The name of the compound is the name of the first element, *phosphorus*, prefixed by *tetra-* to indicate four, followed by the base name of the second element, *sulf*, prefixed by *deca-* to indicate ten and given the suffix *-ide*.

$$P_4S_{10} \quad \text{tetraphosphorus decasulfide}$$

**FOR PRACTICE 3.8**   Name the compound $N_2O_5$.

**FOR MORE PRACTICE 3.8**   Write the formula for phosphorus tribromide.

---

**Nomenclature**  The compound $NCl_3$ is nitrogen trichloride, but $AlCl_3$ is simply aluminum chloride. Why?

CONCEPTUAL
**CONNECTION 3.6**

---

## Naming Acids

We can define acids in a number of ways, as we will discuss in Chapter 16. For now we define **acids** as molecular compounds that release hydrogen ions ($H^+$) when dissolved in water. Acids are composed of hydrogen, usually written first in their formula, and one or more nonmetals, written second. For example, HCl is a molecular compound that, when dissolved in water, forms $H^+(aq)$ and $Cl^-(aq)$ ions, where *aqueous (aq)* means *dissolved in water*. Therefore, HCl is an acid when dissolved in water. To distinguish between gaseous HCl (which is named hydrogen monochloride because it is a molecular compound) and HCl in solution (which is named hydrochloric acid because it is an acid), we write the former as HCl(*g*) and the latter as HCl(*aq*).

Acids are characterized by their sour taste and their ability to dissolve many metals. For example, hydrochloric acid is present in stomach fluids, and its sour taste becomes painfully obvious during vomiting. Hydrochloric acid also dissolves some metals. For example, if we put a strip of zinc into a test tube of hydrochloric acid, it slowly dissolves as the $H^+(aq)$ ions convert the zinc metal into $Zn^{2+}(aq)$ cations (Figure 3.10▼).

Acids are present in foods such as lemons and limes and are used in household products such as toilet bowl cleaner and Lime-A-Way. In this section, we discuss how to name them; in Chapter 16 you will learn more about their properties. We categorize acids into two types: binary acids and oxyacids.

▲ Many fruits are acidic and have the characteristically sour taste of acids.

**Acids Dissolve Many Metals**

Zn atoms

Zn metal

$H_2$ gas

HCl solution

$Zn^{2+}$  $H_3O^+$  $Cl^-$  $H_2O$

▲ **FIGURE 3.10  Hydrochloric Acid Dissolving Zinc Metal**  The zinc atoms are ionized to zinc ions, which dissolve in the water. The HCl forms $H_2$ gas, which is responsible for the bubbles you can see in the test tube.

### Naming Binary Acids

**Binary acids** are composed of hydrogen and a nonmetal. Names for binary acids have the form:

hydro | base name of nonmetal + *-ic* | acid

For example, HCl(*aq*) is hydro*chlor*ic acid and HBr(*aq*) is hydro*brom*ic acid.

HCl(*aq*)    hydrochloric acid            HBr(*aq*)    hydrobromic acid

### Example 3.9    Naming Binary Acids

Name the acid HI(*aq*).

**SOLUTION**

The base name of I is *iod*, so HI(*aq*) is hydroiodic acid.

HI(*aq*)        hydroiodic acid

**FOR PRACTICE 3.9**    Name the acid HF(*aq*).

## Naming Oxyacids

**Oxyacids** contain hydrogen and an oxyanion (an anion containing a nonmetal and oxygen). The common oxyanions are listed in the table of polyatomic ions (Table 3.4). For example, $HNO_3$(*aq*) contains the nitrate ($NO_3^-$) ion, $H_2SO_3$(*aq*) contains the sulfite ($SO_3^{2-}$) ion, and $H_2SO_4$(*aq*) contains the sulfate ($SO_4^{2-}$) ion. Oxyacids are a combination of one or more $H^+$ ions with an oxyanion. The number of $H^+$ ions depends on the

---

# Chemistry IN THE Environment | Acid Rain

Certain pollutants—such as NO, $NO_2$, and $SO_2$—form acids when mixed with water. NO and $NO_2$, primarily emitted in vehicular exhaust, combine with atmospheric oxygen and water to form nitric acid, $HNO_3$(*aq*). $SO_2$, emitted primarily from coal-powered electricity generation, combines with atmospheric oxygen and water to form sulfuric acid, $H_2SO_4$(*aq*). Both $HNO_3$(*aq*) and $H_2SO_4$(*aq*) result in acidic rainwater. The problem is greatest in the northeastern United States where pollutants from midwestern electrical power plants combine with rainwater to produce rain that is up to ten times more acidic than normal.

Acid rain can fall or flow into lakes and streams, making these bodies of water more acidic. Some species of aquatic animals—such as trout, bass, snails, salamanders, and clams—cannot tolerate the increased acidity and die. This in turn disturbs the ecosystem of the lake, resulting in imbalances that may lead to the death of other aquatic species. Acid rain also weakens trees by dissolving and washing away nutrients in the soil and by damaging leaves. Appalachian red spruce trees have been the hardest hit, with many forests showing significant acid rain damage.

In addition, acid rain degrades building materials because acids dissolve iron, the main component of steel, and $CaCO_3$ (limestone), a main component of marble and concrete. Consequently, acid rain has damaged many statues, buildings, and bridges in the northeastern United States.

Acid rain has been a problem for many years, but legislation passed toward the end of the last century has begun to address this issue. In 1990, Congress passed several amendments to the Clean Air Act that included provisions requiring electrical utilities to lower $SO_2$ emissions. Since then, $SO_2$ emissions have decreased and rain in the northeastern United States has become less acidic. With time, and with continued enforcement of the acid rain regulation, lakes, streams, and forests damaged by acid rain should recover.

**QUESTION**

*Name each compound: NO, $NO_2$, $SO_2$, $H_2SO_4$, $HNO_3$, $CaCO_3$.*

◀ A forest damaged by acid rain.

▶ Acid rain damages building materials including the limestone that composes many statues.

charge of the oxyanion; the formula is always charge-neutral. The names of oxyacids depend on the ending of the oxyanion and take the following forms:

oxyanions ending with *-ate*

oxyanions ending with *-ite*

For example, $HNO_3(aq)$ is nitric acid (oxyanion is nitrate), and $H_2SO_3(aq)$ is sulfurous acid (oxyanion is sulfite).

$HNO_3(aq)$  nitric acid   $H_2SO_3(aq)$  sulfurous acid

## Example 3.10 Naming Oxyacids

Name the acid $HC_2H_3O_2(aq)$.

**SOLUTION**

The oxyanion is acetate, which ends in *-ate*; therefore, the name of the acid is *acetic acid*.

$HC_2H_3O_2(aq)$   acetic acid

**FOR PRACTICE 3.10**   Name the acid $HNO_2(aq)$.

**FOR MORE PRACTICE 3.10**   Write the formula for perchloric acid.

## 3.7 Summary of Inorganic Nomenclature

In Section 3.5 and Section 3.6, we discussed naming inorganic compounds, specifically ionic compounds, molecular compounds, and acids. However, we often have to name a compound without initially knowing the category into which it falls. In other words, real-life nomenclature is a bit messier than the categorized nomenclature we just worked through. Figure 3.11▼ summarizes inorganic nomenclature in a flowchart that will help you to tackle nomenclature from beginning to end.

▼ **FIGURE 3.11** **Inorganic Nomenclature Flowchart** The chart summarizes how to name inorganic compounds. Begin by determining if the compound is ionic, molecular, or an acid. Then follow the flowchart for that category from top to bottom until you arrive at a name for the compound.

## Inorganic Nomenclature Flow Chart

**▲ FIGURE 3.12 Flowchart Path for MgCl₂**

To use the flowchart, begin by determining what type of compound you are trying to name. For example, to name the compound $MgCl_2$, you need to decide if the compound is ionic, molecular, or an acid. In this case, since $MgCl_2$ is composed of a metal and nonmetal, it is ionic. Therefore, you begin at the box labeled "IONIC" at the far left side of the flowchart.

Next, decide whether the metal in the compound forms only one type of ion or more than one type. You can determine this by looking for the metal (in this case magnesium) in Figure 3.7. Since magnesium is listed in the figure, it forms only one type of ion; therefore, you take the left branch in the flowchart as shown in Figure 3.12◄.

Finally, name the compound according to the blocks at the end of the path in the flowchart. In this case, write the name of the cation (the metal) followed by the base name of the anion (the nonmetal) appended with the ending *-ide*. Its full name is magnesium chloride.

---

## Example 3.11   Using the Nomenclature Flowchart to Name Compounds   ▶

Use the flowchart in Figure 3.11 to name each compound.

**(a)** $SO_2$          **(b)** $HClO_4(aq)$          **(c)** $CoF_2$

**SOLUTION**

**(a)** $SO_2$

Begin by determining whether the compound is ionic, molecular, or an acid. $SO_2$ contains only nonmetals; therefore it is molecular.

Name the compound as the name of the first element, *sulfur* (no prefix since the prefix is dropped for mono), followed by the base name of the second element, *ox*, prefixed by *di-* to indicate two, and given the suffix *-ide*.

**(b)** $HClO_4(aq)$

Begin by determining whether the compound is ionic, molecular, or an acid. Since $HClO_4(aq)$ contains H and one more nonmetal and is designated as aqueous, it is an acid.

Next determine whether the acid contains oxygen. Since $HClO_4$ contains oxygen, it is an oxyacid.

Then determine whether the name of the oxyanion ends in *-ate* or *-ite*. Since the oxyanion is perchlor*ate*, it ends in *-ate*.

Finally, name the acid as the base name of the oxyanion, *perchlor*, with the ending *-ic*, followed by the word *acid*.

**(c)** $CoF_2$

Begin by determining whether the compound is ionic, molecular, or an acid. Since $CoF_2$ contains a metal and a nonmetal, it is ionic.

Next refer to Figure 3.7 to determine whether the metal forms one type of ion or more than one type. Since Co is not listed in Figure 3.7, it must form more than one type of ion.

Name the compound as the name of the cation, *cobalt*, followed by the charge of the cation in parentheses *(II)*, and the base name of the anion, *fluor*, with the ending *-ide*.

**FOR PRACTICE 3.11**   Use the flowchart in Figure 3.11 to name $H_2SO_3(aq)$.

# 3.8  Formula Mass and the Mole Concept for Compounds

In Chapter 2, we defined the average mass of an atom of an element as its *atomic mass*. Similarly, we now define the average mass of a molecule (or a formula unit) of a compound as its **formula mass**. (The common terms *molecular mass* and *molecular weight* are synonymous with formula mass.) For any compound, the formula mass is the sum of the atomic masses of all the atoms in its chemical formula.

> Remember, ionic compounds do not contain individual molecules. In casual language, the smallest electrically neutral collection of ions is sometimes called a molecule but is more correctly called a formula unit.

$$\text{Formula mass} = \left(\begin{array}{c}\text{Number of atoms}\\\text{of 1st element in}\\\text{chemical formula}\end{array} \times \begin{array}{c}\text{Atomic mass}\\\text{of}\\\text{1st element}\end{array}\right) + \left(\begin{array}{c}\text{Number of atoms}\\\text{of 2nd element in}\\\text{chemical formula}\end{array} \times \begin{array}{c}\text{Atomic mass}\\\text{of}\\\text{2nd element}\end{array}\right) + \dots$$

For example, the formula mass of carbon dioxide, $CO_2$, is:

*Multiply by 2 because formula has two oxygen atoms.*

$$\text{Formula mass} = 12.01 \text{ amu} + 2(16.00 \text{ amu})$$
$$= 44.01 \text{ amu}$$

and that of sodium oxide, $Na_2O$, is:

*Multiply by 2 because formula has two sodium atoms.*

$$\text{Formula mass} = 2(22.99 \text{ amu}) + 16.00 \text{ amu}$$
$$= 61.98 \text{ amu}$$

---

## Example 3.12  Calculating Formula Mass

Calculate the formula mass of glucose, $C_6H_{12}O_6$.

**SOLUTION**

To find the formula mass, add the atomic masses of each atom in the chemical formula.

$$\text{Formula mass} = 6 \times (\text{atomic mass C}) + 12 \times (\text{atomic mass H}) + 6 \times (\text{atomic mass O})$$

$$= 6(12.01 \text{ amu}) \qquad + 12(1.008 \text{ amu}) \qquad + 6(16.00 \text{ amu})$$

$$= 180.16 \text{ amu}$$

**FOR PRACTICE 3.12**   Calculate the formula mass of calcium nitrate.

---

## Molar Mass of a Compound

In Chapter 2 (Section 2.9), we saw that an element's molar mass—the mass in grams of one mole of its atoms—is numerically equivalent to its atomic mass. We then used the molar mass in combination with Avogadro's number to determine the number of atoms in a given mass of the element. We can apply the same concept to compounds. The *molar mass of a compound*—the mass in grams of 1 mol of its molecules or formula units—is numerically equivalent to its formula mass. For example, we just calculated the formula mass of $CO_2$ to be 44.01 amu. The molar mass is, therefore:

$$CO_2 \text{ molar mass} = 44.01 \text{ g/mol}$$

## Using Molar Mass to Count Molecules by Weighing

The molar mass of $CO_2$ is a conversion factor between mass (in grams) and amount (in moles) of $CO_2$. Suppose we want to find the number of $CO_2$ molecules in a sample of dry ice (solid $CO_2$) with a mass of 10.8 g. This calculation is analogous to Example 2.8, where we found the number of atoms in a sample of copper of a given mass. We begin with the mass of 10.8 g and use the molar mass to convert to the amount in moles. Then we use Avogadro's number to convert to number of molecules. The conceptual plan is as follows:

**Conceptual Plan**

To solve the problem, we follow the conceptual plan, beginning with 10.8 g $CO_2$, converting to moles, and then to molecules.

**Solution**

$$10.8 \text{ g } CO_2 \times \frac{1 \text{ mol } CO_2}{44.01 \text{ g } CO_2} \times \frac{6.022 \times 10^{23} \, CO_2 \text{ molecules}}{1 \text{ mol } CO_2}$$

$$= 1.48 \times 10^{23} \, CO_2 \text{ molecules}$$

---

## Example 3.13    The Mole Concept—Converting between Mass and Number of Molecules    ▶

An aspirin tablet contains 325 mg of acetylsalicylic acid ($C_9H_8O_4$). How many acetylsalicylic acid molecules does it contain?

| | |
|---|---|
| **SORT** You are given the mass of acetyl-salicylic acid and asked to find the number of molecules. | **GIVEN:** 325 mg $C_9H_8O_4$ <br><br> **FIND:** number of $C_9H_8O_4$ molecules |
| **STRATEGIZE** First convert to moles (using the molar mass of the compound) and then to number of molecules (using Avogadro's number). You need both the molar mass of acetylsalicylic acid and Avogadro's number as conversion factors. You also need the conversion factor between g and mg. | **CONCEPTUAL PLAN** <br><br>  <br><br> **RELATIONSHIPS USED** <br> $C_9H_8O_4$ molar mass = 9(12.01) + 8(1.008) + 4(16.00) <br> $\qquad\qquad\qquad = 180.15$ g/mol <br><br> $6.022 \times 10^{23} = 1$ mol <br> $1 \text{ mg} = 10^{-3} \text{ g}$ |
| **SOLVE** Follow the conceptual plan to solve the problem. | **SOLUTION** <br><br> $325 \text{ mg } C_9H_8O_4 \times \dfrac{10^{-3} \text{ g}}{1 \text{ mg}} \times \dfrac{1 \text{ mol } C_9H_8O_4}{180.15 \text{ g } C_9H_8O_4} \times$ <br><br> $\dfrac{6.022 \times 10^{23} \, C_9H_8O_4 \text{ molecules}}{1 \text{ mol } C_9H_8O_4} = 1.09 \times 10^{21} \, C_9H_8O_4 \text{ molecules}$ |

**CHECK** The units of the answer, $C_9H_8O_4$ molecules, are correct. The magnitude is smaller than Avogadro's number, as expected, since you have less than 1 molar mass of acetylsalicylic acid.

**FOR PRACTICE 3.13**    Find the number of ibuprofen molecules in a tablet containing 200.0 mg of ibuprofen ($C_{13}H_{18}O_2$).

**FOR MORE PRACTICE 3.13**    Determine the mass of a sample of water containing $3.55 \times 10^{22}$ $H_2O$ molecules.

**Molecular Models and the Size of Molecules** Throughout this book, you will find space-filling molecular models to represent molecules. Which number is the best estimate for the scaling factor used in these models? In other words, by approximately what number would you have to multiply the radius of an actual oxygen atom to get the radius of the sphere used to represent the oxygen atom in the water molecule shown to the right.

**(a)** 10      **(b)** $10^4$      **(c)** $10^8$      **(d)** $10^{16}$

## 3.9  Composition of Compounds

A chemical formula, in combination with the molar masses of its constituent elements, indicates the relative quantities of each element in a compound, which is extremely useful information. For example, about 35 years ago, scientists began to suspect that synthetic compounds known as chlorofluorocarbons (or CFCs) were destroying ozone ($O_3$) in Earth's upper atmosphere. Upper atmospheric ozone is important because it acts as a shield, protecting life on Earth from the sun's harmful ultraviolet light.

CFCs are chemically inert compounds used primarily as refrigerants and industrial solvents. Over time, CFCs accumulated in the atmosphere. In the upper atmosphere, sunlight breaks bonds within CFCs, releasing chlorine atoms. The chlorine atoms react with ozone, converting it into $O_2$. So the harmful part of CFCs is the chlorine atoms that they carry. How can we determine the mass of chlorine in a given mass of a CFC?

One way to express how much of an element is in a given compound is to use the element's mass percent composition for that compound. The **mass percent composition** or **mass percent** of an element is that element's percentage of the compound's total mass. We calculate the mass percent of element X in a compound from the chemical formula as follows:

$$\text{Mass percent of element X} = \frac{\text{mass of element X in 1 mol of compound}}{\text{mass of 1 mol of the compound}} \times 100\%$$

Suppose, for example, that we want to calculate the mass percent composition of Cl in the chlorofluorocarbon $CCl_2F_2$. The mass percent Cl is given by:

▲ The chlorine in chlorofluorocarbons caused the ozone hole over Antarctica. The dark blue color indicates depressed ozone levels.

$CCl_2F_2$

$$\text{Mass percent Cl} = \frac{2 \times \text{molar mass Cl}}{\text{molar mass } CCl_2F_2} \times 100\%$$

We multiply the molar mass of Cl by 2 because the chemical formula has a subscript of 2 for Cl, indicating that 1 mol of $CCl_2F_2$ contains 2 mol of Cl atoms. We calculate the molar mass of $CCl_2F_2$ as follows:

$$\text{Molar mass} = 12.01 \text{ g/mol} + 2(35.45 \text{ g/mol}) + 2(19.00 \text{ g/mol})$$

$$= 120.91 \text{ g/mol}$$

So the mass percent of Cl in $CCl_2F_2$ is:

$$\text{Mass percent Cl} = \frac{2 \times \text{molar mass Cl}}{\text{molar mass } CCl_2F_2} \times 100\%$$

$$= \frac{2 \times 35.45 \text{ g/mol}}{120.91 \text{ g/mol}} \times 100\%$$

$$= 58.64\%$$

## Example 3.14   Mass Percent Composition

Calculate the mass percent of Cl in Freon-112 ($C_2Cl_4F_2$), a CFC refrigerant.

| | |
|---|---|
| **SORT** You are given the molecular formula of Freon-112 and asked to find the mass percent of Cl. | **GIVEN:** $C_2Cl_4F_2$ <br><br> **FIND:** mass percent Cl |
| **STRATEGIZE** The molecular formula tells you that there are 4 mol of Cl in each mole of Freon-112. Find the mass percent composition from the chemical formula by using the equation that defines mass percent. The conceptual plan shows you how to use the mass of Cl in 1 mol of $C_2Cl_4F_2$ and the molar mass of $C_2Cl_4F_2$ to find the mass percent of Cl. | **CONCEPTUAL PLAN** <br><br> $$\text{Mass \% Cl} = \frac{4 \times \text{molar mass Cl}}{\text{molar mass } C_2Cl_4F_2} \times 100\%$$ <br><br> **RELATIONSHIPS USED** <br><br> $$\text{Mass percent of element X} = \frac{\text{mass of element X in 1 mol of compound}}{\text{mass of 1 mol of compound}} \times 100\%$$ |
| **SOLVE** Calculate the necessary parts of the equation and substitute the values into the equation to find mass percent Cl. | **SOLUTION** <br> $4 \times$ molar mass Cl $= 4(35.45 \text{ g/mol}) = 141.8 \text{ g/mol}$ <br> Molar mass $C_2Cl_4F_2 = 2(12.01 \text{ g/mol}) + 4(35.45 \text{ g/mol}) + 2(19.00 \text{ g/mol})$ <br> $\qquad = 24.02 \text{ g/mol} + 141.8 \text{ g/mol} + 38.00 \text{ g/mol} = 203.8 \text{ g/mol}$ <br><br> $$\text{Mass \% Cl} = \frac{4 \times \text{molar mass Cl}}{\text{molar mass } C_2Cl_4F_2} \times 100\%$$ <br> $$= \frac{141.8 \text{ g/mol}}{203.8 \text{ g/mol}} \times 100\%$$ <br> $$= 69.58\%$$ |

**CHECK** The units of the answer (%) are correct. The magnitude is reasonable because it is between 0 and 100% and chlorine is the heaviest atom in the molecule and there are four atoms of it.

**FOR PRACTICE 3.14** Acetic acid ($C_2H_4O_2$) is the active ingredient in vinegar. Calculate the mass percent composition of oxygen in acetic acid.

**FOR MORE PRACTICE 3.14** Calculate the mass percent composition of sodium in sodium oxide.

---

CONCEPTUAL
**CONNECTION 3.8**

**Chemical Formula and Mass Percent Composition** Without doing any calculations, list the elements in $C_6H_6O$ in order of decreasing mass percent composition.

---

## Mass Percent Composition as a Conversion Factor

The mass percent composition of an element in a compound is a conversion factor between mass of the element and mass of the compound. For example, we saw that the mass percent composition of Cl in $CCl_2F_2$ is 58.64%. Since percent means *per hundred*, there are 58.64 g Cl *per hundred* grams $CCl_2F_2$, which can be expressed as the ratio:

$$58.64 \text{ g Cl} : 100 \text{ g } CCl_2F_2$$

or, in fractional form:

$$\frac{58.64 \text{ g Cl}}{100 \text{ g } CCl_2F_2} \quad \text{or} \quad \frac{100 \text{ g } CCl_2F_2}{58.64 \text{ g Cl}}$$

These ratios can function as conversion factors between grams of Cl and grams of $CCl_2F_2$. For example, to calculate the mass of Cl in 1.00 kg $CCl_2F_2$, we use the following conceptual plan:

**Conceptual Plan**

Notice that the mass percent composition acts as a conversion factor between grams of the compound and grams of the constituent element. To calculate grams Cl, we follow the conceptual plan.

**Solution**

$$1.00 \text{ kg } CCl_2F_2 \times \frac{1000 \text{ g}}{1 \text{ kg}} \times \frac{58.64 \text{ g Cl}}{100 \text{ g } CCl_2F_2} = 5.86 \times 10^2 \text{ g Cl}$$

## Example 3.15    Using Mass Percent Composition as a Conversion Factor

The U.S. Food and Drug Administration (FDA) recommends that an adult consume less than 2.4 g of sodium per day. What mass of sodium chloride (in grams) can you consume and still be within the FDA guidelines? Sodium chloride is 39% sodium by mass.

| | |
|---|---|
| **SORT** You are given a mass of sodium and the mass percent of sodium in sodium chloride. You are asked to find the mass of NaCl that contains the given mass of sodium. | **GIVEN:** 2.4 g Na<br><br>**FIND:** g NaCl |
| **STRATEGIZE** Convert between mass of a constituent element and mass of a compound by using mass percent composition as a conversion factor. | **CONCEPTUAL PLAN**<br><br>g Na → g NaCl<br>$\dfrac{100 \text{ g NaCl}}{39 \text{ g Na}}$<br><br>**RELATIONSHIPS USED**<br>39 g Na: 100 g NaCl |
| **SOLVE** Follow the conceptual plan to solve the problem. | **SOLUTION**<br><br>$2.4 \text{ g Na} \times \dfrac{100 \text{ g NaCl}}{39 \text{ g Na}} = 6.2 \text{ g NaCl}$<br><br>You can consume 6.2 g NaCl and still be within the FDA guidelines. |

**CHECK** The units of the answer are correct. The magnitude seems reasonable because it is larger than the amount of sodium, as expected, because sodium is only one of the elements in NaCl.

**FOR PRACTICE 3.15** What mass (in grams) of iron(III) oxide contains 58.7 g of iron? Iron(III) oxide is 69.94% iron by mass.

**FOR MORE PRACTICE 3.15** If someone consumes 22 g of sodium chloride per day, what mass (in grams) of sodium does that person consume? Sodium chloride is 39% sodium by mass.

◀ 12.5 packets of salt contain 6.2 g of NaCl.

# Conversion Factors from Chemical Formulas

Mass percent composition is one way to understand how much chlorine is in a particular chlorofluorocarbon or, more generally, how much of a constituent element is present in a given mass of any compound. However, we can also approach this type of problem in a different way. Chemical formulas contain within them inherent relationships between atoms (or moles of atoms) and molecules (or moles of molecules). For example, the formula for $CCl_2F_2$ tells us that 1 mol of $CCl_2F_2$ contains 2 mol of Cl atoms. We write the ratio as:

$$1 \text{ mol } CCl_2F_2 : 2 \text{ mol } Cl$$

With ratios such as these—which come from the chemical formula—we can directly determine the amounts of the constituent elements present in a given amount of a compound without having to calculate mass percent composition. For example, we calculate the number of moles of Cl in 38.5 mol of $CCl_2F_2$ as follows:

**Conceptual Plan**

**Solution**

$$38.5 \text{ mol } CCl_2F_2 \times \frac{2 \text{ mol } Cl}{1 \text{ mol } CCl_2F_2} = 77.0 \text{ mol } Cl$$

As we have seen, however, we often want to know, not the *amount in moles* of an element in a certain number of moles of compound, but the *mass in grams* (or other units) of a constituent element in a given *mass* of the compound. Suppose we want to know the mass (in grams) of Cl in 25.0 g $CCl_2F_2$. The relationship inherent in the chemical formula (2 mol Cl : 1 mol $CCl_2F_2$) applies to the amount in moles, not to mass. Therefore, we first convert the mass of $CCl_2F_2$ to moles $CCl_2F_2$. *Then* we use the conversion factor from the chemical formula to convert to moles Cl. Finally, we use the molar mass of Cl to convert to grams Cl.

**Conceptual Plan**

**Solution**

$$25.0 \text{ g } CCl_2F_2 \times \frac{1 \text{ mol } CCl_2F_2}{120.91 \text{ g } CCl_2F_2} \times \frac{2 \text{ mol } Cl}{1 \text{ mol } CCl_2F_2} \times \frac{35.45 \text{ g } Cl}{1 \text{ mol } Cl} = 14.7 \text{ g } Cl$$

Notice that we must convert from g $CCl_2F_2$ to mol $CCl_2F_2$ *before* we can use the chemical formula as a conversion factor. *Always remember that the chemical formula indicates the relationship between the amounts (in moles) of substances, not between the masses (in grams) of them.*

The general form for solving problems in which we need to find the mass of an element present in a given mass of a compound is:

Mass compound $\longrightarrow$ moles compound $\longrightarrow$ moles element $\longrightarrow$ mass element

We use the atomic or molar mass to convert between mass and moles, and we use relationships inherent in the chemical to convert between moles and moles.

**Example** 3.16    **Chemical Formulas as Conversion Factors**

Hydrogen may be used in the future to replace gasoline as a fuel. Most major automobile companies are developing vehicles that run on hydrogen. These cars have the potential to be less environmentally harmful than our current vehicles because their only emission is water vapor. One way to obtain hydrogen for fuel is to use an emission-free energy source such as wind power to form elemental hydrogen from water. What mass of hydrogen (in grams) is contained in 1.00 gallon of water? (The density of water is 1.00 g/mL.)

| | |
|---|---|
| **SORT** You are given a volume of water and asked to find the mass of hydrogen it contains. You are also given the density of water. | **GIVEN:** 1.00 gal $H_2O$ <br> $d_{H_2O} = 1.00$ g/mL <br> **FIND:** g H |
| **STRATEGIZE** The first part of the conceptual plan shows how to convert the units of volume from gallons to liters and then to mL. It also shows how to use the density to convert mL to g. <br><br> The second part of the conceptual plan is the basic sequence: mass → moles → moles → mass. Convert between moles and mass using the appropriate molar masses, and convert from mol $H_2O$ to mol H using the conversion factor derived from the molecular formula. | **CONCEPTUAL PLAN** <br>  <br>  <br> **RELATIONSHIPS USED** <br> 3.785 L = 1 gal (Table 1.3) <br> 1000 mL = 1 L <br> 1.00 g $H_2O$ = 1 mL $H_2O$ (density of $H_2O$) <br> Molar mass $H_2O$ = 2(1.008) + 16.00 = 18.02 g/mol <br> 2 mol H : 1 mol $H_2O$ <br> 1.008 g H = 1 mol H |
| **SOLVE** Follow the conceptual plan to solve the problem. | **SOLUTION** <br> $1.00 \text{ gal } H_2O \times \dfrac{3.785 \text{ L}}{1 \text{ gal}} \times \dfrac{1000 \text{ mL}}{1 \text{ L}} \times \dfrac{1.00 \text{ g}}{\text{mL}} = 3.785 \times 10^3 \text{ g } H_2O$ <br><br> $3.785 \times 10^3 \text{ g } H_2O \times \dfrac{1 \text{ mol } H_2O}{18.02 \text{ g } H_2O} \times \dfrac{2 \text{ mol H}}{1 \text{ mol } H_2O}$ <br> $\times \dfrac{1.008 \text{ g H}}{1 \text{ mol H}} = 4.23 \times 10^2 \text{ g H}$ |

**CHECK** The units of the answer (g H) are correct. Since a gallon of water is about 3.8 L, its mass is about 3.8 kg. H is a light atom, so its mass should be significantly less than 3.8 kg, which it is in the answer.

**FOR PRACTICE 3.16** Determine the mass of oxygen in a 7.2-g sample of $Al_2(SO_4)_3$.

**FOR MORE PRACTICE 3.16** Butane ($C_4H_{10}$) is the liquid fuel in lighters. How many grams of carbon are present within a lighter containing 7.25 mL of butane? (The density of liquid butane is 0.601 g/mL.)

**Chemical Formulas and Elemental Composition** The molecular formula for water is $H_2O$. Which ratio can be correctly derived from this formula? Explain.
   **(a)** 2 g H : 1 g $H_2O$    **(b)** 2 mL H : 1 mL $H_2O$    **(c)** 2 mol H : 1 mol $H_2O$

CONCEPTUAL CONNECTION **3.9**

# Chemistry AND Medicine | Methylmercury in Fish

In the last decade, the U.S. Environmental Protection Agency (EPA) has grown increasingly concerned about mercury levels in fish. Mercury—which is present in fish as methylmercury—affects the central nervous system of humans who eat the fish, especially children and developing fetuses. In a developing fetus, excessive mercury exposure can result in slowed mental development and even retardation. Some lakes now have warnings about eating too much fish caught in the lakes.

Recent regulations force fish vendors to alert customers about the dangers of eating too much of certain kinds of commercial fish, including shark, tuna, and mackerel. These fish tend to contain high levels of methylmercury and therefore should be eaten in moderation, especially by children and pregnant women. The U.S. Food and Drug Administration (FDA) *action level*—the level below which the FDA claims the food has no adverse health effects—for methylmercury in fish is 1.0 ppm or 1.0 g of methylmercury per million grams of fish. However, a number of environmental advocacy groups, including the EPA, have suggested that, while this level may be safe for adults, it is too high for children and pregnant women. Consequently, the FDA suggests that pregnant women limit their intake of fish to 12 ounces per week.

### QUESTION

*The levels of methylmercury in fish are normally tested by laboratory techniques that measure only the mercury (Hg). Suppose a lab analyzes a 14.5 g sample of fish and finds that it contains $1.03 \times 10^{-5}$ g of mercury. How much methylmercury ($HgCH_3Cl$) is in the fish in parts per million (ppm)? Is this above the FDA action level?*

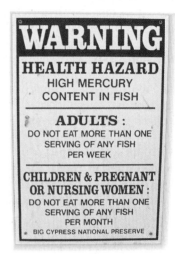

▲ Lakes containing mercury—either from natural sources or from pollution—often have posted limits for the number of fish from the lake that can be eaten safely.

## 3.10 Determining a Chemical Formula from Experimental Data

In Section 3.8, we calculated mass percent composition from a chemical formula. Can we also do the reverse? Can we calculate a chemical formula from mass percent composition? This question is important because many laboratory analyses of compounds list the relative masses of each element present in the compound. For example, if we decompose water into hydrogen and oxygen in the laboratory, we can measure the masses of hydrogen and oxygen produced. Can we determine a chemical formula from this data? The answer is a qualified yes. We can determine a chemical formula, but it is an empirical formula (not a molecular formula). To get a molecular formula, we need additional information, such as the molar mass of the compound.

Suppose we decompose a sample of water in the laboratory and find that it produces 0.857 g of hydrogen and 6.86 g of oxygen. How do we determine an empirical formula from these data? We know that an empirical formula represents a ratio of atoms or a ratio of moles of atoms, *not a ratio of masses*. So the first thing we must do is convert our data from mass (in grams) to amount (in moles). How many moles of each element are present in the sample? To convert to moles, we divide each mass by the molar mass of that element:

$$\text{Moles H} = 0.857 \ \cancel{\text{g H}} \times \frac{1 \ \text{mol H}}{1.01 \ \cancel{\text{g H}}} = 0.849 \ \text{mol H}$$

$$\text{Moles O} = 6.86 \ \cancel{\text{g O}} \times \frac{1 \ \text{mol O}}{16.00 \ \cancel{\text{g O}}} = 0.429 \ \text{mol O}$$

From these data, we know there are 0.849 mol H for every 0.429 mol O. We can now write a *pseudoformula* for water:

$$H_{0.849}O_{0.429}$$

To get the smallest whole-number subscripts in our formula, we divide all the subscripts by the smallest one, in this case 0.429:

$$H_{\frac{0.849}{0.429}}O_{\frac{0.429}{0.429}} = H_{1.98}O = H_2O$$

Our empirical formula for water, which also happens to be the molecular formula, is $H_2O$. You can use the procedure shown here to obtain the empirical formula of any compound from experimental data giving the relative masses of the constituent elements. The left column outlines the procedure, and the center and right columns contain two examples of how to apply the procedure.

## Procedure for...

| **Obtaining an Empirical Formula from Experimental Data** | **Example 3.17** **Obtaining an Empirical Formula from Experimental Data** A compound containing nitrogen and oxygen is decomposed in the laboratory. It produces 24.5 g nitrogen and 70.0 g oxygen. Calculate the empirical formula of the compound. | **Example 3.18** ▶ **Obtaining an Empirical Formula from Experimental Data** A laboratory analysis of aspirin determines the following mass percent composition: C 60.00%; H 4.48%; O 35.52% Find the empirical formula. |
|---|---|---|
| 1. Write down (or calculate) as *given* the masses of each element present in a sample of the compound. If you are given mass percent composition, assume a 100-g sample and calculate the masses of each element from the given percentages. | **GIVEN:** 24.5 g N, 70.0 g O **FIND:** empirical formula | **GIVEN:** In a 100-g sample: 60.00 g C, 4.48 g H, 35.52 g O **FIND:** empirical formula |
| 2. Convert each of the masses in step 1 to moles by using the appropriate molar mass for each element as a conversion factor. | $24.5 \text{ g N} \times \dfrac{1 \text{ mol N}}{14.01 \text{ g N}} = 1.75 \text{ mol N}$ $70.0 \text{ g O} \times \dfrac{1 \text{ mol O}}{16.00 \text{ g O}} = 4.38 \text{ mol O}$ | $60.00 \text{ g C} \times \dfrac{1 \text{ mol C}}{12.01 \text{ g C}} = 4.996 \text{ mol C}$ $4.48 \text{ g H} \times \dfrac{1 \text{ mol H}}{1.008 \text{ g H}} = 4.44 \text{ mol H}$ $35.52 \text{ g O} \times \dfrac{1 \text{ mol O}}{16.00 \text{ g O}} = 2.220 \text{ mol O}$ |
| 3. Write down a pseudoformula for the compound using the number of moles of each element (from step 2) as subscripts. | $N_{1.75}O_{4.38}$ | $C_{4.996}H_{4.44}O_{2.220}$ |
| 4. Divide all the subscripts in the formula by the smallest subscript. | $N_{\frac{1.75}{1.75}}O_{\frac{4.38}{1.75}} \longrightarrow N_1O_{2.5}$ | $C_{\frac{4.996}{2.220}}H_{\frac{4.44}{2.220}}O_{\frac{2.220}{2.220}} \longrightarrow C_{2.25}H_2O_1$ |
| 5. If the subscripts are not whole numbers, multiply all the subscripts by a small whole number (see table) to get whole-number subscripts. | $N_1O_{2.5} \times 2 \longrightarrow N_2O_5$ The correct empirical formula is $N_2O_5$. | $C_{2.25}H_2O_1 \times 4 \longrightarrow C_9H_8O_4$ The correct empirical formula is $C_9H_8O_4$. |

| Fractional Subscript | Multiply by This |
|---|---|
| 0.20 | 5 |
| 0.25 | 4 |
| 0.33 | 3 |
| 0.40 | 5 |
| 0.50 | 2 |
| 0.66 | 3 |
| 0.75 | 4 |
| 0.80 | 5 |

**FOR PRACTICE 3.17**
A sample of a compound is decomposed in the laboratory and produces 165 g carbon, 27.8 g hydrogen, and 220.2 g oxygen. Calculate the empirical formula of the compound.

**FOR PRACTICE 3.18**
Ibuprofen has the following mass percent composition:
C 75.69%, H 8.80%, O 15.51%.
What is the empirical formula of ibuprofen?

## Determining Molecular Formulas for Compounds

We can find the molecular formula of a compound from the empirical formula if we also know the molar mass of the compound. Recall from Section 3.3 that the molecular formula is always a whole-number multiple of the empirical formula:

$$\text{Molecular formula} = \text{empirical formula} \times n, \text{ where } n = 1, 2, 3,...$$

Suppose we want to find the molecular formula for fructose (a sugar found in fruit) from its empirical formula, $CH_2O$, and its molar mass, 180.2 g/mol. We know that the molecular formula is a whole-number multiple of $CH_2O$:

$$\text{Molecular formula} = (CH_2O) \times n$$
$$= C_nH_{2n}O_n$$

We also know that the molar mass is a whole-number multiple of the **empirical formula molar mass**, the sum of the masses of all the atoms in the empirical formula.

$$\text{Molar mass} = \text{empirical formula molar mass} \times n$$

For a particular compound, the value of $n$ in both cases is the same. Therefore, we can find $n$ by calculating the ratio of the molar mass to the empirical formula molar mass:

$$n = \frac{\text{molar mass}}{\text{empirical formula molar mass}}$$

For fructose, the empirical formula molar mass is:

Empirical formula molar mass

$$= 12.01 \text{ g/mol} + 2(1.01 \text{ g/mol}) + 16.00 \text{ g/mol} = 30.03 \text{ g/mol}$$

Therefore, $n$ is:

$$n = \frac{180.2 \text{ g/mol}}{30.03 \text{ g/mol}} = 6$$

We can then use this value of $n$ to find the molecular formula:

$$\text{Molecular formula} = (CH_2O) \times 6 = C_6H_{12}O_6$$

---

**Example 3.19** | **Determining a Molecular Formula from an Empirical Formula and Molar Mass**

Butanedione—the component responsible for the smell and taste of butter and cheese—contains the elements carbon, hydrogen, and oxygen. The empirical formula of butanedione is $C_2H_3O$, and its molar mass is 86.09 g/mol. Determine its molecular formula.

| | |
|---|---|
| **SORT** You are given the empirical formula and molar mass of butanedione and asked to find the molecular formula. | **GIVEN:** Empirical formula = $C_2H_3O$<br>molar mass = 86.09 g/mol<br><br>**FIND:** Molecular formula |
| **STRATEGIZE** A molecular formula is always a whole-number multiple of the empirical formula. Divide the molar mass by the empirical formula molar mass to find the whole number. | Molecular formula = empirical formula $\times$ $n$<br><br>$n = \dfrac{\text{molar mass}}{\text{empirical formula molar mass}}$ |
| **SOLVE** Calculate the empirical formula mass.<br><br>Divide the molar mass by the empirical formula mass to find $n$.<br><br>Multiply the empirical formula by $n$ to obtain the molecular formula. | Empirical formula molar mass<br>$\quad = 2(12.01 \text{ g/mol}) + 3(1.008 \text{ g/mol}) + 16.00 \text{ g/mol} = 43.04 \text{ g/mol}$<br><br>$n = \dfrac{\text{molar mass}}{\text{empirical formula molar mass}} = \dfrac{86.09 \text{ g/mol}}{43.04 \text{ g/mol}} = 2$<br><br>Molecular formula = $C_2H_3O \times 2$<br>$\quad\quad\quad\quad\quad\quad = C_4H_6O_2$ |

CHECK  Check the answer by calculating the molar mass of the formula as follows:

4(12.01 g/mol) + 6(1.008 g/mol) + 2(16.00 g/mol) = 86.09 g/mol

The calculated molar mass is in agreement with the given molar mass.

**FOR PRACTICE 3.19**   A compound has the empirical formula CH and a molar mass of 78.11 g/mol. What is its molecular formula?

**FOR MORE PRACTICE 3.19**   Determine the molecular formula for the compound with a molar mass of 60.10 g/mol and the following percent composition:

C, 39.97%

H, 13.41%

N, 46.62%

## Combustion Analysis

In the previous section, we discussed how to determine the empirical formula of a compound from the relative masses of its constituent elements. Another common (and related) way to obtain empirical formulas for unknown compounds, especially those containing carbon and hydrogen, is **combustion analysis**. In combustion analysis, the unknown compound undergoes combustion (or burning) in the presence of pure oxygen, as shown in Figure 3.13▼. When the sample burns, all of the carbon converts to $CO_2$, and all of the hydrogen converts to $H_2O$. The $CO_2$ and $H_2O$ are weighed. With these masses, we can use the numerical relationships between moles inherent in the formulas for $CO_2$ and $H_2O$ (1 mol $CO_2$ : 1 mol C and 1 mol $H_2O$ : 2 mol H) to determine the amounts of C and H in the original sample. We can determine the amounts of any other elemental constituents, such as O, Cl, or N, by subtracting the sum of the masses of C and H from the original mass of the sample. Examples 20 and 21 illustrate how to perform these calculations for a sample containing only C and H and for a sample containing C, H, and O.

*Combustion is a type of chemical reaction. We discuss chemical reactions and their representation in Section 3.11.*

### Combustion Analysis

Unknown compound is burned in oxygen.

Water and carbon dioxide produced are isolated and weighed.

Oxygen

Other substances not absorbed

Furnace with sample

$H_2O$ absorber

$CO_2$ absorber

▲ **FIGURE 3.13  Combustion Analysis Apparatus**   The sample is placed in a furnace and burned in oxygen. The water and carbon dioxide produced are absorbed into separate containers and weighed.

## *Procedure for...*

| | **Example** 3.20 | **Example** 3.21 ▶ |
|---|---|---|
| **Determining an Empirical Formula from Combustion Analysis** | **Determining an Empirical Formula from Combustion Analysis** | **Determining an Empirical Formula from Combustion Analysis** |
| | Upon combustion, a compound containing only carbon and hydrogen produces 1.83 g $CO_2$ and 0.901 g $H_2O$. Find the empirical formula of the compound. | Upon combustion, a 0.8233-g sample of a compound containing only carbon, hydrogen, and oxygen produces 2.445 g $CO_2$ and 0.6003 g $H_2O$. Find the empirical formula of the compound. |
| **1.** Write down as *given* the masses of each combustion product and the mass of the sample (if given). | GIVEN: 1.83 g $CO_2$, 0.901 g $H_2O$<br><br>FIND: empirical formula | GIVEN: 0.8233-g sample, 2.445 g $CO_2$, 0.6003 g $H_2O$<br><br>FIND: empirical formula |
| **2.** Convert the masses of $CO_2$ and $H_2O$ from step 1 to moles by using the appropriate molar mass for each compound as a conversion factor. | $1.83 \text{ g } CO_2 \times \dfrac{1 \text{ mol } CO_2}{44.01 \text{ g } CO_2}$<br>$= 0.0416 \text{ mol } CO_2$<br><br>$0.901 \text{ g } H_2O \times \dfrac{1 \text{ mol } H_2O}{18.02 \text{ g } H_2O}$<br>$= 0.0500 \text{ mol } H_2O$ | $2.445 \text{ g } CO_2 \times \dfrac{1 \text{ mol } CO_2}{44.01 \text{ g } CO_2}$<br>$= 0.05556 \text{ mol } CO_2$<br><br>$0.6003 \text{ g } H_2O \times \dfrac{1 \text{ mol } H_2O}{18.02 \text{ g } H_2O}$<br>$= 0.03331 \text{ mol } H_2O$ |
| **3.** Convert the moles of $CO_2$ and moles of $H_2O$ from step 2 to moles of C and moles of H using the conversion factors inherent in the chemical formulas of $CO_2$ and $H_2O$. | $0.0416 \text{ mol } CO_2 \times \dfrac{1 \text{ mol C}}{1 \text{ mol } CO_2}$<br>$= 0.0416 \text{ mol C}$<br><br>$0.0500 \text{ mol } H_2O \times \dfrac{2 \text{ mol H}}{1 \text{ mol } H_2O}$<br>$= 0.100 \text{ mol H}$ | $0.05556 \text{ mol } CO_2 \times \dfrac{1 \text{ mol C}}{1 \text{ mol } CO_2}$<br>$= 0.05556 \text{ mol C}$<br><br>$0.03331 \text{ mol } H_2O \times \dfrac{2 \text{ mol H}}{1 \text{ mol } H_2O}$<br>$= 0.06662 \text{ mol H}$ |
| **4.** If the compound contains an element other than C and H, find the mass of the other element by subtracting the sum of the masses of C and H from the mass of the sample. Finally, convert the mass of the other element to moles. | The sample contains no elements other than C and H, so proceed to the next step. | Mass C = $0.05556 \text{ mol C} \times \dfrac{12.01 \text{ g C}}{\text{mol C}}$<br>$= 0.6673 \text{ g C}$<br><br>Mass H = $0.06662 \text{ mol H} \times \dfrac{1.008 \text{ g H}}{\text{mol H}}$<br>$= 0.06715 \text{ g H}$<br><br>Mass O = 0.8233 g<br>$-(0.6673 \text{ g} + 0.06715 \text{ g})$<br>$= 0.0889 \text{ g}$<br><br>Mol O = $0.0889 \text{ g O} \times \dfrac{\text{mol O}}{16.00 \text{ g O}}$<br>$= 0.00556 \text{ mol O}$ |
| **5.** Write down a pseudoformula for the compound using the number of moles of each element (from steps 3 and 4) as subscripts. | $C_{0.0416}H_{0.100}$ | $C_{0.05556}H_{0.06662}O_{0.00556}$ |
| **6.** Divide all the subscripts in the formula by the smallest subscript. (Round all subscripts that are within 0.1 of a whole number.) | $C_{\frac{0.0416}{0.0416}}H_{\frac{0.100}{0.0416}} \longrightarrow C_1H_{2.4}$ | $C_{\frac{0.05556}{0.00556}}H_{\frac{0.0662}{0.00556}}O_{\frac{0.00556}{0.00556}} \longrightarrow C_{10}H_{12}O_1$ |

| 7. If the subscripts are not whole numbers, multiply all the subscripts by a small whole number to get whole-number subscripts. | $C_1H_{2.4} \times 5 \longrightarrow C_5H_{12}$<br>The correct empirical formula is $C_5H_{12}$. | The subscripts are whole numbers; no additional multiplication is needed. The correct empirical formula is $C_{10}H_{12}O$. |
| --- | --- | --- |
| | **FOR PRACTICE 3.20**<br>Upon combustion, a compound containing only carbon and hydrogen produces 1.60 g $CO_2$ and 0.819 g $H_2O$. Find the empirical formula of the compound. | **FOR PRACTICE 3.21**<br>Upon combustion, a 0.8009-g sample of a compound containing only carbon, hydrogen, and oxygen produces 1.6004 g $CO_2$ and 0.6551 g $H_2O$. Find the empirical formula of the compound. |

## 3.11 Writing and Balancing Chemical Equations

 **Writing and Balancing Chemical Equations**

Combustion analysis (which we just examined in Section 3.10) employs a **chemical reaction**, a process in which one or more substances are converted into one or more different ones. Compounds form and change through chemical reactions. For example, water is formed by the reaction of hydrogen with oxygen. A **combustion reaction** is a particular type of chemical reaction in which a substance combines with oxygen to form one or more oxygen-containing compounds. Combustion reactions also emit heat. The heat produced in a number of combustion reactions is critical to supplying our society's energy needs. For example, the heat from the combustion of gasoline expands the gaseous combustion products in a car engine's cylinders, which push the pistons and propel the car. We use the heat released by the combustion of *natural gas* to cook food and to heat our homes.

We represent a chemical reaction with a **chemical equation**. For example, we represent the combustion of natural gas with the equation:

$$CH_4 + O_2 \longrightarrow CO_2 + H_2O$$
reactants          products

The substances on the left side of the equation are the **reactants**, and the substances on the right side are the **products**. We often specify the states of each reactant or product in parentheses next to the formula as follows:

$$CH_4(g) + O_2(g) \longrightarrow CO_2(g) + H_2O(g)$$

The (g) indicates that these substances are gases in the reaction. Table 3.5 summarizes the common states of reactants and products and their symbols used in chemical equations.

The equation just presented for the combustion of natural gas is not complete, however. If we look closely, we can immediately spot a problem.

$$CH_4(g) + O_2(g) \longrightarrow CO_2(g) + H_2O(g)$$

| **2 O atoms** | **2 O atoms + 1 O atom =**<br>**3 O atoms** |

The left side of the equation has two oxygen atoms, while the right side has three. The reaction as written, therefore, violates the law of conservation of mass because an oxygen atom formed out of nothing. Notice also that the left side has four hydrogen atoms, while the right side has only two.

$$CH_4(g) + O_2(g) \longrightarrow CO_2(g) + H_2O(g)$$

| **4 H atoms** | **2 H atoms** |

Two hydrogen atoms have vanished, again violating mass conservation. To correct these problems—that is, to write an equation that more closely represents *what actually happens*—we must **balance** the equation. We need to change the coefficients (the

**TABLE 3.5** States of Reactants and Products in Chemical Equations

| Abbreviation | State |
| --- | --- |
| (g) | Gas |
| (l) | Liquid |
| (s) | Solid |
| (aq) | Aqueous (water solution) |

We cannot change the subscripts when balancing a chemical equation because changing the subscripts changes the substance itself, while changing the coefficients changes the number of molecules of the substance. For example, 2 $H_2O$ is simply two water molecules, but $H_2O_2$ is hydrogen peroxide, a drastically different compound.

numbers *in front of* the chemical formulas), not the subscripts (the numbers within the chemical formulas), to ensure that the number of each type of atom on the left side of the equation is equal to the number on the right side. New atoms do not form during a reaction, nor do atoms vanish—matter must be conserved.

When we add coefficients to the reactants and products to balance an equation, we change the number of molecules in the equation but not the *kind of* molecules. To balance the equation for the combustion of methane, we put the coefficient 2 before $O_2$ in the reactants, and the coefficient 2 before $H_2O$ in the products.

$$CH_4(g) + 2\,O_2(g) \longrightarrow CO_2(g) + 2\,H_2O(g)$$

The equation is now balanced because the numbers of each type of atom on either side of the equation are equal. The balanced equation tells us that one $CH_4$ molecule reacts with two $O_2$ molecules to form one $CO_2$ molecule and two $H_2O$ molecules. We verify that the equation is balanced by summing the number of each type of atom on each side of the equation.

$$CH_4(g) + 2\,O_2(g) \longrightarrow CO_2(g) + 2\,H_2O(g)$$

| Reactants | Products |
|---|---|
| 1 C atom (1 × $\underline{C}H_4$) | 1 C atom (1 × $\underline{C}O_2$) |
| 4 H atoms (1 × $C\underline{H}_4$) | 4 H atoms (2 × $H_2O$) |
| 4 O atoms (2 × $\underline{O}_2$) | 4 O atoms (1 × $C\underline{O}_2$ + 2 × $H_2\underline{O}$) |

The number of each type of atom on both sides of the equation is now equal—the equation is balanced.

We can balance many chemical equations simply by trial and error. However, some guidelines are useful. For example, balancing the atoms in the most complex substances first and the atoms in the simplest substances (such as pure elements) last often makes the process shorter. The following examples of how to balance chemical equations are presented in a two- or three-column format. The general guidelines are shown on the left, with the application of the guidelines on the right. This procedure is meant only as a flexible guide, not a rigid set of steps.

## Procedure for...

| Balancing Chemical Equations | **Example 3.22** — Balancing Chemical Equations — Write a balanced equation for the reaction between solid cobalt(III) oxide and solid carbon to produce solid cobalt and carbon dioxide gas. | **Example 3.23** — Balancing Chemical Equations — Write a balanced equation for the combustion of gaseous butane ($C_4H_{10}$), a fuel used in portable stoves and grills, in which it combines with gaseous oxygen to form gaseous carbon dioxide and gaseous water. |
|---|---|---|
| 1. Write a skeletal equation by writing chemical formulas for each of the reactants and products. Review Sections 3.5 and 3.6 for nomenclature rules. (If a skeletal equation is provided, go to step 2.) | $Co_2O_3(s) + C(s) \longrightarrow Co(s) + CO_2(g)$ | $C_4H_{10}(g) + O_2(g) \longrightarrow CO_2(g) + H_2O(g)$ |

**2.** Balance atoms that occur in more complex substances first. Always balance atoms in compounds before atoms in pure elements.

**Begin with O:**

$Co_2O_3(s) + C(s) \longrightarrow$
$$Co(s) + CO_2(g)$$

3 O atoms $\longrightarrow$ 2 O atoms

To balance O, put a 2 before $Co_2O_3(s)$ and a 3 before $CO_2(g)$.

$2\,Co_2O_3(s) + C(s) \longrightarrow$
$$Co(s) + 3\,CO_2(g)$$

6 O atoms $\longrightarrow$ 6 O atoms

**Begin with C:**

$C_4H_{10}(g) + O_2(g) \longrightarrow$
$$CO_2(g) + H_2O(g)$$

4 C atoms $\longrightarrow$ 1 C atom

To balance C, put a 4 before $CO_2(g)$.

$C_4H_{10}(g) + O_2(g) \longrightarrow$
$$4\,CO_2(g) + H_2O(g)$$

4 C atoms $\longrightarrow$ 4 C atoms

**Balance H:**

$C_4H_{10}(g) + O_2(g) \longrightarrow$
$$4\,CO_2(g) + H_2O(g)$$

10 H atoms $\longrightarrow$ 2 H atoms

To balance H, put a 5 before $H_2O(g)$:

$C_4H_{10}(g) + O_2(g) \longrightarrow$
$$4\,CO_2(g) + 5\,H_2O(g)$$

10 H atoms $\longrightarrow$ 10 H atoms

**3.** Balance atoms that occur as free elements on either side of the equation last. Balance free elements by adjusting their coefficients.

**Balance Co:**

$2\,Co_2O_3(s) + C(s) \longrightarrow$
$$Co(s) + 3\,CO_2(g)$$

4 Co atoms $\longrightarrow$ 1 Co atom

To balance Co, put a 4 before $Co(s)$.

$2\,Co_2O_3(s) + C(s) \longrightarrow$
$$4\,Co(s) + 3\,CO_2(g)$$

4 Co atoms $\longrightarrow$ 4 Co atoms

**Balance C:**

$2\,Co_2O_3(s) + C(s) \longrightarrow$
$$4\,Co(s) + 3\,CO_2(g)$$

1 C atom $\longrightarrow$ 3 C atoms

To balance C, put a 3 before $C(s)$.

$2\,Co_2O_3(s) + 3\,C(s) \longrightarrow$
$$4\,Co(s) + 3\,CO_2(g)$$

**Balance O:**

$C_4H_{10}(g) + O_2(g) \longrightarrow$
$$4\,CO_2(g) + 5\,H_2O(g)$$

2 O atoms $\longrightarrow$ 8 O + 5 O = 13 O atom

To balance O, put a 13/2 before $O_2(g)$.

$C_4H_{10}(g) + 13/2\,O_2(g) \longrightarrow$
$$4\,CO_2(g) + 5\,H_2O(g)$$

13 O atoms $\longrightarrow$ 13 O atoms

**4.** If the balanced equation contains coefficient fractions, clear these by multiplying each of the coefficients in the entire equation by the denominator of the fraction.

This step is not necessary in this example. Proceed to step 5.

$[C_4H_{10}(g) + 13/2\,O_2(g) \longrightarrow$
$$4\,CO_2(g) + 5\,H_2O(g)] \times 2$$
$2\,C_4H_{10}(g) + 13\,O_2(g) \longrightarrow$
$$8\,CO_2(g) + 10\,H_2O(g)$$

**5.** Check to make certain the equation is balanced by summing the total number of each type of atom on both sides of the equation.

$2\,Co_2O_3(s) + 3\,C(s) \longrightarrow$
$$4\,Co(s) + 3\,CO_2(g)$$

| Left | Right |
|---|---|
| 4 Co atoms | 4 Co atoms |
| 6 O atoms | 6 O atoms |
| 3 C atoms | 3 C atoms |

The equation is balanced.

$2\,C_4H_{10}(g) + 13\,O_2(g) \longrightarrow$
$$8\,CO_2(g) + 10\,H_2O(g)$$

| Left | Right |
|---|---|
| 8 C atoms | 8 C atoms |
| 20 H atoms | 20 H atoms |
| 26 O atoms | 26 O atoms |

The equation is balanced.

**FOR PRACTICE 3.22**
Write a balanced equation for the reaction between solid silicon dioxide and solid carbon to produce solid silicon carbide and carbon monoxide gas.

**FOR PRACTICE 3.23**
Write a balanced equation for the combustion of gaseous ethane ($C_2H_6$), a minority component of natural gas, in which it combines with gaseous oxygen to form gaseous carbon dioxide and gaseous water.

**Balanced Chemical Equations** Which quantity or quantities must always be the same on both sides of a chemical equation?

(a) the number of atoms of each kind
(b) the number of molecules of each kind
(c) the number of moles of each kind of molecule
(d) the sum of masses of all substances involved

---

**Example** 3.24 | **Balancing Chemical Equations Containing Ionic Compounds with Polyatomic Ions**

Write a balanced equation for the reaction between aqueous strontium chloride and aqueous lithium phosphate to form solid strontium phosphate and aqueous lithium chloride.

**SOLUTION**

| | |
|---|---|
| 1. Write a skeletal equation by writing chemical formulas for each of the reactants and products. Review Sections 3.5 and 3.6 for naming rules. (If a skeletal equation is provided, go to step 2.) | $SrCl_2(aq) + Li_3PO_4(aq) \longrightarrow Sr_3(PO_4)_2(s) + LiCl(aq)$ |
| 2. Balance metal ions (cations) first. If a polyatomic cation exists on both sides of the equation, balance it as a unit. | **Begin with Sr$^{2+}$:**<br>$SrCl_2(aq) + Li_3PO_4(aq) \longrightarrow Sr_3(PO_4)_2(s) + LiCl(aq)$<br>$\quad$ 1 Sr$^{2+}$ ion $\longrightarrow$ 3 Sr$^{2+}$ ions<br><br>To balance Sr$^{2+}$, put a 3 before $SrCl_2(aq)$.<br>$3\,SrCl_2(aq) + Li_3PO_4(aq) \longrightarrow Sr_3(PO_4)_2(s) + LiCl(aq)$<br>$\quad$ 3 Sr$^{2+}$ ions $\longrightarrow$ 3 Sr$^{2+}$ ions<br><br>**Balance Li$^+$:**<br>$3\,SrCl_2(aq) + Li_3PO_4(aq) \longrightarrow Sr_3(PO_4)_2(s) + LiCl(aq)$<br>$\quad$ 3 Li$^+$ ions $\longrightarrow$ 1 Li$^+$ ion<br><br>To balance Li$^+$, put a 3 before $LiCl(aq)$.<br>$3\,SrCl_2(aq) + Li_3PO_4(aq) \longrightarrow Sr_3(PO_4)_2(s) + 3\,LiCl(aq)$<br>$\quad$ 3 Li$^+$ ions $\longrightarrow$ 3 Li$^+$ ions |
| 3. Balance nonmetal ions (anions) second. If a polyatomic anion exists on both sides of the equation, balance it as a unit. | **Balance PO$_4^{3-}$:**<br>$3\,SrCl_2(aq) + Li_3PO_4(aq) \longrightarrow Sr_3(PO_4)_2(s) + 3\,LiCl(aq)$<br>$\quad$ 1 PO$_4^{3-}$ ions $\longrightarrow$ 2 PO$_4^{3-}$ ions<br><br>To balance PO$_4^{3-}$, put a 2 before $Li_3PO_4(aq)$.<br>$3\,SrCl_2(aq) + 2\,Li_3PO_4(aq) \longrightarrow Sr_3(PO_4)_2(s) + 3\,LiCl(aq)$<br>$\quad$ 2 PO$_4^{3-}$ ions $\longrightarrow$ 2 PO$_4^{3-}$ ions<br><br>**Balance Cl$^-$:**<br>$3\,SrCl_2(aq) + 2\,Li_3PO_4(aq) \longrightarrow Sr_3(PO_4)_2(s) + 3\,LiCl(aq)$<br>$\quad$ 6 Cl$^-$ ions $\longrightarrow$ 3 Cl$^-$ ion<br><br>To balance Cl$^-$, replace the 3 before $LiCl(aq)$ with a 6. This also corrects the balance for Li$^+$, which was thrown off in the previous step.<br>$3\,SrCl_2(aq) + 2\,Li_3PO_4(aq) \longrightarrow Sr_3(PO_4)_2(s) + 6\,LiCl(aq)$<br>$\quad$ 6 Cl$^-$ ions $\longrightarrow$ 6 Cl$^-$ ions |

Check to make certain the equation is balanced by summing the total number of each type of ion on both sides of the equation.

$$3\ SrCl_2(aq) + 2\ Li_3PO_4(aq) \longrightarrow Sr_3(PO_4)_2(s) + 6\ LiCl(aq)$$

| Left | Right |
|------|-------|
| 3 $Sr^{2+}$ ions | 3 $Sr^{2+}$ ions |
| 6 $Li^+$ ions | 6 $Li^+$ ions |
| 2 $PO_4^{3-}$ ions | 2 $PO_4^{3-}$ ions |
| 6 $Cl^-$ ions | 6 $Cl^-$ ions |

The equation is balanced.

**FOR PRACTICE 3.24** Write a balanced equation for the reaction between aqueous lead(II) nitrate and aqueous potassium chloride to form solid lead(II) chloride and aqueous potassium nitrate.

## 3.12 Organic Compounds

Early chemists divided compounds into two types: organic and inorganic. They designated organic compounds as those that originate from living things. Sugar—from sugar-cane or the sugar beet—is a common example of an organic compound. Inorganic compounds, on the other hand, originate from the earth. Salt—mined from the ground or from the ocean—is a common example of an inorganic compound.

Not only did early chemists view organic and inorganic compounds as different in their origin, but also they recognized organic and inorganic compounds to be different in their properties. Organic compounds are easily decomposed. Inorganic compounds, however, are typically more difficult to decompose. Eighteenth-century chemists could synthesize inorganic compounds in the laboratory, but they could not synthesize organic compounds. This was considered to be another great difference between the two different types of compounds. Today, chemists can synthesize both organic and inorganic compounds, and even though organic chemistry is a subfield of chemistry, the differences between organic and inorganic compounds are now viewed as primarily organizational (not fundamental).

Organic compounds are common in everyday substances. Many smells—such as those in perfumes, spices, and foods—are caused by organic compounds. When you sprinkle cinnamon onto your French toast, some cinnamaldehyde—an organic compound present in cinnamon—evaporates into the air. As you inhale cinnamaldehyde molecules, you experience the unique smell of cinnamon. Organic compounds are the major components of living organisms. They are also the main components of most fuels, such as gasoline, oil, and natural gas, and they are the active ingredients in most pharmaceuticals, such as aspirin and ibuprofen.

**Organic compounds** are composed of carbon and hydrogen and a few other elements, including nitrogen, oxygen, and sulfur. The key element in organic chemistry, however, is carbon. In its compounds, carbon always forms four bonds. The simplest organic compound is methane, or $CH_4$.

▲ The organic compound cinnamaldehyde is largely responsible for the taste and smell of cinnamon.

Structural formula    Space-filling model

Methane, $CH_4$

The chemistry of carbon is unique and complex because carbon frequently bonds to itself to form chain, branched, and ring structures.

Propane ($C_3H_8$)                 Isobutane ($C_4H_{10}$)                 Cyclohexane ($C_6H_{12}$)

Carbon can also form double bonds and triple bonds with itself and with other elements.

Ethene ($C_2H_4$)          Ethyne ($C_2H_2$)          Acetic acid ($CH_3COOH$)

This versatility allows carbon to serve as the backbone of millions of different chemical compounds, which is why a general survey of organic chemistry is a yearlong course.

## Hydrocarbons

We can begin to scratch the surface of organic chemistry by categorizing organic compounds into types: hydrocarbons and functionalized hydrocarbons.

▲ Gasoline is composed mostly of hydrocarbons.

**Hydrocarbons** are organic compounds that contain only carbon and hydrogen. Hydrocarbons compose common fuels such as oil, gasoline, liquid propane gas, and natural gas. Hydrocarbons containing only single bonds are **alkanes**, while those containing double or triple bonds are **alkenes** and **alkynes**, respectively. The names of simple, straight-chain hydrocarbons consist of a base name, which is determined by the number of carbon atoms in the chain, and a suffix, determined by whether the hydrocarbon is an alkane (-*ane*), alkene (-*ene*), or alkyne (-*yne*).

The base names for a number of hydrocarbons are listed here:

$$
\begin{array}{ll}
\text{meth} = 1 & \text{hex} = 6 \\
\text{eth} = 2 & \text{hept} = 7 \\
\text{prop} = 3 & \text{oct} = 8 \\
\text{but} = 4 & \text{non} = 9 \\
\text{pent} = 5 & \text{dec} = 10
\end{array}
$$

Table 3.6 lists some common hydrocarbons, their names, and their uses.

**TABLE 3.6 Common Hydrocarbons**

| Name | Molecular Formula | Structural Formula | Space-filling Model | Common Uses |
|---|---|---|---|---|
| Methane | $CH_4$ | | | Primary component of natural gas |
| Propane | $C_3H_8$ | | | LP gas for grills and outdoor stoves |
| *n*-Butane* | $C_4H_{10}$ | | | Common fuel for lighters |
| *n*-Pentane* | $C_5H_{12}$ | | | Component of gasoline |
| Ethene | $C_2H_4$ | | | Ripening agent in fruit |
| Ethyne | $C_2H_2$ | $H-C\equiv C-H$ | | Fuel for welding torches |

*The "*n*" in the names of these hydrocarbons stands for "normal," which means straight chain.

## Functionalized Hydrocarbons

Functionalized hydrocarbons are hydrocarbons in which a **functional group**—a characteristic atom or group of atoms—is incorporated into the hydrocarbon. For example, **alcohols** are organic compounds that have an —OH functional group. We designate the hydrocarbon portion of a molecule as "R," and we write the general formula for an alcohol

The term *functional group* derives from the functionality or chemical character that a specific atom or group of atoms imparts to an organic compound. Even a carbon–carbon double bond can justifiably be called a "functional group."

▲ Rubbing alcohol is isopropyl alcohol.

as R—OH. Some examples of alcohols include methanol (also known as methyl alcohol or wood alcohol) and isopropanol (also known as isopropyl alcohol or rubbing alcohol).

Hydrocarbon (R) group →

$CH_3OH$ ←
Methanol      OH functional group

Hydrocarbon (R) group →

$CH_3$
|
$CH_3CHOH$ ←
Isopropanol
(2-propanol)      OH functional group

A group of organic compounds with the same functional group forms a **family**. Methanol and isopropyl alcohol are both members of the alcohol family of compounds.

The addition of a functional group to a hydrocarbon usually alters the properties of the compound significantly. Take *methanol*, which can be thought of as methane with an —OH group substituted for one of the hydrogen atoms. Methanol is a liquid at room temperature, whereas *methane* is a gas. Although each member of a family is unique, the common functional group bestows some chemical similarities on members of the same family. The names of functional groups have suffixes or endings unique to that functional group. Alcohols, for example, always have names that end in *-ol*. Table 3.7 provides examples of some common functional groups, their general formulas, and their characteristic suffixes or endings.

## TABLE 3.7 Families of Organic Compounds

| Family | Name Ending | General Formula | Example | Name | Occurrence/Use |
|---|---|---|---|---|---|
| Alcohols | -ol | R—OH | $CH_3CH_2$—OH | Ethanol (ethyl alcohol) | Alcohol in fermented beverages |
| Ethers | ether | R—O—R′ | $CH_3CH_2$—O—$CH_2CH_3$ | Diethyl ether | Anesthetic; laboratory solvent |
| Aldehydes | -al | $\overset{\displaystyle O}{\overset{\|}{R-C-H}}$ | $\overset{\displaystyle O}{\overset{\|}{H_3C-C-H}}$ | Ethanal (acetaldehyde) | Perfumes; flavors |
| Ketones | -one | $\overset{\displaystyle O}{\overset{\|}{R-C-R'}}$ | $\overset{\displaystyle O}{\overset{\|}{H_3C-C-CH_3}}$ | Propanone (acetone) | Fingernail polish remover |
| Carboxylic acids | acid | $\overset{\displaystyle O}{\overset{\|}{R-C-OH}}$ | $\overset{\displaystyle O}{\overset{\|}{H_3C-C-OH}}$ | Acetic acid | Vinegar |
| Esters | -ate | $\overset{\displaystyle O}{\overset{\|}{R-C-OR'}}$ | $\overset{\displaystyle O}{\overset{\|}{H_3C-C-OCH_3}}$ | Methyl acetate | Laboratory solvent |
| Amines | amine | $RNH_2$ | $CH_3CH_2-\overset{\displaystyle H}{\overset{\|}{N}}-H$ | Ethyl amine | Smell of rotten fish |

# Self-Assessment Quiz

**Q1.** What is the empirical formula of a compound with the molecular formula $C_{10}H_8$?
a) $C_5H_3$   b) $C_2H_4$   c) $C_5H_4$   d) CH

**Q2.** Which substance is an ionic compound?
a) $SrI_2$   b) $N_2O_4$   c) He   d) $CCl_4$

**Q3.** What is the correct formula for the compound formed between calcium and sulfur?
a) CaS   b) $Ca_2S$   c) $CaS_2$   d) $CaS_3$

**Q4.** Name the compound $SrI_2$.
a) strontium iodide
b) strontium diiodide
c) strontium(II) iodide
d) strontium(II) diiodide

**Q5.** What is the formula for manganese(IV) oxide?
a) $Mn_4O$   b) $MnO_4$   c) $Mn_2O$   d) $MnO_2$

**Q6.** Name the compound $Pb(C_2H_3O_2)_2$.
a) lead(II) carbonate
b) lead(II) acetate
c) lead bicarbonate
d) lead diacetate

**Q7.** Name the compound $P_2I_4$.
a) phosphorus iodide
b) phosphorus diiodide
c) phosphorus(II) iodide
d) diphosphorus tetraiodide

**Q8.** Name the compound $HNO_2(aq)$.
a) hydrogen nitrogen dioxide
b) hydrogen nitrate
c) nitric acid
d) nitrous acid

**Q9.** Determine the number of $CH_2Cl_2$ molecules in 25.0 g $CH_2Cl_2$.
a) 0.294 molecules   b) $1.77 \times 10^{23}$ molecules
c) $1.28 \times 10^{27}$ molecules   d) $1.51 \times 10^{25}$ molecules

**Q10.** List the elements in the compound $CF_2Cl_2$ in order of decreasing mass percent composition.
a) C > F > Cl   b) F > Cl > C
c) Cl > C > F   d) Cl > F > C

**Q11.** Determine the mass of potassium in 35.5 g of KBr.
a) 17.4 g   b) 0.298 g   c) 11.7 g   d) 32.9 g

**Q12.** A compound is 52.14% C, 13.13% H, and 34.73% O by mass. What is the empirical formula of the compound?
a) $C_2H_8O_3$   b) $C_2H_6O$   c) $C_4HO_3$   d) $C_3HO_6$

**Q13.** A compound has the empirical formula $CH_2O$ and a formula mass of 120.10 amu. What is the molecular formula of the compound?
a) $CH_2O$   b) $C_2H_4O_2$   c) $C_3H_6O_3$   d) $C_4H_8O_4$

**Q14.** Combustion of 30.42 g of a compound containing only carbon, hydrogen, and oxygen produces 35.21 g $CO_2$ and 14.42 g $H_2O$. What is the empirical formula of the compound?
a) $C_4H_8O_6$   b) $C_2H_4O_3$
c) $C_2H_2O_3$   d) $C_6HO_{12}$

**Q15.** What are the correct coefficients (reading from left to right) when the chemical equation is balanced?
$$\_PCl_3(l) + \_H_2O(l) \longrightarrow \_H_3PO_3(aq) + \_HCl(aq)$$
a) 1, 3, 1, 3   b) 1, 2, 1, 1
c) 1, 3, 2, 1   d) 3, 6, 1, 9

**Answers: 1.** (c) **2.** (a) **3.** (a) **4.** (a) **5.** (d) **6.** (b) **7.** (d) **8.** (d) **9.** (b) **10.** (d) **11.** (c) **12.** (b) **13.** (d) **14.** (b) **15.** (a)

# Chapter 3 IN REVIEW

## Key Terms

### Section 3.2
ionic bond (90)
ionic compound (90)
covalent bond (90)
molecular compound (90)

### Section 3.3
chemical formula (90)
empirical formula (90)
molecular formula (90)
structural formula (91)
ball-and-stick molecular model (92)
space-filling molecular model (92)

### Section 3.4
atomic element (92)
molecular element (92)
formula unit (94)
polyatomic ion (95)

### Section 3.5
common name (97)
systematic name (97)
binary compound (98)
oxyanion (100)
hydrate (101)

### Section 3.6
acid (103)
binary acid (103)
oxyacid (104)

### Section 3.8
formula mass (107)

### Section 3.9
mass percent composition (mass percent) (109)

### Section 3.10
empirical formula molar mass (116)

combustion analysis (117)

### Section 3.11
chemical reaction (119)
combustion reaction (119)
chemical equation (119)
reactants (119)
products (119)
balanced chemical equation (119)

### Section 3.12
organic compound (123)
hydrocarbon (124)
alkane (124)
alkene (124)
alkyne (124)
functional group (125)
alcohol (125)
family (126)

# Key Concepts

## Chemical Bonds (3.2)

▶ Chemical bonds, the forces that hold atoms together in compounds, arise from the interactions between nuclei and electrons in atoms.

▶ In an ionic bond, one or more electrons are *transferred* from one atom to another, forming a cation (positively charged) and an anion (negatively charged). The two ions are drawn together by the attraction between the opposite charges.

▶ In a covalent bond, one or more electrons are *shared* between two atoms. The atoms are held together by the attraction between their nuclei and the shared electrons.

## Representing Molecules and Compounds (3.3, 3.4)

▶ A compound is represented with a chemical formula, which indicates the elements present and the number of atoms of each.

▶ An empirical formula gives only the *relative* number of atoms, while a molecular formula gives the *actual* number of atoms present in the molecule.

▶ Structural formulas show how atoms are bonded together, while molecular models portray the geometry of the molecule.

▶ Compounds can be divided into two types: molecular compounds, formed between two or more covalently bonded nonmetals, and ionic compounds, usually formed between a metal ionically bonded to one or more nonmetals. The smallest identifiable unit of a molecular compound is a molecule, and the smallest identifiable unit of an ionic compound is a formula unit: the smallest electrically neutral collection of ions.

▶ Elements can also be divided into two types: molecular elements, which occur as (mostly diatomic) molecules, and atomic elements, which occur as individual atoms.

## Naming Inorganic Ionic and Molecular Compounds and Acids (3.5–3.7)

▶ A flowchart for naming simple inorganic compounds is provided in Section 3.7.

## Formula Mass and Mole Concept for Compounds (3.8)

▶ The formula mass of a compound is the sum of the atomic masses of all the atoms in the chemical formula. Like the atomic masses of elements, the formula mass characterizes the average mass of a molecule (or a formula unit).

▶ The mass of one mole of a compound is the molar mass of that compound and equals its formula mass (in grams).

## Chemical Composition (3.9, 3.10)

▶ The mass percent composition of a compound indicates each element's percentage of the total compound's mass. We can determine the mass percent composition from the compound's chemical formula and the molar masses of its elements.

▶ The chemical formula of a compound provides the relative number of atoms (or moles) of each element in a compound, and we can therefore use it to determine numerical relationships between moles of the compound, and moles of its constituent elements. We can extend this relationship to mass by using the molar masses of the compound and its constituent elements.

▶ If the mass percent composition and molar mass of a compound are known, we can determine its empirical and molecular formulas.

## Writing and Balancing Chemical Equations (3.11)

▶ In chemistry, we represent chemical reactions with chemical equations. The substances on the left-hand side of a chemical equation are the reactants, and the substances on the right-hand side are the products.

▶ Chemical equations are balanced when the number of each type of atom on the left side of the equation is equal to the number on the right side.

## Organic Compounds (3.12)

▶ Organic compounds are composed of carbon, hydrogen, and a few other elements such as nitrogen, oxygen, and sulfur.

▶ The simplest organic compounds are hydrocarbons, compounds composed of only carbon and hydrogen.

▶ Hydrocarbons are categorized into three types based on the bonds they contain: alkanes contain single bonds, alkenes contain double bonds, and alkynes contain triple bonds.

▶ All other organic compounds can be thought of as hydrocarbons with one or more functional groups—characteristic atoms or groups of atoms.

▶ Common functionalized hydrocarbons include alcohols, ethers, aldehydes, ketones, carboxylic acids, esters, and amines.

# Key Equations and Relationships

Formula Mass (**3.8**)

$$\left(\begin{array}{c}\text{No. of atoms of 1st element}\\ \text{in chemical formula}\end{array} \times \begin{array}{c}\text{atomic mass}\\ \text{of 1st element}\end{array}\right) + \left(\begin{array}{c}\text{No. of atoms of 2nd element}\\ \text{in chemical formula}\end{array} \times \begin{array}{c}\text{atomic mass}\\ \text{of 2nd element}\end{array}\right) + \cdots$$

Mass Percent Composition (**3.9**)

$$\text{Mass \% of element X} = \frac{\text{mass of X in 1 mol compound}}{\text{mass of 1 mol compound}} \times 100\%$$

Empirical Formula Molar Mass (**3.10**)

$$\text{Molecular formula} = n \times (\text{empirical formula})$$

$$n = \frac{\text{molar mass}}{\text{empirical formula molar mass}}$$

# Key Learning Outcomes

| CHAPTER OBJECTIVES | ASSESSMENT |
|---|---|
| Writing Molecular and Empirical Formulas (**3.3**) | Example 3.1   For Practice 3.1   Exercises 4, 23–26 |
| Classifying Substances as Atomic Elements, Molecular Elements, Molecular Compounds, or Ionic Compounds (**3.4**) | Example 3.2   For Practice 3.2   Exercises 27–32 |
| Writing Formulas for Ionic Compounds (**3.5**) | Examples 3.3, 3.4   For Practice 3.3, 3.4   Exercises 33–36, 43, 44 |
| Naming Ionic Compounds (**3.5**) | Examples 3.5, 3.6   For Practice 3.5, 3.6   For More Practice 3.5, 3.6   Exercises 37–40 |
| Naming Ionic Compounds Containing Polyatomic Ions (**3.5**) | Example 3.7   For Practice 3.7   For More Practice 3.7   Exercises 41–44 |
| Naming Molecular Compounds (**3.6**) | Example 3.8   For Practice 3.8   For More Practice 3.8   Exercises 47–50 |
| Naming Acids (**3.6**) | Examples 3.9, 3.10   For Practice 3.9, 3.10   For More Practice 3.10   Exercises 51–54 |
| Naming Uncategorized Inorganic Compounds (**3.7**) | Example 3.11   For Practice 3.11   Exercises 55–58 |
| Calculating Formula Mass (**3.8**) | Example 3.12   For Practice 3.12   Exercises 59, 60 |
| Using Formula Mass to Count Molecules by Weighing (**3.8**) | Example 3.13   For Practice 3.13   For More Practice 3.13   Exercises 65–70 |
| Calculating Mass Percent Composition (**3.9**) | Example 3.14   For Practice 3.14   For More Practice 3.14   Exercises 71–76 |
| Using Mass Percent Composition as a Conversion Factor (**3.9**) | Example 3.15   For Practice 3.15   For More Practice 3.15   Exercises 77, 78 |
| Using Chemical Formulas as Conversion Factors (**3.9**) | Example 3.16   For Practice 3.16   For More Practice 3.16   Exercises 83, 84 |
| Obtaining an Empirical Formula from Experimental Data (**3.10**) | Examples 3.17, 3.18   For Practice 3.17, 3.18   Exercises 87–94 |
| Calculating a Molecular Formula from an Empirical Formula and Molar Mass (**3.10**) | Example 3.19   For Practice 3.19   For More Practice 3.19   Exercises 95–96 |
| Obtaining an Empirical Formula from Combustion Analysis (**3.10**) | Examples 3.20, 3.21   For Practice 3.20, 3.21   Exercises 97–100 |
| Balancing Chemical Equations (**3.11**) | Examples 3.22, 3.23, 3.24   For Practice 3.22, 3.23, 3.24   Exercises 101–112 |

# EXERCISES

## Review Questions

**1.** How do the properties of compounds compare to the properties of the elements from which the compounds are composed?

**2.** What is a chemical bond? Explain the difference between an ionic bond and a covalent bond.

**3.** Explain the different ways to represent compounds. Why are there so many?

**4.** What is the difference between an empirical formula and a molecular formula?

**5.** Define and provide an example for each of the following: atomic element, molecular element, ionic compound, molecular compound.

6. Explain how to write a formula for an ionic compound given the names of the metal and nonmetal (or polyatomic ion) in the compound.

7. Explain how to name binary ionic compounds. How do you name an ionic compound if it contains a polyatomic ion?

8. Why do the names of some ionic compounds include the charge of the metal ion while others do not?

9. Explain how to name molecular inorganic compounds.

10. How many atoms are specified by each of these prefixes: *mono-, di-, tri-, tetra-, penta-, hexa-*?

11. Explain how to name binary and oxyacids.

12. What is the formula mass for a compound? Why is it useful?

13. Explain how you can use the information in a chemical formula to determine how much of a particular element is present in a given amount of a compound. Provide some examples of why this might be important.

14. What is mass percent composition? Why is it useful?

15. What kinds of conversion factors are inherent in chemical formulas? Provide an example.

16. What kind of chemical formula can be obtained from experimental data showing the relative masses of the elements in a compound?

17. How can a molecular formula be obtained from an empirical formula? What additional information is required?

18. What is combustion analysis? What is it used for?

19. Which elements are normally present in organic compounds?

20. What is the difference between an alkane, an alkene, and an alkyne?

21. What are functionalized hydrocarbons? Cite an example of a functionalized hydrocarbon.

22. Write a generic formula for each of the families of organic compounds.

   a. alcohols        e. carboxylic acids
   b. ethers          f. esters
   c. aldehydes       g. amines
   d. ketones

# Problems by Topic

*Note: Answers to all odd-numbered Problems, numbered in blue, can be found in Appendix III. Exercises in the Problems by Topic section are paired, with each odd-numbered problem followed by a similar even-numbered problem. Exercises in the Cumulative Problems section are also paired, but somewhat more loosely. (Challenge Problems and Conceptual Problems, because of their nature, are unpaired.)*

## Chemical Formulas and Molecular View of Elements and Compounds

23. Determine the number of each type of atom in each formula.
   a. $Mg_3(PO_4)_2$   b. $BaCl_2$   c. $Fe(NO_2)_2$   d. $Ca(OH)_2$

24. Determine the number of each type of atom in each formula.
   a. $Ca(NO_2)_2$   b. $CuSO_4$   c. $Al(NO_3)_3$   d. $Mg(HCO_3)_2$

25. Write a chemical formula for each molecular model. (See Appendix IIA for color codes.)

   (a)              (b)              (c)

26. Write a chemical formula for each molecular model. (See Appendix IIA for color codes.)

   (a)              (b)              (c)

27. Classify each element as atomic or molecular.
   a. neon   b. fluorine   c. potassium   d. nitrogen

28. Identify the elements that have molecules as their basic units.
   a. hydrogen   b. iodine   c. lead   d. oxygen

29. Classify each compound as ionic or molecular.
   a. $CO_2$   b. $NiCl_2$   c. $NaI$   d. $PCl_3$

30. Classify each compound as ionic or molecular.
   a. $CF_2Cl_2$   b. $CCl_4$   c. $PtO_2$   d. $SO_3$

31. Based on the molecular views, classify each substance as an atomic element, a molecular element, an ionic compound, or a molecular compound.

   (a)              (b)              (c)

32. Based on the molecular views, classify each substance as an atomic element, a molecular element, an ionic compound, or a molecular compound.

   (a)                        (b)

   (c)

## Formulas and Names for Ionic Compounds

**33.** Write a formula for the ionic compound that forms between each pair of elements.
a. calcium and oxygen
b. zinc and sulfur
c. rubidium and bromine
d. aluminum and oxygen

**34.** Write a formula for the ionic compound that forms between each pair of elements.
a. silver and chlorine
b. sodium and sulfur
c. aluminum and sulfur
d. potassium and chlorine

**35.** Write a formula for the compound that forms between calcium and each polyatomic ion.
a. hydroxide
b. chromate
c. phosphate
d. cyanide

**36.** Write a formula for the compound that forms between potassium and each polyatomic ion.
a. carbonate
b. phosphate
c. hydrogen phosphate
d. acetate

**37.** Name each ionic compound.
a. $Mg_3N_2$
b. $KF$
c. $Na_2O$
d. $Li_2S$
e. $CsF$
f. $KI$

**38.** Name each ionic compound.
a. $SnCl_4$
b. $PbI_2$
c. $Fe_2O_3$
d. $CuI_2$
e. $HgBr_2$
f. $CrCl_2$

**39.** Give each ionic compound an appropriate name.
a. $SnO$
b. $Cr_2S_3$
c. $RbI$
d. $BaBr_2$

**40.** Give each ionic compound an appropriate name.
a. $BaS$
b. $FeCl_3$
c. $PbI_4$
d. $SrBr_2$

**41.** Name each ionic compound containing a polyatomic ion.
a. $CuNO_2$
b. $Mg(C_2H_3O_2)_2$
c. $Ba(NO_3)_2$
d. $Pb(C_2H_3O_2)_2$

**42.** Name each ionic compound containing a polyatomic ion.
a. $Ba(OH)_2$
b. $NH_4I$
c. $NaBrO_4$
d. $Fe(OH)_3$

**43.** Write the formula for each ionic compound.
a. sodium hydrogen sulfite
b. lithium permanganate
c. silver nitrate
d. potassium sulfate
e. rubidium hydrogen sulfate
f. potassium hydrogen carbonate

**44.** Write the formula for each ionic compound.
a. copper(II) chloride
b. copper(I) iodate
c. lead(II) chromate
d. calcium fluoride
e. potassium hydroxide
f. iron(II) phosphate

**45.** Write the name from the formula or the formula from the name for each hydrated ionic compound.
a. $CoSO_4 \cdot 7\,H_2O$
b. iridium(III) bromide tetrahydrate
c. $Mg(BrO_3)_2 \cdot 6\,H_2O$
d. potassium carbonate dihydrate

**46.** Write the name from the formula or the formula from the name for each hydrated ionic compound.
a. cobalt(II) phosphate octahydrate
b. $BeCl_2 \cdot 2\,H_2O$
c. chromium(III) phosphate trihydrate
d. $LiNO_2 \cdot H_2O$

## Formulas and Names for Molecular Compounds and Acids

**47.** Name each molecular compound.
a. $CO$
b. $NI_3$
c. $SiCl_4$
d. $N_4Se_4$

**48.** Name each molecular compound.
a. $SO_3$
b. $SO_2$
c. $BrF_5$
d. $NO$

**49.** Write the formula for each molecular compound.
a. phosphorus trichloride
b. chlorine monoxide
c. disulfur tetrafluoride
d. phosphorus pentafluoride

**50.** Write the formula for each molecular compound.
a. boron tribromide
b. dichlorine monoxide
c. xenon tetrafluoride
d. carbon tetrabromide

**51.** Name each acid.
a. $HI(aq)$
b. $HNO_3(aq)$
c. $H_2CO_3(aq)$

**52.** Name each acid.
a. $HCl(aq)$
b. $HClO_2(aq)$
c. $H_2SO_4(aq)$

**53.** Write the formula for each acid.
a. hydrofluoric acid
b. hydrobromic acid
c. sulfurous acid

**54.** Write the formula for each acid.
a. phosphoric acid
b. hydrocyanic acid
c. chlorous acid

## Using the Nomenclature Flowchart

**55.** Refer to the nomenclature flowchart (Figure 3.11) to name each compound.
a. $SrCl_2$
b. $SnO_2$
c. $P_2S_5$
d. $HC_2H_3O_2(aq)$

**56.** Refer to the nomenclature flowchart (Figure 3.11) to name each compound.
a. $HNO_2(aq)$
b. $B_2Cl_2$
c. $BaCl_2$
d. $CrCl_3$

**57.** Refer to the nomenclature flowchart (Figure 3.11) to name each compound.
a. $KClO_3$
b. $I_2O_5$
c. $PbSO_4$

**58.** Refer to the nomenclature flowchart (Figure 3.11) to name each compound.
a. $XeO_3$
b. $KClO$
c. $CoSO_4$

## Formula Mass and the Mole Concept for Compounds

**59.** Calculate the formula mass for each compound.
a. $NO_2$
b. $C_4H_{10}$
c. $C_6H_{12}O_6$
d. $Cr(NO_3)_3$

**60.** Calculate the formula mass for each compound.
a. $MgBr_2$
b. $HNO_2$
c. $CBr_4$
d. $Ca(NO_3)_2$

**61.** Calculate the number of moles in each sample.
a. 72.5 g $CCl_4$
b. 12.4 g $C_{12}H_{22}O_{11}$
c. 25.2 kg $C_2H_2$
d. 12.3 g dinitrogen monoxide

**62.** Calculate the mass of each sample.
a. 15.7 mol $HNO_3$
b. $1.04 \times 10^{-3}$ mol $H_2O_2$
c. 72.1 mmol $SO_2$
d. 1.23 mol xenon difluoride

**63.** Determine the number of moles (of molecules or formula units) in each sample.
a. 25.5 g $NO_2$
b. 1.25 kg $CO_2$
c. 38.2 g $KNO_3$
d. 155.2 kg $Na_2SO_4$

**64.** Determine the number of moles (of molecules or formula units) in each sample.
a. 55.98 g $CF_2Cl_2$
b. 23.6 kg $Fe(NO_3)_2$
c. 0.1187 g $C_8H_{18}$
d. 195 kg $CaO$

**65.** How many molecules are in each sample?
a. 6.5 g $H_2O$
b. 389 g $CBr_4$
c. 22.1 g $O_2$
d. 19.3 g $C_8H_{10}$

**66.** How many molecules (or formula units) are in each sample?
a. 85.26 g $CCl_4$
b. 55.93 kg $NaHCO_3$
c. 119.78 g $C_4H_{10}$
d. $4.59 \times 10^5$ g $Na_3PO_4$

**67.** Calculate the mass (in g) of each sample.
 **a.** $5.94 \times 10^{20}$ $SO_3$ molecules
 **b.** $2.8 \times 10^{22}$ $H_2O$ molecules
 **c.** 1 glucose molecule ($C_6H_{12}O_6$)

**68.** Calculate the mass (in g) of each sample.
 **a.** $4.5 \times 10^{25}$ $O_3$ molecules
 **b.** $9.85 \times 10^{19}$ $CCl_2F_2$ molecules
 **c.** 1 water molecule

**69.** A sugar crystal contains approximately $1.8 \times 10^{17}$ sucrose ($C_{12}H_{22}O_{11}$) molecules. What is its mass in mg?

**70.** A salt crystal has a mass of 0.12 mg. How many NaCl formula units does it contain?

## Composition of Compounds

**71.** Calculate the mass percent composition of carbon in each carbon-containing compound.
 **a.** $CH_4$    **b.** $C_2H_6$    **c.** $C_2H_2$    **d.** $C_2H_5Cl$

**72.** Calculate the mass percent composition of nitrogen in each nitrogen-containing compound.
 **a.** $N_2O$    **b.** NO    **c.** $NO_2$    **d.** $HNO_3$

**73.** Most fertilizers consist of nitrogen-containing compounds such as $NH_3$, $CO(NH_2)_2$, $NH_4NO_3$, and $(NH_4)_2SO_4$. Plants use the nitrogen content in these compounds for protein synthesis. Calculate the mass percent composition of nitrogen in each of the fertilizers listed. Which fertilizer has the highest nitrogen content?

**74.** Iron in the earth is in the form of iron ore. Common ores include $Fe_2O_3$ (hematite), $Fe_3O_4$ (magnetite), and $FeCO_3$ (siderite). Calculate the mass percent composition of iron for each of these iron ores. Which ore has the highest iron content?

**75.** Copper(II) fluoride contains 37.42% F by mass. Calculate the mass of fluorine (in g) in 55.5 g of copper(II) fluoride.

**76.** Silver chloride, often used in silver plating, contains 75.27% Ag by mass. Calculate the mass of silver chloride required to plate 155 mg of pure silver.

**77.** The iodide ion is a dietary mineral essential to good nutrition. In countries where potassium iodide is added to salt, iodine deficiency (or goiter) has been almost completely eliminated. The recommended daily allowance (RDA) for iodine is 150 $\mu$g/day. How much potassium iodide (76.45% I) should you consume if you want to meet the RDA?

**78.** The American Dental Association recommends that an adult female should consume 3.0 mg of fluoride ($F^-$) per day to prevent tooth decay. If the fluoride is consumed in the form of sodium fluoride (45.24% F), what amount of sodium fluoride contains the recommended amount of fluoride?

**79.** Write a ratio showing the relationship between the molar amounts of each element for each compound. (See Appendix IIA for color codes.)

 **(a)**          **(b)**          **(c)**

**80.** Write a ratio showing the relationship between the molar amounts of each element for each compound. (See Appendix IIA for color codes.)

 **(a)**          **(b)**          **(c)**

**81.** Determine the number of moles of hydrogen atoms in each sample.
 **a.** 0.0885 mol $C_4H_{10}$    **b.** 1.3 mol $CH_4$
 **c.** 2.4 mol $C_6H_{12}$    **d.** 1.87 mol $C_8H_{18}$

**82.** Determine the number of moles of oxygen atoms in each sample.
 **a.** 4.88 mol $H_2O_2$    **b.** 2.15 mol $N_2O$
 **c.** 0.0237 mol $H_2CO_3$    **d.** 24.1 mol $CO_2$

**83.** Calculate mass (in grams) of sodium in 8.5 g of each sodium-containing food additive.
 **a.** NaCl (table salt)
 **b.** $Na_3PO_4$ (sodium phosphate)
 **c.** $NaC_7H_5O_2$ (sodium benzoate)
 **d.** $Na_2C_6H_6O_7$ (sodium hydrogen citrate)

**84.** Calculate the mass (in kilograms) of chlorine in 25 kg of each chlorofluorocarbon (CFC).
 **a.** $CF_2Cl_2$    **b.** $CFCl_3$    **c.** $C_2F_3Cl_3$    **d.** $CF_3Cl$

**85.** How many fluorine atoms are present in 5.85 g of $C_2F_4$?

**86.** How many bromine atoms are present in 35.2 g of $CH_2Br_2$?

## Chemical Formulas from Experimental Data

**87.** A chemist decomposes samples of several compounds; the masses of their constituent elements are listed. Calculate the empirical formula for each compound.
 **a.** 1.651 g Ag, 0.1224 g O
 **b.** 0.672 g Co, 0.569 g As, 0.486 g O
 **c.** 1.443 g Se, 5.841 g Br

**88.** A chemist decomposes samples of several compounds; the masses of their constituent elements are listed. Calculate the empirical formula for each compound.
 **a.** 1.245 g Ni, 5.381 g I
 **b.** 2.677 g Ba, 3.115 g Br
 **c.** 2.128 g Be, 7.557 g S, 15.107 g O

**89.** Calculate the empirical formula for each stimulant based on its elemental mass percent composition.
 **a.** nicotine (found in tobacco leaves): C 74.03%, H 8.70%, N 17.27%
 **b.** caffeine (found in coffee beans): C 49.48%, H 5.19%, N 28.85%, O 16.48%

**90.** Calculate the empirical formula for each natural flavor based on its elemental mass percent composition.
 **a.** methyl butyrate (component of apple taste and smell): C 58.80%, H 9.87%, O 31.33%
 **b.** vanillin (responsible for the taste and smell of vanilla): C 63.15%, H 5.30%, O 31.55%

**91.** The elemental mass percent composition of ibuprofen (a nonsteroidal anti-inflammatory drug [NSAID]) is 75.69% C, 8.80% H, and 15.51% O. Determine the empirical formula of ibuprofen.

**92.** The elemental mass percent composition of ascorbic acid (vitamin C) is 40.92% C, 4.58% H, and 54.50% O. Determine the empirical formula of ascorbic acid.

93. A 0.77-mg sample of nitrogen reacts with chlorine to form 6.61 mg of the chloride. Determine the empirical formula of nitrogen chloride.

94. A 45.2-mg sample of phosphorus reacts with selenium to form 131.6 mg of the selenide. Determine the empirical formula of phosphorus selenide.

95. From the given empirical formula and molar mass, find the molecular formula of each compound.
    a. $C_6H_7N$, 186.24 g/mol      b. $C_2HCl$, 181.44 g/mol
    c. $C_5H_{10}NS_2$, 296.54 g/mol

96. From the given molar mass and empirical formula of several compounds, find the molecular formula of each compound.
    a. $C_4H_9$, 114.22 g/mol      b. $CCl$, 284.77 g/mol
    c. $C_3H_2N$, 312.29 g/mol

97. Combustion analysis of a hydrocarbon produces 33.01 g $CO_2$ and 13.51 g $H_2O$. Calculate the empirical formula of the hydrocarbon.

98. Combustion analysis of naphthalene, a hydrocarbon used in mothballs, produces 8.80 g $CO_2$ and 1.44 g $H_2O$. Calculate the empirical formula of naphthalene.

99. The foul odor of rancid butter is due largely to butyric acid, a compound containing carbon, hydrogen, and oxygen. Combustion analysis of a 4.30-g sample of butyric acid produces 8.59 g $CO_2$ and 3.52 g $H_2O$. Determine the empirical formula of butyric acid.

100. Tartaric acid is the white, powdery substance that coats tart candies such as Sour Patch Kids. Combustion analysis of a 12.01-g sample of tartaric acid—which contains only carbon, hydrogen, and oxygen—produces 14.08 g $CO_2$ and 4.32 g $H_2O$. Determine the empirical formula of tartaric acid.

## Writing and Balancing Chemical Equations

101. Sulfuric acid is a component of acid rain formed when gaseous sulfur dioxide pollutant reacts with gaseous oxygen and liquid water to form aqueous sulfuric acid. Write the balanced chemical equation this reaction. (Note: this is a simplified representation of this reaction.)

102. Nitric acid is a component of acid rain that forms when gaseous nitrogen dioxide pollutant reacts with gaseous oxygen and liquid water to form aqueous nitric acid. Write the balanced chemical equation for this reaction. (Note: this is a simplified representation of this reaction.)

103. In a popular classroom demonstration, solid sodium is added to liquid water and reacts to produce hydrogen gas and aqueous sodium hydroxide. Write the balanced chemical equation for this reaction.

104. When iron rusts, solid iron reacts with gaseous oxygen to form solid iron(III) oxide. Write the balanced chemical equation for this reaction.

105. Write the balanced chemical equation for the fermentation of sucrose ($C_{12}H_{22}O_{11}$) by yeasts in which the aqueous sugar reacts with water to form aqueous ethanol ($C_2H_5OH$) and carbon dioxide gas.

106. Write the balanced equation for the photosynthesis reaction in which gaseous carbon dioxide and liquid water react in the presence of chlorophyll to produce aqueous glucose ($C_6H_{12}O_6$) and oxygen gas.

107. Write the balanced chemical equation for each reaction.
    a. Solid lead(II) sulfide reacts with aqueous hydrobromic acid to form solid lead(II) bromide and dihydrogen monosulfide gas.
    b. Gaseous carbon monoxide reacts with hydrogen gas to form gaseous methane ($CH_4$) and liquid water.
    c. Aqueous hydrochloric acid reacts with solid manganese(IV) oxide to form aqueous manganese(II) chloride, liquid water, and chlorine gas.
    d. Liquid pentane ($C_5H_{12}$) reacts with gaseous oxygen to form carbon dioxide and liquid water.

108. Write the balanced chemical equation for each reaction.
    a. Solid copper reacts with solid sulfur to form solid copper(I) sulfide.
    b. Solid iron(III) oxide reacts with hydrogen gas to form solid iron and liquid water.
    c. Sulfur dioxide gas reacts with oxygen gas to form sulfur trioxide gas.
    d. Gaseous ammonia ($NH_3$) reacts with gaseous oxygen to form gaseous nitrogen monoxide and gaseous water.

109. Write the balanced chemical equation for the reaction of aqueous sodium carbonate with aqueous copper(II) chloride to form solid copper(II) carbonate and aqueous sodium chloride.

110. Write the balanced chemical equation for the reaction of aqueous potassium hydroxide with aqueous iron(III) chloride to form solid iron(III) hydroxide and aqueous potassium chloride.

111. Balance each chemical equation.
    a. $CO_2(g) + CaSiO_3(s) + H_2O(l) \longrightarrow$
    $$SiO_2(s) + Ca(HCO_3)_2(aq)$$
    b. $Co(NO_3)_3(aq) + (NH_4)_2S(aq) \longrightarrow Co_2S_3(s) + NH_4NO_3(aq)$
    c. $Cu_2O(s) + C(s) \longrightarrow Cu(s) + CO(g)$
    d. $H_2(g) + Cl_2(g) \longrightarrow HCl(g)$

112. Balance each chemical equation.
    a. $Na_2S(aq) + Cu(NO_3)_2(aq) \longrightarrow NaNO_3(aq) + CuS(s)$
    b. $N_2H_4(l) \longrightarrow NH_3(g) + N_2(g)$
    c. $HCl(aq) + O_2(g) \longrightarrow H_2O(l) + Cl_2(g)$
    d. $FeS(s) + HCl(aq) \longrightarrow FeCl_2(aq) + H_2S(g)$

## Organic Compounds

113. Classify each compound as organic or inorganic.
    a. $CaCO_3$      b. $C_4H_8$      c. $C_4H_6O_6$      d. $LiF$

114. Classify each compound as organic or inorganic.
    a. $C_8H_{18}$      b. $CH_3NH_2$      c. $CaO$      d. $FeCO_3$

115. Classify each hydrocarbon as an alkane, alkene, or alkyne.
    a. $H_2C{=}CH{-}CH_3$
    b. $H_3C{-}CH_2{-}CH_3$
    c. $HC{\equiv}C{-}CH_3$
    d. $H_3C{-}CH_2{-}CH_2{-}CH_3$

116. Classify each hydrocarbon as an alkane, alkene, or alkyne.
    a. $HC{\equiv}CH$      b. $H_3C{-}CH{=}CH{-}CH_3$
    c.
    $$\begin{matrix} & CH_3 \\ & | \\ H_3C{-}CH & {-}CH_3 \end{matrix}$$
    d. $H_3C{-}C{\equiv}C{-}CH_3$

117. Write the formula based on the name, or the name based on the formula, for each hydrocarbon.
    a. propane      b. $CH_3CH_2CH_3$
    c. octane      d. $CH_3CH_2CH_2CH_2CH_3$

118. Write the formula based on the name, or the name based on the formula, for each hydrocarbon.
    a. $CH_3CH_3$      b. pentane
    c. $CH_3CH_2CH_2CH_2CH_2CH_3$      d. heptane

**119.** Classify each organic compound as a hydrocarbon or a functionalized hydrocarbon. For functionalized hydrocarbons, identify the compound's family.

a. $H_3C—CH_2OH$

b. $H_3C—CH_3$

c.

$$H_3C—\overset{\overset{\textstyle O}{\|}}{C}—CH_2—CH_3$$

d. $H_3C—NH_2$

**120.** Classify each organic compound as a hydrocarbon or a functionalized hydrocarbon. For functionalized hydrocarbons, identify the compound's family.

a.

$$H_3C—CH_2—\overset{\overset{\textstyle O}{\|}}{C}—OH$$

b.

$$H_3C—\overset{\overset{\textstyle O}{\|}}{CH}$$

c.

$$H_3C—\overset{\overset{\textstyle CH_3}{|}}{\underset{\underset{\textstyle CH_3}{|}}{C}}—CH_3$$

d. $H_3C—CH_2—O—CH_3$

# Cumulative Problems

**121.** How many molecules of ethanol ($C_2H_5OH$) (the alcohol in alcoholic beverages) are present in 145 mL of ethanol? The density of ethanol is 0.789 g/cm³.

**122.** A drop of water has a volume of approximately 0.05 mL. How many water molecules does it contain? The density of water is 1.0 g/cm³.

**123.** Determine the chemical formula of each compound and then use it to calculate the mass percent composition of each constituent element.

a. potassium chromate    b. lead(II) phosphate
c. sulfurous acid    d. cobalt(II) bromide

**124.** Determine the chemical formula of each compound and then use it to calculate the mass percent composition of each constituent element.

a. perchloric acid
b. phosphorus pentachloride
c. nitrogen triiodide
d. carbon dioxide

**125.** A Freon leak in the air-conditioning system of an old car releases 25 g of $CF_2Cl_2$ per month. What mass of chlorine does this car emit into the atmosphere each year?

**126.** A Freon leak in the air-conditioning system of a large building releases 12 kg of $CHF_2Cl$ per month. If the leak is allowed to continue, how many kilograms of Cl will be emitted into the atmosphere each year?

**127.** A metal (M) forms a compound with the formula $MCl_3$. If the compound contains 65.57% Cl by mass, what is the identity of the metal?

**128.** A metal (M) forms an oxide with the formula $M_2O$. If the oxide contains 16.99% O by mass, what is the identity of the metal?

**129.** Estradiol is a female sexual hormone that is responsible for the maturation and maintenance of the female reproductive system. Elemental analysis of estradiol gives the following mass percent composition: C 79.37%, H 8.88%, O 11.75%. The molar mass of estradiol is 272.37 g/mol. Find the molecular formula of estradiol.

**130.** Fructose is a common sugar found in fruit. Elemental analysis of fructose gives the following mass percent composition: C 40.00%, H 6.72%, O 53.28%. The molar mass of fructose is 180.16 g/mol. Find the molecular formula of fructose.

**131.** Combustion analysis of a 13.42-g sample of equilin (which contains only carbon, hydrogen, and oxygen) produces 39.61 g $CO_2$ and 9.01 g $H_2O$. The molar mass of equilin is 268.34 g/mol. Find its molecular formula.

**132.** Estrone, which contains only carbon, hydrogen, and oxygen, is a female sexual hormone in the urine of pregnant women. Combustion analysis of a 1.893-g sample of estrone produces 5.545 g of $CO_2$ and 1.388 g $H_2O$. The molar mass of estrone is 270.36 g/mol. Find its molecular formula.

**133.** Epsom salts is a hydrated ionic compound with the following formula: $MgSO_4 \cdot x\, H_2O$. A 4.93-g sample of Epsom salts is heated to drive off the water of hydration. The mass of the sample after complete dehydration is 2.41 g. Find the number of waters of hydration ($x$) in Epsom salts.

**134.** A hydrate of copper(II) chloride has the following formula: $CuCl_2 \cdot x\, H_2O$. The water in a 3.41-g sample of the hydrate is driven off by heating. The remaining sample has a mass of 2.69 g. Find the number of waters of hydration ($x$) in the hydrate.

**135.** A compound of molar mass 177 g/mol contains only carbon, hydrogen, bromine, and oxygen. Analysis reveals that the compound contains eight times as much carbon as hydrogen by mass. Find the molecular formula.

**136.** Researchers obtained the following data from experiments to find the molecular formula of benzocaine, a local anesthetic, which contains only carbon, hydrogen, nitrogen, and oxygen. Complete combustion of a 3.54-g sample of benzocaine with excess $O_2$ forms 8.49 g of $CO_2$ and 2.14 g $H_2O$. Another 2.35-g sample contains 0.199 g of N. The molar mass of benzocaine is 165 g/mol. Find the molar formula of benzocaine.

**137.** Find the total number of atoms in a sample of cocaine hydrochloride, $C_{17}H_{22}ClNO_4$, of mass 23.5 mg.

**138.** Vanadium forms four different oxides in which the percent by mass of vanadium is, respectively, (a) 76%, (b) 68%, (c) 61%, and (d) 56%. Determine the formula and the name of each oxide.

**139.** The chloride of an unknown metal is believed to have the formula $MCl_3$. A 2.395-g sample of the compound contains $3.606 \times 10^{-2}$ mol Cl. Find the atomic mass of M.

**140.** Write the structural formulas of three different compounds that each have the molecular formula $C_5H_{12}$.

**141.** A chromium-containing compound has the formula $Fe_xCr_yO_4$ and is 28.59% oxygen by mass. Find $x$ and $y$.

**142.** A phosphorus compound that contains 34.00% phosphorus by mass has the formula $X_3P_2$. Identify the element X.

**143.** A particular brand of beef jerky contains 0.0552% sodium nitrite by mass and is sold in an 8.00-oz bag. What mass of sodium does the sodium nitrite contribute to the sodium content of the bag of beef jerky?

**144.** Phosphorus is obtained primarily from ores containing calcium phosphate. If a particular ore contains 57.8% calcium phosphate, what minimum mass of the ore must be processed to obtain 1.00 kg of phosphorus?

# Challenge Problems

**145.** A mixture of NaCl and NaBr has a mass of 2.00 g and contains 0.75 g of Na. What is the mass of NaBr in the mixture?

**146.** Three pure compounds form when 1.00-g samples of element X combine with, respectively, 0.472 g, 0.630 g, and 0.789 g of element Z. The first compound has the formula $X_2Z_3$. Find the empirical formulas of the other two compounds.

**147.** A mixture of $CaCO_3$ and $(NH_4)_2CO_3$ is 61.9% $CO_3$ by mass. Find the mass percent of $CaCO_3$ in the mixture.

**148.** A mixture of 50.0 g of S and $1.00 \times 10^2$ g of $Cl_2$ reacts completely to form $S_2Cl_2$ and $SCl_2$. Find the mass of $S_2Cl_2$ formed.

**149.** Because of increasing evidence of damage to the ozone layer, chlorofluorocarbon (CFC) production was banned in 1996. However, many older cars still have air conditioners that use CFC-12 ($CF_2Cl_2$). These air conditioners are recharged from stockpiled supplies of CFC-12. Suppose that 100 million automobiles each contain 1.1 kg of CFC-12 and leak 25% of their CFC-12 into the atmosphere per year. How much chlorine, in kg, is added to the atmosphere each year due to these air conditioners? (Assume two significant figures in your calculations.)

**150.** A particular coal contains 2.55% sulfur by mass. When the coal is burned, it produces $SO_2$ emissions, which combine with rainwater to produce sulfuric acid. Use the formula of sulfuric acid to calculate the mass percent of S in sulfuric acid. Then determine how much sulfuric acid (in metric tons) is produced by the combustion of 1.0 metric ton of this coal. (A metric ton is 1000 kg.)

**151.** Lead is found in Earth's crust as several different lead ores. Suppose a certain rock is 38.0% PbS (galena), 25.0% $PbCO_3$ (cerussite), and 17.4% $PbSO_4$ (anglesite). The remainder of the rock is composed of substances containing no lead. How much of this rock (in kg) must be processed to obtain 5.0 metric tons of lead? (A metric ton is 1000 kg.)

**152.** A 2.52-g sample of a compound containing only carbon, hydrogen, nitrogen, oxygen, and sulfur is burned in excess oxygen to yield 4.23 g of $CO_2$ and 1.01 g of $H_2O$. Another sample of the same compound, of mass 4.14 g, yields 2.11 g of $SO_3$. A third sample, of mass 5.66 g, yields 2.27 g of $HNO_3$. Calculate the empirical formula of the compound.

**153.** A compound of molar mass 229 g/mol contains only carbon, hydrogen, iodine, and sulfur. Analysis shows that a sample of the compound contains six times as much carbon as hydrogen, by mass. Calculate the molecular formula of the compound.

**154.** The elements X and Y form a compound that is 40% X and 60% Y by mass. The atomic mass of X is twice that of Y. What is the empirical formula of the compound?

**155.** A compound of X and Y is $\frac{1}{3}$ X by mass. The atomic mass of element X is $\frac{1}{3}$ the atomic mass of element Y. Find the empirical formula of the compound.

**156.** A mixture of carbon and sulfur has a mass of 9.0 g. Complete combustion with excess $O_2$ gives 23.3 g of a mixture of $CO_2$ and $SO_2$. Find the mass of sulfur in the original mixture.

# Conceptual Problems

**157.** When molecules are represented by molecular models, what does each sphere represent? How big is the nucleus of an atom in comparison to the sphere used to represent an atom in a molecular model?

**158.** Without doing any calculations, determine which element in each compound has the highest mass percent composition.
  a. CO                b. $N_2O$
  c. $C_6H_{12}O_6$        d. $NH_3$

**159.** Explain the problem with the following statement and correct it: "The chemical formula for ammonia ($NH_3$) indicates that ammonia contains three grams of hydrogen for each gram of nitrogen."

**160.** Explain the problem with the following statement and correct it. "When a chemical equation is balanced, the number of molecules of each type on both sides of the equation is equal."

**161.** Without doing any calculations, arrange the elements in $H_2SO_4$ in order of decreasing mass percent composition.

**162.** Element A is an atomic element, and element B is a diatomic molecular element. Using circles to represent atoms of A and squares to represent atoms of B, draw molecular-level views of each element.

*Active Classroom Learning*

# Questions for Group Work

Discuss these questions with the group and record your consensus answer.

**163.** With group members playing the roles of nuclei and electrons, demonstrate the formation of an ionic bond between Na and Cl. Demonstrate the formation of the covalent bonds in $H_2O$.

**164.** Create a flowchart with a series of simple questions that can be used to determine whether a chemical formula is that of an atomic element, a molecular element, a molecular compound, or an ionic compound. Use your flowchart to identify the correct category for $P_4$, KCl, $CH_4$, Ne, and $NH_4NO_3$.

**165.** Have each member of your group list one similarity or difference between the naming conventions for ionic and molecular compounds.

**166.** A compound isolated from the rind of lemons is found to be 88.14% carbon and 11.86% hydrogen by mass. How many grams of C and H are there in a 100.0-g sample of this substance? How many moles of C and H? What is the empirical formula? The molar mass is determined to be 136.26 g/mol. What is the molecular formula? Which step of the process just described does your group understand the least? Which step will be hardest for the members of your group to remember?

**167.** Octane ($C_8H_{18}$), a component of gasoline, reacts with oxygen to form carbon dioxide and water. Write the balanced chemical reaction for this process by passing a single piece of paper around your group, having each group member complete the next logical step. As you each complete your step, explain your reasoning to the group.

# Data Interpretation and Analysis

## Carbon and Fossil Fuel Consumption

168. Since the 1800s, the concentration of carbon dioxide in Earth's atmosphere has been increasing (Figure a ▼). The scientific consensus is that the increase in $CO_2$ is due to fossil fuel combustion.

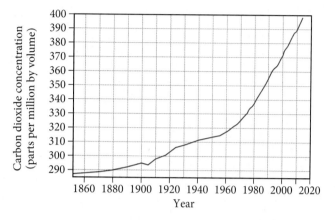

▲ **FIGURE a** **Carbon Dioxide Concentration in the Atmosphere**

Use the information provided in the figure to answer the following questions (Note that 1 ppmv = 1 gas volume/1 × $10^6$ air volumes = 1 $\mu L$ gas/1 L air):

a. Globally, in 2010 and 2012, respectively, there were about 9.20 billion metric tons and 9.60 billion metric tons of carbon in Earth's atmosphere.

i. Determine the increase in carbon in the atmosphere in units of metric tons between 2010 and 2012.

ii. Determine the increase of carbon dioxide in the atmosphere in units of metric tons between 2010 and 2012.

b. Figure a lists the concentration of $CO_2$ in air in terms of parts per million by volume (ppmv).

i. Compare the volume (in mL) of $CO_2$ in 1.00 $m^3$ of air in 1880 to the volume (in mL) of $CO_2$ in 1.00 $m^3$ of air in 2014.

ii. Determine the increase in volume (in mL) of $CO_2$ in 1.00 $m^3$ of air between 1880 and 2014.

iii. What is the percent increase of $CO_2$ in 1.00 $m^3$ of air between 1880 and 2014?

iv. Assuming that the total volume of air in the atmosphere around Earth is 5.1 × $10^9$ $km^3$ and that the concentration of $CO_2$ is uniform throughout the atmosphere, estimate the volume of $CO_2$ in the atmosphere in 2014 in units of $km^3$.

c. Based on the information in Figure a, determine the average yearly increase in the concentration of $CO_2$ in the atmosphere for the period 1960 to 2014. Estimate the concentration of $CO_2$ in the atmosphere, in units of ppmv, in 2030.

# Answers to Conceptual Connections

### Types of Chemical Bonds

**3.1**   The bond is covalent because it is forming between two non-metals.

### Structural Formulas

**3.2**   H—O—H

### Representing Molecules

**3.3**   The spheres represent the electron cloud of the atom. It would be nearly impossible to draw a nucleus to scale on any of the space-filling molecular models in this book—the nucleus would be too small to see.

### A Molecular View of Elements and Compounds

**3.4**

$A_2B$

CD

## Ionic and Molecular Compounds

**3.5** Choice **(a)** best describes the difference between ionic and molecular compounds. Answer **(b)** is incorrect because there are no "new" forces in bonding (just rearrangements that result in lower potential energy) and because ions do not group together in pairs in the solid phase. Answer **(c)** is incorrect because the main difference between ionic and molecular compounds is the way that the atoms bond. Answer **(d)** is incorrect because ionic compounds do not contain molecules.

## Nomenclature

**3.6** This conceptual connection addresses one of the main errors you can make in nomenclature: the failure to correctly categorize the compound. Remember that you must first determine whether the compound is an ionic compound, a molecular compound, or an acid, and then you must name it accordingly. $NCl_3$ is a molecular compound (two or more nonmetals), and therefore in its name prefixes indicate the number of each type of atom—so $NCl_3$ is nitrogen trichloride. The compound $AlCl_3$, however, is an ionic compound (metal and nonmetal), and therefore does not require prefixes—so $AlCl_3$ is aluminum chloride.

## Molecular Models and the Size of Molecules

**3.7** **(c)** Atomic radii range in the hundreds of picometers, while the spheres in these models have radii of about a centimeter. The scaling factor is therefore about $10^8$ (100 million).

## Chemical Formula and Mass Percent Composition

**3.8** $C > O > H$ Since carbon and oxygen differ in atomic mass by only 4 amu, and since there are six carbon atoms in the formula, we can conclude that carbon constitutes the greatest fraction of the mass. Oxygen is next because its mass is 16 times that of hydrogen and there are only six hydrogen atoms for every one oxygen atom.

## Chemical Formulas and Elemental Composition

**3.9** **(c)** The chemical formula for a compound gives relationships between *atoms* or *moles of atoms*. The chemical formula for water states that water molecules contain two H atoms to every one O atom or 2 mol H to every 1 mol $H_2O$. This *does not* imply a two-to-one relationship between *masses* of hydrogen and oxygen because these atoms have different masses. It also does not imply a two-to-one relationship between volumes.

## Balanced Chemical Equations

**3.10** Both **(a)** and **(d)** are correct. When the number of atoms of each type is balanced, the sum of the masses of the substances will be the same on both sides of the equation. Since molecules change during a chemical reaction, the number of molecules is not the same on both sides, nor is the number of moles necessarily the same.

# 4

# Chemical Quantities and Aqueous Reactions

> I feel sorry for people who don't know anything about chemistry. They are missing an important source of happiness.
>
> —LINUS PAULING (1901–1994)

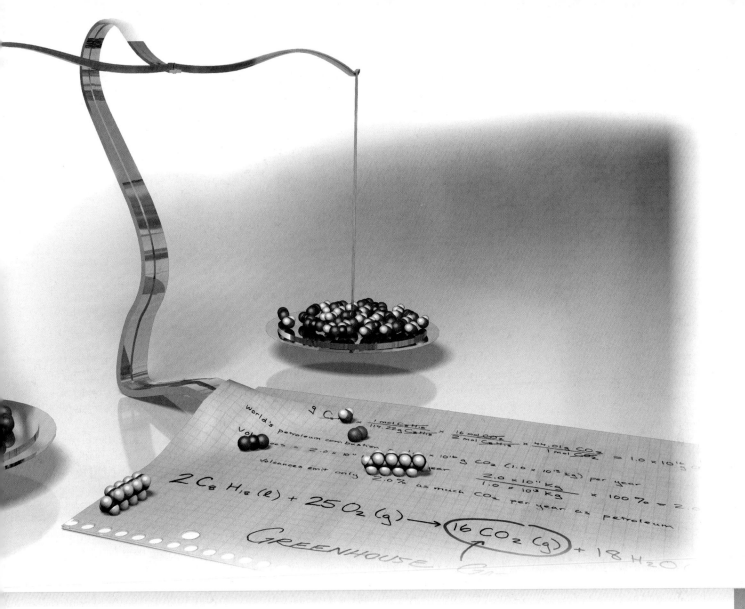

The amount of product formed in a chemical reaction is related to the amount of reactant that reacts. This concept makes sense intuitively, but how can we describe and understand this relationship more fully? The first half of this chapter focuses on chemical stoichiometry—the numerical relationships between the amounts of reactants and products in chemical reactions. In Chapter 3, you learned how to write balanced chemical equations for chemical reactions. Here we examine more closely the meaning of those balanced equations. In the second half of this chapter, we turn to describing chemical reactions that occur in water. You have probably witnessed many of these types of reactions in your daily life because they are so common. Have you ever mixed baking soda with vinegar and observed the subsequent bubbling or noticed the hard water deposits that form on plumbing fixtures? These reactions—and many others, including those that occur within the watery environment of living cells—are aqueous chemical reactions, the subject of the second half of this chapter.

▲ The molecular models on this balance represent the reactants and products in the combustion of octane, a component of petroleum. One of the products, carbon dioxide, is the main greenhouse gas implicated in global climate change.

## 4.1 Climate Change and the Combustion of Fossil Fuels

The temperature outside my office today is a cool 48°F, lower than normal for this time of year on the California coast. However, today's "chill" pales in comparison with how cold it would be without the presence of *greenhouse gases* in the atmosphere.

► FIGURE 4.1 **The Greenhouse Effect** Greenhouse gases in the atmosphere act as a one-way filter. They allow visible light to pass through and warm Earth's surface, but they prevent heat energy from radiating back out into space.

## The Greenhouse Effect

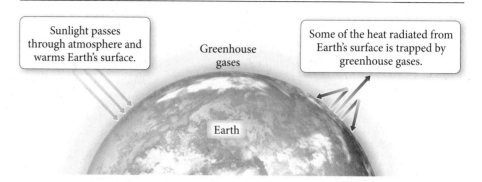

Sunlight passes through atmosphere and warms Earth's surface.

Greenhouse gases

Some of the heat radiated from Earth's surface is trapped by greenhouse gases.

Earth

These gases act like the glass of a greenhouse, allowing sunlight to enter the atmosphere and warm Earth's surface but preventing some of the heat generated by the sunlight from escaping, as shown in Figure 4.1▲. The balance between incoming and outgoing energy from the sun determines Earth's average temperature.

If the greenhouse gases in the atmosphere were not present, more heat energy would escape, and Earth's average temperature would be about 60°F colder than it currently is. The temperature outside of my office today would be below 0°F, and even the sunniest U.S. cities would most likely be covered with snow. However, if the concentration of greenhouse gases in the atmosphere were to increase, Earth's average temperature would rise.

The extremely cold temperatures of Mars are as much a result of its lack of atmosphere as its greater distance from the sun than Earth. Conversely, Venus is an inferno partly because its thick atmosphere is rich in greenhouse gases.

In recent decades scientists have become increasingly concerned because the quantity of atmospheric carbon dioxide ($CO_2$)—Earth's most significant greenhouse gas—is rising. More $CO_2$ enhances the atmosphere's ability to hold heat and is believed to lead to *global warming,* an increase in Earth's average temperature. Since 1860, atmospheric $CO_2$ levels have risen by 38% (Figure 4.2▼), and Earth's average temperature has risen by about 1.2°F (0.7°C) (as shown in Figure 4.3▼).

Most scientists believe that the primary cause of rising atmospheric $CO_2$ concentration is the burning of fossil fuels (natural gas, petroleum, and coal), which provides 90% of our society's energy. Some people, however, have suggested that fossil fuel combustion does not significantly contribute to global warming and climate change. They argue that the amount of carbon dioxide emitted into the atmosphere by natural sources, such as volcanic eruptions, far exceeds that from fossil fuel combustion. Which group is right? One way to judge the validity of the naysayers' argument is by calculating how much carbon dioxide is emitted by fossil fuel combustion and comparing that amount to the amount released by volcanic eruptions.

## Atmospheric Carbon Dioxide

▲ FIGURE 4.2 **Carbon Dioxide Concentrations in the Atmosphere** The rise in carbon dioxide levels is due largely to fossil fuel combustion.

## Global Temperature

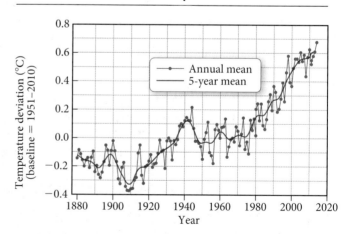

▲ FIGURE 4.3 **Global Temperature** Average temperatures worldwide have risen by about 0.7°C since 1880.

# 4.2 Reaction Stoichiometry: How Much Carbon Dioxide?

The balanced chemical equations for fossil fuel combustion reactions provide the exact relationships between the amount of fossil fuel burned and the amount of carbon dioxide emitted. In this discussion, we use octane (a component of gasoline) as a representative fossil fuel. The balanced equation for the combustion of octane is:

$$2\,C_8H_{18}(l) + 25\,O_2(g) \longrightarrow 16\,CO_2(g) + 18\,H_2O(g)$$

The balanced equation shows that 16 $CO_2$ molecules are produced for every two molecules of octane burned. We can extend this numerical relationship between molecules to the amounts in moles as follows:

**The coefficients in a chemical equation specify the relative amounts in moles of each of the substances involved in the reaction.**

In other words, from the equation, we know that 16 *moles* of $CO_2$ are produced for every two *moles* of octane burned. The numerical relationships between chemical amounts in a balanced chemical equation are called reaction **stoichiometry**. Stoichiometry allows us to predict the amounts of products that will form in a chemical reaction based on the amounts of reactants that react. Stoichiometry also allows us to determine the amount of reactants necessary to form a given amount of product. These calculations are central to chemistry, allowing chemists to plan and carry out chemical reactions to obtain products in the desired quantities.

Stoichiometry is pronounced stoy-kee-AHM-e-tree.

## Making Pizza: The Relationships among Ingredients

The concepts of stoichiometry are similar to a cooking recipe. Calculating the amount of carbon dioxide produced by the combustion of a given amount of a fossil fuel is analogous to calculating the number of pizzas that can be made from a given amount of cheese. For example, suppose we use the following pizza recipe:

1 crust + 5 ounces tomato sauce + 2 cups cheese ⟶ 1 pizza

The recipe contains the numerical relationships between the pizza ingredients. It says that if we have two cups of cheese—and enough of everything else—we can make one pizza. We can write this relationship as a ratio between the cheese and the pizza:

2 cups cheese : 1 pizza

What if we have six cups of cheese? Assuming that we have enough of everything else, we can use the ratio as a conversion factor to calculate the number of pizzas:

$$6\,\text{cups cheese} \times \frac{1\,\text{pizza}}{2\,\text{cups cheese}} = 3\,\text{pizzas}$$

Six cups of cheese are sufficient to make three pizzas. The pizza recipe contains numerical ratios between other ingredients as well, including the crust and the tomato sauce:

1 crust : 1 pizza

5 ounces tomato sauce : 1 pizza

## Making Molecules: Mole-to-Mole Conversions

In a balanced chemical equation, we have a "recipe" for how reactants combine to form products. From our balanced equation for the combustion of octane, for example, we can write the following stoichiometric ratio:

2 mol $C_8H_{18}$ : 16 mol $CO_2$

We can use this ratio to determine how many moles of $CO_2$ form when a given number of moles of $C_8H_{18}$ burns. Suppose that we burn 22.0 moles of $C_8H_{18}$; how many moles of

$CO_2$ form? We use the ratio from the balanced chemical equation in the same way that we used the ratio from the pizza recipe. The ratio acts as a conversion factor between the amount in moles of the reactant ($C_8H_{18}$) and the amount in moles of the product ($CO_2$):

$$22.0 \ \text{mol} \ C_8H_{18} \times \frac{16 \ \text{mol} \ CO_2}{2 \ \text{mol} \ C_8H_{18}} = 176 \ \text{mol} \ CO_2$$

The combustion of 22.0 moles of $C_8H_{18}$ adds 176 moles of $CO_2$ to the atmosphere.

## Making Molecules: Mass-to-Mass Conversions

According to the U.S. Department of Energy, the world burned $3.3 \times 10^{10}$ barrels of petroleum in 2013, the equivalent of approximately $3.7 \times 10^{15}$ g of gasoline. We can estimate the mass of $CO_2$ emitted into the atmosphere from burning this much gasoline using the combustion of $3.7 \times 10^{15}$ g octane as the representative reaction. This calculation is similar to the one we just did, except that we are now given the *mass* of octane instead of the *amount* of octane in moles. Consequently, we must first convert the mass (in grams) to the amount (in moles). The general conceptual plan for calculations in which we are given the mass of a reactant or product in a chemical reaction and asked to find the mass of a different reactant or product takes the form:

$$\boxed{\text{Mass A}} \longrightarrow \boxed{\begin{array}{c}\text{Amount A}\\(\text{in moles})\end{array}} \longrightarrow \boxed{\begin{array}{c}\text{Amount B}\\(\text{in moles})\end{array}} \longrightarrow \boxed{\text{Mass B}}$$

where A and B are two different substances involved in the reaction.

We use the molar mass of A to convert from the mass of A to the amount of A (in moles). We use the appropriate ratio from the balanced chemical equation to convert from the amount of A (in moles) to the amount of B (in moles). And finally, we use the molar mass of B to convert from the amount of B (in moles) to the mass of B. To calculate the mass of $CO_2$ emitted upon the combustion of $3.7 \times 10^{15}$ g of octane, we use the following conceptual plan:

### Conceptual Plan

### Relationships Used

2 mol $C_8H_{18}$ : 16 mol $CO_2$ (from the chemical equation)

molar mass $C_8H_{18}$ = 114.22 g/mol

molar mass $CO_2$ = 44.01 g/mol

### Solution

We follow the conceptual plan to solve the problem, beginning with g $C_8H_{18}$ and canceling units to arrive at g $CO_2$:

$$3.7 \times 10^{15} \ \text{g} \ C_8H_{18} \times \frac{1 \ \text{mol} \ C_8H_{18}}{114.22 \ \text{g} \ C_8H_{18}} \times$$

$$\frac{16 \ \text{mol} \ CO_2}{2 \ \text{mol} \ C_8H_{18}} \times \frac{44.01 \ \text{g} \ CO_2}{1 \ \text{mol} \ CO_2} = 1.1 \times 10^{16} \ \text{g} \ CO_2$$

The percentage of $CO_2$ emitted by volcanoes relative to all fossil fuels is even less than 1.8% because the combustion of coal and natural gas also emits $CO_2$.

The world's petroleum combustion produces $1.1 \times 10^{16}$ g $CO_2$ ($1.1 \times 10^{13}$ kg) per year. In comparison, volcanoes produce about $2 \times 10^{11}$ kg $CO_2$ per year.* In other words, volcanoes emit only $\dfrac{2 \times 10^{11} \ \text{kg}}{1.1 \times 10^{13} \ \text{kg}} \times 100\% = 1.8\%$ as much $CO_2$ per year as

*Gerlach, T. M., Present-day $CO_2$ emissions from volcanoes. *Eos, Transactions, American Geophysical Union.* **1991** *72*, 249, 254–255.

petroleum combustion. The argument that volcanoes emit more carbon dioxide than fossil fuel combustion is clearly mistaken. Examples 4.1 and 4.2 provide additional practice with stoichiometric calculations.

---

### Example 4.1    Stoichiometry

During photosynthesis, plants convert carbon dioxide and water into glucose ($C_6H_{12}O_6$) according to the reaction:

$$6\,CO_2(g) + 6\,H_2O(l) \xrightarrow{\text{sunlight}} 6\,O_2(g) + C_6H_{12}O_6(aq)$$

Suppose that a particular plant consumes 37.8 g of $CO_2$ in one week. Assuming that there is more than enough water present to react with all of the $CO_2$, what mass of glucose (in grams) can the plant synthesize from the $CO_2$?

| | |
|---|---|
| **SORT** The problem provides the mass of carbon dioxide and asks you to find the mass of glucose that can be produced. | **GIVEN:** 37.8 g $CO_2$<br><br>**FIND:** g $C_6H_{12}O_6$ |

| | |
|---|---|
| **STRATEGIZE** The conceptual plan follows the general pattern of mass A → amount A (in moles) → amount B (in moles) → mass B. From the chemical equation, deduce the relationship between moles of carbon dioxide and moles of glucose. Use the molar masses to convert between grams and moles. | **CONCEPTUAL PLAN** <br><br>**RELATIONSHIPS USED**<br>molar mass $CO_2$ = 44.01 g/mol<br>6 mol $CO_2$ : 1 mol $C_6H_{12}O_6$<br>molar mass $C_6H_{12}O_6$ = 180.2 g/mol |

| | |
|---|---|
| **SOLVE** Follow the conceptual plan to solve the problem. Begin with g $CO_2$ and use the conversion factors to arrive at g $C_6H_{12}O_6$. | **SOLUTION**<br><br>$37.8\ \cancel{\text{g CO}_2} \times \dfrac{1\ \cancel{\text{mol CO}}}{44.01\ \cancel{\text{g CO}_2}} \times \dfrac{1\ \cancel{\text{mol C}_6\text{H}_{12}\text{O}_6}}{6\ \cancel{\text{mol CO}_2}} \times \dfrac{180.2\ \text{g C}_6\text{H}_{12}\text{O}_6}{1\ \cancel{\text{mol C}_6\text{H}_{12}\text{O}_6}} = 25.8\ \text{g C}_6\text{H}_{12}\text{O}_6$ |

**CHECK** The units of the answer are correct. The magnitude of the answer (25.8 g) is less than the initial mass of $CO_2$ (37.8 g). This is reasonable because each carbon in $CO_2$ has two oxygen atoms associated with it, while in $C_6H_{12}O_6$ each carbon has only one oxygen atom associated with it and two hydrogen atoms, which are much lighter than oxygen. Therefore, the mass of glucose produced should be less than the mass of carbon dioxide for this reaction.

**FOR PRACTICE 4.1** Magnesium hydroxide, the active ingredient in milk of magnesia, neutralizes stomach acid, primarily HCl, according to the reaction:

$$Mg(OH)_2(aq) + 2\,HCl(aq) \longrightarrow 2\,H_2O(l) + MgCl_2(aq)$$

What mass of HCl, in grams, is neutralized by a dose of milk of magnesia containing 3.26 g $Mg(OH)_2$?

---

### Example 4.2    Stoichiometry

Sulfuric acid ($H_2SO_4$) is a component of acid rain that forms when $SO_2$, a pollutant, reacts with oxygen and water according to the simplified reaction:

$$2\,SO_2(g) + O_2(g) + 2\,H_2O(l) \longrightarrow 2\,H_2SO_4(aq)$$

The generation of the electricity used by a medium-sized home produces about 25 kg of $SO_2$ per year. Assuming that there is more than enough $O_2$ and $H_2O$, what mass of $H_2SO_4$, in kg, can form from this much $SO_2$?

*—Continued on the next page*

*—Continued*

| SORT The problem gives the mass of sulfur dioxide and asks you to find the mass of sulfuric acid. | GIVEN: 25 kg $SO_2$ <br><br> FIND: kg $H_2SO_4$ |
|---|---|

**STRATEGIZE** The conceptual plan follows the standard format of mass → amount (in moles) → amount (in moles) → mass. Since the original quantity of $SO_2$ is given in kilograms, you must first convert to grams. You can deduce the relationship between moles of sulfur dioxide and moles of sulfuric acid from the chemical equation. Since the final quantity is requested in kilograms, convert to kilograms at the end.

**CONCEPTUAL PLAN**

$$\boxed{\text{kg } SO_2} \xrightarrow{\dfrac{1000\text{ g}}{1\text{ kg}}} \boxed{\text{g } SO_2} \xrightarrow{\dfrac{1\text{ mol } SO_2}{64.07\text{ g } SO_2}} \boxed{\text{mol } SO_2} \xrightarrow{\dfrac{2\text{ mol } H_2SO_4}{2\text{ mol } SO_2}}$$

$$\boxed{\text{mol } H_2SO_4} \xrightarrow{\dfrac{98.09\text{ g } H_2SO_4}{1\text{ mol } H_2SO_4}} \boxed{\text{g } H_2SO_4} \xrightarrow{\dfrac{1\text{ kg}}{1000\text{ g}}} \boxed{\text{kg } H_2SO_4}$$

**RELATIONSHIPS USED**

1 kg = 1000 g          2 mol $SO_2$ : 2 mol $H_2SO_4$

molar mass $SO_2$ = 64.07 g/mol          molar mass $H_2SO_4$ = 98.09 g/mol

**SOLVE** Follow the conceptual plan to solve the problem. Begin with the given amount of $SO_2$ in kilograms and use the conversion factors to arrive at kg $H_2SO_4$.

**SOLUTION**

$$25 \text{ kg } SO_2 \times \frac{1000 \text{ g}}{1 \text{ kg}} \times \frac{1 \text{ mol } SO_2}{64.07 \text{ g } SO_2} \times \frac{2 \text{ mol } H_2SO_4}{2 \text{ mol } SO_2}$$

$$\times \frac{98.09 \text{ g } H_2SO_4}{1 \text{ mol } H_2SO_4} \times \frac{1 \text{ kg}}{1000 \text{ g}} = 38 \text{ kg } H_2SO_4$$

**CHECK** The units of the final answer are correct. The magnitude of the final answer (38 kg $H_2SO_4$) is larger than the amount of $SO_2$ given (25 kg). This is reasonable because in the reaction each $SO_2$ molecule "gains weight" by reacting with $O_2$ and $H_2O$.

**FOR PRACTICE 4.2** Another component of acid rain is nitric acid, which forms when $NO_2$, also a pollutant, reacts with oxygen and water according to the simplified equation:

$$4\,NO_2(g) + O_2(g) + 2\,H_2O(l) \longrightarrow 4\,HNO_3(aq)$$

The generation of the electricity used by a medium-sized home produces about 16 kg of $NO_2$ per year. Assuming that there is adequate $O_2$ and $H_2O$, what mass of $HNO_3$, in kg, can form from this amount of $NO_2$ pollutant?

---

CONCEPTUAL
**CONNECTION 4.1**

**Stoichiometry** Under certain conditions sodium reacts with oxygen to form sodium oxide according to the reaction:

$$4\,Na(s) + O_2(g) \longrightarrow 2\,Na_2O(s)$$

A flask contains the amount of oxygen represented by the diagram shown at left.

Which image below best represents the amount of sodium required to completely react with all of the oxygen in the flask according to the equation?

      (a)           (b)           (c)           (d)

**Stoichiometry II** Consider the generic chemical equation
$A + 3B \rightarrow 2C$. Let circles represent molecules of A, squares
represent molecules of B, and triangles represent molecules of C.
If the diagram shown here represents the amount of B available
for reaction, draw similar diagrams showing (a) the amount of A
necessary to completely react with B, and (b) the amount of C
that forms if B completely reacts.

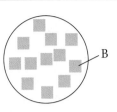

## 4.3 Limiting Reactant, Theoretical Yield, and Percent Yield

KCV  Limiting Reactant, Theoretical Yield, and Percent Yield

Let's return to our pizza analogy to understand three more important concepts in reaction stoichiometry: *limiting reactant, theoretical yield,* and *percent yield.* Recall our pizza recipe from Section 4.2:

$$1 \text{ crust} + 5 \text{ ounces tomato sauce} + 2 \text{ cups cheese} \longrightarrow 1 \text{ pizza}$$

Suppose that we have four crusts, ten cups of cheese, and 15 ounces of tomato sauce. How many pizzas can we make?

**We have enough crusts to make:**

$$4 \text{ crusts} \times \frac{1 \text{ pizza}}{1 \text{ crust}} = 4 \text{ pizzas}$$

**We have enough cheese to make:**

$$10 \text{ cups cheese} \times \frac{1 \text{ pizza}}{2 \text{ cups cheese}} = 5 \text{ pizzas}$$

**We have enough tomato sauce to make:**

$$15 \text{ ounces tomato sauce} \times \frac{1 \text{ pizza}}{5 \text{ ounces tomato sauce}} = 3 \text{ pizzas}$$

Limiting reactant ↗    ↖ Smallest number of pizzas

We have enough crusts for four pizzas and enough cheese for five pizzas, but enough tomato sauce for only three pizzas. Consequently, unless we get more ingredients, we can make only three pizzas. The tomato sauce *limits* how many pizzas we can make. If the pizza recipe were a chemical reaction, the tomato sauce would be the **limiting reactant,** the reactant that limits the amount of product in a chemical reaction. Notice that the limiting reactant is the reactant that makes *the least amount of product.*

The term *limiting reagent* is sometimes used in place of *limiting reactant.*

4 crusts → 4 pizzas

10 cups cheese → 5 pizzas

Limiting reactant → 15 ounces tomato sauce → 3 pizzas ← Least amount of product

◀ The ingredient that makes the least amount of pizza determines how many pizzas you can make.

The reactants that *do not* limit the amount of product—such as the crusts and the cheese in this example—are said to be *in excess*. If this were a chemical reaction, three pizzas would be the **theoretical yield**, the maximum amount of product that can be made in a chemical reaction based on the amount of limiting reactant.

Let us carry this analogy one step further. Suppose we go on to cook our pizzas and accidentally burn one of them. Even though we theoretically have enough ingredients for three pizzas, we end up with only two. If this were a chemical reaction, the two pizzas would be our **actual yield**, the amount of product actually produced by a chemical reaction. (The actual yield is always equal to or less than the theoretical yield because a small amount of product is usually lost to other reactions or does not form during a reaction.) Finally, our **percent yield**, the percentage of the theoretical yield that was actually attained, is calculated as the ratio of the actual yield to the theoretical yield:

Actual yield

$$\% \text{ yield } = \frac{2 \text{ pizzas}}{3 \text{ pizzas}} \times 100\% = 67\%$$

Theoretical yield

Since one of our pizzas burned, our percent yield for pizzas is 67%.

*Summarizing Limiting Reactant and Yield:*

▶ **The limiting reactant** (or **limiting reagent**) is the reactant that is completely consumed in a chemical reaction and limits the amount of product.

▶ **The reactant in excess** is any reactant that occurs in a quantity greater than is required to completely react with the limiting reactant.

▶ **The theoretical yield** is the amount of product that can be made in a chemical reaction based on the amount of limiting reactant.

▶ **The actual yield** is the amount of product actually produced by a chemical reaction.

▶ **The percent yield** is calculated as $\dfrac{\text{actual yield}}{\text{theoretical yield}} \times 100\%$.

## Calculating Limiting Reactant, Theoretical Yield, and Percent Yield

We can apply these concepts to a chemical reaction. Recall from Section 3.11 our balanced equation for the combustion of methane:

$$CH_4(g) + 2\,O_2(g) \longrightarrow CO_2(g) + 2\,H_2O(l)$$

If we start out with five $CH_4$ molecules and eight $O_2$ molecules, what is our limiting reactant? What is our theoretical yield of carbon dioxide molecules? First we calculate the number of $CO_2$ molecules that can be made from five $CH_4$ molecules:

$$5\,CH_4 \times \frac{1\,CO_2}{1\,CH_4} = 5\,CO_2$$

We then calculate the number of $CO_2$ molecules that can be made from eight $O_2$ molecules:

$$8 \, O_2 \times \frac{1 \, CO_2}{2 \, O_2} = 4 \, CO_2$$

Limiting reactant

Least amount of product

We have enough $CH_4$ to make five $CO_2$ molecules and enough $O_2$ to make four $CO_2$ molecules; therefore, $O_2$ is the limiting reactant, and four $CO_2$ molecules is the theoretical yield. The $CH_4$ is in excess.

An alternative way to calculate the limiting reactant (which we describe here but do not use again in this book) is to pick any reactant and determine how much of the *other reactant* is necessary to completely react with it. For the reaction we just examined, we have five $CH_4$ molecules and eight $O_2$ molecules. Let's pick the five $CH_4$ molecules and determine how many $O_2$ molecules are necessary to completely react with them:

$$5 \, CH_4 \times \frac{2 \, O_2}{1 \, CH_4} = 10 \, O_2$$

Since we need ten $O_2$ molecules to completely react with the five $CH_4$ molecules, and since we have only eight $O_2$ molecules, we know that the $O_2$ is the limiting reactant. The same method can be applied by comparing the amounts of reactants in moles.

**Limiting Reactant and Theoretical Yield**  Nitrogen and hydrogen gas react to form ammonia according to the reaction:

CONCEPTUAL CONNECTION **4.3**

$$N_2(g) + 3 \, H_2(g) \longrightarrow 2 \, NH_3(g)$$

If a flask contains a mixture of reactants represented by the image shown at right, which image below best represents the mixture in the flask after the reactants have reacted as completely as possible? What is the limiting reactant? Which reactant is in excess?

(a)          (b)          (c)

## Calculating Limiting Reactant, Theoretical Yield, and Percent Yield from Initial Reactant Masses

When working in the laboratory, we normally measure the initial quantities of reactants in grams, not in number of molecules. To find the limiting reactant and theoretical yield from initial masses, we must first convert the masses to amounts in moles. Consider the reaction:

$$2 \, Mg(s) + O_2(g) \longrightarrow 2 \, MgO(s)$$

A reaction mixture contains 42.5 g Mg and 33.8 g $O_2$; what is the limiting reactant and theoretical yield?

To solve this problem, we must determine which of the reactants makes the least amount of product.

### Conceptual Plan

We can find the limiting reactant by calculating how much product can be made from each reactant. However, we are given the initial quantities in grams, and stoichiometric relationships are between moles, so we must first convert to moles. We then convert from moles of the reactant to moles of product. The reactant that makes the *least amount of product* is the limiting reactant. The conceptual plan is:

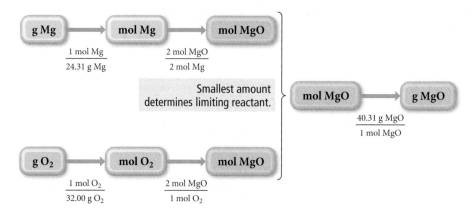

In this conceptual plan, we compare the number of moles of MgO made by each reactant and convert only the smaller amount to grams. (Alternatively, we can convert both quantities to grams and determine the limiting reactant based on the mass of the product.)

### Relationships Used

molar mass Mg = 24.31 g/mol

molar mass $O_2$ = 32.00 g/mol

2 mol Mg : 2 mol MgO

1 mol $O_2$ : 2 mol MgO

molar mass MgO = 40.31 g/mol

### Solution

Beginning with the masses of each reactant, we follow the conceptual plan to calculate how much product can be made from each:

$$42.5 \text{ g Mg} \times \frac{1 \text{ mol Mg}}{24.31 \text{ g Mg}} \times \frac{2 \text{ mol MgO}}{2 \text{ mol Mg}} = 1.7483 \text{ mol MgO}$$

Limiting reactant

Least amount of product

$$1.7483 \text{ mol MgO} \times \frac{40.31 \text{ g MgO}}{1 \text{ mol MgO}} = 70.5 \text{ g MgO}$$

$$33.8 \text{ g } O_2 \times \frac{1 \text{ mol } O_2}{32.00 \text{ g } O_2} \times \frac{2 \text{ mol MgO}}{1 \text{ mol } O_2} = 2.1125 \text{ mol MgO}$$

Since Mg makes the least amount of product, it is the limiting reactant, and $O_2$ is in excess. Notice that the limiting reactant is not necessarily the reactant with the least mass. In this case, the mass of $O_2$ is less than the mass of Mg, yet Mg is the limiting reactant because it makes the least amount of MgO. The theoretical yield is 70.5 g of MgO, the mass of product possible based on the limiting reactant.

Suppose that after the synthesis, the actual yield of MgO is 55.9 g. What is the percent yield? We calculate the percent yield as follows:

$$\% \text{ yield} = \frac{\text{actual yield}}{\text{theoretical yield}} \times 100\% = \frac{55.9 \text{ g}}{70.5 \text{ g}} \times 100\% = 79.3\%$$

## Example 4.3 ▸ Limiting Reactant and Theoretical Yield

Ammonia, $NH_3$, can be synthesized by the reaction:

$$2\,NO(g) + 5\,H_2(g) \longrightarrow 2\,NH_3(g) + 2\,H_2O(g)$$

Starting with 86.3 g NO and 25.6 g $H_2$, find the theoretical yield of ammonia in grams.

| | |
|---|---|
| **SORT** You are given the mass of each reactant in grams and asked to find the theoretical yield of a product. | **GIVEN:** 86.3 g NO, 25.6 g $H_2$ <br><br> **FIND:** theoretical yield of $NH_3(g)$ |

| | |
|---|---|
| **STRATEGIZE** Determine which reactant makes the least amount of product by converting from grams of each reactant to moles of the reactant to moles of the product. Use molar masses to convert between grams and moles and use the stoichiometric relationships (from the chemical equation) to convert between moles of reactant and moles of product. Remember that the reactant that makes *the least amount of product* is the limiting reactant. Convert the number of moles of product obtained using the limiting reactant to grams of product. | **CONCEPTUAL PLAN** <br><br> 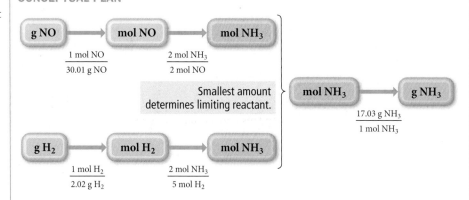 <br><br> **RELATIONSHIPS USED** <br> molar mass NO = 30.01 g/mol <br> molar mass $H_2$ = 2.02 g/mol <br> 2 mol NO : 2 mol $NH_3$  (from chemical equation) <br> 5 mol $H_2$ : 2 mol $NH_3$  (from chemical equation) <br> molar mass $NH_3$ = 17.03 g/mol |

| | |
|---|---|
| **SOLVE** Beginning with the given mass of each reactant, calculate the amount of product that can be made in moles. Convert the amount of product made by the limiting reactant to grams—this is the theoretical yield. | **SOLUTION** <br><br> $86.3\ \text{g NO} \times \dfrac{1\ \text{mol NO}}{30.01\ \text{g NO}} \times \dfrac{2\ \text{mol NH}_3}{2\ \text{mol NO}} = 2.8\underline{7}57\ \text{mol NH}_3$ <br><br> Limiting reactant        Least amount of product <br><br> $25.6\ \text{g H}_2 \times \dfrac{1\ \text{mol H}_2}{2.02\ \text{g H}_2} \times \dfrac{2\ \text{mol NH}_3}{5\ \text{mol H}_2} = 5.0\underline{6}93\ \text{mol NH}_3$ <br><br> $2.8\underline{7}57\ \text{mol NH}_3 \times \dfrac{17.03\ \text{g NH}_3}{\text{mol NH}_3}$ <br><br> $= 49.0\ \text{g NH}_3$ <br><br> Since NO makes the least amount of product, it is the limiting reactant, and the theoretical yield of ammonia is 49.0 g. |

**CHECK** The units of the answer (g $NH_3$) are correct. The magnitude (49.0 g) seems reasonable given that 86.3 g NO is the limiting reactant. NO contains one oxygen atom per nitrogen atom and $NH_3$ contains three hydrogen atoms per nitrogen atom. Since three hydrogen atoms have less mass than one oxygen atom, it is reasonable that the mass of $NH_3$ obtained is less than the mass of NO.

**FOR PRACTICE 4.3** Ammonia can also be synthesized by the reaction:

$$3\,H_2(g) + N_2(g) \longrightarrow 2\,NH_3(g)$$

What is the theoretical yield of ammonia, in kg, that we can synthesize from 5.22 kg of $H_2$ and 31.5 kg of $N_2$?

## Example 4.4 | Limiting Reactant and Theoretical Yield

We can obtain titanium metal from its oxide according to the following balanced equation:

$$TiO_2(s) + 2\,C(s) \longrightarrow Ti(s) + 2\,CO(g)$$

When 28.6 kg of C reacts with 88.2 kg of $TiO_2$, 42.8 kg of Ti is produced. Find the limiting reactant, theoretical yield (in kg), and percent yield.

| | |
|---|---|
| **SORT** You are given the mass of each reactant and the mass of product formed. You are asked to find the limiting reactant, theoretical yield, and percent yield. | **GIVEN:** 28.6 kg C, 88.2 kg $TiO_2$, 42.8 kg Ti produced<br>**FIND:** limiting reactant, theoretical yield, % yield |

**STRATEGIZE** Determine which of the reactants makes the least amount of product by converting from kilograms of each reactant to moles of product. Convert between grams and moles using molar mass. Convert between moles of reactant and moles of product using the stoichiometric relationships derived from the chemical equation. Remember that the reactant that makes the *least amount of product* is the limiting reactant.

Determine the theoretical yield (in kilograms) by converting the number of moles of product obtained with the limiting reactant to kilograms of product.

**CONCEPTUAL PLAN**

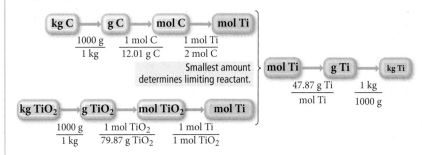

**RELATIONSHIPS USED**

1000 g = 1 kg

molar mass of C = 12.01 g/mol

molar mass of $TiO_2$ = 79.87 g/mol

1 mol $TiO_2$ : 1 mol Ti

2 mol C : 1 mol Ti

molar mass of Ti = 47.87 g/mol

**SOLVE** Beginning with the actual amount of each reactant, calculate the amount of product that can be made in moles. Convert the amount of product made by the limiting reactant to kilograms—this is the theoretical yield.

**SOLUTION**

$$28.6 \text{ kg C} \times \frac{1000\,\text{g}}{1\,\text{kg}} \times \frac{1\,\text{mol C}}{12.01\,\text{g C}} \times \frac{1\,\text{mol Ti}}{2\,\text{mol C}} = 1.1907 \times 10^3 \text{ mol Ti}$$

Limiting reactant

Least amount of product

$$88.2 \text{ kg TiO}_2 \times \frac{1000\,\text{g}}{1\,\text{kg}} \times \frac{1\,\text{mol TiO}_2}{79.87\,\text{g TiO}_2} \times \frac{1\,\text{mol Ti}}{1\,\text{mol TiO}_2} = 1.1043 \times 10^3 \text{ mol Ti}$$

$$1.1043 \times 10^3 \text{ mol Ti} \times \frac{47.87\,\text{g Ti}}{1\,\text{mol Ti}} \times \frac{1\,\text{kg}}{1000\,\text{g}} = 52.9 \text{ kg Ti}$$

Calculate the percent yield by dividing the actual yield (42.8 kg Ti) by the theoretical yield.

Since $TiO_2$ makes the least amount of product, it is the limiting reactant, and 52.9 kg Ti is the theoretical yield.

$$\% \text{ yield} = \frac{\text{actual yield}}{\text{theoretical yield}} \times 100\% = \frac{42.8\,\text{kg}}{52.9\,\text{kg}} \times 100\% = 80.9\%$$

**CHECK** The theoretical yield has the correct units (kg Ti) and has a reasonable magnitude compared to the mass of $TiO_2$. Since Ti has a lower molar mass than $TiO_2$, the amount of Ti made from $TiO_2$ should have a lower mass. The percent yield is reasonable (under 100% as it should be).

**FOR PRACTICE 4.4** Mining companies use this reaction to obtain iron from iron ore:

$$Fe_2O_3(s) + 3\,CO(g) \longrightarrow 2\,Fe(s) + 3\,CO_2(g)$$

The reaction of 167 g $Fe_2O_3$ with 85.8 g CO produces 72.3 g Fe. Determine the limiting reactant, theoretical yield, and percent yield.

**Reactant in Excess**  Nitrogen dioxide reacts with water to form nitric acid and nitrogen monoxide according to the equation:

$$3\,NO_2(g) + H_2O(l) \longrightarrow 2\,HNO_3(l) + NO(g)$$

Suppose that 5 mol $NO_2$ and 1 mol $H_2O$ combine and react completely. How many moles of the reactant in excess are present after the reaction has completed?

## 4.4  Solution Concentration and Solution Stoichiometry

The reactions that occur in lakes, streams, and oceans, as well as the reactions that occur in every cell within our bodies, take place in water. Chemical reactions involving reactants dissolved in water are among the most common and important. A homogeneous mixture of two substances—such as salt and water—is a **solution**. The majority component of the mixture is the **solvent**, and the minority component is the **solute**. An **aqueous solution** is one in which water acts as the solvent. In this section, we first examine how to quantify the concentration of a solution (the amount of solute relative to solvent) and then turn to applying the principles of stoichiometry, which we discussed in the previous section, to reactions occurring in solution.

### Solution Concentration

The amount of solute in a solution is variable. For example, we can add just a little salt to water to make a **dilute solution**, one that contains a small amount of solute relative to the solvent, or we can add a lot of salt to water to make a **concentrated solution**, one that contains a large amount of solute relative to the solvent (Figure 4.4▼). A common way to express solution concentration is **molarity (M)**, the amount of solute (in moles) divided by the volume of solution (in liters):

$$\text{Molarity (M)} = \frac{\text{amount of solute (in mol)}}{\text{volume of solution (in L)}}$$

Notice that molarity is a ratio of the amount of solute per liter of *solution,* not per liter of solvent. To make an aqueous solution of a specified molarity, we usually put the solute into a flask and then add water to reach the desired volume of solution. For example,

**Concentrated and Dilute Solutions**

Concentrated solution        Dilute solution

◄ **FIGURE 4.4  Concentrated and Dilute Solutions**   A concentrated solution contains a relatively large amount of solute relative to solvent. A dilute solution contains a relatively small amount of solute relative to solvent.

▶ FIGURE 4.5 **Preparing a**
**1 Molar NaCl Solution**

**Preparing a Solution of Specified Concentration**

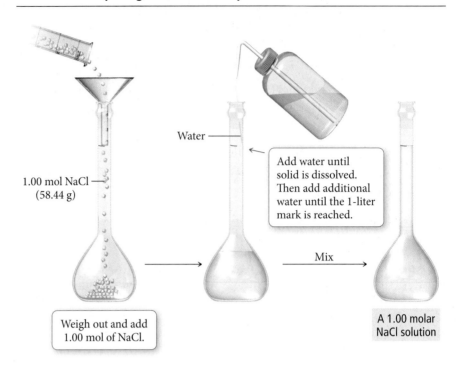

1.00 mol NaCl
(58.44 g)

Water

Add water until
solid is dissolved.
Then add additional
water until the 1-liter
mark is reached.

Mix

Weigh out and add
1.00 mol of NaCl.

A 1.00 molar
NaCl solution

to make 1 L of a 1 M NaCl solution, we add 1 mol of NaCl to a flask and then add enough
water to make 1 L of solution (Figure 4.5▲). We *do not* combine 1 mol of NaCl with 1 L
of water because the resulting solution would have a total volume different from 1 L
and therefore a molarity different than 1 M. To calculate molarity, divide the amount of
the solute in moles by the volume of the solution (solute *and* solvent) in liters, as shown
in Example 4.5.

**Example 4.5**    **Calculating Solution Concentration**

What is the molarity of a solution containing 25.5 g KBr dissolved in enough water to make 1.75 L of solution?

| SORT You are given the mass of KBr and the volume of a solution and asked to find its molarity. | GIVEN: 25.5 g KBr, 1.75 L of solution |
| --- | --- |
| | FIND: molarity (M) |

STRATEGIZE When formulating the conceptual plan, think about the definition of molarity, the amount of solute *in moles* per liter of solution.

You are given the mass of KBr, so first use the molar mass of KBr to convert from g KBr to mol KBr.

Then use the number of moles of KBr and liters of solution to find the molarity.

**CONCEPTUAL PLAN**

g KBr ⟶ mol KBr

$$\frac{1 \text{ mol}}{119.00 \text{ g}}$$

mol KBr, L solution ⟶ Molarity

$$\text{Molarity (M)} = \frac{\text{amount of solute (in mol)}}{\text{volume of solution (in L)}}$$

**RELATIONSHIPS USED**
molar mass of KBr = 119.00 g/mol

| SOLVE Follow the conceptual plan. Begin with g KBr and convert to mol KBr; then use mol KBr and L solution to calculate molarity. | SOLUTION<br><br>$25.5 \text{ g KBr} \times \dfrac{1 \text{ mol KBr}}{119.00 \text{ g KBr}} = 0.21\underline{4}29 \text{ mol KBr}$<br><br>$\text{molarity (M)} = \dfrac{\text{amount of solute (in mol)}}{\text{volume of solution (in L)}}$<br><br>$= \dfrac{0.21\underline{4}29 \text{ mol KBr}}{1.75 \text{ L solution}}$<br><br>$= 0.122 \text{ M}$ |
|---|---|

CHECK The units of the answer (M) are correct. The magnitude is reasonable since common solutions range in concentration from 0 to about 18 M. Concentrations significantly above 18 M are suspect and should be double-checked.

FOR PRACTICE 4.5    Calculate the molarity of a solution made by adding 45.4 g of $NaNO_3$ to a flask and dissolving it with water to create a total volume of 2.50 L.

FOR MORE PRACTICE 4.5    What mass of KBr (in grams) do you need to make 250.0 mL of a 1.50 M KBr solution?

## Using Molarity in Calculations

We can use the molarity of a solution as a conversion factor between moles of the solute and liters of the solution. For example, a 0.500 M NaCl solution contains 0.500 moles NaCl for every liter of solution:

$$\dfrac{0.500 \text{ mol NaCl}}{\text{L solution}} \quad converts \quad \boxed{\text{L solution}} \longrightarrow \boxed{\text{mol NaCl}}$$

This conversion factor converts from liters solution to mole NaCl. If we want to convert in the other direction, we invert the conversion factor:

$$\dfrac{\text{L solution}}{0.500 \text{ mol NaCl}} \quad converts \quad \boxed{\text{mol NaCl}} \longrightarrow \boxed{\text{L solution}}$$

Example 4.6 illustrates how to use molarity in this way.

**Example** 4.6    **Using Molarity in Calculations**

How many liters of a 0.125 M NaOH solution contain 0.255 mol of NaOH?

| SORT You are given the concentration of a NaOH solution. You are asked to find the volume of the solution that contains a given amount (in moles) of NaOH. | GIVEN: 0.125 M NaOH solution, 0.255 mol NaOH<br><br>FIND: volume of NaOH solution (in L) |
|---|---|
| STRATEGIZE The conceptual plan begins with mol NaOH and shows the conversion to L of solution using molarity as a conversion factor. | CONCEPTUAL PLAN<br><br>$\boxed{\text{mol NaOH}} \longrightarrow \boxed{\text{L solution}}$<br><br>$\dfrac{1 \text{ L solution}}{0.125 \text{ mol NaOH}}$<br><br>RELATIONSHIPS USED  $0.125 \text{ M NaOH} = \dfrac{0.125 \text{ mol NaOH}}{1 \text{ L solution}}$ |

—Continued on the next page

*—Continued*

| SOLVE Follow the conceptual plan. Begin with mol NaOH and convert to L solution. | SOLUTION $$0.255 \text{ mol NaOH} \times \frac{1 \text{ L solution}}{0.125 \text{ mol NaOH}} = 2.04 \text{ L solution}$$ |
|---|---|

CHECK The units of the answer (L) are correct. The magnitude is reasonable because the solution contains 0.125 mol per liter. Therefore, roughly 2 L contains the given amount of moles (0.255 mol).

FOR PRACTICE 4.6   How many grams of sucrose ($C_{12}H_{22}O_{11}$) are in 1.55 L of 0.758 M sucrose solution?

FOR MORE PRACTICE 4.6   How many mL of a 0.155 M KCl solution contain 2.55 g KCl?

CONCEPTUAL
CONNECTION **4.5**

**Solutions** If we dissolve 25 g of salt in 251 g of water, what is the mass of the resulting solution?

    **(a)** 251 g      **(b)** 276 g      **(c)** 226 g

> When diluting acids, always add the concentrated acid to the water. Never add water to concentrated acid solutions, as the heat generated may cause the concentrated acid to splatter and burn your skin.

## Solution Dilution

To save space in storerooms, laboratories often store solutions in concentrated forms called **stock solutions**. For example, hydrochloric acid is frequently stored as a 12 M stock solution. However, many lab procedures call for much less concentrated hydrochloric acid solutions, so we must dilute the stock solution to the required concentration. How do we know how much of the stock solution to use? The easiest way to solve dilution problems is to use the following dilution equation:

$$M_1V_1 = M_2V_2 \qquad [4.1]$$

where $M_1$ and $V_1$ are the molarity and volume of the initial concentrated solution, and $M_2$ and $V_2$ are the molarity and volume of the final diluted solution. This equation works because the molarity multiplied by the volume gives the number of moles of solute, which is the same in both solutions.

$$M_1V_1 = M_2V_2$$

$$\text{mol}_1 = \text{mol}_2$$

In other words, the number of moles of solute does not change when we dilute a solution.

For example, suppose a laboratory procedure calls for 3.00 L of a 0.500 M $CaCl_2$ solution. How should we prepare this solution from a 10.0 M stock solution? We solve Equation 4.1 for $V_1$, the volume of the stock solution required for the dilution, and then substitute in the correct values to calculate it.

$$M_1V_1 = M_2V_2$$

$$V_1 = \frac{M_2V_2}{M_1}$$

$$= \frac{0.500 \text{ mol/L} \times 3.00 \text{ L}}{10.0 \text{ mol/L}}$$

$$= 0.150 \text{ L}$$

Consequently, we make the solution by diluting 0.150 L of the stock solution to a total volume of 3.00 L ($V_2$). The resulting solution will be 0.500 M in $CaCl_2$ (Figure 4.6▶).

### Diluting a Solution

Measure 0.150 L of 10.0 M stock solution.

Dilute with water to total volume of 3.00 L.

0.150 L of 10.0 M stock solution

0.500 M CaCl$_2$

$$M_1 V_1 = M_2 V_2$$

$$\frac{10.0 \text{ mol}}{\cancel{L}} \times 0.150 \, \cancel{L} = \frac{0.500 \text{ mol}}{\cancel{L}} \times 3.00 \, \cancel{L}$$

$$1.50 \text{ mol} = 1.50 \text{ mol}$$

◀ **FIGURE 4.6  Preparing 3.00 L of 0.500 M CaCl$_2$ from a 10.0 M Stock Solution**

---

## Example 4.7    Solution Dilution

To what volume should you dilute 0.200 L of a 15.0 M NaOH solution to obtain a 3.00 M NaOH solution?

| | |
|---|---|
| **SORT**  You are given the initial volume, initial concentration, and final concentration of a solution. You need to determine the final volume. | **GIVEN:**<br>$V_1 = 0.200 \, \text{L}$<br>$M_1 = 15.0 \, \text{M}$<br>$M_2 = 3.00 \, \text{M}$<br><br>**FIND:** $V_2$ |
| **STRATEGIZE**  Equation 4.1 relates the initial and final volumes and concentrations for solution dilution problems. You are asked to find $V_2$. The other quantities ($V_1$, $M_1$, and $M_2$)  are all given in the problem. | **CONCEPTUAL PLAN**<br><br>$\boxed{V_1, M_1, M_2} \longrightarrow \boxed{V_2}$<br>$\qquad\qquad M_1 V_1 = M_2 V_2$<br><br>**RELATIONSHIPS USED**<br>$M_1 V_1 = M_2 V_2$ |

—Continued on the next page

*—Continued*

| **SOLVE** Begin with the solution dilution equation and solve it for $V_2$.<br><br>Substitute in the required quantities and calculate $V_2$.<br><br>Make the solution by diluting 0.200 L of the stock solution to a total volume of 1.00 L ($V_2$). The resulting solution will have a concentration of 3.00 M. | **SOLUTION** $M_1V_1 = M_2V_2$<br><br>$V_2 = \dfrac{M_1V_1}{M_2}$<br><br>$= \dfrac{15.0 \text{ mol/L} \times 0.200 \text{ L}}{3.00 \text{ mol/L}}$<br><br>$= 1.00 \text{ L}$ |
|---|---|

**CHECK** The final units (L) are correct. The magnitude of the answer is reasonable because the solution is diluted from 15.0 M to 3.00 M, a factor of five. Therefore, the volume should increase by a factor of five.

**FOR PRACTICE 4.7** To what volume (in mL) should you dilute 100.0 mL of a 5.00 M $CaCl_2$ solution to obtain a 0.750 M $CaCl_2$ solution?

**FOR MORE PRACTICE 4.7** What volume of a 6.00 M $NaNO_3$ solution should you use to make 0.525 L of a 1.20 M $NaNO_3$ solution?

---

CONCEPTUAL
**CONNECTION 4.6**

**Solution Dilution** The image shown at left represents a small volume within 500 mL of aqueous ethanol ($CH_3CH_2OH$) solution. (The water molecules have been omitted for clarity.)

Which image below best represents the same volume of the solution after we add an additional 500 mL of water?

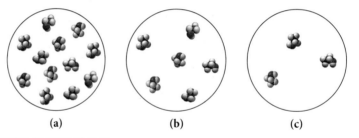

(a)          (b)          (c)

---

## Solution Stoichiometry

In Section 4.2 we discussed how we can use the coefficients in chemical equations as conversion factors between the amounts of reactants (in moles) and the amounts of products (in moles). In aqueous reactions, quantities of reactants and products are often specified in terms of volumes and concentrations. We can use the volume and concentration of a reactant or product to calculate its amount in moles. We can then use the stoichiometric coefficients in the chemical equation to convert to the amount of another reactant or product in moles. The general conceptual plan for these kinds of calculations begins with the volume of a reactant or product:

We make the conversions between solution volumes and amounts of solute in moles using the molarities of the solutions. We make the conversions between amounts in moles of A and B using the stoichiometric coefficients from the balanced chemical equation. Example 4.8 demonstrates solution stoichiometry.

**Example** 4.8 | Solution Stoichiometry

What volume (in L) of 0.150 M KCl solution will completely react with 0.150 L of a 0.175 M Pb(NO₃)₂ solution according to the following balanced chemical equation?

$$2\ KCl(aq) + Pb(NO_3)_2(aq) \longrightarrow PbCl_2(s) + 2\ KNO_3(aq)$$

| | |
|---|---|
| **SORT** You are given the volume and concentration of a Pb(NO₃)₂ solution. You are asked to find the volume of KCl solution (of a given concentration) required to react with it. | **GIVEN:** 0.150 L of Pb(NO₃)₂ solution, 0.175 M Pb(NO₃)₂ solution, 0.150 M KCl solution<br><br>**FIND:** volume KCl solution (in L) |
| **STRATEGIZE** The conceptual plan has the form: volume A → amount A (in moles) → amount B (in moles) → volume B. Use the molar concentrations of the KCl and Pb(NO₃)₂ solutions as conversion factors between the number of moles of reactants in these solutions and their volumes. Use the stoichiometric coefficients from the balanced equation to convert between number of moles of Pb(NO₃)₂ and number of moles of KCl. | **CONCEPTUAL PLAN**<br><br>**RELATIONSHIPS USED**<br>$M\ Pb(NO_3)_2 = \dfrac{0.175\ mol\ Pb(NO_3)_2}{1\ L\ Pb(NO_3)_2\ solution}$<br>$2\ mol\ KCl : 1\ mol\ Pb(NO_3)_2$<br>$M\ KCl = \dfrac{0.150\ mol\ KCl}{1\ L\ KCl\ solution}$ |
| **SOLVE** Begin with L Pb(NO₃)₂ solution and follow the conceptual plan to arrive at L KCl solution. | **SOLUTION**<br>$0.150\ L\ Pb(NO_3)_2\ solution \times \dfrac{0.175\ mol\ Pb(NO_3)_2}{1\ L\ Pb(NO_3)_2\ solution}$<br>$\times \dfrac{2\ mol\ KCl}{1\ mol\ Pb(NO_3)_2} \times \dfrac{1\ L\ KCl\ solution}{0.150\ mol\ KCl} = 0.350\ L\ KCl\ solution$ |

**CHECK** The final units (L KCl solution) are correct. The magnitude (0.350 L) is reasonable because the reaction stoichiometry requires 2 mol of KCl per mole of Pb(NO₃)₂. Since the concentrations of the two solutions are not very different (0.150 M compared to 0.175 M), the volume of KCl required is roughly two times the 0.150 L of Pb(NO₃)₂ given in the problem.

**FOR PRACTICE 4.8** What volume (in mL) of a 0.150 M HNO₃ solution will completely react with 35.7 mL of a 0.108 M Na₂CO₃ solution according to the following balanced chemical equation?

$$Na_2CO_3(aq) + 2\ HNO_3(aq) \longrightarrow 2\ NaNO_3(aq) + CO_2(g) + H_2O(l)$$

**FOR MORE PRACTICE 4.8** In the previous reaction, what mass (in grams) of carbon dioxide forms?

**Solution Stoichiometry** Consider the reaction:

$$2\ A(aq) + B(aq) \longrightarrow C(aq)$$

What is the limiting reactant if you mix equal volumes of a 1 M solution of A and a 1 M solution of B?

CONCEPTUAL
**CONNECTION** 4.7

## Solute and Solvent Interactions

▲ **FIGURE 4.7 Solute and Solvent Interactions** When a solid is put into a solvent, the interactions between solvent and solute particles compete with the interactions between the solute particles themselves.

▲ **FIGURE 4.8 Charge Distribution in a Water Molecule** An uneven distribution of electrons within the water molecule causes the oxygen side of the molecule to have a partial negative charge and the hydrogen side to have a partial positive charge.

## 4.5 Types of Aqueous Solutions and Solubility

Consider two familiar aqueous solutions: salt water and sugar water. Salt water is a homogeneous mixture of NaCl and $H_2O$, and sugar water is a homogeneous mixture of $C_{12}H_{22}O_{11}$ and $H_2O$. You may have made these solutions yourself by adding table salt or sugar to water. As you stir either of these two substances into the water, the substance seems to disappear. However, you know that the original substance is still present because the mixture tastes salty or sweet. How do solids such as salt and sugar dissolve in water?

When a solid is put into a liquid solvent, the attractive forces that hold the solid together (the solute–solute interactions) compete with the attractive forces between the solvent molecules and the particles that compose the solid (the solvent–solute interactions), as shown in Figure 4.7◀. For example, when sodium chloride is put into water, there is a competition between the attraction of $Na^+$ cations and $Cl^-$ anions to each other (due to their opposite charges) and the attraction of $Na^+$ and $Cl^-$ to water molecules. The attraction of $Na^+$ and $Cl^-$ to water is based on the *polar nature* of the water molecule. For reasons we discuss later in this book (Section 9.6), the oxygen atom in water is electron-rich, giving it a partial negative charge ($\delta^-$), as shown in Figure 4.8◀. The hydrogen atoms, in contrast, are electron-poor, giving them a partial positive charge ($\delta^+$). As a result, the positively charged sodium ions are strongly attracted to the oxygen side of the water molecule (which has a partial negative charge), and the negatively charged chloride ions are attracted to the hydrogen side of the water molecule (which has a partial positive charge), as shown in Figure 4.9▼. In the case of NaCl, the attraction between the separated ions and the water molecules overcomes the attraction of sodium and chloride ions to each other, and the sodium chloride dissolves in the water (Figure 4.10▼).

## Electrolyte and Nonelectrolyte Solutions

As Figure 4.11▶ illustrates, a salt solution conducts electricity while a sugar solution does not. The difference between the ways that salt (an ionic compound) and sugar (a molecular compound) dissolve in water illustrates a fundamental difference between types of

## Interactions in a Sodium Chloride Solution

▲ **FIGURE 4.9 Solute and Solvent Interactions in a Sodium Chloride Solution** When sodium chloride is put into water, the attraction of $Na^+$ and $Cl^-$ ions to water molecules competes with the attraction between the oppositely charged ions themselves.

## Dissolution of an Ionic Compound

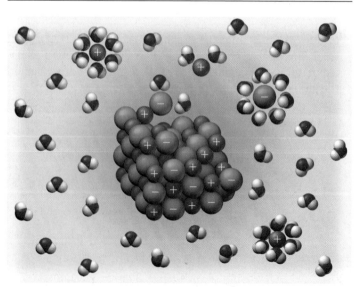

▲ **FIGURE 4.10 Sodium Chloride Dissolving in Water** The attraction between water molecules and the ions of sodium chloride causes NaCl to dissolve in the water.

## Electrolyte and Nonelectrolyte Solutions

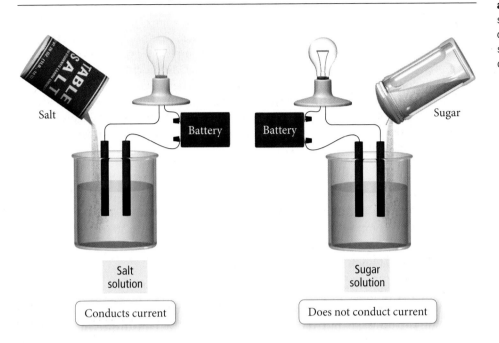

solutions. Ionic compounds such as the sodium chloride in the previous example dissociate into their component ions when they dissolve in water. An NaCl solution, represented as $NaCl(aq)$, does not contain any NaCl units, but rather dissolved $Na^+$ ions and $Cl^-$ ions. The dissolved ions act as charge carriers, allowing the solution to conduct electricity. Substances that dissolve in water to form solutions that conduct electricity are **electrolytes**. Substances such as sodium chloride that completely dissociate into ions when they dissolve in water are **strong electrolytes**, and the resulting solutions are *strong electrolyte solutions*.

In contrast to sodium chloride, sugar is a molecular compound. Most molecular compounds—with the important exception of acids, which we discuss shortly—dissolve in water as intact molecules. Sugar dissolves because the attraction between sugar molecules and water molecules (shown in Figure 4.12▼) overcomes the attraction of sugar molecules to each other (Figure 4.13▼). So unlike a sodium chloride solution

$NaCl(aq)$

Strong electrolyte

## Interactions between Sugar and Water Molecules

▲ **FIGURE 4.12** **Sugar and Water Interactions** Partial charges on sugar molecules and water molecules (which we will discuss more fully in Chapter 11) result in attractions between the sugar molecules and water molecules.

## Sugar Solution

▲ **FIGURE 4.13** **A Sugar Solution** Sugar dissolves because the attractions between sugar molecules and water molecules, which both contain a distribution of electrons that results in partial positive and partial negative charges, overcome the attractions between sugar molecules to each other.

HCl(*aq*)

Strong acid

Unlike soluble ionic compounds, which contain ions and therefore *dissociate* in water, acids are molecular compounds that *ionize* in water.

HC$_2$H$_3$O$_2$(*aq*)

Weak acid

(which is composed of dissociated ions), a sugar solution is composed of intact C$_{12}$H$_{22}$O$_{11}$ molecules homogeneously mixed with the water molecules. Compounds such as sugar that do not dissociate into ions when dissolved in water are called **nonelectrolytes**, and the resulting solutions—called *nonelectrolyte solutions*—do not conduct electricity.

Acids, which we introduced in Section 3.6, are molecular compounds, but they do ionize—form ions—when they dissolve in water. Hydrochloric acid (HCl) is a molecular compound that ionizes into H$^+$ and Cl$^-$ when it dissolves in water. HCl is an example of a **strong acid**, one that completely ionizes in solution. Since strong acids completely ionize in solution, they are also strong electrolytes. We represent the complete ionization of a strong acid with a single reaction arrow between the acid and its ionized form:

$$HCl(aq) \longrightarrow H^+(aq) + Cl^-(aq)$$

Many acids are **weak acids**; they do not completely ionize in water. For example, acetic acid (HC$_2$H$_3$O$_2$), the acid in vinegar, is a weak acid. A solution of a weak acid is composed mostly of the nonionized acid—only a small percentage of the acid molecules ionize. We represent the partial ionization of a weak acid with opposing half arrows between the reactants and products:

$$HC_2H_3O_2(aq) \rightleftharpoons H^+(aq) + C_2H_3O_2^-(aq)$$

Weak acids are **weak electrolytes**, and the resulting solutions—called *weak electrolyte solutions*—conduct electricity only weakly. Figure 4.14▼ summarizes the electrolytic properties of solutions.

## The Solubility of Ionic Compounds

As we have just discussed, when an ionic compound dissolves in water, the resulting solution contains not the intact ionic compound itself, but its component ions dissolved in water. However, not all ionic compounds dissolve in water. If we add AgCl to

▼ **FIGURE 4.14 Electrolytic Properties of Solutions**

### Electrolytic Properties of Solutions

C$_{12}$H$_{22}$O$_{11}$(*aq*)

HC$_2$H$_3$O$_2$(*aq*)

NaCl(*aq*)

Nonelectrolyte

Weak electrolyte

Strong electrolyte

water, for example, it remains solid and appears as a white powder at the bottom of the water.

In general, a compound is termed **soluble** if it dissolves in water and **insoluble** if it does not. However, these classifications are a bit of an oversimplification. (In reality, solubility is a continuum and even "insoluble" compounds dissolve to some extent, though usually orders of magnitude less than soluble compounds.) For example, silver nitrate is soluble. If we mix solid $AgNO_3$ with water, it dissolves and forms a strong electrolyte solution. Silver chloride, as we just saw, is almost completely insoluble. If we mix solid AgCl with water, virtually all of it remains as a solid within the liquid water.

Whether a particular compound is soluble or insoluble depends on several factors. In Section 12.3, we will examine more closely the energy changes associated with solution formation. For now, we can follow a set of empirical rules that chemists have inferred from observations on many ionic compounds. Table 4.1 summarizes these *solubility rules*.

The solubility rules state that compounds containing the sodium ion are soluble. That means that compounds such as NaBr, $NaNO_3$, $Na_2SO_4$, NaOH, and $Na_2CO_3$ all dissolve in water to form strong electrolyte solutions. Similarly, the solubility rules state that compounds containing the $NO_3^-$ ion are soluble. That means that compounds such as $AgNO_3$, $Pb(NO_3)_2$, $NaNO_3$, $Ca(NO_3)_2$, and $Sr(NO_3)_2$ all dissolve in water to form strong electrolyte solutions.

Notice that when compounds containing polyatomic ions such as $NO_3^-$ dissolve, the polyatomic ions dissolve as intact units.

The solubility rules also state that, with some exceptions, compounds containing the $CO_3^{2-}$ ion are insoluble. Therefore, compounds such as $CuCO_3$, $CaCO_3$, $SrCO_3$, and $FeCO_3$ do not dissolve in water. Note that the solubility rules contain many exceptions. For example, compounds containing $CO_3^{2-}$ *are* soluble when paired with $Li^+$, $Na^+$, $K^+$, or $NH_4^+$. Thus $Li_2CO_3$, $Na_2CO_3$, $K_2CO_3$, and $(NH_4)_2CO_3$ are all soluble.

AgCl(s)    AgNO₃(aq)
Insoluble    Soluble

▲ AgCl does not dissolve in water; it remains as a white powder at the bottom of the beaker.

| TABLE 4.1 Solubility Rules for Ionic Compounds in Water | |
|---|---|
| **Compounds Containing the Following Ions Are Generally Soluble** | **Exceptions** |
| $Li^+$, $Na^+$, $K^+$, and $NH_4^+$ | None |
| $NO_3^-$ and $C_2H_3O_2^-$ | None |
| $Cl^-$, $Br^-$, and $I^-$ | When these ions pair with $Ag^+$, $Hg_2^{2+}$, or $Pb^{2+}$, the resulting compounds are insoluble. |
| $SO_4^{2-}$ | When $SO_4^{2-}$ pairs with $Sr^{2+}$, $Ba^{2+}$, $Pb^{2+}$, $Ag^+$, or $Ca^{2+}$, the resulting compound is insoluble. |
| **Compounds Containing the Following Ions Are Generally Insoluble** | **Exceptions** |
| $OH^-$ and $S^{2-}$ | When these ions pair with $Li^+$, $Na^+$, $K^+$, or $NH_4^+$, the resulting compounds are soluble. |
| | When $S^{2-}$ pairs with $Ca^{2+}$, $Sr^{2+}$, or $Ba^{2+}$, the resulting compound is soluble. |
| | When $OH^-$ pairs with $Ca^{2+}$, $Sr^{2+}$, or $Ba^{2+}$, the resulting compound is slightly soluble. |
| $CO_3^{2-}$ and $PO_4^{3-}$ | When these ions pair with $Li^+$, $Na^+$, $K^+$, or $NH_4^+$, the resulting compounds are soluble. |

**Example 4.9** | **Predicting Whether an Ionic Compound Is Soluble**

Predict whether each compound is soluble or insoluble.

**(a)** $PbCl_2$    **(b)** $CuCl_2$    **(c)** $Ca(NO_3)_2$    **(d)** $BaSO_4$

**SOLUTION**

**(a)** Insoluble. Compounds containing $Cl^-$ are normally soluble, but $Pb^{2+}$ is an exception.

**(b)** Soluble. Compounds containing $Cl^-$ are normally soluble and $Cu^{2+}$ is not an exception.

**(c)** Soluble. Compounds containing $NO_3^-$ are always soluble.

**(d)** Insoluble. Compounds containing $SO_4^{2-}$ are normally soluble, but $Ba^{2+}$ is an exception.

**FOR PRACTICE 4.9**   Predict whether each compound is soluble or insoluble.

**(a)** $NiS$    **(b)** $Mg_3(PO_4)_2$    **(c)** $Li_2CO_3$    **(d)** $NH_4Cl$

 **Reactions in Solution**

# 4.6 Precipitation Reactions

Have you ever taken a bath in hard water? Hard water contains dissolved ions such as $Ca^{2+}$ and $Mg^{2+}$ that diminish the effectiveness of soap. These ions react with soap to form a gray curd that may appear as a "bathtub ring" after you drain the tub. Hard water is particularly troublesome when washing clothes. Consider how your white shirt would look covered with the gray curd from the bathtub and you can understand the problem. Consequently, most laundry detergents include substances designed to remove $Ca^{2+}$ and $Mg^{2+}$ from the laundry mixture. The most common substance used for this purpose is sodium carbonate, which dissolves in water to form sodium cations ($Na^+$) and carbonate ($CO_3^{2-}$) anions:

$$Na_2CO_3(aq) \longrightarrow 2\,Na^+(aq) + CO_3^{2-}(aq)$$

Sodium carbonate is soluble, but calcium carbonate and magnesium carbonate are not (see the solubility rules in Table 4.1). Consequently, the carbonate anions react with dissolved $Mg^{2+}$ and $Ca^{2+}$ ions in hard water to form solids that *precipitate* from (or come out of) solution:

$$Mg^{2+}(aq) + CO_3^{2-}(aq) \longrightarrow MgCO_3(s)$$
$$Ca^{2+}(aq) + CO_3^{2-}(aq) \longrightarrow CaCO_3(s)$$

▲ The reaction of ions in hard water with soap produces a gray curd that is visible after you drain the bathwater.

The precipitation of these ions prevents their reaction with the soap, eliminating curd and preventing white shirts from turning gray.

The reactions between $CO_3^{2-}$ and $Mg^{2+}$ and $Ca^{2+}$ are examples of **precipitation reactions**, reactions in which a solid or **precipitate** forms when we mix two solutions. Precipitation reactions are common in chemistry. As another example, consider potassium iodide and lead(II) nitrate, which form colorless, strong electrolyte solutions when dissolved in water separately. When the two solutions are combined, however, a brilliant yellow precipitate forms (Figure 4.15▶). We describe this precipitation reaction with the following chemical equation:

$$2\,KI(aq) + Pb(NO_3)_2(aq) \longrightarrow PbI_2(s) + 2\,KNO_3(aq)$$

Precipitation reactions do not always occur when two aqueous solutions are mixed. For example, if we combine solutions of $KI(aq)$ and $NaCl(aq)$, nothing happens (Figure 4.16▶, on page 164):

$$KI(aq) + NaCl(aq) \longrightarrow NO\ REACTION$$

The key to predicting precipitation reactions is to understand that *only insoluble compounds form precipitates*. In a precipitation reaction, two solutions containing soluble compounds combine and an insoluble compound precipitates. Consider the precipitation reaction described previously:

$$\underset{\text{soluble}}{2\,KI(aq)} + \underset{\text{soluble}}{Pb(NO_3)_2(aq)} \longrightarrow \underset{\text{insoluble}}{PbI_2(s)} + \underset{\text{soluble}}{2\,KNO_3(aq)}$$

## Precipitation Reaction

$$2 \text{ KI}(aq) + \text{Pb(NO}_3)_2(aq) \longrightarrow \text{PbI}_2(s) + 2 \text{ KNO}_3(aq)$$
(soluble)     (soluble)         (insoluble)    (soluble)

When a potassium iodide solution is mixed with a lead(II) nitrate solution, a yellow lead(II) iodide precipitate forms.

2 KI(*aq*)
(soluble)

+

Pb(NO$_3$)$_2$(*aq*)
(soluble)

2 KNO$_3$(*aq*)
(soluble)

+

PbI$_2$(*s*)
(insoluble)

KI and Pb(NO$_3$)$_2$ are both soluble, but the precipitate, PbI$_2$, is insoluble. Before mixing, KI(*aq*) and Pb(NO$_3$)$_2$(*aq*) are both dissociated in their respective solutions:

The instant that the solutions come into contact, all four ions are present:

Now, new compounds—one or both of which might be insoluble—are possible. Specifically, the cation from either compound can pair with the anion from the other to form possibly insoluble products:

If the possible products are both soluble, no reaction occurs and no precipitate forms. If one or both of the possible products are insoluble, a precipitation reaction occurs. In this case, $KNO_3$ is soluble, but $PbI_2$ is insoluble. Consequently, $PbI_2$ precipitates.

To predict whether a precipitation reaction will occur when two solutions are mixed and to write an equation for the reaction, we use the procedure that follows. The steps are outlined in the left column, and two examples illustrating how to apply the procedure are shown in the center and right columns.

$PbI_2(s)$ and $KNO_3(aq)$

## No Reaction

KI(aq)

+

NaCl(aq)

No reaction

▶ **FIGURE 4.16 No Precipitation** When a potassium iodide solution is mixed with a sodium chloride solution, no reaction occurs.

## Procedure for...

| Writing Equations for Precipitation Reactions | **Example 4.10**<br>**Writing Equations for Precipitation Reactions**<br>Write an equation for the precipitation reaction that occurs (if any) when solutions of potassium carbonate and nickel(II) chloride are mixed. | **Example 4.11**<br>**Writing Equations for Precipitation Reactions**<br>Write an equation for the precipitation reaction that occurs (if any) when solutions of sodium nitrate and lithium sulfate are mixed. |
|---|---|---|
| 1. Write the formulas of the two compounds being mixed as reactants in a chemical equation. | $K_2CO_3(aq) + NiCl_2(aq) \longrightarrow$ | $NaNO_3(aq) + Li_2SO_4(aq) \longrightarrow$ |
| 2. Below the equation, write the formulas of the products that could form from the reactants. Obtain these by combining the cation from each reactant with the anion from the other. Make sure to write correct formulas for these ionic compounds using the procedure demonstrated in Section 3.5. | $K_2CO_3(aq) + NiCl_2(aq) \longrightarrow$<br><br>Possible products<br>KCl        $NiCO_3$ | $NaNO_3(aq) + Li_2SO_4(aq) \longrightarrow$<br><br>Possible products<br>$LiNO_3$    $Na_2SO_4$ |
| 3. Refer to the solubility rules to determine whether any of the possible products are insoluble. | KCl is soluble. (Compounds containing $Cl^-$ are usually soluble and $K^+$ is not an exception.)<br><br>$NiCO_3$ is insoluble. (Compounds containing $CO_3^{2-}$ are usually insoluble and $Ni^{2+}$ is not an exception.) | $LiNO_3$ is soluble. (Compounds containing $NO_3^-$ are soluble and $Li^+$ is not an exception.)<br><br>$Na_2SO_4$ is soluble. (Compounds containing $SO_4^{2-}$ are generally soluble and $Na^+$ is not an exception.) |
| 4. If all of the possible products are soluble, there is no precipitate. Write "NO REACTION" after the arrow. | Since this example has an insoluble product, we proceed to the next step. | Since this example has no insoluble product, there is no reaction.<br><br>$NaNO_3(aq) + Li_2SO_4(aq) \longrightarrow$<br>          NO REACTION |
| 5. If any of the possible products are insoluble, write their formulas as the products of the reaction, using (s) to indicate solid. Include an (aq) to indicate aqueous after any soluble products. | $K_2CO_3(aq) + NiCl_2(aq) \longrightarrow$<br>          $NiCO_3(s) + KCl(aq)$ | |
| 6. Balance the equation. Remember to adjust only coefficients, not subscripts. | $K_2CO_3(aq) + NiCl_2(aq) \longrightarrow$<br>          $NiCO_3(s) + 2\,KCl(aq)$ | |
| | **FOR PRACTICE 4.10** Write an equation for the precipitation reaction that occurs (if any) when solutions of ammonium chloride and iron(III) nitrate mix. | **FOR PRACTICE 4.11** Write an equation for the precipitation reaction that occurs (if any) when solutions of sodium hydroxide and copper(II) bromide mix. |

**Precipitation Reactions** Consider the generic ionic compounds with the formulas AX and BY and the following solubility rules:

AX soluble; BY soluble; AY soluble; BX insoluble

Let circles represent $A^+$ ions; squares represent $B^+$ ions; triangles represent $X^-$ ions; and diamonds represent $Y^-$ ions. We represent solutions of the two compounds (AX and BY) as follows:

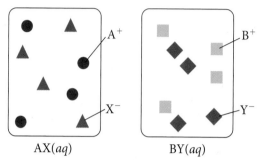

Draw a molecular-level representation showing the result of mixing the two solutions and write an equation to represent the reaction.

## 4.7  Representing Aqueous Reactions: Molecular, Ionic, and Complete Ionic Equations

Consider the following equation for a precipitation reaction:

$$Pb(NO_3)_2(aq) + 2\ KCl(aq) \longrightarrow PbCl_2(s) + 2\ KNO_3(aq)$$

This equation is a **molecular equation**, an equation showing the complete neutral formulas for each compound in the reaction as if they existed as molecules. In actual solutions of soluble ionic compounds, dissolved substances are present as ions. We can write equations for reactions occurring in aqueous solution in a way that better shows the dissociated nature of dissolved ionic compounds. For example, we can rewrite the previous equation as:

$$Pb^{2+}(aq) + 2\ NO_3^-(aq) + 2\ K^+(aq) + 2\ Cl^-(aq) \longrightarrow PbCl_2(s) + 2\ K^+(aq) + 2\ NO_3^-(aq)$$

Equations such as this, which list all of the ions present as either reactants or products in a chemical reaction, are **complete ionic equations**. Strong electrolytes are always represented as their component ions in ionic equations—weak electrolytes are not.

Notice that in the complete ionic equation, some of the ions in solution appear unchanged on both sides of the equation. These ions are called **spectator ions** because they do not participate in the reaction.

$$Pb^{2+}(aq) + 2\ NO_3^-(aq) + 2\ K^+(aq) + 2\ Cl^-(aq) \longrightarrow$$

$$PbCl_2(s) + 2\ K^+(aq) + 2\ NO_3^-(aq)$$

Spectator ions

To simplify the equation and to show more clearly what is happening, we can omit spectator ions:

$$Pb^{2+}(aq) + 2\ Cl^-(aq) \longrightarrow PbCl_2(s)$$

Equations that show only the species that actually change during the reaction are **net ionic equations**.

As another example, consider the reaction between $HCl(aq)$ and $KOH(aq)$:

$$HCl(aq) + KOH(aq) \longrightarrow H_2O(l) + KCl(aq)$$

Since HCl, KOH, and KCl all exist in solution primarily as independent ions, the complete ionic equation is:

$$H^+(aq) + Cl^-(aq) + K^+(aq) + OH^-(aq) \longrightarrow H_2O(l) + K^+(aq) + Cl^-(aq)$$

To write the net ionic equation, we remove the spectator ions, those that are unchanged on both sides of the equation:

$$H^+(aq) + \cancel{Cl^-(aq)} + \cancel{K^+(aq)} + OH^-(aq) \longrightarrow H_2O(l) + \cancel{K^+(aq)} + \cancel{Cl^-(aq)}$$

Spectator ions

The net ionic equation is $H^+(aq) + OH^-(aq) \longrightarrow H_2O(l)$.

*Summarizing Aqueous Equations:*

▶ A **molecular equation** is a chemical equation showing the complete, neutral formulas for every compound in a reaction.

▶ A **complete ionic equation** is a chemical equation showing all of the species as they are actually present in solution.

▶ A **net ionic equation** is an equation showing only the species that actually change during the reaction.

---

**Example 4.12   Writing Complete Ionic and Net Ionic Equations**

Write complete ionic and net ionic equations for each reaction.

**(a)** $3 SrCl_2(aq) + 2 Li_3PO_4(aq) \longrightarrow Sr_3(PO_4)_2(s) + 6 LiCl(aq)$

**(b)** $HC_2H_3O_2(aq) + KOH(aq) \longrightarrow H_2O(l) + KC_2H_3O_2(aq)$

| SOLUTION | |
|---|---|
| **(a)** Write the complete ionic equation by separating aqueous ionic compounds into their constituent ions. The $Sr_3(PO_4)_2(s)$, precipitating as a solid, remains as one unit. | **Complete ionic equation:**<br>$3 Sr^{2+}(aq) + 6 Cl^-(aq) + 6 Li^+(aq) + 2 PO_4{}^{3-}(aq) \longrightarrow$<br>$\qquad\qquad Sr_3(PO_4)_2(s) + 6 Li^+(aq) + 6 Cl^-(aq)$ |
| Write the net ionic equation by eliminating the spectator ions, those that do not change from one side of the reaction to the other. | **Net ionic equation:**<br>$3 Sr^{2+}(aq) + 2 PO_4{}^{3-}(aq) \longrightarrow Sr_3(PO_4)_2(s)$ |
| **(b)** Write the complete ionic equation by separating aqueous ionic compounds into their constituent ions. Do not separate $HC_2H_3O_2(aq)$ because it is a weak electrolyte. | **Complete ionic equation:**<br>$HC_2H_3O_2(aq) + K^+(aq) + OH^-(aq) \longrightarrow$<br>$\qquad\qquad H_2O(l) + K^+(aq) + C_2H_3O_2{}^-(aq)$ |
| Write the net ionic equation by eliminating the spectator ions. | **Net ionic equation:**<br>$HC_2H_3O_2(aq) + OH^-(aq) \longrightarrow H_2O(l) + C_2H_3O_2{}^-(aq)$ |

**FOR PRACTICE 4.12**   Write the complete ionic equation and net ionic equation for the following reaction:

$$2 HI(aq) + Ba(OH)_2(aq) \longrightarrow 2 H_2O(l) + BaI_2(aq)$$

---

## 4.8   Acid–Base and Gas-Evolution Reactions

Two other important classes of reactions that occur in aqueous solution are acid–base reactions and gas-evolution reactions. In an **acid–base reaction** (also called a **neutralization reaction**), an acid reacts with a base and the two neutralize each other, producing water (or in some cases a weak electrolyte). In a **gas-evolution reaction**, a gas forms, resulting in bubbling. In both cases, as in precipitation reactions, the

▲ Gas-evolution reactions, such as the reaction of hydrochloric acid (HCl) with limestone ($CaCO_3$), typically produce $CO_2$; bubbling occurs as the gas is released.

$H_3O^+$

▲ **FIGURE 4.17** **The Hydronium Ion** Protons normally associate with water molecules in solution to form $H_3O^+$ ions, which in turn interact with other water molecules.

▲ These household substances all contain acids.

reactions occur when the anion from one reactant combines with the cation of the other. Many gas-evolution reactions are also acid–base reactions.

## Acid–Base Reactions

Our stomachs contain hydrochloric acid (HCl), which acts in the digestion of food. Certain foods or stress, however, can increase the stomach's acidity to uncomfortable levels, causing acid stomach or heartburn. Antacids are over-the-counter medicines that work by reacting with and neutralizing stomach acid. Antacids employ different *bases*—substances that produce hydroxide ($OH^-$) ions in water—as neutralizing agents. Milk of magnesia, for example, contains $Mg(OH)_2$ and Mylanta contains $Al(OH)_3$. All antacids, regardless of the base they employ, have the same effect of neutralizing stomach acid and relieving heartburn through *acid–base reactions*.

Recall from Chapter 3 that an acid forms $H^+$ ions in solution, and we just saw that a base is a substance that produces $OH^-$ ions in solution. More formally:

- Acid: Substance that produces $H^+$ ions in aqueous solution
- Base: Substance that produces $OH^-$ ions in aqueous solution

These definitions of acids and bases, called the **Arrhenius definitions**, are named after Swedish chemist Svante Arrhenius (1859–1927). In Chapter 15, we will learn more general definitions of acids and bases, but these definitions are sufficient to describe neutralization reactions.

According to the Arrhenius definition, HCl is an acid because it produces $H^+$ ions in solution:

$$HCl(aq) \longrightarrow H^+(aq) + Cl^-(aq)$$

An $H^+$ ion is a bare proton. In solution bare protons normally associate with water molecules to form **hydronium ions** (Figure 4.17 ◀):

$$H^+(aq) + H_2O(l) \longrightarrow H_3O^+(aq)$$

Chemists use $H^+(aq)$ and $H_3O^+(aq)$ interchangeably to mean the same thing—a hydronium ion. The chemical equation for the ionization of HCl and other acids is often written to show the association of the proton with a water molecule to form the hydronium ion:

$$HCl(aq) + H_2O(l) \longrightarrow H_3O^+(aq) + Cl^-(aq)$$

As we discussed in Section 4.5, some acids are weak acids—they do not completely ionize in solution. We represent the partial ionization of a weak acid with opposing half arrows.

$$HC_2H_3O_2(aq) \rightleftharpoons H^+(aq) + C_2H_3O_2^-(aq)$$

Some acids—called **polyprotic acids**—contain more than one ionizable proton and release them sequentially. For example, sulfuric acid, $H_2SO_4$, is a **diprotic acid**. It is strong in its first ionizable proton, but weak in its second:

$$H_2SO_4(aq) \longrightarrow H^+(aq) + HSO_4^-(aq)$$
$$HSO_4^-(aq) \rightleftharpoons H^+(aq) + SO_4^{2-}(aq)$$

According to the Arrhenius definition, NaOH is a base because it produces $OH^-$ ions in solution:

$$NaOH(aq) \longrightarrow Na^+(aq) + OH^-(aq)$$

In analogy to diprotic acids, some bases such as $Sr(OH)_2$ produce two moles of $OH^-$ per mole of the base:

$$Sr(OH)_2(aq) \longrightarrow Sr^{2+}(aq) + 2 OH^-(aq)$$

Table 4.2 lists common acids and bases. You can find acids and bases in many everyday substances. Foods such as citrus fruits and vinegar contain acids. Soap, baking soda, and milk of magnesia all contain bases.

When we mix an acid and a base, the $H^+(aq)$ from the acid—whether it is weak or strong—combines with the $OH^-(aq)$ from the base to form $H_2O(l)$ (Figure 4.18 ▶). Consider the reaction between hydrochloric acid and sodium hydroxide:

$$HCl(aq) + NaOH(aq) \longrightarrow H_2O(l) + NaCl(aq)$$

Acid      Base                Water      Salt

**TABLE 4.2 Some Common Acids and Bases**

| Name of Acid | Formula | Name of Base | Formula |
|---|---|---|---|
| Hydrochloric acid | HCl | Sodium hydroxide | NaOH |
| Hydrobromic acid | HBr | Lithium hydroxide | LiOH |
| Hydroiodic acid | HI | Potassium hydroxide | KOH |
| Nitric acid | $HNO_3$ | Calcium hydroxide | $Ca(OH)_2$ |
| Sulfuric acid | $H_2SO_4$ | Barium hydroxide | $Ba(OH)_2$ |
| Perchloric acid | $HClO_4$ | Ammonia* | $NH_3$ (weak base) |
| Formic acid | $HCHO_2$ (weak acid) | | |
| Acetic acid | $HC_2H_3O_2$ (weak acid) | | |
| Hydrofluoric acid | HF (weak acid) | | |

*Ammonia does not contain $OH^-$, but it produces $OH^-$ in a reaction with water that occurs only to a small extent: $NH_3(aq) + H_2O(l) \rightleftharpoons NH_4^+(aq) + OH^-(aq)$.

▲ Many common household products contain bases.

## Acid–Base Reaction

$$HCl(aq) + NaOH(aq) \longrightarrow H_2O(l) + NaCl(aq)$$

The reaction between hydrochloric acid and sodium hydroxide forms water and a salt, sodium chloride, which remains dissolved in the solution.

$H_3O^+$

$Cl^-$

HCl($aq$)

+

$Na^+$

$OH^-$

NaOH($aq$)

$Cl^-$

$Na^+$

$H_2O(l)$

+

NaCl($aq$)

◄ **FIGURE 4.18 Acid–Base Reaction** The reaction between hydrochloric acid and sodium hydroxide forms water and a salt, sodium chloride, which remains dissolved in the solution.

The word *salt* in this sense applies to any ionic compound and is therefore more general than the common usage, which refers only to table salt (NaCl).

Acid–base reactions generally form water and an ionic compound—called a **salt**—that usually remains dissolved in the solution. The net ionic equation for acid–base reactions involving a strong acid is:

$$H^+(aq) + OH^-(aq) \longrightarrow H_2O(l)$$

However, if the acid is a weak acid, the net ionic equation is slightly different. For example, consider the acid–base equation between hydrofluoric acid and sodium hydroxide:

$$\underset{acid}{HF(aq)} + \underset{base}{NaOH(aq)} \longrightarrow \underset{water}{H_2O(l)} + \underset{salt}{NaF(aq)}$$

The complete ionic equation and the net ionic equation for this reaction are:

$$HF(aq) + Na^+(aq) + OH^-(aq) \longrightarrow H_2O(l) + Na^+(aq) + F^-(aq) \qquad \text{(Complete ionic equation)}$$

$$HF(aq) + OH^-(aq) \longrightarrow H_2O(l) + F^-(aq) \qquad \text{(Net ionic equation)}$$

Notice that, since HF is a weak acid, we do not show it as ionized in the ionic equations.

Another example of an acid–base reaction is the reaction between sulfuric acid and potassium hydroxide:

$$\underset{acid}{H_2SO_4(aq)} + \underset{base}{2\,KOH(aq)} \longrightarrow \underset{water}{2\,H_2O(l)} + \underset{salt}{K_2SO_4(aq)}$$

Again, notice the pattern of acid and base reacting to form water and a salt.

**Acid + Base $\longrightarrow$ Water + Salt** (acid–base reactions)

When writing equations for acid–base reactions, write the formula of the salt using the procedure for writing formulas of ionic compounds demonstrated in Section 3.5.

**Example** 4.13 | **Writing Equations for Acid–Base Reactions Involving a Strong Acid**

Write a molecular and net ionic equation for the reaction between aqueous HI and aqueous Ba(OH)$_2$.

| | |
|---|---|
| **SOLUTION** First identify these substances as an acid and a base. Begin by writing the unbalanced equation in which the acid and the base combine to form water and a salt. | $\underset{acid}{HI(aq)} + \underset{base}{Ba(OH)_2(aq)} \longrightarrow \underset{water}{H_2O(l)} + \underset{salt}{BaI_2(aq)}$ |
| Next, balance the equation; this is the molecular equation. | **Molecular equation:** $2\,HI(aq) + Ba(OH)_2(aq) \longrightarrow 2\,H_2O(l) + BaI_2(aq)$ |
| Write the net ionic equation by removing the spectator ions. | **Net ionic equation:** $2\,H^+(aq) + 2\,OH^-(aq) \longrightarrow 2\,H_2O(l)$ or simply $H^+(aq) + OH^-(aq) \longrightarrow H_2O(l)$ |

**FOR PRACTICE 4.13** Write a molecular and a net ionic equation for the reaction that occurs between aqueous HBr and aqueous LiOH.

**Example** 4.14 | **Writing Equations for Acid–Base Reactions Involving a Weak Acid**

Write a molecular equation, ionic equation, and net ionic equation for the reaction between aqueous acetic acid (HC$_2$H$_3$O$_2$) and aqueous potassium hydroxide (KOH).

| | |
|---|---|
| **SOLUTION** Begin by writing the molecular equation in which the acid and the base combine to form water and a salt. (The equation is already balanced.) | **Molecular equation:** $HC_2H_3O_2(aq) + KOH(aq) \longrightarrow H_2O(l) + KC_2H_3O_2(aq)$ |
| Write the complete ionic equation by separating aqueous ionic compounds into their constituent ions. Do not separate HC$_2$H$_3$O$_2$(aq) because it is a weak acid (and a weak electrolyte). | **Complete ionic equation:** $HC_2H_3O_2(aq) + K^+(aq) + OH^-(aq) \longrightarrow$ $\qquad\qquad H_2O(l) + K^+(aq) + C_2H_3O_2^-(aq)$ |

| Write the net ionic equation by eliminating the spectator ions. | **Net ionic equation:**<br>$HC_2H_3O_2(aq) + OH^-(aq) \longrightarrow H_2O(l) + C_2H_3O_2^-(aq)$ |

**FOR PRACTICE 4.14**   Write the net ionic equation for the reaction between $HCHO_2$ (a weak acid) and NaOH.

## Acid–Base Titrations

We can apply the principles of acid–base neutralization and stoichiometry to a common laboratory procedure called a *titration*. In a **titration**, a substance in a solution of known concentration is reacted with another substance in a solution of unknown concentration. For example, consider the following acid–base reaction:

$$HCl(aq) + NaOH(aq) \longrightarrow H_2O(l) + NaCl(aq)$$

The net ionic equation for this reaction eliminates the spectator ions:

$$H^+(aq) + OH^-(aq) \longrightarrow H_2O(l)$$

Suppose we have an HCl solution represented by the molecular diagram shown at right (we have omitted the $Cl^-$ ions and the $H_2O$ molecules not involved in the reaction from this representation for clarity).

In titrating this sample, we slowly add a solution of known $OH^-$ concentration, as shown in the molecular diagrams in Figure 4.19▼. As we add the $OH^-$, it reacts with and neutralizes the $H^+$, forming water. At the **equivalence point**—the point in the titration when the number of moles of $OH^-$ added equals the number of moles of $H^+$ initially in solution—the titration is complete. The equivalence point is typically signaled by an **indicator**, a dye whose color depends on the acidity or basicity of the solution (Figure 4.20▶, on the next page).

$-H^+$

### Acid–Base Titration

$OH^-(aq)$

$H^+(aq)$

$$H^+(aq) + OH^-(aq) \longrightarrow H_2O(l)$$

Beginning of titration

Equivalence point

◀ **FIGURE 4.19  Acid–Base Titration**

▶ FIGURE 4.20 **Titration** In this titration, NaOH is added to a dilute HCl solution. When the NaOH and HCl reach stoichiometric proportions (the equivalence point), the phenolphthalein indicator changes color to pink.

**Indicator in Titration**

We cover acid–base titrations and indicators in more detail in Chapter 17. In most laboratory titrations, the concentration of one of the reactant solutions is unknown, and the concentration of the other is precisely known. By carefully measuring the volume of each solution required to reach the equivalence point, we can determine the concentration of the unknown solution, as demonstrated in Example 4.15.

---

## Example 4.15 | Acid–Base Titration

The titration of 10.00 mL of HCl solution of unknown concentration requires 12.54 mL of a 0.100 M NaOH solution to reach the equivalence point. What is the concentration of the unknown HCl solution in M?

| | |
|---|---|
| **SORT** You are given the volume and concentration of NaOH solution required to titrate a given volume of HCl solution. You are asked to find the concentration of the HCl solution. | **GIVEN:** 12.54 mL of NaOH solution, 0.100 M NaOH solution, 10.00 mL of HCl solution<br><br>**FIND:** concentration of HCl solution |

**STRATEGIZE** Since this problem involves an acid–base neutralization reaction between HCl and NaOH, start by writing the balanced equation, using the techniques covered earlier in this section.

The first part of the conceptual plan has the form volume A → moles A → moles B. The concentration of the NaOH solution is a conversion factor between moles and volume of NaOH. The balanced equation provides the relationship between number of moles of NaOH and number of moles of HCl.

In the second part of the conceptual plan, use the number of moles of HCl (from the first part) and the volume of HCl solution (given) to calculate the molarity of the HCl solution.

$HCl(aq) + NaOH(aq) \longrightarrow H_2O(l) + NaCl(aq)$

**CONCEPTUAL PLAN**

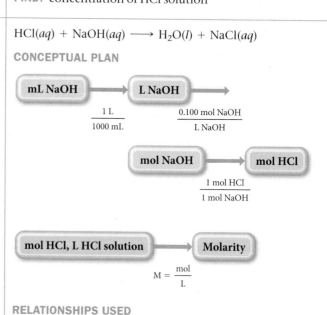

**RELATIONSHIPS USED**

$1\ L = 1000\ mL$

$M\ (NaOH) = \dfrac{0.100\ mol\ NaOH}{L\ NaOH}$

$1\ mol\ HCl : 1\ mol\ NaOH$

$Molarity\ (M) = \dfrac{moles\ of\ solute\ (mol)}{volume\ of\ solution\ (L)}$

| SOLVE In the first part of the solution, determine the number of moles of HCl in the unknown solution. | SOLUTION $12.54 \text{ mL NaOH} \times \dfrac{1 \text{ L}}{1000 \text{ mL}} \times \dfrac{0.100 \text{ mol NaOH}}{\text{L NaOH}}$ $\times \dfrac{1 \text{ mol HCl}}{1 \text{ mol NaOH}} = 1.25 \times 10^{-3} \text{ mol HCl}$ |
|---|---|
| In the second part of the solution, divide the number of moles of HCl by the volume of the HCl solution in L. 10.00 mL is equivalent to 0.01000 L. | $\text{Molarity} = \dfrac{1.25 \times 10^{-3} \text{ mol HCl}}{0.01000 \text{ L}} = 0.125 \text{ M HCl}$ |

CHECK  The units of the answer (M HCl) are correct. The magnitude of the answer (0.125 M) is reasonable because it is similar to the molarity of the NaOH solution, as expected from the reaction stoichiometry (1 mol HCl reacts with 1 mol NaOH) and the similar volumes of NaOH and HCl.

FOR PRACTICE 4.15   The titration of a 20.0-mL sample of an $H_2SO_4$ solution of unknown concentration requires 22.87 mL of a 0.158 M KOH solution to reach the equivalence point. What is the concentration of the unknown $H_2SO_4$ solution?

FOR MORE PRACTICE 4.15   What volume (in mL) of 0.200 M NaOH do we need to titrate 35.00 mL of 0.140 M HBr to the equivalence point?

## Gas-Evolution Reactions

In a *gas-evolution reaction*, two aqueous solutions mix to form a gaseous product that bubbles out of solution. Some gas-evolution reactions form a gaseous product directly when the cation of one reactant combines with the anion of the other. For example, when sulfuric acid reacts with lithium sulfide, dihydrogen sulfide gas forms:

$$H_2SO_4(aq) + Li_2S(aq) \longrightarrow \underset{\text{gas}}{H_2S(g)} + Li_2SO_4(aq)$$

Other gas-evolution reactions often form an intermediate product that then decomposes (breaks down into simpler substances) to form a gas. For example, when aqueous hydrochloric acid is mixed with aqueous sodium bicarbonate, the following reaction occurs (Figure 4.21▶, on the next page):

$$HCl(aq) + NaHCO_3(aq) \longrightarrow \underset{\text{intermediate product}}{H_2CO_3(aq)} + NaCl(aq) \longrightarrow H_2O(l) + \underset{\text{gas}}{CO_2(g)} + NaCl(aq)$$

Many gas-evolution reactions such as this one are also acid–base reactions. In Chapter 16 we will learn how ions such as $CO_3^{2-}$ act as bases in aqueous solution.

The intermediate product, $H_2CO_3$, is not stable and decomposes into $H_2O$ and gaseous $CO_2$. Other important gas-evolution reactions form either $H_2SO_3$ or $NH_4OH$ as intermediate products:

$$HCl(aq) + NaHSO_3(aq) \longrightarrow \underset{\text{intermediate product}}{H_2SO_3(aq)} + NaCl(aq) \longrightarrow H_2O(l) + \underset{\text{gas}}{SO_2(g)} + NaCl(aq)$$

$$NH_4Cl(aq) + NaOH(aq) \longrightarrow \underset{\text{intermediate product}}{NH_4OH(aq)} + NaCl(aq) \longrightarrow H_2O(l) + \underset{\text{gas}}{NH_3(g)} + NaCl(aq)$$

The intermediate product $NH_4OH$ provides a convenient way to think about this reaction, but the extent to which it actually forms is debatable.

Table 4.3 lists the main types of compounds that form gases in aqueous reactions, as well as the gases formed.

**TABLE 4.3  Types of Compounds That Undergo Gas-Evolution Reactions**

| Reactant Type | Intermediate Product | Gas Evolved | Example |
|---|---|---|---|
| Sulfides | None | $H_2S$ | $2 \text{ HCl}(aq) + K_2S(aq) \longrightarrow H_2S(g) + 2 \text{ KCl}(aq)$ |
| Carbonates and bicarbonates | $H_2CO_3$ | $CO_2$ | $2 \text{ HCl}(aq) + K_2CO_3(aq) \longrightarrow H_2O(l) + CO_2(g) + 2 \text{ KCl}(aq)$ |
| Sulfites and bisulfites | $H_2SO_3$ | $SO_2$ | $2 \text{ HCl}(aq) + K_2SO_3(aq) \longrightarrow H_2O(l) + SO_2(g) + 2 \text{ KCl}(aq)$ |
| Ammonium | $NH_4OH$ | $NH_3$ | $NH_4Cl(aq) + KOH(aq) \longrightarrow H_2O(l) + NH_3(g) + KCl(aq)$ |

► **FIGURE 4.21** **Gas-Evolution Reaction** When aqueous hydrochloric acid is mixed with aqueous sodium bicarbonate, gaseous $CO_2$ bubbles out of the reaction mixture.

## Gas-Evolution Reaction

$$NaHCO_3(aq) + HCl(aq) \longrightarrow H_2O(l) + NaCl(aq) + CO_2(g)$$

When aqueous sodium bicarbonate is mixed with aqueous hydrochloric acid, gaseous $CO_2$ bubbles are the result of the reaction.

NaHCO$_3$(aq)

+

HCl(aq)

H$_2$O(l)

+

NaCl(aq)

+

CO$_2$(g)

## Example 4.16  Writing Equations for Gas-Evolution Reactions

Write a molecular equation for the gas-evolution reaction that occurs when you mix aqueous nitric acid and aqueous sodium carbonate.

| | |
|---|---|
| Begin by writing an unbalanced equation in which the cation of each reactant combines with the anion of the other. | $HNO_3(aq) + Na_2CO_3(aq) \longrightarrow$ $H_2CO_3(aq) + NaNO_3(aq)$ |
| You must then recognize that $H_2CO_3(aq)$ decomposes into $H_2O(l)$ and $CO_2(g)$ and write these products into the equation. | $HNO_3(aq) + Na_2CO_3(aq) \longrightarrow H_2O(l) + CO_2(g) + NaNO_3(aq)$ |
| Finally, balance the equation. | $2\,HNO_3(aq) + Na_2CO_3(aq) \longrightarrow H_2O(l) + CO_2(g) + 2\,NaNO_3(aq)$ |

**FOR PRACTICE 4.16** Write a molecular equation for the gas-evolution reaction that occurs when you mix aqueous hydrobromic acid and aqueous potassium sulfite.

**FOR MORE PRACTICE 4.16** Write a net ionic equation for the reaction that occurs when you mix hydroiodic acid with calcium sulfide.

# 4.9   Oxidation–Reduction Reactions

**Oxidation–reduction reactions** or **redox reactions** are reactions in which electrons transfer from one reactant to the other. The rusting of iron, the bleaching of hair, and the production of electricity in batteries involve redox reactions. Many redox reactions involve the reaction of a substance with oxygen (Figure 4.22▼).

$$4\,Fe(s) + 3\,O_2(g) \longrightarrow 2\,Fe_2O_3(s)$$     (rusting of iron)

$$2\,C_8H_{18}(l) + 25\,O_2(g) \longrightarrow 16\,CO_2(g) + 18\,H_2O(g)$$     (combustion of octane)

$$2\,H_2(g) + O_2(g) \longrightarrow 2\,H_2O(g)$$     (combustion of hydrogen)

> Oxidation–reduction reactions are covered in more detail in Chapter 19.

## Oxidation–Reduction Reaction

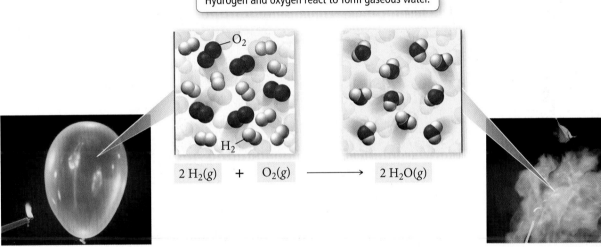

$$2\,H_2(g) + O_2(g) \longrightarrow 2\,H_2O(g)$$

Hydrogen and oxygen react to form gaseous water.

$$2\,H_2(g) \quad + \quad O_2(g) \longrightarrow 2\,H_2O(g)$$

▲ **FIGURE 4.22  Oxidation–Reduction Reaction**   The hydrogen in the balloon reacts with oxygen upon ignition to form gaseous water (which is dispersed in the flame).

However, redox reactions need not involve oxygen. Consider, for example, the reaction between sodium and chlorine to form sodium chloride (NaCl), depicted in Figure 4.23▶, on the next page:

$$2\,Na(s) + Cl_2(g) \longrightarrow 2\,NaCl(s)$$

This reaction is similar to the reaction between sodium and oxygen, which forms sodium oxide:

$$4\,Na(s) + O_2(g) \longrightarrow 2\,Na_2O(s)$$

> The reaction between sodium and oxygen forms other oxides as well.

In both cases, a metal (which has a tendency to lose electrons) reacts with a nonmetal (which has a tendency to gain electrons). And in both cases, metal atoms lose electrons to nonmetal atoms. A fundamental definition of **oxidation** is the loss of electrons, and a fundamental definition of **reduction** is the gain of electrons.

> Helpful Mnemonic: O I L R I G— Oxidation Is Loss; Reduction Is Gain.

The transfer of electrons does not need to be a *complete* transfer (as occurs in the formation of an ionic compound) for the reaction to qualify as oxidation–reduction. For example, consider the reaction between hydrogen gas and chlorine gas:

$$H_2(g) + Cl_2(g) \longrightarrow 2\,HCl(g)$$

> The ability of an element to attract electrons in a chemical bond is called electronegativity. We cover electronegativity in more detail in Section 9.6.

Even though hydrogen chloride is a molecular compound with a covalent bond, and even though the hydrogen has not completely transferred its electron to chlorine during

## Oxidation–Reduction Reaction without Oxygen

$$2\,Na(s)\ +\ Cl_2(g)\ \longrightarrow\ 2\,NaCl(s)$$

Electrons are transferred from sodium to chlorine, forming sodium chloride.
Sodium is oxidized and chlorine is reduced.

Na

Na(s)

Electron transfer

Na$^+$
Cl$^-$

+

Cl$_2$

Cl$_2$(g)

NaCl(s)

▲ **FIGURE 4.23 Oxidation–Reduction without Oxygen** When sodium reacts with chlorine, electrons transfer from the sodium to the chlorine, resulting in the formation of sodium chloride. In this redox reaction, sodium is oxidized and chlorine is reduced.

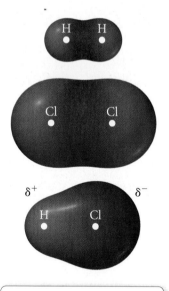

H    H

Cl        Cl

$\delta^+$                    $\delta^-$

H        Cl

Hydrogen loses electron density (oxidation) and chlorine gains electron density (reduction).

▲ **FIGURE 4.24 Redox with Partial Electron Transfer** When hydrogen bonds to chlorine, the electrons are unevenly shared, resulting in an increase of electron density (reduction) for chlorine and a decrease in electron density (oxidation) for hydrogen.

the reaction, we can see from the electron density diagrams (Figure 4.24 ◀) that hydrogen has lost some of its electron density—it has *partially* transferred its electron to chlorine. In the reaction, hydrogen is oxidized and chlorine is reduced and, therefore, this is a redox reaction.

## Oxidation States

Identifying whether or not a reaction between a metal and a nonmetal is a redox reaction is fairly straightforward because of ion formation. But how do we identify redox reactions that occur between nonmetals? Chemists have devised a scheme to track electrons before and after a chemical reaction. In this scheme—which is like bookkeeping for electrons—each shared electron is assigned to the atom that attracts the electrons most strongly. Then a number, called the **oxidation state** or **oxidation number**, is given to each atom based on the electron assignments. In other words, the oxidation number of an atom in a compound is the "charge" it would have if all shared electrons were assigned to the atom with the greatest attraction for those electrons.

For example, consider HCl. Since chlorine attracts electrons more strongly than hydrogen, we assign the two shared electrons in the bond to chlorine; then H (which has lost an electron in our assignment) has an oxidation state of +1, and Cl (which has gained one electron in our assignment) has an oxidation state of −1. Notice that in contrast to ionic charges, which are usually written with the sign of the charge *after* the magnitude (1+ and 1−, for example), oxidation states are written with the sign of the charge *before* the magnitude (+1 and −1, for example). We use the following rules to assign oxidation states to atoms in elements, ions, and compounds.

| **Rules for Assigning Oxidation States** | **Examples** |

(These rules are hierarchical. If any two rules conflict, follow the rule that is higher on the list.)

**1.** The oxidation state of an atom in a free element is 0.

$$\underset{\text{0 ox state}}{Cu} \quad \underset{\text{0 ox state}}{Cl_2}$$

**2.** The oxidation state of a monoatomic ion is equal to its charge.

$$\underset{\text{+2 ox state}}{Ca^{2+}} \quad \underset{\text{−1 ox state}}{Cl^{-}}$$

Do not confuse oxidation state with ionic charge. Unlike ionic charge—which is a real property of an ion—the oxidation state of an atom is merely a theoretical (but useful) construct.

**3.** The sum of the oxidation states of all atoms in:

- A neutral molecule or formula unit is 0.

$$\underset{\text{2(H ox state) + 1(O ox state) = 0}}{H_2O}$$

- An ion is equal to the charge of the ion.

$$\underset{\text{1(N ox state) + 3(O ox state) = −1}}{NO_3^{-}}$$

**4.** In their compounds, metals have positive oxidation states.

- Group 1A metals *always* have an oxidation state of +1.

$$\underset{\text{+1 ox state}}{NaCl}$$

- Group 2A metals *always* have an oxidation state of +2.

$$\underset{\text{+2 ox state}}{CaF_2}$$

**5.** In their compounds, nonmetals are assigned oxidation states according to the table shown here. Entries at the top of the table take precedence over entries at the bottom of the table.

| Nonmetal | Oxidation State | Example |
|---|---|---|
| Fluorine | −1 | $MgF_2$ <br> −1 ox state |
| Hydrogen | +1 | $H_2O$ <br> +1 ox state |
| Oxygen | −2 | $CO_2$ <br> −2 ox state |
| Group 7A | −1 | $CCl_4$ <br> −1 ox state |
| Group 6A | −2 | $H_2S$ <br> −2 ox state |
| Group 5A | −3 | $NH_3$ <br> −3 ox state |

When assigning oxidation states, keep these points in mind:

- The oxidation state of any given element generally depends on what other elements are present in the compound. (The exceptions are the group 1A and 2A metals, which are *always* +1 and +2, respectively.)

- Rule 3 must always be followed. Therefore, when following the hierarchy shown in rule 5, give priority to the element(s) highest on the list and then assign the oxidation state of the element lowest on the list using rule 3.

- When assigning oxidation states to elements that are not covered by rules 4 and 5 (such as carbon), use rule 3 to deduce their oxidation state once all other oxidation states have been assigned.

---

**Example** 4.17 | **Assigning Oxidation States**

Assign an oxidation state to each atom in each element, ion, or compound.
**(a)** $Cl_2$ **(b)** $Na^+$ **(c)** KF **(d)** $CO_2$ **(e)** $SO_4^{2-}$ **(f)** $K_2O_2$

**SOLUTION**

| | |
|---|---|
| Since $Cl_2$ is a free element, the oxidation state of both Cl atoms is 0 (rule 1). | **(a)** $Cl_2$ <br> $\underset{0 \quad 0}{ClCl}$ |
| Since $Na^+$ is a monoatomic ion, the oxidation state of the $Na^+$ ion is +1 (rule 2). | **(b)** $Na^+$ <br> $\underset{+1}{Na^+}$ |
| The oxidation state of K is +1 (rule 4). The oxidation state of F is −1 (rule 5). Since this is a neutral compound, the sum of the oxidation states is 0. | **(c)** KF <br> $\underset{+1 \; -1}{KF}$ <br> sum: +1−1 = 0 |

—*Continued on the next page*

*—Continued*

| | |
|---|---|
| The oxidation state of oxygen is −2 (rule 5). The oxidation state of carbon must be deduced using rule 3, which says that the sum of the oxidation states of all the atoms must be 0. | **(d)** $CO_2$<br>(C ox state) + 2(O ox state) = 0<br>(C ox state) + 2(−2) = 0<br>(C ox state) = +4<br>$CO_2$<br>+4 −2<br>sum: +4+2(−2)=0 |
| The oxidation state of oxygen is −2 (rule 5). You would ordinarily expect the oxidation state of S to be −2 (rule 5). However, if that were the case, the sum of the oxidation states would not equal the charge of the ion. Since O is higher on the list than S, it takes priority and you calculate the oxidation state of sulfur by setting the *sum* of all of the oxidation states equal to −2 (the charge of the ion). | **(e)** $SO_4^{2-}$<br>(S ox state) + 4(O ox state) = −2<br>(S ox state) + 4(−2) = −2<br>S ox state = +6<br>$SO_4^{2-}$<br>+6 −2<br>sum: +6+4(−2)=−2 |
| The oxidation state of potassium is +1 (rule 4). You would ordinarily expect the oxidation state of O to be −2 (rule 5), but rule 4 takes priority, and you deduce the oxidation state of O by setting the sum of all of the oxidation states equal to 0. | **(f)** $K_2O_2$<br>2(K ox state) + 2(O ox state) = 0<br>2(+1) + 2(O ox state) = 0<br>O ox state = −1<br>$K_2O_2$<br>+1 −1<br>sum: 2(+1)+2(−1)=0 |

**FOR PRACTICE 4.17** Assign an oxidation state to each atom in each element, ion, or compound.

**(a)** Cr    **(b)** $Cr^{3+}$    **(c)** $CCl_4$    **(d)** $SrBr_2$    **(e)** $SO_3$    **(f)** $NO_3^-$

In most cases, oxidation states are positive or negative integers; however, on occasion an atom within a compound can have a fractional oxidation state. Consider $KO_2$. The oxidation states are assigned as follows:

$$KO_2$$
$$+1 \quad -\tfrac{1}{2}$$
$$\text{sum: } +1+2\left(-\tfrac{1}{2}\right) = 0$$

In $KO_2$, oxygen has a $-\tfrac{1}{2}$ oxidation state. Although this seems unusual, it is accepted because oxidation states are merely an imposed electron bookkeeping scheme, not an actual physical quantity.

**CONCEPTUAL CONNECTION 4.9**

**Oxidation Numbers in Polyatomic Ions** Which statement best describes the *difference* between the *charge* of a polyatomic ion and the *oxidation states* of its constituent atoms? (For example, the charge of $NO_3^-$ is 1−, and the oxidation states of its atoms are +5 for the nitrogen atom and −2 for each oxygen atom.)

**(a)** The charge of a polyatomic ion is a property of the entire ion, while the oxidation states are assigned to each individual atom.

**(b)** The oxidation state of the ion is the same as its charge.

**(c)** The charge of a polyatomic ion is not a real physical property, while the oxidation states of atoms are actual physical properties.

## Identifying Redox Reactions

We can use oxidation states to identify redox reactions, even between nonmetals. For example, is the following reaction between carbon and sulfur a redox reaction?

$$C + 2S \longrightarrow CS_2$$

If so, what element is oxidized? What element is reduced? We can use the oxidation state rules to assign oxidation states to all elements on both sides of the equation.

$$C + 2S \longrightarrow CS_2$$

Oxidation states: 0      0      +4 −2

Reduction

Oxidation

Carbon changes from an oxidation state of 0 to an oxidation state of +4. In terms of our electron bookkeeping scheme (the assigned oxidation state), carbon *loses electrons* and is *oxidized*. Sulfur changes from an oxidation state of 0 to an oxidation state of −2. In terms of our electron bookkeeping scheme, sulfur *gains electrons* and is *reduced*. In terms of oxidation states, oxidation and reduction are defined as follows:

- Oxidation: An increase in oxidation state
- Reduction: A decrease in oxidation state

> Remember that a reduction is a *reduction* in oxidation state.

## Example 4.18    Using Oxidation States to Identify Oxidation and Reduction

Use oxidation states to identify the element that is oxidized and the element that is reduced in the following redox reaction:

$$Mg(s) + 2 H_2O(l) \longrightarrow Mg(OH)_2(aq) + H_2(g)$$

**SOLUTION**

Begin by assigning oxidation states to each atom in the reaction.

$$Mg(s) + 2 H_2O(l) \longrightarrow Mg(OH)_2(aq) + H_2(g)$$

Oxidation states: 0      +1 −2      +2 −2 +1      0

Reduction

Oxidation

Since Mg increased in oxidation state, it was oxidized. Since H decreased in oxidation state, it was reduced.

**FOR PRACTICE 4.18**    Use oxidation states to identify the element that is oxidized and the element that is reduced in the following redox reaction:

$$Sn(s) + 4 HNO_3(aq) \longrightarrow SnO_2(s) + 4 NO_2(g) + 2 H_2O(g)$$

**FOR MORE PRACTICE 4.18**    Determine whether or not each reaction is a redox reaction. If the reaction is a redox reaction, identify which element is oxidized and which is reduced.

**(a)** $Hg_2(NO_3)_2(aq) + 2 KBr(aq) \longrightarrow Hg_2Br_2(s) + 2 KNO_3(aq)$
**(b)** $4 Al(s) + 3 O_2(g) \longrightarrow 2 Al_2O_3(s)$
**(c)** $CaO(s) + CO_2(g) \longrightarrow CaCO_3(s)$

Notice that *oxidation and reduction must occur together*. If one substance loses electrons (oxidation), then another substance must gain electrons (reduction). A substance that causes the oxidation of another substance is an **oxidizing agent**. Oxygen, for example, is an excellent oxidizing agent. In a redox reaction, *the oxidizing agent is always reduced*. A substance that causes the reduction of another substance is a **reducing agent**. Hydrogen, for example, and group 1A and group 2A metals (because of their tendency to lose electrons) are excellent reducing agents. In a redox reaction, *the reducing agent is always oxidized*.

In Section 19.2 you will learn more about redox reactions, including how to balance them. For now, be able to identify redox reactions, as well as oxidizing and reducing agents, according to the following guidelines.

*Redox reactions:*
- Any reaction in which there is a change in the oxidation states of atoms in going from reactants to products.

*In a redox reaction:*
- The oxidizing agent oxidizes another substance (and is itself reduced).
- The reducing agent reduces another substance (and is itself oxidized).

**Example 4.19**    **Identifying Redox Reactions, Oxidizing Agents, and Reducing Agents**

Determine whether each reaction is an oxidation–reduction reaction. For each oxidation–reduction reaction, identify the oxidizing agent and the reducing agent.

**(a)**  $2 \, Mg(s) + O_2(g) \longrightarrow 2 \, MgO(s)$

**(b)**  $2 \, HBr(aq) + Ca(OH)_2(aq) \longrightarrow 2 \, H_2O(l) + CaBr_2(aq)$

**(c)**  $Zn(s) + Fe^{2+}(aq) \longrightarrow Zn^{2+}(aq) + Fe(s)$

**SOLUTION**

| | |
|---|---|
| *This is a redox reaction* because magnesium increases in oxidation number (oxidation) and oxygen decreases in oxidation number (reduction). | **(a)**  $2 \, Mg(s) + O_2(g) \longrightarrow 2 \, MgO(s)$<br><br>Oxidizing agent: $O_2$<br>Reducing agent: Mg |
| *This is not a redox reaction* because none of the atoms undergo a change in oxidation number. | **(b)**  $2 \, HBr(aq) + Ca(OH)_2(aq) \rightarrow 2 \, H_2O(l) + CaBr_2(aq)$ |
| *This is a redox reaction* because zinc increases in oxidation number (oxidation) and iron decreases in oxidation number (reduction). | **(c)**  $Zn(s) + Fe^{2+}(aq) \longrightarrow Zn^{2+}(aq) + Fe(s)$<br><br>Oxidizing agent: $Fe^{2+}$<br>Reducing agent: Zn |

**FOR PRACTICE 4.19**    Determine whether or not each reaction is a redox reaction. For each redox reaction, identify the oxidizing agent and the reducing agent.

**(a)**  $2 \, Li(s) + Cl_2(g) \longrightarrow 2 \, LiCl(s)$

**(b)**  $2 \, Al(s) + 3 \, Sn^{2+}(aq) \longrightarrow 2 \, Al^{3+}(aq) + 3 \, Sn(s)$

**(c)**  $Pb(NO_3)_2(aq) + 2 \, LiCl(aq) \longrightarrow PbCl_2(s) + 2 \, LiNO_3(aq)$

**(d)**  $C(s) + O_2(g) \longrightarrow CO_2(g)$

**CONCEPTUAL CONNECTION 4.10**    **Oxidation and Reduction**    Which statement is true?

**(a)**  A redox reaction involves *either* the transfer of an electron *or* a change in the oxidation state of an element.

**(b)**  If any of the reactants or products in a reaction contain oxygen, the reaction is a redox reaction.

**(c)**  In a reaction, oxidation can occur independently of reduction.

**(d)**  In a redox reaction, any increase in the oxidation state of a reactant must be accompanied by a decrease in the oxidation state of a reactant.

# Chemistry IN YOUR DAY | Bleached Blonde

Have you ever bleached your hair? Most home kits for hair bleaching contain hydrogen peroxide ($H_2O_2$), an excellent oxidizing agent. When applied to hair, hydrogen peroxide oxidizes melanin, the dark pigment that gives hair its color. Once melanin is oxidized, it no longer imparts a dark color, leaving the hair with the familiar bleached look. Hydrogen peroxide also oxidizes other components of hair. For example, protein molecules in hair contain —SH groups called thiols. Hydrogen peroxide oxidizes these thiol groups to sulfonic acid groups, —$SO_3H$. The oxidation of thiol groups to sulfonic acid groups causes changes in the proteins that compose hair, making the hair more brittle and more likely to tangle. Consequently, people with heavily bleached hair generally use conditioners, which contain compounds that form thin, lubricating coatings on individual hair shafts. These coatings prevent tangling and make hair softer and more manageable.

### QUESTION

*The following is a reaction of hydrogen peroxide with an alkene:*

$$H_2O_2 + C_2H_4 \longrightarrow C_2H_4O + H_2O$$

*Can you see why this reaction is a redox reaction? Can you identify the oxidizing and reducing agents?*

▲ The bleaching of hair involves a redox reaction in which melanin—the main pigment in hair—is oxidized.

## Combustion Reactions

We encountered combustion reactions, a type of redox reaction, in the opening section of this chapter. Combustion reactions are important because most of our society's energy is derived from them (Figure 4.25 ▼).

As we discussed in Section 4.1, *combustion reactions* are characterized by the reaction of a substance with $O_2$ to form one or more oxygen-containing compounds, often including water. Combustion reactions also emit heat. For example, as we saw earlier in this chapter, natural gas ($CH_4$) reacts with oxygen to form carbon dioxide and water:

$$CH_4(g) + 2\,O_2(g) \longrightarrow CO_2(g) + 2\,H_2O(g)$$

Oxidation state: $-4 +1$      $0$      $+4 -2$      $+1 -2$

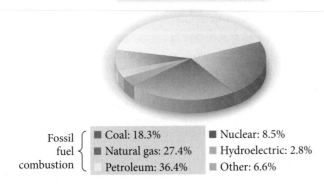

U.S. Energy Use by Source, 2012

Fossil fuel combustion
- Coal: 18.3%
- Natural gas: 27.4%
- Petroleum: 36.4%
- Nuclear: 8.5%
- Hydroelectric: 2.8%
- Other: 6.6%

◀ **FIGURE 4.25 U.S. Energy Consumption** About 82% of the energy the United States uses is produced by combustion reactions. Source: U.S. Energy Information Administration. *Annual Energy Review.* 2014.

In this reaction, carbon is oxidized and oxygen is reduced. Ethanol, the alcohol in alcoholic beverages, also reacts with oxygen in a combustion reaction to form carbon dioxide and water:

$$C_2H_5OH(l) + 3 O_2(g) \longrightarrow 2 CO_2(g) + 3 H_2O(g)$$

Compounds containing carbon and hydrogen—or carbon, hydrogen, and oxygen—always form carbon dioxide and water upon complete combustion. Other combustion reactions include the reaction of carbon with oxygen to form carbon dioxide:

$$C(s) + O_2(g) \longrightarrow CO_2(g)$$

and the reaction of hydrogen with oxygen to form water:

$$2 H_2(g) + O_2(g) \longrightarrow 2 H_2O(g)$$

---

**Example** 4.20 | **Writing Equations for Combustion Reactions**

Write a balanced equation for the combustion of liquid methyl alcohol ($CH_3OH$).

**SOLUTION**

| Begin by writing an unbalanced equation showing the reaction of $CH_3OH$ with $O_2$ to form $CO_2$ and $H_2O$. | $CH_3OH(l) + O_2(g) \longrightarrow CO_2(g) + H_2O(g)$ |
|---|---|
| Balance the equation using the guidelines in Section 3.10. | $2 CH_3OH(l) + 3 O_2(g) \longrightarrow 2 CO_2(g) + 4 H_2O(g)$ |

**FOR PRACTICE 4.20** Write a balanced equation for the complete combustion of liquid $C_2H_5SH$.

---

# Self-Assessment Quiz

**Q1.** Manganese(IV) oxide reacts with aluminum to form elemental manganese and aluminum oxide:

$$3 MnO_2 + 4 Al \longrightarrow 3 Mn + 2 Al_2O_3$$

What mass of Al is required to completely react with 25.0 g $MnO_2$?

a) 7.76 g Al   b) 5.82 g Al
c) 33.3 g Al   d) 10.3 g Al

**Q2.** Sodium and chlorine react to form sodium chloride:

$$2 Na(s) + Cl_2(g) \longrightarrow 2 NaCl(s)$$

What is the theoretical yield of sodium chloride for the reaction of 55.0 g Na with 67.2 g $Cl_2$?

a) $1.40 \times 10^2$ g NaCl   b) 111 g NaCl
c) 55.4 g NaCl   d) 222 g NaCl

**Q3.** Sulfur and fluorine react to form sulfur hexafluoride:

$$S(s) + 3 F_2(g) \longrightarrow SF_6(g)$$

If 50.0 g S is allowed to react as completely as possible with 105.0 g $F_2$, what mass of the excess reactant is left?

a) 20.5 g S   b) 45.7 g $F_2$
c) 15.0 g S   d) 36.3 g $F_2$

**Q4.** A reaction has a theoretical yield of 45.8 g. When the reaction is carried out, 37.2 g of the product is obtained. What is the percent yield?

a) 55.1%   b) 44.8%
c) 123%   d) 81.2%

**Q5.** What is the molarity of a solution containing 55.8 g of $MgCl_2$ dissolved in 1.00 L of solution?

a) 55.8 M   b) 1.71 M
c) 0.586 M   d) 0.558 M

**Q6.** What mass (in grams) of $Mg(NO_3)_2$ is present in 145 mL of a 0.150 M solution of $Mg(NO_3)_2$?

a) 3.23 g   b) 0.022 g
c) 1.88 g   d) 143 g

**Q7.** What volume of a 1.50 M HCl solution should you use to prepare 2.00 L of a 0.100 M HCl solution?

a) 0.300 L   b) 0.133 L
c) 30.0 L   d) 2.00 L

**Q8.** Potassium iodide reacts with lead(II) nitrate in the following precipitation reaction:

$$2 KI(aq) + Pb(NO_3)_2(aq) \longrightarrow 2 KNO_3(aq) + PbI_2(s)$$

What minimum volume of 0.200 M potassium iodide solution is required to completely precipitate all of the lead in 155.0 mL of a 0.112 M lead(II) nitrate solution?

a) 348 mL   b) 86.8 mL
c) 174 mL   d) 43.4 mL

**Q9.** Which solution forms a precipitate when mixed with a solution of aqueous $Na_2CO_3$?

a) $KNO_3(aq)$        b) $NaBr(aq)$

c) $NH_4Cl(aq)$       d) $CuCl_2(aq)$

**Q10.** What is the net ionic equation for the reaction that occurs when aqueous solutions of KOH and $SrCl_2$ are mixed?

a) $K^+(aq) + Cl^-(aq) \longrightarrow KCl(s)$

b) $Sr^{2+}(aq) + 2\,OH^-(aq) \longrightarrow Sr(OH)_2(s)$

c) $H^+(aq) + OH^-(aq) \longrightarrow H_2O(l)$

d) None of the above because no reaction occurs

**Q11.** What is the net ionic equation for the reaction that occurs when aqueous solutions of KOH and $HNO_3$ are mixed?

a) $K^+(aq) + NO_3^-(aq) \longrightarrow KNO_3(s)$

b) $NO_3^-(aq) + OH^-(aq) \longrightarrow NO_3OH(s)$

c) $H^+(aq) + OH^-(aq) \longrightarrow H_2O(l)$

d) None of the above because no reaction occurs.

**Q12.** What is the net ionic equation for the reaction that occurs when aqueous solutions of $KHCO_3$ and HBr are mixed?

a) $K^+(aq) + C_2H_3O_2^-(aq) \longrightarrow KC_2H_3O_2(s)$

b) $H^+(aq) + HCO_3^-(aq) \longrightarrow CO_2(g) + H_2O(l)$

c) $H^+(aq) + OH^-(aq) \longrightarrow H_2O(l)$

d) None of the above because no reaction occurs.

**Q13.** What is the oxidation state of carbon in $CO_3{}^{2-}$?

a) $+4$        b) $+3$

c) $-3$        d) $-2$

**Q14.** Sodium reacts with water according to the reaction:

$$2\,Na(s) + 2\,H_2O(l) \longrightarrow 2\,NaOH(aq) + H_2(g)$$

Identify the oxidizing agent.

a) $Na(s)$       b) $H_2O(l)$

c) $NaOH(aq)$     d) $H_2(aq)$

**Q15.** Identify the correct balanced equation for the combustion of propane ($C_3H_8$).

a) $C_3H_8(g) \longrightarrow 4\,H_2(g) + 3\,C(s)$

b) $C_3H_8(g) + 5\,O_2(g) \longrightarrow 4\,H_2O(g) + 3\,CO_2(g)$

c) $C_3H_8(g) + 3\,O_2(g) \longrightarrow 4\,H_2O(g) + 3\,CO_2(g)$

d) $2\,C_3H_8(g) + 9\,O_2(g) \longrightarrow 6\,H_2CO_3(g) + 2\,H_2(g)$

**Answers: 1.** (d) **2.** (b) **3.** (a) **4.** (d) **5.** (c) **6.** (a) **7.** (b) **8.** (c) **9.** (d) **10.** (b) **11.** (c) **12.** (b) **13.** (a) **14.** (b) **15.** (b)

# Chapter 4 IN REVIEW

## Key Terms

### Section 4.2
stoichiometry (141)

### Section 4.3
limiting reactant (145)
theoretical yield (146)
actual yield (146)
percent yield (146)

### Section 4.4
solution (151)
solvent (151)
solute (151)
aqueous solution (151)
dilute solution (151)
concentrated solution (151)
molarity (M) (151)
stock solution (154)

### Section 4.5
electrolyte (159)
strong electrolyte (159)

nonelectrolyte (159)
strong acid (160)
weak acid (160)
weak electrolyte (160)
soluble (160)
insoluble (160)

### Section 4.6
precipitation reaction (162)
precipitate (162)

### Section 4.7
molecular equation (166)
complete ionic equation (166)
spectator ion (166)
net ionic equation (166)

### Section 4.8
acid–base reaction (neutralization reaction) (167)
gas-evolution reaction (167)
Arrhenius definitions (168)

hydronium ion (168)
polyprotic acid (168)
diprotic acid (168)
salt (170)
titration (171)
equivalence point (171)
indicator (171)

### Section 4.9
oxidation–reduction (redox) reaction (175)
oxidation (175)
reduction (175)
oxidation state (oxidation number) (176)
oxidizing agent (179)
reducing agent (179)

# Key Concepts

## Climate Change and the Combustion of Fossil Fuels (4.1)

▶ Greenhouse gases warm Earth by trapping some of the sunlight that penetrates Earth's atmosphere. Global warming, resulting from rising atmospheric carbon dioxide levels, is potentially harmful.

▶ The largest atmospheric carbon dioxide source is the burning of fossil fuels. This can be verified by reaction stoichiometry.

## Reaction Stoichiometry (4.2)

▶ Reaction stoichiometry refers to the numerical relationships between the reactants and products in a balanced chemical equation.

▶ Reaction stoichiometry allows us to predict, for example, the amount of product that can be formed for a given amount of reactant, or how much of one reactant is required to react with a given amount of another.

## Limiting Reactant, Theoretical Yield, and Percent Yield (4.3)

▶ When a chemical reaction actually occurs, the reactants are usually not present in the exact stoichiometric ratios specified by the balanced chemical equation. The limiting reactant is the one that is available in the smallest stoichiometric quantity—it will be completely consumed in the reaction, and it limits the amount of product that can be made.

▶ Any reactant that does not limit the amount of product is in excess.

▶ The amount of product that can be made from the limiting reactant is the theoretical yield.

▶ The actual yield—always equal to or less than the theoretical yield—is the amount of product that is actually made when the reaction is carried out.

▶ The percentage of the theoretical yield that is actually produced when the reaction is carried out is the percent yield.

## Solution Concentration and Stoichiometry (4.4)

▶ An aqueous solution is a homogeneous mixture of water (the solvent) with another substance (the solute).

▶ We express the concentration of a solution in molarity, the number of moles of solute per liter of solution.

▶ We can use the molarities and volumes of reactant solutions to predict the amount of product that forms in an aqueous reaction.

## Aqueous Solutions and Precipitation Reactions (4.5, 4.6)

▶ Solutes that completely dissociate (or completely ionize in the case of the strong acids) to ions in solution are strong electrolytes and their solutions are good conductors of electricity.

▶ Solutes that only partially dissociate (or partially ionize) are weak electrolytes.

▶ Solutes that do not dissociate (or ionize) are nonelectrolytes.

▶ A substance that dissolves in water to form a solution is soluble.

▶ In a precipitation reaction, we mix two aqueous solutions and a solid (precipitate) forms.

▶ The solubility rules are an empirical set of guidelines that help predict the solubilities of ionic compounds; these rules are especially useful when determining whether or not a precipitate will form.

## Equations for Aqueous Reactions (4.7)

▶ We can represent an aqueous reaction with a molecular equation, which shows the complete neutral formula for each compound in the reaction.

▶ We can also represent an aqueous reaction with a complete ionic equation, which shows the dissociated nature of the aqueous ionic compounds.

▶ A third representation of an aqueous reaction is the net ionic equation, in which the spectator ions—those that do not change in the course of the reaction—are left out of the equation.

## Acid–Base and Gas-Evolution Reactions (4.8)

▶ In an acid–base reaction, an acid, a substance that produces $H^+$ in solution, reacts with a base, a substance that produces $OH^-$ in solution, and the two neutralize each other, producing water (or in some cases a weak electrolyte).

▶ An acid–base titration is a laboratory procedure in which a reaction is carried to its equivalence point—the point at which the reactants are in exact stoichiometric proportions; titrations are useful in determining the concentrations of unknown solutions.

▶ In gas-evolution reactions, two aqueous solutions combine and a gas is produced.

## Oxidation–Reduction Reactions (4.9)

▶ In oxidation–reduction reactions, one substance transfers electrons to another substance.

▶ In oxidation–reduction reactions, the substance that loses electrons is oxidized and the substance that gains them is reduced.

▶ An oxidation state is a fictitious charge given to each atom in an oxidation–reduction reaction by assigning all shared electrons to the atom with the greater attraction for those electrons. An oxidation state is an imposed electronic bookkeeping scheme, not an actual physical state.

▶ The oxidation state of an atom increases upon oxidation and decreases upon reduction.

▶ A combustion reaction is a specific type of oxidation–reduction reaction in which a substance reacts with oxygen, emitting heat and forming one or more oxygen-containing products.

# Key Equations and Relationships

Mass-to-Mass Conversion: Stoichiometry (**4.2**)

mass A $\longrightarrow$ amount A (in moles) $\longrightarrow$
$\qquad$ amount B (in moles) $\longrightarrow$ mass B

Percent Yield (**4.3**)

$$\% \text{ yield} = \frac{\text{actual yield}}{\text{theoretical yield}} \times 100\%$$

Molarity (M): Solution Concentration (**4.4**)

$$M = \frac{\text{amount of solute (in mol)}}{\text{volume of solution (in L)}}$$

Solution Dilution (**4.4**)

$$M_1 V_1 = M_2 V_2$$

Solution Stoichiometry (**4.4**)

volume A $\longrightarrow$ amount A (in moles) $\longrightarrow$
$\qquad$ amount B (in moles) $\longrightarrow$ volume B

# Key Learning Outcomes

| CHAPTER OBJECTIVES | ASSESSMENT |
|---|---|
| Calculations Involving the Stoichiometry of a Reaction (**4.2**) | Examples 4.1, 4.2   For Practice 4.1, 4.2   Exercises 31–36 |
| Determining the Limiting Reactant and Calculating Theoretical and Percent Yield (**4.3**) | Examples 4.3, 4.4   For Practice 4.3, 4.4   Exercises 41–51 |
| Calculating and Using Molarity as a Conversion Factor (**4.4**) | Examples 4.5, 4.6   For Practice 4.5, 4.6   For More Practice 4.5, 4.6   Exercises 53–60 |
| Determining Solution Dilutions (**4.4**) | Example 4.7   For Practice 4.7   For More Practice 4.7   Exercises 63, 64 |
| Using Solution Stoichiometry to Find Volumes and Amounts (**4.4**) | Example 4.8   For Practice 4.8   For More Practice 4.8   Exercises 65–67 |
| Predicting Whether a Compound Is Soluble (**4.5**) | Example 4.9   For Practice 4.9   Exercises 73, 74 |
| Writing Equations for Precipitation Reactions (**4.6**) | Examples 4.10, 4.11   For Practice 4.10, 4.11   Exercises 75–78 |
| Writing Complete Ionic and Net Ionic Equations (**4.7**) | Example 4.12   For Practice 4.12   For More Practice 4.12   Exercises 79, 80 |
| Writing Equations for Acid–Base Reactions (**4.8**) | Example 4.13   For Practice 4.13   Exercises 83, 84 |
| Calculations Involving Acid–Base Titrations (**4.8**) | Example 4.15   For Practice 4.15   For More Practice 4.15   Exercises 89, 90 |

$K^+$

$NO_3^-$

$PbI_2$

| Writing Equations for Gas-Evolution Reactions (**4.8**)  | Example 4.16   For Practice 4.16   For More Practice 4.16<br>Exercises 91, 92 |
| Assigning Oxidation States (**4.9**)  | Example 4.17   For Practice 4.17   Exercises 93–96 |
| Identifying Redox Reactions, Oxidizing Agents, and Reducing Agents Using Oxidation States (**4.9**) | Examples 4.18, 4.19   For Practice 4.18, 4.19<br>For More Practice 4.18   Exercises 97, 98 |
| Writing Equations for Combustion Reactions (**4.9**) | Example 4.20   For Practice 4.20   Exercises 99, 100 |

# EXERCISES
## Review Questions

1. What is reaction stoichiometry? What is the significance of the coefficients in a balanced chemical equation?

2. In a chemical reaction, what is the limiting reactant? The theoretical yield? The percent yield? What do we mean when we say a reactant is in excess?

3. We typically calculate the percent yield using the actual yield and theoretical yield in units of mass (grams or kilograms). Would the percent yield be different if the actual yield and theoretical yield were in units of amount (moles)?

4. What is an aqueous solution? What is the difference between the solute and the solvent?

5. What is molarity? How is it useful?

6. Explain how a strong electrolyte, a weak electrolyte, and a non-electrolyte differ.

7. Explain the difference between a strong acid and a weak acid.

8. What does it mean for a compound to be soluble? Insoluble?

9. What are the solubility rules? How are they useful?

10. What are the cations and anions whose compounds are usually soluble? What are the exceptions? What are the anions whose compounds are mostly insoluble? What are the exceptions?

11. What is a precipitation reaction? Give an example.

12. How can you predict whether a precipitation reaction will occur upon mixing two aqueous solutions?

13. Explain how a molecular equation, a complete ionic equation, and a net ionic equation differ.

14. What is the Arrhenius definition of an acid? A base?

15. What is an acid–base reaction? Give an example.

16. Explain the principles behind an acid–base titration. What is an indicator?

17. What is a gas-evolution reaction? Give an example.

18. What reactant types give rise to gas-evolution reactions?

19. What is an oxidation–reduction reaction? Give an example.

20. What are oxidation states?

21. How can oxidation states be used to identify redox reactions?

22. What happens to a substance when it becomes oxidized? Reduced?

23. In a redox reaction, which reactant is the oxidizing agent? The reducing agent?

24. What is a combustion reaction? Why are they important? Give an example.

## Problems by Topic

### Reaction Stoichiometry

25. Consider the unbalanced equation for the combustion of hexane:

$$C_6H_{14}(g) + O_2(g) \longrightarrow CO_2(g) + H_2O(g)$$

Balance the equation and determine how many moles of $O_2$ are required to react completely with 7.2 moles of $C_6H_{14}$.

26. Consider the unbalanced equation for the neutralization of acetic acid:

$$HC_2H_3O_2(aq) + Ba(OH)_2(aq) \longrightarrow H_2O(l) + Ba(C_2H_3O_2)_2(aq)$$

Balance the equation and determine how many moles of $Ba(OH)_2$ are required to completely neutralize 0.461 mole of $HC_2H_3O_2$.

27. Calculate how many moles of $NO_2$ form when each quantity of reactant completely reacts.

$$2 N_2O_5(g) \longrightarrow 4 NO_2(g) + O_2(g)$$

   a. 2.5 mol $N_2O_5$        b. 6.8 mol $N_2O_5$
   c. 15.2 g $N_2O_5$         d. 2.87 kg $N_2O_5$

28. Calculate how many moles of $NH_3$ form when each quantity of reactant completely reacts.

$$3 N_2H_4(l) \longrightarrow 4 NH_3(g) + N_2(g)$$

   a. 2.6 mol $N_2H_4$        b. 3.55 mol $N_2H_4$
   c. 65.3 g $N_2H_4$         d. 4.88 kg $N_2H_4$

**29.** Consider the balanced equation:

$$SiO_2(s) + 3\,C(s) \longrightarrow SiC(s) + 2\,CO(g)$$

Complete the table showing the appropriate number of moles of reactants and products. If the number of moles of a *reactant* is provided, fill in the required amount of the other reactant, as well as the moles of each product that forms. If the number of moles of a *product* is provided, fill in the required amount of each reactant to make that amount of product, as well as the amount of the other product that forms.

| Mol SiO$_2$ | Mol C | Mol SiC | Mol CO |
|---|---|---|---|
| 3 | ___ | ___ | ___ |
| ___ | 6 | ___ | ___ |
| ___ | ___ | ___ | 10 |
| 2.8 | ___ | ___ | ___ |
| ___ | 1.55 | ___ | ___ |

**30.** Consider the balanced equation:

$$2\,N_2H_4(g) + N_2O_4(g) \longrightarrow 3\,N_2(g) + 4\,H_2O(g)$$

Complete the table showing the appropriate number of moles of reactants and products. If the number of moles of a *reactant* is provided, fill in the required amount of the other reactant, as well as the moles of each product that forms. If the number of moles of a *product* is provided, fill in the required amount of each reactant to make that amount of product, as well as the amount of the other product that forms.

| Mol N$_2$H$_4$ | Mol N$_2$O$_4$ | Mol N$_2$ | Mol H$_2$O |
|---|---|---|---|
| 2 | ___ | ___ | ___ |
| ___ | 5 | ___ | ___ |
| ___ | ___ | ___ | 10 |
| 2.5 | ___ | ___ | ___ |
| ___ | 4.2 | ___ | ___ |
| ___ | ___ | 11.8 | ___ |

**31.** Hydrobromic acid dissolves solid iron according to the reaction:

$$Fe(s) + 2\,HBr(aq) \longrightarrow FeBr_2(aq) + H_2(g)$$

What mass of HBr (in g) do you need to dissolve a 3.2-g pure iron bar on a padlock? What mass of $H_2$ would the complete reaction of the iron bar produce?

**32.** Sulfuric acid dissolves aluminum metal according to the reaction:

$$2\,Al(s) + 3\,H_2SO_4(aq) \longrightarrow Al_2(SO_4)_3(aq) + 3\,H_2(g)$$

Suppose you want to dissolve an aluminum block with a mass of 15.2 g. What minimum mass of $H_2SO_4$ (in g) do you need? What mass of $H_2$ gas (in g) does the complete reaction of the aluminum block produce?

**33.** For each of the reactions, calculate the mass (in grams) of the product that forms when 3.67 g of the underlined reactant completely reacts. Assume that there is more than enough of the other reactant.
a. <u>Ba</u>(s) + Cl$_2$(g) $\longrightarrow$ BaCl$_2$(s)
b. <u>CaO</u>(s) + CO$_2$(g) $\longrightarrow$ CaCO$_3$(s)
c. 2 <u>Mg</u>(s) + O$_2$(g) $\longrightarrow$ 2 MgO(s)
d. 4 <u>Al</u>(s) + 3 O$_2$(g) $\longrightarrow$ 2 Al$_2$O$_3$(s)

**34.** For each of the reactions, calculate the mass (in grams) of the product that forms when 15.39 g of the underlined reactant completely reacts. Assume that there is more than enough of the other reactant.
a. 2 K(s) + <u>Cl$_2$</u>(g) $\longrightarrow$ 2 KCl(s)
b. 2 K(s) + <u>Br$_2$</u>(l) $\longrightarrow$ 2 KBr(s)
c. 4 Cr(s) + 3 <u>O$_2$</u>(g) $\longrightarrow$ 2 Cr$_2$O$_3$(s)
d. 2 <u>Sr</u>(s) + O$_2$(g) $\longrightarrow$ 2 SrO(s)

**35.** For each of the acid–base reactions, calculate the mass (in grams) of each acid necessary to completely react with and neutralize 4.85 g of the base.
a. HCl(aq) + NaOH(aq) $\longrightarrow$ H$_2$O(l) + NaCl(aq)
b. 2 HNO$_3$(aq) + Ca(OH)$_2$(aq) $\longrightarrow$ 2 H$_2$O(l) + Ca(NO$_3$)$_2$(aq)
c. H$_2$SO$_4$(aq) + 2 KOH(aq) $\longrightarrow$ 2 H$_2$O(l) + K$_2$SO$_4$(aq)

**36.** For each precipitation reaction, calculate how many grams of the first reactant are necessary to completely react with 55.8 g of the second reactant.
a. 2 KI(aq) + Pb(NO$_3$)$_2$(aq) $\longrightarrow$ PbI$_2$(s) + 2 KNO$_3$(aq)
b. Na$_2$CO$_3$(aq) + CuCl$_2$(aq) $\longrightarrow$ CuCO$_3$(s) + 2 NaCl(aq)
c. K$_2$SO$_4$(aq) + Sr(NO$_3$)$_2$(aq) $\longrightarrow$ SrSO$_4$(s) + 2 KNO$_3$(aq)

## Limiting Reactant, Theoretical Yield, and Percent Yield

**37.** Find the limiting reactant for each initial amount of reactants.

$$2\,Na(s) + Br_2(g) \longrightarrow 2\,NaBr(s)$$

a. 2 mol Na, 2 mol Br$_2$    b. 1.8 mol Na, 1.4 mol Br$_2$
c. 2.5 mol Na, 1 mol Br$_2$    d. 12.6 mol Na, 6.9 mol Br$_2$

**38.** Find the limiting reactant for each initial amount of reactants.

$$4\,Al(s) + 3\,O_2(g) \longrightarrow 2\,Al_2O_3(s)$$

a. 1 mol Al, 1 mol O$_2$    b. 4 mol Al, 2.6 mol O$_2$
c. 16 mol Al, 13 mol O$_2$    d. 7.4 mol Al, 6.5 mol O$_2$

**39.** Consider the reaction:

$$4\,HCl(g) + O_2(g) \longrightarrow 2\,H_2O(g) + 2\,Cl_2(g)$$

Each molecular diagram represents an initial mixture of reactants. Which mixture produces the greatest amount of products? How many molecules of Cl$_2$ form from the reaction mixture that produces the greatest amount of products?

    (a)           (b)           (c)

**40.** Consider the reaction:

$$2\,CH_3OH(g) + 3\,O_2(g) \longrightarrow 2\,CO_2(g) + 4\,H_2O(g)$$

Each of the molecular diagrams represents an initial mixture of the reactants. Which reaction mixture produces the greatest amount of products? How many CO$_2$ molecules form from the reaction mixture that produces the greatest amount of products?

    (a)           (b)           (c)

**41.** Calculate the theoretical yield of the product (in moles) for each initial amount of reactants.

$$Ti(s) + 2\,Cl_2(g) \longrightarrow TiCl_4(s)$$

**a.** 4 mol Ti, 4 mol $Cl_2$      **b.** 7 mol Ti, 17 mol $Cl_2$
**c.** 12.4 mol Ti, 18.8 mol $Cl_2$

**42.** Calculate the theoretical yield of product (in moles) for each initial amount of reactants.

$$3\,Mn(s) + 2\,O_2(g) \longrightarrow Mn_3O_4(s)$$

**a.** 3 mol Mn, 3 mol $O_2$      **b.** 4 mol Mn, 7 mol $O_2$
**c.** 27.5 mol Mn, 43.8 mol $O_2$

**43.** Zinc sulfide reacts with oxygen according to the reaction:

$$2\,ZnS(s) + 3\,O_2(g) \longrightarrow 2\,ZnO(s) + 2\,SO_2(g)$$

A reaction mixture initially contains 4.2 mol ZnS and 6.8 mol $O_2$. Once the reaction has occurred as completely as possible, what amount (in moles) of the excess reactant remains?

**44.** Iron(II) sulfide reacts with hydrochloric acid according to the reaction:

$$FeS(s) + 2\,HCl(aq) \longrightarrow FeCl_2(s) + H_2S(g)$$

A reaction mixture initially contains 0.223 mol FeS and 0.652 mol HCl. Once the reaction has occurred as completely as possible, what amount (in moles) of the excess reactant remains?

**45.** For the reaction shown, calculate the theoretical yield of product (in grams) for each initial amount of reactants.

$$2\,Al(s) + 3\,Cl_2(g) \longrightarrow 2\,AlCl_3(s)$$

**a.** 2.0 g Al, 2.0 g $Cl_2$      **b.** 7.5 g Al, 24.8 g $Cl_2$
**c.** 0.235 g Al, 1.15 g $Cl_2$

**46.** For the reaction shown, calculate the theoretical yield of the product (in grams) for each initial amount of reactants.

$$Ti(s) + 2\,F_2(g) \longrightarrow TiF_4(s)$$

**a.** 5.0 g Ti, 5.0 g $F_2$      **b.** 2.4 g Ti, 1.6 g $F_2$
**c.** 0.233 g Ti, 0.288 g $F_2$

**47.** Iron(III) oxide reacts with carbon monoxide according to the equation:

$$Fe_2O_3(s) + 3\,CO(g) \longrightarrow 2\,Fe(s) + 3\,CO_2(g)$$

A reaction mixture initially contains 22.55 g $Fe_2O_3$ and 14.78 g CO. Once the reaction has occurred as completely as possible, what mass (in g) of the excess reactant remains?

**48.** Elemental phosphorus reacts with chlorine gas according to the equation:

$$P_4(s) + 6\,Cl_2(g) \longrightarrow 4\,PCl_3(l)$$

A reaction mixture initially contains 45.69 g $P_4$ and 131.3 g $Cl_2$. Once the reaction has occurred as completely as possible, what mass (in g) of the excess reactant remains?

**49.** Lead ions can be precipitated from solution with KCl according to the reaction:

$$Pb^{2+}(aq) + 2\,KCl(aq) \longrightarrow PbCl_2(s) + 2\,K^+(aq)$$

When 28.5 g KCl is added to a solution containing 25.7 g $Pb^{2+}$, a $PbCl_2$ precipitate forms. The precipitate is filtered and dried and found to have a mass of 29.4 g. Determine the limiting reactant, theoretical yield of $PbCl_2$, and percent yield for the reaction.

**50.** Magnesium oxide can be made by heating magnesium metal in the presence of oxygen. The balanced equation for the reaction is:

$$2\,Mg(s) + O_2(g) \longrightarrow 2\,MgO(s)$$

When 10.1 g of Mg reacts with 10.5 g $O_2$, 11.9 g MgO is collected. Determine the limiting reactant, theoretical yield, and percent yield for the reaction.

**51.** Urea ($CH_4N_2O$) is a common fertilizer that is synthesized by the reaction of ammonia ($NH_3$) with carbon dioxide:

$$2\,NH_3(aq) + CO_2(aq) \longrightarrow CH_4N_2O(aq) + H_2O(l)$$

In an industrial synthesis of urea, a chemist combines 136.4 kg of ammonia with 211.4 kg of carbon dioxide and obtains 168.4 kg of urea. Determine the limiting reactant, theoretical yield of urea, and percent yield for the reaction.

**52.** Many computer chips are manufactured from silicon, which occurs in nature as $SiO_2$. When $SiO_2$ is heated to melting, it reacts with solid carbon to form liquid silicon and carbon monoxide gas. In an industrial preparation of silicon, 155.8 kg of $SiO_2$ reacts with 78.3 kg of carbon to produce 66.1 kg of silicon. Determine the limiting reactant, theoretical yield, and percent yield for the reaction.

## Solution Concentration and Solution Stoichiometry

**53.** Calculate the molarity of each solution.
   **a.** 3.25 mol of LiCl in 2.78 L solution
   **b.** 28.33 g $C_6H_{12}O_6$ in 1.28 L of solution
   **c.** 32.4 mg NaCl in 122.4 mL of solution

**54.** Calculate the molarity of each solution.
   **a.** 0.38 mol of $LiNO_3$ in 6.14 L of solution
   **b.** 72.8 g $C_2H_6O$ in 2.34 L of solution
   **c.** 12.87 mg KI in 112.4 mL of solution

**55.** What is the molarity of $NO_3^-$ in each solution?
   **a.** 0.150 M $KNO_3$      **b.** 0.150 M $Ca(NO_3)_2$
   **c.** 0.150 M $Al(NO_3)_3$

**56.** What is the molarity of $Cl^-$ in each solution?
   **a.** 0.200 M NaCl      **b.** 0.150 M $SrCl_2$
   **c.** 0.100 M $AlCl_3$

**57.** How many moles of KCl are contained in each solution?
   **a.** 0.556 L of a 2.3 M KCl solution
   **b.** 1.8 L of a 0.85 M KCl solution
   **c.** 114 mL of a 1.85 M KCl solution

**58.** What volume of 0.200 M ethanol solution contains each amount in moles of ethanol?
   **a.** 0.45 mol ethanol      **b.** 1.22 mol ethanol
   **c.** $1.2 \times 10^{-2}$ mol ethanol

**59.** A laboratory procedure calls for making 400.0 mL of a 1.1 M $NaNO_3$ solution. What mass of $NaNO_3$ (in g) is needed?

**60.** A chemist wants to make 5.5 L of a 0.300 M $CaCl_2$ solution. What mass of $CaCl_2$ (in g) should the chemist use?

**61.** If 123 mL of a 1.1 M glucose solution is diluted to 500.0 mL, what is the molarity of the diluted solution?

**62.** If 3.5 L of a 4.8 M $SrCl_2$ solution is diluted to 45 L, what is the molarity of the diluted solution?

**63.** To what volume should you dilute 50.0 mL of a 12 M stock $HNO_3$ solution to obtain a 0.100 M $HNO_3$ solution?

**64.** To what volume should you dilute 25 mL of a 10.0 M $H_2SO_4$ solution to obtain a 0.150 M $H_2SO_4$ solution?

**65.** Consider the precipitation reaction:

$$2\,Na_3PO_4(aq) + 3\,CuCl_2(aq) \longrightarrow Cu_3(PO_4)_2(s) + 6\,NaCl(aq)$$

What volume of 0.175 M $Na_3PO_4$ solution is necessary to completely react with 95.4 mL of 0.102 M $CuCl_2$?

**66.** Consider the reaction:

$$Li_2S(aq) + Co(NO_3)_2(aq) \longrightarrow 2\,LiNO_3(aq) + CoS(s)$$

What volume of 0.150 M $Li_2S$ solution is required to completely react with 125 mL of 0.150 M $Co(NO_3)_2$?

**67.** What is the minimum amount of 6.0 M $H_2SO_4$ necessary to produce 25.0 g of $H_2(g)$ according to the reaction between aluminum and sulfuric acid?

$$2\,Al(s) + 3\,H_2SO_4(aq) \longrightarrow Al_2(SO_4)_3(aq) + 3\,H_2(g)$$

**68.** What is the molarity of $ZnCl_2$ that forms when 25.0 g of zinc completely reacts with $CuCl_2$ according to the following reaction? Assume a final volume of 275 mL.

$$Zn(s) + CuCl_2(aq) \longrightarrow ZnCl_2(aq) + Cu(s)$$

**69.** A 25.0-mL sample of a 1.20 M potassium chloride solution is mixed with 15.0 mL of a 0.900 M lead(II) nitrate solution and this precipitation reaction occurs:

$$2\,KCl(aq) + Pb(NO_3)_2(aq) \longrightarrow PbCl_2(s) + 2\,KNO_3(aq)$$

The solid $PbCl_2$ is collected, dried, and found to have a mass of 2.45 g. Determine the limiting reactant, the theoretical yield, and the percent yield.

**70.** A 55.0-mL sample of a 0.102 M potassium sulfate solution is mixed with 35.0 mL of a 0.114 M lead(II) acetate solution and this precipitation reaction occurs:

$$K_2SO_4(aq) + Pb(C_2H_3O_2)_2(aq) \longrightarrow 2\,KC_2H_3O_2(aq) + PbSO_4(s)$$

The solid $PbSO_4$ is collected, dried, and found to have a mass of 1.01 g. Determine the limiting reactant, the theoretical yield, and the percent yield.

## Types of Aqueous Solutions and Solubility

**71.** For each compound (all water soluble), would you expect the resulting aqueous solution to conduct electrical current?
   **a.** CsCl     **b.** $CH_3OH$     **c.** $Ca(NO_2)_2$   **d.** $C_6H_{12}O_6$

**72.** Classify each compound as a strong electrolyte or nonelectrolyte.
   **a.** $MgBr_2$     **b.** $C_{12}H_{22}O_{11}$     **c.** $Na_2CO_3$     **d.** KOH

**73.** Determine whether each compound is soluble or insoluble. If the compound is soluble, list the ions present in solution.
   **a.** $AgNO_3$     **b.** $Pb(C_2H_3O_2)_2$     **c.** $KNO_3$     **d.** $(NH_4)_2S$

**74.** Determine whether each compound is soluble or insoluble. If the compound is soluble, list the ions present in solution.
   **a.** AgI     **b.** $Cu_3(PO_4)_2$     **c.** $CoCO_3$     **d.** $K_3PO_4$

## Precipitation Reactions

**75.** Complete and balance each equation. If no reaction occurs, write "NO REACTION."
   **a.** $LiI(aq) + BaS(aq) \longrightarrow$
   **b.** $KCl(aq) + CaS(aq) \longrightarrow$
   **c.** $CrBr_2(aq) + Na_2CO_3(aq) \longrightarrow$
   **d.** $NaOH(aq) + FeCl_3(aq) \longrightarrow$

**76.** Complete and balance each equation. If no reaction occurs, write "NO REACTION."
   **a.** $NaNO_3(aq) + KCl(aq) \longrightarrow$
   **b.** $NaCl(aq) + Hg_2(C_2H_3O_2)_2(aq) \longrightarrow$
   **c.** $(NH_4)_2SO_4(aq) + SrCl_2(aq) \longrightarrow$
   **d.** $NH_4Cl(aq) + AgNO_3(aq) \longrightarrow$

**77.** Write a molecular equation for the precipitation reaction that occurs (if any) when each pair of aqueous solutions is mixed. If no reaction occurs, write "NO REACTION."

   **a.** potassium carbonate and lead(II) nitrate
   **b.** lithium sulfate and lead(II) acetate
   **c.** copper(II) nitrate and magnesium sulfide
   **d.** strontium nitrate and potassium iodide

**78.** Write a molecular equation for the precipitation reaction that occurs (if any) when each pair of aqueous solutions is mixed. If no reaction occurs, write "NO REACTION."
   **a.** sodium chloride and lead(II) acetate
   **b.** potassium sulfate and strontium iodide
   **c.** cesium chloride and calcium sulfide
   **d.** chromium(III) nitrate and sodium phosphate

## Ionic and Net Ionic Equations

**79.** Write balanced complete ionic and net ionic equations for each reaction.
   **a.** $HCl(aq) + LiOH(aq) \longrightarrow H_2O(l) + LiCl(aq)$
   **b.** $CaS(aq) + CuCl_2(aq) \longrightarrow CuS(s) + CaCl_2(aq)$
   **c.** $NaOH(aq) + HC_2H_3O_2(aq) \longrightarrow H_2O(l) + NaC_2H_3O_2(aq)$
   **d.** $Na_3PO_4(aq) + NiCl_2(aq) \longrightarrow Ni_3(PO_4)_2(s) + NaCl(aq)$

**80.** Write balanced complete ionic and net ionic equations for each reaction.
   **a.** $K_2SO_4(aq) + CaI_2(aq) \longrightarrow CaSO_4(s) + KI(aq)$
   **b.** $NH_4Cl(aq) + NaOH(aq) \longrightarrow H_2O(l) + NH_3(g) + NaCl(aq)$
   **c.** $AgNO_3(aq) + NaCl(aq) \longrightarrow AgCl(s) + NaNO_3(aq)$
   **d.** $HC_2H_3O_2(aq) + K_2CO_3(aq) \longrightarrow$
   $$H_2O(l) + CO_2(g) + KC_2H_3O_2(aq)$$

**81.** Mercury(I) ions ($Hg_2^{2+}$) can be removed from solution by precipitation with $Cl^-$. Suppose that a solution contains aqueous $Hg_2(NO_3)_2$. Write complete ionic and net ionic equations for the reaction of aqueous $Hg_2(NO_3)_2$ with aqueous sodium chloride to form solid $Hg_2Cl_2$ and aqueous sodium nitrate.

**82.** Lead(II) ions can be removed from solution by precipitation with sulfate ions. Suppose that a solution contains lead(II) nitrate. Write complete ionic and net ionic equations for the reaction of aqueous lead(II) nitrate with aqueous potassium sulfate to form solid lead(II) sulfate and aqueous potassium nitrate.

## Acid–Base and Gas-Evolution Reactions

**83.** Write balanced molecular and net ionic equations for the reaction between hydrobromic acid and potassium hydroxide.

**84.** Write balanced molecular and net ionic equations for the reaction between nitric acid and calcium hydroxide.

**85.** Complete and balance each acid–base equation.
   **a.** $H_2SO_4(aq) + Ca(OH)_2(aq) \longrightarrow$
   **b.** $HClO_4(aq) + KOH(aq) \longrightarrow$
   **c.** $H_2SO_4(aq) + NaOH(aq) \longrightarrow$

**86.** Complete and balance each acid–base equation.
   **a.** $HI(aq) + LiOH(aq) \longrightarrow$
   **b.** $HC_2H_3O_2(aq) + Ca(OH)_2(aq) \longrightarrow$
   **c.** $HCl(aq) + Ba(OH)_2(aq) \longrightarrow$

**87.** Write balanced complete ionic and net ionic equations for each acid–base reaction.
   **a.** $HBr(aq) + NaOH(aq) \longrightarrow$
   **b.** $HF(aq) + NaOH(aq) \longrightarrow$
   **c.** $HC_2H_3O_2(aq) + RbOH(aq) \longrightarrow$

**88.** Write balanced complete ionic and net ionic equations for each acid–base reaction.
   **a.** $HI(aq) + RbOH(aq) \longrightarrow$
   **b.** $HCHO_2(aq) + NaOH(aq) \longrightarrow$
   **c.** $HC_2H_3O_2(aq) + LiOH(aq) \longrightarrow$

**89.** A 25.00-mL sample of an unknown $HClO_4$ solution requires titration with 22.62 mL of 0.2000 M NaOH to reach the equivalence point. What is the concentration of the unknown $HClO_4$ solution? The neutralization reaction is

$$HClO_4(aq) + NaOH(aq) \longrightarrow H_2O(l) + NaClO_4(aq)$$

**90.** A 30.00-mL sample of an unknown $H_3PO_4$ solution is titrated with a 0.100 M NaOH solution. The equivalence point is reached when 26.38 mL of NaOH solution is added. What is the concentration of the unknown $H_3PO_4$ solution? The neutralization reaction is

$$H_3PO_4(aq) + 3\,NaOH(aq) \longrightarrow 3\,H_2O(l) + Na_3PO_4(aq)$$

**91.** Complete and balance each gas-evolution equation.
  **a.** $HBr(aq) + NiS(s) \longrightarrow$
  **b.** $NH_4I(aq) + NaOH(aq) \longrightarrow$
  **c.** $HBr(aq) + Na_2S(aq) \longrightarrow$
  **d.** $HClO_4(aq) + Li_2CO_3(aq) \longrightarrow$

**92.** Complete and balance each gas-evolution equation.
  **a.** $HNO_3(aq) + Na_2SO_3(aq) \longrightarrow$
  **b.** $HCl(aq) + KHCO_3(aq) \longrightarrow$
  **c.** $HC_2H_3O_2(aq) + NaHSO_3(aq) \longrightarrow$
  **d.** $(NH_4)_2SO_4(aq) + Ca(OH)_2(aq) \longrightarrow$

## Oxidation–Reduction and Combustion

**93.** Assign oxidation states to each atom in each element, ion, or compound.
  **a.** Ag     **b.** $Ag^+$     **c.** $CaF_2$
  **d.** $H_2S$     **e.** $CO_3^{2-}$     **f.** $CrO_4^{2-}$

**94.** Assign oxidation states to each atom in each element, ion, or compound.
  **a.** $Cl_2$     **b.** $Fe^{3+}$     **c.** $CuCl_2$
  **d.** $CH_4$     **e.** $Cr_2O_7^{2-}$     **f.** $HSO_4^-$

**95.** What is the oxidation state of Cr in each compound?
  **a.** CrO     **b.** $CrO_3$     **c.** $Cr_2O_3$

**96.** What is the oxidation state of Cl in each ion?
  **a.** $ClO^-$     **b.** $ClO_2^-$     **c.** $ClO_3^-$     **d.** $ClO_4^-$

**97.** Determine whether each reaction is a redox reaction. For each redox reaction, identify the oxidizing agent and the reducing agent.
  **a.** $4\,Li(s) + O_2(g) \longrightarrow 2\,Li_2O(s)$
  **b.** $Mg(s) + Fe^{2+}(aq) \longrightarrow Mg^{2+}(aq) + Fe(s)$
  **c.** $Pb(NO_3)_2(aq) + Na_2SO_4(aq) \longrightarrow PbSO_4(s) + 2\,NaNO_3(aq)$
  **d.** $HBr(aq) + KOH(aq) \longrightarrow H_2O(l) + KBr(aq)$

**98.** Determine whether each reaction is a redox reaction. For each redox reaction, identify the oxidizing agent and the reducing agent.
  **a.** $Al(s) + 3\,Ag^+(aq) \longrightarrow Al^{3+}(aq) + 3\,Ag(s)$
  **b.** $SO_3(g) + H_2O(l) \longrightarrow H_2SO_4(aq)$
  **c.** $Ba(s) + Cl_2(g) \longrightarrow BaCl_2(s)$
  **d.** $Mg(s) + Br_2(l) \longrightarrow MgBr_2(s)$

**99.** Complete and balance each combustion reaction equation.
  **a.** $S(s) + O_2(g) \longrightarrow$        **b.** $C_3H_6(g) + O_2(g) \longrightarrow$
  **c.** $Ca(s) + O_2(g) \longrightarrow$        **d.** $C_5H_{12}S(l) + O_2(g) \longrightarrow$

**100.** Complete and balance each combustion reaction equation:
  **a.** $C_4H_6(g) + O_2(g) \longrightarrow$        **b.** $C(s) + O_2(g) \longrightarrow$
  **c.** $CS_2(s) + O_2(g) \longrightarrow$        **d.** $C_3H_8O(l) + O_2(g) \longrightarrow$

# Cumulative Problems

**101.** The density of a 20.0% by mass ethylene glycol ($C_2H_6O_2$) solution in water is 1.03 g/mL. Find the molarity of the solution.

**102.** Find the percent by mass of sodium chloride in a 1.35 M NaCl solution. The density of the solution is 1.05 g/mL.

**103.** People sometimes use sodium bicarbonate as an antacid to neutralize excess hydrochloric acid in an upset stomach. What mass of hydrochloric acid (in grams) can 2.5 g of sodium bicarbonate neutralize? (Hint: Begin by writing a balanced equation for the reaction between aqueous sodium bicarbonate and aqueous hydrochloric acid.)

**104.** Toilet bowl cleaners often contain hydrochloric acid, which dissolves the calcium carbonate deposits that accumulate within a toilet bowl. What mass of calcium carbonate (in grams) can 3.8 g of HCl dissolve? (Hint: Begin by writing a balanced equation for the reaction between hydrochloric acid and calcium carbonate.)

**105.** The combustion of gasoline produces carbon dioxide and water. Assume gasoline to be pure octane ($C_8H_{18}$) and calculate the mass (in kg) of carbon dioxide that is added to the atmosphere per 1.0 kg of octane burned. (Hint: Begin by writing a balanced equation for the combustion reaction.)

**106.** Many home barbeques are fueled with propane gas ($C_3H_8$). What mass of carbon dioxide (in kg) is produced upon the complete combustion of 18.9 L of propane (approximate contents of one 5-gallon tank)? Assume that the density of the liquid propane in the tank is 0.621 g/mL. (Hint: Begin by writing a balanced equation for the combustion reaction.)

**107.** Aspirin can be made in the laboratory by reacting acetic anhydride ($C_4H_6O_3$) with salicylic acid ($C_7H_6O_3$) to form aspirin ($C_9H_8O_4$) and acetic acid ($C_2H_4O_2$). The balanced equation is:

$$C_4H_6O_3 + C_7H_6O_3 \longrightarrow C_9H_8O_4 + C_2H_4O_2$$

In a laboratory synthesis, a student begins with 3.00 mL of acetic anhydride (density = 1.08 g/mL) and 1.25 g of salicylic acid. Once the reaction is complete, the student collects 1.22 g of aspirin. Determine the limiting reactant, theoretical yield of aspirin, and percent yield for the reaction.

**108.** The combustion of liquid ethanol ($C_2H_5OH$) produces carbon dioxide and water. After 4.62 mL of ethanol (density = 0.789 g/mL) is allowed to burn in the presence of 15.55 g of oxygen gas, 3.72 mL of water (density = 1.00 g/mL) is collected. Determine the limiting reactant, theoretical yield of $H_2O$, and percent yield for the reaction. (Hint: Write a balanced equation for the combustion of ethanol.)

**109.** A loud classroom demonstration involves igniting a hydrogen-filled balloon. The hydrogen within the balloon reacts explosively with oxygen in the air to form water. If the balloon is filled with a mixture of hydrogen and oxygen, the explosion is even louder than if the balloon is filled only with hydrogen—the intensity of the explosion depends on the relative amounts of oxygen and hydrogen within the balloon. Look at the molecular views representing different amounts of hydrogen and oxygen in four different balloons. Based on the balanced chemical equation, which balloon will make the loudest explosion?

**(a)**          **(b)**

**(c)**          **(d)**

**110.** A hydrochloric acid solution will neutralize a sodium hydroxide solution. Look at the molecular views showing one beaker of HCl and four beakers of NaOH. Which NaOH beaker will just neutralize the HCl beaker? Begin by writing a balanced chemical equation for the neutralization reaction.

$Cl^-$
$H^+$

$Na^+$
$OH^-$

**(a)**     **(b)**     **(c)**     **(d)**

**111.** Predict the products and write a balanced molecular equation for each reaction. If no reaction occurs, write "NO REACTION."
   **a.** $HCl(aq) + Hg_2(NO_3)_2(aq) \longrightarrow$
   **b.** $KHSO_3(aq) + HNO_3(aq) \longrightarrow$
   **c.** aqueous ammonium chloride and aqueous lead(II) nitrate
   **d.** aqueous ammonium chloride and aqueous calcium hydroxide

**112.** Predict the products and write a balanced molecular equation for each reaction. If no reaction occurs, write "NO REACTION."
   **a.** $H_2SO_4(aq) + HNO_3(aq) \longrightarrow$
   **b.** $Cr(NO_3)_3(aq) + LiOH(aq) \longrightarrow$
   **c.** liquid pentanol ($C_5H_{12}O$) and gaseous oxygen
   **d.** aqueous strontium sulfide and aqueous copper(II) sulfate

**113.** Hard water often contains dissolved $Ca^{2+}$ and $Mg^{2+}$ ions. One way to soften water is to add phosphates. The phosphate ion forms insoluble precipitates with calcium and magnesium ions, removing them from solution. A solution is 0.050 M in calcium chloride and 0.085 M in magnesium nitrate. What mass of sodium phosphate would you add to 1.5 L of this solution to completely eliminate the hard water ions? Assume complete reaction.

**114.** An acid solution is 0.100 M in HCl and 0.200 M in $H_2SO_4$. What volume of a 0.150 M KOH solution would completely neutralize all the acid in 500.0 mL of this solution?

**115.** Find the mass of barium metal (in grams) that must react with $O_2$ to produce enough barium oxide to prepare 1.0 L of a 0.10 M solution of $OH^-$.

**116.** A solution contains $Cr^{3+}$ ions and $Mg^{2+}$ ions. The addition of 1.00 L of 1.51 M NaF solution causes the complete precipitation of these ions as $CrF_3(s)$ and $MgF_2(s)$. The total mass of the precipitate is 49.6 g. Find the mass of $Cr^{3+}$ in the original solution.

**117.** The nitrogen in sodium nitrate and in ammonium sulfate is available to plants as fertilizer. Which is the more economical source of nitrogen, a fertilizer containing 30.0% sodium nitrate by weight and costing $9.00 per 100 lb or one containing 20.0% ammonium sulfate by weight and costing $8.10 per 100 lb?

**118.** Find the volume of 0.110 M hydrochloric acid necessary to react completely with 1.52 g $Al(OH)_3$.

**119.** Treatment of gold metal with $BrF_3$ and KF produces $Br_2$ and $KAuF_4$, a salt of gold. Identify the oxidizing agent and the reducing agent in this reaction. Find the mass of the gold salt that forms when a 73.5-g mixture of equal masses of all three reactants is prepared.

**120.** We prepare a solution by mixing 0.10 L of 0.12 M sodium chloride with 0.23 L of a 0.18 M $MgCl_2$ solution. What volume of a 0.20 M silver nitrate solution do we need to precipitate all the $Cl^-$ ion in the solution as AgCl?

**121.** A solution contains one or more of the following ions: $Ag^+$, $Ca^{2+}$, and $Cu^{2+}$. When you add sodium chloride to the solution, no precipitate forms. When you add sodium sulfate to the solution, a white precipitate forms. You filter off the precipitate and add sodium carbonate to the remaining solution, producing another precipitate. Which ions were present in the original solution? Write net ionic equations for the formation of each of the precipitates observed.

**122.** A solution contains one or more of the following ions: $Hg_2^{2+}$, $Ba^{2+}$, and $Fe^{2+}$. When you add potassium chloride to the solution, a precipitate forms. The precipitate is filtered off, and you add potassium sulfate to the remaining solution, producing no precipitate. When you add potassium carbonate to the remaining solution, a precipitate forms. Which ions were present in the original solution? Write net ionic equations for the formation of each of the precipitates observed.

**123.** The reaction of $NH_3$ and $O_2$ forms NO and water. The NO can be used to convert $P_4$ to $P_4O_6$, forming $N_2$ in the process. The $P_4O_6$ can be treated with water to form $H_3PO_3$, which forms $PH_3$ and $H_3PO_4$ when heated. Find the mass of $PH_3$ that forms from the reaction of 1.00 g of $NH_3$.

**124.** An important reaction that takes place in a blast furnace during the production of iron is the formation of iron metal and $CO_2$ from $Fe_2O_3$ and CO. Determine the mass of $Fe_2O_3$ required to form 910 kg of iron. Determine the amount of $CO_2$ that forms in this process.

**125.** A liquid fuel mixture contains 30.35% hexane ($C_6H_{14}$), 15.85% heptane ($C_7H_{16}$), and the rest octane ($C_8H_{18}$). What maximum mass of carbon dioxide is produced by the complete combustion of 10.0 kg of this fuel mixture?

**126.** Titanium occurs in the magnetic mineral ilmenite ($FeTiO_3$), which is often found mixed with sand. The ilmenite can be separated from the sand with magnets. The titanium can then be extracted from the ilmenite by the following set of reactions:

$$FeTiO_3(s) + 3 Cl_2(g) + 3 C(s) \longrightarrow 3 CO(g) + FeCl_2(s) + TiCl_4(g)$$
$$TiCl_4(g) + 2 Mg(s) \longrightarrow 2 MgCl_2(l) + Ti(s)$$

Suppose that an ilmenite–sand mixture contains 22.8% ilmenite by mass and that the first reaction is carried out with a 90.8% yield. If the second reaction is carried out with an 85.9% yield, what mass of titanium can be obtained from 1.00 kg of the ilmenite–sand mixture?

# Challenge Problems

**127.** A mixture of $C_3H_8$ and $C_2H_2$ has a mass of 2.0 g. It is burned in excess $O_2$ to form a mixture of water and carbon dioxide that contains 1.5 times as many moles of $CO_2$ as of water. Find the mass of $C_2H_2$ in the original mixture.

**128.** A mixture of 20.6 g of P and 79.4 g of $Cl_2$ reacts completely to form $PCl_3$ and $PCl_5$ as the only products. Find the mass of $PCl_3$ that forms.

**129.** A solution contains $Ag^+$ and $Hg^{2+}$ ions. The addition of 0.100 L of 1.22 M NaI solution is just enough to precipitate all the ions as AgI and $HgI_2$. The total mass of the precipitate is 28.1 g. Find the mass of AgI in the precipitate.

**130.** The water in lakes that have been acidified by acid rain ($HNO_3$ and $H_2SO_4$) can be neutralized by a process called liming, in which limestone ($CaCO_3$) is added to the acidified water. What mass of limestone (in kg) would completely neutralize a 15.2 billion-liter lake that is $1.8 \times 10^{-5}$ M in $H_2SO_4$ and $8.7 \times 10^{-6}$ M in $HNO_3$?

**131.** Recall from Section 4.6 that sodium carbonate is often added to laundry detergents to soften hard water and make the detergent more effective. Suppose that a particular detergent mixture is designed to soften hard water that is $3.5 \times 10^{-3}$ M in $Ca^{2+}$ and $1.1 \times 10^{-3}$ M in $Mg^{2+}$ and that the average capacity of a washing machine is 19.5 gallons of water. If the detergent requires using 0.65 kg detergent per load of laundry, what percentage (by mass) of the detergent should be sodium carbonate in order to completely precipitate all of the calcium and magnesium ions in an average load of laundry water?

**132.** Lead poisoning is a serious condition resulting from the ingestion of lead in food, water, or other environmental sources. It affects the central nervous system, leading to a variety of symptoms such as distractibility, lethargy, and loss of motor coordination. Lead poisoning is treated with chelating agents, substances that bind to metal ions, allowing them to be eliminated in the urine. A modern chelating agent used for this purpose is succimer ($C_4H_6O_4S_2$). Suppose you are trying to determine the appropriate dose for succimer treatment of lead poisoning. What minimum mass of succimer (in mg) is needed to bind all of the lead in a patient's bloodstream? Assume that patient blood lead levels are $45\ \mu g/dL$, that total blood volume is 5.0 L, and that 1 mol of succimer binds 1 mol of lead.

**133.** A particular kind of emergency breathing apparatus—often placed in mines, caves, or other places where oxygen might become depleted or where the air might become poisoned—works via the following chemical reaction:

$$4\,KO_2(s) + 2\,CO_2(g) \rightarrow 2\,K_2CO_3(s) + 3\,O_2(g)$$

Notice that the reaction produces $O_2$, which can be breathed, and absorbs $CO_2$, a product of respiration. Suppose you work for a company interested in producing a self-rescue breathing apparatus (based on the given reaction) that would allow the user to survive for 10 minutes in an emergency situation. What are the important chemical considerations in designing such a unit? Estimate how much $KO_2$ would be required for the apparatus. (Find any necessary additional information—such as human breathing rates—from appropriate sources. Assume that normal air is 20% oxygen.)

**134.** Metallic aluminum reacts with $MnO_2$ at elevated temperatures to form manganese metal and aluminum oxide. A mixture of the two reactants is 67.2% mole percent Al. Find the theoretical yield (in grams) of manganese from the reaction of 250 g of this mixture.

**135.** Hydrolysis of the compound $B_5H_9$ forms boric acid, $H_3BO_3$. Fusion of boric acid with sodium oxide forms a borate salt, $Na_2B_4O_7$. Without writing complete equations, find the mass (in grams) of $B_5H_9$ required to form 151 g of the borate salt by this reaction sequence.

# Conceptual Problems

**136.** Consider the reaction:

$$4\,K(s) + O_2(g) \longrightarrow 2\,K_2O(s)$$

The molar mass of K is 39.10 g/mol, and that of $O_2$ is 32.00 g/mol. Without doing any calculations, pick the conditions under which potassium is the limiting reactant and explain your reasoning.
**a.** 170 g K, 31 g $O_2$  **b.** 16 g K, 2.5 g $O_2$
**c.** 165 kg K, 28 kg $O_2$  **d.** 1.5 g K, 0.38 g $O_2$

**137.** Consider the reaction:

$$2\,NO(g) + 5\,H_2(g) \longrightarrow 2\,NH_3(g) + 2\,H_2O(g)$$

A reaction mixture initially contains 5 moles of NO and 10 moles of $H_2$. Without doing any calculations, determine which set of amounts best represents the mixture after the reactants have reacted as completely as possible. Explain your reasoning.
**a.** 1 mol NO, 0 mol $H_2$, 4 mol $NH_3$, 4 mol $H_2O$
**b.** 0 mol NO, 1 mol $H_2$, 5 mol $NH_3$, 5 mol $H_2O$
**c.** 3 mol NO, 5 mol $H_2$, 2 mol $NH_3$, 2 mol $H_2O$
**d.** 0 mol NO, 0 mol $H_2$, 4 mol $NH_3$, 4 mol $H_2O$

**138.** The following circle represents 1.0 liter of a solution with a solute concentration of 1 M:

Explain what you would add (the amount of solute or volume of solvent) to the solution to obtain a solution represented by each diagram:

(a)          (b)          (c)

**139.** Consider the reaction:

$$2\,N_2H_4(g) + N_2O_4(g) \longrightarrow 3\,N_2(g) + 4\,H_2O(g)$$

Consider also this representation of an initial mixture of $N_2H_4$ and $N_2O_4$:

Which diagram best represents the reaction mixture after the reactants have reacted as completely as possible?

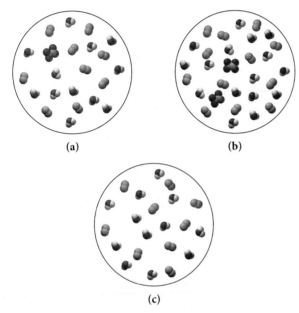

(a)     (b)

(c)

**140.** Consider the generic ionic compounds with the formulas $A_2X$ and $BY_2$ and the following solubility rules:

$A_2X$ soluble; $BY_2$ soluble; AY insoluble; BX soluble.

Assume $A^+$ ions are circles, $B^{2+}$ ions are squares, $X^{2-}$ ions are triangles, and $Y^-$ ions are diamonds. Solutions of the two compounds ($A_2X$ and $BY_2$) are represented as follows:

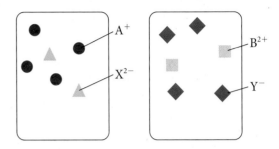

Draw a molecular-level representation showing the result of mixing the two given solutions and write an equation to represent the reaction.

*Active Classroom Learning*

# Questions for Group Work

*Discuss these questions with the group and record your consensus answer.*

**141.** Imagine you mix 16.05 g of methane ($CH_4$) gas and 96.00 g of oxygen ($O_2$) gas and then ignite the mixture. After a bright flash and a loud bang, some water droplets form on the inside of the reaction vessel.

 **a.** Write the balanced chemical reaction for the combustion of methane.

 **b.** Depict the process that occurred in the vessel using circles to represent atoms. Represent carbon with black circles, hydrogen with white circles, and oxygen with gray circles. Let one circle (or one molecule made of circles bonded together) represent exactly one mole.

 **c.** How many moles of water can you make? How many moles of carbon dioxide?

 **d.** Will anything be left over? What? How much?

 **e.** Identify the following: limiting reagent, excess reagent, and theoretical yield.

**142.** Write a detailed set of instructions for making two solutions: (1) 100 mL of 12 M NaOH from solid sodium hydroxide and (2) 1.00 L of 0.1 M NaOH from your first solution. You have in your lab: volumetric flasks marked to contain 100.0 mL and 1.000 L, a graduated cylinder, and a balance.

**143.** Review the solubility rules. Without referring back to the rules, have each group member list two ionic compounds that are expected to be soluble and two that are expected to be insoluble. Include at least one exception. Check the work of the other members of your group.

**144.** Define and give an example of each of the following classes of reactions: precipitation, acid–base, gas evolution, redox (noncombustion), and combustion. Each group member can do one, and then present his or her reaction to the group.

**145.** Using group members to represent atoms, ions, or electrons, act out the reaction $Zn(s) + Fe^{2+}(aq) \longrightarrow Zn^{2+}(aq) + Fe(s)$. Which group member is oxidized? Which is reduced? Which is the oxidizing agent? Which is the reducing agent?

# Data Interpretation and Analysis

## Fluoride in Drinking Water

146. In most communities in the United States, fluoride ions are added to public drinking water to help reduce dental caries (cavities in teeth). Fluoride ions in water react with calcium hydroxyapatite, a chemical component of teeth, to form hydroxyfluorapatite according to this reaction:

$$Ca_{10}(PO_4)_6(OH)_2 + F^- \longrightarrow Ca_{10}(PO_4)_6FOH + OH^-$$

Hydroxyfluorapatite reduces the rate of demineralization of tooth enamel; demineralization can lead to dental caries. However, too much fluoride in drinking water can cause dental fluorosis and/or enamel fluorosis, which stain teeth. Figure a ▼ lists the 1962 recommended and maximum fluoride ion concentration for drinking water in the United States as a function of annual average air temperature.

| Annual average of maximum daily air temperatures (°C) | Recommended fluoride Concentration (mg/L) | | | Maximum allowable fluoride concentration (mg/L) |
|---|---|---|---|---|
| | Lower | Optimum | Upper | |
| 10.0–12.0 | 0.9 | 1.2 | 1.7 | 2.4 |
| 12.1–14.6 | 0.8 | 1.1 | 1.5 | 2.2 |
| 14.7–17.7 | 0.8 | 1.0 | 1.3 | 2.0 |
| 17.8–21.4 | 0.7 | 0.9 | 1.2 | 1.8 |
| 21.5–26.3 | 0.7 | 0.8 | 1.0 | 1.6 |
| 26.4–31.5 | 0.6 | 0.7 | 0.8 | 1.4 |

Source: http://water.epa.gov/drink/contaminants/basicinformation/fluoride.cfm

▲ **FIGURE a   Recommended and Maximum Allowable Fluoride Concentration for U.S. Drinking Water**

A 1994 World Health Organization committee recommended an optimal level of fluoride in drinking water from 0.5 to 1.0 mg/L (milligrams per liter), depending on climate region. Municipalities typically use three commercially available compounds to add fluoride ions to public drinking water. Figure b ▼ lists the three compounds containing fluoride ions along with their molar mass, percent fluoride in the pure compound, and percent pure compound in the commercial product.

| | | Molar Mass | % F⁻ | % Pure Compound in the Commercial Product |
|---|---|---|---|---|
| NaF | Sodium fluoride | 42 | 45 | 98 |
| Na₂SiF₆ | Sodium fluorosilicate | 188 | 61 | 98 |
| H₂SiF₆ | Fluorosilicic acid | 144 | 79 | 25 |

▲ **FIGURE b   Percentage (by mass) of Fluoride Ion in Fluoride Compounds**

Use the information provided in the figures to answer the following questions:

a. Sodium fluoride (NaF) and sodium fluorosilicate (Na₂SiF₆) are commercially available in 100.0-lb bags. Calculate the mass, in kg, of fluoride ion in a 100.0-lb bag for each of these two compounds.

b. The density of fluorosilicic acid ($H_2SiF_6$) is related to its percent by weight as shown in Figure c ▼.

▲ **FIGURE c   Density of Fluorosilicic Acid at 20 °C**
Source: http://www.ml2rconsultancy.com/Pages/H2SiF6DATA.aspx

  i. Calculate the mass, in kg, of fluoride ion in 100.0 lbs of 25.0% (weight/weight) commercial fluorosilicic acid. For a mixture of two components, A and B, the % (weight/weight) means the weight of component A (lbs.) in the total weight (lbs.) of the contents − weight of A + weight B.

  ii. Calculate the molarity of the dilute (25.0% weight/weight) fluorosilicic acid.

c. A public water treatment plant located in a region with an annual average maximum temperature of 16.0 °C uses fluorosilicic acid to treat 50,000 gallons of water per day (5.00 × 10⁴ gallons).

  i. Calculate the amount of fluoride ions that must be added to the water, in kg, in order to meet the optimum amount of fluoride.

  ii. How much 25.0 % commercial fluorosilicic acid in kg must be added to 50,000 gallons of water in order to fluoridate at the optimum level?

d. The U.S. Institute of Medicine has established a maximum daily intake of fluoride of 4 mg F⁻/day for adults. Fluoridated drinking water provides about 75% of an adult's daily fluoride, assuming the adult is consuming mostly public water.

  i. If a community is located in a cold climate (10.0–12.0 °C) and the water has the optimum level of fluoride ions, calculate the mg of fluoride ion present in one glass of water. (Note that: 1 glass = 12.0 ounces of water = 0.355 liters.)

  ii. A public water treatment plant located in a hot climate has a lower fluoride ion concentration in the drinking water compared to the water from a public water treatment plant located in a colder climate. Both meet the recommended daily F⁻ guideline. Explain how this is possible.

e. If you were in charge of fluoridating the public water supply in your community, which compound would you use, how much would you need, and what would it cost? Refer to Figure d▼ to answer this question. Assume you need to treat 50,000 gallons of water per day.

| | | 2015 Price | % Pure Compound in the Commercial Product |
|---|---|---|---|
| NaF | Sodium fluoride | $85.60/kg | 98 |
| $Na_2SiF_6$ | Sodium fluorosilicate | $89.60/kg | 98 |
| $H_2SiF_6$ | Fluorosilicic acid | $52.80/L | 25 |

▲ **FIGURE d** **Commercial Price of Fluoride Compounds**

f. Under certain conditions, fluoride ions react with the calcium hydroxyapatite in teeth according to this reaction:

$$Ca_{10}(PO_4)_6(OH)_2 + 2\,F^- \longrightarrow Ca_{10}(PO_4)_6F_2 + 2\,OH^-$$

If this reaction occurs, how would this influence the amount of fluoridation required in public drinking water and the cost?

g. On April 27, 2015, the U.S. Department of Health and Human Services recommended decreasing the amount of fluoride added to drinking water by about 40% for all regions of the country. For example, communities located in the 17.8–21.4 °C annual average air temperature range should now have a maximum of 0.7 milligram of fluoride per liter of water instead of the long-standing recommended maximum of 1.2 milligrams of fluoride per liter. Explain how this new standard impacts your answers to parts e and f.

# Answers to Conceptual Connections

## Stoichiometry

**4.1** **(c)** Since each $O_2$ molecule reacts with 4 Na atoms, 12 Na atoms are required to react with 3 $O_2$ molecules.

## Stoichiometry II

**4.2**

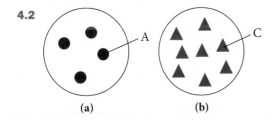

(a)   (b)

## Limiting Reactant and Theoretical Yield

**4.3** **(c)** Nitrogen is the limiting reactant, and there is enough nitrogen to make four $NH_3$ molecules. Hydrogen is in excess, and two hydrogen molecules remain after the reactants have reacted as completely as possible.

## Reactant in Excess

**4.4** The limiting reactant is the 1 mol $H_2O$, which is completely consumed. The 1 mol of $H_2O$ requires 3 mol of $NO_2$ to completely react; therefore, 2 mol $NO_2$ remain after the reaction is complete.

## Solutions

**4.5** **(b)** The mass of a solution is equal to the mass of the solute plus the mass of the solvent. Although the solute seems to disappear, it really does not, and its mass becomes part of the mass of the solution, in accordance with the law of mass conservation.

## Solution Dilution

**4.6** **(c)** Since the volume has doubled, the concentration is halved, so the same volume should contain half as many solute molecules.

## Solution Stoichiometry

**4.7** A is the limiting reactant. You have equal amounts of both reactants, but the reaction requires twice as much of A as B.

## Precipitation Reactions

**4.8**

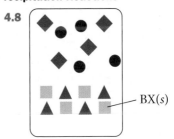

$$AX(aq) + BY(aq) \longrightarrow BX(s) + AY(aq)$$

## Oxidation Numbers in Polyatomic Ions

**4.9** **(a)** The charge of a polyatomic ion is the charge associated with the ion *as a whole*. The oxidation states of the individual atoms must sum to the charge of the ion, but they are assigned to *the individual atoms themselves.* Answer **(b)** is incorrect because oxidation state and charge *are not identical,* even though the charge of a *monoatomic* ion is equal to its oxidation state. Answer **(c)** is incorrect because charge *is* a physical property of ions. Conversely, the oxidation states of atoms are *not* real physical properties, but an imposed electron bookkeeping scheme.

## Oxidation and Reduction

**4.10** **(d)** Since oxidation and reduction must occur together, an increase in the oxidation state of a reactant is always accompanied by a decrease in the oxidation state of a reactant.

CHAPTER

# 5

# Gases

> So many of the properties of matter, especially when in the gaseous form, can be deduced from the hypothesis that their minute parts are in rapid motion, the velocity increasing with the temperature, that the precise nature of this motion becomes a subject of rational curiosity.

—JAMES CLERK MAXWELL (1831–1879)

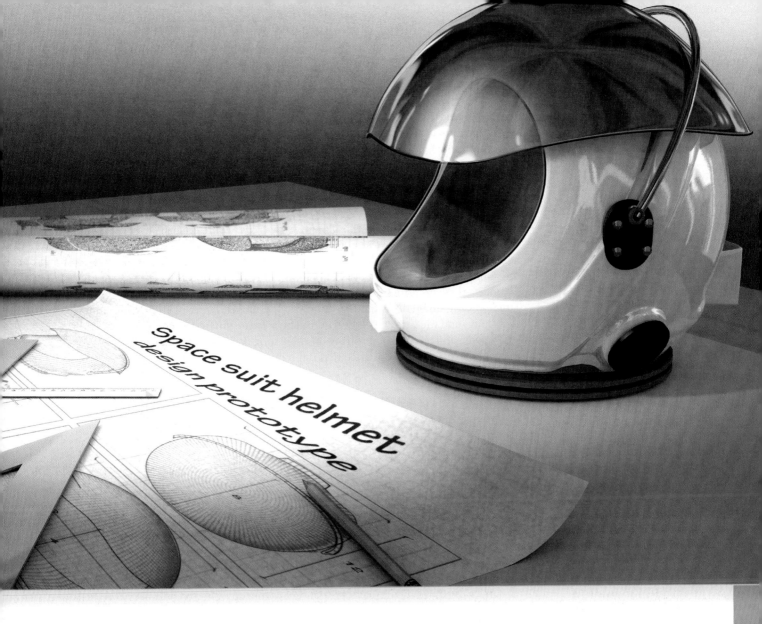

Space suit helmet
design prototype

Human beings can survive for weeks without food, days without water, but only minutes without air. Fortunately, we live at the bottom of a vast ocean of air, held to Earth by gravity. We inhale a lungful of this air every few seconds, keep some of the molecules for our own uses, add some molecules that our bodies no longer need, and exhale the mixture back into the surrounding air. The air around us is matter in the gaseous state. What are the fundamental properties of these gases? What laws describe their behavior? What theory explains these properties and laws? Recall that the scientific approach (see Section 1.2) proceeds in this way—from observations to laws to theories—exactly the way we will proceed in this chapter. The gaseous state is the simplest and best-understood state of matter. In this chapter, we examine that state.

▲ A pressurized suit protected *Felix Baumgartner from the vacuum of space during his record-breaking skydive. Any significant damage to the suit carried the risk of uncontrolled decompression, which would likely result in Baumgartner's death.*

## 5.1 Supersonic Skydiving and the Risk of Decompression

On October 14, 2012, just after midday in New Mexico, Austrian daredevil Felix Baumgartner stepped into the dark void of space 24 miles (38.6 km) above the Earth's surface. Baumgartner's 20-minute journey back to the desert floor broke the sound barrier and shattered the previous skydiving record of 19.5 miles (31.4 km).

When Baumgartner stepped into space, he was protected from the surrounding vacuum by a pressurized suit. The suit contained air at a pressure similar to that found on

Gas molecules

Collisions with surfaces create pressure.

▲ **FIGURE 5.1   Gas Pressure**
Pressure is the force per unit area exerted by gas molecules colliding with the surfaces around them.

**Pressure and Density**

Lower pressure          Higher pressure

▲ **FIGURE 5.2   Pressure and Particle Density**   A low density of gas particles results in low pressure. A high density of gas particles results in high pressure.

▶ Pressure variations in Earth's atmosphere create wind and weather. The H on this map indicates a region of high pressure, usually associated with clear weather. The L indicates a region of low pressure, usually associated with unstable weather.

the surface of Earth. **Pressure** is the force exerted per unit area by gas particles as they strike the surfaces around them (Figure 5.1 ◀). Just as a ball exerts a force when it bounces against a wall, so a gaseous atom or molecule exerts a force when it collides with a surface. The result of these collisions is pressure—a constant force on the surfaces exposed to any gas. The total pressure exerted by a gas depends on several factors, including the concentration of gas particles in the sample; the lower the concentration, the lower the pressure. At 24 miles (38.6 km) above the Earth's surface, the concentration of gas particles is much lower than it is at sea level; consequently, the pressure at that high altitude is very low. Without the pressurized suit, Baumgartner could not have survived the space-like conditions.

One of the primary risks that Baumgartner faced while in space was uncontrolled decompression. Any significant damage to his suit could have caused the air within the suit to escape, resulting in a large pressure drop. While the effects of a large pressure drop are sometimes exaggerated—for example, one urban myth claims that a person can explode—they are nonetheless lethal. If a pressure drop had occurred fast enough, a large pressure difference would have quickly developed between the air in Baumgartner's lungs and the surrounding vacuum. The pressure difference would have caused his lungs to expand too much, resulting in severe lung damage. Fortunately, Baumgartner's suit worked just as it was designed to do, and he plunged safely back to Earth.

## 5.2   Pressure: The Result of Molecular Collisions

Air can hold up a jumbo jet or knock down a building. How? As we discussed in Section 5.1, air contains gaseous atoms and molecules in constant motion. The particles collide with each other and with the surfaces around them. Each collision exerts only a small force, but when the forces of the many particles are summed, they quickly add up. The result of the constant collisions between the atoms or molecules in a gas and the surfaces around them is *pressure*. Because of pressure, we can drink from straws, inflate basketballs, and breathe. Variation in pressure in Earth's atmosphere creates wind, and changes in pressure help us to predict weather. Pressure is all around us and inside of us.

The pressure that a gas sample exerts is the *force* that results from the collisions of gas particles divided by the *area* of the surface with which they collide:

$$\text{Pressure} = \frac{\text{force}}{\text{area}} = \frac{F}{A} \tag{5.1}$$

The pressure exerted by a gas sample, therefore, depends on the number of gas particles in a given volume—the fewer the gas particles, the lower the force per unit area and the lower the pressure (Figure 5.2 ◀). Since the number of gas particles in a given volume

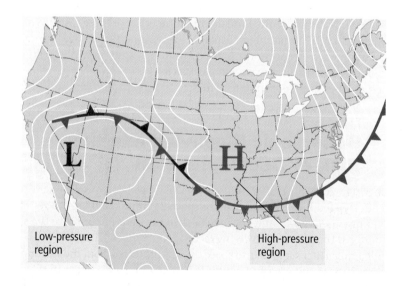

Low-pressure region                     High-pressure region

generally decreases with increasing altitude, *pressure decreases with increasing altitude.* Above 30,000 ft, for example, where most commercial airplanes fly, the pressure is so low that a person could pass out due to a lack of oxygen. For this reason, most airplane cabins are artificially pressurized.

You may sometimes feel the effect of a drop in pressure as a brief pain in your ears. This pain arises within the air-containing cavities in your ear (Figure 5.3▶). When you ascend to a higher altitude, the external pressure (the pressure that surrounds you) drops, while the pressure within your ear cavities (the internal pressure) remains the same. This creates an imbalance—the greater internal pressure forces your eardrum to bulge outward, causing pain. With time, and with the help of a yawn or two, the excess air within your ear's cavities escapes, equalizing the internal and external pressure and relieving the pain.

## Pressure Units

We measure pressure in several different units. A common unit of pressure, the **millimeter of mercury (mmHg)**, originates from how pressure is measured with a **barometer** (Figure 5.4▼). A barometer is an evacuated glass tube, the tip of which is submerged in a pool of mercury (Hg). Atmospheric pressure on the liquid mercury's surface forces the mercury upward into the evacuated tube. Because mercury is so dense (13.5 times more dense than water), atmospheric pressure can support a column of Hg that is only about 0.760 m or 760 mm (about 30 in) tall. (By contrast, atmospheric pressure can support a column of water that is about 10.3 m tall.) This makes a column of mercury a convenient way to measure pressure.

In a barometer, when the atmospheric pressure rises, the height of the mercury column rises as well. Similarly, when atmospheric pressure falls, the height of the column falls. The unit *millimeter of mercury* is often called a **torr**, after the Italian physicist Evangelista Torricelli (1608–1647) who invented the barometer:

$$1 \text{ mmHg} = 1 \text{ torr}$$

A second unit of pressure is the **atmosphere (atm)**, the average pressure at sea level. Since one atmosphere of pressure pushes a column of mercury to a height of 760 mm, 1 atm and 760 mmHg are equal:

$$1 \text{ atm} = 760 \text{ mmHg}$$

A fully inflated mountain bike tire has a pressure of about 6 atm, and the pressure at the top of Mount Everest is about 0.31 atm.

The SI unit of pressure is the **pascal (Pa)**, defined as 1 newton (N) per square meter:

$$1 \text{ Pa} = 1 \text{ N/m}^2$$

The pascal is a much smaller unit of pressure than the atmosphere:

$$1 \text{ atm} = 101{,}325 \text{ Pa}$$

Other common units of pressure include inches of mercury (in Hg) and pounds per square inch (psi):

$$1 \text{ atm} = 29.92 \text{ in Hg} \qquad 1 \text{ atm} = 14.7 \text{ psi}$$

Table 5.1, on the next page, summarizes common pressure units.

### Pressure Imbalance

Eardrum

Reduced external pressure

Normal pressure

Eardrum bulges outward, causing pain.

▲ **FIGURE 5.3 Pressure Imbalance** The discomfort you may feel in your ears upon ascending a mountain is caused by a pressure imbalance between the cavities in your ears and the outside air.

### The Mercury Barometer

Vacuum

Glass tube

Atmospheric pressure

760 mm (29.92 in)

Mercury

▲ **FIGURE 5.4 The Mercury Barometer** Average atmospheric pressure at sea level can support a column of mercury 760 mm in height.

| TABLE 5.1 Common Units of Pressure | | |
|---|---|---|
| Unit | Abbreviation | Average Air Pressure at Sea Level |
| Pascal (1 N/m²) | Pa | 101,325 Pa |
| Pounds per square inch | psi | 14.7 psi |
| Torr (1 mmHg) | torr | 760 torr (exact) |
| Inches of mercury | in Hg | 29.92 in Hg |
| Atmosphere | atm | 1 atm |

## Example 5.1  Converting between Pressure Units

A high-performance road bicycle tire is inflated to a total pressure of 132 psi. What is this pressure in mmHg?

**SORT** The problem gives a pressure in psi and asks you to convert the units to mmHg.

**GIVEN:** 132 psi

**FIND:** mmHg

**STRATEGIZE** Since Table 5.1 does not have a direct conversion factor between psi and mmHg but does provide relationships between both of these units and atmospheres, convert to atm as an intermediate step.

**CONCEPTUAL PLAN**

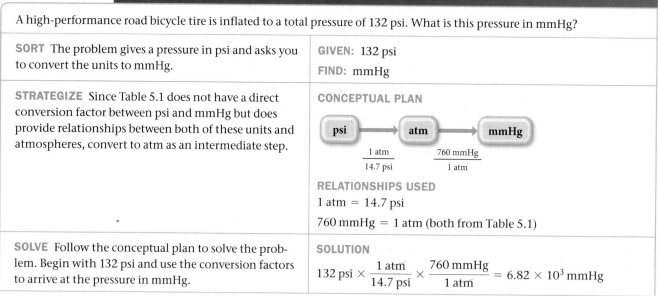

$$\text{psi} \xrightarrow{\dfrac{1\ \text{atm}}{14.7\ \text{psi}}} \text{atm} \xrightarrow{\dfrac{760\ \text{mmHg}}{1\ \text{atm}}} \text{mmHg}$$

**RELATIONSHIPS USED**

1 atm = 14.7 psi

760 mmHg = 1 atm (both from Table 5.1)

**SOLVE** Follow the conceptual plan to solve the problem. Begin with 132 psi and use the conversion factors to arrive at the pressure in mmHg.

**SOLUTION**

$$132\ \text{psi} \times \frac{1\ \text{atm}}{14.7\ \text{psi}} \times \frac{760\ \text{mmHg}}{1\ \text{atm}} = 6.82 \times 10^3\ \text{mmHg}$$

**CHECK** The units of the answer are correct. The magnitude of the answer ($6.82 \times 10^3$ mmHg) is greater than the given pressure in psi. This is reasonable since mmHg is a much smaller unit than psi.

**FOR PRACTICE 5.1** Your local weather report announces that the barometric pressure is 30.44 in Hg. Convert this pressure to psi.

**FOR MORE PRACTICE 5.1** Convert a pressure of 23.8 in Hg to kPa.

### The Manometer

Atmospheric pressure

$h$

High-pressure gas

Height difference ($h$) indicates pressure of gas relative to atmospheric pressure.

▲ **FIGURE 5.5 The Manometer** A manometer measures the pressure exerted by a sample of gas.

## The Manometer: A Way to Measure Pressure in the Laboratory

We can measure the pressure of a gas sample in the laboratory with a **manometer**. A manometer is a U-shaped tube containing a dense liquid, usually mercury, as shown in Figure 5.5◄. In this manometer, one end of the tube is open to atmospheric pressure and the other is attached to a flask containing the gas sample. If the pressure of the gas sample is exactly equal to atmospheric pressure, then the mercury levels on both sides of the tube are the same. If the pressure of the gas sample is *greater than* atmospheric pressure, the mercury level on the left side of the tube is *higher than* the level on the right. If the pressure of the sample is *less than* atmospheric pressure, the mercury level on the left side is *lower than* the level on the right. This type of manometer always measures the pressure of the gas sample relative to atmospheric pressure. The difference in height between the two levels is equal to the difference between the sample's pressure and atmospheric pressure. To accurately calculate the absolute pressure of the sample, we also need a barometer to measure atmospheric pressure (which can vary from day to day).

---

## Chemistry AND **Medicine** | **Blood Pressure**

**B**lood pressure is the force within our arteries that drives the circulation of blood throughout the body. Blood pressure in the body is analogous to water pressure in a plumbing system. Just as water pressure pushes water through the pipes to faucets and fixtures throughout a house, blood pressure pushes blood to muscles and other tissues throughout the body. However, unlike the water pressure in a plumbing system—which is typically nearly constant—blood pressure varies with each heartbeat. When the heart muscle contracts, blood pressure increases; between contractions it decreases. Systolic blood pressure is the peak pressure during a contraction, and diastolic blood pressure is the lowest pressure between contractions. Just as excessively high water pressure in a plumbing system can damage pipes, blood pressure that is too high in a circulatory system can damage the heart and arteries, resulting in increased risk of stroke and heart attack.

Medical professionals usually measure blood pressure with an instrument called a sphygmomanometer—an inflatable cuff equipped with a pressure gauge—and a stethoscope. The cuff is wrapped around the patient's arm and inflated with air. As air is pumped into the cuff, the pressure in the cuff increases. The cuff tightens around the arm and compresses the artery, momentarily stopping blood flow. The person measuring the blood pressure listens to the artery through the stethoscope while slowly releasing the air pressure in the cuff. When the air pressure in the cuff equals the systolic blood pressure (the peak pressure), a pulse is heard through the stethoscope. The pulse is the sound of blood getting through the compressed artery during a contraction of the heart. The pressure reading at that exact moment is the systolic blood pressure. As the pressure in the cuff continues to decrease, the blood can flow through the compressed artery even between contractions, so the pulsing sound stops. The pressure reading when the pulsing sound stops is the diastolic blood pressure (the lowest pressure).

A blood pressure measurement is usually reported as two pressures, in mmHg, separated by a slash. For example, a blood pressure measurement of 122/84 indicates that the systolic blood pressure is 122 mmHg and the diastolic blood pressure is 84 mmHg. Although the value of blood pressure can vary throughout the day, healthy (or normal) values are

▲ A health care provider measures blood pressure with an inflatable cuff that compresses the main artery in the arm. A stethoscope is used to listen for blood flowing through the artery with each heartbeat.

usually considered to be below 120 mmHg for systolic and below 80 mmHg for diastolic (Table 5.2). High blood pressure, also called hypertension, entails the health risks mentioned previously.

Risk factors for hypertension include obesity, high salt (sodium) intake, high alcohol intake, lack of exercise, stress, a family history of high blood pressure, and age (blood pressure tends to increase as we get older). Mild hypertension can be managed with diet and exercise. Moderate to severe cases require doctor-prescribed medication.

**TABLE 5.2 Blood Pressure Ranges**

| Blood Pressure | Systolic (mmHg) | Diastolic (mmHg) |
|---|---|---|
| Hypotension | <100 | <60 |
| Normal | 100–119 | 60–79 |
| Prehypertension | 120–139 | 80–89 |
| Hypertension Stage 1 | 140–159 | 90–99 |
| Hypertension Stage 2 | >160 | >100 |

---

## 5.3 The Simple Gas Laws: Boyle's Law, Charles's Law, and Avogadro's Law

 **Simple Gas Laws and Ideal Gas Law**

We have learned about pressure and its characteristics in Section 5.2. We now broaden our discussion to include the four basic properties of a gas sample: pressure (*P*), volume (*V*), temperature (*T*), and amount in moles (*n*). These properties are interrelated—when one changes, it affects the others. The *simple gas laws* describe the relationships between pairs of these properties. For example, one simple gas law describes how *volume* varies with *pressure* at constant temperature and amount of gas; another law describes how volume varies with *temperature* at constant pressure and amount of gas. These laws were deduced from observations in which two of the four basic properties were held constant in order to elucidate the relationship between the other two.

## Boyle's Law: Volume and Pressure

In the early 1660s, the pioneering English scientist Robert Boyle (1627–1691) and his assistant Robert Hooke (1635–1703) used a J-tube (Figure 5.6▼) to measure the volume of a sample of gas at different pressures. They trapped a sample of air in the J-tube and added mercury to increase the pressure on the gas. Boyle and Hooke observed an *inverse relationship* between volume and pressure—an increase in one causes a decrease in the other—as shown in Figure 5.7▼. This relationship is now known as **Boyle's law**.

> Boyle's law assumes constant temperature and constant amount of gas.

$$\text{Boyle's law:} \quad V \propto \frac{1}{P} \quad \text{(constant } T \text{ and } n\text{)}$$

Boyle's law follows from the idea that pressure results from the collisions of the gas particles with the walls of their container. When the volume of a gas sample decreases, the same number of gas particles is crowded into a smaller volume, resulting in more collisions with the walls and therefore an increase in pressure (Figure 5.8▶).

Scuba divers learn about Boyle's law during certification because it explains why a diver should not ascend toward the surface without continuous breathing. For every 10 m of depth that a diver descends in water, she experiences an additional 1 atm of pressure due to the weight of the water above her (Figure 5.9▶). The pressure regulator used in scuba diving delivers air into the diver's lungs at a pressure that matches the external pressure; otherwise the diver could not inhale the air (see *Chemistry in Your Day: Extra-Long Snorkels* on page 205). For example, when a diver is 20 m below the surface, the regulator delivers air at a pressure of 3 atm to match the 3 atm of pressure around the diver (1 atm due to normal atmospheric pressure and 2 additional atmospheres due to the weight of the water at 20 m). Suppose that a diver inhaled a lungful of air at a pressure of 3 atm and swam quickly to the surface (where the pressure is 1 atm) while holding her breath. What would happen to the volume of air in her lungs?

### The J-Tube

▲ **FIGURE 5.6  The J-Tube**  In a J-tube, a column of mercury traps a sample of gas. The pressure on the gas can be increased by increasing the height (*h*) of mercury in the column.

▲ **FIGURE 5.7  Volume versus Pressure**  A plot of the volume of a gas sample versus pressure. The plot shows that volume and pressure are inversely related.

## Volume versus Pressure: A Molecular View

◀ **FIGURE 5.8 Molecular Interpretation of Boyle's Law**   As the volume of a gas sample decreases, gas molecules collide with surrounding surfaces more frequently, resulting in greater pressure.

Since the pressure decreases by a factor of 3, the volume of the air in her lungs would increase by a factor of 3—a dangerous situation. The volume increase in the diver's lungs would be so great that she would not be able to hold her breath all the way to the surface—the air would force itself out of her mouth but probably not before the expanded air severely damaged her lungs, possibly killing her. Consequently, the most important rule in diving is *never hold your breath*. To avoid catastrophic results, divers must ascend slowly and breathe continuously, allowing the regulator to bring the air pressure in their lungs back to 1 atm by the time they reach the surface.

We can use Boyle's law to calculate the volume of a gas following a pressure change or the pressure of a gas following a volume change *as long as the temperature and the amount of gas remain constant*. For these types of calculations, we write Boyle's law in a slightly different way.

Since $V \propto \dfrac{1}{P}$,    then   $V = (\text{constant}) \times \dfrac{1}{P}$   or   $V = \dfrac{(\text{constant})}{P}$

> If two quantities are proportional, then one is equal to the other multiplied by a constant.

If we multiply both sides by $P$, we find that the product of $P$ and $V$ is equal to a constant:

$$PV = \text{constant}$$

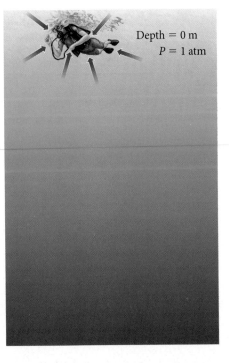

◀ **FIGURE 5.9 Increase in Pressure with Depth**   For every 10 m of depth, a diver experiences approximately 1 additional atmosphere of pressure due to the weight of the surrounding water. At 20 m, for example, the diver experiences approximately 3 atm of pressure (1 atm of normal atmospheric pressure plus an additional 2 atm due to the weight of the water).

This relationship indicates that if the pressure increases, the volume decreases, but the product $P \times V$ always equals the same constant. For two different sets of conditions, we can say that:

$$P_1 V_1 = \text{constant} = P_2 V_2$$

or:

$$P_1 V_1 = P_2 V_2 \qquad [5.2]$$

where $P_1$ and $V_1$ are the initial pressure and volume of the gas and $P_2$ and $V_2$ are the final volume and pressure.

---

### Example 5.2 | Boyle's Law

A woman has an initial lung volume of 2.75 L, which is filled with air at an atmospheric pressure of 1.02 atm. If she increases her lung volume to 3.25 L without inhaling any additional air, what is the pressure in her lungs?

| | |
|---|---|
| To solve the problem, first solve Boyle's law (Equation 5.2) for $P_2$ and then substitute the given quantities to calculate $P_2$. | **SOLUTION** $$P_1 V_1 = P_2 V_2$$ $$P_2 = \frac{V_1}{V_2} P_1$$ $$= \frac{2.75 \text{ L}}{3.25 \text{ L}} 1.02 \text{ atm}$$ $$= 0.863 \text{ atm}$$ |

**FOR PRACTICE 5.2** A snorkeler takes a syringe filled with 16 mL of air from the surface, where the pressure is 1.0 atm, to an unknown depth. The volume of the air in the syringe at this depth is 7.5 mL. What is the pressure at the unknown depth? If the pressure increases by 1 atm for every additional 10 m of depth, how deep is the snorkeler?

---

## Charles's Law: Volume and Temperature

Suppose we keep the pressure of a gas sample constant and measure its volume at a number of different temperatures. Figure 5.10▼ shows the results of several such measurements.

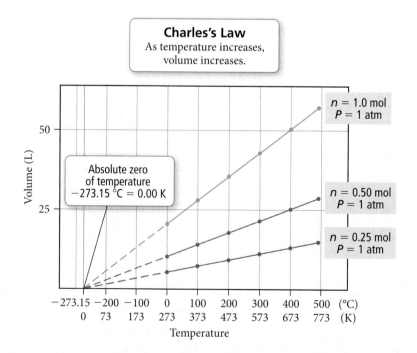

► **FIGURE 5.10 Volume versus Temperature** The volume of a fixed amount of gas at a constant pressure increases linearly with increasing temperature in kelvins. (The extrapolated lines cannot be measured experimentally because all gases condense into liquids before −273.15 °C is reached.)

# Chemistry IN YOUR **DAY** | Extra-Long Snorkels

Several episodes of *The Flintstones* cartoon featured Fred Flintstone and Barney Rubble snorkeling. Their snorkels, however, were not the modern kind, but long reeds that stretched from the surface of the water down to many meters of depth. Fred and Barney swam around in deep water while breathing air provided to them by these extra-long snorkels. Would this work? Why do people bother with scuba diving equipment if they could instead simply use 10-meter snorkels as Fred and Barney did?

When we breathe, we expand the volume of our chest cavity, reducing the pressure in our lungs to less than 1 atm (Boyle's law). Because of this pressure differential, air from outside our lungs flows into them. Extra-long snorkels do not work because the pressure exerted by water at depth is too great. A diver at 10 m experiences an external pressure of 2 atm. This is more than the muscles of the chest cavity can overcome. Instead, the chest cavity and lungs are compressed, resulting in an air pressure within them of more than 1 atm. If the diver had a snorkel that went to the surface—where the air pressure is 1 atm—air would flow out of the diver's lungs (*from* greater pressure *to* less pressure), not into them. It would be impossible for the diver to breathe.

### QUESTION

*A diver takes a balloon with a volume of 2.5 L from the surface, where the pressure is 1.0 atm, to a depth of 20 m, where the pressure is 3.0 atm. What happens to the volume of the balloon? What if the end of the submerged balloon is on a long pipe that goes to the surface and is attached to another balloon? Which way would air flow as the diver descended?*

▲ In the popular cartoon *The Flintstones*, cavemen used long reeds to breathe surface air while swimming at depth. This would not work because the increased pressure at depth would force air out of their lungs; the pressure would not allow them to inhale.

▲ If two balloons were joined by a long tube and one end was submerged in water, what would happen to the volumes of the two balloons?

---

From the plot we can see a relationship between volume and temperature: the volume of a gas increases with increasing temperature. Looking at the plot more closely reveals more—volume and temperature are *linearly related*. If two variables are linearly related, plotting one against the other produces a straight line.

Another interesting feature emerges if we extend or *extrapolate* each line in the plot in Figure 5.10 backward from the lowest measured temperature. The dotted extrapolated lines show that the gas should have a zero volume at −273.15 °C. Recall from Chapter 1 that −273.15 °C corresponds to 0 K (zero on the Kelvin scale), the coldest possible temperature. The extrapolated lines indicate that below −273.15 °C, the gas would have a negative volume, which is physically impossible. For this reason, we refer to 0 K as *absolute zero*—colder temperatures do not exist.

The first person to carefully quantify the relationship between the volume of a gas and its temperature was J. A. C. Charles (1746–1823), a French mathematician and physicist. Charles was interested in gases and was among the first people to ascend in a hydrogen-filled balloon. The direct proportionality between volume and temperature is named **Charles's law** after him.

Charles's law:   $V \propto T$   (constant $P$ and $n$)

| Charles's law assumes constant pressure and constant amount of gas.

When the temperature of a gas sample increases, the gas particles move faster; collisions with the walls are more frequent, and the force exerted with each collision is greater. The only way for the pressure (the force per unit area) to remain constant is for

▲ A hot-air balloon floats because the hot air within the balloon is less dense than the surrounding cold air.

▲ If we place a balloon into liquid nitrogen (77 K), it shrivels up as the air within it cools and occupies less volume at the same external pressure.

## Volume versus Temperature: A Molecular View

▲ **FIGURE 5.11** **Molecular Interpretation of Charles's Law** If we move a balloon from an ice water bath to a boiling water bath, its volume expands as the gas particles within the balloon move faster (due to the increased temperature) and collectively occupy more space.

the gas to occupy a larger volume so that collisions become less frequent and occur over a larger area (Figure 5.11▲).

Charles's law explains why the second floor of a house is usually warmer than the ground floor. According to Charles's law, when air is heated, its volume increases, resulting in a lower density. The warm, less dense air tends to rise in a room filled with colder, denser air. Similarly, Charles's law explains why a hot-air balloon can take flight. The gas that fills a hot-air balloon is warmed with a burner, increasing its volume, lowering its density, and causing it to float in the colder, denser surrounding air.

You can experience Charles's law directly by holding a partially inflated balloon over a warm toaster. As the air in the balloon warms, you can feel the balloon expanding. Alternatively, you can put an inflated balloon into liquid nitrogen and watch it become smaller as it cools.

We can use Charles's law to calculate the volume of a gas following a temperature change or the temperature of a gas following a volume change *as long as the pressure and the amount of gas are constant*. For these calculations, we rearrange Charles's law as follows:

$$\text{Since } V \propto T, \quad \text{then } V = \text{constant} \times T$$

If we divide both sides by $T$, we find that $V/T$ is equal to a constant:

$$V/T = \text{constant}$$

If the temperature increases, the volume increases in direct proportion so that the quotient, $V/T$, is always equal to the same constant. So, for two different measurements, we can say that:

$$V_1/T_1 = \text{constant} = V_2/T_2$$

or

$$\frac{V_1}{T_1} = \frac{V_2}{T_2} \qquad [5.3]$$

where $V_1$ and $T_1$ are the initial volume and temperature of the gas and $V_2$ and $T_2$ are the final volume and temperature. *The temperatures must always be expressed in kelvins (K)* because, as shown in Figure 5.10, the volume of a gas is directly proportional to its *absolute* temperature, not its temperature in °C. For example, doubling the temperature of a gas sample from 1 °C to 2 °C does not double its volume, but doubling the temperature from 200 K to 400 K does.

## Example 5.3  Charles's Law

A sample of gas has a volume of 2.80 L at an unknown temperature. When the sample is submerged in ice water at $T = 0.00\,°C$, its volume decreases to 2.57 L. What was its initial temperature (in K and in °C)?

| | |
|---|---|
| To solve the problem, first solve Charles's law for $T_1$. | **SOLUTION** $$\frac{V_1}{T_2} = \frac{V_2}{T_2}$$ $$T_1 = \frac{V_1}{V_2}T_2$$ |
| Before you substitute the numerical values to calculate $T_1$, convert the temperature to kelvins (K). *Remember, gas law problems must always be worked with Kelvin temperatures.* | $T_2(K) = 0.00 + 273.15 = 273.15\ K$ |
| Substitute $T_2$ and the other given quantities to calculate $T_1$. | $$T_1 = \frac{V_1}{V_2}T_2$$ $$= \frac{2.80\ L}{2.57\ L}\,273.15 K$$ $$= 297.6\ K$$ |
| Calculate $T_1$ in °C by subtracting 273.15 from the value in kelvins. | $T_1(°C) = 297.6 - 273.15 = 24\,°C$ |

**FOR PRACTICE 5.3**  A gas in a cylinder with a moveable piston has an initial volume of 88.2 mL. If we heat the gas from 35 °C to 155 °C, what is its final volume (in mL)?

---

**Boyle's Law and Charles's Law**  The pressure exerted on a sample of a fixed amount of gas is doubled at constant temperature, and then the temperature of the gas in kelvins is doubled at constant pressure. What is the final volume of the gas?

CONCEPTUAL
**CONNECTION 5.1**

(a) The final volume is twice the initial volume.
(b) The final volume of the gas is four times the initial volume.
(c) The final volume of the gas is one-half the initial volume.
(d) The final volume of the gas is one-fourth the initial volume.
(e) The final volume of the gas is the same as the initial volume.

---

## Avogadro's Law: Volume and Amount (in Moles)

So far, we have discussed the relationships between volume and pressure, and volume and temperature, but we have considered only a constant amount of a gas. What happens when the amount of gas changes? The volume of a gas sample (at constant temperature and pressure) as a function of the amount of gas (in moles) in the sample is shown in Figure 5.12▶. We can see that the relationship between volume and amount is linear. As we might expect, extrapolation to zero moles shows zero volume. This relationship, first stated formally by Amedeo Avogadro (1776–1856), is **Avogadro's law**.

Avogadro's law:    $V \propto n$    (constant $T$ and $P$)

When the amount of gas in a sample increases at constant temperature and pressure, its volume increases in direct proportion because the greater number of gas particles fill more space.

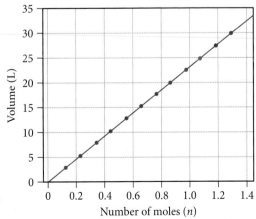

**Avogadro's Law**
As amount of gas increases, volume increases.

▶ **FIGURE 5.12  Volume versus Number of Moles**  The volume of a gas sample increases linearly with the number of moles of gas in the sample.

You experience Avogadro's law when you inflate a balloon. With each exhaled breath, you add more gas particles to the inside of the balloon, increasing its volume. We can use Avogadro's law to calculate the volume of a gas following a change in the amount of the gas *as long as the pressure and temperature of the gas are constant*. For these types of calculations, we express Avogadro's law as:

Avogadro's law assumes constant temperature and constant pressure and is independent of the nature of the gas.

$$\frac{V_1}{n_1} = \frac{V_2}{n_2}$$  [5.4]

where $V_1$ and $n_1$ are the initial volume and number of moles of the gas and $V_2$ and $n_2$ are the final volume and number of moles. In calculations, we use Avogadro's law in a manner similar to the other gas laws, as demonstrated in Example 5.4.

## Example 5.4  Avogadro's Law

A male athlete in a kinesiology research study has a lung volume of 6.15 L during a deep inhalation. At this volume, his lungs contain 0.254 mol of air. During exhalation, his lung volume decreases to 2.55 L. How many moles of gas did the athlete exhale? Assume constant temperature and pressure.

| | |
|---|---|
| To solve the problem, first solve Avogadro's law for the number of moles of gas left in the athlete's lungs after exhalation, $n_2$. Then substitute the given quantities to calculate $n_2$. | **SOLUTION** $$\frac{V_1}{n_1} = \frac{V_2}{n_2}$$ $$n_2 = \frac{V_2}{V_1} n_1$$ $$= \frac{2.55\ \text{L}}{6.15\ \text{L}} 0.254\ \text{mol}$$ $$= 0.105\ \text{mol}$$ |
| Since the lungs initially contained 0.254 mol of air, calculate the amount of air exhaled by subtracting the result from 0.254 mol. (In Chapter 1, we introduced the practice of underlining the least [rightmost] significant digit of intermediate answers but not rounding the final answer until the very end of the calculation. We continue that practice throughout the book. However, in order to avoid unnecessary notation, we will not carry additional digits in cases, such as this one, where doing so does not affect the final answer.) | moles exhaled = 0.254 mol − 0.105 mol $$= 0.149\ \text{mol}$$ |

**FOR PRACTICE 5.4**  A chemical reaction occurring in a cylinder equipped with a moveable piston produces 0.621 mol of a gaseous product. If the cylinder contained 0.120 mol of gas before the reaction and had an initial volume of 2.18 L, what is its volume after the reaction? (Assume constant pressure and temperature and that the initial amount of gas completely reacts.)

 **Simple Gas Laws and Ideal Gas Law**

## 5.4  The Ideal Gas Law

The relationships that we have discussed so far can be combined into a single law that encompasses all of them. So far, we have shown that:

$$V \propto \frac{1}{P} \qquad \text{(Boyle's law)}$$

$$V \propto T \qquad \text{(Charles's law)}$$

$$V \propto n \qquad \text{(Avogadro's law)}$$

Combining these three expressions, we find that $V$ is proportional to $nT/P$:

$$V \propto \frac{nT}{P}$$

The volume of a gas is directly proportional to the number of moles of gas and to the temperature of the gas, but is inversely proportional to the pressure of the gas.

We can replace the proportionality sign with an equals sign by incorporating $R$, a proportionality constant called the *ideal gas constant*:

$$V = \frac{RnT}{P}$$

Rearranging, we get the equation:

$$PV = nRT \qquad [5.5]$$

This equation is the **ideal gas law**, and a hypothetical gas that exactly follows this law is an **ideal gas**. The value of $R$, the **ideal gas constant**, is the same for all gases and has the value:

$$R = 0.08206 \frac{L \cdot atm}{mol \cdot K}$$

L = liters
atm = atmospheres
mol = moles
K = kelvins

The ideal gas law contains within it the simple gas laws that we have discussed. For example, recall that Boyle's law states that $V \propto 1/P$ when the amount of gas ($n$) and the temperature of the gas ($T$) are kept constant. We can rearrange the ideal gas law as follows:

$$PV = nRT$$

First, divide both sides by $P$:

$$V = \frac{nRT}{P}$$

Then put the variables that are constant, along with $R$, in parentheses:

$$V = (nRT)\frac{1}{P}$$

Since $n$ and $T$ are constant in this case, and since $R$ is always a constant, we can write:

$$V \propto (\text{constant}) \times \frac{1}{P}$$

which means that $V \propto 1/P$.

The ideal gas law also shows how other pairs of variables are related. For example, from Charles's law we know that $V \propto T$ at constant pressure and constant number of moles. But what if we heat a sample of gas at constant *volume* and constant number of moles? This question applies to the warning labels on aerosol cans such as hair spray or deodorants. These labels warn against excessive heating or incineration of the can, even after the contents are used up. Why? An "empty" aerosol can is not really empty but contains a fixed amount of gas trapped in a fixed volume. What would happen if we were to heat the can? We can rearrange the ideal gas law to clearly see the relationship between pressure and temperature at constant volume and constant number of moles:

$$PV = nRT$$

$$P = \frac{nRT}{V} = \left(\frac{nR}{V}\right)T$$

Since $n$ and $V$ are constant and since $R$ is always a constant:

$$P = (\text{constant}) \times T$$

This relationship between pressure and temperature is also known as *Gay-Lussac's law*. As the temperature of a fixed amount of gas in a fixed volume increases, the pressure increases. In an aerosol can, this pressure increase can blow the can apart, which is why aerosol cans should not be heated or incinerated. They might explode.

We can use the ideal gas law to determine the value of any one of the four variables ($P$, $V$, $n$, or $T$) given the other three. To do so, we must express each of the quantities in the ideal gas law in the units within $R$:

- pressure ($P$) in atm
- volume ($V$) in L
- moles ($n$) in mol
- temperature ($T$) in K

**Ideal Gas Law**

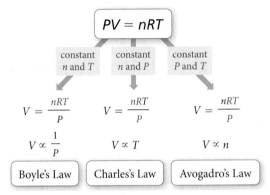

▲ The ideal gas law contains the simple gas laws within it.

| Divide both sides by $V$.

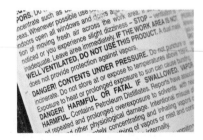

▲ The labels on most aerosol cans warn against incineration. Since the volume of the can is constant, an increase in temperature causes an increase in pressure and possibly an explosion.

## Example 5.5    Ideal Gas Law I

Calculate the volume occupied by 0.845 mol of nitrogen gas at a pressure of 1.37 atm and a temperature of 315 K.

| | |
|---|---|
| **SORT** The problem gives you the number of moles of nitrogen gas, the pressure, and the temperature. You are asked to find the volume. | **GIVEN:** $n = 0.845$ mol, $P = 1.37$ atm, $T = 315$ K <br><br> **FIND:** $V$ |
| **STRATEGIZE** You are given three of the four variables ($P$, $T$, and $n$) in the ideal gas law and asked to find the fourth ($V$). The conceptual plan shows how the ideal gas law provides the relationship between the known quantities and the unknown quantity. | **CONCEPTUAL PLAN** <br><br>  <br><br> **RELATIONSHIPS USED** <br> $PV = nRT$ (ideal gas law) |
| **SOLVE** To solve the problem, first solve the ideal gas law for $V$. <br><br><br><br> Then substitute the given quantities to calculate $V$. | **SOLUTION** <br> $PV = nRT$ <br> $$V = \frac{nRT}{P}$$ <br><br> $$V = \frac{0.845 \text{ mol} \times 0.08206 \dfrac{\text{L} \cdot \text{atm}}{\text{mol} \cdot \text{K}} \times 315 \text{ K}}{1.37 \text{ atm}}$$ <br> $= 15.9$ L |

**CHECK** The units of the answer are correct. The magnitude of the answer (15.9 L) makes sense because, as you will see in the next section, one mole of an ideal gas under standard temperature and pressure (273 K and 1 atm) occupies 22.4 L. Although this is not standard temperature and pressure, the conditions are close enough for a ballpark check of the answer. Since this gas sample contains 0.845 mol, a volume of 15.9 L is reasonable.

**FOR PRACTICE 5.5** An 8.50-L tire contains 0.552 mol of gas at a temperature of 305 K. What is the pressure (in atm and psi) of the gas in the tire?

## Example 5.6    Ideal Gas Law II

Calculate the number of moles of gas in a 3.24-L basketball inflated to a *total pressure* of 24.3 psi at 25 °C. (Note: The *total pressure* is not the same as the pressure read on a pressure gauge such as the type used for checking a car or bicycle tire. That pressure, called the *gauge pressure*, is the *difference* between the total pressure and atmospheric pressure. In this case, if atmospheric pressure is 14.7 psi, the gauge pressure would be 9.6 psi. However, for calculations involving the ideal gas law, you must use the *total pressure* of 24.3 psi.)

| | |
|---|---|
| **SORT** The problem gives you the pressure, the volume, and the temperature. You are asked to find the number of moles of gas. | **GIVEN:** $P = 24.3$ psi, $V = 3.24$ L, $T(°C) = 25 °C$ <br><br> **FIND:** $n$ |
| **STRATEGIZE** The conceptual plan shows how the ideal gas law provides the relationship between the given quantities and the quantity to be found. | **CONCEPTUAL PLAN** <br><br> P, V, T $\longrightarrow$ n <br> $PV = nRT$ <br><br> **RELATIONSHIPS USED** $PV = nRT$ (ideal gas law) |

| SOLVE To solve the problem, first solve the ideal gas law for $n$. | SOLUTION<br>$PV = nRT$<br>$n = \dfrac{PV}{RT}$ |
|---|---|
| Before substituting into the equation, convert $P$ and $T$ into the correct units. | $P = 24.3 \text{ psi} \times \dfrac{1 \text{ atm}}{14.7 \text{ psi}} = 1.65\underline{3}1 \text{ atm}$<br><br>(Since rounding the intermediate answer would result in a slightly different final answer, mark the least significant digit in the intermediate answer but don't round until the end.)<br><br>$T(\text{K}) = 25 + 273 = 298 \text{ K}$ |
| Finally, substitute into the equation and calculate $n$. | $n = \dfrac{1.65\underline{3}1 \text{ atm} \times 3.24 \text{ L}}{0.08206 \dfrac{\text{L} \cdot \text{atm}}{\text{mol} \cdot \text{K}} \times 298 \text{ K}} = 0.219 \text{ mol}$ |

CHECK The units of the answer are correct. The magnitude of the answer (0.219 mol) makes sense because, as you will see in the next section, one mole of an ideal gas under standard temperature and pressure (273 K and 1 atm) occupies 22.4 L. At a pressure that is 65% higher than standard pressure, the volume of 1 mol of gas is proportionally lower. Since this gas sample occupies 3.24 L, the answer of 0.219 mol is reasonable.

FOR PRACTICE 5.6    What volume does 0.556 mol of gas occupy at a pressure of 715 mmHg and a temperature of 58 °C?

FOR MORE PRACTICE 5.6    Determine the pressure in mmHg of a 0.133-g sample of helium gas in a 648-mL container at 32 °C.

# 5.5 Applications of the Ideal Gas Law: Molar Volume, Density, and Molar Mass of a Gas

> The molar volume of 22.4 L only applies at STP.

We just examined how we can use the ideal gas law to calculate one of the variables ($P$, $V$, $T$, or $n$) given the other three. We now turn to three other applications of the ideal gas law: molar volume, density, and molar mass.

## Molar Volume at Standard Temperature and Pressure

The volume occupied by one mole of a substance is its **molar volume**. For gases, we often specify the molar volume under conditions known as **standard temperature** ($T = 0$ °C or 273 K) **and pressure** ($P = 1.00$ atm), abbreviated as **STP**. Applying the ideal gas law, we can determine that the molar volume of an ideal gas at STP is:

$$V = \frac{nRT}{P}$$

$$= \frac{1.00 \text{ mol} \times 0.08206 \dfrac{\text{L} \cdot \text{atm}}{\text{mol} \cdot \text{K}} \times 273 \text{ K}}{1.00 \text{ atm}}$$

$$= 22.4 \text{ L}$$

1 mol He($g$) at STP     1 mol Xe($g$) at STP     1 mol CH$_4$($g$) at STP

▲ One mole of any gas occupies approximately 22.4 L at standard temperature (273 K) and pressure (1.0 atm).

The molar volume of an ideal gas at STP is useful because—as we saw in the *Check* steps of Examples 5.5 and 5.6—it gives us a way to approximate the volume of an ideal gas under conditions that are close to STP.

**Molar Volume** Assuming ideal behavior, which of these gas samples has the greatest volume at STP?

**(a)** 1 g of $H_2$

**(b)** 1 g of $O_2$

**(c)** 1 g of Ar

## Density of a Gas

Since one mole of an ideal gas occupies 22.4 L under standard temperature and pressure, we can readily calculate the density of an ideal gas under these conditions. Because density is mass/volume, and because the mass of one mole of a gas is simply its molar mass, the *density of a gas* is its molar mass divided by its molar volume:

$$Density = \frac{molar\ mass}{molar\ volume}$$

We can calculate the density of a gas at STP by using 22.4 L as the molar volume. For example, the densities of helium and nitrogen gas at STP are:

$$d_{He} = \frac{4.00\ g/mol}{22.4\ L/mol} = 0.179\ g/L \qquad d_{N_2} = \frac{28.02\ g/mol}{22.4\ L/mol} = 1.25\ g/L$$

Notice that *the density of a gas is directly proportional to its molar mass.* The greater the molar mass of a gas, the more dense the gas. For this reason, a gas with a molar mass lower than that of air tends to rise in air. For example, both helium and hydrogen gas (molar masses of 4.00 and 2.01 g/mol, respectively) have molar masses that are lower than the average molar mass of air (approximately 28.8 g/mol). Therefore, a balloon filled with either helium or hydrogen gas floats in air.

We can calculate the density of a gas more generally (under any conditions) by using the ideal gas law. To do so, we can arrange the ideal gas law as:

$$PV = nRT$$

$$\frac{n}{V} = \frac{P}{RT}$$

Since the left-hand side of this equation has units of moles/liter, it represents the *molar density.* We obtain the density in grams/liter from the molar density by multiplying by the molar mass ($\mathcal{M}$):

$$\frac{moles}{liter} \times \frac{grams}{mole} = \frac{grams}{liter}$$

Molar density    Molar mass    Density in grams/liter

Therefore,

$$d = \frac{P\mathcal{M}}{RT} \tag{5.6}$$

Notice that, as expected, density increases with increasing molar mass. Notice also that, as we discussed in Section 5.3, density decreases with increasing temperature.

The primary components of air are nitrogen (about four-fifths) and oxygen (about one-fifth). We discuss the detailed composition of air in Section 5.6.

Density

$$\underbrace{\frac{n}{V}}_{Molar\ density} \quad \mathcal{M} = \underbrace{\frac{P\mathcal{M}}{RT}}_{Molar\ mass}$$

$$d = \frac{P\mathcal{M}}{RT}$$

## Example 5.7    Density

Calculate the density of nitrogen gas at 125 °C and a pressure of 755 mmHg.

| | |
|---|---|
| **SORT** The problem gives you the temperature and pressure of nitrogen gas and asks you to find its density. | **GIVEN:** $T(°C) = 125 °C$, $P = 755$ mmHg <br><br> **FIND:** $d$ |
| **STRATEGIZE** Equation 5.6 provides the relationship between the density of a gas and its temperature, pressure, and molar mass. The temperature and pressure are given. You can calculate the molar mass from the formula of the gas, which you know is $N_2$. | **CONCEPTUAL PLAN** <br><br>  <br> $d = \dfrac{P\mathcal{M}}{RT}$ <br><br> **RELATIONSHIPS USED** <br> $d = \dfrac{P\mathcal{M}}{RT}$ (density of a gas) <br> Molar mass $N_2 = 28.02$ g/mol |
| **SOLVE** To solve the problem, gather each of the required quantities in the correct units. Convert the temperature to kelvins and the pressure to atmospheres. <br><br> Substitute the quantities into the equation to calculate density. | **SOLUTION** <br> $T(K) = 125 + 273 = 398$ K <br> $P = 755 \text{ mmHg} \times \dfrac{1 \text{ atm}}{760 \text{ mmHg}} = 0.99\underline{3}42$ atm <br><br> $d = \dfrac{P\mathcal{M}}{RT}$ <br><br> $= \dfrac{0.99\underline{3}42 \text{ atm}\left(28.02\dfrac{\text{g}}{\text{mol}}\right)}{0.08206\dfrac{\text{L} \cdot \text{atm}}{\text{mol} \cdot \text{K}}(398 \text{ K})}$ <br><br> $= 0.852$ g/L |

**CHECK** The units of the answer are correct. The magnitude of the answer (0.852 g/L) makes sense because earlier you calculated the density of nitrogen gas at STP as 1.25 g/L. Since the temperature is higher than standard temperature, it follows that the density is lower.

**FOR PRACTICE 5.7**    Calculate the density of xenon gas at a pressure of 742 mmHg and a temperature of 45 °C.

**FOR MORE PRACTICE 5.7**    The density of a gas is 1.43 g/L at a temperature of 23 °C and a pressure of 0.789 atm. Calculate the molar mass of the gas.

**Density of a Gas** Arrange the following gases in order of increasing density at STP: Ne, $Cl_2$, $F_2$, and $O_2$.

CONCEPTUAL
CONNECTION **5.3**

## Molar Mass of a Gas

We can use the ideal gas law in combination with mass measurements to calculate the molar mass of an unknown gas. First, we measure the mass and volume of an unknown gas under conditions of known pressure and temperature. Then, we determine the amount of the gas in moles from the ideal gas law. Finally, we calculate the molar mass by dividing the mass (in grams) by the amount (in moles) as demonstrated in Example 5.8.

**Example 5.8**   **Molar Mass of a Gas**

A sample of gas has a mass of 0.311 g. Its volume is 0.225 L at a temperature of 55 °C and a pressure of 886 mmHg. Find its molar mass.

| SORT The problem gives you the mass of a gas sample, along with its volume, temperature, and pressure. You are asked to find the molar mass. | GIVEN: $m = 0.331$ g, $V = 0.225$ L, $T(°C) = 55 °C$, $P = 886$ mmHg  FIND: molar mass (g/mol) |
|---|---|

STRATEGIZE The conceptual plan has two parts. In the first part, use the ideal gas law to find the number of moles of gas.

In the second part, use the definition of molar mass to determine the molar mass.

CONCEPTUAL PLAN

$$P, V, T \longrightarrow n$$
$$PV = nRT$$

$$n, m \longrightarrow \text{molar mass}$$
$$\text{molar mass} = \frac{\text{mass } (m)}{\text{moles } (n)}$$

RELATIONSHIPS USED
$$PV = nRT$$
$$\text{Molar mass} = \frac{\text{mass } (m)}{\text{moles } (n)}$$

SOLVE To find the number of moles, first solve the ideal gas law for $n$.

Before substituting into the equation for $n$, convert the pressure to atm and the temperature to K.

Then substitute into the equation and calculate $n$, the number of moles.

Finally, use the number of moles ($n$) and the given mass ($m$) to calculate the molar mass.

SOLUTION
$$PV = nRT$$
$$n = \frac{PV}{RT}$$

$$P = 886 \text{ mmHg} \times \frac{1 \text{ atm}}{760 \text{ mmHg}} = 1.1658 \text{ atm}$$

$$T(K) = 55 + 273 = 328 \text{ K}$$

$$n = \frac{1.1658 \text{ atm} \times 0.225 \text{ L}}{0.08206 \frac{\text{L} \cdot \text{atm}}{\text{mol} \cdot \text{K}} \times 328 \text{ K}}$$
$$= 9.7454 \times 10^{-3} \text{ mol}$$

$$\text{molar mass} = \frac{\text{mass } (m)}{\text{moles } (n)}$$
$$= \frac{0.311 \text{ g}}{9.7454 \times 10^{-3} \text{ mol}}$$
$$= 31.9 \text{ g/mol}$$

CHECK The units of the answer are correct. The magnitude of the answer (31.9 g/mol) is a reasonable number for a molar mass. If you calculated a very small number (such as any number smaller than 1) or a very large number, you solved the problem incorrectly. Most gases have molar masses between one and several hundred grams per mole.

FOR PRACTICE 5.8   The mass of a sample of gas is 827 mg. Its volume is 0.270 L at a temperature of 88 °C and a pressure of 975 mmHg. Find its molar mass.

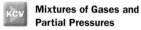
**Mixtures of Gases and Partial Pressures**

## 5.6   Mixtures of Gases and Partial Pressures

Many gas samples are not pure, but are mixtures of gases. Dry air, for example, is a mixture containing nitrogen, oxygen, argon, carbon dioxide, and a few other gases in trace amounts (Table 5.3).

Because the molecules in an ideal gas do not interact (as we will discuss further in Section 5.8), each of the components in an ideal gas mixture acts independently of the others. For example, the nitrogen molecules in air exert a certain pressure—78% of the total pressure—that is independent of the other gases in the mixture. Likewise, the oxygen molecules in air exert a certain pressure—21% of the total pressure—that is also independent of the other gases in the mixture. The pressure due to any individual component in a gas mixture is its **partial pressure** ($P_n$). We can calculate partial pressure from the ideal gas law by assuming that each gas component acts independently.

$$P_n = n_n \frac{RT}{V}$$

For a multicomponent gas mixture, we calculate the partial pressure of each component from the ideal gas law and the number of moles of that component ($n_n$) as follows:

$$P_a = n_a \frac{RT}{V}; \quad P_b = n_b \frac{RT}{V}; \quad P_c = n_c \frac{RT}{V}; \ldots \quad [5.7]$$

The sum of the partial pressures of the components in a gas mixture equals the total pressure:

$$P_{total} = P_a + P_b + P_c + \ldots \quad [5.8]$$

where $P_{total}$ is the total pressure and $P_a, P_b, P_c, \ldots$ are the partial pressures of the components. This relationship is known as **Dalton's law of partial pressures**.

Combining Equations 5.7 and 5.8, we get:

$$
\begin{aligned}
P_{total} &= P_a + P_b + P_c + \ldots \\
&= n_a \frac{RT}{V} + n_b \frac{RT}{V} + n_c \frac{RT}{V} + \ldots \\
&= (n_a + n_b + n_c + \ldots) \frac{RT}{V} \quad [5.9] \\
&= (n_{total}) \frac{RT}{V}
\end{aligned}
$$

The total number of moles in the mixture, when substituted into the ideal gas law, indicates the total pressure of the sample.

If we divide Equation 5.7 by Equation 5.9, we get:

$$\frac{P_a}{P_{total}} = \frac{n_a(RT/V)}{n_{total}(RT/V)} = \frac{n_a}{n_{total}} \quad [5.10]$$

The quantity $n_a/n_{total}$, the number of moles of a component in a mixture divided by the total number of moles in the mixture, is the **mole fraction** ($\chi_a$):

$$\chi_a = \frac{n_a}{n_{total}} \quad [5.11]$$

Rearranging Equation 5.10 and substituting the definition of mole fraction gives:

$$\frac{P_a}{P_{total}} = \frac{n_a}{n_{total}}$$

$$P_a = \frac{n_a}{n_{total}} P_{total} = \chi_a P_{total}$$

or simply

$$P_a = \chi_a P_{total} \quad [5.12]$$

The partial pressure of a component in a gaseous mixture is its mole fraction multiplied by the total pressure. For gases, the mole fraction of a component is equivalent to its percent by volume divided by 100%. Therefore, based on Table 5.3, we calculate the partial pressure of nitrogen ($P_{N_2}$) in air at 1.00 atm as:

$$P_{N_2} = 0.78 \times 1.00 \text{ atm}$$

$$= 0.78 \text{ atm}$$

**TABLE 5.3 Composition of Dry Air**

| Gas | Percent by Volume (%) |
|---|---|
| Nitrogen ($N_2$) | 78 |
| Oxygen ($O_2$) | 21 |
| Argon (Ar) | 0.9 |
| Carbon dioxide ($CO_2$) | 0.04 |

■ $N_2$    ■ $O_2$
▨ Ar    ■ $CO_2$

Likewise, the partial pressure of oxygen in air at 1.00 atm is 0.21 atm and the partial pressure of Ar in air is 0.009 atm. Applying Dalton's law of partial pressures to air at 1.00 atm, we can find the total pressure:

$$P_{total} = P_{N_2} + P_{O_2} + P_{Ar}$$
$$P_{total} = 0.78 \text{ atm} + 0.21 \text{ atm} + 0.009 \text{ atm}$$
$$= 1.00 \text{ atm}$$

For these purposes, we can ignore the contribution of the $CO_2$ and other trace gases because they are so small.

CONCEPTUAL
**CONNECTION 5.4**

**Partial Pressures** A gas mixture contains an equal number of moles of He and Ne. The total pressure of the mixture is 3.0 atm. What are the partial pressures of He and Ne?

---

**Example 5.9** | **Total Pressure and Partial Pressures**

A 1.00-L mixture of helium, neon, and argon has a total pressure of 662 mmHg at 298 K. If the partial pressure of helium is 341 mmHg and the partial pressure of neon is 112 mmHg, what mass of argon is present in the mixture?

| | |
|---|---|
| **SORT** The problem gives you the partial pressures of two of the three components in a gas mixture, along with the total pressure, the volume, and the temperature, and asks you to find the mass of the third component. | **GIVEN:** $P_{He} = 341$ mmHg, $\qquad P_{Ne} = 112$ mmHg, $\qquad P_{total} = 662$ mmHg, $\qquad V = 1.00$ L, $T = 298$ K <br><br> **FIND:** $m_{Ar}$ |
| **STRATEGIZE** You can find the mass of argon from the number of moles of argon, which you can calculate from the partial pressure of argon and the ideal gas law. Begin by using Dalton's law to determine the partial pressure of argon. <br><br> Then use the partial pressure of argon together with the volume of the sample and the temperature to find the number of moles of argon. <br><br> Finally, use the molar mass of argon to calculate the mass of argon from the number of moles of argon. | **CONCEPTUAL PLAN** <br><br>  <br> $P_{tot} = P_{He} + P_{Ne} + P_{Ar}$ <br><br>  <br> $PV = nRT$ <br><br> $n_{Ar} \longrightarrow m_{Ar}$ <br> $\dfrac{39.95 \text{ g Ar}}{1 \text{ mol Ar}}$ <br><br> **RELATIONSHIPS USED** <br> $P_{total} = P_{He} + P_{Ne} + P_{Ar}$ (Dalton's law) <br> $PV = nRT$ (ideal gas law) <br> molar mass Ar = 39.95 g/mol |
| **SOLVE** Follow the conceptual plan. To find the partial pressure of argon, solve the equation for $P_{Ar}$ and substitute the values of the other partial pressures to calculate $P_{Ar}$. | **SOLUTION** <br> $P_{total} = P_{He} + P_{Ne} + P_{Ar}$ <br> $P_{Ar} = P_{total} - P_{He} - P_{Ne}$ <br> $\qquad = 662 \text{ mmHg} - 341 \text{ mmHg} - 112 \text{ mmHg}$ <br> $\qquad = 209 \text{ mmHg}$ |
| Convert the partial pressure from mmHg to atm and use it in the ideal gas law to calculate the amount of argon in moles. | $209 \text{ mmHg} \times \dfrac{1 \text{ atm}}{760 \text{ mmHg}} = 0.275 \text{ atm}$ <br><br> $n = \dfrac{PV}{RT} = \dfrac{0.275 \text{ atm} (1.00 \text{ L})}{0.08206 \dfrac{\text{L} \cdot \text{atm}}{\text{mol} \cdot \text{K}} (298 \text{ K})} = 1.125 \times 10^{-2} \text{ mol Ar}$ |
| Use the molar mass of argon to convert from amount of argon in moles to mass of argon. | $1.125 \times 10^{-2} \text{ mol Ar} \times \dfrac{39.95 \text{ g Ar}}{1 \text{ mol Ar}} = 0.449 \text{ g Ar}$ |

**CHECK** The units of the answer are correct. The magnitude of the answer makes sense because the volume is 1.0 L, which at STP would contain about 1/22 mol. Since the partial pressure of argon in the mixture is about 1/3 of the total pressure, you can roughly estimate about 1/66 of one molar mass of argon, which is fairly close to your answer.

**FOR PRACTICE 5.9**   A sample of hydrogen gas is mixed with water vapor. The mixture has a total pressure of 755 torr, and the water vapor has a partial pressure of 24 torr. What amount (in moles) of hydrogen gas is contained in 1.55 L of this mixture at 298 K?

## Deep-Sea Diving and Partial Pressures

Our lungs have evolved to breathe oxygen at a partial pressure of $P_{O_2} = 0.21$ atm. If the total pressure decreases—when we climb a mountain, for example—the partial pressure of oxygen also decreases. On top of Mount Everest, where the total pressure is 0.311 atm, the partial pressure of oxygen is only 0.065 atm. Low oxygen levels produce a physiological condition called **hypoxia** or oxygen starvation (Figure 5.13▶). Mild hypoxia causes dizziness, headache, and shortness of breath. Severe hypoxia, which occurs when $P_{O_2}$ drops below 0.1 atm, may result in unconsciousness or even death. For this reason, climbers hoping to make the summit of Mount Everest usually carry oxygen to breathe.

Though not as dangerous as a lack of oxygen, too much oxygen can also cause physiological problems. Recall from Section 5.3 that scuba divers breathe pressurized air. At 30 m, a scuba diver breathes air at a total pressure of 4.0 atm, making $P_{O_2}$ about 0.84 atm. This elevated partial pressure of oxygen raises the density of oxygen molecules in the lungs, resulting in a higher concentration of oxygen in body tissues. When $P_{O_2}$ increases beyond 1.4 atm, the increased oxygen concentration in body tissues causes a condition called **oxygen toxicity**, which results in muscle twitching, tunnel vision, and convulsions. Divers who venture too deep without proper precautions have drowned because of oxygen toxicity. A second problem associated with breathing pressurized air is the increase of nitrogen in the lungs. At 30 m, a scuba diver breathes nitrogen at $P_{N_2} = 3.12$ atm, which increases the nitrogen concentration in body tissues and fluids. When $P_{N_2}$ increases beyond about 4 atm, a condition called **nitrogen narcosis** or *rapture of the deep* results. Divers describe this condition as feeling inebriated or drunk. A diver breathing compressed air at 60 m feels as if he has consumed too much wine.

To avoid oxygen toxicity and nitrogen narcosis, deep-sea divers—those who descend beyond 50 m—breathe specialized mixtures of gases. One common mixture is heliox, a mixture of helium and oxygen. These mixtures usually contain a smaller percentage of oxygen than would be found in air, thereby lowering the risk of oxygen toxicity. Heliox also contains helium instead of nitrogen, eliminating the risk of nitrogen narcosis.

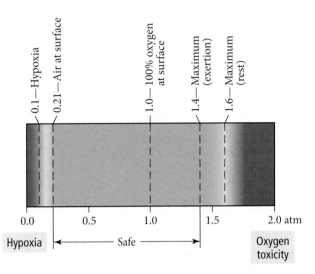

▲ **FIGURE 5.13  Oxygen Partial Pressure Limits**   The partial pressure of oxygen in air at sea level is 0.21 atm. Partial pressures of oxygen below 0.1 atm and above 1.4 atm are dangerous to humans.

| Surface | |
|---|---|
| $P_{tot}$ | = 1 atm |
| $P_{N_2}$ | = 0.78 atm |
| $P_{O_2}$ | = 0.21 atm |

| 30 m | |
|---|---|
| $P_{tot}$ | = 4 atm |
| $P_{N_2}$ | = 3.12 atm |
| $P_{O_2}$ | = 0.84 atm |

▶ When a diver breathes compressed air, the abnormally high partial pressure of oxygen in the lungs leads to an elevated concentration of oxygen in body tissues.

## Example 5.10   Partial Pressures and Mole Fractions

A 12.5-L scuba diving tank contains a helium–oxygen (heliox) mixture made up of 24.2 g of He and 4.32 g of $O_2$ at 298 K. Calculate the mole fraction and partial pressure of each component in the mixture and the total pressure of the mixture.

| | |
|---|---|
| **SORT** The problem gives the masses of two gases in a mixture and the volume and temperature of the mixture. You are to find the mole fraction and partial pressure of each component, as well as the total pressure. | **GIVEN:** $m_{He} = 24.2$ g, $m_{O_2} = 4.32$ g, $V = 12.5$ L, $T = 298$ K <br><br> **FIND:** $\chi_{He}, \chi_{O_2}, P_{He}, P_{O_2}, P_{total}$ |

**STRATEGIZE** The conceptual plan has several parts. To calculate the mole fraction of each component, you must first find the number of moles of each component. Therefore, in the first part of the conceptual plan, convert the masses to moles using the molar masses.

**CONCEPTUAL PLAN**

$$m_{He} \longrightarrow n_{He} \qquad m_{O_2} \longrightarrow n_{O_2}$$
$$\frac{1 \text{ mol He}}{4.00 \text{ g He}} \qquad\qquad \frac{1 \text{ mol } O_2}{32.00 \text{ g } O_2}$$

In the second part, calculate the mole fraction of each component using the mole fraction definition.

$$\chi_{He} = \frac{n_{He}}{n_{He} + n_{O_2}}; \quad \chi_{O_2} = \frac{n_{O_2}}{n_{He} + n_{O_2}}$$

To calculate *partial pressures* calculate the *total pressure* and then use the mole fractions from the previous part to calculate the partial pressures. Calculate the total pressure from the sum of the moles of both components. (Alternatively, you can calculate the partial pressures of the components individually, using the number of moles of each component. Then you can sum them to obtain the total pressure.)

$$P_{total} = \frac{(n_{He} + n_{O_2})RT}{V}$$
$$P_{He} = \chi_{He}P_{total}; \quad P_{O_2} = \chi_{O_2}P_{total}$$

**RELATIONSHIPS USED**

$\chi_a = n_a/n_{total}$ (mole fraction definition)

$P_{total}V = n_{total}RT$

Last, use the mole fractions of each component and the total pressure to calculate the partial pressure of each component.

$P_a = \chi_a P_{total}$

**SOLVE** Follow the plan to solve the problem. Begin by converting each of the masses to amounts in moles.

**SOLUTION**

$$24.2 \text{ g He} \times \frac{1 \text{ mol He}}{4.00 \text{ g He}} = 6.05 \text{ mol He}$$

$$4.32 \text{ g } O_2 \times \frac{1 \text{ mol } O_2}{32.00 \text{ g } O_2} = 0.135 \text{ mol } O_2$$

Calculate each of the mole fractions.

$$\chi_{He} = \frac{n_{He}}{n_{He} + n_{O_2}} = \frac{6.05}{6.05 + 0.135} = 0.97817$$

$$\chi_{O_2} = \frac{n_{O_2}}{n_{He} + n_{O_2}} = \frac{0.135}{6.05 + 0.135} = 0.021827$$

Calculate the total pressure.

$$P_{total} = \frac{(n_{He} + n_{O_2})RT}{V}$$

$$= \frac{(6.05 \text{ mol} + 0.135 \text{ mol})\left(0.08206\dfrac{\text{L} \cdot \text{atm}}{\text{mol} \cdot \text{K}}\right)(298 \text{ K})}{12.5 \text{ L}}$$

$$= 12.099 \text{ atm}$$

Finally, calculate the partial pressure of each component.

$$P_{He} = \chi_{He} P_{total} = 0.97817 \times 12.099 \text{ atm} = 11.8 \text{ atm}$$

$$P_{O_2} = \chi_{O_2} P_{total} = 0.021827 \times 12.099 \text{ atm}$$
$$= 0.264 \text{ atm}$$

**CHECK** The units of the answers are correct, and the magnitudes are reasonable.

**FOR PRACTICE 5.10**   A diver breathes a heliox mixture with an oxygen mole fraction of 0.050. What must the total pressure be for the partial pressure of oxygen to be 0.21 atm?

# Collecting Gases over Water

When the desired product of a chemical reaction is a gas, the gas is often collected by the displacement of water. For example, suppose we use the reaction of zinc with hydrochloric acid as a source of hydrogen gas:

$$Zn(s) + 2\,HCl(aq) \longrightarrow ZnCl_2(aq) + H_2(g)$$

To collect the gas, we can set up an apparatus like the one shown in Figure 5.14▼. As the hydrogen gas forms, it bubbles through the water and gathers in the collection flask. The hydrogen gas collected in this way is not pure, however. It is mixed with water vapor because some water molecules evaporate and combine with the hydrogen molecules.

The partial pressure of water in the mixture, called its **vapor pressure**, depends on temperature (Table 5.4). Vapor pressure increases with increasing temperature because higher temperatures cause more water molecules to evaporate. We will discuss vapor pressure more thoroughly in Chapter 11.

Appendix IIE includes a more complete table of the vapor pressure of water versus temperature.

Suppose we collect hydrogen gas over water at a total pressure of 758.2 mmHg and a temperature of 25 °C. What is the partial pressure of the hydrogen gas? We know that the total pressure is 758.2 mmHg and that the partial pressure of water is 23.78 mmHg (its vapor pressure at 25 °C):

$$P_{total} = P_{H_2} + P_{H_2O}$$

$$758.2 \text{ mmHg} = P_{H_2} + 23.78 \text{ mmHg}$$

Therefore,

$$P_{H_2} = 758.2 \text{ mmHg} - 23.78 \text{ mmHg}$$

$$= 734.4 \text{ mmHg}$$

The partial pressure of the hydrogen in the mixture is 734.4 mmHg.

**TABLE 5.4 Vapor Pressure of Water versus Temperature**

| Temperature (°C) | Pressure (mmHg) | Temperature (°C) | Pressure (mmHg) |
|---|---|---|---|
| 0 | 4.58 | 55 | 118.2 |
| 5 | 6.54 | 60 | 149.6 |
| 10 | 9.21 | 65 | 187.5 |
| 15 | 12.79 | 70 | 233.7 |
| 20 | 17.55 | 75 | 289.1 |
| 25 | 23.78 | 80 | 355.1 |
| 30 | 31.86 | 85 | 433.6 |
| 35 | 42.23 | 90 | 525.8 |
| 40 | 55.40 | 95 | 633.9 |
| 45 | 71.97 | 100 | 760.0 |
| 50 | 92.6 | | |

**Collecting a Gas over Water**

▶ **FIGURE 5.14 Collecting a Gas over Water** When the gaseous product of a chemical reaction is collected over water, the product molecules (in this case $H_2$) mix with water molecules. The pressure of water in the final mixture is equal to the vapor pressure of water at the temperature at which the gas is collected. The partial pressure of the product is the total pressure minus the partial pressure of water.

## Example 5.11 Collecting Gases over Water

In order to determine the rate of photosynthesis (the conversion by plants of carbon dioxide and water into glucose and oxygen), the oxygen gas emitted by an aquatic plant is collected over water at a temperature of 293 K and a total pressure of 755.2 mmHg. Over a specific time period, a total of 1.02 L of gas is collected. What mass of oxygen gas (in grams) forms?

| | |
|---|---|
| **SORT** The problem gives the volume of gas collected over water as well as the temperature and the pressure. You are to find the mass in grams of oxygen that forms. | **GIVEN:** $V = 1.02$ L, $P_{total} = 755.2$ mmHg, $T = 293$ K<br><br>**FIND:** g $O_2$ |
| **STRATEGIZE** You can determine the mass of oxygen from the amount of oxygen in moles, which you can calculate from the ideal gas law if you know the partial pressure of oxygen. Since the oxygen is mixed with water vapor, you find the partial pressure of oxygen in the mixture by subtracting the partial pressure of water at 293 K (20 °C) from the total pressure. | **CONCEPTUAL PLAN**<br><br>$P_{O_2} = P_{total} - P_{H_2O}\ (20\,°C)$ |
| Next, use the ideal gas law to determine the number of moles of oxygen from its partial pressure, volume, and temperature. | <br>$P_{O_2}V = n_{O_2}RT$ |
| Finally, use the molar mass of oxygen to convert the number of moles to grams. | <br>$\dfrac{32.00\ \text{g } O_2}{\text{mol } O_2}$<br><br>**RELATIONSHIPS USED**<br>$P_{total} = P_a + P_b + P_c + \ldots$ (Dalton's law)<br>$PV = nRT$ (ideal gas law) |
| **SOLVE** Follow the conceptual plan to solve the problem. Begin by calculating the partial pressure of oxygen in the oxygen/water mixture. You can find the partial pressure of water at 20 °C in Table 5.4. | **SOLUTION**<br>$P_{O_2} = P_{total} - P_{H_2O}\ (20\,°C)$<br>$\quad = 755.2\ \text{mmHg} - 17.55\ \text{mmHg}$<br>$\quad = 737.\underline{6}5\ \text{mmHg}$ |
| Next, solve the ideal gas law for number of moles. | $n_{O_2} = \dfrac{P_{O_2}V}{RT}$ |
| Before substituting into the ideal gas law, convert the partial pressure of oxygen from mmHg to atm. | $737.\underline{6}5\ \text{mmHg} \times \dfrac{1\ \text{atm}}{760\ \text{mmHg}} = 0.97\underline{0}59\ \text{atm}$ |
| Substitute into the ideal gas law to find the number of moles of oxygen. | $n_{O_2} = \dfrac{P_{O_2}V}{RT} = \dfrac{0.97\underline{0}59\ \text{atm}\ (1.02\ \text{L})}{0.08206\ \dfrac{\text{L} \cdot \text{atm}}{\text{mol} \cdot \text{K}}\ (293\ \text{K})}$<br>$\quad = 4.1\underline{1}75 \times 10^{-2}\ \text{mol}$ |
| Finally, use the molar mass of oxygen to convert to grams of oxygen. | $4.1\underline{1}75 \times 10^{-2}\ \text{mol } O_2 \times \dfrac{32.00\ \text{g } O_2}{1\ \text{mol } O_2} = 1.32\ \text{g } O_2$ |

**CHECK** The answer is in the correct units. You can quickly check the magnitude of the answer by using molar volume. Under STP one liter is about 1/22 of one mole. Therefore the answer should be about 1/22 the molar mass of oxygen (1/22 × 32 = 1.45). The magnitude of the answer seems reasonable.

**FOR PRACTICE 5.11**   A common way to make hydrogen gas in the laboratory is to place a metal such as zinc in hydrochloric acid. The hydrochloric acid reacts with the metal to produce hydrogen gas, which is then collected over water. Suppose a student carries out this reaction and collects a total of 154.4 mL of gas at a pressure of 742 mmHg and a temperature of 25 °C. What mass of hydrogen gas (in mg) does the student collect?

# 5.7   Gases in Chemical Reactions: Stoichiometry Revisited

In Chapter 4, we discussed how to use the coefficients in chemical equations as conversion factors between number of moles of reactants and number of moles of products in a chemical reaction. We can use these conversion factors to determine, for example, the mass of product obtained in a chemical reaction based on a given mass of reactant, or the mass of one reactant needed to react completely with a given mass of another reactant. The general conceptual plan for these kinds of calculations is:

where A and B are two different substances involved in the reaction and the conversion factor between amounts (in moles) of each comes from the stoichiometric coefficients in the balanced chemical equation.

In reactions involving *gaseous* reactant or products, we often specify the quantity of a gas in terms of its volume at a given temperature and pressure. As we have seen, stoichiometry involves relationships between amounts in moles. For stoichiometric calculations involving gases, we can use the ideal gas law to determine the amounts in moles from the volumes, or to determine the volumes from the amounts in moles.

$$n = \frac{PV}{RT} \qquad V = \frac{nRT}{P}$$

| The pressures here could also be partial pressures.

The general conceptual plan for these kinds of calculations is:

Examples 5.12 and 5.13 demonstrate this kind of calculation.

---

**Example** 5.12   **Gases in Chemical Reactions**   ▶

Methanol ($CH_3OH$) can be synthesized by the reaction:

$$CO(g) + 2\,H_2(g) \longrightarrow CH_3OH(g)$$

What volume (in liters) of hydrogen gas, at a temperature of 355 K and a pressure of 738 mmHg, is needed to synthesize 35.7 g of methanol?

| **SORT**  You are given the mass of methanol, the product of the chemical reaction. You are asked to find the required volume of one of the reactants (hydrogen gas) at a specified temperature and pressure. | **GIVEN:** 35.7 g $CH_3OH$, $T = 355$ K, $P = 738$ mmHg <br><br> **FIND:** $V_{H_2}$ |

*—Continued on the next page*

*—Continued*

**STRATEGIZE** You can calculate the required volume of hydrogen gas from the number of moles of hydrogen gas, which you can obtain from the number of moles of methanol via the stoichiometry of the reaction.

First, find the number of moles of methanol from its mass by using the molar mass.

Then use the stoichiometric relationship from the balanced chemical equation to find the number of moles of hydrogen you need to form that quantity of methanol.

Finally, substitute the number of moles of hydrogen together with the pressure and temperature into the ideal gas law to find the volume of hydrogen.

**CONCEPTUAL PLAN**

**RELATIONSHIPS USED**
$PV = nRT$ (ideal gas law)
2 mol $H_2$ : 1 mol $CH_3OH$ (from balanced chemical equation)
molar mass $CH_3OH$ = 32.04 g/mol

**SOLVE** Follow the conceptual plan to solve the problem. Begin by using the mass of methanol to determine the number of moles of methanol.

Next, convert the number of moles of methanol to moles of hydrogen.

Finally, use the ideal gas law to calculate the volume of hydrogen. Before substituting into the equation, you need to convert the pressure to atmospheres.

**SOLUTION**

$$35.7 \text{ g } CH_3OH \times \frac{1 \text{ mol } CH_3OH}{32.04 \text{ g } CH_3OH} = 1.1142 \text{ mol } CH_3OH$$

$$1.1142 \text{ mol } CH_3OH \times \frac{2 \text{ mol } H_2}{1 \text{ mol } CH_3OH} = 2.2284 \text{ mol } H_2$$

$$V_{H_2} = \frac{n_{H_2}RT}{P}$$

$$P = 738 \text{ mmHg} \times \frac{1 \text{ atm}}{760 \text{ mmHg}} = 0.97105 \text{ atm}$$

$$V_{H_2} = \frac{(2.2284 \text{ mol})\left(0.08206\frac{\text{L}\cdot\text{atm}}{\text{mol}\cdot\text{K}}\right)(355 \text{ K})}{0.97105 \text{ atm}}$$

$$= 66.9 \text{ L}$$

**CHECK** The units of the answer are correct. The magnitude of the answer (66.9 L) seems reasonable. You are given slightly more than one molar mass of methanol, which is therefore slightly more than one mole of methanol. From the equation you can see that you need 2 mol hydrogen to make 1 mol methanol, so the answer must be slightly greater than 2 mol hydrogen. Under standard temperature and pressure, slightly more than 2 mol hydrogen occupies slightly more than 2 × 22.4 L = 44.8 L. At a temperature greater than standard temperature, the volume would be even greater; therefore, your answer is reasonable.

**FOR PRACTICE 5.12** In the reaction shown here, 4.58 L of $O_2$ forms at $P$ = 745 mmHg and $T$ = 308 K. How many grams of $Ag_2O$ decomposed?

$$2 Ag_2O(s) \longrightarrow 4 Ag(s) + O_2(g)$$

**FOR MORE PRACTICE 5.12** In the previous reaction, what mass of $Ag_2O(s)$ (in grams) is required to form 388 mL of oxygen gas at $P$ = 734 mmHg and 25.0 °C?

## Molar Volume and Stoichiometry

In Section 5.5 we saw that under standard temperature and pressure, 1 mol of an ideal gas occupies 22.4 L. Consequently, if a reaction occurs at or near standard temperature and pressure, we can use 1 mol = 22.4 L as a conversion factor in stoichiometric calculations, as demonstrated in Example 5.13.

| **Example** 5.13 | **Using Molar Volume in Gas Stoichiometric Calculations** |
|---|---|

How many grams of water form when 1.24 L of $H_2$ gas at STP completely reacts with $O_2$?

$$2 H_2(g) + O_2(g) \longrightarrow 2 H_2O(g)$$

| **SORT** You are given the volume of hydrogen gas (a reactant) at STP and asked to determine the mass of water that forms upon complete reaction. | **GIVEN:** 1.24 L $H_2$ <br><br> **FIND:** g $H_2O$ |
|---|---|
| **STRATEGIZE** Since the reaction occurs under standard temperature and pressure, you can convert directly from the volume (in liters) of hydrogen gas to the amount in moles. Then use the stoichiometric relationship from the balanced equation to find the number of moles of water formed. Finally, use the molar mass of water to obtain the mass of water formed. | **CONCEPTUAL PLAN** <br><br>  <br><br> **RELATIONSHIPS USED** <br> 1 mol = 22.4 L (at STP) <br> 2 mol $H_2$ : 2 mol $H_2O$ (from balanced equation) <br> molar mass $H_2O$ = 18.02 g/mol |
| **SOLVE** Follow the conceptual plan to solve the problem. | $1.24 \text{ L } H_2 \times \dfrac{1 \text{ mol } H_2}{22.4 \text{ L } H_2} \times \dfrac{2 \text{ mol } H_2O}{2 \text{ mol } H_2} \times \dfrac{18.02 \text{ g } H_2O}{1 \text{ mol } H_2O} = 0.998 \text{ g } H_2O$ |

**CHECK** The units of the answer are correct. The magnitude of the answer (0.998 g) is about 1/18 of the molar mass of water, roughly equivalent to the approximately 1/22 of a mole of hydrogen gas given, as expected for the 1:1 stoichiometric relationship between number of moles of hydrogen and number of moles of water.

**FOR PRACTICE 5.13** How many liters of oxygen (at STP) are required to form 10.5 g of $H_2O$?

$$2 H_2(g) + O_2(g) \longrightarrow 2 H_2O(g)$$

**Pressure and Number of Moles** Nitrogen and hydrogen react to form ammonia according to the following equation:

CONCEPTUAL
CONNECTION **5.5**

$$N_2(g) + 3 H_2(g) \rightleftharpoons 2 NH_3(g)$$

Consider the following representations of the initial mixture of reactants and the resulting mixture after the reaction has been allowed to react for some time:

If the volume is kept constant, and nothing is added to the reaction mixture, what happens to the total pressure during the course of the reaction?

**(a)** the pressure increases

**(b)** the pressure decreases

**(c)** the pressure does not change

**KCV** **Kinetic Molecular Theory**

**Kinetic Molecular Theory**

▲ **FIGURE 5.15 A Model for Gas Behavior** In the kinetic molecular theory of gases, a gas sample is modeled as a collection of particles in constant straight-line motion. The size of each particle is negligibly small, and the particles' collisions are elastic.

# 5.8 Kinetic Molecular Theory: A Model for Gases

In Chapter 1, we discussed how the scientific approach proceeds from observations to laws and eventually to theories. Remember that laws summarize behavior—for example, Charles's law summarizes *how* the volume of a gas depends on temperature—whereas theories give the underlying reasons for the behavior. A theory of gas behavior explains, for example, *why* the volume of a gas increases with increasing temperature.

The simplest model for the behavior of gases is the **kinetic molecular theory**. In this theory, a gas is modeled as a collection of particles (either molecules or atoms, depending on the gas) in constant motion (Figure 5.15 ◄). A single particle moves in a straight line until it collides with another particle (or with the wall of the container). Kinetic molecular theory has three basic postulates (or assumptions):

1. **The size of a particle is negligibly small.** Kinetic molecular theory assumes that the particles themselves occupy negligible volume, even though they have mass. This postulate is justified because, under normal pressures, the space between atoms or molecules in a gas is very large compared to the size of the atoms or molecules themselves. For example, in a sample of argon gas at STP, only about 0.01% of the volume is occupied by atoms, and the average distance from one argon atom to another is 3.3 nm. In comparison, the atomic radius of argon is 97 pm. If an argon atom were the size of a golf ball, its nearest neighbor would be, on average, over 4 ft (1.2 m) away at STP.

2. **The average kinetic energy of a particle is proportional to the temperature in kelvins.** The motion of atoms or molecules in a gas is due to thermal energy, which distributes itself among the particles in the gas. At any given moment, some particles are moving faster than others—there is a distribution of velocities—but the higher the temperature, the faster the overall motion and the greater the average kinetic energy. Notice that *kinetic energy* $(\frac{1}{2}mv^2)$—not *velocity*—is proportional to temperature. The atoms in a sample of helium and a sample of argon at the same temperature have the same average *kinetic energy* but not the same average *velocity*. Because the helium atoms are lighter, they must move faster to have the same kinetic energy as argon atoms.

3. **The collision of one particle with another (or with the walls of its container) is completely elastic.** When two particles collide, they may *exchange energy*, but there is no overall *loss of energy*. Any kinetic energy lost by one particle is completely gained by the other. In other words, the particles have no "stickiness," and they are not deformed by the collision. An encounter between two particles in kinetic molecular theory is more like the collision between two billiard balls than the collision between two lumps of clay (Figure 5.16 ▼). Between collisions, the particles do not exert any forces on one another.

Elastic collision

Inelastic collision

▲ **FIGURE 5.16 Elastic versus Inelastic Collisions** When two billiard balls collide, the collision is elastic—the total kinetic energy of the colliding bodies is the same before and after the collision. When two lumps of clay collide, the collision is inelastic—the kinetic energy of the colliding bodies dissipates in the form of heat during the collision.

If we start with the postulates of kinetic molecular theory, we can mathematically derive the ideal gas law (as we demonstrate later). Since the ideal gas law follows directly from kinetic molecular theory, we have confidence that the assumptions of the theory are valid, at least under conditions where the ideal gas law applies.

## How Kinetic Molecular Theory Explains Pressure and the Simple Gas Laws

Let's examine how the concept of pressure as well as each of the gas laws we have examined in this chapter follow conceptually from kinetic molecular theory.

### The Nature of Pressure

In Section 5.2, we defined pressure as force divided by area:

$$P = \frac{F}{A}$$

From Newton's second law, the force ($F$) associated with an individual collision is given by $F = ma$, where $m$ is the mass of the particle and $a$ is its acceleration as it changes its direction of travel due to the collision.

According to kinetic molecular theory, a gas is a collection of particles in constant motion. The motion results in collisions between the particles and the surfaces around them. As each particle collides with a surface, it exerts a force upon that surface. The result of many particles in a gas sample exerting forces on the surfaces around them is a constant pressure.

### Boyle's Law

Boyle's law states that for a constant number of particles at constant temperature, the volume of a gas is inversely proportional to its pressure. According to kinetic molecular theory, if we decrease the volume of a gas, we force the gas particles to occupy a smaller space. As long as the temperature remains the same, the number of collisions with the surrounding surfaces (per unit surface area) must necessarily increase, resulting in a greater pressure.

### Charles's Law

Charles's law states that for a constant number of particles at constant pressure, the volume of a gas is proportional to its temperature. According to kinetic molecular theory, when we increase the temperature of a gas, the average speed, and thus the average kinetic energy, of the particles increases. Since this greater kinetic energy results in more frequent collisions and more force per collision, the pressure of the gas increases if its volume is held constant (Gay-Lussac's law). The only way for the pressure to remain constant is for the volume to increase. The greater volume spreads the collisions out over a greater surface area so that the pressure (defined as force per unit area) is unchanged.

### Avogadro's Law

Avogadro's law states that at constant temperature and pressure, the volume of a gas is proportional to the number of particles. According to kinetic molecular theory, when we increase the number of particles in a gas sample, the number of collisions with the surrounding surfaces increases. The greater number of collisions results in a greater overall force on surrounding surfaces; the only way for the pressure to remain constant is for the volume to increase so that the number of particles per unit volume (and thus the number of collisions) remains constant.

### Dalton's Law

Dalton's law states that the total pressure of a gas mixture is the sum of the partial pressures of its components. According to Dalton's law, the components in a gas mixture act identically to, and independently of, one another. According to kinetic molecular theory, the particles have negligible size and they do not interact. Consequently, the only property that distinguishs one type of particle from another is its mass.

However, even particles of different masses have the same average kinetic energy at a given temperature, so they exert the same force upon collision with a surface. Consequently, adding components to a gas mixture—even different *kinds* of gases—has the same effect as simply adding more particles. The partial pressures of all the components sum to the overall pressure.

**Kinetic Molecular Theory I** As described by kinetic molecular theory, depict a gas sample containing equal molar amounts of argon and xenon. Use red dots to represent argon atoms and blue dots to represent xenon atoms. Give each atom a "tail" to represent its velocity relative to the others in the mixture.

## Kinetic Molecular Theory and the Ideal Gas Law

We have just seen how each of the gas laws conceptually follows from kinetic molecular theory. We can also *derive* the ideal gas law from the postulates of kinetic molecular theory. In other words, the kinetic molecular theory is a quantitative model that *implies* $PV = nRT$. Let's explore this derivation.

The pressure on a wall of a container (Figure 5.17▼) occupied by particles in constant motion is the total force on the wall (due to the collisions) divided by the area of the wall.

$$P = \frac{F_{\text{total}}}{A} \tag{5.13}$$

Recall that the force ($F$) associated with an individual collision is given by $F = ma$, where $m$ is the mass of the particle and $a$ is its acceleration as it changes its direction of travel due to the collision. The acceleration for each collision is the change in velocity ($\Delta v$) divided by the time interval ($\Delta t$), so the force imparted for each collision is:

$$F_{\text{collision}} = m\frac{\Delta v}{\Delta t} \tag{5.14}$$

If a particle collides elastically with the wall, it bounces off the wall with no loss of energy. For a straight-line collision, the change in velocity is $2v$ (the particle's velocity was $v$ before the collision and $-v$ after the collision; therefore, the change is $2v$). The force per collision is given by:

$$F_{\text{collision}} = m\frac{2v}{\Delta t} \tag{5.15}$$

The total number of collisions in the time interval $\Delta t$ on a wall of surface area $A$ is proportional to the number of particles that can reach the wall in this time interval—in other words, all particles within a distance of $v\Delta t$ of the wall. These particles occupy a

**Calculating Gas Pressure: A Molecular View**

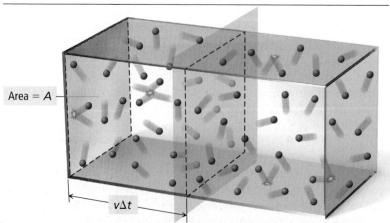

► **FIGURE 5.17 The Pressure on the Wall of a Container** The pressure on the wall of a container can be calculated by determining the total force due to collisions of the particles with the wall.

volume given by $v \Delta t \times A$, and their total number is equal to this volume multiplied by the density of particles in the container $(n/V)$:

$$\text{Number of collisions} \propto \text{number of particles within } v\,\Delta t$$

$$\propto \boxed{v\,\Delta t \times A} \times \boxed{\frac{n}{V}}$$

Volume    Density of particles    [5.16]

The *total force* on the wall is equal to the force per collision multiplied by the number of collisions:

$$F_{\text{total}} = F_{\text{collision}} \times \text{number of collisions}$$

$$\propto m\frac{2v}{\Delta t} \times v\Delta t \times A \times \frac{n}{V}$$    [5.17]

$$\propto mv^2 \times A \times \frac{n}{V}$$

The pressure on the wall is equal to the total force divided by the surface area of the wall:

$$P = \frac{F_{\text{total}}}{A}$$

$$\propto \frac{mv^2 \times A \times \dfrac{n}{V}}{A}$$    [5.18]

$$P \propto mv^2 \times \frac{n}{V}$$

Notice that Equation 5.18 contains within it Boyle's law ($P \propto 1/V$) and Avogadro's law ($V \propto n$). We can get the complete ideal gas law from postulate 2 of the kinetic molecular theory, which states that the average kinetic energy ($\frac{1}{2}mv^2$) is proportional to the temperature in kelvins ($T$):

$$mv^2 \propto T$$    [5.19]

By combining Equations 5.18 and 5.19, we get:

$$P \propto \frac{T \times n}{V}$$

$$PV \propto nT$$    [5.20]

The proportionality can be replaced by an equals sign if we provide the correct constant, $R$:

$$PV = nRT$$    [5.21]

In other words, the kinetic molecular theory (a model for how gases behave) predicts behavior that is consistent with our observations and measurements of gases—the theory agrees with the experiment. Recall from Chapter 1 that a scientific theory is the most powerful kind of scientific knowledge. In the kinetic molecular theory, we have a model for what a gas is like. Although the model is not perfect—indeed, it breaks down under certain conditions, as we shall see later in this chapter—it predicts a great deal about the behavior of gases. Therefore, the model is a good approximation of what a gas is actually like. A careful examination of the conditions under which the model breaks down (see Section 5.10) gives us even more insight into the behavior of gases.

## Temperature and Molecular Velocities

According to kinetic molecular theory, particles of different masses have the same average kinetic energy at a given temperature. The kinetic energy of a particle depends on its mass and velocity according to the equation:

$$\text{KE} = \tfrac{1}{2}mv^2$$

The only way for particles of different masses to have the same kinetic energy is for them to have different velocities as we saw in Conceptual Connection 5.6. *In a gas mixture at a given temperature, lighter particles travel faster (on average) than heavier ones.*

In kinetic molecular theory, we define the root mean square velocity ($u_{rms}$) of a particle as:

$$u_{rms} = \sqrt{\overline{u^2}}$$ [5.22]

where $\overline{u^2}$ is the average of the squares of the particle velocities. Even though the root mean square velocity of a collection of particles is not identical to the average velocity, the two are close in value and conceptually similar. Root mean square velocity is a special *type* of average. The average kinetic energy of one mole of gas particles is given by the equation:

$$KE_{avg} = \tfrac{1}{2}N_A m\overline{u^2}$$ [5.23]

where $N_A$ is Avogadro's number, and $m$ is the mass of the particle.

Postulate 2 of the kinetic molecular theory states that the average kinetic energy is proportional to the temperature in kelvins. The constant of proportionality in this relationship is $(3/2) R$:

$$KE_{avg} = (3/2)RT$$ [5.24]

> The $(3/2) R$ proportionality constant comes from a derivation that is beyond the scope of this textbook.

where $R$ is the gas constant, but in different units ($R = 8.314\,\text{J/mol} \cdot \text{K}$) than those we use in the ideal gas law. If we combine Equations 5.23 and 5.24, and solve for $\overline{u^2}$, we get:

$$(1/2)\,N_A m\overline{u^2} = (3/2)RT$$

$$\overline{u^2} = \frac{(3/2)RT}{(1/2)N_A m} = \frac{3RT}{N_A m}$$

> The joule (J) is a unit of energy that we will discuss in more detail in Section 6.2.
>
> $$\left(1\,\text{J} = 1\,\text{kg}\frac{\text{m}^2}{\text{s}^2}\right)$$

Taking the square root of both sides, we get:

$$\sqrt{\overline{u^2}} = u_{rms} = \sqrt{\frac{3RT}{N_A m}}$$ [5.25]

In Equation 5.25, $m$ is the mass of a particle in kg and $N_A$ is Avogadro's number. The product $N_A m$, then, is the molar mass in kg/mol. If we call this quantity $\mathcal{M}$, then the expression for mean square velocity as a function of temperature becomes the following important result:

$$u_{rms} = \sqrt{\frac{3RT}{\mathcal{M}}}$$ [5.26]

The root mean square velocity of a collection of gas particles is proportional to the square root of the temperature in kelvins and inversely proportional to the square root of the molar mass of the particles (which because of the units of $R$ must be in kilograms per mole). The root mean square velocity of nitrogen molecules at 25 °C, for example, is 515 m/s (1152 mi/hr). The root mean square velocity of hydrogen molecules at room temperature is 1920 m/s (4295 mi/hr). Notice that the lighter molecules move much faster at a given temperature.

CONCEPTUAL
**CONNECTION 5.7**

**Root Mean Square Velocity** How does the approximate root mean square velocity of neon compare to that of krypton at the same temperature?
  (a)  The root mean square velocity of Ne is approximately four times that of Kr.
  (b)  The root mean square velocity of Ne is approximately two times that of Kr.
  (c)  The root mean square velocity of Ne is approximately one-half that of Kr.
  (d)  The root mean square velocity of Ne is approximately one-fourth that of Kr.

## Variation of Velocity Distribution with Molar Mass

▲ **FIGURE 5.18 Velocity Distribution for Several Gases at 25 °C**  At a given temperature, there is a distribution of velocities among the particles in a sample of gas. The exact shape and peak of the distribution vary with the molar mass of the gas.

## Variation of Velocity Distribution with Temperature

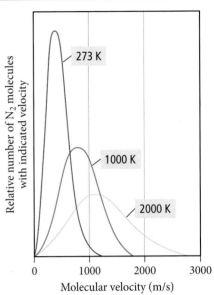

▲ **FIGURE 5.19 Velocity Distribution for Nitrogen at Several Temperatures**  As the temperature of a gas sample increases, the velocity distribution of the molecules shifts toward higher velocity and becomes less sharply peaked.

The root mean square velocity, as we have seen, is a kind of average velocity. Some particles move faster and some move more slowly than this average. The velocities of all the particles in a gas sample form distributions like those shown in Figure 5.18▲. We can see from these distributions that some particles indeed travel at the root mean square velocity. However, many particles travel faster and many travel more slowly than the root mean square velocity. For lighter particles, such as helium and hydrogen, the velocity distribution is shifted toward higher velocities and the curve becomes broader, indicating a wider range of velocities. Figure 5.19▶ illustrates the velocity distribution for nitrogen at different temperatures. As the temperature increases, the root mean square velocity increases and the distribution becomes broader.

## Example 5.14  Root Mean Square Velocity

Calculate the root mean square velocity of oxygen molecules at 25 °C.

| | |
|---|---|
| **SORT**  You are given the kind of molecule and the temperature and asked to find the root mean square velocity. | **GIVEN:** $O_2, t = 25\ °C$<br>**FIND:** $u_{rms}$ |
| **STRATEGIZE**  The conceptual plan for this problem shows how you can use the molar mass of oxygen and the temperature (in kelvins) with the equation that defines the root mean square velocity to calculate root mean square velocity. | **CONCEPTUAL PLAN**<br><br>$\mathcal{M}, T \longrightarrow u_{rms}$<br><br>$$u_{rms} - \sqrt{\frac{3RT}{\mathcal{M}}}$$<br><br>**RELATIONSHIP USED**<br>$u_{rms} = \sqrt{\dfrac{3RT}{\mathcal{M}}}$ (Equation 5.26) |
| **SOLVE**  First, gather the required quantities in the correct units. Note that molar mass must be in kg/mol. | **SOLUTION**<br>$T = 25 + 273 = 298\ K$<br><br>$$\mathcal{M} = \frac{32.00\ \text{g}\ O_2}{1\ \text{mol}\ O_2} \times \frac{1\ \text{kg}}{1000\ \text{g}} = \frac{32.00 \times 10^{-3}\ \text{kg}\ O_2}{1\ \text{mol}\ O_2}$$ |

—Continued on the next page

*—Continued*

| Substitute the quantities into the equation to calculate root mean square velocity. Note that $1\,J\ =\ 1\,kg\cdot m^2/s^2$. | $u_{rms} = \sqrt{\dfrac{3RT}{\mathcal{M}}}$ |
|---|---|
| | $= \sqrt{\dfrac{3\left(8.314\dfrac{J}{mol\cdot K}\right)(298\ K)}{\dfrac{32.00\times 10^{-3}\ kg\ O_2}{1\ mol\ O_2}}}$ |
| | $= \sqrt{2.32\times 10^5\dfrac{J}{kg}}$ |
| | $= \sqrt{2.32\times 10^5\dfrac{\dfrac{kg\cdot m^2}{s^2}}{kg}} = 482\ m/s$ |

**CHECK**   The units of the answer (m/s) are correct. The magnitude of the answer seems reasonable because oxygen is slightly heavier than nitrogen and should therefore have a slightly lower root mean square velocity at the same temperature. Recall from earlier that the root mean square velocity of nitrogen is 515 m/s at 25 °C.

**FOR PRACTICE 5.14**   Calculate the root mean square velocity of gaseous xenon atoms at 25 °C.

CONCEPTUAL
**CONNECTION 5.8**

**Kinetic Molecular Theory II**   Which sample of an ideal gas has the greatest pressure? Assume that the mass of each particle is proportional to its size and that all the gas samples are at the same temperature.

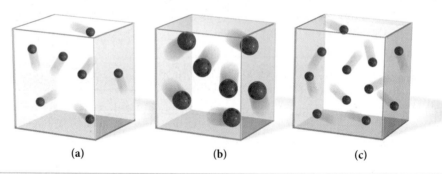

(a)                                (b)                                (c)

## 5.9   Mean Free Path, Diffusion, and Effusion of Gases

We have just seen that the root mean square velocity of gas molecules at room temperature is in the range of hundreds of meters per second. However, suppose that your roommate just put on too much perfume in the bathroom only 6 ft (1.8 m) away. Why does it take a minute or two before you can smell the fragrance? Although most molecules in a perfume bottle have higher molar masses than nitrogen, their velocities are still hundreds of meters per second, so why the delay? The answer is that even though gaseous particles travel at tremendous speeds, they also travel in haphazard paths (Figure 5.20▶). To a perfume molecule, the path from the perfume bottle in the bathroom to your nose 6 ft away is much like a bargain hunter's path through a busy shopping mall during a clearance sale. The molecule travels only a short distance before it collides with another molecule and changes direction, only to collide again, and so on. In fact, at room temperature and atmospheric pressure, a molecule in the air experiences several billion collisions per second. The average distance that a molecule travels between collisions is its **mean free path**. At room temperature and atmospheric pressure, the mean free path of a nitrogen molecule with a molecular diameter of 300 pm (four times the covalent radius) is 93 nm, or about 310 molecular diameters. If the nitrogen molecule were the size of a golf ball, it would travel about 40 ft between collisions. Mean free path increases with *decreasing* pressure. Under conditions of ultrahigh vacuum ($10^{-10}$ torr), the mean free path of a nitrogen molecule is hundreds of kilometers.

The process by which gas molecules spread out in response to a concentration gradient is **diffusion**, and even though the particles undergo many collisions, the root mean square velocity still influences the rate of diffusion. Heavier molecules diffuse more slowly than lighter ones, so the first molecules you smell from a perfume mixture (in a room with no air currents) are the lighter ones.

A process related to diffusion is **effusion**, the process by which a gas escapes from a container into a vacuum through a small hole (Figure 5.21 ▼). The rate of effusion is also related to root mean square velocity—heavier molecules effuse more slowly than lighter ones. The rate of effusion—the amount of gas that effuses in a given time—is inversely proportional to the square root of the molar mass of the gas as follows:

$$\text{rate} \propto \frac{1}{\sqrt{\mathcal{M}}}$$

The ratio of effusion rates of two different gases is given by **Graham's law of effusion**, named after Thomas Graham (1805–1869):

$$\frac{\text{rate}_A}{\text{rate}_B} = \sqrt{\frac{\mathcal{M}_B}{\mathcal{M}_A}}$$   [5.27]

In this expression, $\text{rate}_A$ and $\text{rate}_B$ are the effusion rates of gases A and B, and $\mathcal{M}_A$ and $\mathcal{M}_B$ are their molar masses.

Graham's law explains, in part, why helium balloons only float for a day or so. Because helium has such a low molar mass, it escapes from the balloon quite quickly. A balloon filled with air, by contrast, remains inflated longer.

## Typical Gas Molecule Path

The average distance between collisions is the mean free path.

▲ **FIGURE 5.20   Mean Free Path**   A molecule in a volume of gas follows a haphazard path, involving many collisions with other molecules.

In a ventilated room, air currents also enhance the transport of gas molecules.

## Effusion

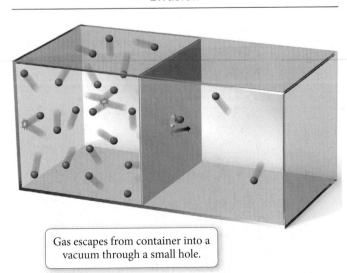

Gas escapes from container into a vacuum through a small hole.

◄ **FIGURE 5.21   Effusion**
Effusion is the escape of a gas from a container into a vacuum through a small hole.

## Example 5.15    Graham's Law of Effusion

An unknown gas effuses at a rate that is 0.462 times that of nitrogen gas (at the same temperature). Calculate the molar mass of the unknown gas in g/mol.

| | |
|---|---|
| **SORT** You are given the ratio of effusion rates for the unknown gas and nitrogen and asked to find the molar mass of the unknown gas. | **GIVEN:** $\dfrac{\text{rate}_{\text{unk}}}{\text{rate}_{N_2}} = 0.462$ <br><br> **FIND:** $\mathcal{M}_{\text{unk}}$ |
| **STRATEGIZE** The conceptual plan uses Graham's law of effusion. You are given the ratio of rates and you know the molar mass of the nitrogen. You can use Graham's law to determine the molar mass of the unknown gas. | **CONCEPTUAL PLAN** <br><br> 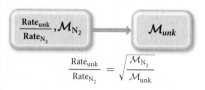 <br><br> $\dfrac{\text{Rate}_{\text{unk}}}{\text{Rate}_{N_2}} = \sqrt{\dfrac{\mathcal{M}_{N_2}}{\mathcal{M}_{\text{unk}}}}$ <br><br> **RELATIONSHIPS USED** <br><br> $\dfrac{\text{rate}_A}{\text{rate}_B} = \sqrt{\dfrac{\mathcal{M}_B}{\mathcal{M}_A}}$  (Graham's law) |
| **SOLVE** Solve the equation for $\mathcal{M}_{\text{unk}}$ and substitute the correct values and calculate. | **SOLUTION** <br><br> $\dfrac{\text{rate}_{\text{unk}}}{\text{rate}_{N_2}} = \sqrt{\dfrac{\mathcal{M}_{N_2}}{\mathcal{M}_{\text{unk}}}}$ <br><br> $\mathcal{M}_{\text{unk}} = \dfrac{\mathcal{M}_{N_2}}{\left(\dfrac{\text{rate}_{\text{unk}}}{\text{rate}_{N_2}}\right)^2}$ <br><br> $= \dfrac{28.02 \text{ g/mol}}{(0.462)^2}$ <br><br> $= 131 \text{ g/mol}$ |

**CHECK** The units of the answer are correct. The magnitude of the answer seems reasonable for the molar mass of a gas. From the answer you can even conclude that the gas is probably xenon, which has a molar mass of 131.29 g/mol.

**FOR PRACTICE 5.15**  Find the ratio of effusion rates of hydrogen gas and krypton gas.

## 5.10  Real Gases: The Effects of Size and Intermolecular Forces

One mole of an ideal gas has a volume of 22.41 L at STP. Figure 5.22▼ shows the molar volume of several real gases at STP. Note that most of these gases have a volume that is

### Molar Volume

► **FIGURE 5.22  Molar Volumes of Real Gases**  The molar volumes of several gases at STP are all close to 22.414 L, indicating that their departures from ideal behavior are small.

very close to 22.41 L, meaning that they act very nearly as ideal gases. Gases behave ideally when both of the following are true: (a) the volume of the gas particles is small compared to the space between them, and (b) the forces between the gas particles are not significant. At STP, these assumptions are valid for most common gases. However, these assumptions break down at higher pressures or lower temperatures.

## The Effect of the Finite Volume of Gas Particles

The finite volume of gas particles—that is, their actual *sizes*—becomes important at high pressure because the volume of the particles themselves occupies a significant portion of the total gas volume (Figure 5.23▶). We can see the effect of particle volume by comparing the molar volume of argon to the molar volume of an ideal gas as a function of pressure at 500 K as shown in Figure 5.24▼. At low pressures, the molar volume of argon is nearly identical to that of an ideal gas. But as the pressure increases, the molar volume of argon becomes *greater than* that of an ideal gas. At the higher pressures, the argon atoms themselves occupy a significant portion of the gas volume, making the actual volume greater than that predicted by the ideal gas law.

▲ **FIGURE 5.23 Particle Volume and Ideal Behavior** As a gas is compressed, the gas particles themselves begin to occupy a significant portion of the total gas volume, leading to deviations from ideal behavior.

**Nonideal Behavior: The effect of particle volume**

At high pressure, volume is higher than predicted.

$T = 500$ K

Argon

Ideal gas

◀ **FIGURE 5.24 The Effect of Particle Volume** At high pressures, 1 mol of argon occupies a larger volume than 1 mol of an ideal gas because of the volume of the argon atoms themselves. (This example was chosen to minimize the effects of intermolecular forces, which are very small in argon at 500 K, thereby isolating the effect of particle volume.)

In 1873, Johannes van der Waals (1837–1923) modified the ideal gas equation to fit the behavior of real gases. From the graph for argon in Figure 5.24 we can see that the ideal gas law predicts a volume that is too small. Van der Waals suggested a small correction factor that accounts for the volume of the gas particles themselves:

**Ideal behavior** $$V = \frac{nRT}{P}$$

**Corrected for volume of gas particles** $$V = \frac{nRT}{P} + nb \qquad [5.28]$$

The correction adds the quantity $nb$ to the volume, where $n$ is the number of moles and $b$ is a constant that depends on the gas (see Table 5.5). We can rearrange the corrected equation as follows:

$$(V - nb) = \frac{nRT}{P} \qquad [5.29]$$

| **TABLE 5.5 Van der Waals Constants for Common Gases** | | |
|---|---|---|
| Gas | $a$ (L$^2 \cdot$ atm/mol$^2$) | $b$ (L/mol) |
| He | 0.0342 | 0.02370 |
| Ne | 0.211 | 0.0171 |
| Ar | 1.35 | 0.0322 |
| Kr | 2.32 | 0.0398 |
| Xe | 4.19 | 0.0511 |
| $H_2$ | 0.244 | 0.0266 |
| $N_2$ | 1.39 | 0.0391 |
| $O_2$ | 1.36 | 0.0318 |
| $Cl_2$ | 6.49 | 0.0562 |
| $H_2O$ | 5.46 | 0.0305 |
| $CH_4$ | 2.25 | 0.0428 |
| $CO_2$ | 3.59 | 0.0427 |
| $CCl_4$ | 20.4 | 0.1383 |

## The Effect of Intermolecular Forces

Intermolecular forces, which we will discuss in more detail in Chapter 11, are attractions between the atoms or molecules that compose any substance. These attractions are typically small in gases and therefore do not matter much at low pressure because the molecules are too far apart to "feel" the attractions. They also do not matter much at high temperatures because the molecules have a lot of kinetic energy. Therefore a weak attraction between them does not affect the collision much. At lower temperatures, however, the collisions occur with less kinetic energy, and weak attractions can affect the collisions. We can understand this difference with an analogy to billiard balls. Imagine two billiard balls coated with a substance that makes them slightly sticky. If they collide when moving at high velocities, the stickiness will not have much of an effect—the balls bounce off one another as if the sticky substance was not even there. However, if the two billiard balls collide when moving very slowly (say barely rolling), the sticky substance would have an effect. The billiard balls might even stick together and not bounce off one another.

The effect of these weak attractions between particles is a decrease in the number of collisions with the surfaces of the container and a corresponding decrease in the pressure compared to that of an ideal gas. We can see the effect of intermolecular forces when we compare the pressure of 1.0 mol of xenon gas to the pressure of 1.0 mol of an ideal gas as a function of temperature and at a fixed volume of 1.0 L, as shown in Figure 5.25▼. At high temperature, the pressure of the xenon gas is nearly identical to that of an ideal gas. But at lower temperatures, the pressure of xenon is *less than* that of an ideal gas. At the lower temperatures, the xenon atoms spend more time interacting with each other and less time colliding with the walls, making the actual pressure less than that predicted by the ideal gas law.

From the graph for xenon shown in Figure 5.25 we can see that the ideal gas law predicts a pressure that is too large at low temperatures. Van der Waals suggested a small correction factor that accounts for the intermolecular forces between gas particles:

**Ideal behavior**    $$P = \frac{nRT}{V}$$

**Corrected for intermolecular forces**    $$P = \frac{nRT}{V} - a\left(\frac{n}{V}\right)^2 \qquad [5.30]$$

The correction subtracts the quantity $a(n/V)^2$ from the pressure, where $n$ is the number of moles, $V$ is the volume, and $a$ is a constant that depends on the gas (see Table 5.5). Notice that the correction factor increases as $n/V$ (the number of moles of particles per

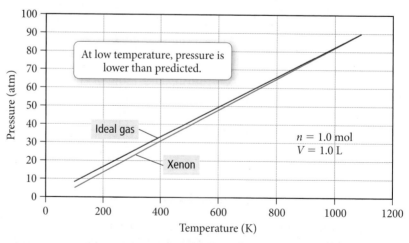

**Nonideal Behavior: The effect of intermolecular forces**

At low temperature, pressure is lower than predicted.

Ideal gas

Xenon

$n = 1.0$ mol
$V = 1.0$ L

Pressure (atm) / Temperature (K)

▲ **FIGURE 5.25  The Effect of Intermolecular Forces**  At low temperatures, the pressure of xenon is less than an ideal gas exerts because interactions among xenon molecules reduce the number of collisions with the walls of the container.

unit volume) increases because a greater concentration of particles makes it more likely that the particles will interact with one another. We can rearrange the corrected equation as:

$$P + a\left(\frac{n}{V}\right)^2 = \frac{nRT}{V} \qquad [5.31]$$

## Van der Waals Equation

We can combine the effects of particle volume (Equation 5.29) and particle intermolecular forces (Equation 5.31) into one equation that describes nonideal gas behavior:

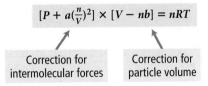

$$[P + a(\tfrac{n}{V})^2] \times [V - nb] = nRT$$

Correction for intermolecular forces

Correction for particle volume

This equation is the **van der Waals equation** and we can use it to calculate the properties of a gas under nonideal conditions.

## Real Gases

We can see the combined effects of particle volume and intermolecular forces by examining a plot of $PV/RT$ versus $P$ for 1 mol of a number of real gases (Figure 5.26 ▼). For an ideal gas, $PV/RT = n$, the number of moles of gas. Therefore, for 1 mol of an ideal gas, $PV/RT$ is equal to 1, as shown in the plot. For real gases, $PV/RT$ deviates from 1, but the deviations are not uniform. For example, water displays a large negative deviation from $PV/RT$ because, for water, the effect of intermolecular forces on lowering the pressure (relative to an ideal gas) is far greater than the effect of particle size on increasing the volume. Notice from Table 5.5 that water has a high value of $a$, the constant that corrects for intermolecular forces, but a moderate value of $b$, the constant that corrects for particle size. Therefore, $PV/RT$ for water is lower than predicted from the ideal gas law.

By contrast, consider the behavior of helium, which displays a positive deviation from the ideal behavior. This is because helium has very weak intermolecular forces and the effect of these forces on lowering the pressure (relative to ideal gas) is small compared to the effect of particle size on increasing the volume. Therefore $PV/RT$ for helium is greater than predicted from the ideal gas law.

**The Behavior of Real Gases**

◄ **FIGURE 5.26  Real versus Ideal Behavior**   For 1 mol of an ideal gas, $PV/RT$ is equal to 1. The combined effects of the volume of gas particles and the interactions among them cause each real gas to deviate from ideal behavior in a slightly different way. These curves were calculated at a temperature of 500 K.

CONCEPTUAL
CONNECTION **5.9**

**Real Gases** The graph shows $PV/RT$ for carbon dioxide at three different temperatures. Rank the curves in order of increasing temperature.

# Self-Assessment Quiz

**Q1.** A gas sample has an initial pressure of 547 mmHg and an initial volume of 0.500 L. What is the pressure (in atm) when the volume of the sample is decreased to 225 mL? (Assume constant temperature and constant number of moles of gas.)

  a) $1.60 \times 10^{-3}$ atm      b) 1.60 atm
  c) 0.324 atm                d) 1.22 atm

**Q2.** A gas sample has a volume of 178 mL at 0.00 °C. The temperature is raised (at constant pressure) until the volume is 211 mL. What is the temperature of the gas sample in °C at this volume?

  a) 0.00 °C               b) 324 °C
  c) −43 °C                d) 51 °C

**Q3.** What is the pressure of 1.78 g of nitrogen gas confined to a volume of 0.118 L at 25 °C?

  a) 13.2 atm            b) 369 atm
  c) 1.10 atm            d) 26.3 atm

**Q4.** What is the density of a sample of argon gas at 55 °C and 765 mmHg?

  a) 2.99 g/L
  b) $1.13 \times 10^3$ g/L
  c) 1.49 g/L
  d) 8.91 g/L

**Q5.** Which gas sample has the greatest volume at STP?

  a) 10.0 g Ar
  b) 10.0 g Kr
  c) 10.0 g Xe
  d) None of the above (They all have the same volume.)

**Q6.** A 1.25-g gas sample occupies 663 mL at 25 °C and 1.00 atm. What is the molar mass of the gas?

  a) 0.258 g/mol
  b) 0.0461 g/mol
  c) 3.87 g/mol
  d) 46.1 g/mol

**Q7.** A 255-mL gas sample contains argon and nitrogen at a temperature of 65 °C. The total pressure of the sample is 725 mmHg, and the partial pressure of argon is 231 mmHg. What mass of nitrogen is present in the sample?

  a) 0.324 g nitrogen        b) 0.167 g nitrogen
  c) 0.0837 g nitrogen      d) 0.870 g nitrogen

**Q8.** A gas mixture in a 1.55-L container at 298 K contains 10.0 g of Ne and 10.0 g of Ar. Calculate the partial pressure (in atm) of Ne and Ar in the container.

  a) $P_{Ne} = 10.5$ atm, $P_{Ar} = 5.29$ atm
  b) $P_{Ne} = 5.83$ atm, $P_{Ar} = 2.95$ atm
  c) $P_{Ne} = 5.88$ atm, $P_{Ar} = 5.88$ atm
  d) $P_{Ne} = 7.82$ atm, $P_{Ar} = 3.95$ atm

**Q9.** A gas sample at STP contains 1.15 g oxygen and 1.55 g nitrogen. What is the volume of the gas sample?

  a) 1.26 L     b) 2.04 L     c) 4.08 L     d) 61.0 L

**Q10.** Aluminum reacts with chlorine gas to form aluminum chloride.

$$2\,Al(s) \;+\; 3\,Cl_2(g) \longrightarrow 2\,AlCl_3(s)$$

What minimum volume of chlorine gas (at 298 K and 225 mmHg) is required to completely react with 7.85 g of aluminum?

  a) 36.0 L    b) 24.0 L    c) 0.0474 L    d) 16.0 L

**Q11.** Calculate the root mean square velocity of $I_2(g)$ at 373 K.

  a) 19.0 m/s           b) 191 m/s
  c) 6.05 m/s           d) 99.1 m/s

**Q12.** Which gas has the greatest kinetic energy at STP?

  a) He
  b) Ne
  c) Ar
  d) None of the above (All have the same kinetic energy.)

**Q13.** A sample of Xe takes 75 seconds to effuse out of a container. An unknown gas takes 37 seconds to effuse out of the identical container under identical conditions. What is the most likely identity of the unknown gas?

  a) He     b) $O_2$     c) $Br_2$     d) Kr

**Q14.** Consider the generic reaction: $2 A(g) + B(g) \longrightarrow 2 C(g)$. If a flask initially contains 1.0 atm of A and 1.0 atm of B, what will be the pressure in the flask if the reaction proceeds to completion? (Assume constant volume and temperature.)

a) 1.0 atm    b) 1.5 atm    c) 2.0 atm    d) 3.0 atm

**Q15.** Rank the gases Ar, $N_2$, $CH_4$, and $C_2H_6$ in order of increasing density at STP.

a) $CH_4 < C_2H_6 < N_2 < Ar$

b) $CH_4 < N_2 < Ar < C_2H_6$

c) $Ar < C_2H_6 < N_2 < CH_4$

d) $CH_4 < N_2 < C_2H_6 < Ar$

**Answers: 1.** (b) **2.** (d) **3.** (a) **4.** (c) **5.** (a) **6.** (d) **7.** (b) **8.** (d) **9.** (b) **10.** (a) **11.** (b) **12.** (d) **13.** (b) **14.** (b) **15.** (d)

# Chapter 5 IN REVIEW

## Key Terms

**Section 5.1**
pressure (198)

**Section 5.2**
millimeter of mercury (mmHg) (199)
barometer (199)
torr (199)
atmosphere (atm) (199)
pascal (Pa) (199)
manometer (200)

**Section 5.3**
Boyle's law (202)
Charles's law (205)
Avogadro's law (207)

**Section 5.4**
ideal gas law (209)
ideal gas (209)
ideal gas constant (209)

**Section 5.5**
molar volume (211)
standard temperature and
  pressure (STP) (211)

**Section 5.6**
partial pressure (215)
Dalton's law of partial pressures (215)
mole fraction ($\chi_a$) (215)
hypoxia (217)

oxygen toxicity (217)
nitrogen narcosis (217)
vapor pressure (219)

**Section 5.8**
kinetic molecular theory (224)

**Section 5.9**
mean free path (231)
diffusion (231)
effusion (231)
Graham's law of effusion (231)

**Section 5.10**
van der Waals equation (235)

## Key Concepts

### Pressure (5.1, 5.2)

▶ Gas pressure is the force per unit area that results from gas particles colliding with the surfaces around them. We use various units to measure pressure including mmHg, torr, Pa, psi, in Hg, and atm.

### The Simple Gas Laws (5.3)

▶ The simple gas laws express relationships between pairs of variables when other variables are constant.

▶ Boyle's law states that the volume of a gas is inversely proportional to its pressure.

▶ Charles's law states that the volume of a gas is directly proportional to its temperature.

▶ Avogadro's law states that the volume of a gas is directly proportional to the amount (in moles).

### The Ideal Gas Law and Its Applications (5.4, 5.5)

▶ The ideal gas law, $PV = nRT$, describes the relationship among all four gas variables and contains the simple gas laws within it.

▶ We can use the ideal gas law to find one of the four variables given the other three. We can use it to calculate the molar volume of an ideal gas, which is 22.4 L at STP, and to calculate the density and molar mass of a gas.

### Mixtures of Gases and Partial Pressures (5.6)

▶ In a mixture of gases, each gas acts independently of the others so that any overall property of the mixture is the sum of the properties of the individual components.

▶ The pressure of any individual component is its partial pressure.

### Gas Stoichiometry (5.7)

▶ In reactions involving gaseous reactants and products, we often report quantities in volumes at specified pressures and temperatures. We can convert these quantities to amounts (in moles) using the ideal gas law. Then we can use the stoichiometric coefficients from the balanced equation to determine the stoichiometric amounts of other reactants or products.

▶ The general form for these types of calculations is: volume A → amount A (in moles) → amount B (in moles) → quantity of B (in desired units).

▶ In cases where the reaction is carried out at STP, we can use the molar volume at STP (22.4 L = 1 mol) to convert between volume in liters and amount in moles.

## Kinetic Molecular Theory and Its Applications (**5.8**, **5.9**)

▶ Kinetic molecular theory is a quantitative model for gases. The theory has three main assumptions: (1) gas particles are negligibly small, (2) the average kinetic energy of a gas particle is proportional to the temperature in kelvins, and (3) the collision of one gas particle with another is completely elastic (the particles do not stick together). The gas laws all follow from the kinetic molecular theory.

▶ We can use kinetic molecular theory to derive the expression for the root mean square velocity of gas particles. This velocity is inversely proportional to the molar mass of the gas, and therefore—at a given temperature—smaller gas particles are (on average) moving more quickly than larger ones.

▶ The kinetic molecular theory also allows us to predict the mean free path of a gas particle (the distance it travels between collisions) and relative rates of diffusion or effusion.

## Real Gases (**5.10**)

▶ Real gases differ from ideal gases to the extent that they do not always fit the assumptions of kinetic molecular theory.

▶ These assumptions tend to break down at high pressures where the volume is higher than predicted for an ideal gas because the particles are no longer negligibly small compared to the space between them.

▶ The assumptions also break down at low temperatures where the pressure is lower than predicted because the attraction between molecules combined with low kinetic energies causes partially inelastic collisions.

▶ The van der Waals equation predicts gas properties under nonideal conditions.

# Key Equations and Relationships

Relationship between Pressure (*P*), Force (*F*), and Area (*A*) (**5.2**)

$$P = \frac{F}{A}$$

Boyle's Law: Relationship between Pressure (*P*) and Volume (*V*) (**5.3**)

$$V \propto \frac{1}{P}$$
$$P_1 V_1 = P_2 V_2$$

Charles's Law: Relationship between Volume (*V*) and Temperature (*T*) (**5.3**)

$$V \propto T \ \ (\text{in K})$$
$$\frac{V_1}{T_1} = \frac{V_2}{T_2}$$

Avogadro's Law: Relationship between Volume (*V*) and Amount in Moles (*n*) (**5.3**)

$$V \propto n$$
$$\frac{V_1}{n_1} = \frac{V_2}{n_2}$$

Ideal Gas Law: Relationship between Volume (*V*), Pressure (*P*), Temperature (*T*), and Amount (*n*) (**5.4**)

$$PV = nRT$$

Dalton's Law: Relationship between Partial Pressures $P_n$ in Mixture of Gases and Total Pressure ($P_{total}$) (**5.6**)

$$P_{total} = P_a + P_b + P_c + \dots$$
$$P_a = \frac{n_a RT}{V} \quad P_b = \frac{n_b RT}{V} \quad P_c = \frac{n_c RT}{V}$$

Mole Fraction ($\chi_a$) (**5.6**)

$$\chi_a = \frac{n_a}{n_{total}}$$
$$P_a = \chi_a P_{total}$$

Average Kinetic Energy ($KE_{avg}$) (**5.8**)

$$KE_{avg} = \frac{3}{2}RT$$

Relationship between Root Mean Square Velocity ($u_{rms}$) and Temperature (*T*) (**5.8**)

$$u_{rms} = \sqrt{\frac{3RT}{\mathcal{M}}}$$

Relationship of Effusion Rates of Two Different Gases (**5.9**)

$$\frac{\text{rate A}}{\text{rate B}} = \sqrt{\frac{\mathcal{M}_B}{\mathcal{M}_A}}$$

Van der Waals Equation: The Effects of Volume and Intermolecular Forces on Nonideal Gas Behavior (**5.10**)

$$\left[ P + a\left(\frac{n}{V}\right)^2 \right] \times (V - nb) = nRT$$

# Key Learning Outcomes

| CHAPTER OBJECTIVES | ASSESSMENT |
|---|---|
| Converting between Pressure Units (**5.2**) | Example 5.1   For Practice 5.1   For More Practice 5.1 Exercises 25–28 |
| Relating Volume and Pressure: Boyle's Law (**5.3**)  | Example 5.2   For Practice 5.2   Exercises 31, 32 |

| Relating Volume and Temperature: Charles's Law (**5.3**) | Example 5.3    For Practice 5.3    Exercises 33, 34 |
|---|---|
| Relating Volume and Moles: Avogadro's Law (**5.3**) | Example 5.4    For Practice 5.4    Exercises 35, 36 |
| Determining *P*, *V*, *n*, or *T* Using the Ideal Gas Law (**5.4**) | Examples 5.5, 5.6    For Practice 5.5, 5.6    For More Practice 5.6 Exercises 37–46, 51, 52 |
| Relating the Density of a Gas to Its Molar Mass (**5.5**) | Example 5.7    For Practice 5.7    For More Practice 5.7 Exercises 55, 56 |
| Calculating the Molar Mass of a Gas with the Ideal Gas Law (**5.5**) | Example 5.8    For Practice 5.8    Exercises 57–60 |
| Calculating Total Pressure, Partial Pressures, and Mole Fractions of Gases in a Mixture (**5.6**) | Examples 5.9, 5.10, 5.11    For Practice 5.9, 5.10, 5.11 Exercises 61, 62, 65, 67, 68, 70 |
| Relating the Amounts of Reactants and Products in Gaseous Reactions: Stoichiometry (**5.7**) | Examples 5.12, 5.13    For Practice 5.12, 5.13    For More Practice 5.12 Exercises 71–77 |
| Calculating the Root Mean Square Velocity of a Gas (**5.8**) | Example 5.14    For Practice 5.14    Exercises 83, 84 |
| Calculating the Effusion Rate or the Ratio of Effusion Rates of Two Gases (**5.9**) | Example 5.15    For Practice 5.15    Exercises 85–88 |

# EXERCISES

## Review Questions

1. What is pressure? What causes pressure?

2. Explain what happens when a person inhales. What forces air into the lungs?

3. Explain what happens when a person exhales. What forces air out of the lungs?

4. What are the common units of pressure? List them in order of smallest to largest unit.

5. What is a manometer? How does it measure the pressure of a sample of gas?

6. Summarize each of the simple gas laws (Boyle's law, Charles's law, and Avogadro's law). For each, explain the relationship between the two variables and also state which variables must be kept constant.

7. Explain why people may experience ear pain after a rapid change in altitude.

8. Explain why scuba divers should never hold their breath when they ascend to the surface.

9. Why is it impossible to breathe air through an extra-long snorkel (longer than a couple of meters) while swimming under water?

10. Explain why hot-air balloons float above the ground and why the second story of a two-story home is often warmer than the ground story.

11. What is the ideal gas law? Why is it useful?

12. Explain how the ideal gas law contains within it the simple gas laws (show an example).

13. Define molar volume and lists its value for a gas at STP.

14. How does the density of a gas depend on temperature? Pressure? How does it depend on the molar mass of the gas?

15. What is partial pressure? What is the relationship between the partial pressures of each gas in a sample and the total pressure of gas in the sample?

16. Why do deep-sea divers breathe a mixture of helium and oxygen?

17. When a gas is collected over water, is the gas pure? Why or why not? How can the partial pressure of the collected gas be determined?

18. If a reaction occurs in the gas phase at STP, we can determine the mass of a product from the volumes of reactants. Explain.

19. What are the basic postulates of kinetic molecular theory? How does the concept of pressure follow from kinetic molecular theory?

20. Explain how Boyle's law, Charles's law, Avogadro's law, and Dalton's law all follow from kinetic molecular theory.

21. How is the kinetic energy of a gas related to temperature? How is the root mean square velocity of a gas related to its molar mass?

22. Describe how the molecules in a perfume bottle travel from the bottle to your nose. What is mean free path?

**23.** Explain the difference between diffusion and effusion. How is the effusion rate of a gas related to its molar mass?

**24.** Deviations from the ideal gas law are observed at high pressure and low temperature. Explain this in light of kinetic molecular theory.

# Problems by Topic

## Converting between Pressure Units

**25.** The pressure in Denver, Colorado (elevation 5280 ft), averages about 24.9 in Hg. Convert this pressure to each indicated unit.
**a.** atm      **b.** mmHg      **c.** psi      **d.** Pa

**26.** The pressure on top of Mount Everest (29,029 ft) averages about 235 mmHg. Convert this pressure to each indicated unit.
**a.** torr      **b.** psi      **c.** in Hg      **d.** atm

**27.** The North American record for highest recorded barometric pressure is 31.85 in Hg, set in 1989 in Northway, Alaska. Convert this pressure to each indicated unit.
**a.** mmHg      **b.** atm      **c.** torr      **d.** kPa (kilopascals)

**28.** The world record for lowest pressure (at sea level) was 652.5 mmHg recorded inside Typhoon Tip on October 12, 1979, in the western Pacific Ocean. Convert this pressure to each indicated unit.
**a.** torr      **b.** atm      **c.** in Hg      **d.** psi

**29.** Given a barometric pressure of 762.4 mmHg, calculate the pressure of each gas sample as indicated by the manometer.

**30.** Given a barometric pressure of 751.5 mmHg, calculate the pressure of each gas sample as indicated by the manometer.

## Simple Gas Laws

**31.** A sample of gas has an initial volume of 5.6 L at a pressure of 735 mmHg. If the volume of the gas is increased to 9.4 L, what is its pressure?

**32.** A sample of gas has an initial volume of 13.9 L at a pressure of 1.22 atm. If the sample is compressed to a volume of 10.3 L, what is its pressure?

**33.** A 48.3-mL sample of gas in a cylinder is warmed from 22 °C to 87 °C. What is its volume at the final temperature?

**34.** A syringe containing 1.55 mL of oxygen gas is cooled from 95.3 °C to 0.0 °C. What is the final volume of oxygen gas?

**35.** A balloon contains 0.158 mol of gas and has a volume of 2.46 L. If an additional 0.113 mol of gas is added to the balloon (at the same temperature and pressure), what is its final volume?

**36.** A cylinder with a moveable piston contains 0.553 mol of gas and has a volume of 253 mL. What is its volume if an additional 0.365 mol of gas is added to the cylinder? (Assume constant temperature and pressure.)

## Ideal Gas Law

**37.** What volume is occupied by 0.118 mol of helium gas at a pressure of 0.97 atm and a temperature of 305 K? Would the volume be different if the gas was argon (under the same conditions)?

**38.** What volume is occupied by 12.5 g of argon gas at a pressure of 1.05 atm and a temperature of 322 K? Would the volume be different if the sample were 12.5 g of helium (under identical conditions)?

**39.** What is the pressure in a 10.0-L cylinder filled with 0.448 mol of nitrogen gas at a temperature of 315 K?

**40.** What is the pressure in a 15.0-L cylinder filled with 32.7 g of oxygen gas at a temperature of 302 K?

**41.** A cylinder contains 28.5 L of oxygen gas at a pressure of 1.8 atm and a temperature of 298 K. How much gas (in moles) is in the cylinder?

**42.** What is the temperature of 0.52 mol of gas at a pressure of 1.3 atm and a volume of 11.8 L?

**43.** An automobile tire has a maximum rating of 38.0 psi (gauge pressure). The tire is inflated (while cold) to a volume of 11.8 L and a gauge pressure of 36.0 psi at a temperature of 12.0 °C. On a hot day, the tire warms to 65.0 °C, and its volume expands to 12.2 L. Does the pressure in the tire exceed its maximum rating? (Note: The *gauge pressure* is the *difference* between the total pressure and atmospheric pressure. In this case, assume that atmospheric pressure is 14.7 psi.)

**44.** A weather balloon is inflated to a volume of 28.5 L at a pressure of 748 mmHg and a temperature of 28.0 °C. The balloon rises in the atmosphere to an altitude of approximately 25,000 ft, where the pressure is 385 mmHg and the temperature is −15.0 °C. Assuming the balloon can freely expand, calculate the volume of the balloon at this altitude.

**45.** A piece of dry ice (solid carbon dioxide) with a mass of 28.8 g sublimes (converts from solid to gas) into a large balloon. Assuming that all of the carbon dioxide ends up in the balloon, what is the volume of the balloon at 22 °C and a pressure of 742 mmHg?

**46.** A 1.0-L container of liquid nitrogen is kept in a closet measuring 1.0 m by 1.0 m by 2.0 m. Assuming that the container is completely full, that the temperature is 25.0 °C, and that the atmospheric pressure is 1.0 atm, calculate the percent (by volume) of air that is displaced if all of the liquid nitrogen evaporates. (Liquid nitrogen has a density of 0.807 g/mL.)

**47.** A wine-dispensing system uses argon canisters to pressurize and preserve wine in the bottle. An argon canister for the system has a volume of 55.0 mL and contains 26.0 g of argon. Assuming ideal gas behavior, what is the pressure in the canister at 295 K? When the argon is released from the canister, it expands to fill the wine bottle. How many 750.0-mL wine bottles can be purged with the argon in the canister at a pressure of 1.20 atm and a temperature of 295 K?

**48.** Cyclists sometimes use pressurized carbon dioxide inflators to inflate a bicycle tire in the event of a flat. These inflators use metal cartridges that contain 16.0 g of carbon dioxide. At 298 K, to what pressure (in psi) can the carbon dioxide in the cartridge inflate a 3.45-L mountain bike tire? (Note: The *gauge pressure* is the *difference* between the total pressure and atmospheric pressure. In this case, assume that atmospheric pressure is 14.7 psi.)

**49.** Which gas sample has the greatest pressure? Assume that all the samples are at the same temperature. Explain.

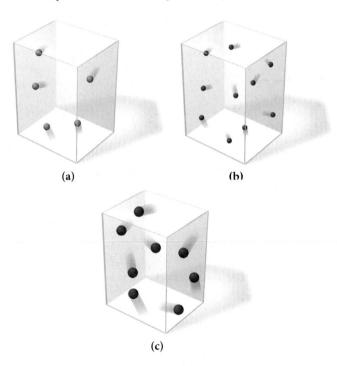

(a)        (b)

(c)

**50.** This picture represents a sample of gas at a pressure of 1 atm, a volume of 1 L, and a temperature of 25 °C. Draw a similar picture showing what would happen to the sample if the volume were reduced to 0.5 L  and the temperature were increased to 250 °C. What would happen to the pressure?

**51.** Aerosol cans carry clear warnings against incineration because of the high pressures that can develop upon heating. Suppose that a can contains a residual amount of gas at a pressure of 755 mmHg and a temperature of 25 °C. What would the pressure be if the can were heated to 1155 °C?

**52.** A sample of nitrogen gas in a 1.75-L container exerts a pressure of 1.35 atm at 25 °C. What is the pressure if the volume of the container is maintained constant and the temperature is raised to 355 °C?

## Molar Volume, Density, and Molar Mass of a Gas

**53.** Use the molar volume of a gas at STP to determine the volume (in L) occupied by 33.6 g of neon at STP.

**54.** Use the molar volume of a gas at STP to calculate the density (in g/L) of nitrogen gas at STP.

**55.** What is the density (in g/L) of hydrogen gas at 20.0 °C and a pressure of 1655 psi?

**56.** A sample of $N_2O$ gas has a density of 2.85 g/L at 298 K. What is the pressure of the gas (in mmHg)?

**57.** A 248-mL gas sample has a mass of 0.433 g at a pressure of 745 mmHg and a temperature of 28 °C. What is the molar mass of the gas?

**58.** A 113-mL gas sample has a mass of 0.171 g at a pressure of 721 mmHg and a temperature of 32 °C. What is the molar mass of the gas?

**59.** A sample of gas has a mass of 38.8 mg. Its volume is 224 mL at a temperature of 55 °C and a pressure of 886 torr. Find the molar mass of the gas.

**60.** A sample of gas has a mass of 0.555 g. Its volume is 117 mL at a temperature of 85 °C and a pressure of 753 mmHg. Find the molar mass of the gas.

## Partial Pressure

**61.** A gas mixture contains each of the following gases at the indicated partial pressures: $N_2$, 215 torr; $O_2$, 102 torr; and He, 117 torr. What is the total pressure of the mixture? What mass of each gas is present in a 1.35-L sample of this mixture at 25.0 °C?

**62.** A gas mixture with a total pressure of 745 mmHg contains each of the following gases at the indicated partial pressures: $CO_2$, 125 mmHg; Ar, 214 mmHg; and $O_2$, 187 mmHg. The mixture also contains helium gas. What is the partial pressure of the helium gas? What mass of helium gas is present in a 12.0-L sample of this mixture at 273 K?

**63.** A 1.20-g sample of dry ice is added to a 755 mL flask containing nitrogen gas at a temperature of 25.0 °C and a pressure of 725 mmHg. The dry ice sublimes (converts from solid to gas), and the mixture returns to 25.0 °C. What is the total pressure in the flask?

**64.** A 275-mL flask contains pure helium at a pressure of 752 torr. A second flask with a volume of 475 mL contains pure argon at a pressure of 722 torr. If we connect the two flasks through a stopcock and we open the stopcock, what is the partial pressure of each gas and the total pressure?

**65.** A gas mixture contains 1.25 g $N_2$ and 0.85 g $O_2$ in a 1.55 L container at 18 °C. Calculate the mole fraction and partial pressure of each component in the gas mixture.

**66.** What is the mole fraction of oxygen gas in air (see Table 5.3)? What volume of air contains 10.0 g of oxygen gas at 273 K and 1.00 atm?

**67.** The hydrogen gas formed in a chemical reaction is collected over water at 30.0 °C at a total pressure of 732 mmHg. What is the partial pressure of the hydrogen gas collected in this way? If the total volume of gas collected is 722 mL, what mass of hydrogen gas is collected?

**68.** The air in a bicycle tire is bubbled through water and collected at 25 °C. If the total volume of gas collected is 5.45 L at a temperature of 25 °C and a pressure of 745 torr, how many moles of gas were in the bicycle tire?

**69.** The zinc in a copper-plated penny will dissolve in hydrochloric acid if the copper coating is filed down in several spots (so that the hydrochloric acid can get to the zinc). The reaction between the acid and the zinc is $2 H^+(aq) + Zn(s) \longrightarrow H_2(g) + Zn^{2+}(aq)$. When the zinc in a certain penny dissolves, the total volume of gas collected over water at 25 °C is 0.951 L at a total pressure of 748 mmHg. What mass of hydrogen gas is collected?

**70.** A heliox deep-sea diving mixture contains 2.0 g of oxygen to every 98.0 g of helium. What is the partial pressure of oxygen when this mixture is delivered at a total pressure of 8.5 atm?

## Reaction Stoichiometry Involving Gases

**71.** Consider the chemical reaction:

$$C(s) + H_2O(g) \longrightarrow CO(g) + H_2(g)$$

How many liters of hydrogen gas are formed from the complete reaction of 15.7 g C? Assume that the hydrogen gas is collected at a pressure of 1.0 atm and a temperature of 355 K.

**72.** Consider the chemical reaction:

$$2 H_2O(l) \longrightarrow 2 H_2(g) + O_2(g)$$

What mass of $H_2O$ is required to form 1.4 L of $O_2$ at a temperature of 315 K and a pressure of 0.957 atm?

**73.** $CH_3OH$ can be synthesized by the reaction:

$$CO(g) + 2 H_2(g) \longrightarrow CH_3OH(g)$$

What volume of $H_2$ gas (in L), at 748 mmHg and 86 °C, is required to synthesize 25.8 g $CH_3OH$? How many liters of CO gas, measured under the same conditions, are required?

**74.** Oxygen gas reacts with powdered aluminum according to the reaction:

$$4 Al(s) + 3 O_2(g) \longrightarrow 2 Al_2O_3(s)$$

What volume of $O_2$ gas (in L), measured at 782 mmHg and 25 °C, completely reacts with 53.2 g Al?

**75.** Automobile air bags inflate following a serious impact. The impact triggers the chemical reaction:

$$2 NaN_3(s) \longrightarrow 2 Na(s) + 3 N_2(g)$$

If an automobile air bag has a volume of 11.8 L, what mass of $NaN_3$ (in g) is required to fully inflate the air bag upon impact? Assume STP conditions.

**76.** Lithium reacts with nitrogen gas according to the reaction:

$$6 Li(s) + N_2(g) \longrightarrow 2 Li_3N(s)$$

What mass of lithium (in g) reacts completely with 58.5 mL of $N_2$ gas at STP?

**77.** Hydrogen gas (a potential future fuel) can be formed by the reaction of methane with water according to the equation:

$$CH_4(g) + H_2O(g) \longrightarrow CO(g) + 3 H_2(g)$$

In a particular reaction, 25.5 L of methane gas (measured at a pressure of 732 torr and a temperature of 25 °C) mixes with 22.8 L of water vapor (measured at a pressure of 702 torr and a temperature of 125 °C). The reaction produces 26.2 L of hydrogen gas at STP. What is the percent yield of the reaction?

**78.** Ozone is depleted in the stratosphere by chlorine from $CF_3Cl$ according to this set of equations:

$$CF_3Cl + UV \text{ light} \longrightarrow CF_3 + Cl$$
$$Cl + O_3 \longrightarrow ClO + O_2$$
$$O_3 + UV \text{ light} \longrightarrow O_2 + O$$
$$ClO + O \longrightarrow Cl + O_2$$

What total volume of ozone at a pressure of 25.0 mmHg and a temperature of 225 K is destroyed when all of the chlorine from 15.0 g of $CF_3Cl$ goes through 10 cycles of the given reactions?

**79.** Chlorine gas reacts with fluorine gas to form chlorine trifluoride.

$$Cl_2(g) + 3 F_2(g) \longrightarrow 2 ClF_3(g)$$

A 2.00-L reaction vessel, initially at 298 K, contains chlorine gas at a partial pressure of 337 mmHg and fluorine gas at a partial pressure of 729 mmHg. Identify the limiting reactant and determine the theoretical yield of $ClF_3$ in grams.

**80.** Carbon monoxide gas reacts with hydrogen gas to form methanol.

$$CO(g) + 2 H_2(g) \longrightarrow CH_3OH(g)$$

A 1.50-L reaction vessel, initially at 305 K, contains carbon monoxide gas at a partial pressure of 232 mmHg and hydrogen gas at a partial pressure of 397 mmHg. Identify the limiting reactant and determine the theoretical yield of methanol in grams.

## Kinetic Molecular Theory

**81.** Consider a 1.0-L sample of helium gas and a 1.0-L sample of argon gas, both at room temperature and atmospheric pressure.
   **a.** Do the atoms in the helium sample have the same *average kinetic energy* as the atoms in the argon sample?
   **b.** Do the atoms in the helium sample have the same *average velocity* as the atoms in the argon sample?
   **c.** Do the argon atoms, because they are more massive, exert a greater pressure on the walls of the container? Explain.
   **d.** Which gas sample has the faster rate of effusion?

**82.** A flask at room temperature contains exactly equal amounts (in moles) of nitrogen and xenon.
   **a.** Which of the two gases exerts the greater partial pressure?
   **b.** The molecules or atoms of which gas have the greater average velocity?
   **c.** The molecules of which gas have the greater average kinetic energy?
   **d.** If a small hole were opened in the flask, which gas effuses more quickly?

**83.** Calculate the root mean square velocity and kinetic energy of $F_2$, $Cl_2$, and $Br_2$ at 298 K. Rank these three halogens with respect to their rate of effusion.

**84.** Calculate the root mean square velocity and kinetic energy of CO, $CO_2$, and $SO_3$ at 298 K. Which gas has the greatest velocity? The greatest kinetic energy? The greatest effusion rate?

**85.** We separate U-235 from U-238 by fluorinating a sample of uranium to form $UF_6$ (which is a gas) and then taking advantage of the different rates of effusion and diffusion for compounds containing the two isotopes. Calculate the ratio of effusion rates for $^{238}UF_6$ and $^{235}UF_6$. The atomic mass of U-235 is 235.054 amu and that of U-238 is 238.051 amu.

**86.** Calculate the ratio of effusion rates for Ar and Kr.

**87.** A sample of neon effuses from a container in 76 seconds. The same amount of an unknown noble gas requires 155 seconds. Identify the second gas.

**88.** A sample of $N_2O$ effuses from a container in 42 seconds. How long will it take the same amount of gaseous $I_2$ to effuse from the same container under identical conditions?

**89.** The graph shows the distribution of molecular velocities for two different molecules (A and B) at the same temperature. Which molecule has the higher molar mass? Which molecule has the higher rate of effusion?

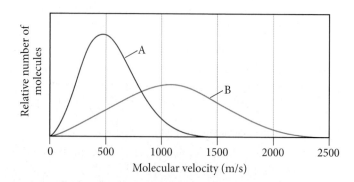

**90.** The graph shows the distribution of molecular velocities for the same molecule at two different temperatures ($T_1$ and $T_2$). Which temperature is greater? Explain.

**Real Gases**

**91.** Which postulate of the kinetic molecular theory breaks down under conditions of high pressure? Explain.

**92.** Which postulate of the kinetic molecular theory breaks down under conditions of low temperature? Explain.

**93.** Use the van der Waals equation and the ideal gas equation to calculate the volume of 1.000 mol of neon at a pressure of 500.0 atm and a temperature of 355.0 K. Explain why the two values are different. (Hint: One way to solve the van der Waals equation for $V$ is to use successive approximations. Use the ideal gas law to get a preliminary estimate for $V$.)

**94.** Use the van der Waals equation and the ideal gas equation to calculate the pressure exerted by 1.000 mol of $Cl_2$ in a volume of 5.000 L at a temperature of 273.0 K. Explain why the two values are different.

# Cumulative Problems

**95.** Modern pennies are composed of zinc coated with copper. A student determines the mass of a penny to be 2.482 g and then makes several scratches in the copper coating (to expose the underlying zinc). The student puts the scratched penny in hydrochloric acid, where the following reaction occurs between the zinc and the HCl (the copper remains undissolved):

$$Zn(s) + 2\,HCl(aq) \longrightarrow H_2(g) + ZnCl_2(aq)$$

The student collects the hydrogen produced over water at 25 °C. The collected gas occupies a volume of 0.899 L at a total pressure of 791 mmHg. Calculate the percent zinc (by mass) in the penny. (Assume that all the Zn in the penny dissolves.)

**96.** A 2.85-g sample of an unknown chlorofluorocarbon decomposes and produces 564 mL of chlorine gas at a pressure of 752 mmHg and a temperature of 298 K. What is the percent chlorine (by mass) in the unknown chlorofluorocarbon?

**97.** The mass of an evacuated 255 mL flask is 143.187 g. The mass of the flask filled with 267 torr of an unknown gas at 25 °C is 143.289 g. Calculate the molar mass of the unknown gas.

**98.** A 118-mL flask is evacuated and found to have a mass of 97.129 g. When the flask is filled with 768 torr of helium gas at 35 °C, it has a mass of 97.171 g. Was the helium gas pure?

**99.** A gaseous hydrogen- and carbon-containing compound is decomposed and found to contain 82.66% carbon and 17.34% hydrogen by mass. The mass of 158 mL of the gas, measured at 556 mmHg and 25 °C, was 0.275 g. What is the molecular formula of the compound?

**100.** A gaseous hydrogen- and carbon-containing compound is decomposed and found to contain 85.63% C and 14.37% H by mass. The mass of 258 mL of the gas, measured at STP, was 0.646 g. What is the molecular formula of the compound?

**101.** Consider the reaction:

$$2\,NiO(s) \longrightarrow 2\,Ni(s) + O_2(g)$$

If $O_2$ is collected over water at 40.0 °C and a total pressure of 745 mmHg, what volume of gas is collected for the complete reaction of 24.78 g of NiO?

**102.** Consider the reaction:

$$2\,Ag_2O(s) \longrightarrow 4\,Ag(s) + O_2(g)$$

If this reaction produces 15.8 g of Ag(s), what total volume of gas can be collected over water at a temperature of 25 °C and a total pressure of 752 mmHg?

**103.** When hydrochloric acid is poured over potassium sulfide, 42.9 mL of hydrogen sulfide gas is produced at a pressure of 752 torr and 25.8 °C. Write an equation for the gas-evolution reaction and determine how much potassium sulfide (in grams) reacted.

**104.** Consider the reaction:

$$2\,SO_2(g) + O_2(g) \longrightarrow 2\,SO_3(g)$$

**a.** If 285.5 mL of $SO_2$ reacts with 158.9 mL of $O_2$ (both measured at 315 K and 50.0 mmHg), what is the limiting reactant and the theoretical yield of $SO_3$?

**b.** If 187.2 mL of $SO_3$ is collected (measured at 315 K and 50.0 mmHg), what is the percent yield for the reaction?

**105.** Ammonium carbonate decomposes upon heating according to the balanced equation:

$$(NH_4)_2CO_3(s) \longrightarrow 2\,NH_3(g) + CO_2(g) + H_2O(g)$$

Calculate the total volume of gas produced at 22 °C and 1.02 atm by the complete decomposition of 11.83 g of ammonium carbonate.

**106.** Ammonium nitrate decomposes explosively upon heating according to the balanced equation:

$$2\,NH_4NO_3(s) \longrightarrow 2\,N_2(g) + O_2(g) + 4\,H_2O(g)$$

Calculate the total volume of gas (at 125 °C and 748 mmHg) produced by the complete decomposition of 1.55 kg of ammonium nitrate.

**107.** Olympic cyclists fill their tires with helium to make them lighter. Calculate the mass of air in an air-filled tire and the mass of helium in a helium-filled tire. What is the mass difference between the two? Assume that the volume of the tire is 855 mL, that it is filled to a total pressure of 125 psi, and that the temperature is 25 °C. Also, assume an average molar mass for air of 28.8 g/mol.

**108.** In a common classroom demonstration, a balloon is filled with air and submerged in liquid nitrogen. The balloon contracts as the gases within the balloon cool. Suppose a balloon initially contains 2.95 L of air at a temperature of 25.0 °C and a pressure of 0.998 atm. Calculate the expected volume of the balloon upon cooling to −196 °C (the boiling point of liquid nitrogen). When the demonstration is carried out, the actual volume of the balloon decreases to 0.61 L. How does the observed volume of the balloon compare to your calculated value? Explain the difference.

**109.** Gaseous ammonia is injected into the exhaust stream of a coal-burning power plant to reduce the pollutant NO to $N_2$ according to the reaction:

$$4 NH_3(g) + 4 NO(g) + O_2(g) \longrightarrow 4 N_2(g) + 6 H_2O(g)$$

Suppose that the exhaust stream of a power plant has a flow rate of 335 L/s at a temperature of 955 K, and that the exhaust contains a partial pressure of NO of 22.4 torr. What should be the flow rate of ammonia delivered at 755 torr and 298 K into the stream to react completely with the NO if the ammonia is 65.2% pure (by volume)?

**110.** The emission of $NO_2$ by fossil fuel combustion can be prevented by injecting gaseous urea into the combustion mixture. The urea reduces NO (which oxidizes in air to form $NO_2$) according to the reaction:

$$2 CO(NH_2)_2(g) + 4 NO(g) + O_2(g) \longrightarrow$$
$$4 N_2(g) + 2 CO_2(g) + 4 H_2O(g)$$

Suppose that the exhaust stream of an automobile has a flow rate of 2.55 L/s at 655 K and contains a partial pressure of NO of 12.4 torr. What total mass of urea is necessary to react completely with the NO formed during 8.0 hours of driving?

**111.** An ordinary gasoline can measuring 30.0 cm by 20.0 cm by 15.0 cm is evacuated with a vacuum pump. Assuming that virtually all of the air can be removed from inside the can and that atmospheric pressure is 14.7 psi, what is the total force (in pounds) on the surface of the can? Do you think that the can could withstand the force?

**112.** Twenty-five milliliters of liquid nitrogen (density = 0.807 g/mL) is poured into a cylindrical container with a radius of 10.0 cm and a length of 20.0 cm. The container initially contains only air at a pressure of 760.0 mmHg (atmospheric pressure) and a temperature of 298 K. If the liquid nitrogen completely vaporizes, what is the total force (in lb) on the interior of the container at 298 K?

**113.** A 160.0-L helium tank contains pure helium at a pressure of 1855 psi and a temperature of 298 K. How many 3.5-L helium balloons will the helium in the tank fill? (Assume an atmospheric pressure of 1.0 atm and a temperature of 298 K.)

**114.** An 11.5-mL sample of liquid butane (density = 0.573 g/mL) is evaporated in an otherwise empty container at a temperature of 28.5 °C. The pressure in the container following evaporation is 892 torr. What is the volume of the container?

**115.** A scuba diver creates a spherical bubble with a radius of 2.5 cm at a depth of 30.0 m where the total pressure (including atmospheric pressure) is 4.00 atm. What is the radius of the bubble when it reaches the surface of the water? (Assume that the atmospheric pressure is 1.00 atm and the temperature is 298 K.)

**116.** A particular balloon can be stretched to a maximum surface area of 1257 $cm^2$. The balloon is filled with 3.0 L of helium gas at a pressure of 755 torr and a temperature of 298 K. The balloon is then allowed to rise in the atmosphere. If the atmospheric temperature is 273 K, at what pressure will the balloon burst? (Assume the balloon is the shape of a sphere.)

**117.** A catalytic converter in an automobile uses a palladium or platinum catalyst (a substance that increases the rate of a reaction without being consumed by the reaction) to convert carbon monoxide gas to carbon dioxide according to the reaction:

$$2 CO(g) + O_2(g) \longrightarrow 2 CO_2(g)$$

A chemist researching the effectiveness of a new catalyst combines a 2.0:1.0 mole ratio mixture of carbon monoxide and oxygen gas (respectively) over the catalyst in a 2.45-L flask at a total pressure of 745 torr and a temperature of 552 °C. When the reaction is complete, the pressure in the flask has dropped to 552 torr. What percentage of the carbon monoxide was converted to carbon dioxide?

**118.** A quantity of $N_2$ occupies a volume of 1.0 L at 300 K and 1.0 atm. The gas expands to a volume of 3.0 L as the result of a change in both temperature and pressure. Find the density of the gas at these new conditions.

**119.** A mixture of CO(g) and $O_2(g)$ in a 1.0-L container at $1.0 \times 10^3$ K has a total pressure of 2.2 atm. After some time, the total pressure falls to 1.9 atm as the result of the formation of $CO_2$. Determine the mass (in grams) of $CO_2$ that forms.

**120.** The radius of a xenon atom is $1.3 \times 10^{-8}$ cm. A 100-mL flask is filled with Xe at a pressure of 1.0 atm and a temperature of 273 K. Calculate the fraction of the volume that is occupied by Xe atoms. (Hint: The atoms are spheres.)

**121.** A natural gas storage tank is a cylinder with a moveable top. Its volume can change only as its height changes, and its radius remains fixed. The height of the cylinder is 22.6 m on a day when the temperature is 22 °C. The next day the height of the cylinder increases to 23.8 m when the gas expands because of a heat wave. Determine the temperature on the second day, assuming that the pressure and amount of gas in the storage tank have not changed.

**122.** A mixture of 8.0 g $CH_4$ and 8.0 g Xe is placed in a container and the total pressure is found to be 0.44 atm. Determine the partial pressure of $CH_4$.

**123.** A steel container of volume 0.35 L can withstand pressures up to 88 atm before exploding. What mass of helium can be stored in this container at 299 K?

**124.** Binary compounds of alkali metals and hydrogen react with water to liberate $H_2(g)$. The $H_2$ from the reaction of a sample of NaH with an excess of water fills a volume of 0.490 L above the water. The temperature of the gas is 35 °C and the total pressure is 758 mmHg. Determine the mass of $H_2$ liberated and the mass of NaH that reacted.

**125.** In a given diffusion apparatus, 15.0 mL of HBr gas diffuses in 1.0 min. In the same apparatus and under the same conditions, 20.3 mL of an unknown gas diffuses in 1.0 min. The unknown gas is a hydrocarbon. Find its molecular formula.

**126.** A sample of $N_2O_3(g)$ has a pressure of 0.017 atm. The temperature (in K) is doubled and the $N_2O_3$ undergoes complete decomposition to $NO_2(g)$ and NO(g). Find the total pressure of the mixture of gases assuming constant volume and no additional temperature change.

**127.** When 0.583 g of neon is added to an 800-cm$^3$ bulb containing a sample of argon, the total pressure of the gases is 1.17 atm at a temperature of 295 K. Find the mass of the argon in the bulb.

**128.** A gas mixture composed of helium and argon has a density of 0.670 g/L at a 755 mmHg and 298 K. What is the composition of the mixture by volume?

**129.** A gas mixture contains 75.2% nitrogen and 24.8% krypton by mass. What is the partial pressure of krypton in the mixture if the total pressure is 745 mmHg?

# Challenge Problems

**130.** A 10-L container is filled with 0.10 mol of H$_2(g)$ and heated to 3000 K, causing some of the H$_2(g)$ to decompose into H$(g)$. The pressure is found to be 3.0 atm. Find the partial pressure of the H$(g)$ that forms from H$_2$ at this temperature. (Assume two significant figures for the temperature.)

**131.** A mixture of NH$_3(g)$ and N$_2$H$_4(g)$ is placed in a sealed container at 300 K. The total pressure is 0.50 atm. The container is heated to 1200 K at which time both substances decompose completely according to the equations 2 NH$_3(g) \longrightarrow$ N$_2(g)$ + 3 H$_2(g)$; N$_2$H$_4(g) \longrightarrow$ N$_2(g)$ + 2 H$_2(g)$. After decomposition is complete, the total pressure at 1200 K is found to be 4.5 atm. Find the percent of N$_2$H$_4(g)$ in the original mixture. (Assume two significant figures for the temperature.)

**132.** A quantity of CO gas occupies a volume of 0.48 L at 1.0 atm and 275 K. The pressure of the gas is lowered, and its temperature is raised until its volume is 1.3 L. Determine the density of the CO under the new conditions.

**133.** When CO$_2(g)$ is put in a sealed container at 701 K and a pressure of 10.0 atm and is heated to 1401 K, the pressure rises to 22.5 atm. Some of the CO$_2$ decomposes to CO and O$_2$. Calculate the mole percent of CO$_2$ that decomposes.

**134.** The world burns approximately $3.7 \times 10^{12}$ kg of fossil fuel per year. Use the combustion of octane as the representative reaction and determine the mass of carbon dioxide (the most significant greenhouse gas) formed per year. The current concentration of carbon dioxide in the atmosphere is approximately 399 ppm (by volume). By what percentage does the concentration increase each year due to fossil fuel combustion? Approximate the average properties of the entire atmosphere by assuming that the atmosphere extends from sea level to 15 km and that it has an average pressure of 381 torr and average temperature of 275 K. Assume Earth is a perfect sphere with a radius of 6371 km.

**135.** The atmosphere slowly oxidizes hydrocarbons in a number of steps that eventually convert the hydrocarbon into carbon dioxide and water. The overall reaction of a number of such steps for methane gas is

CH$_4(g)$ + 5 O$_2(g)$ + 5 NO$(g) \longrightarrow$
$$CO_2(g) + H_2O(g) + 5 NO_2(g) + 2 OH(g)$$

Suppose that an atmospheric chemist combines 155 mL of methane at STP, 885 mL of oxygen at STP, and 55.5 mL of NO at STP in a 2.0 L flask. The flask is allowed to stand for several weeks at 275 K. If the reaction reaches 90.0% of completion (90.0% of the limiting reactant is consumed), what is the partial pressure of each of the reactants and products in the flask at 275 K? What is the total pressure in the flask?

**136.** Two identical balloons are filled to the same volume, one with air and one with helium. The next day, the volume of the air-filled balloon has decreased by 5.0%. By what percent has the volume of the helium-filled balloon decreased? (Assume that the air is four-fifths nitrogen and one-fifth oxygen and that the temperature did not change.)

**137.** A mixture of CH$_4(g)$ and C$_2$H$_6(g)$ has a total pressure of 0.53 atm. Just enough O$_2(g)$ is added to the mixture to bring about its complete combustion to CO$_2(g)$ and H$_2$O$(g)$. The total pressure of the two product gases is 2.2 atm. Assuming constant volume and temperature, find the mole fraction of CH$_4$ in the mixture.

**138.** A sample of C$_2$H$_2(g)$ has a pressure of 7.8 kPa. After some time a portion of it reacts to form C$_6$H$_6(g)$. The total pressure of the mixture of gases is then 3.9 kPa. Assume the volume and the temperature do not change. What fraction of C$_2$H$_2(g)$ has undergone reaction?

# Conceptual Problems

**139.** When the driver of an automobile applies the brakes, the passengers are pushed toward the front of the car, but a helium balloon is pushed toward the back of the car. Upon forward acceleration, the passengers are pushed toward the back of the car, but the helium balloon is pushed toward the front of the car. Why?

**140.** Suppose that a liquid is 10 times denser than water. If you were to sip this liquid at sea level using a straw, what is the maximum length your straw would be?

**141.** This reaction occurs in a closed container:
$$A(g) + 2 B(g) \longrightarrow 2 C(g)$$

A reaction mixture initially contains 1.5 L of A and 2.0 L of B. Assuming that the volume and temperature of the reaction mixture remain constant, what is the percent change in pressure if the reaction goes to completion?

**142.** One mole of nitrogen and one mole of neon are combined in a closed container at STP. How big is the container?

**143.** Exactly equal amounts (in moles) of gas A and gas B are combined in a 1-L container at room temperature. Gas B has a molar mass that is twice that of gas A. Which statement is true for the mixture of gases and why?
a. The molecules of gas B have greater kinetic energy than those of gas A.
b. Gas B has a greater partial pressure than gas A.
c. The molecules of gas B have a greater average velocity than those of gas A.
d. Gas B makes a greater contribution to the average density of the mixture than gas A.

**144.** Which gas would you expect to deviate most from ideal behavior under conditions of low temperature: F$_2$, Cl$_2$, or Br$_2$? Explain.

145. The volume of a sample of a fixed amount of gas is decreased from 2.0 L to 1.0 L. The temperature of the gas in kelvins is then doubled. What is the final pressure of the gas in terms of the initial pressure?

146. Which gas sample has the greatest volume at STP?
    a. 10.0 g Kr    b. 10.0 g Xe    c. 10.0 g He

# Questions for Group Work

Discuss these questions with the group and record your consensus answer.

148. Assign one of the three simple gas laws to each member of your group. For the assigned gas law, have each member write two equations, draw a graph, and describe it in a complete sentence. Have each group member present his or her law to the group.

149. Review the ideal gas law. Without referring back to the text, use algebra to write the ideal gas law and solve for each of the individual variables it contains. Have each group member solve for a different variable and present answers to the group.

150. Hydrogen peroxide ($H_2O_2$) decomposes in the presence of a catalyst to form water and oxygen. The catalyst is added to 5.00 mL of a hydrogen peroxide solution at 25.0 °C, and 49.5 mL of gas is collected over water at a total pressure of 763.8 mmHg.
    a. Write and balance the chemical reaction (note: catalysts do not appear in balanced chemical equations).
    b. Look up the vapor pressure of water under these conditions.
    c. What is the partial pressure of oxygen collected over the water?
    d. How many moles of oxygen are collected?

147. Draw a depiction of a gas sample, as described by kinetic molecular theory, containing equal molar amounts of helium, neon, and krypton. Use different color dots to represent each element. Give each atom a "tail" to represent its velocity relative to the others in the mixture.

*Active Classroom Learning*

    e. How many grams of hydrogen peroxide were in the original sample?
    f. What is the concentration (in mol/L) of the hydrogen peroxide solution?
    g. Which part of this process is conceptually most difficult for your group?

151. A box contains equal amounts of helium, argon, and krypton (all gases) at 25 °C. Using complete sentences, describe the temperatures, masses, average velocities, and average kinetic energy of the three kinds of gas in the mixture. What do they have in common? What are the differences? How are these properties related?

152. Calculate the pressure exerted by 1 mol of an ideal gas in a box that is 0.500 L and 298 K. Have each group member calculate the pressure of 1 mol of the following gases in the same box at the same temperature: He, Ne, $H_2$, $CH_4$, and $CO_2$. Compare group members' answers as well as all answers with the pressure of an ideal gas. Assuming that the van der Waals equation predictions are accurate, account for why the pressure of each gas is higher or lower than that predicted for an ideal gas.

# Data Interpretation and Analysis

## Nitrogen Oxide from Automobiles

153. Air contains about 78.08% nitrogen gas, 20.95% oxygen gas, and 0.97% other gases by volume. When air is combusted in automobile engines, nitrogen oxide gases form. One of these gases, nitric oxide (NO), rapidly oxidizes in air to form nitrogen dioxide ($NO_2$).

Scientists are concerned about the level of $NO_x$ gas ($NO_x$ is the term used to refer collectively to nitrogen oxide gases) being emitted by automobile exhaust. People breathing air with elevated levels of $NO_2$ gas are susceptible to increased incidence of asthma attacks, respiratory infection, and inflammation of the breathing airways, which leads to coughing and wheezing. As a result, government and state regulations require new cars to emit less nitrogen oxide gas than new cars did in the past and older cars to have yearly inspections to check the efficiency of their engines. One purpose of a three-way catalytic converter is to convert the $NO_x$ gases produced as a by-product of the combustion of gasoline with air to $N_2$ and $O_2$.

The EPA mandates that the overall average concentration of nitrogen dioxide in the air not exceed 0.053 ppmv and not exceed a one-hour average of 0.100 ppmv. Monitoring gas pollutants in cities allows engineers and scientists to caution the public about possible health risks. Figure a▶ charts the average concentration of nitrogen dioxide ($NO_2$) and ozone ($O_3$), gases over seven days from nine gas detection stations located equal distances from each other in a city. The particular gas detection system employed in this city can measure the concentration of $NO_2$, $O_3$, and CO gases but does not measure the concentration of NO.

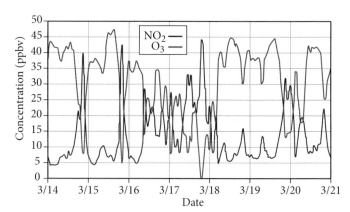

▲ **FIGURE a** **Average Concentration of Nitrogen Dioxide, NO₂, and Ozone, O₃, over Seven Days**
Source: http://www.cas.manchester.ac.uk/resprojects/holmemoss/results/fig2/

Parts-per-billion-volume (ppbv) or parts-per-million-volume (ppmv) units are used in gas measurements. This unit is based on volume-to-volume ratio and is not the same as parts-per-billion (or parts-per-million) used in aqueous measurements, which is based on a mass-per-mass ratio. The following formula converts ppmv (parts-per-million-volume) to mg/m³ for a gas at a pressure of 1 atm:

$$\text{mg/m}^3 = \text{ppmv} \cdot \frac{\mathcal{M}}{(0.08206 \cdot T)}$$

where $\mathcal{M}$ is the molar mass of a gas in units of grams per mole, $T$ is the temperature in Kelvin, and 0.08206 is the ideal gas constant in units of L·atm/mol·K.

Use the information provided in the figure to answer the following questions:

a. What type of relationship exists between nitrogen dioxide and ozone between March 14 and March 16?

b. What number of moles of $NO_2$ in 1.00 m³ of air was produced on March 14? Assume an average temperature of 25.0 °C and a pressure of 1.00 atm. Note that the moles of $NO_2$ produced is the difference between the existing amount at the start of the day and the peak amount.

c. Calculate the number of moles of $O_3$ in 1.00 m³ consumed on March 14. Note that the moles of $O_3$ consumed is the difference between the existing amount at the start of the day and the minimum amount.

d. Determine the mole-to-mole ratio of $O_3$ consumed to $NO_2$ produced.

e. The following chemical equations model the interactions of nitrogen dioxide gas and ozone gas. Which set of equations best accounts for the trends observed in the graph? Explain your answer.

  i. $N_2 + O_2 \longrightarrow 2\,NO$
     $NO + O_3 \longrightarrow NO_2 + O_2 + \text{light}$

  ii. $4\,NH_3 + 5\,O_2 \longrightarrow 4\,NO + 6\,H_2O$
      $2\,NO + O_2 \longrightarrow 2\,NO_2$
      $NO_2 + O_3 \longrightarrow NO_3 + O_2$

  iii. $N_2 + 2\,O_3 \longrightarrow 2\,NO + 2\,O_2$
       $NO + H_2O \longrightarrow NO_2 + H_2$
       $2\,NO_2 + 2\,ClO_2 + 2\,O_3 \longrightarrow 2\,NO_2ClO_4 + O_2$

f. What were inhabitants doing in the city between 9 am and 9 pm on March 14 and 15 that might account for the rise and fall of the concentration of both nitrogen dioxide and ozone? Most likely, what day of the week was March 18?

g. Assume you have a 1.00-m³ sample of air collected in the city during the peak production of $NO_2$ on March 15. The gas sample has a total pressure of 1.00 atm and a temperature of 25.00 °C. Calculate the partial pressures of $NO_2$ and $O_3$ in this sample of gas in units of atmospheres.

h. On average, what was the highest $NO_2$ gas concentration produced in one day in this city during the time period illustrated in Figure a?

i. Did the concentration of $NO_2$ gas exceed either the 0.053-ppmv level standard or the 100.0-ppb average over the period shown in Figure a?

j. Assume the nine gas detection stations cover a volume of 2.0 mile × 2.0 mile × 0.15 mile. If the $NO_2$ gas concentration is uniform throughout this volume at the highest concentration, how many grams of $NO_2$ gas are in the air?

# Answers to Conceptual Connections

### Boyle's Law and Charles's Law

**5.1 (e)** The final volume of the gas is the same as the initial volume because doubling the pressure *decreases* the volume by a factor of two, but doubling the temperature *increases* the volume by a factor of two. The two changes in volume are equal in magnitude but opposite in sign, resulting in a final volume that is equal to the initial volume.

### Molar Volume

**5.2 (a)** Since one gram of $H_2$ contains the greatest number of moles (due to $H_2$ having the lowest molar mass of the listed gases), and since one mole of *any* ideal gas occupies the same volume, the $H_2$ occupies the greatest volume.

### Density of a Gas

**5.3** $Ne < O_2 < F_2 < Cl_2$

### Partial Pressures

**5.4** $P_{He} = 1.5$ atm; $P_{Ne} = 1.5$ atm. Since the number of moles of each gas is equal, the mole fraction of each gas is 0.50 and the partial pressure of each gas is $0.50 \times P_{tot}$.

### Pressure and Number of Moles

**5.5 (b)** Since the total number of gas molecules decreases, the total pressure—the sum of all the partial pressures—must also decrease.

### Kinetic Molecular Theory I

**5.6** Although the velocity "tails" have different lengths, the average length of the tails on the argon atoms in your drawing should be longer than the average length of the tails on the xenon atoms. Because the argon atoms are  lighter, they must on average move faster than the xenon atoms to have the same kinetic energy.

### Root Mean Square Velocity

**5.7 (b)** The molar mass of neon (20.18 g/mol) is approximately one-quarter the molar mass of krypton (83.80 g/mol), so the root mean square velocity of neon is $\sqrt{\frac{1}{1/4}}$ times that of kypton.

### Kinetic Molecular Theory II

**5.8 (c)** Since the temperature and the volume are both constant, the ideal gas law tells us that the pressure depends solely on the number of particles. Sample (c) has the greatest number of particles per unit volume and therefore has the greatest pressure. The pressures of samples (a) and (b) at a given temperature are identical. Even though the particles in (b) are more massive than those in (a), they have the same average kinetic energy at a given temperature. The particles in (b) move more slowly than those in (a), and so exert the same pressure as the particles in (a).

### Real Gases

**5.9** A < B < C. Curve A is the lowest temperature curve because it deviates the most from ideality. The tendency for the intermolecular forces in carbon dioxide to lower the pressure (relative to that of an ideal gas) is greatest at low temperature (because the molecules are moving more slowly and are therefore less able to overcome the intermolecular forces). As a result, the curve that dips the lowest must correspond to the lowest temperature.

CHAPTER

# 6

# Thermochemistry

There is a fact, or if you wish, a law, governing all natural phenomena that are known to date. There is no known exception to this law—it is exact as far as we know. The law is called the conservation of energy. It states that there is a certain quantity, which we call energy, that does not change in the manifold changes which nature undergoes.

—RICHARD P. FEYNMAN (1918–1988)

We have spent the first few chapters of this book examining one of the two major components of our universe—matter. We now turn our attention to the other major component—energy. As far as we know, matter and energy—which can be interchanged but not destroyed—make up the physical universe. Unlike matter, energy is not something we can touch or hold in our hand, but we experience it in many ways. The warmth of sunlight, the feel of wind on our faces, and the force that presses us back when a car accelerates are all manifestations of energy and its interconversions. And of course energy is critical to society and to the world. The standard of living around the globe is strongly correlated with the access to and use of energy resources. Most of those resources, as we shall see, are chemical ones, and we can understand their advantages as well as their drawbacks in terms of chemistry.

▲ *A chemical hand warmer contains substances that react to emit heat.*

## 6.1 Chemical Hand Warmers

My family loves to snowboard. However, my wife hates being cold (with a passion), especially in her hands and toes. Her solution is the chemical hand warmer, a small pouch that comes sealed in a plastic package. She opens the package and places the pouch in her glove or boot. The pouch slowly warms up and keeps her hand (or foot) warm all day long.

Warming your hands with chemical hand warmers involves many of the principles of **thermochemistry**, the study of the relationships between chemistry and energy. When you open the package that contains the hand warmer, the contents are exposed to air, and an *exothermic reaction* occurs. Most hand warmers use the oxidation of iron as the exothermic reaction:

$$4 \, \text{Fe}(s) + 3 \, \text{O}_2(g) \longrightarrow 2 \, \text{Fe}_2\text{O}_3(s) + \text{heat}$$

The useful product of this reaction is not a substance—it is *heat*. We'll define heat more carefully later, but heat is what you feel when you touch something that is warmer than your hand (in this case, the hand warmer). Although some of the heat is lost through the minute openings in your gloves (which is why my wife prefers mittens, which have fewer openings), most of it is transferred to your hands and to the pocket of air surrounding your hands, resulting in a temperature increase. The magnitude of the temperature increase depends on the size of the hand warmer and the size of your glove (as well as some other details). But in general, the size of the temperature increase is proportional to the amount of heat released by the reaction.

In this chapter, we examine the relationship between chemical reactions and energy. Specifically, we look at how chemical reactions can *exchange* energy with their surroundings and how we can quantify the magnitude of those exchanges. These kinds of calculations are important, not only for chemical hand warmers, but also for many other important processes such as the production of energy.

**(a)**

**(b)**

**(c)**

▲ **(a)** A rolling billiard ball has energy due to its motion. **(b)** When the ball collides with a second ball it does work, transferring energy to the second ball. **(c)** The second ball has energy as it rolls away from the collision.

Breaking chemical bonds always takes energy. The chemical reactions that produce energy involve breaking weak bonds (which absorbs small amounts of energy) and forming stronger bonds (which releases large amounts of energy).

## 6.2 The Nature of Energy: Key Definitions

Recall that we briefly examined energy in Section 1.5. We defined **energy** as the capacity to do work and defined **work** as the result of a force acting through a distance. When you push a box across the floor, you have done work. Consider another example of work: a billiard ball rolling across a billiard table and colliding straight on with a second, stationary billiard ball. The rolling ball has *energy* due to its motion. When it collides with another ball it does *work*, resulting in the *transfer* of energy from one ball to the other. The second billiard ball absorbs the energy and begins to roll across the table.

As we just saw with chemical hand warmers, energy can also be transferred through **heat**, the flow of energy caused by a temperature difference. For example, if you hold a cup of coffee in your hand, energy is transferred, in the form of heat, from the hot coffee to your cooler hand. Think of *energy* as something that an object or set of objects possesses. Think of *heat* and *work* as ways that objects or sets of objects *exchange* energy.

### Types of Energy

The energy contained in a rolling billiard ball is an example of **kinetic energy**, the energy associated with the *motion* of an object. The energy contained in a hot cup of coffee is **thermal energy**, the energy associated with the *temperature* of an object. Thermal energy is actually a type of kinetic energy because it arises from the motions of atoms or molecules within a substance. If you raise a billiard ball off the table, you increase its **potential energy**, the energy associated with the *position* or *composition* of an object. The potential energy of the billiard ball, for example, is a result of its position in Earth's gravitational field. Raising the ball off the table, against Earth's gravitational pull, gives the ball more potential energy. Another example of potential energy is the energy contained in a compressed spring. When you compress a spring, you push against the forces that tend to maintain the spring's uncompressed shape, storing energy as potential energy. **Chemical energy**, the energy associated with the relative positions of electrons and nuclei in atoms and molecules, is also a form of potential energy. Some chemical compounds, such as the methane in natural gas or the iron in a chemical hand warmer, are like compressed springs—they contain potential energy, and a chemical reaction can release that potential energy. Figure 6.1 ◄ summarizes these different kinds of energy.

▲ **FIGURE 6.1 The Different Manifestations of Energy**

**Energy Transformation I**

(a)                                    (b)

▲ **FIGURE 6.2** **Energy Transformation: Potential and Kinetic Energy I** **(a)** A billiard ball held above the ground has gravitational potential energy. **(b)** When the ball is released, the potential energy is transformed into kinetic energy, the energy of motion.

## Energy Conservation and Energy Transfer

The **law of conservation of energy** states that *energy can be neither created nor destroyed*. However, energy can be transferred from one object to another, and it can assume different forms. For example, if you drop a raised billiard ball, some of its potential energy becomes kinetic energy as the ball falls toward the table, as shown in Figure 6.2▲. If you release a compressed spring, the potential energy becomes kinetic energy as the spring expands outward, as shown in Figure 6.3▼. When iron reacts with oxygen within a chemical hand warmer, the chemical energy of the iron and oxygen becomes thermal energy that increases the temperature of your hand and glove.

A good way to understand and track energy changes is to define the **system** under investigation. For example, the system may be the chemicals in a beaker, or it may be the iron reacting in a hand warmer. The system's **surroundings** are everything with which the system can exchange energy. If we define the chemicals in a beaker as the system, the surroundings may include the water in which the chemicals are dissolved (for aqueous solutions), the beaker itself, the lab bench on which the beaker sits, the air in the room, and so on. For the iron in the hand warmer, the surroundings include your hand, your glove, the air in the glove, and even the air outside of the glove.

In an energy exchange, energy transfers between the system and the surroundings, as shown in Figure 6.4▶ on the next page. If the system loses energy, the surroundings gain the same exact amount of energy, and vice versa. When the iron within the chemical hand warmer reacts, the system loses energy to the surroundings, producing the desired temperature increase within your gloves.

Einstein showed that it is mass–energy that is conserved; one can be converted into the other. This equivalence becomes important in nuclear reactions, discussed in Chapter 20. In ordinary chemical reactions, however, the interconversion of mass and energy is not a significant factor, and we can regard mass and energy as independently conserved.

**Energy Transformation II**

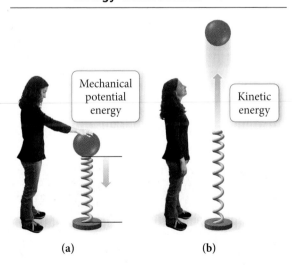

(a)                                    (b)

▲ **FIGURE 6.3** **Energy Transformation: Potential and Kinetic Energy II** **(a)** A compressed spring has potential energy. **(b)** When the spring is released, the potential energy is transformed into kinetic energy.

## Units of Energy

We can deduce the units of energy from the definition of kinetic energy. An object of mass $m$, moving at velocity $v$, has a kinetic energy KE given by the equation:

$$KE = \frac{1}{2}mv^2 \qquad [6.1]$$

kg   m/s

▶ **FIGURE 6.4 Energy Transfer**   If a system and surroundings had energy gauges (which measure energy content in the way a fuel gauge measures fuel content), an energy transfer in which the system transfers energy to the surroundings would result in a decrease in the energy content of the system and an increase in the energy content of the surroundings. The total amount of energy, however, must be conserved.

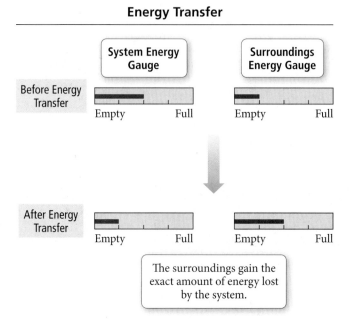

**Energy Transfer**

The surroundings gain the exact amount of energy lost by the system.

$3.6 \times 10^5$ J or 0.10 kWh used in 1 hour

▲ A watt (W) is 1 J/s, so a 100 W lightbulb uses 100 J every second or $3.6 \times 10^5$ J every hour.

The "calorie" referred to on all nutritional labels (regardless of the capitalization) is always the capital C Calorie.

The SI unit of mass is the kg and the unit of velocity is m/s. The SI unit of energy is therefore $kg \cdot m^2/s^2$, defined as the **joule (J)**, named after the English scientist James Joule (1818–1889).

$$1 \, kg \frac{m^2}{s^2} = 1 \, J$$

One joule is a relatively small amount of energy—for example, a 100-watt lightbulb uses $3.6 \times 10^5$ J in 1 hour. Therefore, we often use the kilojoule (kJ) in our energy discussions and calculations (1 kJ = 1000 J). A second commonly used unit of energy is the **calorie (cal)**, originally defined as the amount of energy required to raise the temperature of 1 g of water by 1 °C. The current definition is 1 cal = 4.184 J (exact); a calorie is a larger unit than a joule. A related energy unit is the nutritional, or uppercase "C" **Calorie (Cal)**, equivalent to 1000 lowercase "c" calories. The Calorie is the same as a kilocalorie (kcal): 1 Cal = 1 kcal = 1000 cal. Electricity bills typically are based on another, even larger, energy unit, the **kilowatt-hour (kWh)**: 1 kWh = $3.60 \times 10^6$ J. Electricity costs $0.08–$0.15 per kWh. Table 6.1 lists various energy units and their conversion factors. Table 6.2 shows the amount of energy required for various processes.

| TABLE 6.1  Energy Conversion Factors* | |
|---|---|
| 1 calorie (cal) | = 4.184 joules (J) |
| 1 Calorie (Cal) or kilocalorie (kcal) | = 1000 cal = 4184 J |
| 1 kilowatt-hour (kWh) | = $3.60 \times 10^6$ J |

*All conversion factors in this table are exact.

| TABLE 6.2  Energy Uses in Various Units | | | | |
|---|---|---|---|---|
| Unit | Amount Required to Raise Temperature of 1 g of Water by 1 °C | Amount Required to Light 100 W Bulb for 1 Hour | Amount Used by Human Body in Running 1 Mile (Approximate) | Amount Used by Average U.S. Citizen in 1 Day |
| joule (J) | 4.18 | $3.60 \times 10^5$ | $4.2 \times 10^5$ | $9.0 \times 10^8$ |
| calorie (cal) | 1.00 | $8.60 \times 10^4$ | $1.0 \times 10^5$ | $2.2 \times 10^8$ |
| Calorie (Cal) | 0.00100 | 86.0 | 100 | $2.2 \times 10^5$ |
| kilowatt-hour (kWh) | $1.16 \times 10^{-6}$ | 0.100 | 0.12 | $2.5 \times 10^2$ |

## 6.3   The First Law of Thermodynamics: There Is No Free Lunch

KCV   **The First Law of Thermodynamics**

**Thermodynamics** is the general study of energy and its interconversions. The laws of thermodynamics are among the most fundamental in all of science, governing virtually every process that involves change. The **first law of thermodynamics** is the law of energy conservation, which we state as follows:

**The total energy of the universe is constant.**

In other words, since energy is neither created nor destroyed, and since the universe does not exchange energy with anything else, its energy content does not change.

The first law has many implications: the most important one is that with energy, you cannot get something for nothing. The best you can do with energy is break even—there is no free lunch. According to the first law, a device that would continually produce energy with no energy input, sometimes known as a *perpetual motion machine*, cannot exist. Occasionally, the media report or speculate on the discovery of a machine that can produce energy without the need for energy input. For example, you may have heard someone propose an electric car that recharges itself while driving, or a new motor that can create additional usable electricity as well as the electricity to power itself. Although some hybrid (electric and gasoline powered) vehicles can capture energy from braking and use that energy to recharge their batteries, they could never run indefinitely without additional fuel. As for the motor that powers an external load as well as itself—no such thing exists. Our society has a continual need for energy, and as our current energy resources dwindle, new energy sources will be required. But those sources, whatever they may be, must follow the first law of thermodynamics—energy is always conserved.

### Internal Energy

The **internal energy (E)** of a system is *the sum of the kinetic and potential energies of all of the particles that compose the system.* Internal energy is a **state function**, which means that its value depends *only on the state of the system*, not on how the system arrived at that state.

---

## Chemistry IN YOUR DAY | Redheffer's Perpetual Motion Machine

In 1812, a man named Charles Redheffer appeared in Philadelphia with a machine that he claimed could run forever without any energy input—a perpetual motion machine. He set up the machine on the edge of town and charged admission to view it. He also appealed to the city for money to build a larger version of the machine. When city commissioners came out to inspect the machine, Redheffer did his best to keep them from viewing it too closely. Nonetheless, one of the commissioners noticed something suspicious: the gears that supposedly ran to an external driveshaft were cut in the wrong direction. The driveshaft that the machine was allegedly powering was instead powering the machine. The city commissioners hired a local engineer and clockmaker named Isaiah Lukens to make a similar machine to expose Redheffer's deception. Lukens's machine was even more ingenious than Redheffer's, and Redheffer left Philadelphia exposed as a fraud.

Redheffer was persistent, however, and took his machine to New York. In 1813, during a public display of the machine, the famous mechanical engineer Robert Fulton (1765–1815)—who had demonstrated the first successful steamboat six years earlier—noticed a rhythm to the machine's motion. It seemed to speed up and slow down at regular intervals. Fulton knew that such rhythmic motion is indicative of motion generated by a manual crank. He knocked away some boards in a wall next to the machine and discovered a long belt that led to an enclosed room where, indeed, an old man sat turning a crank. Redheffer's machine—like many other perpetual motion machines throughout history—was again exposed as a hoax.

**QUESTION**

*Can you think of any recent claims of perpetual motion or limitless free energy?*

► **FIGURE 6.5 Altitude as a State Function** The change in altitude during a climb depends only on the difference between the final and initial altitudes, not on the route traveled.

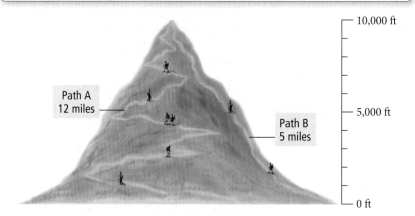

**A State Function**
Change in altitude depends only on the difference between the initial and final values, not on the path taken.

Path A
12 miles

Path B
5 miles

10,000 ft

5,000 ft

0 ft

The state of a chemical system is specified by parameters such as temperature, pressure, concentration, and physical state (solid, liquid, or gas). Consider the mountain-climbing analogy depicted in Figure 6.5▲. The elevation at any point during a mountain climb is analogous to a state function. For example, when we reach 10,000 ft, our elevation is 10,000 ft, no matter how we got there. The distance we traveled to get there, by contrast, is not a state function; we could have climbed the mountain by any number of routes, each requiring us to cover a different distance.

Since state functions depend only on the state of the system, the value of a *change* in a state function is always the difference between its final and initial values. If we start climbing a mountain at an elevation of 3000 ft and reach the summit at 10,000 ft, then our elevation change is 7000 ft (10,000 ft – 3000 ft), regardless of what path we traveled.

Like an altitude change, an internal energy change ($\Delta E$) is determined by the difference in internal energy between the final and initial states:

$$\Delta E = E_{final} - E_{initial}$$

In a chemical system, the reactants constitute the initial state and the products constitute the final state. So $\Delta E$ is the difference in internal energy between the products and the reactants:

$$\Delta E = E_{products} - E_{reactants} \qquad [6.2]$$

For example, consider the reaction between carbon and oxygen to form carbon dioxide:

$$C(s) + O_2(g) \longrightarrow CO_2(g)$$

Just as we can portray the changes that occur when climbing a mountain with an *altitude* diagram that depicts the *altitude* before and after the climb (see Figure 6.5), we can portray the energy changes that occur during a reaction with an *energy* diagram, which compares the *internal energy* of the reactants and the products:

Internal energy

$C(s), O_2(g)$    (reactants)

$\Delta E < 0$ (negative)

$CO_2(g)$    (product)

The vertical axis of the diagram is internal *energy*, which increases as we move up on the diagram. For this reaction, the reactants are *higher* on the diagram than the products because they have higher internal energy. As the reaction occurs, the reactants become products, which have lower internal energy. Therefore, energy is given off by the reaction and $\Delta E$ (that is, $E_{products} - E_{reactants}$) is *negative*.

Where does the energy lost by the reactants (as they transform to products) go? If we define the thermodynamic *system* as the reactants and products of the reaction, then energy flows *out of the system* and *into the surroundings*.

$\Delta E_{sys} < 0$ (negative)    $\Delta E_{surr} > 0$ (positive)

According to the first law, energy must be conserved. Therefore, the amount of energy lost by the system must exactly equal the amount gained by the surroundings:

$$\Delta E_{sys} = -\Delta E_{surr} \qquad [6.3]$$

Now, suppose the reaction is reversed:

$$CO_2(g) \longrightarrow C(s) + O_2(g)$$

The energy-level diagram is nearly identical, with one important difference: $CO_2(g)$ is now the reactant and $C(s)$ and $O_2(g)$ are the products. Instead of decreasing in energy as the reaction occurs, the system increases in energy as shown at right. In this reversed reaction $\Delta E$ is *positive* and energy flows *into the system* and *out of the surroundings*:

$\Delta E_{sys} > 0$ (positive)    $\Delta E_{surr} < 0$ (negative)

***Summarizing Energy Flow:***

▶ If the reactants have a higher internal energy than the products, $\Delta E_{sys}$ is negative and energy flows out of the system into the surroundings.

▶ If the reactants have a lower internal energy than the products, $\Delta E_{sys}$ is positive and energy flows into the system from the surroundings.

We can think of the internal energy of the system in the same way we think about the balance in a checking account. Energy flowing *out of* the system is like a withdrawal and therefore carries a negative sign. Energy flowing *into* the system is like a deposit and carries a positive sign.

**System and Surroundings**  Consider these fictitious internal energy gauges for a chemical system and its surroundings:

CONCEPTUAL
CONNECTION **6.1**

Which diagram best represents the energy gauges for the same system and surroundings following an energy exchange in which $\Delta E_{sys}$ is negative?

## Heat and Work

As we discussed in Section 6.2, a system can exchange energy with its surroundings through *heat* and *work*:

According to the first law of thermodynamics, the change in the internal energy of the system ($\Delta E$) is the sum of the heat transferred ($q$) and the work done ($w$):

$$\Delta E = q + w \qquad [6.4]$$

In Equation 6.4, and from this point forward, we follow the standard convention that $\Delta E$ (with no subscript) refers to the internal energy change of the *system*. As shown in Table 6.3, energy entering the system through heat or work carries a positive sign, and energy leaving the system through heat or work carries a negative sign. Recall our checking account analogy. The system is like the checking account—withdrawals are negative and deposits are positive.

| **TABLE 6.3**  **Sign Conventions for $q$, $w$, and $\Delta E$** | | |
|---|---|---|
| $q$ (heat) | + system *gains* thermal energy | − system *loses* thermal energy |
| $w$ (work) | + work done *on* the system | − work done *by* the system |
| $\Delta E$ (change in internal energy) | + energy flows *into* the system | − energy flows *out* of the system |

We can define our system as the previously discussed billiard ball rolling across a pool table. The rolling ball has a certain initial amount of kinetic energy. When it reaches the other end of the table, the rolling ball collides head-on with a second ball. Assume that the first ball loses all of its kinetic energy so that it remains completely still (it has no kinetic energy) at the point of collision. The total change in internal energy ($\Delta E$) for the first ball is the difference between its initial kinetic energy and its final kinetic energy (which is zero); the first billiard ball lost all of its energy. What happened to that energy? According to the first law, it must have been transferred to the surroundings. In fact, the energy lost by the system must *exactly equal* the amount gained by the surroundings:

$$\Delta E_{\text{sys}} = -\Delta E_{\text{surr}}$$

The surroundings include both the pool table and the second ball. The pool table absorbs some of the ball's kinetic energy as the ball rolls down the table. Minute bumps on the table surface cause friction, which slows the ball down by converting kinetic energy to heat ($q$). The second ball absorbs some of the ball's kinetic energy in the form of work ($w$) upon collision.

Although it is always the case that $\Delta E_{\text{sys}} = -\Delta E_{\text{surr}}$, the exact amount of *work* done on the second ball depends on the quality of the billiard table. On a smooth, high-quality billiard table, the amount of energy lost to friction is relatively small, as illustrated in Figure 6.6(a)▶. The speed of the first ball is not reduced by much as it travels across the table, and a great deal of its original kinetic energy is available to perform work when it collides with the second ball. In contrast, on a rough, poor-quality table, the ball loses much of its initial kinetic energy as heat, leaving only a relatively small amount available for work, as illustrated in Figure 6.6(b)▶.

Notice that the respective amounts of energy converted to heat and work depend on the details of the pool table and the path taken, while the change in internal energy of the rolling ball does not. Because internal energy is a state function, the value of $\Delta E$ for the process in which the ball moves across the table and collides with another ball depends

**(a) Smooth table**

**(b) Rough table**

▲ **FIGURE 6.6 Energy, Work, and Heat** **(a)** On a smooth table, most of the first billiard ball's initial kinetic energy is transferred to the second ball as work. Only a small amount is lost to heat. **(b)** On a rough table, most of the first billiard ball's initial kinetic energy is lost to heat. Only a small amount is left to do work on the second billiard ball.

only on the ball's initial and final kinetic energy. Work and heat, however, are *not* state functions; therefore, the values of $q$ and $w$ depend on the details of the ball's journey across the table. On the smooth table, $w$ is greater in magnitude than $q$; on the rough table, $q$ is greater in magnitude than $w$. However, $\Delta E$ (the sum of $q$ and $w$) is constant.

CONCEPTUAL
**CONNECTION 6.2**

**Heat and Work**  Identify each energy exchange as heat or work and determine whether the sign of heat or work (relative to the system) is positive or negative.
  **(a)**  An ice cube melts and cools the surrounding beverage. (The ice cube is the system.)
  **(b)**  A metal cylinder is rolled up a ramp. (The metal cylinder is the system.)
  **(c)**  Steam condenses on skin, causing a burn. (The condensing steam is the system.)

**Example 6.1** | **Internal Energy, Heat, and Work**

A potato cannon provides a good example of the heat and work associated with a chemical reaction. In a potato cannon, a potato is stuffed into a long cylinder that is capped on one end and open at the other. Some kind of fuel is introduced under the potato at the capped end—usually through a small hole—and ignited. The potato shoots out of the cannon, sometimes flying hundreds of feet, and the cannon emits heat to the surroundings. If the burning of the fuel performs 855 J of work on the potato and produces 1422 J of heat, what is $\Delta E$ for the burning of the fuel? (Note: A potato cannon can be dangerous and should not be constructed without proper training and experience.)

**SOLUTION** To solve the problem, substitute the values of $q$ and $w$ into the equation for $\Delta E$. Since work is done by the system on the surroundings, $w$ is negative. Similarly, because heat is released by the system to the surroundings, $q$ is also negative.

$$\Delta E = q + w$$
$$= -1422\,\text{J} - 855\,\text{J}$$
$$= -2277\,\text{J}$$

**FOR PRACTICE 6.1** A cylinder and piston assembly (defined as the system) is warmed by an external flame. The contents of the cylinder expand, doing work on the surroundings by pushing the piston outward against the external pressure. If the system absorbs 559 J of heat and does 488 J of work during the expansion, what is the value of $\Delta E$?

 **KCV** Heat Capacity

# 6.4 Quantifying Heat and Work

In the previous section, we calculated $\Delta E$ based on *given values of q and w*. We now turn to *calculating q (heat) and w (work)* based on changes in temperature and volume.

## Heat

Recall from Section 6.2 that *heat* is the exchange of thermal energy between a system and its surroundings caused by a temperature difference. Notice the distinction between heat and temperature. Temperature is a *measure* of the thermal energy within a sample of matter. Heat is the *transfer* of thermal energy. Thermal energy always flows from matter at higher temperatures to matter at lower-temperatures. For example, a hot cup of coffee transfers thermal energy—as heat—to the lower-temperature surroundings as it cools down. Imagine a world where the cooler surroundings actually got colder as they transferred thermal energy to the hot coffee, which got hotter. Such a world exists only in our imaginations (or in the minds of science fiction writers), because the spontaneous transfer of heat from a hotter object to a colder one is a fundamental principle of our universe—no exception has ever been observed. The thermal energy in the molecules that compose the hot coffee distributes itself to the molecules in the surroundings. The heat transfer from the coffee to the surroundings stops when the two reach the same temperature, a condition called **thermal equilibrium**. At thermal equilibrium, there is no additional net transfer of heat.

The reason for this one-way transfer is related to the second law of thermodynamics, which we discuss in Chapter 18.

### Temperature Changes and Heat Capacity

When a system absorbs heat ($q$), its temperature changes by $\Delta T$:

Heat ($q$) → System

$\Delta T$

Experiments show that the heat absorbed by a system and its corresponding temperature change are directly proportional: $q \propto \Delta T$. The constant of proportionality between $q$ and $\Delta T$ is the system's *heat capacity* ($C$), a measure of the system's ability to absorb thermal energy without undergoing a large change in temperature.

$$q = C \times \Delta T \qquad [6.5]$$

Heat capacity

Notice that the higher the heat capacity of a system, the smaller the change in temperature for a given amount of absorbed heat. We define the **heat capacity (*C*)** of a system as the quantity of heat required to change its temperature by 1 °C. As we can see by solving Equation 6.5 for heat capacity, the units of heat capacity are those of heat (typically J) divided by those of temperature (typically °C):

$$C = \frac{q}{\Delta T} = \frac{J}{°C}$$

In order to understand two important concepts related to heat capacity, imagine putting a steel saucepan on a kitchen flame. The saucepan's temperature rises rapidly as it absorbs heat from the flame. However, if you add some water to the saucepan, the temperature rises more slowly. Why? The first reason is that when you add the water, the same amount of heat must now warm more matter, so the temperature rises more slowly. In other words, heat capacity is an extensive property—*it depends on the amount of matter being heated* (see Section 1.6). The second (and more fundamental) reason is that *water is more resistant to temperature change than steel*—water has an intrinsically higher capacity to absorb heat without undergoing a large temperature change. The measure of the *intrinsic capacity* of a substance to absorb heat is its **specific heat capacity (*C*ₛ)**, the amount of heat required to raise the temperature of *1 gram* of the substance by 1 °C. The units of specific heat capacity (also called *specific heat*) are J/g·°C. Table 6.4 lists the values of the specific heat capacity for several substances. Heat capacity is sometimes reported as **molar heat capacity**, the amount of heat required to raise the temperature of *1 mole* of a substance by 1 °C. The units of molar heat capacity are J/mol·°C. We can see from these definitions that *specific* heat capacity and *molar* heat capacity are intensive properties—they depend on the *kind* of substance being heated, not on the amount.

Notice that water has the highest specific heat capacity of all the substances in Table 6.4—changing the temperature of water requires a lot of heat. If you have ever experienced the drop in temperature that occurs when traveling from an inland region to the coast during the summer, you have experienced the effects of water's high specific heat capacity. On a summer's day in California, for example, the temperature difference between Sacramento (an inland city) and San Francisco (a coastal city) may be as large as 18 °C (30 °F)—San Francisco enjoys a cool 20 °C (68 °F), while Sacramento bakes at nearly 38 °C (100 °F). Yet the intensity of sunlight falling on these two cities is the same. Why the large temperature difference? San Francisco sits on a peninsula, surrounded by the water of the Pacific Ocean. Water, with its high heat capacity, absorbs much of the sun's heat without undergoing a large increase in temperature, keeping San Francisco cool. Sacramento, by contrast, is about 160 km (100 mi) inland. The land surrounding Sacramento, with its low heat capacity, undergoes a large increase in temperature as it absorbs a similar amount of heat.

Similarly, only two U.S. states have never recorded a temperature above 100 °F. One of them is obvious: Alaska. It is too far north to get that hot. The other one, however, may come as a surprise. It is Hawaii. The high heat capacity of the water that surrounds the only island state moderates the temperature, preventing Hawaii from ever getting too hot.

We can use the specific heat capacity of a substance to quantify the relationship between the amount of heat added to a given amount of the substance and the corresponding temperature increase. The following equation relates these quantities:

Heat (J) ⟶ $q = m \times C_s \times \Delta T$ ⟵ Temperature change (°C)  $\qquad [6.6]$

Mass (g)   Specific heat capacity J/g · °C

| TABLE 6.4 Specific Heat Capacities of Some Common Substances | |
|---|---|
| **Substance** | **Specific Heat Capacity, $C_s$ (J/g · °C)\*** |
| **Elements** | |
| Lead | 0.128 |
| Gold | 0.128 |
| Silver | 0.235 |
| Copper | 0.385 |
| Iron | 0.449 |
| Aluminum | 0.903 |
| **Compounds** | |
| Ethanol | 2.42 |
| Water | 4.18 |
| **Materials** | |
| Glass (Pyrex) | 0.75 |
| Granite | 0.79 |
| Sand | 0.84 |

*At 298 K.

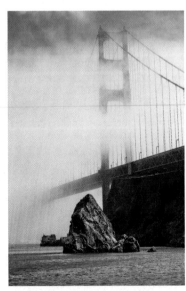

▲ The high heat capacity of the water surrounding San Francisco results in relatively cool summer temperatures.

$\Delta T$ in °C is equal to $\Delta T$ in K, but not equal to $\Delta T$ in °F (Section 1.6).

where $q$ is the amount of heat in J, $m$ is the mass of the substance in g, $C_s$ is the specific heat capacity in J/g·°C, and $\Delta T$ is the temperature change in °C. Example 6.2 demonstrates the use of this equation.

## Example 6.2 | Temperature Changes and Heat Capacity

Suppose you find a penny (minted before 1982, when pennies were almost entirely copper) in the snow. How much heat is absorbed by the penny as it warms from the temperature of the snow, which is −8.0 °C, to the temperature of your body, 37.0 °C? Assume the penny is pure copper and has a mass of 3.10 g.

| | |
|---|---|
| **SORT** You are given the mass of copper as well as its initial and final temperature. You are asked to find the heat required for the given temperature change. | **GIVEN:** $m = 3.10$ g copper<br>$T_i = -8.0\,°C$<br>$T_f = 37.0\,°C$<br>**FIND:** $q$ |
| **STRATEGIZE** The equation $q = m \times C_s \times \Delta T$ gives the relationship between the amount of heat ($q$) and the temperature change ($\Delta T$). | **CONCEPTUAL PLAN**<br><br>$\boxed{C_s, m, \Delta T} \longrightarrow \boxed{q}$<br>$q = m \times C_s \times \Delta T$<br><br>**RELATIONSHIPS USED**<br>$q = m \times C_s \times \Delta T$ (Equation 6.6)<br>$C_s = 0.385$ J/g·°C (Table 6.4) |
| **SOLVE** Gather the necessary quantities for the equation in the correct units and substitute into the equation to calculate $q$. | **SOLUTION**<br>$\Delta T = T_f - T_i = 37.0\,°C - (-8.0\,°C) = 45.0\,°C$<br>$q = m \times C_s \times \Delta T$<br><br>$= 3.10\,g \times 0.385\,\dfrac{J}{g \cdot °C} \times 45.0\,°C = 53.7\,J$ |

**CHECK** The units (J) are correct for heat. The sign of $q$ is *positive,* as it should be because the penny *absorbed* heat from the surroundings.

**FOR PRACTICE 6.2** To determine whether a shiny gold-colored rock is actually gold, a chemistry student decides to measure its heat capacity. She first weighs the rock and finds it has a mass of 4.7 g. She then finds that upon absorption of 57.2 J of heat, the temperature of the rock rises from 25 °C to 57 °C. Find the specific heat capacity of the substance composing the rock and determine whether the value is consistent with the rock being pure gold.

**FOR MORE PRACTICE 6.2** A 55.0-g aluminum block initially at 27.5 °C absorbs 725 J of heat. What is the final temperature of the aluminum?

CONCEPTUAL
**CONNECTION 6.3**

**The Heat Capacity of Water** Suppose you are cold-weather camping and decide to heat some objects to bring into your sleeping bag for added warmth. You place a large water jug and a rock of equal mass near the fire. Over time, both the rock and the water jug warm to about 38 °C (100 °F). If you could bring only one into your sleeping bag, which one should you choose to keep you the warmer? Why?

## Thermal Energy Transfer

As we noted earlier, when two substances of different temperature are combined, thermal energy flows as heat from the hotter substance to the cooler one. If we assume that the two substances are thermally isolated from everything else, then the heat lost by one substance exactly equals the heat gained by the other (according to the law of energy conservation). If we define one substance as the system and the other as the surroundings, we can quantify the heat exchange as:

$$q_{sys} = -q_{surr}$$

Suppose a block of metal initially at 55 °C is submerged into water initially at 25 °C. Thermal energy transfers as heat from the metal to the water:

$$q_{metal} = -q_{water}$$

The metal becomes colder and the water becomes warmer until the two substances reach the same temperature (thermal equilibrium). The exact temperature change that occurs depends on the masses of the metal and the water and on their specific heat capacities. Since $q = m \times C_s \times \Delta T$, we can arrive at the following relationship:

$$q_{metal} = -q_{water}$$
$$m_{metal} \times C_{s,\,metal} \times \Delta T_{metal} = -m_{water} \times C_{s,\,water} \times \Delta T_{water}$$

Example 6.3 shows how to work with thermal energy transfer.

---

**Example 6.3** | **Thermal Energy Transfer**

A 32.5 g cube of aluminum initially at 45.8 °C is submerged into 105.3 g of water at 15.4 °C. What is the final temperature of both substances at thermal equilibrium? (Assume that the aluminum and the water are thermally isolated from everything else.)

| | |
|---|---|
| **SORT** You are given the masses of aluminum and water and their initial temperatures. You are asked to find the final temperature. | **GIVEN:** $m_{Al} = 32.5$ g $\quad m_{H_2O} = 105.3$ g <br> $\qquad T_{i,\,Al} = 45.8\,°C; \; T_{i,\,H_2O} = 15.4\,°C$ <br> **FIND:** $T_f$ |

| | |
|---|---|
| **STRATEGIZE** The heat lost by the aluminum ($q_{Al}$) equals the heat gained by the water ($q_{H_2O}$). <br><br> Use the relationship between $q$ and $\Delta T$ and the given variables to find a relationship between $\Delta T_{Al}$ and $\Delta T_{H_2O}$. <br><br><br><br> Use the relationship between $\Delta T_{Al}$ and $\Delta T_{H_2O}$ (that you just found) along with the initial temperatures of the aluminum and the water to determine the final temperature. Note that at thermal equilibrium, the final temperature of the aluminum and the water is the same, that is, $T_{f,\,Al} = T_{f,\,H_2O} = T_f$. | **CONCEPTUAL PLAN** $q_{Al} = -q_{H_2O}$ <br><br> $\boxed{m_{Al},\, C_{s,\,Al},\, m_{H_2O}, C_{s,\,H_2O}} \longrightarrow \boxed{\Delta T_{Al} = \text{constant} \times \Delta T_{H_2O}}$ <br> $m_{Al} \times C_{s,\,Al} \times \Delta T_{Al} = -m_{H_2O} \times C_{s,\,H_2O} \times \Delta T_{H_2O}$ <br><br><br><br> $\boxed{T_{i,\,Al},\, T_{i,\,H_2O}} \longrightarrow \boxed{T_f}$ <br> $\Delta T_{Al} = \text{constant} \times \Delta T_{H_2O}$ <br><br> **RELATIONSHIPS USED** <br> $C_{s,\,H_2O} = 4.18$ J/g · °C; $C_{s,\,Al} = 0.903$ J/g · °C (Table 6.4) <br> $q = m \times C_s \times \Delta T$ (Equation 6.6) <br><br> *—Continued on the next page* |

*—Continued*

| SOLVE | SOLUTION |
|---|---|
| Write the equation for the relationship between the heat lost by the aluminum ($q_{Al}$) and the heat gained by the water $q_{Al} = -q_{H_2O}$ and substitute $q = m \times C_s \times \Delta T$ for each substance. | $q_{Al} = -q_{H_2O}$<br><br>$m_{Al} \times C_{s,\,Al} \times \Delta T_{Al} = -m_{H_2O} \times C_{s,H_2O} \times \Delta T_{H_2O}$ |
| Substitute the values of $m$ (given) and $C_s$ (from Table 6.4) for each substance and solve the equation for $\Delta T_{Al}$. (Alternatively, you can solve the equation for $\Delta T_{H_2O}$.) | $32.5 \text{ g} \times \dfrac{0.903 \text{ J}}{\text{g} \cdot {}^\circ\text{C}} \cdot \Delta T_{Al} = -105.3 \text{ g} \times \dfrac{4.18 \text{ J}}{\text{g} \cdot {}^\circ\text{C}} \cdot \Delta T_{H_2O}$<br><br>$29.3\underline{4}8 \cdot \Delta T_{Al} = -440.1\underline{5} \cdot \Delta T_{H_2O}$<br><br>$\Delta T_{Al} = -14.9\underline{9}8 \cdot \Delta T_{H_2O}$ |
| Substitute the initial temperatures of aluminum and water into the relationship from the previous step and solve the expression for the final temperature ($T_f$). Remember that the final temperature for both substances will be the same. | $T_f - T_{i,\,Al} = -14.9\underline{9}8(T_f - T_{i,\,H_2O})$<br><br>$T_f = -14.9\underline{9}8 \cdot T_f + 14.9\underline{9}8 \cdot T_{i,\,H_2O} + T_{i,\,Al}$<br><br>$T_f + 14.9\underline{9}8 \cdot T_f = 14.9\underline{9}8 \cdot T_{i,\,H_2O} + T_{i,\,Al}$<br><br>$15.9\underline{9}8 \cdot T_f = 14.9\underline{9}8 \cdot T_{i,\,H_2O} + T_{i,\,Al}$<br><br>$T_f = \dfrac{14.9\underline{9}8 \cdot T_{i,\,H_2O} + T_{i,\,Al}}{15.9\underline{9}8} = \dfrac{14.9\underline{9}8 \cdot 15.4\,^\circ\text{C} + 45.8\,^\circ\text{C}}{15.9\underline{9}8}$<br><br>$= 17.3\,^\circ\text{C}$ |

**CHECK**  The units °C are correct. The final temperature of the mixture is closer to the initial temperature of the *water* than the *aluminum*. This makes sense for two reasons: (1) water has a higher specific heat capacity than aluminum, and (2) there is more water than aluminum. Because the aluminum loses the same amount of heat that is gained by the water, the greater mass and specific heat capacity of the water make the temperature change in the water *less than* the temperature change in the aluminum.

**FOR PRACTICE 6.3**   A block of copper of unknown mass has an initial temperature of 65.4 °C. The copper is immersed in a beaker containing 95.7 g of water at 22.7 °C. When the two substances reach thermal equilibrium, the final temperature is 24.2 °C. What is the mass of the copper block?

**CONCEPTUAL**
**CONNECTION 6.4**

**Thermal Energy Transfer**  Substances A and B, initially at different temperatures, come in contact with each other and reach thermal equilibrium. The mass of substance A is twice the mass of substance B. The specific heat capacity of substance B is twice the specific heat capacity of substance A. Which statement is true about the final temperature of the two substances once thermal equilibrium is reached?

**(a)**  The final temperature is closer to the initial temperature of substance A than it is to the initial temperature of substance B.

**(b)**  The final temperature is closer to the initial temperature of substance B than it is to the initial temperature of substance A.

**(c)**  The final temperature is exactly midway between the initial temperatures of substances A and B.

## Work: Pressure–Volume Work

We have seen that energy transfer between a system and its surroundings can occur via heat ($q$) or work ($w$). We just calculated the *heat* associated with an observed *temperature* change. We now turn to calculating the *work* associated with an observed *volume* change. Although a chemical reaction can do several different types of work, for now we limit our discussion to **pressure–volume work**. We have already defined work as a force acting through a distance. Pressure–volume work occurs when the force is caused by a volume change against an external pressure. For example, pressure–volume work occurs in the cylinder of an automobile engine. The combustion of gasoline causes gases within the cylinders to expand, pushing the piston outward and ultimately moving the wheels of the car.

We can derive an equation for the value of pressure–volume work from the definition of work as a force ($F$) acting through a distance ($D$):

$$w = F \times D \qquad [6.7]$$

When the volume of a cylinder increases (Figure 6.7▼), it pushes against an external force. That external force is the external pressure ($P_{ext}$), which is defined as force ($F$) divided by area ($A$):

$$P_{ext} = \frac{F}{A} \quad \text{or} \quad F = P_{ext} \times A$$

If we substitute this expression for force into the definition of work given in Equation 6.7, we arrive at the expression:

$$w = F \times D$$
$$= P_{ext} \times A \times D$$

The distance through which the force acts is the change in the height of the piston as it moves during the expansion ($\Delta h$). Substituting $\Delta h$ for $D$, we get:

$$w = P_{ext} \times A \times \Delta h$$

The volume of a cylinder is the area of its base times its height, so $A \times \Delta h$ is actually the change in volume ($\Delta V$) that occurs during the expansion. Thus, the expression for work becomes the product of the pressure times the change in volume:

$$w = P_{ext} \Delta V$$

Still missing from the equation is the *sign* of the work done by the expanding gases. As the volume of the cylinder increases, work is done *on* the surroundings *by* the system, so $w$ should be negative. However, upon expansion, $V_2$ (the final volume) is greater than $V_1$ (the initial volume) so $\Delta V$ is positive. In order for $w$ to be negative for a positive expansion, we need to add a negative sign to our equation. In other words, $w$ and $\Delta V$ must be opposite in sign:

$$w = -P_{ext} \Delta V \qquad [6.8]$$

So the work caused by an expansion of volume is the negative of the pressure that the volume expands against multiplied by the change in volume that occurs during the expansion. The units of the work obtained by using this equation are those of pressure (usually atm) multiplied by those of volume (usually L). To convert between L·atm and J, we use the conversion factor 101.3 J = 1 L·atm.

Combustion

▲ The combustion of gasoline within an engine's cylinders does pressure–volume work that ultimately results in the motion of the car.

The force in this equation must be a constant force.

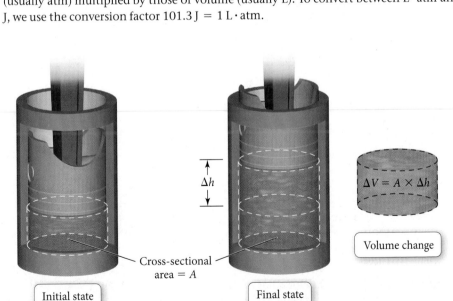

$\Delta h$

Cross-sectional area = $A$

$\Delta V = A \times \Delta h$

Volume change

Initial state

Final state

▲**FIGURE 6.7 Piston Moving within a Cylinder against an External Pressure**

## Example 6.4    Pressure–Volume Work

To inflate a balloon you must do pressure–volume work on the surroundings. If you inflate a balloon from a volume of 0.100 L to 1.85 L against an external pressure of 1.00 atm, how much work is done (in joules)?

| | |
|---|---|
| **SORT** You know the initial and final volumes of the balloon and the pressure against which it expands. The balloon and its contents are the system. | **GIVEN:** $V_1 = 0.100$ L<br>$V_2 = 1.85$ L<br>$P = 1.00$ atm<br><br>**FIND:** $w$ |
| **STRATEGIZE** The equation $w = -P_{ext} \Delta V$ specifies the amount of work done during a volume change against an external pressure. | **CONCEPTUAL PLAN**<br><br><br>$$w = -P_{ext} \Delta V$$ |
| **SOLVE** To solve the problem, calculate the value of $\Delta V$ and substitute it, together with $P_{ext}$, into the equation. | **SOLUTION**<br>$\Delta V = V_2 - V_1$<br>$\quad = 1.85$ L $- 0.100$ L<br>$\quad = 1.75$ L<br>$w = -P_{ext} \Delta V$<br>$\quad = -1.00$ atm $\times 1.75$ L<br>$\quad = -1.75$ L·atm |
| Convert the units of the answer (L·atm) to joules using 101.3 J = 1 L·atm. | $-1.75 \ \text{L·atm} \times \dfrac{101.3 \ \text{J}}{1 \ \text{L·atm}} = -177$ J |

**CHECK** The units (J) are correct for work. The sign of the work is negative, as it should be for an expansion: work is done on the surroundings by the expanding balloon.

**FOR PRACTICE 6.4**    A cylinder equipped with a piston expands against an external pressure of 1.58 atm. If the initial volume is 0.485 L and the final volume is 1.245 L, how much work (in J) is done?

**FOR MORE PRACTICE 6.4**    When fuel is burned in a cylinder equipped with a piston, the volume expands from 0.255 L to 1.45 L against an external pressure of 1.02 atm. In addition, 875 J is emitted as heat. What is $\Delta E$ for the burning of the fuel?

## 6.5    Measuring $\Delta E$ for Chemical Reactions: Constant-Volume Calorimetry

We now have a complete picture of how a system exchanges energy with its surroundings via heat and pressure–volume work:

$$\text{Heat } (q) = m \times C_s \times \Delta T$$

System ⟷ Surroundings

$$\text{Work } (w) = -P_{ext} \Delta V$$

Recall from Section 6.3 that the change in internal energy that occurs during a chemical reaction ($\Delta E$) is a measure of *all of the energy* (heat and work) exchanged with the surroundings ($\Delta E = q + w$). Therefore, we can measure the changes in temperature (to calculate heat) and the changes in volume (to calculate work) that occur during a chemical reaction, and then sum them together to calculate $\Delta E$. However, an easier way to obtain the value of $\Delta E$ for a chemical reaction is to force all of the energy change associated with a reaction to manifest itself as heat rather than work. We can then measure the temperature change caused by the heat flow.

Recall that $\Delta E = q + w$ and that $w = -P_{ext}\Delta V$. If a reaction is carried out at constant volume, then $\Delta V = 0$ and $w = 0$. The heat evolved (or given off), called the *heat at constant volume* ($q_v$), is then equal to $\Delta E_{rxn}$:

$$\Delta E_{rxn} = q_v + w \quad \overset{\text{Equals zero}}{\underset{\text{at constant volume}}{}}$$

$$\Delta E_{rxn} = q_v \qquad [6.9]$$

We can measure the heat evolved in a chemical reaction using *calorimetry*. In **calorimetry**, we measure the thermal energy exchanged between the reaction (defined as the system) and the surroundings by observing the change in temperature of the surroundings:

The magnitude of the temperature change in the surroundings depends on the magnitude of $\Delta E$ for the reaction and on the heat capacity of the surroundings.

Figure 6.8▶ shows a **bomb calorimeter**, a piece of equipment that measures $\Delta E$ for combustion reactions. In a bomb calorimeter, the reaction occurs in a sealed container called a *bomb*, which ensures that the reaction occurs at constant volume. To use a bomb calorimeter, we put the sample to be burned (of known mass) into a cup equipped with an ignition wire. We then seal the cup into the bomb, which is filled with oxygen gas, and place the bomb in a water-filled, insulated container. The container is equipped with a stirrer and a thermometer. Finally, we ignite the sample with a wire coil and monitor the temperature with the thermometer. The temperature change ($\Delta T$) is related to the heat absorbed by the entire calorimeter assembly ($q_{cal}$) by the equation:

$$q_{cal} = C_{cal} \times \Delta T \qquad [6.10]$$

where $C_{cal}$ is the heat capacity of the entire calorimeter assembly (which is usually determined in a separate measurement involving the burning of a substance that gives off a known amount of heat). If no heat escapes from the calorimeter, the amount of heat *gained by* the calorimeter exactly equals that *released by* the reaction (the two are equal in magnitude but opposite in sign):

$$q_{cal} = -q_{rxn} \qquad [6.11]$$

Since the reaction occurs under conditions of constant volume, $q_{rxn} = q_v = \Delta E_{rxn}$. This measured quantity is the change in the internal energy of the reaction for the specific amount of reactant burned. To get $\Delta E_{rxn}$ per mole of a particular reactant—a more general quantity—we divide by the number of moles that actually reacted, as demonstrated in Example 6.5.

## The Bomb Calorimeter

▲ **FIGURE 6.8** **The Bomb Calorimeter** A bomb calorimeter measures changes in internal energy for combustion reactions.

The heat capacity of the calorimeter, $C_{cal}$, has units of energy over temperature; its value accounts for all of the heat absorbed by all of the components within the calorimeter (including the water).

**Example 6.5** | **Measuring $\Delta E_{rxn}$ in a Bomb Calorimeter**

When 1.010 g of sucrose ($C_{12}H_{22}O_{11}$) undergoes combustion in a bomb calorimeter, the temperature rises from 24.92 °C to 28.33 °C. Find $\Delta E_{rxn}$ for the combustion of sucrose in kJ/mol sucrose. The heat capacity of the bomb calorimeter, determined in a separate experiment, is 4.90 kJ/°C. (You can ignore the heat capacity of the small sample of sucrose because it is negligible compared to the heat capacity of the calorimeter.)

| | |
|---|---|
| **SORT** You are given the mass of sucrose, the heat capacity of the calorimeter, and the initial and final temperatures. You are asked to find the change in internal energy for the reaction. | **GIVEN:** 1.010 g $C_{12}H_{22}O_{11}$ $T_i = 24.92$ °C $T_f = 28.33$ °C $C_{cal} = 4.90$ kJ/°C **FIND:** $\Delta E_{rxn}$ |

**STRATEGIZE** The conceptual plan has three parts. In the first part, use the temperature change and the heat capacity of the calorimeter to find $q_{cal}$.

In the second part, use $q_{cal}$ to get $q_{rxn}$ (which just involves changing the sign). Since the bomb calorimeter ensures constant volume, $q_{rxn}$ is equivalent to $\Delta E_{rxn}$ for the amount of sucrose burned.

In the third part, divide $q_{rxn}$ by the number of moles of sucrose to determine $\Delta E_{rxn}$ per mole of sucrose.

**CONCEPTUAL PLAN**

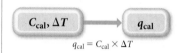

$$q_{cal} = C_{cal} \times \Delta T$$

$$q_{rxn} = -q_{cal}$$

$$\Delta E_{rxn} = \frac{q_{rxn}}{mol\ C_{12}H_{22}O_{11}}$$

**RELATIONSHIPS USED**
$$q_{cal} = C_{cal} \times \Delta T = -q_{rxn}$$
molar mass $C_{12}H_{22}O_{11} = 342.3$ g/mol

**SOLVE** Gather the necessary quantities in the correct units and substitute these into the equation to calculate $q_{cal}$.

**SOLUTION**
$$\Delta T = T_f - T_i$$
$$= 28.33 \text{ °C} - 24.92 \text{ °C} = 3.41 \text{ °C}$$
$$q_{cal} = C_{cal} \times \Delta T$$
$$q_{cal} = 4.90 \frac{kJ}{\text{°C}} \times 3.41 \text{ °C} = 16.7 \text{ kJ}$$

Find $q_{rxn}$ by taking the negative of $q_{cal}$.
$$q_{rxn} = -q_{cal} = -16.7 \text{ kJ}$$

Find $\Delta E_{rxn}$ per mole of sucrose by dividing $q_{rxn}$ by the number of moles of sucrose (calculated from the given mass of sucrose and its molar mass).
$$\Delta E_{rxn} = \frac{q_{rxn}}{mol\ C_{12}H_{22}O_{11}}$$
$$= \frac{-16.7 \text{ kJ}}{1.010 \text{ g } C_{12}H_{22}O_{11} \times \frac{1 \text{ mol } C_{12}H_{22}O_{11}}{342.3 \text{ g } C_{12}H_{22}O_{11}}}$$
$$= -5.66 \times 10^3 \text{ kJ/mol } C_{12}H_{22}O_{11}$$

**CHECK** The units of the answer (kJ) are correct for a change in internal energy. The sign of $\Delta E_{rxn}$ is negative, as it should be for a combustion reaction that gives off energy.

**FOR PRACTICE 6.5** When 1.550 g of liquid hexane ($C_6H_{14}$) undergoes combustion in a bomb calorimeter, the temperature rises from 25.87 °C to 38.13 °C. Find $\Delta E_{rxn}$ for the reaction in kJ/mol hexane. The heat capacity of the bomb calorimeter, determined in a separate experiment, is 5.73 kJ/°C.

**FOR MORE PRACTICE 6.5** The combustion of toluene has a $\Delta E_{rxn}$ of $-3.91 \times 10^3$ kJ/mol. When 1.55 g of toluene ($C_7H_8$) undergoes combustion in a bomb calorimeter, the temperature rises from 23.12 °C to 37.57 °C. Find the heat capacity of the bomb calorimeter.

# 6.6 Enthalpy: The Heat Evolved in a Chemical Reaction at Constant Pressure

 **The Change in Enthalpy for a Chemical Reaction**

We have just seen that when a chemical reaction occurs in a sealed container under conditions of constant volume, the energy evolves only as heat. However, when a chemical reaction occurs open to the atmosphere under conditions of constant pressure—such as a reaction occurring in an open beaker or the burning of natural gas on a stove—the energy can evolve as both heat and work. As we have also seen, $\Delta E_{rxn}$ is a measure of the *total energy change* (both heat and work) that occurs during the reaction. However, in many cases, we are interested only in the heat exchanged, not the work done. For example, when we burn natural gas on a stove to cook food, we do not really care how much work the combustion reaction does on the atmosphere by expanding against it—we just want to know how much heat is given off to cook the food. Under conditions of constant pressure, a thermodynamic quantity called *enthalpy* represents exactly this.

We define the **enthalpy (H)** of a system as the sum of its internal energy and the product of its pressure and volume:

$$H = E + PV \qquad [6.12]$$

Since internal energy, pressure, and volume are all state functions, enthalpy is also a state function. The *change in enthalpy* ($\Delta H$) for any process occurring under constant pressure is given by the expression:

$$\Delta H = \Delta E + P\,\Delta V \qquad [6.13]$$

To better understand this expression, we can interpret the two terms on the right with the help of relationships already familiar to us. We saw previously that $\Delta E = q + w$. If we represent the heat at constant pressure as $q_p$, then the change in internal energy at constant pressure is $\Delta E = q_p + w$. In addition, from our definition of pressure–volume work, we know that $P_{ext}\,\Delta V = -w$. Substituting these expressions into the expression for $\Delta H$ gives us the following expression:

$$
\begin{aligned}
\Delta H &= \Delta E + P\,\Delta V \\
&= (q_p + w) + P\,\Delta V \qquad [6.14]\\
&= q_p + w - w \\
\Delta H &= q_p
\end{aligned}
$$

We can see that $\Delta H$ is equal to $q_p$, the heat at constant pressure.

Conceptually (and often numerically), $\Delta H$ and $\Delta E$ are similar: they both represent changes in a state function for the system. However, $\Delta E$ is a measure of *all of the energy* (heat and work) exchanged with the surroundings, while $\Delta H$ is a measure of only the heat exchanged under conditions of constant pressure. For chemical reactions that do not exchange much work with the surroundings—that is, those that do not cause a large change in reaction volume as they occur—$\Delta H$ and $\Delta E$ are nearly identical in value. For chemical reactions that produce or consume large amounts of gas, and therefore result in large volume changes, $\Delta H$ and $\Delta E$ may differ slightly in value.

---

**The Difference between $\Delta H$ and $\Delta E$** Lighters are usually fueled by butane ($C_4H_{10}$). When 1 mol of butane burns at constant pressure, it produces 2658 kJ of heat and does 3 kJ of work. What are the values of $\Delta H$ and $\Delta E$ for the combustion of one mole of butane?

 CONCEPTUAL CONNECTION **6.5**

---

The signs of $\Delta H$ and $\Delta E$ follow the same conventions. A positive $\Delta H$ indicates that heat flows into the system as the reaction occurs. A chemical reaction with a positive $\Delta H$, called an **endothermic reaction**, absorbs heat from its surroundings.

▶ The reaction that occurs in a chemical cold pack is endothermic—it absorbs energy from the surroundings. The combustion of natural gas is an exothermic reaction—it releases energy to the surroundings.

Endothermic

Exothermic

The reaction that occurs in a chemical cold pack, often used to ice athletic injuries, is a good example of an endothermic reaction. When you break the barrier separating the reactants in a chemical cold pack, the substances mix, react, and absorb heat from the surroundings. The surroundings—including, say, your bruised wrist—get *colder* because they *lose* energy as the cold pack absorbs it.

A chemical reaction with a negative $\Delta H$, called an **exothermic reaction**, gives off heat to its surroundings. The reaction that occurs in the chemical hand warmer discussed in Section 6.1 is a good example of an exothermic reaction. When the reaction takes place, heat is given off into the surroundings (including your hand and glove), making them warmer. The burning of natural gas is another exothermic reaction. As the gas burns, it gives off energy, raising the temperature of its surroundings.

*Summarizing Enthalpy:*

▶ The value of $\Delta H$ for a chemical reaction is the amount of heat absorbed or evolved in the reaction under conditions of constant pressure.

▶ An endothermic reaction has a *positive* $\Delta H$ and absorbs heat from the surroundings. An endothermic reaction feels cold to the touch.

▶ An exothermic reaction has a *negative* $\Delta H$ and gives off heat to the surroundings. An exothermic reaction feels warm to the touch.

### Example 6.6 | Exothermic and Endothermic Processes

Identify each process as endothermic or exothermic and indicate the sign of $\Delta H$.

**(a)** sweat evaporating from skin

**(b)** water freezing in a freezer

**(c)** wood burning in a fire

**SOLUTION**

**(a)** Sweat evaporating from skin cools the skin and is therefore endothermic, with a positive $\Delta H$. The skin must supply heat to the perspiration in order for it to continue to evaporate.

**(b)** Water freezing in a freezer releases heat and is therefore exothermic, with a negative $\Delta H$. The refrigeration system in the freezer must remove this heat for the water to continue to freeze.

**(c)** Wood burning in a fire releases heat and is therefore exothermic, with a negative $\Delta H$.

**FOR PRACTICE 6.6** Identify each process as endothermic or exothermic and indicate the sign of $\Delta H$.

**(a)** an ice cube melting

**(b)** nail polish remover quickly evaporating after it is accidentally spilled on the skin

**(c)** gasoline burning within the cylinder of an automobile engine

## Exothermic and Endothermic Processes: A Molecular View

When a chemical system undergoes a change in enthalpy, where does the energy come from or go to? For example, we just saw that an exothermic chemical reaction gives off *thermal energy*—what is the source of that energy?

First, we know that the emitted thermal energy *does not* come from the original thermal energy of the system. Recall from Section 6.2 that the thermal energy of a system is the total kinetic energy of the atoms and molecules that compose the system. This kinetic energy *cannot* be the source of the energy given off in an exothermic reaction because if the atoms and molecules that compose the system were to lose kinetic energy, their temperature would necessarily fall—the system would get colder. Yet we know that in exothermic reactions, the temperature of the system and the surroundings rises. So there must be some other source of energy.

Recall also from Section 6.2 that the internal energy of a chemical system is the sum of its kinetic energy and its *potential energy*. This potential energy is the source in an exothermic chemical reaction. Under normal circumstances, chemical potential energy (or simply chemical energy) arises primarily from the electrostatic forces between the protons and electrons that compose the atoms and molecules within the system. In an exothermic reaction, some bonds break and new ones form, and the nuclei and electrons reorganize into an arrangement with lower potential energy. As the molecules rearrange, their potential energy converts into thermal energy, the heat emitted in the reaction. Note that breaking bonds always *absorbs* energy. In an exothermic reaction, weak bonds break and stronger bonds form. It only takes a small amount of energy to break the weak bonds and a greater amount is given off when the stronger bonds form, resulting in net energy production. In an endothermic reaction, the opposite happens: strong bonds break and weak ones form. The nuclei and electrons reorganize into an arrangement with higher potential energy, absorbing thermal energy in the process.

**Exothermic and Endothermic Reactions**  If an endothermic reaction absorbs heat, then why does it feel cold to the touch?

CONCEPTUAL
**CONNECTION 6.6**

## Stoichiometry Involving Δ*H*: Thermochemical Equations

The enthalpy change for a chemical reaction, abbreviated $\Delta H_{rxn}$, is also called the **enthalpy of reaction** or **heat of reaction** and depends on the amount of material undergoing the reaction. In other words, the amount of heat generated or absorbed when a chemical reaction occurs depends on the *amounts* of reactants that actually react. We usually specify $\Delta H_{rxn}$ in combination with the balanced chemical equation for the reaction. *The magnitude of $\Delta H_{rxn}$ reflects the stoichiometric amounts of reactants and products for the reaction as written.* For example, consider the balanced equation and $\Delta H_{rxn}$ for the combustion of propane, the main component of LP gas:

$$C_3H_8(g) + 5\,O_2(g) \longrightarrow 3\,CO_2(g) + 4\,H_2O(g) \quad \Delta H_{rxn} = -2044\ \text{kJ}$$

The equation tells us that when 1 mol of $C_3H_8$ reacts with 5 mol of $O_2$ to form 3 mol of $CO_2$ and 4 mol of $H_2O$, 2044 kJ of heat is emitted. We can write these relationships in the same way that we expressed stoichiometric relationships in Chapter 4, as ratios between two quantities. For example, for the reactants, we write the following ratios:

$$1\ \text{mol}\ C_3H_8 : -2044\ \text{kJ} \quad \text{or} \quad 5\ \text{mol}\ O_2 : -2044\ \text{kJ}$$

The ratios indicate that 2044 kJ of heat evolves when 1 mol of $C_3H_8$ and 5 mol of $O_2$ completely react. We can use these ratios to construct conversion factors between amounts of reactants or products and the quantity of heat emitted (for exothermic reactions) or absorbed (for endothermic reactions). To find out how much heat is emitted upon the combustion of a certain mass in grams of $C_3H_8$, we work with the following conceptual plan:

We use the molar mass to convert between grams and moles, and the stoichiometric relationship between moles of $C_3H_8$ and the heat of reaction to convert between moles and kilojoules, as demonstrated in Example 6.7.

## Example 6.7   Stoichiometry Involving $\Delta H$

An LP gas tank in a home barbeque contains 13.2 kg of propane, $C_3H_8$. Calculate the heat (in kJ) associated with the complete combustion of all of the propane in the tank.

$$C_3H_8(g) + 5\,O_2(g) \longrightarrow 3\,CO_2(g) + 4\,H_2O(g) \qquad \Delta H_{rxn} = -2044 \text{ kJ}$$

| | |
|---|---|
| **SORT** You are given the mass of propane and asked to find the heat evolved in its combustion. | **GIVEN:** 13.2 kg $C_3H_8$ <br> **FIND:** $q$ |
| **STRATEGIZE** Starting with kg $C_3H_8$, convert to g $C_3H_8$ and then use the molar mass of $C_3H_8$ to find the number of moles. Next, use the stoichiometric relationship between mol $C_3H_8$ and kJ to determine the heat evolved. | **CONCEPTUAL PLAN** <br><br>  <br><br> **RELATIONSHIPS USED** <br> 1000 g = 1 kg <br> molar mass $C_3H_8$ = 44.09 g/mol <br> 1 mol $C_3H_8$ : $-2044$ kJ (from balanced equation) |
| **SOLVE** Follow the conceptual plan to solve the problem. Begin with 13.2 kg $C_3H_8$ and multiply by the appropriate conversion factors to arrive at kJ. | **SOLUTION** <br><br> $13.2 \text{ kg } C_3H_8 \times \dfrac{1000\,g}{1 \text{ kg}} \times \dfrac{1 \text{ mol } C_3H_8}{44.09 \text{ g } C_3H_8} \times \dfrac{-2044 \text{ kJ}}{1 \text{ mol } C_3H_8}$ <br><br> $= -6.12 \times 10^5 \text{ kJ}$ |

**CHECK** The units of the answer (kJ) are correct for energy. The answer is negative, as it should be for heat evolved by the reaction.

**FOR PRACTICE 6.7**   Ammonia reacts with oxygen according to the equation:

$$4\,NH_3(g) + 5\,O_2(g) \longrightarrow 4\,NO(g) + 6\,H_2O(g) \qquad \Delta H_{rxn} = -906 \text{ kJ}$$

Calculate the heat (in kJ) associated with the complete reaction of 155 g of $NH_3$.

**FOR MORE PRACTICE 6.7**   What mass of butane in grams is necessary to produce $1.5 \times 10^3$ kJ of heat? What mass of $CO_2$ is produced?

$$C_4H_{10}(g) + 13/2\,O_2(g) \longrightarrow 4\,CO_2(g) + 5\,H_2O(g) \qquad \Delta H_{rxn} = -2658 \text{ kJ}$$

## 6.7   Constant-Pressure Calorimetry: Measuring $\Delta H_{rxn}$

For many aqueous reactions, we can measure $\Delta H_{rxn}$ fairly simply using a **coffee-cup calorimeter** shown in Figure 6.9▶. The calorimeter consists of two Styrofoam coffee cups, one inserted into the other, to provide insulation from the laboratory environment.

The calorimeter is equipped with a thermometer and a stirrer. The reaction occurs in a specifically measured quantity of solution within the calorimeter, so that the mass of the solution is known. During the reaction, the heat evolved (or absorbed) causes a temperature change in the solution, which the thermometer measures. If we know the specific heat capacity of the solution, normally assumed to be that of water, we can calculate $q_{soln}$, the heat absorbed by or lost from the solution (which is acting as the surroundings) using the equation:

$$q_{soln} = m_{soln} \times C_{s,\,soln} \times \Delta T$$

The insulated calorimeter prevents heat from escaping, so we assume that the heat gained by the solution equals that lost by the reaction (or vice versa):

$$q_{rxn} = -q_{soln}$$

Because the reaction happens under conditions of constant pressure (open to the atmosphere), $q_{rxn} = q_p = \Delta H_{rxn}$. This measured quantity is the heat of reaction for the specific amount (which is measured ahead of time) of reactants that react. To find $\Delta H_{rxn}$ per mole of a particular reactant—a more general quantity—we divide by the number of moles that actually reacted, as demonstrated in Example 6.8.

*Summarizing Calorimetry:*

▶ Bomb calorimetry occurs at constant *volume* and measures $\Delta E$ for a reaction.

▶ Coffee-cup calorimetry occurs at constant *pressure* and measures $\Delta H$ for a reaction.

## The Coffee-Cup Calorimeter

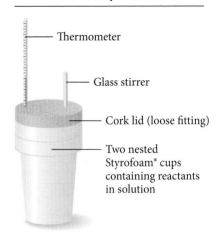

▲ FIGURE 6.9 **The Coffee-Cup Calorimeter** A coffee-cup calorimeter measures enthalpy changes for chemical reactions in solution.

The equation, $q_{rxn} = -q_{soln}$ assumes that no heat is lost to the calorimeter itself. If heat absorbed by the calorimeter is accounted for, the equation becomes $q_{rxn} = -(q_{soln} + q_{cal})$.

## Example 6.8  Measuring $\Delta H_{rxn}$ in a Coffee-Cup Calorimeter

Magnesium metal reacts with hydrochloric acid according to the balanced equation:

$$Mg(s) + 2\,HCl(aq) \longrightarrow MgCl_2(aq) + H_2(g)$$

In an experiment to determine the enthalpy change for this reaction, 0.158 g of Mg metal is combined with enough HCl to make 100.0 mL of solution in a coffee-cup calorimeter. The HCl is sufficiently concentrated so that the Mg completely reacts. The temperature of the solution rises from 25.6 °C to 32.8 °C as a result of the reaction. Find $\Delta H_{rxn}$ for the reaction as written. Use 1.00 g/mL as the density of the solution and $C_{s,\,soln} = 4.18\,\text{J/g}\cdot°\text{C}$ as the specific heat capacity of the solution.

| | |
|---|---|
| **SORT** You are given the mass of magnesium, the volume of solution, the initial and final temperatures, the density of the solution, and the heat capacity of the solution. You are asked to find the change in enthalpy for the reaction. | **GIVEN:** 0.158 g Mg<br>100.0 mL soln<br>$T_i = 25.6\,°\text{C}$  $T_f = 32.8\,°\text{C}$<br>$d = 1.00\,\text{g/mL}$  $C_{s,\,soln} = 4.18\,\text{J/g}\cdot°\text{C}$<br><br>**FIND:** $\Delta H_{rxn}$ |

**STRATEGIZE** The conceptual plan has three parts. In the first part, use the temperature change and the other given quantities, together with the equation $q = m \times C_s \times \Delta T$, to find $q_{soln}$.

In the second part, use $q_{soln}$ to get $q_{rxn}$ (which simply involves changing the sign). Because the pressure is constant, $q_{rxn}$ is equivalent to $\Delta H_{rxn}$ for the amount of magnesium that reacts.

In the third part, divide $q_{rxn}$ by the number of moles of magnesium to determine $\Delta H_{rxn}$ per mole of magnesium.

**CONCEPTUAL PLAN**

$$\Delta H_{rxn} = \frac{q_{rxn}}{\text{mol Mg}}$$

**RELATIONSHIPS USED**
$q = m \times C_s \times \Delta T$
$q_{rxn} = -q_{soln}$

—Continued on the next page

*—Continued*

| | |
|---|---|
| **SOLVE** Gather the necessary quantities in the correct units for the equation $q = m \times C_s \times \Delta T$ and substitute these into the equation to calculate $q_{soln}$. Notice that the sign of $q_{soln}$ is positive, indicating that the solution absorbed heat from the reaction. | **SOLUTION** $C_{s,\,soln} = 4.18\,J/g\cdot°C$ $m_{soln} = 100.0\,mL\,soln \times \dfrac{1.00\,g}{1\,mL\,soln} = 1.00 \times 10^2\,g$ $\Delta T = T_f - T_i$ $\quad = 32.8\,°C - 25.6\,°C = 7.2\,°C$ $q_{soln} = m_{soln} \times C_{s,\,soln} \times \Delta T$ $\quad = 1.00 \times 10^2\,g \times 4.18\dfrac{J}{g\cdot°C} \times 7.2\,°C = 3.0 \times 10^3\,J$ |
| Find $q_{rxn}$ by taking the negative of $q_{soln}$. Notice that $q_{rxn}$ is negative, as you expect for an exothermic reaction. | $q_{rxn} = -q_{soln} = -3.0 \times 10^3\,J$ |
| Finally, find $\Delta H_{rxn}$ per mole of magnesium by dividing $q_{rxn}$ by the number of moles of magnesium that reacts. Find the number of moles of magnesium from the given mass of magnesium and its molar mass. | $\Delta H_{rxn} = \dfrac{q_{rxn}}{mol\ Mg}$ $\quad = \dfrac{-3.0 \times 10^3\,J}{0.158\,g\,Mg \times \dfrac{1\,mol\ Mg}{24.31\,g\,Mg}}$ $\quad = -4.6 \times 10^5\,J/mol\ Mg$ |
| Since the stoichiometric coefficient for magnesium in the balanced chemical equation is 1, the calculated value represents $\Delta H_{rxn}$ for the reaction as written. | $Mg(s) + 2\,HCl(aq) \longrightarrow MgCl_2(aq) + H_2(g)$ $\hspace{5cm} \Delta H_{rxn} = -4.6 \times 10^5\,J$ |

**CHECK** The units of the answer (J) are correct for the change in enthalpy of a reaction. The sign is negative, as you expect for an exothermic reaction.

**FOR PRACTICE 6.8** The addition of hydrochloric acid to a silver nitrate solution precipitates silver chloride according to the reaction:

$$AgNO_3(aq) + HCl(aq) \longrightarrow AgCl(s) + HNO_3(aq)$$

When 50.0 mL of 0.100 M AgNO₃ is combined with 50.0 mL of 0.100 M HCl in a coffee-cup calorimeter, the temperature changes from 23.40 °C to 24.21 °C. Calculate $\Delta H_{rxn}$ for the reaction as written. Use 1.00 g/mL as the density of the solution and $C = 4.18\,J/g\cdot°C$ as the specific heat capacity.

CONCEPTUAL
CONNECTION **6.7**

**Constant-Pressure versus Constant-Volume Calorimetry** The same reaction, with exactly the same amount of reactant, is conducted in a bomb calorimeter and in a coffee-cup calorimeter. In one of the calorimeters, $q_{rxn} = -12.5\,kJ$ and in the other $q_{rxn} = -11.8\,kJ$. Which value was obtained in the bomb calorimeter? (Assume that the reaction has a positive $\Delta V$ in the coffee-cup calorimeter.)

## 6.8 Relationships Involving $\Delta H_{rxn}$

The change in enthalpy for a reaction is always associated with a *particular* reaction. If we change the reaction in well-defined ways, then $\Delta H_{rxn}$ also changes in well-defined ways. We now turn our attention to three quantitative relationships between a chemical equation and $\Delta H_{rxn}$.

1. **If a chemical equation is multiplied by some factor, then $\Delta H_{rxn}$ is also multiplied by the same factor.**

Recall from Section 6.6 that $\Delta H_{rxn}$ is an extensive property; it depends on the quantity of reactants undergoing reaction. Recall also that $\Delta H_{rxn}$ is usually reported for a reaction involving stoichiometric amounts of reactants. For example, for a reaction $A + 2B \longrightarrow C$, $\Delta H_{rxn}$ is typically reported as the amount of heat emitted or absorbed when 1 mol A reacts with 2 mol B to form 1 mol C. Therefore, if a chemical equation is multiplied by a factor, then $\Delta H_{rxn}$ is also multiplied by the same factor. For example,

$$A + 2B \longrightarrow C \qquad \Delta H_1$$

$$2A + 4B \longrightarrow 2C \qquad \Delta H_2 = 2 \times \Delta H_1$$

2. **If a chemical equation is reversed, then $\Delta H_{rxn}$ changes sign.**

Recall from Section 6.6 that $\Delta H_{rxn}$ is a state function, which means that its value depends only on the initial and final states of the system.

$$\Delta H = H_{final} - H_{initial}$$

When a reaction is reversed, the final state becomes the initial state and vice versa. Consequently, $\Delta H_{rxn}$ changes sign:

$$A + 2B \longrightarrow C \qquad \Delta H_1$$

$$C \longrightarrow A + 2B \qquad \Delta H_2 = -\Delta H_1$$

3. **If a chemical equation can be expressed as the sum of a series of steps, then $\Delta H_{rxn}$ for the overall equation is the sum of the heats of reaction for each step.**

This last relationship, known as **Hess's law**, also follows from the enthalpy of reaction being a state function. Since $\Delta H_{rxn}$ is dependent only on the initial and final states, and not on the pathway the reaction follows, $\Delta H$ obtained from summing the individual steps that lead to an overall reaction must be the same as $\Delta H$ for that overall reaction. For example,

$$A + 2B \longrightarrow C \qquad \Delta H_1$$
$$C \longrightarrow 2D \qquad \Delta H_2$$
$$\overline{A + 2B \longrightarrow 2D \qquad \Delta H_3 = \Delta H_1 + \Delta H_2}$$

We illustrate Hess's law with the energy-level diagram in Figure 6.10▶.

These three quantitative relationships make it possible to determine $\Delta H$ for a reaction without directly measuring it in the laboratory. (For some reactions, direct measurement can be difficult.) If we can find related reactions (with known $\Delta H$'s) that sum to the reaction of interest, we can find $\Delta H$ for the reaction of interest. For example, the following reaction between $C(s)$ and $H_2O(g)$ is an industrially important method of generating hydrogen gas:

$$C(s) + H_2O(g) \longrightarrow CO(g) + H_2(g) \qquad \Delta H_{rxn} = ?$$

We can find $\Delta H_{rxn}$ from the following reactions that have known $\Delta H$'s:

$$C(s) + O_2(g) \longrightarrow CO_2(g) \qquad \Delta H = -393.5 \text{ kJ}$$

$$2CO(g) + O_2(g) \longrightarrow 2CO_2(g) \qquad \Delta H = -566.0 \text{ kJ}$$

$$2H_2(g) + O_2(g) \longrightarrow 2H_2O(g) \qquad \Delta H = -483.6 \text{ kJ}$$

We just have to determine how to sum these reactions to get the overall reaction of interest. We do this by manipulating the reactions with known $\Delta H$'s in such a way as to get the reactants of interest on the left, the products of interest on the right, and other species to cancel.

**Hess's Law**
The change in enthalpy for a stepwise process is the sum of the enthalpy changes of the steps.

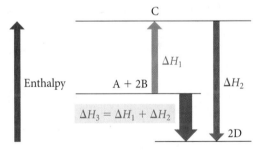

▲ **FIGURE 6.10  Hess's Law**  The change in enthalpy for a stepwise process is the sum of the enthalpy changes of the steps.

Since the first reaction has C(s) as a reactant, and the reaction of interest also has C(s) as a reactant, we write the first reaction unchanged.

$$C(s) + O_2(g) \longrightarrow CO_2(g) \quad \Delta H = -393.5 \text{ kJ}$$

The second reaction has 2 mol of CO(g) as a reactant. However, the reaction of interest has 1 mol of CO(g) as a product. Therefore, we reverse the second reaction, change the sign of $\Delta H$, and multiply the reaction and $\Delta H$ by $^1/_2$.

$$^1/_2 \times [2\,CO_2(g) \longrightarrow 2\,CO(g) + O_2(g)] \quad \Delta H = ^1/_2 \times (+566.0 \text{ kJ})$$

In the third reaction $H_2(g)$ is a reactant. In the reaction of interest, however, $H_2(g)$ is a product. Therefore, we reverse the equation and change the sign of $\Delta H$. In addition, to obtain coefficients that match the reaction of interest, and to cancel $O_2$, we must multiply the reaction and $\Delta H$ by $^1/_2$.

$$^1/_2 \times [2\,H_2O(g) \longrightarrow 2\,H_2(g) + O_2(g)] \quad \Delta H = ^1/_2 \times (+483.6 \text{ kJ})$$

Lastly, we rewrite the three reactions after multiplying through by the indicated factors and show how they sum to the reaction of interest. $\Delta H$ for the reaction of interest is the sum of the $\Delta H$'s for the steps.

$$
\begin{aligned}
C(s) + O_2(g) &\longrightarrow CO_2(g) & \Delta H &= -393.5 \text{ kJ} \\
CO_2(g) &\longrightarrow CO(g) + ^1/_2\,O_2(g) & \Delta H &= +283.0 \text{ kJ} \\
H_2O(g) &\longrightarrow H_2(g) + ^1/_2\,O_2(g) & \Delta H &= +241.8 \text{ kJ} \\
\hline
C(s) + H_2O(g) &\longrightarrow CO(g) + H_2(g) & \Delta H_{rxn} &= +131.3 \text{ kJ}
\end{aligned}
$$

## Example 6.9   Hess's Law

Find $\Delta H_{rxn}$ for the reaction:

$$3\,C(s) + 4\,H_2(g) \longrightarrow C_3H_8(g)$$

Use these reactions with known $\Delta H$'s:

$$
\begin{aligned}
C_3H_8(g) + 5\,O_2(g) &\longrightarrow 3\,CO_2(g) + 4\,H_2O(g) & \Delta H &= -2043 \text{ kJ} \\
C(s) + O_2(g) &\longrightarrow CO_2(g) & \Delta H &= -393.5 \text{ kJ} \\
2\,H_2(g) + O_2(g) &\longrightarrow 2\,H_2O(g) & \Delta H &= -483.6 \text{ kJ}
\end{aligned}
$$

**SOLUTION**

To work this and other Hess's law problems, manipulate the reactions with known $\Delta H$'s in such a way as to get the reactants of interest on the left, the products of interest on the right, and other species to cancel.

| | |
|---|---|
| The first reaction has $C_3H_8$ as a reactant, and the reaction of interest has $C_3H_8$ as a product, so reverse the first reaction and change the sign of $\Delta H$. | $3\,CO_2(g) + 4\,H_2O(g) \longrightarrow C_3H_8(g) + 5\,O_2(g) \quad \Delta H = +2043 \text{ kJ}$ |
| The second reaction has C as a reactant and $CO_2$ as a product, just as required in the reaction of interest. However, the coefficient for C is 1, and in the reaction of interest, the coefficient for C is 3. You need to multiply this equation and its $\Delta H$ by 3. | $3 \times [C(s) + O_2(g) \longrightarrow CO_2(g)] \quad \Delta H = 3 \times (-393.5 \text{ kJ})$ |
| The third reaction has $H_2(g)$ as a reactant, as required. However, the coefficient for $H_2$ is 2, and in the reaction of interest, the coefficient for $H_2$ is 4. Multiply this reaction and its $\Delta H$ by 2. | $2 \times [2\,H_2(g) + O_2(g) \longrightarrow 2\,H_2O(g)] \quad \Delta H = 2 \times (-483.6 \text{ kJ})$ |
| Rewrite the three reactions after multiplying through by the indicated factors and show how they sum to the reaction of interest. $\Delta H$ for the reaction of interest is the sum of the $\Delta H$'s for the steps. | $\begin{aligned} 3\,CO_2(g) + 4\,H_2O(g) &\longrightarrow C_3H_8(g) + 5\,O_2(g) & \Delta H &= +2043 \text{ kJ} \\ 3\,C(s) + 3\,O_2(g) &\longrightarrow 3\,CO_2(g) & \Delta H &= -1181 \text{ kJ} \\ 4\,H_2(g) + 2\,O_2(g) &\longrightarrow 4\,H_2O(g) & \Delta H &= -967.2 \text{ kJ} \\ \hline 3\,C(s) + 4\,H_2(g) &\longrightarrow C_3H_8(g) & \Delta H_{rxn} &= -105 \text{ kJ} \end{aligned}$ |

**FOR PRACTICE 6.9**   Find $\Delta H_{rxn}$ for the reaction:

$$N_2O(g) + NO_2(g) \longrightarrow 3\,NO(g)$$

Use these reactions with known $\Delta H$'s:

$$
\begin{aligned}
2\,NO(g) + O_2(g) &\longrightarrow 2\,NO_2(g) & \Delta H &= -113.1 \text{ kJ} \\
N_2(g) + O_2(g) &\longrightarrow 2\,NO(g) & \Delta H &= +182.6 \text{ kJ} \\
2\,N_2O(g) &\longrightarrow 2\,N_2(g) + O_2(g) & \Delta H &= -163.2 \text{ kJ}
\end{aligned}
$$

**FOR MORE PRACTICE 6.9**   Find $\Delta H_{rxn}$ for the reaction:

$$3\,H_2(g) + O_3(g) \longrightarrow 3\,H_2O(g)$$

Use these reactions with known $\Delta H$'s:

$$2\,H_2(g) + O_2(g) \longrightarrow 2\,H_2O(g) \quad \Delta H = -483.6 \text{ kJ}$$
$$3\,O_2(g) \longrightarrow 2\,O_3(g) \quad\quad\quad\quad \Delta H = +285.4 \text{ kJ}$$

# 6.9 Determining Enthalpies of Reaction from Standard Enthalpies of Formation

We have examined two ways to determine $\Delta H$ for a chemical reaction: experimentally through calorimetry and inferentially through Hess's law. We now turn to a third and more convenient way to determine $\Delta H$ for a large number of chemical reactions: from tabulated *standard enthalpies of formation*.

## Standard States and Standard Enthalpy Changes

Recall that $\Delta H$ is the *change* in enthalpy for a chemical reaction—the difference in enthalpy between the products and the reactants. Because we are interested in changes in enthalpy (and not in absolute values of enthalpy itself), we are free to define the *zero* of enthalpy as conveniently as possible. Returning to our mountain-climbing analogy, a change in altitude (like a change in enthalpy) is an absolute quantity. Altitude itself (like enthalpy), however, is a relative quantity, defined relative to some standard (such as sea level in the case of altitude). We define a similar, albeit slightly more complex, standard for enthalpy. This standard has three parts: the **standard state**, the **standard enthalpy change ($\Delta H°$)**, and the **standard enthalpy of formation ($\Delta H_f°$)**.

1. **Standard State**
   - *For a Gas:*  The standard state for a gas is the pure gas at a pressure of exactly 1 atm.
   - *For a Liquid or Solid:*  The standard state for a liquid or solid is the pure substance in its most stable form at a pressure of 1 atm and at the temperature of interest (often taken to be 25 °C).
   - *For a Substance in Solution:*  The standard state for a substance in solution is a concentration of exactly 1 M.

2. **Standard Enthalpy Change ($\Delta H°$)**
   - The change in enthalpy for a process when all reactants and products are in their standard states. The degree sign indicates standard states.

3. **Standard Enthalpy of Formation ($\Delta H_f°$)**
   - *For a Pure Compound:*  The change in enthalpy when 1 mol of the compound forms from its constituent elements in their standard states.
   - *For a Pure Element in Its Standard State:*  $\Delta H_f° = 0$.

> The standard state was changed in 1997 to a pressure of 1 bar, which is very close to 1 atm (1 atm = 1.013 bar). Both standards are now in common use.

The standard enthalpy of formation is also called the **standard heat of formation**.

Assigning the value of zero to the standard enthalpy of formation for an element in its standard state is the equivalent of assigning an altitude of zero to sea level. Once we assume sea level is zero, we can then measure all subsequent changes in altitude relative to sea level. Similarly, we can measure all changes in enthalpy relative to those of pure elements in their standard states. For example, consider the standard enthalpy of formation of methane gas at 25 °C:

$$C(s, \text{graphite}) + 2\,H_2(g) \longrightarrow CH_4(g) \quad \Delta H_f° = -74.6 \text{ kJ/mol}$$

> The carbon in this equation must be graphite (the most stable form of carbon at 1 atm and 25 °C).

For methane, as with most compounds, $\Delta H_f°$ is negative. Continuing our analogy, if we think of pure elements in their standard states as being at *sea level*, then most compounds lie *below sea level*. The chemical equation for the enthalpy of formation of a compound is always written to form 1 mol of the compound, so $\Delta H_f°$ has the units of kJ/mol. Table 6.5 lists $\Delta H_f°$ values for some selected compounds. A more complete list is in Appendix IIB.

**TABLE 6.5  Standard Enthalpies (or Heats) of Formation, $\Delta H_f^\circ$, at 298 K**

| Formula | $\Delta H_f^\circ$(kJ/mol) | Formula | $\Delta H_f^\circ$(kJ/mol) | Formula | $\Delta H_f^\circ$(kJ/mol) |
|---|---|---|---|---|---|
| *Bromine* | | $C_3H_8O(l$, isopropanol) | −318.1 | *Oxygen* | |
| $Br(g)$ | 111.9 | $C_6H_6(l)$ | 49.1 | $O_2(g)$ | 0 |
| $Br_2(l)$ | 0 | $C_6H_{12}O_6(s$, glucose) | −1273.3 | $O_3(g)$ | 142.7 |
| $HBr(g)$ | −36.3 | $C_{12}H_{22}O_{11}(s$, sucrose) | −2226.1 | $H_2O(g)$ | −241.8 |
| *Calcium* | | *Chlorine* | | $H_2O(l)$ | −285.8 |
| $Ca(s)$ | 0 | $Cl(g)$ | 121.3 | *Silver* | |
| $CaO(s)$ | −634.9 | $Cl_2(g)$ | 0 | $Ag(s)$ | 0 |
| $CaCO_3(s)$ | −1207.6 | $HCl(g)$ | −92.3 | $AgCl(s)$ | −127.0 |
| *Carbon* | | *Fluorine* | | *Sodium* | |
| $C(s$, graphite) | 0 | $F(g)$ | 79.38 | $Na(s)$ | 0 |
| $C(s$, diamond) | 1.88 | $F_2(g)$ | 0 | $Na(g)$ | 107.5 |
| $CO(g)$ | −110.5 | $HF(g)$ | −273.3 | $NaCl(s)$ | −411.2 |
| $CO_2(g)$ | −393.5 | *Hydrogen* | | $Na_2CO_3(s)$ | −1130.7 |
| $CH_4(g)$ | −74.6 | $H(g)$ | 218.0 | $NaHCO_3(s)$ | −950.8 |
| $CH_3OH(l)$ | −238.6 | $H_2(g)$ | 0 | *Sulfur* | |
| $C_2H_2(g)$ | 227.4 | *Nitrogen* | | $S_8(s$, rhombic) | 0 |
| $C_2H_4(g)$ | 52.4 | $N_2(g)$ | 0 | $S_8(s$, monoclinic) | 0.3 |
| $C_2H_6(g)$ | −84.68 | $NH_3(g)$ | −45.9 | $SO_2(g)$ | −296.8 |
| $C_2H_5OH(l)$ | −277.6 | $NH_4NO_3(s)$ | −365.6 | $SO_3(g)$ | −395.7 |
| $C_3H_8(g)$ | −103.85 | $NO(g)$ | 91.3 | $H_2SO_4(l)$ | −814.0 |
| $C_3H_6O(l$, acetone) | −248.4 | $N_2O(g)$ | 81.6 | | |

## Example 6.10   Standard Enthalpies of Formation

Write equations for the formation of (a) $MgCO_3(s)$ and (b) $C_6H_{12}O_6(s)$ from their respective elements in their standard states. Include the value of $\Delta H_f^\circ$ for each equation.

**SOLUTION**

**a.** $MgCO_3(s)$

Write the equation with the elements in $MgCO_3$ in their standard states as the reactants and 1 mol of $MgCO_3$ as the product.

$$Mg(s) + C(s, graphite) + O_2(g) \longrightarrow MgCO_3(s)$$

Balance the equation and look up $\Delta H_f^\circ$ in Appendix IIB. (Use fractional coefficients so that the product of the reaction is 1 mol of $MgCO_3$.)

$$Mg(s) + C(s, graphite) + \frac{3}{2}O_2(g) \longrightarrow MgCO_3(s)$$
$$\Delta H_f^\circ = -1095.8 \text{ kJ/mol}$$

**b.** $C_6H_{12}O_6(s)$

Write the equation with the elements in $C_6H_{12}O_6$ in their standard states as the reactants and 1 mol of $C_6H_{12}O_6$ as the product.

$$C(s, graphite) + H_2(g) + O_2(g) \longrightarrow C_6H_{12}O_6(s)$$

Balance the equation and look up $\Delta H_f^\circ$ in Appendix IIB.

$$6\,C(s, graphite) + 6\,H_2(g) + 3\,O_2(g) \longrightarrow C_6H_{12}O_6(s)$$
$$\Delta H_f^\circ = -1273.3 \text{ kJ/mol}$$

**FOR PRACTICE 6.10**   Write equations for the formation of (a) $NaCl(s)$ and (b) $Pb(NO_3)_2(s)$ from their respective elements in their standard states. Include the value of $\Delta H_f^\circ$ for each equation.

# Calculating the Standard Enthalpy Change for a Reaction

We have just seen that the standard enthalpy of formation corresponds to the *formation* of a compound from its constituent elements in their standard states:

$$\text{elements} \longrightarrow \text{compound} \qquad \Delta H_f^\circ$$

Therefore, the *negative* of the standard enthalpy of formation corresponds to the *decomposition* of a compound into its constituent elements in their standard states:

$$\text{compound} \longrightarrow \text{elements} \qquad -\Delta H_f^\circ$$

We can use these two concepts—the decomposition of a compound into its elements and the formation of a compound from its elements—to calculate the enthalpy change of any reaction by mentally taking the reactants through two steps. In the first step we *decompose the reactants* into their constituent elements in their standard states; in the second step we *form the products* from the constituent elements in their standard states:

$$\text{reactants} \longrightarrow \text{elements} \qquad \Delta H_1 = -\Sigma \Delta H_f^\circ \text{ (reactants)}$$
$$\underline{\text{elements} \longrightarrow \text{products} \qquad \Delta H_2 = +\Sigma \Delta H_f^\circ \text{ (products)}}$$
$$\text{reactants} \longrightarrow \text{products} \qquad \Delta H_{rxn}^\circ = \Delta H_1 + \Delta H_2$$

In these equations, $\Sigma$ means "the sum of" so that $\Delta H_1$ is the sum of the negatives of the heats of formation of the reactants and $\Delta H_2$ is the sum of the heats of formation of the products.

We can demonstrate this procedure by calculating the standard enthalpy change ($\Delta H_{rxn}^\circ$) for the combustion of methane:

$$CH_4(g) + 2\,O_2(g) \longrightarrow CO_2(g) + 2\,H_2O(g) \qquad \Delta H_{rxn}^\circ = ?$$

The energy changes associated with the decomposition of the reactants and the formation of the products are shown in Figure 6.11 ▼. The first step (1) is the decomposition of 1 mol of methane into its constituent elements in their standard states. We can obtain the change in enthalpy for this step by reversing the enthalpy of formation equation for methane and changing the sign of $\Delta H_f^\circ$:

$$(1)\ CH_4(g) \longrightarrow C(s, \text{graphite}) + 2\,H_2(g) \qquad -\Delta H_f^\circ = +74.6 \text{ kJ/mol}$$

The second step, the formation of the products from their constituent elements, has two parts: (a) the formation of 1 mol $CO_2$ and (b) the formation of 2 mol $H_2O$. Because part (b) forms 2 mol $H_2O$, we multiply the $\Delta H_f^\circ$ for that step by 2:

$$(2a)\ C(s, \text{graphite}) + O_2(g) \longrightarrow CO_2(g) \qquad \Delta H_f^\circ = -393.5 \text{ kJ/mol}$$
$$(2b)\ 2 \times [H_2(g) + {}^1\!/_2 O_2(g) \longrightarrow H_2O(g)] \qquad 2 \times \Delta H_f^\circ = 2 \times (-241.8 \text{ kJ/mol})$$

## Calculating the Enthalpy Change for the Combustion of Methane

◀ **FIGURE 6.11  Calculating the Enthalpy Change for the Combustion of Methane**

As we know from Hess's law, the enthalpy of reaction for the overall reaction is the sum of the enthalpies of reaction of the individual steps:

(1) $CH_4(g) \longrightarrow C(s, \text{graphite}) + 2 H_2(g)$ $\qquad -\Delta H_f^\circ = +74.6 \text{ kJ/mol}$

(2a) $C(s, \text{graphite}) + O_2(g) \longrightarrow CO_2(g)$ $\qquad \Delta H_f^\circ = -393.5 \text{ kJ/mol}$

(2b) $2 H_2(g) + O_2(g) \longrightarrow 2 H_2O(g)$ $\qquad 2 \times \Delta H_f^\circ = -483.6 \text{ kJ/mol}$

$\overline{CH_4(g) + 2 O_2(g) \longrightarrow CO_2(g) + 2 H_2O(g)} \qquad \Delta H_{rxn}^\circ = -802.5 \text{ kJ/mol}$

We can streamline and generalize this process as follows:

> **To calculate $\Delta H_{rxn}^\circ$, subtract the enthalpies of formation of the reactants multiplied by their stoichiometric coefficients from the enthalpies of formation of the products multiplied by their stoichiometric coefficients.**

In the form of an equation:

$$\Delta H_{rxn}^\circ = \sum n_p \Delta H_f^\circ (\text{products}) - \sum n_r \Delta H_f^\circ (\text{reactants}) \qquad [6.15]$$

In this equation, $n_p$ represents the stoichiometric coefficients of the products, $n_r$ represents the stoichiometric coefficients of the reactants, and $\Delta H_f^\circ$ represents the standard enthalpies of formation. Keep in mind when using this equation that elements in their standard states have $\Delta H_f^\circ = 0$. Examples 6.11 and 6.12 demonstrate this process.

---

**Example 6.11**   **$\Delta H_{rxn}^\circ$ and Standard Enthalpies of Formation**

Use the standard enthalpies of formation to determine $\Delta H_{rxn}^\circ$ for the reaction:

$$4 NH_3(g) + 5 O_2(g) \longrightarrow 4 NO(g) + 6 H_2O(g)$$

| | |
|---|---|
| **SORT** You are given the balanced equation and asked to find the enthalpy of reaction. | **GIVEN:** $4 NH_3(g) + 5 O_2(g) \longrightarrow 4 NO(g) + 6 H_2O(g)$<br><br>**FIND:** $\Delta H_{rxn}^\circ$ |
| **STRATEGIZE** To calculate $\Delta H_{rxn}^\circ$ from standard enthalpies of formation, subtract the heats of formation of the reactants multiplied by their stoichiometric coefficients from the heats of formation of the products multiplied by their stoichiometric coefficients. | **CONCEPTUAL PLAN**<br>$\Delta H_{rxn}^\circ = \Sigma n_p \Delta H_f^\circ (\text{products}) - \Sigma n_r \Delta H_f^\circ (\text{reactants})$ |

**SOLVE** Begin by looking up (in Appendix IIB) the standard enthalpy of formation for each reactant and product. Remember that the standard enthalpy of formation of pure elements in their standard state is zero. Calculate $\Delta H_{rxn}^\circ$ by substituting into the equation.

**SOLUTION**

| Reactant or product | $\Delta H_f^\circ$ (kJ/mol, from Appendix IIB) |
|---|---|
| $NH_3(g)$ | −45.9 |
| $O_2(g)$ | 0.0 |
| $NO(g)$ | +91.3 |
| $H_2O(g)$ | −241.8 |

$$
\begin{aligned}
\Delta H_{rxn}^\circ &= \Sigma n_p \Delta H_f^\circ (\text{products}) - \Sigma n_r \Delta H_f^\circ (\text{reactants}) \\
&= [4(\Delta H_{f,\, NO(g)}^\circ) + 6(\Delta H_{f,\, H_2O(g)}^\circ)] - [4(\Delta H_{f,\, NH_3(g)}^\circ) + 5(\Delta H_{f,\, O_2(g)}^\circ)] \\
&= [4(+91.3 \text{ kJ}) + 6(-241.8 \text{ kJ})] - [4(-45.9 \text{ kJ}) + 5(0.0 \text{ kJ})] \\
&= -1085.6 \text{ kJ} - (-183.6 \text{ kJ}) \\
&= -902.0 \text{ kJ}
\end{aligned}
$$

CHECK   The units of the answer (kJ) are correct. The answer is negative, which means that the reaction is exothermic.

FOR PRACTICE 6.11   The thermite reaction, in which powdered aluminum reacts with iron oxide, is highly exothermic.

$$2\,Al(s) + Fe_2O_3(s) \longrightarrow Al_2O_3(s) + 2\,Fe(s)$$

Use standard enthalpies of formation to determine $\Delta H^\circ_{rxn}$ for the thermite reaction.

▶ The reaction of powdered aluminum with iron oxide, known as the thermite reaction, releases a large amount of heat.

## Example 6.12   $\Delta H^\circ_{rxn}$ and Standard Enthalpies of Formation

A city of 100,000 people uses approximately $1.0 \times 10^{11}$ kJ of energy per day. Suppose all of that energy comes from the combustion of liquid octane ($C_8H_{18}$) to form gaseous water and gaseous carbon dioxide. Use standard enthalpies of formation to calculate $\Delta H^\circ_{rxn}$ for the combustion of octane and then determine the number of kilograms of octane necessary to provide this amount of energy.

| | |
|---|---|
| **SORT** You are given the amount of energy used and asked to find the mass of octane required to produce the energy. | **GIVEN:** $1.0 \times 10^{11}$ kJ <br><br> **FIND:** kg $C_8H_{18}$ |

**STRATEGIZE** The conceptual plan has three parts. In the first part, write a balanced equation for the combustion of octane.

**CONCEPTUAL PLAN**

(1) Write the balanced equation.

In the second part, calculate $\Delta H^\circ_{rxn}$ from the $\Delta H^\circ_f$'s of the reactants and products.

(2) $\boxed{\Delta H^\circ_f\text{'s}} \longrightarrow \boxed{\Delta H^\circ_{rxn}}$

$\Delta H^\circ_{rxn} = \Sigma\, n_p \Delta H^\circ_f \text{ (products)} - \Sigma\, n_r \Delta H^\circ_f \text{ (reactants)}$

In the third part, convert from kilojoules of energy to moles of octane using the conversion factor found in step 2, and then convert from moles of octane to mass of octane using the molar mass.

(3) $\boxed{kJ} \longrightarrow \boxed{mol\ C_8H_{18}} \longrightarrow \boxed{g\ C_8H_{18}} \longrightarrow \boxed{kg\ C_8H_{18}}$

       $\dfrac{114.22\ g\ C_8H_{18}}{mol\ C_8H_{18}}$    $\dfrac{1\ kg}{1000\ g}$

Conversion factor to be determined from steps 1 and 2

**RELATIONSHIPS USED**

molar mass $C_8H_{18}$ = 114.22 g/mol

1 kg = 1000 g

**SOLVE** Begin by writing the balanced equation for the combustion of octane. For convenience, do not clear the 25/2 fraction in order to keep the coefficient on octane as 1.

**SOLUTION STEP 1**

$$C_8H_{18}(l) + \frac{25}{2}O_2(g) \longrightarrow 8\,CO_2(g) + 9\,H_2O(g)$$

**SOLUTION STEP 2**

| Reactant or product | $\Delta H^\circ_f$ (kJ/mol, from Appendix IIB) |
|---|---|
| $C_8H_{18}(l)$ | −250.1 |
| $O_2(g)$ | 0.0 |
| $CO_2(g)$ | −393.5 |
| $H_2O(g)$ | −241.8 |

Look up (in Appendix IIB) the standard enthalpy of formation for each reactant and product and then calculate $\Delta H^\circ_{rxn}$.

*—Continued on the next page*

—*Continued*

$$\Delta H^\circ_{rxn} = \Sigma n_p \Delta H^\circ_f \text{ (products)} - \Sigma n_r \Delta H^\circ_f \text{ (reactants)}$$
$$= [8(\Delta H^\circ_{f,\,CO_2(g)}) + 9(\Delta H^\circ_{f,\,H_2O(g)})]$$
$$- \left[1(\Delta H^\circ_{f,\,C_8H_{18}(l)}) + \frac{25}{2}(\Delta H^\circ_{f,\,O_2(g)})\right]$$
$$= [8(-393.5 \text{ kJ}) + 9(-241.8 \text{ kJ})] - \left[1(-250.1 \text{ kJ}) + \frac{25}{2}(0.0 \text{ kJ})\right]$$
$$= -5324.2 \text{ kJ} - (-250.1 \text{ kJ})$$
$$= -5074.1 \text{ kJ}$$

| | |
|---|---|
| From steps 1 and 2 build a conversion factor between mol $C_8H_{18}$ and kJ. | **SOLUTION STEP 3**<br>1 mol $C_8H_{18}$ : $-5074.1$ kJ |
| Follow step 3 of the conceptual plan. Begin with $-1.0 \times 10^{11}$ kJ (since the city uses this much energy, the reaction must emit it, and therefore the sign is negative) and follow the steps to determine kilograms of octane. | $-1.0 \times 10^{11} \text{ kJ} \times \dfrac{1 \text{ mol } C_8H_{18}}{-5074.1 \text{ kJ}} \times \dfrac{114.22 \text{ g } C_8H_{18}}{1 \text{ mol } C_8H_{18}} \times \dfrac{1 \text{ kg}}{1000 \text{ g}}$ $= 2.3 \times 10^6 \text{ kg } C_8H_{18}$ |

**CHECK** The units of the answer (kg $C_8H_{18}$) are correct. The answer is positive, as it should be for mass. The magnitude is fairly large, as you would expect since this amount of octane is supposed to provide the energy for an entire city.

**FOR PRACTICE 6.12** The chemical hand warmers described in Section 6.1 produce heat when they are removed from their airtight plastic wrappers. Recall that they utilize the oxidation of iron to form iron oxide according to the reaction $4 \text{ Fe}(s) + 3 \text{ O}_2(g) \longrightarrow 2 \text{ Fe}_2\text{O}_3(s)$. Calculate $\Delta H^\circ_{rxn}$ for this reaction and calculate how much heat is produced from a hand warmer containing 15.0 g of iron powder.

## 6.10 Energy Use and the Environment

In this chapter, we explored the relationship between chemical reactions and energy changes. As noted earlier, our society derives the majority of its energy from the energy changes associated with burning fossil fuels. Fossil fuels have traditionally been regarded as convenient sources of energy due to their abundance and portability and because they undergo combustion reactions that have large negative enthalpies of reaction (the reactions are highly exothermic). However, the burning of fossil fuels also has some serious environmental impacts.

### Energy Consumption

According to the U.S. Department of Energy, the United States currently consumes close to 100 quads (1 quad = 1 quadrillion British thermal units $= 1.06 \times 10^{18}$ J) of energy per year in the categories shown in this chart.

This corresponds to over 100,000 kWh of energy use per person per year. If we used physical laborers to do the equivalent amount of work, each person would need about 120 people. In other words, the average American employs the work output of 120 people, day and night, all year long! For this reason, Americans enjoy one of the highest standards of living in the world. However, our consumption of energy has significant environmental consequences.

■ Residential: 21.0%
■ Commercial: 18.2%
▢ Industrial: 33.4%
■ Transportation: 27.4%

Source: U.S. Energy Information Administration. *Monthly Energy Review*, February, 2015.

Most U.S. energy comes from the combustion of fossil fuels, as shown in Figure 6.12▶. Fossil fuels include petroleum, natural gas, and coal, all of which have been considered convenient fuels because they are relatively abundant in Earth's crust (and therefore relatively inexpensive), they are easily transportable, and their combustion is highly exothermic.

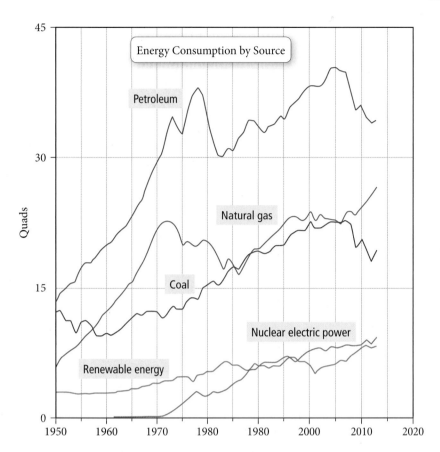

◀ **FIGURE 6.12** **Energy Consumption by Source**
Source: U.S. Energy Information Administration. *Monthly Energy Review*, February, 2015.

The reactions for the combustion of the main or representative components of several fossil fuels, and the associated enthalpies of reaction, are:

Coal: $C(s) + O_2(g) \longrightarrow CO_2(g)$    $\Delta H^\circ_{rxn} = -393.5 \text{ kJ}$

Natural gas: $CH_4(g) + 2\,O_2(g) \longrightarrow CO_2(g) + 2\,H_2O(g)$    $\Delta H^\circ_{rxn} = -802.3 \text{ kJ}$

Petroleum: $C_8H_{18}(l) + 25/2\,O_2(g) \longrightarrow 8\,CO_2(g) + 9\,H_2O(g)$    $\Delta H^\circ_{rxn} = -5074.1 \text{ kJ}$

## Environmental Problems Associated with Fossil Fuel Use

One of the main problems associated with the burning of fossil fuels is that even though they are abundant in Earth's crust, fossil fules are also finite. Fossil fuels originate from ancient plant and animal life and are a nonrenewable energy source—once they are all burned, they cannot be replenished. At current rates of consumption, oil and natural gas supplies will be depleted in 50 to 100 years. Although there is enough coal to last much longer, it is a dirtier fuel (it produces more pollution) and, because it is a solid, is less convenient (more difficult to transport and use) than petroleum and natural gas.

The other major problems associated with fossil fuel use stem from the products of combustion. The chemical equations we show in this section of the chapter for fossil fuel combustion all produce carbon dioxide and water. However, these equations represent the reactions under ideal conditions and do not account for impurities in the fuel, side reactions, and incomplete combustion. When these are taken into account, we can identify three major environmental problems associated with the emissions of fossil fuel combustion: air pollution, acid rain, and global climate change. We discussed acid rain in Chapter 3 (see *Chemistry in the Environment: Acid Rain* in Section 3.6). Here we address air pollution and global climate change, which we first touched on in Section 4.1.

## Air Pollution

The air in all major cities in the world is polluted. Pollution comes from a number of sources, including electricity generation, motor vehicle emissions, and industrial waste. There are many different kinds of air pollutants.

### Sulfur Oxides (SO$_x$)

Sulfur oxides include SO$_2$ and SO$_3$, produced primarily during coal-fired electricity generation and industrial metal refining. The sulfur oxides are lung and eye irritants that affect the respiratory system and are the main precursors of acid rain (see Section 3.6).

### Carbon Monoxide (CO)

Carbon monoxide forms during the incomplete combustion of fossil fuels (petroleum, natural gas, and coal). It is emitted mainly by motor vehicles. In humans and other animals, carbon monoxide displaces oxygen in the blood, forcing the heart and lungs to work harder. At high levels, CO can cause sensory impairment, decreased thinking ability, unconsciousness, and even death.

### Nitrogen Oxides (NO$_x$)

The nitrogen oxides include NO and NO$_2$, emitted by motor vehicles, fossil fuel–based electricity generation plants, and any high-temperature combustion process occurring in air. Nitrogen dioxide is an orange-brown gas that causes the dark haze over polluted cities. Nitrogen oxides are eye and lung irritants and precursors of acid rain.

### Ozone (O$_3$)

Ozone is produced when some of the products of fossil fuel combustion, especially nitrogen oxides and unburned volatile organic compounds (VOCs), react in the presence of sunlight. The products of this reaction, which include ozone, are called *photochemical smog*. Ozone produced in this way—sometimes called ground-level ozone—should not be confused with upper atmospheric or *stratospheric* ozone. Although ozone is always the same molecule (O$_3$), stratospheric ozone is a natural part of our environment that protects Earth from harmful ultraviolet light. Stratospheric ozone does not harm humans because we are not directly exposed to it. Ground-level ozone, on the other hand, is a pollutant to which we are directly exposed; it is an eye and lung irritant, and prolonged exposure can cause permanent lung damage.

In the United States, the U.S. Environmental Protection Agency (EPA) has established limits on these pollutants. Beginning in the 1970s, the U.S. Congress passed the Clean Air Act and its amendments, requiring U.S. cities to reduce their pollution and maintain levels below the standards set by the EPA. As a result of this legislation, pollutant levels in U.S. cities have decreased over the last 30 years, even as the number of vehicles has increased. According to the EPA, the levels of the major pollutants in the air of U.S. cities decreased significantly during the period 1980–2013, as shown in Table 6.6.

Although the levels of pollutants (especially ozone) in some cities are still above what the EPA considers safe, real progress has been made. These trends demonstrate that legislation can improve our environment.

**TABLE 6.6 Changes in National Average Pollutant Levels, 1980–2013**

| Pollutant | Change (%) in Average Level |
|---|---|
| SO$_2$ | −81 |
| CO | −84 |
| NO$_2$ | −60 |
| O$_3$ | −33 |

*Source: EPA's National Air Trends

**Atmospheric Carbon Dioxide**

▲ FIGURE 6.13 **The Rise in Atmospheric Carbon Dioxide** Atmospheric carbon dioxide levels have been steadily increasing as a result of fossil fuel combustion. Source: Dr. Pieter Tans, NOAA/ESRL (www.esrl.noaa.gov/gmd/ccgg/trends/) and Dr. Ralph Keeling, Scripps Institution of Oceanography (www.scrippsco2.ucsd.edu/)

## Global Climate Change

One of the main products of fossil fuel combustion is carbon dioxide. Carbon dioxide is a greenhouse gas; it allows visible light from the sun to enter Earth's atmosphere but prevents heat (in the form of infrared light) from escaping. In doing so, carbon dioxide acts as a blanket, keeping Earth warm, which, in moderation, is a very good thing and allows life as we know it to exist on our planet. However, because of fossil fuel combustion, carbon dioxide levels in the atmosphere have been steadily increasing, as shown in Figure 6.13◄. This increase is expected to change the global climate and further raise Earth's average temperature. Current observations suggest that Earth has already warmed by about 0.7 °C in the last century due to an approximately 38% increase in atmospheric carbon dioxide. Computer models suggest that the warming could worsen if carbon dioxide emissions are not curbed. The possible effects of this warming include heightened storm severity, increasing numbers of floods and droughts, major shifts in agricultural zones, rising sea levels and coastal flooding, and profound changes in habitats that could result in the extinction of some plant and animal species.

| **Example 6.13** | **Fossil Fuels and Climate Change** |
|---|---|

One way to evaluate fuels with respect to global warming is to determine how much heat they release during combustion relative to how much $CO_2$ they produce. The greater the heat relative to the amount of $CO_2$, the better the fuel. Use the combustion reactions of carbon, natural gas, and octane, in combination with the enthalpy of combustion for each reaction (all given earlier), to calculate the heat (in kJ) released by each fuel per 1.00 kg of $CO_2$ produced.

| | |
|---|---|
| **SORT** You are given the mass of $CO_2$ emitted and asked to find the energy output for three different fuels. | **GIVEN:** 1.00 kg $CO_2$ <br><br> **FIND:** kJ |
| **STRATEGIZE** First write the thermochemical equations for the combustion of each fuel given in Section 6.10. | $C(s) + O_2(g) \longrightarrow CO_2(g) \qquad \Delta H°_{rxn} = -393.5$ kJ <br> $CH_4(g) + 2\,O_2(g) \longrightarrow CO_2(g) + 2\,H_2O(g)$ <br> $\qquad\qquad\qquad\qquad\qquad\qquad \Delta H°_{rxn} = -802.3$ kJ <br> $C_8H_{18}(l) + 25/2\,O_2(g) \longrightarrow 8\,CO_2(g) + 9\,H_2O(g)$ <br> $\qquad\qquad\qquad\qquad\qquad\qquad \Delta H°_{rxn} = -5074.1$ kJ |
| The conceptual plan has two parts. In the first part, use the molar mass of $CO_2$ to convert from mass of $CO_2$ to moles of $CO_2$. This step is the same for each fuel. <br><br><br> In the second part, use the stoichiometric relationship between moles of $CO_2$ produced and kilojoules of energy released to calculate the energy output. Repeat the second part for each fuel using the appropriate stoichiometric relationship from the balanced equations. | **CONCEPTUAL PLAN** <br><br> kg $CO_2$ $\longrightarrow$ g $CO_2$ $\longrightarrow$ mol $CO_2$ <br> $\dfrac{1000\text{ g}}{1\text{ kg}}$ $\qquad$ $\dfrac{1\text{ mol}}{44.01\text{ g}}$ <br><br> mol $CO_2$ $\longrightarrow$ kJ <br> $\dfrac{\text{kJ}}{1\text{ mol fuel}}$ — This conversion factor, based on the following stoichiometric relationships, is different for each fuel. <br><br> **STOICHIOMETRIC RELATIONSHIPS** <br> For C: $\qquad$ 1 mol $CO_2$ : $-393.5$ kJ <br> For $CH_4$: $\qquad$ 1 mol $CO_2$ : $-802.3$ kJ <br> For $C_8H_{18}$: $\qquad$ 8 mol $CO_2$ : $-5074.1$ kJ <br><br> **OTHER RELATIONSHIPS USED** <br> 1 kg = 1000 g <br> molar mass $CO_2$ = 44.01 g/mol |
| **SOLVE** Begin by converting kg $CO_2$ to mol $CO_2$. <br><br><br><br> Then, for each fuel, convert mol $CO_2$ to kJ. <br><br><br><br><br><br><br><br><br><br> As you can see from the heat released in the production of 1 kg $CO_2$, natural gas, $CH_4$, provides the most energy per kg $CO_2$; therefore, it is the best fuel with respect to climate change. | **SOLUTION** <br><br> $1.00\text{ kg } CO_2 \times \dfrac{1000\text{ g}}{1\text{ kg}} \times \dfrac{1\text{ mol } CO_2}{44.01\text{ g } CO_2} = 22.72\text{ mol } CO_2$ <br><br> For C: <br> $22.72\text{ mol } CO_2 \times \dfrac{-393.5\text{ kJ}}{1\text{ mol } CO_2} = -8.94 \times 10^3\text{ kJ}$ <br><br> For $CH_4$: <br> $22.72\text{ mol } CO_2 \times \dfrac{-802.3\text{ kJ}}{1\text{ mol } CO_2} = -1.82 \times 10^4\text{ kJ}$ <br><br> For $C_8H_{18}$: <br> $22.72\text{ mol } CO_2 \times \dfrac{-5074.1\text{ kJ}}{8\text{ mol } CO_2} = -1.44 \times 10^4\text{ kJ}$ |

—Continued on the next page

*—Continued*

**CHECK** Each answer is in kJ, as it should be for heat produced. Each answer is negative, as expected for exothermic combustion reactions.

**FOR PRACTICE 6.13** What mass of $CO_2$ (in kilograms) does the combustion of a 15-gallon tank of gasoline release into the atmosphere? Assume the gasoline is pure octane ($C_8H_{18}$) and that it has a density of 0.70 g/mL.

---

# Chemistry IN THE Environment | Renewable Energy

**B**ecause of their finite supply and environmental impacts, fossil fuels will not be our major source of energy in the future. What will replace them? Although the answer is not clear, several alternative energy technologies are beginning to emerge. Unlike fossil fuels, these technologies are renewable, and we can use them indefinitely.

Our planet's greatest source of renewable energy is the sun. If we could capture and harness just a small fraction of the total sunlight falling on Earth, we could meet our energy needs several times over. The main problem with solar energy, however, is diffuseness—the sun's energy falls over an enormous area. How do we concentrate and store it? In California, some of the state's electricity is generated by parabolic troughs, solar power towers, and dish/engines.

These devices use reflective surfaces to focus the sun's energy and produce enough heat to generate electricity. Although the direct cost of generating electricity this way is higher than using fossil fuels, the benefits to the environment are obvious. In addition, with time, the costs are expected to fall.

Another way to capture the sun's energy is in chemical bonds. For example, solar energy could be used to drive the decomposition of water:

$$H_2O(l) \longrightarrow H_2(g) + \tfrac{1}{2} O_2(g) \quad \Delta H^\circ_{rxn} = +285.8 \text{ kJ}$$

The hydrogen gas produced could be stored until needed to provide energy by re-forming water in the reverse reaction:

$$H_2(g) + \tfrac{1}{2} O_2(g) \longrightarrow H_2O(l) \quad \Delta H^\circ_{rxn} = -285.8 \text{ kJ}$$

This reaction can be carried out in an electrochemical device called a fuel cell. In a fuel cell, hydrogen and oxygen gases combine to form water and produce electricity. In 2015, Toyota introduced the Mirai (which means *future* in Japanese), the first mass-produced fuel cell vehicle. The Mirai has a 300-mile range and produces 153 horsepower. Unlike most electric vehicles (such as the Tesla Model S), a fuel cell vehicle can be refueled instead of recharged.

The refueling process only takes a few minutes (compared to recharging, which can take several hours). California has earmarked $200 million to build a "hydrogen highway," a series of 100 hydrogen refueling stations to be operational by 2020.

▲ The Toyota Mirai, a fuel cell vehicle that runs on hydrogen gas and produces only water as exhaust.

Other renewable energy sources are hydroelectric power and wind power. Hydroelectric power plants—which generate approximately 7% of U.S. electricity—harness the gravitational potential energy of water held behind a dam. Water is released at a controlled rate. As the released water falls, it acquires kinetic energy that is used to spin turbines, generating electricity. Wind power plants—which produce nearly 5% of U.S. electricity—consist of hundreds of turbines that are spun by the wind to generate electricity. Both of these technologies are cost competitive with fossil fuels, have no emissions, and are completely renewable.

Our energy future will probably involve a combination of these technologies and some new ones, combined with a focus on greater efficiency and conservation. One thing, however, is clear—the future of fossil fuels is limited.

▲ Wind turbines such as these generate nearly 5% of U.S. electricity.

Parabolic troughs

Solar power tower

Dish/engine

▲ The sun's energy, concentrated by reflective surfaces in various arrangements, can produce enough heat to generate electricity.

# Self-Assessment Quiz

**Q1.** A chemical system produces 155 kJ of heat and does 22 kJ of work. What is $\Delta E$ for the *surroundings*?

a) 177 kJ  b) −177 kJ  c) 133 kJ  d) −133 kJ

**Q2.** Which sample is most likely to undergo the smallest change in temperature upon the absorption of 100 kJ of heat?

a) 15 g water  b) 15 g lead
c) 50 g water  d) 50 g lead

**Q3.** How much heat must be absorbed by a 15.0-g sample of water to raise its temperature from 25.0 °C to 55.0 °C? (For water, $C_s = 4.18 \, \text{J/g °C}$.)

a) 1.57 kJ  b) 1.88 kJ  c) 3.45 kJ  d) 107 J

**Q4.** A 12.5-g sample of granite initially at 82.0 °C is immersed into 25.0 g of water initially at 22.0 °C. What is the final temperature of both substances when they reach thermal equilibrium? (For water, $C_s = 4.18 \, \text{J/g °C}$ and for granite, $C_s = 0.790 \, \text{J/g °C}$.)

a) 52.0 °C  b) $1.55 \times 10^3$ °C
c) 15.7 °C  d) 27.2 °C

**Q5.** A cylinder with a moving piston expands from an initial volume of 0.250 L against an external pressure of 2.00 atm. The expansion does 288 J of work on the surroundings. What is the final volume of the cylinder?

a) 1.42 L  b) 1.17 L  c) 144 L  d) 1.67 L

**Q6.** When a 3.80-g sample of liquid octane ($C_8H_{18}$) is burned in a bomb calorimeter, the temperature of the calorimeter rises by 27.3 °C. The heat capacity of the calorimeter, measured in a separate experiment, is 6.18 kJ/°C. Determine the $\Delta E$ for octane combustion in units of kJ/mol octane.

a) $-5.07 \times 10^3$ kJ/mol
b) $5.07 \times 10^3$ kJ/mol
c) $-44.4 \times 10^3$ kJ/mol
d) $-16.7 \times 10^3$ kJ/mol

**Q7.** Hydrogen gas reacts with oxygen to form water:

$$2\,H_2(g) + O_2(g) \longrightarrow 2\,H_2O(g) \qquad \Delta H = -483.5 \text{ kJ}$$

Determine the minimum mass of hydrogen gas required to produce 226 kJ of heat.

a) 8.63 g  b) 1.88 g  c) 0.942 g  d) 0.935 g

**Q8.** Manganese reacts with hydrochloric acid to produce manganese(II) chloride and hydrogen gas:

$$Mn(s) + 2\,HCl(aq) \longrightarrow MnCl_2(aq) + H_2(g)$$

When 0.625 g Mn is combined with enough hydrochloric acid to make 100.0 mL of solution in a coffee-cup calorimeter, all of the Mn reacts, raising the temperature of the solution from 23.5 °C to 28.8 °C. Find $\Delta H_{rxn}$ for the reaction as written. (Assume that the specific heat capacity of the solution is 4.18 J/g °C and the density is 1.00 g/mL.)

a) −195 kJ  b) −3.54 kJ
c) −1.22 kJ  d) −2.21 kJ

**Q9.** Consider the reactions:

$$A \longrightarrow 2\,B \qquad \Delta H_1$$
$$A \longrightarrow 3\,C \qquad \Delta H_2$$

What is $\Delta H$ for the reaction $2\,B \longrightarrow 3\,C$?

a) $\Delta H_1 + \Delta H_2$  b) $\Delta H_1 - \Delta H_2$
c) $\Delta H_2 - \Delta H_1$  d) $2 \times (\Delta H_1 + \Delta H_2)$

**Q10.** Use standard enthalpies of formation to determine $\Delta H^\circ_{rxn}$ for the reaction:

$$Fe_2O_3(s) + 3\,CO(g) \longrightarrow 2\,Fe(s) + 3\,CO_2(g)$$

a) −541.2 kJ  b) −2336 kJ
c) 541.2 kJ  d) −24.8 kJ

**Q11.** Two substances, A and B, of equal mass but at different temperatures come into thermal contact. The specific heat capacity of substance A is twice the specific heat capacity of substance B. Which statement is true of the temperature of the two substances when they reach thermal equilibrium? (Assume no heat loss other than the thermal transfer between the substances.)

a) The final temperature of both substances is closer to the initial temperature of substance A than the initial temperature of substance B.
b) The final temperature of both substances is closer to the initial temperature of substance B than the initial temperature of substance A.
c) The final temperature of both substances is exactly midway between the initial temperatures of substance A and substance B.
d) The final temperature of substance B is greater than the final temperature of substance A.

**Q12.** Which process is endothermic?

a) the evaporation of water from the skin
b) the burning of candle wax
c) the oxidation of iron in a chemical hand warmer
d) the combustion of natural gas in a stove

**Q13.** Which fuel is not a fossil fuel?

a) coal  b) hydrogen
c) natural gas  d) petroleum

**Q14.** The standard enthalpy of formation for glucose [$C_6H_{12}O_6(s)$] is −1273.3 kJ/mol. What is the correct formation equation corresponding to this $\Delta H^\circ_f$?

a) $6\,C(s, \text{graphite}) + 6\,H_2O(g) \longrightarrow C_6H_{12}O_6(s, \text{glucose})$
b) $6\,C(s, \text{graphite}) + 6\,H_2O(l) \longrightarrow C_6H_{12}O_6(s, \text{glucose})$
c) $6\,C(s, \text{graphite}) + 6\,H_2(l) + 3\,O_2(l) \longrightarrow$
$C_6H_{12}O_6(s, \text{glucose})$
d) $6\,C(s, \text{graphite}) + 6\,H_2(g) + 3\,O_2(g) \longrightarrow$
$C_6H_{12}O_6(s, \text{glucose})$

**Q15.** Natural gas burns in air to form carbon dioxide and water, releasing heat.

$$CH_4(g) + O_2(g) \longrightarrow CO_2(g) + H_2O(g)$$
$$\Delta H^\circ_{rxn} = -802.3 \text{ kJ}$$

What minimum mass of $CH_4$ is required to heat 55 g of water by 25 °C? (Assume 100% heating efficiency.)

a) 0.115 g  b) $2.25 \times 10^3$ g
c) 115 g  d) 8.70 g

**Answers: 1.** (a) **2.** (c) **3.** (b) **4.** (d) **5.** (d) **6.** (a) **7.** (b) **8.** (a) **9.** (c) **10.** (d) **11.** (a) **12.** (a) **13.** (b) **14.** (d) **15.** (a)

# Chapter 6 IN REVIEW

## Key Terms

**Section 6.1**
thermochemistry (249)

**Section 6.2**
energy (250)
work (250)
heat (250)
kinetic energy (250)
thermal energy (250)
potential energy (250)
chemical energy (250)
law of conservation of energy (251)
system (251)
surroundings (251)
joule (J) (252)
calorie (cal) (252)
Calorie (Cal) (252)
kilowatt-hour (kWh) (252)

**Section 6.3**
thermodynamics (253)
first law of thermodynamics (253)
internal energy ($E$) (253)
state function (253)

**Section 6.4**
thermal equilibrium (258)
heat capacity ($C$) (259)
specific heat capacity ($C_s$) (259)
molar heat capacity (259)
pressure–volume work (262)

**Section 6.5**
calorimetry (265)
bomb calorimeter (265)

**Section 6.6**
enthalpy ($H$) (267)

endothermic reaction (267)
exothermic reaction (268)
enthalpy (heat) of reaction
($\Delta H_{rxn}$) (269)

**Section 6.7**
coffee-cup calorimeter (270)

**Section 6.8**
Hess's law (273)

**Section 6.9**
standard state (275)
standard enthalpy change
($\Delta H°$) (275)
standard enthalpy of formation
($\Delta H_f°$) (275)
standard heat of formation (275)

## Key Concepts

### The Nature of Energy and Thermodynamics (6.2, 6.3)

▶ Energy, which is measured in the SI unit of joules (J), is the capacity to do work.

▶ Work is the result of a force acting through a distance.

▶ Many different kinds of energy exist, including kinetic energy, thermal energy, potential energy, and chemical energy, a type of potential energy associated with the relative positions of electrons and nuclei in atoms and molecules.

▶ The first law of thermodynamics states that energy can be converted from one form to another, but the total amount of energy is always conserved.

▶ The internal energy ($E$) of a system is the sum of all of its kinetic and potential energy. Internal energy is a state function, which means that it depends only on the state of the system and not on the pathway by which it got to that state.

▶ A chemical system exchanges energy with its surroundings through heat (the transfer of thermal energy caused by a temperature difference) or work. The total change in internal energy is the sum of these two quantities.

### Heat and Work (6.4)

▶ We quantify heat with the equation $q = m \times C_s \times \Delta T$. In this expression, $C_s$ is the specific heat capacity, the amount of heat required to change the temperature of 1 g of the substance by 1 °C. Compared to most substances, water has a very high heat capacity—it takes a lot of heat to change its temperature.

▶ The type of work most characteristic of chemical reactions is pressure–volume work, which occurs when a gas expands against an external pressure. Pressure–volume work can be quantified with the equation $w = -P_{ext}\Delta V$.

▶ The change in internal energy ($\Delta E$) that occurs during a chemical reaction is the sum of the heat ($q$) exchanged and the work ($w$) done: $\Delta E = q + w$.

### Enthalpy (6.6)

▶ The heat evolved in a chemical reaction occurring at constant pressure is the change in enthalpy ($\Delta H$) for the reaction. Like internal energy, enthalpy is a state function.

▶ An endothermic reaction has a positive enthalpy of reaction; an exothermic reaction has a negative enthalpy of reaction.

▶ We can use the enthalpy of reaction to stoichiometrically determine the heat evolved when a specific amount of reactant reacts.

### Calorimetry (6.5, 6.7)

▶ Calorimetry is a method of measuring $\Delta E$ or $\Delta H$ for a reaction.

▶ In bomb calorimetry, the reaction is carried out under conditions of constant volume, so $\Delta E = q_v$. We can therefore use the temperature change of the calorimeter to calculate $\Delta E$ for the reaction.

▶ When a reaction takes place at constant pressure, energy may be released both as heat and as work.

▶ In coffee-cup calorimetry, a reaction is carried out under atmospheric pressure in a solution, so $q = \Delta H$. We use the temperature change of the solution to calculate $\Delta H$ for the reaction.

### Calculating $\Delta H_{rxn}$ (6.8, 6.9)

▶ We can calculate the enthalpy of reaction ($\Delta H_{rxn}$) from known thermochemical data using the following relationships: (a) when a reaction is multiplied by a factor, $\Delta H_{rxn}$ is multiplied by the same factor; (b) when a reaction is reversed, $\Delta H_{rxn}$ changes sign; and (c) if a chemical reaction can be expressed as a sum of two or more steps,

$\Delta H_{rxn}$ is the sum of the $\Delta H$'s for the individual steps (Hess's law). We can use these relationships to determine the enthalpy change of an unknown reaction from reactions with known enthalpy changes.

▶ A second method to calculate $\Delta H_{rxn}$ from known thermochemical data involves using tabulated standard enthalpies of formation for the reactants and products of the reaction. These are usually tabulated for substances in their standard states, and the enthalpy of reaction is called the standard enthalpy of reaction ($\Delta H^\circ_{rxn}$). For any reaction, we obtain $\Delta H^\circ_{rxn}$ by subtracting the sum of the enthalpies of formation of the reactants multiplied by their stoichiometric coefficients from the sum of the enthalpies of formation of the products multiplied by their stoichiometric coefficients.

## Environmental Problems Associated with Fossil Fuel Use (**6.10**)

▶ Fossil fuels are nonrenewable fuels; once humans consume them, they cannot be replaced.

▶ At current rates of consumption, natural gas and petroleum reserves will be depleted in 50–100 years.

▶ In addition to their limited supply, the products of the combustion of fossil fuels—directly or indirectly formed—contribute to environmental problems including air pollution, acid rain, and global climate change, which involves an increase in Earth's average temperature caused by $CO_2$ emission.

## Key Equations and Relationships

Kinetic Energy (**6.2**)

$$KE = \frac{1}{2}mv^2$$

Change in Internal Energy ($\Delta E$) of a Chemical System (**6.3**)

$$\Delta E = E_{products} - E_{reactants}$$

Energy Flow between System and Surroundings (**6.3**)

$$\Delta E_{system} = -\Delta E_{surroundings}$$

Relationship between Internal Energy ($\Delta E$), Heat ($q$), and Work ($w$) (**6.3**)

$$\Delta E = q + w$$

Relationship between Heat ($q$), Temperature ($T$), and Heat Capacity ($C$) (**6.4**)

$$q = C \times \Delta T$$

Relationship between Heat ($q$), Mass ($m$), Temperature ($T$), and Specific Heat Capacity of a Substance ($C_s$) (**6.4**)

$$q = m \times C_s \times \Delta T$$

Relationship between Work ($w$), Force ($F$), and Distance ($D$) (**6.4**)

$$w = F \times D$$

Relationship between Work ($w$), Pressure ($P$), and Change in Volume ($\Delta V$) (**6.4**)

$$w = -P_{ext} \Delta V$$

Change in Internal Energy ($\Delta E$) of System at Constant Volume (**6.5**)

$$\Delta E = q_v$$

Heat of a Bomb Calorimeter ($q_{cal}$) (**6.5**)

$$q_{cal} = C_{cal} \times \Delta T$$

Heat Exchange between a Calorimeter and a Reaction (**6.5**)

$$q_{cal} = -q_{rxn}$$

Relationship between Enthalpy ($\Delta H$), Internal Energy ($\Delta E$), Pressure ($P$), and Volume ($V$) (**6.6**)

$$\Delta H = \Delta E + P\Delta V$$

$$\Delta H = q_p$$

Relationship between Enthalpy of a Reaction ($\Delta H^\circ_{rxn}$) and the Heats of Formation ($\Delta H^\circ_f$) (**6.9**)

$$\Delta H^\circ_{rxn} = \sum n_p \Delta H^\circ_f \text{ (products)} - \sum n_r \Delta H^\circ_f \text{ (reactants)}$$

## Key Learning Outcomes

| CHAPTER OBJECTIVES | ASSESSMENT |
|---|---|
| Calculating Internal Energy from Heat and Work (**6.3**)  | Example 6.1  For Practice 6.1  Exercises 41–44 |
| Finding Heat from Temperature Changes (**6.4**) | Example 6.2  For Practice 6.2  For More Practice 6.2 Exercises 47–50 |
| Determining Quantities in Thermal Energy Transfer (**6.4**) | Example 6.3  For Practice 6.3  Exercises 65–70 |
| Finding Work from Volume Changes (**6.4**)  Combustion | Example 6.4  For Practice 6.4  For More Practice 6.4 Exercises 51, 52 |

| Using Bomb Calorimetry to Calculate $\Delta E_{rxn}$ (**6.5**) | Example 6.5    For Practice 6.5    For More Practice 6.5   Exercises 53, 54 |
|---|---|
| Predicting Endothermic and Exothermic Processes (**6.6**) | Example 6.6    For Practice 6.6    Exercises 57, 58 |
| Determining Heat from $\Delta H$ and Stoichiometry (**6.6**) | Examples 6.7, 6.13    For Practice 6.7, 6.13    For More Practice 6.7   Exercises 59–64 |
| Finding $\Delta H_{rxn}$ Using Calorimetry (**6.7**) | Example 6.8    For Practice 6.8    Exercises 75, 76 |
| Finding $\Delta H_{rxn}$ Using Hess's Law (**6.8**) | Example 6.9    For Practice 6.9    For More Practice 6.9   Exercises 79–82 |
| Finding $\Delta H_{rxn}^{\circ}$ Using Standard Enthalpies of Formation (**6.9**) | Examples 6.10, 6.11, 6.12    For Practice 6.10, 6.11, 6.12   Exercises 83–92 |

# EXERCISES

## Review Questions

1. What is thermochemistry? Why is it important?

2. What is energy? What is work? List some examples of each.

3. What is kinetic energy? What is potential energy? List some examples of each.

4. State the law of conservation of energy. How does it relate to energy exchanges between a thermodynamic system and its surroundings?

5. What is the SI unit of energy? List some other common units of energy.

6. State the first law of thermodynamics. What are its implications?

7. A friend claims to have constructed a machine that creates electricity but requires no energy input. Explain why you should be suspicious of your friend's claim.

8. What is a state function? List some examples of state functions.

9. What is internal energy? Is internal energy a state function?

10. If energy flows out of a chemical system and into the surroundings, what is the sign of $\Delta E_{system}$?

11. If the internal energy of the products of a reaction is higher than the internal energy of the reactants, what is the sign of $\Delta E$ for the reaction? In which direction does energy flow?

12. What is heat? Explain the difference between heat and temperature.

13. How is the change in internal energy of a system related to heat and work?

14. Explain how the sum of heat and work can be a state function, even though heat and work are themselves not state functions.

15. What is heat capacity? Explain the difference between heat capacity and specific heat capacity.

16. Explain how the high specific heat capacity of water can affect the weather in coastal regions.

17. If two objects, A and B, of different temperature come into direct contact, what is the relationship between the heat lost by one object and the heat gained by the other? What is the relationship between the temperature changes of the two objects? (Assume that the two objects do not lose any heat to anything else.)

18. What is pressure–volume work? How is it calculated?

19. What is calorimetry? Explain the difference between a coffee-cup calorimeter and a bomb calorimeter. What is each designed to measure?

20. What is the change in enthalpy ($\Delta H$) for a chemical reaction? How is $\Delta H$ different from $\Delta E$?

21. Explain the difference between an exothermic and an endothermic reaction. Give the sign of $\Delta H$ for each type of reaction.

22. From a molecular viewpoint, where does the energy emitted in an exothermic chemical reaction come from? Why does the reaction mixture undergo an increase in temperature even though energy is emitted?

23. From a molecular viewpoint, where does the energy absorbed in an endothermic chemical reaction go? Why does the reaction mixture undergo a decrease in temperature even though energy is absorbed?

24. Is the change in enthalpy for a reaction an extensive property? Explain the relationship between $\Delta H$ for a reaction and the amounts of reactants and products that undergo reaction.

25. Explain how the value of $\Delta H$ for a reaction changes upon each operation.
   a. multiplying the reaction by a factor
   b. reversing the reaction
   Why do these relationships hold?

26. What is Hess's law? Why is it useful?

27. What is a standard state? What is the standard enthalpy change for a reaction?

28. What is the standard enthalpy of formation for a compound? For a pure element in its standard state?

29. How do you calculate $\Delta H_{rxn}^{\circ}$ from tabulated standard enthalpies of formation?

30. What are the main sources of the energy consumed in the United States?

31. What are the main environmental problems associated with fossil fuel use?

32. Explain global climate change. What causes global warming? What is the evidence that global warming is occurring?

# Problems by Topic

## Energy Units

**33.** Convert between energy units.
a. 534 kWh to J
b. 215 kJ to Cal
c. 567 Cal to J
d. $2.85 \times 10^3$ J to cal

**34.** Convert between energy units.
a. 231 cal to kJ
b. $132 \times 10^4$ kJ to kcal
c. $4.99 \times 10^3$ kJ to kWh
d. $2.88 \times 10^4$ J to Cal

**35.** Suppose that a person eats 2387 Calories per day. Convert this amount of energy into each unit.
a. J
b. kJ
c. kWh

**36.** A particular frost-free refrigerator uses about 745 kWh of electrical energy per year. Express this amount of energy in each unit.
a. J
b. kJ
c. Cal

## Internal Energy, Heat, and Work

**37.** Which statement is true of the internal energy of a system and its surroundings during an energy exchange with a negative $\Delta E_{sys}$?
a. The internal energy of the system increases and the internal energy of the surroundings decreases.
b. The internal energy of both the system and the surroundings increases.
c. The internal energy of both the system and the surroundings decreases.
d. The internal energy of the system decreases and the internal energy of the surroundings increases.

**38.** During an energy exchange, a chemical system absorbs energy from its surroundings. What is the sign of $\Delta E_{sys}$ for this process? Explain.

**39.** Identify each energy exchange as primarily heat or work and determine whether the sign of $\Delta E$ is positive or negative for the system.
a. Sweat evaporates from skin, cooling the skin. (The evaporating sweat is the system.)
b. A balloon expands against an external pressure. (The contents of the balloon is the system.)
c. An aqueous chemical reaction mixture is warmed with an external flame. (The reaction mixture is the system.)

**40.** Identify each energy exchange as primarily heat or work and determine whether the sign of $\Delta E$ is positive or negative for the system.
a. A rolling billiard ball collides with another billiard ball. The first billiard ball (defined as the system) stops rolling after the collision.
b. A book falls to the floor. (The book is the system).
c. A father pushes his daughter on a swing. (The daughter and the swing are the system).

**41.** A system releases 622 kJ of heat and does 105 kJ of work on the surroundings. What is the change in internal energy of the system?

**42.** A system absorbs 196 kJ of heat and the surroundings do 117 kJ of work on the system. What is the change in internal energy of the system?

**43.** The gas in a piston (defined as the system) warms and absorbs 655 J of heat. The expansion performs 344 J of work on the surroundings. What is the change in internal energy for the system?

**44.** The air in an inflated balloon (defined as the system) warms over a toaster and absorbs 115 J of heat. As it expands, it does 77 kJ of work. What is the change in internal energy for the system?

## Heat, Heat Capacity, and Work

**45.** We pack two identical coolers for a picnic, placing 24 12-ounce soft drinks and five pounds of ice in each. However, the drinks that we put into cooler A were refrigerated for several hours before they were packed in the cooler, while the drinks that we put into cooler B were at room temperature. When we open the two coolers three hours later, most of the ice in cooler A is still present, while nearly all of the ice in cooler B has melted. Explain this difference.

**46.** A kilogram of aluminum metal and a kilogram of water are each warmed to 75 °C and placed in two identical insulated containers. One hour later, the two containers are opened and the temperature of each substance is measured. The aluminum has cooled to 35 °C, while the water has cooled only to 66 °C. Explain this difference.

**47.** How much heat is required to warm 1.50 L of water from 25.0 °C to 100.0 °C? (Assume a density of 1.0 g/mL for the water.)

**48.** How much heat is required to warm 1.50 kg of sand from 25.0 °C to 100.0 °C?

**49.** Suppose that 25 g of each substance is initially at 27.0 °C. What is the final temperature of each substance upon absorbing 2.35 kJ of heat?
a. gold
b. silver
c. aluminum
d. water

**50.** An unknown mass of each substance, initially at 23.0 °C, absorbs $1.95 \times 10^3$ J of heat. The final temperature is recorded. Find the mass of each substance.
a. Pyrex glass ($T_f = 55.4$ °C)
b. sand ($T_f = 62.1$ °C)
c. ethanol ($T_f = 44.2$ °C)
d. water ($T_f = 32.4$ °C)

**51.** How much work (in J) is required to expand the volume of a pump from 0.0 L to 2.5 L against an external pressure of 1.1 atm?

**52.** The average human lung expands by about 0.50 L during each breath. If this expansion occurs against an external pressure of 1.0 atm, how much work (in J) is done during the expansion?

**53.** The air within a piston equipped with a cylinder absorbs 565 J of heat and expands from an initial volume of 0.10 L to a final volume of 0.85 L against an external pressure of 1.0 atm. What is the change in internal energy of the air within the piston?

**54.** A gas is compressed from an initial volume of 5.55 L to a final volume of 1.22 L by an external pressure of 1.00 atm. During the compression the gas releases 124 J of heat. What is the change in internal energy of the gas?

## Enthalpy and Thermochemical Stoichiometry

**55.** When 1 mol of a fuel burns at constant pressure, it produces 3452 kJ of heat and does 11 kJ of work. What are $\Delta E$ and $\Delta H$ for the combustion of the fuel?

**56.** The change in internal energy for the combustion of 1.0 mol of octane at a pressure of 1.0 atm is 5084.3 kJ. If the change in enthalpy is 5074.1 kJ, how much work is done during the combustion?

**57.** Determine whether each process is exothermic or endothermic and indicate the sign of $\Delta H$.
a. natural gas burning on a stove
b. isopropyl alcohol evaporating from skin
c. water condensing from steam

**58.** Determine whether each process is exothermic or endothermic and indicate the sign of $\Delta H$.
a. dry ice evaporating
b. a sparkler burning
c. the reaction that occurs in a chemical cold pack used to ice athletic injuries

**59.** Consider the thermochemical equation for the combustion of acetone ($C_3H_6O$), the main ingredient in nail polish remover:

$$C_3H_6O(l) + 4\,O_2(g) \longrightarrow 3\,CO_2(g) + 3\,H_2O(g)$$
$$\Delta H_{rxn}^\circ = -1790 \text{ kJ}$$

If a bottle of nail polish remover contains 177 mL of acetone, how much heat is released by its complete combustion? The density of acetone is 0.788 g/mL.

**60.** What mass of natural gas ($CH_4$) must burn to emit 267 kJ of heat?

$$CH_4(g) + 2\,O_2(g) \longrightarrow CO_2(g) + 2\,H_2O(g) \quad \Delta H_{rxn}^\circ = -802.3 \text{ kJ}$$

**61.** Nitromethane ($CH_3NO_2$) burns in air to produce significant amounts of heat.

$$2\,CH_3NO_2(l) + \tfrac{3}{2}O_2(g) \longrightarrow 2\,CO_2(g) + 3\,H_2O(l) + N_2(g)$$
$$\Delta H_{rxn}^\circ = -1418 \text{ kJ}$$

How much heat is produced by the complete reaction of 5.56 kg of nitromethane?

**62.** Titanium reacts with iodine to form titanium(III) iodide, emitting heat.

$$2\,Ti(s) + 3\,I_2(g) \longrightarrow 2\,TiI_3(s) \quad \Delta H_{rxn}^\circ = -839 \text{ kJ}$$

Determine the masses of titanium and iodine that react if $1.55 \times 10^3$ kJ of heat is emitted by the reaction.

**63.** The propane fuel ($C_3H_8$) used in gas barbeques burns according to the thermochemical equation:

$$C_3H_8(g) + 5\,O_2(g) \longrightarrow 3\,CO_2(g) + 4\,H_2O(g)$$
$$\Delta H_{rxn}^\circ = -2044 \text{ kJ}$$

If a pork roast must absorb $1.6 \times 10^3$ kJ to fully cook, and if only 10% of the heat produced by the barbeque is actually absorbed by the roast, what mass of $CO_2$ is emitted into the atmosphere during the grilling of the pork roast?

**64.** Charcoal is primarily carbon. Determine the mass of $CO_2$ produced by burning enough carbon (in the form of charcoal) to produce $5.00 \times 10^2$ kJ of heat.

$$C(s) + O_2(g) \longrightarrow CO_2(g) \quad \Delta H_{rxn}^\circ = -393.5 \text{ kJ}$$

## Thermal Energy Transfer

**65.** A silver block, initially at 58.5 °C, is submerged into 100.0 g of water at 24.8 °C, in an insulated container. The final temperature of the mixture upon reaching thermal equilibrium is 26.2 °C. What is the mass of the silver block?

**66.** A 32.5-g iron rod, initially at 22.7 °C, is submerged into an unknown mass of water at 63.2 °C, in an insulated container. The final temperature of the mixture upon reaching thermal equilibrium is 59.5 °C. What is the mass of the water?

**67.** A 31.1-g wafer of pure gold, initially at 69.3 °C, is submerged into 64.2 g of water at 27.8 °C in an insulated container. What is the final temperature of both substances at thermal equilibrium?

**68.** A 2.85-g lead weight, initially at 10.3 °C, is submerged in 7.55 g of water at 52.3 °C in an insulated container. What is the final temperature of both substances at thermal equilibrium?

**69.** Two substances, A and B, initially at different temperatures, come into contact and reach thermal equilibrium. The mass of substance A is 6.15 g and its initial temperature is 20.5 °C. The mass of substance B is 25.2 g and its initial temperature is 52.7 °C. The final temperature of both substances at thermal equilibrium is 46.7 °C. If the specific heat capacity of substance B is 1.17 J/g·°C, what is the specific heat capacity of substance A?

**70.** A 2.74-g sample of a substance suspected of being pure gold is warmed to 72.1 °C and submerged into 15.2 g of water initially at 24.7 °C. The final temperature of the mixture is 26.3 °C. What is the heat capacity of the unknown substance? Could the substance be pure gold?

## Calorimetry

**71.** Exactly 1.5 g of a fuel burns under conditions of constant pressure and then again under conditions of constant volume. In measurement A the reaction produces 25.9 kJ of heat, and in measurement B the reaction produces 23.3 kJ of heat. Which measurement (A or B) corresponds to conditions of constant pressure? Which one corresponds to conditions of constant volume? Explain.

**72.** In order to obtain the largest possible amount of heat from a chemical reaction in which there is a large increase in the number of moles of gas, should you carry out the reaction under conditions of constant volume or constant pressure? Explain.

**73.** When 0.514 g of biphenyl ($C_{12}H_{10}$) undergoes combustion in a bomb calorimeter, the temperature rises from 25.8 °C to 29.4 °C. Find $\Delta E_{rxn}$ for the combustion of biphenyl in kJ/mol biphenyl. The heat capacity of the bomb calorimeter, determined in a separate experiment, is 5.86 kJ/°C.

**74.** Mothballs are composed primarily of the hydrocarbon naphthalene ($C_{10}H_8$). When 1.025 g of naphthalene burns in a bomb calorimeter, the temperature rises from 24.25 °C to 32.33 °C. Find $\Delta E_{rxn}$ for the combustion of naphthalene. The heat capacity of the bomb calorimeter, determined in a separate experiment, is 5.11 kJ/°C.

**75.** Zinc metal reacts with hydrochloric acid according to the balanced equation:

$$Zn(s) + 2\,HCl(aq) \longrightarrow ZnCl_2(aq) + H_2(g)$$

When 0.103 g of $Zn(s)$ is combined with enough HCl to make 50.0 mL of solution in a coffee-cup calorimeter, all of the zinc reacts, raising the temperature of the solution from 22.5 °C to 23.7 °C. Find $\Delta H_{rxn}$ for this reaction as written. (Use 1.0 g/mL for the density of the solution and 4.18 J/g·°C as the specific heat capacity.)

**76.** Instant cold packs used to ice athletic injuries on the field contain ammonium nitrate and water separated by a thin plastic divider. When the divider is broken, the ammonium nitrate dissolves according to the endothermic reaction:

$$NH_4NO_3(s) \longrightarrow NH_4^+(aq) + NO_3^-(aq)$$

In order to measure the enthalpy change for this reaction, 1.25 g of $NH_4NO_3$ is dissolved in enough water to make 25.0 mL of solution. The initial temperature is 25.8 °C and the final temperature (after the solid dissolves) is 21.9 °C. Calculate the change in enthalpy for the reaction in kJ. (Use 1.0 g/mL as the density of the solution and 4.18 J/g·°C as the specific heat capacity.)

## Quantitative Relationships Involving $\Delta H$ and Hess's Law

**77.** For each generic reaction, determine the value of $\Delta H_2$ in terms of $\Delta H_1$.

**a.** $A + B \longrightarrow 2\,C \qquad\qquad \Delta H_1$
   $2\,C \longrightarrow A + B \qquad\qquad \Delta H_2 = ?$

**b.** $A + \tfrac{1}{2}B \longrightarrow C \qquad\qquad \Delta H_1$
   $2\,A + B \longrightarrow 2\,C \qquad\qquad \Delta H_2 = ?$

**c.** $A \longrightarrow B + 2\,C \qquad\qquad \Delta H_1$
   $\tfrac{1}{2}B + C \longrightarrow \tfrac{1}{2}A \qquad\qquad \Delta H_2 = ?$

**78.** Consider the generic reaction:

$$A + 2\,B \longrightarrow C + 3\,D \qquad \Delta H = 155 \text{ kJ}$$

Determine the value of $\Delta H$ for each related reaction.

**a.** $3\,A + 6\,B \longrightarrow 3\,C + 9\,D$
**b.** $C + 3\,D \longrightarrow A + 2\,B$
**c.** $\tfrac{1}{2}C + \tfrac{3}{2}D \longrightarrow \tfrac{1}{2}A + B$

**79.** Calculate $\Delta H_{rxn}$ for the reaction:

$$Fe_2O_3(s) + 3\,CO(g) \longrightarrow 2\,Fe(s) + 3\,CO_2(g)$$

Use the following reactions and given $\Delta H$'s:

$2\,Fe(s) + \frac{3}{2}\,O_2(g) \longrightarrow Fe_2O_3(s) \quad \Delta H = -824.2\ kJ$

$CO(g) + \frac{1}{2}\,O_2(g) \longrightarrow CO_2(g) \quad \Delta H = -282.7\ kJ$

**80.** Calculate $\Delta H_{rxn}$ for the reaction:

$$CaO(s) + CO_2(g) \longrightarrow CaCO_3(s)$$

Use the following reactions and given $\Delta H$'s:

$Ca(s) + CO_2(g) + \frac{1}{2}\,O_2(g) \longrightarrow CaCO_3(s) \quad \Delta H = -812.8\ kJ$

$2\,Ca(s) + O_2(g) \longrightarrow 2\,CaO(s) \quad \Delta H = -1269.8\ kJ$

**81.** Calculate $\Delta H_{rxn}$ for the reaction:

$$5\,C(s) + 6\,H_2(g) \longrightarrow C_5H_{12}(l)$$

Use the following reactions and given $\Delta H$'s:

$C_5H_{12}(l) + 8\,O_2(g) \longrightarrow 5\,CO_2(g) + 6\,H_2O(g)$
$\quad\quad \Delta H = -3244.8\ kJ$

$C(s) + O_2(g) \longrightarrow CO_2(g) \quad\quad \Delta H = -393.5\ kJ$

$2\,H_2(g) + O_2(g) \longrightarrow 2\,H_2O(g) \quad\quad \Delta H = -483.5\ kJ$

**82.** Calculate $\Delta H_{rxn}$ for the reaction:

$$CH_4(g) + 4\,Cl_2(g) \longrightarrow CCl_4(g) + 4\,HCl(g)$$

Use the following reactions and given $\Delta H$'s:

$C(s) + 2\,H_2(g) \longrightarrow CH_4(g) \quad \Delta H = -74.6\ kJ$

$C(s) + 2\,Cl_2(g) \longrightarrow CCl_4(g) \quad \Delta H = -95.7\ kJ$

$H_2(g) + Cl_2(g) \longrightarrow 2\,HCl(g) \quad \Delta H = -92.3\ kJ$

### Enthalpies of Formation and $\Delta H$

**83.** Write an equation for the formation of each compound from its elements in their standard states, and find $\Delta H_f^\circ$ for each in Appendix IIB.
  **a.** $NH_3(g)$  **b.** $CO_2(g)$  **c.** $Fe_2O_3(s)$  **d.** $CH_4(g)$

**84.** Write an equation for the formation of each compound from its elements in their standard states, and find $\Delta H_{rxn}^\circ$ for each in Appendix IIB.
  **a.** $NO_2(g)$  **b.** $MgCO_3(s)$
  **c.** $C_2H_4(g)$  **d.** $CH_3OH(l)$

**85.** Hydrazine ($N_2H_4$) is a fuel used by some spacecraft. It is normally oxidized by $N_2O_4$ according to the equation:

$$N_2H_4(l) + N_2O_4(g) \longrightarrow 2\,N_2O(g) + 2\,H_2O(g)$$

Calculate $\Delta H_{rxn}^\circ$ for this reaction using standard enthalpies of formation.

**86.** Pentane ($C_5H_{12}$) is a component of gasoline that burns according to the following balanced equation:

$$C_5H_{12}(l) + 8\,O_2(g) \longrightarrow 5\,CO_2(g) + 6\,H_2O(g)$$

Calculate $\Delta H_{rxn}^\circ$ for this reaction using standard enthalpies of formation. (The standard enthalpy of formation of liquid pentane is $-146.8\ kJ/mol$.)

**87.** Use standard enthalpies of formation to calculate $\Delta H_{rxn}^\circ$ for each reaction.
  **a.** $C_2H_4(g) + H_2(g) \longrightarrow C_2H_6(g)$
  **b.** $CO(g) + H_2O(g) \longrightarrow H_2(g) + CO_2(g)$
  **c.** $3\,NO_2(g) + H_2O(l) \longrightarrow 2\,HNO_3(aq) + NO(g)$
  **d.** $Cr_2O_3(s) + 3\,CO(g) \longrightarrow 2\,Cr(s) + 3\,CO_2(g)$

**88.** Use standard enthalpies of formation to calculate $\Delta H_{rxn}^\circ$ for each reaction.
  **a.** $2\,H_2S(g) + 3\,O_2(g) \longrightarrow 2\,H_2O(l) + 2\,SO_2(g)$
  **b.** $SO_2(g) + \frac{1}{2}\,O_2(g) \longrightarrow SO_3(g)$
  **c.** $C(s) + H_2O(g) \longrightarrow CO(g) + H_2(g)$
  **d.** $N_2O_4(g) + 4\,H_2(g) \longrightarrow N_2(g) + 4\,H_2O(g)$

**89.** During photosynthesis, plants use energy from sunlight to form glucose ($C_6H_{12}O_6$) and oxygen from carbon dioxide and water. Write a balanced equation for photosynthesis and calculate $\Delta H_{rxn}^\circ$.

**90.** Ethanol ($C_2H_5OH$) can be made from the fermentation of crops and has been used as a fuel additive to gasoline. Write a balanced equation for the combustion of ethanol and calculate $\Delta H_{rxn}^\circ$.

**91.** Top fuel dragsters and funny cars burn nitromethane as fuel according to the balanced combustion equation:

$$2\,CH_3NO_2(l) + \frac{3}{2}\,O_2(g) \longrightarrow 2\,CO_2(g) + 3\,H_2O(l) + N_2(g)$$
$$\Delta H_{rxn}^\circ = -1418\ kJ$$

The enthalpy of combustion for nitromethane is $-709.2\ kJ/mol$. Calculate the standard enthalpy of formation ($\Delta H_f^\circ$) for nitromethane.

**92.** The explosive nitroglycerin ($C_3H_5N_3O_9$) decomposes rapidly upon ignition or sudden impact according to the balanced equation:

$$4\,C_3H_5N_3O_9(l) \longrightarrow 12\,CO_2(g) + 10\,H_2O(g) + 6\,N_2(g) + O_2(g)$$
$$\Delta H_{rxn}^\circ = -5678\ kJ$$

Calculate the standard enthalpy of formation ($\Delta H_f^\circ$) for nitroglycerin.

### Energy Use and the Environment

**93.** Determine the mass of $CO_2$ produced by burning enough of each fuel to produce $1.00 \times 10^2\ kJ$ of heat. Which fuel contributes least to global warming per kJ of heat produced?
  **a.** $CH_4(g) + 2\,O_2(g) \longrightarrow CO_2(g) + 2\,H_2O(g)$
  $$\Delta H_{rxn}^\circ = -802.3\ kJ$$
  **b.** $C_3H_8(g) + 5\,O_2(g) \longrightarrow 3\,CO_2(g) + 4\,H_2O(g)$
  $$\Delta H_{rxn}^\circ = -2043\ kJ$$
  **c.** $C_8H_{18}(l) + \frac{25}{2}\,O_2(g) \longrightarrow 8\,CO_2(g) + 9\,H_2O(g)$
  $$\Delta H_{rxn}^\circ = -5074.1\ kJ$$

**94.** Methanol ($CH_3OH$) has been suggested as a fuel to replace gasoline. Write a balanced equation for the combustion of methanol, find $\Delta H_{rxn}^\circ$, and determine the mass of carbon dioxide emitted per kJ of heat produced. Use the information from the previous exercise to calculate the same quantity for octane, $C_8H_{18}$. How does methanol compare to octane with respect to global warming?

**95.** The citizens of the world burn the fossil fuel equivalent of $7 \times 10^{12}$ kg of petroleum per year. Assume that all of this petroleum is in the form of octane ($C_8H_{18}$) and calculate how much $CO_2$ (in kg) the world produces from fossil fuel combustion per year. (Hint: Begin by writing a balanced equation for the combustion of octane.) If the atmosphere currently contains approximately $3 \times 10^{15}$ kg of $CO_2$, how long will it take for the world's fossil fuel combustion to double the amount of atmospheric carbon dioxide?

**96.** In a sunny location, sunlight has a power density of about $1\ kW/m^2$. Photovoltaic solar cells can convert this power into electricity with 15% efficiency. If a typical home uses 385 kWh of electricity per month, how many square meters of solar cells are required to meet its energy requirements? Assume that electricity can be generated from the sunlight for 8 hours per day.

▲ What area of solar cells do you need to power a home?

# Cumulative Problems

**97.** The kinetic energy of a rolling billiard ball is given by $KE = \frac{1}{2}mv^2$. Suppose a 0.17-kg billiard ball is rolling down a pool table with an initial speed of 4.5 m/s. As it travels, it loses some of its energy as heat. The ball slows down to 3.8 m/s and then collides head-on with a second billiard ball of equal mass. The first billiard ball completely stops and the second one rolls away with a velocity of 3.8 m/s. Assume the first billiard ball is the system. Calculate $w$, $q$, and $\Delta E$ for the process.

**98.** A 100-W lightbulb is placed in a cylinder equipped with a moveable piston. The lightbulb is turned on for 0.015 hour, and the assembly expands from an initial volume of 0.85 L to a final volume of 5.88 L against an external pressure of 1.0 atm. Use the wattage of the lightbulb and the time it is on to calculate $\Delta E$ in joules (assume that the cylinder and lightbulb assembly is the system and assume two significant figures). Calculate $w$ and $q$.

**99.** Evaporating sweat cools the body because evaporation is an endothermic process:

$$H_2O(l) \longrightarrow H_2O(g) \qquad \Delta H^\circ_{rxn} = +44.01 \text{ kJ}$$

Estimate the mass of water that must evaporate from the skin to cool the body by 0.50 °C. Assume a body mass of 95 kg and assume that the specific heat capacity of the body is 4.0 J/g·°C.

**100.** LP gas burns according to the exothermic reaction:

$$C_3H_8(g) + 5 O_2(g) \longrightarrow 3 CO_2(g) + 4 H_2O(g)$$
$$\Delta H^\circ_{rxn} = -2044 \text{ kJ}$$

What mass of LP gas is necessary to heat 1.5 L of water from room temperature (25.0 °C) to boiling (100.0 °C)? Assume that during heating, 15% of the heat emitted by the LP gas combustion goes to heat the water. The rest is lost as heat to the surroundings.

**101.** Use standard enthalpies of formation to calculate the standard change in enthalpy for the melting of ice. (The $\Delta H^\circ_f$ for $H_2O(s)$ is −291.8 kJ/mol.) Use this value to calculate the mass of ice required to cool 355 mL of a beverage from room temperature (25.0 °C) to 0.0 °C. Assume that the specific heat capacity and density of the beverage are the same as those of water.

**102.** Dry ice is solid carbon dioxide. Instead of melting, solid carbon dioxide sublimes according to the equation:

$$CO_2(s) \longrightarrow CO_2(g)$$

◄ When carbon dioxide sublimes, the gaseous $CO_2$ is cold enough to cause water vapor in the air to condense, forming fog.

When dry ice is added to warm water, heat from the water causes the dry ice to sublime more quickly. The evaporating carbon dioxide produces a dense fog often used to create special effects. In a simple dry ice fog machine, dry ice is added to warm water in a Styrofoam cooler. The dry ice produces fog until it evaporates away, or until the water gets too cold to sublime the dry ice quickly enough. Suppose that a small Styrofoam cooler holds 15.0 L of water heated to 85 °C. Use standard enthalpies of formation to calculate the change in enthalpy for dry ice sublimation, and calculate the mass of dry ice that should be added to the water so that the dry ice completely sublimes away when the water reaches 25 °C. Assume no heat loss to the surroundings. (The $\Delta H^\circ_f$ for $CO_2(s)$ is −427.4 kJ/mol.)

**103.** A 25.5-g aluminum block is warmed to 65.4 °C and plunged into an insulated beaker containing 55.2 g water initially at 22.2 °C. The aluminum and the water are allowed to come to thermal equilibrium. Assuming that no heat is lost, what is the final temperature of the water and aluminum?

**104.** If 50.0 mL of ethanol (density = 0.789 g/mL) initially at 7.0 °C is mixed with 50.0 mL of water (density = 1.0 g/mL) initially at 28.4 °C in an insulated beaker, and assuming that no heat is lost, what is the final temperature of the mixture?

**105.** Palmitic acid ($C_{16}H_{32}O_2$) is a dietary fat found in beef and butter. The caloric content of palmitic acid is typical of fats in general. Write a balanced equation for the complete combustion of palmitic acid and calculate the standard enthalpy of combustion. What is the caloric content of palmitic acid in Cal/g? Do the same calculation for table sugar (sucrose, $C_{12}H_{22}O_{11}$). Which dietary substance (sugar or fat) contains more Calories per gram? The standard enthalpy of formation of palmitic acid is −208 kJ/mol and that of sucrose is −2226.1 kJ/mol. [Use $H_2O(l)$ in the balanced chemical equations because the metabolism of these compounds produces liquid water.]

**106.** Hydrogen and methanol have both been proposed as alternatives to hydrocarbon fuels. Write balanced reactions for the complete combustion of hydrogen and methanol and use standard enthalpies of formation to calculate the amount of heat released per kilogram of the fuel. Which fuel contains the most energy in the least mass? How does the energy of these fuels compare to that of octane ($C_8H_{18}$)?

**107.** Derive a relationship between $\Delta H$ and $\Delta E$ for a process in which the temperature of a fixed amount of an ideal gas changes.

**108.** Under certain nonstandard conditions, oxidation by $O_2(g)$ of 1 mol of $SO_2(g)$ to $SO_3(g)$ absorbs 89.5 kJ. The enthalpy of formation of $SO_3(g)$ is −204.2 kJ under these conditions. Find the enthalpy of formation of $SO_2(g)$.

**109.** One tablespoon of peanut butter has a mass of 16 g. It is combusted in a calorimeter whose heat capacity is 120.0 kJ/°C. The temperature of the calorimeter rises from 22.2 °C to 25.4 °C. Find the food caloric content of peanut butter.

**110.** A mixture of 2.0 mol of $H_2(g)$ and 1.0 mol of $O_2(g)$ is placed in a sealed evacuated container made of a perfect insulating material at 25 °C. The mixture is ignited with a spark and reacts to form liquid water. Determine the temperature of the water.

**111.** A 20.0-L volume of an ideal gas in a cylinder with a piston is at a pressure of 3.0 atm. Enough weight is suddenly removed from the piston to lower the external pressure to 1.5 atm. The gas then expands at constant temperature until its pressure is 1.5 atm. Find $\Delta E$, $\Delta H$, $q$, and $w$ for this change in state.

**112.** When 10.00 g of phosphorus is burned in $O_2(g)$ to form $P_4O_{10}(s)$, enough heat is generated to raise the temperature of 2950 g of water from 18.0 °C to 38.0 °C. Calculate the enthalpy of formation of $P_4O_{10}(s)$ under these conditions.

**113.** The $\Delta H$ for the oxidation of sulfur in the gas phase to $SO_3$ is −204 kJ/mol and for the oxidation of $SO_2$ to $SO_3$ is 89.5 kJ/mol. Find the enthalpy of formation of $SO_2$ under these conditions.

**114.** The $\Delta H^\circ_f$ of $TiI_3(s)$ is −328 kJ/mol and the $\Delta H^\circ$ for the reaction $2 Ti(s) + 3 I_2(g) \longrightarrow 2 TiI_3(s)$ is −839 kJ. Calculate the $\Delta H$ of sublimation of $I_2(s)$, which is a solid at 25 °C.

**115.** A gaseous fuel mixture contains 25.3% methane ($CH_4$), 38.2% ethane ($C_2H_6$), and the rest propane ($C_3H_8$) by volume. When the fuel mixture contained in a 1.55 L tank, stored at 755 mmHg and 298 K, undergoes complete combustion, how much heat is emitted? (Assume that the water produced by the combustion is in the gaseous state.)

**116.** A gaseous fuel mixture stored at 745 mmHg and 298 K contains only methane ($CH_4$) and propane ($C_3H_8$). When 11.7 L of this fuel mixture is burned, it produces 769 kJ of heat. What is the mole fraction of methane in the mixture? (Assume that the water produced by the combustion is in the gaseous state.)

**117.** A copper cube with an edge measuring 1.55 cm and an aluminum cube with an edge measuring 1.62 cm are both heated to 55.0 °C and submerged in 100.0 mL of water at 22.2 °C. What is the final temperature of the water when equilibrium is reached? (Assume a density of 0.998 g/mL for water.)

**118.** A pure gold ring and a pure silver ring have a total mass of 14.9 g. The two rings are heated to 62.0 °C and dropped into 15.0 mL of water at 23.5 °C. When equilibrium is reached, the temperature of the water is 25.0 °C. What is the mass of each ring? (Assume a density of 0.998 g/mL for water.)

# Challenge Problems

**119.** A typical frostless refrigerator uses 655 kWh of energy per year in the form of electricity. Suppose that all of this electricity is generated at a power plant that burns coal containing 3.2% sulfur by mass and that all of the sulfur is emitted as $SO_2$ when the coal is burned. If all of the $SO_2$ goes on to react with rainwater to form $H_2SO_4$, what mass of $H_2SO_4$ does the annual operation of the refrigerator produce? (Hint: Assume that the remaining percentage of the coal is carbon and begin by calculating $\Delta H^\circ_{rxn}$ for the combustion of carbon.)

**120.** A large sport utility vehicle has a mass of $2.5 \times 10^3$ kg. Calculate the mass of $CO_2$ emitted into the atmosphere upon accelerating the SUV from 0.0 mph to 65.0 mph. Assume that the required energy comes from the combustion of octane with 30% efficiency. (Hint: Use KE = $\frac{1}{2}mv^2$ to calculate the kinetic energy required for the acceleration.)

**121.** Combustion of natural gas (primarily methane) occurs in most household heaters. The heat given off in this reaction is used to raise the temperature of the air in the house. Assuming that all the energy given off in the reaction goes to heating up only the air in the house, determine the mass of methane required to heat the air in a house by 10.0 °C. Assume that the house dimensions are 30.0 m × 30.0 m × 3.0 m, specific heat capacity of air is 30 J/K·mol, and 1.00 mol of air occupies 22.4 L for all temperatures concerned.

**122.** When backpacking in the wilderness, hikers often boil water to sterilize it for drinking. Suppose that you are planning a backpacking trip and will need to boil 35 L of water for your group. What volume of fuel should you bring? Assume that the fuel has an average formula of $C_7H_{16}$, 15% of the heat generated from combustion goes to heat the water (the rest is lost to the surroundings), the density of the fuel is 0.78 g/mL, the initial temperature of the water is 25.0 °C, and the standard enthalpy of formation of $C_7H_{16}$ is −224.4 kJ/mol.

**123.** An ice cube of mass 9.0 g is added to a cup of coffee. The coffee's initial temperature is 90.0 °C and the cup contains 120.0 g of liquid. Assume the specific heat capacity of the coffee is the same as that of water. The heat of fusion of ice (the heat associated with ice melting) is 6.0 kJ/mol. Find the temperature of the coffee after the ice melts.

**124.** Find $\Delta H$, $\Delta E$, $q$, and $w$ for the freezing of water at −10.0 °C. The specific heat capacity of ice is 2.04 J/g·°C and its heat of fusion (the quantity of heat associated with melting) is −332 J/g.

**125.** Starting from the relationship between temperature and kinetic energy for an ideal gas, find the value of the molar heat capacity of an ideal gas when its temperature is changed at constant volume. Find its molar heat capacity when its temperature is changed at constant pressure.

**126.** An amount of an ideal gas expands from 12.0 L to 24.0 L at a constant pressure of 1.0 atm. Then the gas is cooled at a constant volume of 24.0 L back to its original temperature. Then it contracts back to its original volume. Find the total heat flow for the entire process.

**127.** The heat of vaporization of water at 373 K is 40.7 kJ/mol. Find $q$, $w$, $\Delta E$, and $\Delta H$ for the evaporation of 454 g of water at this temperature at 1 atm.

**128.** Find $\Delta E$, $\Delta H$, $q$, and $w$ for the change in state of 1.0 mol $H_2O(l)$ at 80 °C to $H_2O(g)$ at 110 °C. The heat capacity of $H_2O(l)$ at = 75.3 J/mol K, heat capacity of $H_2O(g)$ = 25.0 J/mol K, and the heat of vaporization of $H_2O$ is 40.7 × $10^3$ J/mol at 100 °C.

**129.** The heat of combustion of liquid octane ($C_8H_{18}$) to carbon dioxide and liquid water at 298 K is −1303 kJ/mol. Find $\Delta E$ for this reaction.

**130.** Find $\Delta H$ for the combustion of ethanol ($C_2H_6O$) to carbon dioxide and liquid water from the following data. The heat capacity of the bomb calorimeter is 34.65 kJ/K and the combustion of 1.765 g of ethanol raises the temperature of the calorimeter from 294.33 K to 295.84 K.

# Conceptual Problems

**131.** Which statement is true of the internal energy of the system and its surroundings following a process in which $\Delta E_{sys}$ = +65 kJ? Explain.
   **a.** The system and the surroundings both lose 65 kJ of energy.
   **b.** The system and the surroundings both gain 65 kJ of energy.
   **c.** The system loses 65 kJ of energy and the surroundings gain 65 kJ of energy.
   **d.** The system gains 65 kJ of energy and the surroundings lose 65 kJ of energy.

**132.** The internal energy of an ideal gas depends only on its temperature. Which statement is true of an isothermal (constant-temperature) expansion of an ideal gas against a constant external pressure? Explain.
   **a.** $\Delta E$ is positive       **b.** $w$ is positive
   **c.** $q$ is positive             **d.** $\Delta E$ is negative

**133.** Which expression describes the heat evolved in a chemical reaction when the reaction is carried out at constant pressure? Explain.
   **a.** $\Delta E - w$    **b.** $\Delta E$    **c.** $\Delta E - q$

**134.** Two identical new refrigerators are plugged in for the first time. Refrigerator A is empty (except for air) and refrigerator B is filled with jugs of water. The compressors of both refrigerators immediately turn on and begin cooling the interiors of the refrigerators. After two hours, the compressor of refrigerator A turns off while the compressor of refrigerator B continues to run. The next day, the compressor of refrigerator A can be heard turning on and off every few minutes, while the compressor of refrigerator B turns off and on every hour or so (and stays on longer each time). Explain these observations.

135. A 1-kg cylinder of aluminum and 1-kg jug of water, both at room temperature, are put into a refrigerator. After 1 hour, the temperature of each object is measured. One of the objects is much cooler than the other. Which one is cooler and why?

136. Two substances A and B, initially at different temperatures, are thermally isolated from their surroundings and allowed to come into thermal contact. The mass of substance A is twice the mass of substance B, but the specific heat capacity of substance B is four times the specific heat capacity of substance A. Which substance will undergo a larger change in temperature?

137. When 1 mol of a gas burns at constant pressure, it produces 2418 J of heat and does 5 J of work. Determine $\Delta E$, $\Delta H$, $q$, and $w$ for the process.

138. In an exothermic reaction, the reactants lose energy and the reaction feels hot to the touch. Explain why the reaction feels hot even though the reactants are losing energy. Where does the energy come from?

139. Which statement is true of a reaction in which $\Delta V$ is positive? Explain.
   a. $\Delta H = \Delta E$   b. $\Delta H > \Delta E$   c. $\Delta H < \Delta E$

---

*Active Classroom Learning*

## Questions for Group Work

Discuss these questions with the group and record your consensus answer.

140. Have each group member write a problem involving the transfer of heat from one material in Table 6.4 to another material in the table. Working as a group, solve each problem. The group member who wrote each problem may act as the group facilitator when the group is working on his or her problem. What do all of your problems have in common? How do they differ?

141. Classify each process as endothermic or exothermic. What is the sign of $\Delta H$ for each process? Explain your answers.
   a. gasoline burning in an engine
   b. steam condensing on a mirror
   c. water boiling in a pot

   Provide at least two additional examples of exothermic processes and two additional examples of endothermic processes. Have each member of your group provide an example.

142. A propane tank on a home barbeque contains $10.4 \times 10^3$ g of propane.
   a. Write the balanced chemical reaction for the combustion of gaseous propane ($C_3H_8$) to form water vapor and gaseous carbon dioxide.

   b. Use the value for $\Delta H_{rxn}$ provided in the text to calculate the total amount of heat produced when the entire contents of the tank of propane is burned.
   c. What mass of water could be warmed from 25 °C to 100 °C with this much heat?

143. Consider the decomposition of liquid hydrogen peroxide ($H_2O_2$) to form water and oxygen.
   a. What is the heat of formation for hydrogen peroxide?
   b. What is the heat of formation for liquid water?
   c. What is the heat of formation for gaseous oxygen? Why?
   d. Write the balanced chemical equations that correspond to the $\Delta H$ values you looked up for parts a, b, and c.
   e. Write the balanced chemical equation for the decomposition of hydrogen peroxide to form water and oxygen. (Write the equation such that the coefficient on oxygen is 1.)
   f. What is the heat of reaction for the process in part e?
   g. Draw a scale diagram of this reaction (1 cm = 100 kJ) that shows the relative energies of reactants (on the left), products (on the right), and the elements in their most stable states (in the middle). Label all the energies you know.

## Data Interpretation and Analysis

### Heat Content of Fuels

144. To compare the energy content of fuels, we can use the quantities known as *higher heating value* (*HHV*) and *lower heating value* (*LHV*). The HHV assumes that the reaction product water is formed in the liquid state. The HHV is the same as the change in enthalpy of combustion, $\Delta H_{comb}$. The LHV assumes that water is formed in the gas state. The LHV is therefore the sum of the HHV (which is negative) and the heat of vaporization of water (which is positive).

   The structures shown here represent molecules of $C_1$, $C_2$, and $C_3$ straight-chain alkanes and straight-chain alcohols.

Methane    Ethane    Propane

Methanol    Ethanol    1-Propanol

Figure a▼ lists the changes in enthalpy when straight-chain alkanes and straight-chain alcohols undergo combustion to form gaseous carbon dioxide and liquid water.

| Alkane | $\Delta H_{comb}$(kJ/mol) | Alcohol | $\Delta H_{comb}$(kJ/mol) |
|---|---|---|---|
| $CH_4(g)$ | −890 | $CH_3OH(l)$ | −726 |
| $C_2H_6(g)$ | −1560 | $CH_3CH_2OH(l)$ | −1367 |
| $C_3H_8(g)$ | −2219 | $CH_3(CH_2)_2OH(l)$ | −2021 |
| $C_4H_{10}(g)$ | −2877 | $CH_3(CH_2)_3OH(l)$ | −2676 |
| $C_5H_{12}(l)$ | −3509 | $CH_3(CH_2)_4OH(l)$ | −3329 |
| $C_6H_{14}(l)$ | −4163 | $CH_3(CH_2)_5OH(l)$ | −3984 |
| $C_7H_{16}(l)$ | −4817 | $CH_3(CH_2)_6OH(l)$ | −4638 |
| $C_8H_{18}(l)$ | −5470 | $CH_3(CH_2)_7OH(l)$ | −5294 |

▲ **FIGURE a** **Change in Enthalpy Values for Combustion of Straight-Chain Alkanes and Straight-Chain Alcohols**

Figure b▼ plots $\Delta H_{comb}$ versus the number of carbon atoms for alkanes and alcohols. Note that there is an almost linear relationship between $\Delta H_{comb}$ and the number of carbon atoms.

◄ **FIGURE b** **Enthalpies of Combustion of Linear Alkanes and Alcohols**
Source: http://www.docbrown.info/page07/delta1Hd.htm

Use the information provided in the figures to answer the following questions:

a. Complete and balance the combustion equations using whole-number coefficients for carbon dioxide and $H_2O$, assuming liquid water forms:

   i. $CH_3CH_2CH_3(g) + O_2(g) \longrightarrow$ _____ + _____

   ii. $CH_3CH_2CH_2OH(l) + O_2(g) \longrightarrow$ _____ + _____

b. Given $\Delta H_{vap}(H_2O)$ is +44 kJ/mol, if the reactions in part a were written to produce $H_2O(g)$ instead of $H_2O(l)$, would you expect the magnitude (or absolute value) of $\Delta H_{comb}$ to increase, decrease, or stay the same? Explain.

c. Calculate $\Delta H$ for the combustion of propane gas, $C_3H_8$, assuming $H_2O(g)$ is produced instead of $H_2O(l)$. Compare your value to the value of $\Delta H$ for propane in Figure a.

d. When one mole of propane gas (used to cook food on a outdoor grill) undergoes complete combustion at constant pressure, is the amount of heat released the HHV or the LHV? Explain.

e. For each –$CH_2$– unit added to the alkanes, what is the change in enthalpy of combustion, $\Delta H_{comb}$?

f. Given $\Delta H_{vap}$ for pentane ($C_5H_{12}$) is +36 kJ/mol, explain why the change in enthalpy of combustion, $\Delta H_{comb}$ between gaseous methane and gaseous ethane is larger than the change in enthalpy of combustion, $\Delta H_{comb}$ between gaseous butane, $C_4H_{10}$, and liquid pentane.

# Answers to Conceptual Connections

### System and Surroundings

**6.1** The correct answer is **(a)**. When $\Delta E_{sys}$ is negative, energy flows out of the system and into the surroundings. The energy increase in the surroundings must exactly match the decrease in the system.

### Heat and Work

**6.2** **(a)** heat, sign is positive **(b)** work, sign is positive **(c)** heat, sign is negative

### The Heat Capacity of Water

**6.3** Bring the water into your sleeping bag; it has the higher heat capacity and will therefore release more heat as it cools.

### Thermal Energy Transfer

**6.4** **(c)** The specific heat capacity of substance B is twice that of A, but since the mass of B is half that of A, the quantity $m \times C_s$ is identical for both substances so that the final temperature is exactly midway between the two initial temperatures.

### The Difference between $\Delta H$ and $\Delta E$

**6.5** $\Delta H$ represents only the heat exchanged; therefore, $\Delta H = -2658$ kJ. $\Delta E$ represents the heat *and* work exchanged; therefore, $\Delta E = -2661$ kJ. The signs of both $\Delta H$ and $\Delta E$ are negative because heat and work are flowing out of the system and into the surroundings. Notice that the values of $\Delta H$ and $\Delta E$ are similar in magnitude, as is the case in many chemical reactions.

### Exothermic and Endothermic Reactions

**6.6** An endothermic reaction feels cold to the touch because the reaction (acting here as the system) absorbs heat from the surroundings. When you touch the vessel in which the reaction occurs, you, being part of the surroundings, lose heat to the system (the reaction), which makes you feel cold. The heat absorbed by the reaction (from your body, in this case) does not increase its temperature, but rather becomes potential energy stored in chemical bonds.

### Constant-Pressure versus Constant-Volume Calorimetry

**6.7** The value of $q_{rxn}$ with the greater magnitude (−12.5 kJ) must have come from the bomb calorimeter. Recall that $\Delta E_{rxn} = q_{rxn} + w_{rxn}$. In a bomb calorimeter, the energy change that occurs in the course of the reaction all takes the form of heat ($q$). In a coffee-cup calorimeter, the amount of energy released as heat may be smaller because some of the energy may be used to do work ($w$).

# 7

# The Quantum-Mechanical Model of the Atom

Anyone who is not shocked by quantum mechanics has not understood it.

—NIELS BOHR (1885–1962)

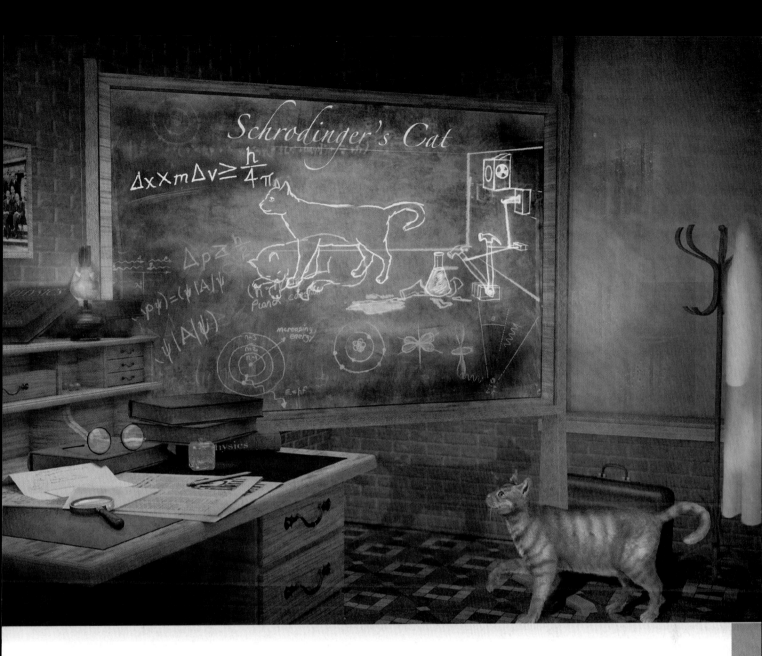

$$\Delta x \times m \Delta v \geq \frac{h}{4\pi}$$

Schrödinger's Cat

The early part of the twentieth century brought changes that revolutionized how we think about physical reality, especially in the atomic realm. Before then, all descriptions of the behavior of matter had been deterministic—the present conditions completely determining the future. The new theory of quantum mechanics suggested that for subatomic particles—electrons, neutrons, and protons—the present does NOT completely determine the future. For example, if you shoot one electron down a path and measure where it lands, a second electron shot down the same path under the same conditions will most likely land in a different place! Several unusually gifted scientists, including Albert Einstein, Niels Bohr, Louis de Broglie, Max Planck, Werner Heisenberg, P. A. M. Dirac, and Erwin Schrödinger, developed quantum-mechanical theory. These scientists did not necessarily feel comfortable with their own theory. Bohr said, "Anyone who is not shocked by quantum mechanics has not understood it." Schrödinger wrote, "I don't like it, and I'm sorry I ever had anything to do with it." Albert Einstein disbelieved the very theory he helped develop, stating, "God does not play dice with the universe." In fact, Einstein attempted to disprove quantum mechanics—without success—until he died. Today, quantum mechanics forms the foundation of chemistry—explaining the periodic table and the behavior of the elements in chemical bonding—as well as providing the practical basis for lasers, computers, and countless other applications.

▲ *The thought experiment known as Schrödinger's cat is intended to show that the strangeness of the quantum world does not transfer to the macroscopic world.*

## 7.1 Schrödinger's Cat

Atoms and the particles that compose them are unimaginably small. Electrons have a mass of less than a trillionth of a trillionth of a gram, a size so small that it is immeasurable. A single speck of dust contains more electrons than the number of people who have existed on Earth over all the centuries of time. Electrons are *small* in the absolute sense of the word—they are among the smallest particles that make up matter. And yet, as we have seen in the preceding chapters, an atom's electrons determine many of its chemical and physical properties. If we are to understand these properties, we must try to understand electrons.

In the early twentieth century, scientists discovered that the *absolutely small* (or *quantum*) world of the electron behaves differently than the *large* (or *macroscopic*) world that we are used to observing. Chief among these differences is the idea that, when unobserved, *absolutely small particles like electrons can be in two different states at the same time*. For example, through a process called radioactive decay (see Chapter 20) an atom can emit small (that is, *absolutely* small) energetic particles from its nucleus. In the macroscopic world, something either emits an energetic particle or it doesn't. In the quantum world, however, the unobserved atom can be in a state in which it is doing both—emitting the particle and not emitting the particle—simultaneously. At first, this seems absurd. The absurdity resolves itself, however, upon observation. When we set out to measure the emitted particle, the act of measurement actually forces the atom into one state or the other.

Early twentieth-century physicists struggled with this idea. Austrian physicist Erwin Schrödinger (1887–1961), in an attempt to demonstrate that this quantum strangeness could never transfer itself to the macroscopic world, published a paper in 1935 that contained a thought experiment about a cat, now known as Schrödinger's cat. In the thought experiment, the imaginary cat is put into a steel chamber that contains radioactive atoms such as the atom described in the previous paragraph. The chamber is equipped with a mechanism that, upon the emission of an energetic particle by one of the radioactive atoms, causes a hammer to break a flask of hydrocyanic acid, a poison. If the flask breaks, the poison is released and the cat dies.

Now here comes the absurdity: if the steel chamber is closed, the whole system remains unobserved, and the radioactive atom is in a state in which it has both emitted the particle and not emitted the particle (with equal probability). Therefore, the cat is both dead and undead. Schrödinger put it this way: "[the steel chamber would have] *in it the living and dead cat (pardon the expression) mixed or smeared out in equal parts*." When the chamber is opened, the act of observation forces the entire system into one state or the other: the cat is either dead or alive, not both. However, while unobserved, the cat is both dead and alive. The absurdity of the both dead and undead cat in Schrödinger's thought experiment was meant to demonstrate how quantum strangeness does not transfer to the macroscopic world.

In this chapter, we examine the **quantum-mechanical model** of the atom, a model that explains the strange behavior of electrons. In particular, we focus on how the model describes electrons as they exist within atoms, and how those electrons determine the chemical and physical properties of elements. You have already learned much about those properties. You know, for example, that some elements are metals and that others are nonmetals. You know that the noble gases are chemically inert and that the alkali metals are chemically reactive. You know that sodium tends to form 1+ ions and that fluorine tends to form 1− ions. But we have not explored *why*. The quantum-mechanical model explains why. In doing so, it explains the modern periodic table and provides the basis for our understanding of chemical bonding.

KCV **The Nature of Light**

## 7.2    **The Nature of Light**

Before we explore electrons and their behavior within the atom, we must understand some of the properties of light. As quantum mechanics developed, light was (surprisingly) found to have many characteristics in common with electrons. Chief among these is the *wave–particle duality* of light. Certain properties of light are best described by thinking of it as a wave, while other properties are best described by thinking of it as a particle. In this section, we first explore the wave behavior of light and its particle behavior. We then turn to electrons to see how they display the same wave–particle duality.

## The Wave Nature of Light

Light is **electromagnetic radiation**, a type of energy embodied in oscillating electric and magnetic fields. A *magnetic field* is a region of space where a magnetic particle experiences a force (think of the space around a magnet). An *electric field* is a region of space where an electrically charged particle experiences a force. A proton, for example, has an electric field around it. If you bring another charged particle into that field, that particle will experience a force.

We can describe electromagnetic radiation as a wave composed of oscillating, mutually perpendicular electric and magnetic fields propagating through space, as shown in Figure 7.1▼. In a vacuum, these waves move at a constant speed of $3.00 \times 10^8$ m/s (186,000 mi/s)—fast enough to circle Earth in one-seventh of a second. This great speed is the reason for the delay between the moment when you see a firework in the sky and the moment when you hear the sound of its explosion. The light from the exploding firework reaches your eye almost instantaneously. The sound, traveling much more slowly (340 m/s), takes longer. The same thing happens in a thunderstorm—you see the flash of lightning immediately, but the sound of thunder takes a few seconds to reach you. (The sound of thunder is delayed by five seconds for each mile between you and its origin.)

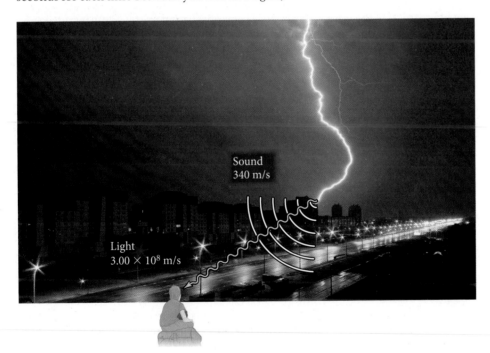

◀ Because light travels nearly a million times faster than sound, the flash of lightning reaches your eyes before the roll of thunder reaches your ears.

### Electromagnetic Radiation

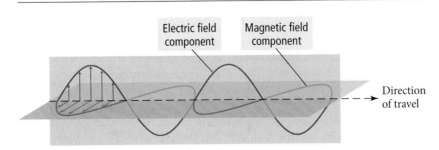

◀ **FIGURE 7.1 Electromagnetic Radiation** We describe electromagnetic radiation as a wave composed of oscillating electric and magnetic fields. The fields oscillate in perpendicular planes.

▶ **FIGURE 7.2 Wavelength and Amplitude** Wavelength and amplitude are independent properties. The wavelength of light determines its color. The amplitude, or intensity, determines its brightness.

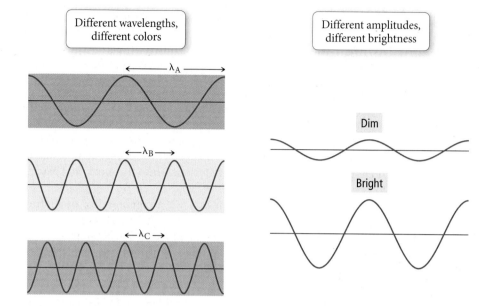

The symbol $\lambda$ is the Greek letter lambda, pronounced "lamb-duh."

We characterize a wave by its *amplitude* and its *wavelength*. In the graphical representation shown here, the **amplitude** of the wave is the vertical height of a crest (or depth of a trough). The amplitude of the electric and magnetic field waves in light determines the light's *intensity* or brightness—the greater the amplitude, the greater the intensity. The **wavelength** ($\lambda$) of the wave is the distance between adjacent crests (or any two analogous points) and is measured in units such as meters, micrometers, or nanometers. Amplitude and wavelength can vary independently of one another, as shown in Figure 7.2▲.

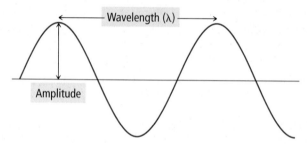

The symbol $\nu$ is the Greek letter nu, pronounced "noo."

Like all waves, light is also characterized by its **frequency ($\nu$)**, the number of cycles (or wave crests) that pass through a stationary point in a given period of time. The units of frequency are cycles per second (cycle/s) or simply $s^{-1}$. An equivalent unit of frequency is the hertz (Hz), defined as 1 cycle/s. The frequency of a wave is directly proportional to the speed at which the wave is traveling—the faster the wave, the greater the number of crests that pass a fixed location per unit time. Frequency is also *inversely* proportional to the wavelength ($\lambda$)—the farther apart the crests, the lower the number of crests that pass a fixed location per unit time. For light, therefore, we can write the equation:

$$\nu = \frac{c}{\lambda} \qquad [7.1]$$

where the speed of light, $c$, and the wavelength, $\lambda$, are both expressed in the same unit of distance. Wavelength and frequency represent different ways of specifying the same information—if we know one, we can readily calculate the other.

For *visible light*—light that can be seen by the human eye—wavelength (or, alternatively, frequency) determines color. White light, produced by the sun or by a lightbulb, contains a spectrum of wavelengths and therefore a spectrum of colors. We see these colors—red, orange, yellow, green, blue, indigo, and violet—in a rainbow or when white light passes through a prism (Figure 7.3▶). Red light, with a wavelength of about 750 nanometers (nm), has the longest wavelength of visible light; violet light, with a wavelength of about 400 nm, has the shortest. The presence of a variety of wavelengths in white light is responsible for the way we perceive colors in objects. When a substance

nano = $10^{-9}$

▲ **FIGURE 7.3 Components of White Light** We can pass white light through a prism and decompose it into its constituent colors, each with a different wavelength. The array of colors makes up the spectrum of visible light.

▲ **FIGURE 7.4 The Color of an Object** A red shirt is red because it reflects predominantly red light while absorbing most other colors.

absorbs some colors while reflecting others, it appears colored. For example, a red shirt appears red because it reflects predominantly red light while absorbing most other colors (Figure 7.4▲). Our eyes see only the reflected light, making the shirt appear red.

---

**Example 7.1    Wavelength and Frequency**

Calculate the wavelength (in nm) of the red light emitted by a barcode scanner that has a frequency of $4.62 \times 10^{14}\,s^{-1}$.

**SOLUTION**

You are given the frequency of the light and asked to find its wavelength. Use Equation 7.1, which relates frequency to wavelength. You can convert the wavelength from meters to nanometers by using the conversion factor between the two (1 nm = $10^{-9}$ m).

$$\nu = \frac{c}{\lambda}$$

$$\lambda = \frac{c}{\nu} = \frac{3.00 \times 10^8\,\text{m/s}}{4.62 \times 10^{14}/\text{s}}$$
$$= 6.49 \times 10^{-7}\,\text{m}$$

$$= 6.49 \times 10^{-7}\,\text{m} \times \frac{1\,\text{nm}}{10^{-9}\,\text{m}} = 649\,\text{nm}$$

**FOR PRACTICE 7.1** A laser dazzles the audience in a rock concert by emitting green light with a wavelength of 515 nm. Calculate the frequency of the light.

---

## The Electromagnetic Spectrum

Visible light makes up only a tiny portion of the entire **electromagnetic spectrum**, which includes all wavelengths of electromagnetic radiation. Figure 7.5▶ on the next page shows the main regions of the electromagnetic spectrum, ranging in wavelength from $10^{-15}$ m (gamma rays) to $10^5$ m (radio waves). In Figure 7.5, short-wavelength, high-frequency radiation is on the right and long-wavelength, low-frequency radiation is on the left.

We will see later in this section that short-wavelength light inherently has greater energy than long-wavelength light (because of the particle nature of light). The most energetic forms of electromagnetic radiation have the shortest wavelengths. The form of electromagnetic radiation with the shortest wavelength is the **gamma ($\gamma$) ray**. Gamma rays are produced by the sun, other stars, and certain unstable atomic nuclei on Earth. Excessive exposure to gamma rays is dangerous to humans because the high energy of gamma rays can damage biological molecules.

Next on the electromagnetic spectrum, with longer wavelengths than gamma rays, are **X-rays**, familiar to us from their medical use. X-rays pass through many substances

We will discuss gamma rays in more detail in Chapter 20.

## The Electromagnetic Spectrum

▲ **FIGURE 7.5 The Electromagnetic Spectrum** The right side of the spectrum consists of high-energy, high-frequency, short-wavelength radiation. The left side consists of low-energy, low-frequency, long-wavelength radiation. Visible light constitutes a small segment in the middle.

▲ To produce a medical X-ray, short-wavelength electromagnetic radiation passes through the skin and creates an image of bones and internal organs.

▲ Warm objects emit infrared light, which is invisible to the eye but can be captured on film or by detectors to produce an infrared photograph.

that block visible light and are therefore used to image bones and internal organs. Like gamma rays, X-rays are sufficiently energetic to damage biological molecules. Although several annual medical X-rays are relatively harmless, too much exposure to X-rays increases cancer risk.

Sandwiched between X-rays and visible light in the electromagnetic spectrum is **ultraviolet (UV) radiation**, most familiar to us as the component of sunlight that produces a sunburn or suntan. Though not as energetic as gamma rays or X-rays, ultraviolet light still carries enough energy to damage biological molecules. Excessive exposure to ultraviolet light increases the risk of skin cancer and cataracts and causes premature wrinkling of the skin.

Next on the spectrum is **visible light**, ranging from violet (shorter wavelength, higher energy) to red (longer wavelength, lower energy). Visible light—at low to moderate intensity—does not carry enough energy to damage biological molecules. It does, however, cause certain molecules in our eyes to change their shape, sending a signal to our brains that results in our ability to see.

Beyond visible light lies **infrared (IR) radiation**. The heat you feel when you place your hand near a hot object is infrared radiation. All warm objects, including human bodies, emit infrared light. Although infrared light is invisible to our eyes, infrared sensors can detect it and are often employed in night vision technology to help people "see" in the dark.

Beyond infrared light, at longer wavelengths still, are **microwaves**, used for radar and in microwave ovens. Although microwave radiation has longer wavelengths and therefore lower energies than visible or infrared light, it is efficiently absorbed by water and can therefore heat substances that contain water. The longest wavelengths on the electromagnetic spectrum are those of **radio waves**, which are used to transmit the signals responsible for AM and FM radio, cellular telephone, television, and other forms of communication.

CONCEPTUAL
**CONNECTION 7.1**

**Frequency of Light** Which group of types of light is listed in order of increasing frequency?
**(a)** visible, ultraviolet, infrared
**(b)** infrared, visible, ultraviolet
**(c)** ultraviolet, visible, infrared
**(d)** infrared, ultraviolet, visible

# Chemistry AND Medicine | Radiation Treatment for Cancer

X-rays and gamma rays are sometimes called ionizing radiation because their short wavelengths correspond to high energies that can ionize atoms and molecules. When ionizing radiation interacts with biological molecules, it can permanently change or even destroy them. Consequently, we normally try to limit our exposure to ionizing radiation. However, doctors use ionizing radiation to destroy molecules within unwanted cells such as cancer cells.

In radiation therapy (also called radiotherapy) doctors aim X-ray or gamma-ray beams at cancerous tumors (groups of cells that divide uncontrollably and invade surrounding healthy tissue). The ionizing radiation damages the molecules within the tumor's cells that carry genetic information—information necessary for the cell to grow and divide. Consequently, the cell dies or stops dividing. Ionizing radiation also damages molecules in healthy cells, but cancerous cells divide more quickly than normal cells, making them more susceptible to genetic damage. Nonetheless, harm to healthy tissues during radiation therapy treatments can result in side effects such as fatigue, skin lesions, hair loss, and organ damage. Medical workers try to reduce such effects by appropriate shielding (of healthy tissue) and by targeting the tumor from multiple directions, minimizing the exposure of healthy cells while maximizing the exposure of cancerous cells.

Another side effect of exposing healthy cells to radiation is that the healthy cells may become cancerous. If a treatment for cancer may cause cancer, why do we continue to use it? In radiation therapy, as in most other disease therapies, there is an associated risk. We take risks all the time, many of them for lesser reasons. For example, every time we fly in an airplane or drive in a car, we risk injury or even death. Why? Because we perceive the benefit—the convenience of being able to travel a significant distance in a short time—to be worth the relatively small risk. The situation is similar in cancer therapy, or any other medical therapy for that matter. The benefit of cancer therapy (possibly curing a cancer that might otherwise kill you) is worth the risk (a slight increase in the chance of developing a future cancer).

### QUESTION

*Why is visible light (by itself) not used to destroy cancerous tumors?*

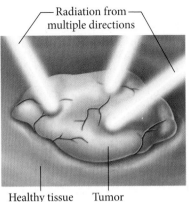

Radiation from multiple directions

Healthy tissue    Tumor

▲ During radiation therapy, a tumor is targeted from multiple directions in order to minimize the exposure of healthy cells while maximizing the exposure of cancerous cells.

▲ In radiation therapy, highly energetic gamma rays are aimed at cancerous tumors.

# Interference and Diffraction

Waves, including electromagnetic waves, interact with each other in a characteristic way called **interference**: they cancel each other out or build each other up, depending on their alignment upon interaction. For example, if two waves of equal amplitude are *in phase* when they interact—that is, they align with overlapping crests—a wave with twice the amplitude results. This is called **constructive interference**.

▲ When a reflected wave meets an incoming wave near the shore, the two waves interfere constructively for an instant, producing a large-amplitude spike.

Waves in phase → Constructive interference

On the other hand, if two waves are completely *out of phase* when they interact—that is, they align so that the crest from one source overlaps with the trough from the other source—the waves cancel by **destructive interference**.

Waves out of phase → Destructive interference

► FIGURE 7.6 **Diffraction** In this view from above, we can see how a wave bends, or diffracts, when it encounters an obstacle or slit with a size comparable to its wavelength. When a wave passes through a small opening, it spreads out. Particles, by contrast, do not diffract; they simply pass through the opening.

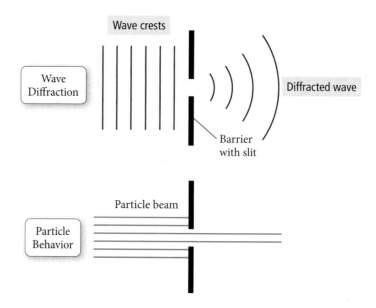

Waves also exhibit a characteristic behavior called **diffraction** (Figure 7.6▲). When a wave encounters an obstacle or a slit that is comparable in size to its wavelength, it bends (or *diffracts*) around it. The diffraction of light through two slits separated by a distance comparable to the wavelength of the light, coupled with interference, results in an *interference pattern*, illustrated in Figure 7.7▼. Each slit acts

## Interference from Two Slits

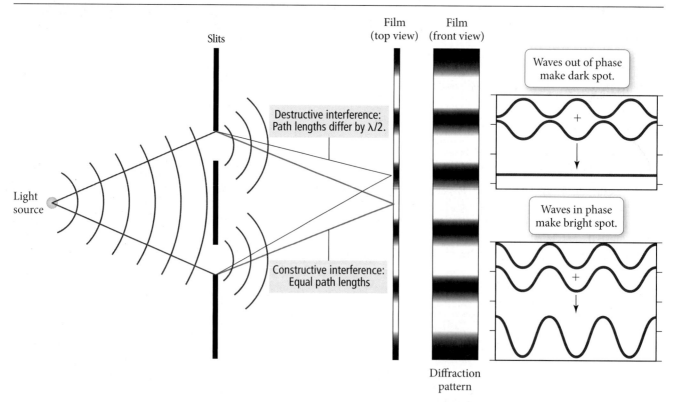

▲ FIGURE 7.7 **Interference from Two Slits** When a beam of light passes through two small slits, the two resulting waves interfere with each other. Whether the interference is constructive or destructive at any given point depends on the difference in the path lengths traveled by the waves. The resulting interference pattern appears as a series of bright and dark lines on a screen.

as a new wave source, and the two new waves interfere with each other. The resulting pattern is a series of bright and dark lines that can be viewed on a screen (or recorded on a film) placed a short distance behind the slits. At the center of the screen, the two waves travel equal distances and interfere constructively to produce a bright line. A small distance away from the center in either direction, the two waves travel slightly different distances, so they are out of phase. At the point where the difference in distance is one-half of one wavelength, the interference is destructive and a dark line appears on the screen. A bit further away from the center we see constructive interference again because the difference between the paths is one whole wavelength. The end result is the interference pattern shown by the light and dark bars in Figure 7.7. Notice that interference results from the ability of a wave to diffract through two slits—an inherent property of waves.

## The Particle Nature of Light

Prior to the early 1900s, and especially after the discovery of the diffraction of light, light was thought to be purely a wave phenomenon. Its behavior was described adequately by classical electromagnetic theory, which treated the electric and magnetic fields that constitute light as waves propagating through space. However, a number of discoveries brought the classical view into question. Chief among these discoveries was the *photoelectric effect*.

The **photoelectric effect** is the observation that many metals emit electrons when light shines upon them, as shown in Figure 7.8 ▼. Classical electromagnetic theory attributed this effect to the transfer of energy from the light to an electron in the metal, which resulted in the dislodgment of the electron. According to this explanation, only the amplitude (intensity) of the light affects the emission of electrons, not the wavelength. In other words, according to the classical description, the rate at which electrons leave the metal due to the photoelectric effect increases with increasing intensity of the light. A dim light was expected to result in a *lag time* between the initial shining of the light and the subsequent emission of an electron. The lag time would be the

The term *classical*, as in classical electromagnetic theory or classical mechanics, refers to descriptions of matter and energy before the advent of quantum mechanics.

**The Photoelectric Effect**

(a)  (b)

▲ **FIGURE 7.8 The Photoelectric Effect** **(a)** When sufficiently energetic light shines on a metal surface, the surface emits electrons. **(b)** The emitted electrons can be measured as an electrical current.

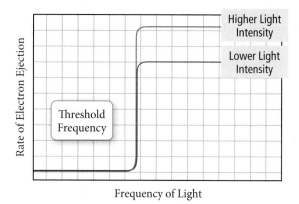

Rate of Electron Ejection

Frequency of Light

▲ **FIGURE 7.9** **The Photoelectric Effect** A plot of the electron ejection rate versus frequency of light for the photoelectric effect.

Einstein was not the first to suggest that energy was quantized. Max Planck (1858–1947) used the idea in 1900 to account for certain characteristics of radiation from hot bodies.

The energy of a photon is directly proportional to its frequency and inversely proportional to its wavelength.

minimum amount of time required for the dim light to transfer sufficient energy to the electron to dislodge it.

The experimental results, however, did not support the classical prediction. Scientists found that a high-frequency, low-intensity light produces electrons *without* the predicted lag time. Furthermore, the light used to dislodge electrons in the photoelectric effect exhibits a *threshold frequency*, below which no electrons are emitted from the metal, no matter how long the light shines on the metal. Figure 7.9 ◀ is a graph of the rate of electron ejection from the metal versus the frequency of light used. Notice that increasing the intensity of the light does not change the threshold frequency. In other words, low-frequency (long-wavelength) light *does not* eject electrons from a metal regardless of its intensity or its duration. But high-frequency (short-wavelength) light *does* eject electrons, even if its intensity is low. What could explain this odd behavior?

In 1905, Albert Einstein (1879–1955) proposed a bold explanation for the photoelectric effect: *light energy must come in packets.* According to Einstein, the amount of energy ($E$) in a light packet depends on its frequency ($\nu$) according to the following equation:

$$E = h\nu \qquad [7.2]$$

where $h$, called *Planck's constant*, has the value $h = 6.626 \times 10^{-34}$ J·s. A *packet* of light is called a **photon** or a **quantum** of light. Since $\nu = c/\lambda$, the energy of a photon can also be expressed in terms of wavelength as follows:

$$E = \frac{hc}{\lambda} \qquad [7.3]$$

Unlike classical electromagnetic theory, in which light was viewed purely as a wave whose intensity was *continuously variable*, Einstein suggested that light was *lumpy*. From this perspective, a beam of light is *not* a wave propagating through space, but a shower of particles (photons), each with energy $h\nu$.

## Example 7.2 | Photon Energy

A nitrogen gas laser pulse with a wavelength of 337 nm contains 3.83 mJ of energy. How many photons does it contain?

| | |
|---|---|
| **SORT** You are given the wavelength and total energy of a light pulse and asked to find the number of photons it contains. | **GIVEN:** $E_{pulse} = 3.83$ mJ<br>$\lambda = 337$ nm<br><br>**FIND:** number of photons |
| **STRATEGIZE** In the first part of the conceptual plan, calculate the energy of an individual photon from its wavelength. | **CONCEPTUAL PLAN**<br><br>$\lambda \longrightarrow E_{photon}$<br><br>$E = \frac{hc}{\lambda}$ |
| In the second part, divide the total energy of the pulse by the energy of a photon to determine the number of photons in the pulse. | $\dfrac{E_{pulse}}{E_{photon}} =$ number of photons<br><br>**RELATIONSHIPS USED** $E = hc/\lambda$ (Equation 7.3) |
| **SOLVE** To execute the first part of the conceptual plan, convert the wavelength to meters and substitute it into the equation to calculate the energy of a 337-nm photon. | **SOLUTION**<br><br>$\lambda = 337 \text{ nm} \times \dfrac{10^{-9} \text{ m}}{1 \text{ nm}} = 3.37 \times 10^{-7} \text{ m}$<br><br>$E_{photon} = \dfrac{hc}{\lambda} = \dfrac{(6.626 \times 10^{-34} \text{ J·s})\left(3.00 \times 10^{8} \frac{\text{m}}{\text{s}}\right)}{3.37 \times 10^{-7} \text{ m}}$<br><br>$= 5.8985 \times 10^{-19} \text{ J}$ |

| To execute the second part of the conceptual plan, convert the energy of the pulse from mJ to J. Then divide the energy of the pulse by the energy of a photon to obtain the number of photons. | $3.83 \text{ mJ} \times \dfrac{10^{-3} \text{ J}}{1 \text{ mJ}} = 3.83 \times 10^{-3} \text{ J}$ <br><br> $\text{number of photons} = \dfrac{E_{\text{pulse}}}{E_{\text{photon}}} = \dfrac{3.83 \times 10^{-3} \text{ J}}{5.8985 \times 10^{-19} \text{ J}}$ <br> $= 6.49 \times 10^{15} \text{ photons}$ |
|---|---|

**CHECK** The units of the answer, photons, are correct. The magnitude of the answer ($10^{15}$) is reasonable. Photons are small particles, and any macroscopic collection must contain a large number of them.

**FOR PRACTICE 7.2** A 100-watt lightbulb radiates energy at a rate of 100 J/s. (The watt, a unit of power, or energy over time, is defined as 1 J/s.) If all of the light emitted has a wavelength of 525 nm, how many photons are emitted per second? (Assume three significant figures in this calculation.)

**FOR MORE PRACTICE 7.2** The energy required to dislodge electrons from sodium metal via the photoelectric effect is 275 kJ/mol. What wavelength in nm of light has sufficient energy per photon to dislodge an electron from the surface of sodium?

---

## Example 7.3 | Wavelength, Energy, and Frequency

Arrange these three types of electromagnetic radiation—visible light, X-rays, and microwaves—in order of increasing

**(a)** wavelength.         **(b)** frequency.         **(c)** energy per photon.

**SOLUTION**

| Examine Figure 7.5 and note that X-rays have the shortest wavelength, followed by visible light and then microwaves. | **(a)** wavelength <br> X-rays < visible < microwaves |
|---|---|
| Since frequency and wavelength are inversely proportional—the longer the wavelength, the shorter the frequency—the ordering with respect to frequency is the reverse of the ordering with respect to wavelength. | **(b)** frequency <br> microwaves < visible < X-rays |
| Energy per photon decreases with increasing wavelength but increases with increasing frequency; therefore, the ordering with respect to energy per photon is the same as for frequency. | **(c)** energy per photon <br> microwaves < visible < X-rays |

**FOR PRACTICE 7.3** Arrange these three colors of visible light—green, red, and blue—in order of increasing

(a) wavelength.             (b) frequency.             (c) energy per photon.

---

Einstein's idea that light is quantized elegantly explains the photoelectric effect. The emission of electrons from the metal depends on whether or not a single photon has sufficient energy (as given by $h\nu$) to dislodge a single electron. For an electron bound to the metal with binding energy $\phi$, the threshold frequency is reached when the energy of the photon is equal to $\phi$.

The symbol $\phi$ is the Greek letter phi, pronounced "fi."

Threshold frequency condition

$$h\nu = \phi$$

Energy of photon      Binding energy of emitted electron

Low-frequency light does not eject electrons because no single photon has the minimum energy necessary to dislodge the electron. We can draw an analogy between a photon ejecting an electron from a metal surface and a ball breaking a glass window. In this analogy, low-frequency photons are like ping-pong balls—a ping-pong ball thrown at a glass window does not break it (just as a low-frequency photon does not eject an electron). Increasing the *intensity* of low-frequency light is like increasing the number of ping-pong balls thrown at the window—doing so simply increases the number of low-energy photons but does not produce any single photon with sufficient energy. In contrast, increasing the *frequency* of the light, even at low intensity, *increases the energy of each photon*. In our analogy, a high-frequency photon is like a baseball—a baseball thrown at a glass window breaks it (just as a high-frequency photon dislodges an electron with no lag time).

As the frequency of the light increases past the threshold frequency, the excess energy of the photon (beyond what is needed to dislodge the electron) transfers to the electron in the form of kinetic energy. The kinetic energy (KE) of the ejected electron, therefore, is the difference between the energy of the photon (*hv*), and the binding energy of the electron, as given by the equation:

$$KE = hv - \phi$$

Although the quantization of light explained the photoelectric effect, the wave explanation of light continued to have explanatory power as well, depending on the circumstances of the particular observation. So the principle that slowly emerged (albeit with some measure of resistance) is what we now call the *wave–particle duality of light*. Sometimes light appears to behave like a wave, at other times like a particle. Which behavior we observe depends on the particular experiment.

CONCEPTUAL
**CONNECTION 7.2**

**The Photoelectric Effect** We shine light of three different wavelengths—325 nm, 455 nm, and 632 nm—on a metal surface. We record the following observations for each wavelength, labeled A, B, and C:

Observation A: No photoelectrons were observed.

Observation B: Photoelectrons with a kinetic energy of 155 kJ/mol were observed.

Observation C: Photoelectrons with a kinetic energy of 51 kJ/mol were observed.

Which observation corresponds to which wavelength of light?

# 7.3 Atomic Spectroscopy and the Bohr Model

The discovery of the particle nature of light began to break down the division that existed in nineteenth-century physics between electromagnetic radiation, which was thought of as a wave phenomenon, and the small particles (protons, neutrons, and electrons) that compose atoms, which were thought to follow Newton's laws of motion. Just as the photoelectric effect suggested the particle nature of light, so certain observations about atoms began to suggest a wave nature for particles. The most important of these observations came from *atomic spectroscopy*, the study of the electromagnetic radiation absorbed and emitted by atoms.

When an atom absorbs energy—in the form of heat, light, or electricity—it often reemits that energy as light. For example, a neon sign is composed of one or more glass tubes filled with neon gas. When an electric current is passed through the tube, the neon atoms absorb some of the electrical energy and reemit it as the familiar red light of a neon sign. If the atoms in the tube are different (that is, not neon), they emit light of a different color. In other words, atoms of each element emit light of a characteristic color. Mercury atoms, for example, emit light that appears blue, helium atoms emit light that appears violet, and hydrogen atoms emit light that appears reddish (Figure 7.10▶).

Remember that the color of visible light is determined by its wavelength. Closer inspection of the light emitted by various atoms reveals that it contains several distinct wavelengths. We can separate the light emitted by a single element in a glass tube into its constituent wavelengths by passing it through a prism (just like we separate the white

▲ The familiar red light from a neon sign is emitted by neon atoms that have absorbed electrical energy, which the atoms reemit as visible radiation.

light from a lightbulb), as shown in Figure 7.11▼. The result is a series of bright lines called an **emission spectrum**. The emission spectrum of a particular element is always the same—it consists of the same bright lines at the same characteristic wavelengths. We can use the emission spectrum to identify the element. For example, light arriving from a distant star contains the emission spectra of the elements that compose the star. Analysis of the light allows us to identify the elements present in the star.

Notice the differences between a white light spectrum and the emission spectra of hydrogen, helium, and barium (shown in Figure 7.11). The white light spectrum is *continuous,* meaning that there are no sudden interruptions in the intensity of the light as a function of wavelength—the spectrum consists of light of all wavelengths. The emission spectra of hydrogen, helium, and barium, however, are not continuous—they consist of bright lines at specific wavelengths, with complete darkness in between. That is, only certain discrete wavelengths of light are present. Classical physics could not explain why these spectra consisted of discrete lines. In fact, according to classical physics, an atom composed of an electron orbiting a nucleus should emit a continuous white light spectrum. Even more problematic, the electron should lose energy as it emits the light and spirals into the nucleus. According to classical physics, an atom should not even be stable.

Johannes Rydberg (1854–1919), a Swedish mathematician, analyzed many atomic spectra and developed an equation that predicts the wavelengths of the hydrogen emission spectrum. However, his equation (shown in the margin) gives little insight into *why* atomic spectra are discrete, *why* atoms are stable, or *why* his equation works.

The Danish physicist Niels Bohr (1885–1962) attempted to develop a model for the atom that explained atomic spectra. In his model, electrons travel around the nucleus in

▲ **FIGURE 7.10 Mercury, Helium, and Hydrogen** Each element emits a characteristic color.

The Rydberg equation is $1/\lambda = R(1/m^2 - 1/n^2)$, where $R$ is the Rydberg constant $(1.097 \times 10^7 \, m^{-1})$ and $m$ and $n$ are integers.

## Emission Spectra

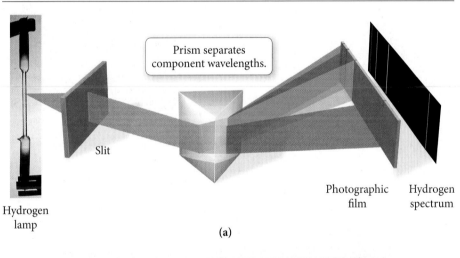

Prism separates component wavelengths.

Slit

Photographic film

Hydrogen spectrum

Hydrogen lamp

**(a)**

Helium spectrum

Barium spectrum

White light spectrum

**(b)**

▲ **FIGURE 7.11 Emission Spectra** **(a)** The light emitted from a hydrogen, helium, or barium lamp consists of specific wavelengths that can be separated by passing the light through a prism. **(b)** The resulting bright lines constitute an emission spectrum characteristic of the element that produced it.

## The Bohr Model and Emission Spectra

▲ **FIGURE 7.12 The Bohr Model and Emission Spectra** According to the Bohr model, each spectral line is produced when an electron falls from one stable orbit, or stationary state, to another of lower energy.

circular orbits (analogous to those of the planets around the sun). However, in contrast to planetary orbits—which theoretically can exist at any distance from the sun—Bohr's orbits exist only at specific, fixed distances from the nucleus. The energy of each Bohr orbit is also fixed, or *quantized*. Bohr called these orbits *stationary states* and suggested that, although they obey the laws of classical mechanics, they also possess "a peculiar, mechanically unexplainable, stability." Scientists now know that the stationary states were really manifestations of the wave nature of the electron, which we will expand upon shortly. Bohr further proposed that, in contradiction to classical electromagnetic theory, an electron orbiting the nucleus in a stationary state emits no radiation. It is only when an electron jumps, or makes a *transition*, from one stationary state to another that radiation is emitted or absorbed (Figure 7.12◄).

---

## Chemistry IN YOUR DAY | Atomic Spectroscopy, a Bar Code for Atoms

When you check out of the grocery store, a laser scanner reads the bar code on the items that you buy. Each item has a unique code that identifies the item and its price. Similarly, each element in the periodic table has a spectrum unlike that of any other element. For example, Figure 7.13▼ shows the emission spectra of oxygen and neon. (In Figure 7.11, we saw the emission spectra of hydrogen, helium, and barium.) Notice that each spectrum is unique and, as such, can be used to identify the substance.

The presence of intense lines in the spectra of a number of metals is the basis for *flame tests*, simple tests used to identify elements in ionic compounds in the absence of a precise analysis of a compound's spectrum. For example, the emission spectrum of sodium features two closely spaced, bright yellow lines. When a crystal of a sodium salt (or a drop of a solution containing a sodium salt) is put into a flame, the flame glows bright yellow (Figure 7.14►). As Figure 7.14 shows, other metals exhibit similarly characteristic colors in flame tests. Each color represents an especially bright spectral emission line (or a combination of two or more such lines). Similar emissions form the basis of the colors seen in fireworks.

Although the *emission* of light from elements is easier to detect, the *absorption* of light by elements is even more commonly used for purposes of identification. Whereas emission spectra consist of bright lines on a dark background, absorption spectra consist of dark lines on a bright background (Figure 7.15►). An absorption spectrum is measured by passing white light through a sample and observing what wavelengths are *missing* due to absorption by the sample. Notice that, in the spectrum of mercury shown in Figure 7.15, the absorption lines are at the same wavelengths as the emission lines. This is because the processes that produce them are mirror images. In emission, an electron makes a transition from a higher-energy level to a lower-energy one. In absorption, the transition is between the same two energy levels, but from the lower level to the higher one.

▲ Fireworks typically contain the salts of such metals as sodium, calcium, strontium, barium, and copper. Emissions from these elements produce the brilliant colors of pyrotechnic displays.

Oxygen

Neon

▲ **FIGURE 7.13 Emission Spectra of Oxygen and Neon** The emission spectrum of each element is unique, and we can use it to identify the element.

◀ **FIGURE 7.14 Flame Tests (from left to right) for Sodium, Potassium, Lithium, and Barium** We can identify elements by the characteristic color of the light they produce when heated. The colors derive from especially bright lines in their emission spectra.

Absorption spectrometers, found in most chemistry laboratories, typically plot the intensity of absorption as a function of wavelength. Such plots are useful both for identifying substances (qualitative analysis) and for determining the concentration of substances (quantitative analysis). Quantitative analysis is possible because the amount of light absorbed by a sample depends on the concentration of the absorbing substance within the sample. For example, we can determine the concentration of $Ca^{2+}$ in a hard water sample by measuring the quantity of light absorbed by the calcium ion at its characteristic wavelength.

In spite of its initial success in explaining the line spectrum of hydrogen, the Bohr model left many unanswered questions. It did, however, serve as an intermediate model between a classical view of the electron and a fully quantum-mechanical view, and therefore has great historical and conceptual importance. Nonetheless, it was ultimately replaced by a more complete quantum-mechanical theory that fully incorporated the wave nature of the electron.

◀ **FIGURE 7.15 Emission and Absorption Spectrum of Mercury** Elements absorb light of the same wavelengths that they radiate when heated. When these wavelengths are subtracted from a beam of white light, the result is a pattern of dark lines corresponding to an absorption spectrum.

The transitions between stationary states in a hydrogen atom are quite unlike transitions in the macroscopic world. The electron is *never* observed *between states*; it is observed only in one state or another. The emission spectrum of an atom consists of discrete lines because the stationary states exist only at specific, fixed energies. The energy of the photon emitted when an electron makes a transition from one stationary state to another is the energy difference between the two stationary states. Transitions between stationary states that are closer together, therefore, produce light of lower energy (longer wavelength) than transitions between stationary states that are farther apart.

# 7.4 The Wave Nature of Matter: The de Broglie Wavelength, the Uncertainty Principle, and Indeterminacy

 **The Wave Nature of Matter**

The Davisson–Germer experiment of 1927 provided the first evidence of electron wave properties. In this experiment, electrons were observed to undergo diffraction by a metal crystal.

The heart of quantum-mechanical theory is the wave nature of the electron, first proposed by Louis de Broglie (1892–1987) in 1924 and confirmed by experiments in 1927. It seemed incredible at the time, but electrons—which were then thought of only as particles and known to have mass—also have a wave nature. The wave nature of

For interference to occur, the spacing of the slits has to be on the order of atomic dimensions.

the electron is seen most clearly in its diffraction. If an electron beam is aimed at two closely spaced slits, and a series (or array) of detectors is arranged to detect the electrons after they pass through the slits, an interference pattern similar to that observed for light (see Figure 7.6) is recorded behind the slits (Figure 7.16(a)▼). The detectors at the center of the array (midway between the two slits) detect a large number of electrons—exactly the opposite of what you would expect for particles (Figure 7.16(b)▼). Moving outward from this center spot, the detectors alternately detect small numbers of electrons and then large numbers again and so on, forming an interference pattern characteristic of waves.

Counter to what might be our initial intuition about electron interference, the interference pattern is *not caused by pairs of electrons interfering with each other, but rather by single electrons interfering with themselves.* If the electron source is turned down to a very low level, so that electrons come out only one at a time, *the interference pattern remains.* In other words, we can design an experiment in which electrons come out of the source singly. We can then record where each electron strikes the detector after it has

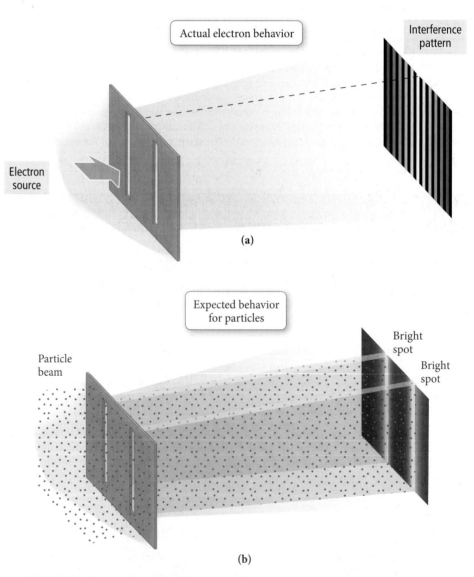

▲ FIGURE 7.16 **Electron Diffraction** When a beam of electrons goes through two closely spaced slits **(a)**, they create an interference pattern, as if the electrons were waves. By contrast, a beam of particles passing through two slits **(b)** produces two smaller beams of particles. Notice that particle beams produce two bright stripes with darkness in between but waves produce the brightest strip directly in the center of the screen.

passed through the slits. If we record the positions of thousands of electrons over a long period of time, we find the same interference pattern shown in Figure 7.16(a).

This observation leads us to an important conclusion: *the wave nature of the electron is an inherent property of individual electrons*. Recall from Section 7.1 that unobserved electrons can simultaneously occupy two different states. In this case, the unobserved electron goes through both slits—it exists in two states simultaneously, just like Schrödinger's cat—and interferes with itself. This wave nature explains the existence of stationary states (in the Bohr model) and prevents the electrons in an atom from crashing into the nucleus as predicted by classical physics. We now turn to three important manifestations of the electron's wave nature: the de Broglie wavelength, the uncertainty principle, and indeterminacy.

## The de Broglie Wavelength

As we have seen, a single electron traveling through space has a wave nature; its wavelength is related to its kinetic energy (the energy associated with its motion). The faster the electron is moving, the higher its kinetic energy and the shorter its wavelength. The wavelength ($\lambda$) of an electron of mass $m$ moving at velocity $v$ is given by the **de Broglie relation**:

$$\lambda = \frac{h}{mv} \quad \text{de Broglie relation} \qquad [7.4]$$

where $h$ is Planck's constant. *Notice that the velocity of a moving electron is related to its wavelength—knowing one is equivalent to knowing the other.*

> The mass of an object ($m$) times its velocity ($v$) is its momentum. Therefore, the wavelength of an electron is inversely proportional to its momentum.

## Example 7.4   De Broglie Wavelength

Calculate the wavelength of an electron traveling with a speed of $2.65 \times 10^6$ m/s.

**SORT** You are given the speed of an electron and asked to calculate its wavelength.

**GIVEN:** $v = 2.65 \times 10^6$ m/s

**FIND:** $\lambda$

**STRATEGIZE** The conceptual plan shows how the de Broglie relation relates the wavelength of an electron to its mass and velocity.

**CONCEPTUAL PLAN**

$$\boxed{v} \longrightarrow \boxed{\lambda}$$

$$\lambda = \frac{h}{mv}$$

**RELATIONSHIPS USED**
$\lambda = h/mv$ (de Broglie relation, Equation 7.4)

**SOLVE** Substitute the velocity, Planck's constant, and the mass of an electron into Equation 7.4 to calculate the electron's wavelength. To correctly cancel the units, break down the J in Planck's constant into its SI base units (1 J = 1 kg·m²/s²).

**SOLUTION**

$$\lambda = \frac{h}{mv} = \frac{6.626 \times 10^{-34} \dfrac{\text{kg} \cdot \text{m}^2}{\text{s}^2} \text{s}}{(9.11 \times 10^{-31} \text{ kg})\left(2.65 \times 10^6 \dfrac{\text{m}}{\text{s}}\right)}$$

$$= 2.74 \times 10^{-10} \text{ m}$$

**CHECK** The units of the answer (m) are correct. The magnitude of the answer is very small, as expected for the wavelength of an electron.

**FOR PRACTICE 7.4** What is the velocity of an electron that has a de Broglie wavelength approximately the length of a chemical bond? Assume the length of a chemical bond is $1.2 \times 10^{-10}$ m.

CONCEPTUAL
**CONNECTION 7.3**

**The de Broglie Wavelength of Macroscopic Objects** Since quantum-mechanical theory is universal, it applies to all objects, regardless of size. Therefore, according to the de Broglie relation, a thrown baseball should also exhibit wave properties. Why don't we observe such properties at the ballpark?

## The Uncertainty Principle

The wave nature of the electron is difficult to reconcile with its particle nature. How can a single entity behave as both a wave and a particle? We can begin to address this question by returning to the single-electron diffraction experiment. How does a single electron aimed at a double slit produce an interference pattern? We saw previously that the electron travels through both slits and interferes with itself. This idea is testable. We simply have to observe the single electron as it travels through both of the slits. If it travels through both slits simultaneously, our hypothesis is correct. But here is where nature gets tricky.

Any experiment designed to observe the electron as it travels through the slits results in the detection of an electron "particle" traveling through a single slit and no interference pattern. Recall from Section 7.1 that an *unobserved* electron can occupy two different states; however, the act of observation forces it into one state or the other. Similarly, the act of observing the electron as it travels through both slits forces it to go through only one slit. The electron diffraction experiment illustrated here is designed to observe which slit the electron travels through by using a laser beam placed directly behind the slits:

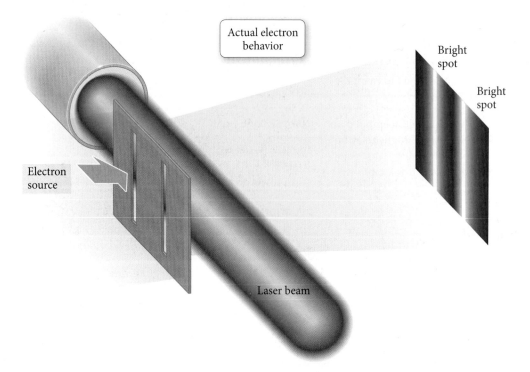

An electron that crosses the laser beam produces a tiny "flash" when a single photon is scattered at the point of crossing. If a flash shows up behind a particular slit, that indicates an electron is passing through that slit. When the experiment is performed, the flash always originates either from one slit *or* the other, but *never* from both at once. Furthermore, the interference pattern, which was present without the laser, is absent when the electrons are observed going through the slits. With the laser on, the electrons hit positions directly behind each slit, as if they were ordinary particles; their wavelike behavior is no longer manifested.

As it turns out, no matter how hard we try or whatever method we set up, *we can never both see the interference pattern and simultaneously determine which hole the electron goes through*. It has never been done, and most scientists agree that it never will. In the words of theoretical physicist P. A. M. Dirac (1902–1984),

> There is a limit to the fineness of our powers of observation and the smallness of the accompanying disturbance—a limit which is inherent in the nature of things and can never be surpassed by improved technique or increased skill on the part of the observer.

The single electron diffraction experiment demonstrates that we cannot simultaneously observe both the wave nature and the particle nature of the electron. When we try to observe which hole the electron goes through (associated with the particle nature of the electron), we lose the interference pattern (associated with the wave nature of the electron). When we try to observe the interference pattern, we cannot determine which hole the electron goes through. The wave nature and particle nature of the electron are said to be **complementary properties**. Complementary properties exclude one another—the more we know about one, the less we know about the other. Which of two complementary properties we observe depends on the experiment we perform—in quantum mechanics, the observation of an event affects its outcome.

As we just saw in the de Broglie relation, the *velocity* of an electron is related to its *wave nature*. The *position* of an electron, however, is related to its *particle nature*. (Particles have well-defined position, but waves do not.) Consequently, our inability to observe the electron simultaneously as both a particle and a wave means that *we cannot simultaneously measure its position and its velocity*. Werner Heisenberg (1901–1976) formalized this idea with the equation:

$$\Delta x \times m\Delta v \geq \frac{h}{4\pi} \text{ Heisenberg's uncertainty principle} \qquad [7.5]$$

where $\Delta x$ is the uncertainty in the position, $\Delta v$ is the uncertainty in the velocity, $m$ is the mass of the particle, and $h$ is Planck's constant. **Heisenberg's uncertainty principle** states that the product of $\Delta x$ and $m\Delta v$ must be greater than or equal to a finite number ($h/4\pi$). In other words, the more accurately we know the position of an electron (the smaller $\Delta x$), the less accurately we can know its velocity (the bigger $\Delta v$) and vice versa. The complementarity of the wave nature and particle nature of the electron results in the complementarity of velocity and position.

Although Heisenberg's uncertainty principle may seem puzzling, it actually solves a great puzzle. Without the uncertainty principle, we are left with a paradox: how can something be *both* a particle and a wave? Saying that an object is both a particle and a wave is like saying that an object is both a circle and a square, a contradiction. Heisenberg solved the contradiction by introducing complementarity—an electron is observed as *either* a particle or a wave, but never both at once. This idea was captured by Schrödinger's thought experiment about the cat, which we discussed in Section 7.1: when observed, the cat is either dead or alive, not both.

▲ Werner Heisenberg, a pioneer of quantum mechanics

## Indeterminacy and Probability Distribution Maps

According to classical physics, and in particular Newton's laws of motion, particles move in a *trajectory* (or path) that is determined by the particle's velocity (the speed and direction of travel), its position, and the forces acting on it. Even if you are not familiar with Newton's laws, you probably have an intuitive sense of them. For example, when you chase a baseball in the outfield, you visually predict where the ball will land by observing its path. You do this by noting its initial position and velocity, watching how these are affected by the forces acting on it (gravity, air resistance, wind), and then inferring its trajectory, as shown in Figure 7.17▶, on the next page. If you knew only the ball's velocity, or only its position (imagine a still photo of the baseball in the air), you could not predict its landing spot. In classical mechanics, both position and velocity are required to predict a trajectory.

▶ FIGURE 7.17 **The Concept of Trajectory** In classical mechanics, the position and velocity of a particle determine its future trajectory, or path. Thus, an outfielder can catch a baseball by observing its position and velocity, allowing for the effects of forces acting on it, such as gravity, and estimating its trajectory. (For simplicity, air resistance and wind are not shown.)

## The Classical Concept of Trajectory

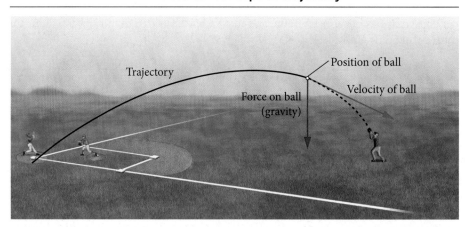

Trajectory

Position of ball

Velocity of ball

Force on ball (gravity)

| Remember that velocity includes speed as well as direction of travel.

Newton's laws of motion are **deterministic**—the present *determines* the future. This means that if two baseballs are hit consecutively with the same velocity from the same position under identical conditions, they will land in exactly the same place. The same is not true of electrons. We have just seen that we cannot simultaneously know the position and velocity of an electron; therefore, we cannot know its trajectory. In quantum mechanics, trajectories are replaced with *probability distribution maps* (Figure 7.18▼). A probability distribution map is a statistical map that shows where an electron is likely to be found under a given set of conditions.

Classical trajectory

Quantum-mechanical probability distribution map

▲ FIGURE 7.18 **Trajectory versus Probability** In quantum mechanics, we cannot calculate deterministic trajectories. Instead, it is necessary to think in terms of probability maps: statistical pictures of where a quantum-mechanical particle, such as an electron, is most likely to be found. In this hypothetical map, darker shading indicates greater probability.

To understand the concept of a probability distribution map, let us return to baseball. Imagine a baseball thrown from the pitcher's mound to a catcher behind home plate (Figure 7.19◀). The catcher can watch the baseball's path, predict exactly where it will cross home plate, and place his mitt in the correct place to catch it. As we have seen, the same predictions cannot be made for an electron. If an electron were thrown from the pitcher's mound to home plate, it would land in a different place every time, even if it were thrown in exactly the same way. This behavior is called **indeterminacy**. A baseball's future path is *determined* by its position and velocity when it leaves the pitcher's hand. In contrast, the future path of an electron is indeterminate and can only be described statistically.

In the quantum-mechanical world of the electron, the catcher cannot know where the electron will cross the plate for any given throw. However, if he were to record hundreds of identical electron throws, the catcher would observe a reproducible, *statistical pattern* of where the electron crosses the plate. He could even draw a map of the strike zone

◀ FIGURE 7.19 **Trajectory of a Macroscopic Object** A baseball follows a well-defined trajectory from the hand of the pitcher to the mitt of the catcher.

The Quantum-Mechanical Strike Zone

Number of pitches

Distance from center of strike zone

◀ **FIGURE 7.20 The Quantum-Mechanical Strike Zone** An electron does not have a well-defined trajectory. However, we can construct a probability distribution map to show the relative probability of it crossing home plate at different points.

showing the probability of an electron crossing a certain area, as shown in Figure 7.20▲. This is a probability distribution map. In the sections that follow, we discuss quantum-mechanical electron *orbitals*, which are essentially probability distribution maps for electrons as they exist within atoms.

# 7.5 Quantum Mechanics and the Atom

KCV Quantum Mechanics and the Atom: Orbitals and Quantum Numbers

As we have seen, the position and velocity of the electron are complementary properties—if we know one accurately, the other becomes indeterminate. Since velocity is directly related to energy (recall that kinetic energy equals $\frac{1}{2}mv^2$), position and *energy* are also complementary properties—the more we know about one, the less we know about the other. Many of the properties of an element, however, depend on the energies of its electrons. For example, whether an electron is transferred from one atom to another to form an ionic bond depends in part on the relative energies of the electron in the two atoms.

In the following paragraphs, we describe the probability distribution maps for electron states in which the electron has well-defined energy but not well-defined position. In other words, for each of these states, we can specify the *energy* of the electron precisely, but not its location at a given instant. Instead, the electron's position is described in terms of an **orbital**, a probability distribution map showing where the electron is likely to be found. Since chemical bonding often involves the sharing of electrons between atoms to form covalent bonds, the spatial distribution of atomic electrons is important to bonding.

> These states are known as energy *eigenstates*.

The mathematical derivation of energies and orbitals for electrons in atoms comes from solving the Schrödinger equation for the atom of interest. The general form of the Schrödinger equation is:

$$\mathcal{H}\psi = E\psi \qquad [7.6]$$

The symbol $\mathcal{H}$ is the Hamiltonian operator, a set of mathematical operations that represents the total energy (kinetic and potential) of the electron within the atom. The symbol $E$ is the actual energy of the electron. The symbol $\psi$ is the **wave function**, a mathematical function that describes the wavelike nature of the electron. A plot of the wave function squared ($\psi^2$) represents an orbital, a position probability distribution map of the electron.

> An operator is different from a normal algebraic entity. In general, an operator transforms a mathematical function into another mathematical function. For example, $d/dx$ is an operator that means "take the derivative of." When $d/dx$ operates on a function (such as $x^2$), it returns another function ($2x$).

> The symbol $\psi$ is the Greek letter psi, pronounced "sigh."

## Solutions to the Schrödinger Equation for the Hydrogen Atom

When the Schrödinger equation is solved, it yields many solutions—many possible wave functions. The wave functions themselves are fairly complicated mathematical functions, and we do not examine them in detail in this book. Instead, we introduce graphical representations (or plots) of the orbitals that correspond to the wave functions. Each orbital is specified by three interrelated **quantum numbers: *n***, the **principal quantum number**; ***l***, the **angular momentum quantum number** (sometimes called the *azimuthal quantum number*); and $m_l$, the **magnetic quantum number**. These quantum numbers all have integer values, as had been hinted at by both the Rydberg equation and Bohr's model. A fourth quantum number, $m_s$, the **spin quantum number**, specifies the orientation of the spin of the electron. We examine each of these quantum numbers individually.

### The Principal Quantum Number ($n$)

The principal quantum number is an integer that determines the overall size and energy of an orbital. Its possible values are $n = 1, 2, 3, \ldots$ and so on. For the hydrogen atom, the energy of an electron in an orbital with quantum number $n$ is given by the equation:

$$E_n = -2.18 \times 10^{-18}\,J\left(\frac{1}{n^2}\right) \quad (n = 1, 2, 3, \ldots) \tag{7.7}$$

The energy is negative because the electron's energy is lowered (made more negative) by its interaction with the nucleus (see the description of Coulomb's law in Section 8.3). The constant, $2.18 \times 10^{-18}\,J$, is known as the Rydberg constant for hydrogen ($R_H$). Notice that orbitals with higher values of $n$ have greater (less negative) energies, as shown in the energy-level diagram at left. Notice also that, as $n$ increases, the spacing between the energy levels decreases.

On the energy diagram at left:

Energy (vertical axis)

$n = 4$ ———— $E_4 = -1.36 \times 10^{-19}\,J$
$n = 3$ ———— $E_3 = -2.42 \times 10^{-19}\,J$

$n = 2$ ———— $E_2 = -5.45 \times 10^{-19}\,J$

$n = 1$ ———— $E_1 = -2.18 \times 10^{-18}\,J$

### The Angular Momentum Quantum Number ($l$)

The angular momentum quantum number is an integer that determines the shape of the orbital. We will consider these shapes in Section 7.6. The possible values of $l$ are $0, 1, 2, \ldots, (n - 1)$. In other words, for a given value of $n$, $l$ can be any integer (including 0) up to $n - 1$. For example, if $n = 1$, then the only possible value of $l$ is 0; if $n = 2$, the possible values of $l$ are 0 and 1. To avoid confusion between $n$ and $l$, values of $l$ are often assigned letters as follows:

The values of $l$ beyond 3 are designated with letters in alphabetical order so that $l = 4$ is designated $g$, $l = 5$ is designated $h$, and so on.

| Value of $l$ | Letter Designation |
|---|---|
| $l = 0$ | s |
| $l = 1$ | p |
| $l = 2$ | d |
| $l = 3$ | f |

**CONCEPTUAL CONNECTION 7.4**

**The Relationship between $n$ and $l$** What values of $l$ are possible for $n = 3$?
(a) 0 (or $s$)
(b) 0 and 1 (or $s$ and $p$)
(c) 0, 1, and 2 (or $s$, $p$, and $d$)
(d) 0, 1, 2, and 3 (or $s$, $p$, $d$, and $f$)

### The Magnetic Quantum Number ($m_l$)

The magnetic quantum number is an integer that specifies the orientation of the orbital. We will consider these orientations in Section 7.6. The possible values of $m_l$ are the integer values (including zero) ranging from $-l$ to $+l$. For example, if $l = 0$, then the only possible value of $m_l$ is 0; if $l = 1$, the possible values of $m_l$ are $-1$, 0, and $+1$.

**CONCEPTUAL CONNECTION 7.5**

**The Relationship between $l$ and $m_l$** What values of $m_l$ are possible for $l = 2$?
(a) 0, 1, and 2
(b) 0
(c) $-1$, 0, and $+1$
(d) $-2$, $-1$, 0, $+1$, and $+2$

## The Spin Quantum Number ($m_s$)

The spin quantum number specifies the orientation of the *spin* of the electron. Electron spin is a fundamental property of an electron (like its negative charge). One electron does not have more or less spin than another—all electrons have the same amount of spin. The orientation of the electron's spin is quantized, with only two possibilities: that we can call spin up ($m_s = +1/2$) and spin down ($m_s = -1/2$). The spin quantum number becomes important in Section 8.3 when we begin to consider how electrons occupy orbitals. Here, we focus only on the first three quantum numbers.

Each specific combination of the first three quantum numbers ($n$, $l$, and $m_l$) specifies one atomic orbital. For example, the orbital with $n = 1$, $l = 0$, and $m_l = 0$ is known as the 1s orbital. The 1 in 1s is the value of $n$, and the s specifies that $l = 0$. There is only one 1s orbital in an atom, and its $m_l$ value is zero. Orbitals with the same value of $n$ are said to be in the same **principal level (or principal shell)**. Orbitals with the same value of $n$ and $l$ are said to be in the same **sublevel (or subshell)**. The following diagram shows all of the orbitals (each represented by a small square) in the first three principal levels:

> The idea of a "spinning" electron is something of a metaphor. A more correct way to express the same idea is to say that an electron has inherent angular momentum.

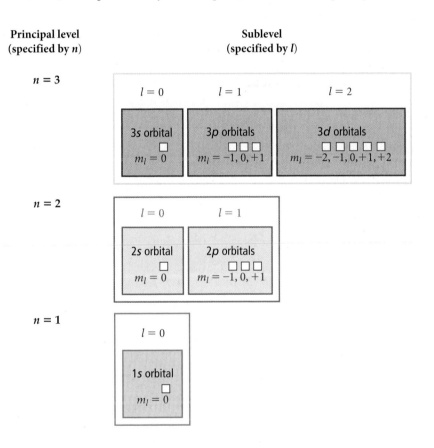

For example, the $n = 2$ level contains the $l = 0$ and $l = 1$ sublevels. Within the $n = 2$ level, the $l = 0$ sublevel—called the 2s sublevel—contains only one orbital (the 2s orbital), with $m_l = 0$. The $l = 1$ sublevel—called the 2p sublevel—contains three 2p orbitals, with $m_l = -1, 0, +1$.

### *Summarizing Orbitals:*

▶ The number of sublevels in any level is equal to $n$, the principal quantum number. Therefore, the $n = 1$ level has one sublevel, the $n = 2$ level has two sublevels, and so on.

▶ The number of orbitals in any sublevel is equal to $2l + 1$. Therefore, the s sublevel ($l = 0$) has one orbital, the p sublevel ($l = 1$) has three orbitals, the d sublevel ($l = 2$) has five orbitals, and so on.

▶ The number of orbitals in a level is equal to $n^2$. Therefore, the $n = 1$ level has one orbital, the $n = 2$ level has four orbitals, the $n = 3$ level has nine orbitals, and so on.

## Example 7.5 Quantum Numbers I

What are the quantum numbers and names (for example, $2s$, $2p$) of the orbitals in the $n = 4$ principal level? How many $n = 4$ orbitals exist?

**SOLUTION**

| | |
|---|---|
| First, determine the possible values of $l$ (from the given value of $n$). For a given value of $n$, the possible values of $l$ are $0, 1, 2, \ldots, (n - 1)$. | $n = 4$; therefore, $l = 0, 1, 2,$ and $3$ |

Next, determine the possible values of $m_l$ for each value of $l$. For a given value of $l$, the possible values of $m_l$ are the integer values, including zero, ranging from $-l$ to $+l$. The name of an orbital is its principal quantum number ($n$) followed by the letter corresponding to the value $l$. The total number of orbitals is given by $n^2$.

| $l$ | Possible $m_l$ Values | Orbital Name |
|---|---|---|
| 0 | 0 | $4s$ (1 orbital) |
| 1 | $-1, 0, +1$ | $4p$ (3 orbitals) |
| 2 | $-2, -1, 0, +1, +2$ | $4d$ (5 orbitals) |
| 3 | $-3, -2, -1, 0, +1, +2, +3$ | $4f$ (7 orbitals) |

Total number of orbitals $= 4^2 = 16$

**FOR PRACTICE 7.5** List the quantum numbers associated with all of the $5d$ orbitals. How many $5d$ orbitals exist?

## Example 7.6 Quantum Numbers II

These sets of quantum numbers are each supposed to specify an orbital. One set, however, is erroneous. Which one and why?

**(a)** $n = 3$; $l = 0$; $m_l = 0$      **(b)** $n = 2$; $l = 1$; $m_l = -1$

**(c)** $n = 1$; $l = 0$; $m_l = 0$      **(d)** $n = 4$; $l = 1$; $m_l = -2$

**SOLUTION**

Choice **(d)** is erroneous because for $l = 1$, the possible values of $m_l$ are only $-1, 0$, and $+1$.

**FOR PRACTICE 7.6** Each set of quantum numbers is supposed to specify an orbital. However, each set contains one quantum number that is not allowed. Replace the quantum number that is not allowed with one that is allowed.

(a) $n = 3$; $l = 3$; $m_l = +2$      (b) $n = 2$; $l = 1$; $m_l = -2$

(c) $n = 1$; $l = 1$; $m_l = 0$

## Atomic Spectroscopy Explained

Quantum theory explains the atomic spectra of atoms we discussed in Section 7.3. Each wavelength in the emission spectrum of an atom corresponds to an electron *transition* between quantum-mechanical orbitals. When an atom absorbs energy, an electron in a lower-energy orbital is *excited* or promoted to a higher-energy orbital, as shown in Figure 7.21▼. In this new configuration, however, the atom is unstable,

**Excitation and Radiation**

▶ **FIGURE 7.21 Excitation and Radiation** When an atom absorbs energy, an electron can be excited from an orbital in a lower-energy level to an orbital in a higher-energy level. The electron in this "excited state" is unstable, however, and relaxes to a lower-energy level, releasing energy in the form of electromagnetic radiation.

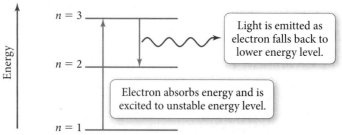

and the electron quickly falls back or *relaxes* to a lower-energy orbital. As it does so, it releases a photon of light containing an amount of energy precisely equal to the energy difference between the two energy levels. We saw previously (see Equation 7.7) that the energy of an orbital in a hydrogen atom with principal quantum number $n$ is given by $E_n = -2.18 \times 10^{-18} \, \text{J} \, (1/n^2)$, where $n = 1, 2, 3, \ldots$ The *difference* in energy between two levels $n_{\text{initial}}$ and $n_{\text{final}}$ is given by $\Delta E = E_{\text{final}} - E_{\text{initial}}$. If we substitute the expression for $E_n$ into the expression for $\Delta E$, we get the following important expression for the change in energy that occurs in a hydrogen atom when an electron changes energy levels:

$$\Delta E = E_{\text{final}} - E_{\text{initial}}$$

$$= -2.18 \times 10^{-18} \, \text{J} \left(\frac{1}{n_f^2}\right) - \left[-2.18 \times 10^{-18} \, \text{J} \left(\frac{1}{n_i^2}\right)\right]$$

$$\Delta E = -2.18 \times 10^{-18} \, \text{J} \left(\frac{1}{n_f^2} - \frac{1}{n_i^2}\right) \qquad [7.8]$$

For example, suppose that an electron in a hydrogen atom relaxes from an orbital in the $n = 3$ level to an orbital in the $n = 2$ level. We determine $\Delta E$, the energy difference corresponding to the transition from $n = 3$ to $n = 2$, as follows:

$$\Delta E_{\text{atom}} = E_2 - E_3$$

$$= -2.18 \times 10^{-18} \, \text{J} \left(\frac{1}{2^2}\right) - \left[-2.18 \times 10^{-18} \, \text{J} \left(\frac{1}{3^2}\right)\right]$$

$$= -2.18 \times 10^{-18} \, \text{J} \left(\frac{1}{2^2} - \frac{1}{3^2}\right)$$

$$= -3.03 \times 10^{-19} \, \text{J}$$

The energy carries a negative sign because the atom *emits* the energy as it relaxes from $n = 3$ to $n = 2$. Since energy must be conserved, the exact amount of energy emitted by the atom is carried away by the photon:

$$\Delta E_{\text{atom}} = -E_{\text{photon}}$$

This energy determines the frequency and wavelength of the photon. Since the wavelength of the photon is related to its energy as $E = hc/\lambda$, we calculate the wavelength of the photon as:

$$\lambda = \frac{hc}{E}$$

$$= \frac{(6.626 \times 10^{-34} \, \text{J} \cdot \text{s})(3.00 \times 10^8 \, \text{m/s})}{3.03 \times 10^{-19} \, \text{J}}$$

$$= 6.56 \times 10^{-7} \, \text{m} \quad \text{or} \quad 656 \, \text{nm}$$

The Rydberg equation, $1/\lambda = R(1/m^2 - 1/n^2)$, can be derived from the relationships just covered. We leave this derivation to an exercise (see Problem 7.98).

Consequently, the light emitted by an excited hydrogen atom as it relaxes from an orbital in the $n = 3$ level to an orbital in the $n = 2$ level has a wavelength of 656 nm (red). Similarly, we can calculate the light emitted due to a transition from $n = 4$ to $n = 2$, which is 486 nm (green). Notice that transitions between orbitals that are further apart in energy produce light that is higher in energy, and therefore shorter in wavelength, than transitions between orbitals that are closer together. Figure 7.22► on the next page shows several of the transitions in the hydrogen atom and their corresponding wavelengths.

**Emission Spectra** Which transition emits light with the shortest wavelength?

    **(a)** $n = 5 \longrightarrow n = 4$     **(b)** $n = 4 \longrightarrow n = 3$     **(c)** $n = 3 \longrightarrow n = 2$

CONCEPTUAL
CONNECTION **7.6**

▶ **FIGURE 7.22** **Hydrogen Energy Transitions and Radiation** An atomic energy level diagram for hydrogen, showing some possible electron transitions between levels and the corresponding wavelengths of emitted light.

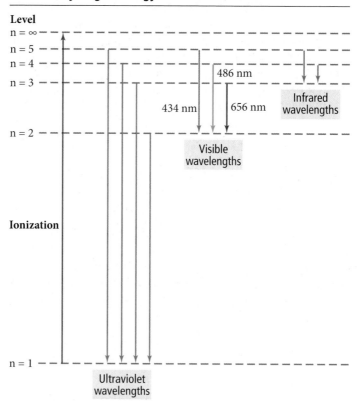

**Hydrogen Energy Transitions and Radiation**

## Example 7.7    Wavelength of Light for a Transition in the Hydrogen Atom

Determine the wavelength of light emitted when an electron in a hydrogen atom makes a transition from an orbital in $n = 6$ to an orbital in $n = 5$.

| | |
|---|---|
| **SORT** You are given the energy levels of an atomic transition and asked to find the wavelength of emitted light. | **GIVEN:** $n = 6 \longrightarrow n = 5$ <br><br> **FIND:** $\lambda$ |
| **STRATEGIZE** In the first part of the conceptual plan, calculate the energy of the electron in the $n = 6$ and $n = 5$ orbitals using Equation 7.7 and subtract to find $\Delta E_{atom}$. | **CONCEPTUAL PLAN** <br><br>  <br> $$\Delta E_{atom} = E_5 - E_6$$ |
| In the second part, find $E_{photon}$ by taking the negative of $\Delta E_{atom}$ and then calculate the wavelength corresponding to a photon of this energy using Equation 7.3. (The difference in sign between $E_{photon}$ and $\Delta E_{atom}$ applies only to emission. *The energy of a photon must always be positive.*) | $$\boxed{\Delta E_{atom}} \longrightarrow \boxed{E_{photon}} \longrightarrow \boxed{\lambda}$$ <br> $\Delta E_{atom} = -E_{photon} \qquad E = \dfrac{hc}{\lambda}$ <br><br> **RELATIONSHIPS USED** <br> $E_n = -2.18 \times 10^{-18}\,\text{J}(1/n^2)$ <br> $E = hc/\lambda$ |
| **SOLVE** Follow the conceptual plan. Begin by calculating $\Delta E_{atom}$. | **SOLUTION** <br> $\Delta E_{atom} = E_5 - E_6$ <br><br> $\quad = -2.18 \times 10^{-18}\,\text{J}\left(\dfrac{1}{5^2}\right) - \left[-2.18 \times 10^{-18}\text{J}\left(\dfrac{1}{6^2}\right)\right]$ <br><br> $\quad = -2.18 \times 10^{-18}\,\text{J}\left(\dfrac{1}{5^2} - \dfrac{1}{6^2}\right)$ <br><br> $\quad = -2.6644 \times 10^{-20}\,\text{J}$ |

| Calculate $E_{photon}$ by changing the sign of $\Delta E_{atom}$. | $E_{photon} = -\Delta E_{atom} = +2.6644 \times 10^{-20}\,J$ |
|---|---|
| Solve the equation relating the energy of a photon to its wavelength for $\lambda$. Substitute the energy of the photon and calculate $\lambda$. | $E = \dfrac{hc}{\lambda}$ <br><br> $\lambda = \dfrac{hc}{E}$ <br><br> $= \dfrac{(6.626 \times 10^{-34}\,J \cdot s)\,(3.00 \times 10^{8}\,m/s)}{2.6644 \times 10^{-20}\,J}$ <br><br> $= 7.46 \times 10^{-6}\,m \text{ or } 7460\,nm$ |

**CHECK** The units of the answer (m) are correct for wavelength. The magnitude is reasonable because $10^{-6}$ m is in the infrared region of the electromagnetic spectrum. You know that transitions from $n = 3$ or $n = 4$ to $n = 2$ lie in the visible region, so it makes sense that a transition between levels of higher $n$ value (which are energetically closer to one another) results in light of longer wavelength.

**FOR PRACTICE 7.7** Determine the wavelength of the light absorbed when an electron in a hydrogen atom makes a transition from an orbital in the $n = 2$ level to an orbital in the $n = 7$ level.

**FOR MORE PRACTICE 7.7** An electron in the $n = 6$ level of the hydrogen atom relaxes to a lower-energy level, emitting light of $\lambda = 93.8$ nm. Find the principal level to which the electron relaxed.

## 7.6 The Shapes of Atomic Orbitals

As we noted previously, the shapes of atomic orbitals are important because covalent chemical bonds depend on the sharing of the electrons that occupy these orbitals. In one model of chemical bonding, for example, a bond consists of the overlap of atomic orbitals on adjacent atoms. The shapes of the overlapping orbitals determine the shape of the molecule. Although we limit ourselves in this chapter to the orbitals of the hydrogen atom, we will see in Chapter 8 that the orbitals of all atoms can be approximated as being hydrogen-like and therefore having very similar shapes to those of hydrogen.

The shape of an atomic orbital is determined primarily by $l$, the angular momentum quantum number. Recall that each value of $l$ is assigned a letter that corresponds to particular orbitals. For example, the orbitals with $l = 0$ are called $s$ orbitals; those with $l = 1$, $p$ orbitals; those with $l = 2$, $d$ orbitals, and so on. We now examine the shape of each of these orbitals.

### s Orbitals ($l = 0$)

The lowest energy orbital is the spherically symmetrical $1s$ orbital shown in Figure 7.23(a)▼. This image is actually a three-dimensional plot of the wave function squared ($\psi^2$), which represents **probability density**, the probability (per unit volume) of finding the electron at a point in space.

$$\psi^2 = \text{probability density} = \frac{\text{probability}}{\text{unit volume}}$$

▼ **FIGURE 7.23 The 1s Orbital: Two Representations** In **(a)** the dot density is proportional to the electron probability density. In **(b)**, the height of the curve is proportional to the electron probability density. The x-axis is $r$, the distance from the nucleus.

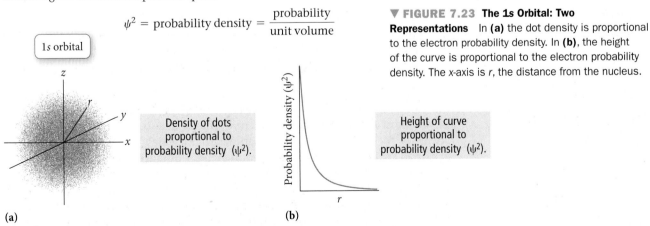

1s orbital

Density of dots proportional to probability density ($\psi^2$).

Height of curve proportional to probability density ($\psi^2$).

Probability density ($\psi^2$)

(a)

(b)

The magnitude of $\psi^2$ in this plot is proportional to the density of the dots shown in the image. The high dot density near the nucleus (at the very center of the plot) indicates a higher probability density for the electron. As you move away from the nucleus, the probability density decreases. Figure 7.23(b) on the previous page hows a plot of probability density ($\psi^2$) versus $r$, the distance from the nucleus. The plot represents a slice through the three-dimensional plot of $\psi^2$ and illustrates how the probability density decreases as $r$ increases.

We can better understand probability density with the help of a thought experiment. Imagine an electron in the 1s orbital located within the volume surrounding the nucleus. Imagine also taking a photograph of the electron every second for 10 or 15 minutes. In one photograph, the electron is very close to the nucleus, in another it is farther away, and so on. Each photo has a dot showing the electron's position relative to the nucleus when the photo was taken. Remember that we can never predict where the electron will be for any one photo. However, if we took hundreds of photos and superimposed all of them, we would have a plot similar to Figure 7.23(a)—a statistical representation of how likely the electron is to be found at each point.

This thought experiment can result in a possible misunderstanding: that the electron is moving around (like a moth near a flame) between photographs. However, in the quantum-mechanical model, that is not the case. Between photographs, the location of the electron is uncertain—in a sense its location is spread out over the entire volume of the orbital. Only when the photograph is taken (that is, when a measurement of the electron's location is made) does the location of the electron become localized to one spot. Between measurements, the electron has no single location. Recall from Section 7.1 that the measurement affects the outcome of any quantum system.

We can also represent an atomic orbital as a geometrical shape that encompasses the volume where the electron is likely to be found most frequently—typically, 90% of the time. For example, we represent the 1s orbital as the three-dimensional sphere shown in Figure 7.24 ◄. If we were to superimpose the dot-density representation of the 1s orbital on the shape representation, 90% of the dots would be within the sphere, meaning that when the electron is in the 1s orbital it has a 90% chance of being found within the sphere.

The plots we have just seen in Figures 7.23 and 7.24 represent probability *density*. However, they are a bit misleading because they seem to imply that the electron is most likely to be found *at the nucleus*. To get a better idea of where the electron is most likely to be found, we use a plot called the **radial distribution function**, shown in Figure 7.25 ◄ for the 1s orbital. The radial distribution function represents the *total probability of finding the electron within a thin spherical shell at a distance r from the nucleus*:

$$\text{Total radial probability (at a given } r) = \frac{\text{probability}}{\text{unit volume}} \times \text{volume of shell at } r$$

The radial distribution function represents, not probability density *at a point r*, but total probability *at a radius r*. In contrast to probability density, which has a maximum at the nucleus, the radial distribution function has a value of *zero* at the nucleus. It increases to a maximum at 52.9 pm and then decreases again with increasing $r$.

The shape of the radial distribution function is the result of multiplying together two functions with opposite trends in $r$. The two functions are:

1. the probability density function ($\psi^2$), which is the probability per unit volume, has a maximum at the nucleus, and decreases with increasing $r$; and

2. the volume of the thin shell, which is zero at the nucleus and increases with increasing $r$.

At the nucleus ($r = 0$), for example, the probability *density* is at a maximum; however, the volume of a thin spherical shell is zero, so the radial distribution function is zero. As $r$ increases, the volume of the thin spherical shell increases. We can understand this by making an analogy to an onion. A spherical shell at a distance $r$ from the nucleus is like a layer in an onion at a distance $r$ from its center. If the layers of the onion all have the same thickness, then the volume of any one layer—think of this as the total amount of onion in the layer—is greater as $r$ increases. Similarly, the volume of any one spherical shell in the radial distribution function increases with increasing distance from the

▲ **FIGURE 7.24 The 1s Orbital Surface** In this representation, the surface of the sphere encompasses the volume where the electron is found 90% of the time when the electron is in the 1s orbital.

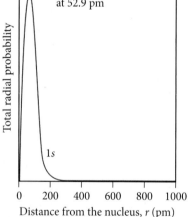

▲ **FIGURE 7.25 The Radial Distribution Function for the 1s Orbital** The curve shows the total probability of finding the electron within a thin shell at a distance $r$ from the nucleus.

1 pm = $10^{-12}$ m

nucleus, resulting in a greater total probability of finding the electron within that shell. Close to the nucleus, this increase in volume with increasing *r* outpaces the decrease in probability density, reaching a maximum at 52.9 pm. Farther out, however, the density tapers off faster than the volume increases.

The maximum in the radial distribution function, 52.9 pm, turns out to be the very same radius that Bohr had predicted for the innermost orbit of the hydrogen atom. However, there is a significant conceptual difference between the two radii. In the Bohr model, every time you probe the atom (in its lowest energy state), you would find the electron at a radius of 52.9 pm. In the quantum-mechanical model, you would generally find the electron at various radii, with 52.9 pm having the greatest probability.

The probability densities and radial distribution functions for the 2*s* and 3*s* orbitals are shown in Figure 7.26▼. Like the 1*s* orbital, these orbitals are spherically symmetric.

▼ **FIGURE 7.26** **Probability Densities and Radial Distribution Functions for the 2s and 3s Orbitals.**

## The 2*s* and 3*s* Orbitals

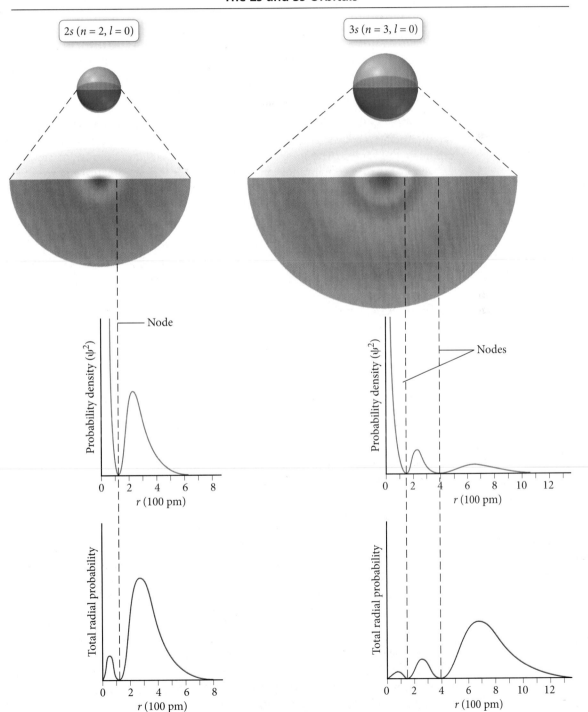

These orbitals are larger in size than the 1s orbital, and, unlike the 1s orbital, they contain at least one *node*. A **node** is a point where the wave function ($\psi$), and therefore the probability density ($\psi^2$) and radial distribution function, all pass through zero. A node in a wave function is much like a node in a standing wave on a vibrating string. We can see nodes in an orbital most clearly by looking at a slice through the orbital. Plots of probability density and the radial distribution function as a function of *r* both reveal the presence of nodes. The probability of finding the electron at a node is zero.

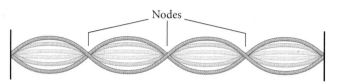

Nodes

▲ The nodes in quantum-mechanical atomic orbitals are three-dimensional analogs of the nodes on a vibrating string.

## *p* Orbitals (*l* = 1)

Each principal level with $n = 2$ or greater contains three *p* orbitals ($m_l = -1, 0, +1$). The three 2*p* orbitals and their radial distribution functions are shown in Figure 7.27 ▼. The *p* orbitals are not spherically symmetric like the *s* orbitals, but they have two *lobes* of electron density on either side of the nucleus and a node located at the nucleus. The three *p* orbitals differ only in their orientation and are orthogonal (mutually perpendicular) to one another. It is convenient to define an *x*-, *y*-, and *z*-axis system and then label each *p* orbital as $p_x$, $p_y$, and $p_z$. The 3*p*, 4*p*, 5*p*, and higher *p* orbitals are all similar in shape to the 2*p* orbitals, but they contain additional nodes (like the higher *s* orbitals) and are progressively larger in size.

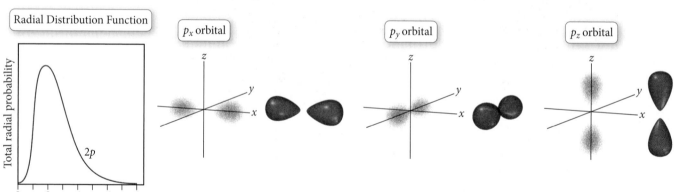

Radial Distribution Function

Total radial probability

2p

0   2   4   6   8
*r* (along lobe), 100 ppm

$p_x$ orbital

$p_y$ orbital

$p_z$ orbital

▲ **FIGURE 7.27 The 2*p* Orbitals and Their Radial Distribution Function** The radial distribution function is the same for each of the three 2*p* orbitals when the *x*-axis of the graph is taken as the axis containing the lobes of the orbital.

## *d* Orbitals (*l* = 2)

Each principal level with $n = 3$ or greater contains five *d* orbitals ($m_l = -2, -1, 0, +1, +2$). The five 3*d* orbitals are shown in Figure 7.28▶. Four of these orbitals have a cloverleaf shape, with four lobes of electron density around the nucleus and two perpendicular nodal planes. The $d_{xy}$, $d_{xz}$, and $d_{yz}$ orbitals are oriented along the *xy*, *xz*, and *yz* planes, respectively, and their lobes are oriented *between* the corresponding axes. The four lobes of the $d_{x^2-y^2}$ orbital are oriented along the *x*- and *y*-axes. The $d_{z^2}$ orbital is different in shape from the other four, having two lobes oriented along the *z*-axis and a donut-shaped ring along the *xy* plane. The 4*d*, 5*d*, 6*d*, ..., orbitals are all similar in shape to the 3*d* orbitals, but they contain additional nodes and are progressively larger in size.

A nodal plane is a plane where the electron probability density is zero. For example, in the $d_{xy}$ orbitals, the nodal planes lie in the *xz* and *yz* planes.

## *f* Orbitals (*l* = 3)

Each principal level with $n = 4$ or greater contains seven *f* orbitals ($m_l = -3, -2$ $-2, -1, 0, +1, +2, +3$), as shown in Figure 7.29▶. These *f* orbitals have more lobes and nodes than *d* orbitals.

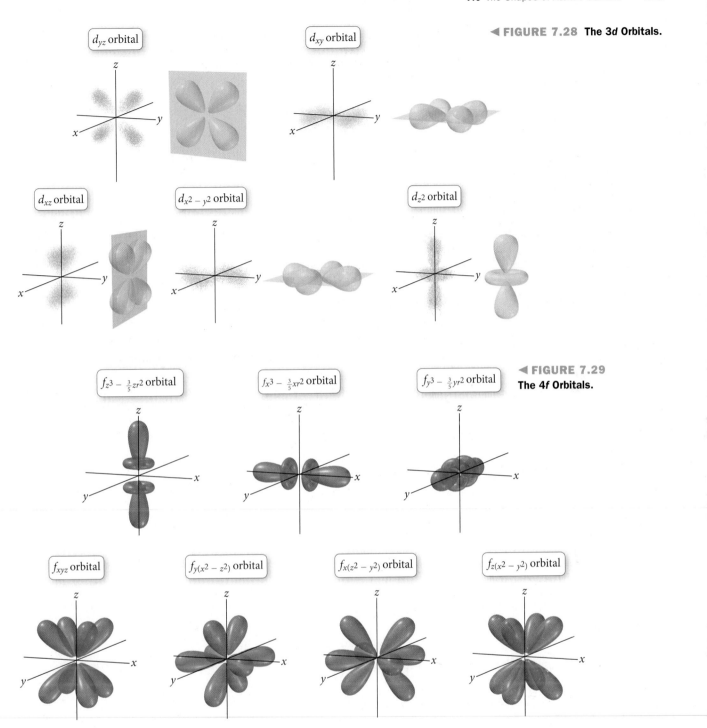

◀ **FIGURE 7.28** **The 3*d* Orbitals.**

◀ **FIGURE 7.29**
**The 4*f* Orbitals.**

## The Phase of Orbitals

The orbitals we have just shown are three-dimensional waves. We can understand an important property of these orbitals by analogy to one-dimensional waves. Consider the following one-dimensional waves:

The wave on the left has a positive amplitude over its entire length, whereas the wave on the right has a positive amplitude over half of its length and a negative amplitude over the other half. The sign of the amplitude of a wave—positive or negative—is known as its **phase**. In these images, blue indicates positive phase and red indicates negative phase. The phase of a wave determines how it interferes with another wave as we saw in Section 7.2.

Just as a one-dimensional wave has a phase, so does a three-dimensional wave. We often represent the phase of a quantum-mechanical orbital with color. For example, the phases of a 1s and 2p orbital are represented as follows:

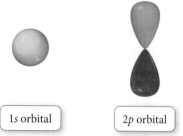

1s orbital          2p orbital

In these depictions, blue represents positive phase and red represents negative phase. The 1s orbital is all one phase, whereas the 2p orbital exhibits two different phases. The phase of quantum-mechanical orbitals is important in bonding, as we shall see in Chapter 10.

## The Shape of Atoms

If some orbitals are shaped like dumbbells and three-dimensional cloverleaves, and if most of the volume of an atom is empty space diffusely occupied by electrons in these orbitals, then why do we often depict atoms as spheres? Atoms are drawn as spheres because most atoms contain many electrons occupying a number of different orbitals. Therefore, the shape of an atom is obtained by superimposing all of its orbitals. If we superimpose the *s*, *p*, and *d* orbitals, we get a roughly spherical shape, as shown in Figure 7.30 ◄.

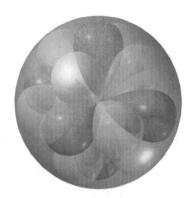

▲ **FIGURE 7.30 Why Atoms Are Spherical** Atoms are depicted as roughly spherical because all the orbitals together make up a roughly spherical shape.

# Self-Assessment Quiz

**Q1.** Which wavelength of light has the highest frequency?

a) 10 nm    b) 10 mm    c) 1 nm    d) 1 mm

**Q2.** Which kind of electromagnetic radiation contains the greatest energy per photon?

a) microwaves          b) gamma rays
c) X-rays              d) visible light

**Q3.** How much energy (in J) is contained in 1.00 mole of 552-nm photons?

a) $3.60 \times 10^{-19}$ J        b) $2.17 \times 10^5$ J
c) $3.60 \times 10^{-28}$ J        d) $5.98 \times 10^{-43}$ J

**Q4.** Light from three different lasers (A, B, and C), each with a different wavelength, was shined onto the same metal surface. Laser A produced no photoelectrons. Lasers B and C both produced photoelectrons, but the photoelectrons produced by laser B had a greater velocity than those produced by laser C. Arrange the lasers in order of increasing wavelength.

a) A < B < C          b) B < C < A
c) C < B < A          d) A < C < B

**Q5.** Calculate the wavelength of an electron traveling at $1.85 \times 10^7$ m/s.

a) $2.54 \times 10^{13}$ m
b) $3.93 \times 10^{-14}$ m
c) $2.54 \times 10^{10}$ m
d) $3.93 \times 10^{-11}$ m

**Q6.** Which set of three quantum numbers *does not* specify an orbital in the hydrogen atom?

a) $n = 2; l = 1; m_l = -1$    b) $n = 3; l = 3; m_l = -2$
c) $n = 2; l = 0; m_l = 0$     d) $n = 3; l = 2; m_l = 2$

**Q7.** Calculate the wavelength of light emitted when an electron in the hydrogen atom makes a transition from an orbital with $n = 5$ to an orbital with $n = 3$.

a) $1.28 \times 10^{-6}$ m          b) $6.04 \times 10^{-7}$ m
c) $2.28 \times 10^{-6}$ m          d) $1.55 \times 10^{-19}$ m

**Q8.** Which electron transition produces light of the highest frequency in the hydrogen atom?

a) $5p \longrightarrow 1s$          b) $4p \longrightarrow 1s$
c) $3p \longrightarrow 1s$          d) $2p \longrightarrow 1s$

**Q9.** How much time (in seconds) does it take light in a vacuum to travel 1.00 billion km?

a) $3.00 \times 10^{17}$ s          b) 3.33 s
c) $3.33 \times 10^3$ s             d) $3.00 \times 10^{20}$ s

**Q10.** Which orbital is a *d* orbital?

a)

b)

c)

d) None of the above

# Chapter 7 IN REVIEW

## Key Terms

**Section 7.1**
quantum-mechanical model (298)

**Section 7.2**
electromagnetic radiation (299)
amplitude (299)
wavelength ($\lambda$) (300)
frequency ($\nu$) (300)
electromagnetic spectrum (301)
gamma rays (301)
X-rays (301)
ultraviolet (UV) radiation (302)
visible light (302)
infrared (IR) radiation (302)
microwaves (302)
radio waves (302)
interference (303)

constructive interference (303)
destructive interference (303)
diffraction (304)
photoelectric effect (305)
photon (quantum) (306)

**Section 7.3**
emission spectrum (309)

**Section 7.4**
de Broglie relation (313)
complementary properties (315)
Heisenberg's uncertainty
    principle (315)
deterministic (316)
indeterminacy (316)

**Section 7.5**
orbital (317)
wave function (317)
quantum number (317)
principal quantum number ($n$) (317)
angular momentum quantum
    number ($l$) (317)
magnetic quantum number ($m_l$) (318)
spin quantum number ($m_s$) (318)
principal level (shell) (319)
sublevel (subshell) (319)

**Section 7.6**
probability density (323)
radial distribution function (324)
node (326)
phase (327)

## Key Concepts

### The Realm of Quantum Mechanics (7.1)

▶ The theory of quantum mechanics explains the behavior of particles, such as photons (particles of light) and electrons, in the atomic and subatomic realms.

▶ Since the electrons of an atom determine many of its chemical and physical properties, quantum mechanics is foundational to understanding chemistry.

### The Nature of Light (7.2)

▶ Light is a type of electromagnetic radiation—a form of energy embodied in oscillating electric and magnetic fields that travels through space at $3.00 \times 10^8$ m/s.

▶ The wave nature of light is characterized by its wavelength—the distance between wave crests—and its ability to experience interference (constructive or destructive) and diffraction.

▶ The particle nature of light is characterized by the specific quantity of energy carried in each photon.

▶ The electromagnetic spectrum includes all wavelengths of electromagnetic radiation from gamma rays (high energy per photon, short wavelength) to radio waves (low energy per photon, long wavelength). Visible light is a tiny sliver in the middle of the electromagnetic spectrum.

### Atomic Spectroscopy (7.3)

▶ Atomic spectroscopy is the study of the light absorbed and emitted by atoms when an electron makes a transition from one energy level to another.

▶ The wavelengths absorbed or emitted in atomic spectra depend on the energy differences between the levels involved in the transition; large energy differences result in short wavelengths, and small energy differences result in long wavelengths.

### The Wave Nature of Matter (7.4)

▶ Electrons have a wave nature with an associated wavelength; the de Broglie relation quantifies the wavelength of an electron.

▶ The wave nature and particle nature of matter are complementary—the more we know of one, the less we know of the other.

▶ The wave–particle duality of electrons is quantified in Heisenberg's uncertainty principle, which states that there is a limit to how well we can know both the position of an electron (associated with the electron's particle nature) and the velocity times the mass of an electron (associated with the electron's wave nature)—the more accurately one is measured, the greater the uncertainty in measurement of the other.

▶ The inability to simultaneously know both the position and the velocity of an electron results in indeterminacy, the inability to predict a trajectory for an electron. Consequently, electron behavior is described differently than the behavior of everyday-sized particles.

▶ The trajectory we normally associate with macroscopic objects is replaced, for electrons, with statistical descriptions that show not the electron's path, but the region where it is most likely to be found.

### The Quantum-Mechanical Model of the Atom (7.5, 7.6)

▶ The most common way to describe electrons in atoms according to quantum mechanics is to solve the Schrödinger equation for the energy states of the electrons within the atom. When the electron is in these states, its energy is well defined but its position is not. The position of an electron is described by a probability distribution map called an orbital.

▶ The solutions to the Schrödinger equation (including the energies and orbitals) are characterized by four quantum numbers: $n$, $l$, $m_l$, and $m_s$.

▶ The principal quantum number ($n$) determines the energy of the electron and the size of the orbital; the angular momentum quantum number ($l$) determines the shape of the orbital; the magnetic quantum number ($m_l$) determines the orientation of the orbital; and the spin quantum number ($m_s$) specifies the orientation of the spin of the electron.

# Key Equations and Relationships

Relationship between Frequency ($\nu$), Wavelength ($\lambda$), and the Speed of Light ($c$) (**7.2**)

$$\nu = \frac{c}{\lambda}$$

Relationship between Energy ($E$), Frequency ($\nu$), Wavelength ($\lambda$), and Planck's Constant ($h$) (**7.2**)

$$E = h\nu$$

$$E = \frac{hc}{\lambda}$$

De Broglie Relation: Relationship between Wavelength ($\lambda$), Mass ($m$), and Velocity ($v$) of a Particle (**7.4**)

$$\lambda = \frac{h}{mv}$$

Heisenberg's Uncertainty Principle: Relationship between a Particle's Uncertainty in Position ($\Delta x$) and Uncertainty in Velocity ($\Delta v$) (**7.4**)

$$\Delta x \times m\Delta v \geq \frac{h}{4\pi}$$

Energy of an Electron in an Orbital with Quantum Number $n$ in a Hydrogen Atom (**7.5**)

$$E_n = -2.18 \times 10^{-18} \text{ J}\left(\frac{1}{n^2}\right) \quad (n = 1, 2, 3, \ldots)$$

Change in Energy That Occurs in an Atom When It Undergoes a Transition between Levels $n_{\text{initial}}$ and $n_{\text{final}}$ (**7.5**)

$$\Delta E = -2.18 \times 10^{-18} \text{ J}\left(\frac{1}{n_f^2} - \frac{1}{n_i^2}\right)$$

# Key Learning Outcomes

| CHAPTER OBJECTIVES | ASSESSMENT |
|---|---|
| Calculating the Wavelength and Frequency of Light (**7.2**)  | Example 7.1   For Practice 7.1   Exercises 39, 40 |
| Calculating the Energy of a Photon (**7.2**) | Example 7.2   For Practice 7.2   For More Practice 7.4   Exercises 41–46 |
| Relating Wavelength, Energy, and Frequency to the Electromagnetic Spectrum (**7.2**) | Example 7.3   For Practice 7.3   Exercises 37, 42 |
| Using the de Broglie Relation to Calculate Wavelength (**7.4**) | Example 7.4   For Practice 7.4   Exercises 49–54 |
| Relating Quantum Numbers to One Another and to Their Corresponding Orbitals (**7.5**) | Examples 7.5, 7.6   For Practice 7.5, 7.6   Exercises 59–62 |
| Relating the Wavelength of Light to Transitions in the Hydrogen Atom (**7.5**) | Example 7.7   For Practice 7.7   For More Practice 7.7   Exercises 69–72 |

# EXERCISES

## Review Questions

1. Why is the quantum-mechanical model of the atom important for understanding chemistry?

2. What is light? How fast does it travel in a vacuum?

3. Define the wavelength and amplitude of a wave. How are these related to the energy of the wave?

4. Define the frequency of electromagnetic radiation. How is frequency related to wavelength?

5. What determines the color of light? Describe the difference between red light and blue light.

6. What determines the color of a colored object? Explain why grass appears green.

7. Give an approximate range of wavelengths for each type of electromagnetic radiation and summarize the characteristics and/or the uses of each.
   a. gamma rays
   b. X-rays
   c. ultraviolet radiation
   d. visible light
   e. infrared radiation
   f. microwave radiation
   g. radio waves

8. Explain the wave behavior known as interference. Explain the difference between constructive and destructive interference.

9. Explain the wave behavior known as diffraction. Draw the diffraction pattern that occurs when light travels through two slits comparable in size and separation to the light's wavelength.

**10.** Describe the photoelectric effect. How did experimental observations of this phenomenon differ from the predictions of classical electromagnetic theory?

**11.** How did the photoelectric effect lead Einstein to propose that light is quantized?

**12.** What is a photon? How is the energy of a photon related to its wavelength? Its frequency?

**13.** What is an emission spectrum? How does an emission spectrum of a gas in a discharge tube differ from a white light spectrum?

**14.** Describe the Bohr model for the atom. How did the Bohr model account for the emission spectra of atoms?

**15.** Explain electron diffraction.

**16.** What is the de Broglie wavelength of an electron? What determines the value of the de Broglie wavelength for an electron?

**17.** What are complementary properties? How does electron diffraction demonstrate the complementarity of the wave nature and particle nature of the electron?

**18.** Explain Heisenberg's uncertainty principle. What paradox is at least partially solved by the uncertainty principle?

**19.** What is a trajectory? What kind of information do you need to predict the trajectory of a particle?

**20.** Why does the uncertainty principle make it impossible to predict a trajectory for the electron?

**21.** Newton's laws of motion are *deterministic*. Explain this statement.

**22.** An electron behaves in ways that are at least partially indeterminate. Explain this statement.

**23.** What is a probability distribution map?

**24.** For each solution to the Schrödinger equation, what can be precisely specified: the electron's energy or its position? Explain.

**25.** What is a quantum-mechanical orbital?

**26.** What is the Schrödinger equation? What is a wave function? How is a wave function related to an orbital?

**27.** What are the possible values of the principal quantum number $n$? What does the principal quantum number determine?

**28.** What are the possible values of the angular momentum quantum number $l$? What does the angular momentum quantum number determine?

**29.** What are the possible values of the magnetic quantum number $m_l$? What does the magnetic quantum number determine?

**30.** List all the orbitals in each principal level. Specify the three quantum numbers for each orbital.
   **a.** $n = 1$   **b.** $n = 2$   **c.** $n = 3$   **d.** $n = 4$

**31.** Explain the difference between a plot showing the probability density for an orbital and one showing the radial distribution function.

**32.** Make sketches of the general shapes of the $s$, $p$, and $d$ orbitals.

**33.** List the four different sublevels. Given that only a maximum of two electrons can occupy an orbital, determine the maximum number of electrons that can exist in each sublevel.

**34.** Why are atoms usually portrayed as spheres when most orbitals are not spherically shaped?

# Problems by Topic

## Electromagnetic Radiation

**35.** The distance from the sun to Earth is $1.496 \times 10^8$ km. How long does it take light to travel from the sun to Earth?

**36.** The nearest star to our sun is Proxima Centauri, at a distance of 4.3 light-years from the sun. A light-year is the distance that light travels in one year (365 days). How far away, in km, is Proxima Centauri from the sun?

**37.** List these types of electromagnetic radiation in order of (i) increasing wavelength and (ii) increasing energy per photon.
   **a.** radio waves
   **b.** microwaves
   **c.** infrared radiation
   **d.** ultraviolet radiation

**38.** List these types of electromagnetic radiation in order of (i) increasing frequency and (ii) decreasing energy per photon.
   **a.** gamma rays
   **b.** radio waves
   **c.** microwaves
   **d.** visible light

**39.** Calculate the frequency of each wavelength of electromagnetic radiation.
   **a.** 632.8 nm (wavelength of red light from helium–neon laser)
   **b.** 503 nm (wavelength of maximum solar radiation)
   **c.** 0.052 nm (a wavelength contained in medical X-rays)

**40.** Calculate the wavelength of each frequency of electromagnetic radiation.
   **a.** 100.2 MHz (typical frequency for FM radio broadcasting)
   **b.** 1070 kHz (typical frequency for AM radio broadcasting) (assume four significant figures)

   **c.** 835.6 MHz (common frequency used for cell phone communication)

**41.** Calculate the energy of a photon of electromagnetic radiation at each of the wavelengths indicated in Problem 39.

**42.** Calculate the energy of a photon of electromagnetic radiation at each of the frequencies indicated in Problem 40.

**43.** A laser pulse with wavelength 532 nm contains 3.85 mJ of energy. How many photons are in the laser pulse?

**44.** A heat lamp produces 32.8 watts of power at a wavelength of $6.5 \, \mu m$. How many photons are emitted per second? (1 watt = 1 J/s)

**45.** Determine the energy of 1 mol of photons for each kind of light. (Assume three significant figures.)
   **a.** infrared radiation (1500 nm)
   **b.** visible light (500 nm)
   **c.** ultraviolet radiation (150 nm)

**46.** How much energy is contained in 1 mol of each?
   **a.** X-ray photons with a wavelength of 0.135 nm
   **b.** $\gamma$-ray photons with a wavelength of $2.15 \times 10^{-5}$ nm

## The Wave Nature of Matter and the Uncertainty Principle

**47.** Sketch the interference pattern that results from the diffraction of electrons passing through two closely spaced slits.

**48.** What happens to the interference pattern described in Problem 47 if the rate of electrons going through the slits is decreased to one electron per hour? What happens to the pattern if we try to determine which slit the electron goes through by using a laser placed directly behind the slits?

49. The resolution limit of a microscope is roughly equal to the wavelength of light used in producing the image. Electron microscopes use an electron beam (in place of photons) to produce much higher resolution images, about 0.20 nm in modern instruments. Assuming that the resolution of an electron microscope is equal to the de Broglie wavelength of the electrons used, to what speed must the electrons be accelerated to obtain a resolution of 0.20 nm?

50. The smallest atoms can themselves exhibit quantum-mechanical behavior. Calculate the de Broglie wavelength (in pm) of a hydrogen atom traveling at 475 m/s.

51. What is the de Broglie wavelength of an electron traveling at $1.35 \times 10^5$ m/s?

52. A proton in a linear accelerator has a de Broglie wavelength of 122 pm. What is the speed of the proton?

53. Calculate the de Broglie wavelength of a 143-g baseball traveling at 95 mph. Why is the wave nature of matter not important for a baseball?

54. A 0.22-caliber handgun fires a 1.9-g bullet at a velocity of 765 m/s. Calculate the de Broglie wavelength of the bullet. Is the wave nature of matter significant for bullets?

55. An electron has an uncertainty in its position of 552 pm. What is the uncertainty in its velocity?

56. An electron traveling at $3.7 \times 10^5$ m/s has an uncertainty in its velocity of $1.88 \times 10^5$ m/s. What is the uncertainty in its position?

## Orbitals and Quantum Numbers

57. Which electron is, on average, closer to the nucleus: an electron in a 2s orbital or an electron in a 3s orbital?

58. Which electron is, on average, farther from the nucleus: an electron in a 3p orbital or an electron in a 4p orbital?

59. What are the possible values of $l$ for each value of $n$?
    **a.** 1      **b.** 2      **c.** 3      **d.** 4

60. What are the possible values of $m_l$ for each value of $l$?
    **a.** 0      **b.** 1      **c.** 2      **d.** 3

61. Which set of quantum numbers *cannot* specify an orbital?
    **a.** $n = 2, l = 1, m_l = -1$      **b.** $n = 3, l = 2, m_l = 0$
    **c.** $n = 3, l = 3, m_l = 2$      **d.** $n = 4, l = 3, m_l = 0$

62. Which combinations of $n$ and $l$ represent real orbitals, and which do not exist?
    **a.** 1s      **b.** 2p      **c.** 4s      **d.** 2d

63. Sketch the 1s and 2p orbitals. How do the 2s and 3p orbitals differ from the 1s and 2p orbitals?

64. Sketch the 3d orbitals. How do the 4d orbitals differ from the 3d orbitals?

## Atomic Spectroscopy

65. An electron in a hydrogen atom is excited with electrical energy to an excited state with $n = 2$. The atom then emits a photon. What is the value of $n$ for the electron following the emission?

66. Determine whether each transition in the hydrogen atom corresponds to absorption or emission of energy.
    **a.** $n = 3 \longrightarrow n = 1$
    **b.** $n = 2 \longrightarrow n = 4$
    **c.** $n = 4 \longrightarrow n = 3$

67. According to the quantum-mechanical model for the hydrogen atom, which electron transitions produces light with the longer wavelength: $2p \longrightarrow 1s$ or $3p \longrightarrow 1s$?

68. According to the quantum-mechanical model for the hydrogen atom, which electron transition produces light with the longer wavelength: $3p \longrightarrow 2s$ or $4p \longrightarrow 3p$?

69. Calculate the wavelength of the light emitted when an electron in a hydrogen atom makes each transition and indicate the region of the electromagnetic spectrum (infrared, visible, ultraviolet, etc.) where the light is found.
    **a.** $n = 2 \longrightarrow n = 1$      **b.** $n = 3 \longrightarrow n = 1$
    **c.** $n = 4 \longrightarrow n = 2$      **d.** $n = 5 \longrightarrow n = 2$

70. Calculate the frequency of the light emitted when an electron in a hydrogen atom makes each transition.
    **a.** $n = 4 \longrightarrow n = 3$      **b.** $n = 5 \longrightarrow n = 1$
    **c.** $n = 5 \longrightarrow n = 4$      **d.** $n = 6 \longrightarrow n = 5$

71. An electron in the $n = 7$ level of the hydrogen atom relaxes to a lower-energy level, emitting light of 397 nm. What is the value of $n$ for the level to which the electron relaxed?

72. An electron in a hydrogen atom relaxes to the $n = 4$ level, emitting light of 114 THz. What is the value of $n$ for the level in which the electron originated?

# Cumulative Problems

73. Ultraviolet radiation and radiation of shorter wavelengths can damage biological molecules because these kinds of radiation carry enough energy to break bonds within the molecules. A typical carbon–carbon bond requires 348 kJ/mol to break. What is the longest wavelength of radiation with enough energy to break carbon–carbon bonds?

74. The human eye contains a molecule called 11-*cis*-retinal that changes shape when struck with light of sufficient energy. The change in shape triggers a series of events that results in an electrical signal being sent to the brain that results in vision. The minimum energy required to change the conformation of 11-*cis*-retinal within the eye is about 164 kJ/mol. Calculate the longest wavelength visible to the human eye.

75. An argon ion laser puts out 5.0 W of continuous power at a wavelength of 532 nm. The diameter of the laser beam is 5.5 mm. If the laser is pointed toward a pinhole with a diameter of 1.2 mm, how many photons travel through the pinhole per second? Assume that the light intensity is equally distributed throughout the entire cross-sectional area of the beam. (1 W = 1 J/s)

76. A green leaf has a surface area of 2.50 cm². If solar radiation is 1000 W/m², how many photons strike the leaf every second? Assume three significant figures and an average wavelength of 504 nm for solar radiation.

77. In a technique used for surface analysis called Auger electron spectroscopy (AES), electrons are accelerated toward a metal surface. These electrons cause the emissions of secondary electrons—called Auger electrons—from the metal surface. The kinetic energy of the Auger electrons depends on the composition of the surface. The presence of oxygen atoms on the surface results in

Auger electrons with a kinetic energy of approximately 506 eV. What is the de Broglie wavelength of one of these electrons?

$$[KE = \tfrac{1}{2}mv^2; 1 \text{ electron volt (eV)} = 1.602 \times 10^{-19} \text{ J}]$$

78. An X-ray photon of wavelength 0.989 nm strikes a surface. The emitted electron has a kinetic energy of 969 eV. What is the binding energy of the electron in kJ/mol?

$$[KE = \tfrac{1}{2}mv^2; 1 \text{ electron volt (eV)} = 1.602 \times 10^{-19} \text{ J}]$$

79. Ionization involves completely removing an electron from an atom. How much energy is required to ionize a hydrogen atom in its ground (or lowest energy) state? What wavelength of light contains enough energy in a single photon to ionize a hydrogen atom?

80. The energy required to ionize sodium is 496 kJ/mol. What minimum frequency of light is required to ionize sodium?

81. Suppose that in an alternate universe, the possible values of $l$ are the integer values from 0 to $n$ (instead of 0 to $n - 1$). Assuming no other differences between this imaginary universe and ours, how many orbitals exist in each level in the alternate universe?
    **a.** $n = 1$    **b.** $n = 2$    **c.** $n = 3$

82. Suppose that, in an alternate universe, the possible values of $m_l$ are the integer values including 0 ranging from $-l - 1$ to $l + 1$ (instead of simply $-l$ to $+l$). How many orbitals exist in each sublevel in the alternate universe?
    **a.** $s$ sublevel    **b.** $p$ sublevel    **c.** $d$ sublevel

83. An atomic emission spectrum of hydrogen shows three wavelengths: 1875 nm, 1282 nm, and 1093 nm. Assign these wavelengths to transitions in the hydrogen atom.

84. An atomic emission spectrum of hydrogen shows three wavelengths: 121.5 nm, 102.6 nm, and 97.23 nm. Assign these wavelengths to transitions in the hydrogen atom.

85. The binding energy of electrons in a metal is 193 kJ/mol. Find the threshold frequency of the metal.

86. In order for a thermonuclear fusion reaction of two deuterons ($^2_1\text{H}^+$) to take place, the deuterons must collide and each must have a velocity of about $1 \times 10^6$ m/s. Find the wavelength of such a deuteron.

87. The speed of sound in air is 344 m/s at room temperature. The lowest frequency of a large organ pipe is 30 s$^{-1}$ and the highest frequency of a piccolo is $1.5 \times 10^4$ s$^{-1}$. Find the difference in wavelength between these two sounds.

88. The distance from Earth to the sun is $1.5 \times 10^8$ km. Find the number of crests in a light wave of frequency $1.0 \times 10^{14}$ s$^{-1}$ traveling from the sun to Earth.

89. The iodine molecule can be photodissociated (broken apart with light) into iodine atoms in the gas phase with light of wavelengths shorter than about 792 nm. A 100.0-mL glass tube contains 55.7 mtorr of gaseous iodine at 25.0 °C. What minimum amount of light energy must be absorbed by the iodine in the tube to dissociate 15.0% of the molecules?

90. A 5.00-mL ampule of a 0.100-M solution of naphthalene in hexane is excited with a flash of light. The naphthalene emits 15.5 J of energy at an average wavelength of 349 nm. What percentage of the naphthalene molecules emitted a photon?

91. A laser produces 20.0 mW of red light. In 1.00 hr, the laser emits $2.29 \times 10^{20}$ photons. What is the wavelength of the laser?

92. A particular laser consumes 150.0 watts of electrical power and produces a stream of $1.33 \times 10^{19}$ 1064-nm photons per second. What is the percent efficiency of the laser in converting electrical power to light?

93. The quantum yield of light-induced chemical reactions (called photochemical reactions) measures the efficiency of the process. The quantum yield, $\phi$, is defined as:
    $$\phi = \frac{\text{number of reaction events}}{\text{number of photons absorbed}}.$$ Suppose the quantum yield for the reaction $CH_3X \longrightarrow CH_3 + X$ is $\phi = 0.24$. A cuvette containing a solution of $CH_3X$ is irradiated with 280-nm light with a power of 885 mW for 10.0 minutes. Assuming total absorption of the light by the sample, what is the maximum amount (in moles) of $CH_3X$ that breaks apart?

94. A student is studying the photodissociation (dissociation with light) of $I_2 \longrightarrow 2I$. When a sample of $I_2$ is irradiated with a power of 255 mW at 590 nm for 35 seconds, 0.0256 mmol of I forms. Assuming complete absorption of the incident radiation, what is the quantum yield, $\phi$, of the reaction? (See Problem 93 for definition of quantum yield.)

# Challenge Problems

95. An electron confined to a one-dimensional box has energy levels given by the equation:
    $$E_n = n^2h^2/8\,mL^2$$
    where $n$ is a quantum number with possible values of $1, 2, 3, \ldots$, $m$ is the mass of the particle, and $L$ is the length of the box.
    **a.** Calculate the energies of the $n = 1$, $n = 2$, and $n = 3$ levels for an electron in a box with a length of 155 pm.
    **b.** Calculate the wavelength of light required to make a transition from $n = 1 \longrightarrow n = 2$ and from $n = 2 \longrightarrow n = 3$. In what region of the electromagnetic spectrum do these wavelengths lie?

96. The energy of a vibrating molecule is quantized much like the energy of an electron in the hydrogen atom. The energy levels of a vibrating molecule are given by the equation:
    $$E_n = \left(n + \frac{1}{2}\right)h\nu$$

where $n$ is a quantum number with possible values of $1, 2, \ldots$, and $\nu$ is the frequency of vibration. The vibration frequency of HCl is approximately $8.85 \times 10^{13}$ s$^{-1}$. What minimum energy is required to excite a vibration in HCl? What wavelength of light is required to excite this vibration?

97. The wave functions for the 1s and 2s orbitals are as follows:
    $$1s \quad \psi = (1/\pi)^{1/2}(1/a_0^{3/2})\exp(-r/a_0)$$
    $$2s \quad \psi = (1/32\pi)^{1/2}(1/a_0^{3/2})(-2r/a_0)\exp(-r/a_0)$$

    where $a_0$ is a constant ($a_0 = 53$ pm) and $r$ is the distance from the nucleus. Use a spreadsheet to make a plot of each of these wave functions for values of $r$ ranging from 0 pm to 200 pm. Describe the differences in the plots and identify the node in the 2s wave function.

**98.** Before quantum mechanics was developed, Johannes Rydberg developed an equation that predicted the wavelengths ($\lambda$) in the atomic spectrum of hydrogen:

$$1/\lambda = R(1/m^2 - 1/n^2)$$

In this equation $R$ is a constant and $m$ and $n$ are integers. Use the quantum-mechanical model for the hydrogen atom to derive the Rydberg equation.

**99.** Find the velocity of an electron emitted by a metal with a threshold frequency of $2.25 \times 10^{14}\,\text{s}^{-1}$ when it is exposed to visible light of wavelength $5.00 \times 10^{-7}\,\text{m}$.

**100.** Water is exposed to infrared radiation of wavelength $2.8 \times 10^{-4}$ cm. Assume that all the radiation is absorbed and converted to heat. How many photons are required to raise the temperature of 2.0 g of water by 2.0 K?

**101.** The 2005 Nobel Prize in Physics was given, in part, to scientists who had made ultrashort pulses of light. These pulses are important in making measurements involving very short time periods. One challenge in making such pulses is the uncertainty principle, which can be stated with respect to energy and time as $\Delta E \cdot \Delta t \geq h/4\pi$. What is the energy uncertainty ($\Delta E$) associated with a short pulse of laser light that lasts for only 5.0 femtoseconds (fs)? Suppose the low-energy end of the pulse had a wavelength of 722 nm. What is the wavelength of the high-energy end of the pulse that is limited only by the uncertainty principle?

**102.** A metal with a threshold frequency of $6.71 \times 10^{14}\,\text{s}^{-1}$ emits an electron with a velocity of $6.95 \times 10^5$ m/s when radiation of $1.01 \times 10^{15}\,\text{s}^{-1}$ strikes the metal. Calculate the mass of the electron.

**103.** Find the longest wavelength of a wave that can travel around in a circular orbit of radius 1.8 m.

**104.** The heat of fusion of ice is 6.00 kJ/mol. Find the number of photons of wavelength $= 6.42 \times 10^{-6}$ m that must be absorbed to melt 1.00 g of ice.

# Conceptual Problems

**105.** Explain the difference between the Bohr model for the hydrogen atom and the quantum-mechanical model. Is the Bohr model consistent with Heisenberg's uncertainty principle?

**106.** The light emitted from one of the following electronic transitions ($n = 4 \longrightarrow n = 3$ or $n = 3 \longrightarrow n = 2$) in the hydrogen atom causes the photoelectric effect in a particular metal, while light from the other transition does not. Which transition causes the photoelectric effect and why?

**107.** Determine whether an interference pattern is observed on the other side of the slits in each experiment.
   **a.** An electron beam is aimed at two closely spaced slits. The beam produces only one electron per minute.
   **b.** An electron beam is aimed at two closely spaced slits. A light beam is placed at each slit to determine when an electron goes through the slit.

   **c.** A high-intensity light beam is aimed at two closely spaced slits.
   **d.** A gun is fired at a solid wall containing two closely spaced slits. (Will the bullets that pass through the slits form an interference pattern on the other side of the solid wall?)

**108.** Which transition in the hydrogen atom produces emitted light with the longest wavelength?
   **a.** $n = 4 \longrightarrow n = 3$
   **b.** $n = 2 \longrightarrow n = 1$
   **c.** $n = 3 \longrightarrow n = 2$

# Questions for Group Work

Active Classroom Learning

Discuss these questions with the group and record your consensus answer.

**109.** Discuss the nature of light with your group. Ask each member of your group to transcribe one complete sentence about the physical nature of light.

**110.** How are electrons like baseballs? How are they unlike baseballs?

**111.** What are all the possible values of $m_l$ if $l = 0$ (an $s$ orbital)? If $l = 1$ (a $p$ orbital)? If $l = 2$ (a $d$ orbital)? How many possible values of $m_l$ would there be if $l = 20$? Write an equation to determine the number of possible values of $m_l$ from the value of $l$.

**112.** Have each group member choose a set of quantum numbers for an electron in a hydrogen atom. Calculate the wavelength of light produced if an electron moves from your state to each state of the other group members. Make a table comparing all possible combinations, and list all wavelengths in order of increasing energy.

**113.** How many nodes are there in the $1s$, $2p$, and $3d$ orbitals? How many nodes are in a $4f$ orbital?

# Data Interpretation and Analysis

## UV Radiation and Sunscreen

**114.** On average sunlight shines on the surface of Earth with an intensity of 910 Watts/m$^2$. Some of this energy is in the form of ultraviolet (UV) radiation. The compounds in chemical sunscreens absorb ultraviolet rays and release the energy as lower-energy infrared (IR) rays, thereby preventing the skin-damaging UV rays from reaching the skin. You may have noticed references to UV-A (320–400 nm) and UV-B (280–320 nm) protection in sunscreen packaging and advertisements. Figure a▶ lists the chemical formula and molar mass of two compounds, 2-ethylhexyl $p$-methoxycinnamate (2-EHMC), a UV-B sunscreen, and terephthalylidenedicamphor sulfonic acid (TDSA), a UV-A sunscreen.

| Sunscreen Compound | Chemical Formula | $\mathcal{M}$ (g/mol) |
|---|---|---|
| terephthalylidenedicamphor sulfonic acid (TDSA) | $C_{28}H_{34}O_8S_2$ | 562.70 |
| 2-ethylhexyl p-methoxycinnamate (2-EHMC) | $C_{18}H_{26}O_3$ | 290.39 |

▲ **FIGURE a   Chemical Formula and Molar Mass ($\mathcal{M}$) of Chemicals Commonly Found in Sunscreens**

Organic molecules have groups of atoms bonded together that can vibrate and rotate, making more energy states available. As a result, instead of absorbing exact wavelengths of radiation as elements do, organic molecules absorb over a broad range of wavelengths. Figure b▼ plots absorbance versus wavelength for 2-EHMC and TDSA. Greater absorbance indicates that the compound absorbs more light.

▲ **FIGURE b   Sunscreen Absorption**
http://mycpss.com/critical-wavelength-broad-spectrum-uv-protection/

Use the information provided in the figures to answer the following questions:

a. Calculate the energy absorbed by TDSA at its maximum absorption for one photon of UV radiation.

b. Calculate the energy absorbed by 2-EHMC at its maximum absorption for one photon of UV radiation.

c. Which absorbs more energy, 2-EHMC or TDSA, at their respective maximum wavelengths?

d. Why are these two compounds commonly included in sunscreen products?

e. A person does not apply sunscreen and has 0.42 m² of skin exposed to the sun for 1.00 hour. Calculate the total energy (in Joules) absorbed by the skin. Assume there are $3.066 \times 10^{21}$ photons/meter²/sec. Assume an average wavelength of 330 nm in order to calculate the energy of a UV photon. Assume half of the photons that strike the skin are reflected.

f. In order for a sunscreen product with equal amounts of TDSA and 2-EHMC to provide adequate protection from UV-A and UV-B rays for 1.00 hour, how many molecules of TDSA and 2-EHMC must be present on the skin? Assume one molecule of TDSA and one molecule 2-EHMC can each absorb 10,000 photons per hour. Assume the photon rate is the same as in part e.

g. Calculate the mass of TDSA and 2-EHMC in grams that must be present on the exposed skin referenced in parts e and f in order to protect the skin.

# Answers to Conceptual Connections

### Frequency of Light

**7.1 (b)** Infrared light has a lower frequency range than visible light, which in turn has a lower frequency range than ultraviolet light.

### The Photoelectric Effect

**7.2** Observation A corresponds to 632 nm, observation B corresponds to 325 nm, and observation C corresponds to 455 nm. The shortest wavelength of light (highest energy per photon) corresponds to the photoelectrons with the greatest kinetic energy. The longest wavelength of light (lowest energy per photon) corresponds to the instance in which no photoelectrons were observed.

### The de Broglie Wavelength of Macroscopic Objects

**7.3** Because of the baseball's large mass, its de Broglie wavelength is minuscule. (For a 150-g baseball, $\lambda$ is on the order of $10^{-34}$ m.) This minuscule wavelength is insignificant compared to the size of the baseball itself, and therefore its effects are not observable.

### The Relationship between n and l

**7.4 (c)** Since $l$ can have a maximum value of $n - 1$ and since $n = 3$, $l$ can have a maximum value of 2.

### The Relationship between l and $m_l$

**7.5 (d)** Since $m_l$ can have the integer values (including 0) between $-l$ and $+l$ and since $l = 2$, the possible values of $m_l$ are $-2, -1, 0, +1,$ and $+2$.

### Emission Spectra

**7.6 (c)** The energy difference between $n = 3$ and $n = 2$ is greatest because the energy differences become closer together with increasing $n$. The greater energy difference results in an emitted photon of greater energy and therefore shorter wavelength.

CHAPTER

# Periodic Properties of the Elements

> Beginning students of chemistry often think of the science as a mere collection of disconnected data to be memorized by brute force. Not at all! Just look at it properly and everything hangs together and makes sense.
>
> —ISAAC ASIMOV (1920–1992)

Great advances in science occur not only when a scientist sees something new, but also when a scientist sees what everyone else has seen in a new way. In other words, great scientists often see patterns where others have seen only disjointed facts. Such was the case in 1869 when Dmitri Mendeleev (1834–1907), a Russian chemist, saw a pattern in the properties of elements. Mendeleev's insight led to the periodic table, arguably the single most important tool for chemists. Recall that scientists devise theories that explain the underlying reasons for observations. Mendeleev's periodic table is a compact way to summarize a large number of observations; quantum mechanics (covered in Chapter 7) is the theory that explains the underlying reasons for the periodic table. Quantum mechanics explains how electrons are arranged in atoms, which in turn determines the element's properties. Since the periodic table is organized according to those properties, quantum mechanics beautifully accounts for Mendeleev's periodic table. In this chapter, we see a continuation of the theme we have been developing since page one of this book—the properties of macroscopic substances (in this case, the elements in the periodic table) are explained by the properties of the particles that compose them (in this case, atoms and their electrons).

▲ *In order for a nerve cell to transmit a signal, sodium and potassium ions must flow in opposite directions through specific ion channels in the cell membrane.*

## 8.1 Nerve Signal Transmission

As you sit reading this book, tiny pumps in the membranes of your cells are working hard to transport ions—especially sodium ($Na^+$) and potassium ($K^+$)—through those membranes. Amazingly, the ions are pumped in opposite directions. Sodium ions are pumped *out of cells*, while potassium ions are pumped *into cells*. The result is a *chemical gradient* for each ion: the concentration of sodium is higher outside the cell than within, whereas just the opposite is true for potassium. These ion pumps are analogous to the water pumps in a high-rise building that pump water against the force of gravity to a tank on the roof. Other structures within the membrane, called ion channels, are like the building's faucets. When these open, sodium and potassium ions flow back down their gradients—sodium flowing in and potassium flowing out. This movement of ions is the basis for the transmission of nerve signals in the brain and throughout the body. Every move you make, every thought you have, and every sensation you experience is mediated by these ion movements.

How do the pumps and channels differentiate between sodium and potassium ions to selectively move one out of the cell and the other into the cell? To answer this question, we must examine the ions more closely. Both ions are cations of group 1A metals. All group 1A metals tend to lose one electron to form cations with a 1+ charge, so that cannot be the decisive factor. Potassium (atomic number 19) lies directly below sodium in the periodic table (atomic number 11), indicating that it has more protons, neutrons, and electrons than sodium. How do these additional subatomic particles affect the properties of potassium? As we will see in this chapter, although a higher atomic number does not always result in a larger ion (or atom), it does in the case of potassium (relative to sodium). The potassium ion has a radius of 133 pm whereas the sodium ion has a radius of 95 pm. (Recall that 1 pm = $10^{-12}$ m.) The pumps and channels within cell membranes are so sensitive that they distinguish between the sizes of these two ions and selectively allow only one or the other to pass.

$r = 95$ pm        $r = 133$ pm

The relative size of sodium and potassium ions is an example of a **periodic property**: one that is predictable based on an element's position within the periodic table. In this chapter, we examine several periodic properties of elements, including atomic radius, ionization energy, and electron affinity. We will see that these properties, as well as the overall arrangement of the periodic table, are explained by quantum-mechanical theory, which we examined in Chapter 7. The arrangement of elements in the periodic table—originally based on similarities in the properties of the elements—reflects how electrons fill quantum-mechanical orbitals.

| 3 |
|---|
| **Li** |
| 6.94 |

| 11 |
|---|
| **Na** |
| 22.99 |

| 19 |
|---|
| **K** |
| 39.10 |

| 37 |
|---|
| **Rb** |
| 85.47 |

| 55 |
|---|
| **Cs** |
| 132.91 |

| 87 |
|---|
| **Fr** |
| (223.02) |

▲ The group 1A metals. Potassium is directly beneath sodium in the periodic table.

## 8.2 The Development of the Periodic Table

Prior to the 1700s, the number of known elements was relatively small, consisting mostly of the metals that had long been used for coinage, jewelry, and weapons. From the early 1700s to the mid-1800s, however, chemists discovered over 50 new elements. The first attempt to organize these elements according to similarities in their properties was made by the German chemist Johann Döbereiner (1780–1849), who grouped elements into *triads*: three elements with similar properties. For example, Döbereiner formed a triad out of barium, calcium, and strontium, three fairly reactive metals. About 50 years later, English chemist John Newlands (1837–1898) organized elements into *octaves*, in analogy to musical notes. When arranged this way, the properties of every eighth element were similar, much as every eighth note in the musical scale is similar. Newlands endured some ridicule for drawing an analogy between chemistry and music, including the derisive comments of one colleague who asked Newlands if he had ever tried ordering the elements according to the first letters of their names.

The modern periodic table is credited primarily to the Russian chemist Dmitri Mendeleev, even though a similar organization had been suggested by the German chemist Julius Lothar Meyer (1830–1895). Recall from Chapter 2 that Mendeleev's table is based

▲ Dmitri Mendeleev is credited with the arrangement of the periodic table.

| Gallium (eka-aluminum) | | |
| --- | --- | --- |
|  | Mendeleev's predicted properties | Actual properties |
| Atomic mass | About 68 amu | 69.72 amu |
| Melting point | Low | 29.8 °C |
| Density | 5.9 g/cm³ | 5.90 g/cm³ |
| Formula of oxide | $X_2O_3$ | $Ga_2O_3$ |
| Formula of chloride | $XCl_3$ | $GaCl_3$ |

| Germanium (eka-silicon) | | |
| --- | --- | --- |
| | Mendeleev's predicted properties | Actual properties |
| Atomic mass | About 72 amu | 72.64 amu |
| Density | 5.5 g/cm³ | 5.35 g/cm³ |
| Formula of oxide | $XO_2$ | $GeO_2$ |
| Formula of chloride | $XCl_4$ | $GeCl_4$ |

▲ **FIGURE 8.1 Eka-aluminum and Eka-silicon** Mendeleev's arrangement of elements in the periodic table allowed him to predict the existence of these elements, now known as gallium and germanium, and anticipate their properties.

on the periodic law, which states that when elements are arranged in order of increasing mass, certain properties recur periodically. Mendeleev arranged the elements in a table in which mass increases from left to right and elements with similar properties fall in the same columns.

Mendeleev's arrangement was a huge success, allowing him to predict the existence and properties of yet undiscovered elements such as eka-aluminum (later discovered and named gallium) and eka-silicon (later discovered and named germanium). The properties of these two elements are summarized in Figure 8.1▲. (As noted in Chapter 2, *eka* means "the one beyond" or "the next one" in a family of elements.) However, Mendeleev did encounter some difficulties. For example, according to accepted values of atomic masses, tellurium (with higher mass) should come *after* iodine. But based on their properties, Mendeleev placed tellurium *before* iodine and suggested that the mass of tellurium was erroneous. The mass was correct; later work by the English physicist Henry Moseley (1887–1915) showed that listing elements according to *atomic number*, rather than atomic mass, resolves this problem and results in even better correlation with elemental properties.

Notice the scientific approach in practice in the history of the periodic table. A number of related observations led to a scientific law—the periodic law. Mendeleev's table, an expression of the periodic law, had predictive power as laws usually do. However, it did not explain *why* the properties of elements recurred or *why* certain elements had similar properties. Recall from Chapter 1 that laws *summarize* behavior while theories *explain* behavior. The theory that explains the reasons behind the periodic law is quantum-mechanical theory. In this chapter, we explore the connection between the periodic table and quantum-mechanical theory.

## 8.3 Electron Configurations: How Electrons Occupy Orbitals

 Electron Configurations

Quantum-mechanical theory describes the behavior of electrons in atoms. Since chemical bonding involves the transfer or sharing of electrons, quantum-mechanical theory helps us understand and describe chemical behavior. As we saw in Chapter 7, electrons in atoms exist within orbitals. An **electron configuration** for an atom shows the particular orbitals that electrons occupy for that atom. For example, consider the **ground state**—or lowest energy state—electron configuration for a hydrogen atom:

H  $1s^1$ ←— Number of electrons in orbital

Orbital

The electron configuration indicates that hydrogen's one electron is in the 1s orbital. Electrons generally occupy the lowest energy orbitals available. Since the 1s orbital is the lowest energy orbital in hydrogen (see Section 7.5), hydrogen's electron occupies that orbital. If we could write electron configurations for all the elements, we could see how the arrangements of the electrons within their atoms correlate with the element's chemical properties. However, the Schrödinger equation solutions (the atomic orbitals and their energies) that we described in Chapter 7 are for the hydrogen atom. What do the atomic orbitals of *other atoms* look like? What are their relative energies?

The Schrödinger equation for multielectron atoms has terms to account for the interactions of the electrons with one another that make it too complicated to solve exactly. However, approximate solutions indicate that the orbitals in multielectron atoms are hydrogen-like—they are similar to the *s*, *p*, *d*, and *f* orbitals that we examined in Chapter 7. In order to see how the electrons in multielectron atoms occupy these hydrogen-like orbitals, we must examine two additional concepts: *the effects of electron spin*, a fundamental property of all electrons that affects the number of electrons allowed in any one orbital; and *sublevel energy splitting*, which determines the order of orbital filling within a level.

## Electron Spin and the Pauli Exclusion Principle

We can represent the electron configuration of hydrogen ($1s^1$) in a slightly different way with an **orbital diagram**, which gives similar information but symbolizes the electron as an arrow and the orbital as a box. The orbital diagram for a hydrogen atom is:

H  ⊡
1s

In an orbital diagram, the direction of the arrow (pointing up or pointing down) represents the orientation of the *electron's spin*. Recall from Section 7.5 that the orientation of the electron's spin is quantized, with only two possibilities—spin up ($m_s = +\frac{1}{2}$) and spin down ($m_s = -\frac{1}{2}$). In an orbital diagram, $m_s = +\frac{1}{2}$ is represented with a half-arrow pointing up ( ↑ ) and $m_s = -\frac{1}{2}$ is represented with a half-arrow pointing down ( ↓ ). In a collection of hydrogen atoms, the electrons in about half of the atoms are spin up and the electrons in the other half are spin down. Since no additional electrons are present within the hydrogen atom, we conventionally represent the hydrogen atom electron configuration with its one electron as spin up.

Helium is the first element on the periodic table that contains two electrons. Its two electrons occupy the 1s orbital:

He     $1s^2$

How do the spins of the two electrons in helium align relative to each other? The answer to this question is addressed by the **Pauli exclusion principle**, formulated by Wolfgang Pauli (1900–1958) in 1925:

> **Pauli exclusion principle: no two electrons in an atom can have the same four quantum numbers.**

Since two electrons occupying the same orbital have three identical quantum numbers ($n$, $l$, and $m_l$), they must have different spin quantum numbers. Because there are only two possible spin quantum numbers ($+\frac{1}{2}$ and $-\frac{1}{2}$), the Pauli exclusion principle implies that *each orbital can have a maximum of only two electrons, with opposing spins.*

By applying the exclusion principle, we write an electron configuration and orbital diagram for helium as follows:

Electron configuration     Orbital diagram

He     $1s^2$

1s

The following table shows the four quantum numbers for each of the two electrons in helium:

| $n$ | $l$ | $m_l$ | $m_s$ |
|-----|-----|-------|-------|
| 1 | 0 | 0 | $+\dfrac{1}{2}$ |
| 1 | 0 | 0 | $-\dfrac{1}{2}$ |

The two electrons have three quantum numbers in common (because they are in the same orbital) but have different spin quantum numbers (as indicated by the opposing half-arrows in the orbital diagram).

## Sublevel Energy Splitting in Multielectron Atoms

A major difference in the (approximate) solutions to the Schrödinger equation for multi-electron atoms compared to the solutions for the hydrogen atom is the energy ordering of the orbitals. In the hydrogen atom, the energy of an orbital depends only on $n$, the principal quantum number. For example, the 3$s$, 3$p$, and 3$d$ orbitals (which are empty for hydrogen in its lowest energy state) all have the same energy—they are **degenerate**. The orbitals within a principal level of a *multielectron atom*, in contrast, are not degenerate—their energy depends on the value of $l$. We say that the energies of the sublevels are *split*. In general, the lower the value of $l$ *within a principal level*, the lower the energy ($E$) of the corresponding orbital. Thus, for a given value of $n$:

$$E(s \text{ orbital}) < E(p \text{ orbital}) < E(d \text{ orbital}) < E(f \text{ orbital})$$

In order to understand why the sublevels split in this way, we must examine three key concepts associated with the energy of an electron in the vicinity of a nucleus: (1) Coulomb's law, which describes the interactions between charged particles; (2) shielding, which describes how one electron can shield another electron from the full charge of the nucleus; and (3) penetration, which describes how one atomic orbital can overlap spatially with another, thus penetrating into a region that is close to the nucleus (and therefore less shielded from nuclear charge). We then discuss how these concepts, together with the spatial distributions of electron probability for each orbital, result in the energy ordering that we just saw.

### Coulomb's Law

The attractions and repulsions between charged particles, first introduced in Section 2.4, are described by **Coulomb's law**, which states that the potential energy ($E$) of two charged particles depends on their charges ($q_1$ and $q_2$) and on their separation ($r$):

$$E = \frac{1}{4\pi\varepsilon_0} \frac{q_1 q_2}{r} \tag{8.1}$$

In this equation, $\varepsilon_0$ is a constant ($\varepsilon_0 = 8.85 \times 10^{-12}\,\text{C}^2/\text{J} \cdot \text{m}$). The potential energy is positive for charges of the same sign (plus $\times$ plus, or minus $\times$ minus) and negative for charges of opposite sign (plus $\times$ minus, or minus $\times$ plus). The *magnitude* of the potential energy depends inversely on the separation between the charged particles. We can draw three important conclusions from Coulomb's law:

- For like charges, the potential energy ($E$) is positive and decreases as the particles get *farther apart* (as $r$ increases). Since systems tend toward lower potential energy, like charges repel each other (in much the same way that like poles of two magnets repel each other).

- For opposite charges, the potential energy is negative and becomes more negative as the particles get *closer together* (as $r$ decreases). Therefore, opposite charges (like opposite poles on a magnet) *attract each other.*

- The *magnitude* of the interaction between charged particles increases as the charges of the particles increase. Consequently, an electron with a charge of 1− is more strongly attracted to a nucleus with a charge of 2+ than it it is to a nucleus with a charge of 1+.

**Coulomb's Law** According to Coulomb's law, what happens to the potential energy of two oppositely charged particles as they get closer together?

**(a)** Their potential energy decreases.

**(b)** Their potential energy increases.

**(c)** Their potential energy does not change.

## Shielding

In multielectron atoms, any one electron experiences both the positive charge of the nucleus (which is attractive) and the negative charges of the other electrons (which are repulsive). We can think of the repulsion of one electron by other electrons as *screening* or **shielding** that electron from the full effects of the nuclear charge. For example, consider a lithium ion (Li$^+$). The lithium ion contains two electrons, so its electron configuration is identical to that of helium:

$$\text{Li}^+ \quad 1s^2$$

Imagine bringing a third electron toward the lithium ion. When the third electron is far from the nucleus, it experiences the 3+ charge of the nucleus through the *screen* or *shield* of the 2− charge of the two 1s electrons, as shown in Figure 8.2(a)▼. We can think of the third electron as experiencing an **effective nuclear charge ($Z_{eff}$)** of approximately 1+ (3+ from the nucleus and 2− from the electrons, for a net charge of 1+). The inner electrons in effect *shield* the outer electron from the full nuclear charge.

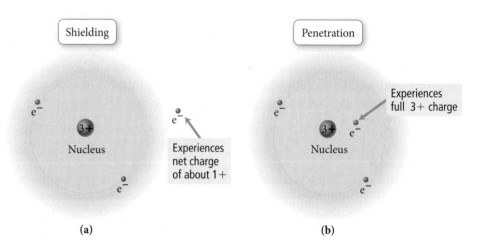

▲ **FIGURE 8.2 Shielding and Penetration** **(a)** An electron far from the nucleus is partly shielded by the electrons in the 1s orbital, reducing the effective net nuclear charge that it experiences. **(b)** An electron that penetrates the electron cloud of the 1s orbital experiences more of the nuclear charge.

## Penetration

Now imagine allowing this third electron to come closer to the nucleus. As the third electron *penetrates* the electron cloud of the 1s electrons, it begins to experience the 3+ charge of the nucleus more fully because it is less shielded by the intervening electrons. If the electron could somehow get closer to the nucleus than the 1s electrons, it would experience the full 3+ charge, as shown in Figure 8.2(b)▲. In other words, as the outer electron undergoes **penetration** into the region occupied by the inner electrons, it experiences a greater nuclear charge and therefore (according to Coulomb's law) a lower energy.

## Electron Spatial Distributions and Sublevel Splitting

We now have examined the concepts we need to understand the energy splitting of the sublevels within a principal level. The splitting is a result of the spatial distributions of electrons within a sublevel. Recall from Section 7.6 that the radial distribution function for an atomic orbital shows the total probability of finding the electron within a thin spherical shell at a distance *r* from the nucleus.

Figure 8.3► shows the radial distribution functions of the 2*s* and 2*p* orbitals superimposed on one another (the radial distribution function of the 1*s* orbital is also shown). Notice that, in general, an electron in a 2*p* orbital has a greater probability of being found closer to the nucleus than an electron in a 2*s* orbital. We might initially expect, therefore, that the 2*p* orbital would be lower in energy. However, exactly the opposite is true—the 2*s* orbital is actually lower in energy, *but only when the 1s orbital is occupied*. (When the 1*s* orbital is empty, the 2*s* and 2*p* orbitals are degenerate.) Why? The reason is the bump near *r* = 0 (near the nucleus) for the 2*s* orbital. This bump represents a significant probability of the electron being found very close to the nucleus. Even more importantly, this area of the probability penetrates into the 1*s* orbital—it gets into the region where shielding by the 1*s* electrons is less effective. In contrast, most of the probability in the radial distribution function of the 2*p* orbital lies *outside* the radial distribution function of the 1*s* orbital. Consequently, almost all of the 2*p* orbital is shielded from nuclear charge by the 1*s* orbital. The end result is that the 2*s* orbital—since it experiences more of the nuclear charge due to its greater *penetration*—is lower in energy than the 2*p* orbital. The results are similar when we compare the 3*s*, 3*p*, and 3*d* orbitals. The *s* orbitals penetrate more fully than the *p* orbitals, which in turn penetrate more fully than the *d* orbitals, as shown in Figure 8.4►.

Figure 8.5▼ shows the energy ordering of a number of orbitals in multi-electron atoms. Notice these features of Figure 8.5:

- Because of penetration, the sublevels of each principal level are *not* degenerate for multielectron atoms.

- In the fourth and fifth principal levels, the effects of penetration become so important that the 4*s* orbital lies lower in energy than the 3*d* orbitals and the 5*s* orbital lies lower in energy than the 4*d* orbitals.

- The energy separations between one set of orbitals and the next become smaller for 4*s* orbitals and beyond, and the relative energy ordering of these orbitals can actually vary among elements. These variations result in irregularities in the electron configurations of the transition metals and their ions (as we shall see later).

▲ **FIGURE 8.3** **Radial Distribution Functions for the 1s, 2s, and 2p Orbitals.**

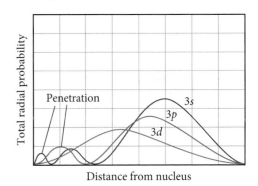

▲ **FIGURE 8.4** **Radial Distribution Functions for the 3s, 3p, and 3d Orbitals** The 3s electrons penetrate most deeply into the inner orbitals, are least shielded, and experience the "greatest" effective nuclear charge. The 3d electrons penetrate least. This accounts for the energy ordering of the sublevels: s < p < d.

## General Energy Ordering of Orbitals for Multielectron Atoms

▲ **FIGURE 8.5** **General Energy Ordering of Orbitals for Multielectron Atoms.**

**Penetration and Shielding** Which statement is true?

**(a)** An orbital that penetrates into the region occupied by core electrons is more shielded from nuclear charge than an orbital that does not penetrate and therefore has a higher energy.

**(b)** An orbital that penetrates into the region occupied by core electrons is less shielded from nuclear charge than an orbital that does not penetrate and therefore has a higher energy.

**(c)** An orbital that penetrates into the region occupied by core electrons is less shielded from nuclear charge than an orbital that does not penetrate and therefore has a lower energy.

**(d)** An orbital that penetrates into the region occupied by core electrons is more shielded from nuclear charge than an orbital that does not penetrate and therefore has a lower energy.

## Electron Configurations for Multielectron Atoms

Unless otherwise specified, we use the term *electron configuration* to mean the ground state (or lowest energy) configuration.

Now that we know the energy ordering of orbitals in multielectron atoms, we can determine ground state electron configurations for the rest of the elements. Since we know that electrons occupy the lowest energy orbitals available when the atom is in its ground state and that only two electrons (with opposing spins) are allowed in each orbital, we can systematically build up the electron configurations for the elements. This pattern of orbital filling is known as the **aufbau principle** (the German word *aufbau* means "build up"). For lithium, with three electrons, the electron configuration and orbital diagram are:

Remember that the number of electrons in a neutral atom is equal to its atomic number.

Electron configuration    Orbital diagram

Li      $1s^2 2s^1$

| 1s | 2s |
|----|----|
| ↑↓ | ↑ |

For carbon, which has six electrons, the electron configuration and orbital diagram are:

Electrons with parallel spins have correlated motion that minimizes their mutual repulsion.

Electron configuration      Orbital diagram

C      $1s^2 2s^2 2p^2$

| 1s | 2s | 2p |   |
|----|----|----|----|
| ↑↓ | ↑↓ | ↑ | ↑ |

Notice that the 2p electrons occupy the p orbitals (of equal energy) singly, rather than pairing in one orbital. This way of filling orbitals is known as **Hund's rule**, which states that *when filling degenerate orbitals, electrons fill them singly first, with parallel spins*. Hund's rule is a result of an atom's tendency to find the lowest energy state possible. When two electrons occupy separate orbitals of equal energy, the repulsive interaction between them is lower than when they occupy the same orbital because the electrons are spread out over different regions of space. By convention we denote these parallel spins with half arrows pointing up.

*Summarizing Orbital Filling:*

▶ Electrons occupy orbitals so as to minimize the energy of the atom; therefore, lower energy orbitals fill before higher energy orbitals. Orbitals fill in the following order: 1s 2s 2p 3s 3p 4s 3d 4p 5s 4d 5p 6s.

▶ Orbitals can hold no more than two electrons each. When two electrons occupy the same orbital, their spins are opposite. This is another way of expressing the Pauli exclusion principle (no two electrons in one atom can have the same four quantum numbers).

▶ When orbitals of identical energy are available, electrons first occupy these orbitals singly with parallel spins rather than in pairs. Once the orbitals of equal energy are half full, the electrons start to pair (Hund's rule).

Consider the electron configurations and orbital diagrams for elements with atomic numbers 3–10:

| Symbol | Number of electrons | Electron configuration | Orbital diagram |
|---|---|---|---|
| Li | 3 | $1s^2 2s^1$ | |
| Be | 4 | $1s^2 2s^2$ | |
| B | 5 | $1s^2 2s^2 2p^1$ | |
| C | 6 | $1s^2 2s^2 2p^2$ | |

Notice that, as a result of Hund's rule, the *p* orbitals fill with single electrons before the electrons pair:

| | | | |
|---|---|---|---|
| N | 7 | $1s^2 2s^2 2p^3$ | |
| O | 8 | $1s^2 2s^2 2p^4$ | |
| F | 9 | $1s^2 2s^2 2p^5$ | |
| Ne | 10 | $1s^2 2s^2 2p^6$ | |

The electron configuration of neon represents the complete filling of the $n = 2$ principal level. When writing electron configurations for elements beyond neon, or beyond any other noble gas, we can abbreviate the electron configuration of the previous noble gas—sometimes called the *inner electron configuration*—by the symbol for the noble gas in square brackets. For example, the electron configuration of sodium is:

$$Na \quad 1s^2 2s^2 2p^6 3s^1$$

We can write this configuration more compactly by using [Ne] to represent the inner electrons:

$$Na \quad [Ne]\, 3s^1$$

[Ne] represents $1s^2 2s^2 2p^6$, the electron configuration for neon.

To write an electron configuration for an element, first find its atomic number from the periodic table—this number equals the number of electrons. Then use the order of filling to distribute the electrons in the appropriate orbitals. Remember that each orbital can hold a maximum of two electrons. Consequently,

- the *s* sublevel has only one orbital and can therefore hold only 2 electrons.
- the *p* sublevel has three orbitals and can hold 6 electrons.
- the *d* sublevel has five orbitals and can hold 10 electrons.
- the *f* sublevel has seven orbitals and can hold 14 electrons.

**Example 8.1** | **Electron Configurations**

Write electron configurations for each element.
**(a)** Mg    **(b)** P    **(c)** Br    **(d)** Al

**SOLUTION**

| | |
|---|---|
| **(a)** Mg<br><br>Magnesium has 12 electrons. Distribute 2 of these into the 1s orbital, 2 into the 2s orbital, 6 into the 2p orbitals, and 2 into the 3s orbital. | Mg    $1s^22s^22p^63s^2$   or   [Ne] $3s^2$ |
| **(b)** P<br><br>Phosphorus has 15 electrons. Distribute 2 of these into the 1s orbital, 2 into the 2s orbital, 6 into the 2p orbitals, 2 into the 3s orbital, and 3 into the 3p orbitals. | P    $1s^22s^22p^63s^23p^3$   or   [Ne] $3s^23p^3$ |
| **(c)** Br<br><br>Bromine has 35 electrons. Distribute 2 of these into the 1s orbital, 2 into the 2s orbital, 6 into the 2p orbitals, 2 into the 3s orbital, 6 into the 3p orbitals, 2 into the 4s orbital, 10 into the 3d orbitals, and 5 into the 4p orbitals. | Br    $1s^22s^22p^63s^23p^64s^23d^{10}4p^5$   or   [Ar] $4s^23d^{10}4p^5$ |
| **(d)** Al<br><br>Aluminum has 13 electrons. Distribute 2 of these into the 1s orbital, 2 into the 2s orbital, 6 into the 2p orbitals, 2 into the 3s orbital, and 1 into the 3p orbital. | Al    $1s^22s^22p^63s^23p^1$   or   [Ne] $3s^23p^1$ |

**FOR PRACTICE 8.1**   Write electron configurations for each element.
**(a)** Cl    **(b)** Si    **(c)** Sr    **(d)** O

---

**Example 8.2** | **Writing Orbital Diagrams**

Write the orbital diagram for sulfur and determine the number of unpaired electrons.

**SOLUTION**

| | |
|---|---|
| Sulfur's atomic number is 16; it therefore has 16 electrons and the electron configuration $1s^22s^22p^63s^23p^4$. Draw a box for each orbital, putting the lowest energy orbital (1s) on the far left and proceeding to orbitals of higher energy to the right. | ☐   ☐   ☐☐☐   ☐   ☐☐☐<br>1s    2s     2p     3s     3p |
| Distribute the 16 electrons into the boxes representing the orbitals, allowing a maximum of two electrons per orbital and remembering Hund's rule. You can see from the diagram that sulfur has two unpaired electrons. | ⇅   ⇅   ⇅ ⇅ ⇅   ⇅   ⇅ ↑ ↑<br>1s    2s     2p     3s     3p<br><br>Two unpaired electrons |

**FOR PRACTICE 8.2**   Write the orbital diagram for Ar and determine the number of unpaired electrons.

---

CONCEPTUAL
**CONNECTION 8.3**

**Electron Configurations and Quantum Numbers**   What are the four quantum numbers for each of the two electrons in a 4s orbital?

# 8.4  Electron Configurations, Valence Electrons, and the Periodic Table

 **Writing an Electron Configuration based on an Element's Position on the Periodic Table**

Recall that Mendeleev arranged the periodic table so that elements with similar chemical properties lie in the same column. We can begin to make the connection between an element's properties and its electron configuration by superimposing the electron configurations of the first 18 elements onto a partial periodic table, as shown in Figure 8.6▶. As we move to the right across a row, the orbitals fill in the correct order. With each subsequent row, the highest principal quantum number increases by one. Notice that as we move down a column, *the number of electrons in the outermost principal energy level (highest n value) remains the same.* The key connection between the macroscopic world (an element's chemical properties) and the atomic world (an atom's electronic structure) lies in these outermost electrons.

An atom's **valence electrons** are those that are important in chemical bonding. *For main-group elements, the valence electrons are those in the outermost principal energy level.* For transition elements, we also count the outermost *d* electrons among the valence electrons (even though they are not in an outermost principal energy level). The chemical properties of an element depend on its valence electrons, which are instrumental in bonding because they are held most loosely (and are therefore the easiest to lose or share). We can now see *why* the elements in a column of the periodic table have similar chemical properties: *they have the same number of valence electrons.*

We distinguish between valence electrons and all the other electrons in an atom, which are called **core electrons**. For example, silicon, with the electron configuration $1s^2 2s^2 2p^6 3s^2 3p^2$, has four valence electrons (those in the $n = 3$ principal level) and ten core electrons as shown in the right margin.

### Outer Electron Configurations of Elements 1–18

| 1A | | | | | | | 8A |
|---|---|---|---|---|---|---|---|
| 1<br>**H**<br>$1s^1$ | 2A | 3A | 4A | 5A | 6A | 7A | 2<br>**He**<br>$1s^2$ |
| 3<br>**Li**<br>$2s^1$ | 4<br>**Be**<br>$2s^2$ | 5<br>**B**<br>$2s^2 2p^1$ | 6<br>**C**<br>$2s^2 2p^2$ | 7<br>**N**<br>$2s^2 2p^3$ | 8<br>**O**<br>$2s^2 2p^4$ | 9<br>**F**<br>$2s^2 2p^5$ | 10<br>**Ne**<br>$2s^2 2p^6$ |
| 11<br>**Na**<br>$3s^1$ | 12<br>**Mg**<br>$3s^2$ | 13<br>**Al**<br>$3s^2 3p^1$ | 14<br>**Si**<br>$3s^2 3p^2$ | 15<br>**P**<br>$3s^2 3p^3$ | 16<br>**S**<br>$3s^2 3p^4$ | 17<br>**Cl**<br>$3s^2 3p^5$ | 18<br>**Ar**<br>$3s^2 3p^6$ |

▲ **FIGURE 8.6  Outer Electron Configurations of the First 18 Elements in the Periodic Table.**

Si    $1s^2 2s^2 2p^6 3s^2 3p^2$

Core electrons        Valence electrons

---

## Example 8.3    Valence Electrons and Core Electrons

Write the electron configuration for Ge. Identify the valence electrons and the core electrons.

**SOLUTION**

| | |
|---|---|
| Determine the total number of electrons from germanium's atomic number (32). Distribute the electrons into the appropriate orbitals according to the information in *Summarizing Orbital Filling* on p. 344. | Ge    $1s^2 2s^2 2p^6 3s^2 3p^6 4s^2 3d^{10} 4p^2$ |
| Since germanium is a main-group element, its valence electrons are those in the outermost principal energy level. For germanium the $n = 4$ principal level is outermost. Consequently, the $n = 4$ electrons are valence electrons and the rest are core electrons.<br><br>*Note: In this book, we write electron configurations with the orbitals in the order of filling. However, electron configurations are sometimes written in order of increasing principal quantum number. The electron configuration of germanium written in order of increasing principal quantum number is Ge $1s^2 2s^2 2p^6 3s^2 3p^6 3d^{10} 4s^2 4p^2$.* | 4 valence electrons<br><br>Ge    $1s^2 2s^2 2p^6 3s^2 3p^6 4s^2 3d^{10} 4p^2$<br><br>28 core electrons |

**FOR PRACTICE 8.3**  Write the electron configuration for phosphorus. Identify the valence electrons and core electrons.

## Orbital Blocks in the Periodic Table

A pattern similar to what we just saw for the first 18 elements exists for the entire periodic table, as shown in Figure 8.7 ▼. Note that because of the filling order of orbitals, we can divide the periodic table into blocks representing the filling of particular sublevels. The first two columns on the left side of the periodic table comprise the $s$ block, with outer electron configurations of $ns^1$ (the alkali metals) and $ns^2$ (the alkaline earth metals). The six columns on the right side of the periodic table comprise the $p$ block, with outer electron configurations of $ns^2 np^1$, $ns^2 np^2$, $ns^2 np^3$, $ns^2 np^4$, $ns^2 np^5$ (halogens), and $ns^2 np^6$ (noble gases). The transition elements comprise the $d$ block, and the lanthanides and actinides (also called the inner transition elements) comprise the $f$ block. (For compactness, the $f$ block is normally printed below the $d$ block instead of being embedded within it.)

You can see that *the number of columns in a block corresponds to the maximum number of electrons that can occupy the particular sublevel of that block.* The $s$ block has two columns (corresponding to one $s$ orbital holding a maximum of two electrons); the $p$ block has six columns (corresponding to three $p$ orbitals with two electrons each); the $d$ block has 10 columns (corresponding to five $d$ orbitals with two electrons each); and the $f$ block has 14 columns (corresponding to seven $f$ orbitals with two electrons each).

Notice also that, except for helium, *the number of valence electrons in any main-group element is equal to its lettered group number.* For example, we know that chlorine has seven valence electrons because it is in group 7A.

Lastly, note that, for main-group elements, *the row number in the periodic table is equal to the number (or n value) of the highest occupied principal level.* For example, because chlorine is in row 3, its highest principal level is the $n = 3$ level.

> Helium is an exception. Even though it lies in the column with an outer electron configuration of $ns^2np^6$, its electron configuration is simply $1s^2$.

## Orbital Blocks of the Periodic Table

▲ **FIGURE 8.7** The *s, p, d,* and *f* Blocks of the Periodic Table.

Groups — 1 (1A) … 18 (8A)

| Period | 1 | 2 | 3 | 4 | 5 | 6 | 7 | 8 | 9 | 10 | 11 | 12 | 13 | 14 | 15 | 16 | 17 | 18 |
|---|---|---|---|---|---|---|---|---|---|---|---|---|---|---|---|---|---|---|
| | 1A | 2A | 3B | 4B | 5B | 6B | 7B | 8B | 8B | 8B | 1B | 2B | 3A | 4A | 5A | 6A | 7A | 8A |
| 1 | 1 H $1s^1$ | | | | | | | | | | | | | | | | | 2 He $1s^2$ |
| 2 | 3 Li $2s^1$ | 4 Be $2s^2$ | | | | | | | | | | | 5 B $2s^2 2p^1$ | 6 C $2s^2 2p^2$ | 7 N $2s^2 2p^3$ | 8 O $2s^2 2p^4$ | 9 F $2s^2 2p^5$ | 10 Ne $2s^2 2p^6$ |
| 3 | 11 Na $3s^1$ | 12 Mg $3s^2$ | | | | | | | | | | | 13 Al $3s^2 3p^1$ | 14 Si $3s^2 3p^2$ | 15 P $3s^2 3p^3$ | 16 S $3s^2 3p^4$ | 17 Cl $3s^2 3p^5$ | 18 Ar $3s^2 3p^6$ |
| 4 | 19 K $4s^1$ | 20 Ca $4s^2$ | 21 Sc $4s^2 3d^1$ | 22 Ti $4s^2 3d^2$ | 23 V $4s^2 3d^3$ | 24 Cr $4s^1 3d^5$ | 25 Mn $4s^2 3d^5$ | 26 Fe $4s^2 3d^6$ | 27 Co $4s^2 3d^7$ | 28 Ni $4s^2 3d^8$ | 29 Cu $4s^1 3d^{10}$ | 30 Zn $4s^2 3d^{10}$ | 31 Ga $4s^2 4p^1$ | 32 Ge $4s^2 4p^2$ | 33 As $4s^2 4p^3$ | 34 Se $4s^2 4p^4$ | 35 Br $4s^2 4p^5$ | 36 Kr $4s^2 4p^6$ |
| 5 | 37 Rb $5s^1$ | 38 Sr $5s^2$ | 39 Y $5s^2 4d^1$ | 40 Zr $5s^2 4d^2$ | 41 Nb $5s^1 4d^4$ | 42 Mo $5s^1 4d^5$ | 43 Tc $5s^2 4d^5$ | 44 Ru $5s^1 4d^7$ | 45 Rh $5s^1 4d^8$ | 46 Pd $4d^{10}$ | 47 Ag $5s^1 4d^{10}$ | 48 Cd $5s^2 4d^{10}$ | 49 In $5s^2 5p^1$ | 50 Sn $5s^2 5p^2$ | 51 Sb $5s^2 5p^3$ | 52 Te $5s^2 5p^4$ | 53 I $5s^2 5p^5$ | 54 Xe $5s^2 5p^6$ |
| 6 | 55 Cs $6s^1$ | 56 Ba $6s^2$ | 57 La $6s^2 5d^1$ | 72 Hf $6s^2 5d^2$ | 73 Ta $6s^2 5d^3$ | 74 W $6s^2 5d^4$ | 75 Re $6s^2 5d^5$ | 76 Os $6s^2 5d^6$ | 77 Ir $6s^2 5d^7$ | 78 Pt $6s^1 5d^9$ | 79 Au $6s^1 5d^{10}$ | 80 Hg $6s^2 5d^{10}$ | 81 Tl $6s^2 6p^1$ | 82 Pb $6s^2 6p^2$ | 83 Bi $6s^2 6p^3$ | 84 Po $6s^2 6p^4$ | 85 At $6s^2 6p^5$ | 86 Rn $6s^2 6p^6$ |
| 7 | 87 Fr $7s^1$ | 88 Ra $7s^2$ | 89 Ac $7s^2 6d^1$ | 104 Rf $7s^2 6d^2$ | 105 Db $7s^2 6d^3$ | 106 Sg $7s^2 6d^4$ | 107 Bh | 108 Hs | 109 Mt | 110 Ds | 111 Rg | 112 Cn | 113 ** | 114 Fl | 115 ** | 116 Lv | 117 ** | 118 ** |

s-block elements · p-block elements · d-block elements · f-block elements

Lanthanides:

| 58 Ce $6s^2 4f^1 5d^1$ | 59 Pr $6s^2 4f^3$ | 60 Nd $6s^2 4f^4$ | 61 Pm $6s^2 4f^5$ | 62 Sm $6s^2 4f^6$ | 63 Eu $6s^2 4f^7$ | 64 Gd $6s^2 4f^7 5d^1$ | 65 Tb $6s^2 4f^9$ | 66 Dy $6s^2 4f^{10}$ | 67 Ho $6s^2 4f^{11}$ | 68 Er $6s^2 4f^{12}$ | 69 Tm $6s^2 4f^{13}$ | 70 Yb $6s^2 4f^{14}$ | 71 Lu $6s^2 4f^{14} 5d^1$ |
|---|---|---|---|---|---|---|---|---|---|---|---|---|---|

Actinides:

| 90 Th $7s^2 6d^2$ | 91 Pa $7s^2 5f^2 6d^1$ | 92 U $7s^2 5f^3 6d^1$ | 93 Np $7s^2 5f^4 6d^1$ | 94 Pu $7s^2 5f^6$ | 95 Am $7s^2 5f^7$ | 96 Cm $7s^2 5f^7 6d^1$ | 97 Bk $7s^2 5f^9$ | 98 Cf $7s^2 5f^{10}$ | 99 Es $7s^2 5f^{11}$ | 100 Fm $7s^2 5f^{12}$ | 101 Md $7s^2 5f^{13}$ | 102 No $7s^2 5f^{14}$ | 103 Lr $7s^2 5f^{14} 6d^1$ |
|---|---|---|---|---|---|---|---|---|---|---|---|---|---|

Recall from Chapter 2 that main-group elements are those in the two far left columns (groups 1A, 2A) and the six far right columns (groups 3A–8A) of the periodic table.

***Summarizing Periodic Table Organization:***

▶ The periodic table is divisible into four blocks corresponding to the filling of the four quantum sublevels (*s*, *p*, *d*, and *f*).

▶ The group number of a main-group element is equal to the number of valence electrons in that element.

▶ The row number of a main-group element is equal to the highest principal quantum number of that element.

**Valence Electrons and Group Number**  Without writing an electron configuration, determine the number of valence electrons in nitrogen.

CONCEPTUAL
CONNECTION **8.4**

# Writing an Electron Configuration for an Element from Its Position in the Periodic Table

The organization of the periodic table allows us to write the electron configuration for an element based on its position in the periodic table. For example, suppose we want to write an electron configuration for Cl. The *inner electron configuration* of Cl is that of the noble gas that precedes it in the periodic table, Ne. So we can represent the inner electron configuration with [Ne]. We can determine the *outer electron configuration*—the configuration of the electrons beyond the previous noble gas—by tracing the elements between Ne and Cl and assigning electrons to the appropriate orbitals, as shown here. Remember that the highest *n* value is indicated by the row number (in this case, 3 for chlorine).

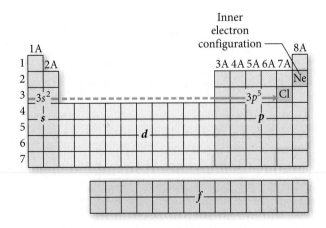

We begin with [Ne], then add in the two 3*s* electrons as we trace across the *s* block, followed by five 3*p* electrons as we trace across the *p* block to Cl, which is in the fifth column of the *p* block. The electron configuration is:

$$\text{Cl} \qquad [\text{Ne}]3s^2 3p^5$$

Notice that Cl is in column 7A and therefore has seven valence electrons and an outer electron configuration of $ns^2 np^5$.

## **Example** 8.4   **Writing Electron Configurations from the Periodic Table** ▶

Refer to the periodic table to write the electron configuration for selenium (Se).

**SOLUTION**

The atomic number of Se is 34. The noble gas that precedes Se in the periodic table is argon, so the inner electron configuration is [Ar]. Obtain the outer electron configuration by tracing the elements between Ar and Se and assigning electrons to the appropriate orbitals. Begin with [Ar]. Because Se is in row 4, add two 4s electrons as you trace across the s block ($n$ = row number). Next, add ten 3d electrons as you trace across the d block ($n$ = row number − 1). Lastly, add four 4p electrons as you trace across the p block to Se, which is in the fourth column of the p block ($n$ = row number).

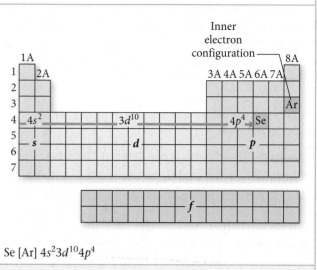

Se [Ar] $4s^2 3d^{10} 4p^4$

**FOR PRACTICE 8.4**   Refer to the periodic table to write the electron configuration of bismuth (Bi).

**FOR MORE PRACTICE 8.4**   Refer to the periodic table to write the electron configuration for iodine (I).

## The Transition and Inner Transition Elements

The electron configurations of the transition elements (*d* block) and inner transition elements (*f* block) exhibit trends that differ somewhat from those of the main-group elements. As we move to the right across a row in the *d* block, the *d* orbitals fill as shown here:

| 21<br>**Sc**<br>$4s^2 3d^1$ | 22<br>**Ti**<br>$4s^2 3d^2$ | 23<br>**V**<br>$4s^2 3d^3$ | 24<br>**Cr**<br>$4s^1 3d^5$ | 25<br>**Mn**<br>$4s^2 3d^5$ | 26<br>**Fe**<br>$4s^2 3d^6$ | 27<br>**Co**<br>$4s^2 3d^7$ | 28<br>**Ni**<br>$4s^2 3d^8$ | 29<br>**Cu**<br>$4s^1 3d^{10}$ | 30<br>**Zn**<br>$4s^2 3d^{10}$ |
|---|---|---|---|---|---|---|---|---|---|
| 39<br>**Y**<br>$5s^2 4d^1$ | 40<br>**Zr**<br>$5s^2 4d^2$ | 41<br>**Nb**<br>$5s^1 4d^4$ | 42<br>**Mo**<br>$5s^1 4d^5$ | 43<br>**Tc**<br>$5s^2 4d^5$ | 44<br>**Ru**<br>$5s^1 4d^7$ | 45<br>**Rh**<br>$5s^1 4d^8$ | 46<br>**Pd**<br>$4d^{10}$ | 47<br>**Ag**<br>$5s^1 4d^{10}$ | 48<br>**Cd**<br>$5s^2 4d^{10}$ |

Notice that *the principal quantum number of the d orbitals that fill across each row in the transition series is equal to the row number minus one.* In the fourth row, the 3d orbitals fill; in the fifth row, the 4d orbitals fill; and so on. This happens because, as we discussed in Section 8.3, the 4s orbital is generally lower in energy than the 3d orbital (because it more efficiently penetrates into the region occupied by the core electrons). The result is that the 4s orbital fills before the 3d orbital, even though the 4s orbital's principal quantum number ($n$ = 4) is higher.

Keep in mind, however, that the 4s and the 3d orbitals are extremely close to each other in energy so their relative energy ordering depends on the exact species under consideration; this causes some irregular behavior in the transition metals. For example, in the first transition series of the *d* block, the outer configuration is $4s^2 3d^x$ with two exceptions: Cr is $4s^1 3d^5$ and Cu is $4s^1 3d^{10}$. This behavior is related to the closely spaced 3d and 4s energy levels and the stability associated with a half-filled (as in Cr) or completely filled (as in Cu) sublevel. Actual electron configurations are definitively determined experimentally (through spectroscopy) and do not always conform to the general pattern. Nonetheless, the patterns described here allow us to accurately predict electron configurations for most of the elements in the periodic table.

As we move across the *f* block (the inner transition series), the *f* orbitals fill. For these elements, the principal quantum number of the *f* orbitals that fill across each row is the row number *minus two*. (In the sixth row, the 4*f* orbitals fill, and in the seventh row, the 5*f* orbitals fill.) In addition, within the inner transition series, the close energy spacing of the 5*d* and 4*f* orbitals sometimes causes an electron to enter a 5*d* orbital instead of the expected 4*f* orbital. For example, the electron configuration of gadolinium is [Xe] $6s^2 4f^7 5d^1$ (instead of the expected [Xe] $6s^2 4f^8$).

## 8.5 The Explanatory Power of the Quantum-Mechanical Model

We can now see how the quantum-mechanical model accounts for the chemical properties of the elements, such as the inertness of helium or the reactivity of hydrogen, and (more generally) how it accounts for the periodic law. *The chemical properties of elements are largely determined by the number of valence electrons they contain.* Their properties are periodic because the number of valence electrons is periodic.

Because elements within a column in the periodic table have the same number of valence electrons, they have similar chemical properties. The noble gases, for example, all have eight valence electrons, except for helium, which has two. Although we do not cover the quantitative (or numerical) aspects of the quantum-mechanical model in this book, calculations of the overall energy of the electrons within atoms with eight valence electrons (or two, in the case of helium) show that they are particularly stable. In other words, when *s* and *p* quantum sublevels are completely full, the overall energy of the electrons that occupy that level is particularly low. Therefore, those electrons *cannot* lower their energy by reacting with other atoms or molecules, so the corresponding atom is relatively unreactive or inert. The noble gases are the most chemically stable and relatively unreactive family in the periodic table.

Elements with electron configurations *close* to those of the noble gases are the most reactive because they can attain noble gas electron configurations by losing or gaining a small number of electrons. For example, alkali metals (group 1A) are among the most reactive metals because their outer electron configuration ($ns^1$) is one electron beyond a noble gas configuration. They readily react to lose the $ns^1$ electron, obtaining a noble gas configuration. This explains why—as we saw in Chapter 2—the group 1A metals tend to form 1+ cations. Similarly, alkaline earth metals, with an outer electron configuration of $ns^2$, also tend to be reactive metals, losing their $ns^2$ electrons to form 2+ cations. This does not mean that forming an ion with a noble gas configuration is in itself energetically favorable. In fact, forming cations always *requires energy*. But when the cation formed has a noble gas configuration, the energy cost of forming the cation is often less than the energy payback that occurs when that cation forms ionic bonds with anions, as we shall see in Chapter 9.

On the right side of the periodic table, halogens are among the most reactive nonmetals because of their $ns^2np^5$ electron configurations. They are only one electron short of a noble gas configuration and tend to react to gain that one electron, forming 1− ions. Figure 8.8▼, which is similar to Figure 2.14 in Chapter 2,

| 8A | 1A |
|---|---|
| 2 **He** $1s^2$ | 3 **Li** $2s^1$ |
| 10 **Ne** $2s^2 2p^6$ | 11 **Na** $3s^1$ |
| 18 **Ar** $3s^2 3p^6$ | 19 **K** $4s^1$ |
| 36 **Kr** $4s^2 4p^6$ | 37 **Rb** $5s^1$ |
| 54 **Xe** $5s^2 5p^6$ | 55 **Cs** $6s^1$ |
| 86 **Rn** $6s^2 6p^6$ | 87 **Fr** $7s^1$ |
| Noble gases | Alkali metals |

| 2A | 7A |
|---|---|
| 4 **Be** $2s^2$ | 9 **F** $2s^2 2p^5$ |
| 12 **Mg** $3s^2$ | 17 **Cl** $3s^2 3p^5$ |
| 20 **Ca** $4s^2$ | 35 **Br** $4s^2 4p^5$ |
| 38 **Sr** $5s^2$ | 53 **I** $5s^2 5p^5$ |
| 56 **Ba** $6s^2$ | 85 **At** $6s^2 6p^5$ |
| 88 **Ra** $7s^2$ | Halogens |
| Alkaline earth metals | |

### Elements That Form Ions with Predictable Charges

| | 1A | 2A | 3B | 4B | 5B | 6B | 7B | 8B | | 1B | 2B | 3A | 4A | 5A | 6A | 7A | 8A |
|---|---|---|---|---|---|---|---|---|---|---|---|---|---|---|---|---|---|
| 1 | $Li^+$ | | | | | | | | | | | | | $N^{3-}$ | $O^{2-}$ | $F^-$ | |
| 2 | $Na^+$ | $Mg^{2+}$ | | | | | | | | | | $Al^{3+}$ | | | $S^{2-}$ | $Cl^-$ | |
| 3 | $K^+$ | $Ca^{2+}$ | | | | | | | | | | | | | $Se^{2-}$ | $Br^-$ | |
| 4 | $Rb^+$ | $Sr^{2+}$ | | | | | | | | | | | | | $Te^{2-}$ | $I^-$ | |
| 5 | $Cs^+$ | $Ba^{2+}$ | | | | | | | | | | | | | | | |

▲ **FIGURE 8.8 Elements That Form Ions with Predictable Charges** Notice that each ion has a noble gas electron configuration.

shows the elements that form predictable ions. The charges of these ions reflect their electron configurations—in their reactions, these elements form ions with noble gas electron configurations.

van der Waals radius

2 × Krypton radius

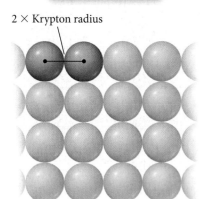

Krypton solid

▲ The van der Waals radius of an atom is one-half the distance between adjacent nuclei in the atomic solid.

Covalent radius

Br          Br

228 pm

$$\text{Br radius} = \frac{228 \text{ pm}}{2} = 114 \text{ pm}$$

▲ The covalent radius of bromine is one-half the distance between two bonded bromine atoms.

The bonding radii of some elements, such as helium and neon, must be approximated because they do not form chemical bonds or metallic crystals.

# 8.6  Periodic Trends in the Size of Atoms and Effective Nuclear Charge

In previous chapters, we saw that the volume of an atom is taken up primarily by its electrons (Chapter 2) occupying quantum-mechanical orbitals (Chapter 7). We also discussed how these orbitals do not have a definite boundary but represent only a statistical probability distribution for where the electron is found. So how do we define the size of an atom? One way to define atomic radii is to consider the distance between *non-bonding* atoms that are in direct contact. For example, krypton can be frozen into a solid in which the krypton atoms are touching each other but are not bonded together. The distance between the centers of adjacent krypton atoms—which we can determine from the solid's density—is then twice the radius of a krypton atom. An atomic radius determined in this way is called the **nonbonding atomic radius** or the **van der Waals radius**. The van der Waals radius represents the radius of an atom when it is not bonded to another atom.

Another way to define the size of an atom—the **bonding atomic radius** or **covalent radius**—is defined differently for nonmetals and metals, as follows:

Nonmetals: one-half the distance between two of the atoms bonded together

Metals: one-half the distance between two of the atoms next to each other in a crystal of the metal

For example, the distance between Br atoms in $Br_2$ is 228 pm; therefore, the Br covalent radius is one-half of 228 pm or 114 pm.

Using this method, we can assign radii to all elements in the periodic table that form chemical bonds or form metallic crystals. A more general term, the **atomic radius**, refers to a set of average bonding radii determined from measurements on a large number of elements and compounds. The atomic radius represents the radius of an atom when it is bonded to another atom and is always smaller than the van der Waals radius. The approximate bond length of any two covalently bonded atoms is the sum of their atomic radii. For example, the approximate bond length for ICl is iodine's atomic radius (133 pm) plus chlorine's atomic radius (99 pm), a bond length of 232 pm. (The actual experimentally measured bond length in ICl is 232.07 pm.)

Figure 8.9▶ plots atomic radius as a function of atomic number for the first 57 elements in the periodic table. Notice the periodic trend in the radii. Atomic radii peak with each alkali metal. Figure 8.10▶ is a relief map of atomic radii for most of the elements in the periodic table. The general trends in the atomic radii of main-group elements, which are the same as trends observed in van der Waals radii, are:

1. As we move down a column (or family) in the periodic table, atomic radius increases.
2. As we move to the right across a period (or row) in the periodic table, atomic radius decreases.

We can understand the observed trend in radii as we move down a column based on the trends in the sizes of atomic orbitals. Atomic radius is largely determined by the valence electrons, the electrons farthest from the nucleus. As we move down a column in the periodic table, the highest principal quantum number ($n$) of the valence electrons increases. Consequently, the valence electrons occupy larger orbitals, resulting in larger atoms.

The observed trend in atomic radii as we move to the right across a row, however, is a bit more complex. To understand this trend, we must revisit some concepts from Section 8.3, including effective nuclear charge and shielding.

## Atomic Radii

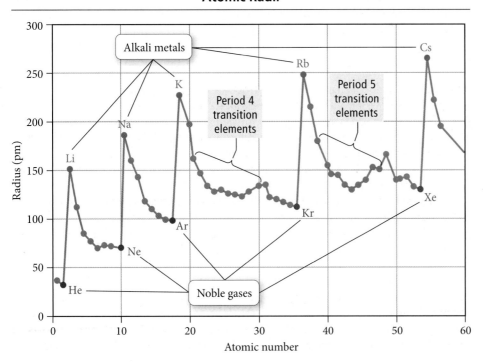

▲ **FIGURE 8.9 Atomic Radius versus Atomic Number** Notice the periodic trend in atomic radius, starting at a peak with each alkali metal and falling to a minimum with each noble gas.

## Trends in Atomic Radius

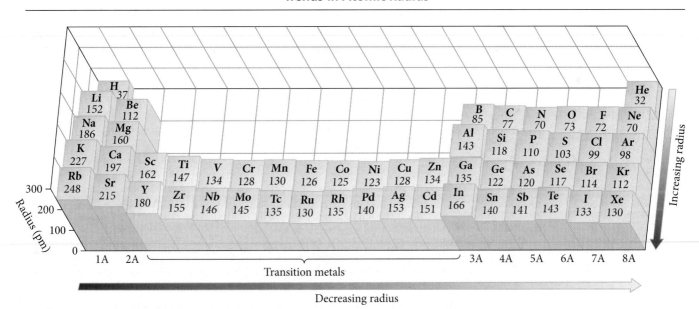

▲ **FIGURE 8.10 Trends in Atomic Radius** In general, atomic radii increase as we move down a column and decrease as we move to the right across a period in the periodic table.

# Effective Nuclear Charge

The trend in atomic radius as we move to the right across a row in the periodic table is determined by the inward pull of the nucleus on the electrons in the outermost principal energy level (highest *n* value). According to Coulomb's law, the attraction between a nucleus and an electron increases with increasing magnitude of nuclear charge. For example, compare the H atom to the He$^+$ ion:

$$H \quad 1s^1$$

$$He^+ \; 1s^1$$

It takes 1312 kJ/mol of energy to remove the $1s$ electron from hydrogen but 5251 kJ/mol of energy to remove it from He$^+$. Why? Although each electron is in a $1s$ orbital, the electron in the helium ion is attracted to the nucleus by a 2+ charge, while the electron in the hydrogen atom is attracted to the nucleus by only a 1+ charge. Therefore, the electron in the helium ion is held more tightly (it has lower potential energy according to Coulomb's law), making it more difficult to remove and making the helium ion smaller than the hydrogen atom.

As we saw in Section 8.3, any one electron in a multielectron atom experiences both the positive charge of the nucleus (which is attractive) and the negative charges of the other electrons (which are repulsive). Consider again the outermost electron in the lithium atom:

$$Li \quad 1s^2 2s^1$$

Even though the $2s$ orbital penetrates into the $1s$ orbital to some degree, the majority of the $2s$ orbital is outside of the $1s$ orbital. Therefore the electron in the $2s$ orbital is partially *screened* or *shielded* from the 3+ charge of the nucleus by the 2− charge of the $1s$ (core) electrons. This reduces the net charge experienced by the $2s$ electron as shown in Figure 8.11 ▼.

### Shielding and Effective Nuclear Charge

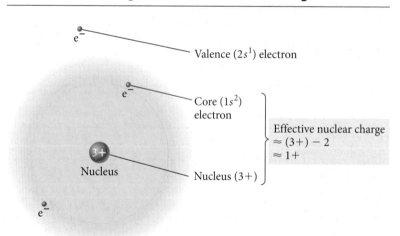

**Lithium**

▲ **FIGURE 8.11 Shielding and Effective Nuclear Charge** The valence electron in lithium experiences the 3+ charge of the nucleus through the screen of the 2− charge of the core electrons. The effective nuclear charge acting on the valence electron is approximately 1+.

As we have seen, we can define the average or net charge experienced by an electron as the *effective nuclear charge*. The effective nuclear charge experienced by a particular electron in an atom is the *actual nuclear charge (Z)* minus *the charge shielded by other electrons (S)*:

$$Z_{\text{eff}} = Z - S$$

Effective nuclear charge

Actual nuclear charge

Charge screened by other electrons

For lithium, we estimate that the two core electrons shield the valence electron from the nuclear charge with high efficiency ($S$ is nearly 2). The effective nuclear charge experienced by lithium's valence electron is therefore slightly greater than 1+.

Now consider the valence electrons in beryllium (Be), with atomic number 4. Its electron configuration is:

$$\text{Be}\quad 1s^2 2s^2$$

To estimate the effective nuclear charge experienced by the $2s$ electrons in beryllium, we must distinguish between two different types of shielding: (1) the shielding of the outermost electrons by the core electrons and (2) the shielding of the outermost electrons by *each other*. The key to understanding the trend in atomic radius is the difference between these two types of shielding:

> **Core electrons efficiently shield electrons in the outermost principal energy level from nuclear charge, but outermost electrons do not efficiently shield one another from nuclear charge.**

For example, the two outermost electrons in beryllium experience the 4+ charge of the nucleus through the shield of the two $1s$ core electrons without shielding each other from that charge very much. We estimate that the shielding ($S$) experienced by any one of the outermost electrons due to the core electrons is nearly 2, but that the shielding due to the other outermost electron is nearly 0. The effective nuclear charge experienced by beryllium's outermost electrons is therefore slightly greater than 2+.

The effective nuclear charge experienced by *beryllium*'s outermost electrons is greater than that experienced by *lithium*'s outermost electron. Consequently, beryllium's outermost electrons are held more tightly than lithium's, resulting in a smaller atomic radius for beryllium. The effective nuclear charge experienced by an atom's outermost electrons continues to become more positive as we move to the right across the rest of the second row in the periodic table, resulting in successively smaller atomic radii. The same trend is generally observed in all main-group elements.

*Summarizing Atomic Radii for Main-Group Elements:*

▶ As we move down a column in the periodic table, the principal quantum number ($n$) of the electrons in the outermost principal energy level increases, resulting in larger orbitals and therefore larger atomic radii.

▶ As we move to the right across a row in the periodic table, the effective nuclear charge ($Z_{\text{eff}}$) experienced by the electrons in the outermost principal energy level increases, resulting in a stronger attraction between the outermost electrons and the nucleus, and smaller atomic radii.

---

**Effective Nuclear Charge**  Which electrons experience the greatest effective nuclear charge?

    **(a)** the valence electrons in Mg

    **(b)** the valence electrons in Al

    **(c)** the valence electrons in S

CONCEPTUAL
**CONNECTION 8.5**

---

## Atomic Radii and the Transition Elements

In Figure 8.10, we can see that as we move down the first two rows of a column within the transition metals, the elements follow the same general trend in atomic radii as the main-group elements (the radii get larger). However, with the exception of the first couple of elements in each transition series, the atomic radii of the transition elements *do not* follow the same trend as the main-group elements as we move to the right across a row. Instead of decreasing in size, *the radii of transition elements stay roughly constant across each row*. Why? The difference is that, across a row of transition elements, the number of electrons in the outermost principal energy level (highest $n$ value) is nearly constant (recall from Section 8.3, for example, that the 4*s* orbital fills before the 3*d*). As another proton is added to the nucleus of each successive element, another electron is added as well, but the electron goes into an $n_{highest} - 1$ orbital. The number of outermost electrons stays constant, and they experience a roughly constant effective nuclear charge, keeping the radius approximately constant.

---

## Example 8.5     Atomic Size

On the basis of periodic trends, choose the larger atom in each pair (if possible). Explain your choices.

**(a)** C or F

**(b)** C or Ge

**(c)** N or Al

**(d)** Al or Ge

**SOLUTION**

| | |
|---|---|
| **(a)** C atoms are larger than F atoms because as you trace the path between C and F on the periodic table, you move to the right within the same period. As you move to the right across a period, the effective nuclear charge experienced by the outermost electrons increases, resulting in a smaller radius. |  |
| **(b)** Ge atoms are larger than C atoms because as you trace the path between C and Ge on the periodic table, you move down a column. Atomic size increases as you move down a column because the outermost electrons occupy orbitals with a higher principal quantum number that are therefore larger, resulting in a larger atom. |  |
| **(c)** Al atoms are larger than N atoms because as you trace the path between N and Al on the periodic table, you move down a column (atomic size increases) and then to the left across a period (atomic size increases). These effects add together for an overall increase. |  |
| **(d)** Based on periodic trends alone, you cannot tell which atom is larger because as you trace the path between Al and Ge you move to the right across a period (atomic size decreases) and then down a column (atomic size increases). These effects tend to counter each other, and it is not easy to tell which will predominate. |  |

---

**FOR PRACTICE 8.5**     On the basis of periodic trends, choose the larger atom in each pair (if possible):

**(a)** Sn or I

**(b)** Ge or Po

**(c)** Cr or W

**(d)** F or Se

**FOR MORE PRACTICE 8.5**     Arrange the elements in order of decreasing radius: S, Ca, F, Rb, Si.

# 8.7 Ions: Electron Configurations, Magnetic Properties, Ionic Radii, and Ionization Energy

Recall that ions are atoms (or groups of atoms) that have lost or gained electrons. In this section, we examine periodic trends in ionic electron configurations, magnetic properties, ionic radii, and ionization energies.

## Electron Configurations and Magnetic Properties of Ions

We can deduce the electron configuration of a main-group monoatomic ion from the electron configuration of the neutral atom and the charge of the ion. For anions, we *add* the number of electrons indicated by the magnitude of the charge of the anion. For example, the electron configuration of fluorine (F) is $1s^2 2s^2 2p^5$ and that of the fluoride ion (F$^-$) is $1s^2 2s^2 2p^6$.

We determine the electron configuration of cations by *subtracting* the number of electrons indicated by the magnitude of the charge. For example, the electron configuration of lithium (Li) is $1s^2 2s^1$ and that of the lithium ion (Li$^+$) is $1s^2 2s^0$ (or simply $1s^2$). For main-group cations, we remove the required number of electrons in the reverse order of filling. However, for transition metal cations, the trend is different. When writing the electron configuration of a transition metal cation, we *remove the electrons in the highest n-value orbitals first, even if this does not correspond to the reverse order of filling.* For example, the electron configuration of vanadium is:

$$V \quad [Ar] \, 4s^2 3d^3$$

The V$^{2+}$ ion, however, has the following electron configuration:

$$V^{2+} \quad [Ar] \, 4s^0 3d^3$$

In other words, for transition metal cations, the order in which electrons are removed upon ionization is *not* the reverse of the filling order. During filling, the $4s$ orbital normally fills before the $3d$ orbital. When a fourth period transition metal ionizes, however, it loses its $4s$ electrons before its $3d$ electrons. Why this unexpected behavior? The full answer to this question is beyond our scope, but the following two factors contribute to this phenomenon:

*The order of removing electrons is not the reverse of the filling order because the energy levels in a cation shift relative to the neutral atom.*

- As discussed previously, the $ns$ and $(n - 1)d$ orbitals are extremely close in energy and, depending on the exact configuration, can vary in relative energy ordering.

- As the $(n - 1)d$ orbitals begin to fill in the first transition series, the increasing nuclear charge stabilizes the $(n - 1)d$ orbitals relative to the $ns$ orbitals. This happens because the $(n - 1)d$ orbitals are not the outermost (or highest $n$) orbitals and are therefore not effectively shielded from the increasing nuclear charge by the $ns$ orbitals.

The bottom-line experimental observation is that an $ns^0(n - 1)d^x$ configuration is lower in energy than an $ns^2(n - 1)d^{x-2}$ configuration for transition metal ions. Therefore, we remove the $ns$ electrons before the $(n - 1)d$ electrons when we write electron configurations for transition metal ions.

The magnetic properties of transition metal ions support these assignments. Recall from Section 8.3 that an unpaired electron generates a magnetic field due to its spin. Consequently, an atom or ion that contains unpaired electrons is attracted to an external magnetic field, and we say that the atom or ion is **paramagnetic**. For example, consider the electron configuration of silver:

Ag    [Kr]  $5s^1 4d^{10}$    | ↑ |    | ↑↓ | ↑↓ | ↑↓ | ↑↓ | ↑↓ |
  
$5s$        $4d$

Silver's unpaired $5s$ electron causes silver to be paramagnetic. In fact, an early demonstration of electron spin—called the Stern–Gerlach experiment—involved the interaction of a beam of silver atoms with a magnetic field.

An atom or ion in which all electrons are paired is not attracted to an external magnetic field—it is instead slightly repelled—and we say that the atom or ion is **diamagnetic**. The zinc atom is diamagnetic:

Zn    [Ar]  $4s^2 3d^{10}$  ⊞⊞ | ⊞⊞ ⊞⊞ ⊞⊞ ⊞⊞ ⊞⊞
                              4s          3d

The magnetic properties of the zinc ion provide confirmation that the 4s electrons are indeed lost before 3d electrons in the ionization of zinc. If zinc lost two 3d electrons upon ionization, then the $Zn^{2+}$ would be paramagnetic (because the two electrons would come out of two different filled d orbitals, leaving each of them with one unpaired electron). That does not happen and the zinc ion, like the zinc atom, is diamagnetic because the 4s electrons are lost instead:

$Zn^{2+}$    [Ar]  $4s^0 3d^{10}$  ☐ | ⊞⊞ ⊞⊞ ⊞⊞ ⊞⊞ ⊞⊞
                                   4s          3d

Observations in other transition metals confirm that the ns electrons are lost before the $(n - 1)d$ electrons upon ionization.

## Example 8.6    Electron Configurations and Magnetic Properties for Ions

Write the electron configuration and orbital diagram for each ion and determine whether each is diamagnetic or paramagnetic.

**(a)** $Al^{3+}$        **(b)** $S^{2-}$        **(c)** $Fe^{3+}$

**SOLUTION**

**(a)** $Al^{3+}$

Begin by writing the electron configuration of the neutral atom.

This ion has a 3+ charge, so you remove three electrons to write the electron configuration of the ion. Write the orbital diagram by drawing half-arrows to represent each electron in boxes representing the orbitals. Because there are no unpaired electrons, $Al^{3+}$ is diamagnetic.

Al      [Ne] $3s^2 3p^1$

$Al^{3+}$   [Ne] or [He] $2s^2 2p^6$

$Al^{3+}$   [He]  ⊞⊞ | ⊞⊞ ⊞⊞ ⊞⊞
                  2s        2p

Diamagnetic

**(b)** $S^{2-}$

Begin by writing the electron configuration of the neutral atom.

Because this ion has a 2− charge, add two electrons to write the electron configuration of the ion. Write the orbital diagram by drawing half-arrows to represent each electron in boxes representing the orbitals. Because there are no unpaired electrons, $S^{2-}$ is diamagnetic.

S      [Ne] $3s^2 3p^4$

$S^{2-}$    [Ne] $3s^2 3p^6$

$S^{2-}$    [Ne]  ⊞⊞ | ⊞⊞ ⊞⊞ ⊞⊞
                  3s        3p

Diamagnetic

**(c)** $Fe^{3+}$

Begin by writing the electron configuration of the neutral atom.

Since this ion has a 3+ charge, remove three electrons to write the electron configuration of the ion. Because it is a transition metal, remove the electrons from the 4s orbital before removing electrons from the 3d orbitals. Write the orbital diagram by drawing half-arrows to represent each electron in boxes representing the orbitals. There are unpaired electrons, so $Fe^{3+}$ is paramagnetic.

Fe      [Ar] $4s^2 3d^6$

$Fe^{3+}$   [Ar] $4s^0 3d^5$

$Fe^{3+}$   [Ar]  ☐ | ↑ ↑ ↑ ↑ ↑
                  4s        3d

Paramagnetic

**FOR PRACTICE 8.6** Write the electron configuration and orbital diagram for each ion and predict whether each is paramagnetic or diamagnetic.

**(a)** $Co^{2+}$        **(b)** $N^{3-}$        **(c)** $Ca^{2+}$

## Ionic Radii

What happens to the radius of an atom when it becomes a cation? An anion? Consider, for example, the difference between the Na atom and the $Na^+$ ion. Their electron configurations are:

$$Na \quad [Ne]\, 3s^1$$
$$Na^+ \quad [Ne]\, 3s^0$$

The sodium atom has an outer $3s$ electron and a neon core. Because the $3s$ electron is the outermost electron and it is shielded from the nuclear charge by the core electrons, it contributes greatly to the size of the sodium atom. The sodium cation, having lost the outermost $3s$ electron, has only the neon core and carries a charge of 1+. Without the $3s$ electron, the sodium cation (ionic radius = 95 pm) is much smaller than the sodium atom (covalent radius = 186 pm). The trend is the same with all cations and their atoms, as shown in Figure 8.12▼:

**Cations are much smaller than their corresponding atoms.**

### Radii of Atoms and Their Cations (pm)

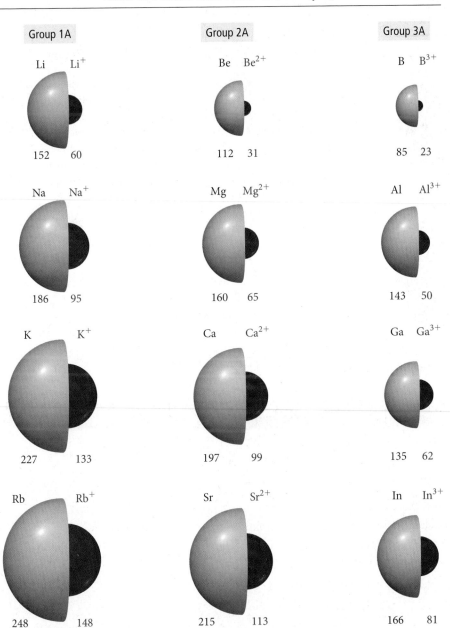

◄ **FIGURE 8.12 Sizes of Atoms and Their Cations** Atomic and ionic radii (pm) for the first three columns of main-group elements.

## Radii of Atoms and Their Anions (pm)

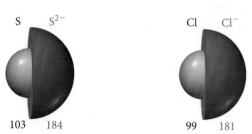

| Group 6A | | Group 7A | |
|---|---|---|---|
| O | O²⁻ | F | F⁻ |
| 73 | 140 | 72 | 136 |

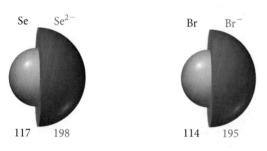

| S | S²⁻ | Cl | Cl⁻ |
|---|---|---|---|
| 103 | 184 | 99 | 181 |

| Se | Se²⁻ | Br | Br⁻ |
|---|---|---|---|
| 117 | 198 | 114 | 195 |

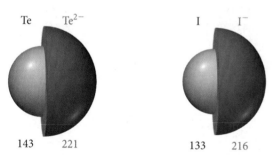

| Te | Te²⁻ | I | I⁻ |
|---|---|---|---|
| 143 | 221 | 133 | 216 |

What about anions? Consider the difference between Cl and Cl⁻. Their electron configurations are:

$$\text{Cl} \quad [\text{Ne}]\, 3s^2 3p^5$$

$$\text{Cl}^- \quad [\text{Ne}]\, 3s^2 3p^6$$

The chlorine anion has one additional outermost electron but no additional proton to increase the nuclear charge. The extra electron increases the repulsions among the outermost electrons, resulting in a chloride anion that is larger than the chlorine atom. The trend is the same with all anions and their atoms, as shown in Figure 8.13◄.

**Anions are much larger than their corresponding atoms.**

We can observe an interesting trend in ionic size when we examine the radii of an *isoelectronic* series of ions—ions with the same number of electrons. Consider the following ions and their radii:

| S²⁻ (184 pm) | Cl⁻ (181 pm) | K⁺ (133 pm) | Ca²⁺ (99 pm) |
|---|---|---|---|
| 18 electrons | 18 electrons | 18 electrons | 18 electrons |
| 16 protons | 17 protons | 19 protons | 20 protons |

All of these ions have 18 electrons in exactly the same orbitals, but the radius of each ion gets successively smaller. Why? The reason is the progressively greater number of protons. The S²⁻ ion has 16 protons and therefore a charge of 16+ pulling on 18 electrons. The Ca²⁺ ion, in contrast, has 20 protons, and therefore a charge of 20+ pulling on the same 18 electrons. The result is a much smaller radius for the calcium ion. For a given number of electrons, a greater nuclear charge results in a smaller atom or ion.

◄ **FIGURE 8.13 Sizes of Atoms and Their Anions** Atomic and ionic radii for groups 6A and 7A in the periodic table.

## Example 8.7 | Ion Size

Choose the larger atom or ion from each pair.

**(a)** S or S²⁻  **(b)** Ca or Ca²⁺  **(c)** Br⁻ or Kr

**SOLUTION**

**(a)** The S²⁻ ion is larger than an S atom because anions are larger than the atoms from which they are formed.

**(b)** A Ca atom is larger than Ca²⁺ because cations are smaller than the atoms from which they are formed.

**(c)** A Br⁻ ion is larger than a Kr atom because, although they are isoelectronic, Br⁻ has one fewer proton than Kr, resulting in less pull on the electrons and therefore a larger radius.

**FOR PRACTICE 8.7** Choose the larger atom or ion from each pair.

**(a)** K or K⁺  **(b)** F or F⁻  **(c)** Ca²⁺ or Cl⁻

**Ions, Isotopes, and Atomic Size** In the previous sections, we have seen how the number of electrons and the number of protons affect the size of an atom or ion. However, we have not considered how the number of neutrons affects the size of an atom. Why not? Would you expect isotopes—for example, C-12 and C-13—to have different atomic radii?

## Ionization Energy

The **ionization energy (IE)** of an atom or ion is the energy required to remove an electron from the atom or ion in the gaseous state. Ionization energy is always positive because removing an electron always takes energy. (The process is similar to an endothermic reaction, which absorbs heat and therefore has a positive $\Delta H$.) The energy required to remove the first electron is the *first ionization energy* ($IE_1$). For example, we represent the first ionization of sodium with the equation:

$$Na(g) \longrightarrow Na^+(g) + 1\,e^- \qquad IE_1 = 496 \text{ kJ/mol}$$

The energy required to remove the second electron is the *second ionization energy* ($IE_2$), the energy required to remove the third electron is the *third ionization energy* ($IE_3$), and so on. We represent the second ionization energy of sodium as:

$$Na^+(g) \longrightarrow Na^{2+}(g) + 1\,e^- \qquad IE_2 = 4560 \text{ kJ/mol}$$

Notice that the second ionization energy is not the energy required to remove *two* electrons from sodium (that quantity is the sum of $IE_1$ and $IE_2$), but rather the energy required to remove one electron from $Na^+$. We look at trends in $IE_1$ and $IE_2$ separately.

## Trends in First Ionization Energy

The first ionization energies of the elements through Xe are shown in Figure 8.14▼. Notice the periodic trend in ionization energy, peaking at each noble gas and bottoming at each alkali metal. Based on what we have learned about electron configurations and effective nuclear charge, how can we account for the observed trend? As we have seen,

### First Ionization Energies

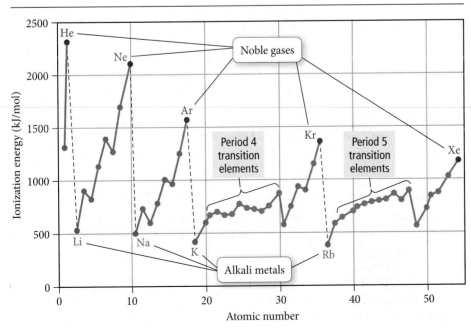

◀ **FIGURE 8.14 First Ionization Energy versus Atomic Number for the Elements through Xenon** First ionization energy starts at a minimum with each alkali metal and rises to a peak with each noble gas.

the principal quantum number, *n*, increases as we move down a column. Within a given sublevel, orbitals with higher principal quantum numbers are larger than orbitals with smaller principal quantum numbers. Consequently, electrons in the outermost principal level are farther away from the positively charged nucleus—and are therefore held less tightly—as we move down a column. This results in a lower ionization energy as we move down a column, as shown in Figure 8.15 ▼.

## Trends in First Ionization Energy

▲ **FIGURE 8.15** **Trends in Ionization Energy** First ionization energy increases as we move to the right across a period and decreases as we move down a column in the periodic table.

What about the trend as we move to the right across a row? For example, would it take more energy to remove an electron from Na or from Cl, two elements on either end of the third row in the periodic table? We know that Na has an outer electron configuration of $3s^1$ and Cl has an outer electron configuration of $3s^2 3p^5$. As discussed previously, the outermost electrons in chlorine experience a higher effective nuclear charge than the outermost electrons in sodium (which is why chlorine has a smaller atomic radius than sodium). Consequently, we would expect chlorine to have a higher ionization energy than sodium, which is indeed the case. We can make a similar argument for the other main-group elements: first ionization energy generally increases as we move to the right across a row in the periodic table, as shown in Figure 8.15.

### Summarizing First Ionization Energy for Main-Group Elements:

▶ First ionization energy generally *decreases* as we move down a column (or family) in the periodic table because electrons in the outermost principal level are increasingly farther away from the positively charged nucleus and are therefore held less tightly.

▶ First ionization energy generally *increases* as we move to the right across a row (or period) in the periodic table because electrons in the outermost principal energy level generally experience a greater effective nuclear charge ($Z_{eff}$).

**Example 8.8** **First Ionization Energy**

On the basis of periodic trends, determine which element in each pair has the higher first ionization energy (if possible).

**(a)** Al or S  **(b)** As or Sb  **(c)** N or Si  **(d)** O or Cl

**SOLUTION**

**(a)** Al or S

S has a higher ionization energy than Al because as you trace the path between Al and S on the periodic table, you move to the right within the same row. Ionization energy increases as you go to the right due to increasing effective nuclear charge.

**(b)** As or Sb

As has a higher ionization energy than Sb because as you trace the path between As and Sb on the periodic table, you move down a column. Ionization energy decreases as you go down a column as a result of the increasing size of orbitals with increasing $n$.

**(c)** N or Si

N has a higher ionization energy than Si because as you trace the path between N and Si on the periodic table, you move down a column (ionization energy decreases) and then to the left across a row (ionization energy decreases). These effects sum together for an overall decrease.

**(d)** O or Cl

Based on periodic trends alone, it is impossible to tell which has a higher ionization energy because, as you trace the path between O and Cl, you go to the right across a row (ionization energy increases) and then down a column (ionization energy decreases). These effects tend to counter each other, and it is not obvious which will dominate.

**FOR PRACTICE 8.8** On the basis of periodic trends, determine the element in each pair with the higher first ionization energy (if possible).

**(a)** Sn or I  **(b)** Ca or Sr  **(c)** C or P  **(d)** F or S

**FOR MORE PRACTICE 8.8** Arrange the following elements in order of decreasing first ionization energy: S, Ca, F, Rb, Si.

## Exceptions to Trends in First Ionization Energy

If we carefully examine Figure 8.15, we can see some exceptions to the trends in first ionization energies. For example, boron has a smaller ionization energy than beryllium, even though it lies to the right of beryllium in the same row. This exception is caused by the difference between the $s$ block and the $p$ block. Recall from Section 8.3 that the $2p$ orbital penetrates into the nuclear region *less than* the $2s$ orbital. Consequently, the $1s$ electrons shield the electron in the $2p$ orbital from nuclear charge more than they shield the electrons in the $2s$ orbital. The result, as we saw in Section 8.3, is that the $2p$ orbitals are higher in energy, and therefore the electron is easier to remove (it has a lower first ionization energy). Similar exceptions occur for aluminum and gallium, both directly below boron in group 3A.

Another exception occurs between nitrogen and oxygen. Although oxygen is to the right of nitrogen in the same row, it has a lower first ionization energy. This exception is caused by the repulsion between electrons when they occupy the same orbital. Examine the electron configurations and orbital diagrams of nitrogen and oxygen shown here:

N    $1s^2 2s^2 2p^3$

O    $1s^2 2s^2 2p^4$

Nitrogen has three electrons in three $p$ orbitals, whereas oxygen has four. In nitrogen, the $2p$ orbitals are half-filled (which makes the configuration particularly stable). One of oxygen's four $2p$ electrons must pair with another electron, making it easier to remove (and less stable). Exceptions for similar reasons occur for S and Se, directly below oxygen in group 6A.

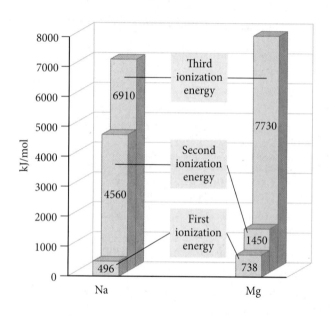

# Trends in Second and Successive Ionization Energies

Notice the trends in the first, second, and third ionization energies of sodium (group 1A) and magnesium (group 2A), as shown at left.

For sodium, there is a huge jump between the first and second ionization energies. For magnesium, the ionization energy roughly doubles from the first to the second, but then a huge jump occurs between the second and third ionization energies. What is the reason for these jumps?

We can understand these trends by examining the electron configurations of sodium and magnesium:

Na    [Ne] $3s^1$

Mg    [Ne] $3s^2$

The first ionization of sodium involves removing the valence electron in the $3s$ orbital. Recall that these valence electrons are held more loosely than the core electrons and that the resulting ion has a noble gas configuration, which is particularly stable. Consequently, the first ionization energy is fairly low. The second ionization of sodium, however, involves removing a core electron from an ion with a noble gas configuration. This requires a tremendous amount of energy, making the value of IE$_2$ very high.

As with sodium, the first ionization of magnesium involves removing a valence electron in the $3s$ orbital. This requires a bit more energy than the corresponding ionization of sodium because of the trends in $Z_{eff}$ that we discussed earlier ($Z_{eff}$ increases as we move to the right across a row). The second ionization of magnesium also involves removing an outer electron in the $3s$ orbital, but this time from an ion with a 1+ charge (instead of from a neutral atom). This requires roughly twice the energy as removing the electron from the neutral atom. The third ionization of magnesium is analogous to the second ionization of sodium—it requires removing a core electron from an ion with a noble gas configuration. This requires a tremendous amount of energy, making the value of IE$_3$ very high.

As shown in Table 8.1, similar trends exist for the successive ionization energies of many elements. The ionization energy increases fairly uniformly with each successive removal of an outermost electron, but then takes a large jump with the removal of the first core electron.

| Element | IE₁ | IE₂ | IE₃ | IE₄ | IE₅ | IE₆ | IE₇ |
|---------|-----|-----|-----|-----|-----|-----|-----|
**TABLE 8.1** Successive Ionization Energies for the Elements Sodium through Argon (kJ/mol)

| Element | IE$_1$ | IE$_2$ | IE$_3$ | IE$_4$ | IE$_5$ | IE$_6$ | IE$_7$ |
|---------|-----|-----|-----|-----|-----|-----|-----|
| Na | 496 | 4560 | | | Core electrons | | |
| Mg | 738 | 1450 | 7730 | | | | |
| Al | 578 | 1820 | 2750 | 11,600 | | | |
| Si | 786 | 1580 | 3230 | 4360 | 16,100 | | |
| P | 1012 | 1900 | 2910 | 4960 | 6270 | 22,200 | |
| S | 1000 | 2250 | 3360 | 4560 | 7010 | 8500 | 27,100 |
| Cl | 1251 | 2300 | 3820 | 5160 | 6540 | 9460 | 11,000 |
| Ar | 1521 | 2670 | 3930 | 5770 | 7240 | 8780 | 12,000 |

**Ionization Energies and Chemical Bonding** Based on what you just learned about ionization energies, explain why valence electrons are more important than core electrons in determining the reactivity and bonding in atoms.

CONCEPTUAL
**CONNECTION 8.7**

## 8.8 Electron Affinities and Metallic Character

Electron affinity and metallic character also exhibit periodic trends. Electron affinity is a measure of how easily an atom will accept an additional electron; it is crucial to chemical bonding because bonding involves the transfer or sharing of electrons. Metallic character is important because of the high proportion of metals in the periodic table and the large role they play in our lives. Of the roughly 110 elements, 87 are metals. We examine each of these periodic properties individually in this section.

### Electron Affinity

The **electron affinity (EA)** of an atom (or ion) is the energy change associated with the gaining of an electron by the atom in the gaseous state. Electron affinity is usually—though not always—negative because an atom or ion usually releases energy when it gains an electron. (The process is analogous to an exothermic reaction, which releases heat and therefore has a negative $\Delta H$.) In other words, the coulombic attraction between the nucleus of an atom and the incoming electron usually results in the release of energy as the electron is gained. For example, we represent the electron affinity of chlorine with the equation:

$$Cl(g) + 1\,e^- \longrightarrow Cl^-(g) \quad EA = -349 \text{ kJ/mol}$$

Figure 8.16▶ displays the electron affinities for a number of main-group elements. As you can see from this figure, the trends in electron affinity are not as regular as trends in other properties we examined in Sections 8.6 and 8.7. For instance, we might expect electron affinities to become relatively more positive (so that the addition of an electron is less exothermic) as we move down a column because the electron is located in orbitals with successively higher principal quantum numbers and therefore is farther from the

**Electron Affinities (kJ/mol)**

| 1A | | | | | | | 8A |
|----|----|----|----|----|----|----|----|
| **H** −73 | 2A | 3A | 4A | 5A | 6A | 7A | **He** >0 |
| **Li** −60 | **Be** >0 | **B** −27 | **C** −122 | **N** >0 | **O** −141 | **F** −328 | **Ne** >0 |
| **Na** −53 | **Mg** >0 | **Al** −43 | **Si** −134 | **P** −72 | **S** −200 | **Cl** −349 | **Ar** >0 |
| **K** −48 | **Ca** −2 | **Ga** −30 | **Ge** −119 | **As** −78 | **Se** −195 | **Br** −325 | **Kr** >0 |
| **Rb** −47 | **Sr** −5 | **In** −30 | **Sn** −107 | **Sb** −103 | **Te** −190 | **I** −295 | **Xe** >0 |

▲ **FIGURE 8.16** Electron Affinities of Selected Main-Group Elements.

nucleus. This trend applies to the group 1A metals but does not hold for the other columns in the periodic table.

A more regular trend in electron affinity occurs as we move to the right across a row, however. Based on the periodic properties we have learned so far, would you expect more energy to be released by Na or Cl when an electron is gained? We know that Na has an outer electron configuration of $3s^1$ and Cl has an outer electron configuration of $3s^2 3p^5$. Since adding an electron to chlorine gives it a noble gas configuration and adding an electron to sodium does not, and since the outermost electrons in chlorine experience a higher $Z_{eff}$ than the outermost electrons in sodium, we would expect chlorine to have a more negative electron affinity—the process should be more exothermic for chlorine. This is in fact the case. For main-group elements, electron affinity generally becomes more negative (more exothermic) as we move to the right across a row in the periodic table. The halogens (group 7A) therefore have the most negative electron affinities. But exceptions do occur. For example, notice that nitrogen and the other group 5A elements do not follow the general trend. These elements have $ns^2 np^3$ outer electron configurations. When an electron is added to this configuration, it must pair with another electron in an already occupied $p$ orbital. The repulsion between two electrons occupying the same orbital causes the electron affinity to be more positive than for elements in the previous column.

### *Summarizing Electron Affinity for Main-Group Elements:*

▶ Most groups (columns) of the periodic table do not exhibit any definite trend in electron affinity. Among the group 1A metals, however, electron affinity becomes more positive as we move down a column (adding an electron becomes less exothermic).

▶ Electron affinity generally becomes more negative (adding an electron becomes more exothermic) as we move to the right across a period (row) in the periodic table.

## Metallic Character

As we discussed in Chapter 2, metals are good conductors of heat and electricity: they can be pounded into flat sheets (malleability), they can be drawn into wires (ductility), they are often shiny, and they tend to lose electrons in chemical reactions. Nonmetals, in contrast, have more varied physical properties; some are solids at room temperature, others are gases. However, in general, nonmetals are typically poor conductors of heat and electricity, and they all tend to gain electrons in chemical reactions.

As we move to the right across a row in the periodic table, ionization energy increases and electron affinity becomes more negative; therefore, elements on the left side of the periodic table are more likely to lose electrons than elements on the right side (which are more likely to gain them). The other properties associated with metals follow the same general trend (even though we do not quantify them here). Consequently, as shown in Figure 8.17▶:

> **As we move to the right across a row (or period) in the periodic table, metallic character decreases.**

As we move down a column in the periodic table, ionization energy decreases, making electrons more likely to be lost in chemical reactions. Therefore,

> **As we move down a column (or family) in the periodic table, metallic character increases.**

These trends, based on the quantum-mechanical model, explain the distribution of metals and nonmetals in the periodic table discussed in Chapter 2. Metals are found on the left side and toward the center and nonmetals on the upper right side. The change in chemical behavior from metallic to nonmetallic can be seen most clearly as we proceed to the right across period 3, or down along group 5A as we can see in Figure 8.18▶.

## Trends in Metallic Character

Metallic character decreases →

Metallic character increases ↓

▲ **FIGURE 8.17  Trends in Metallic Character I**  Metallic character decreases as we move to the right across a period and increases as we move down a column in the periodic table.

## Trends in Metallic Character

Period 3

Group 5A

Metallic character increases

Metallic character decreases →

▲ **FIGURE 8.18  Trends in Metallic Character II**  As we move down group 5A in the periodic table, metallic character increases. As we move across period 3, metallic character decreases.

## Example 8.9 | Metallic Character

On the basis of periodic trends, choose the more metallic element from each pair (if possible).

**(a)** Sn or Te

**(b)** P or Sb

**(c)** Ge or In

**(d)** S or Br

SOLUTION

**(a)** Sn or Te

Sn is more metallic than Te because as you trace the path between Sn and Te on the periodic table, you move to the right within the same period. Metallic character decreases as we move to the right.

**(b)** P or Sb

Sb is more metallic than P because as you trace the path between P and Sb on the periodic table, you move down a column. Metallic character increases as we move down a column.

**(c)** Ge or In

In is more metallic than Ge because as you trace the path between Ge and In on the periodic table, you move down a column (metallic character increases) and then to the left across a period (metallic character increases). These effects add together for an overall increase.

**(d)** S or Br

Based on periodic trends alone, we cannot tell which is more metallic because as you trace the path between S and Br, you move to the right across a period (metallic character decreases) and then down a column (metallic character increases). These effects tend to counter each other, and it is not easy to tell which will predominate.

**FOR PRACTICE 8.9** On the basis of periodic trends, choose the more metallic element from each pair (if possible).

**(a)** Ge or Sn

**(b)** Ga or Sn

**(c)** P or Bi

**(d)** B or N

**FOR MORE PRACTICE 8.9** Arrange the following elements in order of increasing metallic character: Si, Cl, Na, Rb.

CONCEPTUAL
**CONNECTION 8.8**

**Periodic Trends** Use the trends in ionization energy and electron affinity to explain why sodium chloride has the formula NaCl and not $Na_2Cl$ or $NaCl_2$.

# 8.9 Some Examples of Periodic Chemical Behavior: The Alkali Metals, the Halogens, and the Noble Gases

In this section, we explore some of the properties and chemical reactions of three families in the periodic table: the alkali metals, the halogens, and the noble gases. These families exemplify the connection between chemical behavior and electron configuration.

The alkali metals (group 1A) have $ns^1$ outer electron configurations. The single valence electron that keeps these metals from having noble gas configurations is easily removed (the metals have low ionization energies), making these elements the most active metals in the periodic table. The halogens (group 7A) have $ns^2np^5$ outer electron configurations. The one electron needed to attain noble gas configurations is easily acquired (the halogens have highly negative electron affinities), making these elements among the most active nonmetals in the periodic table. The noble gases (group 8A) have electron configurations with full outer $s$ and $p$ sublevels ($ns^2np^6$) and so are the most chemically inert family in the periodic table.

In this section of the chapter, we examine the properties of each of these groups separately. (Even though hydrogen is often listed in group 1A, it behaves like a nonmetal because of its high ionization energy: 1312 kJ/mol. We therefore do not include hydrogen in our discussion of the group 1A metals.)

## The Alkali Metals (Group 1A)

Table 8.2 lists some selected properties of the alkali metals. Notice that, in general, the properties of the alkali metals vary fairly regularly as we proceed down the column. As expected from periodic trends, the atomic radius increases steadily while the first ionization energy decreases steadily.

With the exception of potassium, density increases as we move down the column. This is a general trend that occurs in other columns within the periodic table. As we move down a column, the increase in mass (due to the additional protons and neutrons) outpaces the increase in volume caused by greater atomic radii. The result is successively greater densities. The melting points of the alkali metals as a group are anomalously low for metals, and they steadily decrease as we move down the column. (This is not a general trend for the rest of the periodic table, which shows more irregular patterns in melting points.)

Because of their generally low first ionization energies, the alkali metals are excellent reducing agents—they are readily oxidized, losing electrons to other substances. Consequently, the alkali metals exist naturally in their oxidized state, either in compounds or as dissolved ions in seawater. Since first ionization energy *decreases* as we go down the column, the relative reactivities of the alkali metals tend to *increase* as we move down the column. In other words, the lower the first ionization energy of an alkali metal, the greater tendency it will have to lose its electron and the more reactive it is.

| TABLE 8.2 Properties of the Alkali Metals* | | | | | |
|---|---|---|---|---|---|
| Element | Electron Configuration | Atomic Radius (pm) | $IE_1$ (kJ/mol) | Density at 25 °C (g/cm³) | Melting Point (°C) |
| Li | [He] 2s¹ | 152 | 520 | 0.535 | 181 |
| Na | [Ne] 3s¹ | 186 | 496 | 0.968 | 102 |
| K | [Ar] 4s¹ | 227 | 419 | 0.856 | 98 |
| Rb | [Kr] 5s¹ | 248 | 403 | 1.532 | 39 |
| Cs | [Xe] 6s¹ | 265 | 376 | 1.879 | 29 |

*Francium is omitted because it has no stable isotopes.

▲ **FIGURE 8.19** **Reaction of Sodium and Chlorine to Form Sodium Chloride.**

The reactions of the alkali metals with nonmetals are vigorous. For example, the alkali metals (M) react with halogens (X) according to the reaction:

$$2\,M + X_2 \longrightarrow 2\,MX$$

The reaction of sodium and chlorine to form sodium chloride is typical:

$$2\,Na(s) + Cl_2(g) \longrightarrow 2\,NaCl(s)$$

This reaction emits heat and sparks as it occurs (Figure 8.19◄). Each successive alkali metal reacts even more vigorously with chlorine.

The alkali metals also react with water to form the dissolved alkali metal ion, the hydroxide ion, and hydrogen gas:

$$2\,M(s) + 2\,H_2O(l) \longrightarrow 2\,M^+(aq) + 2\,OH^-(aq) + H_2(g)$$

The reaction is highly exothermic and can be explosive because the heat from the reaction can ignite the hydrogen gas. The reaction becomes more explosive as we move down the column from one metal to the next, as shown in Figure 8.20▼.

### Reactions of the Alkali Metals with Water

► **FIGURE 8.20** **Reactions of the Alkali Metals with Water** The reactions become progressively more vigorous as we move down the group. (The rate of the alkali metal reaction with water, and therefore its vigor, are enhanced by the successively lower melting points of the alkali metals as we move down the column. The low melting points of the heavier metals allow the emitted heat to actually melt the metal, increasing the reaction rate.)

Lithium

Sodium

Potassium

## The Halogens (Group 7A)

Table 8.3 lists selected properties of the first four halogens. Notice that the properties of the halogens, like those of the alkali metals, vary fairly regularly as we proceed down the column. As expected from periodic trends, the atomic radius and the density increase for each successive halogen. We can see from the melting and boiling points that fluorine and chlorine are both gases at room temperature, bromine is a liquid, and iodine is a solid.

All of the halogens are powerful oxidizing agents—they are readily reduced, gaining electrons from other substances in their reactions. Fluorine is the most powerful oxidizing agent of the group—reacting with almost everything, including the heavier noble gases—and iodine is the least. The halogens all react with metals to form *metal halides* according to the equation:

$$2\,M + n\,X_2 \longrightarrow 2\,MX_n$$

**TABLE 8.3** **Properties of the Halogens\***

| Element | Electron Configuration | Atomic Radius (pm) | EA (kJ/mol) | Melting Point (°C) | Boiling Point (°C) | Density of Liquid (g/cm³) |
|---|---|---|---|---|---|---|
| F | [He] $2s^2 2p^5$ | 72 | −328 | −219 | −188 | 1.51 |
| Cl | [Ne] $3s^2 3p^5$ | 99 | −349 | −101 | −34 | 2.03 |
| Br | [Ar] $4s^2 4p^5$ | 114 | −325 | −7 | 59 | 3.19 |
| I | [Kr] $5s^2 5p^5$ | 133 | −295 | 114 | 184 | 3.96 |

\*At is omitted because it is rare and radioactive.

where M is the metal, X is the halogen, and MX$_n$ is the metal halide. For example, chlorine reacts with iron according to the equation:

$$2\,Fe(s) + 3\,Cl_2(g) \longrightarrow 2\,FeCl_3(s)$$

Since metals tend to lose electrons and the halogens tend to gain them, the metal halides—like all compounds that form between metals and nonmetals—contain ionic bonds.

The halogens also react with hydrogen to form *hydrogen halides* according to the equation:

$$H_2(g) + X_2 \longrightarrow 2\,HX(g)$$

The hydrogen halides—like all compounds that form between two nonmetals—contain covalent bonds. All of the hydrogen halides form acidic solutions when combined with water.

The halogens also react with each other to form *interhalogen compounds*. For example, bromine reacts with fluorine according to the equation:

$$Br_2(l) + F_2(g) \longrightarrow 2\,BrF(g)$$

Again, like all compounds that form between two nonmetals, the interhalogen compounds contain covalent bonds.

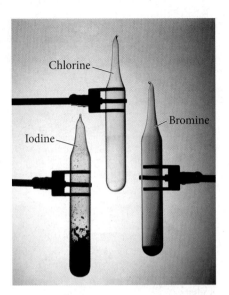

Chlorine

Bromine

Iodine

## Example 8.10 | Alkali Metal and Halogen Reactions

Write a balanced chemical equation for each reaction.
- **(a)** the reaction between potassium metal and bromine gas
- **(b)** the reaction between rubidium metal and liquid water
- **(c)** the reaction between gaseous chlorine and solid iodine

**SOLUTION**

| | |
|---|---|
| **(a)** Alkali metals react with halogens to form metal halides. Write the formulas for the reactants and the metal halide product (making sure to write the correct ionic chemical formula for the metal halide, as outlined in Section 3.5), and then balance the equation. | $2\,K(s) + Br_2(g) \longrightarrow 2\,KBr(s)$ |
| **(b)** Alkali metals react with water to form the dissolved metal ion, the hydroxide ion, and hydrogen gas. Write the skeletal equation including each of these and then balance it. | $2\,Rb(s) + 2\,H_2O(l) \longrightarrow 2\,Rb^+(aq) + 2\,OH^-(aq) + H_2(g)$ |
| **(c)** Halogens react with each other to form interhalogen compounds. Write the skeletal equation with each of the halogens as the reactants and the interhalogen compound as the product and balance the equation. | $Cl_2(g) + I_2(s) \longrightarrow 2\,ICl(g)$ |

**FOR PRACTICE 8.10**  Write a balanced chemical equation for each reaction.
- **(a)** the reaction between aluminum metal and chlorine gas
- **(b)** the reaction between lithium metal and liquid water
- **(c)** the reaction between gaseous hydrogen and liquid bromine

## The Noble Gases (Group 8A)

Table 8.4 lists selected properties of the noble gases. Notice that the properties of the noble gases, like those of the alkali metals and halogens, vary fairly regularly as we proceed down the column. As expected from periodic trends, the atomic radius and the density increase

**TABLE 8.4 Properties of the Noble Gases***

| Element | Electron Configuration | Atomic Radius (pm)** | IE$_1$(kJ/mol) | Boiling Point (K) | Density of Gas (g/L at STP) |
|---------|------------------------|----------------------|----------------|-------------------|------------------------------|
| He | $1s^2$ | 32 | 2372 | 4.2 | 0.18 |
| Ne | [He] $2s^22p^6$ | 70 | 2081 | 27.1 | 0.90 |
| Ar | [Ne] $3s^23p^6$ | 98 | 1521 | 87.3 | 1.78 |
| Kr | [Ar] $4s^24p^6$ | 112 | 1351 | 119.9 | 3.74 |
| Xe | [Kr] $5s^25p^6$ | 130 | 1170 | 165.1 | 5.86 |

*Radon is omitted because it is radioactive.

**Since only the heavier noble gases form compounds, covalent radii for the smaller noble gases are estimated.

▲ Liquid helium, a cryogenic liquid, cools substances to temperatures as low as 1.2 K.

for each successive noble gas, and the first ionization energy decreases. As their boiling points indicate, all of the noble gases are gases at room temperature and must be cooled to extremely low temperatures before they liquefy. For this reason, some noble gases can be cryogenic liquids—liquids used to cool other substances to low temperatures. For example, researchers often submerse samples of interest in boiling liquid helium to cool them down to 4.2 K (or lower) and study their properties at this extremely low temperature.

The high first ionization energies of the noble gases and their completely full *s* and *p* sublevels make them exceptionally unreactive. In fact, before the 1960s, no noble gas compounds were known. Since then, two of the noble gases have been shown to react with fluorine (the most reactive nonmetal on the periodic table) under fairly extreme conditions. For instance, krypton reacts with fluorine to form $KrF_2$:

$$Kr + F_2 \longrightarrow KrF_2$$

Similarly, Xe reacts with fluorine to form three different xenon fluorides:

$$Xe + F_2 \longrightarrow XeF_2$$
$$Xe + 2 F_2 \longrightarrow XeF_4$$
$$Xe + 3 F_2 \longrightarrow XeF_6$$

Xenon can also be forced to react with oxygen to form $XeO_3$ and $XeO_4$.

The inertness of the noble gases has led to their use in situations where reactions are undesirable. For example, argon is used in lightbulbs to prevent the hot tungsten filament from oxidizing, and helium is part of the mixture breathed by deep-sea divers to prevent the toxicity caused by too much oxygen and nitrogen under high pressures. (The helium replaces some of the oxygen and nitrogen in the tank, lowering the concentrations of oxygen and nitrogen in the blood.)

# Self-Assessment Quiz

**Q1.** According to Coulomb's law, if the separation between two particles of the same charge is doubled, what happens to the potential energy of the two particles?

a) It is twice as high as it was before the distance separation.
b) It is one-half as high as it was before the separation.
c) It does not change.
d) It is one-fourth as high as it was before the separation.

**Q2.** Which electron in sulfur is most shielded from nuclear charge?

a) an electron in the $1s$ orbital
b) an electron in a $2p$ orbital
c) an electron in a $3p$ orbital
d) none of the above (All of these electrons are equally shielded from nuclear charge.)

**Q3.** Choose the correct electron configuration for Se.

a) $1s^22s^22p^63s^23p^4$
b) $1s^22s^22p^63s^23p^64s^23d^{10}4p^4$
c) $1s^22s^22p^63s^23p^64s^24p^4$
d) $1s^22s^22p^63s^23p^64s^23d^4$

**Q4.** Choose the correct orbital diagram for vanadium.

**Q5.** Which set of four quantum numbers corresponds to an electron in a 4*p* orbital?

a) $n = 4, l = 1, m_l = 0, m_s = \frac{1}{2}$

b) $n = 4, l = 3, m_l = 3, m_s = -\frac{1}{2}$

c) $n = 4, l = 2, m_l = 0, m_s = \frac{1}{2}$

d) $n = 4, l = 4, m_l = 3, m_s = -\frac{1}{2}$

**Q6.** Which element has the smallest atomic radius?

a) C        b) Si        c) Be        d) F

**Q7.** Which statement is true about electron shielding of nuclear charge?

a) Outermost electrons efficiently shield one another from nuclear charge.

b) Core electrons efficiently shield one another from nuclear charge.

c) Outermost electrons efficiently shield core electrons from nuclear charge.

d) Core electrons efficiently shield outermost electrons from nuclear charge.

**Q8.** Which statement is true about effective nuclear charge?

a) Effective nuclear charge *decreases* as you move to the right across a row in the periodic table.

b) Effective nuclear charge *increases* as you move to the right across a row in the periodic table.

c) Effective nuclear charge *remains constant* as you move to the right across a row in the periodic table.

d) Effective nuclear charge *increases and decreases* at regular intervals as you move to the right across a row in the periodic table.

**Q9.** Which is the correct electron configuration for $Fe^{2+}$?

a) $[Ar]4s^2 3d^6$        b) $[Ar]4s^2 3d^4$

c) $[Ar]4s^0 3d^6$        d) $[Ar]4s^2 3d^8$

**Q10.** Which species is diamagnetic?

a) $Cr^{2+}$     b) Zn     c) Mn     d) C

**Q11.** Arrange these atoms and ions in order of increasing radius: $Cs^+, Ba^{2+}, I^-$.

a) $I^- < Ba^{2+} < Cs^+$        b) $Cs^+ < Ba^{2+} < I^-$

c) $Ba^{2+} < Cs^+ < I^-$        d) $I^- < Cs^+ < Ba^{2+}$

**Q12.** Arrange these elements in order of increasing first ionization energy: Cl, Sn, Si.

a) Cl < Si < Sn

b) Sn < Si < Cl

c) Si < Cl < Sn

d) Sn < Cl < Si

**Q13.** The ionization energies of an unknown third period element are listed here. Identify the element. $IE_1 = 786$ kJ/mol; $IE_2 = 1580$ kJ/mol; $IE_3 = 3230$ kJ/mol; $IE_4 = 4360$ kJ/mol; $IE_5 = 16,100$ kJ/mol

a) Mg        b) Al        c) Si        d) P

**Q14.** Which statement is true about trends in metallic character?

a) Metallic character *increases* as you move to the right across a row in the periodic table and *increases* as you move down a column.

b) Metallic character *decreases* as you move to the right across a row in the periodic table and *increases* as you move down a column.

c) Metallic character *decreases* as you move to the right across a row in the periodic table and *decreases* as you move down a column.

d) Metallic character *increases* as you move to the right across a row in the periodic table and *decreases* as you move down a column.

**Q15.** For which element is the gaining of an electron most exothermic?

a) Li        b) N

c) F        d) B

**Answers: 1.** (b) **2.** (c) **3.** (b) **4.** (d) **5.** (a) **6.** (d) **7.** (d) **8.** (b) **9.** (c) **10.** (b) **11.** (b) **12.** (c) **13.** (c) **14.** (b) **15.** (c)

# Chapter 8 IN REVIEW

## Key Terms

### Section 8.1
periodic property (338)

### Section 8.3
electron configuration (339)
ground state (339)
orbital diagram (340)
Pauli exclusion principle (340)
degenerate (341)
Coulomb's law (341)
shielding (342)
effective nuclear charge ($Z_{eff}$) (342)
penetration (342)

aufbau principle (344)
Hund's rule (344)

### Section 8.4
valence electrons (347)
core electrons (347)

### Section 8.6
van der Waals radius (nonbonding atomic radius) (352)
covalent radius (bonding atomic radius) (352)
atomic radius (352)

### Section 8.7
paramagnetic (357)
diamagnetic (358)
ionization energy (IE) (361)

### Section 8.8
electron affinity (EA) (365)

# Key Concepts

## Periodic Properties and the Development of the Periodic Table (8.1, 8.2)

▶ In the nineteenth century, Dmitri Mendeleev arranged the elements in an early version of the periodic table so that atomic mass increased from left to right in a row and elements with similar properties fell in the same columns.

▶ Periodic properties are predictable based on an element's position within the periodic table. Periodic properties include atomic radius, ionization energy, electron affinity, density, and metallic character.

▶ Quantum mechanics explains the periodic table by describing how electrons fill the quantum-mechanical orbitals within the atoms that compose the elements.

## Electron Configurations (8.3)

▶ An electron configuration for an atom shows which quantum-mechanical orbitals the atom's electrons occupy. For example, the electron configuration of helium ($1s^2$) indicates that helium's two electrons occupy the $1s$ orbital.

▶ The order of filling quantum-mechanical orbitals in multielectron atoms is $1s\ 2s\ 2p\ 3s\ 3p\ 4s\ 3d\ 4p\ 5s\ 4d\ 5p\ 6s$.

▶ According to the Pauli exclusion principle, each orbital can hold a maximum of two electrons and those electrons have opposing spins.

▶ According to Hund's rule, orbitals of the same energy first fill singly with electrons having parallel spins before pairing.

## Electron Configurations and the Periodic Table (8.4, 8.5)

▶ Because quantum-mechanical orbitals fill sequentially with increasing atomic number, we can infer the electron configuration of an element from its position in the periodic table.

▶ The most stable electron configurations are those with completely full $s$ and $p$ sublevels. Therefore, the most stable and unreactive elements—those with the lowest energy electron configurations—are the noble gases.

▶ Elements with one or two valence electrons are among the most active metals, readily losing their valence electrons to attain noble gas configurations.

▶ Elements with six or seven valence electrons are among the most active nonmetals, readily gaining enough electrons to attain a noble gas configuration.

## Effective Nuclear Charge and Periodic Trends in Atomic Size (8.6)

▶ The size of an atom is largely determined by its outermost electrons. As we move down a column in the periodic table, the principal quantum number ($n$) of the outermost electrons increases, resulting in successively larger orbitals and therefore larger atomic radii.

▶ As we move across a row in the periodic table, atomic radii decrease because the effective nuclear charge—the net or average charge experienced by the atom's outermost electrons—increases.

▶ The atomic radii of the transition elements stay roughly constant as we move across each row because electrons are added to the $n_{highest} - 1$ orbitals while the number of highest $n$ electrons stays roughly constant.

## Ion Properties (8.7)

▶ We determine the electron configuration of an ion by adding or subtracting the corresponding number of electrons to the electron configuration of the neutral atom.

▶ For main-group ions, the order in which electrons are removed is the same as the order in which they are added in building up the electron configuration.

▶ For transition metal atoms, the $ns$ electrons are removed before the $(n - 1)d$ electrons.

▶ The radius of a cation is much *smaller* than that of the corresponding atom, and the radius of an anion is much *larger* than that of the corresponding atom.

▶ The first ionization energy—the energy required to remove the first electron from an atom in the gaseous state—generally decreases as we move down a column in the periodic table and increases when we move to the right across a row.

▶ Successive ionization energies increase smoothly from one valence electron to the next, but the ionization energy increases dramatically for the first core electron.

## Electron Affinities and Metallic Character (8.8)

▶ Electron affinity—the energy associated with an atom in its gaseous state gaining an electron—does not show a general trend as we move down a column in the periodic table, but it generally becomes more negative (more exothermic) to the right across a row.

▶ Metallic character—the tendency to lose electrons in a chemical reaction—generally increases down a column in the periodic table and decreases to the right across a row.

## The Alkali Metals, Halogens, and Noble Gases (8.9)

▶ The most active metals are the alkali metals (group 1A), and the most active nonmetals are the halogens (group 7A).

▶ The alkali metals are powerful reducing agents, reacting with many nonmetals—including the halogens and water—to form ionic compounds.

▶ The halogens are powerful oxidizing agents, reacting with many metals to form ionic compounds. The halogens also react with many nonmetals to form covalent compounds.

▶ The noble gases are relatively unreactive; only krypton and xenon form compounds, typically only with fluorine, the most reactive element in the periodic table.

# Key Equations and Relationships

Order of Filling Quantum-Mechanical Orbitals (8.3)

$$1s\ 2s\ 2p\ 3s\ 3p\ 4s\ 3d\ 4p\ 5s\ 4d\ 5p\ 6s$$

# Key Learning Outcomes

| CHAPTER OBJECTIVES | ASSESSMENT |
|---|---|
| Writing Electron Configurations (**8.3**) | Example 8.1  For Practice 8.1  Exercises 41, 42 |
| Writing Orbital Diagrams (**8.3**)  | Example 8.2  For Practice 8.2  Exercises 43, 44 |
| Valence Electrons and Core Electrons (**8.4**) | Example 8.3  For Practice 8.3  Exercises 51, 52 |
| Electron Configurations from the Periodic Table (**8.4**) | Example 8.4  For Practice 8.4  For More Practice 8.4  Exercises 45, 46 |
| Using Periodic Trends to Predict Atomic Size (**8.6**) | Example 8.5  For Practice 8.5  For More Practice 8.5  Exercises 61–64 |
| Writing Electron Configurations for Ions (**8.7**) | Example 8.6  For Practice 8.6  Exercises 65, 66 |
| Using Periodic Trends to Predict Ion Size (**8.7**) | Example 8.7  For Practice 8.7  Exercises 69–72 |
| Using Periodic Trends to Predict Relative Ionization Energies (**8.7**) | Example 8.8  For Practice 8.8  For More Practice 8.8  Exercises 73–76 |
| Predicting Metallic Character Based on Periodic Trends (**8.8**) | Example 8.9  For Practice 8.9  For More Practice 8.9  Exercises 81–84 |
| Writing Reactions for Alkali Metal and Halogen Reactions (**8.9**) | Example 8.10  For Practice 8.10  Exercises 85–90 |

# EXERCISES

## Review Questions

1. What are periodic properties?

2. Which periodic property is particularly important to nerve signal transmission? Why?

3. Explain the contributions of Johann Döbereiner and John Newlands to the organization of elements according to their properties.

4. Who is credited with arranging the periodic table? How are the elements arranged in the modern periodic table?

5. Explain the contributions of Meyer and Moseley to the periodic table.

6. The periodic table is a result of the periodic law. What observations led to the periodic law? What theory explains the underlying reasons for the periodic law?

7. What is an electron configuration? Give an example.

8. What is Coulomb's law? Explain how the potential energy of two charged particles depends on the distance between the charged particles and on the magnitude and sign of their charges.

9. What is shielding? In an atom, which electrons tend to do the most shielding (core electrons or valence electrons)?

10. What is penetration? How does the penetration of an orbital into the region occupied by core electrons affect the energy of an electron in that orbital?

11. Why are the sublevels within a principal level split into different energies for multielectron atoms but not for the hydrogen atom?

12. What is an orbital diagram? Provide an example.

13. Why is electron spin important when writing electron configurations? Explain in terms of the Pauli exclusion principle.

14. What are degenerate orbitals? According to Hund's rule, how are degenerate orbitals occupied?

15. List all orbitals from 1s through 5s according to increasing energy for multielectron atoms.

16. What are valence electrons? Why are they important?

17. Copy this blank periodic table onto a sheet of paper and label each of the blocks within the table: s block, p block, d block, and f block.

18. Explain why the s block in the periodic table has only two columns while the p block has six.

**19.** Why do the rows in the periodic table get progressively longer as you move down the table? For example, the first row contains 2 elements, the second and third rows each contain 8 elements, and the fourth and fifth rows each contain 18 elements. Explain.

**20.** Describe the relationship between a main-group element's lettered group number (the number of the element's column) and its valence electrons.

**21.** Describe the relationship between an element's row number in the periodic table and the highest principal quantum number in the element's electron configuration. How does this relationship differ for main-group elements, transition elements, and inner transition elements?

**22.** Which of the transition elements in the first transition series have anomalous electron configurations?

**23.** Describe how to write the electron configuration for an element based on its position in the periodic table.

**24.** Describe the relationship between the properties of an element and the number of valence electrons that it contains.

**25.** List the number of valence electrons for each family, and explain the relationship between the number of valence electrons and the resulting chemistry of the elements in the family.
  **a.** alkali metals        **b.** alkaline earth metals
  **c.** halogens             **d.** oxygen family

**26.** Define atomic radius. For main-group elements, describe the observed trends in atomic radius as you move:
  **a.** across a period in the periodic table
  **b.** down a column in the periodic table

**27.** What is effective nuclear charge? What is shielding?

**28.** Use the concepts of effective nuclear charge, shielding, and $n$ value of the valence orbital to explain the trend in atomic radius as you move across a period in the periodic table.

**29.** For transition elements, describe and explain the observed trends in atomic radius as you move:
  **a.** across a period in the periodic table
  **b.** down a column in the periodic table

**30.** How is the electron configuration of an anion different from that of the corresponding neutral atom? How is the electron configuration of a cation different?

**31.** Describe how to write an electron configuration for a transition metal cation. Is the order of electron removal upon ionization simply the reverse of electron addition upon filling? Why or why not?

**32.** Describe the relationship between
  **a.** the radius of a cation and that of the atom from which forms
  **b.** the radius of an anion and that of the atom from which it forms

**33.** What is ionization energy? What is the difference between first ionization energy and second ionization energy?

**34.** What is the general trend in first ionization energy as you move down a column in the periodic table? As you move across a row?

**35.** What are the exceptions to the periodic trends in first ionization energy? Why do they occur?

**36.** Examination of the first few successive ionization energies for a given element usually reveals a large jump between two ionization energies. For example, the successive ionization energies of magnesium show a large jump between $IE_2$ and $IE_3$. The successive ionization energies of aluminum show a large jump between $IE_3$ and $IE_4$. Explain why these jumps occur and how you might predict them.

**37.** What is electron affinity? What are the observed periodic trends in electron affinity?

**38.** What is metallic character? What are the observed periodic trends in metallic character?

**39.** Write a general equation for the reaction of an alkali metal with each substance.
  **a.** a halogen        **b.** water

**40.** Write a general equation for the reaction of a halogen with each substance.
  **a.** a metal        **b.** hydrogen        **c.** another halogen

# Problems by Topic

## Electron Configurations

**41.** Write the full electron configuration for each element.
  **a.** Si    **b.** O    **c.** K    **d.** Ne

**42.** Write the full electron configuration for each element.
  **a.** C    **b.** P    **c.** Ar    **d.** Na

**43.** Write the full orbital diagram for each element.
  **a.** N    **b.** F    **c.** Mg    **d.** Al

**44.** Write the full orbital diagram for each element.
  **a.** S    **b.** Ca    **c.** Ne    **d.** He

**45.** Use the periodic table to write an electron configuration for each element. Represent core electrons with the symbol of the previous noble gas in brackets.
  **a.** P    **b.** Ge    **c.** Zr    **d.** I

**46.** Use the periodic table to determine the element corresponding to each electron configuration.
  **a.** $[Ar]\,4s^2 3d^{10} 4p^6$        **b.** $[Ar]\,4s^2 3d^2$
  **c.** $[Kr]\,5s^2 4d^{10} 5p^2$        **d.** $[Kr]\,5s^2$

**47.** Use the periodic table to determine each quantity.
  **a.** the number of $2s$ electrons in Li
  **b.** the number of $3d$ electrons in Cu
  **c.** the number of $4p$ electrons in Br
  **d.** the number of $4d$ electrons in Zr

**48.** Use the periodic table to determine each quantity.
  **a.** the number of $3s$ electrons in Mg
  **b.** the number of $3d$ electrons in Cr
  **c.** the number of $4d$ electrons in Y
  **d.** the number of $6p$ electrons in Pb

**49.** Name an element in the fourth period (row) of the periodic table with the following:
  **a.** five valence electrons
  **b.** four $4p$ electrons
  **c.** three $3d$ electrons
  **d.** full $s$ and $p$ sublevels

**50.** Name an element in the third period (row) of the periodic table with the following:
  **a.** three valence electrons
  **b.** four $3p$ electrons
  **c.** six $3p$ electrons
  **d.** two $3s$ electrons and no $3p$ electrons

## Valence Electrons and Simple Chemical Behavior from the Periodic Table

**51.** Determine the number of valence electrons in an atom of each element.
  **a.** Ba    **b.** Cs    **c.** Ni    **d.** S

**52.** Determine the number of valence electrons in an atom of each element. Which elements do you expect to lose electrons in their chemical reactions? Which do you expect to gain electrons?
**a.** Al    **b.** Sn    **c.** Br    **d.** Se

**53.** Which outer electron configuration would you expect to belong to a reactive metal? To a reactive nonmetal?
**a.** $ns^2$    **b.** $ns^2np^6$    **c.** $ns^2np^5$    **d.** $ns^2np^2$

**54.** Which outer electron configurations would you expect to belong to a noble gas? To a metalloid?
**a.** $ns^2$    **b.** $ns^2np^6$    **c.** $ns^2np^5$    **d.** $ns^2np^2$

## Coulomb's Law and Effective Nuclear Charge

**55.** According to Coulomb's law, which pair of charged particles has the lowest potential energy?
**a.** a particle with a 1− charge separated by 150 pm from a particle with a 2+ charge
**b.** a particle with a 1− charge separated by 150 pm from a particle with a 1+ charge
**c.** a particle with a 1− charge separated by 100 pm from a particle with a 3+ charge

**56.** According to Coulomb's law, rank the interactions between charged particles from lowest potential energy to highest potential energy.
**a.** a 1+ charge and a 1− charge separated by 100 pm
**b.** a 2+ charge and a 1− charge separated by 100 pm
**c.** a 1+ charge and a 1+ charge separated by 100 pm
**d.** a 1+ charge and a 1− charge separated by 200 pm

**57.** Which experience a greater effective nuclear charge: the valence electrons in beryllium or the valence electrons in nitrogen? Why?

**58.** Arrange the atoms according to decreasing effective nuclear charge experienced by their valence electrons: S, Mg, Al, Si.

**59.** If core electrons completely shielded valence electrons from nuclear charge (i.e., if each core electron reduced nuclear charge by 1 unit) and if valence electrons did not shield one another from nuclear charge at all, what would be the effective nuclear charge experienced by the valence electrons of each atom?
**a.** K    **b.** Ca    **c.** O    **d.** C

**60.** In Section 8.6, we estimated the effective nuclear charge on beryllium's valence electrons to be slightly greater than 2+. What would a similar process predict for the effective nuclear charge on boron's valence electrons? Would you expect the effective nuclear charge to be different for boron's 2s electrons compared to its 2p electron? In what way? (Hint: Consider the shape of the 2p orbital compared to that of the 2s orbital.)

## Atomic Radius

**61.** Choose the larger atom from each pair.
**a.** Al or In    **b.** Si or N    **c.** P or Pb    **d.** C or F

**62.** Choose the larger atom from each pair, if possible.
**a.** Sn or Si    **b.** Br or Ga    **c.** Sn or Bi    **d.** Se or Sn

**63.** Arrange these elements in order of increasing atomic radius: Ca, Rb, S, Si, Ge, F.

**64.** Arrange these elements in order of decreasing atomic radius: Cs, Sb, S, Pb, Se.

## Ionic Electron Configurations, Ionic Radii, Magnetic Properties, and Ionization Energy

**65.** Write the electron configuration for each ion.
**a.** $O^{2-}$    **b.** $Br^-$    **c.** $Sr^{2+}$    **d.** $Co^{3+}$    **e.** $Cu^{2+}$

**66.** Write the electron configuration for each ion.
**a.** $Cl^-$    **b.** $P^{3-}$    **c.** $K^+$    **d.** $Mo^{3+}$    **e.** $V^{3+}$

**67.** Write orbital diagrams for each ion and indicate whether the ion is diamagnetic or paramagnetic.
**a.** $V^{5+}$    **b.** $Cr^{3+}$    **c.** $Ni^{2+}$    **d.** $Fe^{3+}$

**68.** Write orbital diagrams for each ion and indicate whether the ion is diamagnetic or paramagnetic.
**a.** $Cd^{2+}$    **b.** $Au^+$    **c.** $Mo^{3+}$    **d.** $Zr^{2+}$

**69.** Which is the larger species in each pair?
**a.** Li or $Li^+$    **b.** $I^-$ or $Cs^+$    **c.** Cr or $Cr^{3+}$    **d.** O or $O^{2-}$

**70.** Which is the larger species in each pair?
**a.** Sr or $Sr^{2+}$    **b.** N or $N^{3-}$    **c.** Ni or $Ni^{2+}$    **d.** $S^{2-}$ or $Ca^{2+}$

**71.** Arrange this isoelectronic series in order of decreasing radius: $F^-, O^{2-}, Mg^{2+}, Na^+$.

**72.** Arrange this isoelectronic series in order of increasing atomic radius: $Se^{2-}, Sr^{2+}, Rb^+, Br^-$.

**73.** Choose the element with the higher first ionization energy from each pair.
**a.** Br or Bi    **b.** Na or Rb    **c.** As or At    **d.** P or Sn

**74.** Choose the element with the higher first ionization energy from each pair.
**a.** P or I    **b.** Si or Cl    **c.** P or Sb    **d.** Ga or Ge

**75.** Arrange these elements in order of increasing first ionization energy: Si, F, In, N.

**76.** Arrange these elements in order of decreasing first ionization energy: Cl, S, Sn, Pb.

**77.** For each element, predict where the "jump" occurs for successive ionization energies. (For example, does the jump occur between the first and second ionization energies, the second and third, or the third and fourth?)
**a.** Be    **b.** N    **c.** O    **d.** Li

**78.** Consider this set of ionization energies.

$$IE_1 = 578 \text{ kJ/mol}$$
$$IE_2 = 1820 \text{ kJ/mol}$$
$$IE_3 = 2750 \text{ kJ/mol}$$
$$IE_4 = 11,600 \text{ kJ/mol}$$

To which third-period element do these ionization values belong?

## Electron Affinities and Metallic Character

**79.** Choose the element with the more negative (more exothermic) electron affinity from each pair.
**a.** Na or Rb    **b.** B or S    **c.** C or N    **d.** Li or F

**80.** Choose the element with the more negative (more exothermic) electron affinity from each pair.
**a.** Mg or S    **b.** K or Cs    **c.** Si or P    **d.** Ga or Br

**81.** Choose the more metallic element from each pair.
**a.** Sr or Sb    **b.** As or Bi    **c.** Cl or O    **d.** S or As

**82.** Choose the more metallic element from each pair.
**a.** Sb or Pb    **b.** K or Ge    **c.** Ge or Sb    **d.** As or Sn

**83.** Arrange these elements in order of increasing metallic character: Fr, Sb, In, S, Ba, Se.

**84.** Arrange these elements in order of decreasing metallic character: Sr, N, Si, P, Ga, Al.

## Chemical Behavior of the Alkali Metals and the Halogens

**85.** Write the balanced chemical equation for the reaction of solid strontium with iodine gas.

**86.** Based on the ionization energies of the alkali metals, which alkali metal would you expect to undergo the most exothermic reaction with chlorine gas? Write a balanced chemical equation for the reaction.

87. Write the balanced chemical equation for the reaction of solid lithium with liquid water.

88. Write the balanced chemical equation for the reaction of solid potassium with liquid water.

89. Write the balanced equation for the reaction of hydrogen gas with bromine gas.

90. Write the balanced equation for the reaction of chlorine gas with fluorine gas.

# Cumulative Problems

91. Bromine is a highly reactive liquid while krypton is an inert gas. Explain this difference based on their electron configurations.

92. Potassium is a highly reactive metal while argon is an inert gas. Explain this difference based on their electron configurations.

93. Both vanadium and its 3+ ion are paramagnetic. Refer to their electron configurations to explain this statement.

94. Refer to their electron configurations to explain why copper is paramagnetic while its 1+ ion is not.

95. Suppose you were trying to find a substitute for $K^+$ in nerve signal transmission. Where would you begin your search? What ions would be most like $K^+$? For each ion you propose, explain the ways in which it would be similar to $K^+$ and the ways it would be different. Refer to periodic trends in your discussion.

96. Suppose you were trying to find a substitute for $Na^+$ in nerve signal transmission. Where would you begin your search? What ions would be most like $Na^+$? For each ion you propose, explain the ways in which it would be similar to $Na^+$ and the ways it would be different. Use periodic trends in your discussion.

97. Life on Earth evolved based on the element carbon. Based on periodic properties, what two or three elements would you expect to be most like carbon?

98. Which pair of elements would you expect to have the most similar atomic radii, and why?
    a. Si and Ga
    b. Si and Ge
    c. Si and As

99. Consider these elements: N, Mg, O, F, Al.
    a. Write the electron configuration for each element.
    b. Arrange the elements in order of decreasing atomic radius.
    c. Arrange the elements in order of increasing ionization energy.
    d. Use the electron configurations in part a to explain the differences between your answers to parts b and c.

100. Consider these elements: P, Ca, Si, S, Ga.
    a. Write the electron configuration for each element.
    b. Arrange the elements in order of decreasing atomic radius.
    c. Arrange the elements in order of increasing ionization energy.
    d. Use the electron configurations in part a to explain the differences between your answers to parts b and c.

101. Explain why atomic radius decreases as you move to the right across a period for main-group elements but not for transition elements.

102. Explain why vanadium (radius = 134 pm) and copper (radius = 128 pm) have nearly identical atomic radii, even though the atomic number of copper is about 25% higher than that of vanadium. What would you predict about the relative densities of these two metals? Look up the densities in a reference book, periodic table, or on the Internet. Are your predictions correct?

103. The lightest noble gases, such as helium and neon, are completely inert—they do not form any chemical compounds whatsoever. The heavier noble gases, in contrast, do form a limited number of compounds. Explain this difference in terms of trends in fundamental periodic properties.

104. The lightest halogen is also the most chemically reactive, and reactivity generally decreases as you move down the column of halogens in the periodic table. Explain this trend in terms of periodic properties.

105. Write general outer electron configurations ($ns^x np^y$) for groups 6A and 7A in the periodic table. The electron affinity of each group 7A element is more negative than that of each corresponding group 6A element. Use the electron configurations to explain why this is so.

106. The electron affinity of each group 5A element is more positive than that of each corresponding group 4A element. Use the outer electron configurations for these columns to suggest a reason for this observation.

107. The elements with atomic numbers 35 and 53 have similar chemical properties. Based on their electronic configurations, predict the atomic number of a heavier element that also should share these chemical properties.

108. Write the electron configurations of the six cations that form from sulfur by the loss of one to six electrons. For those cations that have unpaired electrons, write orbital diagrams.

109. You have cracked a secret code that uses elemental symbols to spell words. The code uses numbers to designate the elemental symbols. Each number is the sum of the atomic number and the highest principal quantum number of the highest occupied orbital of the element whose symbol is to be used. The message may be written forward or backward. Decode the following messages:
    a. 10, 12, 58, 11, 7, 44, 63, 66
    b. 9, 99, 30, 95, 19, 47, 79

110. The electron affinity of sodium is lower than that of lithium, while the electron affinity of chlorine is higher than that of fluorine. Suggest an explanation for this observation.

111. Use Coulomb's law to calculate the ionization energy in kJ/mol of an atom composed of a proton and an electron separated by 100.00 pm. What wavelength of light has sufficient energy to ionize the atom?

112. The first ionization energy of sodium is 496 kJ/mol. Use Coulomb's law to estimate the average distance between the sodium nucleus and the 3s electron. How does this distance compare to the atomic radius of sodium? Explain the difference.

113. Consider the elements: B, C, N, O, F.
    a. Which element has the highest first ionization energy?
    b. Which element has the largest atomic radius?
    c. Which element is most metallic?
    d. Which element has three unpaired electrons?

114. Consider the elements: Na, Mg, Al, Si, P.
    a. Which element has the highest second ionization energy?
    b. Which element has the smallest atomic radius?
    c. Which element is least metallic?
    d. Which element is diamagnetic?

# Challenge Problems

**115.** Consider the densities and atomic radii of the noble gases at 25 °C:

| Element | Atomic Radius (pm) | Density (g/L) |
|---------|--------------------|-----------------| 
| He | 32 | 0.18 |
| Ne | 70 | 0.90 |
| Ar | 98 | – |
| Kr | 112 | 3.75 |
| Xe | 130 | – |
| Rn | – | 9.73 |

    **a.** Estimate the densities of argon and xenon by interpolation from the data.

    **b.** Estimate the density of the element with atomic number 118 by extrapolation from the data.

    **c.** Use the molar mass of neon to estimate the mass of a neon atom. Then use the atomic radius of neon to calculate the average density of a neon atom. How does this density compare to the density of neon gas? What does this comparison suggest about the nature of neon gas?

    **d.** Use the densities and molar masses of krypton and neon to calculate the number of atoms of each element found in a volume of 1.0 L. Use these values to estimate the number of atoms present in 1.0 L of Ar. Now use the molar mass of argon to estimate the density of Ar. How does this estimate compare to that in part a?

**116.** As you have seen, the periodic table is a result of empirical observation (i.e., the periodic law), but quantum-mechanical theory explains *why* the table is so arranged. Suppose that, in another universe, quantum theory was such that there were one *s* orbital but only two *p* orbitals (instead of three) and only three *d* orbitals (instead of five). Draw out the first four periods of the periodic table in this alternative universe. Which elements would be the equivalent of the noble gases? Halogens? Alkali metals?

**117.** Consider the metals in the first transition series. Use periodic trends to predict a trend in density as you move to the right across the series.

**118.** Imagine a universe in which the value of $m_s$ can be $+\frac{1}{2}$, 0, and $-\frac{1}{2}$. Assuming that all the other quantum numbers can take only the values possible in our world and that the Pauli exclusion principle applies, determine:

    **a.** the new electronic configuration of neon

    **b.** the atomic number of the element with a completed $n = 2$ shell

    **c.** the number of unpaired electrons in fluorine

**119.** A carbon atom can absorb radiation of various wavelengths with resulting changes in its electron configuration. Write orbital diagrams for the electron configuration of carbon that results from absorption of the three longest wavelengths of radiation it can absorb.

**120.** Only trace amounts of the synthetic element darmstadtium, atomic number 110, have been obtained. The element is so highly unstable that no observations of its properties have been possible. Based on its position in the periodic table, propose three different reasonable valence electron configurations for this element.

**121.** What is the atomic number of the as yet undiscovered element in which the 8*s* and 8*p* electron energy levels fill? Predict the chemical behavior of this element.

**122.** The trend in second ionization energy for the elements from lithium to fluorine is not a regular one. Predict which of these elements has the highest second ionization energy and which has the lowest and explain. Of the elements N, O, and F, O has the highest and N the lowest second ionization energy. Explain.

**123.** Unlike the elements in groups 1A and 2A, those in group 3A do not show a regular decrease in first ionization energy as you move down the column. Explain the irregularities.

**124.** Using the data in Figures 8.15 and 8.16, calculate $\Delta E$ for the reaction $Na(g) + Cl(g) \longrightarrow Na^+(g) + Cl^-(g)$.

**125.** Even though adding two electrons to O or S forms an ion with a noble gas electron configuration, the second electron affinity of both of these elements is positive. Explain.

**126.** In Section 2.7 we discussed the metalloids, which form a diagonal band separating the metals from the nonmetals. There are other instances in which elements such as lithium and magnesium that are diagonal to each other have comparable metallic character. Suggest an explanation for this observation.

**127.** The heaviest known alkaline earth metal is radium, atomic number 88. Find the atomic numbers of the as yet undiscovered next two members of the series.

**128.** Predict the electronic configurations of the first two excited states (next higher-energy states beyond the ground state) of Pd.

**129.** Table 8.2 does not include francium because none of francium's isotopes are stable. Predict the values of the entries for Fr in Table 8.2. Predict the nature of the products of the reaction of Fr with: (a) water, (b) oxygen, and (c) chlorine.

**130.** From its electronic configuration, predict which of the first 10 elements would be most similar in chemical behavior to the as yet undiscovered element 165.

# Conceptual Problems

**131.** Imagine that in another universe atoms and elements are identical to ours, except that atoms with six valence electrons have particular stability (in contrast to our universe where atoms with eight valence electrons have particular stability). Give an example of an element in the alternative universe that corresponds to each of the following:

    **a.** a noble gas     **b.** a reactive nonmetal

    **c.** a reactive metal

**132.** The outermost valence electron in atom A experiences an effective nuclear charge of 2+ and is on average 225 pm from the nucleus. The outermost valence electron in atom B experiences an effective nuclear charge of 1+ and is on average 175 pm from the nucleus. Which atom (A or B) has the higher first ionization energy? Explain.

133. Determine whether each statement regarding penetration and shielding is true or false. (Assume that all lower-energy orbitals are fully occupied.)
    a. An electron in a 3s orbital is more shielded than an electron in a 2s orbital.
    b. An electron in a 3s orbital penetrates into the region occupied by core electrons more than electrons in a 3p orbital penetrates into the region occupied by core electrons.
    c. An electron in an orbital that penetrates closer to the nucleus always experiences more shielding than an electron in an orbital that does not penetrate as far.
    d. An electron in an orbital that penetrates close to the nucleus tends to experience a higher effective nuclear charge than an electron in an orbital that does not penetrate close to the nucleus.

134. Give a combination of four quantum numbers that could be assigned to an electron occupying a 5p orbital. Do the same for an electron occupying a 6d orbital.

135. Use the trends in ionization energy and electron affinity to explain why calcium fluoride has the formula $CaF_2$ and not $Ca_2F$ or $CaF$.

*Active Classroom Learning*

# Questions for Group Work

*Discuss these questions with the group and record your consensus answer.*

136. In a complete sentence describe the relationship between shielding and penetration.

137. Play a game to memorize the order in which orbitals fill. Have each group member in turn state the name of the next orbital to fill and the maximum number of electrons it can hold (for example, "1s two," "2s two," "2p six," etc.). If a member gets stuck, other group members can help, consulting Figure 8.5 and the accompanying text summary if necessary. However, when a member gets stuck, the next player starts back at "1s two." Keep going until each group member can list all the orbitals in order up to "6s two."

138. Sketch a periodic table (without element symbols). Include the correct number of rows and columns in the s, p, d, and f blocks. Shade in the squares for elements that have irregular electron configurations.

139. In complete sentences, explain: (a) why $Se^{2-}$ and $Br^-$ are about the same size; (b) why $Br^-$ is slightly smaller than $Se^{2-}$; and (c) which singly charged cation you would expect to be approximately the same size as $Se^{2-}$ and $Br^-$ and why.

140. Have each member of your group sketch a periodic table indicating a periodic trend (atomic size, first ionization energy, metallic character, etc.). Have each member present his or her table to the rest of the group and explain the trend based on concepts such as orbital size or effective nuclear charge.

# Data Interpretation and Analysis

### Periodic Properties of Period 3 Elements

141. Figure a▼ plots the first ionization energies of the elements in period 3. Figure b▶ lists some of the binary compounds that form between fluorine and the elements in period 3. Figure c▶ plots the electron affinities of the elements in period 3.

| Element | Na | Mg | Al | Si | P | S | Cl | Ar |
|---|---|---|---|---|---|---|---|---|
| Compound | NaF | $MgF_2$ | $AlF_3$ | | $PF_3$ | $SF_2$ | ClF | |

▲ **FIGURE b** **Binary Compounds That Form between Fluorine and Period 3 Elements**

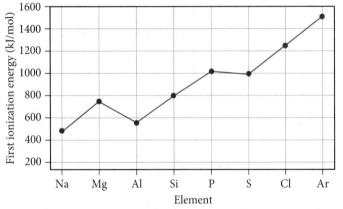

▲ FIGURE a **First Ionization Energies of Period 3 Elements**

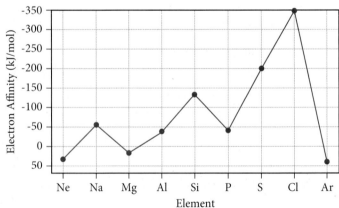

▲ FIGURE c **Electron Affinities of Period 3 Elements**

Use the information provided in the figures to answer the following questions:

a. In general, first ionization energy increases as we move from left to right across the third period. Explain.

b. There is a decrease in first ionization from Mg to Al. There is an increase in electron affinity from Mg to Al. Explain.

c. The first ionization energy of S is slightly less than that of P. Explain.

d. After reviewing the information in Figures a, b, and c, Student X offers the following hypothesis: The number of fluorine atoms that bond to a metal atom is equal to the number of valence electrons of the metal. The number of fluorine atoms that bond to a nonmetal atom is equal to 8 minus the number of valence electrons of the nonmetal. Student Y offers the following hypothesis: In general, there is an inverse relationship between the electron affinities of the period 3 elements and the number of fluorine atoms that bond to the metal or nonmetal atoms.

Do you agree with Student X's hypothesis, with Student Y's hypothesis, both of them, or neither of them? Explain.

e. Write the formula for a compound that forms between silicon and flourine. Does this compound support Student X's hypothesis, Student Y's hypothesis, or neither hypothesis? Explain.

# Answers to Conceptual Connections

### Coulomb's Law

**8.1** **(a)** Since the charges are opposite, the potential energy of the interaction is negative. As the charges get closer together, $r$ becomes smaller and the potential energy decreases (becomes more negative).

### Penetration and Shielding

**8.2** **(c)** Penetration results in less shielding from nuclear charge and therefore lower energy.

### Electron Configurations and Quantum Numbers

**8.3** $n = 4, l = 0, m_l = 0, m_s = +\frac{1}{2}; n = 4, l = 0, m_l = 0, m_s = -\frac{1}{2}$

### Valence Electrons and Group Number

**8.4** Nitrogen has five valence electrons. Since nitrogen is a main-group element, it has the same number of valence electrons as its group number in the periodic table.

### Effective Nuclear Charge

**8.5** **(c)** Since $Z_{\text{eff}}$ increases from left to right across a row in the periodic table, the valence electrons in S experience a greater effective nuclear charge than the valence electrons in Al or in Mg.

### Ions, Isotopes, and Atomic Size

**8.6** The isotopes of an element all have the same radius for two reasons: (1) neutrons are negligibly small compared to the size of an atom and therefore extra neutrons do not increase atomic size, and (2) neutrons have no charge and therefore do not attract electrons in the way that protons do.

### Ionization Energies and Chemical Bonding

**8.7** As you can see from the successive ionization energies of any element, valence electrons are held most loosely and can therefore be transferred or shared most easily. Core electrons, on the other hand, are held tightly and are not easily transferred or shared. Consequently, valence electrons are most important to chemical bonding.

### Periodic Trends

**8.8** The 3s electron in sodium has a relatively low first ionization energy (496 kJ/mol) because it is a valence electron. The energetic cost for sodium to lose a second electron is extraordinarily high (4560 kJ/mol) because the next electron to be lost is a core electron (2p). Similarly, the electron affinity of chlorine to gain one electron ($-349$ kJ/mol) is highly exothermic because the added electron completes chlorine's valence shell. The gain of a second electron by the negatively charged chlorine anion is not so favorable. Therefore, we expect sodium and chlorine to combine in a 1:1 ratio.

CHAPTER

# 9

# Chemical Bonding I: The Lewis Model

Theories are nets cast to catch what we call "the world": to rationalize, to explain, and to master it. We endeavor to make the mesh ever finer and finer.

—KARL POPPER (1902–1994)

Chemical bonding is at the heart of chemistry. The bonding theories that we are about to examine are—as Karl Popper, a philosopher of science, eloquently states in the chapter-opening quote—nets cast to understand the world. In the next two chapters, we will examine three theories with successively finer "meshes." The first is the Lewis model, a simple model of chemical bonding, which can be sketched out on the back of an envelope. With just a few dots, dashes, and chemical symbols, the Lewis model can help us to understand and predict many chemical observations. The second is valence bond theory, which treats electrons in a more quantum-mechanical manner but stops short of viewing them as belonging to the entire molecule. The third is molecular orbital theory, essentially a full quantum-mechanical treatment of the molecule and its electrons as a whole. Molecular orbital theory has great predictive power, but at the expense of great complexity and intensive computational requirements. Which theory is "correct"? Remember that theories are models that help us understand and predict behavior. All three of these theories are extremely useful, depending on exactly what aspect of chemical bonding we want to predict or understand.

▲ *The AIDS drug Indinavir—shown here as the missing piece in a puzzle depicting the protein HIV-protease—was developed with the help of chemical bonding theories.*

## 9.1 Bonding Models and AIDS Drugs

In 1989, researchers using X-ray crystallography—a technique in which X-rays are scattered from crystals of the molecule of interest—determined the structure of a molecule called HIV-protease. HIV-protease is a protein (a class of large biological molecules) synthesized by

We discuss X-ray crystallography in more detail in Section 12.2.

the human immunodeficiency virus (HIV). This particular protein is crucial to the virus's ability to multiply and cause acquired immune deficiency syndrome, or AIDS. Without HIV-protease, HIV cannot spread in the human body because the virus cannot replicate. In other words, without HIV-protease, AIDS can't develop.

With knowledge of the HIV-protease structure, pharmaceutical companies set out to create a molecule that would disable HIV-protease by attaching to the working part of the molecule, called the active site. To design such a molecule, researchers used *bonding theories*—models that predict how atoms bond together to form molecules—to simulate the shape of potential drug molecules and determine how they would interact with the protease molecule. By the early 1990s, pharmaceutical companies had developed several drug molecules that seemed to work. Since these molecules inhibit the action of HIV-protease, they were named *protease inhibitors*. Protease inhibitors, when given in combination with other drugs, decrease the viral count in HIV-infected individuals to undetectable levels. Although protease inhibitors do not cure AIDS, many AIDS patients are still alive today because of these drugs.

Bonding theories are central to chemistry because they explain how atoms bond together to form molecules. They explain why some combinations of atoms are stable and others are not. For example, bonding theories explain why table salt is NaCl and not $NaCl_2$ and why water is $H_2O$ and not $H_3O$. Bonding theories also predict the shapes of molecules—a topic in the next chapter—which in turn determine many of the physical and chemical properties of compounds.

The bonding model we examine in this chapter is the **Lewis model**, named after the American chemist G. N. Lewis (1875–1946). In the Lewis model, valence electrons are represented as dots, and we draw **Lewis electron-dot structures** (or simply **Lewis structures**) to depict molecules. These structures, which are fairly simple to draw, have tremendous predictive power. With minimal computation, the Lewis model can be used to predict whether a particular set of atoms will form a stable molecule and what that molecule might look like. Although we will also examine more advanced theories in the following chapter, the Lewis model remains the simplest model for making quick, everyday predictions about most molecules.

We discuss proteins in more detail in Chapter 22.

▲ G. N. Lewis

## 9.2 Types of Chemical Bonds

We begin our discussion of chemical bonding by asking why bonds form in the first place. This seemingly simple question is vitally important. Imagine our universe without chemical bonding. Such a universe would contain only 91 different kinds of substances (the 91 naturally occurring elements). With such a poor diversity of substances, life would be impossible, and we would not be around to wonder why. The *answer* to the question of why bonds form, however, is not simple and involves not only quantum mechanics but also some thermodynamics that we do not introduce until Chapter 18. Nonetheless, we can address an important *aspect* of the answer now: *Chemical bonds form because they lower the potential energy between the charged particles that compose atoms.*

As you already know, atoms are composed of particles with positive charges (the protons in the nucleus) and negative charges (the electrons). When two atoms approach each other, the electrons of one atom are attracted to the nucleus of the other according to Coulomb's law (see Section 8.3) and vice versa. However, at the same time, the electrons of each atom repel the electrons of the other, and the nucleus of each atom repels the nucleus of the other. The result is a complex set of interactions among a potentially large number of charged particles. If these interactions lead to an overall net reduction of energy between the charged particles, a chemical bond forms. Bonding theories help us to predict the circumstances under which bonds form and also the properties of the resultant molecules.

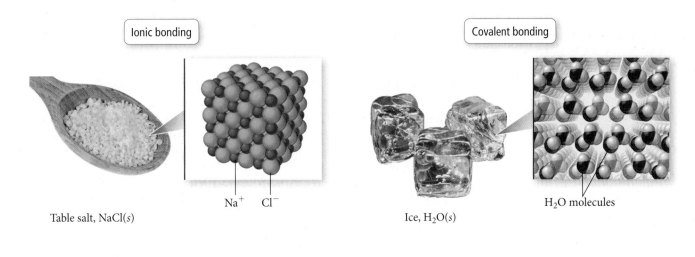

Ionic bonding

Covalent bonding

Table salt, NaCl(s)

Na$^+$    Cl$^-$

Ice, H$_2$O(s)

H$_2$O molecules

Metallic bonding

Sodium metal, Na(s)

e$^-$ sea

Na$^+$

▲ **FIGURE 9.1  Ionic, Covalent, and Metallic Bonding**

We broadly classify chemical bonds into three types depending on the kind of atoms involved in the bonding (Figure 9.1▲):

| Types of Atoms | Type of Bond | Characteristic of Bond |
|---|---|---|
| Metal and nonmetal | Ionic | Electrons transferred |
| Nonmetal and nonmetal | Covalent | Electrons shared |
| Metal and metal | Metallic | Electrons pooled |

Recall from Chapter 8 that metals tend to have low ionization energies (their electrons are relatively easy to remove) and that nonmetals tend to have negative electron affinities (they readily gain electrons). When a metal atom bonds with a nonmetal atom, it transfers one or more electrons to the nonmetal. The metal atom becomes a cation and the nonmetal atom an anion. These oppositely charged ions attract one another, lowering their overall potential energy as described by Coulomb's law. The resulting bond is an **ionic bond**.

We also discussed in Chapter 8 that nonmetals tend to have high ionization energies (their electrons are relatively difficult to remove). Therefore, when a nonmetal bonds with another nonmetal, neither atom transfers electrons to the other. Instead, the two atoms *share* some electrons. The shared electrons interact with the nuclei of both of the bonding atoms, lowering their potential energy in accordance with Coulomb's law. The resulting bond is a **covalent bond**.

Recall from Section 3.2 that we can understand the stability of a covalent bond by considering the most stable arrangement (the one with the lowest potential energy) of two positively charged particles separated by a small distance and a negatively charged particle. As we can see in Figure 9.2▶ on the next page, the arrangement in which the negatively charged particle lies *between* the two positively charged ones has the lowest potential energy because in this arrangement, the negatively charged particle interacts

► **FIGURE 9.2 Possible Configurations of One Negatively Charged Particle and Two Positively Charged Ones**

Lowest potential energy (most stable)

most strongly with *both of the positively charged ones*. In a sense, the negatively charged particle holds the two positively charged ones together. Similarly, shared electrons in a covalent chemical bond *hold* the bonding atoms together by attracting the positive charges of their nuclei.

A third type of bonding, **metallic bonding**, occurs in metals. Since metals have low ionization energies, they tend to lose electrons easily. In the simplest model for metallic bonding—called the *electron sea* model—all of the atoms in a metal lattice pool their valence electrons. These pooled electrons are no longer localized on a single atom, but delocalized over the entire metal. The positively charged metal atoms are then attracted to the sea of electrons, holding the metal together. We discuss metallic bonding in more detail in Section 9.11.

CONCEPTUAL
**CONNECTION 9.1**

**Bond Types** Which compound is most likely to contain ionic bonds?
**(a)** $CH_4$     **(b)** $N_2O$     **(c)** $MgF_2$

## 9.3 Representing Valence Electrons with Dots

In Chapter 8, we saw that, for main-group elements, valence electrons are those electrons in the outermost principal energy level. Because valence electrons are held most loosely, and chemical bonding involves the transfer or sharing of electrons between two or more atoms, valence electrons are most important in bonding, so the Lewis model focuses on these. In a **Lewis symbol**, we represent the valence electrons of main-group elements as dots surrounding the element symbol. For example, the electron configuration of O is:

> Remember, the number of valence electrons for any main group is equal to the group number of the element (except for helium, which is in group 8A but has only two valence electrons).

$$1s^2 2s^2 2p^4$$
6 valence electrons

The Lewis symbol for oxygen is:

·Ö: — 6 dots representing valence electrons

Each dot in a Lewis symbol represents a valence electron. The dots are placed around the element's symbol with a maximum of two dots per side. We draw the Lewis symbols for all of the period 2 elements similarly.

> Although the exact location of dots is not critical, in this book we first place dots singly before pairing (except for helium, which always has two paired dots signifying its duet).

Li·  ·Be·  ·B·  ·Ċ·  ·N̈·  ·Ö:  :F̈:  :N̈e:

Lewis symbols provide a simple way to visualize the number of valence electrons in a main-group atom. Notice that atoms with eight valence electrons—which are particularly stable because they have full s and p sublevels—are easily identified because they have eight dots, an **octet**.

Helium is somewhat of an exception. Its electron configuration and Lewis symbol are:

$$1s^2 \quad He:$$

The Lewis symbol of helium contains only two dots (a **duet**). For helium, a duet represents a stable electron configuration because the $n = 1$ quantum level fills with only two electrons.

In the Lewis model, a **chemical bond** is the sharing or transfer of electrons to attain stable electron configurations for the bonding atoms. If electrons are transferred, as occurs between a metal and a nonmetal, the bond is an *ionic bond*. If the electrons are shared, as occurs between two nonmetals, the bond is a *covalent bond*. In either case, the bonding atoms attain stable electron configurations; since the stable configuration is usually eight electrons in the outermost shell, this is known as the **octet rule**.

When applying the Lewis model, we do not try to calculate the energies associated with the attractions and repulsions between electrons and nuclei on neighboring atoms. The energy changes that occur because of these interactions are central to chemical bonding (as we saw in Section 9.2), yet the Lewis model ignores them because calculating these energy changes is extremely complicated. Instead the Lewis model uses the simple octet rule, a practical approach that accurately predicts what we see in nature for a large number of compounds—hence the success and longevity of the Lewis model.

## 9.4 Ionic Bonding: Lewis Symbols and Lattice Energies

Although the Lewis model's strength is in modeling covalent bonding, it can also be applied to ionic bonding. To represent ionic bonding, we move electron dots from the Lewis symbol of the metal to the Lewis symbol of the nonmetal and allow the resultant ions to form a crystalline lattice composed of alternating cations and anions.

### Ionic Bonding and Electron Transfer

Consider potassium and chlorine, which have the following Lewis symbols:

$$\text{K} \cdot \quad : \overset{\cdot\cdot}{\underset{\cdot\cdot}{\text{Cl}}} :$$

When these atoms bond, potassium transfers its valence electron to chlorine:

$$\text{K} \cdot \; + \; : \overset{\cdot\cdot}{\underset{\cdot\cdot}{\text{Cl}}} : \; \longrightarrow \; \text{K}^+ \left[ : \overset{\cdot\cdot}{\underset{\cdot\cdot}{\text{Cl}}} : \right]^-$$

The transfer of the electron gives chlorine an octet (shown as eight dots around chlorine) and leaves potassium without any valence electrons but with an octet in the *previous* principal energy level (which is now the outermost level):

$$\text{K} \qquad 1s^2 2s^2 2p^6 3s^2 3p^6 4s^1$$
$$\text{K}^+ \qquad 1s^2 2s^2 2p^6 3s^2 \underbrace{3p^6}\; 4s^0$$

Octet in previous level

The potassium, having lost an electron, becomes positively charged (a cation), while the chlorine, which has gained an electron, becomes negatively charged (an anion). We typically write the Lewis symbol of an anion within brackets with the charge in the upper right-hand corner, outside the brackets. The positive and negative charges attract one another, resulting in the compound KCl.

This example shows how we can use the Lewis model to predict the correct chemical formulas for ionic compounds. For the compound that forms between K and Cl, for example, the Lewis model predicts a ratio of one potassium cation to every one chloride anion, KCl. In nature, when we examine the compound formed between potassium and chlorine, we indeed find one potassium ion to every chloride ion.

As another example, consider the ionic compound formed between sodium and sulfur. The Lewis symbols for sodium and sulfur are:

$$\text{Na} \cdot \quad \cdot \overset{\cdot\cdot}{\underset{\cdot\cdot}{\text{S}}} :$$

Sodium must lose its one valence electron in order to have an octet (in the previous principal shell), while sulfur must gain two electrons to attain an octet. Consequently, the compound that forms between sodium and sulfur requires a ratio of two sodium atoms to every one sulfur atom:

$$2\,\text{Na}^+ \left[ : \overset{\cdot\cdot}{\underset{\cdot\cdot}{\text{S}}} : \right]^{2-}$$

Recall that solid ionic compounds do not contain distinct molecules; they are composed of alternating positive and negative ions in a three-dimensional crystalline array.

The two sodium atoms each lose their one valence electron, while the sulfur atom gains two electrons and attains an octet. The Lewis model predicts that the correct chemical formula is $Na_2S$, which is exactly what we see in nature.

| **Example 9.1** | **Using Lewis Symbols to Predict the Chemical Formula of an Ionic Compound** |
|---|---|

Use Lewis symbols to predict the formula for the compound that forms between calcium and chlorine.

**SOLUTION**

| Draw Lewis symbols for calcium and chlorine based on their respective numbers of valence electrons, which you can determine from their group number in the periodic table. | ·Ca·    ·Cl: |
|---|---|
| Calcium must lose its two valence electrons (to end up with an octet in its previous principal shell), while chlorine only needs to gain one electron to attain an octet. Draw two chlorine anions, each with an octet and a 1– charge, and one calcium cation with a 2+ charge. Place brackets around the chlorine anions and indicate the charges on each ion. | $Ca^{2+} 2[:\ddot{Cl}:]^-$ |
| Finally, write the formula with subscripts to indicate the number of each ion. | $CaCl_2$ |

**FOR PRACTICE 9.1**   Use Lewis symbols to predict the formula for the compound that forms between magnesium and nitrogen.

## Lattice Energy: The Rest of the Story

The formation of an ionic compound from its constituent elements is usually quite exothermic. For example, when sodium chloride (table salt) forms from elemental sodium and chlorine, 411 kJ of heat evolves in the following violent reaction:

$$Na(s) + \tfrac{1}{2}Cl_2(g) \longrightarrow NaCl(s) \qquad \Delta H_f^\circ = -411 \text{ kJ/mol}$$

Where does this energy come from? It might seem that it comes solely from the tendency of metals to lose electrons and nonmetals to gain electrons—but it does not. In fact, the transfer of an electron from sodium to chlorine—by itself—actually *absorbs* energy. The first ionization energy of sodium is +496 kJ/mol, and the electron affinity of Cl is only −349 kJ/mol. Based only on these energies, the reaction should be *endothermic* by +147 kJ/mol. So why is the reaction so *exothermic*?

The answer lies in the **lattice energy**—the energy associated with the formation of a crystalline lattice of alternating cations and anions from the gaseous ions. Because the sodium ions are positively charged and the chlorine ions are negatively charged, the potential energy decreases—as prescribed by Coulomb's law—when these ions come together to form a lattice. That energy is emitted as heat when the lattice forms, as shown in Figure 9.3▶. The exact value of the lattice energy, however, is not simple to determine because it involves a large number of interactions among many charged particles in a lattice. The most straightforward way to calculate lattice energy is with the *Born–Haber cycle*.

## The Born–Haber Cycle

The **Born–Haber cycle** is a hypothetical series of steps that represents the formation of an ionic compound from its constituent elements. The steps are chosen so that the change in enthalpy of each step is known except for the last one, which is the lattice energy. The change in enthalpy for the overall process is also known. Using Hess's law (see Section 6.8), we can therefore determine the enthalpy change for the unknown last step, the lattice energy.

Consider the formation of NaCl from its constituent elements in their standard states. The enthalpy change for the overall reaction is simply the standard enthalpy of formation of NaCl(s):

$$Na(s) + \tfrac{1}{2}Cl_2(g) \longrightarrow NaCl(s) \qquad \Delta H_f^\circ = -411 \text{ kJ/mol}$$

Recall that Hess's law states that the change in the overall enthalpy of a stepwise process is the sum of the enthalpy changes of the steps.

## Lattice Energy of an Ionic Compound

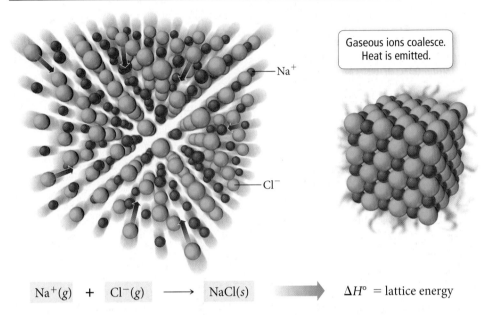

Gaseous ions coalesce.
Heat is emitted.

— Na$^+$

— Cl$^-$

$$\mathrm{Na^+(\textit{g})} \quad + \quad \mathrm{Cl^-(\textit{g})} \quad \longrightarrow \quad \mathrm{NaCl(\textit{s})} \qquad \Delta H° = \text{lattice energy}$$

▲ **FIGURE 9.3 Lattice Energy** The lattice energy of an ionic compound is the energy associated with the formation of a crystalline lattice of the compound from the gaseous ions.

Now consider the following set of steps—the Born–Haber cycle—from which NaCl(*s*) can also be made from Na(*s*) and Cl$_2$(*g*):

- The first step is the formation of gaseous sodium from solid sodium.

$$\mathrm{Na(\textit{s})} \longrightarrow \mathrm{Na(\textit{g})} \qquad \Delta H°_{\text{step 1}} \text{ (sublimation energy of Na)} = +108 \text{ kJ}$$

- The second step is the formation of a chlorine atom from a chlorine molecule.

$$\tfrac{1}{2}\mathrm{Cl_2\,(\textit{g})} \longrightarrow \mathrm{Cl(\textit{g})} \qquad \Delta H°_{\text{step 2}} \text{ (bond energy of Cl}_2 \times \tfrac{1}{2}) = +122 \text{ kJ}$$

- The third step is the ionization of gaseous sodium. The enthalpy change for this step is the ionization energy of sodium.

$$\mathrm{Na(\textit{g})} \longrightarrow \mathrm{Na^+(\textit{g})} + \mathrm{e^-} \qquad \Delta H°_{\text{step 3}} \text{ (ionization energy of Na)} = +496 \text{ kJ}$$

- The fourth step is the addition of an electron to gaseous chlorine. The enthalpy change for this step is the electron affinity of chlorine.

$$\mathrm{Cl(\textit{g})} + \mathrm{e^-} \longrightarrow \mathrm{Cl^-(\textit{g})} \qquad \Delta H°_{\text{step 4}} \text{ (electron affinity of Cl)} = -349 \text{ kJ}$$

- The fifth and final step is the formation of the crystalline solid from the gaseous ions. The enthalpy change for this step is the lattice energy, the unknown quantity.

$$\mathrm{Na^+(\textit{g})} + \mathrm{Cl^-(\textit{g})} \longrightarrow \mathrm{NaCl(\textit{s})} \qquad \Delta H°_{\text{step 5}} = \Delta H°_{\text{lattice}} = ?$$

Figure 9.4▶, on the next page, illustrates the entire Born–Haber cycle for NaCl.

Since the overall reaction obtained by summing the steps in the Born–Haber cycle is equivalent to the formation of NaCl from its constituent elements, we use Hess's law to set the overall enthalpy of formation for NaCl(*s*) equal to the sum of the steps in the Born–Haber cycle:

Lattice energy

$$\Delta H°_{\mathrm{f}} = \Delta H°_{\text{step 1}} + \Delta H°_{\text{step 2}} + \Delta H°_{\text{step 3}} + \Delta H°_{\text{step 4}} + \Delta H°_{\text{step 5}}$$

## Born–Haber Cycle for Production of NaCl from Na(*s*) and Cl₂(*g*)

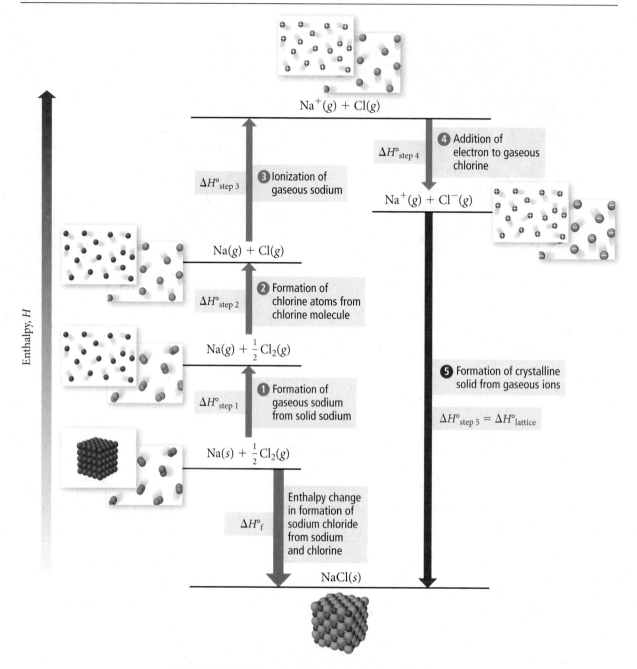

▲ **FIGURE 9.4 Born–Haber Cycle for Sodium Chloride** The sum of the steps is the formation of NaCl from elemental Na and Cl₂. The enthalpy change of the last step (step 5) is the lattice energy.

We then solve this equation for $\Delta H°_{\text{step 5}}$, which is $\Delta H°_{\text{lattice}}$, and substitute the appropriate values to calculate the lattice energy:

$$\Delta H°_{\text{lattice}} = \Delta H°_{\text{step 5}} = \Delta H°_{\text{f}} - (\Delta H°_{\text{step 1}} + \Delta H°_{\text{step 2}} + \Delta H°_{\text{step 3}} + \Delta H°_{\text{step 4}})$$

$$= -411\ \text{kJ} - (+108\ \text{kJ} + 122\ \text{kJ} + 496\ \text{kJ} - 349\ \text{kJ})$$

$$= -788\ \text{kJ}$$

Some books define the lattice energy as the energy associated with separating one mole of an ionic lattice into its gaseous constituents, which makes the sign of the lattice energy positive instead of negative.

The value of the lattice energy is a large negative number. The formation of the crystalline NaCl lattice from sodium cations and chloride anions is highly exothermic and more than compensates for the endothermicity of the electron transfer process.

In other words, the formation of ionic compounds is not exothermic because sodium "wants" to lose electrons and chlorine "wants" to gain them; rather, it is exothermic because of the large amount of heat released when sodium and chlorine ions coalesce to form a crystalline lattice.

## Trends in Lattice Energies: Ion Size

Consider the lattice energies of the following alkali metal chlorides:

| Metal Chloride | Lattice Energy (kJ/mol) |
|---|---|
| LiCl | −834 |
| NaCl | −788 |
| KCl | −701 |
| CsCl | −657 |

Why does the magnitude of the lattice energy decrease as we move down the column? We know from the periodic trends discussed in Chapter 8 that ionic radius increases as we move down a column in the periodic table (see Section 8.7). We also know, from our discussion of Coulomb's law in Section 8.3, that the potential energy of oppositely charged ions becomes less negative (or more positive) as the distance between the ions increases. As the size of the alkali metal ions increases down the column, so does the distance between the metal cations and the chloride anions. The magnitude of each successive lattice energy of the chlorides decreases accordingly, making the formation of the chlorides less exothermic as you move down the column. In other words, *as the ionic radii increase as we move down the column, the ions cannot get as close to each other and therefore do not release as much energy when the lattice forms.*

▲ Bond lengths of the group 1A metal chlorides.

## Trends in Lattice Energies: Ion Charge

Consider the lattice energies of the following two compounds:

| Compound | Lattice Energy (kJ/mol) |
|---|---|
| NaF | −910 |
| CaO | −3414 |

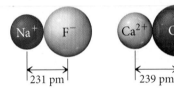

Why is the magnitude of the lattice energy of CaO so much greater than the lattice energy of NaF? $Na^+$ has a radius of 95 pm and $F^-$ has a radius of 136 pm, resulting in a distance between ions of 231 pm. $Ca^{2+}$ has a radius of 99 pm and $O^{2-}$ has a radius of 140 pm, resulting in a distance between ions of 239 pm. Even though the separation between the calcium and oxygen is slightly greater (which would tend to lower the lattice energy), the lattice energy for CaO is almost four times *greater*.

The explanation lies in the charges of the ions. Recall from Coulomb's law that the magnitude of the potential energy of two interacting charges depends not only on the distance between the charges, but also on the product of the charges:

$$E = \frac{1}{4\pi\varepsilon_0}\frac{q_1 q_2}{r}$$

For NaF, $E$ is proportional to $(1+)(1-) = 1-$, while for CaO, $E$ is proportional to $(2+)(2-) = 4-$, so the relative stabilization for CaO relative to NaF is roughly four times greater, as observed in the lattice energy.

***Summarizing Trends in Lattice Energies:***

▶ Lattice energies become less exothermic (less negative) with increasing ionic radius.

▶ Lattice energies become more exothermic (more negative) with increasing magnitude of ionic charge.

## Example 9.2     Predicting Relative Lattice Energies

Arrange these ionic compounds in order of increasing *magnitude* of lattice energy: CaO, KBr, KCl, SrO.

**SOLUTION**

KBr and KCl have lattice energies of smaller magnitude than CaO and SrO because of their lower ionic charges (1+, 1− compared to 2+, 2−). When you compare KBr and KCl, you expect KBr to have a lattice energy of lower magnitude due to the larger ionic radius of the bromide ion relative to the chloride ion. Between CaO and SrO, you expect SrO to have a lower magnitude lattice energy due to the larger ionic radius of the strontium ion relative to the calcium ion.

Order of increasing *magnitude* of lattice energy:

$$KBr < KCl < SrO < CaO$$

Actual lattice energy values:

| Compound | Lattice Energy (kJ/mol) |
|----------|-------------------------|
| KBr | −671 |
| KCl | −701 |
| SrO | −3217 |
| CaO | −3414 |

**FOR PRACTICE 9.2**     Arrange the following in order of increasing magnitude of lattice energy: LiBr, KI, and CaO.

**FOR MORE PRACTICE 9.2**     Which compound has a higher magnitude lattice energy: NaCl or $MgCl_2$?

## Ionic Bonding: Models and Reality

In this section, we have developed a model for ionic bonding. The value of a model is in how well it accounts for what we see in nature (through experiments). Does our ionic bonding model explain the properties of ionic compounds, including their high melting and boiling points, their tendency *not to conduct* electricity as solids, and their tendency *to conduct* electricity when dissolved in water?

We modeled an ionic solid as a lattice of individual ions held together by coulombic forces that are *nondirectional* (which means that, as we move away from the center of an ion, the forces are equally strong in all directions). To melt the solid, these forces must be overcome, which requires a significant amount of heat. Therefore, our model accounts for the high melting points of ionic solids. In the model, electrons transfer from the metal to the nonmetal, but the transferred electrons remain localized on one atom. In other words, our model does not include any free electrons that might conduct electricity (the movement or flow of electrons or other charged particles in response to an electric potential, or voltage, is electrical current). In addition, the ions themselves are fixed in place; therefore, our model accounts for the nonconductivity of ionic solids. When our idealized ionic solid dissolves in water, however, the cations and anions dissociate, forming free ions in solution. These ions can move in response to electrical forces, creating an electrical current. Thus, our model predicts that solutions of ionic compounds conduct electricity (which in fact they do).

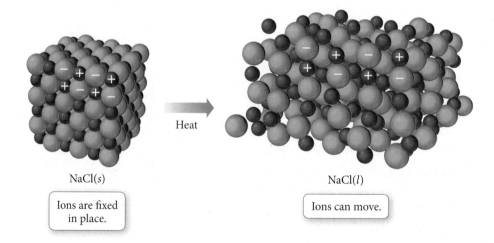

▶ The melting of solid ionic compounds such as sodium chloride requires enough heat to overcome the electrical forces holding the anions and cations together in a lattice. Thus, the melting points of ionic compounds are relatively high.

NaCl(*s*)

Ions are fixed in place.

Heat

NaCl(*l*)

Ions can move.

NaCl(*s*)

▲ Solid sodium chloride does not conduct electricity.

NaCl(*aq*)

▲ Aqueous sodium chloride conducts electricity.

**Melting Points of Ionic Solids**  Use the ionic bonding model to determine which has the higher melting point: NaCl or MgO. Explain the relative ordering.

CONCEPTUAL
**CONNECTION** **9.2**

---

## Chemistry AND **Medicine**  |  **Ionic Compounds in Medicine**

Although most drugs are molecular compounds, a number of ionic compounds have medical uses. Consider the following partial list of ionic compounds used in medicine. Notice that many of these compounds contain polyatomic ions. The bonding between a metal and a polyatomic ion is ionic. However, the bonding within a polyatomic ion is covalent, the topic of our next section.

| Formula | Name | Medical Use |
|---|---|---|
| $AgNO_3$ | Silver nitrate | Topical anti-infective agent; in solution, used to treat and prevent eye infection, especially in newborn infants |
| $BaSO_4$ | Barium sulfate | Given as a contrast medium—or image enhancer—in X-rays |
| $CaSO_4$ | Calcium sulfate | Used to make plaster casts |
| $KMnO_4$ | Potassium permanganate | Topical anti-infective agent; often used to treat fungal infections on the feet |
| $KI$ | Potassium iodide | Antiseptic and disinfectant; given orally to prevent radiation sickness |
| $Li_2CO_3$ | Lithium carbonate | Used to treat bipolar (manic–depressive) disorders |
| $MgSO_4$ | Magnesium sulfate | Used to treat eclampsia (a condition that can occur during pregnancy in which elevated blood pressure leads to convulsions) |
| $Mg(OH)_2$ | Magnesium hydroxide | Antacid and mild laxative |
| $NaHCO_3$ | Sodium bicarbonate | Oral antacid used to treat heartburn and acid stomach; injected into blood to treat severe acidosis (acidification of the blood) |
| $NaF$ | Sodium fluoride | Used to strengthen teeth |
| $ZnO$ | Zinc oxide | Used as protection from ultraviolet light in sun |

**The Lewis Model for Chemical Bonding**

# 9.5 Covalent Bonding: Lewis Structures

The Lewis model provides us with a simple and useful model for covalent bonding. In this model, we represent molecular compounds with *Lewis structures*, which depict neighboring atoms as sharing some (or all) of their valence electrons in order to attain octets (or duets for hydrogen).

## Single Covalent Bonds

To see how covalent bonding is conceived in terms of the Lewis model, consider hydrogen and oxygen, which have the following Lewis symbols:

$$\text{H}\cdot \quad \cdot\overset{..}{\underset{..}{\text{O}}}:$$

In water, these atoms share their unpaired valence electrons so that each hydrogen atom gets a duet and the oxygen atom gets an octet as represented with this Lewis structure:

$$\text{H}:\overset{..}{\underset{..}{\text{O}}}:\text{H}$$

The shared electrons—those that appear in the space between the two atoms—count toward the octets (or duets) of *both of the atoms* (shown at left).

A shared pair of electrons is called a **bonding pair**, whereas a pair that is associated with only one atom—and therefore not involved in bonding—is a **lone pair** (shown at left). Lone pair electrons are also called **nonbonding electrons**.

We often represent a bonding pair of electrons by a dash to emphasize that it constitutes a chemical bond:

$$\text{H}-\overset{..}{\underset{..}{\text{O}}}-\text{H}$$

The Lewis model also shows why the halogens form diatomic molecules. Consider the Lewis symbol for chlorine:

$$:\overset{..}{\underset{..}{\text{Cl}}}\cdot$$

If two Cl atoms pair together, they can each have an octet:

$$:\overset{..}{\underset{..}{\text{Cl}}}:\overset{..}{\underset{..}{\text{Cl}}}: \quad or \quad :\overset{..}{\underset{..}{\text{Cl}}}-\overset{..}{\underset{..}{\text{Cl}}}:$$

Elemental chlorine does indeed exist as a diatomic molecule in nature, just as the Lewis model predicts. The same is true for the other halogens.

Similarly, the Lewis model predicts that hydrogen, which has the Lewis symbol:

$$\text{H}\cdot$$

should exist as $H_2$. When two hydrogen atoms share their valence electrons, each gets a duet, a stable configuration for hydrogen:

$$\text{H}:\text{H} \quad or \quad \text{H}-\text{H}$$

Again, the Lewis model predicts correctly. In nature, elemental hydrogen exists as $H_2$ molecules.

## Double and Triple Covalent Bonds

In the Lewis model, two atoms may share more than one electron pair to get octets. For example, two oxygen atoms share two electron pairs in order for each oxygen atom to have an octet:

$$\cdot\overset{..}{\underset{..}{\text{O}}}: + \cdot\overset{..}{\underset{..}{\text{O}}}:$$

$$\downarrow$$

$$:\overset{..}{\underset{..}{\text{O}}}::\overset{..}{\underset{..}{\text{O}}}: \quad or \quad :\overset{..}{\text{O}}=\overset{..}{\text{O}}:$$

Keep in mind that one dash always stands for *two* electrons (a single bonding pair).

Each oxygen atom has an octet because *the additional bonding pair counts toward the octet of both oxygen atoms.* When two atoms share two electron pairs, the resulting bond is a **double bond**. In general, double bonds are shorter and stronger than single bonds. Atoms can also share three electron pairs. Consider the Lewis structure of $N_2$. Since each N atom has five valence electrons, the Lewis structure for $N_2$ has 10 electrons. Both nitrogen atoms attain octets by sharing three electron pairs:

$$:\text{N}:::\text{N}: \quad or \quad :\text{N}\equiv\text{N}:$$

H:O:H

Duet / Octet / Duet

Bonding pair

H:O:H

Lone pair

The bond is a **triple bond**. Triple bonds are even shorter and stronger than double bonds. When we examine nitrogen in nature, we find that it exists as a diatomic molecule with a very strong bond between the two nitrogen atoms. The bond is so strong that it is difficult to break, making $N_2$ a relatively unreactive molecule.

We explore the characteristics of multiple bonds more fully in Section 9.10.

## Covalent Bonding: Models and Reality

The Lewis model predicts the properties of molecular compounds in many ways. First, it accounts for why particular combinations of atoms form molecules and others do not. For example, why is water $H_2O$ and not $H_3O$? We can write a good Lewis structure for $H_2O$, but not for $H_3O$.

$$H—\overset{..}{\underset{..}{O}}—H \qquad H—\overset{\overset{\textstyle H}{|}}{\underset{..}{O}}—H$$

Oxygen has nine electrons
(one electron beyond an octet)

In this way, the Lewis model predicts that $H_2O$ should be stable, while $H_3O$ should not be, and that is in fact the case. However, if we remove an electron from $H_3O$, we get $H_3O^+$, which should be stable (according to the Lewis model) because, when we remove the extra electron, oxygen attains an octet:

$$\left[ H—\overset{\overset{\textstyle H}{|}}{\underset{..}{O}}—H \right]^+$$

This ion, called the hydronium ion, is in fact stable in aqueous solutions (see Section 4.8).

The Lewis model predicts other possible combinations for hydrogen and oxygen as well. For example, we can write a Lewis structure for $H_2O_2$ as follows:

$$H—\overset{..}{\underset{..}{O}}—\overset{..}{\underset{..}{O}}—H$$

Indeed, $H_2O_2$, or hydrogen peroxide, exists and is often used as a disinfectant and a bleach.

The Lewis model also accounts for why covalent bonds are highly *directional*. The attraction between two covalently bonded atoms is due to the sharing of one or more electron pairs in the space between them. Thus, each bond links just one specific pair of atoms—*in contrast to ionic bonds, which are nondirectional and hold together an entire array of ions*. As a result, the fundamental units of covalently bonded compounds are individual molecules. These molecules can interact with one another in a number of different ways that we cover in Chapter 11. However, in covalently bonded molecular compounds the interactions *between* molecules (intermolecular forces) are generally much weaker than the bonding interactions within a molecule (intramolecular forces), as shown in Figure 9.5 ▼.

## Molecular Compound

$C_5H_{12}(g)$

$C_5H_{12}(l)$

Strong covalent bonds *within* molecules

Weaker intermolecular forces *between* molecules

◀ **FIGURE 9.5 Intermolecular and Intramolecular Forces** The covalent bonds between atoms of a molecule are much stronger than the interactions between molecules. To boil a molecular substance, we simply have to overcome the relatively weak intermolecular forces, so molecular compounds generally have low boiling points.

When a molecular compound melts or boils, the molecules themselves remain intact—only the relatively weak interactions between molecules must be overcome. Consequently, molecular compounds tend to have lower melting and boiling points than ionic compounds.

---

**CONCEPTUAL CONNECTION 9.3**

**Energy and the Octet Rule** What is wrong with the following statement? *Atoms form bonds in order to satisfy the octet rule.*

---

## 9.6 Electronegativity and Bond Polarity

We know from Chapter 7 that representing electrons with dots, as we do in the Lewis model, is a drastic oversimplification. As we have already discussed, this does not invalidate the Lewis model—which is an extremely useful theory—but we must recognize and compensate for its inherent limitations. One limitation of representing electrons as dots, and covalent bonds as two dots shared between two atoms, is that the shared electrons always appear to be *equally* shared. Such is not the case.

For example, consider the Lewis structure of hydrogen fluoride:

$$H : \ddot{\underset{\cdot\cdot}{F}} :$$

The two shared electron dots sitting between the H and the F atoms appear to be equally shared between hydrogen and fluorine. However, based on laboratory measurements, we know they are not. When HF is put in an electric field, the molecules orient as shown in Figure 9.6▼. From this observation, we know that the hydrogen side of the molecule

▶ **FIGURE 9.6 Orientation of Gaseous Hydrogen Fluoride in an Electric Field** Because one side of the HF molecule has a slight positive charge and the other side a slight negative charge, the molecules align themselves with an external electric field.

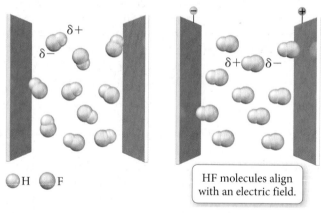

$\delta+$

$\delta-$

$\delta+$ $\delta-$

○ H ● F

HF molecules align with an electric field.

must have a slight positive charge and the fluorine side of the molecule must have a slight negative charge. We represent this partial separation of charge as follows:

$$\overset{+\longrightarrow}{H—F} \quad or \quad \overset{\delta+ \quad \delta-}{H—F}$$

The red arrow on the left, with a positive sign on the tail, indicates that the left side of the molecule has a partial positive charge and that the right side of the molecule (the side the arrow is pointing *toward*) has a partial negative charge. Similarly, the $\delta+$ (delta plus) represents a partial positive charge and the $\delta-$ (delta minus) represents a partial negative charge. Does this make the bond ionic? No. In an ionic bond, the electron is essentially *transferred* from one atom to another. In HF, the electron is *unequally shared*. In other words, even though the Lewis structure of HF portrays the bonding electrons as residing *between* the two atoms, in reality the electron density is greater on the fluorine atom than on the hydrogen atom (Figure 9.7◀). The bond is said to be *polar*—having a positive pole and a negative pole. A **polar covalent bond** is intermediate in nature between a pure covalent bond and an ionic bond. In fact, the categories of pure covalent and ionic are really two extremes within a broad continuum. Most covalent bonds between dissimilar atoms are actually *polar covalent*—somewhere between the two extremes.

▲ **FIGURE 9.7 Electrostatic Potential Map for the HF Molecule** The F end of the molecule, with its partial negative charge, is yellow/pink; the H end, with its partial positive charge, is blue.

# Electronegativity

The ability of an atom to attract electrons to itself in a chemical bond (which results in polar and ionic bonds) is called **electronegativity**. We say that fluorine is more *electronegative* than hydrogen because it takes a greater share of the electron density in HF.

The American chemist Linus Pauling (1901–1994) quantified electronegativity in his classic book, *The Nature of the Chemical Bond*. Pauling compared the bond energy—the energy required to break a bond—of a heteronuclear diatomic molecule such as HF with the bond energies of its homonuclear counterparts, in this case $H_2$ and $F_2$. The bond energies of $H_2$ and $F_2$ are 436 kJ/mol and 155 kJ/mol, respectively. Pauling reasoned that if the HF bond were purely covalent—that is, if the electrons were shared exactly equally—the bond energy of HF should simply be an average of the bond energies of $H_2$ and $F_2$, which would be 296 kJ/mol. However, the bond energy of HF is experimentally measured to be 565 kJ/mol. Pauling suggested that the additional bond energy was due to the *ionic character* of the bond. Based on many such comparisons of bond energies, and by arbitrarily assigning an electronegativity of 4.0 to fluorine (the most electronegative element on the periodic table), Pauling developed the electronegativity values shown in Figure 9.8▼.

For main-group elements, notice the following periodic trends in electronegativity from Figure 9.8:

- Electronegativity generally increases across a period in the periodic table.
- Electronegativity generally decreases down a column in the periodic table.
- Fluorine is the most electronegative element.
- Francium is the least electronegative element (sometimes called the most *electropositive*).

We cover the concept of bond energy in more detail in Section 9.10.

Pauling's "average" bond energy was actually calculated a little bit differently than the normal average shown here. He took the square root of the product of the bond energies of the homonuclear counterparts as the "average."

▼ **FIGURE 9.8 Electronegativities of the Elements** Electronegativity generally increases as we move across a row in the periodic table and decreases as we move down a column.

## Trends in Electronegativity

| | 0.7–1.1 |
| | 1.2–1.6 |
| | 1.7–2.1 |
| | 2.2–2.6 |
| | 2.7–4.0 |

The periodic trends in electronegativity are consistent with other periodic trends we have seen. In general, electronegativity is inversely related to atomic size—the larger the atom, the less ability it has to attract electrons to itself in a chemical bond.

CONCEPTUAL
CONNECTION **9.4**

**Periodic Trends in Electronegativity** Arrange these elements in order of decreasing electronegativity: P, Na, N, Al.

## Bond Polarity, Dipole Moment, and Percent Ionic Character

The degree of polarity in a chemical bond depends on the electronegativity difference (sometimes abbreviated $\Delta$EN) between the two bonding atoms. The greater the electronegativity difference, the more polar the bond. If two atoms with identical electronegativities form a covalent bond, they share the electrons equally, and the bond is purely covalent or *nonpolar*. For example, the chlorine molecule, composed of two chlorine atoms (which necessarily have identical electronegativities), has a covalent bond in which electrons are evenly shared:

If there is a large electronegativity difference between the two atoms in a bond, such as normally occurs between a metal and a nonmetal, the electron from the metal is almost completely transferred to the nonmetal, and the bond is ionic. For example, sodium and chlorine form an ionic bond:

If there is an intermediate electronegativity difference between the two atoms, such as between two different nonmetals, then the bond is polar covalent. For example, HCl has a polar covalent bond:

Although all attempts to divide the bond polarity continuum into specific regions are necessarily arbitrary, it is helpful to classify bonds as covalent, polar covalent, and ionic, based on the electronegativity difference between the bonding atoms as shown in Table 9.1 and Figure 9.9▶.

| TABLE 9.1 The Effect of Electronegativity Difference on Bond Type | | |
| --- | --- | --- |
| **Electronegativity Difference ($\Delta$EN)** | **Bond Type** | **Example** |
| Small (0–0.4) | Covalent | $Cl_2$ |
| Intermediate (0.4–2.0) | Polar covalent | HCl |
| Large (2.0+) | Ionic | NaCl |

We quantify the polarity of a bond by the size of its dipole moment. A **dipole moment** ($\mu$) occurs any time there is a separation of positive and negative charge. The magnitude of a dipole moment created by separating two particles of equal but opposite charges of magnitude $q$ by a distance $r$ is given by the equation:

$$\mu = qr \qquad [9.1]$$

## The Continuum of Bond Types

We can get a sense for the dipole moment of a completely ionic bond by calculating the dipole moment that results from separating a proton and an electron ($q = 1.6 \times 10^{-19}$ C) by a distance of $r = 130$ pm (the approximate length of a short chemical bond):

$$\mu = qr$$
$$= (1.6 \times 10^{-19}\,\text{C})(130 \times 10^{-12}\,\text{m})$$
$$= 2.1 \times 10^{-29}\,\text{C} \cdot \text{m}$$
$$= 6.2\,\text{D}$$

The debye (D) is the unit commonly used for reporting dipole moments ($1\,\text{D} = 3.34 \times 10^{-30}\,\text{C} \cdot \text{m}$). Based on this calculation, we would expect the dipole moment of completely ionic bonds with bond lengths close to 130 pm to be about 6 D. The smaller the magnitude of the charge separation, and the smaller the distance between the charges, the smaller the dipole moment. Table 9.2 lists the dipole moments of several molecules along with the electronegativity differences of their atoms.

By comparing the *actual* dipole moment of a bond to what the dipole moment would be if the electron were completely transferred from one atom to the other, we can get a sense of the degree to which the electron is transferred (or the degree to which the bond is ionic). The **percent ionic character** is the ratio of a bond's actual dipole moment to the dipole moment it would have if the electron were completely transferred from one atom to the other, multiplied by 100%:

$$\text{Percent ionic character} = \frac{\text{measured dipole moment of bond}}{\text{dipole moment if electron were completely transferred}} \times 100\%$$

For example, suppose a diatomic molecule with a bond length of 130 pm has a dipole moment of 3.5 D. We previously calculated that separating a proton and an electron by 130 pm results in a dipole moment of 6.2 D. Therefore, the percent ionic character of the bond is 56%:

$$\text{Percent ionic character} = \frac{3.5\,\text{D}}{6.2\,\text{D}} \times 100\%$$
$$= 56\%$$

A bond in which an electron is completely transferred from one atom to another would have 100% ionic character (although even the most ionic bonds do not reach this ideal). Figure 9.10▶ shows the percent ionic character of a number of diatomic gas-phase molecules plotted against the electronegativity difference between the bonding atoms. As expected, the percent ionic character generally increases as the electronegativity difference increases. However, as we can see, no bond is 100% ionic. In general, bonds with greater than 50% ionic character are referred to as ionic bonds.

| TABLE 9.2 Dipole Moments of Several Molecules in the Gas Phase | | |
|---|---|---|
| **Molecule** | **ΔEN** | **Dipole Moment (D)** |
| $Cl_2$ | 0 | 0 |
| ClF | 1.0 | 0.88 |
| HF | 1.9 | 1.82 |
| LiF | 3.0 | 6.33 |

▼ **FIGURE 9.10** Percent Ionic Character versus Electronegativity Difference for Some Compounds

**Example** 9.3 | **Classifying Bonds as Pure Covalent, Polar Covalent, or Ionic**

Classify the bond formed between each pair of atoms as covalent, polar covalent, or ionic.

**(a)** Sr and F          **(b)** N and Cl          **(c)** N and O

**SOLUTION**

**(a)** In Figure 9.8, find the electronegativity of Sr (1.0) and of F (4.0). The electronegativity difference ($\Delta$EN) is $\Delta$EN = 4.0 − 1.0 = 3.0. Using Table 9.1, classify this bond as ionic.

**(b)** In Figure 9.8, find the electronegativity of N (3.0) and of Cl (3.0). The electronegativity difference ($\Delta$EN) is $\Delta$EN = 3.0 − 3.0 = 0. Using Table 9.1, classify this bond as covalent.

**(c)** In Figure 9.8, find the electronegativity of N (3.0) and of O (3.5). The electronegativity difference ($\Delta$EN) is $\Delta$EN = 3.5 − 3.0 = 0.5. Using Table 9.1, classify this bond as polar covalent.

**FOR PRACTICE 9.3** Determine if the bond formed between each pair of atoms is pure covalent, polar covalent, or ionic.

**(a)** I and I          **(b)** Cs and Br          **(c)** P and O

**CONCEPTUAL**
**CONNECTION 9.5**

**Percent Ionic Character** The HCl(*g*) molecule has a bond length of 127 pm and a dipole moment of 1.08 D. Without doing detailed calculations, identify the best estimate for its percent ionic character.

**(a)** 5%          **(b)** 15%          **(c)** 50%          **(d)** 80%

 **Writing Lewis Structures for Molecular Compound**

## 9.7 Lewis Structures of Molecular Compounds and Polyatomic Ions

We now turn to the basic sequence of steps involved in actually writing Lewis structures for given combinations of atoms.

### Writing Lewis Structures for Molecular Compounds

To write a Lewis structure for a molecular compound, follow these steps:

Often, chemical formulas provide clues to how the atoms are bonded together. For example, CH₃OH indicates that three hydrogen atoms and the oxygen atom are bonded to the carbon atom, but the fourth hydrogen atom is bonded to the oxygen atom.

There are a few exceptions to this rule, such as diborane (B₂H₆), which contains *bridging hydrogens*, but these are rare and cannot be adequately addressed by the Lewis model.

1. **Write the correct skeletal structure for the molecule.** The Lewis structure of a molecule must have the atoms in the correct positions. For example, you could not write a Lewis structure for water if you started with the hydrogen atoms next to each other and the oxygen atom at the end (H H O). In nature, oxygen is the central atom and the hydrogen atoms are *terminal* (at the ends). The correct skeletal structure is H O H. The only way to determine the skeletal structure of a molecule with absolute certainty is to examine its structure experimentally. However, you can write likely skeletal structures by remembering two guidelines. First, *hydrogen atoms are always terminal*. Hydrogen does not ordinarily occur as a central atom because central atoms must form at least two bonds, and hydrogen, which has only a single valence electron to share and requires only a duet, can form just one bond. Second, *put the more electronegative elements in terminal positions* and the less electronegative elements (other than hydrogen) in the central position. Later in this section, you will learn how to distinguish between competing skeletal structures by applying the concept of formal charge.

2. **Calculate the total number of electrons for the Lewis structure by summing the valence electrons of each atom in the molecule.** Remember that the number of valence electrons for any main-group element is equal to its group number in the periodic table. *If you are writing a Lewis structure for a polyatomic ion, you must consider the charge of the ion when calculating the total number of electrons.* Add one electron for each negative charge and subtract one electron for each positive charge. Don't worry about which electron comes from which atom—only the total number is important.

3. **Distribute the electrons among the atoms, giving octets (or duets in the case of hydrogen) to as many atoms as possible.** Begin by placing two electrons between every two atoms. These represent the minimum number of bonding electrons. Then distribute the remaining electrons as lone pairs, first to terminal atoms and then to the central atom, giving octets (or duets for hydrogen) to as many atoms as possible.

4. **If any atoms lack an octet, form double or triple bonds as necessary to give them octets.** Do this by moving lone electron pairs from terminal atoms into the bonding region with the central atom.

> Sometimes distributing all the remaining electrons to the central atom results in more than an octet. This is called an expanded octet, which we discuss in Section 9.9.

In the left column that follows, there is an abbreviated version of the procedure for writing Lewis structures; the center and right columns feature two examples of applying the procedure.

## Procedure for ...

| Writing Lewis Structures for Molecular Compounds | Example 9.4 ▶ **Writing Lewis Structures** Write the Lewis structure for $CO_2$. | Example 9.5 **Writing Lewis Structures** Write the Lewis structure for $NH_3$. |
|---|---|---|
| 1. Write the correct skeletal structure for the molecule. | **SOLUTION** Because carbon is the less electronegative atom, put it in the central position. <br> O C O | **SOLUTION** Since hydrogen is always terminal, put nitrogen in the central position. <br> H N H <br> H |
| 2. Calculate the total number of electrons for the Lewis structure by summing the valence electrons of each atom in the molecule. | Total number of electrons for Lewis structure = $\left(\begin{array}{c}\text{number of}\\ \text{valence}\\ \text{e}^-\text{ for C}\end{array}\right) + 2\left(\begin{array}{c}\text{number of}\\ \text{valence}\\ \text{e}^-\text{ for O}\end{array}\right)$ <br> $= 4 + 2(6) = 16$ | Total number of electrons for Lewis structure = $\left(\begin{array}{c}\text{number of}\\ \text{valence}\\ \text{e}^-\text{ for N}\end{array}\right) + 3\left(\begin{array}{c}\text{number of}\\ \text{valence}\\ \text{e}^-\text{ for H}\end{array}\right)$ <br> $= 5 + 3(1) = 8$ |
| 3. Distribute the electrons among the atoms, giving octets (or duets for hydrogen) to as many atoms as possible. Begin with the bonding electrons, and then proceed to lone pairs on terminal atoms, and finally to lone pairs on the central atom. | Bonding electrons are first. <br> O:C:O <br> (4 of 16 electrons used) <br><br> Lone pairs on terminal atoms are next. <br> :Ö:C:Ö: <br><br> (16 of 16 electrons used) | Bonding electrons are first. <br> H:N:H <br> Ḧ <br> (6 of 8 electrons used) <br> Lone pairs on terminal atoms are next, but none is needed on hydrogen. <br> Lone pairs on the central atom are last. <br><br> H—N̈—H <br>     &#124; <br>     H <br> (8 of 8 electrons used) |
| 4. If any atom lacks an octet, form double or triple bonds as necessary until all atoms have octets. | Since carbon lacks an octet, move lone pairs from the oxygen atoms to bonding regions to form double bonds. <br><br> :Ö:C:Ö: <br> ↓ <br> :O̤=C̈=O̤: | Since all of the atoms have octets (or duets for hydrogen), the Lewis structure for $NH_3$ is complete as shown in the previous step. |
| | **FOR PRACTICE 9.4** Write the Lewis structure for CO. | **FOR PRACTICE 9.5** Write the Lewis structure for $H_2CO$. |

## Writing Lewis Structures for Polyatomic Ions

Write Lewis structures for polyatomic ions by following the same procedure, but pay special attention to the charge of the ion when calculating the number of electrons for the Lewis structure. Add one electron for each negative charge and subtract one electron for each positive charge. Write the Lewis structure for a polyatomic ion within brackets with the charge of the ion in the upper right-hand corner, outside the bracket.

---

**Example 9.6**   | **Writing Lewis Structures for Polyatomic Ions** ▶

Write the Lewis structure for the $NH_4^+$ ion.

**SOLUTION**

| | |
|---|---|
| Begin by writing the skeletal structure. Since hydrogen is always terminal, put the nitrogen atom in the central position. | H<br>H N H<br>H |
| Calculate the total number of electrons for the Lewis structure by summing the number of valence electrons for each atom and subtracting 1 for the 1+ charge. | Total number of electrons for Lewis structure<br>= (number of valence $e^-$ in N) + 4 (number of valence $e^-$ in H) − 1<br>= 5 + 4(1) − 1<br>= 8          Subtract 1 $e^-$ to account<br>for $1^+$ charge of ion. |
| Place two bonding electrons between every two atoms. Since all of the atoms have complete octets, no double bonds are necessary. | H<br>H:N:H<br>H<br><br>(8 of 8 electrons used) |
| Lastly, write the Lewis structure in brackets with the charge of the ion in the upper right-hand corner. | $$\left[\begin{array}{c} H \\ | \\ H\!-\!N\!-\!H \\ | \\ H \end{array}\right]^+$$ |

**FOR PRACTICE 9.6**   Write the Lewis structure for the hypochlorite ion, $ClO^-$.

---

 **Resonance and Formal Charge**

## 9.8   Resonance and Formal Charge

We need two additional concepts to write the best possible Lewis structures for a large number of compounds. The concepts are *resonance*, which is relevant when two or more valid Lewis structures can be drawn for the same compound, and *formal charge*, an electron bookkeeping system that allows us to discriminate between alternative Lewis structures.

### Resonance

For some molecules, we can write more than one valid Lewis structure. For example, consider writing a Lewis structure for $O_3$. The following two Lewis structures, with the double bond on alternate sides, are equally correct:

$$\ddot{O}\!=\!\ddot{O}\!-\!\ddot{O}\!: \quad :\ddot{O}\!-\!\ddot{O}\!=\!\ddot{O}$$

In cases such as this—where there are two or more valid Lewis structures for the same molecule—we find that, in nature, the molecule is an *average* of the two Lewis structures. Both of the two Lewis structures for $O_3$ predict that $O_3$ contains two different bonds (one double bond and one single bond). However, when we experimentally examine the structure of $O_3$, we find that the bonds in the $O_3$ molecule are equivalent and each is intermediate in strength and length between a double bond and single bond. We account for this by representing the molecule with both structures, called *resonance structures*, with a double-headed arrow between them:

$$:\ddot{O}{=}\ddot{O}{-}\ddot{O}: \longleftrightarrow :\ddot{O}{-}\ddot{O}{=}\ddot{O}:$$

A **resonance structure** is one of two or more Lewis structures that have the same skeletal formula (the atoms are in the same locations), but different electron arrangements. The actual structure of the molecule is intermediate between the two (or more) resonance structures and is called a **resonance hybrid**. The term *hybrid* comes from breeding and means the offspring of two animals or plants of different varieties or breeds. If we breed a Labrador retriever with a German shepherd, we get a *hybrid* that is intermediate between the two breeds (Figure 9.11(a)▼). Similarly, the actual structure of a resonance hybrid is intermediate between the two (or more) resonance structures (Figure 9.11(b)▼). The only structure that actually exists is the hybrid structure—the individual resonance structures do not exist and are merely a way to describe the actual structure. Notice that the actual structure of ozone has two equivalent bonds and a bent geometry (we discuss molecular geometries in Chapter 10).

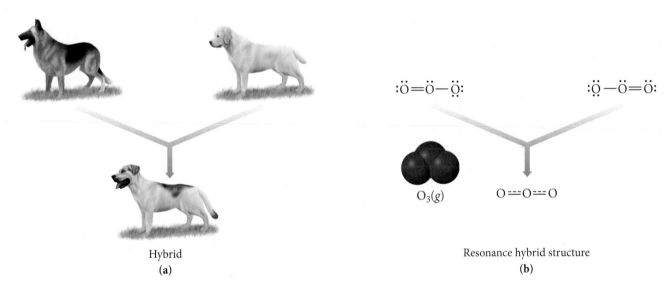

▲ **FIGURE 9.11 Hybridization**   Just as the offspring of two different dog breeds is a hybrid that is intermediate between the two breeds (**a**), the structure of a resonance hybrid is intermediate between that of the contributing resonance structures (**b**).

The concept of resonance is an adaptation of the Lewis model that helps account for the complexity of actual molecules. In the Lewis model, electrons are *localized* either on one atom (lone pair) or between atoms (bonding pair). However, in nature, the electrons in molecules are often *delocalized* over several atoms or bonds. The delocalization of electrons lowers their energy; it stabilizes them (for reasons that are beyond the scope of this book). Resonance depicts two or more structures with the electrons in different places in an attempt to more accurately reflect the delocalization of electrons. In the real hybrid structure, an average between the resonance structures, the electrons are more spread out (or delocalized) than in any of the resonance structures. The resulting stabilization of the electrons (that is, the lowering of their potential energy due to delocalization) is sometimes called *resonance stabilization*. Resonance stabilization makes an important contribution to the stability of many molecules.

## Example 9.7 Writing Resonance Structures ▶

Write a Lewis structure for the $NO_3^-$ ion. Include resonance structures.

SOLUTION

| | |
|---|---|
| Begin by writing the skeletal structure. Since nitrogen is the least electronegative atom, put it in the central position. | O<br>O N O |
| Calculate the total number of electrons for the Lewis structure by summing the number of valence electrons for each atom and adding 1 for the 1− charge. | Total number of electrons for Lewis structure<br>= (number of valence e⁻ in N) + 3 (number of valence e⁻ in O) + 1<br>= 5 + 3(6) + 1<br>= 24　　　　　　　　　　　　　　　↑<br>　　　　　　　　　　　　Add 1 e⁻ to account for 1− charge of ion. |
| Place two bonding electrons between each pair of atoms. | O<br>O:N:O<br><br>(6 of 24 electrons used) |
| Distribute the remaining electrons, first to terminal atoms. There are not enough electrons to complete the octet on the central atom. | :Ö:<br>:Ö:N:Ö:<br><br>(24 of 24 electrons used) |
| Form a double bond by moving a lone pair from one of the oxygen atoms into the bonding region with nitrogen. Enclose the structure in brackets and include the charge. | $\left[ \begin{array}{c} \text{:Ö:} \\ \text{:Ö:N::Ö:} \end{array} \right]^-$<br>*or*<br>$\left[ \begin{array}{c} \text{:Ö:} \\ \text{\|} \\ \text{:Ö—N=Ö:} \end{array} \right]^-$ |
| Since the double bond can form equally well with any of the three oxygen atoms, write all three structures as resonance structures. (The actual space-filling model of $NO_3^-$ is shown here for comparison. Note that all three bonds are equal in length.) | $\left[ \begin{array}{c} \text{:Ö:} \\ \text{\|} \\ \text{:Ö—N=Ö:} \end{array} \right]^- \longleftrightarrow \left[ \begin{array}{c} \text{:O:} \\ \text{\|\|} \\ \text{:Ö—N—Ö:} \end{array} \right]^- \longleftrightarrow \left[ \begin{array}{c} \text{:Ö:} \\ \text{\|} \\ \text{:Ö=N—Ö:} \end{array} \right]^-$<br><br>$NO_3^-$ |

FOR PRACTICE 9.7　Write a Lewis structure for the $NO_2^-$ ion. Include resonance structures.

In the examples of resonance hybrids that we have examined so far, the contributing structures have been equivalent (or equally valid) Lewis structures. In these cases, the true structure is an equally weighted average of the resonance structures. In some cases, however, we can write resonance structures that are not equivalent. For reasons that we cover in the material that follows—such as formal charge—one resonance structure may be somewhat better than another. In such cases, the true structure is still an average of the resonance structures, but the better resonance structure contributes more to the true structure. In other words, multiple nonequivalent resonance structures may be weighted differently in their contributions to the true overall structure of a molecule (see Example 9.8).

## Formal Charge

*Formal charge* is a fictitious charge assigned to each atom in a Lewis structure that helps us to distinguish among competing Lewis structures. The **formal charge** of an atom in a Lewis structure is *the charge it would have if all bonding electrons were shared equally between the bonded atoms.* In other words, formal charge is the calculated charge for an atom if we completely ignore the effects of electronegativity. For example, we know that because fluorine is more electronegative than hydrogen, HF has a dipole moment—the hydrogen atom has a slight positive charge and the fluorine atom has a slight negative charge. However, the *formal charges* of hydrogen and fluorine in HF (the calculated charges if we ignore their differences in electronegativity) are both zero:

$$H\!:\!\ddot{\underset{..}{F}}\!:$$

Formal charge = 0    Formal charge = 0

We can calculate the formal charge on any atom as the difference between the number of valence electrons in the atom and the number of electrons that it "owns" in a Lewis structure. An atom in a Lewis structure can be thought of as "owning" all of its nonbonding electrons and one-half of its bonding electrons:

Formal charge = number of valence electrons −
  (number of nonbonding electrons + $\frac{1}{2}$ number of bonding electrons)

So the formal charge of hydrogen in HF is 0:

$$\text{Formal charge} = 1 - \left[0 + \tfrac{1}{2}(2)\right] = 0$$

Number of valence electrons for H

Number of electrons that H "owns" in the Lewis structure

Similarly, we calculate the formal charge of fluorine in HF as 0:

$$\text{Formal charge} = 7 - \left[6 + \tfrac{1}{2}(2)\right] = 0$$

Number of valence electrons for F

Number of electrons that F "owns" in the Lewis structure

The concept of formal charge is useful because it can help us distinguish between competing skeletal structures or competing resonance structures. In general, when we calculate formal charges, these four rules apply:

1. The sum of all formal charges in a neutral molecule must be zero.
2. The sum of all formal charges in an ion must equal the charge of the ion.
3. Small (or zero) formal charges on individual atoms are better than large ones.
4. When formal charge cannot be avoided, negative formal charge should reside on the most electronegative atom.

We can use formal charge to determine which of the competing skeletal structures for hydrogen cyanide shown here is best (or most stable). Notice that both skeletal structures satisfy the octet rule equally. The formal charge of each atom in the structure is calculated below it.

|  | Structure A | | | Structure B | | |
|---|---|---|---|---|---|---|
|  | H — | C ≡ | N: | H — | N ≡ | C: |
| number of valence e⁻ | 1 | 4 | 5 | 1 | 5 | 4 |
| −number of nonbonding e⁻ | −0 | −0 | −2 | −0 | −0 | −2 |
| −$\frac{1}{2}$(number of bonding e⁻) | −$\frac{1}{2}$(2) | −$\frac{1}{2}$(8) | −$\frac{1}{2}$(6) | −$\frac{1}{2}$(2) | −$\frac{1}{2}$(8) | −$\frac{1}{2}$(6) |
| **Formal charge** | **0** | **0** | **0** | **0** | **+1** | **−1** |

As required, the sum of the formal charges for each of these structures is zero (as it always must be for neutral molecules). However, structure B has formal charges on both the N atom and the C atom, while structure A has no formal charges on any atom. Furthermore, in structure B, the negative formal charge is not on the most electronegative element (nitrogen is more electronegative than carbon). Consequently, structure A is the better skeletal structure. Since atoms in the middle of a molecule tend to have more bonding electrons and fewer nonbonding electrons, they also tend to have more positive formal charges. Consequently, the best skeletal structure usually has the least electronegative atom in the central position, as we learned in step 1 of our procedure for writing Lewis structures.

Both HCN and HNC exist, but—as predicted by formal charge—HCN is more stable than HNC.

## Example 9.8 | Assigning Formal Charges

Assign formal charges to each atom in the resonance structures of the cyanate ion (OCN⁻). Which resonance structure is likely to contribute most to the correct structure of OCN⁻?

$$\begin{array}{ccc} A & B & C \\ [:\ddot{O}-C\equiv N:]^- & [:\ddot{O}=C=\ddot{N}:]^- & [:O\equiv C-\ddot{N}:]^- \end{array}$$

**SOLUTION**

Calculate the formal charge on each atom by finding the number of valence electrons and subtracting the number of nonbonding electrons and one-half the number of bonding electrons.

|  | A $[:\ddot{O}-C\equiv N:]^-$ | | | B $[:\ddot{O}=C=\ddot{N}:]^-$ | | | C $[:O\equiv C-\ddot{N}:]^-$ | | |
|---|---|---|---|---|---|---|---|---|---|
| Number of valence e⁻ | 6 | 4 | 5 | 6 | 4 | 5 | 6 | 4 | 5 |
| −number of nonbonding e⁻ | −6 | −0 | −2 | −4 | −0 | −4 | −2 | −0 | −6 |
| −½(number of bonding e⁻) | −1 | −4 | −3 | −2 | −4 | −2 | −3 | −4 | −1 |
| **Formal charge** | **−1** | **0** | **0** | **0** | **0** | **−1** | **+1** | **0** | **−2** |

The sum of all formal charges for each structure is −1, as it should be for a 1− ion. Structures A and B have the fewest non-zero formal charges and are preferred over structure C. Structure A is preferable to B because it has the negative formal charge on the more electronegative atom. You therefore expect structure A to make the biggest contribution to the resonance forms of the cyanate ion.

**FOR PRACTICE 9.8** Assign formal charges to each atom in the resonance structures of $N_2O$. Which resonance structure is likely to contribute most to the correct structure of $N_2O$?

$$\begin{array}{ccc} A & B & C \\ :\ddot{N}=N=\ddot{O}: & :N\equiv N-\ddot{O}: & :\ddot{N}-N\equiv O: \end{array}$$

**FOR MORE PRACTICE 9.8** Assign formal charges to each of the atoms in the nitrate ion ($NO_3^-$). The Lewis structure for the nitrate ion is shown in Example 9.7.

## Example 9.9 | Drawing Resonance Structures and Assigning Formal Charge for Organic Compounds

Draw the Lewis structure (including resonance structures) for nitromethane ($CH_3NO_2$). For each resonance structure, assign formal charges to all atoms that have formal charge.

**SOLUTION**

Begin by writing the skeletal structure. For organic compounds, the condensed structural formula (in this case $CH_3NO_2$) indicates how the atoms are connected.

$$\begin{array}{cccc} & H & O & \\ H & C & N & O \\ & H & & \end{array}$$

| Calculate the total number of electrons for the Lewis structure by summing the number of valence electrons for each atom. | Total number of electrons for Lewis structure<br><br>= (# valence e$^-$ in C) + 3(# valence e$^-$ in H)<br><br>$\qquad$ + (# valence e$^-$ in N) + 2(# valence e$^-$ in O)<br><br>= 4 + 3(1) + 5 + 2(6)<br><br>= 24 |
|---|---|
| Place a dash between each pair of atoms to indicate a bond. Each dash represents two electrons. | <br><br>(12 of 24 electrons used) |
| Distribute the remaining electrons, first to terminal atoms then to interior atoms. | <br><br>(24 of 24 electrons used) |
| If there are not enough electrons to complete the octets on the interior atoms, form double bonds by moving lone pair electrons from terminal atoms into the bonding region with interior atoms. | |
| Draw any necessary resonance structures by moving only electron dots. (In this case, you can form a double bond between the nitrogen atom and the other oxygen atom.) | |
| Assign formal charges (FC) to each atom.<br><br>FC = # valence e$^-$<br>$\qquad$ − (# nonbonding e$^-$ + $\frac{1}{2}$ # bonding e$^-$) | <br><br>Carbon, hydrogen, and the double-bonded oxygen atoms have formal charges of zero. Nitrogen has a +1 formal charge [$5 - \frac{1}{2}(8)$] and the singly bonded oxygen atom in each resonance structure has a −1 formal charge [$6 - (6 + \frac{1}{2}(2))$]. |

**FOR PRACTICE 9.9** Draw the Lewis structure (including resonance structures) for diazomethane ($CH_2N_2$). For each resonance structure, assign formal charges to all atoms that have formal charge.

# 9.9 Exceptions to the Octet Rule: Odd-Electron Species, Incomplete Octets, and Expanded Octets

The octet rule in the Lewis model has some exceptions, which we examine in this section of the chapter. They include (1) *odd-electron species,* molecules or ions with an odd number of electrons; (2) *incomplete octets,* molecules or ions with *fewer than eight electrons* around an atom; and (3) *expanded octets,* molecules or ions with *more than eight electrons* around an atom.

## Odd-Electron Species

Molecules and ions with an odd number of electrons in their Lewis structures are **free radicals** (or simply *radicals*). For example, nitrogen monoxide—a pollutant found in motor vehicle exhaust—has 11 electrons. If we try to write a Lewis structure for nitrogen monoxide, we can't achieve octets for both atoms:

$$\dot{N}\!:\!\ddot{O}: \quad or \quad :\dot{N}\!=\!\ddot{O}:$$

The unpaired electron in nitrogen monoxide is put on the nitrogen rather than the oxygen in order to minimize formal charges.

The nitrogen atom in nitrogen monoxide does not have an octet, so this Lewis structure does not satisfy the octet rule. Yet, nitrogen monoxide exists, especially in polluted air. Why? As with any simple theory, the Lewis model is not sophisticated enough to handle every single case. It is impossible to write good Lewis structures for free radicals; nevertheless, some of these molecules exist in nature. Perhaps it is a testament to the Lewis model, however, that *relatively few* such molecules exist and that, in general, they tend to be somewhat unstable and reactive. NO, for example, reacts with oxygen in the air to form $NO_2$, another odd-electron molecule represented with the following 17-electron resonance structures:

$$:\ddot{O}\!=\!\dot{N}\!-\!\ddot{O}: \longleftrightarrow :\ddot{O}\!-\!\dot{N}\!=\!\ddot{O}:$$

In turn, $NO_2$ reacts with water to form nitric acid (a component of acid rain) and also reacts with other atmospheric pollutants to form peroxyacetylnitrate (PAN), an active component of photochemical smog. For free radicals, such as NO and $NO_2$, we simply write the best Lewis structure that we can.

CONCEPTUAL
**CONNECTION 9.6**

**Odd-Electron Species**   Which molecule would you expect to be a free radical?
**(a)** CO
**(b)** $CO_2$
**(c)** $N_2O$
**(d)** $NO_2$

## Incomplete Octets

Another significant exception to the octet rule involves elements that tend to form *incomplete octets*. The most important of these is boron, which forms compounds with only six electrons around B, rather than eight. For example, $BF_3$ and $BH_3$ lack an octet for B.

$$:\ddot{F}:\!\ddot{B}:\!\ddot{F}: \qquad H\!:\!\ddot{B}:\!H$$

Beryllium compounds, such as $BeH_2$, also have incomplete octets.

You might be wondering why we don't just form double bonds to increase the number of electrons around B. For $BH_3$, we can't because there are no additional electrons to move into the bonding region. For $BF_3$, however, we could attempt to give B an octet by moving a lone pair from an F atom into the bonding region with B:

$$:\ddot{F}\!-\!B\!=\!\ddot{F}:$$

# Chemistry IN THE Environment | Free Radicals and the Atmospheric Vacuum Cleaner

Free radicals play a key role in much of the chemistry of the atmosphere. The free radical that is most important to atmospheric reactions is the hydroxyl radical:

$$:\ddot{O}-H$$

▲ $NO_2(g)$ is a pollutant found in urban air.

Many free radical structures are abbreviated by writing a single dot with the formula. Thus, the hydroxyl radical is often abbreviated as:

$$\cdot OH$$

In the atmosphere, the hydroxyl radical forms when excited oxygen atoms—formed from the photodecomposition of ozone—react with water vapor:

$$O_3 \xrightarrow{\text{UV light}} O_2 + O^*$$

$$O^* + H_2O \longrightarrow 2 \cdot OH$$

The * next to the O indicates that the oxygen atom has excess energy.

The resulting hydroxyl radical reacts with a wide variety of molecules from both natural sources and from air pollution that are present in the atmosphere. For example, the hydroxyl radical reacts with carbon monoxide, an atmospheric pollutant that we first encountered in Chapter 1, in the following two-step process:

$$CO + \cdot OH \longrightarrow HOCO \cdot$$

$$HOCO \cdot + O_2 \longrightarrow CO_2 + HOO \cdot$$

We can see from this reaction that the hydroxyl radical converts toxic CO into relatively nontoxic $CO_2$.

The HOO· free radical generated by the second reaction is converted back into the hydroxyl radical when it reacts with other atmospheric substances, and the process repeats itself. Therefore, a single hydroxyl radical can convert a lot of CO into $CO_2$.

Do you ever wonder what happens to the hydrocarbons you accidentally spill when filling a car's gas tank or to the natural gas that is released into the atmosphere as you light a kitchen stove? Hydrocarbons released into the atmosphere are converted to $CO_2$ and $H_2O$ in a series of steps initiated by the hydroxyl free radical. Consider the following representative reaction of methane, the main hydrocarbon in natural gas:

$$CH_4 + 5\,O_2 + NO\cdot + 2\cdot OH \xrightarrow{\text{UV light}}$$
$$CO_2 + H_2O + NO_2\cdot + 4\,HOO\cdot$$

Notice the similarity between this reaction and the direct combustion (or burning) of methane:

$$CH_4 + 2\,O_2 \longrightarrow CO_2 + 2\,H_2O$$

As you can see, the free radical reaction initiates a slow "burning" of $CH_4$ in a series of steps that produces carbon dioxide and water and some additional free radicals. The hydroxyl radical initiates similar reactions with other pollutants as well as undesirable naturally occurring atmospheric gases. Without the hydroxyl free radical—sometimes called *the atmospheric vacuum cleaner*—our atmosphere would be a much dirtier place.

▲ Hydrocarbons such as octane evaporate into the atmosphere when a motor vehicle is fueled. What happens to them?

### QUESTION

*Draw the best possible Lewis structures for the free radicals important in atmospheric chemistry: NO, $NO_2$, HOO, OH, $CH_3$.*

This Lewis structure has octets for all atoms, including boron. However, when we assign formal charges to this structure, we get a negative formal charge on B and a positive formal charge on F:

$$\overset{+1}{\ddot{\text{F}}:} \\ \| \\ ^0:\ddot{\text{F}}-\underset{-1}{\text{B}}-\ddot{\text{F}}:^0$$

The positive formal charge on fluorine—the most electronegative element in the periodic table—makes this an unfavorable structure. This leaves us with some questions. Do we complete the octet on B at the expense of giving fluorine a positive formal charge? Or do we leave B without an octet in order to avoid the positive formal charge on fluorine? The answers to these kinds of questions are not always clear because we are pushing the limits of the Lewis model. In the case of boron, we usually accept the incomplete octet as the better Lewis structure. However, doing so does not rule out the possibility that the double-bonded Lewis structure might be a minor contributing resonance structure. The ultimate answers to these kinds of questions must be determined from experiments. Experimental measurements of the B—F bond length in $BF_3$ suggest that the bond may be slightly shorter than expected for a single B—F bond, indicating that it may indeed have a small amount of double-bond character.

$BF_3$ can complete its octet in another way—via a chemical reaction. The Lewis model predicts that $BF_3$ might react in ways that would complete its octet, and indeed it does. For example, $BF_3$ reacts with $NH_3$ as follows:

When nitrogen bonds to boron, the nitrogen atom provides both of the electrons. This kind of bond is called a *coordinate covalent bond*, which we discuss in Chapter 25.

The product has complete octets for all atoms in the structure.

## Expanded Octets

Elements in the third row of the periodic table and beyond often exhibit *expanded octets* of up to 12 (and occasionally 14) electrons. Consider the Lewis structures of arsenic pentafluoride and sulfur hexafluoride:

In $AsF_5$ arsenic has an expanded octet of 10 electrons, and in $SF_6$ sulfur has an expanded octet of 12 electrons. Both of these compounds exist and are stable. In third-period elements and beyond, 10- and 12-electron expanded octets are common because the *d* orbitals in these elements are energetically accessible (they are not much higher in energy than the orbitals occupied by the valence electrons) and can accommodate the extra electrons (see Section 8.3). Expanded octets *never* occur in second-period elements.

In some Lewis structures, we must decide whether or not to expand an octet in order to lower formal charge. For example, consider the Lewis structure of $H_2SO_4$:

Notice that two of the oxygen atoms have a −1 formal charge and that sulfur has a +2 formal charge. While this amount of formal charge is acceptable, especially since the negative formal charge resides on the more electronegative atom, it is possible to eliminate the formal charge by expanding the octet on sulfur:

$$
\begin{array}{c}
H \\
| \\
:\ddot{O}: \\
| \\
:\ddot{O}\!=\!\underset{0}{S}\!-\!\ddot{O}\!-\!H \\
\| \\
:\underset{0}{\ddot{O}}:
\end{array}
$$

Which of these two Lewis structures for $H_2SO_4$ is better? Again, the answer is not straight-forward. Experiments show that the sulfur–oxygen bond lengths in the two sulfur–oxygen bonds without the hydrogen atoms are shorter than expected for sulfur–oxygen single bonds, indicating that the double-bonded Lewis structure plays an important role in describing the bonding in $H_2SO_4$. In general, we expand octets in third-row (or beyond) elements in order to lower formal charge. However, we should *never* expand the octets of second-row elements. Second-row elements do not have energetically accessible *d* orbitals and therefore never exhibit expanded octets.

| **Example 9.10** | **Writing Lewis Structures for Compounds Having Expanded Octets** |  |
|---|---|---|

Write the Lewis structure for $XeF_2$.

**SOLUTION**

| | |
|---|---|
| Begin by writing the skeletal structure. Since xenon is the less electronegative atom, put it in the central position. | F Xe F |
| Calculate the total number of electrons for the Lewis structure by summing the number of valence electrons for each atom. | Total number of electrons for Lewis structure = (number of valence $e^-$ in Xe) + 2(number of valence $e^-$ in F) <br> = 8 + 2(7) <br> = 22 |
| Place two bonding electrons between the atoms of each pair of atoms. | F:Xe:F <br> (4 of 22 electrons used) |
| Distribute the remaining electrons to give octets to as many atoms as possible, beginning with terminal atoms and finishing with the central atom. Arrange additional electrons around the central atom, giving it an expanded octet of up to 12 electrons. | :$\ddot{F}$: Xe :$\ddot{F}$: <br> (16 of 22 electrons used) <br> :$\ddot{F}$ :$\ddot{X}e$:$\ddot{F}$: or :$\ddot{F}$ — $\ddot{X}e$ — $\ddot{F}$: <br> (22 of 22 electrons used) |

**FOR PRACTICE 9.10**   Write the Lewis structure for $XeF_4$.

**FOR MORE PRACTICE 9.10**   Write the Lewis structure for $H_3PO_4$. If necessary, expand the octet on any appropriate atoms to lower formal charge.

**Expanded Octets**   Which molecule could have an expanded octet?
   **(a)** $H_2CO_3$      **(b)** $H_3PO_4$      **(c)** $HNO_2$

CONCEPTUAL
CONNECTION **9.7**

## 9.10   **Bond Energies and Bond Lengths**

In Chapter 6, we discussed how to calculate the standard enthalpy change for a chemical reaction ($\Delta H^\circ_{rxn}$) from tabulated standard enthalpies of formation. However, sometimes we may not easily find standard enthalpies of formation for all of the reactants and products of a reaction. In such cases, we can use individual *bond energies* to estimate enthalpy changes of reaction. In this section, we examine the concept of bond energy and how we can use bond energies to calculate enthalpy changes of reaction. We also look at average bond lengths for a number of commonly encountered bonds.

## Bond Energy

Bond energy is also called bond enthalpy or bond dissociation energy.

The **bond energy** of a chemical bond is the energy required to break 1 mole of the bonds in the gas phase. For example, the bond energy of the $Cl-Cl$ bond in $Cl_2$ is 243 kJ/mol.

$$Cl_2(g) \longrightarrow 2\ Cl(g) \qquad \Delta H = 243\ \text{kJ}$$

The bond energy of HCl is 431 kJ/mol.

$$HCl(g) \longrightarrow H(g) + Cl(g) \qquad \Delta H = 431\ \text{kJ}$$

Bond energies are always positive because it always takes energy to break a bond. We say that the HCl bond is *stronger* than the $Cl_2$ bond because it requires more energy to break it. In general, compounds with stronger bonds tend to be more chemically stable, and therefore less chemically reactive, than compounds with weaker bonds. The triple bond in $N_2$ has a bond energy of 946 kJ/mol.

$$N_2(g) \longrightarrow N(g) + N(g) \qquad \Delta H = 946\ \text{kJ}$$

It is a very strong and stable bond, which explains nitrogen's relative inertness.

The bond energy of a particular bond in a polyatomic molecule is a little more difficult to determine because a particular type of bond can have different bond energies in different molecules. For example, consider the $C-H$ bond. In $CH_4$, the energy required to break one $C-H$ bond is 438 kJ/mol.

$$H_3C-H(g) \longrightarrow H_3C(g) + H(g) \qquad \Delta H = 438\ \text{kJ}$$

However, the energy required to break a $C-H$ bond in other molecules varies slightly, as shown here:

$$F_3C-H(g) \longrightarrow F_3C(g) + H(g) \qquad \Delta H = 446\ \text{kJ}$$

$$Br_3C-H(g) \longrightarrow Br_3C(g) + H(g) \qquad \Delta H = 402\ \text{kJ}$$

$$Cl_3C-H(g) \longrightarrow Cl_3C(g) + H(g) \qquad \Delta H = 401\ \text{kJ}$$

We can calculate an *average bond energy* for a chemical bond, which is an average of the bond energies for that bond in a large number of compounds. For the four compounds just discussed, we calculate an average $C-H$ bond energy of 422 kJ/mol. Table 9.3 lists average bond energies for a number of common chemical bonds averaged over a large number of compounds. Notice that the $C-H$ bond energy listed is 414 kJ/mol, which is not too different from the value we calculated from our limited number of compounds. Notice also that bond energies depend, not only on the kind of atoms involved in the bond, but also on the type of bond: single, double, or triple. In general, for a given pair of atoms, triple bonds are stronger than double bonds, which are, in turn, stronger than single bonds. For example, consider the bond energies of carbon–carbon triple, double, and single bonds listed in Table 9.3.

### TABLE 9.3 Average Bond Energies

| Bond | Bond Energy (kJ/mol) | Bond | Bond Energy (kJ/mol) | Bond | Bond Energy (kJ/mol) |
|------|----------------------|------|----------------------|------|----------------------|
| H—H | 436 | N—N | 163 | Br—F | 237 |
| H—C | 414 | N=N | 418 | Br—Cl | 218 |
| H—N | 389 | N≡N | 946 | Br—Br | 193 |
| H—O | 464 | N—O | 222 | I—Cl | 208 |
| H—S | 368 | N=O | 590 | I—Br | 175 |
| H—F | 565 | N—F | 272 | I—I | 151 |
| H—Cl | 431 | N—Cl | 200 | Si—H | 323 |
| H—Br | 364 | N—Br | 243 | Si—Si | 226 |
| H—I | 297 | N—I | 159 | Si—C | 301 |
| C—C | 347 | O—O | 142 | S—O | 265 |
| C=C | 611 | O=O | 498 | Si=O | 368 |
| C≡C | 837 | O—F | 190 | S=O | 523 |
| C—N | 305 | O—Cl | 203 | Si—Cl | 464 |
| C=N | 615 | O—I | 234 | S=S | 418 |
| C≡N | 891 | F—F | 159 | S—F | 327 |
| C—O | 360 | Cl—F | 253 | S—Cl | 253 |
| C=O | 736* | Cl—Cl | 243 | S—Br | 218 |
| C≡O | 1072 | | | S—S | 266 |
| C—Cl | 339 | | | | |

*799 in $CO_2$.

# Using Average Bond Energies to Estimate Enthalpy Changes for Reactions

We can use average bond energies to *estimate* the enthalpy change of a reaction. For example, consider the reaction between methane and chlorine:

$$H_3C-H(g) + Cl-Cl(g) \longrightarrow H_3C-Cl(g) + H-Cl(g)$$

We can imagine this reaction occurring by the breaking of a C—H bond and a Cl—Cl bond and the forming of a C—Cl bond and an H—Cl bond. We know that when bonds break, the process is endothermic (positive bond energy) and when bonds form, the process is exothermic (negative bond energy). So we can calculate the overall enthalpy change as a sum of the enthalpy changes associated with breaking the required bonds in the reactants and forming the required bonds in the products, as shown in Figure 9.12▼.

$$H_3C-H(g) + Cl-Cl(g) \longrightarrow H_3C-Cl(g) + H-Cl(g)$$

| Bonds Broken | | Bonds Formed | |
|---|---|---|---|
| C—H break | +414 kJ | C—Cl form | −339 kJ |
| Cl—Cl break | +243 kJ | H—Cl form | −431 kJ |
| *Sum ($\Sigma$)$\Delta H$'s bonds broken:* +657 kJ | | *Sum ($\Sigma$)$\Delta H$'s bonds formed:* −770 kJ | |

$$\begin{aligned}
\Delta H_{rxn} &= \Sigma(\Delta H\text{'s bonds broken}) + \Sigma(\Delta H\text{'s bonds formed}) \\
&= +657 \text{ kJ} - 770 \text{ kJ} \\
&= -113 \text{ kJ}
\end{aligned}$$

We find that $\Delta H_{rxn} = -113$ kJ. Calculating $\Delta H_{rxn}^\circ$ from tabulated enthalpies of formation—the method we learned in Chapter 6—gives $\Delta H_{rxn}^\circ = -101$ kJ, fairly close to the value we obtained from average bond energies.

In general, we can calculate $\Delta H_{rxn}$ from average bond energies by summing the changes in enthalpy for all of the bonds that are broken and adding the sum of the enthalpy changes for all of the bonds that are formed. Remember that $\Delta H$ is positive for breaking bonds and negative for forming them:

$$\Delta H_{rxn} = \underbrace{\Sigma(\Delta H\text{'s bonds broken})}_{\text{Positive}} + \underbrace{\Sigma(\Delta H\text{'s bonds formed})}_{\text{Negative}}$$

As we can see from the equation:

- A reaction is *exothermic* when weak bonds break and strong bonds form.

- A reaction is *endothermic* when strong bonds break and weak bonds form.

Scientists sometimes say that "energy is stored in chemical bonds or in a chemical compound," which may make it sound as if breaking the bonds in the compound releases energy. For example, in biology we often hear that energy is stored in glucose or in ATP. However, *breaking a chemical bond always requires energy.* When we say that energy is stored in a compound, or that a compound is energy rich, we mean that the compound can undergo a reaction in which weak bonds break and strong bonds form, thereby releasing energy in the overall process. However, *it is always the forming of chemical bonds that releases energy.*

▼ **FIGURE 9.12 Estimating $\Delta H_{rxn}$ from Bond Energies** We can approximate the enthalpy change of a reaction by summing up the enthalpy changes involved in breaking old bonds and forming new ones.

## Estimating the Enthalpy Change of a Reaction from Bond Energies

CONCEPTUAL
**CONNECTION 9.8**

**Bond Energies and $\Delta H_{rxn}$** The reaction between hydrogen and oxygen to form water is highly exothermic. Which statement is true of the energies of the bonds that break and form during the reaction?

(a) The energy needed to break the required bonds is greater than the energy released when the new bonds form.

(b) The energy needed to break the required bonds is less than the energy released when the new bonds form.

(c) The energy needed to break the required bonds is about the same as the energy released when the new bonds form.

---

**Example 9.11** | **Calculating $\Delta H_{rxn}$ from Bond Energies** ▶

Hydrogen gas, a potential fuel, can be made by the reaction of methane gas and steam.

$$CH_4(g) + 2 H_2O(g) \longrightarrow 4 H_2(g) + CO_2(g)$$

Use bond energies to calculate $\Delta H_{rxn}$ for this reaction.

**SOLUTION**

| | |
|---|---|
| Begin by rewriting the reaction using the Lewis structures of the molecules involved. | H—C—H + 2 H—Ö—H ⟶ 4 H—H + Ö=C=Ö |
| Determine which bonds are broken in the reaction and sum the bond energies of these. | H—C—H + 2 H—Ö—H<br><br>$\Sigma(\Delta H$'s bonds broken)<br>$= 4 (C—H) + 4 (O—H)$<br>$= 4 (414 \text{ kJ}) + 4 (464 \text{ kJ})$<br>$= 3512 \text{ kJ}$ |
| Determine which bonds are formed in the reaction and add the negatives of their bond energies. | 4 H—H + Ö=C=Ö<br>$\Sigma(\Delta H$'s bonds formed)<br>$= -4 (H—H) - 2 (C=O)$<br>$= -4 (436 \text{ kJ}) - 2 (799 \text{ kJ})$<br>$= -3342 \text{ kJ}$ |
| Find $\Delta H_{rxn}$ by summing the results of the previous two steps. | $\Delta H_{rxn} = \Sigma(\Delta H$'s bonds broken) + $\Sigma(\Delta H$'s bonds formed)<br>$= 3512 \text{ kJ} - 3342 \text{ kJ}$<br>$= 1.70 \times 10^2 \text{ kJ}$ |

**FOR PRACTICE 9.11** Another potential future fuel is methanol ($CH_3OH$). Write a balanced equation for the combustion of gaseous methanol and use bond energies to calculate the enthalpy of combustion of methanol in kJ/mol.

**FOR MORE PRACTICE 9.11** Use bond energies to calculate $\Delta H_{rxn}$ for this reaction: $N_2(g) + 3 H_2(g) \longrightarrow 2 NH_3(g)$.

---

## Bond Lengths

Just as we can tabulate average bond energies, which represent the average energy of a bond between two particular atoms in a large number of compounds, we can tabulate average bond lengths (Table 9.4). The average **bond length** represents the average

**TABLE 9.4 Average Bond Lengths**

| Bond | Bond Length (pm) | Bond | Bond Length (pm) | Bond | Bond Length (pm) |
|------|------------------|------|------------------|------|------------------|
| H—H  | 74  | C—C   | 154 | N—N   | 145 |
| H—C  | 110 | C=C   | 134 | N=N   | 123 |
| H—N  | 100 | C≡C   | 120 | N≡N   | 110 |
| H—O  | 97  | C—N   | 147 | N—O   | 136 |
| H—S  | 132 | C=N   | 128 | N=O   | 120 |
| H—F  | 92  | C≡N   | 116 | O—O   | 145 |
| H—Cl | 127 | C—O   | 143 | O=O   | 121 |
| H—Br | 141 | C=O   | 120 | F—F   | 143 |
| H—I  | 161 | C—Cl  | 178 | Cl—Cl | 199 |
|      |     |       |     | Br—Br | 228 |
|      |     |       |     | I—I   | 266 |

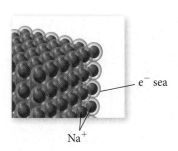

Bond Lengths

$F_2$

143 pm

$Cl_2$

199 pm

$Br_2$

228 pm

$I_2$

266 pm

▲ Bond lengths in the diatomic halogen molecules.

length of a bond between two particular atoms in a large number of compounds. Like bond energies, bond lengths depend not only on the kind of atoms involved in the bond, but also on the type of bond: single, double, or triple. In general, for a particular pair of atoms, triple bonds are shorter than double bonds, which are in turn shorter than single bonds. For example, consider the bond lengths (shown here with bond strengths) of carbon–carbon triple, double, and single bonds:

| Bond | Bond Length (pm) | Bond Strength (kJ/mol) |
|------|------------------|------------------------|
| C≡C  | 120 pm | 837 kJ/mol |
| C=C  | 134 pm | 611 kJ/mol |
| C—C  | 154 pm | 347 kJ/mol |

Notice that, as the bond gets longer, it also becomes weaker. This relationship between the length of a bond and the strength of a bond does not necessarily hold true for all bonds. Consider the following series of nitrogen–halogen single bonds:

| Bond | Bond Length (pm) | Bond Strength (kJ/mol) |
|------|------------------|------------------------|
| N—F  | 139 | 272 |
| N—Cl | 191 | 200 |
| N—Br | 214 | 243 |
| N—I  | 222 | 159 |

Although the bonds generally get weaker as they get longer, the trend is not a smooth one.

## 9.11 Bonding in Metals: The Electron Sea Model

So far, we have explored simple models for bonding between a metal and a nonmetal (ionic bonding) and for bonding between two nonmetals (covalent bonding). We have seen how these models account for and predict the properties of ionic and molecular compounds. The last type of bonding that we examine in this chapter is metallic bonding, which occurs between metals.

As we have already discussed, metals have a tendency to lose electrons, which means that they have relatively low ionization energies. When metal atoms bond together to form a solid lattice, each metal atom donates one or more electrons to an *electron sea*. For example, we can think of sodium metal as an array of positively charged $Na^+$ ions immersed in a sea of negatively charged electrons ($e^-$), as shown in Figure 9.13►.

$e^-$ sea

$Na^+$

▲ **FIGURE 9.13 The Electron Sea Model for Sodium** In this model of metallic bonding, $Na^+$ ions are immersed in a "sea" of electrons.

# Chemistry IN THE Environment | The Lewis Structure of Ozone

Ozone is a form of oxygen in which three oxygen atoms bond together. Its Lewis structure consists of the following resonance structures:

$$:\ddot{O}=\ddot{O}-\ddot{O}: \longleftrightarrow :\ddot{O}-\ddot{O}=\ddot{O}:$$

Compare the Lewis structure of ozone to the Lewis structure of $O_2$:

$$:\ddot{O}=\ddot{O}:$$

Since double bonds are stronger and shorter than single bonds, $O_2$ must have a stronger bond because it is a double bond. $O_3$, on the other hand, has bonds that are intermediate between single and double, which are weaker bonds. The effects of this are significant. As we discussed in Section 6.10, $O_3$ absorbs harmful ultraviolet light entering Earth's atmosphere. Ozone is ideally suited to do this because photons at wavelengths of 280–320 nm (the most harmful components of sunlight) are just strong enough to break the bonds in the $O_3$ molecule:

$$:\ddot{O}=\ddot{O}-\ddot{O}: + \text{ UV light} \longrightarrow :\ddot{O}=\ddot{O}: + \cdot\ddot{O}:$$

In this process, the photon is absorbed. $O_2$ and $O$ then recombine to re-form $O_3$, which can in turn absorb more UV light. The same wavelengths of UV light, however, do not have sufficient energy to break the stronger double bond of $O_2$. No other molecules in our atmosphere can do the job that ozone does. Consequently, we should continue, and even strengthen, the ban on ozone-depleting compounds.

### QUESTION

*Calculate the average bond energy of one $O_3$ bond. What wavelength of light has just the right amount of energy to break this bond?*

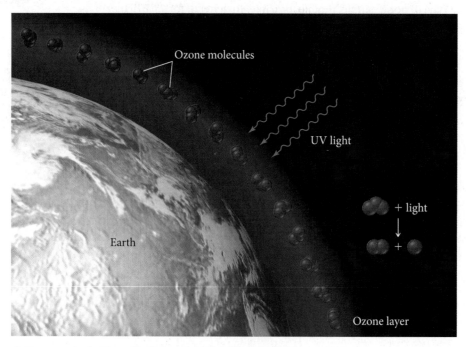

▲ Ozone protects life on Earth from harmful ultraviolet light.

▲ Copper can easily be drawn into fine strands like those used in household electrical cords.

Each sodium atom donates its one valence electron to the "sea" and becomes a sodium ion. The sodium cations are held together by their attraction to the sea of electrons.

Although the electron sea model is simple, it accounts for many of the properties of metals. For example, metals conduct electricity because—in contrast to ionic solids, in which electrons are localized on an ion—the electrons in a metal are free to move. The movement or flow of electrons in response to an electric potential (or voltage) is an electric current. Metals are also excellent conductors of heat, again because of the highly mobile electrons, which help to disperse thermal energy throughout the metal.

The electron sea model also accounts for the *malleability* of metals (their capacity to be pounded into sheets) and the *ductility* of metals (their capacity to be drawn into wires). Since there are no localized or specific "bonds" in a metal, we can deform it relatively easily by forcing the metal ions to slide past one another. The electron sea easily accommodates deformations by flowing into the new shape.

# Self-Assessment Quiz

**Q1.** Which pair of elements is most likely to form an ionic bond?

a) nitrogen and oxygen    b) carbon and hydrogen
c) sulfur and oxygen    d) calcium and oxygen

**Q2.** Which set of elements is arranged in order of increasing electronegativity?

a) $O < S < As < Ge$    b) $Ge < As < S < O$
c) $S < O < As < Ge$    d) $As < O < Ge < S$

**Q3.** Which is the correct Lewis structure for magnesium bromide?

a) $2\,Mg^{2+}\left[:\ddot{B}r:\right]^{-}$     b) $Mg:\ddot{B}r:$

c) $Mg^{2+}2\left[:\ddot{B}r:\right]^{-}$     d) $:\dot{M}g:\ddot{B}r:$

**Q4.** Which compound is likely to have an incomplete octet?

a) $NH_3$   b) $SO_3$   c) $N_2O$   d) $BH_3$

**Q5.** Which compound has the lattice energy with the greatest magnitude?

a) MgS   b) CaS   c) SrS   d) BaS

**Q6.** Which set of compounds is arranged in order of increasing magnitude of lattice energy?

a) $CsI < NaCl < MgS$    b) $NaCl < CsI < MgS$
c) $MgS < NaCl < CsI$    d) $CsI < MgS < NaCl$

**Q7.** Which pair of atoms forms the most polar bond?

a) N and O   b) C and O   c) C and F   d) N and F

**Q8.** Which pair of atoms forms a nonpolar covalent bond?

a) C and S   b) C and O   c) B and O   d) Na and Cl

**Q9.** Which is the correct Lewis structure for nitrogen trifluoride?

a) $:\ddot{F}=\ddot{N}-\ddot{F}:$     b) $:\ddot{F}-N-\ddot{F}:$
   $\quad\quad\;\;|$                 $|$
   $\quad\quad\;\;:\ddot{F}:$              $:\ddot{F}:$

c) $:\ddot{F}-N-\ddot{F}:$     d) $:\ddot{F}-N-\ddot{F}-\ddot{F}:$
   $\quad\quad\;\;|$
   $\quad\quad\;\;:\ddot{F}:$

**Q10.** Which is the correct Lewis structure for $CO_3{}^{2-}$?

**Q11.** Determine the formal charge of nitrogen in the structure shown here:

a) +1
b) +2
c) −1
d) −2

**Q12.** A Lewis structure for the acetate ion is shown here:

Which structure is the best resonance structure for the acetate ion?

**Q13.** Use formal charge to choose the best Lewis structure for $CH_3SOCH_3$.

*Continued—*

**Q14.** Use bond energies from Table 9.3 to determine $\Delta H_{rxn}$ for the reaction between ethanol and hydrogen chloride.

$$CH_3CH_2OH(g) + HCl(g) \longrightarrow CH_3CH_2Cl(g) + H_2O(g)$$

a)  −1549 kJ        b) 1549 kJ
c)  −12 kJ          d) 12 kJ

**Q15.** Consider the halogenation of ethene, where X is a generic halogen:

$$H_2C{=}CH_2(g) + X_2(g) \longrightarrow H_2XC{-}CH_2X(g)$$

Use bond energies to determine which halogen produces the most exothermic halogenation reaction with ethene. The C—F, C—Br, and C—I bond energies are 552 kJ/mol, 280 kJ/mol, and 209 kJ/mol, respectively. Look up all other necessary bond energies in Table 9.3.

a) fluorine        b) chlorine
c) bromine         d) iodine

**Answers: 1.** (d) **2.** (b) **3.** (c) **4.** (d) **5.** (a) **6.** (a) **7.** (c) **8.** (a) **9.** (c) **10.** (b) **11.** (a) **12.** (d) **13.** (b) **14.** (c) **15.** (a)

# Chapter 9 IN REVIEW

## Key Terms

**Section 9.1**
Lewis model (384)
Lewis electron-dot structures
    (Lewis structures) (384)

**Section 9.2**
ionic bond (385)
covalent bond (385)
metallic bonding (386)

**Section 9.3**
Lewis symbol (386)
octet (386)
duet (386)
chemical bond (387)
octet rule (387)

**Section 9.4**
lattice energy (388)
Born–Haber cycle (388)

**Section 9.5**
bonding pair (394)
lone pair (394)
nonbonding electrons (394)
double bond (394)
triple bond (395)

**Section 9.6**
polar covalent bond (396)
electronegativity (397)
dipole moment ($\mu$) (398)
percent ionic character (399)

**Section 9.8**
resonance structures (403)
resonance hybrid (403)
formal charge (404)

**Section 9.9**
free radical (408)

**Section 9.10**
bond energy (412)
bond length (414)

## Key Concepts

### Bonding Models and AIDS Drugs (9.1)

▶ Theories that predict how and why atoms bond together are central to chemistry because they explain compound stability and molecule shape.

▶ Bonding theories have been useful in combating HIV because they help in the design of molecules that bind to the active site of a protein crucial for the development of AIDS.

### Types of Chemical Bonds (9.2)

▶ We divide chemical bonds into three general types: ionic bonds, which occur between a metal and a nonmetal; covalent bonds, which occur between two nonmetals; and metallic bonds, which occur within metals.

▶ In an ionic bond, an electron transfers from the metal to the nonmetal and the resultant ions attract each other by coulombic forces.

▶ In a covalent bond, nonmetals share electrons that interact with the nuclei of both atoms via coulombic forces, holding the atoms together.

▶ In a metallic bond, the atoms form a lattice in which each metal loses electrons to an "electron sea." The attraction of the positively charged metal ions to the electron sea holds the metal together.

### The Lewis Model and Electron Dots (9.3)

▶ In the Lewis model, chemical bonds form when atoms transfer (ionic bonding) or share (covalent bonding) valence electrons to attain noble gas electron configurations.

▶ The Lewis model represents valence electrons as dots surrounding the symbol for an element. When two or more elements bond together, the dots are transferred or shared so that every atom has eight dots, an octet (or two dots, a duet, in the case of hydrogen).

### Ionic Lewis Structures and Lattice Energy (9.4)

▶ In an ionic Lewis structure involving main-group metals, the metal transfers its valence electrons (dots) to the nonmetal.

▶ The formation of most ionic compounds is exothermic because of lattice energy, the energy released when metal cations and nonmetal anions coalesce to form the solid. The smaller the radius of the ions and the greater their charge, the more exothermic the lattice energy.

## Covalent Lewis Structures, Electronegativity, and Polarity (9.5, 9.6, 9.7)

▶ In a covalent Lewis structure, neighboring atoms share valence electrons to attain octets (or duets).

▶ A single shared electron pair constitutes a single bond, while two or three shared pairs constitute double or triple bonds, respectively.

▶ The shared electrons in a covalent bond are not always *equally* shared; when two dissimilar nonmetals form a covalent bond, the electron density is greater on the more electronegative element. The result is a polar bond, in which one element has a partial positive charge and the other a partial negative charge.

▶ Electronegativity—the ability of an atom to attract electrons to itself in chemical bonding—increases as we move to the right across a period in the periodic table and decreases as we move down a column.

▶ Elements with very dissimilar electronegativities form ionic bonds, those with very similar electronegativities form nonpolar covalent bonds, and those with intermediate electronegativity differences form polar covalent bonds.

## Resonance and Formal Charge (9.8)

▶ We can best represent some molecules not by a single Lewis structure, but by two or more resonance structures. The actual structure of these molecules is a resonance hybrid: a combination or average of the contributing structures.

▶ The formal charge of an atom in a Lewis structure is the charge the atom would have if all bonding electrons were shared equally between bonding atoms.

▶ In general, the best Lewis structures have the fewest atoms with formal charge, and any negative formal charges are on the most electronegative atom.

## Exceptions to the Octet Rule (9.9)

▶ Although the octet rule normally applies when we draw Lewis structures, some exceptions occur.

▶ These exceptions include odd-electron species, which necessarily have Lewis structures with only seven electrons around an atom. Such molecules, called free radicals, tend to be unstable and chemically reactive.

▶ Other exceptions to the octet rule include molecules with incomplete octets—usually totaling six electrons (especially important in compounds containing boron)—and molecules with expanded octets—usually 10 or 12 electrons (which can occur in compounds containing elements from the third row of the periodic table and below). Expanded octets never occur in second-period elements.

## Bond Energies and Bond Lengths (9.10)

▶ The bond energy of a chemical bond is the energy required to break one mole of the bonds in the gas phase.

▶ Average bond energies for a number of different bonds are tabulated, and we can use them to calculate enthalpies of reaction.

▶ Average bond lengths are also tabulated.

▶ In general, triple bonds are shorter and stronger than double bonds, which are in turn shorter and stronger than single bonds.

## Bonding in Metals (9.11)

▶ When metal atoms bond together to form a solid, each metal atom donates one or more electrons to an *electron sea*. The metal cations are held together by their attraction to the sea of electrons.

▶ The *electron sea* model accounts for the electrical conductivity, thermal conductivity, malleability, and ductility of metals.

# Key Equations and Relationships

Dipole Moment ($\mu$): Separation of Two Particles of Equal but Opposite Charge of Magnitude $q$ by a Distance $r$ (9.6)

$$\mu = qr$$

Percent Ionic Character (9.6)

Percent ionic character =

$$\frac{\text{measured dipole moment of bond}}{\text{dipole moment if electron were completely transferred}} \times 100\%$$

Formal Charge (9.8)

Formal charge = number of valence electrons −

(number of nonbonding electrons + $\frac{1}{2}$ number of shared electrons)

Enthalpy Change of a Reaction ($\Delta H_{rxn}$): Relationship of Bond Energies (9.10)

$$\Delta H_{rxn} = \Sigma(\Delta H\text{'s bonds broken}) + \Sigma(\Delta H\text{'s bonds formed})$$

# Key Learning Outcomes

| CHAPTER OBJECTIVES | ASSESSMENT |
|---|---|
| Predicting Chemical Formulas of an Ionic Compound (9.4) | Example 9.1   For Practice 9.1   Exercises 41–42 |
| Predicting Relative Lattice Energies (9.4) | Example 9.2   For Practice 9.2   For More Practice 9.2   Exercise 46 |
| Classifying Bonds: Pure Covalent, Polar Covalent, or Ionic (9.6) | Example 9.3   For Practice 9.3   Exercises 55–56 |
| Writing Lewis Structures for Covalent Compounds (9.7) | Examples 9.4–9.5   For Practice 9.4–9.5   Exercises 51–52 |

| Writing Lewis Structures for Polyatomic Ions (**9.7**) | Example 9.6  For Practice 9.6  Exercises 59–64 |
|---|---|
| Writing Resonance Lewis Structures (**9.8**) | Example 9.7  For Practice 9.7  Exercises 63–64 |
| Assigning Formal Charges to Assess Competing Resonance Structures (**9.8**) | Example 9.8  For Practice 9.8  For More Practice 9.8  Exercises 65–66 |
| Drawing Resonance Structures and Assigning Formal Charge for Organic Compounds (**9.8**) | Example 9.9  For Practice 9.9  Exercises 69–72 |
| Writing Lewis Structures for Compounds Having Expanded Octets (**9.9**) | Example 9.10  For Practice 9.10  For More Practice 9.10  Exercises 77–78 |
| Calculating $\Delta H_{rxn}$ from Bond Energies (**9.10**) | Example 9.11  For Practice 9.11  For More Practice 9.11  Exercises 81–83 |

# EXERCISES
## Review Questions

**1.** Why are bonding theories important? Provide some examples of what bonding theories can predict.

**2.** Why do chemical bonds form? What basic forces are involved in bonding?

**3.** What are the three basic types of chemical bonds? What happens to electrons in the bonding atoms in each type?

**4.** How do you determine how many dots to put around the Lewis symbol of an element?

**5.** Describe the octet rule in the Lewis model.

**6.** According to the Lewis model, what is a chemical bond?

**7.** How do you draw an ionic Lewis structure?

**8.** How can Lewis structures be used to determine the formula of ionic compounds? Give an example.

**9.** What is lattice energy?

**10.** Why is the formation of solid sodium chloride from solid sodium and gaseous chlorine exothermic, even though it takes more energy to form the $Na^+$ ion than the amount of energy released upon formation of $Cl^-$ ?

**11.** What is the Born–Haber cycle? List each step in the cycle and show how the cycle is used to calculate lattice energy.

**12.** How does lattice energy relate to ionic radii? To ion charge?

**13.** How does the ionic bonding model explain the relatively high melting points of ionic compounds?

**14.** How does the ionic bonding model explain the nonconductivity of ionic solids, and at the same time the conductivity of ionic solutions?

**15.** In a covalent Lewis structure, what is the difference between lone pair and bonding pair electrons?

**16.** In what ways are double and triple covalent bonds different from single covalent bonds?

**17.** How does the Lewis model for covalent bonding account for why certain combinations of atoms are stable while others are not?

**18.** How does the Lewis model for covalent bonding account for the relatively low melting and boiling points of molecular compounds (compared to ionic compounds)?

**19.** What is electronegativity? What are the periodic trends in electronegativity?

**20.** How do a pure covalent bond, a polar covalent bond, and an ionic bond differ?

**21.** Explain percent ionic character of a bond. Do any bonds have 100% ionic character?

**22.** What is a dipole moment?

**23.** What is the magnitude of the dipole moment formed by separating a proton and an electron by 100 pm? 200 pm?

**24.** What is the basic procedure for writing a covalent Lewis structure?

**25.** How do you determine the number of electrons in the Lewis structure of a molecule? A polyatomic ion?

**26.** What are resonance structures? What is a resonance hybrid?

**27.** Do resonance structures always contribute equally to the overall structure of a molecule? Explain.

**28.** What is formal charge? How is formal charge calculated? How is it helpful?

**29.** Why does the octet rule have exceptions? List the three major categories of exceptions and an example of each.

**30.** Which elements can have expanded octets? Which elements should never have expanded octets?

**31.** What is bond energy? How can you use average bond energies to calculate enthalpies of reaction?

**32.** Explain the difference between endothermic reactions and exothermic reactions with respect to the bond energies of the bonds broken and formed.

**33.** What is the electron sea model for bonding in metals?

**34.** How does the electron sea model explain the conductivity of metals? The malleability and ductility of metals?

# Problems by Topic

## Valence Electrons and Dot Structures

35. Write the electron configuration for N. Then write the Lewis symbol for N and show which electrons from the electron configuration are included in the Lewis symbol.

36. Write the electron configuration for Ne. Then write the Lewis symbol for Ne and show which electrons from the electron configuration are included in the Lewis symbol.

37. Write the Lewis symbol for each atom or ion.
    **a.** Al    **b.** $Na^+$    **c.** Cl    **d.** $Cl^-$

38. Write the Lewis symbol for each atom or ion.
    **a.** $S^{2-}$    **b.** Mg    **c.** $Mg^{2+}$    **d.** P

## Ionic Lewis Symbols and Lattice Energy

39. Write the Lewis symbols for the ions in each ionic compound.
    **a.** NaF    **b.** CaO    **c.** $SrBr_2$    **d.** $K_2O$

40. Write the Lewis symbols for the ions in each ionic compound.
    **a.** SrO    **b.** $Li_2S$    **c.** $CaI_2$    **d.** RbF

41. Use Lewis symbols to determine the formula for the compound that forms between each pair of elements.
    **a.** Sr and Se    **b.** Ba and Cl    **c.** Na and S    **d.** Al and O

42. Use Lewis symbols to determine the formula for the compound that forms between each pair of elements:
    **a.** Ca and N    **b.** Mg and I    **c.** Ca and S    **d.** Cs and F

43. Explain the trend in the lattice energies of the alkaline earth metal oxides.

| Metal Oxide | Lattice Energy (kJ/mol) |
|---|---|
| MgO | −3795 |
| CaO | −3414 |
| SrO | −3217 |
| BaO | −3029 |

44. Rubidium iodide has a lattice energy of −617 kJ/mol, while potassium bromide has a lattice energy of −671 kJ/mol. Why is the lattice energy of potassium bromide more exothermic than the lattice energy of rubidium iodide?

45. The lattice energy of CsF is −744 kJ/mol, whereas that of BaO is −3029 kJ/mol. Explain this large difference in lattice energy.

46. Arrange these compounds in order of increasing magnitude of lattice energy: KCl, SrO, RbBr, CaO.

47. Use the Born–Haber cycle and data from Appendix IIB, Chapter 8 and this chapter to calculate the lattice energy of KCl. ($\Delta H_{sub}$ for potassium is 89.0 kJ/mol.)

48. Use the Born–Haber cycle and data from Appendix IIB and Table 9.3 to calculate the lattice energy of CaO. ($\Delta H_{sub}$ for calcium is 178 kJ/mol; $IE_1$ and $IE_2$ for calcium are 590 kJ/mol and 1145 kJ/mol, respectively; $EA_1$ and $EA_2$ for O are −141 kJ/mol and 744 kJ/mol, respectively.)

## Simple Covalent Lewis Structures, Electronegativity, and Bond Polarity

49. Use covalent Lewis structures to explain why each element (or family of elements) occurs as diatomic molecules.
    **a.** hydrogen    **b.** the halogens    **c.** oxygen    **d.** nitrogen

50. Use covalent Lewis structures to explain why the compound that forms between nitrogen and hydrogen has the formula $NH_3$. Show why $NH_2$ and $NH_4$ are not stable.

51. Write the Lewis structure for each molecule.
    **a.** $PH_3$    **b.** $SCl_2$    **c.** HI    **d.** $CH_4$

52. Write the Lewis structure for each molecule.
    **a.** $NF_3$    **b.** HBr    **c.** $SBr_2$    **d.** $CCl_4$

53. Write the Lewis structure for each molecule.
    **a.** $SF_2$            **b.** $SiH_4$
    **c.** HCOOH (both O bonded to C)    **d.** $CH_3SH$ (C and S central)

54. Write the Lewis structure for each molecule.
    **a.** $CH_2O$    **b.** $C_2Cl_4$
    **c.** $CH_3NH_2$    **d.** $CFCl_3$ (C central)

55. Determine if a bond between each pair of atoms would be pure covalent, polar covalent, or ionic.
    **a.** Br and Br    **b.** C and Cl    **c.** C and S    **d.** Sr and O

56. Determine if a bond between each pair of atoms would be pure covalent, polar covalent, or ionic.
    **a.** C and N    **b.** N and S    **c.** K and F    **d.** N and N

57. Draw the Lewis structure for CO with an arrow representing the dipole moment. Refer to Figure 9.10 to estimate the percent ionic character of the CO bond.

58. Draw the Lewis structure for BrF with an arrow representing the dipole moment. Refer to Figure 9.10 to estimate the percent ionic character of the BrF bond.

## Covalent Lewis Structures, Resonance, and Formal Charge

59. Write the Lewis structure for each molecule or ion.
    **a.** $CI_4$    **b.** $N_2O$    **c.** $SiH_4$    **d.** $Cl_2CO$

60. Write the Lewis structure for each molecule or ion.
    **a.** $H_3COH$    **b.** $OH^-$    **c.** $BrO^-$    **d.** $O_2^{2-}$

61. Write the Lewis structure for each molecule or ion.
    **a.** $N_2H_2$    **b.** $N_2H_4$    **c.** $C_2H_2$    **d.** $C_2H_4$

62. Write the Lewis structure for each molecule or ion.
    **a.** $H_3COCH_3$    **b.** $CN^-$    **c.** $NO_2^-$    **d.** $ClO^-$

63. Write a Lewis structure that obeys the octet rule for each molecule or ion. Include resonance structures if necessary and assign formal charges to each atom.
    **a.** $SeO_2$    **b.** $CO_3^{2-}$    **c.** $ClO^-$    **d.** $NO_2^-$

64. Write a Lewis structure that obeys the octet rule for each ion. Include resonance structures if necessary and assign formal charges to each atom.
    **a.** $ClO_3^-$    **b.** $ClO_4^-$    **c.** $NO_3^-$    **d.** $NH_4^+$

65. Use formal charge to identify the better Lewis structure.

66. Use formal charges to identify the better Lewis structure.

67. How important is the resonance structure shown here to the overall structure of carbon dioxide? Explain.

**68.** In $N_2O$, nitrogen is the central atom and the oxygen atom is terminal. In $OF_2$, however, oxygen is the central atom. Use formal charges to explain why.

**69.** Draw the Lewis structure (including resonance structures) for the acetate ion ($CH_3COO^-$). For each resonance structure, assign formal charges to all atoms that have formal charge.

**70.** Draw the Lewis structure (including resonance structures) for methyl azide ($CH_3N_3$). For each resonance structure, assign formal charges to all atoms that have formal charge.

**71.** What are the formal charges of the atoms shown in red?

$$CH_3-N-\ddot{\underset{\displaystyle CH_3}{\overset{\displaystyle CH_3}{N}}}-\ddot{O}:$$

**72.** What are the formal charges of the atoms shown in red?

$$CH_3-\underset{\displaystyle \ddot{..}}{\overset{\displaystyle :\ddot{O}:}{S}}-CH_3$$

## Odd-Electron Species, Incomplete Octets, and Expanded Octets

**73.** Write the Lewis structure for each molecule (octet rule not followed).
  **a.** $BCl_3$   **b.** $NO_2$   **c.** $BH_3$

**74.** Write the Lewis structure for each molecule (octet rule not followed).
  **a.** $BBr_3$   **b.** $NO$   **c.** $ClO_2$

**75.** Write the Lewis structure for each ion. Include resonance structures if necessary and assign formal charges to all atoms. If necessary, expand the octet on the central atom to lower formal charge.
  **a.** $PO_4^{3-}$   **b.** $CN^-$   **c.** $SO_3^{2-}$   **d.** $ClO_2^-$

**76.** Write Lewis structures for each molecule or ion. Include resonance structures if necessary and assign formal charges to all atoms. If necessary, expand the octet on the central atom to lower formal charge.
  **a.** $SO_4^{2-}$   **b.** $HSO_4^-$   **c.** $SO_3$   **d.** $BrO_2^-$

**77.** Write Lewis structures for each molecule or ion. Use expanded octets as necessary.
  **a.** $PF_5$   **b.** $I_3^-$   **c.** $SF_4$   **d.** $GeF_4$

**78.** Write Lewis structures for each molecule or ion. Use expanded octets as necessary.
  **a.** $ClF_5$   **b.** $AsF_6^-$   **c.** $Cl_3PO$   **d.** $IF_5$

## Bond Energies and Bond Lengths

**79.** Order these compounds in order of increasing carbon–carbon bond strength and in order of decreasing carbon–carbon bond length: HCCH, $H_2CCH_2$, $H_3CCH_3$.

**80.** Which compound shown here has the stronger nitrogen–nitrogen bond? The shorter nitrogen–nitrogen bond?
$$H_2NNH_2, \ HNNH$$

**81.** Hydrogenation reactions are used to add hydrogen across double bonds in hydrocarbons and other organic compounds. Use average bond energies to calculate $\Delta H_{rxn}$ for the hydrogenation reaction.
$$H_2C{=}CH_2(g) + H_2(g) \longrightarrow H_3C{-}CH_3(g)$$

**82.** Ethanol is a possible fuel. Use average bond energies to calculate $\Delta H_{rxn}$ for the combustion of ethanol.
$$CH_3CH_2OH(g) + 3\,O_2(g) \longrightarrow 2\,CO_2(g) + 3\,H_2O(g)$$

**83.** Hydrogen, a potential future fuel, can be produced from carbon (from coal) and steam by this reaction:
$$C(s) + 2\,H_2O(g) \longrightarrow H_2(g) + CO_2(g)$$
Use average bond energies to calculate $\Delta H_{rxn}$ for the reaction.

**84.** In the *Chemistry and the Environment* box on free radicals in this chapter, we discussed the importance of the hydroxyl radical in reacting with and eliminating many atmospheric pollutants. However, the hydroxyl radical does not clean up everything. For example, chlorofluorocarbons—which destroy stratospheric ozone—are not attacked by the hydroxyl radical. Consider the hypothetical reaction by which the hydroxyl radical might react with a chlorofluorocarbon:
$$OH(g) + CF_2Cl_2(g) \longrightarrow HOF(g) + CFCl_2(g)$$
Use bond energies to explain why this reaction is improbable. (The $C{-}F$ bond energy is 552 kJ/mol.)

# Cumulative Problems

**85.** Write an appropriate Lewis structure for each compound. Make certain to distinguish between ionic and molecular compounds.
  **a.** $BI_3$   **b.** $K_2S$
  **c.** HCFO   **d.** $PBr_3$

**86.** Write an appropriate Lewis structure for each compound. Make certain to distinguish between ionic and molecular compounds.
  **a.** $Al_2O_3$   **b.** $ClF_5$
  **c.** $MgI_2$   **d.** $XeO_4$

**87.** Each compound contains both ionic and covalent bonds. Write ionic Lewis structures for each of them, including the covalent structure for the ion in brackets. Write resonance structures if necessary.
  **a.** $BaCO_3$   **b.** $Ca(OH)_2$
  **c.** $KNO_3$   **d.** LiIO

**88.** Each compound contains both ionic and covalent bonds. Write ionic Lewis structures for each of them, including the covalent structure for the ion in brackets. Write resonance structures if necessary.
  **a.** $RbIO_2$   **b.** $NH_4Cl$
  **c.** KOH   **d.** $Sr(CN)_2$

**89.** Carbon ring structures are common in organic chemistry. Draw a Lewis structure for each carbon ring structure, including any necessary resonance structures.
  **a.** $C_4H_8$   **b.** $C_4H_4$   **c.** $C_6H_{12}$   **d.** $C_6H_6$

**90.** Amino acids are the building blocks of proteins. The simplest amino acid is glycine ($H_2NCH_2COOH$). Draw a Lewis structure for glycine. (Hint: The central atoms in the skeletal structure are nitrogen and the two carbon atoms. Each oxygen atom is bonded directly to the right-most carbon atom.)

**91.** Formic acid is responsible for the sting of ant bites. By mass, formic acid is 26.10% C, 4.38% H, and 69.52% O. The molar mass of formic acid is 46.02 g/mol. Determine the molecular formula of formic acid and draw its Lewis structure.

**92.** Diazomethane is a highly poisonous, explosive compound because it readily evolves $N_2$. Diazomethane has the following composition by mass: 28.57% C; 4.80% H; and 66.64% N. The molar mass of diazomethane is 42.04 g/mol. Find the molecular formula of diazomethane, draw its Lewis structure, and assign formal charges to each atom. Why is diazomethane not very stable? Explain.

93. The reaction of $Fe_2O_3(s)$ with $Al(s)$ to form $Al_2O_3(s)$ and $Fe(s)$ is called the thermite reaction and is highly exothermic. What role does lattice energy play in the exothermicity of the reaction?

94. NaCl has a lattice energy of $-787$ kJ/mol. Consider a hypothetical salt XY. $X^{3+}$ has the same radius of $Na^+$ and $Y^{3-}$ has the same radius as $Cl^-$. Estimate the lattice energy of XY.

95. Draw the Lewis structure for nitric acid (the hydrogen atom is attached to one of the oxygen atoms). Include all three resonance structures by alternating the double bond among the three oxygen atoms. Use formal charge to determine which of the resonance structures is most important to the structure of nitric acid.

96. Phosgene ($Cl_2CO$) is a poisonous gas used as a chemical weapon during World War I. It is a potential agent for chemical terrorism today. Draw the Lewis structure of phosgene. Include all three resonance forms by alternating the double bond among the three terminal atoms. Which resonance structure is the best?

97. The cyanate ion ($OCN^-$) and the fulminate ion ($CNO^-$) share the same three atoms but have vastly different properties. The cyanate ion is stable, while the fulminate ion is unstable and forms explosive compounds. The resonance structures of the cyanate ion are explored in Example 9.8. Draw Lewis structures for the fulminate ion—including possible resonance forms—and use formal charge to explain why the fulminate ion is less stable (and therefore more reactive) than the cyanate ion.

98. Draw the Lewis structure for each organic compound from its condensed structural formula.
    a. $C_3H_8$
    b. $CH_3OCH_3$
    c. $CH_3COCH_3$
    d. $CH_3COOH$
    e. $CH_3CHO$

99. Draw the Lewis structure for each organic compound from its condensed structural formula.
    a. $C_2H_4$
    b. $CH_3NH_2$
    c. $HCHO$
    d. $CH_3CH_2OH$
    e. $HCOOH$

100. Use Lewis structures to explain why $Br_3^-$ and $I_3^-$ are stable, while $F_3^-$ is not.

101. Draw the Lewis structure for $HCSNH_2$. (The carbon and nitrogen atoms are bonded together, and the sulfur atom is bonded to the carbon atom.) Label each bond in the molecule as polar or nonpolar.

102. Draw the Lewis structure for urea, $H_2NCONH_2$, one of the compounds responsible for the smell of urine. (The central carbon atom is bonded to both nitrogen atoms and to the oxygen atom.) Does urea contain polar bonds? Which bond in urea is most polar?

103. Some theories of aging suggest that free radicals cause certain diseases and perhaps aging in general. As you know from the Lewis model, such molecules are not chemically stable and will quickly react with other molecules. According to certain theories, free radicals may attack molecules within the cell, such as DNA, changing them and causing cancer or other diseases. Free radicals may also attack molecules on the surfaces of cells, making them appear foreign to the body's immune system. The immune system then attacks the cells and destroys them, weakening the body. Draw Lewis structures for each free radical implicated in this theory of aging.

a. $O_2^-$
b. $O^-$
c. $OH$
d. $CH_3OO$ (unpaired electron on terminal oxygen)

104. Free radicals are important in many environmentally significant reactions (see the *Chemistry in the Environment* box on free radicals in this chapter). For example, photochemical smog—smog that results from the action of sunlight on air pollutants—forms in part by these two steps:

$$NO_2 \xrightarrow{\text{UV light}} NO + O$$

$$O + O_2 \longrightarrow O_3$$

The product of this reaction, ozone, is a pollutant in the lower atmosphere. (Upper atmospheric ozone is a natural part of the atmosphere that protects life on Earth from ultraviolet light.) Ozone is an eye and lung irritant and also accelerates the weathering of rubber products. Rewrite the given reactions using the Lewis structure of each reactant and product. Identify the free radicals.

105. If hydrogen were used as a fuel, it could be burned according to this reaction:

$$H_2(g) + \tfrac{1}{2}O_2(g) \longrightarrow H_2O(g)$$

Use average bond energies to calculate $\Delta H_{rxn}$ for this reaction and also for the combustion of methane ($CH_4$). Which fuel yields more energy per mole? Per gram?

106. Calculate $\Delta H_{rxn}$ for the combustion of octane ($C_8H_{18}$), a component of gasoline, by using average bond energies and then calculate it using enthalpies of formation from Appendix IIB. What is the percent difference between your results? Which result would you expect to be more accurate?

107. Draw the Lewis structure for each compound.
    a. $Cl_2O_7$ (no Cl—Cl bond)
    b. $H_3PO_3$ (two OH bonds)
    c. $H_3AsO_4$

108. The azide ion, $N_3^-$, is a symmetrical ion, all of whose contributing resonance structures have formal charges. Draw three important contributing structures for this ion.

109. List the following gas-phase ion pairs in order of the quantity of energy released when they form from separated gas-phase ions. List the pair that releases the least energy first. $Na^+F^-$, $Mg^{2+}F^-$, $Na^+O^{2-}$, $Mg^{2+}O^{2-}$, $Al^{3+}O^{2-}$.

110. Calculate $\Delta H°$ for the reaction $H_2(g) + Br_2(g) \longrightarrow 2\,HBr(g)$ using the bond energy values. The $\Delta H_f°$ of $HBr(g)$ is not equal to one-half of the value calculated. Account for the difference.

111. The heat of atomization is the heat required to convert a molecule in the gas phase into its constituent atoms in the gas phase. The heat of atomization is used to calculate average bond energies. Without using any tabulated bond energies, calculate the average C—Cl bond energy from the following data: the heat of atomization of $CH_4$ is 1660 kJ/mol, and the heat of atomization of $CH_2Cl_2$ is 1495 kJ/mol.

112. Calculate the heat of atomization (see previous problem) of $C_2H_3Cl$, using the average bond energies in Table 9.3.

113. A compound composed of only carbon and hydrogen is 7.743% hydrogen by mass. Propose a Lewis structure for the compound.

114. A compound composed of only carbon and chlorine is 85.5% chlorine by mass. Propose a Lewis structure for the compound.

# Challenge Problems

**115.** The main component of acid rain ($H_2SO_4$) forms from $SO_2$ pollutant in the atmosphere via these steps:

$$SO_2 + OH \cdot \longrightarrow HSO_3 \cdot$$
$$HSO_3 \cdot + O_2 \longrightarrow SO_3 + HOO \cdot$$
$$SO_3 + H_2O \longrightarrow H_2SO_4$$

Draw the Lewis structure for each of the species in these steps and use bond energies and Hess's law to estimate $\Delta H_{rxn}$ for the overall process. (Use 265 kJ/mol for the S—O single bond energy.)

**116.** A 0.167-g sample of an unknown acid requires 27.8 mL of 0.100 M NaOH to titrate to the equivalence point. Elemental analysis of the acid gives the following percentages by mass: 40.00% C, 6.71% H, 53.29% O. Determine the molecular formula, molar mass, and Lewis structure of the unknown acid.

**117.** Use the dipole moments of HF and HCl (given at the end of the problem) together with the percent ionic character of each bond (Figure 9.10) to estimate the bond length in each molecule. How well does your estimated bond length agree with the bond length in Table 9.4?

$$HCl\ \mu = 1.08\ D$$
$$HF\ \mu = 1.82\ D$$

**118.** Use average bond energies together with the standard enthalpy of formation of $C(g)$ (718.4 kJ/mol) to estimate the standard enthalpy of formation of gaseous benzene, $C_6H_6(g)$. (Remember that average bond energies apply to the gas phase only.) Compare the value you obtain using average bond energies to the actual standard enthalpy of formation of gaseous benzene, 82.9 kJ/mol. What does the difference between these two values tell you about the stability of benzene?

**119.** The standard state of phosphorus at 25 °C is $P_4$. This molecule has four equivalent P atoms, no double or triple bonds, and no expanded octets. Draw its Lewis structure.

**120.** The standard heat of formation of $CaBr_2$ is −675 kJ/mol. The first ionization energy of Ca is 590 kJ/mol, and its second ionization energy is 1145 kJ/mol. The heat of sublimation of $Ca[Ca(s) \longrightarrow Ca(g)]$ is 178kJ/mol. The bond energy of $Br_2$ is 193 kJ/mol, the heat of vaporization of $Br_2(l)$ is 31 kJ/mol, and the electron affinity of Br is −325 kJ/mol. Calculate the lattice energy of $CaBr_2$.

**121.** The standard heat of formation of $PI_3(s)$ is −24.7 kJ/mol, and the PI bond energy in this molecule is 184 kJ/mol. The standard heat of formation of $P(g)$ is 334 kJ/mol, and that of $I_2(g)$ is 62 kJ/mol. The $I_2$ bond energy is 151 kJ/mol. Calculate the heat of sublimation of $PI_3[PI_3(s) \longrightarrow PI_3(g)]$.

**122.** A compound has the formula $C_8H_8$ and does not contain any double or triple bonds. All the carbon atoms are chemically identical, and all the hydrogen atoms are chemically identical. Draw the Lewis structure for this molecule.

**123.** Find the oxidation number of each sulfur in the molecule $H_2S_4$, which has a linear arrangement of its atoms.

**124.** Ionic solids of the $O^-$ and $O^{3-}$ anions do not exist, while ionic solids of the $O^{2-}$ anion are common. Explain.

**125.** The standard state of sulfur is solid rhombic sulfur. Use the appropriate standard heats of formation given in Appendix II to find the average bond energy of the S=O in $SO_2$.

# Conceptual Problems

**126.** Which statement is true of an endothermic reaction?
 **a.** Strong bonds break and weak bonds form.
 **b.** Weak bonds break and strong bonds form.
 **c.** The bonds that break and those that form are of approximately the same strength.

**127.** When a firecracker explodes, energy is obviously released. The compounds in the firecracker can be viewed as being "energy rich." What does this mean? Explain the source of the energy in terms of chemical bonds.

**128.** A fundamental difference between compounds containing ionic bonds and those containing covalent bonds is the existence of molecules. Explain why molecules exist in solid covalent compounds but not in solid ionic compounds.

**129.** In the very first chapter of this book, we described the scientific approach and put a special emphasis on scientific models or theories. In this chapter, we looked carefully at a model for chemical bonding (the Lewis model). Why is this theory successful? What are some of the limitations of the theory?

*Active Classroom Learning*

# Questions for Group Work

Discuss these questions with the group and record your consensus answer.

**130.** Have each member of your group represent an atom of a metal or an atom of a nonmetal. Each group member holds a coin to represent an electron. Which group members are most reluctant to give up their electrons? Which group members are most willing to give up their electrons? Determine which kind of bond could form between each pair of group members. Tabulate your results.

**131.** Spend a few minutes reviewing the Lewis dot symbols for the atoms H through Ne. Form a circle and have each group member ask the group member on his or her right to draw the Lewis symbol for a specific atom. Keep going around until each group member can write all the Lewis dot symbols for the atoms H through Ne. Determine the formal charge for each symbol. In a complete sentence or two, describe why they are all the same.

**132.** Draw the Lewis dot symbols for the atoms Al and O. Use the Lewis model to determine the formula for the compound formed from these two atoms.

**133.** Draft a list of step-by-step instructions for writing the correct Lewis dot structure for any molecule or polyatomic ion.

**134.** Pass a piece of paper around the group and ask each group member in turn to perform the next step in the process of determining a correct Lewis structure (including formal charges on all atoms and resonance structures, if appropriate) for the following molecules and ions: $N_2H_4$, $CCl_4$, $CO_3^{2-}$, and $NH_4^+$.

# Data Interpretation and Analysis

## Lattice Energies of Period 4 Elements

**135.** Evidence for the additional stabilization of certain electron configurations comes from the experimental lattice energies of the metal fluorides–$MF_2$. Figure a▼ plots lattice energy for the 2+ metal cations for period 4 elements. Figure b▶ plots the ionic radii of the 2+ metal cations of period 4 elements versus atomic number.

▲ **FIGURE b  Ionic Radius Versus Atomic Number of 2+ Metal Cations for Period 4 Elements**

▲ **FIGURE a  Lattice Energy of Period 4 Metal Fluorides**
*Source: After C.S.G. Phillips, R. J. P Williams, Inorganic Chemistry, Volume 1, p. 179.*

Use the information provided in the figures to answer the following questions:

a. Explain the general trend in lattice energy observed in Figure a.

b. Is there a correlation between ionic radius and lattice energy? Explain.

c. What could account for the decrease in lattice energy between $CrF_2$ and $MnF_2$?

d. Which has the higher lattice energy: $VF_2$ or $VCl_2$? Explain.

# Answers to Conceptual Connections

## Bond Types

**9.1** **(c)** Because $MgF_2$ is composed of a metal and a nonmetal, it is most likely to contain ionic bonds.

## Melting Points of Ionic Solids

**9.2** You would expect MgO to have the higher melting point because, in our bonding model, the magnesium and oxygen ions are held together in a crystalline lattice by charges of 2+ for magnesium and 2− for oxygen. In contrast, the NaCl lattice is held together by charges of 1+ for sodium and 1− for chlorine. The experimentally measured melting points of these compounds are 801 °C for NaCl and 2852 °C for MgO, in accordance with our model.

## Energy and the Octet Rule

**9.3** The reasons that atoms form bonds are complex. One contributing factor is the lowering of their potential energy. The octet rule is just a handy way to predict the combinations of atoms that have a lower potential energy when they bond.

## Periodic Trends in Electronegativity

**9.4** $N > P > Al > Na$

## Percent Ionic Character

**9.5** **(b)** You are given that the dipole moment of the HCl bond is about 1 D and that the bond length is 127 pm. Previously, we calculated the dipole moment for a 130-pm bond that is 100% ionic to be about 6.2 D. You can therefore estimate the bond's ionic character as $1/6 \times 100$, which is closest to 15%.

## Odd-Electron Species

**9.6** **(d)** $NO_2$ because the sum of the valence electrons of its atoms is an odd number.

## Expanded Octets

**9.7** **(b)** The only molecule in this group that could have an expanded octet is $H_3PO_4$ because phosphorus is a third-period element. Expanded octets *never* occur in second-period elements such as carbon and nitrogen.

## Bond Energies and $\Delta H_{rxn}$

**9.8** **(b)** In a highly exothermic reaction, the energy needed to break bonds is less than the energy released when the new bonds form, resulting in a net release of energy.

# Chemical Bonding II: Molecular Shapes, Valence Bond Theory, and Molecular Orbital Theory

> No theory ever solves all the puzzles with which it is confronted at a given time; nor are the solutions already achieved often perfect.
>
> —THOMAS KUHN (1922–1996)

▲ Similarities in the shape of sucrose and aspartame give both molecules the ability to stimulate a sweet taste sensation.

In Chapter 9, we examined a simple model for chemical bonding called the Lewis model. We saw how this model helps us to explain and predict the combinations of atoms that form stable molecules. When we combine the Lewis model with the idea that valence electron groups repel one another—the basis of an approach known as VSEPR theory—we can predict the general shape of a molecule from its Lewis structure. We address molecular shape and its importance in the first part of this chapter. We then move on to explore two additional bonding theories—called valence bond theory and molecular orbital theory—that are progressively more sophisticated, but at the cost of being more complex, than the Lewis model. As you work through this chapter, our second on chemical bonding, keep in mind the importance of this topic. In our universe, elements join together to form compounds, and that makes many things possible, including our own existence.

## 10.1 Artificial Sweeteners: Fooled by Molecular Shape

Artificial sweeteners, such as aspartame (NutraSweet), taste sweet but have few or no calories. Why? *Because taste and caloric value are independent properties of foods.* The caloric value of a food depends on the amount of energy released when your body metabolizes the food. For example, sucrose (table sugar) is metabolized by oxidation to carbon dioxide and water:

$$C_{12}H_{22}O_{11} + 12\,O_2 \longrightarrow 12\,CO_2 + 11\,H_2O \qquad \Delta H^\circ_{rxn} = -5644\ \text{kJ}$$

When your body metabolizes a mole of sucrose, it obtains 5644 kJ of energy. Some artificial sweeteners, such as saccharin, for example, are not metabolized at all—they just pass through the body unchanged—and therefore have no caloric value. Other artificial sweeteners, such as aspartame, are metabolized but have a much lower caloric content (for a given amount of sweetness) than sucrose.

The *taste* of a food, however, is independent of its metabolism. The sensation of taste originates in the tongue, where specialized taste cells act as highly sensitive and specific molecular detectors. These cells discern sugar molecules from the thousands of different types of molecules present in a mouthful of food. The main factors for this discrimination are the sugar molecule's shape and charge distribution.

The surface of a taste cell contains specialized protein molecules called taste receptors. A particular *tastant*—a molecule that we can taste—fits snugly into a special pocket (just as a key fits into a lock) on the taste receptor protein called the *active site*. For example, a sugar molecule precisely fits into the active site of the sugar receptor protein called T1r3. When the sugar molecule (the key) enters the active site (the lock), the different subunits of the T1r3 protein split apart. This split opens ion channels in the cell membrane, resulting in nerve signal transmission (see Section 8.1) that reaches the brain and causes the sensation of sweetness.

Artificial sweeteners taste sweet because they fit into the receptor pocket that normally binds sucrose. In fact, both aspartame and saccharin bind to the active site in the T1r3 protein more strongly than sugar! For this reason, artificial sweeteners are "sweeter than sugar." Aspartame, for example, is 200 times sweeter than sugar, meaning that it takes 200 times as much sugar as aspartame to trigger the same amount of nerve signal transmission from taste cells.

The lock-and-key fit between the active site of a protein and a particular molecule is important not only to taste but to many other biological functions as well. Immune response, the sense of smell, and many types of drug action all depend on shape-specific interactions between molecules and proteins. Our ability to determine the shapes of key biological molecules is largely responsible for the revolution in biology research that has occurred over the last 50 years.

In this chapter, we look at ways to predict and account for the shapes of molecules. The molecules we examine are much smaller than the protein molecules we just discussed, but the same principles apply to both. The simple model we examine to account for molecular shape is *valence shell electron pair repulsion* (VSEPR) theory, and we will use it in conjunction with the Lewis model. We then proceed to explore two additional bonding theories: valence bond theory and molecular orbital theory. These bonding theories are more complex, but also more powerful, than the Lewis model. They predict and account for molecular shape as well as other properties of molecules.

## 10.2  VSEPR Theory: The Five Basic Shapes

**KCV**  **VSEPR Theory**

Electron groups

Central atom

Repulsions

▲ **FIGURE 10.1  Repulsion between Electron Groups** The basic idea of VSEPR theory is that repulsions between electron groups determine molecular geometry.

**Valence shell electron pair repulsion (VSEPR) theory** is based on the simple idea that **electron groups**—which we define as lone pairs, single bonds, multiple bonds, and even single electrons—repel one another through coulombic forces. The electron groups are also attracted to the nucleus (otherwise the molecule would fall apart), but VSEPR theory focuses on the repulsions. According to VSEPR theory, the repulsions between electron groups on *interior atoms* of a molecule determine the geometry of the molecule (Figure 10.1 ◀). The preferred geometry of a molecule is the one in which the electron groups have the maximum separation (and therefore the minimum repulsion energy) possible. Consequently, for molecules having just one interior atom (the central atom), molecular geometry depends on (a) the number of electron groups around the central atom and (b) how many of those electron groups are bonding groups and how many are lone pairs. We first look at the molecular geometries associated with two to six electron groups around the central atom when all of those groups are bonding groups (single or multiple bonds). The resulting geometries constitute the five basic shapes of molecules. We then consider how these basic shapes are modified if one or more of the electron groups are lone pairs.

# Two Electron Groups: Linear Geometry

Consider the Lewis structure of $BeCl_2$, which has two electron groups (two single bonds) about the central atom:

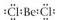

According to VSEPR theory, the geometry of $BeCl_2$ is determined by the repulsion between these two electron groups, which maximize their separation by assuming a 180° bond angle or a **linear geometry**. Experimental measurements of the geometry of $BeCl_2$ indicate that the molecule is indeed linear, as predicted by the theory.

Molecules that form only two single bonds, with no lone pairs, are rare because they do not follow the octet rule. However, the same geometry is observed in all molecules that have two electron groups (and no lone pairs). Consider the Lewis structure of $CO_2$, which has two electron groups (the double bonds) around the central carbon atom:

$$:\ddot{O}=C=\ddot{O}:$$

According to VSEPR theory, the two double bonds repel each other (just as the two single bonds in $BeCl_2$ repel each other), resulting in a linear geometry for $CO_2$. Experimental observations confirm that $CO_2$ is indeed a linear molecule.

Beryllium often forms incomplete octets, as it does in this structure.

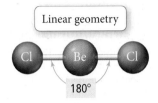

Linear geometry

A double bond counts as one electron group.

Linear geometry

# Three Electron Groups: Trigonal Planar Geometry

The Lewis structure of $BF_3$ (another molecule with an incomplete octet) has three electron groups around the central atom:

These three electron groups can maximize their separation by assuming 120° bond angles in a plane—a **trigonal planar geometry**. Experimental observations of the structure of $BF_3$ again confirm the predictions of VSEPR theory.

Another molecule with three electron groups, formaldehyde (HCHO), has one double bond and two single bonds around the central atom:

$$
\begin{array}{c}
:O: \\
\parallel \\
H-C-H
\end{array}
$$

Because formaldehyde has three electron groups around the central atom, we initially predict that the bond angles should also be 120°. However, experimental observations show that the HCO bond angles are 121.9° and the HCH bond angle is 116.2°. These bond angles are close to the idealized 120° that we originally predicted, but the HCO bond angles are slightly greater than the HCH bond angle because the double bond contains more electron density than the single bond and therefore exerts a slightly greater repulsion on the single bonds. In general, *different types of electron groups exert slightly different repulsions*—the resulting bond angles reflect these differences.

Trigonal planar geometry

**Electron Groups and Molecular Geometry**   In determining electron geometry, why do we consider only the electron groups on the central atom? In other words, why don't we consider electron groups on terminal atoms?

CONCEPTUAL
CONNECTION **10.1**

# Four Electron Groups: Tetrahedral Geometry

The VSEPR geometries of molecules with two or three electron groups around the central atom are two-dimensional, and we can therefore easily visualize and represent them on paper. For molecules with four or more electron groups around the central atom, the geometries are three-dimensional and therefore more difficult to imagine and draw. We can understand these basic shapes by analogy to balloons tied together.

▶ **FIGURE 10.2 Representing Electron Geometry with Balloons** **(a)** The bulkiness of balloons causes them to assume a linear arrangement when two of them are tied together. Similarly, the repulsion between two electron groups produces a linear geometry. **(b)** Like three balloons tied together, three electron groups adopt a trigonal planar geometry.

**(a)** Linear geometry     **(b)** Trigonal planar geometry

In this analogy, each electron group around a central atom is like a balloon tied to a central point. The bulkiness of the balloons causes them to spread out as much as possible, much as the repulsion between electron groups causes them to position themselves as far apart as possible.

For example, if you tie two balloons together, they assume a roughly linear arrangement, as shown in Figure 10.2(a)▲, analogous to the linear geometry of $BeCl_2$ that we just examined. Notice that the balloons do not represent atoms, but *electron groups*. Similarly, if you tie three balloons together—in analogy to three electron groups—they assume a trigonal planar geometry, as shown in Figure 10.2(b)▲, much like the $BF_3$ molecule. If you tie *four* balloons together, however, they assume a three-dimensional **tetrahedral geometry** with 109.5° angles between the balloons. That is, the balloons point toward the vertices of a *tetrahedron*—a geometrical shape with four identical faces, each an equilateral triangle, as shown here:

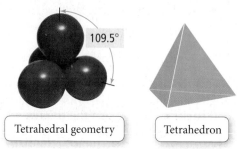

Tetrahedral geometry     Tetrahedron

Methane is an example of a molecule with four electron groups around the central atom:

Tetrahedral geometry

For four electron groups, the tetrahedron is the three-dimensional shape that allows the maximum separation among the groups. The repulsions among the four electron groups in the C—H bonds cause the molecule to assume the tetrahedral shape. When we write the Lewis structure of $CH_4$ on paper, it may seem that the molecule should be square planar, with bond angles of 90°. However, in three dimensions, the electron groups can get farther away from each other by forming the tetrahedral geometry, as illustrated by our balloon analogy.

**Molecular Geometry**   What is the geometry of the HCN molecule? The Lewis structure of HCN is H—C≡N:

   **(a)** linear      **(b)** trigonal planar      **(c)** tetrahedral

## Five Electron Groups: Trigonal Bipyramidal Geometry

Five electron groups around a central atom assume a **trigonal bipyramidal geometry**, like five balloons tied together. In this structure, three of the groups lie in a single plane, as in the trigonal planar configuration, whereas the other two are positioned above and below this plane. The angles in the trigonal bipyramidal structure are not all the same. The angles between the *equatorial positions* (the three bonds in the trigonal plane) are 120°, whereas the angle between the *axial positions* (the two bonds on either side of the trigonal plane) and the trigonal plane is 90°. PCl₅ is an example of a molecule with five electron groups around the central atom:

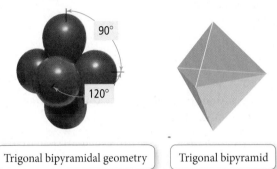

Trigonal bipyramidal geometry       Trigonal bipyramid

Trigonal bipyramidal geometry

The three equatorial chlorine atoms are separated by 120° bond angles, and the two axial chlorine atoms are separated from the equatorial atoms by 90° bond angles.

## Six Electron Groups: Octahedral Geometry

Six electron groups around a central atom assume an **octahedral geometry**, like six balloons tied together. In this structure—named after the eight-sided geometrical shape called the octahedron—four of the groups lie in a single plane, with a fifth group above the plane and the sixth below it. The angles in this geometry are all 90°. SF₆ is a molecule with six electron groups around the central atom:

Octahedral geometry       Octahedron

Octahedral geometry

The structure of this molecule is highly symmetrical; all six bonds are equivalent.

## Example 10.1 | VSEPR Theory and the Basic Shapes

Determine the molecular geometry of $NO_3^-$.

**SOLUTION**

| | |
|---|---|
| Determine the molecular geometry of $NO_3^-$ by counting the number of electron groups around the central atom (N). Begin by drawing a Lewis structure for $NO_3^-$. | $NO_3^-$ has $5 + 3(6) + 1 = 24$ valence electrons. The Lewis structure has three resonance structures: <br><br>  <br><br> The hybrid structure is intermediate between these three and has three equivalent bonds. |
| Use any one of the resonance structures to determine the number of electron groups around the central atom. | The nitrogen atom has three electron groups: two single bonds and one double bond. |
| Based on the number of electron groups, determine the geometry that minimizes the repulsions between the groups. | The electron geometry that minimizes the repulsions between three electron groups is trigonal planar. <br><br> The three bonds are equivalent (because of the resonance structures), so they each exert the same repulsion on the other two and the molecule has three equal bond angles of 120°. |

**FOR PRACTICE 10.1** Determine the molecular geometry of $CCl_4$.

 **VSEPR Theory: The Effect of Lone Pairs**

## 10.3 VSEPR Theory: The Effect of Lone Pairs

Each of the examples we have just seen has only bonding electron groups around the central atom. What happens in molecules that have lone pairs around the central atom as well? The lone pairs also repel other electron groups, as we see in the examples that follow.

### Four Electron Groups with Lone Pairs

Consider the Lewis structure of ammonia:

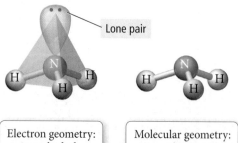

Electron geometry: tetrahedral

Molecular geometry: trigonal pyramidal

The central nitrogen atom has four electron groups (one lone pair and three bonding pairs) that repel one another. If we do not distinguish between bonding electron groups and lone pairs, we find that the **electron geometry**—the geometrical arrangement of the *electron groups*—is still tetrahedral, as we expect for four electron groups. However, the **molecular geometry**—the geometrical arrangement of the atoms—is **trigonal pyramidal**, as shown at left.

Notice that although the electron geometry and the molecular geometry are different, *the electron geometry is relevant to the molecular geometry*. The lone pair exerts its influence on the bonding pairs.

As we noted previously, different kinds of electron groups generally result in different amounts of repulsion. Lone pair electrons typically exert slightly greater repulsions than bonding electrons. If all four electron groups in $NH_3$ exerted equal repulsions on one another, the bond angles in the molecule would all be the ideal tetrahedral angle, 109.5°. However, the actual angle between N—H bonds in ammonia is slightly smaller, 107°. A lone electron pair is more spread out in space than a bonding electron pair because a lone pair is attracted to only one nucleus while a bonding pair is attracted to two (Figure 10.3▼). The lone pair occupies more of the angular space around a nucleus, exerting a greater repulsive force on neighboring electrons and compressing the N—H bond angles.

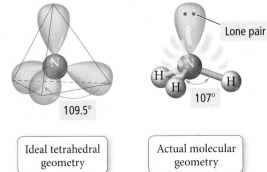

Lone pair

109.5°

Ideal tetrahedral geometry

107°

Actual molecular geometry

Bonding electron pair

Nuclei

Lone pair

Nucleus

◀ **FIGURE 10.3  Lone Pair versus Bonding Electron Pairs**  A lone electron pair occupies more space than a bonding pair.

A water molecule's Lewis structure is:

$$H—\overset{..}{\underset{..}{O}}—H$$

Because water has four electron groups (two bonding pairs and two lone pairs), its *electron geometry* is also tetrahedral, but its *molecular geometry* is **bent**, as shown at right. As in $NH_3$, the bond angles in $H_2O$ are smaller (104.5°) than the ideal tetrahedral bond angles because of the greater repulsion exerted by the lone pair electrons. The bond angle in $H_2O$ is even smaller than in $NH_3$ because $H_2O$ has *two* lone pairs of electrons on the central oxygen atom. These lone pairs compress the $H_2O$ bond angle to a greater extent than in $NH_3$.

In general, electron group repulsions compare as follows:

Lone pair–lone pair > Lone pair–bonding pair > Bonding pair–bonding pair

Most repulsive                    Least repulsive

We see the effects of this ordering in the progressively smaller bond angles of $CH_4$, $NH_3$, and $H_2O$, shown in Figure 10.4▼. The relative ordering of repulsions also helps to determine the geometry of molecules with five and six electron groups when one or more of those groups are lone pairs.

Lone pair

Lone pair

Electron geometry: tetrahedral

Molecular geometry: bent

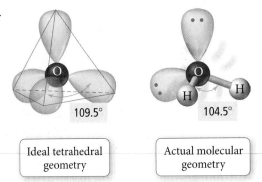

109.5°

Ideal tetrahedral geometry

104.5°

Actual molecular geometry

**Effect of Lone Pairs on Molecular Geometry**

No lone pairs

One lone pair

Two lone pairs

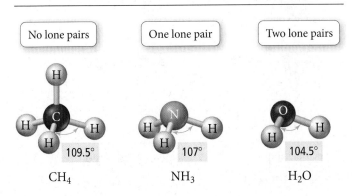

109.5°

$CH_4$

107°

$NH_3$

104.5°

$H_2O$

◀ **FIGURE 10.4  The Effect of Lone Pairs on Molecular Geometry**  The bond angles get progressively smaller as the number of lone pairs on the central atom increases from zero in $CH_4$ to one in $NH_3$ to two in $H_2O$.

# Five Electron Groups with Lone Pairs

Consider the Lewis structure of $SF_4$:

$$:\ddot{F}: $$
$$:\ddot{F}—\ddot{S}—\ddot{F}:$$
$$:\ddot{F}:$$

The central sulfur atom has five electron groups (one lone pair and four bonding pairs). The *electron geometry*, due to the five electron groups, is trigonal bipyramidal.

    To determine the molecular geometry for sulfur tetrafluoride, notice that the lone pair could occupy either an equatorial position or an axial position within the trigonal bipyramidal electron geometry. Which position is more favorable? To answer this question, recall that, as we have just seen, lone pair–bonding pair repulsions are greater than bonding pair–bonding pair repulsions. Therefore, the lone pair occupies the position that minimizes its interaction with the bonding pairs. If the lone pair were in an axial position, it would have three 90° interactions with bonding pairs. In an equatorial position, however, it has only two 90° interactions. Consequently, the lone pair occupies an equatorial position. The resulting molecular geometry is called **seesaw** because it resembles a seesaw (or teeter-totter).

> The seesaw molecular geometry is sometimes called an *irregular tetrahedron*.

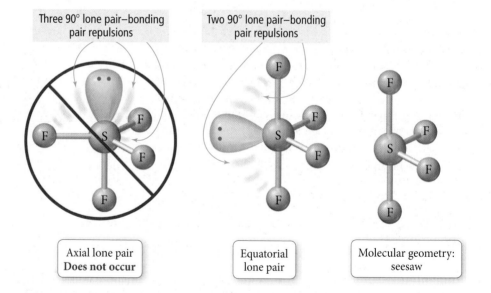

Three 90° lone pair–bonding pair repulsions

Two 90° lone pair–bonding pair repulsions

| Axial lone pair **Does not occur** | Equatorial lone pair | Molecular geometry: seesaw |

    When two of the five electron groups around the central atom are lone pairs, as in $BrF_3$, the lone pairs occupy two of the three equatorial positions—again minimizing 90° interactions with bonding pairs and also avoiding a lone pair–lone pair 90° repulsion. The resulting molecular geometry is **T-shaped**.

$$:\ddot{F}:$$
$$|$$
$$:\ddot{Br}—\ddot{F}:$$
$$|$$
$$:\ddot{F}:$$

| Electron geometry: trigonal bipyramidal | Molecular geometry: T-shaped |

When three of the five electron groups around the central atom are lone pairs, as in $XeF_2$, the lone pairs occupy all three of the equatorial positions, and the resulting molecular geometry is linear.

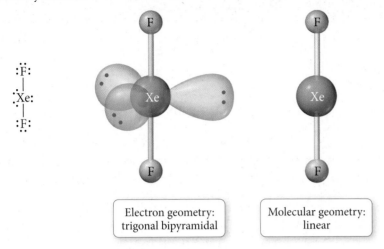

Electron geometry:
trigonal bipyramidal

Molecular geometry:
linear

## Six Electron Groups with Lone Pairs

Consider the Lewis structure of $BrF_5$ shown here. The central bromine atom has six electron groups (one lone pair and five bonding pairs). The electron geometry, due to the six electron groups, is octahedral. Since all six positions in the octahedral geometry are equivalent, the lone pair can be situated in any one of these positions. The resulting molecular geometry is **square pyramidal**.

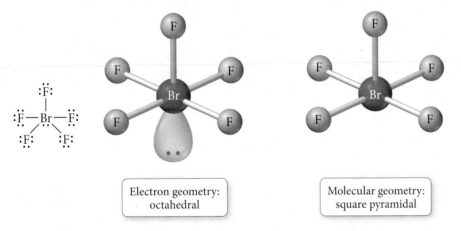

Electron geometry:
octahedral

Molecular geometry:
square pyramidal

When two of the six electron groups around the central atom are lone pairs, as in $XeF_4$, the lone pairs occupy positions across from one another (to minimize lone pair–lone pair repulsions), and the resulting molecular geometry is **square planar**.

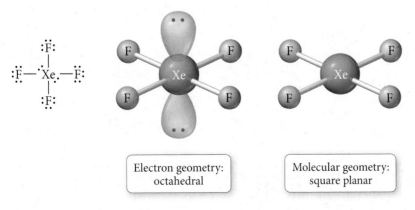

Electron geometry:
octahedral

Molecular geometry:
square planar

**TABLE 10.1 Electron and Molecular Geometries**

| Electron Groups* | Bonding Groups | Lone Pairs | Electron Geometry | Molecular Geometry | Approximate Bond Angles | Example | |
|---|---|---|---|---|---|---|---|
| 2 | 2 | 0 | Linear | Linear | 180° | $\ddot{O}=C=\ddot{O}$ | |
| 3 | 3 | 0 | Trigonal planar | Trigonal planar | 120° | $F{-}B{-}F$ with $F$ above | |
| 3 | 2 | 1 | Trigonal planar | Bent | <120° | $\ddot{O}=\ddot{S}{-}\ddot{O}$ | |
| 4 | 4 | 0 | Tetrahedral | Tetrahedral | 109.5° | $H{-}C{-}H$ (with H above and below) | |
| 4 | 3 | 1 | Tetrahedral | Trigonal pyramidal | <109.5° | $H{-}N{-}H$ with H below | |
| 4 | 2 | 2 | Tetrahedral | Bent | <109.5° | $H{-}\ddot{O}{-}H$ | |
| 5 | 5 | 0 | Trigonal bipyramidal | Trigonal bipyramidal | 120° (equatorial) 90° (axial) | $PCl_5$ structure | |
| 5 | 4 | 1 | Trigonal bipyramidal | Seesaw | <120° (equatorial) <90° (axial) | $SF_4$ structure | |
| 5 | 3 | 2 | Trigonal bipyramidal | T-shaped | <90° | $BrF_3$ structure | |
| 5 | 2 | 3 | Trigonal bipyramidal | Linear | 180° | $F{-}Xe{-}F$ | |
| 6 | 6 | 0 | Octahedral | Octahedral | 90° | $SF_6$ structure | |
| 6 | 5 | 1 | Octahedral | Square pyramidal | <90° | $BrF_5$ structure | |
| 6 | 4 | 2 | Octahedral | Square planar | 90° | $XeF_4$ structure | |

*Count only electron groups around the central atom. Each of the following is considered one electron group: a lone pair, a single bond, a double bond, a triple bond, or a single electron.

**Lone Pair Electrons and Molecular Geometry** Imagine that a molecule with six electron groups is confined to two dimensions and therefore has a hexagonal planar electron geometry. If two of the six groups are lone pairs, where are they located?

(a) positions 1 and 2
(b) positions 1 and 3
(c) positions 1 and 4

*Summarizing VSEPR Theory:*

▶ The geometry of a molecule is determined by the number of electron groups on the central atom (or on all interior atoms, if there is more than one).

▶ The number of electron groups is determined from the Lewis structure of the molecule. If the Lewis structure contains resonance structures, use any one of the resonance structures to determine the number of electron groups.

▶ Each of the following counts as a single electron group: a lone pair, a single bond, a double bond, a triple bond, or a single electron (as in a free radical).

▶ The geometry of the electron groups is determined by their repulsions as summarized in Table 10.1. In general, electron group repulsions vary as follows:

  Lone pair–lone pair > lone pair–bonding pair > bonding pair–bonding pair

▶ Bond angles can vary from the idealized angles because double and triple bonds occupy more space than single bonds (they are bulkier even though they are shorter), and lone pairs occupy more space than bonding groups. The presence of lone pairs usually makes bond angles smaller than the ideal angles for the particular geometry.

**Molecular Geometry and Electron Group Repulsions** Which statement is *always* true according to VSEPR theory?

(a) The shape of a molecule is determined only by repulsions among bonding electron groups.
(b) The shape of a molecule is determined only by repulsions among nonbonding electron groups.
(c) The shape of a molecule is determined by the polarity of its bonds.
(d) The shape of a molecule is determined by repulsions among all electron groups on the central atom (or interior atoms, if there is more than one).

## 10.4  VSEPR Theory: Predicting Molecular Geometries

To determine the geometry of a molecule, follow the procedure presented on the following page. As in other examples, we provide the steps in the left column and two examples of applying the steps in the center and right columns.

## Procedure for...

| Predicting Molecular Geometries | **Example 10.2** ▶ **Predicting Molecular Geometries** Predict the geometry and bond angles of $PCl_3$. | **Example 10.3** **Predicting Molecular Geometries** Predict the geometry and bond angles of $ICl_4^-$. |
|---|---|---|
| 1. **Draw the Lewis structure for the molecule.** | $PCl_3$ has 26 valence electrons. :Cl:  ⎪  :Cl—P—Cl: | $ICl_4^-$ has 36 valence electrons. [ :Cl:  :Cl—I—Cl:  :Cl: ]⁻ |
| 2. **Determine the total number of electron groups around the central atom.** Lone pairs, single bonds, double bonds, triple bonds, and single electrons each count as one group. | The central atom (P) has four electron groups. | The central atom (I) has six electron groups. |
| 3. **Determine the number of bonding groups and the number of lone pairs around the central atom.** These should sum to your result from step 2. Bonding groups include single bonds, double bonds, and triple bonds. | :Cl:  ⎪  :Cl—P—Cl:  Lone pair ⌐ Three of the four electron groups around P are bonding groups, and one is a lone pair. | Lone pairs [ :Cl:  :Cl—I—Cl:  :Cl: ]⁻ Four of the six electron groups around I are bonding groups, and two are lone pairs. |
| 4. **Refer to Table 10.1 to determine the electron geometry and molecular geometry.** If no lone pairs are present around the central atom, the bond angles will be that of the ideal geometry. If lone pairs are present, the bond angles may be smaller than the ideal geometry. | The electron geometry is tetrahedral (four electron groups), and the molecular geometry—the shape of the molecule—is *trigonal pyramidal* (three bonding groups and one lone pair). Because of the presence of a lone pair, the bond angles are less than 109.5°. | The electron geometry is octahedral (six electron groups), and the molecular geometry—the shape of the molecule—is *square planar* (four bonding groups and two lone pairs). Even though lone pairs are present, the bond angles are 90° because the lone pairs are symmetrically arranged and do not compress the I—Cl bond angles. |

<109.5°

Trigonal pyramidal

90°

Square planar

**FOR PRACTICE 10.2**  Predict the molecular geometry and bond angle of ClNO.

**FOR PRACTICE 10.3**  Predict the molecular geometry of $I_3^-$.

# Representing Molecular Geometries on Paper

Since molecular geometries are three-dimensional, they are often difficult to represent on two-dimensional paper. Many chemists use the notation shown here for bonds to indicate three-dimensional structures on two-dimensional paper.

| | | |
|---|---|---|
| *Straight line* Bond in plane of paper | *Hatched wedge* Bond going into the page | *Solid wedge* Bond coming out of the page |

Some examples of the molecular geometries used in this book are shown below using this notation.

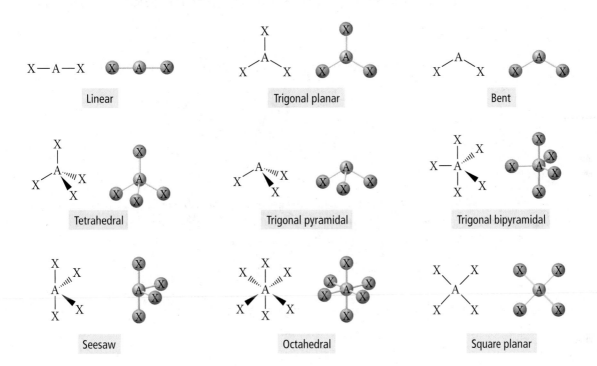

Linear     Trigonal planar     Bent

Tetrahedral     Trigonal pyramidal     Trigonal bipyramidal

Seesaw     Octahedral     Square planar

# Predicting the Shapes of Larger Molecules

Larger molecules may have two or more *interior* atoms. When predicting the shapes of these molecules, we apply the principles we just covered to each interior atom. Consider glycine, an amino acid found in many proteins (such as those involved in taste discussed in Section 10.1). Glycine, shown in the margin, contains four interior atoms: one nitrogen atom, two carbon atoms, and an oxygen atom. To determine the shape of glycine, we determine the geometry about each interior atom as follows:

Four interior atoms

Glycine

| Atom | Number of Electron Groups | Number of Lone Pairs | Molecular Geometry |
|---|---|---|---|
| Nitrogen | 4 | 1 | Trigonal pyramidal |
| Leftmost carbon | 4 | 0 | Tetrahedral |
| Rightmost carbon | 3 | 0 | Trigonal planar |
| Oxygen | 4 | 2 | Bent |

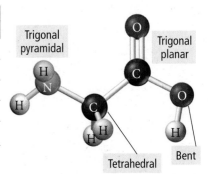

Considering the geometries of each of these, we can determine the entire three-dimensional shape of the molecule as shown to the right.

**Example** 10.4 | **Predicting the Shape of Larger Molecules**

Predict the geometry about each interior atom in methanol ($CH_3OH$) and make a sketch of the molecule.

**SOLUTION**

Begin by drawing the Lewis structure of $CH_3OH$. $CH_3OH$ contains two interior atoms: one carbon atom and one oxygen atom. To determine the shape of methanol, determine the geometry about each interior atom as follows:

| Atom | Number of Electron Groups | Number of Lone Pairs | Molecular Geometry |
|---|---|---|---|
| Carbon | 4 | 0 | Tetrahedral |
| Oxygen | 4 | 2 | Bent |

Considering these interior atom geometries, draw a three-dimensional sketch of the molecule:

**FOR PRACTICE 10.4** Predict the geometry about each interior atom in acetic

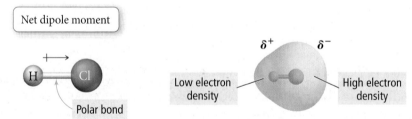

acid ($H_3C$—C—H) and make a sketch of the molecule.

---

CONCEPTUAL
CONNECTION 10.5

**The Shape of Larger Molecules** What is the molecular geometry about nitrogen in $CH_3NH_2$?

---

## 10.5 Molecular Shape and Polarity

In Chapter 9, we discussed polar bonds. Entire molecules can also be polar, depending on their shape and the nature of their bonds. For example, if a diatomic molecule has a polar bond, the molecule as a whole is polar:

Net dipole moment

$\delta^+$        $\delta^-$

H — Cl

Low electron density        High electron density

Polar bond

The figure on the right is an electrostatic potential map of HCl. In these maps, red areas indicate electron-rich regions in the molecule and the blue areas indicate electron-poor regions. Yellow indicates moderate electron density. Notice that the region around the more electronegative atom (chlorine) is more electron-rich than the region around the hydrogen atom. Thus the molecule itself is polar. If the bond in a diatomic molecule is *nonpolar*, the molecule as a whole is *nonpolar*.

In polyatomic molecules, the presence of polar bonds may or may not result in a polar molecule, depending on the molecular geometry. If the molecular geometry is such that the dipole moments of individual polar bonds sum together to a net dipole moment, then the molecule is polar. But, if the molecular geometry is such that the dipole moments of the individual polar bonds cancel each other (that is, sum to zero), then the molecule is nonpolar. It all depends on the geometry of the molecule. Consider carbon dioxide:

$$\ddot{\text{O}}=\text{C}=\ddot{\text{O}}\!:$$

Each $C=O$ bond in $CO_2$ is polar because oxygen and carbon have significantly different electronegativities (3.5 and 2.5, respectively). However, since $CO_2$ is a linear molecule, the polar bonds directly oppose one another and the dipole moment of one bond exactly opposes the dipole moment of the other—the two dipole moments sum to zero and the *molecule* is nonpolar.

Dipole moments can cancel each other because they are *vector quantities*; they have both a magnitude and a direction. Think of each polar bond as a vector, pointing in the direction of the more electronegative atom. The length of the vector is proportional to the electronegativity difference between the bonding atoms. In $CO_2$, we have two identical vectors pointing in exactly opposite directions—the vectors sum to zero, much as +1 and −1 sum to zero:

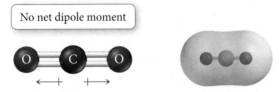

Notice that the electrostatic potential map shows regions of moderately high electron density (yellow with slight red) positioned symmetrically on either end of the molecule with a region of low electron density (blue) located in the middle.

In contrast, consider water:

$$\text{H}-\ddot{\text{O}}-\text{H}$$

The $O-H$ bonds in water are also polar; oxygen and hydrogen have electronegativities of 3.5 and 2.1, respectively. However, the water molecule is not linear but bent, so the two dipole moments do not sum to zero as they do in carbon dioxide. If we imagine each bond in water as a vector pointing toward oxygen (the more electronegative atom), we see that, because of the angle between the vectors, they do not cancel but sum to an overall vector or a net dipole moment (shown by the dashed arrow):

The electrostatic potential map reveals an electron-rich region at the oxygen end of the molecule. Consequently, water is a polar molecule. Table 10.2 (on page 443) summarizes common geometries and molecular polarity.

*Summarizing the Process to Determine Molecular Shape and Polarity:*

▶ **Draw the Lewis structure for the molecule and determine its molecular geometry.**

▶ **Determine whether the molecule contains polar bonds.**    A bond is polar if the two bonding atoms have sufficiently different electronegativities (see Figure 9.8). If the molecule contains polar bonds, superimpose a vector, pointing toward the more electronegative atom, on each bond. Draw the length of the vector proportional to the electronegativity difference between the bonding atoms.

▶ **Determine whether the polar bonds add together to form a net dipole moment.**    Sum the vectors corresponding to the polar bonds together. If the vectors sum to zero, the molecule is nonpolar. If the vectors sum to a net vector, the molecule is polar.

## Vector Addition

We determine whether a molecule is polar by summing the vectors associated with the dipole moments of all the polar bonds in the molecule. If the vectors sum to zero, the molecule is nonpolar. If they sum to a net vector, the molecule is polar. In this section, we demonstrate how to add vectors together in one dimension and in two or more dimensions.

### One Dimension

To add two vectors that lie on the same line, assign one direction as positive. Vectors pointing in that direction have positive magnitudes. Consider vectors pointing in the opposite direction to have negative magnitudes. Then sum the vectors (always remembering to include their signs), as shown in Examples 1–3.

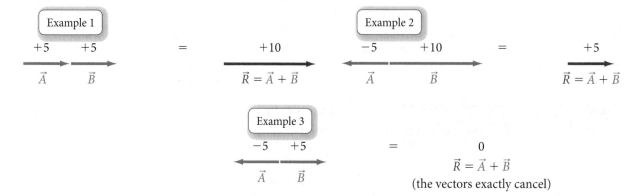

### Two or More Dimensions

To add two vectors, draw a parallelogram in which the two vectors form two adjacent sides. Draw the other two sides of the parallelogram parallel to and the same length as the two original vectors. Draw the resultant vector beginning at the origin and extending to the far corner of the parallelogram as shown in Examples 4 and 5.

To add three or more vectors, add two of them together first, and then add the third vector to the result as shown in Examples 6 and 7.

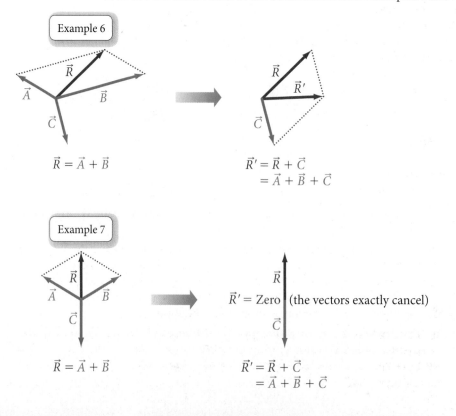

**TABLE 10.2 Common Cases of Adding Dipole Moments to Determine Whether a Molecule Is Polar**

Linear

Nonpolar

The dipole moments of two identical polar bonds pointing in opposite directions cancel. The molecule is nonpolar.

Bent

Polar

The dipole moments of two polar bonds with an angle of less than 180° between them do not cancel. The resultant dipole moment vector is shown in red. The molecule is polar.

Trigonal planar

Nonpolar

The dipole moments of three identical polar bonds at 120° from each other cancel. The molecule is nonpolar.

Tetrahedral

Nonpolar

The dipole moments of four identical polar bonds in a tetrahedral arrangement (109.5° from each other) cancel. The molecule is nonpolar.

Trigonal pyramidal

Polar

The dipole moments of three polar bonds in a trigonal pyramidal arrangement do not cancel. The resultant dipole moment vector is shown in red. The molecule is polar.

Note: In all cases in which the dipoles of two or more polar bonds cancel, the bonds are assumed to be identical. If one or more of the bonds are different from the other(s), the dipoles will not cancel and the molecule will be polar.

**Example 10.5**   **Determining Whether a Molecule Is Polar**

Determine whether $NH_3$ is polar.

SOLUTION

| | |
|---|---|
| **Draw the Lewis structure for the molecule and determine its molecular geometry.** | $$H-\overset{\overset{\textstyle H}{\textstyle \mid}}{\underset{\cdot\cdot}{N}}-H$$ The Lewis structure has three bonding groups and one lone pair about the central atom. Therefore the molecular geometry is trigonal pyramidal. |
| **Determine whether the molecule contains polar bonds.** Sketch the molecule and superimpose a vector for each polar bond. The relative length of each vector is proportional to the electronegativity difference between the atoms forming each bond. The vectors point in the direction of the more electronegative atom. |  The electronegativities of nitrogen and hydrogen are 3.0 and 2.1, respectively. Therefore, the bonds are polar. |
| **Determine whether the polar bonds add together to form a net dipole moment.** Examine the symmetry of the vectors (representing dipole moments) and determine whether they cancel each other or sum to a net dipole moment. |  The three dipole moments sum to a net dipole moment. The molecule is polar. |

**FOR PRACTICE 10.5**   Determine whether $CF_4$ is polar.

Opposite magnetic poles attract one another.

Opposite partial charges on molecules attract one another.

▲ **FIGURE 10.5** **Interaction of Polar Molecules** The north pole of one magnet attracts the south pole of another magnet. In an analogous way, the positively charged end of one molecule attracts the negatively charged end of another (although the forces involved are different). As a result of this electrical attraction, polar molecules interact strongly with one another.

Polar and nonpolar molecules have different properties. Water and oil do not mix, for example, because water molecules are polar and the molecules that compose oil are generally nonpolar. Polar molecules interact strongly with other polar molecules because the positive end of one molecule is attracted to the negative end of another, just as the south pole of a magnet is attracted to the north pole of another magnet (Figure 10.5 ◄). A mixture of polar and nonpolar molecules is similar to a mixture of small magnetic particles and nonmagnetic ones. The magnetic particles (which are like polar molecules) clump together, excluding the nonmagnetic particles (which are like nonpolar molecules) and separating into distinct regions.

Oil is nonpolar.

Water is polar.

▲ Oil and water do not mix because water molecules are polar and the molecules that compose oil are nonpolar.

▲ A mixture of polar and nonpolar molecules is analogous to a mixture of magnetic marbles (opaque) and nonmagnetic marbles (transparent). As with the magnetic marbles, mutual attraction causes polar molecules to clump together, excluding the nonpolar molecules.

# Chemistry IN YOUR DAY | How Soap Works

Imagine eating a greasy cheeseburger with both hands without a napkin. By the end of the meal, grease and oil coat your hands. If you try to wash them with only water, they remain greasy. However, if you add a little soap, the grease washes away. Why? As we just learned, water molecules are polar and the molecules that compose grease and oil are nonpolar. As a result, water and grease do not mix.

The molecules that compose soap, however, have a special structure that allows them to interact strongly with both water and grease. One end of a soap molecule is polar and the other end is nonpolar.

The nonpolar end is a long hydrocarbon chain. Hydrocarbons are always nonpolar because the electronegativity difference between carbon and hydrogen is small and because the tetrahedral arrangement about each carbon atom tends to cancel any small dipole moments of individual bonds. The polar head of a soap molecule—usually, though not always, ionic—strongly attracts water molecules, while the nonpolar tail interacts more strongly with grease and oil molecules (we examine the nature of these interactions in Chapter 11). Thus, soap acts as a sort of molecular liaison, one end interacting with water and the other end interacting with grease. Soap allows water and grease to mix, removing the grease from your hands and washing it down the drain.

**QUESTION**

*Consider the detergent molecule shown at right. Which end do you think is polar? Which end is nonpolar?*

Polar head attracts water.

Nonpolar tail interacts with grease.

$CH_3(CH_2)_{11}OCH_2CH_2OH$

# 10.6 Valence Bond Theory: Orbital Overlap as a Chemical Bond

In the Lewis model, we use dots to represent electrons. We know from quantum-mechanical theory, however, that such a treatment is an oversimplification. More advanced bonding theories treat electrons in a quantum-mechanical manner. These theories are actually extensions of quantum mechanics, applied to molecules. Although a detailed quantitative treatment of these theories is beyond the scope of this book, we introduce them in a *qualitative* manner in the sections that follow. Keep in mind, however, that modern *quantitative* approaches to chemical bonding using these theories accurately predict many of the properties of molecules—such as bond lengths, bond strengths, molecular geometries, and dipole moments—that we discuss in this book.

The simpler of the two more advanced bonding theories is **valence bond theory**. According to valence bond theory, electrons reside in quantum-mechanical orbitals localized on individual atoms. In many cases, these orbitals are simply the standard *s*, *p*, *d*, and *f* atomic orbitals that we learned about in Chapter 7. In other cases, these orbitals are *hybridized atomic orbitals*, a kind of blend or combination of two or more standard atomic orbitals.

When two atoms approach each other, the electrons and nucleus of one atom interact with the electrons and nucleus of the other atom. In valence bond theory, we calculate how these interactions affect the energies of the electrons in the atomic orbitals. If the energy of the system is lowered because of the interactions, then a chemical bond forms. If the energy of the system is raised by the interactions, then a chemical bond does not form.

The interaction energy is usually calculated as a function of the internuclear distance between the two bonding atoms. For example, Figure 10.6▼ shows the calculated interaction energy between two hydrogen atoms as a function of the distance between them. The *y*-axis of the graph is the potential energy of the interaction between the electron and nucleus of one hydrogen atom and the electron and nucleus of the other hydrogen atom. The *x*-axis is the separation (or internuclear distance) between the two atoms. As we can see from the graph, when the atoms are far apart (right side of the graph, labeled 1), the interaction energy is nearly zero because the two atoms do not interact to any significant extent. As the atoms get closer (labeled 2 and 3 on the graph), the interaction energy becomes negative. The lowering of the interaction energy is a net stabilization that attracts one hydrogen atom to the other. If the atoms get too close

Valence bond theory is an application of a general quantum-mechanical approximation method called *perturbation theory*. In perturbation theory, a complex system (such as a molecule) is viewed as a simpler system (such as two atoms) that is slightly altered or perturbed by some additional force or interaction (such as the interaction between the two atoms).

## Interaction Energy of Two Hydrogen Atoms

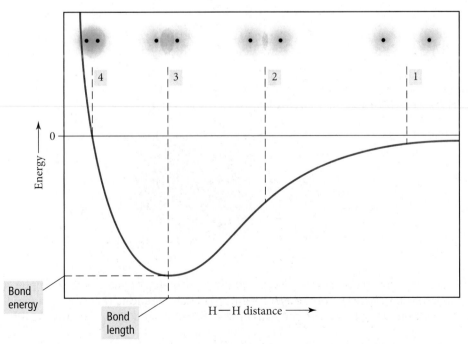

◀ **FIGURE 10.6 Interaction Energy Diagram for H₂** The potential energy of two hydrogen atoms is lowest when they are separated by a distance that allows their 1s orbitals a substantial degree of overlap without too much repulsion between their nuclei. This distance, at which the system is most stable, is the bond length of the H₂ molecule (labeled 3 on this graph).

(labeled 4 on the graph), however, the interaction energy begins to rise, primarily because of the mutual repulsion of the two positively charged nuclei. The most stable point on the curve occurs at the minimum of the interaction energy—this is the equilibrium bond length (labeled 3 on the graph). At this distance, the two atomic $1s$ orbitals have a significant amount of overlap, and the electrons spend time in the internuclear region where they can interact with both nuclei. The value of the interaction energy at the equilibrium bond distance is the bond energy.

When we apply valence bond theory to a number of atoms and their corresponding molecules, we arrive at the following general observation: *the interaction energy is usually negative (or stabilizing) when the interacting atomic orbitals contain a total of two electrons that can spin-pair (orient with opposing spins).* Most commonly, the two electrons come from two half-filled orbitals, but in some cases, the two electrons come from one filled orbital overlapping with a completely empty orbital (this is called a coordinate covalent bond, and we will cover it in more detail in Chapter 25). In other words, when two atoms with half-filled orbitals approach each other, the half-filled orbitals overlap—parts of the orbitals occupy the same space—and the electrons occupying them align with opposite spins. This results in a net energy stabilization that constitutes a covalent chemical bond. The resulting geometry of the molecule emerges from the geometry of the overlapping orbitals.

| When *completely filled* orbitals overlap, the interaction energy is positive (or destabilizing) and no bond forms.

### *Summarizing Valence Bond Theory:*

▶ The valence electrons of the atoms in a molecule reside in quantum-mechanical atomic orbitals. The orbitals can be the standard $s$, $p$, $d$, and $f$ orbitals, or they may be hybrid combinations of these.

▶ A chemical bond results from the overlap of two half-filled orbitals with spin-pairing of the two valence electrons (or less commonly the overlap of a completely filled orbital with an empty orbital).

▶ The geometry of the overlapping orbitals determines the shape of the molecule.

We can apply the general concepts of valence bond theory to explain bonding in hydrogen sulfide, $H_2S$. The valence electron configurations of the atoms in the molecule are as follows:

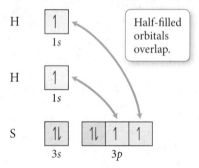

The hydrogen atoms each have one half-filled orbital, and the sulfur atom has two half-filled orbitals. The half-filled orbitals on each hydrogen atom overlap with the two half-filled orbitals on the sulfur atom, forming two chemical bonds:

To show the spin-pairing of the electrons in the overlapping orbitals, we superimpose a half-arrow for each electron in each half-filled orbital to indicate that, within a bond, the electrons are spin-paired (one half-arrow is pointing up and the other is pointing down). We also superimpose paired half-arrows in the filled sulfur *s* and *p* orbitals to represent the lone pair electrons in those orbitals. (Since those orbitals are full, they are not involved in bonding.)

A quantitative calculation of $H_2S$ using valence bond theory yields bond energies, bond lengths, and bond angles. In our qualitative treatment, we simply show how orbital overlap leads to bonding and make a rough sketch of the molecule based on the overlapping orbitals. Notice that, because the overlapping orbitals on the central atom (sulfur) are *p* orbitals and because *p* orbitals are oriented at 90° to one another, the predicted bond angle is 90°. The actual bond angle in $H_2S$ is 92°. In the case of $H_2S$, a simple valence bond treatment matches well with the experimentally measured bond angle (in contrast to VSEPR theory, which predicts a bond angle of less than 109.5°).

---

**What Is a Chemical Bond? Part I**  The answer to the question "What is a chemical bond?" depends on the bonding model. Answer these three questions:
  **(a)** What is a covalent chemical bond according to the Lewis model?
  **(b)** What is a covalent chemical bond according to valence bond theory?
  **(c)** Why are the answers different?

CONCEPTUAL
CONNECTION **10.6**

---

## 10.7 Valence Bond Theory: Hybridization of Atomic Orbitals

**KCV**  Valence Bond Theory: Hybridization

Although the overlap of half-filled *standard* atomic orbitals adequately explains the bonding in $H_2S$, it cannot adequately explain the bonding in many other molecules. For example, suppose we try to explain the bonding between hydrogen and carbon using the same approach. The valence electron configurations of H and C are as follows:

H
     1s

C
     2s      2p

Theoretical prediction

Carbon has only two half-filled orbitals and should therefore form only two bonds with two hydrogen atoms. We would therefore predict that carbon and hydrogen should form a molecule with the formula $CH_2$ and with a bond angle of 90° (corresponding to the angle between any two *p* orbitals).

However, from experiments, we know that the stable compound formed from carbon and hydrogen is $CH_4$ (methane), which has bond angles of 109.5°. The experimental reality is different from our simple prediction in two ways. First, carbon forms bonds to four hydrogen atoms, not two. Second, the bond angles are much larger than the angle between two *p* orbitals. Valence bond theory accounts for the bonding in $CH_4$ and many other polyatomic molecules by incorporating an additional concept called *orbital hybridization*.

So far, we have assumed that the overlapping orbitals that form chemical bonds are simply the standard *s*, *p*, or *d* atomic orbitals. Valence bond theory treats the electrons in a molecule as if they occupied these standard atomic orbitals, but this is a major oversimplification. The concept of hybridization in valence bond theory is essentially a step toward recognizing that *the orbitals in a molecule are not necessarily the same as the orbitals in an atom.* **Hybridization** is a mathematical procedure in which the standard atomic orbitals are combined to form new atomic orbitals called **hybrid orbitals** that

Observed reality

In Section 10.8, we examine another theory called *molecular orbital theory*, which treats electrons in a molecule as occupying orbitals that belong to the molecule as a whole.

As we saw in Chapter 9, the word *hybrid* comes from breeding. A *hybrid* is an offspring of two animals or plants of different standard races or breeds. Similarly, a hybrid orbital is a product of mixing two or more standard atomic orbitals.

correspond more closely to the actual distribution of electrons in chemically bonded atoms. Hybrid orbitals are still localized on individual atoms, but their shapes and energies differ from those of standard atomic orbitals.

Why do we hypothesize that electrons in some molecules occupy hybrid orbitals? In valence bond theory, a chemical bond is the overlap of two orbitals that together contain two electrons. The greater the overlap, the stronger the bond and the lower the energy. In hybrid orbitals, the electron probability density is more concentrated in a single directional lobe, allowing greater overlap with the orbitals of other atoms. Hybrid orbitals *minimize* the energy of the molecule by *maximizing* the orbital overlap in a bond.

Hybridization, however, is not a free lunch—in most cases it actually costs some energy. So hybridization occurs only to the degree that the energy payback through bond formation is large. In general, therefore, the more bonds that an atom forms, the greater the tendency of its orbitals to hybridize. Central or interior atoms, which form the most bonds, have the greatest tendency to hybridize. Terminal atoms, which form the fewest bonds, have the least tendency to hybridize. *In this book, we focus on the hybridization of interior atoms and assume that all terminal atoms—those bonding to only one other atom—are unhybridized.* Hybridization is particularly important in carbon, which tends to form four bonds in its compounds and therefore always hybridizes.

In a more nuanced treatment, hybridization is not an all-or-nothing process—it can occur to varying degrees that are not always easy to predict. We saw earlier, for example, that sulfur does not hybridize very much in forming $H_2S$.

Although we cannot examine the procedure for obtaining hybrid orbitals in mathematical detail here, we can make the following general statements regarding hybridization:

▶ The *number of standard atomic orbitals* added together always equals the *number of hybrid orbitals* formed. The total number of orbitals is conserved.

▶ The *particular combination* of standard atomic orbitals added together determines the *shapes and energies* of the hybrid orbitals formed.

▶ The *particular type of hybridization that occurs* is the one that yields the *lowest overall energy for the molecule*. Since actual energy calculations are beyond the scope of this book, we use electron geometries as determined by VSEPR theory to predict the type of hybridization.

## $sp^3$ Hybridization

We can account for the tetrahedral geometry in $CH_4$ by the hybridization of the one $2s$ orbital and the three $2p$ orbitals on the carbon atom. The four new orbitals that result, called $sp^3$ hybrids, are shown in the following energy diagram:

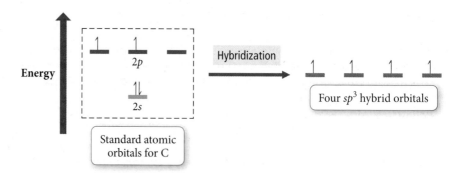

The notation $sp^3$ indicates that the hybrid orbitals are mixtures of one $s$ orbital and three $p$ orbitals. Notice that the hybrid orbitals all have the same energy—they are degenerate. The shapes of the $sp^3$ hybrid orbitals are shown in Figure 10.7 ▶. The four hybrid orbitals are arranged in a tetrahedral geometry with 109.5° angles between them.

We can write an orbital diagram for carbon using these hybrid orbitals:

## Formation of *sp*³ Hybrid Orbitals

One *s* orbital and three *p* orbitals combine to form four *sp*³ orbitals.

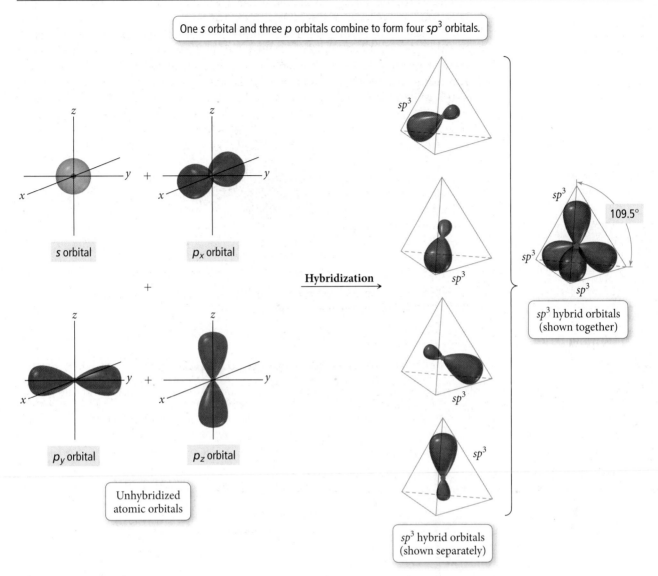

▲ **FIGURE 10.7** *sp*³ **Hybridization**   One s orbital and three p orbitals combine to form four *sp*³ hybrid orbitals.

Carbon's four valence electrons occupy the orbitals singly with parallel spins as dictated by Hund's rule. With this electron configuration, carbon has four half-filled orbitals and can form four bonds with four hydrogen atoms:

The geometry of the *overlapping orbitals* (the hybrids) is tetrahedral, with angles of 109.5° between the orbitals, so the *resulting geometry of the molecule* is tetrahedral, with 109.5° bond angles. This agrees with the experimentally measured geometry of $CH_4$ and with the predicted VSEPR geometry.

Hybridized orbitals readily form chemical bonds because they tend to maximize overlap with other orbitals. However, if the central atom of a molecule contains lone pairs, hybrid orbitals can also accommodate them. For example, the nitrogen orbitals in ammonia are $sp^3$ hybrids. Three of the hybrids are involved in bonding with three hydrogen atoms, but the fourth hybrid contains a lone pair. The presence of the lone pair lowers the tendency of nitrogen's orbitals to hybridize. (Remember that the tendency to hybridize increases with the number of bonds formed.) Therefore the bond angle in $NH_3$ is 107°, a bit closer to the unhybridized $p$ orbital bond angle of 90°.

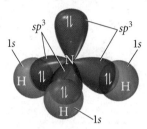

CONCEPTUAL
**CONNECTION 10.7**

**Number of Hybrid Orbitals Part I** How many $sp^3$ hybrid orbitals result from the hybridization of $s$ and $p$ orbitals?

## $sp^2$ Hybridization and Double Bonds

Hybridization of one $s$ and two $p$ orbitals results in three $sp^2$ hybrids and one leftover unhybridized $p$ orbital:

In valence bond theory, the particular hybridization scheme to follow ($sp^2$ versus $sp^3$, for example) for a given molecule is determined computationally, which is beyond the scope of this text. In this book, we determine the particular hybridization scheme from the VSEPR geometry of the molecule, as shown later in this section.

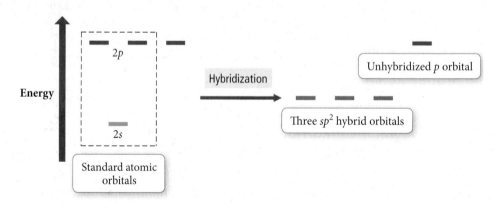

The notation $sp^2$ indicates that the hybrids are mixtures of one $s$ orbital and two $p$ orbitals. Figure 10.8▶ illustrates the shapes of the $sp^2$ hybrid orbitals. Notice that the three hybrid orbitals have a trigonal planar geometry with 120° angles between them. The unhybridized $p$ orbital is perpendicular to the three hybridized orbitals.

As an example of a molecule with $sp^2$ hybrid orbitals, consider $H_2CO$. The unhybridized valence electron configurations of each of the atoms are as follows:

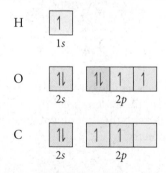

## Formation of $sp^2$ Hybrid Orbitals

One s orbital and two p orbitals combine to form three $sp^2$ orbitals.

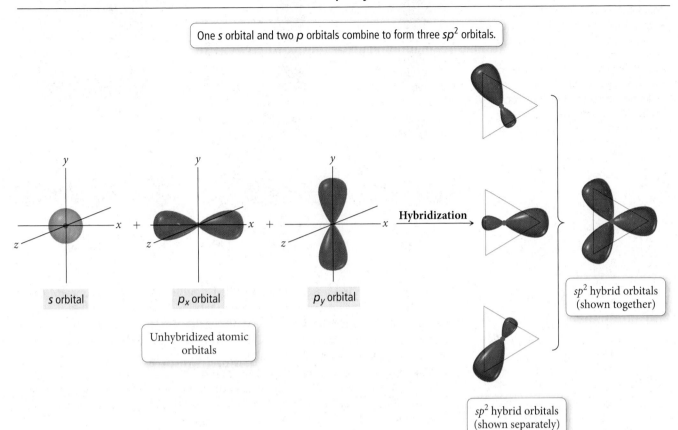

s orbital    $p_x$ orbital    $p_y$ orbital

Unhybridized atomic orbitals

Hybridization

$sp^2$ hybrid orbitals (shown together)

$sp^2$ hybrid orbitals (shown separately)

▲ **FIGURE 10.8** $sp^2$ **Hybridization** One s orbital and two p orbitals combine to form three $sp^2$ hybrid orbitals. One p orbital (not shown) remains unhybridized.

Carbon is the central atom and the hybridization of its orbitals is $sp^2$:

Each of the $sp^2$ orbitals is half filled. The remaining electron occupies the leftover p orbital, even though it is slightly higher in energy. We can now see that the carbon atom has four half-filled orbitals and can therefore form four bonds: two with two hydrogen atoms and two (a double bond) with the oxygen atom. We draw the molecule and the overlapping orbitals as follows:

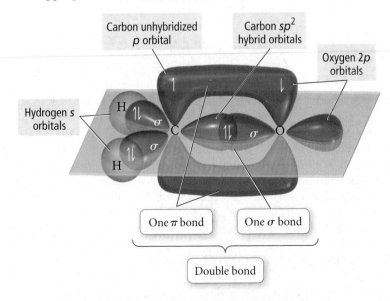

Carbon unhybridized p orbital

Carbon $sp^2$ hybrid orbitals

Oxygen 2p orbitals

Hydrogen s orbitals

One π bond    One σ bond

Double bond

▶ **FIGURE 10.9 Sigma and Pi Bonding** When orbitals overlap side by side, the result is a pi ($\pi$) bond. When orbitals overlap end to end, they form a sigma ($\sigma$) bond. Two atoms can form only one sigma bond. A single bond is a sigma bond, a double bond consists of a sigma bond and a pi bond, and a triple bond consists of a sigma bond and two pi bonds.

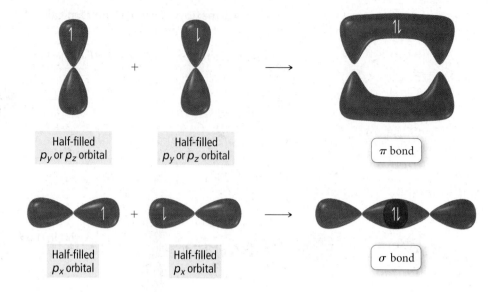

Notice the overlap between the half-filled *p* orbitals on the carbon and oxygen atoms. When *p* orbitals overlap this way (side by side) the resulting bond is a **pi ($\pi$) bond**, and the electron density is above and below the internuclear axis. When orbitals overlap end to end, as is the case in all of the rest of the bonds in the molecule, the resulting bond is a **sigma ($\sigma$) bond** (Figure 10.9▲). Even though we represent the two electrons in a $\pi$ bond as two half-arrows in the upper lobe, they are actually spread out over both the upper and lower lobes (this is one of the limitations we encounter when we try to represent electrons with arrows). We can label all the bonds in the molecule using a notation that specifies the type of bond ($\sigma$ or $\pi$) as well as the type of overlapping orbitals. We have included this notation, as well as the Lewis structure of $H_2CO$ for comparison, in the bonding diagram for $H_2CO$:

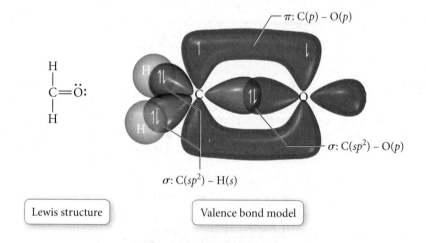

| Lewis structure | Valence bond model |

Notice the correspondence between valence bond theory and the Lewis model. In both models, the central carbon atom forms four bonds: two single bonds and one double bond. However, valence bond theory gives us more insight into the bonds. The double bond between carbon and oxygen according to valence bond theory consists of two different kinds of bonds—one $\sigma$ and one $\pi$—whereas in the Lewis model the two bonds within the double bond appear identical. *A double bond in the Lewis model always corresponds to one $\sigma$ and one $\pi$ bond in valence bond theory.* In general, $\pi$ bonds are weaker than $\sigma$ bonds because side-to-side orbital overlap tends to be less efficient than end-to-end orbital overlap. Consequently, the $\pi$ bond in a double bond is generally easier to break than the $\sigma$ bond. Valence bond theory, as you can see, gives us more insight into the nature of a double bond than the Lewis model.

One—and only one—$\sigma$ bond forms between any two atoms. Additional bonds must be $\pi$ bonds.

Valence bond theory shows us the types of orbitals involved in the bonding and their shapes. For example, in $H_2CO$, the $sp^2$ hybrid orbitals on the central atom are trigonal planar with 120° angles between them, so the resulting predicted geometry of the molecule is trigonal planar with 120° bond angles. The experimentally measured bond angles in $H_2CO$, as discussed previously, are 121.9° for the HCO bond and 116.2° for the HCH bond angle, close to the predicted values.

Valence bond theory allows us to see why the rotation about a double bond is severely restricted. Because of the side-by-side overlap of the $p$ orbitals, the $\pi$ bond must essentially break for rotation to occur (see Chemistry in Your Day: *The Chemistry of Vision*).

Although rotation about a double bond is highly restricted, rotation about a single bond is relatively unrestricted. Consider, for example, the structures of two chlorinated hydrocarbons, 1,2-dichloroethane and 1,2-dichloroethene:

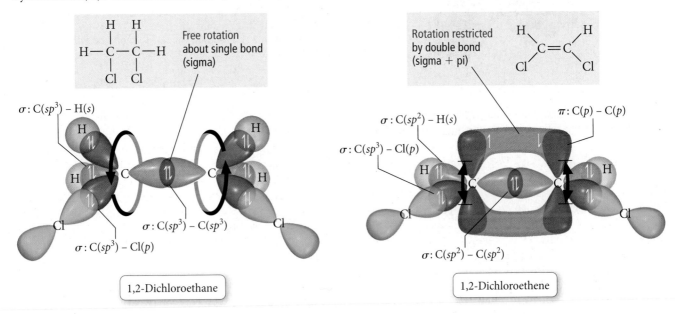

| 1,2-Dichloroethane | 1,2-Dichloroethene |

The hybridization of the carbon atoms in 1,2-dichloroethane is $sp^3$, resulting in relatively free rotation about the sigma single bond. Consequently, there is no difference between the following two structures at room temperature because they quickly interconvert:

In contrast, rotation about the double bond (sigma + pi) in 1,2-dichloroethene is restricted; therefore at room temperature, 1,2-dichloroethene exists in two forms:

*cis*-1,2-Dichloroethene        *trans*-1,2-Dichloroethene

## Chemistry IN YOUR DAY | **The Chemistry of Vision**

In the human eye, light is detected by a chemical switch involving the breaking and re-forming of a π bond. The back portion of the eye, the retina, is coated with millions of light-sensitive cells called rods and cones. Each of these cells contains proteins that bind a compound called 11-*cis*-retinal, which has the following structure:

11-*cis*-Retinal

When a photon of sufficient energy strikes a rod or cone, it causes the isomerization of 11-*cis*-retinal to all-*trans*-retinal:

all-*trans*-Retinal

The isomerization occurs because visible light contains enough energy to break the π bond between the eleventh and twelfth carbon atom in 11-*cis*-retinal. The σ bond, which is stronger, does not break, allowing the molecule to freely rotate about that bond. The π bond then re-forms with the molecule in the *trans* conformation. The different shape of the resultant all-*trans*-retinal causes conformational changes in the protein to which it is bound. These changes cause an electrical signal to be transmitted to the brain.

**QUESTION**

*What is the hybridization of the eleventh and twelfth carbon atoms in retinal?*

These two forms of 1,2-dichloroethene are different compounds with different properties. We distinguish between them with the designations *cis* (meaning "same side") and *trans* (meaning "opposite sides"). Compounds such as these, with the same molecular formula but different structures or different spatial arrangement of atoms, are called *isomers*. Nature can—and does—make different compounds out of the same atoms by arranging the atoms in different ways. Isomerism is common throughout chemistry and especially important in organic chemistry, as we will discuss in Chapter 21.

**CONCEPTUAL CONNECTION 10.8**

**Single and Double Bonds**  In Section 9.10 we learned that double bonds are stronger and shorter than single bonds. For example, a C—C single bond has an average bond energy of 347 kJ/mole, whereas a C=C double bond has an average bond energy of 611 kJ/mole. Use valence bond theory to explain why a double bond is *not* simply twice as strong as a single bond.

## *sp* Hybridization and Triple Bonds

Hybridization of one *s* and one *p* orbital results in two *sp* hybrid orbitals and two leftover unhybridized *p* orbitals:

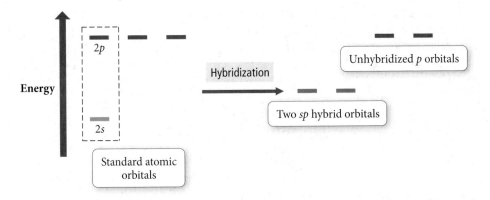

Figure 10.10▼ shows the shapes of the *sp* hybrid orbitals. Notice that the two *sp* hybrid orbitals are arranged in a linear geometry with a 180° angle between them. The unhybridized *p* orbitals are oriented in the plane that is perpendicular to the hybridized *sp* orbitals.

### Formation of *sp* Hybrid Orbitals

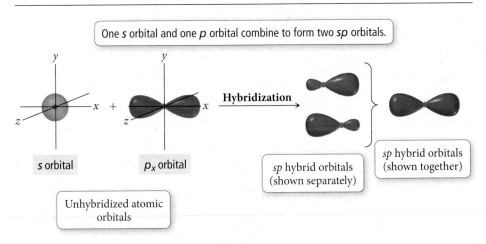

◀ **FIGURE 10.10**

***sp* Hybridization**  One s orbital and one *p* orbital combine to form two *sp* hybrid orbitals. Two *p* orbitals (not shown) remain unhybridized.

The acetylene molecule, HC≡CH, has *sp* hybrid orbitals. The four valence electrons of carbon distribute themselves among the two *sp* hybrid orbitals and the two *p* orbitals:

Each carbon atom then has four half-filled orbitals and can form four bonds: one with a hydrogen atom and three (a triple bond) with the other carbon atom. We draw the molecule and the overlapping orbitals as follows:

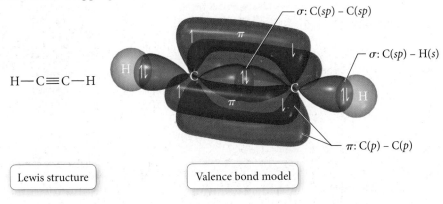

Lewis structure          Valence bond model

Notice that the triple bond between the two carbon atoms consists of two $\pi$ bonds (overlapping $p$ orbitals) and one $\sigma$ bond (overlapping $sp$ orbitals). The $sp$ orbitals on the carbon atoms are linear with 180° between them, so the resulting geometry of each carbon on the molecule is linear with 180° bond angles, in agreement with the experimentally measured geometry of HC≡CH, and also in agreement with the prediction of VSEPR theory.

CONCEPTUAL
**CONNECTION 10.9**

**Number of Hybrid Orbitals Part II** How many $sp$ hybrid orbitals result from the hybridization of $s$ and $p$ orbitals?

## $sp^3d$ and $sp^3d^2$ Hybridization

Recall that, according to the Lewis model, elements occurring in the third row of the periodic table (or below) can exhibit expanded octets (see Section 9.9). The equivalent concept in valence bond theory is hybridization involving the $d$ orbitals. For third-period elements, the $3d$ orbitals are involved in hybridization because their energies are close to the energies of the $3s$ and $3p$ orbitals. The hybridization of one $s$ orbital, three $p$ orbitals, and one $d$ orbital results in $sp^3d$ hybrid orbitals (Figure 10.11(a)▼). The five $sp^3d$ hybrid orbitals have a trigonal bipyramidal arrangement (Figure 10.11(b)▼). As an example of $sp^3d$ hybridization, consider arsenic pentafluoride, $AsF_5$. The arsenic atom bonds to five fluorine atoms by overlap between the $sp^3d$ hybrid orbitals on arsenic and $p$ orbitals on the fluorine atoms, as shown here:

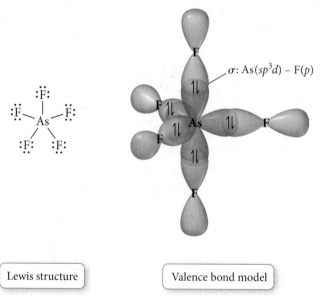

$\sigma$: As($sp^3d$) – F($p$)

Lewis structure

Valence bond model

**▼ FIGURE 10.11 $sp^3d$**
**Hybridization** One s orbital, three $p$ orbitals, and one $d$ orbital combine to form five $sp^3d$ hybrid orbitals.

The $sp^3d$ orbitals on the arsenic atom are trigonal bipyramidal, so the molecular geometry is trigonal bipyramidal.

(a)

(b)

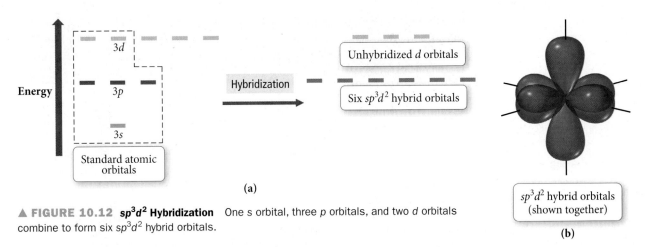

▲ **FIGURE 10.12** $sp^3d^2$ **Hybridization** One s orbital, three $p$ orbitals, and two $d$ orbitals combine to form six $sp^3d^2$ hybrid orbitals.

$sp^3d^2$ hybrid orbitals (shown together)

**(b)**

The hybridization of one $s$ orbital, three $p$ orbitals, and *two d orbitals* results in $sp^3d^2$ hybrid orbitals (Figure 10.12(a)▲). The six $sp^3d^2$ hybrid orbitals have an octahedral geometry (Figure 10.12(b)▲). As an example of $sp^3d^2$ hybridization, consider sulfur hexa-fluoride, $SF_6$. The sulfur atom bonds to six fluorine atoms by overlap between the $sp^3d^2$ hybrid orbitals on sulfur and $p$ orbitals on the fluorine atoms:

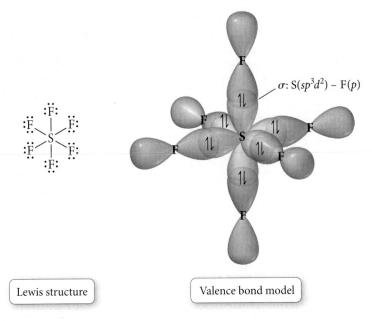

Lewis structure

Valence bond model

The $sp^3d^2$ orbitals on the sulfur atom are octahedral, so the molecular geometry is octa-hedral, again in agreement with VSEPR theory and with the experimentally observed geometry.

## Writing Hybridization and Bonding Schemes

We have now studied examples of the five main types of atomic orbital hybridization. *But how do we know which hybridization scheme best describes the orbitals of a specific atom in a specific molecule?* In computational valence bond theory, the energy of the mole-cule is calculated using a computer; the degree of hybridization as well as the type of hybridization are varied to find the combination that gives the molecule the lowest overall energy. For our purposes, we can assign a hybridization scheme from the electron geometry—determined using VSEPR theory—of the central atom (or interior atoms) of

**TABLE 10.3 Hybridization Scheme from Electron Geometry**

| Number of Electron Groups | Electron Geometry (from VSEPR Theory) | Hybridization Scheme | |
|---|---|---|---|
| 2 | Linear | $sp$ | |
| 3 | Trigonal planar | $sp^2$ | |
| 4 | Tetrahedral | $sp^3$ | |
| 5 | Trigonal bipyramidal | $sp^3d$ | |
| 6 | Octahedral | $sp^3d^2$ | |

the molecule. The five VSEPR electron geometries and the corresponding hybridization schemes are shown in Table 10.3. For example, if the electron geometry of the central atom is tetrahedral, then the hybridization is $sp^3$, and if the electron geometry is octahedral, then the hybridization is $sp^3d^2$, and so on. Although this method of determining the hybridization scheme is not 100% accurate (for example, it predicts that $H_2S$ should be $sp^3$ when in fact $H_2S$ is largely unhybridized), it is the best we can do without more complex computer-based calculations.

We are now ready to put the Lewis model and valence bond theory together to describe bonding in molecules. In the procedure and examples that follow, you will learn how to write a *hybridization and bonding scheme* for a molecule. This involves drawing the Lewis structure for the molecule, determining its geometry using VSEPR theory, determining the hybridization of the interior atoms, drawing the molecule with its overlapping orbitals, and labeling each bond with the $\sigma$ and $\pi$ notation followed by the type of overlapping orbitals. As you can see, this procedure involves virtually everything you have learned about bonding in this chapter and Chapter 9. The procedure for writing a hybridization and bonding scheme is shown in the left column, with two examples of how to apply the procedure in the columns to the right.

# Procedure for...

| Hybridization and Bonding Scheme | **Example 10.6**  **Hybridization and Bonding Scheme**  Write a hybridization and bonding scheme for bromine trifluoride, $BrF_3$. | **Example 10.7**  **Hybridization and Bonding Scheme**  Write a hybridization and bonding scheme for acetaldehyde,  |
|---|---|---|
| 1. **Write the Lewis structure for the molecule.** | SOLUTION  $BrF_3$ has 28 valence electrons and the following Lewis structure:  | SOLUTION  Acetaldehyde has 18 valence electrons and the following Lewis structure:  |
| 2. **Use VSEPR theory to predict the electron geometry about the central atom (or interior atoms).** | The bromine atom has five electron groups and therefore has a trigonal bipyramidal electron geometry. | The leftmost carbon atom has four electron groups and a tetrahedral electron geometry. The rightmost carbon atom has three electron groups and a trigonal planar geometry. |
| 3. **Select the correct hybridization for the central atom (or interior atoms) based on the electron geometry (see Table 10.3).** | A trigonal bipyramidal electron geometry corresponds to $sp^3d$ hybridization. | The leftmost carbon atom is $sp^3$ hybridized, and the rightmost carbon atom is $sp^2$ hybridized. |
| 4. **Sketch the molecule, beginning with the central atom and its orbitals. Show overlap with the appropriate orbitals on the terminal atoms.** | | |
| 5. **Label all bonds using $\sigma$ or $\pi$ notation followed by the type of overlapping orbitals.** |  $\sigma$: Br($sp^3d$) – F($p$)  Lone pairs in $sp^3d$ orbitals |  $\sigma$: C($sp^3$) – H($s$)   $\pi$: C($p$) – O($p$)  $\sigma$: C($sp^2$) – H($s$)  $\sigma$: C($sp^3$) – C($sp^2$)   $\sigma$: C($sp^2$) – O($p$) |
| | **FOR PRACTICE 10.6**  Write the hybridization and bonding scheme for $XeF_4$. | **FOR PRACTICE 10.7**  Write the hybridization and bonding scheme for HCN. |

**Example** 10.8    **Hybridization and Bonding Scheme**    ▶

Use valence bond theory to write a hybridization and bonding scheme for ethene, $H_2C$=$CH_2$.

**SOLUTION**

| | |
|---|---|
| **1.** Write the Lewis structure for the molecule. | $$\begin{array}{c} \ \ \ \ \ \text{H}\ \ \ \ \text{H} \\ \ \ \ \ \ \| \ \ \ \ \ \| \\ \text{H---C}\!=\!\text{C---H} \end{array}$$ |
| **2.** Apply VSEPR theory to predict the electron geometry about the central atom (or interior atoms). | The molecule has two interior atoms. Because each atom has three electron groups (one double bond and two single bonds), the electron geometry about each atom is trigonal planar. |
| **3.** Select the correct hybridization for the central atom (or interior atoms) based on the electron geometry (see Table 10.3). | A trigonal planar geometry corresponds to $sp^2$ hybridization. |
| **4.** Sketch the molecule, beginning with the central atom and its orbitals. Show overlap with the appropriate orbitals on the terminal atoms. |  |
| **5.** Label all bonds using $\sigma$ or $\pi$ notation followed by the type of overlapping orbitals. | 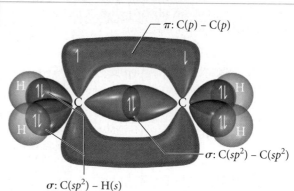 $\pi$: C($p$) – C($p$) $\sigma$: C($sp^2$) – C($sp^2$) $\sigma$: C($sp^2$) – H($s$) |

**FOR PRACTICE 10.8**    Use valence bond theory to write a hybridization and bonding scheme for $CO_2$.

**FOR MORE PRACTICE 10.8**    What is the hybridization of the central iodine atom in $I_3^-$?

## 10.8  Molecular Orbital Theory: Electron Delocalization

Valence bond theory can explain many aspects of chemical bonding—such as the rigidity of a double bond—but it also has limitations. In valence bond theory, we treat electrons as if they reside in the quantum-mechanical orbitals that we calculated *for atoms*. This is a significant oversimplification that we partially compensate for by hybridization. Nevertheless, we can do better.

In Chapter 7, we stated that the  mathematical derivation of energies and orbitals for electrons *in atoms* comes from solving the Schrödinger equation for the atom of interest. For a molecule, we can theoretically do the same thing. The resulting orbitals would be the actual *molecular* orbitals of the molecule as a whole (in contrast to valence bond theory, in which the orbitals are those of individual atoms). As it turns out, however, solving the Schrödinger equation exactly for even the simplest molecules is impossible without making some approximations.

In **molecular orbital (MO) theory**, we do not actually solve the Schrödinger equation for a molecule directly. Instead, we use a trial function, an "educated guess" as to what the solution might be. In other words, rather than mathematically solving the Schrödinger equation, which would give us a mathematical function describing an orbital, we start with a trial mathematical function for the orbital. We then test the trial function to see how well it "works."

We can understand this process by analogy to solving an algebraic equation. Suppose we want to know $x$ in the equation $4x + 5 = 70$ without actually solving the equation. For an easy equation like this one, we might first estimate that $x = 16$. We can then determine how well our estimate works by substituting $x = 16$ into the equation. If the estimate did not work, we could try again until we found the right value of $x$. (In this case, we can quickly see that $x$ must be a little more than 16.)

In molecular orbital theory, the estimating procedure is analogous. However, we need to add one more important concept to get at the heart of molecular orbital theory. In order to determine how well a trial function for an orbital "works" in molecular orbital theory, we calculate its energy. No matter how good our trial function, *we can never do better than nature at minimizing the energy of the orbital.* In other words, we can devise any trial function for an orbital in a molecule and calculate its energy. The energy we calculate for the devised orbital will always be greater than or (at best) equal to the energy of the actual orbital.

How does this help us? The best possible orbital is therefore the one with the minimum energy. In modern molecular orbital theory, computer programs are designed to try many different variations of a guessed orbital and compare the energies of each one. The variation with the lowest energy is the best approximation for the actual molecular orbital.

> Molecular orbital theory is a specific application of a more general quantum-mechanical approximation technique called the variational method. In the variational method, the energy of a trial function within the Schrödinger equation is minimized.

> We calculate the energy of an estimated orbital by substituting it into the Schrödinger equation and solving for the energy.

## Linear Combination of Atomic Orbitals (LCAOs)

The simplest trial functions that work reasonably well in molecular orbital theory turn out to be linear combinations of atomic orbitals, or LCAOs. An LCAO molecular orbital is a *weighted linear sum—analogous to a weighted average—of the valence atomic orbitals* of the atoms in the molecule. At first glance, this concept might seem very similar to that of hybridization in valence bond theory. However, in valence bond theory, hybrid orbitals are weighted linear sums of the valence atomic orbitals of a *particular atom*, and the hybrid orbitals remain *localized* on that atom. In molecular orbital theory, the molecular orbitals are weighted linear sums of the valence atomic orbitals of *all the atoms* in a molecule, and many of the molecular orbitals are *delocalized* over the entire molecule.

Consider the $H_2$ molecule. One of the molecular orbitals for $H_2$ is simply an equally weighted sum of the $1s$ orbital from one atom and the $1s$ orbital from the other. We represent this pictorially and energetically as follows:

> When molecular orbitals are calculated mathematically, it is actually the *wave functions* corresponding to the orbitals that are combined.

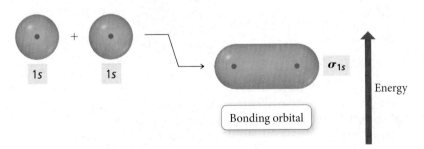

The name of this molecular orbital is $\sigma_{1s}$. The $\sigma$ comes from the shape of the orbital, which looks like a $\sigma$ bond in valence bond theory, and the $1s$ comes from its formation by a linear sum of $1s$ orbitals. The $\sigma_{1s}$ orbital is lower in energy than either of the two $1s$ atomic orbitals from which it was formed. For this reason, this orbital is called a **bonding orbital**. When electrons occupy bonding molecular orbitals, the energy of the electrons is lower than it would be if they were occupying atomic orbitals.

We can think of a molecular orbital in a molecule in much the same way that we think about an atomic orbital in an atom. Electrons seek the lowest energy molecular orbital available, but just as an atom has more than one atomic orbital (and some may be empty), a molecule has more than one molecular orbital (and some may be empty). The next molecular orbital of $H_2$ is approximated by summing the $1s$ orbital on one hydrogen atom with the *negative* (opposite phase) of the $1s$ orbital on the other hydrogen atom:

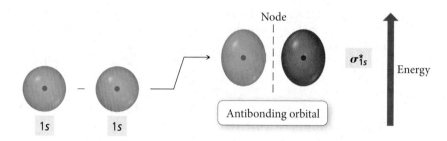

The different phases of the orbitals result in *destructive* interference between them. The resulting molecular orbital therefore has a node between the two atoms. The different colors (red and blue) on either side of the node represent the different phases of the orbital (see Section 7.6). The name of this molecular orbital is $\sigma_{1s}^*$. The star indicates that this orbital is an **antibonding orbital**. Electrons in antibonding orbitals have higher energies than they did in their respective atomic orbitals and therefore tend to raise the energy of the system (relative to the unbonded atoms).

In general, when two atomic orbitals are added together to form molecular orbitals, one of the resultant molecular orbitals is lower in energy (the bonding orbital) than the atomic orbitals and the other is higher in energy (the antibonding orbital). Remember that electrons in orbitals behave like waves. The bonding molecular orbital arises out of constructive interference between the overlapping atomic orbitals because both orbitals have the same phase. The antibonding orbital arises out of destructive interference between the overlapping atomic orbitals because *subtracting* one from the other means the two interacting orbitals have opposite phases (Figure 10.13 ▼).

For this reason, the bonding orbital has an increased electron density in the internuclear region, whereas the antibonding orbital has a node in the internuclear region. Bonding orbitals have greater electron density in the internuclear region, thereby lowering their energy compared to the orbitals in nonbonded atoms. Antibonding orbitals have less electron density in the internuclear region, and their energies are generally higher than in the orbitals of nonbonded atoms.

▶ **FIGURE 10.13 Formation of Bonding and Antibonding Orbitals** Constructive interference between two atomic orbitals gives rise to a molecular orbital that is lower in energy than the atomic orbitals. This is the bonding orbital. Destructive interference between two atomic orbitals gives rise to a molecular orbital that is higher in energy than the atomic orbitals. This is the antibonding orbital.

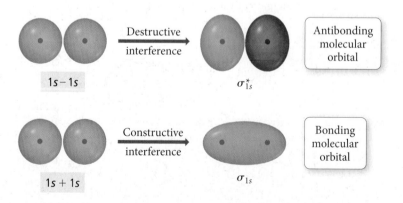

We put all of this together in the molecular orbital energy diagram for $H_2$:

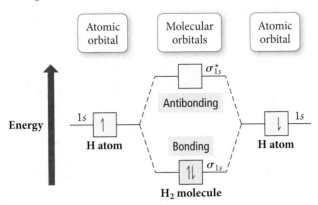

We can represent the molecular orbital energy diagram with a molecular orbital electron configuration (which is analogous to the electron configurations we wrote for elements in Section 8.3):

$$H_2 \qquad (\sigma_{1s})^2$$

In this notation, the $\sigma_{1s}$ represents the molecular orbital, and the superscript 2 represents the two electrons in the orbital.

The molecular orbital (MO) energy diagram shows that two hydrogen atoms can lower their overall energy by forming $H_2$ because the electrons can move from higher-energy atomic orbitals into the lower-energy $\sigma_{1s}$ bonding molecular orbital. In molecular orbital theory, we define the **bond order** of a diatomic molecule such as $H_2$ as follows:

$$\text{Bond order} = \frac{(\text{number of electrons in bonding MOs}) - (\text{number of electrons in antibonding MOs})}{2}$$

For $H_2$, the bond order is 1:

$$H_2 \text{ bond order} = \frac{2 - 0}{2} = 1$$

A positive bond order means that there are more electrons in bonding molecular orbitals than in antibonding molecular orbitals. The electrons therefore have lower energy than they did in the orbitals of the isolated atoms, and a chemical bond forms. In general, the higher the bond order, the stronger the bond. A negative or zero bond order indicates that a bond does *not* form between the atoms. For example, consider the MO diagram for $He_2$:

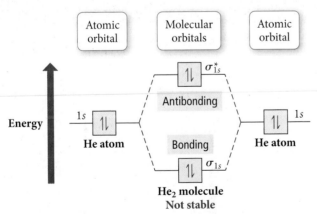

Notice that the two additional electrons must go into the higher-energy antibonding orbital. There is no net stabilization by joining two helium atoms to form a helium molecule, as indicated by the bond order:

$$He_2 \text{ bond order} = \frac{2 - 2}{2} = 0$$

So according to MO theory, $He_2$ should not exist as a stable molecule, and in fact it does not. Another interesting case is the helium–helium ion, $He_2^+$, with the following MO diagram:

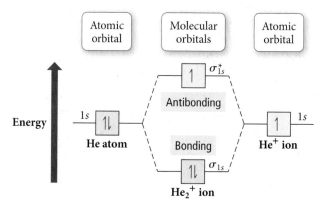

The bond order is $\frac{1}{2}$, indicating that $He_2^+$ should exist, and indeed it does.

### Summarizing LCAO–MO Theory:

▶ We can approximate molecular orbitals (MOs) as a linear combination of atomic orbitals (AOs). The total number of MOs formed from a particular set of AOs always equals the number of AOs in the set.

▶ When two AOs combine to form two MOs, one MO is lower in energy (the bonding MO) and the other is higher in energy (the antibonding MO).

▶ When assigning the electrons of a molecule to MOs, fill the lowest energy MOs first with a maximum of two spin-paired electrons per orbital.

▶ When assigning electrons to two MOs of the same energy, follow Hund's rule—fill the orbitals singly first, with parallel spins, before pairing.

▶ The bond order in a diatomic molecule is the number of electrons in bonding MOs minus the number in antibonding MOs divided by two. Stable bonds are associated with a positive bond order (more electrons in bonding MOs than in antibonding MOs).

Notice the power of the molecular orbital approach. Every electron that enters a bonding molecular orbital stabilizes the molecule or polyatomic ion, and every electron that enters an antibonding molecular orbital destabilizes it. The emphasis on electron pairs has been removed. One electron in a bonding molecular orbital stabilizes half as much as two, so a bond order of one-half is nothing mysterious.

## Example 10.9 | Bond Order

Use molecular orbital theory to predict the bond order in $H_2^-$. Is the $H_2^-$ bond a stronger or weaker bond than the $H_2$ bond?

#### SOLUTION

The $H_2^-$ ion has three electrons. Assign the three electrons to the molecular orbitals, filling lower-energy orbitals first and proceeding to higher-energy orbitals.

Calculate the bond order by subtracting the number of electrons in antibonding orbitals from the number in bonding orbitals and dividing the result by two.

$$H_2^- \text{ bond order} = \frac{2-1}{2} = +\frac{1}{2}$$

Since the bond order is positive, $H_2^-$ should be stable. However, the bond order of $H_2^-$ is lower than the bond order of $H_2$ (which is 1); therefore, the bond in $H_2^-$ is weaker than in $H_2$.

**FOR PRACTICE 10.9** Use molecular orbital theory to predict the bond order in $H_2^+$. Is the $H_2^+$ bond a stronger or weaker bond than the $H_2$ bond?

## Period Two Homonuclear Diatomic Molecules

Homonuclear diatomic molecules (molecules made up of two atoms of the same kind) formed from second-period elements have between 2 and 16 valence electrons. To explain bonding in these molecules, we must consider the next set of higher-energy molecular orbitals, which can be approximated by linear combinations of the valence atomic orbitals of the period 2 elements.

We begin with $Li_2$. Even though lithium is normally a metal, we can use MO theory to predict whether or not the $Li_2$ molecule should exist in the gas phase. We approximate the molecular orbitals in $Li_2$ as linear combinations of the $2s$ atomic orbitals. The resulting molecular orbitals look much like those of the $H_2$ molecule. The MO diagram for $Li_2$ therefore looks a lot like the MO diagram for $H_2$:

> The core electrons can be ignored (as they are in other models for bonding) because these electrons are held too tightly to individual nuclei to contribute significantly to chemical bonding.

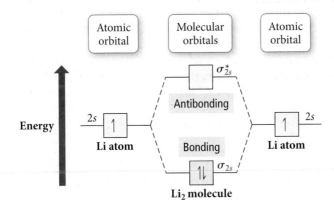

The two valence electrons of $Li_2$ occupy a bonding molecular orbital. We would predict that the $Li_2$ molecule is stable with a bond order of 1. Experiments confirm this prediction. In contrast, consider the MO diagram for $Be_2$:

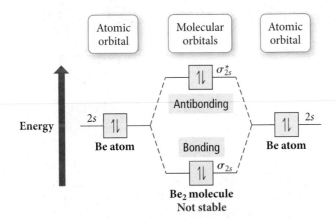

The four valence electrons of $Be_2$ occupy one bonding MO and one antibonding MO. The bond order is 0 and we predict that $Be_2$ should not be stable, which is again consistent with experimental findings.

The next homonuclear molecule composed of second row elements is $B_2$, which has six total valence electrons to accommodate. We can approximate the next higher-energy molecular orbitals for $B_2$ and the rest of the period 2 diatomic molecules as linear combinations of the $2p$ orbitals taken pairwise. Since the three $2p$ orbitals orient along three orthogonal axes, we must assign similar axes to the molecule. In this book, we assign the internuclear axis to be the $x$ direction. Then we represent the LCAO–MOs that result from combining the $2p_x$ orbitals—the ones that lie along the internuclear axis—from each atom as follows:

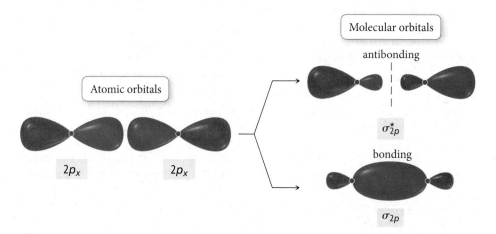

The bonding MO in this pair looks something like candy in a wrapper, with increased electron density in the internuclear region due to constructive interference between the two $2p$ atomic orbitals. It has the characteristic $\sigma$ shape (it is cylindrically symmetric about the bond axis) and is therefore called the $\sigma_{2p}$ bonding orbital. The antibonding orbital, called $\sigma_{2p}^*$, has a node between the two nuclei (due to destructive interference between the two $2p$ orbitals) and is higher in energy than either of the $2p_x$ orbitals.

The LCAO–MOs that result from combining the $2p_z$ orbitals from each atom are represented as follows:

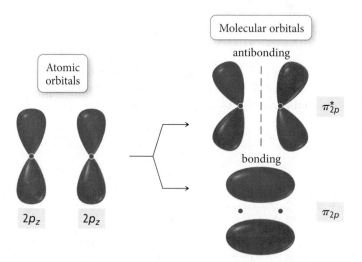

Notice that in this case the $p$ orbitals are added together in a side-by-side orientation (in contrast to the $2p_x$ orbitals, which were oriented end to end). The resultant molecular orbitals consequently have a different shape. The electron density in the bonding molecular orbital is above and below the internuclear axis with a nodal plane that includes the internuclear axis. This orbital resembles the electron density distribution of a $\pi$ bond in valence bond theory. We call this orbital the $\pi_{2p}$ orbital. The corresponding antibonding orbital has an additional node *between* the nuclei (perpendicular to the internuclear axis) and is called the $\pi_{2p}^*$ orbital.

The LCAO–MOs that result from combining the $2p_y$ orbitals from each atom are represented as follows:

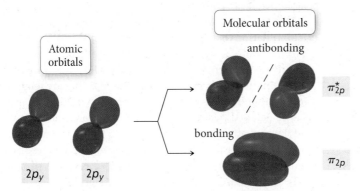

The only difference between the $2p_y$ and the $2p_z$ atomic orbitals is a 90° rotation about the internuclear axis. Consequently, the only difference between the resulting MOs is a 90° rotation about the internuclear axis. The energies and the names of the bonding and antibonding MOs obtained from the combination of the $2p_y$ AOs are identical to those obtained from the combination of the $2p_z$ AOs.

Before we can draw MO diagrams for $B_2$ and the other second-period diatomic molecules, we must determine the relative energy ordering of the MOs obtained from the $2p$ AO combinations. This is not a simple task. The relative ordering of MOs obtained from LCAO–MO theory is usually determined computationally. There is no single order that works for all molecules. For second-period diatomic molecules, computations reveal that the energy ordering for $B_2$, $C_2$, and $N_2$ is slightly different than that for $O_2$, $F_2$, and $Ne_2$ as follows:

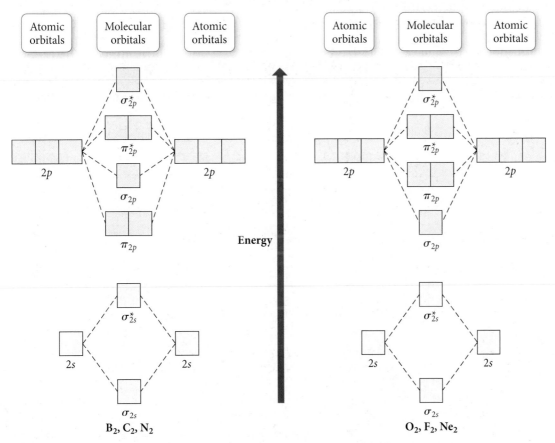

▲ Molecular orbital energy diagrams for second-period diatomic molecules show that the energy ordering of the $\pi_{2p}$ and $\sigma_{2p}$ molecular orbitals can vary.

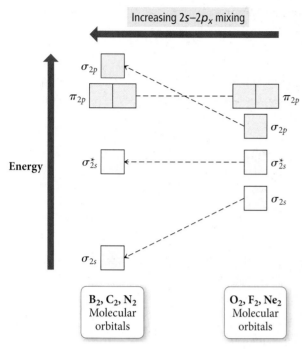

Increasing $2s$–$2p_x$ mixing

Energy

| B₂, C₂, N₂ Molecular orbitals | O₂, F₂, Ne₂ Molecular orbitals |

▲ **FIGURE 10.14 The Effects of 2s–2p Mixing** The degree of mixing between two orbitals decreases with increasing energy difference between them. Mixing of the 2s and $2p_x$ orbitals is therefore greater in B₂, C₂, and N₂ than in O₂, F₂, and Ne₂ because in B, C, and N the energy levels of the atomic orbitals are more closely spaced than in O, F, and Ne. This mixing produces a change in energy ordering for the $\pi_{2p}$ and $\sigma_{2p}$ molecular orbitals.

The difference in energy ordering can only be explained by revisiting our LCAO–MO model. In our simplified treatment, we assumed that the MOs that result from the second-period AOs could be calculated pairwise. In other words, we took the linear combination of a 2s from one atom with the 2s from another, a $2p_x$ from one atom with the $2p_x$ from the other, and so on. However, in a more comprehensive treatment, the MOs are formed from linear combinations that include all of the AOs that are relatively close to each other in energy and of the correct symmetry. Specifically, in a more detailed treatment, the two 2s orbitals and the two $2p_x$ orbitals should all be combined to form a total of four molecular orbitals. The extent to which we include this type of mixing affects the energy levels of the corresponding MOs, as shown in Figure 10.14 ◀. The bottom line is that s–p mixing is significant in B₂, C₂, and N₂ but not in O₂, F₂, and Ne₂. The result of the mixing is a different energy ordering for some of these diatomic molecules.

Figure 10.15 ▼ shows the MO energy diagrams for the rest of the second-period homonuclear diatomic molecules, as well as their bond orders, bond energies, and bond lengths. Notice that as bond order increases, the bond gets stronger (greater bond energy) and shorter (smaller bond length). For B₂, with six electrons, the bond order is 1. For C₂, the bond order is 2, and for N₂, the bond order reaches a maximum with a value of 3. Recall that the Lewis structure for N₂ has a triple bond, so both the Lewis model and molecular orbital theory predict a strong bond for N₂, which is experimentally observed.

▶ **FIGURE 10.15 Molecular Orbital Energy Diagrams for Second-Row p-block Homonuclear Diatomic Molecules**

| | Large 2s–2pₓ interaction | | | | Small 2s–2pₓ interaction | | |
|---|---|---|---|---|---|---|---|
| | **B₂** | **C₂** | **N₂** | | **O₂** | **F₂** | **Ne₂** |
| $\sigma_{2p}^*$ | | | | $\sigma_{2p}^*$ | | | ⇅ |
| $\pi_{2p}^*$ | | | | $\pi_{2p}^*$ | ↑ ↑ | ⇅ ⇅ | ⇅ ⇅ |
| $\sigma_{2p}$ | | | ⇅ | $\pi_{2p}$ | ⇅ ⇅ | ⇅ ⇅ | ⇅ ⇅ |
| $\pi_{2p}$ | ↑ ↑ | ⇅ ⇅ | ⇅ ⇅ | $\sigma_{2p}$ | ⇅ | ⇅ | ⇅ |
| $\sigma_{2s}^*$ | ⇅ | ⇅ | ⇅ | $\sigma_{2s}^*$ | ⇅ | ⇅ | ⇅ |
| $\sigma_{2s}$ | ⇅ | ⇅ | ⇅ | $\sigma_{2s}$ | ⇅ | ⇅ | ⇅ |
| Bond order | 1 | 2 | 3 | | 2 | 1 | 0 |
| Bond energy (kJ/mol) | 290 | 620 | 946 | | 498 | 159 | — |
| Bond length (pm) | 159 | 131 | 110 | | 121 | 143 | — |

CONCEPTUAL
**CONNECTION 10.10**

**Bond Order** What is the bond order for a second-period diatomic molecule containing three electrons in antibonding molecular orbitals and six electrons in bonding molecular orbitals?

In $O_2$, the two additional electrons occupy antibonding orbitals and the bond order is 2. These two electrons are unpaired—they occupy the $\pi_{2p}^*$ orbitals *singly with parallel spins*, as indicated by Hund's rule. The presence of unpaired electrons in the molecular orbital diagram of oxygen is significant because oxygen is known from experiment to be *paramagnetic*—it is attracted to a magnetic field (see Section 8.7). The paramagnetism of oxygen can be demonstrated by suspending liquid oxygen between the poles of a magnet. This magnetic property is the direct result of *unpaired electrons,* whose spin and movement around the nucleus (more accurately known as orbital angular momentum) generate tiny magnetic fields. When a paramagnetic substance is placed in an external magnetic field, the magnetic fields of each atom or molecule align with the external field, creating the attraction (much as two magnets attract each other when properly oriented). In contrast, when the electrons in an atom or molecule are all *paired*, the magnetic fields caused by electron spin and orbital angular momentum tend to cancel each other, resulting in diamagnetism. A *diamagnetic* substance is not attracted to a magnetic field (and is, in fact, slightly repelled).

In the Lewis structure of $O_2$, as well as in the valence bond model of $O_2$, all of the electrons seem to be paired:

▲ Liquid oxygen can be suspended between the poles of a magnet because it is paramagnetic. It contains unpaired electrons (depicted here in the inset) that generate tiny magnetic fields, which align with and interact with the external field.

The s orbital on each oxygen atom contains two electrons, but for clarity neither the s orbitals nor the electrons that occupy them are shown.

In the MO diagram for $O_2$, however, we can see the unpaired electrons. Molecular orbital theory is the more powerful theory in that it can account for the paramagnetism of $O_2$—it gives us a picture of bonding that more closely corresponds to what we see in experiments. Continuing along the second-row homonuclear diatomic molecules, we see that $F_2$ has a bond order of 1 and $Ne_2$ has a bond order of 0, again consistent with experiment since $F_2$ exists and $Ne_2$ does not.

---

## Example 10.10 | **Molecular Orbital Theory**

Draw an MO energy diagram and determine the bond order for the $N_2^-$ ion. Do you expect the bond in the $N_2^-$ ion to be stronger or weaker than the bond in the $N_2$ molecule? Is $N_2^-$ diamagnetic or paramagnetic?

**SOLUTION**

Write an energy-level diagram for the molecular orbitals in $N_2^-$. Use the energy ordering for $N_2$.

$\sigma_{2p}^*$

$\pi_{2p}^*$

$\sigma_{2p}$

$\pi_{2p}$

$\sigma_{2s}^*$

$\sigma_{2s}$

—Continued on the next page

*—Continued*

| | |
|---|---|
| The $N_2^-$ ion has 11 valence electrons (five for each nitrogen atom plus one for the negative charge). Assign the electrons to the molecular orbitals beginning with the lowest energy orbitals and following Hund's rule. | $\sigma_{2p}^*$  $\pi_{2p}^*$  $\sigma_{2p}$  $\pi_{2p}$  $\sigma_{2s}^*$  $\sigma_{2s}$ |
| Calculate the bond order by subtracting the number of electrons in antibonding orbitals from the number in bonding orbitals and dividing the result by 2. | $N_2^-$ bond order $= \dfrac{8-3}{2} = +2.5$ |

The bond order is 2.5, which is a lower bond order than in the $N_2$ molecule (bond order $= 3$); therefore, the bond is weaker. The MO diagram shows that the $N_2^-$ ion has one unpaired electron and is therefore paramagnetic.

**FOR PRACTICE 10.10** Draw an MO energy diagram and determine the bond order for the $N_2^+$ ion. Do you expect the bond in the $N_2^+$ ion to be stronger or weaker than the bond in the $N_2$ molecule? Is $N_2^+$ diamagnetic or paramagnetic?

**FOR MORE PRACTICE 10.10** Apply molecular orbital theory to determine the bond order of $Ne_2$.

## Second-Period Heteronuclear Diatomic Molecules

We can also apply molecular orbital theory to heteronuclear diatomic molecules (two different atoms). For example, we can draw an MO diagram for NO as follows:

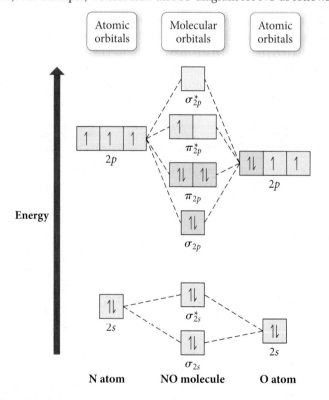

Oxygen is more electronegative than nitrogen, so its atomic orbitals are lower in energy than nitrogen's atomic orbitals. When two atomic orbitals are identical and of equal energy, the weighting of each orbital in forming a molecular orbital is identical. However, when two atomic orbitals are different, the weighting of each orbital in forming a molecular orbital may be different. More specifically, when a molecular orbital is approximated as a linear combination of atomic orbitals of different energies, the lower-energy atomic orbital makes a greater contribution to the bonding molecular orbital and the higher-energy atomic orbital makes a greater contribution to the antibonding molecular orbital. For example, notice that the $\sigma_{2s}$ bonding orbital is closer in energy to the oxygen 2s orbital than to the nitrogen 2s orbital. We can also see this unequal weighting in the shape of the resultant molecular orbital, in which the electron density is concentrated on the oxygen atom, as shown in Figure 10.16►.

As another example of a heteronuclear diatomic molecule, consider the MO diagram for HF:

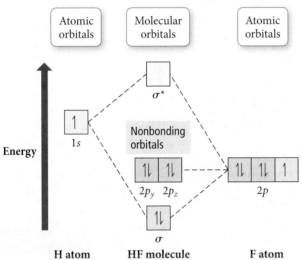

▲ FIGURE 10.16 **Shape of** $\sigma_{2s}$ **bonding orbital in NO** The molecular orbital shows more electron density at the oxygen end of the molecule because the atomic orbitals of oxygen, the more electronegative element, are lower in energy than those of nitrogen. They therefore contribute more to the bonding molecular orbital.

Fluorine is so electronegative that all of its atomic orbitals are lower in energy than hydrogen's atomic orbitals. In fact, fluorine's 2s orbital is so low in energy compared to hydrogen's 1s orbital that it does not contribute appreciably to the molecular orbitals. The molecular orbitals in HF are approximated by the linear combination of the fluorine $2p_x$ orbital and the hydrogen 1s orbital. The other 2p orbitals remain localized on the fluorine and appear in the energy diagram as **nonbonding orbitals**. The electrons in the nonbonding orbitals remain localized on the fluorine atom.

**Example** 10.11 | **Molecular Orbital Theory Applied to Heteronuclear Diatomic Molecules and Ions**

Use molecular orbital theory to determine the bond order of the $CN^-$ ion. Is the ion paramagnetic or diamagnetic?

SOLUTION

| | |
|---|---|
| Determine the number of valence electrons in the molecule or ion. | Number of valence electrons<br>= 4 (from C) + 5 (from N) +<br>1 (from negative charge) = 10 |
| Write an energy-level diagram using Figure 10.15 as a guide. Fill the orbitals beginning with the lowest energy orbital and progressing upward until all electrons have been assigned to an orbital. Remember to allow no more than two electrons (with paired spins) per orbital and to fill degenerate orbitals with single electrons (with parallel spins) before pairing. | $\sigma_{2p}^*$<br><br>$\pi_{2p}^*$<br><br>↿⇂ $\sigma_{2p}$<br><br>↿⇂ ↿⇂ $\pi_{2p}$<br><br>↿⇂ $\sigma_{2s}^*$<br><br>↿⇂ $\sigma_{2s}$ |

—*Continued on the next page*

*—Continued*

| Calculate the bond order using the appropriate formula:<br><br>Bond order =<br>$$\frac{(\text{number of } e^- \text{ in bonding MOs}) - (\text{number of } e^- \text{ in antibonding MOs})}{2}$$ | $CN^-$ bond order $= \dfrac{8-2}{2} = +3$ |
| --- | --- |
| If the MO diagram has unpaired electrons, the molecule or ion is paramagnetic. If the electrons are all paired, the molecule or ion is diamagnetic. | The MO diagram has no unpaired electrons, so the ion is diamagnetic. |

**FOR PRACTICE 10.11** Apply molecular orbital theory to determine the bond order of NO. (Use the energy ordering of $O_2$.) Is the molecule paramagnetic or diamagnetic?

## Polyatomic Molecules

With the aid of computers, molecular orbital theory can be applied to polyatomic molecules and ions, yielding results that correlate very well with experimental measurements. These applications are beyond the scope of this text. However, the delocalization of electrons over an entire molecule is an important contribution of molecular orbital theory to our basic understanding of chemical bonding. For example, consider the Lewis structure and valence bond diagram of ozone:

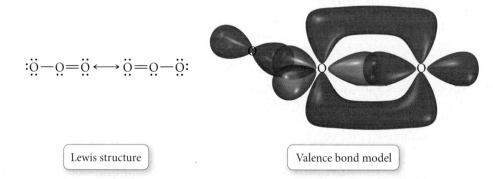

Lewis structure                Valence bond model

In the Lewis model, we use resonance forms to represent the two equivalent bonds. In valence bond theory, it appears that the two oxygen–oxygen bonds should be different. In molecular orbital theory, however, the $\pi$ molecular orbitals in ozone are formed from a linear combination of the three oxygen $2p$ orbitals and are delocalized over the entire molecule. The lowest energy $\pi$ bonding molecular orbital is shown in the left margin.

When we examine ozone in nature, we indeed find two equivalent bonds. A similar situation occurs with benzene ($C_6H_6$). In the Lewis model, we represent the structure with two resonance forms:

In molecular orbital theory, the $\pi$ molecular orbitals in benzene are formed from a linear combination of the six carbon $2p$ orbitals and are delocalized over the entire molecule. The lowest energy $\pi$ bonding molecular orbital is shown in the left margin.

Benzene is in fact a highly symmetric molecule with six identical carbon–carbon bonds. The best picture of the $\pi$ electrons in benzene is one in which the electrons occupy roughly circular-shaped orbitals above and below the plane of the molecule, as depicted in the molecular orbital theory approach.

**What Is a Chemical Bond? Part II**   We have learned that the Lewis model portrays a chemical bond as the transfer or sharing of electrons represented as dots. Valence bond theory portrays a chemical bond as the overlap of two half-filled atomic orbitals. What is a chemical bond according to molecular orbital theory?

CONCEPTUAL
**CONNECTION 10.11**

# Self-Assessment Quiz

**Q1.** Determine the molecular geometry of $CBr_4$.

  a) linear
  b) trigonal planar
  c) tetrahedral
  d) trigonal pyramidal

**Q2.** Determine the molecular geometry of $SeF_4$.

  a) tetrahedral
  b) trigonal bipyramidal
  c) T-shaped
  d) seesaw

**Q3.** Predict the relative bond angles in $BF_3$ and $SO_2$.

  a) $BF_3$ bond angles $>$ $SO_2$ bond angle
  b) $SO_2$ bond angle $>$ $BF_3$ bond angles
  c) $BF_3$ bond angles $=$ $SO_2$ bond angle
  d) Relative bond angles cannot be predicted.

**Q4.** Predict the molecular geometry about nitrogen in the molecule $CH_3NHCH_3$.

  a) linear
  b) trigonal planar
  c) trigonal pyramidal
  d) bent

**Q5.** Which molecule is polar?

  a) $SF_2$     b) $BH_3$
  c) $PF_5$     d) $CS_2$

**Q6.** Determine the hybridization about oxygen in $CH_3OH$.

  a) $sp$          b) $sp^2$
  c) $sp^3$        d) $sp^3d$

**Q7.** Determine the hybridization about carbon in $H_2CO$.

  a) $sp$
  b) $sp^2$
  c) $sp^3$
  d) $sp^3d$

**Q8.** According to valence bond theory, which kind of orbitals overlap to form the P—Cl bonds in $PCl_5$?

  a) $P(sp^3)$ - $Cl(p)$
  b) $P(sp^3d)$ - $Cl(s)$
  c) $P(sp^3)$ - $Cl(s)$
  d) $P(sp^3d)$ - $Cl(p)$

**Q9.** Apply molecular orbital theory to determine the bond order in $C_2$.

  a) 0      b) 1      c) 2      d) 3

**Q10.** Apply molecular orbital theory to predict which species has the strongest bond.

  a) $N_2$
  b) $N_2^-$
  c) $N_2^+$
  d) All bonds are equivalent according to molecular orbital theory.

**Q11.** Apply molecular orbital theory to determine which molecule is diamagnetic.

  a) CO
  b) $B_2$
  c) $O_2$
  d) None of the above (all are paramagnetic)

**Q12.** Which hybridization scheme occurs about nitrogen when nitrogen forms a double bond?

  a) $sp$
  b) $sp^2$
  c) $sp^3$
  d) $sp^3d$

**Q13.** Which molecular geometry results when a central atom has five total electron groups, with three of those being bonding groups and two being lone pairs?

  a) trigonal bipyramidal
  b) seesaw
  c) T-shaped
  d) bent

**Q14.** Determine the correct molecular geometry (from left to right) about each interior atom in $CH_3CH_2OH$.

  a) 1st C tetrahedral; 2nd C trigonal planar; O linear
  b) 1st C trigonal planar; 2nd C bent; O linear
  c) 1st C trigonal planar; 2nd C trigonal pyramidal; O bent
  d) 1st C tetrahedral; 2nd C tetrahedral; O bent

**Q15.** The central atom in a molecule has a bent molecular geometry. Determine the hybridization of the orbitals in the atom.

  a) $sp$     b) $sp^2$     c) $sp^3$
  d) Hybridization cannot be determined from the information given.

**Answers: 1.** (c) **2.** (d) **3.** (a) **4.** (c) **5.** (a) **6.** (c) **7.** (b) **8.** (d) **9.** (c) **10.** (a) **11.** (a) **12.** (b) **13.** (c) **14.** (d) **15.** (d)

# Chapter 10 IN REVIEW

## Key Terms

**Section 10.2**
valence shell electron pair repulsion (VSEPR) theory (428)
electron groups (428)
linear geometry (429)
trigonal planar geometry (429)
tetrahedral geometry (430)
trigonal bipyramidal geometry (431)
octahedral geometry (431)

**Section 10.3**
electron geometry (432)
molecular geometry (432)
trigonal pyramidal geometry (432)
bent geometry (433)
seesaw geometry (434)
T-shaped geometry (434)
square pyramidal geometry (435)
square planar geometry (435)

**Section 10.6**
valence bond theory (445)

**Section 10.7**
hybridization (447)
hybrid orbitals (447)
pi ($\pi$) bond (452)
sigma ($\sigma$) bond (452)

**Section 10.8**
molecular orbital (MO) theory (461)
bonding orbital (462)
antibonding orbital (462)
bond order (463)
nonbonding orbitals (472)

## Key Concepts

### Molecular Shape and VSEPR Theory (10.1–10.4)

▶ The properties of molecules are directly related to their shapes. In VSEPR theory, molecular geometries are determined by the repulsions between electron groups on the central atom. An electron group can be a single bond, double bond, triple bond, lone pair, or even a single electron.

▶ The five basic molecular shapes are linear (two electron groups), trigonal planar (three electron groups), tetrahedral (four electron groups), trigonal bipyramidal (five electron groups), and octahedral (six electron groups).

▶ When lone pairs are present on the central atom, the *electron* geometry is still one of the five basic shapes, but one or more positions are occupied by lone pairs. The *molecular* geometry is therefore different from the electron geometry. Lone pairs are positioned so as to minimize repulsions with other lone pairs and with bonding pairs.

### Polarity (10.5)

▶ The polarity of a polyatomic molecule containing polar bonds depends on its geometry. If the dipole moments of the polar bonds are aligned in such a way that they cancel one another, the molecule is not polar. If they are aligned in such a way as to sum together, the molecule is polar.

▶ Highly symmetric molecules tend to be nonpolar, whereas asymmetric molecules containing polar bonds tend to be polar. The polarity of a molecule dramatically affects its properties.

### Valence Bond Theory (10.6–10.7)

▶ In contrast to the Lewis model, in which a covalent chemical bond is the sharing of electrons represented by dots, in valence bond theory a chemical bond is the overlap of half-filled atomic orbitals (or in some cases the overlap between a completely filled orbital

and an empty one).

▶ The overlapping orbitals may be the standard atomic orbitals, such as $1s$ or $2p$, or they may be hybridized atomic orbitals, which are mathematical combinations of the standard orbitals on a single atom. The basic hybridized orbitals are $sp$, $sp^2$, $sp^3$, $sp^3d$, and $sp^3d^2$.

▶ The geometry of the molecule is determined by the geometry of the overlapping orbitals.

▶ In our treatment of valence bond theory, we use the electron geometry determined by VSEPR theory to determine the correct hybridization scheme.

▶ In valence bond theory, we distinguish between two types of bonds, $\sigma$ (sigma) and $\pi$ (pi). In a $\sigma$ bond, the orbital overlap occurs in the region that lies directly between the two bonding atoms. In a $\pi$ bond, formed from the side-by-side overlap of $p$ orbitals, the overlap occurs above and below the region that lies directly between the two bonding atoms.

▶ Rotation about a $\sigma$ bond is relatively free, whereas rotation about a $\pi$ bond is restricted.

### Molecular Orbital Theory (10.8)

▶ The simplest molecular orbitals are linear combinations of atomic orbitals (LCAOs), weighted averages of the atomic orbitals of the different atoms in the molecule.

▶ When two atomic orbitals combine to form molecular orbitals, they form one molecular orbital of lower energy (the bonding orbital) and one of higher energy (the antibonding orbital).

▶ A set of molecular orbitals fills in much the same way as atomic orbitals.

▶ The stability of the molecule and the strength of the bond depend on the number of electrons in bonding orbitals compared to the number in antibonding orbitals.

## Key Equations and Relationships

Bond Order of a Diatomic Molecule (**10.8**)

$$\text{Bond order} = \frac{(\text{number of electrons in bonding MOs}) - (\text{number of electrons in antibonding MOs})}{2}$$

## Key Learning Outcomes

| CHAPTER OBJECTIVES | ASSESSMENT |
|---|---|
| Using VSEPR Theory to Predict the Basic Shapes of Molecules (**10.2**) 180° | Example 10.1    For Practice 10.1    Exercises 31–32 |
| Predicting Molecular Geometries Using VSEPR Theory and the Effects of Lone Pairs (**10.4**) | Examples 10.2–10.3    For Practice 10.2–10.3    Exercises 35–36 |
| Predicting the Shapes of Larger Molecules (**10.4**) | Example 10.4    For Practice 10.4    Exercises 41–42, 45–46 |
| Using Molecular Shape to Determine Polarity of a Molecule (**10.5**) | Example 10.5    For Practice 10.5    Exercises 49–52 |
| Writing Hybridization and Bonding Schemes Using Valence Bond Theory (**10.7**) | Examples 10.6–10.8    For Practice 10.6–10.8 For More Practice 10.8    Exercises 61–66 |
| Drawing Molecular Orbital Diagrams to Predict Bond Order and Magnetism of a Diatomic Molecule (**10.8**) | Examples 10.9–10.11    For Practice 10.9–10.11 For More Practice 10.10    Exercises 71–72, 75–78, 81–82 |

# EXERCISES

## Review Questions

1. Why is molecular geometry important? Cite some examples.

2. According to VSEPR theory, what determines the geometry of a molecule?

3. Name and sketch the five basic electron geometries, and state the number of electron groups corresponding to each. What constitutes an *electron group*?

4. Explain the difference between electron geometry and molecular geometry. Under what circumstances are they not the same?

5. Give the correct electron and molecular geometries that correspond to each set of electron groups around the central atom of a molecule.
   a. four electron groups overall; three bonding groups and one lone pair
   b. four electron groups overall; two bonding groups and two lone pairs
   c. five electron groups overall; four bonding groups and one lone pair
   d. five electron groups overall; three bonding groups and two lone pairs
   e. five electron groups overall; two bonding groups and three lone pairs
   f. six electron groups overall; five bonding groups and one lone pair
   g. six electron groups overall; four bonding groups and two lone pairs

6. How do you apply VSEPR theory to predict the shape of a molecule with more than one interior atom?

7. How do you determine whether a molecule is polar? Why is polarity important?

8. What is a chemical bond according to valence bond theory?

9. In valence bond theory, what determines the geometry of a molecule?

10. In valence bond theory, the interaction energy between the electrons and nucleus of one atom with the electrons and nucleus of another atom is usually negative (stabilizing) when _____.

11. What is hybridization? Why is hybridization necessary in valence bond theory?

12. How does hybridization of the atomic orbitals in the central atom of a molecule help lower the overall energy of the molecule?

13. How is the *number* of hybrid orbitals related to the number of standard atomic orbitals that are hybridized?

14. Sketch each set of hybrid orbitals.
   a. $sp$    b. $sp^2$    c. $sp^3$
   d. $sp^3d$    e. $sp^3d^2$

15. In the Lewis model, the two bonds in a double bond look identical. However, valence bond theory shows that they are not. Describe a double bond according to valence bond theory. Explain why rotation is restricted about a double bond but not about a single bond.

16. Name the hybridization scheme that corresponds to each electron geometry.
   a. linear                b. trigonal planar
   c. tetrahedral           d. trigonal bipyramidal
   e. octahedral

17. What is a chemical bond according to molecular orbital theory?

18. Explain the difference between hybrid atomic orbitals in valence bond theory and LCAO molecular orbitals in molecular orbital theory.

19. What is a bonding molecular orbital?

20. What is an antibonding molecular orbital?

21. What is the role of wave interference in determining whether a molecular orbital is bonding or antibonding?

**22.** In molecular orbital theory, what is bond order? Why is it important?

**23.** How is the number of molecular orbitals approximated by a linear combination of atomic orbitals related to the number of atomic orbitals used in the approximation?

**24.** Sketch each molecular orbital.

   **a.** $\sigma_{2s}$    **b.** $\sigma_{2s}^*$    **c.** $\sigma_{2p}$
   **d.** $\sigma_{2p}^*$    **e.** $\pi_{2p}$    **f.** $\pi_{2p}^*$

**25.** Draw an energy diagram for the molecular orbitals of period 2 diatomic molecules. Show the difference in ordering for $B_2$, $C_2$, and $N_2$ compared to $O_2$, $F_2$, and $Ne_2$.

**26.** Why does the energy ordering of the molecular orbitals of the period 2 diatomic molecules change in going from $N_2$ to $O_2$?

**27.** Explain the difference between a paramagnetic species and a diamagnetic one.

**28.** When applying molecular orbital theory to heteronuclear diatomic molecules, the atomic orbitals used may be of different energies. If two atomic orbitals of different energies make two molecular orbitals, how are the energies of the molecular orbitals related to the energies of the atomic orbitals? How is the shape of the resultant molecular orbitals related to the shape of the atomic orbitals?

**29.** In molecular orbital theory, what is a nonbonding orbital?

**30.** Write a short paragraph describing chemical bonding according to the Lewis model, valence bond theory, and molecular orbital theory. Indicate how the theories differ in their description of a chemical bond and indicate the strengths and weaknesses of each theory. Which theory is correct?

# Problems by Topic

## VSEPR Theory and Molecular Geometry

**31.** A molecule with the formula $AB_3$ has a trigonal pyramidal geometry. How many electron groups are on the central atom (A)?

**32.** A molecule with the formula $AB_3$ has a trigonal planar geometry. How many electron groups are on the central atom?

**33.** For each molecular geometry, list the number of total electron groups, the number of bonding groups, and the number of lone pairs on the central atom.

   **(a)**       **(b)**       **(c)**

**34.** For each molecular geometry, list the number of total electron groups, the number of bonding groups, and the number of lone pairs on the central atom.

   **(a)**       **(b)**       **(c)**

**35.** Determine the electron geometry, molecular geometry, and idealized bond angles for each molecule. In which cases do you expect deviations from the idealized bond angle?

   **a.** $PF_3$    **b.** $SBr_2$    **c.** $CHCl_3$    **d.** $CS_2$

**36.** Determine the electron geometry, molecular geometry, and idealized bond angles for each molecule. In which cases do you expect deviations from the idealized bond angle?

   **a.** $CF_4$    **b.** $NF_3$    **c.** $OF_2$    **d.** $H_2S$

**37.** Which species has the smaller bond angle, $H_3O^+$ or $H_2O$? Explain.

**38.** Which species has the smaller bond angle, $ClO_4^-$ or $ClO_3^-$? Explain.

**39.** Determine the molecular geometry and sketch each molecule or ion using the bond conventions shown in "Representing Molecular Geometries on Paper" in Section 10.4.

   **a.** $SF_4$    **b.** $ClF_3$    **c.** $IF_2^-$    **d.** $IBr_4^-$

**40.** Determine the molecular geometry and sketch each molecule or ion, using the bond conventions shown in "Representing Molecular Geometries on Paper" in Section 10.4.

   **a.** $BrF_5$    **b.** $SCl_6$    **c.** $PF_5$    **d.** $IF_4^+$

**41.** Determine the molecular geometry about each interior atom and sketch each molecule.

   **a.** $C_2H_2$ (skeletal structure HCCH)
   **b.** $C_2H_4$ (skeletal structure $H_2CCH_2$)
   **c.** $C_2H_6$ (skeletal structure $H_3CCH_3$)

**42.** Determine the molecular geometry about each interior atom and sketch each molecule.

   **a.** $N_2$    **b.** $N_2H_2$ (skeletal structure HNNH)
   **c.** $N_2H_4$ (skeletal structure $H_2NNH_2$)

**43.** Each ball-and-stick model shows the electron and molecular geometry of a generic molecule. Explain what is wrong with each molecular geometry and provide the correct molecular geometry, given the number of lone pairs and bonding groups on the central atom.

   **(a)**       **(b)**       **(c)**

**44.** Each ball-and-stick model shows the electron and molecular geometry of a generic molecule. Explain what is wrong with each molecular geometry and provide the correct molecular geometry, given the number of lone pairs and bonding groups on the central atom.

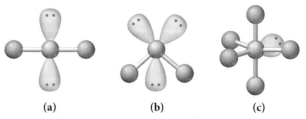

   **(a)**       **(b)**       **(c)**

**45.** Determine the geometry about each interior atom in each molecule and sketch the molecule. (Skeletal structure is indicated in parentheses.)

   **a.** $CH_3OH$ ($H_3COH$)       **b.** $CH_3OCH_3$ ($H_3COCH_3$)
   **c.** $H_2O_2$ (HOOH)

**46.** Determine the geometry about each interior atom in each molecule and sketch the molecule. (Skeletal structure is indicated in parentheses.)
**a.** $CH_3NH_2$ ($H_3CNH_2$)
**b.** $CH_3CO_2CH_3$ ($H_3CCOOCH_3$ One O atom attached to 2nd C atom; the other O atom is bonded to the 2nd and 3rd C atom)
**c.** $NH_2CO_2H$ ($H_2NCOOH$ both O atoms attached to C)

## Molecular Shape and Polarity

**47.** Explain why $CO_2$ and $CCl_4$ are both nonpolar even though they contain polar bonds.

**48.** $CH_3F$ is a polar molecule, even though the tetrahedral geometry often leads to nonpolar molecules. Explain.

**49.** Determine whether each molecule in Exercise 35 is polar or nonpolar.

**50.** Determine whether each molecule in Exercise 36 is polar or nonpolar.

**51.** Determine whether each molecule is polar or nonpolar.
**a.** $SCl_2$     **b.** $SCl_4$     **c.** $BrCl_5$

**52.** Determine whether each molecule is polar or nonpolar.
**a.** $SiCl_4$     **b.** $CF_2Cl_2$     **c.** $SeF_6$     **d.** $IF_5$

## Valence Bond Theory

**53.** The valence electron configurations of several atoms are shown here. How many bonds can each atom make without hybridization?
**a.** Be $2s^2$     **b.** P $3s^23p^3$     **c.** F $2s^22p^5$

**54.** The valence electron configurations of several atoms are shown here. How many bonds can each atom make without hybridization?
**a.** B $2s^22p^1$     **b.** N $2s^22p^3$     **c.** O $2s^22p^4$

**55.** Write orbital diagrams (boxes with arrows in them) to represent the electron configurations—without hybridization—for all the atoms in $PH_3$. Circle the electrons involved in bonding. Draw a three-dimensional sketch of the molecule and show orbital overlap. What bond angle do you expect from the unhybridized orbitals? How well does valence bond theory agree with the experimentally measured bond angle of 93.3°?

**56.** Write orbital diagrams (boxes with arrows in them) to represent the electron configurations—without hybridization—for all the atoms in $SF_2$. Circle the electrons involved in bonding. Draw a three-dimensional sketch of the molecule and show orbital overlap. What bond angle do you expect from the unhybridized orbitals? How well does valence bond theory agree with the experimentally measured bond angle of 98.2°?

**57.** Write orbital diagrams (boxes with arrows in them) to represent the electron configuration of carbon before and after $sp^3$ hybridization.

**58.** Write orbital diagrams (boxes with arrows in them) to represent the electron configurations of carbon before and after $sp$ hybridization.

**59.** Which hybridization scheme allows the formation of at least one $\pi$ bond?
$$sp^3, sp^2, sp^3d^2$$

**60.** Which hybridization scheme allows the central atom to form more than four bonds?
$$sp^3, sp^3d, sp^2$$

**61.** Write a hybridization and bonding scheme for each molecule. Sketch the molecule, including overlapping orbitals, and label all bonds using the notation shown in Examples 10.6 and 10.7.
**a.** $CCl_4$     **b.** $NH_3$     **c.** $OF_2$     **d.** $CO_2$

**62.** Write a hybridization and bonding scheme for each molecule. Sketch the molecule, including overlapping orbitals, and label all bonds using the notation shown in Examples 10.6 and 10.7.
**a.** $CH_2Br_2$     **b.** $SO_2$     **c.** $NF_3$     **d.** $BF_3$

**63.** Write a hybridization and bonding scheme for each molecule or ion. Sketch the structure, including overlapping orbitals, and label all bonds using the notation shown in Examples 10.6 and 10.7.
**a.** $COCl_2$ (carbon is the central atom)
**b.** $BrF_5$
**c.** $XeF_2$
**d.** $I_3^-$

**64.** Write a hybridization and bonding scheme for each molecule or ion. Sketch the structure, including overlapping orbitals, and label all bonds using the notation shown in Examples 10.6 and 10.7.
**a.** $SO_3^{2-}$     **b.** $PF_6^-$     **c.** $BrF_3$     **d.** HCN

**65.** Write a hybridization and bonding scheme for each molecule that contains more than one interior atom. Indicate the hybridization about each interior atom. Sketch the structure, including overlapping orbitals, and label all bonds using the notation shown in Examples 10.6 and 10.7.
**a.** $N_2H_2$ (skeletal structure HNNH)
**b.** $N_2H_4$ (skeletal structure $H_2NNH_2$)
**c.** $CH_3NH_2$ (skeletal structure $H_3CNH_2$)

**66.** Write a hybridization and bonding scheme for each molecule that contains more than one interior atom. Indicate the hybridization about each interior atom. Sketch the structure, including overlapping orbitals, and label all bonds using the notation shown in Examples 10.6 and 10.7.
**a.** $C_2H_2$ (skeletal structure HCCH)
**b.** $C_2H_4$ (skeletal structure $H_2CCH_2$)
**c.** $C_2H_6$ (skeletal structure $H_3CCH_3$)

**67.** Consider the structure of the amino acid alanine. Indicate the hybridization about each interior atom.

**68.** Consider the structure of the amino acid aspartic acid. Indicate the hybridization about each interior atom.

## Molecular Orbital Theory

69. Sketch the bonding molecular orbital that results from the linear combination of two $1s$ orbitals. Indicate the region where interference occurs and state the kind of interference (constructive or destructive).

70. Sketch the antibonding molecular orbital that results from the linear combination of two $1s$ orbitals. Indicate the region where interference occurs and state the kind of interference (constructive or destructive).

71. Draw an MO energy diagram and predict the bond order of $Be_2^+$ and $Be_2^-$. Do you expect these molecules to exist in the gas phase?

72. Draw an MO energy diagram and predict the bond order of $Li_2^+$ and $Li_2^-$. Do you expect these molecules to exist in the gas phase?

73. Sketch the bonding and antibonding molecular orbitals that result from linear combinations of the $2p_x$ atomic orbitals in a homonuclear diatomic molecule. (The $2p_x$ orbitals are those whose lobes are oriented along the bonding axis.)

74. Sketch the bonding and antibonding molecular orbitals that result from linear combinations of the $2p_z$ atomic orbitals in a homonuclear diatomic molecule. (The $2p_z$ orbitals are those whose lobes are oriented perpendicular to the bonding axis.) How do these molecular orbitals differ from those obtained from linear combinations of the $2p_y$ atomic orbitals? (The $2p_y$ orbitals are also oriented perpendicular to the bonding axis, but also perpendicular to the $2p_z$ orbitals.)

75. Using the molecular orbital energy ordering for second-row homonuclear diatomic molecules in which the $\pi_{2p}$ orbitals lie at *lower* energy than the $\sigma_{2p}$, draw MO energy diagrams and predict the bond order in a molecule or ion with each number of total valence electrons. Will the molecule or ion be diamagnetic or paramagnetic?
    **a.** 4        **b.** 6        **c.** 8        **d.** 9

76. Using the molecular orbital energy ordering for second-row homonuclear diatomic molecules in which the $\pi_{2p}$ orbitals lie at *higher* energy than the $\sigma_{2p}$, draw MO energy diagrams and predict the bond order in a molecule or ion with each number of total valence electrons. Will the molecule or ion be diamagnetic or paramagnetic?
    **a.** 10        **b.** 12        **c.** 13        **d.** 14

77. Use molecular orbital theory to predict if each molecule or ion exists in a relatively stable form.
    **a.** $H_2^{2-}$        **b.** $Ne_2$        **c.** $He_2^{2+}$        **d.** $F_2^{2-}$

78. Use molecular orbital theory to predict if each molecule or ion exists in a relatively stable form.
    **a.** $C_2^{2+}$        **b.** $Li_2$        **c.** $Be_2^{2+}$        **d.** $Li_2^{2-}$

79. According to MO theory, which molecule or ion has the highest bond order? Highest bond energy? Shortest bond length?
$$C_2, C_2^+, C_2^-$$

80. According to MO theory, which molecule or ion has the highest bond order? Highest bond energy? Shortest bond length?
$$O_2, O_2^-, O_2^{2-}$$

81. Draw an MO energy diagram for CO. (Use the energy ordering of $O_2$.) Predict the bond order and make a sketch of the lowest energy bonding molecular orbital.

82. Draw an energy diagram for HCl. Predict the bond order and make a sketch of the lowest energy bonding molecular orbital.

# Cumulative Problems

83. For each compound, draw the Lewis structure, determine the geometry using VSEPR theory, determine whether the molecule is polar, identify the hybridization of all interior atoms, and make a sketch of the molecule, according to valence bond theory, showing orbital overlap.
    **a.** $COF_2$ (carbon is the central atom)
    **b.** $S_2Cl_2$ (ClSSCl)
    **c.** $SF_4$

84. For each compound, draw the Lewis structure, determine the geometry using VSEPR theory, determine whether the molecule is polar, identify the hybridization of all interior atoms, and make a sketch of the molecule, according to valence bond theory, showing orbital overlap.
    **a.** $IF_5$        **b.** $CH_2CHCH_3$        **c.** $CH_3SH$

85. Amino acids are biological compounds that link together to form proteins, the workhorse molecules in living organisms. The skeletal structures of several simple amino acids are shown here. For each skeletal structure, complete the Lewis structure, determine the geometry and hybridization about each interior atom, and make a sketch of the molecule, using the bond conventions of Section 10.4.

(a) serine

(b) asparagine

(c) cysteine

86. The genetic code is based on four different bases with the structures shown here. Assign a geometry and hybridization to each interior atom in these four bases.
    **a.** cytosine        **b.** adenine        **c.** thymine        **d.** guanine

(a)

(b)

(c)

(d)

**87.** The structure of caffeine, present in coffee and many soft drinks, is shown here. How many pi bonds are present in caffeine? How many sigma bonds? Insert the lone pairs in the molecule. What kinds of orbitals do the lone pairs occupy?

**88.** The structure of acetylsalicylic acid (aspirin) is shown here. How many pi bonds are present in acetylsalicylic acid? How many sigma bonds? What parts of the molecule are free to rotate? What parts are rigid?

**89.** Most vitamins can be classified as either fat soluble, which results in their tendency to accumulate in the body (so that taking too much can be harmful), or water soluble, which results in their tendency to be quickly eliminated from the body in urine. Examine the structural formulas and space-filling models of these vitamins and determine whether each one is fat soluble (mostly nonpolar) or water soluble (mostly polar).

**(a)** vitamin C

**(b)** vitamin A

**(c)** niacin (vitamin B$_3$)

**(d)** vitamin E

**90.** Water does not easily remove grease from dishes or hands because grease is nonpolar and water is polar. The addition of soap to water, however, allows the grease to dissolve. Study the structure of sodium stearate (a soap) and describe how it works.

**91.** Draw a molecular orbital energy diagram for ClF. (Assume that the $\sigma_p$ orbitals are lower in energy than the $\pi$ orbitals.) What is the bond order in ClF?

**92.** Draw Lewis structures and MO diagrams for $CN^+$, CN, and $CN^-$. According to the Lewis model, which species is most stable? According to MO theory, which species is most stable? Do the two theories agree?

**93.** Bromine can form compounds or ions with any number of fluorine atoms from one to five. Write the formulas of all five of these species, assign a hybridization, and describe their electron and molecular geometry.

**94.** The compound $C_3H_4$ has two double bonds. Describe its bonding and geometry, using a valence bond approach.

**95.** Draw the structure of a molecule with the formula $C_4H_6Cl_2$ that has a dipole moment of 0.

**96.** Draw the structures of two compounds that have the composition $CH_3NO_2$ and have all three H atoms bonded to the C. Predict which compound has the larger ONO bond angle.

**97.** How many types of hybrid orbitals do we use to describe each molecule?
**a.** $N_2O_5$
**b.** $C_2H_5NO$ (four C — H bonds and one O — H bond)
**c.** BrCN (no formal charges)

**98.** Indicate which orbitals overlap to form the $\sigma$ bonds in each molecule.
**a.** $BeBr_2$      **b.** $HgCl_2$      **c.** ICN

# Challenge Problems

**99.** In VSEPR theory, which uses the Lewis model to determine molecular geometry, the trend of decreasing bond angles in $CH_4$, $NH_3$, and $H_2O$ is accounted for by the greater repulsion of lone pair electrons compared to bonding pair electrons. How would this trend be accounted for in valence bond theory?

**100.** The results of a molecular orbital calculation for $H_2O$ are shown here. Examine each of the orbitals and classify them as bonding, antibonding, or nonbonding. Assign the correct number of electrons to the energy diagram. According to this energy diagram, is $H_2O$ stable? Explain.

**101.** The results of a molecular orbital calculation for $NH_3$ are shown here. Examine each of the orbitals and classify them as bonding, antibonding, or nonbonding. Assign the correct number of electrons to the energy diagram. According to this energy diagram, is $NH_3$ stable? Explain.

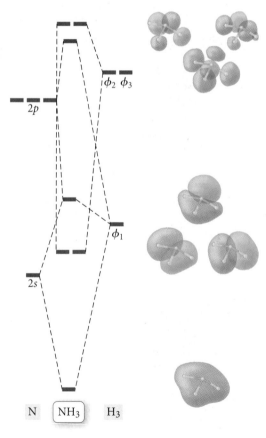

**102.** *cis*-2-Butene isomerizes to *trans*-2-butene via the reaction shown here.

    **a.** If isomerization requires breaking the π bond, what minimum energy is required for isomerization in J/mol? In J/molecule?

    **b.** If the energy for isomerization came from light, what minimum frequency of light would be required? In what portion of the electromagnetic spectrum does this frequency lie?

**103.** The species $NO_2$, $NO_2^+$, and $NO_2^-$ in which N is the central atom have very different bond angles. Predict what these bond angles might be with respect to the ideal angles and justify your prediction.

**104.** The bond angles increase steadily in the series $PF_3$, $PCl_3$, $PBr_3$, and $PI_3$. After consulting the data on atomic radii in Chapter 8, provide an explanation for this observation.

**105.** The ion $CH_5^+$ can form under very special high-energy conditions in the vapor phase in a mass spectrometer. Propose a hybridization for the carbon atom and predict the geometry.

**106.** Neither the VSEPR model nor the hybridization model is able to account for the experimental observation that the F — Ba — F bond angle in gaseous $BaF_2$ is 108° rather than the predicted 180°. Suggest some possible explanations for this observation.

**107.** Draw the Lewis structure for acetamide ($CH_3CONH_2$), an organic compound, and determine the geometry about each interior atom. Experiments show that the geometry about the nitrogen atom in acetamide is nearly planar. What resonance structure can account for the planar geometry about the nitrogen atom?

**108.** Use VSEPR theory to predict the geometry (including bond angles) about each interior atom of methyl azide ($CH_3N_3$), and make a sketch of the molecule. Would you expect the bond angle between the two interior nitrogen atoms to be the same or different? Would you expect the two nitrogen–nitrogen bond lengths to be the same or different?

## Conceptual Problems

**109.** Which statement best captures the fundamental idea behind VSEPR theory? Explain what is wrong with each of the other statements.

    **a.** The angle between two or more bonds is determined primarily by the repulsions between the electrons within those bonds and other (lone pair) electrons on the central atom of a molecule. Each of these electron groups (bonding electrons or lone pair electrons) will lower its potential energy by maximizing its separation from other electron groups, thus determining the geometry of the molecule.

    **b.** The angle between two or more bonds is determined primarily by the repulsions between the electrons within those bonds. Each of these bonding electrons will lower its potential energy by maximizing its separation from other electron groups, thus determining the geometry of the molecule.

    **c.** The geometry of a molecule is determined by the shapes of the overlapping orbitals that form the chemical bonds. Therefore, to determine the geometry of a molecule, you must determine the shapes of the orbitals involved in bonding.

**110.** Suppose that a molecule has four bonding groups and one lone pair on the central atom. Suppose further that the molecule is confined to two dimensions (this is a purely hypothetical assumption for the sake of understanding the principles behind VSEPR theory). Make a sketch of the molecule and estimate the bond angles.

**111.** How does each of the three major bonding theories (the Lewis model, valence bond theory, and molecular orbital theory) define a single chemical bond? A double bond? A triple bond? How are these definitions similar? How are they different?

**112.** The most stable forms of the nonmetals in groups 4A, 5A, and 6A of the second period are molecules with multiple bonds. Beginning with the third period, the most stable forms of the nonmetals of these groups are molecules without multiple bonds. Propose an explanation for this observation based on valence bond theory.

*Active Classroom Learning*

## Questions for Group Work

Discuss these questions with the group and record your consensus answer.

**113.** In complete sentences, describe why someone might expect the bond angles in methane ($CH_4$) to be 90° even though the bond angles are actually 109.5°.

**114.** At least two different numbers of electron groups can result in a linear molecule. What are they? What are the numbers of bonding groups and lone pairs in each case? Provide an example of a linear molecule in each case.

**115.** Have each member of your group select one of the molecules shown below and complete steps a–d. Each member should then present his or her results to the rest of the group, explaining the reasoning used to determine the answers.

    $CS_2$    $NCl_3$    $CF_4$    $CH_2F_2$

    **a.** Draw the Lewis dot structure.

    **b.** Determine the molecular geometry and draw it accurately.

    **c.** Indicate the polarity of any polar bonds within the structure.

    **d.** Classify the molecule as polar or nonpolar.

**116.** How many atomic orbitals form a set of $sp^3$ hybrid orbitals? A set of $sp^2$ hybrid orbitals? A set of $sp$ hybrid orbitals? What is the relationship between these numbers and the number of electron groups around the central atom?

**117.** Use molecular orbital theory to explain in detail why $N_2^+$ and $N_2^-$ have similar bond strengths and both are very different from neutral $N_2$.

# Data Interpretation and Analysis

## Using Distances between Atoms to Predict Bond Angles

**118.** The distance between two atoms that are attached to a common central atom is approximately equal to the sum of their "target distance" atomic radii. For example, consider the structures of $CF_4$ and $OCF_2$ in Figure a ▼. Given that the target distance radius of fluorine when it is bonded to carbon is 108 pm, the target distance between the two fluorine atoms

▲ **FIGURE a** Lewis Structures and Target Distances for $CF_4$ and $OCF_2$

(represented as dashed lines in the figure) is about 216 pm (108 pm + 108 pm) in both $CF_4$ and $OCF_2$. If we know the C—F bond length to be 132.3 pm, we can use trigonometry (Figure b▶) to determine the F—C—F bond angle in each molecule.

$$\sin\theta = \frac{\text{opposite}}{\text{hypotenuse}} = \frac{108 \text{ pm}}{132.3 \text{ pm}} = 0.816$$

$$\theta = 54.7°$$
$$\theta + \theta = 109.4°$$

▲ **FIGURE b** Using Trigonometry to Determine the F—C—F Bond Angles in $OCF_2$

Use the relation $\sin\theta = \text{opposite/hypotenuse}$ to complete the information in the table:

| | C—Cl Bond Length (pm) | Cl—C—Cl Bond Angle (°) | Cl—Cl target distance (pm) |
|---|---|---|---|
| | C—Cl | | |
| $CCl_4$ | 177.1 | | 290 |
| $CF_2Cl_2$ | 174.4 | | 290 |
| $CCl_3F$ | 176 | | 290 |

# Answers to Conceptual Connections

## Electron Groups and Molecular Geometry

**10.1** The geometry of a molecule is determined by how the terminal atoms are arranged around the central atom, which is in turn determined by how the electron groups are arranged around the *central* atom. The electron groups on the terminal atoms do not affect this arrangement.

## Molecular Geometry

**10.2** **(a)** Linear. HCN has two electron groups (the single bond and the triple bond) resulting in a linear geometry.

## Lone Pair Electrons and Molecular Geometry

**10.3** **(c)** Positions 1 and 4 would put the greatest distance between the lone pairs and minimize lone pair–lone pair repulsions.

## Molecular Geometry and Electron Group Repulsions

**10.4** **(d)** All electron groups on the central atom (or interior atoms, if there is more than one) determine the shape of a molecule according to VSEPR theory.

## The Shape of Larger Molecules

**10.5** The nitrogen atom has four electron groups and one lone pair, as the Lewis structure shown to the right illustrates; therefore, the molecular geometry about the nitrogen atom is trigonal pyramidal.

## What Is a Chemical Bond? Part I

**10.6** **(a)** In the Lewis model, a covalent chemical bond is the sharing of electrons (represented by dots). (b) In valence bond theory, a covalent chemical bond is the overlap of half-filled atomic orbitals. (c) The answers are different because the Lewis model and valence bond theory are different models for chemical bonding. They both make useful and often similar predictions, but the assumptions of each model are different, and so are their respective descriptions of a chemical bond.

## Number of Hybrid Orbitals Part I

**10.7** The number of $sp^3$ hybrid orbitals is 4. You can determine the number of hybrid orbitals by summing the superscripts in the hybridization notation. For $sp^3$, you add the implied 1 superscript on the $s$ to the 3 superscript on the $p$ and you get $1 + 3 = 4$.

## Single and Double Bonds

**10.8** Applying valence bond theory, we see that a double bond is actually composed of two different kinds of bonds, one $\sigma$

and one $\pi$. The orbital overlap in the $\pi$ bond is side to side between two $p$ orbitals and consequently different from the end-to-end overlap in a $\sigma$ bond. Since the bonds are different types, the bond energy of the double bond is not just twice the bond energy of the single bond.

### Number of Hybrid Orbitals Part II

**10.9** The number of $sp$ hybrid orbitals is 2. You can determine the number of hybrid orbitals by summing the superscripts in the hybridization notation. For $sp$, you add the implied 1 superscript on the $s$ to the implied 1 superscript on the $p$ and you get $1 + 1 = 2$.

### Bond Order

**10.10** The bond order is 1.5. Remember that bond order is equal to the number of bonding electrons minus the number of antibonding electrons, all divided by two, so you get that the bond order $= (6 - 3)/2 = 1.5$.

### What Is a Chemical Bond? Part II

**10.11** In MO theory, atoms will join together (or bond) when the electrons in the atoms can lower their energy by occupying the molecular orbitals of the resultant molecule. Unlike the Lewis model or valence bond theory, the chemical "bonds" in MO theory are not localized between atoms, but spread throughout the entire molecule.

# CHAPTER 11

# Liquids, Solids, and Intermolecular Forces

It's a wild dance floor there at the molecular level.

—ROALD HOFFMANN (1937–)

Recall from Chapter 1 that matter exists primarily in three states (or phases): solid, liquid, and gas. In Chapter 5, we examined the gas state. In this chapter we turn to the solid and liquid states, known collectively as the *condensed* states. The solid and liquid states are more similar to each other than they are to the gas state. In the gas state, constituent particles—atoms or molecules—are separated by large distances and do not interact with each other very much. In the condensed states, constituent particles are close together and exert moderate to strong attractive forces on one another. Whether a substance is a solid, liquid, or gas at a given temperature depends on the structure of the particles that compose the substance. Remember the theme we have emphasized since Chapter 1 of this book: the properties of matter are determined by the properties of molecules and atoms. In this chapter, we see how the structure of a particular atom or molecule determines the state in which it will exist at a given temperature.

▲ In the absence of gravity (such as in a space station), a sample of spilled water forms a perfect sphere. This behavior is a direct result of intermolecular forces—attractive forces that exist among the particles that compose matter.

## 11.1 Water, No Gravity

In the space station there are no spills. When an astronaut squeezes a full water bottle, the water squirts out like it does on Earth, but instead of falling to the floor and forming a puddle, the water molecules stick together to form a floating, oscillating blob. Over time, the blob stops oscillating and forms a nearly perfect sphere. Why?

The reason is the main topic of this chapter: *intermolecular forces*, the attractive forces that exist among the particles that compose matter. The molecules that compose water are attracted to one another, much like a collection of small magnets are attracted to each other. These attractions hold the water together as a liquid (instead of a gas) at

485

room temperature. These forces also cause samples of water to clump together into a blob, which is clearly seen in the absence of gravity. Over time, irregularities in the shape of the blob smooth out, and the blob becomes a sphere. The sphere is the geometrical shape with the lowest surface area to volume ratio. By forming a sphere, the water molecules maximize their interaction with one another because the sphere results in the minimum number of molecules being at the surface of the liquid, where fewer interactions occur (compared to the interior of the liquid).

Intermolecular forces exist, not only among water molecules, but among all particles that compose matter. You can see the effect of these attractive forces in the image to the left, which shows an astronaut touching a floating blob of water in the absence of gravity. Notice how the water sticks to the astronaut's finger. The water molecules experience an attractive force to the molecules that compose skin. This attractive force deforms the entire blob of water. *Intermolecular forces exist among all the particles that compose matter.*

Intermolecular forces are responsible for the very existence of condensed states. The state of a sample of matter—solid, liquid, or gas—depends on the magnitude of intermolecular forces among the constituent particles relative to the amount of thermal energy in the sample. Recall from Chapter 6 that the molecules and atoms composing matter are in constant random motion that increases with increasing temperature. The energy associated with this motion is *thermal energy*. When thermal energy is high relative to intermolecular forces, matter tends to be gaseous. When thermal energy is low relative to intermolecular forces, matter tends to be liquid or solid.

## 11.2  Solids, Liquids, and Gases: A Molecular Comparison

We are all familiar with solids and liquids. Water, gasoline, rubbing alcohol, and nail polish remover are common liquids. Ice, dry ice, and diamond are familiar solids. What makes these states so different?

### Differences between States of Matter

To illustrate the differences between the three common states of matter, Table 11.1 lists the density and molar volume of water in its three different states, along with molecular representations of each state. Notice that the densities of the solid and liquid states are

| TABLE 11.1  The Three States of Water | | | | |
|---|---|---|---|---|
| Phase | Temperature (°C) | Density (g/cm$^3$, at 1 atm) | Molar Volume | Molecular View |
| Gas (steam) | 100 | $5.90 \times 10^{-4}$ | 30.5 L | |
| Liquid (water) | 20 | 0.998 | 18.0 mL | |
| Solid (ice) | 0 | 0.917 | 19.6 mL | |

much greater than the density of the gas state. Notice also that the solid and liquid states are more similar in density and molar volume to one another than they are to the gas state. The molecular representations show the reason for these differences. The molecules in liquid water and ice are in close contact with one another—essentially touching— whereas those in gaseous water are separated by large distances. The molecular representation of gaseous water in Table 11.1 is actually out of proportion—the water molecules in the figure should be much farther apart for their size. (Only a fraction of a molecule could be included in the figure if it were drawn to scale.) From the molar volumes, we know that 18.0 mL of liquid water (slightly more than a tablespoon) occupies 30.5 L when converted to gas at 100 °C (at atmospheric pressure). The low density of gaseous water is a direct result of this large separation between molecules.

Notice also that, for water, the solid is slightly less dense than the liquid. This is *atypical* behavior. Most solids are slightly denser than their corresponding liquids because the molecules move closer together upon freezing. As we will discuss in Section 11.9, ice is less dense than liquid water because the unique crystal structure of ice results in water molecules moving slightly farther apart upon freezing.

A major difference between liquids and solids is the freedom of movement of the constituent molecules or atoms. Even though the atoms or molecules in a liquid are in close contact, thermal energy partially overcomes the attractions between them, allowing them to move around one another. This is not the case in solids; the atoms or molecules in a solid are virtually locked in their positions, only vibrating back and forth about a fixed point. Table 11.2 summarizes the properties of liquids and solids, as well as the properties of gases for comparison.

| TABLE 11.2 Properties of the States of Matter | | | | |
|---|---|---|---|---|
| **State** | **Density** | **Shape** | **Volume** | **Strength of Intermolecular Forces (Relative to Thermal Energy)** |
| Gas | Low | Indefinite | Indefinite | Weak |
| Liquid | High | Indefinite | Definite | Moderate |
| Solid | High | Definite | Definite | Strong |

Liquids assume the shape of their containers because the atoms or molecules that compose liquids are free to flow (or move around one another). When we pour water into a flask, the water flows and assumes the shape of the flask (Figure 11.1▶). Liquids are not easily compressed because the molecules or atoms that compose them are already in close contact—they cannot be pushed much closer together. The molecules in a gas, by contrast, have a great deal of space between them and are easily forced into a smaller volume by an increase in external pressure (Figure 11.2▼).

▲ **FIGURE 11.1  Liquids Assume the Shapes of Their Containers**  When we pour water into a flask, it assumes the shape of the flask because water molecules are free to flow.

Molecules closely spaced — not easily compressible

Molecules widely spaced — highly compressible

Liquid

Gas

▲ **FIGURE 11.2  Gases Are Compressible**  Molecules in a liquid are closely spaced and are not easily compressed. Molecules in a gas have a great deal of space between them, making gases compressible.

According to some definitions, an amorphous solid is considered a unique state, different from the normal solid state because it lacks any long-range order.

Solids have a definite shape because, in contrast to liquids and gases, the molecules or atoms that compose solids are fixed in place—each molecule or atom in a solid merely vibrates about a fixed point. Like liquids, solids have a definite volume and generally cannot be compressed because the molecules or atoms composing them are already in close contact. Solids may be **crystalline**, in which case the atoms or molecules that compose them are arranged in a well-ordered three-dimensional array, or they may be **amorphous**, in which case the atoms or molecules that compose them have no long-range order (Figure 11.3▼).

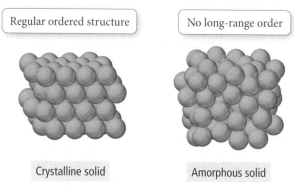

Regular ordered structure | No long-range order

Crystalline solid | Amorphous solid

▶ **FIGURE 11.3 Crystalline and Amorphous Solids** In a crystalline solid, the arrangement of the particles displays long-range order. In an amorphous solid, the arrangement of the particles has no long-range order.

## Changes between States

We can transform one state of matter to another by changing the temperature, pressure, or both. For example, we can convert solid ice to liquid water by heating, and liquid water to solid ice by cooling. The following diagram shows the three states of matter and the changes in conditions that commonly induce transitions between the states:

Heat | Heat or reduce pressure

Cool | Cool or increase pressure

Solid | Liquid | Gas

$C_3H_8$ (g)

We can induce a transition between the liquid and gas state, not only by heating and cooling, but also by changing the pressure. In general, increases in pressure favor the denser state, so increasing the pressure of a gas sample results in a transition to the liquid state.

The most familiar example of this phenomenon occurs in the LP (liquefied petroleum) gas used as a fuel for outdoor grills and lanterns. LP gas is composed primarily of propane, a gas at room temperature and atmospheric pressure. However, it liquefies at pressures exceeding about 2.7 atm. The propane you buy in a tank is under pressure and therefore in the liquid form. When you open the tank, some of the propane escapes as a gas, lowering the pressure in the tank for a brief moment. Immediately, however, some of the liquid propane evaporates, replacing the gas that escaped. Storing gases like propane as liquids is efficient because, in their liquid form, they occupy much less space.

$C_3H_8$ (l)

◀ The propane in an LP gas tank is in the liquid state. When you open the tank, some propane vaporizes and escapes as a gas.

**State Changes** This molecular diagram shows a sample of liquid water:

Which of the diagrams below best depicts the vapor emitted from a pot of boiling water?

(a)

(b)

(c)

# 11.3 Intermolecular Forces: The Forces That Hold Condensed States Together

**KCV**   Intermolecular Forces

The structure of the particles that compose a substance determines the strength of the intermolecular forces that hold the substance together, which in turn determines whether the substance is a solid, liquid, or gas at a given temperature. At room temperature, moderate to strong intermolecular forces tend to result in liquids and solids (high melting and boiling points), and weak intermolecular forces tend to result in gases (low melting and boiling points).

Intermolecular forces originate from the interactions among charges, partial charges, and temporary charges on molecules (or atoms and ions), much as bonding forces originate from interactions among charged particles in atoms. Recall from Section 8.3 that according to Coulomb's law, the potential energy ($E$) of two oppositely charged particles (with charges $q_1$ and $q_2$) decreases (becomes more negative) with increasing magnitude of charge and with decreasing separation ($r$):

$$E = \frac{1}{4\pi\varepsilon_0} \frac{q_1 q_2}{r} \quad \text{(When } q_1 \text{ and } q_2 \text{ are opposite in sign, } E \text{ is negative.)}$$

Therefore, as we have seen, protons and electrons are attracted to each other because their potential energy decreases as they get closer together. Similarly, molecules with partial or temporary charges are attracted to each other because *their* potential energy decreases as they get closer together. However, intermolecular forces, even the strongest ones, are generally *much weaker* than bonding forces.

The reason for the relative weakness of intermolecular forces compared to bonding forces is also related to Coulomb's law. Bonding forces are the result of large charges (the charges on protons and electrons) interacting at very close distances. Intermolecular forces are the result of smaller charges (as we shall see in the following discussion) interacting at greater distances. For example, consider the interaction between two water molecules in liquid water:

Intermolecular
Force

←96 pm→

←——— 300 pm ———→

The length of an O—H bond in liquid water is 96 pm; however, the average distance between water molecules in liquid water is about 300 pm. The larger distances between molecules, as well as the smaller charges involved (the partial charges on the hydrogen and oxygen atoms), result in weaker forces. To break the O—H bonds in water, we have to heat the water to thousands of degrees Celsius. However, to completely overcome the intermolecular forces *between* water molecules, we have to heat water only to its boiling point, 100 °C (at sea level).

In this section we examine several different types of intermolecular forces, including dispersion forces, dipole–dipole forces, hydrogen bonding, and ion–dipole forces. The first three of these can potentially occur in all substances; the last one occurs only in mixtures.

## Dispersion Force

The one intermolecular force present between all molecules and atoms is the **dispersion force** (also called the London force). Dispersion forces are the result of fluctuations in the electron distribution within molecules or atoms. Since all atoms and molecules have electrons, they all exhibit dispersion forces. The electrons in an atom or molecule may, *at any one instant*, be unevenly distributed. Imagine a frame-by-frame movie of a helium atom in which each "frame" captures the position of the helium atom's two electrons:

The nature of dispersion forces was first recognized by Fritz W. London (1900–1954), a German-American physicist.

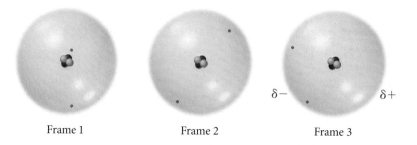

Frame 1            Frame 2            Frame 3

In any one frame, the electrons may not be symmetrically arranged around the nucleus. In frame 3, for example, helium's two electrons are on the left side of the helium atom. At that instant, the left side has a slightly negative charge ($\delta-$). The right side of the atom, which temporarily has no electrons, has a slightly positive charge ($\delta+$) because of the charge of the nucleus. This fleeting charge separation is called an *instantaneous dipole* or a *temporary dipole*. As shown in Figure 11.4▼, an instantaneous dipole on one helium atom induces an instantaneous dipole on its neighboring atoms because the positive end of the instantaneous dipole attracts electrons in the neighboring atoms. The neighboring atoms then attract one another—the positive end of one instantaneous dipole attracting the negative end of another. This attraction is the dispersion force.

► **FIGURE 11.4  Dispersion Interactions**   The temporary dipole in one helium atom induces a temporary dipole in its neighbor. The resulting attraction between the positive and negative charges creates the dispersion force.

## Dispersion Force

An instantaneous dipole on any one helium atom induces instantaneous dipoles on neighboring atoms, which then attract one another.

To polarize means to form a dipole moment (see Section 9.6).

The *magnitude* of the dispersion force depends on how easily the electrons in the atom or molecule can move or *polarize* in response to an instantaneous dipole, which in turn depends on the size (or volume) of the electron cloud. A larger electron cloud results in a greater dispersion force because the electrons are held less tightly by the nucleus and therefore polarize more easily. If all other variables are constant, the dispersion force increases with increasing molar mass because molecules or atoms of higher molar mass generally have more electrons dispersed over a greater volume. For example, consider the boiling points of the noble gases displayed in Table 11.3. As the molar masses and electron cloud volumes of the noble gases increase, the greater dispersion forces result in increasing boiling points (because the atoms are more strongly attracted to one another).

Molar mass alone, however, does not always exactly correlate with the magnitude of the dispersion force. Compare the molar masses and boiling points of *n*-pentane and neopentane:

| TABLE 11.3 Boiling Points of the Noble Gases | | | |
|---|---|---|---|
| Noble Gas | | Molar Mass (g/mol) | Boiling Point (K) |
| He | | 4.00 | 4.2 |
| Ne | | 20.18 | 27 |
| Ar | | 39.95 | 87 |
| Kr | | 83.80 | 120 |
| Xe | | 131.30 | 165 |

**n-Pentane**
molar mass = 72.15 g/mol
boiling point = 36.1 °C

**Neopentane**
molar mass = 72.15 g/mol
boiling point = 9.5 °C

These molecules have identical molar masses, but *n*-pentane has a higher boiling point than neopentane. Why? Because the two molecules have different shapes. The *n*-pentane molecules are long and can interact with one another along their entire length, as shown in Figure 11.5(a)▼. In contrast, the bulky, round shape of neopentane molecules results in a smaller area of interaction between neighboring molecules, as shown in Figure 11.5(b)▼. The result is a lower boiling point for neopentane.

Large area for interaction

Small area for interaction

(a) *n*-Pentane

(b) Neopentane

▲ **FIGURE 11.5 Dispersion Force and Molecular Shape** **(a)** The straight shape of *n*-pentane molecules allows them to interact with one another along the entire length of the molecules. **(b)** The nearly spherical shape of neopentane molecules allows for only a small area of interaction. Thus, dispersion forces are weaker in neopentane than in *n*-pentane, resulting in a lower boiling point.

Although we must always consider molecular shape and other factors in predicting the magnitude of dispersion forces, molar mass can act as a guide when comparing dispersion forces within a family of similar elements or compounds as illustrated in Figure 11.6▶ on the next page.

▶ **FIGURE 11.6** **Boiling Points of the *n*-Alkanes** The boiling points of the *n*-alkanes rise with increasing molar mass and the consequent stronger dispersion forces.

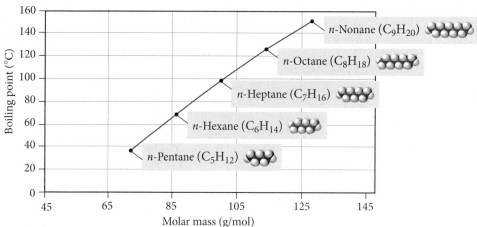

**CONCEPTUAL CONNECTION 11.2**

**Dispersion Forces** Which halogen has the highest boiling point?
**(a)** $Cl_2$  **(b)** $Br_2$  **(c)** $I_2$

## Dipole–Dipole Force

See Section 9.6 and 10.5 to review how to determine if a molecule is polar.

The **dipole–dipole force** exists between all molecules that are polar. Polar molecules have electron-rich regions (which have a partial negative charge) and electron-deficient regions (which have a partial positive charge). For example, consider acetone:

| Structural formula | Space-filling model | Electrostatic potential map |

The image on the right is an electrostatic potential map of acetone; we first saw these kinds of maps in Sections 9.6 and 10.5. Recall that the red areas indicate electron-rich regions in the molecule and that the blue areas indicate electron-poor regions.

Notice that acetone has an electron-rich region surrounding the oxygen atom (because oxygen is more electronegative than the rest of the molecule) and electron-poorer regions surrounding the carbon and hydrogen atoms. The result is that acetone has a **permanent dipole** that can interact with other acetone molecules as shown in Figure 11.7▼. The positive end of one permanent dipole attracts the negative end of

**Dipole–Dipole Interaction**

The positive end of a polar molecule is attracted to the negative end of its neighbor.

▶ **FIGURE 11.7** **Dipole–Dipole Interaction** Molecules with permanent dipoles, such as acetone, are attracted to one another via dipole–dipole interactions.

Space-filling model          Electrostatic potential map

another; this attraction is the dipole–dipole force. Polar molecules, therefore, have higher melting and boiling points than nonpolar molecules of similar molar mass.

Remember that all molecules (including polar ones) have dispersion forces. Polar molecules have, *in addition*, dipole–dipole forces. This additional attractive force raises their melting and boiling points relative to nonpolar molecules of similar molar mass. For example, consider formaldehyde and ethane:

| Name | Formula | Molar Mass (g/mol) | Structure | bp (°C) | mp (°C) |
|------|---------|--------------------|-----------|---------|---------|
| Formaldehyde | $CH_2O$ | 30.03 | | −19.5 | −92 |
| Ethane | $C_2H_6$ | 30.07 | | −88 | −172 |

Formaldehyde is polar, and has a higher melting point and boiling point than nonpolar ethane, even though the two compounds have the same molar mass. Figure 11.8▼ shows the boiling points of a series of molecules with similar molar mass but progressively greater dipole moments. Notice that the boiling points increase with increasing dipole moment.

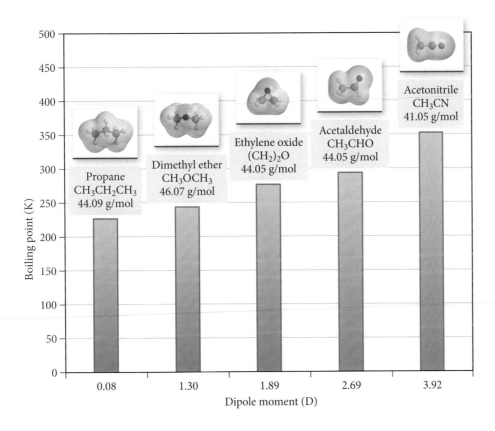

◄ **FIGURE 11.8 Dipole Moment and Boiling Point** The molecules shown here all have similar molar masses but different dipole moments. The boiling points increase with increasing dipole moment.

The polarity of molecules is also important in determining the **miscibility**—the ability to mix without separating into two states—of liquids. This is especially true for water, which is highly polar. In general, polar liquids tend to be miscible with water, while nonpolar liquids are not. For example, pentane ($C_5H_{12}$) is nonpolar and does not mix with water (Figure 11.9► on the next page). Similarly, water and oil (nonpolar) do not mix. Consequently, oily hands or oily stains on clothes cannot be washed with plain water (see *Chemistry in Your Day: How Soap Works* in Section 10.5).

▶ **FIGURE 11.9** **Polar and Nonpolar Substances** Water and pentane do not mix because water molecules are polar and pentane molecules are nonpolar.

$C_5H_{12}(l)$

$H_2O(l)$

## Example 11.1 Dipole–Dipole Forces

Which molecules have dipole–dipole forces?

**(a)** $CO_2$

**(b)** $CH_2Cl_2$

**(c)** $CH_4$

**SOLUTION**

A molecule has dipole–dipole forces if it is polar. To determine if a molecule is polar: (1) *determine if the molecule contains polar bonds* (Section 9.6); and (2) *determine if the polar bonds add together to form a net dipole moment* (Section 10.5).

| | |
|---|---|
| **(a)** $CO_2$<br>**(1)** Since the electronegativity of carbon is 2.5 and that of oxygen is 3.5 (see Figure 9.8), $CO_2$ has polar bonds.<br>**(2)** The geometry of $CO_2$ is linear. Consequently, the dipoles of the polar bonds cancel, so the molecule is *not polar* and does not have dipole–dipole forces. | $O{=}C{=}O$<br><br>No dipole forces present |
| **(b)** $CH_2Cl_2$<br>**(1)** The electronegativity of C is 2.5, that of H is 2.1, and that of Cl is 3.0. Consequently, $CH_2Cl_2$ has two polar bonds (C—Cl) and two bonds that are nearly nonpolar (C—H).<br>**(2)** The geometry of $CH_2Cl_2$ is tetrahedral. Because the C—Cl bonds and the C—H bonds are different, their dipoles do not cancel but sum to a net dipole moment. The molecule is polar and has dipole–dipole forces. | $CH_2Cl_2$<br><br>Dipole forces present |
| **(c)** $CH_4$<br>**(1)** The electronegativity of C is 2.5 and that of hydrogen is 2.1, so the C—H bonds are nearly nonpolar.<br>**(2)** In addition, since the geometry of the molecule is tetrahedral, any slight polarities that the bonds might have cancel. $CH_4$ is therefore nonpolar and does not have dipole–dipole forces. | $CH_4$<br><br>No dipole forces present |

**FOR PRACTICE 11.1** Which molecules have dipole–dipole forces?

**(a)** $CI_4$     **(b)** $CH_3Cl$     **(c)** HCl

**Dipole–Dipole Interaction** An electrostatic potential map for acetonitrile ($CH_3CN$), which is polar, is shown here: From this map, determine the geometry for how two acetonitrile molecules would interact with each other. Draw structural formulas, using the three-dimensional bond notation introduced in Section 10.4, to illustrate the geometry of the interaction.

CONCEPTUAL CONNECTION **11.3**

## Hydrogen Bonding

Polar molecules containing hydrogen atoms bonded directly to small electronegative atoms—most importantly fluorine, oxygen, or nitrogen—exhibit an intermolecular force called **hydrogen bonding**. HF, $NH_3$, and $H_2O$, for example, all exhibit hydrogen bonding. The hydrogen bond is a sort of *super* dipole–dipole force. The large electronegativity difference between hydrogen and any of these electronegative elements causes the hydrogen atom to have a fairly large partial positive charge ($\delta+$) within the bond, while the F, O, or N atom has a fairly large partial negative charge ($\delta-$). In addition, since these atoms are all quite small, the H atom on one molecule can approach the F, O, or N atom on an adjacent molecule very closely. The result is a strong attraction between the H atom on one molecule and the F, O, or N on its neighbor—an attraction called a **hydrogen bond**.

For example, in HF, the hydrogen atom in one molecule is strongly attracted to the fluorine atom on a neighboring molecule (Figure 11.10▼). The electrostatic potential maps in Figure 11.10 illustrate the large differences in electron density that result in unusually large partial charges.

### Hydrogen Bonding

When H bonds directly to F, O, or N, the bonding atoms acquire relatively large partial charges, giving rise to strong dipole–dipole attractions between neighboring molecules.

Space-filling model        Electrostatic potential map

◀ **FIGURE 11.10 Hydrogen Bonding in HF** The hydrogen of one HF molecule, with its partial positive charge, is attracted to the fluorine of its neighbor, with its partial negative charge. This dipole–dipole interaction is an example of a hydrogen bond.

*Hydrogen bonds should not be confused with chemical bonds.* Chemical bonds occur *between individual atoms within a molecule,* whereas hydrogen bonds—like dispersion forces and dipole–dipole forces—are intermolecular forces that occur *between molecules.* A typical hydrogen bond is only 2–5% as strong as a typical covalent chemical bond. Hydrogen bonds are, however, the strongest of the three *intermolecular* forces we have discussed so far. Substances composed of molecules that form hydrogen bonds have higher melting and boiling points than comparable substances composed of molecules that do not form hydrogen bonds.

For example, consider ethanol and dimethyl ether:

| Name | Formula | Molar Mass (g/mol) | Structure | bp (°C) | mp (°C) |
|------|---------|--------------------|-----------|---------|---------|
| Ethanol | $C_2H_6O$ | 46.07 | $CH_3CH_2OH$ | 78.3 | −114.1 |
| Dimethyl ether | $C_2H_6O$ | 46.07 | $CH_3OCH_3$ | −22.0 | −138.5 |

► **FIGURE 11.11 Hydrogen Bonding in Ethanol** The left side shows the space-filling models, and the right side shows the electrostatic potential maps.

## Hydrogen Bonding in Ethanol

The partial positive charge on H is strongly attracted to the partial negative charge on O.

Space-filling model    Electrostatic potential map

## Hydrogen Bonding in Water

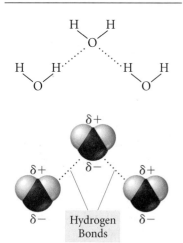

▲ **FIGURE 11.12 Hydrogen Bonding in Water**

Since ethanol contains hydrogen bonded directly to oxygen, ethanol molecules form hydrogen bonds with each other as shown in Figure 11.11▲. The hydrogen that is directly bonded to oxygen in an individual ethanol molecule is strongly attracted to the oxygen on neighboring molecules. This strong attraction (i.e., hydrogen bond) makes the boiling point of ethanol 78.3 °C. Consequently, ethanol is a liquid at room temperature. In contrast, dimethyl ether has an identical molar mass to ethanol but does not exhibit hydrogen bonding because in the dimethyl ether molecule, the oxygen atom is not bonded directly to hydrogen; this results in lower boiling and melting points, and dimethyl ether is a gas at room temperature.

Water is another good example of a molecule with hydrogen bonding (Figure 11.12◄). Figure 11.13▼ plots the boiling points of the simple hydrogen compounds of the group 4A and group 6A elements. In general, boiling points increase with increasing molar mass, as expected based on increasing dispersion forces. However, because of hydrogen bonding, the boiling point of water (100 °C) is much higher than expected based on its molar mass (18.0 g/mol). Without hydrogen bonding, all the water on our planet would be gaseous.

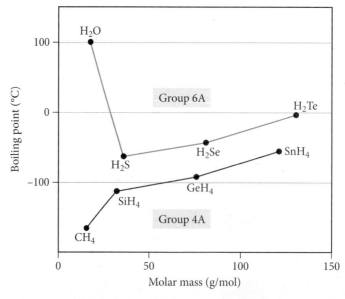

▲ **FIGURE 11.13 Boiling Points of Group 4A and 6A Compounds** Because of hydrogen bonding, the boiling point of water is anomalous compared to the boiling points of other hydrogen-containing compounds.

## Example 11.2    Hydrogen Bonding

One of these compounds is a liquid at room temperature. Which one and why?

$$
\begin{array}{ccc}
\text{O} & \text{H} & \\
\| & | & \\
\text{H—C—H} & \text{H—C—F} & \text{H—O—O—H} \\
& | & \\
& \text{H} & \\
\end{array}
$$

Formaldehyde          Fluoromethane          Hydrogen peroxide

**SOLUTION**

The three compounds have similar molar masses:

| | |
|---|---|
| Formaldehyde | 30.03 g/mol |
| Fluoromethane | 34.03 g/mol |
| Hydrogen peroxide | 34.02 g/mol |

As a result, the strengths of their dispersion forces are similar. All three compounds are also polar, so they have dipole–dipole forces. Hydrogen peroxide, however, is the only one of these compounds that also contains H bonded directly to F, O, or N. Therefore, it also has hydrogen bonding and is likely to have the highest boiling point of the three. Since the example stated that only one of the compounds was a liquid, it is safe to assume that hydrogen peroxide is the liquid. Note that, although fluoromethane *contains* both H and F, H is not *directly bonded* to F, so fluoromethane does not have hydrogen bonding as an intermolecular force. Similarly, formaldehyde *contains* both H and O, but H is not *directly bonded* to O, so formaldehyde does not exhibit hydrogen bonding either.

**FOR PRACTICE 11.2**    Which has the higher boiling point, HF or HCl? Why?

## Ion–Dipole Force

The **ion–dipole force** occurs when an ionic compound is mixed with a polar compound; it is especially important in aqueous solutions of ionic compounds. For example, when sodium chloride is mixed with water, the sodium and chloride ions interact with water molecules via ion–dipole forces, as shown in Figure 11.14 ▶. The positive sodium ions interact with the negative poles of water molecules, while the negative chloride ions interact with the positive poles. Ion–dipole forces are the strongest of the types of intermolecular forces discussed and are responsible for the ability of ionic substances to form solutions with water. We discuss aqueous solutions more thoroughly in Chapter 13.

*Summarizing Intermolecular Forces (as shown in Table 11.4 on the next page):*

▶ Dispersion forces are present in all molecules and atoms and generally increase with increasing molar mass. These forces are always weak in small molecules but can be significant in molecules with high molar masses.

▶ Dipole–dipole forces are present in polar molecules.

▶ Hydrogen bonds, the strongest of the intermolecular forces that can occur in pure substances (second only to ion–dipole forces in general), are present in molecules containing hydrogen bonded directly to fluorine, oxygen, or nitrogen.

▶ Ion–dipole forces are present in mixtures of ionic compounds and polar compounds. These forces are very strong and are especially important in aqueous solutions of ionic compounds.

### Ion–Dipole Forces

The positively charged end of a polar molecule such as $H_2O$ is attracted to negative ions, and the negatively charged end of the molecule is attracted to positive ions.

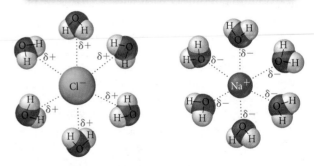

▲ **FIGURE 11.14  Ion–Dipole Forces**  Ion–dipole forces exist between $Na^+$ and the negative ends of $H_2O$ molecules and between $Cl^-$ and the positive ends of $H_2O$ molecules.

**TABLE 11.4 Types of Intermolecular Forces**

| Type | Present In | Molecular Perspective | Strength |
|---|---|---|---|
| Dispersion* | All molecules and atoms | $\delta-$ $\delta+$ $\cdots$ $\delta-$ $\delta+$ | 0.05–20+ kJ/mol |
| Dipole–dipole | Polar molecules | $\delta+$ $\delta-$ $\cdots$ $\delta+$ $\delta-$ | 3–20+ kJ/mol |
| Hydrogen bonding | Molecules containing H bonded to F, O, or N | $\delta+$ $\delta+$ $\delta-$ $\delta-$ $\delta-$ | 10–40 kJ/mol |
| Ion–dipole | Mixtures of ionic compounds and polar compounds | $\delta-$ $\delta-$ $\delta-$ $+$ $\delta-$ $\delta-$ $\delta-$ | 30–100+ kJ/mol |

*The dispersion force can become very strong (as strong and even stronger than the others) for molecules of high molar mass.

CONCEPTUAL
**CONNECTION 11.4**

**Intermolecular Forces and Boiling Point** Which substance has the highest boiling point?

**(a)** $CH_3OH$     **(b)** CO     **(c)** $N_2$

---

## Chemistry AND Medicine | Hydrogen Bonding in DNA

DNA is a long, chainlike molecule that acts as a blueprint for each living organism. Copies of DNA are passed from parent to offspring, which is how we inherit traits from our parents. A DNA molecule is composed of thousands of repeating units called *nucleotides* (Figure 11.15▼). Each nucleotide contains one of four different *organic bases:* adenine, thymine, cytosine, and guanine (abbreviated A, T, C,

and G). The sequence of these bases along DNA encodes the information that determines the nature of the proteins that are made in the body (proteins are the molecules that do most of the work in living organisms). Our proteins in turn determine many of our characteristics, including how we look, what diseases we are at risk of developing, and even our behavior.

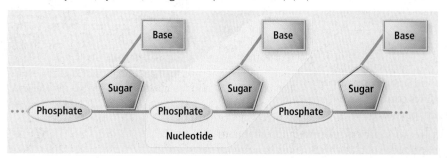

◄ **FIGURE 11.15 Nucleotides** The individual units in a DNA polymer are nucleotides. Each nucleotide contains one of four bases: adenine, thymine, cytosine, and guanine (abbreviated A, T, C, and G).

The replicating mechanism of DNA is related to its structure, which was discovered in 1953 by James Watson (1928–) and Francis Crick (1916–2004). DNA consists of two *complementary* strands wrapped around each other in the now famous double helix and linked by hydrogen bonds between the bases on each strand. Each base (A, T, C, and G) has a complementary partner with which it forms hydrogen bonds (Figure 11.16►): adenine (A) with thymine (T) and cytosine (C) with guanine (G). The hydrogen bonding is so specific that each base will pair only with its complementary partner. When a cell is going to divide, enzymes unzip the DNA molecule across the hydrogen bonds that join its two strands (Figure 11.17▼). Then new bases, complementary to the bases in each strand, are added along each of the original strands, forming hydrogen bonds with their complements. The result is two identical copies of the original DNA.

► **FIGURE 11.16**
**Complementary Base Pairing via Hydrogen Bonds** The individual bases in DNA interact with one another via specific hydrogen bonds that form between A and T and between C and G.

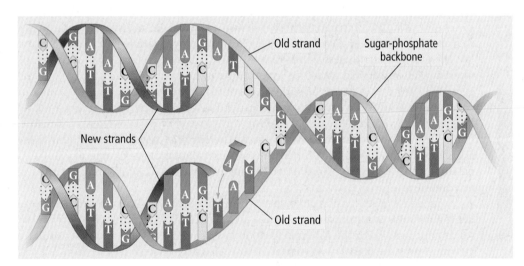

**QUESTION**

*Why would dispersion forces not work as a way to hold the two strands of DNA together? Why would covalent bonds not work?*

◄ **FIGURE 11.17**
**Copying DNA** The two strands of the DNA molecule "unzip" by breaking the hydrogen bonds that join the base pairs. New bases complementary to the bases of each strand are assembled and joined together. The result is two molecules, each identical to the original one.

# 11.4 Intermolecular Forces in Action: Surface Tension, Viscosity, and Capillary Action

The most important manifestation of intermolecular forces is the very existence of liquids and solids. In liquids, we also observe several other manifestations of intermolecular forces, including surface tension, viscosity, and capillary action.

## Surface Tension

A fly fisherman delicately casts a small fishing fly (a metal hook with a few feathers and strings attached to make it look like an insect) onto the surface of a moving stream. The fly floats on the surface of the water—even though the metal composing the hook is

▲ A trout fly can float on water because of surface tension.

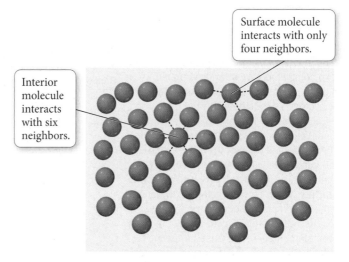

▲ **FIGURE 11.18** **The Origin of Surface Tension** Molecules at the liquid surface have a higher potential energy than those in the interior. As a result, a liquid tends to minimize its surface area; its surface behaves like a "skin."

Recall from Section 11.3 that the interactions between molecules lower their potential energy in much the same way that the interaction between protons and electrons lowers their potential energy, in accordance with Coulomb's law.

▲ **FIGURE 11.19** **Surface Tension in Action** A paper clip floats on water because of surface tension.

▲ **FIGURE 11.20** **Spherical Water Droplets** Tiny water droplets are not distorted much by gravity and form nearly perfect spheres held together by intermolecular forces between water molecules.

denser than water—and attracts trout. Why? The hook floats because of *surface tension*, the tendency of liquids to minimize their surface area.

Figure 11.18▲ depicts the intermolecular forces experienced by a molecule at the surface of the liquid compared to those experienced by a molecule in the interior. Notice that a molecule at the surface has relatively fewer neighbors with which to interact, and it is therefore inherently less stable—it has higher potential energy—than those in the interior. (Remember that attractive interactions with other molecules lower potential energy.) In order to increase the surface area of the liquid, molecules from the interior have to be moved to the surface, and, because molecules at the surface have a higher potential energy than those in the interior, this movement requires energy. Therefore, liquids tend to minimize their surface area. The **surface tension** of a liquid is the energy required to increase the surface area by a unit amount. For example, at room temperature, water has a surface tension of 72.8 mJ/m$^2$—it takes 72.8 mJ to increase the surface area of water by one square meter.

Why does surface tension allow the fly fisherman's hook to float on water? The tendency for liquids to minimize their surface area creates a kind of skin at the surface that resists penetration. For the fisherman's hook to sink into the water, the water's surface area must increase slightly—an increase that is resisted by the surface tension. We can observe surface tension by carefully placing a paper clip on the surface of water (Figure 11.19◄). The paper clip, even though it is denser than water, floats on the surface of the water. A slight tap on the clip provides the energy necessary to overcome the surface tension and causes the clip to sink.

Surface tension decreases as intermolecular forces decrease. You can't float a paper clip on benzene, for example, because the dispersion forces among the molecules composing benzene are significantly weaker than the hydrogen bonds among water molecules. The surface tension of benzene is only 28 mJ/m$^2$—just 40% that of water.

Surface tension is also the reason for the behavior of water that we discussed in the chapter-opening section on water and gravity. You can see this behavior on Earth by looking at small water droplets (those not large enough to be distorted by gravity) on a leaf or a spider's web (Figure 11.20◄). Just as larger samples of water form spheres in the space station, so smaller samples form spheres on Earth. Why? In the same way that gravity pulls the matter of a planet or star inward to form a sphere, intermolecular forces among collections of water molecules pull the water into a sphere. As we discussed in Section 11.1, the sphere is the geometrical shape with the smallest ratio of surface area to volume; therefore, the formation of a sphere minimizes the number of molecules at the surface, minimizing the potential energy of the system.

## Viscosity

Another manifestation of intermolecular forces is **viscosity**, the resistance of a liquid to flow. Motor oil, for example, is more viscous than gasoline, and maple syrup is more viscous than water. Viscosity is measured in a unit called the poise (P), defined as 1 g/cm·s. The viscosity of water at room temperature is approximately one centipoise (cP). Viscosity is greater in substances with stronger intermolecular forces because if molecules are more strongly attracted to each other, they do not flow around each other as freely. Viscosity also depends on molecular shape, increasing in longer molecules that can interact over a greater area and possibly become entangled. Table 11.5 lists the viscosity of several hydrocarbons. Notice the increase in viscosity with increasing molar mass (and therefore increasing magnitude of dispersion forces) and with increasing length (and therefore increasing potential for molecular entanglement).

Viscosity also depends on temperature because thermal energy partially overcomes the intermolecular forces, allowing molecules to flow past each other more easily. Table 11.6 lists the viscosity of water as a function of temperature. Nearly all liquids become less viscous as temperature increases.

**TABLE 11.5** Viscosity of Several Hydrocarbons at 20 °C

| Hydrocarbon | Molar Mass (g/mol) | Formula | Viscosity (cP) |
|---|---|---|---|
| *n*-Pentane | 72.15 | $CH_3CH_2CH_2CH_2CH_3$ | 0.240 |
| *n*-Hexane | 86.17 | $CH_3CH_2CH_2CH_2CH_2CH_3$ | 0.326 |
| *n*-Heptane | 100.2 | $CH_3CH_2CH_2CH_2CH_2CH_2CH_3$ | 0.409 |
| *n*-Octane | 114.2 | $CH_3CH_2CH_2CH_2CH_2CH_2CH_2CH_3$ | 0.542 |
| *n*-Nonane | 128.3 | $CH_3CH_2CH_2CH_2CH_2CH_2CH_2CH_2CH_3$ | 0.711 |

**TABLE 11.6** Viscosity of Liquid Water at Several Temperatures

| Temperature (°C) | Viscosity (cP) |
|---|---|
| 20 | 1.002 |
| 40 | 0.653 |
| 60 | 0.467 |
| 80 | 0.355 |
| 100 | 0.282 |

## Chemistry IN YOUR DAY | Viscosity and Motor Oil

Viscosity is an important property of the motor oil you put into your car. The oil must be thick enough to adequately coat engine surfaces to lubricate them, but also thin enough to be pumped easily into all the required engine compartments. Motor oil viscosity is usually reported on the SAE scale (named after the Society of Automotive Engineers). The higher the SAE rating, the more viscous the oil. The thinnest motor oils have SAE ratings of 5 or 10, whereas the thickest have SAE ratings of up to 50.

Before the 1950s, most automobile owners changed the oil in their engine to accommodate seasonal changes in weather—a higher SAE rating was required in the summer months and a lower rating in the winter. Today, the advent of multigrade oils allows car owners in many climates to keep the same oil all year long. Multigrade oils, such as the 10W-40 oil shown at right, contain polymers (long molecules made up of repeating structural units) that coil at low temperatures but unwind at high temperatures. At low temperatures, the coiled polymers—because of their compact shape—do not contribute very much to the oil's viscosity. As the temperature increases, however, the molecules unwind and their long shape results in intermolecular forces and molecular entanglements that prevent the oil's viscosity from decreasing as much as it would normally. The result is an oil whose viscosity is less temperature-dependent than it would be otherwise, allowing the same oil to be used over a wider range of temperatures. The 10W-40 designation indicates that the oil has an SAE rating of 10 at low temperatures and 40 at high temperatures.

## Capillary Action

Medical technicians take advantage of **capillary action**—the ability of a liquid to flow against gravity up a narrow tube—when taking a blood sample. The technician pokes the patient's finger with a pin, squeezes some blood out of the puncture, and

▲ Blood is drawn into a capillary tube by capillary action.

▲ **FIGURE 11.21 Capillary Action** The attraction of water molecules to the glass surface draws the liquid around the edge of the tube up the walls. The water in the rest of the column is pulled along by the attraction of water molecules to one another. As can be seen in this figure, the narrower the tube, the higher the liquid will rise.

then collects the blood with a thin tube. When the tube's tip comes into contact with the blood, the blood is drawn into the tube by capillary action. The same force helps trees and plants draw water from the soil.

Capillary action results from a combination of two forces: the attraction between molecules in a liquid, called *cohesive forces*, and the attraction between these molecules and the surface of the tube, called *adhesive forces*. The adhesive forces cause the liquid to spread out over the surface of the tube, whereas the cohesive forces cause the liquid to stay together. If the adhesive forces are greater than the cohesive forces (as is the case for water in a glass tube), the attraction to the surface draws the liquid up the tube and the cohesive forces pull along those molecules not in direct contact with the tube walls (Figure 11.21◀). The water rises up the tube until the force of gravity balances the capillary action—the thinner the tube, the higher the rise. If the adhesive forces are less than the cohesive forces (as is the case for liquid mercury), the liquid does not rise up the tube at all (and in fact drops to a level below the level of the surrounding liquid).

We can see the result of the differences in the relative magnitudes of cohesive and adhesive forces by comparing the meniscus of water to the meniscus of mercury (Figure 11.22▶). (The meniscus is the curved shape of a liquid surface within a tube.) The meniscus of water is concave (rounded inward) because the *adhesive forces* are greater than the cohesive forces, causing the edges of the water to creep up the sides of the tube a bit, forming the familiar cupped shape. The meniscus of mercury is convex (rounded outward) because the *cohesive forces*—due to metallic bonding between the atoms—are greater than the adhesive forces. The mercury atoms crowd toward the interior of the liquid to maximize their interactions with each other, resulting in the upward bulge at the center of the surface.

▲ **FIGURE 11.22 Meniscuses of Water and Mercury** The meniscus of water is concave because water molecules are more strongly attracted to the glass wall than to one another. The meniscus of mercury is convex because mercury atoms are more strongly attracted to one another than to the glass walls.

H₂O(*g*)

H₂O(*l*)

▲ **FIGURE 11.23 Vaporization of Water** Some molecules in an open beaker have enough kinetic energy to vaporize from the surface of the liquid.

# 11.5 Vaporization and Vapor Pressure

We now turn our attention to *vaporization*, the process by which thermal energy can overcome intermolecular forces and produce a state change from liquid to gas. We first discuss the process of vaporization itself, then the energetics of vaporization, and finally the concepts of *vapor pressure*, *dynamic equilibrium*, and *critical point*. Vaporization is a common occurrence that we experience every day and even depend on to maintain proper body temperature.

## The Process of Vaporization

Imagine water molecules in a beaker at room temperature and open to the atmosphere (Figure 11.23◀). The molecules are in constant motion due to thermal energy. If we could actually see the molecules at the surface, we would witness what theoretical chemist Roald Hoffmann describes as a "wild dance floor" (see the chapter-opening quote) because of all the vibrating, jostling, and molecular movement. *The higher the temperature, the greater the average energy of the collection of molecules.* However, at any one time, some molecules have more thermal energy than the average and some have less.

The distributions of thermal energies for the molecules in a sample of water at two different temperatures are shown in Figure 11.24▶. The molecules at the high end of the distribution curve have enough energy to break free from the surface—where

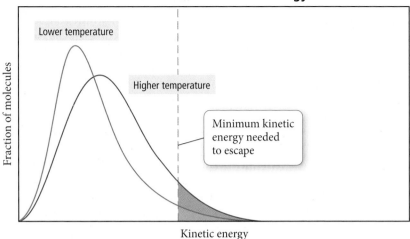

**Distribution of Thermal Energy**

◀ **FIGURE 11.24 Distribution of Thermal Energy** The thermal energies of the molecules in a liquid are distributed over a range. The peak energy increases with increasing temperature.

molecules are held less tightly than they are in the interior due to fewer neighbor–neighbor interactions—and into the gas state. This transition, from liquid to gas, is known as **vaporization**. Some of the water molecules in the gas state, at the low end of the energy distribution curve for the gaseous molecules, plunge back into the water and are captured by intermolecular forces. This transition, from gas to liquid, is the opposite of vaporization and is called **condensation**.

Although both vaporization and condensation occur in a beaker open to the atmosphere, under normal conditions (such as relatively dry air at room temperature) evaporation takes place at a greater rate because most of the newly evaporated molecules escape into the surrounding atmosphere and never come back. The result is a noticeable decrease in the water level within an open beaker over time (usually several days).

What happens if we increase the temperature of the water within the beaker? Because of the shift in the energy distribution to higher energies (see Figure 11.24), more molecules now have enough energy to break free and evaporate, so vaporization occurs more quickly. What happens if we spill the water on the table or floor? The same amount of water is now spread over a wider area, resulting in more molecules at the surface of the liquid. Since molecules at the surface have the greatest tendency to evaporate—because they are held less tightly—vaporization also occurs more quickly in this case. You probably know from experience that water in a beaker or glass may take many days to evaporate completely, whereas the same amount of water spilled on a table or floor typically evaporates within a few hours (depending on the exact conditions).

What happens if the liquid in the beaker is not water, but some other substance with weaker intermolecular forces, such as acetone (the main component in nail polish remover)? The weaker intermolecular forces allow more molecules to evaporate at a given temperature, increasing the rate of vaporization. We call liquids that vaporize easily **volatile** and those that do not vaporize easily **nonvolatile**. Acetone is more volatile than water. Motor oil is virtually nonvolatile at room temperature.

*Summarizing the Process of Vaporization:*

▶ The rate of vaporization increases with increasing temperature.

▶ The rate of vaporization increases with increasing surface area.

▶ The rate of vaporization increases with decreasing strength of intermolecular forces.

## The Energetics of Vaporization

To understand the energetics of vaporization, consider again a beaker of water from the molecular point of view, except now imagine that the beaker is thermally insulated so that heat from the surroundings cannot enter the beaker. What happens to the temperature of the water left in the beaker as molecules evaporate? To answer this question,

See Chapter 6 to review endothermic and exothermic processes.

think about the energy distribution curve again (see Figure 11.24). The molecules that leave the beaker are the ones at the high end of the energy curve—the most energetic. If no additional heat enters the beaker, the average energy of the entire collection of molecules decreases—much as the class average on an exam goes down if you eliminate the highest-scoring students. So vaporization is an *endothermic* process; it takes energy to vaporize the molecules in a liquid. Another way to understand the endothermicity of vaporization is to remember that vaporization requires overcoming the intermolecular forces that hold liquids together. Since energy is needed to pull the molecules away from one another, the process is endothermic.

Our bodies use the endothermic nature of vaporization for cooling. When we overheat, we sweat and our skin is covered with liquid water. As this water evaporates, it absorbs heat from the body, cooling the skin. A fan makes us feel cooler because it blows newly vaporized water away from our skin, allowing more sweat to vaporize and causing even more cooling. High humidity, in contrast, slows down the net rate of evaporation, preventing cooling. When the air already contains large amounts of water vapor, the sweat evaporates more slowly, making the body's cooling system less efficient.

*Condensation*, the opposite of vaporization, is exothermic—heat is released when a gas condenses to a liquid. If you have ever accidentally put your hand above a steaming kettle or opened a bag of microwaved popcorn too soon, you may have experienced a *steam burn*. As the steam condenses to a liquid on your skin, it releases a lot of heat, causing the burn. The condensation of water vapor is also the reason that winter overnight temperatures in coastal regions, which tend to have water vapor in the air, do not get as low as in deserts, which tend to have dry air. As the air temperature in a coastal area drops, water condenses out of the air, releasing heat and preventing the temperature from dropping further. In deserts, the air contains almost no moisture to condense, so the temperature drop is more extreme.

The amount of heat required to vaporize one mole of a liquid to gas is its **heat (or enthalpy) of vaporization ($\Delta H_{vap}$).** The heat of vaporization of water at its normal boiling point of 100 °C is +40.7 kJ/mol:

$$H_2O(l) \longrightarrow H_2O(g) \quad \Delta H_{vap} = +40.7 \text{ kJ/mol}$$

The heat of vaporization is always positive because the process is endothermic—energy must be absorbed to vaporize a substance. The heat of vaporization is somewhat temperature dependent. For example, at 25 °C the heat of vaporization of water is +44.0 kJ/mol, slightly more than it is at 100 °C because the water contains less thermal energy at 25 °C. Table 11.7 lists the heats of vaporization of several liquids at their boiling points and at 25 °C.

When a substance condenses from a gas to a liquid, the same amount of heat is involved, but the heat is emitted rather than absorbed:

$$H_2O(g) \longrightarrow H_2O(l) \quad \Delta H = -\Delta H_{vap} = -40.7 \text{ kJ (at 100 °C)}$$

When one mole of water condenses, it releases 40.7 kJ of heat. The sign of $\Delta H$ in this case is negative because the process is exothermic.

We can use the heat of vaporization of a liquid to calculate the amount of energy required to vaporize a given mass of the liquid (or the amount of heat given off by the condensation of a given mass of liquid), using concepts similar to those covered in Section 6.6 (stoichiometry of $\Delta H$). The heat of vaporization is like a conversion factor between the number of moles of a liquid and the amount of heat required to vaporize it (or the amount of heat emitted when it condenses), as demonstrated in Example 11.3.

▲ When we sweat, water evaporates from the skin. Since evaporation is endothermic, the result is a cooling effect.

The normal boiling point of a substance is its boiling point at a pressure of 1 atm, as discussed later in this section.

The sign conventions of $\Delta H$ were introduced in Chapter 6.

| TABLE 11.7 Heats of Vaporization of Several Liquids at Their Boiling Points and at 25 °C | | | | |
|---|---|---|---|---|
| Liquid | Chemical Formula | Normal Boiling Point (°C) | $\Delta H_{vap}$(kJ/mol) at Boiling Point | $\Delta H_{vap}$(kJ/mol) at 25 °C |
| Water | $H_2O$ | 100 | 40.7 | 44.0 |
| Rubbing alcohol (isopropyl alcohol) | $C_3H_8O$ | 82.3 | 39.9 | 45.4 |
| Acetone | $C_3H_6O$ | 56.1 | 29.1 | 31.0 |
| Diethyl ether | $C_4H_{10}O$ | 34.6 | 26.5 | 27.1 |

**Example 11.3** **Using the Heat of Vaporization in Calculations** ▶

Calculate the mass of water (in g) that can be vaporized at its boiling point with 155 kJ of heat.

| | |
|---|---|
| **SORT** You are given a certain amount of heat in kilojoules and asked to find the mass of water that can be vaporized. | **GIVEN:** 155 kJ<br><br>**FIND:** g $H_2O$ |
| **STRATEGIZE** The heat of vaporization gives the relationship between heat absorbed and moles of water vaporized. Begin with the given amount of heat (in kJ) and convert to moles of water that can be vaporized. Then use the molar mass as a conversion factor to convert from moles of water to mass of water. | **CONCEPTUAL PLAN**<br><br>$$\boxed{kJ} \longrightarrow \boxed{mol\ H_2O} \longrightarrow \boxed{g\ H_2O}$$<br><br>$\dfrac{1\ mol\ H_2O}{40.7\ kJ}$ $\qquad$ $\dfrac{18.02\ g\ H_2O}{1\ mol\ H_2O}$<br><br>**RELATIONSHIPS USED**<br>$\Delta H_{vap} = 40.7$ kJ/mol (at 100 °C)<br>$18.02$ g $H_2O = 1$ mol $H_2O$ |
| **SOLVE** Follow the conceptual plan to solve the problem. | **SOLUTION**<br><br>$155\ kJ \times \dfrac{1\ mol\ H_2O}{40.7\ kJ} \times \dfrac{18.02\ g\ H_2O}{1\ mol\ H_2O} = 68.6\ g\ H_2O$ |

**FOR PRACTICE 11.3** Calculate the amount of heat (in kJ) required to vaporize 2.58 kg of water at its boiling point.

**FOR MORE PRACTICE 11.3** Suppose that 0.48 g of water at 25 °C condenses on the surface of a 55-g block of aluminum that is initially 25 °C. If the heat released during condensation goes only toward heating the metal, what is the final temperature (in °C) of the metal block? (The specific heat capacity of aluminum is 0.903 J/g °C.)

## Vapor Pressure and Dynamic Equilibrium

We have already seen that if a container of water is left uncovered at room temperature, the water slowly evaporates away. What happens if the container is sealed? Imagine a sealed evacuated flask—one from which the air has been removed—containing liquid water, as shown in Figure 11.25▼. Initially, the water molecules evaporate, as

Dynamic equilibrium:
Rate of evaporation =
rate of condensation

(a)   (b)   (c)

▲ **FIGURE 11.25 Vaporization in a Sealed Flask** **(a)** When water is in a sealed container, water molecules begin to vaporize. **(b)** As water molecules build up in the gas state, they begin to recondense into the liquid. **(c)** When the rate of evaporation equals the rate of condensation, dynamic equilibrium is reached.

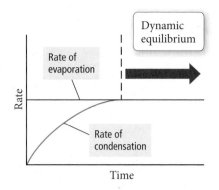

▲ **FIGURE 11.26 Dynamic Equilibrium** Dynamic equilibrium occurs when the rate of condensation is equal to the rate of evaporation.

Boyle's law is discussed in Section 5.3.

they did in the open beaker. But because of the seal, the evaporated molecules cannot escape into the atmosphere. As water molecules enter the gas state, some start condensing back into the liquid. As the concentration (or partial pressure) of gaseous water molecules increases, the rate of condensation also increases. However, as long as the water remains at a constant temperature, the rate of evaporation remains constant. Eventually, the rate of condensation and the rate of vaporization become equal—**dynamic equilibrium** has been reached (Figure 11.26◀). Condensation and vaporization continue at equal rates, and the concentration of water vapor above the liquid is constant.

The pressure of a gas in dynamic equilibrium with its liquid is its **vapor pressure**. The vapor pressure of a particular liquid depends on the intermolecular forces present in the liquid and the temperature. Weak intermolecular forces result in volatile (easily vaporizable) substances with high vapor pressures at a given temperature because the intermolecular forces are easily overcome by thermal energy. Strong intermolecular forces result in nonvolatile substances with low vapor pressures at a given temperature.

A liquid in dynamic equilibrium with its vapor is a balanced system that tends to return to equilibrium if disturbed. For example, consider a sample of *n*-pentane (a component of gasoline) at 25 °C in a cylinder equipped with a moveable piston (Figure 11.27(a)▼). The cylinder contains no other gases except *n*-pentane vapor in dynamic equilibrium with the liquid. Since the vapor pressure of *n*-pentane at 25 °C is 510 mmHg, the pressure in the cylinder is 510 mmHg.

Now, what happens when the piston is moved upward to expand the volume within the cylinder? Initially, the pressure in the cylinder drops below 510 mmHg, in accordance with Boyle's law. Then, however, more liquid vaporizes until equilibrium is reached once again (Figure 11.27(b)▼). If the volume of the cylinder is expanded again, the same thing happens—the pressure initially drops and more *n*-pentane vaporizes to bring the system back into equilibrium. Further expansion causes the same result *as long as some liquid* n-*pentane remains in the cylinder.*

Conversely, what happens if the piston is lowered, decreasing the volume in the cylinder? Initially, the pressure in the cylinder rises above 510 mmHg, but then some of the gas condenses into liquid until equilibrium is reached again (Figure 11.27(c)▼).

(a)  (b)  (c)

▲ **FIGURE 11.27 Dynamic Equilibrium in *n*-Pentane** **(a)** Liquid *n*-pentane is in dynamic equilibrium with its vapor. **(b)** When the volume is increased, the pressure drops and some liquid converts to gas to bring the pressure back up. **(c)** When the volume is decreased, the pressure increases and some gas converts to liquid to bring the pressure back down.

We describe the tendency of a system in dynamic equilibrium to return to equilibrium with the following general statement:

**When a system in dynamic equilibrium is disturbed, the system responds so as to minimize the disturbance and return to a state of equilibrium.**

If the pressure above a liquid–vapor system in equilibrium decreases, some of the liquid evaporates, restoring the equilibrium pressure. If the pressure increases, some of the vapor condenses, bringing the pressure back down to the equilibrium pressure. This basic principle—Le Châtelier's principle—is applicable to any chemical system in equilibrium, as we shall see in Chapter 15.

**CONCEPTUAL CONNECTION 11.5**

**Vapor Pressure**   What happens to the vapor pressure of a substance when its surface area is increased at constant temperature?
  **(a)**  The vapor pressure increases.
  **(b)**  The vapor pressure remains the same.
  **(c)**  The vapor pressure decreases.

## Temperature Dependence of Vapor Pressure and Boiling Point

When the temperature of a liquid increases, its vapor pressure rises because the higher thermal energy increases the number of molecules that have enough energy to vaporize (see Figure 11.24). Because of the shape of the thermal energy distribution curve, a small change in temperature makes a large difference in the number of molecules that have enough energy to vaporize, which results in a large increase in vapor pressure. For example, the vapor pressure of water at 25 °C is 23.3 torr, while at 60 °C the vapor pressure is 149.4 torr. Figure 11.28 ▼ plots the vapor pressure of water and several other liquids as a function of temperature.

The **boiling point** of a liquid is *the temperature at which the liquid's vapor pressure equals the external pressure.* When a liquid reaches its boiling point, the thermal energy is enough for molecules in the interior of the liquid (not just those at the surface) to break

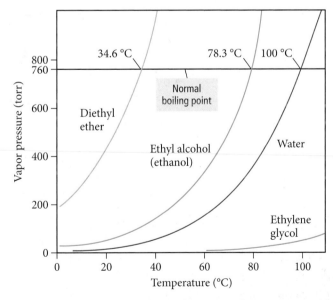

▲ **FIGURE 11.28  Vapor Pressure of Several Liquids at Different Temperatures**   At higher temperatures, more molecules have enough thermal energy to escape into the gas state, so vapor pressure increases with increasing temperature.

► FIGURE 11.29 **Boiling** A liquid boils when thermal energy is high enough to cause molecules in the interior of the liquid to become gaseous, forming bubbles that rise to the surface.

Sometimes you see bubbles begin to form in hot water below 100 °C. These bubbles are dissolved air—not gaseous water—leaving the liquid. Dissolved air comes out of water as you heat it because the solubility of a gas in a liquid decreases with increasing temperature (as we will see in Chapter 13).

free of their neighbors and enter the gas state (Figure 11.29▲). The bubbles in boiling water are pockets of gaseous water that form within the liquid water. The bubbles float to the surface and leave as gaseous water or steam.

The **normal boiling point** of a liquid is *the temperature at which its vapor pressure equals 1 atm.* The normal boiling point of pure water is 100 °C. However, at a lower pressure, water boils at a lower temperature. In Denver, Colorado, where the altitude is around 1600 meters (5200 feet) above sea level, for example, the average atmospheric pressure is about 83% of what it is at sea level, and water boils at approximately 94 °C. For this reason, it takes slightly longer to cook food in boiling water in Denver than in San Francisco (which is at sea level). Table 11.8 lists the boiling point of water at several locations of varied altitudes.

**TABLE 11.8 Boiling Points of Water at Several Locations of Varied Altitudes**

| Location | Elevation (ft) | Approximate Pressure (atm)* | Approximate Boiling Point of Water (°C) |
|---|---|---|---|
| Mount Everest, Tibet (highest mountain peak on Earth) | 29,035 | 0.32 | 78 |
| Mount McKinley (Denali), Alaska (highest mountain peak in North America) | 20,320 | 0.46 | 83 |
| Mount Whitney, California (highest mountain peak in 48 contiguous U.S. states) | 14,495 | 0.60 | 87 |
| Denver, Colorado (mile high city) | 5,280 | 0.83 | 94 |
| Boston, Massachusetts (sea level) | 20 | 1.0 | 100 |

*The atmospheric pressure in each of these locations is subject to weather conditions and can vary significantly from these values.

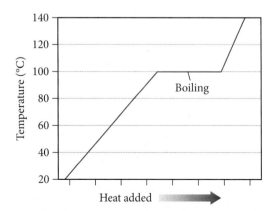

▲ FIGURE 11.30 **The Temperature during Boiling** The temperature of water during boiling remains at 100 °C.

Once the boiling point of a liquid is reached, additional heating only causes more rapid boiling; it does not raise the temperature of the liquid above its boiling point, as shown in the *heating curve* in Figure 11.30◄. Therefore, at 1 atm, boiling water always has a temperature of 100 °C. *As long as liquid water is present, its temperature cannot rise above its boiling point.* After all the water has been converted to steam, the temperature of the steam can continue to rise beyond 100 °C.

### The Clausius–Clapeyron Equation

Now, let's return our attention to Figure 11.28. As we can see from the graph, the vapor pressure of a liquid increases with increasing temperature. However, *the relationship is not linear.* In other words, doubling the temperature results in more than doubling the vapor pressure.

The relationship between vapor pressure and temperature is exponential, and we can express it as follows:

$$P_{vap} = \beta \exp\left(\frac{-\Delta H_{vap}}{RT}\right) \qquad [11.1]$$

In this expression $P_{vap}$ is the vapor pressure, $\beta$ is a constant that depends on the gas, $\Delta H_{vap}$ is the heat of vaporization, $R$ is the gas constant ($8.314\,J/mol \cdot K$), and $T$ is the temperature in kelvins. We can rearrange Equation 11.1 by taking the natural logarithm of both sides:

$$\ln P_{vap} = \ln\left[\beta \exp\left(\frac{-\Delta H_{vap}}{RT}\right)\right] \qquad [11.2]$$

Because $\ln AB = \ln A + \ln B$, we can rearrange the right side of Equation 11.2:

$$\ln P_{vap} = \ln \beta + \ln\left[\exp\left(\frac{-\Delta H_{vap}}{RT}\right)\right] \qquad [11.3]$$

Because $\ln e^x = x$ (see Appendix IB), we can simplify Equation 11.3:

$$\ln P_{vap} = \ln \beta + \frac{-\Delta H_{vap}}{RT} \qquad [11.4]$$

A slight additional rearrangement gives the following important result:

$$\ln P_{vap} = \frac{-\Delta H_{vap}}{R}\left(\frac{1}{T}\right) + \ln \beta \qquad \text{Clausius–Clapeyron equation}$$

$$y = m(x) + b \qquad \text{(equation for a line)}$$

Notice the parallel relationship between the **Clausius–Clapeyron equation** and the equation for a straight line. Just as a plot of $y$ versus $x$ yields a straight line with slope $m$ and intercept $b$, so a plot of $\ln P_{vap}$ (equivalent to $y$) versus $1/T$ (equivalent to $x$) gives a straight line with slope $-\Delta H_{vap}/R$ (equivalent to $m$) and $y$-intercept $\ln \beta$ (equivalent to $b$), as shown in Figure 11.31 ▼. The Clausius–Clapeyron equation gives a linear relationship—not between the vapor pressure and the temperature (which have an exponential relationship)—but between the *natural log* of the vapor pressure and the *inverse* of temperature. This is a common technique in the analysis of chemical data. If two variables are not linearly related, it is often convenient to find ways to graph *functions of those variables* that are linearly related.

Using the Clausius–Clapeyron equation in this way ignores the relatively small temperature dependence of $\Delta H_{vap}$.

Slope $= -3478\,K$
$\quad = -\Delta H_{vap}/R$
$\Delta H_{vap} = -\text{slope} \times R$
$\quad = 3478\,K \times 8.314\,J/mol \cdot K$
$\quad = 28.92 \times 10^3\,J/mol$

◀ **FIGURE 11.31  A Clausius–Clapeyron Plot for Diethyl Ether ($CH_3CH_2OCH_2CH_3$)**   A plot of the natural log of the vapor pressure versus the inverse of the temperature in K yields a straight line with slope $-\Delta H_{vap}/R$.

The Clausius–Clapeyron equation leads to a convenient way to determine the heat of vaporization in the laboratory. We simply measure the vapor pressure of a liquid as a function of temperature and create a plot of the natural log of the vapor pressure versus the inverse of the temperature. We then determine the slope of the line to find the heat of vaporization, as demonstrated in Example 11.4.

**Example** 11.4 | **Using the Clausius–Clapeyron Equation to Determine Heat of Vaporization from Experimental Measurements of Vapor Pressure**

The vapor pressure of dichloromethane is measured as a function of temperature, and the results are tabulated. Determine the heat of vaporization of dichloromethane from the results.

| Temperature (K) | Vapor Pressure (torr) |
| --- | --- |
| 200 | 0.8 |
| 220 | 4.5 |
| 240 | 21 |
| 260 | 71 |
| 280 | 197 |
| 300 | 391 |

**SOLUTION**

To find the heat of vaporization, use an Excel spreadsheet or a graphing calculator to plot the natural log of vapor pressure ($\ln P$) as a function of the inverse of the temperature in kelvins ($1/T$). Then fit the points to a line and determine the slope of the line. The slope of the best-fitting line is $-3773$ K. Since the slope equals $-\Delta H_{vap}/R$, we find the heat of vaporization as follows:

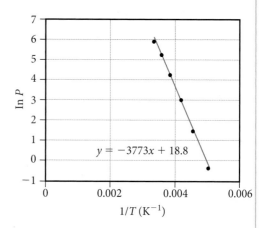

$$\text{slope} = -\Delta H_{vap}/R$$

$$\Delta H_{vap} = -\text{slope} \times R$$

$$= -(-3773 \text{ K})(8.314 \text{ J/mol} \cdot \text{K})$$

$$= 3.14 \times 10^4 \text{ J/mol}$$

$$= 31.4 \text{ kJ/mol}$$

**FOR PRACTICE 11.4** The vapor pressure of carbon tetrachloride is measured as a function of the temperature and the results are tabulated. Determine the heat of vaporization of carbon tetrachloride from the results.

| Temperature (K) | Vapor Pressure (torr) |
| --- | --- |
| 255 | 11.3 |
| 265 | 21.0 |
| 275 | 36.8 |
| 285 | 61.5 |
| 295 | 99.0 |
| 300 | 123.8 |

The Clausius–Clapeyron equation can also be expressed in a two-point form that we can use with just two measurements of vapor pressure and temperature to determine the heat of vaporization:

$$\ln \frac{P_2}{P_1} = \frac{-\Delta H_{vap}}{R}\left(\frac{1}{T_2} - \frac{1}{T_1}\right) \qquad \text{Clausius–Clapeyron equation (two-point form)}$$

The two-point method is generally inferior to plotting multiple points because fewer data points result in more chance for error.

We can use this form of the equation to predict the vapor pressure of a liquid at any temperature if we know the enthalpy of vaporization and the normal boiling point (or the vapor pressure at some other temperature), as demonstrated in Example 11.5.

**Example 11.5** | **Using the Two-Point Form of the Clausius–Clapeyron Equation to Predict the Vapor Pressure at a Given Temperature**

Methanol has a normal boiling point of 64.6 °C and a heat of vaporization ($\Delta H_{vap}$) of 35.2 kJ/mol. What is the vapor pressure of methanol at 12.0 °C?

| | |
|---|---|
| **SORT** You are given the normal boiling point of methanol (the temperature at which the vapor pressure is 760 mmHg) and the heat of vaporization. You are asked to find the vapor pressure at a specified temperature that is also given. | **GIVEN:** $T_1(°C) = 64.6\,°C$ <br> $P_1 = 760\ torr$ <br> $\Delta H_{vap} = 35.2\ kJ/mol$ <br> $T_2(°C) = 12.0\,°C$ <br> **FIND:** $P_2$ |
| **STRATEGIZE** The conceptual plan is essentially the Clausius–Clapeyron equation, which relates the given and find quantities. | **CONCEPTUAL PLAN** <br> $$\ln\frac{P_2}{P_1} = \frac{-\Delta H_{vap}}{R}\left(\frac{1}{T_2} - \frac{1}{T_1}\right)$$ <br> (Clausius–Clapeyron equation, two-point form) |
| **SOLVE** First, convert $T_1$ and $T_2$ from °C to K. | **SOLUTION** <br> $T_1(K) = T_1(°C) + 273.15$ <br> $= 64.6 + 273.15$ <br> $= 337.8\ K$ <br><br> $T_2(K) = T_2(°C) + 273.15$ <br> $= 12.0 + 273.15$ <br> $= 285.2\ K$ |
| Then, substitute the required values into the Clausius–Clapeyron equation and solve for $P_2$. | $$\ln\frac{P_2}{P_1} = \frac{-\Delta H_{vap}}{R}\left(\frac{1}{T_2} - \frac{1}{T_1}\right)$$ <br> $$\ln\frac{P_2}{P_1} = \frac{-35.2 \times 10^3\dfrac{J}{mol}}{8.314\dfrac{J}{mol \cdot K}}\left(\frac{1}{285.2\ K} - \frac{1}{337.8\ K}\right)$$ <br> $= -2.31$ |
| In order to isolate $P_2$, you must take the inverse natural log (or inv ln) of $\ln(P_2/P_1)$. Remember that inv ln $(\ln x) = x$. | $$\frac{P_2}{P_1} = e^{-2.31}$$ <br> $P_2 = P_1(e^{-2.31})$ <br> $= 760\ torr(0.0993)$ <br> $= 75.4\ torr$ |

**CHECK** The units of the answer are correct. The magnitude of the answer makes sense because vapor pressure should be significantly lower at the lower temperature.

**FOR PRACTICE 11.5** Propane has a normal boiling point of −42.0 °C and a heat of vaporization ($\Delta H_{vap}$) of 19.04 kJ/mol. What is the vapor pressure of propane at 25.0 °C?

## The Critical Point: The Transition to an Unusual State of Matter

We have considered the vaporization of a liquid in a container open to the atmosphere with and without heating, and the vaporization of a liquid in a *sealed* container without heating. We now examine the vaporization of a liquid in a *sealed* container *during heating*.

Consider liquid *n*-pentane in equilibrium with its vapor in a sealed container initially at 25 °C. At this temperature, the vapor pressure of *n*-pentane is 0.67 atm.

What happens if we heat the liquid? As the temperature rises, more *n*-pentane vaporizes and the pressure within the container increases. At 100 °C, the pressure is 5.5 atm, and at 190 °C the pressure is 29 atm. As the temperature and pressure increase, more and more gaseous *n*-pentane is forced into the same amount of space, and the density of the *gas* gets higher and higher. At the same time, the increasing temperature causes the density of the *liquid* to become lower and lower. At 197 °C, the meniscus between the liquid and gaseous *n*-pentane disappears, and the gas and liquid states commingle to form a *supercritical fluid* (Figure 11.32▼). For any substance, the *temperature* at which this transition occurs is the **critical temperature (T_c)**. The liquid cannot exist (regardless of pressure) above this temperature. The *pressure* at which this transition occurs is the **critical pressure (P_c)**.

Gas

Liquid

Super-
critical
fluid

$T < T_c$ — Two States

$T > T_c$ — One State

Increasing temperature

▲ **FIGURE 11.32  Critical Point Transition**   As *n*-pentane is heated in a sealed container, it undergoes a transition to a supercritical fluid. At the critical point, the meniscus separating the liquid and gas disappears, and the fluid becomes supercritical—neither a liquid nor a gas.

Researchers are interested in supercritical fluids because of their unique properties. A supercritical fluid has properties of both liquids and gases—it is in some sense intermediate between the two. Supercritical fluids can act as good solvents, selectively dissolving a number of compounds. For example, supercritical carbon dioxide is used as a solvent to extract caffeine from coffee beans. The caffeine dissolves in the supercritical carbon dioxide, but other substances—such as those responsible for the flavor of coffee—do not. Consequently, the caffeine is removed and the flavor of the coffee is not substantially altered. The supercritical carbon dioxide is easily removed from the mixture by simply lowering the pressure below the critical pressure, at which point the carbon dioxide evaporates, leaving no residue.

## 11.6  Sublimation and Fusion

In Section 11.5, we examined a beaker of liquid water at room temperature from the molecular viewpoint. Now, let's examine a block of ice at −10 °C from the same molecular perspective, paying close attention to two common processes: sublimation and fusion.

### Sublimation

Even though a block of ice is solid, the water molecules in the ice have thermal energy, which causes each one to vibrate about a fixed point. The motion is much less vigorous than in a liquid, but it is significant nonetheless. As in liquids, at any instant some molecules in the block of ice have more thermal energy than the average and some have less. The molecules with high enough thermal energy can break free from the ice surface—where, as in liquids, molecules are held less tightly than in the interior due to fewer neighbor–neighbor interactions—and go directly into the gas state (Figure 11.33◀). This process is **sublimation**, the transition from solid to gas. Some of the water molecules in the gas state (those at the low end of the energy distribution curve for the gaseous

$H_2O$ (*g*)

$H_2O$ (*s*)

▲ **FIGURE 11.33  The Sublimation of Ice**   The water molecules at the surface of an ice cube can sublime directly into the gas state.

molecules) collide with the surface of the ice and are captured by the intermolecular forces with other molecules. This process—the opposite of sublimation—is **deposition**, the transition from gas to solid. As is the case with liquids, the pressure of a gas in dynamic equilibrium with its solid is the vapor pressure of the solid.

Although both sublimation and deposition occur on the surface of an ice block open to the atmosphere at $-10\,°C$, sublimation usually occurs at a greater rate than deposition because most of the newly sublimed molecules escape into the surrounding atmosphere and never come back. The result is a noticeable decrease in the size of the ice block over time (even though the temperature is below the melting point).

If you live in a cold climate, you may have noticed the disappearance of ice and snow from the ground even though the temperature remains below $0\,°C$. Similarly, ice cubes left in the freezer for a long time slowly shrink, even though the freezer is always below $0\,°C$. In both cases, the ice is *subliming*, turning directly into water vapor. Ice also sublimes out of frozen foods. You may have noticed, for example, the gradual growth of ice crystals on the *inside* of airtight plastic food-storage bags in a freezer. The ice crystals are composed of water that has sublimed out of the food and redeposited on the surface of the bag or on the surface of the food. For this reason, food that remains frozen for too long becomes dried out. Such dehydration can be avoided to some degree by freezing foods to colder temperatures, a process called deep-freezing. The colder temperature lowers the vapor pressure of ice and preserves the food longer. Freezer burn on meats is another common manifestation of sublimation. When you improperly store meat (for example, in a container that is not airtight) sublimation continues unabated. The result is the dehydration of the surface of the meat, which becomes discolored and loses flavor and texture.

A substance commonly associated with sublimation is solid carbon dioxide or dry ice, which does not melt under atmospheric pressure no matter what the temperature is. However, at $-78\,°C$ the $CO_2$ molecules have enough energy to leave the surface of the dry ice and become gaseous through sublimation.

▲ Dry ice (solid $CO_2$) sublimes but does not melt at atmospheric pressure.

## Fusion

Let's return to our ice block and examine what happens at the molecular level as we increase its temperature. The increasing thermal energy causes the water molecules to vibrate faster and faster. At the **melting point** ($0\,°C$ for water), the molecules have enough thermal energy to overcome the intermolecular forces that hold the molecules at their stationary points, and the solid turns into a liquid. This process is **melting** or **fusion**, the transition from solid to liquid. The opposite of melting is **freezing**, the transition from liquid to solid. Once the melting point of a solid is reached, additional heating only causes more rapid melting; it does not raise the temperature of the solid above its melting point (Figure 11.34▶). Only after all of the ice has melted will additional heating raise the temperature of the liquid water past $0\,°C$. A mixture of water *and* ice always has a temperature of $0\,°C$ (at 1 atm pressure).

▲ **FIGURE 11.34 Temperature during Melting** The temperature of water during melting remains at 0.0 °C as long as both solid and liquid water remain.

## Energetics of Melting and Freezing

The most common way to cool a beverage quickly is to drop several ice cubes into it. As the ice melts, the drink cools because melting is endothermic—the melting ice absorbs heat from the liquid. The amount of heat required to melt 1 mol of a solid is the **heat of fusion ($\Delta H_{fus}$)**. The heat of fusion for water is 6.02 kJ/mol:

$$H_2O(s) \longrightarrow H_2O(l) \quad \Delta H_{fus} = 6.02\ kJ/mol$$

The heat of fusion is positive because melting is endothermic.

Freezing, the opposite of melting, is exothermic—heat is released when a liquid freezes into a solid. For example, as water in the freezer turns into ice, it releases heat, which must be removed by the refrigeration system of the freezer. If the refrigeration system did not remove the heat, the water would not completely freeze into ice. The heat

The term *fusion* is used for melting because if we heat several crystals of a solid, they fuse into a continuous liquid upon melting.

released as the water began to freeze would warm the freezer, preventing further freezing. The change in enthalpy for freezing has the same magnitude as the heat of fusion but the opposite sign:

$$H_2O(l) \longrightarrow H_2O(s) \quad \Delta H = -\Delta H_{fus} = -6.02 \, kJ/mol$$

Different substances have different heats of fusion as listed in Table 11.9.

**TABLE 11.9 Heats of Fusion of Several Substances**

| Liquid | Chemical Formula | Melting Point (°C) | $\Delta H_{fus}$(kJ/mol) |
|---|---|---|---|
| Water | $H_2O$ | 0.00 | 6.02 |
| Rubbing alcohol (isopropyl alcohol) | $C_3H_8O$ | −89.5 | 5.37 |
| Acetone | $C_3H_6O$ | −94.8 | 5.69 |
| Diethyl ether | $C_4H_{10}O$ | −116.3 | 7.27 |

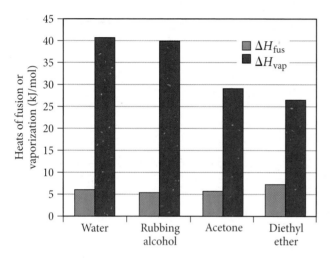

▲ **FIGURE 11.35 Heat of Fusion and Heat of Vaporization** Typical heats of fusion are significantly less than heats of vaporization.

In general, the heat of fusion for a substance is significantly less than its heat of vaporization (Figure 11.35◄). We have already seen that the solid and liquid states are closer to each other in many ways than they are to the gas state. It takes less energy to melt 1 mol of ice into liquid than it does to vaporize 1 mol of liquid water into gas because vaporization requires complete separation of molecules from one another, so the intermolecular forces must be completely overcome. Melting, in contrast, requires that intermolecular forces be only partially overcome, allowing molecules to move around one another while still remaining in contact.

The amount of heat required to sublime one mole of a solid to a gas is its **heat (or enthalpy) of sublimation** **($\Delta H_{sub}$)**. Since enthalpy is a state function, the change in enthalpy of any two-step process is simply the sum of the changes in enthalpies of each step. Because the sublimation of a substance is equivalent to its fusion followed by its vaporization, the heat of sublimation is the sum of the heat of fusion and the heat of vaporization. For water, we calculate the heat of sublimation at 0 °C as follows:

$$\Delta H_{sub} = \Delta H_{fus} + \Delta H_{vap}$$

$$= 6.01 \, kJ/mol + 40.7 \, kJ/mol$$

$$= 46.7 \, kJ/mol$$

 **Heating Curve for Water**

## 11.7 Heating Curve for Water

We can combine and build on the concepts from Sections 11.5 and 11.6 by examining the *heating curve* for 1.00 mol of water at 1.00 atm pressure shown in Figure 11.36►. The *y*-axis of the heating curve represents the temperature of the water sample. The *x*-axis represents the amount of heat added (in kilojoules) during heating. In the diagram, we divide the process into five segments: (1) ice warming; (2) ice melting into liquid water; (3) liquid water warming; (4) liquid water vaporizing into steam; and (5) steam warming.

In two of these segments (2 and 4) the temperature is constant as heat is added because the added heat goes into producing the transition between states, not into increasing the temperature. The two states are in equilibrium during the transition, and the temperature remains constant. The amount of heat required to achieve the state change is given by $q = n\Delta H$.

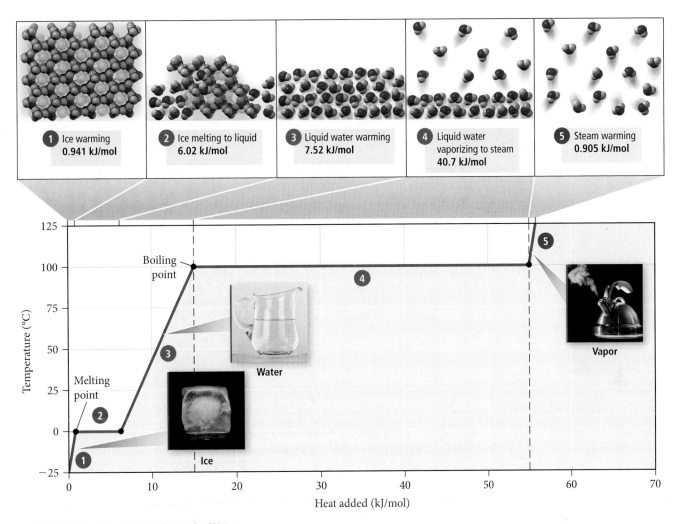

① Ice warming
0.941 kJ/mol

② Ice melting to liquid
6.02 kJ/mol

③ Liquid water warming
7.52 kJ/mol

④ Liquid water vaporizing to steam
40.7 kJ/mol

⑤ Steam warming
0.905 kJ/mol

▲ **FIGURE 11.36  Heating Curve for Water**

In the other three segments (1, 3, and 5), temperature increases linearly. These segments represent the heating of a single state in which the deposited heat raises the temperature in accordance with the substance's heat capacity ($q = mC_s \Delta T$). We examine each of these segments individually.

### Segment 1

In segment 1, solid ice is warmed from $-25\,°C$ to $0\,°C$. Since no transition between states occurs here, the amount of heat required to heat the solid ice is given by $q = mC_s \Delta T$ (see Section 6.4), where $C_s$ is the specific heat capacity of ice ($C_{s,\,ice} = 2.09\,J/g \cdot °C$). For 1.00 mol of water (18.0 g), we calculate the amount of heat as follows:

$$q = mC_{s,\,ice}\,\Delta T$$

$$= 18.0\,g\left(2.09\frac{J}{g \cdot °C}\right)[0.0\,°C - (-25.0\,°C)]$$

$$= 941\,J = 0.941\,kJ$$

So in segment 1, 0.941 kJ of heat is added to the ice, warming it from $-25°\,C$ to $0\,°C$.

### Segment 2

In segment 2, the added heat does not change the temperature of the ice and water mixture because the heat is absorbed by the transition from solid to liquid. The amount of heat required to convert the ice to liquid water is given by $q = n \, \Delta H_{fus}$, where $n$ is the number of moles of water and $\Delta H_{fus}$ is the heat of fusion (see Section 11.6):

$$q = n \, \Delta H_{fus}$$

$$= 1.00 \, mol \left( \frac{6.02 \, kJ}{mol} \right)$$

$$= 6.02 \, kJ$$

In segment 2, 6.02 kJ is added to the ice, melting it into liquid water. Notice that the temperature does not change during melting. The liquid and solid coexist at 0 °C as the melting occurs.

### Segment 3

In segment 3, the liquid water warms from 0 °C to 100 °C. Since no transition between states occurs here, the amount of heat required to heat the liquid water is given by $q = mC_s \Delta T$, as in segment 1. However, now we must use the heat capacity of liquid water (not ice) for the calculation. For 1.00 mol of water (18.0 g), we calculate the amount of heat as follows:

$$q = mC_{s, \, liq} \, \Delta T$$

$$= 18.0 \, g \left( 4.18 \frac{J}{g \cdot °C} \right) (100.0 \, °C - 0.0 \, °C)$$

$$= 7.52 \times 10^3 \, J = 7.52 \, kJ$$

So in segment 3, 7.52 kJ of heat is added to the liquid water, warming it from 0 °C to 100 °C.

### Segment 4

In segment 4, the water undergoes a second transition between states, this time from liquid to gas. The amount of heat required to convert the liquid to gas is given by $q = n \, \Delta H_{vap}$, where $n$ is the number of moles and $\Delta H_{vap}$ is the heat of vaporization (see Section 11.5):

$$q = n \, \Delta H_{vap}$$

$$= 1.00 \, mol \left( \frac{40.7 \, kJ}{mol} \right)$$

$$= 40.7 \, kJ$$

Thus, in segment 4, 40.7 kJ is added to the water, vaporizing it into steam. Notice that the temperature does not change during boiling. The liquid and gas coexist at 100 °C as the boiling occurs.

### Segment 5

In segment 5, the steam warms from 100 °C to 125 °C. Since no transition between states occurs here, the amount of heat required to heat the steam is given by $q = mC_s \Delta T$ (as in segments 1 and 3) except that we must use the heat capacity of steam (2.01 J/g · °C):

$$q = mC_{s, \, steam} \, \Delta T$$

$$= 18.0 \, g \left( 2.01 \frac{J}{g \cdot °C} \right) (125.0 \, °C - 100.0 \, °C)$$

$$= 905 \, J = 0.905 \, kJ$$

So in segment 5, 0.905 kJ of heat is added to the steam, warming it from 100 °C to 125 °C.

CONCEPTUAL
CONNECTION **11.6**

**Cooling of Water with Ice** You just saw that the heat capacity of ice is $C_{s, ice} = 2.09\,J/g \cdot {}^\circ C$ and that the heat of fusion of ice is 6.02 kJ/mol. When a small ice cube at $-10\,{}^\circ C$ is put into a cup of water at room temperature, which of the following plays a greater role in cooling the liquid water: the warming of the ice from $-10\,{}^\circ C$ to $0\,{}^\circ C$, or the melting of the ice?

# 11.8  Phase Diagrams

**KCV**  Phase Diagrams

Throughout most of this chapter, we have examined how the state of a substance changes in response to changes in temperature and pressure. We can combine both the temperature dependence and pressure dependence of the state of a particular substance in a graph called a *phase diagram*. A **phase diagram** is a map of the state or *phase* of a substance as a function of pressure (on the *y*-axis) and temperature (on the *x*-axis). In this section of the chapter, we first examine the major features of a phase diagram, then turn to navigating within a phase diagram, and finally examine and compare the phase diagrams of selected substances.

## The Major Features of a Phase Diagram

We can introduce the major features of a phase diagram by examining the phase diagram for water as an example (Figure 11.37 ▼). The *y*-axis displays the pressure in torr and the *x*-axis displays the temperature in degrees Celsius. We categorize the main features of the phase diagram as regions, lines, and points.

### Phase Diagram for Water

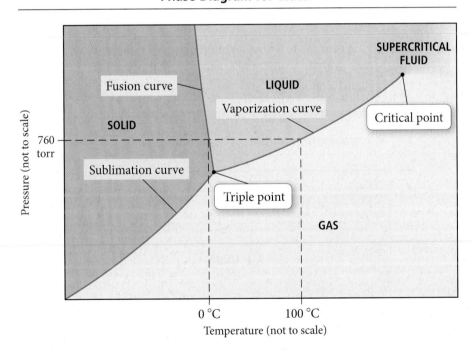

◀ **FIGURE 11.37** **Phase Diagram for Water**

## Regions

*Each of the three main regions—solid, liquid, and gas—in the phase diagram represents conditions under which that particular state is stable.* For example, under any of the temperatures and pressures within the liquid region in the phase diagram of water, the liquid is the stable state. Notice that the point 25 °C and 760 torr falls within the liquid region, as we know from everyday experience. In general, low temperature and high pressure favor the solid state, high temperature and low pressure favor the gas state, and intermediate conditions favor the liquid state. A sample of matter that is not in the state indicated by its phase diagram for a given set of conditions converts to that state when those conditions are imposed. For example, steam that is cooled to room temperature at 1 atm condenses to liquid.

### Lines

*Each of the lines (or curves) in the phase diagram represents a set of temperatures and pressures at which the substance is in equilibrium between the two states on either side of the line.* In the phase diagram for water shown in Figure 11.37, consider the curved line beginning just beyond 0 °C separating the liquid from the gas. This line is the vaporization curve (also called the vapor pressure curve) for water that we examined in Section 11.5. At any of the temperatures and pressures that fall along this line, the liquid and gas states of water are equally stable and in equilibrium. For example, at 100 °C and 760 torr pressure, water and its vapor are in equilibrium—they are equally stable and coexist. The other two major lines in a phase diagram are the *sublimation curve* (separating the solid and the gas) and the *fusion curve* (separating the solid and the liquid).

### The Triple Point

The triple point *in a phase diagram represents the unique set of conditions at which the three states are equally stable and in equilibrium.* In the phase diagram for water, the triple point occurs at 0.0098 °C and 4.58 torr. Under these unique conditions (and only under these conditions), the solid, liquid, and gas states of water are equally stable and coexist in equilibrium.

The triple point of a substance such as water can be reproduced anywhere to calibrate a thermometer or pressure gauge with a known temperature and pressure.

### The Critical Point

As we discussed in Section 11.5, at the critical temperature and pressure, the liquid and gas states coalesce into a *supercritical fluid.* The **critical point** *in a phase diagram represents the temperature and pressure above which a supercritical fluid exists.*

## Navigation within a Phase Diagram

We can represent changes in the temperature or pressure of a sample of water as movement within the phase diagram. For example, suppose we heat a block of ice initially at 1.0 atm and −25 °C. We represent the change in temperature at constant pressure as movement along the line marked A in Figure 11.38 ▼. As the temperature rises, we move to the right along the line. At the fusion curve, the temperature stops rising and melting

**Navigation within a Phase Diagram**

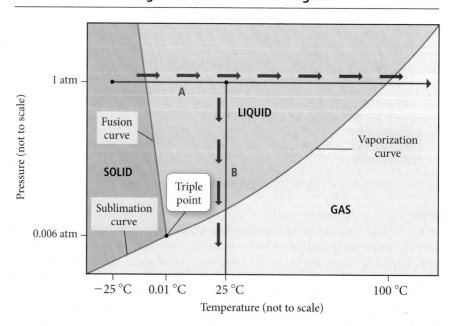

► **FIGURE 11.38 Navigation on the Phase Diagram for Water**

occurs until the solid ice is completely converted to liquid water. Crossing the fusion curve requires the complete transition from solid to liquid. Once the ice has completely melted, the temperature of the liquid water can begin to rise until the vaporization curve is reached. At this point, the temperature again stops rising and boiling occurs until all the liquid is converted to gas.

We represent a change in pressure with a vertical line on the phase diagram. For example, suppose we lower the pressure above a sample of water initially at 1.0 atm and 25 °C. We represent the change in pressure at constant temperature as movement along the line marked B in Figure 11.38. As the pressure drops, we move down the line and approach the vaporization curve. At the vaporization curve, the pressure stops dropping and vaporization occurs until the liquid completely converts to vapor. Crossing the vaporization curve requires the complete transition from liquid to gas. Only after the liquid has all vaporized can the pressure continue to drop.

## The Phase Diagrams of Other Substances

Next, let's examine the phase diagrams of iodine and carbon dioxide in Figure 11.39▼. The phase diagrams are similar to the diagram of water in most of their general features, but some significant differences exist.

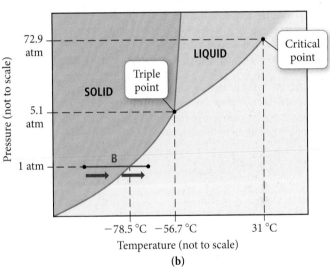

▲ **FIGURE 11.39  Phase Diagrams for Other Substances**   (a) Iodine, (b) Carbon dioxide.

Note that the fusion curves for both carbon dioxide and iodine have a positive slope—as the temperature increases, the pressure also increases—in contrast to the fusion curve for water, which has a negative slope. The behavior of water is atypical. The fusion curve within the phase diagrams for most substances has a positive slope because increasing pressure favors the denser state, which for most substances is the solid state.

For example, suppose we increase the pressure on a sample of iodine from 1 atm to 100 atm at 184 °C, as shown by line A in Figure 11.39(a). Notice that this change crosses the fusion curve, converting the liquid into a solid. In contrast, a pressure increase from 1 atm to 100 atm at −0.1 °C in water causes a state transition from solid to liquid. Unlike most substances, the liquid state of water is actually denser than the solid state.

Both water and iodine have stable solid, liquid, and gaseous states at a pressure of 1 atm. However, notice that carbon dioxide has no stable liquid state at a pressure of 1 atm. If we increase the temperature of a block of solid carbon dioxide (dry ice) at 1 atm, as indicated by line B in Figure 11.39(b), we cross the sublimation curve at −78.5 °C. At this temperature, the solid sublimes to a gas, which is one reason that dry ice is useful (it does not melt into a liquid at atmospheric pressure). Carbon dioxide forms a liquid only above pressures of 5.1 atm.

**Phase Diagrams** A substance has a triple point at −24.5 °C and 225 mmHg. What is most likely to happen to a solid sample of the substance as it is warmed from −35 °C to 0 °C at a pressure of 220 mmHg?

**(a)** The solid melts into a liquid.

**(b)** The solid sublimes into a gas.

**(c)** Nothing (the solid remains solid).

## 11.9  Water: An Extraordinary Substance

Water is the most common and important liquid on Earth. It fills our oceans, lakes, and streams. In its solid form, it caps our mountains, and in its gaseous form, it humidifies our air. We drink water, we sweat water, and we excrete bodily wastes dissolved in water. Indeed, the majority of our body mass *is* water. Life is impossible without water, and in most places on Earth where liquid water exists, life exists. Recent evidence for the past existence of water on Mars has fueled hopes of finding life or evidence of past life there. And though it may not be obvious to us (because we take water for granted), this familiar substance has many remarkable properties.

Among liquids, water is unique. It has a low molar mass (18.02 g/mol), yet it is a liquid at room temperature. Other main-group hydrides have higher molar masses but lower boiling points, as shown in Figure 11.40▼. No other substance of similar molar mass (except HF) comes close to being a liquid at room temperature.

▲ The Mars Curiosity Rover has found evidence that liquid water (which could possibly sustain life) once existed on the surface of Mars.

▶ **FIGURE 11.40  Boiling Points of Main-Group Hydrides** Water is the only common main-group hydride that is a liquid at room temperature.

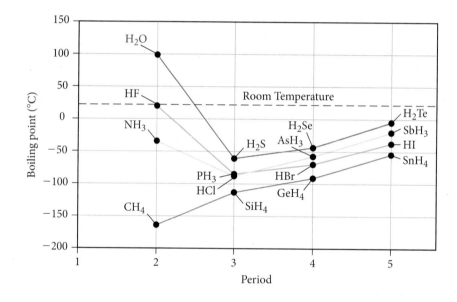

We can understand water's high boiling point (in spite of its low molar mass) by examining its molecular structure. The bent geometry of the water molecule and the highly polar nature of the O—H bonds result in a molecule with a significant dipole moment. Water's two O—H bonds (hydrogen directly bonded to oxygen) allow a water molecule to form strong hydrogen bonds with four other water molecules (Figure 11.41◀), resulting in a relatively high boiling point.

Water's high polarity allows it to dissolve many other polar and ionic compounds and even a number of nonpolar gases such as oxygen and carbon dioxide (by inducing a dipole moment in their molecules). Consequently, water is the main solvent within living organisms, transporting nutrients and other important compounds throughout the body. Water is the main solvent in our environment as well, allowing aquatic animals, for example, to survive by breathing dissolved oxygen and allowing aquatic plants to survive by using dissolved carbon dioxide for photosynthesis.

—H bonds

▲ **FIGURE 11.41  Hydrogen Bonding in Water** A water molecule can form four strong hydrogen bonds with four other water molecules.

Recall from Section 6.4 that water has an exceptionally high specific heat capacity, which has a moderating effect on the climate of coastal cities. In some cities, such as San Francisco, for example, the daily fluctuation in temperature can be less than 10 °C. This same moderating effect occurs over the entire planet, two-thirds of which is covered by water. Without water, the daily temperature fluctuations on our planet might be more like those on Mars, where temperatures fluctuate 63 °C (113 °F) between midday and early morning. Imagine awakening to below-freezing temperatures, only to bake at summer desert temperatures in the afternoon! The presence of water on Earth and water's uniquely high specific heat capacity are largely responsible for our planet's much smaller daily fluctuations.

As we have discussed, the way water freezes is also unique. Unlike other substances, which contract upon freezing, water expands upon freezing. Consequently, ice is less dense than liquid water, which is why ice floats. This seemingly trivial property has significant consequences. The frozen layer of ice at the surface of a winter lake insulates the water in the lake from further freezing. If this ice layer sank, it would kill bottom-dwelling aquatic life and possibly allow the lake to freeze solid, eliminating virtually all life in the lake.

The expansion of water upon freezing, however, is one reason that most organisms do not survive freezing. When the water within a cell freezes, it expands and often ruptures the cell, just as water freezing within a pipe bursts the pipe. Many foods, especially those with high water content, do not survive freezing very well either. Have you ever tried, for example, to freeze your own vegetables? If you put lettuce or spinach in the freezer, it will be limp and damaged upon thawing. The frozen-food industry gets around this problem by *flash freezing* vegetables and other foods. In this process, foods are frozen nearly instantaneously, which prevents water molecules from settling into their preferred crystalline structure. Consequently, the water does not expand very much, and the food remains largely undamaged.

▲ When lettuce freezes, the water within its cells expands, rupturing them.

# Chemistry IN THE Environment | Water Pollution

Water quality is critical to human health. Many human diseases—especially in developing nations—are caused by poor water quality. Several kinds of pollutants, including biological and chemical contaminants, can enter water supplies.

◀ Uncontaminated, sanitary water supplies are critical to human health.

Biological contaminants are microorganisms that cause diseases such as hepatitis, cholera, dysentery, and typhoid. They get into drinking water primarily when human or animal waste is dumped into bodies of water. Drinking water in developed nations is usually chemically treated to kill microorganisms. Water containing biological contaminants poses an immediate danger to human health and

should not be consumed. Boiling eliminates most biological contaminants from untreated water.

Chemical contaminants enter drinking water supplies as a result of industrial dumping, pesticide and fertilizer use, and household dumping. These contaminants include organic compounds, such as carbon tetrachloride and dioxin, and inorganic elements and compounds, such as mercury, lead, and nitrates. Because many chemical contaminants are neither volatile nor alive (like biological contaminants are), they are usually not eliminated through boiling.

The U.S. Environmental Protection Agency (EPA), under the Safe Drinking Water Act of 1974 and its amendments, sets standards that specify the maximum contamination level (MCL) of nearly 100 biological and chemical contaminants in water. Water providers that serve more than 25 people must periodically test the water they deliver to their consumers for these contaminants. If levels exceed the standards set by the EPA, the water provider must notify consumers and take appropriate measures to remove the contaminant from the water. According to the EPA, water from providers that serve more than 25 people should be safe to consume over a lifetime. If it is not safe to drink for a short period of time, providers must notify consumers.

## QUESTION

*Why doesn't boiling eliminate nonvolatile contaminants such as lead?*

# Self-Assessment Quiz

**Q1.** Which state of matter is compressible?

a) gas

b) liquid

c) solid

d) none of the above

**Q2.** Liquid nitrogen boils at 77 K. The image shown below depicts a sample of liquid nitrogen.

Which image best depicts the nitrogen after it has boiled?

a)

b)

c)

d)

**Q3.** Based on the expected intermolecular forces, which halogen has the highest boiling point?

a) $F_2$

b) $Cl_2$

c) $Br_2$

d) $I_2$

**Q4.** Which substance experiences dipole–dipole forces?

a) $CCl_4$

b) $NF_3$

c) $CS_2$

d) $SO_3$

**Q5.** One of these substances is a liquid at room temperature. Which one?

a) $CH_3OH$    b) $CF_4$    c) $SiH_4$    d) $CO_2$

**Q6.** Which property of a liquid increases with increasing temperature?

a) surface tension

b) viscosity

c) vapor pressure

d) none of the above

**Q7.** Determine the amount of heat (in kJ) required to vaporize 1.55 kg of water at its boiling point. For water, $\Delta H_{vap} = 40.7$ kJ/mol (at 100 °C).

a) $3.50 \times 10^3$ kJ

b) $1.14 \times 10^6$ kJ

c) 2.11 kJ

d) 686 kJ

**Q8.** The vapor pressure of a substance is measured over a range of temperatures. A plot of the natural log of the vapor pressure versus the inverse of the temperature (in Kelvin) produces a straight line with a slope of $-3.46 \times 10^3$ K. Find the enthalpy of vaporization of the substance.

a) $2.40 \times 10^{-3}$ kJ/mol

b) 28.8 kJ/mol

c) 0.416 kJ/mol

d) 3.22 kJ/mol

**Q9.** Acetic acid has a normal boiling point of 118 °C and a $\Delta H_{vap}$ of 23.4 kJ/mol. What is the vapor pressure (in mmHg) of acetic acid at 25 °C?

a) $2.92 \times 10^{-39}$ mmHg

b) $7.16 \times 10^3$ mmHg

c) 758 mmHg

d) 80.6 mmHg

**Q10.** Consider the phase diagram shown here. A sample of the substance in the phase diagram is initially at 175 °C and 925 mmHg. What phase transition occurs when the pressure is decreased to 760 mmHg at constant temperature?

a) solid to liquid

b) liquid to gas

c) solid to gas

d) liquid to solid

**Answers: 1.** (a) **2.** (d) **3.** (d) **4.** (b) **5.** (a) **6.** (c) **7.** (a) **8.** (b) **9.** (d) **10.** (b)

# Chapter 11 IN REVIEW

## Key Terms

**Section 11.2**
crystalline (488)
amorphous (488)

**Section 11.3**
dispersion force (490)
dipole–dipole force (492)
permanent dipole (492)
miscibility (493)
hydrogen bonding (495)
hydrogen bond (495)
ion–dipole force (497)

**Section 11.4**
surface tension (500)
viscosity (501)
capillary action (501)

**Section 11.5**
vaporization (503)
condensation (503)
volatile (503)
nonvolatile (503)
heat (or enthalpy) of vaporization ($\Delta H_{vap}$) (504)
dynamic equilibrium (506)
vapor pressure (506)
boiling point (507)
normal boiling point (508)
Clausius–Clapeyron equation (509)
critical temperature ($T_c$) (512)
critical pressure ($P_c$) (512)

**Section 11.6**
sublimation (512)
deposition (513)
melting point (513)
melting (fusion) (513)
freezing (513)
heat of fusion ($\Delta H_{fus}$) (513)
heat (or enthalpy) of sublimation ($\Delta H_{sub}$) (514)

**Section 11.8**
phase diagram (517)
triple point (518)
critical point (518)

## Key Concepts

### Solids, Liquids, and Intermolecular Forces (11.1, 11.2, 11.3)

▶ The forces that hold molecules or atoms together in a liquid or solid are intermolecular forces. The strength of the intermolecular forces in a substance determines its state.

▶ Dispersion forces are present in all elements and compounds; they arise from the fluctuations in electron distribution within atoms and molecules. These are the weakest intermolecular forces, but they are significant in molecules with high molar masses.

▶ Dipole–dipole forces, generally stronger than dispersion forces, are present in all polar molecules.

▶ Hydrogen bonding occurs in polar molecules that contain hydrogen atoms bonded directly to fluorine, oxygen, or nitrogen. These are among the strongest intermolecular forces.

▶ Ion–dipole forces occur when ionic compounds are mixed with polar compounds, and they are especially important in aqueous solutions.

### Surface Tension, Viscosity, and Capillary Action (11.4)

▶ Surface tension results from the tendency of liquids to minimize their surface area in order to maximize the interactions between their constituent particles, thus lowering potential energy. Surface tension causes water droplets to form spheres and allows insects and paper clips to "float" on the surface of water.

▶ Viscosity is the resistance of a liquid to flow. Viscosity increases with increasing strength of intermolecular forces and decreases with increasing temperature.

▶ Capillary action is the ability of a liquid to flow against gravity up a narrow tube. It is the result of adhesive forces, the attraction between the molecules and the surface of the tube, and cohesive forces, the attraction between the molecules in the liquid.

### Vaporization and Vapor Pressure (11.5, 11.7)

▶ Vaporization, the transition from liquid to gas, occurs when thermal energy overcomes the intermolecular forces present in a liquid. The opposite process is condensation. Vaporization is endothermic and condensation is exothermic.

▶ The rate of vaporization increases with increasing temperature, increasing surface area, and decreasing strength of intermolecular forces.

▶ The heat of vaporization ($\Delta H_{vap}$) is the heat required to vaporize one mole of a liquid.

▶ In a sealed container, a liquid and its vapor come into dynamic equilibrium, at which point the rate of vaporization equals the rate of condensation. The pressure of a gas that is in dynamic equilibrium with its liquid is its vapor pressure.

▶ The vapor pressure of a substance increases with increasing temperature and with decreasing strength of its intermolecular forces.

▶ The boiling point of a liquid is the temperature at which its vapor pressure equals the external pressure.

▶ The Clausius–Clapeyron equation expresses the relationship between the vapor pressure of a substance and its temperature and can be used to calculate the heat of vaporization from experimental measurements.

▶ When a liquid is heated in a sealed container, it eventually forms a supercritical fluid, which has properties intermediate between a liquid and a gas. This occurs at critical temperature and critical pressure.

## Fusion and Sublimation (11.6, 11.7)

▶ Sublimation is the transition from solid to gas. The opposite process is deposition.

▶ Fusion, or melting, is the transition from solid to liquid. The opposite process is freezing.

▶ The heat of fusion ($\Delta H_{fus}$) is the amount of heat required to melt one mole of a solid. Fusion is endothermic.

▶ The heat of fusion is generally less than the heat of vaporization because intermolecular forces do not have to be completely overcome for melting to occur.

## Phase Diagrams (11.8)

▶ A phase diagram is a map of the states of a substance as a function of its pressure ($y$-axis) and temperature ($x$-axis).

▶ The regions in a phase diagram represent conditions under which a single stable state (solid, liquid, gas) exists.

▶ The lines represent conditions under which two states are in equilibrium.

▶ The triple point represents the conditions under which all three states coexist.

▶ The critical point is the temperature and pressure above which a supercritical fluid exists.

## The Uniqueness of Water (11.9)

▶ Water is a liquid at room temperature despite its low molar mass. Water forms strong hydrogen bonds, resulting in its high boiling point.

▶ The polarity of water enables it to dissolve many polar and ionic compounds and even nonpolar gases.

▶ Water expands upon freezing, so ice is less dense than liquid water. Water is critical both to the existence of life and to human health.

# Key Equations and Relationships

Clausius–Clapeyron Equation: Relationship between Vapor Pressure ($P_{vap}$), the Heat of Vaporization ($\Delta H_{vap}$), and Temperature ($T$) (**11.5**)

$$\ln P_{vap} = \frac{-\Delta H_{vap}}{RT} + \ln \beta \ (\beta \text{ is a constant})$$

$$\ln \frac{P_2}{P_1} = \frac{-\Delta H_{vap}}{R}\left(\frac{1}{T_2} - \frac{1}{T_1}\right)$$

# Key Learning Outcomes

| CHAPTER OBJECTIVES | ASSESSMENT |
|---|---|
| Determining Whether a Molecule Has Dipole–Dipole Forces (**11.3**) | Example 11.1   For Practice 11.1   Exercises 35–46 |
| Determining Whether a Molecule Displays Hydrogen Bonding (**11.3**) | Example 11.2   For Practice 11.2   Exercises 35–46 |
| Using the Heat of Vaporization in Calculations (**11.5**) | Example 11.3   For Practice 11.3   For More Practice 11.3   Exercises 57–60 |
| Using the Clausius–Clapeyron Equation (**11.5**) | Examples 11.4, 11.5   For Practice 11.4, 11.5   Exercises 61–66 |

# EXERCISES

## Review Questions

1. Explain why water drops are spherical in the absence of gravity.

2. Why are intermolecular forces important?

3. What are the main properties of liquids (in contrast to gases and solids)?

4. What are the main properties of solids (in contrast to liquids and gases)?

5. What is the fundamental difference between an amorphous solid and a crystalline solid?

6. What factors cause transitions between the solid and liquid state? The liquid and gas state?

7. Describe the relationship between the state of a substance, its temperature, and the strength of its intermolecular forces.

8. From what kinds of interactions do intermolecular forces originate?

9. Why are intermolecular forces generally much weaker than bonding forces?

10. What is the dispersion force? What does the magnitude of the dispersion force depend on? How can you predict the magnitude of the dispersion force for closely related elements or compounds?

11. What is the dipole–dipole force? How can you predict the presence of dipole–dipole forces in a compound?

**12.** How is the polarity of a liquid generally related to its miscibility with water?

**13.** What is hydrogen bonding? How can you predict the presence of hydrogen bonding in a compound?

**14.** What is the ion–dipole force? Why is it important?

**15.** What is surface tension? How does surface tension result from intermolecular forces? How is it related to the strength of intermolecular forces?

**16.** What is viscosity? How does viscosity depend on intermolecular forces? What other factors affect viscosity?

**17.** What is capillary action? How does it depend on the relative strengths of adhesive and cohesive forces?

**18.** Explain what happens in the processes of vaporization and condensation. Why does the rate of vaporization increase with increasing temperature and surface area?

**19.** Why is vaporization endothermic? Why is condensation exothermic?

**20.** How is the volatility of a substance related to the intermolecular forces present within the substance?

**21.** What is the heat of vaporization for a liquid and why is it useful?

**22.** Explain the process of dynamic equilibrium. How is dynamic equilibrium related to vapor pressure?

**23.** What happens to a system in dynamic equilibrium when it is disturbed in some way?

**24.** How is vapor pressure related to temperature? What happens to the vapor pressure of a substance when the temperature is increased? Decreased?

**25.** Define the terms *boiling point* and *normal boiling point*.

**26.** What is the Clausius–Clapeyron equation and why is it important?

**27.** Explain what happens to a substance when it is heated in a closed container to its critical temperature.

**28.** What is sublimation? Give a common example of sublimation.

**29.** What is fusion? Is fusion exothermic or endothermic? Why?

**30.** What is the heat of fusion and why is it important?

**31.** Examine the heating curve for water in Section 11.7 (Figure 11.36). Explain why the curve has two segments in which heat is added to the water but the temperature does not rise.

**32.** Examine the heating curve for water in Section 11.7 (Figure 11.36). Explain the significance of the slopes of each of the three rising segments. Why are the slopes different?

**33.** What is a phase diagram? Draw a generic phase diagram and label its important features. What is the significance of crossing a line in a phase diagram?

**34.** How do the properties of water differ from those of most other substances?

# Problems by Topic

## Intermolecular Forces

**35.** Determine the kinds of intermolecular forces that are present in each element or compound.
a. $N_2$      b. $NH_3$      c. CO      d. $CCl_4$

**36.** Determine the kinds of intermolecular forces that are present in each element or compound.
a. Kr      b. $NCl_3$      c. $SiH_4$      d. HF

**37.** Determine the kinds of intermolecular forces that are present in each element or compound.
a. HCl      b. $H_2O$      c. $Br_2$      d. He

**38.** Determine the kinds of intermolecular forces that are present in each element or compound.
a. $PH_3$
b. HBr
c. $CH_3OH$
d. $I_2$

**39.** Arrange these compounds in order of increasing boiling point. Explain your reasoning.
a. $CH_4$
b. $CH_3CH_3$
c. $CH_3CH_2Cl$
d. $CH_3CH_2OH$

**40.** Arrange these compounds in order of increasing boiling point. Explain your reasoning.
a. $H_2S$      b. $H_2Se$      c. $H_2O$

**41.** In each pair of compounds, pick the one with the higher boiling point. Explain your reasoning.
a. $CH_3OH$ or $CH_3SH$
b. $CH_3OCH_3$ or $CH_3CH_2OH$
c. $CH_4$ or $CH_3CH_3$

**42.** In each pair of compounds, pick the one with the higher boiling point. Explain your reasoning.
a. $NH_3$ or $CH_4$
b. $CS_2$ or $CO_2$
c. $CO_2$ or $NO_2$

**43.** In each pair of compounds, pick the one with the higher vapor pressure at a given temperature. Explain your reasoning.
a. $Br_2$ or $I_2$
b. $H_2S$ or $H_2O$
c. $NH_3$ or $PH_3$

**44.** In each pair of compounds, pick the one with the higher vapor pressure at a given temperature. Explain your reasoning.
a. $CH_4$ or $CH_3Cl$
b. $CH_3CH_2CH_2OH$ or $CH_3OH$
c. $CH_3OH$ or $H_2CO$

**45.** Determine whether each pair of compounds forms a homogeneous solution when combined. For those that form homogeneous solutions, indicate the type of forces that are involved.
a. $CCl_4$ and $H_2O$
b. KCl and $H_2O$
c. $Br_2$ and $CCl_4$
d. $CH_3CH_2OH$ and $H_2O$

**46.** Determine whether each pair of compounds forms a homogeneous solution when combined. For those that form homogeneous solutions, indicate the type of forces that are involved.
a. $CH_3CH_2CH_2CH_2CH_3$ and $CH_3CH_2CH_2CH_2CH_2CH_3$
b. $CBr_4$ and $H_2O$
c. $LiNO_3$ and $H_2O$
d. $CH_3OH$ and $CH_3CH_2CH_2CH_2CH_3$

## Surface Tension, Viscosity, and Capillary Action

**47.** Which compound would you expect to have greater surface tension: acetone [$(CH_3)_2CO$] or water ($H_2O$)? Explain.

**48.** Water (a) "wets" some surfaces and beads up on others. Mercury (b), in contrast, beads up on almost all surfaces. Explain this difference.

(a)          (b)

**49.** The structures of two isomers of heptane are shown. Which of these two compounds would you expect to have the greater viscosity?

Compound A

Compound B

**50.** Explain why the viscosity of multigrade motor oils is less temperature-dependent than that of single-grade motor oils.

**51.** Water in a glass tube that contains grease or oil residue displays a flat meniscus (left), whereas water in a clean glass tube displays a concave meniscus (right). Explain this observation.

**52.** When a thin glass tube is put into water, the water rises 1.4 cm. When the same tube is put into hexane, the hexane rises only 0.4 cm. Explain.

## Vaporization and Vapor Pressure

**53.** Which evaporates more quickly: 55 mL of water in a beaker with a diameter of 4.5 cm or 55 mL of water in a dish with a diameter of 12 cm? Is the vapor pressure of the water different in the two containers? Explain.

**54.** Which evaporates more quickly: 55 mL of water ($H_2O$) in a beaker or 55 mL of acetone [$(CH_3)_2CO$] in an identical beaker under identical conditions? Is the vapor pressure of the two substances different? Explain.

**55.** Spilling room-temperature water over your skin on a hot day cools you down. Spilling room-temperature vegetable oil over your skin on a hot day does not. Explain the difference.

**56.** Why is the heat of vaporization of water greater at room temperature than it is at its boiling point?

**57.** The human body obtains 915 kJ of energy from a candy bar. If this energy were used to vaporize water at 100.0 °C, how much water (in liters) could be vaporized? (Assume the density of water is 1.00 g/mL.)

**58.** A 100.0-mL sample of water is heated to its boiling point. How much heat (in kJ) is required to vaporize it? (Assume a density of 1.00 g/mL.)

**59.** Suppose that 0.95 g of water condenses on a 75.0-g block of iron that is initially at 22 °C. If the heat released during condensation goes only to warming the iron block, what is the final temperature (in °C) of the iron block? (Assume a constant enthalpy of vaporization for water of 44.0 kJ/mol.)

**60.** Suppose that 1.15 g of rubbing alcohol ($C_3H_8O$) evaporates from a 65.0-g aluminum block. If the aluminum block is initially at 25 °C, what is the final temperature of the block after the evaporation of the alcohol? Assume that the heat required for the vaporization of the alcohol comes only from the aluminum block and that the alcohol vaporizes at 25 °C.

**61.** This table displays the vapor pressure of ammonia at several different temperatures. Use the data to determine the heat of vaporization and normal boiling point of ammonia.

| Temperature (K) | Pressure (torr) |
|---|---|
| 200 | 65.3 |
| 210 | 134.3 |
| 220 | 255.7 |
| 230 | 456.0 |
| 235 | 597.0 |

**62.** This table displays the vapor pressure of nitrogen at several different temperatures. Use the data to determine the heat of vaporization and normal boiling point of nitrogen.

| Temperature (K) | Pressure (torr) |
|---|---|
| 65 | 130.5 |
| 70 | 289.5 |
| 75 | 570.8 |
| 80 | 1028 |
| 85 | 1718 |

**63.** Ethanol has a heat of vaporization of 38.56 kJ/mol and a normal boiling point of 78.4 °C. What is the vapor pressure of ethanol at 15 °C?

**64.** Benzene has a heat of vaporization of 30.72 kJ/mol and a normal boiling point of 80.1 °C. At what temperature does benzene boil when the external pressure is 445 torr?

65. Carbon disulfide has a vapor pressure of 363 torr at 25 °C and a normal boiling point of 46.3 °C. Find $\Delta H_{vap}$ for carbon disulfide.

66. Methylamine has a vapor pressure of 344 torr at −25 °C and a boiling point of −6.4 °C. Find $\Delta H_{vap}$ for methylamine.

## Sublimation and Fusion

67. How much energy is released when 65.8 g of water freezes?

68. Calculate the amount of heat required to completely sublime 50.0 g of solid dry ice ($CO_2$) at its sublimation temperature. The heat of sublimation for carbon dioxide is 32.3 kJ/mol.

69. An 8.5-g ice cube is placed into 255 g of water. Calculate the temperature change in the water upon the complete melting of the ice. Assume that all of the energy required to melt the ice comes from the water.

70. How much ice (in grams) would have to melt to lower the temperature of 352 mL of water from 25 °C to 5 °C? (Assume the density of water is 1.0 g/mL.)

71. How much heat (in kJ) is required to warm 10.0 g of ice, initially at −10.0 °C, to steam at 110.0 °C? The heat capacity of ice is 2.09 J/g·°C, and that of steam is 2.01 J/g·°C.

72. How much heat (in kJ) is evolved in converting 1.00 mol of steam at 145 °C to ice at −50 °C? The heat capacity of steam is 2.01 J/g·°C, and that of ice is 2.09 J/g·°C.

## Phase Diagrams

73. Consider the phase diagram shown here. Identify the states present at points *a* through *g*.

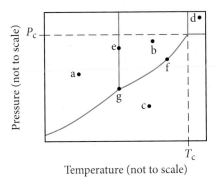

74. Consider the phase diagram for iodine shown here.
    a. What is the normal boiling point for iodine?
    b. What is the melting point for iodine at 1 atm?
    c. What state is present at room temperature and normal atmospheric pressure?
    d. What state is present at 186 °C and 1.0 atm?

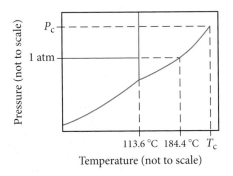

75. Nitrogen has a normal boiling point of 77.3 K and a melting point (at 1 atm) of 63.1 K. Its critical temperature is 126.2 K and its critical pressure is $2.55 \times 10^4$ torr. It has a triple point at 63.1 K and 94.0 torr. Sketch the phase diagram for nitrogen. Does nitrogen have a stable liquid state at 1 atm?

76. Argon has a normal boiling point of 87.2 K and a melting point (at 1 atm) of 84.1 K. Its critical temperature is 150.8 K and its critical pressure is 48.3 atm. It has a triple point at 83.7 K and 0.68 atm. Sketch the phase diagram for argon. Which has the greater density, solid argon or liquid argon?

77. The phase diagram for sulfur is shown here. The rhombic and monoclinic states are two solid states with different structures.
    a. Below what pressure does solid sulfur sublime?
    b. Which of the two solid states of sulfur is more dense?

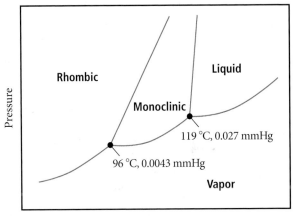

78. The high-pressure phase diagram of ice is shown here. Notice that, under high pressure, ice can exist in several different solid forms. What three forms of ice are present at the triple point marked O? How does the density of ice II compare to ice I (the familiar form of ice)? Would ice III sink or float in liquid water?

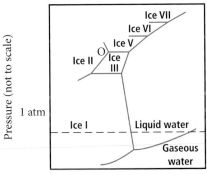

## The Uniqueness of Water

79. Water has a high boiling point given its relatively low molar mass. Why?

80. Water is a good solvent for many substances. What is the molecular basis for this property, and why is it significant?

81. Explain the role of water in moderating Earth's climate.

82. How is the density of solid water compared to that of liquid water atypical among substances? Why is this significant?

# Cumulative Problems

83. Explain the observed trend in the melting points of the hydrogen halides.

| | |
|---|---|
| HI | $-50.8\,°C$ |
| HBr | $-88.5\,°C$ |
| HCl | $-114.8\,°C$ |
| HF | $-83.1\,°C$ |

84. Explain the observed trend in the boiling points of these compounds.

| | |
|---|---|
| $H_2Te$ | $-2\,°C$ |
| $H_2Se$ | $-41.5\,°C$ |
| $H_2S$ | $-60.7\,°C$ |
| $H_2O$ | $100\,°C$ |

85. The vapor pressure of water at 25 °C is 23.76 torr. If 1.25 g of water is enclosed in a 1.5-L container, will any liquid be present? If so, what mass of liquid?

86. The vapor pressure of $CCl_3F$ at 300 K is 856 torr. If 11.5 g of $CCl_3F$ is enclosed in a 1.0-L container, will any liquid be present? If so, what mass of liquid?

87. Examine the phase diagram for iodine shown in Figure 11.39(a). What state transitions occur as we uniformly increase the pressure on a gaseous sample of iodine from 0.010 atm at 185 °C to 100 atm at 185 °C? Make a graph, analogous to the heating curve for water shown in Figure 11.36. Plot pressure versus time during the pressure increase.

88. Carbon tetrachloride displays a triple point at 249.0 K and a melting point (at 1 atm) of 250.3 K. Which state of carbon tetrachloride is more dense, the solid or the liquid? Explain.

89. Four ice cubes at exactly 0 °C with a total mass of 53.5 g are combined with 115 g of water at 75 °C in an insulated container. If no heat is lost to the surroundings, what is the final temperature of the mixture?

90. A sample of steam with a mass of 0.552 g and at a temperature of 100 °C condenses into an insulated container holding 4.25 g of water at 5.0 °C. Assuming that no heat is lost to the surroundings, what is the final temperature of the mixture?

91. Draw a heating curve (such as the one in Figure 11.36) for 1 mole of methanol beginning at 170 K and ending at 350 K. Assume that the values given here are constant over the relevant temperature ranges.

| | |
|---|---|
| Melting point | 176 K |
| Boiling point | 338 K |
| $\Delta H_{fus}$ | 2.2 kJ/mol |
| $\Delta H_{vap}$ | 35.2 kJ/mol |
| $C_{s,solid}$ | 105 J/mol·K |
| $C_{s,liquid}$ | 81.3 J/mol·K |
| $C_{s,gas}$ | 48 J/mol·K |

92. Draw a heating curve (such as the one in Figure 11.36) for 1 mol of benzene beginning at 0 °C and ending at 100 °C. Assume that the values given here are constant over the relevant temperature ranges.

| | |
|---|---|
| Melting point | 5.4 °C |
| Boiling point | 80.1 °C |
| $\Delta H_{fus}$ | 9.9 kJ/mol |
| $\Delta H_{vap}$ | 30.7 kJ/mol |
| $C_{s,solid}$ | 118 J/mol·K |
| $C_{s,liquid}$ | 135 J/mol·K |
| $C_{s,gas}$ | 104 J/mol·K |

93. Air conditioners not only cool air, but dry it as well. A room in a home measures 6.0 m × 10.0 m × 2.2 m. If the outdoor temperature is 30 °C and the partial pressure of water in the air is 85% of the vapor pressure of water at this temperature, what mass of water must be removed from the air each time the volume of air in the room is cycled through the air conditioner? (Assume that all of the water must be removed from the air.) The vapor pressure for water at 30 °C is 31.8 torr.

94. A sealed flask contains 0.55 g of water at 28 °C. The vapor pressure of water at this temperature is 28.35 mmHg. What is the minimum volume of the flask in order that no liquid water be present in the flask?

95. Based on the phase diagram of $CO_2$ shown in Figure 11.39(b), describe the state changes that occur when the temperature of $CO_2$ is increased from 190 K to 350 K at a constant pressure of (a) 1 atm, (b) 5.1 atm, (c) 10 atm, (d) 100 atm.

96. Consider a planet where the pressure of the atmosphere at sea level is 2500 mmHg. Does water behave in a way that can sustain life on the planet?

# Challenge Problems

97. Liquid nitrogen can be used as a cryogenic substance to obtain low temperatures. Under atmospheric pressure, liquid nitrogen boils at 77 K, allowing low temperatures to be reached. However, if the nitrogen is placed in a sealed, insulated container connected to a vacuum pump, even lower temperatures can be reached. Why? If the vacuum pump has sufficient capacity and is left on for an extended period of time, the liquid nitrogen will start to freeze. Explain.

98. Given that the heat of fusion of water is $-6.02\,kJ/mol$, the heat capacity of $H_2O(l)$ is 75.2 J/mol·K, and the heat capacity of $H_2O(s)$ is 37.7 J/mol·K, calculate the heat of fusion of water at $-10\,°C$.

99. The heat of combustion of $CH_4$ is 890.4 kJ/mol, and the heat capacity of $H_2O$ is 75.2 J/mol·K. Find the volume of methane measured at 298 K and 1.00 atm required to convert 1.00 L of water at 298 K to water vapor at 373 K.

**100.** Two liquids, A and B, have vapor pressures at a given temperature of 24 mmHg and 36 mmHg, respectively. We prepare solutions of A and B at a given temperature and measure the total pressures above the solutions. We obtain this data:

| Solution | Amt A (mol) | Amt B (mol) | P (mmHg) |
|----------|-------------|-------------|----------|
| 1 | 1 | 1 | 30 |
| 2 | 2 | 1 | 28 |
| 3 | 1 | 2 | 32 |
| 4 | 1 | 3 | 33 |

Predict the total pressure above a solution of 5 mol A and 1 mol B.

**101.** Three 1.0-L flasks, maintained at 308 K, are connected to each other with stopcocks. Initially, the stopcocks are closed. One of the flasks contains 1.0 atm of $N_2$; the second, 2.0 g of $H_2O$; and the third, 0.50 g of ethanol, $C_2H_6O$. The vapor pressure of $H_2O$ at 308 K is 42 mmHg and that of ethanol is 102 mmHg. The stopcocks are then opened and the contents mix freely. What is the pressure?

**102.** Butane ($C_4H_{10}$) has a heat of vaporization of 22.44 kJ/mol and a normal boiling point of −0.4 °C. A 250-mL sealed flask contains 0.55 g of butane at −22 °C. How much butane is present as a liquid? If the butane is warmed to 25 °C, how much is present as a liquid?

# Conceptual Problems

**103.** The following image is an electrostatic potential map for ethylene oxide, $(CH_2)_2O$, a polar molecule. Use the electrostatic potential map to predict the geometry for how one ethylene oxide molecule interacts with another. Draw structural formulas, using the 3D bond notation introduced in Section 10.4, to show the geometry of the interaction.

**104.** One prediction of global warming is the melting of global ice, which may result in coastal flooding. A criticism of this prediction is that the melting of icebergs does not increase ocean levels any more than the melting of ice in a glass of water increases the level of liquid in the glass. Is this a valid criticism? Does the melting of an ice cube in a cup of water raise the level of the liquid in the cup? Why or why not? In response to this criticism, scientists have asserted that they are not worried about melting icebergs, but rather the melting of ice sheets that sit on the continent of Antarctica. Would the melting of this ice increase ocean levels? Why or why not?

**105.** The rate of vaporization depends on the surface area of the liquid. However, the vapor pressure of a liquid does not depend on the surface area. Explain.

**106.** Substance A has a smaller heat of vaporization than substance B. Which of the two substances will undergo a larger change in vapor pressure for a given change in temperature?

**107.** The density of a substance is greater in its solid state than in its liquid state. If the triple point in the phase diagram of the substance is below 1.0 atm, which will necessarily be at a lower temperature, the triple point or the normal melting point?

**108.** A substance has a heat of vaporization of $\Delta H_{vap}$ and a heat of fusion of $\Delta H_{fus}$. Express the heat of sublimation in terms of $\Delta H_{vap}$ and $\Delta H_{fus}$.

**109.** Examine the heating curve for water in Section 11.7 (Figure 11.36). If heat is added to the water at a constant rate, which of the three segments in which temperature is rising will have the least steep slope? Why?

**110.** A root cellar is an underground chamber used to store fruits, vegetables, and even meats. In extreme cold, farmers put large vats of water into the root cellar to prevent the fruits and vegetables from freezing. Explain why this works.

**111.** Suggest an explanation for the observation that the heat of fusion of a substance is always smaller than its heat of vaporization.

**112.** Refer to Figure 11.36 to answer each question.

a. A sample of steam begins on the line segment labeled 5 on the graph. Is heat absorbed or released in moving from the line segment labeled 5 to the line segment labeled 3? What is the sign of $q$ for this change?

b. In moving from left to right along the line segment labeled 2 on the graph, heat is absorbed, but the temperature remains constant. Where does the heat go?

c. How would the graph change if it were for another substance (other than water)?

*Active Classroom Learning*

# Questions for Group Work

Discuss these questions with the group and record your consensus answer.

**113.** The boiling points of three compounds are tabulated here.

| | Molar Mass | Boiling Point |
|---|------------|---------------|
| 2-hexanone | 100.16 | 128 °C |
| heptane | 100.20 | 98 °C |
| 1-hexanol | 102.17 | 156 °C |

Answer the following questions without looking up the structures for these molecules: Which compound experiences hydrogen bonding? Which compound is polar but is unable to experience hydrogen bonding? Which is neither polar nor capable of hydrogen bonding? Explain your answers.

**114.** The vapor pressure for pure water and pure acetone is measured as a function of temperature. In each case, a graph of the log of the vapor pressure versus $1/T$ is found to be a straight line. The slope of the line for water is $-4895$ K and the slope of the line for acetone is $-3765$ K. Determine $\Delta H_{vap}$ for each substance. Account for the difference by discussing the molecular structure of the two molecules.

**115.** Based on the heating curve for water, does it take more energy to melt a mole of water or to boil a mole of water? Does it take more energy to warm the solid, the liquid, or the gas by 10 °C? Explain your answers clearly.

**116.** Sketch the phase diagram for carbon dioxide. If you have carbon dioxide at 1.0 atm and 25 °C, could you make it a liquid by cooling it down? How could you make it a liquid at 25 °C? If you increase the pressure of carbon dioxide that is at body temperature (37 °C), will it ever liquefy?

# Data Interpretation and Analysis

## Intermolecular Forces of Group 6A Hydrides

**117.** Students in a chemistry laboratory course are investigating the following question: "How do the molar mass and structure influence the boiling point of the group 6A hydrides?" The students review Figure a▼, which is a graph of the boiling point versus molar mass of the group 6A hydrides.

▲ FIGURE a  **Boiling Point and Molar Mass of Group 6A Hydrides**

The molar mass, boiling points, dipole moments, and polarizability of some group 6A compounds are listed in Table a. (Polarizability is the ability of an atom's or a molecule's outer electron cloud to be distorted from its normal shape by an electric field. The units of polarizability are in $\mathring{A}^3 = 10^{-24}\ cm^3$.)

| Compound | Molar Mass | n-Boiling Point (°C) | Dipole Moment (D) | Polarizability ($\times 10^{-24}\ cm^3$) |
|---|---|---|---|---|
| $H_2O$ | 18.01 | 100 | 1.85 | 1.45 |
| $H_2S$ | 34.08 | −60 | 1.10 | 3.81 |
| $H_2Se$ | 80.98 | −42.2 | 0.41 | 4.71 |
| $H_2Te$ | 129.6 | −2.2 | 0.22 | 5.01 |

TABLE a  **Relationship among Boiling Points, Dipole Moments, and Polarizability of Some Binary Hydrides**

Use the information provided in the figure and the table to answer the following questions:

a. Does molar mass correlate with the trend in boiling points for the group 6A hydrides?

b. Which intermolecular force correlates with dipole moment?

c. Which intermolecular force correlates with polarizability?

d. Which is the predominant intermolecular force responsible for the boiling-point trend of $H_2S$, $H_2Se$ and $H_2Te$. Justify your answer using the information presented in Table a.

e. Is your answer to part d in agreement with the information presented in Table 11.4? Explain.

f. Explain which intermolecular force is primarily responsible for the unusually high boiling point of water. Use the data presented in Table a to justify your answer. Compare the strengths of intermolecular forces of attraction present in a sample of liquid water to the strengths of intermolecular forces of attraction present in liquid samples of $H_2S$, $H_2Se$, and $H_2Te$.

g. Starting with a diagram of six $H_2O$ molecules in the liquid state, draw a diagram showing what happens at the molecular level when water boils (a change from the liquid phase to the gas phase).

# Answers to Conceptual Connections

**State Changes**

**11.1** **(a)** When water boils, it simply changes state from liquid to gas. Water molecules do not decompose during boiling.

**Dispersion Forces**

**11.2** **(c)** $I_2$ has the highest boiling point because it has the highest molar mass. Since the halogens are all similar in other ways, we expect $I_2$ to have the greatest dispersion forces and therefore the highest boiling point (and in fact it does).

**Dipole–Dipole Interaction**

**11.3**

**Intermolecular Forces and Boiling Point**

**11.4** **(a)** $CH_3OH$. The compounds all have similar molar masses, so the dispersion forces are similar in all three. CO is polar, but because $CH_3OH$ contains H directly bonded to O, it has hydrogen bonding, resulting in the highest boiling point.

**Vapor Pressure**

**11.5** **(b)** Although the *rate of vaporization* increases with increasing surface area, the *vapor pressure* of a liquid is independent of surface area. An increase in surface area increases both the rate of vaporization and the rate of condensation—the effects exactly cancel and the vapor pressure does not change.

**Cooling of Water with Ice**

**11.6** The warming of the ice from $-10\,°C$ to $0\,°C$ absorbs only $20.9\,J/g$ of ice. The melting of the ice, however, absorbs about $334\,J/g$ of ice. (You can obtain this value by dividing the heat of fusion of water by its molar mass.) Therefore, the melting of the ice produces a larger temperature decrease in the water than does the warming of the ice.

**Phase Diagrams**

**11.7** **(b)** The solid will sublime into a gas. Since the pressure is below the triple point, the liquid state is not stable.

CHAPTER

# 12

# Solids and Modern Materials

> Should we not suppose that in the formation of a crystal, the particles are not only established in rows and columns set in regular figures, but also by means of some polar property have turned identical sides in identical directions?
>
> —ISAAC NEWTON (1642–1727)

In this chapter, we focus on the solid state of matter. We first examine the structures of solids, keeping in mind that these structures determine the properties of solids. For example, the repeating hexagonal pattern of water molecules in crystalline ice determines the hexagonal shape of a snowflake, and the repeating cubic pattern of sodium and chloride ions in sodium chloride determines the cubic shape of salt grains. We then turn our attention to the study and development of solids with unique and useful properties, a field known as *materials science*. The ceramics that compose your coffee cups, the semiconductors in your electronic devices, and the plastics that are most likely all around you even at this moment are materials developed to have specific properties that serve specific purposes. In this chapter, we take a brief look at each of these kinds of materials.

▲ Graphene is the thinnest material known–it is only one atom thick.

## 12.1 Friday Night Experiments: The Discovery of Graphene

In 2010, Andre Geim (1958–) and Konstantin Novoselov (1974–) were awarded the Nobel Prize in Physics for the discovery of a new material—**graphene**. Graphene is the thinnest known material (only one atom thick); it is also the strongest. It conducts heat and electricity, it is transparent, and it is completely impermeable to all substances, including helium. Although its many possible applications are yet to be realized, graphene may one day be used to make faster computers, foldable touchscreens, ultrathin light panels, and super-strong plastics to construct satellites, airplanes, ships, and cars. And how did these scientists discover this wonder material? Through *Friday night experiments*.

Most successful modern scientists work in a narrow field of study, chipping away at a problem about which they have become experts. Andre Geim also worked this way—but not all of the time. Geim often ventured into what he calls *lateral ideas*, which were explorations into fields far from his immediate area of expertise. Over a 15-year period, he explored about two dozen such ideas. These explorations became known as *Friday night experiments* (even though they took much longer than just one night). Most of Geim's Friday night experiments were complete failures, but some succeeded. For example, in one Friday night experiment, Geim was able to suspend an entire frog in a magnetic field. In another, he developed a new kind of tape based on mimicking the toes of a gecko. However, his most successful Friday night experiment was the discovery of graphene.

In 2002, Geim assigned his brand-new graduate student a Friday night experiment as a way to give the student time to get acquainted with the laboratory. His assignment was to make films of graphite "as thin as possible." The graduate student tackled the task by attempting to polish a tablet of graphite down to almost nothingness. In spite of his best efforts, the thinnest sheet he could attain by this method was still pretty thick (about 10 $\mu$m).

In a breakthrough moment, a colleague brought Geim a piece of scotch tape that had been adhered to a graphite surface and then removed. The tape had remnants of graphite stuck to it. When Geim examined the graphite flakes under a microscope, he realized that these flakes where much thinner than those his graduate student was able to produce through polishing. In fact, some of the flakes were just one layer thick! Over the next several months, Geim and his colleagues were able to isolate these flakes and measure some of their properties. In 2004, Geim and his colleagues published their results—the discovery of atomically thin carbon—in *Nature*. In 2010, Geim and his closest collaborator on the project, Konstantin Novoselov, were awarded the Nobel Prize in Physics for "groundbreaking experiments regarding the two-dimensional material graphene."

Graphene has been right under our noses for some time. Every time you write with a pencil, layers of graphite flake onto the paper, leaving the familiar gray trail. That trail undoubtedly also contains flakes of graphene—too small to see. Geim and Novoselov were able to see what others had missed and as a result, discovered a new wonder material. These kinds of discoveries often have huge impacts on society—although it may take decades to realize and refine their applications.

In this chapter, we look into the world of solids and materials and see how they permanently changed societies, just as graphene may someday change our present society. For example, the discovery of cement led to a revolution in building, the discovery of porcelain led to a revolution in dishware, and the discovery of semiconducting materials made possible the computer and Internet revolution which is still unfolding today. The development of new materials continues to be an active and growing area of research. We begin by examining a technique to determine the structure of solids.

▲ The hexagonal shape of a snowflake derives from the hexagonal arrangement of water molecules in crystalline ice.

## 12.2 X-Ray Crystallography

In Section 11.2, we discussed the differences between solids, liquids, and gases. Recall that solids can be *crystalline* (composed of atoms or molecules arranged in structures with long-range order) or *amorphous* (composed of atoms or molecules with no long-range order). If you have ever visited the mineral section of a natural history museum and seen crystals with smooth faces and well-defined angles, or if you have carefully observed the hexagonal shapes of snowflakes, you have witnessed some of the effects of the underlying long-range order in crystalline solids. The beautiful geometric shapes that are visible on the macroscopic scale are the result of specific structural patterns on the molecular and atomic scales. But how do we study these patterns? How do we look into the atomic and molecular world to determine the arrangement of the atoms and measure the distances between them? In this section, we examine **X-ray diffraction**, a powerful laboratory technique that enables us to do exactly that.

In Section 7.2 we saw that electromagnetic waves (light waves) interact with each other in a characteristic way called *interference*: they can cancel each other out or reinforce each other, depending on the alignment of their crests and troughs. *Constructive interference* occurs when two waves interact with their crests and troughs in alignment. *Destructive interference* occurs when two waves interact with the crests of one aligning with the troughs of the other. Recall also that when light encounters two slits separated by a distance comparable to the wavelength of the light, constructive and destructive interference produces a characteristic *interference pattern*, consisting of alternating bright and dark lines.

Atomic planes within crystals are separated by about $10^2$ pm. As a result, light of similar wavelength (which happens to fall in the X-ray region of the electromagnetic spectrum) forms interference patterns or *diffraction patterns* when it interacts with those atoms. The exact pattern of diffraction reveals the spacing between planes of atoms.

Consider two planes of atoms within a crystalline lattice separated by a distance $d$, as illustrated in Figure 12.1▼. If two rays of light with wavelength $\lambda$ that are initially in phase (that is, the crests of one wave are aligned with the crests of the other) diffract from the two planes, the diffracted rays may interfere with each other constructively or destructively, depending on the difference between the path lengths traveled by each ray. If the difference between the two path lengths ($2a$) is an integral number ($n$) of wavelengths, then the interference is constructive:

$$n\lambda = 2a \quad \text{(criterion for constructive interference)} \quad [12.1]$$

Constructive interference

Destructive interference

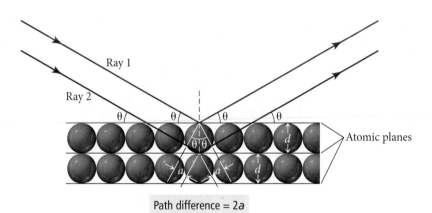

Path difference = 2a

◀ **FIGURE 12.1 Diffraction from a Crystal** When X-rays strike parallel planes of atoms in a crystal, constructive interference occurs if the difference in path length between beams reflected from adjacent planes is an integral number of wavelengths.

Using trigonometry, we can determine that constructive interference occurs when the angle of reflection ($\theta$) is related to the distance $a$ and the separation between layers ($d$) by the following relation:

$$\sin\theta = \frac{a}{d} \quad [12.2]$$

Rearranging, we get:

$$a = d\sin\theta \quad [12.3]$$

By substituting Equation 12.3 into Equation 12.1, we arrive at the following important relationship:

$$n\lambda = 2d\sin\theta \quad \text{Bragg's law}$$

This equation is *Bragg's law*. For a given wavelength of light incident on atoms arranged in layers, we can measure the angle that produces constructive interference (which appears as a bright spot on the X-ray diffraction pattern) and then calculate $d$, the distance between the atomic layers:

$$d = \frac{n\lambda}{2\sin\theta} \quad [12.4]$$

▶ **FIGURE 12.2 X-Ray Diffraction Analysis**
In X-ray crystallography, an X-ray beam is passed
through a sample, which is rotated to allow
diffraction from different crystalline planes. The
resulting patterns, representing constructive
interference from various planes, are analyzed to
determine crystalline structure.

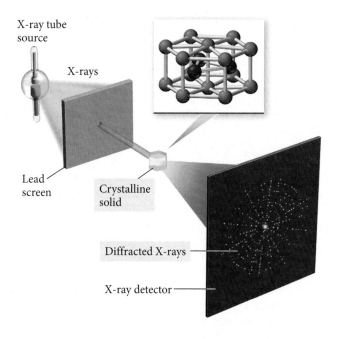

X-ray tube
source

X-rays

Lead
screen

Crystalline
solid

Diffracted X-rays

X-ray detector

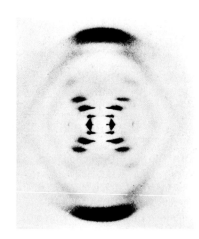

In a modern X-ray diffractometer (Figure 12.2▲), a computer collects and analyzes
the diffraction pattern from a crystal. By rotating the crystal and collecting the resulting
diffraction patterns at different angles, the distances between various crystalline planes
can be measured, eventually yielding the entire crystalline structure. This process is
**X-ray crystallography**.

Researchers use X-ray crystallography to determine not only the structures of simple atomic lattices, but also the structures of proteins, DNA, and other biologically
important molecules. For example, the famous X-ray diffraction photograph shown in
the margin, obtained by Rosalind Franklin (1920–1958) and Maurice Wilkins (1916–
2004), helped Watson and Crick determine the double-helical structure of DNA. Recall
from Section 9.1 that researchers also used X-ray diffraction to determine the structure
of HIV protease, a protein critical to the reproduction of HIV and the development of
AIDS. Knowledge of that structure was then employed to design drug molecules to
inhibit the action of HIV protease and halt the advance of the disease.

## Example 12.1    Using Bragg's Law

When an X-ray beam of $\lambda = 154$ pm is incident on the surface of an iron crystal,
it produces a maximum diffraction at an angle of $\theta = 32.6°$. Assuming $n = 1$,
calculate the separation between layers of iron atoms in the crystal.

**SOLUTION**

To solve this problem, use Bragg's law in the form
given by Equation 12.4. The distance, $d$, is the separation between layers in the crystal.

$$d = \frac{n\lambda}{2 \sin \theta}$$

$$= \frac{154 \text{ pm}}{2 \sin(32.6°)}$$

$$= 143 \text{ pm}$$

**FOR PRACTICE 12.1**    The spacing between layers of molybdenum atoms is 157
pm. Calculate the angle at which 154-pm X-rays produce a maximum reflection
for $n = 1$.

# 12.3  Unit Cells and Basic Structures

X-ray crystallography allows us to determine the regular arrangements of atoms within a crystalline solid. This arrangement is called the **crystalline lattice**. The crystalline lattice of any solid is nature's way of aggregating the particles to minimize their energy. We represent the crystalline lattice with a small collection of atoms, ions, or molecules called the **unit cell**. When the unit cell is repeated over and over—like the tiles of a floor or the pattern in a wallpaper design, but in three dimensions—the entire lattice is reproduced. For example, consider the two-dimensional crystalline lattice shown here:

The unit cell for this lattice is the dark-colored square. Each circle represents an atom, ion, or molecule. Repeating the pattern in the square throughout the two-dimensional space generates the entire lattice.

## Cubic Unit Cells

Many different unit cells exist, and we often classify unit cells by their symmetry. Figure 12.3▼ shows the different fundamental types of unit cells. In this book, we focus primarily on *cubic unit cells* (although we look at one hexagonal unit cell). Cubic unit cells are characterized by equal edge lengths and 90° angles at their corners. Figure 12.4▶ on the next page lists three cubic unit cells—simple cubic, body-centered cubic, and face-centered cubic—along with some of their basic characteristics. The two colors in the images in this table make it easier to visualize the different positions of the atoms; the colors *do not* represent different *kinds* of atoms. For these unit cells, *each atom in any one structure is identical to the other atoms in that structure.*

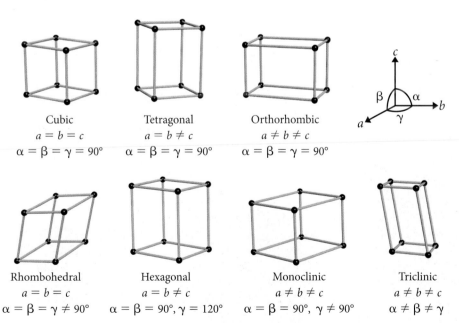

**▲ FIGURE 12.3  The Seven Fundamental Types of Unit Cells**   In this book, we focus on cubic unit cells.

| Cubic Cell Name | Atoms per Unit Cell | Structure | Coordination Number | Edge Length in terms of $r$ | Packing Efficiency (fraction of volume occupied) |
|---|---|---|---|---|---|
| Simple Cubic | 1 | | 6 | $2r$ | 52% |
| Body-Centered Cubic | 2 | | 8 | $\dfrac{4r}{\sqrt{3}}$ | 68% |
| Face-Centered Cubic | 4 | | 12 | $2\sqrt{2}r$ | 74% |

▲ **FIGURE 12.4 Cubic Crystalline Lattices** The different colors used for the atoms in this figure are for clarity only. All atoms within each structure are identical. Unit cells, such as the cubic ones shown here, are customarily portrayed with "whole" atoms, even though only a part of the whole atom may actually be in the unit cell.

The **simple cubic** unit cell (Figure 12.5▼) consists of a cube with one atom at each corner. As we can see in the figure at the left, the atoms touch along each edge of the cube, so the edge length is twice the radius of the atoms ($l = 2r$).

Even though it may seem like the unit cell contains eight atoms, it actually contains only one as close examination of Figure 12.5 makes evident. Each corner atom is shared by eight other unit cells. In other words, any one unit cell actually contains only one-eighth of each of the eight atoms at its corners, for a total of only one atom per unit cell.

Simple cubic

$l = 2r$

▲ In the simple cubic lattice, the atoms touch along each edge so that the edge length is $2r$.

## Simple Cubic Unit Cell

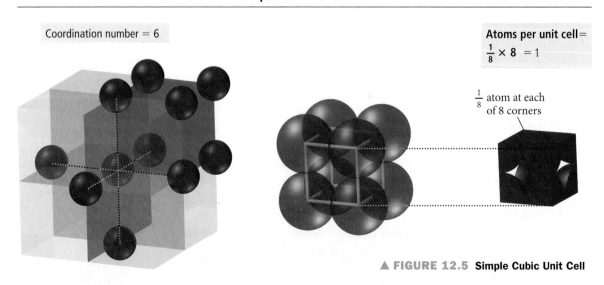

Coordination number = 6

Atoms per unit cell = $\dfrac{1}{8} \times 8 = 1$

$\dfrac{1}{8}$ atom at each of 8 corners

▲ **FIGURE 12.5 Simple Cubic Unit Cell**

A characteristic feature of any unit cell is the **coordination number**, the number of atoms with which each atom is in *direct contact*. The coordination number is the number of atoms with which a particular atom can strongly interact. The simple cubic unit cell has a coordination number of 6; any one atom touches only six others, as you can see in Figure 12.5. A quantity closely related to the coordination number is the **packing efficiency**, the percentage of the volume of the unit cell occupied by the spheres. The higher the coordination number is, the greater the packing efficiency is. The simple cubic unit cell has a packing efficiency of 52%—it contains a lot of empty space.

## Example 12.2 | Calculating the Packing Efficiency of a Unit Cell

Calculate the packing efficiency of the simple cubic unit cell.

**SOLUTION**

| | |
|---|---|
| The packing efficiency is defined as the ratio of the volume of the unit cell occupied by atoms to the volume of the unit cell itself all multiplied by 100%. | $\text{Packing Efficiency} = \dfrac{V_{\text{atom}}}{V_{\text{unit cell}}} \times 100\%$ |
| First, determine the volume occupied by atoms. Each unit cell contains one atom, so the volume of the unit cell occupied by atoms is simply the volume of one atom. | $V_{\text{atom}} = \dfrac{4}{3}\pi r^3$ |
| Next, calculate the volume of the cube in terms of the radius of the atom. The volume of a cube is its edge length cubed, and the edge length of the simple cubic unit cell is $2r$. | $\begin{aligned} V_{\text{unit cell}} &= l^3 \\ &= (2r)^3 \\ &= 8r^3 \end{aligned}$ |
| Calculate the packing efficiency by dividing the volume of the atom by the volume of the unit cell and multiplying by 100%. | $\begin{aligned} \text{Packing Efficiency} &= \dfrac{V_{\text{atom}}}{V_{\text{unit cell}}} \times 100\% \\[2mm] &= \dfrac{\frac{4}{3}\pi r^3}{8r^3} \times 100\% \\[2mm] &= \dfrac{\frac{4}{3}\pi r^3}{8r^3} \times 100\% \\[2mm] &= \dfrac{4}{24}\pi \times 100\% = 52.36\% \end{aligned}$ |

**FOR PRACTICE 12.2** Calculate the packing efficiency of the two-dimensional lattice shown here:

The **body-centered cubic** unit cell (Figure 12.6▶, on the next page) consists of a cube with one atom at each corner and one atom (of the same kind) in the very center of the cube. Note that in the body-centered unit cell, the atoms *do not* touch along each edge of the cube, but instead touch along the diagonal line that runs from one corner,

## Body-Centered Cubic Unit Cell

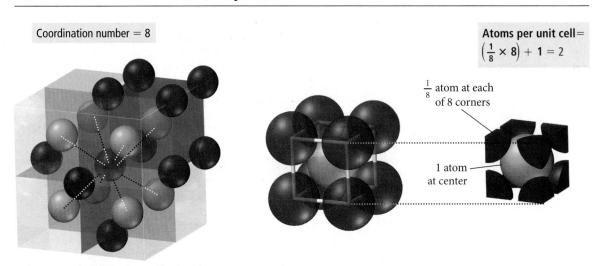

Coordination number = 8

Atoms per unit cell=
$\left(\frac{1}{8} \times 8\right) + 1 = 2$

$\frac{1}{8}$ atom at each
of 8 corners

1 atom
at center

▲ **FIGURE 12.6 Body-Centered Cubic Crystal Structure**   The different colors used for the atoms in this figure are for clarity only. All atoms within the structure are identical.

► In the body-centered cubic lattice, the atoms touch only along the cube diagonal. The edge length is $4r/\sqrt{3}$.

through the middle of the cube, to the opposite corner. The edge length in terms of the atomic radius is therefore $l = 4r/\sqrt{3}$ as shown here:

Body-centered cubic

$$c^2 = b^2 + l^2 \qquad b^2 = l^2 + l^2$$
$$c = 4r \qquad\qquad b^2 = 2l^2$$
$$(4r)^2 = 2l^2 + l^2$$
$$(4r)^2 = 3l^2$$
$$l^2 = \frac{(4r)^2}{3}$$
$$l = \frac{4r}{\sqrt{3}}$$

The body-centered unit cell contains two atoms per unit cell because the center atom is not shared with any other neighboring cells. The coordination number of the body-centered cubic unit cell is 8, which we can see by examining the atom in the very center of the cube, which touches eight atoms at the corners. The packing efficiency for the body-centered unit cell is 68%, significantly higher than for the simple cubic unit cell. Each atom in this structure strongly interacts with more atoms than each atom in the simple cubic unit cell.

## Example 12.3   Relating Unit Cell Volume, Edge Length, and Atomic Radius

A body-centered cubic unit cell has a volume of $4.32 \times 10^{-23}$ cm³. Find the radius of the atom in pm.

| | |
|---|---|
| **SORT**  You are given the volume of a unit cell and asked to find the radius of the atom. | **GIVEN:** $V = 4.32 \times 10^{-23}$ cm³   **FIND:** $r$ (in pm) |
| **STRATEGIZE**  Use the given volume to find the edge length of the unit cell.    Then use the edge length to determine the radius of the atom. | **CONCEPTUAL PLAN**   $V \longrightarrow l$    $V = l^3$    $l \longrightarrow r$    $l = \frac{4r}{\sqrt{3}}$ |

| | **RELATIONSHIPS USED:** |
|---|---|
| | $V = l^3$ (Volume of a cube) |
| | $l = \dfrac{4r}{\sqrt{3}}$ (Edge length of body-centered cubic unit cell) |
| **SOLVE** | **SOLUTION** |
| Solve the equation for the volume of a cube for $l$ and substitute in the given value for $V$ to find $l$. | $V = l^3$ |
| | $l = \sqrt[3]{V} = \sqrt[3]{4.32 \times 10^{-23}\ cm^3} = 3.5\underline{0}88 \times 10^{-8}\ cm$ |
| Solve the equation for the edge length of a body-centered cubic unit cell for $r$ and substitute in the value of $l$ (from the previous step) to find $r$. | $l = \dfrac{4r}{\sqrt{3}}$ |
| | $r = \dfrac{\sqrt{3}\,l}{4} = \dfrac{\sqrt{3}\,(3.5\underline{0}88 \times 10^{-8}\ cm)}{4} = 1.5\underline{1}93 \times 10^{-8}\ cm$ |
| Convert $r$ from cm to m and then to pm. | $1.5\underline{1}93 \times 10^{-8}\ \cancel{cm} \times \dfrac{0.01\ \cancel{m}}{1\ \cancel{cm}} \times \dfrac{1\ pm}{10^{-12}\ \cancel{m}} = 152\ pm$ |

**CHECK** The units of the answer (pm) are correct. The magnitude is reasonable because atomic radii range roughly from 50 to 200 pm.

**FOR PRACTICE 12.3** An atom has radius of 138 pm and crystallizes in the body-centered cubic unit cell. What is the volume of the unit cell in $cm^3$?

---

The **face-centered cubic** unit cell (Figure 12.7▼) is a cube with one atom at each corner and one atom (of the same kind) in the center of each cube face. Note that in the face-centered unit cell (like the body-centered unit cell), the atoms *do not* touch along each edge of the cube. Instead, the atoms touch *along the face diagonal*. The edge length in terms of the atomic radius is therefore $l = 2\sqrt{2}r$, as shown here.

$$b^2 = l^2 + l^2 = 2l^2$$
$$b = 4r$$
$$(4r)^2 = 2l^2$$
$$l^2 = \frac{(4r)^2}{2}$$
$$l = \frac{4r}{\sqrt{2}}$$
$$= 2\sqrt{2}r$$

▶ In the face-centered cubic lattice, the atoms touch along a face diagonal. The edge length is $2\sqrt{2}r$.

Face-centered cubic

## Face-Centered Cubic Unit Cell

Coordination number = 12

Atoms/unit $= \left(\dfrac{1}{8} \times 8\right) + \left(\dfrac{1}{2} \times 6\right) = 4$

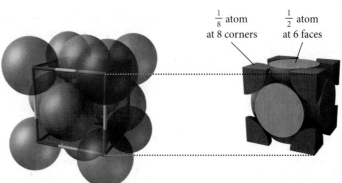

$\frac{1}{8}$ atom at 8 corners

$\frac{1}{2}$ atom at 6 faces

▲ **FIGURE 12.7 Face-Centered Cubic Crystal Structure** The different colors used on the atoms in this figure are for clarity only. All atoms within the structure are identical.

The face-centered unit cell contains four atoms per unit cell because the center atoms on each of the six faces are shared between two unit cells. There are $\frac{1}{2} \times 6 = 3$ face-centered atoms plus $\frac{1}{8} \times 8 = 1$ corner atom, for a total of four atoms per unit cell. The coordination number of the face-centered cubic unit cell is 12, and its packing efficiency is 74%. In this structure, any one atom strongly interacts with more atoms than the atoms in either the simple cubic unit cell or the body-centered cubic unit cell.

**Example** 12.4     **Relating Density to Crystal Structure**

Aluminum crystallizes with a face-centered cubic unit cell. The radius of an aluminum atom is 143 pm. Calculate the density of solid crystalline aluminum in g/cm³.

| | |
|---|---|
| **SORT**  You are given the radius of an aluminum atom and its crystal structure. You are asked to find the density of solid aluminum. | **GIVEN:** $r = 143$ pm, face-centered cubic<br><br>**FIND:** $d$ |
| **STRATEGIZE**  The conceptual plan is based on the definition of density.<br>Since the unit cell has the physical properties of the entire crystal, you can find the mass and volume of the unit cell and use these to calculate its density. | **CONCEPTUAL PLAN**<br>$d = m/V$<br>$m$ = mass of unit cell<br>    = number of atoms in unit cell × mass of each atom<br>$V$ = volume of unit cell<br>    = (edge length)³ |
| **SOLVE**  Begin by finding the mass of the unit cell. Determine the mass of an aluminum atom from its molar mass. Because the face-centered cubic unit cell contains four atoms per unit cell, multiply the mass of aluminum by 4 to get the mass of a unit cell. | **SOLUTION**<br>$m(\text{Al atom}) = 26.98 \dfrac{\text{g}}{\text{mol}} \times \dfrac{1 \text{ mol}}{6.022 \times 10^{23} \text{ atoms}}$<br>$\qquad = 4.480 \times 10^{-23}$ g/atom<br>$m(\text{unit cell}) = 4 \text{ atoms}(4.480 \times 10^{-23} \text{ g/atom})$<br>$\qquad = 1.792 \times 10^{-22}$ g |
| Next, calculate the edge length ($l$) of the unit cell (in m) from the atomic radius of aluminum. For the face-centered cubic structure, $l = 2\sqrt{2}r$. | $l = 2\sqrt{2}r$<br>$\quad = 2\sqrt{2}(143 \text{ pm})$<br>$\quad = 2\sqrt{2}(143 \times 10^{-12} \text{ m})$<br>$\quad = 4.045 \times 10^{-10}$ m |
| Calculate the volume of the unit cell (in cm) by converting the edge length to cm and cubing the edge length. (Use centimeters because you want to report the density in units of g/cm³.) | $V = l^3$<br>$\quad = \left(4.045 \times 10^{-10} \text{ m} \times \dfrac{1 \text{ cm}}{10^{-2} \text{ m}}\right)^3$<br>$\quad = 6.618 \times 10^{-23}$ cm³ |
| Finally, calculate the density by dividing the mass of the unit cell by the volume of the unit cell. | $d = \dfrac{m}{V} = \dfrac{1.792 \times 10^{-22} \text{ g}}{6.618 \times 10^{-23} \text{ cm}^3}$<br>$\quad = 2.71$ g/cm³ |

**CHECK**  The units of the answer are correct. The magnitude of the answer is reasonable because the density is greater than 1 g/cm³ (as you would expect for metals), but still not too high (because aluminum is a low-density metal).

**FOR PRACTICE 12.4**   Chromium crystallizes with a body-centered cubic unit cell. The radius of a chromium atom is 125 pm. Calculate the density of solid crystalline chromium in g/cm³.

CONCEPTUAL
**CONNECTION 12.1**      **Cubic Structures**  How many atoms are in the unit cell in the face-centered cubic structure?

# Closest-Packed Structures

Another way to envision crystal structures, especially useful for metals, in which bonds are not usually directional, is to think of the atoms as stacking in layers, much as fruit is stacked at the grocery store. For example, we can envision the simple cubic structure as one layer of atoms arranged in a square pattern, with the next layer stacking directly over the first, so that the atoms in one layer align exactly on top of the atoms in the layer beneath it, as shown here:

As we saw previously, this crystal structure has a great deal of empty space—only 52% of the volume is occupied by the spheres, and the coordination number is 6.

More space-efficient packing can be achieved by aligning neighboring rows of atoms in a pattern, with one row offset from the next by one-half a sphere, as shown here:

In this way, the atoms pack more closely to each other in any one layer. We can further increase the packing efficiency by placing the next layer *not directly on top of the first*, but again offset so that any one atom actually sits in the indentation formed by three atoms in the layer beneath it, as shown here:

This kind of packing leads to two different crystal structures called *closest-packed structures*, both of which have a packing efficiency of 74% and a coordination number of 12. In the first of these two closest-packed structures—called **hexagonal closest packing**—the third layer of atoms aligns exactly on top of the first, as shown here:

Hexagonal closest packing

## Hexagonal Closest Packing

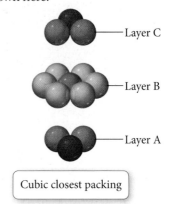

▲ **FIGURE 12.8 Hexagonal Closest-Packing Crystal Structure** The unit cell is outlined in bold.

The pattern from one layer to the next is ABAB, with the third layer aligning exactly above the first. Notice that the central atom in layer B of this structure is touching six atoms in its own layer, three atoms in the layer above it, and three atoms in the layer below, for a coordination number of 12. The unit cell for this crystal structure is not a cubic unit cell, but a hexagonal one, shown in Figure 12.8 ◀.

In the second of the two closest-packed structures—called **cubic closest packing**—the third layer of atoms is offset from the first, as shown here:

Layer C

Layer B

Layer A

Cubic closest packing

The pattern from one layer to the next is ABCABC, with every fourth layer aligning with the first. Although not simple to visualize, the unit cell for cubic closest packing is the face-centered cubic unit cell, shown in Figure 12.9 ▼. The cubic closest-packed structure is identical to the face-centered cubic unit cell structure.

## Cubic Closest Packed = Face-Centered Cubic

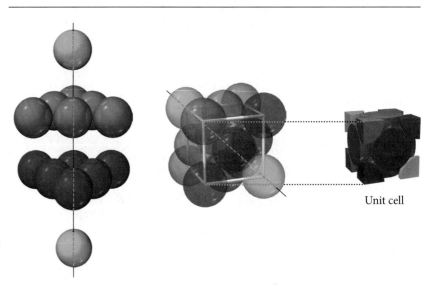

Unit cell

▶ **FIGURE 12.9 Cubic Closest-Packing Crystal Structure** The unit cell of the cubic closest-packed structure is a face-centered cubic.

## 12.4 The Fundamental Types of Crystalline Solids

We examined some of the basic *structures* of crystalline solids in Section 12.3. In this section, we turn to the different *types* of crystalline solids. We can classify crystalline solids into three categories—molecular, ionic, and atomic—based on the individual particles that compose the solid. Atomic solids can themselves be classified into three categories—nonbonded, metallic, and network covalent—depending on the types of interactions between atoms within the solid. Figure 12.10▶ shows the different categories of crystalline solids.

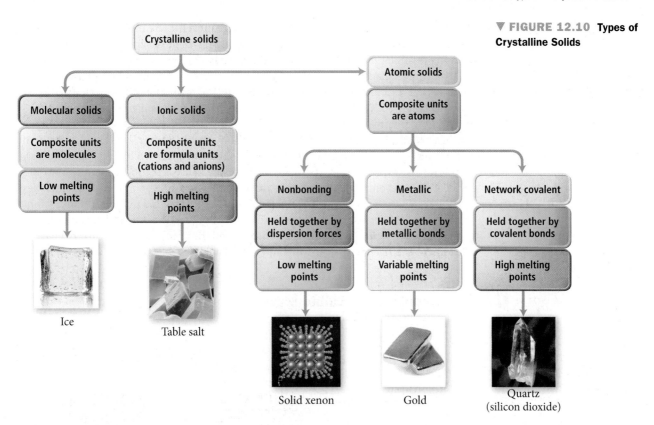

▼ **FIGURE 12.10 Types of Crystalline Solids**

## Molecular Solids

**Molecular solids** are solids in which the composite particles are *molecules*. The unit cells in a crystalline molecular solid are therefore occupied by molecules. Ice (solid $H_2O$) and dry ice (solid $CO_2$) are examples of molecular solids. Molecular solids are held together by the kinds of intermolecular forces—dispersion forces, dipole–dipole forces, and hydrogen bonding—discussed in Chapter 11. Although molecular solids as a whole tend to have low to moderately low melting points, their specific properties depend on the types of intermolecular forces between the molecules as well as the structure of the molecule and the crystal structure of the solid. For example, benzene and toluene have similar molar masses and similar intermolecular forces (dispersion forces):

| Compound | Structure | Molar Mass | Type of Intermolecular Force | Melting Point of Solid |
|---|---|---|---|---|
| Benzene ($C_6H_6$) | | 78.11 g/mol | Dispersion | 5.5 °C |
| Toluene ($C_7H_8$) | | 92.14 g/mol | Dispersion | −95.5 °C |

However, the extra —$CH_3$ group on toluene prevents the molecules from packing as efficiently in the crystal as they do in benzene. The result is that, even though toluene has a higher molar molar mass than benzene, it has a much a lower melting point.

Polymorphism can exist in any crystalline material, not just molecular solids.

Some molecular solids crystallize in different structures called **polymorphs**. The different crystal structures can have different properties, including different melting points and solubilities (see *Chemistry in Your Day: Chocolate, An Edible Material*). Polymorphism is important in pharmaceuticals because different forms of a drug can have different physiological activities. For example, ritonavir was among the first drugs found to be effective in treating HIV and AIDS. When it was first discovered and approved for use, only one polymorph was known. However, soon after its widespread use, a second polymorph was discovered. This second polymorph was more stable, less water-soluble, and much less physiologically active. Even worse, small amounts of this second polymorph in a sample could convert the active polymorph into the less active one. The drug manufacturer had to halt production of ritonavir until researchers solved the problem by switching from a capsule form to a refrigerated gelcap form and later to a stable tablet form.

# Chemistry IN YOUR DAY | Chocolate, An Edible Material

Chocolate is an edible and delicious material designed to please your senses. Good chocolate has a smooth and shiny surface, snaps when you break it, and melts in your mouth, releasing its delicious combination of flavors and texture. Like all materials, chocolate's properties depend on its structure. The main structural ingredient in chocolate is cocoa butter, a triglyceride. We examine triglycerides more carefully in Chapter 22, but for now, just know that triglycerides are long organic molecules, which means that chocolate is a crystalline molecular solid. Many crystalline molecular solids only have one crystalline structure; however, some (including cocoa butter) have several possible crystalline structures called polymorphs, each with different properties. Cocoa butter has six possible polymorphs, often called Types I–VI, as shown in Table 12.1. For reference, body temperature is 37 °C.

The most desirable crystalline form of cocoa butter in chocolate is Type V. Types I and II are soft and crumbly and show noticeable blooming. *Blooming* is the white residue that forms on the outside of chocolate. If you have ever let a chocolate bar melt and then re-solidify, you have probably noticed this undesirable residue. Types III and IV are firmer, but they don't have a sharp snap when you break or bite into them. Type V is firm, displays a sharp snap, has a smooth and shiny surface, and tends to melt in your mouth rather than your hand. For this reason, chocolatiers aim for this

▲ The white residue on the surface of the chocolate bon bon on the left is called bloom. The chocolate bon bon on the right has a smooth, shiny surface, characteristic of Type V cocoa butter.

crystal structure. However, this structure is not easy to achieve. If you simply melt chocolate and let it re-solidify, you get a mix of Types I–V with all of their undesirable qualities. Chocolatiers "temper" their chocolate in their quest for primarily Type V characteristics. (*Continued on next page.*)

**TABLE 12.1 Crystalline Forms of Cocoa Butter**

| Crystalline Structure Type (Polymorph) | Melting Point (°C) | Description | Increasing Density and Stability |
|---|---|---|---|
| I | 17.3 | Soft, crumbly, noticeable blooming, melts in the hand. | |
| II | 23.3 | Soft, crumbly, noticeable blooming, melts in the hand. | |
| III | 25.5 | Firm, poor snap, some blooming, melts in the hand. | |
| IV | 27.3 | Firm, poor snap, some blooming, melts in the hand. | |
| V | 33.8 | Firm, good snap, smooth and shiny surface, melts in the mouth, most desirable. | |
| VI | 36.3 | Hard, some blooming, cannot be formed from melt, takes several months to form. | |

Tempering chocolate requires a specific and carefully controlled heating and cooling sequence (Figure 12.11▼). First chocolatiers heat the chocolate to melt it. Then they allow it to cool slowly. The slow cooling favors the formation of Type V crystals; however, the other types form as well. The final step is to reheat the chocolate to a temperature just below the melting point of the Type V crystals. This step melts all of the Type I–IV crystals, leaving only the Type V crystals to act as "seed crystals."* A seed crystal is a crystal upon which more of the crystal can grow. Seed crystals have the effect of imposing their own structure on the growing crystalline solid. When the properly tempered chocolate cools, the Type V seed crystals cause the majority of the chocolate to crystallize in the Type V structure, resulting in silky smooth, melt-in-your-mouth, chocolate.

▶ **FIGURE 12.11 Chocolate Tempering** In order to temper chocolate, chocolate makers heat it carefully. After the tempering sequence is completed, the chocolate can be poured into molds or used as coatings. When the tempered chocolate cools, it is primarily Type V and has a shiny smooth surface.

Chocolate Tempering Sequence

*Sometimes solid bits of tempered chocolate are added at this point to act as additional seed crystals.

## Ionic Solids

The composite particles of **ionic solids** are ions. Table salt (NaCl) and calcium fluoride ($CaF_2$) are examples of ionic solids. The forces holding ionic solids together are strong coulombic forces (or ionic bonds), and because these forces are much stronger than intermolecular forces, ionic solids tend to have much higher melting points than molecular solids. For example, sodium chloride melts at 801 °C, while carbon disulfide ($CS_2$)—a molecular solid with a higher molar mass—melts at −110 °C. We examine the structure of ionic solids in more detail in Section 12.5.

## Atomic Solids

In **atomic solids** the composite particles are individual atoms. Solid xenon (Xe), iron (Fe), and silicon dioxide ($SiO_2$) are examples of atomic solids. We classify atomic solids into three categories—*nonbonding atomic solids, metallic atomic solids, and network covalent atomic solids*—with each type held together by a different kind of force.

### Nonbonding Atomic Solids

**Nonbonding atomic solids** are held together by relatively weak dispersion forces. In order to maximize these interactions, nonbonding atomic solids form closest-packed structures, maximizing their coordination numbers and minimizing the distance between composite units. Nonbonding atomic solids have very low melting points that increase uniformly with molar mass. The only nonbonding atomic solids are noble gases in their solid form. Argon, for example, has a melting point of −189 °C, and xenon has a melting point of −112 °C.

### Metallic Atomic Solids

**Metallic atomic solids**, such as iron or gold, are held together by *metallic bonds*, which in the simplest model are represented by the interaction of metal cations with the sea of electrons that surrounds them, as described in Section 9.11 (Figure 12.12▶).

Because metallic bonds are not directional, many metals tend to form closest-packed crystal structures. For example, nickel crystallizes in the cubic closest-packed structure and zinc crystallizes in the hexagonal closest-packed structure (Figure 12.13▶, on the next page). Metallic bonds have varying strengths. Some metals, such as mercury,

▲ **FIGURE 12.12 The Electron Sea Model** In the electron sea model for metals, the metal cations exist in a "sea" of electrons.

▶ **FIGURE 12.13 Closest-Packed Crystal Structures in Metals** Nickel crystallizes in the cubic closest-packed structure. Zinc crystallizes in the hexagonal closest-packed structure.

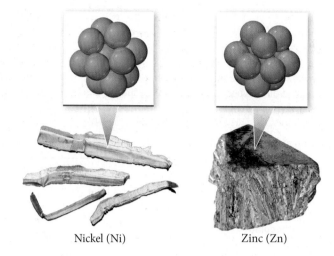

Nickel (Ni)     Zinc (Zn)

have melting points below room temperature, whereas other metals, such as iron, have relatively high melting points (iron melts at 1538 °C).

## Network Covalent Solids

Covalent bonds hold together **network covalent atomic solids**, such as diamond, graphite, and silicon dioxide. The crystal structures of these solids are more restricted by the geometrical constraints of the covalent bonds (which tend to be more directional than intermolecular forces, ionic bonds, or metallic bonds) so they *do not* tend to form closest-packed structures. Network covalent atomic solids have very high melting points because the crystalline solid is held together by covalent bonds. We examine some examples of this class of solids in Section 12.6.

| **Example 12.5** | **Classifying Crystalline Solids** |
| --- | --- |

Classify each crystalline solid as molecular, ionic, or atomic.

**(a)** Au(s)          **(b)** $CH_3CH_2OH(s)$          **(c)** $CaF_2(s)$

**SOLUTION**

**(a)** Gold (Au) is a metal and is therefore an atomic solid.

**(b)** Ethanol ($CH_3CH_2OH$) is a molecular compound. Solid ethanol is therefore a molecular solid.

**(c)** Calcium fluoride ($CaF_2$) is an ionic compound, so it is an ionic solid.

**FOR PRACTICE 12.5**     What type of atomic solid is Au(s)?

**CONCEPTUAL CONNECTION 12.2**

**Crystalline Solid Types and Melting Points** Which solid would you expect to have the highest melting point?

**(a)** MgO(s)          **(b)** $I_2(s)$          **(c)** Kr(s)

## 12.5 The Structures of Ionic Solids

Many ionic solids have crystalline structures that are closely related to the unit cells that we examined in Section 12.3. However, because ionic compounds necessarily contain both cations and anions, their structures must accommodate the two different types of ions. In an ionic solid, the cations and anions attract one another. The coordination number of the unit cell represents the number of close cation–anion interactions. Because these interactions lower potential energy, the crystal structure of a particular ionic compound is the structure that maximizes the coordination number while accommodating both charge neutrality (each unit cell must be charge neutral) and the different sizes of the cations and anions that compose the particular compound. In general, the more similar the radii of the cation and the anion, the higher the coordination number.

Cesium chloride (CsCl) is a good example of an ionic compound with cations and anions of similar size ($Cs^+$ radius = 167 pm; $Cl^-$ radius = 181 pm). In the cesium chloride structure, the chloride ions occupy the lattice sites of a simple cubic cell and one cesium ion lies in the very center of the cell, as shown in Figure 12.14▶. (Note that in this and subsequent figures of ionic crystal structures, the different colored spheres represent different ions.) The coordination number for cesium chloride is 8, meaning that each cesium ion is in direct contact with eight chloride ions (and vice versa). The cesium chloride unit cell contains one chloride anion ($8 \times 1/8 = 1$) and one cesium cation for a ratio of Cs to Cl of 1:1, as the formula for the compound indicates. (Note that complete chloride ions are shown in Figure 12.14, even though only 1/8 of each ion is in the unit cell.) Calcium sulfide (CaS) has the same structure as cesium chloride.

Sodium chloride is an example of an ionic compound with different ion sizes: $Na^+$ (radius = 97 pm) and $Cl^-$ (radius = 181 pm). The crystal structure must accommodate the disproportionate sizes. If ion size were the only consideration, the larger chloride anion could theoretically fit many of the smaller sodium cations around it, but charge neutrality requires that each sodium cation be surrounded by an equal number of chloride anions. Therefore, the coordination number is limited by the number of chloride anions that can fit around the relatively small sodium cation. The structure that minimizes the energy is shown in Figure 12.15▶ and has a coordination number of 6 (each chloride anion is surrounded by six sodium cations and vice versa). We can visualize this structure, called the *rock salt* structure, as chloride anions occupying the lattice sites of a face-centered cubic structure with the smaller sodium cations occupying the holes between the anions. (Alternatively, we can visualize this structure as the *sodium cations* occupying the lattice sites of a face-centered cubic structure with the *larger chloride anions* occupying the spaces between the cations.) Each unit cell contains four chloride anions [$(8 \times \frac{1}{8}) + (6 \times \frac{1}{2}) = 4$] and four sodium cations [$(12 \times \frac{1}{4}) + 1 = 4$], resulting in a ratio of 1:1, as the formula of the compound specifies. Other compounds exhibiting the sodium chloride structure include LiF, KCl, KBr, AgCl, MgO, and CaO.

When there is a greater disproportion between the sizes of the cations and anions in a compound, a coordination number of even 6 is physically impossible. For example, in ZnS ($Zn^{2+}$ radius = 74 pm; $S^{2-}$ radius = 184 pm) the crystal structure, shown in Figure 12.16▶, has a coordination number of only 4. We can visualize this structure, called the *zinc blende* structure, as sulfide anions occupying the lattice sites of a face-centered cubic structure with the smaller zinc cations occupying four of the eight tetrahedral holes located directly beneath each corner atom. A tetrahedral hole is the empty space that lies in the center of a tetrahedral arrangement of four atoms, as shown in the figure below. Each unit cell contains four sulfide anions [$(8 \times \frac{1}{8}) + (6 \times \frac{1}{2}) = 4$] and four zinc cations (each of the four zinc cations is completely contained within the unit cell), resulting in a ratio of 1:1, just as the formula of the compound indicates. Other compounds that exhibit the zinc blende structure include CuCl, AgI, and CdS.

◀ A tetrahedral hole

When the ratio of cations to anions is not 1:1, the crystal structure must accommodate the unequal number of cations and anions. Many compounds that contain a cation-to-anion ratio of 1:2 adopt the *fluorite (CaF$_2$) structure* shown in Figure 12.17▶, on the next page. We can visualize this structure as calcium cations occupying the lattice sites of

Cesium chloride (CsCl)

▲ **FIGURE 12.14 Cesium Chloride Unit Cell** The different colored spheres in this figure represent the different ions in the compound.

Sodium chloride (NaCl)

▲ **FIGURE 12.15 Sodium Chloride Unit Cell** The different colored spheres in this figure represent the different ions in the compound.

Zinc blende (ZnS)

▲ **FIGURE 12.16 Zinc Sulfide (Zinc Blende) Unit Cell** The different colored spheres in this figure represent the different ions in the compound.

Calcium fluoride (CaF$_2$)

a face-centered cubic structure with the larger fluoride anions occupying all eight of the tetrahedral holes located directly beneath each corner atom. Each unit cell contains four calcium cations $[(8 \times \frac{1}{8}) + (6 \times \frac{1}{2}) = 4]$ and eight fluoride anions (each of the eight fluoride anions is completely contained within the unit cell), resulting in a cation to anion ratio of 1:2, just as in the formula of the compound. Other compounds exhibiting the fluorite structure include PbF$_2$, SrF$_2$, and BaCl$_2$. Compounds with a cation-to-anion ratio of 2:1 often exhibit the *antifluorite structure*, in which the anions occupy the lattice sites of a face-centered cubic structure and the cations occupy the tetrahedral holes beneath each corner atom.

◀ **FIGURE 12.17 Calcium Fluoride (or Fluorite) Unit Cell** The different colored spheres in this figure represent the different ions in the compound.

---

**CONCEPTUAL CONNECTION 12.3**

**Ionic Crystalline Solid Unit Cells** Which compound is most likely to crystallize in the zinc blende structure?

(a) RbCl (Rb$^+$ radius = 148 pm; Cl$^-$ radius = 181 pm)

(b) MgCl$_2$ (Mg$^{2+}$ radius = 65 pm; Cl$^-$ radius = 181 pm)

(c) CuI (Cu$^+$ radius = 96 pm; I$^-$ radius = 216 pm)

---

## 12.6 Network Covalent Atomic Solids: Carbon and Silicates

As we saw in Section 12.4, network covalent atomic solids are composed of atoms held together by covalent bonds. Because the covalent bonds have to be broken for the solid to melt, network covalent atomic solids have some of the highest melting points of all substances. In this section, we examine two different families of network covalent atomic solids: carbon and silicates.

**(a) Graphite**　　　　**(b) Diamond**

▲ **FIGURE 12.18 Network Covalent Atomic Solids** **(a)** In graphite, carbon atoms are arranged in sheets. Within each sheet, the atoms are covalently bonded to one another by a network of sigma and pi bonds. Neighboring sheets are held together by dispersion forces. **(b)** In diamond, each carbon atom forms four covalent bonds to four other carbon atoms in a tetrahedral geometry.

### Carbon

Elemental carbon exists in several different forms (or allotropes) including graphene, which we discussed in Section 12.1. Two well-known naturally occurring crystalline forms of carbon are **graphite** and **diamond**. Graphite's structure, shown in Figure 12.18a ◀, consists of flat sheets of carbon atoms covalently bonded together as interconnected hexagonal rings. The bond length between carbon atoms *within a sheet* is 142 pm. However, the forces *between* sheets are much different. There are no covalent bonds between sheets, only relatively weak dispersion forces, and the separation between sheets is 341 pm. Consequently, the sheets slide past each other relatively easily, which explains the slippery feel of graphite and its extensive use as a lubricant. The electrons in the extended pi bonding network within a sheet make graphite a good electrical conductor in the direction of the plane of the sheets. Because of its relative stability and electrical properties, graphite is used for electrodes in electrochemical applications and for heating elements in furnaces.

The density of graphite is 2.2 g/cm$^3$. Under high pressure, the carbon atoms in graphite rearrange to form diamond, which has a higher density of 3.5 g/cm$^3$.

Diamond forms naturally when carbon is exposed to high pressures deep underground. Through movements in Earth's crust, diamond rises toward the surface. Most diamond is found in Africa, mainly in the Congo region and in South Africa. The first synthetic diamonds were produced in the 1940s, using pressures of 50,000 atm and a temperature of 1600 °C. The unit cell for diamond is shown here:

Zinc blende (ZnS)

▲ Diamond Unit Cell

▲ Zinc Blende Unit Cell

You can imagine the unit cell for diamond as a face-centered cubic structure with an additional carbon atom in half of the eight tetrahedral holes located directly beneath each corner atom. Notice the similarity between the diamond unit cell and the zinc blende unit cell that we discussed in Section 12.5. Notice also that each carbon atom is covalently bonded to four other carbon atoms at the corners of a tetrahedron. This structure extends throughout the entire crystal, so that a diamond crystal can be thought of as a giant molecule, held together by these covalent bonds.

Because covalent bonds are very strong, diamond has a very high melting point (it is estimated to melt at about 3800 °C). The electrons in diamond are confined to the covalent bonds and are not free to flow. Therefore, diamond does not conduct electricity. Diamond is very hard and is an excellent conductor of heat. Consequently, the largest use of diamonds is for abrasives and cutting tools. Small diamonds are used at the cutting edge of tools, making the edges much harder and giving them a longer life. Natural diamonds are valued as gems for their brilliance and relative inertness.

**Phase Changes and Pressure**  Why do high pressures favor the formation of diamond from graphite?

CONCEPTUAL
CONNECTION **12.4**

In the 1980s, researchers discovered a new form of carbon when they aimed a powerful laser at a graphite surface. This new form of carbon occurs as soccer-ball-shaped clusters of 60 carbon atoms ($C_{60}$). The atoms form five- and six-membered carbon rings wrapped into a 20-sided icosahedral structure (Figure 12.19▼). The compound was named *buckminsterfullerene*, honoring R. Buckminster Fuller (1895–1983), a twentieth-century engineer and architect who advocated the construction of buildings using a structurally strong geodesic dome shape that he patented.

$C_{60}$

◀ **FIGURE 12.19** $C_{60}$ **and a Geodesic Dome**  The $C_{60}$ structure resembles Buckminster Fuller's geodesic dome.

Researchers have since identified carbon clusters similar to $C_{60}$ containing from 36 to over 100 carbon atoms. As a class, these carbon clusters are called **fullerenes** and nicknamed *buckyballs*. At room temperature, fullerenes are black solids similar to graphite—the individual clusters are held to one another by dispersion forces. Fullerenes are somewhat soluble in nonpolar solvents, and the different fullerenes form solutions of different colors.

Researchers have also synthesized long carbon structures called **nanotubes**, which consist of sheets of interconnected $C_6$ rings that assume the shape of a cylinder (like a roll of chicken wire). The first nanotubes discovered consisted of tubes with double walls of $C_6$ rings with closed ends. The ends of the tubes can be opened when they are heated under the proper conditions. Today, two general types of nanotubes can be produced: (1) *single-walled nanotubes (SWNT)* that have one layer of interconnected $C_6$ rings forming the walls and (2) *multiwalled nanotubes (MWNT)* that have concentric layers of interconnected $C_6$ rings forming the walls. In addition, researchers have been able to form *nanoribbons* by slicing open nanotubes. As we discussed in Section 12.1, carbon can also form graphene, long, extended sheets of carbon atoms just one atom thick. These carbon structures are all shown in Figure 12.20▼.

**(a) Single-walled nanotube (SWNT)**

**(b) Multiwalled nanotube (MWNT)**

**(c) Graphene nanoribbon**

**(d) Graphene sheet**

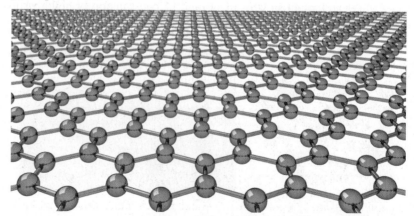

▶ **FIGURE 12.20 Carbon Structures include: (a)** a single-walled nanotube; **(b)** a multiwalled nanotube; **(c)** a graphene nanoribbon; **(d)** a graphene sheet.

Nanotubes are 100 times stronger than steel and only 1/16 as dense. Consequently, we use carbon nanotubes commercially for lightweight applications that require strength, such as golf clubs and bicycle frames. When the nanotubes are lined up parallel to one another, a bundle of the tubes forms a "wire" with very low electrical resistance. These tiny wires raise the possibility of making incredibly small electronic devices.

## Silicates

The **silicates** (extended arrays of silicon and oxygen) are the most common network covalent atomic solids. Geologists estimate that 90% of Earth's crust is composed of silicates. Silicon and oxygen form a network covalent structure in which a silicon atom bonds to four oxygen atoms, assuming a tetrahedral shape with the silicon atom in the middle and the four oxygen atoms at the corners of the tetrahedron (Figure 12.21▶).

In this structure, the silicon atom bonds to each oxygen atom with a single covalent sigma bond. In contrast to carbon, which often bonds to oxygen with a double bond (one sigma and one pi bond), silicon typically forms only a single bond with oxygen because the silicon atom is too large to allow substantial overlap between the *p*-orbitals on the two atoms. The silicon atom in this structure, by bonding to four oxygen atoms, obtains a complete octet. However, each oxygen atom is one electron short of an octet. Therefore, each O atom forms a second covalent bond to a different Si atom, resulting in the three-dimensional structure of **quartz**. Quartz has a formula unit of $SiO_2$ and is generally called **silica**. In silica, each Si atom is in a tetrahedron surrounded by four O atoms, and each O atom acts as a bridge connecting the corners of two tetrahedrons (Figure 12.22▶). See Section 23.3 for a more extended discussion of silicate structures.

▲ **FIGURE 12.21** **SiO₄ Tetrahedron** In an SiO₄ tetrahedron, silicon occupies the center of the tetrahedron, and one oxygen atom occupies each corner.

▲ **FIGURE 12.22** **Structure of Quartz** In the quartz structure, each Si atom is in a tetrahedron surrounded by four O atoms, and each O atom is a bridge connecting the corners of two tetrahedrons.

## 12.7 Ceramics, Cement, and Glass

The silicate structures we examined in Section 12.6 are common in ceramics, cement, and glass. These substances are used in buildings, electrical devices, pottery, and kitchenware. If you look around wherever you are sitting at this moment, you are likely to see examples of these materials. In this section of the chapter, we examine each of these amazing classes of materials individually.

## Ceramics

**Ceramics** are traditionally defined as inorganic nonmetallic solids that are prepared from powders usually mixed with water, formed into the desired shape and then heated. Ceramics are extensively used to make bricks, tiles, pottery, dishware, and insulating elements in electrical devices. They display a range of properties but are usually hard, strong, nonconductive, and brittle. The word "ceramics" originates from the Greek *keramikos*. Keramikos is an area of Athens that was the potters' quarter of the city, and the Greeks were among the first accomplished potters. However, pottery itself predates recorded history. The oldest known example is the Venus of Dolni, a figurine that dates to 29,000–25,000 BCE. Ceramics can be categorized into three types: silicate, oxide, and nonoxide.

### Silicate Ceramics

Many ceramics are composed of aluminosilicates, a class of minerals in which some of the silicon atoms in the silicate structure are replaced by aluminum (see Section 23.3). The weathering of naturally occurring aluminosilicates produces **clays**, which are essentially powdered forms of the minerals mixed with water. When heated, reactions occur that transform the clay into the ceramic substance.

For example, the clay kaolinite, $Al_2Si_2O_5(OH)_4$, undergoes irreversible chemical and structural changes when heated (or fired) above 1500 °C. These changes transform the clay into a white ceramic solid containing an extended network of Si-O and Al-O tetrahedra. Kaolinite is the most important component of porcelain, a ceramic substance that

▲ The Venus of Dolni, the oldest known ceramic object, dates to 29,000–25,000 BCE. It was discovered in 1925 in the Czech Republic.

▲ Ceramic electrical insulator

▲ Aluminum oxide crucible, a container in which substances can be subjected to high temperatures.

originated in China during the Han dynasty about 2000 years ago. Chinese porcelain was superior to previous ceramics and became highly valued and prized for its beauty and durability, especially when formed into cups for drinking tea. The Chinese profitably exported their porcelain wares to the Middle East and the West and kept their recipe for making it a secret for hundreds of years. The term *china* today still refers to the fine plates, saucers, and cups made from porcelain and related ceramic materials. Silicate ceramics also find extensive use as insulators in electrical applications.

### Oxide Ceramics

Among the most common oxide ceramics are $Al_2O_3$ and MgO. These materials demonstrate the advantages and disadvantages of ceramic materials compared to metals. $Al_2O_3$ melts at 2072 °C, compared to aluminum metal, which melts at 660 °C. MgO melts at an even higher temperature (2852 °C). Both are physically and chemically stable at high temperatures. The high melting point and chemical stability make $Al_2O_3$ and MgO outstanding *refractory materials*—materials that can be used in high-temperature applications.

For example, aluminum oxide and magnesium oxide are used in industrial furnaces, high-speed cutting tools, crucibles, heating elements, and fire proofing. Engineers have long desired to take advantage of the high-temperature stability of ceramics to make ceramic engines. A ceramic engine can run at a higher temperature, which makes it more efficient and would require less cooling. However, certain less desirable characteristics of ceramics have made that pursuit difficult. Ceramic materials tend to be brittle and subject to thermal shock (they can crack upon fast changes in temperature). Metals can bend and stretch under stress, but ceramics break. Even though ceramics have found some limited use as components of engines, a completely ceramic mass-produced engine remains elusive.

### Nonoxide Ceramics

The nonoxide ceramics include substances such as $Si_3N_4$, BN, and SiC. Silicon nitride is a network covalent solid with a structure similar to silica; the silicon atoms sit in the center of nitrogen tetrahedra that are linked. Silicon nitride is used extensively in engine parts and nonmetallic ball bearings. Boron nitride is isoelectronic with $C_2$ (BN and $C_2$ have the same number of valence electrons) and forms structures similar to carbon. For example, BN can form layered sheets (similar to graphite), which are covalently bound within the sheets but with only dispersion forces between the sheets. As a result, this form of BN makes a good high-temperature lubricant. Another form of boron nitride has the diamond structure, which results in a hard, strong substance similar to diamond. This form of boron nitride is used as an abrasive and to make cutting tools. Boron nitride can even form nanotubes similar to those formed by carbon. Silicon carbide also has a diamond structure, but in silicon carbide, half of the carbon atoms are replaced by silicon atoms. It too finds applications as an abrasive, and it is also used as a refractory material and an additive to steel.

### Cement

Cement was first discovered by the Romans, who used lime, volcanic ash, and clay to make a pourable slurry that hardened into a rocklike substance. The Romans used cement to construct the 43.3-m-diameter dome of the Pantheon, a 2000-year-old edifice. This dome remains the largest unreinforced concrete dome in existence.

The majority of the cement used today is Portland cement. **Portland cement** is a powdered mixture consisting mostly of limestone ($CaCO_3$) and silica ($SiO_2$), with smaller amounts of alumina ($Al_2O_3$), iron(III) oxide ($Fe_2O_3$), and gypsum ($CaSO_4 \cdot 2H_2O$). The powdered mixture reacts with water in a number of complex reactions that produce a final rocklike substance. Unlike clays, which lose water upon setting, Portland cement reacts *with* water as it hardens. The hardening process involves the formation of Si-O-Si bridges that produce fibrous structures. These structures bond strongly to each other (and most other substances). The original Portland cement was patented in 1824 by an English bricklayer named Joseph Aspidin (1788–1855). The hardened cement resembles a prestigious building stone from the Isle of Portland in Southwest England and derives its name from this resemblance.

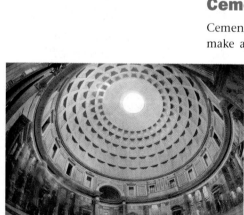

▲ The Roman Pantheon, still a popular tourist attraction, sits in the center of the ancient heart of Rome.

Portland cement is combined with sand and pebbles to make **concrete**, the most widely used building material in the world. Concrete dramatically revolutionized construction. Before concrete, buildings were made exclusively by arranging the materials piece by piece. With the development of concrete, buildings could literally be poured into place. Concrete is used extensively to make foundations, walls, buildings, bridges, aqueducts, roads, and dams. About half of all the structures humans construct are made of concrete.

## Glass

Silica melts when heated above 1500 °C. After melting, if cooled quickly, silica does not crystallize back into the quartz structure. Instead, the Si atoms and O atoms form a randomly ordered amorphous structure called a **glass**. Silicate glass is transparent, impervious to water, and an outstanding material for making windows and drinking vessels. The Egyptians and Greeks likely made glass objects; however, the Romans in the first century BCE were the first to extensively develop glass making. They discovered that adding sodium carbonate to silica dramatically lowers its melting point, allowing glass to form at much lower temperatures. In addition, they developed glass-blowing, which involves melting glass and then using a tube to blow the glass into spherical shapes.

When $SiO_2$ is made into a glass, the result is called **vitreous silica** or **fused silica**. This type of glass is hard, resists high temperatures, has a low thermal expansion, and is transparent to both visible light and ultraviolet light. Vitreous silica (because of the high temperatures required to produce it) is too expensive for most common applications. The most common modern glass is **soda-lime glass**, which is also referred to as *window glass*. Soda-lime glass is about 70% $SiO_2$; the balance is mostly $Na_2O$ and $CaO$. This type of glass is transparent to visible light (not ultraviolet) and has a high thermal expansion, but it is less expensive to make and form into desired shapes than vitreous silica.

One disadvantage of soda-lime glass is its tendency to crack under thermal shock. Adding boric oxide ($B_2O_3$) to the glass mixture instead of $CaO$ produces **borosilicate glass** (also known as **Pyrex**®), which expands less when heated. As a result, vessels made of Pyrex® can withstand heating and cooling cycles that would shatter soda-lime glass. The beakers, flasks, and other glassware found in most chemistry labs are made of borosilicate glass.

**Leaded glass** (often called *crystal*, even though it is not a crystal) results when PbO is mixed with $SiO_2$ and a couple of other minor components. This type of glass has a higher index of refraction (it bends light more than ordinary glass), which results in glassware with a more brilliant appearance than soda-lime glass. It also makes a ringing sound when tapped, which is a common test to distinguish lead crystal from ordinary glass. Recent concerns about the toxicity of lead in leaded glass have led to the development of lead-free alternatives (often called lead-free crystal) for stemware and drinking glasses, which has properties similar to those of leaded glass, but without the negative health risks.

▲ Glass blowing involves blowing air into a hot piece of glass to form spherical shapes.

▲ Leaded glass has a higher index of refraction than ordinary glass, making it appear more brilliant.

The amount of lead that leaches into a drink from leaded glass over a short period of time is likely inconsequential. However, storing drinks in leaded glass for long periods of time can lead to elevated lead levels in the liquid.

## 12.8 Semiconductors and Band Theory

Recall that in Section 12.4 we discussed a model for bonding in metals called the *electron sea model*, which we first introduced and explained in Section 9.11. We now turn to a model for bonding in solids that is both more sophisticated and more broadly applicable—it applies to both metallic solids and covalent solids. The model is **band theory**, and it grows out of molecular orbital theory, first discussed in Section 10.8.

### Molecular Orbitals and Energy Bands

Recall that according to molecular orbital theory, the atomic orbitals of the atoms combine within a molecule to form molecular orbitals. These molecular orbitals are not localized on individual atoms but *delocalized over the entire molecule*. Similarly, in band theory, the atomic orbitals of the atoms within a solid crystal combine to form orbitals that are not localized on individual atoms, but delocalized over the entire *crystal*. In some sense, the crystal is like a very large molecule, and its valence electrons occupy the molecular orbitals formed from the atomic orbitals of each atom in the crystal.

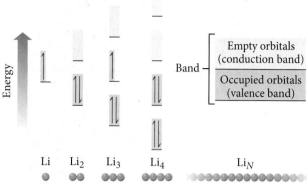

▲ **FIGURE 12.23 Energy Levels of Molecular Orbitals in Lithium Molecules** When many Li atoms are present, the energy levels of the molecular orbitals are so closely spaced that they fuse to form a band. Half of the orbitals are bonding orbitals and contain valence electrons; the other half are antibonding orbitals and are empty.

Recall from Section 2.7 that a semiconductor is a substance of intermediate (and highly temperature-dependent) electrical conductivity.

▶ **FIGURE 12.24 Band Gap** In a conductor, there is no energy gap between the valence band and the conduction band. In semiconductors there is a small energy gap, and in insulators there is a large energy gap.

Consider a series of molecules constructed from individual lithium atoms. The energy levels of the atomic orbitals and resulting molecular orbitals for Li, $Li_2$, $Li_3$, $Li_4$, and $Li_N$ (where $N$ is a large number on the order of $10^{23}$) are shown in Figure 12.23◀. The lithium atom has a single electron in a single $2s$ atomic orbital. The $Li_2$ molecule contains two electrons and two molecular orbitals. The electrons occupy the lower-energy bonding orbital—the higher-energy, or antibonding, molecular orbital is empty. The $Li_4$ molecule contains four electrons and four molecular orbitals. The electrons occupy the two bonding molecular orbitals—the two antibonding orbitals are completely empty.

The $Li_N$ molecule contains $N$ electrons and $N$ molecular orbitals. However, because there are so many molecular orbitals, the energy spacings between them are infinitesimally small; they are no longer discrete energy levels, but rather they form a *band* of energy levels. One half of the orbitals in the band ($N/2$) are bonding molecular orbitals, and (at 0 K) these orbitals contain the $N$ valence electrons; this band is called the **valence band**. The other $N/2$ molecular orbitals are antibonding and (at 0 K) are completely empty; this band is called the **conduction band**. If the atoms composing a solid have $p$ orbitals available, then the same process leads to another set of bands at higher energies.

In lithium metal, the highest occupied molecular orbital lies in the middle of a band of orbitals, and the energy difference between it and the next higher-energy orbital is infinitesimally small. Therefore, above 0 K, electrons can easily make the transition from the valence band to the conduction band. Because electrons in the conduction band are mobile, lithium, like all metals, is a good electrical conductor. Mobile electrons in the conduction band are also responsible for the thermal conductivity of metals. When a metal is heated, electrons are excited to higher-energy molecular orbitals. These electrons quickly transport the thermal energy throughout the crystal lattice.

In metals, the valence band and conduction band are energetically continuous—the energy difference between the top of the valence band and the bottom of the conduction band is infinitesimally small. In semiconductors and insulators, however, an energy gap, called the **band gap**, exists between the valence band and conduction band as shown in Figure 12.24▼. Band gaps are usually specified in units of electron volts abbreviated as eV (1 eV = $1.602 \times 10^{-19}$ J). In insulators, the band gap is large, and electrons are not promoted into the conduction band at ordinary temperatures, resulting in no electrical conductivity. In semiconductors, the band gap is smaller. As a result, the conductivity of semiconductors increases with increasing temperature because heat allows a greater number of electrons to be thermally promoted into the conduction band.

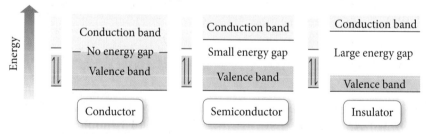

The Group 4A elements exhibit a range of band gaps as shown in Table 12.2.

| TABLE 12.2 Band Gap of Group 4A Elements | | | |
|---|---|---|---|
| **Group 4A Element** | **Atomic Radius (pm)** | **Band Gap** | **Classification** |
| Carbon (diamond) | 77 | 5.5 eV | Insulator |
| Silicon | 118 | 1.11 eV | Semiconductor |
| Germanium | 122 | 0.67 eV | Semiconductor |
| Tin | 140 | 0.08 eV | Metal |
| Lead | 180 | none | Metal |

Notice that the band gap decreases as you move down the column of elements. Why? We know from our discussion of periodic properties (see Section 8.6) that atomic radius increases as we move down a column in the periodic table. The increasing radius reduces the overlap between orbitals on neighboring atoms, which in turn reduces the energy difference between the antibonding orbitals (the conduction band) and the bonding orbitals (the valence band). This results in a decreasing band gap as you move down the column of elements.

## Doping: Controlling the Conductivity of Semiconductors

Doped semiconductors contain minute amounts of impurities that result in additional electrons in the conduction band or electron "holes" in the valence band. For example, silicon is a group 4A semiconductor. Its valence electrons just fill its valence band. The band gap in silicon is large enough that only a few electrons are promoted into the conduction band at room temperature; therefore, silicon is a poor electrical conductor. However, silicon can be doped with phosphorus, a group 5A element with five valence electrons, to increase its conductivity. The phosphorus atoms are incorporated into the silicon crystal structure, and each phosphorous atom brings with it one additional electron. Since the valence band is completely full, the additional electrons must go into the conduction band. These electrons are then mobile and can conduct electrical current. This type of semiconductor is an **n-type semiconductor** because the charge carriers are *negatively* charged electrons in the conduction band.

Silicon can also be doped with a group 3A element, such as gallium, which has only three valence electrons. When gallium is incorporated into the silicon crystal structure, gallium atoms trap some of the electrons in silicon's valence band, resulting in electron "holes," or empty molecular orbitals, in the valence band. The presence of holes also allows for the movement of electrical current because electrons in the valence band can move between holes. In this way, the holes move in the opposite direction as the electrons. This type of semiconductor is a **p-type semiconductor** because each hole acts as a *positive* charge.

At the heart of most modern electronic devices are silicon chips containing millions of **p–n junctions**, tiny spots that are p-type on one side and n-type on the other. These junctions can serve a number of functions, including acting as **diodes** (circuit elements that allow the flow of electrical current in only one direction) or *amplifiers* (elements that amplify a small electrical current into a larger one).

**Semiconductor Type**  Which element would you dope into germanium to create a p-type semiconductor?

**(a)** indium          **(b)** tin          **(c)** arsenic

CONCEPTUAL
**CONNECTION 12.5**

## 12.9  Polymers and Plastics

**Polymers** are long, chainlike molecules composed of repeating units called **monomers**. In Chapter 22, we discuss natural polymers such as starches, proteins, and DNA, which play important roles in living organisms. In this section, we learn about synthetic polymers, which we encounter daily in plastic products such as PVC tubing, styrofoam coffee cups, nylon rope, and plexiglass windows. Polymeric materials are common in our everyday lives, found in everything from computers to toys to packaging materials. Most polymers are durable, partly because of the length of their molecules. In general, the longer a molecule, the greater the intermolecular forces between molecules and the higher the melting point and boiling point of the substance. Since breaking or tearing a polymeric material involves either overcoming the intermolecular forces between chains or actually breaking the covalent bonds between monomers, polymers tend to be durable materials.

One of the simplest synthetic polymers is polyethylene. The polyethylene monomer is ethene (also called ethylene):

$$H_2C = CH_2$$

Monomer

Ethene or ethylene

Ethene monomers react with each other, breaking the double bond between carbons and bonding together to form a long polymer chain:

$$\cdots CH_2 - CH_2 - CH_2 - CH_2 - CH_2 - CH_2 - CH_2 - CH_2 - CH_2 \cdots$$

Polymer

Polyethylene

► **FIGURE 12.25 Polyethylene**
Plastic milk bottles are made from polyethylene.

Polyethylene is an **addition polymer**, a polymer in which the monomers link together without the elimination of any atoms. Polyethylene is the plastic that is used for milk jugs, juice containers, and garbage bags. The properties of polyethylene depend on the exact structure of the ethylene chain and the resulting structure of the material. Although polymers are generally amorphous, they can have regions of crystallinity. For polyethylene, the degree of crystallinity depends in part on the degree of branching in the chain. High-density polyethylene (HDPE), for example, has little branching, so that the chains can align closer to one another, resulting in high crystallinity, high density, and good strength and heat resistance. Common plastic milk jugs are made of HDPE (Figure 12.25◄). Low-density polyethylene (LDPE), in contrast, has more highly branched chains. This branching prevents the chains from interacting as efficiently, resulting in lower crystallinity, density, and strength and heat resistance. Many common plastic bags are made of LDPE.

Substituted polyethylenes make up an entire class of polymers (Table 12.3). For example, polyvinyl chloride (PVC)—the plastic used to make certain kinds of pipes and plumbing fixtures—is composed of monomers in which a chlorine atom has been substituted for one of the hydrogen atoms in ethene (Figure 12.26▼). These monomers, which are shown here, react to form PVC:

$$\cdots CH - CH_2 - CH - CH_2 - CH - CH_2 - CH - CH_2 - CH \cdots$$
$$\ \ \ |\qquad\qquad |\qquad\qquad |\qquad\qquad |\qquad\qquad |$$
$$\ \ \ Cl\qquad\quad Cl\qquad\quad Cl\qquad\quad Cl\qquad\quad Cl$$

Polymer

Polyvinyl chloride (PVC)

$$HC = CH_2$$
$$\ \ |$$
$$\ \ Cl$$

Monomer

Chloroethene

► **FIGURE 12.26 Polyvinyl Chloride**
Polyvinyl chloride is used for many plastic plumbing supplies, such as pipes and connectors.

Some polymers—called *copolymers*—consist of two different kinds of monomers. For example, the monomers that compose nylon 6,6 are hexamethylenediamine and adipic acid. These two monomers add together via a condensation reaction as follows:

| TABLE 12.3 | Polymers of Commercial Importance | |
|---|---|---|
| **Polymer** | **Structure** | **Uses** |
| **Addition Polymers** | | |
| Polyethylene | $-(CH_2-CH_2)_n$ | Films, packaging, bottles |
| Polypropylene | $\left[ CH_2-CH_2 \atop \quad\quad CH_3 \right]_n$ | Kitchenware, fibers, appliances |
| Polystyrene | $\left[ CH_2-CH \atop \text{(phenyl)} \right]_n$ | Packaging, disposable food containers, insulation |
| Polyvinyl chloride | $\left[ CH_2-CH \atop \quad\quad Cl \right]_n$ | Pipe fittings, clear film for meat packaging |
| **Condensation Polymers** | | |
| Polyurethane | $\left[ \underset{O}{C}-NH-R-NH-\underset{O}{C}-O-R'-O \right]_n$  R, R' = $-CH_2-CH_2-$ (*for example*) | "Foam" furniture stuffing, spray-on insulation, automotive parts, footwear, water-protective coatings |
| Polyethylene terephthalate (a polyester) | $\left[ O-CH_2-CH_2-O-\underset{O}{C}-\bigcirc-\underset{O}{C} \right]_n$ | Tire cord, magnetic tape, apparel, soda bottles |
| Nylon 6,6 | $\left[ NH-(CH_2)_6-NH-\underset{O}{C}-(CH_2)_4-\underset{O}{C} \right]_n$ | Home furnishings, apparel, carpet fibers, fish line, polymer blends |

The product that forms between the reaction of two monomers is a **dimer**. The polymer (nylon 6,6) forms as the dimer continues to add more monomers. Polymers that eliminate an atom or a small group of atoms during polymerization are **condensation polymers**. Nylon 6,6 and other similar nylons are drawn into fibers and used to make consumer products such as panty hose, carpet fibers, and fishing line (see *Chemistry in Your Day: Kevlar*).

---

## Chemistry IN YOUR DAY | Kevlar

In 1965, Stephanie Kwolek, working for DuPont to develop new polymer fibers, noticed an odd cloudy product from a polymerization reaction. Some researchers might have rejected the product, but Kwolek insisted on examining its properties more carefully. The results were astonishing—when the polymer was spun into a fiber, it was stronger than any fiber known before. Kwolek had discovered Kevlar, a material that is pound for pound five times stronger than steel.

Kevlar is a condensation polymer featuring aromatic rings and amide linkages:

The polymeric chains within Kevlar crystallize in a parallel arrangement (like dry spaghetti noodles in a box), with strong cross-linking between neighboring chains resulting from hydrogen bonding. The hydrogen bonding occurs between the N—H groups on one chain and the C=O groups on neighboring chains:

This structure is responsible for Kevlar's high strength and its other properties, including fire resistance and chemical resistance (for example, resistance to acids).

Today, DuPont sells hundreds of millions of dollars worth of Kevlar every year. Kevlar is particularly well known for its use in bulletproof vests. This application of Kwolek's discovery has saved thousands of lives. In addition, Kevlar is used to make helmets, radial tires, brake pads, racing sails, suspension bridge cables, skis, and high-performance hiking and camping gear.

▲ Kevlar is used to make bulletproof vests.

---

# Self-Assessment Quiz

**Q1.** An X-ray beam of $\lambda = 154$ pm is incident on the surface of an unknown metallic crystal. It produces a maximum diffraction at an angle of $\theta = 29.1°$. Assuming $n = 1$, calculate the separation between layers of metal atoms in the crystal.

a) 316 pm        b) 105 pm

c) 154 pm        d) 158 pm

**Q2.** How many atoms are in the body-centered cubic unit cell?

a) 1      b) 2      c) 4      d) 5

**Q3.** What is the coordination number of an atom in a face-centered unit cell?

   a) 4      b) 6      c) 8      d) 12

**Q4.** What is the packing efficiency of the face-centered cubic unit cell?

   a) 52%      b) 68%      c) 74%      d) 88%

**Q5.** What is the edge length of a body-centered cubic unit cell made up of atoms that each has a radius of 168 pm?

   a) 388 pm      b) 336 pm      c) 475 pm      d) 84 pm

**Q6.** Rhodium crystallizes in a face-centered cubic unit cell. The radius of a rhodium atom is 135 pm. Determine the density of rhodium in $g/cm^3$.

   a) $3.07 \, g/cm^3$      b) $12.4 \, g/cm^3$
   c) $278 \, g/cm^3$      d) $0.337 \, g/cm^3$

**Q7.** Which type of solid is dry ice (solid carbon dioxide)?

   a) ionic      b) molecular
   c) atomic      d) none of the above

**Q8.** Which solid has the highest melting point?

   a) $CS_2(s)$      b) $CaF_2(s)$
   c) $C_8H_{18}(s)$      d) $Ar(s)$

**Q9.** Which material is a common oxide ceramic?

   a) $N_2O$      b) $Al_2O_3$      c) $Fe_2O_3$      d) $CuO$

**Q10.** What is the main component of glass?

   a) $Al_2O_3$      b) $SiO_2$      c) $CaF_2$      d) $SiC$

**Q11.** Which substance has the greatest band gap?

   a) silicon      b) germanium
   c) lead      d) gold

**Q12.** Which element produces a p-type semiconductor when doped into silicon?

   a) Ge      b) As
   c) Ga      d) Sb

**Q13.** Which monomer produces the addition polymer shown here?

$$\cdots CH-CH_2-CH-CH_2-CH\cdots$$
$$\ \ \ |\ \ \ \ \ \ \ \ \ \ \ \ \ \ |\ \ \ \ \ \ \ \ \ \ \ \ \ \ |$$
$$\ \ \ F\ \ \ \ \ \ \ \ \ \ \ \ \ F\ \ \ \ \ \ \ \ \ \ \ \ F$$

  **(a)** $H_2C{=}CH_2$

  **(b)** $HC{=}CH_2$ with F below

  **(c)** $HC{=}CH$ with F, F below

  **(d)** $FC{=}C{-}F$ with F, F below

# Chapter 12 IN REVIEW

## Key Terms

### Section 12.1
graphene (533)

### Section 12.2
X-ray diffraction (534)
X-ray crystallography (536)

### Section 12.3
crystalline lattice (537)
unit cell (537)
simple cubic (538)
coordination number (539)
packing efficiency (539)
body-centered cubic (540)
face-centered cubic (541)
hexagonal closest packing (543)
cubic closest packing (544)

### Section 12.4
molecular solids (545)
polymorphs (546)

ionic solids (547)
atomic solids (547)
nonbonding atomic solids (547)
metallic atomic solids (547)
network covalent atomic solids (548)

### Section 12.6
graphite (550)
diamond (550)
fullerenes (552)
nanotubes (552)
silicates (553)
quartz (553)
silica (553)

### Section 12.7
ceramics (553)
clay (553)
Portland cement (554)
concrete (555)
glass (555)

vitreous silica (fused silica) (555)
soda-lime glass (555)
borosilicate glass (Pyrex®) (555)
leaded glass (555)

### Section 12.8
band theory (555)
valence band (556)
conduction band (556)
band gap (556)
n-type semiconductor (557)
p-type semiconductor (557)
p–n junctions (557)
diodes (557)

### Section 12.9
polymer (557)
monomer (557)
addition polymer (558)
dimer (560)
condensation polymer (560)

# Key Concepts

## Crystalline Structures (12.2–12.4)

▶ X-ray crystallography uses the diffraction pattern of X-rays to determine the crystal structure of solids.

▶ The crystal lattice is represented by a unit cell, a structure that reproduces the entire lattice when repeated in all three dimensions.

▶ Three basic cubic unit cells are the simple cubic, the body-centered cubic, and the face-centered cubic.

▶ Some crystal lattices can also be depicted as closest-packed structures, including the hexagonal closest-packing structure (not cubic) and the cubic closest-packing structure (which has a face-centered cubic unit cell).

▶ The basic types of crystal solids are molecular, ionic, and atomic solids. We divide atomic solids into three different types: non-bonded, metallic, and network covalent.

## Ionic Solids (12.5)

▶ Unit cells in the structures of ionic solids are charge neutral. The unit cells must also accommodate the difference in size between the anion and cation.

▶ Common unit cells for ionic compounds include the sodium chloride unit cell, the zinc blende unit cell, and the calcium fluoride unit cell.

## Network Covalent Atomic Solids (12.6)

▶ Network covalent solids are composed of extended arrays of atoms held together by covalent bonds.

▶ Common network covalent solids include diamond, graphite, graphene, and silica.

## Ceramics, Cement, and Glass (12.7)

▶ Ceramics are inorganic nonmetallic solids that are prepared from powders typically mixed with water, formed into the desired shape, and then heated. Ceramics are usually hard, strong, nonconductive, and brittle.

▶ Cement is a powdered mixture consisting mostly of limestone ($CaCO_3$) and silica ($SiO_2$), with smaller amounts of other substances. When cement is mixed with water, it reacts to form a hard, stone-like substance.

▶ Glass is primarily amorphous $SiO_2$ or silica. Silicate glass is transparent, impervious to water, and an outstanding material for making windows and drinking vessels.

## Semiconductors and Band Theory (12.8)

▶ Band theory is a model for bonding in solids in which the atomic orbitals of the atoms are combined and delocalized over the entire crystal solid. In band theory, solids form energy bands that are occupied by electrons.

▶ In metals, the valence band (composed of bonding molecular orbitals) is continuous with the conduction band (composed of antibonding molecular orbitals).

▶ Semiconductors have a small energy gap–called the band gap–between the valence band and the conduction band.

▶ Semiconductors can be doped with small amounts of impurities to modify their conductivity.

## Polymers and Plastics (12.9)

▶ Polymers are long, chainlike molecules that consist of repeating units called monomers. They can be natural or synthetic.

▶ Polyethylene is an addition polymer, a polymer formed without the elimination of any atoms.

▶ Condensation polymers, such as nylon, are formed by the elimination of small groups of atoms.

# Key Equations and Relationships

Bragg's Law: Relationship between Light Wavelength ($\lambda$), Angle of Reflection ($\theta$), and Distance ($d$) between the Atomic Layers (**12.2**)

$$n\lambda = 2d \sin \theta \ (n = \text{integer})$$

# Key Learning Outcomes

| CHAPTER OBJECTIVES | ASSESSMENT |
|---|---|
| Using Bragg's Law in X-Ray Diffraction Calculations (**12.2**)  | Example 12.1 For Practice 12.1 Exercises 27, 28 |

| Calculating the Packing Efficiency of a Unit Cell (**12.3**) | Example 12.2   For Practice 12.2   Exercises 31, 32 |
|---|---|
| Relating Unit Cell Volume, Edge Length, and Atomic Radius (**12.3**) | Example 12.3   For Practice 12.3   Exercises 33–36 |
| Relating Density to Crystal Structure (**12.3**) | Example 12.4   For Practice 12.4   Exercises 37, 38 |
| Classifying Crystalline Solids (**12.4**) | Example 12.5   For Practice 12.5   Exercises 49, 50 |

# EXERCISES

## Review Questions

1. What is graphene? Why is graphene unique?

2. Explain the basic principles involved in X-ray crystallography. Include Bragg's law in your explanation.

3. What is a crystalline lattice? How is the lattice represented with the unit cell?

4. Make a drawing of each unit cell: simple cubic, body-centered cubic, and face-centered cubic.

5. For each of the cubic cells in the previous problem, give the coordination number, edge length in terms of $r$, and number of atoms per unit cell.

6. What is the difference between hexagonal closest packing and cubic closest packing? What are the unit cells for each of these structures?

7. What are the three basic types of solids and the composite units of each? What types of forces hold each type of solid together?

8. What are the three categories of atomic solids?

9. What kinds of forces hold each of the three basic categories of atomic solids together?

10. What is a polymorph?

11. In an ionic compound, how are the relative sizes of the cation and anion related to the coordination number of the crystal structure?

12. Show how the cesium chloride, sodium chloride, and zinc blende unit cells each contain a cation-to-anion ratio of 1:1.

13. Show how the fluorite structure accommodates a cation-to-anion ratio of 1:2.

14. Name and describe the different allotropes of carbon.

15. What are silicates? What is quartz?

16. What is the definition of a *ceramic*? What are the three categories of ceramics?

17. List the major and minor components of Portland cement. What is the difference between the hardening process of Portland cement and the hardening process of clays?

18. How is concrete made from Portland cement? What advantage does concrete have for building compared to the construction methods that predated the development of concrete?

19. Describe what happens on the molecular level when silica is heated and then cooled to make glass.

20. Describe the difference between vitreous silica and soda-lime glass. What are some advantages and disadvantages of each of these types of glass?

21. In band theory of bonding for solids, what is a *band*? What is the difference between the *valence band* and the *conduction band?*

22. In band theory of bonding for solids, what is a band gap? How does the band gap differ in metals, semiconductors, and insulators?

23. Explain how doping can increase the conductivity of a semiconductor. What is the difference between an n-type semiconductor and a p-type semiconductor?

24. What is the trend in the size of the band gap as you move down the column of the group 4A elements?

25. What is a polymer? What is the difference between a polymer and a copolymer?

26. How do an addition polymer and a condensation polymer differ from each other?

## Problems by Topic

### Types of Solids and Their Structures

27. An X-ray beam with $\lambda = 154$ pm incident on the surface of a crystal produced a maximum reflection at an angle of $\theta = 28.3°$. Assuming $n = 1$, calculate the separation between layers of atoms in the crystal.

28. An X-ray beam of unknown wavelength is diffracted from a NaCl surface. If the interplanar distance in the crystal is 286 pm, and the angle of maximum reflection is found to be 7.23°, what is the wavelength of the X-ray beam? (Assume $n = 1$.)

**29.** Determine the number of atoms per unit cell for each metal.

(a) Polonium      (b) Tungsten

(c) Nickel

**30.** Determine the coordination number for each structure.

(a) Gold

(b) Ruthenium

(c) Chromium

**31.** Calculate the packing efficiency of the body-centered cubic unit cell. Show your work.

**32.** Calculate the packing efficiency of the face-centered cubic unit cell. Show your work.

**33.** Platinum crystallizes with the face-centered cubic unit cell. The radius of a platinum atom is 139 pm. Calculate the edge length of the unit cell and the density of platinum in g/cm$^3$.

**34.** Molybdenum crystallizes with the body-centered unit cell. The radius of a molybdenum atom is 136 pm. Calculate the edge length of the unit cell and the density of molybdenum.

**35.** Rhodium has a density of 12.41 g/cm$^3$ and crystallizes with the face-centered cubic unit cell. Calculate the radius of a rhodium atom.

**36.** Barium has a density of 3.59 g/cm$^3$ and crystallizes with the body-centered cubic unit cell. Calculate the radius of a barium atom.

**37.** Polonium crystallizes with a simple cubic structure. It has a density of 9.3 g/cm$^3$, a radius of 167 pm, and a molar mass of 209 g/mol. Use these data to calculate Avogadro's number (the number of atoms in one mole).

**38.** Palladium crystallizes with a face-centered cubic structure. It has a density of 12.0 g/cm$^3$, a radius of 138 pm, and a molar mass of 106.42 g/mol. Use these data to calculate Avogadro's number.

**39.** Identify each solid as molecular, ionic, or atomic.
  **a.** Ar($s$)
  **b.** H$_2$O($s$)
  **c.** K$_2$O($s$)
  **d.** Fe($s$)

**40.** Identify each solid as molecular, ionic, or atomic.
  **a.** CaCl$_2$($s$)
  **b.** CO$_2$($s$)
  **c.** Ni($s$)
  **d.** I$_2$($s$)

**41.** Which solid has the highest melting point? Why?

$$\text{Ar}(s), \text{CCl}_4(s), \text{LiCl}(s), \text{CH}_3\text{OH}(s)$$

**42.** Which solid has the highest melting point? Why?

$$\text{C}(s, \text{diamond}), \text{Kr}(s), \text{NaCl}(s), \text{H}_2\text{O}(s)$$

**43.** Which solid in each pair has the higher melting point and why?
  **a.** TiO$_2$($s$) or HOOH($s$)
  **b.** CCl$_4$($s$) or SiCl$_4$($s$)
  **c.** Kr($s$) or Xe($s$)
  **d.** NaCl($s$) or CaO($s$)

**44.** Which solid in each pair has the higher melting point and why?
  **a.** Fe($s$) or CCl$_4$($s$)
  **b.** KCl($s$) or HCl($s$)
  **c.** Ti($s$) or Ne($s$)
  **d.** H$_2$O($s$) or H$_2$S($s$)

**45.** An oxide of titanium crystallizes with the unit cell shown here (titanium = gray; oxygen = red). What is the formula of the oxide?

**46.** An oxide of rhenium crystallizes with the unit cell shown here (rhenium=gray; oxygen=red). What is the formula of the oxide?

**47.** The unit cells for cesium chloride and barium chloride are shown here. Show that the ratio of cations to anions in each unit cell corresponds to the ratio of cations to anions in the formula of each compound.

Cesium chloride          Barium chloride

**48.** The unit cells for lithium oxide and silver iodide are shown here. Show that the ratio of cations to anions in each unit cell corresponds to the ratio of cations to anions in the formula of each compound.

Lithium oxide          Silver iodide

**49.** Identify the structure of each of the two unit cells shown in Problem 47 as the rock salt structure, zinc blende structure, fluorite structure, antifluorite structure, or none of these.

**50.** Identify the structure of each of the two unit cells shown in Problem 48 as the rock salt structure, zinc blende structure, fluorite structure, antifluorite structure, or none of these.

**51.** Consider the rock salt structure in Figure 12.15. What type of structure would result if all the anions were somehow removed, leaving only cations?

**52.** Consider the zinc blende structure in Figure 12.16. What type of structure would result if the remaining tetrahedral sites in the unit cell were also filled with cations?

### Ceramics, Cement, and Glass

**53.** Classify each of the following as a component of a silicate ceramic, an oxide ceramic, or a nonoxide ceramic.
  **a.** $B_4C$      **b.** $Mg_2SiO_4$      **c.** $MoSi_2$

**54.** Classify each of the following as a component of a silicate ceramic, an oxide ceramic, or a nonoxide ceramic.
  **a.** $TiB_2$
  **b.** $ZrO_2$
  **c.** $NaAlSi_3O_8$

**55.** What are the name and formula of the compound commonly used in the manufacture of glass to reduce its tendency to crack or shatter under thermal shock?

**56.** What are the name and formula of the compound commonly used in the manufacture of glass to increase the index of refraction?

**57.** One of the key components in the manufacture of Portland cement is $Ca_3SiO_5$, a compound that is obtained by firing the reactants in a kiln at 1400–1500 °C. Assign an oxidation state to each element in this compound.

**58.** Replacement of aluminum ions in kaolinite with magnesium ions yields a compound with the formula $Mg_3Si_2O_5(OH)_4$. Assign an oxidation state to each element in this compound.

### Semiconductors and Band Theory

**59.** Which solid would you expect to have little or no band gap?
  **a.** $Zn(s)$
  **b.** $Si(s)$
  **c.** $As(s)$

**60.** Which solid would you expect to have the largest band gap?
  **a.** $As(s)$
  **b.** $Sb(s)$
  **c.** $Bi(s)$

**61.** How many molecular orbitals are present in the conduction band of a lithium crystal with a mass of 11.2 g?

**62.** How many molecular orbitals are present in the valence band of a sodium crystal with a mass of 5.45 g?

**63.** A substance has a band gap of 6.9 eV at 273 K. Is this substance best classified as an insulator, a semiconductor, or a metal?

**64.** A substance has a band gap of 0.85 eV at 273 K. Is this substance best classified as an insulator, a semiconductor, or a metal?

**65.** Indicate if each solid forms an n-type or a p-type semiconductor.
  **a.** germanium doped with gallium
  **b.** silicon doped with arsenic

**66.** Indicate if each solid forms an n-type or a p-type semiconductor.
  **a.** silicon doped with gallium
  **b.** germanium doped with antimony

**67.** Does a photon of red light with a frequency of $4.29 \times 10^{14}$ Hz have sufficient energy to promote an electron from the valence band to the conduction band in a sample of silicon (the band gap in silicon is 1.11 eV)?

**68.** Which wavelength of light (in nm) is emitted if an electron moves from the conduction band to the valence band in a sample of diamond (diamond has a band gap of 5.5 eV)?

### Polymers

**69.** Teflon is an addition polymer formed from the monomer shown here. Draw the structure of the polymer.

$$F_2C=CF_2$$

**70.** Saran, the polymer used to make saran wrap, is an addition polymer formed from two monomers—vinylidene chloride and vinyl chloride. Draw the structure of the polymer. (Hint: The monomers alternate.)

Vinylidene chloride      Vinyl chloride

**71.** One kind of polyester is a condensation copolymer formed from terephthalic acid and ethylene glycol. Draw the structure of the dimer. [Hint: Water (circled) is eliminated when the bond between the monomers forms.]

Terephthalic acid            Ethylene glycol

**72.** Nomex, a condensation copolymer used by firefighters because of its flame-resistant properties, forms from isophthalic acid and *m*-aminoaniline. Draw the structure of the dimer. (Hint: Water is eliminated when the bond between the monomers forms.)

Isophthalic acid            *m*-Aminoaniline

**73.** Polyacetylene is an addition polymer with the structure shown here. Draw the structure of the monomer.

**74.** Polyacrylonitrile (PAN) is an addition polymer with the structure shown here. Draw the structure of the monomer.

**75.** The polycarbonate Lexan is a condensation polymer and has the structure shown here

Assuming this polymer was formed from the monomer shown here plus another monomer, provide a possible structure for the other monomer that would be required to form Lexan.

**76.** Consider the condensation polymer with the structure shown here:

Assuming this polymer was formed from the monomer shown below plus another monomer, provide the structure of the other monomer that would be required to form this polymer.

# Cumulative Problems

**77.** Silver iodide crystallizes in the zinc blende structure. The separation between nearest-neighbor cations and anions is approximately 325 pm, and the melting point is 558 °C. Cesium chloride, by contrast, crystallizes in the structure shown in Figure 12.14. Even though the separation between nearest-neighbor cations and anions is greater (348 pm), the melting point of cesium chloride is higher (645 °C). Explain.

**78.** Copper iodide crystallizes in the zinc blende structure. The separation between nearest neighbor cations and anions is approximately 311 pm, and the melting point is 606 °C. Potassium chloride, by contrast, crystallizes in the rock salt structure. Even though the separation between nearest-neighbor cations and anions is greater (319 pm), the melting point of potassium chloride is higher (776 °C). Explain.

**79.** Consider the face-centered cubic structure shown here:

**a.** What is the length of the line (labeled *c*) that runs diagonally across one of the faces of the cube in terms of *r* (the atomic radius)?

**b.** Use the answer to part a and the Pythagorean theorem to derive the expression for the edge length (*l*) in terms of *r*.

**80.** Consider the body-centered cubic structure shown here:

a. What is the length of the line (labeled $c$) that runs from one corner of the cube diagonally through the center of the cube to the other corner in terms of $r$ (the atomic radius)?

b. Use the Pythagorean theorem to derive an expression for the length of the line (labeled $b$) that runs diagonally across one of the faces of the cube in terms of the edge length ($l$).

c. Use the answer to parts a and b along with the Pythagorean theorem to derive the expression for the edge length ($l$) in terms of $r$.

**81.** The volume of a unit cell of diamond is $0.0454 \text{ nm}^3$, and the density of diamond is $3.52 \text{ g/cm}^3$. Find the number of carbon atoms in a unit cell of diamond.

**82.** The density of an unknown metal is $12.3 \text{ g/cm}^3$, and its atomic radius is 0.134 nm. It has a face-centered cubic lattice. Find the atomic mass of this metal.

**83.** An unknown metal is found to have a density of $7.8748 \text{ g/cm}^3$ and to crystallize in a body-centered cubic lattice. The edge of the unit cell is 0.28664 nm. Calculate the atomic mass of the metal.

**84.** When spheres of radius $r$ are packed in a body-centered cubic arrangement, they occupy 68.0% of the available volume.

Use the fraction of occupied volume to calculate the value of $a$, the length of the edge of the cube, in terms of $r$.

**85.** Potassium chloride crystallizes in the rock salt structure. Estimate the density of potassium chloride using the ionic radii provided in Chapter 8.

**86.** Calculate the fraction of empty space in cubic closest packing to five significant figures.

**87.** A tetrahedral site in a closest-packed lattice is formed by four spheres at the corners of a regular tetrahedron. This is equivalent to placing the spheres at alternate corners of a cube. In such a closest-packed arrangement the spheres are in contact, and if the spheres have a radius $r$, the diagonal of the face of the cube is $2r$. The tetrahedral hole is inside the middle of the cube. Find the length of the body diagonal of this cube and then find the radius of the tetrahedral hole.

**88.** X-ray diffractometers often use metals that have had their core electrons excited as a source of X-rays. Consider the $2p \longrightarrow 1s$ transition for copper, which is called the $K\alpha$ transition. Calculate the wavelength of X-rays (in Å) given off by the $K\alpha$ transition if the energy given off by a mole of copper atoms is $7.77 \times 10^5 \text{ kJ}$. ($1 \text{ Å} = 10^{-10} \text{ m}$)

**89.** Why is it necessary to use the $K\alpha$ transition ($2p \longrightarrow 1s$) in copper (see Problem 88) to generate X-rays? Why not use, for example, the $4s \longrightarrow 3p$ transition?

**90.** In certain cases where X-ray diffraction is unsuitable for determining the structure of a crystal, *neutron diffraction* can be used. Instead of X-rays, a beam of neutrons is used to analyze the sample. Calculate the velocity of a beam of neutrons with a wavelength of 2.00 Å.

## Challenge Problems

**91.** The structure of the addition polymer polypropylene is shown in Table 12.3.

a. Draw the structure of the monomer.

b. Show how an alternate version of the polypropylene polymer (with a different arrangement) could be formed from the monomer you found in part a.

**92.** Perovskite is a compound with a cubic unit cell and has a strontium atom at the center of the cell, titanium atoms at the corners of the unit cell, and oxygen atoms at the centers of each face of the unit cell.

a. What is the formula of perovskite?

b. What is the coordination number of strontium in the perovskite structure?

c. If the edge length of the unit cell is 3.905 Å, calculate the density of perovskite in $\text{g/cm}^3$.

**93.** A compound with the formula $Rb_3C_{60}$ has been shown to demonstrate superconductivity below 30.0 K. Given that the $C_{60}$ molecules have a face-centered cubic arrangement, which of the tetrahedral and octahedral sites are occupied by Rb atoms?

**94.** Despite Dalton's laws, it is now known that many ionic compounds do not always contain atoms in small integer ratios. For example, a sample of iron(II) oxide may, in fact, contain a significant amount of $Fe^{3+}$ in addition to $Fe^{2+}$. This is an example of a *nonstoichiometric* compound. A sample of iron(II) oxide is found to be 75.65% iron by mass. Determine the percentage of $Fe^{3+}$ ions in the sample.

## Conceptual Problems

**95.** During the glass manufacturing process, the liquid must be cooled relatively quickly to form the glass. Why?

**96.** Why are X-rays used for crystallography? Why not use some other, more accessible type of electromagnetic radiation such as ultraviolet light?

**97.** Compare the crystal structure of diamond (C) and zinc blende (ZnS). What are the key similarities and differences between the two structures?

**98.** In X-ray crystallography, a small amount of oil is often used to mount the crystal sample on a glass fiber. Explain why neither

the oil nor the glass fiber interferes with the diffraction pattern of the crystal.

**99.** Which is not likely to lead to an increase in electrical conductivity?

a. Increasing the temperature of a semiconductor

b. Choosing a semiconductor with a smaller band gap

c. Doping the semiconductor

d. All of the above would likely lead to an increase in electrical conductivity.

# Questions for Group Work

Discuss these questions with the group and record your consensus answer.

**100.** Have each group member select one of the cubic crystalline lattices. Learn everything you can about your lattice and present it to the group.

**101.** Using group members to play the role of atoms or molecules, demonstrate the main structural features of the following categories of crystalline solids: molecular solid, ionic solid, nonbonded solid, metallic solid, network covalent solid. Clearly identify what you represent and how you are representing various interactions between particles (e.g., covalent bond, ionic bond, etc.).

**102.** Make a list of questions you would need to ask in order to classify a solid into one of the categories of crystalline solids (molecular solid, ionic solid, nonbonded solid, metallic solid, and network covalent solid). Determine a good order to ask them. (You may

need a branching decision tree.) Once you have agreed on a good set of questions, have each group member choose a substance from the chapter and then have the other group members ask the questions in turn until the correct classification is reached. You may agree to edit your questions if you discover ways to improve them when you are using your decision tree.

**103.** Have each group member select and study a material from the section on ceramics, cement, and glass. Take turns describing your material to the group, and see if they can identify the type of material based on your description (without consulting the text).

**104.** Describe how a common object or toy (e.g., a train, building blocks, or beads on a string) could represent the structure of a polymer. Describe how the following terms would be represented using your model: monomer, dimer, addition polymer, condensation polymer, branching.

# Data Interpretation and Analysis

## Magnesium Niobium Hydride as a Source of Hydrogen Fuel

**105.** The use of hydrogen as a fuel to power cars and trucks is an active research area. One challenge is storing the hydrogen, which is problematic because the steel fuel tanks are heavy and the hydrogen gas can explode in a crash.

Some metals have the ability to absorb hydrogen gas to form metal hydrides. For example, niobium and magnesium each react with hydrogen gas to form niobium hydride and magnesium hydride, respectively:

$$Nb(s) + H_2(g) \longrightarrow NbH_2(s) \quad \Delta H = -127 \text{ kJ/mol } H_2$$

$$Mg(s) + H_2(g) \longrightarrow MgH_2(s) \quad \Delta H = -77 \text{ kJ/mol } H_2$$

Metal hydrides make it possible to store a large amount of hydrogen in a small space. Once each of these reactions occurs, the hydrogen is stored in the solid metal hydride. After a metal hydride is formed, the process can be reversed.

Not all elements form hydrides. Table a lists some elements that form hydrides and some that do not.

Figure a ▶ shows the relationship of the reaction enthalpies as a function of hydrogen content for several hydrides.

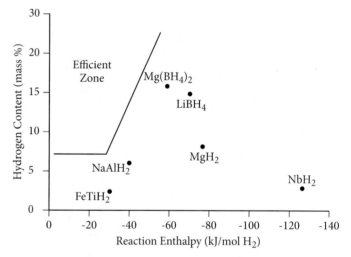

▲ **FIGURE a** **Hydrogen Content versus Reaction Enthalpy for Several Hydrides (at Room Temperature)**

**Elements that Form Hydrides**

| | | | | |
|---|---|---|---|---|
| Li | | | | |
| Na | Mg | | | |
| K | Ca | Sc | Ti | V |
| Rb | Sr | | | |

**Elements that do not Form Hydrides**

| | | | | |
|---|---|---|---|---|
| Cr | Fe | Co | Cu | Zn |
| Mo | Ru | Rh | Ag | Cd |

**TABLE a** **Some Elements That Form Hydrides and Some That Do Not.**

Recently, a research group synthesized magnesium niobium hydride ($Mg_{(0.75)}Nb_{(0.25)}H_2$). The hydrogen absorption measurements of magnesium niobium hydride yielded an enthalpy of hydride formation of $-53$ kJ/mol $H_2$.

Use the information provided in the figure and table to answer the following questions:

a. Identify a pattern for which elements form hydrides and which elements do not. (Hint: There are some exceptions to the pattern, but think about how you would name ionic compounds containing these metals.)

b. Refer to your knowledge about the size of atoms to offer an explanation why the research group decided to use two metals (magnesium and niobium) to form a complex hydride capable of storing hydrogen.

c. Calculate the mass percentage of hydrogen in magnesium niobium hydride ($Mg_{(0.75)}Nb_{(0.25)}H_2$).

   **d.** Does the hydrogen content and reaction enthalpy for magnesium niobium hydride fall in the efficient zone according to Figure a? Explain your answer.

   **e.** Would the metal hydrides discussed in this problem need to be heated or cooled to release hydrogen?

## Answers to Conceptual Connections

**Cubic Structures**

**12.1** The face-centered cubic structure has four atoms per unit cell.

**Crystalline Solid Types and Melting Points**

**12.2** **(a)** MgO, an ionic compound, has the highest melting point because the solid is held together by ionic bonds. $I_2$ is a molecular compound with only dispersion forces, and Kr is a nonbonding atomic solid with only dispersion forces. As a result, these two substances have much lower melting points.

**Ionic Crystalline Solid Unit Cells**

**12.3** **(c)** The zinc blende structure occurs in ionic compounds with a one-to-one cation-to-anion ratio in which the cation is much smaller than the anion.

**Phase Changes and Pressure**

**12.4** An increase in pressure favors the denser phase, in this case diamond.

**Semiconductor Type**

**12.5** **(a)** Since indium has three valence electrons and germanium has four, the presence of indium would introduce holes into the valence band, resulting in a p-type semiconductor.

# 13

# Solutions

> One molecule of nonsaline substance (held in the solvent) dissolved in 100 molecules of any volatile liquid decreases the vapor pressure of this liquid by a nearly constant fraction, nearly 0.0105.
>
> —FRANÇOIS-MARIE RAOULT (1830–1901)

We learned in Chapter 1 that most of the matter we encounter is in the form of mixtures. In this chapter, we focus on homogeneous mixtures, known as solutions. Solutions are mixtures in which atoms and molecules intermingle on the molecular and atomic scale. Common examples of solutions include ocean water, gasoline, and air. Why do solutions form? How are their properties different from the properties of the pure substances that compose them? As you read this chapter, keep in mind the large number of solutions that surround you at every moment, including those that exist within your own body.

▲ *Drinking seawater causes dehydration because seawater draws water out of body tissues.*

## 13.1 Thirsty Solutions: Why You Shouldn't Drink Seawater

In the popular novel *Life of Pi* by Yann Martel, the main character (whose name is Pi) is stranded on a lifeboat with a Bengal tiger in the middle of the Pacific Ocean for 227 days. Pi survives in part by rigging a solar still to distill seawater for drinking. However, in the first three days of his predicament (before he rigs the still) he becomes severely dehydrated from lack of water. He is surrounded by seawater, but drinking *that* water would only have made his condition worse. Why? Seawater actually draws water *out of the body* as it passes through the stomach and intestines, resulting in diarrhea and further dehydration. We can think of seawater as a *thirsty solution*—one that draws more water to itself. Consequently, seawater should never be consumed as drinking water.

In some cases, the concepts of solute and solvent are not useful. For example, a homogeneous mixture of water and ethanol can contain equal amounts of both components, and neither component can then be identified as the solvent.

Seawater is a **solution**, a homogeneous mixture of two or more substances or components as shown in Figure 13.1▼. The majority component in a solution is typically called the **solvent**, and the minority component is called the **solute**. In seawater, water is the solvent, and sodium chloride is the primary solute. Solutions form in part because of the intermolecular forces we discussed in Chapter 11. In most solutions, the particles of the solute interact with the particles of the solvent through intermolecular forces.

▶ **FIGURE 13.1** **A Typical Solution** In seawater, sodium chloride is the primary solute. Water is the solvent.

H$_2$O **Solvent**

Na$^+$
Cl$^-$ **Solute**

The reason that seawater draws water to itself is related to nature's tendency toward spontaneous mixing, which we discuss in more detail later in this chapter and in Chapter 18. For now, we simply observe that unless the mixing process is highly unfavorable energetically, substances tend to combine into uniform mixtures, not separate into pure substances. For example, suppose pure water and a sodium chloride solution are in separate compartments with a removable barrier between them, as shown in Figure 13.2(a)▼. If we remove the barrier, the two liquids spontaneously mix, eventually forming a more dilute sodium chloride solution of uniform concentration, as shown in Figure 13.2(b)▼. The tendency toward mixing results in a uniform concentration of the final solution.

Seawater is a *thirsty* solution because of this tendency toward mixing. As seawater moves through the intestine, it flows past cells that line the digestive tract. These cells consist of largely fluid interiors surrounded by membranes. Cellular fluids themselves contain dissolved ions, including sodium and chloride, but the fluids in the cells are more dilute than seawater. Nature's tendency toward mixing (which tends to produce solutions of uniform concentration), together with the selective permeability of the cell

### Spontaneous Mixing

When the barrier is removed, spontaneous mixing occurs, producing a solution of uniform concentration.

Concentration difference

Uniform concentration

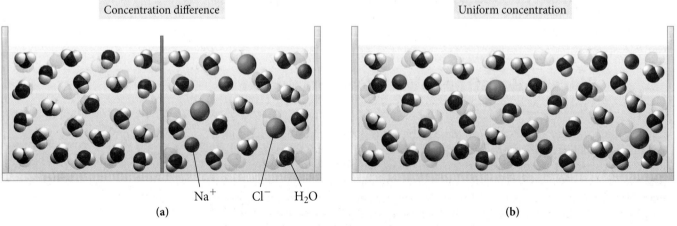

Na$^+$     Cl$^-$   H$_2$O

(a)

(b)

▲ **FIGURE 13.2** **The Tendency to Mix** **(a)** A barrier separates pure water and a sodium chloride solution. **(b)** When the barrier is removed, the two liquids spontaneously mix, producing a single solution of uniform concentration.

membranes (which allows water to flow in and out, but restricts the flow of dissolved solids), cause a *flow of solvent out of the body's cells into the seawater*. In this way, the two solutions become more similar in concentration (as though they had mixed). The solution in the intestine becomes somewhat more dilute than it was, and the solution in the cells becomes somewhat more concentrated. The accumulation of extra fluid in the intestines causes diarrhea, and the decreased fluid in the cells causes dehydration. If Pi had drunk the seawater instead of constructing the solar still, neither he nor his companion, the large Bengal tiger, would have survived their ordeal.

Interior of body cells
–dilute solution

Seawater in intestinal tract
–concentrated solution

Direction of water flow

◀ Seawater is a more concentrated solution than the fluids in body cells. As a result, when seawater flows through the digestive tract, it draws water out of the surrounding tissues.

# 13.2 Types of Solutions and Solubility

A solution may be composed of a solid and a liquid (such as the salt and water that are the primary components of seawater), but it may also be composed of a gas and a liquid, two different liquids, or other combinations (see Table 13.1). In **aqueous solutions**, water is the solvent, and a solid, liquid, or gas is the solute. For example, sugar water and salt water are both aqueous solutions. Similarly, ethyl alcohol—the alcohol in alcoholic beverages—readily mixes with water to form a solution, and carbon dioxide dissolves in water to form the aqueous solution that we know as club soda.

$CO_2$    $H_2O$

▲ Club soda is a solution of carbon dioxide and water.

| TABLE 13.1 Common Types of Solutions | | | |
|---|---|---|---|
| Solution Phase | Solute Phase | Solvent Phase | Example |
| Gaseous solution | Gas | Gas | Air (mainly oxygen and nitrogen) |
| Liquid solution | Gas | Liquid | Club soda ($CO_2$ and water) |
| | Liquid | Liquid | Vodka (ethanol and water) |
| | Solid | Liquid | Seawater (salt and water) |
| Solid solution | Solid | Solid | Brass (copper and zinc) and other alloys |

You probably know from experience that a particular solvent, such as water, does not dissolve all possible solutes. For example, you cannot clean your greasy hands with water alone because the water does not dissolve the grease. However, another solvent, such as paint thinner, can easily dissolve the grease. The grease is *insoluble* in water but *soluble* in the paint thinner. The **solubility** of a substance is the amount of the substance that will dissolve in a given amount of solvent. The solubility of sodium chloride in water at 25 °C is 36 g NaCl per 100 g water, whereas the solubility of grease in water is nearly zero. The solubility of one substance in another depends both on the tendency toward mixing that we discussed in Section 13.1 and on the types of intermolecular forces that we discussed in Chapter 11.

The general solubilities of a number of ionic compounds are described by the solubility rules in Section 4.5.

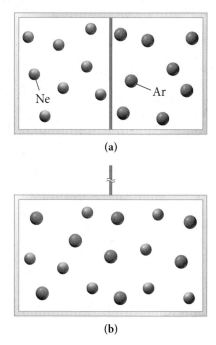

**(a)**

**(b)**

▲ **FIGURE 13.3 Spontaneous Mixing of Two Ideal Gases** **(a)** A barrier separates neon and argon. **(b)** When the barrier is removed, the two gases spontaneously mix to form a uniform solution.

The potential energy of the gas particles does not change upon mixing because the gas is ideal; the particles do not exert any attractive or repulsive forces on each other, so mixing does not change their potential energy.

# Nature's Tendency toward Mixing: Entropy

So far in this book, we have seen that many physical systems tend toward lower *potential energy*. For example, two particles with opposite charges (such as a proton and an electron or a cation and an anion) move toward each other because their potential energy decreases as their separation decreases according to Coulomb's law. The formation of a solution, however, *does not necessarily* lower the potential energy of its constituent particles. The clearest example of this phenomenon is the formation of a homogeneous mixture (a *solution*) of two ideal gases. Suppose that we enclose neon and argon in a container with a removable barrier between them, as shown in Figure 13.3(a)◄. As soon as we remove the barrier, the neon and argon mix together to form a solution, as shown in Figure 13.3(b)◄. *Why?*

At low pressures and moderate temperatures, both neon and argon behave as ideal gases—they do not interact with each other in any way (that is, there are no significant intermolecular forces between their constituent particles). When the barrier is removed, the two gases mix, but their potential energy remains unchanged. In other words, *we cannot think of the mixing of two ideal gases as lowering their potential energy*. Rather, the tendency to mix is related to a concept called *entropy*.

**Entropy** is a measure of *energy randomization* or *energy dispersal* in a system. Recall that a gas at any temperature above 0 K has kinetic energy due to the motion of its atoms. When neon and argon are confined to their individual compartments, their kinetic energies are also confined to those compartments. When the barrier between the compartments is removed, each gas—along with its kinetic energy—becomes *spread out* or *dispersed* over a larger volume. Thus, the mixture of the two gases has greater energy dispersal, or greater *entropy*, than the separated components.

*The pervasive tendency for energy to spread out, or disperse, whenever it is not restrained from doing so* is the reason that two ideal gases mix. Another common example of the tendency toward energy dispersal is the transfer of thermal energy from hot to cold. If we heat one end of an iron rod, the thermal energy deposited at the end of the rod will spontaneously spread along the entire length of the rod. In contrast to the mixing of two ideal gases—where the kinetic energy of the particles becomes dispersed over a larger volume because the particles themselves become dispersed—the thermal energy in the rod, initially concentrated in relatively fewer particles, becomes dispersed by being distributed over a larger number of particles. The tendency for energy to disperse is why thermal energy flows from the hot end of the rod to the cold one, and not the other way around. Imagine a metal rod that became spontaneously hotter on one end and ice cold on the other—this does not happen because energy does not spontaneously concentrate itself. In Chapter 18, we will see that the dispersal of energy is actually the fundamental criterion that ultimately determines the spontaneity of any process.

# The Effect of Intermolecular Forces

We have just seen that, in the absence of intermolecular forces, two substances spontaneously mix to form a homogeneous solution. We know from Chapter 11, however, that solids and liquids exhibit a number of different types of intermolecular forces including dispersion forces, dipole–dipole forces, hydrogen bonding, and ion–dipole forces (Figure 13.4►). These forces may promote the formation of a solution or prevent it, depending on the nature of the forces in the particular combination of solute and solvent.

Intermolecular forces exist between (a) the solvent and solute particles, (b) the solvent particles themselves, and (c) the solute particles themselves, as shown in Figure 13.5►.

| | |
|---|---|
| *Solvent–solute interactions*: | The interactions between a solvent particle and a solute particle. |
| *Solvent–solvent interactions*: | The interactions between a solvent particle and another solvent particle. |
| *Solute–solute interactions*: | The interactions between a solute particle and another solute particle. |

## Intermolecular Forces

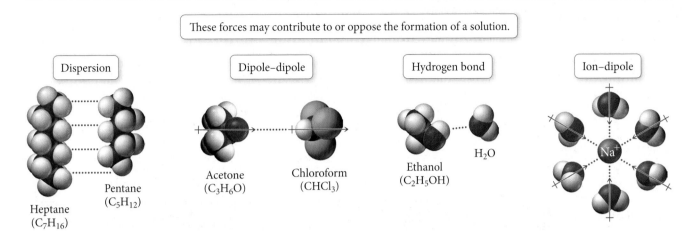

These forces may contribute to or oppose the formation of a solution.

Dispersion

Heptane
($C_7H_{16}$)

Pentane
($C_5H_{12}$)

Dipole–dipole

Acetone
($C_3H_6O$)

Chloroform
($CHCl_3$)

Hydrogen bond

Ethanol
($C_2H_5OH$)

$H_2O$

Ion–dipole

$Na^+$

▲ **FIGURE 13.4 Intermolecular Forces Involved in Solutions**

As Table 13.2 indicates, a solution always forms if the solvent–solute interactions are comparable to, or stronger than, the solvent–solvent interactions and the solute–solute interactions. For example, consider mixing the hydrocarbons pentane ($C_5H_{12}$) and heptane ($C_7H_{16}$). The intermolecular forces present *within* both pentane and heptane are dispersion forces. Similarly, the intermolecular forces present *between* heptane and pentane are also dispersion forces. All three interactions are of similar magnitude, so the two substances are soluble in each other in all proportions—they are said to be **miscible**. The formation of the solution is driven by the tendency toward mixing, or toward greater entropy, that we just discussed.

**TABLE 13.2 Relative Interactions and Solution Formation**

| Solvent–solute interactions | > | Solvent–solvent and solute–solute interactions | Solution forms |
|---|---|---|---|
| Solvent–solute interactions | = | Solvent–solvent and solute–solute interactions | Solution forms |
| Solvent–solute interactions | < | Solvent–solvent and solute–solute interactions | Solution may or may not form, depending on relative disparity |

If solvent–solute interactions are weaker than solvent–solvent and solute–solute interactions—in other words, if solvent molecules and solute molecules each interact more strongly with molecules of their own kind than with molecules of the other kind—then a solution may still form, depending on the relative disparities between the interactions. If the disparity is small, the tendency to mix results in the formation of a solution even though the process is energetically uphill. If the disparity is large, however, a solution does not form. For example, consider mixing hexane and water. The water molecules have strong hydrogen-bonding attractions to each other but cannot form hydrogen bonds with hexane. The energy required to pull water molecules away

## Solution Interactions

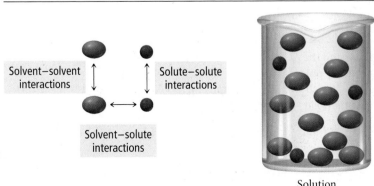

Solvent–solvent interactions

Solute–solute interactions

Solvent–solute interactions

Solution

◀ **FIGURE 13.5 Forces in a Solution** The relative strengths of these three interactions determine whether a solution forms.

## TABLE 13.3 Common Laboratory Solvents

| Common Polar Solvents | Common Nonpolar Solvents |
|---|---|
| Water ($H_2O$) | Hexane ($C_6H_{14}$) |
| Acetone ($CH_3COCH_3$) | Diethyl ether ($CH_3CH_2OCH_2CH_3$)* |
| Methanol ($CH_3OH$) | Toluene ($C_7H_8$) |
| Ethanol ($CH_3CH_2OH$) | Carbon tetrachloride ($CCl_4$) |

*Diethyl ether has a small dipole moment and can be considered intermediate between polar and nonpolar.

from one another is too great, and too little energy is returned when the water molecules interact with hexane molecules. As a result, a solution does not form when we mix hexane and water. Although the tendency to mix is strong, it cannot overcome the large energy disparity between the powerful solvent–solvent interactions and the weak solvent–solute interactions.

In general, we can use the rule of thumb that *like dissolves like* when predicting the formation of solutions. Similar kinds of solvents dissolve similar kinds of solutes. Polar solvents, such as water, tend to dissolve many polar or ionic solutes, and nonpolar solvents, such as hexane, tend to dissolve many nonpolar solutes. Table 13.3 lists some common polar and nonpolar laboratory solvents.

## Example 13.1 Solubility

Vitamins are often categorized as either fat-soluble or water-soluble. Water-soluble vitamins, on one hand, dissolve in body fluids and are easily eliminated in the urine, so there is little danger of overconsumption. Fat-soluble vitamins, on the other hand, can accumulate in the body's fatty deposits. Overconsumption of a fat-soluble vitamin can be dangerous to your health. Examine the structure of the vitamins shown here and classify each as either fat-soluble or water-soluble.

**(a)** Vitamin C

**(b)** Vitamin $K_3$

**(c)** Vitamin A

**(d)** Vitamin $B_5$

### SOLUTION

**(a)** The four —OH bonds in vitamin C make it highly polar and allow it to hydrogen-bond with water. Vitamin C is water-soluble.

**(b)** The C—C bonds in vitamin $K_3$ are nonpolar and the C—H bonds are nearly so. The C=O bonds are polar, but the bond dipoles oppose and largely cancel each other, so the molecule is dominated by the nonpolar bonds. Vitamin $K_3$ is fat-soluble.

**(c)** The C—C bonds in vitamin A are nonpolar and the C—H bonds are nearly so. The one polar —OH bond may increase its water solubility slightly, but overall vitamin A is nonpolar and therefore fat-soluble.

**(d)** The three —OH bonds and one —NH bond in vitamin $B_5$ make it highly polar and allow it to hydrogen-bond with water. Vitamin $B_5$ is water-soluble.

**FOR PRACTICE 13.1**   Determine whether each compound is soluble in hexane.

**(a)** water ($H_2O$)

**(b)** propane ($CH_3CH_2CH_3$)

**(c)** ammonia ($NH_3$)

**(d)** hydrogen chloride (HCl)

**Solubility**   Consider the table listing the solubilities of several alcohols in water and in hexane. Explain the observed trend in terms of intermolecular forces.

CONCEPTUAL
**CONNECTION 13.1**

| Alcohol | Space-Filling Model | Solubility in $H_2O$ (mol alcohol/100 g $H_2O$) | Solubility in Hexane ($C_6H_{14}$) (mol alcohol/100 g $C_6H_{14}$) |
|---|---|---|---|
| Methanol ($CH_3OH$) | | Miscible | 0.12 |
| Ethanol ($CH_3CH_2OH$) | | Miscible | Miscible |
| Propanol ($CH_3CH_2CH_2OH$) | | Miscible | Miscible |
| Butanol ($CH_3CH_2CH_2CH_2OH$) | | 0.11 | Miscible |
| Pentanol ($CH_3CH_2CH_2CH_2CH_2OH$) | | 0.030 | Miscible |

# 13.3   Energetics of Solution Formation

In Chapter 6, we examined the energy changes associated with chemical reactions. Similar energy changes can occur when a solution forms, depending on the magnitude of the interactions between the solute and solvent particles. For example, when we dissolve sodium hydroxide in water, heat is evolved—the solution process is *exothermic*. In contrast, when we dissolve ammonium nitrate ($NH_4NO_3$) in water, heat is absorbed—this solution process is *endothermic*. Other solutions, such as sodium chloride in water, barely absorb or evolve heat upon formation. What causes these different behaviors?

## Energy Changes in Solution Formation

We can understand the energy changes associated with solution formation by envisioning the process as occurring in three steps, each with an associated change in enthalpy:

**1.** Separating the solute into its constituent particles.

This step is always endothermic (positive $\Delta H$) because energy is required to overcome the forces that hold the solute particles together.

**2.** Separating the solvent particles from each other to make room for the solute particles.

This step is also endothermic because energy is required to overcome the intermolecular forces among the solvent particles.

**3.** Mixing the solute particles with the solvent particles.

This step is exothermic because energy is released as the solute particles interact (through intermolecular forces) with the solvent particles.

According to Hess's law, the overall enthalpy change upon solution formation, called the **enthalpy of solution ($\Delta H_{soln}$)**, is the sum of the changes in enthalpy for each step:

$$\Delta H_{soln} = \underset{\text{endothermic (+)}}{\Delta H_{solute}} + \underset{\text{endothermic (+)}}{\Delta H_{solvent}} + \underset{\text{exothermic (−)}}{\Delta H_{mix}}$$

Since the first two terms are endothermic (positive $\Delta H$) and the third term is exothermic (negative $\Delta H$), the overall sign of $\Delta H_{soln}$ depends on the magnitudes of the individual terms, as shown in Figure 13.6▶.

**1.** *If the sum of the endothermic terms is approximately equal in magnitude to the exothermic term, then $\Delta H_{soln}$ is about zero.* The increasing entropy upon mixing drives the formation of a solution while the overall energy of the system remains nearly constant.

**2.** *If the sum of the endothermic terms is smaller in magnitude than the exothermic term, then $\Delta H_{soln}$ is negative and the solution process is exothermic.* In this case, both the tendency toward lower energy and the tendency toward greater entropy drive the formation of a solution.

**3.** *If the sum of the endothermic terms is greater in magnitude than the exothermic term, then $\Delta H_{soln}$ is positive and the solution process is endothermic.* In this case, as long as $\Delta H_{soln}$ is not too large, the tendency toward greater entropy still drives the formation of a solution. If, on the other hand, $\Delta H_{soln}$ is too large, a solution does not form.

## Energetics of Solution Formation

(a) Exothermic

(b) Endothermic

▲ **FIGURE 13.6 Energetics of the Solution Process** **(a)** When $\Delta H_{mix}$ is greater in magnitude than the sum of $\Delta H_{solute}$ and $\Delta H_{solvent}$, the heat of solution is negative (exothermic). **(b)** When $\Delta H_{mix}$ is smaller in magnitude than the sum of $\Delta H_{solute}$ and $\Delta H_{solvent}$, the heat of solution is positive (endothermic).

## Aqueous Solutions and Heats of Hydration

Many common solutions, such as the seawater mentioned in Section 13.1, contain an ionic compound dissolved in water. In these aqueous solutions, $\Delta H_{solvent}$ and $\Delta H_{mix}$ can be combined into a single term called the **heat of hydration ($\Delta H_{hydration}$)** (Figure 13.7▼).

## Heat of Hydration

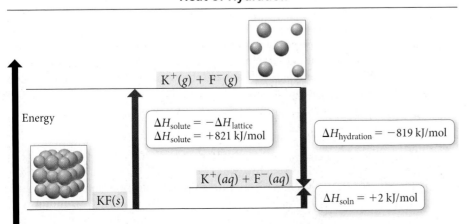

◀ **FIGURE 13.7 Heat of Hydration and Heat of Solution** The heat of hydration is the heat emitted when 1 mol of gaseous solute ions is dissolved in water. The sum of the negative of the lattice energy (which is $\Delta H_{solute}$) and the heat of hydration is the heat of solution.

## Ion–Dipole Interactions

KF solution

The heat of hydration is the enthalpy change that occurs when 1 mol of the gaseous solute ions is dissolved in water. Because the ion–dipole interactions that occur between a dissolved ion and the surrounding water molecules (Figure 13.8▲) are much stronger than the hydrogen bonds in water, $\Delta H_{\text{hydration}}$ is always largely negative (exothermic) for ionic compounds. Using the heat of hydration, we can write the enthalpy of solution as a sum of just two terms, one endothermic and one exothermic:

$$\Delta H_{\text{soln}} = \Delta H_{\text{solute}} + \underbrace{\Delta H_{\text{solvent}} + \Delta H_{\text{mix}}}$$

$$\Delta H_{\text{soln}} = \underset{\substack{\text{endothermic}\\(\text{positive})}}{\Delta H_{\text{solute}}} + \underset{\substack{\text{exothermic}\\(\text{negative})}}{\Delta H_{\text{hydration}}}$$

For ionic compounds, $\Delta H_{\text{solute}}$, the energy required to separate the solute into its constituent particles, is the negative of the solute's lattice energy ($\Delta H_{\text{solute}} = -\Delta H_{\text{lattice}}$), discussed in Section 9.4. For ionic aqueous solutions, then, the overall enthalpy of solution depends on the relative magnitudes of $\Delta H_{\text{solute}}$ and $\Delta H_{\text{hydration}}$, with three possible scenarios (in each case we refer to the *magnitude (absolute value)* of $\Delta H$):

1. **$|\Delta H_{\text{solute}}| < |\Delta H_{\text{hydration}}|$**. The amount of energy required to separate the solute into its constituent ions is less than the energy given off when the ions are hydrated. $\Delta H_{\text{soln}}$ is therefore negative, and the solution process is exothermic. Solutes with negative enthalpies of solution include lithium bromide and potassium hydroxide. When these solutes dissolve in water, the resulting solutions feel warm to the touch:

$$\text{LiBr}(s) \xrightarrow{\text{H}_2\text{O}} \text{Li}^+(aq) + \text{Br}^-(aq) \quad \Delta H_{\text{soln}} = -48.78 \text{ kJ/mol}$$
$$\text{KOH}(s) \xrightarrow{\text{H}_2\text{O}} \text{K}^+(aq) + \text{OH}^-(aq) \quad \Delta H_{\text{soln}} = -57.56 \text{ kJ/mol}$$

2. **$|\Delta H_{\text{solute}}| > |\Delta H_{\text{hydration}}|$**. The amount of energy required to separate the solute into its constituent ions is greater than the energy given off when the ions are hydrated. $\Delta H_{\text{soln}}$ is therefore positive, and the solution process is endothermic (if a solution forms at all). Solutes that form aqueous solutions with positive enthalpies of solution include ammonium nitrate and silver nitrate. When these solutes dissolve in water, the resulting solutions feel cool to the touch:

$$\text{NH}_4\text{NO}_3(s) \xrightarrow{\text{H}_2\text{O}} \text{NH}_4^+(aq) + \text{NO}_3^-(aq) \quad \Delta H_{\text{soln}} = +25.67 \text{ kJ/mol}$$
$$\text{AgNO}_3(s) \xrightarrow{\text{H}_2\text{O}} \text{Ag}^+(aq) + \text{NO}_3^-(aq) \quad \Delta H_{\text{soln}} = +36.91 \text{ kJ/mol}$$

3. **$|\Delta H_{\text{solute}}| \approx |\Delta H_{\text{hydration}}|$**. The amount of energy required to separate the solute into its constituent ions is roughly equal to the energy given off when the ions are hydrated. $\Delta H_{\text{soln}}$ is therefore approximately zero, and the solution process is

neither appreciably exothermic nor appreciably endothermic. Solutes with enthalpies of solution near zero include sodium chloride and sodium fluoride. When these solutes dissolve in water, the resulting solutions do not undergo a noticeable change in temperature:

$$NaCl(s) \xrightarrow{H_2O} Na^+(aq) + Cl^-(aq) \quad \Delta H_{soln} = +3.88 \text{ kJ/mol}$$

$$NaF(s) \xrightarrow{H_2O} Na^+(aq) + F^-(aq) \quad \Delta H_{soln} = +0.91 \text{ kJ/mol}$$

**Energetics of Aqueous Solution Formation**   The enthalpy of solution for cesium fluoride is −36.8 kJ/mol. What can you conclude about the relative magnitudes of $\Delta H_{solute}$ and $\Delta H_{hydration}$?

CONCEPTUAL
CONNECTION **13.2**

# 13.4  Solution Equilibrium and Factors Affecting Solubility

KCV   **Solution Equilibrium and the Factors Affecting Solubility**

The dissolution of a solute in a solvent is an equilibrium process similar to the equilibrium process associated with a phase change (discussed in Chapter 11). Imagine, from a molecular viewpoint, the dissolution of a solid solute such as sodium chloride in a liquid solvent such as water (Figure 13.9▼). Initially, water molecules rapidly solvate sodium cations and chloride anions, resulting in a noticeable decrease in the amount of solid sodium chloride in the water. Over time, however, the concentration of dissolved

## Solution Equilibrium

| NaCl(s) | NaCl(s) $\longrightarrow$ Na$^+$(aq) + Cl$^-$(aq) | NaCl(s) $\rightleftharpoons$ Na$^+$(aq) + Cl$^-$(aq) |
| --- | --- | --- |
| When sodium chloride is first added to water, sodium and chloride ions begin to dissolve into the water. | As the solution becomes more concentrated, some of the sodium and chloride ions can begin to recrystallize as solid sodium chloride. | When the rate of dissolution equals the rate of recrystallization, dynamic equilibrium has been reached. |

**(a) Initial**

Rate of dissolution > Rate of recrystallization
**(b) Dissolving**

Rate of dissolution = Rate of recrystallization
**(c) Dynamic equilibrium**

▲ **FIGURE 13.9  Dissolution of NaCl   (a)** When sodium chloride is first added to water, sodium and chloride ions dissolve into the water. **(b)** As the solution becomes more concentrated, some of the sodium and chloride ions recrystallize as solid sodium chloride. **(c)** When the rate of dissolution equals the rate of recrystallization, dynamic equilibrium is reached.

▶ **FIGURE 13.10** **Precipitation from a Supersaturated Solution** When a small piece of solid sodium acetate is added to a supersaturated sodium acetate solution, the excess solid precipitates out of the solution.

sodium chloride in the solution increases. This dissolved sodium chloride then begins to recrystallize as solid sodium chloride. Initially the rate of dissolution far exceeds the rate of recrystallization, but as the concentration of dissolved sodium chloride increases, the rate of recrystallization also increases. Eventually, the rates of dissolution and recrystallization become equal—**dynamic equilibrium** has been reached.

$$NaCl(s) \rightleftharpoons Na^+(aq) + Cl^-(aq)$$

A solution in which the dissolved solute is in dynamic equilibrium with the solid (undissolved) solute is a **saturated solution**. *If you add additional solute to a saturated solution, it will not dissolve.* A solution containing less than the equilibrium amount of solute is an **unsaturated solution**. *If you add additional solute to an unsaturated solution, it will dissolve.*

Under certain circumstances, a **supersaturated solution**—one containing more than the equilibrium amount of solute—may form. Supersaturated solutions are unstable and the excess solute normally precipitates out of the solution. However, in some cases, if left undisturbed, a supersaturated solution can exist for an extended period of time. For example, in a common classroom demonstration, a tiny piece of solid sodium acetate is added to a supersaturated solution of sodium acetate. This triggers the precipitation of the solute, which crystallizes out of solution in a dramatic and often beautiful way (Figure 13.10▲).

## The Temperature Dependence of the Solubility of Solids

In the case of sugar dissolving in water, the higher temperature increases both *how fast* the sugar dissolves and *how much* sugar dissolves.

The solubility of solids in water can be highly dependent on temperature. Have you ever noticed how much more sugar you can dissolve in hot tea than in cold tea? Although exceptions exist, *the solubility of most solids in water increases with increasing temperature*, as shown in Figure 13.11▼. For example, the solubility of potassium nitrate ($KNO_3$) at

▶ **FIGURE 13.11** **Solubility and Temperature** The solubility of most solids increases with increasing temperature.

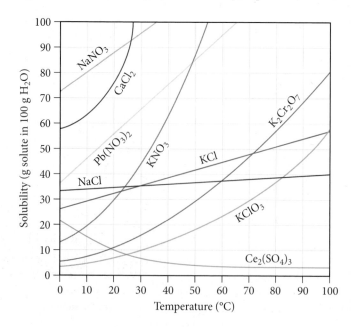

room temperature is about 37 g $KNO_3$ per 100 g of water. At 50 °C, the solubility rises to 88 g $KNO_3$ per 100 g of water.

A common way to purify a solid is a technique called **recrystallization**. In this technique, enough solid is added to water (or some other solvent) to create a saturated solution at an elevated temperature. As the solution cools, it becomes supersaturated and the excess solid precipitates out of solution. If the solution cools slowly, the solid forms crystals as it comes out of solution. The crystalline structure tends to reject impurities, resulting in a purer solid.

You can use the temperature dependence of the solubility of solids to make rock candy. Prepare a saturated sucrose (table sugar) solution at an elevated temperature, and allow a string or stick to dangle into the solution for several days. As the solution cools and the solvent evaporates, the solution becomes supersaturated and sugar crystals grow on the string or stick. After several days, beautiful edible crystals or "rocks" of sugar cover the string.

▲ Rock candy is formed by the recrystallization of sugar.

## Factors Affecting the Solubility of Gases in Water

Solutions of gases dissolved in water are common. Club soda, for example, is a solution of carbon dioxide and water, and most liquids exposed to air contain dissolved gases from air. Fish and other aquatic animals depend on the oxygen dissolved in lake or ocean water for life, and our blood contains dissolved nitrogen, oxygen, and carbon dioxide. Even tap water contains dissolved gases. The solubility of a gas in a liquid is affected by both temperature and pressure.

### The Effect of Temperature

You can observe the effect of temperature on the solubility of a gas in water by heating ordinary tap water on a stove. Before the water reaches its boiling point, small bubbles develop in the water. These bubbles are the dissolved air (mostly nitrogen and oxygen) coming out of solution. (Once the water boils, the bubbling becomes more vigorous— these larger bubbles are composed of water vapor.) The dissolved air comes out of the heated solution because—unlike solids, whose solubility generally increases with increasing temperature—*the solubility of gases in water decreases with increasing temperature.*

The inverse relationship between gas solubility and temperature is the reason that warm soda pop bubbles more than cold soda pop when you open it and warm beer goes flat faster than cold beer. More carbon dioxide comes out of these solutions at room temperature than at a lower temperature because the gas is less soluble at room temperature. The decreasing solubility of gases with increasing temperature is also the reason that fish don't bite much in a warm lake. The warm temperature results in a lower oxygen concentration. With lower oxygen levels, the fish become lethargic and do not strike as aggressively at the lure or bait you cast their way.

| Cold soda pop | Warm soda pop |

▲ Warm soda pop bubbles more than cold soda pop because carbon dioxide is less soluble in the warm solution.

**Solubility and Temperature**  A solution is saturated in both nitrogen gas and potassium bromide at 75 °C. When the solution is cooled to room temperature, what is most likely to happen?

  **(a)** Some nitrogen gas bubbles out of solution.
  **(b)** Some potassium bromide precipitates out of solution.
  **(c)** Some nitrogen gas bubbles out of solution, *and* some potassium bromide precipitates out of solution.
  **(d)** Nothing happens.

CONCEPTUAL
**CONNECTION 13.3**

### The Effect of Pressure

The solubility of gases also depends on pressure. The higher the pressure of a gas above a liquid, the more soluble the gas is in the liquid. In a sealed can of soda pop, for example, the carbon dioxide is maintained in solution by a high pressure of carbon dioxide within the can. When the can is opened, this pressure is released and the solubility of carbon dioxide decreases, resulting in bubbling (Figure 13.12▶, on the next page).

▶ **FIGURE 13.12** **Soda Fizz**
The bubbling that occurs when a can of soda is opened results from the reduced pressure of carbon dioxide over the liquid. At lower pressure, the carbon dioxide is less soluble and bubbles out of solution.

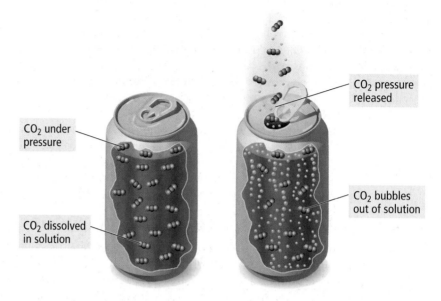

CO$_2$ pressure released

CO$_2$ under pressure

CO$_2$ bubbles out of solution

CO$_2$ dissolved in solution

The increased solubility of a gas in a liquid can be understood by considering cylinders containing water and carbon dioxide gas:

Equilibrium

Pressure is increased. More CO$_2$ dissolves.

Equilibrium is restored.

The first cylinder represents an equilibrium between gaseous and dissolved carbon dioxide—the rate of carbon dioxide molecules entering solution exactly equals the rate of molecules leaving the solution. Now imagine decreasing the volume, as we have done in the second cylinder. The pressure of carbon dioxide increases, causing the rate of molecules entering the solution to rise. The number of molecules in solution increases until equilibrium is established again, as shown in the third cylinder. However, the amount of carbon dioxide in solution in the third cylinder is greater.

We quantify the solubility of gases with increasing pressure with **Henry's law**:

$$S_{gas} = k_H P_{gas}$$

where $S_{gas}$ is the solubility of the gas (usually in M), $k_H$ is a constant of proportionality (called the *Henry's law constant*) that depends on the specific solute and solvent and also on temperature, and $P_{gas}$ is the partial pressure of the gas (usually in atm). The equation shows that the solubility of a gas in a liquid is directly proportional to the pressure of the gas above the liquid. Table 13.4 lists the Henry's law constants for several common gases.

**TABLE 13.4** **Henry's Law Constants for Several Gases in Water at 25 °C**

| Gas | $k_H$ (M/atm) |
|---|---|
| O$_2$ | $1.3 \times 10^{-3}$ |
| N$_2$ | $6.1 \times 10^{-4}$ |
| CO$_2$ | $3.4 \times 10^{-2}$ |
| NH$_3$ | $5.8 \times 10^{1}$ |
| He | $3.7 \times 10^{-4}$ |

**Henry's Law** Examine the Henry's law constants in Table 13.4. Why is the constant for ammonia larger than the others?

---

**Example** 13.2    **Henry's Law**

What pressure of carbon dioxide is required to keep the carbon dioxide concentration in a bottle of club soda at 0.12 M at 25 °C?

| | |
|---|---|
| **SORT** You are given the desired solubility of carbon dioxide and asked to find the pressure required to achieve this solubility. | **GIVEN:** $S_{CO_2} = 0.12\,M$ <br><br> **FIND:** $P_{CO_2}$ |
| **STRATEGIZE** Use Henry's law to find the required pressure from the solubility. You will need the Henry's law constant for carbon dioxide, which is listed in Table 13.4. | **CONCEPTUAL PLAN** <br><br> $\boxed{S_{CO_2}} \longrightarrow \boxed{P_{CO_2}}$ <br> $S_{CO_2} = k_{H,CO_2}P_{CO_2}$ <br><br> **RELATIONSHIPS USED** <br> $S_{gas} = k_H P_{gas}$ (Henry's law) <br> $k_{H,\,CO_2} = 3.4 \times 10^{-2}\,M/atm$ (from Table 13.4) |
| **SOLVE** Solve the Henry's law equation for $P_{CO_2}$ and substitute the other quantities to calculate it. | **SOLUTION** <br> $S_{CO_2} = k_{H,\,CO_2}P_{CO_2}$ <br><br> $P_{CO_2} = \dfrac{S_{CO_2}}{k_{H,\,CO_2}}$ <br><br> $= \dfrac{0.12\,M}{3.4 \times 10^{-2}\dfrac{M}{atm}}$ <br><br> $= 3.5\,atm$ |

**CHECK** The answer is in the correct units and seems reasonable. A small answer (for example, less than 1 atm) would be suspect because you know that the soda is under a pressure greater than atmospheric pressure when you open it. A very large answer (for example, over 100 atm) would be suspect because an ordinary can or bottle probably could not sustain such high pressures without bursting.

**FOR PRACTICE 13.2** Determine the solubility of oxygen in water at 25 °C exposed to air at 1.0 atm. Assume a partial pressure for oxygen of 0.21 atm.

---

# 13.5   Expressing Solution Concentration

Solution Concentration: Molarity, Molality, Parts by Mass and Volume, Mole Fraction

As we have seen, the amount of solute in a solution is an important property of the solution. For example, the amount of sodium chloride in a solution determines whether or not the solution will cause dehydration if consumed. A **dilute solution** contains small quantities of solute relative to the amount of solvent. Drinking a dilute sodium chloride solution does not cause dehydration. A **concentrated solution** contains large quantities of solute relative to the amount of solvent. Drinking a concentrated sodium chloride solution does cause dehydration. Common ways of reporting solution concentration include molarity, molality, parts by mass, parts by volume, mole fraction,

# Chemistry IN THE Environment | Lake Nyos

Most people living near Lake Nyos in the West African country of Cameroon began August 22, 1986, like any other day. Unfortunately, their day ended in tragedy. On that evening, a large cloud of carbon dioxide gas burped up from the depths of Lake Nyos, killing over 1700 people and about 3000 cattle. Two years before, a similar tragedy had occurred in Lake Monoun, just 60 miles away, killing 37 people. Today, scientists have taken steps to prevent these lakes from accumulating the carbon dioxide that caused the disasters.

◄ Lake Nyos, in Cameroon, has a deceptively peaceful appearance; in the summer of 1986, more than 1700 people died around its shores.

Lake Nyos is a water-filled volcanic crater. Some 50 miles beneath the surface of the lake, molten volcanic rock (magma) produces carbon dioxide gas that seeps into the lake. The carbon dioxide forms a solution with the lake water. The high pressure at the bottom of the deep lake allows the solution to become highly concentrated in carbon dioxide. The 1986 disaster occurred because over time—either because of the high concentration itself or because of some other natural trigger such as a landslide or small earthquake—some gaseous carbon dioxide escaped. The rising bubbles disrupted the stratified layers of lake water, causing water at the bottom of the lake to rise to a region of lower pressure. The drop in pressure decreased the solubility of the carbon dioxide, so more carbon dioxide bubbles formed. This in turn caused more churning and still more carbon dioxide release. The result was a massive cloud of carbon dioxide gas that escaped from the lake. Since carbon dioxide is heavier than air, the carbon dioxide traveled down the sides of the volcano and into the nearby valley, displacing air and asphyxiating many of the local residents.

In an effort to keep these events from recurring, scientists have constructed a piping system that slowly vents carbon dioxide from the lake bottom, preventing the buildup that led to the tragedy.

◄ A plumbing system slowly vents carbon dioxide from Lake Nyos.

**QUESTION**

*Suppose that the water pressure at the bottom of Lake Nyos is 25 atm. What is the solubility of carbon dioxide at that depth?*

and mole percent, as summarized in Table 13.5. We have seen two of these units before: molarity in Section 4.4, and mole fraction in Section 5.6. In the following section, we review the terms we have already covered and introduce the new ones.

## Molarity

The **molarity (M)** of a solution is the amount of solute (in moles) divided by the volume of solution (in liters):

$$\text{Molarity (M)} = \frac{\text{amount solute (in mol)}}{\text{volume solution (in L)}}$$

| TABLE 13.5 Solution Concentration Terms | | |
|---|---|---|
| **Unit** | **Definition** | **Units** |
| Molarity (M) | $\dfrac{\text{amount solute (in mol)}}{\text{volume solution (in L)}}$ | $\dfrac{\text{mol}}{\text{L}}$ |
| Molality (*m*) | $\dfrac{\text{amount solute (in mol)}}{\text{mass solvent (in kg)}}$ | $\dfrac{\text{mol}}{\text{kg}}$ |
| Mole fraction ($\chi$) | $\dfrac{\text{amount solute (in mol)}}{\text{total amount of solute and solvent (in mol)}}$ | None |
| Mole percent (mol %) | $\dfrac{\text{amount solute (in mol)}}{\text{total amount of solute and solvent (in mol)}} \times 100\%$ | % |
| Parts by mass | $\dfrac{\text{mass solute}}{\text{mass solution}} \times \text{multiplication factor}$ | |
| Percent by mass (%) | Multiplication factor = 100 | % |
| Parts per million by mass (ppm) | Multiplication factor = $10^6$ | ppm |
| Parts per billion by mass (ppb) | Multiplication factor = $10^9$ | ppb |
| Parts by volume (%, ppm, ppb) | $\dfrac{\text{volume solute}}{\text{volume solution}} \times \text{multiplication factor*}$ | |

*Multiplication factors for parts by volume are identical to those for parts by mass.

Note that molarity is moles of solute per liter of *solution*, not per liter of solvent. To make a solution of a specified molarity, we usually put the solute into a flask and then add water (or another solvent) to the desired volume of solution, as shown in Figure 13.13▼. Molarity is a convenient unit to use when making, diluting, and transferring solutions because it specifies the amount of solute per unit of solution.

Weigh out 1.00 mol NaCl (58.44 g).

Add water until solid is dissolved. Then add additional water until the 1 L mark is reached.

Mix

A 1.00 molar NaCl solution

▲ **FIGURE 13.13 Preparing a Solution of Known Concentration**　To make a 1 M NaCl solution, we add 1 mol of the solid to a flask and dilute with water to make 1 L of solution.

Molarity depends on volume, and because volume varies with temperature, molarity also varies with temperature. For example, a 1 M aqueous solution at room temperature is slightly less than 1 M at an elevated temperature because the volume of the solution is greater at the elevated temperature.

## Molality

A concentration unit that is independent of temperature is **molality (*m*)**, the amount of solute (in moles) divided by the mass of solvent (in kilograms):

$$\text{Molality } (m) = \frac{\text{amount solute (in mol)}}{\text{mass solvent (in kg)}}$$

Notice that molality is defined with respect to kilograms *solvent*, not kilograms solution. Molality is particularly useful when we need to compare concentrations over a range of different temperatures.

**CONCEPTUAL CONNECTION 13.5**

**Molality**  If you combine 25.0 g of a solute that has a mass of 25.0 g/mol with 100.0 g of a solvent, what is the molality of the resulting solution?

## Parts by Mass and Parts by Volume

It is often convenient to report a concentration as a ratio of masses. A **parts by mass** concentration is the ratio of the mass of the solute to the mass of the solution, all multiplied by a multiplication factor:

$$\frac{\text{Mass solute}}{\text{Mass solution}} \times \text{multiplication factor}$$

The concentration of the solution determines the best parts by mass unit (and therefore the multiplication factor) to use. For example, for **percent by mass** the multiplication factor is 100:

$$\text{Percent by mass } = \frac{\text{mass solute}}{\text{mass solution}} \times 100\%$$

*Percent* means *per hundred*; a solution with a concentration of 14% by mass contains 14 g of solute per 100 g of solution.

For more dilute solutions, we might use **parts per million (ppm)**, which has a multiplication factor of $10^6$, or **parts per billion (ppb)**, which has a multiplication factor of $10^9$.

$$\text{ppm} = \frac{\text{mass solute}}{\text{mass solution}} \times 10^6$$

$$\text{ppb} = \frac{\text{mass solute}}{\text{mass solution}} \times 10^9$$

A solution with a concentration of 15 ppm by mass, for example, contains 15 g of solute per $10^6$ g of solution.

Sometimes, we report concentrations as a ratio of volumes, especially for solutions in which both the solute and solvent are liquids. A **parts by volume** concentration is usually the ratio of the volume of the solute to the volume of the solution, all multiplied by a multiplication factor:

$$\frac{\text{Volume solute}}{\text{Volume solution}} \times \text{multiplication factor}$$

The multiplication factors are identical to those just described for parts by mass concentrations. For example, a 22% ethanol solution by volume contains 22 mL of ethanol for every 100 mL of solution.

## Using Parts by Mass (or Parts by Volume) in Calculations

We can use the parts by mass (or parts by volume) concentration of a solution as a conversion factor between mass (or volume) of the solute and mass (or volume) of the solution. For example, for a solution containing 3.5% sodium chloride by mass, we write the following conversion factor:

$$\frac{3.5 \text{ g NaCl}}{100 \text{ g solution}}$$    converts    **g solution** → **g NaCl**

This conversion factor converts from grams solution to grams NaCl. To convert the other way, we invert the conversion factor:

$$\frac{100 \text{ g solution}}{3.5 \text{ g NaCl}}$$    converts    **g NaCl** → **g solution**

## Example 13.3    Using Parts by Mass in Calculations

What volume (in mL) of a soft drink that is 10.5% sucrose ($C_{12}H_{22}O_{11}$) by mass contains 78.5 g of sucrose? (The density of the solution is 1.04 g/mL.)

| SORT You are given a mass of sucrose and the concentration and density of a sucrose solution, and you are asked to find the volume of solution containing the given mass. | GIVEN: 78.5 g $C_{12}H_{22}O_{11}$<br>10.5% $C_{12}H_{22}O_{11}$ by mass<br>density = 1.04 g/mL<br><br>FIND: mL |
|---|---|
| STRATEGIZE Begin with the mass of sucrose in grams. Use the mass percent concentration of the solution (written as a ratio, as shown under Relationships Used) to find the number of grams of solution containing this quantity of sucrose. Then use the density of the solution to convert grams to milliliters of solution. | CONCEPTUAL PLAN<br><br><br>**g $C_{12}H_{22}O_{11}$** → **g soln** → **mL soln**<br>$\frac{100 \text{ g soln}}{10.5 \text{ g } C_{12}H_{22}O_{11}}$    $\frac{1 \text{ mL}}{1.04 \text{ g}}$<br><br>RELATIONSHIPS USED<br>$\frac{10.5 \text{ g } C_{12}H_{22}O_{11}}{100 \text{ g soln}}$ (percent by mass written as ratio)<br><br>$\frac{1.04 \text{ g}}{1 \text{ mL}}$ (given density of the solution) |
| SOLVE Begin with 78.5 g $C_{12}H_{22}O_{11}$ and multiply by the conversion factors to arrive at the volume of solution. | SOLUTION<br>$78.5 \text{ g } C_{12}H_{22}O_{11} \times \dfrac{100 \text{ g soln}}{10.5 \text{ g } C_{12}H_{22}O_{11}} \times \dfrac{1 \text{ mL}}{1.04 \text{ g}} = 719 \text{ mL soln}$ |

CHECK The units of the answer are correct. The magnitude seems correct because the solution is approximately 10% sucrose by mass. Since the density of the solution is approximately 1 g/mL, the volume containing 78.5 g sucrose should be roughly 10 times larger, as calculated ($719 \approx 10 \times 78.5$).

**FOR PRACTICE 13.3**    What mass of sucrose ($C_{12}H_{22}O_{11}$), in g, is contained in 355 mL (12 ounces) of a soft drink that is 11.5% sucrose by mass? (Assume a density of 1.04 g/mL.)

**FOR MORE PRACTICE 13.3**    A water sample is found to contain the pollutant chlorobenzene with a concentration of 15 ppb (by mass). What volume of this water contains $5.00 \times 10^2$ mg of chlorobenzene? (Assume a density of 1.00 g/mL.)

## Mole Fraction and Mole Percent

For some applications, especially those in which the ratio of solute to solvent can vary widely, the most useful way to express concentration is the amount of solute (in moles) divided by the total amount of solute and solvent (in moles). This ratio is the **mole fraction ($\chi_{solute}$)**:

The mole fraction can also be defined for the solvent:

$$\chi_{solvent} = \frac{n_{solvent}}{n_{solute} + n_{solvent}}$$

$$\chi_{solute} = \frac{\text{amount solute (in mol)}}{\text{total amount of solute and solvent (in mol)}} = \frac{n_{solute}}{n_{solute} + n_{solvent}}$$

Also in common use is the **mole percent (mol %)**, which is simply the mole fraction × 100 percent:

$$mol\% = \chi_{solute} \times 100\%$$

# Chemistry IN THE Environment | The Dirty Dozen

A number of potentially harmful chemicals—such as DDT, dioxin, and polychlorinated biphenyls (PCBs)—can make their way into water sources from industrial dumping, atmospheric emissions, agricultural use, and household dumping. Since crops, livestock, and fish consume water, they too can accumulate these chemicals. Human consumption of food or water contaminated with harmful chemicals can lead to a number of diseases and adverse health effects such as increased cancer risk, liver damage, and central nervous system damage. Governments around the world have joined forces to ban the production of a number of these kinds of chemicals—called persistent organic pollutants or POPs. The original treaty named the Stockholm Convention on Persistent Organic Pollutants was signed in 2001. It targeted 12 such substances referred to as the "dirty dozen" (Table 13.6).

### TABLE 13.6 The Dirty Dozen

| | |
|---|---|
| 1. Aldrin—insecticide | 7. Furan—industrial by-product |
| 2. Chlordane—insecticide | 8. Heptachlor—insecticide |
| 3. DDT—insecticide | 9. Hexachlorobenzene—fungicide, industrial by-product |
| 4. Dieldrin—insecticide | 10. Mirex—insecticide, fire retardant |
| 5. Dioxin—industrial by-product | 11. Polychlorinated biphenyls (PCBs)—electrical insulators |
| 6. Eldrin—insecticide | 12. Toxaphene—insecticide |

▲ Potentially harmful chemicals make their way into water sources in many ways.

One problem common to all of these chemicals is their persistence in the environment. These compounds are fairly stable and do not break down under normal environmental conditions. Once they get into the environment, they stay there for a long time. A second problem with these chemicals is their contribution to a process called *bioamplification*. Because they are nonpolar, these chemicals are stored and concentrated in the fatty tissues of the organisms that consume them. As larger organisms eat smaller ones, they consume more of the stored chemicals. The result is an increase in the concentrations of these chemicals as they move up the food chain.

◄ The bald eagle was nearly driven to extinction by human use of DDT. Bald eagles eat fish, which became contaminated due to bioamplification. The DDT consumed with the fish caused the eagle's egg shells to become fragile. The DDT ban has allowed the bald eagle population to recover.

For example, the use of DDT nearly drove the bald eagle (whose egg shells were greatly weakened by eating fish contaminated with DDT) to extinction. In 1963, the population dwindled to less than 500 nesting pairs. Fortunately, nearly all intentional production of DDT (and the other chemicals on this list) has been banned. The bald eagle population has recovered to an estimated 10,000 nesting pairs. The recovery has been so strong that the bald eagle has been removed from both the endangered and threatened species lists.

In the United States, the presence of these contaminants in water supplies is monitored by the Environmental Protection Agency (EPA). The EPA has set limits, called maximum contaminant levels (MCLs), for each of these contaminants in food and drinking water. Table 13.7 lists some MCLs for selected compounds in water supplies. Notice the units that the EPA uses to express the concentration of the contaminants, mg/L. This unit is a conversion factor between liters of water consumed and the mass (in mg) of the pollutant. According to the EPA, as long as the contaminant concentrations are below these levels, the water is safe to drink.

### TABLE 13.7 EPA Maximum Contaminant Level (MCL) for Several "Dirty Dozen" Chemicals

| | |
|---|---|
| Chlordane | 0.002 mg/L |
| Dioxin | 0.00000003 mg/L |
| Heptachlor | 0.0004 mg/L |
| Hexachlorobenzene | 0.001 mg/L |

### QUESTION

*Calculate how much of each chemical in Table 13.7 at the MCL is present in 715 L of water, the approximate amount of water consumed by an adult in one year.*

**Example** 13.4    **Calculating Concentrations**

A solution is prepared by dissolving 17.2 g of ethylene glycol ($C_2H_6O_2$) in 0.500 kg of water. The final volume of the solution is 515 mL. Calculate the concentration of the solution in each unit.

**(a)** molarity

**(b)** molality

**(c)** percent by mass

**(d)** mole fraction

**(e)** mole percent

**SOLUTION**

| | | |
|---|---|---|
| **(a)** | To calculate molarity, first find the amount of ethylene glycol in moles from the mass and molar mass. Then divide the amount in moles by the volume of the solution in liters. | $\text{mol } C_2H_6O_2 = 17.2 \text{ g } C_2H_6O_2 \times \dfrac{1 \text{ mol g } C_2H_6O_2}{62.07 \text{ g } C_2H_6O_2} = 0.27\underline{7}1 \text{ mol } C_2H_6O_2$ <br><br> $\text{Molarity (M)} = \dfrac{\text{amount solute (in mol)}}{\text{volume solution (in L)}}$ <br><br> $= \dfrac{0.27\underline{7}1 \text{ mol } C_2H_6O}{0.515 \text{ L solution}}$ <br><br> $= 0.538 \text{ M}$ |
| **(b)** | To calculate molality, use the amount of ethylene glycol in moles from part (a), and divide by the mass of the water in kilograms. | $\text{Molality } (m) = \dfrac{\text{amount solute (in mol)}}{\text{mass solvent (in kg)}}$ <br><br> $= \dfrac{0.27\underline{7}1 \text{ mol } C_2H_6O}{0.500 \text{ kg } H_2O}$ <br><br> $= 0.554 \; m$ |
| **(c)** | To calculate percent by mass, divide the mass of the solute by the sum of the masses of the solute and solvent and multiply the ratio by 100%. | $\text{Percent by mass} = \dfrac{\text{mass solute}}{\text{mass solution}} \times 100\%$ <br><br> $= \dfrac{17.2 \text{ g}}{17.2 \text{ g} + 5.00 \times 10^2 \text{ g}} \times 100\%$ <br><br> $= 3.33\%$ |
| **(d)** | To calculate mole fraction, first determine the amount of water in moles from the mass of water and its molar mass. Then divide the amount of ethylene glycol in moles (from part (a)) by the total number of moles. | $\text{mol } H_2O = 5.00 \times 10^2 \text{ g } H_2O \times \dfrac{1 \text{ mol } H_2O}{18.02 \text{ g } H_2O}$ <br><br> $= 27.\underline{7}5 \text{ mol } H_2O$ <br><br> $\chi_{\text{solute}} = \dfrac{n_{\text{solute}}}{n_{\text{solute}} + n_{\text{solvent}}}$ <br><br> $= \dfrac{0.27\underline{7}1 \text{ mol}}{0.27\underline{7}1 \text{ mol} + 27.\underline{7}5 \text{ mol}}$ <br><br> $= 9.89 \times 10^{-3}$ |
| **(e)** | To calculate mole percent, multiply the mole fraction by 100%. | $\text{mol } \% = \chi_{\text{solute}} \times 100\%$ <br><br> $= 0.989\%$ |

**FOR PRACTICE 13.4**   A solution is prepared by dissolving 50.4 g sucrose ($C_{12}H_{22}O_{11}$) in 0.332 kg of water. The final volume of the solution is 355 mL. Calculate the concentration of the solution in each unit.

**(a)** molarity

**(b)** molality

**(c)** percent by mass

**(d)** mole fraction

**(e)** mole percent

## Example 13.5    Converting between Concentration Units ▶

What is the molarity of a 6.56% by mass glucose ($C_6H_{12}O_6$) solution? (The density of the solution is 1.03 g/mL.)

| | |
|---|---|
| **SORT** You are given the concentration of a glucose solution in percent by mass and the density of the solution. Find the concentration of the solution in molarity. | **GIVEN:** 6.56% $C_6H_{12}O_6$<br><br>density = 1.03 g/mL<br><br>**FIND:** M |

**STRATEGIZE** Begin with the mass percent concentration of the solution written as a ratio, and separate the numerator from the denominator. Convert the numerator from g $C_6H_{12}O_6$ to mol $C_6H_{12}O_6$. Convert the denominator from g soln to mL of solution and then to L solution. Divide the numerator (now in mol) by the denominator (now in L) to obtain molarity.

**CONCEPTUAL PLAN**

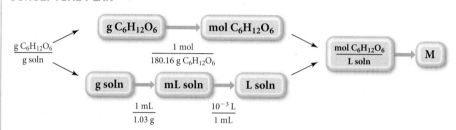

**RELATIONSHIPS USED**

$$\frac{6.56 \text{ g } C_6H_{12}O_6}{100 \text{ g soln}} \text{ (percent by mass written as ratio)}$$

$$\frac{1 \text{ mol}}{180.16 \text{ g } C_6H_{12}O_6} \text{ (from molar mass of glucose)}$$

$$\frac{1 \text{ mL}}{1.03 \text{ g}} \text{ (from given density of the solution)}$$

**SOLVE** Begin with the numerator (6.56 g $C_6H_{12}O_6$) and use the molar mass to convert to mol $C_6H_{12}O_6$.

**SOLUTION**

$$6.56 \text{ g } C_6H_{12}O_6 \times \frac{1 \text{ mol } C_6H_{12}O_6}{180.16 \text{ g } C_6H_{12}O_6} = 0.036\underline{4}12 \text{ mol } C_6H_{12}O_6$$

Convert the denominator (100 g solution) into mL of solution (using the density) and then to L of solution.

$$100 \text{ g soln} \times \frac{1 \text{ mL}}{1.03 \text{ g}} \times \frac{10^{-3} \text{ L}}{\text{mL}} = 0.097\underline{0}87 \text{ L soln}$$

Finally, divide mol $C_6H_{12}O_6$ by L solution to arrive at molarity.

$$\frac{0.036\underline{4}12 \text{ mol } C_6H_{12}O_6}{0.097\underline{0}87 \text{ L soln}} = 0.375 \text{ M } C_6H_{12}O_6$$

**CHECK** The units of the answer are correct. The magnitude seems correct. Very high molarities (especially above 25 M) should immediately appear suspect. One liter of water contains about 55 moles of water molecules, so molarities higher than 55 M are physically impossible.

**FOR PRACTICE 13.5**   What is the molarity of a 10.5% by mass glucose ($C_6H_{12}O_6$) solution? (The density of the solution is 1.03 g/mL.)

**FOR MORE PRACTICE 13.5**   What is the molality of a 10.5% by mass glucose ($C_6H_{12}O_6$) solution? (The density of the solution is 1.03 g/mL.)

# 13.6 Colligative Properties: Vapor Pressure Lowering, Freezing Point Depression, Boiling Point Elevation, and Osmotic Pressure

**KCV** **Colligative Properties**

Have you ever wondered why you add salt to the ice in an ice-cream maker? Or why salt is scattered on icy roads in cold climates? Salt lowers the temperature at which a saltwater solution freezes. A salt and water solution remains liquid even below 0 °C. When salt is added to ice in the ice-cream maker, an ice/water/salt mixture forms that can reach a temperature of about −10 °C, at which point the cream freezes. On the winter road, the salt allows the ice to melt when the ambient temperature is below freezing.

▲ In winter, salt is often applied to roads so that the ice will melt at lower temperatures.

The depression of the freezing point of water by salt is an example of a **colligative property**, a property that depends on the number of particles dissolved in solution, not on the type of particle. In this section, we examine four colligative properties: vapor pressure lowering, freezing point depression, boiling point elevation, and osmotic pressure.

Since these properties depend on the *number* of dissolved particles, nonelectrolytes are treated slightly differently than electrolytes when determining colligative properties. When 1 mol of a nonelectrolyte dissolves in water, it forms 1 mol of dissolved particles. When 1 mol of an electrolyte dissolves in water, however, it normally forms more than 1 mol of dissolved particles (as shown in Figure 13.14▶). For example, when 1 mol of NaCl dissolves in water, it forms 1 mol of dissolved $Na^+$ ions and 1 mol of dissolved $Cl^-$ ions. Therefore, the resulting solution has 2 mol of dissolved particles. The colligative properties of electrolyte solutions reflect this higher concentration of dissolved particles. In this section we examine colligative properties of nonelectrolyte solutions; we expand the concept to include electrolyte solutions in Section 13.7.

$Cl^-$　$Na^+$

4 formula units

8 ions

◀ **FIGURE 13.14** When sodium chloride is dissolved in water, each mole of NaCl produces 2 mol of particles: 1 mol of $Na^+$ cations and 1 mol of $Cl^-$ anions.

## Vapor Pressure Lowering

Recall from Section 11.5 that the vapor pressure of a liquid is the pressure of the gas above the liquid when the two are in dynamic equilibrium (that is, when the rate of vaporization equals the rate of condensation). How does a nonvolatile nonelectrolyte solute affect the vapor pressure of the liquid into which it dissolves? The basic answer to this question is that *the vapor pressure of the solution is lower than the vapor pressure of the pure solvent*. We can understand why this happens in two different ways.

The simplest explanation for why the vapor pressure of a solution is lower than that of the pure solvent is related to the concept of dynamic equilibrium itself. Consider the following representation of a liquid in dynamic equilibrium with its vapor. Here the rate of vaporization is equal to the rate of condensation:

Dynamic equilibrium

When a nonvolatile solute is added, however, the solute particles (shown in red) interfere with the ability of the solvent particles (blue) to vaporize. The rate of vaporization is thus diminished compared to that of the pure solvent:

Rate of vaporization
reduced by solute

The change in the rate of vaporization creates an imbalance in the rates; the rate of condensation is now *greater* than the rate of vaporization. The net effect is that some of the molecules that were in the gas state condense into the liquid state. As they condense, the reduced number of molecules in the gas state causes the rate of condensation to decrease. Eventually the two rates become equal again, but only after the concentration of solvent molecules in the gas state has decreased:

Equilibrium reestablished
but with fewer molecules
in gas phase

The result is a lower vapor pressure for the solution compared to the pure solvent.

A more fundamental explanation of why the vapor pressure of a solution is lower than that of the pure solvent is related to the tendency toward mixing (toward greater entropy) that we discussed in Sections 13.1 and 13.2. Recall that a concentrated solution is a *thirsty* solution—it has the ability to draw solvent to itself. We can see a dramatic demonstration of this tendency by placing a concentrated solution of a nonvolatile solute and a beaker of the pure solvent in a sealed container. Over time, the level of the pure solvent will drop, and the level of the solution will rise as molecules vaporize out of the pure solvent and condense into the solution.

Notice the similarity between this process and the dehydration caused by drinking seawater. In both cases, a concentrated solution has the ability to draw solvent to itself. The reason is nature's tendency to mix. If a pure solvent and concentrated solution are combined in a beaker, they naturally form a mixture in which the concentrated solution becomes less concentrated than it was initially. Similarly, if a pure solvent and concentrated solution are combined in a sealed container—even though they are in separate beakers—the two mix so that the concentrated solution becomes less concentrated.

The net transfer of solvent from the beaker containing pure solvent to the one containing the solution shows that the vapor pressure of the solution is lower than that of the pure solvent. As solvent molecules vaporize, the vapor pressure in the sealed container rises. Before dynamic equilibrium can be attained, however, the pressure exceeds the vapor pressure of the solution, causing molecules to condense into the solution (the beaker on the right). Therefore, molecules constantly vaporize from the pure solvent, but the solvent's vapor pressure is never reached because molecules are constantly condensing into the solution. The result is a continuous transfer of solvent molecules from the pure solvent to the solution.

We can quantify the vapor pressure of a solution with **Raoult's law**:

$$P_{solution} = \chi_{solvent} P^{\circ}_{solvent}$$

In this equation, $P_{solution}$ is the vapor pressure of the solution, $\chi_{solvent}$ is the mole fraction of the solvent, and $P^{\circ}_{solvent}$ is the vapor pressure of the pure solvent at the same temperature. For example, suppose a water sample at 25 °C contains 0.900 mol of water and 0.100 mol of a nonvolatile solute such as sucrose. The pure water has a vapor pressure of 23.8 torr. We calculate the vapor pressure of the solution by substituting into Raoult's law:

$$P_{solution} = \chi_{H_2O} P^{\circ}_{H_2O}$$

$$= 0.900(23.8 \text{ torr})$$

$$= 21.4 \text{ torr}$$

Notice that the vapor pressure of the solution is directly proportional to the amount of the solvent in the solution. Since the solvent particles compose 90% of all of the particles in the solution, the vapor pressure of the solution is 90% of the vapor pressure of the pure solvent.

To arrive at an equation that shows how much the vapor pressure is lowered by a solute, we define the **vapor pressure lowering ($\Delta P$)** as the difference in vapor pressure between the pure solvent and the solution:

$$\Delta P = P^{\circ}_{solvent} - P_{solution}$$

Then, for a two-component solution, we can substitute $\chi_{solvent} = 1 - \chi_{solute}$ into Raoult's law:

$$P_{solution} = \chi_{solvent} P^{\circ}_{solvent}$$

$$P_{solution} = (1 - \chi_{solute}) P^{\circ}_{solvent}$$

$$P^{\circ}_{solvent} - P_{solution} = \chi_{solute} P^{\circ}_{solvent}$$

$$\Delta P = \chi_{solute} P^{\circ}_{solvent}$$

This last equation indicates that the lowering of the vapor pressure is directly proportional to the mole fraction of the solute.

## Example 13.6   Calculating the Vapor Pressure of a Solution Containing a Nonelectrolyte and Nonvolatile Solute

Calculate the vapor pressure at 25 °C of a solution containing 99.5 g sucrose ($C_{12}H_{22}O_{11}$) and 300.0 mL water. The vapor pressure of pure water at 25 °C is 23.8 torr. Assume the density of water is 1.00 g/mL.

| | |
|---|---|
| **SORT** You are given the mass of sucrose and volume of water in a solution. You are also given the vapor pressure and density of pure water and asked to find the vapor pressure of the solution. | **GIVEN:** 99.5 g $C_{12}H_{22}O_{11}$<br>300.0 mL $H_2O$<br>$P^{\circ}_{H_2O}$ = 23.8 torr at 25 °C<br>$d_{H_2O}$ = 1.00 g/mL<br><br>**FIND:** $P_{solution}$ |

| | |
|---|---|
| **STRATEGIZE** Raoult's law relates the vapor pressure of a solution to the mole fraction of the solvent and the vapor pressure of the pure solvent. Begin by calculating the amount in moles of sucrose and water. | **CONCEPTUAL PLAN**<br><br> |
| Calculate the mole fraction of the solvent from the calculated amounts of solute and solvent. | <br><br>$$\chi_{H_2O} = \frac{n_{H_2O}}{n_{H_2O} + n_{C_{12}H_{22}O_{11}}}$$ |
| Then use Raoult's law to calculate the vapor pressure of the solution. | $$\boxed{\chi_{H_2O}, P^{\circ}_{H_2O}} \longrightarrow \boxed{P_{solution}}$$<br><br>$$P_{solution} = \chi_{H_2O} P^{\circ}_{H_2O}$$ |

| | |
|---|---|
| **SOLVE** Calculate the number of moles of each solution component. | **SOLUTION**<br><br>$$99.5 \text{ g } C_{12}H_{22}O_{11} \times \frac{1 \text{ mol } C_{12}H_{22}O_{11}}{342.30 \text{ g } C_{12}H_{22}O_{11}} = 0.2907 \text{ mol } C_{12}H_{22}O_{11}$$<br><br>$$300.0 \text{ mL } H_2O \times \frac{1.00 \text{ g}}{1 \text{ mL}} \times \frac{1 \text{ mol } H_2O}{18.02 \text{ g } H_2O} = 16.65 \text{ mol } H_2O$$ |
| Use the number of moles of each component to calculate the mole fraction of the solvent ($H_2O$). | $$\chi_{H_2O} = \frac{n_{H_2O}}{n_{C_{12}H_{22}O_{11}} + n_{H_2O}}$$<br><br>$$= \frac{16.65 \text{ mol}}{0.2907 \text{ mol} + 16.65 \text{ mol}}$$<br><br>$$= 0.9828$$ |
| Use the mole fraction of water and the vapor pressure of pure water to calculate the vapor pressure of the solution. | $$P_{solution} = \chi_{H_2O} P^{\circ}_{H_2O}$$<br><br>$$= 0.9828 (23.8 \text{ torr})$$<br><br>$$= 23.4 \text{ torr}$$ |

**CHECK** The units of the answer are correct. The magnitude of the answer seems right because the calculated vapor pressure of the solution is just below that of the pure liquid, as you would expect for a solution with a large mole fraction of solvent.

**FOR PRACTICE 13.6** Calculate the vapor pressure at 25 °C of a solution containing 55.3 g ethylene glycol (HOCH$_2$CH$_2$OH) and 285.2 g water. The vapor pressure of pure water at 25 °C is 23.8 torr.

**FOR MORE PRACTICE 13.6** A solution containing ethylene glycol and water has a vapor pressure of 7.88 torr at 10 °C. Pure water has a vapor pressure of 9.21 torr at 10 °C. What is the mole fraction of ethylene glycol in the solution?

# Vapor Pressures of Solutions Containing a Volatile (Nonelectrolyte) Solute

Some solutions contain not only a volatile solvent, but also a volatile *solute*. In this case, *both* the solvent and the solute contribute to the overall vapor pressure of the solution. A solution like this may be an **ideal solution** (in which case its behavior follows Raoult's law at all concentrations for both the solvent and the solute), or it may be nonideal (in which case it does not follow Raoult's law).

An ideal solution is similar in concept to an ideal gas. Just as an ideal gas follows the ideal gas law exactly, so an ideal solution follows Raoult's law exactly. In an ideal solution, the solute–solvent interactions are similar in magnitude to the solute–solute and solvent–solvent interactions. In this type of solution, the solute simply dilutes the solvent and ideal behavior is observed. The vapor pressure of each solution component is described by Raoult's law throughout the entire composition range of the solution. For a two-component solution containing liquids A and B, we write:

> Over a complete range of composition of a solution, it no longer makes sense to designate a solvent and solute, so we simply label the two components A and B.

$$P_A = \chi_A P_A^\circ$$
$$P_B = \chi_B P_B^\circ$$

The total pressure above such a solution is the sum of the partial pressures of the components:

$$P_{tot} = P_A + P_B$$

Figure 13.15(a)▼ is a plot of vapor pressure versus solution composition for an ideal two-component solution.

In a nonideal solution, the solute–solvent interactions are either stronger or weaker than the solvent–solvent interactions. If the solute–solvent interactions are stronger, then the solute tends to prevent the solvent from vaporizing as readily as it would otherwise. If the solution is sufficiently dilute, then the effect will be small and Raoult's law works as an approximation. However, if the solution is not dilute, the effect will be significant and the vapor pressure of the solution will be *less than* that predicted by Raoult's law, as shown in Figure 13.15(b)▼.

If, however, the solute–solvent interactions are weaker than solvent–solvent interactions, then the solute tends to allow more vaporization than would occur with just the solvent. If the solution is not dilute, the effect will be significant and the vapor pressure of the solution will be *greater than* predicted by Raoult's law, as shown in Figure 13.15(c)▼.

## Deviations from Raoult's Law

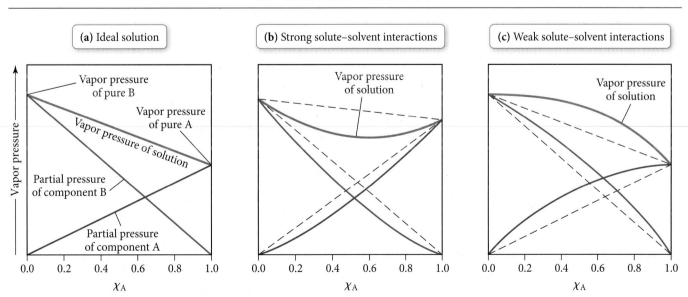

▲ **FIGURE 13.15** **Behavior of Ideal and Nonideal Solutions**   **(a)** An ideal solution follows Raoult's law for both components. **(b)** A solution with particularly strong solute–solvent interactions displays negative deviations from Raoult's law. **(c)** A solution with particularly weak solute–solvent interactions displays positive deviations from Raoult's law. (The dashed lines in parts b and c represent ideal behavior.)

## Example 13.7 Calculating the Vapor Pressure of a Two-Component Solution

A solution contains 3.95 g of carbon disulfide ($CS_2$) and 2.43 g of acetone ($CH_3COCH_3$). At 35 °C the vapor pressures of pure carbon disulfide and pure acetone are 515 torr and 332 torr, respectively. Assuming ideal behavior, calculate the vapor pressures of each component and the total vapor pressure above the solution. The experimentally measured total vapor pressure of the solution at 35 °C is 645 torr. Is the solution ideal? If not, what can you say about the relative strength of carbon disulfide–acetone interactions compared to the acetone–acetone and carbon disulfide–carbon disulfide interactions?

| | |
|---|---|
| **SORT** You are given the masses and vapor pressures of carbon disulfide and acetone and are asked to find the vapor pressures of each component in the mixture and the total pressure assuming ideal behavior. | **GIVEN:** 3.95 g $CS_2$ <br><br> 2.43 g $CH_3COCH_3$ <br><br> $P^\circ_{CS_2}$ = 515 torr (at 35 °C) <br><br> $P^\circ_{CH_3COCH_3}$ = 332 torr (at 35 °C) <br><br> $P_{tot}(exp)$ = 645 torr (at 35 °C) <br><br> **FIND:** $P_{CS_2}$, $P_{CH_3COCH_3}$, $P_{tot}(ideal)$ |

**STRATEGIZE** This problem requires that you use Raoult's law to calculate the partial pressures of each component. In order to use Raoult's law, first calculate the mole fractions of the two components. Convert the masses of each component to moles and then use the definition of mole fraction to calculate the mole fraction of carbon disulfide. You can then find the mole fraction of acetone because the mole fractions of the two components add up to 1.

**CONCEPTUAL PLAN**

$$3.95 \text{ g CS}_2 \longrightarrow \text{mol CS}_2$$
$$\frac{1 \text{ mol CS}_2}{76.15 \text{ g CS}_2}$$

$$2.43 \text{ g CH}_3\text{COCH}_3 \longrightarrow \text{mol CH}_3\text{COCH}_3$$
$$\frac{1 \text{ mol CH}_3\text{COCH}_3}{58.08 \text{ g CH}_3\text{COCH}_3}$$

$$\text{mol CS}_2, \text{mol CH}_3\text{OCH}_3 \longrightarrow \chi_{CS_2}, \chi_{CH_3COCH_3}$$
$$\chi_{CS_2} = \frac{n_{CS_2}}{n_{CS_2} + n_{CH_3COCH_3}}$$

Use the mole fraction of each component along with Raoult's law to calculate the partial pressure of each component. The total pressure is the sum of the partial pressures.

$$P_{CS_2} = \chi_{CS_2} P^\circ_{CS_2}$$
$$P_{CH_3COCH_3} = \chi_{CH_3COCH_3} P^\circ_{CH_3COCH_3}$$
$$P_{tot} = P_{CS_2} + P_{CH_3COCH_3}$$

**RELATIONSHIPS USED**

$$\chi_A = \frac{n_A}{n_A + n_B} \text{ (mole fraction definition)}$$
$$P_A = \chi_A P^\circ_A \text{ (Raoult's law)}$$

**SOLVE** Begin by converting the mass of each component to the amounts in moles.

**SOLUTION**

$$3.95 \text{ g CS}_2 \times \frac{1 \text{ mol CS}_2}{76.15 \text{ g CS}_2} = 0.05187 \text{ mol CS}_2$$

$$2.43 \text{ g CH}_3\text{COCH}_3 \times \frac{1 \text{ mol CH}_3\text{COCH}_3}{58.08 \text{ g CH}_3\text{COCH}_3} = 0.04184 \text{ mol CH}_3\text{COCH}_3$$

| Then calculate the mole fraction of carbon disulfide. | $$\chi_{CS_2} = \frac{n_{CS_2}}{n_{CS_2} + n_{CH_3COCH_3}}$$ $$= \frac{0.05187 \text{ mol}}{0.05187 \text{ mol} + 0.04184 \text{ mol}}$$ $$= 0.5535$$ |
|---|---|
| Calculate the mole fraction of acetone by subtracting the mole fraction of carbon disulfide from 1. | $$\chi_{CH_3COCH_3} = 1 - 0.5535$$ $$= 0.4465$$ |
| Calculate the partial pressures of carbon disulfide and acetone by using Raoult's law and the given values of the vapor pressures of the pure substances. | $$P_{CS_2} = \chi_{CS_2} P^\circ_{CS_2}$$ $$= 0.5535 \, (515 \text{ torr})$$ $$= 285 \text{ torr}$$ $$P_{CH_3COCH_3} = \chi_{CH_3COCH_3} P^\circ_{CH_3COCH_3}$$ $$= 0.4465 \, (332 \text{ torr})$$ $$= 148 \text{ torr}$$ |
| Calculate the total pressure by summing the partial pressures. | $$P_{tot}(\text{ideal}) = 285 \text{ torr} + 148 \text{ torr}$$ $$= 433 \text{ torr}$$ |
| Lastly, compare the calculated total pressure for the ideal case to the experimentally measured total pressure. Since the experimentally measured pressure is greater than the calculated pressure, we can conclude that the interactions between the two components are weaker than the interactions between the components themselves. | $P_{tot}(\text{exp}) = 645 \text{ torr}$ $P_{tot}(\text{exp}) > P_{tot}(\text{ideal})$ The solution is not ideal and shows positive deviations from Raoult's law. Therefore, carbon disulfide–acetone interactions must be weaker than acetone–acetone and carbon disulfide–carbon disulfide interactions. |

**CHECK** The units of the answer (torr) are correct. The magnitude seems reasonable given the partial pressures of the pure substances.

**FOR PRACTICE 13.7** A solution of benzene ($C_6H_6$) and toluene ($C_7H_8$) is 25.0% benzene by mass. At 25 °C the vapor pressures of pure benzene and pure toluene are 94.2 torr and 28.4 torr, respectively. Assuming ideal behavior, calculate the following:

**(a)** The vapor pressure of each of the solution components in the mixture.

**(b)** The total pressure above the solution.

**(c)** The composition of the vapor in mass percent.

Why is the composition of the vapor different from the composition of the solution?

**CONCEPTUAL CONNECTION 13.6**

**Raoult's Law** A solution contains equal amounts (in moles) of liquid components A and B. The vapor pressure of pure A is 100 mmHg and that of pure B is 200 mmHg. The experimentally measured vapor pressure of the solution is 120 mmHg. What are the relative strengths of the solute–solute, solute–solvent, and solvent–solvent interactions in this solution?

# Freezing Point Depression and Boiling Point Elevation

Vapor pressure lowering occurs at all temperatures. We can see the effect of vapor pressure lowering over a range of temperatures by comparing the phase diagrams for a pure solvent and for a solution containing a nonvolatile solute:

▶ A nonvolatile solute lowers the vapor pressure of a solution, resulting in a lower freezing point and an elevated boiling point.

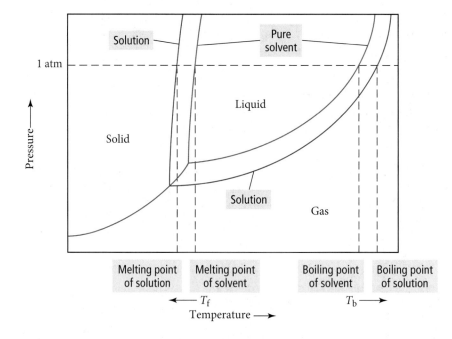

Notice that the vapor pressure for the solution is shifted downward compared to that of the pure solvent. Consequently, the vapor pressure curve intersects the solid–gas curve at a lower temperature. The net effect is that the solution has a *lower melting point* and a *higher boiling point* than the pure solvent. These effects are called **freezing point depression** and **boiling point elevation**, respectively, both of which are colligative properties (like vapor pressure lowering).

The freezing point of a solution containing a nonvolatile solute is lower than the freezing point of the pure solvent. For example, antifreeze, used to prevent the freezing of engine blocks in cold climates, is an aqueous solution of ethylene glycol ($C_2H_6O_2$). The more concentrated the solution, the lower the freezing point becomes.

The amount that the freezing point decreases is given by the following equation:

$$\Delta T_f = m \times K_f$$

▲ Antifreeze is an aqueous solution of ethylene glycol. The solution has a lower freezing point and higher boiling point than pure water.

where

- $\Delta T_f$ is the change in temperature of the freezing point in Celsius degrees (relative to the freezing point of the pure solvent), usually reported as a positive number;
- $m$ is the molality of the solution in moles solute per kilogram solvent; and
- $K_f$ is the freezing point depression constant for the solvent.

For water,

$$K_f = 1.86\,°C/m$$

When an aqueous solution containing a dissolved solid solute freezes slowly, the ice that forms does not normally contain much of the solute. For example, when ice forms in ocean water, the ice is not salt water, but freshwater. As the ice forms, the crystal structure of the ice tends to exclude the solute particles. You can verify this yourself by partially freezing a saltwater solution in the freezer. Take out the newly formed ice, rinse it several times, and taste it. Compare its taste to the taste of the original solution. The ice is much less salty.

**TABLE 13.8 Freezing Point Depression and Boiling Point Elevation Constants for Several Liquid Solvents**

| Solvent | Normal Freezing Point (°C) | $K_f$ (°C/m) | Normal Boiling Point (°C) | $K_b$ (°C/m) |
|---|---|---|---|---|
| Benzene ($C_6H_6$) | 5.5 | 5.12 | 80.1 | 2.53 |
| Carbon tetrachloride ($CCl_4$) | −22.9 | 29.9 | 76.7 | 5.03 |
| Chloroform ($CHCl_3$) | −63.5 | 4.70 | 61.2 | 3.63 |
| Ethanol ($C_2H_5OH$) | −114.1 | 1.99 | 78.3 | 1.22 |
| Diethyl ether ($C_4H_{10}O$) | −116.3 | 1.79 | 34.6 | 2.02 |
| Water ($H_2O$) | 0.00 | 1.86 | 100.0 | 0.512 |

Table 13.8 provides freezing point depression and boiling point elevation constants for several liquids. Calculating the freezing point of a solution involves substituting into the freezing point depression equation as Example 13.8 demonstrates.

| **Example** 13.8 | **Freezing Point Depression** |
|---|---|

Calculate the freezing point of a 1.7 $m$ aqueous ethylene glycol solution.

| **SORT** You are given the molality of a solution and asked to find its freezing point. | **GIVEN:** 1.7 $m$ solution<br><br>**FIND:** freezing point (from $\Delta T_f$) |
|---|---|
| **STRATEGIZE** To solve this problem, use the freezing point depression equation. | **CONCEPTUAL PLAN**<br><br>$$\boxed{m} \longrightarrow \boxed{\Delta T_f}$$<br>$$\Delta T_f = m \times K_f$$ |
| **SOLVE** Substitute into the equation to calculate $\Delta T_f$. The actual freezing point is the freezing point of pure water (0.00 °C) $- \Delta T_f$. | **SOLUTION**<br><br>$\Delta T_f = m \times K_f$<br>$\quad = 1.7\,m \times 1.86\ °C/m$<br>$\quad = 3.2\ °C$<br><br>Freezing point $= 0.00\ °C - 3.2\ °C$<br>$\qquad\qquad\quad = -3.2\ °C$ |

**CHECK** The units of the answer are correct. The magnitude seems about right. The expected range for freezing points of an aqueous solution is anywhere from −10 °C to just below 0 °C. Any answers out of this range would be suspect.

**FOR PRACTICE 13.8**  Calculate the freezing point of a 2.6 $m$ aqueous sucrose solution.

The boiling point of a solution containing a nonvolatile solute is higher than the boiling point of the pure solvent. In automobiles, antifreeze not only prevents the freezing of water within engine blocks in cold climates, it also prevents the boiling of water within engine blocks in hot climates. The amount that boiling points rise in solutions is given by the equation:

$$\Delta T_b = m \times K_b$$

where

- $\Delta T_b$ is the change in temperature of the boiling point in Celsius degrees (relative to the boiling point of the pure solvent);
- $m$ is the molality of the solution in moles solute per kilogram solvent; and
- $K_b$ is the boiling point elevation constant for the solvent.

For water,

$$K_b = 0.512 \,°C/m$$

Calculating the boiling point of a solution involves substituting into the boiling point elevation equation, as Example 13.9 demonstrates.

CONCEPTUAL
**CONNECTION 13.7**

**Boiling Point Elevation** Solution A is a 1.0 M solution with a nonionic solute and water as the solvent. Solution B is a 1.0 M solution with the same nonionic solute and ethanol as the solvent. Which solution has the greater increase in its boiling point (relative to the pure solvent)?

**Example 13.9** | **Boiling Point Elevation** ▶

What mass of ethylene glycol ($C_2H_6O_2$), in grams, must be added to 1.0 kg of water to produce a solution that boils at 105.0 °C?

| | |
|---|---|
| **SORT** You are given the desired boiling point of an ethylene glycol solution containing 1.0 kg of water and asked to find the mass of ethylene glycol you need to add to raise the boiling point. | **GIVEN:** $\Delta T_b = 5.0$ °C, 1.0 kg $H_2O$ <br><br> **FIND:** g $C_2H_6O_2$ |
| **STRATEGIZE** To solve this problem, use the boiling point elevation equation to calculate the desired molality of the solution from $\Delta T_b$. <br><br> Then use that molality to determine how many moles of ethylene glycol are needed per kilogram of water. Finally, calculate the molar mass of ethylene glycol and use it to convert from moles of ethylene glycol to mass of ethylene glycol. | **CONCEPTUAL PLAN**  <br> $\Delta T_b = m \times K_b$ <br><br>  <br><br> **RELATIONSHIPS USED** $C_2H_6O_2$ molar mass = 62.07 g/mol <br> $\Delta T_b = m \times K_b$ (boiling point elevation) |
| **SOLVE** Begin by solving the boiling point elevation equation for molality and substituting the required quantities to calculate $m$. | **SOLUTION** <br> $\Delta T_b = m \times K_b$ <br><br> $m = \dfrac{\Delta T_b}{K_b} = \dfrac{5.0 \,°C}{0.512 \,\dfrac{°C}{m}}$ <br><br> $= 9.7\underline{7} \, m$ <br><br> $1.0 \text{ kg } H_2O \times \dfrac{9.7\underline{7} \text{ mol } C_2H_6O_2}{\text{kg } H_2O} \times \dfrac{62.07 \text{ g } C_2H_6O_2}{1 \text{ mol } C_2H_6O_2} = 6.1 \times 10^2 \text{ g } C_2H_6O_2$ |

**CHECK** The units of the answer are correct. The magnitude might seem a little high initially, but the boiling point elevation constant is so small that a lot of solute is required to raise the boiling point by a small amount.

**FOR PRACTICE 13.9** Calculate the boiling point of a 3.60 $m$ aqueous sucrose solution.

# Chemistry IN YOUR DAY | Antifreeze in Frogs

Wood frogs (*Rana sylvatica*) look like most other frogs. They are a few inches long and have characteristic greenish-brown skin. However, wood frogs survive cold winters in a remarkable way—they partially freeze. In its partially frozen state, a wood frog has no heartbeat, no blood circulation, no breathing, and no brain activity. Within 1–2 hours of thawing, however, these vital functions return and the frog hops off to find food. How does the wood frog do this?

Most cold-blooded animals cannot survive freezing temperatures because the water within their cells freezes. As we learned in Section 11.9, when water freezes, it expands, irreversibly damaging cells. When the wood frog hibernates for the winter, however, it produces large amounts of glucose that is secreted into its bloodstream and fills the interior of its cells. When the temperature drops below freezing, extracellular body fluids, such as those in the frog's abdominal cavity, freeze solid. Fluids within cells, however, remain liquid because the high glucose concentration lowers their freezing point. In other words, the concentrated glucose solution within the frog's cells acts as antifreeze, preventing the water within the cells from freezing and allowing the frog to survive.

▲ The wood frog survives winter by partially freezing. It protects its cells by flooding them with glucose, which acts as an antifreeze.

### QUESTION

*The wood frog can survive at body temperatures as low as −8.0 °C. Calculate the molality of a glucose solution ($C_6H_{12}O_6$) required to lower the freezing point of water to −8.0 °C.*

## Osmotic Pressure

The process by which seawater causes dehydration (discussed in Section 13.1) is *osmosis*. **Osmosis** is the flow of solvent from a solution of lower solute concentration to one of higher solute concentration. Concentrated solutions draw solvent from more dilute solutions because of nature's tendency to mix.

Figure 13.16▼ illustrates an osmosis cell. The left side of the cell contains a concentrated saltwater solution and the right side of the cell contains pure water. A **semipermeable membrane**—a membrane that selectively allows some substances to pass through but not

### Osmosis and Osmotic Pressure

▲ **FIGURE 13.16 An Osmosis Cell** In an osmosis cell, water flows from the pure-water side of the cell through the semipermeable membrane to the saltwater side.

others—separates the two halves of the cell. Water flows by osmosis from the pure-water side of the cell through the semipermeable membrane and into the saltwater side. Over time, the water level on the left side of the cell rises, while the water level on the right side of the cell falls. If external pressure is applied to the water in the left side, this process can be opposed and even reversed. The pressure required to stop the osmotic flow, called the **osmotic pressure**, is given by the following equation:

$$\Pi = MRT$$

where M is the molarity of the solution, $T$ is the temperature (in Kelvin), and $R$ is the ideal gas constant ($0.08206 \ \text{L} \cdot \text{atm/mol} \cdot \text{K}$).

## Example 13.10 Osmotic Pressure

The osmotic pressure of a solution containing 5.87 mg of an unknown protein per 10.0 mL of solution is 2.45 torr at 25 °C. Find the molar mass of the unknown protein.

| **SORT** You are told that a solution of an unknown protein contains 5.87 mg of the protein per 10.0 mL of solution. You are also given the osmotic pressure of the solution at a particular temperature and asked to find the molar mass of the unknown protein. | **GIVEN:** 5.87 mg protein<br>10.0 mL solution<br>$\Pi = 2.45$ torr<br>$T = 25\,°\text{C}$<br><br>**FIND:** molar mass of protein (g/mol) |
|---|---|

| **STRATEGIZE**<br>Step 1: Use the given osmotic pressure and temperature to find the molarity of the protein solution. | **CONCEPTUAL PLAN**<br><br>$\Pi = MRT$ |
|---|---|
| Step 2: Use the molarity calculated in step 1 to find the number of moles of protein in 10 mL of solution. |  |
| Step 3: Finally, use the number of moles of the protein calculated in step 2 and the given mass of the protein in 10.0 mL of solution to find the molar mass. | $\text{Molar mass} = \dfrac{\text{mass protein}}{\text{moles protein}}$<br><br>**RELATIONSHIPS USED** $\Pi = MRT$ (osmotic pressure equation) |

| **SOLVE** Step 1: Begin by solving the osmotic pressure equation for molarity and substituting in the required quantities in the correct units to calculate M. | **SOLUTION**<br>$\Pi = MRT$<br><br>$M = \dfrac{\Pi}{RT} = \dfrac{2.45 \ \text{torr} \times \dfrac{1 \ \text{atm}}{760 \ \text{torr}}}{0.08206 \ \dfrac{\text{L} \cdot \text{atm}}{\text{mol} \cdot \text{K}}(298 \ \text{K})}$<br><br>$= 1.3\underline{1}8 \times 10^{-4} \ \text{M}$ |
|---|---|
| Step 2: Begin with the given volume, convert to liters, then use the molarity to find the number of moles of protein. | $10.0 \ \text{mL} \times \dfrac{1 \ \text{L}}{1000 \ \text{mL}} \times \dfrac{1.3\underline{1}8 \times 10^{-4} \ \text{mol}}{\text{L}}$<br><br>$\qquad\qquad\qquad = 1.3\underline{1}8 \times 10^{-6} \ \text{mol}$ |
| Step 3: Use the given mass and the number of moles from step 2 to calculate the molar mass of the protein. | $\text{Molar mass} = \dfrac{\text{mass protein}}{\text{moles protein}}$<br><br>$= \dfrac{5.87 \times 10^{-3} \ \text{g}}{1.3\underline{1}8 \times 10^{-6} \ \text{mol}} = 4.45 \times 10^{3} \ \text{g/mol}$ |

**CHECK**  The units of the answer are correct. The magnitude might seem a little high initially, but proteins are large molecules and therefore have high molar masses.

**FOR PRACTICE 13.10**  Calculate the osmotic pressure (in atm) of a solution containing 1.50 g ethylene glycol ($C_2H_6O_2$) in 50.0 mL of solution at 25 °C.

# 13.7  Colligative Properties of Strong Electrolyte Solutions

At the beginning of Section 13.6, we saw that colligative properties depend on the number of dissolved particles and that we must therefore treat electrolytes slightly differently than nonelectrolytes when determining colligative properties. For example, the freezing point depression of a 0.10 *m* sucrose solution is $\Delta T_f = 0.186$ °C. However, the freezing point depression of a 0.10 *m* sodium chloride solution is nearly twice this large. Why? Because 1 mol of sodium chloride dissociates into nearly 2 mol of ions in solution. The ratio of moles of particles in solution to moles of formula units dissolved is called the **van't Hoff factor (*i*)**:

$$i = \frac{\text{moles of particles in solution}}{\text{moles of formula units dissolved}}$$

**TABLE 13.9 Van't Hoff Factors at 0.05 *m* Concentration in Aqueous Solution**

| Solute | *i* Expected | *i* Measured |
|---|---|---|
| Nonelectrolyte | 1 | 1 |
| NaCl | 2 | 1.9 |
| MgSO$_4$ | 2 | 1.3 |
| MgCl$_2$ | 3 | 2.7 |
| K$_2$SO$_4$ | 3 | 2.6 |
| FeCl$_3$ | 4 | 3.4 |

Since 1 mol of NaCl produces 2 mol of particles in solution, we expect the van't Hoff factor for NaCl to be exactly 2. In reality, this expected factor only occurs in very dilute solutions. For example, the van't Hoff factor for a 0.10 *m* NaCl solution is 1.87, and for a 0.010 *m* NaCl solution the van't Hoff factor is 1.94. The van't Hoff factor approaches the expected value at infinite dilution (as the concentration approaches zero). Table 13.9 lists the actual and expected van't Hoff factors for a number of solutes.

The reason that van't Hoff factors do not exactly equal expected values is that some ions effectively pair in solution. We expect the dissociation of an ionic compound to be complete in solution. In reality, however, the dissociation is not complete—at any moment, some cations pair with anions (Figure 13.17▶), slightly reducing the number of particles in solution.

Ion pairing

◀ **FIGURE 13.17  Ion Pairing** Hydrated anions and cations may get close enough together to effectively pair, lowering the concentration of particles below what would be expected.

To calculate freezing point depression, boiling point elevation, and osmotic pressure of ionic solutions, we use the van't Hoff factor in each equation as follows:

$$\Delta T_f = im \times K_f \text{ (freezing point depression)}$$

$$\Delta T_b = im \times K_b \text{ (boiling point elevation)}$$

$$\Pi = iMRT \text{ (osmotic pressure)}$$

**Colligative Properties**  Which solution has the highest boiling point?

    **(a)**  0.50 M C$_{12}$H$_{22}$O$_{11}$
    **(b)**  0.50 M NaCl
    **(c)**  0.50 M MgCl$_2$

CONCEPTUAL
**CONNECTION 13.8**

## Example 13.11 | Van't Hoff Factor and Freezing Point Depression

The freezing point of an aqueous 0.050 $m$ CaCl$_2$ solution is $-0.27$ °C. What is the van't Hoff factor ($i$) for CaCl$_2$ at this concentration? How does it compare to the expected value of $i$?

| | |
|---|---|
| **SORT** You are given the molality of a solution and its freezing point. You are asked to find the value of $i$, the van't Hoff factor, and compare it to the expected value. | **GIVEN:** 0.050 $m$ CaCl$_2$ solution, $\Delta T_f = 0.27$ °C <br><br> **FIND:** $i$ |
| **STRATEGIZE** To solve this problem, use the freezing point depression equation including the van't Hoff factor. | **CONCEPTUAL PLAN** <br> $\Delta T_f = im \times K_f$ |
| **SOLVE** Solve the freezing point depression equation for $i$, substituting in the given quantities to calculate its value. <br><br> The expected value of $i$ for CaCl$_2$ is 3 because calcium chloride forms 3 mol of ions for each mole of calcium chloride that dissolves. Because of ion pairing, the experimental value is slightly less than 3. | **SOLUTION** <br> $\Delta T_f = im \times K_f$ <br><br> $i = \dfrac{\Delta T_f}{m \times K_f}$ <br><br> $= \dfrac{0.27\,°\text{C}}{0.050\,m \times \dfrac{1.86\,°\text{C}}{m}}$ <br><br> $= 2.9$ |

**CHECK** The answer has no units, as expected since $i$ is a ratio. The magnitude is about right since it is close to the value you would expect upon complete dissociation of CaCl$_2$.

**FOR PRACTICE 13.11** Calculate the freezing point of an aqueous 0.10 $m$ FeCl$_3$ solution using a van't Hoff factor of 3.2.

## Strong Electrolytes and Vapor Pressure

Just as the freezing point depression of a solution containing an electrolyte solute is greater than that of a solution containing the same concentration of a nonelectrolyte solute, so the vapor pressure lowering is greater (for the same reasons). The vapor pressure for a sodium chloride solution, for example, is lowered about twice as much as it is for a nonelectrolyte solution of the same concentration. To calculate the vapor pressure of a solution containing an ionic solute, we need to account for the dissociation of the solute when we calculate the mole fraction of the solvent, as we demonstrate in Example 13.12.

## Example 13.12 | Calculating the Vapor Pressure of a Solution Containing an Ionic Solute

A solution contains 0.102 mol Ca(NO$_3$)$_2$ and 0.927 mol H$_2$O. Calculate the vapor pressure of the solution at 55 °C. The vapor pressure of pure water at 55 °C is 118.1 torr. (Assume that the solute completely dissociates.)

| | |
|---|---|
| **SORT** You are given the number of moles of each component of a solution and asked to find the vapor pressure of the solution. You are also given the vapor pressure of pure water at the appropriate temperature. | **GIVEN:** 0.102 mol Ca(NO$_3$)$_2$ <br><br> 0.927 mol H$_2$O <br><br> $P^\circ_{H_2O} = 118.1$ torr (at 55 °C) <br><br> **FIND:** $P_{solution}$ |
| **STRATEGIZE** To solve this problem, use Raoult's law as you did in Example 13.6. Calculate $\chi_{solvent}$ from the given amounts of solute and solvent. | **CONCEPTUAL PLAN** <br><br> $\boxed{\chi_{H_2O},\ P^\circ_{H_2O}} \longrightarrow \boxed{P_{solution}}$ <br><br> $P_{solution} = \chi_{H_2O} P^\circ_{H_2O}$ |

**SOLVE** The key to this problem is to understand the dissociation of calcium nitrate. Write the equation showing the dissociation.

**SOLUTION**

$$Ca(NO_3)_2(s) \longrightarrow Ca^{2+}(aq) + 2\,NO_3^-(aq)$$

Since 1 mol of calcium nitrate dissociates into 3 mol of dissolved particles, multiply the number of moles of calcium nitrate by 3 when calculating the mole fraction.

$$\chi_{H_2O} = \frac{n_{H_2O}}{3 \times n_{Ca(NO_3)_2} + n_{H_2O}}$$

$$= \frac{0.927\ \text{mol}}{3\,(0.102)\ \text{mol} + 0.927\ \text{mol}}$$

$$= 0.751\underline{8}$$

Use the mole fraction of water and the vapor pressure of pure water to calculate the vapor pressure of the solution.

$$P_{solution} = \chi_{H_2O}P^\circ_{H_2O}$$

$$= 0.751\underline{8}\,(118.1\ \text{torr})$$

$$= 88.8\ \text{torr}$$

**CHECK** The units of the answer are correct. The magnitude also seems right because the calculated vapor pressure of the solution is significantly less than that of the pure solvent, as you would expect for a solution with a significant amount of solute.

**FOR PRACTICE 13.12** A solution contains 0.115 mol $H_2O$ and an unknown number of moles of sodium chloride. The vapor pressure of the solution at 30 °C is 25.7 torr. The vapor pressure of pure water at 30 °C is 31.8 torr. Calculate the number of moles of sodium chloride in the solution.

## Colligative Properties and Medical Solutions

Doctors and others health care workers often administer solutions to patients. The osmotic pressure of these solutions is controlled for the desired effect on the patient. Solutions having osmotic pressures greater than those of body fluids are called *hyperosmotic*. These solutions take water out of cells and tissues. When a human cell is placed in a hyperosmotic solution, it tends to shrivel as it loses water to the surrounding solution (Figure 13.18(b)▼). Solutions having osmotic pressures less than those of body fluids are called *hyposmotic*. These solutions pump water into cells. When a human cell is placed in a hyposmotic solution—such as pure water, for example—water enters the cell, sometimes causing it to burst (Figure 13.18(c)▼).

Isosmotic solution

Hyperosmotic solution

Hyposmotic solution

**(a)** Normal red blood cells

**(b)** Shriveled red blood cells

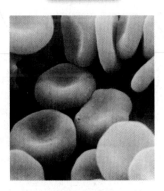

**(c)** Swollen red blood cells

▲ **FIGURE 13.18** **Red Blood Cells and Osmosis** **(a)** In an isosmotic solution, red blood cells have the normal shape shown here. In a hyperosmotic solution **(b)**, they lose water and shrivel. In a hyposmotic solution **(c)**, they swell up and may burst as water flows into the cell.

Intravenous solutions—those that are administered directly into a patient's veins—must have osmotic pressures equal to those of body fluids. These solutions are called *isosmotic* (or *isotonic*). When a patient is given an IV in a hospital, the majority of the fluid is usually an isosmotic saline solution—a solution containing 0.9 g NaCl per 100 mL of solution. In medicine and in other health-related fields, solution concentrations are often reported in units that indicate the mass of the solute per given volume of solution. Also common is *percent mass to volume*—which is the mass of the solute in grams divided by the volume of the solution in milliliters times 100%. In these units, the concentration of an isotonic saline solution is 0.9% mass/volume.

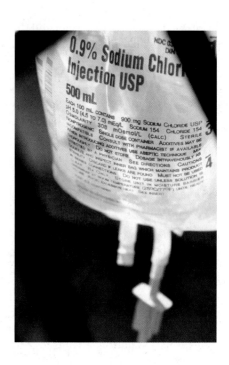

▶ Fluids used for intravenous transfusion must be isosmotic with bodily fluids—that is, they must have the same osmotic pressure as body fluids.

## 13.8 Colloids

When you mix water and soap together, the resulting mixture has a distinctive haze (Figure 13.19◀). Soapy water is hazy because soap and water form a *colloidal dispersion* rather than a true solution. A **colloidal dispersion**, or more simply a **colloid**, is a mixture in which a dispersed substance (which is solute-like) is finely divided in a dispersing medium (which is solvent-like). Examples of colloids including fog, smoke, whipped cream, and milk are listed in Table 13.10.

Whether or not a mixture is a colloid is determined by the size of the particles it contains. If the particles are small (for example, individual small molecules), then the mixture is a solution. If the particles have a diameter greater than 1 μm (for example, grains of sand), then the mixture is a heterogeneous mixture. Sand stirred into water slowly settles out of the water. *If the particles are between 1 nm and 1000 nm in size, the mixture is a colloid.* Colloidal particles are small enough that they stay dispersed throughout the dispersing medium by collisions with other molecules or atoms. When you view a colloidal particle dispersed in a liquid under a microscope, you can witness its jittery motion, which proceeds along a random path, as shown in Figure 13.20▶. This motion, called Brownian motion, is caused by collisions with molecules in the liquid. In the beginning of the twentieth century, Brownian motion was a decisive factor in confirming the molecular and atomic nature of matter.

Soap forms a colloid because of its unique structure, shown in Figure 13.21▶. One end of the molecule is ionic and therefore interacts strongly with water molecules via ion–dipole interactions. The other end of the soap molecule is a long, nonpolar, hydrocarbon tail.

▲ **FIGURE 13.19 A Colloid** Soapy water is an example of a colloidal dispersion. The haze is due to the scattering of light by the colloidal particles.

**TABLE 13.10 Types of Colloidal Dispersions**

| Classification | Dispersing Substance (Solute-like) | Dispersing Medium (Solvent-like) | | Example |
|---|---|---|---|---|
| Aerosol | Liquid | Gas | 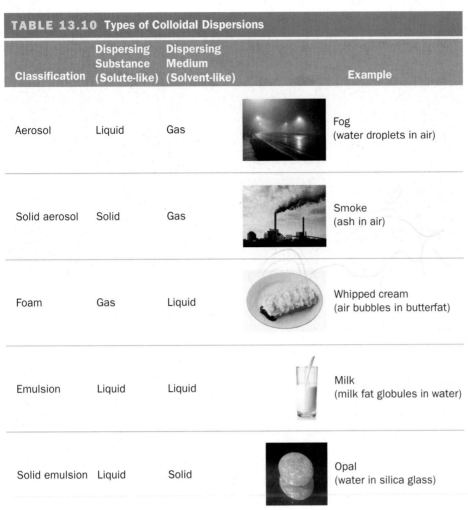 | Fog (water droplets in air) |
| Solid aerosol | Solid | Gas | | Smoke (ash in air) |
| Foam | Gas | Liquid | | Whipped cream (air bubbles in butterfat) |
| Emulsion | Liquid | Liquid | | Milk (milk fat globules in water) |
| Solid emulsion | Liquid | Solid | | Opal (water in silica glass) |

When enough soap is added to water, the soap molecules aggregate in structures called *micelles* (Figure 13.22▼). In a micelle, the nonpolar hydrocarbon tails crowd into the center of a sphere to maximize their interactions with one another. The ionic heads orient toward the surface of the sphere where they can interact with water molecules. The micelle structures are responsible for the haze seen in soapy water—they are too small to be seen by the naked eye, but they still scatter light (as particles in other types of colloids do).

▲ **FIGURE 13.20 Brownian Motion** A colloidal particle exhibits Brownian motion, moving in a jerky, haphazard path as it undergoes collisions with molecules in the liquid.

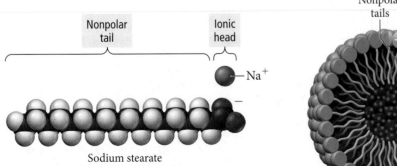

▲ **FIGURE 13.21 Soap Molecule Structure** A soap molecule has a charged ionic head and a long nonpolar hydrocarbon tail.

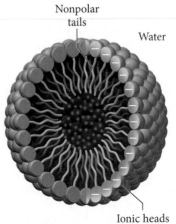

◄ **FIGURE 13.22 Micelle Structure** In a micelle, the nonpolar tails of soap molecules (or of other molecules that have properties that are similar to soap) are oriented inward (where they can interact with one another), and the ionic heads are oriented outward (where they can interact with the polar water molecules).

▲ **FIGURE 13.23** **The Tyndall Effect** When a light beam passes through a colloidal suspension (left), it is visible because the colloid particles scatter some of the light. The beam is not visible in a noncolloidal solution (right), nor would it be visible in pure water.

▲ Light beams are invisible when they are not scattered by colloidally dispersed particles such as dust or mist in the air.

This scattering of light by a colloidal dispersion is known as the **Tyndall effect** (Figure 13.23▲). You can observe the Tyndall effect in other colloids such as fog (water droplets dispersed in air) or dusty air. In fact, you can use the Tyndall effect as a test to determine whether a mixture is a solution or a colloid, since solutions contain completely dissolved solute molecules that are too small to scatter light.

Colloidal suspensions of micelles are kept stable by electrostatic repulsions that occur at their surfaces. For example, in soap, the ionic heads of the soap molecules compose the surface of the spherical particle (Figure 13.24▼). These ionic heads interact strongly with water molecules but repel other colloid particles. Heating a colloid composed of micelles can destroy the micelles because collisions occur with enough force to overcome the electrostatic repulsions and allow the molecules within the micelles to coalesce with those in other micelles. Similarly, adding an electrolyte to a colloidal suspension of micelles can also disrupt the electrostatic repulsions that occur between the particles and thus destroy the colloid. For this reason, soap does not work well in a saltwater solution.

The particles in a colloid need not be clusters of molecules. Some colloids, such as many protein solutions, contain dispersed macromolecules. For example, blood contains dispersed hemoglobin. The hemoglobin molecules are so large that they scatter light; thus, blood is considered a colloid.

▲ **FIGURE 13.24** **Micelle Repulsions** Micelles do not coalesce because the charged surface of one micelle repels the charged surface of another.

# Self-Assessment Quiz

**Q1.** Which compound do you expect to be soluble in octane ($C_8H_{18}$)?

a) $CH_3OH$    b) $CBr_4$    c) $H_2O$    d) $NH_3$

**Q2.** An aqueous solution is saturated in both potassium chlorate and carbon dioxide gas at room temperature. What happens when the solution is warmed to 85 °C?

a) Potassium chlorate precipitates out of solution.

b) Carbon dioxide bubbles out of solution.

c) Potassium chlorate precipitates out of solution and carbon dioxide bubbles out of solution.

d) Nothing happens; all of the potassium chloride and the carbon dioxide remain dissolved in solution.

**Q3.** A 500.0-mL sample of pure water is allowed to come to equilibrium with pure oxygen gas at a pressure of 755 mmHg. What mass of oxygen gas dissolves in the water? (The Henry's law constant for oxygen gas is $1.3 \times 10^{-3}$ M/atm.)

a) 15.7 g    b) $6.5 \times 10^{-3}$ g

c) 0.041 g    d) 0.021 g

**Q4.** A potassium bromide solution is 7.55% potassium bromide by mass and its density is 1.03 g/mL. What mass of potassium bromide is contained in 35.8 mL of the solution?

a) 2.78 g    b) 2.70 g    c) 4.88 g    d) 2.62 g

**Q5.** A solution contains 22.4 g glucose ($C_6H_{12}O_6$) dissolved in 0.500 L of water. What is the molality of the solution? (Assume a density of 1.00 g/mL for water.)

a) 0.238 $m$    b) 44.8 $m$    c) 0.249 $m$    d) 4.03 $m$

**Q6.** A sodium nitrate solution is 12.5% $NaNO_3$ by mass and has a density of 1.02 g/mL. Calculate the molarity of the solution.

a) 1.44 M    b) 12.8 M

c) 6.67 M    d) 1.50 M

**Q7.** Determine the vapor pressure at 25 °C of an aqueous ethylene glycol ($C_2H_6O_2$) solution that is 14.8% $C_2H_6O_2$ by mass. The vapor pressure of pure water at 25 °C is 23.8 torr.

a) 3.52 torr    b) 22.7 torr    c) 1.14 torr    d) 20.3 torr

**Q8.** A solution contains a mixture of substance A and substance B, both of which are volatile. The mole fraction of substance A is 0.35. At 32 °C the vapor pressure of pure A is 87 mmHg and the vapor pressure of pure B is 122 mmHg. What is the total vapor pressure of the solution at this temperature?

a) 110 mmHg    b) 209 mmHg

c) 99.3 mmHg    d) 73.2 mmHg

**Q9.** What mass of glucose ($C_6H_{12}O_6$) should be dissolved in 10.0 kg of water to obtain a solution with a freezing point of −4.2 °C?

a) 0.023 kg    b) 4.1 kg

c) 0.41 kg    d) 14.1 kg

**Q10.** Which of these aqueous solutions has the highest boiling point?

a) 1.25 M $C_6H_{12}O_6$

b) 1.25 M $KNO_3$

c) 1.25 M $Ca(NO_3)_2$

d) None of the above (they all have the same boiling point)

**Q11.** The osmotic pressure of a solution containing 22.7 mg of an unknown protein in 50.0 mL of solution is 2.88 mmHg at 25 °C. Determine the molar mass of the protein.

a) 246 g/mol

b) 3.85 g/mol

c) $2.93 \times 10^3$ g/mol

d) 147 g/mol

**Q12.** The enthalpy of solution for NaOH is −44.46 kJ/mol. What can you conclude about the relative magnitudes of the absolute values of $\Delta H_{solute}$ and $\Delta H_{hydration}$, where $\Delta H_{solute}$ is the heat associated with separating the solute particles and $\Delta H_{hydration}$ is the heat associated with dissolving the solute particles in water?

a) $|\Delta H_{solute}| > |\Delta H_{hydration}|$

b) $|\Delta H_{solute}| < |\Delta H_{hydration}|$

c) $|\Delta H_{solute}| = |\Delta H_{hydration}|$

d) None of the above (nothing can be concluded about the relative magnitudes)

**Q13.** A 2.4 $m$ aqueous solution of an ionic compound with the formula $MX_2$ has a boiling point of 103.4 °C. Calculate the van't Hoff factor ($i$) for $MX_2$ at this concentration.

a) 2.8    b) 83    c) 0.73    d) 1.0

**Q14.** A solution is an equimolar mixture of two volatile components A and B. Pure A has a vapor pressure of 50 torr and pure B has a vapor pressure of 100 torr. The vapor pressure of the mixture is 85 torr. What can you conclude about the relative strengths of the intermolecular forces between particles of A and B (relative to those between particles of A and those between particles of B)?

a) The intermolecular forces between particles A and B are *weaker* than those between particles of A and those between particles of B.

b) The intermolecular forces between particles A and B are *stronger* than those between particles of A and those between particles of B.

c) The intermolecular forces between particles A and B are *the same as* those between particles of A and those between particles of B.

d) Nothing can be concluded about the relative strengths of intermolecular forces from this observation.

**Q15.** An aqueous solution is in equilibrium with a gaseous mixture containing an equal number of moles of oxygen, nitrogen, and helium. Rank the relative concentrations of each gas in the aqueous solution from highest to lowest.

a) $[O_2] > [N_2] > [He]$

b) $[He] > [N_2] > [O_2]$

c) $[N_2] > [He] > [O_2]$

d) $[N_2] > [O_2] > [He]$

**Answers: 1.** (b) **2.** (b) **3.** (d) **4.** (a) **5.** (c) **6.** (d) **7.** (b) **8.** (a) **9.** (b) **10.** (c) **11.** (c) **12.** (b) **13.** (a) **14.** (a) **15.** (a)

# Chapter 13 IN REVIEW

## Key Terms

### Section 13.1
solution (572)
solvent (572)
solute (572)

### Section 13.2
aqueous solution (573)
solubility (573)
entropy (574)
miscible (575)

### Section 13.3
enthalpy of solution ($\Delta H_{soln}$) (578)
heat of hydration ($\Delta H_{hydration}$) (579)

### Section 13.4
dynamic equilibrium (582)
saturated solution (582)
unsaturated solution (582)

supersaturated solution (582)
recrystallization (583)
Henry's law (584)

### Section 13.5
dilute solution (585)
concentrated solution (585)
molarity (M) (588)
molality ($m$) (588)
parts by mass (588)
percent by mass (588)
parts per million (ppm) (588)
parts per billion (ppb) (588)
parts by volume (588)
mole fraction ($\chi_{solute}$) (589)
mole percent (mol%) (590)

### Section 13.6
colligative property (593)
Raoult's law (595)
vapor pressure lowering ($\Delta P$) (595)
ideal solution (597)
freezing point depression (600)
boiling point elevation (600)
osmosis (603)
semipermeable membrane (603)
osmotic pressure (604)

### Section 13.7
van't Hoff factor ($i$) (605)

### Section 13.8
colloidal dispersion (colloid) (608)
Tyndall effect (610)

## Key Concepts

### Solutions (13.1, 13.2)
▶ A solution is a homogeneous mixture of two or more substances. In a solution, the majority component is the solvent and the minority component is the solute.
▶ The tendency toward greater entropy (or greater energy dispersal) is the driving force for solution formation.
▶ In aqueous solutions, water is a solvent, and a solid, liquid, or gas is the solute.

### Solubility and Energetics of Solution Formation (13.2, 13.3)
▶ The solubility of a substance is the amount of the substance that dissolves in a given amount of solvent. The solubility of one substance in another depends on the types of intermolecular forces that exist *between* the substances as well as *within* each substance.
▶ We can determine the overall enthalpy change upon solution formation by adding the enthalpy changes for the three steps of solution formation: (1) separation of the solute particles, (2) separation of the solvent particles, and (3) mixing of the solute and solvent particles. The first two steps are both endothermic, whereas the last is exothermic.
▶ In aqueous solutions of an ionic compound, the combined change in enthalpy for steps 2 and 3 is the heat of hydration ($\Delta H_{hydration}$), which is always negative.

### Solution Equilibrium (13.4)
▶ Dynamic equilibrium in a solution occurs when the rates of dissolution and recrystallization in a solution are equal. A solution in this state is saturated. Solutions containing less than or more than the equilibrium amount of solute are unsaturated or supersaturated, respectively.

▶ The solubility of most solids in water increases with increasing temperature.
▶ The solubility of gases in water generally decreases with increasing temperature, but it increases with increasing pressure.

### Concentration Units (13.5)
▶ Common units to express solution concentration include molarity (M), molality ($m$), mole fraction ($\chi$), mole percent (mol %), percent (%) by mass or volume, parts per million (ppm) by mass or volume, and parts per billion (ppb) by mass or volume. Table 13.5 summarizes these units.

### Vapor Pressure Lowering, Freezing Point Depression, Boiling Point Elevation, and Osmosis (13.6, 13.7)
▶ The presence of a nonvolatile solute in a liquid results in a lower vapor pressure of the solution relative to the vapor pressure of the pure liquid. Raoult's law for an ideal solution predicts this lower vapor pressure.
▶ If the solute–solvent interactions are particularly strong, the actual vapor pressure is lower than that predicted by Raoult's law.
▶ If the solute–solvent interactions are particularly weak, the actual vapor pressure is higher than that predicted by Raoult's law.
▶ The addition of a nonvolatile solute to a liquid results in a solution with a lower freezing point and a higher boiling point than those of the pure solvent.
▶ The flow of solvent from a solution of lower concentration to a solution of higher concentration is osmosis.
▶ All of these phenomena (vapor pressure lowering, freezing point depression, boiling point elevation, and osmosis) are colligative properties and depend only on the number of solute particles added, not the type of solute particles.

▶ Electrolyte solutes have a greater effect on these properties than the corresponding amount of a nonelectrolyte solute as specified by the van't Hoff factor.

## Colloids (13.8)

▶ A colloid is a mixture in which a substance is finely divided in a dispersing medium.

▶ Colloidal mixtures occur when the dispersed substance ranges in size from 1 nm to 1000 nm.

▶ One way to identify colloidal mixtures is by their tendency to scatter light, known as the Tyndall effect.

## Key Equations and Relationships

Henry's Law: Solubility of Gases with Increasing Pressure (**13.4**)

$$S_{gas} = k_H P_{gas} \ (k_H \text{ is Henry's law constant})$$

Molarity (M) of a Solution (**13.5**)

$$(M) = \frac{\text{amount solute (in mol)}}{\text{volume solution (in L)}}$$

Molality (m) of a Solution (**13.5**)

$$(m) = \frac{\text{amount solute (in mol)}}{\text{mass solvent (in kg)}}$$

Concentration of a Solution in Parts by Mass and Parts by Volume (**13.5**)

$$\text{Percent by mass} = \frac{\text{mass solute} \times 100\%}{\text{mass solution}}$$

$$\text{Parts per million (ppm)} = \frac{\text{mass solute} \times 10^6}{\text{mass solution}}$$

$$\text{Parts per billion (ppb)} = \frac{\text{mass solute} \times 10^9}{\text{mass solution}}$$

$$\text{Parts by volume} = \frac{\text{volume solute} \times \text{multiplication factor}}{\text{volume solution}}$$

Concentration of a Solution in Mole Fraction ($\chi$) and Mole Percent (**13.5**)

$$\chi_{solute} = \frac{n_{solute}}{n_{solute} + n_{solvent}}$$
$$\text{Mol \%} = \chi \times 100\%$$

Raoult's Law: Relationship between the Vapor Pressure of a Solution ($P_{solution}$), the Mole Fraction of the Solvent ($\chi_{solvent}$), and the Vapor Pressure of the Pure Solvent ($P^\circ_{solvent}$) (**13.6**)

$$P_{solution} = \chi_{solvent} P^\circ_{solvent}$$

The Vapor Pressure of a Solution Containing Two Volatile Components (**13.6**)

$$P_A = \chi_A P^\circ_A$$
$$P_B = \chi_B P^\circ_B$$
$$P_{tot} = P_A + P_B$$

Relationship between Freezing Point Depression ($\Delta T_f$), Molality (m), and Freezing Point Depression Constant ($K_f$) (**13.6**)

$$\Delta T_f = m \times K_f$$

Relationship between Boiling Point Elevation ($\Delta T_b$), Molality (m), and Boiling Point Elevation Constant ($K_b$) (**13.6**)

$$\Delta T_b = m \times K_b$$

Relationship between Osmotic Pressure ($\Pi$), Molarity (M), the Ideal Gas Constant (R), and Temperature (T, in K) (**13.6**)

$$\Pi = MRT \ (R = 0.08206 \ \text{L} \cdot \text{atm}/\text{mol} \cdot \text{K})$$

van't Hoff Factor (i): Ratio of Moles of Particles in Solution to Moles of Formula Units Dissolved (**13.7**)

$$i = \frac{\text{moles of particles in solution}}{\text{moles of formula units dissolved}}$$

## Key Learning Outcomes

| CHAPTER OBJECTIVES | ASSESSMENT |
|---|---|
| Determining Whether a Solute Is Soluble in a Solvent (**13.2**) | Example 13.1    For Practice 13.1    Exercises 31–34 |
| Using Henry's Law to Predict the Solubility of Gases with Increasing Pressure (**13.4**) | Example 13.2    For Practice 13.2    Exercises 49, 50 |
| Calculating Concentrations of Solutions (**13.5**)  $\dfrac{3.5 \text{ g NaCl}}{100 \text{ g solution}}$ converts **g solution** → **g NaCl** | Examples 13.3, 13.4    For Practice 13.3, 13.4    For More Practice 13.3    Exercises 51–56, 63, 64 |
| Converting between Concentration Units (**13.5**) | Example 13.5    For Practice 13.5    For More Practice 13.5    Exercises 65–68 |
| Determining the Vapor Pressure of a Solution Containing a Nonelectrolyte and Nonvolatile Solute (**13.6**) | Example 13.6    For Practice 13.6    For More Practice 13.6    Exercises 71, 72 |
| Determining the Vapor Pressure of a Two-Component Solution (**13.6**) | Example 13.7    For Practice 13.7    Exercises 73–76 |
| Calculating Freezing Point Depression (**13.6**) | Example 13.8    For Practice 13.8    Exercises 77–82, 87, 88 |
| Calculating Boiling Point Elevation (**13.6**) | Example 13.9    For Practice 13.9    Exercises 77–80, 87, 88 |
| Determining the Osmotic Pressure (**13.6**) | Example 13.10    For Practice 13.10    Exercises 83–86 |

| Determining and Using the van't Hoff Factor (**13.7**) | Example 13.11 For Practice 13.11 Exercises 91–96 |
| --- | --- |
| Determining the Vapor Pressure of a Solution Containing an Ionic Solute (**13.7**)  | Example 13.12 For Practice 13.12 Exercises 97, 98 |

# EXERCISES

## Review Questions

1. Explain why drinking seawater results in dehydration.

2. What is a solution? What are the solute and solvent?

3. What does it mean to say that a substance is soluble in another substance? Which units are used in reporting solubility?

4. Why do two ideal gases thoroughly mix when combined? What drives the mixing?

5. What is entropy? Why is entropy important in discussing the formation of solutions?

6. What kinds of intermolecular forces are involved in solution formation?

7. Explain how the relative strengths of solute–solute interactions, solvent–solvent interactions, and solvent–solute interactions affect solution formation.

8. What does the statement *like dissolves like* mean with respect to solution formation?

9. What are the three steps involved in evaluating the enthalpy changes associated with solution formation?

10. What is the heat of hydration ($\Delta H_{hydration}$)? How does the enthalpy of solution depend on the relative magnitudes of $\Delta H_{solute}$ and $\Delta H_{hydration}$?

11. Explain dynamic equilibrium with respect to solution formation. What is a saturated solution? An unsaturated solution? A supersaturated solution?

12. How does the solubility of a solid in a liquid depend on temperature? How is this temperature dependence exploited to purify solids through recrystallization?

13. How does the solubility of a gas in a liquid depend on temperature? How does this temperature dependence affect the amount of oxygen available for fish and other aquatic animals?

14. How does the solubility of a gas in a liquid depend on pressure? How does this pressure dependence account for the bubbling that occurs upon opening a can of soda?

15. What is Henry's law? For what kinds of calculations is Henry's law useful?

16. What are the common units for expressing solution concentration?

17. How are parts by mass and parts by volume used in calculations?

18. What is the effect of a nonvolatile solute on the vapor pressure of a liquid? Why is the vapor pressure of a solution different from the vapor pressure of the pure liquid solvent?

19. What is Raoult's law? For what kind of calculations is Raoult's law useful?

20. Explain the difference between an ideal and a nonideal solution.

21. What is the effect on vapor pressure of a solution with particularly *strong* solute–solvent interactions? With particularly *weak* solute–solvent interactions?

22. Explain why the lower vapor pressure for a solution containing a nonvolatile solute results in a higher boiling point and lower melting point compared to the pure solvent.

23. What are colligative properties?

24. What is osmosis? What is osmotic pressure?

25. Explain the meaning of the van't Hoff factor and its role in determining the colligative properties of solutions containing ionic solutes.

26. Describe a colloidal dispersion. What is the difference between a colloidal dispersion and a true solution?

27. What is the Tyndall effect, and how can it be used to help identify colloidal dispersions?

28. What keeps the particles in a colloidal dispersion from coalescing?

## Problems by Topic

### Solubility

29. Pick an appropriate solvent from Table 13.3 to dissolve each substance. State the kind of intermolecular forces that would occur between the solute and solvent in each case.
   a. motor oil (nonpolar)
   b. ethanol (polar, contains an OH group)
   c. lard (nonpolar)
   d. potassium chloride (ionic)

30. Pick an appropriate solvent from Table 13.3 to dissolve each substance. State the kind of intermolecular forces that would occur between the solute and solvent in each case.
   a. isopropyl alcohol (polar, contains an OH group)
   b. sodium chloride (ionic)
   c. vegetable oil (nonpolar)
   d. sodium nitrate (ionic)

31. Which molecule would you expect to be more soluble in water: $CH_3CH_2CH_2OH$ or $HOCH_2CH_2CH_2OH$?

**32.** Which molecule would you expect to be more soluble in water: $CCl_4$ or $CH_2Cl_2$?

**33.** For each compound, would you expect greater solubility in water or in hexane? Indicate the kinds of intermolecular forces that occur between the solute and the solvent in which the molecule is most soluble.

**a.** glucose

**b.** naphthalene

**c.** dimethyl ether

**d.** alanine (an amino acid)

**34.** For each compound, would you expect greater solubility in water or in hexane? Indicate the kinds of intermolecular forces that would occur between the solute and the solvent in which the molecule is most soluble.

**a.** toluene

**b.** sucrose (table sugar)

**c.** isobutene

**d.** ethylene glycol

## Energetics of Solution Formation

**35.** When ammonium chloride ($NH_4Cl$) is dissolved in water, the solution becomes colder.
  **a.** Is the dissolution of ammonium chloride endothermic or exothermic?
  **b.** What can you conclude about the relative magnitudes of the lattice energy of ammonium chloride and its heat of hydration?
  **c.** Sketch a qualitative energy diagram similar to Figure 13.7 for the dissolution of $NH_4Cl$.
  **d.** Why does the solution form? What drives the process?

**36.** When lithium iodide (LiI) is dissolved in water, the solution becomes hotter.
  **a.** Is the dissolution of lithium iodide endothermic or exothermic?
  **b.** What can you conclude about the relative magnitudes of the lattice energy of lithium iodide and its heat of hydration?
  **c.** Sketch a qualitative energy diagram similar to Figure 13.7 for the dissolution of LiI.
  **d.** Why does the solution form? What drives the process?

**37.** Silver nitrate has a lattice energy of $-820$ kJ/mol and a heat of solution of 22.6 kJ/mol. Calculate the heat of hydration for silver nitrate.

**38.** Use the data to calculate the heats of hydration of lithium chloride and sodium chloride. Which of the two cations, lithium or sodium, has stronger ion–dipole interactions with water? Why?

| Compound | Lattice Energy (kJ/mol) | $\Delta H_{soln}$ (kJ/mol) |
|---|---|---|
| LiCl | −834 | −37.0 |
| NaCl | −769 | +3.88 |

**39.** Lithium iodide has a lattice energy of $-7.3 \times 10^2$ kJ/mol and a heat of hydration of $-793$ kJ/mol. Find the heat of solution for lithium iodide and determine how much heat is evolved or absorbed when 15.0 g of lithium iodide completely dissolves in water.

**40.** Potassium nitrate has a lattice energy of $-163.8$ kcal/mol and a heat of hydration of $-155.5$ kcal/mol. How much potassium nitrate has to dissolve in water to absorb $1.00 \times 10^2$ kJ of heat?

## Solution Equilibrium and Factors Affecting Solubility

**41.** A solution contains 25 g of NaCl per 100.0 g of water at 25 °C. Is the solution unsaturated, saturated, or supersaturated? (Use Figure 13.11.)

**42.** A solution contains 32 g of $KNO_3$ per 100.0 g of water at 25 °C. Is the solution unsaturated, saturated, or supersaturated? (Use Figure 13.11.)

**43.** A $KNO_3$ solution containing 45 g of $KNO_3$ per 100.0 g of water is cooled from 40 °C to 0 °C. What happens during cooling? (Use Figure 13.11.)

**44.** A KCl solution containing 42 g of KCl per 100.0 g of water is cooled from 60 °C to 0 °C. What happens during cooling? (Use Figure 13.11.)

**45.** Some laboratory procedures involving oxygen-sensitive reactants or products call for using water that has been boiled (and then cooled). Explain.

**46.** A person preparing a fish tank fills the tank with water that has been boiled (and then cooled). When the person puts fish into the tank, they die. Explain.

**47.** Scuba divers breathing air at increased pressure can suffer from nitrogen narcosis—a condition resembling drunkenness—when the partial pressure of nitrogen exceeds about 4 atm. What property of gas/water solutions causes this to happen? How can a diver reverse this effect?

**48.** Scuba divers breathing air at increased pressure can suffer from oxygen toxicity—too much oxygen in their bloodstream—when the partial pressure of oxygen exceeds about 1.4 atm. What happens to the amount of oxygen in a diver's bloodstream when he or she breathes oxygen at elevated pressures? How can this be reversed?

**49.** Calculate the mass of nitrogen dissolved at room temperature in an 80.0-L home aquarium. Assume a total pressure of 1.0 atm and a mole fraction for nitrogen of 0.78.

**50.** Use Henry's law to determine the molar solubility of helium at a pressure of 1.0 atm and 25 °C.

### Concentrations of Solutions

**51.** An aqueous NaCl solution is made using 112 g of NaCl diluted to a total solution volume of 1.00 L. Calculate the molarity, molality, and mass percent of the solution. (Assume a density of 1.08 g/mL for the solution.)

**52.** An aqueous $KNO_3$ solution is made using 72.5 g of $KNO_3$ diluted to a total solution volume of 2.00 L. Calculate the molarity, molality, and mass percent of the solution. (Assume a density of 1.05 g/mL for the solution.)

**53.** To what volume should you dilute 50.0 mL of a 5.00 M KI solution so that 25.0 mL of the diluted solution contains 3.05 g of KI?

**54.** To what volume should you dilute 125 mL of an 8.00 M $CuCl_2$ solution so that 50.0 mL of the diluted solution contains 4.67 g $CuCl_2$?

**55.** Silver nitrate solutions are often used to plate silver onto other metals. What is the maximum amount of silver (in grams) that can be plated out of 4.8 L of an $AgNO_3$ solution containing 3.4% Ag by mass? Assume that the density of the solution is 1.01 g/mL.

**56.** A dioxin-contaminated water source contains 0.085% dioxin by mass. How much dioxin is present in 2.5 L of this water? Assume a density of 1.00 g/mL.

**57.** A hard water sample contains 0.0085% Ca by mass (in the form of $Ca^{2+}$ ions). How much water (in grams) contains 1.2 g of Ca? (1.2 g of Ca is the recommended daily allowance of calcium for adults between 19 and 24 years old.)

**58.** Lead is a toxic metal that affects the central nervous system. A Pb-contaminated water sample contains 0.0011% Pb by mass. How much of the water (in mL) contains 150 mg of Pb? (Assume a density of 1.0 g/mL.)

**59.** You can purchase nitric acid in a concentrated form that is 70.3% $HNO_3$ by mass and has a density of 1.41 g/mL. Describe exactly how you would prepare 1.15 L of 0.100 M $HNO_3$ from the concentrated solution.

**60.** You can purchase hydrochloric acid in a concentrated form that is 37.0% HCl by mass and that has a density of 1.20 g/mL. Describe exactly how to prepare 2.85 L of 0.500 M HCl from the concentrated solution.

**61.** Describe how to prepare each solution from the dry solute and the solvent.
   **a.** $1.00 \times 10^2$ mL of 0.500 M KCl
   **b.** $1.00 \times 10^2$ g of 0.500 *m* KCl
   **c.** $1.00 \times 10^2$ g of 5.0% KCl solution by mass

**62.** Describe how to prepare each solution from the dry solute and the solvent.
   **a.** 125 mL of 0.100 M $NaNO_3$
   **b.** 125 g of 0.100 *m* $NaNO_3$
   **c.** 125 g of 1.0% $NaNO_3$ solution by mass

**63.** A solution is prepared by dissolving 28.4 g of glucose ($C_6H_{12}O_6$) in 355 g of water. The final volume of the solution is 378 mL. For this solution, calculate the concentration in each unit.
   **a.** molarity                **b.** molality
   **c.** percent by mass          **d.** mole fraction
   **e.** mole percent

**64.** A solution is prepared by dissolving 20.2 mL of methanol ($CH_3OH$) in 100.0 mL of water at 25 °C. The final volume of the solution is 118 mL. The densities of methanol and water at this temperature are 0.782 g/mL and 1.00 g/mL, respectively. For this solution, calculate the concentration in each unit.
   **a.** molarity                **b.** molality
   **c.** percent by mass          **d.** mole fraction
   **e.** mole percent

**65.** Household hydrogen peroxide is an aqueous solution containing 3.0% hydrogen peroxide by mass. What is the molarity of this solution? (Assume a density of 1.01 g/mL.)

**66.** One brand of laundry bleach is an aqueous solution containing 4.55% sodium hypochlorite (NaOCl) by mass. What is the molarity of this solution? (Assume a density of 1.02 g/mL.)

**67.** An aqueous solution contains 36% HCl by mass. Calculate the molality and mole fraction of the solution.

**68.** An aqueous solution contains 5.0% NaCl by mass. Calculate the molality and mole fraction of the solution.

### Vapor Pressure of Solutions

**69.** A beaker contains 100.0 mL of pure water. A second beaker contains 100.0 mL of seawater. The two beakers are left side by side on a lab bench for 1 week. At the end of the week, the liquid level in both beakers has decreased. However, the level has decreased more in one of the beakers than in the other. Which one and why?

**70.** Which solution has the highest vapor pressure?
   **a.** 20.0 g of glucose ($C_6H_{12}O_6$) in 100.0 mL of water
   **b.** 20.0 g of sucrose ($C_{12}H_{22}O_{11}$) in 100.0 mL of water
   **c.** 10.0 g of potassium acetate $KC_2H_3O_2$ in 100.0 mL of water

**71.** Calculate the vapor pressure of a solution containing 24.5 g of glycerin ($C_3H_8O_3$) in 135 mL of water at 30.0 °C. The vapor pressure of pure water at this temperature is 31.8 torr. Assume that glycerin is not volatile and dissolves molecularly (i.e., it is not ionic), and use a density of 1.00 g/mL for the water.

**72.** A solution contains naphthalene ($C_{10}H_8$) dissolved in hexane ($C_6H_{14}$) at a concentration of 12.35% naphthalene by mass. Calculate the vapor pressure at 25 °C of hexane above the solution. The vapor pressure of pure hexane at 25 °C is 151 torr.

**73.** A solution contains 50.0 g of heptane ($C_7H_{16}$) and 50.0 g of octane ($C_8H_{18}$) at 25 °C. The vapor pressures of pure heptane and pure octane at 25 °C are 45.8 torr and 10.9 torr, respectively. Assuming ideal behavior, answer the following:
   **a.** What is the vapor pressure of each of the solution components in the mixture?
   **b.** What is the total pressure above the solution?
   **c.** What is the composition of the vapor in mass percent?
   **d.** Why is the composition of the vapor different from the composition of the solution?

**74.** A solution contains a mixture of pentane and hexane at room temperature. The solution has a vapor pressure of 258 torr. Pure pentane and hexane have vapor pressures of 425 torr and 151 torr, respectively, at room temperature. What is the mole fraction composition of the mixture? (Assume ideal behavior.)

**75.** A solution contains 4.08 g of chloroform ($CHCl_3$) and 9.29 g of acetone ($CH_3COCH_3$). The vapor pressures at 35 °C of pure chloroform and pure acetone are 295 torr and 332 torr, respectively. Assuming ideal behavior, calculate the vapor pressures of each of the components and the total vapor pressure above the solution. The experimentally measured total vapor pressure of the solution at 35 °C is 312 torr. Is the solution ideal? If not, what can you say about the relative strength of chloroform–acetone interactions compared to the acetone–acetone and chloroform–chloroform interactions?

**76.** A solution of methanol and water has a mole fraction of water of 0.312 and a total vapor pressure of 211 torr at 39.9 °C. The vapor pressures of pure methanol and pure water at this temperature are 256 torr and 55.3 torr, respectively. Is the solution ideal? If not, what can you say about the relative strengths of the solute–solvent interactions compared to the solute–solute and solvent–solvent interactions?

## Freezing Point Depression, Boiling Point Elevation, and Osmosis

**77.** A glucose solution contains 55.8 g of glucose ($C_6H_{12}O_6$) in 455 g of water. Determine the freezing point and boiling point of the solution.

**78.** An ethylene glycol solution contains 21.2 g of ethylene glycol ($C_2H_6O_2$) in 85.4 mL of water. Determine the freezing point and boiling point of the solution. (Assume a density of 1.00 g/mL for water.)

**79.** Calculate the freezing point and boiling point of a solution containing 10.0 g of naphthalene ($C_{10}H_8$) in 100.0 mL of benzene. Benzene has a density of 0.877 g/$cm^3$.

**80.** Calculate the freezing point and boiling point of a solution containing 7.55 g of ethylene glycol ($C_2H_6O_2$) in 85.7 mL of ethanol. Ethanol has a density of 0.789 g/$cm^3$.

**81.** An aqueous solution containing 17.5 g of an unknown molecular (nonelectrolyte) compound in 100.0 g of water has a freezing point of −1.8 °C. Calculate the molar mass of the unknown compound.

**82.** An aqueous solution containing 35.9 g of an unknown molecular (nonelectrolyte) compound in 150.0 g of water has a freezing point of −1.3 °C. Calculate the molar mass of the unknown compound.

**83.** Calculate the osmotic pressure of a solution containing 24.6 g of glycerin ($C_3H_8O_3$) in 250.0 mL of solution at 298 K.

**84.** What mass of sucrose ($C_{12}H_{22}O_{11}$) would you combine with $5.00 \times 10^2$ g of water to make a solution with an osmotic pressure of 8.55 atm at 298 K? (Assume a density of 1.0 g/mL for the solution.)

**85.** A solution containing 27.55 mg of an unknown protein per 25.0 mL solution was found to have an osmotic pressure of 3.22 torr at 25 °C. What is the molar mass of the protein?

**86.** Calculate the osmotic pressure of a solution containing 18.75 mg of hemoglobin in 15.0 mL of solution at 25 °C. The molar mass of hemoglobin is $6.5 \times 10^4$ g/mol.

**87.** Calculate the freezing point and boiling point of each aqueous solution, assuming complete dissociation of the solute.
   **a.** 0.100 $m$ $K_2S$
   **b.** 21.5 g of $CuCl_2$ in $4.50 \times 10^2$ g water
   **c.** 5.5% $NaNO_3$ by mass (in water)

**88.** Calculate the freezing point and boiling point in each solution, assuming complete dissociation of the solute.
   **a.** 10.5 g $FeCl_3$ in $1.50 \times 10^2$ g water
   **b.** 3.5% KCl by mass (in water)
   **c.** 0.150 m $MgF_2$

**89.** What mass of salt (NaCl) should you add to 1.00 L of water in an ice cream maker to make a solution that freezes at −10.0 °C? Assume complete dissociation of the NaCl and density of 1.00 g/mL for water.

**90.** Determine the required concentration (in percent by mass) for an aqueous ethylene glycol ($C_2H_6O_2$) solution to have a boiling point of 104.0 °C.

**91.** Use the van't Hoff factors in Table 13.9 to calculate each colligative property:
   **a.** the melting point of a 0.100 $m$ iron(III) chloride solution
   **b.** the osmotic pressure of a 0.085 M potassium sulfate solution at 298 K
   **c.** the boiling point of a 1.22% by mass magnesium chloride solution

**92.** Using the van't Hoff factors in Table 13.9, calculate the mass of solute required to make each aqueous solution:
   **a.** a sodium chloride solution containing $1.50 \times 10^2$ g of water that has a melting point of −1.0 °C
   **b.** $2.50 \times 10^2$ mL of a magnesium sulfate solution that has an osmotic pressure of 3.82 atm at 298 K
   **c.** an iron(III) chloride solution containing $2.50 \times 10^2$ g of water that has a boiling point of 102 °C

**93.** A 1.2 $m$ aqueous solution of an ionic compound with the formula $MX_2$ has a boiling point of 101.4 °C. Calculate the van't Hoff factor ($i$) for $MX_2$ at this concentration.

**94.** A 0.95 $m$ aqueous solution of an ionic compound with the formula MX has a freezing point of −3.0 °C. Calculate the van't Hoff factor ($i$) for MX at this concentration.

**95.** A 0.100 M ionic solution has an osmotic pressure of 8.3 atm at 25 °C. Calculate the van't Hoff factor ($i$) for this solution.

**96.** A solution contains 8.92 g of KBr in 500.0 mL of solution and has an osmotic pressure of 6.97 atm at 25 °C. Calculate the van't Hoff factor ($i$) for KBr at this concentration.

**97.** Calculate the vapor pressure at 25 °C of an aqueous solution that is 5.50% NaCl by mass. (Assume complete dissociation of the solute.)

**98.** An aqueous $CaCl_2$ solution has a vapor pressure of 81.6 mmHg at 50 °C. The vapor pressure of pure water at this temperature is 92.6 mmHg. What is the concentration of $CaCl_2$ in mass percent? (Assume complete dissociation of the solute.)

# Cumulative Problems

**99.** The solubility of carbon tetrachloride ($CCl_4$) in water at 25 °C is 1.2 g/L. The solubility of chloroform ($CHCl_3$) at the same temperature is 10.1 g/L. Why is chloroform almost ten times more soluble in water than carbon tetrachloride?

**100.** The solubility of phenol in water at 25 °C is 87 g/L. The solubility of naphthol at the same temperature is only 0.74 g/L. Examine the structures of phenol and naphthol shown here and explain why phenol is so much more soluble than naphthol.

Phenol          Naphthol

**101.** Potassium perchlorate ($KClO_4$) has a lattice energy of −599 kJ/mol and a heat of hydration of −548 kJ/mol. Find the heat of solution for potassium perchlorate and determine the temperature change that occurs when 10.0 g of potassium perchlorate is dissolved with enough water to make 100.0 mL of solution. (Assume a heat capacity of 4.05 J/g · °C for the solution and a density of 1.05 g/mL.)

**102.** Sodium hydroxide (NaOH) has a lattice energy of −887 kJ/mol and a heat of hydration of −932 kJ/mol. How much solution could be heated to boiling by the heat evolved by the dissolution of 25.0 g of NaOH? (For the solution, assume a heat capacity of 4.0 J/g · °C, an initial temperature of 25.0 °C, a boiling point of 100.0 °C, and a density of 1.05 g/mL.)

**103.** A saturated solution forms when 0.0537 L of argon, at a pressure of 1.0 atm and temperature of 25 °C, is dissolved in 1.0 L of water. Calculate the Henry's law constant for argon.

**104.** A gas has a Henry's law constant of 0.112 M/atm. What total volume of solution is needed to completely dissolve 1.65 L of the gas at a pressure of 725 torr and a temperature of 25 °C?

**105.** The Safe Drinking Water Act (SDWA) sets a limit for mercury—a toxin to the central nervous system—at 0.0020 ppm by mass. Water suppliers must periodically test their water to ensure that mercury levels do not exceed this limit. Suppose water becomes contaminated with mercury at twice the legal limit (0.0040 ppm). How much of this water would a person have to consume to ingest 50.0 mg of mercury?

**106.** Water softeners often replace calcium ions in hard water with sodium ions. Since sodium compounds are soluble, the presence of sodium ions in water does not cause the white, scaly residues caused by calcium ions. However, calcium is more beneficial to human health than sodium because calcium is a necessary part of the human diet, while high levels of sodium intake are linked to increases in blood pressure. The U.S. Food and Drug Administration (FDA) recommends that adults ingest less than 2.4 g of sodium per day. How many liters of softened water, containing a sodium concentration of 0.050% sodium by mass, would a person have to consume to exceed the FDA recommendation? (Assume a water density of 1.0 g/mL.)

**107.** An aqueous solution contains 12.5% NaCl by mass. What mass of water (in grams) is contained in 2.5 L of the vapor above this solution at 55 °C? The vapor pressure of pure water at 55 °C is 118 torr. (Assume complete dissociation of NaCl.)

**108.** The vapor above an aqueous solution contains 19.5 mg water per liter at 25 °C. Assuming ideal behavior, what is the concentration of the solute within the solution in mole percent?

**109.** What is the freezing point of an aqueous solution that boils at 106.5 °C?

**110.** What is the boiling point of an aqueous solution that has a vapor pressure of 20.5 torr at 25 °C? (Assume a nonvolatile solute.)

**111.** An isotonic solution contains 0.90% NaCl mass to volume. Calculate the percent mass to volume for isotonic solutions containing each solute at 25 °C. Assume a van't Hoff factor of 1.9 for all *ionic* solutes.
   **a.** KCl          **b.** NaBr          **c.** glucose ($C_6H_{12}O_6$)

**112.** Magnesium citrate, $Mg_3(C_6H_5O_7)_2$, belongs to a class of laxatives called *hyperosmotics*, which cause rapid emptying of the bowel. When a concentrated solution of magnesium citrate is consumed, it passes through the intestines, drawing water and promoting diarrhea, usually within 6 hours. Calculate the osmotic pressure of a magnesium citrate laxative solution containing 28.5 g of magnesium citrate in 235 mL of solution at 37 °C (approximate body temperature). Assume complete dissociation of the ionic compound.

**113.** A solution is prepared from 4.5701 g of magnesium chloride and 43.238 g of water. The vapor pressure of water above this solution is 0.3624 atm at 348.0 K. The vapor pressure of pure water at this temperature is 0.3804 atm. Find the value of the van't Hoff factor (*i*) for magnesium chloride in this solution.

**114.** When $HNO_2$ is dissolved in water, it partially dissociates according to the equation $HNO_2 \rightleftharpoons H^+ + NO_2^-$. A solution is prepared that contains 7.050 g of $HNO_2$ in 1.000 kg of water. Its freezing point is −0.2929 °C. Calculate the fraction of $HNO_2$ that has dissociated.

**115.** A solution of a nonvolatile solute in water has a boiling point of 375.3 K. Calculate the vapor pressure of water above this solution at 338 K. The vapor pressure of pure water at this temperature is 0.2467 atm.

**116.** The density of a 0.438 M solution of potassium chromate ($K_2CrO_4$) at 298 K is 1.063 g/mL. Calculate the vapor pressure of water above the solution. The vapor pressure of pure water at this temperature is 0.0313 atm. (Assume complete dissociation of the solute.)

**117.** The vapor pressure of carbon tetrachloride, $CCl_4$, is 0.354 atm, and the vapor pressure of chloroform, $CHCl_3$, is 0.526 atm at 316 K. A solution is prepared from equal masses of these two compounds at this temperature. Calculate the mole fraction of the chloroform in the vapor above the solution. If the vapor above the original solution is condensed and isolated into a separate flask, what will the vapor pressure of chloroform be above this new solution?

**118.** Distillation is a method of purification based on successive separations and recondensations of vapor above a solution. Use the result of the previous problem to calculate the mole fraction of chloroform in the vapor above a solution obtained by three successive separations and condensations of the vapors above the original solution of carbon tetrachloride and chloroform. Show how this result explains the use of distillation as a separation method.

**119.** A solution of 49.0% $H_2SO_4$ by mass has a density of 1.39 g/cm³ at 293 K. A 25.0-cm³ sample of this solution is mixed with enough water to increase the volume of the solution to 99.8 cm³. Find the molarity of sulfuric acid in this solution.

**120.** Find the mass of urea ($CH_4N_2O$) needed to prepare 50.0 g of a solution in water in which the mole fraction of urea is 0.0770.

**121.** A solution contains 10.05 g of unknown compound dissolved in 50.0 mL of water. (Assume a density of 1.00 g/mL for water.) The freezing point of the solution is $-3.16 \, °C$. The mass percent composition of the compound is 60.97% C, 11.94% H, and the rest is O. What is the molecular formula of the compound?

**122.** The osmotic pressure of a solution containing 2.10 g of an unknown compound dissolved in 175.0 mL of solution at 25 °C is 1.93 atm. The combustion of 24.02 g of the unknown compound produced 28.16 g $CO_2$ and 8.64 g $H_2O$. What is the molecular formula of the compound (which contains only carbon, hydrogen, and oxygen)?

**123.** A 100.0-mL aqueous sodium chloride solution is 13.5% NaCl by mass and has a density of 1.12 g/mL. What would you add (solute or solvent) and what mass of it to make the boiling point of the solution 104.4 °C? (Use $i = 1.8$ for NaCl.)

**124.** A 50.0-mL solution is initially 1.55% $MgCl_2$ by mass and has a density of 1.05 g/mL. What is the freezing point of the solution after you add an additional 1.35 g $MgCl_2$? (Use $i = 2.5$ for $MgCl_2$.)

# Challenge Problems

**125.** The small bubbles that form on the bottom of a water pot that is being heated (before boiling) are due to dissolved air coming out of solution. Use Henry's law and the solubilities given to calculate the total volume of nitrogen and oxygen gas that should bubble out of 1.5 L of water upon warming from 25 °C to 50 °C. Assume that the water is initially saturated with nitrogen and oxygen gas at 25 °C and a total pressure of 1.0 atm. Assume that the gas bubbles out at a temperature of 50 °C. The solubility of oxygen gas at 50 °C is 27.8 mg/L at an oxygen pressure of 1.00 atm. The solubility of nitrogen gas at 50 °C is 14.6 mg/L at a nitrogen pressure of 1.00 atm. Assume that the air above the water contains an oxygen partial pressure of 0.21 atm and a nitrogen partial pressure of 0.78 atm.

**126.** The vapor above a mixture of pentane and hexane at room temperature contains 35.5% pentane by mass. What is the mass percent composition of the solution? Pure pentane and hexane have vapor pressures of 425 torr and 151 torr, respectively, at room temperature.

**127.** A 1.10-g sample contains only glucose ($C_6H_{12}O_6$) and sucrose ($C_{12}H_{22}O_{11}$). When the sample is dissolved in water to a total solution volume of 25.0 mL, the osmotic pressure of the solution is 3.78 atm at 298 K. What is the mass percent composition of glucose and sucrose in the sample?

**128.** A solution is prepared by mixing 631 mL of methanol with 501 mL of water. The molarity of methanol in the resulting solution is 14.29 M. The density of methanol at this temperature is 0.792 g/mL. Calculate the difference in volume between this solution and the total volume of water and methanol that were mixed to prepare the solution.

**129.** Two alcohols, isopropyl alcohol and propyl alcohol, have the same molecular formula, $C_3H_8O$. A solution of the two that is two-thirds by mass isopropyl alcohol has a vapor pressure of 0.110 atm at 313 K. A solution that is one-third by mass isopropyl alcohol has a vapor pressure of 0.089 atm at 313 K. Calculate the vapor pressure of each pure alcohol at this temperature. Explain the difference given that the formula of propyl alcohol is $CH_3CH_2CH_2OH$ and that of isopropyl alcohol is $(CH_3)_2CHOH$.

**130.** A metal, M, of atomic mass 96 amu reacts with fluorine to form a salt that can be represented as $MF_x$. In order to determine $x$ and therefore the formula of the salt, a boiling point elevation experiment is performed. A 9.18-g sample of the salt is dissolved in 100.0 g of water, and the boiling point of the solution is found to be 374.38 K. Find the formula of the salt. (Assume complete dissociation of the salt in solution.)

**131.** Sulfuric acid in water dissociates completely into $H^+$ and $HSO_4^-$ ions. The $HSO_4^-$ ion dissociates to a limited extent into $H^+$ and $SO_4^{2-}$. The freezing point of a 0.1000 $m$ solution of sulfuric acid in water is 272.76 K. Calculate the molality of $SO_4^{2-}$ in the solution, assuming ideal solution behavior.

**132.** A solution of 75.0 g of benzene ($C_6H_6$) and 75.0 g of toluene ($C_7H_8$) has a total vapor pressure of 80.9 mmHg at 303 K. Another solution of 100.0 g benzene and 50.0 g toluene has a total vapor pressure of 93.9 mmHg at this temperature. Find the vapor pressure of pure benzene and pure toluene at 303 K.

**133.** A solution is prepared by dissolving 11.60 g of a mixture of sodium carbonate and sodium bicarbonate in 1.00 L of water. A 300.0 $cm^3$ sample of the solution is treated with excess $HNO_3$ and boiled to remove all the dissolved gas. A total of 0.940 L of dry $CO_2$ is collected at 298 K and 0.972 atm. Find the molarity of the carbonate and bicarbonate in the solution.

# Conceptual Problems

**134.** Substance A is a nonpolar liquid and has only dispersion forces among its constituent particles. Substance B is also a nonpolar liquid and has about the same magnitude of dispersion forces among its constituent particles as substance A. When substance A and substance B are combined, they spontaneously mix.
   **a.** Why do the two substances mix?
   **b.** Predict the sign and magnitude of $\Delta H_{soln}$.
   **c.** Determine the signs and relative magnitudes of $\Delta H_{solute}$, $\Delta H_{solvent}$, and $\Delta H_{mix}$.

**135.** A power plant built on a river uses river water as a coolant. The water is warmed as it is used in heat exchangers within the plant. Should the warm water be immediately cycled back into the river? Why or why not?

**136.** The vapor pressure of a 1 M ionic solution is different from the vapor pressure of a 1 M nonelectrolyte solution. In both cases, the solute is nonvolatile. Which set of diagrams best represents the differences between the two solutions and their vapors?

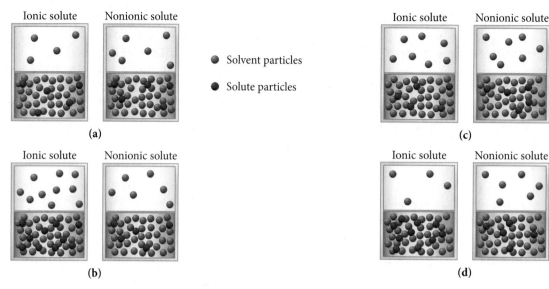

- Solvent particles
- Solute particles

**137.** If each substance listed here costs the same amount per kilogram, which would be most cost-effective as a way to lower the freezing point of water? (Assume complete dissociation for all ionic compounds.) Explain.
 a. $HOCH_2CH_2OH$     b. NaCl     c. KCl
 d. $MgCl_2$     e. $SrCl_2$

**138.** A helium balloon inflated on one day will fall to the ground by the next day. The volume of the balloon decreases somewhat overnight but not by enough to explain why it no longer floats. (If you inflate a new balloon with helium to the same size as the balloon that fell to the ground, the newly inflated balloon floats.) Explain.

*Active Classroom Learning*

# Questions for Group Work

Discuss these questions with the group and record your consensus answer.

**139.** Explain why 1-propanol ($CH_3CH_2CH_2OH$) is miscible in both water ($H_2O$) and hexane ($C_6H_6$) when hexane and water are barely soluble in each other.

**140.** Have each group member make a flashcard with one of the following on the front: $\Delta H_{soln}$, $\Delta H_{lattice}$, $\Delta H_{solvent}$, $\Delta H_{mix}$, and $\Delta H_{hydration}$. On the back of the card, each group member should describe (in words) the $\Delta H$ process his or her card lists and how that $\Delta H$ relates to other $\Delta H$ values mathematically. Each member presents his or her $\Delta H$ to the group. After everyone has presented, members should trade cards and quiz each other.

**141.** Complete the following table by adding *increases, decreases,* or *no effect*:

|  | Increasing Temperature | Increasing Pressure |
|---|---|---|
| solubility of gas in water |  |  |
| solubility of a solid in water |  |  |

**142.** When 13.62 g (about one tablespoon) of table sugar (sucrose, $C_{12}H_{22}O_{11}$) is dissolved in 241.5 mL of water (density 0.997 g/mL), the final volume is 250.0 mL (about one cup). Have each group member calculate one of the following for the solution and present his or her answer to the group:
 a. mass percent
 b. molarity
 c. molality

**143.** Calculate the expected boiling and freezing point for the solution in the previous problem. If you had to bring this syrup to the boiling point for a recipe, would you expect it to take much more time than it takes to boil the same amount of pure water? Why or why not? Would the syrup freeze in a typical freezer (−18 °C)? Why or why not?

# Data Interpretation and Analysis

## Freezing Point of Salt Water

**144.** The coldest seawater ever recorded (in a liquid state) was in a stream under an Antarctic glacier in 2010; it measured −2.6 °C (27.3 °F).

 On average, water in the world's oceans has a salinity of about 3.5% (35 g/L, or 599 mM). This means that every kilogram (roughly one liter by volume) of seawater contains approximately 35 grams (1.2 ounces) of dissolved salts (predominantly sodium ($Na^+$) and chloride ($Cl^-$) ions). Figure a▶ provides information on the amounts of various cations and anions in seawater with an average density of 1.025 g/ml.

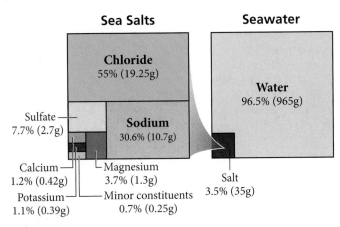

▲ **FIGURE a** Cations and Anions in Ocean Water

Figure b▼ illustrates the salinity of the oceans on Earth in grams of dissolved material (solute) per kilogram of seawater (solution) or parts per thousand (ppt).

▲ **FIGURE b** Salinity of Seawater in the Earth's Oceans

Because most of the dissolved minerals in seawater is sodium chloride, we can make the assumption that seawater

having a salinity of 35 g/kg contains 35 grams of NaCl per one kilogram of seawater.

Use the information provided in the figures to answer the following questions:

a. Calculate the freezing point of a sample of seawater having the highest salinity. Calculate the freezing point of a sample of seawater having the lowest salinity. Use a Van't Hoff Factor of 1.9 in your calculation. Assume the two samples contain water as the solvent and NaCl as the solute.

b. Figure c▼ is a proposed graph of the freezing temperature of saltwater versus the amount of salinity. A student claims that saltwater having a salinity of 280 g NaCl/kg of saltwater freezes at −20°C. Is this claim accurate? Explain.

▲ **FIGURE c** Salinity of Saltwater Versus the Freezing Point of Seawater

c. Use the information provided in Figure b and redraw the graph in Figure c using a range of salinity values from six different oceans. The y-axis should be the freezing point of water, and the x-axis should be the salinity of seawater in units of grams of NaCl/kg of seawater. Does Figure c provide an accurate view of the freezing point of the oceans?

# Answers to Conceptual Connections

## Solubility

**13.1** The first alcohol on the list is methanol, which is highly polar and forms hydrogen bonds with water. It is miscible in water and has only limited solubility in hexane, which is nonpolar. However, as the carbon chain gets longer in the series of alcohols, the OH group becomes less important relative to the growing nonpolar carbon chain. Therefore, the alcohols become progressively less soluble in water and more soluble in hexane. This table demonstrates the rule of thumb *like dissolves like*. Methanol is like water and therefore dissolves in water. It is unlike hexane and therefore has limited solubility in hexane. As you move down the list, the alcohols become increasingly like hexane and increasingly unlike water and therefore become increasingly soluble in hexane and increasingly insoluble in water.

## Energetics of Aqueous Solution Formation

**13.2** You can conclude that $|\Delta H_{solute}| < |\Delta H_{hydration}|$. Since $\Delta H_{soln}$ is negative, the absolute value of the negative term ($\Delta H_{hydration}$) must be greater than the absolute value of the positive term ($\Delta H_{solute}$).

## Solubility and Temperature

**13.3** **(b)** Some potassium bromide precipitates out of solution. The solubility of most solids decreases with decreasing temperature. However, the solubility of gases increases with decreasing

temperature. Therefore, the nitrogen becomes more soluble and will not bubble out of solution.

## Henry's Law

**13.4** Ammonia is the only compound on the list that is polar, so we would expect its solubility in water to be greater than the solubilities of the other gases (which are all nonpolar).

## Molality

**13.5** The solution has a molality of 10.0 *m*. You combined 1.00 mol of solute with 0.100 kg of solvent, so the molality is 1.00 mol/ 0.100 kg = 10.0 *m*.

## Raoult's Law

**13.6** The solute–solvent interactions must be stronger than the solute–solute and solvent–solvent interactions. The stronger interactions lower the vapor pressure from the expected ideal value of 150 mmHg.

## Boiling Point Elevation

**13.7** Solution B because $K_b$ for ethanol is greater than $K_b$ for water.

## Colligative Properties

**13.8** **(c)** The 0.50 M $MgCl_2$ solution has the highest boiling point because it has the highest concentration of particles. We expect 1 mol of $MgCl_2$ to form 3 mol of particles in solution (although it effectively forms slightly fewer).

CHAPTER

# 14

# Chemical Kinetics

Nobody, I suppose, could devote many years to the study of chemical kinetics without being deeply conscious of the fascination of time and change: this is something that goes outside science into poetry. . . .

—SIR CYRIL N. HINSHELWOOD (1897–1967)

In this chapter-opening quotation, Oxford chemistry professor Sir Cyril Hinshelwood calls attention to an aspect of chemistry often overlooked by the casual observer—the mystery of change with time. Since the opening chapter of this book, you have learned that the goal of chemistry is to understand the macroscopic world by examining the molecular one. In this chapter, we focus on understanding how this molecular world changes with time, an area of study called chemical kinetics. The molecular world is anything but static. Thermal energy produces constant molecular motion, causing molecules to repeatedly collide with one another. In a tiny fraction of these collisions, something extraordinary happens—the electrons on one molecule or atom are attracted to the nuclei of another. Some bonds weaken and new bonds form—a chemical reaction occurs. Chemical kinetics is the study of how these kinds of changes occur in time.

▲ *Pouring ice water on a lizard slows it down, making it easier to catch.*

## 14.1 Catching Lizards

The children who live in my neighborhood (including my own kids) have a unique way of catching lizards. Armed with cups of ice water, they chase one of these cold-blooded reptiles into a corner, and take aim and pour the cold water directly onto the lizard's body. The lizard's body temperature drops and it becomes virtually immobilized—easy prey for little hands. The kids scoop up the lizard and place it in a tub filled with sand and leaves. They then watch as the lizard warms back up and becomes active again. They usually release the lizard back into the yard within hours. I guess you could call them catch-and-release lizard hunters.

Unlike mammals, which actively regulate their body temperature through metabolic activity, lizards are *ectotherms*—their body temperature depends on their surroundings. When splashed with cold water, a lizard's body simply gets colder. The drop in body temperature immobilizes the lizard because its movement depends on chemical reactions that occur within its muscles, and the *rates* of those reactions—how fast they occur—are highly sensitive to temperature. In other words, when the temperature drops, the reactions that produce movement in the lizard occur more slowly; therefore, the movement itself slows down. When reptiles get cold, they become lethargic, unable to move very quickly. For this reason, reptiles try to maintain their body temperature within a narrow range by moving between sun and shade.

The rates of chemical reactions, and especially the ability to *control* those rates, are important not just in reptile movement but in many other phenomena as well. For example, a successful rocket launch depends on the rate at which fuel burns—too quickly and the rocket can explode, too slowly and it will not leave the ground. Chemists must always consider reaction rates when synthesizing compounds. No matter how stable a compound might be, its synthesis is impossible if the rate at which it forms is too slow. As we have seen with reptiles, reaction rates are important to life. In fact, the human body's ability to switch a specific reaction on or off at a specific time is achieved largely by controlling the rate of that reaction through the use of enzymes (biological molecules that we explore more fully in Section 14.7).

The first person to measure the rate of a chemical reaction carefully was Ludwig Wilhelmy (1812–1864). In 1850, he measured how fast sucrose, upon treatment with acid, hydrolyzed (broke up with the addition of water) into glucose and fructose. This reaction occurred over several hours, and Wilhelmy was able to show how the rate depended on the initial amount of sugar present—the greater the initial amount, the faster the initial rate. Today we can measure the rates of reactions that occur in times as short as several femtoseconds (femto = $10^{-15}$). The knowledge of reaction rates is not only practically important—giving us the ability to control how fast a reaction occurs—but also theoretically important. As we will discuss in Section 14.6, the rate of a reaction can tell us much about how the reaction occurs on the molecular scale.

## 14.2  The Rate of a Chemical Reaction

The rate of a chemical reaction is a measure of how fast the reaction occurs, as shown in Figure 14.1 ◀. If a chemical reaction has a fast rate, a large fraction of molecules react to form products in a given period of time. If a chemical reaction has a slow rate, only a relatively small fraction of molecules react to form products in a given period of time.

### Definition of Reaction Rate

When we measure how fast something occurs, or more specifically the *rate* at which it occurs, we usually express the measurement as a change in some

▲ **FIGURE 14.1  The Rate of a Chemical Reaction**

quantity per unit of time. For example, we measure the speed of a car—the rate at which it travels—in *miles per hour*, and we measure how quickly (or slowly) people lose weight in *pounds per week*. We report these rates in units that represent the change in what we are measuring (distance or weight) divided by the change in time:

$$\text{Speed} = \frac{\text{change in distance}}{\text{change in time}} = \frac{\Delta x}{\Delta t} \qquad \text{Weight loss} = \frac{\text{change in weight}}{\text{change in time}} = \frac{\Delta \text{ weight}}{\Delta t}$$

Similarly, the rate of a chemical reaction is measured as a change in the amounts of reactants or products (usually in concentration units) divided by the change in time. For example, consider the gas-phase reaction between $H_2(g)$ and $I_2(g)$:

$$H_2(g) + I_2(g) \longrightarrow 2\,HI(g)$$

We can define the rate of this reaction in the time interval $t_1$ to $t_2$ as follows:

$$\text{Rate} = -\frac{\Delta[H_2]}{\Delta t} = -\frac{[H_2]_{t_2} - [H_2]_{t_1}}{t_2 - t_1} \qquad [14.1]$$

> Recall that [A] means the concentration of A in M (mol/L).

In this expression, $[H_2]_{t_2}$ is the hydrogen concentration at time $t_2$ and $[H_2]_{t_1}$ is the hydrogen concentration at time $t_1$. The reaction rate is defined as *the negative* of the change in concentration of a reactant divided by the change in time. The negative sign is part of the definition when we define the reaction rate with respect to a reactant because reactant concentrations decrease as a reaction proceeds; therefore, *the change in the concentration of a reactant is negative*. The negative sign thus makes the overall *rate* positive. (By convention, we report reaction rates as positive quantities.)

Similarly, we can define the reaction rate with respect to the other reactant:

$$\text{Rate} = -\frac{\Delta[I_2]}{\Delta t} \qquad [14.2]$$

Since 1 mol of $H_2$ reacts with 1 mol of $I_2$, we define the rate in the same way. We can also define the rate with respect to the *product* of the reaction:

$$\text{Rate} = +\frac{1}{2}\frac{\Delta[HI]}{\Delta t} \qquad [14.3]$$

Because product concentrations *increase* as the reaction proceeds, the change in concentration of a product is positive. Therefore, when we define the rate with respect to a product, we do not include a negative sign in the definition—the rate is naturally positive. The factor of $\frac{1}{2}$ in this definition is related to the stoichiometry of the reaction. In order to have a single rate for the entire reaction, the definition of the rate with respect to each reactant and product must reflect the stoichiometric coefficients of the reaction. For this particular reaction, 2 mol of HI is produced from 1 mol of $H_2$ and 1 mol of $I_2$:

The concentration of HI increases at twice the rate that the concentration of $H_2$ or $I_2$ decreases. In other words, if 100 $I_2$ molecules react per second, then 200 HI molecules form per second. In order for the overall rate to have the same value when defined with respect to any of the reactants or products, we must multiply the change in HI concentration by a factor of one-half.

Consider the graph in Figure 14.2▶, which represents the changes in concentration for $H_2$ (one of the reactants) and HI (the product) versus time. Let's examine several features of this graph individually.

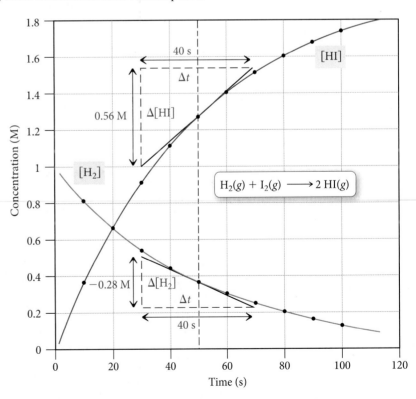

▲ **FIGURE 14.2 Reactant and Product Concentrations as a Function of Time** The graph shows the concentration of one of the reactants ($H_2$) and the product (HI) as a function of time. The other reactant ($I_2$) is omitted for clarity.

### Change in Reactant and Product Concentrations

The reactant concentration, as expected, *decreases* with time because reactants are consumed in a reaction. The product concentration *increases* with time because products are formed in a reaction. The increase in HI concentration occurs at exactly twice the rate of the decrease in $H_2$ concentration because of the stoichiometry of the reaction: 2 mol of HI form for every 1 mol of $H_2$ consumed.

### The Average Rate of the Reaction

We can calculate the average rate of the reaction for any time interval using Equation 14.1 for $H_2$. The following table lists $H_2$ concentration ($[H_2]$) at various times, the change in $H_2$ concentration for each interval ($\Delta[H_2]$), the change in time for each interval ($\Delta t$), and the rate for each interval ($-\Delta[H_2]/\Delta t$). The rate is the average rate within the given time interval. For example, the average rate of the reaction in the time interval between 10 and 20 seconds is 0.0149 M/s, whereas the average rate in the time interval between 20 and 30 seconds is 0.0121 M/s.

| Time (s) | $[H_2]$ (M) | $\Delta[H_2]$ (M) | $\Delta t$ (s) | Rate $= -\Delta[H_2]/\Delta t$ (M/s) |
|---|---|---|---|---|
| 0.000 | 1.000 | | | |
| 10.000 | 0.819 | −0.181 | 10.000 | 0.0181 |
| 20.000 | 0.670 | −0.149 | 10.000 | 0.0149 |
| 30.000 | 0.549 | −0.121 | 10.000 | 0.0121 |
| 40.000 | 0.449 | −0.100 | 10.000 | 0.0100 |
| 50.000 | 0.368 | −0.081 | 10.000 | 0.0081 |
| 60.000 | 0.301 | −0.067 | 10.000 | 0.0067 |
| 70.000 | 0.247 | −0.054 | 10.000 | 0.0054 |
| 80.000 | 0.202 | −0.045 | 10.000 | 0.0045 |
| 90.000 | 0.165 | −0.037 | 10.000 | 0.0037 |
| 100.000 | 0.135 | −0.030 | 10.000 | 0.0030 |

Notice that the average rate *decreases* as the reaction progresses. In other words, the reaction slows down as it proceeds. We discuss this further in the next section, where we will see that, for most reactions, the rate depends on the concentrations of the reactants. As the reactants transform to products, their concentrations decrease, and the reaction slows down.

### The Instantaneous Rate of the Reaction

The instantaneous rate of the reaction is the rate at any one point in time and is represented by the instantaneous slope of the curve at that point. We can determine the instantaneous rate by calculating the slope of the tangent to the curve at the point of interest. In Figure 14.2, we have drawn the tangent lines for both $[H_2]$ and $[HI]$ at 50 seconds. We calculate the instantaneous rate at 50 seconds as follows:

Using $[H_2]$

$$\text{Instantaneous rate (at 50 s)} = -\frac{\Delta[H_2]}{\Delta t} = -\frac{-0.28 \text{ M}}{40 \text{ s}} = 0.0070 \text{ M/s}$$

Using $[HI]$

$$\text{Instantaneous rate (at 50 s)} = +\frac{1}{2}\frac{\Delta[HI]}{\Delta t} = +\frac{1}{2}\frac{0.56 \text{ M}}{40 \text{ s}} = 0.0070 \text{ M/s}$$

As we would expect, the rate is the same whether we use one of the reactants or the product for the calculation. Notice that the instantaneous rate at 50 seconds (0.0070 M/s) is between the average rates calculated for the 10-second intervals just before and just after 50 seconds.

We can generalize our definition of reaction rate for the generic reaction:

$$a\text{A} + b\text{B} \longrightarrow c\text{C} + d\text{D} \qquad\qquad [14.4]$$

where A and B are reactants, C and D are products, and $a$, $b$, $c$, and $d$ are the stoichiometric coefficients. We define the rate of the reaction as follows:

$$\text{Rate} = -\frac{1}{a}\frac{\Delta[\text{A}]}{\Delta t} = -\frac{1}{b}\frac{\Delta[\text{B}]}{\Delta t} = +\frac{1}{c}\frac{\Delta[\text{C}]}{\Delta t} = +\frac{1}{d}\frac{\Delta[\text{D}]}{\Delta t} \qquad\qquad [14.5]$$

From the definition, we can see that knowing the rate of change in the concentration of any one reactant or product at a point in time allows us to determine the rate of change in the concentration of any other reactant or product at that point in time (from the balanced equation). *However, predicting the rate at some future time is not possible from just the balanced equation.*

---

## Example 14.1   Expressing Reaction Rates

Consider this balanced chemical equation:

$$\text{H}_2\text{O}_2(aq) + 3\,\text{I}^-(aq) + 2\,\text{H}^+(aq) \longrightarrow \text{I}_3^-(aq) + 2\,\text{H}_2\text{O}(l)$$

In the first 10.0 seconds of the reaction, the concentration of $\text{I}^-$ drops from 1.000 M to 0.868 M.

**(a)** Calculate the average rate of this reaction in this time interval.

**(b)** Determine the rate of change in the concentration of $\text{H}^+$ (that is, $\Delta[\text{H}^+]/\Delta t$) during this time interval.

**SOLUTION**

| | |
|---|---|
| **(a)** Use Equation 14.5 to calculate the average rate of the reaction. | $\text{Rate} = -\dfrac{1}{3}\dfrac{\Delta[\text{I}^-]}{\Delta t}$ <br><br> $= -\dfrac{1}{3}\dfrac{(0.868\,\text{M} - 1.000\,\text{M})}{10.0\,\text{s}}$ <br><br> $= 4.40 \times 10^{-3}\,\text{M/s}$ |
| **(b)** Use Equation 14.5 again for the relationship between the rate of the reaction and $\Delta[\text{H}^+]/\Delta t$. After solving for $\Delta[\text{H}^+]/\Delta t$, substitute the calculated rate from part (a) and calculate $\Delta[\text{H}^+]/\Delta t$. | $\text{Rate} = -\dfrac{1}{2}\dfrac{\Delta[\text{H}^+]}{\Delta t}$ <br><br> $\dfrac{\Delta[\text{H}^+]}{\Delta t} = -2(\text{rate})$ <br><br> $= -2(4.40 \times 10^{-3}\,\text{M/s})$ <br><br> $= -8.80 \times 10^{-3}\,\text{M/s}$ |

**FOR PRACTICE 14.1**   For the reaction shown in Example 14.1, predict the rate of change in concentration of $\text{H}_2\text{O}_2$ ($\Delta[\text{H}_2\text{O}_2]/\Delta t$) and $\text{I}_3^-$ ($\Delta[\text{I}_3^-]/\Delta t$) during this time interval.

---

**Reaction Rates**   For the reaction $\text{A} + 2\text{B} \longrightarrow \text{C}$ under a given set of conditions, the initial rate is 0.100 M/s. What is $\Delta[\text{B}]/\Delta t$ under the same conditions?

  **(a)** $-0.0500\,\text{M/s}$       **(b)** $-0.100\,\text{M/s}$       **(c)** $-0.200\,\text{M/s}$

CONCEPTUAL
CONNECTION **14.1**

---

## Measuring Reaction Rates

In order to study the kinetics of a reaction, we must have an experimental way to measure the concentration of at least one of the reactants or products as a function of time. For example, Ludwig Wilhelmy, whose experiment on the rate of the conversion of sucrose to glucose and fructose we discussed briefly in Section 14.1, took advantage of sucrose's ability to rotate polarized light. (Polarized light is light with an electric field oriented along one plane.) When a beam of polarized light is passed through a sucrose solution, the polarization of the light is rotated clockwise. In contrast, the products of the reaction (glucose and fructose) rotate polarized light counterclockwise. By measuring the degree of polarization of light passing through a reacting solution—a technique known as polarimetry—Wilhelmy was able to determine the relative concentrations of the reactants and products as a function of time.

Perhaps the most common way to study the kinetics of a reaction is through spectroscopy (see Section 7.3). For example, the reaction of $H_2$ and $I_2$ to form HI can be followed spectroscopically because $I_2$ is violet and $H_2$ and HI are colorless. As $I_2$ reacts with $H_2$ to form HI, the violet color of the reaction mixture fades. We can monitor the fading color with a spectrometer, a device that passes light through a sample and measures how strongly the light is absorbed (Figure 14.3▼). If the sample contains the reacting mixture, the intensity of the light absorption will decrease as the reaction proceeds, providing a direct measure of the concentration of $I_2$ as a function of time. Because light travels so fast and because experimental techniques can produce very short pulses of light, spectroscopy can be used to measure reactions that happen on time scales as short as several femtoseconds.

▼ **FIGURE 14.3 The Spectrometer** In a spectrometer, light of a specific wavelength is passed through the sample, and the intensity of the transmitted light—which depends on how much light is absorbed by the sample—is measured and recorded.

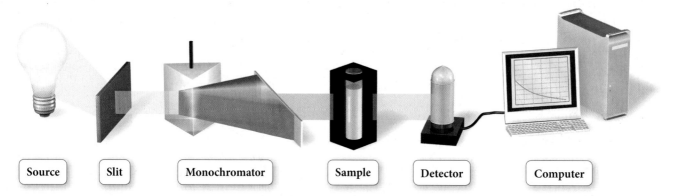

| Source | Slit | Monochromator | Sample | Detector | Computer |

By measuring changes in pressure, we can also monitor reactions in which the number of moles of gaseous reactants and products changes as the reaction proceeds. Consider the reaction in which dinitrogen monoxide reacts to form nitrogen and oxygen gas:

$$2\,N_2O(g) \longrightarrow 2\,N_2(g) + O_2(g)$$

For every 2 mol of $N_2O$ that reacts, the reaction vessel contains one additional mole of gas. As the reaction proceeds and the amount of gas increases, the pressure steadily rises. We can use the rise in pressure to determine the relative concentrations of reactants and products as a function of time.

We can use all of the three techniques mentioned here—polarimetry, spectroscopy, and pressure measurement—to monitor a reaction as it occurs in a reaction vessel. Some reactions occur slowly enough that samples, or *aliquots*, can be periodically withdrawn from the reaction vessel and analyzed to determine the progress of the reaction. We can use instrumental techniques such as gas chromatography (Figure 14.4◄) or mass spectrometry, as well as wet chemical techniques such as titration, to measure the relative amounts of reactants or products in the aliquot. By taking aliquots at regular time intervals, we can determine the relative amounts of reactants and products as a function of time.

▲ **FIGURE 14.4 The Gas Chromatograph** In a gas chromatograph (GC), a sample of the reaction mixture, or aliquot, is injected into a specially constructed column. Because of their characteristic physical and chemical properties, different components of the mixture pass through the column at different rates and thus exit at different times. As each component leaves the column, it is detected electronically and a chromatogram is recorded. The area under each peak in the chromatogram is proportional to the amount of one particular component in the sample mixture.

# 14.3 The Rate Law: The Effect of Concentration on Reaction Rate

The rate of a reaction often depends on the concentration of one or more of the reactants. As we have already seen, Ludwig Wilhelmy noticed this effect for the hydrolysis of sucrose. For simplicity, let's consider a reaction in which a single reactant, A, decomposes into products:

$$A \longrightarrow products$$

As long as the rate of the reverse reaction (in which the products return to reactants) is negligibly slow, we can express the relationship between the rate of the reaction and the concentration of the reactant—called the **rate law**—as follows:

$$Rate = k[A]^n \qquad [14.6]$$

where $k$ is a constant of proportionality called the **rate constant** and $n$ is the **reaction order**. The value of $n$ (usually an integer) determines how the rate depends on the concentration of the reactant:

- If $n = 0$, the reaction is *zero order* and the rate is independent of the concentration of A.

- If $n = 1$, the reaction is *first order* and the rate is directly proportional to the concentration of A.

- If $n = 2$, the reaction is *second order* and the rate is proportional to the square of the concentration of A.

Although other orders are possible, including noninteger (or fractional) orders, these three are the most common.

> By definition, $[A]^0 = 1$, so the rate is equal to $k$ regardless of $[A]$.

## The Three Common Reaction Orders ($n = 0$, 1, and 2)

Figure 14.5▼ shows three plots illustrating how the *concentration of A changes with time* for the three common reaction orders with identical numerical values for the rate constant ($k$) and identical initial concentrations. Figure 14.6▼ has three plots showing the *rate of the reaction* (the slope of the lines in Figure 14.5) *as a function of the reactant concentration* for each reaction order.

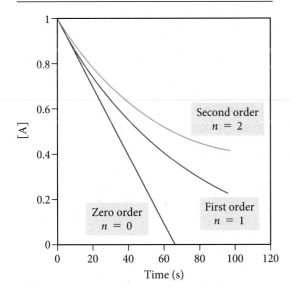

**Reactant Concentration versus Time**

▲ FIGURE 14.5 **Reactant Concentration as a Function of Time for Different Reaction Orders**

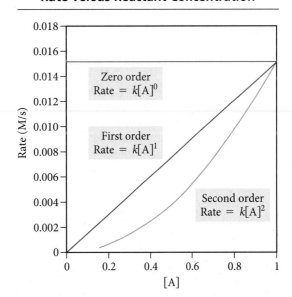

**Rate versus Reactant Concentration**

▲ FIGURE 14.6 **Reaction Rate as a Function of Reactant Concentration for Different Reaction Orders**

## Sublimation Is Zero Order

Ice in glass tube

> When one layer of particles sublimes, the next layer is exposed. The number of particles available to sublime remains constant.

▲ **FIGURE 14.7** **Sublimation** When a layer of particles sublimes, another identical layer is just below it. Consequently, the number of particles available to sublime at any one time does not change with the total number of particles in the sample, and the process is zero order.

## Zero-Order Reaction

In a zero-order reaction, the rate of the reaction is independent of the concentration of the reactant:

$$\text{Rate} = k[A]^0 = k \qquad [14.7]$$

Consequently, for a zero-order reaction, the concentration of the reactant decreases linearly with time, as shown in Figure 14.5. The slope of the line is constant, indicating a constant rate. The rate is constant because the reaction does not slow down as the concentration of A decreases. The graph in Figure 14.6 shows that the rate of a zero-order reaction is the same at any concentration of A.

Zero-order reactions occur under conditions where the amount of reactant actually *available for reaction* is unaffected by changes in the *overall quantity of reactant*. For example, sublimation is normally zero order because only molecules at the surface of a substance can sublime, and the concentration of the surface molecules does not change as the amount of subliming substance decreases (Figure 14.7◄).

## First-Order Reaction

In a first-order reaction, the rate of the reaction is directly proportional to the concentration of the reactant:

$$\text{Rate} = k[A]^1 \qquad [14.8]$$

For a first-order reaction, the rate slows down as the reaction proceeds because the concentration of the reactant decreases. We can see this in Figure 14.5—the slope of the curve (the rate) becomes less steep (slower) with time. Figure 14.6 shows the rate as a function of the concentration of A. Notice the linear relationship—the rate is directly proportional to the concentration.

## Second-Order Reaction

In a second-order reaction, the rate of the reaction is proportional to the square of the concentration of the reactant:

$$\text{Rate} = k[A]^2 \qquad [14.9]$$

Consequently, for a second-order reaction, the rate is even more sensitive to the reactant concentration. We can see this in Figure 14.5—the slope of the curve (the rate) flattens out more quickly than it does for a first-order reaction. Figure 14.6 shows the rate as a function of the concentration of A. Notice the quadratic relationship—the rate is proportional to the square of the concentration.

## Determining the Order of a Reaction

*The order of a reaction can be determined only by experiment.* A common way to determine reaction order is the *method of initial rates.* In this method, the initial rate—the rate for a short period of time at the beginning of the reaction—is measured by running the reaction several times with different initial reactant concentrations to determine the effect of the concentration on the rate. For example, let's return to our simple reaction in which a single reactant, A, decomposes into products:

$$A \longrightarrow \text{products}$$

In an experiment, the initial rate is measured at several different initial concentrations with the following results:

| [A] (M) | Initial Rate (M/s) |
|---------|--------------------|
| 0.10    | 0.015              |
| 0.20    | 0.030              |
| 0.40    | 0.060              |

In this data set, when the concentration of A doubles, the rate doubles—the initial rate is directly proportional to the initial concentration. The reaction is therefore first order in A, and the rate law takes the first-order form:

$$\text{Rate} = k[A]^1$$

We can determine the value of the rate constant, $k$, by solving the rate law for $k$ and substituting the concentration and the initial rate from any one of the three measurements. Here we use the first measurement:

$$\text{Rate} = k[A]^1$$

$$k = \frac{\text{rate}}{[A]} = \frac{0.015 \text{ M/s}}{0.10 \text{ M}} = 0.15 \text{ s}^{-1}$$

Notice that the rate constant for a first-order reaction has units of $\text{s}^{-1}$.

The following two data sets show how measured initial rates are different for zero-order and for second-order reactions having the same initial rate at $[A] = 0.10$ M:

**Zero Order ($n = 0$)**

| [A] (M) | Initial Rate (M/s) |
|---------|--------------------|
| 0.10 | 0.015 |
| 0.20 | 0.015 |
| 0.40 | 0.015 |

**Second Order ($n = 2$)**

| [A] (M) | Initial Rate (M/s) |
|---------|--------------------|
| 0.10 | 0.015 |
| 0.20 | 0.060 |
| 0.40 | 0.240 |

For a zero-order reaction, the initial rate is independent of the reactant concentration—the rate is the same at all measured initial concentrations. For a second-order reaction, the initial rate quadruples for a doubling of the reactant concentration—the relationship between concentration and rate is quadratic. If we are unsure about how the initial rate is changing with the initial reactant concentration, or if the numbers are not as obvious as they are in these examples, we can substitute any two initial concentrations and the corresponding initial rates into a ratio of the rate laws to determine the order ($n$):

$$\frac{\text{rate 2}}{\text{rate 1}} = \frac{k[A]_2^n}{k[A]_1^n}$$

For example, we can substitute the last two measurements in the data set just given for the second-order reaction as follows:

$$\frac{0.240 \text{ M/s}}{0.060 \text{ M/s}} = \frac{k(0.40 \text{ M})^n}{k(0.20 \text{ M})^n}$$

$$4.0 = \left(\frac{0.40}{0.20}\right)^n = 2^n$$

$$\log 4.0 = \log(2^n)$$

$$= n \log 2$$

$$n = \frac{\log 4}{\log 2}$$

$$= 2$$

> Remember that $\dfrac{x^n}{y^n} = \left(\dfrac{x}{y}\right)^n$.

> Remember that $\log x^n = n \log x$.

The rate constants for zero- and second-order reactions have different units than for first-order reactions. The rate constant for a zero-order reaction has units of $M \cdot s^{-1}$, and the rate constant for a second-order reaction has units of $M^{-1} \cdot s^{-1}$.

> The general expression for the units of the rate constant for an $n$th-order reaction is $M^{(1-n)} s^{-1}$.

**Order of Reaction** The reaction $A \longrightarrow B$ has been experimentally determined to be second order. The initial rate is 0.0100 M/s at an initial concentration of A of 0.100 M. What is the initial rate at $[A] = 0.500$ M?

   **(a)** 0.00200 M/s    **(b)** 0.0100 M/s    **(c)** 0.0500 M/s    **(d)** 0.250 M/s

CONCEPTUAL
**CONNECTION 14.2**

## Reaction Order for Multiple Reactants

So far, we have considered a simple reaction with only one reactant. How do we define the rate law for reactions with more than one reactant? Consider the generic reaction:

$$a\text{A} + b\text{B} \longrightarrow c\text{C} + d\text{D}$$

As long as the reverse reaction is negligibly slow, the rate law is proportional to the concentration of [A] raised to the $m$ power multiplied by the concentration of [B] raised to the $n$ power:

$$\text{Rate} = k[\text{A}]^m[\text{B}]^n \qquad [14.10]$$

where $m$ is the reaction order with respect to A and $n$ is the reaction order with respect to B. The **overall order** is the sum of the exponents ($m + n$). For example, the reaction between hydrogen and iodine has been experimentally determined to be first order with respect to hydrogen, first order with respect to iodine, and thus second order overall:

$$\text{H}_2(g) + \text{I}_2(g) \longrightarrow 2\,\text{HI}(g) \quad \text{Rate} = k[\text{H}_2]^1[\text{I}_2]^1$$

Similarly, the reaction between hydrogen and nitrogen monoxide has been experimentally determined to be first order with respect to hydrogen, second order with respect to nitrogen monoxide, and thus third order overall:

$$2\,\text{H}_2(g) + 2\,\text{NO}(g) \longrightarrow \text{N}_2(g) + 2\,\text{H}_2\text{O}(g) \quad \text{Rate} = k[\text{H}_2]^1[\text{NO}]^2$$

*The rate law for any reaction must always be determined by experiment*, often by the method of initial rates described previously. We can't simply look at a chemical equation and determine the rate law for the reaction. When the reaction has two or more reactants, the concentration of each reactant is usually varied independently of the others to determine the dependence of the rate on the concentration of that reactant. Example 14.2 demonstrates how to use the method of initial rates to determine the order of a reaction with multiple reactants.

**Example 14.2** **Determining the Order and Rate Constant of a Reaction**

Consider the reaction between nitrogen dioxide and carbon monoxide:

$$\text{NO}_2(g) + \text{CO}(g) \longrightarrow \text{NO}(g) + \text{CO}_2(g)$$

The initial rate of the reaction is measured at several different concentrations of the reactants, and the tabulated results are shown here.

From the data, determine:

**(a)** the rate law for the reaction

**(b)** the rate constant ($k$) for the reaction

| [NO$_2$] (M) | [CO] (M) | Initial Rate (M/s) |
|---|---|---|
| 0.10 | 0.10 | 0.0021 |
| 0.20 | 0.10 | 0.0082 |
| 0.20 | 0.20 | 0.0083 |
| 0.40 | 0.10 | 0.033 |

**SOLUTION**

**(a)** Begin by examining how the rate changes for each change in concentration. Between the first two experiments, the concentration of NO$_2$ doubles, the concentration of CO stays constant, and the rate quadruples, suggesting that the reaction is second order in NO$_2$.

Between the second and third experiments, the concentration of NO$_2$ stays constant, the concentration of CO doubles, and the rate remains constant (the small change in the least significant figure is simply experimental error), suggesting that the reaction is zero order in CO.

Between the third and fourth experiments, the concentration of NO$_2$ again doubles and the concentration of CO halves, yet the rate quadruples again, confirming that the reaction is second order in NO$_2$ and zero order in CO.

| [NO$_2$] (M) | [CO] (M) | Initial Rate (M/s) |
|---|---|---|
| 0.10 | 0.10 | 0.0021 |
| ↓ × 2 | ↓ constant | ↓ × 4 |
| 0.20 | 0.10 | 0.0082 |
| ↓ constant | ↓ × 2 | ↓ × 1 |
| 0.20 | 0.20 | 0.0083 |
| ↓ × 2 | ↓ × $\frac{1}{2}$ | ↓ × 4 |
| 0.40 | 0.10 | 0.033 |

Write the overall rate expression.

$$\text{Rate} = k[\text{NO}_2]^2[\text{CO}]^0 = k[\text{NO}_2]^2$$

**(a) ALTERNATIVE APPROACH**

If the relationship between the changes in concentrations and the changes in initial rates is not immediately obvious, you can determine the reaction order for any reactant by substituting any two initial rates and the corresponding initial concentrations into a ratio of the rate laws to determine the order ($n$).

$$\frac{\text{rate } 2}{\text{rate } 1} = \frac{k[A]_2^n}{k[A]_1^n}$$

For $NO_2$ use the first and second concentrations and rates (because $[NO_2]$ changes here, but $[CO]$ is constant). Substitute the rates and concentrations into the expression for the ratio of the rate constants.

For $NO_2$

$$\frac{\text{rate } 2}{\text{rate } 1} = \frac{k[NO_2]_2^n}{k[NO_2]_1^n}$$

$$\frac{0.0082 \text{ M/s}}{0.0021 \text{ M/s}} = \frac{k(0.20 \text{ M})^n}{k(0.10 \text{ M})^n}$$

$$3.9 = \left(\frac{0.20}{0.10}\right)^n = 2^n$$

Take the log of both sides of the equation and solve for $n$.

$$\log 3.9 = \log 2^n$$
$$= n \log 2$$

$$n = \frac{\log 3.9}{\log 2}$$
$$= 1.96 = 2$$

For CO, use the second and third concentrations and rates (because $[CO]$ changes here, but $[NO_2]$ is constant).

For CO

$$\frac{\text{rate } 3}{\text{rate } 2} = \frac{k[CO]_3^n}{k[CO_2]_2^n}$$

Substitute the rates and concentrations into the expression for the ratio of the rate laws.

$$\frac{0.0083 \text{ M/s}}{0.0082 \text{ M/s}} = \frac{k(0.20 \text{ M})^n}{k(0.10 \text{ M})^n}$$

$$1.01 = \left(\frac{0.20}{0.10}\right)^n = 2^n$$

Take the log of both sides of the equation and solve for $n$.

$$\log 1.01 = \log 2^n$$
$$= n \log 2$$

$$n = \frac{\log 1.01}{\log 2}$$
$$= 0.01 = 0$$

Write the overall rate expression from the order of each reactant.

$$\text{Rate} = k[NO_2]^2[CO]^0 = k[NO_2]^2$$

**(b)** To determine the rate constant for the reaction, solve the rate law for $k$ and substitute the concentration and the initial rate from any one of the four measurements. In this case, we use the first measurement.

$$\text{Rate} = k[NO_2]^2$$
$$k = \frac{\text{rate}}{[NO_2]^2} = \frac{0.0021 \text{ M/s}}{(0.10 \text{ M})^2} = 0.21 \text{ M}^{-1} \cdot \text{s}^{-1}$$

**FOR PRACTICE 14.2** Consider the equation:

$$CHCl_3(g) + Cl_2(g) \longrightarrow CCl_4(g) + HCl(g)$$

The initial rate of reaction is measured at several different concentrations of the reactants with the tabulated results shown here.

| $[CHCl_3]$ (M) | $[Cl_2]$ (M) | Initial Rate (M/s) |
|---|---|---|
| 0.010 | 0.010 | 0.0035 |
| 0.020 | 0.010 | 0.0069 |
| 0.020 | 0.020 | 0.0098 |
| 0.040 | 0.040 | 0.027 |

From the data, determine:

**(a)** the rate law for the reaction     **(b)** the rate constant ($k$) for the reaction

**Rate and Concentration** This reaction was experimentally determined to be first order with respect to $O_2$ and second order with respect to NO:

$$O_2(g) + 2\,NO(g) \longrightarrow 2\,NO_2(g)$$

The diagrams shown here represent reaction mixtures in which the number of each type of molecule represents its relative initial concentration. Which mixture has the fastest initial rate?

(a)                    (b)                    (c)

 The Integrated Rate Law

## 14.4 The Integrated Rate Law: The Dependence of Concentration on Time

The rate laws we have examined so far show the relationship between *the rate of a reaction and the concentration of a reactant*. But we often want to know the relationship between *the concentration of a reactant and time*. For example, the presence of chlorofluorocarbons (CFCs) in the atmosphere threatens the ozone layer. One of the reasons that CFCs pose such a significant threat is that the reactions that consume them are so slow (Table 14.1). Legislation has significantly reduced CFC emissions, but even if humans were to completely stop adding CFCs to the atmosphere, their concentration would decrease only very slowly. Nonetheless, we would like to know how their concentration changes with time. How much will be left in 20 years? In 50 years?

**TABLE 14.1 Atmospheric Lifetimes of Several CFCs**

| CFC Name | Structure | Atmospheric Lifetime* |
|---|---|---|
| CFC-11 ($CCl_3F$) Trichlorofluoromethane | | 45 years |
| CFC-12 ($CCl_2F_2$) Dichlorodifluoromethane | | 100 years |
| CFC-113 ($C_2F_3Cl_3$) 1,1,2-Trichloro-1,2,2-trifluoroethane | | 85 years |
| CFC-114 ($C_2F_4Cl_2$) 1,2-Dichlorotetrafluoroethane | | 300 years |
| CFC-115 ($C_2F_5Cl$) Monochloropentafluoroethane | | 1700 years |

*Data taken from EPA site (under section 602 of Clean Air Act).

### The Integrated Rate Law

The **integrated rate law** for a chemical reaction is a relationship between the concentrations of the reactants and time. For simplicity, we return to a single reactant decomposing into products:

$$A \longrightarrow products$$

The integrated rate law for this reaction depends on the order of the reaction; let's examine each of the common reaction orders individually.

### First-Order Integrated Rate Law

If our simple reaction is first order, the rate is directly proportional to the concentration of A:

$$Rate = k[A]$$

Since Rate $= -\Delta[A]/\Delta t$, we can write:

$$-\frac{\Delta[A]}{\Delta t} = k[A] \qquad\qquad [14.11]$$

In this form, the rate law is also known as the *differential rate law*.

We can use calculus (see margin) to integrate the differential rate law and obtain the first-order *integrated rate law*:

$$\ln[A]_t = -kt + \ln[A]_0 \qquad\qquad [14.12]$$

We can also rearrange the integrated rate law by subtracting $\ln[A]_0$ from both sides of Equation 14.12:

$$\ln[A]_t = -kt + \ln[A]_0$$

$$\ln[A]_t - \ln[A]_0 = -kt$$

Since $\ln A - \ln B = \ln(A/B)$, we can rearrange this equation to get:

$$\ln\frac{[A]_t}{[A]_0} = -kt \qquad\qquad [14.13]$$

where $[A]_t$ is the concentration of A at any time $t$, $k$ is the rate constant, and $[A]_0$ is the initial concentration of A.

Notice that the integrated rate law shown in Equation 14.12 has the form of an equation for a straight line:

$$\ln[A]_t = -kt + \ln[A]_0$$

$$y = mx + b$$

For a first-order reaction, a plot of the natural log of the reactant concentration as a function of time yields a straight line with a slope of $-k$ and a $y$-intercept of $\ln[A]_0$, as shown in Figure 14.8▶. (Note that the slope is negative but that the rate constant is always positive.)

$$-\frac{d[A]}{dt} = k[A]$$

$$\frac{d[A]}{[A]} = -kdt$$

$$\int_{[A]_0}^{[A]}\frac{d[A]}{[A]} = -\int_0^t kdt$$

$$[\ln[A]]_{[A]_0}^{[A]} = -k[t]_0^t$$

$$\ln[A] - \ln[A]_0 = -kt$$

$$\ln[A] = -kt + \ln[A]_0$$

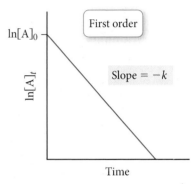

▲ FIGURE 14.8 **First-Order Integrated Rate Law** For a first-order reaction, a plot of the natural log of the reactant concentration as a function of time yields a straight line. The slope of the line is equal to $-k$ and the $y$-intercept is $\ln[A]_0$.

---

**Example 14.3** | **The First-Order Integrated Rate Law: Using Graphical Analysis of Reaction Data**

Consider the equation for the decomposition of $SO_2Cl_2$:

$$SO_2Cl_2(g) \longrightarrow SO_2(g) + Cl_2(g)$$

The concentration of $SO_2Cl_2$ is monitored at a fixed temperature as a function of time during the decomposition reaction, and the following data are tabulated:

| Time (s) | [SO₂Cl₂] (M) | Time (s) | [SO₂Cl₂] (M) |
|----------|--------------|----------|--------------|
| 0 | 0.100 | 800 | 0.0793 |
| 100 | 0.0971 | 900 | 0.0770 |
| 200 | 0.0944 | 1000 | 0.0748 |
| 300 | 0.0917 | 1100 | 0.0727 |
| 400 | 0.0890 | 1200 | 0.0706 |
| 500 | 0.0865 | 1300 | 0.0686 |
| 600 | 0.0840 | 1400 | 0.0666 |
| 700 | 0.0816 | 1500 | 0.0647 |

Show that the reaction is first order, and determine the rate constant for the reaction.

—Continued on the next page

—*Continued*

**SOLUTION**

In order to show that the reaction is first order, prepare a graph of ln [$SO_2Cl_2$] versus time as shown:

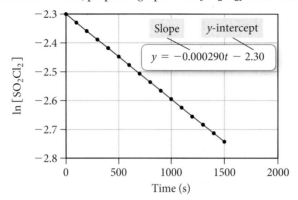

The plot is linear, confirming that the reaction is indeed first order. To obtain the rate constant, fit the data to a line. The slope of the line is equal to $-k$. Since the slope of the best-fitting line (which is most easily determined on a graphing calculator or with spreadsheet software such as Microsoft Excel) is $-2.90 \times 10^{-4}\,s^{-1}$, the rate constant is $+2.90 \times 10^{-4}\,s^{-1}$.

**FOR PRACTICE 14.3** Use the graph and the best-fitting line in Example 14.3 to predict the concentration of $SO_2Cl_2$ at 1900 s.

---

**Example 14.4** | **The First-Order Integrated Rate Law: Determining the Concentration of a Reactant at a Given Time**

In Example 14.3, you determined that the decomposition of $SO_2Cl_2$ (under the given reaction conditions) is first order and has a rate constant of $+2.90 \times 10^{-4}\,s^{-1}$. If the reaction is carried out at the same temperature and the initial concentration of $SO_2Cl_2$ is 0.0225 M, what is the $SO_2Cl_2$ concentration after 865 s?

| | |
|---|---|
| **SORT** You are given the rate constant of a first-order reaction and the initial concentration of the reactant and asked to find the concentration at 865 seconds. | **GIVEN:** $k = +2.90 \times 10^{-4}\,s^{-1}$ <br> $[SO_2Cl_2]_0 = 0.0225$ M <br><br> **FIND:** $[SO_2Cl_2]$ at $t = 865$ s |
| **STRATEGIZE** Refer to the first-order integrated rate law to determine the $SO_2Cl_2$ concentration at $t = 865$ s. | **EQUATION** $\ln [A]_t = -kt + \ln [A]_0$ |
| **SOLVE** Substitute the rate constant, the initial concentration, and the time into the integrated rate law. <br> Solve the integrated rate law for the concentration of $[SO_2Cl_2]_t$. | **SOLUTION** <br> $\ln[SO_2Cl_2]_t = -kt + \ln[SO_2Cl_2]_0$ <br> $\ln[SO_2Cl_2]_t = -(2.90 \times 10^{-4}\,s^{-1})865\,s + \ln(0.0225)$ <br> $\ln[SO_2Cl_2]_t = -0.251 - 3.79$ <br> $[SO_2Cl_2]_t = e^{-4.04}$ <br> $\phantom{[SO_2Cl_2]_t} = 0.0176$ M |

**CHECK** The concentration is smaller than the original concentration as expected. If the concentration were larger than the initial concentration, this would indicate a mistake in the signs of one of the quantities on the right-hand side of the equation.

**FOR PRACTICE 14.4** Cyclopropane rearranges to form propene in the gas phase.

$$\underset{H_2C-CH_2}{\overset{CH_2}{\triangle}} \longrightarrow CH_3-CH=CH_2$$

The reaction is first order in cyclopropane and has a measured rate constant of $3.36 \times 10^{-5}\,s^{-1}$ at 720 K. If the initial cyclopropane concentration is 0.0445 M, what is the cyclopropane concentration after 235.0 minutes?

## Second-Order Integrated Rate Law

If our simple reaction (A $\longrightarrow$ products) is second order, the rate is proportional to the square of the concentration of A:

$$\text{Rate} = k[A]^2$$

Since Rate $= -\Delta[A]/\Delta t$, we can write the differential rate law:

$$-\frac{\Delta[A]}{\Delta t} = k[A]^2 \qquad [14.14]$$

Again, we can use calculus to integrate the second-order differential rate law. We leave this derivation to an exercise (see Exercise 115 at the end of this chapter). The derivation results in the *second-order integrated rate law*:

$$\frac{1}{[A]_t} = kt + \frac{1}{[A]_0} \qquad [14.15]$$

The second-order integrated rate law is also in the form of an equation for a straight line:

$$\frac{1}{[A]_t} = kt + \frac{1}{[A]_0}$$
$$y = mx + b$$

However, you must plot the inverse of the concentration of the reactant as a function of time. The plot yields a straight line with a slope of $k$ and an intercept of $1/[A]_0$, as shown in Figure 14.9▶.

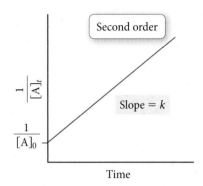

▲ **FIGURE 14.9 Second-Order Integrated Rate Law** For a second-order reaction, a plot of the inverse of the reactant concentration as a function of time yields a straight line. The slope of the line is equal to $k$ and the $y$-intercept is $1/[A]_0$.

---

**Example 14.5** | **The Second-Order Integrated Rate Law: Using Graphical Analysis of Reaction Data**

Consider the equation for the decomposition of $NO_2$:

$$NO_2(g) \longrightarrow NO(g) + O(g)$$

The concentration of $NO_2$ is monitored at a fixed temperature as a function of time during the decomposition reaction and the data tabulated in the margin at right. Show by graphical analysis that the reaction is not first order and that it is second order. Determine the rate constant for the reaction.

**SOLUTION**

In order to show that the reaction is *not* first order, prepare a graph of $\ln[NO_2]$ versus time as shown:

The plot is *not* linear (the straight line does not fit the data points), confirming that the reaction is not first order. In order to show that the reaction is second order, prepare a graph of $1/[NO_2]$ versus time as shown:

*—Continued on the next page*

| Time (s) | [NO$_2$] (M) |
|---|---|
| 0 | 0.01000 |
| 50 | 0.00887 |
| 100 | 0.00797 |
| 150 | 0.00723 |
| 200 | 0.00662 |
| 250 | 0.00611 |
| 300 | 0.00567 |
| 350 | 0.00528 |
| 400 | 0.00495 |
| 450 | 0.00466 |
| 500 | 0.00440 |
| 550 | 0.00416 |
| 600 | 0.00395 |
| 650 | 0.00376 |
| 700 | 0.00359 |
| 750 | 0.00343 |
| 800 | 0.00329 |
| 850 | 0.00316 |
| 900 | 0.00303 |
| 950 | 0.00292 |
| 1000 | 0.00282 |

*—Continued*

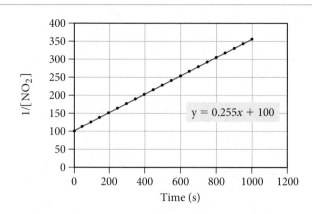

This graph is linear (the data points fit well to a straight line), confirming that the reaction is indeed second order. To obtain the rate constant, determine the slope of the best-fitting line. The slope is $0.255 \, M^{-1} \cdot s^{-1}$; therefore, the rate constant is $0.255 \, M^{-1} \cdot s^{-1}$.

**FOR PRACTICE 14.5** Use the graph and the best-fitting line in Example 14.5 to predict the concentration of $NO_2$ at 2000 s.

## Zero-Order Integrated Rate Law

If our simple reaction is zero order, the rate is proportional to a constant:

$$\text{Rate} = k[A]^0 = k$$

Since $\text{Rate} = -\Delta[A]/\Delta t$, we can write:

$$-\frac{\Delta[A]}{\Delta t} = k \qquad [14.16]$$

We can integrate the zero-order differential rate law (see margin) to obtain the *zero-order integrated rate law*:

$$[A]_t = -kt + [A]_0 \qquad [14.17]$$

The zero-order integrated rate law in Equation 14.17 is also in the form of an equation for a straight line. A plot of the concentration of the reactant as a function of time yields a straight line with a slope of $-k$ and an intercept of $[A]_0$, shown in Figure 14.10 ◄.

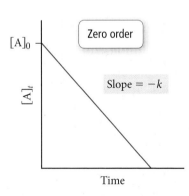

**▲ FIGURE 14.10 Zero-Order Integrated Rate Law** For a zero-order reaction, a plot of the reactant concentration as a function of time yields a straight line. The slope of the line is equal to $-k$ and the *y*-intercept is $[A]_0$.

## The Half-Life of a Reaction

The **half-life ($t_{1/2}$)** of a reaction is the time required for the concentration of a reactant to fall to one-half of its initial value. For example, if a reaction has a half-life of 100 seconds, and if the initial concentration of the reactant is 1.0 M, the concentration will fall to 0.50 M in 100 s. The half-life expression—which defines the dependence of half-life on the rate constant and the initial concentration—is different for different reaction orders.

### First-Order Reaction Half-Life

From the definition of half-life and from the integrated rate law, we can derive an expression for the half-life. For a first-order reaction, the integrated rate law is:

$$\ln \frac{[A]_t}{[A]_0} = -kt$$

At a time equal to the half-life ($t = t_{1/2}$), the concentration is exactly half of the initial concentration: ($[A]_{t_{1/2}} = \frac{1}{2}[A]_0$). Therefore, when $t = t_{1/2}$ we can write the following expression:

$$\ln \frac{\frac{1}{2}[A]_0}{[A]_0} = \ln \frac{1}{2} = -kt_{1/2} \qquad [14.18]$$

Solving for $t_{1/2}$, and substituting $-0.693$ for $\ln\frac{1}{2}$, we arrive at the expression for the half-life of a first-order reaction:

$$t_{1/2} = \frac{0.693}{k} \qquad [14.19]$$

Notice that, for a first-order reaction, $t_{1/2}$ *is independent of the initial concentration*. For example, if $t_{1/2}$ is 100 s, and if the initial concentration is 1.0 M, the concentration falls to 0.50 M in 100 s, to 0.25 M in another 100 s, to 0.125 M in another 100 s, and so on (Figure 14.11▶). Even though the concentration is changing as the reaction proceeds, the half-life (how long it takes for the concentration to halve) is constant. A constant half-life is unique to first-order reactions, making the concept of half-life particularly useful for first-order reactions.

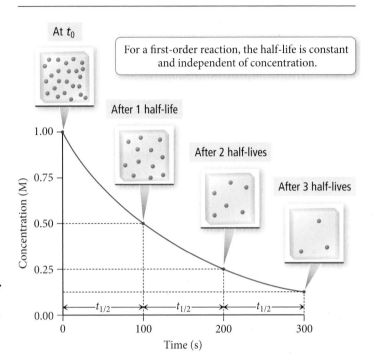

**Half-Life for a First-Order Reaction**

At $t_0$

For a first-order reaction, the half-life is constant and independent of concentration.

After 1 half-life

After 2 half-lives

After 3 half-lives

▶ **FIGURE 14.11  Half-Life: Concentration versus Time for a First-Order Reaction**   For this reaction, the concentration decreases by one-half every 100 seconds ($t_{1/2} = 100$ s). The blue spheres represent reactant molecules (the products are omitted for clarity).

**Half-Life Part I**   The images shown here depict the first-order reaction A $\longrightarrow$ B at various times during the reaction process. The black circles represent reactant A and the red circles represent product B. What is the half-life of the reaction?

CONCEPTUAL
**CONNECTION 14.4**

$t = 0$

$t = 30$ s

$t = 60$ s

$t = 90$ s

$t = 120$ s

**Half-Life Part II**   A first-order reaction (A $\longrightarrow$ B) has a half-life of 25 minutes. If the initial concentration of A is 0.300 M, what is the concentration of B after 50 minutes? (Do not use a calculator to solve this problem.)

CONCEPTUAL
**CONNECTION 14.5**

## Example 14.6 Half-Life

Molecular iodine dissociates at 625 K with a first-order rate constant of $0.271 \text{ s}^{-1}$. What is the half-life of this reaction?

**SOLUTION**

| | |
|---|---|
| Because the reaction is first order, the half-life is given by Equation 14.19. Substitute the value of $k$ into the expression and calculate $t_{1/2}$. | $t_{1/2} = \dfrac{0.693}{k}$ <br><br> $= \dfrac{0.693}{0.271 \text{ s}^{-1}} = 2.56 \text{ s}$ |

**FOR PRACTICE 14.6** A first-order reaction has a half-life of 26.4 seconds. How long does it take for the concentration of the reactant in the reaction to fall to one-eighth of its initial value?

### Second-Order Reaction Half-Life

For a second-order reaction, the integrated rate law is:

$$\frac{1}{[A]_t} = kt + \frac{1}{[A]_0}$$

At a time equal to the half-life ($t = t_{1/2}$), the concentration is exactly one-half of the initial concentration ($[A]_t = \frac{1}{2}[A]_0$). We can therefore write the following expression at $t = t_{1/2}$:

$$\frac{1}{\frac{1}{2}[A]_0} = kt_{1/2} + \frac{1}{[A]_0} \qquad [14.20]$$

And then solve for $t_{1/2}$:

$$kt_{1/2} = \frac{1}{\frac{1}{2}[A]_0} - \frac{1}{[A]_0}$$

$$kt_{1/2} = \frac{2}{[A]_0} - \frac{1}{[A]_0}$$

$$t_{1/2} = \frac{1}{k[A]_0} \qquad [14.21]$$

Notice that, for a second-order reaction, the half-life depends on the initial concentration. So if the initial concentration of a reactant in a second-order reaction is 1.0 M, and the half-life is 100 s, the concentration falls to 0.50 M in 100 s. However, the time it takes for the concentration to fall to 0.25 M is now *longer than 100 s* because the initial concentration has decreased. Thus, the half-life continues to get longer as the concentration decreases.

### Zero-Order Reaction Half-Life

For a zero-order reaction, the integrated rate law is:

$$[A]_t = -kt + [A]_0$$

Making the substitutions ($t = t_{1/2}$; $[A]_t = \frac{1}{2}[A]_0$), we can write the expression at $t = t_{1/2}$:

$$\frac{1}{2}[A]_0 = -kt_{1/2} + [A]_0 \qquad [14.22]$$

We then solve for $t_{1/2}$:

$$t_{1/2} = \frac{[A]_0}{2k} \qquad [14.23]$$

Notice that, for a zero-order reaction, the half-life also depends on the initial concentration; however, unlike in the second-order case, the two are directly proportional—the half-life gets shorter as the concentration decreases.

*Summarizing Basic Kinetic Relationships (see Table 14.2):*

▶ The reaction order and rate law must be determined experimentally.

▶ The rate law relates the *rate* of the reaction to the *concentration* of the reactant(s).

▶ The integrated rate law (which is mathematically derived from the rate law) relates the *concentration* of the reactant(s) to *time*.

▶ The half-life is the time it takes for the concentration of a reactant to fall to one-half of its initial value.

▶ The half-life of a first-order reaction is independent of the initial concentration.

▶ The half-lives of zero-order and second-order reactions depend on the initial concentrations.

**Rate Law and Integrated Rate Law**  A decomposition reaction, with a rate that is observed to slow down as the reaction proceeds, has a half-life that depends on the initial concentration of the reactant. Which statement is most likely true for this reaction?

CONCEPTUAL
CONNECTION **14.6**

(a) A plot of the natural log of the concentration of the reactant as a function of time is linear.

(b) The half-life of the reaction increases as the initial concentration increases.

(c) A doubling of the initial concentration of the reactant results in a quadrupling of the rate.

**TABLE 14.2 Rate Law Summary Table**

| Order | Rate Law | Units of $k$ | Integrated Rate Law | Straight-Line Plot | Half-Life Expression |
|---|---|---|---|---|---|
| 0 | $\text{Rate} = k[A]^0$ | $M \cdot s^{-1}$ | $[A]_t = -kt + [A]_0$ | 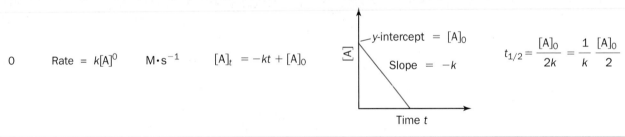 | $t_{1/2} = \dfrac{[A]_0}{2k} = \dfrac{1}{k}\dfrac{[A]_0}{2}$ |
| 1 | $\text{Rate} = k[A]^1$ | $s^{-1}$ | $\ln[A]_t = -kt + \ln[A]_0$ $\ln\dfrac{[A]_t}{[A]_0} = -kt$ | | $t_{1/2} = \dfrac{0.693}{k} = \dfrac{1}{k}(0.693)$ |
| 2 | $\text{Rate} = k[A]^2$ | $M^{-1} \cdot s^{-1}$ | $\dfrac{1}{[A]_t} = kt + \dfrac{1}{[A]_0}$ | 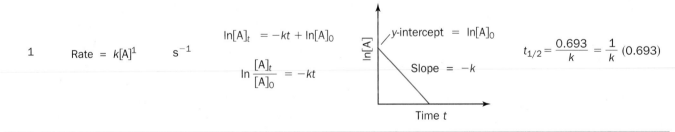 | $t_{1/2} = \dfrac{1}{k[A]_0} = \dfrac{1}{k}\dfrac{1}{[A]_0}$ |

For order 0 plot: $y$-intercept $= [A]_0$; Slope $= -k$; axes [A] vs Time $t$.

For order 1 plot: $y$-intercept $= \ln[A]_0$; Slope $= -k$; axes $\ln[A]$ vs Time $t$.

For order 2 plot: Slope $= k$; $y$-intercept $= 1/[A]_0$; axes $1/[A]$ vs Time $t$.

# 14.5 The Effect of Temperature on Reaction Rate

In the opening section of this chapter, we discussed how lizards become lethargic when their body temperature drops and the chemical reactions that control their muscle movement slow. The rates of chemical reactions are, in general, highly sensitive to temperature. For example, at around room temperature, a 10 °C increase in temperature increases the rate of a typical biological reaction by two or three times. How do we explain this highly sensitive temperature dependence?

## The Arrhenius Equation

Recall that the rate law for a reaction is Rate = $k[A]^n$. *The temperature dependence of the reaction rate is contained in the rate constant, k* (which is actually a constant only when the temperature remains constant). An increase in temperature generally results in an increase in $k$, which results in a faster rate. In 1889, Swedish chemist Svante Arrhenius (1859–1927) wrote a paper quantifying the temperature dependence of the rate constant.

The modern form of the **Arrhenius equation** shows the relationship between the rate constant ($k$) and the temperature in kelvin ($T$):

$$k = Ae^{\frac{-E_a}{RT}}$$

Frequency factor     Activation energy     Exponential factor

[14.24]

In this equation, $R$ is the gas constant (8.314 J/mol · K), $A$ is a constant called the *frequency factor* (or the *pre-exponential factor*), and $E_a$ is the *activation energy* (or *activation barrier*).

## Activation Energy

$$2\,H_2(g) + O_2(g) \rightleftharpoons 2\,H_2O(g)$$

▲ **FIGURE 14.12 The Activation Energy Barrier** Even though the reaction is energetically favorable (the energy of the products is lower than that of the reactants), an input of energy is needed for the reaction to take place.

The **activation energy ($E_a$)** is an energy barrier or hump that must be surmounted in order for the reactants to transform into products (Figure 14.12▲). We examine the frequency factor more closely in the next section; for now, we can think of the **frequency factor (A)** as the number of times that the reactants approach the activation barrier per unit time.

# The Activation Energy, Frequency Factor, and Exponential Factor

To understand each of these quantities better, consider the simple reaction in which $CH_3NC$ (methyl isonitrile) rearranges to form $CH_3CN$ (acetonitrile):

Let's examine the physical meaning of the activation energy, frequency factor, and exponential factor for this reaction.

## The Activation Energy

Figure 14.13▶ shows the energy of a molecule as the reaction proceeds. The *x*-axis represents the progress of the reaction from left (reactant) to right (product). To get from the reactant to the product, the molecule must go through a high-energy state called the **activated complex** or **transition state.** Even though the overall reaction is energetically downhill (exothermic), it must first go uphill to reach the activated complex because energy is required to initially weaken the $H_3C-N$ bond and allow the NC group to begin to rotate:

Bond weakens

NC group begins to rotate

▲ **FIGURE 14.13  The Activated Complex**  The reaction pathway includes a transitional state—the activated complex—that has a higher energy than either the reactant or the product.

The energy required to reach the activated complex is the *activation energy. The higher the activation energy, the slower the reaction rate (at a given temperature).*

## The Frequency Factor

Recall that the frequency factor represents the number of approaches to the activation barrier per unit time. Any time that the NC group begins to rotate, it approaches the activation barrier. For this reaction, the frequency factor represents the rate at which the NC part of the molecule wags (vibrates side to side). With each wag, the reactant approaches the activation barrier:

Each wag is an approach to the activation barrier.

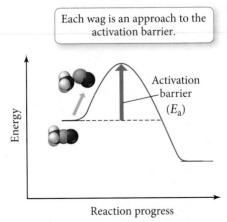

However, approaching the activation barrier is not equivalent to surmounting it. Most of the approaches do not have enough total energy to make it over the activation barrier.

## The Exponential Factor

The **exponential factor** is a number between 0 and 1 that represents the fraction of molecules that have enough energy to make it over the activation barrier on a given approach. The exponential factor is the fraction of approaches that are actually successful and result in the product. For example, if the frequency factor is $10^9/s$ and the exponential factor is $10^{-7}$ at a certain temperature, then the overall rate constant at that temperature is $10^9/s \times 10^{-7} = 10^2/s$. In this case, the CN group wags at a rate of $10^9/s$. With each wag, the activation barrier is approached. However, for a given wag only 1 in $10^7$ molecules has sufficient energy to actually make it over the activation barrier.

The exponential factor depends on both the temperature ($T$) and the activation energy ($E_a$) of the reaction:

$$\text{Exponential factor} = e^{-E_a/RT}$$

A low activation energy and a high temperature make the negative exponent small, so the exponential factor approaches one. For example, if the activation energy is zero, then the exponent is zero, and the exponential factor is exactly one ($e^{-0} = 1$)—every approach to the activation barrier is successful. By contrast, a large activation energy and a low temperature make the exponent a very large negative number, so the exponential factor becomes very small. For example, as the temperature approaches 0 K, the exponent approaches an infinitely large negative number, and the exponential factor approaches zero ($e^{-\infty} = 0$).

As the temperature increases, the number of molecules having enough thermal energy to surmount the activation barrier increases. At any given temperature, a sample of molecules has a distribution of energies, as shown in Figure 14.14 ◄. Under common circumstances, only a small fraction of the molecules has enough energy to make it over the activation barrier. Because of the shape of the energy distribution curve, however, a small change in temperature results in a large difference in the number of molecules having enough energy to surmount the activation barrier. This explains the sensitivity of reaction rates to temperature.

### Thermal Energy Distribution

As temperature increases, the fraction of molecules with enough energy to surmount the activation energy barrier also increases.

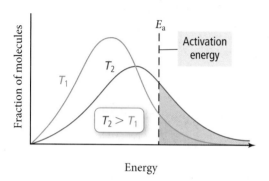

▲ **FIGURE 14.14 Thermal Energy Distribution** At any given temperature, the atoms or molecules in a gas sample have a range of energies. The higher the temperature, the wider the energy distribution and the greater the average energy. The fraction of molecules with enough energy to surmount the activation energy barrier and react (shaded regions) increases sharply as the temperature rises.

*Summarizing Temperature and Reaction Rate:*

▶ The frequency factor is the number of times that the reactants approach the activation barrier per unit time.

▶ The exponential factor is the fraction of the approaches that are successful in surmounting the activation barrier and forming products.

▶ The exponential factor increases with increasing temperature but decreases with increasing activation energy.

## Arrhenius Plots: Experimental Measurements of the Frequency Factor and the Activation Energy

The frequency factor and activation energy are important quantities in understanding the kinetics of any reaction. To see how we measure these factors in the laboratory, consider again Equation 14.24: $k = Ae^{-E_a/RT}$. Taking the natural log of both sides of this equation, we get the following result:

| Remember that $\ln(AB) = \ln A + \ln B$.

| Remember that $\ln e^x = x$.

| In an Arrhenius analysis, the pre-exponential factor ($A$) is assumed to be independent of temperature. Although the pre-exponential factor does depend on temperature to some degree, its temperature dependence is much less than that of the exponential factor and is often ignored.

$$\ln k = \ln(Ae^{-E_a/RT})$$

$$\ln k = \ln A + \ln e^{-E_a/RT}$$

$$\ln k = \ln A - \frac{E_a}{RT} \qquad [14.25]$$

$$\ln k = -\frac{E_a}{R}\left(\frac{1}{T}\right) + \ln A \qquad [14.26]$$

$$y = \quad mx \quad + \quad b$$

Equation 14.26 is in the form of a straight line. *A plot of the natural log of the rate constant (ln k) versus the inverse of the temperature in kelvins (1/T) yields a straight line with a slope of $-E_a/R$ and a y-intercept of ln A.* Such a plot is called an **Arrhenius plot** and is commonly used in the analysis of kinetic data, as shown in Example 14.7.

---

**Example 14.7**    **Using an Arrhenius Plot to Determine Kinetic Parameters**

The decomposition of ozone shown here is important to many atmospheric reactions:

$$O_3(g) \longrightarrow O_2(g) + O(g)$$

A study of the kinetics of the reaction results in the following data:

| Temperature (K) | Rate Constant ($M^{-1} \cdot s^{-1}$) | Temperature (K) | Rate Constant ($M^{-1} \cdot s^{-1}$) |
|---|---|---|---|
| 600 | $3.37 \times 10^3$ | 1300 | $7.83 \times 10^7$ |
| 700 | $4.85 \times 10^4$ | 1400 | $1.45 \times 10^8$ |
| 800 | $3.58 \times 10^5$ | 1500 | $2.46 \times 10^8$ |
| 900 | $1.70 \times 10^6$ | 1600 | $3.93 \times 10^8$ |
| 1000 | $5.90 \times 10^6$ | 1700 | $5.93 \times 10^8$ |
| 1100 | $1.63 \times 10^7$ | 1800 | $8.55 \times 10^8$ |
| 1200 | $3.81 \times 10^7$ | 1900 | $1.19 \times 10^9$ |

Determine the value of the frequency factor and activation energy for the reaction.

**SOLUTION**

To determine the frequency factor and activation energy, prepare a graph of the natural log of the rate constant (ln k) versus the inverse of the temperature (1/T).

The plot is linear, as expected for Arrhenius behavior. The line that fits best has a slope of $-1.12 \times 10^4$ K and a y-intercept of 26.8. Calculate the activation energy from the slope by setting the slope equal to $-E_a/R$ and solving for $E_a$:

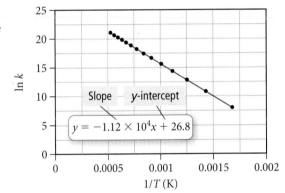

$$-1.12 \times 10^4 \text{ K} = \frac{-E_a}{R}$$

$$E_a = 1.12 \times 10^4 \text{ K} \left( 8.314 \frac{\text{J}}{\text{mol} \cdot \text{K}} \right)$$

$$= 9.31 \times 10^4 \text{ J / mol}$$

$$= 93.1 \text{ kJ/mol}$$

Calculate the frequency factor (A) by setting the intercept equal to ln A.

$$26.8 = \ln A$$

$$A = e^{26.8}$$

$$= 4.36 \times 10^{11}$$

Since the rate constants are measured in units of $M^{-1} \cdot s^{-1}$, the frequency factor is in the same units. Consequently, you can conclude that the reaction has an activation energy of 93.1 kJ/mol and a frequency factor of $4.36 \times 10^{11} \, M^{-1} \cdot s^{-1}$.

**FOR PRACTICE 14.7**    For the decomposition of ozone reaction in Example 14.7, use the results of the Arrhenius analysis to predict the rate constant at 298 K.

In some cases, when either data are limited or plotting capabilities are absent, we can calculate the activation energy if we know the rate constant at just two different temperatures. We can apply the Arrhenius expression in Equation 14.26 to the two different temperatures as follows:

$$\ln k_2 = -\frac{E_a}{R}\left(\frac{1}{T_2}\right) + \ln A$$

$$\ln k_1 = -\frac{E_a}{R}\left(\frac{1}{T_1}\right) + \ln A$$

We can then subtract $\ln k_1$ from $\ln k_2$:

$$\ln k_2 - \ln k_1 = \left[-\frac{E_a}{R}\left(\frac{1}{T_2}\right) + \ln A\right] - \left[-\frac{E_a}{R}\left(\frac{1}{T_1}\right) + \ln A\right]$$

Rearranging, we get the two-point form of the Arrhenius equation:

$$\ln\frac{k_2}{k_1} = \frac{E_a}{R}\left(\frac{1}{T_1} - \frac{1}{T_2}\right) \qquad\qquad [14.27]$$

Example 14.8 demonstrates how to use this equation to calculate the activation energy from experimental measurements of the rate constant at two different temperatures.

---

**Example 14.8**   **Using the Two-Point Form of the Arrhenius Equation**

Consider the reaction between nitrogen dioxide and carbon monoxide:

$$NO_2(g) + CO(g) \longrightarrow NO(g) + CO_2(g)$$

The rate constant at 701 K is measured as $2.57\ \text{M}^{-1}\cdot\text{s}^{-1}$ and that at 895 K is measured as $567\ \text{M}^{-1}\cdot\text{s}^{-1}$. Find the activation energy for the reaction in kJ/mol.

| | |
|---|---|
| **SORT** You are given the rate constant of a reaction at two different temperatures and asked to find the activation energy. | **GIVEN:** $T_1 = 701\ \text{K}$, $k_1 = 2.57\ \text{M}^{-1}\cdot\text{s}^{-1}$<br>$T_2 = 895\ \text{K}$, $k_2 = 567\ \text{M}^{-1}\cdot\text{s}^{-1}$<br><br>**FIND:** $E_a$ |
| **STRATEGIZE** Use the two-point form of the Arrhenius equation, which relates the activation energy to the given information and $R$ (a constant). | **EQUATION** $\ln\dfrac{k_2}{k_1} = \dfrac{E_a}{R}\left(\dfrac{1}{T_1} - \dfrac{1}{T_2}\right)$ |
| **SOLVE** Substitute the two rate constants and the two temperatures into the equation.<br><br><br><br><br>Solve the equation for $E_a$, the activation energy, and convert to kJ/mol. | **SOLUTION**<br>$\ln\dfrac{567\ \text{M}^{-1}\cdot\text{s}^{-1}}{2.57\ \text{M}^{-1}\cdot\text{s}^{-1}} = \dfrac{E_a}{R}\left(\dfrac{1}{701\ \text{K}} - \dfrac{1}{895\ \text{K}}\right)$<br><br>$5.40 = \dfrac{E_a}{R}\left(\dfrac{3.09 \times 10^{-4}}{\text{K}}\right)$<br><br>$E_a = 5.40\left(\dfrac{\text{K}}{3.09 \times 10^{-4}}\right)R$<br><br>$= 5.40\left(\dfrac{\text{K}}{3.09 \times 10^{-4}}\right)8.314\ \dfrac{\text{J}}{\text{mol}\cdot\text{K}}$<br><br>$= 1.45 \times 10^5\ \text{J/mol}$<br><br>$= 1.5 \times 10^2\ \text{kJ/mol}$ |

**CHECK** The magnitude of the answer is reasonable. Activation energies for most reactions range from tens to hundreds of kilojoules per mole.

**FOR PRACTICE 14.8** Use the results from Example 14.8 and the given rate constant of the reaction at either of the two temperatures to predict the rate constant for this reaction at 525 K.

**Temperature Dependence of Reaction Rate**  Reaction A and reaction B have identical frequency factors. However, reaction B has a higher activation energy than reaction A. Which reaction has a greater rate constant at room temperature?

## The Collision Model: A Closer Look at the Frequency Factor

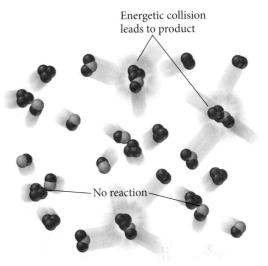

▲ **FIGURE 14.15  The Collision Model**  In the collision model, two molecules react after a sufficiently energetic collision with the correct orientation brings the reacting groups together.

Recall that the frequency factor in the Arrhenius equation represents the number of approaches to the activation barrier per unit time. Let's refine that idea for a reaction involving two gas-phase reactants:

$$A(g) + B(g) \longrightarrow products$$

In the **collision model**, a chemical reaction occurs after a sufficiently energetic collision between two reactant molecules (Figure 14.15▶). In collision theory, therefore, each approach to the activation barrier is a collision between the reactant molecules. Consequently, the value of the frequency factor should simply be the number of collisions that occur per second. However, the frequency factors of most (though not all) gas-phase chemical reactions tend to be smaller than the number of collisions that occur per second. Why?

In the collision model, we can separate the frequency factor into two distinct parts, as shown in the following equations:

$$k = Ae^{\frac{-E_a}{RT}}$$
$$= pze^{\frac{-E_a}{RT}}$$

Orientation factor    Collision frequency

where $p$ is the **orientation factor** and $z$ is the **collision frequency**. The collision frequency is the number of collisions that occur per unit time, which we calculate for a gas-phase reaction from the pressure of the gases and the temperature of the reaction mixture. Under typical conditions, a single molecule undergoes on the order of $10^9$ collisions every second. The orientation factor is a number, usually between 0 and 1, which represents the fraction of collisions with an orientation that allows the reaction to occur.

To better understand the orientation factor, consider the reaction represented by the following equation:

$$NOCl(g) + NOCl(g) \longrightarrow 2 NO(g) + Cl_2(g)$$

In order for the reaction to occur, two NOCl molecules must collide with sufficient energy. However, not all collisions with sufficient energy lead to products because the reactant molecules must also be properly oriented.

Consider the three possible collision orientations of the reactant molecules shown here:

Ineffective collision

Ineffective collision

Effective collision

The first two collisions, even if they occur with sufficient energy, do not result in a reaction because the reactant molecules are not oriented in a way that allows the chlorine atoms to bond. In other words, if two molecules are to react with each other, they must collide in such a way that allows the necessary bonds to break and form. For the reaction of $NOCl(g)$, the orientation factor is $p = 0.16$. This means that only 16 out of 100 sufficiently energetic collisions are actually successful in forming the products.

Some reactions have orientation factors that are much smaller than one. Consider the reaction between hydrogen and ethene:

$$H_2(g) + CH_2{=}CH_2(g) \longrightarrow CH_3{-}CH_3(g)$$

The orientation factor for this reaction is $1.7 \times 10^{-6}$, which means that fewer than two out of each million sufficiently energetic collisions actually form products. The small orientation factor indicates that the orientational requirements for this reaction are very stringent—the molecules must be aligned in a *very specific way* for the reaction to occur.

Reactions between *individual atoms* usually have orientation factors of approximately one because atoms are spherically symmetric and thus any orientation can lead to the formation of products. A few reactions have orientation factors greater than one. Consider the reaction between potassium and bromine:

$$K(g) + Br_2(g) \longrightarrow KBr(g) + Br(g)$$

This reaction has an orientation factor of $p = 4.8$. In other words, there are more reactions than collisions—the reactants do not even have to collide to react! Apparently, through a process dubbed *the harpoon mechanism*, a potassium atom can actually transfer an electron to a bromine molecule without a collision. The resulting positive charge on the potassium and the negative charge on the bromine cause the two species to attract each other and form a bond. The potassium atom essentially *harpoons* a passing bromine molecule with an electron and *reels it in* through the coulombic attraction between unlike charges.

We can picture a sample of reactive gases as a frenzy of collisions between the reacting atoms or molecules. At normal temperatures, the vast majority of these collisions do not have sufficient energy to overcome the activation barrier, and the atoms or molecules simply bounce off one another. Of the collisions having sufficient energy to overcome the activation barrier, most do not have the proper orientation for the reaction to occur (for the majority of common reactions). When two molecules with sufficient energy *and* the correct orientation collide, something extraordinary happens. The electrons on one of the atoms or molecules are attracted to the nuclei of the other; some bonds begin to weaken while other bonds begin to form and, if all goes well, the reactants go through the transition state and are transformed into the products. This is how a chemical reaction occurs.

CONCEPTUAL
**CONNECTION 14.8**

**Collision Theory** Which reaction do you expect to have the smallest orientation factor?

(a) $H(g) + I(g) \longrightarrow HI(g)$
(b) $H_2(g) + I_2(g) \longrightarrow 2\,HI(g)$
(c) $HCl(g) + HCl(g) \longrightarrow H_2(g) + Cl_2(g)$

## 14.6 Reaction Mechanisms

Most chemical reactions do not occur in a single step but rather over the course of several steps. When we write a chemical equation to represent a chemical reaction, *we usually represent the overall reaction, not the series of individual steps by which the reaction occurs.* Consider the reaction in which hydrogen gas reacts with iodine monochloride:

$$H_2(g) + 2\,ICl(g) \longrightarrow 2\,HCl(g) + I_2(g)$$

The overall equation shows only the substances present at the beginning of the reaction and the substances formed by the reaction—it does not show the intermediate steps. The **reaction mechanism** is the series of individual chemical steps by which an overall chemical reaction occurs. For example, the proposed mechanism for the reaction between hydrogen and iodine monochloride contains two steps:

> **Step 1**   $H_2(g) + ICl(g) \longrightarrow HI(g) + HCl(g)$
>
> **Step 2**   $HI(g) + ICl(g) \longrightarrow HCl(g) + I_2(g)$

In the first step, an $H_2$ molecule collides with an ICl molecule and forms an HI molecule and an HCl molecule. In the second step, the HI molecule formed in the first step collides with a second ICl molecule to form another HCl molecule and an $I_2$ molecule. Each step in a reaction mechanism is an **elementary step**. Elementary steps cannot be broken down into simpler steps—they occur as written (they represent the exact species that are colliding in the reaction).

One of the requirements for a valid reaction mechanism is that the individual steps in the mechanism add to the overall reaction. For example, the mechanism just shown sums to the overall reaction as shown here:

$$\begin{array}{r} H_2(g) + ICl(g) \longrightarrow HI(g) + HCl(g) \\ HI(g) + ICl(g) \longrightarrow HCl(g) + I_2(g) \\ \hline H_2(g) + 2\,ICl(g) \longrightarrow 2\,HCl(g) + I_2(g) \end{array}$$

Notice that the HI molecule appears in the reaction mechanism but not in the overall reaction equation. We call species such as HI *reaction intermediates*. A **reaction intermediate** forms in one elementary step and is consumed in another. The reaction mechanism is a complete, detailed description of the reaction at the molecular level—it specifies the individual collisions and reactions that result in the overall reaction. As such, reaction mechanisms are highly sought-after pieces of chemical knowledge.

How do we determine reaction mechanisms? Recall from the opening section of this chapter that chemical kinetics are not only practically important (because they allow us to control the rate of a particular reaction), but they are also theoretically important because they can help us determine the mechanism of the reaction. We can piece together possible reaction mechanisms by measuring the kinetics of the overall reaction and working backward to write a mechanism consistent with the measured kinetics.

An elementary step represents an actual interaction between the reactant molecules in the step. An overall reaction equation shows only the starting substances and the ending substances, not the path between them.

## Rate Laws for Elementary Steps

We characterize elementary steps by their **molecularity**, the number of reactant particles involved in the step. The most common molecularities are **unimolecular** and **bimolecular**:

| | |
|---|---|
| $A \longrightarrow$ products | **Unimolecular** |
| $A + A \longrightarrow$ products | **Bimolecular** |
| $A + B \longrightarrow$ products | **Bimolecular** |

Elementary steps in which three reactant particles collide, called **termolecular** steps, are very rare because the probability of three particles simultaneously colliding is small.

Although we cannot deduce the rate law for an *overall chemical reaction* from the balanced chemical equation, we can deduce the rate law for an *elementary step* from its equation. Since we know that an elementary step occurs through the collision of the reactant particles, the rate is proportional to the product of the concentrations of those particles. For example, the rate for the bimolecular elementary step in which A reacts with B is proportional to the concentration of A multiplied by the concentration of B:

$$A + B \longrightarrow \text{products} \qquad \text{Rate} = k[A][B]$$

Similarly, the rate law for the bimolecular step in which A reacts with A is proportional to the square of the concentration of A:

$$A + A \longrightarrow \text{products} \qquad \text{Rate} = k[A]^2$$

Table 14.3 summarizes the rate laws for the common elementary steps, as well as those for the rare termolecular step. Notice that the molecularity of the elementary step is equal to the overall order of the step.

| TABLE 14.3 Rate Laws for Elementary Steps | | |
| --- | --- | --- |
| Elementary Step | Molecularity | Rate Law |
| A $\longrightarrow$ products | 1 | Rate = $k[A]$ |
| A + A $\longrightarrow$ products | 2 | Rate = $k[A]^2$ |
| A + B $\longrightarrow$ products | 2 | Rate = $k[A][B]$ |
| A + A + A $\longrightarrow$ products | 3 (rare) | Rate = $k[A]^3$ |
| A + A + B $\longrightarrow$ products | 3 (rare) | Rate = $k[A]^2[B]$ |
| A + B + C $\longrightarrow$ products | 3 (rare) | Rate = $k[A][B][C]$ |

## Rate-Determining Steps and Overall Reaction Rate Laws

In most chemical reactions, one of the elementary steps—called the **rate-determining step**—is much slower than the others. The rate-determining step in a chemical reaction is analogous to the narrowest section on a freeway. If a section of a freeway narrows from four lanes to two lanes, for even a short distance, the rate at which cars travel along the freeway is limited by the rate at which they can travel through the narrow section (even though the rate might be much faster along the four-lane section). Similarly, the rate-determining step in a reaction mechanism limits the overall rate of the reaction (even though the other steps occur much faster) and therefore determines *the rate law for the overall reaction.*

▶ The rate-limiting step in a reaction mechanism limits the overall rate of the reaction just as the narrowest section of a highway limits the rate at which traffic can pass.

Rate-limiting section

As an example, consider the reaction between nitrogen dioxide gas and carbon monoxide gas:

$$NO_2(g) + CO(g) \longrightarrow NO(g) + CO_2(g)$$

The experimentally determined rate law for this reaction is Rate = $k[NO_2]^2$. We can see from this rate law that the reaction must not be a single-step reaction—otherwise the rate law would be Rate = $k[NO_2][CO]$. A possible mechanism for this reaction involves two steps:

$$NO_2(g) + NO_2(g) \longrightarrow NO_3(g) + NO(g) \qquad \text{Slow}$$

$$NO_3(g) + CO(g) \longrightarrow NO_2(g) + CO_2(g) \qquad \text{Fast}$$

Figure 14.16▶ shows the energy diagram for this mechanism. The first step has a much larger activation energy than the second step. This greater activation energy results in a much smaller rate constant for the first step compared to the second step. The first step determines the overall rate of the reaction, and the predicted *overall* rate law is therefore Rate = $k[NO_2]^2$, which is consistent with the observed experimental rate law.

## Energy Diagram for a Two-Step Mechanism

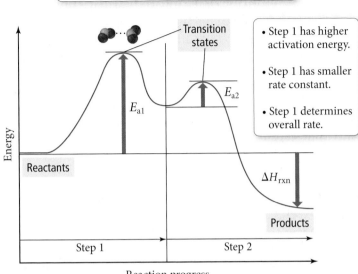

Because $E_a$ for Step 1 > $E_a$ for Step 2, Step 1 has the smaller rate constant and is rate limiting.

Transition states

$E_{a1}$

$E_{a2}$

- Step 1 has higher activation energy.
- Step 1 has smaller rate constant.
- Step 1 determines overall rate.

Reactants

$\Delta H_{rxn}$

Products

Step 1    Step 2

Reaction progress

For a proposed reaction mechanism, such as the one shown here for $NO_2$ and CO, to be *valid* (mechanisms can only be validated, not proven), two conditions must be met:

1. **The elementary steps in the mechanism must sum to the overall reaction.**
2. **The rate law predicted by the mechanism must be consistent with the experimentally observed rate law.**

We have already seen that the rate law predicted by the earlier mechanism is consistent with the experimentally observed rate law. We can check whether the elementary steps sum to the overall reaction by adding them together:

$$NO_2(g) + NO_2(g) \longrightarrow NO_3(g) + NO(g) \qquad \text{Slow}$$
$$\underline{NO_3(g) + CO(g) \longrightarrow NO_2(g) + CO_2(g) \qquad \text{Fast}}$$
$$NO_2(g) + CO(g) \longrightarrow NO(g) + CO_2(g) \qquad \text{Overall}$$

The mechanism fulfills both of the given requirements and is therefore valid. A valid mechanism is not a *proven* mechanism (because other mechanisms may also fulfill both of the given requirements). We can only say that a given mechanism is consistent with kinetic observations of the reaction and is therefore possible. Other types of data—such as the experimental evidence for a proposed intermediate—are necessary to further strengthen the validity of a proposed mechanism.

## Mechanisms with a Fast Initial Step

When the proposed mechanism for a reaction has a slow initial step—such as the one shown previously for the reaction between $NO_2$ and CO—the rate law predicted by the mechanism normally contains only reactants involved in the overall reaction. However, when a mechanism begins with a fast initial step, some other subsequent step in the mechanism is the rate-limiting step. In these cases, the rate law predicted by the rate-limiting step may contain reaction intermediates. Because reaction intermediates do not appear in the overall reaction equation, a rate law containing intermediates cannot generally correspond to the experimental rate law. Fortunately, however, we can often express the concentration of intermediates in terms of the concentrations of the reactants of the overall reaction.

In a multistep mechanism where the first step is fast, the products of the first step may build up because the rate at which they are consumed is limited by some slower step further down the line. As those products build up, they can begin to react with one another to re-form the reactants. As long as the first step is fast enough compared to the rate-limiting step, the first-step reaction will reach equilibrium. We indicate equilibrium as follows:

$$\text{Reactants} \underset{k_{-1}}{\overset{k_1}{\rightleftharpoons}} \text{products}$$

The double arrows indicate that both the forward reaction and the reverse reaction occur. If equilibrium is reached, then the rate of the forward reaction equals the rate of the reverse reaction.

As an example, consider the reaction by which hydrogen reacts with nitrogen monoxide to form water and nitrogen gas:

$$2\,H_2(g) + 2\,NO(g) \longrightarrow 2\,H_2O(g) + N_2(g)$$

The experimentally observed rate law is Rate = $k[H_2][NO]^2$. The reaction is first order in hydrogen and second order in nitrogen monoxide. The proposed mechanism has a slow second step:

$$2\,NO(g) \underset{k_{-1}}{\overset{k_1}{\rightleftharpoons}} N_2O_2(g) \qquad\qquad \text{Fast}$$

$$H_2(g) + N_2O_2(g) \overset{k_2}{\longrightarrow} H_2O(g) + N_2O(g) \qquad \text{Slow (rate limiting)}$$

$$N_2O(g) + H_2(g) \overset{k_3}{\longrightarrow} N_2(g) + H_2O(g) \qquad \text{Fast}$$

$$\overline{2\,H_2(g) + 2\,NO(g) \longrightarrow 2\,H_2O(g) + N_2(g)} \qquad \text{Overall}$$

To determine whether the mechanism is valid, we must find whether the two conditions described previously are met. As we can see, the steps do indeed sum to the overall reaction, so the first condition is met.

The second condition is that the rate law predicted by the mechanism must be consistent with the experimentally observed rate law. Because the second step is rate limiting, we write the following expression for the rate law:

$$\text{Rate} = k_2[H_2][N_2O_2] \qquad\qquad\qquad [14.28]$$

This rate law contains an intermediate ($N_2O_2$) and therefore cannot be consistent with the experimentally observed rate law (which does not contain intermediates). Because of the equilibrium in the first step, however, *we can express the concentration of the intermediate in terms of the reactants of the overall equation.* Since the first step reaches equilibrium, the rate of the forward reaction in the first step equals the rate of the reverse reaction:

$$\text{Rate (forward)} = \text{Rate (reverse)}$$

The rate of the forward reaction is given by the rate law:

$$\text{Rate} = k_1[NO]^2$$

The rate of the reverse reaction is given by the rate law:

$$\text{Rate} = k_{-1}[N_2O_2]$$

Since these two rates are equal at equilibrium, we write the expression:

$$k_1[NO]^2 = k_{-1}[N_2O_2]$$

Rearranging, we get:

$$[N_2O_2] = \frac{k_1}{k_{-1}}[NO]^2$$

We can now substitute this expression into Equation 14.28, the rate law obtained from the slow step:

$$\text{Rate} = k_2[H_2][N_2O_2]$$

$$= k_2[H_2]\frac{k_1}{k_{-1}}[NO]^2$$

$$= \frac{k_2k_1}{k_{-1}}[H_2][NO]^2$$

If we combine the individual rate constants into one overall rate constant, we get the predicted rate law:

$$\text{Rate} = k[H_2][NO]^2 \qquad\qquad\qquad [14.29]$$

Since this rate law is consistent with the experimentally observed rate law, condition 2 is met and the proposed mechanism is valid.

**Example 14.9**   **Reaction Mechanisms**

Ozone naturally decomposes to oxygen by this reaction:

$$2\,O_3(g) \longrightarrow 3\,O_2(g)$$

The experimentally observed rate law for this reaction is:

$$\text{Rate} = k[O_3]^2[O_2]^{-1}$$

Show that this proposed mechanism is consistent with the experimentally observed rate law.

$$O_3(g) \underset{k_{-1}}{\overset{k_1}{\rightleftharpoons}} O_2(g) + O(g) \qquad \text{Fast} \qquad O_3(g) + O(g) \xrightarrow{k_2} 2\,O_2(g) \qquad \text{Slow}$$

**SOLUTION**

| | |
|---|---|
| To determine whether the mechanism is valid, you must first determine whether the steps sum to the overall reaction. The steps do indeed sum to the overall reaction, so the first condition is met. | $O_3(g) \underset{k_{-1}}{\overset{k_1}{\rightleftharpoons}} O_2(g) + O(g)$ <br> $\underline{O_3(g) + O(g) \xrightarrow{k_2} 2\,O_2(g)}$ <br> $2\,O_3(g) \longrightarrow 3\,O_2(g)$ |
| The second condition is that the rate law predicted by the mechanism is consistent with the experimentally observed rate law. Because the second step is rate limiting, write the rate law based on the second step. | $\text{Rate} = k_2[O_3][O]$ |
| Because the rate law contains an intermediate (O), you must express the concentration of the intermediate in terms of the concentrations of the reactants of the overall reaction. To do this, set the rates of the forward reaction and the reverse reaction of the first step equal to each other. Solve the expression from the previous step for [O], the concentration of the intermediate. | $\text{Rate (forward)} = \text{Rate (backward)}$ <br> $k_1[O_3] = k_{-1}[O_2][O]$ <br> $[O] = \dfrac{k_1[O_3]}{k_{-1}[O_2]}$ |
| Finally, substitute [O] into the rate law predicted by the slow step. | $\begin{aligned} \text{Rate} &= k_2[O_3][O] \\ &= k_2[O_3]\dfrac{k_1[O_3]}{k_{-1}[O_2]} \\ &= k_2\dfrac{k_1}{k_{-1}}\dfrac{[O_3]^2}{[O_2]} = k[O_3]^2[O_2]^{-1} \end{aligned}$ |

**CHECK** Because the two steps in the proposed mechanism sum to the overall reaction, and because the rate law obtained from the proposed mechanism is consistent with the experimentally observed rate law, the proposed mechanism is valid. The −1 reaction order with respect to [O$_2$] indicates that the rate slows down as the concentration of oxygen increases—oxygen inhibits, or slows down, the reaction.

**FOR PRACTICE 14.9**   Predict the overall reaction and rate law that results from the following two-step mechanism:

$$2\,A \longrightarrow A_2 \qquad \text{Slow}$$
$$A_2 + B \longrightarrow A_2B \qquad \text{Fast}$$

# 14.7 Catalysis

Throughout this chapter, we have learned about ways to control the rates of chemical reactions. We can speed up the rate of a reaction by increasing the concentration of the reactants or by increasing the temperature. However, these approaches are not always feasible. There are limits to how concentrated we can make a reaction mixture, and increases in temperature may allow unwanted reactions—such as the decomposition of a reactant—to occur.

Alternatively, we can increase reaction rates by using a **catalyst**, a substance that increases the rate of a chemical reaction but is not consumed by the reaction. A catalyst works by providing an alternative mechanism for the reaction—one in which the rate-determining step has a lower activation energy.

For example, consider the noncatalytic destruction of ozone in the upper atmosphere, discussed in Section 6.10, which happens according to this reaction:

$$O_3(g) + O(g) \longrightarrow 2\,O_2(g)$$

In this reaction, an ozone molecule collides with an oxygen atom to form two oxygen molecules in a single elementary step.

Earth has a protective ozone layer in the upper atmosphere because the activation energy for this reaction is fairly high; the reaction, therefore, proceeds at a fairly slow rate, and the ozone layer does not rapidly decompose into $O_2$. However, the addition of Cl atoms (which come from the photodissociation of human-made chlorofluorocarbons) to the upper atmosphere makes available another pathway by which $O_3$ can be destroyed. The first step in this pathway—called the catalytic destruction of ozone—is the reaction of Cl with $O_3$ to form ClO and $O_2$:

$$Cl + O_3 \longrightarrow ClO + O_2$$

This is followed by a second step in which ClO reacts with O, regenerating Cl:

$$ClO + O \longrightarrow Cl + O_2$$

If we add the two reactions, the overall reaction is identical to the noncatalytic reaction:

$$\begin{array}{c} \cancel{Cl} + O_3 \longrightarrow \cancel{ClO} + O_2 \\ \cancel{ClO} + O \longrightarrow \cancel{Cl} + O_2 \\ \hline O_3 + O \longrightarrow 2\,O_2 \end{array}$$

However, the activation energy for the rate-limiting step in this pathway is much smaller than for the first, uncatalyzed pathway (as shown in Figure 14.17◄), and therefore the reaction occurs at a much faster rate. Note that the Cl is not consumed in the overall reaction—this is characteristic of a catalyst.

In the case of the catalytic destruction of ozone, the catalyst speeds up a reaction that we *do not* want to happen. Most of the time, however, catalysts are used to speed up reactions that we *do* want to happen. For example, your car most likely has a catalytic converter in its exhaust system. The catalytic converter contains solid catalysts, such as platinum, rhodium, or palladium, dispersed on an underlying high-surface-area ceramic structure. These catalysts convert exhaust pollutants such as nitrogen monoxide and carbon monoxide into less harmful substances:

$$2\,NO(g) + 2\,CO(g) \xrightarrow{\text{catalyst}} N_2(g) + 2\,CO_2(g)$$

The catalytic converter also promotes the complete combustion of any fuel fragments present in the exhaust:

$$CH_3CH_2CH_3(g) + 5\,O_2(g) \xrightarrow{\text{catalyst}} 3\,CO_2(g) + 4\,H_2O(g)$$

Fuel fragment

Photodissociation means *light-induced* dissociation. The energy from a photon of light can break chemical bonds and therefore dissociate, or break apart, a molecule.

## Energy Diagram for Catalyzed and Uncatalyzed Pathways

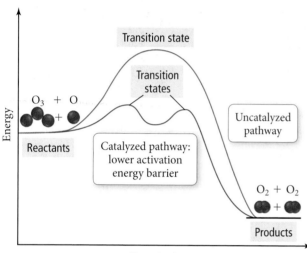

▲ **FIGURE 14.17 Catalyzed and Uncatalyzed Decomposition of Ozone** In the catalytic destruction of ozone (red), the activation barrier for the rate-limiting step is much lower than in the uncatalyzed process (blue).

▶ The catalytic converter in the exhaust system of a car helps eliminate pollutants in the exhaust.

Fuel fragments in exhaust are harmful because they lead to the formation of ozone. Recall from Section 6.10 that although ozone is a natural part of our *upper* atmosphere that protects us from excess exposure to ultraviolet light, it is a pollutant in the *lower* atmosphere, interfering with cardiovascular function and acting as an eye and lung irritant. The use of catalytic converters in motor vehicles has resulted in lower levels of these pollutants over most U.S. cities in the last 30 years, even though the number of cars on the roadways has dramatically increased (see Table 14.4).

## Homogeneous and Heterogeneous Catalysis

We categorize catalysts into two types: homogeneous and heterogeneous (Figure 14.18▼). In **homogeneous catalysis**, the catalyst exists in the same phase (or state) as the reactants. The catalytic destruction of ozone by Cl is an example of homogeneous catalysis—the chlorine atoms exist in the gas phase with the gas-phase reactants. In **heterogeneous catalysis**, the catalyst exists in a different phase than the reactants. The solid catalysts used in catalytic converters are examples of heterogeneous catalysts—they are solids while the reactants are gases. The use of solid catalysts with gas-phase or solution-phase reactants is the most common type of heterogeneous catalysis.

**TABLE 14.4 Change in Pollutant Levels**

| Pollutant | Change 1980–2013 |
|---|---|
| $NO_2$ | −60% |
| $O_3$ | −33% |
| CO | −84% |

Source: *EPA's National Air Trends*

▲ Polar stratospheric clouds contain ice particles that catalyze reactions by which chlorine is released from its atmospheric chemical reservoirs.

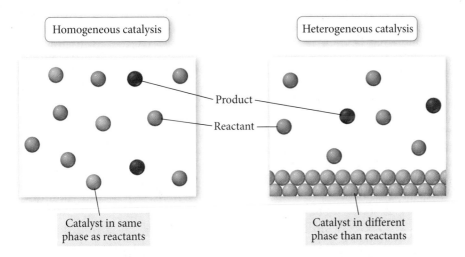

▲ **FIGURE 14.18 Homogeneous and Heterogeneous Catalysis** A homogeneous catalyst exists in the same phase as the reactants. A heterogeneous catalyst exists in a different phase than the reactants. Often, a heterogeneous catalyst provides a solid surface on which the reaction can take place.

Research has shown that heterogeneous catalysis is most likely responsible for the annual formation of an ozone hole over Antarctica. After the discovery of the Antarctic ozone hole in 1985, scientists wondered why there was such a dramatic drop in ozone over Antarctica but not over the rest of the planet. After all, the chlorine from chlorofluorocarbons that catalyzes ozone destruction is evenly distributed throughout the entire atmosphere.

As it turns out, most of the chlorine that enters the atmosphere from chlorofluorocarbons gets bound up in chlorine *reservoirs*, which are substances such as $ClONO_2$ that hold chlorine and prevent it from catalyzing ozone destruction. The unique conditions over Antarctica—especially the cold isolated air mass that exists during the long dark winter—result in clouds that contain solid ice particles. These unique clouds are called polar stratospheric clouds (or PSCs), and the surfaces of the ice particles within these clouds appear to catalyze the release of chlorine from their reservoirs:

$$ClONO_2 + HCl \xrightarrow[\text{PSCs}]{} Cl_2 + HNO_3$$

When the sun rises in the Antarctic spring, the sunlight dissociates the chlorine molecules into chlorine atoms:

$$Cl_2 \xrightarrow[\text{light}]{} 2\,Cl$$

The chlorine atoms then catalyze the destruction of ozone by the mechanism discussed previously. This continues until the sun melts the stratospheric clouds, allowing chlorine

May    September

▲ **FIGURE 14.19** **Ozone Depletion in the Antarctic Spring**  The concentration of ozone over Antarctica drops sharply during the months of September and October due to the catalyzed destruction of ozone by chlorine. The image on the left shows the average ozone levels in May 2014, and the image on the right shows the average levels from September 2014. (The lowest ozone levels are represented in purple.) Source: NASA Ozone Watch, OMI instrument (KNMI/NASA) onboard the Aura satellite.

atoms to be reincorporated into their reservoirs. The result is an ozone hole that forms every spring and lasts about six to eight weeks (Figure 14.19◄).

A second example of heterogeneous catalysis involves the **hydrogenation** of double bonds within alkenes. Consider the reaction between ethene and hydrogen, which is relatively slow at normal temperatures:

$$H_2C{=}CH_2(g) + H_2(g) \longrightarrow H_3C{-}CH_3(g)$$

Slow at room temperature

However, in the presence of finely divided platinum, palladium, or nickel, the reaction happens rapidly. The catalysis occurs by the four-step process depicted in Figure 14.20▼:

1. Adsorption: the reactants are adsorbed onto the metal surface.
2. Diffusion: the reactants diffuse on the surface until they approach each other.
3. Reaction: the reactants react to form the products.
4. Desorption: the products desorb from the surface into the gas phase.

The large activation energy of the hydrogenation reaction—due primarily to the strength of the hydrogen–hydrogen bond in $H_2$—is greatly lowered when the reactants adsorb onto the surface.

**Heterogeneous Catalysis**

Hydrogen    Ethene

Adsorption    Diffusion

Reaction    Desorption

▲ **FIGURE 14.20** **Catalytic Hydrogenation of Ethene**

## Enzymes: Biological Catalysts

The strategies used to speed up chemical reactions in the laboratory—high temperatures, high pressures, strongly acidic or alkaline conditions—are not available to living organisms, since they would be fatal to cells.

We find perhaps the best example of chemical catalysis in living organisms. Most of the thousands of reactions that must occur for an organism to survive are too slow at normal temperatures. So living organisms rely on **enzymes**, biological catalysts that increase the rates of biochemical reactions. Enzymes are usually large protein molecules with complex three-dimensional structures. Within each enzyme's structure is a specific area called the **active site**. The properties and shape of the active site are just right to bind the reactant molecule, usually called the **substrate**. The substrate fits into the active site in a manner that is analogous to a key fitting into a lock (Figure 14.21 ▼). When the substrate binds to the active site of the enzyme—through intermolecular forces such as hydrogen bonding and dispersion forces, or even covalent bonds—the activation energy of the reaction is greatly lowered, allowing the reaction to occur at a much faster rate. The general mechanism by which an enzyme (E) binds a substrate (S) and then reacts to form the products (P) is:

$$E + S \rightleftharpoons ES \qquad \text{Fast}$$
$$ES \longrightarrow E + P \qquad \text{Slow, rate limiting}$$

### Enzyme–Substrate Binding

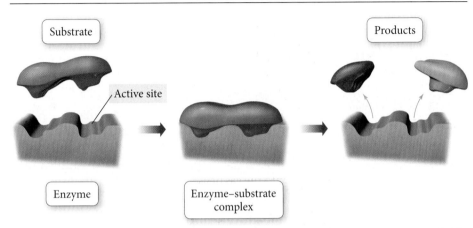

◄ **FIGURE 14.21 Enzyme–Substrate Binding** A substrate (or reactant) fits into the active site of an enzyme much as a key fits into a lock. Intermolecular forces hold the substrate in place and form an enzyme–substrate complex. (Sometimes temporary covalent bonding may also be involved.) After the reaction occurs, the products are released from the active site.

Sucrase, for example, is an enzyme that catalyzes the breaking up of sucrose (table sugar) into glucose and fructose within the body. At body temperature, sucrose does not

$$C_{12}H_{22}O_{11} + H_2O \longrightarrow C_6H_{12}O_6 + C_6H_{12}O_6$$

Sucrose                          Glucose        Fructose

◄ Sucrose breaks up into glucose and fructose during digestion.

break into glucose and fructose because the activation energy is high, resulting in a slow reaction rate. However, when a sucrose molecule binds to the active site within sucrase, the bond between the glucose and fructose units weakens because glucose is forced into a geometry that stresses the bond (Figure 14.22▶). Weakening of this bond lowers the activation energy for the reaction, increasing the reaction rate. The reaction can then proceed toward equilibrium—which favors the products—at a much lower temperature.

By allowing otherwise slow reactions to occur at reasonable rates, enzymes give living organisms tremendous control over which reactions occur and when they occur. Enzymes are extremely specific (each enzyme catalyzes only a single reaction) and efficient, speeding up reaction rates by factors of as much as a billion.

To turn a particular reaction on, a living organism produces or activates the correct enzyme to catalyze that reaction. Because organisms are so dependent on the reactions enzymes catalyze,

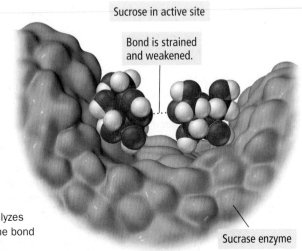

Sucrose in active site

Bond is strained and weakened.

Sucrase enzyme

▶ **FIGURE 14.22 An Enzyme-Catalyzed Reaction** Sucrase catalyzes the conversion of sucrose into glucose and fructose by weakening the bond that joins the two rings.

many substances that inhibit the action of enzymes are highly toxic. Locking up a single enzyme molecule can stop the reaction of billions of substrates, much as one motorist stalled at a tollbooth can paralyze an entire highway full of cars. (For another example of enzyme action, see the *Chemistry and Medicine* box on the role of chymotrypsin in digestion.)

# Chemistry AND Medicine | Enzyme Catalysis and the Role of Chymotrypsin in Digestion

When we eat foods containing proteins—such as meats, eggs, beans, and nuts—the proteins must be digested. Proteins are large biological molecules composed of individual units called amino acids. (We discuss the structure of proteins and other biologically important molecules more fully in Chapter 22.) The amino acids in proteins are linked together via peptide bonds, as shown in Figure 14.23▼.

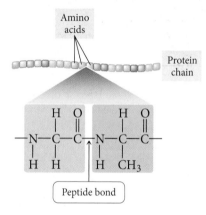

▲ **FIGURE 14.23** **The Structure of a Protein** Proteins are chains of amino acids linked together by peptide bonds.

During digestion, the protein must be broken up into individual amino acids (Figure 14.24▼), which can pass through the walls of the small intestine and into the bloodstream. However, the peptide bonds that link amino acids together are relatively stable, and under ordinary conditions the reaction is slow.

◀ **FIGURE 14.24** **Protein Digestion** During digestion, a protein is broken down into its component amino acids.

The pancreas secretes an enzyme called chymotrypsin (Figure 14.25▶) into the small intestine. Like many enzymes, chymotrypsin is highly selective in its action—it operates only on peptide bonds between certain kinds of amino acids. When a protein molecule containing such a pair of amino acids is attached to the active site of chymotrypsin, the peptide bond between them is weakened as the chymotrypsin forms a covalent bond with the carbon in the peptide bond. A water molecule can then come in and cleave the bond, with

◀ **FIGURE 14.25**
**Chymotrypsin, a Digestive Enzyme** This model of chymotrypsin shows a section of a protein substrate (red) in the active site.

an —OH from the water binding to the carbon atom and the remaining —H bonding to the nitrogen (Figure 14.26▼).

The amino acid chain is thus clipped at the peptide bond. The products of the reaction leave the active site, another protein binds to the chymotrypsin, and the process is repeated. Other digestive enzymes cleave protein chains between other pairs of amino acids. Together, these enzymes eventually reduce the entire protein to its constituent amino acids.

▲ **FIGURE 14.26** **The Action of Chymotrypsin**

# Self-Assessment Quiz

**Q1.** This graph shows the concentration of the reactant A in the reaction A ⟶ B. Determine the average rate of the reaction between 0 and 10 seconds.

a) 0.07 M/s  b) 0.007 M/s  c) 0.86 M/s  d) 0.014 M/s

**Q2.** Dinitrogen monoxide decomposes into nitrogen and oxygen when heated. The initial rate of the reaction is 0.022 M/s. What is the initial rate of change of the concentration of $N_2O$ (that is, $\Delta[N_2O]/\Delta t$)?

$$2\,N_2O(g) \longrightarrow 2\,N_2(g) + O_2(g)$$

a) $-0.022$ M/s    b) $-0.011$ M/s
c) $-0.044$ M/s    d) $+0.022$ M/s

**Q3.** This plot shows the rate of the decomposition of $SO_2Cl_2$ into $SO_2$ and $Cl_2$ as a function of the concentration of $SO_2Cl_2$. What is the order of the reaction?

a) first order    b) second order    c) zero order
d) Order cannot be determined without more information.

**Q4.** For the reaction $2\,A + B \longrightarrow C$, the initial rate is measured at several different reactant concentrations. From the resulting tabulated data, determine the rate law for the reaction.

| [A] (M) | [B] (M) | Initial Rate (M/s) |
|---------|---------|--------------------|
| 0.05    | 0.05    | 0.035              |
| 0.10    | 0.05    | 0.070              |
| 0.20    | 0.10    | 0.56               |

a) Rate = $k[A][B]$          b) Rate = $k[A]^2[B]$
c) Rate = $k[A][B]^2$        d) Rate = $k[A]^2[B]^2$

**Q5.** What is the rate constant for the reaction in Question 4?

a) $2.8 \times 10^2\,M^{-2}\cdot s^{-1}$       b) $14\,M^{-2}\cdot s^{-1}$
c) $1.4 \times 10^2\,M^{-2}\cdot s^{-1}$       d) $1.4 \times 10^3\,M^{-2}\cdot s^{-1}$

**Q6.** The decomposition of $Br_2$ is followed as a function of time; two different plots of the data are shown. Determine the order and rate constant for the reaction.

a) first order; $0.030\,s^{-1}$
b) first order; $33.3\,s^{-1}$
c) second order; $0.045\,M^{-1}\cdot s^{-1}$
d) second order; $22.2\,M^{-1}\cdot s^{-1}$

*Continued—*

**Q7.** The reaction X $\longrightarrow$ products is second order in X and has a rate constant of 0.035 $M^{-1}s^{-1}$. If a reaction mixture is initially 0.45 M in X, what is the concentration of X after 155 seconds?

a) 7.6 M

b) $2.0 \times 10^{-3}$ M

c) 0.13 M

d) 0.00 M

**Q8.** A decomposition reaction has a half-life that does not depend on the initial concentration of the reactant. What is the order of the reaction?

a) zero order

b) first order

c) second order

d) Order cannot be determined without more information.

**Q9.** The rate constant of a reaction is measured at different temperatures. A plot of the natural log of the rate constant as a function of the inverse of the temperature (in kelvins) yields a straight line with a slope of $-8.55 \times 10^3\,K^{-1}$. What is the activation energy ($E_a$) for the reaction?

a) −71 kJ

b) 71 kJ

c) 1.0 kJ

d) −1.0 kJ

**Q10.** The rate constant for a reaction at 25.0 °C is 0.010 $s^{-1}$, and its activation energy is 35.8 kJ. Find the rate constant at 50.0 °C.

a) 0.021 $s^{-1}$

b) 0.010 $s^{-1}$

c) 0.0033 $s^{-1}$

d) 0.031 $s^{-1}$

**Q11.** The mechanism shown is proposed for the gas phase reaction, $2\,N_2O_5 \longrightarrow 4\,NO_2 + O_2$. What rate law does the mechanism predict?

$$N_2O_5 \underset{k_{-1}}{\overset{k_1}{\rightleftharpoons}} NO_2 + NO_3 \qquad \text{Fast}$$

$$NO_2 + NO_3 \longrightarrow NO_2 + O_2 + NO \qquad \text{Slow}$$

$$NO + N_2O_5 \longrightarrow 3\,NO_2 \qquad \text{Fast}$$

a) Rate = $k[N_2O_5]$

b) Rate = $k[N_2O_5]^2$

c) Rate = $k[N_2O_5]^0$

d) Rate = $k[NO_2][NO_3]$

**Q12.** Which statement is true regarding the function of a catalyst in a chemical reaction?

a) A catalyst increases the rate of a reaction.

b) A catalyst provides an alternate mechanism for the reaction.

c) A catalyst is not consumed by the reaction.

d) All of the above are true.

**Q13.** These images represent the first-order reaction A $\longrightarrow$ B initially and at some later time. The rate law for the reaction is Rate = 0.010 $s^{-1}$ [A]. How much time has passed between the two images?

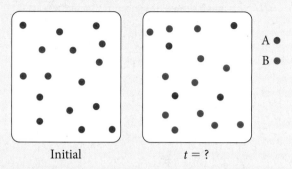

Initial          $t = ?$

A •

B •

a) 69 s    b) 139 s    c) 60 s    d) 12.5 s

**Q14.** Use collision theory to determine which single-step reaction has the smallest orientation factor.

a) $H + H \longrightarrow H_2$

b) $I + HI \longrightarrow I_2 + H$

c) $H_2 + H_2C=CH_2 \longrightarrow H_3C-CH_3$

d) All of these reactions have the same orientation factor.

**Q15.** Carbon monoxide and chlorine gas react to form phosgene ($COCl_2$) according to the equation:

$$CO(g) + Cl_2(g) \longrightarrow COCl_2(g)$$

The rate law for the reaction is Rate = $k[Cl_2]^{3/2}[CO]$. Which representation of a mixture of chlorine gas and carbon monoxide gas has the fastest initial rate?

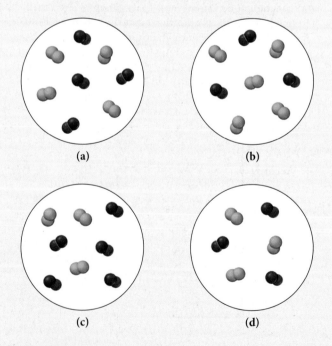

(a)          (b)

(c)          (d)

**Answers: 1.** (d) **2.** (c) **3.** (a) **4.** (c) **5.** (a) **6.** (a) **7.** (c) **8.** (b) **9.** (b) **10.** (d) **11.** (a) **12.** (d) **13.** (b) **14.** (c) **15.** (b)

# Chapter 14 IN REVIEW

## Key Terms

**Section 14.3**
rate law (629)
rate constant (*k*) (629)
reaction order (*n*) (629)
overall order (632)

**Section 14.4**
integrated rate law (634)
half-life ($t_{1/2}$) (638)

**Section 14.5**
Arrhenius equation (642)
activation energy ($E_a$) (642)

frequency factor (*A*) (642)
activated complex (transition state) (643)
exponential factor (644)
Arrhenius plot (645)
collision model (647)
orientation factor (647)
collision frequency (647)

**Section 14.6**
reaction mechanism (649)
elementary step (649)
reaction intermediate (649)
molecularity (649)

unimolecular (649)
bimolecular (649)
termolecular (649)
rate-determining step (650)

**Section 14.7**
catalyst (653)
homogeneous catalysis (655)
heterogeneous catalysis (655)
hydrogenation (656)
enzyme (656)
active site (656)
substrate (656)

## Key Concepts

### Reaction Rates, Orders, and Rate Laws (14.1–14.3)

▶ The rate of a chemical reaction is a measure of how fast a reaction occurs. The rate reflects the change in the concentration of a reactant or product per unit time and is usually reported in units of M/s.

▶ Reaction rates generally depend on the concentration of the reactants. The rate of a first-order reaction is directly proportional to the concentration of the reactant, the rate of a second-order reaction is proportional to the square of the concentration of the reactant, and the rate of a zero-order reaction is independent of the concentration of the reactant.

▶ For a reaction with more than one reactant, the order with respect to each reactant is, in general, independent of the order with respect to other reactants. The rate law shows the relationship between the rate and the concentrations of each reactant and must be determined experimentally.

### Integrated Rate Laws and Half-Life (14.4)

▶ The rate law for a reaction describes the relationship between the rate of the reaction and the concentrations of the reactants.

▶ The integrated rate law for a reaction describes the relationship between the concentration of a reactant and time.

▶ The integrated rate law for a zero-order reaction shows that the concentration of the reactant varies linearly with time. For a first-order reaction, the *natural log* of the concentration of the reactant varies linearly with time, and for a second-order reaction, the *inverse* of the concentration of the reactant varies linearly with time.

▶ The half-life of a reaction can be derived from the integrated rate law and represents the time required for the concentration of a reactant to fall to one-half of its initial value. The half-life of a first-order reaction is *independent* of initial concentration of the reactant. The half-life of a zero-order or second-order reaction *depends* on the initial concentration of reactant.

### The Effect of Temperature on Reaction Rate (14.5)

▶ The rate constant of a reaction generally depends on temperature and can be expressed by the Arrhenius equation, which consists of a frequency factor and an exponential factor.

▶ The frequency factor represents the number of times that the reactants approach the activation barrier per unit time. The exponential factor is the fraction of approaches that are successful in surmounting the activation barrier and forming products.

▶ The exponential factor depends on both the temperature and the activation energy, a barrier that the reactants must overcome to become products. The exponential factor increases with increasing temperature but decreases with increasing activation energy.

▶ We can determine the frequency factor and activation energy for a reaction by measuring the rate constant at different temperatures and constructing an Arrhenius plot.

▶ For reactions in the gas phase, Arrhenius behavior can be modeled with the collision model. In this model, reactions occur as a result of sufficiently energetic collisions. The colliding molecules must be oriented in such a way that the reaction can occur. The frequency factor contains two terms: *p*, which represents the fraction of collisions that have the proper orientation, and *z*, which represents the number of collisions per unit time.

### Reaction Mechanisms (14.6)

▶ Most chemical reactions occur not in a single step, but in several steps. The series of individual steps by which a reaction occurs is the reaction mechanism.

▶ In order for a proposed reaction mechanism to be valid, it must fulfill two conditions: (a) the steps must sum to the overall reaction, and (b) the mechanism must predict the experimentally observed rate law.

▶ For mechanisms with a slow initial step, we derive the rate law from the slow step.

▶ For mechanisms with a fast initial step, we first write the rate law based on the slow step but then assume that the fast steps reach equilibrium, so we can write concentrations of intermediates in terms of the reactants.

### Catalysis (14.7)

▶ A catalyst is a substance that increases the rate of a chemical reaction by providing an alternative mechanism that has a lower activation energy for the rate-determining step.

▶ Catalysts can be homogeneous or heterogeneous. A homogeneous catalyst exists in the same phase as the reactants and forms a homogeneous mixture with them. A heterogeneous catalyst generally exists in a different phase than the reactants.

▶ Enzymes are biological catalysts capable of increasing the rate of specific biochemical reactions by many orders of magnitude.

# Key Equations and Relationships

**The Rate of Reaction (14.2)**

For a reaction, $aA + bB \longrightarrow cC + dD$, the rate is defined as:

$$\text{Rate} = -\frac{1}{a}\frac{\Delta[A]}{\Delta t} = -\frac{1}{b}\frac{\Delta[B]}{\Delta t} = +\frac{1}{c}\frac{\Delta[C]}{\Delta t} = +\frac{1}{d}\frac{\Delta[D]}{\Delta t}$$

**The Rate Law (14.3)**

$$\text{Rate} = k[A]^n \qquad \text{(single reactant)}$$
$$\text{Rate} = k[A]^m[B]^n \qquad \text{(multiple reactants)}$$

**Integrated Rate Laws and Half-Life (14.4)**

| Reaction Order | Integrated Rate Law | Units of k | Half-Life Expression |
|---|---|---|---|
| 0 | $[A]_t = -kt + [A]_0$ | $M \cdot s^{-1}$ | $t_{1/2} = \dfrac{[A]_0}{2k}$ |
| 1 | $\ln[A]_t = -kt + \ln[A]_0$ | $s^{-1}$ | $t_{1/2} = \dfrac{0.693}{k}$ |
| 2 | $\dfrac{1}{[A]_t} = kt + \dfrac{1}{[A]_0}$ | $M^{-1} \cdot s^{-1}$ | $t_{1/2} = \dfrac{1}{k[A]_0}$ |

**Arrhenius Equation (14.5)**

$$k = Ae^{-E_a/RT}$$

$$\ln k = -\frac{E_a}{R}\left(\frac{1}{T}\right) + \ln A \qquad \text{(linearized form)}$$

$$\ln\frac{k_2}{k_1} = \frac{E_a}{R}\left(\frac{1}{T_1} - \frac{1}{T_2}\right) \qquad \text{(two-point form)}$$

$$k = pze^{-E_a/RT} \qquad \text{(collision theory)}$$

**Rate Laws for Elementary Steps (14.6)**

| Elementary Step | Molecularity | Rate Law |
|---|---|---|
| $A \longrightarrow$ products | 1 | Rate = $k[A]$ |
| $A + A \longrightarrow$ products | 2 | Rate = $k[A]^2$ |
| $A + B \longrightarrow$ products | 2 | Rate = $k[A][B]$ |
| $A + A + A \longrightarrow$ products | 3 (rare) | Rate = $k[A]^3$ |
| $A + A + B \longrightarrow$ products | 3 (rare) | Rate = $k[A]^2[B]$ |
| $A + B + C \longrightarrow$ products | 3 (rare) | Rate = $k[A][B][C]$ |

# Key Learning Outcomes

| CHAPTER OBJECTIVES | ASSESSMENT |
|---|---|
| Expressing Reaction Rates (14.2) | Example 14.1   For Practice 14.1   Exercises 25–34 |
| Determining the Order, Rate Law, and Rate Constant of a Reaction (14.3) | Example 14.2   For Practice 14.2   Exercises 41–46 |
| Using Graphical Analysis of Reaction Data to Determine Reaction Order and Rate Constants (14.4) | Examples 14.3, 14.5   For Practice 14.3, 14.5   Exercises 49–54 |
| Determining the Concentration of a Reactant at a Given Time (14.4) | Example 14.4   For Practice 14.4   Exercises 53–56 |
| Working with the Half-Life of a Reaction (14.4) | Example 14.6   For Practice 14.6   Exercises 53–58 |
| Using the Arrhenius Equation to Determine Kinetic Parameters (14.5) | Examples 14.7, 14.8   For Practice 14.7, 14.8   Exercises 61–72 |
| Determining whether a Reaction Mechanism Is Valid (14.6) | Example 14.9   For Practice 14.9   Exercises 75–78 |

# EXERCISES

## Review Questions

1. Explain why lizards become sluggish in cold weather. How is this phenomenon related to chemistry?

2. Why are reaction rates important (both practically and theoretically)?

3. What units are typically used to express the rate of a reaction?

4. Why is the reaction rate for reactants defined as the *negative* of the change in reactant concentration with respect to time, whereas for products it is defined as the change in reactant concentration with respect to time (with a positive sign)?

5. Explain the difference between the average rate of reaction and the instantaneous rate of reaction.

6. Consider a simple reaction in which reactant A forms products:

   $$A \longrightarrow products$$

   What is the rate law if the reaction is zero order with respect to A? First order? Second order? For each case, explain how a doubling of the concentration of A would affect the rate of reaction.

7. How is the order of a reaction generally determined?

8. For a reaction with multiple reactants, how is the overall order of the reaction defined?

9. Explain the difference between the rate law for a reaction and the integrated rate law for a reaction. What relationship does each kind of rate law express?

10. Write integrated rate laws for zero-order, first-order, and second-order reactions of the form A $\longrightarrow$ products.

11. What does the term *half-life* mean? Write the expressions for the half-lives of zero-order, first-order, and second-order reactions.

12. How do reaction rates typically depend on temperature? What part of the rate law is temperature dependent?

13. Explain the meaning of each term within the Arrhenius equation: activation energy, frequency factor, and exponential factor. Use these terms and the Arrhenius equation to explain why small changes in temperature can result in large changes in reaction rates.

14. What is an Arrhenius plot? Explain the significance of the slope and intercept of an Arrhenius plot.

15. Explain how a chemical reaction occurs according to the collision model. Explain the meaning of the orientation factor in this model.

16. Explain the difference between a normal chemical equation for a chemical reaction and the mechanism of that reaction.

17. In a reaction mechanism, what is an elementary step? Write down the three most common elementary steps and the corresponding rate law for each one.

18. What are the two requirements for a proposed mechanism to be valid for a given reaction?

19. What is an intermediate within a reaction mechanism?

20. What is a catalyst? How does a catalyst increase the rate of a chemical reaction?

21. Explain the difference between homogeneous catalysis and heterogeneous catalysis.

22. What are the four basic steps involved in heterogeneous catalysis?

23. What are enzymes? What is the active site of an enzyme? What is a substrate?

24. What is the general two-step mechanism by which most enzymes work?

## Problems by Topic

### Reaction Rates

25. Consider the reaction:

    $$2\,HBr(g) \longrightarrow H_2(g) + Br_2(g)$$

    a. Express the rate of the reaction in terms of the change in concentration of each of the reactants and products.

    b. In the first 25.0 s of this reaction, the concentration of HBr dropped from 0.600 M to 0.512 M. Calculate the average rate of the reaction during this time interval.

    c. If the volume of the reaction vessel in part b was 1.50 L, what amount of $Br_2$ (in moles) was formed during the first 15.0 s of the reaction?

26. Consider the reaction:

    $$2\,N_2O(g) \longrightarrow 2\,N_2(g) + O_2(g)$$

    a. Express the rate of the reaction in terms of the change in concentration of each of the reactants and products.

    b. In the first 15.0 s of the reaction, 0.015 mol of $O_2$ is produced in a reaction vessel with a volume of 0.500 L. What is the average rate of the reaction during this time interval?

    c. Predict the rate of change in the concentration of $N_2O$ during this time interval. In other words, what is $\Delta[N_2O]/\Delta t$?

27. For the reaction $2\,A(g) + B(g) \longrightarrow 3\,C(g)$,

    a. determine the expression for the rate of the reaction in terms of the change in concentration of each of the reactants and products.

    b. when A is decreasing at a rate of 0.100 M/s, how fast is B decreasing? How fast is C increasing?

28. For the reaction $A(g) + \frac{1}{2}\,B(g) \longrightarrow 2\,C(g)$,

    a. determine the expression for the rate of the reaction in terms of the change in concentration of each of the reactants and products.

    b. when C is increasing at a rate of 0.025 M/s, how fast is B decreasing? How fast is A decreasing?

29. Consider the reaction:

    $$Cl_2(g) + 3\,F_2(g) \longrightarrow 2\,ClF_3(g)$$

    Complete the table.

    | $\Delta[Cl_2]/\Delta t$ | $\Delta[F_2]/\Delta t$ | $\Delta[ClF_3]/\Delta t$ | Rate |
    |---|---|---|---|
    | −0.012 M/s | | | |

30. Consider the reaction:

    $$8\,H_2S(g) + 4\,O_2(g) \longrightarrow 8\,H_2O(g) + S_8(g)$$

    Complete the table.

    | $\Delta[H_2S]/\Delta t$ | $\Delta[O_2]/\Delta t$ | $\Delta[H_2O]/\Delta t$ | $\Delta[S_8]/\Delta t$ | Rate |
    |---|---|---|---|---|
    | −0.080 M/s | | | | |

**31.** Consider the reaction:
$$C_4H_8(g) \longrightarrow 2\,C_2H_4(g)$$
The tabulated data were collected for the concentration of $C_4H_8$ as a function of time:

| Time (s) | $[C_4H_8]$ (M) |
|---|---|
| 0 | 1.000 |
| 10 | 0.913 |
| 20 | 0.835 |
| 30 | 0.763 |
| 40 | 0.697 |
| 50 | 0.637 |

**a.** What is the average rate of the reaction between 0 and 10 s? Between 40 and 50 s?

**b.** What is the rate of formation of $C_2H_4$ between 20 and 30 s?

**32.** Consider the reaction:
$$NO_2(g) \longrightarrow NO(g) + \tfrac{1}{2}O_2(g)$$
The tabulated data were collected for the concentration of $NO_2$ as a function of time:

| Time (s) | $[NO_2]$ (M) |
|---|---|
| 0 | 1.000 |
| 10 | 0.951 |
| 20 | 0.904 |
| 30 | 0.860 |
| 40 | 0.818 |
| 50 | 0.778 |
| 60 | 0.740 |
| 70 | 0.704 |
| 80 | 0.670 |
| 90 | 0.637 |
| 100 | 0.606 |

**a.** What is the average rate of the reaction between 10 and 20 s? Between 50 and 60 s?

**b.** What is the rate of formation of $O_2$ between 50 and 60 s?

**33.** Consider the reaction:
$$H_2(g) + Br_2(g) \longrightarrow 2\,HBr(g)$$
The graph shows the concentration of $Br_2$ as a function of time.

**a.** Use the graph to calculate each quantity:
  (i) the average rate of the reaction between 0 and 25 s
  (ii) the instantaneous rate of the reaction at 25 s
  (iii) the instantaneous rate of formation of HBr at 50 s

**b.** Make a rough sketch of a curve representing the concentration of HBr as a function of time. Assume that the initial concentration of HBr is zero.

**34.** Consider the reaction:
$$2\,H_2O_2(aq) \longrightarrow 2\,H_2O(l) + O_2(g)$$
The graph shows the concentration of $H_2O_2$ as a function of time.

Use the graph to calculate each quantity:

**a.** the average rate of the reaction between 10 and 20 s

**b.** the instantaneous rate of the reaction at 30 s

**c.** the instantaneous rate of formation of $O_2$ at 50 s

**d.** If the initial volume of the $H_2O_2$ is 1.5 L, what total amount of $O_2$ (in moles) is formed in the first 50 s of reaction?

## The Rate Law and Reaction Orders

**35.** This graph shows a plot of the rate of a reaction versus the concentration of the reactant A for the reaction A $\longrightarrow$ products.

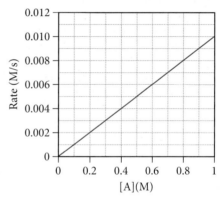

**a.** What is the order of the reaction with respect to A?

**b.** Make a rough sketch of a plot of [A] versus *time*.

**c.** Write a rate law for the reaction including an estimate for the value of $k$.

**36.** This graph shows a plot of the rate of a reaction versus the concentration of the reactant.

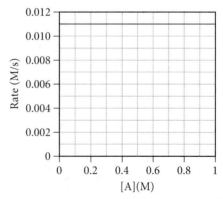

**a.** What is the order of the reaction with respect to A?

**b.** Make a rough sketch of a plot of [A] versus *time*.

**c.** Write a rate law for the reaction including the value of $k$.

37. What are the units of k for each type of reaction?
    a. first-order reaction
    b. second-order reaction
    c. zero-order reaction

38. This reaction is first order in $N_2O_5$:
$$N_2O_5(g) \longrightarrow NO_3(g) + NO_2(g)$$
The rate constant for the reaction at a certain temperature is 0.053/s.
    a. Calculate the rate of the reaction when $[N_2O_5] = 0.055$ M.
    b. What would the rate of the reaction be at the concentration indicated in part a if the reaction were second order? Zero order? (Assume the same *numerical* value for the rate constant with the appropriate units.)

39. A reaction in which A, B, and C react to form products is first order in A, second order in B, and zero order in C.
    a. Write a rate law for the reaction.
    b. What is the overall order of the reaction?
    c. By what factor does the reaction rate change if [A] is doubled (and the other reactant concentrations are held constant)?
    d. By what factor does the reaction rate change if [B] is doubled (and the other reactant concentrations are held constant)?
    e. By what factor does the reaction rate change if [C] is doubled (and the other reactant concentrations are held constant)?
    f. By what factor does the reaction rate change if the concentrations of all three reactants are doubled?

40. A reaction in which A, B, and C react to form products is zero order in A, one-half order in B, and second order in C.
    a. Write a rate law for the reaction.
    b. What is the overall order of the reaction?
    c. By what factor does the reaction rate change if [A] is doubled (and the other reactant concentrations are held constant)?
    d. By what factor does the reaction rate change if [B] is doubled (and the other reactant concentrations are held constant)?
    e. By what factor does the reaction rate change if [C] is doubled (and the other reactant concentrations are held constant)?
    f. By what factor does the reaction rate change if the concentrations of all three reactants are doubled?

41. Consider the data showing the initial rate of a reaction (A $\longrightarrow$ products) at several different concentrations of A. What is the order of the reaction? Write a rate law for the reaction including the value of the rate constant, k.

| [A] (M) | Initial Rate (M/s) |
|---------|--------------------|
| 0.100   | 0.053              |
| 0.200   | 0.210              |
| 0.300   | 0.473              |

42. Consider the data showing the initial rate of a reaction (A $\longrightarrow$ products) at several different concentrations of A. What is the order of the reaction? Write a rate law for the reaction including the value of the rate constant, k.

| [A] (M) | Initial Rate (M/s) |
|---------|--------------------|
| 0.15    | 0.008              |
| 0.30    | 0.016              |
| 0.60    | 0.032              |

43. Consider the tabulated data showing the initial rate of a reaction (A $\longrightarrow$ products) at several different concentrations of A. What is the order of the reaction? Write a rate law for the reaction including the value of the rate constant, k.

| [A] (M) | Initial Rate (M/s) |
|---------|--------------------|
| 0.12    | 0.0078             |
| 0.16    | 0.0104             |
| 0.20    | 0.0130             |

44. Consider the tabulated data showing the initial rate of a reaction (A $\longrightarrow$ products) at several different concentrations of A. What is the order of the reaction? Write a rate law for the reaction including the value of the rate constant, k.

| [A] (M) | Initial Rate (M/s) |
|---------|--------------------|
| 0.12    | $3.89 \times 10^{-4}$ |
| 0.18    | $8.75 \times 10^{-4}$ |
| 0.28    | $2.12 \times 10^{-3}$ |

45. The tabulated data were collected for this reaction:
$$2\,NO_2(g) + F_2(g) \longrightarrow 2\,NO_2F(g)$$

| [$NO_2$] | [$F_2$] (M) | Initial Rate (M/s) |
|----------|-------------|--------------------|
| 0.100    | 0.100       | 0.026              |
| 0.200    | 0.100       | 0.051              |
| 0.200    | 0.200       | 0.103              |
| 0.400    | 0.400       | 0.411              |

Write an expression for the reaction rate law and calculate the value of the rate constant, k. What is the overall order of the reaction?

46. The tabulated data were collected for this reaction:
$$CH_3Cl(g) + 3\,Cl_2(g) \longrightarrow CCl_4(g) + 3\,HCl(g)$$

| [$CH_3Cl$] (M) | [$Cl_2$] (M) | Initial Rate (M/s) |
|----------------|--------------|--------------------|
| 0.050          | 0.050        | 0.014              |
| 0.100          | 0.050        | 0.029              |
| 0.100          | 0.100        | 0.041              |
| 0.200          | 0.200        | 0.115              |

Write an expression for the reaction rate law and calculate the value of the rate constant, k. What is the overall order of the reaction?

## The Integrated Rate Law and Half-Life

47. Indicate the order of reaction consistent with each observation.
    a. A plot of the concentration of the reactant versus time yields a straight line.
    b. The reaction has a half-life that is independent of initial concentration.
    c. A plot of the inverse of the concentration versus time yields a straight line.

48. Indicate the order of reaction consistent with each observation.
    a. The half-life of the reaction gets shorter as the initial concentration is increased.
    b. A plot of the natural log of the concentration of the reactant versus time yields a straight line.
    c. The half-life of the reaction gets longer as the initial concentration is increased.

**49.** The tabulated data show the concentration of AB versus time for this reaction:

$$AB(g) \longrightarrow A(g) + B(g)$$

| Time (s) | [AB] (M) |
|---|---|
| 0 | 0.950 |
| 50 | 0.459 |
| 100 | 0.302 |
| 150 | 0.225 |
| 200 | 0.180 |
| 250 | 0.149 |
| 300 | 0.128 |
| 350 | 0.112 |
| 400 | 0.0994 |
| 450 | 0.0894 |
| 500 | 0.0812 |

Determine the order of the reaction and the value of the rate constant. Predict the concentration of AB at 25 s.

**50.** The tabulated data show the concentration of $N_2O_5$ versus time for this reaction:

$$N_2O_5(g) \longrightarrow NO_3(g) + NO_2(g)$$

| Time (s) | [$N_2O_5$] (M) |
|---|---|
| 0 | 1.000 |
| 25 | 0.822 |
| 50 | 0.677 |
| 75 | 0.557 |
| 100 | 0.458 |
| 125 | 0.377 |
| 150 | 0.310 |
| 175 | 0.255 |
| 200 | 0.210 |

Determine the order of the reaction and the value of the rate constant. Predict the concentration of $N_2O_5$ at 250 s.

**51.** The tabulated data show the concentration of cyclobutane ($C_4H_8$) versus time for this reaction:

$$C_4H_8 \longrightarrow 2\,C_2H_4$$

| Time (s) | [$C_4H_8$] (M) |
|---|---|
| 0 | 1.000 |
| 10 | 0.894 |
| 20 | 0.799 |
| 30 | 0.714 |
| 40 | 0.638 |
| 50 | 0.571 |
| 60 | 0.510 |
| 70 | 0.456 |
| 80 | 0.408 |
| 90 | 0.364 |
| 100 | 0.326 |

Determine the order of the reaction and the value of the rate constant. What is the rate of reaction when [$C_4H_8$] = 0.25 M?

**52.** The reaction A $\longrightarrow$ products was monitored as a function of time. The results are shown here.

| Time (s) | [A] (M) |
|---|---|
| 0 | 1.000 |
| 25 | 0.914 |
| 50 | 0.829 |
| 75 | 0.744 |
| 100 | 0.659 |
| 125 | 0.573 |
| 150 | 0.488 |
| 175 | 0.403 |
| 200 | 0.318 |

Determine the order of the reaction and the value of the rate constant. What is the rate of reaction when [A] = 0.10 M?

**53.** This reaction was monitored as a function of time:

$$A \longrightarrow B + C$$

A plot of ln[A] versus time yields a straight line with slope $-0.0045/s$.

a. What is the value of the rate constant ($k$) for this reaction at this temperature?
b. Write the rate law for the reaction.
c. What is the half-life?
d. If the initial concentration of A is 0.250 M, what is the concentration after 225 s?

**54.** This reaction was monitored as a function of time:

$$AB \longrightarrow A + B$$

A plot of 1/[AB] versus time yields a straight line with a slope of $+0.55/M \cdot s$.

a. What is the value of the rate constant ($k$) for this reaction at this temperature?
b. Write the rate law for the reaction.
c. What is the half-life when the initial concentration is 0.55 M?
d. If the initial concentration of AB is 0.250 M, and the reaction mixture initially contains no products, what are the concentrations of A and B after 75 s?

**55.** The decomposition of $SO_2Cl_2$ is first order in $SO_2Cl_2$ and has a rate constant of $1.42 \times 10^{-4}\,s^{-1}$ at a certain temperature.
a. What is the half-life for this reaction?
b. How long will it take for the concentration of $SO_2Cl_2$ to decrease to 25% of its initial concentration?
c. If the initial concentration of $SO_2Cl_2$ is 1.00 M, how long will it take for the concentration to decrease to 0.78 M?
d. If the initial concentration of $SO_2Cl_2$ is 0.150 M, what is the concentration of $SO_2Cl_2$ after $2.00 \times 10^2\,s$? After $5.00 \times 10^2\,s$?

**56.** The decomposition of XY is second order in XY and has a rate constant of $7.02 \times 10^{-3}\,M^{-1} \cdot s^{-1}$ at a certain temperature.
a. What is the half-life for this reaction at an initial concentration of 0.100 M?
b. How long will it take for the concentration of XY to decrease to 12.5% of its initial concentration when the initial concentration is 0.100 M? When the initial concentration is 0.200 M?
c. If the initial concentration of XY is 0.150 M, how long will it take for the concentration to decrease to 0.062 M?
d. If the initial concentration of XY is 0.050 M, what is the concentration of XY after $5.0 \times 10^1\,s$? After $5.50 \times 10^2\,s$?

57. The half-life for the radioactive decay of U-238 is 4.5 billion years and is independent of initial concentration. How long will it take for 10% of the U-238 atoms in a sample of U-238 to decay? If a sample of U-238 initially contained $1.5 \times 10^{18}$ atoms when the universe was formed 13.8 billion years ago, how many U-238 atoms does it contain today?

58. The half-life for the radioactive decay of C-14 is 5730 years and is independent of the initial concentration. How long does it take for 25% of the C-14 atoms in a sample of C-14 to decay? If a sample of C-14 initially contains 1.5 mmol of C-14, how many millimoles are left after 2255 years?

## The Effect of Temperature and the Collision Model

59. The diagram shows the energy of a reaction as the reaction progresses. Label each blank box in the diagram.

**a.** reactants
**b.** products
**c.** activation energy ($E_a$)
**d.** enthalpy of reaction ($\Delta H_{rxn}$)

60. A chemical reaction is endothermic and has an activation energy that is twice the value of the enthalpy change of the reaction. Draw a diagram depicting the energy of the reaction as it progresses. Label the position of the reactants and products and indicate the activation energy and enthalpy of reaction.

61. The activation energy of a reaction is 56.8 kJ/mol and the frequency factor is $1.5 \times 10^{11}$/s. Calculate the rate constant of the reaction at 25 °C.

62. The rate constant of a reaction at 32 °C is 0.055/s. If the frequency factor is $1.2 \times 10^{13}$/s, what is the activation barrier?

63. The rate constant ($k$) for a reaction was measured as a function of temperature. A plot of ln $k$ versus $1/T$ (in K) is linear and has a slope of −7445 K. Calculate the activation energy for the reaction.

64. The rate constant ($k$) for a reaction was measured as a function of temperature. A plot of ln $k$ versus $1/T$ (in K) is linear and has a slope of $−1.01 \times 10^4$ K. Calculate the activation energy for the reaction.

65. The data shown here were collected for the first-order reaction:

$$N_2O(g) \longrightarrow N_2(g) + O(g)$$

Use an Arrhenius plot to determine the activation barrier and frequency factor for the reaction.

| Temperature (K) | Rate Constant (1/s) |
| --- | --- |
| 800 | $3.24 \times 10^{-5}$ |
| 900 | 0.00214 |
| 1000 | 0.0614 |
| 1100 | 0.955 |

66. The tabulated data show the rate constant of a reaction measured at several different temperatures. Use an Arrhenius plot to determine the activation barrier and frequency factor for the reaction.

| Temperature (K) | Rate Constant (1/s) |
| --- | --- |
| 300 | 0.0134 |
| 310 | 0.0407 |
| 320 | 0.114 |
| 330 | 0.303 |
| 340 | 0.757 |

67. The tabulated data were collected for the second-order reaction:

$$Cl(g) + H_2(g) \longrightarrow HCl(g) + H(g)$$

Use an Arrhenius plot to determine the activation barrier and frequency factor for the reaction.

| Temperature (K) | Rate Constant (L/mol · s) |
| --- | --- |
| 90 | 0.00357 |
| 100 | 0.0773 |
| 110 | 0.956 |
| 120 | 7.781 |

68. The tabulated data show the rate constant of a reaction measured at several different temperatures. Use an Arrhenius plot to determine the activation barrier and frequency factor for the reaction.

| Temperature (K) | Rate Constant (1/s) |
| --- | --- |
| 310 | 0.00434 |
| 320 | 0.0140 |
| 330 | 0.0421 |
| 340 | 0.118 |
| 350 | 0.316 |

69. A reaction has a rate constant of 0.0117/s at 400.0 K and 0.689/s at 450.0 K.
   **a.** Determine the activation barrier for the reaction.
   **b.** What is the value of the rate constant at 425 K?

70. A reaction has a rate constant of 0.000122/s at 27 °C and 0.228/s at 77 °C.
   **a.** Determine the activation barrier for the reaction.
   **b.** What is the value of the rate constant at 17 °C?

71. If a temperature increase from 10.0 °C to 20.0 °C doubles the rate constant for a reaction, what is the value of the activation barrier for the reaction?

72. If a temperature increase from 20.0 °C to 35.0 °C triples the rate constant for a reaction, what is the value of the activation barrier for the reaction?

73. Consider these two gas-phase reactions:
   **a.** $AA(g) + BB(g) \longrightarrow 2\,AB(g)$
   **b.** $AB(g) + CD(g) \longrightarrow AC(g) + BD(g)$

   If the reactions have identical activation barriers and are carried out under the same conditions, which one would you expect to have the faster rate?

74. Which of these two reactions would you expect to have the smaller orientation factor? Explain.
   **a.** $O(g) + N_2(g) \longrightarrow NO(g) + N(g)$
   **b.** $NO(g) + Cl_2(g) \longrightarrow NOCl(g) + Cl(g)$

## Reaction Mechanisms

**75.** Consider this overall reaction, which is experimentally observed to be second order in AB and zero order in C:

$$AB + C \longrightarrow A + BC$$

Is the following mechanism valid for this reaction?

$$AB + AB \xrightarrow{k_1} AB_2 + A \quad \text{Slow}$$

$$AB_2 + C \xrightarrow{k_2} AB + BC \quad \text{Fast}$$

**76.** Consider this overall reaction, which is experimentally observed to be second order in X and first order in Y:

$$X + Y \longrightarrow XY$$

**a.** Does the reaction occur in a single step in which X and Y collide?

**b.** Is this two-step mechanism valid?

$$2X \underset{k_2}{\overset{k_1}{\rightleftharpoons}} X_2 \qquad \text{Fast}$$

$$X_2 + Y \xrightarrow{k_3} XY + X \quad \text{Slow}$$

**77.** Consider this three-step mechanism for a reaction:

$$Cl_2(g) \underset{k_2}{\overset{k_1}{\rightleftharpoons}} 2\,Cl(g) \qquad\qquad\qquad \text{Fast}$$

$$Cl(g) + CHCl_3(g) \xrightarrow{k_3} HCl(g) + CCl_3(g) \quad \text{Slow}$$

$$Cl(g) + CCl_3(g) \xrightarrow{k_4} CCl_4(g) \qquad\qquad \text{Fast}$$

**a.** What is the overall reaction?

**b.** Identify the intermediates in the mechanism.

**c.** What is the predicted rate law?

**78.** Consider this two-step mechanism for a reaction:

$$NO_2(g) + Cl_2(g) \xrightarrow{k_1} ClNO_2(g) + Cl(g) \quad \text{Slow}$$

$$NO_2(g) + Cl(g) \xrightarrow{k_2} ClNO_2(g) \qquad\qquad \text{Fast}$$

**a.** What is the overall reaction?

**b.** Identify the intermediates in the mechanism.

**c.** What is the predicted rate law?

## Catalysis

**79.** Many heterogeneous catalysts are deposited on high-surface-area supports. Why?

**80.** Suppose that the reaction A $\longrightarrow$ products is exothermic and has an activation barrier of 75 kJ/mol. Sketch an energy diagram showing the energy of the reaction as a function of the progress of the reaction. Draw a second energy curve showing the effect of a catalyst.

**81.** Suppose that a catalyst lowers the activation barrier of a reaction from 125 kJ/mol to 55 kJ/mol. By what factor would you expect the reaction rate to increase at 25 °C? (Assume that the frequency factors for the catalyzed and uncatalyzed reactions are identical.)

**82.** The activation barrier for the hydrolysis of sucrose into glucose and fructose is 108 kJ/mol. If an enzyme increases the rate of the hydrolysis reaction by a factor of 1 million, how much lower must the activation barrier be when sucrose is in the active site of the enzyme? (Assume that the frequency factors for the catalyzed and uncatalyzed reactions are identical and a temperature of 25 °C.)

# Cumulative Problems

**83.** The tabulated data were collected for this reaction at 500 °C:

$$CH_3CN(g) \longrightarrow CH_3NC(g)$$

| Time (h) | [CH₃CN] (M) |
|----------|-------------|
| 0.0 | 1.000 |
| 5.0 | 0.794 |
| 10.0 | 0.631 |
| 15.0 | 0.501 |
| 20.0 | 0.398 |
| 25.0 | 0.316 |

**a.** Determine the order of the reaction and the value of the rate constant at this temperature.

**b.** What is the half-life for this reaction (at the initial concentration)?

**c.** How long will it take for 90% of the CH₃CN to convert to CH₃NC?

**84.** The tabulated data were collected for this reaction at a certain temperature:

$$X_2Y \longrightarrow 2X + Y$$

| Time (h) | [X₂Y] (M) |
|----------|-----------|
| 0.0 | 0.100 |
| 1.0 | 0.0856 |
| 2.0 | 0.0748 |
| 3.0 | 0.0664 |
| 4.0 | 0.0598 |
| 5.0 | 0.0543 |

**a.** Determine the order of the reaction and the value of the rate constant at this temperature.

**b.** What is the half-life for this reaction (at the initial concentration)?

**c.** What is the concentration of X after 10.0 hours?

**85.** Consider the reaction:

$$A + B + C \longrightarrow D$$

The rate law for this reaction is:

$$\text{Rate} = k\frac{[A][C]^2}{[B]^{1/2}}$$

Suppose the rate of the reaction at certain initial concentrations of A, B, and C is 0.0115 M/s. What is the rate of the reaction if the concentrations of A and C are doubled and the concentration of B is tripled?

**86.** Consider the reaction:

$$2\,O_3(g) \longrightarrow 3\,O_2(g)$$

The rate law for this reaction is:

$$\text{Rate} = k\frac{[O_3]^2}{[O_2]}$$

Suppose that a 1.0-L reaction vessel initially contains 1.0 mol of O₃ and 1.0 mol of O₂. What fraction of the O₃ will have reacted when the rate falls to one-half of its initial value?

**87.** At 700 K acetaldehyde decomposes in the gas phase to methane and carbon monoxide. The reaction is:

$$CH_3CHO(g) \longrightarrow CH_4(g) + CO(g)$$

A sample of CH₃CHO is heated to 700 K and the pressure is measured as 0.22 atm before any reaction takes place. The kinetics of the reaction are followed by measurements of total pressure and these data are obtained:

| t (s) | 0 | 1000 | 3000 | 7000 |
|-------|------|------|------|------|
| P_Total (atm) | 0.22 | 0.24 | 0.27 | 0.31 |

Find the rate law, the rate constant, and the total pressure after $2.00 \times 10^4$ s.

**88.** At 400 K oxalic acid decomposes according to the reaction:

$$H_2C_2O_4(g) \longrightarrow CO_2(g) + HCOOH(g)$$

In three separate experiments, the initial pressure of oxalic acid and final total pressure after 20,000 s are measured.

| Experiment | 1 | 2 | 3 |
|---|---|---|---|
| $P_{H_2C_2O_4}$ at $t = 0$ | 65.8 | 92.1 | 111 |
| $P_{Total}$ at $t = 20,000$ s | 94.6 | 132 | 160 |

Find the rate law of the reaction and its rate constant.

**89.** Dinitrogen pentoxide decomposes in the gas phase to form nitrogen dioxide and oxygen gas. The reaction is first order in dinitrogen pentoxide and has a half-life of 2.81 h at 25 °C. If a 1.5-L reaction vessel initially contains 745 torr of $N_2O_5$ at 25 °C, what partial pressure of $O_2$ is present in the vessel after 215 minutes?

**90.** Cyclopropane ($C_3H_6$) reacts to form propene ($C_3H_6$) in the gas phase. The reaction is first order in cyclopropane and has a rate constant of $5.87 \times 10^{-4}$/s at 485 °C. If a 2.5-L reaction vessel initially contains 722 torr of cyclopropane at 485 °C, how long will it take for the partial pressure of cyclopropane to drop to below $1.00 \times 10^2$ torr?

**91.** Iodine atoms combine to form $I_2$ in liquid hexane solvent with a rate constant of $1.5 \times 10^{10}$ L/mol·s. The reaction is second order in I. Since the reaction occurs so quickly, the only way to study the reaction is to create iodine atoms almost instantaneously, usually by photochemical decomposition of $I_2$. Suppose a flash of light creates an initial [I] concentration of 0.0100 M. How long will it take for 95% of the newly created iodine atoms to recombine to form $I_2$?

**92.** The hydrolysis of sucrose ($C_{12}H_{22}O_{11}$) into glucose and fructose in acidic water has a rate constant of $1.8 \times 10^{-4}$ s$^{-1}$ at 25 °C. Assuming the reaction is first order in sucrose, determine the mass of sucrose that is hydrolyzed when 2.55 L of a 0.150 M sucrose solution is allowed to react for 195 minutes.

**93.** The reaction $AB(aq) \longrightarrow A(g) + B(g)$ is second order in AB and has a rate constant of 0.0118 M$^{-1}$·s$^{-1}$ at 25.0 °C. A reaction vessel initially contains 250.0 mL of 0.100 M AB that is allowed to react to form the gaseous product. The product is collected over water at 25.0 °C. How much time is required to produce 200.0 mL of the products at a barometric pressure of 755.1 mmHg? (The vapor pressure of water at this temperature is 23.8 mmHg.)

**94.** The reaction $2 H_2O_2(aq) \longrightarrow 2 H_2O(l) + O_2(g)$ is first order in $H_2O_2$ and under certain conditions has a rate constant of 0.00752 s$^{-1}$ at 20.0 °C. A reaction vessel initially contains 150.0 mL of 30.0% $H_2O_2$ by mass solution (the density of the solution is 1.11 g/mL). The gaseous oxygen is collected over water at 20.0 °C as it forms. What volume of $O_2$ forms in 85.0 seconds at a barometric pressure of 742.5 mmHg? (The vapor pressure of water at this temperature is 17.5 mmHg.)

**95.** Consider this energy diagram:

**a.** How many elementary steps are involved in this reaction?
**b.** Label the reactants, products, and intermediates.

**c.** Which step is rate limiting?
**d.** Is the overall reaction endothermic or exothermic?

**96.** Consider the reaction in which HCl adds across the double bond of ethene:

$$HCl + H_2C{=}CH_2 \longrightarrow H_3C{-}CH_2Cl$$

The following mechanism, with the accompanying energy diagram, has been suggested for this reaction:

***Step 1*** $\quad HCl + H_2C{=}CH_2 \longrightarrow H_3C{=}CH_2{}^+ + Cl^-$

***Step 2*** $\quad H_3C{=}CH_2{}^+ + Cl^- \longrightarrow H_3C{-}CH_2Cl$

**a.** Based on the energy diagram, determine which step is rate limiting.
**b.** What is the expected order of the reaction based on the proposed mechanism?
**c.** Is the overall reaction exothermic or endothermic?

**97.** The desorption (leaving of the surface) of a single molecular layer of *n*-butane from a single crystal of aluminum oxide is found to be first order with a rate constant of 0.128/s at 150 K.
**a.** What is the half-life of the desorption reaction?
**b.** If the surface is initially completely covered with *n*-butane at 150 K, how long will it take for 25% of the molecules to desorb (leave the surface)? For 50% to desorb?
**c.** If the surface is initially completely covered, what fraction will remain covered after 10 s? After 20 s?

**98.** The evaporation of a 120-nm film of *n*-pentane from a single crystal of aluminum oxide is zero order with a rate constant of $1.92 \times 10^{13}$ molecules/cm$^2$·s at 120 K.
**a.** If the initial surface coverage is $8.9 \times 10^{16}$ molecules/cm$^2$, how long will it take for one-half of the film to evaporate?
**b.** What fraction of the film is left after 10 s? Assume the same initial coverage as in part a.

**99.** The kinetics of this reaction were studied as a function of temperature. (The reaction is first order in each reactant and second order overall.)

$$C_2H_5Br(aq) + OH^-(aq) \longrightarrow C_2H_5OH(l) + Br^-(aq)$$

| Temperature (°C) | $k$ (L/mol·s) |
|---|---|
| 25 | $8.81 \times 10^{-5}$ |
| 35 | 0.000285 |
| 45 | 0.000854 |
| 55 | 0.00239 |
| 65 | 0.00633 |

**a.** Determine the activation energy and frequency factor for the reaction.
**b.** Determine the rate constant at 15 °C.
**c.** If a reaction mixture is 0.155 M in $C_2H_5Br$ and 0.250 M in $OH^-$, what is the initial rate of the reaction at 75 °C?

**100.** The reaction $2 N_2O_5 \longrightarrow 2 N_2O_4 + O_2$ takes place at around room temperature in solvents such as $CCl_4$. The rate constant at 293 K is found to be $2.35 \times 10^{-4}\,s^{-1}$, and at 303 K the rate constant is found to be $9.15 \times 10^{-4}\,s^{-1}$. Calculate the frequency factor for the reaction.

**101.** This reaction has an activation energy of zero in the gas phase:

$$CH_3 + CH_3 \longrightarrow C_2H_6$$

   **a.** Would you expect the rate of this reaction to change very much with temperature?
   **b.** Why might the activation energy be zero?
   **c.** What other types of reactions would you expect to have little or no activation energy?

**102.** Consider the two reactions:

$$O + N_2 \longrightarrow NO + N \qquad E_a = 315\ kJ/mol$$
$$Cl + H_2 \longrightarrow HCl + H \qquad E_a = 23\ kJ/mol$$

   **a.** Why is the activation barrier for the first reaction so much higher than that for the second?
   **b.** The frequency factors for these two reactions are very close to each other in value. Assuming that they are the same, calculate the ratio of the reaction rate constants for these two reactions at 25 °C.

**103.** Anthropologists can estimate the age of a bone or other sample of organic matter by its carbon-14 content. The carbon-14 in a living organism is constant until the organism dies, after which carbon-14 decays with first-order kinetics and a half-life of 5730 years. Suppose a bone from an ancient human contains 19.5% of the C-14 found in living organisms. How old is the bone?

**104.** Geologists can estimate the age of rocks by their uranium-238 content. The uranium is incorporated in the rock as it hardens and then decays with first-order kinetics and a half-life of 4.5 billion years. A rock contains 83.2% of the amount of uranium-238 that it contained when it was formed. (The amount that the rock contained when it was formed can be deduced from the presence of the decay products of U-238.) How old is the rock?

**105.** Consider the gas-phase reaction:

$$H_2(g) + I_2(g) \longrightarrow 2 HI(g)$$

The reaction was experimentally determined to be first order in $H_2$ and first order in $I_2$. Consider the proposed mechanisms.

Proposed mechanism I:
$$H_2(g) + I_2(g) \longrightarrow 2 HI(g) \qquad \text{Single step}$$

Proposed mechanism II:
$$I_2(g) \underset{k_{-1}}{\overset{k_1}{\rightleftharpoons}} 2 I(g) \qquad \text{Fast}$$
$$H_2(g) + 2 I(g) \xrightarrow{k_2} 2 HI(g) \qquad \text{Slow}$$

   **a.** Show that both of the proposed mechanisms are valid.
   **b.** What kind of experimental evidence might lead you to favor mechanism II over mechanism I?

**106.** Consider the reaction:

$$2 NH_3(aq) + OCl^-(aq) \longrightarrow N_2H_4(aq) + H_2O(l) + Cl^-(aq)$$

This three-step mechanism is proposed:

$$NH_3(aq) + OCl^-(aq) \underset{k_2}{\overset{k_1}{\rightleftharpoons}} NH_2Cl(aq) + OH^-(aq) \quad \text{Fast}$$
$$NH_2Cl(aq) + NH_3(aq) \xrightarrow{k_3} N_2H_5^+(aq) + Cl^-(aq) \quad \text{Slow}$$
$$N_2H_5^+(aq) + OH^-(aq) \xrightarrow{k_4} N_2H_4(aq) + H_2O(l) \quad \text{Fast}$$

   **a.** Show that the mechanism sums to the overall reaction.
   **b.** What is the rate law predicted by this mechanism?

**107.** The proposed mechanism for the formation of hydrogen bromide can be written in a simplified form as:

$$Br_2(g) \underset{k_{-1}}{\overset{k_1}{\rightleftharpoons}} 2 Br(g) \qquad \text{Fast}$$
$$Br(g) + H_2(g) \xrightarrow{k_2} HBr(g) + H(g) \qquad \text{Slow}$$
$$H(g) + Br_2(g) \xrightarrow{k_3} HBr(g) + Br(g) \qquad \text{Fast}$$

What rate law corresponds to this mechanism?

**108.** A proposed mechanism for the formation of hydrogen iodide can be written in simplified form as

$$I_2 \underset{k_{-1}}{\overset{k_1}{\rightleftharpoons}} 2 I \qquad \text{Fast}$$
$$I + H_2 \underset{k_{-2}}{\overset{k_2}{\rightleftharpoons}} H_2I \qquad \text{Fast}$$
$$H_2I + I \xrightarrow{k_3} 2 HI \qquad \text{Slow}$$

What rate law corresponds to this mechanism?

**109.** A certain substance X decomposes. Fifty percent of X remains after 100 minutes. How much X remains after 200 minutes if the reaction order with respect to X is (a) zero order, (b) first order, (c) second order?

**110.** The half-life for radioactive decay (a first-order process) of plutonium-239 is 24,000 years. How many years does it take for one mole of this radioactive material to decay until just one atom remains?

**111.** The energy of activation for the decomposition of 2 mol of HI to $H_2$ and $I_2$ in the gas phase is 185 kJ. The heat of formation of $HI(g)$ from $H_2(g)$ and $I_2(g)$ is $-5.65\ kJ/mol$. Find the energy of activation for the reaction of 1 mol of $H_2$ and 1 mol of $I_2$ to form 2 mol of HI in the gas phase.

**112.** Ethyl chloride vapor decomposes by the first-order reaction:

$$C_2H_5Cl \longrightarrow C_2H_4 + HCl$$

The activation energy is 249 kJ/mol, and the frequency factor is $1.6 \times 10^{14}\,s^{-1}$. Find the value of the rate constant at 710 K. What fraction of the ethyl chloride decomposes in 15 minutes at this temperature? Find the temperature at which the rate of the reaction would be twice as fast.

# Challenge Problems

**113.** In this chapter we have seen a number of reactions in which a single reactant forms products. For example, consider the following first-order reaction:

$$CH_3NC(g) \longrightarrow CH_3CN(g)$$

However, we also learned that gas-phase reactions occur through collisions.

   **a.** One possible explanation for how this reaction occurs is that two molecules of $CH_3NC$ collide with each other and form two molecules of the product in a single elementary step. If that were the case, what reaction order would you expect?
   **b.** Another possibility is that the reaction occurs through more than one step. For example, a possible mechanism involves one step in which the two $CH_3NC$ molecules collide, resulting in the "activation" of one of them. In a second step, the activated molecule goes on to form the product. Write down this mechanism and determine which step must be rate

determining in order for the kinetics of the reaction to be first order. Show explicitly how the mechanism predicts first-order kinetics.

**114.** The first-order *integrated* rate law for the reaction A ⟶ products is derived from the rate law using calculus:

$$\text{Rate} = k[A] \text{ (first-order rate law)}$$

$$\text{Rate} = -\frac{d[A]}{dt}$$

$$\frac{d[A]}{dt} = -k[A]$$

The equation just given is a first-order, separable differential equation that can be solved by separating the variables and integrating:

$$\frac{d[A]}{[A]} = -kdt$$

$$\int_{[A]_0}^{[A]} \frac{d[A]}{[A]} = -\int_0^t kdt$$

In the integral just given, $[A]_0$ is the initial concentration of A. We then evaluate the integral:

$$[\ln[A]]_{[A]_0}^{[A]} = -k[t]_0^t$$

$$\ln[A] - \ln[A]_0 = -kt$$

$$\ln[A] = -kt + \ln[A]_0 \text{ (integrated rate law)}$$

a. Use a procedure similar to the one just shown to derive an integrated rate law for a reaction A ⟶ products, which is one-half order in the concentration of A (that is, Rate = $k[A]^{1/2}$).

b. Use the result from part a to derive an expression for the half-life of a one-half-order reaction.

**115.** The previous exercise shows how the first-order integrated rate law is derived from the first-order differential rate law. Begin with the second-order differential rate law and derive the second-order integrated rate law.

**116.** The rate constant for the first-order decomposition of $N_2O_5(g)$ to $NO_2(g)$ and $O_2(g)$ is $7.48 \times 10^{-3}\,s^{-1}$ at a given temperature.

a. Find the length of time required for the total pressure in a system containing $N_2O_5$ at an initial pressure of 0.100 atm to rise to 0.145 atm.

b. To 0.200 atm.

c. Find the total pressure after 100 s of reaction.

**117.** Phosgene ($Cl_2CO$), a poison gas used in World War I, is formed by the reaction of $Cl_2$ and CO. The proposed mechanism for the reaction is:

| | |
|---|---|
| $Cl_2 \rightleftharpoons 2\,Cl$ | (fast, equilibrium) |
| $Cl + CO \rightleftharpoons ClCO$ | (fast, equilibrium) |
| $ClCO + Cl_2 \longrightarrow Cl_2CO + Cl$ | (slow) |

What rate law is consistent with this mechanism?

**118.** The rate of decomposition of $N_2O_3(g)$ to $NO_2(g)$ and $NO(g)$ is followed by measuring $[NO_2]$ at different times. The following data are obtained.

| $[NO_2]$(mol/L) | 0 | 0.193 | 0.316 | 0.427 | 0.784 |
|---|---|---|---|---|---|
| $t$ (s) | 0 | 884 | 1610 | 2460 | 50,000 |

The reaction follows a first-order rate law. Calculate the rate constant. Assume that after 50,000 s all $N_2O_3(g)$ had decomposed.

**119.** At 473 K, for the elementary reaction $2\,NOCl(g) \underset{k_{-1}}{\overset{k_1}{\rightleftharpoons}} 2\,NO(g) + Cl_2(g)$

$$k_1 = 7.8 \times 10^{-2}\,\text{L/mol s} \quad \text{and}$$

$$k_{-1} = 4.7 \times 10^2\,\text{L}^2/\text{mol}^2\,\text{s}$$

A sample of NOCl is placed in a container and heated to 473 K. When the system comes to equilibrium, [NOCl] is found to be 0.12 mol/L. What are the concentrations of NO and $Cl_2$?

# Conceptual Problems

**120.** Consider the reaction:

$$CHCl_3(g) + Cl_2(g) \longrightarrow CCl_4(g) + HCl(g)$$

The reaction is first order in $CHCl_3$ and one-half order in $Cl_2$. Which reaction mixture would you expect to have the fastest initial rate?

(a)                     (b)

(c)

**121.** The accompanying graph shows the concentration of a reactant as a function of time for two different reactions. One of the reactions is first order and the other is second order. Which of the two reactions is first order? Second order? How would you change each plot to make it linear?

**122.** A particular reaction, A ⟶ products, has a rate that slows down as the reaction proceeds. The half-life of the reaction is found to depend on the initial concentration of A. Determine whether each statement is likely to be true or false for this reaction.

a. A doubling of the concentration of A doubles the rate of the reaction.

b. A plot of $1/[A]$ versus time is linear.

c. The half-life of the reaction gets longer as the initial concentration of A increases.

d. A plot of the concentration of A versus time has a constant slope.

—*Active Classroom Learning*

# Questions for Group Work

Discuss these questions with the group and record your consensus answer.

**123.** A student says, "The initial concentration of a reactant was doubled, and the rate doubled. Therefore the reaction is second order in that reactant." Why might the student say that? What is wrong with the statement? What is the actual order with respect to the reactant? Explain your reasoning clearly.

**124.** A certain compound, A, reacts to form products according to the reaction A ⟶ P. The amount of A is measured as a function of time under a variety of different conditions and the tabulated results are shown here:

| Time (s) | 25.0 °C [A] (M) | 35.0 °C [A] (M) | 45.0 °C [A] (M) |
|---|---|---|---|
| 0 | 1.000 | 1.000 | 1.000 |
| 10 | 0.779 | 0.662 | 0.561 |
| 20 | 0.591 | 0.461 | 0.312 |
| 30 | 0.453 | 0.306 | 0.177 |
| 40 | 0.338 | 0.208 | 0.100 |
| 50 | 0.259 | 0.136 | 0.057 |
| 60 | 0.200 | 0.093 | 0.032 |

Have one group member make a graph of [A] vs. *t*, one group member make a graph of ln[A] vs. *t*, and one group member make a graph of 1/[A] vs. *t* using the data for 25 °C. Additional group members can make similar graphs for the other temperatures. What is the order of the reaction with respect to A? Explain your answer.

a. Use the data to determine the rate constant at each temperature.
b. What is the activation energy for this reaction?
c. The same reaction is conducted in the presence of a catalyst, and the following data are obtained:

| Time (s) | 25.0 °C [A] (M) | 35.0 °C [A] (M) | 45.0 °C [A] (M) |
|---|---|---|---|
| 0 | 1.000 | 1.000 | 1.000 |
| 0.1 | 0.724 | 0.668 | 0.598 |
| 0.2 | 0.511 | 0.433 | 0.341 |
| 0.3 | 0.375 | 0.291 | 0.202 |
| 0.4 | 0.275 | 0.190 | 0.119 |
| 0.5 | 0.198 | 0.122 | 0.071 |
| 0.6 | 0.141 | 0.080 | 0.043 |

What effect does a catalyst have on the rate of the reaction? What is the activation energy for this reaction in the presence of the catalyst? How does it compare with the activation energy for the reaction when the catalyst isn't present?

# Data Interpretation and Analysis

## Kinetics of Color Change in Marionberry Preserves

**125.** The color of food is an important component in consumer decisions. Ultraviolet rays from the sun and other chemical processes can break chemical bonds and cause food color to fade. The time that it takes for the color of food to fade depends on the kinetics of the chemical reactions occurring in the food.

A company makes marionberry preserves and places the food in sealed glass jars. In one experiment, food scientists place some jars near a window where the preserves are exposed to light and the temperature is kept constant at 25.0 °C. The scientists periodically measure the color intensity of the preserves during a 200-day period. Table a displays the change in color intensity of the preserves (as a percentage of the original) in the presence of light at 25.0 °C. Figure a ▼ plots the data in Table a as a graph.

| Time (days) | % of Initial Color | Time (days) | % of Initial Color |
|---|---|---|---|
| 0 | 100.0 | 24 | 72.18 |
| 2 | 97.32 | 26 | 70.25 |
| 4 | 94.71 | 30 | 66.53 |
| 6 | 92.17 | 40 | 58.08 |
| 8 | 89.70 | 50 | 50.71 |
| 10 | 87.30 | 60 | 44.27 |
| 12 | 84.96 | 70 | 38.65 |
| 14 | 82.68 | 80 | 33.74 |
| 16 | 80.47 | 90 | 29.45 |
| 18 | 78.31 | 100 | 25.71 |
| 20 | 76.21 | 150 | 13.04 |
| 22 | 74.17 | 200 | 6.61 |

**TABLE a   Fading of Preserves Color as a Function of Time in the Presence of Light (Days) at 25.0 °C**

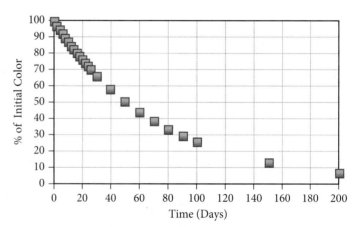

▲ **FIGURE a   Fading in Preserves Color in the Presence of Light as a Function of Time at 25.0 °C**

In another experiment, scientists place several jars of the preserves in a dark room. Tables b-d list the color intensity (as a percentage of the original) of the preserves stored in the absence of light at three different temperatures (5.0 °C, 25.0 °C, and 40.0 °C) over a 200-day period.

| Time (days) | % of Initial Color | Time (days) | % of Initial Color |
|---|---|---|---|
| 0 | 100 | 24 | 79.08 |
| 2 | 98.06 | 26 | 77.55 |
| 4 | 96.16 | 30 | 74.58 |
| 6 | 94.30 | 40 | 67.64 |
| 8 | 92.48 | 50 | 61.34 |
| 10 | 90.68 | 60 | 55.63 |
| 12 | 88.93 | 70 | 50.45 |
| 14 | 87.21 | 80 | 45.75 |
| 16 | 85.52 | 90 | 41.49 |
| 18 | 83.86 | 100 | 37.62 |
| 20 | 82.24 | 150 | 23.08 |
| 22 | 80.65 | 200 | 14.16 |

TABLE b Fading of Preserves Color in the Dark as a Function of Time at 25.0 °C

| Time (days) | % of Initial Color | Time (days) | % of Initial Color |
|---|---|---|---|
| 0 | 100 | 24 | 92.58 |
| 2 | 99.36 | 26 | 91.99 |
| 4 | 98.72 | 30 | 90.82 |
| 6 | 98.09 | 40 | 87.95 |
| 8 | 97.46 | 50 | 85.17 |
| 10 | 96.84 | 60 | 82.48 |
| 12 | 96.22 | 70 | 79.88 |
| 14 | 95.61 | 80 | 77.35 |
| 16 | 94.99 | 90 | 74.91 |
| 18 | 94.38 | 100 | 72.54 |
| 20 | 93.78 | 150 | 61.78 |
| 22 | 93.18 | 200 | 52.62 |

TABLE c Fading of Preserves Color in the Dark as a Function of Time at 5.0 °C

| Time (days) | % of Initial Color | Time (days) | % of Initial Color |
|---|---|---|---|
| 0 | 100 | 24 | 64.30 |
| 2 | 96.38 | 26 | 61.98 |
| 4 | 92.90 | 30 | 57.58 |
| 6 | 89.55 | 40 | 47.90 |
| 8 | 86.31 | 50 | 39.85 |
| 10 | 83.19 | 60 | 33.15 |
| 12 | 80.19 | 70 | 27.58 |
| 14 | 77.29 | 80 | 22.95 |
| 16 | 74.49 | 90 | 19.09 |
| 18 | 71.80 | 100 | 15.88 |
| 20 | 69.21 | 150 | 6.32 |
| 22 | 66.71 | 200 | 2.52 |

TABLE d Fading of Preserves Color in the Dark as a Function of Time at 40.0 °C

Use the information provided in the figure and the tables to answer the following questions (assume the % initial color is directly proportional to concentration of color-causing compounds in the preserves):

a. Which order of reaction does the degradation of the preserve color follow? Explain and justify your answer.

b. Determine $k$, the rate constant for the preserves stored in the presence of light at 25.0 °C and the rate constant for the preserves stored in darkness at 25.0 °C. Which has the lower rate constant? Assuming color preservation is the goal, is it better to store the preserves in the dark or in the presence of light? Explain.

c. Does temperature (5.0 °C, 25.0 °C, and 40.0 °C) influence the degradation of color for the preserves stored in the dark? Explain.

d. When storing the preserves in the dark, which temperature (5.0 °C, 25.0 °C, or 40.0 °C) has the lowest rate constant? In the dark, is it better to store the preserves at a cool or warm temperature? Explain.

e. Determine the activation energy of the degradation reaction in the dark, $E_a$, in units of kJ/mol.

# Answers to Conceptual Connections

**Reaction Rates**

14.1 (c) The rate at which B changes is twice the rate of the reaction because its coefficient is 2, and it is negative because B is a reactant.

**Order of Reaction**

14.2 (d) Since the reaction is second order, increasing the concentration of A by a factor of 5 causes the rate to increase by $5^2$ or 25.

**Rate and Concentration**

14.3 All three mixtures have the same total number of molecules, but mixture (c) has the greatest number of NO molecules. Since the reaction is second order in NO and only first order in $O_2$, mixture (c) has the fastest initial rate.

**Half-Life Part I**

14.4 The half-life is 90 s, because it takes 90 s for the reactant to fall to one-half of its intitial concentration.

**Half-Life Part II**

14.5 The concentration of B after 50 seconds (two half-lives) is 0.225 M. See table.

| Time (min) | [A] (M) | [B] (M) |
|---|---|---|
| 0 | 0.300 | 0.0 |
| 25 | 0.150 | 0.150 |
| 50 | 0.075 | 0.225 |

**Rate Law and Integrated Rate Law**

14.6 (c) The reaction is most likely second order because its rate depends on the concentration (therefore it cannot be zero order), and its half-life depends on the initial concentration (therefore it cannot be first order). For a second-order reaction, a doubling of the initial concentration results in the quadrupling of the rate.

**Temperature Dependence of Reaction Rate**

14.7 Reaction A has a faster rate because it has a lower activation energy; therefore, the exponential factor is larger at a given temperature, making the rate constant larger. (With a larger rate constant and the same initial concentration, the rate is faster).

**Collision Theory**

14.8 (c) Since the reactants in part (a) are atoms, the orientation factor should be about one. The reactants in parts (b) and (c) are both molecules, so we expect orientation factors of less than one. Since the reactants in (b) are symmetrical, we would not expect the collision to have as specific an orientation requirement as in (c), where the reactants are asymmetrical and must therefore collide in such way that a hydrogen atom is in close proximity to another hydrogen atom. Therefore, we expect (c) to have the smallest orientation factor.

> Every system in chemical equilibrium, under the influence of a change of any one of the factors of equilibrium, undergoes a transformation . . . [that produces a change] . . . in the opposite direction of the factor in question.
>
> —HENRI LE CHÂTELIER (1850–1936)

In Chapter 14, we examined *how fast* a chemical reaction occurs. In this chapter we examine *how far* a chemical reaction goes. The *speed* of a chemical reaction is determined by kinetics. The *extent* of a chemical reaction is determined by thermodynamics. In this chapter, we focus on describing and quantifying how far a chemical reaction goes based on an experimentally measurable quantity called *the equilibrium constant*. A reaction with a large equilibrium constant proceeds nearly to completion—almost all the reactants react to form products. A reaction with a small equilibrium constant barely proceeds at all—almost all the reactants remain as reactants, hardly forming any products. In this chapter we simply accept the equilibrium constant as an experimentally measurable quantity and learn how to use it to predict and quantify the extent of a reaction. In Chapter 18, we will explore the reasons underlying the magnitude of equilibrium constants.

▲ *A developing fetus obtains oxygen from the mother's blood because the reaction between oxygen and fetal hemoglobin has a larger equilibrium constant than the reaction between oxygen and maternal hemoglobin.*

## 15.1  Fetal Hemoglobin and Equilibrium

Have you ever wondered how a baby in the womb gets oxygen? Unlike you and me, a fetus does not breathe air. Yet, like you and me, a fetus needs oxygen. Where does that oxygen come from? After we are born, we inhale air into our lungs and that air diffuses into capillaries, where it comes into contact with our blood. Within our red blood cells, a protein called hemoglobin (Hb) reacts with oxygen according to the chemical equation:

$$Hb + O_2 \rightleftharpoons HbO_2$$

▲ Hemoglobin is the oxygen-carrying protein in red blood cells. Oxygen binds to iron atoms, which are depicted here in white.

The double arrows in this equation indicate that the reaction can occur in both the forward and reverse directions and can reach chemical *equilibrium*. We encountered this term in Chapters 11 and 13, and we define it more carefully in the next section. For now, understand that the relative concentrations of the reactants and products in a reaction at equilibrium are described by the *equilibrium constant, K*. A large value of $K$ means that the reaction lies far to the right at equilibrium—a high concentration of products and a low concentration of reactants. A small value of $K$ means that the reaction lies far to the left at equilibrium—a high concentration of reactants and a low concentration of products. In short, the value of $K$ is a measure of how far a reaction proceeds—the larger the value of $K$, the more the reaction proceeds toward the products.

The equilibrium constant for the reaction between hemoglobin and oxygen is such that hemoglobin efficiently binds oxygen at typical lung oxygen concentrations, but it can also release oxygen under the appropriate conditions. Any system at equilibrium, including the hemoglobin–oxygen system, responds to changes in ways that maintain equilibrium. If any of the concentrations of the reactants or products change, the reaction shifts to counteract that change. For the hemoglobin system, as blood flows through the lungs where oxygen concentrations are high, the equilibrium shifts to the right—hemoglobin binds oxygen:

Lungs:
high $[O_2]$

$$Hb + O_2 \rightleftharpoons HbO_2$$

Reaction shifts right.

In our bodies, as blood flows out of the lungs and into muscles and organs where oxygen concentrations have been depleted (because muscles and organs use oxygen), the equilibrium shifts to the left—hemoglobin releases oxygen:

Muscles:
low $[O_2]$

$$Hb + O_2 \rightleftharpoons HbO_2$$

Reaction shifts left.

In other words, to maintain equilibrium, *hemoglobin binds oxygen when the surrounding oxygen concentration is high, but it releases oxygen when the surrounding oxygen concentration is low.* In this way, hemoglobin transports oxygen from the lungs to all parts of the body that use oxygen.

A fetus has its own circulatory system. The mother's blood never flows into the fetus's body, and the fetus cannot get any air in the womb. How, then, does the fetus get oxygen? The answer lies in the properties of fetal hemoglobin (HbF), which is slightly different from adult hemoglobin. Like adult hemoglobin, fetal hemoglobin is in equilibrium with oxygen:

$$HbF + O_2 \rightleftharpoons HbFO_2$$

However, the equilibrium constant for fetal hemoglobin is larger than the equilibrium constant for adult hemoglobin, meaning that the reaction tends to go farther in the direction of the product. Consequently, fetal hemoglobin loads oxygen at a lower oxygen concentration than does adult hemoglobin. In the placenta, fetal blood flows in close proximity to maternal blood. Although the two never mix, because of the different equilibrium constants, the maternal hemoglobin releases oxygen that the fetal hemoglobin then binds and carries into its own circulatory system (Figure 15.1 ▶). Nature has evolved a chemical system through which the mother's hemoglobin can in effect *hand off* oxygen to the hemoglobin of the fetus.

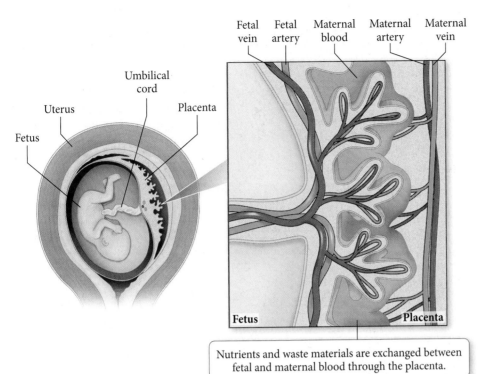

Fetal vein • Fetal artery • Maternal blood • Maternal artery • Maternal vein

Uterus • Umbilical cord • Placenta • Fetus

Fetus • Placenta

Nutrients and waste materials are exchanged between fetal and maternal blood through the placenta.

◀ **FIGURE 15.1 Oxygen Exchange between the Maternal and Fetal Circulation** In the placenta, the blood of the fetus comes into close proximity with the blood of the mother, although the two do not mix directly. Because the reaction of fetal hemoglobin with oxygen has a larger equilibrium constant than the reaction of maternal hemoglobin with oxygen, the fetus receives oxygen from the mother's blood.

## 15.2 The Concept of Dynamic Equilibrium

Recall from the previous chapter that reaction rates generally increase with increasing concentration of the reactants and decrease with decreasing concentration of the reactants (unless the reaction order is zero). With this in mind, consider the reaction between hydrogen and iodine:

$$H_2(g) + I_2(g) \rightleftharpoons 2\,HI(g)$$

In this reaction, $H_2$ and $I_2$ react to form 2 HI molecules, but the 2 HI molecules can also react to re-form $H_2$ and $I_2$. A reaction such as this one—that can proceed in both the forward and reverse directions—is **reversible**. Suppose we begin with only $H_2$ and $I_2$ in a container (Figure 15.2a▶ on the next page). What happens? Initially $H_2$ and $I_2$ begin to react to form HI (Figure 15.2b). However, as $H_2$ and $I_2$ react, their concentrations decrease, which in turn *decreases the rate of the forward reaction*. At the same time, HI begins to form. As the concentration of HI increases, the reverse reaction begins to occur at a faster and faster rate. Eventually the rate of the reverse reaction (which has been increasing) equals the rate of the forward reaction (which has been decreasing). At that point, **dynamic equilibrium** is reached (Figure 15.2c, d▶):

> **Dynamic equilibrium for a chemical reaction is the condition in which the rate of the forward reaction equals the rate of the reverse reaction.**

Dynamic equilibrium is "dynamic" because the forward and reverse reactions are still occurring; however, they are occurring at the same rate. When dynamic equilibrium is reached, the concentrations of $H_2$, $I_2$, and HI no longer change. They remain constant because the reactants and products form at the same rate that they are depleted. Note that just because the concentrations of reactants and products no longer change at equilibrium *does not mean that the concentrations of reactants and products are equal to one another* at equilibrium. Some reactions reach equilibrium only after most of the reactants have formed products. Others reach equilibrium when only a small fraction of the reactants have formed products. It depends on the reaction.

Nearly all chemical reactions are at least theoretically reversible. In many cases, however, the reversibility is so small that it can be ignored.

## Dynamic Equilibrium

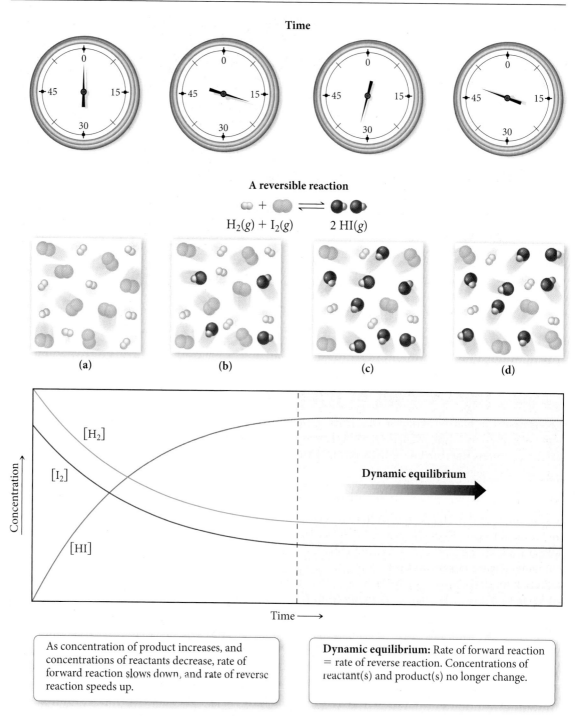

**▲ FIGURE 15.2 Dynamic Equilibrium** Equilibrium is reached in a chemical reaction when the concentrations of the reactants and products no longer change. The molecular images depict the progress of the reaction $H_2(g) + I_2(g) \rightleftharpoons 2\,HI(g)$. The graph shows the concentrations of $H_2$, $I_2$, and HI as a function of time. When the reaction reaches equilibrium, both the forward and reverse reactions continue, but at equal rates, so the concentrations of the reactants and products remain constant.

We can better understand dynamic equilibrium with a simple analogy. Imagine two neighboring countries (A and B) with a closed border between them (Figure 15.3▶). Country A is overpopulated and Country B is underpopulated. One day, the border between the two countries opens, and people immediately begin to leave Country A for Country B:

$$\text{Country A} \longrightarrow \text{Country B}$$

### Dynamic Equilibrium: An Analogy

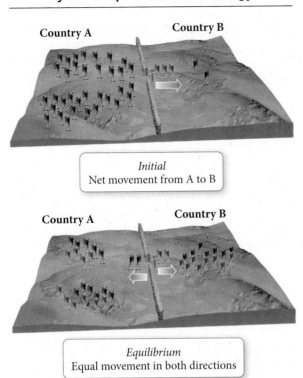

▲ **FIGURE 15.3  A Population Analogy for Chemical Equilibrium**   Because Country A is initially overpopulated, people migrate from Country A to Country B. As the population of Country A falls and that of Country B rises, the rate of migration from Country A to Country B decreases and the rate of migration from Country B to Country A increases. Eventually the two rates become equal. Equilibrium has been reached.

The population of Country A goes down as the population of Country B goes up. As people leave Country A, however, the *rate* at which they leave slows down because as Country A becomes less populated, the pool of potential emigrants gets smaller. (In other words, the rate of emigration is directly proportional to the population—as the population decreases, the emigration rate goes down.) On the other hand, as people move into Country B, it gets more crowded and some people begin to move from Country B to Country A:

$$\text{Country A} \longleftarrow \text{Country B}$$

As the population of Country B continues to grow, the rate of people moving out of Country B gets faster. Eventually the *rate* of people moving out of Country A (which has been slowing down as people leave) equals the *rate* of people moving out of Country B (which has been increasing as Country B gets more crowded). Dynamic equilibrium has been reached:

$$\text{Country A} \rightleftharpoons \text{Country B}$$

Notice that when the two countries reach dynamic equilibrium, their populations no longer change because the number of people moving out of either country equals the number of people moving in. However, one country—because of its charm or the availability of good jobs or lower taxes, or for whatever other reason—may have a higher population than the other country, even when dynamic equilibrium is reached.

Similarly, when a chemical reaction reaches dynamic equilibrium, the rate of the forward reaction equals the rate of the reverse reaction, and the concentrations of reactants and products become *constant*. But the concentrations of reactants and products will not necessarily be *equal* at equilibrium, just as the populations of the two countries are not necessarily equal at equilibrium.

# 15.3 The Equilibrium Constant (*K*)

We have just seen that the *concentrations of reactants and products* are not equal at equilibrium—rather, the *rates of the forward and reverse reactions* are equal. So what about the concentrations? What can we know about them? The concentrations, as we can see by reexamining Figure 15.2, become constant; they don't change once equilibrium is reached (as long as the temperature is constant). We quantify the relative concentrations of reactants and products at equilibrium with a quantity called the *equilibrium constant (K)*. Consider an equation for a generic chemical reaction:

$$aA + bB \rightleftharpoons cC + dD$$

We distinguish between the equilibrium constant (*K*) and the Kelvin unit of temperature (K) by italicizing the equilibrium constant.

where A and B are reactants, C and D are products, and *a*, *b*, *c*, and *d* are the respective stoichiometric coefficients in the chemical equation. The **equilibrium constant (*K*)** for the reaction is defined as the ratio—*at equilibrium*—of the concentrations of the products raised to their stoichiometric coefficients divided by the concentrations of the reactants raised to their stoichiometric coefficients:

Products

$$K = \frac{[C]^c [D]^d}{[A]^a [B]^b} \qquad \boxed{\text{The Law of Mass Action}}$$

Reactants

In this notation, [A] represents the molar concentration of A.

Why is *this* particular ratio of concentrations at equilibrium—and not some other ratio—defined as the equilibrium constant? *Because this particular ratio is always a constant when the reactants and products are at equilibrium* (at constant temperature). As we can see from the expression, the equilibrium constant *quantifies* the relative concentrations of reactants and products *at equilibrium*. The relationship between the balanced chemical equation and the expression of the equilibrium constant is the **law of mass action**.

## Expressing Equilibrium Constants for Chemical Reactions

To express an equilibrium constant for a chemical reaction, we examine the balanced chemical equation and apply the law of mass action. For example, suppose we want to express the equilibrium constant for the reaction:

$$2 N_2O_5(g) \rightleftharpoons 4 NO_2(g) + O_2(g)$$

The equilibrium constant is $[NO_2]$ raised to the fourth power multiplied by $[O_2]$ raised to the first power divided by $[N_2O_5]$ raised to the second power:

$$K = \frac{[NO_2]^4 [O_2]}{[N_2O_5]^2}$$

Notice that the *coefficients* in the chemical equation become the *exponents* in the expression of the equilibrium constant.

## Example 15.1 | Expressing Equilibrium Constants for Chemical Equations

Express the equilibrium constant for the chemical equation:

$$CH_3OH(g) \rightleftharpoons CO(g) + 2\,H_2(g)$$

**SOLUTION**

The equilibrium constant is the equilibrium concentrations of the products raised to their stoichiometric coefficients divided by the equilibrium concentrations of the reactants raised to their stoichiometric coefficients.

$$K = \frac{[CO][H_2]^2}{[CH_3OH]}$$

**FOR PRACTICE 15.1**   Express the equilibrium constant for the combustion of propane:

$$C_3H_8(g) + 5\,O_2(g) \rightleftharpoons 3\,CO_2(g) + 4\,H_2O(g)$$

## The Significance of the Equilibrium Constant

You now know how to express the equilibrium constant, but what does it mean? What, for example, does a large equilibrium constant ($K \gg 1$) imply about a reaction? A large equilibrium constant indicates that the numerator (which specifies the amounts of products at equilibrium) is larger than the denominator (which specifies the amounts of reactants at equilibrium). Therefore, when the equilibrium constant is large, the forward reaction is favored. For example, consider the reaction:

$$H_2(g) + Br_2(g) \rightleftharpoons 2\,HBr(g) \qquad K = 1.9 \times 10^{19} \text{ (at 25 °C)}$$

The equilibrium constant is large, indicating that the equilibrium point for the reaction lies far to the right—high concentrations of products, low concentrations of reactants (Figure 15.4▶). Remember that the equilibrium constant says nothing about *how fast* a reaction reaches equilibrium, only *how far* the reaction has proceeded once equilibrium is reached. A reaction with a large equilibrium constant may be kinetically very slow and take a long time to reach equilibrium.

Conversely, what does a *small* equilibrium constant ($K \ll 1$) mean? It indicates that the reverse reaction is favored and that there will be more reactants than products when equilibrium is reached. For example, consider the reaction:

$$N_2(g) + O_2(g) \rightleftharpoons 2\,NO(g) \qquad K = 4.1 \times 10^{-31} \text{ (at 25 °C)}$$

The equilibrium constant is very small, indicating that the equilibrium point for the reaction lies far to the left—high concentrations of reactants, low concentrations of products (Figure 15.5▶). This is fortunate because $N_2$ and $O_2$ are the main components of air. If this equilibrium constant were large, much of the $N_2$ and $O_2$ in air would react to form NO, a toxic gas.

*Summarizing the Significance of the Equilibrium Constant:*

▶ $K \ll 1$ Reverse reaction is favored; forward reaction does not proceed very far.

▶ $K \approx 1$ Neither direction is favored; forward reaction proceeds about halfway.

▶ $K \gg 1$ Forward reaction is favored; forward reaction proceeds essentially to completion.

▼ **FIGURE 15.4  The Meaning of a Large Equilibrium Constant**   If the equilibrium constant for a reaction is large, the equilibrium point of the reaction lies far to the right—the concentration of products is large and the concentration of reactants is small.

$$H_2(g) + Br_2(g) \rightarrow 2\,HBr(g)$$

$$K = \frac{[HBr]^2}{[H_2][Br_2]} = \text{large number}$$

$$N_2(g) + O_2(g) \rightleftharpoons 2\,NO(g)$$

$$K = \frac{[NO]^2}{[N_2][O_2]} = \text{small number}$$

▲ **FIGURE 15.5  The Meaning of a Small Equilibrium Constant** If the equilibrium constant for a reaction is small, the equilibrium point of the reaction lies far to the left—the concentration of products is small and the concentration of reactants is large.

## CONCEPTUAL
## CONNECTION 15.1

**The Magnitude of the Equilibrium Constant** Consider the reaction A(*g*) ⇌ B(*g*). The images shown here illustrate equilibrium mixtures of A (red) and B (black) at three different temperatures. At which temperature is the equilibrium constant the largest?

T₁

T₂

T₃

## CONCEPTUAL
## CONNECTION 15.2

**Equilibrium Constants and Equilibrium Concentrations**
The equilibrium constant for the reaction A(*g*) ⇌ B(*g*) is 10. A reaction mixture initially contains [A] = 1.1 M and [B] = 0.0 M. Which statement is true at equilibrium?

**(a)** The reaction mixture contains [A] = 1.0 M and [B] = 0.1 M.
**(b)** The reaction mixture contains [A] = 0.1 M and [B] = 1.0 M.
**(c)** The reaction mixture contains equal concentrations of A and B.

---

# Chemistry AND Medicine | Life and Equilibrium

Have you ever tried to define life? If you have, you probably know that a definition is elusive. How are living things different from nonliving things? You may try to define living things as those things that can move. But of course many living things do not move (most plants, for example), and some nonliving things, such as glaciers and Earth itself, do move. So motion is neither unique to nor definitive of life. You may try to define living things as those things that can reproduce. But again, many living things, such as mules or sterile humans, cannot reproduce; yet they are alive. In addition, some nonliving things, such as crystals, for example, reproduce (in some sense). So what is unique about living things?

One definition of life involves the concept of equilibrium—living things *are not* in equilibrium with their surroundings. Our body temperature, for example, is not the same as the temperature of our surroundings. If we jump into a swimming pool, the

◀ What makes these cells alive?

acidity of our blood does not become the same as the acidity of the surrounding water. Living things, even the simplest ones, maintain some measure of *disequilibrium* with their environment.

We must add one more concept, however, to complete our definition of life with respect to equilibrium. A cup of hot water is in disequilibrium with its environment with respect to temperature, yet it is not alive. The cup of hot water has no control over its disequilibrium, however, and will slowly come to equilibrium with its environment. In contrast, living things—as long as they are alive—maintain and *control* their disequilibrium. Your body temperature, for example, is not *only* in disequilibrium with your surroundings—it is in *controlled disequilibrium*. Your body maintains your temperature within a specific range that is not in equilibrium with the surrounding temperature.

So, one criterion for life is that living things are in *controlled disequilibrium* with their environment. Maintaining disequilibrium is a main activity of living organisms, requiring energy obtained from their environment. Plants derive that energy from sunlight; animals eat plants (or other animals that have eaten plants), and thus they too ultimately derive their energy from the sun. A living thing comes into equilibrium with its surroundings only after it dies.

# Relationships between the Equilibrium Constant and the Chemical Equation

If a chemical equation is modified in some way, then the equilibrium constant for the equation changes because of the modification. The following three modifications are common:

1. **If you reverse the equation, invert the equilibrium constant.** For example, consider this equilibrium equation:

$$A + 2B \rightleftharpoons 3C$$

The expression for the equilibrium constant of this reaction is:

$$K_{forward} = \frac{[C]^3}{[A][B]^2}$$

If you reverse the equation:

$$3C \rightleftharpoons A + 2B$$

then, according to the law of mass action, the expression for the equilibrium constant becomes:

$$K_{reverse} = \frac{[A][B]^2}{[C]^3} = \frac{1}{K_{forward}}$$

2. **If you multiply the coefficients in the equation by a factor, raise the equilibrium constant to the same factor.** Consider again this chemical equation and corresponding expression for the equilibrium constant:

$$A + 2B \rightleftharpoons 3C \qquad K = \frac{[C]^3}{[A][B]^2}$$

If you multiply the equation by $n$, you get:

$$nA + 2nB \rightleftharpoons 3nC$$

Applying the law of mass action, the expression for the equilibrium constant becomes:

$$K' = \frac{[C]^{3n}}{[A]^n[B]^{2n}} = \left(\frac{[C]^3}{[A][B]^2}\right)^n = K^n$$

If $n$ is a fractional quantity, raise $K$ to the same fractional quantity.

Remember that $(x^a)^b = x^{ab}$.

3. **If you add two or more individual chemical equations to obtain an overall equation, multiply the corresponding equilibrium constants by each other to obtain the overall equilibrium constant.** Consider these two chemical equations and their corresponding equilibrium constant expressions:

$$A \rightleftharpoons 2B \qquad K_1 = \frac{[B]^2}{[A]}$$

$$2B \rightleftharpoons 3C \qquad K_2 = \frac{[C]^3}{[B]^2}$$

The two equations sum as follows:

$$\begin{array}{c} A \rightleftharpoons 2B \\ 2B \rightleftharpoons 3C \\ \hline A \rightleftharpoons 3C \end{array}$$

According to the law of mass action, the equilibrium constant for this overall equation is then:

$$K_{overall} = \frac{[C]^3}{[A]}$$

Notice that $K_{overall}$ is the product of $K_1$ and $K_2$:

$$K_{overall} = K_1 \times K_2$$
$$= \frac{[B]^2}{[A]} \times \frac{[C]^3}{[B]^2}$$
$$= \frac{[C]^3}{[A]}$$

CONCEPTUAL
**CONNECTION 15.3**

**The Equilibrium Constant and the Chemical Equation** The reaction $A(g) \rightleftharpoons 2 B(g)$ has an equilibrium constant of $K = 0.010$. What is the equilibrium constant for the reaction $B(g) \rightleftharpoons \frac{1}{2}A(g)$?

**(a)** 1   **(b)** 10   **(c)** 100   **(d)** 0.0010

---

**Example 15.2** | **Manipulating the Equilibrium Constant to Reflect Changes in the Chemical Equation**

Consider the chemical equation and equilibrium constant for the synthesis of ammonia at 25 °C:

$$N_2(g) + 3 H_2(g) \rightleftharpoons 2 NH_3(g) \qquad K = 5.6 \times 10^5$$

Calculate the equilibrium constant for the following reaction at 25 °C:

$$NH_3(g) \rightleftharpoons \frac{1}{2} N_2(g) + \frac{3}{2} H_2(g) \qquad K' = ?$$

**SOLUTION**

You want to manipulate the given reaction and value of $K$ to obtain the desired reaction and value of $K$. You can see that the given reaction is the reverse of the desired reaction, and its coefficients are twice those of the desired reaction.

| | |
|---|---|
| Begin by reversing the given reaction and taking the inverse of the value of $K$. | $N_2(g) + 3 H_2(g) \rightleftharpoons 2 NH_3(g) \qquad K = 5.6 \times 10^5$ <br><br> $2 NH_3(g) \rightleftharpoons N_2(g) + 3 H_2(g) \qquad K_{reverse} = \dfrac{1}{5.6 \times 10^5}$ |
| Next, multiply the reaction by $\frac{1}{2}$ and raise the equilibrium constant to the $\frac{1}{2}$ power. | $NH_3(g) \rightleftharpoons \frac{1}{2} N_2(g) + \frac{3}{2} H_2(g)$ <br><br> $K' = K_{reverse}^{1/2} = \left(\dfrac{1}{5.6 \times 10^5}\right)^{1/2}$ |
| Calculate the value of $K'$. | $K' = 1.3 \times 10^{-3}$ |

**FOR PRACTICE 15.2** Consider the following chemical equation and equilibrium constant at 25 °C:

$$2 COF_2(g) \rightleftharpoons CO_2(g) + CF_4(g) \qquad K = 2.2 \times 10^6$$

Calculate the equilibrium constant for the following reaction at 25 °C:

$$2 CO_2(g) + 2 CF_4(g) \rightleftharpoons 4 COF_2(g) \qquad K' = ?$$

**FOR MORE PRACTICE 15.2** Predict the equilibrium constant for the first reaction shown here given the equilibrium constants for the second and third reactions:

$$CO_2(g) + 3 H_2(g) \rightleftharpoons CH_3OH(g) + H_2O(g) \qquad K_1 = ?$$
$$CO(g) + H_2O(g) \rightleftharpoons CO_2(g) + H_2(g) \qquad K_2 = 1.0 \times 10^5$$
$$CO(g) + 2 H_2(g) \rightleftharpoons CH_3OH(g) \qquad K_3 = 1.4 \times 10^7$$

---

## 15.4 Expressing the Equilibrium Constant in Terms of Pressure

So far, we have expressed the equilibrium constant only in terms of the *concentrations* of the reactants and products. For gaseous reactions, the partial pressure of a particular gas is proportional to its concentration. Therefore, we can also express the equilibrium constant in terms of the *partial pressures* of the reactants and products. Consider the gaseous reaction:

$$2 SO_3(g) \rightleftharpoons 2 SO_2(g) + O_2(g)$$

From this point on, we designate $K_c$ as the equilibrium constant with respect to concentration in molarity. For the reaction just given, we can express $K_c$ using the law of mass action:

$$K_c = \frac{[SO_2]^2[O_2]}{[SO_3]^2}$$

We now designate $K_p$ as the equilibrium constant with respect to partial pressures in atmospheres. *The expression for $K_p$ takes the form of the expression for $K_c$, except that we use the partial pressure of each gas in place of its concentration.* For the $SO_3$ reaction, we write $K_p$ as:

$$K_p = \frac{(P_{SO_2})^2 P_{O_2}}{(P_{SO_3})^2}$$

where $P_A$ is simply the partial pressure of gas A in units of atmospheres.

## Relationship Between $K_p$ and $K_c$

Since the partial pressure of a gas in atmospheres is not the same as its concentration in molarity, the value of $K_p$ for a reaction is not necessarily equal to the value of $K_c$. However, as long as the gases are behaving ideally, we can derive a relationship between the two constants. The concentration of an ideal gas A is the number of moles of A ($n_A$) divided by its volume ($V$) in liters:

$$[A] = \frac{n_A}{V}$$

From the ideal gas law, we can relate the quantity $n_A/V$ to the partial pressure of A as follows:

$$P_A V = n_A RT$$

$$P_A = \frac{n_A}{V} RT$$

Since $[A] = n_A/V$, we can write:

$$P_A = [A]RT \quad \text{or} \quad [A] = \frac{P_A}{RT} \tag{15.1}$$

Now consider the following general equilibrium chemical equation:

$$aA + bB \rightleftharpoons cC + dD$$

According to the law of mass action, we write $K_c$ as follows:

$$K_c = \frac{[C]^c[D]^d}{[A]^a[B]^b}$$

Substituting $[X] = P_X/RT$ for each concentration term, we get:

$$K_c = \frac{\left(\dfrac{P_C}{RT}\right)^c\left(\dfrac{P_D}{RT}\right)^d}{\left(\dfrac{P_A}{RT}\right)^a\left(\dfrac{P_B}{RT}\right)^b} = \frac{P_C^c P_D^d \left(\dfrac{1}{RT}\right)^{c+d}}{P_A^a P_B^b \left(\dfrac{1}{RT}\right)^{a+b}} = \frac{P_C^c P_D^d}{P_A^a P_B^b}\left(\frac{1}{RT}\right)^{c+d-(a+b)}$$

$$= K_p\left(\frac{1}{RT}\right)^{c+d-(a+b)}$$

Rearranging,

$$K_p = K_c(RT)^{c+d-(a+b)}$$

Finally, if we let $\Delta n = c + d - (a + b)$, which is the sum of the stoichiometric coefficients of the gaseous products minus the sum of the stoichiometric coefficients of the gaseous reactants, we get the following general result:

$$K_p = K_c(RT)^{\Delta n} \tag{15.2}$$

Notice that if the total number of moles of gas is the same after the reaction as before, then $\Delta n = 0$, and $K_p$ is equal to $K_c$.

In the equation $K_p = K_c(RT)^{\Delta n}$, the quantity $\Delta n$ represents the difference between the number of moles of gaseous products and gaseous reactants.

## Example 15.3 | Relating $K_p$ and $K_c$

Nitrogen monoxide, a pollutant in automobile exhaust, is oxidized to nitrogen dioxide in the atmosphere according to the equation:

$$2\,NO(g) + O_2(g) \rightleftharpoons 2\,NO_2(g) \qquad K_p = 2.2 \times 10^{12} \text{ at } 25\,°C$$

Find $K_c$ for this reaction.

| | |
|---|---|
| **SORT** You are given $K_p$ for the reaction and asked to find $K_c$. | **GIVEN:** $K_p = 2.2 \times 10^{12}$ <br><br> **FIND:** $K_c$ |
| **STRATEGIZE** Use Equation 15.2 to relate $K_p$ and $K_c$. | **EQUATION** $K_p = K_c(RT)^{\Delta n}$ |
| **SOLVE** Solve the equation for $K_c$. <br><br><br> Calculate $\Delta n$. <br><br> Substitute the required quantities to calculate $K_c$. The temperature must be in kelvins. The units are dropped when reporting $K_c$ as described later in this section. | **SOLUTION** <br> $$K_c = \frac{K_p}{(RT)^{\Delta n}}$$ <br> $$\Delta n = 2 - 3 = -1$$ <br> $$K_c = \frac{2.2 \times 10^{12}}{\left(0.08206 \dfrac{L \cdot atm}{mol \cdot K} \times 298\ K\right)^{-1}}$$ <br> $$= 5.4 \times 10^{13}$$ |

**CHECK** The easiest way to check this answer is to substitute it back into Equation 15.2 and confirm that you get the original value for $K_p$.

$$K_p = K_c(RT)^{\Delta n}$$

$$= 5.4 \times 10^{13}\left(0.08206 \frac{L \cdot atm}{mol \cdot K} \times 298\ K\right)^{-1}$$

$$= 2.2 \times 10^{12}$$

**FOR PRACTICE 15.3** Consider the following reaction and corresponding value of $K_c$:

$$H_2(g) + I_2(g) \rightleftharpoons 2\,HI(g) \qquad K_c = 6.2 \times 10^2 \text{ at } 25\,°C$$

What is the value of $K_p$ at this temperature?

## Units of $K$

Throughout this book, we express concentrations and partial pressures within the equilibrium constant expression in units of molarity and atmospheres, respectively. When expressing the value of the equilibrium constant, however, we have not included the units. Formally, the values of concentration or partial pressure that we substitute into the equilibrium constant expression are ratios of the concentration or pressure to a reference concentration (exactly 1 M) or a reference pressure (exactly 1 atm).

For example, within the equilibrium constant expression, a pressure of 1.5 atm becomes:

$$\frac{1.5\ atm}{1\ atm} = 1.5$$

Similarly, a concentration of 1.5 M becomes:

$$\frac{1.5\ M}{1\ M} = 1.5$$

As long as concentration units are expressed in molarity for $K_c$ and pressure units are expressed in atmospheres for $K_p$, we can skip this formality and enter the quantities directly into the equilibrium expression, dropping their corresponding units.

**The Relationship between $K_p$ and $K_c$** Under which circumstances are $K_p$ and $K_c$ equal for the reaction shown here?

$$aA(g) + bB(g) \rightleftharpoons cC(g) + dD(g)$$

**(a)** If $a + b = c + d$.
**(b)** If the reaction is reversible.
**(c)** If the equilibrium constant is small.

CONCEPTUAL CONNECTION **15.4**

## 15.5 Heterogeneous Equilibria: Reactions Involving Solids and Liquids

Many chemical reactions involve pure solids or pure liquids as reactants or products. Consider, for example, the reaction:

$$2\,CO(g) \rightleftharpoons CO_2(g) + C(s)$$

We might expect the expression for the equilibrium constant to be:

$$K_c = \frac{[CO_2][C]}{[CO]^2} \quad \text{(incorrect)}$$

However, since carbon is a solid, its concentration is constant (if you double the amount of carbon its *concentration* remains the same). The concentration of a solid does not change because a solid does not expand to fill its container. Its concentration, therefore, depends only on its density, which is constant as long as *some* solid is present (Figure 15.6▼). Consequently, pure solids—those reactants or products labeled in the chemical equation with an (*s*)—are not included in the equilibrium expression (because their constant value is incorporated into the value of *K*). The correct equilibrium expression for this reaction is therefore:

$$K_c = \frac{[CO_2]}{[CO]^2}$$

Similarly, the concentration of a pure liquid does not change. So, pure liquids—reactants or products labeled in the chemical equation with an (ℓ)—are also excluded from the equilibrium expression. For example, consider the equilibrium expression for the reaction between carbon dioxide and water:

$$CO_2(g) + H_2O(\ell) \rightleftharpoons H^+(aq) + HCO_3^-(aq)$$

Since $H_2O(\ell)$ is pure liquid, it is omitted from the equilibrium expression:

$$K_c = \frac{[H^+][HCO_3^-]}{[CO_2]}$$

### A Heterogeneous Equilibrium

Same $[CO_2]$ and $[CO]$ at equilibrium

Same temperature

C(*s*)  C(*s*)

$$2\,CO(g) \rightleftharpoons CO_2(g) + C(s)$$

◄ **FIGURE 15.6 Heterogeneous Equilibrium** The concentration of solid carbon (the number of atoms per unit volume) is constant as long as some solid carbon is present. The same is true for pure liquids. For this reason, the concentrations of solids and pure liquids are not included in equilibrium constant expressions.

**Example 15.4** | **Writing Equilibrium Expressions for Reactions Involving a Solid or a Liquid**

Write an expression for the equilibrium constant ($K_c$) for this chemical equation:

$$CaCO_3(s) \rightleftharpoons CaO(s) + CO_2(g)$$

**SOLUTION**

| Since $CaCO_3(s)$ and $CaO(s)$ are both solids, omit them from the equilibrium expression. | $K_c = [CO_2]$ |
|---|---|

**FOR PRACTICE 15.4** Write an equilibrium expression ($K_c$) for the equation:

$$4 HCl(g) + O_2(g) \rightleftharpoons 2 H_2O(\ell) + 2 Cl_2(g)$$

**CONCEPTUAL**
**CONNECTION 15.5**

**Heterogeneous Equilibria, $K_p$ and $K_c$** For which reaction does $K_p = K_c$?

**(a)** $2 Na_2O_2(s) + 2 CO_2(g) \rightleftharpoons 2 Na_2CO_3(s) + O(g)$
**(b)** $Fe_2O_3(s) + 3 CO(g) \rightleftharpoons 2 Fe(s) + 3 CO_2(g)$
**(c)** $NH_4NO_3(s) \rightleftharpoons N_2O(g) + 2 H_2O(g)$

## 15.6 Calculating the Equilibrium Constant from Measured Equilibrium Concentrations

The most direct way to obtain an experimental value for the equilibrium constant of a reaction is to measure the concentrations of the reactants and products in a reaction mixture at equilibrium. Consider the following reaction:

$$H_2(g) + I_2(g) \rightleftharpoons 2 HI(g)$$

Since equilibrium constants depend on temperature, many equilibrium problems state the temperature even though it has no formal part in the calculation.

Suppose a mixture of $H_2$ and $I_2$ is allowed to come to equilibrium at 445 °C. The measured equilibrium concentrations are $[H_2] = 0.11$ M, $[I_2] = 0.11$ M, and $[HI] = 0.78$ M. What is the value of the equilibrium constant at this temperature?

We can write the expression for $K_c$ from the balanced equation:

$$K_c = \frac{[HI]^2}{[H_2][I_2]}$$

To calculate the value of $K_c$, we substitute the correct equilibrium concentrations into the expression for $K_c$:

$$K_c = \frac{[HI]^2}{[H_2][I_2]}$$

$$= \frac{(0.78)^2}{(0.11)(0.11)}$$

$$= 5.0 \times 10^1$$

The concentrations within $K_c$ should always be written in moles per liter (M); however, as noted in Section 15.4, we do not normally include the units when expressing the value of the equilibrium constant, so $K_c$ is unitless.

For any reaction, the equilibrium *concentrations* of the reactants and products depend on the initial concentrations (and in general vary from one set of initial concentrations to another). However, the equilibrium *constant* is always the same at a given temperature, regardless of the initial concentrations. For example, Table 15.1 shows several different equilibrium concentrations of $H_2$, $I_2$, and HI, each from a different set of

**TABLE 15.1  Initial and Equilibrium Concentrations for the Reaction**
$H_2(g) + I_2(g) \rightleftharpoons 2 HI(g)$ **at 445 °C**

| Initial Concentrations | | | Equilibrium Concentrations | | | Equilibrium Constant |
|---|---|---|---|---|---|---|
| $[H_2]$ | $[I_2]$ | $[HI]$ | $[H_2]$ | $[I_2]$ | $[HI]$ | $K_c = \dfrac{[HI^2]}{[H_2][I_2]}$ |
| 0.50 | 0.50 | 0.0 | 0.11 | 0.11 | 0.78 | $\dfrac{(0.78)^2}{(0.11)(0.11)} = 50$ |
| 0.0 | 0.0 | 0.50 | 0.055 | 0.055 | 0.39 | $\dfrac{(0.39)^2}{(0.055)(0.055)} = 50$ |
| 0.50 | 0.50 | 0.50 | 0.165 | 0.165 | 1.17 | $\dfrac{(1.17)^2}{(0.165)(0.165)} = 50$ |
| 1.0 | 0.50 | 0.0 | 0.53 | 0.033 | 0.934 | $\dfrac{(0.934)^2}{(0.53)(0.033)} = 50$ |
| 0.50 | 1.0 | 0.0 | 0.033 | 0.53 | 0.934 | $\dfrac{(0.934)^2}{(0.033)(0.53)} = 50$ |

initial concentrations. Notice that the equilibrium constant is always the same, regardless of the initial concentrations. Whether we start with only reactants or only products, the reaction reaches equilibrium concentrations at which the equilibrium constant is the same. No matter what the initial concentrations are, the reaction always goes in a direction that ensures that the equilibrium concentrations—when substituted into the equilibrium expression—give the same constant, $K$.

So far, we have calculated equilibrium constants from values of the equilibrium concentrations of all the reactants and products. In most cases, however, we need only know the initial concentrations of the reactant(s) and the equilibrium concentration of any *one* reactant or product. We can deduce the other equilibrium concentrations from the stoichiometry of the reaction.

For example, consider the simple reaction:

$$A(g) \rightleftharpoons 2 B(g)$$

Suppose that we have a reaction mixture in which the initial concentration of A is 1.00 M and the initial concentration of B is 0.00 M. When equilibrium is reached, the concentration of A is 0.75 M. Since [A] has changed by −0.25 M, we can deduce (based on the stoichiometry) that [B] must have changed by 2 × (+0.25 M) or +0.50 M. We summarize the initial conditions, the changes, and the equilibrium conditions in the following table:

| | [A] | [B] |
|---|---|---|
| Initial | 1.00 | 0.00 |
| Change | −0.25 | +2(0.25) |
| Equilibrium | 0.75 | 0.50 |

We refer to this type of table as an ICE table (I = initial, C = change, E = equilibrium). To calculate the equilibrium constant, we use the balanced equation to write an expression for the equilibrium constant and then substitute the equilibrium concentrations from the ICE table:

$$K = \frac{[B]^2}{[A]} = \frac{(0.50)^2}{(0.75)} = 0.33$$

In Examples 15.5 and 15.6, we show the general procedure for solving these kinds of equilibrium problems in the left column and work two examples exemplifying the procedure in the center and right columns.

## Procedure for...

| **Finding Equilibrium Constants from Experimental Concentration Measurements** | **Example 15.5** ▶ Finding Equilibrium Constants from Experimental Concentration Measurements | **Example 15.6** Finding Equilibrium Constants from Experimental Concentration Measurements |
|---|---|---|
| To solve these types of problems, follow the given procedure. | Consider the following reaction:<br><br>$CO(g) + 2\,H_2(g) \rightleftharpoons CH_3OH(g)$<br><br>A reaction mixture at 780 °C initially contains [CO] = 0.500 M and [H$_2$] = 1.00 M. At equilibrium, the CO concentration is found to be 0.15 M. What is the value of the equilibrium constant? | Consider the following reaction:<br><br>$2\,CH_4(g) \rightleftharpoons C_2H_2(g) + 3\,H_2(g)$<br><br>A reaction mixture at 1700 °C initially contains [CH$_4$] = 0.115 M. At equilibrium, the mixture contains [C$_2$H$_2$] = 0.035 M. What is the value of the equilibrium constant? |

**1. Using the balanced equation as a guide, prepare an ICE table showing the known initial concentrations and equilibrium concentrations of the reactants and products.** Leave space in the middle of the table for determining the changes in concentration that occur during the reaction.

$CO(g) + 2\,H_2(g) \rightleftharpoons CH_3OH(g)$

|         | [CO]  | [H₂] | [CH₃OH] |
|---------|-------|------|---------|
| Initial | 0.500 | 1.00 | 0.00    |
| Change  |       |      |         |
| Equil   | 0.15  |      |         |

$2\,CH_4(g) \rightleftharpoons C_2H_2(g) + 3\,H_2(g)$

|         | [CH₄] | [C₂H₂] | [H₂] |
|---------|-------|--------|------|
| Initial | 0.115 | 0.00   | 0.00 |
| Change  |       |        |      |
| Equil   |       | 0.035  |      |

**2. For the reactant or product whose concentration is known both initially and at equilibrium, calculate the change in concentration that occurs.**

$CO(g) + 2\,H_2(g) \rightleftharpoons CH_3OH(g)$

|         | [CO]  | [H₂] | [CH₃OH] |
|---------|-------|------|---------|
| Initial | 0.500 | 1.00 | 0.00    |
| Change  | −0.35 |      |         |
| Equil   | 0.15  |      |         |

$2\,CH_4(g) \rightleftharpoons C_2H_2(g) + 3\,H_2(g)$

|         | [CH₄] | [C₂H₂] | [H₂] |
|---------|-------|--------|------|
| Initial | 0.115 | 0.00   | 0.00 |
| Change  |       | +0.035 |      |
| Equil   |       | 0.035  |      |

**3. Use the change you calculated in step 2 and the stoichiometric relationships from the balanced chemical equation to determine the changes in concentration of all other reactants and products.** Since reactants are consumed during the reaction, the changes in their concentrations are negative. Since products are formed, the changes in their concentrations are positive.

$CO(g) + 2\,H_2(g) \rightleftharpoons CH_3OH(g)$

|         | [CO]  | [H₂]      | [CH₃OH] |
|---------|-------|-----------|---------|
| Initial | 0.500 | 1.00      | 0.00    |
| Change  | −0.35 | −2(0.35)  | +0.35   |
| Equil   | 0.15  |           |         |

$2\,CH_4(g) \rightleftharpoons C_2H_2(g) + 3\,H_2(g)$

|         | [CH₄]      | [C₂H₂] | [H₂]      |
|---------|------------|--------|-----------|
| Initial | 0.115      | 0.00   | 0.00      |
| Change  | −2(0.035)  | +0.035 | +3(0.035) |
| Equil   |            | 0.035  |           |

**4. Sum each column for each reactant and product to determine the equilibrium concentrations.**

|         | [CO]  | [H₂]  | [CH₃OH] |
|---------|-------|-------|---------|
| Initial | 0.500 | 1.00  | 0.00    |
| Change  | −0.35 | −0.70 | +0.35   |
| Equil   | 0.15  | 0.30  | 0.35    |

|         | [CH₄]  | [C₂H₂] | [H₂]   |
|---------|--------|--------|--------|
| Initial | 0.115  | 0.00   | 0.00   |
| Change  | −0.070 | +0.035 | +0.105 |
| Equil   | 0.045  | 0.035  | 0.105  |

**5. Use the balanced equation to write an expression for the equilibrium constant and substitute the equilibrium concentrations to calculate K.**

$$K_c = \frac{[CH_3OH]}{[CO][H_2]^2}$$
$$= \frac{0.35}{(0.15)(0.30)^2}$$
$$= 26$$

$$K_c = \frac{[C_2H_2][H_2]^3}{[CH_4]^2}$$
$$= \frac{(0.035)(0.105)^3}{(0.045)^2}$$
$$= 0.020$$

**FOR PRACTICE 15.5** The reaction in Example 15.5 between CO and $H_2$ is carried out at a different temperature with initial concentrations of [CO] = 0.27 M and [$H_2$] = 0.49 M. At equilibrium, the concentration of $CH_3OH$ is 0.11 M. Find the equilibrium constant at this temperature.

**FOR PRACTICE 15.6** The reaction of $CH_4$ in Example 15.6 is carried out at a different temperature with an initial concentration of [$CH_4$] = 0.087 M. At equilibrium, the concentration of $H_2$ is 0.012 M. Find the equilibrium constant at this temperature.

# 15.7 The Reaction Quotient: Predicting the Direction of Change

When the reactants of a chemical reaction mix, they generally react to form products—we say that the reaction proceeds to the right (toward the products). The amount of products that forms when a reaction reaches equilibrium depends on the magnitude of the equilibrium constant, as we have seen. However, what if a reaction mixture that is not at equilibrium contains both reactants *and products*? Can we predict the direction of change for such a mixture?

To gauge the progress of a reaction relative to equilibrium, we use a quantity called the *reaction quotient*. The definition of the reaction quotient takes the same form as the definition of the equilibrium constant, except that the reaction need not be at equilibrium. So, for the general reaction:

$$aA + bB \rightleftharpoons cC + dD$$

we define the **reaction quotient ($Q_c$)** as the ratio—at any point in the reaction—of the concentrations of the products raised to their stoichiometric coefficients divided by the concentrations of the reactants raised to their stoichiometric coefficients. For gases with amounts measured in atmospheres, the reaction quotient uses the partial pressures in place of concentrations and is called $Q_p$:

$$Q_c = \frac{[C]^c[D]^d}{[A]^a[B]^b} \qquad Q_p = \frac{P_C^c P_D^d}{P_A^a P_B^b}$$

The difference between the reaction quotient and the equilibrium constant is that, at a given temperature, the equilibrium constant has only one value and it specifies the relative amounts of reactants and products *at equilibrium*. The reaction quotient, by contrast, depends on the current state of the reaction and has many different values as the reaction proceeds. For example, in a reaction mixture containing only reactants, the reaction quotient is zero ($Q_c = 0$):

$$Q_c = \frac{[0]^c[0]^d}{[A]^a[B]^b} = 0$$

In a reaction mixture containing only products, the reaction quotient is infinite ($Q_c = \infty$):

$$Q_c = \frac{[C]^c[D]^d}{[0]^a[0]^b} = \infty$$

In a reaction mixture containing both reactants and products, each at a concentration of 1 M, the reaction quotient is one ($Q_c = 1$):

$$Q_c = \frac{(1)^c(1)^d}{(1)^a(1)^b} = 1$$

The reaction quotient is useful because *the value of Q relative to K is a measure of the progress of the reaction toward equilibrium. At equilibrium, the reaction quotient is*

## Q, K, and the Direction of a Reaction

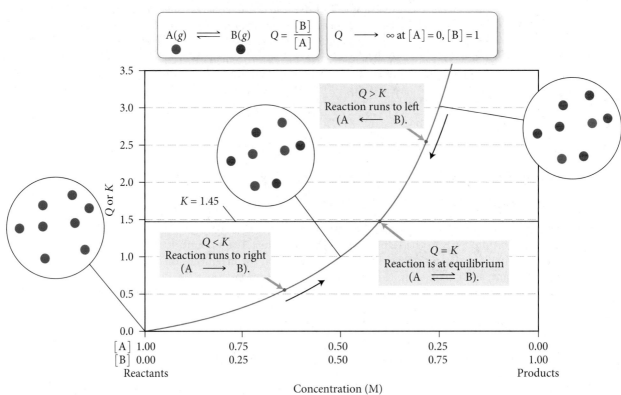

$$A(g) \rightleftharpoons B(g) \qquad Q = \frac{[B]}{[A]}$$

$$Q \longrightarrow \infty \text{ at } [A] = 0, [B] = 1$$

$K = 1.45$

$Q > K$
Reaction runs to left
$(A \longleftarrow B)$.

$Q < K$
Reaction runs to right
$(A \longrightarrow B)$.

$Q = K$
Reaction is at equilibrium
$(A \rightleftharpoons B)$.

Q or K

| | | | | |
|---|---|---|---|---|
| [A] 1.00 | 0.75 | 0.50 | 0.25 | 0.00 |
| [B] 0.00 | 0.25 | 0.50 | 0.75 | 1.00 |
| Reactants | | | | Products |

Concentration (M)

▲ **FIGURE 15.7** *Q, K, and the Direction of a Reaction*   The graph shows a plot of Q as a function of the concentrations of the reactants and products in a simple reaction A $\rightleftharpoons$ B, in which K = 1.45 and the sum of the reactant and product concentrations is 1 M. The far left of the graph represents pure reactant and the far right represents pure product. The midpoint of the graph represents an equal mixture of A and B. When Q is less than K, the reaction moves in the forward direction (A $\longrightarrow$ B). When Q is greater than K, the reaction moves in the reverse direction (A $\longleftarrow$ B). When Q is equal to K, the reaction is at equilibrium.

*equal to the equilibrium constant.* Figure 15.7▲ shows a plot of Q as a function of the concentrations of A and B for the simple reaction A(g) $\rightleftharpoons$ B(g), which has an equilibrium constant of K = 1.45. The following points are representative of three possible conditions:

| Q | K | Predicted Direction of Reaction |
|---|---|---|
| 0.55 | 1.45 | To the right (toward products) |
| 2.55 | 1.45 | To the left (toward reactants) |
| 1.45 | 1.45 | No change (at equilibrium) |

For the first set of values in the table, Q is less than K and must therefore get larger as the reaction proceeds toward equilibrium. Q becomes larger as the reactant concentration decreases and the product concentration increases—the reaction proceeds to the right. For the second set of values, Q is greater than K and must therefore get smaller as the reaction proceeds toward equilibrium. Q gets smaller as the reactant concentration increases and the product concentration decreases—the reaction proceeds to the left. In the third set of values, Q = K, implying that the reaction is at equilibrium—the reaction will not proceed in either direction.

***Summarizing Direction of Change Predictions:***
The reaction quotient (Q) relative to the equilibrium constant (K) is a measure of the progress of a reaction toward equilibrium.

▶   Q < K Reaction goes to the right (toward products).

▶   Q > K Reaction goes to the left (toward reactants).

▶   Q = K Reaction is at equilibrium.

**Example** 15.7 | **Predicting the Direction of a Reaction by Comparing Q and K**

Consider the reaction and its equilibrium constant at 25.0 °C:

$$I_2(g) + Cl_2(g) \rightleftharpoons 2\,ICl(g) \qquad K_p = 81.9$$

A reaction mixture contains $P_{I_2} = 0.114$ atm, $P_{Cl_2} = 0.102$ atm, and $P_{ICl} = 0.355$ atm. Is the reaction mixture at equilibrium? If not, in which direction will the reaction proceed?

**SOLUTION**

| | |
|---|---|
| To determine the progress of the reaction relative to the equilibrium state, first calculate Q. | $$Q_p = \left(\frac{P_{ICl}^2}{P_{I_2}P_{Cl_2}}\right)$$ $$= \frac{(0.355)^2}{(0.114)(0.102)}$$ $$= 10.8$$ |
| Compare Q to K. | $Q_p = 10.8;\ K_p = 81.9$ <br><br> Since $Q_p < K_p$, the reaction is not at equilibrium and will proceed to the right. |

**FOR PRACTICE 15.7**  Consider the reaction and its equilibrium constant:

$$N_2O_4(g) \rightleftharpoons 2\,NO_2(g) \qquad K_c = 5.85 \times 10^{-3}\ \text{(at some temperature)}$$

A reaction mixture contains $[NO_2] = 0.0255$ M and $[N_2O_4] = 0.0331$ M. Calculate $Q_c$ and determine the direction in which the reaction will proceed.

**Q and K**  For the reaction $N_2O_4(g) \rightleftharpoons 2\,NO_2(g)$, a reaction mixture at a certain temperature initially contains both $N_2O_4$ and $NO_2$ in their standard states (see the definition of standard state in Section 6.9). If $K_p = 0.15$, which statement is true of the reaction mixture before any reaction occurs?

CONCEPTUAL
**CONNECTION 15.6**

   **(a)**  $Q = K$; the reaction is at equilibrium.
   **(b)**  $Q < K$; the reaction will proceed to the right.
   **(c)**  $Q > K$; the reaction will proceed to the left.

## 15.8 Finding Equilibrium Concentrations

 **Finding Equilibrium Concentrations from Initial Concentrations**

In Section 15.6, we discussed how to calculate an equilibrium constant from the equilibrium concentrations of the reactants and products. Just as commonly we will want to calculate equilibrium concentrations of reactants or products from the equilibrium constant. These kinds of calculations are important because they allow us to calculate the amount of a reactant or product at equilibrium. For example, in a synthesis reaction, we might want to know how much of the product forms when the reaction reaches equilibrium. Or for the hemoglobin–oxygen equilibrium discussed in Section 15.1, we might want to know the concentration of oxygenated hemoglobin present under certain oxygen concentrations within the lungs or muscles.

We divide these types of problems into two categories: (1) finding equilibrium concentrations when we know the equilibrium constant and all but one of the equilibrium concentrations of the reactants and products, and (2) finding equilibrium concentrations when we know the equilibrium constant and only initial concentrations. The second category of problem is more difficult than the first. Let's examine each separately.

### Finding Equilibrium Concentrations from the Equilibrium Constant and All but One of the Equilibrium Concentrations of the Reactants and Products

We can use the equilibrium constant to calculate the equilibrium concentration of one of the reactants or products, given the equilibrium concentrations of the others. To solve this type of problem, we follow our general problem-solving procedure.

**Example 15.8** **Finding Equilibrium Concentrations When You Know the Equilibrium Constant and All but One of the Equilibrium Concentrations of the Reactants and Products**

Consider the following reaction:

$$2\,COF_2(g) \rightleftharpoons CO_2(g) + CF_4(g) \qquad K_c = 2.00 \text{ at } 1000\,^{\circ}C$$

In an equilibrium mixture, the concentration of $COF_2$ is 0.255 M and the concentration of $CF_4$ is 0.118 M. What is the equilibrium concentration of $CO_2$ ?

| | |
|---|---|
| **SORT** You are given the equilibrium constant of a chemical reaction, together with the equilibrium concentrations of the reactant and one product. You are asked to find the equilibrium concentration of the other product. | **GIVEN:** $[COF_2] = 0.255$ M<br><br>$[CF_4] = 0.118$ M<br><br>$K_c = 2.00$<br><br>**FIND:** $[CO_2]$ |
| **STRATEGIZE** Calculate the concentration of the product using the given quantities and the expression for $K_c$. | **CONCEPTUAL PLAN** 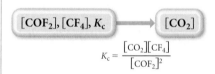 |
| **SOLVE** Solve the equilibrium expression for $[CO_2]$ and then substitute in the appropriate values to calculate it. | **SOLUTION**<br><br>$[CO_2] = K_c \dfrac{[COF_2]^2}{[CF_4]}$<br><br>$[CO_2] = 2.00\left(\dfrac{(0.255)^2}{0.118}\right) = 1.10$ M |

**CHECK** Check your answer by mentally substituting the given values of $[COF_2]$ and $[CF_4]$ as well as your calculated value for $CO_2$ back into the equilibrium expression:

$$K_c = \frac{[CO_2][CF_4]}{[COF_2]^2}$$

$[CO_2]$ is roughly equal to 1. $[COF_2]^2 \approx 0.06$ and $[CF_4] \approx 0.12$. Therefore $K_c$ is approximately 2, as given in the problem.

**FOR PRACTICE 15.8** Diatomic iodine $[I_2]$ decomposes at high temperature to form I atoms according to the reaction:

$$I_2(g) \rightleftharpoons 2\,I(g) \qquad K_c = 0.011 \text{ at } 1200\,^{\circ}C$$

In an equilibrium mixture, the concentration of $I_2$ is 0.10 M. What is the equilibrium concentration of I?

# Finding Equilibrium Concentrations from the Equilibrium Constant and Initial Concentrations or Pressures

More commonly, we know the equilibrium constant and only initial concentrations of reactants and need to find the *equilibrium concentrations* of the reactants or products. These kinds of problems are generally more involved than those we just examined and require a specific procedure to solve them. The procedure has some similarities to the one used in Examples 15.5 and 15.6 in that we set up an ICE table showing the initial concentrations, the changes, and the equilibrium concentrations. However, unlike in Examples 15.5 and 15.6, here the changes in concentration are not known and are represented with the variable $x$.

For example, consider again the simple reaction:

$$A(g) \rightleftharpoons 2\,B(g)$$

Suppose that, as before (see Section 15.6), we have a reaction mixture in which the initial concentration of A is 1.0 M and the initial concentration of B is 0.00 M. However, now we know the equilibrium constant, $K = 0.33$, and want to find the equilibrium concentrations. We know that because $Q = 0$, the reaction proceeds to the right (towards the products). We set up the ICE table with the given initial concentrations and then *represent the unknown change in [A] with the variable x* as follows:

| | [A] | [B] | |
|---------|---------|------|---|
| Initial | 1.0 | 0.00 | Represent changes from initial conditions with the variable $x$. |
| Change | $-x$ | $+2x$ | |
| Equil | $1.0 - x$ | $2x$ | |

Notice that, due to the stoichiometry of the reaction, the change in [B] must be $+2x$. As before, each *equilibrium* concentration is the sum of the two entries above it in the ICE table. In order to find the equilibrium concentrations of A and B, we must find the value of the variable $x$. Since we know the value of the equilibrium constant, we can use the equilibrium constant expression to set up an equation in which $x$ is the only variable:

$$K = \frac{[B]^2}{[A]} = \frac{(2x)^2}{1.0 - x} = 0.33$$

or more simply:

$$\frac{4x^2}{1.0 - x} = 0.33$$

This equation is a *quadratic* equation—it contains the variable $x$ raised to the second power. In general, we can solve quadratic equations with the quadratic formula (see Appendix IC), which we introduce in Example 15.10. If the quadratic equation is a perfect square, however, we can solve it by simpler means, as shown in Example 15.9. For both of these examples, we give the general procedure in the left column and apply the procedure to the two different example problems in the center and right columns. Later in this section, we see that quadratic equations can often be simplified by making some approximations based on our chemical knowledge.

## Procedure for...

| **Finding Equilibrium Concentrations from Initial Concentrations and the Equilibrium Constant** | **Example 15.9** ▶ Finding Equilibrium Concentrations from Initial Concentrations and the Equilibrium Constant | **Example 15.10** Finding Equilibrium Concentrations from Initial Concentrations and the Equilibrium Constant |
|---|---|---|
| To solve these types of problems, follow the given procedure. | Consider the reaction:<br><br>$N_2(g) + O_2(g) \rightleftharpoons 2\,NO(g)$<br><br>$K_c = 0.10$ (at 2000 °C)<br><br>A reaction mixture at 2000 °C initially contains $[N_2] = 0.200$ M and $[O_2] = 0.200$ M. Find the equilibrium concentrations of the reactants and product at this temperature. | Consider the reaction:<br><br>$N_2O_4(g) \rightleftharpoons 2\,NO_2(g)$<br><br>$K_c = 0.36$ (at 100 °C)<br><br>A reaction mixture at 100 °C initially contains $[NO_2] = 0.100$ M. Find the equilibrium concentrations of $NO_2$ and $N_2O_4$ at this temperature. |

**1. Using the balanced equation as a guide, prepare a table showing the known initial concentrations of the reactants and products.** Leave room in the table for the changes in concentrations and for the equilibrium concentrations.

$N_2(g) + O_2(g) \rightleftharpoons 2\,NO(g)$

|  | $[N_2]$ | $[O_2]$ | $[NO]$ |
|---|---|---|---|
| Initial | 0.200 | 0.200 | 0.00 |
| Change |  |  |  |
| Equil |  |  |  |

$N_2O_4(g) \rightleftharpoons 2\,NO_2(g)$

|  | $[N_2O_4]$ | $[NO_2]$ |
|---|---|---|
| Initial | 0.00 | 0.100 |
| Change |  |  |
| Equil |  |  |

**2. Use the initial concentrations to calculate the reaction quotient ($Q$) for the initial concentrations. Compare $Q$ to $K$ to predict the direction in which the reaction proceeds.**

$$Q_c = \frac{[NO]^2}{[N_2][O_2]} = \frac{(0.00)^2}{(0.200)(0.200)}$$
$$= 0$$

$Q < K$; therefore, the reaction will proceed to the right.

$$Q_c = \frac{[NO_2]^2}{[N_2O_4]} = \frac{(0.100)^2}{0.00}$$
$$= \infty$$

$Q > K$; therefore, the reaction will proceed to the left.

**3. Represent the change in the concentration of one of the reactants or products with the variable $x$. Define the changes in the concentrations of the other reactants or products in terms of $x$.** It is usually most convenient to let $x$ represent the change in concentration of the reactant or product with the smallest stoichiometric coefficient.

$N_2(g) + O_2(g) \rightleftharpoons 2\,NO(g)$

|  | $[N_2]$ | $[O_2]$ | $[NO]$ |
|---|---|---|---|
| Initial | 0.200 | 0.200 | 0.00 |
| Change | $-x$ | $-x$ | $+2x$ |
| Equil |  |  |  |

$N_2O_4(g) \rightleftharpoons 2\,NO_2(g)$

|  | $[N_2O_4]$ | $[NO_2]$ |
|---|---|---|
| Initial | 0.00 | 0.100 |
| Change | $+x$ | $-2x$ |
| Equil |  |  |

**4. Sum each column for each reactant and each product to determine the equilibrium concentrations in terms of the initial concentrations and the variable $x$.**

$N_2(g) + O_2(g) \rightleftharpoons 2\,NO(g)$

|  | $[N_2]$ | $[O_2]$ | $[NO]$ |
|---|---|---|---|
| Initial | 0.200 | 0.200 | 0.00 |
| Change | $-x$ | $-x$ | $+2x$ |
| Equil | $0.200 - x$ | $0.200 - x$ | $2x$ |

$N_2O_4(g) \rightleftharpoons 2\,NO_2(g)$

|  | $[N_2O_4]$ | $[NO_2]$ |
|---|---|---|
| Initial | 0.00 | 0.100 |
| Change | $+x$ | $-2x$ |
| Equil | $x$ | $0.100 - 2x$ |

5. **Substitute the expressions for the equilibrium concentrations (from step 4) into the expression for the equilibrium constant. Using the given value of the equilibrium constant, solve the expression for the variable $x$.** In some cases, such as Example 15.9, you can take the square root of both sides of the expression to solve for $x$. In other cases, such as Example 15.10, you must solve a quadratic equation to find $x$.

Remember the quadratic formula:

$$ax^2 + bx + c = 0$$

$$x = \frac{-b \pm \sqrt{b^2 - 4ac}}{2a}$$

| | |
|---|---|
| $K_c = \dfrac{[NO]^2}{[N_2][O_2]}$ | $K_c = \dfrac{[NO_2]^2}{[N_2O_4]}$ |
| $\quad = \dfrac{(2x)^2}{(0.200 - x)(0.200 - x)}$ | $\quad = \dfrac{(0.100 - 2x)^2}{x}$ |
| $0.10 = \dfrac{(2x)^2}{(0.200 - x)^2}$ | $0.36 = \dfrac{0.0100 - 0.400x + 4x^2}{x}$ |
| $\sqrt{0.10} = \dfrac{2x}{0.200 - x}$ | $0.36x = 0.0100 - 0.400x + 4x^2$ |
| $\sqrt{0.10}\,(0.200 - x) = 2x$ | $4x^2 - 0.76x + 0.0100 = 0 \; (quadratic)$ |
| $\sqrt{0.10}\,(0.200) - \sqrt{0.10}\,x = 2x$ | $x = \dfrac{-b \pm \sqrt{b^2 - 4ac}}{2a}$ |
| $0.063 = 2x + \sqrt{0.10}\,x$ | $\quad = \dfrac{-(-0.76) \pm \sqrt{(-0.76)^2 - 4(4)(0.0100)}}{2(4)}$ |
| $0.063 = 2.3x$ | $\quad = \dfrac{0.76 \pm 0.65}{8}$ |
| $x = 0.027$ | $x = 0.176 \quad or \quad x = 0.014$ |

6. **Substitute $x$ into the expressions for the equilibrium concentrations of the reactants and products (from step 4) and calculate the concentrations.** In cases where you solved a quadratic and have two values for $x$, choose the value for $x$ that gives a physically realistic answer. For example, reject the value of $x$ that results in any negative concentrations.

| | |
|---|---|
| $[N_2] = 0.200 - 0.027$ | We reject the root $x = 0.176$ because it gives a negative concentration for $NO_2$. Using $x = 0.014$, we get the following concentrations: |
| $\quad = 0.173\,M$ | |
| $[O_2] = 0.200 - 0.027$ | |
| $\quad = 0.173\,M$ | $[NO_2] = 0.100 - 2x$ |
| $[NO] = 2(0.027)$ | $\quad = 0.100 - 2(0.014) = 0.072\,M$ |
| $\quad = 0.054\,M$ | $[N_2O_4] = x$ |
| | $\quad = 0.014\,M$ |

7. **Check your answer by substituting the calculated equilibrium values into the equilibrium expression. The calculated value of $K$ should match the given value of $K$.** Note that rounding errors could cause a difference in the least significant digit when comparing values of the equilibrium constant.

| | |
|---|---|
| $K_c = \dfrac{[NO]^2}{[N_2][O_2]}$ | $K_c = \dfrac{[NO_2]^2}{[N_2O_4]} = \dfrac{(0.072)^2}{0.014}$ |
| $\quad = \dfrac{(0.054)^2}{(0.173)(0.173)} = 0.097$ | $\quad = 0.37$ |
| Since the calculated value of $K_c$ matches the given value (to within one digit in the least significant figure), the answer is valid. | Since the calculated value of $K_c$ matches the given value (to within one digit in the least significant figure), the answer is valid. |

**FOR PRACTICE 15.9**

The reaction in Example 15.9 is carried out at a different temperature at which $K_c = 0.055$. This time, however, the reaction mixture starts with only the product, $[NO] = 0.0100\,M$, and no reactants. Find the equilibrium concentrations of $N_2$, $O_2$, and NO at equilibrium.

**FOR PRACTICE 15.10**

The reaction in Example 15.10 is carried out at the same temperature, but this time the reaction mixture initially contains only the reactant, $[N_2O_4] = 0.0250\,M$, and no $NO_2$. Find the equilibrium concentrations of $N_2O_4$ and $NO_2$.

When the initial conditions are given in terms of partial pressures (instead of concentrations) and the equilibrium constant is given as $K_p$ instead of $K_c$, use the same procedure, but substitute partial pressures for concentrations, as shown in Example 15.11.

**Example 15.11** | **Finding Equilibrium Partial Pressures When You Are Given the Equilibrium Constant and Initial Partial Pressures**

Consider the reaction:

$$I_2(g) + Cl_2(g) \rightleftharpoons 2\,ICl(g) \qquad K_p = 81.9 \text{ (at 25 °C)}$$

A reaction mixture at 25 °C initially contains $P_{I_2} = 0.100$ atm, $P_{Cl_2} = 0.100$ atm, and $P_{ICl} = 0.100$ atm. Find the equilibrium partial pressures of $I_2$, $Cl_2$, and $ICl$ at this temperature.

**SOLUTION**

Follow the procedure used in Examples 15.5 and 15.6 (using partial pressures in place of concentrations) to solve the problem.

| | | |
|---|---|---|
| 1. Using the balanced equation as a guide, prepare an ICE table showing the known initial partial pressures of the reactants and products. | $I_2(g) + Cl_2(g) \rightleftharpoons 2\,ICl(g)$ <br><br> | |

| | $P_{I_2}$ (atm) | $P_{Cl_2}$ (atm) | $P_{ICl}$ (atm) |
|---|---|---|---|
| Initial | 0.100 | 0.100 | 0.100 |
| Change | | | |
| Equil | | | |

2. Use the initial partial pressures to calculate the reaction quotient ($Q$). Compare $Q$ to $K$ to predict the direction in which the reaction will proceed.

$$Q_p = \frac{(P_{ICl})^2}{P_{I_2}P_{Cl_2}} = \frac{(0.100)^2}{(0.100)(0.100)} = 1$$

$K_p = 81.9$ (given)

$Q < K$; therefore, the reaction will proceed to the right.

3. Represent the change in the partial pressure of one of the reactants or products with the variable $x$. Define the changes in the partial pressures of the other reactants or products in terms of $x$.

$I_2(g) + Cl_2(g) \rightleftharpoons 2\,ICl(g)$

| | $P_{I_2}$(atm) | $P_{Cl_2}$(atm) | $P_{ICl}$(atm) |
|---|---|---|---|
| Initial | 0.100 | 0.100 | 0.100 |
| Change | $-x$ | $-x$ | $+2x$ |
| Equil | | | |

4. Sum each column for each reactant and product to determine the equilibrium partial pressures in terms of the initial partial pressures and the variable $x$.

$I_2(g) + Cl_2(g) \rightleftharpoons 2\,ICl(g)$

| | $P_{I_2}$ (atm) | $P_{Cl_2}$ (atm) | $P_{ICl}$ (atm) |
|---|---|---|---|
| Initial | 0.100 | 0.100 | 0.100 |
| Change | $-x$ | $-x$ | $+2x$ |
| Equil | $0.100 - x$ | $0.100 - x$ | $0.100 + 2x$ |

5. Substitute the expressions for the equilibrium partial pressures (from step 4) into the expression for the equilibrium constant. Use the given value of the equilibrium constant to solve the expression for the variable $x$.

$$K_p = \frac{(P_{ICl})^2}{P_{I_2}P_{Cl_2}} = \frac{(0.100 + 2x)^2}{(0.100 - x)(0.100 - x)}$$

$$81.9 = \frac{(0.100 + 2x)^2}{(0.100 - x)^2} \quad \text{(perfect square)}$$

$$\sqrt{81.9} = \frac{(0.100 + 2x)}{(0.100 - x)}$$

$$\sqrt{81.9}\,(0.100 - x) = 0.100 + 2x$$

$$\sqrt{81.9}\,(0.100) - \sqrt{81.9}\,x = 0.100 + 2x$$

$$\sqrt{81.9}\,(0.100) - 0.100 = 2x + \sqrt{81.9}\,x$$

$$0.805 = 11.05x$$

$$x = 0.0729$$

6. Substitute $x$ into the expressions for the equilibrium partial pressures of the reactants and products (from step 4) and calculate the partial pressures.

$P_{I_2} = 0.100 - 0.0729 = 0.027$ atm

$P_{Cl_2} = 0.100 - 0.0729 = 0.027$ atm

$P_{ICl} = 0.100 + 2(0.0729) = 0.246$ atm

| 7. Check your answer by substituting the calculated equilibrium partial pressures into the equilibrium expression. The calculated value of $K$ should match the given value of $K$. | $K_p = \dfrac{(P_{ICl})^2}{P_{I_2}P_{Cl_2}} = \dfrac{(0.246)^2}{(0.027)(0.027)} = 83$ <br><br> Since the calculated value of $K_p$ matches the given value (within the uncertainty indicated by the significant figures), the answer is valid. |
|---|---|

**FOR PRACTICE 15.11** The reaction between $I_2$ and $Cl_2$ in Example 15.11 is carried out at the same temperature, but with these initial partial pressures: $P_{I_2} = 0.150$ atm, $P_{Cl_2} = 0.150$ atm, $P_{ICl} = 0.00$ atm. Find the equilibrium partial pressures of all three substances.

## Simplifying Approximations in Working Equilibrium Problems

For some equilibrium problems of the type shown in Examples 15.9, 15.10, and 15.11, we can make an approximation that simplifies the calculations without any significant loss of accuracy. For example, if the equilibrium constant is relatively small, the reaction will not proceed very far to the right. Therefore, if the initial reactant concentration is relatively large, we can make the assumption that $x$ is small relative to the initial concentration of reactant.

To see how this approximation works, consider again the simple reaction A $\rightleftharpoons$ 2 B. Suppose that, as before, we have a reaction mixture in which the initial concentration of A is 1.0 M and the initial concentration of B is 0.0 M and that we want to find the equilibrium concentrations. However, suppose that in this case the equilibrium constant is much smaller, say $K = 3.3 \times 10^{-5}$. The ICE table is identical to the one we set up previously:

|         | [A]       | [B]  |
|---------|-----------|------|
| Initial | 1.0       | 0.0  |
| Change  | $-x$      | $+2x$ |
| Equil   | $1.0 - x$ | $2x$ |

Except for the value of $K$, we end up with the exact quadratic equation that we had before:

$$K = \frac{[B]^2}{[A]} = \frac{(2x)^2}{1.0 - x} = 3.3 \times 10^{-5}$$

or more simply:

$$\frac{4x^2}{1.0 - x} = 3.3 \times 10^{-5}$$

We can rearrange this quadratic equation into standard form and solve it using the quadratic formula. But since $K$ is small, the reaction will not proceed very far toward products and, therefore, $x$ will also be small. If $x$ is much smaller than 1.0, then $(1.0 - x)$ (the quantity in the denominator) can be approximated by (1.0):

$$\frac{4x^2}{(1.0 - \cancel{x})} = 3.3 \times 10^{-5}$$

This approximation greatly simplifies the equation, which we can then solve for $x$ as follows:

$$\frac{4x^2}{1.0} = 3.3 \times 10^{-5}$$

$$4x^2 = 3.3 \times 10^{-5}$$

$$x = \sqrt{\frac{3.3 \times 10^{-5}}{4}} = 0.0029$$

We can check the validity of this approximation by comparing the calculated value of $x$ to the number it was subtracted from. The ratio of $x$ to the number it is subtracted from should be less than 0.05 (or 5%) for the approximation to be valid. In this case, we subtracted $x$ from 1.0, and we calculate the ratio of the value of $x$ to 1.0 as follows:

$$\frac{0.0029}{1.0} \times 100\% = 0.29\%$$

The approximation is therefore valid.

In Examples 15.12 and 15.13, we treat two nearly identical problems—the only difference is the initial concentration of the reactant. In Example 15.12, the initial concentration of the reactant is relatively large, the equilibrium constant is small, and the *x is small* approximation works well. In Example 15.13, however, the initial concentration of the reactant is much smaller, and even though the equilibrium constant is the same, the *x is small* approximation does not work (because the initial concentration is also small). In cases such as this, we have a couple of options to solve the problem. We can either solve the equation exactly (using the quadratic formula, for example), or we can use the *method of successive approximations*, which is introduced in Example 15.13. In this method, we essentially solve for *x* as if it were small, and then substitute the value obtained back into the equation (where *x* was initially neglected) to solve for *x* again. We repeat this process until the calculated value of *x* stops changing with each iteration, an indication that we have arrived at an acceptable value for *x*.

Note that the *x is small* approximation does not imply that *x is zero*. If that were the case, the reactant and product concentrations would not change from their initial values. The *x is small* approximation just means that when *x* is added or subtracted to another number, it does not change that number by very much. For example, we can calculate the value of the difference $1.0 - x$ when $x = 3.0 \times 10^{-4}$:

$$1.0 - x = 1.0 - 3.0 \times 10^{-4} = 0.9997 = 1.0$$

Since the value of 1.0 is known only to two significant figures, subtracting the small *x* does not change the value at all. This situation is similar to weighing yourself on a bathroom scale with and without a penny in your pocket. Unless your scale is unusually precise, removing the penny from your pocket will not change the reading on the scale. This does not imply that the penny is weightless, only that its weight is small when compared to your body weight. You can neglect the weight of the penny in reading your weight with no detectable loss in accuracy.

## Procedure for...

| Finding Equilibrium Concentrations from Initial Concentrations in Cases with a Small Equilibrium Constant | Example 15.12   Finding Equilibrium Concentrations from Initial Concentrations in Cases with a Small Equilibrium Constant | Example 15.13  Finding Equilibrium Concentrations from Initial Concentrations in Cases with a Small Equilibrium Constant |
|---|---|---|
| To solve these types of problems, follow the given procedure. | Consider the reaction for the decomposition of hydrogen disulfide:  $$2\,H_2S(g) \rightleftharpoons 2\,H_2(g) + S_2(g)$$  $$K_c = 1.67 \times 10^{-7} \text{ at } 800\,°C$$  A 0.500-L reaction vessel initially contains 0.0125 mol of $H_2S$ at 800 °C. Find the equilibrium concentrations of $H_2$ and $S_2$. | Consider the reaction for the decomposition of hydrogen disulfide:  $$2\,H_2S(g) \rightleftharpoons 2\,H_2(g) + S_2(g)$$  $$K_c = 1.67 \times 10^{-7} \text{ at } 800\,°C$$  A 0.500-L reaction vessel initially contains $1.25 \times 10^{-4}$ mol of $H_2S$ at 800 °C. Find the equilibrium concentrations of $H_2$ and $S_2$. |
| **1.** Using the balanced equation as a guide, prepare a table showing the known initial concentrations of the reactants and products. (In these examples, you first calculate the concentration of $H_2S$ from the given number of moles and volume.) | $$[H_2S] = \frac{0.0125 \text{ mol}}{0.500 \text{ L}} = 0.0250 \text{ M}$$  $$2\,H_2S(g) \rightleftharpoons 2\,H_2(g) + S_2(g)$$ | $$[H_2S] = \frac{1.25 \times 10^{-4} \text{ mol}}{0.500 \text{ L}}$$ $$= 2.50 \times 10^{-4} \text{ M}$$  $$2\,H_2S(g) \rightleftharpoons 2\,H_2(g) + S_2(g)$$ |

|  | **[H₂S]** | **[H₂]** | **[S₂]** |
|---|---|---|---|
| Initial | 0.0250 | 0.00 | 0.00 |
| Change |  |  |  |
| Equil |  |  |  |

|  | **[H₂S]** | **[H₂]** | **[S₂]** |
|---|---|---|---|
| Initial | 2.50 × 10⁻⁴ | 0.00 | 0.00 |
| Change |  |  |  |
| Equil |  |  |  |

| | | |
|---|---|---|
| 2. Use the initial concentrations to calculate the reaction quotient (Q). Compare Q to K to predict the direction in which the reaction proceeds. | By inspection, $Q = 0$; the reaction proceeds to the right. | By inspection, $Q = 0$; the reaction proceeds to the right. |

3. Represent the change in the concentration of one of the reactants or products with the variable $x$. Define the changes in the concentrations of the other reactants or products with respect to $x$.

$$2\,H_2S(g) \rightleftharpoons 2\,H_2(g) + S_2(g)$$

| | [H₂S] | [H₂] | [S₂] |
|---|---|---|---|
| Initial | 0.0250 | 0.00 | 0.00 |
| Change | $-2x$ | $+2x$ | $+x$ |
| Equil | | | |

$$2\,H_2S(g) \rightleftharpoons 2\,H_2(g) + S_2(g)$$

| | [H₂S] | [H₂] | [S₂] |
|---|---|---|---|
| Initial | $2.50 \times 10^{-4}$ | 0.00 | 0.00 |
| Change | $-2x$ | $+2x$ | $+x$ |
| Equil | | | |

4. Sum each column for each reactant and product to determine the equilibrium concentrations in terms of the initial concentrations and the variable $x$.

$$2\,H_2S(g) \rightleftharpoons 2\,H_2(g) + S_2(g)$$

| | [H₂S] | [H₂] | [S₂] |
|---|---|---|---|
| Initial | 0.0250 | 0.00 | 0.00 |
| Change | $-2x$ | $+2x$ | $+x$ |
| Equil | $0.0250 - 2x$ | $2x$ | $x$ |

$$2\,H_2S(g) \rightleftharpoons 2\,H_2(g) + S_2(g)$$

| | [H₂S] | [H₂] | [S₂] |
|---|---|---|---|
| Initial | $2.50 \times 10^{-4}$ | 0.00 | 0.00 |
| Change | $-2x$ | $+2x$ | $+x$ |
| Equil | $2.50 \times 10^{-4} - 2x$ | $2x$ | $x$ |

5. Substitute the expressions for the equilibrium concentrations (from step 4) into the expression for the equilibrium constant. Use the given value of the equilibrium constant to solve the resulting equation for the variable $x$. In this case, the resulting equation is cubic in $x$. Although cubic equations can be solved, the solutions are not usually simple. However, since the equilibrium constant is small, we know that the reaction does not proceed very far to the right. Therefore, $x$ is a small number and can be dropped from any quantities in which it is added to or subtracted from another number (as long as the number itself is not too small).

$$K_c = \frac{[H_2]^2[S_2]}{[H_2S]^2}$$

$$= \frac{(2x)^2 x}{(0.0250 - 2x)^2}$$

$$1.67 \times 10^{-7} = \frac{4x^3}{(0.0250 - 2x)^2}$$

$x$ is small.

$$1.67 \times 10^{-7} = \frac{4x^3}{(0.0250 - 2\cancel{x})^2}$$

$$1.67 \times 10^{-7} = \frac{4x^3}{6.25 \times 10^{-4}}$$

$$6.25 \times 10^{-4}(1.67 \times 10^{-7}) = 4x^3$$

$$x^3 = \frac{6.25 \times 10^{-4}(1.67 \times 10^{-7})}{4}$$

$$x = 2.97 \times 10^{-4}$$

$$K_c = \frac{[H_2]^2[S_2]}{[H_2S]^2}$$

$$= \frac{(2x)^2 x}{(2.50 \times 10^{-4} - 2x)^2}$$

$$1.67 \times 10^{-7} = \frac{4x^3}{(2.50 \times 10^{-4} - 2x)^2}$$

$x$ is small.

$$1.67 \times 10^{-7} = \frac{4x^3}{(2.50 \times 10^{-4} - 2\cancel{x})^2}$$

$$1.67 \times 10^{-7} = \frac{4x^3}{6.25 \times 10^{-8}}$$

$$6.25 \times 10^{-8}(1.67 \times 10^{-7}) = 4x^3$$

$$x^3 = \frac{6.25 \times 10^{-8}(1.67 \times 10^{-7})}{4}$$

$$x = 1.38 \times 10^{-5}$$

Check whether your approximation is valid by comparing the calculated value of $x$ to the number it was added to or subtracted from. The ratio of the two numbers should be less than 0.05 (or 5%) for the approximation to be valid. If the approximation is not valid, proceed to step 5a.

Checking the *x is small* approximation:

$$\frac{2.97 \times 10^{-4}}{0.0250} \times 100\% = 1.19\%$$

The *x is small* approximation is valid; proceed to step 6.

Checking the *x is small* approximation:

$$\frac{1.38 \times 10^{-5}}{2.50 \times 10^{-4}} \times 100\% = 5.52\%$$

The approximation does not satisfy the <5% rule (although it is close).

—Continued on the next page

—*Continued*

**5a.** If the approximation is not valid, you can either solve the equation exactly (by hand or with your calculator), or use the method of successive approximations. In this example, the method of successive approximations is used.

$$1.67 \times 10^{-7} = \frac{4x^3}{(2.50 \times 10^{-4} - 2x)^2}$$

$$x = 1.38 \times 10^{-5}$$

$$1.67 \times 10^{-7} = \frac{4x^3}{(2.50 \times 10^{-4} - 2.76 \times 10^{-5})^2}$$

$$x = 1.27 \times 10^{-5}$$

Substitute the value obtained for $x$ in step 5 back into the original cubic equation, but only at the exact spot where you assumed $x$ was negligible, and then solve the equation for $x$ again. Continue this procedure until the value of $x$ you obtain from solving the equation is the same as the one that you substituted into the equation.

If we substitute this value of $x$ back into the cubic equation and solve it, we get $x = 1.28 \times 10^{-5}$, which is nearly identical to $1.27 \times 10^{-5}$. Therefore, we have arrived at the best approximation for $x$.

**6.** Substitute $x$ into the expressions for the equilibrium concentrations of the reactants and products (from step 4) and calculate the concentrations.

$$[H_2S] = 0.0250 - 2(2.97 \times 10^{-4})$$
$$= 0.0244 \, M$$
$$[H_2] = 2(2.97 \times 10^{-4})$$
$$= 5.94 \times 10^{-4} \, M$$
$$[S_2] = 2.97 \times 10^{-4} \, M$$

$$[H_2S] = 2.50 \times 10^{-4} - 2(1.28 \times 10^{-5})$$
$$= 2.24 \times 10^{-4} \, M$$
$$[H_2] = 2(1.28 \times 10^{-5})$$
$$= 2.56 \times 10^{-5} \, M$$
$$[S_2] = 1.28 \times 10^{-5} \, M$$

**7.** Check your answer by substituting the calculated equilibrium values into the equilibrium expression. The calculated value of $K$ should match the given value of $K$. Note that the approximation method and rounding errors could cause a difference of up to about 5% when comparing values of the equilibrium constant.

$$K_c = \frac{(5.94 \times 10^{-4})^2 (2.97 \times 10^{-4})}{(0.0244)^2}$$
$$= 1.76 \times 10^{-7}$$

The calculated value of $K$ is close enough to the given value when you consider the uncertainty introduced by the approximation. Therefore the answer is valid.

$$K_c = \frac{(2.56 \times 10^{-5})^2 (1.28 \times 10^{-5})}{(2.24 \times 10^{-4})^2}$$
$$= 1.67 \times 10^{-7}$$

The calculated value of $K$ is equal to the given value. Therefore the answer is valid.

**FOR PRACTICE 15.12**
The reaction in Example 15.12 is carried out at the same temperature with the following initial concentrations: $[H_2S] = 0.100 \, M$, $[H_2] = 0.100 \, M$, and $[S_2] = 0.00 \, M$. Find the equilibrium concentration of $[S_2]$.

**FOR PRACTICE 15.13**
The reaction in Example 15.13 is carried out at the same temperature with the following initial concentrations: $[H_2S] = 1.00 \times 10^{-4} \, M$, $[H_2] = 0.00 \, M$, and $[S_2] = 0.00 \, M$. Find the equilibrium concentration of $[S_2]$.

CONCEPTUAL
**CONNECTION 15.7**

**The *x is small* Approximation**  For the generic reaction, $A(g) \rightleftharpoons B(g)$, consider each value of $K$ and initial concentration of A. For which set of values does the *x is small* approximation most likely apply?

**(a)** $K = 1.0 \times 10^{-5}$; $[A] = 0.250 \, M$      **(b)** $K = 1.0 \times 10^{-2}$; $[A] = 0.250 \, M$

**(c)** $K = 1.0 \times 10^{-5}$; $[A] = 0.00250 \, M$    **(d)** $K = 1.0 \times 10^{-2}$; $[A] = 0.00250 \, M$

# 15.9  Le Châtelier's Principle: How a System at Equilibrium Responds to Disturbances

 **Le Châtelier's Principle**

We have seen that a chemical system not in equilibrium tends to progress toward equilibrium and that the relative concentrations of the reactants and products at equilibrium are characterized by the equilibrium constant, $K$. What happens, however, when a chemical system already at equilibrium is disturbed? **Le Châtelier's principle** states that the chemical system responds to minimize the disturbance.

| Pronounced "Le-sha-te-lyay"

> **Le Châtelier's principle: When a chemical system at equilibrium is disturbed, the system shifts in a direction that minimizes the disturbance.**

In other words, a system at equilibrium tends to maintain that equilibrium—it bounces back when disturbed.

We can understand Le Châtelier's principle by returning to our two neighboring countries analogy. Suppose the populations of Country A and Country B are at equilibrium. This means that the rate of people moving out of Country A and into Country B is equal to the rate of people moving into Country A and out of Country B, and the populations of the two countries are stable:

$$\text{Country A} \rightleftharpoons \text{Country B}$$

Now imagine disturbing the balance (Figure 15.8▼). Suppose there is a notable increase in the birthrate in Country B. What happens? After Country B becomes more crowded, the rate of people leaving Country B increases. The net flow of people is out of Country B and into Country A. Equilibrium is disturbed by the addition of more people to Country B, and people leave Country B in response. In effect, the system responded by shifting in the direction that minimized the disturbance.

On the other hand, what happens if there is a baby boom in Country A? As Country A gets more crowded, the rate of people leaving Country A increases. The net flow of people is out of Country A and into Country B. The number of people in Country A initially increases, and the system responds; people move out of Country A.

The two-country analogy should help you see the effects of disturbing a system in equilibrium—it should not be taken as an exact parallel.

### Le Châtelier's Principle: An Analogy

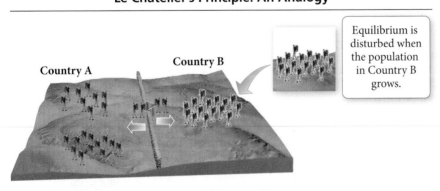

Country A    Country B

Equilibrium is disturbed when the population in Country B grows.

Country A    Country B

System responds to minimize disturbance.

◀ **FIGURE 15.8  A Population Analogy for Le Châtelier's Principle**  A baby boom in Country B shifts the equilibrium to the left. People leave Country B (because it has become too crowded) and migrate to Country A until equilibrium is reestablished.

Chemical systems behave similarly: when their equilibrium is disturbed, they react to counter the disturbance. We can disturb a system in chemical equilibrium in several different ways, including changing the concentration of a reactant or product, changing the volume or pressure, and changing the temperature. We consider each of these separately.

## The Effect of a Concentration Change on Equilibrium

Consider the following reaction in chemical equilibrium:

$$N_2O_4(g) \rightleftharpoons 2\,NO_2(g)$$

Suppose we disturb the equilibrium by adding $NO_2$ to the equilibrium mixture (Figure 15.9▼). In other words, we increase the concentration of $NO_2$, the product. What happens? According to Le Châtelier's principle, the system shifts in a direction to minimize the disturbance. The reaction goes to the left (it proceeds in the reverse direction), consuming some of the added $NO_2$ and thus bringing its concentration back down, as shown graphically in Figure 15.10a▶:

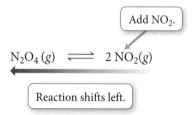

The reaction shifts to the left because the value of $Q$ changes as follows:

- Before addition of $NO_2$: $Q = K$.
- Immediately after addition of $NO_2$: $Q > K$.
- Reaction shifts to the left to reestablish equilibrium.

On the other hand, what happens if we add extra $N_2O_4$ (the reactant), increasing its concentration? In this case, the reaction shifts to the right, consuming some of

**Le Châtelier's Principle: Changing Concentration**

▲ **FIGURE 15.9  Le Châtelier's Principle: The Effect of a Concentration Change**  Adding $NO_2$ causes the reaction to shift left, consuming some of the added $NO_2$ and forming more $N_2O_4$.

## Le Châtelier's Principle: Graphical Representation

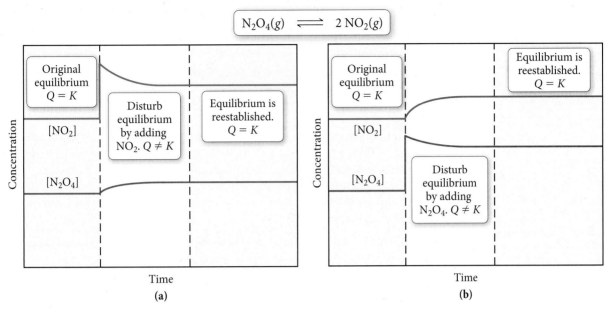

$$N_2O_4(g) \rightleftharpoons 2\ NO_2(g)$$

**(a)**

**(b)**

▲ **FIGURE 15.10  Le Châtelier's Principle: Changing Concentration** The graph shows the concentrations of $NO_2$ and $N_2O_4$ for the reaction $N_2O_4(g) \longrightarrow 2\ NO_2(g)$ as a function of time in three distinct stages of the reaction: initially at equilibrium (left), upon disturbance of the equilibrium by addition of more $NO_2$ **(a)** or $N_2O_4$ **(b)** to the reaction mixture (center), and upon reestablishment of equilibrium (right).

the added $N_2O_4$ and bringing *its* concentration back down, as shown graphically in Figure 15.10b▲:

$$N_2O_4(g) \rightleftharpoons 2\ NO_2(g)$$

The reaction shifts to the right because the value of $Q$ changes as follows:

- Before addition of $N_2O_4$ : $Q = K$.
- Immediately after addition of $N_2O_4$ : $Q < K$.
- Reaction shifts to the right to reestablish equilibrium.

In both of these cases (either adding $NO_2$ or adding $N_2O_4$), the system shifts in a direction that minimizes the disturbance. Lowering the concentration of a reactant (which makes $Q > K$) causes the system to shift in the direction of the reactants to minimize the disturbance. Lowering the concentration of a product (which makes $Q < K$) causes the system to shift in the direction of products.

***Summarizing the Effect of a Concentration Change on Equilibrium:***
If a chemical system is at equilibrium:

▶ *Increasing* the concentration of one or more of the *reactants* (which makes $Q < K$) causes the reaction to *shift to the right* (in the direction of the products).

▶ *Increasing* the concentration of one or more of the *products* (which makes $Q > K$) causes the reaction to *shift to the left* (in the direction of the reactants).

▶ *Decreasing* the concentration of one or more of the *reactants* (which makes $Q > K$) causes the reaction to *shift to the left* (in the direction of the reactants).

▶ *Decreasing* the concentration of one or more of the *products* (which makes $Q < K$) causes the reaction to *shift to the right* (in the direction of the products).

**Example** 15.14 | **The Effect of a Concentration Change on Equilibrium**

Consider the following reaction at equilibrium:

$$CaCO_3(s) \rightleftharpoons CaO(s) + CO_2(g)$$

What is the effect of adding additional $CO_2$ to the reaction mixture? What is the effect of adding additional $CaCO_3$?

**SOLUTION**

Adding additional $CO_2$ increases the concentration of $CO_2$ and causes the reaction to shift to the left. Adding additional $CaCO_3$, however, does *not* increase the concentration of $CaCO_3$ because $CaCO_3$ is a solid and therefore has a constant concentration. Thus, adding additional $CaCO_3$ has no effect on the position of the equilibrium. (Note that, as we saw in Section 15.5, solids are not included in the equilibrium expression.)

**FOR PRACTICE 15.14** Consider the following reaction in chemical equilibrium:

$$2\,BrNO(g) \rightleftharpoons 2\,NO(g) + Br_2(g)$$

What is the effect of adding additional $Br_2$ to the reaction mixture? What is the effect of adding additional BrNO?

## The Effect of a Volume (or Pressure) Change on Equilibrium

In considering the effect of a change in volume, we are assuming that the change in volume is carried out at constant temperature.

How does a system in chemical equilibrium respond to a volume change? Recall from Chapter 5 that changing the volume of a gas (or a gas mixture) results in a change in pressure. Remember also that pressure and volume are inversely related: a *decrease* in volume causes an *increase* in pressure, and an *increase* in volume causes a *decrease* in pressure. So, if the volume of a reaction mixture at chemical equilibrium is changed, the pressure changes and the system shifts in a direction to minimize that change.

For example, consider the following reaction at equilibrium in a cylinder equipped with a moveable piston:

$$N_2(g) + 3\,H_2(g) \rightleftharpoons 2\,NH_3(g)$$

What happens if we push down on the piston, lowering the volume and raising the pressure (Figure 15.11▶)? How can the chemical system respond to bring the pressure back down? Look carefully at the reaction coefficients. If the reaction shifts to the right, 4 mol of gas particles are converted to 2 mol of gas particles. From the ideal gas law ($PV = nRT$), we know that decreasing the number of moles of a gas ($n$) results in a lower pressure ($P$). Therefore, the system shifts to the right, decreasing the number of gas molecules and bringing the pressure back down, minimizing the disturbance.

Consider the same reaction mixture at equilibrium again. What happens if, this time, we pull *up* on the piston, *increasing* the volume (Figure 15.11b▶)? The higher volume results in a lower pressure and the system responds to bring the pressure back up. It does this by shifting to the left, converting 2 mol of gas particles into 4 mol of gas particles, increasing the pressure and minimizing the disturbance.

Consider again the same reaction mixture at equilibrium. What happens if, this time, we keep the volume the same but increase the pressure *by adding an inert gas* to the mixture? Although the overall pressure of the mixture increases, the partial pressures of the reactants and products do not change. Consequently, there is no effect, and the reaction does not shift in either direction.

*Summarizing the Effect of Volume Change on Equilibrium:*

If a chemical system is at equilibrium:

▶ *Decreasing* the volume causes the reaction to shift in the direction that has *the fewer moles of gas particles*.

▶ *Increasing* the volume causes the reaction to shift in the direction that has *the greater number of moles of gas particles*.

▶ If a reaction has an equal number of moles of gas on both sides of the chemical equation, then a change in volume produces no effect on the equilibrium.

▶ Adding an inert gas to the mixture at a fixed volume has no effect on the equilibrium.

## Le Châtelier's Principle: Changing Pressure

◀ **FIGURE 15.11 Le Châtelier's Principle: The Effect of a Pressure Change** **(a)** Decreasing the volume increases the pressure, causing the reaction to shift to the right (fewer moles of gas, lower pressure). **(b)** Increasing the volume reduces the pressure, causing the reaction to shift to the left (more moles of gas, higher pressure).

$N_2(g) + 3 H_2(g) \rightleftharpoons 2 NH_3(g)$

4 mol of gas     2 mol of gas

$N_2(g) + 3 H_2(g) \rightleftharpoons 2 NH_3(g)$

4 mol of gas     2 mol of gas

**Reaction shifts right** (toward side with fewer moles of gas particles).

(a)

**Reaction shifts left** (toward side with more moles of gas particles).

(b)

## Example 15.15    The Effect of a Volume Change on Equilibrium

Consider the following reaction at chemical equilibrium:

$$2 KClO_3(s) \rightleftharpoons 2 KCl(s) + 3 O_2(g)$$

What is the effect of decreasing the volume of the reaction mixture? Increasing the volume of the reaction mixture? Adding an inert gas at constant volume?

**SOLUTION**

The chemical equation has 3 mol of gas on the right and zero moles of gas on the left. Decreasing the volume of the reaction mixture increases the pressure and causes the reaction to shift to the left (toward the side with fewer moles of gas particles). Increasing the volume of the reaction mixture decreases the pressure and causes the reaction to shift to the right (toward the side with more moles of gas particles.) Adding an inert gas has no effect.

**FOR PRACTICE 15.15**    Consider the following reaction at chemical equilibrium:

$$2 SO_2(g) + O_2(g) \rightleftharpoons 2 SO_3(g)$$

What is the effect of decreasing the volume of the reaction mixture? Increasing the volume of the reaction mixture?

## The Effect of a Temperature Change on Equilibrium

In considering the effect of a change in temperature, we are assuming that the heat is added (or removed) at constant pressure.

When a system at equilibrium is disturbed by a change in concentration or a change in volume, the equilibrium shifts to counter the change, but *the equilibrium constant does not change*. In other words, changes in volume or concentration generally change $Q$, not $K$, and the system responds by shifting so that $Q$ becomes equal to $K$. In contrast, a change in temperature changes the actual value of the equilibrium constant. Nonetheless, we can use Le Châtelier's principle to predict the effects of a temperature change.

If we increase the temperature of a reaction mixture at equilibrium, the reaction shifts in the direction that tends to decrease the temperature and vice versa. Recall from Chapter 6 that an exothermic reaction (negative $\Delta H$) emits heat:

$$\text{Exothermic reaction:} \quad A + B \rightleftharpoons C + D + \text{heat}$$

We can think of heat as a product in an exothermic reaction. In an endothermic reaction (positive $\Delta H$), the reaction absorbs heat:

$$\text{Endothermic reaction:} \quad A + B + \text{heat} \rightleftharpoons C + D$$

We can think of heat as a reactant in an endothermic reaction.

At constant pressure, raising the temperature of an *exothermic* reaction—think of this as adding heat—is similar to adding more product, causing the reaction to shift left. For example, the reaction of nitrogen with hydrogen to form ammonia is exothermic:

Add heat

$$N_2(g) + 3\,H_2(g) \rightleftharpoons 2\,NH_3(g) + \text{heat}$$

Reaction shifts left.
Smaller $K$

Raising the temperature of an equilibrium mixture of these three gases causes the reaction to shift left, absorbing some of the added heat and forming less products and more reactants. Note that, unlike adding additional $NH_3$ to the reaction mixture (which does *not* change the value of the equilibrium constant), *changing the temperature does change the value of the equilibrium constant*. The new equilibrium mixture will have more reactants and fewer products and therefore a smaller value of $K$.

Conversely, lowering the temperature causes the reaction to shift right, releasing heat and producing more products because the value of $K$ has increased:

Remove heat

$$N_2(g) + 3\,H_2(g) \rightleftharpoons 2\,NH_3(g) + \text{heat}$$

Reaction shifts right.
Larger $K$

In contrast, for an *endothermic* reaction, raising the temperature (adding heat) causes the reaction to shift right to absorb the added heat. For example, the following reaction is endothermic:

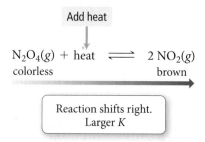

Add heat

$$\underset{\text{colorless}}{N_2O_4(g)} + \text{heat} \rightleftharpoons \underset{\text{brown}}{2\,NO_2(g)}$$

Reaction shifts right.
Larger $K$

## Le Châtelier's Principle: Changing Temperature

$$N_2O_4(g) + \text{heat} \rightleftharpoons 2\,NO_2(g)$$

colorless                    brown

Lower temperature:
$N_2O_4$ favored

Higher temperature:
$NO_2$ favored

▲ FIGURE 15.12  **Le Châtelier's Principle: The Effect of a Temperature Change**
Because the reaction is endothermic, raising the temperature causes a shift to the right,
toward the formation of brown $NO_2$.

Raising the temperature of an equilibrium mixture of these two gases causes the reaction
to shift right, absorbing some of the added heat and producing more products because
the value of $K$ has increased. Because $N_2O_4$ is colorless and $NO_2$ is brown, the effects of
changing the temperature of this reaction are easily seen (Figure 15.12▲).

On the other hand, lowering the temperature (removing heat) of a reaction mixture
of these two gases causes the reaction to shift left, releasing heat, forming less products,
and lowering the value of $K$:

Remove heat

$$N_2O_4(g) + \text{heat} \rightleftharpoons 2\,NO_2(g)$$

colorless                    brown

Reaction shifts left.
Smaller $K$

### *Summarizing the Effect of a Temperature Change on Equilibrium:*
In an *exothermic* chemical reaction, heat is a product.

▶ *Increasing* the temperature causes an exothermic reaction to *shift left* (in the
   direction of the reactants); the value of the equilibrium constant decreases.

▶ *Decreasing* the temperature causes an exothermic reaction to *shift right* (in the
   direction of the products); the value of the equilibrium constant increases.

In an *endothermic* chemical reaction, heat is a reactant.

▶ *Increasing* the temperature causes an endothermic reaction to *shift right* (in the
   direction of the products); the equilibrium constant increases.

▶ *Decreasing* the temperature causes an endothermic reaction to *shift left* (in the
   direction of the reactants); the equilibrium constant decreases.

Adding heat favors the endothermic
direction. Removing heat favors the
exothermic direction.

**Example** 15.16 | **The Effect of a Temperature Change on Equilibrium**

The following reaction is endothermic:

$$CaCO_3(s) \rightleftharpoons CaO(s) + CO_2(g)$$

What is the effect of increasing the temperature of the reaction mixture? Decreasing the temperature?

**SOLUTION**

Since the reaction is endothermic, we can think of heat as a reactant:

$$Heat + CaCO_3(s) \rightleftharpoons CaO(s) + CO_2(g)$$

Raising the temperature is like adding a reactant, causing the reaction to shift to the right. Lowering the temperature is equivalent to removing a reactant, causing the reaction to shift to the left.

**FOR PRACTICE 15.16** The following reaction is exothermic:

$$2\,SO_2(g) + O_2(g) \rightleftharpoons 2\,SO_3(g)$$

What is the effect of increasing the temperature of the reaction mixture? Decreasing the temperature?

# Self-Assessment Quiz

**Q1.** What is the correct expression for the equilibrium constant ($K_c$) for the reaction between carbon and hydrogen gas to form methane shown here?

$$C(s) + 2\,H_2(g) \rightleftharpoons CH_4(g)$$

a) $K_c = \dfrac{[CH_4]}{[H_2]}$   b) $K_c = \dfrac{[CH_4]}{[C][H_2]}$

c) $K_c = \dfrac{[CH_4]}{[C][H_2]^2}$   d) $K_c = \dfrac{[CH_4]}{[H_2]^2}$

**Q2.** The equilibrium constant for the reaction shown here is $K_c = 1.0 \times 10^3$. A reaction mixture at equilibrium contains $[A] = 1.0 \times 10^{-3}$ M. What is the concentration of B in the mixture?

$$A(g) \rightleftharpoons B(g)$$

a) $1.0 \times 10^{-3}$ M   b) 1.0 M
c) 2.0 M   d) $1.0 \times 10^3$ M

**Q3.** Use the data shown here to find the equilibrium constant ($K_c$) for the reaction $A(g) \rightleftharpoons 2\,B(g) + C(g)$.

$$A(g) \rightleftharpoons 2\,X(g) + C(g) \quad K_c = 1.55$$
$$B(g) \rightleftharpoons X(g) \quad K_c = 25.2$$

a) 984   b) 26.8   c) $6.10 \times 10^{-4}$   d) $2.44 \times 10^{-3}$

**Q4.** The reaction shown here has a $K_p = 4.5 \times 10^2$ at 825 K. Find $K_c$ for the reaction at this temperature.

$$CH_4(g) + CO_2(g) \rightleftharpoons 2\,CO(g) + 2\,H_2(g)$$

a) 0.098   b) $2.1 \times 10^6$   c) 6.6   d) $4.5 \times 10^{-2}$

**Q5.** Consider the reaction between NO and $Cl_2$ to form NOCl:

$$2\,NO(g) + Cl_2(g) \rightleftharpoons 2\,NOCl(g)$$

A reaction mixture at a certain temperature initially contains only $[NO] = 0.50$ M and $[Cl_2] = 0.50$ M.

After the reaction comes to equilibrium, the concentration of NOCl is 0.30 M. Find the value of the equilibrium constant ($K_c$) at this temperature.

a) 11   b) 4.3
c) 6.4   d) 0.22

**Q6.** For the reaction $2\,A(g) \rightleftharpoons B(g)$, the equilibrium constant is $K_p = 0.76$. A reaction mixture initially contains 2.0 atm of each gas ($P_A = 2.0$ atm and $P_B = 2.0$ atm). Which statement is true of the reaction mixture?

a) The reaction mixture is at equilibrium.
b) The reaction mixture will proceed toward products.
c) The reaction mixture will proceed toward reactants.
d) It is not possible to determine from the information given the future direction of the reaction mixture.

**Q7.** Consider the reaction between iodine gas and chlorine gas to form iodine monochloride:

$$I_2(g) + Cl_2(g) \rightleftharpoons 2\,ICl(g) \quad K_p = 81.9 \text{ (at 298 K)}$$

A reaction mixture at 298 K initially contains $P_{I_2} = 0.25$ atm and $P_{Cl_2} = 0.25$ atm. What is the partial pressure of iodine monochloride when the reaction reaches equilibrium?

a) 0.17 atm   b) 0.64 atm
c) 0.41 atm   d) 2.3 atm

**Q8.** Consider the reaction of A to form B:

$$2\,A(g) \rightleftharpoons B(g) \quad K_c = 1.8 \times 10^{-5} \text{ (at 298 K)}$$

A reaction mixture at 298 K initially contains $[A] = 0.50$ M. What is the concentration of B when the reaction reaches equilibrium?

a) $9.0 \times 10^{-6}$ M   b) 0.060 M
c) 0.030 M   d) $4.5 \times 10^{-6}$ M

**Q9.** The decomposition of $NH_4HS$ is endothermic:

$$NH_4HS(s) \rightleftharpoons NH_3(g) + H_2S(g)$$

Which change to an equilibrium mixture of this reaction results in the formation of more $H_2S$?

a) a decrease in the volume of the reaction vessel (at constant temperature)

b) an increase in the amount of $NH_4HS$ in the reaction vessel

c) an increase in temperature

d) all of the above

**Q10.** The solid XY decomposes into gaseous X and Y:

$$XY(s) \rightleftharpoons X(g) + Y(g) \quad K_p = 4.1 \text{ (at 0 °C)}$$

If the reaction is carried out in a 22.4-L container, which initial amounts of X and Y result in the formation of solid XY?

a) 5 mol X; 0.5 mol Y

b) 2.0 mol X; 2.0 mol Y

c) 1 mol X; 1 mol Y

d) none of the above

**Q11.** What is the effect of adding helium gas (at constant volume) to an equilibrium mixture of the reaction:

$$CO(g) + Cl_2(g) \rightleftharpoons COCl_2(g)$$

a) The reaction shifts toward the products.

b) The reaction shifts toward the reactants.

c) The reaction does not shift in either direction.

d) The reaction slows down.

**Q12.** The reaction $X_2(g) \rightleftharpoons 2 X(g)$ occurs in a closed reaction vessel at constant volume and temperature. Initially, the vessel contains only $X_2$ at a pressure of 1.55 atm. After the reaction reaches equilibrium, the total pressure is 2.85 atm. What is the value of the equilibrium constant, $K_p$, for the reaction?

a) 27    b) 10    c) 5.2    d) 32

**Answers: 1.** (d) **2.** (b) **3.** (d) **4.** (a) **5.** (c) **6.** (b) **7.** (c) **8.** (d) **9.** (c) **10.** (d) **11.** (c) **12.** (a)

# Chapter 15 IN REVIEW

## Key Terms

**Section 15.2**
reversible (677)
dynamic equilibrium (677)

**Section 15.3**
equilibrium constant (K) (680)
law of mass action (680)

**Section 15.7**
reaction quotient ($Q_c$) (691)

**Section 15.9**
Le Châtelier's principle (703)

## Key Concepts

### The Equilibrium Constant (15.1)

▶ The relative concentrations of the reactants and the products at equilibrium are expressed by the equilibrium constant, $K$.

▶ The equilibrium constant measures how far a reaction proceeds toward products: a large $K$ (greater than 1) indicates a high concentration of products at equilibrium, and a small $K$ (less than 1) indicates a low concentration of products at equilibrium.

### Dynamic Equilibrium (15.2)

▶ Most chemical reactions are reversible; they can proceed in either the forward or the reverse direction.

▶ When a chemical reaction is in dynamic equilibrium, the rate of the forward reaction equals the rate of the reverse reaction, so the net concentrations of reactants and products do not change. However, this does *not* imply that the concentrations of the reactants and the products are equal at equilibrium.

### The Equilibrium Constant Expression (15.3)

▶ The equilibrium constant expression is given by the law of mass action and is equal to the concentrations of the products, raised to their stoichiometric coefficients, divided by the concentrations of the reactants, raised to their stoichiometric coefficients.

▶ When the equation for a chemical reaction is reversed, multiplied, or added to another equation, $K$ must be modified accordingly.

### The Equilibrium Constant, K (15.4)

▶ The equilibrium constant can be expressed in terms of concentrations ($K_c$) or in terms of partial pressures ($K_p$). These two constants are related by Equation 15.2. Concentration must always be expressed in units of molarity for $K_c$. Partial pressures must always be expressed in units of atmospheres for $K_p$.

### States of Matter and the Equilibrium Constant (15.5)

▶ The equilibrium constant expression contains only partial pressures or concentrations of reactants and products that exist as gases or solutes dissolved in solution. Pure liquids and solids are not included in the expression for the equilibrium constant.

### Calculating K (15.6)

▶ We can calculate the equilibrium constant from equilibrium concentrations or partial pressures by substituting measured values into the expression for the equilibrium constant (as obtained from the law of mass action).

▶ In most cases, we can calculate the equilibrium concentrations of the reactants and products—and therefore the value of the equilibrium constant—from the initial concentrations of the reactants and products and the equilibrium concentration of *just one* reactant or product.

## The Reaction Quotient, Q (15.7)

▶ The reaction quotient, Q, is the ratio of the concentrations (or partial pressures) of products raised to their stoichiometric coefficients to the concentrations of reactants raised to their stoichiometric coefficients *at any point in the reaction.*

▶ Like K, Q can be expressed in terms of concentrations ($Q_c$) or partial pressures ($Q_p$).

▶ At equilibrium, Q is equal to K; therefore, we can determine the direction in which a reaction proceeds by comparing Q to K. If Q < K, the reaction moves in the direction of the products; if Q > K, the reaction moves in the reverse direction.

## Finding Equilibrium Concentrations (15.8)

▶ There are two general types of problems in which K is given and we can determine one (or more) equilibrium concentrations:

(1) We are given K and all but one equilibrium concentration.
(2) We are given K and *only* initial concentrations.

▶ We solve the first type by rearranging the law of mass action and substituting the given values.

▶ We solve the second type by creating an ICE table and using a variable x to represent the change in concentration.

## Le Châtelier's Principle (15.9)

▶ When a system at equilibrium is disturbed—by a change in the amount of a reactant or product, a change in volume, or a change in temperature—the system shifts in the direction that minimizes the disturbance.

# Key Equations and Relationships

Expression for the Equilibrium Constant, $K_c$ (**15.3**)

$$aA + bB \rightleftharpoons cC + dD$$

$$K = \frac{[C]^c[D]^d}{[A]^a[B]^b} \quad \text{(equilibrium concentrations only)}$$

Relationship between the Equilibrium Constant and the Chemical Equation (**15.3**)

1. If you reverse the equation, invert the equilibrium constant.
2. If you multiply the coefficients in the equation by a factor, raise the equilibrium constant to the same factor.
3. If you add two or more individual chemical equations to obtain an overall equation, multiply the corresponding equilibrium constants by each other to obtain the overall equilibrium constant.

Expression for the Equilibrium Constant, $K_p$ (**15.4**)

$$aA + bB \rightleftharpoons cC + dD$$

$$K_p = \frac{P_C^c P_D^d}{P_A^a P_B^b} \quad \text{(equilibrium partial pressures only)}$$

Relationship between the Equilibrium Constants, $K_c$ and $K_p$ (**15.4**)

$$K_p = K_c(RT)^{\Delta n}$$

The Reaction Quotient, $Q_c$ (**15.7**)

$$aA + bB \rightleftharpoons cC + dD$$

$$Q_c = \frac{[C]^c[D]^d}{[A]^a[B]^b} \quad \text{(concentrations at any point in the reaction)}$$

The Reaction Quotient, $Q_p$ (**15.7**)

$$aA + bB \rightleftharpoons cC + dD$$

$$Q_P = \frac{P_C^c P_D^d}{P_A^a P_B^b} \quad \text{(partial pressures at any point in the reaction)}$$

Relationship of Q to the Direction of the Reaction (**15.7**)

$$Q < K \quad \text{Reaction goes to the right.}$$

$$Q > K \quad \text{Reaction goes to the left.}$$

$$Q = K \quad \text{Reaction is at equilibrium.}$$

# Key Learning Outcomes

| CHAPTER OBJECTIVES | | ASSESSMENT |
|---|---|---|
| Expressing Equilibrium Constants for Chemical Equations (**15.3**) | **Law of Mass Action** $$K = \frac{\overbrace{[C]^c[D]^d}^{\text{Products}}}{\underbrace{[A]^a[B]^b}_{\text{Reactants}}}$$ | Example 15.1   For Practice 15.1   Exercises 21, 22 |
| Manipulating the Equilibrium Constant to Reflect Changes in the Chemical Equation (**15.3**) | | Example 15.2   For Practice 15.2   For More Practice 15.2 Exercises 27–30 |
| Relating $K_p$ and $K_c$ (**15.4**) | | Example 15.3   For Practice 15.3   Exercises 31, 32 |
| Writing Equilibrium constant Expressions for Reactions Involving a Solid or a Liquid (**15.5**) | | Example 15.4   For Practice 15.4   Exercises 33, 34 |

Add heat

$$N_2(g) + 3\,H_2(g) \rightleftharpoons 2\,NH_3(g) + \text{heat}$$

Reaction shifts left.
Smaller $K$

# EXERCISES

## Review Questions

**1.** How does a developing fetus get oxygen in the womb?

**2.** What is dynamic equilibrium? Why is it called *dynamic*?

**3.** Give the general expression for the equilibrium constant of the following generic reaction:
$$aA + bB \rightleftharpoons cC + dD$$

**4.** What is the significance of the equilibrium constant? What does a large equilibrium constant tell us about a reaction? A small one?

**5.** What happens to the value of the equilibrium constant for a reaction if the reaction equation is reversed? Multiplied by a constant?

**6.** If two reactions sum to an overall reaction, and the equilibrium constants for the two reactions are $K_1$ and $K_2$, what is the equilibrium constant for the overall reaction?

**7.** Explain the difference between $K_c$ and $K_p$. For a given reaction, how are the two constants related?

**8.** What units should be used when expressing concentrations or partial pressures in the equilibrium constant? What are the units of $K_p$ and $K_c$? Explain.

**9.** Why are the concentrations of solids and liquids omitted from equilibrium expressions?

**10.** Does the value of the equilibrium constant depend on the initial concentrations of the reactants and products? Do the equilibrium concentrations of the reactants and products depend on their initial concentrations? Explain.

**11.** Explain how you might deduce the equilibrium constant for a reaction in which you know the initial concentrations of the reactants and products and the equilibrium concentration of only one reactant or product.

**12.** What is the definition of the reaction quotient ($Q$) for a reaction? What does $Q$ measure?

**13.** What is the value of $Q$ when each reactant and product is in its standard state? (See Section 6.9 for the definition of standard states.)

**14.** In what direction will a reaction proceed for each condition: (a) $Q < K$; (b) $Q > K$; and (c) $Q = K$?

**15.** Many equilibrium calculations involve finding the equilibrium concentrations of reactants and products given their initial concentrations and the equilibrium constant. Outline the general procedure used in solving these kinds of problems.

**16.** In equilibrium problems involving equilibrium constants that are small relative to the initial concentrations of reactants, we can often assume that the quantity $x$ (which represents how far the reaction proceeds toward products) is small. When this assumption is made, we can ignore the quantity $x$ when it is subtracted from a large number but not when it is multiplied by a large number. In other words, $2.5 - x \approx 2.5$, but $2.5x \neq 2.5$.

Explain why we can ignore a small $x$ in the first case, but not in the second.

**17.** What happens to a chemical system at equilibrium when that equilibrium is disturbed?

**18.** What is the effect of a change in concentration of a reactant or product on a chemical reaction initially at equilibrium?

**19.** What is the effect of a change in volume on a chemical reaction (that includes gaseous reactants or products) initially at equilibrium?

**20.** What is the effect of a temperature change on a chemical reaction initially at equilibrium? How does the effect differ for an exothermic reaction compared to an endothermic one?

# Problems by Topic

## Equilibrium and the Equilibrium Constant Expression

**21.** Write an expression for the equilibrium constant of each chemical equation.
**a.** $SbCl_5(g) \rightleftharpoons SbCl_3(g) + Cl_2(g)$
**b.** $2 BrNO(g) \rightleftharpoons 2 NO(g) + Br_2(g)$
**c.** $CH_4(g) + 2 H_2S(g) \rightleftharpoons CS_2(g) + 4 H_2(g)$
**d.** $2 CO(g) + O_2(g) \rightleftharpoons 2 CO_2(g)$

**22.** Find and fix each mistake in the equilibrium constant expressions.
**a.** $2 H_2S(g) \rightleftharpoons 2 H_2(g) + S_2(g)$   $K = \dfrac{[H_2][S_2]}{[H_2S]}$

**b.** $CO(g) + Cl_2(g) \rightleftharpoons COCl_2(g)$   $K = \dfrac{[CO][Cl_2]}{[COCl_2]}$

**23.** When this reaction comes to equilibrium, will the concentrations of the reactants or products be greater? Does the answer to this question depend on the initial concentrations of the reactants and products?

$$A(g) + B(g) \rightleftharpoons 2 C(g) \quad K_c = 1.4 \times 10^{-5}$$

**24.** Ethene ($C_2H_4$) can be halogenated by this reaction:

$$C_2H_4(g) + X_2(g) \rightleftharpoons C_2H_4X_2(g)$$

where $X_2$ can be $Cl_2$ (green), $Br_2$ (brown), or $I_2$ (purple). Examine the three figures representing equilibrium concentrations in this reaction at the same temperature for the three different halogens. Rank the equilibrium constants for the three reactions from largest to smallest.

a.     b.     c.

**25.** $H_2$ and $I_2$ are combined in a flask and allowed to react according to the reaction:
$$H_2(g) + I_2(g) \rightleftharpoons 2 HI(g)$$
Examine the figures (sequential in time) and answer the questions:
**a.** Which figure represents the point at which equilibrium is reached?
**b.** How would the series of figures change in the presence of a catalyst?
**c.** Would there be different amounts of reactants and products in the final figure (vi) in the presence of a catalyst?

(i)     (ii)

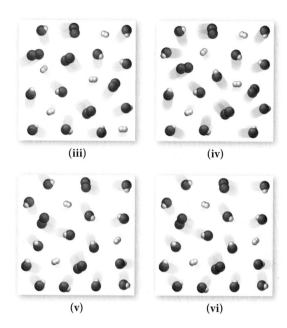

(iii)     (iv)

(v)     (vi)

**26.** A chemist trying to synthesize a particular compound attempts two different synthesis reactions. The equilibrium constants for the two reactions are 23.3 and $2.2 \times 10^4$ at room temperature. However, upon carrying out both reactions for 15 minutes, the chemist finds that the reaction with the smaller equilibrium constant produces more of the desired product. Explain how this might be possible.

**27.** This reaction has an equilibrium constant of $K_p = 2.26 \times 10^4$ at 298 K.

$$CO(g) + 2 H_2(g) \rightleftharpoons CH_3OH(g)$$

Calculate $K_p$ for each reaction and predict whether reactants or products will be favored at equilibrium.
**a.** $CH_3OH(g) \rightleftharpoons CO(g) + 2 H_2(g)$
**b.** $\frac{1}{2} CO(g) + H_2(g) \rightleftharpoons \frac{1}{2} CH_3OH(g)$
**c.** $2 CH_3OH(g) \rightleftharpoons 2 CO(g) + 4 H_2(g)$

**28.** This reaction has an equilibrium constant of $K_p = 2.2 \times 10^6$ at 298 K.

$$2 COF_2(g) \rightleftharpoons CO_2(g) + CF_4(g)$$

Calculate $K_p$ for each reaction and predict whether reactants or products will be favored at equilibrium.
**a.** $COF_2(g) \rightleftharpoons \frac{1}{2} CO_2(g) + \frac{1}{2} CF_4(g)$
**b.** $6 COF_2(g) \rightleftharpoons 3 CO_2(g) + 3 CF_4(g)$
**c.** $2 CO_2(g) + 2 CF_4(g) \rightleftharpoons 4 COF_2(g)$

**29.** Consider the reactions and their respective equilibrium constants:

$NO(g) + \frac{1}{2} Br_2(g) \rightleftharpoons NOBr(g)$        $K_p = 5.3$
$2 NO(g) \rightleftharpoons N_2(g) + O_2(g)$        $K_p = 2.1 \times 10^{30}$

Use these reactions and their equilibrium constants to predict the equilibrium constant for the following reaction:

$$N_2(g) + O_2(g) + Br_2(g) \rightleftharpoons 2 NOBr(g)$$

**30.** Use the reactions and their equilibrium constants to predict the equilibrium constant for the reaction $2 A(s) \rightleftharpoons 3 D(g)$.
$A(s) \rightleftharpoons \frac{1}{2} B(g) + C(g)$        $K_1 = 0.0334$
$3 D(g) \rightleftharpoons B(g) + 2 C(g)$        $K_2 = 2.35$

## $K_p$, $K_c$, and Heterogeneous Equilibria

**31.** Calculate $K_c$ for each reaction.
  **a.** $I_2(g) \rightleftharpoons 2\,I(g)$    $K_p = 6.26 \times 10^{-22}$ (at 298 K)
  **b.** $CH_4(g) + H_2O(g) \rightleftharpoons CO(g) + 3\,H_2(g)$
                                        $K_p = 7.7 \times 10^{24}$ (at 298 K)
  **c.** $I_2(g) + Cl_2(g) \rightleftharpoons 2\,ICl(g)$      $K_p = 81.9$ (at 298 K)

**32.** Calculate $K_p$ for each reaction.
  **a.** $N_2O_4(g) \rightleftharpoons 2\,NO_2(g)$        $K_c = 5.9 \times 10^{-3}$ (at 298 K)
  **b.** $N_2(g) + 3\,H_2(g) \rightleftharpoons 2\,NH_3(g)$    $K_c = 3.7 \times 10^{8}$ (at 298 K)
  **c.** $N_2(g) + O_2(g) \rightleftharpoons 2\,NO(g)$    $K_c = 4.10 \times 10^{-31}$ (at 298 K)

**33.** Write an equilibrium expression for each chemical equation involving one or more solid or liquid reactants or products.
  **a.** $CO_3^{2-}(aq) + H_2O(l) \rightleftharpoons HCO_3^{-}(aq) + OH^{-}(aq)$
  **b.** $2\,KClO_3(s) \rightleftharpoons 2\,KCl(s) + 3\,O_2(g)$
  **c.** $HF(aq) + H_2O(l) \rightleftharpoons H_3O^{+}(aq) + F^{-}(aq)$
  **d.** $NH_3(aq) + H_2O(l) \rightleftharpoons NH_4^{+}(aq) + OH^{-}(aq)$

**34.** Find and fix the mistake in the equilibrium expression.

$$PCl_5(g) \rightleftharpoons PCl_3(l) + Cl_2(g) \qquad K_c = \frac{[PCl_3][Cl_2]}{[PCl_5]}$$

## Relating the Equilibrium Constant to Equilibrium Concentrations and Equilibrium Partial Pressures

**35.** Consider the reaction:

$$CO(g) + 2\,H_2(g) \rightleftharpoons CH_3OH(g)$$

An equilibrium mixture of this reaction at a certain temperature has $[CO] = 0.105$ M, $[H_2] = 0.114$ M, and $[CH_3OH] = 0.185$ M. What is the value of the equilibrium constant ($K_c$) at this temperature?

**36.** Consider the reaction:

$$NH_4HS(s) \rightleftharpoons NH_3(g) + H_2S(g)$$

An equilibrium mixture of this reaction at a certain temperature has $[NH_3] = 0.278$ M and $[H_2S] = 0.355$ M. What is the value of the equilibrium constant ($K_c$) at this temperature?

**37.** Consider the reaction:

$$N_2(g) + 3\,H_2(g) \rightleftharpoons 2\,NH_3(g)$$

Complete the table. Assume that all concentrations are equilibrium concentrations in M.

| T (K) | [N₂] | [H₂] | [NH₃] | Kc |
|---|---|---|---|---|
| 500 | 0.115 | 0.105 | 0.439 | _____ |
| 575 | 0.110 | _____ | 0.128 | 9.6 |
| 775 | 0.120 | 0.140 | _____ | 0.0584 |

**38.** Consider the following reaction:

$$H_2(g) + I_2(g) \rightleftharpoons 2\,HI(g)$$

Complete the table. Assume that all concentrations are equilibrium concentrations in M.

| T (°C) | [H₂] | [I₂] | [HI] | Kc |
|---|---|---|---|---|
| 25 | 0.0355 | 0.0388 | 0.922 | _____ |
| 340 | _____ | 0.0455 | 0.387 | 9.6 |
| 445 | 0.0485 | 0.0468 | _____ | 50.2 |

**39.** Consider the reaction:

$$2\,NO(g) + Br_2(g) \rightleftharpoons 2\,NOBr(g) \qquad K_p = 28.4 \text{ at } 298 \text{ K}$$

In a reaction mixture at equilibrium, the partial pressure of NO is 108 torr and that of $Br_2$ is 126 torr. What is the partial pressure of NOBr in this mixture?

**40.** Consider the reaction:

$$SO_2Cl_2(g) \rightleftharpoons SO_2(g) + Cl_2(g) \quad K_p = 2.91 \times 10^{3} \text{ at } 298 \text{ K}$$

In a reaction at equilibrium, the partial pressure of $SO_2$ is 137 torr and that of $Cl_2$ is 285 torr. What is the partial pressure of $SO_2Cl_2$ in this mixture?

**41.** For the reaction $A(g) \rightleftharpoons 2\,B(g)$, a reaction vessel initially contains only A at a pressure of $P_A = 1.32$ atm. At equilibrium, $P_A = 0.25$ atm. Calculate the value of $K_p$. (Assume no changes in volume or temperature.)

**42.** For the reaction $2\,A(g) \rightleftharpoons B(g) + 2\,C(g)$, a reaction vessel initially contains only A at a pressure of $P_A = 255$ mmHg. At equilibrium, $P_A = 55$ mmHg. Calculate the value of $K_p$. (Assume no changes in volume or temperature.)

**43.** Consider the reaction:

$$Fe^{3+}(aq) + SCN^{-}(aq) \rightleftharpoons FeSCN^{2+}(aq)$$

A solution is made containing an initial $[Fe^{3+}]$ of $1.0 \times 10^{-3}$ M and an initial $[SCN^{-}]$ of $8.0 \times 10^{-4}$ M. At equilibrium, $[FeSCN^{2+}] = 1.7 \times 10^{-4}$ M. Calculate the value of the equilibrium constant ($K_c$).

**44.** Consider the reaction:

$$SO_2Cl_2(g) \rightleftharpoons SO_2(g) + Cl_2(g)$$

A reaction mixture is made containing an initial $[SO_2Cl_2]$ of 0.020 M. At equilibrium, $[Cl_2] = 1.2 \times 10^{-2}$ M. Calculate the value of the equilibrium constant ($K_c$).

**45.** Consider the reaction:

$$H_2(g) + I_2(g) \rightleftharpoons 2\,HI(g)$$

A reaction mixture in a 3.67-L flask at a certain temperature initially contains 0.763 g $H_2$ and 96.9 g $I_2$. At equilibrium, the flask contains 90.4 g HI. Calculate the equilibrium constant ($K_c$) for the reaction at this temperature.

**46.** Consider the reaction:

$$CO(g) + 2\,H_2(g) \rightleftharpoons CH_3OH(g)$$

A reaction mixture in a 5.19-L flask at a certain temperature contains 26.9 g CO and 2.34 g $H_2$. At equilibrium, the flask contains 8.65 g $CH_3OH$. Calculate the equilibrium constant ($K_c$) for the reaction at this temperature.

## The Reaction Quotient and Reaction Direction

**47.** Consider the reaction:

$$NH_4HS(s) \rightleftharpoons NH_3(g) + H_2S(g)$$

At a certain temperature, $K_c = 8.5 \times 10^{-3}$. A reaction mixture at this temperature containing solid $NH_4HS$ has $[NH_3] = 0.166$ M and $[H_2S] = 0.166$ M. Will more of the solid form or will some of the existing solid decompose as equilibrium is reached?

**48.** Consider the reaction:

$$2\,H_2S(g) \rightleftharpoons 2\,H_2(g) + S_2(g) \quad K_p = 2.4 \times 10^{-4} \text{ at } 1073 \text{ K}$$

A reaction mixture contains 0.112 atm of $H_2$, 0.055 atm of $S_2$, and 0.445 atm of $H_2S$. Is the reaction mixture at equilibrium? If not, in what direction will the reaction proceed?

**49.** Silver sulfate dissolves in water according to the reaction:

$$Ag_2SO_4(s) \rightleftharpoons 2\,Ag^{+}(aq) + SO_4^{2-}(aq)$$
$$K_c = 1.1 \times 10^{-5} \text{ at } 298 \text{ K}$$

A 1.5-L solution contains 6.55 g of dissolved silver sulfate. If additional solid silver sulfate is added to the solution, will it dissolve?

**50.** Nitrogen dioxide dimerizes according to the reaction:

$$2\,NO_2(g) \rightleftharpoons N_2O_4(g) \qquad K_p = 6.7 \text{ at } 298 \text{ K}$$

A 2.25-L container contains 0.055 mol of $NO_2$ and 0.082 mol of $N_2O_4$ at 298 K. Is the reaction at equilibrium? If not, in what direction will the reaction proceed?

## Finding Equilibrium Concentrations from Initial Concentrations and the Equilibrium Constant

51. Consider the reaction and the associated equilibrium constant:

$$a A(g) \rightleftharpoons b B(g) \qquad K_c = 4.0$$

Find the equilibrium concentrations of A and B for each value of $a$ and $b$. Assume that the initial concentration of A in each case is 1.0 M and that no B is present at the beginning of the reaction.
**a.** $a = 1; b = 1$   **b.** $a = 2; b = 2$   **c.** $a = 1; b = 2$

52. Consider the reaction and the associated equilibrium constant:

$$a A(g) + b B(g) \rightleftharpoons c C(g) \qquad K_c = 5.0$$

Find the equilibrium concentrations of A, B, and C for each value of $a$, $b$, and $c$. Assume that the initial concentrations of A and B are each 1.0 M and that no product is present at the beginning of the reaction.
**a.** $a = 1; b = 1; c = 2$
**b.** $a = 1; b = 1; c = 1$
**c.** $a = 2; b = 1; c = 1$ (set up equation for $x$; don't solve)

53. For the reaction shown here, $K_c = 0.513$ at 500 K.

$$N_2O_4(g) \rightleftharpoons 2\,NO_2(g)$$

If a reaction vessel initially contains an $N_2O_4$ concentration of 0.0500 M at 500 K, what are the equilibrium concentrations of $N_2O_4$ and $NO_2$ at 500 K?

54. For the reaction shown here, $K_c = 255$ at 1000 K.

$$CO(g) + Cl_2(g) \rightleftharpoons COCl_2(g)$$

If a reaction mixture initially contains a CO concentration of 0.1500 M and a $Cl_2$ concentration of 0.175 M at 1000 K, what are the equilibrium concentrations of CO, $Cl_2$, and $COCl_2$ at 1000 K?

55. Consider the reaction:

$$NiO(s) + CO(g) \rightleftharpoons Ni(s) + CO_2(g)$$
$$K_c = 4.0 \times 10^3 \text{ at } 1500 \text{ K}$$

If a mixture of solid nickel(II) oxide and 0.20 M carbon monoxide comes to equilibrium at 1500 K, what is the equilibrium concentration of $CO_2$?

56. Consider the reaction:

$$CO(g) + H_2O(g) \rightleftharpoons CO_2(g) + H_2(g) \quad K_c = 102 \text{ at } 500 \text{ K}$$

If a reaction mixture initially contains 0.110 M CO and 0.110 M $H_2O$, what will the equilibrium concentration of each of the reactants and products be?

57. Consider the reaction:

$$HC_2H_3O_2(aq) + H_2O(l) \rightleftharpoons H_3O^+(aq) + C_2H_3O_2^-(aq)$$
$$K_c = 1.8 \times 10^{-5} \text{ at } 25\,^\circ\text{C}$$

If a solution initially contains 0.210 M $HC_2H_3O_2$, what is the equilibrium concentration of $H_3O^+$ at 25 °C?

58. Consider the reaction:

$$SO_2Cl_2(g) \rightleftharpoons SO_2(g) + Cl_2(g)$$
$$K_c = 2.99 \times 10^{-7} \text{ at } 227\,^\circ\text{C}$$

If a reaction mixture initially contains 0.175 M $SO_2Cl_2$, what is the equilibrium concentration of $Cl_2$ at 227 °C?

59. Consider the reaction:

$$Br_2(g) + Cl_2(g) \rightleftharpoons 2\,BrCl(g) \quad K_p = 1.11 \times 10^{-4} \text{ at } 150 \text{ K}$$

A reaction mixture initially contains a $Br_2$ partial pressure of 755 torr and a $Cl_2$ partial pressure of 735 torr at 150 K. Calculate the equilibrium partial pressure of BrCl.

60. Consider the reaction:

$$CO(g) + H_2O(g) \rightleftharpoons CO_2(g) + H_2(g)$$
$$K_p = 0.0611 \text{ at } 2000 \text{ K}$$

A reaction mixture initially contains a CO partial pressure of 1344 torr and a $H_2O$ partial pressure of 1766 torr at 2000 K. Calculate the equilibrium partial pressures of each of the products.

61. Consider the reaction:

$$A(g) \rightleftharpoons B(g) + C(g)$$

Find the equilibrium concentrations of A, B, and C for each value of $K_c$. Assume that the initial concentration of A in each case is 1.0 M and that the reaction mixture initially contains no products. Make any appropriate simplifying assumptions.
**a.** $K_c = 1.0$
**b.** $K_c = 0.010$
**c.** $K_c = 1.0 \times 10^{-5}$

62. Consider the reaction:

$$A(g) \rightleftharpoons 2\,B(g)$$

Find the equilibrium partial pressures of A and B for each value of $K$. Assume that the initial partial pressure of B in each case is 1.0 atm and that the initial partial pressure of A is 0.0 atm. Make any appropriate simplifying assumptions.
**a.** $K_p = 1.0$
**b.** $K_p = 1.0 \times 10^{-4}$
**c.** $K_p = 1.0 \times 10^5$

## Le Châtelier's Principle

63. Consider this reaction at equilibrium:

$$CO(g) + Cl_2(g) \rightleftharpoons COCl_2(g)$$

Predict whether the reaction will shift left, shift right, or remain unchanged after each disturbance.
**a.** $COCl_2$ is added to the reaction mixture.
**b.** $Cl_2$ is added to the reaction mixture.
**c.** $COCl_2$ is removed from the reaction mixture.

64. Consider this reaction at equilibrium:

$$2\,BrNO(g) \rightleftharpoons 2\,NO(g) + Br_2(g)$$

Predict whether the reaction will shift left, shift right, or remain unchanged after each disturbance.
**a.** NO is added to the reaction mixture.
**b.** BrNO is added to the reaction mixture.
**c.** $Br_2$ is removed from the reaction mixture.

65. Consider this reaction at equilibrium:

$$2\,KClO_3(s) \rightleftharpoons 2\,KCl(s) + 3\,O_2(g)$$

Predict whether the reaction will shift left, shift right, or remain unchanged after each disturbance.
**a.** $O_2$ is removed from the reaction mixture.
**b.** KCl is added to the reaction mixture.
**c.** $KClO_3$ is added to the reaction mixture.
**d.** $O_2$ is added to the reaction mixture.

66. Consider this reaction at equilibrium:

$$C(s) + H_2O(g) \rightleftharpoons CO(g) + H_2(g)$$

Predict whether the reaction will shift left, shift right, or remain unchanged after each disturbance.
**a.** C is added to the reaction mixture.
**b.** $H_2O$ is condensed and removed from the reaction mixture.
**c.** CO is added to the reaction mixture.
**d.** $H_2$ is removed from the reaction mixture.

67. Each reaction is allowed to come to equilibrium, and then the volume is changed as indicated. Predict the effect (shift right, shift left, or no effect) of the indicated volume change.
**a.** $I_2(g) \rightleftharpoons 2\,I(g)$ (volume is increased)
**b.** $2\,H_2S(g) \rightleftharpoons 2\,H_2(g) + S_2(g)$ (volume is decreased)
**c.** $I_2(g) + Cl_2(g) \rightleftharpoons 2\,ICl(g)$ (volume is decreased)

**68.** Each reaction is allowed to come to equilibrium, and then the volume is changed as indicated. Predict the effect (shift right, shift left, or no effect) of the indicated volume change.

a. $CO(g) + H_2O(g) \rightleftharpoons CO_2(g) + H_2(g)$ (volume is decreased)
b. $PCl_3(g) + Cl_2(g) \rightleftharpoons PCl_5(g)$ (volume is increased)
c. $CaCO_3(s) \rightleftharpoons CaO(s) + CO_2(g)$ (volume is increased)

**69.** This reaction is endothermic.

$$C(s) + CO_2(g) \rightleftharpoons 2\,CO(g)$$

Predict the effect (shift right, shift left, or no effect) of increasing and decreasing the reaction temperature. How does the value of the equilibrium constant depend on temperature?

**70.** This reaction is exothermic.

$$C_6H_{12}O_6(s) + 6\,O_2(g) \rightleftharpoons 6\,CO_2(g) + 6\,H_2O(g)$$

Predict the effect (shift right, shift left, or no effect) of increasing and decreasing the reaction temperature. How does the value of the equilibrium constant depend on temperature?

**71.** Coal, which is primarily carbon, can be converted to natural gas, primarily $CH_4$, by the exothermic reaction:

$$C(s) + 2\,H_2(g) \rightleftharpoons CH_4(g)$$

Which disturbance will favor $CH_4$ at equilibrium?
a. adding more C to the reaction mixture
b. adding more $H_2$ to the reaction mixture
c. raising the temperature of the reaction mixture
d. lowering the volume of the reaction mixture
e. adding a catalyst to the reaction mixture
f. adding neon gas to the reaction mixture

**72.** Coal can be used to generate hydrogen gas (a potential fuel) by the endothermic reaction:

$$C(s) + H_2O(g) \rightleftharpoons CO(g) + H_2(g)$$

If this reaction mixture is at equilibrium, predict whether each disturbance will result in the formation of additional hydrogen gas, the formation of less hydrogen gas, or have no effect on the quantity of hydrogen gas.
a. adding more C to the reaction mixture
b. adding more $H_2O$ to the reaction mixture
c. raising the temperature of the reaction mixture
d. increasing the volume of the reaction mixture
e. adding a catalyst to the reaction mixture
f. adding an inert gas to the reaction mixture

# Cumulative Problems

**73.** Carbon monoxide replaces oxygen in oxygenated hemoglobin according to the reaction:

$$HbO_2(aq) + CO(aq) \rightleftharpoons HbCO(aq) + O_2(aq)$$

a. Use the reactions and associated equilibrium constants at body temperature given here to find the equilibrium constant for the reaction just shown.

$$Hb(aq) + O_2(aq) \rightleftharpoons HbO_2(aq) \qquad K_c = 1.8$$
$$Hb(aq) + CO(aq) \rightleftharpoons HbCO(aq) \qquad K_c = 306$$

b. Suppose that an air mixture becomes polluted with carbon monoxide at a level of 0.10% (by volume). Assuming the air contains 20.0% oxygen, and that the oxygen and carbon monoxide ratios that dissolve in the blood are identical to the ratios in the air, what is the ratio of $HbCO$ to $HbO_2$ in the bloodstream? Comment on the toxicity of carbon monoxide.

**74.** Nitrogen monoxide is a pollutant in the lower atmosphere that irritates the eyes and lungs and leads to the formation of acid rain. Nitrogen monoxide forms naturally in atmosphere according to the endothermic reaction:

$$N_2(g) + O_2(g) \rightleftharpoons 2\,NO(g) \qquad K_p = 4.1 \times 10^{-31} \text{ at } 298 \text{ K}$$

Use the ideal gas law to calculate the concentrations of nitrogen and oxygen present in air at a pressure of 1.0 atm and a temperature of 298 K. Assume that nitrogen composes 78% of air by volume and that oxygen composes 21% of air. Find the "natural" equilibrium concentration of NO in air in units of molecules/$cm^3$. How would you expect this concentration to change in an automobile engine in which combustion is occurring?

**75.** The reaction $CO_2(g) + C(s) \rightleftharpoons 2\,CO(g)$ has $K_p = 5.78$ at 1200 K.

a. Calculate the total pressure at equilibrium when 4.45 g of $CO_2$ is introduced into a 10.0-L container and heated to 1200 K in the presence of 2.00 g of graphite.
b. Repeat the calculation of part a in the presence of 0.50 g of graphite.

**76.** A mixture of water and graphite is heated to 600 K. When the system comes to equilibrium, it contains 0.13 mol of $H_2$, 0.13 mol of CO, 0.43 mol of $H_2O$, and some graphite. Some $O_2$ is added to the

system, and a spark is applied so that the $H_2$ reacts completely with the $O_2$. Find the amount of CO in the flask when the system returns to equilibrium.

**77.** At 650 K, the reaction $MgCO_3(s) \rightleftharpoons MgO(s) + CO_2(g)$ has $K_p = 0.026$. A 10.0-L container at 650 K has 1.0 g of $MgO(s)$ and $CO_2$ at $P = 0.0260$ atm. The container is then compressed to a volume of 0.100 L. Find the mass of $MgCO_3$ that is formed.

**78.** A system at equilibrium contains $I_2(g)$ at a pressure of 0.21 atm and $I(g)$ at a pressure of 0.23 atm. The system is then compressed to half its volume. Find the pressure of each gas when the system returns to equilibrium.

**79.** Consider the exothermic reaction:

$$C_2H_4(g) + Cl_2(g) \rightleftharpoons C_2H_4Cl_2(g)$$

If you were trying to maximize the amount of $C_2H_4Cl_2$ produced, which tactic might you try? Assume that the reaction mixture reaches equilibrium.
a. increasing the reaction volume
b. removing $C_2H_4Cl_2$ from the reaction mixture as it forms
c. lowering the reaction temperature
d. adding $Cl_2$

**80.** Consider the endothermic reaction:

$$C_2H_4(g) + I_2(g) \rightleftharpoons C_2H_4I_2(g)$$

If you were trying to maximize the amount of $C_2H_4I_2$ produced, which tactic might you try? Assume that the reaction mixture reaches equilibrium.
a. decreasing the reaction volume
b. removing $I_2$ from the reaction mixture
c. raising the reaction temperature
d. adding $C_2H_4$ to the reaction mixture

**81.** Consider the reaction:

$$H_2(g) + I_2(g) \rightleftharpoons 2\,HI(g)$$

A reaction mixture at equilibrium at 175 K contains $P_{H_2} = 0.958$ atm, $P_{I_2} = 0.877$ atm, and $P_{HI} = 0.020$ atm. A second reaction mixture, also at 175 K, contains $P_{H_2} = P_{I_2} = 0.621$ atm and $P_{HI} = 0.101$ atm. Is the second reaction at equilibrium? If not, what will be the partial pressure of HI when the reaction reaches equilibrium at 175 K?

**82.** Consider the reaction:
$$2 H_2S(g) + SO_2(g) \rightleftharpoons 3 S(s) + 2 H_2O(g)$$
A reaction mixture initially containing 0.500 M $H_2S$ and 0.500 M $SO_2$ contains 0.0011 M $H_2O$ at equilibrium at a certain temperature. A second reaction mixture at the same temperature initially contains $[H_2S] = 0.250$ M and $[SO_2] = 0.325$ M. Calculate the equilibrium concentration of $H_2O$ in the second mixture at this temperature.

**83.** Ammonia can be synthesized according to the reaction:
$$N_2(g) + 3 H_2(g) \rightleftharpoons 2 NH_3(g) \quad K_p = 5.3 \times 10^{-5} \text{ at } 725 \text{ K}$$
A 200.0-L reaction container initially contains 1.27 kg of $N_2$ and 0.310 kg of $H_2$ at 725 K. Assuming ideal gas behavior, calculate the mass of $NH_3$ (in g) present in the reaction mixture at equilibrium. What is the percent yield of the reaction under these conditions?

**84.** Hydrogen can be extracted from natural gas according to the reaction:
$$CH_4(g) + CO_2(g) \rightleftharpoons 2 CO(g) + 2 H_2(g)$$
$$K_p = 4.5 \times 10^2 \text{ at } 825 \text{ K}$$
An 85.0-L reaction container initially contains 22.3 kg of $CH_4$ and 55.4 kg of $CO_2$ at 825 K. Assuming ideal gas behavior, calculate the mass of $H_2$ (in g) present in the reaction mixture at equilibrium. What is the percent yield of the reaction under these conditions?

**85.** The system described by the reaction: $CO(g) + Cl_2(g) \rightleftharpoons COCl_2(g)$ is at equilibrium at a given temperature when $P_{CO} = 0.30$ atm, $P_{Cl_2} = 0.10$ atm, and $P_{COCl_2} = 0.60$ atm. An additional pressure of $Cl_2(g) = 0.40$ atm is added. Find the pressure of CO when the system returns to equilibrium.

**86.** A reaction vessel at 27 °C contains a mixture of $SO_2$ ($P = 3.00$ atm) and $O_2$ ($P = 1.00$ atm). When a catalyst is added, this reaction takes place: $2 SO_2(g) + O_2(g) \rightleftharpoons 2 SO_3(g)$. At equilibrium, the total pressure is 3.75 atm. Find the value of $K_c$.

**87.** At 70 K, $CCl_4$ decomposes to carbon and chlorine. The $K_p$ for the decomposition is 0.76. Find the starting pressure of $CCl_4$ at this temperature that will produce a total pressure of 1.0 atm at equilibrium.

**88.** The equilibrium constant for the reaction $SO_2(g) + NO_2(g) \rightleftharpoons SO_3(g) + NO(g)$ is $K_c = 3.0$. Find the amount of $NO_2$ that must be added to 2.4 mol of $SO_2$ in order to form 1.2 mol of $SO_3$ at equilibrium.

**89.** A sample of $CaCO_3(s)$ is introduced into a sealed container of volume 0.654 L and heated to 1000 K until equilibrium is reached. The $K_p$ for the reaction $CaCO_3(s) \rightleftharpoons CaO(s) + CO_2(g)$ is $3.9 \times 10^{-2}$ at this temperature. Calculate the mass of $CaO(s)$ that is present at equilibrium.

**90.** An equilibrium mixture contains $N_2O_4$ ($P = 0.28$ atm) and $NO_2$ ($P = 1.1$ atm) at 350 K. The volume of the container is doubled at constant temperature. Write a balanced chemical equation for the reaction and calculate the equilibrium pressures of the two gases when the system reaches a new equilibrium.

**91.** Carbon monoxide and chlorine gas react to form phosgene:
$$CO(g) + Cl_2(g) \rightleftharpoons COCl_2(g) \quad K_p = 3.10 \text{ at } 700 \text{ K}$$
If a reaction mixture initially contains 215 torr of CO and 245 torr of $Cl_2$, what is the mole fraction of $COCl_2$ when equilibrium is reached?

**92.** Solid carbon can react with gaseous water to form carbon monoxide gas and hydrogen gas. The equilibrium constant for the reaction at 700.0 K is $K_p = 1.60 \times 10^{-3}$. If a 1.55-L reaction vessel initially contains 145 torr of water at 700.0 K in contact with excess solid carbon, find the percent by mass of hydrogen gas of the gaseous reaction mixture at equilibrium.

# Challenge Problems

**93.** Consider the reaction:
$$2 NO(g) + O_2(g) \rightleftharpoons 2 NO_2(g)$$
**a.** A reaction mixture at 175 K initially contains 522 torr of NO and 421 torr of $O_2$. At equilibrium, the total pressure in the reaction mixture is 748 torr. Calculate $K_p$ at this temperature.
**b.** A second reaction mixture at 175 K initially contains 255 torr of NO and 185 torr of $O_2$. What is the equilibrium partial pressure of $NO_2$ in this mixture?

**94.** Consider the reaction:
$$2 SO_2(g) + O_2(g) \rightleftharpoons 2 SO_3(g) \quad K_p = 0.355 \text{ at } 950 \text{ K}$$
A 2.75-L reaction vessel at 950 K initially contains 0.100 mol of $SO_2$ and 0.100 mol of $O_2$. Calculate the total pressure (in atmospheres) in the reaction vessel when equilibrium is reached.

**95.** Nitric oxide reacts with chlorine gas according to the reaction:
$$2 NO(g) + Cl_2(g) \rightleftharpoons 2 NOCl(g)$$
$$K_p = 0.27 \text{ at } 700 \text{ K}$$
A reaction mixture initially contains equal partial pressures of NO and $Cl_2$. At equilibrium, the partial pressure of NOCl is 115 torr. What were the initial partial pressures of NO and $Cl_2$?

**96.** At a given temperature, a system containing $O_2(g)$ and some oxides of nitrogen can be described by these reactions:
$$2 NO(g) + O_2(g) \rightleftharpoons 2 NO_2(g) \quad K_p = 10^4$$
$$2 NO_2(g) \rightleftharpoons N_2O_4(g) \quad K_p = 0.10$$
A pressure of 1 atm of $N_2O_4(g)$ is placed in a container at this temperature. Predict which, if any, component (other than $N_2O_4$) will be present at a pressure greater than 0.2 atm at equilibrium.

**97.** A sample of pure $NO_2$ is heated to 337 °C, at which temperature it partially dissociates according to the equation:
$$2 NO_2(g) \rightleftharpoons 2 NO(g) + O_2(g)$$
At equilibrium the density of the gas mixture is 0.520 g/L at 0.750 atm. Calculate $K_c$ for the reaction.

**98.** When $N_2O_5(g)$ is heated, it dissociates into $N_2O_3(g)$ and $O_2(g)$ according to the reaction:
$$N_2O_5(g) \rightleftharpoons N_2O_3(g) + O_2(g)$$
$$K_c = 7.75 \text{ at a given temperature}$$
The $N_2O_3(g)$ dissociates to give $N_2O(g)$ and $O_2(g)$ according the reaction:
$$N_2O_3(g) \rightleftharpoons N_2O(g) + O_2(g)$$
$$K_c = 4.00 \text{ at the same temperature}$$
When 4.00 mol of $N_2O_5(g)$ is heated in a 1.00-L reaction vessel to this temperature, the concentration of $O_2(g)$ at equilibrium is 4.50 mol/L. Find the concentrations of all the other species in the equilibrium system.

**99.** A sample of $SO_3$ is introduced into an evacuated sealed container and heated to 600 K. The following equilibrium is established:
$$2 SO_3(g) \rightleftharpoons 2 SO_2(g) + O_2(g)$$
The total pressure in the system is 3.0 atm and the mole fraction of $O_2$ is 0.12. Find $K_p$.

# Conceptual Problems

**100.** A reaction $A(g) \rightleftharpoons B(g)$ has an equilibrium constant of $1.0 \times 10^{-4}$. For which of the initial reaction mixtures is the *x is small* approximation most likely to apply?
   **a.** [A] = 0.0010 M; [B] = 0.00 M
   **b.** [A] = 0.00 M; [B] = 0.10 M
   **c.** [A] = 0.10 M; [B] = 0.10 M
   **d.** [A] = 0.10 M; [B] = 0.00 M

**101.** The reaction $A(g) \rightleftharpoons 2 B(g)$ has an equilibrium constant of $K_c = 1.0$ at a given temperature. If a reaction vessel contains equal initial amounts (in moles) of A and B, does the direction in which the reaction proceeds depend on the volume of the reaction vessel? Explain.

**102.** A particular reaction has an equilibrium constant of $K_p = 0.50$. A reaction mixture is prepared in which all the reactants and products are in their standard states. In which direction does the reaction proceed?

**103.** Consider the reaction:

$$aA(g) \rightleftharpoons bB(g)$$

Each of the entries in the following table represents equilibrium partial pressures of A and B under different initial conditions. What are the values of *a* and *b* in the reaction?

| $P_A$ (atm) | $P_B$ (atm) |
|---|---|
| 4.0 | 2.0 |
| 2.0 | 1.4 |
| 1.0 | 1.0 |
| 0.50 | 0.71 |
| 0.25 | 0.50 |

**104.** Consider the simple one-step reaction:

$$A(g) \rightleftharpoons B(g)$$

Since the reaction occurs in a single step, the forward reaction has a rate of $k_{for}[A]$ and the reverse reaction has a rate of $k_{rev}[B]$. What happens to the rate of the forward reaction when we increase the concentration of A? How does this explain the reason behind Le Châtelier's principle?

**105.** Consider the reaction: $A(g) \rightleftharpoons 2 B(g)$. The graph plots the concentrations of A and B as a function of time at a constant temperature. What is the equilibrium constant for this reaction at this temperature?

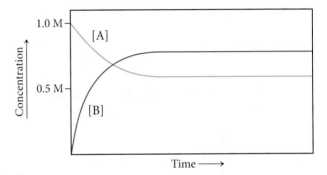

---

*Active Classroom Learning*

# Questions for Group Work

Discuss these questions with the group and record your consensus answer.

**106.** The reactions shown here can be combined to make the overall reaction $C(s) + H_2O(g) \longrightarrow CO(g) + H_2(g)$ by reversing some and/or dividing all the coefficients by a number. As a group, determine how the reactions need to be modified to sum to the overall process. Then have each group member determine the value of *K* for one of the reactions to be combined. Finally, combine all the values of *K* to determine the value of *K* for the overall reaction.
   **a.** $C(s) + O_2(g) \longrightarrow CO_2(g)$ $\quad K = 1.363 \times 10^{69}$
   **b.** $2 H_2(g) + O_2(g) \longrightarrow 2 H_2O(g)$ $\quad K = 1.389 \times 10^{80}$
   **c.** $2 CO(g) + O_2(g) \longrightarrow 2 CO_2(g)$ $\quad K = 1.477 \times 10^{90}$

**107.** Consider the reaction: $N_2(g) + 3 H_2(g) \longrightarrow 2 NH_3(g)$.
   **a.** Write the equilibrium constant expression for this reaction. If some hydrogen is added, before the reaction shifts,
   **b.** How will the numerator and denominator of the expression in part a compare to the value at equilibrium?
   **c.** Will Q be larger or smaller than K? Why?
   **d.** Will the reaction have to shift forward or backward to retain equilibrium? Explain.
   **e.** Are your answers above consistent with Le Châtelier's principle? Explain.

**108.** For the reaction $A \longrightarrow B$, the ratio of products to reactants at equilibrium is always the same number, no matter how much A or B is initially present. Interestingly, in contrast, the ratio of products to reactants for the reaction $C \longrightarrow 2 D$ does depend on how much of C and D you have initially. Explain this observation. Which ratio is independent of the starting amounts of C and D? Answer in complete sentences.

**109.** Solve each of the expressions for *x* using the quadratic formula and the *x is small* approximation. In which of the following expressions is the *x is small* approximation valid?
   **a.** $x^2/(0.2 - x) = 1.3 \times 10^4$
   **b.** $x^2/(0.2 - x) = 1.3$
   **c.** $x^2/(0.2 - x) = 1.3 \times 10^{-4}$
   **d.** $x^2/(0.01 - x) = 1.3 \times 10^{-4}$

   In a complete sentence, describe the factor(s) that tend to make the *x is small* approximation valid in an expression.

**110.** Have each group member explain to the group what happens if a system at equilibrium is subject to one of the following changes and why:
   **a.** the concentration of a reactant is increased
   **b.** a solid product is added
   **c.** the volume is decreased
   **d.** the temperature is raised

# Data Interpretation and Analysis

## Chemical Equilibrium on Venus

**111.** The atmosphere of Venus is currently composed of 96.5% carbon dioxide, 3.5% nitrogen, trace amounts of sulfur dioxide, and clouds of sulfuric acid. The average temperature on this planet is 740 K and the average pressure is 94 atm. Research suggests that, due to different pressure and temperature conditions, carbon dioxide may have existed in different states on Venus in the past. A phase diagram of carbon dioxide (Figure a▶) can help estimate the state of carbon dioxide.

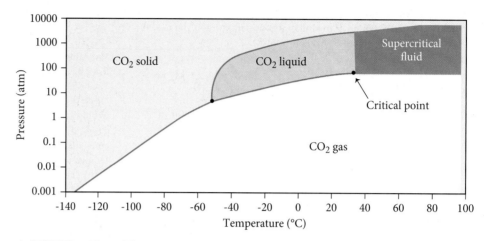

▲ FIGURE a  Phase Diagram of CO₂.

Because of the large quantities of $CO_2$ on Venus, the concentration of $CO_2$ in the atmosphere could depend on one or more equilibrium systems with carbonate ions in minerals. For example, carbon dioxide gas might react with calcium silicate ($CaSiO_3$) according to the following equilibrium:

$$CaSiO_3(s) + CO_2(g) \rightleftharpoons CaCO_3(s) + SiO_2(s) \text{ (Reaction 1)}$$

An alternative viewpoint suggests that the concentration of $CO_2$ gas in the atmosphere of Venus depends on the following equilibrium:

$$MgSiO_3(s) + CaCO_3() + SiO_2(s) \rightleftharpoons CaMgSi_2O_6(s) + CO_2(g)$$
$$\text{(Reaction 2)}$$

Temperature plays an important role in equilibrium systems. The graph in Figure b▶ plots temperature in units of Kelvin versus the log of the partial pressure of $CO_2$ gas for each of the two possible reactions. The area marked as region 1 corresponds to the equilibrium system for Reaction 1, whereas the area marked as region 2 corresponds to the equilibrium system for Reaction 2.

For the questions that follow, choose a point in the middle of the colored band (which represents the range of possible values). For example, according to the graph, on Earth, at a temperature of 298 K, the log $P_{CO_2}$ = −3.90 for Reaction 2. This corresponds to a partial pressure of $1.3 \times 10^{-4}$ atm.

Use the information provided in the figures to answer the following questions:

a. If the partial pressure of $CO_2$ on Venus is 91 atm, calculate the equilibrium constant ($K_p$) for Reactions 1 and 2.

**b.** Determine the partial pressure of $CO_2$ on Venus if the temperature cooled from 750 K to 400 K using Reaction 1? Using Reaction 2? Choose values in the middle of colored bands of the graph in Figure b.

c. Refer to the phase diagram in Figure a. At what minimum pressure can supercritical $CO_2$ form? If the pressure of the Venusian atmosphere in the distant past was much higher than it is today, could supercritical carbon dioxide have existed on the surface of Venus?

▲ FIGURE b  **A Representation of Two Types of Equilibrium Systems Involving Carbon Dioxide on Venus.**

# Answers to Conceptual Connections

**The Magnitude of the Equilibrium Contstant**

**15.1** The equilibrium constant is largest for temperature $T_3$ (because the reaction mixture has the most products relative to reactants at that temperature).

**Equilibrium Constants and Equilibrium Concentrations**

**15.2** **(b)** The reaction mixture will contain $[A] = 0.1\,M$ and $[B] = 1.0\,M$ so that $[B]/[A] = 10$.

**The Equilibrium Constant and the Chemical Equation**

**15.3** **(b)** The reaction is reversed and divided through by two. Therefore, you invert the equilibrium constant and take the square root of the result. $K = (1/0.010)^{1/2} = 10$.

**The Relationship between $K_p$ and $K_c$**

**15.4** **(a)** When $a + b = c + d$, the quantity $\Delta n$ is zero so that $K_p = K_c(RT)^0$. Since $(RT)^0$ is equal to 1, $K_p = K_c$.

**Heterogeneous Equilibria, $K_p$, and $K_c$**

**15.5** **(b)** Since $\Delta n$ for gaseous reactants and products is zero, $K_p = K_c$.

**Q and K**

**15.6** **(c)** Because $N_2O_4$ and $NO_2$ are both in their standard states, they each have a partial pressure of 1.0 atm. Consequently, $Q_p = 1$. Since $K_p = 0.15$, $Q_p > K_p$, and the reaction proceeds to the left.

**The *x is small* Approximation**

**15.7** **(a)** The *x is small* approximation is most likely to apply to a reaction with a small equilibrium constant and an initial concentration of reactant that is not too small. The bigger the equilibrium constant and the smaller the initial concentration of reactant, the less likely that the *x is small* approximation will apply.

CHAPTER

# 16

# Acids and Bases

The differences between the various acid–base concepts are not concerned with which is right, but which is most convenient to use in a particular situation.

—JAMES E. HUHEEY (1935–)

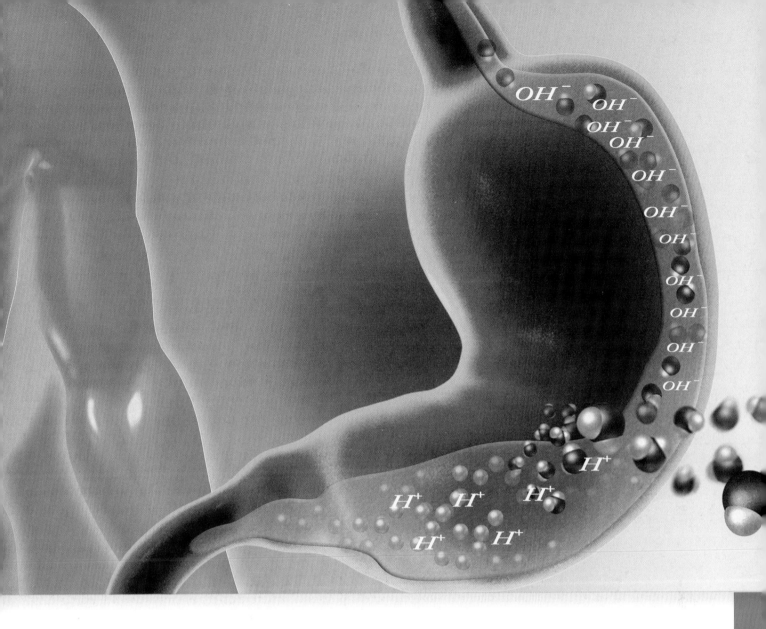

In this chapter, we apply the equilibrium concepts learned in the previous chapter to acid–base phenomena. Acids are common in many foods, such as limes, lemons, and vinegar, and in a number of consumer products, such as toilet cleaners and batteries. Bases are less common in foods but are key ingredients in consumer products such as drain openers and antacids. We will examine three different models for acid–base behavior, all of which define that behavior differently. In spite of their differences, the three models coexist, each being useful at explaining a particular range of acid–base phenomena. We also examine how to calculate the acidity or basicity of solutions and define a useful scale, called the pH scale, to quantify acidity and basicity. These types of calculations often involve solving the kind of equilibrium problems that we explored in Chapter 15.

▲ *Milk of magnesia contains a base that can neutralize stomach acid and relieve heartburn.*

## 16.1 Heartburn

Heartburn is a painful burning sensation in the esophagus (the tube that joins the throat to the stomach) just below the chest. The pain is caused by hydrochloric acid (HCl), which the stomach excretes to kill microorganisms and to activate enzymes that break down food. Hydrochloric acid sometimes backs up out of the stomach and into the esophagus, a phenomenon called *acid reflux*. Recall from Section 4.8 that acids are substances that—by one definition that we elaborate on shortly—produce $H^+$ ions in solution. When hydrochloric acid from the stomach comes in contact with the lining of the esophagus, the $H^+$ ions irritate the tissues, resulting in the burning sensation. Some of the acid can work its way into the lower throat and even the mouth, producing pain in the throat and a sour taste (characteristic of acids) in the mouth. Almost everyone experiences heartburn at some time, most commonly after a large meal when the stomach is full. Strenuous activity or lying in a horizontal position after a large meal increases the likelihood of stomach acid reflux and the resulting heartburn.

The simplest way to relieve mild heartburn is to swallow repeatedly. Saliva contains the bicarbonate ion ($HCO_3^-$), which acts as a base and, when swallowed, neutralizes some of the acid in the esophagus. Later in this chapter, we will see how bicarbonate acts as a base. You can also treat heartburn with antacids such as Tums, milk of magnesia, or Mylanta. These over-the-counter medications contain more base than saliva does and therefore are effective at neutralizing esophageal acid. We look at the bases in these medicines more carefully later (see *Chemistry and Medicine: What's in My Antacid?* in Section 16.7).

For some people, heartburn becomes a chronic problem. Gastroesophageal reflux disease (GERD) is the medical condition associated with chronic heartburn. In patients with GERD, the band of muscles (called the esophageal sphincter) at the bottom of the esophagus just above the stomach does not close tightly enough, allowing stomach contents to leak back into the esophagus on a regular basis. Medical researchers have developed a wireless sensor to help diagnose and evaluate treatment of GERD. Using a tube that goes down through the throat, a physician attaches the sensor to tissues in the patient's esophagus. The sensor reads pH—a measure of acidity that we discuss in Section 16.5—and transmits the readings to a recorder worn on the patient's body. The patient goes about his or her normal activities for the next few days while the recorder monitors esophageal pH. The physician then reads the record of esophageal pH to make a diagnosis or evaluate treatment.

In this chapter, we examine acid and base behavior. Acids and bases are not only important to our health (as we have just seen), but are also found in many household products, foods, medicines, and of course in nearly every chemistry laboratory. Acid–base chemistry is central to much of biochemistry and molecular biology. The building blocks of proteins, for example, are acids (called amino acids) and the molecules that carry the genetic code in DNA are bases.

## 16.2 The Nature of Acids and Bases

*Acids* have the following general properties: a sour taste, the ability to dissolve many metals, the ability to turn blue litmus paper red, and the ability to neutralize bases. Table 16.1 lists some common acids.

| Bases were first defined in Section 4.8. |
| The bicarbonate ion neutralizes stomach acid according to the reaction: $H^+(aq) + HCO_3^-(aq) \longrightarrow H_2O(l) + CO_2(g)$. |
| The concentration of stomach acid, $[H_3O^+]$, varies from about 0.01 to 0.1 M. |
| For a review of acid naming, see Section 3.6. |
| Litmus paper contains dyes that change color in the presence of acids and bases. |
| The formula for acetic acid can also be written as $CH_3COOH$. |

**TABLE 16.1 Some Common Acids**

| Name | Occurrence/Uses |
|------|-----------------|
| Hydrochloric acid (HCl) | Metal cleaning; food preparation; ore refining; primary component of stomach acid |
| Sulfuric acid ($H_2SO_4$) | Fertilizer and explosives manufacturing; dye and glue production; in automobile batteries; electroplating of copper |
| Nitric acid ($HNO_3$) | Fertilizer and explosives manufacturing; dye and glue production |
| Acetic acid ($HC_2H_3O_2$) | Plastic and rubber manufacturing; food preservation; active component of vinegar |
| Citric acid ($H_3C_6H_5O_7$) | In citrus fruits such as lemons and limes; used to adjust pH in foods and beverages |
| Carbonic acid ($H_2CO_3$) | In carbonated beverages due to the reaction of carbon dioxide with water |
| Hydrofluoric acid (HF) | Metal cleaning; glass frosting and etching |
| Phosphoric acid ($H_3PO_4$) | Fertilizer manufacturing; biological buffering; beverage preservation |

You can find hydrochloric acid in most chemistry laboratories. In industry, it is used to clean metals, to prepare and process some foods, and to refine metal ores. As we discussed in Section 16.1, hydrochloric acid is also the main component of stomach acid.

HCl

Hydrochloric acid

Sulfuric acid and nitric acid are also common in the laboratory. They play major roles in the manufacture of fertilizers, explosives, dyes, and glues. Sulfuric acid, produced in larger quantities than any other industrial chemical, is used in most automobile batteries.

$H_2SO_4$

Sulfuric acid

$HNO_3$

Nitric acid

$HC_2H_3O_2$

Acetic acid

You can probably find acetic acid in your home—it is the active component of vinegar. It is also produced in improperly stored wines. The word *vinegar* originates from the French words *vin aigre*, which means "sour wine." Wine experts consider the presence of vinegar in wines a serious fault, since it makes the wine taste like salad dressing.

Acetic acid is a **carboxylic acid**, an acid that contains the following grouping of atoms:

Carboxylic acid group

Carboxylic acids are often found in substances derived from living organisms. Other examples of carboxylic acids are citric acid, found in lemons and limes, and malic acid, found in apples, grapes, and wine.

$H_3C_6H_5O_7$

Citric acid

$H_2C_4H_4O_5$

Malic acid

▲ Acetic acid makes vinegar taste sour.

▲ Citrus fruits, apples, and grapes all contain acids.

▲ Many common household products and remedies contain bases.

Coffee is acidic overall, but bases present in coffee—such as caffeine—and other compounds impart a bitter flavor.

**TABLE 16.2 Common Bases**

| Name | Occurrence/Uses |
|------|-----------------|
| Sodium hydroxide (NaOH) | Petroleum processing; soap and plastic manufacturing |
| Potassium hydroxide (KOH) | Cotton processing; electroplating; soap production; batteries |
| Sodium bicarbonate (NaHCO$_3$) | Sold as baking soda; antacid; source of CO$_2$ |
| Sodium carbonate (Na$_2$CO$_3$) | Glass and soap manufacturing; general cleanser; water softener |
| Ammonia (NH$_3$) | Detergent; fertilizer and explosives manufacturing; synthetic fiber production |

*Bases* have the following general properties: a bitter taste, a slippery feel, the ability to turn red litmus paper blue, and the ability to neutralize acids. Because of their bitterness, bases are less common in foods than are acids. Our aversion to the taste of bases is probably an evolutionary adaptation to warn us against **alkaloids**, organic bases found in plants that are often poisonous. (For example, the active component of hemlock—the poisonous plant that killed the Greek philosopher Socrates—is the alkaloid coniine.) Nonetheless, some foods, such as coffee and chocolate (especially dark chocolate), contain bitter flavors. Many people enjoy the bitterness, but only after acquiring the taste over time.

Bases feel slippery because they react with oils on the skin to form soap-like substances. Some household cleaning solutions, such as ammonia, are basic and have the characteristic slippery feel of a base. Bases turn red litmus paper blue; in the laboratory, litmus paper is routinely used to test the basicity of solutions.

Some common bases are listed in Table 16.2. You can find sodium hydroxide and potassium hydroxide in most chemistry laboratories. They are used in petroleum and cotton processing and in soap and plastic manufacturing. Sodium hydroxide is the active ingredient in products such as Drāno that unclog drains. In many homes, you can find sodium bicarbonate in the medicine cabinet (it is an active ingredient in some antacids) as well as in the kitchen (labeled as baking soda).

**Definitions of Acids and Bases**

# 16.3 Definitions of Acids and Bases

What are the key characteristics of the molecules and ions that exhibit acid and base behavior? In this chapter, we examine three different definitions: the Arrhenius definition, the Brønsted–Lowry definition, and the Lewis definition. Why three definitions, and which one is correct? As inorganic chemist and author James Huheey notes in the quotation that opens this chapter, no single definition is "correct." Rather, each definition is useful in a given instance. We discuss the Lewis definition of acids and bases in Section 16.11; here we discuss the other two.

## The Arrhenius Definition

In the 1880s, Swedish chemist Svante Arrhenius (1859–1927) proposed the following molecular definitions of acids and bases:

**Acid:** A substance that produces H$^+$ ions in aqueous solution.

**Base:** A substance that produces OH$^-$ ions in aqueous solution.

According to the **Arrhenius definition**, HCl is an acid because it produces H$^+$ ions in solution (Figure 16.1 ◀):

$$HCl(aq) \longrightarrow H^+(aq) + Cl^-(aq)$$

Hydrogen chloride (HCl) is a covalent compound and does not contain ions. However, in water it *ionizes* completely to form H$^+(aq)$ ions and Cl$^-(aq)$ ions. The H$^+$ ions are highly reactive. In aqueous solution, the H$^+$ ions bond to water to form H$_3$O$^+$:

Arrhenius Acid

HCl

$$HCl(aq) \longrightarrow H^+(aq) + Cl^-(aq)$$

▲ **FIGURE 16.1 Arrhenius Acid**
An Arrhenius acid produces H$^+$ ions in solution.

$$H^+ + \quad :\!\overset{\displaystyle H}{\underset{\displaystyle \cdot\cdot}{O}}\!:\!H \quad \longrightarrow \quad \left[ \overset{\displaystyle H}{H\!:\!\underset{\displaystyle \cdot\cdot}{O}\!:\!H} \right]^+$$

The $H_3O^+$ ion is the **hydronium ion**. In water, $H^+$ ions *always* associate with $H_2O$ molecules to form hydronium ions and other associated species with the general formula $H(H_2O)_n^+$. For example, an $H^+$ ion can associate with two water molecules to form $H(H_2O)_2^+$, with three to form $H(H_2O)_3^+$, and so on. Chemists often use $H^+(aq)$ and $H_3O^+(aq)$ interchangeably to mean the same thing—an $H^+$ ion that has been solvated (or dissolved) in water.

According to the Arrhenius definition, NaOH is a base because it produces $OH^-$ ions in solution (Figure 16.2▶):

$$NaOH(aq) \longrightarrow Na^+(aq) + OH^-(aq)$$

NaOH is an ionic compound and therefore contains $Na^+$ and $OH^-$ ions. When NaOH is added to water, it *dissociates* (breaks apart) into its component ions.

Under the Arrhenius definition, acids and bases combine to form water, neutralizing each other in the process:

$$H^+(aq) + OH^-(aq) \longrightarrow H_2O(l)$$

$$NaOH(aq) \longrightarrow Na^+(aq) + OH^-(aq)$$

▲ **FIGURE 16.2 Arrhenius Base** An Arrhenius base produces $OH^-$ ions in solution.

## The Brønsted–Lowry Definition

A second, more widely applicable definition of acids and bases, the **Brønsted–Lowry definition**, was introduced in 1923. This definition focuses on the *transfer of $H^+$ ions* in an acid–base reaction. Since an $H^+$ ion is a proton—a hydrogen atom without its electron—this definition focuses on the idea of a proton donor and a proton acceptor:

**Acid:**   proton ($H^+$ ion) *donor*

**Base:**   proton ($H^+$ ion) *acceptor*

According to this definition, HCl is an acid because, in solution, it donates a proton to water:

$$HCl(aq) + H_2O(l) \longrightarrow H_3O^+(aq) + Cl^-(aq)$$

This definition clearly describes what happens to the $H^+$ ion from an acid—it associates with a water molecule to form $H_3O^+$ (a hydronium ion). The Brønsted–Lowry definition also applies nicely to bases (such as $NH_3$) that do not inherently contain $OH^-$ ions but still produce $OH^-$ ions in solution. According to the Brønsted–Lowry definition, $NH_3$ is a base because it accepts a proton from water:

$$NH_3(aq) + H_2O(l) \rightleftharpoons NH_4^+(aq) + OH^-(aq)$$

According to the Brønsted–Lowry definition, acids (proton donors) and bases (proton acceptors) always occur together in an acid–base reaction. In the reaction between HCl and $H_2O$, HCl is the proton donor (acid) and $H_2O$ is the proton acceptor (base):

$$\underset{\substack{\text{acid} \\ \text{(proton donor)}}}{HCl(aq)} + \underset{\substack{\text{base} \\ \text{(proton acceptor)}}}{H_2O(l)} \longrightarrow H_3O^+(aq) + Cl^-(aq)$$

All Arrhenius acids and bases are acids and bases under the Brønsted–Lowry definition. However, some Brønsted–Lowry acids and bases cannot be classified as Arrhenius acids and bases.

In the reaction between $NH_3$ and $H_2O$, $H_2O$ is the proton donor (acid) and $NH_3$ is the proton acceptor (base):

$$\underset{\substack{\text{base} \\ \text{(proton acceptor)}}}{NH_3(aq)} + \underset{\substack{\text{acid} \\ \text{(proton donor)}}}{H_2O(l)} \rightleftharpoons NH_4^+(aq) + OH^-(aq)$$

According to the Brønsted–Lowry definition, some substances—such as water in the previous two equations—can act as acids *or* bases. Substances that can act as acids or bases are **amphoteric**. Notice what happens when we reverse an equation representing a Brønsted–Lowry acid–base reaction:

$$\underset{\substack{\text{acid} \\ \text{(proton donor)}}}{NH_4^+(aq)} + \underset{\substack{\text{base} \\ \text{(proton acceptor)}}}{OH^-(aq)} \rightleftharpoons NH_3(aq) + H_2O(l)$$

▶ **FIGURE 16.3 Conjugate Acid–Base Pairs** A conjugate acid–base pair consists of two substances related to each other by the transfer of a proton.

$$NH_3 \quad \xrightarrow{\text{Add } H^+} \quad NH_4^+ \qquad\qquad H_2O \quad \xrightarrow{\text{Remove } H^+} \quad OH^-$$

NH₃ (base) → NH₄⁺ (conjugate acid)    H₂O (acid) → OH⁻ (conjugate base)

Conjugate acid–base pair          Conjugate acid–base pair

In this reaction, $NH_4^+$ is the proton donor (acid) and $OH^-$ is the proton acceptor (base). The substance that was the base ($NH_3$) has become the acid ($NH_4^+$) and vice versa. $NH_4^+$ and $NH_3$ are often referred to as a **conjugate acid–base pair**, two substances related to each other by the transfer of a proton (Figure 16.3▲). A **conjugate acid** is any base to which a proton has been added, and a **conjugate base** is any acid from which a proton has been removed. Going back to the original forward reaction, we can identify the conjugate acid–base pairs:

$$NH_3(aq) + H_2O(l) \rightleftharpoons NH_4^+(aq) + OH^-(aq)$$

Base          Acid          Conjugate          Conjugate
                            acid                base

***Summarizing the Brønsted-Lowry Definition of Acids and Bases:***

▶ A base accepts a proton and becomes a conjugate acid.

▶ An acid donates a proton and becomes a conjugate base.

---

**Example 16.1**   **Identifying Brønsted–Lowry Acids and Bases and Their Conjugates**

In each reaction, identify the Brønsted–Lowry acid, the Brønsted–Lowry base, the conjugate acid, and the conjugate base.

**(a)** $H_2SO_4(aq) + H_2O(l) \longrightarrow HSO_4^-(aq) + H_3O^+(aq)$

**(b)** $HCO_3^-(aq) + H_2O(l) \rightleftharpoons H_2CO_3(aq) + OH^-(aq)$

**SOLUTION**

**(a)** Because $H_2SO_4$ donates a proton to $H_2O$ in this reaction, it is the acid (proton donor). After $H_2SO_4$ donates the proton, it becomes $HSO_4^-$, the conjugate base. Because $H_2O$ accepts a proton, it is the base (proton acceptor). After $H_2O$ accepts the proton, it becomes $H_3O^+$, the conjugate acid.

$$H_2SO_4(aq) + H_2O(l) \longrightarrow HSO_4^-(aq) + H_3O^+(aq)$$

Acid          Base          Conjugate          Conjugate
                            base                acid

**(b)** Because $H_2O$ donates a proton to $HCO_3^-$ in this reaction, it is the acid (proton donor). After $H_2O$ donates the proton, it becomes $OH^-$, the conjugate base. Because $HCO_3^-$ accepts a proton, it is the base (proton acceptor). After $HCO_3^-$ accepts the proton, it becomes $H_2CO_3$, the conjugate acid.

$$HCO_3^-(aq) + H_2O(l) \rightleftharpoons H_2CO_3(aq) + OH^-(aq)$$

Base          Acid          Conjugate          Conjugate
                            acid                base

**FOR PRACTICE 16.1**   In each reaction, identify the Brønsted–Lowry acid, the Brønsted–Lowry base, the conjugate acid, and the conjugate base.

**(a)** $C_5H_5N(aq) + H_2O(l) \rightleftharpoons C_5H_5NH^+(aq) + OH^-(aq)$

**(b)** $HNO_3(aq) + H_2O(l) \rightleftharpoons H_3O^+(aq) + NO_3^-(aq)$

**Conjugate Acid–Base Pairs** Which pair is *not* a conjugate acid–base pair?
   **(a)** $(CH_3)_3 N; (CH_3)_3 NH^+$    **(b)** $H_2SO_4; H_2SO_3$    **(c)** $HNO_2; NO_2^-$

## 16.4   Acid Strength and the Acid Ionization Constant ($K_a$)

The strength of an electrolyte, first discussed in Section 4.5, depends on the extent of its dissociation into its component ions in solution. A *strong electrolyte* completely dissociates into ions in solution, whereas a *weak electrolyte* only partially dissociates. We define strong and weak acids accordingly. A **strong acid** completely ionizes in solution, whereas a **weak acid** only partially ionizes. In other words, the strength of an acid depends on the equilibrium:

$$HA(aq) + H_2O(l) \rightleftharpoons H_3O^+(aq) + A^-(aq)$$

In the preceding equation, HA is a generic formula for an acid. If the equilibrium lies far to the right, the acid is strong—it completely ionizes. If the equilibrium lies to the left, the acid is weak—only a small percentage of the acid molecules ionize. Of course, the range of acid strength is continuous, but for most purposes, the categories of strong and weak are useful.

### Strong Acids

Hydrochloric acid (HCl) is an example of a strong acid:

> Single arrow indicates
> complete ionization.

$$HCl(aq) + H_2O(l) \longrightarrow H_3O^+(aq) + Cl^-(aq)$$

An HCl solution contains virtually no intact HCl; the HCl has essentially all ionized to form $H_3O^+(aq)$ and $Cl^-(aq)$ (Figure 16.4▼). A 1.0 M HCl solution has an $H_3O^+$ concentration of 1.0 M. Abbreviating the concentration of $H_3O^+$ as $[H_3O^+]$, we say that a 1.0 M HCl solution has $[H_3O^+]$ = 1.0 M.

    Table 16.3 lists the six important strong acids. The first five acids in the table are **monoprotic acids**, acids containing only one ionizable proton. Sulfuric acid is an example of a **diprotic acid**, an acid containing two ionizable protons.

> An ionizable proton is one that ionizes in solution. We discuss polyprotic acids in more detail in Section 16.9.

| TABLE 16.3 Strong Acids | |
|---|---|
| Hydrochloric acid (HCl) | Nitric acid (HNO$_3$) |
| Hydrobromic acid (HBr) | Perchloric acid (HClO$_4$) |
| Hydriodic acid (HI) | Sulfuric acid (H$_2$SO$_4$) (*diprotic*) |

**A Strong Acid**

When HCl dissolves in water, it ionizes completely.

HCl         ●—— HCl

Cl⁻   $H_3O^+$
     Cl⁻
Cl⁻   $H_3O^+$   Cl⁻
$H_3O^+$   $H_3O^+$

—— Cl⁻

—— $H_3O^+$

◀ **FIGURE 16.4 Ionization of a Strong Acid** When HCl dissolves in water, it completely ionizes to form $H_3O^+$ and $Cl^-$. The solution contains virtually no intact HCl.

## A Weak Acid

When HF dissolves in water, only a fraction of the molecules ionize.

▲ **FIGURE 16.5 Ionization of a Weak Acid** When HF dissolves in water, only a fraction of the dissolved molecules ionize to form $H_3O^+$ and $F^-$. The solution contains many intact HF molecules.

## Weak Acids

In contrast to HCl, HF is a weak acid, one that does not completely ionize in solution:

Equilibrium arrow indicates partial ionization.

$$HF(aq) + H_2O(l) \rightleftharpoons H_3O^+(aq) + F^-(aq)$$

An HF solution contains a large number of intact (or un-ionized) HF molecules; it also contains some $H_3O^+(aq)$ and $F^-(aq)$ (Figure 16.5▲). In other words, a 1.0 M HF solution has an $[H_3O^+]$ that is much less than 1.0 M because only some of the HF molecules ionize to form $H_3O^+$.

The degree to which an acid is strong or weak depends on the attraction between the anion of the acid (the conjugate base) and the hydrogen ion, relative to the attractions of these ions to water. Recall that HA is a generic formula for an acid. The degree to which the following reaction proceeds in the forward direction depends on the strength of the attraction between $H^+$ and $A^-$:

$$\underset{\text{acid}}{HA(aq)} + H_2O(l) \rightleftharpoons H_3O^+(aq) + \underset{\text{conjugate base}}{A^-(aq)}$$

If the attraction between $H^+$ and $A^-$ is *weak*, then the reaction favors the forward direction and the acid is *strong*. If the attraction between $H^+$ and $A^-$ is *strong*, then the reaction favors the reverse direction and the acid is *weak*, as shown in Figure 16.6◄.

For example, in HCl, the conjugate base ($Cl^-$) has a relatively weak attraction to $H^+$—the reverse reaction does not occur to any significant extent. In HF, on the other hand, the conjugate base ($F^-$) has a greater attraction to $H^+$—the reverse reaction occurs to a significant degree. *In general, the stronger the acid, the weaker the conjugate base and vice versa.* If the forward reaction (that of the acid) has a high tendency to occur, then the reverse reaction (that of the conjugate base) has a low tendency to occur. Table 16.4 lists some common weak acids.

Notice that two of the weak acids in Table 16.4 are diprotic, meaning that they have two ionizable protons, and one is a **triprotic acid** (three ionizable protons). We discuss polyprotic acids in more detail in Section 16.9.

The terms *strong* and *weak* acids are often confused with the terms *concentrated* and *dilute* acids. Can you articulate the difference between these terms?

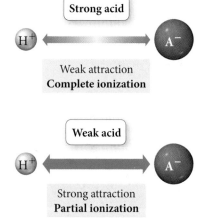

▲ **FIGURE 16.6 Ionic Attraction and Acid Strength** In a strong acid, the attraction between $H^+$ and $A^-$ is weak, resulting in complete ionization. In a weak acid, the attraction between $H^+$ and $A^-$ is strong, resulting in only partial ionization.

Alternatively, we can write the formulas for acetic acid and formic acid as $CH_3COOH$ and $HCOOH$, respectively, to indicate that in these compounds the only H that ionizes is the one attached to an oxygen atom.

| TABLE 16.4 Some Weak Acids | |
| --- | --- |
| Hydrofluoric acid (HF) | Sulfurous acid ($H_2SO_3$) (*diprotic*) |
| Acetic acid ($HC_2H_3O_2$) | Carbonic acid ($H_2CO_3$) (*diprotic*) |
| Formic acid ($HCHO_2$) | Phosphoric acid ($H_3PO_4$) (*triprotic*) |

# The Acid Ionization Constant ($K_a$)

We quantify the relative strength of a weak acid with the **acid ionization constant ($K_a$)**, which is the equilibrium constant for the ionization reaction of the weak acid. As we saw in Section 15.3, for the two equivalent reactions:

$$HA(aq) + H_2O(l) \rightleftharpoons H_3O^+(aq) + A^-(aq)$$
$$HA(aq) \rightleftharpoons H^+(aq) + A^-(aq)$$

the equilibrium constant is:

$$K_a = \frac{[H_3O^+][A^-]}{[HA]} = \frac{[H^+][A^-]}{[HA]}$$

Since $[H_3O^+]$ is equivalent to $[H^+]$, both forms of the expression are equal. Although the ionization constants for all weak acids are relatively small (otherwise the acid would not be a weak acid), they do vary in magnitude. The smaller the constant, the less the acid ionizes, and the weaker the acid. Table 16.5 lists the acid ionization constants for a number of common weak acids in order of decreasing acid strength. Appendix II C contains a more complete list.

> Sometimes $K_a$ is referred to as the acid dissociation constant.

> Recall from Chapter 15 that the concentrations of pure solids or pure liquids are not included in the expression for $K_c$; therefore, $H_2O(l)$ is not included in the expression for $K_a$.

**TABLE 16.5** Acid Ionization Constants ($K_a$) for Some Monoprotic Weak Acids at 25 °C

| Acid | Formula | Structural Formula | Ionization Reaction | $K_a$ | $pK_a$ |
|---|---|---|---|---|---|
| Chlorous acid | $HClO_2$ | H—O—Cl=O | $HClO_2(aq) + H_2O(l) \rightleftharpoons H_3O^+(aq) + ClO_2^-(aq)$ | $1.1 \times 10^{-2}$ | 1.96 |
| Nitrous acid | $HNO_2$ | H—O—N=O | $HNO_2(aq) + H_2O(l) \rightleftharpoons H_3O^+(aq) + NO_2^-(aq)$ | $4.6 \times 10^{-4}$ | 3.34 |
| Hydrofluoric acid | HF | H—F | $HF(aq) + H_2O(l) \rightleftharpoons H_3O^+(aq) + F^-(aq)$ | $6.8 \times 10^{-4}$ | 3.17 |
| Formic acid | $HCHO_2$ | H—O—C(=O)—H | $HCHO_2(aq) + H_2O(l) \rightleftharpoons H_3O^+(aq) + CHO_2^-(aq)$ | $1.8 \times 10^{-4}$ | 3.74 |
| Benzoic acid | $HC_7H_5O_2$ | H—O—C(=O)—C$_6$H$_5$ | $HC_7H_5O_2(aq) + H_2O(l) \rightleftharpoons H_3O^+(aq) + C_7H_5O_2^-(aq)$ | $6.5 \times 10^{-5}$ | 4.19 |
| Acetic acid | $HC_2H_3O_2$ | H—O—C(=O)—CH$_3$ | $HC_2H_3O_2(aq) + H_2O(l) \rightleftharpoons H_3O^+(aq) + C_2H_3O_2^-(aq)$ | $1.8 \times 10^{-5}$ | 4.74 |
| Hypochlorous acid | HClO | H—O—Cl | $HClO(aq) + H_2O(l) \rightleftharpoons H_3O^+(aq) + ClO^-(aq)$ | $2.9 \times 10^{-8}$ | 7.54 |
| Hydrocyanic acid | HCN | H—C≡N | $HCN(aq) + H_2O(l) \rightleftharpoons H_3O^+(aq) + CN^-(aq)$ | $4.9 \times 10^{-10}$ | 9.31 |
| Phenol | $HC_6H_5O$ | H—O—C$_6$H$_5$ | $HC_6H_5O(aq) + H_2O(l) \rightleftharpoons H_3O^+(aq) + C_6H_5O^-(aq)$ | $1.3 \times 10^{-10}$ | 9.89 |

**The Magnitude of the Acid Ionization Constant** Consider the three generic weak acids HA, HB, and HC. The images shown here represent the ionization of each acid at room temperature. Which acid has the largest $K_a$?

HA                HB                HC

**Relative Strengths of Weak Acids** Consider these two acids and their $K_a$ values:

$$\text{HF} \qquad K_a = 6.8 \times 10^{-4}$$
$$\text{HClO} \qquad K_a = 2.9 \times 10^{-8}$$

Which acid is stronger?

## 16.5 Autoionization of Water and pH

We saw previously that water acts as a base when it reacts with HCl and as an acid when it reacts with $NH_3$:

$$HCl(aq) + H_2O(l) \longrightarrow H_3O^+(aq) + Cl^-(aq)$$

**Acid** **Base**
(proton donor) (proton acceptor)

$$NH_3(aq) + H_2O(l) \rightleftharpoons NH_4^+(aq) + OH^-(aq)$$

**Base** **Acid**
(proton acceptor) (proton donor)

Water is *amphoteric*; it can act as either an acid or a base. Even when pure, water acts as an acid and a base with itself, a process called **autoionization**:

$$H_2O(l) + H_2O(l) \rightleftharpoons H_3O^+(aq) + OH^-(aq)$$

**Acid** **Base**
(proton donor) (proton acceptor)

We can write the autoionization reaction as:

$$H_2O(l) \rightleftharpoons H^+(aq) + OH^-(aq)$$

The equilibrium constant expression for this reaction is the product of the concentration of the two ions:

$$K_w = [H_3O^+][OH^-] = [H^+][OH^-]$$

This equilibrium constant is called the **ion product constant for water ($K_w$)** (sometimes called the *dissociation constant for water*). At 25 °C, $K_w = 1.0 \times 10^{-14}$. In pure water, since $H_2O$ is the only source of these ions, the concentrations of $H_3O^+$ and $OH^-$ are equal, and the solution is **neutral**. Since the concentrations of the two ions are equal, we can easily calculate them from $K_w$.

$$[H_3O^+] = [OH^-] = \sqrt{K_w} = 1.0 \times 10^{-7} \quad \text{(in pure water at 25 °C)}$$

As you can see, in pure water, the concentrations of $H_3O^+$ and $OH^-$ are *very small* ($1.0 \times 10^{-7}$ M) at room temperature.

An **acidic solution** contains an acid that creates additional $H_3O^+$ ions, causing $[H_3O^+]$ to increase. However, the *ion product constant still applies*:

$$[H_3O^+][OH^-] = K_w = 1.0 \times 10^{-14}$$

*The concentration of $H_3O^+$ times the concentration of $OH^-$ is always $1.0 \times 10^{-14}$ at 25 °C.* If $[H_3O^+]$ increases, then $[OH^-]$ must decrease for the ion product constant to remain $1.0 \times 10^{-14}$. For example, if $[H_3O^+] = 1.0 \times 10^{-3}$, then we can find $[OH^-]$ by solving the ion product constant expression for $[OH^-]$:

$$(1.0 \times 10^{-3})[OH^-] = 1.0 \times 10^{-14}$$

$$[OH^-] = \frac{1.0 \times 10^{-14}}{1.0 \times 10^{-3}} = 1.0 \times 10^{-11} \text{ M}$$

In an acidic solution $[H_3O^+] > [OH^-]$.

A **basic solution** contains a base that creates additional $OH^-$ ions, causing $[OH^-]$ to increase and $[H_3O^+]$ to decrease, but again the *ion product constant still applies*. Suppose $[OH^-] = 1.0 \times 10^{-2}$; we can find $[H_3O^+]$ by solving the ion product constant expression for $[H_3O^+]$:

$$[H_3O^+](1.0 \times 10^{-2}) = 1.0 \times 10^{-14}$$

$$[H_3O^+] = \frac{1.0 \times 10^{-14}}{1.0 \times 10^{-2}} = 1.0 \times 10^{-12} \text{ M}$$

In a basic solution $[OH^-] > [H_3O^+]$.

Notice that changing $[H_3O^+]$ in an aqueous solution produces an inverse change in $[OH^-]$ and vice versa.

*Summarizing $K_w$:*

▶ A *neutral solution* contains $[H_3O^+] = [OH^-] = 1.0 \times 10^{-7}$ M (at 25 °C).

▶ An *acidic solution* contains $[H_3O^+] > [OH^-]$.

▶ A *basic solution* contains $[OH^-] > [H_3O^+]$.

▶ In *all aqueous solutions* both $H_3O^+$ and $OH^-$ are present, with $[H_3O^+][OH^-] = K_w = 1.0 \times 10^{-14}$ (at 25 °C).

> The ion product constant $K_w$ depends on temperature. For all calculations in this book, assume 25 °C unless otherwise noted.

## Example 16.2    Using $K_w$ in Calculations

Calculate $[OH^-]$ at 25 °C for each solution and determine if the solution is acidic, basic, or neutral.

**(a)** $[H_3O^+] = 7.5 \times 10^{-5}$ M        **(b)** $[H_3O^+] = 1.5 \times 10^{-9}$ M

**(c)** $[H_3O^+] = 1.0 \times 10^{-7}$ M

**SOLUTION**

**(a)** To find $[OH^-]$ use the ion product constant. Substitute the given value for $[H_3O^+]$ and solve the equation for $[OH^-]$.

Since $[H_3O^+] > [OH^-]$, the solution is acidic.

$[H_3O^+][OH^-] = K_w = 1.0 \times 10^{-14}$

$(7.5 \times 10^{-5})[OH^-] = 1.0 \times 10^{-14}$

$[OH^-] = \dfrac{1.0 \times 10^{-14}}{7.5 \times 10^{-5}} = 1.3 \times 10^{-10} \text{ M}$

Acidic solution

*—Continued on the next page*

*Continued—*

| | |
|---|---|
| **(b)** Substitute the given value for $[H_3O^+]$ and solve the acid ionization equation for $[OH^-]$.<br><br>Since $[H_3O^+] < [OH^-]$, the solution is basic. | $(1.5 \times 10^{-9})[OH^-] = 1.0 \times 10^{-14}$<br><br>$[OH^-] = \dfrac{1.0 \times 10^{-14}}{1.5 \times 10^{-9}} = 6.7 \times 10^{-6}\,M$<br>Basic solution |
| **(c)** Substitute the given value for $[H_3O^+]$ and solve the acid ionization equation for $[OH^-]$.<br><br>Since $[H_3O^+] = 1.0 \times 10^{-7}$ and $[OH^-] = 1.0 \times 10^{-7}$, the solution is neutral. | $(1.0 \times 10^{-7})[OH^-] = 1.0 \times 10^{-14}$<br><br>$[OH^-] = \dfrac{1.0 \times 10^{-14}}{1.0 \times 10^{-7}} = 1.0 \times 10^{-7}\,M$<br>Neutral solution |

**FOR PRACTICE 16.2** Calculate $[H_3O^+]$ at 25 °C for each solution and determine if the solution is acidic, basic, or neutral.
**(a)** $[OH^-] = 1.5 \times 10^{-2}\,M$     **(b)** $[OH^-] = 1.0 \times 10^{-7}\,M$     **(c)** $[OH^-] = 8.2 \times 10^{-10}\,M$

## The pH Scale: A Way to Quantify Acidity and Basicity

The pH scale is a compact way to specify the acidity of a solution. We define **pH** as the negative of the log of the hydronium ion concentration:

$$pH = -\log[H_3O^+]$$

A solution with $[H_3O^+] = 1.0 \times 10^{-3}\,M$ (acidic) has a pH of:

$$pH = -\log[H_3O^+]$$
$$= -\log(1.0 \times 10^{-3})$$
$$= -(-3.00)$$
$$= 3.00$$

Notice that we report the pH to two *decimal places* in this example. This is because only the numbers to the right of the decimal point are significant in a logarithm. Because our original value for the concentration had two significant figures, the log of that number has two decimal places:

2 significant digits     2 decimal places

$$\log \underset{}{(1.0)} \times 10^{-3} = 3.\underset{}{00}$$

If the original number has three significant digits, we report the log to three decimal places:

3 significant digits     3 decimal places

$$\log \underset{}{(1.00)} \times 10^{-3} = 3.\underset{}{000}$$

A solution with $[H_3O^+] = 1.0 \times 10^{-7}\,M$ (neutral at 25 °C) has a pH of:

$$pH = -\log[H_3O^+]$$
$$= -\log(1.0 \times 10^{-7})$$
$$= -(-7.00)$$
$$= 7.00$$

In general, at 25 °C:

- pH < 7 The solution is *acidic*.
- pH > 7 The solution is *basic*.
- pH = 7 The solution is *neutral*.

Table 16.6 lists the pH of some common substances. As we discussed in Section 16.2, many foods, especially fruits, are acidic and have low pH values. Relatively few foods are basic. The foods with the lowest pH values are limes and lemons, and they are among the sourest. Because the pH scale is a *logarithmic scale*, a change of 1 pH unit corresponds to a 10-fold change in $H_3O^+$ concentration (Figure 16.7▶). For example, a lime with a pH of 2.0 is 10 times more acidic than a plum with a pH of 3.0 and 100 times more acidic than a cherry with a pH of 4.0.

*The log of a number is the exponent to which 10 must be raised to obtain that number. Thus, $\log 10^1 = 1$; $\log 10^2 = 2$; $\log 10^{-1} = -1$; $\log 10^{-2} = -2$, etc. (see Appendix I).*

*When we take the log of a quantity, the result has the same number of decimal places as the number of significant figures in the original quantity.*

*Concentrated acid solutions can have negative pH. For example, if $[H_3O^+] = 2.0\,M$, the pH is $-0.30$.*

**TABLE 16.6 The pH of Some Common Substances**

| Substance | pH |
|---|---|
| Gastric juice (human stomach) | 1.0–3.0 |
| Limes | 1.8–2.0 |
| Lemons | 2.2–2.4 |
| Soft drinks | 2.0–4.0 |
| Plums | 2.8–3.0 |
| Wines | 2.8–3.8 |
| Apples | 2.9–3.3 |
| Peaches | 3.4–3.6 |
| Cherries | 3.2–4.0 |
| Beers | 4.0–5.0 |
| Rainwater (unpolluted) | 5.6 |
| Human blood | 7.3–7.4 |
| Egg whites | 7.6–8.0 |
| Milk of magnesia | 10.5 |
| Household ammonia | 10.5–11.5 |
| 4% NaOH solution | 14 |

## The pH Scale

| 0 | 1 | 2 | 3 | 4 | 5 | 6 | 7 | 8 | 9 | 10 | 11 | 12 | 13 | 14 |

Acidic                                     pH                                      Basic

| $10^{-0}$ | $10^{-1}$ | $10^{-2}$ | $10^{-3}$ | $10^{-4}$ | $10^{-5}$ | $10^{-6}$ | $10^{-7}$ | $10^{-8}$ | $10^{-9}$ | $10^{-10}$ | $10^{-11}$ | $10^{-12}$ | $10^{-13}$ | $10^{-14}$ |

$[H^+]$

▲ **FIGURE 16.7 The pH Scale** An increase of 1 on the pH scale corresponds to a factor of 10 decrease in $[H_3O^+]$.

---

**Example 16.3**   **Calculating pH from $[H_3O^+]$ or $[OH^-]$**   ▶

Calculate the pH of each solution at 25 °C and indicate whether the solution is acidic or basic.

**(a)** $[H_3O^+] = 1.8 \times 10^{-4}$ M          **(b)** $[OH^-] = 1.3 \times 10^{-2}$ M

**SOLUTION**

**(a)** To calculate pH, substitute the given $[H_3O^+]$ into the pH equation. Since pH < 7, this solution is acidic.

$$\begin{aligned} pH &= -\log[H_3O^+] \\ &= -\log(1.8 \times 10^{-4}) \\ &= -(-3.74) \\ &= 3.74 \text{ (acidic)} \end{aligned}$$

**(b)** First use $K_w$ to find $[H_3O^+]$ from $[OH^-]$.

$$\begin{aligned} [H_3O^+][OH^-] &= K_w = 1.0 \times 10^{-14} \\ [H_3O^+](1.3 \times 10^{-2}) &= 1.0 \times 10^{-14} \\ [H_3O^+] &= \frac{1.0 \times 10^{-14}}{1.3 \times 10^{-2}} = 7.7 \times 10^{-13} \text{ M} \end{aligned}$$

Then substitute $[H_3O^+]$ into the pH expression to find pH.

$$\begin{aligned} pH &= -\log[H_3O^+] \\ &= -\log(7.7 \times 10^{-13}) \\ &= -(-12.11) \\ &= 12.11 \text{ (basic)} \end{aligned}$$

Since pH > 7, this solution is basic.

**FOR PRACTICE 16.3**   Calculate the pH of each solution and indicate whether the solution is acidic or basic.

**(a)** $[H_3O^+] = 9.5 \times 10^{-9}$ M          **(b)** $[OH^-] = 7.1 \times 10^{-3}$ M

---

**Example 16.4**   **Calculating $[H_3O^+]$ from pH**

Calculate $[H_3O^+]$ for a solution with a pH of 4.80.

**SOLUTION**

To find $[H_3O^+]$ from pH, start with the equation that defines pH. Substitute the given value of pH and then solve for $[H_3O^+]$. Since the given pH value was reported to two decimal places, report the $[H_3O^+]$ to two significant figures. (Remember that $10^{\log x} = x$ (see Appendix I). Some calculators use an inv log key to represent this function.)

$$\begin{aligned} pH &= -\log[H_3O^+] \\ 4.80 &= -\log[H_3O^+] \\ -4.80 &= \log[H_3O^+] \\ 10^{-4.80} &= 10^{\log[H_3O^+]} \\ 10^{-4.80} &= [H_3O^+] \\ [H_3O^+] &= 1.6 \times 10^{-5} \text{ M} \end{aligned}$$

**FOR PRACTICE 16.4**   Calculate $[H_3O^+]$ for a solution with a pH of 8.37.

---

## pOH and Other p Scales

The pOH scale is analogous to the pH scale but is defined with respect to $[OH^-]$ instead of $[H_3O^+]$:

$$pOH = -\log[OH^-]$$

| 0.0 | 1.0 | 2.0 | 3.0 | 4.0 | 5.0 | 6.0 | 7.0 | 8.0 | 9.0 | 10.0 | 11.0 | 12.0 | 13.0 | 14.0 |

Acidic           pH           Basic

| 14.0 | 13.0 | 12.0 | 11.0 | 10.0 | 9.0 | 8.0 | 7.0 | 6.0 | 5.0 | 4.0 | 3.0 | 2.0 | 1.0 | 0.0 |

pOH

▲ **FIGURE 16.8 pH and pOH**

Notice that p is the mathematical operator $-\log$; thus, $pX = -\log X$.

A solution with an $[OH^-]$ of $1.0 \times 10^{-3}$ M (basic) has a pOH of 3.00. On the pOH scale, a pOH less than 7 is basic and a pOH greater than 7 is acidic. A pOH of 7 is neutral (Figure 16.8▲). We can derive a relationship between pH and pOH at 25 °C from the expression for $K_w$:

$$[H_3O^+][OH^-] = 1.0 \times 10^{-14}$$

Taking the log of both sides, we get:

$$\log([H_3O^+][OH^-]) = \log(1.0 \times 10^{-14})$$
$$\log[H_3O^+] + \log[OH^-] = -14.00$$
$$-\log[H_3O^+] - \log[OH^-] = 14.00$$
$$pH + pOH = 14.00$$

The sum of pH and pOH is always equal to 14.00 at 25 °C. Therefore, a solution with a pH of 3 has a pOH of 11.

Another common p scale is the $pK_a$ scale defined as:

$$pK_a = -\log K_a$$

The $pK_a$ of a weak acid is another way to quantify its strength. The smaller the $pK_a$, the stronger the acid. For example, chlorous acid, with a $K_a$ of $1.1 \times 10^{-2}$, has a $pK_a$ of 1.96; and formic acid, with a $K_a$ of $1.8 \times 10^{-4}$, has a $pK_a$ of 3.74 (See Table 16.5).

CONCEPTUAL
**CONNECTION 16.4**

**pH and Acidity** As the pH of a solution increases (gets higher), what happens to the acidity of the solution?

**(a)** The acidity increases.      **(b)** The acidity decreases.

**(c)** The acidity remains constant.

# Chemistry AND Medicine | Ulcers

An ulcer is a lesion on the wall of the stomach or small intestine. Under normal circumstances, a thick layer of mucus lines the stomach wall and protects it from the hydrochloric acid and other gastric juices in the stomach. If that mucous layer is damaged, however, stomach juices come in direct contact with the stomach wall and begin to digest it, creating an ulcer. The main symptom of an ulcer is a burning or gnawing pain in the stomach.

Acidic drugs, such as aspirin, and acidic foods, such as citrus fruits and pickling fluids, irritate ulcers. When consumed, these substances increase the acidity of the stomach juices, exacerbating the irritation to the stomach wall. On the other hand, antacids—which contain bases—relieve ulcers. Common antacids include Tums and milk of magnesia.

The causes of ulcers are manifold. For many years, a stressful lifestyle and a rich diet were blamed. More recent research, however, has shown that a bacterial infection of the stomach lining is responsible for many ulcers. (The 2005 Nobel Prize in Physiology or Medicine was awarded to Australians Barry J. Marshall and J. Robin Warren, for their discovery of the bacterial cause of ulcers.) Long-term use of some over-the-counter pain relievers, such as aspirin, is also believed to produce ulcers.

▲ An ulcer is a lesion in the stomach wall.

## QUESTION

*Which dessert is less likely to irritate an ulcer, key lime pie or meringue (made of egg whites)?*

# 16.6 Finding the [H₃O⁺] and pH of Strong and Weak Acid Solutions

A solution containing a strong or weak acid has two potential sources of $H_3O^+$: the ionization of the acid itself and the autoionization of water. If we let HA be a strong or weak acid, the ionization equations are:

$$HA(aq) + H_2O(l) \rightleftharpoons H_3O^+(aq) + A^-(aq) \qquad \text{Strong or Weak Acid}$$

$$H_2O(l) + H_2O(l) \rightleftharpoons H_3O^+(aq) + OH^-(aq) \qquad K_w = 1.0 \times 10^{-14}$$

Except in extremely dilute acid solutions, the autoionization of water contributes a negligibly small amount of $H_3O^+$ compared to the ionization of the strong or weak acid. Recall from Section 16.5 that autoionization in pure water at 25 °C produces an $H_3O^+$ concentration of $1.0 \times 10^{-7}$ M. In a strong or weak acid solution, the additional $H_3O^+$ from the acid causes the autoionization of water equilibrium to shift left (as described by Le Châtelier's principle). Consequently, in most strong or weak acid solutions, the autoionization of water produces even less $H_3O^+$ than in pure water and can be ignored. Therefore we can focus exclusively on the amount of $H_3O^+$ produced by the acid.

The only exceptions would be extremely dilute ($<10^{-5}$ M) strong acid solutions.

## Strong Acids

Because strong acids, by definition, completely ionize in solution, and because we can (in nearly all cases) ignore the contribution of the autoionization of water, *the concentration of $H_3O^+$ in a strong acid solution is equal to the concentration of the strong acid*. For example, a 0.10 M HCl solution has an $H_3O^+$ concentration of 0.10 M and a pH of 1.00:

$$0.10 \text{ M HCl} \Rightarrow [H_3O^+] = 0.10 \text{ M} \Rightarrow pH = -\log(0.10) = 1.00$$

## Weak Acids

Finding the pH of a weak acid solution is more complicated because the concentration of $H_3O^+$ is *not equal* to the concentration of the weak acid. For example, if we make a solution of 0.10 M HCl (a strong acid) and a solution of 0.10 M acetic acid (a weak acid) in the laboratory and measure the pH of each, we get the following results:

| | |
|---|---|
| 0.10 M HCl | pH = 1.00 |
| 0.10 M HC₂H₃O₂ | pH = 2.87 |

The pH of the acetic acid solution is higher (it is less acidic) because acetic acid only partially ionizes. Calculating the $[H_3O^+]$ formed by the ionization of a weak acid requires solving an equilibrium problem similar to those in Chapter 15. Consider, for example, a 0.10 M solution of the generic weak acid HA with an acid ionization constant $K_a$. Since we can ignore the contribution of the autoionization of water, we only have to determine the concentration of $H_3O^+$ formed by the following equilibrium:

$$HA(aq) + H_2O(l) \rightleftharpoons H_3O^+(aq) + A^-(aq) \qquad K_a$$

We can summarize the initial conditions, the changes, and the equilibrium conditions in the following ICE table:

ICE tables were first introduced in Section 15.6. The reactant $H_2O(l)$ is a pure liquid and is therefore not included either in the equilibrium constant or in the ICE table (see Section 15.5).

| | [HA] | [H₃O⁺] | [A⁻] |
|---|---|---|---|
| Initial | 0.10 | ≈0.00 | 0.00 |
| Change | −x | +x | +x |
| Equilibrium | 0.10 − x | x | x |

The initial $H_3O^+$ concentration is listed as *approximately* zero because of the negligibly small contribution of $H_3O^+$ due to the autoionization of water (discussed previously). The variable $x$ represents the amount of HA that ionizes. As discussed in Chapter 15, each *equilibrium* concentration is the sum of the two entries above it in the ICE table. In order to find the equilibrium concentration of $H_3O^+$, we must find the value of the variable $x$. We can use the equilibrium expression to set up an equation in which $x$ is the only variable:

$$K_a = \frac{[H_3O^+][A^-]}{[HA]}$$

$$= \frac{x^2}{0.10 - x}$$

As is often the case with equilibrium problems, we arrive at a quadratic equation in $x$, which we can solve using the quadratic formula (see Appendix IC). However, in many cases we can apply the *x is small* approximation (first discussed in Section 15.8). In Examples 16.5 and 16.6, we examine the general procedure for solving weak acid equilibrium problems. In both of these examples, the *x is small* approximation works well. In Example 16.7, we solve a problem in which the *x is small* approximation does not work. In such cases, we can solve the quadratic equation explicitly, or apply the method of successive approximations (also discussed in Section 15.8). Finally, in Example 16.8, we work a problem in which we find the equilibrium constant of a weak acid from its pH.

# Procedure for...

| Finding the pH (or $[H_3O^+]$) of a Weak Acid Solution | **Example 16.5** ▶ Finding the $[H_3O^+]$ of a Weak Acid Solution | **Example 16.6** Finding the pH of a Weak Acid Solution |
|---|---|---|
| To solve these types of problems, follow the outlined procedure. | Find the $[H_3O^+]$ of a 0.100 M HCN solution. | Find the pH of a 0.200 M $HNO_2$ solution. |

1. **Write the balanced equation for the ionization of the acid and use it as a guide to prepare an ICE table showing the given concentration of the weak acid as its initial concentration.** Leave room in the table for the changes in concentrations and for the equilibrium concentrations. (Note that $[H_3O^+]$ is listed as approximately zero because the autoionization of water produces a negligibly small amount of $H_3O^+$.)

$HCN(aq) + H_2O(l) \rightleftharpoons$
$\qquad\qquad H_3O^+(aq) + CN^-(aq)$

| | [HCN] | [H₃O⁺] | [CN⁻] |
|---|---|---|---|
| Initial | 0.100 | ≈0.00 | 0.00 |
| Change | | | |
| Equil | | | |

$HNO_2(aq) + H_2O(l) \rightleftharpoons$
$\qquad\qquad H_3O^+(aq) + NO_2^-(aq)$

| | [HNO₂] | [H₃O⁺] | [NO₂⁻] |
|---|---|---|---|
| Initial | 0.200 | ≈0.00 | 0.00 |
| Change | | | |
| Equil | | | |

2. **Represent the change in the concentration of $H_3O^+$ with the variable $x$. Define the changes in the concentrations of the other reactants and products in terms of $x$. Always keep in mind the stoichiometry of the reaction.**

$HCN(aq) + H_2O(l) \rightleftharpoons$
$\qquad\qquad H_3O^+(aq) + CN^-(aq)$

| | [HCN] | [H₃O⁺] | [CN⁻] |
|---|---|---|---|
| Initial | 0.100 | ≈0.00 | 0.00 |
| Change | −x | +x | +x |
| Equil | | | |

$HNO_2(aq) + H_2O(l) \rightleftharpoons$
$\qquad\qquad H_3O^+(aq) + NO_2^-(aq)$

| | [HNO₂] | [H₃O⁺] | [NO₂⁻] |
|---|---|---|---|
| Initial | 0.200 | ≈0.00 | 0.00 |
| Change | −x | +x | +x |
| Equil | | | |

3. **Sum each column to determine the equilibrium concentrations in terms of the initial concentrations and the variable $x$.**

$HCN(aq) + H_2O(l) \rightleftharpoons$
$\qquad\qquad H_3O^+(aq) + CN^-(aq)$

| | [HCN] | [H₃O⁺] | [CN⁻] |
|---|---|---|---|
| Initial | 0.100 | ≈0.00 | 0.00 |
| Change | −x | +x | +x |
| Equil | 0.100 − x | x | x |

$HNO_2(aq) + H_2O(l) \rightleftharpoons$
$\qquad\qquad H_3O^+(aq) + NO_2^-(aq)$

| | [HNO₂] | [H₃O⁺] | [NO₂⁻] |
|---|---|---|---|
| Initial | 0.200 | ≈0.00 | 0.00 |
| Change | −x | +x | +x |
| Equil | 0.200 − x | x | x |

| | | |
|---|---|---|
| 4. **Substitute the expressions for the equilibrium concentrations (from step 3) into the expression for the acid ionization constant ($K_a$).** In many cases, you can make the approximation that $x$ is small (as discussed in Section 15.8). **Substitute the value of the acid ionization constant (from Table 16.5) into the $K_a$ expression and solve for $x$.** Confirm that the $x$ is small approximation is valid by calculating the ratio of $x$ to the number it was subtracted from in the approximation. The ratio should be less than 0.05 (or 5%). | $K_a = \dfrac{[H_3O^+][CN^-]}{[HCN]}$<br><br>$= \dfrac{x^2}{0.100 - x}$ $\quad$ (x is small)<br><br>$4.9 \times 10^{-10} = \dfrac{x^2}{0.100}$<br><br>$\sqrt{4.9 \times 10^{-10}} = \sqrt{\dfrac{x^2}{0.100}}$<br><br>$x = \sqrt{(0.100)(4.9 \times 10^{-10})}$<br><br>$= 7.0 \times 10^{-6}$<br><br>$\dfrac{7.0 \times 10^{-6}}{0.100} \times 100\% = 7.0 \times 10^{-3}\%$<br><br>Therefore the approximation is valid. | $K_a = \dfrac{[H_3O^+][NO_2^-]}{[HNO_2]}$<br><br>$= \dfrac{x^2}{0.200 - x}$ $\quad$ (x is small)<br><br>$4.6 \times 10^{-4} = \dfrac{x^2}{0.200}$<br><br>$\sqrt{4.6 \times 10^{-4}} = \sqrt{\dfrac{x^2}{0.200}}$<br><br>$x = \sqrt{(0.200)(4.6 \times 10^{-4})}$<br><br>$= 9.6 \times 10^{-3}$<br><br>$\dfrac{9.6 \times 10^{-3}}{0.200} \times 100\% = 4.8\%$<br><br>Therefore the approximation is valid (but barely so). |
| 5. **Determine [H₃O⁺] from the calculated value of $x$ and calculate the pH if necessary.** | $[H_3O^+] = 7.0 \times 10^{-6}$ M<br>(pH was not asked for in this problem.) | $[H_3O^+] = 9.6 \times 10^{-3}$ M<br>$pH = -\log [H_3O^+]$<br>$= -\log(9.6 \times 10^{-3})$<br>$= 2.02$ |
| 6. **Check your answer by substituting the calculated equilibrium values into the acid ionization expression.** The calculated value of $K_a$ should match the given value of $K_a$. Note that rounding errors and the $x$ is small approximation may result in a difference in the least significant digit when comparing values of $K_a$. | $K_a = \dfrac{[H_3O^+][CN^-]}{[HCN]} = \dfrac{(7.0 \times 10^{-6})^2}{0.100}$<br><br>$= 4.9 \times 10^{-10}$<br>Since the calculated value of $K_a$ matches the given value, the answer is valid. | $K_a = \dfrac{[H_3O^+][NO_2^-]}{[HNO_2]} = \dfrac{(9.6 \times 10^{-3})^2}{0.200}$<br><br>$= 4.6 \times 10^{-4}$<br>Since the calculated value of $K_a$ matches the given value, the answer is valid. |
| | **FOR PRACTICE 16.5**<br>Find the H₃O⁺ concentration of a 0.250 M hydrofluoric acid solution. | **FOR PRACTICE 16.6**<br>Find the pH of a 0.0150 M acetic acid solution. |

---

**Example 16.7** **Finding the pH of a Weak Acid Solution in Cases Where the x is small Approximation Does Not Work**

Find the pH of a 0.100 M HClO₂ solution.

**SOLUTION**

| | |
|---|---|
| 1. Write the balanced equation for the ionization of the acid and use it as a guide to prepare an ICE table showing the given concentration of the weak acid as its initial concentration. (Note that the H₃O⁺ concentration is listed as approximately zero. Although a little H₃O⁺ is present from the autoionization of water, this amount is negligibly small compared to the amount of HClO₂ or H₃O⁺ formed by the acid.) | $HClO_2(aq) + H_2O(l) \rightleftharpoons H_3O^+(aq) + ClO_2^-(aq)$<br><br>| | [HClO₂] | [H₃O⁺] | [ClO₂⁻] |<br>|---|---|---|---|<br>| Initial | 0.100 | ≈0.00 | 0.00 |<br>| Change | | | |<br>| Equil | | | | |
| 2. Represent the change in [H₃O⁺] with the variable $x$. Define the changes in the concentrations of the other reactants and products in terms of $x$. | $HClO_2(aq) + H_2O(l) \rightleftharpoons H_3O^+(aq) + ClO_2^-(aq)$<br><br>| | [HClO₂] | [H₃O⁺] | [ClO₂⁻] |<br>|---|---|---|---|<br>| Initial | 0.100 | ≈0.00 | 0.00 |<br>| Change | −x | +x | +x |<br>| Equil | | | | |

—Continued on the next page

*—Continued*

| 3. | Sum each column to determine the equilibrium concentrations in terms of the initial concentrations and the variable $x$. |
|---|---|

$$HClO_2(aq) + H_2O(l) \rightleftharpoons H_3O^+(aq) + ClO_2^-(aq)$$

| | $[HClO_2]$ | $[H_3O^+]$ | $[ClO_2^-]$ |
|---|---|---|---|
| Initial | 0.100 | $\approx 0.00$ | 0.00 |
| Change | $-x$ | $+x$ | $+x$ |
| Equil | $0.100 - x$ | $x$ | $x$ |

| 4. | Substitute the expressions for the equilibrium concentrations (from step 3) into the expression for the acid ionization constant ($K_a$). Make the *x is small* approximation and substitute the value of the acid ionization constant (from Table 16.5) into the $K_a$ expression. Solve for $x$. |
|---|---|

$$K_a = \frac{[H_3O^+][ClO_2^-]}{[HClO_2]}$$

$$= \frac{x^2}{0.100 - x} \quad (x \text{ is small})$$

$$0.011 = \frac{x^2}{0.100}$$

$$\sqrt{0.011} = \sqrt{\frac{x^2}{0.100}}$$

$$x = \sqrt{(0.100)(0.011)}$$

$$= 0.033$$

| | Check to see if the *x is small* approximation is valid by calculating the ratio of $x$ to the number it was subtracted from in the approximation. The ratio should be less than 0.05 (or 5%). |
|---|---|

$$\frac{0.033}{0.100} \times 100\% = 33\%$$

Therefore, the *x is small* approximation is *not valid*.

| 4a. | If the *x is small* approximation is not valid, solve the quadratic equation explicitly or use the method of successive approximations to find $x$. In this case, you solve the quadratic equation. |
|---|---|

$$0.011 = \frac{x^2}{0.100 - x}$$

$$0.011(0.100 - x) = x^2$$

$$0.0011 - 0.011x = x^2$$

$$x^2 + 0.011x - 0.0011 = 0$$

$$x = \frac{-b \pm \sqrt{b^2 - 4ac}}{2a}$$

$$= \frac{-(0.011) \pm \sqrt{(0.011)^2 - 4(1)(-0.0011)}}{2(1)}$$

$$= \frac{-0.011 \pm 0.0672}{2}$$

$$x = -0.039 \quad or \quad x = 0.028$$

Since $x$ represents the concentration of $H_3O^+$, and since concentrations cannot be negative, you reject the negative root.

$$x = 0.028$$

| 5. | Determine the $H_3O^+$ concentration from the calculated value of $x$ and calculate the pH (if necessary). |
|---|---|

$$[H_3O^+] = 0.028 \text{ M}$$
$$pH = -\log[H_3O^+]$$
$$= -\log 0.028$$
$$= 1.55$$

| 6. | Check your answer by substituting the calculated equilibrium values into the acid ionization expression. The calculated value of $K_a$ should match the given value of $K_a$. Note that rounding errors could result in a difference in the least significant digit when comparing values of $K_a$. |
|---|---|

$$K_a = \frac{[H_3O^+][ClO_2^-]}{[HClO_2]} = \frac{0.028^2}{0.100 - 0.028}$$

$$= 0.011$$

Since the calculated value of $K_a$ matches the given value, the answer is valid.

**FOR PRACTICE 16.7** Find the pH of a 0.010 M $HNO_2$ solution.

**Example 16.8**    **Finding the Equilibrium Constant from pH**

A 0.100 M weak acid (HA) solution has a pH of 4.25. Find $K_a$ for the acid.

**SOLUTION**

| | |
|---|---|
| Use the given pH to find the equilibrium concentration of $H_3O^+$. Then write the balanced equation for the ionization of the acid and use it as a guide to prepare an ICE table showing all known concentrations. | $pH = -\log[H_3O^+]$ <br> $4.25 = -\log[H_3O^+]$ <br> $[H_3O^+] = 5.6 \times 10^{-5}\,M$ <br> $HA(aq) + H_2O(l) \rightleftharpoons H_3O^+(aq) + A^-(aq)$ |

|  | [HA] | [H₃O⁺] | [A⁻] |
|---|---|---|---|
| Initial | 0.100 | ≈0.00 | 0.00 |
| Change | | | |
| Equil | | 5.6 × 10⁻⁵ | |

| | |
|---|---|
| Use the equilibrium concentration of $H_3O^+$ and the stoichiometry of the reaction to predict the changes and equilibrium concentration for all species. For most weak acids, the initial and equilibrium concentrations of the weak acid (HA) are effectively equal because the amount that ionizes is usually very small compared to the initial concentration. | $HA(aq) + H_2O(l) \rightleftharpoons H_3O^+(aq) + A^-(aq)$ |

|  | [HA] | [H₃O⁺] | [A⁻] |
|---|---|---|---|
| Initial | 0.100 | ≈0.00 | 0.00 |
| Change | −5.6 × 10⁻⁵ | +5.6 × 10⁻⁵ | +5.6 × 10⁻⁵ |
| Equil | (0.100−5.6 × 10⁻⁵) ≈0.100 | 5.6 × 10⁻⁵ | 5.6 × 10⁻⁵ |

| | |
|---|---|
| Substitute the equilibrium concentrations into the expression for $K_a$ and calculate its value. | $K_a = \dfrac{[H_3O^+][A^-]}{[HA]}$ <br> $= \dfrac{(5.6 \times 10^{-5})(5.6 \times 10^{-5})}{0.100}$ <br> $= 3.1 \times 10^{-8}$ |

**FOR PRACTICE 16.8**    A 0.175 M weak acid solution has a pH of 3.25. Find $K_a$ for the acid.

**The *x is small* Approximation**  The initial concentration and $K_a$ of several weak acid (HA) solutions are listed. For which solution is the *x is small* approximation *least* likely to work in finding the pH?

CONCEPTUAL CONNECTION **16.5**

(a) initial [HA] = 0.100 M; $K_a = 1.0 \times 10^{-5}$
(b) initial [HA] = 1.00 M; $K_a = 1.0 \times 10^{-6}$
(c) initial [HA] = 0.0100 M; $K_a = 1.0 \times 10^{-3}$
(d) initial [HA] = 1.0 M; $K_a = 1.5 \times 10^{-3}$

**Strong and Weak Acids**  Which solution is most acidic (that is, which one has the lowest pH)?

CONCEPTUAL CONNECTION **16.6**

(a) 0.10 M HCl
(b) 0.10 M HF
(c) 0.20 M HF

## Percent Ionization of a Weak Acid

We can quantify the ionization of a weak acid according to the percentage of acid molecules that actually ionizes. We define the **percent ionization** of a weak acid as the ratio of the ionized acid concentration to the initial acid concentration, multiplied by 100%:

$$\text{Percent ionization} = \frac{\text{concentration of ionized acid}}{\text{initial concentration of acid}} \times 100\% = \frac{[H_3O^+]_{equil}}{[HA]_{init}} \times 100\%$$

Since the concentration of ionized acid is equal to the $H_3O^+$ concentration at equilibrium (for a monoprotic acid), we can use $[H_3O^+]_{equil}$ and $[HA]_{init}$ in the formula to calculate the percent ionization. For example, in Example 16.6, we found that a 0.200 M $HNO_2$ solution contains $9.6 \times 10^{-3}$ M $H_3O^+$. The 0.200 M $HNO_2$ solution therefore has the following percent ionization:

$$\% \text{ ionization} = \frac{[H_3O^+]_{equil}}{[HA]_{init}} \times 100\%$$

$$= \frac{9.6 \times 10^{-3} \text{ M}}{0.200 \text{ M}} \times 100\%$$

$$= 4.8\%$$

As you can see, the percent ionization is relatively small. In this case, even though $HNO_2$ has the second largest $K_a$ in Table 16.5, fewer than 5 molecules out of 100 ionize. For most other weak acids (with smaller $K_a$ values) the percent ionization is even less.

In Example 16.9, we calculate the percent ionization of a more concentrated $HNO_2$ solution. In the example, notice that the calculated $H_3O^+$ concentration is much greater (as we would expect for a more concentrated solution), but the *percent ionization* is actually smaller.

---

**Example 16.9    Finding the Percent Ionization of a Weak Acid**

Find the percent ionization of a 2.5 M $HNO_2$ solution.

**SOLUTION**

| | | |
|---|---|---|
| To find the percent ionization, you must find the equilibrium concentration of $H_3O^+$. Follow the procedure in Example 16.5, shown in condensed form here. | $HNO_2(aq) + H_2O(l) \rightleftharpoons H_3O^+(aq) + NO_2^-(aq)$ | |

| | $[HNO_2]$ | $[H_3O^+]$ | $[NO_2^-]$ |
|---|---|---|---|
| Initial | 2.5 | $\approx 0.00$ | 0.00 |
| Change | $-x$ | $+x$ | $+x$ |
| Equil | $2.5 - x$ | $x$ | $x$ |

$$K_a = \frac{[H_3O^+][NO_2^-]}{[HNO_2]} = \frac{x^2}{2.5 - x} \quad (x \text{ is small})$$

$$4.6 \times 10^{-4} = \frac{x^2}{2.5}$$

$$x = 0.034$$

Therefore, $[H_3O^+] = 0.034$ M.

| | |
|---|---|
| Use the definition of percent ionization to calculate it. (Since the percent ionization is less than 5%, the *x is small* approximation is valid.) | $\% \text{ ionization} = \frac{[H_3O^+]_{equil}}{[HA]_{init}} \times 100\%$ $= \frac{0.034 \text{ M}}{2.5 \text{ M}} \times 100\%$ $= 1.4\%$ |

**FOR PRACTICE 16.9**    Find the percent ionization of a 0.250 M $HC_2H_3O_2$ solution at 25 °C.

We can now compare the results of Examples 16.6 and 16.9. From the table shown here, we can see that the more concentrated $HNO_2$ solution has a lower percent ionization. This trend applies to all weak acid solutions.

| [HNO₂] | [H₃O⁺] | Percent Ionization |
|--------|--------|--------------------|
| 0.200  | 0.0096 | 4.8%               |
| 2.5    | 0.034  | 1.4%               |

*Summarizing [H₃O⁺] and Percent Ionization:*

▶ The *equilibrium H₃O⁺ concentration* of a weak acid *increases* with increasing initial concentration of the acid.

▶ The *percent ionization* of a weak acid *decreases* with increasing concentration of the acid.

In other words, as the concentration of a weak acid solution increases, the concentration of the hydronium ion also increases, but the increase is not linear. The $H_3O^+$ concentration increases less than the concentration of the acid because as the acid concentration increases, a smaller fraction of weak acid molecules ionize.

We can understand this behavior by analogy with Le Châtelier's principle. Consider the following weak acid ionization equilibrium:

$$HA(aq) \rightleftharpoons H^+(aq) + A^-(aq)$$

1 mol dissolved particles    2 mol dissolved particles

If we dilute a weak acid solution initially at equilibrium, the system (according to Le Châtelier's principle) responds to minimize the disturbance. The equilibrium shifts to the right because the right side of the equation contains more particles in solution (2 mol versus 1 mol) than the left side. If the system shifts to the right, the percent ionization is greater in the more dilute solution, which is what we observe.

**CONCEPTUAL CONNECTION 16.7**

**Percent Ionization**  Which weak acid solution has the greatest percent ionization? Which solution has the lowest (most acidic) pH?

**(a)**  0.100 M $HC_2H_3O_2$
**(b)**  0.500 M $HC_2H_3O_2$
**(c)**  0.0100 M $HC_2H_3O_2$

## Mixtures of Acids

Finding the pH of a mixture of acids may seem difficult at first. However, in many cases, the relative strengths of the acids in the mixture allow us to neglect the weaker acid and focus only on the stronger one. Here, we consider two possible acid mixtures: a strong acid with a weak acid and a weak acid with another weak acid.

### A Strong Acid and a Weak Acid

Consider a mixture that is 0.10 M in HCl and 0.10 M in $HCHO_2$. There are three sources of $H_3O^+$ ions: the strong acid (HCl), the weak acid ($HCHO_2$), and the autoionization of water:

$$HCl(aq) + H_2O(l) \longrightarrow H_3O^+(aq) + Cl^-(aq) \qquad \text{Strong}$$
$$HCHO_2(aq) + H_2O(l) \rightleftharpoons H_3O^+(aq) + CHO_2^-(aq) \qquad K_a = 1.8 \times 10^{-4}$$
$$H_2O(l) + H_2O(l) \rightleftharpoons H_3O^+(aq) + OH^-(aq) \qquad K_w = 1.0 \times 10^{-14}$$

Since HCl is strong, we know that it completely ionizes to produce a significant concentration of $H_3O^+$(0.10 M). The $H_3O^+$ formed by HCl then *suppresses* the formation of additional $H_3O^+$ formed by the ionization of $HCHO_2$ or the autoionization of water. In other words, according to Le Châtelier's principle, the formation of $H_3O^+$ by the strong acid causes the weak acid to ionize even less than it would in the absence of the strong acid. To see this clearly, we calculate $[H_3O^+]$ and $[CHO_2^-]$ in this solution.

In an initial estimate of $[H_3O^+]$, we can neglect the contribution of $HCHO_2$ and $H_2O$. The concentration of $H_3O^+$ is then equal to the initial concentration of HCl:

$$[H_3O^+] = [HCl] = 0.10\,M$$

To find $[CHO_2^-]$ we must solve an equilibrium problem. However, the initial concentration of $H_3O^+$ in this case is not negligible (as it has been in all the other weak acid equilibrium problems that we have worked so far) because HCl has formed a significant amount of $H_3O^+$. The concentration of $H_3O^+$ formed by HCl becomes the *initial* concentration of $H_3O^+$ in the ICE table for $HCHO_2$ as shown here:

$$HCHO_2(aq) + H_2O(l) \rightleftharpoons H_3O^+(aq) + CHO_2^-(aq)$$

|         | $[HCHO_2]$ | $[H_3O^+]$ | $[CHO_2^-]$ |
|---------|-----------|-----------|------------|
| Initial | 0.10      | 0.10      | 0.00       |
| Change  | $-x$      | $+x$      | $+x$       |
| Equil   | $0.10 - x$ | $0.10 + x$ | $x$      |

We then use the equilibrium expression to set up an equation in which $x$ is the only variable:

$$K_a = \frac{[H_3O^+][CHO_2^-]}{[HCHO_2]}$$

$$= \frac{(0.10 + x)x}{0.10 - x}$$

Since the equilibrium constant is small relative to the initial concentration of the acid, we can make the *x is small* approximation:

$$K_a = \frac{(0.10 + \cancel{x})x}{0.10 - \cancel{x}}$$

$$1.8 \times 10^{-4} = \frac{(0.10)x}{0.10}$$

$$x = 1.8 \times 10^{-4}$$

Checking the *x is small* approximation:

$$\frac{1.8 \times 10^{-4}}{0.10} \times 100\% = 0.18\%$$

We find that the approximation is valid. Therefore, $[CHO_2^-] = 1.8 \times 10^{-4}\,M$. We can now see that we can completely ignore the ionization of the weak acid ($HCHO_2$) in calculating $[H_3O^+]$ for the mixture. The contribution to the concentration of $H_3O^+$ by the weak acid must necessarily be equal to the concentration of $CHO_2^-$ that we just calculated (because of the stoichiometry of the ionization reaction). Therefore, we have the following contributions to $[H_3O^+]$:

HCl contributes 0.10 M

$HCHO_2$ contributes $1.8 \times 10^{-4}\,M$ or 0.00018 M

Total $[H_3O^+] = 0.10\,M + 0.00018\,M = 0.10\,M$

As we can see, since the significant figure rules for addition limit the answer to two decimal places, the amount of $H_3O^+$ contributed by $HCHO_2$ is completely negligible. The amount of $H_3O^+$ contributed by the autoionization of water is even smaller and therefore similarly negligible.

## A Mixture of Two Weak Acids

When two weak acids are mixed, we again have three potential sources of $H_3O^+$ to consider: each of the two weak acids and the autoionization of water. However, if the values of $K_a$ for the two weak acids are sufficiently different in magnitude (if they differ by more than a factor of several hundred), then as long as the concentrations of the two acids are similar in magnitude (or the concentration of the stronger one is greater than that of the weaker), we can assume that the weaker acid will not make a significant contribution to the concentration of $H_3O^+$. We make this assumption for the same reason that we made a similar assumption for a mixture of a strong acid and a weak one: the $H_3O^+$ formed by the stronger acid suppresses the ionization of the weaker one, in accordance with Le Châtelier's principle. Example 16.10 shows how to calculate the concentration of $H_3O^+$ in a mixture of two weak acids.

---

**Example** 16.10 | **Mixtures of Weak Acids**

Find the pH of a mixture that is 0.300 M in HF and 0.100 M in HClO.

SOLUTION

| | |
|---|---|
| The three possible sources of $H_3O^+$ ions are HF, HClO, and $H_2O$. Write the ionization equations for the three sources and their corresponding equilibrium constants. Since the equilibrium constant for the ionization of HF is about 24,000 times larger than that for the ionization of HClO, the contribution of HF to [$H_3O^+$] is by far the greatest. You can therefore just calculate the [$H_3O^+$] formed by HF and neglect the other two potential sources of $H_3O^+$. | $HF(aq) + H_2O(l) \rightleftharpoons H_3O^+(aq) + F^-(aq)$    $K_a = 6.8 \times 10^{-4}$ <br><br> $HClO(aq) + H_2O(l) \rightleftharpoons H_3O^+(aq) + ClO^-(aq)$    $K_a = 2.9 \times 10^{-8}$ <br><br> $H_2O(l) + H_2O(l) \rightleftharpoons H_3O^+(aq) + OH^-(aq)$    $K_w = 1.0 \times 10^{-14}$ |

| | |
|---|---|
| Write the balanced equation for the ionization of HF and use it as a guide to prepare an ICE table. | $HF(aq) + H_2O(l) \rightleftharpoons H_3O^+(aq) + F^-(aq)$ |

|  | [HF] | [$H_3O^+$] | [$F^-$] |
|---|---|---|---|
| Initial | 0.300 | $\approx 0.00$ | 0.00 |
| Change | $-x$ | $+x$ | $+x$ |
| Equil | $0.300 - x$ | $x$ | $x$ |

| | |
|---|---|
| Substitute the expressions for the equilibrium concentrations into the expression for the acid ionization constant ($K_a$). Since the equilibrium constant is small relative to the initial concentration of HF, you can make the *x is small* approximation. Substitute the value of the acid ionization constant (from Table 16.5) into the $K_a$ expression and solve for $x$. | $K_a = \dfrac{[H_3O^+][F^-]}{[HF]} = \dfrac{x^2}{0.300 - x}$ (x is small) <br><br> $6.8 \times 10^{-4} = \dfrac{x^2}{0.300}$ <br><br> $\sqrt{(0.300)(6.8 \times 10^{-4})} = \sqrt{x^2}$ <br><br> $x = 1.4\underline{2}8 \times 10^{-2}$ |
| Confirm that the *x is small* approximation is valid by calculating the ratio of $x$ to the number it was subtracted from in the approximation. The ratio should be less than 0.05 (or 5%). | $\dfrac{1.4\underline{2}8 \times 10^{-2}}{0.300} \times 100\% = 4.8\%$ <br><br> Therefore, the approximation is valid (though barely so). |

| | |
|---|---|
| Determine the $H_3O^+$ concentration from the calculated value of $x$ and find the pH. | $[H_3O^+] = 1.4 \times 10^{-2}$ M <br> $pH = -\log(1.4 \times 10^{-2}) = 1.85$ |

FOR PRACTICE 16.10   Find the $ClO^-$ concentration of the mixture of HF and HClO discussed in Example 16.10.

---

**Judging Relative pH**   Which solution is most acidic (that is, has the lowest pH)?

CONCEPTUAL
CONNECTION 16.8

(a) 1.0 M HCl

(b) 2.0 M HF

(c) A solution that is 1.0 M in HF and 1.0 M in HClO

## 16.7 Base Solutions

We have seen that bases are solutions that produce $OH^-$ in solution (Arrhenius definition) or accept protons (Brønsted–Lowry definition). In analogy to acids, which can be strong or weak, bases can also be strong or weak.

### Strong Bases

Just as we define a strong acid as one that completely ionizes in solution, analogously we define a **strong base** as a base that completely dissociates in solution. NaOH, for example, is a strong base:

$$NaOH(aq) \longrightarrow Na^+(aq) + OH^-(aq)$$

An NaOH solution contains no intact NaOH—it has all dissociated to form $Na^+(aq)$ and $OH^-(aq)$ (Figure 16.9▼). In other words, a 1.0 M NaOH solution has $[OH^-] = 1.0$ M and $[Na^+] = 1.0$ M. Table 16.7 lists the common strong bases.

► **FIGURE 16.9** **Ionization of a Strong Base** When NaOH dissolves in water, it dissociates completely into $Na^+$ and $OH^-$. The solution contains virtually no intact NaOH.

### A Strong Base

| TABLE 16.7 Strong Bases | |
|---|---|
| Lithium hydroxide (LiOH) | Strontium hydroxide [Sr(OH)$_2$] |
| Sodium hydroxide (NaOH) | Calcium hydroxide [Ca(OH)$_2$] |
| Potassium hydroxide (KOH) | Barium hydroxide [Ba(OH)$_2$] |

As we can see in the table, most strong bases are group 1A or group 2A metal hydroxides. The group 1A metal hydroxides are highly soluble in water and can form concentrated base solutions. The group 2A metal hydroxides, however, are only slightly soluble, a useful property for some applications (see *Chemistry and Medicine: What's in My Antacid?* in this section). Notice that the general formula for the group 2A metal hydroxides is M(OH)$_2$. When they dissolve in water, they produce 2 mol of $OH^-$ per mole of the base. For example, Sr(OH)$_2$ dissociates as follows:

$$Sr(OH)_2(aq) \longrightarrow Sr^{2+}(aq) + 2\,OH^-(aq)$$

Unlike diprotic acids, which ionize in two steps, bases containing two $OH^-$ ions dissociate in one step.

### Weak Bases

The behavior of a **weak base** is analogous to that of a weak acid. Unlike strong bases that contain $OH^-$ and *dissociate* in water, the most common weak bases produce $OH^-$ by accepting a proton from water, ionizing water to form $OH^-$ according to the general equation:

$$B(aq) + H_2O(l) \rightleftharpoons BH^+(aq) + OH^-(aq)$$

In this equation, B is a generic symbol for a weak base. Ammonia, for example, ionizes water as follows:

$$NH_3(aq) + H_2O(l) \rightleftharpoons NH_4^+(aq) + OH^-(aq)$$

## A Weak Base

The double arrow indicates that the ionization is not complete. An $NH_3$ solution contains mostly $NH_3$ with only some $NH_4^+$ and $OH^-$ (Figure 16.10▲). A 1.0 M $NH_3$ solution will have $[OH^-] < 1.0$ M.

The extent of ionization of a weak base is quantified with the **base ionization constant**, $K_b$. For the general reaction in which a weak base ionizes water, we define $K_b$ as follows:

$$B(aq) + H_2O(l) \rightleftharpoons BH^+(aq) + OH^-(aq) \qquad K_b = \frac{[BH^+][OH^-]}{[B]}$$

By analogy with $K_a$, the smaller the value of $K_b$, the weaker the base. Table 16.8 lists some common weak bases, their ionization reactions, and values for $K_b$. Appendix IIC contains a more complete table. We can also apply the "p" scale to $K_b$, so that $pK_b = -\log K_b$.

All but two of the weak bases listed in Table 16.8 are either ammonia or an amine, which we can think of as ammonia with one or more hydrocarbon groups substituted for one or more hydrogen atoms. All of these bases have a nitrogen atom with a lone pair (Figure 16.11▶). This lone pair acts as the proton acceptor that makes the substance a base, as shown in the reactions for ammonia and methylamine:

$$H-\overset{\cdot\cdot}{\underset{\underset{H}{|}}{N}}-H(aq) + H-\overset{\cdot\cdot}{\underset{\cdot\cdot}{O}}-H(l) \rightleftharpoons H-\overset{\overset{H}{|}}{\underset{\underset{H}{|}}{N^+}}-H(aq) + :\overset{\cdot\cdot}{\underset{\cdot\cdot}{O}}-H$$

$$H-\overset{\overset{H}{|}}{\underset{\underset{H}{|}}{C}}-\overset{\cdot\cdot}{\underset{\underset{H}{|}}{N}}-H(aq) + H-\overset{\cdot\cdot}{\underset{\cdot\cdot}{O}}-H(l) \rightleftharpoons H-\overset{\overset{H}{|}}{\underset{\underset{H}{|}}{C}}-\overset{\overset{H}{|}}{\underset{\underset{H}{|}}{N^+}}-H(aq) + :\overset{\cdot\cdot}{\underset{\cdot\cdot}{O}}-H$$

### TABLE 16.8 Some Common Weak Bases

| Weak Base | Ionization Reaction | $K_b$ (at 25 °C) |
|---|---|---|
| Carbonate ion ($CO_3^{2-}$)* | $CO_3^{2-}(aq) + H_2O(l) \rightleftharpoons HCO_3^-(aq) + OH^-(aq)$ | $1.8 \times 10^{-4}$ |
| Methylamine ($CH_3NH_2$) | $CH_3NH_2(aq) + H_2O(l) \rightleftharpoons CH_3NH_3^+(aq) + OH^-(aq)$ | $4.4 \times 10^{-4}$ |
| Ethylamine ($C_2H_5NH_2$) | $C_2H_5NH_2(aq) + H_2O(l) \rightleftharpoons C_2H_5NH_3^+(aq) + OH^-(aq)$ | $5.6 \times 10^{-4}$ |
| Ammonia ($NH_3$) | $NH_3(aq) + H_2O(l) \rightleftharpoons NH_4^+(aq) + OH^-(aq)$ | $1.76 \times 10^{-5}$ |
| Bicarbonate ion ($HCO_3^-$)* (or hydrogen carbonate) | $HCO_3^-(aq) + H_2O(l) \rightleftharpoons H_2CO_3(aq) + OH^-(aq)$ | $2.3 \times 10^{-8}$ |
| Pyridine ($C_5H_5N$) | $C_5H_5N(aq) + H_2O(l) \rightleftharpoons C_5H_5NH^+(aq) + OH^-(aq)$ | $1.7 \times 10^{-9}$ |
| Aniline ($C_6H_5NH_2$) | $C_6H_5NH_2(aq) + H_2O(l) \rightleftharpoons C_6H_5NH_3^+(aq) + OH^-(aq)$ | $3.9 \times 10^{-10}$ |

*The carbonate and bicarbonate ions must occur with a positively charged ion such as $Na^+$ that serves to balance the charge but does not have any part in the ionization reaction. For example, it is the bicarbonate ion that makes sodium bicarbonate ($NaHCO_3$) basic. We look more closely at ionic bases in Section 16.8.

▶ FIGURE 16.11 **Lone Pairs in Weak Bases** Many weak bases have a nitrogen atom with a lone pair that acts as the proton acceptor.

Ammonia    Methylamine    Pyridine

## Finding the [OH⁻] and pH of Basic Solutions

Finding the [OH⁻] and pH of a strong base solution is relatively straightforward, as shown in Example 16.11. As we did in calculating the [$H_3O^+$] in strong acid solutions, we can neglect the contribution of the autoionization of water to the [OH⁻] and focus solely on the strong base itself.

**Example 16.11** | **Finding the [OH⁻] and pH of a Strong Base Solution**

Determine the OH⁻ concentration and pH in each solution.

**(a)** 0.225 M KOH                    **(b)** 0.0015 M Sr(OH)$_2$

SOLUTION

**(a)** Since KOH is a strong base, it completely dissociates into K⁺ and OH⁻ in solution. The concentration of OH⁻ is therefore the same as the given concentration of KOH.
Use this concentration and $K_w$ to find [$H_3O^+$].

$$KOH(aq) \longrightarrow K^+(aq) + OH^-(aq)$$
$$[OH^-] = 0.225 \text{ M}$$

$$[H_3O^+][OH^-] = K_w = 1.0 \times 10^{-14}$$
$$[H_3O^+](0.225) = 1.0 \times 10^{-14}$$
$$[H_3O^+] = 4.\underline{4}4 \times 10^{-14} \text{ M}$$

Then substitute [$H_3O^+$] into the pH expression to find the pH.

$$pH = -\log[H_3O^+]$$
$$= -\log(4.44 \times 10^{-14})$$
$$= 13.35$$

**(b)** Since Sr(OH)$_2$ is a strong base, 1 mol of Sr(OH)$_2$ completely dissociates into 1 mol of Sr$^{2+}$ and 2 mol of OH⁻ in solution. The concentration of OH⁻ is therefore twice the given concentration of Sr(OH)$_2$.

$$Sr(OH)_2(aq) \longrightarrow Sr^{2+}(aq) + 2\,OH^-(aq)$$
$$[OH^-] = 2(0.0015) \text{ M}$$
$$= 0.0030 \text{ M}$$

Use this concentration and $K_w$ to find [$H_3O^+$].

$$[H_3O^+][OH^-] = K_w = 1.0 \times 10^{-14}$$
$$[H_3O^+](0.0030) = 1.0 \times 10^{-14}$$
$$[H_3O^+] = 3.3 \times 10^{-12} \text{ M}$$

Substitute [$H_3O^+$] into the pH expression to find the pH.

$$pH = -\log[H_3O^+]$$
$$= -\log(3.3 \times 10^{-12})$$
$$= 11.48$$

**FOR PRACTICE 16.11** Find the [OH⁻] and pH of a 0.010 M Ba(OH)$_2$ solution.

Finding the [OH⁻] and pH of a *weak base* solution is analogous to finding the [$H_3O^+$] and pH of a weak acid. Similarly, we neglect the contribution of the autoionization of water to the [OH⁻] and focus solely on the weak base itself. We find the contribution of the weak base by preparing an ICE table showing the relevant concentrations of all species and then use the base ionization constant expression to find the [OH⁻]. Example 16.12 demonstrates how to find the [OH⁻] and pH of a weak base solution.

**Example 16.12**    **Finding the [OH⁻] and pH of a Weak Base Solution**

Find the [OH⁻] and pH of a 0.100 M NH₃ solution.

**SOLUTION**

1. Write the balanced equation for the ionization of water by the base and use it as a guide to prepare an ICE table showing the given concentration of the weak base as its initial concentration. Leave room in the table for the changes in concentrations and for the equilibrium concentrations. (Note that you should list the OH⁻ concentration as approximately zero. Although a little OH⁻ is present from the autoionization of water, this amount is negligibly small compared to the amount of OH⁻ formed by the base.)

$$NH_3(aq) + H_2O(l) \rightleftharpoons NH_4^+(aq) + OH^-(aq)$$

|  | [NH₃] | [NH₄⁺] | [OH⁻] |
|---|---|---|---|
| Initial | 0.100 | 0.00 | ≈0.00 |
| Change |  |  |  |
| Equil |  |  |  |

2. Represent the change in the concentration of OH⁻ with the variable $x$. Define the changes in the concentrations of the other reactants and products in terms of $x$.

$$NH_3(aq) + H_2O(l) \rightleftharpoons NH_4^+(aq) + OH^-(aq)$$

|  | [NH₃] | [NH₄⁺] | [OH⁻] |
|---|---|---|---|
| Initial | 0.100 | 0.00 | ≈0.00 |
| Change | −x | +x | +x |
| Equil |  |  |  |

3. Sum each column to determine the equilibrium concentrations in terms of the initial concentrations and the variable $x$.

$$NH_3(aq) + H_2O(l) \rightleftharpoons NH_4^+(aq) + OH^-(aq)$$

|  | [NH₃] | [NH₄⁺] | [OH⁻] |
|---|---|---|---|
| Initial | 0.100 | 0.00 | ≈0.00 |
| Change | −x | +x | +x |
| Equil | 0.100 − x | x | x |

4. Substitute the expressions for the equilibrium concentrations (from step 3) into the expression for the base ionization constant. In many cases, you can make the approximation that $x$ is small (as discussed in Chapter 15).

Substitute the value of the base ionization constant (from Table 16.8) into the $K_b$ expression and solve for $x$.

$$K_b = \frac{[NH_4^+][OH^-]}{[NH_3]}$$

$$= \frac{x^2}{0.100 - x} \quad (x \text{ is small})$$

$$1.76 \times 10^{-5} = \frac{x^2}{0.100}$$

$$\sqrt{1.76 \times 10^{-5}} = \sqrt{\frac{x^2}{0.100}}$$

$$x = \sqrt{(0.100)(1.76 \times 10^{-5})}$$

$$= 1.33 \times 10^{-3}$$

Confirm that the $x$ is small approximation is valid by calculating the ratio of $x$ to the number it was subtracted from in the approximation. The ratio should be less than 0.05 (or 5%).

$$\frac{1.33 \times 10^{-3}}{0.100} \times 100\% = 1.33\%$$

Therefore, the approximation is valid.

5. Determine the OH⁻ concentration from the calculated value of $x$. Use the expression for $K_w$ to find [H₃O⁺].

$$[OH^-] = 1.33 \times 10^{-3} \, M$$
$$[H_3O^+][OH^-] = K_w = 1.00 \times 10^{-14}$$
$$[H_3O^+](1.33 \times 10^{-3}) = 1.00 \times 10^{-14}$$
$$[H_3O^+] = 7.52 \times 10^{-12} \, M$$

Substitute [H₃O⁺] into the pH equation to find pH.

$$pH = -\log[H_3O^+]$$
$$= -\log(7.52 \times 10^{-12})$$
$$= 11.124$$

**FOR PRACTICE 16.12**    Find the [OH⁻] and pH of a 0.33 M methylamine solution.

# Chemistry AND Medicine | What's in My Antacid?

In the opening section of this chapter, we discussed heartburn and its treatment with antacids. Some common antacids and their active ingredients include the following:

| | |
|---|---|
| Amphogel | $Al(OH)_3$ |
| Milk of magnesia | $Mg(OH)_2$ |
| Maalox | $Mg(OH)_2$ and $Al(OH)_3$ |
| Mylanta | $Mg(OH)_2$ and $Al(OH)_3$ |
| Tums | $CaCO_3$ |

We categorize antacids into three main groups: calcium-based, magnesium-based, and aluminum-based. Calcium-based antacids may cause acid rebound—which means that, although they initially relieve heartburn, they can also cause the stomach to produce more acid, resulting in a quick return of the symptoms. Aluminum- and magnesium-based antacids do not cause acid rebound but have downsides of their own. Aluminum-based antacids tend to cause constipation, and magnesium-based ones tend to cause diarrhea. (In fact, milk of magnesia is sometimes used as a laxative.) A person who takes repeated doses of these antacids should alternate between the two or choose a product that contains both.

Notice the absence of group 1A metal hydroxides, such as KOH or NaOH, in the list of antacids. Why are those substances—which are completely soluble and act as strong bases—not used in antacids? Because a solution that contains sufficient KOH or NaOH to neutralize stomach acid would also burn the mouth and throat. In contrast, $Mg(OH)_2$ and $Al(OH)_3$ are only slightly soluble. Therefore liquid antacids containing these are actually *suspensions* of $Mg(OH)_2$ and $Al(OH)_3$—they are heterogeneous mixtures in which the solid is finely divided into the liquid. As a result, the concentration of $OH^-$ in these suspensions is relatively small compared to what it would be with a group 1A metal hydroxide.

▶ Antacids contain a variety of bases that effectively neutralize excess stomach acid.

Initially, it might seem as though the relatively lower $OH^-$ concentration would make the antacid much less effective. However, the solid $Mg(OH)_2$ or $Al(OH)_3$ continues to dissolve as the $OH^-$ neutralizes stomach acid. For example, a suspension of magnesium hydroxide contains solid $Mg(OH)_2$ in equilibrium with dissolved $Mg^{2+}$ and $OH^-$ ions:

$$Mg(OH)_2(s) \rightleftharpoons Mg^{2+}(aq) + 2\,OH^-(aq)$$

As stomach acid is neutralized, $OH^-$ is used up, and the equilibrium shifts to the right providing additional $OH^-$ ions. In this way, a suspension of $Mg(OH)_2$ provides a steady concentration of dissolved $OH^-$ ions to neutralize stomach acid.

### QUESTION

*Write chemical equations showing the reactions of each of the bases in the antacids discussed here with stomach acid (HCl).*

---

 The Acid–Base Properties of Ions and Salts

## 16.8  The Acid–Base Properties of Ions and Salts

We have already seen that some ions act as bases. For example, the bicarbonate ion acts as a base according to the following equation:

$$HCO_3^-(aq) + H_2O(l) \rightleftharpoons H_2CO_3(aq) + OH^-(aq)$$

The bicarbonate ion, like any ion, does not exist by itself—in order to be charge-neutral, it must pair with a counter ion (in this case a cation) to form an ionic compound, called a *salt*. For example, the sodium salt of bicarbonate is sodium bicarbonate. Like all soluble salts, sodium bicarbonate dissociates in solution to form a sodium cation and bicarbonate anion:

$$NaHCO_3(s) \rightleftharpoons Na^+(aq) + HCO_3^-(aq)$$

The sodium ion has neither acidic nor basic properties (it does not ionize water), as we will see shortly. The bicarbonate ion, by contrast, acts as a weak base, ionizing water as just shown to form a basic solution. Consequently, the pH of a sodium bicarbonate solution is above 7 (the solution is basic). In this section, we consider some of the acid–base properties of salts and the ions they contain. Some salts are pH-neutral when put into water, others are acidic, and still others are basic, depending on their constituent anions and cations. In general, anions tend to form either *basic* or neutral solutions, while cations tend to form either *acidic* or neutral solutions.

# Anions as Weak Bases

We can think of any anion as the conjugate base of an acid. Consider the following anions and their corresponding acids:

| **This anion** | **is the conjugate base of** | **this acid** |
|---|---|---|
| $Cl^-$ | | HCl |
| $F^-$ | | HF |
| $NO_3^-$ | | $HNO_3$ |
| $C_2H_3O_2^-$ | | $HC_2H_3O_2$ |

In general, the anion $A^-$ is the conjugate base of the acid HA. Since every anion can be regarded as the conjugate base of an acid, every anion itself can potentially act as a base. However, *not every anion does act as a base*—it depends on the strength of the corresponding acid. In general:

- An anion that is the conjugate base of a *weak acid* is itself a *weak base*.
- An anion that is the conjugate base of a *strong acid* is pH-*neutral* (forms solutions that are neither acidic nor basic).

For example, the $Cl^-$ anion is the conjugate base of HCl, a strong acid. Therefore the $Cl^-$ anion is pH-neutral (neither acidic nor basic). The $F^-$ anion, however, is the conjugate base of HF, a weak acid. Therefore, the $F^-$ ion is itself a weak base and ionizes water according to the reaction:

$$F^-(aq) + H_2O(l) \rightleftharpoons OH^-(aq) + HF(aq)$$

We can understand why the conjugate base of a weak acid is basic by asking ourselves why an acid is weak to begin with. Hydrofluoric acid is a weak acid because the following reaction lies to the left:

$$HF(aq) + H_2O(l) \rightleftharpoons H_3O^+(aq) + F^-(aq)$$

The equilibrium lies to the left because the $F^-$ ion has a significant affinity for $H^+$ ions. Consequently, when $F^-$ is put into water, its affinity for $H^+$ ions allows it to remove $H^+$ ions from water molecules, thus acting as a weak base. In general, as shown in Figure 16.12▼,

| | | Acid | Base | |
|---|---|---|---|---|
| | | HCl | $Cl^-$ | |
| | **Strong** | $H_2SO_4$ | $HSO_4^-$ | **Neutral** |
| | | $HNO_3$ | $NO_3^-$ | |
| | | $H_3O^+$ | $H_2O$ | |
| | | $HSO_4^-$ | $SO_4^{2-}$ | |
| | | $H_2SO_3$ | $HSO_3^-$ | |
| | | $H_3PO_4$ | $H_2PO_4^-$ | |
| | | HF | $F^-$ | |
| | | $HC_2H_3O_2$ | $C_2H_3O_2^-$ | |
| | | $H_2CO_3$ | $HCO_3^-$ | **Weak** |
| | **Weak** | $H_2S$ | $HS^-$ | |
| | | $HSO_3^-$ | $SO_3^{2-}$ | |
| | | $H_2PO_4^-$ | $HPO_4^{2-}$ | |
| | | HCN | $CN^-$ | |
| | | $NH_4^+$ | $NH_3$ | |
| | | $HCO_3^-$ | $CO_3^{2-}$ | |
| | | $HPO_4^{2-}$ | $PO_4^{3-}$ | |
| | | $H_2O$ | $OH^-$ | |
| | | $HS^-$ | $S^{2-}$ | **Strong** |
| | **Negligible** | $OH^-$ | $O^{2-}$ | |

◀ **FIGURE 16.12  Strength of Conjugate Acid–Base Pairs**  The stronger an acid, the weaker its conjugate base.

the weaker the acid, the stronger the conjugate base (as we saw in Section 16.4). In contrast, the conjugate base of a strong acid, such as $Cl^-$, does not act as a base because this reaction lies far to the right:

$$HCl(aq) + H_2O(l) \longrightarrow H_3O^+(aq) + Cl^-(aq)$$

The reaction lies far to the right because the $Cl^-$ ion has a low affinity for $H^+$ ions. Consequently, when $Cl^-$ is put into water, it does not remove $H^+$ ions from water molecules.

---

**Example 16.13** | **Determining Whether an Anion Is Basic or pH-Neutral**

Classify each anion as a weak base or pH-neutral.

**(a)** $NO_3^-$  **(b)** $NO_2^-$  **(c)** $C_2H_3O_2^-$

**SOLUTION**

**(a)** From Table 16.3, you can see that $NO_3^-$ is the conjugate base of a *strong* acid ($HNO_3$) and is therefore pH-neutral.

**(b)** From Table 16.5 (or from its absence in Table 16.3), you know that $NO_2^-$ is the conjugate base of a weak acid ($HNO_2$) and is therefore a weak base.

**(c)** From Table 16.5 (or from its absence in Table 16.3), you know that $C_2H_3O_2^-$ is the conjugate base of a weak acid ($HC_2H_3O_2$) and is therefore a weak base.

**FOR PRACTICE 16.13** Classify each anion as a weak base or pH-neutral.

**(a)** $CHO_2^-$  **(b)** $ClO_4^-$

---

We can determine the pH of a solution containing an anion that acts as a weak base in a manner similar to how we determine the pH of any weak base solution. However, we need to know $K_b$ for the anion acting as a base, which we can readily determine from $K_a$ of the corresponding acid. Recall from Section 16.4 the expression for $K_a$ for a generic acid HA:

$$HA(aq) + H_2O(l) \rightleftharpoons H_3O^+(aq) + A^-(aq)$$

$$K_a = \frac{[H_3O^+][A^-]}{[HA]}$$

Similarly, the expression for $K_b$ for the conjugate base ($A^-$) is:

$$A^-(aq) + H_2O(l) \rightleftharpoons OH^-(aq) + HA(aq)$$

$$K_b = \frac{[OH^-][HA]}{[A^-]}$$

If we multiply the expressions for $K_a$ and $K_b$, we get $K_w$:

$$K_a \times K_b = \frac{[H_3O^+][A^-]}{[HA]} \frac{[OH^-][HA]}{[A]} = [H_3O^+][OH^-] = K_w$$

Or simply,

$$K_a \times K_b = K_w$$

The product of $K_a$ for an acid and $K_b$ for its conjugate base is $K_w$ ($1.0 \times 10^{-14}$ at 25 °C). Consequently, we can find $K_b$ for an anion acting as a base from the value of $K_a$ for the corresponding acid. For example, for acetic acid ($HC_2H_3O_2$), $K_a = 1.8 \times 10^{-5}$. We calculate $K_b$ for the conjugate base ($C_2H_3O_2^-$) by substituting into the equation:

$$K_a \times K_b = K_w$$

$$K_b = \frac{K_w}{K_a} = \frac{1.0 \times 10^{-14}}{1.8 \times 10^{-5}} = 5.6 \times 10^{-10}$$

Knowing $K_b$, we can find the pH of a solution containing an anion acting as a base, as Example 16.14 demonstrates.

**Example 16.14**    **Determining the pH of a Solution Containing an Anion Acting as a Base**

Find the pH of a 0.100 M $NaCHO_2$ solution. The salt completely dissociates into $Na^+(aq)$ and $CHO_2^-(aq)$, and the $Na^+$ ion has no acid or base properties.

**SOLUTION**

1. Since the $Na^+$ ion does not have any acidic or basic properties, you can ignore it. Write the balanced equation for the ionization of water by the basic anion and use it as a guide to prepare an ICE table showing the given concentration of the weak base as its initial concentration.

$CHO_2^-(aq) + H_2O(l) \rightleftharpoons HCHO_2(aq) + OH^-(aq)$

|         | $[CHO_2^-]$ | $[HCHO_2]$ | $[OH^-]$ |
|---------|-------------|------------|----------|
| Initial | 0.100       | 0.00       | ≈0.00    |
| Change  |             |            |          |
| Equil   |             |            |          |

2. Represent the change in the concentration of $OH^-$ with the variable $x$. Define the changes in the concentrations of the other reactants and products in terms of $x$.

$CHO_2^-(aq) + H_2O(l) \rightleftharpoons HCHO_2(aq) + OH^-(aq)$

|         | $[CHO_2^-]$ | $[HCHO_2]$ | $[OH^-]$ |
|---------|-------------|------------|----------|
| Initial | 0.100       | 0.00       | ≈0.00    |
| Change  | −x          | +x         | +x       |
| Equil   |             |            |          |

3. Sum each column to determine the equilibrium concentrations in terms of the initial concentrations and the variable $x$.

$CHO_2^-(aq) + H_2O(l) \rightleftharpoons HCHO_2(aq) + OH^-(aq)$

|         | $[CHO_2^-]$ | $[HCHO_2]$ | $[OH^-]$ |
|---------|-------------|------------|----------|
| Initial | 0.100       | 0.00       | ≈0.00    |
| Change  | −x          | +x         | +x       |
| Equil   | 0.100 − x   | x          | x        |

4. Find $K_b$ from $K_a$ (for the conjugate acid).

$K_a \times K_b = K_w$

$$K_b = \frac{K_w}{K_a} = \frac{1.0 \times 10^{-14}}{1.8 \times 10^{-4}} = 5.6 \times 10^{-11}$$

Substitute the expressions for the equilibrium concentrations (from step 3) into the expression for $K_b$. In many cases, you can make the approximation that $x$ is small.

$$K_b = \frac{[HCHO_2][OH^-]}{[CHO_2^-]}$$

$$= \frac{x^2}{0.100 - x}$$

Substitute the value of $K_b$ into the $K_b$ expression and solve for $x$.

$$5.6 \times 10^{-11} = \frac{x^2}{0.100}$$

$$x = 2.4 \times 10^{-6}$$

Confirm that the $x$ is small approximation is valid by calculating the ratio of $x$ to the number it was subtracted from in the approximation. The ratio should be less than 0.05 (or 5%).

$$\frac{2.4 \times 10^{-6}}{0.100} \times 100\% = 0.0024\%$$

Therefore, the approximation is valid.

5. Determine the $OH^-$ concentration from the calculated value of $x$.

$[OH^-] = 2.4 \times 10^{-6}$ M

Use the expression for $K_w$ to find $[H_3O^+]$.

$[H_3O^+][OH^-] = K_w = 1.0 \times 10^{-14}$
$[H_3O^+](2.4 \times 10^{-6}) = 1.0 \times 10^{-14}$
$[H_3O^+] = 4.2 \times 10^{-9}$ M

Substitute $[H_3O^+]$ into the pH equation to find pH.

$pH = -\log[H_3O^+]$
$\quad = -\log(4.2 \times 10^{-9})$
$\quad = 8.38$

**FOR PRACTICE 16.14**    Find the pH of a 0.250 M $NaC_2H_3O_2$ solution.

We can also express the relationship between $K_a$ and $K_b$ in terms of $pK_a$ and $pK_b$. By taking the log of both sides of $K_a \times K_b = K_w$, we get:

$$\log(K_a \times K_b) = \log K_w$$

$$\log K_a + \log K_b = \log K_w$$

Since $K_w = 10^{-14}$, we can rearrange the equation to get:

$$\log K_a + \log K_b = \log 10^{-14} = -14$$

Rearranging further:

$$-\log K_a - \log K_b = 14$$

Since $-\log K = pK$, we get:

$$pK_a + pK_b = 14$$

## Cations as Weak Acids

In contrast to anions, which in some cases act as weak bases, cations can in some cases act as *weak acids*. Although some exceptions exist, we can generally divide cations into three categories: cations that are the counterions of strong bases; cations that are the conjugate acids of *weak* bases; and cations that are small, highly charged metals. We examine each individually.

### Cations That Are the Counterions of Strong Bases

Strong bases such as $NaOH$ or $Ca(OH)_2$ generally contain hydroxide ions and a counterion. In solution, a strong base completely dissociates to form $OH^-(aq)$ and the solvated (in solution) counterion. Although these counterions interact with water molecules via ion–dipole forces, they do not ionize water and they do not contribute to the acidity or basicity of the solution. In general *cations that are the counterions of strong bases are themselves pH-neutral* (they form solutions that are neither acidic nor basic). For example, $Na^+$, $K^+$, and $Ca^{2+}$ are the counterions of the strong bases $NaOH$, $KOH$, and $Ca(OH)_2$ and are therefore themselves pH-neutral.

### Cations That Are the Conjugate Acids of Weak Bases

A cation can be formed from any nonionic weak base by adding a proton ($H^+$) to its formula. The cation is the conjugate acid of the base. Consider the following cations and their corresponding weak bases:

| This cation | is the conjugate acid of | this weak base |
|---|---|---|
| $NH_4^+$ | | $NH_3$ |
| $C_2H_5NH_3^+$ | | $C_2H_5NH_2$ |
| $CH_3NH_3^+$ | | $CH_3NH_2$ |

Any of these cations, with the general formula $BH^+$, will act as a weak acid according to the equation:

$$BH^+(aq) + H_2O(aq) \rightleftharpoons H_3O^+(aq) + B(aq)$$

In general, *a cation that is the conjugate acid of a weak base is a weak acid.*

We can calculate the pH of a solution containing the conjugate acid of a weak base just like that of any other weak acidic solution. However, we must derive the value of $K_a$ for the acid from $K_b$ using the previously derived relationship: $K_a \times K_b = K_w$.

### Cations That Are Small, Highly Charged Metals

*Small, highly charged metal cations such as $Al^{3+}$ and $Fe^{3+}$ form weakly acidic solutions.* For example, when $Al^{3+}$ is dissolved in water, it becomes hydrated according to the equation:

$$Al^{3+}(aq) + 6H_2O(l) \longrightarrow Al(H_2O)_6^{3+}(aq)$$

The hydrated form of the ion then acts as a Brønsted–Lowry acid:

$$Al(H_2O)_6{}^{3+}(aq) \;+\; H_2O(aq) \;\rightleftharpoons\; Al(H_2O)_5(OH)^{2+}(aq) + H_3O^+(aq)$$

In effect, the binding of a water molecule to the cation makes the water more acidic. See Section 16.10.

Neither the alkali metal cations nor the alkaline earth metal cations ionize water in this way, but the cations of many other metals do. The smaller and more highly charged the cation, the more acidic its behavior.

---

**Example 16.15** | **Determining Whether a Cation Is Acidic or pH-Neutral**

Classify each cation as a weak acid or pH-neutral.

**(a)** $C_5H_5NH^+$      **(b)** $Ca^{2+}$      **(c)** $Cr^{3+}$

**SOLUTION**

**(a)** The $C_5H_5NH^+$ cation is the conjugate acid of a weak base and is therefore a weak acid.

**(b)** The $Ca^{2+}$ cation is the counterion of a strong base and is therefore pH-neutral (neither acidic nor basic).

**(c)** The $Cr^{3+}$ cation is a small, highly charged metal cation and is therefore a weak acid.

**FOR PRACTICE 16.15**   Classify each cation as a weak acid or pH-neutral.

**(a)** $Li^+$      **(b)** $CH_3NH_3{}^+$      **(c)** $Fe^{3+}$

---

## Classifying Salt Solutions as Acidic, Basic, or Neutral

Since salts contain both a cation and an anion, they can form acidic, basic, or neutral solutions when dissolved in water. The pH of the solution depends on the specific cation and anion involved. We examine the four possibilities individually.

1. **Salts in which neither the cation nor the anion acts as an acid or a base form pH-neutral solutions.**   A salt in which the cation is the counterion of a strong base and in which the anion is the conjugate base of strong acid forms a *neutral* solution. Some salts in this category include:

   | NaCl | Ca(NO₃)₂ | KBr |
   |------|----------|-----|
   | sodium chloride | calcium nitrate | potassium bromide |

   Cations are pH–neutral.　　　Anions are conjugate bases of strong acids.

2. **Salts in which the cation does not act as an acid and the anion acts as a base form basic solutions.**   A salt in which the cation is the counterion of a strong base and in which the anion is the conjugate base of *weak* acid forms a *basic* solution. Salts in this category include:

   | NaF | Ca(C₂H₃O₂)₂ | KNO₂ |
   |-----|-------------|------|
   | sodium fluoride | calcium acetate | potassium nitrite |

   Cations are pH–neutral .　　　Anions are conjugate bases of weak acids.

3. **Salts in which the cation acts as an acid and the anion does not act as a base form acidic solutions.** A salt in which the cation is either the conjugate acid of a weak base or a small, highly charged metal ion and in which the anion is the conjugate base of *strong* acid forms an *acidic* solution. Salts in this category include:

| $FeCl_3$ | $Al(NO_3)_3$ | $NH_4Br$ |
|----------|--------------|----------|
| iron(III) chloride | aluminum nitrate | ammonium bromide |

Cations are conjugate acids of *weak* bases or small, highly charged metal ions.

Anions are conjugate bases of *strong* acids.

4. **Salts in which the cation acts as an acid and the anion acts as a base form solutions in which the pH depends on the relative strengths of the acid and the base.** A salt in which the cation is either the conjugate acid of a weak base or a small, highly charged metal ion and in which the anion is the conjugate base of a *weak* acid forms a solution in which the pH depends on the relative strengths of the acid and base. Salts in this category include:

| $FeF_3$ | $Al(C_2H_3O_2)_3$ | $NH_4NO_2$ |
|---------|-------------------|------------|
| iron(III) fluoride | aluminum acetate | ammonium nitrite |

Cations are conjugate acids of *weak* bases or small, highly charged metal ions.

Anions are conjugate bases of *weak* acids.

We can determine the overall acidity or basicity of a solution containing one of these salts by comparing the $K_a$ of the acid to the $K_b$ of the base—the ion with the higher value of $K$ dominates and determines whether the solution will be acidic or basic, as shown in part (e) of Example 16.16. Table 16.9 summarizes all of these possibilities.

**TABLE 16.9 pH of Salt Solutions**

| | | ANION | |
|---|---|---|---|
| | | **Conjugate base of strong acid** | **Conjugate base of weak acid** |
| **CATION** | **Conjugate acid of weak base** | Acidic | Depends on relative strengths |
| | **Small, highly charged metal ion** | Acidic | Depends on relative strengths |
| | **Counterion of strong base** | Neutral | Basic |

## Example 16.16 Determining the Overall Acidity or Basicity of Salt Solutions

Determine if the solution formed by each salt is acidic, basic, or neutral.

**(a)** $SrCl_2$  **(b)** $AlBr_3$  **(c)** $CH_3NH_3NO_3$
**(d)** $NaCHO_2$  **(e)** $NH_4F$

**SOLUTION**

**(a)** The $Sr^{2+}$ cation is the counterion of a strong base ($Sr(OH)_2$) and is pH-neutral. The $Cl^-$ anion is the conjugate base of a strong acid (HCl) and is pH-neutral as well. The $SrCl_2$ solution is therefore pH-neutral (neither acidic nor basic).

$SrCl_2$

pH-neutral cation    pH-neutral anion

Neutral solution

**(b)** The $Al^{3+}$ cation is a small, highly charged metal ion (that is not an alkali metal or an alkaline earth metal) and is a weak acid. The $Br^-$ anion is the conjugate base of a strong acid (HBr) and is pH-neutral. The $AlBr_3$ solution is therefore acidic.

$AlBr_3$

acidic cation     pH-neutral anion

Acidic solution

**(c)** The $CH_3NH_3^+$ ion is the conjugate acid of a weak base ($CH_3NH_2$) and is acidic. The $NO_3^-$ anion is the conjugate base of a strong acid ($HNO_3$) and is pH-neutral. The $CH_3NH_3NO_3$ solution is therefore acidic.

$CH_3NH_3NO_3$

acidic cation     pH-neutral anion

Acidic solution

**(d)** The $Na^+$ cation is the counterion of a strong base and is pH-neutral. The $CHO_2^-$ anion is the conjugate base of a weak acid and is basic. The $NaCHO_2$ solution is therefore basic.

$NaCHO_2$

pH-neutral cation     basic anion

Basic solution

**(e)** The $NH_4^+$ ion is the conjugate acid of a weak base ($NH_3$) and is acidic. The $F^-$ ion is the conjugate base of a weak acid and is basic. To determine the overall acidity or basicity of the solution, compare the values of $K_a$ for the acidic cation and $K_b$ for the basic anion. Obtain each value of $K$ from the conjugate by using $K_a \times K_b = K_w$.

$NH_4F$

acidic cation     basic anion

$$K_a(NH_4^+) = \frac{K_w}{K_b(NH_3)} = \frac{1.0 \times 10^{-14}}{1.76 \times 10^{-5}}$$
$$= 5.68 \times 10^{-10}$$

$$K_b(F^-) = \frac{K_w}{K_a(HF)} = \frac{1.0 \times 10^{-14}}{6.8 \times 10^{-4}}$$
$$= 1.5 \times 10^{-11}$$

Since $K_a$ is greater than $K_b$, the solution is acidic.

$K_a > K_b$     Acidic solution

**FOR PRACTICE 16.16**   Determine if the solution formed by each salt is acidic, basic, or neutral.

**(a)** $NaHCO_3$   **(b)** $CH_3CH_2NH_3Cl$   **(c)** $KNO_3$   **(d)** $Fe(NO_3)_3$

# 16.9 Polyprotic Acids

In Section 16.4, we discussed that some acids, called polyprotic acids, contain two or more ionizable protons. Recall that sulfurous acid ($H_2SO_3$) is a diprotic acid containing two ionizable protons and phosphoric acid ($H_3PO_4$) is a triprotic acid containing three ionizable protons.

Typically, a **polyprotic acid** ionizes in successive steps, each with its own $K_a$. For example, sulfurous acid ionizes in two steps:

$$H_2SO_3(aq) \rightleftharpoons H^+(aq) + HSO_3^-(aq) \qquad K_{a_1} = 1.6 \times 10^{-2}$$

$$HSO_3^-(aq) \rightleftharpoons H^+(aq) + SO_3^{2-}(aq) \qquad K_{a_2} = 6.4 \times 10^{-8}$$

$K_{a_1}$ is the acid ionization constant for the first step, and $K_{a_2}$ is the acid ionization constant for the second step. Notice that $K_{a_2}$ is much smaller than $K_{a_1}$. This is true for all polyprotic acids and makes sense because the first proton separates from a neutral molecule, while the second must separate from an anion. The negatively charged anion holds the positively charged proton more tightly, making the proton more difficult to remove and resulting in a smaller value of $K_a$.

Table 16.10 lists some common polyprotic acids and their acid ionization constants. Notice that in all cases, the value of $K_a$ for each step becomes successively smaller. The value of $K_{a_1}$ for sulfuric acid is identified as strong because sulfuric acid is strong in the first step and weak in the second.

| TABLE 16.10 Common Polyprotic Acids and Ionization Constants at 25 °C | | | | | |
|---|---|---|---|---|---|
| Name (Formula) | Structure | Space-filling model | $K_{a_1}$ | $K_{a_2}$ | $K_{a_3}$ |
| Sulfuric Acid ($H_2SO_4$) | | | Strong | $1.2 \times 10^{-2}$ | |
| Oxalic Acid ($H_2C_2O_4$) | | | $6.0 \times 10^{-2}$ | $6.1 \times 10^{-5}$ | |
| Sulfurous Acid ($H_2SO_3$) | | | $1.6 \times 10^{-2}$ | $6.4 \times 10^{-8}$ | |
| Phosphoric Acid ($H_3PO_4$) | | | $7.5 \times 10^{-3}$ | $6.2 \times 10^{-8}$ | $4.2 \times 10^{-13}$ |
| Citric Acid ($H_3C_6H_5O_7$) | | | $7.4 \times 10^{-4}$ | $1.7 \times 10^{-5}$ | $4.0 \times 10^{-7}$ |
| Ascorbic Acid ($H_2C_6H_6O_6$) | | | $8.0 \times 10^{-5}$ | $1.6 \times 10^{-12}$ | |
| Carbonic Acid ($H_2CO_3$) | | | $4.3 \times 10^{-7}$ | $5.6 \times 10^{-11}$ | |

## Finding the pH of Polyprotic Acid Solutions

Finding the pH of a polyprotic acid solution is less difficult than we might first imagine because, for most polyprotic acids, $K_{a_1}$ is much larger than $K_{a_2}$ (or $K_{a_3}$ for triprotic acids). Therefore the amount of $H_3O^+$ formed by the first ionization step is much larger than that formed by the second or third ionization step. In addition, the formation of $H_3O^+$ in the first step inhibits the formation of additional $H_3O^+$ in the second step (because of Le Châtelier's principle). Consequently, we treat most polyprotic acid solutions as if the first step were the only one that contributes to the $H_3O^+$ concentration, as illustrated in Example 16.17. A major exception is a dilute solution of sulfuric acid, which we examine in Example 16.18.

## Example 16.17   Finding the pH of a Polyprotic Acid Solution

Find the pH of a 0.100 M ascorbic acid ($H_2C_6H_6O_6$) solution.

**SOLUTION**

To find the pH, you must find the equilibrium concentration of $H_3O^+$. Treat the problem as a weak acid pH problem with a single ionizable proton. The second proton contributes a negligible amount to the concentration of $H_3O^+$ and can be ignored. Follow the procedure from Example 16.6, shown in condensed form here. Use $K_{a_1}$ for ascorbic acid from Table 16.10.

$$H_2C_6H_6O_6(aq) + H_2O(l) \rightleftharpoons H_3O^+(aq) + HC_6H_6O_6^-(aq)$$

|  | $[H_2C_6H_6O_6]$ | $[H_3O^+]$ | $[HC_6H_6O_6^-]$ |
|---|---|---|---|
| Initial | 0.100 | $\approx 0.00$ | 0.00 |
| Change | $-x$ | $+x$ | $+x$ |
| Equil | $0.100 - x$ | $x$ | $x$ |

$$K_{a_1} = \frac{[H_3O^+][HC_6H_6O_6^-]}{[H_2C_6H_6O_6]}$$

$$= \frac{x^2}{0.100 - x} \quad (x \text{ is small})$$

$$8.0 \times 10^{-5} = \frac{x^2}{0.100}$$

$$x = 2.8 \times 10^{-3}$$

Confirm that the *x is small* approximation is valid by calculating the ratio of $x$ to the number it was subtracted from in the approximation. The ratio should be less than 0.05 (or 5%).

$$\frac{2.8 \times 10^{-3}}{0.100} \times 100\% = 2.8\%$$

The approximation is valid. Therefore,

Calculate the pH from $H_3O^+$ concentration.

$$[H_3O^+] = 2.8 \times 10^{-3} \text{ M}.$$

$$pH = -\log(2.8 \times 10^{-3}) = 2.55$$

**FOR PRACTICE 16.17**   Find the pH of a 0.050 M $H_2CO_3$ solution.

## Example 16.18   Dilute $H_2SO_4$ Solutions

Find the pH of a 0.0100 M sulfuric acid ($H_2SO_4$) solution.

**SOLUTION**

Sulfuric acid is strong in its first ionization step and weak in its second. Begin by writing the equations for the two steps. As the concentration of an $H_2SO_4$ solution becomes smaller, the second ionization step becomes more significant because the percent ionization increases (as discussed in Section 16.6). Therefore, for a concentration of 0.0100 M, you can't neglect the $H_3O^+$ contribution from the second step, as you can for other polyprotic acids. You must calculate the $H_3O^+$ contributions from both steps.

$$H_2SO_4(aq) + H_2O(l) \longrightarrow H_3O^+(aq) + HSO_4^-(aq) \quad \text{Strong}$$

$$HSO_4^-(aq) + H_2O(l) \rightleftharpoons H_3O^+(aq) + SO_4^{2-}(aq) \quad K_{a_2} = 0.012$$

*—Continued on the next page*

*Continued—*

| The $[H_3O^+]$ that results from the first ionization step is 0.0100 M (because the first step is strong). To determine the $[H_3O^+]$ formed by the second step, prepare an ICE table for the second step in which the initial concentration of $H_3O^+$ is 0.0100 M. The initial concentration of $HSO_4^-$ must also be 0.0100 M (due to the stoichiometry of the ionization reaction). | |  | $[HSO_4^-]$ | $[H_3O^+]$ | $[SO_4^{2-}]$ |
|---|---|---|---|---|---|
| | | Initial | 0.0100 | $\approx$0.0100 | 0.000 |
| | | Change | $-x$ | $+x$ | $+x$ |
| | | Equil | $0.0100 - x$ | $0.0100 + x$ | $x$ |

| Substitute the expressions for the equilibrium concentrations (from the table just shown) into the expression for $K_{a_2}$. In this case, you cannot make the *x is small* approximation because the equilibrium constant (0.012) *is not small* relative to the initial concentration (0.0100). | $$K_{a_2} = \frac{[H_3O^+][SO_4^{2-}]}{[HSO_4^-]}$$ $$= \frac{(0.0100 + x)x}{0.0100 - x}$$ |
|---|---|
| Substitute the value of $K_{a_2}$ and multiply out the expression to arrive at the standard quadratic form. | $$0.012 = \frac{0.0100x + x^2}{0.0100 - x}$$ $$0.012(0.0100 - x) = 0.0100x + x^2$$ $$0.00012 - 0.012x = 0.0100x + x^2$$ $$x^2 + 0.022x - 0.00012 = 0$$ |

| Solve the quadratic equation using the quadratic formula. | $$x = \frac{-b \pm \sqrt{b^2 - 4ac}}{2a}$$ $$= \frac{-(0.022) \pm \sqrt{(0.022)^2 - 4(1)(-0.00012)}}{2(1)}$$ $$= \frac{-0.022 \pm 0.031}{2}$$ $$x = -0.027 \text{ or } x = 0.0045$$ Since $x$ represents a concentration, and since concentrations cannot be negative, we reject the negative root. $$x = 0.0045$$ |
|---|---|

| Determine the $H_3O^+$ concentration from the calculated value of $x$ and calculate the pH. Notice that the second step produces almost half as much $H_3O^+$ as the first step—an amount that must not be ignored. This is always the case with dilute $H_2SO_4$ solutions. | $$[H_3O^+] = 0.0100 + x$$ $$= 0.0100 + 0.0045$$ $$= 0.0145 \text{ M}$$ $$pH = -\log[H_3O^+]$$ $$= -\log(0.0145)$$ $$= 1.84$$ |
|---|---|

**FOR PRACTICE 16.18**   Find the pH and $[SO_4^{2-}]$ of a 0.0075 M sulfuric acid solution.

## Finding the Concentration of the Anions for a Weak Diprotic Acid Solution

In some cases, we may want to know the concentrations of the anions formed by a polyprotic acid. Consider the following generic polyprotic acid $H_2X$ and its ionization steps:

$$H_2X(aq) + H_2O(l) \rightleftharpoons H_3O^+(aq) + HX^-(aq) \quad K_{a_1}$$
$$HX^-(aq) + H_2O(l) \rightleftharpoons H_3O^+(aq) + X^{2-}(aq) \quad K_{a_2}$$

In Examples 16.17 and 16.18, we illustrated how to find the concentration of $H_3O^+$ for such a solution, which is equal to the concentration of $HX^-$. What if instead we needed to find the concentration of $X^{2-}$? To find the concentration of $X^{2-}$, we use the concentration of $HX^-$ and $H_3O^+$ (from the first ionization step) as the initial concentrations for the second ionization step. We then solve a second equilibrium problem using the second ionization equation and $K_{a_2}$, as demonstrated in Example 16.19.

**Example** 16.19 | **Finding the Concentration of the Anions for a Weak Diprotic Acid Solution**

Find the $[C_6H_6O_6^{2-}]$ of the 0.100 M ascorbic acid ($H_2C_6H_6O_6$) solution in Example 16.17.

**SOLUTION**

To find the $[C_6H_6O_6^{2-}]$, use the concentrations of $[HC_6H_6O_6^-]$ and $H_3O^+$ produced by the first ionization step (as calculated in Example 16.17) as the initial concentrations for the second step. Because of the 1:1 stoichiometry, $[HC_6H_6O_6^-] = [H_3O^+]$. Then solve an equilibrium problem for the second step similar to that of Example 16.5, shown in condensed form here. You can find the value of $K_{a_2}$ for ascorbic acid in Table 16.10.

$$HC_6H_6O_6^-(aq) + H_2O(l) \rightleftharpoons H_3O^+(aq) + C_6H_6O_6^{2-}(aq)$$

|         | $[HC_6H_6O_6^-]$ | $[H_3O^+]$ | $[C_6H_6O_6^{2-}]$ |
|---------|------------------|------------|---------------------|
| Initial | $2.8 \times 10^{-3}$ | $2.8 \times 10^{-3}$ | 0.000 |
| Change  | $-x$ | $+x$ | $+x$ |
| Equil   | $2.8 \times 10^{-3} - x$ | $2.8 \times 10^{-3} + x$ | $x$ |

$$K_{a_2} = \frac{[H_3O^+][C_6H_6O_6^{2-}]}{[HC_6H_6O_6^-]}$$

$$= \frac{(2.8 \times 10^{-3} + x)x}{2.8 \times 10^{-3} - x} \quad (x \text{ is small})$$

$$= \frac{(2.8 \times 10^{-3})x}{2.8 \times 10^{-3}}$$

$$x = K_{a_2} = 1.6 \times 10^{-12} \text{ M.}$$

Since $x$ is much smaller than $2.8 \times 10^{-3}$, the $x$ *is small* approximation is valid. Therefore,

$$[C_6H_6O_6^{2-}] = 1.6 \times 10^{-12} \text{ M.}$$

**FOR PRACTICE 16.19**   Find the $[CO_3^{2-}]$ of the 0.050 M carbonic acid ($H_2CO_3$) solution in For Practice 16.17.

Notice from the results of Example 16.19 that the concentration of $X^{2-}$ for a weak diprotic acid $H_2X$ is equal to $K_{a_2}$. This general result applies to all diprotic acids in which the *x is small* approximation is valid. Notice also that the concentration of $H_3O^+$ produced by the second ionization step of a diprotic acid is very small compared to the concentration produced by the first step, as shown in Figure 16.13▼.

## Dissociation of a Polyprotic Acid

$$H_2C_6H_6O_6(aq) + H_2O(l) \rightleftharpoons H_3O^+(aq) + HC_6H_6O_6^-(aq)$$

$$\left[H_3O^+\right] = 2.8 \times 10^{-3} \text{ M}$$

$$HC_6H_6O_6^-(aq) + H_2O(l) \rightleftharpoons H_3O^+(aq) + C_6H_6O_6^{2-}(aq)$$

$$\left[H_3O^+\right] = 1.6 \times 10^{-12} \text{ M}$$

0.100 M $H_2C_6H_6O_6$

$$\text{Total} \left[H_3O^+\right] = 2.8 \times 10^{-3} \text{ M} + 1.6 \times 10^{-12} \text{ M}$$

$$= 2.8 \times 10^{-3} \text{ M}$$

▲ **FIGURE 16.13 Dissociation of a Polyprotic Acid**   A 0.100 M $H_2C_6H_6O_6$ solution contains an $H_3O^+$ concentration of $2.8 \times 10^{-3}$ M from the first step. The amount of $H_3O^+$ contributed by the second step is only $1.6 \times 10^{-12}$ M, which is insignificant compared to the amount produced by the first step.

# 16.10  Acid Strength and Molecular Structure

We have learned that a Brønsted–Lowry acid is a proton ($H^+$) donor. However, we have not explored why some hydrogen-containing molecules act as acids while others do not, or why some acids are strong and others weak. For example, why is $H_2S$ acidic while $CH_4$ is not? Or why is HF a weak acid while HCl is a strong acid? We divide our discussion about these issues into two categories: binary acids (those containing hydrogen and only one other element) and oxyacids (those containing hydrogen bonded to an oxygen atom that is bonded to another element).

## Binary Acids

Consider the bond between a hydrogen atom and some other generic element (which we call Y):

$$H—Y$$

The factors affecting the ease with which this hydrogen is donated (and therefore will be acidic) are the *polarity* of the bond and the *strength* of the bond.

### Bond Polarity

Using the notation introduced in Chapter 9, the H—Y bond must be polarized with the hydrogen atom as the positive pole in order for HY to be acidic:

$$\overset{+}{\underset{\longrightarrow}{\delta^+ H—Y^{\delta-}}}$$

This requirement makes physical sense because the hydrogen atom must be lost as a positively charged ion ($H^+$). A partial positive charge on the hydrogen atom facilitates its loss. Consider the following three bonds and their corresponding dipole moments:

$$\overset{\overset{+}{\longleftarrow}}{H—Li} \qquad H—C \qquad \overset{\overset{+}{\longrightarrow}}{H—F}$$

| Not acidic | Not acidic | Acidic |

LiH is ionic with *the negative charge on the hydrogen atom*; therefore, LiH is not acidic. The C—H bond is virtually nonpolar because the electronegativities of carbon and hydrogen are similar; therefore C—H is not acidic. In contrast, the H—F bond is polar with the positive charge on the hydrogen atom. As we have discussed in this chapter, HF is an acid. This is because the partial positive charge on the hydrogen atom makes it easier for the hydrogen to be lost as an $H^+$ ion.

### Bond Strength

The strength of the H—Y bond also affects the strength of the corresponding acid. As we might expect, the stronger the bond, the weaker the acid—the more tightly the hydrogen atom is held, the less likely it is to come off. We can see the effect of bond strength by comparing the bond strengths and acidities of the hydrogen halides:

| Acid | Bond Energy (kJ/mol) | Type of Acid |
|------|---------------------|--------------|
| H—F | 565 | Weak |
| H—Cl | 431 | Strong |
| H—Br | 364 | Strong |

HCl and HBr have weaker bonds and are both strong acids. HF, however, has a stronger bond and is therefore a weak acid, despite the greater bond polarity of HF.

### The Combined Effect of Bond Polarity and Bond Strength

We can see the combined effect of bond polarity and bond strength by examining the trends in acidity of the group 6A and 7A hydrides as shown in Figure 16.14 ◄. The hydrides become more acidic from left to right as the H—Y bond becomes more polar. The hydrides also become more acidic from top to bottom as the H—Y bond becomes weaker.

Increasing electronegativity
Increasing acidity
→

Decreasing bond strength
Increasing acidity
↓

| 6A | 7A |
|------|------|
| $H_2O$ | HF |
| $H_2S$ | HCl |
| $H_2Se$ | HBr |
| $H_2Te$ | HI |

▲ **FIGURE 16.14  Acidity of the Group 6A and 7A Hydrides**   From left to right, the hydrides become more acidic because the H—Y bond becomes more polar. From top to bottom, these hydrides become more acidic because the H—Y bond becomes weaker.

# Oxyacids

Oxyacids contain a hydrogen atom bonded to an oxygen atom. The oxygen atom is in turn bonded to another atom (which we call Y):

$$H—O—Y—$$

Oxyacids are sometimes called oxoacids.

Y may or may not be bonded to yet other atoms. The factors affecting the ease with which the hydrogen in an oxyacid is donated (and therefore is acidic) are the *electronegativity of the element Y* and the *number of oxygen atoms attached to the element Y*.

### The Electronegativity of Y

The more electronegative the element Y, the more it weakens and polarizes the H—O bond and the more acidic the oxyacid is. We can see this effect by comparing the electronegativity of Y and the acid ionization constants of the following oxyacids:

| Acid | Electronegativity of Y | $K_a$ |
|------|------------------------|-------|
| H—O—I | 2.5 | $2.3 \times 10^{-11}$ |
| H—O—Br | 2.8 | $2.0 \times 10^{-9}$ |
| H—O—Cl | 3.0 | $2.9 \times 10^{-8}$ |

Chlorine is the most electronegative of the three elements, and the corresponding acid has the greatest $K_a$.

### The Number of Oxygen Atoms Bonded to Y

Oxyacids may contain additional oxygen atoms bonded to the element Y. Because these additional oxygen atoms are electronegative, they draw electron density away from Y, which in turn draws electron density away from the H—O bond, further weakening and polarizing it, and leading to increasing acidity. We can see this effect by comparing the following series of acid ionization constants:

| Acid | Structure | $K_a$ |
|------|-----------|-------|
| $HClO_4$ | $H—O—Cl(=O)(=O)=O$ | Strong |
| $HClO_3$ | $H—O—Cl(=O)=O$ | 1 |
| $HClO_2$ | $H—O—Cl=O$ | $1.1 \times 10^{-2}$ |
| $HClO$ | $H—O—Cl$ | $2.9 \times 10^{-8}$ |

The greater the number of oxygen atoms bonded to Y, the stronger the acid. On this basis we would predict that $H_2SO_4$ is a stronger acid than $H_2SO_3$ and that $HNO_3$ is stronger than $HNO_2$. As we have seen in this chapter, both $H_2SO_4$ and $HNO_3$ are strong acids, while $H_2SO_3$ and $HNO_2$ are weak acids, as predicted.

**Acid Strength and Molecular Structure**   Which of the protons shown in red is more acidic?

CONCEPTUAL
CONNECTION **16.9**

(a)                    (b)

# 16.11 Lewis Acids and Bases

We began our definitions of acids and bases with the Arrhenius model. We then saw how the Brønsted–Lowry model, by introducing the concept of a proton donor and proton acceptor, expands the range of substances that we consider acids and bases. We now introduce a third model, which further broadens the range of substances that we can consider acids. This third model is the *Lewis model*, named after G. N. Lewis, the American chemist who devised the electron-dot representation of chemical bonding (Section 9.1).

While the Brønsted–Lowry model focuses on the transfer of a proton, the Lewis model focuses on the transfer of an electron pair. Consider the simple acid–base reaction between the $H^+$ ion and $NH_3$, shown here with Lewis structures:

$$H^+ \ + \ :NH_3 \ \longrightarrow \ \left[H{:}NH_3\right]^+$$

Brønsted–Lowry model focuses on the proton     Lewis model focuses on the electron pair

According to the Brønsted–Lowry model, the ammonia accepts a proton, thus acting as a base. According to the Lewis model, the ammonia acts as a base by *donating an electron pair*. The general definitions of acids and bases according to the Lewis model focus on the electron pair:

**Lewis acid**: electron pair acceptor

**Lewis base**: electron pair donor

According to the Lewis definition, $H^+$ in the reaction just shown is acting as an acid because it is accepting an electron pair from $NH_3$. $NH_3$ is acting as a Lewis base because it is donating an electron pair to $H^+$.

Although the Lewis model does not significantly expand the substances that can be considered bases—because all proton acceptors must have an electron pair to bind the proton—it does significantly expand the substances that can be considered acids. According to the Lewis model, a substance need not even contain hydrogen to be an acid. For example, consider the gas-phase reaction between boron trifluoride and ammonia shown here:

$$BF_3 \quad + \quad :NH_3 \quad \longrightarrow \quad F_3B{:}NH_3$$

Lewis acid       Lewis base       adduct

Boron trifluoride has an empty orbital that can accept the electron pair from ammonia and form the product (the product of a Lewis acid–base reaction is sometimes called an *adduct*). The reaction just shown demonstrates an important property of Lewis acids:

> *A Lewis acid has an empty orbital (or can rearrange electrons to create an empty orbital) that can accept an electron pair.*

Consequently, the Lewis definition subsumes a whole new class of acids. Next we examine a few examples.

## Molecules That Act as Lewis Acids

Since molecules with incomplete octets have empty orbitals, they can serve as Lewis acids. For example, both $AlCl_3$ and $BCl_3$ have incomplete octets:

These both act as Lewis acids, as shown in the following reactions:

$$\begin{array}{ccc} :\!\ddot{C}l\!: & H \\ | & | \\ :\!\ddot{C}l\!-\!Al\ +\ :\!N\!-\!H \\ | & | \\ :\!\ddot{C}l\!: & H \end{array} \longrightarrow \begin{array}{c} :\!\ddot{C}l\!:\ \ H \\ | \ \ \ \ | \\ :\!\ddot{C}l\!-\!Al\!:\!N\!-\!H \\ | \ \ \ \ | \\ :\!\ddot{C}l\!:\ \ H \end{array}$$

Some molecules that may not initially contain empty orbitals can rearrange their electrons to act as Lewis acids. Consider the reaction between carbon dioxide and water:

Water
**Lewis base**

Carbon dioxide
**Lewis acid**

Carbonic acid

The electrons in the double bond on carbon move to the terminal oxygen atom, allowing carbon dioxide to act as a Lewis acid by accepting an electron pair from water. The molecule then undergoes a rearrangement in which the hydrogen atom shown in red bonds with the terminal oxygen atom instead of the internal one.

## Cations That Act as Lewis Acids

Some cations, because they are positively charged and have lost some electrons, have empty orbitals that allow them to also act as Lewis acids. Consider the hydration process of the $Al^{3+}$ ion discussed in Section 16.8 shown here:

$$Al^{3+}(aq)\ +\ 6\left[\begin{array}{c} H \\ | \\ :\!\ddot{O}\!: \\ | \\ H \end{array}\right](l)\ \longrightarrow\ Al\left[\begin{array}{c} H \\ | \\ :\!\ddot{O}\!: \\ | \\ H \end{array}\right]_6^{3+}(aq)$$

Lewis
acid

Lewis
base

The aluminum ion acts as a Lewis acid, accepting lone pairs from six water molecules to form the hydrated ion. Many other small, highly charged metal ions also act as Lewis acids in this way.

## 16.12  Acid Rain

About 82% of U.S. energy comes from the combustion of fossil fuels, including petroleum, natural gas, and coal (Figure 16.15▶). Some fossil fuels, especially coal, contain small amounts of sulfur impurities. During combustion, these impurities react with oxygen to form $SO_2$. In addition, during combustion of any fossil fuel, nitrogen from the air reacts with oxygen to form $NO_2$. Sulfur dioxide and nitrogen dioxide react with water and diatomic oxygen ($O_2$) in the atmosphere to form sulfuric acid and nitric acid:

$$2\,SO_2\ +\ O_2\ +\ 2\,H_2O\ \longrightarrow\ 2\,H_2SO_4$$
$$4\,NO_2\ +\ O_2\ +\ 2\,H_2O\ \longrightarrow\ 4\,HNO_3$$

U.S. Energy Consumption
by Source, 2013

■ Fossil Fuels: 82%
■ Nuclear: 10%
  Renewable Energy: 8%

▲ **FIGURE 16.15  Sources of U.S. Energy**  About 82% of U.S. energy comes from fossil fuel combustion. Source: U.S. EIA Annual Energy Review 2015

**Hydrogen ion concentration as pH from measurements made at the Central Analytical Laboratory, 2013**

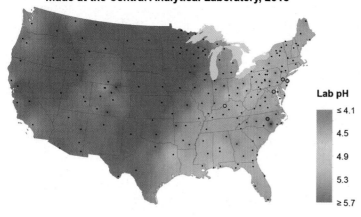

Lab pH

| | |
|---|---|
| | ≤ 4.1 |
| | 4.5 |
| | 4.9 |
| | 5.3 |
| | ≥ 5.7 |

▲ **FIGURE 16.16** **Acid Rain**
Acid rain is a significant problem in the northeastern United States. The map shows the pH of rain that falls across the United States.
Source: National Atmospheric Deposition Program, National Trends Network

These acids combine with rain to form *acid rain*. The problem is greatest in the northeastern portion of the United States, where the rain is significantly acidic because the sulfur and nitrogen oxides produced by coal combustion in the Midwest are carried toward the Northeast by natural air currents.

Even in relatively unpolluted areas, rain is naturally somewhat acidic because of atmospheric carbon dioxide. Carbon dioxide combines with rainwater to form carbonic acid:

$$CO_2 + H_2O \longrightarrow H_2CO_3$$

However, carbonic acid is a relatively weak acid. Rain that is saturated with $CO_2$ has a pH of about 5.6, which is only mildly acidic. When nitric acid and sulfuric acid mix with rain, the pH of the rain can fall below 4.3 (Figure 16.16 ◀). Remember that because of the logarithmic nature of the pH scale, rain with a pH of 4.3 has an $[H_3O^+]$ that is 20 times greater than rain with a pH of 5.6. Rain that is this acidic has a harmful impact.

## Effects of Acid Rain

Acids dissolve metals and acid rain degrades metal structures. Bridges, railroads, and even automobiles can be damaged by acid rain. Since acids react with carbonates, acid rain also harms building materials that contain carbonates ($CO_3^{2-}$), including marble, cement, and limestone. Statues, buildings, and pathways in the Northeast show significant signs of acid rain damage (Figure 16.17a ▼).

Acid rain also accumulates in lakes and rivers and affects aquatic life. Many lakes, especially those that are surrounded by land that contains significant amounts of limestone ($CaCO_3$), have the ability to neutralize acidic rain. In the Midwest, for example, limestone-rich soils prevent most lakes from becoming acidified. In the northeastern United States, however, the lack of limestone makes lakes more susceptible, and over 2000 lakes and streams have increased acidity levels due to acid rain. Aquatic plants, frogs, salamanders, and some species of fish are sensitive to acid levels and cannot survive in the acidified water. Trees can also be affected by acid rain because the acid removes nutrients from the soil, making survival difficult (Figure 16.17b ▼).

▼ **FIGURE 16.17** **The Effects of Acid Rain** **(a)** Buildings, gravestones, and statues damaged by acid rain are a common sight in the northeastern United States and in many other industrialized nations. **(b)** Some species of trees are highly susceptible to the effects of acid rain.

(a)

(b)

## Acid Rain Legislation

In 1990, the U.S. Congress passed amendments to the Clean Air Act specifically targeted at reducing acid rain. These amendments force electrical utilities—which are the most significant source of $SO_2$—to lower their $SO_2$ emissions gradually over time. The result has been a 76% decrease in $SO_2$ pollutant levels in the United States since 1990 (Figure 16.18▼). The acidity of rain over the northeastern United States has begun to decrease, and lakes are beginning to recover. In the early 1990s, scientists categorized 30% of the lakes in the Northeast as being of *acute concern*, which means that a complete loss of fish population is expected. Today, the fraction of lakes in this category has been reduced to less than 18%. The acid rain program has been a dramatic success, and today's environmental legislation—such as legislation aimed at global warming—is using the acid rain legislation as a model for what can work.

### SO₂ AIR QUALITY, 1990–2013

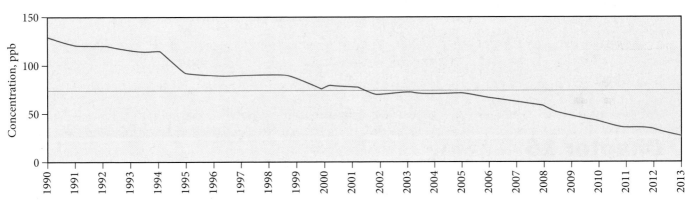

▲ **FIGURE 16.18  U.S. Sulfur Dioxide Pollutant Levels**   As a result of amendments to the Clean Air Act passed in 1990, $SO_2$ levels continue to decrease.
Source: U.S. EPA Air Trends

# Self-Assessment Quiz

**Q1.** Identify the conjugate base in the reaction shown here:

$$HClO_2(aq) + H_2O(aq) \rightleftharpoons H_3O^+(aq) + ClO_2^-(aq)$$

a) $HClO_2$    b) $H_2O$    c) $H_3O^+$    d) $ClO_2^-$

**Q2.** Which pair is a Brønsted–Lowry conjugate acid–base pair?

a) $NH_3$; $NH_4^+$                    b) $H_3O^+$; $OH^-$
c) $HCl$; $HBr$                         d) $ClO_4^-$; $ClO_3^-$

**Q3.** Consider the given acid ionization constants. Identify the strongest conjugate base.

| Acid | $K_a$ |
|---|---|
| $HNO_2(aq)$ | $4.6 \times 10^{-4}$ |
| $HCHO_2(aq)$ | $1.8 \times 10^{-4}$ |
| $HClO(aq)$ | $2.9 \times 10^{-8}$ |
| $HCN(aq)$ | $4.9 \times 10^{-10}$ |

a) $NO_2^-(aq)$                        b) $CHO_2^-(aq)$
c) $ClO^-(aq)$                          d) $CN^-(aq)$

**Q4.** What is the $OH^-$ concentration in an aqueous solution at 25 °C in which $[H_3O^+] = 1.9 \times 10^{-9}$ M?

a) $1.9 \times 10^{-9}$ M              b) $5.3 \times 10^{-6}$ M
c) $5.3 \times 10^6$ M                 d) $1.9 \times 10^{-23}$ M

**Q5.** An $HNO_3(aq)$ solution has a pH of 1.75. What is the molar concentration of the $HNO_3(aq)$ solution?

a) 1.75 M                              b) $5.6 \times 10^{-13}$ M
c) 56 M                                d) 0.018 M

**Q6.** Find the pH of a 0.350 M aqueous benzoic acid solution. For benzoic acid, $K_a = 6.5 \times 10^{-5}$.

a) 4.64                                b) 4.19
c) 2.32                                d) 11.68

**Q7.** Find the pH of a 0.155 M $HClO_2(aq)$ solution. For $HClO_2$, $K_a = 0.011$.

a) 0.92                                b) 1.44
c) 1.39                                d) 0.69

**Q8.** Calculate the percent ionization of 1.45 M aqueous acetic acid solution. For acetic acid, $K_a = 1.8 \times 10^{-5}$.

a) 0.35%                               b) 0.0018%
c) 0.29%                               d) 0.0051%

*Continued—*

**Q9.** Consider two aqueous solutions of nitrous acid ($HNO_2$). Solution A has a concentration of $[HNO_2]$ = 0.55 M and solution B has a concentration of $[HNO_2]$ = 1.25 M. Which statement about the two solutions is true?

a) Solution A has the higher percent ionization and the higher pH.

b) Solution B has the higher percent ionization and the higher pH.

c) Solution A has the higher percent ionization and solution B has the higher pH.

d) Solution B has the higher percent ionization and solution A has the higher pH.

**Q10.** Find the $[OH^-]$ in a 0.200 M solution of ethylamine ($C_2H_5NH_2$). For ethylamine, $K_b$ = 5.6 × $10^{-4}$.

a) 11.52 M    b) 2.48 M    c) 0.033 M    d) 0.011 M

**Q11.** Which ion forms a basic solution when dissolved in water?

a) $Br^-$    b) $NO_3^-$    c) $HSO_4^-$    d) $SO_3^{2-}$

**Q12.** Which compound forms an acidic solution when dissolved in water?

a) $NH_4Cl$    b) NaCl    c) $KNO_2$    d) $Ca(NO_3)_2$

**Q13.** Find the pH of 0.175 M NaCN solution. For HCN, $K_a$ = 4.9 × $10^{-10}$.

a) 5.03    b) 11.28    c) 2.31    d) 8.97

**Q14.** What is the concentration of $X^{2-}$ in a 0.150 M solution of the diprotic acid $H_2X$? For $H_2X$, $K_{a_1}$ = 4.5 × $10^{-6}$ and $K_{a2}$ = 1.2 × $10^{-11}$.

a) 9.9 × $10^{-8}$ M      b) 2.0 × $10^{-9}$ M

c) 8.2 × $10^{-4}$ M      d) 1.2 × $10^{-11}$ M

**Q15.** Which acid has the largest $K_a$: $HClO_2(aq)$, $HBrO_2(aq)$, or $HIO_2(aq)$?

a) $HClO_2(aq)$

b) $HBrO_2(aq)$

c) $HIO_2(aq)$

d) All three acids have the same $K_a$.

**Answers: 1.** (d) **2.** (a) **3.** (d) **4.** (b) **5.** (d) **6.** (c) **7.** (b) **8.** (a) **9.** (a) **10.** (d) **11.** (d) **12.** (a) **13.** (b) **14.** (a) **15.** (a)

# Chapter 16 IN REVIEW

## Key Terms

**Section 16.2**
carboxylic acid (725)
alkaloid (726)

**Section 16.3**
Arrhenius definitions (of acids and bases) (726)
hydronium ion (727)
Brønsted–Lowry definitions (of acids and bases) (727)
amphoteric (727)
conjugate acid–base pair (728)
conjugate acid (728)
conjugate base (728)

**Section 16.4**
strong acid (729)
weak acid (729)
monoprotic acid (729)
diprotic acid (729)
triprotic acid (730)
acid ionization constant ($K_a$) (731)

**Section 16.5**
autoionization (732)
ion product constant for water ($K_w$) (733)
neutral (733)
acidic solution (733)
basic solution (733)
pH (734)

**Section 16.6**
percent ionization (742)

**Section 16.7**
strong base (746)
weak base (746)
base ionization constant ($K_b$) (747)

**Section 16.9**
polyprotic acid (757)

**Section 16.11**
Lewis acid (764)
Lewis base (764)

## Key Concepts

### Heartburn (16.1)

▶ Hydrochloric acid from the stomach sometimes comes in contact with the esophageal lining, resulting in irritation, called heartburn. Heartburn is treated with antacids, bases that neutralize stomach acid.

### The Nature of Acids and Bases (16.2)

▶ Acids generally taste sour, dissolve metals, turn blue litmus paper red, and neutralize bases. Common acids are hydrochloric, sulfuric, nitric, and carboxylic acids.

> Bases generally taste bitter, feel slippery, turn red litmus paper blue, and neutralize acids. Common bases are sodium hydroxide, sodium bicarbonate, and potassium hydroxide.

## Definitions of Acids and Bases (16.3)

> The Arrhenius definition of acids and bases states that in an aqueous solution, an acid produces hydrogen ions and a base produces hydroxide ions.

> The Brønsted–Lowry definition states that an acid is a proton (hydrogen ion) donor and a base is a proton acceptor. According to the Brønsted–Lowry definition, two substances related by the transfer of a proton are a conjugate acid–base pair.

## Acid Strength and the Acid Dissociation Constant, $K_a$ (16.4)

> In a solution, a strong acid completely ionizes but a weak acid only partially ionizes.

> Generally, the stronger the acid, the weaker the conjugate base, and vice versa.

> The extent of dissociation of a weak acid is quantified by the acid dissociation constant, $K_a$, which is the equilibrium constant for the ionization of the weak acid.

## Autoionization of Water and pH (16.5)

> In an acidic solution, the concentration of hydrogen ions is always greater than the concentration of hydroxide ions. $[H_3O^+]$ multiplied by $[OH^-]$ is always constant at a constant temperature.

> There are two types of logarithmic acid–base scales: pH and pOH. At 25 °C, the sum of a solution's pH and pOH is always 14.

## Finding the $[H_3O^+]$ and pH of Strong and Weak Acid Solutions (16.6)

> In a strong acid solution, the hydrogen ion concentration equals the initial concentration of the acid.

> In a weak acid solution, the hydrogen ion concentration—which can be determined by solving an equilibrium problem—is lower than the initial acid concentration.

> The percent ionization of weak acids decreases as the acid (and hydrogen ion) concentration increases.

> In mixtures of two acids with large $K_a$ differences, the concentration of hydrogen ions can usually be determined by considering only the stronger of the two acids.

## Base Solutions (16.7)

> A strong base dissociates completely; a weak base does not.

> Most weak bases produce hydroxide ions through the ionization of water. The base ionization constant, $K_b$, indicates the extent of ionization.

## Ions as Acids and Bases (16.8)

> A cation is a weak acid if it is the conjugate acid of a weak base; it is neutral if it is the counter ion of a strong base.

> An anion is a weak base if it is the conjugate base of a weak acid; it is neutral if it is the conjugate base of a strong acid.

> To calculate the pH of a solution of an acidic cation or basic anion, we determine $K_a$ or $K_b$ from the equation $K_a \times K_b = K_w$.

## Polyprotic Acids (16.9)

> Polyprotic acids contain two or more ionizable protons.

> Generally, polyprotic acids ionize in successive steps, and the value of $K_a$ becomes smaller for each step.

> In many cases, we can determine the $[H_3O^+]$ of a polyprotic acid solution by considering only the first ionization step; then, the concentration of the anion formed in the second ionization step is equivalent to the value of $K_{a_2}$.

## Acid Strength and Molecular Structure (16.10)

> For binary acids, acid strength decreases with increasing bond energy and increases with increasing bond polarity.

> For oxyacids, acid strength increases with the electronegativity of the atoms bonded to the oxygen atom and also increases with the number of oxygen atoms in the molecule.

## Lewis Acids and Bases (16.11)

> A third model of acids and bases, the Lewis model, defines a base as an electron pair donor and an acid as an electron pair acceptor. According to this definition, an acid does not have to contain hydrogen. A Lewis acid can be a compound with an empty orbital—or one that will rearrange to make an empty orbital—or a cation.

## Acid Rain (16.12)

> The combustion of fossil fuels produces oxides of sulfur and nitrogen, which react with oxygen and water to form sulfuric and nitric acids. These acids combine with rain to form acid rain.

> Acid rain corrodes human-made structures and damages aquatic environments and forests. Environmental legislation has helped stabilize the amount of acid rain being produced.

# Key Equations and Relationships

Note: In all of these equations $[H^+]$ is interchangeable with $[H_3O^+]$.

Expression for the Acid Ionization Constant, $K_a$ (16.4)

$$K_a = \frac{[H_3O^+][A^-]}{[HA]}$$

The Ion Product Constant for Water, $K_w$ (16.5)

$$K_w = [H_3O^+][OH^-] = 1.0 \times 10^{-14} \text{ (at 25 °C)}$$

Expression for the pH Scale (16.5)

$$pH = -\log[H_3O^+]$$

Expression for the pOH Scale (16.5)

$$pOH = -\log[OH^-]$$

Relationship between pH and pOH (16.5)

$$pH + pOH = 14.00$$

Expression for the p$K_a$ Scale (16.5)

$$pK_a = -\log K_a$$

Expression for Percent Ionization (16.6)

$$\text{Percent ionization} = \frac{\text{concentration of ionized acid}}{\text{initial concentration of acid}} \times 100\%$$

$$= \frac{[H_3O^+]_{equil}}{[HA]_{init}} \times 100\%$$

Relationship between $K_a$, $K_b$, and $K_w$ (16.8)

$$K_a \times K_b = K_w$$

# Key Learning Outcomes

| CHAPTER OBJECTIVES | ASSESSMENT |
|---|---|
| Identifying Brønsted–Lowry Acids and Bases and Their Conjugates (**16.3**)<br><br>$$H_2SO_4(aq) + H_2O(l) \longrightarrow HSO_4^-(aq) + H_3O^+(aq)$$<br>Acid　　Base　　Conjugate base　　Conjugate acid | Example 16.1　For Practice 16.1　Exercises 35, 36 |
| Using $K_w$ in Calculations (**16.5**) | Example 16.2　For Practice 16.2　Exercises 47, 48 |
| Calculating pH from $[H_3O^+]$ or $[OH^-]$ (**16.5**) | Examples 16.3, 16.4　For Practice 16.3, 16.4　Exercises 49–52 |
| Finding the pH of a Weak Acid Solution (**16.6**) | Examples 16.5, 16.6, 16.7　For Practice 16.5, 16.6, 16.7　Exercises 63–68 |
| Finding the Acid Ionization Constant from pH (**16.6**) | Example 16.8　For Practice 16.8　Exercises 69, 70 |
| Finding the Percent Ionization of a Weak Acid (**16.6**) | Example 16.9　For Practice 16.9　Exercises 71–74 |
| Mixtures of Weak Acids (**16.6**) | Example 16.10　For Practice 16.10　Exercises 79, 80 |
| Finding the $[OH^-]$ and pH of a Strong Base Solution (**16.7**) | Example 16.11　For Practice 16.11　Exercises 81, 82 |
| Finding the $[OH^-]$ and pH of a Weak Base Solution (**16.7**)<br><br>Ammonia　Methylamine　Pyridine | Example 16.12　For Practice 16.12　Exercises 89, 90 |
| Determining Whether an Anion Is Basic or Neutral (**16.8**) | Example 16.13　For Practice 16.13　Exercises 95, 96 |
| Determining the pH of a Solution Containing an Anion Acting as a Base (**16.8**) | Example 16.14　For Practice 16.14　Exercises 97, 98 |
| Determining Whether a Cation Is Acidic or Neutral (**16.8**) | Example 16.15　For Practice 16.15　Exercises 99, 100 |
| Determining the Overall Acidity or Basicity of Salt Solutions (**16.8**) | Example 16.16　For Practice 16.16　Exercises 101, 102 |
| Finding the pH of a Polyprotic Acid Solution (**16.9**) | Example 16.17　For Practice 16.17　Exercises 111, 112 |
| Finding the $[H_3O^+]$ in Dilute $H_2SO_4$ Solutions (**16.9**) | Example 16.18　For Practice 16.18　Exercise 115 |
| Finding the Concentration of the Anions for a Weak Diprotic Acid Solution (**16.9**) | Example 16.19　For Practice 16.19　Exercises 113, 114 |

# EXERCISES

## Review Questions

1. What causes heartburn? What are some possible ways to alleviate heartburn?

2. What are the general physical and chemical properties of acids? Of bases?

3. What is a carboxylic acid? Give an example.

4. What is the Arrhenius definition of an acid? Of a base?

5. What is a hydronium ion? Does $H^+$ exist in solution by itself?

6. What is the Brønsted–Lowry definition of an acid? Of a base?

7. Why is there more than one definition of acid–base behavior? Which definition is the right one?

8. Describe amphoteric behavior and give an example.

9. What is a conjugate acid–base pair? Provide an example.

10. Explain the difference between a strong acid and a weak acid and list one example of each.

11. What are diprotic and triprotic acids? List an example of each.

12. Define the acid ionization constant and explain its significance.

13. Write an equation for the autoionization of water and an expression for the ion product constant for water ($K_w$). What is the value of $K_w$ at 25 °C?

14. What happens to the [OH⁻] of a solution when the [H₃O⁺] is increased? Decreased?

15. Define pH. What pH range is considered acidic? Basic? Neutral? (Assume 25 °C.)

16. Define pOH. What pOH range is considered acidic? Basic? Neutral? (Assume 25 °C.)

17. In most solutions containing a strong or weak acid, the autoionization of water can be neglected when calculating [H₃O⁺]. Explain why this statement is valid.

18. When calculating [H₃O⁺] for weak acid solutions, we can often use the *x is small* approximation. Explain the nature of this approximation and why it is valid.

19. What is the percent ionization of an acid? Explain what happens to the percent ionization of a weak acid as a function of the concentration of the weak acid solution.

20. In calculating [H₃O⁺] for a mixture of a strong acid and weak acid, the weak acid can often be neglected. Explain why this statement is valid.

21. Write a generic equation showing how a weak base ionizes water.

22. How can you determine if an anion will act as a weak base? Write a generic equation showing the reaction by which an anion, A⁻, acts as a weak base.

23. What is the relationship between the acid ionization constant for a weak acid ($K_a$) and the base ionization constant for its conjugate base ($K_b$)?

24. What kinds of cations act as weak acids? List some examples.

25. When calculating the [H₃O⁺] for a polyprotic acid, the second ionization step can often be neglected. Explain why this statement is valid.

26. For a weak diprotic acid H₂X, what is the relationship between [X²⁻] and $K_{a_2}$? Under what conditions does this relationship exist?

27. For a binary acid, H—Y, which factors affect the relative ease with which the acid ionizes?

28. Which factors affect the relative acidity of an oxyacid?

29. What is the Lewis definition of an acid? Of a base?

30. What is a general characteristic of a Lewis acid? Of a Lewis base?

31. What is acid rain? What causes it, and where is the problem the greatest?

32. What are the main detrimental effects of acid rain? What is being done to address the problem of acid rain?

## Problems by Topic

### The Nature and Definitions of Acids and Base

33. Identify each substance as an acid or a base and write a chemical equation showing how it is an acid or a base according to the Arrhenius definition.
    a. HNO₃(*aq*)          b. NH₄⁺(*aq*)
    c. KOH(*aq*)          d. HC₂H₃O₂(*aq*)

34. Identify each substance as an acid or a base and write a chemical equation showing how it is an acid or a base in aqueous solution according to the Arrhenius definition.
    a. NaOH(*aq*)          b. H₂SO₄(*aq*)
    c. HBr(*aq*)          d. Sr(OH)₂(*aq*)

35. In each reaction, identify the Brønsted–Lowry acid, the Brønsted–Lowry base, the conjugate acid, and the conjugate base.
    a. H₂CO₃(*aq*) + H₂O(*l*) ⇌ H₃O⁺(*aq*) + HCO₃⁻(*aq*)
    b. NH₃(*aq*) + H₂O(*l*) ⇌ NH₄⁺(*aq*) + OH⁻(*aq*)
    c. HNO₃(*aq*) + H₂O(*l*) ⟶ H₃O⁺(*aq*) + NO₃⁻(*aq*)
    d. C₅H₅N(*aq*) + H₂O(*l*) ⇌ C₅H₅NH⁺(*aq*) + OH⁻(*aq*)

36. In each reaction, identify the Brønsted–Lowry acid, the Brønsted–Lowry base, the conjugate acid, and the conjugate base.
    a. HI(*aq*) + H₂O(*l*) ⟶ H₃O⁺(*aq*) + I⁻(*aq*)
    b. CH₃NH₂(*aq*) + H₂O(*l*) ⇌ CH₃NH₃⁺(*aq*) + OH⁻(*aq*)
    c. CO₃²⁻(*aq*) + H₂O(*l*) ⇌ HCO₃⁻(*aq*) + OH⁻(*aq*)
    d. HBr(*aq*) + H₂O(*l*) ⟶ H₃O⁺(*aq*) + Br⁻(*aq*)

37. Write the formula for the conjugate base of each acid.
    a. HCl     b. H₂SO₃     c. HCHO₂     d. HF

38. Write the formula for the conjugate acid of each base.
    a. NH₃     b. ClO₄⁻     c. HSO₄⁻     d. CO₃²⁻

39. Both H₂O and H₂PO₄⁻ are amphoteric. Write an equation to show how each substance can act as an acid and another equation to show how each can act as a base.

40. Both HCO₃⁻ and HS⁻ are amphoteric. Write an equation to show how each substance can act as an acid and another equation to show how each can act as a base.

### Acid Strength and $K_a$

41. Classify each acid as strong or weak. If the acid is weak, write an expression for the acid ionization constant ($K_a$).
    a. HNO₃     b. HCl     c. HBr     d. H₂SO₃

42. Classify each acid as strong or weak. If the acid is weak, write an expression for the acid ionization constant ($K_a$).
    a. HF     b. HCHO₂     c. H₂SO₄     d. H₂CO₃

43. The three diagrams represent three different solutions of the binary acid HA. Water molecules have been omitted for clarity, and hydronium ions (H₃O⁺) are represented by hydrogen ions (H⁺). Rank the acids in order of decreasing acid strength.

(a)          (b)          (c)

44. Rank the solutions in order of decreasing [H₃O⁺]: 0.10 M HCl; 0.10 M HF; 0.10 M HClO; 0.10 M HC₆H₅O.

45. Pick the stronger base from each pair.
    a. F⁻ or Cl⁻
    b. NO₂⁻ or NO₃⁻
    c. F⁻ or ClO⁻

46. Pick the stronger base from each pair.
    a. ClO₄⁻ or ClO₂⁻
    b. Cl⁻ or H₂O
    c. CN⁻ or ClO⁻

## Autoionization of Water and pH

**47.** Calculate [OH⁻] in each aqueous solution at 25 °C, and classify the solution as acidic or basic.
  **a.** $[H_3O^+] = 1.2 \times 10^{-8}\,M$    **b.** $[H_3O^+] = 8.5 \times 10^{-5}\,M$
  **c.** $[H_3O^+] = 3.5 \times 10^{-2}\,M$

**48.** Calculate [H₃O⁺] in each aqueous solution at 25 °C, and classify each solution as acidic or basic.
  **a.** $[OH^-] = 1.1 \times 10^{-9}\,M$    **b.** $[OH^-] = 2.9 \times 10^{-2}\,M$
  **c.** $[OH^-] = 6.9 \times 10^{-12}\,M$

**49.** Calculate the pH and pOH of each solution at 25 °C.
  **a.** $[H_3O^+] = 1.7 \times 10^{-8}\,M$    **b.** $[H_3O^+] = 1.0 \times 10^{-7}\,M$
  **c.** $[H_3O^+] = 2.2 \times 10^{-6}\,M$

**50.** Calculate [H₃O⁺] and [OH⁻] for each solution at 25 °C.
  **a.** pH = 8.55    **b.** pH = 11.23    **c.** pH = 2.87

**51.** Complete the table. (All solutions are at 25 °C.)

| [H₃O⁺] | [OH⁻] | pH | Acidic or Basic |
|---|---|---|---|
| | | 3.15 | |
| $3.7 \times 10^{-9}$ | | | |
| | | 11.1 | |
| | $1.6 \times 10^{-11}$ | | |

**52.** Complete the table. (All solutions are at 25 °C.)

| [H₃O⁺] | [OH⁻] | pH | Acidic or Basic |
|---|---|---|---|
| $3.5 \times 10^{-3}$ | | | |
| | $3.8 \times 10^{-7}$ | | |
| $1.8 \times 10^{-9}$ | | | |
| | | 7.15 | |

**53.** Like all equilibrium constants, the value of $K_w$ depends on temperature. At body temperature (37 °C), $K_w = 2.4 \times 10^{-14}$. What are the [H₃O⁺] and pH of pure water at body temperature?

**54.** The value of $K_w$ increases with increasing temperature. Is the autoionization of water endothermic or exothermic?

**55.** Calculate the pH of each acid solution. Explain how the resulting pH values demonstrate that the pH of an acid solution should carry as many digits to the right of the decimal place as the number of significant figures in the concentration of the solution.
  $[H_3O^+] = 0.044\,M$
  $[H_3O^+] = 0.045\,M$
  $[H_3O^+] = 0.046\,M$

**56.** Determine the concentration of H₃O⁺ to the correct number of significant figures in a solution with each pH. Describe how these calculations show the relationship between the number of digits to the right of the decimal place in pH and the number of significant figures in concentration.
  pH = 2.50
  pH = 2.51
  pH = 2.52

## Acid Solutions

**57.** For each strong acid solution, determine [H₃O⁺], [OH⁻], and pH.
  **a.** 0.25 M HCl
  **b.** 0.015 M HNO₃
  **c.** a solution that is 0.052 M in HBr and 0.020 M in HNO₃
  **d.** a solution that is 0.655% HNO₃ by mass (assume a density of 1.01 g/mL for the solution)

**58.** Determine the pH of each solution.
  **a.** 0.048 M HI
  **b.** 0.0895 M HClO₄
  **c.** a solution that is 0.045 M in HClO₄ and 0.048 M in HCl
  **d.** a solution that is 1.09% HCl by mass (assume a density of 1.01 g/mL for the solution)

**59.** What mass of HI must be present in 0.250 L of solution to obtain a solution with each pH value?
  **a.** pH = 1.25    **b.** pH = 1.75    **c.** pH = 2.85

**60.** What mass of HClO₄ must be present in 0.500 L of solution to obtain a solution with each pH value?
  **a.** pH = 2.50    **b.** pH = 1.50    **c.** pH = 0.50

**61.** What is the pH of a solution in which 224 mL of HCl(g), measured at 27.2 °C and 1.02 atm, is dissolved in 1.5 L of aqueous solution?

**62.** What volume of a concentrated HCl solution, which is 36.0% HCl by mass and has a density of 1.179 g/mL, should be used to make 5.00 L of an HCl solution with a pH of 1.8?

**63.** Determine the [H₃O⁺] and pH of a 0.100 M solution of benzoic acid.

**64.** Determine the [H₃O⁺] and pH of a 0.200 M solution of formic acid.

**65.** Determine the pH of an HNO₂ solution of each concentration. In which cases can you *not* make the simplifying assumption that *x is small*?
  **a.** 0.500 M    **b.** 0.100 M    **c.** 0.0100 M

**66.** Determine the pH of an HF solution of each concentration. In which cases can you *not* make the simplifying assumption that *x is small*? ($K_a$ for HF is $6.8 \times 10^{-4}$.)
  **a.** 0.250 M    **b.** 0.0500 M    **c.** 0.0250 M

**67.** If 15.0 mL of glacial acetic acid (pure HC₂H₃O₂) is diluted to 1.50 L with water, what is the pH of the resulting solution? The density of glacial acetic acid is 1.05 g/mL.

**68.** Calculate the pH of a formic acid solution that contains 1.35% formic acid by mass. (Assume a density of 1.01 g/mL for the solution.)

**69.** A 0.185 M solution of a weak acid (HA) has a pH of 2.95. Calculate the acid ionization constant ($K_a$) for the acid.

**70.** A 0.115 M solution of a weak acid (HA) has a pH of 3.29. Calculate the acid ionization constant ($K_a$) for the acid.

**71.** Determine the percent ionization of a 0.125 M HCN solution.

**72.** Determine the percent ionization of a 0.225 M solution of benzoic acid.

**73.** Calculate the percent ionization of an acetic acid solution having the given concentration.
  **a.** 1.00 M    **b.** 0.500 M    **c.** 0.100 M    **d.** 0.0500 M

**74.** Calculate the percent ionization of a formic acid solution having the given concentration.
  **a.** 1.00 M    **b.** 0.500 M    **c.** 0.100 M    **d.** 0.0500 M

**75.** A 0.148 M solution of a monoprotic acid has a percent ionization of 1.55%. Determine the acid ionization constant ($K_a$) for the acid.

**76.** A 0.085 M solution of a monoprotic acid has a percent ionization of 0.59%. Determine the acid ionization constant ($K_a$) for the acid.

**77.** Find the pH and percent ionization of each HF solution. ($K_a$ for HF is $6.8 \times 10^{-4}$.)
  **a.** 0.250 M HF    **b.** 0.100 M HF    **c.** 0.050 M HF

**78.** Find the pH and percent ionization of a 0.100 M solution of a weak monoprotic acid having the given $K_a$ values.
  **a.** $K_a = 1.0 \times 10^{-5}$  **b.** $K_a = 1.0 \times 10^{-3}$  **c.** $K_a = 1.0 \times 10^{-1}$

**79.** Find the pH of each mixture of acids.
   **a.** 0.115 M in HBr and 0.125 M in $HCHO_2$
   **b.** 0.150 M in $HNO_2$ and 0.085 M in $HNO_3$
   **c.** 0.185 M in $HCHO_2$ and 0.225 M in $HC_2H_3O_2$
   **d.** 0.050 M in acetic acid and 0.050 M in hydrocyanic acid

**80.** Find the pH of each mixture of acids.
   **a.** 0.075 M in $HNO_3$ and 0.175 M in $HC_7H_5O_2$
   **b.** 0.020 M in HBr and 0.015 M in $HClO_4$
   **c.** 0.095 M in HF and 0.225 M in $HC_6H_5O$
   **d.** 0.100 M in formic acid and 0.050 M in hypochlorous acid

## Base Solutions

**81.** For each strong base solution, determine $[OH^-]$, $[H_3O^+]$, pH, and pOH.
   **a.** 0.15 M NaOH
   **b.** $1.5 \times 10^{-3}$ M $Ca(OH)_2$
   **c.** $4.8 \times 10^{-4}$ M $Sr(OH)_2$
   **d.** $8.7 \times 10^{-5}$ M KOH

**82.** For each strong base solution, determine $[OH^-]$, $[H_3O^+]$, pH, and pOH.
   **a.** $8.77 \times 10^{-3}$ M LiOH
   **b.** 0.0112 M $Ba(OH)_2$
   **c.** $1.9 \times 10^{-4}$ M KOH
   **d.** $5.0 \times 10^{-4}$ M $Ca(OH)_2$

**83.** Determine the pH of a solution that is 3.85% KOH by mass. Assume that the solution has density of 1.01 g/mL.

**84.** Determine the pH of a solution that is 1.55% NaOH by mass. Assume that the solution has density of 1.01 g/mL.

**85.** What volume of 0.855 M KOH solution is required to make 3.55 L of a solution with pH of 12.4?

**86.** What volume of a 15.0% by mass NaOH solution, which has a density of 1.116 g/mL, should be used to make 5.00 L of an NaOH solution with a pH of 10.8?

**87.** Write equations showing how each weak base ionizes water to form $OH^-$. Also write the corresponding expression for $K_b$.
   **a.** $NH_3$
   **b.** $HCO_3^-$
   **c.** $CH_3NH_2$

**88.** Write equations showing how each weak base ionizes water to form $OH^-$. Also write the corresponding expression for $K_b$.
   **a.** $CO_3^{2-}$
   **b.** $C_6H_5NH_2$
   **c.** $C_2H_5NH_2$

**89.** Determine the $[OH^-]$, pH, and pOH of a 0.15 M ammonia solution.

**90.** Determine the $[OH^-]$, pH, and pOH of a solution that is 0.125 M in $CO_3^{2-}$.

**91.** Caffeine ($C_8H_{10}N_4O_2$) is a weak base with a $pK_b$ of 10.4. Calculate the pH of a solution containing a caffeine concentration of 455 mg/L.

**92.** Amphetamine ($C_9H_{13}N$) is a weak base with a $pK_b$ of 4.2. Calculate the pH of a solution containing an amphetamine concentration of 225 mg/L.

**93.** Morphine is a weak base. A 0.150 M solution of morphine has a pH of 10.7. What is $K_b$ for morphine?

**94.** A 0.135 M solution of a weak base has a pH of 11.23. Determine $K_b$ for the base.

## Acid–Base Properties of Ions and Salts

**95.** Determine if each anion acts as a weak base in solution. For those anions that are basic, write an equation that shows how the anion acts as a base.
   **a.** $Br^-$
   **b.** $ClO^-$
   **c.** $CN^-$
   **d.** $Cl^-$

**96.** Determine whether each anion is basic or neutral. For those anions that are basic, write an equation that shows how the anion acts as a base.
   **a.** $C_7H_5O_2^-$
   **b.** $I^-$
   **c.** $NO_3^-$
   **d.** $F^-$

**97.** Determine the $[OH^-]$ and pH of a solution that is 0.140 M in $F^-$.

**98.** Determine the $[OH^-]$ and pH of a solution that is 0.250 M in $HCO_3^-$.

**99.** Determine whether each cation is acidic or pH-neutral. For those cations that are acidic, write an equation that shows how the cation acts as an acid.
   **a.** $NH_4^+$
   **b.** $Na^+$
   **c.** $Co^{3+}$
   **d.** $CH_2NH_3^+$

**100.** Determine whether each cation is acidic or pH-neutral. For each cation that is acidic, write an equation that shows how the cation acts as an acid.
   **a.** $Sr^{2+}$
   **b.** $Mn^{3+}$
   **c.** $C_5H_5NH^+$
   **d.** $Li^+$

**101.** Determine if each salt will form a solution that is acidic, basic, or pH-neutral.
   **a.** $FeCl_3$
   **b.** NaF
   **c.** $CaBr_2$
   **d.** $NH_4Br$
   **e.** $C_6H_5NH_3NO_2$

**102.** Determine if each salt will form a solution that is acidic, basic, or pH-neutral.
   **a.** $Al(NO_3)_3$
   **b.** $C_2H_5NH_3NO_3$
   **c.** $K_2CO_3$
   **d.** RbI
   **e.** $NH_4ClO$

**103.** Arrange the solutions in order of increasing acidity.
   NaCl, $NH_4Cl$, $NaHCO_3$, $NH_4ClO_2$, NaOH

**104.** Arrange the solutions in order of increasing basicity.
   $CH_3NH_3Br$, KOH, KBr, KCN, $C_5H_5NHNO_2$

**105.** Determine the pH of each solution.
   **a.** 0.10 M $NH_4Cl$
   **b.** 0.10 M $NaC_2H_3O_2$
   **c.** 0.10 M NaCl

**106.** Determine the pH of each solution.
   **a.** 0.20 M $KCHO_2$
   **b.** 0.20 M $CH_3NH_3I$
   **c.** 0.20 M KI

**107.** Calculate the concentration of all species in a 0.15 M KF solution.

**108.** Calculate the concentration of all species in a 0.225 M $C_6H_5NH_3Cl$ solution.

## Polyprotic Acids

**109.** Write chemical equations and corresponding equilibrium expressions for each of the three ionization steps of phosphoric acid.

**110.** Write chemical equations and corresponding equilibrium expressions for each of the two ionization steps of carbonic acid.

**111.** Calculate the $[H_3O^+]$ and pH of each polyprotic acid solution.
   **a.** 0.350 M $H_3PO_4$
   **b.** 0.350 M $H_2C_2O_4$

**112.** Calculate the $[H_3O^+]$ and pH of each polyprotic acid solution.
   **a.** 0.125 M $H_2CO_3$
   **b.** 0.125 M $H_3C_6H_5O_7$

**113.** Calculate the concentration of all species in a 0.500 M solution of $H_2SO_3$.

**114.** Calculate the concentration of all species in a 0.155 M solution of $H_2CO_3$.

**115.** Calculate the $[H_3O^+]$ and pH of each $H_2SO_4$ solution. At approximately what concentration does the *x is small* approximation break down?
   **a.** 0.50 M
   **b.** 0.10 M
   **c.** 0.050 M

**116.** Consider a 0.10 M solution of a weak polyprotic acid ($H_2A$) with the possible values of $K_{a_1}$ and $K_{a_2}$ given here.
   **a.** $K_{a_1} = 1.0 \times 10^{-4}$; $K_{a_2} = 5.0 \times 10^{-5}$
   **b.** $K_{a_1} = 1.0 \times 10^{-4}$; $K_{a_2} = 1.0 \times 10^{-5}$
   **c.** $K_{a_1} = 1.0 \times 10^{-4}$; $K_{a_2} = 1.0 \times 10^{-6}$

   Calculate the contributions to $[H_3O^+]$ from each ionization step. At what point can the contribution of the second step be neglected?

## Molecular Structure and Acid Strength

**117.** Based on their molecular structure, pick the stronger acid from each pair of binary acids. Explain your choice.
  **a.** HF and HCl  **b.** $H_2O$ or HF  **c.** $H_2Se$ or $H_2S$

**118.** Based on molecular structure, arrange the binary compounds in order of increasing acid strength. Explain your choice.
$$H_2Te,\ HI,\ H_2S,\ NaH$$

**119.** Based on their molecular structure, pick the stronger acid from each pair of oxyacids. Explain your choice.
  **a.** $H_2SO_4$ or $H_2SO_3$
  **b.** $HClO_2$ or $HClO$
  **c.** $HClO$ or $HBrO$
  **d.** $CCl_3COOH$ or $CH_3COOH$

**120.** Based on molecular structure, arrange the oxyacids in order of increasing acid strength. Explain your choice.
$$HClO_3,\ HIO_3,\ HBrO_3$$

**121.** Which is a stronger base, $S^{2-}$ or $Se^{2-}$? Explain.

**122.** Which is a stronger base, $PO_4^{3-}$ or $AsO_4^{3-}$? Explain.

## Lewis Acids and Bases

**123.** Classify each species as either a Lewis acid or a Lewis base.
  **a.** $Fe^{3+}$  **b.** $BH_3$
  **c.** $NH_3$  **d.** $F^-$

**124.** Classify each species as either a Lewis acid or a Lewis base.
  **a.** $BeCl_2$  **b.** $OH^-$
  **c.** $B(OH)_3$  **d.** $CN^-$

**125.** Identify the Lewis acid and Lewis base from among the reactants in each equation.
  **a.** $Fe^{3+}(aq) + 6\,H_2O(l) \rightleftharpoons Fe(H_2O)_6^{3+}(aq)$
  **b.** $Zn^{2+}(aq) + 4\,NH_3(aq) \rightleftharpoons Zn(NH_3)_4^{2+}(aq)$
  **c.** $(CH_3)_3N(g) + BF_3(g) \rightleftharpoons (CH_3)_3NBF_3(s)$

**126.** Identify the Lewis acid and Lewis base from among the reactants in each equation.
  **a.** $Ag^+(aq) + 2\,NH_3(aq) \rightleftharpoons Ag(NH_3)_2^+(aq)$
  **b.** $AlBr_3 + NH_3 \rightleftharpoons H_3NAlBr_3$
  **c.** $F^-(aq) + BF_3(aq) \rightleftharpoons BF_4^-(aq)$

# Cumulative Problems

**127.** Based on these molecular views, determine whether each pictured acid is weak or strong.

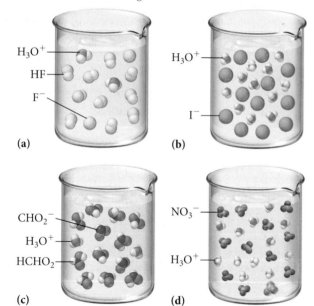

**(a)**  **(b)**  **(c)**  **(d)**

**128.** Based on these molecular views, determine whether each pictured base is weak or strong.

**(a)**  **(b)**

**(c)**  **(d)**

**129.** The binding of oxygen by hemoglobin in the blood involves the equilibrium reaction:
$$HbH^+(aq) + O_2(aq) \rightleftharpoons HbO_2(aq) + H^+(aq)$$

In this equation, Hb is hemoglobin. The pH of normal human blood is highly controlled within a range of 7.35 to 7.45. Given the above equilibrium, why is this important? What would happen to the oxygen-carrying capacity of hemoglobin if blood became too acidic (a dangerous condition known as acidosis)?

**130.** Carbon dioxide dissolves in water according to the equations:
$$CO_2(g) + H_2O(l) \rightleftharpoons H_2CO_3(aq)$$
$$H_2CO_3(aq) + H_2O(l) \rightleftharpoons HCO_3^-(aq) + H_3O^+(aq)$$

Carbon dioxide levels in the atmosphere have increased about 20% over the last century. Given that Earth's oceans are exposed to atmospheric carbon dioxide, what effect might the increased $CO_2$ be having on the pH of the world's oceans? What effect might this change be having on the limestone structures (primarily $CaCO_3$) of coral reefs and marine shells?

**131.** People often take milk of magnesia to reduce the discomfort associated with acid stomach or heartburn. The recommended dose is 1 teaspoon, which contains $4.00 \times 10^2$ mg of $Mg(OH)_2$. What volume of an HCl solution with a pH of 1.3 can be neutralized by one dose of milk of magnesia? If the stomach contains $2.00 \times 10^2$ mL of pH 1.3 solution, is all the acid neutralized? If not, what fraction is neutralized?

**132.** Lakes that have been acidified by acid rain can be neutralized by liming, the addition of limestone ($CaCO_3$). How much limestone (in kg) is required to completely neutralize a 4.3 billion liter lake with a pH of 5.5?

▲ Liming a lake.

**133.** Acid rain over the Great Lakes has a pH of about 4.5. Calculate the $[H_3O^+]$ of this rain and compare that value to the $[H_3O^+]$ of rain over the West Coast that has a pH of 5.4. How many times more concentrated is the acid in rain over the Great Lakes?

**134.** White wines tend to be more acidic than red wines. Find the $[H_3O^+]$ in a Sauvignon Blanc with a pH of 3.23 and a Cabernet Sauvignon with a pH of 3.64. How many times more acidic is the Sauvignon Blanc?

**135.** Common aspirin is acetylsalicylic acid, which has the structure shown here and a $pK_a$ of 3.5.

Calculate the pH of a solution in which one normal adult dose of aspirin ($6.5 \times 10^2$ mg) is dissolved in 8.0 ounces of water.

**136.** The AIDS drug zalcitabine (also known as ddC) is a weak base with a $pK_b$ of 9.8.

What percentage of the base is protonated in an aqueous zalcitabine solution containing 565 mg/L?

**137.** Determine the pH of each solution.
   **a.** 0.0100 M $HClO_4$
   **b.** 0.115 M $HClO_2$
   **c.** 0.045 M $Sr(OH)_2$
   **d.** 0.0852 M KCN
   **e.** 0.155 M $NH_4Cl$

**138.** Determine the pH of each solution.
   **a.** 0.0650 M $HNO_3$
   **b.** 0.150 M $HNO_2$
   **c.** 0.0195 M KOH
   **d.** 0.245 M $CH_3NH_3I$
   **e.** 0.318 M $KC_6H_5O$

**139.** Determine the pH of each two-component solution.
   **a.** 0.0550 M in HI and 0.00850 M in HF
   **b.** 0.112 M in NaCl and 0.0953 M in KF
   **c.** 0.132 M in $NH_4Cl$ and 0.150 M $HNO_3$
   **d.** 0.0887 M in sodium benzoate and 0.225 M in potassium bromide
   **e.** 0.0450 M in HCl and 0.0225 M in $HNO_3$

**140.** Determine the pH of each two-component solution.
   **a.** 0.050 M KOH and 0.015 M $Ba(OH)_2$
   **b.** 0.265 M $NH_4NO_3$ and 0.102 M HCN
   **c.** 0.075 M RbOH and 0.100 M $NaHCO_3$
   **d.** 0.088 M $HClO_4$ and 0.022 M KOH
   **e.** 0.115 M NaClO and 0.0500 M KI

**141.** Write net ionic equations for the reactions that take place when aqueous solutions of the following substances are mixed:
   **a.** sodium cyanide and nitric acid
   **b.** ammonium chloride and sodium hydroxide
   **c.** sodium cyanide and ammonium bromide
   **d.** potassium hydrogen sulfate and lithium acetate
   **e.** sodium hypochlorite and ammonia

**142.** Morphine has the formula $C_{17}H_{19}NO_3$. It is a base and accepts one proton per molecule. It is isolated from opium. A 0.682-g sample of opium is found to require 8.92 mL of a 0.0116 M solution of sulfuric acid for neutralization. Assuming that morphine is the only acid or base present in opium, calculate the percent morphine in the sample of opium.

**143.** The pH of a 1.00 M solution of urea, a weak organic base, is 7.050. Calculate the $K_a$ of protonated urea.

**144.** A solution is prepared by dissolving 0.10 mol of acetic acid and 0.10 mol of ammonium chloride in enough water to make 1.0 L of solution. Find the concentration of ammonia in the solution.

**145.** Lactic acid is a weak acid found in milk. Its calcium salt is a source of calcium for growing animals. A saturated solution of this salt, which we can represent as $Ca(Lact)_2$, has a $[Ca^{2+}] = 0.26$ M and a pH = 8.78. Assuming the salt is completely dissociated, calculate the $K_a$ of lactic acid.

**146.** A solution of 0.23 mol of the chloride salt of protonated quinine ($QH^+$), a weak organic base, in 1.0 L of solution has pH = 4.58. Find the $K_b$ of quinine (Q).

# Challenge Problems

**147.** A student mistakenly calculates the pH of a $1.0 \times 10^{-7}$ M HI solution to be 7.0. Explain why the student is incorrect and calculate the correct pH.

**148.** When 2.55 g of an unknown weak acid (HA) with a molar mass of 85.0 g/mol is dissolved in 250.0 g of water, the freezing point of the resulting solution is $-0.257\,°C$. Calculate $K_a$ for the unknown weak acid.

**149.** Calculate the pH of a solution that is 0.00115 M in HCl and 0.0100 M in $HClO_2$.

**150.** To what volume should you dilute 1 L of a solution of a weak acid HA to reduce the $[H^+]$ to one-half of that in the original solution?

**151.** HA, a weak acid, with $K_a = 1.0 \times 10^{-8}$, also forms the ion $HA_2^-$. The reaction is $HA(aq) + A^-(aq) \rightleftharpoons HA_2^-(aq)$ and its K = 4.0. Calculate the $[H^+]$, $[A^-]$, and $[HA_2^-]$ in a 1.0 M solution of HA.

**152.** Basicity in the gas phase can be defined as the proton affinity of the base, for example, $CH_3NH_2(g) + H^+(g) \rightleftharpoons CH_3NH_3^+(g)$. In the gas phase, $(CH_3)_3N$ is more basic than $CH_3NH_2$, while in solution the reverse is true. Explain this observation.

**153.** Calculate the pH of a solution prepared from 0.200 mol of $NH_4CN$ and enough water to make 1.00 L of solution.

**154.** To 1.0 L of a 0.30 M solution of $HClO_2$ is added 0.20 mol of NaF. Calculate the $[HClO_2]$ at equilibrium.

**155.** A mixture of $Na_2CO_3$ and $NaHCO_3$ has a mass of 82.2 g. It is dissolved in 1.00 L of water, and the pH is found to be 9.95. Find the mass of $NaHCO_3$ in the mixture.

**156.** A mixture of NaCN and $NaHSO_4$ consists of a total of 0.60 mol. When the mixture is dissolved in 1.0 L of water and comes to equilibrium, the pH is found to be 9.9. Find the amount of NaCN in the mixture.

## Conceptual Problems

**157.** Without doing any calculations, determine which solution in each pair is more acidic.
   **a.** 0.0100 M in HCl and 0.0100 M in KOH
   **b.** 0.0100 M in HF and 0.0100 M in KBr
   **c.** 0.0100 M in $NH_4Cl$ and 0.0100 M in $CH_3NH_3Br$
   **d.** 0.100 M in NaCN and 0.100 M in $CaCl_2$

**158.** Without doing any calculations, determine which solution in each pair is more basic.
   **a.** 0.100 M in NaClO and 0.100 M in NaF
   **b.** 0.0100 M in KCl and 0.0100 M in $KClO_2$

   **c.** 0.0100 M in $HNO_3$ and 0.0100 M in NaOH
   **d.** 0.0100 M in $NH_4Cl$ and 0.0100 M in HCN

**159.** Rank the acids in order of increasing acid strength.
   $CH_3COOH$   $CH_2ClCOOH$   $CHCl_2COOH$   $CCl_3COOH$

**160.** Without using a calculator, determine the pH and pOH of each solution. Rank the solutions from most acidic to most basic.
   **a.** $1.0 \times 10^{-2}$ M HCl
   **b.** $1.0 \times 10^{-4}$ M HCl
   **c.** $1.0 \times 10^{-2}$ M NaOH
   **d.** $1.0 \times 10^{-4}$ M NaOH

## Questions for Group Work

*Active Classroom Learning*

Discuss these questions with the group and record your consensus answer.

**161.** Without referring to the text, have each member of your group mention a different property of either an acid or a base, such as "Acids turn blue litmus paper red." Record as many properties as your group can recall without the text.

**162.** Have each group member make two flashcards with an acid or base on one side and its conjugate on the other side. Check each other's cards and quiz each other until each group member is proficient at identifying conjugate pairs.

**163.** Answer the following in a complete sentence or two:
   **a.** How do you know if an acid is strong or weak?
   **b.** How do you calculate the pH of a strong acid solution?

   **c.** How do you calculate the pH of a weak acid solution?
   **d.** If you know the $K_a$ of an acid, how do you determine the $K_b$ of its conjugate base?
   **e.** If you know $[OH^-]$ for a solution, how do you determine $[H_3O^+]$?

**164.** Solve the following problem, taking turns in your group to explain how to do the next step: What is the pH when 5.3 g of sodium acetate, $NaC_2H_3O_2$, is dissolved in 100.0 mL of water? (The $K_a$ of acetic acid, $HC_2H_3O_2$, is $1.8 \times 10^{-5}$.)

**165.** Define each of the following with complete sentences, and provide an example chemical equation: an Arrhenius acid, a Brønsted–Lowry base, and a Lewis acid.

## Data Interpretation and Analysis

### Sulfur Dioxide in Wine

**166.** Sulfur dioxide protects wine from spoiling; it acts as an antioxidant, a bactericide, and a preservative. Potassium or sodium metabisulfite, ($K_2S_2O_5$ or $Na_2S_2O_5$), is a good source of sulfur dioxide. $SO_2$ exists in a pH-dependent equilibrium system:

$$SO_2 + H_2O \rightleftharpoons H^+ + HSO_3^- \rightleftharpoons 2H^+ + SO_3^{2-}$$

$$pK_a = 1.81 \qquad pK_a = 7.2$$

   In this equilibrium system, $SO_2$ is called "molecular $SO_2$". In the $HSO_3^-$ form, it is called "free $SO_2$". Only molecular $SO_2$ is active as a preservative. The amount of molecular $SO_2$ that is

available from a given amount of free $SO_2$ is highly pH-dependent, which is one reason that winemakers carefully measure and adjust pH. The recommended amount of molecular $SO_2$ in white wine is 0.8 ppm, and for red wine it is 0.5 ppm. The bisulfite ion ($HSO_3^-$) or "free $SO_2$" is measured in milligrams $SO_2$ produced per L, or mg/L which is approximately equivalent to 1 ppm. Figure a ▶ shows the recommended amount of free $SO_2$ required to obtain the correct amount of molecular $SO_2$ concentration as a function of pH in white wine and red wine.

| pH | White Wine 0.8 ppm molecular SO₂ | Red Wine 0.5 ppm molecular SO₂ |
|---|---|---|
| 3 | 13 | 8 |
| 3.05 | 15 | 9 |
| 3.1 | 16 | 10 |
| 3.15 | 19 | 12 |
| 3.2 | 21 | 13 |
| 3.25 | 23 | 15 |
| 3.3 | 26 | 16 |
| 3.35 | 29 | 18 |
| 3.4 | 32 | 20 |
| 3.45 | 37 | 23 |
| 3.5 | 40 | 25 |
| 3.55 | 46 | 29 |
| 3.6 | 50 | 31 |
| 3.65 | 57 | 36 |
| 3.7 | 63 | 39 |
| 3.75 | 72 | 45 |
| 3.8 | 79 | 49 |
| 3.85 | 91 | 57 |
| 3.9 | 99 | 62 |

▲ **FIGURE a**  **Amount of "free SO₂" Required to Obtain the Correct Amount of Molecular SO₂ Concentration as a Function of pH in White and Red Wine.**

Figure b ▶ provides information in graph form about the concentration of free $SO_2$ required to maintain 0.8 ppm (0.8 mg/L) of molecular $SO_2$ in white wine at a given pH.

Use the information provided to answer the following questions:

a. Write the Lewis structure of the metabisufite, $S_2O_5^{2-}$ anion, and sketch its VSEPR molecular shape. What is the molecular shape around each sulfur atom? (Hint: The S atoms are bonded to each other.)

b. Write the equilibrium equation to represent how the metabisulfite anion, $S_2O_5^{2-}$, reacts with water to become the bisulfite anion.

c. A batch of white wine has an initial concentration of free $SO_2$ of 22 ppm and pH of 3.70. How much sodium metabisulfite should be added to the wine (per liter) to achieve the proper amount of free $SO_2$. Sodium metabisulfate is 67% free $SO_2$ by mass.

d. How must the graph in Figure b change to represent the concentration of free $SO_2$ required to maintain 0.5 ppm (0.5 mg/L) of molecular $SO_2$ in red wine at a given pH? Should the bars shift upward, downward, or stay the same? Explain your answer.

e. What concentration of free $SO_2$ must be present to protect red wine having a pH of 3.70?

f. In general, white wines tend to be more acidic than red wines. What happens to the equilibrium species present in a sample of white wine when the pH drops from 4.0 to 3.0?

### Free SO₂ Concentration Required to Maintain 0.8 0ppm Molecular SO₂

▲ **FIGURE b**  **Free SO₂ Required to Maintain 0.8 ppm (0.8 mg/L) of Molecular SO₂ in White Wine at a Given pH.**

# Answers to Conceptual Connections

## Conjugate Acid–Base Pairs

**16.1** **(b)** $H_2SO_4$ and $H_2SO_3$ are both acids; this is not a conjugate acid–base pair.

## The Magnitude of the Acid Ionization Constant

**16.2** HB has the largest $K_a$.

## Relative Strengths of Weak Acids

**16.3** HF is stronger because it has a larger acid ionization constant.

## pH and Acidity

**16.4** **(b)** As pH increases, acidity decreases.

## The *x is small* Approximation

**16.5** **(c)** The validity of the *x is small* approximation depends on both the value of the equilibrium constant and the initial concentration—the closer that these are to one another, the less likely the approximation is valid.

## Strong and Weak Acids

**16.6** **(a)** A weak acid solution is usually less than 5% dissociated. Since HCl is a strong acid, the 0.10 M solution is much more acidic than either a weak acid with the same concentration or even a weak acid that is twice as concentrated.

## Percent Ionization

**16.7** Solution **(c)** has the greatest percent ionization because percent ionization increases with decreasing weak acid concentration. Solution **(b)** has the lowest pH because the equilibrium $H_3O^+$ concentration increases with increasing weak acid concentration.

## Judging Relative pH

**16.8** **(a)** A weak acid solution is usually less than 5% dissociated. Therefore, since HCl is the only strong acid, the 1.0 M solution is much more acidic than either a weak acid that is twice as concentrated or a combination of two weak acids with the same concentrations.

## Acid Strength and Molecular Structure

**16.9** **(a)** Since the carbon atom in **(a)** is bonded to another oxygen atom, which draws electron density away from the O—H bond (weakening and polarizing it), and the carbon atom in **(b)** is bonded only to other hydrogen atoms, the proton in structure **(a)** is more acidic.

# Aqueous Ionic Equilibrium

In the strictly scientific sense of the word, insolubility does not exist, and even those substances characterized by the most obstinate resistance to the solvent action of water may properly be designated as extraordinarily difficult of solution, not as insoluble.

—OTTO N. WITT (1853–1915)

We have already seen the importance of aqueous solutions, first in Chapters 4, 13, and 15, and most recently in Chapter 16 on acids and bases. We now turn our attention to two additional topics involving aqueous solutions: buffers (solutions that resist pH change) and solubility equilibria (the extent to which slightly soluble ionic compounds dissolve in water). Buffers are tremendously important in biology because nearly all physiological processes must occur within a narrow pH range. Solubility equilibria are related to the solubility rules that we discussed in Chapter 4. In this chapter, we find a more complicated picture: solids that we considered insoluble under the simple "solubility rules" are actually better described as being only very slightly soluble, as the chapter-opening quotation from Otto Witt suggests. Solubility equilibria are important in predicting not only solubility, but also precipitation reactions that might occur when aqueous solutions are mixed.

▲ *Antifreeze is a dangerous poison that affects the blood's ability to carry oxygen.*

## 17.1  The Danger of Antifreeze

Every year, thousands of dogs and cats die from consuming a common household product: improperly stored or leaked antifreeze. Most types of auto antifreeze are aqueous solutions of ethylene glycol:

Some brands of antifreeze use propylene glycol, which is less toxic than ethylene glycol.

$HOCH_2CH_2OH$

Ethylene glycol has a somewhat sweet taste that can attract curious dogs and cats—and sometimes even young children, who are also vulnerable to this toxic compound. The first stage of ethylene glycol poisoning is a state resembling drunkenness. Because the compound is an alcohol, it affects the brain much as an alcoholic beverage would. Once ethylene glycol starts to be metabolized, however, a second and more deadly stage commences.

In the liver, ethylene glycol is oxidized to glycolic acid ($HOCH_2COOH$), which enters the bloodstream. The acidity of blood is critically important and tightly regulated because many proteins only function in a narrow pH range. In human blood, for example, pH is held between 7.36 and 7.42. This nearly constant blood pH is maintained by *buffers*. We discuss buffers more carefully later in this chapter, but for now know that a buffer is a chemical system that resists pH changes by neutralizing added acid or base. An important buffer in blood is a mixture of carbonic acid ($H_2CO_3$) and the bicarbonate ion ($HCO_3^-$). The carbonic acid neutralizes added base:

$$H_2CO_3(aq) + \underset{\text{added base}}{OH^-(aq)} \longrightarrow H_2O(l) + HCO_3^-(aq)$$

The bicarbonate ion neutralizes added acid:

$$HCO_3^-(aq) + \underset{\text{added acid}}{H^+(aq)} \longrightarrow H_2CO_3(aq)$$

In this way, the carbonic acid and bicarbonate ion buffering system keeps blood pH nearly constant.

When the glycolic acid generated by antifreeze consumption first enters the bloodstream, the acid's tendency to lower blood pH is countered by the buffering action of the bicarbonate ion. However, if the quantities of consumed antifreeze are large enough, the glycolic acid overwhelms the capacity of the buffer (we discuss buffer capacity in Section 17.3), causing blood pH to drop to dangerously low levels.

Low blood pH results in *acidosis*, a condition in which the acid affects the equilibrium between hemoglobin (Hb) and oxygen:

$$\overset{\boxed{\text{Excess H}^+} \downarrow}{HbH^+(aq) + O_2(g)} \underset{\boxed{\text{Shift left}}}{\rightleftharpoons} HbO_2(aq) + H^+(aq)$$

The excess acid causes the equilibrium to shift to the left, reducing the blood's ability to carry oxygen. At this point, the cat or dog may begin hyperventilating in an effort to overcome the acidic blood's lowered oxygen-carrying capacity. If no treatment is administered, the animal will eventually go into a coma and die.

One treatment for ethylene glycol poisoning is the administration of ethyl alcohol (the alcohol found in alcoholic beverages). The two molecules are similar enough that the liver enzyme that catalyzes the metabolism of ethylene glycol also acts on ethyl alcohol, but the enzyme has a higher affinity for ethyl alcohol than for ethylene glycol. Consequently, the enzyme preferentially metabolizes ethyl alcohol, allowing the unmetabolized ethylene glycol to escape through the urine. If administered early, this treatment can save the life of a dog or cat that has consumed ethylene glycol.

## 17.2 Buffers: Solutions That Resist pH Change

 **Buffers**

Most solutions significantly change pH when an acid or base is added to them. As we have just learned, however, a **buffer** resists pH change by neutralizing added acid or added base. A buffer contains either:

1. significant amounts of a weak acid and its conjugate base or
2. significant amounts of a weak base and its conjugate acid.

For example, the buffer in blood is composed of carbonic acid ($H_2CO_3$) and its conjugate base, the bicarbonate ion ($HCO_3^-$). When additional base is added to a buffer, the weak acid reacts with the base, neutralizing it. When additional acid is added to a buffer, the conjugate base reacts with the acid, neutralizing it. In this way, a buffer can maintain a nearly constant pH.

## Formation of a Buffer

Weak acid

Conjugate base

Buffer solution

Acetic acid
$HC_2H_3O_2$

Sodium acetate
$NaC_2H_3O_2$

$C_2H_3O_2^-$    $HC_2H_3O_2$    $H_3O^+$    $Na^+$

◄ **FIGURE 17.1  A Buffer Solution**
A buffer typically consists of a weak acid (which can neutralize added base) and its conjugate base (which can neutralize added acid).

A weak acid by itself, even though it partially ionizes to form some of its conjugate base, does not contain sufficient base to be a buffer. Similarly, a weak base by itself, even though it partially ionizes water to form some of its conjugate acid, does not contain sufficient acid to be a buffer. *A buffer must contain significant amounts of both a weak acid and its conjugate base (or vice versa).* Consider the simple buffer made by dissolving acetic acid ($HC_2H_3O_2$) and sodium acetate ($NaC_2H_3O_2$) in water (Figure 17.1 ▲).

Suppose, on one hand, that we add a strong base, such as NaOH, to this solution. The acetic acid neutralizes the base:

$$NaOH(aq) + HC_2H_3O_2(aq) \longrightarrow H_2O(l) + NaC_2H_3O_2(aq)$$

As long as the amount of added NaOH is less than the amount of $HC_2H_3O_2$ in solution, the buffer neutralizes the added NaOH and the resulting pH change is small.

Suppose, on the other hand, that we add a strong acid, such as HCl, to the solution. In this case, the conjugate base, $NaC_2H_3O_2$, neutralizes the added HCl:

$$HCl(aq) + NaC_2H_3O_2(aq) \longrightarrow HC_2H_3O_2(aq) + NaCl(aq)$$

As long as the amount of added HCl is less than the amount of $NaC_2H_3O_2$ in solution, the buffer neutralizes the added HCl and the resulting pH change is small.

$C_2H_3O_2^-$ is the conjugate base of $HC_2H_3O_2$.

### *Summarizing Buffer Characteristics:*

▶ Buffers resist pH change.

▶ A buffer contains significant amounts of either (1) a weak acid and its conjugate base, or (2) a weak base and its conjugate acid.

▶ The acid neutralizes added base.

▶ The base neutralizes added acid.

CONCEPTUAL
CONNECTION **17.1**

**Buffers**  Which solution is a buffer?

   **(a)**  a solution that is 0.100 M in $HNO_2$ and 0.100 M in HCl

   **(b)**  a solution that is 0.100 M in $HNO_3$ and 0.100 M in $NaNO_3$

   **(c)**  a solution that is 0.100 M in $HNO_2$ and 0.100 M in NaCl

   **(d)**  a solution that is 0.100 M in $HNO_2$ and 0.100 M in $NaNO_2$

**KCV** Finding pH and pH changes in Buffer Solutions

# Calculating the pH of a Buffer Solution

In Chapter 16, we discussed how to calculate the pH of a solution containing either a weak acid or its conjugate base, but not both. How do we calculate the pH of a buffer—a solution containing both? Consider a solution that initially contains $HC_2H_3O_2$ and $NaC_2H_3O_2$, each at a concentration of 0.100 M. The acetic acid ionizes according to the reaction:

$$HC_2H_3O_2(aq) + H_2O(l) \rightleftharpoons H_3O^+(aq) + C_2H_3O_2^-(aq)$$
Initial concentration:   0.100 M                                                     0.100 M

Le Châtelier's principle is discussed in Section 15.9.

However, the ionization of $HC_2H_3O_2$ in the solution is suppressed compared to its ionization in a solution that does not initially contain any $C_2H_3O_2^-$, because the presence of $C_2H_3O_2^-$ *shifts the equilibrium to the left* (as we would expect from Le Châtelier's principle). In other words, the presence of the $C_2H_3O_2^-(aq)$ ion causes the acid to ionize even less than it normally would (Figure 17.2▼), resulting in a less acidic solution (higher pH). This effect is known as the **common ion effect**, so named because the solution contains two substances ($HC_2H_3O_2$ and $NaC_2H_3O_2$) that share a common ion ($C_2H_3O_2^-$). To find the pH of a buffer solution containing common ions, we work an equilibrium problem in which the initial concentrations include both the acid and its conjugate base, as shown in Example 17.1.

▶ **FIGURE 17.2 The Common Ion Effect** The pH of a 0.100 M acetic acid solution is 2.9. The pH of a 0.100 M sodium acetate solution is 8.9. The pH of a solution that is 0.100 M in acetic acid and 0.100 M in sodium acetate is 4.7.

| pH = 2.9 | pH = 8.9 | pH = 4.7 |
|---|---|---|
|  |  |  |
| 0.100 M $HC_2H_3O_2$ | 0.100 M $NaC_2H_3O_2$ | 0.100 M $HC_2H_3O_2$ 0.100 M $NaC_2H_3O_2$ |

## Example 17.1  Calculating the pH of a Buffer Solution

Calculate the pH of a buffer solution that is 0.100 M in $HC_2H_3O_2$ and 0.100 M in $NaC_2H_3O_2$.

**SOLUTION**

**1.** Write the balanced equation for the ionization of the acid and use it as a guide to prepare an ICE table showing the given concentrations of the acid and its conjugate base as the initial concentrations. Leave room in the table for the changes in concentrations and for the equilibrium concentrations.

$$HC_2H_3O_2(aq) + H_2O(l) \rightleftharpoons H_3O^+(aq) + C_2H_3O_2^-(aq)$$

|  | $[HC_2H_3O_2]$ | $[H_3O^+]$ | $[C_2H_3O_2^-]$ |
|---|---|---|---|
| Initial | 0.100 | ≈0.00 | 0.100 |
| Change |  |  |  |
| Equil |  |  |  |

**2.** Represent the change in the concentration of $H_3O^+$ with the variable $x$. Express the changes in the concentrations of the other reactants and products in terms of $x$.

$$HC_2H_3O_2(aq) + H_2O(l) \rightleftharpoons H_3O^+(aq) + C_2H_3O_2^-(aq)$$

|  | $[HC_2H_3O_2]$ | $[H_3O^+]$ | $[C_2H_3O_2^-]$ |
|---|---|---|---|
| Initial | 0.100 | ≈0.00 | 0.100 |
| Change | $-x$ | $+x$ | $+x$ |
| Equil |  |  |  |

| | | |
|---|---|---|
| 3. Sum each column to determine the equilibrium concentrations in terms of the initial concentrations and the variable $x$. | $HC_2H_3O_2(aq) + H_2O(l) \rightleftharpoons H_3O^+(aq) + C_2H_3O_2^-(aq)$ |

| | $[HC_2H_3O_2]$ | $[H_3O^+]$ | $[C_2H_3O_2^-]$ |
|---|---|---|---|
| Initial | 0.100 | $\approx 0.00$ | 0.100 |
| Change | $-x$ | $+x$ | $+x$ |
| Equil | $0.100 - x$ | $x$ | $0.100 + x$ |

| | |
|---|---|
| 4. Substitute the expressions for the equilibrium concentrations (from step 3) into the expression for the acid ionization constant.<br><br>In most cases, you can make the approximation that $x$ *is small*. (See Sections 15.8 and 16.6 to review the *x is small* approximation.)<br><br>Substitute the value of the acid ionization constant (from Table 16.5) into the $K_a$ expression and solve for $x$.<br><br>Confirm that $x$ *is small* by calculating the ratio of $x$ and the number it was subtracted from in the approximation. The ratio should be less than 0.05 (5%). | $K_a = \dfrac{[H_3O^+][C_2H_3O^-]}{[HC_2H_3O_2]}$<br><br>$\phantom{K_a} = \dfrac{x(0.100 + x)}{0.100 - x} \quad$ (*x is small*)<br><br>$1.8 \times 10^{-5} = \dfrac{x(0.100)}{0.100}$<br><br>$x = 1.8 \times 10^{-5}$<br><br>$\dfrac{1.8 \times 10^{-5}}{0.100} \times 100\% = 0.018\%$<br>Therefore the approximation is valid. |
| 5. Determine the $H_3O^+$ concentration from the calculated value of $x$ and substitute into the pH equation to find pH. | $[H_3O^+] = x = 1.8 \times 10^{-5}\,M$<br><br>$pH = -\log[H_3O^+]$<br>$\phantom{pH} = -\log(1.8 \times 10^{-5})$<br>$\phantom{pH} = 4.74$ |

**FOR PRACTICE 17.1**  Calculate the pH of a buffer solution that is 0.200 M in $HC_2H_3O_2$ and 0.100 M in $NaC_2H_3O_2$.

**FOR MORE PRACTICE 17.1**  Calculate the pH of the buffer that results from mixing 60.0 mL of 0.250 M $HCHO_2$ and 15.0 mL of 0.500 M $NaCHO_2$.

## The Henderson–Hasselbalch Equation

We can derive an equation that relates the pH of a buffer solution to the initial concentration of the buffer components, thus simplifying the calculation of the pH of a buffer solution. Consider a buffer containing the generic weak acid HA and its conjugate base $A^-$. The acid ionizes as follows:

$$HA(aq) + H_2O(l) \rightleftharpoons H_3O^+(aq) + A^-(aq)$$

We derive an expression for the concentration of $H_3O^+$ from the acid ionization equilibrium expression by solving the expression for $[H_3O^+]$:

$$K_a = \frac{[H_3O^+][A^-]}{[HA]}$$

$$[H_3O^+] = K_a \frac{[HA]}{[A^-]} \qquad\qquad [17.1]$$

If we make the same *x is small* approximation that we make for weak acid or weak base equilibrium problems, *we can consider the equilibrium concentrations of HA and $A^-$ to be essentially identical to the initial concentrations of HA and $A^-$* (see step 4 of Example 17.1). Therefore, to determine $[H_3O^+]$ for any buffer solution, we multiply $K_a$ by the ratio of

Recall that the variable $x$ in a weak acid equilibrium problem represents the change in the initial acid concentration. The *x is small* approximation is valid because so little of the weak acid ionizes compared to its initial concentration.

the concentrations of the acid and the conjugate base. To find the $[H_3O^+]$ of the buffer in Example 17.1 (a solution that is 0.100 M in $HC_2H_3O_2$ and 0.100 M in $NaC_2H_3O_2$), we substitute the concentrations of $HC_2H_3O_2$ and $C_2H_3O_2^-$ into Equation 17.1:

$$[H_3O^+] = K_a \frac{[HC_2H_3O_2]}{[C_2H_3O_2^-]}$$

$$= K_a \frac{0.100}{0.100}$$

$$= K_a$$

In this buffer solution, as in any in which the acid and conjugate base concentrations are equal, $[H_3O^+]$ is equal to $K_a$.

We can derive an equation for the pH of a buffer by taking the negative logarithm of both sides of Equation 17.1:

$$[H_3O^+] = K_a \frac{[HA]}{[A^-]}$$

Recall that
$\log AB = \log A + \log B$, so $-\log AB$
$= -\log A - \log B$.

$$-\log[H_3O^+] = -\log\left(K_a \frac{[HA]}{[A^-]}\right)$$

$$-\log[H_3O^+] = -\log K_a - \log\frac{[HA]}{[A^-]} \qquad [17.2]$$

We can rearrange Equation 17.2 to get:

Recall that $\log (A/B) = -\log (B/A)$,
so $-\log (A/B) = \log (B/A)$.

$$-\log[H_3O^+] = -\log K_a + \log\frac{[A^-]}{[HA]}$$

Since $pH = -\log[H_3O^+]$ and since $pK_a = -\log K_a$, we obtain the result:

$$pH = pK_a + \log\frac{[A^-]}{[HA]}$$

Since $A^-$ is a weak base and HA is a weak acid, we can generalize the equation:

Note that, as expected, the pH of a buffer increases with an increase in the amount of base relative to the amount of acid.

$$pH = pK_a + \log\frac{[\text{base}]}{[\text{acid}]} \qquad [17.3]$$

where the base is the conjugate base of the acid or the acid is the conjugate acid of the base. This equation, known as the **Henderson–Hasselbalch equation**, allows us to quickly calculate the pH of a buffer solution from the initial concentrations of the buffer components *as long as the x is small approximation is valid*. In Example 17.2, we show how to find the pH of a buffer in two ways: in the left column we solve a common ion effect equilibrium problem using a method similar to the one we used in Example 17.1; in the right column we use the Henderson–Hasselbalch equation.

## Example 17.2 Calculating the pH of a Buffer Solution as an Equilibrium Problem and with the Henderson–Hasselbalch Equation

Calculate the pH of a buffer solution that is 0.050 M in benzoic acid ($HC_7H_5O_2$) and 0.150 M in sodium benzoate ($NaC_7H_5O_2$). For benzoic acid, $K_a = 6.5 \times 10^{-5}$.

SOLUTION

| Equilibrium Approach | Henderson–Hasselbalch Approach |
|---|---|
| Write the balanced equation for the ionization of the acid and use it as a guide to prepare an ICE table. | To find the pH of this solution, determine which component is the acid and which is the base and substitute their concentrations into the Henderson–Hasselbalch equation to calculate pH. |

$$HC_7H_5O_2(aq) + H_2O(l) \rightleftharpoons H_3O^+(aq) + C_7H_5O_2^-(aq)$$

| | $[HC_7H_5O_2]$ | $[H_3O^+]$ | $[C_7H_5O_2^-]$ |
|---|---|---|---|
| Initial | 0.050 | $\approx 0.00$ | 0.150 |
| Change | $-x$ | $+x$ | $+x$ |
| Equil | $0.050 - x$ | $x$ | $0.150 + x$ |

Substitute the expressions for the equilibrium concentrations into the expression for the acid ionization constant. Make the *x is small* approximation and solve for *x*.

$$K_a = \frac{[H_3O^+][C_7H_5O_2^-]}{[HC_7H_5O_2]}$$

$$= \frac{x(0.150 + x)}{0.050 - x} \quad (x \text{ is small})$$

$$6.5 \times 10^{-5} = \frac{x(0.150)}{0.050}$$

$$x = 2.2 \times 10^{-5}$$

Since $[H_3O^+] = x$, we calculate pH as follows:

$$pH = -\log[H_3O^+]$$

$$= -\log(2.2 \times 10^{-5})$$

$$= 4.66$$

$HC_7H_5O_2$ is the acid and $NaC_7H_5O_2$ is the base. Therefore, we calculate the pH as follows:

$$pH = pK_a + \log\frac{[\text{base}]}{[\text{acid}]}$$

$$= -\log(6.5 \times 10^{-5}) + \log\frac{0.150}{0.050}$$

$$= 4.1\underline{8}7 + 0.4\underline{7}7$$

$$= 4.66$$

Confirm that the *x is small* approximation is valid by calculating the ratio of *x* to the number it was subtracted from in the approximation. The ratio should be less than 0.05 (5%). (See Sections 15.8 and 16.6 to review the *x is small* approximation.)

$$\frac{2.2 \times 10^{-5}}{0.050} \times 100\% = 0.044\%$$

The approximation is valid.

Confirm that the *x is small* approximation is valid by calculating the $[H_3O^+]$ from the pH. Since $[H_3O^+]$ is formed by ionization of the acid, the calculated $[H_3O^+]$ has to be less than 0.05 (or 5%) of the initial concentration of the acid in order for the *x is small* approximation to be valid.

$$pH = 4.66 = -\log[H_3O^+]$$

$$[H_3O^+] = 10^{-4.66} = 2.2 \times 10^{-5} \text{ M}$$

$$\frac{2.2 \times 10^{-5}}{0.050} \times 100\% = 0.044\%$$

The approximation is valid.

**FOR PRACTICE 17.2**  Calculate the pH of a buffer solution that is 0.250 M in HCN and 0.170 M in KCN. For HCN, $K_a = 4.9 \times 10^{-10}$ ($pK_a = 9.31$). Use both the equilibrium approach and the Henderson–Hasselbalch approach.

You may be wondering how to decide whether to use the equilibrium approach or the Henderson–Hasselbalch equation when calculating the pH of buffer solutions. The answer depends on the specific problem. In cases where you can make the *x is small* approximation, the Henderson–Hasselbalch equation is adequate. However, as you can see from Example 17.2, checking the *x is small* approximation is not as convenient with the Henderson–Hasselbalch equation (because the approximation is implicit). Thus, the equilibrium approach, though lengthier, gives you a better sense of the important quantities in the problem and the nature of the approximation.

When first working buffer problems, use the equilibrium approach until you get a good sense for when the *x is small* approximation is adequate. Then, you can switch to the more streamlined approach in cases where the approximation applies (and only in those cases). In general, remember that the *x is small* approximation applies to problems in which both of the following are true: (a) the initial concentrations of acids (and/or bases) are not too dilute; and (b) the equilibrium constant is fairly small. Although the exact values depend on the details of the problem, for many buffer problems this means that the initial concentrations of acids and conjugate bases should be at least $10^2$–$10^3$ times greater than the equilibrium constant (depending on the required accuracy).

**pH of Buffer Solutions** A buffer contains the weak acid HA and its conjugate base $A^-$. The weak acid has a $pK_a$ of 4.82 and the buffer has a pH of 4.25. Which statement is true of the relative concentrations of the weak acid and conjugate base in the buffer?

    **(a)** $[HA] > [A^-]$     **(b)** $[HA] < [A^-]$     **(c)** $[HA] = [A^-]$

Which buffer component would you add to change the pH of the buffer to 4.72?

## Calculating pH Changes in a Buffer Solution

When we add acid or base to a buffer, the buffer resists a pH change. Nonetheless, the pH does change by a small amount. Calculating the pH change requires breaking up the problem into two parts:

1. **The stoichiometry calculation** in which we calculate how the addition changes the relative amounts of acid and conjugate base.
2. **The equilibrium calculation** in which we calculate the pH based on the new amounts of acid and conjugate base.

We demonstrate this calculation with a 1.0-L buffer solution that is 0.100 M in the generic acid HA and 0.100 M in its conjugate base $A^-$. Since the concentrations of the weak acid and the conjugate base are equal, the pH of the buffer is equal to $pK_a$. Let's calculate the pH of the solution after we add 0.025 mol of strong acid ($H^+$) (assuming that the change in volume from adding the acid is negligible).

### The Stoichiometry Calculation

As the added acid is neutralized, it converts a stoichiometric amount of the base into its conjugate acid through the neutralization reaction (Figure 17.3a▶, on p. 788):

$$\underset{\text{added acid}}{H^+(aq)} + \underset{\text{weak base in buffer}}{A^-(aq)} \longrightarrow HA(aq)$$

Neutralizing 0.025 mol of the strong acid ($H^+$) requires 0.025 mol of the weak base ($A^-$). Consequently, the amount of $A^-$ *decreases* by 0.025 mol and the amount of HA *increases* by 0.025 mol (because of the 1:1:1 stoichiometry of the neutralization reaction). We can track these changes in tabular form as follows:

It is best to work with amounts in moles instead of concentrations when tracking these changes, as explained later in this chapter.

|  | $H^+(aq)$ | + | $A^-(aq)$ | $\longrightarrow$ | $HA(aq)$ |
|---|---|---|---|---|---|
| **Before addition** | $\approx$0.00 mol | | 0.100 mol | | 0.100 mol |
| **Addition** | +0.025 mol | | — | | — |
| **After addition** | $\approx$0.00 mol | | 0.075 mol | | 0.125 mol |

*Notice that this table is not an ICE table.* This table simply tracks the stoichiometric changes that occur during the neutralization of the added acid. We write $\approx$0.00 mol for the amount of $H^+$ because the amount is so small compared to the amounts of $A^-$ and HA. (Remember that weak acids ionize only to a small extent and that the presence of the common ion further suppresses the ionization.) The amount of $H^+$, of course, is not *exactly* zero, as we can see by completing the equilibrium part of the calculation.

### The Equilibrium Calculation

We have just seen that adding a small amount of acid to a buffer is equivalent to changing the initial concentrations of the acid and conjugate base present in the buffer (in this case, since the volume is 1.0 L, [HA] increased from 0.100 M to 0.125 M and [$A^-$] decreased from 0.100 M to 0.075 M). Knowing these new initial concentrations, we can calculate the new pH in the same way that we calculate the pH of any buffer: either by working a full equilibrium problem or by using the Henderson–Hasselbalch equation (see Examples 17.1 and 17.2).

In this case, we work the full equilibrium problem. We begin by writing the balanced equation for the ionization of the acid and using it as a guide to prepare an ICE table. The initial concentrations for the ICE table are those that we calculated in the stoichiometry part of the calculation:

$$HA(aq) + H_2O(l) \rightleftharpoons H_3O^+(aq) + A^-(aq)$$

| | [HA] | [H₃O⁺] | [A⁻] | |
|---|---|---|---|---|
| Initial | 0.125 | ≈0.00 | 0.075 | From stoichiometry calculation |
| Change | −x | +x | +x | |
| Equil | 0.125 − x | x | 0.075 + x | |

We then substitute the expressions for the equilibrium concentrations into the expression for the acid ionization constant. As long as $K_a$ is sufficiently small relative to the initial concentrations, we can make the *x is small* approximation and solve for x, which is equal to [H₃O⁺]:

$$K_a = \frac{[H_3O^+][A^-]}{[HA]}$$

$$= \frac{x(0.075 + x)}{0.125 - x} \quad (x \text{ is small})$$

$$K_a = \frac{x(0.075)}{0.125}$$

$$x = [H_3O^+] = K_a \frac{0.125}{0.075}$$

Once we calculate [H₃O⁺], we can calculate the pH with the equation pH = −log [H₃O⁺].

Notice that, since the expression for x contains a *ratio* of concentrations [HA]/[A⁻], the *amounts of acid and base in moles* may be substituted in place of concentration because, in a single buffer solution, the volume is the same for both the acid and the base. Therefore, the volumes cancel:

$$[HA]/[A^-] = \frac{\dfrac{n_{HA}}{V}}{\dfrac{n_{A^-}}{V}} = n_{HA}/n_{A^-}$$

The effect of adding a small amount of strong base to the buffer is exactly the opposite of adding acid. The added base converts a stoichiometric amount of the acid into its conjugate base through the neutralization reaction (Figure 17.3b▶, on the next page):

$$\underset{\text{added base}}{OH^-(aq)} + \underset{\text{weak acid in buffer}}{HA(aq)} \longrightarrow H_2O(l) + A^-(aq)$$

If we add 0.025 mol of OH⁻, the amount of A⁻ goes *up* by 0.025 mol and the amount of HA goes *down* by 0.025 mol, as shown in the following table:

| | OH⁻(aq) | + | HA(aq) | ⟶ | H₂O(l) | + | A⁻(aq) |
|---|---|---|---|---|---|---|---|
| **Before addition** | ≈0.00 mol | | 0.100 mol | | | | 0.100 mol |
| **Addition** | +0.025 mol | | — | | | | — |
| **After addition** | ≈0.00 mol | | 0.075 mol | | | | 0.125 mol |

*When you calculate the pH of a buffer after adding small amounts of acid or base, remember the following:*

- Adding a small amount of strong acid to a buffer converts a stoichiometric amount of the base to the conjugate acid and decreases the pH of the buffer (adding acid decreases pH just as we would expect).

- Adding a small amount of strong base to a buffer converts a stoichiometric amount of the acid to the conjugate base and increases the pH of the buffer (adding base increases pH just as we would expect).

The easiest way to remember these changes is relatively simple: adding acid creates more acid; adding base creates more base.

▶ **FIGURE 17.3 Buffering Action**
**(a)** When an acid is added to a buffer, a stoichiometric amount of the weak base is converted to the conjugate acid.
**(b)** When a base is added to a buffer, a stoichiometric amount of the weak acid is converted to the conjugate base.

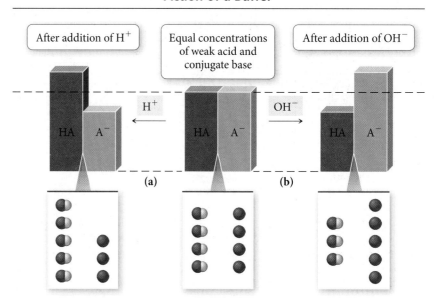

Example 17.3 and the For Practice Problems that follow it involve calculating pH changes in a buffer solution after small amounts of strong acid or strong base are added. As we have seen, these problems generally have two parts:

- Part I. Stoichiometry—use the stoichiometry of the neutralization equation to calculate the changes in the amounts (in moles) of the buffer components upon addition of the acid or base.

- Part II. Equilibrium—use the new amounts of buffer components to work an equilibrium problem to find pH. (For most buffers, this can also be done with the Henderson–Hasselbalch equation.)

CONCEPTUAL
**CONNECTION 17.3**

**Buffering Action** Closely examine Figure 17.3▲. Which image best represents the amount of $OH^-$ added to the buffer in part (b) of the figure?

a.     b.     c.

**Example 17.3** | **Calculating the pH Change in a Buffer Solution after the Addition of a Small Amount of Strong Acid or Base**

A 1.0-L buffer solution contains 0.100 mol $HC_2H_3O_2$ and 0.100 mol $NaC_2H_3O_2$. The value of $K_a$ for $HC_2H_3O_2$ is $1.8 \times 10^{-5}$. Because the initial amounts of acid and conjugate base are equal, the pH of the buffer is equal to $pK_a = -\log(1.8 \times 10^{-5}) = 4.74$. Calculate the new pH after adding 0.010 mol of solid NaOH to the buffer. For comparison, calculate the pH after adding 0.010 mol of solid NaOH to 1.0 L of pure water. (Ignore any small changes in volume that might occur upon addition of the base.)

**SOLUTION**

Part I: Stoichiometry. The addition of the base converts a stoichiometric amount of acid to the conjugate base (adding base creates more base). Write an equation showing the neutralization reaction and then set up a table to track the changes.

| | $OH^-(aq)$ | + $HC_2H_3O_2(aq)$ | ⟶ $H_2O(l)$ + | $C_2H_3O_2^-(aq)$ |
|---|---|---|---|---|
| **Before addition** | ≈0.00 mol | 0.100 mol | | 0.100 mol |
| **Addition** | 0.010 mol | — | | — |
| **After addition** | ≈0.00 mol | 0.090 mol | | 0.110 mol |

| Part II: Equilibrium. Write the balanced equation for the ionization of the acid and use it as a guide to prepare an ICE table. Use the amounts of acid and conjugate base from part I as the initial amounts of acid and conjugate base in the ICE table. | $HC_2H_3O_2(aq) + H_2O(l) \rightleftharpoons H_3O^+(aq) + C_2H_3O_2^-(aq)$ |
|---|---|

|  | $[HC_2H_3O_2]$ | $[H_3O^+]$ | $[C_2H_3O_2^-]$ |
|---|---|---|---|
| Initial | 0.090 | $\approx 0.00$ | 0.110 |
| Change | $-x$ | $+x$ | $+x$ |
| Equil | $0.090 - x$ | $x$ | $0.110 + x$ |

| Substitute the expressions for the equilibrium concentrations of acid and conjugate base into the expression for the acid ionization constant. Make the *x is small* approximation and solve for *x*. Calculate the pH from the value of *x*, which is equal to $[H_3O^+]$. | $K_a = \dfrac{[H_3O^+][C_2H_3O_2^-]}{[HC_2H_3O_2]}$ $= \dfrac{x(0.110 + x)}{0.090 - \cancel{x}}$ (*x* is small) $1.8 \times 10^{-5} = \dfrac{x(0.110)}{0.090}$ $x = [H_3O^+] = 1.\underline{4}7 \times 10^{-5}\,M$ $pH = -\log[H_3O^+]$ $\quad\quad = -\log(1.\underline{4}7 \times 10^{-5})$ $\quad\quad = 4.83$ |
|---|---|
| Confirm that the *x is small* approximation is valid by calculating the ratio of *x* to the smallest number it was subtracted from in the approximation. The ratio should be less than 0.05 (5%). | $\dfrac{1.\underline{4}7 \times 10^{-5}}{0.090} \times 100\% = 0.016\%$ The approximation is valid. |

| Part II: Equilibrium Alternative (using the Henderson–Hasselbalch equation). As long as the *x is small* approximation is valid, you can substitute the quantities of acid and conjugate base after the addition (from part I) into the Henderson–Hasselbalch equation and calculate the new pH. | $pH = pK_a + \log\dfrac{[base]}{[acid]}$ $\quad\quad = -\log(1.8 \times 10^{-5}) + \log\dfrac{0.110}{0.090}$ $\quad\quad = 4.74 + 0.087$ $\quad\quad = 4.83$ |
|---|---|

| The pH of 1.0 L of water after adding 0.010 mol of NaOH is calculated from the $[OH^-]$. For a strong base, $[OH^-]$ is simply the number of moles of $OH^-$ divided by the number of liters of solution. | $[OH^-] = \dfrac{0.010\ mol}{1.0\ L} = 0.010\ M$ $pOH = -\log[OH^-] = -\log(0.010)$ $\quad\quad\quad = 2.00$ $pOH + pH = 14.00$ $\quad\quad\quad pH = 14.00 - pOH$ $\quad\quad\quad\quad\quad = 14.00 - 2.00$ $\quad\quad\quad\quad\quad = 12.00$ |
|---|---|

**CHECK** Notice that the buffer solution changed from pH = 4.74 to pH = 4.83 upon addition of the base (a small fraction of a single pH unit). In contrast, the pure water changed from pH = 7.00 to pH = 12.00, five whole pH units (a factor of $10^5$). Notice also that even the buffer solution got slightly more basic upon addition of a base, as we would expect. To check your answer, always make sure the pH goes in the direction you expect: adding base should make the solution more basic (higher pH); adding acid should make the solution more acidic (lower pH).

**FOR PRACTICE 17.3**   Calculate the pH of the solution in Example 17.3 upon addition of 0.015 mol of NaOH to the original buffer.

**FOR MORE PRACTICE 17.3**   Calculate the pH of the solution in Example 17.3 upon addition of 10.0 mL of 1.00 M HCl to the original buffer.

**Adding Acid or Base to a Buffer** A buffer contains equal amounts of a weak acid and its conjugate base and has a pH of 5.25. Which would be a reasonable value of buffer pH after the addition of a small amount of acid?

(a) 4.15
(b) 5.15
(c) 5.35
(d) 6.35

## Buffers Containing a Base and Its Conjugate Acid

So far, we have seen examples of buffers composed of an acid and its conjugate base (where the conjugate base is an ion). A buffer can also be composed of a base and its conjugate acid (where the conjugate acid is an ion). For example, a solution containing significant amounts of both $NH_3$ and $NH_4Cl$ acts as a buffer (Figure 17.4▼). The $NH_3$ is a weak base that neutralizes small amounts of added acid, and the $NH_4^+$ ion is the conjugate acid that neutralizes small amounts of added base.

We calculate the pH of a solution like this in the same way that we calculated the pH of a buffer containing a weak acid and its conjugate base. When using the Henderson–Hasselbalch equation, however, we must first calculate $pK_a$ for the conjugate acid of the weak base. Recall from Section 16.8 that for a conjugate acid–base pair, $K_a \times K_b = K_w$ and $pK_a + pK_b = 14$. Consequently, we can find $pK_a$ of the conjugate acid by subtracting $pK_b$ of the weak base from 14. Example 17.4 illustrates the procedure for calculating the pH of a buffer composed of a weak base and its conjugate acid.

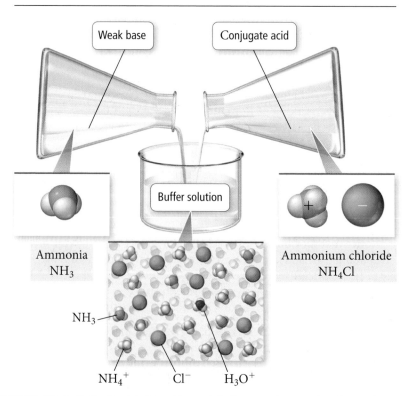

▲ **FIGURE 17.4 Buffer Containing a Base** A buffer can also consist of a weak base and its conjugate acid.

**Example** 17.4 | **Using the Henderson–Hasselbalch Equation to Calculate the pH of a Buffer Solution Composed of a Weak Base and Its Conjugate Acid**

Use the Henderson–Hasselbalch equation to calculate the pH of a buffer solution that is 0.50 M in $NH_3$ and 0.20 M in $NH_4Cl$. For ammonia, $pK_b = 4.75$.

**SOLUTION**

Since $K_b$ for $NH_3$ ($1.76 \times 10^{-5}$) is much smaller than the initial concentrations in this problem, you can use the Henderson–Hasselbalch equation to calculate the pH of the buffer. First calculate $pK_a$ from $pK_b$.

$$pK_a + pK_b = 14$$
$$pK_a = 14 - pK_b$$
$$= 14 - 4.75$$
$$= 9.25$$

Then substitute the given quantities into the Henderson–Hasselbalch equation and calculate pH.

$$pH = pK_a + \log \frac{[\text{base}]}{[\text{acid}]}$$
$$= 9.25 + \log \frac{0.50}{0.20}$$
$$= 9.25 + 0.40$$
$$= 9.65$$

**FOR PRACTICE 17.4**   Calculate the pH of 1.0 L of the solution in Example 17.4 upon addition of 0.010 mol of solid NaOH to the original buffer solution.

**FOR MORE PRACTICE 17.4**   Calculate the pH of 1.0 L of the solution in Example 17.4 upon addition of 30.0 mL of 1.0 M HCl to the original buffer solution.

# 17.3 Buffer Effectiveness: Buffer Range and Buffer Capacity

An effective buffer neutralizes small to moderate amounts of added acid or base. Recall from Section 17.1, however, that a buffer can be destroyed by the addition of too much acid or too much base. What factors influence the effectiveness of a buffer? In this section, we examine two such factors: *the relative amounts of the acid and conjugate base* and *the absolute concentrations of the acid and conjugate base*. We then define the *capacity of a buffer* (how much added acid or base it can effectively neutralize) and the *range of a buffer* (the pH range over which a particular acid and its conjugate base can be effective).

## Relative Amounts of Acid and Base

A buffer is most effective (most resistant to pH changes) when the concentrations of acid and conjugate base are equal. Let's explore this idea by considering the behavior of a generic buffer composed of HA and $A^-$ for which $pK_a = 5.00$ and calculating the percent change in pH upon addition of 0.010 mol of NaOH for two different 1.0-liter solutions of this buffer system. Both solutions have 0.20 mol of *total* acid and conjugate base. However, solution I has equal amounts of acid and conjugate base (0.10 mol of each), while solution II has much more acid than conjugate base (0.18 mol HA and 0.020 mol $A^-$). We can calculate the initial pH values of each solution using the Henderson–Hasselbalch equation. Solution I has an initial pH of 5.00 and solution II has an initial pH of 4.05:

**Solution I: 0.10 mol HA and 0.10 mol A⁻;**
**initial pH = 5.00**

| | $OH^-(aq)$ + | $HA(aq)$ $\longrightarrow$ | $H_2O(l)$ + $A^-(aq)$ |
|---|---|---|---|
| **Before addition** | ≈0.00 mol | 0.100 mol | 0.100 mol |
| **Addition** | 0.010 mol | — | — |
| **After addition** | ≈0.00 mol | 0.090 mol | 0.110 mol |

$$pH = pK_a + \log \frac{[base]}{[acid]}$$
$$= 5.00 + \log \frac{0.110}{0.090}$$
$$= 5.09$$
$$\% \text{ change} = \frac{5.09 - 5.00}{5.00} \times 100\%$$
$$= 1.8\%$$

**Solution II: 0.18 mol HA and 0.020 mol A⁻;**
**initial pH = 4.05**

| | $OH^-(aq)$ + | $HA(aq)$ $\longrightarrow$ | $H_2O(l)$ + $A^-(aq)$ |
|---|---|---|---|
| **Before addition** | ≈0.00 mol | 0.18 mol | 0.020 mol |
| **Addition** | 0.010 mol | — | — |
| **After addition** | ≈0.00 mol | 0.17 mol | 0.030 mol |

$$pH = pK_a + \log \frac{[base]}{[acid]}$$
$$= 5.00 + \log \frac{0.030}{0.17}$$
$$= 4.25$$
$$\% \text{ change} = \frac{4.25 - 4.05}{4.05} \times 100\%$$
$$= 4.9\%$$

As you can see, the buffer with equal amounts of acid and conjugate base is more resistant to pH change and is therefore the more effective buffer. A buffer becomes less effective as the difference in the relative amounts of acid and conjugate base increases. As a guideline, we can say that an effective buffer must have a [base]/[acid] ratio in the range of 0.10 to 10. *In order for a buffer to be reasonably effective, the relative concentrations of acid and conjugate base should not differ by more than a factor of 10.*

## Absolute Concentrations of the Acid and Conjugate Base

*A buffer is most effective (most resistant to pH changes) when the concentrations of acid and conjugate base are high.* Let's explore this idea by again considering a generic buffer composed of HA and A⁻ and a pK_a of 5.00 and calculating the percent change in pH upon addition of 0.010 mol of NaOH for two 1.0-liter solutions of this buffer system. In this case, both the acid and the base in solution I are ten times more concentrated than the acid and base in solution II. Both solutions have equal relative amounts of acid and conjugate base and therefore have the same initial pH of 5.00:

**Solution I: 0.50 mol HA and 0.50 mol A⁻;**
**initial pH = 5.00**

| | $OH^-(aq)$ + | $HA(aq)$ $\longrightarrow$ | $H_2O(l)$ + $A^-(aq)$ |
|---|---|---|---|
| **Before addition** | ≈0.00 mol | 0.50 mol | 0.50 mol |
| **Addition** | 0.010 mol | — | — |
| **After addition** | ≈0.00 mol | 0.49 mol | 0.51 mol |

$$pH = pK_a + \log \frac{[base]}{[acid]}$$
$$= 5.00 + \log \frac{0.51}{0.49}$$
$$= 5.02$$
$$\% \text{ change} = \frac{5.02 - 5.00}{5.00} \times 100\%$$
$$= 0.4\%$$

**Solution II: 0.050 mol HA and 0.050 mol A⁻;**
**initial pH = 5.00**

| | $OH^-(aq)$ + | $HA(aq)$ $\longrightarrow$ | $H_2O(l)$ + $A^-(aq)$ |
|---|---|---|---|
| **Before addition** | ≈0.00 mol | 0.050 mol | 0.050 mol |
| **Addition** | 0.010 mol | — | — |
| **After addition** | ≈0.00 mol | 0.040 mol | 0.060 mol |

$$pH = pK_a + \log \frac{[base]}{[acid]}$$
$$= 5.00 + \log \frac{0.060}{0.040}$$
$$= 5.18$$
$$\% \text{ change} = \frac{5.18 - 5.00}{5.00} \times 100\%$$
$$= 3.6\%$$

As you can see, the buffer with greater amounts of acid and conjugate base is more resistant to pH changes and therefore is the more effective buffer. The more dilute the buffer components, the less effective the buffer.

## Buffer Range

In light of the guideline that the relative concentrations of acid and conjugate base should not differ by more than a factor of 10 in order for a buffer to be reasonably effective, we can calculate the pH range over which a particular acid and its conjugate base make an effective buffer. Because the pH of a buffer is given by the Henderson–Hasselbalch equation, we can calculate the outermost points of the effective range as follows:

*Lowest pH* for effective buffer occurs when the base is one-tenth as concentrated as the acid.

$$pH = pK_a + \log\frac{[base]}{[acid]}$$
$$= pK_a + \log 0.10$$
$$= pK_a - 1$$

*Highest pH* for effective buffer occurs when the base is ten times as concentrated as the acid.

$$pH = pK_a + \log\frac{[base]}{[acid]}$$
$$= pK_a + \log 10$$
$$= pK_a + 1$$

▲ A concentrated buffer contains more of the weak acid and its conjugate base than a weak buffer does. It can therefore neutralize more added acid or added base.

*The effective range for a buffering system is one pH unit on either side of* $pK_a$. For example, we can use a weak acid with a $pK_a$ of 5.0 (and its conjugate base) to prepare a buffer in the range of 4.0–6.0. We can adjust the relative amounts of acid and conjugate base to achieve any pH within this range. As we noted earlier, however, the buffer would be most effective at pH 5.0 because the buffer components would be exactly equal at that pH. Example 17.5 demonstrates how to pick an acid/conjugate base system for a buffer and how to calculate the relative amounts of acid and conjugate base required for a desired pH.

## Example 17.5    Preparing a Buffer

Which acid would you choose to combine with its sodium salt to make a solution buffered at pH 4.25? For the best choice, calculate the ratio of the conjugate base to the acid required to attain the desired pH.

chlorous acid ($HClO_2$)  $pK_a = 1.95$   formic acid ($HCHO_2$)      $pK_a = 3.74$
nitrous acid ($HNO_2$)    $pK_a = 3.34$   hypochlorous acid (HClO)  $pK_a = 7.54$

**SOLUTION**

The best choice is formic acid because its $pK_a$ lies closest to the desired pH. You can calculate the ratio of conjugate base ($CHO_2^-$) to acid ($HCHO_2$) required by using the Henderson–Hasselbalch equation as follows:

$$pH = pK_a + \log\frac{[base]}{[acid]}$$

$$4.25 = 3.74 + \log\frac{[base]}{[acid]}$$

$$\log\frac{[base]}{[acid]} = 4.25 - 3.74$$
$$= 0.51$$

$$\frac{[base]}{[acid]} = 10^{0.51}$$
$$= 3.24$$

**FOR PRACTICE 17.5**   Which acid in Example 17.5 would you choose to create a buffer with pH = 7.35? If you have 500.0 mL of a 0.10 M solution of the acid, what mass of the corresponding sodium salt of the conjugate base do you need to make the buffer?

## Chemistry AND Medicine | Buffer Effectiveness in Human Blood

A s we discussed in Section 17.1, blood contains several buffering systems, the most important of which consists of carbonic acid and the bicarbonate ion. The concentrations of these buffer components in normal blood plasma are $[HCO_3^-] = 0.024$ M and $[H_2CO_3] = 0.0012$ M. The $pK_a$ for carbonic acid at body temperature is 6.1.

If we substitute these quantities into the Henderson–Hasselbalch equation, we can calculate the normal pH of blood:

$$pH = pK_a + \log\frac{[base]}{[acid]}$$

$$= 6.1 + \log\frac{[HCO_3^-]}{[H_2CO_3]}$$

$$= 6.1 + \log\frac{0.024 \text{ M}}{0.0012 \text{ M}}$$

$$= 7.4$$

Normal blood has a pH of 7.4. Notice that the concentration of the bicarbonate ion is 20 times higher than the concentration of carbonic acid and the pH of the buffer is more than one pH unit away from $pK_a$. Why?

The higher bicarbonate ion concentration in blood makes the buffer capacity of blood greater for acid than for base, which is necessary because the products of metabolism that enter blood are mostly acidic. For example, when we exercise, our bodies produce lactic acid ($HC_3H_5O_3$).

The lactic acid enters the bloodstream and must be neutralized. The bicarbonate ion neutralizes the lactic acid according to the equation:

$$HCO_3^-(aq) + HC_3H_5O_3(aq) \longrightarrow H_2CO_3(aq) + C_3H_5O_3^-(aq)$$

An enzyme called carbonic anhydrase then catalyzes the conversion of carbonic acid into carbon dioxide and water:

$$H_2CO_3(aq) \rightleftharpoons CO_2(g) + H_2O(l)$$

We eliminate the carbon dioxide from our blood when we breathe. When large amounts of lactic acid are produced, we must breathe faster to keep up with the need to eliminate carbon dioxide. This is why we pant when we exert ourselves.

**QUESTION**

*A 70-kg person has a total blood volume of about 5.0 L. Given the carbonic acid and bicarbonate concentrations stated, what volume (in mL) of 6.0 M HCl can be neutralized by blood without the blood pH dropping below 7.0 (which would result in death)?*

▶ Normal blood has a pH of 7.4.

## Buffer Capacity

**Buffer capacity** is the amount of acid or base that we can add to a buffer without causing a large change in pH. Given what we just learned about the absolute concentrations of acid and conjugate base in an effective buffer, we can conclude that the *buffer capacity increases with increasing absolute concentrations of the buffer components*. The more concentrated the weak acid and conjugate base that compose the buffer, the higher the buffer capacity. In addition, *overall buffer capacity increases as the relative concentrations of the buffer components become more similar to each other.* As the ratio of the buffer components gets closer to 1, the *overall* capacity of the buffer (the ability to neutralize added acid *and* added base) becomes greater. In some cases, however, a buffer that must neutralize primarily added acid (or primarily added base) may be overweighted in one of the buffer components, as shown in the *Chemistry and Medicine* box in this section.

CONCEPTUAL
**CONNECTION 17.5**

**Buffer Capacity** A 1.0-L buffer solution is 0.10 M in HF and 0.050 M in NaF. Which action destroys the buffer?

**(a)** adding 0.050 mol of HCl

**(b)** adding 0.050 mol of NaOH

**(c)** adding 0.050 mol of NaF

**(d)** none of the above

# 17.4 Titrations and pH Curves

**KCV** The Titration of a Weak Acid and a Strong Base

We first examined acid–base titrations in Section 4.8. In an **acid–base titration**, a basic (or acidic) solution of unknown concentration reacts with an acidic (or basic) solution of known concentration. The known solution is slowly added to the unknown one while the pH is monitored with either a pH meter or an **indicator** (a substance whose color depends on the pH). As the acid and base combine, they neutralize each other. At the **equivalence point**—the point in the titration when the number of moles of base is stoichiometrically equal to the number of moles of acid—the titration is complete. When the equivalence point is reached, neither reactant is in excess and the number of moles of the reactants are related by the reaction stoichiometry (Figure 17.5▼).

In this section, we examine acid–base titrations more closely, concentrating on the pH changes that occur during the titration. A plot of the pH of the solution during a titration is known as a *titration curve* or *pH curve*. Figure 17.6▼ is a pH curve for the titration of HCl with NaOH. Before any base is added to the solution, the pH is low (as we would expect for a solution of HCl). As the NaOH is added, the solution becomes less acidic because the NaOH begins to neutralize the HCl. The point of inflection in the

> The equivalence point is so named because the number of moles of acid and base are stoichiometrically equivalent at this point.

Beginning of titration

Equivalence point

▲ **FIGURE 17.5 Acid–Base Titration** As OH⁻ is added in a titration, it neutralizes the H⁺, forming water. At the equivalence point, the titration is complete.

Strong Acid & Strong Base

Equivalence point

Volume of NaOH added (mL)

▲ **FIGURE 17.6 Titration Curve: Strong Acid + Strong Base** This curve represents the titration of 50.0 mL of 0.100 M HCl with 0.100 M NaOH.

middle of the curve is the equivalence point. Notice that the pH changes very quickly near the equivalence point (small amounts of added base cause large changes in pH). Beyond the equivalence point, the solution is basic because the HCl has been completely neutralized and excess base is being added to the solution. The exact shape of the pH curve depends on several factors, including the strength of the acid or base being titrated. Let's consider several combinations individually.

## The Titration of a Strong Acid with a Strong Base

Consider the titration of 25.0 mL of 0.100 M HCl with 0.100 M NaOH. We begin by calculating the volume of base required to reach the equivalence point and then the pH at several points during the titration.

### Volume of NaOH Required to Reach the Equivalence Point

During the titration, the added sodium hydroxide neutralizes the hydrochloric acid:

$$HCl(aq) + NaOH(aq) \longrightarrow H_2O(l) + NaCl(aq)$$

The equivalence point is reached when the number of moles of base added equals the number of moles of acid initially in solution. We calculate the amount of acid initially in solution from its volume and its concentration:

$$\text{Initial mol HCl} = 0.0250\,\text{L} \times \frac{0.100\,\text{mol}}{1\,\text{L}} = 0.00250\,\text{mol HCl}$$

The amount of NaOH that must be added is 0.00250 mol NaOH. We calculate the volume of NaOH required from its concentration:

$$\text{Volume NaOH solution} = 0.00250\,\text{mol} \times \frac{1\,\text{L}}{0.100\,\text{mol}} = 0.0250\,\text{L}$$

The equivalence point is reached when 25.0 mL of NaOH has been added. In this case, the concentrations of both solutions are identical, so the volume of NaOH solution required to reach the equivalence point is equal to the volume of the HCl solution that is being titrated.

### Initial pH (before Adding Any Base)

The initial pH of the solution is simply the pH of a 0.100 M HCl solution. Since HCl is a strong acid, the concentration of $H_3O^+$ is also 0.100 M and the pH is 1.00:

$$pH = -\log[H_3O^+]$$
$$= -\log(0.100)$$
$$= 1.00$$

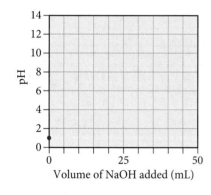

Volume of NaOH added (mL)

### pH after Adding 5.00 mL NaOH

As NaOH is added to the solution, it neutralizes $H_3O^+$:

$$OH^-(aq) + H_3O^+(aq) \longrightarrow 2\,H_2O(l)$$

We calculate the amount of $H_3O^+$ at any given point (before the equivalence point) by using the reaction stoichiometry—1 mol of NaOH neutralizes 1 mol of $H_3O^+$. The initial number of moles of $H_3O^+$ (as just calculated) is 0.00250 mol. We calculate the number of moles of NaOH added at 5.00 mL by multiplying the added volume (in L) by the concentration of the NaOH solution:

$$\text{mol NaOH added} = 0.00500\,\text{L} \times \frac{0.100\,\text{mol}}{1\,\text{L}} = 0.000500\,\text{mol NaOH}$$

The addition of $OH^-$ causes the amount of $H_3O^+$ to decrease as shown in the following table:

|  | $OH^-(aq)$ | $+$ | $H_3O^+(aq) \longrightarrow 2\,H_2O(l)$ |
|---|---|---|---|
| **Before addition** | $\approx 0.00$ mol | | 0.00250 mol |
| **Addition** | 0.000500 mol | | — |
| **After addition** | $\approx 0.00$ mol | | 0.00200 mol |

We calculate the $H_3O^+$ concentration by dividing the number of moles of $H_3O^+$ remaining by the *total volume* (initial volume plus added volume):

$$[H_3O^+] = \frac{0.00200 \text{ mol } H_3O^+}{0.0250 \text{ L} + 0.00500 \text{ L}} = 0.0667 \text{ M}$$

Initial volume     Added volume

The pH is therefore 1.18:

$$pH = -\log 0.0667$$
$$= 1.18$$

### pH after Adding 10.0, 15.0, and 20.0 mL NaOH

As more NaOH is added, it further neutralizes the $H_3O^+$ in the solution. We calculate the pH at each of these points in the same way that we calculated the pH at the 5.00 mL point. The results are tabulated as follows:

| Volume (mL) | pH |
|---|---|
| 10.0 | 1.37 |
| 15.0 | 1.60 |
| 20.0 | 1.95 |

### pH after Adding 25.0 mL NaOH (Equivalence Point)

The pH at the equivalence point of a strong acid–strong base titration will always be 7.00 (at 25 °C). At the equivalence point, the strong base has completely neutralized the strong acid. The only source of hydronium ions then is the ionization of water. The $[H_3O^+]$ at 25 °C from the ionization of water is $1.00 \times 10^{-7}$ M and the pH is 7.00.

### pH after Adding 30.00 mL NaOH

As NaOH is added beyond the equivalence point, it becomes the excess reagent. We calculate the amount of $OH^-$ at any given point (past the equivalence point) by subtracting the initial amount of $H_3O^+$ from the amount of added $OH^-$. The number of moles of $OH^-$ added at 30.00 mL is:

$$\text{mol } OH^- \text{ added} = 0.0300 \text{ L} \times \frac{0.100 \text{ mol}}{1 \text{ L}} = 0.00300 \text{ mol } OH^-$$

The number of moles of $OH^-$ remaining after neutralization is shown in the following table:

| | $OH^-(aq)$ | + $H_3O^+(aq)$ $\longrightarrow$ 2 $H_2O(l)$ |
|---|---|---|
| **Before addition** | $\approx 0.00$ mol | 0.00250 mol |
| **Addition** | 0.00300 mol | — |
| **After addition** | 0.00050 mol | $\approx 0.00$ mol |

We calculate the $OH^-$ concentration by dividing the number of moles of $OH^-$ remaining by the *total volume* (initial volume plus added volume):

$$[OH^-] = \frac{0.000500 \text{ mol } OH^-}{0.0250 \text{ L} + 0.0300 \text{ L}} = 0.00909 \text{ M}$$

We can then calculate the $[H_3O^+]$ and pH:

$$[H_3O^+][OH^-] = 10^{-14}$$

$$[H_3O^+] = \frac{10^{-14}}{[OH^-]} = \frac{10^{-14}}{0.00909}$$

$$= 1.10 \times 10^{-12} \text{M}$$

$$pH = -\log(1.10 \times 10^{-12})$$

$$= 11.96$$

### pH after Adding 35.0, 40.0, and 50.0 mL NaOH

As more NaOH is added, it further increases the basicity of the solution. We calculate the pH at each of these points in the same way that we calculated the pH at the 30.00 mL point. The results are tabulated as follows:

| Volume (mL) | pH |
|---|---|
| 35.0 | 12.22 |
| 40.0 | 12.36 |
| 50.0 | 12.52 |

### The Overall pH Curve

The overall pH curve for the titration of a strong acid with a strong base has the characteristic S-shape we just plotted. The overall curve is as follows:

*Summarizing the Titration of a Strong Acid with a Strong Base:*

▶ The initial pH is simply the pH of the strong acid solution to be titrated.

▶ Before the equivalence point, $H_3O^+$ is in excess. Calculate the $[H_3O^+]$ by subtracting the number of moles of added $OH^-$ from the initial number of moles of $H_3O^+$ and dividing by the *total* volume.

▶ At the equivalence point, neither reactant is in excess and the pH = 7.00.

▶ Beyond the equivalence point, $OH^-$ is in excess. Calculate the $[OH^-]$ by subtracting the initial number of moles of $H_3O^+$ from the number of moles of added $OH^-$ and dividing by the *total* volume.

The pH curve for the titration of a strong base with a strong acid is shown in Figure 17.7▼. Calculating the points along this curve is very similar to calculating the points along the pH curve for the titration of a strong acid with a strong base. The main difference is that the curve starts basic and then becomes acidic after the equivalence point (instead of vice versa).

▶ **FIGURE 17.7** **Titration Curve: Strong Base + Strong Acid** This curve represents the titration of 25.0 mL of 0.100 M NaOH with 0.100 M HCl.

## Example 17.6 | Strong Acid–Strong Base Titration pH Curve

A 50.0-mL sample of 0.200 M sodium hydroxide is titrated with 0.200 M nitric acid. Calculate pH:

**(a)** after adding 30.00 mL of $HNO_3$          **(b)** at the equivalence point

SOLUTION

| | |
|---|---|
| **(a)** Begin by calculating the initial amount of NaOH (in moles) from the volume and molarity of the NaOH solution. Because NaOH is a strong base, it dissociates completely, so the amount of $OH^-$ is equal to the amount of NaOH. | $\text{moles NaOH} = 0.0500 \text{ L} \times \dfrac{0.200 \text{ mol}}{1 \text{ L}}$ <br><br> $= 0.0100 \text{ mol}$ <br> $\text{moles } OH^- = 0.0100 \text{ mol}$ |

Calculate the amount of $HNO_3$ (in moles) added at 30.0 mL from the molarity of the $HNO_3$ solution.

$$\text{mol } HNO_3 \text{ added} = 0.0300 \text{ L} \times \dfrac{0.200 \text{ mol}}{1 \text{ L}}$$
$$= 0.00600 \text{ mol } HNO_3$$

As $HNO_3$ is added to the solution, it neutralizes some of the $OH^-$. Calculate the number of moles of $OH^-$ remaining by setting up a table based on the neutralization reaction that shows the amount of $OH^-$ before the addition, the amount of $H_3O^+$ added, and the amounts left after the addition.

|  | $OH^-(aq)$ | + | $H_3O^+(aq) \longrightarrow 2\ H_2O(l)$ |
|---|---|---|---|
| **Before addition** | 0.0100 mol | | $\approx 0.00$ mol |
| **Addition** | — | | 0.00600 mol |
| **After addition** | 0.0040 mol | | 0.00 mol |

Calculate the $OH^-$ concentration by dividing the amount of $OH^-$ remaining by the *total volume* (initial volume plus added volume).

$$[OH^-] = \dfrac{0.0040 \text{ mol}}{0.0500 \text{ L} + 0.0300 \text{ L}}$$
$$= 0.0500 \text{ M}$$

Calculate the pOH from $[OH^-]$.

$$pOH = -\log(0.0500)$$
$$= 1.30$$

Calculate the pH from the pOH using the equation $pH + pOH = 14$.

$$pH = 14 - pOH$$
$$= 14 - 1.30$$
$$= 12.70$$

**(b)** At the equivalence point, the strong base has completely neutralized the strong acid. The $[H_3O^+]$ at 25 °C from the ionization of water is $1.00 \times 10^{-7}$ M and the pH is therefore 7.00.

$$pH = 7.00$$

FOR PRACTICE 17.6    Calculate the pH in the titration in Example 17.6 after the addition of 60.0 mL of 0.200 M $HNO_3$.

**Titration Equivalence Point**  The amount of strong acid in the flask shown here is to be titrated by a strong base. Which mark on the burette next to the flask indicates the amount of base required to reach the equivalence point?

CONCEPTUAL
CONNECTION **17.6**

**(a)**  A

**(b)**  B

**(c)**  C

**(d)**  D

# The Titration of a Weak Acid with a Strong Base

Consider the titration of 25.0 mL of 0.100 M $HCHO_2$ with 0.100 M NaOH:

$$NaOH(aq) + HCHO_2(aq) \longrightarrow H_2O(l) + NaCHO_2(aq)$$

The concentrations and the volumes here are identical to those in our previous titration, in which we calculated the pH curve for the titration of a *strong* acid with a strong base. The only difference is that $HCHO_2$ is a *weak* acid rather than a strong one. We begin by calculating the volume required to reach the equivalence point of the titration.

## Volume of NaOH Required to Reach the Equivalence Point

From the stoichiometry of the equation, we can see that the equivalence point occurs when the amount (in moles) of added base equals the amount (in moles) of acid initially in solution:

$$\text{Initial mol } HCHO_2 = 0.0250\ L \times \frac{0.100\ \text{mol}}{1\ L} = 0.00250\ \text{mol } HCHO_2$$

The amount of NaOH that must be added is 0.00250 mol NaOH. The volume of NaOH required is therefore:

$$\text{Volume NaOH solution} = 0.00250\ \text{mol} \times \frac{1\ L}{0.100\ \text{mol}} = 0.0250\ L\ \text{NaOH solution}$$

The equivalence point occurs when 25.0 mL of base has been added. Notice that the volume of NaOH required to reach the equivalence point for this weak acid is identical to that required for a strong acid. *The volume at the equivalence point in an acid–base titration does not depend on whether the acid being titrated is a strong acid or a weak acid; it depends only on the amount (in moles) of acid present in solution before the titration begins and on the concentration of the added base.*

## Initial pH (before Adding Any Base)

The initial pH of the solution is the pH of a 0.100 M $HCHO_2$ solution. Since $HCHO_2$ is a weak acid, we calculate the concentration of $H_3O^+$ and the pH by doing an equilibrium problem for the ionization of $HCHO_2$. The procedure for solving weak acid ionization problems is given in Examples 16.5 and 16.6. We show a highly condensed calculation here ($K_a$ for $HCHO_2$ is $1.8 \times 10^{-4}$):

$$HCHO_2(aq) + H_2O(l) \rightleftharpoons H_3O^+(aq) + CHO_2^-(aq)$$

|         | [HCHO₂]     | [H₃O⁺]      | [CHO₂⁻] |
|---------|-------------|-------------|---------|
| Initial | 0.100       | ≈0.00       | 0.00    |
| Change  | −x          | +x          | +x      |
| Equil   | 0.100 − x   | x           | x       |

$$K_a = \frac{[H_3O^+][CHO_2^-]}{[HCHO_2]}$$

$$= \frac{x^2}{0.100 - x} \quad (x \text{ is small})$$

$$1.8 \times 10^{-4} = \frac{x^2}{0.100}$$

$$x = 4.\underline{2}4 \times 10^{-3}$$

Therefore, $[H_3O^+] = 4.\underline{2}4 \times 10^{-3}\ M$.

$$pH = -\log(4.\underline{2}4 \times 10^{-3})$$

$$= 2.37$$

Notice that the pH is initially at a higher value (less acidic) than it is for a strong acid of the same concentration, as we would expect because the acid is weak.

## pH after Adding 5.00 mL NaOH

When titrating a *weak acid* with a strong base, the added NaOH *converts a stoichiometric amount of the acid into its conjugate base.* As we calculated previously, 5.00 mL of the 0.100 M NaOH solution contains 0.000500 mol $OH^-$. When the 0.000500 mol $OH^-$ is added to the weak acid solution, the $OH^-$ reacts stoichiometrically with $HCHO_2$ causing the amount of $HCHO_2$ to *decrease* by 0.000500 mol and the amount of $CHO_2^-$ to *increase* by 0.000500 mol.

This is very similar to what happens when we add strong base to a buffer, and it is summarized in the following table:

| | $OH^-(aq)$ | + | $HCHO_2(aq)$ | $\longrightarrow$ | $H_2O(l)$ | + | $CHO_2^-(aq)$ |
|---|---|---|---|---|---|---|---|
| **Before addition** | $\approx$0.00 mol | | 0.00250 mol | | — | | 0.00 mol |
| **Addition** | 0.000500 mol | | — | | — | | — |
| **After addition** | $\approx$0.00 mol | | 0.00200 mol | | — | | 0.000500 mol |

Notice that, after the addition, the solution contains significant amounts of both an acid ($HCHO_2$) and its conjugate base ($CHO_2^-$)—*the solution is now a buffer.* To calculate the pH of a buffer (when the *x is small* approximation applies as it does here), we can use the Henderson–Hasselbalch equation and $pK_a$ for $HCHO_2$ (which is 3.74):

$$pH = pK_a + \log\frac{[base]}{[acid]}$$

$$= 3.74 + \log\frac{0.000500}{0.00200}$$

$$= 3.74 - 0.60$$

$$= 3.14$$

### pH after Adding 10.0, 12.5, 15.0, and 20.0 mL NaOH

As more NaOH is added, it converts more $HCHO_2$ into $CHO_2^-$. We calculate the relative amounts of $HCHO_2$ and $CHO_2^-$ at each of these volumes using the reaction stoichiometry and then calculate the pH of the resulting buffer using the Henderson–Hasselbalch equation (as we did for the pH at 5.00 mL). The amounts of $HCHO_2$ and $CHO_2^-$ (after addition of the $OH^-$) at each volume and the corresponding values of pH are tabulated as follows:

Half-equivalence point $(pH = pK_a)$

| Volume (mL) | mol $HCHO_2$ | mol $CHO_2^-$ | pH |
|---|---|---|---|
| 10.0 | 0.00150 | 0.00100 | 3.56 |
| 12.5 | 0.00125 | 0.00125 | 3.74 |
| 15.0 | 0.00100 | 0.00150 | 3.92 |
| 20.0 | 0.00050 | 0.00200 | 4.34 |

As the titration proceeds, more of the $HCHO_2$ is converted to the conjugate base ($CHO_2^-$). Notice that an added NaOH volume of 12.5 mL corresponds to one-half of the equivalence point. At this volume, one-half of the initial amount of $HCHO_2$ has been converted to $CHO_2^-$, resulting in *equal amounts of weak acid and conjugate base.* For any buffer in which the amounts of weak acid and conjugate base are equal, the $pH = pK_a$:

$$pH = pK_a + \log\frac{[base]}{[acid]}$$

If [base] = [acid], then [base]/[acid] = 1.

$$pH = pK_a + \log 1$$

$$= pK_a + 0$$

$$= pK_a$$

Since $pH = pK_a$ halfway to the equivalence point, titrations can be used to measure the $pK_a$ of an acid.

### pH after Adding 25.0 mL NaOH (Equivalence Point)

At the equivalence point, 0.00250 mol of $OH^-$ has been added and therefore all of the $HCHO_2$ has been converted into its conjugate base ($CHO_2^-$) as summarized in the following table:

| | $OH^-(aq)$ | + | $HCHO_2(aq)$ | $\longrightarrow$ | $H_2O(l)$ | + | $CHO_2^-(aq)$ |
|---|---|---|---|---|---|---|---|
| **Before addition** | $\approx$0.00 mol | | 0.00250 mol | | — | | 0.00 mol |
| **Addition** | 0.00250 mol | | — | | — | | — |
| **After addition** | $\approx$0.00 mol | | 0.00 mol | | — | | 0.00250 mol |

The solution is no longer a buffer (it no longer contains significant amounts of both a weak acid and its conjugate base). Instead, the solution contains an ion ($CHO_2^-$) acting as a weak base. We discussed how to calculate the pH of solutions such as this in Section 16.8 (see Example 16.14) by solving an equilibrium problem involving the ionization of water by the weak base ($CHO_2^-$):

$$CHO_2^-(aq) + H_2O(l) \rightleftharpoons HCHO_2(aq) + OH^-(aq)$$

We calculate the initial concentration of $CHO_2^-$ for the equilibrium problem by dividing the number of moles of $CHO_2^-$ (0.00250 mol) by the *total* volume at the equivalence point (initial volume plus added volume):

$$[CHO_2^-] = \frac{\overbrace{0.00250 \text{ mol}}^{\text{Moles } CHO_2^- \text{ at equivalence point}}}{\underbrace{0.0250 \text{ L}}_{\text{Initial volume}} + \underbrace{0.0250 \text{ L}}_{\substack{\text{Added volume} \\ \text{at equivalence point}}}} = 0.0500 \text{ M}$$

We then proceed to solve the equilibrium problem as shown in condensed form as follows:

$$CHO_2^-(aq) + H_2O(l) \rightleftharpoons HCHO_2(aq) + OH^-(aq)$$

|  | $[CHO_2^-]$ | $[HCHO_2]$ | $[OH^-]$ |
|---|---|---|---|
| Initial | 0.0500 | 0.00 | $\approx 0.00$ |
| Change | $-x$ | $+x$ | $+x$ |
| Equil | $0.0500 - x$ | $x$ | $x$ |

Before substituting into the expression for $K_b$, we find the value of $K_b$ from $K_a$ for formic acid ($K_a = 1.8 \times 10^{-4}$) and $K_w$:

$$K_a \times K_b = K_w$$

$$K_b = \frac{K_w}{K_a} = \frac{1.0 \times 10^{-14}}{1.8 \times 10^{-4}} = 5.6 \times 10^{-11}$$

Then we substitute the equilibrium concentrations from the previous table into the expression for $K_b$:

$$K_b = \frac{[HCHO_2][OH^-]}{[CHO_2^-]}$$

$$= \frac{x^2}{0.0500 - x} \quad (x \text{ is small})$$

$$5.6 \times 10^{-11} = \frac{x^2}{0.0500}$$

$$x = 1.7 \times 10^{-6}$$

Remember that $x$ represents the concentration of the hydroxide ion. We calculate $[H_3O^+]$ and pH:

$$[OH^-] = 1.7 \times 10^{-6} \text{ M}$$

$$[H_3O^+][OH^-] = K_w = 1.0 \times 10^{-14}$$

$$[H_3O^+](1.7 \times 10^{-6}) = 1.0 \times 10^{-14}$$

$$[H_3O^+] = 5.9 \times 10^{-9} \text{ M}$$

$$pH = -\log[H_3O^+]$$

$$= -\log(5.9 \times 10^{-9})$$

$$= 8.23$$

Notice that the pH at the equivalence point is *not* neutral but basic. *The titration of a weak acid by a strong base always has a basic equivalence point* because, at the equivalence point, all of the acid has been converted into its conjugate base, resulting in a weakly basic solution.

## pH after Adding 30.00 mL NaOH

At this point in the titration, 0.00300 mol of $OH^-$ has been added. NaOH has thus become the excess reagent as shown in the following table:

| | $OH^-(aq)$ | + | $HCHO_2(aq)$ | $\longrightarrow$ | $H_2O(l)$ | + | $CHO_2^-(aq)$ |
|---|---|---|---|---|---|---|---|
| **Before addition** | $\approx 0.00$ mol | | 0.00250 mol | | — | | 0.00 mol |
| **Addition** | 0.00300 mol | | — | | — | | — |
| **After addition** | 0.00050 mol | | $\approx 0.00$ mol | | — | | 0.00250 mol |

The solution is now a mixture of a strong base (NaOH) and a weak base ($CHO_2^-$). The strong base completely overwhelms the weak base, and we can calculate the pH by considering the strong base alone (as we did for the titration of a strong acid and a strong base). We calculate the $OH^-$ concentration by dividing the amount of $OH^-$ remaining by the *total volume* (initial volume plus added volume):

$$[OH^-] = \frac{0.00050 \text{ mol } OH^-}{0.0250 \text{ L} + 0.0300 \text{ L}} = 0.0091 \text{ M}$$

We can then calculate the $[H_3O^+]$ and pH:

$$[H_3O^+][OH^-] = 10^{-14}$$

$$[H_3O^+] = \frac{10^{-14}}{[OH^-]} = \frac{10^{-14}}{0.0091} = 1.1 \times 10^{-12} \text{ M}$$

$$pH = -\log(1.10 \times 10^{-12})$$

$$= 11.96$$

## pH after Adding 35.0, 40.0, and 50.0 mL NaOH

As more NaOH is added, the basicity of the solution increases further. We calculate the pH at each of these volumes in the same way we calculated the pH at 30.00 mL of added NaOH. The results are tabulated as follows:

| Volume (mL) | pH |
|---|---|
| 35.0 | 12.22 |
| 40.0 | 12.36 |
| 50.0 | 12.52 |

## The Overall pH Curve

The overall pH curve for the titration of a weak acid with a strong base has a characteristic S-shape similar to that for the titration of a strong acid with a strong base. The main difference is that the equivalence point pH is basic (not neutral). Notice that calculating the pH in different regions throughout the titration involves working different kinds of acid–base problems, all of which we have encountered before.

*Summarizing Titration of a Weak Acid with a Strong Base:*

▶ The initial pH is that of the weak acid solution to be titrated. Calculate the pH by working an equilibrium problem (similar to Examples 16.5 and 16.6) using the concentration of the weak acid as the initial concentration.

▶ Between the initial pH and the equivalence point, the solution becomes a buffer. Use the reaction stoichiometry to calculate the amounts of each buffer component and then use the Henderson–Hasselbalch equation to calculate the pH (as in Example 17.3).

▶ Halfway to the equivalence point, the buffer components are exactly equal and $pH = pK_a$.

▶ At the equivalence point, the acid has all been converted into its conjugate base. Calculate the pH by working an equilibrium problem for the ionization of water by the ion acting as a weak base (similar to Example 16.14). (Calculate the concentration of the ion acting as a weak base by dividing the number of moles of the ion by the total volume at the equivalence point.)

▶ Beyond the equivalence point, $OH^-$ is in excess. Ignore the weak base and calculate the $[OH^-]$ by subtracting the initial number of moles of the weak acid from the number of moles of added $OH^-$ and dividing by the *total* volume.

## Example 17.7  Weak Acid–Strong Base Titration pH Curve

A 40.0-mL sample of 0.100 M $HNO_2$ is titrated with 0.200 M KOH. Calculate:

**(a)** the volume required to reach the equivalence point

**(b)** the pH after adding 5.00 mL of KOH

**(c)** the pH at one-half the equivalence point

SOLUTION

| | |
|---|---|
| **(a)** The equivalence point occurs when the amount (in moles) of added base equals the amount (in moles) of acid initially in the solution. Begin by calculating the amount (in moles) of acid initially in the solution. The amount (in moles) of KOH that must be added is equal to the amount of the weak acid. | $mol\ HNO_2 = 0.0400\ L \times \dfrac{0.100\ mol}{L}$ <br><br> $= 4.00 \times 10^{-3}\ mol$ <br><br> $mol\ KOH\ required = 4.00 \times 10^{-3}\ mol$ |
| Calculate the volume of KOH required from the number of moles of KOH and the molarity. | $volume\ KOH\ solution = 4.00 \times 10^{-3}\ mol \times \dfrac{1\ L}{0.200\ mol}$ <br><br> $= 0.0200\ L\ KOH\ solution$ <br><br> $= 20.0\ mL\ KOH\ solution$ |

**(b)** Use the concentration of the KOH solution to calculate the amount (in moles) of $OH^-$ in 5.00 mL of the solution.

$$mol\ OH^- = 5.00 \times 10^{-3}\ L \times \frac{0.200\ mol}{1\ L}$$
$$= 1.00 \times 10^{-3}\ mol\ OH^-$$

Prepare a table showing the amounts of $HNO_2$ and $NO_2^-$ before and after the addition of 5.00 mL KOH. The addition of the KOH stoichiometrically reduces the concentration of $HNO_2$ and increases the concentration of $NO_2^-$.

| | $OH^-(aq)$ | $+$ | $HNO_2(aq)$ | $\longrightarrow$ | $H_2O(l)$ | $+$ | $NO_2^-(aq)$ |
|---|---|---|---|---|---|---|---|
| **Before addition** | $\approx 0.00$ mol | | $4.00 \times 10^{-3}$ mol | | — | | $0.00$ mol |
| **Addition** | $1.00 \times 10^{-3}$ mol | | — | | — | | — |
| **After addition** | $\approx 0.00$ mol | | $3.00 \times 10^{-3}$ mol | | — | | $1.00 \times 10^{-3}$ mol |

Since the solution now contains significant amounts of a weak acid and its conjugate base, use the Henderson–Hasselbalch equation and $pK_a$ for $HNO_2$ (which is 3.34) to calculate the pH of the solution.

$$pH = pK_a + \log\frac{[base]}{[acid]}$$
$$= 3.34 + \log\frac{1.00 \times 10^{-3}}{3.00 \times 10^{-3}}$$
$$= 3.34 - 0.48 = 2.86$$

**(c)** At one-half the equivalence point, the amount of added base is exactly one-half the initial amount of acid. The base converts exactly half of the $HNO_2$ into $NO_2^-$, resulting in equal amounts of the weak acid and its conjugate base. The pH is therefore equal to $pK_a$.

| | $OH^-(aq)$ | $+$ | $HNO_2(aq)$ | $\longrightarrow$ | $H_2O(l)$ | $+$ | $NO_2^-(aq)$ |
|---|---|---|---|---|---|---|---|
| **Before addition** | $\approx 0.00$ mol | | $4.00 \times 10^{-3}$ mol | | — | | $0.00$ mol |
| **Addition** | $2.00 \times 10^{-3}$ mol | | — | | — | | — |
| **After addition** | $\approx 0.00$ mol | | $2.00 \times 10^{-3}$ mol | | — | | $2.00 \times 10^{-3}$ mol |

$$pH = pK_a + \log\frac{[base]}{[acid]}$$

$$= 3.34 + \log\frac{2.00 \times 10^{-3}}{2.00 \times 10^{-3}} = 3.34 + 0 = 3.34$$

**FOR PRACTICE 17.7**    Determine the pH at the equivalence point for the titration of $HNO_2$ and KOH in Example 17.7.

# The Titration of a Weak Base with a Strong Acid

The pH curve for the titration of a weak base with a strong acid is shown in Figure 17.8▼. Calculating the points along this curve is very similar to calculating the points along the pH curve for the titration of a weak acid with a strong base (which we just did). The main differences are that the curve starts basic and has an acidic equivalence point. We calculate the pH in the buffer region using $pH = pK_a + \log\frac{[base]}{[acid]}$ where the $pK_a$ corresponds to the conjugate acid of the base being titrated.

**CONCEPTUAL CONNECTION 17.7**

**The Half-Equivalence Point**    What is the pH at the half-equivalence point in the titration of a weak base with a strong acid? The $pK_b$ of the weak base is 8.75.

   **(a)** 8.75        **(b)** 7.0        **(c)** 5.25        **(d)** 4.37

# The Titration of a Polyprotic Acid

When a diprotic acid is titrated with a strong base, if $K_{a_1}$ and $K_{a_2}$ are sufficiently different, the pH curve will have two equivalence points. For example, Figure 17.9▼ shows the pH

▲ **FIGURE 17.8  Titration Curve: Weak Base with Strong Acid** This curve represents the titration of 0.100 M $NH_3$ with 0.100 M HCl.

▲ **FIGURE 17.9  Titration Curve: Diprotic Acid with Strong Base** This curve represents the titration of 25.0 mL of 0.100 M $H_2SO_3$ with 0.100 M NaOH.

curve for the titration of sulfurous acid ($H_2SO_3$) with sodium hydroxide. Recall from Section 16.9 that sulfurous acid ionizes in two steps as follows:

$$H_2SO_3(aq) \rightleftharpoons H^+(aq) + HSO_3^-(aq) \qquad K_{a_1} = 1.6 \times 10^{-2}$$

$$HSO_3^-(aq) \rightleftharpoons H^+(aq) + SO_3^{2-}(aq) \qquad K_{a_2} = 6.4 \times 10^{-8}$$

The first equivalence point in the titration curve represents the titration of the first proton, whereas the second equivalence point represents the titration of the second proton. Notice that the volume required to reach the first equivalence point is identical to the volume required to reach the second point because the number of moles of $H_2SO_3$ in the first step determines the number of moles of $HSO_3^-$ in the second step.

CONCEPTUAL
**CONNECTION 17.8**

**Acid–Base Titrations**  Consider these three titrations:

**(i)**   the titration of 25.0 mL of a 0.100 M monoprotic weak acid with 0.100 M NaOH

**(ii)**   the titration of 25.0 mL of a 0.100 M diprotic weak acid with 0.100 M NaOH

**(iii)**   the titration of 25.0 mL of a 0.100 M strong acid with 0.100 M NaOH

Which statement is most likely to be true?

**(a)**   All three titrations have the same initial pH.

**(b)**   All three titrations have the same pH at their first equivalence points.

**(c)**   All three titrations require the same volume of NaOH to reach their first equivalence points.

## Indicators: pH-Dependent Colors

We can monitor the pH of a titration with either a pH meter or an indicator. The direct monitoring of pH with a meter yields data like the pH curves we have examined previously, allowing determination of the equivalence point from the pH curve itself, as shown in Figure 17.10▼. With an indicator, we rely on the point where the indicator changes color—called the **endpoint**—to determine the equivalence point, as shown

▲ **FIGURE 17.10  Monitoring the pH during a Titration**  A pH meter monitors pH during titration. The inflection point in the resulting pH curve signifies the equivalence point.

## Using an Indicator

in Figure 17.11▲. With the correct indicator, the endpoint of the titration (indicated by the color change) occurs at the equivalence point (when the amount of acid equals the amount of base).

An indicator is itself a weak organic acid that is a different color than its conjugate base. For example, phenolphthalein (whose structure is shown in Figure 17.12▼) is a common indicator whose acid form is colorless and conjugate base form is pink. If we let HIn represent the acid form of a generic indicator and In$^-$ the conjugate base form, we have the following equilibrium:

$$\text{HIn}(aq) + \text{H}_2\text{O}(l) \rightleftharpoons \text{H}_3\text{O}^+(aq) + \text{In}^-(aq)$$

color 1                                color 2

Because its color is intense, only a small amount of indicator is required—an amount that will not affect the pH of the solution or the equivalence point of the neutralization reaction. When the [H$_3$O$^+$] changes during the titration, the equilibrium shifts in response. At low pH, the [H$_3$O$^+$] is high and the equilibrium lies far to the left, resulting in a solution of color 1. As the titration proceeds, the [H$_3$O$^+$] decreases, shifting the equilibrium to the right. Since the pH change is large near the equivalence point of the titration, there is a

## Phenolphthalein, a Common Indicator

$$+2\,\text{OH}^- \rightleftharpoons +2\,\text{H}^+$$

Acidic - Colorless                         Basic - Pink

▲ FIGURE 17.12 **Phenolphthalein** Phenolphthalein, a weakly acidic compound, is colorless. Its conjugate base is pink.

large change in $[H_3O^+]$ near the equivalence point. Provided that the correct indicator is chosen, there will also be a correspondingly significant change in color. For the titration of a strong acid with a strong base, one drop of the base near the endpoint is usually enough to change the indicator from color 1 to color 2.

The color of a solution containing an indicator depends on the relative concentrations of HIn and $In^-$. As a useful guideline, we can assume the following:

If $\dfrac{[In^-]}{[HIn]} = 1$, the indicator solution will be intermediate in color.

If $\dfrac{[In^-]}{[HIn]} > 10$, the indicator solution will be the color of $In^-$.

If $\dfrac{[In^-]}{[HIn]} < 0.1$, the indicator solution will be the color of HIn.

From the Henderson–Hasselbalch equation, we can derive an expression for the ratio of $[In^-]/[HIn]$:

$$pH = pK_a + \log \frac{[base]}{[acid]}$$

$$= pK_a + \log \frac{[In^-]}{[HIn]}$$

$$\log \frac{[In^-]}{[HIn]} = pH - pK_a$$

$$\frac{[In^-]}{[HIn]} = 10^{(pH - pK_a)}$$

Consider the following three pH values relative to $pK_a$ and the corresponding colors of the indicator solution:

| pH (relative to $pK_a$) | $[In^-]/[HIn]$ ratio | Color of Indicator Solution |
|---|---|---|
| pH = $pK_a$ | $\dfrac{[In^-]}{[HIn]} = 10^0 = 1$ | Intermediate color |
| pH = $pK_a + 1$ | $\dfrac{[In^-]}{[HIn]} = 10^1 = 10$ | Color of $In^-$ |
| pH = $pK_a - 1$ | $\dfrac{[In^-]}{[HIn]} = 10^{-1} = 0.10$ | Color of HIn |

### Indicator Color Change: Methyl Red

When the pH of the solution equals the $pK_a$ of the indicator, the solution has an intermediate color. When the pH is 1 unit (or more) above $pK_a$, the indicator is the color of $In^-$, and when the pH is 1 unit (or more) below $pK_a$, the indicator is the color of HIn. As you can see, the indicator changes color within a range of two pH units centered at $pK_a$ (Figure 17.13 ◄). Table 17.1 shows various indicators and their colors as a function of pH.

◄ **FIGURE 17.13 Indicator Color Change** An indicator (in this case, methyl red) generally changes color within a range of two pH units. (The pH for each solution is marked on its test tube.)

**TABLE 17.1 Ranges of Color Changes for Several Acid–Base Indicators**

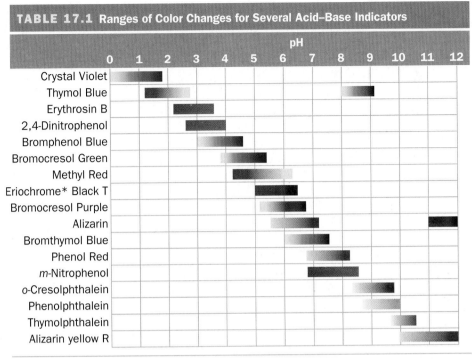

*Trademark of CIBA GEIGY CORP.

## 17.5 Solubility Equilibria and the Solubility Product Constant

Recall from Chapter 4 that a compound is considered *soluble* if it dissolves in water and *insoluble* if it does not. Recall also that, with the *solubility rules* in Table 4.1, we were able to classify ionic compounds simply as soluble or insoluble. Now we have the tools to examine *degrees* of solubility.

We can better understand the solubility of an ionic compound by applying the concept of equilibrium to the process of dissolution. For example, we can represent the dissolution of calcium fluoride in water as an equilibrium:

$$CaF_2(s) \rightleftharpoons Ca^{2+}(aq) + 2\,F^-(aq)$$

The equilibrium constant for a chemical equation representing the dissolution of an ionic compound is the **solubility product constant ($K_{sp}$)**. For $CaF_2$, the expression of the solubility product constant is:

$$K_{sp} = [Ca^{2+}][F^-]^2$$

Notice that, as we discussed in Section 15.5, solids are omitted from the equilibrium expression because the concentration of a solid is constant (it is determined by its density and does not change when more solid is added).

The value of $K_{sp}$ is a measure of the solubility of a compound. Table 17.2 lists the values of $K_{sp}$ for a number of ionic compounds. A more complete list can be found in Appendix IIC.

### $K_{sp}$ and Molar Solubility

Recall from Section 13.2 that the *solubility* of a compound is the quantity of the compound that dissolves in a certain amount of liquid. The **molar solubility** is the solubility in units of moles per liter (mol/L). We can calculate the molar solubility of a compound directly from $K_{sp}$.

Consider silver chloride:

$$AgCl(s) \rightleftharpoons Ag^+(aq) + Cl^-(aq) \qquad K_{sp} = 1.77 \times 10^{-10}$$

**TABLE 17.2 Selected Solubility Product Constants ($K_{sp}$) at 25 °C**

| Compound | Formula | $K_{sp}$ | Compound | Formula | $K_{sp}$ |
|---|---|---|---|---|---|
| Barium fluoride | $BaF_2$ | $2.45 \times 10^{-5}$ | Lead(II) chloride | $PbCl_2$ | $1.17 \times 10^{-5}$ |
| Barium sulfate | $BaSO_4$ | $1.07 \times 10^{-10}$ | Lead(II) bromide | $PbBr_2$ | $4.67 \times 10^{-6}$ |
| Calcium carbonate | $CaCO_3$ | $4.96 \times 10^{-9}$ | Lead(II) sulfate | $PbSO_4$ | $1.82 \times 10^{-8}$ |
| Calcium fluoride | $CaF_2$ | $1.46 \times 10^{-10}$ | Lead(II) sulfide* | $PbS$ | $9.04 \times 10^{-29}$ |
| Calcium hydroxide | $Ca(OH)_2$ | $4.68 \times 10^{-6}$ | Magnesium carbonate | $MgCO_3$ | $6.82 \times 10^{-6}$ |
| Calcium sulfate | $CaSO_4$ | $7.10 \times 10^{-5}$ | Magnesium hydroxide | $Mg(OH)_2$ | $2.06 \times 10^{-13}$ |
| Copper(II) sulfide* | $CuS$ | $1.27 \times 10^{-36}$ | Silver chloride | $AgCl$ | $1.77 \times 10^{-10}$ |
| Iron(II) carbonate | $FeCO_3$ | $3.07 \times 10^{-11}$ | Silver chromate | $Ag_2CrO_4$ | $1.12 \times 10^{-12}$ |
| Iron(II) hydroxide | $Fe(OH)_2$ | $4.87 \times 10^{-17}$ | Silver bromide | $AgBr$ | $5.35 \times 10^{-13}$ |
| Iron(II) sulfide* | $FeS$ | $3.72 \times 10^{-19}$ | Silver iodide | $AgI$ | $8.51 \times 10^{-17}$ |

*Sulfide equilibrium is of the type: $MS(s) + H_2O(l) \rightleftharpoons M^{2+}(aq) + HS^-(aq) + OH^-(aq)$

Notice that $K_{sp}$ is *not* the molar solubility, but the solubility product constant. The solubility product constant has only one value at a given temperature. The solubility, however, can have different values in different kinds of solutions. For example, due to the common ion effect, the solubility of AgCl in pure water is different from its solubility in an NaCl solution, even though the solubility product constant is the same for both solutions. Notice also that the solubility of AgCl is directly related (by the reaction stoichiometry) to the amount of $Ag^+$ or $Cl^-$ present in solution when equilibrium is reached. Consequently, finding molar solubility from $K_{sp}$ involves solving an equilibrium problem.

For AgCl, we set up an ICE table for the dissolution of AgCl into its ions in pure water:

$$AgCl(s) \rightleftharpoons Ag^+(aq) + Cl^-(aq)$$

| | [Ag⁺] | [Cl⁻] |
|---|---|---|
| Initial | 0.00 | 0.00 |
| Change | +S | +S |
| Equil | S | S |

*Alternatively, the variable x can be used in place of S, as it was for other equilibrium calculations.*

We let $S$ represent the concentration of AgCl that dissolves (which is the molar solubility) and then represent the concentrations of the ions formed in terms of $S$. In this case, for every 1 mol of AgCl that dissolves, 1 mol of $Ag^+$ and 1 mol of $Cl^-$ are produced. Therefore, the concentrations of $Ag^+$ or $Cl^-$ present in solution are equal to $S$. Substituting the equilibrium concentrations of $Ag^+$ and $Cl^-$ into the expression for the solubility product constant, we get:

$$K_{sp} = [Ag^+][Cl^-]$$
$$= S \times S$$
$$= S^2$$

Therefore,

$$S = \sqrt{K_{sp}}$$
$$= \sqrt{1.77 \times 10^{-10}}$$
$$= 1.33 \times 10^{-5} \, M$$

The molar solubility of AgCl is $1.33 \times 10^{-5}$ mol per liter.

## Example 17.8 | Calculating Molar Solubility from $K_{sp}$

Calculate the molar solubility of $PbCl_2$ in pure water.

**SOLUTION**

| | |
|---|---|
| Begin by writing the reaction by which solid $PbCl_2$ dissolves into its constituent aqueous ions and write the corresponding expression for $K_{sp}$. | $PbCl_2(s) \rightleftharpoons Pb^{2+}(aq) + 2\,Cl^-(aq)$<br>$K_{sp} = [Pb^{2+}][Cl^-]^2$ |

Refer to the stoichiometry of the reaction and prepare an ICE table, showing the equilibrium concentrations of $Pb^{2+}$ and $Cl^-$ relative to $S$, the amount of $PbCl_2$ that dissolves.

$$PbCl_2(s) \rightleftharpoons Pb^{2+}(aq) + 2\,Cl^-(aq)$$

| | $[Pb^{2+}]$ | $[Cl^-]$ |
|---|---|---|
| Initial | 0.00 | 0.00 |
| Change | +S | +2S |
| Equil | S | 2S |

| | |
|---|---|
| Substitute the equilibrium expressions for $[Pb^{2+}]$ and $[Cl^-]$ from the previous step into the expression for $K_{sp}$. | $K_{sp} = [Pb^{2+}][Cl^-]^2$<br>$\quad = S(2S)^2 = 4S^3$ |
| Solve for $S$ and substitute the numerical value of $K_{sp}$ (from Table 17.2) to calculate $S$. | $S = \sqrt[3]{\dfrac{K_{sp}}{4}}$<br><br>$S = \sqrt[3]{\dfrac{1.17 \times 10^{-5}}{4}} = 1.43 \times 10^{-2}\,M$ |

**FOR PRACTICE 17.8**   Calculate the molar solubility of $Fe(OH)_2$ in pure water.

---

## Chemistry IN YOUR DAY | Hard Water

In many parts of the United States, significant concentrations of $CaCO_3$ and $MgCO_3$ dissolve into rainwater runoff as it flows through soils rich in these compounds. Water containing dissolved calcium and magnesium ions is known as *hard water*. Hard water is not a health hazard because both calcium and magnesium are part of a healthy diet. However, their presence in water can be a nuisance. Because of their relatively low solubility, hard water becomes saturated with $CaCO_3$ and $MgCO_3$ as it evaporates. If evaporation continues, some dissolved ions precipitate out as salts. These precipitates show up as scaly deposits on faucets, sinks, or cookware. Washing cars or dishes with hard water leaves spots of $CaCO_3$ and $MgCO_3$.

Water can be softened with water softeners. These devices replace the $Ca^{2+}$ and $Mg^{2+}$ ions present in hard water with ions such as $K^+$ or $Na^+$. Since potassium and sodium salts are soluble, they do not form scaly deposits in the way that $Ca^{2+}$ and $Mg^{2+}$ ions do. However, when sodium is used to soften drinking water, the resulting water is high in sodium content, a disadvantage to those who must control their sodium intake due to high blood pressure.

▲ The water in reservoirs where the soil is rich in limestone (calcium carbonate) contains $Ca^{2+}$ ions. This so-called hard water contains dissolved $CaCO_3$ and $MgCO_3$. When hard water evaporates, it can leave deposits of these salts on the shores of lakes and reservoirs (at left) and on plumbing fixtures (such as the pipe shown here).

**QUESTION**

Use the $K_{sp}$ values from Table 17.2 to calculate the molar solubility of $CaCO_3$ and $MgCO_3$. What mass of $CaCO_3$ (in grams) is in 5 L of water that is saturated with $CaCO_3$?

## Example 17.9    Calculating $K_{sp}$ from Molar Solubility

The molar solubility of $Ag_2SO_4$ in pure water is $1.4 \times 10^{-2}$ M. Calculate $K_{sp}$.

**SOLUTION**

| | |
|---|---|
| Begin by writing the reaction by which solid $Ag_2SO_4$ dissolves into its constituent aqueous ions; then write the corresponding expression for $K_{sp}$. | $Ag_2SO_4(s) \rightleftharpoons 2\,Ag^+(aq) + SO_4{}^{2-}(aq)$<br><br>$K_{sp} = [Ag^+]^2[SO_4{}^{2-}]$ |

| | |
|---|---|
| Use an ICE table to define $[Ag^+]$ and $[SO_4{}^{2-}]$ in terms of $S$, the amount of $Ag_2SO_4$ that dissolves. | $Ag_2SO_4(s) \rightleftharpoons 2\,Ag^+(aq) + SO_4{}^{2-}(aq)$ |

| | $[Ag^+]$ | $[SO_4{}^{2-}]$ |
|---|---|---|
| Initial | 0.00 | 0.00 |
| Change | +2S | +S |
| Equil | 2S | S |

| | |
|---|---|
| Substitute the expressions for $[Ag^+]$ and $[SO_4{}^{2-}]$ from the previous step into the expression for $K_{sp}$. Substitute the given value *of the molar solubility for S and calculate $K_{sp}$.* | $K_{sp} = [Ag^+]^2[SO_4{}^{2-}]$<br>$\quad = (2S)^2 S$<br>$\quad = 4S^3$<br>$\quad = 4(1.4 \times 10^{-2})^3$<br>$\quad = 1.1 \times 10^{-5}$ |

**FOR PRACTICE 17.9**    The molar solubility of AgBr in pure water is $7.3 \times 10^{-7}$ M. Calculate $K_{sp}$.

## $K_{sp}$ and Relative Solubility

As we have just seen, molar solubility and $K_{sp}$ are related, and each can be calculated from the other; however, we cannot generally use the $K_{sp}$ values of two different compounds to directly compare their relative solubilities.

For example, consider the following compounds, their $K_{sp}$ values, and their molar solubilities:

| Compound | $K_{sp}$ | Solubility |
|---|---|---|
| $Mg(OH)_2$ | $2.06 \times 10^{-13}$ | $3.72 \times 10^{-5}$ M |
| $FeCO_3$ | $3.07 \times 10^{-11}$ | $5.54 \times 10^{-6}$ M |

Magnesium hydroxide has a smaller $K_{sp}$ than iron(II) carbonate, but a higher molar solubility. Why? The relationship between $K_{sp}$ and molar solubility depends on the stoichiometry of the dissociation reaction. Consequently, any direct comparison of $K_{sp}$ values for different compounds can only be made if the compounds have the same dissociation stoichiometry.

Consider the following compounds with the same dissociation stoichiometry, their $K_{sp}$ values, and their molar solubilities:

| Compound | $K_{sp}$ | Solubility |
|---|---|---|
| $Mg(OH)_2$ | $2.06 \times 10^{-13}$ | $3.72 \times 10^{-5}$ M |
| $CaF_2$ | $1.46 \times 10^{-10}$ | $3.32 \times 10^{-4}$ M |

In this case, magnesium hydroxide and calcium fluoride have the same dissociation stoichiometry (1 mol of each compound produces 3 mol of dissolved ions); therefore, the $K_{sp}$ values can be directly compared as a measure of relative solubility.

## The Effect of a Common Ion on Solubility

How is the solubility of an ionic compound affected when the compound is dissolved in a solution that already contains one of its ions? For example, what is the solubility of $CaF_2$ in a solution that is 0.100 M in NaF? We can determine the change in solubility by considering the common ion effect, which we first encountered in Section 17.2. We represent the dissociation of $CaF_2$ in a 0.100 M NaF solution as shown in the right margin.

In accordance with Le Châtelier's principle, the presence of the $F^-$ ion in solution causes the equilibrium to shift to the left (compared to its position in pure water), which means that less $CaF_2$ dissolves—that is, its solubility is decreased.

> **In general, the solubility of an ionic compound is lower in a solution containing a common ion than in pure water.**

We can calculate the exact value of the solubility by working an equilibrium problem in which the concentration of the common ion is accounted for in the initial conditions, as shown in Example 17.10.

Common ion
0.100 M $F^-$(aq)

$$CaF_2(s) \rightleftharpoons Ca^{2+}(aq) + 2\,F^-(aq)$$

Equilibrium shifts left

---

## Example 17.10 Calculating Molar Solubility in the Presence of a Common Ion

What is the molar solubility of $CaF_2$ in a solution containing 0.100 M NaF?

**SOLUTION**

| | |
|---|---|
| Begin by writing the reaction by which solid $CaF_2$ dissolves into its constituent aqueous ions. Write the corresponding expression for $K_{sp}$. | $CaF_2(s) \rightleftharpoons Ca^{2+}(aq) + 2\,F^-(aq)$ <br><br> $K_{sp} = [Ca^{2+}][F^-]^2$ |

Use the stoichiometry of the reaction to prepare an ICE table showing the initial concentration of the common ion. Fill in the equilibrium concentrations of $Ca^{2+}$ and $F^-$ relative to $S$, the amount of $CaF_2$ that dissolves.

$CaF_2(s) \rightleftharpoons Ca^{2+}(aq) + 2\,F^-(aq)$

| | $[Ca^{2+}]$ | $[F^-]$ |
|---|---|---|
| Initial | 0.00 | 0.100 |
| Change | $+S$ | $+2S$ |
| Equil | $S$ | $0.100 + 2S$ |

| | |
|---|---|
| Substitute the equilibrium expressions for $[Ca^{2+}]$ and $[F^-]$ from the previous step into the expression for $K_{sp}$. Because $K_{sp}$ is small, you can make the approximation that $2S$ is much less than 0.100 and will therefore be insignificant when added to 0.100 (this is similar to the $x$ is small approximation in equilibrium problems). | $K_{sp} = [Ca^{2+}][F^-]^2$ <br><br> $\qquad = S(0.100 + 2S)^2 \quad$ (S is small) <br><br> $\qquad = S(0.100)^2$ |

| | |
|---|---|
| Solve for $S$ and substitute the numerical value of $K_{sp}$ (from Table 17.2) to calculate $S$. Note that the calculated value of $S$ is indeed small compared to 0.100; our approximation is valid. | $K_{sp} = S(0.100)^2$ <br><br> $S = \dfrac{K_{sp}}{0.0100} = \dfrac{1.46 \times 10^{-10}}{0.0100} = 1.46 \times 10^{-8}\,M$ |

For comparison, the molar solubility of $CaF_2$ in pure water is $3.32 \times 10^{-4}\,M$, which means $CaF_2$ is over 20,000 times more soluble in water than in the NaF solution. (Confirm this for yourself by calculating its solubility in pure water from the value of $K_{sp}$.)

**FOR PRACTICE 17.10** Calculate the molar solubility of $CaF_2$ in a solution containing 0.250 M $Ca(NO_3)_2$.

---

CONCEPTUAL
CONNECTION **17.9**

**Common Ion Effect** In which solution is $BaSO_4$ most soluble?

    **(a)** a solution that is 0.10 M in $BaNO_3$

    **(b)** a solution that is 0.10 M in $Na_2SO_4$

    **(c)** a solution that is 0.10 M in $NaNO_3$

## The Effect of pH on Solubility

The pH of a solution can affect the solubility of a compound in that solution. For example, consider the dissociation of $Mg(OH)_2$, the active ingredient in milk of magnesia:

$$Mg(OH)_2(s) \rightleftharpoons Mg^{2+}(aq) + 2\,OH^-(aq)$$

The solubility of this compound is highly dependent on the pH of the solution into which it dissolves. If the pH is high, then the concentration of $OH^-$ in the solution is high. In accordance with the common ion effect, this shifts the equilibrium to the left, lowering the solubility:

High $[OH^-]$

$$Mg(OH)_2(s) \rightleftharpoons Mg^{2+}(aq) + 2\,OH^-(aq)$$

Equilibrium shifts left

If the pH is low, then the concentration of $H_3O^+(aq)$ in the solution is high. As the $Mg(OH)_2$ dissolves, these $H_3O^+$ ions neutralize the newly dissolved $OH^-$ ions, driving the reaction to the right:

$H_3O^+$ reacts with $OH^-$

$$Mg(OH)_2(s) \rightleftharpoons Mg^{2+}(aq) + 2\,OH^-(aq)$$

Equilibrium shifts right

Consequently, the solubility of $Mg(OH)_2$ in an acidic solution is higher than in a pH-neutral or basic solution.

**In general, the solubility of an ionic compound with a strongly basic or weakly basic anion increases with increasing acidity (decreasing pH).**

Common basic anions include $OH^-$, $S^{2-}$, and $CO_3^{2-}$. Therefore, hydroxides, sulfides, and carbonates are more soluble in acidic water than in pure water. Since rainwater is naturally acidic due to dissolved carbon dioxide, it can dissolve rocks high in limestone ($CaCO_3$) as it flows through the ground. This process sometimes results in huge underground caverns such as those at Carlsbad Caverns National Park in New Mexico. Dripping water saturated in $CaCO_3$ within the cave creates the dramatic mineral formations known as stalagmites and stalactites.

▲ Stalactites (which hang from the ceiling) and stalagmites (which grow up from the ground) form as calcium carbonate precipitates out of the water evaporating in underground caves.

---

**Example** 17.11 | **The Effect of pH on Solubility**

Determine whether each compound is more soluble in an acidic solution than it is in a neutral solution.

**(a)** $BaF_2$  **(b)** $AgI$  **(c)** $Ca(OH)_2$

**SOLUTION**

**(a)** The solubility of $BaF_2$ is greater in an acidic solution than it is in a neutral solution because the $F^-$ ion is a weak base. ($F^-$ is the conjugate base of the weak acid HF and is therefore a weak base.)

**(b)** The solubility of $AgI$ is not greater in an acidic solution than it is in a neutral solution because the $I^-$ is *not* a base. ($I^-$ is the conjugate base of the *strong* acid HI and is therefore pH-neutral.)

**(c)** The solubility of $Ca(OH)_2$ is greater in an acidic solution than it is in a neutral solution because the $OH^-$ ion is a strong base.

**FOR PRACTICE 17.11**  Which compound, $FeCO_3$ or $PbBr_2$, is more soluble in acid than in base? Why?

# 17.6 Precipitation

In Chapter 4, we learned that a precipitation reaction can occur when we mix two solutions containing ionic compounds when one of the possible cross products—the combination of a cation from one solution and the anion from the other—is insoluble. In this chapter, however, we have seen that the terms *soluble* and *insoluble* are extremes in a continuous range of solubility—many compounds are slightly soluble, and even those that we categorize as insoluble in Chapter 4 actually have some limited degree of solubility (they have very small solubility product constants).

We can better understand precipitation reactions by revisiting a concept from Chapter 15— the reaction quotient ($Q$). The reaction quotient for the reaction by which an ionic compound dissolves is the product of the concentrations of the ionic components raised to their stoichiometric coefficients. For example, consider the reaction by which $CaF_2$ dissolves:

$$CaF_2(s) \rightleftharpoons Ca^{2+}(aq) + 2\,F^-(aq)$$

The reaction quotient for this reaction is:

$$Q = [Ca^{2+}][F^-]^2$$

Na₂CrO₄

AgNO₃
Ag₂CrO₄

The difference between $Q$ and $K_{sp}$ is that $K_{sp}$ is the value of this product *at equilibrium only*, whereas $Q$ is the value of the product under any conditions. We can therefore use the value of $Q$ to compare a solution containing any concentrations of the component ions to a solution that is at equilibrium.

Consider a solution of calcium fluoride in which $Q$ is less than $K_{sp}$. Recall from Chapter 15 that if $Q$ is less than $K_{sp}$, the reaction will proceed to the right (toward products). Consequently, as long as solid $CaF_2$ is present, the $CaF_2$ will continue to dissolve. If all of the solid has already dissolved, the solution will simply remain as it is, containing less than the equilibrium amount of the dissolved ions. Such a solution is an *unsaturated solution*. If more solid is added to an unsaturated solution, it will dissolve, as long as $Q$ remains less than $K_{sp}$.

Now consider a solution in which $Q$ is exactly equal to $K_{sp}$. In this case, the reaction is at equilibrium and does not make progress in either direction. Such a solution most likely contains at least a small amount of the solid in equilibrium with its component ions. However, the amount of solid may be too small to be visible. Such a solution is a *saturated solution*.

Finally, consider a solution in which $Q$ is greater than $K_{sp}$. In this case, the reaction proceeds to the left (toward the reactants) and solid calcium fluoride forms from the dissolved calcium and fluoride ions. In other words, the solid normally precipitates out of a solution in which $Q$ is greater than $K_{sp}$. Under certain circumstances, however, $Q$ can remain greater than $K_{sp}$ for an unlimited period of time. Such a solution, called a *supersaturated solution*, is unstable and forms a precipitate when sufficiently disturbed. Figure 17.14▶ shows a supersaturated solution of sodium acetate. When a small seed crystal of solid sodium acetate is dropped into the solution, it triggers the precipitation reaction.

▼ FIGURE 17.14 **Precipitation from a Supersaturated Solution** The excess solute in a supersaturated solution of sodium acetate precipitates out if a small sodium acetate crystal is added.

***Summarizing the Relationship of Q and K_{sp} in Solutions Containing an Ionic Compound:***

▶ If $Q < K_{sp}$, the solution is unsaturated and more of the solid ionic compound can dissolve in the solution.

▶ If $Q = K_{sp}$, the solution is saturated. The solution is holding the equilibrium amount of the dissolved ions, and additional solid does not dissolve in the solution.

▶ If $Q > K_{sp}$, the solution is supersaturated. Under most circumstances, the excess solid precipitates out of a supersaturated solution.

Seed crystal

Supersaturated solution of sodium acetate

Solid sodium acetate forming

We can use $Q$ to predict whether a precipitation reaction will occur upon the mixing of two solutions containing dissolved ionic compounds. For example, consider mixing a silver nitrate solution with a potassium iodide solution to form a mixture that is 0.010 M in $AgNO_3$ and 0.015 M in KI. Will a precipitate form in the newly mixed solution? From Chapter 4 we know that one of the cross products, $KNO_3$, is soluble and will therefore not precipitate. The other cross product, AgI, *may* precipitate if the concentrations of $Ag^+$ and $I^-$ are high enough in the newly mixed solution: we can compare $Q$ to $K_{sp}$ to determine if a precipitate will form. For AgI, $K_{sp} = 8.51 \times 10^{-17}$. For the newly mixed solution, $[Ag^+] = 0.010$ M and $[I^-] = 0.015$ M. We calculate $Q$ as follows:

$$Q = [Ag^+][I^-] = (0.010)(0.015) = 1.5 \times 10^{-4}$$

The value of $Q$ is much greater than $K_{sp}$; therefore, AgI should precipitate out of the newly mixed solution.

---

**Example 17.12    Predicting Precipitation Reactions by Comparing $Q$ and $K_{sp}$**

A solution containing lead(II) nitrate is mixed with one containing sodium bromide to form a solution that is 0.0150 M in $Pb(NO_3)_2$ and 0.00350 M in NaBr. Does a precipitate form in the newly mixed solution?

**SOLUTION**

| | |
|---|---|
| First, determine the possible cross products and their $K_{sp}$ values (Table 17.2). Any cross products that are soluble do *not* precipitate (see Table 4.1). | Possible cross products:<br>$NaNO_3$   soluble<br>$PbBr_2$   $K_{sp} = 4.67 \times 10^{-6}$ |
| Calculate $Q$ and compare it to $K_{sp}$. A precipitate will only form if $Q > K_{sp}$. | $Q = [Pb^{2+}][Br^-]^2$<br>   $= (0.0150)(0.00350)^2$<br>   $= 1.84 \times 10^{-7}$<br>$Q < K_{sp}$; therefore no precipitate forms. |

**FOR PRACTICE 17.12**   If the original solutions in Example 17.12 are concentrated through evaporation and mixed again to form a solution that is 0.0600 M in $Pb(NO_3)_2$ and 0.0158 M in NaBr, will a precipitate form in this newly mixed solution?

---

## Selective Precipitation

A solution may contain several different dissolved metal cations that can often be separated by **selective precipitation**, a process involving the addition of a reagent that forms a precipitate with one of the dissolved cations but not the others. For example, seawater contains dissolved magnesium and calcium cations with the concentrations $[Mg^{2+}] = 0.059$ M and $[Ca^{2+}] = 0.011$ M. We can separate these ions by adding a reagent that will precipitate one of the ions but not the other. From Table 17.2, we find that $Mg(OH)_2$ has a $K_{sp}$ of $2.06 \times 10^{-13}$ and that $Ca(OH)_2$ has a $K_{sp}$ of $4.68 \times 10^{-6}$, indicating that the hydroxide ion forms a precipitate with magnesium at a much lower concentration than it does with calcium. Consequently, a soluble hydroxide—such as KOH or NaOH—is a good choice for the precipitating reagent. When we add an appropriate amount of KOH or NaOH to seawater, the hydroxide ion causes the precipitation of $Mg(OH)_2$ (the compound with the lowest $K_{sp}$) but not $Ca(OH)_2$. Calculations for this selective precipitation are shown in Examples 17.13 and 17.14. In these calculations, you compare $Q$ to $K_{sp}$ to determine the concentration that triggers precipitation.

The difference in $K_{sp}$ values required for selective precipitation is a factor of at least $10^3$.

## Example 17.13    Finding the Minimum Required Reagent Concentration for Selective Precipitation

The magnesium and calcium ions present in seawater ($[Mg^{2+}] = 0.059$ M and $[Ca^{2+}] = 0.011$ M) can be separated by selective precipitation with KOH. What minimum $[OH^-]$ triggers the precipitation of the $Mg^{2+}$ ion?

**SOLUTION**

| | |
|---|---|
| The precipitation commences when the value of $Q$ for the precipitating compound just equals the value of $K_{sp}$. Set the expression for $Q$ for magnesium hydroxide equal to the value of $K_{sp}$, and solve for $[OH^-]$. This is the concentration above which $Mg(OH)_2$ precipitates. | $Q = [Mg^{2+}][OH^-]^2$ $= (0.059)[OH^-]^2$ When $Q = K_{sp}$, $(0.059)[OH^-]^2 = K_{sp} = 2.06 \times 10^{-13}$ $[OH^-]^2 = \dfrac{2.06 \times 10^{-13}}{0.059}$ $[OH^-] = 1.9 \times 10^{-6}$ M |

**FOR PRACTICE 17.13**    If the concentration of $Mg^{2+}$ in the previous solution was 0.025 M, what minimum $[OH^-]$ triggers precipitation of the $Mg^{2+}$ ion?

## Example 17.14    Finding the Concentrations of Ions Left in Solution after Selective Precipitation

You add potassium hydroxide to the solution in Example 17.13. When the $[OH^-]$ reaches $1.9 \times 10^{-6}$ M (as you just calculated), magnesium hydroxide begins to precipitate out of solution. As you continue to add KOH, the magnesium hydroxide continues to precipitate. However, at some point, the $[OH^-]$ becomes high enough to begin to precipitate the calcium ions as well. What is the concentration of $Mg^{2+}$ when $Ca^{2+}$ begins to precipitate?

**SOLUTION**

| | |
|---|---|
| First, find the $OH^-$ concentration at which $Ca^{2+}$ begins to precipitate by writing the expression for $Q$ for calcium hydroxide and substituting the concentration of $Ca^{2+}$ from Example 17.13. | $Q = [Ca^{2+}][OH^-]^2$ $= (0.011)[OH^-]^2$ |
| Set the expression for $Q$ equal to the value of $K_{sp}$ for calcium hydroxide and solve for $[OH^-]$. This is the concentration above which $Ca(OH)_2$ precipitates. | When $Q = K_{sp}$, $(0.011)[OH^{-2}] = K_{sp} = 4.68 \times 10^{-6}$ $[OH^-]^2 = \dfrac{4.68 \times 10^{-6}}{0.011}$ $[OH^-] = 2.0\underline{6} \times 10^{-2}$ M |
| Find the concentration of $Mg^{2+}$ when $OH^-$ reaches the concentration you just calculated by writing the expression for *Q for magnesium hydroxide* and substituting the concentration of $OH^-$ that you just calculated. Then set the expression for $Q$ equal to the value of $K_{sp}$ for magnesium hydroxide and solve for $[Mg^{2+}]$. This is the concentration of $Mg^{2+}$ that remains when $Ca(OH)_2$ begins to precipitate. | $Q = [Mg^{2+}][OH^-]^2$ $= [Mg^{2+}](2.0\underline{6} \times 10^{-2})^2$ When $Q = K_{sp}$, $[Mg^{2+}](2.0\underline{6} \times 10^{-2})^2 = K_{sp} = 2.06 \times 10^{-13}$ $[Mg^{2+}] = \dfrac{2.06 \times 10^{-13}}{(2.0\underline{6} \times 10^{-2})^2}$ $[Mg^{2+}] = 4.9 \times 10^{-10}$ M |

—Continued on the next page

*Continued—*

As you can see from the results, the selective precipitation worked very well. The concentration of $Mg^{2+}$ dropped from 0.059 M to $4.9 \times 10^{-10}$ M before any calcium began to precipitate, which means that the process separated 99.99% of the magnesium out of the solution.

**FOR PRACTICE 17.14** A solution is 0.085 M in $Pb^{2+}$ and 0.025 M in $Ag^+$. (a) If selective precipitation is to be achieved using NaCl, what minimum concentration of NaCl do you need to begin to precipitate the ion that precipitates first? (b) What is the concentration of each ion left in solution at the point where the second ion begins to precipitate?

## 17.7 Qualitative Chemical Analysis

Selective precipitation as discussed in Section 17.6 can be used in a systematic way to determine which metal ions are present in an unknown solution. This method is known as **qualitative analysis**. The word *qualitative* means *involving quality or kind*. Qualitative analysis involves finding the *kind* of ions present in the solution. This stands in contrast to **quantitative analysis**, which is concerned with quantity, or the amounts of substances in a solution or mixture.

In the past, qualitative analysis by selective precipitation was used extensively to determine the metals present in a sample. This process—dubbed *wet chemistry* because it involves the mixing of many aqueous solutions in the lab—has been replaced by more precise and less time-intensive instrumental techniques. Nonetheless, both for the sake of history and also because of the importance of the principles involved, we now examine a traditional qualitative analysis scheme. You may use such a scheme in your general chemistry laboratory as an exercise in qualitative analysis.

The basic idea behind qualitative analysis is straightforward. A sample containing a mixture of metal cations is subjected to the addition of several precipitating agents. At each step, some of the metal cations—those that form insoluble compounds with the precipitating agent—precipitate from the mixture and are separated out as solids. The remaining aqueous mixture is then subjected to the next precipitating agent, and so on (Figure 17.15▼).

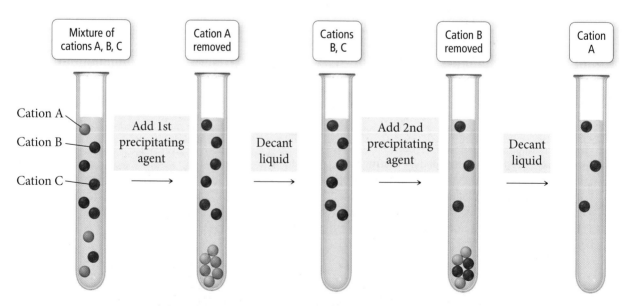

▲ **FIGURE 17.15 Qualitative Analysis** In qualitative analysis, specific ions are precipitated successively by the addition of appropriate reagents.

Figure 17.16▼ is a diagram of a general qualitative analysis scheme. The scheme involves separating a mixture of the common ions into five groups by sequentially adding five different precipitating agents. After each precipitating agent is added, the mixture is put into a centrifuge to separate the solid from the liquid. The liquid is decanted for the next step, and the solid is set aside for subsequent analysis. We examine each group separately.

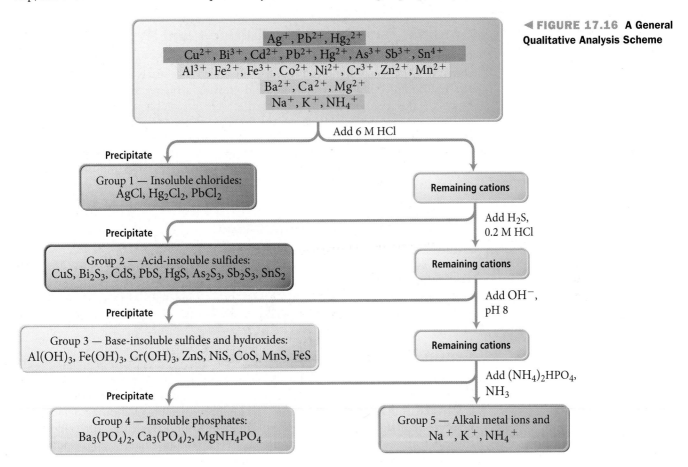

◀ **FIGURE 17.16** A General Qualitative Analysis Scheme

## Group 1: Insoluble Chlorides

In the first step, the aqueous mixture containing the metal cations is treated with 6 M HCl. Since most chlorides are soluble, the chloride ions *do not form* a precipitate with the majority of the cations in the mixture. However, $Ag^+$, $Pb^{2+}$, and $Hg_2^{2+}$ *do form* insoluble chlorides. So, if any of those metal cations are present, they precipitate out. The absence of a precipitate constitutes a negative test for $Ag^+$, $Pb^{2+}$, and $Hg_2^{2+}$. If a precipitate forms, one or more of these ions is present. After the solid is separated from the liquid, the solution is ready for the next step.

## Group 2: Acid-Insoluble Sulfides

In the second step, the acidic aqueous mixture containing the remaining metal cations is treated with $H_2S$, a weak diprotic acid that dissociates in two steps:

$$H_2S \rightleftharpoons H^+ + HS^-$$

$$HS^- \rightleftharpoons H^+ + S^{2-}$$

The concentration of $S^{2-}$ ions in an $H_2S$ solution is pH-dependent. At low pH (high $H^+$ concentration) the equilibria shift left, minimizing the amount of available $S^{2-}$. At high pH (low $H^+$ concentration) the equilbria shift right, maximizing the amount of available $S^{2-}$. At this stage, the solution is acidic (from the addition of HCl in the previous step),

$(NH_4)_2S$   CdS   $Sb_2S_3$   PbS

and the concentration of $S^{2-}$ in solution is relatively low. Only the most insoluble metal sulfides (those with the smallest $K_{sp}$ values) precipitate under these conditions. These include $Hg^{2+}$, $Cd^{2+}$, $Bi^{3+}$, $Cu^{2+}$, $Sn^{4+}$, $As^{3+}$, $Pb^{2+}$ and $Sb^{3+}$. If any of these metal cations are present, they precipitate out as sulfides. After the solid is separated from the liquid, the solution is ready for the next step.

## Group 3: Base-Insoluble Sulfides and Hydroxides

In the third step, additional base and $H_2S$ are added to the acidic aqueous mixture containing the remaining metal cations. The added base reacts with acid, shifting the $H_2S$ ionization equilibria to the right and creating a higher $S^{2-}$ concentration. This causes the precipitation of those sulfides that were too soluble to precipitate out in the previous step but not soluble enough to prevent precipitation with the higher sulfide ion concentration. The ions that precipitate as sulfides at this point (if they are present) are $Fe^{2+}$, $Co^{2+}$, $Zn^{2+}$, $Mn^{2+}$, and $Ni^{2+}$. In addition, the additional base causes $Cr^{3+}$, $Fe^{3+}$, and $Al^{3+}$ to precipitate as hydroxides. After the solid is separated from the liquid, the solution is ready for the next step.

## Group 4: Insoluble Phosphates

At this stage, all of the cations have precipitated except those belonging to the alkali metal family (group 1A in the periodic table) and the alkaline earth metal family (group 2A in the periodic table). The alkaline earth metal cations can be precipitated by adding $(NH_4)_2HPO_4$ to the solution, causing $Mg^{2+}$, $Ca^{2+}$, and $Ba^{2+}$ to precipitate as metal phosphates, which are separated from the liquid.

## Group 5: Alkali Metals and $NH_4^+$

The only dissolved ions that the liquid decanted from the previous step can now contain are $Na^+$, $K^+$, and $NH_4^+$. These cations do not form insoluble compounds with any anions and cannot be precipitated from the solution. Their presence can be determined, however, by other means. Sodium and potassium ions, for example, are usually identified through flame tests. The sodium ion produces a yellow-orange flame, and the potassium ion produces a violet flame, as shown in Figure 17.17 ▼.

By applying this procedure, nearly two dozen metal cations can be separated from a solution initially containing all of them. Each of the groups can be further analyzed to determine which specific ions are present. The procedures for these steps are found in many general chemistry laboratory manuals.

Sodium        Potassium

▶ **FIGURE 17.17 Flame Tests** The sodium ion produces a yellow-orange flame. The potassium ion produces a violet flame.

# 17.8 Complex Ion Equilibria

We have discussed several different types of equilibria so far, including acid–base equilibria and solubility equilibria. We now turn to equilibria of another type, which primarily involve transition metal ions in solution. Transition metal ions tend to be good electron acceptors (good Lewis acids). In aqueous solutions, water molecules can act as electron donors (Lewis bases) to hydrate transition metal ions. For example, silver ions are hydrated by water in solution to form $Ag(H_2O)_2^+(aq)$. Chemists often write $Ag^+(aq)$ as a shorthand notation for the hydrated silver ion, but the bare ion does not really exist by itself in solution.

Species such as $Ag(H_2O)_2^+$ are known as *complex ions*. A **complex ion** contains a central metal ion bound to one or more *ligands*. A **ligand** is a neutral molecule or ion that acts as a Lewis base with the central metal ion. In $Ag(H_2O)_2^+$, water is the ligand. If a stronger Lewis base is put into a solution containing $Ag(H_2O)_2^+$, the stronger Lewis base displaces the water in the complex ion. For example, ammonia reacts with $Ag(H_2O)_2^+$ according to the following reaction:

$$Ag(H_2O)_2^+(aq) + 2\,NH_3(aq) \rightleftharpoons Ag(NH_3)_2^+(aq) + 2\,H_2O(l)$$

For simplicity, we often leave water out of the equation:

$$Ag^+(aq) + 2\,NH_3(aq) \rightleftharpoons Ag(NH_3)_2^+(aq) \quad K_f = 1.7 \times 10^7$$

We cover complex ions in more detail in Chapter 25. Here, we focus on the equilibria associated with their formation.

The equilibrium constant associated with the reaction for the formation of a complex ion, such as the one just shown, is called the **formation constant ($K_f$)**. The expression for $K_f$ is determined by the law of mass action, like any equilibrium constant. For $Ag(NH_3)_2^+$, the expression for $K_f$ is:

$$K_f = \frac{[Ag(NH_3)_2^+]}{[Ag^+][NH_3]^2}$$

Notice that the value of $K_f$ for $Ag(NH_3)_2^+$ is large, indicating that the formation of the complex ion is highly favored. Table 17.3 lists the formation constants for a number of common complex ions. You can see that, in general, values of $K_f$ are very large, indicating that the formation of complex ions is highly favored in each case. Example 17.15 illustrates how to use $K_f$ in calculations.

**TABLE 17.3 Formation Constants of Selected Complex Ions in Water at 25 °C**

| Complex Ion | $K_f$ | Complex Ion | $K_f$ |
|---|---|---|---|
| $Ag(CN)_2^-$ | $1 \times 10^{21}$ | $Cu(NH_3)_4^{2+}$ | $1.7 \times 10^{13}$ |
| $Ag(NH_3)_2^+$ | $1.7 \times 10^7$ | $Fe(CN)_6^{4-}$ | $1.5 \times 10^{35}$ |
| $Ag(S_2O_3)_2^{3-}$ | $2.8 \times 10^{13}$ | $Fe(CN)_6^{3-}$ | $2 \times 10^{43}$ |
| $AlF_6^{3-}$ | $7 \times 10^{19}$ | $Hg(CN)_4^{2-}$ | $1.8 \times 10^{41}$ |
| $Al(OH)_4^-$ | $3 \times 10^{33}$ | $HgCl_4^{2-}$ | $1.1 \times 10^{16}$ |
| $CdBr_4^{2-}$ | $5.5 \times 10^3$ | $HgI_4^{2-}$ | $2 \times 10^{30}$ |
| $CdI_4^{2-}$ | $2 \times 10^6$ | $Ni(NH_3)_6^{2+}$ | $2.0 \times 10^8$ |
| $Cd(CN)_4^{2-}$ | $3 \times 10^{18}$ | $Pb(OH)_3^-$ | $8 \times 10^{13}$ |
| $Co(NH_3)_6^{3+}$ | $2.3 \times 10^{33}$ | $Sn(OH)_3^-$ | $3 \times 10^{25}$ |
| $Co(OH)_4^{2-}$ | $5 \times 10^9$ | $Zn(CN)_4^{2-}$ | $2.1 \times 10^{19}$ |
| $Co(SCN)_4^{2-}$ | $1 \times 10^3$ | $Zn(NH_3)_4^{2+}$ | $2.8 \times 10^9$ |
| $Cr(OH)_4^-$ | $8.0 \times 10^{29}$ | $Zn(OH)_4^{2-}$ | $2 \times 10^{15}$ |
| $Cu(CN)_4^{2-}$ | $1.0 \times 10^{25}$ | | |

**Example 17.15**    **Complex Ion Equilibria**

You mix a 200.0-mL sample of a solution that is $1.5 \times 10^{-3}$ M in $Cu(NO_3)_2$ with a 250.0-mL sample of a solution that is 0.20 M in $NH_3$. After the solution reaches equilibrium, what concentration of $Cu^{2+}(aq)$ remains?

SOLUTION

| | |
|---|---|
| Write the balanced equation for the complex ion equilibrium that occurs and look up the value of $K_f$ in Table 17.3. Since this is an equilibrium problem, you have to create an ICE table, which requires the initial concentrations of $Cu^{2+}$ and $NH_3$. Calculate those concentrations from the given values. | $Cu^{2+}(aq) + 4NH_3(aq) \rightleftharpoons Cu(NH_3)_4{}^{2+}(aq)$ <br> $K_f = 1.7 \times 10^{13}$ <br><br> $[Cu^{2+}]_{initial} = \dfrac{0.200\,L \times \dfrac{1.5 \times 10^{-3}\,mol}{L}}{0.200\,L + 0.250\,L} = 6.7 \times 10^{-4}\,M$ <br><br> $[NH_3]_{initial} = \dfrac{0.250\,L \times \dfrac{0.20\,mol}{1\,L}}{0.200\,L + 0.250\,L} = 0.11\,M$ |

Construct an ICE table for the reaction and write down the initial concentrations of each species.

$Cu^{2+}(aq) + 4NH_3(aq) \rightleftharpoons Cu(NH_3)_4{}^{2+}(aq)$

| | $[Cu^{2+}]$ | $[NH_3]$ | $[Cu(NH_3)_4{}^{2+}]$ |
|---|---|---|---|
| Initial | $6.7 \times 10^{-4}$ | 0.11 | 0.0 |
| Change | | | |
| Equil | | | |

Since the equilibrium constant is large and the concentration of ammonia is much larger than the concentration of $Cu^{2+}$, you can assume that the reaction will be driven to the right so that most of the $Cu^{2+}$ is consumed. Unlike previous ICE tables, where you used $x$ to represent the change in concentration in going to equilibrium, here you let $x$ represent the small amount of $Cu^{2+}$ that remains when equilibrium is reached.

$Cu^{2+}(aq) + 4NH_3(aq) \rightleftharpoons Cu(NH_3)_4{}^{2+}(aq)$

| | $[Cu^{2+}]$ | $[NH_3]$ | $[Cu(NH_3)_4{}^{2+}]$ |
|---|---|---|---|
| Initial | $6.7 \times 10^{-4}$ | 0.11 | 0.0 |
| Change | $\approx(-6.7 \times 10^{-4})$ | $\approx 4(-6.7 \times 10^{-4})$ | $\approx(+6.7 \times 10^{-4})$ |
| Equil | $x$ | 0.11 | $6.7 \times 10^{-4}$ |

Substitute the expressions for the equilibrium concentrations into the expression for $K_f$ and solve for $x$.

$$K_f = \frac{[Cu(NH_3)_4{}^{2+}]}{[Cu^{2+}][NH_3]^4}$$

$$= \frac{6.7 \times 10^{-4}}{x(0.11)^4}$$

$$x = \frac{6.7 \times 10^{-4}}{K_f(0.11)^4}$$

$$= \frac{6.7 \times 10^{-4}}{1.7 \times 10^{13}(0.11)^4}$$

$$= 2.7 \times 10^{-13}$$

| | |
|---|---|
| Confirm that $x$ is indeed small compared to the initial concentration of the metal cation. <br><br> The remaining $Cu^{2+}$ is very small because the formation constant is very large. | Since $x = 2.7 \times 10^{-13} \ll 6.7 \times 10^{-4}$, the approximation is valid. The remaining $[Cu^{2+}] = 2.7 \times 10^{-13}$ M. |

**FOR PRACTICE 17.15**    You mix a 125.0-mL sample of a solution that is 0.0117 M in $NiCl_2$ with a 175.0-mL sample of a solution that is 0.250 M in $NH_3$. After the solution reaches equilibrium, what concentration of $Ni^{2+}(aq)$ remains?

# The Effect of Complex Ion Equilibria on Solubility

Recall from Section 17.5 that the solubility of an ionic compound with a basic anion increases with increasing acidity because the acid reacts with the anion and drives the reaction to the right. Similarly, *the solubility of an ionic compound containing a metal cation that forms complex ions increases in the presence of Lewis bases that complex with the cation*. The most common Lewis bases that increase the solubility of metal cations are $NH_3$, $CN^-$, and $OH^-$.

For example, silver chloride is only slightly soluble in pure water:

$$AgCl(s) \rightleftharpoons Ag^+(aq) + Cl^-(aq) \qquad K_{sp} = 1.77 \times 10^{-10}$$

However, adding ammonia increases its solubility dramatically because, as we saw previously in this section, the ammonia forms a complex ion with the silver cations:

$$Ag^+(aq) + 2\,NH_3(aq) \rightleftharpoons Ag(NH_3)_2^+(aq) \qquad K_f = 1.7 \times 10^7$$

The large value of $K_f$ significantly lowers the concentration of $Ag^+(aq)$ in solution and therefore drives the dissolution of $AgCl(s)$. The two previous reactions can be added together:

$$AgCl(s) \rightleftharpoons Ag^+(aq) + Cl^-(aq) \qquad K_{sp} = 1.77 \times 10^{-10}$$
$$\underline{Ag^+(aq) + 2\,NH_3(aq) \rightleftharpoons Ag(NH_3)_2^+(aq) \qquad\qquad K_f = 1.7 \times 10^7}$$
$$AgCl(s) + 2\,NH_3(aq) \rightleftharpoons Ag(NH_3)_2^+(aq) + Cl^-(aq) \qquad K = K_{sp} \times 3.0 \times 10^{-3}$$

As we learned in Section 15.3, the equilibrium constant for a reaction that is the sum of two other reactions is the product of the equilibrium constants for the two other reactions. Adding ammonia changes the equilibrium constant for the dissolution of $AgCl(s)$ by a factor of $3.0 \times 10^{-3}/1.77 \times 10^{-10} = 1.7 \times 10^7$ (17 million), which makes the otherwise relatively insoluble $AgCl(s)$ quite soluble, as shown in Figure 17.18▼.

## Complex Ion Formation

$$2\,NH_3(aq) + AgCl(s) \rightleftharpoons Ag(NH_3)_2^+(aq) + Cl^-(aq)$$

◄ **FIGURE 17.18 Complex Ion Formation** Normally insoluble AgCl is made soluble by the addition of $NH_3$, which forms a complex ion with $Ag^+$ and dissolves the AgCl.

**Solubility and Complex Ion Equilibria** Which compound, when added to water, is most likely to increase the solubility of CuS?

**(a)** NaCl  **(b)** $KNO_3$  **(c)** NaCN  **(d)** $MgBr_2$

## The Solubility of Amphoteric Metal Hydroxides

Many metal hydroxides are insoluble or only very slightly soluble in pH-neutral water. For example, $Al(OH)_3$ has $K_{sp} = 1.3 \times 10^{-33}$, which means that if we put $Al(OH)_3$ in water, the vast majority of it will settle to the bottom as an undissolved solid. All metal hydroxides, however, have a basic anion ($OH^-$) and therefore become more soluble in acidic solutions (see the previous subsection and Section 17.5). The metal hydroxides become more soluble because they can act as bases and react with $H_3O^+(aq)$. For example, $Al(OH)_3$ dissolves in acid according to the reaction:

$$Al(OH)_3(s) + 3\,H_3O^+(aq) \longrightarrow Al^{3+}(aq) + 6\,H_2O(l)$$

$Al(OH)_3$ acts as a base in this reaction.

> Recall from Section 16.3 that a substance that can act as either an acid or a base is said to be amphoteric.

Interestingly, some metal hydroxides can also act as acids—they are *amphoteric*. The ability of an amphoteric metal hydroxide to act as an acid increases its solubility in basic solution. For example, $Al(OH)_3(s)$ dissolves in basic solution according to the reaction:

$$Al(OH)_3(s) + OH^-(aq) \longrightarrow Al(OH)_4{}^-(aq)$$

$Al(OH)_3$ acts as an acid in this reaction.

$Al(OH)_3$ is soluble at high pH and soluble at low pH but *insoluble* in a pH-neutral solution.

We can observe the whole range of the pH-dependent solubility behavior of $Al^{3+}$ by considering a hydrated aluminum ion in solution, beginning at an acidic pH. We know from Section 16.8 that $Al^{3+}$ in solution is inherently acidic because it complexes with water to form $Al(H_2O)_6{}^{3+}(aq)$. The complex ion then acts as an acid by losing a proton from one of the complexed water molecules according to the reaction:

$$Al(H_2O)_6{}^{3+}(aq) + H_2O(l) \rightleftharpoons Al(H_2O)_5(OH)^{2+}(aq) + H_3O^+(aq)$$

Addition of base to the solution drives the reaction to the right and continues to remove protons from complexed water molecules:

$$Al(H_2O)_5(OH)^{2+}(aq) + OH^-(aq) \rightleftharpoons Al(H_2O)_4(OH)_2{}^+(aq) + H_2O(l)$$

$$Al(H_2O)_4(OH)_2{}^+(aq) + OH^-(aq) \rightleftharpoons Al(H_2O)_3(OH)_3(s) + H_2O(l)$$
equivalent to $Al(OH)_3(s)$

The result of removing three protons from $Al(H_2O)_6{}^{3+}(aq)$ is the solid white precipitate $Al(H_2O)_3(OH)_3(s)$, which is more commonly written as $Al(OH)_3(s)$. The solution is now pH-neutral and the hydroxide is insoluble. Addition of more $OH^-$ makes the solution basic and dissolves the solid precipitate:

$$Al(H_2O)_3(OH)_3(s) + OH^-(aq) \rightleftharpoons Al(H_2O)_2(OH)_4{}^-(aq) + H_2O(l)$$

As the solution goes from acidic to neutral to basic, the solubility of $Al^{3+}$ changes accordingly, as illustrated in Figure 17.19▶.

The extent to which a metal hydroxide dissolves in both acid and base depends on the degree to which it is amphoteric. Cations that form amphoteric hydroxides include $Al^{3+}$, $Cr^{3+}$, $Zn^{2+}$, $Pb^{2+}$, and $Sn^{2+}$. Other metal hydroxides, such as those of $Ca^{2+}$, $Fe^{2+}$, and $Fe^{3+}$, are not amphoteric—they become soluble in acidic solutions, but not in basic ones.

# pH-Dependent Solubility of an Amphoteric Hydroxide

Acidic

$Al(H_2O)_6^{3+}(aq)$

Basic

$Al(H_2O)_2(OH)_4^-(aq)$

pH-neutral

$Al(H_2O)_3(OH)_3(s)$
or $Al(OH)_3(s)$

▲ **FIGURE 17.19  Solubility of an Amphoteric Hydroxide**
Because aluminum hydroxide is amphoteric, its solubility is pH-dependent. At low pH, the formation of $Al(H_2O)_6^{3+}$ drives the dissolution. At neutral pH, insoluble $Al(OH)_3$ precipitates out of solution. At high pH, the formation of $Al(H_2O)_2(OH)_4^-$ drives the dissolution.

# Self-Assessment Quiz

**Q1.** A buffer is 0.100 M in $NH_4Cl$ and 0.100 M in $NH_3$. When a small amount of hydrobromic acid is added to this buffer, which buffer component neutralizes the added acid?

   a) $NH_4^+$      b) $Cl^-$      c) $NH_3$
   d) None of the above (hydrobromic acid will not be neutralized by this buffer).

**Q2.** What is the pH of a buffer that is 0.120 M in formic acid $(HCHO_2)$ and 0.080 M in potassium formate $(KCHO_2)$? For formic acid, $K_a = 1.8 \times 10^{-4}$.

   a) 2.33      b) 3.57      c) 3.74      d) 3.91

**Q3.** A buffer with a pH of 9.85 contains $CH_3NH_2$ and $CH_3NH_3Cl$ in water. What can you conclude about the relative concentrations of $CH_3NH_2$ and $CH_3NH_3Cl$ in this buffer? For $CH_3NH_2$, $pK_b = 3.36$.

   a) $CH_3NH_2 > CH_3NH_3Cl$
   b) $CH_3NH_2 < CH_3NH_3Cl$
   c) $CH_3NH_2 = CH_3NH_3Cl$
   d) Nothing can be concluded about the relative concentrations of $CH_3NH_2$ and $CH_3NH_3Cl$.

**Q4.** A 500.0-mL buffer solution is 0.10 M in benzoic acid and 0.10 M in sodium benzoate and has an initial pH of 4.19. What is the pH of the buffer upon addition of 0.010 mol of NaOH?

   a) 1.70      b) 4.01      c) 4.29      d) 4.37

**Q5.** Consider a buffer composed of the weak acid HA and its conjugate base $A^-$. Which pair of concentrations results in the most effective buffer?

   a) 0.10 M HA; 0.10 M $A^-$
   b) 0.50 M HA; 0.50 M $A^-$
   c) 0.90 M HA; 0.10 M $A^-$
   d) 0.10 M HA; 0.90 M $A^-$

**Q6.** Which combination is the best choice to prepare a buffer with a pH of 9.0?

   a) $NH_3$; $NH_4Cl$ ($pK_b$ for $NH_3$ is 4.75)
   b) $C_5H_5N$; $C_5H_5NHCl$ ($pK_b$ for $C_5H_5N$ is 8.76)
   c) $HNO_2$; $NaNO_2$ ($pK_a$ for $HNO_2$ is 3.33)
   d) $HCHO_2$; $NaCHO_2$ ($pK_a$ for $HCHO_2$ is 3.74)

**Q7.** A 25.0-mL sample of an unknown HBr solution is titrated with 0.100 M NaOH. The equivalence point is reached upon the addition of 18.88 mL of the base. What is the concentration of the HBr solution?

   a) 0.0755 M          b) 0.0376 M
   c) 0.100 M           d) 0.00188 M

**Q8.** A 10.0-mL sample of 0.200 M hydrocyanic acid (HCN) is titrated with 0.0998 M NaOH. What is the pH at the equivalence point? For hydrocyanic acid, $pK_a = 9.31$.

   a) 7.00      b) 8.76      c) 9.31      d) 11.07

**Q9.** A 20.0-mL sample of 0.150 M ethylamine is titrated with 0.0981 M HCl. What is the pH after the addition of 5.0 mL of HCl? For ethylamine, $pK_b = 3.25$.

   a) 10.75      b) 11.04      c) 2.96      d) 11.46

*Continued—*

**Q10.** Three 15.0-mL acid samples—0.10 M HA, 0.10 M HB, and 0.10 M $H_2C$—are all titrated with 0.100 M NaOH. If HA is a weak acid, HB is a strong acid, and $H_2C$ is a diprotic acid, which statement is true of all three titrations?

a) All three titrations have the same pH at the first equivalence point.
b) All three titrations have the same initial pH.
c) All three titrations have the same final pH.
d) All three titrations require the same volume of NaOH to reach the first equivalence point.

**Q11.** A weak unknown monoprotic acid is titrated with a strong base. The titration curve is shown. Find $K_a$ for the unknown acid.

a) $2.5 \times 10^{-3}$    b) $3.2 \times 10^{-5}$
c) $3.2 \times 10^{-7}$    d) $2.5 \times 10^{-9}$

**Q12.** Calculate the molar solubility of lead(II) bromide ($PbBr_2$). For lead(II) bromide, $K_{sp} = 4.67 \times 10^{-6}$.

a) 0.00153 M
b) 0.0105 M
c) 0.0167 M
d) 0.0211 M

**Q13.** Calculate the molar solubility of magnesium fluoride ($MgF_2$) in a solution that is 0.250 M in NaF. For magnesium fluoride, $K_{sp} = 5.16 \times 10^{-11}$.

a) $2.35 \times 10^{-4}$ M
b) $2.06 \times 10^{-10}$ M
c) $2.87 \times 10^{-5}$ M
d) $8.26 \times 10^{-10}$ M

**Q14.** A solution is 0.025 M in $Pb^{2+}$. What minimum concentration of $Cl^-$ is required to begin to precipitate $PbCl_2$? For $PbCl_2$, $K_{sp} = 1.17 \times 10^{-5}$.

a) $1.17 \times 10^{-5}$ M
b) 0.0108 M
c) 0.0216 M
d) $5.41 \times 10^{-4}$ M

**Q15.** Which compound is more soluble in an acidic solution than in a neutral solution?

a) $PbBr_2$    b) CuCl
c) AgI    d) $BaF_2$

**Answers: 1.** (c) **2.** (b) **3.** (b) **4.** (d) **5.** (b) **6.** (a) **7.** (a) **8.** (d) **9.** (d) **10.** (d) **11.** (b) **12.** (b) **13.** (d) **14.** (c) **15.** (d)

# Chapter 17 IN REVIEW

## Key Terms

**Section 17.2**
buffer (780)
common ion effect (782)
Henderson–Hasselbalch equation (784)

**Section 17.3**
buffer capacity (794)

**Section 17.4**
acid–base titration (795)

indicator (795)
equivalence point (795)
endpoint (806)

**Section 17.5**
solubility product constant ($K_{sp}$) (809)
molar solubility (809)

**Section 17.6**
selective precipitation (816)

**Section 17.7**
qualitative analysis (818)
quantitative analysis (818)

**Section 17.8**
complex ion (821)
ligand (821)
formation constant ($K_f$) (821)

## Key Concepts

### The Danger of Antifreeze (17.1)

▶ Although buffers closely regulate the pH of mammalian blood, the capacity of these buffers to neutralize can be overwhelmed.

▶ Ethylene glycol, the main component of antifreeze, is metabolized by the liver into glycolic acid. The resulting acidity can exceed the buffering capacity of blood and cause acidosis, a serious condition that results in oxygen deprivation.

### Buffers: Solutions That Resist pH Change (17.2)

▶ Buffers contain significant amounts of both a weak acid and its conjugate base (or a weak base and its conjugate acid), enabling the buffer to neutralize added acid or added base.

▶ Adding a small amount of acid to a buffer converts a stoichiometric amount of base to the conjugate acid. Adding a small amount of base to a buffer converts a stoichiometric amount of the acid to the conjugate base.

▶ We can determine the pH of a buffer solution by solving an equilibrium problem, focusing on the common ion effect, or by using the Henderson–Hasselbalch equation.

## Buffer Range and Buffer Capacity (17.3)

▶ A buffer works best when the amounts of acid and conjugate base it contains are large and approximately equal.

▶ If the relative amounts of acid and base in a buffer differ by more than a factor of 10, the ability of the buffer to neutralize added acid and added base diminishes. The maximum pH range at which a buffer is effective is one pH unit on either side of the acid's $pK_a$.

## Titrations and pH Curves (17.4)

▶ A titration curve is a graph of the change in pH versus added volume of acid or base during a titration.

▶ This chapter examines three types of titration curves, representing three types of acid–base reactions: a strong acid with a strong base, a weak acid with a strong base (or vice versa), and a polyprotic acid with a strong base.

▶ The equivalence point of a titration can be made visible by an indicator, a compound that changes color over a specific pH range.

## Solubility Equilibria and the Solubility Product Constant (17.5)

▶ The solubility product constant ($K_{sp}$) is an equilibrium constant for the dissolution of an ionic compound in water.

▶ We can determine the molar solubility of an ionic compound from $K_{sp}$ and vice versa. Although the value of $K_{sp}$ is constant at a given temperature, the solubility of an ionic substance can depend on other factors such as the presence of common ions and the pH of the solution.

## Precipitation (17.6)

▶ We can compare the magnitude of $K_{sp}$ to the reaction quotient, $Q$, in order to determine the relative saturation of a solution.

▶ Substances with cations that have sufficiently different values of $K_{sp}$ can be separated by selective precipitation, in which an added reagent forms a precipitate with one of the dissolved cations but not others.

## Qualitative Chemical Analysis (17.7)

▶ Qualitative analysis operates on the principle that a mixture of cations can be separated and analyzed based on the differences in the solubilities of their salts.

▶ In a classic qualitative analysis scheme, an unknown mixture of cations is sequentially treated with different reagents, each of which precipitates a known subgroup of cations.

## Complex Ion Equilibria (17.8)

▶ A complex ion contains a central metal ion coordinated to two or more ligands.

▶ The equilibrium constant for the formation of a complex ion is called a formation constant and is usually quite large.

▶ The solubility of an ionic compound containing a metal cation that forms complex ions increases in the presence of Lewis bases that complex with the cation because the formation of the complex ion drives the dissolution reaction to the right.

▶ All metal hydroxides become more soluble in the presence of acids, but amphoteric metal hydroxides also become more soluble in the presence of bases.

# Key Equations and Relationships

The Henderson–Hasselbalch Equation (**17.2**)

$$pH = pK_a + \log\frac{[\text{base}]}{[\text{acid}]}$$

Effective Buffer Range (**17.3**)

$$pH = \text{range} = pK_a \pm 1$$

The Relation between $Q$ and $K_{sp}$ (**17.6**)

If $Q < K_{sp}$, the solution is unsaturated. More of the solid ionic compound can dissolve in the solution.

If $Q = K_{sp}$, the solution is saturated. The solution is holding the equilibrium amount of the dissolved ions, and additional solid will not dissolve in the solution.

If $Q > K_{sp}$, the solution is supersaturated. Under most circumstances, the solid will precipitate out of a supersaturated solution.

# Key Learning Outcomes

| CHAPTER OBJECTIVES | ASSESSMENT |
|---|---|
| Calculating the pH of a Buffer Solution (**17.2**) | Example 17.1   For Practice 17.1   For More Practice 17.1 Exercises 29, 30, 33, 34 |
| Using the Henderson–Hasselbalch Equation to Calculate the pH of a Buffer Solution (**17.2**) | Example 17.2   For Practice 17.2   Exercises 37–42 |
| Calculating the pH Change in a Buffer Solution after the Addition of a Small Amount of Strong Acid or Base (**17.2**) | Example 17.3   For Practice 17.3   For More Practice 17.3 Exercises 47–50 |
| Using the Henderson–Hasselbalch Equation to Calculate the pH of a Buffer Solution Composed of a Weak Base and Its Conjugate Acid (**17.2**) | Example 17.4   For Practice 17.4   For More Practice 17.4 Exercises 37–42 |
| Preparing a Buffer (**17.3**) | Example 17.5   For Practice 17.5   Exercises 57–58 |

| | |
|---|---|
| Strong Acid–Strong Base Titration pH Curve (**17.4**)  | Example 17.6    For Practice 17.6    Exercises 67–70 |
| Weak Acid–Strong Base Titration pH Curve (**17.4**) | Example 17.7    For Practice 17.7    Exercises 65, 71–72, 75, 77–78 |
| Calculating Molar Solubility from $K_{sp}$ (**17.5**) | Example 17.8    For Practice 17.8    Exercises 87–88 |
| Calculating $K_{sp}$ from Molar Solubility (**17.5**) | Example 17.9    For Practice 17.9    Exercises 89–90, 92, 94 |
| Calculating Molar Solubility in the Presence of a Common Ion (**17.5**) | Example 17.10    For Practice 17.10    Exercises 95–96 |
| Determining the Effect of pH on Solubility (**17.5**) | Example 17.11    For Practice 17.11    Exercises 97–100 |
| Predicting Precipitation Reactions by Comparing $Q$ and $K_{sp}$ (**17.6**)  | Example 17.12    For Practice 17.12    Exercises 101–104 |
| Finding the Minimum Required Reagent Concentration for Selective Precipitation (**17.6**) | Example 17.13    For Practice 17.13    Exercises 105–106 |
| Finding the Concentrations of Ions Left in Solution after Selective Precipitation (**17.6**) | Example 17.14    For Practice 17.14    Exercises 107–108 |
| Working with Complex Ion Equilibria (**17.8**) | Example 17.15    For Practice 17.15    Exercises 109–112 |

# EXERCISES

## Review Questions

1. What is the pH range of human blood? How is human blood maintained in this pH range?

2. What is a buffer? How does a buffer work? How does it neutralize added acid? Added base?

3. What is the common ion effect?

4. What is the Henderson–Hasselbalch equation, and why is it useful?

5. What is the pH of a buffer solution when the concentrations of both buffer components (the weak acid and its conjugate base) are equal? What happens to the pH when the buffer contains more of the weak acid than the conjugate base? More of the conjugate base than the weak acid?

**6.** Suppose that a buffer contains equal amounts of a weak acid and its conjugate base. What happens to the relative amounts of the weak acid and conjugate base when a small amount of strong acid is added to the buffer? What happens when a small amount of strong base is added?

**7.** How do you use the Henderson–Hasselbalch equation to calculate the pH of a buffer containing a base and its conjugate acid? Specifically, how do you determine the correct value for $pK_a$?

**8.** What factors influence the effectiveness of a buffer? What are the characteristics of an effective buffer?

**9.** What is the effective pH range of a buffer (relative to the $pK_a$ of the weak acid component)?

**10.** Describe acid–base titration. What is the equivalence point?

**11.** The pH at the equivalence point of the titration of a strong acid with a strong base is 7.0. However, the pH at the equivalence point of the titration of a *weak* acid with a strong base is above 7.0. Explain.

**12.** The volume required to reach the equivalence point of an acid–base titration depends on the volume and concentration of the acid or base to be titrated and on the concentration of the acid or base used to do the titration. It does not, however, depend on whether or not the acid or base being titrated is strong or weak. Explain.

**13.** In the titration of a strong acid with a strong base, how do you calculate these quantities?
  **a.** initial pH
  **b.** pH before the equivalence point
  **c.** pH at the equivalence point
  **d.** pH beyond the equivalence point

**14.** In the titration of a weak acid with a strong base, how do you calculate these quantities?
  **a.** initial pH
  **b.** pH before the equivalence point
  **c.** pH at one-half the equivalence point
  **d.** pH at the equivalence point
  **e.** pH beyond the equivalence point

**15.** The titration of a polyprotic acid with sufficiently different $pK_a$s displays two equivalence points. Why?

**16.** In the titration of a polyprotic acid, the volume required to reach the first equivalence point is identical to the volume required to reach the second one. Why?

**17.** What is the difference between the endpoint and the equivalence point in a titration?

**18.** What is an indicator? How can an indicator signal the equivalence point of a titration?

**19.** What is the solubility product constant? Write a general expression for the solubility constant of a compound with the general formula $A_mX_n$.

**20.** What is molar solubility? How can you obtain the molar solubility of a compound from $K_{sp}$?

**21.** How does a common ion affect the solubility of a compound? More specifically, how is the solubility of a compound with the general formula AX different in a solution containing one of the common ions ($A^+$ or $X^-$) than it is in pure water? Explain.

**22.** How is the solubility of an ionic compound with a basic anion affected by pH? Explain.

**23.** For a given solution containing an ionic compound, what is the relationship between $Q$, $K_{sp}$, and the relative saturation of the solution?

**24.** What is selective precipitation? Under which conditions does selective precipitation occur?

**25.** What is qualitative analysis? How does *qualitative* analysis differ from *quantitative* analysis?

**26.** What are the main groups in the general qualitative analysis scheme described in this chapter? Describe the steps and reagents necessary to identify each group.

# Problems by Topic

## The Common Ion Effect and Buffers

**27.** In which of these solutions will $HNO_2$ ionize less than it does in pure water?
  **a.** 0.10 M NaCl
  **b.** 0.10 M $KNO_3$
  **c.** 0.10 M NaOH
  **d.** 0.10 M $NaNO_2$

**28.** A formic acid solution has a pH of 3.25. Which of these substances will raise the pH of the solution upon addition? Explain your answer.
  **a.** HCl            **b.** NaBr
  **c.** $NaCHO_2$      **d.** KCl

**29.** Solve an equilibrium problem (using an ICE table) to calculate the pH of each solution.
  **a.** a solution that is 0.20 M in $HCHO_2$ and 0.15 M in $NaCHO_2$
  **b.** a solution that is 0.16 M in $NH_3$ and 0.22 M in $NH_4Cl$

**30.** Solve an equilibrium problem (using an ICE table) to calculate the pH of each solution.
  **a.** a solution that is 0.195 M in $HC_2H_3O_2$ and 0.125 M in $KC_2H_3O_2$
  **b.** a solution that is 0.255 M in $CH_3NH_2$ and 0.135 M in $CH_3NH_3Br$

**31.** Calculate the percent ionization of a 0.15 M benzoic acid solution in pure water and in a solution containing 0.10 M sodium benzoate. Why does the percent ionization differ significantly in the two solutions?

**32.** Calculate the percent ionization of a 0.13 M formic acid solution in pure water and also in a solution containing 0.11 M potassium formate. Explain the difference in percent ionization in the two solutions.

**33.** Solve an equilibrium problem (using an ICE table) to calculate the pH of each solution.
  **a.** 0.15 M HF
  **b.** 0.15 M NaF
  **c.** a mixture that is 0.15 M in HF and 0.15 M in NaF

**34.** Solve an equilibrium problem (using an ICE table) to calculate the pH of each solution.
  **a.** 0.18 M $CH_3NH_2$
  **b.** 0.18 M $CH_3NH_3Cl$
  **c.** a mixture that is 0.18 M in $CH_3NH_2$ and 0.18 M in $CH_3NH_3Cl$

**35.** A buffer contains significant amounts of acetic acid and sodium acetate. Write equations showing how this buffer neutralizes added acid and added base.

**36.** A buffer contains significant amounts of ammonia and ammonium chloride. Write equations showing how this buffer neutralizes added acid and added base.

**37.** Use the Henderson–Hasselbalch equation to calculate the pH of each solution in Problem 29.

**38.** Use the Henderson–Hasselbalch equation to calculate the pH of each solution in Problem 30.

**39.** Use the Henderson–Hasselbalch equation to calculate the pH of each solution.
  **a.** a solution that is 0.135 M in HClO and 0.155 M in KClO
  **b.** a solution that contains 1.05% $C_2H_5NH_2$ by mass and 1.10% $C_2H_5NH_3Br$ by mass
  **c.** a solution that contains 10.0 g of $HC_2H_3O_2$ and 10.0 g of $NaC_2H_3O_2$ in 150.0 mL of solution

**40.** Use the Henderson–Hasselbalch equation to calculate the pH of each solution.
  **a.** a solution that is 0.145 M in propanoic acid and 0.115 M in potassium propanoate
  **b.** a solution that contains 0.785% $C_5H_5N$ by mass and 0.985% $C_5H_5NHCl$ by mass
  **c.** a solution that contains 15.0 g of HF and 25.0 g of NaF in 125 mL of solution

**41.** Calculate the pH of the solution that results from each mixture.
  **a.** 50.0 mL of 0.15 M $HCHO_2$ with 75.0 mL of 0.13 M $NaCHO_2$
  **b.** 125.0 mL of 0.10 M $NH_3$ with 250.0 mL of 0.10 M $NH_4Cl$

**42.** Calculate the pH of the solution that results from each mixture.
  **a.** 150.0 mL of 0.25 M HF with 225.0 mL of 0.30 M NaF
  **b.** 175.0 mL of 0.10 M $C_2H_5NH_2$ with 275.0 mL of 0.20 M $C_2H_5NH_3Cl$

**43.** Calculate the ratio of NaF to HF required to create a buffer with pH = 4.00.

**44.** Calculate the ratio of $CH_3NH_2$ to $CH_3NH_3Cl$ concentration required to create a buffer with pH = 10.24.

**45.** What mass of sodium benzoate should you add to 150.0 mL of a 0.15 M benzoic acid solution to obtain a buffer with a pH of 4.25? (Assume no volume change.)

**46.** What mass of ammonium chloride should you add to 2.55 L of a 0.155 M $NH_3$ to obtain a buffer with a pH of 9.55? (Assume no volume change.)

**47.** A 250.0-mL buffer solution is 0.250 M in acetic acid and 0.250 M in sodium acetate.
  **a.** What is the initial pH of this solution?
  **b.** What is the pH after addition of 0.0050 mol of HCl?
  **c.** What is the pH after addition of 0.0050 mol of NaOH?

**48.** A 100.0-mL buffer solution is 0.175 M in HClO and 0.150 M in NaClO.
  **a.** What is the initial pH of this solution?
  **b.** What is the pH after addition of 150.0 mg of HBr?
  **c.** What is the pH after addition of 85.0 mg of NaOH?

**49.** For each solution, calculate the initial and final pH after adding 0.010 mol of HCl.
  **a.** 500.0 mL of pure water
  **b.** 500.0 mL of a buffer solution that is 0.125 M in $HC_2H_3O_2$ and 0.115 M in $NaC_2H_3O_2$
  **c.** 500.0 mL of a buffer solution that is 0.155 M in $C_2H_5NH_2$ and 0.145 M in $C_2H_5NH_3Cl$

**50.** For each solution, calculate the initial and final pH after adding 0.010 mol of NaOH.
  **a.** 250.0 mL of pure water
  **b.** 250.0 mL of a buffer solution that is 0.195 M in $HCHO_2$ and 0.275 M in $KCHO_2$
  **c.** 250.0 mL of a buffer solution that is 0.255 M in $CH_3CH_2NH_2$ and 0.235 M in $CH_3CH_2NH_3Cl$

**51.** A 350.0-mL buffer solution is 0.150 M in HF and 0.150 M in NaF. What mass of NaOH can this buffer neutralize before the pH rises above 4.00? If the same volume of the buffer were 0.350 M in HF and 0.350 M in NaF, what mass of NaOH could be handled before the pH rises above 4.00?

**52.** A 100.0-mL buffer solution is 0.100 M in $NH_3$ and 0.125 M in $NH_4Br$. What mass of HCl can this buffer neutralize before the pH falls below 9.00? If the same volume of the buffer were 0.250 M in $NH_3$ and 0.400 M in $NH_4Br$, what mass of HCl could be handled before the pH fell below 9.00?

**53.** Determine whether or not the mixing of each pair of solutions results in a buffer.
  **a.** 100.0 mL of 0.10 M $NH_3$; 100.0 mL of 0.15 M $NH_4Cl$
  **b.** 50.0 mL of 0.10 M HCl; 35.0 mL of 0.150 M NaOH
  **c.** 50.0 mL of 0.15 M HF; 20.0 mL of 0.15 M NaOH
  **d.** 175.0 mL of 0.10 M $NH_3$; 150.0 mL of 0.12 M NaOH
  **e.** 125.0 mL of 0.15 M $NH_3$; 150.0 mL of 0.20 M NaOH

**54.** Determine whether or not the mixing of each pair of solutions results in a buffer.
  **a.** 75.0 mL of 0.10 M HF; 55.0 mL of 0.15 M NaF
  **b.** 150.0 mL of 0.10 M HF; 135.0 mL of 0.175 M HCl
  **c.** 165.0 mL of 0.10 M HF; 135.0 mL of 0.050 M KOH
  **d.** 125.0 mL of 0.15 M $CH_3NH_2$; 120.0 mL of 0.25 M $CH_3NH_3Cl$
  **e.** 105.0 mL of 0.15 M $CH_3NH_2$; 95.0 mL of 0.10 M HCl

**55.** Blood is buffered by carbonic acid and the bicarbonate ion. Normal blood plasma is 0.024 M in $HCO_3^-$ and 0.0012 M $H_2CO_3$ ($pK_{a_1}$ for $H_2CO_3$ at body temperature is 6.1).
  **a.** What is the pH of blood plasma?
  **b.** If the volume of blood in a normal adult is 5.0 L, what mass of HCl can be neutralized by the buffering system in blood before the pH falls below 7.0 (which would result in death)?
  **c.** Given the volume from part (b), what mass of NaOH can be neutralized before the pH rises above 7.8?

**56.** The fluids within cells are buffered by $H_2PO_4^-$ and $HPO_4^{2-}$.
  **a.** Calculate the ratio of $HPO_4^{2-}$ to $H_2PO_4^-$ required to maintain a pH of 7.1 within a cell.
  **b.** Could a buffer system employing $H_3PO_4$ as the weak acid and $H_2PO_4^-$ as the weak base be used as a buffer system within cells? Explain.

**57.** Which buffer system is the best choice to create a buffer with pH = 7.20? For the best system, calculate the ratio of the masses of the buffer components required to make the buffer.

$HC_2H_3O_2/KC_2H_3O_2$          $HClO_2/KClO_2$
$NH_3/NH_4Cl$                    $HClO/KClO$

**58.** Which buffer system is the best choice to create a buffer with pH = 9.00? For the best system, calculate the ratio of the masses of the buffer components required to make the buffer.

HF/KF                           $HNO_2/KNO_2$
$NH_3/NH_4Cl$                   HClO/KClO

**59.** A 500.0-mL buffer solution is 0.100 M in $HNO_2$ and 0.150 M in $KNO_2$. Determine if each addition would exceed the capacity of the buffer to neutralize it.
  **a.** 250 mg NaOH        **b.** 350 mg KOH
  **c.** 1.25 g HBr          **d.** 1.35 g HI

**60.** A 1.0-L buffer solution is 0.125 M in $HNO_2$ and 0.145 M in $NaNO_2$. Determine the concentrations of $HNO_2$ and $NaNO_2$ after the addition of each substance:
  **a.** 1.5 g HCl    **b.** 1.5 g NaOH    **c.** 1.5 g HI

## Titrations, pH Curves, and Indicators

**61.** The graphs labeled (a) and (b) show the titration curves for two equal-volume samples of monoprotic acids, one weak and one strong. Both titrations were carried out with the same concentration of strong base.

**(a)**

**(b)**

  **(i)** What is the approximate pH at the equivalence point of each curve?
  **(ii)** Which graph corresponds to the titration of the strong acid and which one to the titration of the weak acid?

**62.** Two 25.0-mL samples, one 0.100 M HCl and the other 0.100 M HF, are titrated with 0.200 M KOH.
  **a.** What is the volume of added base at the equivalence point for each titration?
  **b.** Is the pH at the equivalence point for each titration acidic, basic, or neutral?
  **c.** Which titration curve has the lower initial pH?
  **d.** Sketch each titration curve.

**63.** Two 20.0-mL samples, one 0.200 M KOH and the other 0.200 M $CH_3NH_2$, are titrated with 0.100 M HI.
  **a.** What is the volume of added acid at the equivalence point for each titration?
  **b.** Is the pH at the equivalence point for each titration acidic, basic, or neutral?
  **c.** Which titration curve has the lower initial pH?
  **d.** Sketch each titration curve.

**64.** The graphs labeled (a) and (b) show the titration curves for two equal-volume samples of bases, one weak and one strong. Both titrations were carried out with the same concentration of strong acid.

**(a)**

**(b)**

  **(i)** What is the approximate pH at the equivalence point of each curve?
  **(ii)** Which graph corresponds to the titration of the strong base and which one to the weak base?

**65.** Consider the curve shown here for the titration of a weak monoprotic acid with a strong base and answer each question.

  **a.** What is the pH and what is the volume of added base at the equivalence point?
  **b.** At what volume of added base is the pH calculated by working an equilibrium problem based on the initial concentration and $K_a$ of the weak acid?
  **c.** At what volume of added base does pH = $pK_a$?
  **d.** At what volume of added base is the pH calculated by working an equilibrium problem based on the concentration and $K_b$ of the conjugate base?
  **e.** Beyond what volume of added base is the pH calculated by focusing on the amount of excess strong base added?

**66.** Consider the curve shown here for the titration of a weak base with a strong acid and answer each question.

  **a.** What is the pH and what is the volume of added acid at the equivalence point?
  **b.** At what volume of added acid is the pH calculated by working an equilibrium problem based on the initial concentration and $K_b$ of the weak base?
  **c.** At what volume of added acid does pH = $14 - pK_b$?
  **d.** At what volume of added acid is the pH calculated by working an equilibrium problem based on the concentration and $K_a$ of the conjugate acid?
  **e.** Beyond what volume of added acid is the pH calculated by focusing on the amount of excess strong acid added?

**67.** Consider the titration of a 35.0-mL sample of 0.175 M HBr with 0.200 M KOH. Determine each quantity.
  **a.** the initial pH
  **b.** the volume of added base required to reach the equivalence point
  **c.** the pH at 10.0 mL of added base
  **d.** the pH at the equivalence point
  **e.** the pH after adding 5.0 mL of base beyond the equivalence point

**68.** A 20.0-mL sample of 0.125 M $HNO_3$ is titrated with 0.150 M NaOH. Calculate the pH for at least five different points throughout the titration curve and sketch the curve. Indicate the volume at the equivalence point on your graph.

**69.** Consider the titration of a 25.0-mL sample of 0.115 M RbOH with 0.100 M HCl. Determine each quantity.
   **a.** the initial pH
   **b.** the volume of added acid required to reach the equivalence point
   **c.** the pH at 5.0 mL of added acid
   **d.** the pH at the equivalence point
   **e.** the pH after adding 5.0 mL of acid beyond the equivalence point

**70.** A 15.0-mL sample of 0.100 M $Ba(OH)_2$ is titrated with 0.125 M HCl. Calculate the pH for at least five different points throughout the titration curve and sketch the curve. Indicate the volume at the equivalence point on your graph.

**71.** Consider the titration of a 20.0-mL sample of 0.105 M $HC_2H_3O_2$ with 0.125 M NaOH. Determine each quantity.
   **a.** the initial pH
   **b.** the volume of added base required to reach the equivalence point
   **c.** the pH at 5.0 mL of added base
   **d.** the pH at one-half of the equivalence point
   **e.** the pH at the equivalence point
   **f.** the pH after adding 5.0 mL of base beyond the equivalence point

**72.** A 30.0-mL sample of 0.165 M propanoic acid is titrated with 0.300 M KOH. Calculate the pH at each volume of added base: 0 mL, 5 mL, 10 mL, equivalence point, one-half equivalence point, 20 mL, 25 mL. Sketch the titration curve.

**73.** Consider the titration of a 25.0-mL sample of 0.175 M $CH_3NH_2$ with 0.150 M HBr. Determine each quantity.
   **a.** the initial pH
   **b.** the volume of added acid required to reach the equivalence point
   **c.** the pH at 5.0 mL of added acid
   **d.** the pH at one-half of the equivalence point
   **e.** the pH at the equivalence point
   **f.** the pH after adding 5.0 mL of acid beyond the equivalence point

**74.** A 25.0-mL sample of 0.125 M pyridine is titrated with 0.100 M HCl. Calculate the pH at each volume of added acid: 0 mL, 10 mL, 20 mL, equivalence point, one-half equivalence point, 40 mL, 50 mL. Sketch the titration curve.

**75.** Consider the titration curves (labeled a and b) for two weak acids, both titrated with 0.100 M NaOH.

**(a)**

**(b)**

   **(i)** Which acid solution is more concentrated?
   **(ii)** Which acid has the larger $K_a$?

**76.** Consider the titration curves (labeled a and b) for two weak bases, both titrated with 0.100 M HCl.

**(a)**

**(b)**

   **(i)** Which base solution is more concentrated?
   **(ii)** Which base has the larger $K_b$?

**77.** A 0.229-g sample of an unknown monoprotic acid is titrated with 0.112 M NaOH. The resulting titration curve is shown here. Determine the molar mass and $pK_a$ of the acid.

**78.** A 0.446-g sample of an unknown monoprotic acid is titrated with 0.105 M KOH. The resulting titration curve is shown here. Determine the molar mass and $pK_a$ of the acid.

**79.** A 20.0-mL sample of 0.115 M sulfurous acid ($H_2SO_3$) solution is titrated with 0.1014 M KOH. At what added volume of base solution does each equivalence point occur?

**80.** A 20.0-mL sample of a 0.125 M diprotic acid ($H_2A$) solution is titrated with 0.1019 M KOH. The acid ionization constants for the acid are $K_{a_1} = 5.2 \times 10^{-5}$ and $K_{a_2} = 3.4 \times 10^{-10}$. At what added volume of base does each equivalence point occur?

**81.** Methyl red has a $pK_a$ of 5.0 and is red in its acid form and yellow in its basic form. If several drops of this indicator are placed in a 25.0-mL sample of 0.100 M HCl, what color will the solution appear? If 0.100 M NaOH is slowly added to the HCl sample, in what pH range will the indicator change color?

**82.** Phenolphthalein has a $pK_a$ of 9.7. It is colorless in its acid form and pink in its basic form. For each of the values of pH, calculate $[In^-]/[HIn]$ and predict the color of a phenolphthalein solution.
   **a.** pH = 2.0
   **b.** pH = 5.0
   **c.** pH = 8.0
   **d.** pH = 11.0

83. Referring to Table 17.1, pick an indicator for use in the titration of each acid with a strong base.
    a. HF
    b. HCl
    c. HCN

84. Referring to Table 17.1, pick an indicator for use in the titration of each base with a strong acid.
    a. $CH_3NH_2$     b. NaOH     c. $C_6H_5NH_2$

## Solubility Equilibria

85. Write balanced equations and expressions for $K_{sp}$ for the dissolution of each ionic compound.
    a. $BaSO_4$     b. $PbBr_2$     c. $Ag_2CrO_4$

86. Write balanced equations and expressions for $K_{sp}$ for the dissolution of each ionic compound.
    a. $CaCO_3$     b. $PbCl_2$     c. $AgI$

87. Refer to the $K_{sp}$ values in Table 17.2 to calculate the molar solubility of each compound in pure water.
    a. $AgBr$     b. $Mg(OH)_2$     c. $CaF_2$

88. Refer to the $K_{sp}$ values in Table 17.2 to calculate the molar solubility of each compound in pure water.
    a. MX ($K_{sp} = 1.27 \times 10^{-36}$)
    b. $Ag_2CrO_4$
    c. $Ca(OH)_2$

89. Use the given molar solubilities in pure water to calculate $K_{sp}$ for each compound.
    a. MX; molar solubility $= 3.27 \times 10^{-11}$ M
    b. $PbF_2$; molar solubility $= 5.63 \times 10^{-3}$ M
    c. $MgF_2$; molar solubility $= 2.65 \times 10^{-4}$ M

90. Use the given molar solubilities in pure water to calculate $K_{sp}$ for each compound.
    a. $BaCrO_4$; molar solubility $= 1.08 \times 10^{-5}$ M
    b. $Ag_2SO_3$; molar solubility $= 1.55 \times 10^{-5}$ M
    c. $Pd(SCN)_2$; molar solubility $= 2.22 \times 10^{-8}$ M

91. Two compounds with general formulas AX and $AX_2$ have $K_{sp} = 1.5 \times 10^{-5}$. Which of the two compounds has the higher molar solubility?

92. Consider the compounds with the generic formulas listed and their corresponding molar solubilities in pure water. Which compound has the smallest value of $K_{sp}$?
    a. AX; molar solubility $= 1.35 \times 10^{-4}$ M
    b. $AX_2$; molar solubility $= 2.25 \times 10^{-4}$ M
    c. $A_2X$; molar solubility $= 1.75 \times 10^{-4}$ M

93. Refer to the $K_{sp}$ value from Table 17.2 to calculate the solubility of iron(II) hydroxide in pure water in grams per 100.0 mL of solution.

94. The solubility of copper(I) chloride is 3.91 mg per 100.0 mL of solution. Calculate $K_{sp}$ for CuCl.

95. Calculate the molar solubility of barium fluoride in each liquid or solution.
    a. pure water
    b. $0.10$ M $Ba(NO_3)_2$
    c. $0.15$ M NaF

96. Calculate the molar solubility of MX ($K_{sp} = 1.27 \times 10^{-36}$) in each liquid or solution.
    a. pure water
    b. $0.25$ M $MCl_2$
    c. $0.20$ M $Na_2X$

97. Calculate the molar solubility of calcium hydroxide in a solution buffered at each pH.
    a. pH $= 4$
    b. pH $= 7$
    c. pH $= 9$

98. Calculate the solubility (in grams per $1.00 \times 10^2$ mL of solution) of magnesium hydroxide in a solution buffered at pH $= 10$. How does this compare to the solubility of $Mg(OH)_2$ in pure water?

99. Determine if each compound is more soluble in acidic solution than it is in pure water. Explain.
    a. $BaCO_3$
    b. $CuS$
    c. $AgCl$
    d. $PbI_2$

100. Determine if each compound is more soluble in acidic solution than it is in pure water. Explain.
    a. $Hg_2Br_2$
    b. $Mg(OH)_2$
    c. $CaCO_3$
    d. $AgI$

## Precipitation and Qualitative Analysis

101. A solution containing sodium fluoride is mixed with one containing calcium nitrate to form a solution that is 0.015 M in NaF and 0.010 M in $Ca(NO_3)_2$. Does a precipitate form in the mixed solution? If so, identify the precipitate.

102. A solution containing potassium bromide is mixed with one containing lead acetate to form a solution that is 0.013 M in KBr and 0.0035 M in $Pb(C_2H_3O_2)_2$. Does a precipitate form in the mixed solution? If so, identify the precipitate.

103. Predict whether a precipitate will form if you mix 75.0 mL of a NaOH solution with pOH $= 2.58$ with 125.0 mL of a 0.018 M $MgCl_2$ solution. Identify the precipitate, if any.

104. Predict whether a precipitate will form if you mix 175.0 mL of a 0.0055 M KCl solution with 145.0 mL of a 0.0015 M $AgNO_3$ solution. Identify the precipitate, if any.

105. Potassium hydroxide is used to precipitate each of the cations from their respective solution. Determine the minimum concentration of KOH required for precipitation to begin in each case.
    a. 0.015 M $CaCl_2$
    b. 0.0025 M $Fe(NO_3)_2$
    c. 0.0018 M $MgBr_2$

106. Determine the minimum concentration of the precipitating agent on the right to cause precipitation of the cation from the solution on the left.
    a. 0.035 M $Ba(NO_3)_2$; NaF
    b. 0.085 M $CaI_2$; $K_2SO_4$
    c. 0.0018 M $AgNO_3$; RbCl

107. A solution is 0.010 M in $Ba^{2+}$ and 0.020 M in $Ca^{2+}$.
    a. If sodium sulfate is used to selectively precipitate one of the cations while leaving the other cation in solution, which cation will precipitate first? What minimum concentration of $Na_2SO_4$ will trigger the precipitation of the cation that precipitates first?
    b. What is the remaining concentration of the cation that precipitates first, when the other cation begins to precipitate?

**108.** A solution is 0.022 M in $Fe^{2+}$ and 0.014 M in $Mg^{2+}$.

   **a.** If potassium carbonate is used to selectively precipitate one of the cations while leaving the other cation in solution, which cation will precipitate first? What minimum concentration of $K_2CO_3$ will trigger the precipitation of the cation that precipitates first?

   **b.** What is the remaining concentration of the cation that precipitates first, when the other cation begins to precipitate?

## Complex Ion Equilibria

**109.** A solution is made $1.1 \times 10^{-3}$ M in $Zn(NO_3)_2$ and 0.150 M in $NH_3$. After the solution reaches equilibrium, what concentration of $Zn^{2+}(aq)$ remains?

**110.** A 120.0-mL sample of a solution that is $2.8 \times 10^{-3}$ M in $AgNO_3$ is mixed with a 225.0-mL sample of a solution that is 0.10 M in NaCN. After the solution reaches equilibrium, what concentration of $Ag^+(aq)$ remains?

**111.** Use the appropriate values of $K_{sp}$ and $K_f$ to find the equilibrium constant for the reaction.

$$FeS(s) + 6\,CN^-(aq) \rightleftharpoons Fe(CN)_6{}^{4-}(aq) + S^{2-}(aq)$$

**112.** Use the appropriate values of $K_{sp}$ and $K_f$ to find the equilibrium constant for the reaction.

$$PbCl_2(s) + 3\,OH^-(aq) \rightleftharpoons Pb(OH)_3{}^-(aq) + 2\,Cl^-(aq)$$

# Cumulative Problems

**113.** A 150.0-mL solution contains 2.05 g of sodium benzoate and 2.47 g of benzoic acid. Calculate the pH of the solution.

**114.** A solution is made by combining 10.0 mL of 17.5 M acetic acid with 5.54 g of sodium acetate and diluting to a total volume of 1.50 L. Calculate the pH of the solution.

**115.** A buffer is created by combining 150.0 mL of 0.25 M $HCHO_2$ with 75.0 mL of 0.20 M NaOH. Determine the pH of the buffer.

**116.** A buffer is created by combining 3.55 g of $NH_3$ with 4.78 g of HCl and diluting to a total volume of 750.0 mL. Determine the pH of the buffer.

**117.** A 1.0-L buffer solution initially contains 0.25 mol of $NH_3$ and 0.25 mol of $NH_4Cl$. In order to adjust the buffer pH to 8.75, should you add NaOH or HCl to the buffer mixture? What mass of the correct reagent should you add?

**118.** A 250.0-mL buffer solution initially contains 0.025 mol of $HCHO_2$ and 0.025 mol of $NaCHO_2$. In order to adjust the buffer pH to 4.10, should you add NaOH or HCl to the buffer mixture? What mass of the correct reagent should you add?

**119.** In analytical chemistry, bases used for titrations must often be standardized; that is, their concentration must be precisely determined. Standardization of sodium hydroxide solutions can be accomplished by titrating potassium hydrogen phthalate ($KHC_8H_4O_4$), also known as KHP, with the NaOH solution to be standardized.

   **a.** Write an equation for the reaction between NaOH and KHP.

   **b.** The titration of 0.5527 g of KHP required 25.87 mL of an NaOH solution to reach the equivalence point. What is the concentration of the NaOH solution?

**120.** A 0.5224-g sample of an unknown monoprotic acid was titrated with 0.0998 M NaOH. The equivalence point of the titration occurred at 23.82 mL. Determine the molar mass of the unknown acid.

**121.** A 0.25-mol sample of a weak acid with an unknown $pK_a$ was combined with 10.0 mL of 3.00 M KOH, and the resulting solution was diluted to 1.500 L. The measured pH of the solution was 3.85. What is the $pK_a$ of the weak acid?

**122.** A 5.55-g sample of a weak acid with $K_a = 1.3 \times 10^{-4}$ was combined with 5.00 mL of 6.00 M NaOH, and the resulting solution was diluted to 750.0 mL. The measured pH of the solution was 4.25. What is the molar mass of the weak acid?

**123.** A 0.552-g sample of ascorbic acid (vitamin C) was dissolved in water to a total volume of 20.0 mL and titrated with 0.1103 M KOH. The equivalence point occurred at 28.42 mL. The pH of the solution at 10.0 mL of added base was 3.72. From this data, determine the molar mass and $K_a$ for vitamin C.

**124.** Sketch the titration curve from Problem 123 by calculating the pH at the beginning of the titration, at one-half of the equivalence point, at the equivalence point, and at 5.0 mL beyond the equivalence point. Pick a suitable indicator for this titration from Table 17.1.

**125.** One of the main components of hard water is $CaCO_3$. When hard water evaporates, some of the $CaCO_3$ is left behind as a white mineral deposit. If a hard water solution is saturated with calcium carbonate, what volume of the solution has to evaporate to deposit $1.00 \times 10^2$ mg of $CaCO_3$?

**126.** Gout—a condition that results in joint swelling and pain—is caused by the formation of sodium urate ($NaC_5H_3N_4O_3$) crystals within tendons, cartilage, and ligaments. Sodium urate precipitates out of blood plasma when uric acid levels become abnormally high. This sometimes happens as a result of eating too many rich foods and consuming too much alcohol, which is why gout is sometimes referred to as the "disease of kings." If the sodium concentration in blood plasma is 0.140 M, and $K_{sp}$ for sodium urate is $5.76 \times 10^{-8}$, what minimum concentration of urate would result in precipitation?

**127.** Pseudogout, a condition with symptoms similar to those of gout (see Problem 126), is caused by the formation of calcium diphosphate ($Ca_2P_2O_7$) crystals within tendons, cartilage, and ligaments. Calcium diphosphate will precipitate out of blood plasma when diphosphate levels become abnormally high. If the calcium concentration in blood plasma is 9.2 mg/dL, and $K_{sp}$ for calcium diphosphate is $8.64 \times 10^{-13}$, what minimum concentration of diphosphate results in precipitation?

**128.** Calculate the solubility of silver chloride in a solution that is 0.100 M in $NH_3$.

**129.** Calculate the solubility of CuX in a solution that is 0.150 M in NaCN. $K_{sp}$ for CuX is $1.27 \times 10^{-36}$.

**130.** Aniline, abbreviated $\phi NH_2$, where $\phi$ is $C_6H_5$, is an important organic base used in the manufacture of dyes. It has $K_b = 4.3 \times 10^{-10}$. In a certain manufacturing process it is necessary to keep the concentration of $\phi NH_3{}^+$ (aniline's conjugate acid, the anilinium ion) below $1.0 \times 10^{-9}$ M in a solution that is 0.10 M in aniline. Find the concentration of NaOH required for this process.

131. The $K_b$ of hydroxylamine, $NH_2OH$, is $1.10 \times 10^{-8}$. A buffer solution is prepared by mixing 100.0 mL of a 0.36 M hydroxylamine solution with 50.0 mL of a 0.26 M HCl solution. Determine the pH of the resulting solution.

132. A 0.867-g sample of an unknown acid requires 32.2 mL of a 0.182 M barium hydroxide solution for neutralization. Assuming the acid is diprotic, calculate the molar mass of the acid.

133. A 25.0-mL volume of a sodium hydroxide solution requires 19.6 mL of a 0.189 M hydrochloric acid for neutralization. A 10.0-mL volume of a phosphoric acid solution requires 34.9 mL of the sodium hydroxide solution for complete neutralization. Calculate the concentration of the phosphoric acid solution.

134. Find the mass of sodium formate that must be dissolved in 250.0 cm³ of a 1.4 M solution of formic acid to prepare a buffer solution with pH = 3.36.

135. What relative masses of dimethyl amine and dimethyl ammonium chloride do you need to prepare a buffer solution of pH = 10.43?

136. You are asked to prepare 2.0 L of a HCN/NaCN buffer that has a pH of 9.8 and an osmotic pressure of 1.35 atm at 298 K. What masses of HCN and NaCN should you use to prepare the buffer? (Assume complete dissociation of NaCN.)

137. What should the molar concentrations of benzoic acid and sodium benzoate be in a solution that is buffered at a pH of 4.55 and has a freezing point of −2.0 °C? (Assume complete dissociation of sodium benzoate and a density of 1.01 g/mL for the solution.)

## Challenge Problems

138. Derive an equation similar to the Henderson–Hasselbalch equation for a buffer composed of a weak base and its conjugate acid. Instead of relating pH to $pK_a$ and the relative concentrations of an acid and its conjugate base (as the Henderson–Hasselbalch equation does), the equation should relate pOH to $pK_b$ and the relative concentrations of a base and its conjugate acid.

139. Since soap and detergent action is hindered by hard water, laundry formulations usually include water softeners—called builders—designed to remove hard water ions (especially $Ca^{2+}$ and $Mg^{2+}$) from the water. A common builder used in North America is sodium carbonate. Suppose that the hard water used to do laundry contains 75 ppm $CaCO_3$ and 55 ppm $MgCO_3$ (by mass). What mass of $Na_2CO_3$ is required to remove 90.0% of these ions from 10.0 L of laundry water?

140. A 0.558-g sample of a diprotic acid with a molar mass of 255.8 g/mol is dissolved in water to a total volume of 25.0 mL. The solution is then titrated with a saturated calcium hydroxide solution.
    a. Assuming that the $pK_a$ values for each ionization step are sufficiently different to see two equivalence points, determine the volume of added base for the first and second equivalence points.
    b. The pH after adding 25.0 mL of the base is 3.82. Find the value of $K_{a_1}$.
    c. The pH after adding 20.0 mL past the first equivalence point is 8.25. Find the value of $K_{a_2}$.

141. When excess solid $Mg(OH)_2$ is shaken with 1.00 L of 1.0 M $NH_4Cl$ solution, the resulting saturated solution has pH = 9.00. Calculate the $K_{sp}$ of $Mg(OH)_2$.

142. What amount of solid NaOH must be added to 1.0 L of a 0.10 M $H_2CO_3$ solution to produce a solution with $[H^+] = 3.2 \times 10^{-11}$ M? There is no significant volume change as the result of the addition of the solid.

143. Calculate the solubility of $Au(OH)_3$ in (a) water and (b) 1.0 M nitric acid solution ($K_{sp} = 5.5 \times 10^{-46}$).

144. Calculate the concentration of $I^-$ in a solution obtained by shaking 0.10 M KI with an excess of AgCl(s).

145. What volume of 0.100 M sodium carbonate solution is required to precipitate 99% of the Mg from 1.00 L of 0.100 M magnesium nitrate solution?

146. Find the solubility of CuI in 0.40 M HCN solution. The $K_{sp}$ of CuI is $1.1 \times 10^{-12}$ and the $K_f$ for the $Cu(CN)_2^-$ complex ion is $1 \times 10^{24}$.

147. Find the pH of a solution prepared from 1.0 L of a 0.10 M solution of $Ba(OH)_2$ and excess $Zn(OH)_2(s)$. The $K_{sp}$ of $Zn(OH)_2$ is $3 \times 10^{-15}$ and the $K_f$ of $Zn(OH)_4^{2-}$ is $2 \times 10^{15}$.

148. What amount of HCl gas must be added to 1.00 L of a buffer solution that contains [acetic acid] = 2.0 M and [acetate] = 1.0 M in order to produce a solution with pH = 4.00?

## Conceptual Problems

149. Without doing any calculations, determine if pH = $pK_a$, pH > $pK_a$, or pH < $pK_a$. Assume that HA is a weak monoprotic acid.
    a. 0.10 mol HA and 0.050 mol of $A^-$ in 1.0 L of solution
    b. 0.10 mol HA and 0.150 mol of $A^-$ in 1.0 L of solution
    c. 0.10 mol HA and 0.050 mol of $OH^-$ in 1.0 L of solution
    d. 0.10 mol HA and 0.075 mol of $OH^-$ in 1.0 L of solution

150. A buffer contains 0.10 mol of a weak acid and 0.20 mol of its conjugate base in 1.0 L of solution. Determine whether or not each addition exceeds the capacity of the buffer.
    a. adding 0.020 mol of NaOH
    b. adding 0.020 mol of HCl
    c. adding 0.10 mol of NaOH
    d. adding 0.010 mol of HCl

151. Consider three solutions:
    i. 0.10 M solution of a weak monoprotic acid
    ii. 0.10 M solution of strong monoprotic acid
    iii. 0.10 M solution of a weak diprotic acid

    Each solution is titrated with 0.15 M NaOH. Which quantity is the same for all three solutions?
    a. the volume required to reach the final equivalence point
    b. the volume required to reach the first equivalence point
    c. the pH at the first equivalence point
    d. the pH at one-half the first equivalence point

**152.** Two monoprotic acid solutions (A and B) are titrated with identical NaOH solutions. The volume to reach the equivalence point for solution A is twice the volume required to reach the equivalence point for solution B, and the pH at the equivalence point of solution A is higher than the pH at the equivalence point for solution B. Which statement is true?

**a.** The acid in solution A is more concentrated than in solution B and is also a stronger acid than that in solution B.

**b.** The acid in solution A is less concentrated than in solution B and is also a weaker acid than that in solution B.

**c.** The acid in solution A is more concentrated than in solution B and is also a weaker acid than that in solution B.

**d.** The acid in solution A is less concentrated than in solution B and is also a stronger acid than that in solution B.

**153.** Describe the solubility of $CaF_2$ in each solution compared to its solubility in water.

**a.** in a 0.10 M NaCl solution

**b.** in a 0.10 M NaF solution

**c.** in a 0.10 M HCl solution

**154.** Why does the titration of a weak acid with a strong base always have a basic equivalence point?

*Active Classroom Learning*

# Questions for Group Work

Discuss these questions with the group and record your consensus answer.

**155.** Name a compound that you could add to a solution of each of the compounds to make a buffer. Explain your reasoning in complete sentences.

**a.** acetic acid

**b.** sodium nitrite

**c.** ammonia

**d.** potassium formate

**e.** $Na_2HPO_4$ (two possible answers)

**156.** Derive the Henderson–Hasselbalch equation as a group. Take turns having each group member write and explain the next step in the derivation.

**157.** With group members acting as atoms or ions, act out the reaction that occurs when HCl is added to a buffer solution composed of $HC_2H_3O_2$ and $NaC_2H_3O_2$. Write out a script for a narrator that describes the processes that occur, including how the buffer keeps the pH approximately the same even though a strong acid is added.

**158.** A certain town gets its water from an underground aquifer that contains water in equilibrium with calcium carbonate limestone.

**a.** What is the symbol for the equilibrium constant that describes calcium carbonate dissolving in water? What is the numerical value?

**b.** Calculate the molar solubility of calcium carbonate.

**c.** If an entire coffee cup of water (about 200 mL) evaporated on your desk over spring break, how many grams of calcium carbonate would be left behind?

**d.** If you wanted to clean out your coffee cup, would it be better to use an acidic or basic cleaning solution? Why?

**159.** Have each group member look up the $K_{sp}$ for a different compound. Calculate the molar solubility. Do the numerical values suggest that the compound is soluble or insoluble? Compare answers with the solubility rules from Chapter 4, and have each group member present his or her findings to the group.

# Data Interpretation and Analysis

## Buffers and Hydroponics

**160.** Hydroponics is a method for growing plants in water with added nutrients and oxygen. No soil is used. A deviation from the optimal pH range of 5.5–6.5 can prevent a hydroponic plant from obtaining the necessary minerals and oxygen it needs to grow. To help maintain a pH of 5.5–6.5, growers use a citrate salt buffer system. Citric acid, $H_3C_6H_5O_7$, is a triprotic acid that has three carboxylic acid groups (Figure a ▼).

**▲ FIGURE a  Structure of Citric Acid.**

Citric acid has three acid-dissociation constants shown here:

$$H_3C_6H_5O_7(aq) + H_2O(l) \rightleftharpoons H_2C_6H_5O_7^-(aq) + H_3O^+(aq)$$
$$K_{a1} = 7.4 \times 10^{-4}$$

$$H_2C_6H_5O_7^-(aq) + H_2O(l) \rightleftharpoons HC_6H_5O_7^{2-}(aq) + H_3O^+(aq)$$
$$K_{a2} = 1.7 \times 10^{-5}$$

$$HC_6H_5O_7^{2-}(aq) + H_2O(l) \rightleftharpoons C_6H_5O_7^{3-}(aq) + H_3O^+(aq)$$
$$K_{a3} = 4.0 \times 10^{-7}$$

Use the information provided in the figures and equations to answer the following questions:

**a.** Write a chemical equation that shows how citric acid reacts with a hydroxide ion.

**b.** Which two of the citrate salts listed in Table a would you use to make a buffer with a pH range of 5.5–6.5? Explain.

**c.** Show how your buffer system (from part b) reacts with incoming acid ($H_3O^+$) and incoming base ($OH^-$).

**d.** Figure b▶ represents a titration curve for citric acid ($H_3C_6H_5O_7$) with a strong base. On the graph, indicate the buffer region involving the salts you chose in part b.

**e.** To prepare a buffer with pH = 6.0, what is the proper ratio of the concentrations of the two citrate salts (from part b)? If the buffer must neutralize at least 0.010 M added acid, what should the minimum concentration be for each of the two buffer components?

| Potassium Dihydrogen Citrate | Potassium Hydrogen Citrate | Potassium Citrate |
|---|---|---|
| $KH_2C_6H_5O_7$ | $K_2HC_6H_5O_7$ | $K_3C_6H_5O_7$ |

**TABLE a  Citrate Salt Formulas**

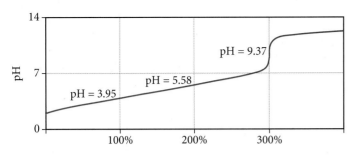

▲ **FIGURE b** **pH as a Function of % Completion of Titration of Citric Acid.**

Nour, V., Trandafir, I., Ionica, M. E. HPLC Organic Acid Analysis in Different Citrus Juices under Reversed Phase Conditions. *Not. Bot. Hort. Agrobot. Cluj* **2010**, 38, 1, pp. 44–48.

Weast, R. C., Ed. *Handbook of Chemistry and Physics,* The Chemical Rubber Company, Ohio, U.S. 1968.

NCIB – National Center for Biotechnology Information. Compound Summary for Citric Acid,http://pubchem.ncbi.nlm.nih.gov/summary/summary.cgi?cid=311#x94 (accessed Oct 2, 2014).

Penniston, K. L., Nakada, S. Y., Holmes, R. P., Assimos, D. G. Quantitative Assessment of Citric Acid in Lemon Juice, Lime Juice, and Commercially-Available Fruit Juice Products. *J Endourol.* Mar 2008; 22, 3, pp. 567–570.

# Answers to Conceptual Connections

## Buffers

**17.1** **(d)** Only this solution contains significant amounts of a weak acid and its conjugate base. (Remember that $HNO_3$ is a strong acid, but $HNO_2$ is a weak acid.)

## pH of Buffer Solutions

**17.2** **(a)** Since the pH of the buffer is less than the $pK_a$ of the acid, the buffer must contain more acid than base ($[HA] > [A^-]$). In order to raise the pH of the buffer from 4.25 to 4.72, you must add more of the weak base (adding a base will make the buffer solution more basic).

## Buffering Action

**17.3** **(a)** Since one HA molecule is converted to one ion of the conjugate base ($A^-$ ion), one $OH^-$ ion was added to the buffer solution.

## Adding Acid or Base to a Buffer

**17.4** **(b)** Since acid is added to the buffer, the pH will become slightly lower (slightly more acidic). Answer **(a)** reflects too large a change in pH for a buffer, and answers **(c)** and **(d)** have the pH changing in the wrong direction.

## Buffer Capacity

**17.5** **(a)** Adding 0.050 mol of HCl destroys the buffer because it will react with all of the NaF, leaving no conjugate base in the buffer mixture.

## Titration Equivalence Point

**17.6** **(d)** Because the flask contains $7 H^+$ ions, the equivalence point is reached when $7 OH^-$ ions have been added.

## The Half-Equivalence Point

**17.7** **(c)** The pH at the half-equivalence point is the $pK_a$ of the conjugate acid, which is equal to $14.00 - 8.75 = 5.25$.

## Acid–Base Titrations

**17.8** **(c)** Since the volumes and concentrations of all three acids are the same, the volume of NaOH required to reach the first equivalence point (and the only equivalence point for titrations i and iii) is the same for all three titrations.

## Common Ion Effect

**17.9** **(c)** The sodium nitrate solution is the only one that has no common ion with barium sulfate. The other two solutions have common ions with barium sulfate; therefore, the solubility of barium sulfate is lower in these solutions.

## Solubility and Complex Ion Equilibria

**17.10** **(c)** Only NaCN contains an anion ($CN^-$) that forms a complex ion with $Cu^{2+}$ [from Table 17.3 we can see that $K_f = 1.0 \times 10^{25}$ for $Cu(CN)_4{}^{2-}$]. Therefore, the presence of $CN^-$ will drive the dissolution reaction of CuS.

# Free Energy and Thermodynamics

Die Energie der Welt ist konstant. Die
Entropie der Welt strebt einem Maximum zu.
(The energy of the world is constant.
The entropy of the world tends towards
a maximum.)

—RUDOLF CLAUSIUS (1822–1888)

▲ In this clever illusion, it seems that the water can perpetually flow through the canal. However, perpetual motion is forbidden under the laws of thermodynamics.

Throughout this book, we have examined chemical and physical changes. We have studied how fast chemical changes occur (kinetics) and how to predict how far they will go (through the use of equilibrium constants). We have learned that acids neutralize bases and that gases expand to fill their containers. We now turn to the following question: why do these changes occur in the first place? What ultimately drives physical and chemical changes in matter? The answer may surprise you. The driving force behind chemical and physical change in the universe is a quantity called *entropy*, which is related to the dispersion (spreading out) of energy. Nature tends toward that state in which energy is spread out to the greatest extent possible. Although it does not seem obvious at first glance, the freezing of water below 0 °C, the dissolving of a solid into a solution, the neutralization of an acid by a base, and even the development of a person from an embryo all increase the entropy in the universe (they all result in greater energy dispersion). In our universe, entropy always increases.

## 18.1 Nature's Heat Tax: You Can't Win and You Can't Break Even

Energy transactions are like gambling—you walk into the casino with your pockets full of cash and (if you keep gambling long enough) you walk out empty-handed. In the long run, you lose money in gambling because the casino takes a cut on each transaction. So it is with energy. Nature takes a cut—sometimes referred to as "nature's heat tax"—on every energy transaction so that, in the end, energy is dissipated.

▲ A rechargeable battery always requires more energy to charge than the energy available for work during discharging because some energy is always lost to the surroundings during the charging/discharging cycle.

See the box entitled *Redheffer's Perpetual Motion Machine* in Section 6.3.

Recall from Chapter 6 that, according to the first law of thermodynamics, energy is conserved in chemical processes. When we burn gasoline to run a car, for example, the amount of energy produced by the chemical reaction does not vanish, nor does any new energy appear that was not present as potential energy (within the gasoline) before the combustion. Some of the energy from the combustion reaction goes toward driving the car forward (about 20%), and the rest is dissipated into the surroundings as heat (feel a car engine after a drive if you doubt this). The total energy given off by the combustion reaction exactly equals the sum of the amount of energy that goes to propel the car and the amount dissipated as heat—energy is conserved. In other words, when it comes to energy, we can't win; we cannot create energy that was not there to begin with.

The picture becomes more interesting, however, when we consider the second law of thermodynamics. The second law—which we examine in more detail throughout this chapter—implies that not only can we not win in an energy transaction, we cannot even break even. For example, consider a rechargeable battery. Suppose that when we use the fully charged battery for some application, the energy from the battery does 100 kJ of work. Recharging the battery to its original state will *necessarily* (according to the second law of thermodynamics) require *more than* 100 kJ of energy. Energy is not destroyed during the cycle of discharging and recharging the battery, but some energy must be lost to the surroundings in order for the process to occur at all. Nature imposes a heat tax, an unavoidable cut of every energy transaction.

The implications of the second law for energy use are significant. First of all, according to the second law, we cannot create a perpetual motion machine (a machine that perpetually moves *without any energy input*). If the machine is to be in motion, it must pay the heat tax with each cycle of its motion—over time, it will therefore run down and stop moving.

Second, in most energy transactions, not only is the heat tax lost to the surroundings, but additional energy is also lost as heat because real-world processes do not achieve the theoretically possible maximum efficiency (Figure 18.1◄). Consequently, the most efficient use of energy generally occurs with the smallest number of transactions. For example, heating your home with natural gas is generally cheaper and more efficient than heating it with electricity (Figure 18.2►). When you heat your home with natural gas, there is only one energy transaction—you burn the gas and the heat from the reaction warms the house. When you heat your home with electricity, several transactions occur. Most electricity is generated from the combustion of fossil fuels; the heat from the burning fuel boils water to create steam. The steam then turns a turbine on a generator to create electricity. The electricity travels from the power plant to your home, and some of the energy is lost as heat during the trip. Finally, the electricity runs the heater that generates heat. With each transaction, energy is lost to the surroundings, resulting in a less efficient use of energy than if you had burned natural gas directly.

100 watts of electrical energy

95 watts dissipated as heat

5 watts emitted as light

◄ FIGURE 18.1 **Energy Loss** In most energy transactions, some energy is lost to the surroundings, so each transaction is only fractionally efficient.

**CONCEPTUAL CONNECTION 18.1**

**Nature's Heat Tax and Diet** Advocates of a vegetarian diet argue that the amount of cropland required for one person to maintain a meat-based diet is about six to ten times greater than the amount required for the same person to maintain a vegetarian diet. Apply the concept of nature's heat tax to explain this assertion.

**Heating with Natural Gas**

**Heating with Electricity**

▲ **FIGURE 18.2 Heating with Gas versus Heating with Electricity** When natural gas heats a home, only a single energy transaction is involved, so the heat tax is minimized. When electricity is used to heat a home, a number of energy transactions are required, each of which involves some loss. The result is a much lower efficiency.

## 18.2 Spontaneous and Nonspontaneous Processes

A fundamental goal of thermodynamics is to predict *spontaneity*. For example, will rust spontaneously form when iron comes into contact with oxygen? Will water spontaneously decompose into hydrogen and oxygen? A **spontaneous process** is one that occurs *without ongoing outside intervention* (such as the performance of work by some external force). For example, when you drop a book in a gravitational field, it spontaneously drops to the floor. When you place a ball on a slope, it spontaneously rolls down the slope.

For simple mechanical systems, such as the dropping of a book or the rolling of a ball, predicting spontaneity is fairly intuitive. A mechanical system tends toward lowest potential energy, which is usually easy to see (at least in *simple* mechanical systems).

◀ Iron spontaneously rusts when it comes in contact with oxygen.

▶ **FIGURE 18.3 Mechanical Potential Energy and Chemical Potential** **(a)** Mechanical potential energy predicts the direction in which a mechanical system will spontaneously move. **(b)** We seek a chemical potential that predicts the direction in which a chemical system will spontaneously move.

## The Concept of Chemical Potential

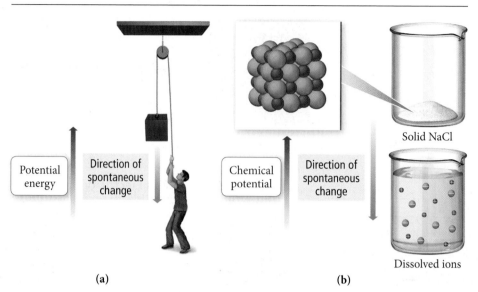

Solid NaCl

Dissolved ions

(a)  (b)

However, the prediction of spontaneity for chemical systems is not so intuitively obvious. To make these predictions, we need to develop a criterion for the spontaneity of chemical systems. In other words, we need to develop a *chemical potential* that predicts the direction of a chemical system, much as mechanical potential energy predicts the direction of a mechanical system (Figure 18.3▲).

We must not confuse the *spontaneity* of a chemical reaction with the *speed* of a chemical reaction. In thermodynamics, we study the *spontaneity* of a reaction—the direction in which and extent to which a chemical reaction proceeds. In kinetics, we study the *speed* of the reaction—how fast a reaction takes place (Figure 18.4▶). A reaction may be thermodynamically spontaneous but kinetically slow at a given temperature. For example, the conversion of diamond to graphite is thermodynamically spontaneous. But your diamonds will not become worthless anytime soon because the process is extremely slow kinetically. Although the rate of a spontaneous process can be increased by the use of a catalyst, a nonspontaneous process cannot be made spontaneous by the use of a catalyst. Catalysts affect only the rate of a reaction, not the spontaneity.

Spontaneous

Slow rate

Diamond  Graphite

▶ Even though graphite is thermodynamically more stable than diamond, the conversion of diamond to graphite is kinetically so slow that it does not occur at any measurable rate.

One more observation about nonspontaneity—a nonspontaneous process is not *impossible*. For example, the extraction of iron metal from iron ore is a nonspontaneous process; it does not happen if the iron ore is left to itself, but that does not mean it is impossible. As we will see later in this chapter, a nonspontaneous process can be made spontaneous by coupling it to another process that is spontaneous or by supplying

◀ **FIGURE 18.4** **Thermodynamics and Kinetics** Thermodynamics deals with the relative chemical potentials of the reactants and products. It enables us to predict whether a reaction will be spontaneous and to calculate how much work it can do. Kinetics deals with the chemical potential of intermediate states and enables us to determine why a reaction is slow or fast.

energy from an external source. Iron can be separated from its ore if external energy is supplied, usually by means of another reaction (that is itself highly spontaneous).

# 18.3 Entropy and the Second Law of Thermodynamics

**Entropy and the Second Law of Thermodynamics**

See Section 6.6 for the definition of enthalpy.

The first candidate in our search for a chemical potential might be enthalpy, which we defined in Chapter 6. Perhaps, just as a mechanical system proceeds in the direction of lowest potential energy, so a chemical system might proceed in the direction of lowest enthalpy. If this were the case, all exothermic reactions would be spontaneous and all endothermic reactions would not. However, although *most* spontaneous processes are exothermic, some spontaneous processes are *endothermic*. For example, above 0 °C, ice spontaneously melts (an endothermic process). So enthalpy must not be the sole criterion for spontaneity.

We can learn more about the driving force behind chemical reactions by considering several processes (like ice melting) that involve an increase in enthalpy. The processes listed here are energetically uphill (they are endothermic), yet they occur spontaneously. What drives them?

- the melting of ice above 0 °C
- the evaporation of liquid water to gaseous water
- the dissolution of sodium chloride in water

Each of these processes is endothermic *and* spontaneous. Do they have anything in common? Notice that, in each process, disorder or randomness increases. In the melting of ice, the arrangement of the water molecules changes from a highly ordered one (in ice) to a somewhat disorderly one (in liquid water).

The use of the word *disorder* here is analogous to our macroscopic notions of disorder. The definition of molecular disorder, which is covered shortly, is very specific.

$H_2O(s)$

Increasing entropy

$H_2O(l)$

◀ When ice melts, the arrangement of water molecules changes from orderly to more disorderly.

During the evaporation of a liquid to a gas, the arrangement changes from a *somewhat* disorderly one (atoms or molecules in the liquid) to a *highly* disorderly one (atoms or molecules in the gas).

▶ When water evaporates, the arrangement of water molecules becomes still more disorderly.

$H_2O(g)$

Increasing entropy

$H_2O(l)$

In the dissolution of a salt into water, the arrangement again changes from an orderly one (in which the ions in the salt occupy regular positions in the crystal lattice) to a more disorderly one (in which the ions are randomly dispersed throughout the liquid water).

$NaCl(aq)$

Increasing entropy

$NaCl(s)$

▶ When salt dissolves in water, the arrangement of the molecules and ions becomes more disorderly.

In all three of these processes, a quantity called *entropy*—related to disorder or randomness at the molecular level—increases.

## Entropy

We have hit upon the criterion for spontaneity in chemical systems: entropy. Informally, we can think of entropy as disorder or randomness. But the concept of disorder or randomness on the macroscopic scale—such as the messiness of a drawer—is only *analogous* to the concept of disorder or randomness on the molecular scale. Formally, **entropy**, abbreviated by the symbol $S$, has the following definition:

> **Entropy ($S$) is a thermodynamic function that increases with the number of *energetically equivalent* ways to arrange the components of a system to achieve a particular state.**

Austrian physicist, Ludwig Boltzmann (1844–1906) expressed this definition mathematically as:

$$S = k \ln W$$

where $k$ is the Boltzmann constant (the gas constant divided by Avogadro's number, $R/N_A = 1.38 \times 10^{-23}$ J/K) and $W$ is the number of energetically equivalent ways to arrange the components of the system. Since $W$ is unitless (it is simply a number), the units of entropy are joules per kelvin (J/K). We talk about the significance of the units shortly. As we can see from the equation, as $W$ increases, entropy increases.

The key to understanding entropy is the quantity $W$. What does $W$—the number of energetically equivalent ways to arrange the components of the system—signify? Imagine a system of particles such as a fixed amount of an ideal gas. A given set of conditions ($P$, $V$, and $T$) defines the *state* (or *macrostate*) of the system. As long as these conditions remain constant, the energy of the system also remains constant. However, exactly where that energy is at any given instant is anything but constant.

At any one instant, a particular gas particle may have a good deal of kinetic energy. However, after a very short period of time, that particle may have very little kinetic energy (because it lost its energy through collisions with other particles). The exact internal energy distribution among the particles at any one instant is sometimes referred to as a *microstate*. We can think of a microstate as a snapshot of the system at a given instant in time. The next instant, the snapshot (the microstate) changes. However, the *macrostate*—defined by $P$, $V$, and $T$—remains constant. A given macrostate can exist as a result of a large number of different microstates. In other words, the snapshot (or microstate) of a given macrostate is generally different from one moment to the next as the energy of the system is constantly redistributing itself among the particles of the system.

We can conceive of $W$ in terms of microstates. The quantity, $W$, is the number of *possible* microstates that can result in a given macrostate. For example, suppose that we have two systems (call them System A and System B) and that each is composed of two particles (one blue and one red). Both systems have a total energy of 4 joules, but System A has only one energy level and System B has two:

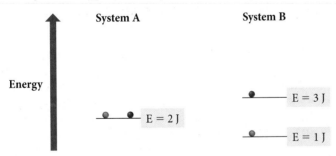

Each system has the same total energy (4 J), but System A has only one possible microstate (the red and the blue particles both occupying the 2 J energy level) while System B has a second possible microstate:

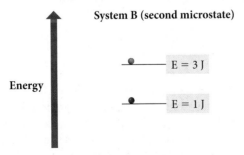

In this second microstate for System B, the blue particle has 3 J and the red one has 1 J (as opposed to System B's first microstate, where the energy of the particles is switched). This second microstate is not possible for System A because it has only one energy level. For System A, $W = 1$, but for System B, $W = 2$. In other words, System B has more microstates that result in the same 4 J macrostate. Since $W$ is larger for System B than for

▲ Boltzmann's equation is engraved on his tombstone.

System A, System B has greater *entropy*; it has more *energetically equivalent ways to arrange the components of the system.*

We can understand an important aspect of entropy by turning our attention to energy for a moment. The entropy of a state increases with the number of *energetically equivalent* ways to arrange the components of the system to achieve that particular state. This implies that *the state with the highest entropy also has the greatest dispersal of energy.* Returning to our previous example, the energy of System B is dispersed over two energy levels instead of being confined to just one. At the heart of entropy is the concept of energy dispersal or energy randomization. *A state in which a given amount of energy is more highly dispersed (or more highly randomized) has more entropy than a state in which the same energy is more highly concentrated.*

We have already alluded to the **second law of thermodynamics**; we now formally define it:

> **For any spontaneous process, the entropy of the *universe* increases ($\Delta S_{univ} > 0$).**

*The criterion for spontaneity is the entropy of the universe.* Processes that increase the entropy of the universe—those that result in greater dispersal or randomization of energy—occur spontaneously. Processes that decrease the entropy of the universe do not occur spontaneously.

See the discussion of state functions in Section 6.3.

Entropy, like enthalpy, is a *state function*—its value depends only on the state of the system, not on how the system arrived at that state. Therefore, for any process, *the change in entropy is the entropy of the final state minus the entropy of the initial state*:

$$\Delta S = S_{final} - S_{initial}$$

Entropy determines the direction of chemical and physical change. *A chemical system proceeds in a direction that increases the entropy of the universe*—it proceeds in a direction that has the largest number of *energetically equivalent* ways to arrange its components.

## The Entropy Change upon the Expansion of an Ideal Gas

See the discussion of work done by an expanding gas in Section 6.4.

To better understand this tendency, let us examine the expansion of an ideal gas into a vacuum (a spontaneous process with no associated change in enthalpy). Consider a flask containing an ideal gas that is connected to another, evacuated, flask by a tube equipped with a stopcock. When the stopcock is opened, the gas spontaneously expands into the evacuated flask. Since the gas is expanding into a vacuum, the pressure against which it expands is zero, and therefore the work ($w = -P_{ext} \Delta V$) is also zero.

However, even though the total energy of the gas does not change during the expansion, the entropy does change. To picture this, consider a simplified system containing only four gas atoms:

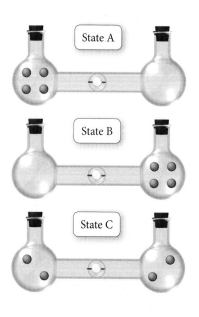

When the stopcock is opened, several possible energetically equivalent final states may result, each with the four atoms distributed in a different way. For example, there could be three atoms in the flask on the left and one in the flask on the right, or vice versa. For simplicity, we consider only the possibilities shown in the figure in the left margin, which are state A, state B, and state C. Since the energy of any one atom is the same in either flask, and since the atoms do not interact, states A, B, and C are energetically equivalent.

Now we ask the following question for each state: how many microstates give rise to the same macrostate? To keep track of the microstates, we label the atoms 1–4. Although they have different numbered labels, since the atoms are all the same, there is externally no difference between them. For states A and B, only one microstate results in the specified macrostate—atoms 1–4 on the left side or the right side, respectively:

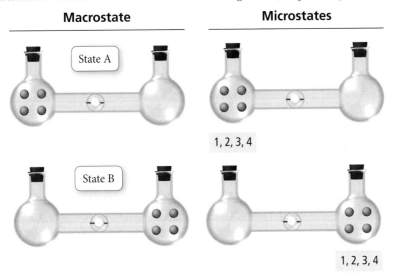

For state C, however, six different possible microstates all result in the same macrostate (two atoms on each side):

This means that if the atoms are just randomly moving between the two flasks, the statistical probability of finding the atoms in state C is six times greater than the probability of finding the atoms in state A or state B. Consequently, even for a simple system consisting of only four atoms, the atoms are most likely to be found in state C. State C has the greatest entropy—it has the greatest number of energetically equivalent ways to distribute its components.

As the number of atoms increases, the number of microstates that leads to the atoms being equally distributed between the two flasks increases dramatically. For example, with 10 atoms, the number of microstates leading to an equal distribution of atoms between two flasks is 252 and with 20 atoms the number of microstates is 184,756. Yet, the number of microstates that leads to all of the atoms being only in the left flask (or all

In these drawings, the exact location of an atom within a flask is insignificant. The significant aspect is whether the atom is in the left flask or the right flask.

For $n$ particles, the number of ways to put $r$ particles in one flask and $n - r$ particles in the other flask is $n!/[(n - r)!r!]$. For 10 atoms, $n = 10$ and $r = 5$.

only in the right flask) does not increase—it is always only 1. In other words, the arrangement in which the atoms are equally distributed between the two flasks has a much larger number of possible microstates and therefore much greater entropy. The system thus tends toward that state.

The *change in entropy* in transitioning from a state in which all of the atoms are in the left flask to the state in which the atoms are evenly distributed between both flasks is *positive* because the final state has a greater entropy than the initial state:

$$\Delta S = S_{final} - S_{initial}$$

Entropy of state in which atoms are distributed between both flasks

Entropy of state in which atoms are all in one flask

Since $S_{final}$ is greater than $S_{initial}$, $\Delta S$ is positive and the process is spontaneous according to the second law. Notice that when the atoms are confined to one flask, their energy is also confined to that one flask; however, when the atoms are evenly distributed between both flasks, their energy is spread out over a greater volume. As the gas expands into the empty flask, energy is dispersed.

The second law explains many phenomena not explained by the first law. In Chapter 6, we learned that heat travels from a substance at higher temperature to one at lower temperature. For example, if we drop an ice cube into water, heat travels from the water to the ice cube—the water cools and the ice warms (and eventually melts). Why? The first law would not prohibit some heat from flowing the other way—from the ice to the water. The ice could lose 10 J of heat (cooling even more) and the water could gain 10 J of heat (warming even more). The first law of thermodynamics would not be violated by such a heat transfer. Imagine putting ice into water only to have the water get warmer as it absorbed thermal energy from the ice! It will never happen because heat transfer from cold to hot violates the second law of thermodynamics. According to the second law, energy is dispersed, not concentrated. The transfer of heat from a substance of higher temperature to one of lower temperature results in greater energy randomization—the energy that was concentrated in the hot substance becomes dispersed between the two substances. The second law accounts for this pervasive tendency.

Heat

CONCEPTUAL
CONNECTION **18.2**

**Entropy** Consider these three changes in the possible distributions of six gaseous particles within three interconnected boxes. Which change has a positive $\Delta S$?

(a)

(b)

(c)

## 18.4 Entropy Changes Associated with State Changes

The entropy of a sample of matter *increases* as it changes state from a solid to a liquid or from a liquid to a gas (Figure 18.5►). We first consider this idea conceptually, and then we turn to actually calculating the value of $\Delta S$ for this change. Recall that, in Chapter 6, we distinguished between a thermodynamic system and its surroundings. This distinction is useful in our discussion of entropy. For a change of state in a substance, we consider the substance to be the system. The surroundings are then the rest of the universe. Here, we

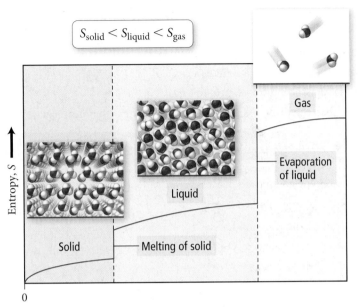

$$S_{solid} < S_{liquid} < S_{gas}$$

Gas

Evaporation of liquid

Liquid

Solid

Melting of solid

Entropy, S

0

Temperature (K)

◀ **FIGURE 18.5** **Entropy and State Change** Entropy increases when matter changes from a solid to a liquid and from a liquid to a gas.

focus on the entropy of the *system* undergoing the change of state (not the surroundings, which we address in Section 18.5).

## Entropy and State Change: The Concept

We can informally think of the increase in entropy in going from a solid to a liquid or from a liquid to a gas by analogy with macroscopic disorder. The gaseous state is more disorderly than the liquid state, which is in turn more disorderly than the solid state. More formally, however, the differences in entropy are related to the number of energetically equivalent ways of arranging the particles in each state—there are more in the gas than in the liquid, and more in the liquid than in the solid.

A gas has more energetically equivalent configurations because it has more ways to distribute its energy than a solid. The energy in a molecular solid consists largely of the vibrations between its molecules. If the same substance is in its gaseous state, however, the energy can take the form of straight-line motions of the molecules (called translational energy) and rotations of the molecules (called rotational energy). In other words, when a solid vaporizes into a gas, there are new "places" to put energy (Figure 18.6▼). The gas thus has more possible microstates (more energetically equivalent configurations) than the solid and therefore a greater entropy.

▼ **FIGURE 18.6** **"Places" for Energy** In the solid state, energy is contained largely in the vibrations between molecules. In the gas state, energy can be contained in both the straight-line motion of molecules (translational energy) and the rotation of molecules (rotational energy).

## Additional "Places" for Energy

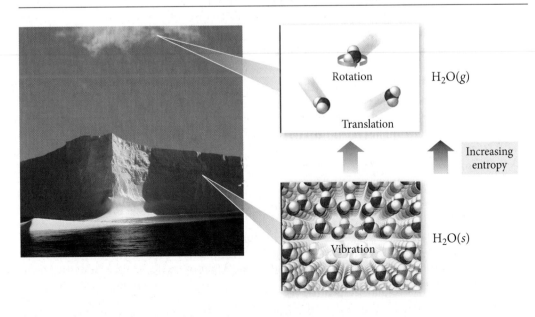

Rotation

$H_2O(g)$

Translation

Increasing entropy

Vibration

$H_2O(s)$

We can now predict the sign of $\Delta S$ for processes involving changes of state (or phase). In general, entropy increases ($\Delta S > 0$) for each of the following:

- the phase transition from a solid to a liquid
- the phase transition from a solid to a gas
- the phase transition from a liquid to a gas
- an increase in the number of moles of a gas during a chemical reaction

**Example 18.1** | **Predicting the Sign of Entropy Change**

Predict the sign of $\Delta S$ for each process.

**(a)** $H_2O(g) \longrightarrow H_2O(l)$

**(b)** Solid carbon dioxide sublimes.

**(c)** $2\,N_2O(g) \longrightarrow 2\,N_2(g) + O_2(g)$

**SOLUTION**

**(a)** Since a gas has a greater entropy than a liquid, the entropy decreases and $\Delta S$ is negative.

**(b)** Since a solid has a lower entropy than a gas, the entropy increases and $\Delta S$ is positive.

**(c)** Since the number of moles of gas increases, the entropy increases and $\Delta S$ is positive.

**FOR PRACTICE 18.1** Predict the sign of $\Delta S$ for each process.

**(a)** the boiling of water

**(b)** $I_2(g) \longrightarrow I_2(s)$

**(c)** $CaCO_3(s) \longrightarrow CaO(s) + CO_2(g)$

## Entropy and State Changes: The Calculation

We have just seen how the entropy of system increases with the transition from a solid to a liquid to a gas. We can calculate the actual changes in entropy that occur in the system during these state transitions. Recall from Sections 11.5 and 11.6 that a change in state is accompanied by an exchange of heat between the system and surroundings. For example, melting and vaporization are endothermic (they absorb heat from the surroundings), while freezing and condensation are exothermic (they give off heat to the surroundings).

The definition of entropy that we introduced previously ($S = k \ln W$) shows how entropy is related to the distribution of energy among the particles that compose matter. On a macroscopic scale, we can define the change in entropy that occurs when a system exchanges a quantity of heat ($q$) with its surroundings at a constant temperature ($T$) with the following equation:

$$\Delta S = \frac{q_{rev}}{T} \text{ (constant temperature)} \qquad [18.1]$$

In this equation, $\Delta S$ is the entropy change of the system, $q_{rev}$ is the heat exchanged with the surroundings in a *reversible* process (more on this to come), and $T$ is the temperature in K. The temperature must be constant throughout the change—the process must be *isothermal*. Let's address two important concepts related to this equation, the units of entropy and reversible processes.

### Units of Entropy

In Section 18.3 we saw that the units of entropy are J/K. We can now better understand these units. Entropy is a measure of energy dispersal per unit temperature and therefore has the units of energy (J) over temperature (K). For example, when a substance melts (an endothermic process), it absorbs a certain amount of energy from the surroundings. That energy becomes dispersed into the system (as the substance changes state from a solid to a liquid) and the entropy of the substance increases. The change in entropy is related to the amount of energy the substance absorbs (units of J) divided by the temperature of the substance (units of K).

## Reversible Processes

A **reversible process** is one that reverses direction upon an infinitesimally small change in some property. The melting of ice when the ice and its surroundings are both at exactly 0 °C is an example of a reversible process. An infinitesimally small *removal* of heat reverses the melting and causes freezing instead. An infinitesimally small *addition* of heat causes the melting to resume. All reversible processes are in a constant state of equilibrium and represent highly idealized conditions. In contrast, the melting of ice that happens when we set an ice cube on a countertop at room temperature is *irreversible*. In this case, the countertop and the surrounding air are at 25.0 °C and an infinitesimally small removal of heat does not reverse the melting.

We can use Equation 18.1 to calculate the change in the entropy of a system that undergoes a state change. For example, when ice melts at its melting point, the heat exchanged with the surroundings is $\Delta H_{fus}$ and the temperatue is 273 K:

$$q_{rev} = \Delta H_{fus} = 6.02 \text{ kJ/mol}$$
$$T = 273 \text{ K}$$

We substitute these quantities into Equation 18.1 to determine the change in entropy of the system when 1 mol of ice melts at its melting point:

$$\Delta S = \frac{q_{rev}}{T}$$

$$= \frac{(1 \text{ mol}) \dfrac{6.02 \times 10^3 \text{ J}}{\text{mol}}}{273 \text{ K}}$$

$$= 22.1 \text{ J/K}$$

The value of $\Delta S$ is positive as we would expect for a transition from solid to liquid.

---

### Example 18.2  Calculating ΔS for a Change of State

Calculate the change in entropy that occurs in the system when 25.0 g of water condenses from a gas to a liquid at the normal boiling point of water (100.0 °C).

**SOLUTION**

| | |
|---|---|
| Because the condensation is occurring at the boiling point of water and the temperature is constant, you can use Equation 18.1 to calculate the change in entropy of the system. | $\Delta S = \dfrac{q_{rev}}{T}$ |
| Gather the necessary quantities. Look up the value of the enthalpy of vaporization for water at its boiling point in Table 11.7. The enthalpy of condensation is the same in value but opposite in sign. <br><br> Calculate the temperature in K. | $\Delta H_{vap} = 40.7 \text{ kJ/mol}$ <br> $\Delta H_{condensation} = -40.7 \text{ kJ/mol}$ <br> $T(K) = T(°C) + 273.15$ <br><br> $= 100.0 + 273.15$ <br> $= 373.15 \text{ K}$ |
| Substitute into Equation 18.1 to calculate the change in entropy for the system. | $\Delta S = \dfrac{q_{rev}}{T}$ <br><br> $= \dfrac{25.0 \text{ g} \times \dfrac{1 \text{ mol}}{18.02 \text{ g}} \times \left( \dfrac{-40.7 \times 10^3 \text{ J}}{\text{mol}} \right)}{373.15 \text{ K}}$ <br><br> $= -151 \text{ J/K}$ |

*—Continued on the next page*

*Continued—*

> CHECK The answer has the correct units for entropy (J/K). The answer is negative as you would expect for the condensation of a gas to a liquid.
>
> FOR PRACTICE 18.2 Calculate the change in entropy that occurs in the system when 10.0 g of acetone ($C_3H_6O$) vaporizes from a liquid to a gas at its normal boiling point (56.1 °C).

## 18.5 Heat Transfer and Changes in the Entropy of the Surroundings

We have now seen that the criterion for spontaneity is an increase in the entropy of the universe. However, you can probably think of several spontaneous processes in which entropy seems to decrease. For example, when water freezes at temperatures below 0 °C, the entropy of the water decreases, yet the process is spontaneous. Similarly, when water vapor in air condenses into fog on a cold night, the entropy of the water also decreases. Why are these processes spontaneous?

To answer this question, we must return to the second law: for any spontaneous process, the entropy *of the universe* increases ($\Delta S_{univ} > 0$). Even though the entropy *of the water* decreases during freezing and condensation, the entropy *of the universe* must somehow increase in order for these processes to be spontaneous. If we define the water as the system, then $\Delta S_{sys}$ is the entropy change for the water itself, $\Delta S_{surr}$ is the entropy change for the surroundings, and $\Delta S_{univ}$ is the entropy change for the universe. The entropy change for the universe is the sum of the entropy changes for the system and the surroundings:

$$\Delta S_{univ} = \Delta S_{sys} + \Delta S_{surr}$$

The second law states that the entropy of the universe must increase ($\Delta S_{univ} > 0$) for a process to be spontaneous. The entropy of the *system* can decrease ($\Delta S_{sys} < 0$) as long as the entropy of the *surroundings* increases by a greater amount ($\Delta S_{surr} > -\Delta S_{sys}$), so that the overall entropy of the *universe* undergoes a net increase.

For liquid water freezing or water vapor condensing, we know that the change in entropy for the system ($\Delta S_{sys}$) is negative as we discussed in Section 18.4. For $\Delta S_{univ}$ to be positive, therefore, $\Delta S_{surr}$ must be positive and greater in absolute value (or magnitude) than $\Delta S_{sys}$ as shown graphically here:

$$\Delta S_{univ} = \Delta S_{sys} + \Delta S_{surr}$$

Even though (as we saw earlier) enthalpy by itself cannot determine spontaneity, the increase in the entropy of the surroundings caused by the release of heat explains why exothermic processes are so *often* spontaneous.

But why does the freezing of ice or the condensation of water increase the entropy of the surroundings? Because both processes are *exothermic*: they give off heat to the surroundings. Because we think of entropy as the dispersal or randomization of energy, *the release of heat energy by the system disperses that energy into the surroundings, increasing the entropy of the surroundings.* The freezing of water below 0 °C and the condensation of water vapor on a cold night both increase the entropy of the universe because the heat given off to the surroundings increases the entropy of the surroundings to a sufficient degree to overcome the entropy decrease in the water.

*Summarizing Entropy Changes in the Surroundings:*

▶ An exothermic process increases the entropy of the surroundings.

▶ An endothermic process decreases the entropy of the surroundings.

## The Temperature Dependence of $\Delta S_{surr}$

We have just seen how the freezing of water increases the entropy of the surroundings by dispersing heat energy into the surroundings. Yet we know that the freezing of water is not spontaneous at all temperatures. The freezing of water becomes *nonspontaneous* above 0 °C. Why? Because the magnitude of the increase in the entropy of the surroundings due to the dispersal of energy into the surroundings is *temperature dependent*.

The greater the temperature, the smaller the increase in entropy for a given amount of energy dispersed into the surroundings. Recall that the units of entropy are joules per kelvin: energy units divided by temperature units. As we have seen, *entropy is a measure of energy dispersal (joules) per unit temperature (kelvins)*. The higher the temperature, the lower the amount of entropy for a given amount of energy dispersed. We can understand the temperature dependence of entropy changes due to heat flow with a simple analogy. Imagine that you have $1000 to give away. If you gave the $1000 to a rich man, the impact on his net worth would be negligible (because he already has so much money). If you gave the same $1000 to a poor man, however, his net worth would change substantially (because he has so little money). Similarly, if you disperse 1000 J of energy into surroundings that are hot, the entropy increase is small (because the impact of the 1000 J is small on surroundings that already contain a lot of energy). If you disperse the same 1000 J of energy into surroundings that are cold, however, the entropy increase is large (because the impact of the 1000 J is great on surroundings that contain little energy). For this same reason, the impact of the heat released to the surroundings by the freezing of water depends on the temperature of the surroundings—the higher the temperature, the smaller the impact.

We can now understand why water spontaneously freezes at low temperature but not at high temperature. For the freezing of liquid water into ice, the change in entropy of the system is negative at all temperatures:

$$\Delta S_{univ} = \Delta S_{sys} + \Delta S_{surr}$$

Negative    Positive and large at low temperature
            Positive and small at high temperature

At low temperatures, on one hand, the decrease in entropy of the system is overcome by the large increase in the entropy of the surroundings (a positive quantity), resulting in a positive $\Delta S_{univ}$ and a spontaneous process. At high temperatures, on the other hand, the decrease in entropy of the system is not overcome by the increase in entropy of the surroundings (because the magnitude of the positive $\Delta S_{surr}$ is smaller at higher temperatures), resulting in a negative $\Delta S_{univ}$; therefore, the freezing of water is not spontaneous at high temperature as shown graphically here:

$$\Delta S_{univ} = \Delta S_{sys} + \Delta S_{surr} \quad \text{(for water freezing)}$$

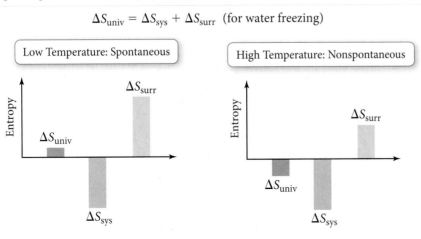

Low Temperature: Spontaneous

High Temperature: Nonspontaneous

## Quantifying Entropy Changes in the Surroundings

We have seen that when a system exchanges heat with the surroundings, it changes the entropy of the surroundings. In Section 18.4, we saw that, at constant temperature, we can use Equation 18.1 ($\Delta S = q_{rev}/T$) to quantify the entropy change in the system. We can use the same expression to quantify entropy changes in the surroundings. In other words, the change in entropy of the surroundings depends on: (a) the amount of heat transferred into or out of the surroundings; and (b) the temperature of the surroundings.

Since the surroundings are usually an infinitely large bath at constant temperature, the heat transferred into or out of that bath under conditions of constant pressure (where only PV work is allowed) is simply $-\Delta H_{sys}$. The negative sign reflects that, according to the first law of thermodynamics, any heat leaving the system must go into the surroundings and vice versa ($q_{sys} = -q_{surr}$). If we incorporate this idea into Equation 18.1, we get the following expression:

$$\Delta S_{surr} = \frac{-\Delta H_{sys}}{T} \qquad \text{(constant } P, \ T) \qquad [18.2]$$

For any chemical or physical process occurring at constant temperature and pressure, the entropy change of the surroundings is equal to the heat dispersed into or out of the surroundings ($-\Delta H_{sys}$) divided by the temperature of the surroundings in kelvins. Notice that:

- A process that emits heat into the surroundings ($\Delta H_{sys}$ negative) *increases* the entropy of the surroundings (positive $\Delta S_{surr}$).

- A process that absorbs heat from the surroundings ($\Delta H_{sys}$ positive) *decreases* the entropy of the surroundings (negative $\Delta S_{surr}$).

- The magnitude of the change in entropy of the surroundings is proportional to the magnitude of $\Delta H_{sys}$.

Equation 18.1 gives us insight into why exothermic processes have a tendency to be spontaneous at low temperatures—they increase the entropy of the surroundings. As temperature increases, however, a given negative $\Delta H_{sys}$ produces a smaller positive $\Delta S_{surr}$; for this reason, exothermicity becomes less of a determining factor for spontaneity as temperature increases.

---

**Example 18.3    Calculating Entropy Changes in the Surroundings**

Consider the combustion of propane gas:
$$C_3H_8(g) + 5\,O_2(g) \longrightarrow 3\,CO_2(g) + 4\,H_2O(g) \qquad \Delta H_{rxn} = -2044 \text{ kJ}$$

**(a)** Calculate the entropy change in the surroundings when this reaction occurs at 25 °C.

**(b)** Determine the sign of the entropy change for the system.

**(c)** Determine the sign of the entropy change for the universe. Is the reaction spontaneous?

**SOLUTION**

| | |
|---|---|
| **(a)** The entropy change of the surroundings is given by Equation 18.2. Substitute the value of $\Delta H_{rxn}$ and the temperature in kelvins and calculate $\Delta S_{surr}$. | $T = 273 + 25 = 298$ K <br><br> $\Delta S_{surr} = \dfrac{-\Delta H_{rxn}}{T}$ <br><br> $\qquad = \dfrac{-(-2044 \text{ kJ})}{298 \text{ K}}$ <br><br> $\qquad = +6.86 \text{ kJ/K}$ <br><br> $\qquad = +6.86 \times 10^3 \text{ J/K}$ |
| **(b)** Determine the number of moles of gas on each side of the reaction. An increase in the number of moles of gas implies a positive $\Delta S_{sys}$. | $C_3H_8(g) + 5\,O_2(g) \longrightarrow 3\,CO_2(g) + 4\,H_2O(g)$ <br> $\qquad\quad$ 6 mol gas $\qquad\qquad\qquad\qquad$ 7 mol gas <br><br> $\Delta S_{sys}$ is positive. |

**(c)** The change in entropy of the universe is the sum of the entropy changes of the system and the surroundings. If the entropy changes of the system and surroundings are both the same sign, the entropy change for the universe also has the same sign.

$$\Delta S_{univ} = \Delta S_{sys} + \Delta S_{surr}$$

Positive    Positive

Therefore, $\Delta S_{univ}$ is positive and the reaction is spontaneous.

---

**FOR PRACTICE 18.3** Consider the reaction between nitrogen and oxygen gas to form dinitrogen monoxide:

$$2\,N_2(g) + O_2(g) \longrightarrow 2\,N_2O(g) \qquad \Delta H_{rxn} = +163.2\,kJ$$

**(a)** Calculate the entropy change in the surroundings when this reaction occurs at 25 °C.

**(b)** Determine the sign of the entropy change for the system.

**(c)** Determine the sign of the entropy change for the universe. Is the reaction spontaneous?

**FOR MORE PRACTICE 18.3** For a reaction, $\Delta H_{rxn} = -107\,kJ$ and $\Delta S_{rxn} = 285\,J/K$. At what temperature is the change in entropy for this reaction equal to the change in entropy for the surroundings?

---

**Entropy and Biological Systems** Do biological systems contradict the second law of thermodynamics? By taking energy from their surroundings and synthesizing large, complex biological molecules, plants and animals tend to concentrate energy, not disperse it. How can this be so?

CONCEPTUAL
**CONNECTION 18.3**

---

## 18.6 Gibbs Free Energy

Equation 18.2 establishes a relationship between the enthalpy change in a system and the entropy change in the surroundings. Recall that for any process the entropy change of the universe is the sum of the entropy change of the system and the entropy change of the surroundings:

$$\Delta S_{univ} = \Delta S_{sys} + \Delta S_{surr} \qquad [18.3]$$

Combining Equation 18.3 with Equation 18.2 gives us the following relationship at constant temperature and pressure:

$$\Delta S_{univ} = \Delta S_{sys} - \frac{\Delta H_{sys}}{T} \qquad [18.4]$$

Using Equation 18.4, we can calculate $\Delta S_{univ}$ while focusing only on the *system*. If we multiply Equation 18.4 by $-T$, we arrive at the expression:

$$-T\Delta S_{univ} = -T\Delta S_{sys} + T\frac{\Delta H_{sys}}{T}$$
$$= \Delta H_{sys} - T\Delta S_{sys} \qquad [18.5]$$

If we drop the subscript *sys*—from now on $\Delta H$ and $\Delta S$ without subscripts mean $\Delta H_{sys}$ and $\Delta S_{sys}$—we get the expression:

$$-T\Delta S_{univ} = \Delta H - T\Delta S \qquad [18.6]$$

The right-hand side of Equation 18.6 represents the change in a thermodynamic function called *Gibbs free energy*. The formal definition of **Gibbs free energy (G)** is:

$$G = H - TS \qquad [18.7]$$

where $H$ is enthalpy, $T$ is the temperature in kelvins, and $S$ is entropy. The *change* in Gibbs free energy, symbolized by $\Delta G$, is expressed as follows (at constant temperature):

$$\Delta G = \Delta H - T\Delta S \qquad [18.8]$$

If we combine Equations 18.6 and 18.8, we have an equation that makes clear the significance of $\Delta G$:

$$\Delta G = -T\Delta S_{univ} \qquad (\text{constant } T, P) \qquad [18.9]$$

*The change in Gibbs free energy for a process occurring at constant temperature and pressure is proportional to the negative of $\Delta S_{univ}$. Since $\Delta S_{univ}$ is a criterion for spontaneity, $\Delta G$ is also a criterion for spontaneity (although opposite in sign). In fact, Gibbs free energy is also called chemical potential because it is analogous to mechanical potential energy discussed earlier. Just as mechanical systems tend toward lower potential energy, so chemical systems tend toward lower Gibbs free energy (toward lower chemical potential) (Figure 18.7▼).*

► **FIGURE 18.7 Gibbs Free Energy** Gibbs free energy is also called chemical potential because it determines the direction of spontaneous change for chemical systems.

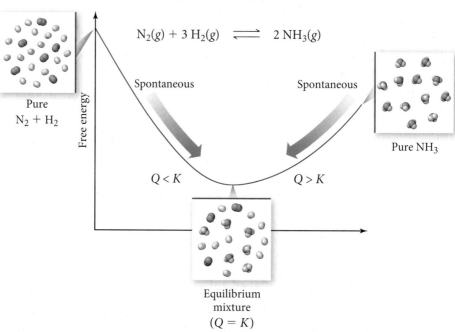

**Gibbs Free Energy Determines the Direction of Spontaneous Change**

$$N_2(g) + 3 H_2(g) \rightleftharpoons 2 NH_3(g)$$

Pure $N_2 + H_2$

Spontaneous

Spontaneous

Pure $NH_3$

Free energy

$Q < K$

$Q > K$

Equilibrium mixture $(Q = K)$

***Summarizing Gibbs Free Energy (at Constant Temperature and Pressure):***

▶ $\Delta G$ is proportional to the negative of $\Delta S_{univ}$.

▶ A decrease in Gibbs free energy ($\Delta G < 0$) corresponds to a spontaneous process.

▶ An increase in Gibbs free energy ($\Delta G > 0$) corresponds to a nonspontaneous process.

Notice that we can calculate changes in Gibbs free energy solely with reference to the system. So, to determine whether a process is spontaneous, we only have to find the change in *entropy* for the system ($\Delta S$) and the change in *enthalpy* for the system ($\Delta H$). We can then predict the spontaneity of the process at any temperature. In Chapter 6, we learned how to calculate changes in enthalpy ($\Delta H$) for chemical reactions. In Section 18.7, we learn how to calculate changes in entropy ($\Delta S$) for chemical reactions. We can then use those two quantities to calculate changes in free energy ($\Delta G$) for chemical reactions and predict their spontaneity (Section 18.8). Before we move on to these topics, let's examine some examples that demonstrate how $\Delta H$, $\Delta S$, and $T$ affect the spontaneity of chemical processes.

 **The Effect of $\Delta H$, $\Delta S$, and $T$ on Reaction Spontaneity**

# The Effect of $\Delta H$, $\Delta S$, and $T$ on Spontaneity

**Case 1: $\Delta H$ Negative, $\Delta S$ Positive** If a reaction is exothermic ($\Delta H < 0$), and if the change in entropy for the reaction is positive ($\Delta S > 0$), then the change in free energy is negative at all temperatures and the reaction is spontaneous at all temperatures:

$$\Delta G = \Delta H - T\,\Delta S$$

Negative at all temperatures

Negative

Positive

As an example, consider the dissociation of $N_2O$ :

$$2\,N_2O(g) \longrightarrow 2\,N_2(g) + O_2(g) \qquad \Delta H^\circ_{rxn} = -163.2\ kJ$$

2 mol gas      3 mol gas

Recall from Chapter 6 that $\Delta H^\circ$ represents the standard enthalpy change. The definition of the standard state was first given in Section 6.9 and is summarized in Section 18.7.

The change in *enthalpy* is negative—heat is emitted, increasing the entropy of the surroundings. The change in *entropy* for the reaction is positive, which means that the entropy of the system increases. (We can see that the change in entropy is positive from the balanced equation—the number of moles of gas increases.) Since the entropy of both the system and the surroundings increases, the entropy of the universe must also increase, making the reaction spontaneous at all temperatures.

**Case 2: $\Delta H$ Positive, $\Delta S$ Negative** If a reaction is endothermic ($\Delta H > 0$), and if the change in entropy for the reaction is negative ($\Delta S < 0$), then the change in free energy is positive at all temperatures and the reaction is nonspontaneous at all temperatures:

$$\Delta G = \Delta H - T\,\Delta S$$

Positive at all temperatures     Positive     Negative

As an example, consider the formation of ozone from oxygen:

$$3\,O_2(g) \longrightarrow 2\,O_3(g) \qquad \Delta H^\circ_{rxn} = +285.4\ kJ$$

3 mol gas      2 mol gas

The change in *enthalpy* is positive—heat is therefore absorbed, *decreasing* the entropy of the surroundings. The change in *entropy* is negative, which means that the entropy of the system decreases. (We can see that the change in entropy is negative from the balanced equation—the number of moles of gas decreases.) Since the entropy of both the system and the surroundings decreases, the entropy of the universe must also decrease, making the reaction nonspontaneous at all temperatures.

**Case 3: $\Delta H$ Negative, $\Delta S$ Negative** If a reaction is exothermic ($\Delta H < 0$), and if the change in entropy for the reaction is negative ($\Delta S < 0$), then the sign of the change in free energy depends on temperature. The reaction is spontaneous at low temperature, but nonspontaneous at high temperature:

$$\Delta G = \Delta H - T\,\Delta S$$

Negative at low temperatures   Positive at high temperatures     Negative     Negative

As an example, consider the freezing of liquid water to form ice:

$$H_2O(l) \longrightarrow H_2O(s) \qquad \Delta H^\circ = -6.01\ kJ$$

The change in *enthalpy* is negative—heat is emitted, increasing the entropy of the surroundings. The change in *entropy* is negative, which means that the entropy of the system decreases. (We can see that the change in entropy is negative from the balanced equation—a liquid turns into a solid.)

Unlike the two previous cases, where the changes in *entropy* of the system and of the surroundings had the *same* sign, the changes here are *opposite* in sign. Therefore, the sign of the change in free energy depends on the relative magnitudes of the two changes. At a low enough temperature, the heat emitted into the surroundings causes a large entropy change in the surroundings, making the process spontaneous. At high temperature, the same amount of heat is dispersed into warmer surroundings, so the positive entropy change in the surroundings is smaller, resulting in a nonspontaneous process.

**Case 4: $\Delta H$ Positive, $\Delta S$ Positive** If a reaction is endothermic ($\Delta H > 0$), and if the change in entropy for the reaction is positive ($\Delta S > 0$), then the sign of the change in free energy again depends on temperature. The reaction is nonspontaneous at low temperature but spontaneous at high temperature:

$$\Delta G = \Delta H - T\Delta S$$

Positive at low temperatures
Negative at high temperatures

Positive

Positive

As an example, consider the vaporizing of liquid water to gaseous water:

$$H_2O(l) \longrightarrow H_2O(g) \qquad \Delta H° = +40.7 \text{ kJ (at } 100\,°C)$$

The change in *enthalpy* is positive—heat is absorbed from the surroundings, so the entropy of the surroundings decreases. The change in *entropy* is positive, which means that the entropy of the system increases. (We can see that the change in entropy is positive from the balanced equation—a liquid turns into a gas.) The changes in entropy of the system and the surroundings again have opposite signs, only this time the entropy of the surroundings decreases while the entropy of the system increases. In cases such as this, high temperature favors spontaneity because the absorption of heat from the surroundings has less effect on the entropy of the surroundings as temperature increases.

Table 18.1 summarizes these effects. Notice that when $\Delta H$ and $\Delta S$ have opposite signs, the spontaneity of the reaction does not depend on temperature. *When $\Delta H$ and $\Delta S$ have the same sign, however, the spontaneity does depend on temperature.* The temperature at which the reaction changes from being spontaneous to being nonspontaneous (or vice versa) is the temperature at which $\Delta G$ changes sign, which we can find by setting $\Delta G = 0$ and solving for $T$, as shown in part b of Example 18.4.

**TABLE 18.1 The Effect of $\Delta H$, $\Delta S$, and $T$ on Spontaneity**

| $\Delta H$ | $\Delta S$ | Low Temperature | High Temperature | Example |
|---|---|---|---|---|
| − | + | Spontaneous ($\Delta G < 0$) | Spontaneous ($\Delta G < 0$) | $2 N_2O(g) \longrightarrow 2 N_2(g) + O_2(g)$ |
| + | − | Nonspontaneous ($\Delta G > 0$) | Nonspontaneous ($\Delta G > 0$) | $3 O_2(g) \longrightarrow 2 O_3(g)$ |
| − | − | Spontaneous ($\Delta G < 0$) | Nonspontaneous ($\Delta G > 0$) | $H_2O(l) \longrightarrow H_2O(s)$ |
| + | + | Nonspontaneous ($\Delta G > 0$) | Spontaneous ($\Delta G < 0$) | $H_2O(l) \longrightarrow H_2O(g)$ |

## Example 18.4 Calculating Gibbs Free Energy Changes and Predicting Spontaneity from $\Delta H$ and $\Delta S$

Consider the reaction for the decomposition of carbon tetrachloride gas:

$$CCl_4(g) \longrightarrow C(s, \text{graphite}) + 2 Cl_2(g) \qquad \Delta H = +95.7 \text{ kJ}; \ \Delta S = +142.2 \text{ J/K}$$

**(a)** Calculate $\Delta G$ at 25 °C and determine whether the reaction is spontaneous.

**(b)** If the reaction is not spontaneous at 25 °C, determine at what temperature (if any) the reaction becomes spontaneous.

**SOLUTION**

**(a)** Use Equation 18.8 to calculate $\Delta G$ from the given values of $\Delta H$ and $\Delta S$. The temperature must be in kelvins. *Be sure to express both $\Delta H$ and $\Delta S$ in the same units (usually joules).*

$T = 273 + 25 = 298 \text{ K}$
$\Delta G = \Delta H - T\Delta S$
$\quad = 95.7 \times 10^3 \text{ J} - (298 \text{ K})142.2 \text{ J/K}$
$\quad = 95.7 \times 10^3 \text{ J} - 42.4 \times 10^3 \text{ J}$
$\quad = +53.3 \times 10^3 \text{ J}$
The reaction is not spontaneous.

**(b)** Since $\Delta S$ is positive, $\Delta G$ becomes more negative with increasing temperature. To determine the temperature at which the reaction becomes spontaneous, use Equation 18.8 to find the temperature at which $\Delta G$ changes from positive to negative (set $\Delta G = 0$ and solve for $T$). The reaction is spontaneous above this temperature.

$\Delta G = \Delta H - T\Delta S$
$0 = 95.7 \times 10^3 \text{ J} - (T)142.2 \text{ J/K}$
$T = \dfrac{95.7 \times 10^3 \text{ J}}{142.2 \text{ J/K}}$
$\quad = 673 \text{ K}$

**FOR PRACTICE 18.4**    Consider the reaction:

$$C_2H_4(g) + H_2(g) \longrightarrow C_2H_6(g) \qquad \Delta H = -137.5 \text{ kJ}; \Delta S = -120.5 \text{ J/K}$$

Calculate $\Delta G$ at 25 °C and determine whether the reaction is spontaneous. Does $\Delta G$ become more negative or more positive as the temperature increases?

---

**CONCEPTUAL
CONNECTION 18.4**

**$\Delta H$, $\Delta S$, and $\Delta G$**    Which statement is true regarding the sublimation of dry ice (solid $CO_2$)?

**(a)**    $\Delta H$ is positive, $\Delta S$ is positive, and $\Delta G$ is positive at low temperatures and negative at high temperatures.

**(b)**    $\Delta H$ is negative, $\Delta S$ is negative, and $\Delta G$ is negative at low temperatures and positive at high temperatures.

**(c)**    $\Delta H$ is negative, $\Delta S$ is positive, and $\Delta G$ is negative at all temperatures.

**(d)**    $\Delta H$ is positive, $\Delta S$ is negative, and $\Delta G$ is positive at all temperatures.

---

## 18.7   Entropy Changes in Chemical Reactions: Calculating $\Delta S^{\circ}_{rxn}$

**KCV**    **Standard Molar Entropies**

In Chapter 6, we learned how to calculate standard changes in enthalpy ($\Delta H^{\circ}_{rxn}$) for chemical reactions. We now turn to calculating standard changes in *entropy* for chemical reactions.

### Defining Standard States and Standard Entropy Changes

Recall from Section 6.9 that the standard enthalpy change for a reaction ($\Delta H^{\circ}_{rxn}$) is the change in enthalpy for a process in which all reactants and products are in their standard states. Recall also the definition of the standard state:

- *For a Gas*: The standard state for a gas is the pure gas at a pressure of exactly 1 atm.

- *For a Liquid or Solid*: The standard state for a liquid or solid is the pure substance in its most stable form at a pressure of 1 atm and at the temperature of interest (often taken to be 25 °C).

- *For a Substance in Solution*: The standard state for a substance in solution is a concentration of 1 M.

We define the **standard entropy change for a reaction ($\Delta S^{\circ}_{rxn}$)** as the change in *entropy* for a process in which all reactants and products are in their standard states. Since entropy is a function of state, the standard change in entropy is therefore the standard entropy of the products minus the standard entropy of the reactants:

$$\Delta S^{\circ}_{rxn} = S^{\circ}_{products} - S^{\circ}_{reactants}$$

But how do we find the standard entropies of the reactants and products? Recall from Chapter 6 that we defined *standard molar enthalpies of formation* ($\Delta H^{\circ}_f$) to use in calculating $\Delta H^{\circ}_{rxn}$. We now need to define **standard molar entropies ($S^{\circ}$)** to use in calculating $\Delta S^{\circ}_{rxn}$.

The standard state has recently been changed to a pressure of 1 bar, which is very close to 1 atm (1 atm = 1.013 bar). Both standards are now in common use.

### Standard Molar Entropies ($S^{\circ}$) and the Third Law of Thermodynamics

In Chapter 6, we defined a *relative* zero for enthalpy. To do this, we assigned a value of zero to the standard enthalpy of formation for an element in its standard state. This was necessary because absolute values of enthalpy cannot be determined. In other words, for

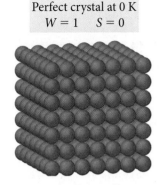

Perfect crystal at 0 K
$W = 1$    $S = 0$

▲ FIGURE 18.8 **Zero Entropy**
A perfect crystal at 0 K has only
one possible way to arrange its
components.

enthalpy, there is no absolute zero against which to measure all other values; therefore, we always have to rely on enthalpy changes from an arbitrarily assigned standard (the elements in their standard states and most stable forms).

For entropy, however, *there is an absolute zero*. The absolute zero of entropy is established by the **third law of thermodynamics**, which states:

**The entropy of a perfect crystal at absolute zero (0 K) is zero.**

A perfect crystal at a temperature of absolute zero has only one possible way ($W = 1$) to arrange its components (Figure 18.8 ◄). Based on Boltzmann's definition of entropy ($S = k \ln W$), its entropy is zero ($S = k \ln 1 = 0$).

We can measure all entropy values against the absolute zero of entropy as defined by the third law. Table 18.2 lists values of standard entropies at 25 °C for selected substances. A more complete list is in Appendix IIB. Standard entropy values are listed in units of joules per mole per kelvin (J/mol·K). The units of mole in the denominator are required because *entropy is an extensive property*—it depends on the amount of the substance.

| TABLE 18.2 Standard Molar Entropy Values ($S°$) for Selected Substances at 298 K | | | | | |
|---|---|---|---|---|---|
| Substance | $S°(J/mol \cdot K)$ | Substance | $S°(J/mol \cdot K)$ | Substance | $S°(J/mol \cdot K)$ |
| Gases | | Liquids | | Solids | |
| $H_2(g)$ | 130.7 | $H_2O(l)$ | 70.0 | $MgO(s)$ | 27.0 |
| $Ar(g)$ | 154.8 | $CH_3OH(l)$ | 126.8 | $Fe(s)$ | 27.3 |
| $CH_4(g)$ | 186.3 | $Br_2(l)$ | 152.2 | $Li(s)$ | 29.1 |
| $H_2O(g)$ | 188.8 | $C_6H_6(l)$ | 173.4 | $Cu(s)$ | 33.2 |
| $N_2(g)$ | 191.6 | | | $Na(s)$ | 51.3 |
| $NH_3(g)$ | 192.8 | | | $K(s)$ | 64.7 |
| $F_2(g)$ | 202.8 | | | $NaCl(s)$ | 72.1 |
| $O_2(g)$ | 205.2 | | | $CaCO_3(s)$ | 91.7 |
| $Cl_2(g)$ | 223.1 | | | $FeCl_3(s)$ | 142.3 |
| $C_2H_4(g)$ | 219.3 | | | | |

At 25 °C, the standard entropy of any substance is the energy dispersed into one mole of that substance at 25 °C, which depends on the number of "places" to put energy within the substance. The factors that affect the number of "places" to put energy—and therefore the standard entropy—include the state of the substance, the molar mass of the substance, the particular allotrope, its molecular complexity, and its extent of dissolution. Let's examine each of these separately.

Some elements exist in two or more forms, called *allotropes*, within the same state.

### Relative Standard Entropies: Gases, Liquids, and Solids

As we saw in Section 18.3, the entropy of a gas is generally greater than the entropy of a liquid, which is in turn greater than the entropy of a solid. We can see these trends in the tabulated values of standard entropies. For example, consider the relative standard entropies of liquid water and gaseous water at 25 °C:

| | $S°(J/mol \cdot K)$ |
|---|---|
| $H_2O(l)$ | 70.0 |
| $H_2O(g)$ | 188.8 |

Gaseous water has a much greater standard entropy because, as we discussed in Section 18.3, it has more energetically equivalent ways to arrange its components, which in turn results in greater energy dispersal at 25 °C.

## Relative Standard Entropies: Molar Mass

Consider the standard entropies of the noble gases at 25 °C :

| | S° (J/mol·K) | |
|---|---|---|
| He(g) | 126.2 | |
| Ne(g) | 146.1 | |
| Ar(g) | 154.8 | |
| Kr(g) | 163.8 | |
| Xe(g) | 169.4 | |

The more massive the noble gas, the greater its entropy at 25 °C. A complete explanation of why entropy increases with increasing molar mass is beyond the scope of this book. Briefly, the energy states associated with the motion of heavy atoms are more closely spaced than those of lighter atoms. The more closely spaced energy states allow for greater dispersal of energy at a given temperature and therefore greater entropy. This trend is true only for elements in the same state. (The effect of a state change—from a liquid to a gas, for example—is far greater than the effect of molar mass.)

## Relative Standard Entropies: Allotropes

As mentioned previously, some elements can exist in two or more forms—called *allotropes*—in the same state of matter. For example, the allotropes of carbon include diamond and graphite—both solid forms of carbon. Because the arrangement of atoms within these forms is different, their standard molar entropies are different:

| | S° (J/mol·K) | |
|---|---|---|
| C(s, diamond) | 2.4 | |
| C(s, graphite) | 5.7 | |

In diamond the atoms are constrained by chemical bonds in a highly restricted three-dimensional crystal structure. In graphite the atoms bond together in sheets, but the sheets have freedom to slide past each other. The less constrained structure of graphite results in more "places" to put energy and therefore greater entropy compared to diamond.

## Relative Standard Entropies: Molecular Complexity

For a given state of matter, entropy generally increases with increasing molecular complexity. For example, consider the standard entropies of argon and nitrogen monoxide gas:

|  | Molar Mass (g/mol) | $S°(J/mol \cdot K)$ |
|---|---|---|
| Ar(g) | 39.948 | 154.8 |
| NO(g) | 30.006 | 210.8 |

Ar has a greater molar mass than NO, yet it has less entropy at 25 °C. Why? Molecules generally have more "places" to put energy than do atoms. In a gaseous sample of argon, on one hand, the only form that energy can take is the translational motion of the atoms. In a gaseous sample of NO, on the other hand, energy can take the form of translational motion, rotational motion, and (at high enough temperatures) vibrational motions of the molecules (Figure 18.9 ▼). Therefore, for a given state, molecules generally have a greater entropy than free atoms.

▶ **FIGURE 18.9** **"Places" for Energy in Gaseous NO** Energy can be contained in translational motion, rotational motion, and (at high enough temperatures) vibrational motion.

| Translational motion | Rotational motion | Vibrational motion |

Similarly, more complex molecules generally have more entropy than simpler ones. For example, consider the standard entropies of carbon monoxide and ethene gas:

|  | Molar Mass (g/mol) | $S°(J/mol \cdot K)$ |
|---|---|---|
| CO(g) | 28.01 | 197.7 |
| $C_2H_4(g)$ | 28.05 | 219.3 |

These two substances have nearly the same molar mass, but the greater complexity of $C_2H_4$ results in a greater molar entropy. When molecular complexity and molar mass both increase (as is often the case), molar entropy also increases, as demonstrated by the oxides of nitrogen:

|  | $S°(J/mol \cdot K)$ |
|---|---|
| NO(g) | 210.8 |
| $NO_2(g)$ | 240.1 |
| $N_2O_4(g)$ | 304.4 |

The increasing molecular complexity as we move down this list, as well as the increasing molar mass, results in more "places" to put energy and therefore greater entropy.

## Relative Standard Entropies: Dissolution

The standard entropies for aqueous solutions are for the solution in its standard state, which is defined as having a concentration of 1 M.

The dissolution of a crystalline solid into solution usually results in an increase in entropy. For example, consider the standard entropies of solid and aqueous potassium chlorate:

|  | $S°(J/mol \cdot K)$ |
|---|---|
| $KClO_3(s)$ | 143.1 |
| $KClO_3(aq)$ | 265.7 |

When solid potassium chlorate dissolves in water, the energy that was concentrated within the crystal becomes dispersed throughout the entire solution. The greater energy dispersal results in greater entropy.

**Standard Entropies** Arrange these gases in order of increasing standard molar entropy: $SO_3$, Kr, $Cl_2$.

## Calculating the Standard Entropy Change ($\Delta S^\circ_{rxn}$) for a Reaction

Since entropy is a state function, and since standard entropies for many common substances are tabulated, we can calculate the standard entropy change for a chemical reaction by calculating the difference in entropy between the products and the reactants. More specifically,

> **To calculate $\Delta S^\circ_{rxn}$, subtract the standard entropies of the reactants multiplied by their stoichiometric coefficients from the standard entropies of the products multiplied by their stoichiometric coefficients. In the form of an equation:**

$$\Delta S^\circ_{rxn} = \sum n_p S^\circ(\text{products}) - \sum n_r S^\circ(\text{reactants}) \qquad [18.10]$$

In Equation 18.10, $n_p$ represents the stoichiometric coefficients of the products, $n_r$ represents the stoichiometric coefficients of the reactants, and $S^\circ$ represents the standard entropies. Keep in mind when using this equation that, *unlike enthalpies of formation, which are zero for elements in their standard states, standard entropies are always nonzero at 25 °C.* Example 18.5 demonstrates the application of Equation 18.10.

---

**Example 18.5** | **Calculating Standard Entropy Changes ($\Delta S^\circ_{rxn}$)**

Calculate $\Delta S^\circ_{rxn}$ for the balanced chemical equation:

$$4\,NH_3(g) + 5\,O_2(g) \longrightarrow 4\,NO(g) + 6\,H_2O(g)$$

**SOLUTION**

Begin by looking up the standard entropy for each reactant and product in Appendix IIB. Always note the correct state—(g), (l), (aq), or (s)—for each reactant and product.

| Reactant or product | S° (in J/mol · K) |
|---|---|
| $NH_3(g)$ | 192.8 |
| $O_2(g)$ | 205.2 |
| $NO(g)$ | 210.8 |
| $H_2O(g)$ | 188.8 |

Calculate $\Delta S^\circ_{rxn}$ by substituting the appropriate values into Equation 18.10. Remember to include the stoichiometric coefficients in your calculation.

$$\begin{aligned}\Delta S^\circ_{rxn} &= \sum n_p S^\circ(\text{products}) - \sum n_r S^\circ(\text{reactants})\\ &= [4(S^\circ_{NO(g)}) + 6(S^\circ_{H_2O(g)})] - [4(S^\circ_{NH_3(g)}) + 5(S^\circ_{O_2(g)})]\\ &= [4(210.8\,J/K) + 6(188.8\,J/K)] - [4(192.8\,J/K) + 5(205.2\,J/K)]\\ &= 1976.0\,J/K - 1797.2\,J/K = 178.8\,J/K\end{aligned}$$

**CHECK** Notice that $\Delta S^\circ_{rxn}$ is positive, as you would expect for a reaction in which the number of moles of gas increases.

**FOR PRACTICE 18.5** Calculate $\Delta S^\circ_{rxn}$ for the balanced chemical equation:

$$2\,H_2S(g) + 3\,O_2(g) \longrightarrow 2\,H_2O(g) + 2\,SO_2(g)$$

---

## 18.8 Free Energy Changes in Chemical Reactions: Calculating $\Delta G^\circ_{rxn}$

In the previous section, we learned how to calculate the standard change in entropy for a chemical reaction ($\Delta S^\circ_{rxn}$). However, the criterion for spontaneity at standard conditions is the **standard change in free energy ($\Delta G^\circ_{rxn}$)**. In this section, we examine three methods to calculate the standard change in free energy for a reaction ($\Delta G^\circ_{rxn}$). In the first method, we calculate $\Delta H°_{rxn}$ and $\Delta S°_{rxn}$ from tabulated values of $\Delta H°_f$ and $S°$, and then use the relationship $\Delta G°_{rxn} = \Delta H°_{rxn} - T\Delta S°_{rxn}$ to calculate $\Delta G°_{rxn}$. In the second method, we use tabulated values of free energies of formation to calculate $\Delta G°_{rxn}$ directly. In the third method, we determine the free energy change for a stepwise reaction from the free energy changes of each of the steps. At the end of this section, we discuss what is "free" about free energy. Remember that $\Delta G°_{rxn}$ is extremely useful because it tells us about the spontaneity of a process at standard conditions. The more negative $\Delta G°_{rxn}$ is, the more spontaneous the process (the further it will go toward products to reach equilibrium).

## Calculating Standard Free Energy Changes with $\Delta G°_{rxn} = \Delta H°_{rxn} - T\Delta S°_{rxn}$

In Chapter 6 (Section 6.9), we used tabulated values of standard enthalpies of formation to calculate $\Delta H°_{rxn}$. In the previous section of this chapter, we used tabulated values of standard entropies to calculate $\Delta S°_{rxn}$. We can use these calculated values of $\Delta H°_{rxn}$ and $\Delta S°_{rxn}$ to determine the standard free energy change for a reaction by using the equation:

$$\Delta G°_{rxn} = \Delta H°_{rxn} - T\Delta S°_{rxn} \qquad [18.11]$$

Since tabulated values of standard enthalpies of formation ($\Delta H°_f$) and standard entropies ($S°$) are usually applicable at 25 °C, the equation should (strictly speaking) be valid only when $T = 298$ K (25 °C). However, the changes in $\Delta H°_{rxn}$ and $\Delta S°_{rxn}$ over a limited temperature range are small when compared to the changes in the value of the temperature itself. Therefore, we can use Equation 18.11 to estimate changes in free energy at temperatures other than 25 °C.

---

**Example 18.6** | **Calculating the Standard Change in Free Energy for a Reaction Using $\Delta G°_{rxn} = \Delta H°_{rxn} - T\Delta S°_{rxn}$**

One of the possible initial steps in the formation of acid rain is the oxidation of the pollutant $SO_2$ to $SO_3$ by the reaction:

$$SO_2(g) + \tfrac{1}{2}O_2(g) \longrightarrow SO_3(g)$$

Calculate $\Delta G°_{rxn}$ at 25 °C and determine whether the reaction is spontaneous.

**SOLUTION**

| Begin by looking up (in Appendix IIB) the standard enthalpy of formation and the standard entropy for each reactant and product. | Reactant or product | $\Delta H°_f$(kJ/mol) | $S°$(J/mol · K) |
|---|---|---|---|
| | $SO_2(g)$ | −296.8 | 248.2 |
| | $O_2(g)$ | 0 | 205.2 |
| | $SO_3(g)$ | −395.7 | 256.8 |

| Calculate $\Delta H°_{rxn}$ using Equation 6.15. | $\Delta H°_{rxn} = \sum n_p \, \Delta H°_f(\text{products}) - \sum n_r \, \Delta H°_f(\text{reactants})$ <br> $= [\Delta H°_{f,\,SO_3(g)}] - [\Delta H°_{f,\,SO_2(g)} + \tfrac{1}{2}(\Delta H°_{f,\,O_2(g)})]$ <br> $= -395.7 \text{ kJ} - (-296.8 \text{ kJ} + 0.0 \text{ kJ}) = -98.9 \text{ kJ}$ |
|---|---|

| Calculate $\Delta S°_{rxn}$ using Equation 18.10. | $\Delta S°_{rxn} = \sum n_p \, S°(\text{products}) - \sum n_r \, S°(\text{reactants})$ <br> $= [S°_{SO_3(g)}] - [S°_{SO_2(g)} + \tfrac{1}{2}(S°_{O_2(g)})]$ <br> $= 256.8 \text{ J/K} - [248.2 \text{ J/K} + \tfrac{1}{2}(205.2 \text{ J/K})] = -94.0 \text{ J/K}$ |
|---|---|

| Calculate $\Delta G°_{rxn}$ using the calculated values of $\Delta H°_{rxn}$ and $\Delta S°_{rxn}$ and Equation 18.11. Convert the temperature to kelvins. | $T = 25 + 273 = 298$ K <br> $\Delta G°_{rxn} = \Delta H°_{rxn} - T\Delta S°_{rxn}$ <br> $= -98.9 \times 10^3 \text{ J} - 298 \text{ K}\,(-94.0 \text{ J/K})$ <br> $= -70.9 \times 10^3 \text{ J} = -70.9 \text{ kJ}$ <br> The reaction is spontaneous at this temperature because $\Delta G°_{rxn}$ is negative. |
|---|---|

---

**FOR PRACTICE 18.6** Consider the oxidation of NO to $NO_2$:

$$NO(g) + \tfrac{1}{2}O_2(g) \longrightarrow NO_2(g)$$

Calculate $\Delta G°_{rxn}$ at 25 °C and determine whether the reaction is spontaneous at standard conditions.

| **Example** 18.7 | **Estimating the Standard Change in Free Energy for a Reaction at a Temperature Other Than 25 °C Using $\Delta G^\circ_{rxn} = \Delta H^\circ_{rxn} - T\Delta S^\circ_{rxn}$** |
|---|---|

For the reaction in Example 18.6, estimate the value of $\Delta G^\circ_{rxn}$ at 125 °C. Is the reaction more or less spontaneous at this elevated temperature; that is, is the value of $\Delta G^\circ_{rxn}$ more negative (more spontaneous) or more positive (less spontaneous)?

**SOLUTION**

| | |
|---|---|
| Estimate $\Delta G^\circ_{rxn}$ at the new temperature using the calculated values of $\Delta H^\circ_{rxn}$ and $\Delta S^\circ_{rxn}$ from Example 18.6. For $T$, convert the given temperature to kelvins. Make sure to use the same units for $\Delta H^\circ_{rxn}$ and $\Delta S^\circ_{rxn}$ (usually joules). | $T = 125 + 273 = 398 \text{ K}$ <br> $\begin{aligned} \Delta G^\circ_{rxn} &= \Delta H^\circ_{rxn} - T\Delta S^\circ_{rxn} \\ &= -98.9 \times 10^3 \text{ J} - 398 \text{ K} (-94.0 \text{ J/K}) \\ &= -61.5 \times 10^3 \text{ J} \\ &= -61.5 \text{ kJ} \end{aligned}$ <br> Since the value of $\Delta G^\circ_{rxn}$ at this elevated temperature is less negative (or more positive) than the value of $\Delta G^\circ_{rxn}$ at 25 °C (which is $-70.9$ kJ), the reaction is less spontaneous. |

**FOR PRACTICE 18.7** For the reaction in For Practice 18.6, calculate the value of $\Delta G^\circ_{rxn}$ at $-55$ °C. Is the reaction more spontaneous (more negative $\Delta G^\circ_{rxn}$) or less spontaneous (more positive $\Delta G^\circ_{rxn}$) at the lower temperature?

## Calculating $\Delta G^\circ_{rxn}$ with Tabulated Values of Free Energies of Formation

Because $\Delta G^\circ_{rxn}$ is the *change* in free energy for a chemical reaction—the difference in free energy between the products and the reactants—and because free energy is a state function, we can calculate $\Delta G^\circ_{rxn}$ by subtracting the free energies of the reactants of the reaction from the free energies of the products of the reaction. Also, since we are interested only in *changes* in free energy (and not in absolute values of free energy), we are free to define the *zero* of free energy as conveniently as possible. By analogy with our definition of enthalpies of formation, we define the **free energy of formation ($\Delta G^\circ_f$)** as follows:

> **The free energy of formation ($\Delta G^\circ_f$) is the change in free energy when 1 mol of a compound in its standard state forms from its constituent elements in their standard states. The free energy of formation of pure elements in their standard states is zero.**

We can measure all changes in free energy relative to pure elements in their standard states. To calculate $\Delta G^\circ_{rxn}$, we subtract the free energies of formation of the reactants multiplied by their stoichiometric coefficients from the free energies of formation of the products multiplied by their stoichiometric coefficients. In the form of an equation:

$$\Delta G^\circ_{rxn} = \sum n_p \Delta G^\circ_f(\text{products}) - \sum n_r \Delta G^\circ_f(\text{reactants}) \qquad [18.12]$$

In Equation 18.12, $n_p$ represents the stoichiometric coefficients of the products, $n_r$ represents the stoichiometric coefficients of the reactants, and $\Delta G^\circ_f$ represents the standard free energies of formation.

Table 18.3 lists $\Delta G^\circ_f$ values for selected substances. You can find a more complete list in Appendix IIB. Notice that, by definition, *elements* have standard free energies of formation of zero. Notice also that most *compounds* have negative standard free energies of formation. This means that those compounds spontaneously form from their elements in their standard states. Compounds with positive free energies of formation do not spontaneously form from their elements and are therefore less common.

Example 18.8 demonstrates the calculation of $\Delta G^\circ_{rxn}$ from $\Delta G^\circ_f$ values. This method of calculating $\Delta G^\circ_{rxn}$ works only at the temperature for which the free energies of formation are tabulated, namely, 25 °C. To estimate $\Delta G^\circ_{rxn}$ at other temperatures we must use $\Delta G^\circ_{rxn} = \Delta H^\circ_{rxn} - T\Delta S^\circ_{rxn}$, as demonstrated previously.

**TABLE 18.3 Standard Molar Free Energies of Formation $\Delta G_f^\circ$ for Selected Substances at 298 K**

| Substance | $\Delta G_f^\circ$ (kJ/mol) | Substance | $\Delta G_f^\circ$ (kJ/mol) |
|---|---|---|---|
| $H_2(g)$ | 0 | $CH_4(g)$ | −50.5 |
| $O_2(g)$ | 0 | $H_2O(g)$ | −228.6 |
| $N_2(g)$ | 0 | $H_2O(l)$ | −237.1 |
| C(s, graphite) | 0 | $NH_3(g)$ | −16.4 |
| C(s, diamond) | 2.900 | NO(g) | +87.6 |
| CO(g) | −137.2 | $NO_2(g)$ | +51.3 |
| $CO_2(g)$ | −394.4 | NaCl(s) | −384.1 |

**Example** 18.8 | **Calculating $\Delta G_{rxn}^\circ$ from Standard Free Energies of Formation**

Ozone in the lower atmosphere is a pollutant that can form by the following reaction involving the oxidation of unburned hydrocarbons:

$$CH_4(g) + 8\,O_2(g) \longrightarrow CO_2(g) + 2\,H_2O(g) + 4\,O_3(g)$$

Use the standard free energies of formation to determine $\Delta G_{rxn}^\circ$ for this reaction at 25 °C.

**SOLUTION**

| | | |
|---|---|---|
| Begin by looking up (in Appendix IIB) the standard free energies of formation for each reactant and product. Remember that the standard free energy of formation of a pure element in its standard state is zero. | **Reactant or product** \  $\Delta G_f^\circ$ **(in kJ/mol)** | |

| Reactant or product | $\Delta G_f^\circ$ (in kJ/mol) |
|---|---|
| $CH_4(g)$ | −50.5 |
| $O_2(g)$ | 0.0 |
| $CO_2(g)$ | −394.4 |
| $H_2O(g)$ | −228.6 |
| $O_3(g)$ | 163.2 |

| | |
|---|---|
| Calculate $\Delta G_{rxn}^\circ$ by substituting into Equation 18.12. | $\Delta G_{rxn}^\circ = \sum n_p \Delta G_f^\circ(\text{products}) - \sum n_r \Delta G_f^\circ(\text{reactants})$ <br><br> $= [\Delta G_{f,\,CO_2(g)}^\circ + 2(\Delta G_{f,\,H_2O(g)}^\circ) + 4(\Delta G_{f,\,O_3(g)}^\circ)] - [\Delta G_{f,\,CH_4(g)}^\circ + 8(\Delta G_{f,\,O_2(g)}^\circ)]$ <br><br> $= [-394.4\text{ kJ} + 2(-228.6\text{ kJ}) + 4(163.2\text{ kJ})] - [-50.5\text{ kJ} + 8(0.0\text{ kJ})]$ <br><br> $= -198.8\text{ kJ} + 50.5\text{ kJ}$ <br><br> $= -148.3\text{ kJ}$ |

**FOR PRACTICE 18.8**    One of the reactions that occurs within a catalytic converter in the exhaust pipe of a car is the simultaneous oxidation of carbon monoxide and reduction of NO (both of which are harmful pollutants):

$$2\,CO(g) + 2\,NO(g) \longrightarrow 2\,CO_2(g) + N_2(g)$$

Use standard free energies of formation to determine $\Delta G_{rxn}^\circ$ for this reaction at 25 °C. Is the reaction spontaneous at standard conditions?

**FOR MORE PRACTICE 18.8**    In For Practice 18.8, you calculated $\Delta G_{rxn}^\circ$ for the simultaneous oxidation of carbon monoxide and reduction of NO using standard free energies of formation. Calculate $\Delta G_{rxn}^\circ$ for that reaction again at 25 °C, only this time use $\Delta G_{rxn}^\circ = \Delta H_{rxn}^\circ - T\Delta S_{rxn}^\circ$. How do the two values compare? Use your results to calculate $\Delta G_{rxn}^\circ$ at 500.0 K and explain why you could not calculate $\Delta G_{rxn}^\circ$ at 500.0 K using tabulated standard free energies of formation.

## Chemistry IN YOUR DAY | Making a Nonspontaneous Process Spontaneous

A process that is nonspontaneous can be made spontaneous by coupling it with another process that is highly spontaneous. For example, hydrogen gas is a potential future fuel because it can be used in a fuel cell (a type of battery in which the reactants are constantly supplied—see Chapter 19) to generate electricity. The main problem with switching to hydrogen is securing a source. Where can we get the huge amounts of hydrogen gas that would be needed to meet our world's energy needs?

Earth's oceans and lakes, of course, contain vast amounts of hydrogen. But that hydrogen is locked up in water molecules, and the decomposition of water into hydrogen and oxygen has a positive $\Delta G_{rxn}^\circ$ and is therefore nonspontaneous:

$$H_2O(g) \longrightarrow H_2(g) + \tfrac{1}{2}O_2(g) \quad \Delta G_{rxn}^\circ = +228.6 \text{ kJ}$$

To obtain hydrogen from water, we need to find another reaction with a highly negative $\Delta G_{rxn}^\circ$ that can couple with the decomposition reaction to give an overall reaction with a negative $\Delta G_{rxn}^\circ$. For example, the oxidation of carbon monoxide to carbon dioxide has a large negative $\Delta G_{rxn}^\circ$ and is highly spontaneous:

$$CO(g) + \tfrac{1}{2}O_2(g) \longrightarrow CO_2(g) \quad \Delta G_{rxn}^\circ = -257.2 \text{ kJ}$$

If we add the two reactions together, we get a negative $\Delta G_{rxn}^\circ$:

Nonspontaneous

$$H_2O(g) \longrightarrow H_2(g) + \tfrac{1}{2}O_2(g) \quad \Delta G_{rxn}^\circ = +228.6 \text{ kJ}$$

$$CO(g) + \tfrac{1}{2}O_2(g) \longrightarrow CO_2(g) \quad \Delta G_{rxn}^\circ = -257.2 \text{ kJ}$$

$$H_2O(g) + CO(g) \longrightarrow H_2(g) + CO_2(g) \quad \Delta G_{rxn}^\circ = -28.6 \text{ kJ}$$

Spontaneous

The reaction between water and carbon monoxide is thus a spontaneous way to generate hydrogen gas.

The coupling of nonspontaneous reactions with highly spontaneous ones is also important in biological systems. The synthesis reactions that create the complex biological molecules (such as proteins and DNA) needed by living organisms, for example, are themselves nonspontaneous. Living systems grow and reproduce by coupling these nonspontaneous reactions to highly spontaneous ones. The main spontaneous reaction that ultimately drives the nonspontaneous ones is the metabolism of food. The oxidation of glucose, for example, is highly spontaneous:

$$C_6H_{12}O_6(s) + 6\, O_2(g) \longrightarrow 6\, CO_2(g) + 6\, H_2O(l)$$
$$\Delta G_{rxn}^\circ = -2880 \text{ kJ}$$

Spontaneous reactions such as these ultimately drive the nonspontaneous reactions necessary to sustain life.

## Calculating $\Delta G_{rxn}^\circ$ for a Stepwise Reaction from the Changes in Free Energy for Each of the Steps

Recall from Section 6.8 that because enthalpy is a state function, we can calculate $\Delta H_{rxn}^\circ$ for a stepwise reaction from the sum of the changes in enthalpy for each step (according to Hess's law). Since free energy is also a state function, the same relationships that we covered in Chapter 6 for enthalpy also apply to free energy:

1. If a chemical equation is multiplied by some factor, then $\Delta G_{rxn}$ is also multiplied by the same factor.

2. If a chemical equation is reversed, then $\Delta G_{rxn}$ changes sign.

3. If a chemical equation can be expressed as the sum of a series of steps, then $\Delta G_{rxn}$ for the overall equation is the sum of the free energies of reactions for each step.

Example 18.9 illustrates the use of these relationships to calculate $\Delta G_{rxn}^\circ$ for a stepwise reaction.

### Example 18.9 | Calculating $\Delta G_{rxn}^\circ$ for a Stepwise Reaction

Find $\Delta G_{rxn}^\circ$ for the reaction:

$$3\, C(s) + 4\, H_2(g) \longrightarrow C_3H_8(g)$$

Use the following reactions with known $\Delta G_{rxn}^\circ$ values:

$$C_3H_8(g) + 5\, O_2(g) \longrightarrow 3\, CO_2(g) + 4\, H_2O(g) \quad \Delta G_{rxn}^\circ = -2074 \text{ kJ}$$
$$C(s) + O_2(g) \longrightarrow CO_2(g) \quad \Delta G_{rxn}^\circ = -394.4 \text{ kJ}$$
$$2\, H_2(g) + O_2(g) \longrightarrow 2\, H_2O(g) \quad \Delta G_{rxn}^\circ = -457.1 \text{ kJ}$$

*—Continued on the next page*

*Continued—*

| SOLUTION | |
|---|---|
| To work this problem, you need to manipulate the given reactions with known values of $\Delta G^\circ_{rxn}$ in such a way as to get the reactants of interest on the left, the products of interest on the right, and other species to cancel. | |
| Since the first reaction has $C_3H_8$ as a reactant, and the reaction of interest has $C_3H_8$ as a product, reverse the first reaction and change the sign of $\Delta G^\circ_{rxn}$. | $3\,CO_2(g) + 4\,H_2O(g) \longrightarrow C_3H_8(g) + 5\,O_2(g)$    $\Delta G^\circ_{rxn} = +2074\ kJ$ |
| The second reaction has C as a reactant and $CO_2$ as a product, as required in the reaction of interest. However, the coefficient for C is 1, and in the reaction of interest, the coefficient for C is 3. Therefore, multiply this equation and its $\Delta G^\circ_{rxn}$ by 3. | $3 \times [C(s) + O_2(g) \longrightarrow CO_2(g)]$    $\Delta G^\circ_{rxn} = 3 \times (-394.4\ kJ)$ $= -1183\ kJ$ |
| In the third reaction $H_2(g)$ is a reactant, as required. However, the coefficient for $H_2$ is 2, and in the reaction of interest, the coefficient for $H_2$ is 4. Multiply this reaction and its $\Delta G^\circ_{rxn}$ by 2. | $2 \times [2\,H_2(g) + O_2(g) \longrightarrow 2\,H_2O(g)]$    $\Delta G^\circ_{rxn} = 2 \times (-457.1\ kJ)$ $= -914.2$ |
| Lastly, rewrite the three reactions after multiplying through by the indicated factors and show how they sum to the reaction of interest. $\Delta G^\circ_{rxn}$ for the reaction of interest is then the sum of the $\Delta G$'s for the steps. | $3\,CO_2(g) + 4\,H_2O(g) \longrightarrow C_3H_8(g) + 5\,O_2(g)$    $\Delta G^\circ_{rxn} = +2074\ kJ$ $3\,C(s) + 3\,O_2(g) \longrightarrow 3\,CO_2(g)$    $\Delta G^\circ_{rxn} = -1183\ kJ$ $4\,H_2(g) + 2\,O_2(g) \longrightarrow 4\,H_2O(g)$    $\Delta G^\circ_{rxn} = -914.2\ kJ$ $\overline{3\,C(s) + 4\,H_2(g) \longrightarrow C_3H_8(g)}$    $\Delta G^\circ_{rxn} = -23\ kJ$ |

**FOR PRACTICE 18.9**

Find $\Delta G^\circ_{rxn}$ for the reaction:

$$N_2O(g) + NO_2(g) \longrightarrow 3\,NO(g)$$

Use the following reactions with known $\Delta G^\circ_{rxn}$ values:

$2\,NO(g) + O_2(g) \longrightarrow 2\,NO_2(g)$    $\Delta G^\circ_{rxn} = -71.2\ kJ$
$N_2(g) + O_2(g) \longrightarrow 2\,NO(g)$    $\Delta G^\circ_{rxn} = +175.2\ kJ$
$2\,N_2O(g) \longrightarrow 2\,N_2(g) + O_2(g)$    $\Delta G^\circ_{rxn} = -207.4\ kJ$

Maximum work = 50.5 kJ

$C(s, graphite) + 2\,H_2(g)$

$CH_4(g)$
$\Delta H^\circ_{rxn} = -74.6\ kJ$

Minimum heat lost to surroundings = 24.1 kJ

▲ **FIGURE 18.10 Free Energy** Although the reaction produces 74.6 kJ of heat, only a maximum of 50.5 kJ is available to do work. The rest of the energy is lost to the surroundings.

# Why Free Energy Is "Free"

We often want to use the energy released by a chemical reaction to do work. For example, in an automobile engine, we use the energy released by the combustion of gasoline to move the car forward. The change in free energy of a chemical reaction represents the maximum amount of energy available, or *free*, to do work (if $\Delta G^\circ_{rxn}$ is negative).

For many reactions, the change in free energy is less than the change in enthalpy for the reaction. Consider the reaction between carbon and hydrogen occurring at 25 °C :

$$C(s, graphite) + 2\,H_2(g) \longrightarrow CH_4(g)$$
$$\Delta H^\circ_{rxn} = -74.6\ kJ$$
$$\Delta S^\circ_{rxn} = -80.8\ J/K$$
$$\Delta G^\circ_{rxn} = -50.5\ kJ$$

The reaction is exothermic and gives off 74.6 kJ of heat energy. However, the maximum amount of energy available for useful work is only 50.5 kJ (Figure 18.10 ◀). Why? We can see that the change in entropy of the *system* is negative. Nevertheless, the reaction is spontaneous. This is possible only if some of the emitted heat goes to increase the entropy of the surroundings by an amount sufficient to make the change in entropy of the *universe* positive. The amount of energy available to do work (the free energy) is what remains after accounting for the heat that must be lost to the surroundings.

## Reversible Process

Weight of sand exactly matches pressure at each increment.

▲ **FIGURE 18.11 A Reversible Process**  In a reversible process, the free energy is drawn out in infinitesimally small increments that exactly match the amount of energy that the process is producing in that increment. In this case, grains of sand are removed one at a time, resulting in a series of small expansions in which the weight of sand almost exactly matches the pressure of the expanding gas. This process is close to reversible—each sand grain would need to have an infinitesimally small mass for the process to be fully reversible.

The change in free energy for a chemical reaction represents a *theoretical limit* as to how much work can be done by the reaction. In a *real* reaction, the amount of energy available to do work is even *less* than $\Delta G^\circ_{rxn}$ because additional energy is lost to the surroundings as heat. The only reactions that reach the theoretical limit are reversible reactions (see Section 18.4 for the definition of reversible processes). A reversible chemical reaction occurs infinitesimally slowly, and the free energy is drawn out in infinitesimally small increments that exactly match the amount of energy that the reaction is producing during that increment (Figure 18.11 ▲).

All real reactions are irreversible and therefore do not achieve the theoretical limit of available free energy. Let's return to the discharging battery from the opening section of this chapter as an example of this concept. A battery contains chemical reactants configured in such a way that, upon spontaneous reaction, they produce an electrical current. We can harness the free energy released by the reaction to do work. For example, an electric motor can be wired to the battery. Flowing electrical current makes the motor turn (Figure 18.12 ▶).

Because of the resistance in the wire, the flowing electrical current also produces some heat, which is lost to the surroundings and is not available to do work. The amount of free energy lost as heat can be decreased by slowing down the rate of current flow. The slower the rate of current flow, the less free energy is lost as heat and the more is available to do work. However, only in the theoretical case of infinitesimally slow current flow is the maximum amount of work (equal to $\Delta G^\circ_{rxn}$) done. Any real rate of current flow results in some loss of energy as heat. This lost energy is the "heat tax" that we discussed in the opening section of this chapter.

Recharging the battery necessarily requires more energy than is obtained as work because some of the energy is lost as heat. In other words, during discharging and recharging, the battery has gone through a cycle in which it returns to its original (charged) state. However, the surroundings have gained some heat during the process, and the entropy of the surroundings has increased (which makes the process spontaneous). Any real (or irreversible) cyclical process such as this suffers the same fate—the system may return to its original state, but the surroundings do not, resulting in the permanent dispersal of energy to the surroundings.

If the change in free energy of a chemical reaction is positive, then $\Delta G^\circ_{rxn}$ represents *the minimum amount of energy required to make the reaction occur*. Again, $\Delta G^\circ_{rxn}$ represents a theoretical limit. Making a real nonspontaneous reaction occur always requires more energy than the theoretical limit.

Applying the definition of reversibility from Section 18.4, a reversible chemical reaction is one that changes direction upon an infinitesimally small change in a variable (such as temperature or pressure) related to the reaction.

▲ **FIGURE 18.12 Energy Loss in a Battery**  When current is drawn from a battery to do work, some energy is lost as heat due to resistance in the wire. Consequently, the quantity of energy required to recharge the battery will be more than the quantity of work done.

### 18.9 Free Energy Changes for Nonstandard States: The Relationship between $\Delta G^{\circ}_{rxn}$ and $\Delta G_{rxn}$

We have learned how to calculate the *standard* free energy change for a reaction ($\Delta G^{\circ}_{rxn}$). However, the standard free energy change applies only to a very narrow set of conditions, namely, those conditions in which the reactants and products are in their standard states.

### Standard versus Nonstandard States

Consider the standard free energy change for the evaporation of liquid water to gaseous water:

$$H_2O(l) \rightleftharpoons H_2O(g) \qquad \Delta G^{\circ}_{rxn} = +8.59 \text{ kJ/mol}$$

The standard free energy change for this process is positive, so the process is nonspontaneous. But you know that if you spill water onto the floor under ordinary conditions, it spontaneously evaporates. Why? *Because under ordinary conditions the reactants and products are not in their standard states* and $\Delta G^{\circ}_{rxn}$ applies only when the reactants and product are in their standard states, sometimes called *standard conditions*. For a gas (such as the water vapor in the reaction just given), the standard state is that the pure gas is present at a partial pressure of 1 atmosphere. In a flask containing liquid water and water vapor in its standard state ($P_{H_2O} = 1$ atm) at 25 °C the water does not vaporize. In fact, since $\Delta G^{\circ}_{rxn}$ is negative for the reverse reaction, the reaction spontaneously occurs in reverse—water vapor condenses.

In open air under ordinary circumstances, the partial pressure of water vapor is much less than 1 atm. The state of the water vapor is not its standard state, and therefore the value of $\Delta G^{\circ}_{rxn}$ does not apply. For the nonstandard state, we must calculate $\Delta G_{rxn}$ (as opposed to $\Delta G^{\circ}_{rxn}$) to predict spontaneity.

▲ Spilled water spontaneously evaporates even though $\Delta G^{\circ}$ for the vaporization of water is positive. Why?

### The Free Energy Change of a Reaction under Nonstandard Conditions

We can calculate the **free energy change of a reaction under nonstandard conditions ($\Delta G_{rxn}$)** from $\Delta G^{\circ}_{rxn}$ using the relationship:

$$\Delta G_{rxn} = \Delta G^{\circ}_{rxn} + RT \ln Q \qquad \text{[18.13]}$$

where $Q$ is the reaction quotient (defined in Section 15.7), $T$ is the temperature in kelvins, and $R$ is the gas constant in the appropriate units (8.314 J/mol·K).

In Equation 18.13 and all subsequent thermodynamic equations, we use $Q_p$ for reactions involving gases, and we use $Q_c$ for reactions involving substances dissolved in solution. We can demonstrate the use of this equation by applying it to the liquid–vapor water equilibrium under several different conditions, as shown in Figure 18.13▶. Note that by the law of mass action, for this equilibrium, $Q = P_{H_2O}$ (where the pressure is expressed in atmospheres):

$$H_2O(l) \longrightarrow H_2O(g) \qquad Q = P_{H_2O}$$

### Standard Conditions

Under standard conditions, $P_{H_2O} = 1$ atm and therefore $Q = 1$. Substituting, we get the expression:

$$\Delta G_{rxn} = \Delta G^{\circ}_{rxn} + RT \ln Q$$
$$= +8.59 \text{ kJ/mol} + RT \ln(1)$$
$$= +8.59 \text{ kJ/mol}$$

$$H_2O(l) \rightleftharpoons H_2O(g)$$

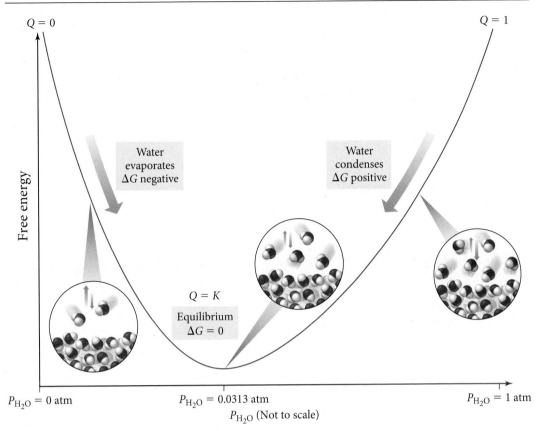

▲ **FIGURE 18.13 Free Energy versus Pressure for Water** The free energy change for the vaporization of water is a function of pressure.

Under standard conditions, $Q$ is always equal to 1, and since $\ln(1) = 0$, the value of $\Delta G_{rxn}$ is therefore equal to $\Delta G^\circ_{rxn}$, as expected. For the liquid–vapor water equilibrium, because $\Delta G^\circ_{rxn} > 0$, the reaction is not spontaneous in the forward direction but is spontaneous in the reverse direction. As stated previously, under standard conditions water vapor condenses into liquid water.

### Equilibrium Conditions

At 25.00 °C, liquid water is in equilibrium with water vapor at a pressure of 0.0313 atm; therefore, $Q = K_p = 0.0313$. Substituting:

$$\Delta G_{rxn} = \Delta G^\circ_{rxn} + RT \ln (0.0313)$$

$$= +8.59 \text{ kJ/mol} + 8.314 \frac{\text{J}}{\text{mol} \cdot \text{K}} (298.15 \text{ K}) \ln (0.0313)$$

$$= +8.59 \text{ kJ/mol} + (-8.59 \times 10^3 \text{ J/mol})$$

$$= +8.59 \text{ kJ/mol} - 8.59 \text{ kJ/mol}$$

$$= 0$$

Under equilibrium conditions, the value of $RT \ln Q$ is always equal in magnitude but opposite in sign to the value of $\Delta G^\circ_{rxn}$. Therefore, the value of $\Delta G_{rxn}$ is zero. Because $\Delta G_{rxn} = 0$, the reaction is not spontaneous in either direction, as expected for a reaction at equilibrium.

### Other Nonstandard Conditions

A water partial pressure of $5.00 \times 10^{-3}$ atm corresponds to a relative humidity of 16% at 25 °C.

To calculate the value of $\Delta G_{rxn}$ under any other set of nonstandard conditions, calculate $Q$ and substitute the value into Equation 18.13. For example, the partial pressure of water vapor in the air on a dry (nonhumid) day is $5.00 \times 10^{-3}$ atm, so $Q = 5.00 \times 10^{-3}$. Substituting:

$$\Delta G_{rxn} = \Delta G^{\circ}_{rxn} + RT \ln (5.00 \times 10^{-3})$$

$$= +8.59 \text{ kJ/mol} + 8.314 \frac{J}{mol \cdot K}(298 \text{ K}) \ln(5.00 \times 10^{-3})$$

$$= +8.59 \text{ kJ/mol} + (-13.1 \times 10^3 \text{ J/mol})$$

$$= +8.59 \text{ kJ/mol} - 13.1 \text{ kJ/mol}$$

$$= -4.5 \text{ kJ/mol}$$

Under these conditions, the value of $\Delta G_{rxn} < 0$, so the reaction is spontaneous in the forward direction, consistent with our experience of water evaporating when spilled on the floor.

---

**Example 18.10** | **Calculating $\Delta G_{rxn}$ under Nonstandard Conditions**

Consider the reaction at 298 K:

$$2 \text{ NO}(g) + \text{O}_2(g) \longrightarrow 2 \text{ NO}_2(g) \qquad \Delta G^{\circ}_{rxn} = -71.2 \text{ kJ}$$

Calculate $\Delta G_{rxn}$ under these conditions:

$$P_{NO} = 0.100 \text{ atm}; \qquad P_{O_2} = 0.100 \text{ atm}; \qquad P_{NO_2} = 2.00 \text{ atm}$$

Is the reaction more or less spontaneous under these conditions than under standard conditions?

**SOLUTION**

| | |
|---|---|
| Use the law of mass action to calculate $Q$. | $Q = \dfrac{(P_{NO_2})^2}{(P_{NO})^2 P_{O_2}} = \dfrac{(2.00)^2}{(0.100)^2(0.100)} = 4.00 \times 10^3$ |
| Substitute $Q$, $T$, and $\Delta G^{\circ}_{rxn}$ into Equation 18.13 to calculate $\Delta G_{rxn}$. (Since the units of $R$ include joules, write $\Delta G^{\circ}_{rxn}$ in joules.) | $\Delta G_{rxn} = \Delta G^{\circ}_{rxn} + RT \ln Q$ <br> $\quad = -71.2 \times 10^3 \text{ J} + 8.314 \dfrac{J}{mol \cdot K}(298 \text{ K}) \ln(4.00 \times 10^3)$ <br> $\quad = -71.2 \times 10^3 \text{ J} + 20.5 \times 10^3 \text{ J}$ <br> $\quad = -50.7 \times 10^3 \text{ J}$ <br> $\quad = -50.7 \text{ kJ}$ <br> The reaction is spontaneous under these conditions, but less spontaneous than it would be under standard conditions (because $\Delta G_{rxn}$ is less negative than $\Delta G^{\circ}_{rxn}$). |

**CHECK** The calculated result is consistent with what you would expect based on Le Châtelier's principle; increasing the concentration of the products and decreasing the concentration of the reactants relative to standard conditions should make the reaction less spontaneous than it was under standard conditions.

**FOR PRACTICE 18.10** Consider the reaction at 298 K:

$$2 \text{ H}_2\text{S}(g) + \text{SO}_2(g) \longrightarrow 3 \text{ S}(s, \text{rhombic}) + 2 \text{ H}_2\text{O}(g) \quad \Delta G^{\circ}_{rxn} = -102 \text{ kJ}$$

Calculate $\Delta G_{rxn}$ under these conditions:

$$P_{H_2S} = 2.00 \text{ atm}; \quad P_{SO_2} = 1.50 \text{ atm}; \quad P_{H_2O} = 0.0100 \text{ atm}$$

Is the reaction more or less spontaneous under these conditions than under standard conditions?

**Free Energy Changes and Le Châtelier's Principle**   According to Le Châtelier's principle and the dependence of free energy on reactant and product concentrations, which statement is true? (Assume that both reactants and products are gaseous.)

**(a)** A high concentration of reactants relative to products results in a more spontaneous reaction than one in which the reactants and products are in their standard states.

**(b)** A high concentration of products relative to reactants results in a more spontaneous reaction than one in which the reactants and products are in their standard states.

**(c)** A reaction in which the reactants are in standard states, but in which no products have formed, has a $\Delta G_{rxn}$ that is more positive than $\Delta G^\circ_{rxn}$.

# 18.10 Free Energy and Equilibrium: Relating $\Delta G^\circ_{rxn}$ to the Equilibrium Constant ($K$)

We have discussed throughout this chapter that $\Delta G^\circ_{rxn}$ determines the spontaneity of a reaction when the reactants and products are in their standard states. In Chapter 15, we learned that the equilibrium constant ($K$) determines how far a reaction goes toward products, a measure of spontaneity. Therefore, as you might expect, the standard free energy change of a reaction and the equilibrium constant are related—the equilibrium constant becomes larger as the standard free energy change becomes more negative. In other words, if, on the one hand, the reactants in a particular reaction undergo a large *negative* free energy change as they become products, then the reaction has a large equilibrium constant, with products strongly favored at equilibrium. If, on the other hand, the reactants in a particular reaction undergo a large *positive* free energy change as they become products, then the reaction has a small equilibrium constant, with reactants strongly favored at equilibrium.

## The Relationship between $\Delta G^\circ_{rxn}$ and $K$

We can derive a relationship between $\Delta G^\circ_{rxn}$ and $K$ from Equation 18.13. We know that at equilibrium $Q = K$ and $\Delta G_{rxn} = 0$. Making these substitutions:

$$\Delta G_{rxn} = \Delta G^\circ_{rxn} + RT \ln Q$$

$$0 = \Delta G^\circ_{rxn} + RT \ln K$$

$$\Delta G^\circ_{rxn} = -RT \ln K \qquad [18.14]$$

In Equation 18.14 and all subsequent thermodynamic equations, we use $K_p$ for reactions involving gases, and we use $K_c$ for reactions involving substances dissolved in solution. We can better understand the relationship between $\Delta G^\circ_{rxn}$ and $K$ by considering the following ranges of values for $K$, as summarized in Figure 18.14▶:

The relationship between $\Delta G^\circ_{rxn}$ and $K$ is logarithmic—small changes in $\Delta G^\circ_{rxn}$ have a large effect on $K$.

- When $K < 1$, $\ln K$ is negative and $\Delta G^\circ_{rxn}$ is positive. Under standard conditions (when $Q = 1$) the reaction is spontaneous in the reverse direction.

- When $K > 1$, $\ln K$ is positive and $\Delta G^\circ_{rxn}$ is negative. Under standard conditions (when $Q = 1$) the reaction is spontaneous in the forward direction.

- When $K = 1$, $\ln K$ is zero and $\Delta G^\circ_{rxn}$ is zero. The reaction happens to be at equilibrium under standard conditions.

### Free Energy and the Equilibrium Constant

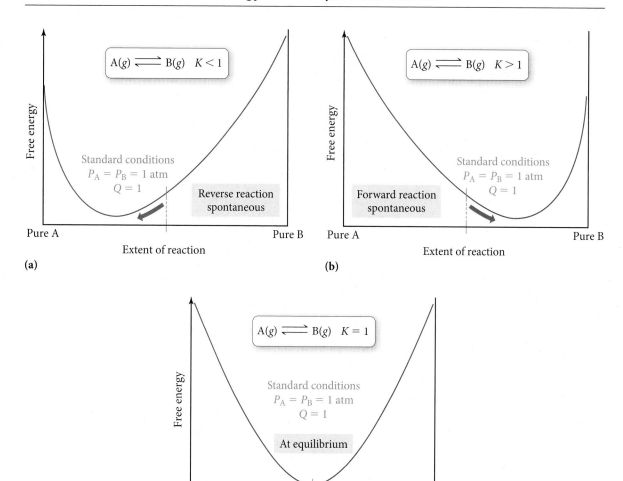

▲ **FIGURE 18.14  Free Energy and the Equilibrium Constant** **(a)** Free energy curve for a reaction with a small equilibrium constant. **(b)** Free energy curve for a reaction with a large equilibrium constant. **(c)** Free energy curve for a reaction in which $K = 1$.

---

**Example 18.11**    **The Equilibrium Constant and $\Delta G^\circ_{rxn}$**    ▶

Use tabulated free energies of formation to calculate the equilibrium constant for the reaction at 298 K:

$$N_2O_4(g) \rightleftharpoons 2\,NO_2(g)$$

**SOLUTION**

| | |
|---|---|
| Look up (in Appendix IIB) the standard free energies of formation for each reactant and product. | |

| Reactant or product | $\Delta G^\circ_f$ (in kJ/mol) |
|---|---|
| $N_2O_4(g)$ | 99.8 |
| $NO_2(g)$ | 51.3 |

| | |
|---|---|
| Calculate $\Delta G^\circ_{rxn}$ by substituting into Equation 18.12. | $\begin{aligned}\Delta G^\circ_{rxn} &= \sum n_p \Delta G^\circ_f(\text{products}) - \sum n_r \Delta G^\circ_f(\text{reactants}) \\ &= 2[\Delta G^\circ_{f,\,NO_2(g)}] - \Delta G^\circ_{f,\,N_2O_4(g)} \\ &= 2(51.3\ \text{kJ}) - 99.8\ \text{kJ} \\ &= 2.8\ \text{kJ}\end{aligned}$ |

Calculate $K$ from $\Delta G^\circ_{rxn}$ by solving Equation 18.14 for $K$ and substituting the values of $\Delta G^\circ_{rxn}$ and temperature.

$$\Delta G^\circ_{rxn} = -RT \ln K$$

$$\ln K = \frac{-\Delta G^\circ_{rxn}}{RT}$$

$$= \frac{-2.8 \times 10^3 \, \text{J/mol}}{8.314 \dfrac{\text{J}}{\text{mol} \cdot \text{K}} (298 \, \text{K})}$$

$$= -1.13$$

$$K = e^{-1.13}$$

$$= 0.32$$

**FOR PRACTICE 18.11**    Calculate $\Delta G^\circ_{rxn}$ at 298 K for the reaction:

$$I_2(g) + Cl_2(g) \rightleftharpoons 2 \, ICl(g) \quad K_p = 81.9$$

**CONCEPTUAL CONNECTION 18.7**

**$K$ and $\Delta G^\circ_{rxn}$**    The reaction $A(g) \rightleftharpoons B(g)$ has an equilibrium constant that is less than one. What can you conclude about $\Delta G^\circ_{rxn}$ for the reaction?

**(a)** $\Delta G^\circ_{rxn} = 0$        **(b)** $\Delta G^\circ_{rxn} < 0$        **(c)** $\Delta G^\circ_{rxn} > 0$

**CONCEPTUAL CONNECTION 18.8**

**The Conditions of Spontaneity**    A reaction is spontaneous under a certain set of conditions. Select all equations that *must be true* for this reaction under these conditions:

$\Delta G_{rxn} < 0$, $\Delta G_{rxn} > 0$, $\Delta G_{rxn} = 0$, $\Delta G^\circ_{rxn} < 0$, $\Delta G^\circ_{rxn} > 0$, $\Delta G^\circ_{rxn} = 0$, $\Delta H_{rxn} < 0$, $\Delta H_{rxn} > 0$, $\Delta H_{rxn} = 0$, $\Delta S_{rxn} < 0$, $\Delta S_{rxn} > 0$, $\Delta S_{rxn} = 0$, $Q = K$, $Q < K$, $Q > K$

## The Temperature Dependence of the Equilibrium Constant

We now have an equation that relates the standard free energy change for a reaction ($\Delta G^\circ_{rxn}$) to the equilibrium constant for a reaction ($K$):

$$\Delta G^\circ_{rxn} = -RT \ln K \quad [18.15]$$

We also have an equation for how the free energy change for a reaction ($\Delta G^\circ_{rxn}$) depends on temperature ($T$):

$$\Delta G^\circ_{rxn} = \Delta H^\circ_{rxn} - T \Delta S^\circ_{rxn} \quad [18.16]$$

We can combine these two equations to obtain an equation for how the equilibrium constant depends on temperature. Combining Equations 18.15 and 18.16, we arrive at:

$$-RT \ln K = \Delta H^\circ_{rxn} - T \Delta S^\circ_{rxn} \quad [18.17]$$

We can then divide both sides of Equation 18.17 by the quantity $RT$:

$$-\ln K = \frac{\Delta H^\circ_{rxn}}{RT} - \frac{T \Delta S^\circ_{rxn}}{RT}$$

Canceling and rearranging, we get this important result:

$$\ln K = -\frac{\Delta H^\circ_{rxn}}{R} \left( \frac{1}{T} \right) + \frac{\Delta S^\circ_{rxn}}{R} \quad [18.18]$$

$$y = mx + b$$

Equation 18.18 is in the form of a straight line. A plot of the natural log of the equilibrium constant ($\ln K$) versus the inverse of the temperature in kelvins ($1/T$) yields a straight line with a slope of $-\Delta H^\circ_{rxn}/R$ and a $y$-intercept of $\Delta S^\circ_{rxn}/R$. Such a plot is useful for obtaining thermodynamic data (namely, $\Delta H^\circ_{rxn}$ and $\Delta S^\circ_{rxn}$) from measurements of $K$ as a function of temperature. However, since $\Delta H^\circ_{rxn}$ and $\Delta S^\circ_{rxn}$ can themselves be slightly temperature dependent, this analysis works only over a relatively limited temperature range.

The equation can also be expressed in a two-point form:

$$\ln \frac{K_2}{K_1} = -\frac{\Delta H^\circ_{rxn}}{R}\left(\frac{1}{T_2} - \frac{1}{T_1}\right)$$

[18.19]

We can use this equation to find $\Delta H^\circ_{rxn}$ from a measurement of the equilibrium constant at two different temperatures or to find the equilibrium constant at some other temperature if we know the equilibrium constant at a given temperature and $\Delta H^\circ_{rxn}$.

# Self-Assessment Quiz

**Q1.** Which reaction is most likely to have a positive $\Delta S_{sys}$?

a) $SiO_2(s) + 3\,C(s) \longrightarrow SiC(s) + 2\,CO(g)$
b) $6\,CO_2(g) + 6\,H_2O(g) \longrightarrow C_6H_{12}O_6(s) + 6\,O_2(g)$
c) $CO(g) + Cl_2(g) \longrightarrow COCl_2(g)$
d) $3\,NO_2(g) + H_2O(l) \longrightarrow 2\,HNO_3(l) + NO(g)$

**Q2.** The sign of $\Delta H_{rxn}$ and $\Delta S_{rxn}$ for several different reactions are given. In which case is the reaction spontaneous at all temperatures?

a) $\Delta H_{rxn} < 0; \Delta S_{rxn} < 0$   b) $\Delta H_{rxn} > 0; \Delta S_{rxn} > 0$
c) $\Delta H_{rxn} < 0; \Delta S_{rxn} > 0$   d) $\Delta H_{rxn} > 0; \Delta S_{rxn} < 0$

**Q3.** Arrange the gases—$F_2$, Ar, and $CH_3F$—in order of increasing standard entropy ($S^\circ$) at 298 K.

a) $F_2 < Ar < CH_3F$   b) $CH_3F < F_2 < Ar$
c) $CH_3F < Ar < F_2$   d) $Ar < F_2 < CH_3F$

**Q4.** Calculate the change in entropy that occurs in the system when 1.00 mol of methanol ($CH_3OH$) vaporizes from a liquid to a gas at its boiling point (64.6 °C). For methanol, $\Delta H_{vap} = 35.21$ kJ/mol.

a) 104 J/K   b) −104 J/K   c) 545 J/K   d) −545 J/K

**Q5.** A reaction has a $\Delta H_{rxn} = 54.2$ kJ. Calculate the change in entropy for the surroundings ($\Delta S_{surr}$) for the reaction at 25.0 °C. (Assume constant pressure and temperature.)

a) $2.17 \times 10^3$ J/K   b) $-2.17 \times 10^3$ J/K
c) −182 J/K   d) 182 J/K

**Q6.** A reaction has $\Delta H^\circ_{rxn} = -255$ kJ and $\Delta S^\circ_{rxn} = 211$ J/K. Calculate $\Delta G^\circ_{rxn}$ at 55 °C.

a) $11.9 \times 10^3$ kJ   b) $69.5 \times 10^3$ kJ
c) −267 kJ   d) −324 kJ

**Q7.** Use standard entropies to calculate $\Delta S^\circ_{rxn}$ for the balanced chemical equation:

$$2\,PCl_3(l) + O_2(g) \longrightarrow 2\,POCl_3(l)$$

| Substance | S° (J/mol · K) |
|---|---|
| POCl₃(*l*) | 222.5 |
| POCl₃(*g*) | 325.5 |
| PCl₃(*l*) | 217.1 |
| PCl₃(*g*) | 311.8 |
| O₂(*g*) | 205.2 |

a) −194.4 J/K   b) −199.8 J/K
c) 10.8 J/K   d) 1084.4 J/K

**Q8.** Use standard free energies of formation to calculate $\Delta G^\circ_{rxn}$ for the balanced chemical equation:

$$Mg(s) + N_2O(g) \longrightarrow MgO(s) + N_2(g)$$

| Substance | $\Delta G^\circ_{rxn}$ (kJ/mol) |
|---|---|
| N₂O(*g*) | 103.7 |
| MgO(s) | −569.3 |

a) 673.0 kJ   b) −673.0 kJ
c) −465.6 kJ   d) 465.6 kJ

**Q9.** Find $\Delta G^\circ_{rxn}$ for the reaction $2\,A + B \longrightarrow 2\,C$ from the given data.

$$A \longrightarrow B \quad \Delta G^\circ_{rxn} = 128 \text{ kJ}$$
$$C \longrightarrow 2\,B \quad \Delta G^\circ_{rxn} = 455 \text{ kJ}$$
$$A \longrightarrow C \quad \Delta G^\circ_{rxn} = -182 \text{ kJ}$$

a) −401 kJ   b) 509 kJ
c) 401 kJ   d) −509 kJ

**Q10.** The given reaction has a $\Delta G^\circ_{rxn} = 9.4$ kJ at 25 °C. Find $\Delta G_{rxn}$ when $P_{NO_2} = 0.115$ atm and $P_{NO} = 9.7$ atm at 25 °C.

$$3\,NO_2(g) + H_2O(l) \longrightarrow 2\,HNO_3(l) + NO(g)$$

a) −12.3 kJ   b) $21.7 \times 10^3$ kJ
c) 31.1 kJ   d) 18.8 kJ

**Q11.** The reaction $A(g) \rightleftharpoons B(g)$ has an equilibrium constant of $K_p = 2.3 \times 10^{-5}$. What can you conclude about the sign of $\Delta G^\circ_{rxn}$ for the reaction?

a) $\Delta G^\circ_{rxn} = 0$
b) $\Delta G^\circ_{rxn}$ is negative
c) $\Delta G^\circ_{rxn}$ is positive
d) Nothing can be concluded about the sign of $\Delta G^\circ_{rxn}$ for the reaction.

**Q12.** A reaction has an equilibrium constant of $K_p = 0.018$ at 25°C. Find $\Delta G^\circ_{rxn}$ for the reaction at this temperature.

a) −835 J   b) −4.32 kJ
c) −9.95 kJ   d) 9.95 kJ

**Q13.** Which distribution of six particles into three interconnected boxes has the highest entropy?

a)

b)

c)

d)

**Q14.** Which process results in the increase in entropy of the universe?

    a) the cooling of a hot cup of coffee in room temperature air

    b) the evaporation of water from a desk at room temperature

    c) the melting of snow above 0 °C

    d) all of the above

**Q15.** Under which set of conditions is $\Delta G_{rxn}$ for the reaction $A(g) \longrightarrow B(g)$ most likely to be negative?

    a) $P_A = 10.0$ atm; $P_B = 10.0$ atm

    b) $P_A = 10.0$ atm; $P_B = 0.010$ atm

    c) $P_A = 0.010$ atm; $P_B = 10.0$ atm

    d) $P_A = 0.010$ atm; $P_B = 0.010$ atm

**Q16.** Which statement is true for the freezing of liquid water below 0 °C?

    a) $\Delta H$ is positive; $\Delta S$ is negative; $\Delta G$ is negative

    b) $\Delta H$ is negative; $\Delta S$ is negative; $\Delta G$ is negative

    c) $\Delta H$ is positive; $\Delta S$ is positive; $\Delta G$ is positive

    d) $\Delta H$ is positive; $\Delta S$ is negative; $\Delta G$ is positive

**Answers: 1.** (a) **2.** (c) **3.** (d) **4.** (a) **5.** (c) **6.** (d) **7.** (a) **8.** (b) **9.** (d) **10.** (c) **11.** (c) **12.** (d) **13.** (b) **14.** (d) **15.** (b) **16.** (b)

# Chapter 18 IN REVIEW

## Key Terms

**Section 18.2**
spontaneous process (841)

**Section 18.3**
entropy (S) (844)
second law of thermodynamics (846)

**Section 18.4**
reversible process (851)

**Section 18.6**
Gibbs free energy (G) (855)

**Section 18.7**
standard entropy change for a reaction ($\Delta S^{\circ}_{rxn}$) (859)
standard molar entropies (S°) (859)
third law of thermodynamics (860)

**Section 18.8**
standard free energy change ($\Delta G^{\circ}_{rxn}$) (863)
free energy of formation ($\Delta G^{\circ}_f$) (865)

**Section 18.9**
free energy change of a reaction under nonstandard conditions ($\Delta G_{rxn}$) (870)

## Key Concepts

### Nature's Heat Tax: You Can't Win and You Can't Break Even (18.1)

▶ The first law of thermodynamics states that energy can be neither created nor destroyed.

▶ According to the second law of thermodynamics, in every energy transaction, some energy is lost to the surroundings; this lost energy is nature's heat tax.

### Spontaneous and Nonspontaneous Processes (18.2)

▶ Both spontaneous and nonspontaneous processes can occur, but only spontaneous processes can take place without outside intervention.

▶ Thermodynamics is the study of the *spontaneity* of reactions, *not* to be confused with kinetics, the study of the *rate* of reactions.

### Entropy and the Second Law of Thermodynamics (18.3)

▶ The second law of thermodynamics states that for *any* spontaneous process, the entropy of the universe increases.

▶ Entropy (S) is proportional to the number of energetically equivalent ways in which the components of a system can be arranged and is a measure of energy dispersal per unit temperature.

### The Entropy Changes Associated with Changes of State (18.4)

▶ The entropy of a substance increases as it changes state from a solid to a liquid and when it changes from a liquid to a gas.

▶ The change in entropy for a system undergoing a reversible change in state is $\Delta S = q_{rev}/T$.

### Heat Transfer and Changes in the Entropy of the Surroundings (18.5)

▶ For a process to be spontaneous, the total entropy of the universe (system plus surroundings) must increase.

▶ The entropy of the surroundings increases when the change in *enthalpy* of the system ($\Delta H_{sys}$) is negative (i.e., for exothermic reactions).

▶ The change in entropy of the surroundings for a given $\Delta H_{sys}$ depends inversely on temperature—the greater the temperature, the smaller the magnitude of $\Delta S_{surr}$.

### Gibbs Free Energy (18.6)

▶ Gibbs free energy, G, is a thermodynamic function that is proportional to the negative of the change in the entropy of the universe. A negative $\Delta G$ represents a spontaneous reaction and a positive $\Delta G$ represents a nonspontaneous reaction.

▶ We can calculate the value of $\Delta G$ for a reaction from the values of $\Delta H$ and $\Delta S$ for the *system* using the equation $\Delta G = \Delta H - T\Delta S$.

### Entropy Changes in Chemical Reactions: Calculating $\Delta S^{\circ}_{rxn}$ (18.7)

▶ We calculate the standard change in entropy for a reaction similarly to the way we calculate the standard change in enthalpy for a reaction: by subtracting the sum of the standard entropies of the reactants multiplied by their stoichiometric coefficients from the sum of the standard entropies of the products multiplied by their stoichiometric coefficients.

▶ Standard entropies are *absolute*; an entropy of zero is established by the third law of thermodynamics as the entropy of a perfect crystal at absolute zero.

▶ The entropy of a substance at a given temperature depends on factors that affect the number of energetically equivalent arrangements of the substance; these include the state, size, and molecular complexity of the substance.

## Free Energy Changes in Chemical Reactions: Calculating $\Delta G_{rxn}^{\circ}$ (18.8)

▶ There are three ways to calculate $\Delta G_{rxn}^{\circ}$: (1) from $\Delta H^{\circ}$ and $\Delta S^{\circ}$, (2) from free energies of formation (only at 25 °C), and (3) from the $\Delta G^{\circ}$'s of reactions that sum to the reaction of interest.

▶ The magnitude of a negative $\Delta G_{rxn}^{\circ}$ represents the theoretical amount of energy available to do work, while a positive $\Delta G_{rxn}^{\circ}$

represents the minimum amount of energy required to make a nonspontaneous process occur.

## Free Energy Changes for Nonstandard States: The Relationship between $\Delta G_{rxn}^{\circ}$ and $\Delta G_{rxn}$ (18.9)

▶ The value of $\Delta G_{rxn}^{\circ}$ applies only to standard conditions, and most real conditions are not standard.

▶ Under nonstandard conditions, we can calculate $\Delta G_{rxn}$ from the equation $\Delta G_{rxn} = \Delta G_{rxn}^{\circ} + RT \ln Q$.

## Free Energy and Equilibrium: Relating $\Delta G_{rxn}^{\circ}$ to the Equilibrium Constant (K) (18.10)

▶ Under standard conditions, the free energy change for a reaction is directly proportional to the negative of the natural log of the equilibrium constant, $K$; the more negative the free energy change (*i.e.*, the more spontaneous the reaction), the larger the equilibrium constant.

▶ We can use the temperature dependence of $\Delta G_{rxn}^{\circ}$, as given by $\Delta G^{\circ} = \Delta H^{\circ} - T\Delta S^{\circ}$, to derive an expression for the temperature dependence of the equilibrium constant.

# Key Equations and Relationships

The Definition of Entropy (**18.3**)
$$S = k \ln W \quad k = 1.38 \times 10^{-23} \text{ J/K}$$

Change in Entropy (**18.3**, **18.4**)
$$\Delta S = S_{final} - S_{initial}$$
$$\Delta S = q_{rev}/T \quad \text{(Isothermal, reversible process)}$$

Change in the Entropy of the Universe (**18.5**)
$$\Delta S_{univ} = \Delta S_{sys} + \Delta S_{surr}$$

Change in the Entropy of the Surroundings (**18.5**)
$$\Delta S_{surr} = \frac{-\Delta H_{sys}}{T} \text{(constant } T, P)$$

Change in Gibbs Free Energy (**18.6**)
$$\Delta G = \Delta H - T\Delta S$$

The Relationship between Spontaneity and $\Delta H$, $\Delta S$, and $T$ (**18.6**)

| ΔH | ΔS | Low Temperature | High Temperature |
|----|----|-----------------|------------------|
| − | + | Spontaneous | Spontaneous |
| + | − | Nonspontaneous | Nonspontaneous |
| − | − | Spontaneous | Nonspontaneous |
| + | + | Nonspontaneous | Spontaneous |

Standard Change in Entropy (**18.7**)
$$\Delta S_{rxn}^{\circ} = \sum n_p S^{\circ}(\text{products}) - \sum n_r S^{\circ}(\text{reactants})$$

Methods for Calculating the Free Energy of Formation ($\Delta G_{rxn}^{\circ}$) (**18.8**)

1. $\Delta G_{rxn}^{\circ} = \Delta H_{rxn}^{\circ} - T\Delta S_{rxn}^{\circ}$

2. $\Delta G_{rxn}^{\circ} = \sum n_p \Delta G_f^{\circ}(\text{products}) - \sum n_r \Delta G_f^{\circ}(\text{reactants})$

3. $\Delta G_{rxn(overall)}^{\circ} = \Delta G_{rxn(step\ 1)}^{\circ} + \Delta G_{rxn(step\ 2)}^{\circ} + \Delta G_{rxn(step\ 3)}^{\circ} + \ldots$

The Relationship between $\Delta G_{rxn}^{\circ}$ and $\Delta G_{rxn}$ (**18.9**)

$$\Delta G_{rxn} = \Delta G_{rxn}^{\circ} + RT \ln Q \quad R = 8.314 \text{ J/mol} \cdot K$$

The Relationship between $\Delta G_{rxn}^{\circ}$ and $K$ (**18.10**)
$$\Delta G_{rxn}^{\circ} = -RT \ln K$$

The Temperature Dependence of the Equilibrium Constant (**18.10**)

$$\ln K = -\frac{\Delta H_{rxn}^{\circ}}{R}\left(\frac{1}{T}\right) + \frac{\Delta S_{rxn}^{\circ}}{R}$$

$$\ln \frac{K_2}{K_1} = -\frac{\Delta H_{rxn}^{\circ}}{R}\left(\frac{1}{T_2} - \frac{1}{T_1}\right)$$

# Key Learning Outcomes

| CHAPTER OBJECTIVES | ASSESSMENT |
|--------------------|------------|
| Predicting the Sign of Entropy Change (**18.4**) | Example 18.1   For Practice 18.1   Exercises 27–28, 35–38, 41–42 |
| Calculating ΔS for a Change of State (**18.4**) | Example 18.2   For Practice 18.2   Exercises 31–34 |
| Calculating Entropy Changes in the Surroundings (**18.4**) $\quad \Delta S_{univ} = \Delta S_{sys} + \Delta S_{surr}$ ↑Positive ↑Positive | Example 18.3   For Practice 18.3   For More Practice 18.3 Exercises 39–42 |

| | |
|---|---|
| Calculating Gibbs Free Energy Changes and Predicting Spontaneity from $\Delta H$ and $\Delta S$ (**18.6**) | Example 18.4  For Practice 18.4  Exercises 43–48 |
| Computing Standard Entropy Changes ($\Delta S^\circ_{rxn}$) (**18.7**) Perfect crystal at 0 K $W = 1$  $S = 0$ | Example 18.5  For Practice 18.5  Exercises 55–58 |
| Calculating the Standard Change in Free Energy for a Reaction using $\Delta G^\circ_{rxn} = \Delta H^\circ_{rxn} - T\Delta S^\circ_{rxn}$ (**18.8**) | Examples 18.6, 18.7  For Practice 18.6, 18.7 Exercises 59–62, 65–66 |
| Calculating $\Delta G^\circ_{rxn}$ from Standard Free Energies of Formation (**18.8**) | Example 18.8  For Practice 18.8  For More Practice 18.8 Exercises 63–64 |
| Determining $\Delta G^\circ_{rxn}$ for a Stepwise Reaction (**18.8**) | Example 18.9  For Practice 18.9  Exercises 67–68 |
| Calculating $\Delta G_{rxn}$ under Nonstandard Conditions (**18.9**) | Example 18.10  For Practice 18.10  Exercises 69–72, 75–76 |
| Relating the Equilibrium Constant and $\Delta G^\circ_{rxn}$ (**18.9**) | Example 18.11  For Practice 18.11  Exercises 73–74, 77–78 |

# EXERCISES
## Review Questions

1. What is the first law of thermodynamics, and how does it relate to energy use?

2. What is nature's heat tax, and how does it relate to energy use?

3. What is a perpetual motion machine? Can such a machine exist given the laws of thermodynamics?

4. Is it more efficient to heat your home with a natural gas furnace or an electric furnace? Explain.

5. What is a spontaneous process? Provide an example.

6. Explain the difference between the spontaneity of a reaction (which depends on thermodynamics) and the speed at which the reaction occurs (which depends on kinetics). Can a catalyst make a nonspontaneous reaction spontaneous?

7. What is the precise definition of entropy? What is the significance of entropy being a state function?

8. Why does the entropy of a gas increase when it expands into a vacuum?

9. Explain the difference between macrostates (external arrangements of particles) and microstates (internal arrangements of particles).

10. Based on its fundamental definition, explain why entropy is a measure of energy dispersion.

11. State the second law of thermodynamics. How does the second law explain why heat travels from a substance at higher temperature to one at lower temperature?

12. What happens to the entropy of a sample of matter when it changes state from a solid to a liquid? From a liquid to a gas?

13. Explain why water spontaneously freezes to form ice below 0 °C even though the entropy of the water decreases during the state transition. Why is the freezing of water not spontaneous above 0 °C?

14. Why do exothermic processes tend to be spontaneous at low temperatures? Why does their tendency toward spontaneity decrease with increasing temperature?

15. What is the significance of the change in Gibbs free energy ($\Delta G$) for a reaction?

16. Predict the spontaneity of a reaction (and the temperature dependence of the spontaneity) for each possible combination of signs for $\Delta H$ and $\Delta S$ (for the system).
    a. $\Delta H$ negative, $\Delta S$ positive     b. $\Delta H$ positive, $\Delta S$ negative
    c. $\Delta H$ negative, $\Delta S$ negative     d. $\Delta H$ positive, $\Delta S$ positive

17. State the third law of thermodynamics and explain its significance.

18. Why is the standard entropy of a substance in the gas state greater than its standard entropy in the liquid state?

19. How does the standard entropy of a substance depend on its molar mass? On its molecular complexity?

20. How can you calculate the standard entropy change for a reaction from tables of standard entropies?

21. What are three different methods to calculate $\Delta G^\circ$ for a reaction? Which method would you choose to calculate $\Delta G^\circ$ for a reaction at a temperature other than 25 °C?

22. Why is free energy "free"?

23. Explain the difference between $\Delta G^\circ$ and $\Delta G$.

24. Why does water spilled on the floor evaporate even though $\Delta G^\circ$ for the evaporation process is positive at room temperature?

25. How do you calculate the change in free energy for a reaction under nonstandard conditions?

26. How does the value of $\Delta G^\circ$ for a reaction relate to the equilibrium constant for the reaction? What does a negative $\Delta G^\circ$ for a reaction imply about $K$ for the reaction? A positive $\Delta G^\circ$?

# Problems by Topic

## Entropy, the Second Law of Thermodynamics, and the Direction of Spontaneous Change

**27.** Which of these processes is spontaneous?
    **a.** the combustion of natural gas
    **b.** the extraction of iron metal from iron ore
    **c.** a hot drink cooling to room temperature
    **d.** drawing heat energy from the ocean's surface to power a ship

**28.** Which of these processes are nonspontaneous? Are the nonspontaneous processes impossible?
    **a.** a bike going up a hill
    **b.** a meteor falling to Earth
    **c.** obtaining hydrogen gas from liquid water
    **d.** a ball rolling down a hill

**29.** Two systems, each composed of two particles represented by circles, have 20 J of total energy. Which system, A or B, has the greater entropy? Why?

**30.** Two systems, each composed of three particles represented by circles, have 30 J of total energy. How many energetically equivalent ways can you distribute the particles in each system? Which system has greater entropy?

**System A**

10 J  •••

**System B**

12 J  •

10 J  •

8 J  •

**31.** Calculate the change in entropy that occurs in the system when 1.00 mole of isopropyl alcohol ($C_3H_8O$) melts at its melting point (−89.5 °C). See Table 11.9 for heats of fusion.

**32.** Calculate the change in entropy that occurs in the system when 1.00 mole of diethyl ether ($C_4H_{10}O$) condenses from a gas to a liquid at its normal boiling point (34.6 °C). See Table 11.7 for heats of vaporization.

**33.** Calculate the change in entropy that occurs in the system when 45.0 g of acetone ($C_3H_6O$) freezes at its melting point (−94.8 °C). See Table 11.9 for heats of fusion.

**34.** Calculate the change in entropy that occurs in the system when 55.0 g of water vaporizes from a liquid to a gas at its boiling point (100.0 °C). See Table 11.7 for heats of vaporization.

**35.** Without doing any calculations, determine the sign of $\Delta S_{sys}$ for each chemical reaction.
    **a.** $2\,KClO_3(s) \longrightarrow 2\,KCl(s) + 3\,O_2(g)$
    **b.** $CH_2{=}CH_2(g) + H_2(g) \longrightarrow CH_3CH_3(g)$
    **c.** $Na(s) + 1/2\,Cl_2(g) \longrightarrow NaCl(s)$
    **d.** $N_2(g) + 3\,H_2(g) \longrightarrow 2\,NH_3(g)$

**36.** Without doing any calculations, determine the sign of $\Delta S_{sys}$ for each chemical reaction.
    **a.** $Mg(s) + Cl_2(g) \longrightarrow MgCl_2(s)$
    **b.** $2\,H_2S(g) + 3\,O_2(g) \longrightarrow 2\,H_2O(g) + 2\,SO_2(g)$
    **c.** $2\,O_3(g) \longrightarrow 3\,O_2(g)$
    **d.** $HCl(g) + NH_3(g) \longrightarrow NH_4Cl(s)$

**37.** Without doing any calculations, determine the signs of $\Delta S_{sys}$ and $\Delta S_{surr}$ for each chemical reaction. In addition, predict under what temperatures (all temperatures, low temperatures, or high temperatures), if any, the reaction is spontaneous.
    **a.** $C_3H_8(g) + 5\,O_2(g) \longrightarrow 3\,CO_2(g) + 4\,H_2O(g)$
          $\Delta H^\circ_{rxn} = -2044$ kJ
    **b.** $N_2(g) + O_2(g) \longrightarrow 2\,NO(g)$  $\Delta H^\circ_{rxn} = +182.6$ kJ
    **c.** $2\,N_2(g) + O_2(g) \longrightarrow 2\,N_2O(g)$  $\Delta H^\circ_{rxn} = +163.2$ kJ
    **d.** $4\,NH_3(g) + 5\,O_2(g) \longrightarrow 4\,NO(g) + 6\,H_2O(g)$
          $\Delta H^\circ_{rxn} = -906$ kJ

**38.** Without doing any calculations, determine the signs of $\Delta S_{sys}$ and $\Delta S_{surr}$ for each chemical reaction. In addition, predict under what temperatures (all temperatures, low temperatures, or high temperatures), if any, the reaction is spontaneous.
    **a.** $2\,CO(g) + O_2(g) \longrightarrow 2\,CO_2(g)$  $\Delta H^\circ_{rxn} = -566.0$ kJ
    **b.** $2\,NO_2(g) \longrightarrow 2\,NO(g) + O_2(g)$  $\Delta H^\circ_{rxn} = +113.1$ kJ
    **c.** $2\,H_2(g) + O_2(g) \longrightarrow 2\,H_2O(g)$  $\Delta H^\circ_{rxn} = -483.6$ kJ
    **d.** $CO_2(g) \longrightarrow C(s) + O_2(g)$  $\Delta H^\circ_{rxn} = +393.5$ kJ

**39.** Calculate $\Delta S_{surr}$ at the indicated temperature for each reaction.
    **a.** $\Delta H^\circ_{rxn} = -385$ kJ; 298 K    **b.** $\Delta H^\circ_{rxn} = -385$ kJ; 77 K
    **c.** $\Delta H^\circ_{rxn} = +114$ kJ; 298 K    **d.** $\Delta H^\circ_{rxn} = +114$ kJ; 77 K

**40.** A reaction has $\Delta H^\circ_{rxn} = -112$ kJ and $\Delta S^\circ_{rxn} = 354$ J/K. At what temperature is the change in entropy for the reaction equal to the change in entropy for the surroundings?

**41.** Given the values of $\Delta H^\circ_{rxn}$, $\Delta S^\circ_{rxn}$, and $T$, determine $\Delta S_{univ}$ and predict whether or not each reaction is spontaneous. (Assume that all reactants and products are in their standard states.)
    **a.** $\Delta H^\circ_{rxn} = +115$ kJ; $\Delta S^\circ_{rxn} = -263$ J/K; $T = 298$ K
    **b.** $\Delta H^\circ_{rxn} = -115$ kJ; $\Delta S^\circ_{rxn} = +263$ J/K; $T = 298$ K
    **c.** $\Delta H^\circ_{rxn} = -115$ kJ; $\Delta S^\circ_{rxn} = -263$ J/K; $T = 298$ K
    **d.** $\Delta H^\circ_{rxn} = -115$ kJ; $\Delta S^\circ_{rxn} = -263$ J/K; $T = 615$ K

**42.** Given the values of $\Delta H^\circ_{rxn}$, $\Delta S^\circ_{rxn}$, and $T$, determine $\Delta S_{univ}$ and predict whether or not each reaction is spontaneous. (Assume that all reactants and products are in their standard states.)
    **a.** $\Delta H^\circ_{rxn} = -95$ kJ; $\Delta S^\circ_{rxn} = -157$ J/K; $T = 298$ K
    **b.** $\Delta H^\circ_{rxn} = -95$ kJ; $\Delta S^\circ_{rxn} = -157$ J/K; $T = 855$ K
    **c.** $\Delta H^\circ_{rxn} = +95$ kJ; $\Delta S^\circ_{rxn} = -157$ J/K; $T = 298$ K
    **d.** $\Delta H^\circ_{rxn} = -95$ kJ; $\Delta S^\circ_{rxn} = +157$ J/K; $T = 398$ K

## Standard Entropy Changes and Gibbs Free Energy

**43.** Calculate the change in Gibbs free energy for each of the sets of $\Delta H_{rxn}$, $\Delta S_{rxn}$, and $T$ given in Problem 41. Predict whether or not each reaction is spontaneous at the temperature indicated. (Assume that all reactants and products are in their standard states.)

**44.** Calculate the change in Gibbs free energy for each of the sets of $\Delta H_{rxn}$, $\Delta S_{rxn}$, and $T$ given in Problem 42. Predict whether or not each reaction is spontaneous at the temperature indicated. (Assume that all reactants and products are in their standard states.)

45. Calculate the free energy change for this reaction at 25 °C. Is the reaction spontaneous? (Assume that all reactants and products are in their standard states.)

$$C_3H_8(g) + 5\,O_2(g) \longrightarrow 3\,CO_2(g) + 4\,H_2O(g)$$
$$\Delta H°_{rxn} = -2217\ kJ;\ \Delta S°_{rxn} = 101.1\ J/K$$

46. Calculate the free energy change for this reaction at 25 °C. Is the reaction spontaneous? (Assume that all reactants and products are in their standard states.)

$$2\,Ca(s) + O_2(g) \longrightarrow 2\,CaO(s)$$
$$\Delta H°_{rxn} = -1269.8\ kJ;\ \Delta S°_{rxn} = -364.6\ J/K$$

47. Fill in the blanks in the table. Both $\Delta H$ and $\Delta S$ refer to the system.

| $\Delta H$ | $\Delta S$ | $\Delta G$ | Low Temperature | High Temperature |
|---|---|---|---|---|
| − | + | − | Spontaneous | _____ |
| − | − | Temperature dependent | _____ | _____ |
| + | + | _____ | _____ | Spontaneous |
| _____ | − | _____ | Nonspontaneous | Nonspontaneous |

48. Predict the conditions (high temperature, low temperature, all temperatures, or no temperatures) under which each reaction is spontaneous.
    a. $H_2O(g) \longrightarrow H_2O(l)$
    b. $CO_2(s) \longrightarrow CO_2(g)$
    c. $H_2(g) \longrightarrow 2\,H(g)$
    d. $2\,NO_2(g) \longrightarrow 2\,NO(g) + O_2(g)$ (endothermic)

49. How does the molar entropy of a substance change with increasing temperature?

50. What is the molar entropy of a pure crystal at 0 K? What is the significance of the answer to this question?

51. For each pair of substances, choose the one that you expect to have the higher standard molar entropy ($S°$) at 25 °C. Explain your choices.
    a. $CO(g);\ CO_2(g)$
    b. $CH_3OH(l);\ CH_3OH(g)$
    c. $Ar(g);\ CO_2(g)$
    d. $CH_4(g);\ SiH_4(g)$
    e. $NO_2(g);\ CH_3CH_2CH_3(g)$
    f. $NaBr(s);\ NaBr(aq)$

52. For each pair of substances, choose the one that you expect to have the higher standard molar entropy ($S°$) at 25 °C. Explain your choices.
    a. $NaNO_3(s);\ NaNO_3(aq)$
    b. $CH_4(g);\ CH_3CH_3(g)$
    c. $Br_2(l);\ Br_2(g)$
    d. $Br_2(g);\ F_2(g)$
    e. $PCl_3(g);\ PCl_5(g)$
    f. $CH_3CH_2CH_2CH_3(g);\ SO_2(g)$

53. Rank each set of substances in order of increasing standard molar entropy ($S°$). Explain your reasoning.
    a. $NH_3(g);\ Ne(g);\ SO_2(g);\ CH_3CH_2OH(g);\ He(g)$
    b. $H_2O(s);\ H_2O(l);\ H_2O(g)$
    c. $CH_4(g);\ CF_4(g);\ CCl_4(g)$

54. Rank each set of substances in order of increasing standard molar entropy ($S°$). Explain your reasoning.
    a. $I_2(g);\ F_2(g);\ Br_2(g);\ Cl_2(g)$
    b. $H_2O(g);\ H_2O_2(g);\ H_2S(g)$
    c. $C(s,\ graphite);\ C(s,\ diamond);\ C(s,\ amorphous)$

55. Use data from Appendix IIB to calculate $\Delta S°_{rxn}$ for each of the reactions. In each case, try to rationalize the sign of $\Delta S°_{rxn}$.
    a. $C_2H_4(g) + H_2(g) \longrightarrow C_2H_6(g)$
    b. $C(s) + H_2O(g) \longrightarrow CO(g) + H_2(g)$
    c. $CO(g) + H_2O(g) \longrightarrow H_2(g) + CO_2(g)$
    d. $2\,H_2S(g) + 3\,O_2(g) \longrightarrow 2\,H_2O(l) + 2\,SO_2(g)$

56. Use data from Appendix IIB to calculate $\Delta S°_{rxn}$ for each of the reactions. In each case, try to rationalize the sign of $\Delta S°_{rxn}$.
    a. $3\,NO_2(g) + H_2O(l) \longrightarrow 2\,HNO_3(aq) + NO(g)$
    b. $Cr_2O_3(s) + 3\,CO(g) \longrightarrow 2\,Cr(s) + 3\,CO_2(g)$
    c. $SO_2(g) + \frac{1}{2}O_2(g) \longrightarrow SO_3(g)$
    d. $N_2O_4(g) + 4\,H_2(g) \longrightarrow N_2(g) + 4\,H_2O(g)$

57. Find $\Delta S°$ for the formation of $CH_2Cl_2(g)$ from its gaseous elements in their standard states. Rationalize the sign of $\Delta S°$.

58. Find $\Delta S°$ for the reaction between nitrogen gas and fluorine gas to form nitrogen trifluoride gas. Rationalize the sign of $\Delta S°$.

59. Methanol ($CH_3OH$) burns in oxygen to form carbon dioxide and water. Write a balanced equation for the combustion of liquid methanol and calculate $\Delta H°_{rxn}$, $\Delta S°_{rxn}$, and $\Delta G°_{rxn}$ at 25 °C. Is the combustion of methanol spontaneous?

60. In photosynthesis, plants form glucose ($C_6H_{12}O_6$) and oxygen from carbon dioxide and water. Write a balanced equation for photosynthesis and calculate $\Delta H°_{rxn}$, $\Delta S°_{rxn}$, and $\Delta G°_{rxn}$ at 25 °C. Is photosynthesis spontaneous?

61. For each reaction, calculate $\Delta H°_{rxn}$, $\Delta S°_{rxn}$, and $\Delta G°_{rxn}$ at 25 °C and state whether or not the reaction is spontaneous. If the reaction is not spontaneous, would a change in temperature make it spontaneous? If so, should the temperature be raised or lowered from 25 °C?
    a. $N_2O_4(g) \longrightarrow 2\,NO_2(g)$
    b. $NH_4Cl(s) \longrightarrow HCl(g) + NH_3(g)$
    c. $3\,H_2(g) + Fe_2O_3(s) \longrightarrow 2\,Fe(s) + 3\,H_2O(g)$
    d. $N_2(g) + 3\,H_2(g) \longrightarrow 2\,NH_3(g)$

62. For each reaction, calculate $\Delta H°_{rxn}$, $\Delta S°_{rxn}$, and $\Delta G°_{rxn}$ at 25 °C and state whether or not the reaction is spontaneous. If the reaction is not spontaneous, would a change in temperature make it spontaneous? If so, should the temperature be raised or lowered from 25 °C?
    a. $2\,CH_4(g) \longrightarrow C_2H_6(g) + H_2(g)$
    b. $2\,NH_3(g) \longrightarrow N_2H_4(g) + H_2(g)$
    c. $N_2(g) + O_2(g) \longrightarrow 2\,NO(g)$
    d. $2\,KClO_3(s) \longrightarrow 2\,KCl(s) + 3\,O_2(g)$

63. Use standard free energies of formation to calculate $\Delta G°$ at 25 °C for each reaction in Problem 61. How do the values of $\Delta G°$ calculated this way compare to those calculated from $\Delta H°$ and $\Delta S°$? Which of the two methods could be used to determine how $\Delta G°$ changes with temperature?

64. Use standard free energies of formation to calculate $\Delta G°$ at 25 °C for each reaction in Problem 62. How well do the values of $\Delta G°$ calculated this way compare to those calculated from $\Delta H°$ and $\Delta S°$? Which of the two methods could be used to determine how $\Delta G°$ changes with temperature?

65. Consider the reaction:

$$2\,NO(g) + O_2(g) \longrightarrow 2\,NO_2(g)$$

Estimate $\Delta G°$ for this reaction at each temperature and predict whether or not the reaction is spontaneous. (Assume that $\Delta H°$ and $\Delta S°$ do not change too much within the given temperature range.)
    a. 298 K   b. 715 K   c. 855 K

66. Consider the reaction:

$$CaCO_3(s) \longrightarrow CaO(s) + CO_2(g)$$

Estimate $\Delta G°$ for this reaction at each temperature and predict whether or not the reaction is spontaneous. (Assume that $\Delta H°$ and $\Delta S°$ do not change too much within the given temperature range.)
    a. 298 K   b. 1055 K   c. 1455 K

**67.** Determine $\Delta G°$ for the reaction:

$$Fe_2O_3(s) + 3\,CO(g) \longrightarrow 2\,Fe(s) + 3\,CO_2(g)$$

Use the following reactions with known $\Delta G°_{rxn}$ values:

| | |
|---|---|
| $2\,Fe(s) + \frac{3}{2}O_2(g) \longrightarrow Fe_2O_3(s)$ | $\Delta G°_{rxn} = -742.2$ kJ |
| $CO(g) + \frac{1}{2}O_2(g) \longrightarrow CO_2(g)$ | $\Delta G°_{rxn} = -257.2$ kJ |

**68.** Calculate $\Delta G°_{rxn}$ for the reaction:

$$CaCO_3(s) \longrightarrow CaO(s) + CO_2(g)$$

Use the following reactions and given $\Delta G°_{rxn}$ values:

| | |
|---|---|
| $Ca(s) + CO_2(g) + \frac{1}{2}O_2(g) \longrightarrow CaCO_3(s)$ | $\Delta G°_{rxn} = -734.4$ kJ |
| $2\,Ca(s) + O_2(g) \longrightarrow 2\,CaO(s)$ | $\Delta G°_{rxn} = -1206.6$ kJ |

## Free Energy Changes, Nonstandard Conditions, and the Equilibrium Constant

**69.** Consider the sublimation of iodine at 25.0 °C :

$$I_2(s) \longrightarrow I_2(g)$$

**a.** Find $\Delta G°_{rxn}$ at 25.0 °C.
**b.** Find $\Delta G_{rxn}$ at 25.0 °C under the following nonstandard conditions:
  **i.** $P_{I_2} = 1.00$ mmHg
  **ii.** $P_{I_2} = 0.100$ mmHg
**c.** Explain why iodine spontaneously sublimes in open air at 25.0 °C.

**70.** Consider the evaporation of methanol at 25.0 °C :

$$CH_3OH(l) \longrightarrow CH_3OH(g)$$

**a.** Find $\Delta G°_r$ at 25.0 °C.
**b.** Find $\Delta G_r$ at 25.0 °C under the following nonstandard conditions:
  **i.** $P_{CH_3OH} = 150.0$ mmHg
  **ii.** $P_{CH_3OH} = 100.0$ mmHg
  **iii.** $P_{CH_3OH} = 10.0$ mmHg
**c.** Explain why methanol spontaneously evaporates in open air at 25.0 °C.

**71.** Consider the reaction:

$$CH_3OH(g) \rightleftharpoons CO(g) + 2\,H_2(g)$$

Calculate $\Delta G$ for this reaction at 25 °C under the following conditions:
  **i.** $P_{CH_3OH} = 0.855$ atm
  **ii.** $P_{CO} = 0.125$ atm
  **iii.** $P_{H_2} = 0.183$ atm

**72.** Consider the reaction:

$$CO_2(g) + CCl_4(g) \rightleftharpoons 2\,COCl_2(g)$$

Calculate $\Delta G$ for this reaction at 25 °C under the following conditions:
  **i.** $P_{CO_2} = 0.112$ atm
  **ii.** $P_{CCl_4} = 0.174$ atm
  **iii.** $P_{COCl_2} = 0.744$ atm

**73.** Use data from Appendix IIB to calculate the equilibrium constants at 25 °C for each reaction.
**a.** $2\,CO(g) + O_2(g) \rightleftharpoons 2\,CO_2(g)$
**b.** $2\,H_2S(g) \rightleftharpoons 2\,H_2(g) + S_2(g)$

**74.** Use data from Appendix IIB to calculate the equilibrium constants at 25 °C for each reaction. $\Delta G°_f$ for BrCl(g) is $-1.0$ kJ/mol.
**a.** $2\,NO_2(g) \rightleftharpoons N_2O_4(g)$
**b.** $Br_2(g) + Cl_2(g) \rightleftharpoons 2\,BrCl(g)$

**75.** Consider the reaction:

$$CO(g) + 2\,H_2(g) \rightleftharpoons CH_3OH(g)$$

$$K_p = 2.26 \times 10^4 \text{ at } 25\,°C$$

Calculate $\Delta G_{rxn}$ for the reaction at 25 °C under each of the following conditions:
**a.** standard conditions
**b.** at equilibrium
**c.** $P_{CH_3OH} = 1.0$ atm; $P_{CO} = P_{H_2} = 0.010$ atm

**76.** Consider the reaction:

$$I_2(g) + Cl_2(g) \rightleftharpoons 2\,ICl(g)$$

$$K_p = 81.9 \text{ at } 25\,°C$$

Calculate $\Delta G_{rxn}$ for the reaction at 25 °C under each of the following conditions:
**a.** standard conditions
**b.** at equilibrium
**c.** $P_{ICl} = 2.55$ atm; $P_{I_2} = 0.325$ atm; $P_{Cl_2} = 0.221$ atm

**77.** Estimate the value of the equilibrium constant at 525 K for each reaction in Problem 73.

**78.** Estimate the value of the equilibrium constant at 655 K for each reaction in Problem 74. ($\Delta H°_f$ for BrCl is 14.6 kJ/mol.)

**79.** Consider the reaction:

$$H_2(g) + I_2(g) \rightleftharpoons 2\,HI(g)$$

The following data show the equilibrium constant for this reaction measured at several different temperatures. Use the data to find $\Delta H°_{rxn}$ and $\Delta S°_{rxn}$ for the reaction.

| Temperature | $K_p$ |
|---|---|
| 150 K | $1.4 \times 10^{-6}$ |
| 175 K | $4.6 \times 10^{-4}$ |
| 200 K | $3.6 \times 10^{-2}$ |
| 225 K | 1.1 |
| 250 K | 15.5 |

**80.** Consider the reaction:

$$2\,NO(g) + O_2(g) \rightleftharpoons 2\,NO_2(g)$$

The following data show the equilibrium constant for this reaction measured at several different temperatures. Use the data to find $\Delta H°_{rxn}$ and $\Delta S°_{rxn}$ for the reaction.

| Temperature | $K_p$ |
|---|---|
| 170 K | $3.8 \times 10^{-3}$ |
| 180 K | 0.34 |
| 190 K | 18.4 |
| 200 K | 681 |

**81.** The change in enthalpy ($\Delta H°_{rxn}$) for a reaction is $-25.8$ kJ/mol. The equilibrium constant for the reaction is $1.4 \times 10^3$ at 298 K. What is the equilibrium constant for the reaction at 655 K?

**82.** A reaction has an equilibrium constant of $8.5 \times 10^3$ at 298 K. At 755 K, the equilibrium constant is 0.65. Find $\Delta H°_{rxn}$ for the reaction.

# Cumulative Problems

**83.** Determine the sign of $\Delta S_{sys}$ for each process.
  **a.** water boiling
  **b.** water freezing
  **c.**

**84.** Determine the sign of $\Delta S_{sys}$ for each process.
  **a.** dry ice subliming
  **b.** dew forming
  **c.**

**85.** Our atmosphere is composed primarily of nitrogen and oxygen, which coexist at 25 °C without reacting to any significant extent. However, the two gases can react to form nitrogen monoxide according to the reaction:

$$N_2(g) + O_2(g) \rightleftharpoons 2\,NO(g)$$

  **a.** Calculate $\Delta G°$ and $K_p$ for this reaction at 298 K. Is the reaction spontaneous?
  **b.** Estimate $\Delta G°$ at 2000 K. Does the reaction become more spontaneous as temperature increases?

**86.** Nitrogen dioxide, a pollutant in the atmosphere, can combine with water to form nitric acid. One of the possible reactions is shown here. Calculate $\Delta G°$ and $K_p$ for this reaction at 25 °C and comment on the spontaneity of the reaction.

$$3\,NO_2(g) + H_2O(l) \longrightarrow 2\,HNO_3(aq) + NO(g)$$

**87.** Ethene ($C_2H_4$) can be halogenated by the reaction:

$$C_2H_4(g) + X_2(g) \rightleftharpoons C_2H_4X_2(g)$$

where $X_2$ can be $Cl_2$, $Br_2$, or $I_2$. Use the thermodynamic data given to calculate $\Delta H°$, $\Delta S°$, $\Delta G°$, and $K_p$ for the halogenation reaction by each of the three halogens at 25 °C. Which reaction is most spontaneous? Least spontaneous? What is the main factor responsible for the difference in the spontaneity of the three reactions? Does higher temperature make the reactions more spontaneous or less spontaneous?

| Compound | $\Delta H_f°$ (kJ/mol) | S° (J/mol · K) |
|---|---|---|
| $C_2H_4Cl_2(g)$ | −129.7 | 308.0 |
| $C_2H_4Br_2(g)$ | +38.3 | 330.6 |
| $C_2H_4I_2(g)$ | +66.5 | 347.8 |

**88.** $H_2$ reacts with the halogens ($X_2$) according to the reaction:

$$H_2(g) + X_2(g) \rightleftharpoons 2\,HX(g)$$

where $X_2$ can be $Cl_2$, $Br_2$, or $I_2$. Use the thermodynamic data in Appendix IIB to calculate $\Delta H°$, $\Delta S°$, $\Delta G°$, and $K_p$ for the reaction between hydrogen and each of the three halogens. Which reaction is most spontaneous? Least spontaneous? What is the main factor responsible for the difference in the spontaneity of the three reactions? Does higher temperature make the reactions more spontaneous or less spontaneous?

**89.** Consider this reaction occurring at 298 K:

$$N_2O(g) + NO_2(g) \rightleftharpoons 3\,NO(g)$$

  **a.** Show that the reaction is not spontaneous under standard conditions by calculating $\Delta G_{rxn}°$.
  **b.** If a reaction mixture contains only $N_2O$ and $NO_2$ at partial pressures of 1.0 atm each, the reaction will be spontaneous until some NO forms in the mixture. What maximum partial pressure of NO builds up before the reaction ceases to be spontaneous?
  **c.** Can the reaction be made more spontaneous by an increase or decrease in temperature? If so, what temperature is required to make the reaction spontaneous under standard conditions?

**90.** Consider this reaction occurring at 298 K:

$$BaCO_3(s) \rightleftharpoons BaO(s) + CO_2(g)$$

  **a.** Show that the reaction is not spontaneous under standard conditions by calculating $\Delta G_{rxn}°$.
  **b.** If $BaCO_3$ is placed in an evacuated flask, what is the partial pressure of $CO_2$ when the reaction reaches equilibrium?
  **c.** Can the reaction be made more spontaneous by an increase or decrease in temperature? If so, at what temperature is the partial pressure of carbon dioxide 1.0 atm?

**91.** Living organisms use energy from the metabolism of food to create an energy-rich molecule called adenosine triphosphate (ATP). The ATP acts as an energy source for a variety of reactions that the living organism must carry out to survive. ATP provides energy through its hydrolysis, which can be symbolized as follows:

$$ATP(aq) + H_2O(l) \longrightarrow ADP(aq) + P_i(aq) \quad \Delta G_{rxn}° = -30.5 \text{ kJ}$$

where ADP represents adenosine diphosphate and $P_i$ represents an inorganic phosphate group (such as $HPO_4{}^{2-}$).
  **a.** Calculate the equilibrium constant, $K$, for the given reaction at 298 K.
  **b.** The free energy obtained from the oxidation (reaction with oxygen) of glucose ($C_6H_{12}O_6$) to form carbon dioxide and water can be used to re-form ATP by driving the given reaction in reverse. Calculate the standard free energy change for the oxidation of glucose and estimate the maximum number of moles of ATP that can be formed by the oxidation of one mole of glucose.

**92.** The standard free energy change for the hydrolysis of ATP was given in Problem 91. In a particular cell, the concentrations of ATP, ADP, and $P_i$ are 0.0031 M, 0.0014 M, and 0.0048 M, respectively. Calculate the free energy change for the hydrolysis of ATP under these conditions. (Assume a temperature of 298 K.)

**93.** These reactions are important in catalytic converters in automobiles. Calculate $\Delta G°$ for each at 298 K. Predict the effect of increasing temperature on the magnitude of $\Delta G°$.
  **a.** $2\,CO(g) + 2\,NO(g) \longrightarrow N_2(g) + 2\,CO_2(g)$
  **b.** $5\,H_2(g) + 2\,NO(g) \longrightarrow 2\,NH_3(g) + 2\,H_2O(g)$
  **c.** $2\,H_2(g) + 2\,NO(g) \longrightarrow N_2(g) + 2\,H_2O(g)$
  **d.** $2\,NH_3(g) + 2\,O_2(g) \longrightarrow N_2O(g) + 3\,H_2O(g)$

**94.** Calculate $\Delta G°$ at 298 K for these reactions and predict the effect on $\Delta G°$ of lowering the temperature.
  **a.** $NH_3(g) + HBr(g) \longrightarrow NH_4Br(s)$
  **b.** $CaCO_3(s) \longrightarrow CaO(s) + CO_2(g)$
  **c.** $CH_4(g) + 3\,Cl_2(g) \longrightarrow CHCl_3(g) + 3\,HCl(g)$
  ($\Delta G_f°$ for $CHCl_3(g)$ is −70.4 kJ/mol.)

**95.** All the oxides of nitrogen have positive values of $\Delta G_f^\circ$ at 298 K, but only one common oxide of nitrogen has a positive $\Delta S_f^\circ$. Identify that oxide of nitrogen without reference to thermodynamic data and explain.

**96.** The values of $\Delta G_f^\circ$ for the hydrogen halides become less negative with increasing atomic number. The $\Delta G_f^\circ$ of HI is slightly positive. However, the trend in $\Delta S_f^\circ$ is to become more positive with increasing atomic number. Explain.

**97.** Consider the reaction $X_2(g) \longrightarrow 2\,X(g)$. When a vessel initially containing 755 torr of $X_2$ comes to equilibrium at 298 K, the equilibrium partial pressure of X is 103 torr. The same reaction is repeated with an initial partial pressure of 748 torr of $X_2$ at 755 K; the equilibrium partial pressure of X is 532 torr. Find $\Delta H^\circ$ for the reaction.

**98.** Dinitrogen tetroxide decomposes to nitrogen dioxide:

$$N_2O_4(g) \longrightarrow 2\,NO_2(g) \quad \Delta H_{rxn}^\circ = 55.3 \text{ kJ}$$

At 298 K, a reaction vessel initially contains 0.100 atm of $N_2O_4$. When equilibrium is reached, 58% of the $N_2O_4$ has decomposed to $NO_2$. What percentage of $N_2O_4$ decomposes at 388 K? Assume that the initial pressure of $N_2O_4$ is the same (0.100 atm).

**99.** Indicate and explain the sign of $\Delta S_{univ}$ for each process.
  **a.** $2\,H_2(g) + O_2(g) \longrightarrow 2\,H_2O(l)$ at 298 K
  **b.** the electrolysis of $H_2O(l)$ to $H_2(g)$ and $O_2(g)$ at 298 K
  **c.** the growth of an oak tree from a little acorn

**100.** The Haber process is very important for agriculture because it converts $N_2(g)$ from the atmosphere into bound nitrogen, which can be taken up and used by plants. The Haber process reaction is $N_2(g) + 3\,H_2(g) \rightleftharpoons 2\,NH_3(g)$. The reaction is exothermic but is carried out at relatively high temperatures. Why?

**101.** A metal salt with the formula $MCl_2$ crystallizes from water to form a solid with the composition $MCl_2 \cdot 6\,H_2O$. The equilibrium vapor pressure of water above this solid at 298 K is 18.3 mmHg. What is the value of $\Delta G$ for the reaction $MCl_2 \cdot 6\,H_2O(s) \rightleftharpoons MCl_2(s) + 6\,H_2O(g)$ when the pressure of water vapor is 18.3 mmHg? When the pressure of water vapor is 760 mmHg?

**102.** The solubility of $AgCl(s)$ in water at 25 °C is $1.33 \times 10^{-5}$ mol/L and its $\Delta H^\circ$ of solution is 65.7 kJ/mol. What is its solubility at 50.0 °C?

## Challenge Problems

**103.** Review the box in this chapter entitled Chemistry in Your Day: Making a Nonspontaneous Process Spontaneous. The hydrolysis of ATP, shown in Problem 91, is often used to drive nonspontaneous processes—such as muscle contraction and protein synthesis—in living organisms. The nonspontaneous process to be driven must be coupled to the ATP hydrolysis reaction. For example, suppose the nonspontaneous process is $A + B \longrightarrow AB$ ($\Delta G^\circ$ positive). The coupling of a nonspontaneous reaction such as this one to the hydrolysis of ATP is often accomplished by the mechanism:

$$A + ATP + H_2O \longrightarrow A\!-\!P_i + ADP$$
$$\underline{A\!-\!P_i + B \longrightarrow AB + P_i}$$
$$A + B + ATP + H_2O \longrightarrow AB + ADP + P_i$$

As long as $\Delta G_{rxn}$ for the nonspontaneous reaction is less than 30.5 kJ, the reaction can be made spontaneous by coupling in this way to the hydrolysis of ATP. Suppose that ATP is to drive the reaction between glutamate and ammonia to form glutamine:

**a.** Calculate $K$ for the reaction between glutamate and ammonia. (The standard free energy change for the reaction is +14.2 kJ/mol. Assume a temperature of 298 K.)

**b.** Write a set of reactions such as those given showing how the glutamate and ammonia reaction can couple with the hydrolysis of ATP. What is $\Delta G_{rxn}^\circ$ and $K$ for the coupled reaction?

**104.** Calculate the entropy of each state and rank the states in order of increasing entropy.

(a)

(b)

**(c)**

**105.** Suppose we redefine the standard state as $P = 2$ atm. Find the new standard $\Delta G_f^\circ$ values of each substance.
  **a.** $HCl(g)$
  **b.** $N_2O(g)$
  **c.** $H(g)$
Explain the results in terms of the relative entropies of reactants and products of each reaction.

**106.** The $\Delta G$ for the freezing of $H_2O(l)$ at $-10\,°C$ is $-210\,J/mol$, and the heat of fusion of ice at this temperature is $5610\,J/mol$. Find the entropy change of the universe when 1 mol of water freezes at $-10\,°C$.

**107.** Consider the reaction that occurs during the Haber process:

$$N_2(g) + 3\,H_2(g) \longrightarrow 2\,NH_3(g)$$

The equilibrium constant is $3.9 \times 10^5$ at 300 K and $1.2 \times 10^{-1}$ at 500 K. Calculate $\Delta H°_{rxn}$ and $\Delta S°_{rxn}$ for this reaction.

**108.** The salt ammonium nitrate can follow three modes of decomposition: (a) to $HNO_3(g)$ and $NH_3(g)$, (b) to $N_2O(g)$ and $H_2O(g)$, and (c) to $N_2(g)$, $O_2(g)$, and $H_2O(g)$. Calculate $\Delta G°_{rxn}$ for each mode of decomposition at 298 K. Explain in light of these results how it is still possible to use ammonium nitrate as a fertilizer and the precautions that should be taken when it is used.

**109.** Given the data, calculate $\Delta S_{vap}$ for each of the first four liquids. ($\Delta S_{vap} = \Delta H_{vap}/T$, where $T$ is in K)

| Compound | Name | BP (°C) | $\Delta H_{vap}$ (kJ/mol) at BP |
|---|---|---|---|
| $C_4H_{10}O$ | Diethyl ether | 34.6 | 26.5 |
| $C_3H_6O$ | Acetone | 56.1 | 29.1 |
| $C_6H_6O$ | Benzene | 79.8 | 30.8 |
| $CHCl_3$ | Chloroform | 60.8 | 29.4 |
| $C_2H_5OH$ | Ethanol | 77.8 | 38.6 |
| $H_2O$ | Water | 100 | 40.7 |

All four values should be close to each other. Predict whether the last two liquids in the table have $\Delta S_{vap}$ in this same range. If not, predict whether it is larger or smaller and explain. Verify your prediction.

# Conceptual Problems

**110.** Which is more efficient, a butane lighter or an electric lighter (such as the ones traditionally found on the dashboard of automobiles)? Explain.

**111.** Which statement is true?
   **a.** A spontaneous reaction is always a fast reaction.
   **b.** A spontaneous reaction is always a slow reaction.
   **c.** The spontaneity of a reaction is not necessarily related to the speed of a reaction.

**112.** Which process is necessarily driven by an increase in the entropy of the surroundings?
   **a.** the condensation of water
   **b.** the sublimation of dry ice
   **c.** the freezing of water

**113.** Consider the changes in the distribution of nine particles into three interconnected boxes shown here. Which has the most negative $\Delta S$?

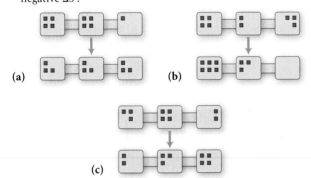

(a)     (b)

(c)

**114.** Which statement is true?
   **a.** A reaction in which the entropy of the system increases can be spontaneous only if it is exothermic.
   **b.** A reaction in which the entropy of the system increases can be spontaneous only if it is endothermic.
   **c.** A reaction in which the entropy of the system decreases can be spontaneous only if it is exothermic.

**115.** Which process is spontaneous at 298 K?
   **a.** $H_2O(l) \longrightarrow H_2O(g, 1\,atm)$
   **b.** $H_2O(l) \longrightarrow H_2O(g, 0.10\,atm)$
   **c.** $H_2O(l) \longrightarrow H_2O(g, 0.010\,atm)$

**116.** The free energy change of the reaction $A(g) \longrightarrow B(g)$ is zero under certain conditions. The *standard* free energy change of the reaction is $-42.5\,kJ$. Which statement must be true about the reaction?
   **a.** The concentration of the product is greater than the concentration of the reactant.
   **b.** The reaction is at equilibrium.
   **c.** The concentration of the reactant is greater than the concentration of the product.

**117.** The reaction $A(g) \rightleftharpoons B(g)$ has an equilibrium constant of 5.8 and under certain conditions has $Q = 336$. What can you conclude about the sign of $\Delta G°_{rxn}$ and $\Delta G_{rxn}$ for this reaction under these conditions?

*Active Classroom Learning*

# Questions for Group Work

Discuss these questions with the group and record your consensus answer.

**118.** Imagine that you roll two dice. Write down all the possible rolls that sum to 2. Write all the possible rolls that sum to 12. Write all the possible rolls that sum to 7. Which configuration has the greatest entropy: 2, 12, or 7?

**119.** If you roll 1 million dice, what will be the average of all the dice? If there is a room with 1 million dice and they all have a 1 on the top face, and there is an earthquake strong enough to roll dice around, what is the likelihood that after the earthquake all the top faces will sum to 1 million? To 6 million? How does this thought experiment illustrate the second law of thermodynamics?

**120.** Not all processes in which the system increases in entropy are spontaneous. How can this observation be consistent with the second law? Provide an example and explain your answer in complete sentences.

**121.** Have each group member look up $\Delta H_f^\circ$ and $S^\circ$ for one substance in the reaction:  $3 \, O_2(g) + 6 \, H_2(g) + 6 \, C(s, \text{graphite}) \longrightarrow$ $C_6H_{12}O_6(s, \text{glucose})$. What is $\Delta H^\circ$ for this reaction? What is $\Delta S^\circ$?

When is $\Delta H_f^\circ$ for a substance equal to zero? When is $S^\circ$ for a substance equal to zero?

**122.** Calculate $\Delta G^\circ$ at 25 °C for the reaction in the previous question. Is this reaction spontaneous under standard conditions? How do you know? What is the determining factor: the change in energy or the change in entropy or both? Explain.

# Data Interpretation and Analysis

## Thermodynamic Functions for Borax

**123.** Borax (sodium tetraboratedecahydrate), a mineral found in dry lakebeds in California, is used as a preservative and in the manufacturing of soap and glass. By determining the $K_{sp}$ of borax at different temperatures, we can determine $\Delta H^\circ$, $\Delta S^\circ$, and $\Delta G^\circ$ for the dissolution of borax:

$Na_2B_4O_5(OH)_4 \cdot 8 \, H_2O(s) \longrightarrow 2 \, Na^+(aq) + B_4O_5(OH)_4{}^{2-}(aq) + 8 \, H_2O(l)$
  (Borax)                                    (Tetraborate)

The relationship:

$$\ln(K_{sp}) = \frac{-\Delta H^\circ}{RT} + \frac{\Delta S^\circ}{R}$$

has the form of a linear equation $y = mx + b$, where $y$ is the $\ln K_{sp}$ and $x$ is $1/T$ ($T$ in Kelvin). The slope is equal to $-\Delta H^\circ/R$, and the $y$ intercept is $\Delta S^\circ/R$, where $R$ is the gas constant, 8.314 J/K mol. Determining $K_{sp}$ at several different temperatures allows us to plot a graph of $\ln K$ versus $1/T$ as shown in Figure a ▼.

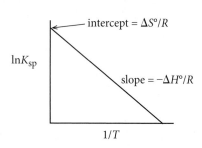

▲ **FIGURE a**  **Plot of $\ln K_{sp}$ versus $1/T$ (Units of Temperature in Kelvin).**

Knowing the values of $\Delta H^\circ$ and $\Delta S^\circ$ at a specific temperature allows the calculation of the change in Gibbs free energy for the reaction $\Delta G^\circ = \Delta H^\circ - T \Delta S^\circ$.

Table a lists $K_{sp}$ values for the dissolution of borax at several different temperatures (°C).

| Temperature (°C) | $K_{sp}$ |
|---|---|
| 40.0 | 0.041 |
| 45.0 | 0.083 |
| 50.0 | 0.264 |
| 55.0 | 0.486 |
| 60.0 | 0.552 |

**TABLE a**  **$K_{sp}$ Values for Dissolution of Borax**

Use the information provided in the figure and table to do the following:

a. Plot a graph of $\ln K_{sp}$ versus $1/T$ ($T$ in Kelvin).

**b.** Determine $\Delta H^\circ$. Is this process endothermic or exothermic?

c. Determine $\Delta S^\circ$.

**d.** Determine $\Delta G^\circ$ at 298 K.

e. Sketch a graph of $\ln K$ versus $1/T$ for an exothermic process.

# Answers to Conceptual Connections

### Nature's Heat Tax and Diet

**18.1** A person subsisting on a vegetarian diet eats fruits and vegetables, metabolizing the energy-containing molecules of these foods directly. A person subsisting on a meat-based diet eats the meat of animals such as cows and metabolizes energy-containing molecules that were part of the animals. Beef cattle synthesize their energy-containing molecules from compounds that they obtain by eating and digesting plants. Since breaking down and resynthesizing biological molecules requires energy—and since the cattle also need to extract some of the energy from their food to live—a meat-based diet necessitates additional energy transactions in the overall process of obtaining energy for life. Due to nature's heat tax, the meat-based diet is less efficient than the vegetarian diet.

### Entropy

**18.2** **(a)** The more spread out the particles are between the three boxes, the greater the entropy. Therefore, the entropy change is positive only in scheme **(a)**.

### Entropy and Biological Systems

**18.3** Biological systems do not violate the second law of thermodynamics. The key to understanding this concept is realizing that entropy changes in the system can be negative as long as the entropy change of the universe is positive. Biological systems can decrease their own entropy, but only at the expense of creating more entropy in the surroundings (which they do primarily by emitting the heat they generate by their metabolic processes). Thus, for any biological process, $\Delta S_{univ}$ is positive.

### $\Delta H$, $\Delta S$, and $\Delta G$

**18.4** **(a)** Sublimation is endothermic (it requires energy to overcome the intermolecular forces that hold solid carbon dioxide together), so $\Delta H$ is positive. The number of moles of gas increases when the solid turns into a gas, so the entropy of the carbon dioxide increases and $\Delta S$ is positive. Since $\Delta G = \Delta H - T\Delta S$, $\Delta G$ is positive at low temperature and negative at high temperature.

### Standard Entropies

**18.5** $Kr < Cl_2 < SO_3$. Because krypton is a monoatomic gas, it has the least entropy. Because $SO_3$ is the most complex molecule, it has the most entropy. The molar masses of the three gases vary slightly, but not enough to overcome the differences in molecular complexity.

### Free Energy Changes and Le Châtelier's Principle

**18.6** **(a)** A high concentration of reactants relative to products will lead to $Q < 1$, making the term $RT \ln Q$ in Equation 18.14 negative. $\Delta G_{rxn}$ is more negative than $\Delta G_{rxn}^{\circ}$ and the reaction is more spontaneous.

### $K$ and $\Delta G_{rxn}^{\circ}$

**18.7** **(c)** Since the equilibrium constant is less than 1, the reaction proceeds toward reactants under standard conditions (when $Q = 1$). Therefore, $\Delta G_{rxn}^{\circ}$ is positive.

### The Conditions of Spontaneity

**18.8** $\Delta G_{rxn} < 0$, $Q < K$

# 19

# Electrochemistry

> **One day sir, you may tax it.**
>
> —MICHAEL FARADAY (1791–1867)
> [In response to Mr. Gladstone, the British Chancellor of the Exchequer, when asked about the practical worth of electricity.]

Anode reaction:
$2 H_2(g) + 4 OH^-(aq) \rightarrow 4 H_2O(l) + 4e^-$

Cathode reaction:
$O_2(g) + 2 H_2O(l) + 4e^- \rightarrow 4 OH^-(aq)$

Overall reaction:
$2 H_2(g) + O_2(g) \rightarrow 2 H_2O$

Fuel Cell

This chapter's opening quote from Michael Faraday illustrates an important aspect of basic research (research for the sake of understanding how nature works). The Chancellor of the Exchequer (the British cabinet minister responsible for all financial matters) wanted to know how Michael Faraday's apparently esoteric investigations of electricity would ever be useful to the empire. Faraday responded in a way that the chancellor would understand—he pointed out the eventual financial payoff. Today electricity is a fundamental form of energy, powering our entire economy. Although basic research does not always lead to useful applications, much of the technology our society relies on has grown out of basic research. The history of modern science shows that we must first understand nature (the goal of basic research) before we can harness its power. In this chapter, we discuss oxidation–reduction reactions (first introduced in Chapter 4) and how we can exploit them to generate electricity. The applications range from the batteries that power flashlights to the fuel cells that may one day power our homes and automobiles.

▲ The smartphone shown here is powered by a hydrogen–oxygen fuel cell, a device that generates electricity from the reaction between hydrogen and oxygen to form water.

## 19.1 Pulling the Plug on the Power Grid

The power grid distributes centrally generated electricity throughout the country to homes and businesses. When you turn on a light or electrical appliance, electricity flows from the grid through the wires in your home to the light or appliance. The electrical energy is converted into light within the lightbulb or into work within the appliance. The average U.S. household consumes about 1000 kilowatt-hours (kWh) of electricity per month. The local electrical utility monitors your electricity use and bills you for it.

The kilowatt-hour is a unit of energy first introduced in Section 6.2.

▲ The energy produced by this fuel cell can power an entire house.

▲ The Toyota Mirai is the first mass-produced fuel-cell vehicle that runs on hydrogen gas and produces only water as exhaust.

In the future, you may have the option of disconnecting from the power grid. Several innovative companies are developing small, fuel-cell power plants—each no bigger than a refrigerator—to sit next to homes and quietly generate enough electricity to meet each household's power needs. The heat produced by a fuel cell's operation can be recaptured and used to heat water or the space within the home, eliminating the need for a hot-water heater or a furnace. Similar fuel cells can also power cars. Fuel cells are highly efficient and, although obstacles to their development and use must yet be overcome, one day they may supply a majority of our power while producing less pollution than fossil fuel combustion.

Fuel cells are based on oxidation–reduction reactions (see Section 4.9). The most common type of fuel cell—the hydrogen–oxygen fuel cell—is based on the reaction between hydrogen and oxygen:

$$2\,H_2(g) + O_2(g) \longrightarrow 2\,H_2O(l)$$

In this reaction, hydrogen and oxygen form covalent bonds with one another. Recall that, according to the Lewis theory, a single covalent bond is a shared electron pair. However, since oxygen is more electronegative than hydrogen, the electron pair in a hydrogen–oxygen bond is not *equally* shared: oxygen gets the larger portion (see Section 9.6). In effect, oxygen has more electrons in $H_2O$ than it does in elemental $O_2$—oxygen has gained electrons in the reaction and has therefore been reduced.

In a reaction between hydrogen and oxygen, oxygen atoms gain electrons directly from hydrogen atoms. In a hydrogen–oxygen fuel cell, this same redox reaction occurs, but the hydrogen and oxygen are separated, forcing electrons to travel through an external wire to get from hydrogen to oxygen. The moving electrons constitute an electrical current. In this way, a fuel cell employs the electron-gaining tendency of oxygen and the electron-losing tendency of hydrogen to force electrons to move through a wire to create the electricity that can provide power for a home or an electric automobile. Smaller fuel cells can also replace batteries and be used to power consumer electronics such as laptop computers, smartphones, and MP3 players. The generation of electricity from spontaneous redox reactions (such as those that occur in a fuel cell) and the use of electricity to drive nonspontaneous redox reactions (such as those that occur in gold or silver plating) are examples of electrochemistry, the subject of this chapter.

## 19.2 Balancing Oxidation–Reduction Equations

Recall from Section 4.9 that *oxidation* is the loss of electrons, and *reduction* is the gain of electrons. Recall also that we can identify oxidation–reduction reactions through changes in oxidation states: *oxidation corresponds to an increase in oxidation state, and reduction corresponds to a decrease in oxidation state.* For example, consider the following reaction between calcium and water:

$$Ca(s) + 2\,H_2O(l) \longrightarrow Ca(OH)_2(aq) + H_2(g)$$

Oxidation / Reduction

Because calcium increases in oxidation state from 0 to +2, it is oxidized. Because hydrogen decreases in oxidation state from +1 to 0, it is reduced.

Balancing redox reactions can be more complicated than balancing other types of reactions because both the mass (or number of each type of atom) and the *charge* must be balanced. We can balance redox reactions occurring in aqueous solutions with a special procedure called the *half-reaction method of balancing*. In this procedure, we break down the overall equation into two half-reactions: one for oxidation and one for reduction. We then balance the half-reactions individually and add them together. The steps differ slightly for reactions occurring in acidic and in basic solution. Examples 19.1 and 19.2 demonstrate the method for an acidic solution, and Example 19.3 demonstrates the method for a basic solution.

Review Section 4.9 on assigning oxidation states.

# Procedure for...

| **Half-Reaction Method of Balancing Aqueous Redox Equations in Acidic Solution** | **Example 19.1**<br>**Half-Reaction Method of Balancing Aqueous Redox Equations in Acidic Solution** | **Example 19.2**<br>**Half-Reaction Method of Balancing Aqueous Redox Equations in Acidic Solution** |
|---|---|---|
| **GENERAL PROCEDURE** | Balance the redox equation:<br>$Al(s) + Cu^{2+}(aq) \longrightarrow Al^{3+}(aq) + Cu(s)$ | Balance the redox equation:<br>$Fe^{2+}(aq) + MnO_4^-(aq) \longrightarrow$ <br> $Fe^{3+}(aq) + Mn^{2+}(aq)$ |
| **Step 1** *Assign oxidation states* to all atoms and identify the substances being oxidized and reduced. | $$Al(s) + Cu^{2+}(aq) \longrightarrow Al^{3+}(aq) + Cu(s)$$<br>0    +2      +3    0<br>Oxidation / Reduction | $$Fe^{2+}(aq) + MnO_4^-(aq) \longrightarrow Fe^{3+}(aq) + Mn^{2+}(aq)$$<br>+2    +7 −2      +3    +2<br>Reduction / Oxidation |
| **Step 2** *Separate the overall reaction into two half-reactions:* one for oxidation and one for reduction. | **Oxidation:** $Al(s) \longrightarrow Al^{3+}(aq)$<br>**Reduction:** $Cu^{2+}(aq) \longrightarrow Cu(s)$ | **Oxidation:** $Fe^{2+}(aq) \longrightarrow Fe^{3+}(aq)$<br>**Reduction:** $MnO_4^-(aq) \longrightarrow Mn^{2+}(aq)$ |
| **Step 3** *Balance each half-reaction with respect to mass* in the following order:<br>• Balance all elements other than H and O.<br>• Balance O by adding $H_2O$.<br>• Balance H by adding $H^+$. | All elements are balanced, so proceed to the next step. | All elements other than H and O are balanced, so proceed to balance H and O.<br>$Fe^{2+}(aq) \longrightarrow Fe^{3+}(aq)$<br>$MnO_4^-(aq) \longrightarrow Mn^{2+}(aq) + 4\,H_2O(l)$<br>$8\,H^+(aq) + MnO_4^-(aq) \longrightarrow$ <br> $Mn^{2+}(aq) + 4\,H_2O(l)$ |
| **Step 4** *Balance each half-reaction with respect to charge* by adding electrons. (Make the sum of the charges on both sides of the equation equal by adding as many electrons as necessary.) | $Al(s) \longrightarrow Al^{3+}(aq) + 3\,e^-$<br>$2\,e^- + Cu^{2+}(aq) \longrightarrow Cu(s)$ | $Fe^{2+}(aq) \longrightarrow Fe^{3+}(aq) + 1\,e^-$<br>$5\,e^- + 8\,H^+(aq) + MnO_4^-(aq) \longrightarrow$ <br> $Mn^{2+}(aq) + 4\,H_2O(l)$ |
| **Step 5** *Make the number of electrons in both half-reactions equal* by multiplying one or both half-reactions by a small whole number. | $2[Al(s) \longrightarrow Al^{3+}(aq) + 3\,e^-]$<br>$2\,Al(s) \longrightarrow 2\,Al^{3+}(aq) + 6\,e^-$<br>$3[2\,e^- + Cu^{2+}(aq) \longrightarrow Cu(s)]$<br>$6\,e^- + 3\,Cu^{2+}(aq) \longrightarrow 3\,Cu(s)$ | $5[Fe^{2+}(aq) \longrightarrow Fe^{3+}(aq) + 1\,e^-]$<br>$5\,Fe^{2+}(aq) \longrightarrow 5\,Fe^{3+}(aq) + 5\,e^-$<br>$5\,e^- + 8\,H^+(aq) + MnO_4^-(aq) \longrightarrow$ <br> $Mn^{2+}(aq) + 4\,H_2O(l)$ |
| **Step 6** *Add the two half-reactions together,* canceling electrons and other species as necessary. | $2\,Al(s) \longrightarrow 2\,Al^{3+}(aq) + \cancel{6\,e^-}$<br>$\cancel{6\,e^-} + 3\,Cu^{2+}(aq) \longrightarrow 3\,Cu(s)$<br>―――――――――<br>$2\,Al(s) + 3\,Cu^{2+}(aq) \longrightarrow$ <br> $2\,Al^{3+}(aq) + 3\,Cu(s)$ | $5\,Fe^{2+}(aq) \longrightarrow 5\,Fe^{3+}(aq) + \cancel{5\,e^-}$<br>$\cancel{5\,e^-} + 8\,H^+(aq) + MnO_4^-(aq) \longrightarrow$ <br> $Mn^{2+}(aq) + 4\,H_2O(l)$<br>―――――――――<br>$5\,Fe^{2+}(aq) + 8\,H^+(aq) + MnO_4^-(aq) \longrightarrow$ <br> $5\,Fe^{3+}(aq) + Mn^{2+}(aq) + 4\,H_2O(l)$<br>*—Continued on the next page* |

*—Continued*

| Step 7 | Verify that the reaction is balanced with respect to both mass and charge. |
|---|---|

| Reactants | Products |
|---|---|
| 2 Al | 2 Al |
| 3 Cu | 3 Cu |
| +6 charge | +6 charge |

| Reactants | Products |
|---|---|
| 5 Fe | 5 Fe |
| 8 H | 8 H |
| 1 Mn | 1 Mn |
| 4 O | 4 O |
| +17 charge | +17 charge |

**FOR PRACTICE 19.1**
Balance the redox reaction in acidic solution:

$$H^+(aq) + Cr(s) \longrightarrow H_2(g) + Cr^{2+}(aq)$$

**FOR PRACTICE 19.2**
Balance the redox reaction in acidic solution:

$$Cu(s) + NO_3^-(aq) \longrightarrow Cu^{2+}(aq) + NO_2(g)$$

When a redox reaction occurs in basic solution, we balance the reaction in a similar manner, except that we add an additional step to neutralize any $H^+$ with $OH^-$. The $H^+$ and the $OH^-$ combine to form $H_2O$ as shown in Example 19.3.

## Example 19.3  Balancing Redox Reactions Occurring in Basic Solution

Balance the equation occurring in basic solution:

$$I^-(aq) + MnO_4^-(aq) \longrightarrow I_2(aq) + MnO_2(s)$$

**SOLUTION**

To balance redox reactions occurring in basic solution, follow the half-reaction method outlined in Examples 19.1 and 19.2, but add an extra step to neutralize the acid with $OH^-$ as shown in step 3.

| | |
|---|---|
| 1. Assign oxidation states. | $$I^-(aq) + MnO_4^-(aq) \longrightarrow I_2(aq) + MnO_2(s)$$ <br> $-1$ ⟶ $+7$ $-2$ ⟶ $0$ ⟶ $+4$ $-2$ <br> ⟵ Reduction ⟶ <br> ⟵ Oxidation ⟶ |
| 2. Separate the overall reaction into two half-reactions. | **Oxidation:** $I^-(aq) \longrightarrow I_2(aq)$ <br> **Reduction:** $MnO_4^-(aq) \longrightarrow MnO_2(s)$ |
| 3. Balance each half-reaction with respect to mass. <br>• Balance all elements other than H and O. <br><br>• Balance O by adding $H_2O$. <br><br>• Balance H by adding $H^+$. <br><br><br>• Neutralize $H^+$ by adding enough $OH^-$ to neutralize each $H^+$. Add the same number of $OH^-$ ions to each side of the equation. | $\begin{cases} 2\,I^-(aq) \longrightarrow I_2(aq) \\ MnO_4^-(aq) \longrightarrow MnO_2(s) \end{cases}$ <br><br> $\begin{cases} 2\,I^-(aq) \longrightarrow I_2(aq) \\ MnO_4^-(aq) \longrightarrow MnO_2(s) + 2\,H_2O(l) \end{cases}$ <br><br> $\begin{cases} 2\,I^-(aq) \longrightarrow I_2(aq) \\ 4\,H^+(aq) + MnO_4^-(aq) \longrightarrow MnO_2(s) + 2\,H_2O(l) \end{cases}$ <br><br> $\begin{cases} 2\,I^-(aq) \longrightarrow I_2(aq) \\ 4\,H^+(aq) + 4\,OH^-(aq) + MnO_4^-(aq) \longrightarrow MnO_2(s) + 2\,H_2O(l) + 4\,OH^-(aq) \end{cases}$ <br> $\underbrace{\qquad\qquad}_{4\,H_2O(l)}$ |

| 4. Balance each half-reaction with respect to charge. | $2\,I^-(aq) \longrightarrow I_2(aq) + 2\,e^-$ |
| | $4\,H_2O(l) + MnO_4^-(aq) + 3\,e^- \longrightarrow MnO_2(s) + 2\,H_2O(l) + 4\,OH^-(aq)$ |
| 5. Make the number of electrons in both half-reactions equal. | $3[2\,I^-(aq) \longrightarrow I_2(aq) + 2\,e^-]$ |
| | $6\,I^-(aq) \longrightarrow 3\,I_2(aq) + 6\,e^-$ |
| | $2[4\,H_2O(l) + MnO_4^-(aq) + 3\,e^- \longrightarrow MnO_2(s) + 2\,H_2O(l) + 4\,OH^-(aq)]$ |
| | $8\,H_2O(l) + 2\,MnO_4^-(aq) + 6\,e^- \longrightarrow 2\,MnO_2(s) + 4\,H_2O(l) + 8\,OH^-(aq)$ |
| 6. Add the half-reactions together. | $6\,I^-(aq) \longrightarrow 3\,I_2(aq) + 6\,e^-$ |
| | $\cancel{4}8\,H_2O(l) + 2\,MnO_4^-(aq) + \cancel{6\,e^-} \longrightarrow 2\,MnO_2(s) + 4\,H_2O(l) + 8\,OH^-(aq)$ |
| | $\overline{6\,I^-(aq) + 4\,H_2O(l) + 2\,MnO_4^-(aq) \longrightarrow 3\,I_2(aq) + 2\,MnO_2(s) + 8\,OH^-(aq)}$ |
| 7. Verify that the reaction is balanced. | |

| Reactants | Products |
| --- | --- |
| 6 I | 6 I |
| 8 H | 8 H |
| 2 Mn | 2 Mn |
| 12 O | 12 O |
| −8 charge | −8 charge |

**FOR PRACTICE 19.3** Balance the following redox reaction occurring in basic solution:

$$ClO^-(aq) + Cr(OH)_4^-(aq) \longrightarrow CrO_4^{2-}(aq) + Cl^-(aq)$$

# 19.3 Voltaic (or Galvanic) Cells: Generating Electricity from Spontaneous Chemical Reactions

**Electrical current** is the flow of electric charge. Electrons flowing through a wire or ions flowing through a solution both constitute electrical current. Since redox reactions involve the transfer of electrons from one substance to another, these reactions have the potential to generate electrical current as we discussed in Section 19.1.

For example, consider the spontaneous redox reaction:

$$Zn(s) + Cu^{2+}(aq) \longrightarrow Zn^{2+}(aq) + Cu(s)$$

When Zn metal is placed in a $Cu^{2+}$ solution, the greater tendency of zinc to lose electrons results in Zn being oxidized and $Cu^{2+}$ being reduced. Electrons are transferred directly from the Zn to the $Cu^{2+}$ (Figure 19.1▶, on the next page). Although the actual process is more complicated, we can imagine that—on the atomic scale—a zinc atom within the zinc metal transfers two electrons to a copper ion in solution. The zinc atom then becomes a zinc ion dissolved in the solution. The copper ion accepts the two electrons and is deposited on the zinc as solid copper.

Suppose we were to separate the zinc atoms and copper ions and force the electron transfer to occur another way—not directly from the zinc atom to the copper ion, but through a wire connecting the two half-reactions. The flowing electrons would constitute an electrical current and could be used to do electrical work.

▶ FIGURE 19.1 **A Spontaneous Oxidation–Reduction Reaction** When zinc is immersed in a solution containing copper ions, the zinc atoms transfer electrons to the copper ions. The zinc atoms are oxidized and dissolve in the solution. The copper ions are reduced and are deposited on the electrode.

**A Spontaneous Redox Reaction: Zn + Cu$^{2+}$**

Zinc strip

Copper(II) sulfate solution

Zn atoms (solid)

Cu$^{2+}$ ions in solution

e$^-$

Zn$^{2+}$ ion

Cu atom

$$Zn(s) + Cu^{2+}(aq) \longrightarrow Zn^{2+}(aq) + Cu(s)$$

## The Voltaic Cell

The generation of electricity through redox reactions is carried out in a device called an **electrochemical cell**. A **voltaic (or galvanic) cell** is an electrochemical cell that *produces* electrical current from a *spontaneous* chemical reaction. A second type of electrochemical cell, called an **electrolytic cell**, *consumes* electrical current to drive a *nonspontaneous* chemical reaction. We discuss voltaic cells in this section and electrolytic cells in Section 19.8.

In the voltaic cell in Figure 19.2▶, a solid strip of zinc is placed in a Zn(NO$_3$)$_2$ solution to form a **half-cell**. A solid strip of copper placed in a Cu(NO$_3$)$_2$ solution forms a second half-cell. The strips act as **electrodes**, conductive surfaces through which electrons can enter or leave the half-cells. Each metal strip reaches equilibrium with its ions in solution according to these half-reactions:

$$Zn(s) \rightleftharpoons Zn^{2+}(aq) + 2\,e^-$$

$$Cu(s) \rightleftharpoons Cu^{2+}(aq) + 2\,e^-$$

The *continual* flow of electrical current in a voltaic cell requires a pathway by which counterions can flow to neutralize charge buildup; this is discussed later.

The idea that one electrode in a voltaic cell becomes more negatively charged relative to the other electrode due to differences in ionization tendencies is central to understanding how a voltaic cell works.

However, the position of these equilibria is not the same for both metals. As we have just seen, the zinc has a greater tendency to ionize than the copper, so the zinc half-reaction lies further to the right. As a result, the zinc electrode becomes negatively charged relative to the copper electrode.

If the two half-cells are connected by a wire running from the zinc—through a light-bulb or other electrical device—to the copper, electrons spontaneously flow from the zinc electrode (which is more negatively charged, and therefore, repels electrons) to the copper electrode. As the electrons flow away from the zinc electrode, the Zn/Zn$^{2+}$ equilibrium shifts to the right (according to Le Châtelier's principle) and oxidation occurs. As electrons flow to the copper electrode, the Cu/Cu$^{2+}$ equilibrium shifts to the left, and reduction occurs. The flowing electrons constitute an electrical current that lights the bulb.

## A Voltaic Cell

**2 e⁻ lost by each Zn atom oxidized**

Zn²⁺

Zn

NO₃⁻    K⁺

Anode
Zn(s)

Salt bridge
containing
KNO₃(aq)

Glass wool
plugs allow
ions to pass

Cathode
Cu(s)

**2 e⁻ gained by each Cu²⁺ ion reduced**

Cu²⁺

Cu

Zn(NO₃)₂(aq)    Cu(NO₃)₂(aq)

| **Oxidation** | **Reduction** |
|---|---|
| $Zn(s) \longrightarrow Zn^{2+}$ ions $+ \ 2 \ e^-$ | $Cu^{2+}(aq) + 2 \ e^- \longrightarrow Cu(s)$ |

▲ **FIGURE 19.2  A Voltaic Cell**   The tendency of zinc to transfer electrons to copper results in a flow of electrons through the wire that lights the bulb. The movement of electrons from the zinc anode to the copper cathode creates a positive charge buildup at the zinc half-cell and a negative charge buildup at the copper half-cell. The flow of ions within the salt bridge neutralizes this charge buildup, allowing the reaction to continue.

## Current and Potential Difference

We can understand electrical current and why it flows by analogy with water current in a stream (Figure 19.3▶). The *rate of electrons flowing* through a wire is analogous to the *rate of water moving* through a stream. Electrical current is measured in units of **amperes (A)**, also called amps. One ampere represents the flow of one coulomb (a measure of electrical charge) per second:

$$1 \ A = 1 \ C/s$$

Because an electron has a charge of $1.602 \times 10^{-19}$ C, 1 A corresponds to the flow of $6.242 \times 10^{18}$ electrons per second.

The *driving force* for electrical current is analogous to the driving force for water current. Water current is driven by a difference in gravitational potential energy (caused by a gravitational field). Streams flow downhill, from higher to lower potential energy. Electrical current is also driven by a difference in potential energy (caused by an electric field resulting from the charge difference on the two electrodes) called **potential difference**. *Potential difference is a measure of the difference in potential energy (usually in joules) per unit of charge (coulombs).* The SI unit of potential difference is the **volt (V)**, which is equal to one joule per coulomb:

$$1 \ V = 1 \ J/C$$

In other words, a potential difference of one volt indicates that a charge of one coulomb experiences an energy difference of one joule between the two electrodes.

A large potential difference corresponds to a large difference in charge between the two electrodes, and therefore, a strong tendency for electron flow (analogous to a steeply descending streambed). Potential difference, since it gives rise to the force that results in the motion of electrons, is also referred to as **electromotive force (emf)**.

The ampere is often abbreviated as *amp*.

▲ **FIGURE 19.3  An Analogy for Electrical Current**   Just as water flows downhill in response to a difference in gravitational potential energy, electrons flow through a conductor in response to an electrical potential difference, creating an electrical current.

In a voltaic cell, the potential difference between the two electrodes is the **cell potential ($E_{cell}$) or cell emf**. The cell potential depends on the relative tendencies of the reactants to undergo oxidation and reduction. Combining the oxidation of a substance with a strong tendency to undergo oxidation and the reduction of a substance with a strong tendency to undergo reduction produces a large difference in charge between the two electrodes, and therefore, a high positive cell potential.

In general, cell potential also depends on the concentrations of the reactants and products in the cell and the temperature (which we will assume to be 25 °C unless otherwise noted). Under standard conditions (1 M concentration for reactants in solution and 1 atm pressure for gaseous reactants), the cell potential is called the **standard cell potential ($E^\circ_{cell}$)** or **standard emf**. For example, the standard cell potential in the zinc and copper cell described previously is 1.10 volts:

$$Zn(s) + Cu^{2+}(aq) \longrightarrow Zn^{2+}(aq) + Cu(s) \qquad E^\circ_{cell} = +1.10 \text{ V}$$

If the zinc is replaced with nickel (which has a lower tendency to be oxidized) the cell potential is lower:

$$Ni(s) + Cu^{2+}(aq) \longrightarrow Ni^{2+}(aq) + Cu(s) \qquad E^\circ_{cell} = +0.62 \text{ V}$$

The cell potential is a measure of the overall tendency of the redox reaction to occur spontaneously—the lower the cell potential, the lower the tendency to occur. A negative cell potential indicates that the forward reaction is not spontaneous.

## Anode, Cathode, and Salt Bridge

In all electrochemical cells, we call the electrode where oxidation occurs the **anode** and the electrode where reduction occurs the **cathode**. In a voltaic cell, the anode is the more negatively charged electrode, and we label it with a negative (−) sign. The cathode of a voltaic cell is the more positively charged electrode, and we label it with a (+) sign. Electrons flow from the anode to the cathode (from negative to positive) through the wires connecting the electrodes.

As electrons flow out of the anode, positive ions ($Zn^{2+}$ in the preceding example) form in the oxidation half-cell, resulting in a buildup of *positive charge* in the *solution*. As electrons flow into the cathode, positive ions ($Cu^{2+}$ in the preceding example) are reduced at the reduction half-cell, resulting in a buildup of *negative charge* in the *solution*.

If the movement of electrons from anode to cathode were the only flow of charge, the buildup of the opposite charge in the solution would stop electron flow almost immediately. The cell needs a pathway by which counterions can flow between the half-cells without the solutions in the half-cells totally mixing. One such pathway is a **salt bridge**, an inverted, U-shaped tube that contains a strong electrolyte such as $KNO_3$ and connects the two half-cells (see Figure 19.2). The electrolyte is usually suspended in a gel and held within the tube by permeable stoppers. The salt bridge allows a flow of ions that neutralizes the charge buildup in the solution. *The negative ions within the salt bridge flow to neutralize the accumulation of positive charge at the anode, and the positive ions flow to neutralize the accumulation of negative charge at the cathode.* In other words, the salt bridge completes the circuit, allowing electrical current to flow.

> Note that the anode and cathode need not actually be negatively and positively charged, respectively. The anode is the electrode with the relatively *more* negative (or less positive) charge.

CONCEPTUAL
**CONNECTION 19.1**

**Voltaic Cells** In a voltaic cell, electrons flow:

(a) from the more negatively charged electrode to the more positively charged electrode.

(b) from the more positively charged electrode to the more negatively charged electrode.

(c) from lower potential energy to higher potential energy.

# Electrochemical Cell Notation

We represent electrochemical cells with a compact notation called a *cell diagram* or *line notation*. For example, we represent the electrochemical cell discussed previously in which Zn is oxidized to $Zn^{2+}$ and $Cu^{2+}$ is reduced to Cu as follows:

$$Zn(s)\,|\,Zn^{2+}(aq)\,\|\,Cu^{2+}(aq)\,|\,Cu(aq)$$

In this representation,

- the oxidation half-reaction is on the left and the reduction is on the right. A double vertical line, indicating the salt bridge, separates the two half-reactions.
- substances in different phases are separated by a single vertical line, which represents the boundary between the phases.
- for some redox reactions, the reactants and products of one or both of the half-reactions may be in the same phase. In these cases (which we explain further next), we separate the reactants and products from each other with a comma in the line diagram. Such cells use an inert electrode, such as platinum (Pt) or graphite, as the anode or cathode (or both).

Consider the redox reaction in which Fe(s) is oxidized and $MnO_4^-(aq)$ is reduced:

$$5\,Fe(s) + 2\,MnO_4^-(aq) + 16\,H^+(aq) \longrightarrow 5\,Fe^{2+}(aq) + 2\,Mn^{2+}(aq) + 8\,H_2O(s)$$

The half-reactions for this overall reaction are:

**Oxidation:** $Fe(s) \longrightarrow Fe^{2+}(aq) + 2\,e^-$

**Reduction:** $MnO_4^-(aq) + 5\,e^- + 8\,H^+(aq) \longrightarrow Mn^{2+}(aq) + 4\,H_2O(l)$

Notice that in the reduction half-reaction the principal species are all in the aqueous phase. In this case, the electron transfer needs an electrode on which to occur. An inert platinum electrode is employed, and the electron transfer takes place at its surface. Using line notation, we represent the electrochemical cell corresponding to the above reaction as:

$$Fe(s)\,|\,Fe^{2+}(aq)\,\|\,MnO_4^-(aq),\,H^+(aq),\,Mn^{2+}(aq)\,|\,Pt(s)$$

The Pt(s) on the far right indicates that an inert platinum electrode acts as the cathode in this reaction, as depicted in Figure 19.4▼.

## Inert Platinum Electrode

**Oxidation**
$Fe(s) \longrightarrow Fe^{2+}(aq) + 2\,e^-$

**Reduction**
$MnO_4^-(aq) + 5\,e^- + 8\,H^+(aq) \longrightarrow Mn^{2+}(aq) + 4\,H_2O(l)$

◀ **FIGURE 19.4 Inert Platinum Electrode** When the participants in a half-reaction are all in the aqueous phase, a conductive surface is needed for electron transfer to take place. In such cases an inert electrode of graphite or platinum is often used. In this electrochemical cell, an iron strip acts as the anode and a platinum strip acts as the cathode. Iron is oxidized at the anode, and $MnO_4^-$ is reduced at the cathode.

# 19.4 Standard Electrode Potentials

As we have just seen, the standard cell potential ($E°_{cell}$) for an electrochemical cell depends on the specific half-reactions occurring in the half-cells and is a measure of the potential energy difference (per unit charge) between the two electrodes. We can think of the electrode in each half-cell as having its own individual potential, called the **standard electrode potential**. The overall standard cell potential ($E°_{cell}$) is the difference between the two standard electrode potentials.

We can better understand this idea with an analogy. Consider two water tanks with different water levels connected by a common pipe, as shown in Figure 19.5 ◄. The water in each tank has its own level and corresponding potential energy. When the tanks are connected, water flows from the tank with the higher level (higher potential energy) to the tank with a lower water level (lower potential energy). Similarly, each half-cell in an electrochemical cell has its own charge and corresponding electrode potential. *When the cells are connected, electrons flow from the electrode with more negative charge (greater potential energy) to the electrode with more positive charge (less potential energy).*

One limitation of this analogy is that, unlike the water level in a tank, we cannot measure the electrode potential in a half-cell directly—we can only measure the overall potential that occurs when two half-cells are combined in a whole cell. However, we can arbitrarily assign a potential of zero to the electrode in a *particular* type of half-cell and then measure all other electrode potentials relative to that zero.

The half-cell electrode that is normally chosen to have a potential of zero is the **standard hydrogen electrode (SHE)**. This half-cell consists of an inert platinum electrode immersed in 1 M HCl with hydrogen gas at 1 atm bubbling through the solution, as shown in Figure 19.6 ▼. When the SHE acts as the cathode, the following half-reaction occurs:

$$2 \, H^+(aq) + 2 \, e^- \longrightarrow H_2(g) \quad E°_{cathode} = 0.00 \, V$$

If we connect the standard hydrogen electrode to an electrode in another half-cell of interest, we can measure the potential difference (or voltage) between the two electrodes. Because we assigned the standard hydrogen electrode zero voltage, we can now determine the electrode potential of the other half-cell.

For example, consider the electrochemical cell in Figure 19.7 ►. In this electrochemical cell, Zn is oxidized to $Zn^{2+}$ and $H^+$ is reduced to $H_2$ under standard conditions (all solutions are 1 M in concentration, and all gases are 1 atm in pressure) and at 25 °C.

▲ **FIGURE 19.5 An Analogy for Electrode Potential**

**Standard Hydrogen Electrode (SHE)**

► **FIGURE 19.6 The Standard Hydrogen Electrode** We arbitrarily assign the standard hydrogen electrode (SHE) an electrode potential of zero. All other electrode potentials are then measured relative to the SHE.

## Measuring Half-Cell Potential with the SHE

Oxidation
$$Zn(s) \longrightarrow Zn^{2+}(aq) + 2\,e^-$$

Reduction
$$2\,H^+(aq) + 2\,e^- \longrightarrow H_2(g)$$

▲ **FIGURE 19.7 Measuring Electrode Potential** Because the electrode potential of the SHE is zero, the electrode potential for the oxidation of Zn is equal to the cell potential.

Electrons travel from the anode (where oxidation occurs) to the cathode (where reduction occurs), so we define $E^\circ_{cell}$ as *the difference in voltage between the cathode (final state) and the anode (initial state)*:

$$E^\circ_{cell} = E^\circ_{final} - E^\circ_{initial}$$
$$= E^\circ_{cathode} - E^\circ_{anode}$$

The measured cell potential for this cell is $+0.76$ V. The anode (in this case, $Zn/Zn^{2+}$) is at a more negative voltage (higher potential energy) than the cathode (in this case, the SHE). Therefore, electrons spontaneously flow from the anode to the cathode. We can diagram the potential energy and the voltage as follows:

Referring back to our water tank analogy, the zinc half-cell is like the water tank with the higher water level, and electrons therefore flow from the zinc electrode to the standard hydrogen electrode.

Because we assigned the SHE a potential of zero (0.00 V), we can determine the electrode potential for the $Zn/Zn^{2+}$ half-cell (the anode) from the measured cell potential ($E^\circ_{cell}$):

$$E^\circ_{cell} = E^\circ_{cathode} - E^\circ_{anode}$$
$$0.76\ V = 0.00\ V - E^\circ_{anode}$$
$$E^\circ_{anode} = -0.76\ V$$

The potential for the $Zn/Zn^{2+}$ electrode is *negative*. The negative potential indicates that an electron at the $Zn/Zn^{2+}$ electrode has greater potential energy than it does at the SHE. *Remember that the more negative the electrode potential is, the greater the potential energy of an electron at that electrode (because negative charge repels electrons).*

What happens if we connect an electrode in which the electron has *more positive* potential (than the standard hydrogen electrode) to the standard hydrogen electrode? That electrode would then have a more *positive* voltage (lower potential energy for an electron).

For example, suppose we connect the standard hydrogen electrode to a Cu electrode immersed in a 1 M $Cu^{2+}$ solution. The measured cell potential for this cell is $-0.34$ V. The anode (defined as $Cu/Cu^{2+}$) is at a more positive voltage (lower potential energy) than the cathode (the SHE). Therefore, electrons will *not* spontaneously flow from the anode to the cathode. We can diagram the potential energy and the voltage of this cell as follows:

The copper half-cell is like the water tank with the lower water level, and electrons *do not* spontaneously flow from the copper electrode to the standard hydrogen electrode.

We can again determine the electrode potential for the $Cu/Cu^{2+}$ half-cell (the anode) from the measured cell potential:

$$E^\circ_{cell} = E^\circ_{cathode} - E^\circ_{anode}$$

$$-0.34 \text{ V} = 0.00 \text{ V} - E^\circ_{anode}$$

$$E^\circ_{anode} = +0.34 \text{ V}$$

The potential for the $Cu/Cu^{2+}$ electrode is *positive*. The positive potential indicates that an electron at the $Cu/Cu^{2+}$ electrode has *lower* potential energy than it does at the SHE. The more positive the electrode potential, the lower the potential energy of an electron at that electrode (because positive charge attracts electrons).

By convention, standard electrode potentials are written for *reduction* half-reactions. We write the standard electrode potentials for the two half-reactions just discussed as:

$$Cu^{2+}(aq) + 2\,e^- \longrightarrow Cu(s) \quad E^\circ = +0.34 \text{ V}$$

$$Zn^{2+}(aq) + 2\,e^- \longrightarrow Zn(s) \quad E^\circ = -0.76 \text{ V}$$

We can see that the $Cu/Cu^{2+}$ electrode is positive relative to the SHE (and therefore, tends to draw electrons *away* from the SHE) and that the $Zn/Zn^{2+}$ electrode is negative relative to the SHE (and therefore, tends to repel electrons toward the SHE). Table 19.1 lists the standard electrode potentials for a number of common half-reactions.

### Summarizing Standard Electrode Potentials:

▶ The electrode potential of the standard hydrogen electrode (SHE) is exactly zero.

▶ The electrode in any half-cell with a greater tendency to undergo reduction is positively charged relative to the SHE, and therefore, has a positive $E^\circ$.

▶ The electrode in any half-cell with a lesser tendency to undergo reduction (or greater tendency to undergo oxidation) is negatively charged relative to the SHE, and therefore, has a negative $E^\circ$.

▶ The cell potential of any electrochemical cell ($E^\circ_{cell}$) is the difference between the electrode potentials of the cathode and the anode ($E^\circ_{cell} = E^\circ_{cat} - E^\circ_{an}$).

▶ $E^\circ_{cell}$ is positive for spontaneous reactions and negative for nonspontaneous reactions.

Multiplying a half-reaction by some constant to balance a redox reaction does not affect the value of $E^\circ$ for the half-reaction.

**TABLE 19.1 Standard Electrode Potentials at 25 °C**

| Reduction Half-Reaction | | $E°(V)$ |
|---|---|---|
| $F_2(g) + 2\ e^-$ | $\longrightarrow 2\ F^-(aq)$ | 2.87 |
| $H_2O_2(aq) + 2\ H^+(aq) + 2\ e^-$ | $\longrightarrow 2\ H_2O(l)$ | 1.78 |
| $PbO_2(s) + 4\ H^+(aq) + SO_4^{2-}(aq) + 2\ e^-$ | $\longrightarrow PbSO_4(s) + 2\ H_2O(l)$ | 1.69 |
| $MnO_4^-(aq) + 4\ H^+(aq) + 3\ e^-$ | $\longrightarrow MnO_2(s) + 2\ H_2O(l)$ | 1.68 |
| $MnO_4^-(aq) + 8\ H^+(aq) + 5\ e^-$ | $\longrightarrow Mn^{2+}(aq) + 4\ H_2O(l)$ | 1.51 |
| $Au^{3+}(aq) + 3\ e^-$ | $\longrightarrow Au(s)$ | 1.50 |
| $PbO_2(s) + 4\ H^+(aq) + 2\ e^-$ | $\longrightarrow Pb^{2+}(aq) + 2\ H_2O(l)$ | 1.46 |
| $Cl_2(g) + 2\ e^-$ | $\longrightarrow 2\ Cl^-(aq)$ | 1.36 |
| $Cr_2O_7^{2-}(aq) + 14\ H^+(aq) + 6\ e^-$ | $\longrightarrow 2\ Cr^{3+}(aq) + 7\ H_2O(l)$ | 1.33 |
| $O_2(g) + 4\ H^+(aq) + 4\ e^-$ | $\longrightarrow 2\ H_2O(l)$ | 1.23 |
| $MnO_2(s) + 4\ H^+(aq) + 2\ e^-$ | $\longrightarrow Mn^{2+}(aq) + 2\ H_2O(l)$ | 1.21 |
| $IO_3^-(aq) + 6\ H^+(aq) + 5\ e^-$ | $\longrightarrow \frac{1}{2}I_2(aq) + 3\ H_2O(l)$ | 1.20 |
| $Br_2(l) + 2\ e^-$ | $\longrightarrow 2\ Br^-(aq)$ | 1.09 |
| $VO_2^+(aq) + 2\ H^+(aq) + e^-$ | $\longrightarrow VO^{2+}(aq) + H_2O(l)$ | 1.00 |
| $NO_3^-(aq) + 4\ H^+(aq) + 3\ e^-$ | $\longrightarrow NO(g) + 2\ H_2O(l)$ | 0.96 |
| $ClO_2(g) + e^-$ | $\longrightarrow ClO_2^-(aq)$ | 0.95 |
| $Ag^+(aq) + e^-$ | $\longrightarrow Ag(s)$ | 0.80 |
| $Fe^{3+}(aq) + e^-$ | $\longrightarrow Fe^{2+}(aq)$ | 0.77 |
| $O_2(g) + 2\ H^+(aq) + 2\ e^-$ | $\longrightarrow H_2O_2(aq)$ | 0.70 |
| $MnO_4^-(aq) + e^-$ | $\longrightarrow MnO_4^{2-}(aq)$ | 0.56 |
| $I_2(s) + 2\ e^-$ | $\longrightarrow 2\ I^-(aq)$ | 0.54 |
| $Cu^+(aq) + e^-$ | $\longrightarrow Cu(s)$ | 0.52 |
| $O_2(g) + 2\ H_2O(l) + 4\ e^-$ | $\longrightarrow 4\ OH^-(aq)$ | 0.40 |
| $Cu^{2+}(aq) + 2\ e^-$ | $\longrightarrow Cu(s)$ | 0.34 |
| $SO_4^{2-}(aq) + 4\ H^+(aq) + 2\ e^-$ | $\longrightarrow H_2SO_3(aq) + H_2O(l)$ | 0.20 |
| $Cu^{2+}(aq) + e^-$ | $\longrightarrow Cu^+(aq)$ | 0.16 |
| $Sn^{4+}(aq) + 2\ e^-$ | $\longrightarrow Sn^{2+}(aq)$ | 0.15 |
| $2\ H^+(aq) + 2\ e^-$ | $\longrightarrow H_2(g)$ | 0 |
| $Fe^{3+}(aq) + 3\ e^-$ | $\longrightarrow Fe(s)$ | −0.036 |
| $Pb^{2+}(aq) + 2\ e^-$ | $\longrightarrow Pb(s)$ | −0.13 |
| $Sn^{2+}(aq) + 2\ e^-$ | $\longrightarrow Sn(s)$ | −0.14 |
| $Ni^{2+}(aq) + 2\ e^-$ | $\longrightarrow Ni(s)$ | −0.23 |
| $Cd^{2+}(aq) + 2\ e^-$ | $\longrightarrow Cd(s)$ | −0.40 |
| $Fe^{2+}(aq) + 2\ e^-$ | $\longrightarrow Fe(s)$ | −0.45 |
| $Cr^{3+}(aq) + e^-$ | $\longrightarrow Cr^{2+}(aq)$ | −0.50 |
| $Cr^{3+}(aq) + 3\ e^-$ | $\longrightarrow Cr(s)$ | −0.73 |
| $Zn^{2+}(aq) + 2\ e^-$ | $\longrightarrow Zn(s)$ | −0.76 |
| $2\ H_2O(l) + 2\ e^-$ | $\longrightarrow H_2(g) + 2\ OH^-(aq)$ | −0.83 |
| $Mn^{2+}(aq) + 2\ e^-$ | $\longrightarrow Mn(s)$ | −1.18 |
| $Al^{3+}(aq) + 3\ e^-$ | $\longrightarrow Al(s)$ | −1.66 |
| $Mg^{2+}(aq) + 2\ e^-$ | $\longrightarrow Mg(s)$ | −2.37 |
| $Na^+(aq) + e^-$ | $\longrightarrow Na(s)$ | −2.71 |
| $Ca^{2+}(aq) + 2\ e^-$ | $\longrightarrow Ca(s)$ | −2.76 |
| $Ba^{2+}(aq) + 2\ e^-$ | $\longrightarrow Ba(s)$ | −2.90 |
| $K^+(aq) + e^-$ | $\longrightarrow K(s)$ | −2.92 |
| $Li^+(aq) + e^-$ | $\longrightarrow Li(s)$ | −3.04 |

Stronger oxidizing agent (↑, left)

Weaker oxidizing agent (bottom left)

Weaker reducing agent (top right)

Stronger reducing agent (↓, bottom right)

Example 19.4 illustrates how to calculate the potential of an electrochemical cell from the standard electrode potentials of the half-reactions.

---

**Example 19.4** **Calculating Standard Potentials for Electrochemical Cells from Standard Electrode Potentials of the Half-Reactions**

Use tabulated standard electrode potentials to calculate the standard cell potential for the following reaction occurring in an electrochemical cell at 25 °C. (The equation is balanced.)

$$Al(s) + NO_3^-(aq) + 4\,H^+(aq) \longrightarrow Al^{3+}(aq) + NO(g) + 2\,H_2O(l)$$

**SOLUTION**

| | |
|---|---|
| Begin by separating the reaction into oxidation and reduction half-reactions. (In this case, you can readily see that Al(s) is oxidized. In cases where it is not so apparent, you may want to assign oxidation states to determine the correct half-reactions.) | ***Oxidation:*** $Al(s) \longrightarrow Al^{3+}(aq) + 3\,e^-$ <br><br> ***Reduction:*** $NO_3^-(aq) + 4\,H^+(aq) + 3\,e^- \longrightarrow NO(g) + 2\,H_2O(l)$ |
| Look up the standard electrode potentials for each half-reaction in Table 19.1. Add the half-cell reactions together to obtain the overall redox equation. Calculate the standard cell potential by subtracting the electrode potential of the anode from the electrode potential of the cathode. | ***Oxidation*** <br> ***(Anode):*** $\quad\quad\quad\quad\quad\quad\quad\quad Al(s) \longrightarrow Al^{3+}(aq) + 3\,e^- \quad E° = -1.66\text{ V}$ <br> ***Reduction*** <br> ***(Cathode):*** $\underline{\quad NO_3^-(aq) + 4\,H^+(aq) + 3\,e^- \longrightarrow NO(g) + 2\,H_2O(l) \quad E° = 0.96\text{ V}}$ <br> $\quad\quad\quad\quad\quad Al(s) + NO_3^-(aq) + 4\,H^+(aq) \longrightarrow Al^{3+}(aq) + NO(g) + 2\,H_2O(l)$ <br><br> $E°_{cell} = E°_{cat} - E°_{an}$ <br> $\quad\quad = 0.96\text{ V} - (-1.66\text{ V})$ <br> $\quad\quad = 2.62\text{ V}$ |

**FOR PRACTICE 19.4** Use tabulated standard electrode potentials to calculate the standard cell potential for the following reaction occurring in an electrochemical cell at 25 °C. (The equation is balanced.)

$$3\,Pb^{2+}(aq) + 2\,Cr(s) \longrightarrow 3\,Pb(s) + 2\,Cr^{3+}(aq)$$

---

**CONCEPTUAL CONNECTION 19.2**

**Standard Electrode Potentials** An electrode has a negative electrode potential. Which statement is correct regarding the potential energy of an electron at this electrode?

**(a)** An electron at this electrode has a lower potential energy than it has at a standard hydrogen electrode.

**(b)** An electron at this electrode has a higher potential energy than it has at a standard hydrogen electrode.

**(c)** An electron at this electrode has the same potential energy as it has at a standard hydrogen electrode.

---

## Predicting the Spontaneous Direction of an Oxidation–Reduction Reaction

To determine the spontaneous direction of an oxidation–reduction reaction, we examine the electrode potentials of the two relevant half-reactions in Table 19.1. The half-reaction with the more *negative* electrode potential tends to lose electrons, and therefore, undergo oxidation. (Remember that negative charge repels electrons.) The half-reaction having the more *positive* electrode potential tends to gain electrons, and therefore, undergo reduction. (Remember that positive charge attracts electrons.)

For example, consider the two reduction half-reactions:

$$Ni^{2+}(aq) + 2\,e^- \longrightarrow Ni(s) \qquad E° = -0.23\,V$$

$$Mn^{2+}(aq) + 2\,e^- \longrightarrow Mn(s) \qquad E° = -1.18\,V$$

Because the manganese half-reaction has a more negative electrode potential, it repels electrons and proceeds in the reverse direction (oxidation). Because the nickel half-reaction has the more positive (or less negative) electrode potential, it attracts electrons and proceeds in the forward direction.

We can confirm this by calculating the standard electrode potential for manganese acting as the anode (oxidation) and nickel acting as the cathode (reduction):

**Oxidation *(Anode)*:**

**Reduction (Cathode):**

$$Mn(s) \longrightarrow Mn^{2+}(aq) + 2\,e^- \qquad E° = -1.18\,V$$
$$Ni^{2+}(aq) + 2\,e^- \longrightarrow Ni(s) \qquad E° = -0.23\,V$$
$$\overline{Ni^{2+}(aq) + Mn(s) \longrightarrow Ni(s) + Mn^{2+}(aq)} \qquad E°_{cell} = E°_{cat} - E°_{an}$$
$$= -0.23\,V - (-1.18\,V)$$
$$= 0.95\,V$$

The overall cell potential is positive, indicating a spontaneous reaction. Consider the electrochemical cell corresponding to this spontaneous redox reaction in Figure 19.8▼. We draw the manganese half-cell on the left as the anode and the nickel half-cell on the right as the cathode. Electrons flow from the anode to the cathode.

Another way to predict the spontaneity of a redox reaction is to note the relative positions of the two half-reactions in Table 19.1. The table lists half-reactions in order of *decreasing* electrode potential, so the half-reactions near the top of the table—those having large *positive* electrode potentials—attract electrons, and therefore, tend to occur in the forward direction. Half-reactions near the bottom of the table—those having large *negative* electrode potentials—repel electrons, and therefore, tend to occur in the reverse direction. In other words, as we move down Table 19.1, the half-reactions become less likely to occur in the forward direction and more likely to occur in the reverse direction. As a result, *any reduction half-reaction listed is spontaneous when paired with the reverse of a half-reaction that appears below it in* Table 19.1.

For example, if we return to our two previous half-reactions involving manganese and nickel, we can see that the manganese half-reaction is below the nickel half-reaction in Table 19.1:

$$Ni^{2+}(aq) + 2\,e^- \longrightarrow Ni(s) \qquad E° = -0.23\,V$$

$$Mn^{2+}(aq) + 2\,e^- \longrightarrow Mn(s) \qquad E° = -1.18\,V$$

Therefore, the nickel reaction occurs in the forward direction (reduction) and the manganese reaction occurs in the reverse direction (oxidation).

The following mnemonics (NIO and PIR) help to predict the spontaneous direction of redox reactions:

N.I.O.—More **N**egative **I**s **O**xidation
**P.I.R.**—More **P**ositive **I**s **R**eduction

◀ **FIGURE 19.8 Mn/Ni²⁺ Electrochemical Cell** The reduction of $Mn^{2+}$ is listed below the reduction of $Ni^{2+}$ in Table 19.1, so the reduction of $Ni^{2+}$ is spontaneous when paired with the oxidation of Mn.

Recall from Section 4.9 that an *oxidizing agent* causes the oxidation of another substance (and is itself reduced) and that a *reducing agent* causes the reduction of another substance (and is itself oxidized).

*Summarizing the Prediction of Spontaneous Direction for Redox Reactions:*

▶ The half-reaction with the more *positive* electrode potential attracts electrons more strongly and undergoes reduction. (Substances listed at the top of Table 19.1 tend to undergo reduction; they are good oxidizing agents.)

▶ The half-reaction with the more *negative* electrode potential repels electrons more strongly and undergoes oxidation. (Substances listed near the bottom of Table 19.1 tend to undergo oxidation; they are good reducing agents.)

▶ Any reduction reaction in Table 19.1 is spontaneous when paired with the *reverse* of any of the reactions listed below it on the table.

### Example 19.5  Predicting Spontaneous Redox Reactions and Sketching Electrochemical Cells

Without calculating $E^\circ_{cell}$, predict whether each of the redox reactions is spontaneous. If the reaction is spontaneous as written, make a sketch of the electrochemical cell in which the reaction could occur. If the reaction is not spontaneous as written, write an equation for the spontaneous direction in which the reaction would occur and sketch the electrochemical cell in which the spontaneous reaction would occur. In your sketches, make sure to label the anode (which should be drawn on the left), the cathode, and the direction of electron flow.

**(a)** $Fe(s) + Mg^{2+}(aq) \longrightarrow Fe^{2+}(aq) + Mg(s)$
**(b)** $Fe(s) + Pb^{2+}(aq) \longrightarrow Fe^{2+}(aq) + Pb(s)$

**SOLUTION**

**(a)** $Fe(s) + Mg^{2+}(aq) \longrightarrow Fe^{2+}(aq) + Mg(s)$
This reaction involves the reduction of $Mg^{2+}$:
$$Mg^{2+}(aq) + 2\,e^- \longrightarrow Mg(s) \qquad E^\circ = -2.37\ V$$
and the oxidation of Fe:
$$Fe(s) \longrightarrow Fe^{2+}(aq) + 2\,e^- \qquad E^\circ = -0.45\ V$$
However, the magnesium half-reaction has the more negative electrode potential, and therefore, repels electrons more strongly and undergoes oxidation. The iron half-reaction has the more positive electrode potential, and therefore, attracts electrons more strongly and undergoes reduction. So the reaction as written is *not* spontaneous. (The reaction pairs the reduction of $Mg^{2+}$ with the reverse of a half-reaction *above it* in Table 19.1—such pairings are not spontaneous.)
However, the reverse reaction is spontaneous:
$$Fe^{2+}(aq) + Mg(s) \longrightarrow Fe(s) + Mg^{2+}(aq)$$
The corresponding electrochemical cell is shown in Figure 19.9◀.

**(b)** $Fe(s) + Pb^{2+}(aq) \longrightarrow Fe^{2+}(aq) + Pb(s)$
This reaction involves the reduction of $Pb^{2+}$:
$$Pb^{2+}(aq) + 2\,e^- \longrightarrow Pb(s) \qquad E^\circ = -0.13\ V$$
and the oxidation of iron:
$$Fe(s) \longrightarrow Fe^{2+}(aq) + 2\,e^- \qquad E^\circ = -0.45\ V$$
The iron half-reaction has the more negative electrode potential, and therefore, repels electrons and undergoes oxidation. The lead half-reaction has the more positive electrode potential, and therefore, attracts electrons and undergoes reduction. Therefore, the reaction *is* spontaneous as written. (The reaction pairs the reduction of $Pb^{2+}$ with the reverse of a half-reaction *below it* in Table 19.1—such pairings are always spontaneous.) The corresponding electrochemical cell is shown in Figure 19.10◀.

▲ **FIGURE 19.9  Mg/Fe²⁺ Electrochemical Cell**

▲ **FIGURE 19.10  Fe/Pb²⁺ Electrochemical Cell**

**FOR PRACTICE 19.5**   Are the following redox reactions spontaneous under standard conditions?
**(a)** $Zn(s) + Ni^{2+}(aq) \longrightarrow Zn^{2+}(aq) + Ni(s)$
**(b)** $Zn(s) + Ca^{2+}(aq) \longrightarrow Zn^{2+}(aq) + Ca(s)$

**Selective Oxidation** A solution contains both NaI and NaBr. Which oxidizing agent could you add to the solution to selectively oxidize $I^-(aq)$ but not $Br^-(aq)$?

   **(a)** $Cl_2$     **(b)** $H_2O_2$     **(c)** $CuCl_2$     **(d)** $HNO_3$

## Predicting Whether a Metal Will Dissolve in Acid

In Chapter 16, we learned that acids dissolve metals. Most acids dissolve metals by the reduction of $H^+$ ions to hydrogen gas and the corresponding oxidation of the metal to its ion. For example, if solid Zn is dropped into hydrochloric acid, the following reaction occurs:

$$2\,H^+(aq) + 2\,e^- \longrightarrow H_2(g)$$
$$\underline{Zn(s) \longrightarrow Zn^{2+}(aq) + 2\,e^-}$$
$$Zn(s) + 2\,H^+(aq) \longrightarrow Zn^{2+}(aq) + H_2(g)$$

We observe the reaction as the dissolving of the zinc and the bubbling of hydrogen gas. The zinc is oxidized and the $H^+$ ions are reduced. Notice that this reaction involves the pairing of a reduction half-reaction (the reduction of $H^+$) with the reverse of a half-reaction that falls below it in Table 19.1. Therefore, this reaction is spontaneous. What happens, however, if we pair the reduction of $H^+$ with the oxidation of Cu? The reaction is not spontaneous because it involves pairing the reduction of $H^+$ with the reverse of a half-reaction that is listed *above it* in the table. Consequently, copper does not react with $H^+$ and does not dissolve in acids such as HCl. In general, *metals whose reduction half-reactions are listed below the reduction of $H^+$ to $H_2$ in Table 19.1 dissolve in acids, while metals listed above it do not.*

An important exception to this rule is nitric acid ($HNO_3$), which can oxidize metals through the reduction half-reaction:

$$NO_3^-(aq) + 4\,H^+(aq) + 3\,e^- \longrightarrow NO(g) + 2\,H_2O(l) \quad E° = 0.96\ V$$

Since this half-reaction is above the reduction of $H^+$ in Table 19.1, $HNO_3$ can oxidize metals (such as copper) that can't be oxidized by HCl.

$$Zn(s) + 2\,H^+(aq) \longrightarrow$$
$$Zn^{2+}(aq) + H_2(g)$$

▲ When zinc is immersed in hydrochloric acid, the zinc is oxidized, forming ions that become solvated in the solution. Hydrogen ions are reduced, forming bubbles of hydrogen gas.

**Metals Dissolving in Acids** Which metal dissolves in $HNO_3$ but not in HCl?

   **(a)** Fe     **(b)** Au     **(c)** Ag

## 19.5 Cell Potential, Free Energy, and the Equilibrium Constant

**KCV** Cell Potential, Free Energy, and the Equilibrium Constant

We have seen that a positive standard cell potential ($E°_{cell}$) corresponds to a spontaneous oxidation–reduction reaction. And we know (from Chapter 18) that the spontaneity of a reaction is determined by the sign of $\Delta G°$. Therefore, $E°_{cell}$ and $\Delta G°$ must be related. We also know from Section 18.10 that $\Delta G°$ for a reaction is related to the equilibrium constant ($K$) for the reaction. Since $E°_{cell}$ and $\Delta G°$ are related, then $E°_{cell}$ and $K$ must also be related.

Before we look at the nature of each of these relationships in detail, let's consider the following generalizations.

For a spontaneous redox reaction (one that proceeds in the forward direction when all reactants and products are in their standard states):

- $\Delta G°$ is negative (<0)
- $E°_{cell}$ is positive (>0)
- $K > 1$

For a nonspontaneous reaction (one that proceeds in the reverse direction when all reactants and products are in their standard states):

- $\Delta G°$ is positive (>0)
- $E°_{cell}$ is negative (<0)
- $K < 1$

## The Relationship between $\Delta G°$ and $E°_{cell}$

We can derive a relationship between $\Delta G°$ and $E°_{cell}$ by briefly returning to the definition of potential difference from Section 19.3—a potential difference is a measure of the difference of potential energy per unit charge ($q$):

$$E = \frac{\text{potential energy difference (in J)}}{\text{charge (in C)}}$$

Since the potential energy difference represents the maximum amount of work that can be done by the system on the surroundings, we can write:

$$w_{max} = -qE°_{cell} \qquad [19.1]$$

The negative sign follows the convention used throughout this book that work done by the system on the surroundings is negative.

We can quantify the charge ($q$) that flows in an electrochemical reaction by using **Faraday's constant ($F$)**, which represents the charge in coulombs of 1 mol of electrons:

$$F = \frac{96{,}485 \text{ C}}{\text{mol e}^-}$$

The total charge is $q = nF$, where $n$ is the number of moles of electrons from the balanced chemical equation and $F$ is Faraday's constant. Substituting $q = nF$ into Equation 19.1:

$$w_{max} = -qE°_{cell}$$
$$= -nFE°_{cell} \qquad [19.2]$$

Finally, recall from Chapter 18 that the standard change in free energy for a chemical reaction ($\Delta G°$) represents the maximum amount of work that can be done by the reaction. Therefore, $w_{max} = \Delta G°$. Making this substitution into Equation 19.2, we arrive at the following important result:

$$\Delta G° = -nFE°_{cell} \qquad [19.3]$$

where $\Delta G°$ is the standard change in free energy for an electrochemical reaction, $n$ is the number of moles of electrons transferred in the balanced equation, $F$ is Faraday's constant, and $E°_{cell}$ is the standard cell potential. Example 19.6 shows how to apply this equation to calculate the standard free energy change for an electrochemical cell.

**Example 19.6** Relating $\Delta G°$ and $E°_{cell}$

Use the tabulated electrode potentials to calculate $\Delta G°$ for the reaction:

$$I_2(s) + 2\,Br^-(aq) \longrightarrow 2\,I^-(aq) + Br_2(l)$$

Is the reaction spontaneous?

| | |
|---|---|
| **SORT** You are given a redox reaction and asked to find $\Delta G°$. | **GIVEN:** $I_2(s) + 2\,Br^-(aq) \longrightarrow 2\,I^-(aq) + Br_2(l)$<br><br>**FIND:** $\Delta G°$ |

| | |
|---|---|
| **STRATEGIZE** Refer to the values of electrode potentials in Table 19.1 to calculate $E°_{cell}$. Then use Equation 19.3 to calculate $\Delta G°$ from $E°_{cell}$. | **CONCEPTUAL PLAN**<br><br><br><br>$\Delta G° = -nFE°_{cell}$ |

**SOLVE** Separate the reaction into oxidation and reduction half-reactions and find the standard electrode potentials for each. Determine $E°_{cell}$ by subtracting $E_{an}$ from $E_{cat}$.

**SOLUTION**

| | | |
|---|---|---|
| **Oxidation (Anode):** | $2\,Br^-(aq) \longrightarrow Br_2(l) + 2\,e^-$ | $E° = 1.09\ V$ |
| **Reduction (Cathode):** | $I_2(s) + 2\,e^- \longrightarrow 2\,I^-(aq)$ | $E° = 0.54\ V$ |
| | $I_2(s) + 2\,Br^-(aq) \longrightarrow 2\,I^-(aq) + Br_2(l)$ | $E°_{cell} = E°_{cat} - E°_{an}$ |
| | | $= -0.55\ V$ |

Calculate $\Delta G°$ from $E°_{cell}$. The value of $n$ (the number of moles of electrons) corresponds to the number of electrons that are canceled in the half-reactions. Remember that 1 V = 1 J/C.

$$\Delta G° = -nFE°_{cell}$$

$$= -2\ \text{mol}\ e^- \left( \frac{96,485\ C}{\text{mol}\ e^-} \right)\left( -0.55\ \frac{J}{C} \right)$$

$$= +1.1 \times 10^5\ J$$

Since $\Delta G°$ is positive, the reaction is not spontaneous under standard conditions.

**CHECK** The answer is in the correct units (joules) and seems reasonable in magnitude ($\approx$ 110 kJ). You have seen (in Chapter 18) that values of $\Delta G°$ typically range from plus or minus tens to hundreds of kilojoules. The sign is positive, as expected for a reaction in which $E°_{cell}$ is negative.

**FOR PRACTICE 19.6** Use tabulated electrode potentials to calculate $\Delta G°$ for the reaction.

$$2\,Na(s) + 2\,H_2O(l) \longrightarrow H_2(g) + 2\,OH^-(aq) + 2\,Na^+(aq)$$

Is the reaction spontaneous?

---

**Periodic Trends and the Direction of Spontaneity for Redox Reactions** Consider the result of Example 19.6. The calculation revealed that the reaction is not spontaneous. Based on conceptual reasoning, which of the following best explains why $I_2$ does not oxidize $Br^-$?

CONCEPTUAL CONNECTION **19.5**

**(a)** Br is more electronegative than I; therefore, you do not expect $Br^-$ to give up an electron to $I_2$.

**(b)** I is more electronegative than Br; therefore, you do not expect $I_2$ to give up an electron to $Br^-$.

**(c)** $Br^-$ is in solution and $I_2$ is a solid. Solids do not gain electrons from substances in solution.

## The Relationship between $E^{\circ}_{cell}$ and $K$

We can derive a relationship between the standard cell potential ($E^{\circ}_{cell}$) and the equilibrium constant for the redox reaction occurring in the cell ($K$) by returning to the relationship between $\Delta G^{\circ}$ and $K$ that we learned in Chapter 18. Recall from Section 18.10 that:

$$\Delta G^{\circ} = -RT \ln K \qquad [19.4]$$

By setting Equations 19.3 and 19.4 equal to each other, we get:

$$-nFE^{\circ}_{cell} = -RT \ln K$$

$$E^{\circ}_{cell} = \frac{RT}{nF} \ln K \qquad [19.5]$$

Equation 19.5 is usually simplified for use at 25 °C by making the following substitutions:

$$R = 8.314 \frac{J}{mol \cdot K}; \; T = 298.15 \text{ K}; \; F = \left(\frac{96,485 \text{ C}}{mol \text{ e}^-}\right); \text{ and } \ln K = 2.303 \log K$$

Substituting into Equation 19.5, we get the following important result:

$$E^{\circ}_{cell} = \frac{0.0592 \text{ V}}{n} \log K \qquad [19.6]$$

where $E^{\circ}_{cell}$ is the standard cell potential, $n$ is the number of moles of electrons transferred in the redox reaction, and $K$ is the equilibrium constant for the balanced redox reaction at 25 °C. Example 19.7 demonstrates how to use Equation 19.6.

---

### Example 19.7    Relating $E^{\circ}_{cell}$ and $K$

Use the tabulated electrode potentials to calculate $K$ for the oxidation of copper by $H^+$ (at 25 °C):

$$Cu(s) + 2 H^+(aq) \longrightarrow Cu^{2+}(aq) + H_2(g)$$

| SORT You are given a redox reaction and asked to find $K$. | **GIVEN:** $Cu(s) + 2 H^+(aq) \longrightarrow Cu^{2+}(aq) + H_2(g)$<br>**FIND:** $K$ |
|---|---|
| STRATEGIZE Refer to the values of electrode potentials in Table 19.1 to calculate $E^{\circ}_{cell}$. Then use Equation 19.6 to calculate $K$ from $E^{\circ}_{cell}$. | **CONCEPTUAL PLAN**<br><br>$$E^{\circ}_{cell} = \frac{0.0592 \text{ V}}{n} \log K$$ |

SOLVE Separate the reaction into oxidation and reduction half-reactions and find the standard electrode potentials for each. Find $E^{\circ}_{cell}$ by subtracting $E_{an}$ from $E_{cat}$.

**SOLUTION**

***Oxidation***
*(Anode):*                   $Cu(s) \longrightarrow Cu^{2+}(aq) + 2 e^-$              $E^{\circ} = 0.34 \text{ V}$

***Reduction***
*(Cathode):*   $2 H^+(aq) + 2 e^- \longrightarrow H_2(g)$              $E^{\circ} = 0.00 \text{ V}$

$\overline{\qquad\qquad Cu(s) + 2 H^+(aq) \longrightarrow Cu^{2+}(aq) + H_2(g)\qquad E^{\circ}_{cell} = E^{\circ}_{cat} - E^{\circ}_{an}}$

$$= -0.34 \text{ V}$$

Calculate $K$ from $E^{\circ}_{cell}$. The value of $n$ (the number of moles of electrons) corresponds to the number of electrons that are canceled in the half-reactions.

$$E^{\circ}_{cell} = \frac{0.0592 \text{ V}}{n} \log K$$

$$\log K = E^{\circ}_{cell} \frac{n}{0.0592 \text{ V}}$$

$$\log K = -0.34 \text{ V} \frac{2}{0.0592 \text{ V}} = -11.486$$

$$K = 10^{-11.486} = 3.3 \times 10^{-12}$$

CHECK The answer has no units, as expected for an equilibrium constant. The magnitude of the answer is small, indicating that the reaction lies far to the left at equilibrium, as expected for a reaction in which $E°_{cell}$ is negative.

FOR PRACTICE 19.7 Use the tabulated electrode potentials to calculate $K$ for the oxidation of iron by $H^+$ (at 25 °C):

$$2 \, Fe(s) + 6 \, H^+(aq) \longrightarrow 2 \, Fe^{3+}(aq) + 3 \, H_2(g)$$

Notice that the fundamental quantity in the given relationships is the standard change in free energy ($\Delta G°$). From that quantity, we can calculate both $E°_{cell}$ and $K$. The relationships between these three quantities is summarized with the following diagram:

$$E°_{cell} = \frac{0.0592 \, V}{n} \log K$$

**Relating $K$, $\Delta G°_{rxn}$, and $E°_{cell}$** A redox reaction has an equilibrium constant of $K = 1.2 \times 10^3$. Which statement is true regarding $\Delta G°_{rxn}$ and $E°_{cell}$ for this reaction?

(a) $E°_{cell}$ is positive and $\Delta G°_{rxn}$ is positive.
(b) $E°_{cell}$ is negative and $\Delta G°_{rxn}$ is negative.
(c) $E°_{cell}$ is positive and $\Delta G°_{rxn}$ is negative.
(d) $E°_{cell}$ is negative and $\Delta G°_{rxn}$ is positive.

CONCEPTUAL
CONNECTION **19.6**

## 19.6 Cell Potential and Concentration

We have learned how to find $E°_{cell}$ under standard conditions. For example, we know that when $[Cu^{2+}] = 1 \, M$ and $[Zn^{2+}] = 1 \, M$, the following reaction produces a potential of 1.10 V:

$$Zn(s) + Cu^{2+}(aq, 1 \, M) \longrightarrow Zn^{2+}(aq, 1 \, M) + Cu(s) \quad E°_{cell} = 1.10 \, V$$

However, what if $[Cu^{2+}] > 1 \, M$ and $[Zn^{2+}] < 1 \, M$? For example, how would the cell potential for the following conditions be different from the potential under standard conditions?

$$Zn(s) + Cu^{2+}(aq, 2 \, M) \longrightarrow Zn^{2+}(aq, 0.010 \, M) + Cu(s) \quad E_{cell} = ?$$

Since the concentration of a reactant is greater than standard conditions, and since the concentration of product is less than standard conditions, we can use Le Châtelier's principle to predict that the reaction has an even stronger tendency to occur in the forward direction and that $E_{cell}$ is therefore greater than +1.10 V (Figure 19.11▶, on the next page).

### Cell Potential under Nonstandard Conditions: The Nernst Equation

We can derive an exact relationship between $E_{cell}$ (under nonstandard conditions) and $E°_{cell}$ by considering the relationship between the change in free energy ($\Delta G$) and the *standard* change in free energy ($\Delta G°$) from Section 18.9:

$$\Delta G = \Delta G° + RT \ln Q \qquad [19.7]$$

▲ **FIGURE 19.11 Cell Potential and Concentration** This figure compares the $Zn/Cu^{2+}$ electrochemical cell under standard and nonstandard conditions. In this case, the nonstandard conditions consist of a higher $Cu^{2+}$ concentration ($[Cu^{2+}] > 1$ M) at the cathode and a lower $Zn^{2+}$ concentration at the anode ($[Zn^{2+}] < 1$ M). According to Le Châtelier's principle, the forward reaction has a greater tendency to occur, resulting in a greater overall cell potential than the cell potential under standard conditions.

where $R$ is the gas constant $(8.314 \text{ J/mol} \cdot \text{K})$, $T$ is the temperature in kelvins, and $Q$ is the reaction quotient corresponding to the nonstandard conditions. Since we know the relationship between $\Delta G$ and $E_{cell}$ (Equation 19.3), we can substitute into Equation 19.7:

$$\Delta G = \Delta G° + RT \ln Q$$

$$-nFE_{cell} = -nFE°_{cell} + RT \ln Q$$

We then divide each side by $-nF$ to arrive at:

$$E_{cell} = E°_{cell} - \frac{RT}{nF} \ln Q \qquad [19.8]$$

As we have seen, $R$ and $F$ are constants; at $T = 25\,°C$, $\frac{RT}{nF} \ln Q = \frac{0.0592 \text{ V}}{n} \log Q$.

Substituting into Equation 19.8, we arrive at the **Nernst equation**:

$$E_{cell} = E°_{cell} - \frac{0.0592 \text{ V}}{n} \log Q \qquad [19.9]$$

where $E_{cell}$ is the cell potential in volts, $E°_{cell}$ is the *standard* cell potential in volts, $n$ is the number of moles of electrons transferred in the redox reaction, and $Q$ is the reaction quotient. Notice that, under standard conditions, $Q = 1$, and (since $\log 1 = 0$) $E_{cell} = E°_{cell}$, as expected. Example 19.8 shows how to calculate the cell potential under nonstandard conditions.

## Example 19.8    Calculating $E_{cell}$ under Nonstandard Conditions

Determine the cell potential for an electrochemical cell based on the following two half-reactions:

**Oxidation:** $Cu(s) \longrightarrow Cu^{2+}(aq, 0.010\ M) + 2\ e^-$

**Reduction:** $MnO_4^-(aq, 2.0\ M) + 4\ H^+(aq, 1.0\ M) + 3\ e^- \longrightarrow MnO_2(s) + 2\ H_2O(l)$

| | |
|---|---|
| **SORT** You are given the half-reactions for a redox reaction and the concentrations of the aqueous reactants and products. You are asked to find the cell potential. | **GIVEN:** $[MnO_4^-] = 2.0\ M;\ [H^+] = 1.0\ M;\ [Cu^{2+}] = 0.010\ M$ <br><br> **FIND:** $E_{cell}$ |
| **STRATEGIZE** Use the tabulated values of electrode potentials to calculate $E°_{cell}$. Then use Equation 19.9 to calculate $E_{cell}$. | **CONCEPTUAL PLAN** <br><br>  <br><br> $E_{cell} = E°_{cell} - \dfrac{0.0592\ V}{n} \log Q$ |

| | |
|---|---|
| **SOLVE** Write the oxidation and reduction half-reactions, multiplying by the appropriate coefficients to cancel the electrons. Find the standard electrode potentials for each half-reaction. Find $E°_{cell}$. | **SOLUTION** <br> **Oxidation** <br> **(Anode):** $\qquad\qquad\qquad\qquad 3[Cu(s) \longrightarrow Cu^{2+}(aq) + 2\ e^-] \qquad E° = 0.34\ V$ <br> **Reduction** <br> **(Cathode):** $\quad 2[MnO_4^-(aq) + 4\ H^+(aq) + 3\ e^- \longrightarrow MnO_2(s) + 2\ H_2O(l)] \quad E° = 1.68\ V$ <br> $\overline{\quad 3\ Cu(s) + 2\ MnO_4^-(aq) + 8\ H^+(aq) \longrightarrow 3\ Cu^{2+}(aq) + 2\ MnO_2(s) + 4\ H_2O(l)\quad}$ <br> $\qquad\qquad\qquad\qquad\qquad\qquad\qquad\qquad\qquad E°_{cell} = E°_{cat} - E°_{an} = 1.34\ V$ |
| Calculate $E_{cell}$ from $E°_{cell}$. The value of $n$ (the number of moles of electrons) corresponds to the number of electrons (six in this case) canceled in the half-reactions. Determine $Q$ based on the overall balanced equation and the given concentrations of the reactants and products. (Note that pure liquid water, solid $MnO_2$, and solid copper are omitted from the expression for $Q$.) | $\begin{aligned} E_{cell} &= E°_{cell} - \dfrac{0.0592\ V}{n} \log Q \\[2mm] &= E°_{cell} - \dfrac{0.0592\ V}{n} \log \dfrac{[Cu^{2+}]^3}{[MnO_4^-]^2[H^+]^8} \\[2mm] &= 1.34\ V - \dfrac{0.0592\ V}{6} \log \dfrac{(0.010)^3}{(2.0)^2(1.0)^8} \\[2mm] &= 1.34\ V - (-0.065\ V) \\[2mm] &= 1.41\ V \end{aligned}$ |

**CHECK** The answer has the correct units (V). The value of $E_{cell}$ is larger than $E°_{cell}$, as expected based on Le Châtelier's principle because one of the aqueous reactants has a concentration greater than standard conditions and the one aqueous product has a concentration less than standard conditions. Therefore, the reaction has a greater tendency to proceed toward products and a greater cell potential.

**FOR PRACTICE 19.8** Determine the cell potential of an electrochemical cell based on the following two half-reactions:

**Oxidation:** $Ni(s) \longrightarrow Ni^{2+}(aq, 2.0\ M) + 2\ e^-$

**Reduction:** $VO_2^+(aq, 0.010\ M) + 2\ H^+(aq, 1.0\ M) + e^- \longrightarrow VO^{2+}(aq, 2.0\ M) + H_2O(l)$

From Equation 19.9, we can conclude the following:

- When a redox reaction within a voltaic cell occurs under standard conditions, $Q = 1$; therefore, $E_{cell} = E°_{cell}$.

$$E_{cell} = E°_{cell} - \frac{0.0592 \text{ V}}{n} \log Q$$

$$= E°_{cell} - \frac{0.0592 \text{ V}}{n} \log 1 \qquad \text{log } 1 = 0$$

$$= E°_{cell}$$

- When a redox reaction within a voltaic cell occurs under conditions in which $Q < 1$, the greater concentration of reactants relative to products drives the reaction to the right, resulting in $E_{cell} > E°_{cell}$.

- When a redox reaction within an electrochemical cell occurs under conditions in which $Q > 1$, the greater concentration of products relative to reactants drives the reaction to the left, resulting in $E_{cell} < E°_{cell}$.

- When a redox reaction reaches equilibrium, $Q = K$. The redox reaction has no tendency to occur in either direction and $E_{cell} = 0$:

$$E_{cell} = E°_{cell} - \frac{0.0592 \text{ V}}{n} \log Q \qquad E°_{cell} \quad \text{(see Equation 19.6)}$$

$$= E°_{cell} - \frac{0.0592 \text{ V}}{n} \log K$$

$$= E°_{cell} - E°_{cell}$$

$$= 0$$

This last point explains why batteries do not last forever—as the reactants are depleted, the reaction proceeds toward equilibrium and the potential tends toward zero.

CONCEPTUAL
CONNECTION **19.7**

**Relating $Q$, $K$, $E_{cell}$, and $E°_{cell}$**  In an electrochemical cell, $Q = 0.0010$ and $K = 0.10$. What can you conclude about $E_{cell}$ and $E°_{cell}$?

**(a)** $E_{cell}$ is positive and $E°_{cell}$ is negative.
**(b)** $E_{cell}$ is negative and $E°_{cell}$ is positive.
**(c)** Both $E_{cell}$ and $E°_{cell}$ are positive.
**(d)** Both $E_{cell}$ and $E°_{cell}$ are negative.

## Concentration Cells

Since cell potential depends not only on the half-reactions occurring in the cell, but also on the *concentrations* of the reactants and products in those half-reactions, we can construct a voltaic cell in which both half-reactions are the same, but in which *a difference in concentration drives the current flow*. For example, consider the electrochemical cell shown in Figure 19.12▶, in which copper is oxidized at the anode and copper ions are reduced at the cathode. The second part of Figure 19.12 depicts this cell under nonstandard conditions, with $[Cu^{2+}] = 2.0$ M in one half-cell and $[Cu^{2+}] = 0.010$ M in the other:

$$Cu(s) + Cu^{2+}(aq, 2.0 \text{ M}) \longrightarrow Cu^{2+}(aq, 0.010 \text{ M}) + Cu(s)$$

## A Concentration Cell

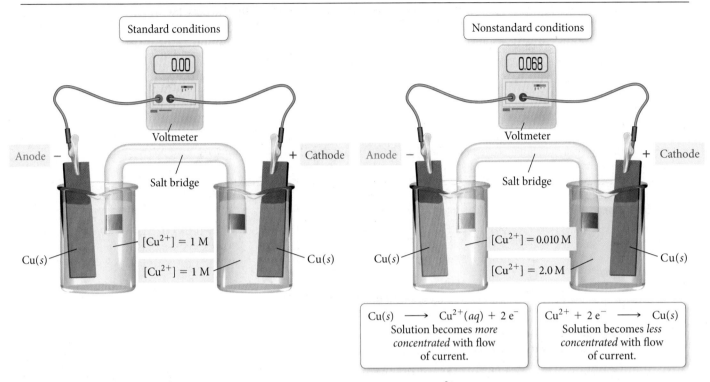

▲ **FIGURE 19.12 Cu/Cu²⁺ Concentration Cell** If two half-cells have the same $Cu^{2+}$ concentration, the cell potential is zero. If one half-cell has a greater $Cu^{2+}$ concentration than the other, a spontaneous reaction occurs. In the reaction, $Cu^{2+}$ ions in the more concentrated cell are reduced (to solid copper), while $Cu^{2+}$ ions in the more dilute cell are formed (from solid copper). The concentration of copper ions in the two half-cells tends toward equality.

The half-reactions are identical, and the *standard* cell potential is therefore zero:

**Reduction** *(Cathode):*  $Cu^{2+}(aq) + 2e^- \longrightarrow Cu(s)$     $E° = 0.34$ V

**Oxidation** *(Anode):*         $Cu(s) \longrightarrow Cu^{2+}(aq) + 2e^-$  $E° = 0.34$ V

$$Cu^{2+}(aq) + Cu(s) \longrightarrow Cu(s) + Cu^{2+}(aq) \quad E°_{cell} = E°_{cat} - E°_{an}$$

$$= +0.00 \text{ V}$$

Because of the different concentrations in the two half-cells, the cell potential must be calculated using the Nernst equation:

$$E_{cell} = E°_{cell} - \frac{0.0592 \text{ V}}{2} \log \frac{0.010}{2.0}$$

$$= 0.000 \text{ V} + 0.068 \text{ V}$$

$$= 0.068 \text{ V}$$

The cell produces a potential of 0.068 V. *Electrons spontaneously flow from the half-cell with the lower copper ion concentration to the half-cell with the higher copper ion concentration.* We can imagine a concentration cell in the same way we think about any concentration gradient. If we mix a concentrated solution of $Cu^{2+}$ with a dilute solution, the $Cu^{2+}$ ions flow from the concentrated solution to the dilute one. Similarly, in a concentration cell, the transfer of electrons *from* the dilute half-cell results in the formation of $Cu^{2+}$ ions in the dilute half-cell. The electrons flow to the concentrated cell, where they react with $Cu^{2+}$ ions and reduce them to $Cu(s)$. Therefore, *the flow of electrons has the effect of increasing the concentration of $Cu^{2+}$ in the dilute cell and decreasing the concentration of $Cu^{2+}$ in the concentrated half-cell.*

# Chemistry AND Medicine | Concentration Cells in Human Nerve Cells

Recall from Section 8.1 that tiny pumps in the membranes of human nerve cells pump ions—especially sodium ($Na^+$) and potassium ($K^+$)—through those membranes, establishing a concentration gradient for each type of ion. The concentration of sodium ions is higher outside the cell than within, while exactly the opposite is true for potassium ions. These concentration gradients result in an electrical potential across the cell membrane, called the resting potential, of about −70 mV. (The interior of the cell is negative with respect to the exterior.)

When the nerve cell is stimulated, certain channels in the membrane open, allowing $Na^+$ ions to rush into the cell and causing the potential to temporarily rise to about +30 mV (Figure 19.13▼). Other channels that allow $K^+$ ions to rush

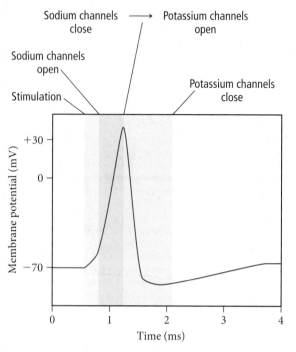

▲ **FIGURE 19.14 Potential Changes across the Nerve Cell Membrane** The changes in ion concentrations that take place when a nerve cell is stimulated result in a spike in the electrochemical potential across the membrane.

▲ **FIGURE 19.13 Concentration Changes in Nerve Cells**

out of the cell open in turn, bringing the potential back down to near its resting potential. The result is a spike in the electrochemical potential across the membrane, which provides

the stimulus for a similar spike in the neighboring segment of the membrane (Figure 19.14▲). In this way, an electrical signal moves down the length of a nerve cell.

When the electrical signal reaches the end of the nerve cell, it triggers the release of a chemical neurotransmitter, which travels to the neighboring nerve cell and stimulates the same kind of electrochemical spike. In this way, neural signals travel throughout the brain and nervous system.

## 19.7 Batteries: Using Chemistry to Generate Electricity

We have seen that we can combine the electron-losing tendency of one substance with the electron-gaining tendency of another to create electrical current in a voltaic cell. Batteries are voltaic cells conveniently packaged to act as portable sources of electricity. The actual oxidation and reduction reactions vary depending on the particular type of battery. In this section, we examine several different types.

### Dry-Cell Batteries

Common batteries, such as the kind in a flashlight, are called **dry-cell batteries** because they do not contain large amounts of liquid water. There are several familiar types of dry-cell batteries. The most inexpensive are composed of a zinc case that acts as the anode (Figure 19.15(a)▶). The zinc is oxidized according to the reaction:

*Oxidation (Anode):* $Zn(s) \longrightarrow Zn^{2+}(aq) + 2\,e^-$

The cathode is a carbon rod immersed in a moist paste of $MnO_2$ that also contains $NH_4Cl$. The $MnO_2$ is reduced to $Mn_2O_3$ according to the reaction:

**_Reduction (Cathode):_** $2 MnO_2(s) + 2 NH_4^+(aq) + 2 e^- \longrightarrow$
$$Mn_2O_3(s) + 2 NH_3(g) + H_2O(l)$$

These two half-reactions produce a voltage of about 1.5 V. Two or more of these batteries can be connected in series (cathode-to-anode connection) to produce higher voltages.

The more common **alkaline batteries** (Figure 19.15(b)▶) employ slightly different half-reactions in a basic medium (therefore the name alkaline). In an alkaline battery, the zinc is oxidized in a basic environment:

**_Oxidation (Anode):_** $Zn(s) + 2 OH^-(aq) \longrightarrow Zn(OH)_2(s) + 2 e^-$

**_Reduction (Cathode):_** $2 MnO_2(s) + 2 H_2O(l) + 2 e^- \longrightarrow$
$$2 MnO(OH)(s) + 2 OH^-(aq)$$

**_Overall reaction:_** $Zn(s) + 2 MnO_2(s) + 2 H_2O(l) \longrightarrow$
$$Zn(OH)_2(s) + 2 MnO(OH)(s)$$

Alkaline batteries have a longer working life and a longer shelf life than their nonalkaline counterparts.

## Lead–Acid Storage Batteries

The batteries in most automobiles are **lead–acid storage batteries**. These batteries consist of six electrochemical cells wired in series (Figure 19.16▼). Each cell produces 2 V for a total of 12 V. Each cell contains a porous lead anode where oxidation occurs and a lead(IV) oxide cathode where reduction occurs according to the reactions:

**_Oxidation (Anode):_** $Pb(s) + HSO_4^-(aq) \longrightarrow PbSO_4(s) + H^+(aq) + 2 e^-$

**_Reduction (Cathode):_** $PbO_2(s) + HSO_4^-(aq) + 2 3 H^+(aq) + 2 e^- \longrightarrow$
$$PbO_4(s) + 2 H_2O(l)$$

**_Overall reaction:_** $Pb(s) + PbO_2(s) + 2 HSO_4^-(aq) + 2 H^+(aq) \longrightarrow$
$$2 PbSO_4(s) + 2 H_2O(l)$$

Both the anode and the cathode are immersed in sulfuric acid ($H_2SO_4$). As electrical current is drawn from the battery, both electrodes become coated with $PbSO_4(s)$. If the battery is run for a long time without recharging, too much $PbSO_4(s)$ develops on the surface of the electrodes and the battery goes dead. The lead–acid storage battery can be recharged by an electrical current (which must come from an external source such as an alternator in a car). The current causes the preceding reaction to occur in reverse, converting the $PbSO_4(s)$ back to $Pb(s)$ and $PbSO_4(s)$.

(a)

(b)

▲ **FIGURE 19.15 Dry-Cell Battery**
**(a)** In a common dry-cell battery, the zinc case acts as the anode and a graphite rod immersed in a moist, slightly acidic paste of $MnO_2$ and $NH_4Cl$ acts as the cathode.
**(b)** Longer-lived alkaline batteries employ a graphite cathode immersed in a paste of $MnO_2$ and a base.

▲ **FIGURE 19.16 Lead–Acid Storage Battery** A lead–acid storage battery consists of six cells wired in series. Each cell contains a porous lead anode and a lead oxide cathode, both immersed in sulfuric acid.

## Other Rechargeable Batteries

The ubiquity of power electronic products such as laptops and smartphones, as well as the growth in popularity of hybrid electric vehicles, drives the need for efficient, long-lasting, rechargeable batteries. The most common types include the **nickel–cadmium (NiCad) battery**, the **nickel–metal hydride (NiMH) battery**, and the **lithium ion battery**.

### The Nickel–Cadmium (NiCad) Battery

Nickel–cadmium batteries consist of a solid cadmium anode and a NiO(OH)(s) cathode. The electrolyte is usually KOH(aq). During operation, the cadmium is oxidized and the NiO(OH) is reduced according to the equations:

*Oxidation (Anode):* $Cd(s) + 2 OH^-(aq) \longrightarrow Cd(OH)_2(s) + 2 e^-$

*Reduction (Cathode):* $2 NiO(OH)(s) + 2 H_2O(l) + 2 e^- \longrightarrow$
$$2 Ni(OH)_2(s) + 2 OH^-(aq)$$

▲ Several types of batteries, including NiCad, NiMH, and lithium ion batteries, are recharged by chargers that use household current.

The overall reaction produces about 1.30 V. As current is drawn from the NiCad battery, solid cadmium hydroxide accumulates on the anode and solid nickel(II) hydroxide accumulates on the cathode. But if current is run in the opposite direction, the reactants can be regenerated from the products. A common problem in recharging NiCad and other rechargeable batteries is knowing when to stop. Once all of the products of the reaction are converted back to reactants, the charging process should ideally terminate. If charging does not stop, the electrical current drives other, usually unwanted, reactions such as the electrolysis of water to form hydrogen and oxygen gas. These reactions typically damage the battery and may sometimes even cause an explosion. Consequently, most commercial battery chargers have sensors that measure when the charging is complete. These sensors rely on the small changes in voltage or increases in temperature that occur once the products have all been converted back to reactants.

### The Nickel–Metal Hydride (NiMH) Battery

Although NiCad batteries were the standard rechargeable battery for many years, they are being replaced by other types of rechargeable batteries, in part because of the toxicity of cadmium and the resulting disposal problems. One of these replacements is the nickel–metal hydride or NiMH battery. The NiMH battery employs the same cathode reaction as the NiCad battery, but it employs a different anode reaction. In the anode of a NiMH battery, hydrogen atoms held in a metal alloy are oxidized. If we let M represent the metal alloy, we can write the half-reactions as follows:

*Oxidation (Anode):* $M \cdot H(s) + OH^-(aq) \longrightarrow M(s) + H_2O(l) + e^-$

*Reduction (Cathode):* $NiO(OH)(s) + H_2O(l) + e^- \longrightarrow Ni(OH)_2(s) + OH^-(aq)$

In addition to being more environmentally friendly than NiCad batteries, NiMH batteries also have a greater energy density (energy content per unit battery mass), as we can see in Table 19.2. In some cases, a NiMH battery can carry twice the energy of a NiCad battery of the same mass, making NiMH batteries the most common choice for hybrid electric vehicles.

### The Lithium Ion Battery

The newest and most expensive common type of rechargeable battery is the lithium ion battery. Since lithium is the least dense metal ($0.53 \text{ g/cm}^3$), lithium batteries have high-energy densities (see Table 19.2). The lithium battery works differently than the other batteries we have examined so far, and the details of its operation are beyond the scope of

| TABLE 19.2 Energy Density and Overcharge Tolerance of Several Rechargeable Batteries | | |
|---|---|---|
| **Battery Type** | **Energy Density (W · h/kg)** | **Overcharge Tolerance** |
| NiCad | 45–80 | Moderate |
| NiMH | 60–120 | Low |
| Li ion | 110–160 | Low |
| Pb storage | 30–50 | High |

our current discussion. Briefly, we can describe operation of the lithium battery as being due primarily to the motion of lithium ions from the anode to the cathode. The anode is composed of graphite into which lithium ions are incorporated between layers of carbon atoms. Upon discharge, the lithium ions spontaneously migrate to the cathode, which consists of a lithium transition metal oxide such as $LiCoO_2$ or $LiMn_2O_4$. The transition metal is reduced during this process. Upon recharging, the transition metal is oxidized, forcing the lithium to migrate back into the graphite (Figure 19.17▶). The flow of lithium ions from the anode to the cathode causes a corresponding flow of electrons in the external circuit. Lithium ion batteries are commonly used in applications where light weight and high-energy density are important. These include smartphones, laptop computers, and digital cameras.

## Fuel Cells

We discussed the potential for *fuel cells* in the opening section of this chapter. Fuel cells may one day replace—or at least work in combination with—centralized power grid electricity. In addition, vehicles powered by fuel cells may one day usurp vehicles powered by internal combustion engines. Fuel cells are like batteries; the key difference is that a battery is self-contained, while in a fuel cell the reactants need to be constantly replenished from an external source. With use, normal batteries lose their ability to generate voltage because the reactants become depleted as electrical current is drawn from the battery. In a **fuel cell**, the reactants—the fuel provided from an external source—constantly flow through the battery, generating electrical current as they undergo a redox reaction.

The most common fuel cell is the hydrogen–oxygen fuel cell (Figure 19.18▼). In this cell, hydrogen gas flows past the anode (a screen coated with platinum catalyst) and undergoes oxidation:

***Oxidation*** *(Anode):*    $2\,H_2(g) + 4\,OH^-(aq) \longrightarrow 4\,H_2O(l) + 4\,e^-$

**Graphite**              **Lithium-transition metal oxide**

**Lithium ions**

Charge ⟵ $Li^+$ ⟶ Discharge

▲ **FIGURE 19.17 Lithium Ion Battery** In the lithium ion battery, the spontaneous flow of lithium ions from the graphite anode to the lithium transition metal oxide cathode causes a corresponding flow of electrons in the external circuit.

## Hydrogen–Oxygen Fuel Cell

- ● Electron
- ○ Proton
- ● Oxygen atom

Electric circuit

$e^-$

Anode          Cathode

$H^+$    $OH^-$

Hydrogen inlet

$H_2$

$H_2O$    $O_2$

Oxygen inlet

**Electrolyte**

| **Oxidation** | **Reduction** |
|---|---|
| $2\,H_2(g) + 4\,OH^-(aq)$ $\longrightarrow\ 4\,H_2O(l) + 4\,e^-$ | $O_2(g) + 2\,H_2O(l) + 4\,e^-$ $\longrightarrow\ 4\,OH^-(aq)$ |

◀ **FIGURE 19.18 Hydrogen–Oxygen Fuel Cell** In this fuel cell, hydrogen and oxygen combine to form water.

# Chemistry IN YOUR DAY | The Fuel-Cell Breathalyzer

Police use a device called a breathalyzer to measure the amount of ethyl alcohol ($CH_3CH_2OH$) in the bloodstream of a person suspected of driving under the influence of alcohol (Figure 19.19 ▼). Breathalyzers work because the quantity of ethyl alcohol in the breath is proportional to the quantity of ethyl alcohol in the bloodstream. One type of breathalyzer employs a fuel cell to measure the quantity of alcohol in the breath. When a suspect blows into the breathalyzer, ethyl alcohol is oxidized to acetic acid at the anode:

*Oxidation (Anode):*  $CH_3CH_2OH(g) + 4\,OH^-(aq) \longrightarrow$
ethyl alcohol

$$HC_2H_3O_2(g) + 3\,H_2O(l) + 4\,e^-$$
acetic acid

At the cathode, oxygen is reduced:

*Reduction (Cathode):*  $O_2(g) + 2\,H_2O(l) + 4\,e^- \longrightarrow$
$$4\,OH^-(aq)$$

The overall reaction is the oxidation of ethyl alcohol to acetic acid and water:

*Overall:*  $CH_3CH_2OH(g) + O_2(g) \longrightarrow$
$$HC_2H_3O_2(g) + H_2O(l)$$

The magnitude of electrical current produced depends on the quantity of alcohol in the breath. A higher current results from a higher blood alcohol level. When calibrated correctly, the fuel-cell breathalyzer can precisely measure the blood alcohol level of a suspected drunk driver.

▲ **FIGURE 19.19  Fuel-Cell Breathalyzer**   The fuel-cell breathalyzer works by oxidizing ethyl alcohol in the breath to acetic acid. The electrical current that is produced is proportional to the concentration of ethyl alcohol in the breath.

Oxygen gas flows past the cathode (a similar screen) and undergoes reduction:

> *Reduction (Cathode):*  $O_2(g) + 2\,H_2O(l) + 4\,e^- \longrightarrow 4\,OH^-(aq)$

The half-reactions sum to the following overall reaction:

> *Overall reaction:*  $2\,H_2(g) + O_2(g) \longrightarrow 2\,H_2O(l)$

Notice that the only product is water. In the space shuttle program, hydrogen–oxygen fuel cells consume hydrogen to provide electricity and astronauts drink the water that is produced by the reaction. In order for hydrogen-powered fuel cells to become more widely used, a more readily available source of hydrogen must be developed.

## 19.8  Electrolysis: Driving Nonspontaneous Chemical Reactions with Electricity

In a voltaic cell, a spontaneous redox reaction produces electrical current. In an *electrolytic cell*, electrical current drives an otherwise nonspontaneous redox reaction through a process called **electrolysis**. We have seen that the reaction of hydrogen with

## Electrolysis of Water

**Anode**
$$2\,H_2O(l) \longrightarrow O_2(g) + 4\,H^+(aq) + 4\,e^-$$

**Cathode**
$$2\,H_2O(l) + 2\,e^- \longrightarrow H_2(g) + 2\,OH^-(aq)$$

◀ **FIGURE 19.20 Electrolysis of Water**
Electrical current can decompose water into hydrogen and oxygen gas.

oxygen to form water is spontaneous and can be used to produce an electrical current in a fuel cell. Conversely, by *supplying* electrical current, we can cause the reverse reaction to occur, separating water into hydrogen and oxygen (Figure 19.20▲):

$$2\,H_2(g) + O_2(g) \longrightarrow 2\,H_2O(l)$$ 
(spontaneous—produces electrical current; occurs in a voltaic cell)

$$2\,H_2O(l) \longrightarrow 2\,H_2(g) + O_2(g)$$ 
(nonspontaneous—consumes electrical current; occurs in an electrolytic cell)

Recall from Section 19.7 that one of the problems associated with the widespread adoption of hydrogen fuel cells is the scarcity of hydrogen. Where will the hydrogen to power these fuel cells come from? One possible solution is to obtain hydrogen from water through solar-powered electrolysis. A solar-powered electrolytic cell can produce hydrogen from water when the sun is shining. The hydrogen made in this way could be converted back to water to generate electricity and could also be used to power fuel-cell vehicles.

Electrolysis also has numerous other applications. For example, most metals are found in Earth's crust as metal oxides. Converting an oxide to a pure metal requires that the metal be reduced, a nonspontaneous process. Electrolysis can be used to produce these metals. Thus, sodium is produced by the electrolysis of molten sodium chloride (discussed in the following subsection).

Electrolysis is also used to plate metals onto other metals. For example, silver can be plated onto a less expensive metal using the electrolytic cell shown in Figure 19.21▶. In this cell, a silver electrode is placed in a solution containing silver ions. An electrical current causes the oxidation of silver at the anode (replenishing the silver ions in solution) and the reduction of silver ions at the cathode (coating the less expensive metal with solid silver):

***Oxidation*** *(Anode):* $\quad Ag(s) \longrightarrow Ag^+(aq) + e^-$

***Reduction (Cathode):*** $\quad Ag^+(aq) + e^- \longrightarrow Ag(s)$

▼ **FIGURE 19.21 Silver Plating**
Silver can be plated from a solution of silver ions onto metallic objects in an electrolytic cell.

## Electrolytic Cell for Silver Plating

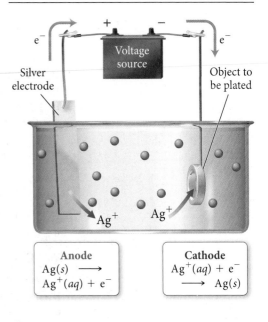

**Anode**
$$Ag(s) \longrightarrow Ag^+(aq) + e^-$$

**Cathode**
$$Ag^+(aq) + e^- \longrightarrow Ag(s)$$

## Voltaic Cell

## Electrolytic Cell

▲ **FIGURE 19.22  Voltaic versus Electrolytic Cells**  In a Zn/Cu²⁺ voltaic cell, the reaction proceeds in the spontaneous direction. In a Zn²⁺/Cu electrolytic cell, electrical current drives the reaction in the nonspontaneous direction.

Since the standard cell potential of the reaction is zero, the reaction is not spontaneous under standard conditions. An external power source can be used to drive current flow and cause the reaction to occur.

The voltage required to cause electrolysis depends on the specific half-reactions. For example, we have seen that the oxidation of zinc and the reduction of $Cu^{2+}$ produce a voltage of 1.10 V under standard conditions:

$$
\begin{aligned}
\textbf{\textit{Reduction (Cathode):}} \quad & Cu^{2+}(aq) + 2\,e^- \longrightarrow Cu(s) & E^\circ = 0.34\ \text{V} \\
\textbf{\textit{Oxidation (Anode):}} \quad & Zn(s) \longrightarrow Zn^{2+}(aq) + 2\,e^- & E^\circ = -0.76\ \text{V} \\
\hline
& Cu^{2+}(aq) + Zn(s) \longrightarrow Cu(s) + Zn^{2+}(aq) & E^\circ_{cell} = E^\circ_{cat} - E^\circ_{an} \\
& & = +1.10\ \text{V}
\end{aligned}
$$

If a power source producing *more than 1.10 V* is inserted into the Zn/Cu²⁺ voltaic cell, electrons can be forced to flow in the opposite direction, causing the reduction of $Zn^{2+}$ and the oxidation of Cu, as shown in Figure 19.22▲. Notice that in the electrolytic cell, the anode has become the cathode (oxidation always occurs at the anode) and the cathode has become the anode.

In a *voltaic cell*, the anode is the source of electrons and is therefore labeled with a negative charge. The cathode draws electrons and is therefore labeled with a positive charge. In an *electrolytic cell*, however, the source of the electrons is the external power source. The external power source must *draw electrons away* from the anode; thus, the anode must be connected to the positive terminal of the battery (as shown in Figure 19.22). Similarly, the power source drives electrons toward the cathode (where they are used in reduction), so the cathode must be connected to the *negative* terminal of the battery. The charge labels (+ and −) on an electrolytic cell are opposite of what they are in a voltaic cell.

*Summarizing Characteristics of Electrochemical Cell Types:*
**In all electrochemical cells**:

▶ Oxidation occurs at the anode.

▶ Reduction occurs at the cathode.

**In voltaic cells**:

▶ The anode is the source of electrons and has a negative charge (anode −).

▶ The cathode draws electrons and has a positive charge (cathode +).

**In electrolytic cells**:

▶ Electrons are drawn away from the anode, which must be connected to the positive terminal of the external power source (anode +).

▶ Electrons are forced to the cathode, which must be connected to the negative terminal of the power source (cathode −).

## Predicting the Products of Electrolysis

Predicting the products of an electrolysis reaction is in some cases relatively straightforward and in other cases more complex. We cover the simpler cases first and follow with the more complex ones.

### Pure Molten Salts

Consider the electrolysis of a molten salt such as sodium chloride, shown in Figure 19.23▶. $Na^+$ and $Cl^-$ are the only species present in the cell. The chloride ion cannot be further reduced (−1 is its lowest oxidation state), so it must be oxidized. The sodium ion cannot be further oxidized (+1 is its highest oxidation state), so it must be reduced. Thus, we can write the half-reactions:

*Oxidation (Anode):*        $2 \, Cl^-(l) \longrightarrow Cl_2(g) + 2 \, e^-$

*Reduction (Cathode):*    $2 \, Na^+(l) + 2 \, e^- \longrightarrow 2 \, Na(s)$

$$2 \, Na^+(l) + 2 \, Cl^-(l) \longrightarrow 2 \, Na(s) + Cl_2(g)$$

Although the reaction as written is not spontaneous, it can be driven to occur in an electrolytic cell by an external power source.

We can generalize as follows:

• In the electrolysis of a pure molten salt, the anion is oxidized and the cation is reduced.

### Mixtures of Cations or Anions

What if a molten salt contains more than one anion or cation? For example, suppose our electrolysis cell contains both NaCl and KCl. Which of the two cations is reduced at the cathode? In order to answer this question, we must determine which of the two cations is more easily reduced.

Although the values of electrode potentials for aqueous solutions given in Table 19.1 do not apply to molten salts, the relative ordering of the electrode potentials does reflect the relative ease with which the metal cations are reduced. We can see from the table that the reduction of $Na^+$ is listed *above* the reduction of $K^+$; that is, $Na^+$ has a more positive electrode potential:

$Na^+(aq) + e^- \longrightarrow Na(s)$      $E° = -2.71 \, V$     (for aqueous solution)

$K^+(aq) + e^- \longrightarrow K(s)$      $E° = -2.92 \, V$     (for aqueous solution)

Therefore, $Na^+$ is easier to reduce than $K^+$. Consequently, in a mixture of NaCl and KCl, $Na^+$ has a greater tendency to be reduced at the cathode.

Similarly, what if a mixture of molten salts contained more than one anion? For example, in a mixture of NaBr and NaCl, which of the two anions is oxidized at the cathode? The answer is similar: the anion that is more easily oxidized (the one with the more negative electrode potential):

$2 \, Cl^-(aq) \longrightarrow Cl_2(g) + 2 \, e^-$     $E° = 1.36 \, V$     (for aqueous solution)

$2 \, Br^-(aq) \longrightarrow Br_2(l) + 2 \, e^-$     $E° = 1.09 \, V$     (for aqueous solution)

Since the electrode potential for the bromine half-reaction is more negative, electrons are more easily extracted from it. The bromide ion is therefore oxidized at the anode.

We can generalize as follows:

• The cation that is most easily reduced (the one with the more positive electrode potential) is reduced first.

• The anion that is most easily oxidized (the one with the more negative electrode potential) is oxidized first.

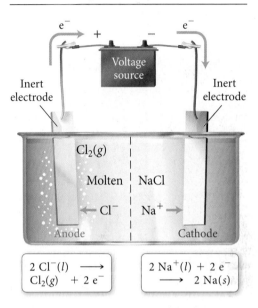

**Electrolysis of a Molten Salt**

$2 \, Cl^-(l) \longrightarrow$
$Cl_2(g) + 2 \, e^-$

$2 \, Na^+(l) + 2 \, e^-$
$\longrightarrow 2 \, Na(s)$

▲ **FIGURE 19.23 Electrolysis of Molten NaCl** In the electrolysis of a pure molten salt, the anion (in this case $Cl^-$) is oxidized and the cation (in this case $Na^+$) is reduced.

Throughout this discussion, "more positive" means the same thing as "less negative."

Throughout this discussion, "more negative" means the same thing as "less positive."

Remember NIO and PIR:
**N.I.O.**—More **N**egative **I**s **O**xidation
**P.I.R.**—More **P**ositive **I**s **R**eduction

## Aqueous Solutions

Electrolysis in an aqueous solution is complicated by the possibility of the electrolysis of water itself. Recall that water can be either oxidized or reduced according to the following half-reactions:

**Oxidation (Anode):**   $2 H_2O(l) \longrightarrow O_2(g) + 4 H^+(aq) + 4 e^-$     $E° = 1.23$ V (standard conditions)
$E = 0.82$ V([$H^+$] = $10^{-7}$ M)

**Reduction (Cathode):**   $2 H_2O(l) + 2 e^- \longrightarrow H_2(g) + 2 OH^-(aq)$   $E° = -0.83$ V (standard conditions)
$E = -0.41$ V([$OH^-$] = $10^{-7}$ M)

▲ Pure water is a poor conductor of electrical current, but the addition of an electrolyte allows electrolysis to take place, producing hydrogen and oxygen gas in a stoichiometric ratio.

The electrode potentials under standard conditions are shown to the right of each half-reaction. However, in pure water at room temperature, the concentrations of $H^+$ and $OH^-$ are not standard. The electrode potentials for [$H^+$] = $10^{-7}$ M and [$OH^-$] = $10^{-7}$ M are shown in blue. Using those electrode potentials, we can calculate $E_{cell}$ for the electrolysis of water as follows:

$$E_{cell} = E_{cat} - E_{an} = -0.41V - 0.82 V = -1.23 V$$

When a battery with a potential of several volts is connected to an electrolysis cell containing pure water, no reaction occurs because the concentration of ions in pure water is too low to conduct any significant electrical current. When an electrolyte such as $Na_2SO_4$ is added to the water, however, electrolysis occurs readily.

In any aqueous solution in which electrolysis is to take place, the electrolysis of water is also possible. For example, consider the electrolysis of a sodium iodide solution, as shown in Figure 19.24▼. For the electrolysis of *molten* NaI, we can readily predict that $I^-$ is oxidized at the anode and that $Na^+$ is reduced at the cathode. In an aqueous solution, however, two different oxidation half-reactions are possible at the anode, the oxidation of $I^-$ and the oxidation of water:

**Oxidation:** $2 I^-(aq) \longrightarrow I_2(aq) + 2 e^-$     $E° = 0.54$ V
**Oxidation:** $2 H_2O(l) \longrightarrow O_2(g) + 4 H^+(aq) + 4 e^-$   $E = 0.82$ V ([$H^+$] = $10^{-7}$ M)

Similarly, two different reduction half-reactions are possible at the cathode, the reduction of $Na^+$ and the reduction of water:

**Reduction:** $2 Na^+(aq) + 2 e^- \longrightarrow 2 Na(s)$     $E° = -2.71$ V
**Reduction:** $2 H_2O(l) + 2 e^- \longrightarrow H_2(g) + 2 OH^-(aq)$
$E = -0.41$ V ([$OH^-$] = $10^{-7}$ M)

How do we know which reactions actually occur? In both cases, the answer is the same: *the half-reaction that occurs more easily.* For oxidation, the half-reaction with the more negative electrode potential is the easier one from which to extract electrons. In this case, therefore, the iodide ion is oxidized at the anode. For reduction, the half-reaction with the more positive electrode potential is the one that accepts electrons more easily. In this case, therefore, water is reduced at the cathode. Notice that $Na^+$ cannot be reduced in an aqueous solution—water is reduced before $Na^+$.

We can make the following generalization:

- The cations of active metals—those that are not easily reduced, such as $Li^+$, $K^+$, $Na^+$, $Mg^{2+}$, $Ca^{2+}$, and $Al^{3+}$—cannot be reduced from aqueous solutions by electrolysis because water is reduced at a lower voltage.

### The Electrolysis of Aqueous Sodium Chloride and Overvoltage

An additional complication that we must consider when predicting the products of electrolysis is *overvoltage*—an additional voltage that must be applied in order to make some nonspontaneous reactions occur.

## Electrolysis of an Aqueous Salt Solution

$2 I^-(aq) \longrightarrow$
$I_2(aq) + 2 e^-$

$2 H_2O(l) + 2 e^- \longrightarrow$
$H_2(g) + 2 OH^-(aq)$

▲ **FIGURE 19.24  Electrolysis of Aqueous NaI**
In this cell, $I^-$ is oxidized to $I_2$ at the anode and $H_2O$ is reduced to $H_2$ at the cathode. Sodium ions are not reduced because their electrode potential is more negative than the electrode potential of water.

We can demonstrate this concept by considering the electrolysis of a sodium chloride solution, shown in Figure 19.25▶. In order to predict the product of the electrolysis, we consider the two possible oxidation half-reactions:

**Oxidation:** $2\,Cl^-(aq) \longrightarrow Cl_2(g) + 2\,e^-$        $E° = 1.36\ V$

**Oxidation:** $2\,H_2O(l) \longrightarrow O_2(g) + 4\,H^+(aq) + 4\,e^-$

$E = 0.82\ V([H^+] = 10^{-7}\ M)$

and the two possible reduction half-reactions:

**Reduction:** $2\,Na^+(aq) + 2\,e^- \longrightarrow 2\,Na(s)$        $E° = -2.71\ V$

**Reduction:** $2\,H_2O(l) + 2\,e^- \longrightarrow H_2(g) + 2\,OH^-(aq)$

$E = -0.41\ V([OH^-] = 10^{-7}\ M)$

Since the oxidation of water has a more negative electrode potential than the oxidation of $Cl^-$, we would initially predict that it would be easier to remove electrons from water, and thus water should be oxidized at the anode. Similarly, since the reduction of water has a more positive electrode potential than the reduction of $Na^+$, we would expect that it would be easier to get water to accept electrons, so water should be reduced at the cathode. In other words, we initially predict that a sodium chloride solution would simply result in the electrolysis of water, producing oxygen gas at the anode and hydrogen gas at the cathode. If we construct such a cell, however, we find that, although hydrogen gas is indeed formed at the cathode (as predicted), oxygen gas is *not* formed at the anode—chlorine gas is formed instead. Why? The answer is that even though the electrode potential for the oxidation of water is 0.82 V, the reaction actually requires a voltage greater than 0.82 V in order to occur. (The reasons for this behavior are related to kinetic factors that are beyond the scope of our current discussion.) This additional voltage, the *overvoltage*, increases the voltage required for the oxidation of water to about 1.4 V. The result is that the chloride ion oxidizes more easily than water and $Cl_2(g)$ is observed at the anode.

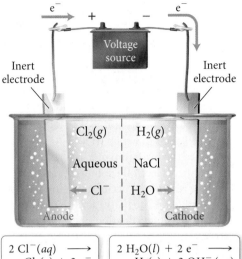

▲ **FIGURE 19.25 Electrolysis of Aqueous NaCl: The Effect of Overvoltage**   Because of overvoltage, the anode reaction of this cell is the oxidation of $Cl^-$ to $Cl_2$ gas rather than the oxidation of water to $H^+$ and $O_2$ gas.

---

**Example 19.9** | **Predicting the Products of Electrolysis Reactions**

Predict the half-reaction occurring at the anode and the half-reaction occurring at the cathode for electrolysis for each reaction.

**(a)** a mixture of molten $AlBr_3$ and $MgBr_2$

**(b)** an aqueous solution of LiI

**SOLUTION**

**(a)** In the electrolysis of a molten salt, the anion is oxidized and the cation is reduced. However, this mixture contains two cations. Start by writing the possible oxidation and reduction half-reactions that might occur. Because $Br^-$ is the only anion, write the equation for its oxidation, which occurs at the anode.

At the cathode, both the reduction of $Al^{3+}$ and the reduction of $Mg^{2+}$ are possible. The one that actually occurs is the one that occurs more easily. Since the reduction of $Al^{3+}$ has a more positive electrode potential in aqueous solution, this ion is more easily reduced. Therefore, the reduction of $Al^{3+}$ occurs at the cathode.

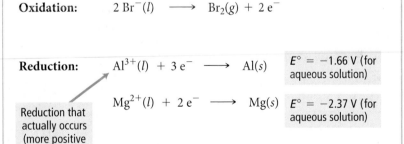

**Oxidation:**        $2\,Br^-(l) \longrightarrow Br_2(g) + 2\,e^-$

**Reduction:**        $Al^{3+}(l) + 3\,e^- \longrightarrow Al(s)$    $E° = -1.66\ V$ (for aqueous solution)

$Mg^{2+}(l) + 2\,e^- \longrightarrow Mg(s)$    $E° = -2.37\ V$ (for aqueous solution)

Reduction that actually occurs (more positive potential)

*—Continued on the next page*

*Continued—*

**(b)** Since LiI is in an aqueous solution, two different oxidation half-reactions are possible at the anode, the oxidation of $I^-$ and the oxidation of water. Write half-reactions for each including the electrode potential. Remember to use the electrode potential of water under conditions in which $[H^+] = 10^{-7}$ M. Since the oxidation of $I^-$ has the more negative electrode potential, it is the half-reaction to occur at the anode. Similarly, write half-reactions for the two possible reduction half-reactions that might occur at the cathode, the reduction of $Li^+$ and the reduction of water. Since the reduction of water has the more positive electrode potential (even when considering overvoltage, which would raise the necessary voltage by about 0.4–0.6 V), it is the half-reaction to occur at the cathode.

Oxidation that actually occurs (more negative potential)

**Oxidation:**
$$2\,I^-(aq) \longrightarrow I_2(l) + 2\,e^-$$
$$E° = 0.54\text{ V}$$

**Oxidation:**
$$2\,H_2O(l) \longrightarrow O_2(g) + 4\,H^+(aq) + 4\,e^-$$
$$E° = 0.82\text{ V } ([H^+] = 10^{-7}\text{ M})$$

**Reduction:**
$$2\,Li^+(aq) + 2\,e^- \longrightarrow 2\,Li(s)$$
$$E° = -3.04\text{ V}$$

**Reduction:**
$$2\,H_2O(l) + 2\,e^- \longrightarrow H_2(g) + 2\,OH^-(aq)$$
$$E = -0.41\text{ V } ([OH^-] = 10^{-7}\text{ M})$$

Reduction that actually occurs (more positive potential)

**FOR PRACTICE 19.9** Predict the half-reactions occurring at the anode and the cathode for the electrolysis of aqueous $Na_2SO_4$.

## Stoichiometry of Electrolysis

**▲ FIGURE 19.26 Electrolytic Cell for Copper Plating** In this cell, copper ions are plated onto other metals. It takes two moles of electrons to plate one mole of copper atoms.

In an electrolytic cell, electrical current is used to drive a particular chemical reaction. In a sense, the electrons act as a reactant, and therefore, have a stoichiometric relationship with the other reactants and products. Unlike ordinary reactants, for which we usually measure quantity as mass, for electrons we measure quantity as charge.

For example, consider an electrolytic cell used to coat copper onto metals, as shown in Figure 19.26 ◀. The half-reaction by which copper is deposited onto the metal is:

$$Cu^{2+}(aq) + 2\,e^- \longrightarrow Cu(s)$$

For every 2 mol of electrons that flow through the cell, 1 mol of solid copper is plated. We can write the stoichiometric relationship:

$$2\text{ mol }e^- : 1\text{ mol }Cu(s)$$

We can determine the number of moles of electrons that have flowed in a given electrolysis cell by measuring the total charge that has flowed through the cell, which in turn depends on the *magnitude* of the current and on the *time* that the current runs. Recall from Section 19.3 that the unit of current is the ampere:

$$1\text{ A} = 1\frac{C}{s}$$

If we multiply the amount of current (in A) flowing through the cell by the time (in s) that the current flowed, we find the total charge that passed through the cell in that time:

$$\text{Current}\left(\frac{C}{s}\right) \times \text{time (s)} = \text{charge (C)}$$

The relationship between charge and the number of moles of electrons is given by Faraday's constant, which, as we saw previously, corresponds to the charge in coulombs of 1 mol of electrons:

$$F = \frac{96{,}485\text{ C}}{\text{mol }e^-}$$

These relationships can be used to solve problems involving the stoichiometry of electrolytic cells, as demonstrated in Example 19.10.

## Example 19.10 | Stoichiometry of Electrolysis

Gold can be plated out of a solution containing $Au^{3+}$ according to the half-reaction:

$$Au^{3+}(aq) + 3\,e^- \longrightarrow Au(s)$$

What mass of gold (in grams) is plated by a 25-minute flow of 5.5 A current?

| | |
|---|---|
| **SORT** You are given the half-reaction for the plating of gold, which shows the stoichiometric relationship between moles of electrons and moles of gold. You are also given the current and duration. You must find the mass of gold that will be deposited in that time. | **GIVEN:** $3\ \text{mol e}^- : 1\ \text{mol Au}$ <br> 5.5 amps <br> 25 min <br><br> **FIND:** g Au |
| **STRATEGIZE** You need to find the amount of gold, which is related stoichiometrically to the number of electrons that have flowed through the cell. Begin with time in minutes and convert to seconds. Then, because current is a measure of charge per unit time, use the given current and the time to find the number of coulombs. Use Faraday's constant to calculate the number of moles of electrons and the stoichiometry of the reaction to find the number of moles of gold. Finally, use the molar mass of gold to convert to mass of gold. | **CONCEPTUAL PLAN** <br><br> 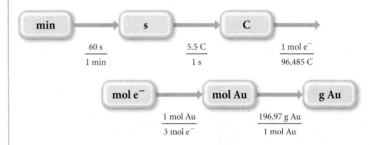 |
| **SOLVE** Follow the conceptual plan to solve the problem, canceling units to arrive at the mass of gold. | **SOLUTION** <br><br> $25\ \text{min} \times \dfrac{60\ \text{s}}{1\ \text{min}} \times \dfrac{5.5\ \text{C}}{1\ \text{s}} \times \dfrac{1\ \text{mol e}^-}{96{,}485\ \text{C}} \times \dfrac{1\ \text{mol Au}}{3\ \text{mol e}^-} \times \dfrac{196.97\ \text{g Au}}{1\ \text{mol Au}} = 5.6\ \text{g Au}$ |

**CHECK** The answer has the correct units (g Au). The magnitude of the answer is reasonable if you consider that 10 amps of current for 1 hour is the equivalent of about 1/3 mol of electrons (check for yourself), which would produce 1/9 mol (or about 20 g) of gold.

**FOR PRACTICE 19.10** Silver can be plated out of a solution containing $Ag^+$ according to the half-reaction:

$$Ag^+(aq) + e^- \longrightarrow Ag(s)$$

How much time (in minutes) does it take to plate 12 g of silver using a current of 3.0 A?

## 19.9 Corrosion: Undesirable Redox Reactions

**Corrosion** is the (usually) gradual, nearly always undesired, oxidation of metals that are exposed to oxidizing agents in the environment. From Table 19.1, we can see that the reduction of oxygen in the presence of water has an electrode potential of +0.40 V:

$$O_2(g) + 2\,H_2O(l) + 4\,e^- \longrightarrow 4\,OH^-(aq) \qquad E° = +0.40\ \text{V}$$

In the presence of acid, the reduction of oxygen has an even more positive electrode potential of +1.23 V:

$$O_2(g) + 4\,H^+(aq) + 4\,e^- \longrightarrow 2\,H_2O(l) \qquad E° = +1.23\ \text{V}$$

The reduction of oxygen, therefore, has a strong tendency to occur and can bring about the oxidation of other substances, especially metals. Notice that the half-reactions for the reduction of most metal ions are listed *below* the half-reactions for the reduction of oxygen in Table 19.1. Consequently, the oxidation (or corrosion) of those metals is spontaneous when paired with the reduction of oxygen. Corrosion is the opposite of the process by which metals are extracted from their ores. In extraction, the free metal is reduced out from its ore. In corrosion, the metal is oxidized.

▲ A metal must usually be reduced to extract it from its ore. In corrosion, the metal is oxidized back to its more natural state.

▲ Aluminum is stable because its oxide forms a protective film over the underlying metal, preventing further oxidation.

## Corrosion of Iron

Given the ease with which metals oxidize in the presence of oxygen, acid, and water, why are metals used so frequently as building materials in the first place? Many metals form oxides that coat the surface of the metal and prevent further corrosion. For example, bare aluminum metal, with an electrode potential of $-1.66$ V, is quickly oxidized in the presence of oxygen. However, the oxide that forms at the surface of aluminum is $Al_2O_3$. In its crystalline form, $Al_2O_3$ is sapphire, a highly inert and structurally solid substance. The $Al_2O_3$ coating acts to protect the underlying aluminum metal, preventing further corrosion.

The oxides of iron, in contrast, are not structurally stable, and they tend to flake away from the underlying metal, exposing it to further corrosion. A significant part of the iron produced each year is used to replace rusted iron. Rusting is a redox reaction in which iron is oxidized according to the following half-reaction:

$$Fe(s) \longrightarrow Fe^{2+}(aq) + 2\,e^- \qquad E° = -0.45 \text{ V}$$

▼ **FIGURE 19.27 Corrosion of Iron: Rusting** The oxidation of iron occurs at anodic regions on the metal surface. The iron ions migrate to cathodic regions, where they react with oxygen and water to form rust.

This oxidation reaction tends to occur at defects on the surface of the iron—known as *anodic regions* because oxidation is occurring at these locations—as shown in Figure 19.27▼. The electrons produced at the anodic region travel through the metal to areas called *cathodic regions* where they react with oxygen and H$^+$ ions dissolved in moisture:

$$O_2(g) + 4\,H^+(aq) + 4\,e^- \longrightarrow 2\,H_2O(l) \qquad E° = +1.23 \text{ V}$$

### The Rusting of Iron

(The H$^+$ ions come from carbonic acid, which naturally forms in water from carbon dioxide in air.) The overall reaction has a cell potential of $+1.68$ V and is highly spontaneous:

$$2\,Fe(s) + O_2(g) + 4\,H^+(aq) \longrightarrow 2\,H_2O(l) + 2\,Fe^{2+}(aq) \qquad E°_{cell} = +1.68 \text{ V}$$

The Fe$^{2+}$ ions formed in the anodic regions can migrate through moisture on the surface of the iron to cathodic regions, where they are further oxidized by reaction with more oxygen:

$$4\,Fe^{2+}(aq) + O_2(g) + (4 + 2n)\,H_2O(l) \longrightarrow \underset{\text{rust}}{2\,Fe_2O_3 \cdot n H_2O(s)} + 8\,H^+(aq)$$

Rust is a hydrated form of iron(III) oxide whose exact composition depends on the conditions under which it forms.

Consider each of the following important aspects of the formation of rust:

- *Moisture must be present for rusting to occur.* The presence of water is necessary because water is a reactant in the last reaction and because charge (either electrons or ions) must be free to flow between the anodic and cathodic regions.

- *Additional electrolytes promote rusting.* The presence of an electrolyte (such as sodium chloride) on the surface of iron promotes rusting because it enhances current flow. This is why cars rust so quickly in cold climates where roads are salted, or in coastal areas where seawater mist is present.

- *The presence of acids promotes rusting.* Because H$^+$ ions are involved in the reduction of oxygen, lower pH enhances the cathodic reaction and leads to faster rusting.

# Preventing the Corrosion of Iron

▲ In galvanized nails, a layer of zinc prevents the underlying iron from rusting. The zinc oxidizes in place of the iron, forming a protective layer of zinc oxide.

Preventing the rusting of iron is a major industry. The most obvious way to prevent rust is to keep iron dry. Without water, the redox reaction cannot occur. Another way to prevent rust is to coat iron with a substance that is impervious to water. Cars, for example, are painted and sealed to prevent rust. A scratch in the paint can lead to rusting of the underlying iron.

Rust can also be prevented by placing a *sacrificial electrode* in electrical contact with the iron. The sacrificial electrode must be composed of a metal that oxidizes more easily than iron (that is, it must be below iron in Table 19.1). The sacrificial electrode oxidizes in place of the iron (just as the more easily oxidizable species in a mixture is the one to oxidize), protecting the iron from oxidation. A related way to protect iron from rusting is to coat it with a metal that oxidizes more easily than iron. Galvanized nails, for example, are coated with a thin layer of zinc. Since zinc has a more negative electrode potential than iron, it will oxidize in place of the underlying iron (just

as a sacrificial electrode does). The oxide of zinc is not crumbly and remains on the nails as a protective coating.

▲ A scratch in paint often allows the underlying iron to rust.

▲ If a metal more active than iron, such as magnesium or aluminum, is in electrical contact with iron, the metal rather than the iron will be oxidized. This principle underlies the use of sacrificial electrodes to prevent the corrosion of iron.

**Sacrificial Electrodes**  Which of these metals does not act as a sacrificial electrode for iron?

Zn, Mg, Mn, Cu

CONCEPTUAL
**CONNECTION 19.8**

# Self-Assessment Quiz

**Q1.** Balance the redox reaction equation (occurring in acidic solution) and choose the correct coefficients for each reactant and product.

__VO₂⁺(aq) + __Sn(s) + __H⁺(aq) ⟶
__VO²⁺(aq) + __Sn²⁺(aq) + __H₂O(l)

a) 2,1,4 ⟶ 2,1,2          b) 1,1,2 ⟶ 1,1,1
c) 2,1,2 ⟶ 2,1,1          d) 2,1,2 ⟶ 2,1,2

**Q2.** Which statement is true for voltaic cells?

a) Electrons flow from the anode to the cathode.
b) Electrons flow from the more negatively charged electrode to the more positively charged electrode.
c) Electrons flow from higher potential energy to lower potential energy.
d) All of the above are true.

**Q3.** Use Table 19.1 to calculate $E°_{cell}$ for the reaction.

$2\ ClO_2(g) + Pb(s) \longrightarrow 2\ ClO_2^-(aq) + Pb^{2+}(aq)$

a) 1.77 V          b) 2.03 V
c) 0.82 V          d) 1.08 V

**Q4.** Use Table 19.1 to determine which statement is true of the voltaic cell pictured here.

a) Sn is the anode; Ag is the cathode; electrons flow from left to right.
b) Sn is the cathode; Ag is the anode; electrons flow from left to right.
c) Sn is the anode; Ag is the cathode; electrons flow from right to left.
d) Sn is the cathode; Ag is the anode; electrons flow from right to left.

*Continued—*

**Q5.** Use Table 19.1 to determine which metal *does not* dissolve in hydrochloric acid (HCl).

a) Zn        b) Cd        c) Cu        d) Fe

**Q6.** The $Zn/Zn^{2+}$ electrode has a standard electrode potential of $E° = -0.76$ V. How does the relative potential energy of an electron at the $Zn/Zn^{2+}$ electrode compare to the potential energy of an electron at the standard hydrogen electrode?

a) An electron at the $Zn/Zn^{2+}$ electrode has a higher potential energy than an electron at the standard hydrogen electrode.

b) An electron at the $Zn/Zn^{2+}$ electrode has a lower potential energy than an electron at the standard hydrogen electrode.

c) An electron at the $Zn/Zn^{2+}$ electrode has the same potential energy as an electron at the standard hydrogen electrode.

d) Nothing can be concluded about the relative potential energy of an electron at the standard electrode potential.

**Q7.** Use Table 19.1 to calculate $\Delta G°$ for the reaction.

$$2 MnO_4^-(aq) + Cd(s) \longrightarrow 2 MnO_4^{2-}(aq) + Cd^{2+}(aq)$$

a) +30.9 kJ    b) −30.9 kJ    c) −185 kJ    d) +185 kJ

**Q8.** A redox reaction has an $E°_{cell} = -0.56$ V. What can you conclude about the equilibrium constant $(K)$ for the reaction?

a) $K < 1$        b) $K > 1$        c) $K = 0$

d) Nothing can be concluded about $K$ from $E°_{cell}$.

**Q9.** Find $E_{cell}$ for an electrochemical cell based on the following reaction with $[MnO_4^-] = 2.0$ M, $[H^+] = 1.0$ M, and $[Ag^+] = 0.010$ M. $E°_{cell}$ for the reaction is +0.88 V.

$$MnO_4^-(aq) + 4 H^+(aq) + 3 Ag(s) \longrightarrow$$
$$MnO_2(s) + 2 H_2O(l) + 3 Ag^+(aq)$$

a) 0.83 V    b) 1.00 V    c) 0.76 V    d) 0.93 V

**Q10.** In an electrochemical cell, $Q = 0.010$ and $K = 855$. What can you conclude about $E_{cell}$ and $E°_{cell}$?

a) $E_{cell}$ is positive and $E°_{cell}$ is negative.

b) $E_{cell}$ is negative and $E°_{cell}$ is positive.

c) $E_{cell}$ and $E°_{cell}$ are both negative.

d) $E_{cell}$ and $E°_{cell}$ are both positive.

**Q11.** Which reaction occurs at the *anode* of a lead storage battery?

a) $Zn(s) + 2 OH^-(aq) \longrightarrow Zn(OH)_2(s) + 2 e^-$

b) $PbO_2(s) + HSO_4^-(aq) + 3 H^+(aq) + 2 e^- \longrightarrow$
$$PbSO_4(s) + 2 H_2O(l)$$

c) $Pb(s) + HSO_4^-(aq) \longrightarrow PbSO_4(s) + H^+(aq) + 2 e^-$

d) $O_2(g) + 2 H_2O(l) + 4 e^- \longrightarrow 4 OH^-(aq)$

**Q12.** Which reaction could be used to generate electricity in a voltaic electrochemical cell?

a) $Pb^{2+}(aq) + Mg(s) \longrightarrow Mg^{2+}(aq) + Pb(s)$

b) $Zn^{2+}(aq) + Sn(s) \longrightarrow Sn^{2+}(aq) + Zn(s)$

c) $NaCl(aq) + AgNO_3(aq) \longrightarrow AgCl(s) + NaNO_3(aq)$

d) None of the above

**Q13.** Which reaction occurs at the cathode of an electrolytic cell containing a mixture of molten KCl and $ZnCl_2$?

a) $K(s) \longrightarrow K^+(l) + e^-$

b) $K^+(l) + e^- \longrightarrow K(s)$

c) $Zn^{2+}(l) + 2 e^- \longrightarrow Zn(s)$

d) $2 Cl^-(l) \longrightarrow Cl_2(g) + 2 e^-$

**Q14.** Copper is plated onto the cathode of an electrolytic cell containing $CuCl_2(aq)$. How long does it take to plate 111 mg of copper with a current of 3.8 A?

a) $1.3 \times 10^3$ s              b) 44 s

c) 89 s                                d) 22 s

**Q15.** Which metal can be used as a sacrificial electrode to prevent the rusting of an iron pipe?

a) Au        b) Ag        c) Cu        d) Mn

**Answers:** 1. (a)  2. (d)  3. (d)  4. (a)  5. (c)  6. (a)  7. (c)  8. (a)  9. (b)  10. (d)  11. (c)  12. (a)  13. (c)  14. (c)  15. (d)

# Chapter 19 IN REVIEW

## Key Terms

**Section 19.3**

electrical current (893)
electrochemical cell (894)
voltaic (galvanic) cell (894)
electrolytic cell (894)
half-cell (894)
electrode (894)
ampere (A) (895)
potential difference (895)
volt (V) (895)
electromotive force (emf) (895)
cell potential (cell emf) $(E_{cell})$ (896)
standard cell potential
(standard emf) $(E°_{cell})$ (896)

anode (896)
cathode (896)
salt bridge (896)

**Section 19.4**

standard electrode potential (898)
standard hydrogen electrode
(SHE) (898)

**Section 19.5**

Faraday's constant $(F)$ (906)

**Section 19.6**

Nernst equation (910)

**Section 19.7**

dry-cell battery (914)
alkaline battery (915)
lead–acid storage battery (915)
nickel–cadmium (NiCad) battery (916)
nickel–metal hydride (NiMH) battery (916)
lithium ion battery (916)
fuel cell (917)

**Section 19.8**

electrolysis (918)

**Section 19.9**

corrosion (925)

# Key Concepts

## Pulling the Plug on the Power Grid (19.1)

▶ Oxidation–reduction reactions are reactions in which electrons are transferred from one reactant to another.

▶ In the most common form of fuel cell, an electrical current is created as hydrogen is oxidized and oxygen is reduced; water is the only product.

## Balancing Oxidation–Reduction Equations (19.2)

▶ Oxidation is the loss of electrons and corresponds to an increase in oxidation state; reduction is the gain of electrons and corresponds to a decrease in oxidation state.

▶ We can balance redox reactions using the half-reaction method, in which the oxidation and reduction reactions are balanced separately and then added. This method differs slightly for redox reactions in acidic and in basic solutions.

## Voltaic (or Galvanic) Cells: Generating Electricity from Spontaneous Chemical Reactions (19.3)

▶ A voltaic electrochemical cell separates the reactants of a spontaneous redox reaction into two half-cells that are connected by a wire and a means to exchange ions so that electricity is generated.

▶ In an electrochemical cell, the electrode where oxidation occurs is the anode and the electrode where reduction occurs is the cathode; electrons flow from the anode to the cathode.

▶ The rate of electrons flowing through a wire is measured in amperes (A), and the cell potential is measured in volts (V).

▶ A salt bridge allows ions to flow between the half-cell solutions and prevent the buildup of charge.

▶ The cell diagram or line notation symbolizes electrochemical cells concisely by separating the components of the reaction using lines or commas.

## Standard Electrode Potentials (19.4)

▶ The electrode potentials of half-cells are measured in relation to that of a standard hydrogen electrode, which is assigned an electrode potential of zero under standard conditions (solution concentrations of 1 M, gas pressures of 1 atm, and a temperature of 25 °C).

▶ A species with a highly positive $E°$ has a strong tendency to attract electrons and undergo reduction (and is therefore an excellent oxidizing agent).

▶ A species with a highly negative $E°$ has a strong tendency to repel electrons and undergo oxidation (and is therefore an excellent reducing agent).

## Cell Potential, Free Energy, and the Equilibrium Constant (19.5)

▶ In a spontaneous reaction, $E°_{cell}$ is positive, the change in free energy ($\Delta G°$) is negative, and the equilibrium constant ($K$) is greater than 1.

▶ In a nonspontaneous reaction, $E°_{cell}$ is negative, $\Delta G°$ is positive, and $K$ is less than 1.

▶ Because $E°_{cell}$, $\Delta G°$, and $K$ all relate to spontaneity, we can derive equations relating all three quantities.

## Cell Potential and Concentration (19.6)

▶ The standard cell potential ($E°_{cell}$) is related to the cell potential ($E_{cell}$) by the Nernst equation, $E_{cell} = E°_{cell} - (0.0592 \text{ V}/n) \log Q$.

▶ As shown by the Nernst equation, $E_{cell}$ is related to the reaction quotient ($Q$); $E_{cell}$ equals zero when $Q$ equals $K$.

▶ In a concentration cell, the reactions at both electrodes are identical and electrons flow because of a difference in concentration. Nerve cells are a biological example of concentration cells.

## Batteries: Using Chemistry to Generate Electricity (19.7)

▶ Batteries are packaged voltaic cells.

▶ Dry-cell batteries, including alkaline batteries, do not contain large amounts of water.

▶ The reactions in rechargeable batteries, such as lead–acid storage, nickel–cadmium, nickel–metal hydride, and lithium ion batteries, can be reversed.

▶ Fuel cells are similar to batteries except that fuel-cell reactants must be continually replenished from an external source.

## Electrolysis: Driving Nonspontaneous Chemical Reactions with Electricity (19.8)

▶ An electrolytic electrochemical cell differs from a voltaic cell in that (1) an electrical charge is used to drive the reaction, and (2) although the anode is still the site of oxidation and the cathode the site of reduction, they are represented with signs opposite those of a voltaic cell (anode +, cathode −).

▶ In electrolysis reactions, the anion is oxidized; if there is more than one anion, the anion with the more negative $E°$ is oxidized.

▶ We can use stoichiometry to calculate the quantity of reactants consumed or products produced in an electrolytic cell.

## Corrosion: Undesirable Redox Reactions (19.9)

▶ Corrosion is the undesired oxidation of metal by environmental oxidizing agents.

▶ When some metals, such as aluminum, oxidize, they form a stable compound that prevents further oxidation. Iron, in contrast, does not form a structurally stable compound when oxidized, and therefore, rust flakes off and exposes more iron to corrosion.

▶ Iron corrosion can be prevented by protecting iron from water, minimizing the presence of electrolytes and acids, or using a sacrificial electrode.

# Key Equations and Relationships

**Definition of an Ampere (19.3)**

$$1 \text{ A} = 1 \text{ C/s}$$

**Definition of a Volt (19.3)**

$$1 \text{ V} = 1 \text{ J/C}$$

**Standard Hydrogen Electrode (19.4)**

$$2 \text{ H}^+(aq) + 2 \text{ e}^- \longrightarrow \text{H}_2(g) \quad E° = 0.00 \text{ V}$$

**Equation for Cell Potential (19.4)**

$$E°_{cell} = E°_{cathode} - E°_{anode}$$

Relating $\Delta G°$ and $E°_{cell}$ (**19.5**)

$$\Delta G° = -nFE°_{cell} \qquad F = \frac{96,485 \text{ C}}{\text{mol e}^-}$$

Relating $E°_{cell}$ and $K$ (**19.5**)

$$E°_{cell} = \frac{0.0592 \text{ V}}{n} \log K \qquad \text{(at 25 °C)}$$

The Nernst Equation (**19.6**)

$$E_{cell} = E°_{cell} - \frac{0.0592 \text{ V}}{n} \log Q \qquad \text{(at 25 °C)}$$

# Key Learning Outcomes

| CHAPTER OBJECTIVES | ASSESSMENT |
|---|---|
| Half-Reaction Method of Balancing Aqueous Redox Equations in Acidic Solution (**19.2**) | Examples 19.1, 19.2   For Practice 19.1, 19.2   Exercises 37–40 |
| Balancing Redox Reactions Occurring in Basic Solution (**19.2**) | Example 19.3   For Practice 19.3   Exercises 41, 42 |
| Calculating Standard Potentials for Electrochemical Cells from Standard Electrode Potentials of the Half-Reactions (**19.4**) | Example 19.4   For Practice 19.4   Exercises 45, 46, 61, 62 |
| Predicting Spontaneous Redox Reactions and Sketching Electrochemical Cells (**19.4**) | Example 19.5   For Practice 19.5   Exercises 43, 44, 47, 48, 51–54 |
| Relating $\Delta G°$ and $E°_{cell}$ (**19.5**) | Example 19.6   For Practice 19.6   Exercises 65, 66 |
| Relating $E°_{cell}$ and $K$ (**19.5**) | Example 19.7   For Practice 19.7   Exercises 67–72 |
| Calculating $E_{cell}$ under Nonstandard Conditions (**19.6**) | Example 19.8   For Practice 19.8   Exercises 73–78 |
| Predicting the Products of Electrolysis Reactions (**19.8**) | Example 19.9   For Practice 19.9   Exercises 91–96 |
| Stoichiometry of Electrolysis (**19.8**) | Example 19.10   For Practice 19.10   Exercises 99–102 |

# EXERCISES

## Review Questions

**1.** In electrochemistry, spontaneous redox reactions are used for what purpose?

**2.** In electrochemistry, what kind of reaction can be driven by electricity?

**3.** Give the basic definitions of oxidation and reduction and explain the basic procedure for balancing redox reactions.

**4.** Explain the difference between a voltaic (or galvanic) electrochemical cell and an electrolytic cell.

**5.** What reaction (oxidation or reduction) occurs at the anode of a voltaic cell? What is the sign of the anode? Do electrons flow toward or away from the anode?

**6.** What reaction (oxidation or reduction) occurs at the cathode of a voltaic cell? What is the sign of the cathode? Do electrons flow toward or away from the cathode?

**7.** Explain the purpose of a salt bridge in an electrochemical cell.

**8.** What unit is used to measure the magnitude of electrical current? What unit is used to measure the magnitude of a potential difference? Explain how electrical current and potential difference differ.

**9.** What is the definition of the standard cell potential ($E°_{cell}$)? What does a large positive standard cell potential imply about the spontaneity of the redox reaction occurring in the cell? What does a negative standard cell potential imply about the reaction?

**10.** Describe the basic features of a cell diagram (or line notation) for an electrochemical cell.

**11.** Why do some electrochemical cells employ inert electrodes such as platinum?

**12.** Describe the standard hydrogen electrode (SHE) and explain its use in determining standard electrode potentials.

13. How is the cell potential of an electrochemical cell ($E^\circ_{cell}$) related to the potentials of the half-cells?

14. Does a large positive electrode potential indicate a strong oxidizing agent or a strong reducing agent? What about a large negative electrode potential?

15. Is a spontaneous redox reaction obtained by pairing any reduction half-reaction with one listed above it or with one listed below it in Table 19.1?

16. How can Table 19.1 be used to predict whether or not a metal will dissolve in HCl? In HNO$_3$?

17. Explain why $E^\circ_{cell}$, $\Delta G^\circ_{rxn}$, and $K$ are all interrelated.

18. Does a redox reaction with a small equilibrium constant ($K < 1$) have a positive or a negative $E^\circ_{cell}$? Does it have a positive or a negative $\Delta G^\circ_{rxn}$?

19. How does $E_{cell}$ depend on the concentrations of the reactants and products in the redox reaction occurring in the cell? What effect does increasing the concentration of a reactant have on $E_{cell}$? Increasing the concentration of a product?

20. Use the Nernst equation to show that $E_{cell} = E^\circ_{cell}$ under standard conditions.

21. What is a concentration electrochemical cell?

22. What are the anode and cathode reactions in a common dry-cell battery? In an alkaline battery?

23. What are the anode and cathode reactions in a lead–acid storage battery? What happens when the battery is recharged?

24. What are the three common types of portable rechargeable batteries, and how does each one work?

25. What is a fuel cell? What is the most common type of fuel cell, and what reactions occur at its anode and cathode?

26. Explain how a fuel-cell breathalyzer works.

27. List some applications of electrolysis.

28. The anode of an electrolytic cell must be connected to which terminal—positive or negative—of the power source?

29. What species is oxidized, and what species is reduced in the electrolysis of a pure molten salt?

30. If an electrolytic cell contains a mixture of species that can be oxidized, how do you determine which species will actually be oxidized? If it contains a mixture of species that can be reduced, how do you determine which one will actually be reduced?

31. Why does the electrolysis of an aqueous sodium chloride solution produce hydrogen gas at the cathode?

32. What is overvoltage in an electrochemical cell? Why is it important?

33. How is the amount of current flowing through an electrolytic cell related to the amount of product produced in the redox reaction?

34. What is corrosion? Why is corrosion only a problem for some metals (such as iron)?

35. Explain the role of each of the following in promoting corrosion: moisture, electrolytes, and acids.

36. How can the corrosion of iron be prevented?

# Problems by Topic

## Balancing Redox Reactions

37. Balance each redox reaction occurring in acidic aqueous solution.
    **a.** $K(s) + Cr^{3+}(aq) \longrightarrow Cr(s) + K^+(aq)$
    **b.** $Al(s) + Fe^{2+}(aq) \longrightarrow Al^{3+}(aq) + Fe(s)$
    **c.** $BrO_3^-(aq) + N_2H_4(g) \longrightarrow Br^-(aq) + N_2(g)$

38. Balance each redox reaction occurring in acidic aqueous solution.
    **a.** $Zn(s) + Sn^{2+}(aq) \longrightarrow Zn^{2+}(aq) + Sn(s)$
    **b.** $Mg(s) + Cr^{3+}(aq) \longrightarrow Mg^{2+}(aq) + Cr(s)$
    **c.** $MnO_4^-(aq) + Al(s) \longrightarrow Mn^{2+}(aq) + Al^{3+}(aq)$

39. Balance each redox reaction occurring in acidic aqueous solution.
    **a.** $PbO_2(s) + I^-(aq) \longrightarrow Pb^{2+}(aq) + I_2(s)$
    **b.** $SO_3^{2-}(aq) + MnO_4^-(aq) \longrightarrow SO_4^{2-}(aq) + Mn^{2+}(aq)$
    **c.** $S_2O_3^{2-}(aq) + Cl_2(g) \longrightarrow SO_4^{2-}(aq) + Cl^-(aq)$

40. Balance each redox reaction occurring in acidic aqueous solution.
    **a.** $I^-(aq) + NO_2^-(aq) \longrightarrow I_2(s) + NO(g)$
    **b.** $ClO_4^-(aq) + Cl^-(aq) \longrightarrow ClO_3^-(aq) + Cl_2(g)$
    **c.** $NO_3^-(aq) + Sn^{2+}(aq) \longrightarrow Sn^{4+}(aq) + NO(g)$

41. Balance each redox reaction occurring in basic aqueous solution.
    **a.** $H_2O_2(aq) + ClO_2(aq) \longrightarrow ClO_2^-(aq) + O_2(g)$
    **b.** $Al(s) + MnO_4^-(aq) \longrightarrow MnO_2(s) + Al(OH)_4^-(aq)$
    **c.** $Cl_2(g) \longrightarrow Cl^-(aq) + ClO^-(aq)$

42. Balance each redox reaction occurring in basic aqueous solution.
    **a.** $MnO_4^-(aq) + Br^-(aq) \longrightarrow MnO_2(s) + BrO_3^-(aq)$
    **b.** $Ag(s) + CN^-(aq) + O_2(g) \longrightarrow Ag(CN)_2^-(aq)$
    **c.** $NO_2^-(aq) + Al(s) \longrightarrow NH_3(g) + AlO_2^-(aq)$

## Voltaic Cells, Standard Cell Potentials, and Direction of Spontaneity

43. Sketch a voltaic cell for each redox reaction. Label the anode and cathode and indicate the half-reaction that occurs at each electrode and the species present in each solution. Also indicate the direction of electron flow.
    **a.** $2 Ag^+(aq) + Pb(s) \longrightarrow 2 Ag(s) + Pb^{2+}(aq)$
    **b.** $2 ClO_2(g) + 2 I^-(aq) \longrightarrow 2 ClO_2^-(aq) + I_2(s)$
    **c.** $O_2(g) + 4 H^+(aq) + 2 Zn(s) \longrightarrow 2 H_2O(l) + 2 Zn^{2+}(aq)$

44. Sketch a voltaic cell for each redox reaction. Label the anode and cathode and indicate the half-reaction that occurs at each electrode and the species present in each solution. Also indicate the direction of electron flow.
    **a.** $Ni^{2+}(aq) + Mg(s) \longrightarrow Ni(s) + Mg^{2+}(aq)$
    **b.** $2 H^+(aq) + Fe(s) \longrightarrow H_2(g) + Fe^{2+}(aq)$
    **c.** $2 NO_3^-(aq) + 8 H^+(aq) + 3 Cu(s) \longrightarrow$
    $$2 NO(g) + 4 H_2O(l) + 3 Cu^{2+}(aq)$$

45. Calculate the standard cell potential for each of the electrochemical cells in Problem 43.

46. Calculate the standard cell potential for each of the electrochemical cells in Problem 44.

**47.** Consider the voltaic cell:

a. Determine the direction of electron flow and label the anode and the cathode.

b. Write a balanced equation for the overall reaction and calculate $E°_{cell}$.

c. Label each electrode as negative or positive.

d. Indicate the direction of anion and cation flow in the salt bridge.

**48.** Consider the voltaic cell:

a. Determine the direction of electron flow and label the anode and the cathode.

b. Write a balanced equation for the overall reaction and calculate $E°_{cell}$.

c. Label each electrode as negative or positive.

d. Indicate the direction of anion and cation flow in the salt bridge.

**49.** Use line notation to represent each electrochemical cell in Problem 43.

**50.** Use line notation to represent each electrochemical cell in Problem 44.

**51.** Make a sketch of the voltaic cell represented by the line notation. Write the overall balanced equation for the reaction and calculate $E°_{cell}$.

$$Sn(s) \,|\, Sn^{2+}(aq) \,\|\, NO(g) \,|\, NO_3^-(aq), H^+(aq) \,|\, Pt(s)$$

**52.** Make a sketch of the voltaic cell represented by the line notation. Write the overall balanced equation for the reaction and calculate $E°_{cell}$.

$$Mn(s) \,|\, Mn^{2+}(aq) \,\|\, ClO_2^-(aq) \,|\, ClO_2(g) \,|\, Pt(s)$$

**53.** Determine whether or not each redox reaction occurs spontaneously in the forward direction.

a. $Ni(s) + Zn^{2+}(aq) \longrightarrow Ni^{2+}(aq) + Zn(s)$

b. $Ni(s) + Pb^{2+}(aq) \longrightarrow Ni^{2+}(aq) + Pb(s)$

c. $Al(s) + 3\,Ag^+(aq) \longrightarrow Al^{3+}(aq) + 3\,Ag(s)$

d. $Pb(s) + Mn^{2+}(aq) \longrightarrow Pb^{2+}(aq) + Mn(s)$

**54.** Determine whether or not each redox reaction occurs spontaneously in the forward direction.

a. $Ca^{2+}(aq) + Zn(s) \longrightarrow Ca(s) + Zn^{2+}(aq)$

b. $2\,Ag^+(aq) + Ni(s) \longrightarrow 2\,Ag(s) + Ni^{2+}(aq)$

c. $Fe(s) + Mn^{2+}(aq) \longrightarrow Fe^{2+}(aq) + Mn(s)$

d. $2\,Al(s) + 3\,Pb^{2+}(aq) \longrightarrow 2\,Al^{3+}(aq) + 3\,Pb(s)$

**55.** Which metal could you use to reduce $Mn^{2+}$ ions but not $Mg^{2+}$ ions?

**56.** Which metal can be oxidized with an $Sn^{2+}$ solution but not with an $Fe^{2+}$ solution?

**57.** Determine whether or not each metal dissolves in 1 M HCl. For those metals that do dissolve, write a balanced redox reaction showing what happens when the metal dissolves.

a. Al      b. Ag      c. Pb

**58.** Determine whether or not each metal dissolves in 1 M HCl. For those metals that do dissolve, write a balanced redox reaction showing what happens when the metal dissolves.

a. Cu      b. Fe      c. Au

**59.** Determine whether or not each metal dissolves in 1 M $HNO_3$. For those metals that do dissolve, write a balanced redox reaction showing what happens when the metal dissolves.

a. Cu      b. Au

**60.** Determine whether or not each metal dissolves in 1 M $HIO_3$. For those metals that do dissolve, write a balanced redox equation for the reaction that occurs.

a. Au      b. Cr

**61.** Calculate $E°_{cell}$ for each balanced redox reaction and determine if the reaction is spontaneous as written.

a. $2\,Cu(s) + Mn^{2+}(aq) \longrightarrow 2\,Cu^+(aq) + Mn(s)$

b. $MnO_2(aq) + 4\,H^+(aq) + Zn(s) \longrightarrow$
$$Mn^{2+}(aq) + 2\,H_2O(l) + Zn^{2+}(aq)$$

c. $Cl_2(g) + 2\,F^-(aq) \longrightarrow F_2(g) + 2\,Cl^-(aq)$

**62.** Calculate $E°_{cell}$ for each balanced redox reaction and determine if the reaction is spontaneous as written.

a. $O_2(g) + 2\,H_2O(l) + 4\,Ag(s) \longrightarrow 4\,OH^-(aq) + 4\,Ag^+(aq)$

b. $Br_2(l) + 2\,I^-(aq) \longrightarrow 2\,Br^-(aq) + I_2(s)$

c. $PbO_2(s) + 4\,H^+(aq) + Sn(s) \longrightarrow$
$$Pb^{2+}(aq) + 2\,H_2O(l) + Sn^{2+}(aq)$$

**63.** Which metal cation is the best oxidizing agent?

a. $Pb^{2+}$      b. $Cr^{3+}$      c. $Fe^{2+}$      d. $Sn^{2+}$

**64.** Which metal is the best reducing agent?

a. Mn      b. Al      c. Ni      d. Cr

## Cell Potential, Free Energy, and the Equilibrium Constant

**65.** Use tabulated electrode potentials to calculate $\Delta G°_{rxn}$ for each reaction at 25 °C.

a. $Pb^{2+}(aq) + Mg(s) \longrightarrow Pb(s) + Mg^{2+}(aq)$

b. $Br_2(l) + 2\,Cl^-(aq) \longrightarrow 2\,Br^-(aq) + Cl_2(g)$

c. $MnO_2(s) + 4\,H^+(aq) + Cu(s) \longrightarrow$
$$Mn^{2+}(aq) + 2\,H_2O(l) + Cu^{2+}(aq)$$

**66.** Use tabulated electrode potentials to calculate $\Delta G°_{rxn}$ for each reaction at 25 °C.

a. $2\,Fe^{3+}(aq) + 3\,Sn(s) \longrightarrow 2\,Fe(s) + 3\,Sn^{2+}(aq)$

b. $O_2(g) + 2\,H_2O(l) + 2\,Cu(s) \longrightarrow 4\,OH^-(aq) + 2\,Cu^{2+}(aq)$

c. $Br_2(l) + 2\,I^-(aq) \longrightarrow 2\,Br^-(aq) + I_2(s)$

**67.** Calculate the equilibrium constant for each of the reactions in Problem 65.

**68.** Calculate the equilibrium constant for each of the reactions in Problem 66.

**69.** Calculate the equilibrium constant for the reaction between $Ni^{2+}(aq)$ and $Cd(s)$ (at 25 °C).

**70.** Calculate the equilibrium constant for the reaction between $Fe^{2+}(aq)$ and $Zn(s)$ (at 25 °C).

**71.** Calculate $\Delta G^\circ_{rxn}$ and $E^\circ_{cell}$ for a redox reaction with $n = 2$ that has an equilibrium constant of $K = 25$ (at 25 °C).

**72.** Calculate $\Delta G^\circ_{rxn}$ and $E^\circ_{cell}$ for a redox reaction with $n = 3$ that has an equilibrium constant of $K = 0.050$ (at 25 °C).

### Nonstandard Conditions and the Nernst Equation

**73.** A voltaic cell employs the following redox reaction:

$$Sn^{2+}(aq) + Mn(s) \longrightarrow Sn(s) + Mn^{2+}(aq)$$

Calculate the cell potential at 25 °C under each set of conditions.
**a.** standard conditions
**b.** $[Sn^{2+}] = 0.0100 \text{ M}; [Mn^{2+}] = 2.00 \text{ M}$
**c.** $[Sn^{2+}] = 2.00 \text{ M}; [Mn^{2+}] = 0.0100 \text{ M}$

**74.** A voltaic cell employs the redox reaction:

$$2 Fe^{3+}(aq) + 3 Mg(s) \longrightarrow 2 Fe(s) + 3 Mg^{2+}(aq)$$

Calculate the cell potential at 25 °C under each set of conditions.
**a.** standard conditions
**b.** $[Fe^{3+}] = 1.0 \times 10^{-3} \text{ M}; [Mg^{2+}] = 2.50 \text{ M}$
**c.** $[Fe^{3+}] = 2.00 \text{ M}; [Mg^{2+}] = 1.5 \times 10^{-3} \text{ M}$

**75.** An electrochemical cell is based on these two half-reactions:

Ox:  $Pb(s) \longrightarrow Pb^{2+}(aq, 0.10 \text{ M}) + 2 e^-$

Red:  $MnO_4^-(aq, 1.50 \text{ M}) + 4 H^+(aq, 2.0 \text{ M}) + 3 e^- \longrightarrow$
$$MnO_2(s) + 2 H_2O(l)$$

Calculate the cell potential at 25 °C.

**76.** An electrochemical cell is based on these two half-reactions:

Ox:  $Sn(s) \longrightarrow Sn^{2+}(aq, 2.00 \text{ M}) + 2 e^-$

Red:  $ClO_2(g, 0.100 \text{ atm}) + e^- \longrightarrow ClO_2^-(aq, 2.00 \text{ M})$

Calculate the cell potential at 25 °C.

**77.** A voltaic cell consists of a $Zn/Zn^{2+}$ half-cell and a $Ni/Ni^{2+}$ half-cell at 25 °C. The initial concentrations of $Ni^{2+}$ and $Zn^{2+}$ are 1.50 M and 0.100 M, respectively.
**a.** What is the initial cell potential?
**b.** What is the cell potential when the concentration of $Ni^{2+}$ has fallen to 0.500 M?
**c.** What are the concentrations of $Ni^{2+}$ and $Zn^{2+}$ when the cell potential falls to 0.45 V?

**78.** A voltaic cell consists of a $Pb/Pb^{2+}$ half-cell and a $Cu/Cu^{2+}$ half-cell at 25 °C. The initial concentrations of $Pb^{2+}$ and $Cu^{2+}$ are 0.0500 M and 1.50 M, respectively.
**a.** What is the initial cell potential?
**b.** What is the cell potential when the concentration of $Cu^{2+}$ has fallen to 0.200 M?
**c.** What are the concentrations of $Pb^{2+}$ and $Cu^{2+}$ when the cell potential falls to 0.35 V?

**79.** Make a sketch of a concentration cell employing two $Zn/Zn^{2+}$ half-cells. The concentration of $Zn^{2+}$ in one of the half-cells is 2.0 M and the concentration in the other half-cell is $1.0 \times 10^{-3}$ M. Label the anode and the cathode and indicate the half-reaction occurring at each electrode. Also indicate the direction of electron flow.

**80.** Consider the concentration cell:

**a.** Label the anode and cathode.
**b.** Indicate the direction of electron flow.
**c.** Indicate what happens to the concentration of $Pb^{2+}$ in each half-cell.

**81.** A concentration cell consists of two $Sn/Sn^{2+}$ half-cells. The cell has a potential of 0.10 V at 25 °C. What is the ratio of the $Sn^{2+}$ concentrations in the two half-cells?

**82.** A $Cu/Cu^{2+}$ concentration cell has a voltage of 0.22 V at 25 °C. The concentration of $Cu^{2+}$ in one of the half-cells is $1.5 \times 10^{-3}$ M. What is the concentration of $Cu^{2+}$ in the other half-cell? (Assume the concentration in the unknown cell is the *lower* of the two concentrations.)

### Batteries, Fuel Cells, and Corrosion

**83.** Determine the optimum mass ratio of Zn to $MnO_2$ in an alkaline battery.

**84.** What mass of lead sulfate is formed in a lead–acid storage battery when 1.00 g of Pb undergoes oxidation?

**85.** Refer to the tabulated values of $\Delta G^\circ_f$ in Appendix IIB to calculate $E^\circ_{cell}$ for a fuel-cell that employs the reaction between methane gas ($CH_4$) and oxygen to form carbon dioxide and gaseous water.

**86.** Refer to the tabulated values of $\Delta G^\circ_f$ in Appendix IIB to calculate $E^\circ_{cell}$ for the fuel-cell breathalyzer, which employs the following reaction. ($\Delta G^\circ_f$ for $HC_2H_3O_2(g) = -374.2$ kJ/mol.)

$$CH_3CH_2OH(g) + O_2(g) \longrightarrow HC_2H_3O_2(g) + H_2O(g)$$

**87.** Determine whether or not each metal, if coated onto iron, would prevent the corrosion of iron.
**a.** Zn
**b.** Sn
**c.** Mn

**88.** Determine whether or not each metal, if coated onto iron, would prevent the corrosion of iron.
**a.** Mg
**b.** Cr
**c.** Cu

## Electrolytic Cells and Electrolysis

89. Consider the electrolytic cell:

a. Label the anode and the cathode and indicate the half-reactions occurring at each.

b. Indicate the direction of electron flow.

c. Label the terminals on the battery as positive or negative and calculate the minimum voltage necessary to drive the reaction.

90. Draw an electrolytic cell in which $Mn^{2+}$ is reduced to Mn and Sn is oxidized to $Sn^{2+}$. Label the anode and cathode, indicate the direction of electron flow, and write an equation for the half-reaction occurring at each electrode. What minimum voltage is necessary to drive the reaction?

91. Write equations for the half-reactions that occur in the electrolysis of molten potassium bromide.

92. What products are obtained in the electrolysis of molten NaI?

93. Write equations for the half-reactions that occur in the electrolysis of a mixture of molten potassium bromide and molten lithium bromide.

94. What products are obtained in the electrolysis of a molten mixture of KI and KBr?

95. Write equations for the half-reactions that occur at the anode and cathode for the electrolysis of each aqueous solution.
a. $NaBr(aq)$
b. $PbI_2(aq)$
c. $Na_2SO_4(aq)$

96. Write equations for the half-reactions that occur at the anode and cathode for the electrolysis of each aqueous solution.
a. $Ni(NO_3)_2(aq)$
b. $KCl(aq)$
c. $CuBr_2(aq)$

97. Make a sketch of an electrolysis cell that electroplates copper onto other metal surfaces. Label the anode and the cathode and indicate the reactions that occur at each.

98. Make a sketch of an electrolysis cell that electroplates nickel onto other metal surfaces. Label the anode and the cathode and indicate the reactions that occur at each.

99. Copper can be electroplated at the cathode of an electrolysis cell by the half-reaction:
$$Cu^{2+}(aq) + 2\,e^- \longrightarrow Cu(s)$$
How much time would it take for 325 mg of copper to be plated at a current of 5.6 A?

100. Silver can be electroplated at the cathode of an electrolysis cell by the half-reaction:
$$Ag^+(aq) + e^- \longrightarrow Ag(s)$$
What mass of silver would plate onto the cathode if a current of 6.8 A flowed through the cell for 72 min?

101. A major source of sodium metal is the electrolysis of molten sodium chloride. What magnitude of current produces 1.0 kg of sodium metal in 1 hour?

102. What mass of aluminum metal can be produced per hour in the electrolysis of a molten aluminum salt by a current of 25 A?

103. Consider the reaction shown here occurring at 25 °C.
$$A(s) + B^{2+}(aq) \longrightarrow A^{2+}(aq) + B(s) \qquad \Delta G^\circ_{rxn} = -14.0 \text{ kJ}$$
Determine the value of $E^\circ_{cell}$ and $K$ for the reaction and complete the table.

| $[B^{2+}]$ | $[A^{2+}]$ | Q | $E_{cell}$ | $\Delta G_{rxn}$ |
|---|---|---|---|---|
| 1.00 | 1.00 | | | |
| 1.00 | $1.00 \times 10^{-4}$ | | | |
| $1.0 \times 10^{-4}$ | 1.0 | | | |
| $3.54 \times 10^{-3}$ | 1.0 | | | |

104. Consider the reaction shown here occurring at 25 °C.
$$Cr(s) + Cd^{2+}(aq) \longrightarrow Cr^{2+}(aq) + Cd(s)$$
Determine $E^\circ_{cell}$, $K$, and $\Delta G^\circ_{rxn}$ for the reaction and complete the table.

| $[Cd^{2+}]$ | $[Cr^{2+}]$ | Q | $E_{cell}$ | $\Delta G_{rxn}$ |
|---|---|---|---|---|
| 1.00 | 1.00 | | | |
| 1.00 | $1.00 \times 10^{-5}$ | | | |
| $1.00 \times 10^{-5}$ | 1.00 | | | |
| $4.18 \times 10^{-4}$ | 1.00 | | | |

# Cumulative Problems

105. Consider the unbalanced redox reaction:
$$MnO_4^-(aq) + Zn(s) \longrightarrow Mn^{2+}(aq) + Zn^{2+}(aq)$$
Balance the equation and determine the volume of a 0.500 M $KMnO_4$ solution required to completely react with 2.85 g of Zn.

106. Consider the unbalanced redox reaction:
$$Cr_2O_7^{2-}(aq) + Cu(s) \longrightarrow Cr^{3+}(aq) + Cu^{2+}(aq)$$
Balance the equation and determine the volume of a 0.850 M $K_2Cr_2O_7$ solution required to completely react with 5.25 g of Cu.

**107.** Consider the molecular views of an Al strip and $Cu^{2+}$ solution. Draw a similar sketch showing what happens to the atoms and ions after the Al strip is submerged in the solution for a few minutes.

**108.** Consider the molecular view of an electrochemical cell involving the overall reaction:

$$Zn(s) + Ni^{2+}(aq) \longrightarrow Zn^{2+}(aq) + Ni(s)$$

Draw a similar sketch of the cell after it has generated a substantial amount of electrical current.

**109.** Determine whether HI can dissolve each metal sample. If it can, write a balanced chemical reaction showing how the metal dissolves in HI and determine the minimum volume of 3.5 M HI required to completely dissolve the sample.
   **a.** 2.15 g Al    **b.** 4.85 g Cu    **c.** 2.42 g Ag

**110.** Determine if $HNO_3$ can dissolve each metal sample. If it can, write a balanced chemical reaction showing how the metal dissolves in $HNO_3$ and determine the minimum volume of 6.0 M $HNO_3$ required to completely dissolve the sample.
   **a.** 5.90 g Au    **b.** 2.55 g Cu    **c.** 4.83 g Sn

**111.** The cell potential of this electrochemical cell depends on the pH of the solution in the anode half-cell.

$$Pt(s) \,|\, H_2(g, 1 \text{ atm}) \,|\, H^+(aq, ? \text{ M}) \,\|\, Cu^{2+}(aq, 1.0 \text{ M}) \,|\, Cu(s)$$

What is the pH of the solution if $E_{cell}$ is 355 mV?

**112.** The cell potential of this electrochemical cell depends on the gold concentration in the cathode half-cell.

$$Pt(s) \,|\, H_2(g, 1.0 \text{ atm}) \,|\, H^+(aq, 1.0 \text{ M}) \,\|\, Au^{3+}(aq, ? \text{ M}) \,|\, Au(s)$$

What is the concentration of $Au^{3+}$ in the solution if $E_{cell}$ is 1.22 V?

**113.** A friend wants you to invest in a new battery she has designed that produces 24 V in a single voltaic cell. Why should you be wary of investing in such a battery?

**114.** What voltage can theoretically be achieved in a battery in which lithium metal is oxidized and fluorine gas is reduced? Why might such a battery be difficult to produce?

**115.** A battery relies on the oxidation of magnesium and the reduction of $Cu^{2+}$. The initial concentrations of $Mg^{2+}$ and $Cu^{2+}$ are $1.0 \times 10^{-4}$ M and 1.5 M, respectively, in 1.0-liter half-cells.
   **a.** What is the initial voltage of the battery?
   **b.** What is the voltage of the battery after delivering 5.0 A for 8.0 h?
   **c.** How long can the battery deliver 5.0 A before going dead?

**116.** A rechargeable battery is constructed based on a concentration cell constructed of two $Ag/Ag^+$ half-cells. The volume of each half-cell is 2.0 L, and the concentrations of $Ag^+$ in the half-cells are 1.25 M and $1.0 \times 10^{-3}$ M.
   **a.** How long can this battery deliver 2.5 A of current before it goes dead?
   **b.** What mass of silver is plated onto the cathode by running at 3.5 A for 5.5 h?
   **c.** Upon recharging, how long would it take to redissolve $1.00 \times 10^2$ g of silver at a charging current of 10.0 amps?

**117.** If a water electrolysis cell operates at a current of 7.8 A, how long will it take to generate 25.0 L of hydrogen gas at a pressure of 25.0 atm and a temperature of 25 °C?

**118.** When a suspected drunk driver blows 188 mL of his breath through the fuel-cell breathalyzer described in Section 19.7, the breathalyzer produces an average of 324 mA of current for 10 s. Assuming a pressure of 1.0 atm and a temperature of 25 °C, what percent (by volume) of the driver's breath is ethanol?

**119.** The $K_{sp}$ of CuI is $1.1 \times 10^{-12}$. Find $E_{cell}$ for the cell:

$$Cu(s) \,|\, CuI(s) \,|\, I^-(aq)(1.0 \text{ M}) \,\|\, Cu^+(aq)(1.0 \text{ M}) \,|\, Cu(s)$$

**120.** The $K_{sp}$ of $Zn(OH)_2$ is $1.8 \times 10^{-14}$. Find $E_{cell}$ for the half-reaction:

$$Zn(OH)_2(s) + 2 e^- \rightleftharpoons Zn(s) + 2 OH^-(aq)$$

**121.** Calculate $\Delta G°_{rxn}$ and $K$ for each reaction.
   **a.** The disproportionation of $Mn^{2+}(aq)$ to Mn(s) and $MnO_2(s)$ in acid solution at 25 °C.
   **b.** The disproportionation of $MnO_2(s)$ to $Mn^{2+}(aq)$ and $MnO_4^-(aq)$ in acid solution at 25 °C.

**122.** Calculate $\Delta G°_{rxn}$ and $K$ for each reaction.
   **a.** The reaction of $Cr^{2+}(aq)$ with $Cr_2O_7^{2-}(aq)$ in acid solution to form $Cr^{3+}(aq)$.
   **b.** The reaction of $Cr^{3+}(aq)$ and Cr(s) to form $Cr^{2+}(aq)$. [The electrode potential of $Cr^{2+}(aq)$ to Cr(s) is −0.91 V.]

**123.** The molar mass of a metal (M) is 50.9 g/mol; it forms a chloride of unknown composition. Electrolysis of a sample of the molten chloride with a current of 6.42 A for 23.6 minutes produces 1.20 g of M at the cathode. Determine the empirical formula of the chloride.

**124.** A metal forms the fluoride $MF_3$. Electrolysis of the molten fluoride by a current of 3.86 A for 16.2 minutes deposits 1.25 g of the metal. Calculate the molar mass of the metal.

**125.** A sample of impure tin of mass 0.535 g is dissolved in strong acid to give a solution of $Sn^{2+}$. The solution is then titrated with a 0.0448 M solution of $NO_3^-$, which is reduced to NO(g). The equivalence point is reached upon the addition of 0.0344 L of the $NO_3^-$ solution. Find the percent by mass of tin in the original sample, assuming that it contains no other reducing agents.

**126.** A 0.0251-L sample of a solution of $Cu^+$ requires 0.0322 L of 0.129 M $KMnO_4$ solution to reach the equivalence point. The products of the reaction are $Cu^{2+}$ and $Mn^{2+}$. What is the concentration of the $Cu^+$ solution?

127. A current of 11.3 A is applied to 1.25 L of a solution of 0.552 M HBr converting some of the $H^+$ to $H_2(g)$, which bubbles out of solution. What is the pH of the solution after 73 minutes?

128. A 215-mL sample of a 0.500 M NaCl solution with an initial pH of 7.00 is subjected to electrolysis. After 15.0 minutes, a 10.0-mL portion (or aliquot) of the solution was removed from the cell and titrated with 0.100 M HCl solution. The end point in the titration was reached upon addition of 22.8 mL of HCl.

Assuming constant current, what was the current (in A) running through the cell?

129. An $MnO_2(s)/Mn^{2+}(aq)$ electrode in which the pH is 10.24 is prepared. Find the $[Mn^{2+}]$ necessary to lower the potential of the half-cell to 0.00 V (at 25 °C).

130. To what pH should you adjust a standard hydrogen electrode to get an electrode potential of −0.122 V? (Assume that the partial pressure of hydrogen gas remains at 1 atm.)

# Challenge Problems

131. Suppose a hydrogen–oxygen fuel-cell generator produces electricity for a house. Use the balanced redox reactions and the standard cell potential to predict the volume of hydrogen gas (at STP) required each month to generate the electricity. Assume the home uses $1.2 \times 10^3$ kWh of electricity per month.

132. A voltaic cell designed to measure $[Cu^{2+}]$ is constructed of a standard hydrogen electrode and a copper metal electrode in the $Cu^{2+}$ solution of interest. If you want to construct a calibration curve for how the cell potential varies with the concentration of copper(II), what do you plot in order to obtain a straight line? What is the slope of the line?

133. The surface area of an object to be gold plated is 49.8 cm$^2$, and the density of gold is 19.3 g/cm$^3$. A current of 3.25 A is applied to a solution that contains gold in the +3 oxidation state. Calculate the time required to deposit an even layer of gold $1.00 \times 10^{-3}$ cm thick on the object.

134. To electrodeposit all the Cu and Cd from a solution of $CuSO_4$ and $CdSO_4$ required 1.20 F of electricity (1 F = 1 mol e$^-$). The mixture of Cu and Cd that was deposited had a mass of 50.36 g. What mass of $CuSO_4$ was present in the original mixture?

135. Sodium oxalate, $Na_2C_2O_4$, in solution is oxidized to $CO_2(g)$ by $MnO_4^-$, which is reduced to $Mn^{2+}$. A 50.1-mL volume of a solution of $MnO_4^-$ is required to titrate a 0.339-g sample of sodium oxalate. This solution of $MnO_4^-$ is used to analyze uranium-containing samples. A 4.62-g sample of a uranium-containing material requires 32.5 mL of the solution for titration. The oxidation of the uranium can be represented by the change $UO^{2+} \longrightarrow UO_2^{2+}$. Calculate the percentage of uranium in the sample.

136. Three electrolytic cells are connected in a series. The electrolytes in the cells are aqueous copper(II) sulfate, gold(III) sulfate, and silver nitrate. A current of 2.33 A is applied, and after some time 1.74 g Cu is deposited. How long was the current applied? What mass of gold and silver was deposited?

137. The cell $Pt(s)|Cu^+(1\ M),\ Cu^{2+}(1\ M)\|Cu^+(1\ M)|Cu(s)$ has $E° = 0.364$ V. The cell $Cu(s)|Cu^{2+}(1\ M)\|Cu^+(1\ M)|Cu(s)$ has $E° = 0.182$ V. Write the cell reaction for each cell and explain the differences in $E°$. Calculate $\Delta G°$ for each cell reaction to help explain these differences.

# Conceptual Problems

138. An electrochemical cell has a positive standard cell potential but a negative cell potential. Which statement is true for the cell?
   a. $K > 1; Q > K$  b. $K < 1; Q > K$
   c. $K > 1; Q < K$  d. $K < 1; Q < K$

139. Which oxidizing agent will oxidize $Br^-$ but not $Cl^-$?
   a. $K_2Cr_2O_7$ (in acid)  b. $KMnO_4$ (in acid)  c. $HNO_3$

140. A redox reaction employed in an electrochemical cell has a negative $\Delta G°_{rxn}$. Which statement is true?
   a. $E°_{cell}$ is positive; $K < 1$  b. $E°_{cell}$ is positive; $K > 1$
   c. $E°_{cell}$ is negative; $K > 1$  d. $E°_{cell}$ is negative; $K < 1$

141. A redox reaction has an equilibrium constant of $K = 0.055$. What is true of $\Delta G°_{rxn}$ and $E°_{cell}$ for this reaction?

*Active Classroom Learning*

# Questions for Group Work

Discuss these questions with the group and record your consensus answer.

142. Balance the redox reactions by following the steps in the text. Rotate through the group, having each group member do the next step in the process and explain that step to the rest of the group.
   a. $I_2(s) + Fe(s) \longrightarrow FeI_2(s)$
   b. $Cl_2(g) + H_2O_2(aq) \longrightarrow Cl^-(aq) + O_2(g)$ (acidic)
   c. $Hg^{2+}(aq) + H_2(g) \longrightarrow Hg(l) + H_2O(l)$ (basic)
   d. $CH_3OH(l) + O_2(g) \longrightarrow CO_2(g) + H_2O(l)$ (acidic)

143. Have each group member select a half-reaction from Table 19.1. Each member should calculate the standard cell potential of an electrochemical cell formed between each member's half-reaction and the half-reaction of each of the other group members. For each pair of half-reactions, write the overall balanced chemical reaction that will be spontaneous.

144. Calculate $\Delta G°$ and $K$ for each reaction the group created in Question 2. For one of the reactions, explain how the sign or magnitude of each quantity ($E°_{cell}$, $\Delta G°$, and $K$) is consistent with the fact that the reaction is spontaneous in the direction written.

145. Design a device that uses an electrochemical cell to determine the amount of $Cu^{2+}$ in a sample of water. Describe, in detail, the construction and the theory of operation of your device. If you are able to measure voltage with one-millivolt accuracy, what will the uncertainty in your measured concentration be?

146. Using a library or the Internet, research a fuel cell that uses methanol for fuel. What is the reaction at the anode? What is the reaction at the cathode? What is the overall reaction? What is the standard cell potential? How many kWh can it generate from 1 L (0.792 kg) of methanol?

# Data Interpretation and Analysis

## Concentration and emf in Voltaic Cells

147. A series of experiments is conducted using a voltaic cell consisting of a $Zn/Zn^{2+}$ half-cell and a $Cu/Cu^{2+}$ half-cell at 25 °C similar to the cell in Figure 19.11. The cell potential is measured using different concentrations of the $Zn^{2+}$ ions and $Cu^{2+}$ ions, and the data are recorded in Table a.

| $[Zn^{2+}]$ (M) | $[Cu^{2+}]$ (M) | $E_{cell}$ (volts) | $\log\{[Zn^{2+}]/[Cu^{2+}]\}$ |
|---|---|---|---|
| 2.00 | 0.00100 | 1.000 | |
| 1.00 | 0.00100 | 1.012 | |
| 0.100 | 0.00100 | 1.042 | |
| 0.0100 | 0.00100 | 1.075 | |
| 0.00100 | 0.00100 | 1.10 | |
| 0.00100 | 2.00 | 1.20 | |
| 0.0100 | 2.00 | 1.17 | |
| 0.100 | 2.00 | 1.14 | |
| 1.00 | 2.00 | 1.11 | |
| 2.00 | 2.00 | 1.10 | |

**TABLE a  Concentrations of $Zn^{2+}$ and $Cu^{2+}$ and Measured Cell Potential**

Use the information provided in the questions and table to answer the following questions:

a. Complete Table a (either by hand or with a spreadsheet program) and graph the cell potential versus $\log\{[Zn^{2+}]/[Cu^{2+}]\}$ in Figure a▶.

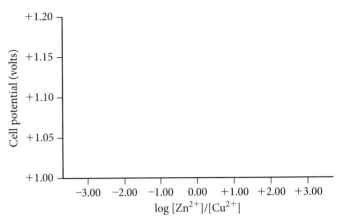

▲ **FIGURE a  Cell Potential as a Function of $\log\{[Zn^{2+}]/[Cu^{2+}]\}$.**

b. To obtain a cell potential of 1.15 volts using a Zn-Cu cell, should the concentration of $Zn^{2+}$ be greater than, less than, or equal to the concentration of $Cu^{2+}$? Explain your answer.

c. Show that there are at least three different combinations of concentrations of $Zn^{2+}$ and $Cu^{2+}$ ions capable of generating a cell potential of +1.15 V.

d. Use your graph from part a to determine the cell potential when $[Zn^{2+}] = 0.050$ M and $[Cu^{2+}] = 1.75$ M.

e. Consider a Zn-Cu cell with the following initial concentrations: $[Cu^{2+}] = 2.00$ M and $[Zn^{2+}] = 0.010$ M. As this cell operates over a period of time, sketch a graph (qualitative) of $[Cu^{2+}]$ and $[Zn^{2+}]$ as a function of time.

# Answers to Conceptual Connections

## Voltaic Cells

**19.1 (a)** Electrons are negatively charged, and therefore, flow away from the more negatively charged electrode and toward the more positively charged electrode.

## Standard Electrode Potentials

**19.2 (b)** A negative electrode potential indicates that an electron at that electrode has greater potential energy than it has at a standard hydrogen electrode.

## Selective Oxidation

**19.3 (d)** The reduction of $HNO_3$ is below the reduction of $Br_2$ and above the reduction of $I_2$ in Table 19.1. Since any reduction half-reaction is spontaneous when paired with the reverse of a half-reaction below it in the table, the reduction of $HNO_3$ is spontaneous when paired with the oxidation of $I^-$ but is not spontaneous when paired with the oxidation of $Br^-$.

## Metals Dissolving in Acids

**19.4 (c)** Ag falls above the half-reaction for the reduction of $H^+$ but below the half-reaction for the reduction of $NO_3^-$ in Table 19.1.

## Periodic Trends and the Direction of Spontaneity for Redox Reactions

**19.5 (a)** Br is more electronegative than I. If the two atoms were in competition for the electron, the electron would go to the more electronegative atom (Br). Therefore, $I_2$ does not spontaneously gain electrons from $Br^-$.

## Relating $K$, $\Delta G^\circ_{rxn}$, and $E^\circ_{cell}$

**19.6 (c)** Since $K > 1$, the reaction is spontaneous under standard conditions (when $Q = 1$, the reaction proceeds toward the products). Therefore, $E^\circ_{cell}$ is positive and $\Delta G^\circ_{rxn}$ is negative.

## Relating $Q$, $K$, $E_{cell}$, and $E^\circ_{cell}$

**19.7 (a)** Since $K < 1$, $E^\circ_{cell}$ is negative (under standard conditions, the reaction is not spontaneous). Since $Q < K$, $E_{cell}$ is positive (the reaction is spontaneous under the nonstandard conditions of the cell).

## Sacrificial Electrodes

**19.8** Cu. The electrode potential for Fe is more negative than that of Cu. Therefore, Fe oxidizes more easily than Cu.

CHAPTER

# 20

# Radioactivity and Nuclear Chemistry

I am among those who think that science has great beauty. A scientist in his laboratory is not only a technician; he is also a child placed before natural phenomena which impress him like a fairy tale.

—MARIE CURIE (1867–1934)

In this chapter, we examine radioactivity and nuclear chemistry, both of which involve changes within the *nuclei* of atoms. Unlike ordinary chemical processes, in which elements retain their identity, nuclear processes often result in one element changing into another, frequently emitting tremendous amounts of energy. Radioactivity has numerous applications, including the diagnosis and treatment of medical conditions such as cancer, thyroid disease, abnormal kidney and bladder function, and heart disease. Naturally occurring radioactivity allows us to estimate the age of fossils, rocks, and ancient artifacts. And radioactivity, perhaps most famously, led to the discovery of nuclear fission, used for electricity generation and nuclear weapons.

## 20.1 Diagnosing Appendicitis

One morning a few years ago I awoke with a dull pain on the lower right side of my abdomen that was worse by early afternoon. Since pain in this area can indicate appendicitis (inflammation of the appendix), and since I know that appendicitis can be dangerous if left untreated, I went to the hospital emergency room. The doctor who examined me recommended a simple blood test to determine my white blood cell count. Patients with appendicitis usually have a high white blood cell count because the body is trying to fight the infection. In my case, the test was negative—I had a normal white blood cell count.

Although my symptoms were consistent with appendicitis, the negative blood test clouded the diagnosis. The doctor said that I could elect to have my appendix removed anyway (even though it might be healthy) or I could submit to another test that might confirm the appendicitis. I chose the additional test, which involved *nuclear medicine*, an area of medical practice that employs *radioactivity* to diagnose and treat disease. **Radioactivity** is the emission of subatomic particles or high-energy electromagnetic radiation by the nuclei of certain atoms. Such atoms are said to be **radioactive**. Most radioactive emissions can pass through many types of matter (such as skin and muscle, in this case).

During the test, antibodies—naturally occurring molecules that fight infection—were labeled with radioactive atoms and then injected into my bloodstream. Because antibodies attack infection, they migrate to areas of the body where infection is present. If my appendix was indeed infected, the antibodies would accumulate there. I waited about an hour, and then a technician took me to a room and laid me on a table. She inserted a photographic film in a panel above me and removed the covering that prevents exposure of the film. Radioactivity is invisible to the eye, but it exposes photographic film. If my appendix had been infected, it would have (by then) contained a high concentration of the radioactively labeled antibodies. The antibodies would emit radiation and expose the film. The test, however, was negative. No radioactivity was emanating from my appendix. After several hours, the pain in my abdomen subsided and I went home. I never did find out what caused the pain.

## 20.2 The Discovery of Radioactivity

Radioactivity was first observed in 1896 by a French scientist named Antoine-Henri Becquerel (1852–1908). Becquerel was interested in the newly discovered X-rays (see Chapter 7), which were a hot topic of physics research in his time. He hypothesized that X-rays were emitted in conjunction with **phosphorescence**, the long-lived *emission* of light that sometimes follows the absorption of light by certain atoms and molecules. Phosphorescence is probably most familiar to you as the *glow* in glow-in-the-dark products (such as toys or stickers). After such a product is exposed to light, it reemits some of that light, usually at slightly longer wavelengths. If you turn off the room lights or put the glow-in-the-dark product in the dark, you see the greenish glow of the emitted light. Becquerel hypothesized that this visible greenish glow was associated with the emission of X-rays (which are invisible).

▲ The greenish light emitted from glow-in-the-dark toys is phosphorescence.

To test his hypothesis, Becquerel placed crystals—composed of potassium uranyl sulfate, a compound known to phosphoresce—on top of a photographic plate wrapped in black cloth. He then exposed the crystals to sunlight. He knew the crystals had phosphoresced because he could see the emitted light when he brought them back into the dark. If the crystals had also emitted X-rays, the X-rays would have passed through the black cloth and exposed the underlying photographic plate. Becquerel performed the experiment several times and always got the same result—the photographic plate showed a dark exposure spot where the crystals had been (Figure 20.1 ◀). Becquerel believed his hypothesis was correct and presented the results—that phosphorescence and X-rays were linked—to the French Academy of Sciences.

Becquerel later retracted his results, however, when he discovered that a photographic plate with the same crystals showed a dark exposure spot even when the plate and the crystals were stored in a drawer and not exposed to sunlight. Becquerel realized that the crystals themselves were constantly emitting something that exposed the photographic plate, regardless of whether or not they phosphoresced. Becquerel concluded that it was the uranium within the crystals that was the source of the emissions, and he named the emissions *uranic rays*.

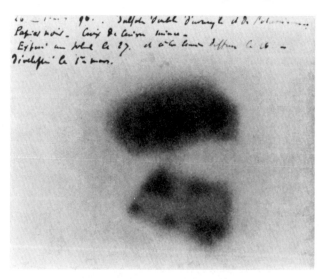

▲ **FIGURE 20.1 The Discovery of Radioactivity** This photographic plate (with Becquerel's original comments at top) played a key role in the discovery of radioactivity. Becquerel placed a uranium-containing compound on the plate (which was wrapped in black cloth to shield it from visible light). The plate was darkened by some unknown form of penetrating radiation that was produced continuously, independently of phosphorescence.

Soon after Becquerel's discovery, a young graduate student named Marie Sklodowska Curie (1867–1934) (one of the first women in France to pursue doctoral work) decided to study uranic rays for her doctoral thesis. Her first task was to determine whether any other substances besides uranium (the heaviest known element at the time) emitted these rays. In her search, Curie discovered two new elements, both of which also emitted uranic rays. Curie named one of her newly discovered elements polonium, after her home country of Poland. The other element she named radium, because of its high level of radioactivity. Radium is so radioactive that it gently glows in the dark and emits significant amounts of heat. Since it was clear that these rays were not unique to uranium, Curie changed the name of uranic rays to radioactivity. In 1903, Curie and her husband, Pierre Curie, as well as Becquerel were all awarded the Nobel Prize in physics for the discovery of radioactivity. In 1911, Curie received a second Nobel Prize, this time in chemistry, for her discovery of the two new elements.

▲ Marie Curie, one of the first women in France to pursue a doctoral degree, was twice awarded the Nobel Prize, in 1903 and 1911. She is seen here with her daughters, in about 1905. Irène (left) became a distinguished nuclear physicist in her own right, winning a Nobel Prize in 1935. Eve (right) wrote a highly acclaimed biography of her mother.

## 20.3 Types of Radioactivity

While Curie focused her work on discovering the different kinds of radioactive elements, Ernest Rutherford and others focused on characterizing the radioactivity itself. These scientists found that the emissions are produced by the nuclei of radioactive atoms. Such nuclei are unstable and spontaneously decompose, emitting small pieces of themselves to gain stability. These fragments are the radioactivity that Becquerel and Curie detected. Natural radioactivity can be categorized into several different types, including *alpha (α) decay*, *beta (β) decay*, *gamma (γ) ray emission*, and *positron emission*. In addition, some unstable atomic nuclei can attain greater stability by absorbing electrons from one of the atoms' own orbitals, a process called *electron capture*.

Element 96 is named curium in honor of Marie Curie and her contributions to our understanding of radioactivity.

In order to understand these different types of radioactivity, we must briefly review the notation for symbolizing isotopes from Section 2.6. Recall that we can represent any isotope with the following notation:

Mass number $(A)$ = the sum of the number of protons and number of neutrons in the nucleus

Atomic number $(Z)$ = the number of protons in the nucleus

Since $A$ represents the sum of the number of protons and neutrons, and since $Z$ represents the number of protons, the number of neutrons in the nucleus $(N)$ is $A - Z$.

$$N = A - Z$$

Number of neutrons

▲ Radium, discovered by Marie Curie, is so radioactive that it glows visibly and emits heat.

 **Types of Radioactivity**

For example, the symbol $^{21}_{10}\text{Ne}$ represents the neon isotope containing 10 protons and 11 neutrons. The symbol $^{20}_{10}\text{Ne}$ represents the neon isotope containing 10 protons and 10 neutrons. Remember that most elements have several different isotopes. When we are discussing nuclear properties, we often refer to a particular isotope (or species) of an element as a **nuclide**.

We represent the main subatomic particles—protons, neutrons, and electrons— with similar notation.

Proton symbol $^1_1\text{p}$   Neutron symbol $^1_0\text{n}$   Electron symbol $^0_{-1}\text{e}$

The 1 in the lower left of the proton symbol represents 1 proton, and the 0 in the lower left corner of the neutron symbol represents 0 protons. The −1 in the lower left corner of the electron symbol is a bit different from the other atomic numbers, but it will make sense to you when we see it in the context of nuclear decay a bit later in this section.

## Alpha Decay

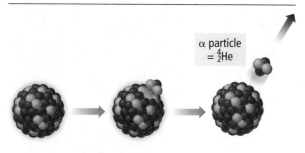

▲ **FIGURE 20.2 Alpha Decay** In alpha decay, a nucleus emits a particle composed of two protons and two neutrons (a helium-4 nucleus).

As we discuss in Section 20.4, nuclei are unstable when they are too large or when they contain an unbalanced ratio of neutrons to protons.

In nuclear chemistry, we are primarily interested in changes within the nucleus; therefore, the 2+ charge that we would normally write for a helium nucleus is omitted for an alpha particle.

## Alpha (α) Decay

**Alpha (α) decay** occurs when an unstable nucleus emits a particle composed of two protons and two neutrons (Figure 20.2◀). Since two protons and two neutrons combined are identical to a helium-4 nucleus, the symbol for alpha radiation is the symbol for helium-4:

**Alpha (α) particle** $\quad {}^{4}_{2}\text{He}$

When an element emits an alpha particle, the number of protons in its nucleus changes, transforming the element into a different element. We symbolize this phenomenon with a **nuclear equation**, an equation that represents nuclear processes such as radioactivity. For example, the nuclear equation for the alpha decay of uranium-238 is:

Parent nuclide        Daughter nuclide

$${}^{238}_{92}\text{U} \longrightarrow {}^{234}_{90}\text{Th} + {}^{4}_{2}\text{He}$$

The original atom is the *parent nuclide*, and the product of the decay is the *daughter nuclide*. In this case, uranium-238 (the parent nuclide) becomes thorium-234 (the daughter nuclide). Unlike a chemical reaction, in which elements retain their identities, in a nuclear reaction elements often change their identities. Like a chemical equation, however, a nuclear equation must be balanced. *The sum of the atomic numbers on both sides of a nuclear equation must be equal, and the sum of the mass numbers on both sides must also be equal:*

$${}^{238}_{92}\text{U} \longrightarrow {}^{234}_{90}\text{Th} + {}^{4}_{2}\text{He}$$

| Reactants | Products |
|---|---|
| Sum of mass numbers = 238 | Sum of mass numbers = 234 + 4 = 238 |
| Sum of atomic numbers = 92 | Sum of atomic numbers = 90 + 2 = 92 |

We can deduce the identity and symbol of the daughter nuclide in any alpha decay from the mass and atomic number of the parent nuclide. During alpha decay, the mass number decreases by 4 and the atomic number decreases by 2, as shown in Example 20.1.

---

**Example 20.1** | **Writing Nuclear Equations for Alpha Decay**

Write the nuclear equation for the alpha decay of Ra-224.

**SOLUTION**

| | |
|---|---|
| Begin with the symbol for Ra-224 on the left side of the equation and the symbol for an alpha particle on the right side. | ${}^{224}_{88}\text{Ra} \longrightarrow {}^{?}_{?}? + {}^{4}_{2}\text{He}$ |
| Equalize the sum of the mass numbers and the sum of the atomic numbers on both sides of the equation by writing the appropriate mass number and atomic number for the unknown daughter nuclide. | ${}^{224}_{88}\text{Ra} \longrightarrow {}^{220}_{86}? + {}^{4}_{2}\text{He}$ |
| Refer to the periodic table to determine the identity of the unknown daughter nuclide from its atomic number and write its symbol. Since the atomic number is 86, the daughter nuclide is radon (Rn). | ${}^{224}_{88}\text{Ra} \longrightarrow {}^{220}_{86}\text{Rn} + {}^{4}_{2}\text{He}$ |

**FOR PRACTICE 20.1** Write the nuclear equation for the alpha decay of Po-216.

---

Alpha radiation is the 18-wheeler truck of radioactivity. The alpha particle is by far the most massive of all particles commonly emitted by radioactive nuclei. Consequently, alpha radiation has the most potential to interact with and damage other molecules, including biological ones. Highly energetic radiation interacts with other molecules and atoms by ionizing them. When radiation ionizes molecules within the cells of living organisms, the cells can usually repair the damage. However, in some cases the cells can die or begin to reproduce abnormally. The ability of radiation to ionize other molecules and atoms is called its **ionizing power**. Of all types of radioactivity, alpha radiation has the highest ionizing power.

However, alpha particles, because of their large size, have the lowest **penetrating power**—the ability to penetrate matter. (Imagine a semi trying to get through a traffic jam.) In order for radiation to damage important molecules within living cells, it must penetrate into the cell. Alpha radiation does not easily penetrate into cells because it can be stopped by a sheet of paper, by clothing, or even by air. Consequently, a low-level alpha emitter that remains outside the body is relatively safe. However, consuming large amounts of an alpha emitter can be dangerous because the alpha particles then have direct access to the molecules that compose organs and tissues.

## Beta ($\beta$) Decay

**Beta ($\beta$) decay** occurs when an unstable nucleus emits an electron (Figure 20.3▶). How does a nucleus, which contains only protons and neutrons, emit an electron? In some unstable nuclei, a neutron changes into a proton and emits an electron:

**Beta decay**        Neutron $\longrightarrow$ proton + emitted electron

The symbol for a beta ($\beta$) particle in a nuclear equation is:

**Beta ($\beta$) particle**        $_{-1}^{0}e$        •

We represent beta decay with this nuclear equation:

$$_{0}^{1}n \longrightarrow {}_{1}^{1}p + {}_{-1}^{0}e$$

The −1 reflects the charge of the electron, which is equivalent to an atomic number of −1 in a nuclear equation. When an atom emits a beta particle, its atomic number increases by 1 because it then has an additional proton. For example, the nuclear equation for the beta decay of radium-228 is:

$$_{88}^{228}Ra \longrightarrow {}_{89}^{228}Ac + {}_{-1}^{0}e$$

Notice that the nuclear equation is balanced—the sum of the mass numbers on both sides is equal, and the sum of the atomic numbers on both sides is equal.

Beta radiation is the four-door sedan of radioactivity. Beta particles are much less massive than alpha particles and consequently have a lower ionizing power. However, because of their smaller size, beta particles have a higher penetrating power and only something as substantive as a sheet of metal or a thick piece of wood stops them. Consequently, a beta emitter outside of the body poses a higher risk than an alpha emitter. If ingested, however, the beta emitter does less damage than an alpha emitter.

## Gamma ($\gamma$) Ray Emission

**Gamma ($\gamma$) ray emission** is significantly different from alpha or beta radiation. Gamma radiation is a form of *electromagnetic* radiation (see Section 7.2). Gamma rays are high-energy (short-wavelength) photons. The symbol for a gamma ray is:

**Gamma ($\gamma$) ray**        $_{0}^{0}\gamma$        〜〜〜〜〜➤

A gamma ray has no charge and no mass. When a gamma-ray photon is emitted from a radioactive atom, it does not change the mass number or the atomic number of the element. Gamma rays are usually emitted from nuclei in excited states or in conjunction with other types of radiation. For example, the alpha emission of U-238 (discussed previously) is also accompanied by the emission of a gamma ray:

$$_{92}^{238}U \longrightarrow {}_{90}^{234}Th + {}_{2}^{4}He + {}_{0}^{0}\gamma$$

The emission of gamma rays in conjunction with other types of radiation is so common that it is often left out of nuclear equations. Nuclear chemists simply understand that the gamma rays are emitted along with the other types of decay. However, in some instances, there can be a delay between the initial decay and the subsequent emission of gamma ray. The initial decay leaves the daughter nucleus in *metastable* state (an unstable state that can exist for a prolonged period of time). The daughter then emits a

**Beta Decay**

▲ **FIGURE 20.3 Beta Decay** In beta decay, a neutron emits an electron and becomes a proton.

This kind of beta radiation is also called beta minus ($\beta^-$) radiation due to its negative charge.

See Section 7.2 for a review of electromagnetic radiation.

gamma particle at a later time. For example, technetium-99 has a gamma-emitting metastable state:

$$^{99m}_{43}\text{Tc} \longrightarrow ^{99}_{43}\text{Tc} + ^{0}_{0}\gamma$$

The m next to the mass number stands for metastable.

Gamma rays are the motorbikes of radioactivity. They have the lowest ionizing power, but the highest penetrating power. (Imagine a motorbike zipping through a traffic jam.) Shielding gamma rays requires several inches of lead or thick slabs of concrete.

## Positron Emission

**Positron Emission**

▲ **FIGURE 20.4  Positron Emission**   In positron emission, a proton emits a positron and becomes a neutron.

**Positron emission** occurs when an unstable nucleus emits a positron (Figure 20.4◀). A **positron** is the *antiparticle* of the electron; it has the same mass as an electron, but the opposite charge. If a positron collides with an electron, the two particles annihilate each other, releasing energy in the form of gamma rays. In positron emission, a proton is converted into a neutron and emits a positron:

**Positron emission**          Proton ⟶ neutron + emitted positron

The symbol for a positron in a nuclear equation is:

**Positron**          $^{0}_{+1}\text{e}$          •

We represent positron emission with this nuclear equation:

$$^{1}_{1}\text{p} \longrightarrow ^{1}_{0}\text{n} + ^{0}_{+1}\text{e}$$

When an atom emits a positron, its atomic number *decreases* by 1 because it has one less proton after emission. Consider the nuclear equation for the positron emission of phosphorus-30 as an example:

$$^{30}_{15}\text{P} \longrightarrow ^{30}_{14}\text{Si} + ^{0}_{+1}\text{e}$$

Positron emission can be thought of as a type of beta emission and is sometimes referred to as beta plus emission ($\beta^+$).

We can determine the identity and symbol of the daughter nuclide in any positron emission in a manner similar to that used for alpha and beta decay, as shown in Example 20.2. Positrons are similar to beta particles in their ionizing and penetrating power.

## Electron Capture

Unlike the forms of radioactive decay that we have discussed so far, electron capture involves a particle being *absorbed by* rather than *emitted from* an unstable nucleus. **Electron capture** occurs when a nucleus assimilates an electron from an inner orbital of its electron cloud. Like positron emission, the net effect of electron capture is the conversion of a proton into a neutron:

**Electron capture**          Proton + electron ⟶ neutron

We represent electron capture with this nuclear equation:

$$^{1}_{1}\text{p} + ^{0}_{-1}\text{e} \longrightarrow ^{1}_{0}\text{n}$$

When an atom undergoes electron capture, its atomic number decreases by 1 because it has one less proton. For example, when Ru-92 undergoes electron capture, its atomic number changes from 44 to 43:

$$^{92}_{44}\text{Ru} + ^{0}_{-1}\text{e} \longrightarrow ^{92}_{43}\text{Tc}$$

Table 20.1 summarizes the different kinds of radiation.

---

**Example 20.2**    **Writing Nuclear Equations for Beta Decay, Positron Emission, and Electron Capture**

Write the nuclear equation for each type of decay.

**(a)** beta decay in Bk-249          **(b)** positron emission in O-15          **(c)** electron capture in I-111

| SOLUTION | |
|---|---|
| **(a)** In beta decay, the atomic number *increases* by 1 and the mass number remains unchanged. <br> The daughter nuclide is element number 98, californium. | $^{249}_{97}\text{Bk} \longrightarrow \, ^{249}_{98}? + \, ^{0}_{-1}\text{e}$ <br><br> $^{249}_{97}\text{Bk} \longrightarrow \, ^{249}_{98}\text{Cf} + \, ^{0}_{-1}\text{e}$ |
| **(b)** In positron emission, the atomic number *decreases* by 1 and the mass number remains unchanged. <br> The daughter nuclide is element number 7, nitrogen. | $^{15}_{8}\text{O} \longrightarrow \, ^{15}_{7}? + \, ^{0}_{+1}\text{e}$ <br><br> $^{15}_{8}\text{O} \longrightarrow \, ^{15}_{7}\text{N} + \, ^{0}_{+1}\text{e}$ |
| **(c)** In electron capture, the atomic number also *decreases* by 1 and the mass number remains unchanged. <br> The daughter nuclide is element number 52, tellurium. | $^{111}_{53}\text{I} + \, ^{0}_{-1}\text{e} \longrightarrow \, ^{111}_{52}?$ <br><br> $^{111}_{53}\text{I} + \, ^{0}_{-1}\text{e} \longrightarrow \, ^{111}_{52}\text{Te}$ |

**FOR PRACTICE 20.2**
**(a)** Write three nuclear equations to represent the nuclear decay sequence that begins with the alpha decay of U-235 followed by a beta decay of the daughter nuclide and then another alpha decay.
**(b)** Write the nuclear equation for the positron emission of Na-22.
**(c)** Write the nuclear equation for electron capture in Kr-76.

**FOR MORE PRACTICE 20.2** Potassium-40 can decay to produce Ar-40. What is the method of decay? Write the nuclear equation for this decay.

**TABLE 20.1 Modes of Radioactive Decay**

| Decay Mode | Process | | Change in: A | Z | N/Z* | Example |
|---|---|---|---|---|---|---|
| $\alpha$ | Parent nuclide — Daughter nuclide — $\alpha$ particle | $^{4}_{2}\text{He}$ | $-4$ | $-2$ | Increase | $^{238}_{92}\text{U} \longrightarrow \, ^{234}_{90}\text{Th} + \, ^{4}_{2}\text{He}$ |
| $\beta$ | Neutron — Neutron becomes a proton $+ \, ^{0}_{-1}\text{e}$ Parent nuclide — Daughter nuclide — $\beta$ particle | | $0$ | $+1$ | Decrease | $^{228}_{88}\text{Ra} \longrightarrow \, ^{228}_{89}\text{Ac} + \, ^{0}_{-1}\text{e}$ |
| $\gamma$ | Excited nuclide — Stable nuclide — Photon $^{0}_{0}\gamma$ | | $0$ | $0$ | None | $^{99m}_{43}\text{Tc} \longrightarrow \, ^{99}_{43}\text{Tc} + \, ^{0}_{0}\gamma$ |
| Positron emission | Proton — Proton becomes a neutron $+ \, ^{0}_{+1}\text{e}$ Parent nuclide — Daughter nuclide — Positron | | $0$ | $-1$ | Increase | $^{30}_{15}\text{P} \longrightarrow \, ^{30}_{14}\text{Si} + \, ^{0}_{+1}\text{e}$ |
| Electron capture | Proton — Proton becomes a neutron $^{0}_{-1}\text{e} +$ Parent nuclide — Daughter nuclide | | $0$ | $-1$ | Increase | $^{92}_{44}\text{Ru} + \, ^{0}_{-1}\text{e} \longrightarrow \, ^{92}_{43}\text{Tc}$ |

* Neutron-to-proton ratio

**Alpha and Beta Decay**   Consider the graphical representation of a series of decays shown here.

The arrow labeled $x$ and the arrow labeled $y$ each correspond to what kind of decay?

**(a)**   $x$ corresponds to alpha decay and $y$ corresponds to positron emission.

**(b)**   $x$ corresponds to positron emission and $y$ corresponds to alpha decay.

**(c)**   $x$ corresponds to alpha decay and $y$ corresponds to beta decay.

**(d)**   $x$ corresponds to beta decay and $y$ corresponds to alpha decay.

## 20.4   The Valley of Stability: Predicting the Type of Radioactivity

So far, we have described various different types of radioactivity. But we have not discussed what causes a particular nuclide to be radioactive in the first place. And why do some nuclides decay via alpha decay, while others decay via beta decay or positron emission? The answers to these questions are not simple, but we can get a basic idea of the factors that influence the stability of the nucleus and the nature of its decay.

A nucleus is a collection of protons (positively charged) and neutrons (uncharged). We know that positively charged particles such as protons repel one another. So what binds the nucleus together? A fundamental force of physics known as the **strong force** binds the nucleus together. All **nucleons**—protons and neutrons—are attracted to one another by the strong force. However, the strong force acts only at very short distances. We can think of the stability of a nucleus as a balance between the *repulsive* coloumbic force among protons and the *attractive* strong force among all nucleons. The neutrons in a nucleus, therefore, play an important role in stabilizing the nucleus because they attract other nucleons (through the strong force) but lack the repulsive force associated with positive charge.

It might seem that adding more neutrons would *always* lead to greater stability, so that the more neutrons the better. This is not the case, however, because protons and neutrons occupy energy levels in a nucleus that are similar to those occupied by electrons in an atom. As you add more neutrons, the neutrons must occupy increasingly higher energy levels within the nucleus. At some point, the energy payback from the strong force is not enough to compensate for the high energy state that the neutrons must occupy.

An important number in determining nuclear stability is the *ratio* of neutrons to protons ($N/Z$). Figure 20.5 ◄ is a plot of the number of neutrons versus the number of protons for all known nuclei. The yellow dots represent unstable nuclei and the green dots represent stable ones. The region of the graph with the green dots (stable nuclei) is known as the *valley* (or *island*) *of stability*. Notice that for the lighter elements, the $N/Z$ ratio of stable nuclei is about 1 (equal numbers of neutrons and protons). For example, the most abundant isotope of carbon ($Z = 6$) is carbon-12, which contains six protons and six neutrons. However, beyond about $Z = 20$, the $N/Z$ ratio of stable nuclei begins to get larger. For example, at $Z = 44$ stable nuclei have $N/Z$ ratios of about 1.275 and at $Z = 80$, the $N/Z$ ratio reaches about 1.5. Above $Z = 83$, stable nuclei do not exist—bismuth ($Z = 83$) is the heaviest element with stable (nonradioactive) isotopes.

**▼ FIGURE 20.5   Stable and Unstable Nuclei**   A plot of $N$ (the number of neutrons) versus $Z$ (the number of protons) for all known nuclei. The green dots represent stable nuclei and the yellow dots represent unstable nuclei. A nucleus with an $N/Z$ ratio that is too high tends to undergo beta decay. A nucleus with an $N/Z$ ratio that is too low tends to undergo positron emission or electron capture.

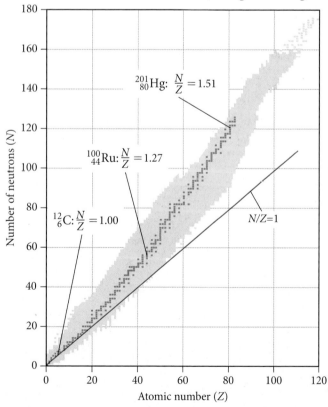

The type of radioactivity emitted by a nuclide depends in part on the $N/Z$ ratio.

**$N/Z$ too high:** Nuclides that lie above the valley of stability have too many neutrons and tend to convert neutrons to protons via beta decay. The process of undergoing beta decay moves the nuclide down in the plot in Figure 20.5 and closer to (or into) the valley of stability.

**$N/Z$ too low:** Nuclides that lie below the valley of stability have too many protons and tend to convert protons to neutrons via positron emission or electron capture. This moves the nuclide up in the plot in Figure 20.5 and closer to (or into) the valley of stability.

One way to decide whether a particular nuclide has an $N/Z$ that is too high, too low, or about right is to consult Figure 20.5. Those nuclides that lie within the valley of stability are stable. Alternatively, we can also compare the mass number of the nuclide to the atomic mass listed in the periodic table for the corresponding element. The atomic mass is an average of the masses of the stable nuclides for an element and thus represents an $N/Z$ that is about right. For example, suppose we want to evaluate $N/Z$ for Ru-112. Ruthenium has an atomic mass of 101.07, so we know that the nuclide with a mass number of 112 contains too many neutrons and therefore has an $N/Z$ that is too high. The following example demonstrates how to apply these considerations in predicting the mode of decay for a nucleus.

---

**Example 20.3** | **Predicting the Type of Radioactive Decay**

Predict whether each nuclide is more likely to decay via beta decay or positron emission.

**(a)** Mg-28   **(b)** Mg-22   **(c)** Mo-102

**SOLUTION**

**(a)** Magnesium-28 has 16 neutrons and 12 protons, so $N/Z = 1.33$. However, for $Z = 12$, you can see from Figure 20.5 that stable nuclei should have an $N/Z$ of about 1. Alternatively, you can see from the periodic table that the atomic mass of magnesium is 24.31. Therefore, a nuclide with a mass number of 28 is too heavy to be stable because the $N/Z$ ratio is too high and Mg-28 undergoes *beta decay*, resulting in the conversion of a neutron to a proton.

**(b)** Magnesium-22 has 10 neutrons and 12 protons, so $N/Z = 0.83$ (too low). Alternatively, you can see from the periodic table that the atomic mass of magnesium is 24.31. A nuclide with a mass number of 22 is too light; the $N/Z$ ratio is too low. Therefore, Mg-22 undergoes *positron emission*, resulting in the conversion of a proton to a neutron. (Electron capture would accomplish the same thing as positron emission, but in Mg-22, positron emission is the only decay mode observed.)

**(c)** Molybdenum-102 has 60 neutrons and 42 protons, so $N/Z = 1.43$. However, for $Z = 42$, you can see from Figure 20.5 that stable nuclei should have $N/Z$ ratios of about 1.3. Alternatively, you can see from the periodic table that the atomic mass of molybdenum is 95.94. A nuclide with a mass number of 102 is too heavy to be stable; the $N/Z$ ratio is too high. Therefore, Mo-102 undergoes *beta decay*, resulting in the conversion of a neutron to a proton.

**FOR PRACTICE 20.3**   Predict whether each nuclide is more likely to decay via beta decay or positron emission.

**(a)** Pb-192   **(b)** Pb-212   **(c)** Xe-114

---

## Magic Numbers

In addition to the $N/Z$ ratio, the *actual number* of protons and neutrons also affects the stability of the nucleus. Table 20.2 shows the number of nuclei with different possible combinations of even or odd nucleons. Notice that a large number of stable nuclides have both an even number of protons and an even number of neutrons. Only five stable nuclides have an odd and odd combination.

**TABLE 20.2 Number of Stable Nuclides with Even and Odd Numbers of Nucleons**

| Z | N | Number of Nuclides |
|------|------|------|
| Even | Even | 157 |
| Even | Odd | 53 |
| Odd | Even | 50 |
| Odd | Odd | 5 |

## A Decay Series

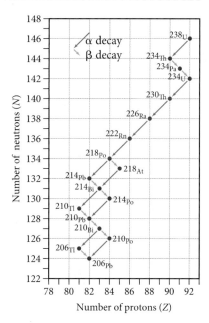

▲ **FIGURE 20.6** **The Uranium-238 Radioactive Decay Series**
Uranium-238 decays via a series of steps ending in Pb-206, a stable element. Each diagonal line to the left represents an alpha decay, and each diagonal line to the right represents a beta decay.

The reason for this is that nucleons occupy energy levels within the nucleus much as electrons occupy energy levels within an atom. Just as atoms with certain numbers of electrons are uniquely stable (in particular, the number of electrons associated with the noble gases: 2, 10, 18, 36, 54, etc.), so atoms with certain numbers of nucleons ($N$ or $Z = 2, 8, 20, 28, 50, 82$, and $N = 126$) are uniquely stable. These numbers are referred to as **magic numbers**. Nuclei containing a magic number of protons or neutrons are particularly stable. Since the magic numbers are even, this accounts in part for the abundance of stable nuclides with even numbers of nucleons. Moreover, nucleons also have a tendency to pair (much as electrons pair). This tendency and the resulting stability of paired nucleons also contribute to the abundance of stable nuclides with even numbers of nucleons.

## Radioactive Decay Series

Atoms with $Z > 83$ are radioactive and decay in one or more steps involving primarily alpha and beta decay (with some gamma decay to carry away excess energy). For example, uranium (atomic number 92) is the heaviest naturally occurring element. Its most common isotope is U-238, an alpha emitter that decays to Th-234:

$$^{238}_{92}\text{U} \longrightarrow {}^{234}_{90}\text{Th} + {}^{4}_{2}\text{He}$$

The daughter nuclide, Th-234, is itself radioactive—it is a beta emitter that decays to Pa-234:

$$^{234}_{90}\text{Th} \longrightarrow {}^{234}_{91}\text{Pa} + {}^{0}_{-1}\text{e}$$

Protactinium-234 is also radioactive, decaying to U-234 via beta emission. Radioactive decay continues until a stable nuclide, Pb-206, is reached. Figure 20.6 ◀ illustrates the entire uranium-238 decay series.

## 20.5 Detecting Radioactivity

The particles emitted by radioactive nuclei are very energetic and therefore can be readily detected. In a radiation detector, the particles are detected through their interactions with atoms or molecules. The most common radiation detectors are **thermoluminescent dosimeters** (Figure 20.7 ◀), which are issued to people working with or near radioactive substances. These dosimeters contain crystals of salts such as calcium fluoride that are excited by the ionizing radiation. The excited electrons are trapped by impurities that are intentionally introduced into the crystals. When the crystals are heated, the electrons relax to their ground state, emitting light. The amount of light emitted is proportional to the radiation exposure. These dosimeters are collected and processed regularly as a way to monitor a person's exposure. We discuss the effects and measurement of exposure in more detail in Section 20.11.

Radioactivity can be instantly detected with devices such as a **Geiger-Müller counter** (Figure 20.8 ▶). In this instrument (commonly referred to as a Geiger counter), particles emitted by radioactive nuclei pass through an argon-filled chamber. The energetic particles create a trail of ionized argon atoms. High voltage applied between a wire within the chamber and the chamber itself causes these newly formed ions to produce an electrical signal that can be displayed on a meter or turned into an audible click. Each click corresponds to a radioactive particle passing through the argon gas chamber. This clicking is the stereotypical sound most people associate with a radiation detector.

A second type of device commonly used to detect radiation instantly is a **scintillation counter**. In a scintillation counter, radioactive emissions pass through a material (such as NaI or CsI) that emits ultraviolet or visible light in response to excitation by high-energy photons. The photons excite the atoms to a higher-energy state. The atoms release this energy as light, which is detected and turned into an electrical signal that is read on a meter.

▲ **FIGURE 20.7**
**Thermoluminescent Dosimeter**
A thermoluminescent dosimeter contains crystals that when heated emit light in proportion to radiation exposure.

▲ **FIGURE 20.8** **Geiger-Müller Counter** When ionizing radiation passes through the argon-filled chamber, it ionizes the argon atoms, giving rise to a brief, tiny pulse of electrical current that is transduced onto a meter or into an audible click.

## 20.6 The Kinetics of Radioactive Decay and Radiometric Dating

Radioactivity is a natural component of our environment. The ground beneath you most likely contains radioactive atoms that emit radiation. The food you eat contains a residual quantity of radioactive atoms that are absorbed into your body fluids and incorporated into tissues. Small amounts of radiation from space make it through our atmosphere to constantly bombard Earth. Humans and other living organisms have evolved in this environment and have adapted to survive in it.

One reason for the radioactivity in our environment is the instability of all atomic nuclei beyond atomic number 83 (bismuth). Every element with more than 83 protons in its nucleus is unstable and therefore radioactive. In addition, some isotopes of elements with fewer than 83 protons are also unstable and radioactive. Radioactive nuclides *persist* in our environment because new radioactive nuclides are constantly being formed and because many of the existing ones decay away only very slowly.

All radioactive nuclei decay via first-order kinetics, so the rate of decay in a particular sample is directly proportional to the number of nuclei present as indicated in the equation:

You may find it useful to review the discussion of first-order kinetics in Section 14.3.

$$\text{Rate} = kN$$

where $N$ is the number of radioactive nuclei and $k$ is the rate constant. Different radioactive nuclides decay into their daughter nuclides with different rate constants. Some nuclides decay quickly (large rate constant), while others decay slowly (small rate constant).

The time it takes for one-half of the parent nuclides in a radioactive sample to decay to the daughter nuclides is the *half-life* and is identical to the concept of half-life for chemical reactions that we discussed in Chapter 14. Thus, the relationship between the half-life of a nuclide and its rate constant is given by the same expression (Equation 14.19) that we derived for a first-order reaction in Section 14.4:

$$t_{1/2} = \frac{0.693}{k} \tag{20.1}$$

Nuclides that decay quickly have short half-lives and large rate constants—they are considered very active (many decay events per unit time). Nuclides that decay slowly have long half-lives and are less active (fewer decay events per unit time). For example, thorium-232 is an alpha emitter with a half-life of $1.4 \times 10^{10}$ years, or 14 billion years. A sample of Th-232 containing 1 million atoms decays to $^1/_2$ million atoms in 14 billion years and then to $^1/_4$ million in another 14 billion years and so on. Notice that a radioactive sample does *not* decay to *zero* atoms in two half-lives—we can't add two half-lives together to get a "whole" life. The amount that remains after one half-life is always one-half of what was present at the start. The amount that remains after two half-lives is one-quarter of what was present at the start, and so on.

| 1 million Th-232 atoms | → | 1/2 million Th-232 atoms | → | 1/4 million Th-232 atoms |

14 billion years          14 billion years

Some nuclides have very short half-lives. For example, radon-220 has a half-life of approximately 1 minute (Figure 20.9▼). A 1-million-atom sample of radon-220 decays to $^1/_4$ million radon-220 atoms in just two minutes and to approximately 1000 atoms in 10 minutes. Table 20.3 lists several nuclides and their half-lives.

| 1 million Rn-220 atoms | → | 1/2 million Rn-220 atoms | → | 1/4 million Rn-220 atoms |

1 minute          1 minute

**TABLE 20.3 Selected Nuclides and Their Half-Lives**

| Nuclide | Half-Life | Type of Decay |
|---------|-----------|---------------|
| $^{232}_{90}$Th | $1.4 \times 10^{10}$ yr | Alpha |
| $^{238}_{92}$U | $4.5 \times 10^{9}$ yr | Alpha |
| $^{14}_{6}$C | 5715 yr | Beta |
| $^{220}_{86}$Rn | 55.6 s | Alpha |
| $^{219}_{90}$Th | $1.05 \times 10^{-6}$ s | Alpha |

▶ **FIGURE 20.9 The Decay of Radon-220** Radon-220 decays with a half-life of approximately 1 minute.

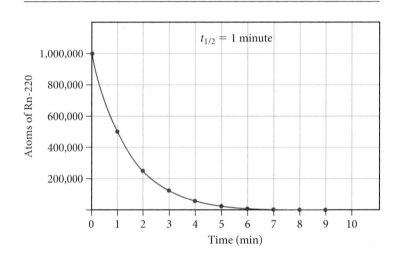

**Decay of Radon-220**

$t_{1/2} = 1$ minute

CONCEPTUAL CONNECTION **20.2**

**Half-Life** Consider this graph representing the decay of a radioactive nuclide.

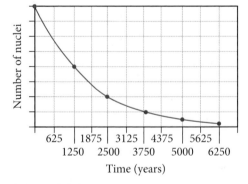

What is the half-life of the nuclide?

(a) 625 years          (b) 1250 years

(c) 2500 years          (d) 3125 years

## The Integrated Rate Law

Recall from Chapter 14 that for first-order chemical reactions, the concentration of a reactant as a function of time is given by the integrated rate law:

$$\ln\frac{[A]_t}{[A]_0} = -kt \qquad\qquad [20.2]$$

Because nuclear decay follows first-order kinetics, we can substitute the number of nuclei for concentration to arrive at the equation:

$$\ln\frac{N_t}{N_0} = -kt \qquad\qquad [20.3]$$

where $N_t$ is the number of radioactive nuclei at time $t$ and $N_0$ is the initial number of radioactive nuclei. Example 20.4 demonstrates the use of this equation.

**Example** 20.4 | **Radioactive Decay Kinetics** ▶

Plutonium-236 is an alpha emitter with a half-life of 2.86 years. If a sample initially contains 1.35 mg of Pu-236, what mass of Pu-236 is present after 5.00 years?

| | |
|---|---|
| **SORT** You are given the initial mass of Pu-236 in a sample and asked to find the mass after 5.00 years. | **GIVEN:** $m_{Pu\text{-}236}(\text{initial}) = 1.35$ mg; $t = 5.00$ yr; $t_{1/2} = 2.86$ yr<br><br>**FIND:** $m_{Pu\text{-}236}$ (final) |
| **STRATEGIZE** Use the integrated rate law (Equation 20.3) to solve this problem. You must determine the value of the rate constant ($k$) from the half-life expression (Equation 20.1). | **CONCEPTUAL PLAN**<br><br>$t_{1/2}$ → $k$<br><br>$t_{1/2} = \dfrac{0.693}{k}$ |
| Use the value of the rate constant, the initial mass of Pu-236, and the time along with integrated rate law to find the final mass of Pu-236. Since the mass of the Pu-236 ($m_{Pu\text{-}236}$) is directly proportional to the number of atoms ($N$), and since the integrated rate law contains the ratio ($N_t/N_0$), you can substitute the initial and final masses for the initial and final number of atoms. | <br>$k, m_{Pu\text{-}236}(\text{initial}), t$ → $m_{Pu\text{-}236}(\text{final})$<br><br>$\ln\dfrac{N_t}{N_0} = -kt$ |
| **SOLVE** Follow your plan. Begin by determining the rate constant from the half-life. | **SOLUTION**<br><br>$t_{1/2} = \dfrac{0.693}{k}$<br><br>$k = \dfrac{0.693}{t_{1/2}} = \dfrac{0.693}{2.86\text{ yr}}$<br><br>$= 0.2423/\text{yr}$ |
| Solve the integrated rate law for $N_t$ and substitute the values of the rate constant, the initial mass of Pu-236, and the time into the solved equation. Calculate the final mass of Pu-236. | $\ln\dfrac{N_t}{N_0} = -kt$<br><br>$\dfrac{N_t}{N_0} = e^{-kt}$<br><br>$N_t = N_0\,e^{-kt}$<br><br>$N_t = 1.35\text{ mg }[e^{-(0.2423/\text{yr})(5.00\text{ yr})}]$<br><br>$N_t = 0.402$ mg |

**CHECK** The units of the answer (mg) are correct. The magnitude of the answer (0.402 mg) is about one-third of the original mass (1.35 mg), which seems reasonable given that the amount of time is between one and two half-lives. (One half-life would result in one-half of the original mass, and two half-lives would result in one-fourth of the original mass.)

**FOR PRACTICE 20.4** How long will it take for the 1.35 mg sample of Pu-236 in Example 20.4 to decay to 0.100 mg?

Since radioactivity is a first-order process, the rate of decay is linearly proportional to the number of nuclei in the sample. Therefore, the initial rate of decay ($rate_0$) and the rate of decay at time $t$ ($rate_t$) can also be used in the integrated rate law:

$$Rate_t = kN_t \qquad Rate_0 = kN_0$$

$$\frac{N_t}{N_0} = \frac{rate_t/k}{rate_0/k} = \frac{rate_t}{rate_0}$$

Substituting into Equation 20.3, we get the following result:

$$\ln \frac{rate_t}{rate_0} = -kt \qquad\qquad [20.4]$$

We can use Equation 20.4 to predict how the rate of decay of a radioactive sample will change with time or how much time has passed based on how the rate has changed (see examples later in this section).

The radioactive isotopes in our environment and their predictable decay with time can therefore be used to estimate the age of rocks or artifacts containing those isotopes. The technique is known as **radiometric dating**, and we examine two different types individually.

CONCEPTUAL
**CONNECTION 20.3**

**Half-Life and the Amount of Radioactive Sample** A sample initially contains 1.6 moles of a radioactive isotope. How much of the sample remains after four half-lives?

**(a)** 0.0 mol **(b)** 0.10 mol **(c)** 0.20 mol **(d)** 0.40 mol

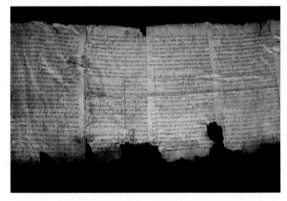

▲ The Dead Sea Scrolls are 2000-year-old biblical manuscripts. Their age was determined by radiocarbon dating.

Libby received the Nobel Prize in 1960 for the development of radiocarbon dating.

A steady state is a set of conditions that results in constant concentrations of the relevant species.

# Radiocarbon Dating: Using Radioactivity to Measure the Age of Fossils and Artifacts

Archaeologists, geologists, anthropologists, and other scientists use **radiocarbon dating**, a technique devised in 1949 by Willard Libby (1908–1980) at the University of Chicago, to estimate the ages of fossils and artifacts. For example, in 1947, young shepherds searching for a stray goat near the Dead Sea (east of Jerusalem) entered a cave and discovered ancient scrolls that had been stuffed into jars. These scrolls—now named the Dead Sea Scrolls—are 2000-year-old texts of the Hebrew Bible, predating other previously discovered manuscripts by almost a thousand years.

The Dead Sea Scrolls, like other ancient artifacts, contain a radioactive signature that reveals their age. This signature results from the presence of carbon-14 (which is radioactive) in the environment. Carbon-14 is constantly formed in the upper atmosphere by the neutron bombardment of nitrogen:

$$^{14}_{7}N + ^{1}_{0}n \longrightarrow ^{14}_{6}C + ^{1}_{1}H$$

After it forms, carbon-14 decays back to nitrogen by beta emission with a half-life of 5715 years:

$$^{14}_{6}C \longrightarrow ^{14}_{7}N + ^{0}_{-1}e \quad t_{1/2} = 5715 \text{ yr}$$

The continuous formation of carbon-14 in the atmosphere and its continuous decay to nitrogen-14 produce a nearly constant steady-state amount of atmospheric carbon-14. The atmospheric carbon-14 is oxidized to carbon dioxide and incorporated into plants by photosynthesis. The C-14 then makes its way up the food chain and ultimately into all living organisms. As a result, the tissues in all living plants, animals, and humans contain the same ratio of carbon-14 to carbon-12 ($^{14}C : ^{12}C$) as that found in the atmosphere. When a living organism dies, however, it stops incorporating new carbon-14 into its tissues. The $^{14}C : ^{12}C$ ratio then begins to decrease with a half-life of 5715 years.

Because many artifacts, including the Dead Sea Scrolls, are made from materials that were once living—such as papyrus, wood, or other plant and animal derivatives—the $^{14}C : ^{12}C$ ratio in these artifacts indicates their age. For example, suppose an ancient artifact has a $^{14}C : {}^{12}C$ ratio that is 25% of that found in living organisms. How old is the artifact? Because it contains one-quarter as much carbon-14 as a living organism, it is two half-lives or 11,460 years old. The maximum age that can be estimated from carbon-14 dating is about 50,000 years—beyond that, the amount of carbon-14 is too low to measure accurately.

We can check the accuracy of carbon-14 dating against objects whose ages are known from historical sources. These kinds of comparisons reveal that ages obtained from C-14 dating deviate from the actual ages by up to about 5%. For a 6000-year-old object, that is an error of about 300 years. The reason for the deviations is the variance of atmospheric C-14 levels over time.

In order to make C-14 dating more accurate, scientists have studied the carbon-14 content of western bristlecone pine trees, which can live up to 5000 years. The tree trunks of the pines contain growth rings corresponding to each year of the tree's life, and the wood laid down in each ring incorporates carbon derived from the carbon dioxide in the atmosphere at that time. The rings thus provide a historical record of the atmospheric carbon-14 content. In addition, the rings of living trees can be correlated with the rings of dead trees (if the lifetimes of the trees overlap), allowing the record to be extended back about 11,000 years. Using the data from the bristlecone pine, we can correct the 5% deviations from historical dates. In this way, the known ages of bristlecone pine trees are used to calibrate C-14 dating, resulting in more accurate results.

▲ Western bristlecone pine trees can live up to 5000 years; scientists can precisely determine the age of a tree by counting the annual rings in its trunk. In this way, the trees can calibrate the time scale for radiocarbon dating.

## Example 20.5    Radiocarbon Dating

A skull believed to belong to an ancient human being has a carbon-14 decay rate of 4.50 disintegrations per minute per gram of carbon (4.50 dis/min · g C). If living organisms have a decay rate of 15.3 dis/min · g C, how old is the skull? (The decay rate is directly proportional to the amount of carbon-14 present.)

| | |
|---|---|
| **SORT** You are given the current rate of decay for the skull and the assumed initial rate. You are asked to find the age of the skull, which is the time that passed in order for the rate to have reached its current value. | **GIVEN:** $\text{rate}_t = 4.50$ dis/min · g C; $\text{rate}_0 = 15.3$ dis/min · g C; <br><br> **FIND:** $t$ |
| **STRATEGIZE** Use the expression for half-life (Equation 20.1) to find the rate constant ($k$) from the half-life for C-14, which is 5715 years (Table 20.3). <br><br><br> Use the value of the rate constant and the initial and current rates to find $t$ from the integrated rate law (Equation 20.4). | **CONCEPTUAL PLAN** <br><br>  $$t_{1/2} = \frac{0.693}{k}$$ <br><br> $\boxed{k, \text{rate}_t, \text{rate}_0} \longrightarrow \boxed{t}$ <br><br> $$\ln\frac{\text{rate}_t}{\text{rate}_0} = -kt$$ |
| **SOLVE** Begin by finding the rate constant from the half-life. <br><br><br><br><br><br><br> Substitute the rate constant and the initial and current rates into the integrated rate law and solve for $t$. | **SOLUTION** <br><br> $$t_{1/2} = \frac{0.693}{k}$$ <br> $$k = \frac{0.693}{t_{1/2}} = \frac{0.693}{5715 \text{ yr}}$$ <br> $$= 1.2\underline{1}3 \times 10^{-4}/\text{yr}$$ <br><br> $$\ln\frac{\text{rate}_t}{\text{rate}_0} = -kt$$ <br><br> $$t = -\frac{\ln\dfrac{\text{rate}_t}{\text{rate}_0}}{k} = -\frac{\ln\dfrac{4.50 \text{ dis/min} \cdot \text{g C}}{15.3 \text{ dis/min} \cdot \text{g C}}}{1.2\underline{1}3 \times 10^{-4}/\text{yr}}$$ <br><br> $$= 1.0 \times 10^4 \text{ yr}$$ |

*—Continued on the next page*

*Continued—*

> **CHECK** The units of the answer (yr) are correct. The magnitude of the answer is about 10,000 years, which is a little less than two half-lives. This value is reasonable given that two half-lives would result in a decay rate of about $3.8 \text{ dis/min} \cdot \text{g C}$.

> **FOR PRACTICE 20.5** A researcher claims that an ancient scroll originated from Greek scholars in about 500 B.C. Its carbon-14 decay rate is 89% of that found in living organisms. How old is the scroll, and could it be authentic?

---

## Chemistry IN YOUR DAY | Radiocarbon Dating and the Shroud of Turin

The Shroud of Turin—kept in the cathedral of Turin in Italy—is a linen cloth bearing a mysterious image. Many people have interpreted the image as that of a man who appears to have been crucified. The image becomes clearer if the shroud is photographed and viewed as a negative. Some claim that the shroud is the original burial cloth of Jesus, miraculously imprinted with his image. In 1988, the Roman Catholic Church chose three independent laboratories to perform radiocarbon dating on the shroud. The laboratories took samples of the cloth and measured the carbon-14 content. The three independent laboratories all arrived at similar results—the shroud was made from linen originating in about A.D. 1325. Although some have disputed the results (and continue to do so), and although no scientific test is 100% reliable, newspapers around the world announced that the Shroud did not date back to biblical times.

▶ The linen cloth known as the Shroud of Turin bears the image of a man believed by some to be Jesus.

---

## Uranium/Lead Dating

Radiocarbon dating can only measure the ages of objects that were once living and that are relatively young (<50,000 years). Other radiometric dating techniques can measure the ages of prehistoric objects that were never alive. The most dependable technique relies on the ratio of uranium-238 to lead-206 within igneous rocks (rocks of volcanic origin). This technique measures the time that has passed since the rock solidified (at which point the "radiometric clock" is reset).

Because U-238 decays into Pb-206 with a half-life of $4.5 \times 10^9$ years, the relative amounts of U-238 and Pb-206 in a uranium-containing rock reveal its age. For example, if a rock originally contained U-238 and currently contains equal amounts of U-238 and Pb-206, it is 4.5 billion years old, assuming that the rock did not contain any Pb-206 when it was formed. The latter assumption can be tested because the lead that results from the decay of uranium has a different isotopic composition than the lead that was deposited in rocks at the time of their formation. Example 20.6 shows how we can use the relative amounts of Pb-206 and U-238 to estimate a rock's age.

---

## Example 20.6 | Using Uranium/Lead Dating to Estimate the Age of a Rock

A meteor contains 0.556 g of Pb-206 to every 1.00 g of U-238. Assuming that the meteor did not contain any Pb-206 at the time of its formation, determine the age of the meteor. Uranium-238 decays to lead-206 with a half-life of 4.5 billion years.

| **SORT** You are given the current masses of Pb-206 and U-238 in a rock and asked to find its age. You are also given the half-life of U-238. | **GIVEN:** $m_{\text{U-238}} = 1.00 \text{ g}; m_{\text{Pb-206}} = 0.556 \text{ g};$ <br> $\quad\quad t_{1/2} = 4.5 \times 10^9 \text{ yr}$ <br><br> **FIND:** $t$ |

**STRATEGIZE** Use the integrated rate law (Equation 20.3) to solve this problem. To do so, first determine the value of the rate constant ($k$) from the half-life expression (Equation 20.1).

Before substituting into the integrated rate law, you need the ratio of the current amount of U-238 to the original amount ($N_t/N_0$). The current mass of uranium is simply 1.00 g. The initial mass includes the current mass (1.00 g) plus the mass that has decayed into lead-206, which you can determine from the current mass of Pb-206.

Use the value of the rate constant and the initial and current amounts of U-238 and the integrated rate law to find $t$.

**CONCEPTUAL PLAN**

$$t_{1/2} = \frac{0.693}{k}$$

$$\frac{1 \text{ mol Pb}}{206 \text{ g Pb}} \qquad \frac{1 \text{ mol U}}{1 \text{ mol Pb}} \qquad \frac{238 \text{ g U}}{1 \text{ mol U}}$$

$$k, N_t, N_0 \longrightarrow t$$

$$\ln \frac{N_t}{N_0} = -kt$$

**SOLVE** Follow your plan. Begin by finding the rate constant from the half-life.

**SOLUTION**

$$t_{1/2} = \frac{0.693}{k}$$

$$k = \frac{0.693}{t_{1/2}} = \frac{0.693}{4.5 \times 10^9 \text{ yr}}$$

$$= 1.5\underline{4} \times 10^{-10}/\text{yr}$$

Determine the mass in grams of U-238 required to form the given mass of Pb-206.

$$0.556 \text{ g Pb-206} \times \frac{1 \text{ mol Pb-206}}{206 \text{ g Pb-206}} \times \frac{1 \text{ mol U-238}}{1 \text{ mol Pb-206}} \times \frac{238 \text{ g U-238}}{1 \text{ mol U-238}}$$

$$= 0.64\underline{2}4 \text{ g U-238}$$

Substitute the rate constant and the initial and current masses of U-238 into the integrated rate law and solve for $t$. (The initial mass of U-238 is the sum of the current mass and the mass required to form the given mass of Pb-206.)

$$\ln \frac{N_t}{N_0} = -kt$$

$$t = -\frac{\ln \dfrac{N_t}{N_0}}{k} = -\frac{\ln \dfrac{1.00 \text{ g}}{1.00 \text{ g} + 0.64\underline{2}4 \text{ g}}}{1.5\underline{4} \times 10^{-10}/\text{yr}}$$

$$= 3.2 \times 10^9 \text{ yr}$$

**CHECK** The units of the answer (yr) are correct. The magnitude of the answer is about 3.2 billion years, which is less than one half-life. This value is reasonable given that less than half of the uranium in the meteor has decayed into lead.

**FOR PRACTICE 20.6** A rock contains a Pb-206 to U-238 mass ratio of 0.145 : 1.00. Assuming that the rock did not contain any Pb-206 at the time of its formation, determine its age.

## The Age of Earth

The uranium/lead radiometric dating technique as well as other radiometric dating techniques (such as the decay of potassium-40 to argon-40) have been widely used to measure the ages of rocks on Earth and have produced highly consistent results. Rocks with ages greater than 3.5 billion years have been found on every continent. The oldest rocks have an age of approximately 4.0 billion years, establishing a lower limit for Earth's age (Earth is at least as old as its oldest rocks). The ages of about 70 meteorites that have struck Earth have also been extensively studied and have been found to be about 4.5 billion years old. Since the meteorites were formed at the same time as our solar system (which includes Earth), the best estimate for Earth's age is therefore about 4.5 billion years. That age is consistent with the estimated age of our universe—about 13.7 billion years.

The age of the universe is estimated from its expansion rate, which can be measured by examining changes in the wavelength of light from distant galaxies.

## 20.7 The Discovery of Fission: The Atomic Bomb and Nuclear Power

The element with atomic number 100 is named fermium in honor of Enrico Fermi.

In the mid-1930s Enrico Fermi (1901–1954), an Italian physicist, attempted to synthesize a new element by bombarding uranium—the heaviest known element at that time—with neutrons. Fermi speculated that if a neutron could be incorporated into the nucleus of a uranium atom, the nucleus might undergo beta decay, converting a neutron into a proton. If that happened, a new element, with atomic number 93, would be synthesized for the first time. The nuclear equation for the process is shown here:

$$\underset{\text{Neutron}}{{}^{238}_{92}\text{U} + {}^{1}_{0}\text{n}} \longrightarrow {}^{239}_{92}\text{U} \longrightarrow \underset{\text{Newly synthesized element}}{{}^{239}_{93}\text{X} + {}^{0}_{-1}\text{e}}$$

Fermi performed the experiment and detected the emission of beta particles. However, his results were inconclusive. Had he synthesized a new element? Fermi never chemically examined the products to determine their composition and therefore could not say with certainty that he had.

The element with atomic number 109 is named meitnerium in honor of Lise Meitner.

Three researchers in Germany—Lise Meitner (1878–1968), Fritz Strassmann (1902–1980), and Otto Hahn (1879–1968)—repeated Fermi's experiments and then performed careful chemical analysis of the products. What they found in the products—several elements *lighter* than uranium—would change the world forever. On January 6, 1939, Meitner, Strassmann, and Hahn reported that the neutron bombardment of uranium resulted in **nuclear fission**—the splitting of the uranium atom. The nucleus of the neutron-bombarded uranium atom was split into barium, krypton, and other smaller products. They also determined that the process emits enormous amounts of energy. A nuclear equation for a fission reaction, showing how uranium breaks apart into the daughter nuclides, is shown here:

$$^{235}_{92}\text{U} + {}^{1}_{0}\text{n} \longrightarrow {}^{140}_{56}\text{Ba} + {}^{93}_{36}\text{Kr} + 3\,{}^{1}_{0}\text{n} + \text{energy}$$

▲ Lise Meitner in Otto Hahn's Berlin laboratory. Together with Hahn and Fritz Strassmann, Meitner determined that U-235 could undergo nuclear fission.

Notice that the initial uranium atom is the U-235 isotope, which constitutes less than 1% of all naturally occurring uranium. U-238, the most abundant uranium isotope, does not undergo fission. Notice also that the process produces three neutrons, which have the potential to initiate fission in three other U-235 atoms.

Scientists quickly realized that a sample rich in U-235 could undergo a **chain reaction** in which neutrons produced by the fission of one uranium nucleus would induce fission in other uranium nuclei (Figure 20.10 ▶). This self-amplifying reaction is capable of producing an enormous amount of energy. This is the energy that is harnessed in an atomic bomb. However, to make a bomb, a **critical mass** of U-235—enough U-235 to produce a self-sustaining reaction—is necessary.

### The Manhattan Project

Fearing that Nazi Germany would develop a bomb based on nuclear fission, several U.S. scientists persuaded Albert Einstein, the most famous scientist of the time, to write a letter to President Franklin Roosevelt warning of this possibility. Einstein wrote, ". . . and it is conceivable—though much less certain—that extremely powerful bombs of a new type may thus be constructed. A single bomb of this type, carried by boat and exploded in a port, might very well destroy the whole port together with some of the surrounding territory."

## Fission Chain Reaction

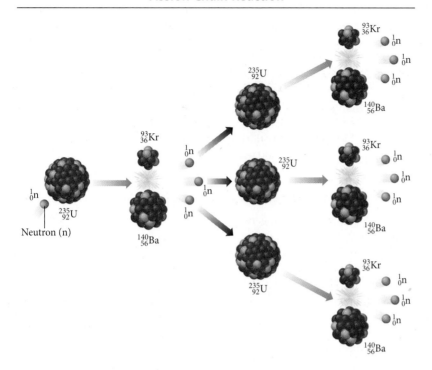

◀ **FIGURE 20.10 A Self-Amplifying Chain Reaction** The fission of one U-235 nucleus emits neutrons that can then initiate fission in other U-235 nuclei, resulting in a chain reaction that releases enormous amounts of energy.

◀ Einstein's letter (part of which is shown here) encouraged Franklin Roosevelt to fund research for a fission bomb.

Albert Einstein
Old Grove Rd.
Nassau Point
Peconic, Long Island

August 2nd, 1939

F.D. Roosevelt,
President of the United States,
White House
Washington, D.C.

Sir:

Some recent work by E.Fermi and L. Szilard, which has been communicated to me in manuscript, leads me to expect that the element uranium may be turned into a new and important source of energy in the immediate future. Certain aspects of the situation which has arisen seem to call for watchfulness and, if necessary, quick action on the part of the Administration. I believe therefore that it is my duty to bring to your attention the following facts and recommendations:

In the course of the last four months it has been made probable - through the work of Joliot in France as well as Fermi and Szilard in America - that it may become possible to set up a nuclear chain reaction in a large mass of uranium,by which vast amounts of power and large quantities of new radium-like elements would be generated. Now it appears almost certain that this could be achieved in the immediate future.

This new phenomenon would also lead to the construction of bombs, and it is conceivable - though much less certain - that extremely powerful bombs of a new type may thus be constructed. A single bomb of this type, carried by boat and exploded in a port, might very well destroy the whole port together with some of the surrounding territory. However, such bombs might very well prove to be too heavy for transportation by air.

Einstein's letter helped persuade Roosevelt, and in 1941 he assembled the resources to begin the costliest scientific project ever attempted. The top-secret endeavor was called the *Manhattan Project* and its primary goal was to build an atomic bomb before the Germans did. Physicist J. R. Oppenheimer (1904–1967) led the project at a high-security research facility in Los Alamos, New Mexico. Four years later, on July 16, 1945, the United States detonated the world's first nuclear weapon at a test site in New Mexico. The first atomic bomb exploded with a force equivalent to 18,000 tons of dynamite. Ironically, the Germans—who had *not* made a successful nuclear bomb—had already been defeated by this time. Instead, the United States used the atomic bomb on Japan. One bomb was dropped on Hiroshima and a second bomb was dropped on Nagasaki. Together, the bombs killed approximately 200,000 people and led to Japan's surrender.

▲ On July 16, 1945, in the New Mexico desert, the world's first atomic bomb was detonated. It had the power of 18,000 tons of dynamite.

## Nuclear Power: Using Fission to Generate Electricity

Nuclear reactions, such as fission, generate enormous amounts of energy. In a nuclear bomb, the energy is released all at once. The energy can also be released more slowly and used for peaceful purposes such as electricity generation. In the United States, nuclear fission generates about 20% of electricity. In some other countries, nuclear fission generates as much as 70% of electricity. To get an idea of the amount of energy released during fission, imagine a hypothetical nuclear-powered car. Suppose the fuel for such a car was a uranium cylinder about the size of a pencil. How often would you have to refuel the car? The energy content of the uranium cylinder is equivalent to about 1000 twenty-gallon tanks of gasoline. If you refuel your gasoline-powered car once a week, your nuclear-powered car could go 1000 weeks—almost 20 years—before refueling.

Similarly, a nuclear-powered electricity generation plant can produce a lot of electricity from a small amount of fuel. Such plants exploit the heat created by fission, using it to boil water and make steam, which then turns the turbine on a generator to produce electricity (Figure 20.11 ▼). The fission reaction occurs in the nuclear core of

## Nuclear Reactor

▲ **FIGURE 20.11 A Nuclear Reactor** The fission of U-235 in the core of a nuclear power plant generates heat that creates steam and turns a turbine on an electrical generator. Control rods are raised or lowered to control the fission reaction. (Note that the water carrying heat away from the reactor core is contained within its own pipes and does not come into direct contact with the steam that drives the turbines.)

the power plant. The core consists of uranium fuel rods—enriched to about 3.5% U-235—interspersed between retractable neutron-absorbing control rods. When the control rods are retracted from the fuel rod assembly, the chain reaction can occur. When the control rods are fully inserted into the fuel assembly, however, they absorb the neutrons that would otherwise induce fission, shutting down the chain reaction. By retracting or inserting the control rods, the operator increases or decreases the rate of fission. In this way, the fission reaction is controlled to produce the right amount of heat needed for electricity generation. In case of a power failure, the control rods automatically drop into the fuel rod assembly, shutting down the fission reaction.

A typical nuclear power plant generates enough electricity for a city of about one million people and uses about 50 kg of fuel per day. In contrast, a coal-burning power plant uses about 2,000,000 kg of fuel to generate the same amount of electricity. Furthermore, a nuclear power plant generates no air pollution and no greenhouse gases. A coal-burning power plant emits pollutants such as carbon monoxide, nitrogen oxides, and sulfur oxides. Coal-burning power plants also emit carbon dioxide, a greenhouse gas.

## Problems with Nuclear Power

Nuclear power generation, however, is not without problems. Foremost among them is the danger of nuclear accidents. In spite of safety precautions, the fission reaction occurring in a nuclear power plant can overheat. The most famous examples of this occurred in Chernobyl, in the former Soviet Union, on April 26, 1986, and at the Fukushima Daiichi Nuclear Power Plant in Japan in March of 2011.

In the Chernobyl incident, operators of the plant were performing an experiment designed to reduce maintenance costs. In order to perform the experiment, they had to disable many of the safety features of the reactor core. The experiment failed with disastrous results. The nuclear core, composed partly of graphite, overheated and began to burn. The accident caused 31 immediate deaths and produced a fire that scattered radioactive debris into the atmosphere, making much of the surrounding land (within about a 32-kilometer radius) uninhabitable. As bad as the accident was, however, it was a not a nuclear detonation. A nuclear power plant *cannot* become a nuclear bomb. The uranium fuel used in electricity generation is not sufficiently enriched in U-235 to produce a nuclear detonation. U.S. nuclear power plants have additional safety features designed to prevent similar accidents. For example, U.S. nuclear power plants have large containment structures to contain radioactive debris in the event of an accident.

In the 2011 Japanese accident, a 9.0 magnitude earthquake triggered a tsunami that flooded the coastal plant and caused the cooling system pumps to fail. Three of the nuclear cores within the plant dramatically overheated and experienced a partial meltdown (in which the fuel becomes so hot that it melts). The accident was intensified by the loss of water in the fuel storage ponds (pools of water normally used to keep spent fuel cool), which caused the fuel stored in the ponds to also overheat. The release of radiation into the environment, however, while significant, was much lower in Japan than at Chernobyl. In fact, no radioactivity-related deaths were reported at the Fukushima plant or the surrounding area. The cleanup of the site, however, will continue for many years.

A second problem associated with nuclear power is waste disposal. Although the amount of nuclear fuel used in electricity generation is small compared to other fuels, some of the products of the reaction are radioactive and have long half-lives. What do we do with this waste? Currently, in the United States, nuclear waste is stored on site at the nuclear power plants. A single permanent disposal site was being developed in Yucca Mountain, Nevada, to store U.S. waste. However, in 2010, the Obama administration halted the development of this project and formed the Blue Ribbon Commission on America's Nuclear Future to explore alternatives. The committee has made several important recommendations. These include the development of both a temporary above-ground storage facility and a permanent underground facility. However, the committee has not made any decisions regarding Yucca Mountain as a potential site for the storage facility.

▲ In 1986, the reactor core at Chernobyl (in what is now Ukraine) overheated, exploded, and destroyed part of the containment structure. The release of radioactive nuclides into the environment forced the government to relocate over 335,000 people.

Reactor cores in the United States are not made of graphite and cannot burn in the way that the Chernobyl core did.

## 20.8 Converting Mass to Energy: Mass Defect and Nuclear Binding Energy

Nuclear fission produces large amounts of energy. But where does the energy come from? We can answer this question by carefully examining the masses of the reactants and products in the fission equation from Section 20.7:

$$^{235}_{92}U + ^{1}_{0}n \longrightarrow ^{140}_{56}Ba + ^{93}_{36}Kr + 3^{1}_{0}n$$

| Mass Reactants | | Mass Products | |
|---|---|---|---|
| $^{235}_{92}U$ | 235.04392 amu | $^{140}_{56}Ba$ | 139.910581 amu |
| $^{1}_{0}n$ | 1.00866 amu | $^{93}_{36}Kr$ | 92.931130 amu |
| | | $3^{1}_{0}n$ | 3(1.00866) amu |
| **Total** | **236.05258 amu** | | **235.86769 amu** |

In a chemical reaction, there are also mass changes associated with the emission or absorption of energy. Because the energy involved in chemical reactions is so much smaller than that of nuclear reactions, however, these mass changes are completely negligible.

Notice that the products of the nuclear reaction have *less mass* than the reactants. The missing mass is converted to energy. In Chapter 1, we learned that matter is conserved in chemical reactions. In nuclear reactions matter can be converted to energy. The relationship between the amount of matter that is lost and the amount of energy formed is given by Einstein's famous equation relating the two quantities:

$$E = mc^2$$

where $E$ is the energy produced, $m$ is the mass lost, and $c$ is the speed of light.

For example, in the fission reaction just shown, we calculate the quantity of energy produced as follows:

$$\text{Mass lost } (m) = 236.05258 \text{ amu} - 235.86769 \text{ amu}$$

$$= 0.18489 \text{ amu} \times \frac{1.66054 \times 10^{-27} \text{ kg}}{1 \text{ amu}}$$

$$= 3.0702 \times 10^{-28} \text{ kg}$$

$$\text{Energy produced } (E) = mc^2$$

$$= 3.0702 \times 10^{-28} \text{ kg } (2.9979 \times 10^8 \text{ m/s})^2$$

$$= 2.7593 \times 10^{-11} \text{ J}$$

The result $(2.7593 \times 10^{-11} \text{ J})$ is the energy produced when one nucleus of U-235 undergoes fission. This may not seem like much energy, but it is only the energy produced by the fission of a *single* nucleus. We can also calculate the energy produced *per mole* of U-235 to compare it to a chemical reaction:

$$2.7593 \times 10^{-11} \frac{\text{J}}{\text{U-235 atom}} \times \frac{6.0221 \times 10^{23} \text{ U-235 atoms}}{1 \text{ mol U-235}}$$

$$= 1.6617 \times 10^{13} \text{ J/mol U-235}$$

The energy produced by the fission of 1 mol of U-235 is about 17 billion kJ. In contrast, a highly exothermic chemical reaction produces 1000 kJ per mole of reactant. Fission produces over a million times more energy per mole than chemical processes.

### Mass Defect and Nuclear Binding Energy

We can examine the formation of a stable nucleus from its component particles as a nuclear reaction in which mass is converted to energy. For example, consider the formation of helium-4 from its components:

$$2^{1}_{1}H + 2^{1}_{0}n \longrightarrow ^{4}_{2}He$$

| Mass Reactants | | Mass Products | |
|---|---|---|---|
| $2^{1}_{1}H$ | 2(1.00783) amu | $^{4}_{2}He$ | 4.00260 amu |
| $2^{1}_{0}n$ | 2(1.00866) amu | | |
| **Total** | **4.03298 amu** | | **4.00260 amu** |

The electrons are contained on the left side in the two $^{1}_{1}H$, and on the right side in $^{4}_{2}He$. If you write the equation using only two protons on the left ($^{1}_{1}p$), you must also add two electrons to the left.

A helium-4 atom has less mass than the sum of the masses of its separate components. This difference in mass, known as the **mass defect**, exists in all stable nuclei. The energy corresponding to the mass defect—obtained by substituting the mass defect into the equation $E = mc^2$—is the **nuclear binding energy**, the amount of energy required to break apart the nucleus into its component nucleons.

Although chemists often report energies in joules, nuclear physicists often use the electron volt (eV) or megaelectron volt (MeV): $1 \text{ MeV} = 1.602 \times 10^{-13} \text{ J}$. Unlike energy in joules, which is typically reported per mole, energy in electron volts is reported per nucleus. A particularly useful conversion for calculating and reporting nuclear binding energies is the relationship between amu (mass units) and MeV (energy units):

> An electron volt is defined as the kinetic energy of an electron that has been accelerated through a potential difference of 1 V.

$$1 \text{ amu} = 931.5 \text{ MeV}$$

A mass defect of 1 amu, when substituted into the equation $E = mc^2$, gives an energy of 931.5 MeV. Using this conversion factor, we can readily calculate the binding energy of the helium nucleus:

$$\text{Mass defect} = 4.03298 \text{ amu} - 4.00260 \text{ amu}$$
$$= 0.03038 \text{ amu}$$

$$\text{Nuclear binding energy} = 0.03038 \text{ amu} \times \frac{931.5 \text{ MeV}}{1 \text{ amu}}$$
$$= 28.30 \text{ MeV}$$

The binding energy of the helium nucleus is 28.30 MeV. In order to compare the binding energy of one nucleus to that of another, we calculate the *binding energy per nucleon*, which is the nuclear binding energy of a nuclide divided by the number of nucleons in the nuclide. For helium-4, we calculate the binding energy per nucleon as follows:

$$\text{Binding energy per nucleon} = \frac{28.30 \text{ MeV}}{4 \text{ nucleons}}$$
$$= 7.075 \text{ MeV per nucleon}$$

We can calculate the binding energy per nucleon for other nuclides in the same way. For example, the nuclear binding energy of carbon-12 is 7.680 MeV per nucleon. Since the binding energy per nucleon of carbon-12 is greater than that of helium-4, we conclude the carbon-12 nuclide is more *stable* (it has lower potential energy).

---

### Example 20.7   Mass Defect and Nuclear Binding Energy

Calculate the mass defect and nuclear binding energy per nucleon (in MeV) for C-16, a radioactive isotope of carbon with a mass of 16.014701 amu.

**SOLUTION**

| | |
|---|---|
| Calculate the mass defect as the difference between the mass of one C-16 atom and the sum of the masses of six hydrogen atoms and 10 neutrons. | $\text{Mass defect} = 6(\text{mass } {}^{1}_{1}\text{H}) + 10(\text{mass } {}^{1}_{0}\text{n}) - \text{mass } {}^{16}_{6}\text{C}$ <br> $= 6(1.00783 \text{ amu}) + 10(1.00866 \text{ amu}) - 16.014701 \text{ amu}$ <br> $= 0.118879 \text{ amu}$ |
| Calculate the nuclear binding energy by converting the mass defect (in amu) into MeV. (Use 1 amu = 931.5 MeV.) | $0.118879 \text{ amu} \times \dfrac{931.5 \text{ MeV}}{\text{amu}} = 110.74 \text{ MeV}$ |
| Determine the nuclear binding energy per nucleon by dividing by the number of nucleons in the nucleus. | $\text{Nuclear binding energy per nucleon} = \dfrac{110.74 \text{ MeV}}{16 \text{ nucleons}}$ <br> $= 6.921 \text{ MeV/nucleon}$ |

**FOR PRACTICE 20.7**   Calculate the mass defect and nuclear binding energy per nucleon (in MeV) for U-238, which has a mass of 238.050784 amu.

▶ **FIGURE 20.12** **Nuclear Binding Energy per Nucleon** The nuclear binding energy per nucleon (a measure of the stability of a nucleus) reaches a maximum at Fe-56. Energy can be obtained either by breaking a heavy nucleus up into lighter ones (fission) or by combining lighter nuclei into heavier ones (fusion).

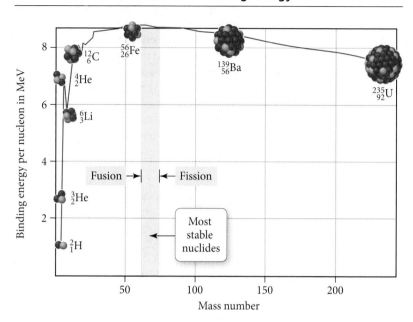

The Curve of Binding Energy

## The Nuclear Binding Energy Curve

Figure 20.12▲ shows the binding energy per nucleon plotted as a function of mass number ($A$). The binding energy per nucleon is relatively low for small mass numbers and increases until about $A = 60$, where it reaches a maximum. Nuclides with mass numbers of about 60, therefore, are among the most stable. Beyond $A = 60$, the binding energy per nucleon decreases again. Figure 20.12 illustrates why nuclear fission is a highly exothermic process. When a heavy nucleus, such as U-235, breaks up into smaller nuclei, such as Ba-140 and Kr-93, the binding energy per nucleon increases. This is analogous to a chemical reaction in which weak bonds break and strong bonds form. In both cases, the process is exothermic. Figure 20.12 also reveals that the *combining* of two lighter nuclei (below $A = 60$) to form a heavier nucleus is exothermic as well. This process is called *nuclear fusion*, which we discuss in the next section of this chapter.

## 20.9 Nuclear Fusion: The Power of the Sun

Nuclear fission is the *splitting* of a heavy nucleus to form two or more lighter ones. **Nuclear fusion**, by contrast, is the *combination* of two light nuclei to form a heavier one. Both fusion and fission emit large amounts of energy because, as we have just seen, they both form daughter nuclides with greater binding energies per nucleon than the parent nuclides. Nuclear fusion is the energy source of stars, including our sun. In stars, hydrogen atoms fuse together to form helium atoms, emitting energy in the process.

Nuclear fusion is also the basis of hydrogen bombs. A modern hydrogen bomb has up to 1000 times the explosive force of the first atomic bombs. These bombs employ the fusion reaction shown here:

$$\,^2_1\text{H} + \,^3_1\text{H} \longrightarrow \,^4_2\text{He} + \,^1_0\text{n}$$

In this reaction, deuterium (the isotope of hydrogen with one neutron) and tritium (the isotope of hydrogen with two neutrons) combine to form helium-4 and a neutron (Figure 20.13◀). Because fusion reactions require two positively charged nuclei (which repel each other) to fuse together, extremely high temperatures are required. In a hydrogen bomb, a small fission bomb is detonated first, creating temperatures and pressures high enough for fusion to proceed.

Nuclear fusion has been intensely investigated as a way to produce electricity. Because of the higher energy density—fusion provides about ten times more energy per gram of fuel than does fission—and because the products of the reaction are

**Deuterium-Tritium Fusion Reaction**

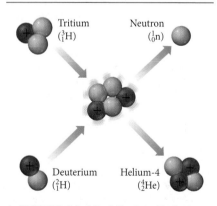

▲ **FIGURE 20.13** **A Nuclear Fusion Reaction** In this reaction, two heavy isotopes of hydrogen, deuterium (hydrogen-2) and tritium (hydrogen-3), fuse to form helium-4 and a neutron.

less problematic than those of fission, fusion holds promise as a future energy source. However, despite concerted efforts, the generation of electricity by fusion remains elusive. One of the main problems is the high temperature required for fusion to occur—no material can withstand those temperatures. Using powerful magnetic fields or laser beams, scientists have succeeded in compressing and heating nuclei to the point where fusion has been initiated and even sustained for brief periods of time (Figure 20.14▶). To date, however, the amount of energy generated by fusion reactions has been less than the amount required to get it to occur. After years of spending billions of dollars on fusion research, the U.S. Congress has reduced funding for these projects. Whether fusion will ever be a viable energy source remains uncertain.

## 20.10 Nuclear Transmutation and Transuranium Elements

One of the goals of the early chemists of the Middle Ages, who were known as *alchemists*, was the transformation of ordinary metals into gold. Many alchemists hoped to turn low-cost metals, such as lead or tin, into precious metals, and in this way become wealthy. These alchemists were never successful because their attempts were merely chemical—they mixed different metals together or tried to get them to react with other substances in order to turn them into gold. In a chemical reaction, an element retains its identity, so a less valuable metal—such as lead—always remains lead, even when it forms a compound with another element.

*Nuclear* reactions, in contrast to chemical reactions, result in the transformation of one element into another, a process known as **transmutation**. We have already seen how this occurs in radioactive decay, in fission, and in fusion. In addition, other nuclear reactions that transmute elements are possible. For example, in 1919 Ernest Rutherford bombarded nitrogen-17 with alpha particles to form oxygen:

$$^{14}_{7}\text{N} + ^{4}_{2}\text{He} \longrightarrow ^{17}_{8}\text{O} + ^{1}_{1}\text{H}$$

Irène Joliot-Curie (daughter of Marie Curie) and her husband Frédéric bombarded aluminum-27 with alpha particles to form phosphorus:

$$^{27}_{13}\text{Al} + ^{4}_{2}\text{He} \longrightarrow ^{30}_{15}\text{P} + ^{1}_{0}\text{n}$$

In the 1930s, scientists began building devices that accelerate particles to high velocities, opening the door to even more possibilities. These devices are generally of two types, the **linear accelerator** and the **cyclotron**.

In a single-stage linear accelerator, a charged particle such as a proton is accelerated in an evacuated tube. The accelerating force is provided by a potential difference (or voltage) between the ends of the tube. In multistage linear accelerators, such as the Stanford Linear Accelerator (SLAC) at Stanford University, a series of tubes of increasing length are connected to a source of alternating voltage, as shown in Figure 20.15▼. The voltage alternates in such a way that, as a positively charged particle leaves a particular tube,

**Tokamak Fusion Reactor**

Coils generate magnetic fields to contain fusing nuclei

Plasma

▲ **FIGURE 20.14 Tokamak Fusion Reactor** A tokamak uses powerful magnetic fields to confine nuclear fuel at the enormous temperatures needed for fusion. The high temperatures produce a *plasma*, a state of matter in which some fraction of the atoms are ionized.

▲ The Joliot-Curies won the 1935 Nobel Prize in chemistry for their work on nuclear transmutation.

◀ **FIGURE 20.15 The Linear Accelerator** In a multistage linear accelerator, the charge on successive tubes is rapidly alternated in such a way that as a positively charged particle leaves a particular tube, that tube becomes positively charged, repelling the particle toward the next tube. At the same time, the tube that the particle is now approaching becomes negatively charged, pulling the particle toward it. This process repeats through a number of tubes until the particle has been accelerated to a high velocity.

Alternating voltage source

Vacuum pump

+/−    +/−

Proton source

1  2  3  4  5  6

Target

Evacuated tube

▲ The Stanford Linear Accelerator (top) is located at Stanford University in California. The Fermi National Accelerator Laboratory complex in Batavia, Illinois (bottom), includes two cyclotrons in a figure-8 configuration.

▲ **FIGURE 20.16  The Cyclotron**  In a cyclotron, two semicircular D-shaped structures are subjected to an alternating voltage. A charged particle, starting from a point between the two, is accelerated back and forth between them, while additional magnets cause the particle to move in a spiral path.

that tube becomes positively charged, repelling the particle to the next tube. At the same time, the tube that the particle is now approaching becomes negatively charged, pulling the particle toward it. This continues throughout the linear accelerator, allowing the particle to be accelerated to velocities up to 90% of the speed of light. When particles of this speed collide with a target, they produce a shower of subatomic particles that can be studied. For example, researchers using the Stanford Linear Accelerator were awarded the 1990 Nobel Prize in physics for discovering evidence that protons and neutrons were composed of still smaller subatomic particles called *quarks*.

In a cyclotron, a similarly alternating voltage is used to accelerate a charged particle, only this time the alternating voltage is applied between the two semicircular halves of the cyclotron (Figure 20.16▲). A charged particle originally in the middle of the two semicircles is accelerated back and forth between them. Additional magnets cause the particle to move in a spiral path. As the charged particle spirals out from the center, it gains speed and eventually exits the cyclotron aimed at the target.

With linear accelerators or cyclotrons, all sorts of nuclear transmutations can be achieved. In this way, scientists have made nuclides that don't normally exist in nature. For example, uranium-238 can be made to collide with carbon-12 to form an element with atomic number 98:

$$\ce{^{238}_{92}U} + \ce{^{12}_{6}C} \longrightarrow \ce{^{244}_{98}Cf} + 6\,\ce{^{1}_{0}n}$$

This element was named californium (Cf) because it was first produced (by a slightly different nuclear reaction) at the University of California at Berkeley. Many other nuclides with atomic numbers larger than that of uranium have been synthesized since the 1940s. These synthetic elements—called transuranium elements—have been added to the periodic table.

Most synthetic elements are unstable and have very short half-lives. Some exist for only fractions of a second after they are made.

CONCEPTUAL
**CONNECTION 20.4**

**Nuclear Transformations**  Californium-252 is bombarded with a boron-10 nucleus to produce another nuclide and six neutrons. Which nuclide forms?

## 20.11  The Effects of Radiation on Life

As we discussed in Section 20.2, the energy associated with radioactivity can ionize molecules. When radiation ionizes important molecules in living cells, problems can develop. The ingestion of radioactive materials—especially alpha and beta emitters—is particularly dangerous because the radioactivity once inside the body can do additional damage. The effects of radiation can be divided into three different types, which we discuss in this section: acute radiation damage, increased cancer risk, and genetic effects.

## Acute Radiation Damage

Acute radiation damage results from exposure to large amounts of radiation in a short period of time. The main sources of this kind of exposure are nuclear bombs and exposed nuclear reactor cores. These high levels of radiation kill large numbers of cells. Rapidly dividing cells, such as those in the immune system and the intestinal lining, are most susceptible to radiation. Consequently, people exposed to high levels of radiation have weakened immune systems and a lowered ability to absorb nutrients from food. In milder cases, recovery is possible with time. In more extreme cases, death results, often from infection.

## Increased Cancer Risk

Lower doses of radiation over extended periods of time can increase cancer risk. Radiation increases cancer risk because it can damage DNA, the molecules in cells that carry instructions for cell growth and replication. When the DNA within a cell is damaged, the cell normally dies. Occasionally, however, changes in DNA cause cells to grow abnormally and to become cancerous. These cancerous cells grow into tumors that can spread and, in some cases, cause death. Cancer risk increases with increasing radiation exposure. However, cancer is so prevalent and has so many convoluted causes that determining an exact threshold for increased cancer risk from radiation exposure is difficult.

We explain DNA in more detail in Chapter 22.

## Genetic Defects

Another possible result of radiation exposure is genetic defects in future generations. If radiation damages the DNA of reproductive cells—such as eggs or sperm—then the offspring that develop from those cells may have genetic abnormalities. Genetic defects of this type have been observed in laboratory animals exposed to high levels of radiation. However, such genetic defects—with a clear causal connection to radiation exposure— have yet to be verified in humans, even in studies of Hiroshima survivors.

## Measuring Radiation Exposure and Dose

We can measure radiation exposure in a number of different ways. One method is to measure **exposure**, the number of decay events to which a person is exposed. The unit used for this type of exposure measurement is the *curie* (Ci), defined as $3.7 \times 10^{10}$ decay events per second. A person exposed to a curie of radiation from an alpha emitter is bombarded by $3.7 \times 10^{10}$ alpha particles per second. However, recall that different kinds of radiation produce different effects. For example, we know that alpha radiation has a much greater ionizing power than beta radiation. Consequently, a certain number of alpha decays occurring within a person's body (due to the ingestion of an alpha emitter) do more damage than the same number of beta decays. If the alpha emitter and beta emitter are external to the body, however, the radiation from the alpha emitter is largely stopped by clothing or the skin (due to the low penetrating power of alpha radiation), whereas the radiation from the beta emitter penetrates the skin and causes more damage. Consequently, the curie is not an effective measure of how much biological tissue damage the radiation actually does.

A better way to assess radiation exposure is to measure radiation **dose**, the amount of energy actually absorbed by body tissue. The units used for this type of exposure measurement are the *gray* (Gy), which corresponds to 1 J of energy absorbed per kilogram of body tissue, and the *rad* (for *r*adiation *a*bsorbed *d*ose), which corresponds to 0.01 Gy:

$$1 \text{ gray (Gy)} = 1 \text{ J/kg body tissue}$$

$$1 \text{ rad} = 0.01 \text{ J/kg body tissue}$$

Although these units measure the actual energy absorbed by bodily tissues, they still do not account for the amount of damage to biological molecules caused by that energy absorption, which differs from one type of radiation to another and from one type of biological tissue to another. For example, when a gamma ray passes through biological tissue, the energy absorbed is spread out over the long distance that the radiation travels through the body, resulting in a low ionization density within the tissue. When an alpha particle passes through biological tissue, in contrast, the energy is absorbed over a much shorter distance, resulting in a much higher ionization density. The higher ionization density results in greater damage, even though the amount of energy absorbed by the tissue might be the same.

Consequently, a correction factor, called the **biological effectiveness factor**, or **RBE** (for *Relative Biological Effectiveness*), is usually multiplied by the dose in rads to obtain the dose in a unit called the **rem** for *roentgen equivalent man*:

Dose in rads × biological effectiveness factor = dose in rems

The biological effectiveness factor for alpha radiation, for example, is much higher than that for gamma radiation.

On average, each adult is exposed to approximately 310 mrem of radiation per year from natural sources shown in Table 20.4. The majority of this exposure comes from radon, one of the products in the uranium decay series. As we can see from Table 20.4, however, some medical procedures also involve exposure levels similar to those received from natural sources. The increased use of computed tomography (CT) scans over the last decade—which have associated exposures of 200–1600 mrem—has raised some concerns about the overuse of that technology in medicine.

It takes much more than the average natural radiation dose or the dose from a medical diagnostic procedure to produce significant health effects in humans. The first measurable effect, a decreased white blood cell count, occurs at instantaneous exposures of approximately 20 rem (Table 20.5). Exposures of 100 rem produce a definite increase in cancer risk, and exposures of over 500 rem often result in death.

A *roentgen* is defined as the amount of radiation that produces $2.58 \times 10^{-1}$ C of charge per kg of air.

The SI unit that corresponds to the rem is the sievert (Sv). However, the rem is still commonly used in the United States. The conversion factor is 1 rem = 0.01 Sv.

| Source | Dose |
|---|---|
| **TABLE 20.4 Radiation Dose by Source for Persons Living in the United States** | |
| **Natural Radiation** | |
| A 5-hour jet airplane ride | 2.5 mrem/trip (0.5 mrem/hr at 39,000 feet) (whole body dose) |
| Cosmic radiation from outer space | 27 mrem/yr (whole body dose) |
| Terrestrial radiation | 28 mrem/yr (whole body dose) |
| Natural radionuclides in the body | 35 mrem/yr (whole body dose) |
| Radon gas | 200 mrem/yr (lung dose) |
| **Diagnostic Medical Procedures** | |
| Chest X-ray | 8 mrem (whole body dose) |
| Dental X-rays (panoramic) | 30 mrem (skin dose) |
| Dental X-rays (two bitewings) | 80 mrem (skin dose) |
| Mammogram | 138 mrem per image |
| Barium enema (X-ray portion only) | 406 mrem (bone marrow dose) |
| Upper gastrointestinal tract test | 244 mrem (X-ray portion only) (bone marrow dose) |
| Thallium heart scan | 500 mrem (whole body dose) |
| Computed tomography (CT) head | 200 mrem (whole body dose) |
| Computed tomography (CT) abdomen and pelvis | 1000 mrem (whole body dose) |
| **Consumer Products** | |
| Building materials | 3.5 mrem/yr (whole body dose) |
| Luminous watches (H-3 and Pm-147) | 0.04–0.1 mrem/yr (whole body dose) |
| Tobacco products (to smokers of 30 cigarettes per day) | 16,000 mrem/yr (bronchial epithelial dose) |

*Source:* Department of Health and Human Services, National Institutes of Health.

| TABLE 20.5 Effects of Instantaneous Radiation Exposure | |
|---|---|
| **Approximate Dose (rem)** | **Probable Outcome** |
| 20–100 | Decreased white blood cell count; possible increase in cancer risk |
| 100–400 | Radiation sickness including vomiting and diarrhea; skin lesions; increase in cancer risk |
| 500 | Death (often within 2 months) |
| 1000 | Death (often within 2 weeks) |
| 2000 | Death (within hours) |

**CONCEPTUAL CONNECTION 20.5**

**Radiation Dose** Suppose a person ingests equal amounts of two nuclides, both of which are beta emitters (of roughly equal energy). Nuclide A has a half-life of 8.5 hours and Nuclide B has a half-life of 15.0 hours. Both nuclides are eliminated from the body within 24 hours of ingestion. Which of the two nuclides produces the greater radiation dose?

# 20.12 Radioactivity in Medicine and Other Applications

Radioactivity is often perceived as dangerous; however, it is also immensely useful to physicians in the diagnosis and treatment of disease and has numerous other valuable applications. The use of radioactivity in medicine can be broadly divided into *diagnostic techniques* (which diagnose disease) and *therapeutic techniques* (which treat disease).

## Diagnosis in Medicine

The use of radioactivity in diagnosis usually involves a **radiotracer**, a radioactive nuclide attached to a compound or introduced into a mixture in order to track the movement of the compound or mixture within the body. Tracers are useful in the diagnosis of disease because of two main factors: (1) the sensitivity with which radioactivity can be detected, and (2) the identical chemical behavior of a radioactive nucleus and its nonradioactive counterpart. For example, the thyroid gland naturally concentrates iodine. When a patient is given small amounts of iodine-131 (a radioactive isotope of iodine), the radioactive iodine accumulates in the thyroid, just as nonradioactive iodine does. However, the radioactive iodine emits radiation, which can be detected with great sensitivity and used to measure the rate of iodine uptake by the thyroid, and thus to image the gland.

Because different radiotracers are taken up preferentially by different organs or tissues, various radiotracers are used to monitor metabolic activity and image a variety of organs and structures, including the kidneys, heart, brain, gallbladder, bones, and arteries, as shown in Table 20.6. Radiotracers can also be employed to locate infections or cancers within the body. To locate an infection, antibodies are labeled (or tagged) with a radioactive nuclide, such as technetium-99m (where "m" means metastable), and administered to the patient. The tagged antibodies aggregate at the infected site, as described in Section 20.1. Cancerous tumors can be detected because they naturally concentrate phosphorus. When a patient is given phosphorus-32 (a radioactive isotope of phosphorus) or a phosphate compound incorporating another radioactive isotope such as Tc-99m, the tumors concentrate the radioactive substance and become sources of radioactivity that can be detected (Figure 20.17▲).

▲ **FIGURE 20.17 A Bone Scan** These images, front and rear views of the human body, were created by the gamma ray emissions of Tc-99m. Such scans are used to locate cancer that has metastasized (spread) to the bones from a primary tumor elsewhere.

| TABLE 20.6 Common Radiotracers | | | |
|---|---|---|---|
| **Nuclide** | **Type of Emission** | **Half-Life** | **Body Part Studied** |
| Technetium-99m | Gamma (primarily) | 6.01 hours | Various organs, bones |
| Iodine-131 | Beta | 8.0 days | Thyroid |
| Iron-59 | Beta | 44.5 days | Blood, spleen |
| Thallium-201 | Electron capture | 3.05 days | Heart |
| Fluorine-18 | Positron emission | 1.83 hours | PET studies of heart, brain |
| Phosphorus-32 | Beta | 14.3 days | Tumors in various organs |

▲ **FIGURE 20.18** **A PET Scan** The colored areas indicate regions of high metabolic activity in the brain of a schizophrenic patient experiencing hallucinations.

A specialized imaging technique known as **positron emission tomography (PET)** employs positron-emitting nuclides, such as fluorine-18, synthesized in cyclotrons. The fluorine-18 is attached to a metabolically active substance such as glucose and administered to the patient. As the glucose travels through the bloodstream and to the heart and brain, it carries the radioactive fluorine, which decays with a half-life of just under two hours. When a fluorine-18 nuclide decays, it emits a positron that immediately combines with any nearby electrons. Since a positron and an electron are antiparticles, they annihilate one other, producing two gamma rays that travel in exactly opposing directions. The gamma rays can be detected by an array of detectors that can locate the point of origin of the rays with great accuracy. The result is a set of highly detailed images that show both the rate of glucose metabolism and structural features of the imaged organ (Figure 20.18 ◄).

## Radiotherapy in Medicine

Because radiation kills cells and is particularly effective at killing rapidly dividing cells, it is often used as a therapy for cancer (cancer cells reproduce much faster than normal cells). Medical technicians focus high-energy photons on internal tumors to kill them. The photon beam is usually moved in a circular path around the tumor (Figure 20.19 ▼), maximizing the exposure of the tumor while minimizing the exposure of the surrounding healthy tissue. Nonetheless, cancer patients receiving such treatment usually develop the symptoms of radiation sickness, which include vomiting, skin burns, and hair loss.

▶ **FIGURE 20.19** **Radiotherapy for Cancer** This treatment involves exposing a malignant tumor to gamma rays generated by nuclides such as cobalt-60. The beam is moved in a circular pattern around the tumor to maximize exposure of the tumor to radiation while minimizing the exposure of healthy tissues.

You may wonder why radiation—which is known to cause cancer—is also used to treat cancer. The answer lies in risk analysis. A cancer patient is normally exposed to radiation doses of about 100 rem. Such a dose increases cancer risk by about 1%. However, if the patient has a 100% chance of dying from the cancer that he already has, such a risk becomes acceptable, especially since there is a significant chance of curing the cancer.

## Other Applications

Radioactivity is often used to kill microorganisms. For example, physicians use radiation to sterilize medical devices that are to be surgically implanted. The radiation kills bacteria that might otherwise lead to infection. Similarly, radiation is used to kill bacteria and parasites in foods. Like the pasteurization of milk, the irradiation of foods makes them safer to consume and gives them a longer shelf life (Figure 20.20▶). The irradiation of raw meat and poultry kills *E. coli* and *Salmonella*, bacteria that can lead to serious illness and even death when consumed. The irradiation of food does not, however, make the food itself radioactive, nor does it decrease the nutritional value of the food. In the United States, the irradiation of many different types of foods—including beef, poultry, potatoes, flour, and fruit—has been approved by the U.S. Food and Drug Administration (FDA) and the U.S. Department of Agriculture (USDA).

Radioactivity is also used to control harmful insect populations. For example, fruit flies can be raised in large numbers in captivity and sterilized with radiation. When these fruit flies are released, they mate with wild fruit flies but do not produce offspring. The efforts of the wild fruit flies, which might otherwise lead to reproduction, are wasted, and the next generation of flies is smaller than it would otherwise have been. Similar strategies have been employed to control the populations of disease-carrying mosquitoes.

▲ **FIGURE 20.20 Irradiation of Food** Irradiation kills microbes that cause food to decay, allowing for longer and safer storage. The food is not made radioactive, and its properties are unchanged in the process. These strawberries were picked at the same time, but those on the bottom were irradiated before storage.

# Self-Assessment Quiz

**Q1.** Which daughter nuclide forms when polonium-214 undergoes alpha decay?

a) $^{218}_{86}Rn$

b) $^{214}_{85}At$

c) $^{214}_{83}Bi$

d) $^{210}_{82}Pb$

**Q2.** Which nuclear equation accurately represents the beta decay of Xe-133?

a) $^{133}_{54}Xe \longrightarrow ^{133}_{55}Cs + ^{0}_{-1}e$

b) $^{133}_{54}Xe \longrightarrow ^{133}_{53}I + ^{0}_{+1}e$

c) $^{133}_{54}Xe + ^{0}_{-1}e \longrightarrow ^{133}_{53}I$

d) $^{133}_{54}Xe \longrightarrow ^{129}_{52}Cs + ^{4}_{2}He$

**Q3.** Which nuclide is most likely to undergo beta decay?

a) Si-22

b) Rb-91

c) Ar-35

d) Co-52

**Q4.** Which form of radioactive decay would you be most likely to detect if it were happening in the room next to the one you are currently in?

a) alpha

b) beta

c) gamma

d) positron emission

**Q5.** The chart shows the mass of a decaying nuclide versus time. What is the half-life of the decay?

a) 15 min

b) 25 min

c) 35 min

d) 70 min

**Q6.** Iron-59 is a beta emitter with a half-life of 44.5 days. If a sample initially contains 132 mg of iron-59, how much iron-59 is left in the sample after 265 days?

a) 0.00 mg     b) 2.13 mg

c) 33.2 mg     d) 66.0 mg

*Continued—*

**Q7.** An artifact has a carbon-14 decay rate of 8.55 disintegrations per minute per gram of carbon (8.55 dis/min · g C). Living organisms have carbon-14 decay rate of 15.3 dis/min · g C. How old is the artifact? (The half-life of carbon-14 is 5715 yr.)

a) $4.81 \times 10^3$ yr      b) $2.10 \times 10^3$ yr
c) $3.20 \times 10^3$ yr      d) $1.21 \times 10^{-4}$ yr

**Q8.** An igneous rock contains a Pb-206/U-238 mass ratio of 0.372. How old is the rock? (U-238 decays into Pb-206 with a half-life of $4.5 \times 10^9$ yr.)

a) $4.50 \times 10^9$ yr      b) $6.42 \times 10^9$ yr
c) $2.05 \times 10^9$ yr      d) $2.32 \times 10^9$ yr

**Q9.** Calculate the nuclear binding energy per nucleon for cobalt-59, the only stable isotope of cobalt. The mass of

cobalt-59 is 58.933198 amu. (The mass of $^1_1$H is 1.00783 amu, and the mass of a neutron is 1.00866 amu.)

a) 517.3 MeV
b) 8.768 MeV
c) 19.16 MeV
d) $1.011 \times 10^{-5}$ MeV

**Q10.** Which problem is not associated with nuclear power generation?

a) danger of overheated nuclear core
b) waste disposal
c) global warming
d) none of the above (All of the above are problems associated with nuclear power generation.)

**Answers: 1.** (d) **2.** (a) **3.** (b) **4.** (c) **5.** (c) **6.** (b) **7.** (a) **8.** (d) **9.** (b) **10.** (c)

# Chapter 20 IN REVIEW

## Key Terms

**Section 20.1**
radioactivity (940)
radioactive (940)

**Section 20.2**
phosphorescence (940)

**Section 20.3**
nuclide (941)
alpha ($\alpha$) decay (942)
alpha ($\alpha$) particle (942)
nuclear equation (942)
ionizing power (943)
penetrating power (943)
beta ($\beta$) decay (943)
beta ($\beta$) particle (943)
gamma ($\gamma$) ray emission (943)
gamma ($\gamma$) ray (943)
positron emission (944)
positron (944)
electron capture (944)

**Section 20.4**
strong force (946)
nucleons (946)
magic numbers (948)

**Section 20.5**
thermoluminescent dosimeter (948)
Geiger-Müller counter (948)
scintillation counter (948)

**Section 20.6**
radiometric dating (952)
radiocarbon dating (952)

**Section 20.7**
nuclear fission (956)
chain reaction (956)
critical mass (956)

**Section 20.8**
mass defect (961)
nuclear binding energy (961)

**Section 20.9**
nuclear fusion (962)

**Section 20.10**
transmutation (963)
linear accelerator (963)
cyclotron (963)

**Section 20.11**
exposure (965)
dose (965)
biological effectiveness factor (RBE) (966)
rem (966)

**Section 20.12**
radiotracer (967)
positron emission tomography (PET) (968)

## Key Concepts

### Diagnosing Appendicitis (20.1)

▶ Radioactivity is the emission of subatomic particles or energetic electromagnetic radiation by the nuclei of certain atoms.

▶ Because some radioactive emissions can pass through matter, radioactivity is useful in medicine and many other areas of study.

### The Discovery of Radioactivity (20.2)

▶ Radioactivity was discovered by Antoine-Henri Becquerel when he found that uranium causes a photographic exposure in the absence of light.

▶ Marie Sklodowska Curie later determined that this phenomenon was not unique to uranium, and she began calling the rays that produced the exposure radioactivity. Curie also discovered two new elements, polonium and radium.

### Types of Radioactivity (20.3)

▶ The major types of natural radioactivity are alpha ($\alpha$) decay, beta ($\beta$) decay, gamma ($\gamma$) ray emission, and positron emission.

▶ Alpha radiation is helium nuclei. Beta particles are electrons. Gamma rays are electromagnetic radiation of very high energy. Positrons are the antiparticles of electrons.

- A nucleus may absorb one of its orbital electrons in a process called electron capture.
- We can represent each radioactive process with a nuclear equation that illustrates how the parent nuclide changes into the daughter nuclide. In a nuclear equation, although the specific types of atoms may not balance, the atomic numbers and mass numbers must.
- Each type of radioactivity has a different ionizing and penetrating power. These values are inversely related; a particle with a higher ionizing power has a lower penetrating power. Alpha particles are the most massive and they have the highest ionizing power, followed by beta particles and positrons, which are equivalent in their ionizing power. Gamma rays have the lowest ionizing power.

## The Valley of Stability: Predicting the Type of Radioactivity (20.4)

- The stability of a nucleus, and therefore the probability that it will undergo radioactive decay, depends largely on two factors. The first is the ratio of neutrons to protons ($N/Z$) because neutrons provide a strong force that overcomes the electromagnetic repulsions between the positive protons. This ratio is one for smaller elements but becomes greater than one for larger elements. The second factor related to nuclei stability is a concept known as magic numbers; certain numbers of nucleons are more stable than others.

## Detecting Radioactivity (20.5)

- Radiation detectors determine the quantity of radioactivity in an area or sample.
- Thermoluminescent dosimeters employ salt crystals for detection of radiation; such detectors do not provide an instantaneous response.
- Two detectors that instantly register the amount of radiation are the Geiger-Müller counter, which uses the ionization of argon by radiation to produce an electrical signal, and the scintillation counter, which uses the emission of light induced by radiation.

## The Kinetics of Radioactive Decay and Radiometric Dating (20.6)

- All radioactive elements decay according to first-order kinetics (Chapter 14); the half-life equation and the integrated rate law for radioactive decay are derived from the first-order rate laws.
- The kinetics of radioactive decay is used to date objects and artifacts. The age of materials that were once part of living organisms is measured by carbon-14 dating. The age of ancient rocks and Earth itself is determined by uranium/lead dating.

## The Discovery of Fission: The Atomic Bomb and Nuclear Power (20.7)

- Fission is the splitting of an atom, such as uranium-235, into two atoms of lesser atomic weight.
- Because the fission of one uranium-235 atom releases enormous amounts of energy and produces neutrons that can split other uranium-235 atoms, the energy from these collective reactions can be harnessed in an atomic bomb or nuclear reactor.
- Nuclear power produces no air pollution and requires little mass to release lots of energy; however, there is always a danger of accidents, and it is difficult to dispose of nuclear waste.

## Converting Mass to Energy: Mass Defect and Nuclear Binding Energy (20.8)

- In a nuclear fission reaction, mass is converted into energy.
- The difference in mass between a nuclide and the individual protons and neutrons that compose it is the mass defect, and the corresponding energy, calculated from Einstein's equation $E = mc^2$, is the nuclear binding energy.
- The stability of a nucleus is determined by the binding energy per nucleon, which increases up to mass number 60 and then decreases.

## Nuclear Fusion: The Power of the Sun (20.9)

- Stars produce their energy by the process that is the opposite of fission: nuclear fusion, the combination of two light nuclei to form a heavier one.
- Modern nuclear weapons employ fusion. Although fusion has been examined as a possible method to produce electricity, experiments with hydrogen fusion have thus far been more costly than productive.

## Nuclear Transmutation and Transuranium Elements (20.10)

- Nuclear transmutation, the changing of one element to another element, was used to create the transuranium elements, elements with atomic numbers greater than that of uranium.
- Two devices are most commonly used to accelerate particles to the high speeds necessary for transmutation reactions: the linear accelerator and the cyclotron. Both use alternating voltage to propel particles by electromagnetic forces.

## The Effects of Radiation on Life (20.11)

- The effects of radiation can be grouped into three categories. Acute radiation damage is caused by a large exposure to radiation for a short period of time. Lower radiation exposures may result in increased cancer risk because of damage to DNA. Genetic defects are caused by damage to the DNA of reproductive cells.
- The most effective unit of measurement for the amount of radiation absorbed is the rem, which takes into account the different penetrating and ionizing powers of the various types of radiation.

## Radioactivity in Medicine (20.12)

- Radioactivity is central to the diagnosis of medical problems by means of radiotracers and positron emission tomography (PET). Both of these techniques can provide data about the appearance and metabolic activity of an organ, or help locate a tumor.
- Radiation is also used to treat cancer because it kills cells. Radiation can kill bacteria in foods and control harmful insect populations.

# Key Equations and Relationships

The First-Order Rate Law (**20.6**)

$$\text{Rate} = kN$$

The Half-Life Equation (**20.6**)

$$t_{1/2} = \frac{0.693}{k} \qquad k = \text{rate constant}$$

The Integrated Rate Law (**20.6**)

$$\ln \frac{N_t}{N_0} = -kt \quad N_t = \text{number of radioactive nuclei at time } t$$

$$N_0 = \text{initial number of radioactive nuclei}$$

Einstein's Energy–Mass Equation (**20.6**)

$$E = mc^2$$

## Key Learning Outcomes

| CHAPTER OBJECTIVES | ASSESSMENT |
|---|---|
| Writing Nuclear Equations for Alpha Decay (**20.3**) | Example 20.1    For Practice 20.1    Exercises 31–36 |
| Writing Nuclear Equations for Beta Decay, Positron Emission, and Electron Capture (**20.3**) | Example 20.2    For Practice 20.2    For More Practice 20.2 Exercises 31–36 |
| Predicting the Type of Radioactive Decay (**20.4**) | Example 20.3    For Practice 20.3    Exercises 41, 42 |
| Using Radioactive Decay Kinetics (**20.6**) | Example 20.4    For Practice 20.4    Exercises 45–52 |
| Using Radiocarbon Dating (**20.6**) | Example 20.5    For Practice 20.5    Exercises 53, 54 |
| Using Uranium/Lead Dating to Estimate the Age of a Rock (**20.6**) | Example 20.6    For Practice 20.6    Exercises 55, 56 |
| Determining the Mass Defect and Nuclear Binding Energy (**20.8**) | Example 20.7    For Practice 20.7    Exercises 65–72 |

# EXERCISES

## Review Questions

**1.** What is radioactivity? Who discovered it? How was it discovered?

**2.** Explain Marie Curie's role in the discovery of radioactivity.

**3.** Define $A$, $Z$, and $X$ in the following notation used to specify a nuclide: $^A_Z X$.

**4.** Use the notation from Question 3 to write symbols for a proton, a neutron, and an electron.

**5.** What is an alpha particle? What happens to the mass number and atomic number of a nuclide that emits an alpha particle?

**6.** What is a beta particle? What happens to the mass number and atomic number of a nuclide that emits a beta particle?

**7.** What is a gamma ray? What happens to the mass number and atomic number of a nuclide that emits a gamma ray?

**8.** What is a positron? What happens to the mass number and atomic number of a nuclide that emits a positron?

**9.** Describe the process of electron capture. What happens to the mass number and atomic number of a nuclide that undergoes electron capture?

**10.** Rank alpha particles, beta particles, positrons, and gamma rays in terms of: (a) increasing ionizing power; (b) increasing penetrating power.

**11.** Explain why the ratio of neutrons to protons ($N/Z$) is important in determining nuclear stability. How can you use the $N/Z$ ratio of a nuclide to predict the kind of radioactive decay that it might undergo?

**12.** What are magic numbers? How are they important in determining the stability of a nuclide?

**13.** Describe the basic way that each device detects radioactivity: (a) thermoluminescent dosimeter; (b) Geiger-Müller counter; and (c) scintillation counter.

**14.** Explain the concept of half-life with respect to radioactive nuclides. What rate law is characteristic of radioactivity?

**15.** Explain the main concepts behind the technique of radiocarbon dating. How can radiocarbon dating be corrected for changes in atmospheric concentrations of C-14? What range of ages can be reliably determined by C-14 dating?

**16.** How is the uranium to lead ratio in a rock used to estimate its age? How does this dating technique provide an estimate for Earth's age? How old is Earth according to this dating method?

**17.** Describe fission. Include the concepts of chain reaction and critical mass in your description. How and by whom was fission discovered? Explain how fission can be used to generate electricity.

**18.** What was the Manhattan Project? Briefly describe its development and culmination.

**19.** Describe the advantages and disadvantages of using fission to generate electricity.

**20.** The products of a nuclear reaction usually have a different mass than the reactants. Why?

**21.** Explain the concepts of mass defect and nuclear binding energy. At what mass number does the nuclear binding energy per nucleon peak? What is the significance of this?

**22.** What is fusion? Why can fusion and fission both produce energy? Explain.

**23.** What are some of the problems associated with using fusion to generate electricity?

**24.** Explain transmutation and provide one or two examples.

**25.** How does a linear accelerator work? For what purpose is it used?

**26.** Explain the basic principles of cyclotron function.

**27.** How does radiation affect living organisms?

**28.** Explain why different kinds of radiation affect biological tissues differently, even though the amount of radiation exposure may be the same.

**29.** Explain the significance of the biological effectiveness factor in measuring radiation exposure. What types of radiation would you expect to have the highest biological effectiveness factor?

**30.** Describe some of the medical uses, both in diagnosis and in treatment of disease, of radioactivity.

# Problems by Topic

## Radioactive Decay and Nuclide Stability

**31.** Write a nuclear equation for the indicated decay of each nuclide.
  **a.** U-234 (alpha)
  **b.** Th-230 (alpha)
  **c.** Pb-214 (beta)
  **d.** N-13 (positron emission)
  **e.** Cr-51 (electron capture)

**32.** Write a nuclear equation for the indicated decay of each nuclide.
  **a.** Po-210 (alpha)
  **b.** Ac-227 (beta)
  **c.** Tl-207 (beta)
  **d.** O-15 (positron emission)
  **e.** Pd-103 (electron capture)

**33.** Write a partial decay series for Th-232 undergoing the sequential decays: $\alpha, \beta, \beta, \alpha$.

**34.** Write a partial decay series for Rn-220 undergoing the sequential decays: $\alpha, \alpha, \beta, \beta$.

**35.** Fill in the missing particles in each nuclear equation.
  **a.** $\underline{\quad} \longrightarrow {}^{217}_{85}\text{At} + {}^{4}_{2}\text{He}$
  **b.** ${}^{241}_{94}\text{Pu} \longrightarrow {}^{241}_{95}\text{Am} + \underline{\quad}$
  **c.** ${}^{19}_{11}\text{Ne} \longrightarrow {}^{19}_{10}\text{Ne} + \underline{\quad}$
  **d.** ${}^{75}_{34}\text{Se} + \underline{\quad} \longrightarrow {}^{75}_{33}\text{As}$

**36.** Fill in the missing particles in each nuclear equation.
  **a.** ${}^{241}_{95}\text{Am} \longrightarrow {}^{237}_{93}\text{Np} + \underline{\quad}$
  **b.** $\underline{\quad} \longrightarrow {}^{233}_{92}\text{U} + {}^{0}_{-1}\text{e}$
  **c.** ${}^{237}_{93}\text{Np} \longrightarrow \underline{\quad} + {}^{4}_{2}\text{He}$
  **d.** ${}^{75}_{35}\text{Br} \longrightarrow \underline{\quad} + {}^{0}_{+1}\text{e}$

**37.** Determine whether or not each nuclide is likely to be stable. State your reasons.
  **a.** Mg-26         **b.** Ne-25
  **c.** Co-51         **d.** Te-124

**38.** Determine whether or not each nuclide is likely to be stable. State your reasons.
  **a.** Ti-48     **b.** Cr-63     **c.** Sn-102     **d.** Y-88

**39.** The first six elements of the first transition series have the following number of stable isotopes:

| Element | Number of Stable Isotopes |
|---------|---------------------------|
| Sc | 1 |
| Ti | 5 |
| V | 1 |
| Cr | 3 |
| Mn | 1 |
| Fe | 4 |

Explain why Sc, V, and Mn each have only one stable isotope while the other elements have several.

**40.** Neon and magnesium each have three stable isotopes while sodium and aluminum each have only one. Explain why this might be so.

**41.** Predict a likely mode of decay for each unstable nuclide.
  **a.** Mo-109         **b.** Ru-90
  **c.** P-27           **d.** Sn-100

**42.** Predict a likely mode of decay for each unstable nuclide.
  **a.** Sb-132         **b.** Te-139
  **c.** Fr-202         **d.** Ba-123

**43.** Which nuclide in each pair would you expect to have the longer half-life?
  **a.** Cs-113 or Cs-125
  **b.** Fe-62 or Fe-70

**44.** Which nuclide in each pair would you expect to have the longer half-life?
  **a.** Cs-149 or Cs-139
  **b.** Fe-45 or Fe-52

## The Kinetics of Radioactive Decay and Radiometric Dating

**45.** One of the nuclides in spent nuclear fuel is U-235, an alpha emitter with a half-life of 703 million years. How long will it take for the amount of U-235 to reach 10.0% of its initial amount?

**46.** A patient is given 0.050 $\mu$g of technetium-99m, a radioactive isotope with a half-life of about 6.0 hours. How long does it take for the radioactive isotope to decay to $1.0 \times 10^{-3}$ $\mu$g? (Assume no excretion of the nuclide from the body.)

**47.** A radioactive sample contains 1.55 g of an isotope with a half-life of 3.8 days. What mass of the isotope remains after 5.5 days?

**48.** At 8:00 A.M., a patient receives a 1.5-$\mu$g dose of I-131 to treat thyroid cancer. If the nuclide has a half-life of eight days, what mass of the nuclide remains in the patient at 5:00 P.M. the next day? (Assume no excretion of the nuclide from the body.)

**49.** A sample of F-18 has an initial decay rate of $1.5 \times 10^5$/s. How long will it take for the decay rate to fall to $2.5 \times 10^3$/s? (F-18 has a half-life of 1.83 hours.)

**50.** A sample of Tl-201 has an initial decay rate of $5.88 \times 10^4$/s. How long will it take for the decay rate to fall to 287/s? (Tl-201 has a half-life of 3.042 days.)

**51.** A wooden boat discovered just south of the Great Pyramid in Egypt has a carbon-14/carbon-12 ratio that is 72.5% of that found in living organisms. How old is the boat?

**52.** A layer of peat beneath the glacial sediments of the last ice age has a carbon-14/carbon-12 ratio that is 22.8% of that found in living organisms. How long ago was this ice age?

**53.** An ancient skull has a carbon-14 decay rate of 0.85 disintegration per minute per gram of carbon (0.85 dis/min·g C). How old is the skull? (Assume that living organisms have a carbon-14 decay rate of 15.3 dis/min·g C and that carbon-14 has a half-life of 5715 yr.)

**54.** A mammoth skeleton has a carbon-14 decay rate of 0.48 disintegration per minute per gram of carbon (0.48 dis/min·g C). When did the mammoth live? (Assume that living organisms have a carbon-14 decay rate of 15.3 dis/min·g C and that carbon-14 has a half-life of 5715 yr.)

**55.** A rock from Australia contains 0.438 g of Pb-206 to every 1.00 g of U-238. Assuming that the rock did not contain any Pb-206 at the time of its formation, how old is the rock?

**56.** A meteor has a Pb-206:U-238 mass ratio of 0.855:1.00. What is the age of the meteor? (Assume that the meteor did not contain any Pb-206 at the time of its formation.)

## Fission, Fusion, and Transmutation

**57.** Write the nuclear reaction for the neutron-induced fission of U-235 to form Xe-144 and Sr-90. How many neutrons are produced in the reaction?

**58.** Write the nuclear reaction for the neutron-induced fission of U-235 to produce Te-137 and Zr-97. How many neutrons are produced in the reaction?

**59.** Write the nuclear equation for the fusion of two H-2 atoms to form He-3 and one neutron.

**60.** Write the nuclear equation for the fusion of H-3 with H-1 to form He-4.

**61.** A breeder nuclear reactor is a reactor in which nonfissionable (nonfissile) U-238 is converted into fissionable (fissile) Pu-239. The process involves bombardment of U-238 by neutrons to form U-239, which then undergoes two sequential beta decays. Write nuclear equations for this process.

**62.** Write the series of nuclear equations to represent the bombardment of Al-27 with a neutron to form a product that subsequently undergoes a beta decay.

**63.** Rutherfordium-257 was synthesized by bombarding Cf-249 with C-12. Write the nuclear equation for this reaction.

**64.** Element 107, now named bohrium, was synthesized by German researchers by colliding bismuth-209 with chromium-54 to form a bohrium isotope and one neutron. Write the nuclear equation to represent this reaction.

## Energetics of Nuclear Reactions, Mass Defect, and Nuclear Binding Energy

**65.** If 1.0 g of matter is converted to energy, how much energy is formed?

**66.** A typical home uses approximately $1.0 \times 10^3$ kWh of energy per month. If the energy came from a nuclear reaction, what mass would have to be converted to energy per year to meet the energy needs of the home?

**67.** Calculate the mass defect and nuclear binding energy per nucleon of each nuclide.
   **a.** O-16 (atomic mass = 15.994915 amu)
   **b.** Ni-58 (atomic mass = 57.935346 amu)
   **c.** Xe-129 (atomic mass = 128.904780 amu)

**68.** Calculate the mass defect and nuclear binding energy per nucleon of each nuclide.
   **a.** Li-7 (atomic mass = 7.016003 amu)
   **b.** Ti-48 (atomic mass = 47.947947 amu)
   **c.** Ag-107 (atomic mass = 106.905092 amu)

**69.** Calculate the quantity of energy produced per gram of U-235 (atomic mass = 235.043922 amu) for the neutron-induced fission of U-235 to form Xe-144 (atomic mass = 143.9385 amu) and Sr-90 (atomic mass = 89.907738 amu) (discussed in Problem 57).

**70.** Calculate the quantity of energy produced per mole of U-235 (atomic mass = 235.043922 amu) for the neutron-induced fission of U-235 to produce Te-137 (atomic mass = 136.9253 amu) and Zr-97 (atomic mass = 96.910950 amu) (discussed in Problem 58).

**71.** Calculate the quantity of energy produced per gram of reactant for the fusion of two H-2 (atomic mass = 2.014102 amu) atoms to form He-3 (atomic mass = 3.016029 amu) and one neutron (discussed in Problem 59).

**72.** Calculate the quantity of energy produced per gram of reactant for the fusion of H-3 (atomic mass = 3.016049 amu) with H-1 (atomic mass = 1.007825 amu) to form He-4 (atomic mass = 4.002603 amu) (discussed in Problem 60).

## Effects and Applications of Radioactivity

**73.** A 75-kg human has a dose of 32.8 rad of radiation. How much energy is absorbed by the person's body? Compare this energy to the amount of energy absorbed by the person's body if he or she jumped from a chair to the floor (assume that the chair is 0.50 m from the ground and that all of the energy from the fall is absorbed by the person).

**74.** If a 55-gram laboratory mouse has a dose of 20.5 rad of radiation, how much energy is absorbed by the mouse's body?

**75.** PET studies require fluorine-18, which is produced in a cyclotron and decays with a half-life of 1.83 hours. Assuming that the F-18 can be transported at 60.0 miles/hour, how close must the hospital be to the cyclotron if 65% of the F-18 produced makes it to the hospital?

**76.** Suppose a patient is given 1.55 μg of I-131, a beta emitter with a half-life of 8.0 days. Assuming that none of the I-131 is eliminated from the person's body in the first 4.0 hours of treatment, what is the exposure (in Ci) during those first four hours?

# Cumulative Problems

**77.** Complete each nuclear equation and calculate the energy change (in J/mol of reactant) associated with each (Be-9 = 9.012182 amu, Bi-209 = 208.980384 amu, He-4 = 4.002603 amu, Li-6 = 6.015122 amu, Ni-64 = 63.927969 amu, Rg-272 = 272.1535 amu, Ta-179 = 178.94593 amu, and W-179 = 178.94707 amu).
   **a.** _____ $+ \, ^9_4\text{Be} \longrightarrow \, ^6_3\text{Li} + \, ^4_2\text{He}$
   **b.** $^{209}_{83}\text{Bi} + \, ^{64}_{28}\text{Ni} \longrightarrow \, ^{272}_{111}\text{Rg} + $ _____
   **c.** $^{179}_{74}\text{W} + $ _____ $\longrightarrow \, ^{179}_{73}\text{Ta}$

**78.** Complete each nuclear equation and calculate the energy change (in J/mol of reactant) associated with each (Al-27 = 26.981538 amu, Am-241 = 241.056822 amu, He-4 = 4.002603 amu, Np-237 = 237.048166 amu, P-30 = 29.981801 amu, S-32 = 31.972071 amu, and Si-29 = 28.976495 amu).
   **a.** $^{27}_{13}\text{Al} + \, ^4_2\text{He} \longrightarrow \, ^{30}_{15}\text{P} + $ _____
   **b.** $^{32}_{16}\text{S} + $ _____ $\longrightarrow \, ^{29}_{14}\text{Si} + \, ^4_2\text{He}$
   **c.** $^{241}_{95}\text{Am} \longrightarrow \, ^{237}_{93}\text{Np} + $ _____

**79.** Write the nuclear equation for the most likely mode of decay for each unstable nuclide.
  **a.** Ru-114
  **b.** Ra-216
  **c.** Zn-58
  **d.** Ne-31

**80.** Write the nuclear equation for the most likely mode of decay for each unstable nuclide.
  **a.** Kr-74
  **b.** Th-221
  **c.** Ar-44
  **d.** Nb-85

**81.** Bismuth-210 is a beta emitter with a half-life of 5.0 days. If a sample contains 1.2 g of Bi-210 (atomic mass = 209.984105 amu), how many beta emissions occur in 13.5 days? If a person's body intercepts 5.5% of those emissions, to what amount of radiation (in Ci) is the person exposed?

**82.** Polonium-218 is an alpha emitter with a half-life of 3.0 minutes. If a sample contains 55 mg of Po-218 (atomic mass = 218.008965 amu), how many alpha emissions occur in 25.0 minutes? If the polonium is ingested by a person, to what amount of radiation (in Ci) is the person exposed?

**83.** Radium-226 (atomic mass = 226.025402 amu) decays to radon-224 (a radioactive gas) with a half-life of $1.6 \times 10^3$ years. What volume of radon gas (at 25.0 °C and 1.0 atm) does 25.0 g of radium produce in 5.0 days? (Report your answer to two significant digits.)

**84.** In one of the neutron-induced fission reactions of U-235 (atomic mass = 235.043922 amu), the products are Ba-140 and Kr-93 (a radioactive gas). What volume of Kr-93 (at 25.0 °C and 1.0 atm) is produced when 1.00 g of U-235 undergoes this fission reaction?

**85.** When a positron and an electron annihilate one another, the resulting mass is completely converted to energy. Calculate the energy associated with this process in kJ/mol.

**86.** A typical nuclear reactor produces about 1.0 MW of power per day. What is the minimum rate of mass loss required to produce this much energy?

**87.** Find the binding energy in an atom of $^3$He, which has a mass of 3.016030 amu.

**88.** The overall hydrogen burning reaction in stars can be represented as the conversion of four protons to one $\alpha$ particle. Use the data for the mass of H-1 and He-4 to calculate the energy released by this process.

**89.** The nuclide $^{247}$Es can be made by bombardment of $^{238}$U in a reaction that emits five neutrons. Identify the bombarding particle.

**90.** The nuclide $^6$Li reacts with $^2$H to form two identical particles. Identify the particles.

**91.** The half-life of $^{238}$U is $4.5 \times 10^9$ yr. A sample of rock of mass 1.6 g produces 29 dis/s. Assuming all the radioactivity is due to $^{238}$U, find the percent by mass of $^{238}$U in the rock.

**92.** The half-life of $^{232}$Th is $1.4 \times 10^{10}$ yr. Find the number of disintegrations per hour emitted by 1.0 mol of $^{232}$Th.

**93.** A 1.50-L gas sample at 745 mm Hg and 25.0 °C contains 3.55% radon-220 by volume. Radon-220 is an alpha emitter with a half-life of 55.6 s. How many alpha particles are emitted by the gas sample in 5.00 minutes?

**94.** A 228-mL sample of an aqueous solution contains 2.35% $MgCl_2$ by mass. Exactly one-half of the magnesium ions are Mg-28, a beta emitter with a half-life of 21 hours. What is the decay rate of Mg-28 in the solution after 4.00 days? (Assume a density of 1.02 g/mL for the solution.)

**95.** When a positron and an electron collide and annihilate each other, two photons of equal energy are produced. Find the wavelength of these photons.

**96.** The half-life of $^{235}$U, an alpha emitter, is $7.1 \times 10^8$ yr. Calculate the number of alpha particles emitted by 1.0 mg of this nuclide in 1.0 minute.

**97.** Given that the energy released in the fusion of two deuterons to a $^3$He and a neutron is 3.3 MeV, and in the fusion to tritium and a proton it is 4.0 MeV, calculate the energy change for the process $^3$He + $^1$n $\longrightarrow$ $^3$H + $^1$p. Suggest an explanation for why this process occurs at much lower temperatures than either of the first two.

**98.** The nuclide $^{18}$F decays by both electron capture and $\beta^+$ decay. Find the difference in the energy released by these two processes. The atomic masses are $^{18}$F = 18.000950 and $^{18}$O = 17.9991598.

# Challenge Problems

**99.** The space shuttle carries about 72,500 kg of solid aluminum fuel, which is oxidized with ammonium perchlorate according to the reaction shown here:

$$10 \, Al(s) + 6 \, NH_4ClO_4(s) \longrightarrow$$

$$4 \, Al_2O_3(s) + 2 \, AlCl_3(s) + 12 \, H_2O(g) + 3 \, N_2(g)$$

The space shuttle also carries about 608,000 kg of oxygen (which reacts with hydrogen to form gaseous water).
  **a.** Assuming that aluminum and oxygen are the limiting reactants, determine the total energy produced by these fuels. ($\Delta H_f^\circ$ for solid ammonium perchlorate is −295 kJ/mol.)
  **b.** Suppose that a future space shuttle is powered by matter–antimatter annihilation. The matter could be normal hydrogen (containing a proton and an electron) and the antimatter could be antihydrogen (containing an antiproton and a positron). What mass of antimatter is required to produce the energy equivalent of the aluminum and oxygen fuel currently carried on the space shuttle?

**100.** Suppose that an 85.0-gram laboratory animal ingests 10.0 mg of a substance that contained 2.55% by mass Pu-239, an alpha emitter with a half-life of 24,110 years.
  **a.** What is the animal's initial radiation exposure in curies?
  **b.** If all of the energy from the emitted alpha particles is absorbed by the animal's tissues, and if the energy of each emission is $7.77 \times 10^{-12}$ J, what is the dose in rads to the animal in the first 4.0 hours following the ingestion of the radioactive material? Assuming a biological effectiveness factor of 20, what is the 4.0-hour dose in rems?

**101.** In addition to the natural radioactive decay series that begins with U-238 and ends with Pb-206, there are natural radioactive decay series that begin with U-235 and Th-232. Both of these series end with nuclides of Pb. Predict the likely end product of each series and the number of $\alpha$ decay steps that occur.

**102.** The hydride of an unstable nuclide of a Group IIA metal, $MH_2(s)$, decays by $\alpha$-emission. A 0.025-mol sample of the hydride is placed in an evacuated 2.0 L container at 298 K. After 82 minutes, the pressure in the container is 0.55 atm. Find the half-life of the nuclide.

**103.** The nuclide $^{38}Cl$ decays by beta emission with a half-life of 37.2 min. A sample of 0.40 mol of $H^{38}Cl$ is placed in a 6.24-L container. After 74.4 min the pressure is 1650 mmHg. What is the temperature of the container?

**104.** When $^{10}BF_3$ is bombarded with neutrons, the boron-10 undergoes an $\alpha$ decay, but the F is unaffected. A 0.20-mol sample of $^{10}BF_3$ contained in a 3.0-L container at 298 K is bombarded with neutrons until half of the $^{10}BF_3$ has reacted. What is the pressure in the container at 298 K?

# Conceptual Problems

**105.** Closely examine the diagram representing the beta decay of fluorine-21 and draw in the missing nucleus.

$^{21}_{9}F$

$? \ + \ ^{0}_{-1}e$

**106.** Approximately how many half-lives must pass for the amount of radioactivity in a substance to decrease to below 1% of its initial level?

**107.** A person is exposed for three days to identical amounts of two different nuclides that emit positrons of roughly equal energy. The half-life of nuclide A is 18.5 days and the half-life of nuclide B is 255 days. Which of the two nuclides poses the greater health risk?

**108.** Identical amounts of two different nuclides, an alpha emitter and a gamma emitter, with roughly equal half-lives are spilled in a building adjacent to your bedroom. Which of the two nuclides is likely to pose the greater health threat to you while you sleep in your bed? If you accidentally wander into the building and ingest equal amounts of the two nuclides, which of the two is likely to pose the greater health threat?

**109.** Drugstores in many areas now carry tablets, under such trade names as Iosat and NoRad, designed to be taken in the event of an accident at a nuclear power plant or a terrorist attack that releases radioactive material. These tablets contain potassium iodide (KI). Can you explain the nature of the protection that they provide? (*Hint:* see the label in the photo.)

# Questions for Group Work

*Active Classroom Learning*

Discuss these questions with the group and record your consensus answer.

**110.** Complete the table of particles involved in radioactive decay.

| Particle Name | Symbol | Mass Number | Atomic Number or Charge |
|---|---|---|---|
| alpha particle | $^{4}_{2}He$ | | |
| | $^{0}_{-1}e$ | | −1 |
| gamma ray | $^{0}_{0}Y$ | | |
| positron | | 0 | |
| | $^{1}_{0}n$ | 1 | |
| proton | $^{1}_{1}p$ | | |

**111.** Have each group member study a different mode of radioactive decay (alpha, beta, gamma, positron emission, or electron capture) and present it to the group. Each presentation should include a description of the process, a description of how the atomic and mass numbers change, and at least one specific example. Presentations should also address the questions: What do all nuclear reactions have in common and how do they differ from each other?

**112.** Two students are discussing whether or not the total mass changes during a nuclear reaction. The first student insists that mass is conserved. The second student says that mass is converted into energy. Explain the context in which each student is correct and how that fact is applied to solve problems.

**113.** Write all the balanced nuclear equation for each step of the nuclear decay sequence that starts with U-238 and ends with U-234. Refer to Figure 20.6 for the decay processes involved.

**114.** Radon-220 undergoes alpha decay with a half-life of 55.6 s. Assume there are 16,000 atoms present initially and make a table showing how many atoms will be present at 0 s, 55.6 s, 111.2 s, 166.8 s, 222.4 s, and 278.0 s (all multiples of the half-life). Now calculate how many atoms will be present at 50 s, 100 s, and 200 s (not multiples of the half-life). Make a graph with number of atoms present on the $y$-axis and total time on the $x$-axis.

# Data Interpretation and Analysis

## The Use of Technetium-99m for Medical Diagnosis

115. Radioisotopes are useful for medical diagnosis because once ingested the isotopes accumulate in various organs such as the liver, lungs, lymph nodes, and bones. The energy emitted from the decay of the isotopes provides a way to image an organ or bone.

The radioactive isotope Technetium-99m ("m" means metastable) is prepared by bombarding Molybdenum-99 with deuterons:

$$^{99}_{42}Mo + \text{deuterons} \longrightarrow {}^{99m}_{43}Tc + {}^{0}_{-1}\beta$$

Once formed, $^{99m}_{43}Tc$ decays to $^{99}_{43}Tc$ by a release of gamma rays:

$$^{99m}_{43}Tc \longrightarrow {}^{99}_{43}Tc + {}^{0}_{0}\gamma$$

A patient is given a dose of 0.060 mg of $^{99m}_{43}Tc$ and data in Table a are obtained:

| Time (h) | Mass of $^{99m}_{43}Tc$ (mg) | Emission (photons/s) |
|---|---|---|
| 0.00 | 0.0600 | 6000 |
| 2.00 | 0.0476 | 4762 |
| 4.00 | 0.0378 | 3781 |
| 6.00 |  | 3003 |
| 8.00 | 0.0239 |  |
| 10.00 |  | 1893 |
| 12.00 |  |  |
| 14.00 | 0.0120 |  |

TABLE a  Technetium-m Emission

Use the information provided in the question and the table to answer the following questions:

a. Given that all radioactive decay follows first-order kinetics, determine the half-life of $^{99m}_{43}Tc$.

b. Determine the decay constant, $k$, for $^{99m}_{43}Tc$.

c. Complete the information in the table.

d. How many atoms of $^{99m}_{43}Tc$ are present at $t = 0.0$ h?

e. How many atoms of $^{99m}_{43}Tc$ are present at $t = 12.06$ h?

# Answers to Conceptual Connections

## Alpha and Beta Decay

20.1 **(c)** The arrow labeled $x$ represents a decrease of two neutrons and two protons, indicative of alpha decay. The arrow labeled $y$ represents a decrease of one neutron and an increase of one proton, indicative of beta decay.

## Half-Life

20.2 **(b)** The half-life is the time it takes for the number of nuclei to decay to one-half of their original number.

## Half-Life and the Amount of Radioactive Sample

20.3 **(b)** 0.10 mol. The sample loses one-half of the number of moles per half-life; so over the course of four half-lives, the amount falls to 0.10 mol.

## Nuclear Transformations

20.4 Lawrencium-256

## Radiation Exposure

20.5 Nuclide A. Because nuclide A has a shorter half-life, more of the nuclides will decay, and therefore produce radiation, before they exit the body.

# 21

# Organic Chemistry

Organic chemistry just now is enough to drive one mad. It gives one the impression of a primeval, tropical forest full of the most remarkable things. . . .

—FRIEDRICH WÖHLER (1800–1882)

Organic chemistry is the study of carbon-containing compounds. Carbon is unique in the sheer number of compounds that it forms. Millions of organic compounds are known, and researchers discover new ones every day. Carbon is also unique in the *diversity* of compounds that it forms. In most cases, a fixed number of carbon atoms can combine with a fixed number of atoms of another element to form many different compounds. For example, 10 carbon atoms and 22 hydrogen atoms form 75 distinctly different compounds. With carbon as the backbone, nature is able to take the same combination of atoms and bond them together in slightly different ways to produce a huge variety of substances. It is not surprising that life is based on the chemistry of carbon because life needs diversity to exist, and organic chemistry is nothing if not diverse. In this chapter, we peer into Friedrich Wöhler's "primeval tropical forest" (see chapter-opening quotation) and discover the most remarkable things.

▲ About half of all men's colognes contain at least some patchouli alcohol ($C_{15}H_{26}O$), an organic compound (pictured here) derived from the patchouli plant. Patchouli alcohol has a pungent, musty, earthy fragrance.

## 21.1 Fragrances and Odors

Have you ever ridden an elevator with someone wearing too much perfume? Or found yourself too close to a skunk? Or caught a whiff of rotting fish? What causes these fragrances and odors? When we inhale certain molecules called odorants, they bind with olfactory receptors in our noses. This interaction sends a nerve signal to the brain that we experience as a smell. Some smells, such as that of perfume, are pleasant (when not overdone). Other smells, such as that of the skunk or rotting fish, are unpleasant. Our sense of smell helps us identify food, people, and other organisms, and it alerts us to dangers such as polluted air or spoiled food. Smell (olfaction) is one way we probe the environment around us.

$$CH_3CH=CHCH_2SH$$
2-Butene-1-thiol

$$CH_3$$
$$|$$
$$CH_3CHCH_2CH_2SH$$
3-Methyl-1-butanethiol

▲ The smell of skunk is due primarily to the molecules shown here.

Odorants, if they are to reach our noses, must be volatile. However, many volatile substances have no scent at all. Nitrogen, oxygen, water, and carbon dioxide molecules, for example, are constantly passing through our noses, yet they produce no smell because they do not bind to olfactory receptors. Most common smells are caused by **organic molecules**, molecules containing carbon combined with several other elements, such as hydrogen, nitrogen, oxygen, and sulfur. Organic molecules are responsible for the smells of roses, vanilla, cinnamon, almond, jasmine, body odor, and rotting fish. When you wander into a rose garden, you experience the sweet smell caused in part by geraniol, an organic compound emitted by roses. Men's colognes often contain patchouli alcohol, an earthy-smelling organic compound extracted from the patchouli plant. If you have been in the vicinity of skunk spray (or have been unfortunate enough to be sprayed yourself), you are familiar with the unpleasant smell of 2-butene-1-thiol and 3-methyl-1-butanethiol, two particularly odoriferous compounds present in the secretion that skunks use to defend themselves.

The study of compounds containing carbon combined with one or more of the elements mentioned previously (hydrogen, nitrogen, oxygen, and sulfur), including their properties and their reactions, is **organic chemistry**. Besides composing much of what we smell, organic compounds are prevalent in foods, drugs, petroleum products, and pesticides. Organic chemistry is also the basis for living organisms. Life has evolved based on carbon-containing compounds, making organic chemistry of utmost importance to any person interested in understanding living organisms.

## 21.2 Carbon: Why It Is Unique

Why did life evolve based on the chemistry of carbon? Why is life not based on some other element? The answer may not be simple, but we do know that life—in order to exist—must entail complexity, and carbon chemistry is clearly complex. The number of compounds containing carbon is greater than the number of compounds containing all of the other elements combined. The reasons for carbon's unique and versatile behavior include its ability to form four covalent bonds, its ability to form double and triple bonds, and its tendency to *catenate* (that is, to form chains).

### Carbon's Tendency to Form Four Covalent Bonds

Carbon—with its four valence electrons—forms four covalent bonds. Consider the Lewis structure and space-filling models of two simple carbon compounds, methane and ethane:

Methane

Ethane

The geometry about a carbon atom forming four single bonds is tetrahedral, as shown in the figure for methane. Carbon's ability to form four bonds, and to form those bonds with a number of different elements, results in the potential to form many different compounds. As you learn to draw structures for organic compounds, always remember to draw carbon with four bonds.

### Carbon's Ability to Form Double and Triple Bonds

Carbon atoms also form double bonds (trigonal planar geometry) and triple bonds (linear geometry), adding even more diversity to the number of compounds that carbon forms. Consider the Lewis structure and space-filling models of the carbon compounds ethene and ethyne:

Ethene

Ethyne

# Chemistry IN YOUR DAY | Vitalism and the Perceived Differences between Organic and Inorganic Compounds

By the end of the eighteenth century, chemists had learned that compounds could be broadly categorized as either organic or inorganic. Organic compounds come from living things, whereas inorganic compounds come from the nonliving things on Earth. Sugar—obtained from sugarcane or the sugar beet—is a common example of an organic compound. Salt—mined from the ground or extracted from ocean water—is a common example of an inorganic compound.

Organic and inorganic compounds are different, not only in their origin, but also in their properties. Organic compounds are easily decomposed. Sugar, for example, readily decomposes into carbon and water when heated. (Think of the last time you burned something sugary in a pan or in the oven.) Inorganic compounds are more difficult to decompose. Salt decomposes only when heated to very high temperatures.

Even more curious to these early chemists was their inability to synthesize a single organic compound in the laboratory. Although they were able to synthesize many inorganic compounds, despite concerted efforts, they were not able to synthesize any organic compounds.

The origin and properties of organic compounds led early chemists to postulate that organic compounds were unique to living organisms. They hypothesized that living organisms contained a *vital force*—a mystical or supernatural power—that allowed them to produce organic compounds. They thought that producing an organic compound outside of a living organism was impossible because the vital force was not present. This belief—which became known as *vitalism*—explained why no chemist had succeeded in synthesizing an organic compound in the laboratory.

An experiment performed in 1828 by German chemist Friedrich Wöhler (1800–1882) marked the beginning of the end of vitalism. Wöhler heated ammonium cyanate (an inorganic compound) and formed urea (an organic compound):

$$\underset{\text{ammonium cyanate}}{NH_4OCN} \xrightarrow{\text{heat}} \underset{\text{urea}}{H_2NCONH_2}$$

Urea was a known organic compound that had previously been isolated only from urine. Although it was not realized at the time, Wöhler's simple experiment was a key step in opening all of life to scientific investigation. He showed that the compounds composing living organisms—like all compounds—follow scientific laws and can be studied and understood.

Today, known organic compounds number in the millions, and modern organic chemistry is a vast field that produces substances as diverse as drugs, petroleum products, and plastics.

Friedrich Wöhler

▶ The synthesis of urea in 1828 by German chemist Friedrich Wöhler marked the beginning of the end for vitalism.

Urea

In contrast, silicon (the element in the periodic table with properties closest to that of carbon) does not readily form double or triple bonds because the greater size of silicon atoms results in a Si—Si bond that is too long for much overlap between nonhybridized *p* orbitals.

## Carbon's Tendency to Catenate

Carbon, more than any other element, can bond to itself to form chain, branched, and ring structures as we can see in propane, isobutane, and cyclohexane:

Propane

Isobutane

Cyclohexane

Although other elements can form chains, none surpasses carbon at this ability. Silicon, for example, can form chains with itself. However, silicon's affinity for oxygen (the Si—O bond is stronger than the Si—Si bond) coupled with the prevalence of oxygen in our atmosphere means that silicon–silicon chains are readily oxidized to form silicates (the silicon–oxygen compounds that compose a significant proportion of minerals). By contrast, the C—C bond (347 kJ/mol) and the C—O bond (359 kJ/mol) are nearly the same strength, allowing carbon chains to exist relatively peacefully in an oxygen-rich environment. Silicon's affinity for oxygen robs it of the rich diversity that catenation provides to carbon.

## 21.3 Hydrocarbons: Compounds Containing Only Carbon and Hydrogen

Hydrocarbons—compounds that contain only carbon and hydrogen—are the simplest organic compounds. However, because of the uniqueness of carbon, many different kinds of hydrocarbons exist. We use hydrocarbons as fuels. Candle wax, oil, gasoline, liquid petroleum (LP) gas, and natural gas are all composed of hydrocarbons. Hydrocarbons are also the starting materials in the synthesis of many different consumer products, including fabrics, soaps, dyes, cosmetics, drugs, plastic, and rubber.

As shown in Figure 21.1 ▼, we classify hydrocarbons into four different types: **alkanes**, **alkenes**, **alkynes**, and **aromatic hydrocarbons**. We differentiate alkanes, alkenes, and alkynes—also called **aliphatic hydrocarbons**—based on the kinds of bonds between carbon atoms. (We discuss aromatic hydrocarbons in detail in Section 21.7.) As Table 21.1 illustrates, alkanes have only single bonds between carbon atoms, alkenes have a double bond, and alkynes have a triple bond.

▶ **FIGURE 21.1 Four Types of Hydrocarbons**

| TABLE 21.1 Alkanes, Alkenes, Alkynes | | | |
|---|---|---|---|
| Type of Hydrocarbon | Type of Bonds | Generic Formula* | Example |
| Alkanes | All single | $C_nH_{2n+2}$ | Ethane |
| Alkenes | One (or more) double | $C_nH_{2n}$ | Ethene |
| Alkynes | One (or more) triple | $C_nH_{2n-2}$ | Ethyne |

* $n$ is the number of carbon atoms. These formulas apply only to noncyclic structures containing no more than one multiple bond.

# Drawing Hydrocarbon Structures

Throughout this book, we have relied primarily on molecular formulas as the simplest way to represent compounds. In organic chemistry molecular formulas are insufficient because, as we have already discussed, the same atoms can bond together in different ways to form different compounds. For example, consider an alkane with four carbon atoms and 10 hydrogen atoms. Two different structures, named butane and isobutane, are possible:

Butane

Isobutane

Butane and isobutane are **structural isomers**, molecules with the same molecular formula but different structures. Because of their different structures, they have different properties—indeed, they are different compounds. Isomerism is ubiquitous in organic chemistry. Butane has two structural isomers. Pentane ($C_5H_{12}$) has three, hexane ($C_6H_{14}$) has five, and decane ($C_{10}H_{22}$) has 75!

We represent the structure of a particular hydrocarbon with a **structural formula**, a formula that shows not only the numbers of each kind of atoms, but also how the atoms are bonded. Organic chemists use several different kinds of structural formulas. For example, we can represent butane and isobutane in each of the following ways:

| Structural formula | Condensed structural formula | Carbon skeleton formula | Ball-and-stick model | Space-filling model |
|---|---|---|---|---|

Butane $\quad$ $CH_3-CH_2-CH_2-CH_3$

Isobutane $\quad$ $CH_3-CH-CH_3$ with $CH_3$ above

The structural formula shows all of the carbon and hydrogen atoms in the molecule and how they are bonded. The *condensed structural formula* groups the hydrogen atoms with the carbon atom to which they are bonded. Condensed structural formulas may show some of the bonds (as the previous examples do) or none at all. We can also write the condensed structural formula for butane as $CH_3CH_2CH_2CH_3$. The *carbon skeleton formula* (also called a line formula) shows the carbon–carbon bonds only as lines. Each end or bend of a line represents a carbon atom bonded to as many hydrogen atoms as necessary to form a total of four bonds. Carbon skeleton formulas allow us to draw complex structures quickly.

A carbon skeleton formula is called a *line* formula because it uses lines to represent a molecule.

Note that structural formulas are generally not three-dimensional representations of molecules—as space-filling or ball-and-stick models are—but rather two-dimensional representations that show how atoms are bonded together. As such, the most important feature of a structural formula is the *connectivity* of the atoms, not the exact way the formula is drawn. For example, consider the two condensed structural formulas for butane and the corresponding space-filling models below them as shown on the right.

$CH_3-CH_2-CH_2-CH_3$ $\qquad$ $CH_3-CH_2-CH_2$ with $CH_3$ below

Same molecule

Since rotation about single bonds is relatively unhindered at room temperature, the two structural formulas are identical, even though they are drawn differently.

We represent double and triple bonds in structural formulas with double or triple lines. For example, we draw the structural formulas for $C_3H_6$ (propene) and $C_3H_4$ (propyne) as follows:

The kind of structural formula we use depends on how much information we want to portray. Example 21.1 illustrates how to write structural formulas for a compound.

## Example 21.1   Writing Structural Formulas for Hydrocarbons

Write the structural formulas and carbon skeleton formulas for the five isomers of $C_6H_{14}$ (hexane).

**SOLUTION**

| | |
|---|---|
| To begin, draw the carbon backbone of the straight-chain isomer. | C—C—C—C—C—C |
| Next, determine the carbon backbone structure of the other isomers by arranging the carbon atoms in four other unique ways. | (structural skeletons) |
| Fill in all the hydrogen atoms so that each carbon forms four bonds. | (structural formulas) |

$$
\begin{array}{c}
\qquad\qquad H \\
H\!-\!C\!-\!H \\
H \quad H \quad | \quad H \\
H\!-\!C\!-\!C\!-\!C\!-\!C\!-\!H \\
H \quad H \quad | \quad H \\
H\!-\!C\!-\!H \\
\qquad\qquad H
\end{array}
\qquad\qquad
\begin{array}{c}
\qquad H \\
H\!-\!C\!-\!H \\
H \quad | \quad H \quad H \\
H\!-\!C\!-\!C\!-\!C\!-\!C\!-\!H \\
H \quad H \quad | \quad H \\
\qquad H\!-\!C\!-\!H \\
\qquad\qquad H
\end{array}
$$

| | |
|---|---|
| Write the carbon skeleton formulas by using lines to represent each carbon–carbon bond. Remember that each end and each bend represents a carbon atom. | |

**FOR PRACTICE 21.1**   Write the structural formulas and carbon skeleton formulas for the three isomers of $C_5H_{12}$ (pentane).

---

**Organic Structures**   Which structure is an *isomer* of the molecule shown here  (and not just the same structure)?

$$
\begin{array}{c}
\qquad\quad CH_3 \\
\qquad\quad | \\
CH_3\!-\!CH\!-\!CH_2\!-\!CH\!-\!CH_3 \\
\qquad\qquad\qquad\quad | \\
\qquad\qquad\qquad\quad CH_3
\end{array}
$$

**(a)**
$$
\begin{array}{c}
CH_3\!-\!CH\!-\!CH_2\!-\!CH\!-\!CH_3 \\
\quad\; | \qquad\qquad\quad | \\
\quad\; CH_3 \qquad\qquad CH_3
\end{array}
$$

**(b)**
$$
\begin{array}{c}
\qquad\qquad\qquad\quad CH_3 \\
\qquad\qquad\qquad\quad | \\
CH_3\!-\!CH\!-\!CH_2\!-\!CH \\
\quad\; | \qquad\qquad\quad | \\
\quad\; CH_3 \qquad\qquad CH_3
\end{array}
$$

**(c)**
$$
\begin{array}{c}
CH_3 \qquad\quad CH_3 \\
| \qquad\qquad\;\; | \\
CH\!-\!CH_2\!-\!CH \\
| \qquad\qquad\;\; | \\
CH_3 \qquad\quad CH_3
\end{array}
$$

**(d)**
$$
\begin{array}{c}
\qquad\quad CH_3 \\
\qquad\quad | \\
CH_3\!-\!CH\!-\!CH_2\!-\!CH_2\!-\!CH_2 \\
\qquad\qquad\qquad\qquad\qquad\;\; | \\
\qquad\qquad\qquad\qquad\qquad\;\; CH_3
\end{array}
$$

---

## Stereoisomerism and Optical Isomerism

**Stereoisomers** are molecules in which the atoms have the same connectivity but a different spatial arrangement. We categorize stereoisomers into two types: geometric (or cis–trans) isomers and optical isomers. We discuss geometric isomers in Section 21.5. **Optical isomers** are two molecules that are nonsuperimposable mirror images of one another. Consider the molecule shown here with its mirror image:

Molecule    Mirror image

The molecule cannot be superimposed onto its mirror image. If we swing the mirror image around to try to superimpose the two, we find that there is no way to get all four substituent atoms to align together. (A substituent is an atom or group of atoms that is substituted for a hydrogen atom in an organic compound.)

The mirror image is not superimposable on original molecule.

Molecule         Mirror image

▲ **FIGURE 21.2  Mirror Images** The left and right hands are nonsuperimposable mirror images, just as are optical isomers.

Optical isomers are similar to your right and left hands (Figure 21.2◄). The two are mirror images of one another, but you cannot superimpose one on the other. For this reason, a right-handed glove does not fit on your left hand and vice versa.

Any carbon atom with four different substituents in a tetrahedral arrangement exhibits optical isomerism. Consider 3-methylhexane:

Chiral center

Optical isomers of 3-methylhexane

The molecules shown here are nonsuperimposable mirror images and are optical isomers of one other.

Optical isomers are also called **enantiomers**. Any molecule, such as 3-methylhexane, that exhibits optical isomerism is said to be **chiral**, from the Greek word *cheir*, which means "hand." Optical isomerism is important, not only to organic chemistry, but also to biology and biochemistry. Most biological molecules are chiral and usually only one or the other enantiomer is active in biological systems. For example, glucose, the primary fuel of cells, is chiral. Only one of the enantiomers of glucose has that familiar sweet taste, and only that enantiomer can fuel cellular functioning; the other enantiomer is not even metabolized by the body.

Some of the physical and chemical properties of enantiomers are indistinguishable from one another. For example, both of the optical isomers of 3-methylhexane have identical freezing points, melting points, and densities. However, the properties of enantiomers differ from one another in two important ways: (1) in the direction in which they rotate polarized light and (2) in their chemical behavior in a chiral environment.

### Rotation of Polarized Light

*Plane-polarized light* is light that is made up of electric field waves that oscillate in only one plane as shown in Figure 21.3►. When a beam of plane-polarized light is directed through a sample containing only one of two optical isomers, the plane of polarization of the light is rotated as shown in Figure 21.4►. One of the two optical isomers rotates the polarization of the light clockwise and is the **dextrorotatory** isomer (or the *d* isomer). The other isomer rotates the polarization of the light counterclockwise and is the **levorotatory** isomer (or the *l* isomer). An equimolar mixture of both optical isomers does not rotate the polarization of light at all and is called a **racemic mixture**.

*Dextrorotatory* means turning clockwise or to the right. *Levorotatory* means turning counterclockwise or to the left.

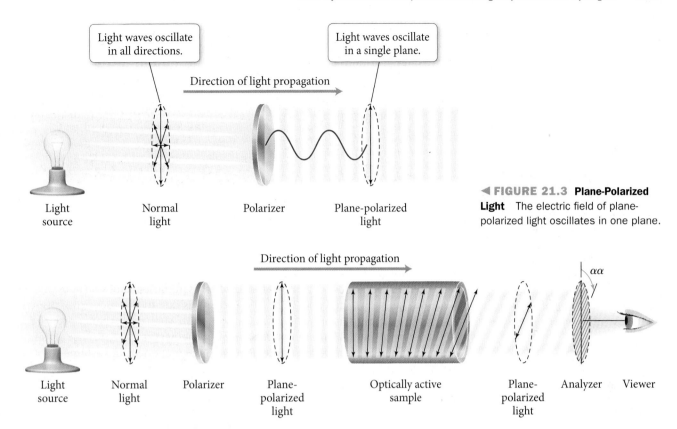

◀ **FIGURE 21.3 Plane-Polarized Light** The electric field of plane-polarized light oscillates in one plane.

▲ **FIGURE 21.4 Rotation of Plane-Polarized Light** Plane-polarized light rotates as it passes through a sample containing only one of two optical isomers.

## Chemical Behavior in a Chiral Environment

Optical isomers also exhibit different chemical behavior when they are in a chiral environment (a chiral environment is one that is not superimposable on its mirror image). Enzymes are large biological molecules that catalyze reactions in living organisms and provide chiral environments. Consider the following simplified picture of two enantiomers in a chiral environment:

One of the enantiomers fits the template, but the other does not, no matter how it is rotated. In this way, an enzyme is able to catalyze the reaction of one enantiomer because that particular enantiomer fits the "template." As we have already seen, most biological molecules are chiral, and usually only one or the other enantiomer is active in biological systems.

**Optical Isomers** Which molecule exhibits optical isomerism?

CONCEPTUAL
CONNECTION **21.2**

## 21.4 Alkanes: Saturated Hydrocarbons

Alkanes (hydrocarbons containing only single bonds) are often referred to as **saturated hydrocarbons** because they are saturated (loaded to capacity) with hydrogen. The simplest hydrocarbons are methane ($CH_4$), the main component of natural gas; ethane ($C_2H_6$), a minority component in natural gas; and propane ($C_3H_8$), the main component of LP gas:

Methane      Ethane      Propane

Alkanes containing four or more carbon atoms may be straight or branched (as we have already seen). The straight-chain isomers are often called normal alkanes, or *n*-alkanes. As the number of carbon atoms increases in the *n*-alkanes, so does their boiling point (as shown in Table 21.2). The increase is due to the increasing dispersion force with increasing molar mass (see Section 11.3). Methane, ethane, propane, and *n*-butane are all gases at room temperature, but the next *n*-alkane in the series, pentane, is a liquid at room temperature. Pentane is a component of gasoline. Table 21.3 summarizes the *n*-alkanes through decane, which contains 10 carbon atoms. Like pentane, hexane through decane are all components of gasoline. Table 21.4 summarizes the uses of hydrocarbons.

### TABLE 21.2 *n*-Alkane Boiling Points

| *n*-Alkane | Boiling Point (°C) |
|---|---|
| Methane | −161.5 |
| Ethane | −88.6 |
| Propane | −42.1 |
| *n*-Butane | −0.5 |
| *n*-Pentane | 36.0 |
| *n*-Hexane | 68.7 |
| *n*-Heptane | 98.5 |
| *n*-Octane | 125.6 |

### TABLE 21.3 *n*-Alkanes

| *n* | Name | Molecular Formula $C_nH_{2n+2}$ | Structural Formula | Condensed Structural Formula |
|---|---|---|---|---|
| 1 | Methane | $CH_4$ | | $CH_4$ |
| 2 | Ethane | $C_2H_6$ | | $CH_3CH_3$ |
| 3 | Propane | $C_3H_8$ | | $CH_3CH_2CH_3$ |
| 4 | *n*-Butane | $C_4H_{10}$ | | $CH_3CH_2CH_2CH_3$ |
| 5 | *n*-Pentane | $C_5H_{12}$ | | $CH_3CH_2CH_2CH_2CH_3$ |

| 6 | *n*-Hexane | $C_6H_{14}$ | | $CH_3CH_2CH_2CH_2CH_2CH_3$ |
|---|---|---|---|---|

| 7 | *n*-Heptane | $C_7H_{16}$ | | $CH_3CH_2CH_2CH_2CH_2CH_2CH_3$ |
|---|---|---|---|---|

| 8 | *n*-Octane | $C_8H_{18}$ | | $CH_3CH_2CH_2CH_2CH_2CH_2CH_2CH_3$ |
|---|---|---|---|---|

| 9 | *n*-Nonane | $C_9H_{20}$ | | $CH_3CH_2CH_2CH_2CH_2CH_2CH_2CH_2CH_3$ |
|---|---|---|---|---|

| 10 | *n*-Decane | $C_{10}H_{22}$ | | $CH_3CH_2CH_2CH_2CH_2CH_2CH_2CH_2CH_2CH_3$ |
|---|---|---|---|---|

**TABLE 21.4 Uses of Hydrocarbons**

| Number of Carbon Atoms | State | Major Uses |
|---|---|---|
| 1–4 | Gas | Heating fuel, cooking fuel |
| 5–7 | Low-boiling liquids | Solvents, gasoline |
| 6–18 | Liquids | Gasoline |
| 12–24 | Liquids | Jet fuel, portable-stove fuel |
| 18–50 | High-boiling liquids | Diesel fuel, lubricants, heating oil |
| 50+ | Solids | Petroleum jelly, paraffin wax |

## Naming Alkanes

Many organic compounds have common names that we can learn only through familiarity. Because of the sheer number of organic compounds, however, we need a systematic method of nomenclature. In this book, we adopt the nomenclature system recommended by the IUPAC (International Union of Pure and Applied Chemistry), which is used throughout the world. In this system, the longest continuous chain of carbon atoms—called the base chain—determines the base name of the compound. The root of the base name depends on the number of carbon atoms in the base chain, as shown in Table 21.5. Base names for alkanes always have the ending *-ane*. Groups of carbon atoms branching off the base chain are alkyl groups and are named as substituents. Remember that a *substituent* is an atom or group of atoms that has been substituted for a hydrogen atom in an organic compound. Common alkyl groups are shown in Table 21.6.

**TABLE 21.5 Prefixes for Base Names of Alkane Chains**

| Number of Carbon Atoms | Prefix |
|---|---|
| 1 | meth- |
| 2 | eth- |
| 3 | prop- |
| 4 | but- |
| 5 | pent- |
| 6 | hex- |
| 7 | hept- |
| 8 | oct- |
| 9 | non- |
| 10 | dec- |

**TABLE 21.6 Common Alkyl Groups**

| Condensed Structural Formula | Name | Condensed Structural Formula | Name |
|---|---|---|---|
| $-CH_3$ | Methyl | $-CHCH_3$ <br> \| <br> $CH_3$ | Isopropyl |
| $-CH_2CH_3$ | Ethyl | $-CH_2CHCH_3$ <br> \| <br> $CH_3$ | Isobutyl |
| $-CH_2CH_2CH_3$ | Propyl | $-CHCH_2CH_3$ <br> \| <br> $CH_3$ | sec-Butyl |
| $-CH_2CH_2CH_2CH_3$ | Butyl | $CH_3$ <br> \| <br> $-CCH_3$ <br> \| <br> $CH_3$ | tert-Butyl |

The procedure in Examples 21.2 and 21.3 will allow you to systematically name many alkanes. The procedure is in the left column and two examples of applying the procedure are in the center and right columns.

## Procedure for...

### Naming Alkanes

**1.** Count the number of carbon atoms in the longest continuous carbon chain to determine the base name of the compound. Locate the prefix corresponding to this number of atoms in Table 21.5 and add the ending *-ane* to form the base name.

---

### Example 21.2
**Naming Alkanes**

Name this alkane:

$$CH_3-CH_2-CH-CH_2-CH_3$$
$$\vert$$
$$CH_2$$
$$\vert$$
$$CH_3$$

**SOLUTION**

This compound has five carbon atoms in its longest continuous chain.

$$CH_3-CH_2-CH-CH_2-CH_3$$
$$\vert$$
$$CH_2$$
$$\vert$$
$$CH_3$$

The correct prefix from Table 21.5 is *pent-*. The base name is pentane.

---

### Example 21.3
**Naming Alkanes**

Name this alkane:

$$CH_3-CH-CH_2-CH-CH_2-CH_2-CH-CH_3$$
$$\vert \qquad\qquad \vert \qquad\qquad\qquad \vert$$
$$CH_3 \qquad CH_2 \qquad\qquad CH_3$$
$$\vert$$
$$CH_3$$

**SOLUTION**

This compound has eight carbon atoms in its longest continuous chain.

$$CH_3-CH-CH_2-CH-CH_2-CH_2-CH-CH_3$$
$$\vert \qquad\qquad \vert \qquad\qquad\qquad \vert$$
$$CH_3 \qquad CH_2 \qquad\qquad CH_3$$
$$\vert$$
$$CH_3$$

The correct prefix from Table 21.5 is *oct-*. The base name is octane.

| | | |
|---|---|---|
| **2.** Consider every branch from the base chain to be a substituent. Name each substituent according to Table 21.6. | This compound has one substituent named *ethyl*.<br><br>$CH_3$—$CH_2$—$CH$—$CH_2$—$CH_3$<br>　　　　　｜<br>　　　　$CH_2$<br>　　　　　｜<br>ethyl →　$CH_3$ | This compound has one substituent named *ethyl* and two named *methyl*.<br><br>$CH_3$—$CH$—$CH_2$—$CH$—$CH_2$—$CH_2$—$CH$—$CH_3$<br>　　　｜　　　　　｜　　　　　　　｜<br>　　$CH_3$　　　$CH_2$ ← ethyl　　$CH_3$<br>　　　　　　　　｜<br>　　　　　　　$CH_3$<br>　　　　　　　methyl |
| **3.** Beginning with the end closest to the branching, number the base chain and assign a number to each substituent. (If two substituents occur at equal distances from each end, go to the next substituent to determine from which end to start numbering.) | Number the base chain as follows:<br><br>$\overset{1}{CH_3}$—$\overset{2}{CH_2}$—$\overset{3}{CH}$—$\overset{4}{CH_2}$—$\overset{5}{CH_3}$<br>　　　　　｜<br>　　　　$CH_2$<br>　　　　　｜<br>　　　　$CH_3$<br><br>Assign the number 3 to the ethyl substituent. | Number the base chain as follows:<br><br>$\overset{1}{CH_3}$—$\overset{2}{CH}$—$\overset{3}{CH_2}$—$\overset{4}{CH}$—$\overset{5}{CH_2}$—$\overset{6}{CH_2}$—$\overset{7}{CH}$—$\overset{8}{CH_3}$<br>　　　　｜　　　　　｜　　　　　　　｜<br>　　　$CH_3$　　　$CH_2$　　　　　$CH_3$<br>　　　　　　　　｜<br>　　　　　　　$CH_3$<br><br>Assign the number 4 to the ethyl substituent and the numbers 2 and 7 to the two methyl substituents. |
| **4.** Write the name of the compound in the following format: (substituent number)-(substituent name)(base name). If there are two or more substituents, give each one a number and list them alphabetically with hyphens between words and numbers. | The name of the compound is:<br><br>3-ethylpentane | The basic form of the name of the compound is:<br><br>4-ethyl-2,7-methyloctane<br><br>List ethyl before methyl because substituents are listed in alphabetical order. |
| **5.** If a compound has two or more identical substituents, indicate the number of identical substituents with the prefix *di-* (2), *tri-* (3), or *tetra-* (4) before the substituent's name.<br><br>Separate the numbers indicating the positions of the substituents relative to each other with a comma. Do not take the prefixes into account when alphabetizing. | Does not apply to this compound. | This compound has two methyl substituents; therefore, the name of the compound is:<br><br>4-ethyl-2,7-dimethyloctane |
| | **FOR PRACTICE 21.2**<br>Name this alkane:<br><br>$CH_3$—$CH_2$—$CH$—$CH_2$—$CH_2$—$CH_3$<br>　　　　　｜<br>　　　　$CH_3$ | **FOR PRACTICE 21.3**<br>Name this alkane:<br><br>$CH_3$—$CH_2$—$CH$—$CH_2$—$CH$—$CH_2$—$CH_3$<br>　　　　　｜　　　　｜<br>　　　　$CH_3$　　$CH_3$ |

## Example 21.4 Naming Alkanes

Name this alkane:   $CH_3-CH-CH_2-CH-CH_3$
                               $|$                $|$
                             $CH_3$            $CH_3$

**SOLUTION**

| | | |
|---|---|---|
| **1.** | The longest continuous carbon chain has five atoms. Therefore, the base name is pentane. | $CH_3-CH-CH_2-CH-CH_3$<br>  $\quad\quad\quad\;\;|\quad\quad\quad\quad|$<br>  $\quad\quad\quad CH_3\quad\quad\;\; CH_3$ |
| **2.** | This compound has two substituents, both of which are named methyl. | $CH_3-CH-CH_2-CH-CH_3$<br>  $\quad\quad\quad\;\;|\quad\quad\quad\quad|$<br>  $\quad\quad\quad CH_3\quad\quad\;\; CH_3$<br>methyl |
| **3.** | Since both substituents are equidistant from the ends, it does not matter from which end you start numbering. | $\overset{1}{C}H_3-\overset{2}{C}H-\overset{3}{C}H_2-\overset{4}{C}H-\overset{5}{C}H_3$<br>  $\quad\quad\quad\;\;|\quad\quad\quad\quad|$<br>  $\quad\quad\quad CH_3\quad\quad\;\; CH_3$ |
| **4, 5.** | Use the general form for the name:<br><br>(substituent number)-(substituent name)(base name)<br><br>Because this compound contains two identical substituents, step 5 from the naming procedure applies and you use the prefix *di-*. Indicate the position of each substituent with a number separated by a comma. | 2,4-dimethylpentane |

**FOR PRACTICE 21.4**   Name this alkane:   $CH_3-CH-CH_2-CH-CH-CH_3$
                                                     $|$              $|\quad\;|$
                                                   $CH_3$          $CH_3\; CH_3$

## 21.5   Alkenes and Alkynes

The general formulas shown here for alkenes and alkynes assume only one multiple bond.

Alkenes are hydrocarbons containing at least one double bond between carbon atoms. Alkynes contain at least one triple bond. Because of the double or triple bond, alkenes and alkynes have fewer hydrogen atoms than the corresponding alkanes and are therefore called **unsaturated hydrocarbons** because they are not loaded to capacity with hydrogen. Recall that noncyclic alkenes have the formula $C_nH_{2n}$ and noncyclic alkynes have the formula $C_nH_{2n-2}$. The simplest alkene is ethene ($C_2H_4$), also called ethylene:

Ethene or ethylene

$C_2H_4$

Formula        Structural formula        Space-filling model

The geometry about each carbon atom in ethene is trigonal planar (see Example 10.8 for the valence bond model of ethene), making ethene a flat, rigid molecule. Ethene is a ripening agent in fruit such as bananas. When a banana within a cluster of bananas begins to ripen, it emits ethene. The ethene then causes other bananas in the cluster to ripen. Banana farmers usually pick bananas green for ease of shipping. When the bananas arrive at their destination, the fruit is often "gassed" with ethene to initiate ripening. Table 21.7 lists the names and structures of several other alkenes. Most of them do not have familiar uses except their presence as minority components in fuels.

**TABLE 21.7** Alkenes

| $n$ | Name | Molecular Formula $C_nH_{2n}$ | Structural Formula | Condensed Structural Formula |
|---|---|---|---|---|
| 2 | Ethene | $C_2H_4$ | | $CH_2$=$CH_2$ |
| 3 | Propene | $C_3H_6$ | | $CH_2$=$CHCH_3$ |
| 4 | 1-Butene* | $C_4H_8$ | | $CH_2$=$CHCH_2CH_3$ |
| 5 | 1-Pentene* | $C_5H_{10}$ | | $CH_2$=$CHCH_2CH_2CH_3$ |
| 6 | 1-Hexene* | $C_6H_{12}$ | | $CH_2$=$CHCH_2CH_2CH_2CH_3$ |

* These alkenes have one or more isomers depending on the position of the double bond. The isomers shown here have the double bond in the 1 position, meaning the first carbon–carbon bond of the chain.

The simplest alkyne is ethyne, $C_2H_2$, also known as acetylene:

Ethyne or    $C_2H_2$    $H$—$C$≡$C$—$H$
acetylene

Formula    Structural formula    Space-filling model

The geometry about each carbon atom in ethyne is linear, making ethyne a linear molecule. Ethyne (or acetylene) is commonly used as fuel for welding torches. Table 21.8 shows the names and structures of several other alkynes. Like alkenes, the alkynes do not have familiar uses other than their presence as minority components of gasoline.

▲ Welding torches burn ethyne in pure oxygen to produce the very hot flame needed for melting metals.

## Naming Alkenes and Alkynes

We name alkenes and alkynes in the same way we name alkanes with the following exceptions:

- The base chain is the longest continuous carbon chain *that contains the double or triple bond.*
- The base name has the ending -*ene* for alkenes and -*yne* for alkynes.
- We number the base chain to give the double or triple bond the lowest possible number.
- We insert a number indicating the position of the double or triple bond (lowest possible number) just before the base name.

For example, the alkene and alkyne shown here are 2-methyl-2-pentene and 1-butyne:

$$CH_3CH_2CH = CCH_3$$
$$| $$
$$CH_3$$
2-Methyl-2-pentene

$$CH \equiv CCH_2CH_3$$
1-Butyne

**TABLE 21.8 Alkynes**

| $n$ | Name | Molecular Formula $C_nH_{2n-2}$ | Structural Formula | Condensed Structural Formula |
|---|---|---|---|---|
| 2 | Ethyne | $C_2H_2$ | $H-C \equiv C-H$ | $CH \equiv CH$ |
| 3 | Propyne | $C_3H_4$ | $H-C \equiv C-\overset{\displaystyle H}{\underset{\displaystyle H}{C}}-H$ | $CH \equiv CCH_3$ |
| 4 | 1-Butyne* | $C_4H_6$ | $H-C \equiv C-\overset{\displaystyle H}{\underset{\displaystyle H}{C}}-\overset{\displaystyle H}{\underset{\displaystyle H}{C}}-H$ | $CH \equiv CCH_2CH_3$ |
| 5 | 1-Pentyne* | $C_5H_8$ | $H-C \equiv C-\overset{\displaystyle H}{\underset{\displaystyle H}{C}}-\overset{\displaystyle H}{\underset{\displaystyle H}{C}}-\overset{\displaystyle H}{\underset{\displaystyle H}{C}}-H$ | $CH \equiv CCH_2CH_2CH_3$ |
| 6 | 1-Hexyne* | $C_6H_{10}$ | $H-C \equiv C-\overset{\displaystyle H}{\underset{\displaystyle H}{C}}-\overset{\displaystyle H}{\underset{\displaystyle H}{C}}-\overset{\displaystyle H}{\underset{\displaystyle H}{C}}-\overset{\displaystyle H}{\underset{\displaystyle H}{C}}-H$ | $CH \equiv CCH_2CH_2CH_2CH_3$ |

* These alkynes have one or more isomers depending on the position of the triple bond. The isomers shown here have the triple bond in the 1 position, meaning the first carbon–carbon bond of the chain.

**Example 21.5  Naming Alkenes and Alkynes**

Name each compound.

$$CH_3$$
$$|$$
$$CH_3-C=C-CH_2-CH_3$$
$$|$$
$$CH_2$$
$$|$$
$$CH_3$$
(a)

$$CH_3$$
$$|$$
$$CH_3-CH$$
$$|$$
$$CH_3-CH-CH-C \equiv CH$$
$$|$$
$$CH_3$$
(b)

**SOLUTION**

(a) 1. The longest continuous carbon chain containing the double bond has six carbon atoms. The base name is therefore *hexene*.

$$CH_3$$
$$|$$
$$CH_3-C=C-CH_2-CH_3$$
$$|$$
$$CH_2$$
$$|$$
$$CH_3$$

| | |
|---|---|
| 2. The two substituents are both methyl. | methyl → CH₃<br>CH₃—C=C—CH₂—CH₃<br>　　　｜<br>　　　CH₂<br>　　　｜<br>　　　CH₃ |
| 3. One of the exceptions listed previously states that, in naming alkenes, you should number the chain so that the *double bond* has the lowest number. In this case, the double bond is equidistant from the ends. Assign the double bond the number 3. The two methyl groups are then at positions 3 and 4. | $\overset{CH_3}{\underset{3\;4\;5\;6}{CH_3-C=C-CH_2-CH_3}}$<br>　　　｜<br>　　2 CH₂<br>　　　｜<br>　　　CH₃<br>　　　1 |
| 4, 5. Use the general form for the name:<br>　　(substituent number)-(substituent name)(base name)<br>Because this compound contains two identical substituents, step 5 of the naming procedure applies. Use the prefix *di-*. In addition, indicate the position of each substituent with a number separated by a comma. Because this compound is an alkene, specify the position of the double bond, isolated by hyphens, just before the base name. | 3,4-dimethyl-3-hexene |
| **(b)** 1. The longest continuous carbon chain containing the triple bond is five carbon atoms long; therefore, the base name is *pentyne*. | CH₃<br>｜<br>CH₃—CH<br>｜<br>CH₃—CH—CH—C≡CH<br>｜<br>CH₃ |
| 2. There are two substituents; one is a methyl group and the other an isopropyl group. | CH₃　　isopropyl<br>｜<br>CH₃—CH<br>｜<br>CH₃—CH—CH—C≡CH<br>｜<br>CH₃ ← methyl |
| 3. Number the base chain, giving the triple bond the lowest number (1). Assign the isopropyl and methyl groups the numbers 3 and 4, respectively. | CH₃<br>｜<br>CH₃—CH<br>｜<br>$\underset{5\;\;\;4\;\;\;3\;\;\;2\;\;1}{CH_3-CH-CH-C\equiv CH}$<br>｜<br>CH₃ |
| 4. Use the general form for the name:<br>　　(substituent number)-(substituent name)(base name)<br>Since there are two substituents, list both of them in alphabetical order. Because this compound is an alkyne, specify the position of the triple bond with a number isolated by hyphens just before the base name. | 3-isopropyl-4-methyl-1-pentyne |

*—Continued on the next page*

*—Continued*

**FOR PRACTICE 21.5** Name each compound.

(a)

$$CH_3-C{\equiv}C-\overset{\overset{\displaystyle CH_3}{|}}{\underset{\underset{\displaystyle CH_3}{|}}{C}}-CH_3$$

(b)

$$CH_3-CH-CH_2-\overset{}{CH}-\overset{\overset{\displaystyle CH_3}{|}}{\underset{\underset{\displaystyle CH_3}{|}}{CH_2}}-CH-CH=CH_2$$
$$\underset{\displaystyle CH_3}{|}\qquad\underset{\displaystyle CH_3}{|}$$

## Geometric (Cis–Trans) Isomerism in Alkenes

A major difference between a single bond and a double bond is the degree to which rotation occurs about the bond. As discussed in Section 10.7, rotation about a double bond is highly restricted due to the overlap between unhybridized *p* orbitals on the adjacent carbon atoms.

Consider the difference between 1,2-dichloroethane and 1,2-dichloroethene:

1,2-Dichloroethane          1,2-Dichloroethene

The hybridization of the carbon atoms in 1,2-dichloroethane is *sp*³, resulting in relatively free rotation about the sigma single bond. Consequently, the two structures are identical at room temperature because they quickly interconvert:

In contrast, rotation about the double bond (sigma + pi) in 1,2-dichloroethene is restricted, so at room temperature, 1,2-dichloroethene exists in two isomeric forms:

*cis*-1,2-Dichloroethene          *trans*-1,2-Dichloroethene

These two forms of 1,2-dichloroethene are different compounds with different properties as shown in Table 21.9. This kind of isomerism is a type of stereoisomerism (see Section 21.3) called **geometric** (or **cis–trans**) **isomerism**. We distinguish between the two isomers with the designations *cis* (meaning "same side") and *trans* (meaning "opposite sides"). Cis–trans isomerism is common in alkenes. As another example, consider *cis*- and *trans*-2-butene. Like the two isomers of 1,2-dichloroethene, these two isomers have different physical properties. For example, *cis*-2-butene boils at 3.7 °C, and *trans*-2-butene boils at 0.9 °C:

*cis*-2-Butene          *trans*-2-Butene

**TABLE 21.9 Physical Properties of *cis*- and *trans*-1,2-Dichloroethene**

| Name | Structure | Space-filling Model | Density (g/mL) | Melting Point (°C) | Boiling Point (°C) |
|---|---|---|---|---|---|
| *cis*-1,2-Dichloroethene | (structure) | (model) | 1.284 | −80.5 | 60.1 |
| *trans*-1,2-Dichloroethene | (structure) | (model) | 1.257 | −49.4 | 47.5 |

# 21.6 Hydrocarbon Reactions

One of the most common hydrocarbon reactions is combustion, the burning of hydrocarbons in the presence of oxygen. Hydrocarbon combustion reactions are highly exothermic and are commonly used to warm homes and buildings, to generate electricity, and to power the engines of cars, ships, and airplanes. It is not an exaggeration to say that hydrocarbon combustion makes our current way of life possible. Approximately 90% of energy produced in the United States is generated by hydrocarbon combustion.

Alkanes, alkenes, and alkynes all undergo combustion. In a combustion reaction, the hydrocarbon reacts with oxygen to form carbon dioxide and water:

$$CH_3CH_2CH_3(g) + 5\,O_2(g) \longrightarrow 3\,CO_2(g) + 4\,H_2O(g) \qquad \text{Alkane combustion}$$

$$CH_2{=}CHCH_2CH_3(g) + 6\,O_2(g) \longrightarrow 4\,CO_2(g) + 4\,H_2O(g) \qquad \text{Alkene combustion}$$

$$CH{\equiv}CCH_3(g) + 4\,O_2(g) \longrightarrow 3\,CO_2(g) + 2\,H_2O(g) \qquad \text{Alkyne combustion}$$

## Reactions of Alkanes

In addition to combustion reactions, alkanes also undergo *substitution* reactions, in which one or more hydrogen atoms on an alkane are replaced by one or more other atoms. The most common substitution reaction is halogen substitution (also referred to as halogenation). For example, methane can react with chlorine gas in the presence of heat or light to form chloromethane:

$$\underset{\text{Methane}}{CH_4(g)} + \underset{\text{Chlorine}}{Cl_2(g)} \xrightarrow{\text{heat or light}} \underset{\text{Chloromethane}}{CH_3Cl(g)} + HCl(g)$$

Ethane reacts with chlorine gas to form chloroethane:

$$\underset{\text{Ethane}}{CH_3CH_3(g)} + \underset{\text{Chlorine}}{Cl_2(g)} \xrightarrow{\text{heat or light}} \underset{\text{Chloromethane}}{CH_3CH_2Cl(g)} + HCl(g)$$

Multiple halogenation reactions can occur because halogens can replace more than one of the hydrogen atoms on an alkane. For example, chloromethane can react with chlorine, and the product of that reaction can react again (and so on):

$$\underset{\text{Chloromethane}}{CH_3Cl(g)} + \underset{\text{Chlorine}}{Cl_2(g)} \xrightarrow{\text{heat or light}} \underset{\substack{\text{Dichloromethane}\\\text{(also known as}\\\text{methylene chloride)}}}{CH_2Cl_2(g)} + HCl(g)$$

$$\underset{\text{Dichloromethane}}{CH_2Cl_2(g)} + \underset{\text{Chlorine}}{Cl_2(g)} \xrightarrow{\text{heat or light}} \underset{\substack{\text{Trichloromethane}\\\text{(also known as}\\\text{chloroform)}}}{CHCl_3(g)} + HCl(g)$$

$$\underset{\text{Trichloromethane}}{CHCl_3(g)} + \underset{\text{Chlorine}}{Cl_2(g)} \xrightarrow{\text{heat or light}} \underset{\substack{\text{Tetrachloromethane}\\\text{(also known as}\\\text{carbon tetrachloride)}}}{CCl_4(g)} + HCl(g)$$

The general form for halogen substitution reactions is:

$$R-H + X_2 \xrightarrow{\text{heat or light}} R-X + HX$$

Alkane  Halogen  Haloalkane Hydrogen halide

Notice that the halogenation of hydrocarbons requires initiation with heat or light, which causes the chlorine–chlorine bond to break:

$$Cl-Cl \xrightarrow{\text{heat or light}} Cl\cdot + Cl\cdot$$

The resulting chlorine atoms are *free radicals* (see Section 9.9), as the dot that represents each chlorine atom's unpaired electron indicates. Chlorine radicals are highly reactive and attack the C—H bond in hydrocarbons. The subsequent reaction proceeds by this mechanism:

$$Cl\cdot + R-H \longrightarrow R\cdot + HCl$$
$$R\cdot + Cl_2 \longrightarrow R-Cl + Cl\cdot$$

Notice that a chlorine free radical is produced as a product of the last step. This free radical can go on to react again, unless it encounters another chlorine free radical, in which case it reacts with it to re-form $Cl_2$.

## Reactions of Alkenes and Alkynes

Alkenes and alkynes undergo addition reactions in which molecules add across (on either side of) the multiple bond. For example, ethene reacts with chlorine gas to form dichloroethane:

$$\underset{H}{\overset{H}{>}}C=C\underset{H}{\overset{H}{<}} + Cl-Cl \longrightarrow H-\underset{Cl}{\overset{H}{\underset{|}{\overset{|}{C}}}}-\underset{Cl}{\overset{H}{\underset{|}{\overset{|}{C}}}}-H$$

The addition of chlorine converts the carbon–carbon double bond into a single bond because each carbon atom bonds to a chlorine atom.

Alkenes and alkynes can also add hydrogen in hydrogenation reactions. For example, in the presence of an appropriate catalyst, propene reacts with hydrogen gas to form propane:

$$H-\underset{H}{\overset{H}{\underset{|}{\overset{|}{C}}}}-C=C\underset{H}{\overset{H}{<}} + H-H \xrightarrow{\text{catalyst}} H-\underset{H}{\overset{H}{\underset{|}{\overset{|}{C}}}}-\underset{H}{\overset{H}{\underset{|}{\overset{|}{C}}}}-\underset{H}{\overset{H}{\underset{|}{\overset{|}{C}}}}-H$$

Hydrogenation reactions convert unsaturated hydrocarbons into saturated hydrocarbons. For example, hydrogenation reactions convert unsaturated vegetable oils into saturated fats. Most vegetable oils are unsaturated because their carbon chains contain double bonds. The double bonds put bends into the carbon chains that result in less efficient packing of molecules; thus vegetable oils are liquids at room temperature, while saturated fats are solids at room temperature. When food manufacturers add hydrogen to the double bonds of vegetable oil, the unsaturated fat is converted into a saturated fat, turning the liquid oil into a solid at room temperature. The words "partially hydrogenated vegetable oil" on a label indicate a food product that contains saturated fats made via hydrogenation reactions.

Alkenes can also add unsymmetrical reagents across the double bond. For example, ethene reacts with hydrogen chloride to form chloroethane:

$$\underset{H}{\overset{H}{>}}C=C\underset{H}{\overset{H}{<}} + HCl \longrightarrow H-\underset{H}{\overset{H}{\underset{|}{\overset{|}{C}}}}-\underset{Cl}{\overset{H}{\underset{|}{\overset{|}{C}}}}-H$$

Chloroethane

We often indicate the presence of a catalyst by adding a label over the reaction arrow.

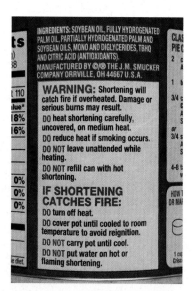

▲ Partially hydrogenated vegetable oil is a saturated fat that is made by hydrogenating unsaturated fats.

If the alkene itself is also unsymmetrical, then the addition of an unsymmetrical reagent leads to the potential for two different products. For example, when HCl reacts with propene, two products are possible:

When this reaction is carried out in the lab, however, only the 2-chloropropane forms. We can predict the product of the addition of an unsymmetrical reagent to an unsymmetrical alkene with Markovnikov's rule, which states the following:

> **When a polar reagent is added to an unsymmetrical alkene, the positive end (the least electronegative part) of the reagent adds to the carbon atom that has the most hydrogen atoms.**

In most reactions of this type, the positive end of the reagent is hydrogen; therefore, the hydrogen atom bonds to the carbon atom that already contains the most hydrogen atoms.

---

**Example 21.6    Alkene Addition Reactions**

Determine the products of each reaction.

(a) $CH_3CH_2CH{=}CH_2 + Br_2 \longrightarrow$    (b) $CH_3CH_2CH{=}CH_2 + HBr \longrightarrow$

**SOLUTION**

(a) The reaction of 1-butene with bromine is an example of a symmetric addition. The bromine adds across the double bond, and each carbon forms a single bond to a bromine atom.

(b) The reaction of 1-butene with hydrogen bromide is an example of an unsymmetrical addition. Apply Markovnikov's rule to determine which carbon the hydrogen bonds with and which carbon the bromine atom bonds with. Markovnikov's rule predicts that the hydrogen bonds to the end carbon in this case.

**FOR PRACTICE 21.6**    Determine the products of each reaction.

## 21.7 Aromatic Hydrocarbons

As you might imagine, determining the structure of organic compounds has not always been easy. During the mid-1800s chemists were working to determine the structure of a particularly stable organic compound named benzene ($C_6H_6$). In 1865, Friedrich August Kekulé (1829–1896) had a dream in which he envisioned chains of carbon atoms as snakes. The snakes danced before him, and one of them twisted around and bit its tail. Based on that vision, Kekulé proposed the following structure for benzene:

This structure has alternating single and double bonds. When we examine the carbon–carbon bond lengths in benzene, however, we find that all the bonds are the same length, which indicates that the following resonance structures are a more accurate representation of benzene:

Recall from Section 9.8 that the actual structure of a molecule represented by resonance structures is intermediate between the two resonance structures and is called a *resonance hybrid*.

The true structure of benzene is a hybrid of the two resonance structures. We often represent benzene with the following carbon skeletal formula (or line formula):

The ring represents the delocalized $\pi$ electrons that occupy the molecular orbital shown superimposed on the ball-and-stick model. When drawing benzene rings, either by themselves or as parts of other compounds, organic chemists use either this diagram or just one of the resonance structures with alternating double bonds. Both representations indicate the same thing—a benzene ring.

The benzene ring structure occurs in many organic compounds. An atom or group of atoms can be substituted for one or more of the six hydrogen atoms on the ring to form compounds referred to as substituted benzenes, such as chlorobenzene and phenol:

Chlorobenzene      Phenol

Because many compounds containing benzene rings have pleasant aromas, benzene rings are also called *aromatic rings*, and compounds containing them are called *aromatic compounds*. Aromatic compounds are responsible for the pleasant smells of cinnamon, vanilla, and jasmine.

## Naming Aromatic Hydrocarbons

Monosubstituted benzenes—benzenes in which only one of the hydrogen atoms has been substituted—are often named as derivatives of benzene:

Ethylbenzene      Bromobenzene

These names take the general form:

(name of substituent)benzene

However, many monosubstituted benzenes have names that can only be learned through familiarity. Some common ones are shown here:

| | | | |
|---|---|---|---|
| ⬡—CH₃ | ⬡—NH₂ | ⬡—OH | ⬡—CH=CH₂ |
| Toluene | Aniline | Phenol | Styrene |

Some substituted benzenes, especially those with large substituents, are named by treating the benzene ring as the substituent. In these cases, the benzene substituent is referred to as a **phenyl group**:

$CH_3-CH_2-CH-CH_2-CH_2-CH_2-CH_3$
3-Phenylheptane

$CH_2=CH-CH_2-CH-CH_2-CH_3$
4-Phenyl-1-hexene

**Disubstituted benzenes**—benzenes in which two hydrogen atoms have been substituted—are numbered, and the substituents are listed alphabetically. We determine the order of numbering on the ring by the alphabetical order of the substituents:

1-Chloro-3-ethylbenzene

1-Bromo-2-chlorobenzene

When the two substituents are identical, we use the prefix *di-*:

1,2-Dichlorobenzene        1,3-Dichlorobenzene        1,4-Dichlorobenzene

Also in common use, in place of numbering, are the prefixes *ortho* (1,2 disubstituted), *meta* (1,3 disubstituted), and *para* (1,4 disubstituted):

*ortho*-Dichlorobenzene
or
*o*-Dichlorobenzene

*meta*-Dichlorobenzene
or
*m*-Dichlorobenzene

*para*-Dichlorobenzene
or
*p*-Dichlorobenzene

Compounds containing fused aromatic rings are called polycyclic aromatic hydrocarbons. Some common examples (shown in Figure 21.5▶, on the next page) include naphthalene, the substance that composes mothballs, and pyrene, a carcinogen found in cigarette smoke.

► **FIGURE 21.5 Polycyclic Aromatic Compounds** The structures of some common polycyclic aromatic compounds contain fused rings.

Naphthalene

Anthracene

Pyrene

Tetracene

## Reactions of Aromatic Compounds

We might expect benzene to react similarly to the alkenes, readily undergoing addition reactions across its double bonds. However, because of electron delocalization around the ring and the resulting greater stability, benzene does not typically undergo addition reactions. Instead, benzene undergoes substitution reactions in which the hydrogen atoms are replaced by other atoms or groups of atoms as shown in the following examples:

The substances shown over the arrows are catalysts needed to increase the rate of the reaction.

Chlorobenzene

Ethylbenzene

# 21.8 Functional Groups

Most other families of organic compounds are hydrocarbons with a *functional group*—a characteristic atom or group of atoms—inserted into the hydrocarbon. A group of organic compounds that all have the same functional group is a *family*. For example, the members of the family of alcohols each have an —OH functional group and the general formula R—OH, where R represents a hydrocarbon group. (That is, we refer to the hydrocarbon group as an "R group.") Some specific examples include methanol and isopropyl alcohol (also known as rubbing alcohol):

Methanol            Isopropyl alcohol

The presence of a functional group in a hydrocarbon alters the properties of the compound significantly. For example, methane is a nonpolar gas. By contrast, methanol—methane with an —OH group substituted for one of the hydrogen atoms—is a polar, hydrogen-bonded liquid at room temperature. Although each member of a family is unique and different, their common functional group causes some similarities in both their physical and chemical properties. Table 21.10 lists some common functional groups, their general formulas, and an example of each.

**TABLE 21.10 Some Common Functional Groups**

| Family | General Formula* | Condensed General Formula | Example | Name |
|---|---|---|---|---|
| Alcohols | R—OH | ROH | $CH_3CH_2OH$ | Ethanol (ethyl alcohol) |
| Ethers | R—O—R | ROR | $CH_3OCH_3$ | Dimethyl ether |
| Aldehydes | R—$\overset{\overset{\displaystyle O}{\|\|}}{C}$—H | RCHO | $CH_3$—$\overset{\overset{\displaystyle O}{\|\|}}{C}$—H | Ethanal (acetaldehyde) |
| Ketones | R—$\overset{\overset{\displaystyle O}{\|\|}}{C}$—R | RCOR | $CH_3$—$\overset{\overset{\displaystyle O}{\|\|}}{C}$—$CH_3$ | Propanone (acetone) |
| Carboxylic acids | R—$\overset{\overset{\displaystyle O}{\|\|}}{C}$—OH | RCOOH | $CH_3$—$\overset{\overset{\displaystyle O}{\|\|}}{C}$—OH | Ethanoic acid (acetic acid) |
| Esters | R—$\overset{\overset{\displaystyle O}{\|\|}}{C}$—OR | RCOOR | $CH_3$—$\overset{\overset{\displaystyle O}{\|\|}}{C}$—$OCH_3$ | Methyl acetate |
| Amines | R—$\overset{\overset{\displaystyle R}{\|}}{N}$—R | $R_3N$ | $CH_3CH_2$—$\overset{\overset{\displaystyle H}{\|}}{N}$—H | Ethylamine |

*In ethers, ketones, esters, and amines, the R groups may be the same or different.

## 21.9 Alcohols

As we discussed in Section 21.8, **alcohols** are organic compounds containing the $-OH$ functional group, or **hydroxyl group**, and they have the general formula $R-OH$. In addition to methanol and isopropyl alcohol, ethanol and 1-butanol (shown here) are also common alcohols:

$$CH_3-CH_2-OH$$
Ethanol

$$CH_3-CH_2-CH_2-CH_2-OH$$
1-Butanol

### Naming Alcohols

The names of alcohols are like the names of alkanes with the following differences:

- The base chain is the longest continuous carbon chain that contains the $-OH$ functional group.
- The base name has the ending -ol.
- We number the base chain to assign the $-OH$ group the lowest possible number.
- We insert a number indicating the position of the $-OH$ group just before the base name. For example:

$$CH_3CH_2CH_2CHCH_3$$
$$|$$
$$OH$$
2-Pentanol

$$CH_2-CH_2-CH-CH_3$$
$$|\qquad\qquad|$$
$$OH\qquad\qquad CH_3$$
3-Methyl-1-butanol

### About Alcohols

The familiar alcohol in alcoholic beverages, ethanol, is most commonly formed by the yeast fermentation of sugars, such as glucose, from fruits and grains:

$$C_6H_{12}O_6 \xrightarrow{\text{yeast}} 2\ CH_3CH_2OH + 2\ CO_2$$
Glucose $\qquad\qquad$ Ethanol

Alcoholic beverages contain ethanol, water, and a few other components that impart flavor and color. Beer usually contains 3–6% ethanol. Wine contains about 12–14% ethanol, and spirits—beverages like whiskey, rum, or tequila—range from 40% to 80% ethanol, depending on their *proof*. The proof of an alcoholic beverage is twice the percentage of its ethanol content, so an 80-proof whiskey contains 40% ethanol. Ethanol is used as a gasoline additive because it increases the octane rating of gasoline and fosters its complete combustion, reducing the levels of certain pollutants such as carbon monoxide and the precursors of ozone.

Isopropyl alcohol (or 2-propanol) is available at any drugstore under the name of rubbing alcohol. It is commonly used as a disinfectant for wounds and to sterilize medical instruments. Isopropyl alcohol should never be consumed internally, as it is highly toxic. Four ounces of isopropyl alcohol can cause death. A third common alcohol is methanol, also called wood alcohol. Methanol is commonly used as a laboratory solvent and as a fuel additive. Like isopropyl alcohol, methanol is toxic and should never be consumed.

### Alcohol Reactions

Alcohols undergo a number of reactions including substitution, elimination (or dehydration), and oxidation. Alcohols also react with active metals to form strong bases.

## Substitution

Alcohols react with acids such as HBr to form halogenated hydrocarbons:

$$ROH + HBr \longrightarrow R{-}Br + H_2O$$

In these reactions, the halogen replaces the hydroxyl group on the alcohol. For example, ethanol reacts with hydrobromic acid to form bromoethane and water:

$$CH_3CH_2OH + HBr \longrightarrow CH_3CH_2Br + H_2O$$

## Elimination (or Dehydration)

In the presence of concentrated acids such as $H_2SO_4$, alcohols react and eliminate water, forming an alkene. For example, ethanol eliminates water to form ethene according to the reaction:

$$
\begin{array}{cc}
CH_2{-}CH_2 \\
\;\,|\qquad\; | \\
\;\,H\qquad OH
\end{array}
\xrightarrow{\;H_2SO_4\;}
CH_2{=}CH_2 + H_2O
$$

## Oxidation

In organic chemistry, we think of oxidation and reduction in terms of the changes to the carbon atoms in the molecule. Thus, oxidation is the gaining of oxygen or the losing of hydrogen by a carbon atom. Reduction is the loss of oxygen or the gaining of hydrogen by a carbon atom. We can draw a series showing relative states of oxidation:

In this view, an alcohol is a partially oxidized hydrocarbon; it can be further oxidized to form an aldehyde or carboxylic acid, or it can be reduced to form a hydrocarbon (but this is rare). For example, ethanol can be oxidized to acetic acid according to the reaction:

$$CH_3CH_2OH \xrightarrow[\;H_2SO_4\;]{\;Na_2Cr_2O_7\;} CH_3COOH$$

## Reaction with Active Metals

Alcohols react with active metals, such as sodium, much as water does. For example, methanol reacts with sodium to form *sodium methoxide* and hydrogen gas:

$$CH_3OH + Na \longrightarrow CH_3ONa + \tfrac{1}{2}H_2$$

Sodium methoxide

The reaction of *water* with sodium produces *sodium hydroxide* and hydrogen gas:

$$H_2O + Na \longrightarrow NaOH + \tfrac{1}{2}H_2$$

In both cases, a strong base forms ($OH^-$ for water and $CH_3O^-$ for methanol).

## Example 21.7 | Alcohol Reactions

Determine the type of reaction (substitution, dehydration, oxidation, or reaction with an active metal) that occurs in each case, and write formulas for the products.

(a) 
$$CH_3-\underset{\underset{CH_3}{|}}{CH}-CH_2-CH_2-OH + HBr \longrightarrow$$

(b) 
$$CH_3-\underset{\underset{CH_3}{|}}{CH}-CH_2-CH_2-OH \xrightarrow[H_2SO_4]{Na_2Cr_2O_7}$$

### SOLUTION

| | |
|---|---|
| (a) An alcohol reacting with an acid is an example of a *substitution reaction*. The product of the substitution reaction is a halogenated hydrocarbon and water. | $CH_3-\underset{\underset{CH_3}{\|}}{CH}-CH_2-CH_2-OH + HBr \longrightarrow$ <br><br> $CH_3-\underset{\underset{CH_3}{\|}}{CH}-CH_2-CH_2-Br + H_2O$ |
| (b) An alcohol in solution with sodium dichromate and acid undergoes an *oxidation reaction*. The product of the oxidation reaction is a carboxylic acid functional group. (We discuss carboxylic acid functional groups in detail in Section 21.11.) | $CH_3-\underset{\underset{CH_3}{\|}}{CH}-CH_2-CH_2-OH \xrightarrow[H_2SO_4]{Na_2Cr_2O_7}$ <br><br> $CH_3-\underset{\underset{CH_3}{\|}}{CH}-CH_2-\overset{\overset{O}{\|}}{C}-OH$ |

**FOR PRACTICE 21.7** Determine the type of reaction (substitution, dehydration, oxidation, or reaction with an active metal) that occurs in each case, and write formulas for the products.

(a) $CH_3CH_2OH + Na \longrightarrow$

(b) $CH_3-\underset{\underset{CH_3}{|}}{CH}-CH_2-OH \xrightarrow{H_2SO_4}$

## 21.10 Aldehydes and Ketones

**Aldehydes** and **ketones** have the following general formulas:

$$\underset{\text{Aldehyde}}{R-\overset{\overset{O}{\|}}{C}-H} \qquad \underset{\text{Ketone}}{R-\overset{\overset{O}{\|}}{C}-R}$$

The condensed structural formula for aldehydes is RCHO and that for ketones is RCOR.

The functional group for both aldehydes and ketones is the **carbonyl group**:

$$\overset{\overset{O}{\|}}{\underset{\diagup^{C}\diagdown}{}}$$

Ketones have an R group attached to both sides of the carbonyl, while aldehydes have one R group and a hydrogen atom. (An exception is formaldehyde, which is an aldehyde with two H atoms bonded to the carbonyl group.)

$$H-\overset{\overset{O}{\|}}{\underset{}{C}}-H$$

Formaldehyde or methanal

Other common aldehydes and ketones are shown in Figure 21.6▼.

| | | | |
|---|---|---|---|
| $CH_3-\overset{\overset{\displaystyle O}{\|\|}}{C}-H$ | $CH_3-CH_2-\overset{\overset{\displaystyle O}{\|\|}}{C}-H$ | $CH_3-\overset{\overset{\displaystyle O}{\|\|}}{C}-CH_3$ | $CH_3-CH_2-\overset{\overset{\displaystyle O}{\|\|}}{C}-CH_3$ |
| Acetaldehyde or ethanal | Propanal | Acetone or propanone | Butanone |

▲ **FIGURE 21.6  Common Aldehydes and Ketones**

## Naming Aldehydes and Ketones

Many aldehydes and ketones have common names that we can learn only by becoming familiar with them, but we can systematically name simple aldehydes according to the number of carbon atoms in the longest continuous carbon chain that contains the carbonyl group. We form the base name from the name of the corresponding alkane by dropping the -*e* and adding the ending –*al*:

$$CH_3-CH_2-CH_2-\overset{\overset{\displaystyle O}{\|\|}}{C}-H \qquad CH_3-CH_2-CH_2-CH_2-\overset{\overset{\displaystyle O}{\|\|}}{C}-H$$

Butanal    Pentanal

We name simple ketones according to the longest continuous carbon chain containing the carbonyl group, forming the base name from the name of the corresponding alkane by dropping the letter -*e* and adding the ending -*one*. For ketones, we number the chain to give the carbonyl group the lowest possible number:

$$CH_3-CH_2-CH_2-\overset{\overset{\displaystyle O}{\|\|}}{C}-CH_3 \qquad CH_3-CH_2-CH_2-\overset{\overset{\displaystyle O}{\|\|}}{C}-CH_2-CH_3$$

2-Pentanone    3-Hexanone

## About Aldehydes and Ketones

The most familiar aldehyde is probably formaldehyde, shown earlier in this section. Formaldehyde is a gas with a pungent odor. It is often mixed with water to make formalin, a preservative and disinfectant. Formaldehyde is also found in wood smoke, which is one reason that smoking is an effective method of food preservation—the formaldehyde kills bacteria. Aromatic aldehydes, those that also contain an aromatic ring, have pleasant aromas. For example, vanillin causes the smell of vanilla, cinnamaldehyde is the sweet-smelling component of cinnamon, and benzaldehyde accounts for the smell of almonds (Figure 21.7▼):

Vanillin          Cinnamaldehyde

Benzaldehyde

▲ **FIGURE 21.7  The Nutty Aroma of Almonds**   Benzaldehyde is partly responsible for the smell of almonds.

Ionone

▲ **FIGURE 21.8 The Fragrance of Raspberries** Ionone is partly responsible for the smell of raspberries.

The most familiar ketone is acetone, the main component of nail polish remover. Other ketones have more pleasant aromas. For example, carvone is largely responsible for the smell of spearmint, 2-heptanone (among other compounds) for the smell of cloves, and ionone for the smell of raspberries (Figure 21.8▲):

Carvone

$CH_3—CH_2—CH_2—CH_2—CH_2—\overset{\displaystyle O}{\overset{\|}{C}}—CH_3$
2-Heptanone

## Aldehyde and Ketone Reactions

Aldehydes and ketones can be formed by the *oxidation* of alcohols. For example, ethanol can be oxidized to ethanal, and 2-propanol can be oxidized to 2-propanone (or acetone):

Ethanol $\xrightarrow{\text{oxidation}}$ Ethanal

2-Propanol $\xrightarrow{\text{oxidation}}$ 2-Propanone

In the reverse reaction, an aldehyde or ketone is reduced to an alcohol. For example, 2-butanone can be reduced to 2-butanol in the presence of a reducing agent:

2-Butanone $\xrightarrow{\text{reduction}}$ 2-Butanol

The carbonyl group in aldehydes and ketones is unsaturated, much like the double bond in an alkene. Because of this feature, the most common reactions of aldehydes and ketones are **addition reactions**. However, in contrast to the carbon–carbon double bond in alkenes, which is nonpolar, the double bond in the carbonyl group is highly polar (Figure 21.9▶). Consequently, additions across the double bond result in the more electronegative part of the reagent bonding to the carbon atom and the less electronegative part (often hydrogen) bonding to the oxygen atom. For example, HCN adds across the carbonyl double bond in formaldehyde:

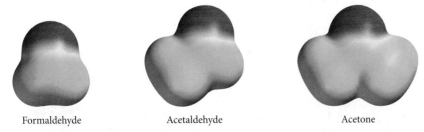

Formaldehyde     Acetaldehyde     Acetone

▲ **FIGURE 21.9** **Charge Density Plots of the Carbonyl Group** The carbonyl group is highly polar, as shown in these plots of electrostatic potential.

## 21.11 Carboxylic Acids and Esters

**Carboxylic acids** and **esters** have the general formulas:

$$\begin{array}{cc} \underset{\text{Carboxylic acid}}{\overset{\displaystyle O \atop \parallel}{R-C-OH}} & \underset{\text{Ester}}{\overset{\displaystyle O \atop \parallel}{R-C-OR}} \end{array}$$

> The condensed structural formula for carboxylic acids is RCOOH; for esters it is RCOOR.

Figure 21.10► shows the structures of some common carboxylic acids and esters.

### Naming Carboxylic Acids and Esters

We systematically name carboxylic acids according to the number of carbon atoms in the longest chain containing the —COOH functional group. We form the base name by dropping the -e from the name of the corresponding alkane and adding the ending -oic acid:

$$\underset{\text{Propanoic acid}}{\overset{\displaystyle O \atop \parallel}{CH_3-CH_2-C-OH}} \qquad \underset{\text{Pentanoic acid}}{\overset{\displaystyle O \atop \parallel}{CH_3-CH_2-CH_2-CH_2-C-OH}}$$

We name esters as if they were derived from a carboxylic acid by replacing the H on the OH with an alkyl group. The R group from the parent acid forms the base name of the compound. We change the -ic on the name of the corresponding carboxylic acid to -ate, and drop acid, naming the R group that replaced the H on the carboxylic acid as an alkyl group with the ending -yl, as shown in the following examples:

$$\underset{\text{Propanoic acid}}{\overset{\displaystyle O \atop \parallel}{CH_3-CH_2-C-OH}} \qquad \underset{\text{Pentanoic acid}}{\overset{\displaystyle O \atop \parallel}{CH_3-CH_2-CH_2-CH_2-C-OH}}$$

$$\underset{\text{Methyl propanoate}}{\overset{\displaystyle O \atop \parallel}{CH_3-CH_2-C-OCH_3}} \qquad \underset{\text{Ethyl pentanoate}}{\overset{\displaystyle O \atop \parallel}{CH_3-CH_2-CH_2-CH_2-C-OCH_2CH_3}}$$

### About Carboxylic Acids and Esters

Like all acids, carboxylic acids taste sour. The most familiar carboxylic acid is ethanoic acid, better known by its common name, acetic acid. Acetic acid is the active ingredient in vinegar. It can form by the oxidation of ethanol, which is why wines left open to air become sour. Some yeasts and bacteria also form acetic acid when they metabolize sugars in bread dough. These are added to bread dough to make sourdough bread. Other common carboxylic acids include methanoic acid (formic acid), present in bee

$$\underset{\text{Ethanoic acid or acetic acid}}{\overset{\displaystyle O \atop \parallel}{CH_3-C-OH}}$$

$$\underset{\text{Butanoic acid}}{\overset{\displaystyle O \atop \parallel}{CH_3-CH_2-CH_2-C-OH}}$$

$$\underset{\text{Methyl butanoate}}{\overset{\displaystyle O \atop \parallel}{CH_3-CH_2-CH_2-C-O-CH_3}}$$

$$\underset{\text{Ethyl propanoate}}{\overset{\displaystyle O \atop \parallel}{CH_3-CH_2-C-O-CH_2-CH_3}}$$

▲ **FIGURE 21.10** **Common Carboxylic Acids and Esters**

Citric acid

▲ **FIGURE 21.11** **The Tart Taste of Limes** Citric acid is partly responsible for the sour taste of limes.

stings and ant bites; lactic acid, which collects in muscles after intense exercise causing soreness; and citric acid, found in limes, lemons, and oranges (Figure 21.11▲):

Formic or methanoic acid          Lactic acid

Esters are best known for their sweet smells. Methyl butanoate is largely responsible for the smell and taste of apples, and ethyl butanoate is partly responsible for the smell and taste of pineapples (see Figure 21.12▼):

$$CH_3—CH_2—CH_2—\overset{\overset{\displaystyle O}{\|}}{C}—O—CH_3$$
Methyl butanoate

$$CH_3—CH_2—CH_2—\overset{\overset{\displaystyle O}{\|}}{C}—O—CH_2—CH_3$$
Ethyl butanoate

▲ **FIGURE 21.12** **The Aroma of Pineapple** Ethyl butanoate is partly responsible for the aroma of pineapples.

## Carboxylic Acid and Ester Reactions

Carboxylic acids act as weak acids in solution:

$$RCOOH(aq) + H_2O(l) \rightleftharpoons H_3O^+(aq) + RCOO^-(aq)$$

Like all acids, carboxylic acids react with strong bases via neutralization reactions. For example, propanoic acid reacts with sodium hydroxide to form sodium propanoate and water:

$$CH_3CH_2COOH(aq) + NaOH(aq) \longrightarrow CH_3CH_2COO^-Na^+(aq) + HOH(l)$$

A carboxylic acid reacts with an alcohol to form an ester via a **condensation reaction**, a reaction in which two (or more) organic compounds join, often with the loss of water (or some other small molecule):

$$R—\overset{\overset{\displaystyle O}{\|}}{C}—OH + HO—R' \xrightarrow{H_2SO_4} R—\overset{\overset{\displaystyle O}{\|}}{C}—O—R' + H_2O$$
Acid          Alcohol                              Ester          Water

An important example of this reaction is the formation of acetylsalicylic acid (aspirin) from ethanoic acid (acetic acid) and salicylic acid (originally obtained from the bark of the willow tree):

Acetic acid          Salicylic acid                    Acetylsalicylic acid

If we subject a carboxylic acid to high temperatures, it undergoes a condensation reaction with itself to form an acid anhydride (anhydride means "without water"):

$$RCOOH(aq) + HOOCR(aq) \longrightarrow RCOOOCR(aq) + HOH(aq)$$
Acid anhydride

We can add water to an acid anhydride to reverse the reaction just shown and regenerate the carboxylic acid molecules.

Conceptual connection box

**Oxidation** Arrange the compounds from least oxidized to most oxidized.

(a) CH₃—C(=O)—CH₃

(b) CH₃—C(=O)—O—CH₃

(c) CH₃—CH₂—CH₃

(d) CH₃—CH(OH)—CH₃

CONCEPTUAL
CONNECTION **21.3**

## 21.12 Ethers

**Ethers** are organic compounds with the general formula ROR. The two R groups may be identical or they may be different. Some common ethers are shown in Figure 21.13▼.

### Naming Ethers

Common names for ethers have the format:
                    (R group 1) (R group 2) ether

CH₃—O—CH₃                    CH₃—O—CH₂—CH₃
Dimethyl ether                 Ethyl methyl ether

CH₃—CH₂—O—CH₂—CH₃
Diethyl ether

▲ FIGURE 21.13 **Ethers**

If the two R groups differ, we use each of their names in alphabetical order. If the two R groups are the same, we use the prefix *di-*. Some examples include:

$$H_3C—CH_2—CH_2—O—CH_2—CH_2—CH_3$$
Dipropyl ether

$$H_3C—CH_2—O—CH_2—CH_2—CH_3$$
Ethyl propyl ether

## About Ethers

The most common ether is diethyl ether. Diethyl ether is a useful laboratory solvent because it can dissolve many organic compounds and it has a low boiling point (34.6 °C). The low boiling point allows for easy removal of the solvent. Diethyl ether was used as a general anesthetic for many years. When inhaled, diethyl ether depresses the central nervous system, causing unconsciousness and insensitivity to pain. Its use as an anesthetic, however, has decreased in recent years because other compounds have the same anesthetic effect with fewer side effects (such as nausea).

## 21.13 Amines

The simplest nitrogen-containing compound is ammonia ($NH_3$). **Amines** are organic compounds containing nitrogen that are derived from ammonia with one or more of the hydrogen atoms replaced by alkyl groups. Like ammonia, amines are weak bases. We systematically name amines according to the hydrocarbon groups attached to the nitrogen and assign the ending *-amine*:

$$CH_3—CH_2—\overset{\displaystyle |}{\underset{\displaystyle |}{N}}—H$$
$$\text{H}$$
Ethylamine

$$CH_3—CH_2—\overset{\displaystyle |}{\underset{\displaystyle |}{N}}—H$$
$$CH_3$$
Ethylmethylamine

Amines are most commonly known for their awful odors. When a living organism dies, the bacteria that feast on its proteins emit amines. For example, trimethylamine causes the smell of rotten fish, and cadaverine causes the smell of decaying animal flesh:

$$\overset{\displaystyle CH_3}{\underset{\displaystyle |}{CH_3—N—CH_3}}$$
Trimethylamine

$$NH_2—CH_2—CH_2—CH_2—CH_2—CH_2—NH_2$$
Cadaverine

## Amine Reactions

Just as carboxylic acids act as weak acids, so amines act as weak bases:

$$RNH_2(aq) + H_2O(l) \rightleftharpoons RNH_3^+(aq) + OH^-(aq)$$

Like all bases, amines react with strong acids to form salts called ammonium salts. For example, methylamine reacts with hydrochloric acid to form methylammonium chloride:

$$CH_3NH_2(aq) + HCl(aq) \longrightarrow CH_3NH_3^+Cl^-(aq)$$
Methylammonium chloride

An important amine reaction that we will see again in Chapter 22 is the condensation reaction between a carboxylic acid and an amine to form an amide:

$$CH_3COOH(aq) + HNHR(aq) \longrightarrow CH_3CONHR(aq) + HOH(l)$$

This reaction is responsible for the formation of proteins from amino acids (see Section 22.4).

# Self-Assessment Quiz

**Q1.** Which property of carbon is related to its ability to form a large number of compounds?

   a) its tendency to form four covalent bonds
   b) its ability to form double and triple bonds
   c) its tendency to catenate
   d) all of the above

**Q2.** What is the correct formula for the alkane (noncyclical) containing eight carbon atoms?

   a) $C_8H_{16}$     b) $C_8H_{18}$     c) $C_8H_{14}$     d) $C_8H_8$

**Q3.** Which structure is not an isomer of $CH_3CH_2CH_2CH_2CH_3$?

   a)
$$CH_3 \qquad CH_3$$
$$CH_2-CH_2-CH_2$$

   b)
$$CH_3-CH-CH_2-CH_3$$
$$\qquad\quad CH_3$$

   c)
$$\qquad\quad CH_3$$
$$CH_3-C-CH_3$$
$$\qquad\quad CH_3$$

   d) None of the above (all are isomers).

**Q4.** Which structure can exhibit optical isomerism?

   a)
$$\begin{array}{ccc} H & H & \\ | & | & \\ H-C-C-Cl \\ | & | & \\ H & H & \end{array}$$

   b)
$$\begin{array}{ccc} H & H & Cl \\ | & | & | \\ H-C-C-C-H \\ | & | & | \\ Cl & H & H \end{array}$$

   c)
$$\begin{array}{ccc} Br & Br & H \\ | & | & | \\ H-C-C-C-H \\ | & | & | \\ H & H & H \end{array}$$

   d)
$$\begin{array}{ccc} Br & H & Br \\ | & | & | \\ H-C-C-C-Br \\ | & | & | \\ Br & H & Br \end{array}$$

**Q5.** Name the compound:

$$\qquad\qquad\qquad\qquad CH_3$$
$$CH_3-CH_2-CH_2-CH-CH-CH_3$$
$$\qquad\qquad\qquad CH_2$$
$$\qquad\qquad\qquad CH_3$$

   a) 4-ethyl-5-methylhexane
   b) 3-ethyl-2-methylhexane
   c) 3-ethyl-2-methylnonane
   d) 4-methyl-5-ethylhexane

**Q6.** Name the compound:

$$CH_3-C\equiv C-CH-CH_3$$
$$\qquad\qquad\qquad CH_3$$

   a) 4-methyl-2-pentyne
   b) 2-methyl-3-pentyne
   c) 2-methyl-3-hexyne
   d) 4-methyl-2-hexyne

**Q7.** Determine the product of the reaction:

$$\qquad\qquad\qquad CH_3$$
$$CH_3-CH=C-CH_3 + HBr \longrightarrow$$

   a)
$$\qquad\qquad\qquad\quad CH_3$$
$$CH_2Br-CH=C-CH_3$$

   b)
$$\qquad\qquad\qquad\quad CH_3$$
$$CH_3-CH_2-CH-CH_3$$

   c)
$$\qquad\qquad\qquad CH_3$$
$$CH_3-CH-CH-CH_3$$
$$\qquad\quad Br$$

   d)
$$\qquad\qquad\qquad CH_3$$
$$CH_3-CH_2-C-CH_3$$
$$\qquad\qquad\qquad Br$$

**Q8.** Determine the product of the reaction:

$$CH_3-CH-CH_2-OH + HCl \longrightarrow$$
$$\qquad\quad CH_3$$

   a)
$$CH_3-CH-CH_3$$
$$\qquad\quad Cl$$

   b)
$$CH_3-CH-CH_3$$
$$\qquad\quad CH_3$$

   c)
$$\qquad\quad Cl$$
$$CH_3-C-CH_2-OH$$
$$\qquad\quad CH_3$$

   d)
$$CH_3-CH-CH_2-Cl$$
$$\qquad\quad CH_3$$

*Continued—*

**Q9.** Which compound is an ester?

a) $CH_3—CH_2—O—CH_3$

b) $CH_3—CH_2—\overset{\displaystyle O}{\overset{\|}{C}}—OH$

c) $CH_3—\overset{\displaystyle O}{\overset{\|}{C}}—CH_3$

d) $CH_3—\overset{\displaystyle O}{\overset{\|}{C}}—O—CH_2CH_3$

**Q10.** Which compound is most likely to have a foul odor?

a) $CH_3—\overset{\displaystyle O}{\overset{\|}{C}}—O—CH_3$

b) $CH_3—CH_2—NH_2$

c) $CH_3—CH_2—O—CH_3$

d) $CH_3—CH_2—CH_2—OH$

**Answers: 1.** (d) **2.** (b) **3.** (a) **4.** (c) **5.** (b) **6.** (a) **7.** (d) **8.** (d) **9.** (d) **10.** (b)

# Chapter 21 IN REVIEW

## Key Terms

**Section 21.1**
organic molecule (980)
organic chemistry (980)

**Section 21.3**
alkane (982)
alkene (982)
alkyne (982)
aromatic hydrocarbon (982)
aliphatic hydrocarbon (982)
structural isomers (983)
structural formula (983)
stereoisomers (985)
optical isomers (985)
enantiomers (986)
chiral (986)
dextrorotatory (986)

levorotatory (986)
racemic mixture (986)

**Section 21.4**
saturated hydrocarbon (988)

**Section 21.5**
unsaturated hydrocarbon (992)
geometric (cis–trans) isomerism (996)

**Section 21.7**
phenyl group (1001)
disubstituted benzene (1001)

**Section 21.9**
alcohol (1004)
hydroxyl group (1004)

**Section 21.10**
aldehyde (1006)
ketone (1006)
carbonyl group (1006)
addition reaction (1008)

**Section 21.11**
carboxylic acid (1009)
esters (1009)
condensation reaction (1010)

**Section 21.12**
ether (1011)

**Section 21.13**
amine (1012)

## Key Concepts

### Fragrances and Odors (21.1)
▶ Organic chemistry is the study of organic compounds, which contain carbon (and other elements including hydrogen, oxygen, and nitrogen).

### Carbon (21.2)
▶ Carbon forms more compounds than all the other elements combined.

▶ Carbon's four valence electrons (in conjunction with its size) allow carbon to form four bonds (in the form of single, double, or triple bonds).
▶ Carbon also has the capacity to catenate, to form long chains, because of the strength of the carbon–carbon bond.

### Hydrocarbons (21.3)
▶ Organic compounds containing only carbon and hydrogen are hydrocarbons, the key components of our world's fuels.

- We categorize hydrocarbons into four different types: alkanes, alkenes, alkynes, and aromatic hydrocarbons.
- Stereoisomers are molecules that feature the same atoms bonded in the same order but arranged differently in space. Optical isomerism, a type of stereoisomerism, occurs when two molecules are nonsuperimposable mirror images of one another.

## Alkanes (21.4)

- Alkanes are saturated hydrocarbons—they contain only single bonds and are therefore represented by the generic formula $C_nH_{2n+2}$. Alkane names always end in -ane.

## Alkenes and Alkynes (21.5)

- Alkenes and alkynes are unsaturated hydrocarbons—they contain double bonds (alkenes) or triple bonds (alkynes) and are represented by the generic formulas $C_nH_{2n}$ and $C_nH_{2n-2}$, respectively.
- Alkene names end in -ene and alkynes end in -yne.
- Because rotation about a double bond is severely restricted, geometric (or cis–trans) isomerism occurs in alkenes.

## Hydrocarbon Reactions (21.6)

- The most common hydrocarbon reaction is probably combustion, in which hydrocarbons react with oxygen to form carbon dioxide and water; this reaction is exothermic and is used to provide most of our society's energy.
- Alkanes can also undergo substitution reactions, where heat or light causes another atom, commonly a halogen such as bromine, to be substituted for a hydrogen atom.
- Unsaturated hydrocarbons undergo addition reactions. If the addition reaction is between two unsymmetrical molecules, Markovnikov's rule predicts that the positive end of the polar reagent adds to the carbon with the most hydrogen atoms.

## Aromatic Hydrocarbons (21.7)

- Aromatic hydrocarbons contain six-membered benzene rings represented with alternating single and double bonds that become equivalent through resonance. These compounds are called aromatic because they often produce pleasant fragrances.
- Because of the stability of the aromatic ring, benzene is more stable than a straight-chain alkene, and it undergoes substitution rather than addition reactions.

## Functional Groups (21.8)

- Characteristic groups of atoms, such as hydroxyl (—OH), are functional groups. Molecules that contain the same functional group have similar chemical and physical properties and are referred to as families.

## Alcohols (21.9)

- The family of alcohols contains the —OH group and is named with the suffix -ol.

- Alcohols are commonly used in gasoline, in alcoholic beverages, and in sterilization procedures.
- Alcohols undergo substitution reactions, in which a substituent such as a halogen replaces the hydroxyl group.
- Alcohols undergo elimination reactions, in which water is eliminated across a bond to form an alkene, and oxidation or reduction reactions.
- Alcohols also react with active metals to form alkoxide ions and hydrogen gas.

## Aldehydes and Ketones (21.10)

- Aldehydes and ketones both contain a carbonyl group (a carbon atom double-bonded to oxygen).
- In aldehydes, the carbonyl group is at the end of a carbon chain, while in ketones it is between two other carbon atoms.
- Aldehydes are named with the suffix -al and ketones with the suffix -one.
- A carbonyl can be formed by the oxidation of an alcohol or reverted to an alcohol by reduction.
- Like alkenes, carbonyls undergo addition reactions; however, because the carbon–oxygen bond is highly polar, the electronegative component of the reagent always adds to the carbon atom, and the less electronegative part adds to the oxygen.

## Carboxylic Acids and Esters (21.11)

- Carboxylic acids contain a carbonyl group and a hydroxide on the same carbon and are named with the suffix -oic acid.
- Esters contain a carbonyl group bonded to an oxygen atom that is in turn bonded to an R group; they are named with the suffix -oate.
- Carboxylic acids taste sour, such as acetic acid in vinegar, while esters smell sweet.
- Carboxylic acids react as weak acids but can also form esters through condensation reactions with alcohols.

## Ethers (21.12)

- The family of ethers contains an oxygen atom between two R groups.
- Ethers are named with the ending -yl ether.

## Amines (21.13)

- Amines are organic compounds that contain nitrogen and are named with the suffix -amine.
- They are known for their terrible odors; the smell of decaying animal flesh is produced by cadaverine.
- Amines act as weak bases and produce a salt when mixed with a strong acid.
- The combination of an amine with a carboxylic acid leads to a condensation reaction; in our bodies, this reaction produces proteins from amino acids.

# Key Equations and Relationships

Halogen Substitution Reactions in Alkanes (21.6)

$$R-H + X_2 \xrightarrow{\text{heat or light}} R-X + HX$$
Alkane   Halogen   Haloalkane   Hydrogen halide

Common Functional Groups (**21.8**)

| Family | General Formula | Condensed General Formula | Example | Name |
|---|---|---|---|---|
| Alcohols | R—OH | ROH | $CH_3CH_2OH$ | Ethanol (ethyl alcohol) |
| Ethers | R—O—R | ROR | $CH_3OCH_3$ | Dimethyl ether |
| Aldehydes | R—C(=O)—H | RCHO | $H_3C$—C(=O)—H | Ethanal (acetaldehyde) |
| Ketones | R—C(=O)—R | RCOR | $H_3C$—C(=O)—$CH_3$ | Propanone (acetone) |
| Carboxylic acids | R—C(=O)—OH | RCOOH | $H_3C$—C(=O)—OH | Acetic acid |
| Esters | R—C(=O)—OR | RCOOR | $H_3C$—C(=O)—$OCH_3$ | Methyl acetate |
| Amines | R—N(R)—R | $R_3N$ | $H_3CH_2C$—N(H)—H | Ethylamine |

Alcohol Reactions (**21.9**)

*Substitution*   $ROH + HBr \longrightarrow R—Br + H_2O$

*Oxidation*

$$R—CH_2—CH_2—OH \xrightarrow[H_2SO_4]{Na_2Cr_2O_7} R—CH_2—C(=O)—OH$$

Alcohol          Aldehyde

*Elimination*

$$R—CH_2—CH_2—OH \xrightarrow{H_2SO_4} R—CH=CH_2 + H_2O$$

Carboxylic Acid Condensation Reactions (**21.11**)

$$R—C(=O)—OH + HO—R' \xrightarrow{H_2SO_4} R—C(=O)—O—R' + H_2O$$

Acid          Alcohol                    Ester          Water

Amine Acid–Base Reactions (**21.13**)

$$RNH_2(aq) + H_2O(l) \longrightarrow RNH_3^+(aq) + OH^-(aq)$$

Amine–Carboxylic Acid Condensation Reactions (**21.13**)

$$CH_3COOH(aq) + HNHR(aq) \longrightarrow CH_3CONHR(aq) + H_2O(l)$$

# Key Learning Outcomes

| CHAPTER OBJECTIVES | ASSESSMENT |
|---|---|
| Writing Structural Formulas for Hydrocarbons (**21.3**) | Example 21.1   For Practice 21.1   Exercises 35, 36 |
| Naming Alkanes (**21.4**) | Examples 21.2, 21.3, 21.4   For Practice 21.2, 21.3, 21.4   Exercises 41, 42 |
| Naming Alkenes and Alkynes (**21.5**) | Example 21.5   For Practice 21.5   Exercises 51–54 |
| Writing Reactions: Addition Reactions (**21.6**) | Example 21.6   For Practice 21.6   Exercises 57–60 |

Writing Reactions: Alcohols (**21.9**)

$$CH_3-CH_2-OH$$
Ethanol

Example 21.7    For Practice 21.7    Exercises 73, 74

# EXERCISES

## Review Questions

1. What kinds of molecules often trigger our sense of smell?

2. What is organic chemistry?

3. What is unique about carbon and carbon-based compounds? Why did life evolve around carbon?

4. Why does carbon form such a large diversity of compounds?

5. Why does silicon exhibit less diversity of compounds than carbon does?

6. Describe the geometry and hybridization about a carbon atom that forms
   a. four single bonds.
   b. two single bonds and one double bond.
   c. one single bond and one triple bond.

7. What are hydrocarbons? What are their main uses?

8. What are the main classifications of hydrocarbons? What are their generic molecular formulas?

9. Explain the differences between a structural formula, a condensed structural formula, a carbon skeleton formula, a ball-and-stick model, and a space-filling model.

10. What are structural isomers? How do the properties of structural isomers differ from one another?

11. What are optical isomers? How do the properties of optical isomers differ from one another?

12. Define each term related to optical isomerism: enantiomers, chiral, dextrorotatory, levorotatory, racemic mixture.

13. What is the difference between saturated and unsaturated hydrocarbons?

14. What are the key differences in the way that alkanes, alkenes, and alkynes are named?

15. Explain geometric isomerism in alkenes. How do the properties of geometric isomers differ from one another?

16. Describe and provide an example of a hydrocarbon combustion reaction.

17. What kinds of reactions are common to alkanes? List an example of each.

18. Describe each kind of reaction.
    a. substitution reaction
    b. addition reaction
    c. elimination reaction

19. What kinds of reactions are common to alkenes? Give an example of each.

20. Explain Markovnikov's rule and give an example of a reaction to which it applies.

21. What is the structure of benzene? What are the different ways in which this structure is represented?

22. What kinds of reactions are common to aromatic compounds? Provide an example of each.

23. What is a functional group? List some examples.

24. What is the generic structure of alcohols? Write the structures of two specific alcohols.

25. Explain oxidation and reduction with respect to organic compounds.

26. What kinds of reactions are common to alcohols? Provide an example of each.

27. What are the generic structures for aldehydes and ketones? Write a structure for a specific aldehyde and ketone.

28. What kinds of reactions are common to aldehydes and ketones? List an example of each.

29. What are the generic structures for carboxylic acids and esters? Write a structure for a specific carboxylic acid and ester.

30. What kinds of reactions are common to carboxylic acids and esters? Provide an example of each.

31. What is the generic structure of ethers? Write the structures of two specific ethers.

32. What is the generic structure of amines? Write the structures of two specific amines.

## Problems by Topic

### Hydrocarbons

33. Based on the molecular formula, determine whether each compound is an alkane, alkene, or alkyne. (Assume that the hydrocarbons are noncyclical and there is no more than one multiple bond.)
    a. $C_5H_{12}$    b. $C_3H_6$
    c. $C_7H_{12}$    d. $C_{11}H_{22}$

34. Based on the molecular formula, determine whether each compound is an alkane, alkene, or alkyne. (Assume that the hydrocarbons are noncyclical and there is no more than one multiple bond.)
    a. $C_8H_{16}$
    b. $C_4H_6$
    c. $C_7H_{16}$
    d. $C_2H_2$

**35.** Write structural formulas for each of the nine structural isomers of heptane.

**36.** Write structural formulas for any 6 of the 18 structural isomers of octane.

**37.** Determine whether each compound exhibits optical isomerism.

a. CCl₄

b. CH₃—CH₂—CH—CH₂—CH₂—CH₂—CH₃
                   |
                  CH₃

c.
$$CH_3-\overset{\displaystyle H}{\underset{\displaystyle NH_2}{C}}-Cl$$

d. CH₃CHClCH₃

**38.** Determine whether each compound exhibits optical isomerism.

a. CH₃CH₂CHClCH₃

b. CH₃CCl₂CH₃

c.

d. CH₃—C—C—OH with CH₃ bearing CH₂—CH₃ and ketone/alkene groups
$$CH_3-\overset{\displaystyle CH_2}{\underset{\displaystyle \overset{\displaystyle CH}{\underset{\displaystyle CH_2}{\parallel}}}{C}}-\overset{\displaystyle O}{\overset{\displaystyle \parallel}{C}}-OH$$

**39.** Determine whether the molecules in each pair are the same or enantiomers.

a. CH₃CH₂⫶⫶C⫶Cl (with CH₃)    Cl⫶C⫶CH₂CH₃ (with CH₃)

b. H₃C⫶C⫶Cl (with HO)    HO⫶C⫶CH₃ (with Cl)

c. (CH₃)₂CH—C...C—CH₃ structures with H₃C, CH₂    H₃C—C...C with HC—CH₃, CH₃, CH₂

**40.** Determine whether the molecules in each pair are the same or enantiomers.

a. H₃C⫶C⫶Cl (with CH₃, HO)    Cl⫶C⫶CH₃ (with CH₃, OH)

b. H₃C⫶C⫶Br (with H, H₂N)    Br⫶C⫶CH₃ (with H, NH₂)

c. 
$$H_3C\underset{\displaystyle CH_3}{\overset{\displaystyle H}{\cdots}}C-CCl_3$$
$$Cl_3C\underset{\displaystyle CH_3}{\overset{\displaystyle HC(CH_3)_2}{C}}$$

## Alkanes

**41.** Name each alkane.

a. CH₃—CH₂—CH₂—CH₂—CH₃

b. CH₃—CH₂—CH—CH₃
                |
              CH₃

c.
$$CH_3-\underset{\displaystyle CH_3}{CH}-CH_2-\underset{\displaystyle \overset{\displaystyle CH-CH_3}{|}}{\overset{\displaystyle CH_3}{CH}}-CH_2-CH_2-CH_3$$

d. CH₃—CH—CH₂—CH—CH₂—CH₃
         |              |
        CH₃       CH₂—CH₃

**42.** Name each alkane.

a. CH₃—CH—CH₃
         |
        CH₃

b. CH₃—CH—CH₂—CH—CH₂
         |           |
       CH₃      CH₃
                           |
                         CH₃

c.
$$CH_3-\underset{\displaystyle CH_3}{\overset{\displaystyle CH_3}{C}}-\underset{\displaystyle CH_3}{\overset{\displaystyle CH_3}{C}}-CH_3$$

d. CH₃—CH—CH₂—CH—CH—CH₂—CH₂—CH₃
with CH₃, CH₂ (CH₃ substituents)
$$CH_3-\underset{\displaystyle CH_3}{CH}-CH_2-CH-\underset{\displaystyle \overset{\displaystyle CH_2}{|}{\underset{\displaystyle CH_3}{}}}{CH}-CH_2-CH_2-CH_3$$

**43.** Draw a structure for each alkane.

a. 3-ethylhexane

b. 3-ethyl-3-methylpentane

c. 2,3-dimethylbutane

d. 4,7-diethyl-2,2-dimethylnonane

**44.** Draw a structure for each alkane.

a. 2,2-dimethylpentane

b. 3-isopropylheptane

c. 4-ethyl-2,2-dimethylhexane

d. 4,4-diethyloctane

**45.** Complete and balance each hydrocarbon combustion reaction.

a. $CH_3CH_2CH_3 + O_2 \longrightarrow$

b. $CH_3CH_2CH{=}CH_2 + O_2 \longrightarrow$

c. $CH{\equiv}CH + O_2 \longrightarrow$

**46.** Complete and balance each hydrocarbon combustion reaction.

a. $CH_3CH_2CH_2CH_3 + O_2 \longrightarrow$

b. $CH_2{=}CHCH_3 + O_2 \longrightarrow$

c. $CH{\equiv}CCH_2CH_3 + O_2 \longrightarrow$

**47.** List all the possible products for each alkane substitution reaction. (Assume monosubstitution.)

a. $CH_3CH_3 + Br_2 \longrightarrow$

b. $CH_3CH_2CH_3 + Cl_2 \longrightarrow$

c. $CH_2Cl_2 + Br_2 \longrightarrow$

d. $CH_3-\underset{\underset{CH_3}{|}}{CH}-CH_3 + Cl_2 \longrightarrow$

**48.** List all the possible products for each alkane substitution reaction. (Assume monosubstitution.)

a. $CH_4 + Cl_2 \longrightarrow$

b. $CH_3CH_2Br + Br_2 \longrightarrow$

c. $CH_3CH_2CH_2CH_3 + Cl_2 \longrightarrow$

d. $CH_3CHBr_2 + Br_2 \longrightarrow$

## Alkenes and Alkynes

**49.** Write structural formulas for each of the possible isomers of *n*-hexene that are formed by moving the position of the double bond.

**50.** Write structural formulas for each of the possible isomers of *n*-pentyne that are formed by moving the position of the triple bond.

**51.** Name each alkene.

a. $CH_2{=}CH-CH_2-CH_3$

b. $CH_3-\underset{\underset{CH_3}{|}}{CH}-\underset{\underset{CH_3}{|}}{C}{=}CH-CH_3$

c. $CH_2{=}HC-\underset{\underset{\underset{\underset{CH_3}{|}}{CH}}{|}}{CH}-CH_2-CH_2-CH_3$  
$\qquad\qquad CH_3-CH$  
$\qquad\qquad\qquad\quad CH_3$

d. $CH_3-\underset{\underset{CH_3}{|}}{CH}-CH{=}\underset{\underset{\underset{\underset{CH_3}{|}}{CH_2}}{|}}{C}-CH_3$

**52.** Name each alkene.

a. $CH_3-CH_2-CH{=}CH-CH_2-CH_3$

b. $CH_3-\underset{\underset{CH_3}{|}}{CH}-CH{=}CH-CH_3$

c. $CH_3-\underset{\underset{CH_3}{|}}{CH}-CH{=}\underset{\underset{\underset{\underset{CH_3}{|}}{CH_2}}{|}}{C}-\underset{}{CH}-CH_3$  with $CH_3$

**d.** $CH_3-\underset{\underset{CH_3}{|}}{\overset{\overset{CH_3}{|}}{C}}-CH{=}\overset{\overset{CH_3}{|}}{C}-CH_2-CH_3$

**53.** Name each alkyne.

a. $CH_3-C{\equiv}C-CH_3$

b. $CH_3-C{\equiv}C-\underset{\underset{CH_3}{|}}{\overset{\overset{CH_3}{|}}{C}}-CH_2-CH_3$

c. $CH{\equiv}C-\underset{\underset{\underset{\underset{CH_3}{|}}{CH-CH_3}}{|}}{CH}-CH_2-CH_2-CH_3$

d. $CH_3-\underset{\underset{\underset{\underset{CH_3}{|}}{CH_2}}{|}}{CH}-C{\equiv}C-\overset{\overset{CH_3}{|}}{CH}-\underset{\underset{\underset{\underset{CH_3}{|}}{CH_2}}{|}}{CH_2}$

**54.** Name each alkyne.

a. $CH{\equiv}C-\underset{\underset{CH_3}{|}}{CH}-CH_3$

b. $CH_3-C{\equiv}C-\underset{\underset{CH_3}{|}}{CH}-\overset{\overset{CH_3}{|}}{CH}-CH_2-CH_3$

c. $CH{\equiv}C-\underset{\underset{\underset{\underset{CH_3}{|}}{CH_2}}{|}}{\overset{\overset{CH_3}{|}}{C}}-CH_2-CH_3$

d. $CH_3-C{\equiv}C-\underset{\underset{\underset{\underset{CH_3}{|}}{CH_2}}{|}}{CH}-\overset{\overset{CH_3}{|}}{\underset{\underset{CH_3}{}}{C}}-CH_3$

**55.** Draw the correct structure for each compound.

a. 4-octyne

b. 3-nonene

c. 3,3-dimethyl-1-pentyne

d. 5-ethyl-3,6-dimethyl-2-heptene

**56.** Draw the correct structure for each compound.

a. 2-hexene

b. 1-heptyne

c. 4,4-dimethyl-2-hexene

d. 3-ethyl-4-methyl-2-pentene

**57.** List the products of each alkene addition reaction.

a. $CH_3-CH=CH-CH_3 + Cl_2 \longrightarrow$

b. $CH_3-\underset{\underset{CH_3}{|}}{CH}-CH=CH-CH_3 + HBr \longrightarrow$

c. $CH_3-CH_2-CH=CH-CH_3 + Br_2 \longrightarrow$

d. $CH_3-\underset{\underset{CH_3}{|}}{CH}-CH=\overset{\overset{CH_3}{|}}{C}-CH_3 + HCl \longrightarrow$

**58.** What are the products of each alkene addition reaction?

a. $CH_3-\underset{\underset{CH_3}{|}}{CH}-CH=CH_2 + Br_2 \longrightarrow$

b. $CH_2=CH-CH_3 + Cl_2 \longrightarrow$

c. $CH_3-\underset{\underset{CH_3}{|}}{\overset{\overset{CH_3}{|}}{C}}-CH=CH_2 + HCl \longrightarrow$

d. $CH_3-\underset{\underset{CH_2-CH_3}{|}}{CH}-CH=\overset{\overset{CH_3}{|}}{C}-CH_3 + HBr \longrightarrow$

**59.** Complete each hydrogenation reaction.

a. $CH_2=CH-CH_3 + H_2 \xrightarrow{\text{catalyst}}$

b. $CH_3-\underset{\underset{CH_3}{|}}{CH}-CH=CH_2 + H_2 \xrightarrow{\text{catalyst}}$

c. $CH_3-\underset{\underset{CH_3}{|}}{CH}-\overset{\overset{CH_3}{|}}{C}=CH_2 + H_2 \xrightarrow{\text{catalyst}}$

**60.** Complete each hydrogenation reaction.

a. $CH_3-CH_2-CH=CH_2 + H_2 \xrightarrow{\text{catalyst}}$

b. $CH_3-CH_2-\underset{\underset{CH_3}{|}}{\overset{\overset{CH_3}{|}}{C}}=\overset{\overset{CH_3}{|}}{C}-CH_3 + H_2 \xrightarrow{\text{catalyst}}$

c. $CH_3-CH_2-\underset{\underset{CH_3}{|}}{C}=CH_2 + H_2 \xrightarrow{\text{catalyst}}$

## Aromatic Hydrocarbons

**61.** Name each monosubstituted benzene.

a. $CH_3$  b. $Br$  c. $Cl$

**62.** Name each monosubstituted benzene.

a. $H_2C-CH_3$  b. $F$  c. $CH_3$ / $H_3C-\overset{\overset{CH_3}{|}}{\underset{|}{C}}-CH_3$

**63.** Name each compound in which the benzene ring is best treated as a substituent.

a. $CH_3-\underset{|}{CH}-CH_2-\underset{\underset{CH_2-CH_3}{|}}{CH}-CH_2-\overset{\overset{\phantom{C}}{|}}{CH}-CH_2-CH_3$ with $CH_3$ and phenyl substituents

b. $CH_3-\underset{|}{CH}-CH=CH-CH_2-CH_2-CH_2-CH_3$ with phenyl substituent

c. $CH_3-C\equiv C-\underset{|}{CH}-\underset{\underset{CH_3}{|}}{CH}-\underset{\underset{CH_3}{|}}{CH}-CH_2-CH_3$ with phenyl substituent

**64.** Name each compound in which the benzene ring is best treated as a substituent.

a. $H_3C-CH_2-\underset{|}{CH}-\underset{\underset{H_3C}{|}}{CH}-CH_2-CH_3$ with phenyl substituent

b. $CH_2-CH_2-CH_2-CH_2-C\equiv C-CH_3$ with phenyl substituent

c. $CH_3-\underset{|}{CH}-\underset{\underset{CH_3}{|}}{CH}-\overset{\overset{CH_3}{|}}{C}=CH-\underset{|}{CH}-CH_2-CH_3$ with phenyl substituent

**65.** Name each disubstituted benzene.

a.

b. CH₂—CH₃ benzene ring CH₂—CH₃

c. F benzene ring Cl

**66.** Name each disubstituted benzene.

a. Br benzene ring Cl

b. Cl benzene ring CH₂—CH₃

c. I benzene ring I

**67.** Draw the structure for each compound.
a. isopropylbenzene
b. *meta*-dibromobenzene
c. 1-chloro-4-methylbenzene

**68.** Draw the structure for each compound.
a. ethylbenzene
b. 1-iodo-2-methylbenzene
c. *para*-diethylbenzene

**69.** What are the products of each aromatic substitution reaction?

a. benzene ring with H's + Br₂ $\xrightarrow{\text{FeBr}_3}$

b. benzene + CH₃—CH—Cl (with CH₃) $\xrightarrow{\text{AlCl}_3}$

**70.** What are the products of each aromatic substitution reaction?

a. benzene ring with H's + Cl₂ $\xrightarrow{\text{FeCl}_3}$

b. benzene + CH₃—C—Cl (with CH₃ top and CH₃ bottom) $\xrightarrow{\text{AlCl}_3}$

**Alcohols**

**71.** Name each alcohol.
a. $CH_3—CH_2—CH_2—OH$

b. $CH_3—CH—CH_2—CH—CH_3$ with CH₂—CH₃ on second carbon and OH on fourth carbon

c. $CH_3—CH—CH_2—CH—CH_2—CH—CH_3$ with CH₃ on second carbon, OH on fourth carbon, CH₃ on sixth carbon

d. $H_3C—CH_2—C—CH_2—CH_3$ with HO above and H₃C below central carbon

**72.** Draw the structure for each alcohol.
a. 2-butanol
b. 2-methyl-1-propanol
c. 3-ethyl-1-hexanol
d. 2-methyl-3-pentanol

**73.** List the products of each alcohol reaction.
a. $CH_3—CH_2—CH_2—OH + HBr \longrightarrow$

b. $CH_3—CH—CH_2—OH$ with CH₃ below $\xrightarrow{\text{H}_2\text{SO}_4}$

c. $CH_3—CH_2—OH + Na \longrightarrow$

d. $CH_3—C—CH_2—CH_2—OH$ with CH₃ above and CH₃ below central carbon $\xrightarrow[\text{H}_2\text{SO}_4]{\text{Na}_2\text{Cr}_2\text{O}_7}$

**74.** List the products of each alcohol reaction.
a. $CH_3—C—OH$ with CH₃ above and CH₃ below $\xrightarrow{\text{H}_2\text{SO}_4}$

b. $CH_3—CH—CH_2—CH_2—OH$ with CH₃ above $\xrightarrow[\text{H}_2\text{SO}_4]{\text{Na}_2\text{Cr}_2\text{O}_7}$

c. $CH_3—CH_2—OH + HCl \longrightarrow$

d. $CH_3—CH—CH_2—OH$ with CH₃ above $+ Na \longrightarrow$

## Aldehydes and Ketones

75. Name each aldehyde or ketone.

a. $CH_3-\overset{\displaystyle O}{\overset{\displaystyle \|}{C}}-CH_2-CH_3$

b. $CH_3-CH_2-CH_2-CH_2-\overset{\displaystyle O}{\overset{\displaystyle \|}{CH}}$

c. $CH_3-\overset{\displaystyle CH_3}{\underset{\displaystyle CH_3}{\overset{\displaystyle |}{\underset{\displaystyle |}{C}}}}-CH_2-\overset{\displaystyle CH_3}{\overset{\displaystyle |}{CH}}-CH_2-\overset{\displaystyle O}{\overset{\displaystyle \|}{C}}-H$

d. $CH_3-\underset{\displaystyle \underset{\displaystyle CH_2-CH_3}{|}}{CH}-CH_2-\overset{\displaystyle O}{\overset{\displaystyle \|}{C}}-CH_3$

76. Draw the structure of each aldehyde or ketone.
   a. hexanal
   b. 2-pentanone
   c. 2-methylbutanal
   d. 4-heptanone

77. Determine the product of the addition reaction.

$CH_3-CH_2-CH_2-\overset{\displaystyle O}{\overset{\displaystyle \|}{CH}} + H-C\equiv N \xrightarrow{\text{NaCN}}$

78. Determine the product of the addition reaction.

$CH_3-\overset{\displaystyle O}{\overset{\displaystyle \|}{C}}-CH_2-CH_3 + HCN \xrightarrow{\text{NaCN}}$

## Carboxylic Acids and Esters

79. Name each carboxylic acid or ester.

a. $CH_3-CH_2-CH_2-\overset{\displaystyle O}{\overset{\displaystyle \|}{C}}-O-CH_3$

b. $CH_3-CH_2-\overset{\displaystyle O}{\overset{\displaystyle \|}{C}}-OH$

c. $CH_3-\underset{\displaystyle \underset{\displaystyle CH_3}{|}}{CH}-CH_2-CH_2-CH_2-\overset{\displaystyle O}{\overset{\displaystyle \|}{C}}-OH$

d. $CH_3-CH_2-CH_2-CH_2-\overset{\displaystyle O}{\overset{\displaystyle \|}{C}}-O-CH_2-CH_3$

80. Draw the structure of each carboxylic acid or ester.
   a. pentanoic acid
   b. methyl hexanoate
   c. 3-ethylheptanoic acid
   d. butyl ethanoate

81. Determine the products of each carboxylic acid reaction.

a. $CH_3-CH_2-CH_2-CH_2-\overset{\displaystyle O}{\overset{\displaystyle \|}{C}}-OH +$
   $CH_3-CH_2-OH \xrightarrow{\text{H}_2\text{SO}_4}$

b. $\underset{\displaystyle CH_2}{\overset{\displaystyle O}{\overset{\displaystyle \|}{\underset{\displaystyle |}{\underset{\displaystyle CH_2}{\underset{\displaystyle \underset{\displaystyle CH_2}{}}{}}}}}}$ ... $\xrightarrow{\text{heat}}$

82. Determine the products of each carboxylic acid reaction.

a. $CH_3-CH_2-\overset{\displaystyle O}{\overset{\displaystyle \|}{C}}-OH + NaOH \longrightarrow$

b. $CH_3-CH_2-CH_2-\overset{\displaystyle O}{\overset{\displaystyle \|}{C}}-OH +$
   $CH_3-CH_2-CH_2-OH \xrightarrow{\text{H}_2\text{SO}_4}$

## Ethers

83. Name each ether.
   a. $CH_3-CH_2-CH_2-O-CH_2-CH_3$
   b. $CH_3-CH_2-CH_2-CH_2-CH_2-O-CH_2-CH_3$
   c. $CH_3-CH_2-CH_2-O-CH_2-CH_2-CH_3$
   d. $CH_3-CH_2-O-CH_2-CH_2-CH_2-CH_3$

84. Draw the structure for each ether.
   a. ethyl propyl ether
   b. dibutyl ether
   c. methyl hexyl ether
   d. dipentyl ether

## Amines

85. Name each amine.
   a. $CH_3-CH_2-\underset{\displaystyle \underset{\displaystyle H}{|}}{N}-CH_2-CH_3$
   b. $CH_3-CH_2-CH_2-\underset{\displaystyle \underset{\displaystyle H}{|}}{N}-CH_3$
   c. $CH_3-CH_2-CH_2-\underset{\displaystyle \underset{\displaystyle CH_3}{|}}{N}-CH_2-CH_2-CH_2-CH_3$

**86.** Draw the structure for each amine.
  **a.** isopropylamine
  **b.** triethylamine
  **c.** butylethylamine

**87.** Classify each amine reaction as acid–base or condensation and list its products.
  **a.** $CH_3NHCH_3 + HCl \longrightarrow$
  **b.** $CH_3CH_2NH_2 + CH_3CH_2COOH \longrightarrow$
  **c.** $CH_3NH_2 + H_2SO_4 \longrightarrow$

**88.** List the products of each amine reaction.
  **a.** $N(CH_2CH_3)_3 + HNO_3 \longrightarrow$

**b.** $CH_3-\overset{\overset{H}{|}}{N}-\underset{\underset{CH_3}{|}}{CH}-CH_3 + HCN \longrightarrow$

**c.** $CH_3-\overset{\overset{H}{|}}{N}-\underset{\underset{CH_3}{|}}{CH}-CH_3 +$ 
$CH_3-\underset{\underset{CH_3}{|}}{CH}-CH_2-\overset{\overset{O}{||}}{C}-OH \longrightarrow$

# Cumulative Problems

**89.** Identify each organic compound as an alkane, alkene, alkyne, aromatic hydrocarbon, alcohol, ether, aldehyde, ketone, carboxylic acid, ester, or amine, and provide a name for the compound.

**a.** $H_3C-\underset{\underset{CH_3}{|}}{HC}-CH_2-\overset{\overset{O}{||}}{C}-O-CH_3$

**b.** $CH_3-CH_2-\underset{\underset{CH_3}{|}}{CH}-CH_2-O-CH_2-CH_3$

**c.** $H_2C-CH_3$ (on benzene ring with $CH_3$)

**d.** $CH_3-C\equiv C-\underset{\underset{CH_3}{|}}{CH}-\overset{\overset{CH_2-CH_3}{|}}{CH}-CH_2-CH_3$

**e.** $H_3C-CH_2-CH_2-\overset{\overset{O}{||}}{CH}$

**f.** $H_3C-\underset{\underset{H_3C}{|}}{CH}-\overset{\overset{OH}{|}}{CH_2}$

**90.** Identify each organic compound as an alkane, alkene, alkyne, aromatic hydrocarbon, alcohol, ether, aldehyde, ketone, carboxylic acid, ester, or amine, and provide a name for the compound.

**a.** $H_3C-HC-\overset{\overset{H_3C}{|}}{C}=\underset{}{C}-CH_3$ with $CH_3$, $CH_3$

**b.** $CH_3-\underset{\underset{CH_3}{|}}{\overset{\overset{CH_3}{|}}{C}}-CH_2-\underset{\underset{}{|}}{CH}-CH_2-CH_3$ with $CH_3$

**c.** $CH_3-CH_2-\underset{\underset{CH_3}{|}}{CH}-CH_2-\overset{\overset{O}{||}}{C}-OH$

**d.** $CH_3-\underset{\underset{CH_3}{|}}{CH}-N-CH_2-CH_2-CH_2-CH_3$ (N with H)

**e.** $CH_3-\underset{\underset{CH_2-CH_3}{|}}{CH}-CH_2-\overset{\overset{CH_2-OH}{|}}{CH}-CH_3$

**f.** $CH_3-CH_2-CH_2-\overset{\overset{O}{||}}{C}-\underset{\underset{CH_3}{|}}{CH}-CH_3$

**91.** Name each compound.

**a.** $CH_3-CH_2-CH-CH_2-CH-CH_2-CH_2-CH_2-CH_3$ with $CH_3$ and $HC-CH_3$, $CH_2$, $CH_3$

**b.** $CH_3-\underset{\underset{CH_3}{|}}{CH}-CH_2-\overset{\overset{O}{||}}{C}-CH_2-CH_3$

**c.** $CH_3-\overset{\overset{OH}{|}}{CH}-\underset{\underset{CH_3}{|}}{CH}-CH_3$

**d.** $CH_3-\underset{\underset{CH_3}{|}}{CH}-CH-CH-C\equiv C-H$ with $CH_2$, $CH_3$ branch ($CH_2$ then $CH_3$)

**92.** Name each compound.

a. CH₃—CH=CH—C—CH—CH₂—CH₃ with CH₃ and CH₃ on top of C and CH, and CH₂—CH₃ below

$$CH_3-CH=CH-\underset{\underset{\underset{CH_3}{|}}{\overset{}{CH_2}}}{\overset{\overset{CH_3}{|}}{C}}-\overset{\overset{CH_3}{|}}{CH}-CH_2-CH_3$$

b.

c.

$$CH_3-CH_2-\overset{\overset{CH_3}{|}}{CH}-CH_2-\overset{\overset{O}{\|}}{C}-O-\overset{\overset{CH_3}{|}}{CH}-CH_3$$

d.

$$CH_3-\underset{\underset{CH_3}{|}}{CH}-CH_2-\overset{\overset{O}{\|}}{CH}$$

**93.** Determine whether the two structures are isomers or the same molecule drawn in two different ways.

a.
$$CH_3-CH_2-\overset{\overset{O}{\|}}{C}-O-CH_3$$
$$CH_3-\overset{\overset{O}{\|}}{C}-O-CH_2-CH_3$$

b.

c.
$$CH_3-HC-CH-CH_3$$ with CH₃ CH₃ below the two central carbons
$$CH_3-HC-\overset{\overset{CH_3}{|}}{CH}-CH_3$$ with CH₃ below HC

**94.** Determine whether the two structures are isomers or the same molecule drawn two different ways.

a. $CH_3-CH_2-\overset{\overset{CH_3}{|}}{CH}-CH_2-CH_3$

$CH_3-CH_2-\overset{\overset{\overset{CH_3}{|}}{CH_2}}{\underset{}{CH}}-CH_3$

**b.**

$$CH_3-\overset{\overset{CH_3}{|}}{CH}-CH_2-O-CH_2-CH_3$$

$$CH_3-CH_2-CH_2-O-CH_2-\overset{\overset{CH_3}{|}}{CH_2}$$

c. $CH_3-\overset{\overset{CH_3}{|}}{CH}-CH_2-\overset{\overset{O}{\|}}{C}-CH_2-CH_3$

$$CH_3-\overset{\overset{}{}}{CH}-CH_2-\overset{\overset{O}{\|}}{C}-\overset{}{CH_2}$$ with CH₃ below CH and CH₃ below CH₂

**95.** What minimum amount of hydrogen gas, in grams, is required to completely hydrogenate 15.5 kg of 2-butene?

**96.** How many kilograms of $CO_2$ does the complete combustion of 3.8 kg of *n*-octane produce?

**97.** Classify each organic reaction as combustion, alkane substitution, alkene addition or hydrogenation, aromatic substitution, or alcohol substitution, elimination, or oxidation.

a. $2\,CH_3CH{=}CH_2 + 9\,O_2 \longrightarrow 6\,CO_2 + 6\,H_2O$

b. $CH_3CH_2CH_3 + Cl_2 \longrightarrow CH_3CH_2CH_2Cl + HCl$

c. $CH_3-CH_2-\overset{\overset{}{}}{CH}-CH_2-OH \xrightarrow{H_2SO_4}$ with CH₃ below CH

$$CH_3-CH_2-\overset{\overset{}{}}{C}=CH_2$$ with CH₃ below C

d.

**98.** Determine the products of each reaction.

a. $CH_3-CH_2-\overset{\overset{}{}}{C}=CH_2 + H_2 \longrightarrow$ with CH₃ below C

b. $CH_3-CH_2-CH_2-CH_2-OH + HCl \longrightarrow$

c. $CH_3-CH_2-\overset{\overset{CH_3}{|}}{CH}-CH_2-\overset{\overset{O}{\|}}{C}-OH +$

$$CH_3CH_2OH \longrightarrow$$

d. $CH_3-CH_2-\overset{\overset{H}{|}}{N}-CH_2-CH_3 + HCl \longrightarrow$

**99.** Draw the structure that corresponds to each name and indicate which structures can exist as stereoisomers.

a. 3-methyl-1-pentene

b. 3,5-dimethyl-2-hexene

c. 3-propyl-2-hexene

**100.** Identify the two compounds that display stereoisomerism and draw their structures.
   **a.** 3-methyl-3-pentanol
   **b.** 2-methyl-2-pentanol
   **c.** 3-methyl-2-pentanol
   **d.** 2-methyl-3-pentanol
   **e.** 2,4-dimethyl-3-pentanol

101. There are 11 structures (ignoring stereoisomerism) with the formula $C_4H_8O$ that have no carbon branches. Draw the structures and identify the functional groups in each.

**102.** There are eight structures with the formula $C_3H_7NO$ in which the O is part of a carbonyl group. Draw the structures and identify the functional groups in each.

103. Explain why carboxylic acids are much stronger acids than alcohols.

**104.** The hydrogen at C-1 of 1-butyne is much more acidic than the one at C-1 in 1-butene. Explain.

# Challenge Problems

105. Determine the one or two steps it takes to get from the starting material to the product using the reactions found in this chapter.

**a.**

**b.**

**c.**

**106.** Given the following synthesis of ethyl 3-chloro-3-methylbutanoate, fill in the missing intermediates or reactants.

107. For the chlorination of propane, the two isomers shown here are possible.

$$CH_3CH_2CH_3 + Cl_2 \longrightarrow$$

1-Chloropropane      2-Chloropropane

Propane has six hydrogen atoms on terminal carbon atoms—called primary (1°) hydrogen atoms—and two hydrogen atoms on the interior carbon atom—called secondary (2°) hydrogen atoms.
   **a.** If the two different types of hydrogen atoms were equally reactive, what ratio of 1-chloropropane to 2-chloropropane would we expect as monochlorination products?
   **b.** The result of a reaction yields 55% 2-chloropropane and 45% 1-chloropropane. What can we conclude about the relative reactivity of the two different kinds of hydrogen atoms? Determine a ratio of the reactivity of one type of hydrogen atom to the other.

**108.** There are two isomers of $C_4H_{10}$. Suppose that each isomer is treated with $Cl_2$ and the products that have the composition $C_4H_8Cl_2$ are isolated. Find the number of different products that form from each of the original $C_4H_{10}$ compounds. Do not consider optical isomerism.

109. Identify the compounds formed in the previous problem that are chiral.

**110.** Nitromethane has the formula $CH_3NO_2$, with the N bonded to the C and without O—O bonds. Draw its two most important contributing structures.
   **a.** What is the hybridization of the C, and how many hybrid orbitals are in the molecule?
   **b.** What is the shortest bond?
   **c.** Between which two atoms is the strongest bond found?
   **d.** Predict whether the HCH bond angles are greater or less than 109.5° and justify your prediction.

111. Free radical fluorination of methane is uncontrollably violent, and free radical iodination of methane is a very poor reaction. Explain these observations in light of bond energies.

**112.** There are two compounds with the formula $C_3H_6$, one of which does not have a multiple bond. Draw its structure and explain why it is much less stable than the isomer with the double bond.

113. Consider molecules that have two carbons and two chlorines. Draw the structures of three of these with no dipole moment and two with a dipole moment.

# Conceptual Problems

**114.** Pick the more oxidized structure in each pair.

a. $CH_3-\overset{\overset{\displaystyle O}{\|}}{CH}$ or $CH_3-CH_2-OH$

b. $CH_3-CH_2-OH$ or $CH_3-CH_3$

c. $CH_3-CH_2-\overset{\overset{\displaystyle O}{\|}}{CH}$ or $CH_3-CH_2-\overset{\overset{\displaystyle O}{\|}}{C}-OH$

**115.** Draw the structure and name a compound with the formula $C_8H_{18}$ that forms only one product with the formula $C_8H_{17}Br$ when it is treated with $Br_2$.

**116.** Determine whether each structure is chiral.

a. $\overset{\overset{\displaystyle Cl}{|}}{\underset{\underset{\displaystyle CH_3}{|}}{HC}}-CH_3$

b. $CH_3-\overset{\overset{\displaystyle }{}}{\underset{\underset{\displaystyle Cl}{|}}{CH}}-\overset{\overset{\displaystyle }{}}{\underset{\underset{\displaystyle CH_3}{|}}{CH}}-CH_3$

c. $CH_3-CH_2-OH$

d. $\overset{\overset{\displaystyle Cl}{|}}{\underset{\underset{\displaystyle CH_3}{|}}{HC}}-\overset{}{\underset{\underset{\displaystyle Br}{|}}{CH_2}}$

# Questions for Group Work

Active Classroom Learning

Discuss these questions with the group and record your consensus answer.

**117.** Have each group member select a different functional group. Individually, draw and name a compound containing your functional group. Pass your drawing to the group member on your right and your compound name to the group member on your left. Name the compound for the drawing you received and draw the compound whose name you received.

**118.** Using complete sentences, compare and contrast each of the following. Mention at least two similarities and two differences for each.
a. salt and sugar
b. methane and 3-methylheptane
c. aldehydes and ketones

**119.** The octane rating for gasoline is a measurement of how readily a fuel combusts compared to 2,2,4-trimethylpentane, an isomer of octane.
a. Draw 2,2,4-trimethylpentane and verify that it is an isomer of octane.
b. Draw four other isomers of octane.
c. Select one of the isomers and draw it so that it looks different on the page, but is still the exact same compound.
d. Name the isomer you drew in part c.
e. Define "isomer" using a complete sentence.

**120.** Working individually, have each group member select from the chapter a reaction characteristic of a particular organic functional group. Individually, write out your reaction with one of the following left as a blank: one reactant, one product, or reaction conditions. Pass your reaction to the group member on your right. Have group members fill in the blank of the reaction they receive.

# Data Interpretation and Analysis

## Preparation of Fruit-Scented Esters

**121.** Ester compounds often have a sweet, pleasant odor. Many characteristic fruit scents are largely due to the natural presence of one or more ester compounds. As such, artificial scents for foods are often composed of complex mixtures of various esters. The exact identities and ratios of ingredients that compose particular scents are closely guarded secrets in the food and fragrance industry.

   Suppose that you are a chemist working for a company that is creating a new line of air fresheners. The company is considering three scents: apple, pear, and pineapple. The project manager has asked you to prepare the ester compounds that are largely responsible for these scents. The carbon skeleton formulas for these ester compounds are shown in Figure a▶.

Apple    $CH_3-CH_2-CH_2-\overset{\overset{\displaystyle O}{\|}}{C}-O-CH_3$

Pear    $CH_3-\overset{\overset{\displaystyle O}{\|}}{C}-O-CH_2-CH_2-CH_3$

Pineapple    $CH_3-CH_2-CH_2-\overset{\overset{\displaystyle O}{\|}}{C}-O-CH_2-CH_3$

▲ **FIGURE a** **Structures of Ester Compounds Responsible for Various Fruit Scents**

To prepare these esters, you have been given the alcohols listed in Table a and an adequate supply of all other necessary reagents, solvents, equipment, and the like.

| Reagent | Molar mass (g/mol) | Density (g/mL) | Cost, per 1.00 L |
|---------|--------------------|-----------------|-------------------|
| methanol | 32.04 | 0.79 | $46.20 |
| ethanol | 46.07 | 0.79 | $112.00 |
| 1-propanol | 60.10 | 0.80 | $72.70 |
| 1-butanol | 74.12 | 0.81 | $72.60 |

**TABLE a  Molar Mass, Density, and Cost of Alcohol Compounds for Ester Compound Preparation**

Use the information provided in the figure and the table to answer the following questions:

a. Provide a name for each ester that you will prepare.
b. Draw the structure of each alcohol listed in Table a.
c. Determine the procedure you will use to prepare each ester, using the reactions found in this chapter. (Hint: Recall that esters are derived from a carboxylic acid and an alcohol.)
d. Calculate the cost to prepare 100.0 g of each ester. Which one will be the most expensive to prepare? Which ester will be the least expensive? (Consider only the cost of the alcohols in Table a and disregard the costs of other reagents. Assume 100% yield for all reactions.)
e. The scent of an apple is due to many different ester compounds including butyl acetate, pentyl acetate, ethyl pentanoate, pentylpentanoate, and pentylhexanoate. Draw structures for each of these ester compounds.

# Answers to Conceptual Connections

### Organic Structures

**21.1 (d)** The others are simply the same structure drawn in slightly different ways.

### Optical Isomers

**21.2 (b)** This structure is the only one that contains a carbon atom (the one on the left) with four different substituent groups attached (a Br atom, a Cl atom, an H atom, and a $CH_3$ group).

### Oxidation

**21.3 (c, d, a, b)** Oxidation includes the gain of oxygen or the loss of hydrogen.

CHAPTER **22**

# Biochemistry

> We've discovered the secret of life.
>
> —FRANCIS H. C. CRICK (1916–2004)

I n Chapter 21, we examined organic chemistry and discussed the different types of organic compounds and their structures and chemistry. In this chapter, we turn to biochemical compounds, those organic compounds important in living organisms. Biochemistry—the area of study at the interface between chemistry and biology that strives to understand living organisms at the molecular level—exploded in the second half of the twentieth century. That explosion began with the discovery of the structure of DNA in 1953 by James D. Watson and Francis H. C. Crick and continues to this day, most recently marked by the 2003 completion of the Human Genome Project, which succeeded in mapping the 3 billion base pairs within the DNA of humans. The benefits of biochemistry to humankind are numerous, ranging from a better understanding of illnesses and better drugs to cure them to a better understanding of ourselves and our origins.

## 22.1 Diabetes and the Synthesis of Human Insulin

Diabetes afflicts over 16 million people in the United States alone. Today, it is a chronic but generally manageable ailment; at the beginning of the twentieth century, however, diabetes was often fatal. The most dangerous form, type 1 diabetes, develops when the pancreas does not make enough *insulin*, a protein that promotes the absorption of glucose from the blood into cells, where glucose is used for energy. Consequently, people with diabetes have high blood-sugar levels that can lead to a number of complications, including heart disease, blindness, and kidney failure. Before 1922, the only option

**1029**

diabetics had was to control their blood-sugar levels through diet, but this was often not enough to overcome the disease.

Important advances throughout the twentieth century dramatically changed the prognosis for diabetics. The initial breakthrough came in 1922, when researchers first injected insulin from animal sources into a diabetic. The insulin worked, resulting in a nearly complete recovery for the patient. Within a year, insulin harvested from the pancreases of slaughtered pigs became widely available, and, for many patients, diabetes became a long-term manageable disease. However, insulin taken from pigs and cattle is not identical to human insulin, and some patients do not tolerate the animal insulin as well as others.

In 1955, Frederick Sanger (1918–2013) discovered the detailed chemical structure of human insulin. As we will see later in this chapter, proteins are biological molecules composed of repeating units called amino acids (of which there are 20 different types in humans). Sanger was able to determine the specific sequence of amino acids in human insulin, work for which he received the 1958 Nobel Prize in Chemistry. Knowing the amino acid sequence allowed researchers to synthesize human insulin in the laboratory by 1963. Unfortunately, they could not make sufficient amounts to meet the needs of diabetics.

The growing field of biotechnology, however, allowed a fledgling company called Genentech to synthesize human insulin on a large scale by the early 1980s. Researchers at Genentech were able to insert the human *gene* for insulin—the blueprint that determines how insulin is synthesized in humans—into the DNA of bacterial cells. When the bacteria reproduced in culture, they made copies of the inserted human insulin gene and passed it on to their offspring. Furthermore, as the growing bacterial culture synthesized the bacterial proteins that it needed to grow and survive, it also synthesized human insulin. In other words, researchers at Genentech were able to get bacteria to produce human insulin for them! The ability to synthesize human insulin in this way revolutionized the treatment of diabetes, resulting in better health and extended lives for hundreds of millions of people who suffer from this disease.

The study of the chemistry occurring in living organisms is **biochemistry**, the topic of this chapter. Many biologically important molecules are very large and complex—they are *macromolecules*. However, understanding their structures is not as difficult as you might imagine because most of them consist of much smaller, simpler components linked together into long chains (*polymers*). In this chapter, we divide our study of biochemistry along the lines of the major chemical components of cells: lipids, carbohydrates, proteins, and nucleic acids.

## 22.2  Lipids

**Lipids** are the chemical components of the cell that are insoluble in water but soluble in nonpolar solvents. Fatty acids, fats, oils, phospholipids, glycolipids, and steroids are all lipids. Their insolubility in water makes it possible for lipids to compose the structural components of cell membranes, which separate the aqueous interior of the cell from its aqueous environment in the body. Lipids also play a role in long-term energy storage and insulation.

### Fatty Acids

One type of lipid is the **fatty acid**, a carboxylic acid (see Section 21.11) with a long hydrocarbon tail. The general structure for a fatty acid contains the carboxylic acid group:

$$\text{R}-\overset{\overset{\displaystyle O}{\|}}{\text{C}}-\text{OH}$$

Fatty acid general structure

The R on the left side of the structure represents a hydrocarbon chain containing 3 to 19 carbon atoms. Fatty acids differ only in their R group.

Myristic acid, found in butterfat and in coconut oil, has the R group $CH_3(CH_2)_{12}$. Myristic acid is a *saturated* fatty acid: its carbon chain has no double bonds:

Myristic acid

Other fatty acids—called *monounsaturated* or *polyunsaturated* fatty acids—have one or more double bonds in their carbon chains. For example, oleic acid—found in olive oil, peanut oil, and human fat—is a monounsaturated fatty acid:

Oleic acid

The long hydrocarbon tails of fatty acids make them insoluble in water. Table 22.1 lists several fatty acids, some common sources for each, and their melting points. Notice that the melting points of the fatty acids *increase* as their carbon chain lengths increase. The longer the chains, the greater the dispersion forces between adjacent molecules in the solid state, and the higher the melting point. Notice also that the melting points *decrease* with the presence of double bonds. For example, stearic acid and oleic acid have the same number of carbon atoms, but stearic acid melts at 70 °C and oleic acid melts at 4 °C. The double bond puts a "kink" in the carbon chain that makes it more difficult for neighboring molecules to interact over the entire length of the chain (Figure 22.1▶, on the next page), thus lowering the melting point.

Dispersion forces are discussed in Section 11.3.

### TABLE 22.1 Fatty Acids

**Saturated Fatty Acids**

| Name | Number of Carbons | mp (°C) | Structure | Sources |
|------|-------------------|---------|-----------|---------|
| Butyric acid | 4 | −7.9 | $CH_3CH_2CH_2COOH$ | Milk fat |
| Caproic acid | 10 | 31 | $CH_3(CH_2)_8COOH$ | Milk fat, whale oil |
| Myristic acid | 14 | 59 | $CH_3(CH_2)_{12}COOH$ | Butterfat, coconut oil |
| Palmitic acid | 16 | 64 | $CH_3(CH_2)_{14}COOH$ | Beef fat, butterfat |
| Stearic acid | 18 | 70 | $CH_3(CH_2)_{16}COOH$ | Beef fat, butterfat |

**Unsaturated Fatty Acids**

| Name | Number of Carbons | Number of Double Bonds | mp (°C) | Structure | Sources |
|------|-------------------|------------------------|---------|-----------|---------|
| Oleic acid | 18 | 1 | 4 | $CH_3(CH_2)_7CH=CH(CH_2)_7COOH$ | Olive oil, peanut oil |
| Linoleic acid | 18 | 2 | −5 | $CH_3(CH_2)_4(CH=CHCH_2)_2(CH_2)_6COOH$ | Linseed oil, corn oil |
| Linolenic acid | 18 | 3 | −11 | $CH_3CH_2(CH=CHCH_2)_3(CH_2)_6COOH$ | Linseed oil, corn oil |

## The Effect of Unsaturation

Stearic acid, mp 70 °C          Oleic acid, mp 4 °C

▲ **FIGURE 22.1 The Effect of Unsaturation** A double bond results in a bend in the carbon chain of a fatty acid or fat. The bend makes it more difficult for neighboring molecules to interact over the entire length of the carbon chain, thus lowering the melting point.

## Fats and Oils

Fats and oils are **triglycerides**, triesters composed of glycerol with three fatty acids attached. Triglycerides form when a glycerol molecule reacts with three fatty acids:

$$
\begin{array}{c}
\text{O} \\
\parallel \\
\text{R—C—OH}
\end{array}
$$

| | | |
|---|---|---|
| H₂C—OH | | |
| HO—CH | | |
| H₂C—OH | | |

Glycerol    +    3 Fatty acids    →    Triglyceride

We discussed the general structure of esters in Section 20.11.

The bonds that join the glycerol to the fatty acids are called **ester linkages**. For example, tristearin—the main component of beef fat—forms from the reaction of glycerol and three stearic acid molecules (Figure 22.2▼). If the fatty acids in a triglyceride are saturated,

Glycerol    + 3    3 Stearic acid molecules

Tristearin

▲ **FIGURE 22.2 The Formation of Tristearin** The reaction between glycerol and stearic acid forms tristearin.

### Tristearin: A Saturated Fat

Tristearin, mp 72 °C

▲ Tristearin is a triglyceride found in lard; it is a saturated fat.

the triglyceride is a **saturated fat** and tends to be solid at room temperature. Triglycerides from warm-blooded animals (for example, lard from pigs) are generally saturated.

If the fatty acids in a triglyceride are unsaturated, the triglyceride is an **unsaturated fat**, or an oil, and tends to be liquid at room temperature. Triglycerides from plants (olive oil, corn oil, canola oil, etc.) or from cold-blooded animals (fish oil) are generally unsaturated.

### Triolein: A Monounsaturated Fat

Triolein, mp −4 °C

Triolein is a triglyceride found in olive oil; it is a monounsaturated fat.

Most of the fats and oils in our diet are triglycerides. During digestion, triglycerides are broken down into fatty acids, glycerol, monoglycerides, and diglycerides. These products pass through the intestinal wall and reassemble into triglycerides before they are absorbed into the blood. This process is slower than the digestion of other food types, and for this reason eating fats and oils gives a lasting feeling of fullness.

The effect of fats and oils on health has been widely debated. Some diets call for a drastic reduction of our daily intake of fats and oils, whereas others actually call for an *increase* in fats and oils. The U.S. Food and Drug Administration (FDA) recommends moderate consumption of fats and oils, less than 30% of total caloric intake.

## Other Lipids

Additional lipids found in cells include phospholipids, glycolipids, and steroids. **Phospholipids** have the same basic structure as triglycerides, except that one of the fatty acid groups is replaced with a phosphate group. Unlike a fatty acid, which is nonpolar, the phosphate group is polar. A phospholipid molecule therefore has a polar region and a nonpolar region. Consider the structure of phosphatidylcholine, a phospholipid found in the cell membranes of higher animals (Figure 22.3▶ on the next page). The polar part of the molecule is *hydrophilic* (has a strong affinity for water), while the nonpolar part is *hydrophobic* (is repelled by water).

**Glycolipids** have similar structures and properties. The nonpolar section of a glycolipid is composed of a fatty acid chain and a hydrocarbon chain. The polar section is a

**Phosphatidylcholine**
Phosphatidylcholine is a
phospholipid. The structure
is similar to a triglyceride
except that one of the fatty
acid groups is replaced with a
phosphate group.

**Phosphatidylcholine: A Phospholipid**

▲ FIGURE 22.4 **Schematic
for Phospholipid or Glycolipid**
We represent a phospholipid or a
glycolipid as a circle (the polar part
of the molecule) with two long tails
(the nonpolar part of the molecule).

Polar head      Nonpolar tails

sugar molecule such as glucose. We often portray phospholipids or glycolipids schemati-
cally as a circle with two long tails (Figure 22.4 ◀). The circle represents the polar, hydrophilic part of the molecule, and the tails represent the nonpolar, hydrophobic parts.
Phospholipids and glycolipids are key components of cell membranes; the polar parts
interact with the aqueous environments inside and outside the cell, and the nonpolar
parts interact with each other, forming a double-layered structure called a **lipid bilayer**
(Figure 22.5 ◀). Lipid bilayers encapsulate cells and many cellular structures.

   **Steroids** are lipids with a four-ring structure:

Cholesterol, testosterone, and β-estradiol are common steroids:

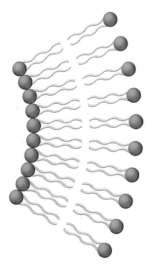

▲ FIGURE 22.5 **Lipid Bilayer**
Lipid bilayers are composed of
phospholipids or glycolipids arranged
in a structure that encapsulates cells
and many cellular structures.

Cholesterol

Testosterone

β-Estradiol

Hormones are chemical messengers
that are released by cells in one part
of the body and cause an effect in
cells in another part of the body.

   Although cholesterol has a bad reputation, it serves many important functions in the
body. Like phospholipids and glycolipids, cholesterol is part of cell membranes. Cholesterol
also serves as a starting material (or precursor) for the body's synthesis of other steroids such
as testosterone, a principal male hormone, and β-estradiol, a principal female hormone.

# 22.3 Carbohydrates

Carbohydrates are responsible for short-term storage of energy in living organisms, and they make up the main structural components of plants. Carbohydrates—as their name, which means carbon and water, implies—often have the general formula $(CH_2O)_n$. Structurally, we identify **carbohydrates** as polyhydroxy aldehydes or ketones. For example, glucose, with the formula $C_6H_{12}O_6$, has the following structure:

> The name should not be taken literally—the hydrogen and oxygen in carbohydrates do not bond together in the same way that they bond in water.

Glucose

Glucose is a six-carbon aldehyde (that is, it contains the —CHO group) with —OH groups on five of the six carbon atoms. The many —OH groups make glucose soluble in water (and therefore in blood), which is crucial to glucose's role as the primary fuel of cells. Glucose is easily transported in the bloodstream and is soluble within the aqueous interior of a cell. We broadly classify carbohydrates as simple carbohydrates (or simple sugars) and complex carbohydrates.

> Recall from Section 21.10 that aldehydes have the general structure RCHO and ketones have the general structure RCOR.

## Simple Carbohydrates: Monosaccharides and Disaccharides

**Monosaccharides**—meaning "one sugar"—are the simplest carbohydrates. Monosaccharides contain between three and eight carbon atoms and have only one aldehyde or ketone functional group. The general names for monosaccharides begin with a prefix that indicates the number of carbon atoms, followed by the suffix *-ose*. The most common carbohydrates in living organisms are pentoses and hexoses.

> 3-carbon sugar: triose
> 4-carbon sugar: tetrose
> 5-carbon sugar: pentose
> 6-carbon sugar: hexose
> 7-carbon sugar: heptose
> 8-carbon sugar: octose

Glucose, the structure of which we saw previously, is an example of a **hexose**, a six-carbon sugar. Glucose is also an example of an **aldose**, a sugar with an aldehyde group. Often, we combine these two ways of designating sugars; glucose is an *aldohexose* (*aldo-* indicates that it is an aldehyde; *-hex-* indicates that it has six carbon atoms; and *-ose* indicates that it is a carbohydrate).

Another common carbohydrate is fructose, a polyhydroxy *ketone* with the following structure:

Fructose

Glucose and fructose are structural isomers—they both have the same formula ($C_6H_{12}O_6$), but they have different structures. Fructose is a **ketose**, a sugar that is a ketone. Since fructose has six carbon atoms, it is a *ketohexose*. Fructose, often called fruit sugar, is in many fruits and vegetables and is a major component of honey.

---

**Example** 22.1 | **Carbohydrates and Optical Isomerism**

Closely examine the structure of glucose shown here. Does glucose exhibit optical isomerism (discussed in Section 21.3)? If so, which carbon atoms are chiral?

Glucose

**SOLUTION**

Any carbon atom with four different substituents attached to it is chiral. Glucose has four chiral carbon atoms (labeled 2, 3, 4, and 5) and therefore exhibits optical isomerism.

Glucose

Variations in the positions of the —OH and —H groups on these carbon atoms result in a number of different possible isomers for glucose. For example, switching the relative positions of the —OH and —H group on the carbon atom closest to the carbonyl group results in mannose, an optical isomer of glucose.

**FOR PRACTICE 22.1**  Examine the structure of fructose (p. 1043). Does fructose exhibit optical isomerism? How many of the carbon atoms in fructose are chiral?

---

▲ **FIGURE 22.6 Intramolecular Reaction of Glucose to Form a Ring**  The alcohol group on C5 in glucose reacts with the carbonyl group (C1) to form a closed ring.

Most five- and six-carbon monosaccharides undergo intramolecular reactions that convert their straight carbon chain into a ring. For example, in glucose, the alcohol group on C5 reacts with the carbonyl group (C1), as shown in Figure 22.6 ◄ to form the ring structure shown here:

In a glucose solution, the vast majority of the molecules are in ring form. However, the molecules in ring form exist in equilibrium with a small fraction in the open-chain form. Other common monosaccharides, in their ring form, include fructose (discussed previously in its straight-chain form) and galactose shown at the right.

Galactose, also known as brain sugar, is a hexose usually found combined with other monosaccharides in disaccharides such as lactose (see next paragraph). Galactose also occurs within the brain and nervous system of most animals. Galactose and glucose differ only in the stereochemistry at C4. Notice that in galactose the —OH group is roughly perpendicular to the plane of the ring while in glucose it is roughly in the same plane as the ring.

Two monosaccharides can link together via a **glycosidic linkage** to form a **disaccharide**, a carbohydrate that can be decomposed into two simpler sugars. For example, glucose and fructose join to form sucrose, commonly known as table sugar (Figure 22.7▶). When we eat disaccharides, the link between individual monosaccharides is broken during digestion by **hydrolysis**, the splitting of a chemical bond with water that results in the addition of H and OH to the products. The resultant monosaccharides readily pass through the intestinal wall and enter the bloodstream to become fuel for cells.

## Complex Carbohydrates

Monosaccharides can also link to form a type of natural polymer (or biopolymer) called a **polysaccharide**, a long, chain-like molecule composed of many bonded monosaccharide units. Polysaccharides are known as **complex carbohydrates** because of their long chains of sugars. The most common polysaccharides are *cellulose*, *starch*, and *glycogen*, all three of which are composed of repeating glucose units. The main difference among them lies in the way the units are bonded to each other. In cellulose, the oxygen atoms are roughly parallel with the planes of the rings. This is referred to as a *β*-glycosidic linkage. In starch and glycogen, the oxygen atoms joining neighboring glucose units point down relative to the planes of the rings. This is an *α*-glycosidic linkage:

▲ FIGURE 22.7 **Formation of a Glycosidic Linkage** Glucose and fructose can join, eliminating water and forming a glycosidic linkage that results in the disaccharide sucrose, commonly known as table sugar.

Cellulose

*β*-Glycosidic linkages

Starch

*α*-Glycosidic linkages

◀ The primary difference between starch and cellulose is in the way the units are bonded together.

## Cellulose

The main structural component of plants, **cellulose**, is the most abundant organic substance on Earth and consists of glucose units bonded together by β-glycosidic linkages. This structure allows neighboring cellulose molecules to form multiple hydrogen bonds with one another, resulting in the rigid and structurally stable properties we associate with wood and fiber. Humans lack the enzyme required to digest cellulose. When we eat cellulose (usually called fiber when it is present in foods), it passes right through our intestines, providing bulk to stools and preventing constipation. In contrast, some bacteria have the enzyme required to metabolize cellulose into its component glucose units. These bacteria are common in the guts of termites and ruminants such as cows, allowing them to extract caloric content from cellulose.

## Starch

The main energy storage medium for plants is **starch**, the soft, pliable substance abundant in potatoes and grains. Starch is composed of two slightly different polysaccharides, *amylose* and *amylopectin*. Both are made up of glucose units bonded together by α-glycosidic linkages, but amylopectin contains branches in the chains. When animals digest starch, the link between individual glucose units is broken by hydrolysis, allowing glucose molecules to pass through the intestinal wall and into the bloodstream.

## Glycogen

The structure of **glycogen** is similar to that of amylopectin, but the chain is even more highly branched. Animals use glycogen to store glucose in the muscles. Glycogen's highly branched structure leaves many end groups that can be quickly hydrolyzed to meet energy needs. When muscles become depleted of glycogen, muscle movement and exercise become much more difficult. Marathon runners often "hit the wall" at about mile 20 because they have depleted most of the glycogen from their muscles.

## 22.4 Proteins and Amino Acids

Proteins are the workhorse molecules in living organisms; they are involved in virtually every facet of cell structure and function. For example, most of the chemical reactions that occur in living organisms are enabled by **enzymes**, proteins that act as catalysts in biochemical reactions. Without enzymes, life would be impossible. Proteins are also the structural elements of muscle, skin, and cartilage. They transport oxygen in the blood, act as antibodies to fight disease, and function as hormones to regulate metabolic processes. Proteins reign supreme as the working molecules of life. Table 22.2 summarizes the functions of some of the important classes of proteins and lists examples of each.

**TABLE 22.2 Protein Functions**

| Class of Protein | Primary Function | Example |
| --- | --- | --- |
| Structural proteins | Compose structures within living organisms | Collagen (skin, tendon, cartilage), keratin (hair, fingernails) |
| Enzymes | Catalyze and control biochemical reactions | DNA polymerase (involved in replication of DNA) |
| Hormones | Regulate metabolic processes | Insulin (regulates glucose metabolism) |
| Transport proteins | Transport substances from one place to another | Hemoglobin (transports oxygen) |
| Storage proteins | Provide source of essential nutrients | Casein (protein in mammalian milk) |
| Contractile and motile proteins | Mediate motion and muscle contraction | Actin and myosin (provide muscle contraction) |
| Protective proteins | Protect and defend cells | Antibodies (neutralize infectious agents) |

# Amino Acids: The Building Blocks of Proteins

Proteins are polymers of amino acids. Each **amino acid** molecule consists of a carbon atom—called the $\alpha$-carbon—bonded to four different groups: an amine group, an R group (also called a side chain), a carboxylic acid group, and a hydrogen atom:

In a protein, the presence of an R group does not necessarily indicate a pure alkyl group. See Table 22.3 for possible R groups.

Amino acid general structure

Amino acids differ from each other in their R groups. For example, the R group in alanine is a methyl group ($CH_3$):

Alanine

Other amino acids are glycine (R = H), phenylalanine (R = $CH_2C_6H_5$), serine (R = $CH_2OH$), aspartic acid (R = $CH_2COOH$), and lysine (R = $CH_2CH_2CH_2CH_2NH_2$):

Glycine

Phenylalanine

Serine

Aspartic acid

Lysine

The R groups, or side chains, differ chemically. For example, phenylalanine has a large nonpolar R group, whereas serine has a polar one. Aspartic acid has an acidic R group, whereas lysine, since it contains nitrogen, has a basic one. When amino acids are strung together to make a protein, the chemical properties of the R groups determine the structure and properties of the protein. Table 22.3 shows the most common amino acids in proteins and their three-letter abbreviations. The diversity of amino acids creates the possibility for an even larger diversity of proteins.

## TABLE 22.3 Common Amino Acids

Glycine (Gly)

Alanine (Ala)

Valine (Val)

Leucine (Leu)

Isoleucine (Ile)

Proline (Pro)

Methionine (Met)

Cysteine (Cys)

Serine (Ser)

Threonine (Thr)

Aspartic acid (Asp)

Glutamic acid (Glu)

Asparagine (Asn)

Glutamine (Gln)

Lysine (Lys)

Arginine (Arg)

Histidine (His)

Phenylalanine (Phe)

Tyrosine (Tyr)

Tryptophan (Trp)

Because all amino acids (except glycine) contain four different groups attached to a tetrahedral carbon (the α-carbon), all amino acids (except glycine) are chiral about that carbon. The amino acids that compose naturally occurring proteins are the L-enantiomers, and they are called L-amino acids. Why life on Earth is based on this enantiomer over the other is an interesting question that remains to be answered. (It seems just as likely that life could have used the D-enantiomer.)

Although we usually write the structures of amino acids as neutral, their actual structure is ionic and depends on pH. In general, amino acids undergo an intramolecular acid–base reaction and form a *dipolar ion*, or *zwitterion*:

Dipolar ion

At room temperature this equilibrium lies far to the right. Since one side of the dipolar ion is positively charged and the other negatively charged, amino acids are highly polar and soluble in water. They also have fairly high melting points (usually >200 °C). In addition, the intramolecular acid–base reaction makes amino acids less acidic and less basic than most carboxylic acids and amines, respectively.

## Peptide Bonding between Amino Acids

Amino acids link together through the reaction of the amine end of one amino acid with the carboxylic end of another:

The resulting bond is a **peptide bond**, and the resulting molecule—two amino acids linked together—is a **dipeptide**. When two or more amino acids link in this way, the molecule they form has two distinct ends: an amino terminal (or N-terminal end) and a carboxyl terminal (or C-terminal end). A *tripeptide* is three amino acids joined by peptide bonds; a *tetrapeptide* is four; and so on. Short chains of amino acids are called *oligopeptides* generally, and longer chains (more than 20) are called **polypeptides**. Functional proteins usually contain one or more polypeptide chains, with each chain consisting of hundreds or even thousands of amino acids joined by peptide bonds.

The formation of a peptide bond is an example of a condensation reaction (see Section 21.11).

---

**Example 22.2    Peptide Bonds**

Show the reaction by which valine, cysteine, and phenylalanine (in that order) link via peptide bonds. Designate valine as N-terminal and label the N-terminal and C-terminal ends in the resulting tripeptide.

Valine          Cysteine          Phenylalanine

*—Continued on the next page*

*Continued—*

**SOLUTION**

Peptide bonds form when the carboxylic end of one amino acid reacts with the amine end of another amino acid.

**FOR PRACTICE 22.2** Show the reaction by which alanine, threonine, and serine (in that order) link via peptide bonds. Designate alanine as the N-terminal and label the N-terminal and C-terminal ends in the resulting tripeptide.

Alanine      Threonine      Serine

---

CONCEPTUAL
**CONNECTION 22.1**

**Peptides** How many different tripeptides can form from the three amino acids listed here? (The amino acids are indicated using the three-letter amino acid abbreviations from Table 22.3.)

Ser, Ala, Gly

---

## 22.5 Protein Structure

A protein's structure is critical to its function. For example, recall from Section 22.1 that insulin is a protein that promotes the absorption of glucose out of the blood and into muscle cells where the glucose is needed for energy. Insulin recognizes muscle cells because muscle cell surfaces contain *insulin receptors*, molecules that fit a specific portion of the insulin protein. If insulin were a different shape, it would not latch onto insulin receptors on muscle cells and could not do its job. Thus, the shape or *conformation* of a protein is crucial to its function.

We can broadly classify proteins into two main structural categories: fibrous proteins and globular proteins (Figure 22.8▶). **Fibrous proteins** tend to have relatively

## Fibrous and Globular Proteins

Fibrous protein

Globular protein

Glycine
X
Y } Other amino acids, often proline or hydroxyproline

Collagen

Hemoglobin

◀ **FIGURE 22.8** **Fibrous and Globular Proteins** Proteins are broadly divided into fibrous proteins (which have relatively simple linear structures) and globular proteins (which have more complex three-dimensional structures).

simple linear structures and be insoluble in aqueous solutions. They serve primarily structural functions within living organisms. Collagen and keratin, for example, are both fibrous proteins (see Table 22.2). **Globular proteins** tend to have more complex structures but are often roughly spherical in overall shape. Globular proteins are generally structured so that polar side chains on amino acids are oriented toward the exterior of the protein, while nonpolar side chains are oriented toward the interior of the protein. Consequently, globular proteins tend to be soluble in water, but they maintain a nonpolar environment within the protein that excludes water. Hemoglobin and insulin are both globular proteins.

Protein structure is analyzed at four levels: primary structure, secondary structure, tertiary structure, and quaternary structure (Figure 22.9▼). We examine each of these categories separately.

▼ **FIGURE 22.9** **Levels of Protein Structure** Protein structure is analyzed at four levels: primary, secondary, tertiary, and quaternary.

## Levels of Protein Structure

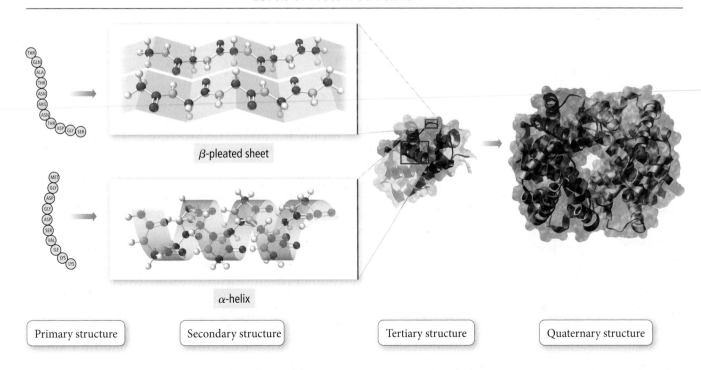

$\beta$-pleated sheet

$\alpha$-helix

Primary structure

Secondary structure

Tertiary structure

Quaternary structure

The genetic disease known as sickle-cell anemia results in red blood cells with a sickle shape. These cells impede circulation of blood, causing damage to major organs.

## Primary Structure

The **primary structure** of a protein is the sequence of amino acids in its chain(s). Primary structure, which determines the other three kinds of structure, is maintained by the covalent peptide bonds between individual amino acids. The primary structure of egg-white lysozyme—a protein that helps fight infection—is shown in Figure 22.10▼. The figure illustrates the amino acid sequence, the N-terminal and C-terminal ends, and the presence of *disulfide linkages*, covalent cross-links between cysteine amino acids in the polymer. We discuss disulfide linkages in more detail later in the section on tertiary structure. Researchers determined the first amino acid sequences for proteins in the 1950s. Today, the amino acid sequences for thousands of proteins are known.

Changes in the amino acid sequence of a protein, even minor ones, can destroy the function of a protein. Hemoglobin, as we saw in Chapter 15, is a protein that transports oxygen in the blood. It is composed of four polypeptide chains made up of a total of 574 amino acid units. If valine is replaced by glutamic acid in just one position on two of these chains, the disease known as sickle-cell anemia results. The red blood cells of people with sickle-cell anemia take on a sickle shape that impedes circulation, causing damage to major organs. In the past, sickle-cell anemia was fatal, often resulting in death before age 30—all due to a change in a few atoms of two amino acids out of 574. Modern therapies have extended the life span of people with sickle-cell anemia so that they now live into their 40s and 50s.

▶ **FIGURE 22.10** **Primary Structure of Egg-White Lysozyme** Primary structure refers to the sequence of amino acids in a protein.

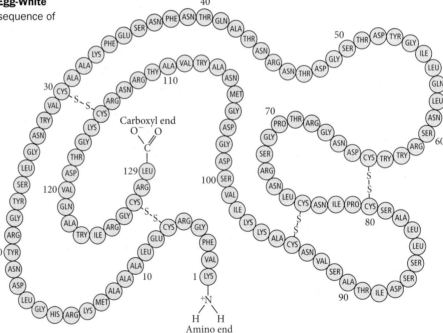

▼ **FIGURE 22.11** **The α-Helix Structure** The α-helix is an example of secondary protein structure.

α-Helix protein structure

## Secondary Structure

The **secondary structure** of a protein refers to certain regular periodic or repeating patterns in the arrangement of protein chains. Secondary structure is maintained by interactions between amino acids that are fairly close together in the linear sequence of the protein chain or that are adjacent to each other on neighboring chains or chains that fold back on themselves. The most common of these patterns is the **α-helix**, shown in Figure 22.11 ◀. In the α-helix structure, the amino acid chain wraps into a tight coil from which side chains extend. The structure is maintained by hydrogen-bonding interactions between NH and CO groups along the peptide backbone of the protein. Some proteins—such as keratin, which composes human hair—have the α-helix pattern throughout their entire chain. Other proteins have very little or no α-helix pattern in their chain.

β-Pleated sheet protein structure

◀ **FIGURE 22.12 The β-Pleated Sheet Structure**   The β-pleated sheet is a secondary protein structure.

A second common pattern in the secondary structure of proteins is the **β-pleated sheet** (Figure 22.12▲). In this structure, the chain is extended (as opposed to coiled) and forms a zigzag pattern. The peptide backbones of neighboring chains interact with one another through hydrogen bonding to form zigzag-shaped sheets. Some proteins—such as silk—have the β-pleated sheet structure throughout their entire chains. Since the protein chains in the β-pleated sheet are fully extended, silk is inelastic. Many proteins have some sections that are β-pleated sheet, other sections that are α-helical, and still other sections that have less regular patterns referred to as **random coils**.

## Tertiary Structure

The **tertiary structure** of a protein consists of the large-scale bends and folds resulting from interactions between the R groups of amino acids that are separated by large distances in the linear sequence of the protein chain. These interactions, shown in Figure 22.13▼, include hydrogen bonding, disulfide linkages (covalent bonds between cysteine amino acids), hydrophobic interactions (attractions between large, nonpolar side chains), and salt bridges (acid–base interactions between acidic and basic side chains). Fibrous proteins generally lack tertiary structure; they simply extend in a long continuous chain with some secondary structure. Globular proteins, by contrast, fold in on themselves, forming complex globular shapes rich in tertiary structure.

### Interactions That Maintain Tertiary Structure

◀ **FIGURE 22.13 Interactions within Proteins**   The tertiary structure of a protein is maintained by interactions between the R groups of amino acids that are separated by large distances in the linear sequence of the protein chain.

## Quaternary Structure

Some proteins—called *monomeric* proteins—are composed of only one polypeptide chain. However, *multimeric* proteins are composed of several polypeptide chains, called subunits. We just saw, for example, that hemoglobin is composed of four such subunits. The way that subunits fit together in a multimeric protein is referred to as the **quaternary structure** of the protein. Quaternary structure is maintained by the same types of interactions that maintain tertiary structure, but the interactions are between amino acids on different subunits.

*Summarizing Protein Structure:*

▶ Primary structure is the amino acid sequence. It is maintained by the peptide bonds that hold amino acids together.

▶ Secondary structure refers to the repeating patterns in the arrangement of protein chains. These are maintained by interactions between the peptide backbones of amino acids that are close together in the chain sequence or adjacent to each other on neighboring chains. Secondary structure is characteristic of fibrous proteins, but globular proteins also frequently feature regions of $\alpha$-helix, $\beta$-pleated sheet, and random coil secondary structure.

▶ Tertiary structure refers to the large-scale twists and folds of globular proteins. These are maintained by interactions between the R groups of amino acids that are separated by long distances in the chain sequence.

▶ Quaternary structure refers to the arrangement of subunits in proteins that have more than one polypeptide chain. Quaternary structure is maintained by interactions between amino acids on different subunits.

## 22.6 Nucleic Acids: Blueprints for Proteins

We have seen that the amino acid sequence in a protein determines that protein's structure and function. If the amino acid sequence is incorrect, the protein is unlikely to function properly. How do cells in living organisms synthesize the many thousands of different required proteins, each with the correct amino acid sequence? The answer lies in **nucleic acids**, molecules that serve as blueprints for protein synthesis. Nucleic acids employ a chemical code to specify the correct amino acid sequences for proteins. Nucleic acids are broadly divided into two types: deoxyribonucleic acid, or DNA, which exists primarily in the nucleus of the cell; and ribonucleic acid, or RNA, which exists throughout the cell.

### The Basic Structure of Nucleic Acids

Like proteins, nucleic acids are polymers. The individual units composing nucleic acids are **nucleotides**. Each nucleotide has three parts: a sugar, a base, and a phosphate group that serves as a link between sugars (Figure 22.14▼).

### DNA: Basic Structure

▲ **FIGURE 22.14 DNA Structure** DNA is composed of repeating units called nucleotides. Each nucleotide contains a sugar, a base, and a phosphate group.

## Sugars

In DNA, the sugar is deoxyribose, whereas in RNA the sugar is ribose:

Deoxyribose                    Ribose

The base attaches to the sugar at C1, and the phosphate attaches to the sugar at C3 and C5. When a base is attached to the sugar, the numbers of the carbon atoms in the sugar ring are primed to distinguish them from the carbon atoms on the bases (which are not primed). For example, C5 becomes C5′ and C3 becomes C3′.

## Bases

Every nucleotide in DNA has the same sugar, but each nucleotide has just one of four different bases. In DNA, the four bases are adenine (A), cytosine (C), guanine (G), and thymine (T). Each of these bases bonds to the sugar via the nitrogen atom circled in the illustrations shown here:

Purine Bases

Adenine                    Guanine

Pyrimidine Bases

Cytosine                    Thymine

Adenine and guanine are called *purine* bases because they resemble the bicyclic compound purine. Cytosine and thymine are called *pyrimidine* bases because they resemble the monocyclic compound pyrimidine:

Purine                    Pyrimidine

In RNA, the base uracil (U), also a pyridimine base, replaces thymine. Uracil and thymine differ only in a methyl ($-CH_3$) group:

No methyl group → on uracil

Uracil

## Base Pairing in DNA

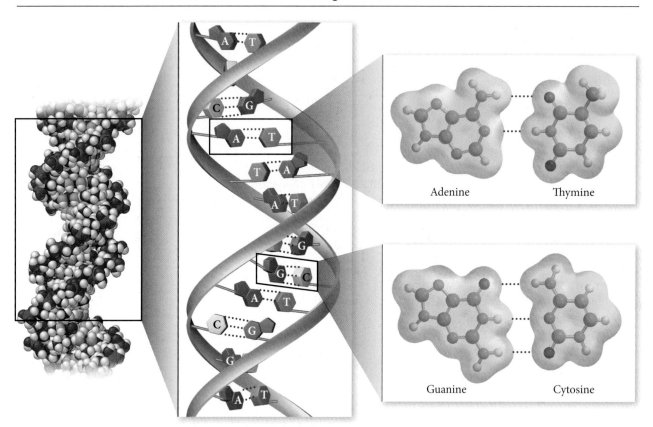

▲ **FIGURE 22.15  Base Pairing in DNA**   The bases in nucleic acids are complementary. Each pyrimidine base pairs with only one purine base (G with C, A with T) via specific hydrogen bonds that occur between the two bases.

The bases in nucleic acids are **complementary**—that is, each one precisely pairs with another base. Each pyrimidine base pairs with only one purine base via specific hydrogen bonds that occur between the two bases. More specifically, guanine pairs only with cytosine through three hydrogen bonds, and adenine pairs only with thymine through two hydrogen bonds as shown in Figure 22.15▲. In other words, guanine is complementary to cytosine and adenine is complementary to thymine. This base pairing is central to DNA replication, as we discuss in Section 22.7.

### Phosphate Links

The sugar units in nucleic acids link together by phosphate groups, which bind to C5′ and C3′ of the sugar as shown in Figure 22.16▶. Note that a nucleic acid molecule has two distinct ends, which we refer to as the 5′ end and the 3′ end.

### The Genetic Code

The order of bases in a nucleic acid chain determines the order of amino acids in a protein. However, since there are only four bases, and since together they must have the ability to code for 20 different amino acids, a single base cannot code for a single amino acid. A

◀ **FIGURE 22.16  Short Strand of DNA**   DNA contains alternating sugar and phosphate groups with bases attached to each sugar. The 3′ end of the molecule is missing an attachment at the number 3 position in the sugar ring, and the 5′ end is missing an attachment at the number 5 position.

**Genetic Structure**

**Chromosome**—structure within cell nucleus that houses DNA.

**Gene**—portion of DNA that codes for a single protein.

**Codon**—sequence of three nucleotides with their associated bases. A codon codes for one amino acid.

**Nucleotide**—individual links in the nucleic acid chain. Nucleotides are composed of a sugar group, a phosphate group, and a base.

▲ **FIGURE 22.17 Genetic Structure** The hierarchical structure of genetic information is chromosome, gene, codon, and nucleotide.

▲ **FIGURE 22.18 Chromosomes** Genes are contained in structures called chromosomes. Most human cells contain 46 chromosomes.

sequence of three nucleotides—called a **codon**—codes for each amino acid (Figure 22.17 ▲). The genetic code—the code that identifies the amino acid specified by a particular codon—was worked out in 1961. It is nearly universal—the same codons specify the same amino acids in nearly all organisms. For example, in DNA the sequence AGT codes for the amino acid serine, and the sequence ACC codes for the amino acid threonine. It does not matter if the DNA is in a rat, a bacterium, or a human—the code is the same.

A **gene** is a sequence of codons within a DNA molecule that codes for a single protein. Since proteins vary in size from a few dozen to thousands of amino acids, genes vary in length from dozens to thousands of codons. For example, egg-white lysozyme (see Figure 22.10) is composed of 129 amino acids. So the lysozyme gene contains 129 codons—one for each amino acid in the lysozyme protein. Each codon is like a three-letter word that specifies one amino acid. String the correct number of codons together in the correct sequence, and you have a gene, the instructions for the amino acid sequence in a protein. Genes are contained in structures called **chromosomes** (Figure 22.18 ▶). A human cell generally contains 46 chromosomes in its nucleus.

In addition to having a codon for each amino acid, genes also contain additional coding that signals, for example, where the gene begins and where it ends.

**The Genetic Code** Assuming you have four different bases, how many amino acids can you code for with two-base sequences? Three-base sequences?

CONCEPTUAL **CONNECTION 22.2**

## 22.7 DNA Replication, the Double Helix, and Protein Synthesis

Most of the cells in our bodies contain all of the genes required to make all of the proteins that we need—the DNA within any one cell is *complete*. However, any particular cell does not express all those genes; it does not synthesize all those proteins. Cells synthesize only the proteins that are important to their function. For example, a pancreatic cell expresses the insulin gene within its nucleus to synthesize insulin. Pancreatic cells do not express the gene for keratin (the protein in hair), even though the keratin gene is also contained in their nuclei. The cells in our scalp, in contrast, which also have both insulin and keratin genes in their nuclei, synthesize keratin but not insulin.

▲ **FIGURE 22.19** **Watson and Crick** James Watson and Francis Crick discovered the structure of DNA, including the double helix and the pairing of complementary bases.

### DNA Replication and the Double Helix

The human body contains on the order of $10^{13}$ cells, most of which have complete copies of the DNA that originally was present in a single cell (the fertilized egg). When a cell divides, it makes complete copies of its DNA for each daughter cell. The ability of DNA to copy itself is related to its structure, discovered in 1953 by James D. Watson and Francis H. C. Crick. Watson and Crick, aided by evidence from X-ray diffraction photos (see Section 12.2), determined that DNA exists as two complementary strands wound around each other in a double helix (Figure 22.19◄). The strands are antiparallel, so one runs $3' \longrightarrow 5'$ while the other runs $5' \longrightarrow 3'$. The bases on each DNA strand are directed toward the interior of the helix, where they hydrogen-bond to their complementary bases on the other strand. For example, if a section of DNA contains the bases:

the complementary strand has the sequence:

As we saw earlier, A pairs only with T, and C pairs only with G. The two complementary strands are tightly wrapped into a helical coil, the famous DNA double helix structure (Figure 22.20▼).

DNA double helix

Nucleotide

► **FIGURE 22.20** **DNA Double Helix** Two complementary strands of DNA wrap around one another to form a double helix.

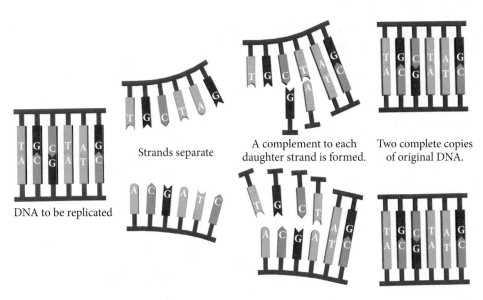

Strands separate

A complement to each daughter strand is formed.

Two complete copies of original DNA.

DNA to be replicated

When a cell is about to divide, the DNA unwinds and the hydrogen bonds joining the complementary bases break (Figure 22.21▲), forming two daughter strands. With the help of the enzyme DNA polymerase, a complement to each daughter strand—with the correct complementary bases in the correct sequence—forms. The hydrogen bonds between the old strands and the newly complementary strands then re-form, resulting in two complete copies of the original DNA, one for each daughter cell.

## Protein Synthesis

Living organisms must continually synthesize thousands of proteins to survive, each when it is needed and in the quantities required. When a cell needs to make a particular protein, the gene—the section of the DNA that codes for that particular protein—unravels. Complementary copies of that gene are then synthesized (or transcribed) as single strands of *messenger RNA* (or mRNA). The mRNA moves out of the cell's nucleus to structures in the cytoplasm called *ribosomes*.

At the ribosome, protein synthesis occurs. The ribosome (which has a large subunit and a small subunit) moves along the mRNA chain that codes for the protein, "reading" the sequence of codons. At each codon, the specified amino acid is brought into place and a peptide bond forms with the previous amino acid (Figure 22.22▼). As the ribosome moves along the mRNA, the protein forms. All of this is orchestrated by enzymes that catalyze the necessary reactions.

Complete protein

Secondary structure begins to form

Growing protein

Large subunit of ribosome

mRNA

Small subunit of ribosome

Ribosomes move along mRNA

Ribosome subunits released

▲ FIGURE 22.22 **Protein Synthesis** A ribosome moves along a strand of mRNA, joining amino acids to form a protein.

# Chemistry AND Medicine | The Human Genome Project

In 1990, the U.S. Department of Energy (DOE) and the National Institutes of Health (NIH) embarked on a 15-year project to map the human genome, all of the genetic material of a human being. Over 2500 researchers from 18 countries contributed to this research, which has been referred to as the Mount Everest of biology. An initial draft of the map was completed in 2001 and the final draft was completed in 2003. Here, we highlight some of what has been learned through this massive undertaking.

- The human genome contains 3165 million nucleotide base pairs.

- The average gene contains about 3000 base pairs. The largest gene is for the protein dystrophin (dystrophin deficiency is the root cause of muscular dystrophy); it contains 2.4 million base pairs.

- The human genome contains about 30,000 genes. The function of over half of these is still unknown. Before the Human Genome Project, researchers had estimated that humans had about 100,000 genes. The number of genes in humans is not much larger than the number found in many simpler organisms. For example, the number of genes in a roundworm is nearly 20,000. Whatever makes humans unique, it is not the number of genes in our genome.

- Less than 2% of human DNA actually consists of genes. These genes are aggregated in seemingly random areas within the genome, with vast expanses of noncoding DNA between the coding regions. This stands in contrast to other organisms, which tend to have more uniform distribution of genes throughout their genome.

- The order of DNA base pairs is 99.9% identical in all humans.

- About 1.4 million single base-pair differences (called SNPs for single-nucleotide polymorphisms) have been identified in the human genome. Understanding SNPs can help in identifying individuals who are susceptible to certain diseases. Knowledge of SNPs may also allow physicians to tailor drugs to match individuals.

Knowledge of the human genome is expected to lead to the development of new therapies in several ways. First, knowledge of genes can lead to smart drug design. Instead of developing drugs by trial and error (the current procedure for many drugs), knowledge of a specific gene will allow scientists to design drugs to carry out a specific function related to that gene or its protein product. Second, human genes themselves can provide the blueprint for the production of certain types of drugs, either in the laboratory or by other organisms. For example, in Section 22.1 we discussed how insulin is made by inserting the insulin gene into bacteria, which then synthesize the needed drug. Intriguingly, it may even be possible to replace abnormal or missing genes in the cells of diseased patients. Such gene therapies are still in the early stages of development, but they may eventually give us a powerful new tool for combating inherited diseases.

Although the completion of the Human Genome Project may seem like the end, it is really just the beginning. Thousands of studies in the coming years will rely on the data obtained through this endeavor.

*Summarizing DNA Coding:*

▶ DNA contains the code for the sequence of amino acids in proteins.

▶ A codon—three nucleotides with their bases—codes for one amino acid.

▶ A gene—a sequence of codons—codes for one protein.

▶ Genes are contained in structures called chromosomes that occur within cells. Humans have 46 chromosomes in the nuclei of their cells.

▶ When a human cell divides, each daughter cell receives a complete copy of the DNA—all 46 chromosomes.

▶ When a cell synthesizes a protein, the base sequence of the gene that codes for that protein is transferred to mRNA. The mRNA then moves to a ribosome, where the amino acids are linked in the correct sequence to synthesize the protein. The general sequence of information flow is:

$$DNA \longrightarrow RNA \longrightarrow PROTEIN$$

# Self-Assessment Quiz

**Q1.** Which compound is a lipid?

a) 

b) $H_3C-(CH_2)_{12}-C\overset{O}{\underset{OH}{\big\|}}$

c) 

d) $H_3C-CH_2-O-CH_2-CH_3$

**Q2.** Which compound is a carbohydrate?

a) 

b) 

c) $H_3C-CH_2-CH_2-CH_2-CH_2-CH_2-CH_2-CH_3$

d) 

**Q3.** Which compound is an amino acid?

a) 

b) $H_2N-CH_2-CH_2-CH_2-CH_2-CH_2-NH_2$

c) 

d)

**Q4.** The sequence of amino acids in a protein chain is an example of which kind of protein structure?

a) primary  b) secondary
c) tertiary  d) quaternary

**Q5.** A section of DNA contains the bases ATTCGGAA. What is the correct sequence of bases in the complementary strand?

a) CGGATTCC  b) ATTCGGAA
c) AAGGCTTA  d) TAAGCCTT

**Q6.** A codon is a sequence within DNA that codes for what?

a) an amino acid  b) a protein
c) a carbohydrate  d) a fat

**Q7.** Which of the functions listed is *not* attributed to proteins?

a) catalyze biochemical reactions
b) transport substances from one place to another
c) mediate muscle contraction
d) none of the above (all of these functions are attributed to proteins)

**Q8.** Which is a primary function of nucleic acids?

a) energy storage
b) coding the amino acid sequence for proteins
c) oxygen transport
d) regulation of metabolic processes

*Continued—*

**Q9.** Peptide bonds occur in which type of biochemical compounds?

    a) lipids            b) carbohydrates

    c) proteins         d) nucleic acids

**Q10.** How many nucleotides are required to code for all of the amino acids in a protein that contains 186 amino acids?

    a) 186     b) 372     c) 558     d) 1

**Answers: 1.** (b) **2.** (a) **3.** (c) **4.** (a) **5.** (d) **6.** (a) **7.** (d) **8.** (b) **9.** (c) **10.** (c)

# Chapter 22 IN REVIEW

## Key Terms

**Section 22.1**

biochemistry (1030)

**Section 22.2**

lipid (1030)
fatty acid (1030)
triglyceride (1032)
ester linkage (1032)
saturated fat (1033)
unsaturated fat (1033)
phospholipid (1033)
glycolipid (1033)
lipid bilayer (1034)
steroid (1034)

**Section 22.3**

carbohydrate (1035)
monosaccharide (1035)

hexose (1035)
aldose (1035)
ketose (1035)
glycosidic linkage (1037)
disaccharide (1037)
hydrolysis (1037)
polysaccharide (1037)
complex carbohydrate (1037)
cellulose (1038)
starch (1038)
glycogen (1038)

**Section 22.4**

enzyme (1038)
amino acid (1039)
peptide bond (1041)
dipeptide (1041)
polypeptide (1041)

**Section 22.5**

fibrous protein (1042)
globular protein (1043)
primary structure (1044)
secondary structure (1044)
$\alpha$-helix (1044)
$\beta$-pleated sheet (1045)
random coil (1045)
tertiary structure (1045)
quaternary structure (1046)

**Section 22.6**

nucleic acid (1046)
nucleotide (1046)
complementary (1048)
codon (1049)
gene (1049)
chromosome (1049)

## Key Concepts

### Diabetes and the Synthesis of Human Insulin (22.1)

▶ Diabetes is a chronic illness that occurs when the pancreas cannot make enough insulin, a protein that promotes the absorption of glucose into cells.

▶ The chemical structure of insulin was discovered in 1955 by Frederick Sanger, who made it possible for insulin to be synthesized in the laboratory.

▶ Eventually, scientists inserted the human gene that codes for the production of insulin into bacteria, which were then able to produce enough insulin to supply diabetics.

### Lipids (22.2)

▶ Lipids are biological chemicals that are nonpolar and, therefore, insoluble in water. In our bodies, lipids compose cell membranes, store energy, and provide insulation.

▶ A type of lipid called a fatty acid is a carboxylic acid with a long hydrocarbon chain. Fatty acids can be saturated, meaning they contain the maximum number of hydrogen atoms, or unsaturated, meaning they contain one or more carbon–carbon double bonds. Saturated fatty acids experience greater intermolecular forces,

making them solid at room temperature, while unsaturated fatty acids are liquids.

▶ Fats and oils are triglycerides, triesters composed of glycerol bonded by ester linkages to three fatty acids. Like fatty acids, triglycerides can be saturated (fats) or unsaturated (oils).

▶ Other lipids include phospholipids, made up of a glycerol bonded to two nonpolar fatty acids and a polar phosphate group, used in animal cell membranes; glycolipids, similar to phospholipids but with a sugar molecule as their polar head; and steroids, four-ringed lipids that include cholesterol and sex hormones.

### Carbohydrates (22.3)

▶ Carbohydrates are polyhydroxy aldehydes or ketones and generally have the formula $(CH_2O)_n$. They are important to short-term energy storage and plant structure composition.

▶ Monosaccharides, the simplest carbohydrates, contain three to eight carbons with one aldehyde or ketone functional group. Glucose, an example of a hexose, can exist both in a linear form and a ring form.

▶ Two monosaccharides combine to form a disaccharide. For example, glucose and fructose combine to form sucrose. The glycosidic linkages that connect the two monosaccharides are broken during digestion by hydrolysis.

▶ Polysaccharides are polymers of monosaccharides known as complex carbohydrates. They include cellulose, also called fiber, the main structural component of plants; starch, an energy storage compound found in potatoes and grains; and glycogen, used by animals to store glucose in the muscles.

## Proteins and Amino Acids (22.4)

▶ Proteins are polymers of amino acids and serve a variety of biological functions including structure composition, metabolic regulation, and muscle contraction. Enzymes are particularly important proteins that catalyze biochemical reactions in cells.

▶ An amino acid contains a carbon atom bonded to an amine group, a carboxylic acid group, a hydrogen atom, and an R group. There are 20 amino acids in humans, and they differ only in their R group.

▶ Amino acids form peptide bonds between the amine end of one amino acid with the carboxylic end of another, creating dipeptides, tripeptides, and so on, or polypeptides, large examples of which are called proteins.

## Protein Structure (22.5)

▶ Protein structure and shape are critical to protein function.

▶ Proteins can be broadly divided into two structural categories. Fibrous proteins are generally linear, insoluble structures that serve structural functions. Globular proteins fold into roughly spherical conformations with nonpolar side chains oriented to the interior and polar side chains oriented to the exterior; this structure makes globular proteins soluble in water.

▶ Protein structure can be analyzed at four levels. The primary structure is the sequence of the amino acid chain. The secondary structure refers to certain regular repeating patterns in the arrangement of protein chains, such as $\alpha$-helix and $\beta$-pleated sheet patterns. The tertiary structure refers to large-scale bends and folds due to interactions between the R groups of amino acids such as hydrogen bonding, disulfide linkages, hydrophobic interactions, and salt bridges.

Quaternary structure shows the way that monomeric subunits fit together in multimeric proteins that have more than one polypeptide chain.

## Nucleic Acids (22.6)

▶ Nucleic acids, such as DNA and RNA, are the chemical blueprints used to synthesize proteins. Nucleic acids are polymers of nucleotides, which are each composed of a sugar, a base, and a phosphate group.

▶ The bases of DNA are adenine (A), cytosine (C), guanine (G), and thymine (T), which are subject to complementary pairing: each pyrimidine base combines with only one purine base.

▶ Phosphate links bind the C5′ carbon of one sugar with the C3′ carbon of another sugar to make the polymeric chain.

▶ The order of the bases in a nucleic acid chain specifies the order of amino acids in a protein.

▶ Each amino acid is coded by a codon, a sequence of three bases. A gene is a sequence of codons that codes for a specific protein. Genes, in turn, are contained in structures called chromosomes.

## DNA Replication, the Double Helix, and Protein Synthesis (22.7)

▶ Though the DNA code is complete in any cell in the body, every cell does not express every gene. Cells only express the genes relevant to their function.

▶ DNA is composed of two complementary strands wound around each other in a double helix. The strands are antiparallel; the bases of each strand face the interior and hydrogen-bond to their complements on the other strand.

▶ In order to replicate, a DNA strand divides and an enzyme called DNA polymerase creates the complement of each of the divided strands, thereby making two copies of the original DNA molecule.

▶ To synthesize proteins, the section of DNA that codes for that gene unravels. Messenger RNA (mRNA) is synthesized as a copy of the gene. The mRNA joins with ribosomes, structures that "read" the mRNA code and synthesize the correct sequence of amino acids.

## Key Learning Outcomes

| CHAPTER OBJECTIVES | ASSESSMENT |
|---|---|
| Recognizing and Working with the Basic Structures of Lipids (22.2) <br><br> Polar head    Nonpolar tails | Exercises 31–34 |
| Recognizing and Working with the Basic Structures of Carbohydrates (22.3) | Exercises 37–40 |
| Identifying Chiral Carbon Atoms in Carbohydrates (22.3) | Example 22.1   For Practice 22.1   Exercises 41, 42, 86 |
| Drawing Structures for Amino Acids and Peptide Bonds (22.4) | Example 22.2   For Practice 22.2   Exercises 49–58 |
| Recognizing Levels of Protein Structure (22.5) | Exercises 59–62 |
| Recognizing Nucleic Acids and Nucleotides (22.6) | Exercises 63, 64 |

# EXERCISES

## Review Questions

1. What is biochemistry? What significant advances in biochemistry have helped diabetics?

2. What is a lipid? What roles do lipids play in living organisms?

3. What is a fatty acid? Draw the general structure of a fatty acid.

4. What effect do double bonds have within the hydrocarbon chain of a fatty acid?

5. What are triglycerides? Draw a general structure of a triglyceride.

6. Explain the difference, both in terms of structure and in terms of properties, between a saturated fat and an unsaturated fat.

7. Describe the basic structure of phospholipids and glycolipids. What functions do these lipids have in living organisms?

8. What is a steroid? List some functions of steroids.

9. What are carbohydrates? What role do they play in living organisms?

10. How do monosaccharides and disaccharides differ? Aldoses and ketoses?

11. How do simple and complex carbohydrates differ?

12. How do cellulose, starch, and glycogen differ? Describe the function of each.

13. What roles do proteins play in living organisms? List specific examples.

14. Describe the basic structure of an amino acid. How are amino acids linked together to form proteins?

15. How do the properties of the R groups in amino acids relate to the properties of proteins?

16. Why are amino acids chiral?

17. Draw the structure of a neutral amino acid and its dipolar ion.

18. Draw the structure of any two amino acids, showing how they link together to form a dipeptide.

19. Why is protein structure important?

20. How do fibrous proteins and globular proteins differ?

21. Describe the various levels of protein structure (primary, secondary, tertiary, and quaternary).

22. What types of interactions or bonds maintain each of the structures listed in the previous problem?

23. Describe the secondary structures known as $\alpha$-helix and $\beta$-pleated sheet.

24. What is the function of nucleic acids in living organisms?

25. What is the general structure of a nucleic acid?

26. The bases in nucleic acids are *complementary*. What does this mean?

27. What is a codon? A gene? A chromosome?

28. Do most cells contain complete copies of an organism's DNA? Do most cells express all of the genes contained in their DNA?

29. Explain the mechanism by which DNA is replicated.

30. Explain the mechanism by which proteins are synthesized from the information contained within DNA.

## Problems by Topic

### Lipids

31. Determine whether or not each molecule is a lipid. If the molecule is a lipid, indicate the kind of lipid. If it is a fatty acid or a triglyceride, classify it as saturated or unsaturated.

a. $CH_3-CH_2-CH_2-CH_2-CH_2-CH_2-CH_2-CH_2-CH_2-CH_3$

b.

c.

d.

32. Determine whether or not each molecule is a lipid. If the molecule is a lipid, state the kind of lipid. If it is a fatty acid or a triglyceride, classify it as saturated or unsaturated.

a.

b. $CH_3-CH_2-CH_2-CH_2-CH_2-CH_2-CH_2-CH_2-CH_2-CH_2-CH_2-CH_3$

c.

d.

33. Determine whether or not each molecule is a fatty acid. If it is a fatty acid, classify it as saturated, monounsaturated, or polyunsaturated.

**a.** [structure of a fatty acid with carboxylic acid group]

**b.** [structure of a hydrocarbon chain]

**c.** [structure with ether linkage]

**d.** [structure of an unsaturated fatty acid with CH=CH and carboxylic acid group]

34. Which fatty acid is most likely to be a solid at room temperature?

**a.** [short chain carboxylic acid structure]

**b.** [longer chain saturated carboxylic acid structure]

**c.** [monounsaturated fatty acid with CH=CH]

**d.** [polyunsaturated fatty acid with two CH=CH groups]

35. Draw structures showing the reaction of glycerol with linoleic acid to form the triglyceride trilinolean. Would you expect this triglyceride to be a fat or an oil?

36. Draw structures showing the reaction of glycerol with myristic acid to form the triglyceride trimyristin. Would you expect this triglyceride to be a fat or an oil?

## Carbohydrates

37. Determine whether or not each structure is a carbohydrate. If the molecule is a carbohydrate, classify it as a monosaccharide, disaccharide, or trisaccharide.

**a.** [open-chain sugar structure with CHO and multiple OH groups]

**b.** [steroid structure with CH₃ and OH groups]

**c.** [disaccharide ring structure]

**d.** [dipeptide structure with amide linkage]

38. Determine whether or not each structure is a carbohydrate. If the molecule is a carbohydrate, classify it as a monosaccharide, disaccharide, or trisaccharide.

**a.** [trisaccharide ring structure]

**b.** [open-chain structure with CHO, OH, CH₂, CH₃]

**c.** [unsaturated fatty acid structure with multiple double bonds]

**d.** [furanose ring structure]

**39.** Classify each saccharide as an aldose or a ketose. Also classify each as a triose, tetrose, pentose, and so on.

a.

b.

c.

d.

**40.** Classify each saccharide as an aldose or a ketose. Also classify each as a triose, tetrose, pentose, and so on.

a.

b.

c.

d.

**41.** How many chiral carbon atoms are in each of the structures in Problem 39?

**42.** How many chiral carbon atoms are in each of the structures in Problem 40?

**43.** Draw structures for the straight-chain and ring forms of glucose.

**44.** Draw structures for the straight-chain and ring forms of fructose.

**45.** Draw the products that result from the hydrolysis of the carbohydrate pictured on the right.

**46.** Draw the products that result from the hydrolysis of the carbohydrate pictured on the right.

**47.** Draw the structure of sucrose. Label the glucose and fructose rings in this disaccharide.

**48.** Lactose is a disaccharide of glucose and galactose. Draw the structure of lactose.

## Amino Acids and Proteins

**49.** Draw each amino acid in its dipolar ion form.
  a. Thr  b. Ala  c. Leu  d. Lys

**50.** Draw each amino acid in its dipolar ion form.
  a. Val  b. Phe  c. Tyr  d. Cys

**51.** Draw the structures of the two enantiomers of alanine.

**52.** Draw the structures of the two enantiomers of cysteine.

**53.** How many different tripeptides can form from serine, glycine, and cysteine? List the amino acid sequence of each one.

**54.** How many dipeptides can form from leucine and serine? List the amino acid sequence for each one.

**55.** Draw the reaction by which serine and tyrosine form a peptide bond.

**56.** Draw the reaction by which valine and asparagine form a peptide bond.

**57.** Draw a structure for each tripeptide.
  a. Gln-Met-Cys  b. Ser-Leu-Cys  c. Cys-Leu-Ser

**58.** Draw a structure for each tetrapeptide.
  a. Ser-Ala-Leu-Cys  b. Gln-Met-Cys-Gly
  c. Gly-Cys-Met-Gln

**59.** A phenylalanine amino acid on a protein strand undergoes hydrophobic interactions with another phenylalanine amino acid that is 26 amino acid units away. The resulting fold in the protein is an example of which kind of structure? (primary, secondary, tertiary, or quaternary)

**60.** An amino acid on a protein strand forms a hydrogen bond to another amino acid that is four amino acid units away. The next amino acid on the chain does the same, hydrogen-bonding to an amino acid that is four amino acids away from it. This pattern repeats itself over a significant part of the protein chain. The resulting pattern in the protein is an example of which kind of structure? (primary, secondary, tertiary, or quaternary)

**61.** The amino acid sequence in one section of a protein is shown here. It represents which kind of structure? (primary, secondary, tertiary, or quaternary)

-Lys-Glu-Thr-Ala-Ala-Ala-Lys-Phe-Glu-

**62.** A dimeric protein is composed of two individual chains of amino acids. The way these two chains fit together is an example of which kind of structure? (primary, secondary, tertiary, or quaternary)

## Nucleic Acids

**63.** Determine whether or not each structure is a nucleotide. For each nucleotide, identify the base as A, T, C, or G.

a.

b.

**c.**    **d.**

**64.** Determine whether or not each structure is a nucleotide. For each nucleotide, identify the base as A, T, C, or G.

**a.**   **b.**

**c.**   **d.**

**65.** Draw the structures of the two purine bases in nucleic acids.

**66.** Draw the structures of the three pyrimidine bases in nucleic acids.

**67.** Draw the DNA strand that is complementary to the DNA strand shown here.

T G T A C G C

**68.** Draw the DNA strand that is complementary to the DNA strand shown here.

A T G A C T G

**69.** A monomeric protein contains 154 amino acids. How many codons code for these amino acids? How many nucleotides?

**70.** A dimeric protein contains 142 amino acids in one strand and 148 in the other. How many codons code for these amino acids? How many nucleotides?

# Cumulative Problems

**71.** Determine the class of biochemical compound that contains each type of linkage.
   **a.** peptide bonds      **b.** glycosidic linkage
   **c.** ester linkage

**72.** Name the type of polymer associated with each monomer.
   **a.** nucleotide      **b.** amino acid
   **c.** saccharide

**73.** What is the difference between a codon and a nucleotide? A codon and a gene?

**74.** What is the difference between a fatty acid and a triglyceride? A triglyceride and a phospholipid?

**75.** The amino acid alanine has the condensed structural formula shown here:

$$NH_2CH(CH_3)COOH$$

Determine the VSEPR geometry about each internal atom and make a three-dimensional sketch of the molecule.

**76.** The amino acid serine has the condensed structural formula shown here:

$$NH_2CH(CH_2OH)COOH$$

Determine the VSEPR geometry about each internal atom and make a three-dimensional sketch of the molecule.

**77.** Which amino acids in Table 22.3 are most likely to be involved in hydrophobic interactions?

**78.** Sickle-cell anemia is caused by a genetic defect that substitutes valine for glutamic acid at one position in two of the four chains of the hemoglobin protein. The result is a decrease in the water solubility of hemoglobin. Examine the structures of valine and glutamic acid and explain why this occurs.

**79.** Determining the amino acid sequence in a protein usually involves treating the protein with various reagents that break up the protein into smaller fragments that can be individually sequenced. Treating a particular 11-amino acid polypeptide with one reagent produced the fragments:

   Ala-Leu-Phe-Gly-Asn-Lys   Trp-Glu-Cys   Gly-Arg

Treating the same polypeptide with a different reagent produced the fragments:

   Glu-Cys   Gly-Asn-Lys-Trp   Gly-Arg-Ala-Leu-Phe

What is the amino acid sequence of the polypeptide?

**80.** Treating a particular polypeptide with one reagent (as described in the previous problem) produced the fragments:

   Gly-Glu-Ser-Lys   Trp-Arg   Leu-Thr-Ala-Trp

Treating the same polypeptide with a different reagent produced the fragments:

   Gly-Glu   Thr-Ala-Trp   Ser-Lys-Trp-Arg-Leu

What is the amino acid sequence of the polypeptide?

**81.** Naturally occurring D-glucose is one of a pair of enantiomers. Its mirror image is L-glucose. Draw the two cyclic six-membered isomers of L-glucose that differ in the configuration around C1 and indicate which is $\alpha$ and which is $\beta$.

**82.** Calculate the mass percent of phosphorus in a strand of DNA that consists of equal amounts of each of the four N-bases.

**83.** The double helical structure of DNA disrupts on heating but reforms on cooling. Use thermodynamic reasoning to account for these observations.

**84.** From the structural formula of cholesterol shown on page 1034, determine (a) the composition, and (b) the number of chiral centers.

# Challenge Problems

**85.** One way to fight viral infections is to prevent a virus from replicating its DNA. Without DNA replication, the virus cannot multiply. Some viral drug therapies cause the introduction of *fake* nucleotides into cells. When the virus uses one of these fake nucleotides in an attempt to replicate its DNA, the fake nucleotide doesn't work and viral DNA replication is halted. For example, azidothymidine (AZT), a drug used to fight the human immunodeficiency virus (HIV) that causes AIDS, results in the introduction of the following fake thymine-containing nucleotide into cells. Examine the structures of the real nucleotide and the AZT fake nucleotide. Propose a mechanism for how this fake nucleotide might halt DNA replication.

AZT nucleotide

Actual nucleotide

**86.** Draw each molecule and identify the chiral centers within them.
   **a.** ribose
   **b.** galactose
   **c.** 5-deoxyribose (*Hint:* The 5 indicates that the oxygen is removed from the 5th carbon.)

**87.** Glucose transport across the red blood cell membranes (erythrocyte membrane) is a well-studied system. One laboratory project obtained the data shown here for glucose transport.

| [Glucose]$_{outside}$ (mM) | Rate of Glucose Entry ($\mu$M/min) |
|:---:|:---:|
| 0.5 | 12 |
| 1.0 | 19 |
| 2.0 | 27 |
| 3.0 | 32 |
| 4.0 | 35 |

The kinetics of glucose transport through the membrane follows the Michaelis–Menten equation:

$$V_0 = \frac{V_{max}[\text{glucose}]}{K_t + [\text{glucose}]}$$

$V_0$ = rate of glucose entry

$V_{max}$ = maximum rate (the point at which addition of glucose has no effect on the rate)

$K_t$ = transport constant

The Michaelis–Menten equation can be rearranged so that a plot $1/V_0$ versus $1/[\text{glucose}]$ produces a straight line. Rearrange the equation and plot the data in order to determine $K_t$ and $V_{max}$ for glucose transport across the erythrocyte membrane.

**88.** Eukaryotic DNA is equipped with special ends called telomers. Telomers are made up of hexanucleotide sequences that repeat at the ends of the DNA. For example, human DNA features repeating AGGGTT sequences. Functionally, telomers protect the ends of chromosomes from being treated as a broken piece of DNA in need of repair. Interestingly, telomers are cut off each time the DNA is replicated, indicating a possible cellular clock that allows only a certain number of cellular replications. Telomerase is the enzyme that catalyzes the synthesis of telomers. Telomerase is present in limited quantities within certain cells such as fetal tissue, adult male germ cells, and stem cells. It is also found in over 85% of tumor cells. Researchers speculate that the telomerase activity may be linked to cancer. Propose an explanation for why telomerase activity could be associated with cancer and speculate on ways in which cancer treatments in the future may capitalize on research on this enzyme.

**89.** Write the major equilibrium that is established in a solution of glycine at pH = 2 and at pH = 10. The p$K_a$ of the COOH group is 2.3 and the p$K_a$ of the NH$_3^+$ group is 9.6. Determine the relative concentrations of each member of the relevant conjugate acid–base pair at pH = 2 and pH = 10. Calculate the pH at which glycine is neutral.

# Conceptual Problems

**90.** How many different tetrapeptides can form from four different amino acids?

**91.** Could the genetic code have been based on just three bases and three-base codons? Explain why or why not. (Assume that the code must accommodate 20 different amino acids.)

**92.** The genetic code is random, which means that a particular codon could have coded for a different amino acid. The genetic code is also nearly universal, meaning that it is the same code in nearly all organisms (and in the few where it differs, it does so only slightly). If scientists ever find life on another planet, they will be curious to know its genetic code. What would a completely different genetic code indicate about the origin of the life forms? What would a genetic code identical to terrestrial life indicate?

*Active Classroom Learning*

# Questions for Group Work

Discuss these questions with the group and record your consensus answer.

**93.** Compare and contrast cornstarch with cotton balls, which are made of almost pure cellulose. Include at least three ways in which they are similar and three ways in which they are different. Mention aspects you have observed in daily life and aspects on the molecular scale.

**94.** With group members acting as atoms, molecules, or molecular fragments, act out the formation of a peptide bond.

**95.** Describe how a common object or toy (e.g., a train, building blocks, or beads on a string) could represent a protein. Describe how amino acids and peptide bonds are represented in your analogy. Also describe the representation of primary, secondary, tertiary, and quaternary structure in your analogy.

**96.** Use the tabulated key to decipher the mRNA encoded message. Group members may split up the message and decode separate parts, but each part must be decoded by at least two group members (to prevent errors). Describe how the process your group uses to decode the message is significant for biochemistry.

| Codon | Amino Acid | Codon | Amino Acid | Codon | Amino Acid | Codon | Amino Acid |
|---|---|---|---|---|---|---|---|
| UUC | A | ACU | H | GAC | O | GGC | V |
| CUA | B | GCG | I | GAA | P | UGA | W |
| AUU | C | UAC | J | UGU | Q | UAA | X |
| AUG | D | CAU | K | UGG | R | UAG | Y |
| GUC | E | CAG | L | CGC | S | GGU | Z |
| UCC | F | AAC | M | AGU | T | | |
| CCG | G | AAG | N | AGG | U | | |

CUAGCGGACAUUACUGUCAACGCGCGCAGUUGGUAG GCGCGC UUCAACUUCGGUGCGAAGCCG!

**97.** Working individually, draw an accurate structure for one of the following (don't tell other group members which one you choose): a monosaccharide, a disaccharide, a polysaccharide, a fatty acid, a triglyceride, a steroid, an amino acid, or a dipeptide. After each group member has completed drawing a structure, take turns showing your structure to the group. How many group members can correctly identify the structure you drew? How many of your group members' structures can you correctly identify?

# Data Interpretation and Analysis

### Digestion of Lactose by the Lactase Enzyme

**98.** Lactose is a disaccharide sugar present in milk. Lactose is composed of two monosaccharides, galactose and glucose, that are joined together by a β-glycosidic bond. The β-glycosidic bond forms between the hydroxyl group on C-1 of galactose and C-4 of glucose.

When lactose is digested, the glycosidic bond between galactose and glucose is hydrolyzed. The enzyme responsible for hydrolyzing lactose is *lactase*, which is found in the small intestine. Millions of people lack sufficient levels of lactase, and as a result, they experience lactose intolerance. If lactose is not hydrolyzed, it remains in the intestines. Bacteria in the gut ferment the lactose, causing many unpleasant symptoms such as indigestion, cramping, and gas. Fortunately, commercial preparations can enable lactose-intolerant individuals to consume and digest milk products without discomfort.

Use the information provided to answer the following questions:

a. Draw the structure of lactose. (The structures of galactose and glucose are shown on page 1037 with their carbon numbering scheme.)

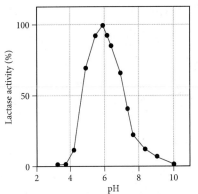

◄ **FIGURE a**
**Lactase Activity and pH**
[Source: *European Journal of Biochemistry*, Vol. 114, p. 653–661 (1981).]

b. Lactase enzyme activity, like most enzymes, is sensitive to pH. Figure a▲ illustrates how lactase activity is affected by the pH of a solution. Based on the data in Figure a, what can you conclude about the pH of the small intestine?

c. A glutamic acid in the active site of lactase is suspected to be involved in the catalytic mechanism. Draw the structure of the glutamic acid side chain in the ionization state that likely exists when lactase is most catalytically active.

# Answers to Conceptual Connections

### Peptides

**22.1** Six possible tripeptides can form from these three amino acids. They are (1) Ser-Gly-Ala; (2) Gly-Ala-Ser; (3) Ala-Ser-Gly; (4) Ala-Gly-Ser; (5) Ser-Ala-Gly; (6) Gly-Ser-Ala. Notice that bonding the amino acids in reverse order results in a different molecule because the N-terminal and the C-terminal ends reside on different amino acids. For example, in Ser-Gly-Ala, the N-terminal amino acid is Ser (conventionally drawn on the left) and the C-terminal side is Ala. In Ala-Gly-Ser, in contrast, the N-terminal amino acid is Ala and the C-terminal one is Ser.

### The Genetic Code

**22.2** The number of unique two-base sequences of four bases is $4^2 = 16$. The number of unique three-base sequences of four bases is $4^3 = 64$. A two-base system could code for 16 amino acids, and a three-base system could code for 64 amino acids.

# 23

# Chemistry of the Nonmetals

What one learns in chemistry is that Nature wrote all the rules of structuring; man does not invent chemical structuring rules; he only discovers the rules. All the chemist can do is find out what Nature permits, and any substances that are thus developed or discovered are inherently natural.

—R. BUCKMINSTER FULLER (1895–1983)

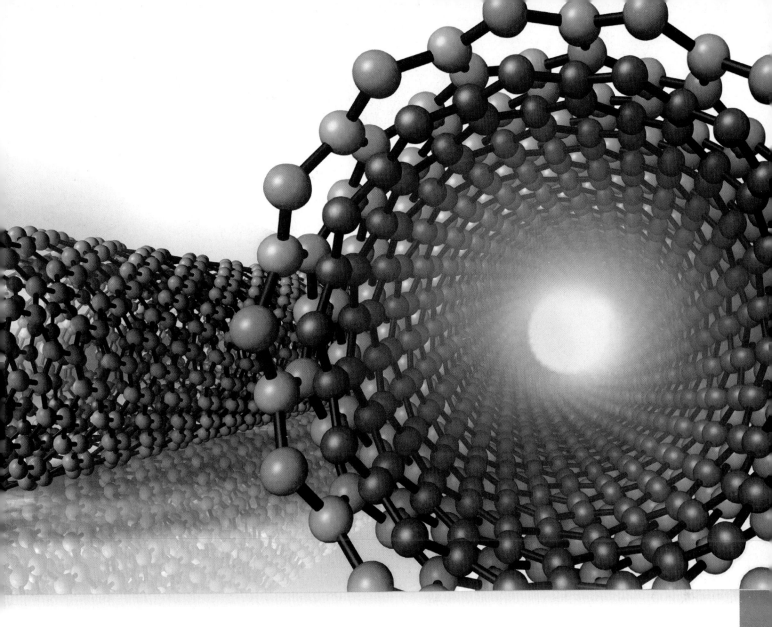

In these last few chapters, we explore the descriptive chemistry of the nonmetals and metals. These descriptions are part of a branch of chemistry called inorganic chemistry. We begin our exploration of descriptive inorganic chemistry by looking at the chemistry of some of the main-group elements. The main-group elements are grouped together because their valence electrons occupy only *s* or *p* orbitals; however, their properties vary greatly. The main-group elements include metals, nonmetals, and metalloids and they may be solids, liquids, or gases at room temperature. This great diversity of properties, bonding, and structures of all the main-group elements cannot be adequately described in a single chapter. Therefore, this chapter focuses on only a few main-group elements (silicon, boron, carbon, nitrogen, phosphorus, oxygen, sulfur, and the halogens) and their compounds in an effort to illustrate the diversity within the group.

▲ Scientists have been able to synthesize carbon nanotubes that have an insulating boron nitride sheath. These structures are like electrical wires 100,000 times thinner than a human hair.

## 23.1  Insulated Nanowires

In 1991, scientists discovered carbon nanotubes, the long, thin, hollow cylinders of carbon atoms that we discussed in Section 12.6. In the late 1990s, scientists discovered that they could make similar tubes from boron nitride. Boron nitride contains BN units that are isoelectronic (having the same number of electrons) with carbon in the sense that each BN unit contains eight valence electrons, or four per atom (just like carbon). The size and electronegativity of a carbon atom are also almost equal to the average of those properties for a boron atom and a nitrogen atom (Table 23.1). As a result, BN forms a number of structures that are similar to those formed by carbon, including nanotubes.

**TABLE 23.1 Properties of BN and C**

| Element | Atomic Radius (pm) | Ionization Energy (kJ/mol) | Electronegativity |
|---|---|---|---|
| B | 85 | 800 | 2 |
| BN | 77.5 average | 1101 average | 2.5 average |
| N | 70 | 1402 | 3 |
| C | 77 | 1086 | 2.5 |

An important difference between boron nitride nanotubes and carbon nanotubes is their conductivity. Carbon nanotubes conduct electrical current, but boron nitride tubes act as insulators. In 2003, scientists were able to combine these two sorts of nanotubes into one structure: a conducting carbon nanotube with an insulating boron nitride sheath, shown in Figure 23.1◄. The result is an insulated conducting wire that is 100,000 times thinner than a human hair. Such thin wires may someday be used in computers and other electronic devices, allowing these devices to continue to become smaller and more efficient.

The more we learn about the structures and reactivities of known materials, the better equipped we are to discover new materials and applications. Even though it may seem that most inorganic compounds have already been discovered and analyzed, new materials, with immense impacts on our society, are constantly being discovered. As Buckminster Fuller stated in this chapter's opening quote, we continue to find "what Nature permits." In some cases, what nature permits turns out to be extremely useful to society.

C nanotube    BN nanotube

▲ **FIGURE 23.1 Boron Nitride Nanotube** The model represents an insulating BN nanotube filled with a conducting carbon nanotube.

## 23.2 The Main-Group Elements: Bonding and Properties

We identify the **main-group elements** by their electron configurations. In this chapter, we focus on groups 3A–7A, the major part of the *p* block in the periodic table. The *p* orbitals fill incrementally across any row of this section of the periodic table; they contain from one electron in group 3A to five electrons in group 7A (the halogens). The physical properties of the elements, such as atomic size and electronegativity, also change across each period, and this affects their reactivity and the types of compounds they form.

Recall from Section 8.6 that the effective nuclear charge of the main-group elements becomes greater as we move to the right across any row in the periodic table. This increasing effective charge results in smaller radii, increasing electronegativity, and increasing ionization energy as we move to the right across the periods. Consequently, as we have seen since the early chapters of this book, the nonmetals on the right side of the periodic table (those toward the right of the *p* block) tend to form anions in ionic compounds. They are easily reduced, gaining electrons to completely fill their *p* orbitals and attain noble gas electron configurations. These elements act in reactions as *oxidizing agents*— they oxidize other substances while they are themselves reduced. The smallest halogens and the elements in the oxygen group are the strongest oxidizing agents in the *p* block.

Elements near the center of the *p* block have fewer *p* electrons and do not usually fill the *p* orbitals by forming anions; instead they share electrons, forming covalent bonds. We see this type of reactivity in the vast array of molecular compounds formed by the smaller elements in the carbon and nitrogen groups. The main-group elements on the far left of the *p* block have only one *p* electron and form cations in ionic compounds and electron-deficient species (species with an incomplete octet) in covalent compounds.

Notice that as we move to the right across any row in the *p* block, the type of bonding changes as the elements become less metallic. Recall from Section 8.8 that metallic character increases as we go down each column. The diagonal group of metalloid elements stretching from boron to astatine divides the main-group elements: to the left of this diagonal, the elements are metals that form cations and metallic compounds; to the right, the elements are nonmetals that form anions and covalent compounds.

The vast range in elemental properties, from those of metallic elements such as thallium and lead that have very low electronegativities of 1.8 and 1.9 (respectively) to those of the nonmetallic elements such as oxygen and fluorine that have the highest electronegativities of 3.5 and 4.0 (respectively), results in the great chemical diversity of the elements in the *p* block. These elements include metals, alloys, simple covalent compounds, enormous covalent network compounds, simple binary ionic compounds, and complex chain and layered ionic compounds.

## Major Divisions of the Periodic Table

## 23.3  Silicates: The Most Abundant Matter in Earth's Crust

The most abundant elements in Earth's crust are oxygen (45–50% by mass) and silicon (about 28% by mass). The few other elements that individually comprise more than 1% of the crust's mass are aluminum, iron, calcium, magnesium, sodium, and potassium (Figure 23.2▼), and most of these are found in silicon and oxygen compounds. In order to understand the matter on Earth's surface, we must understand silicon and oxygen compounds.

■ Oxygen: 46.4%    ■ Ca: 4.15%
■ Silicon: 28.2%    ■ Na: 2.36%
  Al: 8.32%         ■ Mg: 2.33%
■ Fe: 5.63%         ⎰ K: 2.09%
                    ⎱ Ti: 0.57%
                      H: 0.14%

▶ **FIGURE 23.2  Major Elements in Earth's Crust**   The major components of Earth's crust are oxygen and silicon. Only a few other elements compose more than 1% of the crust each.

> A rock is a heterogeneous mixture that may contain one or more minerals.

**Silicates** are covalent atomic solids (see Section 12.6) that contain silicon, oxygen, and various metal atoms. Rocks, clays, and soils contain silicates. Their great diversity illustrates a theme that we have encountered since Chapter 1 of this book: the properties of substances are determined by their atomic and molecular structures. The structures of silicates determine their properties—and since these structures are varied, their properties are also varied. Some silicates form strong three-dimensional materials, whereas others break into sheets, and still others are highly fibrous. Let's examine more closely several of these structures.

▲ **FIGURE 23.3  SiO₄ Tetrahedron**   In a $SiO_4$ tetrahedron, silicon occupies the center of the tetrahedron and one oxygen atom occupies each corner.

### Quartz

Recall from Chapter 12 that silicon and oxygen form a network covalent structure in which a silicon atom bonds to four oxygen atoms, forming a tetrahedral shape with the silicon atom in the middle and the four oxygen atoms at the corners of the tetrahedron (Figure 23.3◀). As we examined in Chapter 12, the silicon atom bonds to each oxygen atom with a single covalent sigma bond. Recall that the silicon atom in this structure bonds to four oxygen atoms and obtains a complete octet. However, each oxygen atom is one electron short of an octet, so each O atom forms a second covalent bond to a different Si atom, forming the three-dimensional structure of **quartz**. Quartz has a formula unit of $SiO_2$ and is generally called **silica**. Each Si atom is in a tetrahedron surrounded by four O atoms, and each O atom acts as a bridge connecting the corners of two tetrahedrons (Figure 23.4▼).

### Aluminosilicates

**Aluminosilicates** are a family of compounds in which aluminum atoms substitute for silicon atoms in some of the lattice sites of the silica structure. Since the aluminum ion has only three valence electrons (in contrast to the four valence electrons of silicon), a $SiO_2$ unit becomes $AlO_2^-$ upon substitution of aluminum. The negative charge is balanced by a positive counterion. A common group of aluminosilicates is the feldspars. The mineral albite is a feldspar. A **mineral** is a homogeneous, crystalline substance that naturally occurs in the Earth's crust. In albite, one-fourth of the Si atoms are replaced by Al atoms. $Na^+$ ions provide the necessary balancing positive charge. The formula for albite is $Na(AlSi_3O_8)$, but it may be written as $Na(AlO_2)(SiO_2)_3$ to illustrate the substitution of Al for Si.

◀ **FIGURE 23.4  Structure of Quartz**   In the quartz structure, each Si atom is in a tetrahedron surrounded by four O atoms, and each O atom is a bridge connecting the corners of two tetrahedrons.

**Example 23.1** | **Determining the Composition of an Aluminosilicate**

Write the formula for anorthite, a crystal in which Al atoms substitute for one-half of the Si atoms and the charge is balanced by $Ca^{2+}$ ions.

**SOLUTION**

The $AlO_2^-$ unit substitutes for one-half of the $SiO_2$ units; therefore, the formula has equal numbers of $AlO_2^-$ and $SiO_2$ units. You must balance every $AlO_2^-$ ion in the formula by a corresponding positive charge. Since $Ca^{2+}$ has a 2+ charge, it can balance two $AlO_2^-$ units. Thus, the formula for anorthite is $Ca(Al_2Si_2O_8)$ or $Ca(AlO_2)_2(SiO_2)_2$.

**FOR PRACTICE 23.1** Orthoclase is a crystal in which $Al^{3+}$ substitutes for one-fourth of the $Si^{4+}$ ions. $K^+$ ions balance the charge. Write the formula for orthoclase.

## Individual Silicate Units, Silicate Chains, and Silicate Sheets

In many silicate compounds, the oxygen atoms are not connected to two silicon atoms to form the neutral compound that is found in quartz. Instead, the oxygen atoms gain electrons from metal atoms and form polyatomic anions, such as $SiO_4^{4-}$. The positively charged metal ions then bond to the negatively charged silicon oxide. In these minerals, the $SiO_4$ tetrahedrons occur singly, in chains, or in sheets.

When a tetrahedron occurs singly (not bonded to other tetrahedrons), it forms the $SiO_4^{4-}$ polyatomic anion (which has four extra electrons that satisfy the octet rule for the four oxygen atoms). These types of silicates are **orthosilicates** (or nesosilicates) and require cations that have a total charge of 4+ to neutralize the negative charge. The cat-ions can be of a single metal, such as $Zn^{2+}$ in $Zn_2SiO_4$ (the mineral wil-lemite), or they can be a mixture of different metals, such as the family of crystals called olivines $[(Mg, Fe)_2SiO_4]$ where the $Mg^{2+}$ and $Fe^{2+}$ ions can exist in variable proportions, providing a total charge of 4+. All of these compounds are held together by ionic bonding between the metal cations and $SiO_4^{4-}$ polyatomic anions.

The silicate tetrahedrons can also form structures called **pyrosilicates** (or sorosilicates) in which two tetrahedrons share one corner (Figure 23.5▶), forming the disilicate ion, which has the for-mula $Si_2O_7^{6-}$. This group requires cations that balance the 6− charge on $Si_2O_7^{6-}$. Again, these cations can be the same metal ions or a mix-ture of different metal ions. For example, in the mineral hardystonite $(Ca_2ZnSi_2O_7)$, two $Ca^{2+}$ ions and one $Zn^{2+}$ ion provide the 6+ charge.

The silicon tetrahedron also forms structures called **pyroxenes** (or inosilicates) in which many of the tetrahe-drons bond together creating chains (Figure 23.6▶). The formula unit for these chains is the $SiO_3^{2-}$ unit, and the repeating unit in the structure is two formula units $(Si_2O_6^{4-})$. Two of the oxygen atoms are bonded to two silicon atoms (and thus to two other tetrahedrons) at two of the four corners of each tetrahedron. The silicate chains are held together by ionic bonding to

Formula unit: $Si_2O_7^{6-}$

▲ **FIGURE 23.5 Pyrosilicate Structure** In pyrosilicates, the silicate tetrahedrons share one corner, forming $Si_2O_7^{6-}$ units. Pyrosilicates are also called sorosilicates.

Repeating unit: $Si_2O_6^{4-}$

Formula unit: $SiO_3^{2-}$

▶ **FIGURE 23.6 Pyroxene Structure** In pyroxenes, chains of silicate tetrahedrons form. Pyroxenes are also called inosilicates.

▶ **FIGURE 23.7** **Amphibole Structure**
Amphiboles contain double chains of silica tetrahedrons.

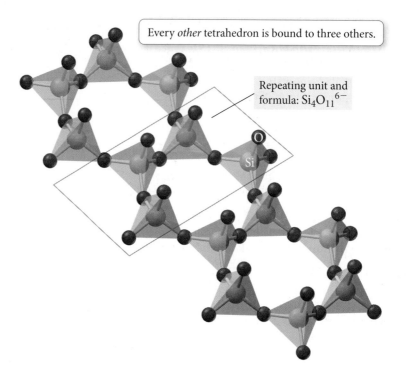

Every *other* tetrahedron is bound to three others.

Repeating unit and formula: $Si_4O_{11}^{6-}$

O
Si

▲ The fibrous texture of asbestos results from the silicate double chains of the amphibole structure.

▲ The flaky texture of mica is due to silicate sheets of the phyllosilicate structure.

metal cations, which lie between the chains. For example, in the crystal diopside, $CaMg(SiO_3)_2$, $Ca^{2+}$ and $Mg^{2+}$ ions bond with the silicate chains.

Some silicate structures have *double chains* in which half of the tetrahedrons of one chain are bonded to tetrahedrons in another chain through oxygen atoms. The minerals with double silicate chains are called amphiboles, and the repeating unit in the crystal is $Si_4O_{11}^{6-}$ (Figure 23.7▲). Half of the tetrahedrons are bonded by two of the four corner O atoms, and half of the tetrahedrons are bonded by three of the four corners, joining the two chains together. The bonding within the double chains is very strong, but the bonding between the double chains is not so strong. This structure often results in fibrous minerals such as asbestos. An example of an asbestos-type mineral is tremolite, $Ca_2(OH)_2Mg_5(Si_4O_{11})_2$. In this crystal, hydroxide ions bonded to some of the metal cations help balance the charge between the cations and the anionic silicate chains.

When three of the four oxygen atoms are bonded between the silicate tetrahedrons, the sheet structure shown in Figure 23.8▼ results. These compounds are called phyllosilicates and have a formula unit of $Si_2O_5^{2-}$. Sheets of tetrahedral silicates are bonded together by metal cations that lie between the sheets. For example, the mineral *talc*, $Mg_3(Si_2O_5)_2(OH)_2$, is a phyllosilicate. The weak interactions between silicate sheets give talc its slippery feel (this is similar to the way the weak interactions between sheets of carbon atoms give graphite its slippery feel). Table 23.2 summarizes the different kinds of silicate structures.

Every tetrahedron is bound to three others.

Repeating unit: $Si_4O_{10}^{4-}$

O
Si

Formula unit: $Si_2O_5^{2-}$

◀ **FIGURE 23.8** **Phyllosilicate Structure**
In phyllosilicates, three of the four oxygens are bonded to two silicon atoms, forming sheets of silica tetrahedrons.

**TABLE 23.2 Types of Silicate Structures**

| Tetrahedrons | Shared Vertices | Formula Unit | Si : O Ratio | Class Name | Example |
|---|---|---|---|---|---|
| Single tetrahedron | 0 | $SiO_4^{4-}$ | 1 : 4 | Orthosilicates, nesosilicates | Olivine, $Mg_2SiO_4$ |
| Double tetrahedron | 1 | $Si_2O_7^{6-}$ | 2 : 7 | Pyrosilicates, sorosilicates | Hardystonite, $Ca_2ZnSi_2O_7$ |
| Single chain | 2 | $SiO_3^{2-}$ | 1 : 3 | Pyroxenes, inosilicates | Jadeite, $NaAl(SiO_3)_2$ |
| Double chain | 2 and 3 | $Si_4O_{11}^{6-}$ | 4 : 11 | Amphiboles | Tremolite, $Ca_2(OH)_2Mg_5(Si_4O_{11})_2$ |
| Sheet | 3 | $Si_2O_5^{2-}$ | 2 : 5 | Phyllosilicates | Talc, $Mg_3(Si_2O_5)_2(OH)_2$ |
| Network covalent | 4 | $SiO_2$ | 1 : 2 | Silicas, tectosilicates | Quartz, $SiO_2$ |
| Network covalent | 4 | $AlSi_3O_8^-$ or $Al_2Si_2O_8^{2-}$ | Variable | Feldspars | Albite, $NaAlSi_3O_8$ |

---

**Example 23.2**   **Composition and Charge Balance of Silicates**

The silicate chrysotile is an amphibole with the formula $Mg_6Si_4O_{11}(OH)_x$. Use charge balancing to calculate the value of $x$ in the formula.

SOLUTION

The silicate unit for amphiboles is $Si_4O_{11}^{6-}$. The formula has six $Mg^{2+}$ ions for a total charge of 12+. To balance, you need to add another 6− charge to the 6− charge for the silicate. Therefore, add six $OH^-$ ions to the formula for chrysotile, giving a formula of $Mg_6Si_4O_{11}(OH)_6$.

FOR PRACTICE 23.2   Use charge balancing to calculate how many hydroxide ions there are in the formula of the mineral pyrophyllite, $Al_2(Si_2O_5)_2(OH)_x$.

---

## Procedure for...

| **Predicting Types of Silicate Structure and Accounting for Charge Balance** | **Example 23.3** **Predicting Silicate Structures** Predict the silicate structure for the mineral spudomene, $LiAlSi_2O_6$, and show that the formula is charge neutral. | **Example 23.4** **Predicting Silicate Structures** Predict the silicate structure for the mineral thortveitite, $Sc_2Si_2O_7$, and show that the formula is charge neutral. |
|---|---|---|
| Determine the ratio of Si to O in the formula. | SOLUTION Si : O = 1 : 3 | SOLUTION Si : O = 2 : 7 |
| Match the Si : O ratio to the type of silicate in Table 23.2. | A 1 : 3 ratio is a single chain, a pyroxene (or inosilicate). | A 2 : 7 ratio is a double tetrahedron, a pyrosilicate (or sorosilicate). |
| Determine the total anion charge. | Each $SiO_3$ group has a charge of 2−, and there are two groups per formula, so the total anion charge is 4−. | Each $Si_2O_7$ group has a charge of 6−, and there is one group per formula, so the total anion charge is 6−. |
| Determine the total cation charge and show that it matches the total anion charge. | The $Li^+$ cation has a 1+ charge and the $Al^{3+}$ cation has a 3+ charge for a total of 4+, which matches the anion charge. | Each scandium cation has a charge of 3+ for a total of 6+, which matches the anion charge. |
| | FOR PRACTICE 23.3   Predict the silicate structure for the mineral phenakite, $Be_2SiO_4$, and show that the formula is charge neutral. | FOR PRACTICE 23.4   Predict the silicate structure for the mineral diopside, $CaMgSi_2O_6$, and show that the formula is charge neutral. |

# 23.4  Boron and Its Remarkable Structures

The group 3A elements each have a filled *s* sublevel and one electron in the *p* sublevel. This electron configuration prevents these main-group elements, especially boron, from easily attaining a full octet. Most of the elements in group 3A are metals; however, because of its small size and higher electronegativity, boron behaves as a semimetal. These characteristics endow boron with some special properties and result in a wide array of structures not common to most elements.

## Elemental Boron

▲ **FIGURE 23.9  $B_{12}$ Icosahedron** An icosahedron contains 20 triangular faces that are connected at 12 vertices. Elemental boron forms several different structures, each based on the basic icosahedral unit.

The structure of elemental boron is complex. Boron has at least five different allotropes (different structures with the same elemental composition). The structure of each allotrope is based on an icosahedron (Figure 23.9◄), a geometrical shape, containing 20 triangular faces joined at 12 vertices, that is very roughly spherical. Twelve boron atoms occupy the 12 vertices. Each different allotrope connects the icosahedrons in different ways. They all have boron atoms outside the icosahedrons that bridge the icosahedrons together.

Boron is rare in Earth's crust, making up less than 0.001% by mass. Yet because it is highly concentrated at various deposits around the world, it can be mined in large quantities. The largest deposit occurs at a mine in Boron, California, which has operated continuously for over 135 years. Naturally occurring boron always occurs in compounds, and it is almost always bonded to oxygen. Among the major sources of boron are the sodium borates, which include borax, $Na_2[B_4O_5(OH)_4] \cdot 8\,H_2O$, and kernite, $Na_2[B_4O_6(OH)_2] \cdot 3\,H_2O$. Another major source of boron is calcium borate, or colemanite, $Ca_2B_6O_{11} \cdot 5\,H_2O$. In all of these compounds, boron is bonded in polyatomic anions.

The primary use for boron today is in glass manufacture. Recall from Section 12.7 that adding boron oxide to silicon oxide glass alters the thermal expansion of the glass, which is important for glassware intended for heating. When glass is heated, the outer edge of the glass warms and expands more quickly than the inner edge, creating stress that can result in cracking. Adding boron oxide to glass reduces its thermal expansion, allowing the glass (called borosilicate glass or Pyrex) to be heated without cracking.

Elemental boron is also used in the nuclear energy industry. Boron readily absorbs neutrons and is used in the control rods of nuclear reactors. When the nuclear reaction needs to be slowed down, the rods are inserted into the reactor to absorb the neutrons (see Section 20.7).

## Boron–Halogen Compounds: Trihalides

Boron forms many covalently bonded compounds in which boron atoms bond to each other. In some ways, these compounds are similar to those in which carbon covalently bonds to itself; however, the structures are different because boron is less electronegative and has only three valence electrons. Recall from earlier in this section that elemental boron has a tendency to form polyhedral cluster structures. Boron also tends to form electron-deficient compounds (compounds in which boron lacks an octet).

Boron halides have the general formula $BX_3$ and have a trigonal planar structure:

$$\ddot{X} \quad 120° \atop :\!\ddot{X}\!\!-\!\!B\!\!-\!\!\ddot{X}\!:$$

The bonds in the boron trihalides are stronger and shorter than a typical single bond, which we can explain using valence bond theory and hybridization. The boron atom uses $sp^2$ hybridized orbitals to form sigma bonds with the three halogen atoms. Because boron's three valence electrons are used to form the sigma bonds, the third *p* orbital of boron is an empty orbital that is perpendicular to the trigonal plane of the molecule. Each halogen atom has a filled *p* orbital, also perpendicular to the trigonal plane of the molecule. The empty *p* orbital on the boron atom can overlap with the full *p* orbitals on the halogens, forming a coordinate-covalent-type second bond. In $BCl_3$, for example, the boron and chlorine are joined by bonds resembling a double bond. Like normal double bonds, the boron–chlorine bond is shorter and stronger than a single bond.

The boron trihalides are strong Lewis acids. For example, $BF_3$ reacts with $NH_3$ according to the following Lewis acid–base reaction:

$$BF_3(g) + :NH_3(g) \longrightarrow F_3B{:}NH_3(s)$$

Boron trihalides act as Lewis acids in many organic reactions, such as those in which alcohols or acids are converted into esters. In water, the trihalides hydrolyze to form acidic solutions according to these reactions:

$$BF_3 + H_2O \longrightarrow BF_3 \cdot H_2O \longrightarrow BF_3OH^-(aq) + H^+(aq)$$

$$4\,BF_3 + 3\,H_2O \longrightarrow 3\,H^+ + 3\,BF_4^-(aq) + B(OH)_3(aq)$$

$$BCl_3 + 3\,H_2O \longrightarrow B(OH)_3(aq) + 3\,H^+(aq) + 3\,Cl^-(aq)$$

## Boron–Oxygen Compounds

Boron forms very strong bonds with oxygen in structures that contain trigonal $BO_3$ structures. The formula for the crystalline structure of boron and oxygen is $B_2O_3$. In this compound, the trigonal $BO_3$ structures hook together to form interlocking $B_6O_6$ hexagonal rings (Figure 23.10 ▶). Each hexagonal ring has a boron atom at each of its six corners and an oxygen atom in the middle of each of the six sides. The compound $B_2O_3$ melts at 450 °C. If the molten $B_2O_3$ cools quickly, it forms a glass (an amorphous solid). The glass still contains many interlocking $B_6O_6$ hexagonal rings but lacks the long-range order of the crystal. Molten boron oxide dissolves many metal oxides and silicon oxide to form glasses of many different compositions.

▲ FIGURE 23.10 **$B_2O_3$ Structure** Crystalline $B_2O_3$ consists of $BO_3$ trigonal structures that form hexagonal rings of $B_6O_6$.

## Boron–Hydrogen Compounds: Boranes

Compounds composed of boron and hydrogen, called **boranes**, form many unique cluster, cagelike, and netlike structures. The **closo-boranes** have the formula $B_nH_n^{2-}$ and form fully closed polyhedrons with triangular sides; Figure 23.11 ▼ shows two of these structures. A boron atom, with an attached hydrogen atom, occupies each of the vertices in the polyhedrons. The *closo*-borane with the formula $B_{12}H_{12}^{2-}$ forms the full icosahedral shape that elemental boron does, but it includes added hydrogen atoms.

If the borane polyhedron is missing one or more boron atoms, extra hydrogen atoms attach to the structure to make the borane neutral. Researchers have identified over 35 structurally different neutral boranes, ranging from $B_2H_6$ to $B_{20}H_{26}$. We classify these neutral boranes on the basis of their different chemical formulas. The **nido-boranes**, named from the Latin word for *nest*, have the formula $B_nH_{n+4}$. They consist of a cage of

### *closo*-Boranes

$B_6H_6^{2-}$          $B_7H_7^{2-}$

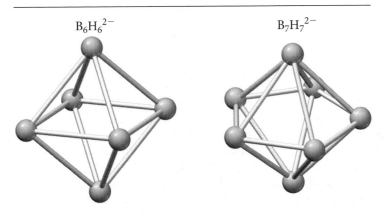

◀ FIGURE 23.11 *closo*-**Borane Structures** *closo*-Borane structures form closed polyhedrons with triangular faces. In this figure, each sphere represents a BH unit. $B_6H_6^{2-}$ has an octahedral (square bipyramidal) shape. $B_7H_7^{2-}$ has a pentagonal bipyramidal shape.

▶ **FIGURE 23.12** *nido-* and *arachno-* **Boranes** The *nido-*borane structure forms a cage missing one boron atom from a corner. The *arachno-*borane structure forms a web missing more than one boron from a corner.

A *nido-*borane        An *arachno-*borane

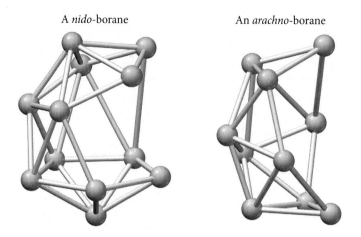

boron atoms missing one corner. The **arachno-boranes**, named from the Greek word for *spiderweb*, have the formula $B_nH_{n+6}$. They consist of a cage of boron atoms that is missing two or three corners. Figure 23.12▲ shows examples of a *nido-* and an *arachno-*borane.

Boranes not only form interesting structures, but they are also valuable as catalysts in organic reactions. For example, adding an alkene to a diborane forms an alkane bonded to the boron atom. The alkane can be cleaved (split) from the boron, resulting in a net hydrogenation reaction such as the one shown here that can be carried out under mild (lower temperature) conditions:

$$B_2H_6(g) + 6\,CH_2=CHCH_3(g) \longrightarrow 2\,B(CH_2CH_2CH_3)_3(l)$$

## 23.5  Carbon, Carbides, and Carbonates

The group 4A elements exhibit the most versatile bonding of all elements. As we saw in Chapters 12 and 21, carbon has the ability to bond with other carbon atoms and with a few other elements to form a great variety of organic compounds. These compounds are the molecules of life, which we examined in Chapter 22. The different forms of elemental crystalline carbon were covered in Chapter 12 (see Section 12.6). Here we focus on amorphous carbon and those compounds of carbon that are known as *inorganic* (rather than organic).

### Amorphous Carbon

Carbon occurs naturally in noncrystalline forms. Noncrystalline **coal** forms from the decomposition of ancient plant material that has been buried for millions of years, during which time it undergoes a process called carbonization. The carbonization reaction, which occurs under high pressure in the presence of water and the absence of air, removes most of the hydrogen and oxygen (which are lost as volatile gases such as methane and water) from the original organic compounds that composed the plant. The resulting coal contains a mixture of various hydrocarbons and carbon-rich particles. It is extensively mined and employed as an energy source throughout the world. Coal types are classified by the amount of carbon and other elements that they contain, as shown in Table 23.3.

Among the types of coal listed in Table 23.3, anthracite has the highest carbon content and consequently yields the most energy per mass when burned. Bituminous coal also contains a relatively high amount of carbon but has in addition high levels of sulfur, which results in increased formation of sulfur oxides when bituminous coal is burned. Sulfur oxides are the pollutants that create acid rain (see Sections 3.6 and 16.12).

Heating coal in the absence of air forms a solid called **coke** that is composed mainly of carbon and ash. Coke is used in the steel industry for the reduction of iron ore to iron.

In a blast furnace, the carbon in the coke is oxidized to form carbon monoxide, which reduces the iron in iron oxide according to these reactions:

$$O_2(g) + \underset{\text{coke}}{C(s)} \longrightarrow CO_2(g)$$

$$CO_2(g) + \underset{\text{coke}}{C(s)} \longrightarrow 2\,CO(g)$$

$$Fe_2O_3(s) + 3\,CO(g) \longrightarrow 2\,Fe(s) + 3\,CO_2(g)$$

Heating wood in the absence of air produces **charcoal**. Like coal, charcoal contains a high amount of amorphous free carbon and is a common fuel for outdoor cooking grills. Charcoal retains the general overall shape of the original wood, but the process creates many voids, resulting in a much lower density. The voids within charcoal create high surface area that makes the charcoal useful for filtration. The impurities in a liquid or gas adsorb (or stick) on the charcoal surface as the liquid or gas flows through the pores in the charcoal.

Very fine carbon particles with high surface areas are called **activated carbon**, or *activated charcoal*. The large surface area of the particles, greater than $10^3$ m$^2$/g, makes activated carbon particles extremely efficient at adsorbing other molecules onto their surfaces. Activated carbon is made by heating amorphous carbon in steam, which breaks the grains into smaller sizes and removes any other materials adsorbed on the surface. Activated carbon is used to filter impurities from gas and as a decolorizing agent, removing impurities that discolor organic products such as sugar or wheat flour.

| TABLE 23.3 Approximate Composition of the Main Types of Coal | | | | | |
|---|---|---|---|---|---|
| Type of Coal | Free C (mol %) | Total C (mol %) | H (mol %) | O (mol %) | S (mol %) |
| Lignite | 22 | 71 | 4 | 23 | 1 |
| Bituminous | 60 | 80 | 6 | 8 | 5 |
| Anthracite | 88 | 93 | 3 | 3 | 1 |

**Soot** is an amorphous form of carbon that forms during the incomplete combustion of hydrocarbons; its presence is indicated by blue or black smoke. Toxic carbon monoxide also forms in the process:

$$H_xC_y(s) + O_2(g) \longrightarrow H_2O(g) + CO_2(g) + CO(g) + \underset{\text{soot}}{C(s)}$$

**Carbon black**, a fine, powdered form of carbon, is a component of soot. Each day, over a million tons of carbon black, a strengthener for rubber tires, is used in manufacturing. The black color of automobile tires is due to the several kilograms of carbon black within each tire, accounting for over 25% of the mass of the typical tire.

▲ Automobile tires are black because of the carbon black that is added to strengthen the tires and maintain flexibility.

## Carbides

Binary compounds composed of carbon combined with less electronegative elements are **carbides**. We classify carbides into three general categories: ionic, covalent, and metallic. All types of carbides have some shared properties; they are extremely hard materials with high melting points.

### Ionic Carbides

Compounds composed of carbon and a low-electronegativity metal such as an alkali metal or an alkaline earth metal are **ionic carbides**. Most ionic carbides contain the dicarbide ion, $C_2^{2-}$, commonly called the *acetylide ion*. For example, calcium carbide has the formula $CaC_2$ and a structure similar to that of NaCl (Figure 23.13▶, on the next page).

Calcium carbide forms by the reaction of calcium oxide with coke in an electric furnace:

$$CaO(s) + 3\,C(s) \longrightarrow CaC_2(s) + CO(g)$$

▶ **FIGURE 23.13  Calcium Carbide Structure**
In the NaCl-type structure for CaC$_2$, the dicarbide ions are in the positions of the chloride ions, making the structure slightly noncubic.

$C_2^{2-}$    $Ca^{2+}$

Ionic carbides react with water, forming acetylene. For example, sodium carbide reacts with water according to the reaction shown here:

$$Na_2C_2(s) + 2\,H_2O(l) \longrightarrow 2\,NaOH(aq) + C_2H_2(g)$$

In the past, calcium carbide was used as a source of acetylene (which is highly flammable) for lighting. The solid CaC$_2$ reacts with water, releasing acetylene gas, which was burned in applications such as automobile headlights and lamps for mines. Transporting solid calcium carbide was more convenient and safer than transporting the flammable gas.

**CONCEPTUAL CONNECTION 23.1**

**Carbide Formulas**   Predict the formula for potassium carbide.

## Covalent Carbides

Compounds composed of carbon and low-electronegativity *nonmetals* or *metalloids* are **covalent carbides**. The most important covalent carbide is silicon carbide (SiC), a very hard material. Over 500,000 tons of silicon carbide are produced annually, mostly for use as an abrasive material in the cutting and polishing of metals. In a process analogous to the formation of calcium carbide, silicon carbide forms by the reaction of silicon oxide with coke at high temperatures:

$$SiO_2(s) + 3\,C(s) \longrightarrow SiC(s) + 2\,CO(g)$$

A gem-quality form of SiC, called *moissanite*, has also been developed. Moissanite is described as being more brilliant than all other gems, including diamonds. Yet moissanite costs much less than diamond and is consequently sold as a diamond substitute (like the more common diamond substitute cubic zirconia, ZrO$_2$). Moissanite was first identified in small particles at the Diablo Canyon meteorite impact crater in Arizona and is sometimes advertised as "a gift from the stars."

## Metallic Carbides

Compounds composed of carbon and metals that have a metallic lattice with holes small enough to fit carbon atoms are **metallic carbides**. Metallic carbides retain many metallic properties, such as high electrical conductivity, but they are stronger, harder, and less malleable than their corresponding metals. Adding carbon to steel, for example,

2500×

▲ This micrograph shows cementite (dark regions) in steel.

increases its hardness by forming regions of cementite ($Fe_3C$) in the steel matrix. Tungsten carbide (WC) is a metallic carbide used in cutting tools.

## Carbon Oxides

Carbon forms two stable oxides, carbon monoxide and carbon dioxide. Earth's atmosphere contains about 0.04% carbon dioxide by volume. Plants use atmospheric carbon dioxide to produce sugars during photosynthesis:

$$6\,CO_2(g) + 6\,H_2O(g) \longrightarrow C_6H_{12}O_6(s) + 6\,O_2(g)$$

Carbon dioxide returns to the atmosphere via animal respiration, plant and animal decay, and (in modern history) fossil fuel combustion. Because carbon dioxide is highly soluble in water, the oceans of the world act as a reservoir for $CO_2$, keeping the amount of $CO_2$ in the atmosphere generally stable. As we saw in Section 6.10, however, the increase in the combustion of fossil fuels in the last century has increased the amount of $CO_2$ in the atmosphere by about 38%.

Recall from Section 11.8 that $CO_2$ has a triple point at $-57\ °C$ and 5.1 atm. At atmospheric pressure, therefore, the liquid phase of $CO_2$ does not exist. Solid carbon dioxide sublimes directly to the gas phase when heated, which is why solid $CO_2$ is called "dry ice."

Carbon monoxide (CO) is a colorless, odorless, and tasteless gas. The boiling point of carbon monoxide is $-192\ °C$ at atmospheric pressure, and CO is only very slightly soluble in water. Carbon monoxide is toxic because it interferes with the ability of hemoglobin to bind oxygen. Unlike carbon dioxide, which is very stable, carbon monoxide is relatively reactive and can be used as a reducing agent.

For example, carbon monoxide reacts with oxygen and metal oxides to form carbon dioxide:

$$2\,CO(g) + O_2(g) \longrightarrow 2\,CO_2(g)$$
$$CO(g) + CuO(s) \longrightarrow CO_2(g) + Cu(s)$$

Carbon monoxide also reduces many nonmetals, producing compounds with the reduced form of the nonmetal:

$$CO(g) + Cl_2(g) \longrightarrow COCl_2(g)$$
$$CO(g) + S(s) \longrightarrow COS(g)$$

The product of the first reaction listed is phosgene ($COCl_2$), also known as carbonyl chloride, a poisonous gas that was used in World War I as a chemical weapon. Phosgene is now an important industrial chemical used in the production of polycarbonates. The product of the second reaction, carbonyl sulfide (COS), is a fungicide (a substance that kills fungi).

## Carbonates

When carbon dioxide dissolves in water, it forms carbonic acid ($H_2CO_3$). As a weak acid, carbonic acid partially ionizes into hydrogen carbonate (or bicarbonate) and carbonate:

$$\underset{\text{carbonic acid}}{CO_2(aq) + H_2O(l) \rightleftharpoons H_2CO_3(aq)}$$

$$H_2CO_3(aq) \rightleftharpoons \underset{\text{hydrogen carbonate}}{H^+(aq) + HCO_3^-(aq)} \rightleftharpoons 2\,H^+(aq) + \underset{\text{carbonate}}{CO_3^{2-}(aq)}$$

Recall from Section 13.4 that the solubility of carbon dioxide, like that of other gases, increases with increasing pressure. Carbon dioxide under high pressure carbonates soft drinks. Under most conditions, less than 0.5% of the dissolved carbon dioxide reacts with water to form carbonic acid. This leaves most of the carbon dioxide as dissolved gas molecules, so the soft drink does not acquire much of a sour acidic taste.

The hydrated crystal of sodium carbonate, $Na_2CO_3 \cdot 10\,H_2O$, is known as **washing soda**. Heating washing soda releases the waters of hydration, forming the stable anhydrous sodium carbonate, $Na_2CO_3$. All of the alkali metal ions form stable carbonates that remain

▲ Alka-Seltzer™ contains sodium bicarbonate mixed with citric acid and aspirin. When put in water, the acid and carbonate react.

stable even when heated. The carbonates all make basic solutions when added to water because the carbonate ions readily ionize water (as described more fully in Section 16.8):

$$Na_2CO_3(s) \longrightarrow 2\,Na^+(aq) + CO_3^{2-}(aq)$$

$$CO_3^{2-}(aq) + H_2O(l) \longrightarrow HCO_3^-(aq) + OH^-(aq)$$

Sodium bicarbonate ($NaHCO_3$) is *baking soda*. When heated, baking soda gives off carbon dioxide gas, which is why its use in baking helps raise dough:

$$2\,NaHCO_3(s) \longrightarrow Na_2CO_3(s) + H_2O(l) + CO_2(g)$$

Baking *powder* is a mixture of $NaHCO_3$ and an acid. The two components of the mixture are kept from reacting by a starch filler. When water is added to the mixture, however, the two components dissolve and react, producing the carbon dioxide that forms pockets of gas in baked products. You can perform a simple test to determine if baking powder is still good (that is, whether the acid has not already slowly reacted with the sodium bicarbonate) by pouring some boiling water over a small sample of the baking powder. If the hot water produces bubbles, then the baking powder is still active. Alka-Seltzer is another common consumer product that employs sodium bicarbonate, in this case mixed with citric acid and aspirin. When immersed in water, the acid and carbonate react to produce carbon dioxide, creating the familiar fizz.

CONCEPTUAL
**CONNECTION 23.2**

**Carbonate Solubility** As we saw in Chapter 4, the carbonates of metal ions other than group 1A are insoluble in water. Which action would increase their solubility?

**(a)** adding acid to the solution

**(b)** adding base to the solution

**(c)** increasing the amount of the solid carbonate in the solution

## 23.6 Nitrogen and Phosphorus: Essential Elements for Life

The group 5A elements range from nonmetallic nitrogen and phosphorus to metallic bismuth. Both nitrogen and phosphorus are nonmetals; they do not conduct electricity, and they form acidic oxides. Both have $s^2p^3$ electron configurations, and yet their chemical properties are very different. Phosphorus is much larger and less electronegative; it also has $d$ orbitals available for bonding.

### Elemental Nitrogen and Phosphorus

Nitrogen was identified in 1772 and phosphorus in 1669. Elemental nitrogen is a diatomic gas that composes about 78% of Earth's atmosphere by volume (see Section 5.6). To obtain elemental nitrogen, air is cooled below $-196\,°C$, which causes it to liquefy. When the liquid air is warmed slightly, the nitrogen boils off, leaving liquid oxygen (which boils at the higher temperature of $-183\,°C$). Passing the vaporized gas over hot copper metal purifies the nitrogen by removing residual oxygen (which reacts with the copper to form CuO). Nitrogen gas can also be separated from the other atmospheric gases by passing air through certain silicate materials called *zeolites,* which have channels of just the right diameter to separate gas molecules of different size. Some mineral sources for nitrogen are saltpeter ($KNO_3$) and Chile saltpeter ($NaNO_3$).

As we first saw in Section 9.5, nitrogen molecules have a triple bond between the two N atoms. The strength of the triple bond makes $N_2$ very stable, and attempts to break the bond have not been commercially successful. When nitrogen gas is heated with oxygen or hydrogen, nitric oxide (NO) or ammonia ($NH_3$), respectively, form with low yields. When nitrogen gas is heated with active metals, metal nitrides form. Aside from this, however, nitrogen gas is relatively unreactive.

The stability of elemental nitrogen makes it useful in creating a protective atmosphere to prevent oxidation in many industrial processes. For example, industrial furnaces employ a nitrogen atmosphere to anneal (hold at elevated temperature below the melting point) products made of metal, and chemical reactions sensitive to oxygen are carried out in a nitrogen atmosphere. Nitrogen is also used to preserve a variety of foods.

Elemental phosphorus was first isolated by accident from urine when Henning Brand (1630–1710), a physician and alchemist from Hamburg, Germany, was distilling urine in an ill-informed attempt to obtain gold from the golden liquid. The elemental form of phosphorus that Brand obtained instead was a white, waxy, flammable solid called **white phosphorus**. White phosphorus is highly toxic to humans. For over a hundred years, urine was the main source for elemental phosphorus. Today, however, phosphorus is obtained from a calcium phosphate mineral called apatite, [$Ca_3(PO_4)_2$]. The mineral is heated with sand and coke in an electric furnace:

$$2\,Ca_3(PO_4)_2(s) + 6\,SiO_2(s) + 10\,C(s) \longrightarrow P_4(g) + 6\,CaSiO_3(l) + 10\,CO(g)$$
$$\quad\text{apatite} \qquad\qquad \text{sand} \qquad \text{coke} \qquad \text{white phosphorus}$$

White phosphorus consists of $P_4$ molecules in a tetrahedral shape, with the phosphorus atoms at the corners of the tetrahedron (Figure 23.14►). The bond angles between the three P atoms on any one face of the tetrahedron are small (60°) and strained, making the $P_4$ molecule unstable and reactive. White phosphorus spontaneously burns in air; it is normally stored under water to prevent contact with air.

When heated to about 300 °C in the absence of air, white phosphorus slightly changes its structure to a different allotrope called **red phosphorus**, which is amorphous. The general structure of red phosphorus is similar to that of white phosphorus, except that one of the bonds between two phosphorus atoms in the tetrahedron is broken (Figure 23.15▼). The two phosphorus atoms link to other phosphorus atoms, making chains that vary in structure.

▲ **FIGURE 23.14 White Phosphorus** The small bond angle of 60° between the phosphorus atoms at the corners of the tetrahedron puts a great strain on the structure and makes the $P_4$ molecule unstable.

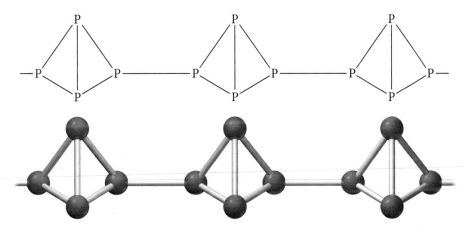

◄ **FIGURE 23.15 Red Phosphorus** Red phosphorus consists of chains of phosphorus atoms that form amorphous structures.

Red phosphorus is neither as reactive nor as toxic as white phosphorus, and even though it is also flammable, it can be stored in air. Red phosphorus is used commercially in applications such as match heads. Rubbing a match head onto a surface produces enough heat (through friction) to ignite the phosphorus. Most strike-anywhere matches feature the phosphorus compound tetraphosphorus trisulfide ($P_4S_3$) and an oxidizing agent, potassium chlorate ($KClO_3$).

A third allotrope of phosphorus is **black phosphorus**. Black phosphorus is obtained by heating white phosphorus under pressure. This form of phosphorus is the most thermodynamically stable form, and therefore the least reactive. Black phosphorus has a layered structure similar to that of graphite.

| TABLE 23.4 Oxidation States of Various Nitrogen Compounds | |
| --- | --- |
| Nitrogen-Containing Compound | Oxidation State |
| $NH_3$ | $-3$ |
| $N_2H_4$ | $-2$ |
| $H_2NOH$ | $-1$ |
| $HN_3$ | $-\frac{1}{3}$ |
| $N_2$ | $0$ |
| $N_2O$ | $+1$ |
| $NO$ | $+2$ |
| $N_2O_3$, $NF_3$ | $+3$ |
| $NO_2$, $N_2O_4$ | $+4$ |
| $N_2O_5$, $HNO_3$ | $+5$ |

# Nitrogen Compounds

Nitrogen, with a valence electron configuration of $2s^2 2p^3$, can gain three electrons or lose five electrons to obtain an octet. Nitrogen forms many covalent compounds with oxidation states from $-3$ to $+5$, as shown in Table 23.4.

## Nitrogen Hydrides

The most common nitrogen hydride is **ammonia** ($NH_3$), the strong-smelling compound in which nitrogen displays its lowest oxidation state ($-3$). Ammonia is important to humans because it reacts with sulfuric acid (or phosphoric acid) to produce ammonium salts for fertilizers:

$$2\,NH_3(g) + H_2SO_4(aq) \longrightarrow (NH_4)_2SO_4(aq)$$

For hundreds of years, natural biological materials such as animal manure were used as nitrogen-containing fertilizers. In the 1800s, however, the nitrogen-bearing nitrate mineral $NaNO_3$ was discovered in Chile (and named Chile saltpeter). This nitrate mineral became an important source of fertilizer and made the country of Chile very wealthy. However, because Chile saltpeter was a limited source, chemists were continually striving to develop a new source.

The obvious *unlimited* source of nitrogen is the atmosphere, but the strong triple bond in elemental nitrogen renders it unusable by plants. In order to be used as fertilizer, elemental nitrogen has to be *fixed*, which means that it has to be converted into a nitrogen-containing compound such as $NH_3$. However, the direct reaction of nitrogen gas with hydrogen gas to form ammonia is very slow and produces low yields of ammonia under normal conditions:

$$N_2(g) + 3\,H_2(g) \rightleftharpoons 2\,NH_3(g)$$

In the early 1900s, German chemist Fritz Haber studied the equilibrium conditions for this reaction and showed that high pressures and lower temperatures favored the product. Carrying out the reaction at a higher pressure, and using a catalyst to increase the reaction rate, the industrial process for producing ammonia from nitrogen gas and hydrogen gas—now called the **Haber–Bosch process**—became practical by the mid-1930s. This process is the main industrial process for making ammonia and fixing nitrogen for many uses, including fertilizers and explosives.

**Hydrazine** ($N_2H_4$) is another nitrogen and hydrogen compound in which nitrogen has a negative oxidation state ($-2$). Hydrazine is the nitrogen analog of hydrogen peroxide; it has a bond between nitrogen atoms that is similar to the bond between oxygen atoms in hydrogen peroxide (Figure 23.16 ◀). Hydrazine, like hydrogen peroxide, is a colorless liquid. However, while hydrogen peroxide is a powerful oxidizing agent, hydrazine is a powerful reducing agent, as shown in the following reactions:

Hydrazine

Hydrogen peroxide

▲ **FIGURE 23.16 Hydrazine and Hydrogen Peroxide** Hydrazine forms a structure similar to hydrogen peroxide with an N—N bond in the place of the O—O bond.

$$N_2H_4(l) + 2\,H_2O_2(l) \longrightarrow N_2(g) + 4\,H_2O(g)$$

$$N_2H_4(aq) + 2\,I_2(aq) \longrightarrow N_2(g) + 4\,HI(aq)$$

$$N_2H_4(aq) + 2\,Pb^{2+}(aq) \longrightarrow N_2(g) + 2\,Pb(s) + 4\,H^+(aq)$$

The oxidation state of each atom appears directly below its symbol. Notice that in each reaction, nitrogen is oxidized and causes the reduction of the other reactant.

**Hydrogen azide** ($HN_3$) is a nitrogen and hydrogen compound with a higher nitrogen-to-hydrogen ratio than ammonia or hydrazine. Ammonia and hydrazine are both basic:

$$N_2H_4(aq) + H_2O(l) \longrightarrow N_2H_5^{2+}(aq) + OH^-(aq)$$

$$NH_3(aq) + H_2O(l) \longrightarrow NH_4^+(aq) + OH^-(aq)$$

Hydrogen azide, by contrast, is acidic, ionizing in water to form the azide ion ($N_3^-$):

$$HN_3(aq) + H_2O(l) \longrightarrow H_3O^+(aq) + N_3^-(aq)$$

The $N_3^-$ ion can be represented with the resonance structures shown here (formal charges indicated in red):

$$\left[:\ddot{N}=N=\ddot{N}:\right]^- \quad \left[:N\equiv N-\ddot{N}:\right]^- \quad \left[\cdot\ddot{N}=\ddot{N}-\ddot{N}\cdot\right]^-$$

Since the rightmost structure has the least amount of formal charge, it contributes most to the hybrid structure.

Hydrogen azide is thermodynamically unstable compared to its constituent elements and reacts explosively to produce hydrogen and nitrogen gas:

$$2\,HN_3(l) \longrightarrow H_2(g) + 3\,N_2(g)$$

The sodium azide salt is a stable solid at room temperature, but at elevated temperatures, or with a spark, it quickly forms elemental sodium and nitrogen gas:

$$2\,NaN_3(s) \longrightarrow 2\,Na(l) + 3\,N_2(g)$$

The large volume of $N_2$ gas that forms from a small volume of $NaN_3(s)$ is the basis for airbags in automobiles. However, pure sodium azide also forms liquid sodium, which is dangerous because of its high reactivity. Therefore, other components, such as $KNO_3$ and $SiO_2$ are added to the mixture in airbags to react with the liquid sodium:

$$10\,Na(l) + 2\,KNO_3(s) \longrightarrow K_2O(s) + 5\,Na_2O(s) + N_2(g)$$

$$2\,K_2O(s) + SiO_2(s) \longrightarrow K_4SiO_4(s)$$

$$2\,Na_2O(s) + SiO_2(s) \longrightarrow Na_4SiO_4(s)$$

The overall reaction produces the large volume of nitrogen gas required to fill an automobile airbag quickly, along with harmless potassium and sodium silicates.

## Nitrogen Oxides

Under certain conditions, especially high temperatures, nitrogen is oxidized by oxygen to form a number of different oxides. For example, during lightning storms, nitrogen monoxide (NO) gas forms in the upper atmosphere:

$$N_2(g) + O_2(g) \xrightarrow{\text{lighting}} 2\,NO(g)$$

Other nitrogen oxides, such as nitrogen dioxide and dinitrogen trioxide, form from the further oxidation of nitrogen monoxide:

$$2\,NO(g) + O_2(g) \longrightarrow 2\,NO_2(g)$$

$$NO(g) + NO_2(g) \longrightarrow N_2O_3(l)$$

All nitrogen oxides are thermodynamically unstable and eventually decompose into their constituent elements or react to form more stable compounds. However, many of these reactions are kinetically slow, so some nitrogen oxides persist for long periods of time.

The most important nitrogen oxide, because of its significance in biological systems, is probably nitrogen monoxide (NO), also called nitric oxide. In 1987, nitrogen monoxide was named molecule of the year by the journal *Science* because of several discoveries related to its biological functions. For example, NO helps control blood pressure through blood vessel dilation, it is important in memory and digestion, and it plays major roles in inducing male erections and female uterine contractions. The ability to adjust NO levels is a key role of the medications that treat erectile dysfunction (such as Viagra).

Earlier in this section, we learned that lightning forms NO in the atmosphere. In Sections 3.6 and 16.12 we saw that NO and $NO_2$, formed as by-products of fossil fuel combustion, are among the important precursors of acid rain. Nitrogen monoxide and nitrogen dioxide are both reactive free radicals (they contain unpaired electrons). At low temperature, two $NO_2$ molecules dimerize to form $N_2O_4$, pairing their two lone electrons. If $N_2O_4$ is heated, it decomposes back to $NO_2$. Consequently, the equilibrium between $NO_2$ and $N_2O_4$ is highly temperature-dependent, as we saw in Section 15.9:

$$N_2O_4(g) \xrightarrow{\text{heat}} 2\,NO_2(g)$$
colorless    reddish brown

Dinitrogen monoxide ($N_2O$), also called nitrous oxide, is a good oxidizing agent. It can support the combustion of active metals:

$$Mg(s) + N_2O(g) \longrightarrow MgO(s) + N_2(g)$$

Dinitrogen monoxide is unstable when heated, decomposing into nitrogen and oxygen gas:

$$2\,N_2O(g) \xrightarrow{\text{heat}} 2\,N_2(g) + O_2(g)$$

Dinitrogen monoxide (often referred to as "nitrous" or laughing gas) is used as an anesthetic by dentists and to pressurize food dispensers (such as whipped-cream dispensers). Commercially, $N_2O$ is produced by the decomposition of ammonium nitrate:

$$NH_4NO_3(aq) \xrightarrow{\text{heat}} N_2O(g) + 2\,H_2O(l)$$

## Nitric Acid, Nitrates, and Nitrides

Nitric acid is an important commercial product of nitrogen. In an electric furnace, nitric acid forms from nitrogen and oxygen gas:

$$2\,N_2(g) + 5\,O_2(g) + 2\,H_2O(g) \longrightarrow 4\,HNO_3(g)$$

This reaction is thermodynamically favored, but it is kinetically slow. (What would happen to our atmosphere if this reaction were thermodynamically *and* kinetically favored?) Because of the slow speed of this reaction, a more efficient process, called the **Ostwald process**, is used to commercially produce nitric acid.

The first step of the Ostwald process involves passing ammonia gas over hot metal gauze at 600–700 °C to form NO gas. The gauze is made of metals such as platinum and rhodium that are good catalysts for this reaction:

$$4\,NH_3(g) + 5\,O_2(g) \xrightarrow{\text{catalyst}} 4\,NO(g) + 6\,H_2O(g)$$

Next, additional oxygen is added to oxidize the NO to $NO_2$ gas, which is then passed through a water spray to form nitric acid:

$$3\,NO_2(g) + H_2O(l) \longrightarrow 2\,HNO_3(l) + NO(g)$$

These steps are similar to the natural process that forms acid rain from NO and $NO_2$ gas in the atmosphere. The NO gas made during this last step can be recycled back to form more $NO_2$ and eventually more $HNO_3$.

Nitric acid is a strong acid that completely ionizes in water. Concentrated nitric acid is 70% nitric acid by mass, or 16 M. A small fraction of the $HNO_3$ in a bottle of concentrated nitric acid will react with water to form $NO_2$, a reddish brown gas that, in small amounts, gives the acid its characteristic pale yellow color:

$$4\,HNO_3(aq) \longrightarrow 4\,NO_2(g) + O_2(g) + 2\,H_2O(l)$$

The main commercial uses of nitric acid are in the production of fertilizers and explosives. Over a million tons of ammonium nitrate fertilizer is produced annually by the reaction between ammonia and nitric acid:

$$NH_3(g) + HNO_3(aq) \longrightarrow NH_4NO_3(aq)$$

Besides being a good fertilizer, ammonium nitrate and some other nitrates are also good explosives. Ammonium nitrate explodes according to the following reaction:

$$2\,NH_4NO_3(s) \xrightarrow{\text{heat}} 2\,N_2O(g) + 4\,H_2O(g) \longrightarrow 2\,N_2(g) + O_2(g) + 4\,H_2O(g)$$

Metal nitrates are responsible for the various colors seen in fireworks displays. The different metal ions emit different colors as the nitrate explodes in air. For example, copper nitrate produces a green-colored light according to the reaction:

$$2\,Cu(NO_3)_2(g) \xrightarrow{\text{heat}} 2\,CuO(s) + 4\,NO_2(g) + O_2(g) + \text{green light}$$

As we learned in Chapter 4, nitrates are very soluble in water. For reactions that need soluble metal cations, a nitrate compound is a good source for the cation without interference from the anion.

Nitrites are compounds containing the nitrite ion ($NO_2^-$). Sodium nitrite is used as a food preservative because it kills *Clostridium botulinum* bacteria, the cause of botulism, and because it keeps meat from discoloring when the meat is exposed to air. Consumers have voiced concerns over this practice, both because it hides the true age of the meat and because the nitrites can react with amines in the meat to form compounds called nitrosamines, which are suspected cancer-causing agents. However, no evidence exists to support the idea that nitrites at levels currently used in meats increase cancer risk in humans.

## Phosphorus Compounds

Phosphorus has a valence electron configuration of $3s^2 3p^3$, similar to that of nitrogen. Phosphorus also forms many compounds, with oxidation states ranging from $-3$ through $+5$. The most stable compounds have the $+5$ oxidation state.

### Phosphine

**Phosphine** ($PH_3$) is a colorless, poisonous gas that smells like decaying fish. In phosphine, phosphorus has an oxidation state of $-3$. Since phosphorus is less electronegative than nitrogen, phosphine is less polar than ammonia. Phosphine forms from the hydrolysis of metal phosphides:

$$Ca_3P_2(s) + 6\,H_2O(l) \longrightarrow 2\,PH_3(g) + 3\,Ca(OH)_2(aq)$$

The disproportionation of white phosphorus in a basic solution can also produce phosphine:

$$2\,P_4 + 3\,OH^- + 9\,H_2O \longrightarrow 5\,PH_3 + 3\,H_2PO_4^-$$

When heated, phosphine decomposes to phosphorus and hydrogen:

$$4\,PH_3(g) \longrightarrow P_4(s) + 6\,H_2(g)$$

Like ammonia, phosphine can form phosphonium compounds such as $PH_4Cl$ and $PH_4I$. Unlike ammonia, phosphine is not basic in aqueous solution.

> Disproportionation is a reaction in which an element is both reduced and oxidized during the same reaction. In this equation the phosphorus in $P_4$ is both oxidized and reduced. Phosphorus has an oxidation number of 0 in $P_4$ and is reduced to $-3$ in $PH_3$ and oxidized to $+5$ in $H_2PO_4^-$.

### Phosphorus Halides

When phosphorus reacts with the halogens, it forms phosphorus halides, the most important of which generally have the formulas $PX_3$ and $PX_5$:

$$P_4(s) + 6\,Cl_2(g) \longrightarrow 4\,PCl_3(l)$$

$$P_4(s) + 10\,Cl_2(g) \longrightarrow 4\,PCl_5(s) \quad \text{(with excess chlorine)}$$

Phosphorus halides react with water to form phosphoric acid and the corresponding hydrogen halide. For example, $PCl_3$ reacts with water:

$$PCl_3(l) + 3\,H_2O(l) \longrightarrow H_3PO_4(aq) + 3\,HCl(aq)$$

Reaction of $PCl_3$ with oxygen at room temperature forms phosphorus oxychloride:

$$2\,PCl_3(l) + O_2(g) \longrightarrow 2\,POCl_3(l)$$

Other phosphorus oxyhalides form as a result of the reactions of $POCl_3$ with metal fluorides or iodides:

$$POCl_3(l) + 3\,NaI(s) \longrightarrow POI_3(g) + 3\,NaCl(s)$$

The phosphorus halides and oxyhalides are important compounds in organic chemistry and serve as starting materials for the production of many phosphorus-containing compounds. Many of the key compounds in pesticides, oil additives, fire retardants for clothing, and surfactants (agents that act at surfaces), for example, are commercially made from phosphorus oxyhalides.

### Phosphorus Oxides

White phosphorus reacts directly with oxygen to form phosphorus oxides, as in the reaction shown here:

$$P_4(s) + 5\,O_2(g) \longrightarrow P_4O_{10}(g)$$

The product depends upon the amount of oxygen. Tetraphosphorus hexaoxide, $P_4O_6(s)$, forms when oxygen is limited, and tetraphosphorus decaoxide, $P_4O_{10}(s)$, forms when greater amounts of oxygen are available.

Phosphorus oxides form interesting cage structures (Figure 23.17▼). We can visualize the $P_4O_6$ structure as a tetrahedron with a phosphorus atom at each of the vertices and an oxygen atom between each pair of phosphorus atoms. The $P_4O_{10}$ structure has four additional oxygen atoms bonded to each phosphorus atom at the vertices of each tetrahedron.

$P_4O_6$                                         $P_4O_{10}$

▲ **FIGURE 23.17** **Tetraphosphorus Hexaoxide and Decaoxide, $P_4O_6$ and $P_4O_{10}$**   The $P_4O_6$ structure has the P atoms at the corners of a tetrahedron and the O atoms on the edges. The $P_4O_{10}$ structure has O atoms also bonded to the P atoms at the corners.

## Phosphoric Acid and Phosphates

Phosphoric acid and phosphates are among the most important phosphorus-containing compounds. Phosphoric acid is a colorless solid that melts at 42 °C. Concentrated phosphoric acid is 85% phosphoric acid by mass, or 14.7 M. Phosphoric acid is produced from the oxidation of white phosphorus to tetraphosphorus decaoxide (see previous reaction), which is then reacted with water:

$$P_4O_{10}(s) + 6\,H_2O(l) \longrightarrow 4\,H_3PO_4(aq)$$

This method produces a very pure phosphoric acid. A less pure product forms from the reaction of calcium phosphate (a mineral source of phosphate) with concentrated sulfuric acid:

$$Ca_3(PO_4)_2(s) + 3\,H_2SO_4(aq) \longrightarrow 3\,CaSO_4(s) + 2\,H_3PO_4(aq)$$

One direct use of phosphoric acid is rust removal. In steel production, thick steel slabs must be heated and rolled into thinner ones. During this process, the hot steel is exposed to air, which oxidizes the surface. To remove this rust, the thin steel sheets pass through phosphoric or hydrochloric acid baths, which dissolve the rust from the metal.

A major use of phosphoric acid is fertilizer production. In the past, phosphorus-containing materials such as fish, bones, and bat guano were used as fertilizer. Sulfuric acid decomposes bones to make phosphorus compounds that are more readily taken up by plants. Today many different phosphorus compounds have been developed specifically as fertilizers for various types of plants.

Detergent manufacturers use sodium phosphate compounds as additives. Compounds such as sodium pyrophosphate ($Na_4P_2O_7$) and sodium tripolyphosphate ($Na_5P_3O_{10}$) remove metal ions such as $Ca^{2+}$ and $Mg^{2+}$ from hard water, increasing the effectiveness of the detergent and preventing scum rings on sinks and tubs. However, phosphate compounds in detergents are being replaced by other compounds because of ecological problems—primarily the overfertilization of algae in bodies of water—associated with the phosphates.

Phosphoric acid and phosphates are also important chemicals in the food industry. Phosphoric acid is a soft drink additive. At a low concentration, phosphoric acid is nontoxic and adds a tart, acidic taste to soft drinks. It also prevents bacterial growth in the soda. Table 23.5 summarizes the uses of phosphates in the food industry.

**TABLE 23.5  Uses of Phosphates in the Food Industry**

| | |
|---|---|
| Phosphoric acid, $H_3PO_4$ | Flavor agent in soda, yeast nutrient |
| Sodium dihydrogen phosphate (also sodium phosphate monobasic), $NaH_2PO_4$ | Emulsifier, pH buffering agent |
| Sodium hydrogen phosphate (also sodium phosphate dibasic), $Na_2HPO_4$ | Baking powder, fermentation auxiliary |
| Sodium hexametaphosphate, $(NaPO_3)_6$ | Preservative, pH buffering agent |
| Sodium trimetaphosphate, $(NaPO_3)_3$ | Starch modifier, juice dispersant |
| Iron(III) pyrophosphate nonahydrate, $Fe_4(P_2O_7)_3 \cdot 9\ H_2O$ | Nutritional supplement |
| Sodium monofluorophosphate, $Na_2PO_3F$ | Fluoride source for toothpaste |
| Pyrophosphate, $P_2O_7{}^{4-}$ | Tartar control for toothpaste |

## 23.7  Oxygen

The group 6A elements have an $s^2p^4$ valence electron configuration and a strong attraction for electrons. They obtain a full octet by gaining only two more electrons. Because of its small size, oxygen is a much stronger oxidizing agent than the rest of the group 6A elements. Oxygen has the second highest electronegativity of any element (3.5), while the rest of the 6A elements range from 2.5 to 2.0. Because of its high abundance (almost half the mass of Earth's crust is composed of oxygen) and its high reactivity, oxygen occurs in many common compounds, including metal oxides, carbonates, silicates, hydrates, and water. Oxygen is also critical for life; the oxidation of biomolecules by oxygen provides energy for most living systems on Earth.

### Elemental Oxygen

Oxygen exists naturally as a colorless, odorless, diatomic, nonpolar gas. It condenses to a pale blue liquid at $-183\,°C$. Oxygen is slightly soluble in water (0.04 g in 1 L or 0.001 M at 25 °C). This rather low concentration of oxygen is enough to support life in aquatic environments. A few types of living systems that dwell deep in the ocean near vents that exude sulfur-containing fumes base their life processes on sulfur chemistry rather than oxygen chemistry.

Today, about 21% of Earth's atmosphere is composed of $O_2$, but this was not always the case. Earth's early atmosphere was reducing (rather than oxidizing) and contained hydrogen, methane, ammonia, and carbon dioxide. About 2.7 billion years ago, cyanobacteria (blue-green algae) began to convert the carbon dioxide and water to oxygen by photosynthesis. It took hundreds of millions of years to reach the current oxygen levels.

Joseph Priestley (1733–1804), an English scientist and minister, is credited with discovering oxygen. In 1774, he isolated oxygen by focusing sunlight on mercury(II) oxide and collecting the gas that was released as the red powder oxide formed liquid mercury. He tested the gas by using it to make a candle burn more brightly. He carried out a number of experiments with oxygen over the years, including bravely breathing his newfound gas. Antoine Lavoisier (1743–1794) is credited with recognizing that oxygen is necessary for combustion. He described combustion as the reaction of a substance with oxygen (and not the loss of a substance). These discoveries and explanations were important steps in the development of modern chemistry.

Oxygen is one of the most abundantly produced industrial chemicals. The major production method is the *fractionation* of air. Fractionation involves cooling air until its components liquefy. Then the air is warmed, and components such as $N_2$ and Ar are separated out, leaving oxygen behind. Most commercial oxygen is stored and transported as a gas in tanks under high pressure. Another method for the production of oxygen is the electrolysis of water. Passing an electric current through water containing a small amount of an electrolyte produces hydrogen gas at the cathode and oxygen gas at the

▲ Black smokers are vents found under the ocean that provide energy based on sulfur chemistry for life dwelling near the vents.

anode (see Section 19.8). Because of the large amount of electricity needed, however, electrolysis is not a cost-efficient method for oxygen production:

$$2\,H_2O(l) \xrightarrow{\text{electrical current}} 2\,H_2(g) + O_2(g)$$

In the laboratory, oxygen can be produced by heating and decomposing metal oxides and other oxygen-containing compounds. The oxides of mercury, silver, and gold lose all their oxygen when heated, while the oxides of other metals, such as barium, lose only some of their oxygen:

$$2\,HgO(s) \xrightarrow{\text{heat}} 2\,Hg(l) + O_2(g)$$

$$2\,BaO_2(s) \xrightarrow{\text{heat}} 2\,BaO(s) + O_2(g)$$

Metal nitrates and chlorates also yield oxygen when heated. A catalyst, such as manganese oxide or iron oxide, can make these reactions very fast and dangerous:

$$2\,NaNO_3(s) \xrightarrow{\text{heat}} 2\,NaNO_2(s) + O_2(g)$$

$$2\,KClO_3(s) \xrightarrow[\text{heat}]{\text{catalyst}} 2\,KCl(s) + 3\,O_2(g)$$

## Uses for Oxygen

The greatest industrial use for oxygen is the enrichment of the air in a blast furnace for the conversion of high-carbon iron to steel. Large quantities of oxygen are also used in oxyhydrogen or oxyacetylene torches for the cutting of metals. In addition, oxygen is used to create artificial air for use underwater, during high-altitude travel, and in safety equipment.

Oxygen plays an important role in the treatment of a number of medical conditions, such as acute and chronic lung diseases and heart disorders. Generally, patients use masks or nasal catheters to receive oxygen from a tank of compressed oxygen. However, portable oxygen concentrators have been developed that use molecular sieves to separate and concentrate oxygen from air. Hyperbaric oxygen therapy is the application of high oxygen levels to patients with skin wounds, such as those with skin grafts or hard-to-heal wounds associated with diabetes. The high oxygen level kills anaerobic bacteria that can infect such wounds.

## Oxides

As a strong oxidizing agent, oxygen reacts with most other elements to form oxides. We classify oxides according to the oxidation state of oxygen in the oxide (Table 23.6). The type of oxide that forms depends on the size and charge of the metal. Regular oxides are more stable for the smaller ions with a higher charge. Superoxides are more stable for the larger ions with a smaller charge.

**TABLE 23.6 Types of Oxides**

| Class | Ion | Oxidation State of O | Example |
|---|---|---|---|
| Oxide | $O^{2-}$ | $-2$ | $Li_2O$, $MgO$ |
| Peroxide | $O_2^{2-}$ | $-1$ | $Na_2O_2$, $BaO_2$ |
| Superoxide | $O_2^{-}$ | $-\frac{1}{2}$ | $RbO_2$, $CsO_2$ |

Oxygen also reacts with many nonmetals to form covalent compounds. Many of these nonmetals form several different binary oxides. For example, we have already seen that carbon forms $CO$ and $CO_2$ and that nitrogen forms $N_2O$, $NO$, $N_2O_3$, $NO_2$, $N_2O_4$, and $N_2O_5$.

## Ozone

**Ozone** ($O_3$), an allotrope of oxygen, is a toxic blue diamagnetic gas with a strong odor. People can detect the smell at levels as low as 0.01 ppm, and the odor is often noticed in electrical storms or near electrical equipment because the gas forms from $O_2$ by an electrical discharge. Ozone is denser than $O_2$ and condenses to a deep blue liquid at −112 °C. Ozone is naturally made by the irradiation of $O_2$ with ultraviolet light in the upper atmosphere:

$$3\,O_2(g) \xrightarrow{\text{UV radiation}} 2\,O_3(g)$$

Ozone is also produced by passing $O_2$ gas through an electric field. The volume of gas decreases as the $O_2$ is converted to $O_3$. Ozone is produced industrially by the electrolysis of cold concentrated sulfuric acid.

Thermodynamically unstable, ozone decomposes spontaneously to oxygen:

$$2\,O_3(g) \longrightarrow 3\,O_2(g)$$

Ozone is used commercially as a strong oxidizing agent. For example, ozone can oxidize $NO_2$ to $N_2O_5$ or PbS to $PbSO_4$:

$$2\,NO_2(g) + O_3(g) \longrightarrow N_2O_5(g) + O_2(g)$$
$$PbS(s) + 4\,O_3(g) \longrightarrow PbSO_4(s) + 4\,O_2(g)$$

Ozone kills bacteria and is an environmentally safe replacement for chlorine in water-purification plants because the only by-product is $O_2$. However, since ozone naturally decomposes, it must constantly be replenished, an economic drawback that limits its use.

The air we breathe contains ozone because it forms as a by-product of fossil fuel combustion. Since it is a strong oxidizing agent, it is a harmful substance. In the lower atmosphere, ozone damages the lungs and skin, stings the eyes, and damages most plant and animal tissues. Ozone also reacts with many types of plastic and rubber materials, causing them to become brittle and to crack.

As we also saw in Section 6.10, the layer of ozone in the upper atmosphere plays an important role in the absorption of harmful ultraviolet radiation from the sun. It absorbs the UV radiation and breaks apart to $O_2$ and O. This oxygen atom will often react with another $O_2$ molecule to re-form as ozone. During this cycle, each ozone molecule absorbs many ultraviolet photons. Recall from Section 3.9 that the ozone layer has been depleted by chlorofluorocarbons (CFCs). However, legislation has banned CFCs in the hope that their reduction will help the ozone layer to recover.

## 23.8  Sulfur: A Dangerous but Useful Element

Like oxygen, sulfur is a nonmetal that belongs to the 6A family. However, sulfur's $3p$ orbitals extend farther out from the nucleus than do oxygen's $2p$ orbitals. Consequently, sulfur is larger and is a much weaker oxidizing agent than oxygen. Unlike oxygen, which forms compounds with *only* negative oxidation states, sulfur forms compounds with both negative and positive oxidation states. Sulfur, selenium, and tellurium generally form covalent compounds with +4 or +6 oxidation states, forming anions only when bonding with highly electropositive metals. Sulfur is also much less abundant than oxygen, yet it still composes about 0.06% of the mass of Earth's crust.

### Elemental Sulfur

Elemental sulfur occurs in a few natural deposits, mostly deep underground in the Gulf Coast area of the United States and in eastern Europe. These deposits are believed to have formed as a by-product of anaerobic bacteria that decomposed sulfur-containing minerals over many years. The **Frasch process**, which allows us to recover sulfur from the earth, is diagrammed in Figure 23.18▶. Superheated water and compressed air are forced down long pipes into the sulfur deposits. The hot water melts the sulfur, which is pushed up to the surface along with

▲ **FIGURE 23.18  The Frasch Process**  The Frasch process extracts molten sulfur from the ground by forcing superheated water into beds of deposited solid sulfur.

▲ **FIGURE 23.19 Sulfur Deposits** Some sulfur deposits can also be found at Earth's surface where volcanic activity has brought the element to the surface.

▲ **FIGURE 23.20 Quenching Liquid Sulfur** When hot liquid sulfur is poured into cold water, the sulfur quenches into an amorphous solid.

▲ Molten sulfur below 150 °C (left) and nearing 180 °C (right).

the hot water through a third pipe. Some sulfur deposits also occur at Earth's surface—often in hot springs—where volcanic activity has made the sulfur accessible (Figure 23.19▲).

Sulfur has several allotropes; the most common naturally occurring allotrope of sulfur is composed of an $S_8$ ring structure called cyclooctasulfur. Most of the different allotropes have ring structures with rings ranging in size from $S_6$ to $S_{20}$. When heated above its melting point of 112 °C, cyclooctasulfur forms a straw-yellow liquid with low viscosity. Above 150 °C, the rings begin to break, and the sulfur becomes a darker, more viscous liquid as the broken rings entangle one another. The color is darkest at 180 °C when the liquid becomes very viscous and pours very slowly, as can be seen in the accompanying photo. Above this temperature, however, the intermolecular forces between the $S_8$ chains weaken, and the liquid becomes less viscous again. If the hot liquid is poured into cold water, the sulfur will quench into an amorphous solid (Figure 23.20▲). Initially, this amorphous material is flexible like a plastic, but it hardens into a brittle solid.

The Frasch process, while important, provides less than one-third of the world's sulfur; sulfur is a by-product of a number of other industrial processes and is thus recovered from them. For example, dihydrogen monosulfide ($H_2S$), which is commonly called hydrogen sulfide, is a component of natural gas. The $H_2S$ is separated from the other components by passing the gas through organic solvents such as ethanolamine. The $H_2S$ dissolves in the organic solvent:

$$HOC_2H_4NH_2(l) + H_2S(g) \longrightarrow HOC_2H_4NH_3^+(\text{solvent}) + HS^-(\text{solute})$$
$$\underset{\text{ethanolamine}}{}$$

The $H_2S$ is recovered and oxidized to elemental sulfur through a two-step process—the **Claus process**—that accounts for over 50% of all sulfur produced:

$$2\,H_2S(g) + 3\,O_2(g) \longrightarrow 2\,SO_2(g) + 2\,H_2O(g)$$

$$4\,H_2S(g) + 2\,SO_2(g) \longrightarrow 6\,S(s) + 4\,H_2O(g)$$

The third major source of sulfur production is metal sulfide minerals, such as the mineral iron(II) disulfide (also known as iron pyrite). Roasting iron pyrite in the absence of air causes the sulfur to disproportionate (react with itself) to form iron(II) sulfide and elemental sulfur:

$$2\,FeS_2(s) \xrightarrow{\text{heat}} 2\,FeS(s) + S_2(g)$$

Alternatively, the metal sulfide can be roasted in air to oxidize the metal sulfide. The sulfur is then removed as sulfur dioxide:

$$2\,ZnS(s) + 3\,O_2(g) \longrightarrow 2\,ZnO(s) + 2\,SO_2(g)$$

# Hydrogen Sulfide and Metal Sulfides

Hydrogen sulfide, a component of natural gas, is toxic and forms by the reactions of anaerobic bacteria on organic substances. Rotting vegetation and bogs are natural sources of $H_2S$ for the atmosphere. Fortunately, we can detect the odor of $H_2S$ (a rotten egg smell) at concentrations as low as 0.02 ppm, which pose no threat to our health. Levels as low as 10 ppm can cause nausea, and 100 ppm can cause death. The smell of $H_2S$ becomes more difficult to detect at high levels because $H_2S$ also has an anesthetic effect that dulls the sense of smell. Consequently, the sudden onset of strong $H_2S$ odor is a reason to move quickly to fresh air.

It might initially seem that hydrogen sulfide ($H_2S$) would share chemical properties with water ($H_2O$), but it does not. Water has a larger bond angle (104.5°) than hydrogen sulfide (92.5°) and is much more polar. Because of its polarity, water forms strong hydrogen bonds, whereas hydrogen sulfide does not. In addition, the O—H bond is much stronger than the S—H bond. These differences result in a far lower boiling point and greater reactivity for hydrogen sulfide compared to water. Water is a stable molecule in the presence of air and oxygen. Hydrogen sulfide burns in air, reacting with oxygen to form elemental sulfur or sulfur oxides.

| **Example** 23.5 | **Balancing of and Assigning Oxidation States to Sulfur Reactions** |
|---|---|

Write a balanced equation for the reaction of $O_2$ and $H_2S$ to form elemental S (in the form of $S_8$). Identify the change of the oxidation state for S.

**SOLUTION**

| Write the skeletal equation. The products are elemental $S_8$ and $H_2O$. | $H_2S(g) + O_2(g) \longrightarrow H_2O(g) + S_8(s)$ |
|---|---|
| Since hydrogen is initially balanced, balance S first, followed by H and O. | $8\,H_2S(g) + 4\,O_2(g) \longrightarrow 8\,H_2O(g) + S_8(s)$ |
| Assign oxidation states to each element (see Section 4.9). | $$8\,H_2S(g) + 4\,O_2(g) \longrightarrow 8\,H_2O(g) + S_8(s)$$ $\underset{+1\ -2}{\phantom{x}} \qquad \underset{0}{\phantom{x}} \qquad\qquad \underset{+1\ -2}{\phantom{x}} \quad \underset{0}{\phantom{x}}$ — Oxidation — The oxidation state of S changes from −2 in $H_2S$ to 0 in $S_8$; therefore, S is oxidized. |

**FOR PRACTICE 23.5** Write a balanced equation for the reaction of oxygen with $H_2S$ to form $SO_2$. Identify the change of the oxidation state for S.

**TABLE 23.7 Common Metal Sulfides**

| Sulfide | Formula | Common Name | $K_{sp}$ (at 25 °C) |
|---|---|---|---|
| Iron(II) sulfide | $FeS_2$ | Pyrite | $3.72 \times 10^{-19}$ |
| Zinc sulfide | ZnS | Sphalerite | $2.0 \times 10^{-25}$ |
| Lead(II) sulfide | PbS | Galena | $9.04 \times 10^{-29}$ |
| Mercury(II) sulfide | HgS | Cinnabar | $1.6 \times 10^{-54}$ |

Hydrogen sulfide forms from the reactions of metal sulfides with hydrochloric acid:

$$FeS(s) + 2\,HCl(aq) \longrightarrow FeCl_2(s) + H_2S(g)$$

Only a few of the metal sulfides, those with group 1A and 2A metals and Al, are very soluble in water. Some common metal sulfides and their solubility product constants are listed in Table 23.7. The low solubility of these sulfides allows the use of $H_2S$ as a good analytical method to determine whether metal ions are present in a solution. Sodium sulfide is used to precipitate toxic metals from industrial waste sources.

Metal sulfides have a number of industrial uses, mostly because they are toxic to bacteria. For example, $SeS_2$ is a shampoo additive that kills bacteria and controls dandruff, and $As_2S_3$ kills parasites.

## Sulfur Dioxide

Sulfur dioxide is another toxic sulfur compound. Under standard conditions, it is a colorless, dense gas that has an acidic taste. The acidic taste results from the reaction of the gas with the water in your mouth:

$$SO_2(g) + H_2O(l) \longrightarrow H_2SO_3(aq)$$

Sulfur dioxide forms naturally during volcanic activity when sulfides oxidize in the high volcanic temperatures. Sulfur dioxide is also a pollutant that forms during many industrial processes, such as coal and oil combustion and metal extraction. As we have seen, when the sulfur dioxide is emitted into the air, it reacts with oxygen and water to produce acid rain:

$$2\,SO_2(g) + 2\,H_2O(g) + O_2(g) \longrightarrow 2\,H_2SO_4(aq)$$

Sulfuric acid as a pollutant is destructive to plants, animals, and man-made structures (see Section 16.12).

To prevent the release of $SO_2$ into the atmosphere, industrial processes "scrub" their emissions to remove the $SO_2$. The exhaust gas flows through stacks lined with calcium carbonate that, when heated, capture sulfur oxides in the form of calcium sulfate dust:

$$CaCO_3(s) \xrightarrow{\text{heat}} CaO(s) + CO_2(g)$$

$$2\,CaO(s) + 2\,SO_2(g) + O_2(g) \longrightarrow 2\,CaSO_4(s)$$

The $CaSO_4$ dust is collected and disposed of. New uses for the tons of waste $CaSO_4$, such as discarded fireproof insulation, are needed.

One use for $SO_2$ capitalizes on its toxicity. Food producers spray fruits and other vegetation with a solution containing $SO_2$ to kill mold and preserve the fruit. As a result, the fruits and vegetables can be shipped throughout the world, making it possible for inhabitants of the Northern Hemisphere to enjoy summer fruits (from the Southern Hemisphere) in the winter.

## Sulfuric Acid

The most important use of sulfur and its compounds is the production of sulfuric acid. In fact, sulfuric acid is the most abundantly produced chemical in the world because it is a strong acid, a strong oxidizing agent, and a good dehydrating agent. It is also plentiful and inexpensive. Sulfuric acid is used in fertilizers, color dyes, petrochemicals, paints, plastics, explosives, battery, steel, and detergent industries, to name just a few.

Pure $H_2SO_4$ melts at 10.4 °C and boils at 337 °C. At room temperature, it is an oily, dense liquid. Sulfuric acid reacts vigorously and exothermically with water. Pure or concentrated $H_2SO_4$ must be added to water slowly to avoid rapid heating, boiling, and splattering:

$$H_2SO_4(l) \xrightarrow{H_2O(l)} H_2SO_4\,(aq) \quad \text{highly exothermic}$$

The strong attraction between sulfuric acid and water makes sulfuric acid a very strong dehydrating agent. As shown in Figure 23.21 ◀, its affinity for water is strong enough to decompose some organic materials, such as sucrose:

$$C_{12}H_{22}O_{11}(s) + H_2SO_4(l) \longrightarrow 12\,C(s) + 11\,H_2O(g) + H_2SO_4(aq)$$

Sulfuric acid is produced industrially by a method known as the **contact process**, developed in the early twentieth century. In this method, elemental sulfur is first heated in air to form $SO_2$ gas, which is then heated in contact with a $V_2O_5$ catalyst to form $SO_3$ gas:

$$S(g) + O_2(g) \xrightarrow{\text{heat}} SO_2(g)$$

$$2\,SO_2(g) + O_2(g) \xrightarrow{V_2O_5 \text{ catalyst}} 2\,SO_3(g)$$

▲ **FIGURE 23.21 Dehydration of Sucrose** Sulfuric acid dehydrates sucrose by removing the hydrogen and oxygen as water molecules and leaving carbon behind. The porous carbon foam forms because the reaction is very exothermic.

The $SO_3$ gas is absorbed into concentrated sulfuric acid, producing a dense form of sulfuric acid called oleum, $H_2S_2O_7$, which produces $H_2SO_4$ when dissolved in water:

$$SO_3(g) + H_2SO_4(l) \longrightarrow H_2S_2O_7(l)$$
$$H_2S_2O_7(l) + H_2O(l) \longrightarrow 2\,H_2SO_4(aq)$$

As we have already seen, sulfuric acid is used in fertilizer production, which consumes a significant amount of the sulfuric acid produced.

## 23.9  Halogens: Reactive Elements with High Electronegativity

The halogens are all one electron short of a noble gas electron configuration. They are the most electronegative elements in their respective periods and are therefore very reactive. They do not naturally occur in their elemental form. The source of most of the halogens (except fluorine) is the dissolved salts present in seawater. The major sources for fluorine are several different minerals, including fluorspar ($CaF_2$) and fluoroapatite $[Ca_{10}F_2(PO_4)_6]$.

We have already seen some of the properties of the halogens, especially those that exhibit periodic trends, in Section 8.9. For example, the atomic radius of the halogens increases regularly from fluorine to iodine, as shown in Table 23.8. Due partly to its small size, fluorine has the highest electronegativity of all elements and is always found in oxidation states of $-1$ or $0$. The other halogens can be found with oxidation states ranging from $-1$ to $+7$. The positive oxidation states occur when a halogen bonds to more electronegative elements such as fluorine or oxygen.

**TABLE 23.8  Selected Properties of the Halogens**

| Element | Melting Point (°C) | Boiling Point (°C) | Atomic Radius (pm) | Electronegativity |
|---------|-------------------|--------------------|--------------------|-------------------|
| Fluorine | −219 | −188 | 72 | 4.0 |
| Chlorine | −101 | −34 | 99 | 3.0 |
| Bromine | −7 | 60 | 113 | 2.8 |
| Iodine | 114 | 185 | 133 | 2.5 |

---

**Example 23.6** | **Determining the Oxidation State of Halogens in Compounds**

Calculate the oxidation state of Cl in each compound.

**(a)** $ClO_3^-$       **(b)** $HClO$       **(c)** $Cl_2$

**SOLUTION**

**(a)** For $ClO_3^-$, each O atom has an oxidation state of $-2$ for a total of $-6$ for all three O atoms. Therefore, the Cl atom has to have an oxidation state of $+5$ for the sum of the oxidation states to equal the charge of the ion $(5 - 6 = -1)$.

**(b)** For $HClO$, the O atom has an oxidation state of $-2$, and the H atom has an oxidation state of $+1$. Therefore the Cl atom has to have an oxidation state of $+1$ to have a neutral charge $(1 + 1 - 2 = 0)$.

**(c)** For $Cl_2$, the Cl atoms are in their elemental state, so they have an oxidation state of $0$.

**FOR PRACTICE 23.6**  Calculate the oxidation state of Cl in $ClO_4^-$ and $Cl^-$.

## Elemental Fluorine and Hydrofluoric Acid

Fluorine is the most reactive element and forms binary compounds with all elements except He, Ne, and Ar. Fluorine even forms compounds with some of the noble gases, producing compounds such as $XeF_2$, $XeF_6$, and $XeOF_4$. The high reactivity of fluorine is related to several factors. First, the F—F bond is among the weakest halogen–halogen bonds, as shown in Table 23.9. In order for a halogen to react with other substances, the halogen–halogen bond must break. The energy required to break that bond is small for $F_2$, so the resulting reaction is more exothermic. Second, the small size of fluorine results in a high lattice energy for the ionic compounds that it forms. The high lattice energy means the compounds are very stable.

| TABLE 23.9 Comparison of Halogen X—X Bond Energy | | | | |
|---|---|---|---|---|
| Halogen | F—F | Cl—Cl | Br—Br | I—I |
| Bond energy (kJ/mol) | 159 | 243 | 193 | 151 |

The high reactivity of fluorine is illustrated by its ability to burn (or quickly oxidize) many substances, such as iron and sulfur, that do not readily burn with oxygen:

$$Fe(s) + F_2(g) \longrightarrow FeF_2(s)$$

$$S(s) + 3 F_2(g) \longrightarrow SF_6(g)$$

Fluorine gas even reacts with asbestos and glass, two materials commonly used as containers for reactive substances. Consequently, fluorine is normally held in metal containers made of iron, copper, or nickel. These metals also initially react with fluorine, but then a thin layer of the product coats the surface of the metal, protecting the underlying metal from further reaction.

Elemental fluorine is produced from the electrolysis of hydrofluoric acid, forming $F_2$ and $H_2$ gases:

*Oxidation:* $\qquad\qquad\qquad 2 F^-(g) \longrightarrow F_2(g) + 2 e^-$

*Reduction:* $\qquad 2 H^+(g) + 2 e^- \longrightarrow H_2(g)$

Gaseous hydrogen fluoride can be obtained from the reaction of the mineral fluorspar $(CaF_2)$ with sulfuric acid:

$$CaF_2(s) + H_2SO_4(l) \longrightarrow 2 HF(g) + CaSO_4(s)$$

In its solid form, HF has a crystal structure that contains zigzag chains of alternating H and F atoms. In aqueous solutions, HF is a weak acid ($K_a = 3.5 \times 10^{-4}$ for HF at 298 K). Like all anions in aqueous solution, the $F^-$ ions from ionized HF are solvated by water molecules. However, the $F^-$ ions can also associate with other HF molecules to form $HF_2^-$.

$$F^-(aq) + HF(aq) \longrightarrow HF_2^-(aq)$$

The structure of $HF_2^-$, shown here, is unique because it contains a bridging hydrogen atom (a hydrogen atom that essentially forms two bonds):

Hydrofluoric acid is a strong oxidizing agent and reacts with glass according to the following reactions:

$$SiO_2(s) + 4 HF(aq) \longrightarrow SiF_4(g) + 2 H_2O(l)$$

or:

$$SiO_2(s) + 6 HF(aq) \longrightarrow SiF_6^{2-}(aq) + 2 H^+(aq) + 2 H_2O(l)$$

As a result, hydrofluoric acid cannot be held in a glass container and is generally stored in plastic. The ability of HF to react with glass makes it useful in etching glass. The parts of the glass to be etched are left exposed, and the rest of the surface is masked with a nonreactive substance such as plastic. The surface is then exposed to hydrofluoric acid and the nonmasked glass etches away, leaving the desired pattern. Hydrofluoric acid is particularly dangerous because it quickly penetrates into tissues, damaging internal organs and bones. Direct exposure of just 2% of body surface area to concentrated hydrofluoric acid can be fatal.

## Elemental Chlorine

Historically, the primary source for chlorine has been seawater. Electrolysis of NaCl in seawater produces $Cl_2$ gas and $H_2$ gas:

$$2\,NaCl(aq) + 2\,H_2O(l) \xrightarrow{\text{electricity}} Cl_2(g) + 2\,NaOH(aq) + H_2(g)$$

Today, much $Cl_2$ gas is produced and collected as a by-product of the various metal processing methods, such as in the reduction of metal chlorides to form metals:

$$MgCl_2(l) \xrightarrow{\text{electricity}} Mg(s) + Cl_2(g)$$

## Halogen Compounds

Halogens form ionic compounds with most metals, they form covalent compounds with many nonmetals, and they can bond with other halogens to form interhalogen compounds. Here we look at some of the compounds halogens form.

### Interhalogen Compounds

Covalent compounds composed of two different halogens are known as **interhalogen compounds** or *interhalides*. The general formula of these compounds is $AB_n$, where A is the larger halogen, B is the smaller halogen, and $n$ is 1, 3, 5, or 7. The smaller halogens surround the larger halogen in the $AB_3$ compounds. The only known $AB_5$ compounds contain fluorine as the smaller halogen (because the other halogens are too large for five of them to fit around another halogen). The compound $IF_7$ is the only known interhalogen compound with $n = 7$. The large size of the iodine atom allows seven small fluorine atoms to surround it. The interhalides can be formed by reacting the elemental halogens with one another. Interhalide polyatomic ions, such as $ICl_2^+$ and $ICl_4^-$, also exist.

---

### Example 23.7 Formation of Interhalogen Compounds

Write the balanced equation for the formation of $ClF_3$ gas from the elemental halogens.

**SOLUTION**

The unbalanced reaction is:

$$Cl_2(g) + F_2(g) \longrightarrow ClF_3(g)$$

At least two $ClF_3$ molecules form from each $Cl_2$ molecule; therefore, add a 2 before $ClF_3$ and then a 3 before $F_2$. The balanced equation is:

$$Cl_2(g) + 3\,F_2(g) \longrightarrow 2\,ClF_3(g)$$

**FOR PRACTICE 23.7** Write a balanced equation for the formation of $IF_5$ gas from the elemental halogens.

---

The geometry of the interhalides can be predicted from the valence shell electron pair repulsion model (VSEPR). Because the halogens do not form double or triple bonds, the shapes of these compounds are relatively straightforward to determine, as shown in Examples 23.8 and 23.9.

## Procedure for...

| Determining the Shape of Interhalogen Compounds | **Example 23.8**<br>Molecular Shapes of Interhalogen Compounds<br>Determine the molecular geometry of $IBr_2^-$. | **Example 23.9**<br>Molecular Shapes of Interhalogen Compounds<br>Determine the molecular geometry of $IF_7$. |
|---|---|---|
| Identify the central atom. | **SOLUTION**<br><br>I | **SOLUTION**<br><br>I |
| Draw the Lewis structure (see Section 9.5). | $$\left[:\ddot{B}r-\ddot{\ddot{I}}-\ddot{B}r:\right]^-$$ | |
| Count the number of bonds and lone pairs on the central atom. | The I atom has two bonds and three lone pairs for a total of five electron groups. | The I atom has seven bonds on the central atom for a total of seven electron groups. |
| Determine the electron geometry from the number of electron groups (see Section 10.4). | With five electron groups, the electron geometry is *trigonal bipyramidal*. | With seven electron groups, the electron geometry is *pentagonal bipyramidal*. |
| Determine the molecular shape from the number of bonds and lone pairs. | With two bonds the molecular geometry is *linear* because the three lone pairs occupy equatorial positions.<br><br> | With seven bonds, the molecular geometry is also *pentagonal bipyramidal*.<br><br> |
| | **FOR PRACTICE 23.8**  Determine the electron geometry and molecular geometry of $ICl_2^+$. | **FOR PRACTICE 23.9**  Determine the electron geometry and molecular geometry of $BrF_5$. |

The only industrially useful interhalide is $ClF_3$, employed in the nuclear energy industry to produce $UF_6(g)$, which is used to separate $^{235}U$ (<1% of naturally occurring uranium) from $^{238}U$. Uranium ores react with gaseous hydrogen fluoride to form $UF_4(s)$, which reacts with $ClF_3(g)$ to form the gaseous uranium compound:

$$UO_2(s) + 4\,HF(g) \longrightarrow UF_4(s) + 2\,H_2O(g)$$

$$UF_4(s) + ClF_3(g) \longrightarrow UF_6(g) + ClF(g)$$

Because of the difference in mass between the two isotopes, $^{238}UF_6$ effuses more slowly than $^{235}UF_6$ (see Section 5.9 for a description of effusion). The mixture of gases is allowed to flow through barriers with very small pores. Since $^{235}UF_6$ effuses more quickly, the initial flow of gas that exits the pores is enriched in $^{235}UF_6$. By repeating the process, the two isotopes can be nearly completely separated.

### Halogen Oxides

Most halogen oxides are unstable, and many are explosive. A unique halogen oxide is $OF_2$; oxygen usually has a negative oxidation state in its compounds but has a +2 oxidation state in $OF_2$ (due to the high electronegativity of fluorine).

A number of different chlorine oxides are known, including $Cl_2O$, $ClO_2$, $Cl_2O_6$, and $Cl_2O_7$. Chlorine dioxide is a powerful oxidizing agent used to bleach flour and wood pulp (to make white paper). Because $ClO_2$ is explosive, the gas is diluted with $CO_2$ or $N_2$ for safety. Some water treatment plants use $ClO_2$ for water disinfection in place of $Cl_2$. $ClO_2$ is produced by the oxdiation of sodium chlorite with $Cl_2$ or by the reduction of sodium chlorate with hydrochloric acid:

$$2\,NaClO_2(aq) + Cl_2(g) \longrightarrow 2\,NaCl(aq) + 2\,ClO_2(g)$$

$$2\,NaClO_3(aq) + 4\,HCl(aq) \longrightarrow 2\,ClO_2(g) + Cl_2(g) + 2\,H_2O(l) + 2\,NaCl(aq)$$

---

**Example 23.10**  |  **Identifying Changes in Oxidation States**

Identify the change of oxidation state for Cl in the production of chlorine dioxide from sodium chlorite. Identify the oxidizing agent and the reducing agent.

**SOLUTION**

First determine the oxidation state of Cl in each compound.

$$2\,NaClO_2(aq) + Cl_2(g) \longrightarrow 2\,NaCl(aq) + 2\,ClO_2(g)$$

$$\underset{+1\ +3\ -2}{\phantom{}} \qquad \underset{0}{\phantom{}} \qquad \underset{+1\ -1}{\phantom{}} \qquad \underset{+4\ -2}{\phantom{}}$$

The Cl in the $NaClO_2$ was oxidized from +3 to +4 by the oxidizing agent $Cl_2$, which was reduced from 0 to −1 by the reducing agent $NaClO_2$.

**FOR PRACTICE 23.10**  Identify the change of oxidation state for Cl in the production of chlorine dioxide from sodium chlorate. Identify the oxidizing agent and the reducing agent.

---

# Self-Assessment Quiz

**Q1.** An aluminosilicate is composed of a crystal in which Al atoms substitute for three-fourths of the Si atoms in the silica structure. The negative charge of the $AlO_2^-$ groups is balanced by $Na^+$ ions. What is the formula of the aluminosilicate?

a) $Na(AlSi_3O_8)$  
b) $Na_2(Al_2Si_2O_8)$  
c) $Na_3(Al_3SiO_8)$  
d) $NaSiO_2$

**Q2.** Use charge balance to determine $x$ in the formula for the sorosilicate bertrandite, $Be_4Si_2O_7(OH)_x$.

a) 1  b) 2  c) 3  d) 4

**Q3.** Which element is important in the manufacturing of glassware intended for heating?

a) carbon  
b) nitrogen  
c) phosphorus  
d) boron

**Q4.** Fertilizers commonly contain compounds of which element?

a) nitrogen  
b) aluminum  
c) fluorine  
d) boron

**Q5.** What is the oxidation state of bromine in the compound $NaBrO_3$?

a) +2  b) −3  c) −4  d) +5

**Q6.** What is the molecular geometry of $ICl_4^-$?

a) tetrahedral  
b) square planar  
c) octahedral  
d) seesaw

**Q7.** Which element is oxidized in the reaction?

$$2\,CaO(s) + 2\,SO_2(g) + O_2(g) \longrightarrow 2\,CaSO_4(s)$$

a) Ca  b) O  c) S  
d) None of the above (no element is oxidized in the reaction)

**Q8.** Which carbon oxide does not exist?

a) CO  b) $CO_2$  c) $CO_3^{2-}$  d) $CO_4$

**Q9.** Which element is extracted from the ground primarily in its elemental form?

a) boron  
b) phosphorus  
c) sulfur  
d) fluorine

**Q10.** Which substance forms when carbon dioxide dissolves in water?

a) CO  b) $H_2CO$  c) $CH_4$  d) $H_2CO_3$

**Answers: 1.** (c) **2.** (b) **3.** (d) **4.** (a) **5.** (d) **6.** (b) **7.** (c) **8.** (d) **9.** (c) **10.** (d)

# Chapter 23 IN REVIEW

## Key Terms

**Section 23.2**
main-group elements (1064)

**Section 23.3**
silicates (1066)
quartz (1066)
silica (1066)
aluminosilicates (1066)
mineral (1068)
orthosilicates (1067)
pyrosilicates (1067)
pyroxenes (1067)

**Section 23.4**
boranes (1071)
*closo*-boranes (1071)
*nido*-boranes (1071)
*arachno*-boranes (1072)

**Section 23.5**
coal (1072)
coke (1072)
charcoal (1073)
activated carbon (1073)
soot (1073)
carbon black (1073)
carbides (1073)
ionic carbides (1073)
covalent carbides (1074)
metallic carbides (1074)
washing soda (1075)

**Section 23.6**
white phosphorus (1077)
red phosphorus (1077)
black phosphorus (1077)
ammonia (1078)

Haber–Bosch process (1078)
hydrazine (1078)
hydrogen azide (1078)
Ostwald process (1080)
phosphine (1081)

**Section 23.7**
ozone (1084)

**Section 23.8**
Frasch process (1085)
Claus process (1086)
contact process (1088)

**Section 23.9**
interhalogen compounds (1091)

## Key Concepts

### Bonding and Properties (23.2)

▶ Main-group elements are defined by their electron configurations and their location on the periodic table.

▶ The properties of the main-group elements show great diversity. Metals, nonmetals, and metalloids are all found among the main-group elements.

▶ Some main-group elements form covalent bonds while others form ionic bonds.

### The Most Common Matter: Silicates (23.3)

▶ Silicates (found in rocks, clays, and soils) are covalent atomic solids that contain silicon, oxygen, and various metal atoms.

▶ Silicate structures consist of four oxygen atoms bonded to silicon, forming a negatively charged polyatomic anion with a tetrahedral shape. Various metal ions within the structure balance the charge of the compound.

▶ The $SiO_4$ tetrahedrons can link to form chains, double chains, sheets, or even extended three-dimensional structures.

▶ The properties of the silicates depend on the connections between the silicate tetrahedrons. Because of the wide variety of combinations of tetrahedron connections and the many different metal ions that fit within the structure, an enormous variety of different silicate minerals exist in nature, making the silicate materials the most common structures found on Earth.

### Boron (23.4)

▶ Because of its small size and high electronegativity, boron behaves as a metalloid.

▶ The structure of elemental boron consists of icosahedron structures bonded together in various ways.

▶ Boron tends to form electron-deficient compounds.

▶ Compounds of boron and hydrogen form cluster compounds resembling spheres, cages, and nets.

### Carbon (23.5)

▶ Organic chemistry is based on the chemistry of carbon, and carbon is also very important in many inorganic compounds and applications. Important types of amorphous elemental carbon include carbon black and coke.

▶ Important types of inorganic carbon compounds include carbides and carbonates. Carbon can form carbides with metallic, covalent, or ionic properties.

▶ Carbon and its oxygen compounds are intimately involved in the functions of life.

### Nitrogen and Phosphorus (23.6)

▶ Nitrogen and phosphorus have been known for over 200 years; both form compounds, with oxidation numbers ranging from $-3$ up to $+5$.

▶ Both nitrogen and phosphorus compounds are critical for plant growth; their most important use is as fertilizers. The strong triple bond between nitrogen atoms in $N_2$ makes nitrogen from the atmosphere inaccessible to most plants, so ingenious chemical processes have been devised to make nitrogen compounds that are more available to plants.

### Oxygen (23.7)

▶ Oxygen is the most common element on Earth. It is found in the atmosphere as the elemental gas and as many oxide gases.

▶ Oxygen is found in ocean water and in Earth's crust as silicate and oxide compounds. Oxygen is a strong oxidizing agent and forms compounds with $-\frac{1}{2}$, $-1$, or $-2$ oxidation states.

▶ Ozone, $O_3$, is a helpful molecule in the upper atmosphere, where it absorbs harmful ultraviolet radiation, but it is a harmful molecule at Earth's surface.

## Sulfur (23.8)

▶ More sulfuric acid is produced than any other chemical; most of it is used to make fertilizers. Other uses take advantage of its strong oxidation and dehydration properties.

▶ Elemental sulfur has several allotropes ranging from ring structures to chain structures and amorphous materials, depending upon temperature.

## Halogens (23.9)

▶ The halogens are the most electronegative elements, so they are always found as compounds, usually ionic. When the halogens

bond with other electronegative elements, however, they can form covalent compounds.

▶ Fluorine has special chemical properties because it is the most electronegative element and is very small, making it a very strong oxidizing agent.

▶ Interhalogen compounds are formed between two halogens, with the larger halogen as the central atom of the structure.

## Key Learning Outcomes

| CHAPTER OBJECTIVES | ASSESSMENT |
|---|---|
| Determining the Composition, Charge Balance, and Type of Silicates and Aluminosilicates (23.3) | Examples 23.1, 23.2, 23.3, 23.4 <br> For Practice 23.1, 23.2, 23.3, 23.4   Exercises 15–24 |
| Writing Equations for Reactions and Assigning Oxidation States (23.8, 23.9) | Examples 23.5, 23.6, 23.10   For Practice 23.5, 23.6, 23.10 <br> Exercises 39–42, 51–54, 69–70, 73–74 |
| Formation of Interhalogen Compounds and Their Structures (23.9) <br> (F)—(H)—(F) | Examples 23.7, 23.8, 23.9   For Practice 23.7, 23.8, 23.9 <br> Exercises 71–72, 76 |

# EXERCISES

## Review Questions

1. Why does BN form compounds similar to those formed by elemental carbon?

2. What is the main characteristic that determines whether or not an element is a main-group element?

3. Does the metallic characteristic of a main-group element increase or decrease as we move down a family? Explain why.

4. Why does silicon form only single bonds with oxygen but carbon, which is in the same family as silicon, forms double bonds with oxygen in many compounds?

5. What is the difference between a rock and a mineral?

6. Briefly define each term.
   a. orthosilicate      b. amphibole
   c. pyroxene           d. pyrosilicate
   e. feldspar

7. Why is boron oxide often added to silica glass?

8. Why does boron form electron-deficient bonds? Provide an example.

9. Explain why solid $CO_2$ is referred to as dry ice.

10. Nitric acid and phosphoric acid are two major chemical products of the chemical industry. Describe some of their uses.

11. What is the typical concentration of oxygen in dry air?

12. Describe how nitrogen can be separated from the other components of air.

13. Earth's atmosphere originally did not contain oxygen. Explain how the atmosphere gained oxygen.

14. Name a benefit, a hazard, and a useful commercial application of ozone.

## Problems by Topic

### Silicates: The Most Abundant Matter in Earth's Crust

15. Silicon bonds to oxygen to form a tetrahedral shape in both the network covalent silica compound, $SiO_2$, and in ionic silicate compounds. What is the oxidation state of Si in each of these structures?
   a. silica compound, $SiO_2$      b. orthosilicates, $SiO_4^{4-}$
   c. pyrosilicates, $Si_2O_7^{6-}$

16. What is the oxidation state of Si in each of these structures?
   a. pyroxenes, $SiO_3^{2-}$      b. amphiboles, $Si_4O_{11}^{6-}$
   c. phyllosilicates, $Si_2O_5^{2-}$

17. In the orthosilicate garnet, the formula unit has three $SiO_4^{4-}$ units and is balanced by $Ca^{2+}$ and $Al^{3+}$ cations. Determine the formula unit of garnet.

18. In the pyroxene kanoite, the formula unit has two $SiO_3^{2-}$ units and is balanced by manganese and magnesium ions. Determine the formula unit of kanoite. Assume that the oxidation state of Mn is +2.

19. Kaolin is a clay material that is a phyllosilicate. Use charge balancing to determine how many hydroxide ions are in the formula for kaolin, $Al_2Si_2O_5(OH)_x$.

20. Tremolite is a double-chain silicate in the amphibole class. Use charge balancing to determine how many hydroxide ions are in the formula for tremolite, $Ca_2Mg_5Si_8O_{22}(OH)_x$.

21. How are the silica tetrahedrons linked for $ZrSiO_4$? Which class of silicates does this compound belong to?

22. How are the silica tetrahedrons linked for $CaSiO_3$? Which class of silicates does this compound belong to?

23. Predict the structure and give the charges on the cations in one of the minerals in the hornblende family, $Ca_2Mg_4FeSi_7AlO_{22}(OH)_2$.

24. Predict the structure and label the charges on the cations in the mineral hedenbergite, $CaFeSi_2O_6$.

## Boron and Its Remarkable Structures

25. A major source of boron is the mineral kernite, $Na_2[B_4O_5(OH)_4] \cdot 3\,H_2O$. Calculate how many grams of boron can be produced from $1.0 \times 10^3$ kg of a kernite-bearing ore if the ore contains 0.98% kernite by mass and the process has a 65% yield.

26. An uncommon mineral of boron is ulexite, $NaCaB_5O_9 \cdot 8\,H_2O$. How many grams of boron can be produced from $5.00 \times 10^2$ kg of ulexite-bearing ore if the ore contains 0.032% ulexite by mass and the process has an 88% yield?

27. Explain why the bond angles in $BCl_3$ and $NCl_3$ are different.

28. Explain why the bond between B and Cl in the molecule $BCl_3$ is shorter than would be expected for a single B—Cl bond.

29. Predict the number of vertices and faces on each *closo*-borane.
   a. $B_6H_6^{2-}$
   b. $B_{12}H_{12}^{2-}$

30. Predict the number of vertices and faces on each *closo*-borane.
   a. $B_4H_4^{2-}$
   b. $B_9H_9^{2-}$

31. Describe the differences among a *closo*-borane, a *nido*-borane, and an *arachno*-borane.

32. Describe how boron is used in the nuclear industry.

## Carbon, Carbides, and Carbonates

33. Describe the difference between regular charcoal and activated charcoal.

34. Explain why the structure of charcoal allows carbon to act as a good filter while the diamond structure does not.

35. Describe the difference between an ionic carbide and a covalent carbide. Which types of atoms will form these carbides with carbon?

36. Silicon carbide is produced by heating silicone polymers, forming methane gas, hydrogen gas, and silicon carbide. Balance the reaction of heating $[(CH_3)_2Si]_8$ to form silicon carbide.

37. Referring to the phase diagram in Section 11.8, describe what happens to the phase of $CO_2$ during each process.
   a. reducing the pressure on solid $CO_2$ that is at $-80\,°C$
   b. decreasing the temperature on $CO_2$ gas that is held at a pressure of 20 atm
   c. increasing the temperature on solid $CO_2$ that is held at a pressure of 0.8 atm

38. Referring to the phase diagram in Section 11.8, describe what happens to the phase of $CO_2$ during each process.
   a. reducing the temperature from the critical point
   b. increasing the pressure on $CO_2$ gas that is held at a temperature of $-50\,°C$
   c. increasing the temperature on solid $CO_2$ that is held at a pressure of 20 atm

39. Predict the products for each reaction and write a balanced equation.
   a. $CO(g) + CuO(s)$
   b. $SiO_2(s) + C(s)$
   c. $S(s) + CO(g)$

40. Predict the products for each reaction and write a balanced equation.
   a. $CO(g) + Cl_2(g)$
   b. $CO_2(g) + Mg(s)$
   c. $S(s) + C(s)$

41. Give the oxidation state for carbon in:
   a. $CO$
   b. $CO_2$
   c. $C_3O_2$

42. Write a balanced reaction for the gas release reaction of Alka-Seltzer, sodium bicarbonate with citric acid, $C_6H_8O_7$. (The acid is a triprotic acid and consists of a chain of three carbon atoms, each with a carboxylic acid group, COOH.)

## Nitrogen and Phosphorus: Essential Elements for Life

43. Explain the process of fixing nitrogen.

44. Explain why the diatomic nitrogen atom is unusable by most plants. Where do plants get nitrogen?

45. Describe the differences in the allotropes of white and red phosphorus. Explain why red phosphorus is more stable.

46. Describe how red and black phosphorus are made from white phosphorus.

47. Saltpeter and Chile saltpeter are two important mineral sources for nitrogen. Calculate the mass percent of nitrogen in each mineral.

48. Apatite is a main mineral source for the production of phosphorus. Calculate the atomic and mass percent of P in the mineral apatite.

49. Referring to the tables in Appendix IIB, determine whether or not hydrogen azide is stable at room temperature compared to its elements, $H_2$ and $N_2$. Is hydrogen azide stable at any temperature?

50. Referring to the tables in Appendix IIB, determine if dinitrogen monoxide is stable at room temperature compared to its elements, $O_2$ and $N_2$. Is dinitrogen monoxide stable at any temperature?

51. Predict the products for each reaction and write a balanced equation.
   a. $NH_4NO_3(aq) +$ heat
   b. $NO_2(g) + H_2O(l)$
   c. $PCl_3(l) + O_2(g)$

52. Predict the products for each reaction and write a balanced equation.
   a. $NO(g) + NO_2(g)$
   b. heating $PH_3$
   c. $P_4(s) + 5\,O_2(g)$

53. Rank the nitrogen ions from the one with N in the highest oxidation state to the one with N in the lowest.

$$N_3^-, N_2H_5^+, NO_3^-, NH_4^+, NO_2^-$$

54. Determine the oxidation state of N in the compounds in the reaction for the formation of nitric acid. Identify the oxidizing agent and the reducing agent.

$$3\,NO_2(g) + H_2O(l) \longrightarrow 2\,HNO_3(l) + NO(g)$$

55. Draw the Lewis structures for the phosphorus halides $PCl_3$ and $PCl_5$. Describe their VSEPR shape.

56. Dinitrogen pentoxide is an ionic compound formed from the ions $NO_2^+$ and $NO_3^-$. Give the oxidation state of N in each ion and the VSEPR shape for each of the ions.

57. Ammonium carbonate is produced from the reaction of urea, $CO(NH_2)_2$ with water. Write a balanced equation for this reaction and determine how much urea is needed to produce 23 g of ammonium carbonate.

58. Explain why phosphine, $PH_3$, is less polar than ammonia.

59. Reacting oxygen with white phosphorus can form either $P_4O_6$ or $P_4O_{10}$. State the conditions that determine which product forms.

**60.** $P_4O_{10}$ is one of the most effective drying agents, having the ability to extract water from other molecules. The $P_4O_{10}$ forms phosphoric acid. Write balanced reactions for the reaction of $P_4O_{10}$ with:
a. $HNO_3$, forming $N_2O_5$    b. $H_2SO_4$, forming $SO_3$

## Oxygen

**61.** Name the major source of the element oxygen and describe how it is produced.

**62.** Explain why either greatly decreasing or increasing the percentage of oxygen in the atmosphere is dangerous.

**63.** Identify each compound as an oxide, peroxide, or superoxide.
a. $LiO_2$        b. $CaO$        c. $K_2O_2$

**64.** Identify each compound as an oxide, peroxide, or superoxide.
a. $MgO$        b. $Na_2O_2$        c. $CsO_2$

## Sulfur: A Dangerous but Useful Element

**65.** Explain why the viscosity of liquid sulfur increases with increasing temperature initially, but then decreases upon further increases in temperature.

**66.** Sulfur dioxide is a toxic sulfur compound. List one natural source and one industrial source that produce $SO_2$.

**67.** Calculate the maximum mass (in grams) of each metal sulfide that will dissolve in 1.0 L of a solution that is $5.00 \times 10^{-5}$ M in $Na_2S$.
a. $PbS$        b. $ZnS$

**68.** A coal source contains 1.1% sulfur by mass. If $2.0 \times 10^4$ kg of coal is burned and forms oxides, calculate the mass of $CaSO_4(s)$ that is produced from "scrubbing" the $SO_2$ pollutant out of the exhaust gas. Assume that all of the sulfur in the coal is converted to calcium sulfate.

**69.** Write the equation for roasting iron pyrite in the absence of air to form elemental sulfur. Calculate the volume of $S_2$ gas that can be produced from roasting 5.5 kg of iron pyrite. Assume that all of the sulfur in the iron pyrite is converted to $S_2$ gas. (Assume STP to calculate the gas volume.)

**70.** Write an overall reaction from the two steps in the Claus process. Calculate the volume of $H_2S$ gas needed to produce 1.0 kg of $S(s)$. (Assume STP to calculate the gas volume.)

## Halogens

**71.** Determine the oxidation state of Xe and give the VSEPR structure for each compound.
a. $XeF_2$        b. $XeF_6$        c. $XeOF_4$

**72.** Describe the shape of each halogen compound.
a. $BrF_4^-$    b. $IF_3$    c. $BrO_2^-$    d. $ClO_4^-$

**73.** When chlorine is bubbled through a colorless aqueous solution containing bromide ions, the solution turns red and the chlorine is reduced. Write a balanced equation for this reaction and identify the oxidizing and reducing agent.

**74.** Carbon tetrachloride is produced by passing chlorine gas over carbon disulfide in the presence of a catalyst. The reaction also produces $S_2Cl_2$. Write the balanced reaction and identify which element is oxidized and which element is reduced.

**75.** If 55 g of $SiO_2(s)$ glass is placed into 111 L of 0.032 M HF, is there enough HF to dissolve all of the glass? Determine which substance is the limiting reagent and calculate how much of the other reagent is left if the reaction proceeds to completion.

**76.** Describe the difference in the types of bonds (single, double, triple) and the shapes of the following two iodine ions: $ICl_4^-$ and $IO_4^-$.

# Cumulative Problems

**77.** From the compositions of lignite and bituminous coal, calculate the mass of sulfuric acid that could potentially form as acid rain from burning $1.00 \times 10^2$ kg of each type of coal.

**78.** Calculate the volume of $CO_2$ released from heating and decomposing 88 g of sodium bicarbonate. (Assume standard pressure and temperature.)

**79.** All of the halogens form oxoacids. The perhalic acids have the general formula of $HXO_4$. Explain why $HClO_4$ is a much stronger acid than $HIO_4$.

**80.** The halogens form oxoacids with different amounts of oxygen. Explain why $HClO_4$ is a stronger acid than $HClO_2$.

**81.** Determine the ratio of effusion rates of HCl compared to each gas.
a. $Cl_2$
b. HF
c. HI

**82.** Calculate the ratio of effusion rates for each pair of gases.
a. $^{238}UF_6$ and ClF
b. $^{238}UF_6$ and $^{235}UF_6$

**83.** Sodium peroxide is a very powerful oxidizing agent. Balance the reaction of sodium peroxide with elemental iron to give sodium oxide and $Fe_3O_4$.

**84.** Sulfur dioxide is a reducing agent. When it is bubbled through an aqueous solution containing $Br_2$, a red-colored solution, it reduces the bromine to colorless bromide ions and forms sulfuric acid. Write a balanced equation for this reaction and identify the oxidizing and reducing agent.

**85.** Using the molecular orbital model for a diatomic molecule, explain the different bond lengths for the ions of oxygen. Also state which ion is diamagnetic.

| Ion | O—O Bond Length (pm) |
|-----|----------------------|
| $O_2^+$ | 112 |
| $O_2$ | 121 |
| $O_2^-$ | 133 |
| $O_2^{2-}$ | 149 |

**86.** The *closo*-borane with the formula $B_6H_6^{2-}$ has the six B atoms at vertices, forming an octahedron structure with eight faces. The formula for the number of sides is $2n - 4$, where $n$ is the number of boron atoms. Determine the number of vertices and faces for each *closo*-borane.
a. $B_4H_4^{2-}$        b. $B_{12}H_{12}^{2-}$

**87.** Find the amount (in moles) of C—C bonds that must be broken when 1.0 mole of $C(g)$ is formed from C(diamond). Calculate the $\Delta H$ of sublimation of diamond from the data in Appendix II, Table B. Then do the calculation using the C—C bond energy in Table 9.3. Suggest a reason for the difference between the two values.

**88.** Breathing air that contains 0.13% CO by volume for 30 minutes will cause death. CO can form by incomplete combustion of carbon-containing compounds. Calculate the minimum volume of octane ($C_8H_{18}$, a component of gasoline, which has a density of 0.70 g/mL) that must burn to produce this composition of CO in a garage of volume 40 m³ at 298 K and 1.0 atm.

**89.** Given that the $\Delta H_f^\circ$ of 1 M $H_2SO_3$ is $-633$ kJ, use the data in Appendix II, Table B to calculate the $\Delta H^\circ$ for the formation of a 1 M solution of $SO_2$ in water from $SO_2(g)$.

**90.** Use the data in Appendix II, Table B to calculate $\Delta H^\circ$ for the formation of a 1 M solution of $H_2SO_4$ from $SO_3(g)$.

# Challenge Problems

**91.** Calculate the standard enthalpy of reaction for reducing the different forms of iron oxide to iron metal and $CO_2$ from the reaction of the oxide with CO. Identify which reaction is the most exothermic per mole of iron and explain why.
   **a.** $Fe_3O_4$
   **b.** FeO
   **c.** $Fe_2O_3$

**92.** Balance the equation for the production of acetylene ($C_2H_2$) from the reaction of calcium carbide with water. If the acetylene is burned to form water and carbon dioxide, how many kilojoules of energy are produced from the complete reaction of 18 g of calcium carbide?

**93.** Carbon suboxide, $C_3O_2$, is a linear molecule with the two oxygen atoms at the end and double bonds between each carbon and oxygen atom.
   **a.** Draw the Lewis structure for $C_3O_2$.
   **b.** State the type of hybridization of each carbon atom.
   **c.** Calculate the heat of reaction for the reaction of carbon suboxide with water to form malonic acid ($HO_2CCH_2CO_2H$). Hint: Each end of malonic acid has a carbon double bonded to an oxygen and bonded to a hydroxide.

**94.** Calcium carbonate is insoluble in water. Yet, it dissolves in an acidic solution. Calculate the standard enthalpy, entropy, and Gibbs free energy change for the reaction between solid calcium carbonate and hydrochloric acid. What drives the reaction, the enthalpy change, or the entropy change?

**95.** When hydrazine is dissolved in water it acts like a base.
$$N_2H_4(aq) + H_2O(l) \rightleftharpoons N_2H_5^+(aq) + OH^-(aq)$$
$$K_{b_1} = 8.5 \times 10^{-7}$$
$$N_2H_5^+(aq) + H_2O(l) \rightleftharpoons N_2H_6^{2+}(aq) + OH^-(aq)$$
$$K_{b_2} = 8.9 \times 10^{-16}$$
   **a.** Calculate the $K_b$ for the overall reaction of hydrazine forming $N_2H_6^{2+}$.
   **b.** Calculate $K_{a_1}$ for $N_2H_5^+$.
   **c.** Calculate the concentration of hydrazine and both cations in a solution buffered at a pH of 8.5 for a solution that was made by dissolving 0.012 mol of hydrazine in 1 L of water.

**96.** Solid fuel in the booster rockets for spacecraft consists of aluminum powder as the fuel and ammonium perchlorate as the oxidizing agent.
$$3 NH_4ClO_4(s) + 3 Al(s) \longrightarrow$$
$$Al_2O_3(s) + AlCl_3(s) + 6 H_2O(g) + 3 NO(g)$$
If a rocket launch burns $2.200 \times 10^3$ kg of aluminum, calculate the energy produced in joules. Calculate the volume of the gas produced assuming it was cooled back to 298 K at 1 atm. The standard enthalpy of formation of solid ammonium perchlorate is $-295.3$ kJ/mol.

**97.** Two known compounds have the formula $H_2N_2O_2$. One of them is a weak acid and one is a weak base. The acid, called hyponitrous acid, has two O—H bonds. The base, called nitramide, has no O—H bonds. Draw Lewis structures for these compounds. Predict whether the acid is stronger or weaker than nitrous acid and whether the base is stronger or weaker than ammonia.

# Conceptual Problems

**98.** Explain why fine particles of activated charcoal can absorb more (as a filter) than large briquettes of charcoal.

**99.** The two major components of the atmosphere are the diatomic molecules of nitrogen and oxygen. Explain why pure nitrogen is used as a protective atmosphere in the laboratory and pure oxygen is much more reactive.

**100.** Explain why nitrogen can form compounds with many different oxidation numbers.

**101.** Describe how sodium dihydrogen phosphate can be used as a pH buffering agent.

**102.** Explain why $H_2S$ has a different bond angle and is much more reactive than $H_2O$.

**103.** Explain why fluorine is found only with the oxidation state of $-1$ or 0, while the other halogens are found in compounds with other oxidation states.

**104.** Why do some substances burn in fluorine gas even if they do not burn in oxygen gas?

**105.** Explain why $SO_2$ is used as a reducing agent but $SO_3$ is not.

**106.** Explain why the interhalogen molecule $BrCl_3$ exists but $ClBr_3$ does not.

*Active Classroom Learning*

# Questions for Group Work

Discuss these questions with the group and record your consensus answer.

**107.** Have each group member select a different element from this list: B, C, N, O, and F. Have all group members look up the following values for their elements: valence electron configuration, covalent atomic radius, effective nuclear charge, first ionization energy, and Pauling electronegativity. All members should compare the values for their elements with those of the group. Repeat with O, S, Se, Te, and Po. Account for the trends you see by referring to quantum mechanical shells. Note and provide a tentative explanation for any exceptions to trends you observe.

**108.** Provide a detailed chemical description of your birthstone or favorite gem. If two members of the group share a birthstone, have one select another gem, so that no two group members work with the same gem. Is your gem a silicate? What type? Share your findings with your group. Does your group agree on which gem has the most interesting structure?

**109.** As a group, create a single schematic drawing of a factory that uses the Haber-Bosch process and the Ostwald process to make $NH_4NO_3$ fertilizer. The factory may only take in $N_2(g)$, $H_2(g)$, and $O_2(g)$. Pass your diagram around the group and have each group member add one step in the process. Where might the factory obtain $H_2$?

**110.** Ask each group member to work individually to write out two or three reactions from one section of the chapter. Replace the symbol for the element that is the topic of that section with a blank or an X. Take turns playing "identify the mystery element" as follows. One person shares their reactions (with the blanks) with the group. The person to the right of the person sharing tries to identify the missing element. If that person can't, the group member to the right tries. Continue until either the element has been identified or all group members have had a chance. The next round starts with the person who was first to guess the element sharing reactions with the group. Continue until all have shared reactions, and all reactions have been correctly identified.

**111.** Working individually, find a sentence in the chapter that uniquely describes a particular main-group compound. As in the previous question, take turns reading your sentences to the group saying "blank" in place of the name of the compound and giving group members a chance to identify the compound.

# Data Interpretation and Analysis

## Hydrogen Sulfide in Well Water

**112.** Hydrogen sulfide ($H_2S$) smells like rotten eggs, and its smell can be detected at concentrations as low as 0.02 ppm. Well water, which is drawn from underground depths of 30–250 meters (100–800 feet), is commonly contaminated with hydrogen sulfide. The contamination arises from bacterial decomposition of sulfur-containing organic and inorganic matter present in the oxygen-deficient underground environment. Some $H_2S$ gas dissolves in the water that is pumped for drinking supplies, where it is detected by taste at concentrations as low as 0.05 ppm. Municipalities remove $H_2S$ to improve water taste and to prevent the accumulation of toxic $H_2S$ gas within the system.

Table a lists hydrogen sulfide solubility in water when the partial pressure of hydrogen sulfide is 1.00 atm. Table b lists solubility product constant data for common metal sulfides.

| Temperature, °C | Mole fraction of hydrogen sulfide in water |
|---|---|
| 15 | $2.335 \times 10^{-3}$ |
| 20 | $2.075 \times 10^{-3}$ |
| 25 | $1.85 \times 10^{-3}$ |
| 30 | $1.66 \times 10^{-3}$ |
| 35 | $1.51 \times 10^{-3}$ |

**TABLE a Hydrogen Sulfide Solubility in Water (Partial Pressure of Hydrogen Sulfide = 1.00 atm)** [Source: *CRC Handbook of Chemistry and Physics*, 85th Edition]

| Sulfide | $K_{sp}$ (at 25 °C) |
|---|---|
| copper(II) sulfide | $6 \times 10^{-37}$ |
| iron(II) sulfide | $6 \times 10^{-19}$ |
| lead(II) sulfide | $9.04 \times 10^{-29}$ |
| zinc sulfide | $2.0 \times 10^{-25}$ |

**TABLE b Solubility Product Constant Data of Common Metal Sulfides**

Use the information provided in the tables to answer the following questions:

a. What can you conclude about the temperature dependence of $H_2S$ solubility in water from the data in Table a?

b. Suppose that water analysis indicates that your well's water contains 1.8 mg $H_2S$ per liter of well water. Using Henry's law (Chapter 13) and the data in Table a, calculate the partial pressure of $H_2S$, in atm, that exists above your well water at 30 °C. (Assume the well water is in a closed container).

c. Air containing $H_2S$ at 20 ppm (by volume) and below is considered safe to breathe. Is the air above the solution in part b safe to breathe? (The total pressure is 1.00 atm.)

d. A water technician recommends that you treat your well water with hydrogen peroxide, $H_2O_2$, in order to convert $H_2S$ to sulfate (which does not have an offensive odor or taste), according to the following unbalanced reaction:

$$H_2O_2(aq) + H_2S(aq) \longrightarrow SO_4{}^{2-}(aq) + H^+(aq) + H_2O(l)$$

What volume of a 3.0% by mass solution of hydrogen peroxide is required to treat 955 liters of your well water? (Assume that no other species are present to react with the $H_2O_2$ and that the density of the hydrogen peroxide solution is 1.0 g/mL.)

e. Hydrogen sulfide is a weak diprotic acid that ionizes in water to form hydrogen sulfide ($HS^-$) and sulfide ($S^{2-}$). Sulfide dissolved in water can be precipitated as an insoluble metal sulfide, according to the following reaction:

$$M^{2+}(aq) + S^{2-}(aq) \longrightarrow MS(s)$$

Which metal in Table b most effectively removes hydrogen sulfide from water?

f. What volume, in mL, of a 0.500 M solution of the metal ion you chose in part e is required to remove the hydrogen sulfide present in 955 liters of your well water (from part b)?

g. What mass, in grams, of the metal sulfide (from part f) precipitates? (Assume 100% yield.)

# Answers to Conceptual Connections

## Carbide Formulas

**23.1** $K_2C_2$

## Carbonate Solubility

**23.2 (a)** Since the carbonate ion is basic, adding acid to the solution drives the dissolution reaction to the right because the acid reacts with the carbonate ion. (Recall from Section 17.5 that the solubility of an ionic compound with a basic anion increases with increasing acidity.)

# Metals and Metallurgy

> The metals are not presented immediately to the hand of man . . . but they are, for the most part, buried in darkness, in the bowels of the earth, where they are so much disguised, by combination and mixture with other substances, that they often appear entirely unlike themselves.
>
> —WILLIAM HENRY (1774–1836)

If you look around your house, classroom, or neighborhood, much of what you see is made of metal or at least has some metal parts. Can you imagine life without metals? Without metals, we would have no skyscrapers, which need the rigid framework of steel beams; no automobiles, which need metal engines and bodies; and no electricity, which needs copper and aluminum wire for transmission. Metallurgy is both a very old and a very new science. We find the roots of metallurgy in the distant past, when our primitive ancestors began to search for and process metals. Of the three prehistoric ages of humankind—the Stone Age, the Bronze Age, and the Iron Age—two are named after our quest for metals. But we do not commonly find metals as elements on Earth; instead we find them in compounds, often scattered within other compounds. The mining, separating, and refining of the vast array of metals is a fairly new science. Many modern metals were not available even as recently as 200 years ago. In the middle of the nineteenth century, gold was cheaper than aluminum because aluminum was so difficult to refine, and titanium, which is important in the aerospace industry, could not even be manufactured. In this chapter, we cover the area of chemistry known as metallurgy.

▲ Oil is sometimes contaminated with vanadium, which is thought to have been the metal used for oxygen transport by some ancient life-forms on Earth.

## 24.1  Vanadium: A Problem and an Opportunity

Companies often seek help from university researchers to solve specific technical problems. In one case, representatives of an oil company sought help from a university's chemistry and engineering departments to address vanadium contamination in oil. Because vanadium is toxic, the U.S. government regulates the quantity of vanadium

▲ Vanadium is a soft white metal.

that can be emitted by industrial processes, and oil companies have to figure out environmentally acceptable ways to remove the vanadium from the oil.

Vanadium is a rare element, making up only about 0.015% of Earth's crust (by mass). It is a soft white metal with high ductility and malleability, which means that it can be easily drawn into narrow wires or rolled into thin sheets. Because of the high reactivity of pure vanadium, this element occurs naturally within compounds (not in its elemental form). However, vanadium compounds are not normally found in the rocks or soils near oil fields, so why does this rare, reactive metal occur in oil? The answer lies in the biology of ancient life-forms. Most modern animals use iron in hemoglobin to transport oxygen through their bloodstreams (see Section 15.1), and a few (such as lobsters) use copper for this function. It appears that some extinct animals used vanadium. Even today, one group of simple marine organisms, the tunicates or sea squirts, are believed to employ vanadium compounds for oxygen transport. The source of vanadium in some crude oil may be the very animals from which the oil was formed.

During refining, crude oil is heated to carry out the reactions that form the different petroleum products. If some of the vanadium-containing oil burns during heating, the vanadium forms vanadate compounds, ionic compounds with vanadium oxide polyatomic ions such as $VO_4^{3-}$ and $VO_3^{-}$. These compounds, which have low melting points, can dissolve the protective oxide coatings on stainless steel containers, causing the steel to corrode and ruining the containers in which the oil is stored. If the vanadium compounds are not removed from the oil before storage, the customer inherits the problem. In addition, vanadium and many of its compounds are toxic, so oil companies cannot simply dump the extracted vanadium compounds into the environment.

The presence of vanadium in crude oil may, however, turn out to be profitable for the oil companies. If vanadium can be economically recovered from oil, then oil could become a major source of vanadium, a valuable metal with several important industrial uses, including the production of iron alloys and sulfuric acid.

The recovery of vanadium from oil involves metallurgy, the topic of this chapter. **Metallurgy** includes all the processes associated with mining, separating, and refining metals and the subsequent production of pure metals and mixtures of metals called *alloys* (which we define in Section 24.4). In this chapter, we explore the general properties and natural distribution of metals; several different categories of metallurgical processes, including pyrometallurgy, hydrometallurgy, electrometallurgy, and powder metallurgy; the structures and alloys formed by metals; and the metallic products and applications for several selected materials.

## 24.2 The General Properties and Natural Distribution of Metals

Metals share several common properties. All metals are opaque (we cannot see through them), and they are good conductors of heat and electricity. They generally have high malleability (the capacity to be bent or hammered into desired forms) and ductility (the capacity to be drawn into wires). We can explain these properties by the bonding theories that we have already examined for metals: the electron sea model and band theory. In the electron sea model, discussed in Section 9.11, each metal atom within a metal sample donates one or more electrons to an *electron sea*, which flows within the metal. In band theory, discussed in Section 12.8, the atomic orbitals of the metal atoms are combined, forming *bands* that are delocalized over the entire crystalline solid; electrons move freely within these bands. The mobile electrons in both of these models endow metals with many of their shared properties.

Each metal, however, is also unique, and even their shared properties may vary within a range. For example, some metals, such as copper, silver, and aluminum, have much higher thermal and electrical conductivities than do other metals. Table 24.1 lists the thermal conductivity and electrical resistivity (a low resistivity corresponds to a high conductivity) of a few metals. Notice that both properties can vary by factors of 10 or more. Most metals are very strong, malleable, and ductile. But lead is a

**TABLE 24.1** Thermal Conductivity and Electrical Resistivity of Several Metals

| Metal | Thermal Conductivity (W/cm · K) | Electrical Resistivity ($\mu\Omega$ · cm) |
|---|---|---|
| Ag, silver | 4.29 | 1.59 |
| Cu, copper | 4.01 | 1.67 |
| Fe, iron | 0.804 | 9.71 |
| V, vanadium | 0.307 | 24.8 |

soft, weak metal, and chromium is a brittle metal that will not bend without breaking. When we think about the shared properties of metals, we must allow for a range of properties as well as some unique ones.

Over 75% of the elements in the periodic table are metals, yet metals make up only about 25% of the mass of Earth's crust (see Figure 23.2 for the composition of Earth's crust). Earth's core is thought to be composed of iron and nickel, but because the core is so far from the surface, these metals are not accessible. The most abundant metal on Earth is aluminum, a main-group metal. Several alkali and alkaline earth metals (calcium, sodium, potassium, and magnesium) make up more than 1% of Earth's crust. Iron, which makes up about 5% of the crust, is the only transition metal that accounts for more than 0.1% of the crust. Of the first-row transition metals, titanium, chromium, iron, nickel, copper, and zinc are all plentiful enough to be important industrial materials.

▲ Gold is one of the few metals that can be found as an element in nature, often in veins like the one shown here.

▲ Most NaCl salt is produced by trapping ocean water in shallow basins and letting the water evaporate.

Only a few metals occur naturally as elements, including nickel, copper, palladium, silver, platinum, and gold. Because of their low reactivity, these metals are often called the *noble metals*. They are usually concentrated within mountainous or volcanic regions in small, isolated veins within a rock matrix.

Most of the rest of the metals occur naturally in positive oxidation states within mineral deposits. **Minerals** (as we saw in Chapter 23) are homogeneous, naturally occurring, crystalline inorganic solids. A rock that contains a high concentration of a specific mineral is called an **ore**. Metallurgical processes separate useful minerals from other, unusable material in the ore.

The main sources for the alkali metals are chloride minerals, such as halite (sodium chloride) and sylvite (potassium chloride). Halite occurs in large deposits in sites that were once ancient oceans, or it can be precipitated by evaporating ocean water. The main source for some other metals are oxide minerals, such as hematite ($Fe_2O_3$), rutile ($TiO_2$), and cassiterite ($SnO_2$). The ores containing these minerals are unevenly distributed throughout Earth. For example, North America has large deposits of hematite but no substantial deposits of cassiterite. Sulfides are the most important minerals for many metals, such as galena (PbS), cinnabar (HgS), sphalerite (ZnS), and molybdenite ($MoS_2$).

The main mineral source for some other metals can be complex. For example, the primary *mineral* sources for vanadium are vanadinite [$Pb_5(VO_4)_3Cl$] and carnotite [$K_2(UO_2)_2(VO_4)_2 \cdot 3 H_2O$], which is also a main mineral source for uranium. Vanadium, as we have seen, is also found in crude oil. No specific minerals contain the metal radium. Yet radium sometimes substitutes for uranium within uranium-containing minerals. The small amount of radium that replaces uranium in carnotite is the major source of radium. Tantalum and niobium, named after the Greek god Tantalus and his daughter Niobe, always occur together in mixed deposits of columbite [$Fe(NbO_3)_2$] and tantalite [$Fe(TaO_3)_2$]. The minerals are mined together, and the elements are later separated by recrystallization (see Section 13.4).

## 24.3 Metallurgical Processes

Mined ores are first physically *separated* into their metal-containing and nonmetal-containing components. Then, the elemental metal is extracted from the compounds in which it is found, a process called **extractive metallurgy**. In this section, we look at several metallurgical processes, including pyrometallurgy, hydrometallurgy, electrometallurgy, and powder metallurgy. After separation and extraction, the metal is **refined**; refining is the process in which the crude material is purified.

### Separation

The first step in processing metal-containing ores is crushing the ore into smaller particles. The particles that contain the minerals are then separated from the undesired material, called **gangue** (pronounced "gang"), usually by physical methods. For example, in some cases, a cyclone of wind is used to lift the gangue away from the metal-containing particles, as shown in Figure 24.1▼. If the minerals are magnetic, magnets can separate the minerals. Electrostatic forces can also separate polar minerals from nonpolar gangue.

Sometimes solutions separate the minerals from the gangue, as shown in Figure 24.2▼. A *wetting agent* (or detergent) that preferentially attaches to the mineral surfaces is added to a mixture of the mined material and water. Air is blown through the mixture, forming bubbles and a froth. Because they are attached to the wetting agent, the minerals segregate into the froth, where they can be collected. The gangue separates into the solution.

▲ **FIGURE 24.1 Separation by Air** An industrial cyclone separates the crushed light particles of gangue from the heavier metal-bearing particles.

▲ **FIGURE 24.2 Separation by Using a Solution** A wetting agent helps the minerals attach to the froth in a bubbled solution. The minerals in the froth are separated from the gangue remaining in the solution.

### Pyrometallurgy

Pyrometallurgy contains the Greek stem *pyro*, meaning fire or heat.

Once the mineral is separated from the gangue, the elemental metal is extracted from the mineral. Several different techniques can achieve this separation. In **pyrometallurgy**, heat is employed to extract a metal from its mineral. Different heating conditions have different effects on the mineral.

**Calcination** is the heating of an ore in order to decompose it and drive off a volatile product. For example, when carbonate minerals are heated, carbon dioxide is driven off, as shown in these examples:

$$PbCO_3(s) \xrightarrow{heat} PbO(s) + CO_2(g)$$

$$4\,FeCO_3(s) + O_2(g) \xrightarrow{heat} 2\,Fe_2O_3(s) + 4\,CO_2(g)$$

Many minerals occur in a hydrated form (that is, they contain water). Calcination can also drive off water:

$$Fe_2O_3 \cdot 2\,Fe(OH)_3(s) \xrightarrow{heat} 2\,Fe_2O_3(s) + 3\,H_2O(g)$$

Hydrated compounds are discussed in Section 3.5.

Heating that causes a chemical reaction between the furnace atmosphere (the gases in the furnace) and the mineral is called **roasting**. Roasting is particularly important in processing sulfide ores. The ores are heated in the presence of oxygen, converting the sulfide into an oxide and emitting sulfur dioxide. For example, the roasting of lead(II) sulfide occurs by the reaction:

$$2\,PbS(s) + 3\,O_2(g) \xrightarrow{heat} 2\,PbO(s) + 2\,SO_2(g)$$

In some cases, especially with the less active metals such as mercury, roasting the sulfide produces the pure metal:

$$HgS(s) + O_2(g) \xrightarrow{heat} Hg(g) + SO_2(g)$$

When roasting forms a liquid product, which makes separation easier, it is called **smelting**. Consider, for example, the smelting of zinc oxide:

$$ZnO(s) + C(s) \xrightarrow{heat} Zn(l) + CO(g)$$

The gaseous carbon monoxide separates from the liquid zinc, allowing the metal to be readily recovered. In some cases, a flux must be added to the mixture during smelting to help separate the two materials. The **flux** is a material that will react with the gangue to form a substance with a low melting point. For example, oxides of silicon within gangue can be liquefied by reaction with calcium carbonate according to the reaction:

$$\underset{\text{gangue}}{SiO_2(s)} + \underset{\text{flux}}{CaCO_3(s)} \longrightarrow CO_2(g) + \underset{\text{slag}}{CaSiO_3(l)}$$

The waste liquid solution that forms from the flux and gangue is usually a silicate material called a **slag**. The liquid metal and the liquid slag have different densities, and therefore separate. Holes tapped at different heights into the side of the container holding the liquid metal and slag allow the more dense liquid to flow out of the lower tap holes and the less dense liquid to flow out of the higher tap holes.

## Hydrometallurgy

The use of an aqueous solution to extract metals from their ores is **hydrometallurgy**. A hydrometallurgy process is used to obtain gold. Gold occurs in its elemental state, but often as very small particles mixed with other substances. The gold can be separated out of the mixture by selectively dissolving it into solution, a process called **leaching**. In this process, solid gold reacts with sodium cyanide to form a soluble gold complex:

$$4\,Au(s) + 8\,CN^-(aq) + O_2(g) + 2\,H_2O(l) \longrightarrow 4\,Au(CN)_2^-(aq) + 4\,OH^-(aq)$$

The impurities are filtered out of the solution, and the gold is reduced back to elemental gold with a reactive metal such as zinc:

$$2\,Au(CN)_2^-(aq) + Zn(s) \longrightarrow Zn(CN)_4^{2-}(aq) + 2\,Au(s)$$

Leaching has been practiced for many years and often results in the contamination of streams and rivers with cyanide. New alternatives using the thiosulfate ion ($S_2O_3^-$) are being investigated to replace leaching.

Different acid, base, and salt solutions are sometimes used to selectively separate out metal-bearing minerals. For example, sulfuric acid is used to separate the copper and iron from the mineral chalcopyrite, $CuFeS_2$, and a sodium chloride solution is used to separate the lead from the insoluble mineral anglesite, $PbSO_4$:

$$2\,CuFeS_2(s) + H_2SO_4(aq) + 4\,O_2(g) \longrightarrow 2\,CuSO_4(aq) + Fe_2O_3(s) + 3\,S(s) + H_2O(l)$$

$$PbSO_4(s) + 4\,NaCl(aq) \longrightarrow Na_2[PbCl_4](aq) + Na_2SO_4(aq)$$

Hydrometallurgy is often more economical than pyrometallurgy because of the high energy costs associated with the elevated temperatures needed for calcination and roasting.

## Electrometallurgy

See Section 19.8 for a description of electrolysis.

In **electrometallurgy**, electrolysis is used to produce metals from their compounds. For example, the *Hall process* is an electrometallurgical process crucial to aluminum production. The main source of aluminum is bauxite, $Al_2O_3 \cdot n\,H_2O$. A hydrometallurgical process, the *Bayer process*, separates bauxite from the iron and silicon dioxide with which it is usually found. In this process, the bauxite is heated in a concentrated aqueous NaOH solution under high pressure. The aluminum oxide dissolves, leaving the other oxides behind in solid form:

$$Al_2O_3 \cdot n\,H_2O(s) + 2\,OH^-(aq) + 2\,H_2O(l) \longrightarrow 2\,Al(OH)_4{}^-(aq)$$

The basic aluminum solution is separated from the oxide solids, and then the aluminum oxide is precipitated out of solution by neutralizing it (recall the pH dependence of the solubility of $Al(OH)_3$ from Section 17.8). Calcination of the precipitate at temperatures greater than 1000 °C yields anhydrous alumina ($Al_2O_3$). Electrolysis is then used to reduce the aluminum out of the aluminum oxide. Because $Al_2O_3$ melts at such a high temperature (greater than 2000 °C), however, the electrolysis is not carried out on molten $Al_2O_3$. Instead, in the Hall process the $Al_2O_3$ is dissolved into molten cryolite ($Na_3AlF_6$), and graphite rods are used as electrodes to carry out the electrolysis in the liquid mixture (Figure 24.3▼). The carbon that composes the graphite electrodes is oxidized by the dissolved oxygen ions in the molten salt and converted to carbon dioxide. The aluminum ions dissolved in the molten salt reduce to molten aluminum, which sinks down to the bottom of the cell and is removed:

**Oxidation:**   $C(s) + 2\,O^{2-}(dissolved) \longrightarrow CO_2(g) + 4\,e^-$

**Reduction:**   $3\,e^- + Al^{3+}(dissolved) \longrightarrow Al(l)$

▶ **FIGURE 24.3 The Hall Process** The Hall process produces aluminum metal by reducing the aluminum ions in alumina. Graphite electrodes act as reducing agents. Dissolved oxygen ions in the molten cryolite oxidize the carbon in the graphite electrodes to form carbon dioxide.

Another important example of metallurgy is the refinement of copper. The most abundant copper source is the mineral chalcopyrite, $CuFeS_2$. First, the chalcopyrite is converted to CuS by roasting. During this process, the iron also forms oxides and sulfides. Silica is added to form an iron silicate slag, which is then removed. The remaining copper(II) sulfide reacts with oxygen to form sulfur dioxide gas and copper metal, but the metal is not very pure. Electrolysis is employed to refine the copper.

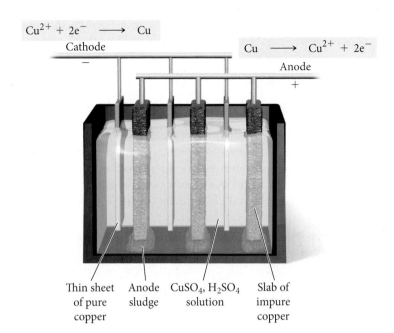

$$Cu^{2+} + 2e^- \longrightarrow Cu$$
Cathode
−

$$Cu \longrightarrow Cu^{2+} + 2e^-$$
Anode
+

Thin sheet of pure copper

Anode sludge

$CuSO_4$, $H_2SO_4$ solution

Slab of impure copper

◀ **FIGURE 24.4 Copper Electrolysis Cell** Copper is refined by electrolysis. The impure copper is oxidized at the anode and then reduced to form pure metal on the cathode. Many precious metals collect in the sludge at the bottom of the electrolysis cell.

In the electrolysis cell used to refine copper, both the anode and the cathode are made of copper (Figure 24.4▲). The anode is the impure copper (that needs to be refined), and the cathode is a thin sheet of pure copper. As the current flows through the cell, the copper from the anode oxidizes and dissolves in a copper sulfate solution. It then plates out as pure copper on the cathode. The impurities in the copper anode separate from the copper during electrolysis because, even though the more active metals also oxidize from the anode, they stay in solution and do not plate out on the cathode. The less active metals do not oxidize at all and simply fall to the bottom of the cell as the copper is dissolved from the anode. The sludge at the bottom of the electrolysis cell contains many precious metals, including gold and silver. About one-quarter of the silver produced in the United States is from the impurities recovered from the refinement of copper.

## Powder Metallurgy

**Powder metallurgy**, first developed in the 1920s, is used to make metallic components from powdered metal. In powder metallurgy, micron-sized metal particles are pressed together under high pressures to form the desired component. The component is then heated (*sintered*). The sintering process occurs below the melting point of the powder but at a temperature high enough to fuse the metal particles together, strengthening the metal and increasing its density.

Originally, iron powder from mill scrap was used in powder metallurgy. The scrap was primarily iron oxide that would fall off the steel as it was being milled. The iron oxide dust was heated in a hydrogen atmosphere to reduce the oxide to iron particles. Manufacturers called the powdered metal *iron sponge* because numerous holes form in the particles when the oxygen escapes.

In the 1960s, the A. O. Smith Company in the United States introduced a new method for the development of powdered metal, called water atomization. In this method, the pure metal is melted and a small stream of the liquid is allowed to flow from the bottom of the container of molten metal. A high-pressure blast of cold water hits the stream of metal, breaking it into small droplets that quickly solidify. Powdered metal particles made in this way are smoother and more dense than the sponge powder particles from oxide scrap. In addition to powdered iron metal, many copper, bronze, carbide, and brass parts are made with powder metallurgical processes.

Powder metallurgy offers several advantages over traditional casting or milling of metal. For example, waste is almost eliminated because the part can be pressed directly into the desired shape. Intricate teeth on gears and multiple holes can be designed into the press, and therefore do not have to be machined after production. Making cast metal

▲ These metal products were all made by powder metallurgy.

objects from metals with high melting points, such as molybdenum and tungsten, can be difficult because of the high temperatures necessary to melt the metal. Using the powder avoids the need for high temperatures.

# 24.4 Metal Structures and Alloys

We can describe the structures of metals as the closest packing of spheres, first discussed in Section 12.3. Elemental metals generally crystallize in one of the basic types of crystal lattices, including face-centered cubic, body-centered cubic, and hexagonal closest packed. The crystal structure of a metal may change, however, as a function of temperature and pressure. Table 24.2 lists the crystal structures for the 3d transition metals.

| TABLE 24.2 The Crystal Structures of the 3d Elements | | |
| --- | --- | --- |
| Metal | Natural Crystal Structure at Room Temperature and Atmospheric Pressure | Other Crystal Structures at Different Temperatures and Pressures |
| Sc | Hexagonal closest packed | Face-centered cubic, body-centered cubic |
| Ti | Hexagonal closest packed | Body-centered cubic above 882 °C |
| V | Body-centered cubic | |
| Cr | Body-centered cubic | Hexagonal closest packed |
| Mn | Alpha complex body-centered cubic form | Beta simple cubic form above 727 °C |
| | | Face-centered cubic above 1095 °C |
| | | Body-centered cubic above 1133 °C |
| Fe | Body-centered cubic | Face-centered cubic above 909 °C |
| | | Body-centered cubic above 1403 °C |
| Co | Hexagonal closest packed | Face-centered cubic above 420 °C |
| Ni | Face-centered cubic | |
| Cu | Face-centered cubic | |
| Zn | Hexagonal closest packed | |

## Alloys

An **alloy** is a metallic material that contains more than one element. Some alloys are simply solid solutions, whereas others are specific compounds with definite ratios of the component elements. Alloys have metallic properties, but they can consist of either two or more metals or a metal and a nonmetal. We broadly classify alloys as substitutional or interstitial. In a **substitutional alloy**, one metal atom substitutes for another in the crystal structure. The crystal structure may either stay the same upon the substitution, or it may change to accommodate the differences between the atoms. In an **interstitial alloy**, small, usually nonmetallic atoms fit in between the metallic atoms of a crystal. The alloy maintains its metallic properties with these interstitial atoms in the structure.

## Substitutional Alloys

For two metals to form a substitutional alloy, the radii of the two metal atoms must be similar, usually within 15% of each other. For example, the atomic radii of copper and nickel are both 135 pm, and both of the elements form the face-centered cubic structure. Therefore, either metal can easily replace the other in the metal crystal structure.

▲ Copper crystallizes in the face-centered structure shown here.

▲ In a nickel and copper alloy, nickel atoms substitute for some of the copper atoms.

Figure 24.5▶ is a phase diagram for a copper and nickel alloy. This phase diagram, called a *binary* phase diagram, is different from the diagrams that we discussed in Section 11.8, which show the phases of a pure substance at different pressures and temperatures. This diagram shows the different phases for a mixture at different *compositions* and temperatures. The x-axis indicates the composition (in this case, the mole percent of nickel in the alloy, with the left side representing pure copper and the right side representing pure nickel). The y-axis indicates the temperature.

Pure copper melts at 1084 °C, as indicated by the change from solid to liquid at 0% nickel. Pure nickel melts at 1455 °C, as indicated by the change from solid to liquid at 100% nickel. The area on the diagram above the line connecting the melting points of copper and nickel represents a liquid solution of the two metals. The area below that line represents a solid solution of the two metals. Any ratios of copper and nickel can form the face-centered cubic structure.

▲ **FIGURE 24.5 Cu–Ni Phase Diagram** Because copper and nickel have similar crystal structures and similar size, together they can form a solid solution. A solution forms at all compositions from pure copper to pure nickel.

## Guidelines for...

| Interpreting a Binary Phase Diagram | **Example 24.1** Determining Alloy Compositions from a Phase Diagram | **Example 24.2** Determining Alloy Compositions from a Phase Diagram |
|---|---|---|
| | Determine the composition and phase present at point A on Figure 24.5. | Determine the composition and phase present at point B on Figure 24.5. |
| Locate the temperature and composition of the alloy on the binary phase diagram in Figure 24.5. | Point A represents 30 mol % Ni at 1300 °C. | Point B represents 75 mol % Ni at 1100 °C. |
| Identify the phase. | The point is above the melting point line, so the phase is liquid. | The point is below the melting point line, so the phase is solid. |
| Identify the amount of copper and nickel in the phase. | This liquid phase is made up of 30 mol % Ni, so it is 70 mol % Cu. | This solid phase is made up of 75 mol % Ni, so it is 25 mol % Cu. |
| | **FOR PRACTICE 24.1** Determine the composition and phase present at point C on Figure 24.5. | **FOR PRACTICE 24.2** Determine the composition and phase present at point D on Figure 24.5. |

The phase diagrams of some alloys are more complex. Consider the phase diagram for chromium and vanadium in Figure 24.6▶, on the next page. Both Cr and V form the body-centered cubic crystal, and the atoms are similar in size. Yet this phase diagram reveals an important difference between this alloy and the Cu and Ni alloy:

▶ **FIGURE 24.6 Cr–V Phase Diagram** In a binary phase diagram of two metals that form a solid solution, an intermediate composition can have either the highest or lowest melting point. In the chromium and vanadium phase diagram an intermediate composition has the lowest melting point.

the melting temperature does not vary in a uniform way from the lower melting point of Cr to the higher melting point of V. Instead we see that the alloys with intermediate compositions melt at lower temperatures than either pure metal. The lowest melting point is 1750 °C at a composition of 30 mol% vanadium. Solid compositions that melt at temperatures lower than either of the pure metals, as well as compositions that melt at temperatures higher than either of the two metals, are common in these types of alloys.

## Alloys with Limited Solubility

Some alloys are composed of metals that each have different crystal structures. For example, nickel crystallizes in the face-centered cubic structure and chromium in the body-centered cubic structure. Because of their different structures, these two metals do not form a miscible solid solution at all compositions. At some intermediate composition, the structure has to change from that of one of the metals to that of the other. Figure 24.7 ◀ is the nickel and chromium phase diagram from 700 °C to 1900 °C. Notice that the diagram has two different solid phases: face-centered cubic and body-centered cubic. From pure nickel (0 mol% chromium) to about 40–50 mol% chromium, the structure is face-centered cubic. In this structure, Cr atoms substitute for Ni atoms in the face-centered cubic structure of nickel. However, beyond a certain percentage of chromium (which depends on temperature), that structure is no longer stable. At 700 °C about 40 mol% Cr can fit in the crystal, and at 1200 °C about 50 mol% Cr can fit in. Adding additional Cr beyond these points results in a different phase.

At the other end of the diagram (nearly pure chromium) and at 700 °C, only a small amount of nickel can be substituted into the body-centered cubic structure of the chromium. As the temperature rises to 1300 °C, however, about 20 mol% nickel can be accommodated into the chromium structure. The region in between the two phases is called a **two-phase region**. At these compositions, the two phases (nickel-rich face-centered cubic and chromium-rich body-centered cubic) exist together. The amount of each phase depends upon the composition of the alloy.

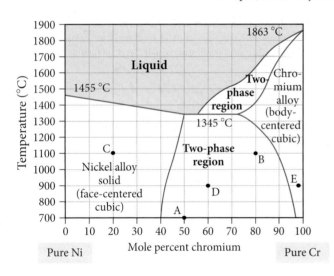

▲ **FIGURE 24.7 Cr–Ni Phase Diagram** Because chromium and nickel form solids with crystal structures that differ from each other, an alloy mixture of the two metals does not form a solid solution at all compositions. A two-phase region exists at compositions between the possible compositions of the two different structures. In the two-phase region, both crystal structures coexist in equilibrium.

We can determine the composition and relative amounts of the two different phases that coexist in a two-phase region from a phase diagram. Point A on Figure 24.7 in the Cr–Ni phase diagram represents 50% composition at 700 °C, and both phases are present. Some of the Cr atoms have substituted into the nickel-rich face-centered cubic structure, but there is too much Cr to all fit into the crystal. The leftover Cr atoms form the chromium-rich body-centered cubic structure with a small number of Ni atoms in the crystal.

The two crystals that exist at point A are (1) the nickel-rich face-centered cubic structure with 40 mol % Cr; and (2) the chromium-rich body-centered cubic structure with 5 mol % Ni. The 50% composition on the phase diagram has just slightly more Cr atoms than can fit into the nickel structure. Thus, most of the crystals in the two-phase region are the nickel-rich face-centered cubic structure with only a small amount of the chromium-rich body-centered cubic structure, as determined by a method called the *lever rule*. The **lever rule** states that in a two-phase region, whichever phase is closer to the composition of the alloy is the more abundant phase. In this example, the 50 mol % Cr composition on the phase diagram is closer to the 40 mol % Cr composition of the nickel-rich face-centered cubic phase than the 95 mol % Cr composition of the chromium-rich body-centered cubic phase, so more face-centered cubic crystals are present than body-centered cubic crystals.

## Guidelines for...

| Interpreting the Phases and Compositions in a Binary Phase Diagram as Described by the Lever Rule | Example 24.3 — Alloy Compositions in a Solid Solution with Limited Solubility | Example 24.4 — Alloy Compositions in a Solid Solution with Limited Solubility |
|---|---|---|
| | Determine the composition, relative amounts, and phases present at point B on Figure 24.7. | Determine the composition, relative amounts, and phases present at point C on Figure 24.7. |
| Locate the temperature and composition of the alloy on the binary phase diagram. | Point B represents 80 mol % Cr at 1100 °C. | Point C represents 20 mol % Cr at 1100 °C. |
| Identify the phases. | Point B is located in the two-phase region, consisting of the nickel-rich face-centered cubic structure and the chromium-rich body-centered cubic structure. | Point C is located in the one-phase region of the nickel-rich face-centered cubic structure. |
| Identify the amount of Ni and Cr in the phases. | The nickel-rich face-centered cubic structure has 45 mol % Cr and 55 mol % Ni, and the chromium-rich body-centered cubic structure has 90 mol % Cr and 10 mol % Ni. | The nickel-rich face-centered cubic structure at this point has 20 mol % Cr and 80 mol % Ni. |
| Identify the relative amounts of the phases. | At a composition of 80 mol % Cr, the composition is closer to the Cr-rich body-centered cubic phase, so there is more of this phase than there is of the Ni-rich face-centered cubic phase. | There is only one phase. It is 100 mol % of the Ni face-centered cubic phase. |
| | **FOR PRACTICE 24.3** Determine the composition, relative amounts, and phases present at point D on Figure 24.7. | **FOR PRACTICE 24.4** Determine the composition, relative amounts, and phases present at point E on Figure 24.7. |

## Interstitial Alloys

Recall that in contrast to substitutional alloys, where one metal atom substitutes for another in the lattice, interstitial alloys contain atoms of one kind that fit into the holes, or interstitial sites, of the crystal structure of the other metal. In metals with the interstitial elements hydrogen, boron, nitrogen, or carbon, the alloy that results retains its metallic properties.

▲ **FIGURE 24.8 Octahedral Holes in Closest-Packed Crystals** Octahedral holes are found in a closest-packed structure. In the closest-packed structure, the octahedral hole is surrounded by six atoms.

Closest-packed crystal structures have two different types of holes between the atoms in the crystalline lattice. An **octahedral hole**, shown in Figure 24.8 ◄, exists in the middle of six atoms on two adjacent closest-packed sheets of metal atoms. The hole is located directly above the center of three closest-packed metal atoms in one sheet and below the three metal atoms in the adjacent sheet. This configuration of metal atoms is identical to a regular octahedral configuration that consists of four atoms in a square plane with one atom above and one atom below the square.

We can calculate the size of an octahedral hole by determining the size of a hole on a square of four atoms in a plane (Figure 24.9 ◄). Any of the four corners of the square is the 90° angle of a right triangle formed from the adjacent two sides and the diagonal line that runs across the center of the square. According to the Pythagorean theorem, the length of the diagonal line ($c$) is related to the length of the sides ($a$ and $b$) as follows:

$$c^2 = a^2 + b^2 \qquad [24.1]$$

If we let $r_m$ equal the metal atom radius, the lengths of the sides are $2r_m$. We can then substitute $2r_m$ for both $a$ and $b$:

$$c^2 = (2r_m)^2 + (2r_m)^2$$

$$c^2 = 8r_m^2$$

$$c = 2.828r_m \qquad [24.2]$$

We can see from Figure 24.9 that the length of the diagonal ($c$) is twice the radius of the atoms plus twice the radius of the hole:

$$c = 2r_m + 2r_{hole} \qquad [24.3]$$

Combining Equations 24.2 and 24.3, we get the important result:

$$2.828r_m = 2r_m + 2r_{hole}$$

$$0.828r_m = 2r_{hole}$$

$$r_{hole} = 0.414r_m \qquad [24.4]$$

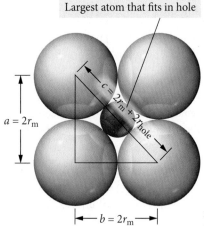

▲ **FIGURE 24.9 A Different View of an Octahedral Hole** An octahedral hole can be viewed as the area in the middle of a square plane of atoms, with one additional atom above the hole and one additional atom below the hole, accounting for the six close atoms. The diagonal of the square is equal to the radius of the two corner atoms plus the diameter of the hole. The Pythagorean theorem relates the length of the diagonal to the lengths of the sides of the square.

The octahedral hole, surrounded by six metal atoms, has a radius that is 41.4% of the metal atom radius. By contrast, the hole in the center of a cube in the *simple cubic* structure has a radius that is 73% of the metal atom radius. The *number* of octahedral holes in a closest-packed structure is equal to the number of metal atoms.

The second type of interstitial hole in a closest-packed structure is a **tetrahedral hole** (first introduced in Section 12.5), which forms directly above the center point of three closest-packed metal atoms in one plane and below a fourth metal atom located directly above the center point in the adjacent plane (Figure 24.10 ▼). The number of tetrahedral holes in a closest-packed structure is equal to *twice* the number of metal atoms. Because this hole is surrounded by only four atoms, the hole is smaller than the octahedral interstitial hole. We can apply geometric considerations similar to those used previously for the octahedral hole to determine that the tetrahedral hole has a radius that is 23% of the metal atom radius.

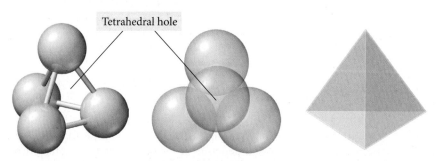

▲ **FIGURE 24.10 A Tetrahedral Hole in a Closest-Packed Crystal** Tetrahedral holes are found in a closest-packed structure. The tetrahedral hole is surrounded by four of the atoms in the closest-packed structure.

| TABLE 24.3 Formulas of Several Interstitial Alloys | | | |
|---|---|---|---|
| **Compound** | **Type of Interstitial Hole Occupied** | **Fraction of Holes Occupied** | **Formula** |
| Titanium carbide | Octahedral | All | TiC |
| Molybdenum nitride | Octahedral | One-half | $Mo_2N$ |
| Tungsten nitride | Octahedral | One-half | $W_2N$ |
| Manganese nitride | Octahedral | One-quarter | $Mn_4N$ |
| Palladium hydride | Tetrahedral | One-quarter | $Pd_2H$ |
| Titanium hydride | Tetrahedral | All | $TiH_2$ |

Interstitial alloys form when small nonmetallic atoms fit within the octahedral or tetrahedral holes of the crystalline lattice of the metal. The formulas for these alloys depend both on the type of hole occupied by the nonmetallic atom and on the fraction of holes occupied. For example, titanium and carbon form an alloy with a closest-packed structure for titanium in which all of the octahedral holes are filled with carbon atoms. Since the number of octahedral holes in a closest-packed structure is equal to the number of atoms in the structure, the ratio of carbon atoms to titanium atoms is 1:1 and the corresponding formula is TiC. In the compound formed between molybdenum and nitrogen, by contrast, only one-half of the octahedral holes in the closest-packed structure of Mo are filled with N. Therefore, the formula for this compound is $Mo_2N$. Table 24.3 lists the formulas and relative number of holes filled for several different interstitial alloys.

**Interstitial Alloys**  An interstitial alloy contains a nonmetal (X) that occupies one-eighth of the tetrahedral holes of the closest-packed structure of the metal (M). What is its formula?

CONCEPTUAL
CONNECTION **24.1**

## 24.5  Sources, Properties, and Products of Some of the 3*d* Transition Metals

In this section we examine major sources, interesting properties, and important products of several of the 3*d* transition metals, specifically, titanium, chromium, manganese, cobalt, copper, nickel, and zinc. We also survey the different metallurgical methods used to separate and refine them. The variety of uses for these metals reflects their varied structures and properties.

### Titanium

Titanium is the ninth most abundant element in Earth's crust and the fourth most abundant metal. Titanium was discovered in 1791, but the pure metal was not isolated until 1910. The principal minerals of titanium are rutile ($TiO_2$) and ilmenite ($FeTiO_3$). Another source of titanium is coal ash, the residue from burned coal. Black shiny ilmenite is found in granite deposits along the North Atlantic coast, often within silica sand. The mineral ilmenite is magnetic due to the presence of $Fe^{2+}$ ions, so it can be separated with a magnet from the nonmagnetic silica. The separated mineral is heated in the presence of carbon under an atmosphere of chlorine gas, forming $TiCl_4$, a volatile gas that can be isolated:

$$FeTiO_3(s) + 3\,Cl_2(g) + 3\,C(s) \longrightarrow 3\,CO(g) + FeCl_2(s) + TiCl_4(g)$$

The $TiCl_4$ gas is reacted with hot magnesium metal turnings (shaved pieces of magnesium), forming elemental titanium—a solid sponge material:

$$TiCl_4(g) + 2\,Mg(s) \longrightarrow 2\,MgCl_2(l) + Ti(s)$$

Titanium is very reactive, readily oxidizing in the presence of oxygen and even nitrogen. Consequently, the elemental titanium is **arc-melted**—a method in which the solid metal is melted with an arc (an electrical discharge) from a high-voltage electric source in a controlled atmosphere to prevent oxidation—and then collected in a water-cooled copper pot. Because of titanium's high reactivity, any further processing must be done in a protective atmosphere of an inert gas, such as Ar, to prevent oxidation.

Despite its reactivity, solid titanium is highly resistant to corrosion in air, acid, and seawater because it quickly reacts with oxygen to form an oxide that coats the surface, preventing further oxidation of the underlying metal. Consequently, titanium is often used in the production of ship components such as propeller shafts and rigging. Titanium is also very strong and light; it is stronger than steel and less than half as dense. It is denser than aluminum but twice as strong. For these reasons, titanium is used in the airline industry for the production of jet engine parts. When titanium is alloyed with 5% aluminum and trace amounts of Fe, Cr, and Mo, the resulting metal retains its strength under higher temperature.

The most common use of titanium, however, is as titanium(IV) oxide ($TiO_2$), which forms a clear crystal but a brilliant white powder. Most white paints include $TiO_2$, which is far less toxic than $PbO_2$, which was used in older paints. Titanium(IV) oxide is made by reacting sulfuric acid with ilmenite, which dissolves the titanium into solution:

$$FeTiO_3(s) + 3\,H_2SO_4(l) \longrightarrow FeSO_4(aq) + Ti(SO_4)_2(aq) + 3\,H_2O(l)$$

Neutralizing the solution with strong base forms titanium(IV) oxide:

$$Ti^{4+}(aq) + 4\,OH^-(aq) \longrightarrow TiO_2(s) + 2\,H_2O(l)$$

Calcination of the oxide dries it to form $TiO_2$ rutile crystals. Large rutile crystals are sometimes used as gems because they resemble diamonds.

## Chromium

The name *chromium* comes from the Greek root *chroma*, which means color. The different compounds of chromium are brightly colored (Table 24.4; Figure 24.11 ◀). The main ore source of chromium is chromite, $FeCr_2O_4$. No appreciable sources of chromium ores exist in the United States. Chromium metal is produced by reducing chromium ore with aluminum:

$$3\,FeCr_2O_4(s) + 8\,Al(s) \longrightarrow 4\,Al_2O_3(s) + 6\,Cr(s) + 3\,Fe(s)$$

Metallic chromium is a white, hard, lustrous, and brittle metal. The metal readily dissolves in acids such as hydrochloric acid and sulfuric acid, but it does not dissolve in concentrated nitric acid (because nitric acid forms an oxide on the surface of the chromium that resists further reaction). The primary use of chromium is in the production of steel alloys called *stainless steels*. Reducing chromite with carbon produces ferrochrome, an alloy that is added to steel:

$$FeCr_2O_4(s) + 4\,C(g) \xrightarrow{\text{heat}} Fe_xCr_y(s) + 4\,CO(g)$$

▲ **FIGURE 24.11 Chromium Compounds** Chromium compounds tend to be brightly colored.

| TABLE 24.4 The Colors of Various Chromium Compounds | |
|---|---|
| **Compound** | **Color** |
| Chromates ($CrO_4{}^{2-}$) | Yellow |
| Chromium(II) iodide | Red-brown |
| Chromium(III) iodide | Green-black |
| Chromium(II) chloride | White |
| Chromium(III) chloride | Violet |
| Dichromates ($Cr_2O_7{}^{2-}$) | Orange |
| Chromium(III) oxide | Deep green |
| Chrome alum | Purple |
| Chromium(VI) oxychloride ($CrO_2Cl_2$) | Dark red |
| Chromium(II) acetate | Red |

The chromium reacts with any oxygen in the steel to protect the iron from rusting. Chromium compounds were also used extensively in metal coatings, such as paints because the chromium helps to rustproof the underlying metal, and in wood preservatives because these compounds kill many of the bacteria and molds that rot wood. Chromium compounds are finding less use today, however, because of their toxicity and potential carcinogenicity. Nonetheless, because of their great corrosion resistance, chromate coatings are still used on large outdoor steel structures such as bridges, and chromate paints are used to mark streets (on the pavement) and on street signs.

Because chromium has the electron configuration $[Ar] 4s^1 3d^5$, with six orbitals available for bonding, it can have oxidation states from +1 to +6. Low-oxidation-state chromium exists primarily as the *cation* in salt compounds such as $Cr(NO_3)_3$ and $CrCl_3$. High-oxidation-state chromium occurs within the polyatomic *anions* of salts. The most important compounds with chromium in the +6 oxidation state are the chromates and dichromates. In an acidic solution (below pH 6), the orange-red dichromate ion $Cr_2O_7^{2-}$ is more stable. In more basic solutions (above pH 6), the yellow chromate ion $CrO_4^{2-}$ dominates. Adding acid to a solution containing the chromate ion produces the dichromate ion:

$$2 H^+(aq) + 2 CrO_4^{2-}(aq) \longrightarrow Cr_2O_7^{2-}(aq) + H_2O(l)$$

The chromate ion has a tetrahedral arrangement of oxygen atoms around a chromium atom. The dichromate ion has one bridging oxygen between the two tetrahedrons surrounding the chromium ions:

16.2 nm   17.7 nm

Chromate ($CrO_4^{2-}$)               Dichromate ($Cr_2O_7^{2-}$)

The high-oxidation-state chromates and dichromates are very strong oxidizing agents (that is, they are easily reduced). Consequently, they are used as coatings on other metal surfaces to prevent oxidation. The chromates react with the atoms on the surface of the other metals, forming a strongly bonded gel-like film. The film is nonmetallic and bonds very effectively with paint and resins that are applied over the film. We can see evidence of the oxidizing power of dichromate in the ability of ammonium dichromate to sustain combustion without any additional oxygen. Once ignited, ammonium dichromate burns with visible flames, giving off a smoke of green chromium oxide dust:

▲ Ammonium dichromate does not require additional oxygen to burn.

$$(NH_4)_2Cr_2O_7(s) \xrightarrow{heat} 4 H_2O(g) + N_2(g) + Cr_2O_3(s)$$

## Manganese

Manganese, with the electron configuration $[Ar]4s^2 3d^5$, exhibits a wide range of oxidation states, from +1 to +7. The most common natural sources of manganese are pyrolusite ($MnO_2$), hausmannite ($Mn_3O_4$), and rhodochrosite ($MnCO_3$) minerals. Calcination of rhodochrosite produces manganese(IV) oxide:

$$MnCO_3(s) + \tfrac{1}{2}O_2(g) \xrightarrow{heat} MnO_2(s) + CO_2(g)$$

The manganese(IV) oxide or pyrolusite minerals can react with active metals such as Al or Na to produce the elemental metal:

$$3 MnO_2(s) + 4 Al(s) \longrightarrow 3 Mn(s) + 2 Al_2O_3(s)$$

Pyrolusite, however, is often found as an impure mineral, containing mixtures of MnO and $Fe_2O_3$. Heating the mineral in the presence of carbon can reduce the mineral, forming an alloy of manganese and iron called ferromanganese:

$$MnO_2(s) + MnO(s) + Fe_2O_3(s) + n\,C(s) \longrightarrow Mn_xFe_yC_z(s) + n\,CO(g)$$

Ferromanganese generally contains about 5–6% carbon and is used as an alloying material in steel production. Manganese is added to steel alloys to change the physical properties of the steel. For example, manganese makes the steel easier to deform at high temperatures. Adding manganese to steel helps the rolling and forging steps of steel production. Steel alloys containing about 12% manganese are used in military armor and industrial applications such as bulldozer blades. Manganese also strengthens copper, aluminum, and magnesium alloys.

Manganese is a reactive metal that dissolves in most acids. When heated in the presence of air, it forms the various manganese oxides, including MnO, $Mn_3O_4$, and $MnO_2$. When heated in pure oxygen, the high-oxidation-state oxide $Mn_2O_7$ also forms. At high-oxidation states, the manganese compounds are good oxidizing agents. Dissolving $MnO_2$ into a solution of hydrochloric acid oxidizes the chloride and produces chlorine gas:

$$MnO_2(s) + 4\,HCl(aq) \longrightarrow Cl_2(g) + 2\,H_2O(l) + MnCl_2(aq)$$

The permanganate ion ($MnO_4^-$) has an oxidation state of +7 for Mn and is also an important oxidizing agent. The permanganate ion can even oxidize hydrogen peroxide, which is itself used as an oxidizing agent:

$$2\,MnO_4^-(aq) + 3\,H_2O_2(aq) + 2\,H^+(aq) \longrightarrow 2\,MnO_2(s) + 3\,O_2(g) + 4\,H_2O(g)$$

The compound $MnO_2$ is a glass additive. By itself, $MnO_2$ is either a brown or black crystal, depending on the degree of hydration. Yet, when added to a silica glass, it imparts a pink color to the glass. The pink color is useful because it counteracts the green color often seen in glass, which is due to small concentrations of impure iron oxides. In other words, $MnO_2$ is added to glass to "decolorize" it. Old manganese-containing glass that has been exposed to UV light for over 100 years develops a slight purple tint. This purple color is due to the oxidation of $MnO_2$ to Mn(VII) oxides.

## Cobalt

Cobalt ore is often found within the ores of other metals, such as iron, nickel, lead, and silver. Cobalt's most common ores are sulfide minerals, such as cobaltite (CoAsS), which is collected as a by-product in the extraction processes of the other metal ores. No large deposits of cobalt ores occur in the United States, but surveys indicate that there are cobalt-rich deposits in the Pacific Ocean near the Hawaiian Islands.

Cobalt, like iron and nickel, is **ferromagnetic** and is important in the production of magnets. The atoms in ferromagnetic materials contain unpaired electrons like the paramagnetic materials described in Section 8.7. In ferromagnetic materials, however, these electrons can all align with their spins oriented in the same direction, creating a permanent magnetic field.

Magnets are increasingly important in industrial and military applications. In the United States, concerns have arisen over the lack of domestic sources of cobalt for these applications. New and stronger magnetic materials that do not require cobalt have been developed, but these require neodymium, which is found mostly in China.

Cobalt is also an important additive for high-strength steels. Carbaloy, a mixture composed primarily of cobalt metal with grains of tungsten carbide, is a very tough material. The strength of the cobalt and the hardness of the carbides make carbaloy a good material for industrial cutting and abrasion. Cobalt also forms many compounds with brilliant blue colors and is used in making pigments and inks. Cobalt compounds are essential for health because cobalt is the metal in vitamin $B_{12}$, which prevents anemia.

# Copper

Copper, which can be found in its elemental form, was one of the first elements to be isolated and used by humans—copper products have been known for over 10,000 years. Ancient civilizations used copper to form tools. The earliest known artifacts produced by the smelting of copper are at a site in Tepe Yahya, Iran, dating from about 3800 B.C.E. The discovery of **bronze**, a copper and tin alloy, improved tool making because copper's bronze alloys are stronger, and they resist wear and corrosion better than pure copper.

The most important copper ores are chalcopyrite ($CuFeS_2$) and malachite [$Cu_2(OH)_2CO_3$]. Copper ores often occur near deposits of elemental copper, providing early humans with an easy way to locate copper ores. The electrometallurgical production of pure copper from chalcopyrite is described in Section 24.3.

The high natural abundance and generally high concentration of copper and copper ores make copper an economical choice for many industrial applications. Today over 40% of copper products are made from recycled copper. The ease and low cost of recycling copper have allowed copper to remain an important industrial metal.

The high conductivity of copper is second only to that of silver, making copper the most important metal for electrical wires. Along with iron in steel, copper is now among the most widely used metals; it is used in electrical motors and devices, the electrical wiring that snakes though the walls of houses, and the electrical transmission network linking power sources to homes and industries all over the world. Because of its high heat conductivity, copper is also used as a heat exchange material—for example, in car radiators. In addition, copper is used to make pipes for water distribution. Copper pipes can be readily connected to each other with watertight seals by soldering. Copper displaced lead for use as water pipes because of the toxicity of lead. Today, because of its lower price and lighter weight, plastic is used for many water pipes.

Copper has a distinctive reddish color that can be polished to a beautiful metallic luster; it has been used in architecture, as a decorative metal in jewelry, and as a material for sculpture. Copper composes the decorative sheathings on many domes and rooftop art works. When exposed to the atmosphere and rain, copper can oxidize, forming a number of different compounds, such as malachite and brochanite ($Cu_4SO_4(OH)_6$), which have beautiful blue or green colors. Copper roofing shingles can be purchased as polished copper sheets or as sheets that have already been oxidized and exhibit a beautiful weathered *patina* (a coating that comes with age and use). Chemical compounds speed up the aging process and produce the desired patina, which takes 20 or more years to develop naturally.

Even though copper has many useful properties, it is not a very strong metal; therefore, alloys of copper with improved strength have been developed. Bronze was one of the first alloys ever produced. This alloy of copper and tin has been used for thousands of years because it could be made even in the low heat of a Stone Age campfire. **Brass**, another widely used alloy, contains copper and zinc. Many brass and bronze alloys also contain other metals to achieve certain physical properties. Some of the most important applications of bronze and brass are for plumbing fixtures, bearings, and art decorations. In addition, the tendency of brass and bronze alloys (unlike those of iron) not to spark when struck makes them useful in applications where sparks could be dangerous.

# Nickel

Most of the world's nickel comes from deposits in Ontario, Canada. These deposits are believed to have formed from a meteorite impact (most meteorites have high nickel content). The nickel occurs as a sulfide compound mixed with copper and iron sulfides. To produce nickel, sulfides are roasted in air to form metal oxides, and then they are reduced to the elemental metals with carbon. The metal mixture is heated in the presence of carbon monoxide, forming nickel carbonyl, which has a boiling point of 43°C and can be collected as a gas:

$$Ni(s) + 4\,CO(g) \xrightarrow{\text{heat}} Ni(CO)_4(g)$$

▲ Copper is used in a variety of applications.

▲ Copper wiring is used to conduct electricity in wires.

When the nickel carbonyl is heated past 200 °C, it decomposes back to nickel metal and carbon monoxide. This method of refining nickel is the *Mond process*.

Nickel metal is fairly unreactive and resistant to corrosion, characteristics that it shares with platinum and palladium. Consequently, nickel is used as an alloying metal in the production of stainless steels. Many nickel alloys are used for applications where corrosion resistance is important. For example, the alloy Monel contains 72% Ni, 25% Cu, and 3% Fe and is resistant to reaction with most chemicals. Monel does not react even with fluorine gas at room temperature. Nickel–steel alloys are used for armor plates, and elemental nickel is often plated onto other metals as a protective coating.

## Zinc

Elemental zinc was officially discovered in Europe in 1746 when calamine (zinc silicate) was reduced with charcoal to produce the metal. However, zinc had been used for many hundreds of years before this discovery because zinc ores and copper ores were used to form copper–zinc brass alloys. The main sources of zinc are ores composed of sphalerite ($ZnS$), smithsonite ($ZnCO_3$), and an oxide mixture of zinc, iron, and manganese called franklinite. These ores are roasted to form the oxides of the metals and then reduced with carbon to produce the elemental metals.

Zinc combines with many different metals to form useful alloys. As we have discussed, the combination of zinc and copper produces the brass family of alloys. The combination of zinc and nickel with copper produces alloys with a silver color called *German* or *silver brass*. Zinc is also used in solder alloys with low melting points.

*Galvanizing* a steel object (such as a nail) involves dipping the object into a molten bath of zinc. The zinc, which is more reactive than the iron in steel, preferentially oxidizes, forming a tough protective coating. Zinc compounds are also used to coat steel before applying other coatings. The zinc compounds, such as zinc phosphate, adhere strongly to the steel surface, forming rough crystals onto which other coatings, such as paint, adhere very well. If a painted steel surface is scratched, the underlying exposed metal is susceptible to rust, but the added zinc ions prevent this by migrating toward the defect and forming a protective zinc oxide coating.

Zinc and zinc compounds have generally been considered safe. Zinc additives in coatings have replaced many of the chromium and lead additives (both of which are toxic) that were previously used. Today, however, even zinc additives in coatings are being studied for environmental hazards. In Europe, all substances that contain zinc compounds must be labeled as potential polluters of environmental water. More environmentally safe organic compounds have been developed to replace the metallic anticorrosion additives, but these compounds are more expensive to produce. Opportunities abound for chemists to develop needed products that are both environmentally safe and economically viable.

▲ Zinc phosphate adheres strongly to steel surfaces, forming rough crystals onto which other coatings, such as paint, can be applied.

# Self-Assessment Quiz

**Q1.** Which metal is naturally found in its elemental state?
   a) gold
   b) titanium
   c) chromium
   d) iron

**Q2.** Which reaction is an example of pyrometallurgy?
   a) $Ti^{4+}(aq) + 4\,OH^-(aq) \rightarrow TiO_2(s) + 2\,H_2O(l)$
   b) $2\,MnCO_3(s) + O_2(g) \xrightarrow{\text{heat}} 2\,MnO_2(s) + 2\,CO_2(g)$
   c) $Ni(s) + 4\,CO(g) \longrightarrow Ni(CO)_4(g)$
   d) $3\,MnO_2(s) + 4\,Al(s) \longrightarrow 3\,Mn(s) + 2\,Al_2O_3(s)$

**Q3.** Based on the data shown here, which pair of metals is likely to form a substitutional alloy?

| Metal | Atomic Radius | Crystal Structure |
|-------|---------------|-------------------|
| Sc | 162 pm | Hexagonal closet packed |
| W | 135 pm | Body-centered cubic |
| Cu | 128 pm | Face-centered cubic |
| Cr | 128 pm | Body-centered cubic |

   a) Sc and W
   b) Cu and Cr
   c) Cu and Sc
   d) Cr and W

**Q4.** Two metals, A and B, form a substitutional alloy with the binary phase diagram shown here. What is the melting point of the alloy at a composition that is 75% A and 25% B?

a) 1000 °C
b) 1300 °C
c) 1450 °C
d) 2000 °C

**Q5.** Two metals, A and B, form a substitutional alloy with the binary phase diagram shown here. Determine the composition and relative amounts of the two phases present at 50% composition and 1500 °C.

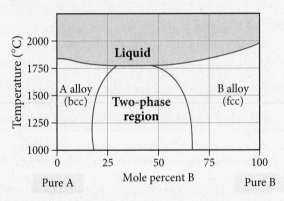

a) The A-rich bcc phase is 80% A and 20% B; the B-rich fcc phase is 37% A and 63% B; the mixture contains more of the A-rich fcc phase.

b) The A-rich bcc phase is 20% A and 80% B; the B-rich fcc phase is 63% A and 37% B; the mixture contains more of the B-rich fcc phase.

c) The A-rich bcc phase is 50% A and 50% B; the B-rich fcc phase is 50% A and 50% B; the mixture contains equal amounts of both phases.

d) The A-rich bcc phase is 80% A and 20% B; the B-rich fcc phase is 37% A and 63% B; the mixture contains more of the B-rich fcc phase.

**Q6.** An interstitial alloy contains a nonmetal (X) that occupies one-fourth of the octahedral holes in the closet-packed lattice of the metal (M). What is the formula for the alloy?

a) $M_4X$
b) $MX_4$
c) $M_2X$
d) $MX_2$

**Q7.** Which metal is commonly used to make electrical wires?

a) zinc
b) cobalt
c) manganese
d) copper

**Q8.** Which metal is commonly used to make parts for aircraft engines?

a) zinc
b) titanium
c) nickel
d) silver

**Q9.** Calculate the mass percent of chromium in the mineral chromite ($FeCr_2O_4$).

a) 83.97%
b) 46.46%
c) 86.78%
d) 36.13%

**Q10.** What is a common use for zinc?

a) galvanization of steel
b) lightweight aircraft parts
c) paint pigments
d) production of magnets

# Chapter 24 IN REVIEW

## Key Terms

**Section 24.1**
metallurgy (1102)

**Section 24.2**
minerals (1103)
ore (1103)

**Section 24.3**
extractive metallurgy (1104)
refining (1104)
gangue (1104)
pyrometallurgy (1104)
calcination (1105)

roasting (1105)
smelting (1105)
flux (1105)
slag (1105)
hydrometallurgy (1105)
leaching (1105)
electrometallurgy (1106)
powder metallurgy (1107)

**Section 24.4**
alloy (1108)
substitutional alloy (1108)

interstitial alloy (1108)
two-phase region (1110)
lever rule (1111)
octahedral hole (1112)
tetrahedral hole (1112)

**Section 24.5**
arc-melting (1114)
ferromagnetic (1116)
bronze (1117)
brass (1117)

## Key Concepts

### General Properties (24.2)

▶ Metals have many common physical properties, such as high conductivity of electricity and heat and high malleability and ductility. These properties vary among the different metals.

### Natural Distribution of Metals (24.2)

▶ Metals are unevenly distributed throughout Earth's crust. All the metals together compose about 25% of the mass of Earth's crust, but just a few key metals are abundant enough to each individually compose more than 1% of the crust.

▶ Few metals exist naturally in their elemental state; most are found in ores, rocks that contain a high concentration of metal-containing minerals.

▶ Most metal-containing minerals are oxides, sulfides, chlorides, carbonates, or more complex compounds.

### Metallurgical Processes (24.3)

▶ To be useful, metals have to be separated from the gangue, the unusable part of the ores, reduced to elemental metals and refined to be more pure.

▶ Extractive metallurgy is the general term for the processes, such as pyrometallurgy, hydrometallurgy, and electrometallurgy, that separate metal from ore.

▶ A method of forming metal components from micron-sized metal particles is powder metallurgy.

### Phase Diagrams (24.4)

▶ Elemental metals tend to crystallize in three crystalline structures: face-centered cubic, body-centered cubic, and hexagonal closest packed.

▶ When two types of metal atoms bond together, they form an alloy.

▶ A binary phase diagram is a graphical representation of the phases and crystal types of alloys present at different compositions and temperatures.

▶ If the two metals in an alloy are similar in size and have the same crystal structure, they tend to form a miscible solid solution, which means that they can form an alloy at any composition ratio. If the two metals are dissimilar in size or crystal structure, the solubility of one atom in the other's crystal structure is often limited. At certain compositions two different crystals can coexist in equilibrium; this is called a two-phase region. The lever rule determines which phase is present in a greater proportion.

### Types of Alloys (24.4)

▶ There are two different types of alloys, substitutional and interstitial.

▶ A substitutional alloy is a mixture in which one type of metal atom replaces another type of metal atom in the crystal structure.

▶ In an interstitial alloy, one type of atom (either a metal or nonmetal) fits into the holes within the crystal structure of the metal. Interstitial alloys can be made with different atoms filling different fractions of the different types of holes.

## Key Equations and Relationships

Lever Rule (24.4)

In a two-phase region on a phase diagram, two different crystal structures coexist in equilibrium. Whichever phase has a composition closer to the overall composition of the alloy is the phase present in the larger relative amount.

# Key Learning Outcomes

| CHAPTER OBJECTIVES | | ASSESSMENT |
|---|---|---|
| Using a Phase Diagram to Determine the Composition, Relative Amounts, and Phases Present (**24.4**) | | Examples 24.1, 24.2, 24.3, 24.4  For Practice 24.1, 24.2, 24.3, 24.4  Exercises 35–38 |
| Determining the Composition of an Interstitial Alloy from the Occupancy of the Interstitial Holes (**24.4**) |  Octahedral hole   60% | Exercises 41, 42 |

# EXERCISES

## Review Questions

1. Why is vanadium present in oil from some sources?

2. List three categories of metallurgical processes.

3. Why is Ni not considered a common metal even though it composes over 2% of the total mass of Earth and only Fe and Mg have a higher percent composition of Earth's total mass?

4. Metal elements are found in both minerals and ores. Describe the difference between a mineral and an ore.

5. Ores contain minerals and gangue. Describe the difference between a mineral and gangue.

6. Calcination, roasting, and smelting are three pyrometallurgical processes. Compare and contrast the processes.

7. What compound has been historically used to leach gold from gold ores? Why is this process being discontinued?

8. Name three advantages of making metal components from the powder metallurgical process.

9. Describe the difference between body-centered cubic and face-centered cubic structures.

10. Describe the difference between a substitutional alloy and an interstitial alloy.

11. Why was copper one of the first metals that humans used?

12. Describe why bronze was one of the first alloys that humans used.

13. Both brass and bronze contain copper. Describe the difference between these two alloys.

14. List the properties of copper that make it appropriate for electric wires and the properties that make it appropriate for water pipes.

## Problems by Topic

### The General Properties and Natural Distribution of Metals

15. Describe three typical properties of metals.

16. Describe whether each property is generally higher or lower for metals compared to nonmetals.
    a. thermal conductivity
    b. electrical resistivity
    c. transparency
    d. ductility

17. List four metal elements that each individually compose more than 1% of Earth's crust.

18. List four metals that occur as elements in their natural state in the earth's crust.

19. List the name and formula of one important mineral source for each of these metals: Fe, Hg, V, and Nb.

20. List the name and formula of one important mineral source for each of these metals: Ti, Zn, U, and Ta.

### Metallurgical Processes

21. Two ores of magnesium are $MgCO_3$ and $Mg(OH)_2$. Write balanced equations for the calcination of these two minerals to form MgO.

22. Two ores of copper are CuO and CuS. Write balanced equations for the roasting of CuO with C to form Cu metal and the roasting of CuS with $O_2$ to form CuO.

23. Give the definition of a flux and identify the flux in the reaction.

$$SiO(s) + MgO(s) \longrightarrow MgSiO_3(l)$$

24. Give the definition of a slag and identify the slag in the reaction.

$$SiO_2(s) + MgO(s) \longrightarrow MgSiO_3(l)$$

25. Provide a general description of how hydrometallurgy is used to extract metals from ores.

26. Provide a general description of how electrometallurgy is used to extract metals from ores.

27. How is $Al_2O_3$ separated from other oxides using the Bayer process? What soluble form of aluminum forms from $Al_2O_3$ during the Bayer process?

28. When copper is purified using an electrochemical cell, which electrode has the pure copper and which electrode has the impure copper? Explain how gold is obtained from this process.

29. Describe the difference between sponge powdered iron and water-atomized powdered iron.

30. Describe the difference in the processing of sponge powdered iron and water-atomized powdered iron.

## Metal Structures and Alloys

31. Determine the composition of each vanadium alloy.
    a. One-half of the V atoms are replaced by Cr atoms.
    b. One-fourth of the V atoms are replaced by Fe atoms.
    c. One-fourth of the V atoms are replaced by Cr atoms, and one-fourth of the V atoms are replaced by Fe atoms.

32. Determine the composition of each cobalt alloy.
    a. One-third of the Co atoms are replaced by Zn atoms.
    b. One-eighth of the Co atoms are replaced by Ti atoms.
    c. One-third of the Co atoms are replaced by Zn atoms, and one-sixth of the Co atoms are replaced by Ti atoms.

33. Referring to Table 24.2, explain why you might expect Cr and Fe to form miscible alloys.

34. Referring to Table 24.2, explain why you might expect Co and Cu to not form miscible alloys.

35. Determine the composition and phases present at points A and B on the Cr–Fe phase diagram.

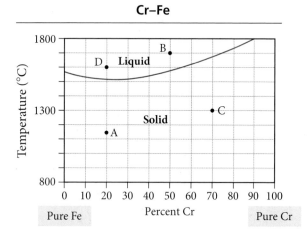

36. Determine the composition and phase present at points C and D on the Cr–Fe phase diagram (see Problem 35 for diagram).

37. Determine the composition, relative amounts, and phases present at points A and B on the Co–Cu phase diagram.

38. Determine the composition, relative amounts, and phases present at points C and D on the Co–Cu phase diagram (see Problem 37 for phase diagram).

39. The elements Mn and Si are added to steel to improve its properties in electric motors, and some C is often found as a detrimental impurity. Which of these elements fills interstitial holes in the Fe lattice, and which substitutes for the Fe in the lattice?

40. The elements Si and P are added to steel to improve its properties in electric motors, and some N is often found as a detrimental impurity. Which of these elements fills interstitial holes in the Fe lattice, and which substitutes for the Fe in the lattice?

41. Determine the formula for each interstitial alloy.
    a. Nitrogen occupies one-half of the octahedral sites of a closest-packed Mo structure.
    b. Hydrogen occupies all of the tetrahedral sites of a Cr closest-packed structure.

42. Determine the formula for each interstitial alloy.
    a. Nitrogen occupies one-fourth of the octahedral sites of a closest-packed Fe structure.
    b. Hydrogen occupies one-half of the tetrahedral sites of a Ti closest-packed structure.

## Sources, Properties, and Products of Some of the 3*d* Transition Metals

43. Identify the metal found in each mineral.
    a. sphalerite
    b. malachite
    c. hausmannite

44. Name at least one important mineral that is a source for each metal.
    a. Fe
    b. Co
    c. Cr

45. Calculate the heat of reaction ($\Delta H^{\circ}_{rxn}$) for the calcination of rhodochrosite. ($\Delta H^{\circ}_{f}$ for rhodochrosite is $-894.1$ kJ/ mol.)

46. The extraction of Mn from pyrolusite with aluminum produces pure Mn metal and $Al_2O_3$. Calculate the heat of reaction ($\Delta H^{\circ}_{rxn}$).

47. Describe the effects of adding Cr to steel and give a use for chromium–steel alloys.

48. Describe the effects of adding Mn to steel and give a use for manganese–steel alloys.

49. Calculate the mole percent and mass percent of Ti in the minerals rutile and ilmenite.

50. Calculate the mole percent and mass percent of Mn in the minerals pyrolusite and rhodochrosite.

51. Why is it important to use an inert atmosphere to surround the metal when arc-melting titanium?

52. Titanium is a very reactive metal. Explain why titanium has a high corrosion resistance to seawater and can be used for the production of ship components.

53. Which compound of Ti is the most important industrial product of titanium metal? Describe an application for this compound.

54. Describe how Zn is used to protect the surface of steel products. What is the name of this process?

55. Describe the Bayer process.

56. Describe the Mond process.

57. Which metals are found in carbaloy steel?

58. Which metals are found in Monel steel?

# Cumulative Problems

**59.** After $2.0 \times 10^4$ kg of an ore that contains 0.051% ilmenite is mined, the percent yield from extracting and refining the metals from the mineral is 87% for the iron and 63% for the titanium. Calculate the mass of iron and titanium produced from the ore.

**60.** Calculate the mass of aluminum metal that is needed to produce Cr metal from $5.00 \times 10^2$ g of chromite. Calculate how many grams of Cr metal are produced from $5.00 \times 10^2$ g of chromite.

**61.** How many lattice atoms surround a tetrahedral hole and an octahedral hole in a closest-packed structure? Describe which hole site is larger and explain why.

**62.** Explain why the crystal structure and atomic size of the two elements are important factors in a two-component phase diagram.

**63.** Why does Mn form compounds with higher oxidation states than Cr?

**64.** Why is $MnO_2$ added to silica glass?

**65.** Co, Fe, and Ni are ferromagnetic. What is the difference between ferromagnetic and paramagnetic?

**66.** Is the chromate ion or dichromate ion more stable in an acidic solution? Which ion has a higher Cr:O ratio? Which ion has a higher oxidation state for Cr?

**67.** The first ionization energies of iron, cobalt, nickel, and the first three platinum metals are all about the same, but the first ionization energies of osmium, iridium, and platinum are substantially greater. Suggest an explanation for this observation.

**68.** Although both the Group IA and IB metals have a half-filled $s$ subshell, the IB metals have markedly higher first ionization energies, densities, and melting points, and markedly lower second and third ionization energies than the corresponding IA metals. Explain these observations.

# Challenge Problems

**69.** Iron powder is placed into a tall cylinder-shaped die and pressed from the top and bottom to make a cylinder-shaped pressed part with a height of 5.62 cm and a radius of 4.00 cm. The density of the iron powder before it was pressed was 2.41 g/mL. The density of a pressed iron part is 6.85 g/mL. The density of pure solid iron is 7.78 g/mL.
   a. Calculate the original height of the powder before it was pressed.
   b. Calculate the theoretical height of the pressed component if it could be pressed to the same density of pure iron.
   c. What percentage of the pressed iron part is composed of voids between the iron particles?

**70.** When a part is made by pressing together powdered metal, no metal is wasted. In contrast, metal is typically scrapped after a metal part is cut from a solid metal plate. If a circular part with a diameter of 10.0 cm is made from an original shape of a square with a side length of 10.0 cm, calculate the percentage of the metal that is thrown away as scrap. If the circular shape also has a circular hole with a diameter of 6.0 cm, calculate the percentage of the metal that is thrown away as scrap.

**71.** Hydrogen can be in both the octahedral and tetrahedral holes for lanthanum. Determine the percentage of the holes that are filled if the formula is $LaH_{2.76}$.

**72.** Tin exists in two allotropic forms. Gray tin has a diamond structure, and white tin has a close-packed structure. Predict which allotrope is (a) denser, (b) a conductor of electricity. Predict the valence electronic configuration of tin in each allotrope.

**73.** Find the equilibrium constant at 298 K for the reaction.

$$[Ag(CN)_2]^-(aq) + Cu(s) \rightleftharpoons [Cu(CN)_2]^-(aq) + Ag(s)$$

The $K_f$ for $[Cu(CN)_2]^- = 1.0 \times 10^{24}$ and the rest of the data needed are in Appendix II.

**74.** Find the equilibrium constant at 298 K for the reaction.

$$2\,[Cu(NH_3)_2]^+(aq) \rightleftharpoons [Cu(NH_3)_4]^{2+}(aq) + Cu(s)$$

The $K_f$ for $[Cu(NH_3)_2]^+ = 6.3 \times 10^{10}$ and the rest of the data needed are in Appendix II.

**75.** Propose a chemical procedure to extract cobalt from the mineral cobaltite, CoAsS. What are some of the hazards of such a procedure?

# Conceptual Problems

**76.** Why are metals such as Ni and Co economical to mine and use in industrial processes, even though they have a very low natural abundance in Earth's crust?

**77.** Explain why metals such as Au and Ag are found in their elemental states in nature, but metals like Na and Ca are always found in compounds in nature.

**78.** What allows some pairs of metals to form alloys that have the same structure at any composition, while other pairs of metals form alloys with structures that depend on the relative composition of the two metals?

# Questions for Group Work

Discuss these questions with the group and record your consensus answer.

**79.** Assign each element in the first row of the transition metals, scandium through zinc, to a group member. Have each group member look up on the Internet or in a reputable chemistry reference book the electrical resistivity for his or her element(s). Agree on the units and temperature. As a group, prepare a graph of resistivity versus atomic number. What is the general trend in resistivity as the atomic number increases? What atomic properties might account for this trend? Are there any notable exceptions?

**80.** Create a table comparing and contrasting pyrometallurgy, hydrometallurgy, and electrometallurgy. Include a definition, an application or two, and any relevant notes for each type of metallurgy. Once your table is complete, pass it around your group asking each group member to read one cell from your table. Have other group members take turns trying to identify the type of metallurgy from the fact that is read aloud.

**81.** Would you expect an alloy of iron and vanadium to be substitutional or interstitial? Explain your choice. Create a phase diagram for the binary alloy of iron and vanadium based on the tabulated data provided.

| % V | Melting Point |
|-----|---------------|
| 0 | 1530 |
| 20 | 1480 |
| 40 | 1470 |
| 60 | 1560 |
| 80 | 1720 |
| 100 | 1900 |

**82.** Assign each group member a different interstitial alloy from Table 24.3. Each group member should look up a few interesting facts for his or her alloy, including how it is used. Present findings, along with a description of each alloy's structure, to the group.

**83.** Working individually, create a fill-in-the-blank question by rewriting a characteristic sentence or chemical reaction from Section 24.5, but leaving out the name of the metal focused on in the content selected. Take turns sharing questions with the rest of the group. Take turns in the group trying to fill in the blank with the correct metal.

# Data Interpretation and Analysis

## Soldering with Tin–Lead Alloys

**84.** Alloys of tin and lead are common solders. A *solder* is a metal alloy that melts at a relatively low temperature and connects two or more metal pieces. The technique of soldering is commonly used by plumbers to fuse copper water pipes and by electricians to affix electrical components to printed circuit boards. Table a lists tensile strength and hardness data for pure tin, pure lead, and two tin–lead alloys. *Tensile strength* is the maximum stress a material that is being stretched or pulled can withstand before breaking. *Hardness* on the Brinell Scale is a measure of a substance's resistance to indentation

when subjected to impact by a steel ball (the higher the number the greater the hardness).

The tin–lead phase diagram in Figure a ▼ contains a *eutectic point*—a point on the phase diagram representing the lowest melting point. The eutectic composition has a sharp melting transition. All other compositions of tin–lead melt over a range of temperatures, beginning at the *solidus* line (the line separating the solid phase and the liquid-solid region) and ending at the *liquidus* line (the line separating the liquid phase and the liquid-solid region). Between these lines, a mixture of liquid and solid

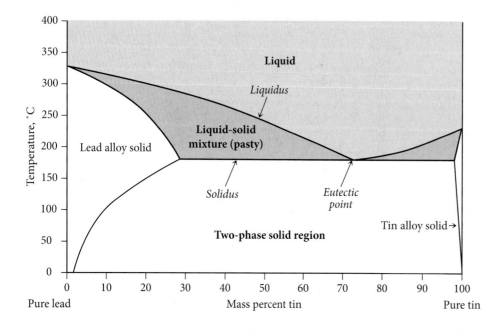

◄ **FIGURE a** **Tin–Lead Phase Diagram.**

exists. This mixture is described as "pasty" because it can be worked into an optimal shape around the connection point of the two base metal pieces.

| Substance | Tensile Strength, MPa | Hardness (Brinell Scale) |
|---|---|---|
| Pure tin | 13.2 | 4 |
| Pure lead | 12.3 | 4 |
| 63% tin–37% lead (mass%) | 51.5 | 17 |
| 50% tin–50% lead (mass%) | 41.2 | 14 |

TABLE a **Strength and Hardness of Tin, Lead, and Tin–Lead alloys**

Use the information provided to answer the following questions:

a. Based on the data provided in Table a, explain why tin–lead alloys are better solders than pure tin or pure lead.

b. Based on the phase diagram shown in Figure a, do tin and lead form a miscible solid solution at all compositions?

c. According to the phase diagram, what are the melting points of pure tin and pure lead?

d. What composition of tin and lead constitutes a eutectic mixture, and what is the melting point of the eutectic mixture?

e. Why is a 63% tin/37% lead (mass %) alloy commonly used for electrical soldering projects?

f. What is the melting range of a 50% tin/50% lead (mass %) alloy?

# Answers to Conceptual Connections

**Interstitial Alloys**

**24.1** $M_4X$. Because there are twice as many tetrahedral holes as metal atoms in a closest-packed structure, and one-eighth of them are occupied by nonmetal atoms, there must be one-quarter as many nonmetal atoms as metal atoms.

CHAPTER

# 25

# Transition Metals and Coordination Compounds

"Chemistry must become the astronomy of the molecular world."

—ALFRED WERNER (1866–1919)

The Dz² orbital

To:
The one in the highest orbital

In this chapter, we examine the chemistry of the transition metals and an important class of transition metal compounds called coordination compounds. We will see that coordination compounds form all of the types of isomers that we have studied so far, as well as some new types. In our examination of the transition metals, we draw on much of what we learned in Chapters 7 and 8 about electronic structure and periodicity. We also briefly revisit valence bond theory to understand bonding in coordination compounds, but we quickly shift to a different theory—called crystal field theory—that better explains many of the properties of these compounds. Transition metals and coordination compounds are important, not only because of their interesting chemistry, but because of their numerous applications. Coordination compounds are the basis for a number of therapeutic drugs, chemical sensors, and coloring agents. In addition, many biological molecules contain transition metals that bond in ways that are similar to coordination compounds. For example, the oxygen-carrying site on hemoglobin is an iron ion bonded partly to an amino acid in the hemoglobin molecule and partly to a flat molecule called a porphyrin. An oxygen molecule reversibly bonds to the iron and is transported throughout the body by blood flow.

▲ The color of ruby is caused by a splitting of the d-orbital energy levels in $Cr^{3+}$ by the host crystal.

## 25.1 The Colors of Rubies and Emeralds

Rubies are deep red and emeralds are brilliant green, yet the color of both gemstones is caused by the same ion, $Cr^{3+}$. The difference lies in the crystal that hosts the ion. Rubies are crystals of aluminum oxide ($Al_2O_3$) in which about 1% of the $Al^{3+}$ ions are replaced by $Cr^{3+}$ ions. Emeralds, by contrast, are crystals of beryllium aluminum silicate

▲ **Ruby and Emerald** The red color of ruby and the green color of emerald are both caused by $Cr^{3+}$.

$[Be_3Al_2(SiO_3)_6]$ in which a similar percentage of the $Al^{3+}$ ions is replaced by $Cr^{3+}$. The embedded $Cr^{3+}$ ion is red in the aluminum oxide crystal but green in the beryllium aluminum silicate crystal. Why?

The answer to this question lies in the effect that the host crystal has on the energies of the atomic orbitals in $Cr^{3+}$. Atoms in the crystal create a field around the ion—sometimes called the *crystal field*—that splits the five normally degenerate *d* orbitals into two or more levels. The color of each gemstone is caused by electron transitions between these levels. In rubies, the crystal field is stronger (and the corresponding splitting of the *d* orbitals greater) than it is in emeralds. Recall from Chapter 7 that the color of a substance depends on the colors *absorbed* by that substance, which in turn depends on the energy differences between the orbitals involved in the absorption. The greater splitting in ruby results in a greater energy difference between the *d* orbitals of $Cr^{3+}$, and consequently the absorption of a different color of light than in emerald.

The colors of several other gemstones are also caused by the splitting of the *d* orbitals in transition metal ions embedded within host crystals. For example, the red in garnet, which has $Mg_3Al_2(SiO_4)_3$ as a host crystal, and the yellow-green of peridot, which has $Mg_2SiO_4$ as a host crystal are both caused by electron transitions between *d* orbitals in $Fe^{2+}$. Similarly, the blue in turquoise, which has $[Al_6(PO_4)_4(OH)_8 \cdot 4\, H_2O]^{2-}$ as a host crystal, is due to transitions between the *d* orbitals in $Cu^{2+}$.

In this chapter, we examine the properties of the transition metals and their ions more closely. We also examine the properties of coordination compounds in some detail. We first discussed this common type of transition metal compound in Chapter 17 (see Section 17.8). In a coordination compound, bonds to a central metal ion split the *d* orbitals much as they are split in the crystals of gemstones. The theory that explains the splitting and the corresponding colors is **crystal field theory**, which we also explore in this chapter.

▲ **Garnet, Peridot, and Turquoise** The red in garnet and the yellow-green of peridot are both caused by $Fe^{2+}$. The blue of turquoise is caused by $Cu^{2+}$.

## 25.2 Properties of Transition Metals

Transition metals, the elements in the *d* block of the periodic table, are a study in similarities and differences. When we compare their properties with the varied properties of the main-group elements, they seem markedly similar. For example, almost all transition metals have moderate to high densities, good electrical conductivity, high melting points, and moderate to extreme hardness. Their similar properties are related to their similar electron configurations: they all have electrons in *d* orbitals that can be involved in metallic bonding. In spite of their similarities, however, each element is also unique, and the transition metals exhibit a wide variety of chemical behavior. Before we examine some of their periodic properties, let's review the electron configurations of these elements, first discussed in Chapter 8.

### Electron Configurations

Recall from Section 8.4 that, as we move to the right across a row of transition elements, electrons are added to $(n - 1)d$ orbitals (where *n* is the row number in the periodic table and also the quantum number of the highest occupied principal level). For example, as we move across the fourth-period transition metals, electrons are added to the 3*d* orbitals, as Table 25.1 shows.

In general, the ground state electron configuration for the first two rows of transition elements is [noble gas] $ns^2(n-1)d^x$ and for the third and fourth rows it is [noble gas] $ns^2(n-2)f^{14}(n-1)d^x$, where $x$ ranges from 1 to 10. Recall from Section 8.4, however, that because the $ns$ and $(n-1)d$ sublevels are close in energy, many exceptions occur. For example, in the first transition series of the $d$ block, the outer configuration is $4s^2 3d^x$ with two exceptions: Cr is $4s^1 3d^5$ and Cu is $4s^1 3d^{10}$. This behavior is related to the closely spaced $3d$ and $4s$ energy levels and the stability associated with a half-filled or completely filled $d$ sublevel.

Recall from Section 8.7 that the transition metals form ions by losing electrons from the $ns$ orbital *before* losing electrons from the $(n-1)d$ orbitals. For example, $Fe^{2+}$ has an electron configuration of [Ar] $3d^6$ because it has lost both of the $4s$ electrons to form the 2+ charge. Examples 25.1 and 25.2 review the steps in writing electron configurations for transition metals and their ions.

**TABLE 25.1 First-Row Transition Metal Orbital Occupancy**

| | 4s | 3d | | | | |
|---|---|---|---|---|---|---|
| Sc | ↑↓ | ↑ | | | | |
| Ti | ↑↓ | ↑ | ↑ | | | |
| V | ↑↓ | ↑ | ↑ | ↑ | | |
| Cr | ↑ | ↑ | ↑ | ↑ | ↑ | ↑ |
| Mn | ↑↓ | ↑ | ↑ | ↑ | ↑ | ↑ |
| Fe | ↑↓ | ↑↓ | ↑ | ↑ | ↑ | ↑ |
| Co | ↑↓ | ↑↓ | ↑↓ | ↑ | ↑ | ↑ |
| Ni | ↑↓ | ↑↓ | ↑↓ | ↑↓ | ↑ | ↑ |
| Cu | ↑ | ↑↓ | ↑↓ | ↑↓ | ↑↓ | ↑↓ |
| Zn | ↑↓ | ↑↓ | ↑↓ | ↑↓ | ↑↓ | ↑↓ |

## Procedure for...

### Writing Electron Configurations

| | **Example 25.1** Writing Electron Configurations for Transition Metals — Write the ground state electron configuration for Zr. | **Example 25.2** Writing Electron Configurations for Transition Metals — Write the ground state electron configuration for $Co^{3+}$. |
|---|---|---|
| Identify the noble gas that precedes the element and write it in square brackets. | **SOLUTION** [Kr] | **SOLUTION** [Ar] |
| Count down the periods to determine the outer principal quantum level—this is the quantum level for the $s$ orbital. Subtract one to obtain the quantum level for the $d$ orbital. If the element is in the third or fourth transition series, include $(n-2)f^{14}$ electrons in the configuration. | Zr is in the fifth period so the orbitals used are: [Kr] $5s4d$ | Co is in the fourth period, so the orbitals used are: [Ar] $4s3d$ |
| Count across the row to see how many electrons are in the neutral atom and fill the orbitals accordingly. | Zr has four more electrons than Kr. [Kr] $5s^2 4d^2$ | Co has nine more electrons than Ar. [Ar] $4s^2 3d^7$ |
| For an ion, remove the required number of electrons, first from the $s$ and then from the $d$ orbitals. | | $Co^{3+}$ has three fewer electrons than the Co atom. [Ar] $4s^0 3d^6$ or [Ar] $3d^6$ |

**FOR PRACTICE 25.1** Write the ground state electron configuration for Os.

**FOR PRACTICE 25.2** Write the ground state electron configuration for $Nb^{2+}$.

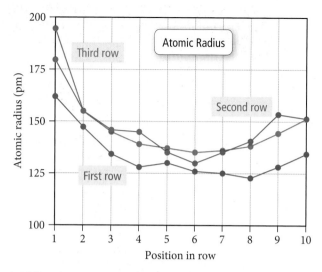

▲ **FIGURE 25.1 Trends in Atomic Radius** With the exception of a decrease in radius from the first to the second element, there is only a small variation in atomic radius across a row. There is a small and expected increase in radius from the first to the second transition row but virtually no difference in radius from the second to the third.

## Atomic Size

As we discussed in Section 8.6, for main-group elements, the size of atoms decreases across a period and increases down a column. For transition metals, however, there is little variation in size across a row (other than for the first two elements in each transition metal row, such as Sc and Ti in the first row), as Figure 25.1◄ illustrates. The reason for the difference is that, across a row of transition elements, the number of electrons in the outermost principal energy level (highest $n$ value) is nearly constant. As another proton is added to the nucleus with each successive element, another electron is added as well, but the electron goes into an $n - 1$ orbital. The number of outermost electrons thus stays the same, and the electrons experience a roughly constant effective nuclear charge as we move across the row, keeping the radii approximately constant.

Looking down a group, we see a small but expected increase in size from the first transition metal row to the second, but the size of elements in the third row is about the same as it is for those in the second row. This pattern is also different from that of the main-group elements, especially when we consider that in any given column, the third transition row has 32 more electrons than the second row. The reason that the third transition row elements are not larger is because 14 of the 32 electrons are in a $(n - 2)f$ sublevel, and while electrons in $f$ orbitals are in lower principal quantum levels, they are not very effective at shielding the outer electrons from nuclear charge. Consequently, the outer electrons are held more tightly by the nucleus, offsetting the typical increase in size between the periods—an effect called the **lanthanide contraction**.

CONCEPTUAL
**CONNECTION 25.1**

**Atomic Size** Which element has the larger atomic radius, Fe or W?

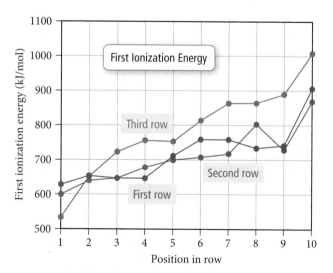

▲ **FIGURE 25.2 Trends in First Ionization Energy** First ionization energy generally increases across a row, following the main-group trend. However, in contrast to the main-group trend, the third transition row has a higher ionization energy than the first and second rows.

## Ionization Energy

The first ionization energies of transition elements follow the expected main-group periodic trend and slowly increase across a row (Figure 25.2◄), but the increase is smaller than for main-group elements. As we move down a group, we see that the third transition row generally has a higher ionization energy than do the first two rows, a trend counter to that observed in the main-group elements. In the transition elements the charge of the nucleus increases substantially from one row to the next, but there is only a small increase in atomic size between the first and second rows, and no increase in size between the second and third row. The outer electrons are therefore held more tightly in the third transition row than in the first two rows.

# Electronegativity

The electronegativity values of the transition metals, like their ionization energies, follow the main-group trend and slowly increase across a row, as shown in Figure 25.3 ▼. The increase is smaller than the increase that occurs in the main-group elements, but we expect that given the similarity in the sizes of the atoms. The trend in electronegativity values down a group (or column) is another example of the transition metals behaving differently from the main-group elements. The electronegativity values generally increase from the first transition row to the second, but there is no further increase for the third row. In the main-group elements, in contrast, we see a *decrease* in electronegativity as we move down a group. The difference is again caused by the relatively small change in atomic size as we move down a column for the transition elements, accompanied by a large increase in nuclear charge. One of the heaviest metals, gold (Au), is also the most electronegative metal. Its electronegativity value (EN = 2.4) is even higher than that of some nonmetals (EN of P is 2.1), and compounds of an $Au^-$ ion have been observed.

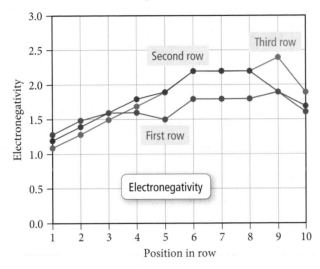

◀ FIGURE 25.3 **Trends in Electronegativity** The electronegativity of the transition elements generally increases across a row, following the main-group trend. However, in contrast to the main-group trend, electronegativity increases from the first transition row to the second. There is little electronegativity difference between the second and third transition rows.

# Oxidation States

Unlike main-group metals, which tend to exhibit only one oxidation state, the transition metals often exhibit a variety of oxidation states (Figure 25.4 ▼). The highest oxidation state for a transition metal is +7 for manganese (Mn). The electron configuration of manganese in this oxidation state corresponds to the loss of all the electrons in the 4s and 3d orbitals, leaving a noble gas electron configuration ([Ar]). This is the same configuration we see for all of the highest oxidation states of the elements to the left of Mn. To the right of manganese, the oxidation states are all lower, mostly +2 or +3. A +2 oxidation state for a transition metal is not surprising, as 4s electrons are readily lost.

> Metals in high oxidation states, such as +7, exist only when the metal is bound to a highly electronegative element such as oxygen; they do not exist as bare ions.

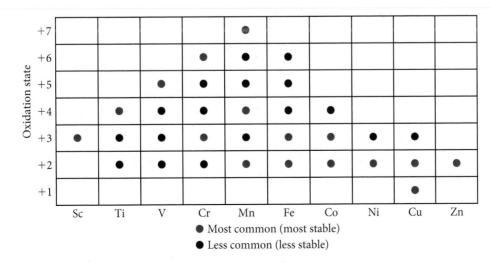

◀ FIGURE 25.4 **First-Row Transition Metal Oxidation States** The transition metals exhibit many more oxidation states than the main-group elements. These oxidation states range from +7 to +1.

# 25.3  Coordination Compounds

We discussed at the end of Chapter 17 that transition metals tend to form *complex ions*. A **complex ion** contains a central metal ion bound to one or more *ligands*. A **ligand** is a Lewis base (or electron donor) that forms a bond with the metal. When a complex ion combines with one or more *counterions* (ions of opposite charge that are not acting as ligands), the resulting neutral compound is a **coordination compound**.

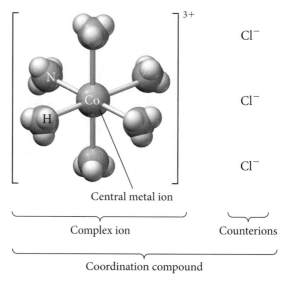

▲ **Complex Ion and Coordination Compound**   A coordination compound contains a complex ion and corresponding counterions. The complex ion contains a central metal atom coordinated to several ligands. The compound shown here is $[Co(NH_3)_6]Cl_3$.

The first coordination compounds were discovered in the early eighteenth century, but their nature was not understood until nearly 200 years later. Swiss chemist Alfred Werner (1866–1919) studied coordination compounds extensively—especially a series of cobalt(III) compounds with ammonia. The formulas for these compounds were then written as $CoCl_3 \cdot 6\,NH_3$, $CoCl_3 \cdot 5\,NH_3$ and, $CoCl_3 \cdot 4\,NH_3$. In 1893, Werner proposed that the central metal ion has two types of interactions that he named **primary valence** and **secondary valence**. The primary valence is the oxidation state on the central metal atom, and the secondary valence is the number of molecules or ions directly bound to the metal atom, called the **coordination number**. In $CoCl_3 \cdot 6\,NH_3$ the primary valence is +3, and it was discovered that the ammonia molecules are directly bound to the central cobalt, giving a coordination number of 6. Today we write the formula of this compound as $[Co(NH_3)_6]Cl_3$ to better represent the coordination compound as the combination of a complex ion, $Co(NH_3)_6{}^{3+}$, and three $Cl^-$ counterions.

We write the formulas of the other cobalt(III) compounds studied by Werner as $[Co(NH_3)_5Cl]Cl_2$ and $[Co(NH_3)_4Cl_2]Cl$. In these two cases, the complex ions are $Co(NH_3)_5Cl^{2+}$ (with two $Cl^-$ counterions) and $Co(NH_3)_4Cl_2{}^+$ (with one $Cl^-$ counterion), respectively. With this series of compounds, Werner demonstrated that the $Cl^-$ can replace $NH_3$ in the secondary valence. In other words, $Cl^-$ can act as a counterion, or it can bond directly to the metal as part of the complex ion.

The complex ion itself contains the metal ion in the center and the ligands—which can be neutral molecules or ions—arranged around it. We can think of the metal–ligand complex as a Lewis acid–base adduct (see Section 16.11) because the bond is formed when the ligand donates a pair of electrons to an empty orbital on the metal. For example, consider the reaction between the silver ion and ammonia:

Coordinate covalent bonds

Lewis acid    Lewis base    Lewis acid–base adduct

A bond of this type, which we first encountered in Section 10.6, is often referred to as a **coordinate covalent bond**. Ligands are therefore good Lewis bases and have at least one pair of electrons to donate to, and bond with, the central metal ion. Table 25.2 contains a number of common ligands.

**TABLE 25.2 Common Ligands**

| Name | Lewis diagram |
|------|---------------|
| Water | |
| Ammonia | |
| Chloride ion | |
| Carbon monoxide | |
| Cyanide ion | |
| Thiocyanate ion | |
| Oxalate ion (ox) | |
| Ethylenediamine (en) | |
| Ethylenediaminetetraacetate (EDTA) | |

Ligands that donate only one electron pair to the central metal are **monodentate**. Some ligands have the ability to donate two pairs of electrons (from two different atoms) to the metal; these are **bidentate**. Examples of bidentate ligands include the oxalate ion (abbreviated ox) and the ethylenediamine molecule (abbreviated en) shown here. The ethylenediamine ligand bonded to $Co^{3+}$ is shown in Figure 25.5(a)▼.

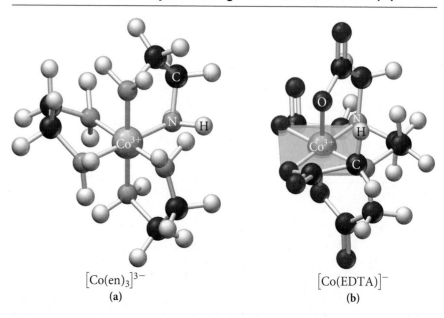

Ethylenediamine

Some ligands, called **polydentate** ligands, can donate even more than two electron pairs (from more than two atoms) to the metal. The most common polydentate ligand is the ethylenediaminetetraacetate ion ($EDTA^{4-}$) shown here:

$EDTA^{4-}$

The EDTA ligand wraps itself completely around the metal, donating up to six pairs of electrons (Figure 25.5b▼). A complex ion that contains either a bidentate or polydentate ligand is a **chelate** (pronounced "key-late"), and the coordinating ligand is known as a **chelating agent.**

A survey of many coordination compounds shows that coordination numbers can vary from as low as 2 to as high as 12. The most common coordination numbers are 6, as occurs in $[Co(NH_3)_6]^{3+}$, and 4, as occurs in $[PdCl_4]^{2-}$. Coordination numbers greater than 6 are rarely observed for the first-row transition metals. Typically, only 1+ metal ions have a coordination number as low as 2, as occurs in $[Ag(NH_3)_2]^+$. Odd coordination numbers exist, but they are rare.

▶ **FIGURE 25.5 Bidentate and Polydentate Ligands Coordinated to Co(III)** (a) Ethylenediamine is a bidentate ligand; (b) EDTA is a hexadentate ligand.

**Bidentate and Polydentate Ligands Coordinated to Co(III)**

$[Co(en)_3]^{3-}$
**(a)**

$[Co(EDTA)]^-$
**(b)**

The common geometries of complex ions, shown in Table 25.3, depend in part on their coordination number. A coordination number of 2 results in a linear geometry, and a coordination number of 6 results in an octahedral geometry. A coordination number of 4 can have either a tetrahedral geometry or a square planar geometry, depending on the number of $d$ electrons in the metal ion. Metal ions with a $d^8$ electron configuration (such as $[PdCl_4]^{2-}$) exhibit square planar geometry, and metal ions with a $d^{10}$ electron configuration (such as $[Zn(NH_3)_4]^{2+}$) exhibit tetrahedral geometry.

**TABLE 25.3  Common Geometries of Complex Ions**

| Coordination Number | Shape | Model | Example |
|---|---|---|---|
| 2 | Linear | | $[Ag(NH_3)_2]^+$ |
| 4 | Square planar | | $[PdCl_4]^{2-}$ |
| 4 | Tetrahedral | | $[Zn(NH_3)_4]^{2+}$ |
| 6 | Octahedral | | $[Fe(H_2O)_6]^{3+}$ |

## Naming Coordination Compounds

To name coordination compounds, we follow a series of general rules based on the system originally proposed by Werner. As with all salts (see Section 3.5), the name of the cation goes before the name of the anion.

| Guidelines for Naming Complex Ions | Examples |
|---|---|
| **1.** Name the ligands.<br>• Name neutral ligands as molecules with the following notable exceptions: | $NH_2CH_2CH_2NH_2$ is ethylenediamine. |
| H₂O (aqua) | $H_2O$ is aqua. |
| NH₃ (ammine) | |
| CO (carbonyl) | |
| • Name anionic ligands with the name of the ion plus an ending modified as follows: | |
| -*ide* becomes -o | $Cl^-$ is chloro. |
| -*ate* becomes -ato | $SO_4^{2-}$ is sulfato. |
| -*ite* becomes -ito | $SO_3^{2-}$ is sulfito. |

Table 25.4 lists the names of some common ligands.

| | |
|---|---|
| **2.** List the names of the ligands in alphabetical order before the name of the metal cation. | Ammine ($NH_3$) is listed before chloro ($Cl^-$), which is listed before nitrito ($NO_2^-$). |

—*Continued on the next page*

| Guidelines for Naming Complex Ions | Examples |
|---|---|
| 3. Use a prefix to indicate the number of ligands (when there is more than one of a particular type): *di-* (2), *tri-* (3), *tetra-* (4), *penta-* (5), or *hexa-* (6). <br><br> If the name of the ligand already contains a prefix, such as ethylenediamine, place parentheses around the ligand name and use *bis-* (2), *tris-* (3), or *tetrakis-* (4) to indicate the number. <br><br> Prefixes do not affect the order in which you list the ligands. | Trichloro indicates three $Cl^-$ ligands. Tetraammine indicates four $NH_3$ ligands. <br><br> Tris(ethylenediamine) indicates three ethylenediamine ligands. |
| 4. Name the metal. <br> a. When the complex ion is a cation, use the name of the metal followed by the oxidation state written with a roman numeral. <br><br><br> b. If the complex ion is an anion, drop the ending of the metal and add *–ate* followed by the oxidation state written with a Roman numeral. Some metals use the Latin root with the *-ate* ending. Table 25.5 lists the names for some common metals in anionic complexes. | In cations: <br> $Co^{3+}$ is cobalt(III). <br> $Pt^{2+}$ is platinum(II). <br> $Cu^+$ is copper(I). <br> In anions: <br> $Co^{3+}$ is cobaltate(III). <br> $Pt^{2+}$ is platinate(II). <br> $Cu^+$ is cuprate(I). |
| 5. Write the entire name of the complex ion by listing the ligands first followed by the metal. | $[Pt(NH_3)_2Cl_4]^{2-}$ is diamminetetrachloroplatinate(II). $[Co(NH_3)_6]^{3+}$ is hexaamminecobalt(III). |

**TABLE 25.4 Names and Formulas of Common Ligands**

| Ligand | Name in Complex Ion |
|---|---|
| **Anions** | |
| Bromide, $Br^-$ | Bromo |
| Chloride, $Cl^-$ | Chloro |
| Hydroxide, $OH^-$ | Hydroxo |
| Cyanide, $CN^-$ | Cyano |
| Nitrite, $NO_2^-$ | Nitro |
| Oxalate, $C_2O_4^{2-}$ (ox) | Oxalato |
| Ethylenediaminetetraacetate (EDTA$^{4-}$) | Ethylenediaminetetraacetato |
| **Neutral molecules** | |
| Water, $H_2O$ | Aqua |
| Ammonia, $NH_3$ | Ammine |
| Carbon monoxide, CO | Carbonyl |
| Ethylenediamine (en) | Ethylenediamine |

**TABLE 25.5 Names of Common Metals when Found in Anionic Complex Ions**

| Metal | Name in Anionic Complex |
|---|---|
| Chromium | Chromate |
| Cobalt | Cobaltate |
| Copper | Cuprate |
| Gold | Aurate |
| Iron | Ferrate |
| Lead | Plumbate |
| Manganese | Manganate |
| Molybdenum | Molybdate |
| Nickel | Nickelate |
| Platinum | Platinate |
| Silver | Argentate |
| Tin | Stannate |
| Zinc | Zincate |

When we write the *formula* of a complex ion, we write the symbol for the metal first, followed by neutral molecules and then anions. If there is more than one anion or neutral molecule acting as a ligand, we list them in alphabetical order based on the chemical symbol.

## Procedure for...

| Naming Coordination Compounds | **Example 25.3** Naming Coordination Compounds | **Example 25.4** Naming Coordination Compounds |
|---|---|---|
| | Name the following compound: $[Cr(H_2O)_5Cl]Cl_2$. | Name the following compound: $K_3[Fe(CN)_6]$. |
| Identify the cation and anion and first name the simple ion (i.e., not the complex one). | **SOLUTION** $[Cr(H_2O)_5Cl]^{2+}$ is a complex cation. $Cl^-$ is chloride. | **SOLUTION** $K^+$ is potassium. $[Fe(CN)_6]^{3-}$ is a complex anion. |
| Give each ligand a name and list them in alphabetical order. | $H_2O$ is aqua. $Cl^-$ is chloro. | $CN^-$ is cyano. |
| Name the metal ion. | $Cr^{3+}$ is chromium(III). | $Fe^{3+}$ is ferrate(III) because the complex is anionic. |
| Name the complex ion by adding prefixes to indicate the number of each ligand followed by the name of each ligand followed by the name of the metal ion. | $[Cr(H_2O)_5Cl]^{2+}$ is pentaaquachlorochromium(III). | $[Fe(CN)_6]^{3-}$ is hexacyanoferrate(III). |
| Name the compound by writing the name of the cation before the anion. The only space is between ion names. | $[Cr(H_2O)_5Cl]Cl_2$ is pentaaquachlorochromium(III) chloride. | $K_3[Fe(CN)_6]$ is potassium hexacyanoferrate(III). |
| | **FOR PRACTICE 25.3** Name the following compound: $[Mn(CO)(NH_3)_5]SO_4$. | **FOR PRACTICE 25.4** Name the following compound: $Na_2[PtCl_4]$. |

## 25.4 Structure and Isomerization

Isomerism is common in coordination compounds. We broadly divide the isomerism observed in coordination compounds into two categories. Each category has subcategories, as shown in Figure 25.6▶, on the next page. In **structural isomers**, atoms are connected to one another in different ways, whereas in **stereoisomers**, atoms are connected in the same way but the ligands have a different spatial arrangement about the metal atom.

### Structural Isomerism

We subdivide the broad category of structural isomers into two types: coordination isomers and linkage isomers. **Coordination isomers** occur when a coordinated ligand exchanges places with the uncoordinated counterion. For example, two different compounds have the general formula $Co(NH_3)_5BrCl$. In one of them, the bromine coordinates to the metal and chloride is a counterion, pentaamminebromocobalt(II) chloride, $[Co(NH_3)_5Br]Cl$; in the other one, the chlorine coordinates to the metal and bromide is the counterion, pentaamminechlorocobalt(II) bromide, $[Co(NH_3)_5Cl]Br$.

**Linkage isomers** have ligands that can coordinate to the metal in different orientations. For example, the nitrite ion ($NO_2^-$) has a lone pair on the N atom as well as lone pairs on the O atoms—either of the two atoms can form coordinate covalent bonds with the metal. When the nitrite ion coordinates through the N atom, it is a *nitro* ligand and

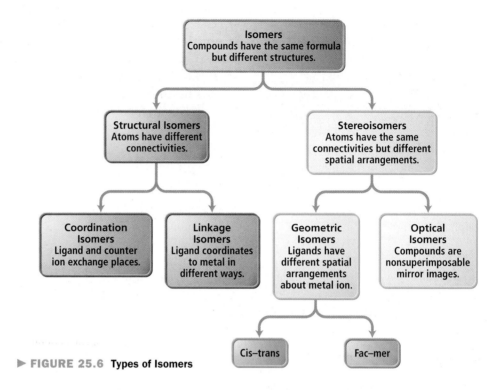

▶ **FIGURE 25.6 Types of Isomers**

is represented as $NO_2^-$, but when it coordinates through the O atom, it is a *nitrito* ligand and is usually represented as $ONO^-$. An example of linkage isomerization can be seen in the yellow-orange complex ion pentaamminenitrocobalt(III), $[Co(NH_3)_5NO_2]^{2+}$, which contrasts with the red-orange complex ion pentaamminenitritocobalt(III), $[Co(NH_3)_5ONO]^{2+}$, as shown in Figure 25.7▼. Table 25.6 lists other ligands capable of linkage isomerization.

N-bond and O-bond $NO_2^-$ ligand

▶ **FIGURE 25.7 Linkage Isomers**
In $[Co(NH_3)_5NO_2]^{2+}$, the $NO_2$ ligand bonds to the central metal atom through the nitrogen atom. In $[Co(NH_3)_5ONO]^{2+}$, the $NO_2$ ligand bonds through the oxygen atoms. The different isomers have different colors.

Nitro isomer                    Nitrito isomer

**TABLE 25.6 Ligands Capable of Linkage Isomerization**

nitro

$$\left[ :\ddot{O}-\ddot{N}=\ddot{O} \right]^{-}$$

nitrito

cyano

$$\left[ :C\equiv N: \right]^{-}$$

isocyano

isocyanato

$$\left[ :\ddot{O}=C=\ddot{N}: \right]^{-}$$

cyanato

isothiocyanato

$$\left[ :\ddot{S}=C=\ddot{N}: \right]^{-}$$

thiocyanato

## Stereoisomerism

We can also subdivide the broad category of stereoisomers into two types: geometric isomers and optical isomers. **Geometric isomers** result when the ligands bonding to the metal have a different spatial arrangement. One type of geometric isomerism, as we saw in Section 21.5, is cis–trans isomerism, which in complex ions occurs in square planar complexes of the general formula $MA_2B_2$ or octahedral complexes of the general formula $MA_4B_2$. For example, cis–trans isomerism occurs in the square planar complex $Pt(NH_3)_2Cl_2$. Figure 25.8(a)▼ shows the two distinct ways in which the ligands can be oriented around the metal. In one complex, the $Cl^-$ ligands are next to each other on one side of the molecule—this is the cis isomer. In the other complex, the $Cl^-$ ligands are on opposite sides of the molecule—this is the trans isomer.

Geometric isomerism also exists in the octahedral complex ion $[Co(NH_3)_4Cl_2]^+$. As shown in Figure 25.8(b)▼, the ligands arrange themselves around the metal in two ways, one with the $Cl^-$ ligands on the same side (the cis isomer) and another with the $Cl^-$ ligands

Cis          Trans

$Pt(NH_3)_2Cl_2$

**(a)**

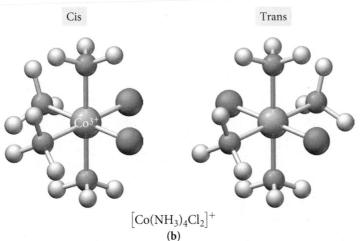

Cis          Trans

$\left[ Co(NH_3)_4Cl_2 \right]^+$

**(b)**

◄ **FIGURE 25.8 Cis–trans Isomerism** **(a)** Cis–trans isomerism in square planar $Pt(NH_3)_2Cl_2$. In the cis isomer, the $Cl^-$ ligands are next to each other on one side of the molecule. In the trans isomer, the $Cl^-$ ligands are on opposite sides of the molecule. **(b)** Cis–trans isomerism in octahedral $[Co(NH_3)_4Cl_2]^+$. In the cis isomer, the $Cl^-$ ligands are on the same side. In the trans isomer, the $Cl^-$ ligands are on opposite sides.

on opposite sides of the metal (the trans isomer). Note that cis–trans isomerism does not occur in tetrahedral complexes because all bond angles around the metal are 109.5°, and each corner of a tetrahedron is considered to be adjacent to all three other corners.

Another type of geometric isomerism is fac–mer isomerism, which occurs in octahedral complexes of the general formula $MA_3B_3$. For example, in $Co(NH_3)_3Cl_3$, the ligands arrange themselves around the metal in two ways (Figure 25.9▼). In the fac isomer the three $Cl^-$ ligands are all on one side of the molecule and make up one face of the octahedron (fac is short for facial). In the mer isomer the three ligands form an arc around the middle of the octahedron (mer is short for meridian).

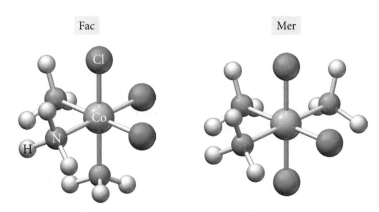

**Fac**                                             **Mer**

▶ **FIGURE 25.9  Fac–Mer Isomerism in $CO(NH_3)_3Cl_3$**  In the fac isomer, all three $Cl^-$ ligands are on one side of the molecule and make up one face of the octahedron. In the mer isomer, the three ligands inscribe an arc around the middle (or meridian) of the octahedron.

## *Procedure for...*

| Identifying and Drawing Geometric Isomers | **Example 25.5** <br> **Identifying and Drawing Geometric Isomers** <br> Draw the structures and label the type of all the isomers of $[Co(en)_2Cl_2]^+$. | **Example 25.6** <br> **Identifying and Drawing Geometric Isomers** <br> Draw the structures and label the type for of the isomers of $[Ni(CN)_2Cl_2]^{2-}$. |
|---|---|---|
| Identify the coordination number and the geometry around the metal. | **SOLUTION** <br> The ethylenediamine (en) ligand is bidentate, so each occupies two coordination sites. Each $Cl^-$ is monodentate, occupying one site. The total coordination number is 6, so this must be an octahedral complex. | **SOLUTION** <br> All the ligands are monodentate, so the total coordination number is 4. $Ni^{2+}$ is a $d^8$ electronic configuration, so we expect a square planar complex. |
| Identify if this is cis–trans or fac–mer isomerism. | With ethylenediamine occupying four sites and $Cl^-$ occupying two sites, it fits the general formula $MA_4B_2$, leading to cis–trans isomers. | Square planar complexes can only have cis–trans isomers. |
| Draw and label the two isomers. | Cis                    Trans | Cis                    Trans |
|  | **FOR PRACTICE 25.5**  Draw the structures and label the type of all the isomers of $[Cr(H_2O)_3Cl_3]^+$. | **FOR PRACTICE 25.6**  Draw the structures and label the type of all the isomers of $[Co(NH_3)_2Cl_2(ox)]^-$. |

The second category of stereoisomerism is optical isomerism. Recall from Section 21.3 that **optical isomers** are nonsuperimposable mirror images of one another. If you hold your right hand up to a mirror, the image in the mirror looks like your left hand.

No matter how you rotate or flip your left hand, you cannot superimpose it on your right hand. Molecules or ions that exhibit this quality are *chiral*. The isomers are *enantiomers*, and they exhibit the property of optical activity (the rotation of polarized light). The complex ion $[Co(en)_3]^{3+}$ is nonsuperimposable on its mirror image, so it is a chiral complex (Figure 25.10 ▼).

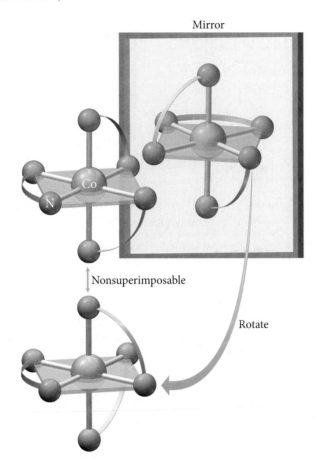

◀ **FIGURE 25.10 Optical Isomerism in $[Co(en)_3]^{3+}$** The mirror images of $[Co(en)_3]^{3+}$ are not superimposable. (The connected nitrogen atoms represent the ethylenediamine ligand.)

## Example 25.7 | Recognizing and Drawing Optical Isomers

Determine whether the cis or trans isomers in Example 25.5 are optically active (demonstrate optical isomerism).

**SOLUTION**

Draw the trans isomer of $[Co(en)_2Cl_2]^+$ and its mirror image. Check to see if they are superimposable by rotating one isomer 180°.

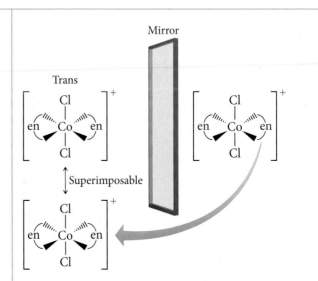

In this case the two are identical, so there is no optical activity.

—*Continued on the next page*

*Continued—*

Draw the cis isomer and its mirror image. Check to see if they are superimposable by rotating one isomer 180°.

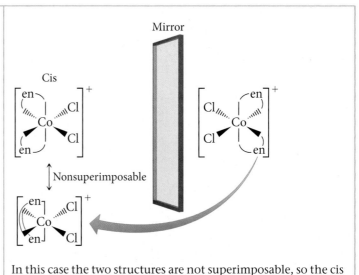

In this case the two structures are not superimposable, so the cis isomer does exhibit optical activity.

**FOR PRACTICE 25.7**    Determine whether the fac or mer isomers of $[Cr(H_2O)_3Cl_3]^+$ are optically active.

The previous examples demonstrate optical isomerism in octahedral complexes. Tetrahedral complexes can also exhibit optical isomerism, but only if all four coordination sites are occupied by different ligands. Square planar complexes do not normally exhibit optical isomerism as they are superimposable on their mirror images.

## 25.5  Bonding in Coordination Compounds

The bonding in complex ions and coordination compounds can be described by valence bond theory, which helps us to understand the geometries of the complex ions. However, other properties of complex ions, such as their colors and their magnetic properties, require a different model called crystal field theory. We examine both models in this section.

| Geometry | Hybridization | Orbitals |
|----------|---------------|----------|
| Linear | $sp$ | |
| Tetrahedral | $sp^3$ | |
| Square planar | $dsp^2$ | |
| Octahedral | $d^2sp^3$ | |

▲ **FIGURE 25.11  Common Hybridization Schemes in Complex Ions**  We can deduce the valence bond model hybridization schemes from the geometry of the complex ion.

### Valence Bond Theory

Recall from Section 10.6 that according to valence bond theory, a coordinate covalent bond is the overlap between a completely filled atomic orbital and an empty atomic orbital. In complex ions, the filled orbital is on the ligand, and the empty orbital is on the metal ion. The metal ion orbitals are hybridized according to the geometry of the complex ion. The common hybridization schemes are shown in Figure 25.11 ◄. An octahedral complex ion requires six empty orbitals in an octahedral arrangement on the metal ion. A full set of $d^2sp^3$ hybrid orbitals results in the exact orbitals needed for this geometry. A set of $sp^3$ hybrid orbitals results in a tetrahedral arrangement of orbitals, a set of $dsp^2$ hybrid orbitals results in a square planar arrangement, and a set of *sp* hybrid orbitals results in a linear arrangement of orbitals. In each case, the coordinate covalent bond is formed by the overlap between the orbitals on the ligands and the hybridized orbitals on the metal ion.

## Crystal Field Theory

Crystal field theory (CFT) is a bonding model for transition metal complexes that helps us to understand their colors and magnetic properties. To illustrate the basic principles of CFT, we examine the central metal atom's *d* orbitals in an octahedral complex.

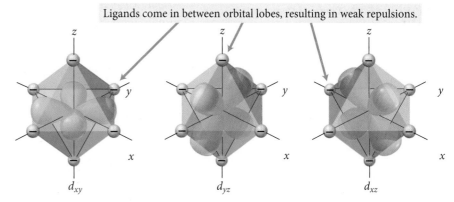

### Octahedral Complexes

The basic premise of CFT is that complex ions form because of attractions between the electrons on the ligands and the positive charge on the metal ion. However, the electrons on the ligands also repel the electrons in the *unhybridized* metal *d* orbitals. CFT focuses on these repulsions. Figure 25.12▶ shows how the ligand positions superimpose on the *d* orbitals in an octahedral complex. Notice that the ligands in an octahedral complex are located in the same space as the lobes of the $d_{x^2-y^2}$

▲ **FIGURE 25.12** **Relative Positions of *d* Orbitals and Ligands in an Octahedral Complex**    The ligands in an octahedral complex (represented here as spheres of negative charge) interact most strongly with the $d_{z^2}$ and $d_{x^2-y^2}$ orbitals.

and $d_{z^2}$ orbitals. The repulsions *between* electron pairs in the ligands and any potential electrons in the *d* orbitals result in an increase in the energies of these orbitals. In contrast, the $d_{xy}$, $d_{xz}$, and $d_{yz}$ orbitals lie *between* the axes and have nodes directly on the axes, which results in less repulsion and lower energies for these three orbitals. In other words, the *d* orbitals—which are degenerate in the bare metal ion—are split into higher- and lower-energy levels because of the spatial arrangement of the ligands (Figure 25.13▼). The difference in energy between these split *d* orbitals is known as the crystal field splitting energy (Δ). The magnitude of the splitting depends on the particular complex. In **strong-field complexes**, the splitting is large, and in **weak-field complexes**, the splitting is small.

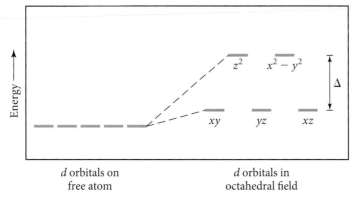

▲ **FIGURE 25.13** ***d* Orbital Splitting in an Octahedral Field**    The otherwise degenerate *d* orbitals are split into two energy levels by the ligands in an octahedral complex ion.

**(a)** **(b)**

▲ FIGURE 25.14 **Colors of Complex Ions**
**(a)** The complex ion $[Fe(CN)_6]^{3-}$ forms a deep red
solution, and **(b)** $[Ni(NH_3)_6]^{2+}$ is blue.

## The Color of Complex Ions and Crystal Field Strength

We discussed in the opening section of this chapter that transition metals in host crystals often show brilliant colors because of the crystal field splitting of their $d$ orbitals. Solutions of complex ions display brilliant colors because of similar splittings. For example, an $[Fe(CN)_6]^{3-}$ solution is deep red, and an $[Ni(NH_3)_6]^{2+}$ solution is blue (Figure 25.14 ◄).

Recall from Chapter 7 that the color of an object is related to the absorption of light energy by its electrons. If a substance absorbs all of the visible wavelengths, it appears black. If it transmits (or reflects) all the wavelengths (absorbs no light), it appears colorless. A substance appears to be a particular color if it absorbs some visible light but also transmits (or reflects) the wavelengths associated with that color. A substance also appears to be a given color if it transmits (or reflects) most wavelengths but absorbs the *complementary color* on a color wheel (Figure 25.15 ▼). For example, a substance that absorbs green light (the complement of red) will appear red. A solution of $[Ti(H_2O)_6]^{3+}$ is purple because it absorbs strongly between 490 and 580 nm, the yellow-green region of the visible spectrum (Figure 25.16(a) ▼).

The easiest way to measure the energy difference between the $d$ orbitals in a complex ion is to use spectroscopy to determine the wavelength of light absorbed when an electron makes a transition from the lower-energy $d$ orbitals to the higher-energy ones. With that information we can calculate the crystal field splitting energy, $\Delta$:

$$E_{photon} = h\nu = \frac{hc}{\lambda} = \Delta$$

Consider the $[Ti(H_2O)_6]^{3+}$ absorption spectrum shown in Figure 25.16(b) ▼. The maximum absorbance is at 498 nm. Using this wavelength, we calculate $\Delta$:

$$\Delta = \frac{hc}{\lambda} = \frac{(6.626 \times 10^{-34}\,J \cdot s)(3.00 \times 10^8\,m/s)}{(498\,nm \times 1 \times 10^{-9}\,m/nm)}$$

$$\Delta = 3.99 \times 10^{-19}\,J$$

This energy corresponds to a single $[Ti(H_2O)_6]^{3+}$ ion. We can convert to kilojoules per mole:

$$\Delta = \left(3.99 \times 10^{-19}\,\frac{J}{ion}\right)\left(6.02 \times 10^{23}\,\frac{ion}{mol}\right)\left(1\,\frac{kJ}{1000\,J}\right) = 240\,kJ/mol$$

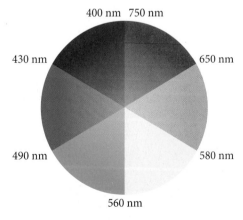

400 nm   750 nm

430 nm                    650 nm

490 nm                    580 nm

560 nm

▲ FIGURE 25.15 **The Color Wheel** Colors
across from one another on the color wheel are
said to be complementary. A substance that
absorbs a color on the wheel will appear to be its
complementary color.

**(a)**

Absorbance

400        500        600        700
Wavelength (nm)

**(b)**

▲ FIGURE 25.16 **The Color and Absorption Spectrum of**
**$[Ti(H_2O)_6]^{3+}$** **(a)** A solution containing $[Ti(H_2O)_6]^{3+}$ is purple.
**(b)** The absorption spectrum of $[Ti(H_2O)_6]^{3+}$ extends across the
green-yellow region of the spectrum.

## Example 25.8 | Crystal Field Splitting Energy

The complex ion $[Cu(NH_3)_6]^{2+}$ is blue in aqueous solution. Estimate the crystal field splitting energy (in kJ/mol) for this ion.

SOLUTION

| | |
|---|---|
| Begin by consulting the color wheel to determine approximately which wavelength is being absorbed. | Because the solution is blue, you can deduce that orange light is absorbed since orange is the complementary color to blue. |
| Estimate the absorbed wavelength. | The color orange ranges from 580 to 650 nm, so you can estimate the average wavelength as 615 nm. |
| Calculate the energy corresponding to this wavelength, using $E = hc/\lambda$. This energy corresponds to $\Delta$. | $E = \dfrac{(6.626 \times 10^{-34}\,J \cdot s)(3.00 \times 10^8\,m/s)}{(615\,nm)(1 \times 10^{-9}\,m/nm)}$ $E = 3.23 \times 10^{-19}\,J = \Delta$ |
| Convert J/ion into kJ/mol. | $E = \dfrac{(3.23 \times 10^{-19}\,J/ion)(6.02 \times 10^{23}\,ion/mol)}{(1000\,J/kJ)}$ $\Delta = 195\,kJ/mol$ |

**FOR PRACTICE 25.8** The complex ion $[Co(NH_3)_5NO_2]^{2+}$ is yellow. Estimate the crystal field splitting energy (in kJ/mol) for this ion.

The magnitude of the crystal field splitting in a complex ion—and, therefore whether it is a strong-field or a weak-field complex—depends in large part on the ligands attached to the central metal ion. Spectroscopic studies of various ligands attached to the same metal allow us to arrange different ligands in order of their ability to split the $d$ orbitals. This list is known as the *spectrochemical series* and is arranged from ligands that result in the largest $\Delta$ to those that result in the smallest:

$$CN^- > NO_2^- > en > NH_3 > H_2O > OH^- > F^- > Cl^- > Br^- > I^-$$

large $\Delta$                                   small $\Delta$

typically strong-field ligands            typically weak-field ligands

Ligands that produce large values of $\Delta$ are *strong-field ligands*, and those that give small values of $\Delta$ are *weak-field ligands*.

The metal ion also has an effect on the magnitude of $\Delta$. If we examine different metal ions with the same ligand, we find that $\Delta$ increases as the charge on the metal ion increases. The greater charge on the metal draws the ligands closer, causing greater repulsion with the $d$ orbitals and therefore a larger $\Delta$. An example of this behavior occurs in the complex ions between $NH_3$ (a ligand in the middle of the spectrochemical series) and the +2 or +3 oxidation states of cobalt. Hexaamminecobalt(II) ion, $[Co(NH_3)_6]^{2+}$, has a weak crystal field (small $\Delta$) and hexaamminecobalt(III) ion, $[Co(NH_3)_6]^{3+}$, has a strong field (large $\Delta$).

**CONCEPTUAL CONNECTION 25.2**

**Weak- and Strong-Field Ligands** Two ligands, A and B, form complexes with a particular metal ion. When the metal ion complexes with ligand A, the resulting solution is red. When the metal ion complexes with ligand B, the resulting solution is yellow. Which of the two ligands produces the larger $\Delta$?

## Magnetic Properties

The strength of the crystal field can affect the magnetic properties of a transition metal complex. Recall that, according to Hund's rule, electrons occupy degenerate orbitals singly as long as an empty orbital is available. When the energies of the $d$ orbitals are split

Recall from Section 8.7 that a paramagnetic species contains unpaired electrons and a diamagnetic one does not.

$[Fe(CN_6)]^{4-}$

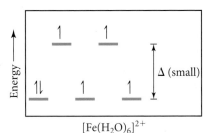

$[Fe(H_2O)_6]^{2+}$

by ligands, the lower-energy orbitals fill first. Once they are half-filled, the next electron can either: (1) pair with an electron in one of the lower-energy half-filled orbitals by overcoming the electron–electron repulsion associated with having two electrons in the same orbital; or (2) go into an empty orbital of higher energy by overcoming the energy difference between the orbitals—in this case, the crystal field splitting energy, $\Delta$. The magnitude of $\Delta$ compared to the electron–electron repulsions determines which of these two actually occurs.

We can compare two iron(II) complexes to see the difference in behavior under strong- and weak-field conditions. $[Fe(CN)_6]^{4-}$ is known to be diamagnetic, and $[Fe(H_2O)_6]^{2+}$ is known to be paramagnetic. Both of these complexes contain $Fe^{2+}$, which has an electron configuration of [Ar] $3d^6$. In the case of $[Fe(CN)_6]^{4-}$, $CN^-$ is a strong-field ligand that generates a large $\Delta$, so it takes more energy to occupy the higher-energy level than it does to pair the electrons in the lower-energy level. The result is that all six electrons are paired and the compound is diamagnetic, as shown in the accompanying figure in the margin.

In $[Fe(H_2O)_6]^{2+}$, $H_2O$ is a weak-field ligand that generates a small $\Delta$, so the electron pairing energy is greater than $\Delta$. Consequently, the first five electrons occupy the five $d$ orbitals singly and only the sixth pairs up, resulting in a paramagnetic compound with four unpaired electrons, as shown in the accompanying figure in the margin.

In general, complexes with strong-field ligands have fewer unpaired electrons relative to the free metal ion and are therefore called **low-spin complexes**. Complexes with weak-field ligands, by contrast, have the same number of unpaired electrons as the free metal ion and are **high-spin complexes**.

When we examine the orbital diagrams of the $d^1$ through $d^{10}$ metal ions in octahedral complexes, we find that only $d^4$, $d^5$, $d^6$, and $d^7$ metal ions have low- and high-spin possibilities. Since there are three lower-energy $d$ orbitals, the $d^1$, $d^2$, and $d^3$ metal ions always have unpaired electrons, independent of $\Delta$. In the $d^8$, $d^9$, and $d^{10}$ metal ions, the three lower-energy orbitals are completely filled, so the remaining electrons fill the two higher orbitals (as expected according to Hund's rule), also independent of $\Delta$.

## Procedure for...

| Determining the Number of Unpaired Electrons in Octahedral Complexes | **Example 25.9**<br>**High- and Low-Spin Octahedral Complexes**<br>How many unpaired electrons are there in the complex ion $[CoF_6]^{3-}$? | **Example 25.10**<br>**High- and Low-Spin Octahedral Complexes**<br>How many unpaired electrons are there in the complex ion $[Co(NH_3)_5NO_2]^{2+}$? |
|---|---|---|
| | **SOLUTION** | **SOLUTION** |
| Begin by determining the charge and number of $d$ electrons on the metal. | The metal is $Co^{3+}$ and has a $d^6$ electronic configuration. | The metal is $Co^{3+}$ and has a $d^6$ electronic configuration. |
| Look at the spectrochemical series to determine whether the ligand is a strong-field or a weak-field ligand. | $F^-$ is a weak-field ligand, so $\Delta$ is relatively small. | $NH_3$ and $NO_2^-$ are both strong-field ligands, so $\Delta$ is relatively large. |
| Decide if the complex is high- or low-spin and draw the electron configuration. | Weak-field ligands yield high-spin configurations. | Strong-field ligands yield low-spin configurations. |

| Count the unpaired electrons. | This configuration has four unpaired electrons. | This configuration has no unpaired electrons. |
|---|---|---|
| | **FOR PRACTICE 25.9** How many unpaired electrons are there in the complex ion $[FeCl_6]^{3-}$? | **FOR PRACTICE 25.10** How many unpaired electrons are there in the complex ion $[Co(CN)_6]^{4-}$? |

### Tetrahedral and Square Planar Complexes

So far, we have examined the $d$ orbital energy changes only for octahedral complexes, but transition metal complexes can have other geometries, such as tetrahedral and square planar. We use crystal field theory to determine the $d$ orbital splitting pattern for these geometries as well. For a tetrahedral complex, the $d$ orbital splitting pattern is the opposite of the octahedral splitting pattern: three $d$ orbitals ($d_{xy}$, $d_{xz}$, and $d_{yz}$) are higher in energy, and two $d$ orbitals ($d_{x^2-y^2}$ and $d_{z^2}$) are lower in energy (Figure 25.17▼). Almost all tetrahedral complexes are high-spin because of reduced ligand–metal interactions. The $d$ orbitals in a tetrahedral complex interact with only four ligands, as opposed to six in the octahedral complex, so the value of $\Delta$ is generally smaller.

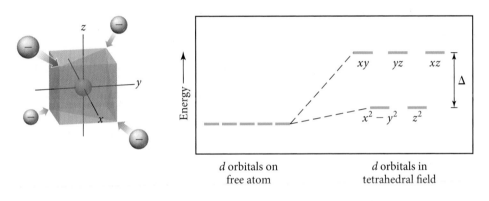

◄ **FIGURE 25.17 Splitting of $d$ Orbitals by a Tetrahedral Ligand Geometry** In tetrahedral complexes, the splitting of the $d$ orbitals has a pattern that is the opposite of the octahedral splitting pattern. The $d_{xy}$, $d_{yz}$, and $d_{xz}$ orbitals are higher in energy than the $d_{z^2}$ and $d_{x^2-y^2}$ orbitals.

A square planar complex gives us the most complex splitting pattern of the three geometries (Figure 25.18▼). As we discussed previously, square planar complexes occur in $d^8$ metal ions, such as $Pt^{2+}$, $Pd^{2+}$, $Ir^+$, or $Au^{3+}$, and they are low-spin in nearly all cases.

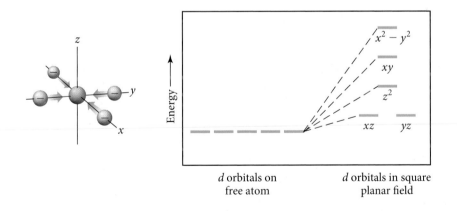

◄ **FIGURE 25.18 Splitting of $d$ Orbitals by a Square Planar Ligand Geometry** Square planar complexes produce the $d$ orbital energy pattern shown here.

## 25.6  Applications of Coordination Compounds

Coordination compounds are found in living systems, in industry, and even in household products. Recall from Chapter 24 how silver and gold metals are extracted from their respective ores using cyanide complexes and how nickel metal is extracted by forming the gaseous carbonyl complex, $Ni(CO)_4$. In this section, we describe a few additional applications of coordination compounds.

## Chelating Agents

In Section 25.3, we introduced the chelating agent ethylenediaminetetraacetate ion (EDTA$^{4-}$). This ligand has lone pairs on six different donor atoms that can interact with a metal ion to form very stable metal complexes. EDTA is used to treat victims of heavy metal poisoning such as lead poisoning. The patient is given [Ca(EDTA)]$^{2-}$ and because the lead complex ($K_f = 2 \times 10^{18}$) is more stable than the calcium complex ($K_f = 4 \times 10^{10}$), the lead displaces the calcium. The body excretes the lead complex and leaves behind the calcium, which is nontoxic (and is in fact a nutrient).

(a)        (b)

▲ **FIGURE 25.19 Chemical Analysis with SCN⁻** **(a)** Blue indicates Co$^{2+}$. **(b)** Red indicates Fe$^{3+}$.

## Chemical Analysis

Some ligands are selective in their binding, preferring specific metal ions; these ligands can be used in chemical analysis. For example, dimethylglyoxime (dmg) is used to chemically analyze a sample for Ni$^{2+}$ or Pd$^{2+}$. In the presence of Ni$^{2+}$, an insoluble red precipitate forms, and in the presence of Pd$^{2+}$, an insoluble yellow precipitate forms. Similarly, the SCN⁻ ligand is used to test for Co$^{2+}$ or Fe$^{3+}$. In the presence of Co$^{2+}$ a blue solution forms, and in the presence of Fe$^{3+}$ a deep red solution forms (Figure 25.19 ◄).

## Coloring Agents

Because of the wide variety of colors found in coordination complexes, they are often used as coloring agents. For example, a commercially available agent, iron blue, is a mixture of the hexacyano complexes of iron(II) and iron(III). Iron blue is used in ink, paint, cosmetics (eye shadow), and blueprints.

## Biomolecules

Living systems contain many molecules based on metal complexes. Hemoglobin (involved in oxygen transport), cytochrome c (involved in electron transport), carbonic anhydrase (involved in respiration), and chlorophyll (involved in photosynthesis) all have coordinated metal ions that are critical to their structure and function. Table 25.7 summarizes the biological significance of many of the other first-row transition metals.

| TABLE 25.7  Transition Metals and Some of Their Functions in the Human Body | |
|---|---|
| **Transition Metal** | **Biological Function** |
| Chromium | Works with insulin to control utilization of glucose |
| Manganese | Fat and carbohydrate synthesis |
| Molybdenum | Involved in hemoglobin synthesis |
| Iron | Oxygen transport |
| Copper | Involved in hemoglobin synthesis |
| Zinc | Involved in cell reproduction and tissue growth; part of more than 70 enzymes; assists in the utilization of carbohydrate, protein, and fat |

## Hemoglobin and Cytochrome C

In hemoglobin and in cytochrome c, an iron complex called a heme is connected to a protein, as shown in the figure below. A heme is an iron ion coordinated to a flat,

Cytochrome c | Heme (H atoms omitted for clarity)

polydentate ligand called a porphyrin (Figure 25.20▼). The porphyrin ligand has a planar ring structure with four nitrogen atoms that can coordinate to the metal ion. Different porphyrins have different substituent groups connected around the outside of the ring.

◀ **FIGURE 25.20 Porphyrin** A porphyrin has four nitrogen atoms that can coordinate to a central metal atom.

In hemoglobin, the iron complex is octahedral, with the four nitrogen atoms of the porphyrin in a square planar arrangement around the metal. A nitrogen atom from a nearby amino acid of the protein occupies the fifth coordination site, and either $O_2$ or $H_2O$ occupies the last coordination site (Figure 25.21▶). In the lungs, where the oxygen content is high, the hemoglobin

Hemoglobin is discussed in Section 15.1.

▶ **FIGURE 25.21 Hemoglobin** In hemoglobin, the iron complex is octahedral, with the four nitrogen atoms of the porphyrin in a square planar arrangement around the metal. A nitrogen atom from a nearby amino acid of the protein occupies the fifth coordination site, and either $O_2$ or $H_2O$ occupies the last coordination site.

Hemoglobin

Heme structure

**▲ FIGURE 25.22 Chlorophyll**
Chlorophyll, involved in photosynthesis in plants, contains magnesium coordinated to a porphyrin.

coordinates to an $O_2$ molecule. The oxygen-rich hemoglobin is carried by the bloodstream to areas throughout the body that are depleted in oxygen, where oxygen is released and replaced by a water molecule. The hemoglobin then travels back to the lungs to repeat the cycle.

## Chlorophyll

Chlorophyll, shown in Figure 25.22◀, is another porphyrin-based biomolecule. In chlorophyll the porphyrin is not surrounded by a protein, and the coordinated metal is magnesium (which is not a transition metal). Chlorophyll is essential for photosynthesis process performed by plants, in which light energy from the sun is converted to chemical energy to fuel the plant's growth.

## Carbonic Anhydrase

In carbonic anhydrase, the zinc ion is bound in a tetrahedral complex, with three of the coordination sites occupied by nitrogen atoms from surrounding amino acids and the fourth site available to bind a water molecule (Figure 25.23▼). Carbonic anhydrase catalyzes the reaction between water and $CO_2$ in respiration:

$$H_2O(l) + CO_2(g) \rightleftharpoons H^+(aq) + HCO_3^-(aq)$$

**▲ FIGURE 25.23 Carbonic Anhydrase** Carbonic anhydrase contains a zinc ion that is bound in a tetrahedral complex, with three of the coordination sites occupied by nitrogen atoms from surrounding amino acids. The fourth site is available to bind a water molecule.

A water molecule alone is not acidic enough to react with a $CO_2$ molecule at a sufficient rate. When the water molecule is bound to the zinc ion in carbonic anhydrase, the positive charge on the metal draws electron density from the O—H bond and the $H_2O$ becomes more acidic—sufficiently so to readily lose a proton. The resulting bound $OH^-$ easily reacts with a $CO_2$ molecule, and the reaction is much faster than the uncatalyzed version.

## Drugs and Therapeutic Agents

In the mid-1960s, researchers found that the platinum(II) complex *cis*-[Pt(NH$_3$)$_2$Cl$_2$], known as cisplatin, is an effective anticancer agent (Figure 25.24◀). Interestingly, the closely related geometric isomer *trans*-[Pt(NH$_3$)$_2$Cl$_2$] has little or no effect on cancer tumors. Cisplatin is believed to function by attaching itself to the cancer cell's DNA and replacing the $Cl^-$ ligands with donor atoms from the DNA strands. The cis arrangement of the $Cl^-$ ligands corresponds to the geometry required to bind to the DNA strands. The trans isomer, though closely related, cannot bind properly due to the arrangement of the $Cl^-$ ligands and is therefore not an effective agent. Cisplatin and other closely related platinum(II) complexes are still used today in chemotherapy for certain types of cancer and are among the most effective anticancer agents available for these cases.

**▲ FIGURE 25.24 Cisplatin**
Cisplatin is an effective anticancer agent.

# Self-Assessment Quiz

**Q1.** What is the electron configuration of the $Cu^+$ ion?

    a) $[Ar]4s^2 3d^9$          b) $[Ar]4s^2 3d^8$

    c) $[Ar]4s^1 3d^9$          d) $[Ar]4s^0 3d^{10}$

**Q2.** Which metal has the highest first ionization energy?

    a) Ti     b) Mn     c) Ru     d) Au

**Q3.** What is the name of the compound $[CoCl(NH_3)_5]Cl_2$?

    a) pentaamminetrichlorocobalt(III) chloride

    b) pentaamminechlorocobalt(III) chloride

    c) pentaamminechlorocobalt(II) chloride

    d) pentaamminetrichlorocobalt(II) chloride

**Q4.** What is the formula of hexaaquamanganese(II) sulfate?

    a) $[Mn(OH)_6]SO_4$          b) $[Mn(H_2O)_6]SO_4$

    c) $[Mn(H_2O)_6]_2SO_4$         d) $[Mn_2(H_2O)_6]SO_4$

**Q5.** Which complex ion can exhibit geometric isomerism? Assume that M is the metal ion, A and B are ligands, and the geometry is octahedral.

    a) $[MA_6]^{2+}$          b) $[MA_5B]^{2+}$

    c) $[MA_4B_2]^{2+}$         d) $[MA_3B_3]^{2+}$

**Q6.** Pick the optical isomer of the complex ion represented here:

**Q7.** According to valence bond theory, what is the hybridization of the central metal ion in an octahedral complex ion?

    a) $sp$     b) $sp^3$     c) $dsp^2$     d) $d^2sp^3$

**Q8.** Estimate the crystal field splitting energy (in kJ/mol) for a complex ion that is red in solution.

    a) 228 kJ/mol

    b) 171 kJ/mol

    c) $2.84 \times 10^{-19}$ kJ/mol

    d) $3.79 \times 10^{-19}$ kJ/mol

**Q9.** Use crystal field theory to determine the number of unpaired electrons in the complex ion $[Fe(CN)_6]^{4-}$.

    a) 0     b) 2     c) 3     d) 4

**Q10.** Which complex ion is diamagnetic?

    a) $[Cr(H_2O)_4Cl_2]^+$         b) $[Fe(H_2O)_6]^{2+}$

    c) $[Co(NH_3)_6]^{3+}$         d) $[CoCl_6]^{3-}$

(a)     (b)

**Answers: 1.** (d) **2.** (d) **3.** (b) **4.** (b) **5.** (c) **6.** (b) **7.** (d) **8.** (a) **9.** (a) **10.** (c)

# Chapter 25 IN REVIEW

## Key Terms

### Section 25.1
crystal field theory (1128)

### Section 25.2
lanthanide contraction (1130)

### Section 25.3
complex ion (1132)
ligand (1132)
coordination compound (1132)
primary valence (1132)
secondary valence (1132)

coordination number (1132)
coordinate covalent bond (1133)
monodentate (1134)
bidentate (1134)
polydentate (1134)
chelate (1134)
chelating agent (1134)

### Section 25.4
structural isomers (1137)
stereoisomers (1137)

coordination isomers (1137)
linkage isomers (1137)
geometric isomers (1139)
optical isomers (1140)

### Section 25.5
strong-field complex (1143)
weak-field complex (1143)
low-spin complex (1146)
high-spin complex (1146)

# Key Concepts

## Electron Configurations (25.2)

▶ As we move across a row of transition elements, electrons add to the $(n-1)d$ orbitals, resulting in a general electron configuration for the first- and second-row transition elements of [noble gas] $ns^2(n-1)d^x$ and for the third and fourth rows of [noble gas] $ns^2(n-2)f^{14}(n-1)d^x$, where $x$ ranges from 1 to 10.

▶ A transition element forms a cation by losing electrons from the $ns$ orbitals before losing electrons from the $(n-1)d$ orbitals.

## Periodic Trends (25.2)

▶ The variations in atomic size, ionization energy, and electronegativity across a row in the periodic table are similar to those of main-group elements (although the trends are less pronounced and less regular). As we move down a group, however, atomic size increases from the first row to the second but stays roughly constant from the second row to the third because of the lanthanide contraction. This contraction results in ionization energy and electronegativity trends as we move down a column that are opposite of the main-group elements.

## Composition and Naming of Coordination Compounds (25.3)

▶ A coordination compound is composed of a complex ion and a counterion.

▶ A complex ion contains a central metal ion bound to one or more ligands. The number of ligands directly bound to the metal ion is the coordination number.

▶ The ligand forms a coordinate covalent bond to the metal ion by donating a pair of electrons to an empty orbital on the metal.

▶ Ligands that donate a single pair of electrons are monodentate. A ligand that donates two pairs of electrons is bidentate, and a ligand that donates more than two pairs is polydentate.

▶ In naming coordination compounds, we use the name of the cation followed by the name of the anion. To name a complex ion, we use the guidelines outlined in Section 25.3.

## Types of Isomers (25.4)

▶ We broadly divide the isomerism observed in coordination compounds into two categories: structural isomers, in which atoms are connected differently to one another, and stereoisomers, in which atoms are connected in the same way but the ligands have a different spatial arrangement about the metal atom.

▶ Structural isomers are either coordination isomers (a coordinated ligand exchanges places with an uncoordinated counterion) or linkage isomers (a particular ligand has the ability to coordinate to the metal in different ways).

▶ Stereoisomers are either geometric isomers (the ligands bonded to the metal have a different spatial arrangement relative to each other, leading to cis–trans or fac–mer isomers) or optical isomers (nonsuperimposable mirror images).

## Crystal Field Theory (25.5)

▶ Crystal field theory is a bonding model for transition metal complex ions. The model describes how the degeneracy of the $d$ orbitals is broken by the repulsive forces between the electrons on the ligands around the metal ion and the $d$ orbitals in the metal ion.

▶ The energy difference between the split $d$ orbitals is the crystal field splitting energy ($\Delta$). The magnitude of $\Delta$ depends in large part on the ligands bound to the metal.

▶ Octahedral complexes with a $d^4$, $d^5$, $d^6$, or $d^7$ metal ion can have two possible electronic configurations with different numbers of unpaired electrons. The first, called high-spin, has the same number of unpaired electrons as the free metal ion and is usually the result of a weak crystal field. The second, called low-spin, has fewer unpaired electrons than the free metal ion and is usually the result of a strong crystal field.

# Key Equations and Relationships

Crystal Field Splitting Energy (25.5)

$\Delta = hc/\lambda$ (where $\lambda$ is the wavelength of maximum absorption)

# Key Learning Outcomes

| CHAPTER OBJECTIVES | ASSESSMENT |
|---|---|
| Writing Electronic Configurations for Transition Metals and Their Ions (25.2) | Examples 25.1, 25.2  For Practice 25.1, 25.2  Exercises 19–20, 57, 58 |
| Naming Coordination Compounds (25.3) | Examples 25.3, 25.4  For Practice 25.3, 25.4  Exercises 23–28 |
| Recognizing and Drawing Geometric Isomers (25.4) | Examples 25.5, 25.6  For Practice 25.5, 25.6  Exercises 5–40, 61–62 |
| Recognizing and Drawing Optical Isomers (25.4) | Example 25.7  For Practice 25.7  Exercises 39–40, 61–62 |

Calculating Crystal Field Splitting Energy (**25.5**)

(a)      (b)

Example 25.8   For Practice 25.8   Exercises 43–46

Recognizing and Predicting High-Spin and Low-Spin
Octahedral Complex Ions (**25.5**)

Examples 25.9, 25.10   For Practice 25.9, 25.10   Exercises 49–52, 65

# EXERCISES

## Review Questions

**1.** When a transition metal atom forms an ion, which electrons are lost first?

**2.** Explain why transition metals exhibit multiple oxidation states instead of a single oxidation state (like most of the main-group metals).

**3.** Why is the +2 oxidation state so common for transition metals?

**4.** Explain why atomic radii of elements in the third row of the transition metals are no larger than those of elements in the second row.

**5.** Gold is the most electronegative transition metal. Explain.

**6.** Briefly define each term.
   **a.** coordination number
   **b.** ligand
   **c.** bidentate and polydentate
   **d.** complex ion
   **e.** chelating agent

**7.** Using the Lewis acid–base definition, how would you categorize a ligand? How would you categorize a transition metal ion?

**8.** Explain the differences between each pair of isomer types.
   **a.** structural isomer and stereoisomer
   **b.** linkage isomer and coordination isomer
   **c.** geometric isomer and optical isomer

**9.** Which complex ion geometry has the potential to exhibit cis–trans isomerism: linear, tetrahedral, square planar, octahedral?

**10.** How can you tell whether a complex ion is optically active?

**11.** Explain the differences between weak-field and strong-field metal complexes.

**12.** Explain why compounds of $Sc^{3+}$ are colorless, but compounds of $Ti^{3+}$ are colored.

**13.** Explain why compounds of $Zn^{2+}$ are white, but compounds of $Cu^{2+}$ are often blue or green.

**14.** Explain the differences between high-spin and low-spin metal complexes.

**15.** Why are almost all tetrahedral complexes high-spin?

**16.** Many transition metal compounds are colored. How does crystal field theory account for this?

## Problems by Topic

### Properties of Transition Metals

**17.** Write the ground state electron configuration for each atom and ion pair.
   **a.** Ni, $Ni^{2+}$
   **b.** Mn, $Mn^{4+}$
   **c.** Y, $Y^+$
   **d.** Ta, $Ta^{2+}$

**18.** Write the ground state electron configuration for each atom and ion pair.
   **a.** Zr, $Zr^{2+}$
   **b.** Co, $Co^{2+}$
   **c.** Tc, $Tc^{3+}$
   **d.** Os, $Os^{4+}$

**19.** Determine the highest possible oxidation state for each element.
   **a.** V      **b.** Re      **c.** Pd

**20.** Which first-row transition metal(s) has the following highest possible oxidation state?
   **a.** +3      **b.** +7      **c.** +4

### Coordination Compounds

**21.** Determine the oxidation state and coordination number of the metal ion in each complex ion.
   **a.** $[Cr(H_2O)_6]^{3+}$
   **b.** $[Co(NH_3)_3Cl_3]^-$
   **c.** $[Cu(CN)_4]^{2-}$
   **d.** $[Ag(NH_3)_2]^+$

**22.** Determine the oxidation state and coordination number of the metal ion in each complex ion.
   **a.** $[Co(NH_3)_5Br]^{2+}$
   **b.** $[Fe(CN)_6]^{4-}$
   **c.** $[Co(ox)_3]^{4-}$
   **d.** $[PdCl_4]^{2-}$

**23.** Name each complex ion or coordination compound.
   **a.** $[Cr(H_2O)_6]^{3+}$
   **b.** $[Cu(CN)_4]^{2-}$
   **c.** $[Fe(NH_3)_5Br]SO_4$
   **d.** $[Co(H_2O)_4(NH_3)(OH)]Cl_2$

**24.** Name each complex ion or coordination compound.
   **a.** $[Cu(en)_2]^{2+}$
   **b.** $[Mn(CO)_3(NO_2)_3]^{2+}$
   **c.** $Na[Cr(H_2O)_2(ox)_2]$
   **d.** $[Co(en)_3][Fe(CN)_6]$

**25.** Write the formula for each complex ion or coordination compound.
   **a.** hexaamminechromium(III)
   **b.** potassium hexacyanoferrate(III)
   **c.** ethylenediaminedithiocyanatocopper(II)
   **d.** tetraaquaplatinum(II) hexachloroplatinate(IV)

**26.** Write the formula for each complex ion or coordination compound.
   **a.** hexaaquanickel(II) chloride
   **b.** pentacarbonylchloromanganese(I)
   **c.** ammonium diaquatetrabromovanadate(III)
   **d.** tris(ethylenediamine)cobalt(III) trioxalatoferrate(III)

27. Write the formula and the name of each complex ion.
    a. a complex ion with $Co^{3+}$ as the central ion and three $NH_3$ molecules and three $CN^-$ ions as ligands
    b. a complex ion with $Cr^{3+}$ as the central ion and a coordination number of 6 with ethylenediamine ligands

28. Write the formula and the name of each complex ion or coordination compound.
    a. a complex ion with four water molecules and two $ONO^-$ ions connected to an Fe(III) ion
    b. a coordination compound made of two complex ions: one a complex of V(III) with two ethylenediamine molecules and two $Cl^-$ ions as ligands and the other a complex of Ni(II) having a coordination number of 4 with $Cl^-$ ions as ligands

## Structure and Isomerism

29. Draw two linkage isomers of $[Mn(NH_3)_5(NO_2)]^{2+}$.

30. Draw two linkage isomers of $[PtCl_3(SCN)]^{2-}$.

31. Write the formulas and names for the coordination isomers of $[Fe(H_2O)_6]Cl_2$.

32. Write the formulas and names for the coordination isomers of $[Co(en)_3][Cr(ox)_3]$.

33. Which complexes exhibit geometric isomerism?
    a. $[Cr(NH_3)_5(OH)]^{2+}$       b. $[Cr(en)_2Cl_2]^+$
    c. $[Cr(H_2O)(NH_3)_3Cl_2]^+$       d. $[Pt(NH_3)Cl_3]^-$
    e. $[Pt(H_2O)_2(CN)_2]$

34. Which complexes exhibit geometric isomerism?
    a. $[Co(H_2O)_2(ox)_2]^-$       b. $[Co(en)_3]^{3+}$
    c. $[Co(H_2O)_2(NH_3)_2(ox)]^+$       d. $[Ni(NH_3)_2(en)]^{2+}$
    e. $[Ni(CO)_2Cl_2]$

35. If W, X, Y, and Z are different monodentate ligands, how many geometric isomers are there for each ion?
    a. square planar $[NiWXYZ]^{2+}$
    b. tetrahedral $[ZnWXYZ]^{2+}$

36. How many geometric isomers are there for each species?
    a. $[Fe(CO)_3Cl_3]$       b. $[Mn(CO)_2Cl_2Br_2]^+$

37. Draw the structures and label the type for all the isomers of each ion.
    a. $[Cr(CO)_3(NH_3)_3]^{3+}$
    b. $[Pd(CO)_2(H_2O)Cl]^+$

38. Draw the structures and label the type for all the isomers of each species.
    a. $[Fe(CO)_4Cl_2]^+$       b. $[Pt(en)Cl_2]$

39. Determine if either isomer of $[Cr(NH_3)_2(ox)_2]^-$ is optically active.

40. Determine if either isomer of $[Fe(CO)_3Cl_3]$ is optically active.

## Bonding in Coordination Compounds

41. Draw the octahedral crystal field splitting diagram for each metal ion.
    a. $Zn^{2+}$       b. $Fe^{3+}$ (high- and low-spin)
    c. $V^{3+}$       d. $Co^{2+}$ (high-spin)

42. Draw the octahedral crystal field splitting diagram for each metal ion.
    a. $Cr^{3+}$
    b. $Cu^{2+}$
    c. $Mn^{3+}$ (high- and low-spin)
    d. $Fe^{2+}$ (low-spin)

43. The $[CrCl_6]^{3-}$ ion has a maximum in its absorption spectrum at 735 nm. Calculate the crystal field splitting energy (in kJ/mol) for this ion.

44. The absorption spectrum of the complex ion $[Rh(NH_3)_6]^{3+}$ has maximum absorbance at 295 nm. Calculate the crystal field splitting energy (in kJ/mol) for this ion.

45. Three complex ions of cobalt(III), $[Co(CN)_6]^{3-}$, $[Co(NH_3)_6]^{3+}$, and $[CoF_6]^{3-}$, absorb light at wavelengths of (in no particular order) 290 nm, 440 nm, and 770 nm. Match each complex ion to the appropriate wavelength absorbed. What color would you expect each solution to be?

46. Three bottles of aqueous solutions are discovered in an abandoned lab. The solutions are green, yellow, and purple. It is known that three complex ions of chromium(III) were commonly used in that lab: $[Cr(H_2O)_6]^{3+}$, $[Cr(NH_3)_6]^{3+}$, and $[Cr(H_2O)_4Cl_2]^+$. Determine the likely identity of each of the colored solutions.

47. The $[Mn(NH_3)_6]^{2+}$ ion is paramagnetic with five unpaired electrons. The $NH_3$ ligand is usually a strong field ligand. Is $NH_3$ acting as a strong field ligand in this case?

48. The complex $[Fe(H_2O)_6]^{2+}$ is paramagnetic. Is the $H_2O$ ligand inducing a strong or weak field?

49. How many unpaired electrons would you expect for each complex ion?
    a. $[RhCl_6]^{3-}$
    b. $[Co(OH)_6]^{4-}$
    c. cis-$[Fe(en)_2(NO_2)_2]^+$

50. How many unpaired electrons would you expect for each complex ion?
    a. $[Cr(CN)_6]^{4-}$
    b. $[MnF_6]^{4-}$
    c. $[Ru(en)_3]^{2+}$

51. How many unpaired electrons would you expect for the complex ion $[CoCl_4]^{2-}$ if it is a tetrahedral shape?

52. The complex ion $[PdCl_4]^{2-}$ is known to be diamagnetic. Use this information to determine if it is a tetrahedral or square planar structure.

## Applications of Coordination Compounds

53. What structural feature do hemoglobin, cytochrome c, and chlorophyll have in common?

54. Identify the central metal atom in each complex.
    a. hemoglobin
    b. carbonic anhydrase
    c. chlorophyll
    d. iron blue

55. Hemoglobin exists in two predominant forms in our bodies. One form, known as oxyhemoglobin, has $O_2$ bound to the iron and the other, known as deoxyhemoglobin, has a water molecule bound instead. Oxyhemoglobin is a low-spin complex that gives arterial blood its red color, and deoxyhemoglobin is a high-spin complex that gives venous blood its darker color. Explain these observations in terms of crystal field splitting. Would you categorize $O_2$ as a strong- or weak-field ligand?

56. Carbon monoxide and the cyanide ion are both toxic because they bind more strongly than oxygen to the iron in hemoglobin (Hb).

$$Hb + O_2 \rightleftharpoons HbO_2 \qquad K = 2 \times 10^{12}$$
$$Hb + CO \rightleftharpoons HbCO \qquad K = 1 \times 10^{14}$$

Calculate the equilibrium constant value for this reaction:

$$HbO_2 + CO \rightleftharpoons HbCO + O_2$$

Does the equilibrium favor reactants or products?

# Cumulative Problems

57. Recall from Chapter 8 that Cr and Cu are exceptions to the normal orbital filling, resulting in a [Ar] $4s^1 3d^x$ configuration. Write the ground state electron configuration for each species.
    a. Cr, $Cr^+$, $Cr^{2+}$, $Cr^{3+}$
    b. Cu, $Cu^+$, $Cu^{2+}$

58. Most of the second row transition metals do not follow the normal orbital filling pattern. Five of them—Nb, Mo, Ru, Rh, and Ag—have a [Kr] $5s^1 4d^x$ configuration and Pd has a [Kr] $4d^{10}$ configuration. Write the ground state electron configuration for each species.
    a. Mo, $Mo^+$, Ag, $Ag^+$
    b. Ru, $Ru^{3+}$
    c. Rh, $Rh^{2+}$
    d. Pd, $Pd^+$, $Pd^{2+}$

59. Draw the Lewis diagrams for each ligand. Indicate the lone pair(s) that may be donated to the metal. Indicate any you expect to be bidentate or polydentate.
    a. $NH_3$
    b. $SCN^-$
    c. $H_2O$

60. Draw the Lewis diagrams for each ligand. Indicate the lone pair(s) that may be donated to the metal. Indicate any you expect to be bidentate or polydentate.
    a. $CN^-$
    b. bipyridyl (bipy), which has the following structure:

    c. $NO_2^-$

61. List all the different formulas for an octahedral complex made from a metal (M) and three different ligands (A, B, and C). Describe any isomers for each complex.

62. Amino acids, such as glycine (gly), form complexes with the trace metal ions found in the bloodstream. Glycine, whose structure is shown here, acts as a bidentate ligand coordinating with the nitrogen atom and one of the oxygen atoms.

    Draw all the possible isomers of:
    a. square planar $[Ni(gly)_2]$
    b. tetrahedral $[Zn(gly)_2]$
    c. octahedral $[Fe(gly)_3]$

63. Oxalic acid solutions remove rust stains. Draw a complex ion that is likely responsible for this effect. Does it have any isomers?

64. W, X, Y, and Z are different monodentate ligands.
    a. Is the square planar $[NiWXYZ]^{2+}$ optically active?
    b. Is the tetrahedral $[ZnWXYZ]^{2+}$ optically active?

65. Hexacyanomanganate(III) ion is a low-spin complex. Draw the crystal field splitting diagram with electrons filled in appropriately. Is this complex paramagnetic or diamagnetic?

66. Determine the color and approximate wavelength absorbed most strongly by each solution.
    a. blue solution
    b. red solution
    c. yellow solution

67. Draw the structures of all the geometric isomers of $[Ru(H_2O)_2(NH_3)_2Cl_2]^+$. Draw the mirror images of any that are chiral.

68. A 0.32 mol amount of $NH_3$ is dissolved in 0.47 L of a 0.38 M silver nitrate solution. Calculate the equilibrium concentrations of all species in the solution.

# Challenge Problems

69. When a solution of $PtCl_2$ reacts with the ligand trimethylphosphine, $P(CH_3)_3$, two compounds are produced. The compounds share the same elemental analysis: 46.7% Pt; 17.0% Cl; 14.8% P; 17.2% C; 4.34% H. Determine the formula, draw the structure, and give the systematic name for each compound.

70. Draw a crystal field splitting diagram for a trigonal planar complex ion. Assume the plane of the molecule is perpendicular to the z axis.

71. Draw a crystal field splitting diagram for a trigonal bipyramidal complex ion. Assume the axial positions are on the z axis.

72. Explain why $[Ni(NH_3)_4]^{2+}$ is paramagnetic, while $[Ni(CN)_4]^{2-}$ is diamagnetic.

73. Sulfide ($S^{2-}$) salts are notoriously insoluble in aqueous solution.
    a. Calculate the molar solubility of nickel(II) sulfide in water. $K_{sp}(NiS) = 3 \times 10^{-16}$
    b. Nickel(II) ions form a complex ion in the presence of ammonia with a formation constant ($K_f$) of $2.0 \times 10^8$: $Ni^{2+} + 6 NH_3 \rightleftharpoons [Ni(NH_3)_6]^{2+}$. Calculate the molar solubility of NiS in 3.0 M $NH_3$.
    c. Explain any differences between the answers to parts **a** and **b**.

74. Calculate the solubility of $Zn(OH)_2(s)$ in 2.0 M NaOH solution. (Hint: you must take into account the formation of $Zn(OH)_4^{2-}$, which has a $K_f = 2 \times 10^{15}$).

75. Halide complexes of metal M of the form $[MX_6]^{3-}$ are found to be stable in aqueous solution. But it is possible that they undergo rapid ligand exchange with water (or other ligands) that is not detectable because the complexes are less stable. This property is referred to as their *lability*. Suggest an experiment to measure the lability of these complexes that does not employ radioactive labels.

76. The $K_f$ for $[Cu(en)_2]^{2+}$ is much larger than the one for $[Cu(NH_3)_4]^{2+}$. This difference is primarily an entropy effect. Explain why and calculate the difference between the $\Delta S°$ values at 298 K for the complete dissociation of the two complex ions. (Hint: the value of $\Delta H$ is about the same for both systems.)

77. When solid $Cd(OH)_2$ is added to a solution of 0.10 M NaI, some of it dissolves. Calculate the pH of the solution at equilibrium.

# Conceptual Problems

**78.** Two ligands, A and B, both form complexes with a particular metal ion. When the metal ion complexes with ligand A, the solution is green. When the metal ion complexes with ligand B, the solution is violet. Which of the two ligands results in the largest Δ?

**79.** Which element has the higher ionization energy, Cu or Au?

**80.** The complexes of $Fe^{3+}$ have magnetic properties that depend on whether the ligands are strong or weak field. Explain why this observation supports the idea that electrons are lost from the $4s$ orbital before the $3d$ orbitals in the transition metals.

*Active Classroom Learning*

# Questions for Group Work

Discuss these questions with the group and record your consensus answer.

**81.** Choose a row of the transition metals in the periodic table. Have each group member look up and graph (where appropriate) a separate trend for the elements in that row, choosing from electron configuration, atomic size, ionization energy, and electronegativity. Each member should present a graph to the group and describe the general trend and any notable exceptions. If possible, form new groups made up of individuals who researched the same property for different rows of the periodic table.

**82.** Have each group member write down the names and formulas for two coordination compounds. Taking turns, show each formula to the group and have the rest of the group name the compound, with each member contributing one step in the process. Once all group members have had their formulas named, repeat the process by showing each formula name to the group and having group members determine the correct formula.

**83.** Working individually, draw a pair of coordination compounds that are isomers. Take turns showing your drawings to the group

and having them identify the type of isomerism and the reasons your drawing demonstrates that type of isomerism. If your group misidentifies your type of isomerism, rather than telling them the correct answer right away, point out the part of the structure that prevents it from being the type of isomerism they thought it was.

**84.** Divide the electron configurations $d^1$ through $d^{10}$ among the group members so that every configuration is assigned to at least two group members. Working individually, draw the orbital diagram for the configurations assigned to you, including both high-spin and low-spin diagrams where possible. Present your diagrams to your group. Combine everyone's diagrams into one set for the group.

**85.** Working individually, review one of the applications of coordination complexes. Without referring to the text or mentioning the key words in the heading of the subsection, describe the application you reviewed. As group members each describe the application they reviewed, take turns attempting to identify the key words from the heading of the subsection they are describing.

# Data Interpretation and Analysis

## Qualitative Test for $Fe^{3+}$ Ions in Solution

**86.** Many aqueous solutions of complex ions display brilliant colors that depend on the identities of the metal ion and ligand(s). Some ligands bind selectively to certain metal ions, producing complex ions with characteristic colors. These distinctive complex ions serve as qualitative indicators of the presence of particular metal ions. For example, $Fe^{3+}$ is identified by the rapid formation of the intensely colored pentaaquathio-cyanatoiron(III) complex ion, $[Fe(H_2O)_5SCN]^{2+}$, when thiocyanate, $SCN^-$, is added to a solution containing hexaaquairon(III), $[Fe(H_2O)_6]^{3+}$, according to the following chemical equation:

$$[Fe(H_2O)_6]^{3+}(aq) + SCN^-(aq) \rightleftharpoons [Fe(H_2O)_5SCN]^{2+}(aq)$$

*pale violet*       *colorless*       *intensely colored*

The absorption spectrum of an aqueous solution of $[Fe(H_2O)_5SCN]^{2+}$ is shown in Figure a▶.

▲ **FIGURE a** **Absorption Spectrum of $[Fe(H_2O)_5SCN]^{2+}$**

Use the information provided to answer the following questions:

a. Determine the color of the aqueous solution of $[Fe(H_2O)_5SCN]^{2+}$ based on the spectrum shown in Figure a.

b. Calculate the crystal field splitting energy, $\Delta$, of $[Fe(H_2O)_5SCN]^{2+}$, in kJ/mol.

c. The hexaaquairon(III) complex ion, $[Fe(H_2O)_6]^{3+}$, produces a pale violet aqueous solution. Is the crystal field splitting

energy, $\Delta$, of $[Fe(H_2O)_6]^{3+}$ smaller or larger than the $\Delta$ of $[Fe(H_2O)_5SCN]^{2+}$?

d. On the basis of your answers to parts b and c, compare the crystal field strengths of water and thiocyanate ligands.

e. The complex ion hexacyanoferrate(III), $[Fe(CN)_6]^{3-}$, is red in aqueous solution. What can you conclude about the relative crystal field splitting energies of $[Fe(CN)_6]^{3-}$ and $[Fe(H_2O)_5SCN]^{2+}$?

# Answers to Conceptual Connections

### Atomic Size

**25.1** The element W has the larger radius because it is in the third transition row and Fe is in the first. Atomic radii increase from the first to the second transition row and stay roughly constant from the second to the third.

### Weak- and Strong-Field Ligands

**25.2** Ligand B forms a yellow solution, which means that the complex absorbs in the violet region. Ligand A forms a red solution, which means that the complex absorbs in the green. Since the violet region of the electromagnetic spectrum is of shorter wavelength (higher energy) than the green region, ligand B produces a higher $\Delta$.

# APPENDIX I  Common Mathematical Operations in Chemistry

## A. Scientific Notation

A number written in scientific notation consists of a **decimal part**, a number that is usually between 1 and 10, and an **exponential part**, 10 raised to an **exponent**, $n$:

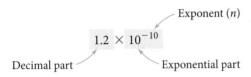

Each of the following numbers is written in both scientific and decimal notation:

$$1.0 \times 10^5 = 100{,}000 \qquad 1.0 \times 10^{-5} = 0.000010$$

$$6.7 \times 10^3 = 6700 \qquad 6.7 \times 10^{-3} = 0.0067$$

A positive exponent ($n$) means 1 multiplied by 10 $n$ times:

$$10^0 = 1$$

$$10^1 = 1 \times 10$$

$$10^2 = 1 \times 10 \times 10 = 100$$

$$10^3 = 1 \times 10 \times 10 \times 10 = 1000$$

A negative exponent ($-n$) means 1 divided by 10 $n$ times:

$$10^{-1} = \frac{1}{10} = 0.1$$

$$10^{-2} = \frac{1}{10 \times 10} = 0.01$$

$$10^{-3} = \frac{1}{10 \times 10 \times 10} = 0.001$$

To convert a number to scientific notation, we move the decimal point to obtain a number between 1 and 10 and then multiply by 10 raised to the appropriate power. For example, to write 5983 in scientific notation, we move the decimal point to the left three places to get 5.983 (a number between 1 and 10) and then multiply by 1000 to make up for moving the decimal point:

$$5983 = 5.983 \times 1000$$

Since 1000 is $10^3$, we write:

$$5983 = 5.983 \times 10^3$$

We can do this in one step by counting how many places we move the decimal point to obtain a number between 1 and 10 and then writing the decimal part multiplied by 10 raised to the number of places we moved the decimal point:

$$5983 = 5.983 \times 10^3$$

If the decimal point is moved to the left, as in the previous example, the exponent is positive. If the decimal is moved to the right, the exponent is negative:

$$0.00034 = 3.4 \times 10^{-4}$$

To express a number in scientific notation:

1.  Move the decimal point to obtain a number between 1 and 10.

2.  Write the result from step 1 multiplied by 10 raised to the number of places you moved the decimal point.

    - *The exponent is positive if you moved the decimal point to the left.*

    - *The exponent is negative if you moved the decimal point to the right.*

Consider the following additional examples:

$$290{,}809{,}000 = 2.90809 \times 10^8$$

$$0.000000000070 \text{ m} = 7.0 \times 10^{-11} \text{ m}$$

## Multiplication and Division

To multiply numbers expressed in scientific notation, multiply the decimal parts and add the exponents:

$$(A \times 10^m)(B \times 10^n) = (A \times B) \times 10^{m+n}$$

To divide numbers expressed in scientific notation, divide the decimal parts and subtract the exponent in the denominator from the exponent in the numerator:

$$\frac{(A \times 10^m)}{(B \times 10^n)} = \left(\frac{A}{B}\right) \times 10^{m-n}$$

Consider the following example involving multiplication:

$$(3.5 \times 10^4)(1.8 \times 10^6) = (3.5 \times 1.8) \times 10^{4+6}$$

$$= 6.3 \times 10^{10}$$

Consider the following example involving division:

$$\frac{(5.6 \times 10^7)}{(1.4 \times 10^3)} = \left(\frac{5.6}{1.4}\right) \times 10^{7-3}$$

$$= 4.0 \times 10^4$$

## Addition and Subtraction

To add or subtract numbers expressed in scientific notation, rewrite all the numbers so that they have the same exponent, then add or subtract the decimal parts of the numbers. The exponents remained unchanged:

$$A \times 10^n$$
$$\pm B \times 10^n$$
$$\overline{(A \pm B) \times 10^n}$$

Notice that the numbers *must have* the same exponent. Consider the following example involving addition:

$$4.82 \times 10^7$$
$$+3.4 \times 10^6$$

First, express both numbers with the same exponent. In this case, you rewrite the lower number and perform the addition as follows:

$$4.82 \times 10^7$$
$$+0.34 \times 10^7$$
$$\overline{5.16 \times 10^7}$$

Consider the following example involving subtraction:

$$7.33 \times 10^5$$
$$-1.9 \times 10^4$$

First, express both numbers with the same exponent. In this case, you rewrite the lower number and perform the subtraction as follows:

$$7.33 \times 10^5$$
$$-0.19 \times 10^5$$
$$\overline{7.14 \times 10^5}$$

## Powers and Roots

To raise a number written in scientific notation to a power, raise the decimal part to the power and multiply the exponent by the power:

$$(4.0 \times 10^6)^2 = 4.0^2 \times 10^{6 \times 2}$$
$$= 16 \times 10^{12}$$
$$= 1.6 \times 10^{13}$$

To take the $n$th root of a number written in scientific notation, take the $n$th root of the decimal part and divide the exponent by the root:

$$(4.0 \times 10^6)^{1/3} = 4.0^{1/3} \times 10^{6/3}$$
$$= 1.6 \times 10^2$$

## B. Logarithms

### Common (or Base 10) Logarithms

The common or base 10 logarithm (abbreviated log) of a number is the exponent to which 10 must be raised to obtain that number. For example, the log of 100 is 2 because 10 must be raised to the second power to get 100. Similarly, the log of 1000 is 3 because 10 must be raised to the third power to get 1000. The logs of several multiples of 10 are shown here:

$$\log 10 = 1$$
$$\log 100 = 2$$
$$\log 1000 = 3$$
$$\log 10{,}000 = 4$$

Because $10^0 = 1$ by definition, $\log 1 = 0$.

The log of a number smaller than one is negative because 10 must be raised to a negative exponent to get a number smaller than one. For example, the log of 0.01 is $-2$ because 10 must be raised to $-2$ to get 0.01. Similarly, the log of 0.001 is $-3$ because 10 must be raised to $-3$ to get 0.001. The logs of several fractional numbers are shown below:

$$\log 0.1 = -1$$
$$\log 0.01 = -2$$
$$\log 0.001 = -3$$
$$\log 0.0001 = -4$$

The logs of numbers that are not multiples of 10 can be computed on your calculator. See your calculator manual for specific instructions.

## Inverse Logarithms

The inverse logarithm or invlog function is exactly the opposite of the log function. For example, the log of 100 is 2 and the inverse log of 2 is 100. The log function and the invlog function undo one another:

$$\log 100 = 2$$
$$\text{invlog } 2 = 100$$
$$\text{invlog}(\log 100) = 100$$

The inverse log of a number is 10 rasied to that number:

$$\text{invlog } x = 10^x$$
$$\text{invlog } 3 = 10^3 = 1000$$

The inverse logs of numbers can be computed on your calculator. See your calculator manual for specific instructions.

## Natural (or Base e) Logarithms

The natural (or base $e$) logarithm (abbreviated ln) of a number is the exponent to which $e$ (which has the value of 2.71828...) must be raised to obtain that number. For example, the ln of 100 is 4.605 because $e$ must be raised to 4.605 to get 100. Similarly, the ln of 10.0 is 2.303 because $e$ must be raised to 2.303 to get 10.0.

The inverse natural logarithm or invln function is exactly the opposite of the ln function. For example, the ln of 100 is

4.605 and the inverse ln of 4.605 is 100. The inverse ln of a number is simply $e$ raised to that number:

$$\text{invln } x = e^x$$

$$\text{invln } 3 = e^3 = 20.1$$

The invln of a number can be computed on your calculator. See your calculator manual for specific instructions.

## Mathematical Operations Using Logarithms

Because logarithms are exponents, mathematical operations involving logarithms are similar to those involving exponents as follows:

$$\log(a \times b) = \log a + \log b \qquad \ln(a \times b) = \ln a + \ln b$$

$$\log \frac{a}{b} = \log a - \log b \qquad \ln \frac{a}{b} = \ln a - \ln b$$

$$\log a^n = n \log a \qquad \ln a^n = n \ln a$$

## C. Quadratic Equations

A quadratic equation contains at least one term in which the variable $x$ is raised to the second power (and no terms in which $x$ is raised to a higher power). A quadratic equation has the following general form:

$$ax^2 + bx + c = 0$$

A quadratic equation can be solved for $x$ using the quadratic formula:

$$x = \frac{-b \pm \sqrt{b^2 - 4ac}}{2a}$$

Quadratic equations are often encountered when solving equilibrium problems. Here we show how to use the quadratic formula to solve a quadratic equation for $x$:

$$3x^2 - 5x + 1 = 0 \quad (\textit{quadratic equation})$$

$$x = \frac{-b \pm \sqrt{b^2 - 4ac}}{2a}$$

$$= \frac{-(-5) \pm \sqrt{(-5)^2 - 4(3)(1)}}{2(3)}$$

$$= \frac{5 \pm 3.6}{6}$$

$$x = 1.43 \quad \text{or} \quad x = 0.233$$

As you can see, the solution to a quadratic equation usually has two values. In any real chemical system, one of the values can be eliminated because it has no physical significance. (For example, it may correspond to a negative concentration, which does not exist.)

## D. Graphs

Graphs are often used to visually show the relationship between two variables. For example, in Chapter 5 we show

the following relationship between the volume of a gas and its pressure:

▲ **Volume versus Pressure**   A plot of the volume of a gas sample—as measured in a J-tube—versus pressure. The plot shows that volume and pressure are inversely related.

The horizontal axis is the $x$-axis and is normally used to show the independent variable. The vertical axis is the $y$-axis and is normally used to show how the other variable (called the dependent variable) varies with a change in the independent variable. In this case, the graph shows that as the pressure of a gas sample increases, its volume decreases.

Many relationships in chemistry are *linear*, which means that if you change one variable by a factor of $n$ the other variable will also change by a factor of $n$. For example, the volume of a gas is linearly related to the number of moles of gas. When two quantities are linearly related, a graph of one versus the other produces a straight line. For example, the graph below shows how the volume of an ideal gas sample depends on the number of moles of gas in the sample:

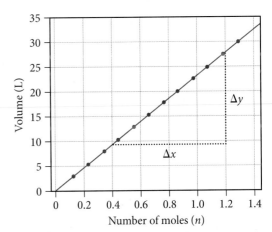

▲ **Volume versus Number of Moles**   The volume of a gas sample increases linearly with the number of moles of gas in the sample.

A linear relationship between any two variables $x$ and $y$ can be expressed by the following equation:

$$y = mx + b$$

where $m$ is the slope of the line and $b$ is the $y$-intercept. The slope is the change in $y$ divided by the change in $x$:

$$m = \frac{\Delta y}{\Delta x}$$

For the graph above, we can estimate the slope by simply estimating the changes in $y$ and $x$ for a given interval. For example, between $x = 0.4\,\text{mol}$ and $1.2\,\text{mol}$, $\Delta x = 0.80\,\text{mol}$ and we can estimate that $\Delta y = 18\,\text{L}$. Therefore the slope is:

$$m = \frac{\Delta y}{\Delta x} = \frac{18\,\text{L}}{0.80\,\text{mol}} = 23\,\text{mol/L}$$

In several places in this book, logarithmic relationships between variables can be plotted in order to obtain a linear relationship. For example, the variables $[A]_t$ and $t$ in the following equation are not linearly related, but the natural logarithm of $[A]_t$ and $t$ are linearly related:

$$\ln[A]_t = -kt + \ln[A]_0$$

$$y = mx + b$$

A plot of $\ln[A]_t$ versus $t$ will therefore produce a straight line with slope $= -k$ and $y$-intercept $= \ln[A]_0$.

# APPENDIX II  Useful Data

## A. Atomic Colors

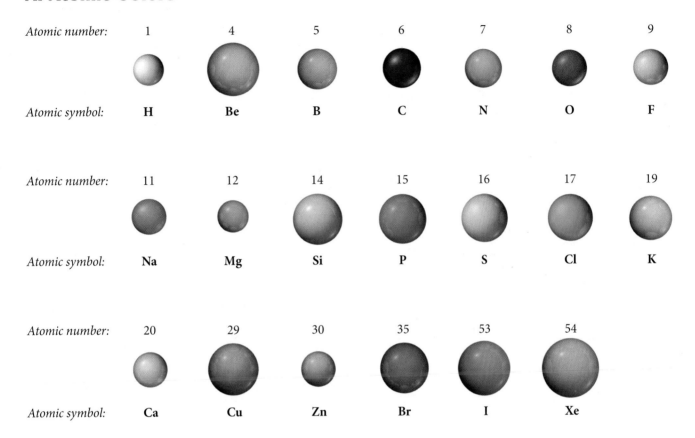

| Atomic number: | 1 | 4 | 5 | 6 | 7 | 8 | 9 |
| --- | --- | --- | --- | --- | --- | --- | --- |
| Atomic symbol: | H | Be | B | C | N | O | F |

| Atomic number: | 11 | 12 | 14 | 15 | 16 | 17 | 19 |
| --- | --- | --- | --- | --- | --- | --- | --- |
| Atomic symbol: | Na | Mg | Si | P | S | Cl | K |

| Atomic number: | 20 | 29 | 30 | 35 | 53 | 54 |
| --- | --- | --- | --- | --- | --- | --- |
| Atomic symbol: | Ca | Cu | Zn | Br | I | Xe |

## B. Standard Thermodynamic Quantities for Selected Substances at 25 °C

| Substance | $\Delta H_f^\circ$(kJ/mol) | $\Delta G_f^\circ$(kJ/mol) | $S^\circ$(J/mol · K) |
| --- | --- | --- | --- |
| **Aluminum** | | | |
| Al(s) | 0 | 0 | 28.32 |
| Al(g) | 330.0 | 289.4 | 164.6 |
| $Al^{3+}$(aq) | −538.4 | −483 | −325 |
| $AlCl_3$(s) | −704.2 | −628.8 | 109.3 |
| $Al_2O_3$(s) | −1675.7 | −1582.3 | 50.9 |
| **Barium** | | | |
| Ba(s) | 0 | 0 | 62.5 |
| Ba(g) | 180.0 | 146.0 | 170.2 |
| $Ba^{2+}$(aq) | −537.6 | −560.8 | 9.6 |
| $BaCO_3$(s) | −1213.0 | −1134.4 | 112.1 |
| $BaCl_2$(s) | −855.0 | −806.7 | 123.7 |
| BaO(s) | −548.0 | −520.3 | 72.1 |
| $Ba(OH)_2$(s) | −944.7 | | |
| $BaSO_4$(s) | −1473.2 | −1362.2 | 132.2 |

| Substance | $\Delta H_f^\circ$(kJ/mol) | $\Delta G_f^\circ$(kJ/mol) | $S^\circ$(J/mol · K) |
| --- | --- | --- | --- |
| **Beryllium** | | | |
| Be(s) | 0 | 0 | 9.5 |
| BeO(s) | −609.4 | −580.1 | 13.8 |
| $Be(OH)_2$(s) | −902.5 | −815.0 | 45.5 |
| **Bismuth** | | | |
| Bi(s) | 0 | 0 | 56.7 |
| $BiCl_3$(s) | −379.1 | −315.0 | 177.0 |
| $Bi_2O_3$(s) | −573.9 | −493.7 | 151.5 |
| $Bi_2S_3$(s) | −143.1 | −140.6 | 200.4 |
| **Boron** | | | |
| B(s) | 0 | 0 | 5.9 |
| B(g) | 565.0 | 521.0 | 153.4 |
| $BCl_3$(g) | −403.8 | −388.7 | 290.1 |
| $BF_3$(g) | −1136.0 | −1119.4 | 254.4 |
| $B_2H_6$(g) | 36.4 | 87.6 | 232.1 |

—(Continued on the next page)

| Substance | $\Delta H_f^\circ$(kJ/mol) | $\Delta G_f^\circ$(kJ/mol) | $S^\circ$(J/mol · K) |
|---|---|---|---|
| $B_2O_3(s)$ | −1273.5 | −1194.3 | 54.0 |
| $H_3BO_3(s)$ | −1094.3 | −968.9 | 90.0 |
| **Bromine** | | | |
| $Br(g)$ | 111.9 | 82.4 | 175.0 |
| $Br_2(l)$ | 0 | 0 | 152.2 |
| $Br_2(g)$ | 30.9 | 3.1 | 245.5 |
| $Br^-(aq)$ | −121.4 | −102.8 | 80.71 |
| $HBr(g)$ | −36.3 | −53.4 | 198.7 |
| **Cadmium** | | | |
| $Cd(s)$ | 0 | 0 | 51.8 |
| $Cd(g)$ | 111.8 | 77.3 | 167.7 |
| $Cd^{2+}(aq)$ | −75.9 | −77.6 | −73.2 |
| $CdCl_2(s)$ | −391.5 | −343.9 | 115.3 |
| $CdO(s)$ | −258.4 | −228.7 | 54.8 |
| $CdS(s)$ | −161.9 | −156.5 | 64.9 |
| $CdSO_4(s)$ | −933.3 | −822.7 | 123.0 |
| **Calcium** | | | |
| $Ca(s)$ | 0 | 0 | 41.6 |
| $Ca(g)$ | 177.8 | 144.0 | 154.9 |
| $Ca^{2+}(aq)$ | −542.8 | −553.6 | −53.1 |
| $CaC_2(s)$ | −59.8 | −64.9 | 70.0 |
| $CaCO_3(s)$ | −1207.6 | −1129.1 | 91.7 |
| $CaCl_2(s)$ | −795.4 | −748.8 | 108.4 |
| $CaF_2(s)$ | −1228.0 | −1175.6 | 68.5 |
| $CaH_2(s)$ | −181.5 | −142.5 | 41.4 |
| $Ca(NO_3)_2(s)$ | −938.2 | −742.8 | 193.2 |
| $CaO(s)$ | −634.9 | −603.3 | 38.1 |
| $Ca(OH)_2(s)$ | −985.2 | −897.5 | 83.4 |
| $CaSO_4(s)$ | −1434.5 | −1322.0 | 106.5 |
| $Ca_3(PO_4)_2(s)$ | −4120.8 | −3884.7 | 236.0 |
| **Carbon** | | | |
| $C(s, graphite)$ | 0 | 0 | 5.7 |
| $C(s, diamond)$ | 1.88 | 2.9 | 2.4 |
| $C(g)$ | 716.7 | 671.3 | 158.1 |
| $CH_4(g)$ | −74.6 | −50.5 | 186.3 |
| $CH_3Cl(g)$ | −81.9 | −60.2 | 234.6 |
| $CH_2Cl_2(g)$ | −95.4 | | 270.2 |
| $CH_2Cl_2(l)$ | −124.2 | −63.2 | 177.8 |
| $CHCl_3(l)$ | −134.1 | −73.7 | 201.7 |
| $CCl_4(g)$ | −95.7 | −62.3 | 309.7 |
| $CCl_4(l)$ | −128.2 | −66.4 | 216.4 |
| $CH_2O(g)$ | −108.6 | −102.5 | 218.8 |
| $CH_2O_2$ (l, formic acid) | −425.0 | −361.4 | 129.0 |
| $CH_3NH_2$ (g, methylamine) | −22.5 | 32.7 | 242.9 |
| $CH_3OH(l)$ | −238.6 | −166.6 | 126.8 |
| $CH_3OH(g)$ | −201.0 | −162.3 | 239.9 |

| Substance | $\Delta H_f^\circ$(kJ/mol) | $\Delta G_f^\circ$(kJ/mol) | $S^\circ$(J/mol · K) |
|---|---|---|---|
| $C_2H_2(g)$ | 227.4 | 209.9 | 200.9 |
| $C_2H_4(g)$ | 52.4 | 68.4 | 219.3 |
| $C_2H_6(g)$ | −84.68 | −32.0 | 229.2 |
| $C_2H_5OH(l)$ | −277.6 | −174.8 | 160.7 |
| $C_2H_5OH(g)$ | −234.8 | −167.9 | 281.6 |
| $C_2H_3Cl$ (g, vinyl chloride) | 37.2 | 53.6 | 264.0 |
| $C_2H_4Cl_2$ (l, dichloroethane) | −166.8 | −79.6 | 208.5 |
| $C_2H_4O$ (g, acetaldehyde) | −166.2 | −133.0 | 263.8 |
| $C_2H_4O_2$ (l, acetic acid) | −484.3 | −389.9 | 159.8 |
| $C_3H_8(g)$ | −103.85 | −23.4 | 270.3 |
| $C_3H_6O$ (l, acetone) | −248.4 | −155.6 | 199.8 |
| $C_3H_7OH$ (l, isopropanol) | −318.1 | | 181.1 |
| $C_4H_{10}(l)$ | −147.3 | −15.0 | 231.0 |
| $C_4H_{10}(g)$ | −125.7 | −15.71 | 310.0 |
| $C_6H_6(l)$ | 49.1 | 124.5 | 173.4 |
| $C_6H_5NH_2$ (l, aniline) | 31.6 | 149.2 | 191.9 |
| $C_6H_5OH$ (s, phenol) | −165.1 | −50.4 | 144.0 |
| $C_6H_{12}O_6$ (s, glucose) | −1273.3 | −910.4 | 212.1 |
| $C_{10}H_8$ (s, naphthalene) | 78.5 | 201.6 | 167.4 |
| $C_{12}H_{22}O_{11}$ (s, sucrose) | −2226.1 | −1544.3 | 360.24 |
| $CO(g)$ | −110.5 | −137.2 | 197.7 |
| $CO_2(g)$ | −393.5 | −394.4 | 213.8 |
| $CO_2(aq)$ | −413.8 | −386.0 | 117.6 |
| $CO_3^{2-}(aq)$ | −677.1 | −527.8 | −56.9 |
| $HCO_3^-(aq)$ | −692.0 | −586.8 | 91.2 |
| $H_2CO_3(aq)$ | −699.7 | −623.2 | 187.4 |
| $CN^-(aq)$ | 151 | 166 | 118 |
| $HCN(l)$ | 108.9 | 125.0 | 112.8 |
| $HCN(g)$ | 135.1 | 124.7 | 201.8 |
| $CS_2(l)$ | 89.0 | 64.6 | 151.3 |
| $CS_2(g)$ | 116.7 | 67.1 | 237.8 |
| $COCl_2(g)$ | −219.1 | −204.9 | 283.5 |
| $C_{60}(s)$ | 2327.0 | 2302.0 | 426.0 |
| **Cesium** | | | |
| $Cs(s)$ | 0 | 0 | 85.2 |
| $Cs(g)$ | 76.5 | 49.6 | 175.6 |
| $Cs^+(aq)$ | −258.0 | −292.0 | 132.1 |
| $CsBr(s)$ | −400 | −387 | 117 |
| $CsCl(s)$ | −438 | −414 | 101.2 |
| $CsF(s)$ | −553.5 | −525.5 | 92.8 |
| $CsI(s)$ | −342 | −337 | 127 |

| Substance | $\Delta H_f^\circ$(kJ/mol) | $\Delta G_f^\circ$(kJ/mol) | $S^\circ$(J/mol · K) |
|---|---|---|---|
| *Chlorine* | | | |
| $Cl(g)$ | 121.3 | 105.3 | 165.2 |
| $Cl_2(g)$ | 0 | 0 | 223.1 |
| $Cl^-(aq)$ | −167.1 | −131.2 | 56.6 |
| $HCl(g)$ | −92.3 | −95.3 | 186.9 |
| $HCl(aq)$ | −167.2 | −131.2 | 56.5 |
| $ClO_2(g)$ | 102.5 | 120.5 | 256.8 |
| $Cl_2O(g)$ | 80.3 | 97.9 | 266.2 |
| *Chromium* | | | |
| $Cr(s)$ | 0 | 0 | 23.8 |
| $Cr(g)$ | 396.6 | 351.8 | 174.5 |
| $Cr^{3+}(aq)$ | −1971 | | |
| $CrO_4^{2-}(aq)$ | −872.2 | −717.1 | 44 |
| $Cr_2O_3(s)$ | −1139.7 | −1058.1 | 81.2 |
| $Cr_2O_7^{2-}(aq)$ | −1476 | −1279 | 238 |
| *Cobalt* | | | |
| $Co(s)$ | 0 | 0 | 30.0 |
| $Co(g)$ | 424.7 | 380.3 | 179.5 |
| $CoO(s)$ | −237.9 | −214.2 | 53.0 |
| $Co(OH)_2(s)$ | −539.7 | −454.3 | 79.0 |
| *Copper* | | | |
| $Cu(s)$ | 0 | 0 | 33.2 |
| $Cu(g)$ | 337.4 | 297.7 | 166.4 |
| $Cu^+(aq)$ | 51.9 | 50.2 | −26 |
| $Cu^{2+}(aq)$ | 64.9 | 65.5 | −98 |
| $CuCl(s)$ | −137.2 | −119.9 | 86.2 |
| $CuCl_2(s)$ | −220.1 | −175.7 | 108.1 |
| $CuO(s)$ | −157.3 | −129.7 | 42.6 |
| $CuS(s)$ | −53.1 | −53.6 | 66.5 |
| $CuSO_4(s)$ | −771.4 | −662.2 | 109.2 |
| $Cu_2O(s)$ | −168.6 | −146.0 | 93.1 |
| $Cu_2S(s)$ | −79.5 | −86.2 | 120.9 |
| *Fluorine* | | | |
| $F(g)$ | 79.38 | 62.3 | 158.75 |
| $F_2(g)$ | 0 | 0 | 202.79 |
| $F^-(aq)$ | −335.35 | −278.8 | −13.8 |
| $HF(g)$ | −273.3 | −275.4 | 173.8 |
| *Gold* | | | |
| $Au(s)$ | 0 | 0 | 47.4 |
| $Au(g)$ | 366.1 | 326.3 | 180.5 |
| *Helium* | | | |
| $He(g)$ | 0 | 0 | 126.2 |
| *Hydrogen* | | | |
| $H(g)$ | 218.0 | 203.3 | 114.7 |
| $H^+(aq)$ | 0 | 0 | 0 |
| $H^+(aq)$ | 1536.3 | 1517.1 | 108.9 |
| $H_2(g)$ | 0 | 0 | 130.7 |

| Substance | $\Delta H_f^\circ$(kJ/mol) | $\Delta G_f^\circ$(kJ/mol) | $S^\circ$(J/mol · K) |
|---|---|---|---|
| *Iodine* | | | |
| $I(g)$ | 106.76 | 70.2 | 180.79 |
| $I_2(s)$ | 0 | 0 | 116.14 |
| $I_2(g)$ | 62.42 | 19.3 | 260.69 |
| $I^-(aq)$ | −56.78 | −51.57 | 106.45 |
| $HI(g)$ | 26.5 | 1.7 | 206.6 |
| *Iron* | | | |
| $Fe(s)$ | 0 | 0 | 27.3 |
| $Fe(g)$ | 416.3 | 370.7 | 180.5 |
| $Fe^{2+}(aq)$ | −87.9 | −84.94 | −113.4 |
| $Fe^{3+}(aq)$ | −47.69 | −10.54 | −293.3 |
| $FeCO_3(s)$ | −740.6 | −666.7 | 92.9 |
| $FeCl_2(s)$ | −341.8 | −302.3 | 118.0 |
| $FeCl_3(s)$ | −399.5 | −334.0 | 142.3 |
| $FeO(s)$ | −272.0 | −255.2 | 60.75 |
| $Fe(OH)_3(s)$ | −823.0 | −696.5 | 106.7 |
| $FeS_2(s)$ | −178.2 | −166.9 | 52.9 |
| $Fe_2O_3(s)$ | −824.2 | −742.2 | 87.4 |
| $Fe_3O_4(s)$ | −1118.4 | −1015.4 | 146.4 |
| *Lead* | | | |
| $Pb(s)$ | 0 | 0 | 64.8 |
| $Pb(g)$ | 195.2 | 162.2 | 175.4 |
| $Pb^{2+}(aq)$ | 0.92 | −24.4 | 18.5 |
| $PbBr_2(s)$ | −278.7 | −261.9 | 161.5 |
| $PbCO_3(s)$ | −699.1 | −625.5 | 131.0 |
| $PbCl_2(s)$ | −359.4 | −314.1 | 136.0 |
| $PbI_2(s)$ | −175.5 | −173.6 | 174.9 |
| $Pb(NO_3)_2(s)$ | −451.9 | | |
| $PbO(s)$ | −217.3 | −187.9 | 68.7 |
| $PbO_2(s)$ | −277.4 | −217.3 | 68.6 |
| $PbS(s)$ | −100.4 | −98.7 | 91.2 |
| $PbSO_4(s)$ | −920.0 | −813.0 | 148.5 |
| *Lithium* | | | |
| $Li(s)$ | 0 | 0 | 29.1 |
| $Li(g)$ | 159.3 | 126.6 | 138.8 |
| $Li^+(aq)$ | −278.47 | −293.3 | 12.24 |
| $LiBr(s)$ | −351.2 | −342.0 | 74.3 |
| $LiCl(s)$ | −408.6 | −384.4 | 59.3 |
| $LiF(s)$ | −616.0 | −587.7 | 35.7 |
| $LiI(s)$ | −270.4 | −270.3 | 86.8 |
| $LiNO_3(s)$ | −483.1 | −381.1 | 90.0 |
| $LiOH(s)$ | −487.5 | −441.5 | 42.8 |
| $Li_2O(s)$ | −597.9 | −561.2 | 37.6 |
| *Magnesium* | | | |
| $Mg(s)$ | 0 | 0 | 32.7 |
| $Mg(g)$ | 147.1 | 112.5 | 148.6 |
| $Mg^{2+}(aq)$ | −467.0 | −455.4 | −137 |
| $MgCl_2(s)$ | −641.3 | −591.8 | 89.6 |
| $MgCO_3(s)$ | −1095.8 | −1012.1 | 65.7 |
| $MgF_2(s)$ | −1124.2 | −1071.1 | 57.2 |

—(Continued on the next page)

| Substance | $\Delta H_f^\circ$(kJ/mol) | $\Delta G_f^\circ$(kJ/mol) | $S^\circ$(J/mol · K) |
|---|---|---|---|
| MgO(s) | −601.6 | −569.3 | 27.0 |
| Mg(OH)$_2$(s) | −924.5 | −833.5 | 63.2 |
| MgSO$_4$(s) | −1284.9 | −1170.6 | 91.6 |
| Mg$_3$N$_2$(s) | −461 | −401 | 88 |
| *Manganese* | | | |
| Mn(s) | 0 | 0 | 32.0 |
| Mn(g) | 280.7 | 238.5 | 173.7 |
| Mn$^{2+}$(aq) | −219.4 | −225.6 | −78.8 |
| MnO(s) | −385.2 | −362.9 | 59.7 |
| MnO$_2$(s) | −520.0 | −465.1 | 53.1 |
| MnO$_4^-$(aq) | −529.9 | −436.2 | 190.6 |
| *Mercury* | | | |
| Hg(l) | 0 | 0 | 75.9 |
| Hg(g) | 61.4 | 31.8 | 175.0 |
| Hg$^{2+}$(aq) | 170.21 | 164.4 | −36.19 |
| Hg$_2^{2+}$(aq) | 166.87 | 153.5 | 65.74 |
| HgCl$_2$(s) | −224.3 | −178.6 | 146.0 |
| HgO(s) | −90.8 | −58.5 | 70.3 |
| HgS(s) | −58.2 | −50.6 | 82.4 |
| Hg$_2$Cl$_2$(s) | −265.4 | −210.7 | 191.6 |
| *Nickel* | | | |
| Ni(s) | 0 | 0 | 29.9 |
| Ni(g) | 429.7 | 384.5 | 182.2 |
| NiCl$_2$(s) | −305.3 | −259.0 | 97.7 |
| NiO(s) | −239.7 | −211.7 | 37.99 |
| NiS(s) | −82.0 | −79.5 | 53.0 |
| *Nitrogen* | | | |
| N(g) | 472.7 | 455.5 | 153.3 |
| N$_2$(g) | 0 | 0 | 191.6 |
| NF$_3$(g) | −132.1 | −90.6 | 260.8 |
| NH$_3$(g) | −45.9 | −16.4 | 192.8 |
| NH$_3$(aq) | −80.29 | −26.50 | 111.3 |
| NH$_4^+$(aq) | −133.26 | −79.31 | 111.17 |
| NH$_4$Br(s) | −270.8 | −175.2 | 113.0 |
| NH$_4$Cl(s) | −314.4 | −202.9 | 94.6 |
| NH$_4$CN(s) | 0.4 | | |
| NH$_4$F(s) | −464.0 | −348.7 | 72.0 |
| NH$_4$HCO$_3$(s) | −849.4 | −665.9 | 120.9 |
| NH$_4$I(s) | −201.4 | −112.5 | 117.0 |
| NH$_4$NO$_3$(s) | −365.6 | −183.9 | 151.1 |
| NH$_4$NO$_3$(aq) | −339.9 | −190.6 | 259.8 |
| HNO$_3$(g) | −133.9 | −73.5 | 266.9 |
| HNO$_3$(aq) | −207 | −110.9 | 146 |
| NO(g) | 91.3 | 87.6 | 210.8 |
| NO$_2$(g) | 33.2 | 51.3 | 240.1 |
| NO$_3^-$(aq) | −206.85 | −110.2 | 146.70 |
| NOBr(g) | 82.2 | 82.4 | 273.7 |
| NOCl(g) | 51.7 | 66.1 | 261.7 |
| N$_2$H$_4$(l) | 50.6 | 149.3 | 121.2 |
| N$_2$H$_4$(g) | 95.4 | 159.4 | 238.5 |

| Substance | $\Delta H_f^\circ$(kJ/mol) | $\Delta G_f^\circ$(kJ/mol) | $S^\circ$(J/mol · K) |
|---|---|---|---|
| N$_2$O(g) | 81.6 | 103.7 | 220.0 |
| N$_2$O$_4$(l) | −19.5 | 97.5 | 209.2 |
| N$_2$O$_4$(g) | 9.16 | 99.8 | 304.4 |
| N$_2$O$_5$(s) | −43.1 | 113.9 | 178.2 |
| N$_2$O$_5$(g) | 13.3 | 117.1 | 355.7 |
| *Oxygen* | | | |
| O(g) | 249.2 | 231.7 | 161.1 |
| O$_2$(g) | 0 | 0 | 205.2 |
| O$_3$(g) | 142.7 | 163.2 | 238.9 |
| OH$^-$(aq) | −230.02 | −157.3 | −10.90 |
| H$_2$O(l) | −285.8 | −237.1 | 70.0 |
| H$_2$O(g) | −241.8 | −228.6 | 188.8 |
| H$_2$O$_2$(l) | −187.8 | −120.4 | 109.6 |
| H$_2$O$_2$(g) | −136.3 | −105.6 | 232.7 |
| *Phosphorus* | | | |
| P(s, white) | 0 | 0 | 41.1 |
| P(s, red) | −17.6 | −12.1 | 22.8 |
| P(g) | 316.5 | 280.1 | 163.2 |
| P$_2$(g) | 144.0 | 103.5 | 218.1 |
| P$_4$(g) | 58.9 | 24.4 | 280.0 |
| PCl$_3$(l) | −319.7 | −272.3 | 217.1 |
| PCl$_3$(g) | −287.0 | −267.8 | 311.8 |
| PCl$_5$(s) | −443.5 | | |
| PCl$_5$(g) | −374.9 | −305.0 | 364.6 |
| PF$_5$(g) | −1594.4 | −1520.7 | 300.8 |
| PH$_3$(g) | 5.4 | 13.5 | 210.2 |
| POCl$_3$(l) | −597.1 | −520.8 | 222.5 |
| POCl$_3$(g) | −558.5 | −512.9 | 325.5 |
| PO$_4^{3-}$(aq) | −1277.4 | −1018.7 | −220.5 |
| HPO$_4^{2-}$(aq) | −1292.1 | −1089.2 | −33.5 |
| H$_2$PO$_4^-$(aq) | −1296.3 | −1130.2 | 90.4 |
| H$_3$PO$_4$(s) | −1284.4 | −1124.3 | 110.5 |
| H$_3$PO$_4$(aq) | −1288.3 | −1142.6 | 158.2 |
| P$_4$O$_6$(s) | −1640.1 | | |
| P$_4$O$_{10}$(s) | −2984 | −2698 | 228.9 |
| *Platinum* | | | |
| Pt(s) | 0 | 0 | 41.6 |
| Pt(g) | 565.3 | 520.5 | 192.4 |
| *Potassium* | | | |
| K(s) | 0 | 0 | 64.7 |
| K(g) | 89.0 | 60.5 | 160.3 |
| K$^+$(aq) | −252.14 | −283.3 | 101.2 |
| KBr(s) | −393.8 | −380.7 | 95.9 |
| KCN(s) | −113.0 | −101.9 | 128.5 |
| KCl(s) | −436.5 | −408.5 | 82.6 |
| KClO$_3$(s) | −397.7 | −296.3 | 143.1 |
| KClO$_4$(s) | −432.8 | −303.1 | 151.0 |
| KF(s) | −567.3 | −537.8 | 66.6 |
| KI(s) | −327.9 | −324.9 | 106.3 |
| KNO$_3$(s) | −494.6 | −394.9 | 133.1 |

| Substance | $\Delta H_f^\circ$(kJ/mol) | $\Delta G_f^\circ$(kJ/mol) | $S^\circ$(J/mol · K) |
|---|---|---|---|
| KOH(s) | −424.6 | −379.4 | 81.2 |
| KOH(aq) | −482.4 | −440.5 | 91.6 |
| KO₂(s) | −284.9 | −239.4 | 116.7 |
| K₂CO₃(s) | −1151.0 | −1063.5 | 155.5 |
| K₂O(s) | −361.5 | −322.1 | 94.14 |
| K₂O₂(s) | −494.1 | −425.1 | 102.1 |
| K₂SO₄(s) | −1437.8 | −1321.4 | 175.6 |
| **Rubidium** | | | |
| Rb(s) | 0 | 0 | 76.8 |
| Rb(g) | 80.9 | 53.1 | 170.1 |
| Rb⁺(aq) | −251.12 | −283.1 | 121.75 |
| RbBr(s) | −394.6 | −381.8 | 110.0 |
| RbCl(s) | −435.4 | −407.8 | 95.9 |
| RbClO₃(s) | −392.4 | −292.0 | 152 |
| RbF(s) | −557.7 | | |
| RbI(s) | −333.8 | −328.9 | 118.4 |
| **Scandium** | | | |
| Sc(s) | 0 | 0 | 34.6 |
| Sc(g) | 377.8 | 336.0 | 174.8 |
| **Selenium** | | | |
| Se(s, gray) | 0 | 0 | 42.4 |
| Se(g) | 227.1 | 187.0 | 176.7 |
| H₂Se(g) | 29.7 | 15.9 | 219.0 |
| **Silicon** | | | |
| Si(s) | 0 | 0 | 18.8 |
| Si(g) | 450.0 | 405.5 | 168.0 |
| SiCl₄(l) | −687.0 | −619.8 | 239.7 |
| SiF₄(g) | −1615.0 | −1572.8 | 282.8 |
| SiH₄(g) | 34.3 | 56.9 | 204.6 |
| SiO₂(s, quartz) | −910.7 | −856.3 | 41.5 |
| Si₂H₆(g) | 80.3 | 127.3 | 272.7 |
| **Silver** | | | |
| Ag(s) | 0 | 0 | 42.6 |
| Ag(g) | 284.9 | 246.0 | 173.0 |
| Ag⁺(aq) | 105.79 | 77.11 | 73.45 |
| AgBr(s) | −100.4 | −96.9 | 107.1 |
| AgCl(s) | −127.0 | −109.8 | 96.3 |
| AgF(s) | −204.6 | −185 | 84 |
| AgI(s) | −61.8 | −66.2 | 115.5 |
| AgNO₃(s) | −124.4 | −33.4 | 140.9 |
| Ag₂O(s) | −31.1 | −11.2 | 121.3 |
| Ag₂S(s) | −32.6 | −40.7 | 144.0 |
| Ag₂SO₄(s) | −715.9 | −618.4 | 200.4 |
| **Sodium** | | | |
| Na(s) | 0 | 0 | 51.3 |
| Na(g) | 107.5 | 77.0 | 153.7 |
| Na⁺(aq) | −240.34 | −261.9 | 58.45 |
| NaBr(s) | −361.1 | −349.0 | 86.8 |
| NaCl(s) | −411.2 | −384.1 | 72.1 |
| NaCl(aq) | −407.2 | −393.1 | 115.5 |

| Substance | $\Delta H_f^\circ$(kJ/mol) | $\Delta G_f^\circ$(kJ/mol) | $S^\circ$(J/mol · K) |
|---|---|---|---|
| NaClO₃(s) | −365.8 | −262.3 | 123.4 |
| NaF(s) | −576.6 | −546.3 | 51.1 |
| NaHCO₃(s) | −950.8 | −851.0 | 101.7 |
| NaHSO₄(s) | −1125.5 | −992.8 | 113.0 |
| NaI(s) | −287.8 | −286.1 | 98.5 |
| NaNO₃(s) | −467.9 | −367.0 | 116.5 |
| NaNO₃(aq) | −447.5 | −373.2 | 205.4 |
| NaOH(s) | −425.8 | −379.7 | 64.4 |
| NaOH(aq) | −470.1 | −419.2 | 48.2 |
| NaO₂(s) | −260.2 | −218.4 | 115.9 |
| Na₂CO₃(s) | −1130.7 | −1044.4 | 135.0 |
| Na₂O(s) | −414.2 | −375.5 | 75.1 |
| Na₂O₂(s) | −510.9 | −447.7 | 95.0 |
| Na₂SO₄(s) | −1387.1 | −1270.2 | 149.6 |
| Na₃PO₄(s) | −1917 | −1789 | 173.8 |
| **Strontium** | | | |
| Sr(s) | 0 | 0 | 55.0 |
| Sr(g) | 164.4 | 130.9 | 164.6 |
| Sr²⁺(aq) | −545.51 | −557.3 | −39 |
| SrCl₂(s) | −828.9 | −781.1 | 114.9 |
| SrCO₃(s) | −1220.1 | −1140.1 | 97.1 |
| SrO(s) | −592.0 | −561.9 | 54.4 |
| SrSO₄(s) | −1453.1 | −1340.9 | 117.0 |
| **Sulfur** | | | |
| S(s, rhombic) | 0 | 0 | 32.1 |
| S(s, monoclinic) | 0.3 | 0.096 | 32.6 |
| S(g) | 277.2 | 236.7 | 167.8 |
| S₂(g) | 128.6 | 79.7 | 228.2 |
| S₈(g) | 102.3 | 49.7 | 430.9 |
| S²⁻(aq) | 41.8 | 83.7 | 22 |
| SF₆(g) | −1220.5 | −1116.5 | 291.5 |
| HS⁻(aq) | −17.7 | 12.4 | 62.0 |
| H₂S(g) | −20.6 | −33.4 | 205.8 |
| H₂S(aq) | −39.4 | −27.7 | 122 |
| SOCl₂(l) | −245.6 | | |
| SO₂(g) | −296.8 | −300.1 | 248.2 |
| SO₃(g) | −395.7 | −371.1 | 256.8 |
| SO₄²⁻(aq) | −909.3 | −744.6 | 18.5 |
| HSO₄⁻(aq) | −886.5 | −754.4 | 129.5 |
| H₂SO₄(l) | −814.0 | −690.0 | 156.9 |
| H₂SO₄(aq) | −909.3 | −744.6 | 18.5 |
| S₂O₃²⁻(aq) | −648.5 | −522.5 | 67 |
| **Tin** | | | |
| Sn(s, white) | 0 | 0 | 51.2 |
| Sn(s, gray) | −2.1 | 0.1 | 44.1 |
| Sn(g) | 301.2 | 266.2 | 168.5 |
| SnCl₄(l) | −511.3 | −440.1 | 258.6 |
| SnCl₄(g) | −471.5 | −432.2 | 365.8 |
| SnO(s) | −280.7 | −251.9 | 57.2 |
| SnO₂(s) | −577.6 | −515.8 | 49.0 |

—(Continued on the next page)

| Substance | $\Delta H_f^\circ$(kJ/mol) | $\Delta G_f^\circ$(kJ/mol) | $S^\circ$(J/mol·K) |
|---|---|---|---|
| **Titanium** | | | |
| Ti(s) | 0 | 0 | 30.7 |
| Ti(g) | 473.0 | 428.4 | 180.3 |
| TiCl$_4$(l) | −804.2 | −737.2 | 252.3 |
| TiCl$_4$(g) | −763.2 | −726.3 | 353.2 |
| TiO$_2$(s) | −944.0 | −888.8 | 50.6 |
| **Tungsten** | | | |
| W(s) | 0 | 0 | 32.6 |
| W(g) | 849.4 | 807.1 | 174.0 |
| WO$_3$(s) | −842.9 | −764.0 | 75.9 |
| **Uranium** | | | |
| U(s) | 0 | 0 | 50.2 |
| U(g) | 533.0 | 488.4 | 199.8 |
| UF$_6$(s) | −2197.0 | −2068.5 | 227.6 |

| Substance | $\Delta H_f^\circ$(kJ/mol) | $\Delta G_f^\circ$(kJ/mol) | $S^\circ$(J/mol·K) |
|---|---|---|---|
| UF$_6$(g) | −2147.4 | −2063.7 | 377.9 |
| UO$_2$(s) | −1085.0 | −1031.8 | 77.0 |
| **Vanadium** | | | |
| V(s) | 0 | 0 | 28.9 |
| V(g) | 514.2 | 754.4 | 182.3 |
| **Zinc** | | | |
| Zn(s) | 0 | 0 | 41.6 |
| Zn(g) | 130.4 | 94.8 | 161.0 |
| Zn$^{2+}$(aq) | −153.39 | −147.1 | −109.8 |
| ZnCl$_2$(s) | −415.1 | −369.4 | 111.5 |
| ZnO(s) | −350.5 | −320.5 | 43.7 |
| ZnS (s, zinc blende) | −206.0 | −201.3 | 57.7 |
| ZnSO$_4$(s) | −982.8 | −871.5 | 110.5 |

# C. Aqueous Equilibrium Constants

## 1. Dissociation Constants for Acids at 25 °C

| Name | Formula | $K_{a_1}$ | $K_{a_2}$ | $K_{a_3}$ |
|---|---|---|---|---|
| Acetic | HC$_2$H$_3$O$_2$ | $1.8 \times 10^{-5}$ | | |
| Acetylsalicylic | HC$_9$H$_7$O$_4$ | $3.3 \times 10^{-4}$ | | |
| Adipic | H$_2$C$_6$H$_8$O$_4$ | $3.9 \times 10^{-5}$ | $3.9 \times 10^{-6}$ | |
| Arsenic | H$_3$AsO$_4$ | $5.5 \times 10^{-3}$ | $1.7 \times 10^{-7}$ | $5.1 \times 10^{-12}$ |
| Arsenous | H$_3$AsO$_3$ | $5.1 \times 10^{-10}$ | | |
| Ascorbic | H$_2$C$_6$H$_6$O$_6$ | $8.0 \times 10^{-5}$ | $1.6 \times 10^{-12}$ | |
| Benzoic | HC$_7$H$_5$O$_2$ | $6.5 \times 10^{-5}$ | | |
| Boric | H$_3$BO$_3$ | $5.4 \times 10^{-10}$ | | |
| Butanoic | HC$_4$H$_7$O$_2$ | $1.5 \times 10^{-5}$ | | |
| Carbonic | H$_2$CO$_3$ | $4.3 \times 10^{-7}$ | $5.6 \times 10^{-11}$ | |
| Chloroacetic | HC$_2$H$_2$O$_2$Cl | $1.4 \times 10^{-3}$ | | |
| Chlorous | HClO$_2$ | $1.1 \times 10^{-2}$ | | |
| Citric | H$_3$C$_6$H$_5$O$_7$ | $7.4 \times 10^{-4}$ | $1.7 \times 10^{-5}$ | $4.0 \times 10^{-7}$ |
| Cyanic | HCNO | $2 \times 10^{-4}$ | | |
| Formic | HCHO$_2$ | $1.8 \times 10^{-4}$ | | |
| Hydrazoic | HN$_3$ | $2.5 \times 10^{-5}$ | | |
| Hydrocyanic | HCN | $4.9 \times 10^{-10}$ | | |
| Hydrofluoric | HF | $6.8 \times 10^{-4}$ | | |
| Hydrogen chromate ion | HCrO$_4^-$ | $3.0 \times 10^{-7}$ | | |
| Hydrogen peroxide | H$_2$O$_2$ | $2.4 \times 10^{-12}$ | | |
| Hydrogen selenate ion | HSeO$_4^-$ | $2.2 \times 10^{-2}$ | | |
| Hydrosulfuric | H$_2$S | $8.9 \times 10^{-8}$ | $1 \times 10^{-19}$ | |
| Hydrotelluric | H$_2$Te | $2.3 \times 10^{23}$ | $1.6 \times 10^{-11}$ | |

| Name | Formula | $K_{a_1}$ | $K_{a_2}$ | $K_{a_3}$ |
|---|---|---|---|---|
| Hypobromous | HBrO | $2.8 \times 10^{-9}$ | | |
| Hypochlorous | HClO | $2.9 \times 10^{-8}$ | | |
| Hypoiodous | HIO | $2.3 \times 10^{-11}$ | | |
| Iodic | HIO$_3$ | $1.7 \times 10^{-1}$ | | |
| Lactic | HC$_3$H$_5$O$_3$ | $1.4 \times 10^{-4}$ | | |
| Maleic | H$_2$C$_4$H$_2$O$_4$ | $1.2 \times 10^{-2}$ | $5.9 \times 10^{-7}$ | |
| Malonic | H$_2$C$_3$H$_2$O$_4$ | $1.5 \times 10^{-3}$ | $2.0 \times 10^{-6}$ | |
| Nitrous | HNO$_2$ | $4.6 \times 10^{-4}$ | | |
| Oxalic | H$_2$C$_2$O$_4$ | $6.0 \times 10^{-2}$ | $6.1 \times 10^{-5}$ | |
| Paraperiodic | H$_5$IO$_6$ | $2.8 \times 10^{-2}$ | $5.3 \times 10^{-9}$ | |
| Phenol | HC$_6$H$_5$O | $1.3 \times 10^{-10}$ | | |
| Phosphoric | H$_3$PO$_4$ | $7.5 \times 10^{-3}$ | $6.2 \times 10^{-8}$ | $4.2 \times 10^{-13}$ |
| Phosphorous | H$_3$PO$_3$ | $5 \times 10^{-2}$ | $2.0 \times 10^{-7}$ | |
| Propanoic | HC$_3$H$_5$O$_2$ | $1.3 \times 10^{-5}$ | | |
| Pyruvic | HC$_3$H$_3$O$_3$ | $4.1 \times 10^{-3}$ | | |
| Pyrophosphoric | H$_4$P$_2$O$_7$ | $1.2 \times 10^{-1}$ | $7.9 \times 10^{-3}$ | $2.0 \times 10^{-7}$ |
| Selenous | H$_2$SeO$_3$ | $2.4 \times 10^{-3}$ | $4.8 \times 10^{-9}$ | |
| Succinic | H$_2$C$_4$H$_4$O$_4$ | $6.2 \times 10^{-5}$ | $2.3 \times 10^{-6}$ | |
| Sulfuric | H$_2$SO$_4$ | Strong acid | $1.2 \times 10^{-2}$ | |
| Sulfurous | H$_2$SO$_3$ | $1.6 \times 10^{-2}$ | $6.4 \times 10^{-8}$ | |
| Tartaric | H$_2$C$_4$H$_4$O$_6$ | $1.0 \times 10^{-3}$ | $4.6 \times 10^{-5}$ | |
| Trichloroacetic | HC$_2$Cl$_3$O$_2$ | $2.2 \times 10^{-1}$ | | |
| Trifluoroacetic acid | HC$_2$F$_3$O$_2$ | $3.0 \times 10^{-1}$ | | |

## 2. Dissociation Constants for Hydrated Metal Ions at 25 °C

| Cation | Hydrated Ion | $K_a$ | Cation | Hydrated Ion | $K_a$ |
|--------|--------------|-------|--------|--------------|-------|
| $Al^{3+}$ | $Al(H_2O)_6^{3+}$ | $1.4 \times 10^{-5}$ | $Fe^{3+}$ | $Fe(H_2O)_6^{3+}$ | $6.3 \times 10^{-3}$ |
| $Be^{2+}$ | $Be(H_2O)_6^{2+}$ | $3 \times 10^{-7}$ | $Ni^{2+}$ | $Ni(H_2O)_6^{2+}$ | $2.5 \times 10^{-11}$ |
| $Co^{2+}$ | $Co(H_2O)_6^{2+}$ | $1.3 \times 10^{-9}$ | $Pb^{2+}$ | $Pb(H_2O)_6^{2+}$ | $3 \times 10^{-8}$ |
| $Cr^{3+}$ | $Cr(H_2O)_6^{3+}$ | $1.6 \times 10^{-4}$ | $Sn^{2+}$ | $Sn(H_2O)_6^{2+}$ | $4 \times 10^{-4}$ |
| $Cu^{2+}$ | $Cu(H_2O)_6^{2+}$ | $3 \times 10^{-8}$ | $Zn^{2+}$ | $Zn(H_2O)_6^{2+}$ | $2.5 \times 10^{-10}$ |
| $Fe^{2+}$ | $Fe(H_2O)_6^{2+}$ | $3.2 \times 10^{-10}$ | | | |

## 3. Dissociation Constants for Bases at 25 °C

| Name | Formula | $K_b$ | Name | Formula | $K_b$ |
|------|---------|-------|------|---------|-------|
| Ammonia | $NH_3$ | $1.76 \times 10^{-5}$ | Ketamine | $C_{13}H_{16}ClNO$ | $3 \times 10^{-7}$ |
| Aniline | $C_6H_5NH_2$ | $3.9 \times 10^{-10}$ | Methylamine | $CH_3NH_2$ | $4.4 \times 10^{-4}$ |
| Bicarbonate ion | $HCO_3^-$ | $2.3 \times 10^{-8}$ | Morphine | $C_{17}H_{19}NO_3$ | $1.6 \times 10^{-6}$ |
| Carbonate ion | $CO_3^{2-}$ | $1.8 \times 10^{-4}$ | Nicotine | $C_{10}H_{14}N_2$ | $1.0 \times 10^{-6}$ |
| Codeine | $C_{18}H_{21}NO_3$ | $1.6 \times 10^{-6}$ | Piperidine | $C_5H_{10}NH$ | $1.33 \times 10^{-3}$ |
| Diethylamine | $(C_2H_5)_2NH$ | $6.9 \times 10^{-4}$ | Propylamine | $C_3H_7NH_2$ | $3.5 \times 10^{-4}$ |
| Dimethylamine | $(CH_3)_2NH$ | $5.4 \times 10^{-4}$ | Pyridine | $C_5H_5N$ | $1.7 \times 10^{-9}$ |
| Ethylamine | $C_2H_5NH_2$ | $5.6 \times 10^{-4}$ | Strychnine | $C_{21}H_{22}N_2O_2$ | $1.8 \times 10^{-6}$ |
| Ethylenediamine | $C_2H_8N_2$ | $8.3 \times 10^{-5}$ | Triethylamine | $(C_2H_5)_3N$ | $5.6 \times 10^{-4}$ |
| Hydrazine | $H_2NNH_2$ | $1.3 \times 10^{-6}$ | Trimethylamine | $(CH_3)_3N$ | $6.4 \times 10^{-5}$ |
| Hydroxylamine | $HONH_2$ | $1.1 \times 10^{-8}$ | | | |

## 4. Solubility Product Constants for Compounds at 25 °C

| Compound | Formula | $K_{sp}$ | Compound | Formula | $K_{sp}$ |
|----------|---------|----------|----------|---------|----------|
| Aluminum hydroxide | $Al(OH)_3$ | $1.3 \times 10^{-33}$ | Calcium hydrogen phosphate | $CaHPO_4$ | $1 \times 10^{-7}$ |
| Aluminum phosphate | $AlPO_4$ | $9.84 \times 10^{-21}$ | Calcium oxalate | $CaC_2O_4$ | $2.32 \times 10^{-9}$ |
| Barium carbonate | $BaCO_3$ | $2.58 \times 10^{-9}$ | Calcium phosphate | $Ca_3(PO_4)_2$ | $2.07 \times 10^{-33}$ |
| Barium chromate | $BaCrO_4$ | $1.17 \times 10^{-10}$ | Calcium sulfate | $CaSO_4$ | $7.10 \times 10^{-5}$ |
| Barium fluoride | $BaF_2$ | $2.45 \times 10^{-5}$ | Chromium(III) hydroxide | $Cr(OH)_3$ | $6.3 \times 10^{-31}$ |
| Barium hydroxide | $Ba(OH)_2$ | $5.0 \times 10^{-3}$ | Cobalt(II) carbonate | $CoCO_3$ | $1.0 \times 10^{-10}$ |
| Barium oxalate | $BaC_2O_4$ | $1.6 \times 10^{-6}$ | Cobalt(II) hydroxide | $Co(OH)_2$ | $5.92 \times 10^{-15}$ |
| Barium phosphate | $Ba_3(PO_4)_2$ | $6 \times 10^{-39}$ | Cobalt(II) sulfide | $CoS$ | $5 \times 10^{-22}$ |
| Barium sulfate | $BaSO_4$ | $1.07 \times 10^{-10}$ | Copper(I) bromide | $CuBr$ | $6.27 \times 10^{-9}$ |
| Cadmium carbonate | $CdCO_3$ | $1.0 \times 10^{-12}$ | Copper(I) chloride | $CuCl$ | $1.72 \times 10^{-7}$ |
| Cadmium hydroxide | $Cd(OH)_2$ | $7.2 \times 10^{-15}$ | Copper(I) cyanide | $CuCN$ | $3.47 \times 10^{-20}$ |
| Cadmium sulfide | $CdS$ | $8 \times 10^{-28}$ | Copper(II) carbonate | $CuCO_3$ | $2.4 \times 10^{-10}$ |
| Calcium carbonate | $CaCO_3$ | $4.96 \times 10^{-9}$ | Copper(II) hydroxide | $Cu(OH)_2$ | $2.2 \times 10^{-20}$ |
| Calcium chromate | $CaCrO_4$ | $7.1 \times 10^{-4}$ | Copper(II) phosphate | $Cu_3(PO_4)_2$ | $1.40 \times 10^{-37}$ |
| Calcium fluoride | $CaF_2$ | $1.46 \times 10^{-10}$ | Copper(II) sulfide | $CuS$ | $1.27 \times 10^{-36}$ |
| Calcium hydroxide | $Ca(OH)_2$ | $4.68 \times 10^{-6}$ | Iron(II) carbonate | $FeCO_3$ | $3.07 \times 10^{-11}$ |

—(Continued on the next page)

| Compound | Formula | $K_{sp}$ |
|---|---|---|
| Iron(II) hydroxide | $Fe(OH)_2$ | $4.87 \times 10^{-17}$ |
| Iron(II) sulfide | $FeS$ | $3.72 \times 10^{-19}$ |
| Iron(III) hydroxide | $Fe(OH)_3$ | $2.79 \times 10^{-39}$ |
| Lanthanum fluoride | $LaF_3$ | $2 \times 10^{-19}$ |
| Lanthanum iodate | $La(IO_3)_3$ | $7.50 \times 10^{-12}$ |
| Lead(II) bromide | $PbBr_2$ | $4.67 \times 10^{-6}$ |
| Lead(II) carbonate | $PbCO_3$ | $7.40 \times 10^{-14}$ |
| Lead(II) chloride | $PbCl_2$ | $1.17 \times 10^{-5}$ |
| Lead(II) chromate | $PbCrO_4$ | $2.8 \times 10^{-13}$ |
| Lead(II) fluoride | $PbF_2$ | $3.3 \times 10^{-8}$ |
| Lead(II) hydroxide | $Pb(OH)_2$ | $1.43 \times 10^{-20}$ |
| Lead(II) iodide | $PbI_2$ | $9.8 \times 10^{-9}$ |
| Lead(II) phosphate | $Pb_3(PO_4)_2$ | $1 \times 10^{-54}$ |
| Lead(II) sulfate | $PbSO_4$ | $1.82 \times 10^{-8}$ |
| Lead(II) sulfide | $PbS$ | $9.04 \times 10^{-29}$ |
| Magnesium carbonate | $MgCO_3$ | $6.82 \times 10^{-6}$ |
| Magnesium fluoride | $MgF_2$ | $5.16 \times 10^{-11}$ |
| Magnesium hydroxide | $Mg(OH)_2$ | $2.06 \times 10^{-13}$ |
| Magnesium oxalate | $MgC_2O_4$ | $4.83 \times 10^{-6}$ |
| Manganese(II) carbonate | $MnCO_3$ | $2.24 \times 10^{-11}$ |
| Manganese(II) hydroxide | $Mn(OH)_2$ | $1.6 \times 10^{-13}$ |
| Manganese(II) sulfide | $MnS$ | $2.3 \times 10^{-13}$ |
| Mercury(I) bromide | $Hg_2Br_2$ | $6.40 \times 10^{-23}$ |
| Mercury(I) carbonate | $Hg_2CO_3$ | $3.6 \times 10^{-17}$ |
| Mercury(I) chloride | $Hg_2Cl_2$ | $1.43 \times 10^{-18}$ |
| Mercury(I) chromate | $Hg_2CrO_4$ | $2 \times 10^{-9}$ |
| Mercury(I) cyanide | $Hg_2(CN)_2$ | $5 \times 10^{-40}$ |

| Compound | Formula | $K_{sp}$ |
|---|---|---|
| Mercury(I) iodide | $Hg_2I_2$ | $5.2 \times 10^{-29}$ |
| Mercury(II) hydroxide | $Hg(OH)_2$ | $3.1 \times 10^{-26}$ |
| Mercury(II) sulfide | $HgS$ | $1.6 \times 10^{-54}$ |
| Nickel(II) carbonate | $NiCO_3$ | $1.42 \times 10^{-7}$ |
| Nickel(II) hydroxide | $Ni(OH)_2$ | $5.48 \times 10^{-16}$ |
| Nickel(II) sulfide | $NiS$ | $3 \times 10^{-20}$ |
| Silver bromate | $AgBrO_3$ | $5.38 \times 10^{-5}$ |
| Silver bromide | $AgBr$ | $5.35 \times 10^{-13}$ |
| Silver carbonate | $Ag_2CO_3$ | $8.46 \times 10^{-12}$ |
| Silver chloride | $AgCl$ | $1.77 \times 10^{-10}$ |
| Silver chromate | $Ag_2CrO_4$ | $1.12 \times 10^{-12}$ |
| Silver cyanide | $AgCN$ | $5.97 \times 10^{-17}$ |
| Silver iodide | $AgI$ | $8.51 \times 10^{-17}$ |
| Silver phosphate | $Ag_3PO_4$ | $8.89 \times 10^{-17}$ |
| Silver sulfate | $Ag_2SO_4$ | $1.20 \times 10^{-5}$ |
| Silver sulfide | $Ag_2S$ | $6 \times 10^{-51}$ |
| Strontium carbonate | $SrCO_3$ | $5.60 \times 10^{-10}$ |
| Strontium chromate | $SrCrO_4$ | $3.6 \times 10^{-5}$ |
| Strontium phosphate | $Sr_3(PO_4)_2$ | $1 \times 10^{-31}$ |
| Strontium sulfate | $SrSO_4$ | $3.44 \times 10^{-7}$ |
| Tin(II) hydroxide | $Sn(OH)_2$ | $5.45 \times 10^{-27}$ |
| Tin(II) sulfide | $SnS$ | $1 \times 10^{-26}$ |
| Zinc carbonate | $ZnCO_3$ | $1.46 \times 10^{-10}$ |
| Zinc hydroxide | $Zn(OH)_2$ | $3 \times 10^{-17}$ |
| Zinc oxalate | $ZnC_2O_4$ | $2.7 \times 10^{-8}$ |
| Zinc sulfide | $ZnS$ | $2 \times 10^{-25}$ |

## 5. Complex Ion Formation Constants in Water at 25 °C

| Complex Ion | $K_f$ |
|---|---|
| $[Ag(CN)_2]^-$ | $1 \times 10^{21}$ |
| $[Ag(EDTA)]^{3-}$ | $2.1 \times 10^7$ |
| $[Ag(en)_2]^+$ | $5.0 \times 10^7$ |
| $[Ag(NH_3)_2]^+$ | $1.7 \times 10^7$ |
| $[Ag(SCN)_4]^{3-}$ | $1.2 \times 10^{10}$ |
| $[Ag(S_2O_3)_2]^{3-}$ | $2.8 \times 10^{13}$ |
| $[Al(EDTA)]^-$ | $1.3 \times 10^{16}$ |
| $[AlF_6]^{3-}$ | $7 \times 10^{19}$ |
| $[Al(OH)_4]^-$ | $3 \times 10^{33}$ |
| $[Al(ox)_3]^{3-}$ | $2 \times 10^{16}$ |
| $[CdBr_4]^{2-}$ | $5.5 \times 10^3$ |
| $[Cd(CN)_4]^{2-}$ | $3 \times 10^{18}$ |

| Complex Ion | $K_f$ |
|---|---|
| $[CdCl_4]^{2-}$ | $6.3 \times 10^2$ |
| $[Cd(en)_3]^{2+}$ | $1.2 \times 10^{12}$ |
| $[CdI_4]^{2-}$ | $2 \times 10^6$ |
| $[Co(EDTA)]^{2-}$ | $2.0 \times 10^{16}$ |
| $[Co(EDTA)]^-$ | $1 \times 10^{36}$ |
| $[Co(en)_3]^{2+}$ | $8.7 \times 10^{13}$ |
| $[Co(en)_3]^{3+}$ | $4.9 \times 10^{48}$ |
| $[Co(NH_3)_6]^{2+}$ | $1.3 \times 10^5$ |
| $[Co(NH_3)_6]^{3+}$ | $2.3 \times 10^{33}$ |
| $[Co(OH)_4]^{2-}$ | $5 \times 10^9$ |
| $[Co(ox)_3]^{4-}$ | $5 \times 10^9$ |
| $[Co(ox)_3]^{3-}$ | $1 \times 10^{20}$ |

| Complex Ion | $K_f$ |
|---|---|
| $[Co(SCN)_4]^{2-}$ | $1 \times 10^3$ |
| $[Cr(EDTA)]^-$ | $1 \times 10^{23}$ |
| $[Cr(OH)_4]^-$ | $8.0 \times 10^{29}$ |
| $[CuCl_3]^{2-}$ | $5 \times 10^5$ |
| $[Cu(CN)_4]^{2-}$ | $1.0 \times 10^{25}$ |
| $[Cu(EDTA)]^{2-}$ | $5 \times 10^{18}$ |
| $[Cu(en)_2]^{2+}$ | $1 \times 10^{20}$ |
| $[Cu(NH_3)_4]^{2+}$ | $1.7 \times 10^{13}$ |
| $[Cu(ox)_2]^{2-}$ | $3 \times 10^8$ |
| $[Fe(CN)_6]^{4-}$ | $1.5 \times 10^{35}$ |
| $[Fe(CN)_6]^{3-}$ | $2 \times 10^{43}$ |
| $[Fe(EDTA)]^{2-}$ | $2.1 \times 10^{14}$ |
| $[Fe(EDTA)]^-$ | $1.7 \times 10^{24}$ |
| $[Fe(en)_3]^{2+}$ | $5.0 \times 10^9$ |
| $[Fe(ox)_3]^{4-}$ | $1.7 \times 10^5$ |
| $[Fe(ox)_3]^{3-}$ | $2 \times 10^{20}$ |
| $[Fe(SCN)]^{2+}$ | $8.9 \times 10^2$ |
| $[Hg(CN)_4]^{2-}$ | $1.8 \times 10^{41}$ |
| $[HgCl_4]^{2-}$ | $1.1 \times 10^{16}$ |
| $[Hg(EDTA)]^{2-}$ | $6.3 \times 10^{21}$ |
| $[Hg(en)_2]^{2+}$ | $2 \times 10^{23}$ |
| $[HgI_4]^{2-}$ | $2 \times 10^{30}$ |

| Complex Ion | $K_f$ |
|---|---|
| $[Hg(ox)_2]^{2-}$ | $9.5 \times 10^6$ |
| $[Ni(CN)_4]^{2-}$ | $2 \times 10^{31}$ |
| $[Ni(EDTA)]^{2-}$ | $3.6 \times 10^{18}$ |
| $[Ni(en)_3]^{2+}$ | $2.1 \times 10^{18}$ |
| $[Ni(NH_3)_6]^{2+}$ | $2.0 \times 10^8$ |
| $[Ni(ox)_3]^{4-}$ | $3 \times 10^8$ |
| $[PbCl_3]^-$ | $2.4 \times 10^1$ |
| $[Pb(EDTA)]^{2-}$ | $2 \times 10^{18}$ |
| $[PbI_4]^{2-}$ | $3.0 \times 10^4$ |
| $[Pb(OH)_3]^-$ | $8 \times 10^{13}$ |
| $[Pb(ox)_2]^{2-}$ | $3.5 \times 10^6$ |
| $[Pb(S_2O_3)_3]^{4-}$ | $2.2 \times 10^6$ |
| $[PtCl_4]^{2-}$ | $1 \times 10^{16}$ |
| $[Pt(NH_3)_6]^{2+}$ | $2 \times 10^{35}$ |
| $[Sn(OH)_3]^-$ | $3 \times 10^{25}$ |
| $[Zn(CN)_4]^{2-}$ | $2.1 \times 10^{19}$ |
| $[Zn(EDTA)]^{2-}$ | $3 \times 10^{16}$ |
| $[Zn(en)_3]^{2+}$ | $1.3 \times 10^{14}$ |
| $[Zn(NH_3)_4]^{2+}$ | $2.8 \times 10^9$ |
| $[Zn(OH)_4]^{2-}$ | $2 \times 10^{15}$ |
| $[Zn(ox)_3]^{4-}$ | $1.4 \times 10^8$ |

# D. Standard Electrode Potentials at 25 °C

| Half-Reaction | $E°(V)$ |
|---|---|
| $F_2(g) + 2\,e^- \longrightarrow 2\,F^-(aq)$ | 2.87 |
| $O_3(g) + 2\,H^+(aq) + 2\,e^- \longrightarrow O_2(g) + H_2O(l)$ | 2.08 |
| $Ag^{2+}(aq) + e^- \longrightarrow Ag^+(aq)$ | 1.98 |
| $Co^{3+}(aq) + e^- \longrightarrow Co^{2+}(aq)$ | 1.82 |
| $H_2O_2(aq) + 2\,H^+(aq) + 2\,e^- \longrightarrow 2\,H_2O(l)$ | 1.78 |
| $PbO_2(s) + 4\,H^+(aq) + SO_4^{2-}aq) + 2\,e^- \longrightarrow$ $PbSO_4(s) + 2\,H_2O(l)$ | 1.69 |
| $MnO_4^-(aq) + 4\,H^+(aq) + 3\,e^- \longrightarrow MnO_2(s) + 2\,H_2O(l)$ | 1.68 |
| $2\,HClO(aq) + 2\,H^+(aq) + 2\,e^- \longrightarrow Cl_2(g) + 2\,H_2O(l)$ | 1.61 |
| $MnO_4^-(aq) + 8\,H^+(aq) + 5\,e^- \longrightarrow Mn^{2+}(aq) + 4\,H_2O(l)$ | 1.51 |
| $Au^{3+}(aq) + 3\,e^- \longrightarrow Au(s)$ | 1.50 |
| $2\,BrO_3^-(aq) + 12\,H^+(aq) + 10\,e^- \longrightarrow Br_2(l) + 6\,H_2O(l)$ | 1.48 |
| $PbO_2(s) + 4\,H^+(aq) + 2\,e^- \longrightarrow Pb^{2+}(aq) + 2\,H_2O(l)$ | 1.46 |
| $Cl_2(g) + 2\,e^- \longrightarrow 2\,Cl^-(aq)$ | 1.36 |

| Half-Reaction | $E°(V)$ |
|---|---|
| $Cr_2O_7^{2-}(aq) + 14\,H^+(aq) + 6\,e^- \longrightarrow$ $2\,Cr^{3+}(aq) + 7\,H_2O(l)$ | 1.33 |
| $O_2(g) + 4\,H^+(aq) + 4\,e^- \longrightarrow 2\,H_2O(l)$ | 1.23 |
| $MnO_2(s) + 4\,H^+(aq) + 2\,e^- \longrightarrow Mn^{2+}(aq) + 2\,H_2O(l)$ | 1.21 |
| $IO_3^-(aq) + 6\,H^+(aq) + 5\,e^- \longrightarrow \frac{1}{2}\,I_2(aq) + 3\,H_2O(l)$ | 1.20 |
| $Br_2(l) + 2\,e^- \longrightarrow 2\,Br^-(aq)$ | 1.09 |
| $AuCl_4^-(aq) + 3\,e^- \longrightarrow Au(s) + 4\,Cl^-(aq)$ | 1.00 |
| $VO_2^+(aq) + 2\,H^+(aq) + e^- \longrightarrow VO^{2+}(aq) + H_2O(l)$ | 1.00 |
| $HNO_2(aq) + H^+(aq) + e^- \longrightarrow NO(g) + 2\,H_2O(l)$ | 0.98 |
| $NO_3^-(aq) + 4\,H^+(aq) + 3\,e^- \longrightarrow NO(g) + 2\,H_2O(l)$ | 0.96 |
| $ClO_2(g) + e^- \longrightarrow ClO_2^-(aq)$ | 0.95 |
| $2\,Hg^{2+}(aq) + 2\,e^- \longrightarrow 2\,Hg_2^{2+}(aq)$ | 0.92 |
| $Ag^+(aq) + e^- \longrightarrow Ag(s)$ | 0.80 |
| $Hg_2^{2+}(aq) + 2\,e^- \longrightarrow 2\,Hg(l)$ | 0.80 |

*—(Continued on the next page)*

| Half-Reaction | E°(V) |
|---|---|
| $Fe^{3+}(aq) + e^- \longrightarrow Fe^{2+}(aq)$ | 0.77 |
| $PtCl_4^{2-}(aq) + 2\,e^- \longrightarrow Pt(s) + 4\,Cl^-(aq)$ | 0.76 |
| $O_2(g) + 2\,H^+(aq) + 2\,e^- \longrightarrow H_2O_2(aq)$ | 0.70 |
| $MnO_4^-(aq) + e^- \longrightarrow MnO_4^{2-}(aq)$ | 0.56 |
| $I_2(s) + 2\,e^- \longrightarrow 2\,I^-(aq)$ | 0.54 |
| $Cu^+(aq) + e^- \longrightarrow Cu(s)$ | 0.52 |
| $O_2(g) + 2\,H_2O(l) + 4\,e^- \longrightarrow 4\,OH^-(aq)$ | 0.40 |
| $Cu^{2+}(aq) + 2\,e^- \longrightarrow Cu(s)$ | 0.34 |
| $BiO^+(aq) + 2\,H^+(aq) + 3\,e^- \longrightarrow Bi(s) + H_2O(l)$ | 0.32 |
| $Hg_2Cl_2(s) + 2\,e^- \longrightarrow 2\,Hg(l) + 2\,Cl^-(aq)$ | 0.27 |
| $AgCl(s) + e^- \longrightarrow Ag(s) + Cl^-(aq)$ | 0.22 |
| $SO_4^{2-}(aq) + 4\,H^+(aq) + 2\,e^- \longrightarrow H_2SO_3(aq) + H_2O(l)$ | 0.20 |
| $Cu^{2+}(aq) + e^- \longrightarrow Cu^+(aq)$ | 0.16 |
| $Sn^{4+}(aq) + 2\,e^- \longrightarrow Sn^{2+}(aq)$ | 0.15 |
| $S(s) + 2\,H^+(aq) + 2\,e^- \longrightarrow H_2S(g)$ | 0.14 |
| $AgBr(s) + e^- \longrightarrow Ag(s) + Br^-(aq)$ | 0.071 |
| $2\,H^+(aq) + 2\,e^- \longrightarrow H_2(g)$ | 0.00 |
| $Fe^{3+}(aq) + 3\,e^- \longrightarrow Fe(s)$ | −0.036 |
| $Pb^{2+}(aq) + 2\,e^- \longrightarrow Pb(s)$ | −0.13 |
| $Sn^{2+}(aq) + 2\,e^- \longrightarrow Sn(s)$ | −0.14 |
| $AgI(s) + e^- \longrightarrow Ag(s) + I^-(aq)$ | −0.15 |

| Half-Reaction | E°(V) |
|---|---|
| $N_2(g) + 5\,H^+(aq) + 4\,e^- \longrightarrow N_2H_5^+(aq)$ | −0.23 |
| $Ni^{2+}(aq) + 2\,e^- \longrightarrow Ni(s)$ | −0.23 |
| $Co^{2+}(aq) + 2\,e^- \longrightarrow Co(s)$ | −0.28 |
| $PbSO_4(s) + 2\,e^- \longrightarrow Pb(s) + SO_4^{2-}(aq)$ | −0.36 |
| $Cd^{2+}(aq) + 2\,e^- \longrightarrow Cd(s)$ | −0.40 |
| $Fe^{2+}(aq) + 2\,e^- \longrightarrow Fe(s)$ | −0.45 |
| $2\,CO_2(g) + 2\,H^+(aq) + 2\,e^- \longrightarrow H_2C_2O_4(aq)$ | −0.49 |
| $Cr^{3+}(aq) + e^- \longrightarrow Cr^{2+}(aq)$ | −0.50 |
| $Cr^{3+}(aq) + 3\,e^- \longrightarrow Cr(s)$ | −0.73 |
| $Zn^{2+}(aq) + 2\,e^- \longrightarrow Zn(s)$ | −0.76 |
| $2\,H_2O(l) + 2\,e^- \longrightarrow H_2(g) + 2\,OH^-(aq)$ | −0.83 |
| $Mn^{2+}(aq) + 2\,e^- \longrightarrow Mn(s)$ | −1.18 |
| $Al^{3+}(aq) + 3\,e^- \longrightarrow Al(s)$ | −1.66 |
| $H_2(g) + 2\,e^- \longrightarrow 2\,H^-(aq)$ | −2.23 |
| $Mg^{2+}(aq) + 2\,e^- \longrightarrow Mg(s)$ | −2.37 |
| $La^{3+}(aq) + 3\,e^- \longrightarrow La(s)$ | −2.38 |
| $Na^+(aq) + e^- \longrightarrow Na(s)$ | −2.71 |
| $Ca^{2+}(aq) + 2\,e^- \longrightarrow Ca(s)$ | −2.76 |
| $Ba^{2+}(aq) + 2\,e^- \longrightarrow Ba(s)$ | −2.90 |
| $K^+(aq) + e^- \longrightarrow K(s)$ | −2.92 |
| $Li^+(aq) + e^- \longrightarrow Li(s)$ | −3.04 |

# E. Vapor Pressure of Water at Various Temperatures

| T (°C) | P (torr) | T (°C) | P (torr) | T (°C) | P (torr) | T (°C) | P (torr) |
|---|---|---|---|---|---|---|---|
| 0 | 4.58 | 21 | 18.65 | 35 | 42.2 | 92 | 567.0 |
| 5 | 6.54 | 22 | 19.83 | 40 | 55.3 | 94 | 610.9 |
| 10 | 9.21 | 23 | 21.07 | 45 | 71.9 | 96 | 657.6 |
| 12 | 10.52 | 24 | 22.38 | 50 | 92.5 | 98 | 707.3 |
| 14 | 11.99 | 25 | 23.76 | 55 | 118.0 | 100 | 760.0 |
| 16 | 13.63 | 26 | 25.21 | 60 | 149.4 | 102 | 815.9 |
| 17 | 14.53 | 27 | 26.74 | 65 | 187.5 | 104 | 875.1 |
| 18 | 15.48 | 28 | 28.35 | 70 | 233.7 | 106 | 937.9 |
| 19 | 16.48 | 29 | 30.04 | 80 | 355.1 | 108 | 1004.4 |
| 20 | 17.54 | 30 | 31.82 | 90 | 525.8 | 110 | 1074.6 |

# APPENDIX III Answers to Selected Exercises

## Chapter 1

**33. a.** theory **b.** observation
**c.** law **d.** observation
**35.** Several answers possible.
**37. a.** mixture, homogeneous
**b.** pure substance, compound
**c.** pure substance, element
**d.** mixture, heterogeneous
**39.**

| Substance | Pure or Mixture | Type |
|---|---|---|
| Aluminum | Pure | Element |
| Apple juice | Mixture | Homogeneous |
| Hydrogen peroxide | Pure | Compound |
| Chicken soup | Mixture | Heterogeneous |

**41. a.** pure substance, compound
**b.** mixture, heterogeneous
**c.** mixture, homogeneous
**d.** pure substance, element
**43.** physical, chemical, physical, physical, physical
**45. a.** chemical **b.** physical
**c.** physical **d.** chemical
**47. a.** chemical **b.** physical
**c.** chemical **d.** chemical
**49. a.** physical **b.** chemical **c.** physical
**51. a.** $0\,°C$ **b.** $-321\,°F$
**c.** $-78.3\,°F$ **d.** $310.2\,K$
**53.** $-62.2\,°C$, $210.9\,K$
**55. a.** $1.2\,nm$ **b.** $22\,fs$ **c.** $1.5\,Gg$ **d.** $3.5\,ML$
**57. a.** $4.5 \times 10^{-9}\,s$ **b.** $1.8 \times 10^{-14}\,s$
**c.** $1.28 \times 10^{-10}\,m$ **d.** $3.5 \times 10^{-5}\,m$
**59.**

| 1245 kg | $1.245 \times 10^6$ g | $1.245 \times 10^9$ mg |
|---|---|---|
| 515 km | $5.15 \times 10^6$ dm | $5.15 \times 10^7$ cm |
| 122.355 s | $1.22355 \times 10^5$ ms | 0.122355 ks |
| 3.345 kJ | $3.345 \times 10^3$ J | $3.345 \times 10^6$ mJ |

**61. a.** $254.998\,km$ **b.** $2.54998 \times 10^{-1}\,Mm$
**c.** $254998 \times 10^3\,mm$ **d.** $254998 \times 10^2\,cm$
**63.** 10,000 1 cm squares
**65.** no
**67.** $1.26\,g/cm^3$
**69. a.** $463\,g$ **b.** $3.7\,L$
**71.** $201. \times 10^3\,g$
**73. a.** $73.7\,mL$ **b.** $88.2\,°C$ **c.** $647\,mL$
**75. a.** $1{,}05\underline{0}{,}501$ **b.** $0.002\underline{0}$
**c.** $0.00000000000000\underline{2}$ **d.** $0.00109\underline{0}$
**77. a.** 3 **b.** ambiguous, without more information assume three significant figures
**c.** 3 **d.** 5
**e.** ambiguous, without more information assume one significant figure

**79. a.** not exact **b.** exact
**c.** not exact **d.** exact
**81. a.** 156.9 **b.** 156.8
**c.** 156.8 **d.** 156.9
**83. a.** 1.84 **b.** 0.033
**c.** 0.500 **d.** 34
**85. a.** 41.4 **b.** 133.5
**c.** 73.0 **d.** 0.42
**87. a.** 391.3 **b.** $1.1 \times 10^4$
**c.** 5.96 **d.** $5.93 \times 10^4$
**89.** $0.74\,g/mL$
**91. a.** $2.78 \times 10^4\,cm^3$ **b.** $1.898 \times 10^{-3}\,kg$
**c.** $1.98 \times 10^7\,cm$
**93. a.** $60.6\,in$ **b.** $3.14 \times 10^3\,g$
**c.** $3.7\,qt$ **d.** $4.29\,in$
**95.** $5.0 \times 10^1\,min$
**97.** $4.0 \times 10^1\,mi/gal$
**99. a.** $1.95 \times 10^{-4}\,km^2$ **b.** $1.95 \times 10^4\,dm^2$
**c.** $1.95 \times 10^6\,cm^2$
**101.** $0.680\,mi^2$
**103.** $0.95\,mL$
**105.** $3.1557 \times 10^7\,s/\text{solar year}$
**107. a.** extensive **b.** intensive
**c.** intensive **d.** intensive
**e.** extensive
**109.** $-34\,°F$
**111.** $F = kg(m/s^2) = N$ (for newton), kN, pN
**113. a.** $2.2 \times 10^{-6}$ **b.** 0.0159 **c.** $6.9 \times 10^4$
**115. a.** mass of can of gold $= 1.9 \times 10^4\,g$
mass of can of sand $= 3.0 \times 10^3\,g$
**b.** Yes, the thief sets off the trap because the can of sand is lighter than the gold cylinder.
**117.** $22\,in^3$
**119.** $7.6\,g/cm^3$
**121.** $3.11 \times 10^5\,lb$
**123.** $3.3 \times 10^2\,km$
**125.** $6.8 \times 10^{-15}$
**127.** $2.4 \times 10^{19}\,km$
**129.** 488 grams
**131.** $0.661\,\Omega$
**133.** 0.492
**135.** 18.2 atm
**137.** $1\,J = 1\,kg\,m^2/s^2$
$m = kg$, $v^2 = (m/s)^2$  $mv = kg\,m^2/s^2$
$P = N/m^2 = kg\,m/s^2/m^2 = kg/m\,s^2$
$V = m^3$  $PV = kg\,m^3/m\,s^2 = kg\,m^2/s^2$
**139.** $9.0 \times 10^1\,mg\,CO$
**141.** 13% increase
**143.** No. Since the container is sealed the atoms and molecules can move around, but they cannot leave. If no atoms or molecules can leave, the mass must be constant.

**145.** 343 1 cm cubes
**147. a.** the dark block          **b.** the light-colored block
      **c.** cannot tell
**149. a.** law       **b.** theory       **c.** observation      **d.** law
**155. a.** ice, 918 kg/m$^3$; water, 999.97 kg/m$^3$; Ice is less
      dense, so it floats on liquid water.
      **c.** 2.44 $\times$ 10$^7$ km$^3$

# Chapter 2

**29.** 13.5 g
**31.** These results are not consistent with the law of definite
      proportions because sample 1 is composed of 11.5 parts
      Cl to 1 part C and sample 2 is composed of 9.05 parts
      Cl to 1 part C. The law of definite proportions states
      that a given compound always contains exactly the
      same proportion of elements by mass.
**33.** 23.8 g
**35.** For the law of multiple proportions to hold, the
      ratio of the masses of O combining with 1 g of O's
      in the compound should be a small whole number.
      0.3369/0.168 = 2.00
**37.** Sample 1: 1.00 g O$_2$/1.00 g S;
      Sample 2: 1.50 g O$_2$/1.00 g S
      Sample 2/sample 1 = 1.50/1.00 = 1.50
      3 O atoms/2 O atoms = 1.5
**39. a.** not consistent
      **b.** consistent: Dalton's atomic theory states that the
          atoms of a given element are identical.
      **c.** consistent: Dalton's atomic theory states that atoms
          combine in simple whole-number ratios to form
          compounds.
      **d.** not consistent
**41. a.** consistent: Rutherford's nuclear model states that
          the atom is largely empty space.
      **b.** consistent: Rutherford's nuclear model states that
          most of the atom's mass is concentrated in a tiny
          region called the nucleus.
      **c.** not consistent          **d.** not consistent
**43.** $-2.3 \times 10^{-19}$ C
**45.** $9.4 \times 10^{13}$ excess electrons, $8.5 \times 10^{-17}$ kg
**47.** a, b, c
**49.** $1.83 \times 10^3$ e$^-$
**51. a.** Ag-107      **b.** Ag-109      **c.** U-238      **d.** H-2
**53. a.** $7_1^1$p and $7_1^0$n          **b.** $11_1^1$p and $12_1^0$n
      **c.** $86_1^1$p and $136_1^0$n      **d.** $82_1^1$p and $126_1^0$n
**55.** $6_1^1$p and $8_0^1$n, $^{14}_6$C
**57. a.** $28_1^1$p and 26 e$^-$          **b.** $16_1^1$p and 18 e$^-$
      **c.** $35_1^1$p and 36 e$^-$          **d.** $24_1^1$p and 21 e$^-$
**59. a.** 2$-$      **b.** 1+      **c.** 3+      **d.** 1+

**61.**

| Symbol | Ion Formed | Number of Electrons in Ion | Number of Protons in Ion |
|--------|-----------|----------------------------|--------------------------|
| Ca | Ca$^{2+}$ | 18 | 20 |
| Be | Be$^{2+}$ | 2 | 4 |
| Se | Se$^{2-}$ | 36 | 34 |
| In | In$^{3+}$ | 46 | 49 |

**63. a.** potassium, metal      **b.** barium, metal
      **c.** iodine, nonmetal      **d.** oxygen, nonmetal
      **e.** antimony, metalloid
**65.** **a** and **b** are main group elements.
**67. a.** alkali metal
      **b.** halogen
      **c.** alkaline earth metal
      **d.** alkaline earth metal
      **e.** noble gas
**69.** Cl and F because they are in the same group or family.
      Elements in the same group or family have similar
      chemical properties.

**71.**

**73.** The fluorine-19 isotope must have a large percent
      abundance, which would make fluorine produce a
      large peak at this mass. Chlorine has two isotopes
      (Cl-35 and Cl-37). The atomic mass is simply the
      weighted average of these two, which means that there
      is no chlorine isotope with a mass of 35.45 amu.
**75.** 121.8 amu, Sb
**77.** Br-79 78.92, amu 50.96%
**79.** 152 amu
**81.** $3.32 \times 10^{24}$ atoms
**83. a.** 0.295 mol Ar          **b.** 0.0543 mol Zn
      **c.** 0.144 mol Ta          **d.** 0.0304 mol Li
**85.** $2.11 \times 10^{22}$ atoms
**87. a.** $1.01 \times 10^{23}$ atoms      **b.** $6.78 \times 10^{21}$ atoms
      **c.** $5.39 \times 10^{21}$ atoms      **d.** $5.6 \times 10^{20}$ atoms
**89. a.** 36 grams          **b.** 0.187 grams
      **c.** 62 grams          **d.** 3.1 grams
**91.** $2.6 \times 10^{21}$ atoms
**93.** $3.239 \times 10^{-22}$ g
**95.** 1.50 g
**97.** C$_2$O$_3$
**99.** $4.82241 \times 10^7$ C/kg
**101.** 207 amu
**103.** $^{237}$Pa, $^{238}$U, $^{239}$Np, $^{240}$Pu, $^{235}$Ac, $^{234}$Ra, etc.
**105.**

| Symbol | Z | A | #p | #e$^-$ | #n | Charge |
|--------|---|---|----|--------|----|--------|
| O | 8 | 16 | 8 | 10 | 8 | 2$-$ |
| Ca$^{2+}$ | 20 | 40 | 20 | 18 | 20 | 2+ |
| Mg$^{2+}$ | 12 | 25 | 12 | 10 | 13 | 2+ |
| N$^{3-}$ | 7 | 14 | 7 | 10 | 7 | 3$-$ |

**107.** $V_n = 8.2 \times 10^{-8}$ pm$^3$, $V_a = 1.4 \times 10^6$ pm$^3$,
      $5.9 \times 10^{-12}$ %
**109.** $6.022 \times 10^{21}$ dollars total, $8.6 \times 10^{11}$ dollars per
      person, billionaires
**111.** 15.985 amu
**113.** $4.76 \times 10^{24}$ atoms

**115.** 3.56 cm
**117.** Li-6 = 7.494%, Li-7 = 92.506%
**119.** 75.0% gold
**121.** 106.91 amu
**123.** $1.66 \times 10^{22}$ gold atoms
**125.** $1 \times 10^{78}$ atoms/universe
**127.** 0.423
**129.** 63.67 g/mol
**131.** 25.06 g/mol

**133.**

     Li-6          Li-7

**135.** If the amu and mole were not based on the same isotope, the numerical values obtained for an atom of material and a mole of material would not be the same. If, for example, the mole were based on the number of particles in C-12 but the amu were changed to a fraction of the mass of an atom of Ne-20, the number of particles and the number of amu that make up one mole of material would no longer be the same. We would no longer have the relationship in which the mass of an atom in amu is numerically equal to the mass of a mole of those atoms in grams.

**137.** The different isotopes of the same element have the same number of protons and electrons, so the attractive forces between the nucleus and the electrons are constant and there is no difference in the radii of the isotopes. Ions, on the other hand, have a different number of electrons than the parent atom from which they are derived. Cations have fewer electrons than the parent atom. The attractive forces are greater because there is a larger positive charge in the nucleus than the negative charge in the electron cloud. So, cations are smaller than the atom they are derived from. Anions have more electrons than the parent. The electron cloud has a greater negative charge than the nucleus, so the anions have larger radii than the parent.

**142. a.** yes
    **c.** The May process because it produces the least Pb, Cd, and Cu and the smallest amount of total heavy metals.
    **e.** 0.148 mg
    **g.**

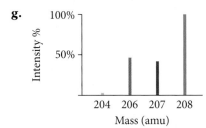

# Chapter 3

**23. a.** 3 Mg, 2 P, 8 O      **b.** 1 Ba, 2 Cl
    **c.** 1 Fe, 2 N, 4 O      **d.** 1 Ca, 2 O, 2 H
**25. a.** $NH_3$     **b.** $C_2H_6$     **c.** $SO_3$
**27. a.** atomic           **b.** molecular
    **c.** atomic           **d.** molecular
**29. a.** molecular       **b.** ionic
    **c.** ionic              **d.** molecular
**31. a.** molecular element    **b.** molecular compound
    **c.** atomic element
**33. a.** CaO    **b.** ZnS    **c.** RbBr    **d.** $Al_2O_3$
**35. a.** $Ca(OH)_2$        **b.** $CaCrO_4$
    **c.** $Ca_3(PO_4)_2$      **d.** $Ca(CN)_2$
**37. a.** magnesium nitride    **b.** potassium fluoride
    **c.** sodium oxide         **d.** lithium sulfide
    **e.** cesium fluoride        **f.** potassium iodide
**39. a.** tin(II) oxide           **b.** chromium(III) sulfide
    **c.** rubidium iodide       **d.** barium bromide
**41. a.** copper(I) nitrite       **b.** magnesium acetate
    **c.** barium nitrate         **d.** lead(II) acetate
**43. a.** $NaHSO_3$          **b.** $LiMnO_4$
    **c.** $AgNO_3$            **d.** $K_2SO_4$
    **e.** $RbHSO_4$          **f.** $KHCO_3$
**45. a.** cobalt(II) sulfate heptahydrate
    **b.** $IrBr_3 \cdot 4\,H_2O$
    **c.** magnesium bromate hexahydrate
    **d.** $K_2CO_3 \cdot 2\,H_2O$
**47. a.** carbon monoxide    **b.** nitrogen triiodide
    **c.** silicon tetrachloride   **d.** tetranitrogen tetraselenide
**49. a.** $PCl_3$             **b.** ClO
    **c.** $S_2F_4$             **d.** $PF_5$
**51. a.** hydroiodic acid     **b.** nitric acid
    **c.** carbonic acid
**53. a.** HF      **b.** HBr      **c.** $H_2SO_3$
**55. a.** strontium chloride
    **b.** tin(IV) oxide
    **c.** diphosphorus pentasulfide
    **d.** acetic acid
**57. a.** potassium chlorate
    **b.** diiodine pentoxide
    **c.** lead(II) sulfate
**59. a.** 46.01 amu        **b.** 58.12 amu
    **c.** 180.16 amu      **d.** 238.03 amu
**61. a.** 0.471 mol        **b.** 0.0362 mol
    **c.** 968 mol          **d.** 0.279 mol
**63. a.** 0.554 mol        **b.** 28.4 mol
    **c.** 0.378 mol        **d.** 1093 mol
**65. a.** $2.2 \times 10^{23}$ molecules
    **b.** $7.06 \times 10^{23}$ molecules
    **c.** $4.16 \times 10^{23}$ molecules
    **d.** $1.09 \times 10^{23}$ molecules
**67. a.** 0.0790 g
    **b.** 0.84 g
    **c.** $2.992 \times 10^{-22}$ g
**69.** 0.10 mg
**71. a.** 74.87% C        **b.** 79.88% C
    **c.** 92.24% C        **d.** 37.23% C

**73.** $NH_3$: 82.27% N
$CO(NH_2)_2$: 46.65% N
$NH_4NO_3$: 35.00% N
$(NH_4)_2SO_4$: 21.20% N
$NH_3$ has the highest N content.

**75.** 20.8 g F

**77.** 196 $\mu$g KI

**79. a.** 2:1    **b.** 4:1    **c.** 6:2:1

**81. a.** 0.885 mol H    **b.** 5.2 mol H
**c.** 29 mol H    **d.** 33.7 mol H

**83. a.** 3.3 g Na    **b.** 3.6 g Na
**c.** 1.4 g Na    **d.** 1.7 g Na

**85.** $1.41 \times 10^{23}$ F atoms

**87. a.** $Ag_2O$    **b.** $Co_3As_2O_8$    **c.** $SeBr_4$

**89. a.** $C_5H_7N$    **b.** $C_4H_5N_2O$

**91.** $C_{13}H_{18}O_2$

**93.** $NCl_3$

**95. a.** $C_{12}H_{14}N_2$    **b.** $C_6H_3Cl_3$    **c.** $C_{10}H_{20}N_2S_4$

**97.** $CH_2$

**99.** $C_2H_4O$

**101.** $2\,SO_2(g) + O_2(g) + 2\,H_2O(l) \longrightarrow 2\,H_2So_4(aq)$

**103.** $2\,Na(s) + 2\,H_2O(l) \longrightarrow H_2(g) + 2\,NaOH(aq)$

**105.** $C_{12}H_{22}O_{11}(s) + H_2O(l) \longrightarrow 4\,C_2H_5OH(aq) + 4\,CO_2(g)$

**107. a.** $PbS(s) + 2\,HBr(aq) \longrightarrow PbBr_2(s) + H_2S(g)$
**b.** $CO(g) + 3\,H_2(g) \longrightarrow CH_4(g) + H_2O(l)$
**c.** $4\,HCl(aq) + MnO_2(s) \longrightarrow$
$\qquad MnCl_2(aq) + 2\,H_2O(l) + Cl_2(g)$
**d.** $C_5H_{12}(l) + 8\,O_2(g) \longrightarrow 5\,CO_2(g) + 6\,H_2O(g)$

**109.** $Na_2CO_3(aq) + CuCl_2(aq)\,CuCO_3(s) + 2\,NaCl(aq)$

**111. a.** $2\,CO_2(g) + CaSiO_3(s) + H_2O(l) \longrightarrow$
$\qquad SiO_2(s) + Ca(HCO_3)_2(aq)$
**b.** $2\,Co(NO_3)_3(aq) + 3\,(NH_4)_2S(aq) \longrightarrow$
$\qquad Co_2S_3(s) + 6\,NH_4NO_3(aq)$
**c.** $Cu_2O(s) + C(s) \longrightarrow 2\,Cu(s) + CO(g)$
**d.** $H_2(g) + Cl_2(g) \longrightarrow 2\,HCl(g)$

**113. a.** inorganic    **b.** organic
**c.** organic    **d.** inorganic

**115. a.** alkene    **b.** alkane
**c.** alkyne    **d.** alkane

**117. a.** $CH_3CH_2CH_3$    **b.** propane
**c.** $CH_3CH_2CH_2CH_2CH_2CH_2CH_2CH_3$    **d.** pentane

**119. a.** functionalized hydrocarbon, alcohol
**b.** hydrocarbon
**c.** functionalized hydrocarbon, ketone
**d.** functionalized hydrocarbon, amine

**121.** $1.50 \times 10^{24}$ molecules EtOH

**123. a.** $K_2CrO_4$, 40.27% K, 26.78% Cr, 32.95% O
**b.** $Pb_3(PO_4)_2$, 76.60% Pb, 7.63% P, 15.77% O
**c.** $H_2SO_3$, 2.46% H, 39.07% S, 58.47% O
**d.** $CoBr_2$, 26.94% Co, 73.06% Br

**125.** $1.80 \times 10^2$ g $Cl_2$/yr

**127.** M = Fe

**129.** estradiol = $C_{18}H_{24}O_2$

**131.** $C_{18}H_{20}O_2$

**133.** $7\,H_2O$

**135.** $C_6H_9BrO$

**137.** $1.87 \times 10^{21}$ atoms

**139.** 92.93 amu

**141.** $x = 1, y = 2$

**143.** 41.7 mg

**145.** 0.224 g

**147.** 22.0% by mass

**149.** $1.6 \times 10^7$ kg Cl

**151.** $7.8 \times 10^3$ kg rock

**153.** $C_5H_{10}SI$

**155.** $X_2Y_3$

**157.** The sphere in the molecular models represents the electron cloud of the atom. On this scale, the nucleus would be too small to see.

**159.** The statement is incorrect because a chemical formula is based on the ratio of atoms combined, not the ratio of grams combined. The statement should read, "The chemical formula for ammonia ($NH_3$) indicates that ammonia contains three hydrogen atoms to each nitrogen atom."

**161.** O, S, H

**168. a. i.** 0.40 billion metric tons;
**ii.** 1.5 billion metric tons
**c.** 1.5 ppm/yr; 423 ppm

# Chapter 4

**25.** $2\,C_6H_{14}(g) + 19\,O_2(g) \longrightarrow$
$\qquad 12\,CO_2(g) + 14\,H_2O(g)$, 68 mol $O_2$

**27. a.** 5.0 mol $NO_2$    **b.** 14. mol $NO_2$
**c.** 0.281 mol $NO_2$    **d.** 53.1 mol $NO_2$

**29.**

| mol $SiO_2$ | mol C | mol SiC | mol CO |
|---|---|---|---|
| 3 | 9 | 3 | 6 |
| 2 | 6 | 2 | 4 |
| 5 | 15 | 5 | 10 |
| 2.8 | 8.4 | 2.8 | 5.6 |
| 0.517 | 1.55 | 0.517 | 1.03 |

**31. a.** 9.3 g HBr, 0.12 g $H_2$

**33. a.** 5.56 g $BaCl_2$    **b.** 6.55 g $CaCO_3$
**c.** 6.09 g MgO    **d.** 6.93 g $Al_2O_3$

**35. a.** 4.42 g HCl    **b.** 8.25 g $HNO_3$
**c.** 4.24 g $H_2SO_4$

**37. a.** Na    **b.** Na    **c.** $Br_2$    **d.** Na

**39.** 3 molecules $Cl_2$

**41. a.** 2 mol    **b.** 7 mol    **c.** 9.40 mol

**43.** 0.5 mol $O_2$

**45. a.** 2.5 g    **b.** 31.1 g    **c.** 1.16 g

**47.** 2.91 grams $CO_2$ remaining

**49.** limiting reactant: $Pb^{2+}$, theoretical yield: 34.5 g $PbCl_2$, percent yield: 85.3%

**51.** limiting reactant: $NH_3$, theoretical yield: 240.5 kg $CH_4N_2O$, percent yield: 70.01%

**53. a.** 1.17 M LiCl
**b.** 0.123 M $C_6H_{12}O_6$
**c.** 0.00453 M NaCl

**55. a.** 0.150 M $NO_3^-$
**b.** 0.300 M $NO_3^-$
**c.** 0.450 M $NO_3^-$

**57. a.** 1.3 mol    **b.** 1.5 mol    **c.** 0.211 mol

**59.** 37 g

**61.** 0.27 M

**63.** 6.0 L

**65.** 37.1 mL

**67.** 2.1 L

**69.** lead nitrate, 3.75 g, 65.3%

**71. a.** yes    **b.** no    **c.** yes    **d.** no

**73. a.** soluble $Ag^+$, $NO_3^-$
   **b.** soluble $Pb^{2+}$, $C_2H_3O_2^-$
   **c.** soluble $K^+$, $NO_3^-$
   **d.** soluble $NH_4^+$, $S^{2-}$

**75. a.** NO REACTION    **b.** NO REACTION
   **c.** $CrBr_2(aq) + Na_2CO_3(aq) \longrightarrow$
$$CrCO_3(s) + 2\,NaBr(aq)$$
   **d.** $3\,NaOH(aq) + FeCl_3(aq) \longrightarrow$
$$Fe(OH)_3(s) + 3\,NaCl(aq)$$

**77. a.** $K_2CO_3(aq) + Pb(NO_3)_2(aq) \longrightarrow$
$$PbCO_3(s) + 2\,KNO_3(aq)$$
   **b.** $Li_2SO_4(aq) + Pb(C_2H_3O_2)_2(aq) \longrightarrow$
$$PbSO_4(s) + 2\,LiC_2H_3O_2(aq)$$
   **c.** $Cu(NO_3)_2(aq) + MgS(aq) \longrightarrow$
$$CuS(s) + Mg(NO_3)_2(aq)$$
   **d.** NO REACTION

**79. a.** Complete:
$$H^+(aq) + Cl^-(aq) + Li^+(aq) + OH^-(aq) \longrightarrow$$
$$H_2O(l) + Li^+(aq) + Cl^-(aq)$$
   Net: $H^+(aq) + OH^-(aq) \longrightarrow H_2O(l)$
   **b.** Complete:
$$Ca^{2+}(aq) + S^{2-}(aq) + Cu^{2+}(aq) + 2\,Cl^-(aq) \longrightarrow$$
$$CuS(s) + Ca^{2+}(aq) + 2\,Cl^-(aq)$$
   Net: $S^{2-}(aq) + Cu^{2+}(aq) \longrightarrow CuS(s)$
   **c.** Complete:
$$Na^+(aq) + OH^-(aq) + HC_2H_3O_2(aq) \longrightarrow$$
$$H_2O(l) + Na^+(aq) + C_2H_3O_2^-(aq)$$
   Net: $OH^-(aq) + HC_2H_3O_2(aq) \longrightarrow$
$$H_2O(l) + C_2H_3O_2^-(aq)$$
   **d.** Complete:
$$6\,Na^+(aq) + 2\,PO_4^{3-}(aq) + 3\,Ni^{2+}(aq) + 6\,Cl^-(aq) \longrightarrow$$
$$Ni_3(PO_4)_2(s) + 6\,Na^+(aq) + 6\,Cl^-(aq)$$
   Net: $3\,Ni^{2+}(aq) + 2\,PO_4^{3-}(aq) \longrightarrow Ni_3(PO_4)_2(s)$

**81.** Complete:
$$Hg_2^{2+}(aq) + 2\,NO_3^-(aq) + 2\,Na^+(aq) + 2\,Cl^-(aq) \longrightarrow$$
$$Hg_2Cl_2(s) + 2\,Na^+(aq) + 2\,NO_3^-(aq)$$
   Net: $Hg_2^{2+}(aq) + 2\,Cl^-(aq) \longrightarrow Hg_2Cl_2(s)$

**83.** Molecular:
$$HBr(aq) + KOH(aq) \longrightarrow H_2O(l) + KBr(aq)$$
   Net ionic: $H^+(aq) + OH^-(aq) \longrightarrow H_2O(l)$

**85. a.** $H_2SO_4(aq) + Ca(OH)_2(aq) \longrightarrow$
$$2\,H_2O(l) + CaSO_4(s)$$
   **b.** $HClO_4(aq) + KOH(aq) \longrightarrow H_2O(l) + KClO_4(aq)$
   **c.** $H_2SO_4(aq) + 2\,NaOH(aq) \longrightarrow$
$$2\,H_2O(l) + Na_2SO_4(aq)$$

**87. a.** Complete ionic:
$$H^+(aq) + Br^-(aq) + Na^+(aq) + OH^-(aq) \longrightarrow$$
$$H_2O(l) + Na^+(aq) + Br^-(aq)$$
   Net ionic: $H^+(aq)\,OH^-(aq) \longrightarrow H_2O(l)$
   **b.** Complete ionic:
$$HF(aq) + Na^+(aq) + OH^-(aq) \longrightarrow$$
$$H_2O(l) + Na^+(aq) + F^-(aq)$$

Net ionic:
$$HF(aq) + OH^-(aq) \longrightarrow H_2O(l) + F^-(aq)$$
   **c.** Complete ionic:
$$HC_2H_3O_2(aq) + Rb^+(aq) + OH^-(aq) \longrightarrow$$
$$H_2O(l) + Rb^+(aq) + C_2H_3O_2^-(aq)$$

Net ionic:
$$HC_2H_3O_2(aq) + OH^-(aq) \longrightarrow H_2O(l) + C_2H_3O_2^-(aq)$$

**89.** 0.1810 M $HClO_4$

**91. a.** $2\,HBr(aq) + NiS(s) \longrightarrow H_2S(g) + NiBr_2(aq)$
   **b.** $NH_4I(aq) + NaOH(aq) \longrightarrow$
$$H_2O(l) + NH_3(g) + NaI(aq)$$
   **c.** $2\,HBr(aq) + Na_2S(aq) \longrightarrow H_2S(g) + 2\,NaBr(aq)$
   **d.** $2\,HClO_4(aq) + Li_2CO_3(aq) \longrightarrow$
$$H_2O(l) + CO_2(g) + 2\,LiClO_4(aq)$$

**93. a.** Ag: 0    **b.** Ag: +1
   **c.** Ca: +2, F: −1    **d.** H: +1, S: −2
   **e.** C: +4, O: −2    **f.** Cr: +6, O: −2

**95. a.** +2    **b.** +6    **c.** +3

**97. a.** redox reaction, oxidizing agent: $O_2$, reducing agent: Li
   **b.** redox reaction, oxidizing agent: $Fe^{2+}$, reducing agent: Mg
   **c.** not a redox reaction
   **d.** not a redox reaction

**99. a.** $S(s) + O_2(g) \longrightarrow SO_2(g)$
   **b.** $2\,C_3H_6(g) + 9\,O_2(g) \longrightarrow 6\,CO_2(g) + 6\,H_2O(g)$
   **c.** $2\,Ca(s) + O_2(g) \longrightarrow 2\,CaO(g)$
   **d.** $C_5H_{12}S(l) + 9\,O_2(g) \longrightarrow$
$$5\,CO_2(g) + SO_2(g) + 6\,H_2O(g)$$

**101.** 3.32 M

**103.** 1.1 g

**105.** 3.1 kg

**107.** limiting reactant: $C_7H_6O_3$, theoretical yield: 1.63 g $C_9H_8O_4$, percent yield: 74.8%

**109. b**

**111. a.** $2\,HCl(aq) + Hg_2(NO_3)_2(aq) \longrightarrow$
$$Hg_2Cl_2(s) + 2\,HNO_3(aq)$$
   **b.** $KHSO_3(aq) + HNO_3(aq) \longrightarrow$
$$H_2O(l) + SO_2(g) + KNO_3(aq)$$
   **c.** $2\,NH_4Cl(aq) + Pb(NO_3)_2(aq) \longrightarrow$
$$PbCl_2(s) + 2\,NH_4NO_3(aq)$$
   **d.** $2\,NH_4Cl(aq) + Ca(OH)_2(aq) \longrightarrow$
$$2\,NH_3(g) + 2\,H_2O(g) + CaCl_2(aq)$$

**113.** 22 g

**115.** 6.9 g

**117.** $NaNO_3$ is more economical.

**119.** Br is the oxidizing agent, Au is the reducing agent, 38.8 g $KAuF_4$.

**121.** $Ca^{2+}$ and $Cu^{2+}$ present in the original solution. Net ionic for first precipitate:
$$Ca^{2+}(aq) + SO_4^{2-}(aq) \longrightarrow CaSO_4(s)$$
Net ionic for second precipitate:
$$Cu^{2+}(aq) + CO_3^{2-}(aq) \longrightarrow CuCO_3(s)$$

**123.** 0.333 g $PH_3$

**125.** 30.8 kg $CO_2$

**127.** 1.6 g $C_2H_2$

**129.** 11.8 g AgI

**131.** 5.5% by mass

**133.** 14 g $KO_2$. In designing the unit, you would need to consider the theoretical yield and % yield of the reaction, how changing the limiting reactant would affect the reaction, and the stoichiometry between $KO_2$ and $O_2$ to determine the mass of $KO_2$ required to produce enough $O_2$ for 10 minutes. You might also consider the speed of the reaction and whether or not the reaction produced heat. Additionally, because your body does not use 100% of the oxygen taken in with each breath, the apparatus would only need to replenish the oxygen used. The percentage of oxygen in air is about 20% and the percentage in exhaled air is about 16%, so you can assume that 4% of the air would need to be replenished with oxygen.

**135.** 37.9 g $B_5H_9$

**137. a.** Since the mol ratio of $H_2$ to NO is 5:2, the 10 mol of $H_2$ will require 4 mole NO, and $H_2$ is the limiting reactant. This eliminates answers b and c. Since there is excess NO, this eliminates d, leaving answer a.

**139. a**

**146. a.** NaF, $2.0 \times 10^1$ kg $F^-$; $Na_2SiF_6$, 27 kg $F^-$
  **c. i.** 0.190 kg $F^-$        **ii.** 0.957 kg $H_2SiF_6$
  **e.** $Na_2SiF_6$; 0.316 kg; $28.31
  **g.** You would require 40% less, and costs would decrease by 40%.

# Chapter 5

**25. a.** 0.832 atm        **b.** 632 mmHg
  **c.** 12.2 psi        **d.** $8.43 \times 10^4$ Pa
**27. a.** 809.0 mmHg        **b.** 1.064 atm
  **c.** 809.0 torr        **d.** 107.9 kPa
**29. a.** 832 mmHg        **b.** 718 mmHg
**31.** $4.4 \times 10^2$ mmHg
**33.** 58.9 mL
**35.** 4.22 L
**37.** 3.0 L. The volume would not be different if the gas was argon.
**39.** 1.16 atm
**41.** 2.1 mol
**43.** Yes, the final gauge pressure is 43.5 psi, which exceeds the maximum rating.
**45.** 16.2 L
**47.** 286 atm, 17.5 bottles purged
**49. b**
**51.** 4.76 atm
**53.** 37.3 L
**55.** 9.43 g/L
**57.** 44.0 g/mol
**59.** 4.00 g/mol
**61.** $P_{tot} = 434$ torr, $mass_{N_2} = 0.437$ g, $mass_{O_2} = 0.237$ g, $mass_{He} = 0.0340$ g
**63.** 1.84 atm
**65.** $\chi_{N_2} = 0.627$, $\chi_{O_2} = 0.373$, $P_{N_2} = 0.687$ atm, $P_{O_2} = 0.409$ atm
**67.** $P_{H_2} = 0.921$ atm, $mass_{H_2} = 0.0539$ g
**69.** $7.47 \times 10^{-2}$ g
**71.** 38 L
**73.** $V_{H_2} = 48.2$ L, $V_{CO} = 24.1$ L

**75.** 22.8 g $NaN_3$
**77.** 60.4%
**79.** $F_2$, 2.84 g $ClF_3$
**81. a.** yes        **b.** no
  **c.** No. Even though the argon atoms are more massive than the helium atoms, both have the same kinetic energy at a given temperature. The argon atoms therefore move more slowly and so exert the same pressure as the helium atoms.
  **d.** He
**83.** $F_2 : u_{rms} = 442$ m/s,  $KE_{avg} = 3.72 \times 10^3$ J;
  $Cl_2 : u_{rms} = 324$ m/s,  $KE_{avg} = 3.72 \times 10^3$ J;
  $Br_2 : u_{rms} = 216$ m/s,  $KE_{avg} = 3.72 \times 10^3$ J;
  rankings : $u_{rms} : Br_2 < Cl_2 < F_2$,
  $KE_{avg} : Br_2 = Cl_2 = F_2$,
  rate of effusion: $Br_2 < Cl_2 < F_2$
**85.** rate $^{238}UF_6$/rate $^{235}UF_6$ = 0.99574
**87.** krypton
**89.** A has the higher molar mass, B has the higher rate of effusion.
**91.** That the volume of gas particles is small compared to the space between them breaks down under conditions of high pressure. At high pressure the particles themselves occupy a significant portion of the total gas volume.
**93.** 0.05826 L (ideal); 0.0708 L (V.D.W.); Difference because of high pressure, at which Ne no longer acts ideally.
**95.** 97.8%
**97.** 27.8 g/mol
**99.** $C_4H_{10}$
**101.** 4.70 L
**103.** 2 HCl($aq$) + $K_2S$($s$) $\longrightarrow$
  $H_2S$($g$) + 2 KCl($aq$), 0.191 g $K_2S$
**105.** 11.7 L
**107.** $mass_{air} = 8.56$ g,  $mass_{He} = 1.20$ g, mass difference = 7.36 g
**109.** 4.76 L/s
**111.** total force $= 6.15 \times 10^3$ pounds; no, the can cannot withstand this force.
**113.** $5.8 \times 10^3$ balloons
**115.** 4.0 cm
**117.** 77.7%
**119.** 0.32 gram
**121.** 311 K
**123.** 5.0 g
**125.** $C_3H_8$
**127.** 0.39 g Ar
**129.** 74.0 mmHg
**131.** 25% $N_2H_4$
**133.** 25%
**135.** $P_{CH_4} = 7.30 \times 10^{-2}$ atm,  $P_{O_2} = 4.20 \times 10^{-1}$ atm,
  $P_{NO} = 2.79 \times 10^{-3}$ atm,  $P_{CO_2} = 5.03 \times 10^{-3}$ atm,
  $P_{H_2O} = 5.03 \times 10^{-3}$ atm,  $P_{NO_2} = 2.51 \times 10^{-2}$ atm,
  $P_{OH} = 1.01 \times 10^{-2}$ atm,  $P_{tot} = 0.542$ atm
**137.** 0.42 atm
**139.** Because helium is less dense than air, the balloon moves in a direction opposite the direction the air inside the car is moving due to the acceleration and deceleration of the car.

**141.** −29%
**143. a.** false    **b.** false    **c.** false    **d.** true
**145.** four times the initial pressure
**147.** Although the velocity "tails" have different lengths, the average length of the tails on the helium atoms is longer than the average length of the tails on the neon atoms, which is in turn longer than the average length of the tails on the krypton atoms. The lighter the atom, the faster the tails must move on average to have the same kinetic energy.

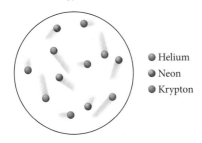

● Helium
● Neon
● Krypton

**153. a.** inverse relationship
**c.** $1.3 \times 10^{-6}$ mol
**e.** Sequence i
**g.** $P_{NO_2} = 4.3 \times 10^{-8}$ atm; $P_{O_3} = 5.0 \times 10^{-9}$ atm

# Chapter 6

**33. a.** $1.92 \times 10^9$ J      **b.** $5.14 \times 10^4$ cal
**c.** $2.37 \times 10^6$ J      **d.** 0.681 Cal
**35. a.** $9.987 \times 10^6$ J      **b.** $9.987 \times 10^3$ kJ
**c.** 2.78 kWh
**37. d**
**39. a.** heat, +      **b.** work, −      **c.** heat, +
**41.** $-7.27 \times 10^2$ kJ
**43.** 311 kJ
**45.** The drinks that went into cooler B had more thermal energy than the refrigerated drinks that went into cooler A. The temperature difference between the drinks in cooler B and the ice was greater than the difference between the drinks and the ice in cooler A. More thermal energy was exchanged between the drinks and the ice in cooler B, which resulted in more melting.
**47.** $4.7 \times 10^5$ J
**49. a.** $7.6 \times 10^2$ °C      **b.** $4.3 \times 10^2$ °C
**c.** $1.3 \times 10^2$ °C      **d.** 49 °C
**51.** $-2.8 \times 10^2$ J
**53.** 489 J
**55.** $\Delta E = -3463$ J, $\Delta H = -3452$ kJ
**57. a.** exothermic, −
**b.** endothermic, +
**c.** exothermic, −
**59.** $-4.30 \times 10^3$ kJ
**61.** $6.46 \times 10^4$ kJ
**63.** 1.0 kg $CO_2$
**65.** mass of silver 77.1 grams
**67.** final temperature 28.4 °C
**69.** specific heat capacity of substance A 1.10 J/g·°C

**71.** Measurement B corresponds to conditions of constant pressure. Measurement A corresponds to conditions of constant volume. When a fuel is burned under constant pressure some of the energy released does work on the atmosphere by expanding against it. Less energy is manifest as heat due to this work. When a fuel is burned under constant volume, all of the energy released by the combustion reaction is evolved as heat.
**73.** $-6.3 \times 10^3$ kJ/mol
**75.** $-1.6 \times 10^5$ J
**77. a.** $-\Delta H_1$      **b.** $2\,\Delta H_1$      **c.** $-\frac{1}{2}\Delta H_1$
**79.** −23.9 kJ
**81.** −173.2 kJ
**83. a.** $N_2(g) + 3 H_2(g) \longrightarrow 2 NH_3(g)$,
$\Delta H_f^\circ = -45.9$ kJ/mol
**b.** $C(s, \text{graphite}) + O_2(g) \longrightarrow CO_2(g)$,
$\Delta H_f^\circ = -393.5$ kJ/mol
**c.** $2 Fe(s) + 3/2\,O_2(g) \longrightarrow Fe_2O_3(s)$,
$\Delta H_f^\circ = -824.2$ kJ/mol
**d.** $C(s, \text{graphite}) + 2 H_2(g) \longrightarrow CH_4(g)$,
$\Delta H_f^\circ = -74.6$ kJ/mol
**85.** −380.2 kJ/mol
**87. a.** −137.1 kJ      **b.** −41.2 kJ
**c.** −137 kJ      **d.** 290.7 kJ
**89.** $6 CO_2(g) + 6 H_2O(l) \longrightarrow C_6H_{12}O_6(s) + 6 O_2(g)$,
$\Delta H_{rxn}^\circ = 2803$ kJ
**91.** −113.0 kJ/mol
**93. a.** 5.49 g $CO_2$      **b.** 6.46 g $CO_2$
**c.** 6.94 g $CO_2$
Natural gas, $CH_4(g)$, contributes the least to global warming by producing the least $CO_2(g)$ per kJ of heat produced.
**95.** $2 \times 10^{13}$ kg $CO_2$ produced per year, 150 years
**97.** $\Delta E = -1.7$ J,  $q = -0.5$ J,  $w = -1.2$ J
**99.** 78 g
**101.** $\Delta H = 6.0$ kJ/mol, $1.1 \times 10^2$ g
**103.** 26.1 °C
**105.** palmitic acid: 9.9378 Cal/g, sucrose: 3.938 Cal/g, fat contains more Cal/g than sugar
**107.** $\Delta H = \Delta E + nR\Delta T$
**109.** 5.7 Cal/g
**111.** $\Delta E = 0, \Delta H = 0, q = -w = 3.0 \times 10^3$ J
**113.** −294 kJ/mol
**115.** 94.0 kJ
**117.** 23.9 °C
**119.** $7.3 \times 10^3$ g $H_2SO_4$
**121.** $7.2 \times 10^2$ g
**123.** 78.2 °C
**125.** $C_v = \frac{3}{2}R, C_p = \frac{5}{2}R$
**127.** $q = 1030$ kJ, $\Delta H = 1030$ kJ, $\Delta E = 952$ kJ, $w = -78$ kJ
**129.** −1292 kJ
**131. d**
**133. a.** At constant pressure, heat can be added and work can be done on the system. $\Delta E = q + w$; therefore, $q = \Delta E - w$.
**135.** The aluminum is cooler because it has a lower heat capacity (specific heat).

**137.** $q = -2418\,J, w = -5\,kJ, \Delta H = -2418\,J/mol,$
$$\Delta E = -2423\,J/mol$$

**139. b.** $\Delta H > \Delta E$

**144. a. i.** $CH_3CH_2CH_3(g) + 5\,O_2(g) \longrightarrow$
$$3\,CO_2(g) + 4\,H_2O(l)$$
**ii.** $2\,CH_3CH_2CH_2OH(l) + 9\,O_2(g) \longrightarrow$
$$6\,CO_2(g) + 8\,H_2O(l)$$
**c.** $-2043.0\,kJ$ compared to $-2219\,kJ$
**e.** about $655\,kJ/CH_2$ group

# Chapter 7

**35.** $499\,s$
**37. i.** d, c, b, a  **ii.** a, b, c, d
**39. a.** $4.74 \times 10^{14}\,Hz$  **b.** $5.96 \times 10^{14}\,Hz$
 **c.** $5.8 \times 10^{18}\,Hz$
**41. a.** $3.14 \times 10^{-19}\,J$
 **b.** $3.95 \times 10^{-19}\,J$
 **c.** $3.8 \times 10^{-15}\,J$
**43.** $1.03 \times 10^{16}$ photons
**45. a.** $79.8\,kJ/mol$  **b.** $239\,kJ/mol$
 **c.** $798\,kJ/mol$
**47.**

**49.** $3.6 \times 10^6\,m/s$
**51.** $5.39\,nm$
**53.** $1.1 \times 10^{-34}\,m$. The wavelength of a baseball is negligible with respect to its size.
**55.** $\Delta v = 1.04 \times 10^5\,m/s$
**57.** $2s$
**59. a.** $l = 0$  **b.** $l = 0, 1$
 **c.** $l = 0, 1, 2$  **d.** $l = 0, 1, 2, 3$
**61.** c
**63.** See Figures 7.25 and 7.26. The $2s$ and $3p$ orbitals would, on average, be farther from the nucleus and have more nodes than the $1s$ and $2p$ orbitals.
**65.** $n = 1$
**67.** $2p \longrightarrow 1s$
**69. a.** $122\,nm$, UV  **b.** $103\,nm$, UV
 **c.** $486\,nm$, visible  **d.** $434\,nm$, visible
**71.** $n = 2$
**73.** $344\,nm$
**75.** $6.4 \times 10^{17}$ photons/s
**77.** $0.0547\,nm$
**79.** $91.2\,nm$
**81. a.** 4  **b.** 9  **c.** 16
**83.** $n = 4 \longrightarrow n = 3, n = 5 \longrightarrow n = 3,$
 $n = 6 \longrightarrow n = 3$, respectively
**85.** $4.84 \times 10^{14}\,s^{-1}$
**87.** $11\,m$
**89.** $6.78 \times 10^{-3}\,J$
**91.** $632\,nm$
**93.** $2.98 \times 10^{-4}\,mol$
**95. a.** $E_1 = 2.51 \times 10^{-18}\,J, E_2 = 1.00 \times 10^{-17}\,J,$
 $E_3 = 2.26 \times 10^{-17}\,J$
 **b.** $26.5\,nm$, UV; $15.8\,nm$, UV

**97.**

$1s$:

$2s$:

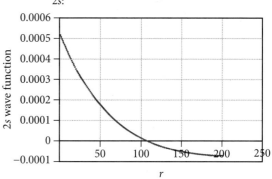

The plot for the $2s$ wave function extends below the $x$-axis. The $x$-intercept represents the radial node of the orbital.

**99.** $7.39 \times 10^5\,m/s$
**101.** $\Delta E = 1.1 \times 10^{-20}\,J, 7.0 \times 10^2\,nm$
**103.** $11\,m$
**105.** In the Bohr model, electrons exist in specific orbits encircling the atom. In the quantum-mechanical model, electrons exist in orbitals that are really probability density maps of where the electron is likely to be found. The Bohr model is inconsistent with Heisenberg's uncertainty principle.
**107. a.** yes  **b.** no  **c.** yes  **d.** no
**114. a.** $5.93 \times 10^{-19}\,J$  **c.** 2-EHMC
 **e.** $1.4 \times 10^6\,J$  **g.** 0.11 g TDSA; 0.055 g EHMC

# Chapter 8

**41. a.** $1s^2 2s^2 2p^6 3s^2 3p^2$  **b.** $1s^2 2s^2 2p^4$
 **c.** $1s^2 2s^2 2p^6 3s^2 3p^6 4s^1$  **d.** $1s^2 2s^2 2p^6$

**43. a.**

**45. a.** [Ne] $3s^2 3p^3$ **b.** [Ar] $4s^2 3d^{10} 4p^2$
  **c.** [Kr] $5s^2 4d^2$ **d.** [Kr] $5s^2 4d^{10} 5p^5$
**47. a.** 1 **b.** 10 **c.** 5 **d.** 2
**49. a.** V, As **b.** Se **c.** V **d.** Kr
**51. a.** 2 **b.** 1 **c.** 10 **d.** 6
**53.** reactive metal: **a**, reactive nonmetal: **c**
**55. c**
**57.** The valence electrons of nitrogen will experience a greater effective nuclear charge. The valence electrons of both atoms are screened by two core electrons, but N has a greater number of protons and therefore a greater net nuclear charge.
**59. a.** 1+ **b.** 2+ **c.** 6+ **d.** 4+
**61. a.** In **b.** Si **c.** Pb **d.** C
**63.** F, S, Si, Ge, Ca, Rb
**65. a.** [Ne] **b.** [Kr] **c.** [Kr]
  **d.** [Ar] $3d^6$ **e.** [Ar] $3d^9$
**67. a.** [Ar] Diamagnetic

  **b.** [Ar]  Paramagnetic
      $3d$

  **c.** [Ar]  Paramagnetic
      $3d$

  **d.** [Ar]  Paramagnetic
      $3d$

**69. a.** Li **b.** I$^-$ **c.** Cr **d.** O$^{2-}$
**71.** O$^{2-}$, F$^-$, Na$^+$, Mg$^{2+}$
**73. a.** Br
  **b.** Na
  **c.** Cannot tell based on periodic trends.
  **d.** P
**75.** In, Si, N, F
**77. a.** second and third **b.** fifth and sixth
  **c.** sixth and seventh **d.** first and second
**79. a.** Na **b.** S **c.** C **d.** F
**81. a.** Sr
  **b.** Bi
  **c.** Cannot tell based on periodic trends.
  **d.** As
**83.** S, Se, Sb, In, Ba, Fr
**85.** Sr$(s)$ + I$_2(g)$ $\longrightarrow$ SrI$_2(s)$
**87.** 2 Li$(s)$ + 2 H$_2$O$(l)$ $\longrightarrow$ 2 Li$^+(aq)$ + 2 OH$^-(aq)$ + H$_2(g)$
**89.** H$_2(g)$ + Br$_2(g)$ $\longrightarrow$ 2 HBr$(g)$
**91.** Br: $1s^2 2s^2 2p^6 3s^2 3p^6 4s^2 3d^{10} 4p^5$
  Kr: $1s^2 2s^2 2p^6 3s^2 3p^6 4s^2 3d^{10} 4p^6$
  Krypton's outer electron shell is filled, giving it chemical stability. Bromine is missing an electron from its outer shell and subsequently has a high electron affinity. Bromine tends to be easily reduced by gaining an electron, giving the bromide ion stability due to the filled $p$ subshell that corresponds to krypton's chemically stable electron configuration.
**93.** V: [Ar] $4s^2 3d^3$
  V$^{3+}$: [Ar] $3d^2$

Both V and V$^{3+}$ contain unpaired electrons in their $3d$ orbitals.
**95.** A substitute for K$^+$ would need to exhibit a 1+ electric charge and have similar mass and atomic radius. Na$^+$ and Rb$^+$ would not be good substitutes because their radii are significantly smaller and larger, respectively. Based on mass, Ca$^+$ and Ar$^+$ are the closest to K$^+$. Because the first ionization energy of Ca$^+$ is closest to that of K$^+$, Ca$^+$ is the best choice for a substitute. The difficulty lies in Ca's low second ionization energy, making it easily oxidized.
**97.** Si, Ge
**99. a.** N: [He] $2s^2 2p^3$, Mg: [Ne] $3s^2$, O: [He] $2s^2 2p^4$,
  F: [He] $2s^2 2p^5$, Al: [Ne] $3s^2 3p^1$
  **b.** Mg, Al, O, F, N
  **c.** Al, Mg, O, N, F
  **d.** Aluminum's first ionization energy is lower than Mg because its $3p$ electron is shielded by the $3s$ orbital. Oxygen's first ionization energy is lower than that of N because its fourth $2p$ electron experiences electron–electron repulsion by the other electron in its orbital.
**101.** For main-group elements, atomic radii decrease across a period because the addition of a proton in the nucleus and an electron in the outermost energy level increases $Z_{eff}$. This does not happen in the transition metals because the electrons are added to the $n_{highest-1}$ orbital and the $Z_{eff}$ stays roughly the same.
**103.** Noble gases are exceptionally unreactive due to the stability of their completely filled outer quantum levels and their high ionization energies. The ionization energies of Kr, Xe, and Rn are low enough to form some compounds.
**105.** 6A: $ns^2 np^4$, 7A: $ns^2 np^5$, group 7A elements require only one electron to achieve a noble gas configuration. Since group 6A elements require two electrons, their affinity for one electron is less negative because one electron will merely give them an $np^5$ configuration.
**107.** 85
**109. a.** One If By Land (O, Ne, I, F, B, Y, La, Nd)
  **b.** Atoms are Fun (N, U, Fe, Ra, S, Mo, Ta backward)
**111.** $1.390 \times 10^3$ kJ/mol, 86.14 nm
**113. a.** F **b.** B **c.** B **d.** O
**115. a.** $d_{Ar} \approx 2$ g/L, $d_{Xe} \approx 6.5$ g/L
  **b.** $d_{118} \approx 13$ g/L
  **c.** mass = $3.35 \times 10^{-23}$ g/Ne atom, density of Ne atom = $2.3 \times 10^4$ g/L. The separation of Ne atoms relative to their size is immense.
  **d.** Kr: $2.69 \times 10^{22}$ atoms/L, Ne: $2.69 \times 10^{22}$ atoms/L. It seems Ar will also have $2.69 \times 10^{22}$ atoms/L. $d_{Ar} = 1.78$ g/L. This corresponds to accepted values.
**117.** Density increases to the right because, though electrons are added successively across the period, they are added to the $3d$ subshell, which is not a part of the outermost principal energy level. As a result, the atomic radius does not increase significantly across the period, while mass does.

**119.**

Longest $\lambda$ :

| 1s | 2s | $2p_x$ | $2p_y$ | $2p_z$ |
|----|----|--------|--------|--------|
| ⇅ | ⇅ | ↑ | ↓ | |

Next longest $\lambda$ :

| 1s | 2s | $2p_x$ | $2p_y$ | $2p_z$ |
|----|----|--------|--------|--------|
| ⇅ | ⇅ | ⇅ | | |

Next longest $\lambda$ :

| 1s | 2s | $2p_x$ | $2p_y$ | $2p_z$ |
|----|----|--------|--------|--------|
| ⇅ | ↑ | ↑ | ↑ | ↑ |

**121.** 168, noble gas
**123.** A relatively high effective nuclear charge is found in gallium with its completed $3d$ subshell and in thallium with its completed $4f$ subshell, accounting for the relatively high first ionization energies of these elements.
**125.** The second electron affinity requires the addition of an electron to something that is already negatively charged. The monoanions of both of these elements have relatively high electron density in a relatively small volume. As we shall see in Chapter 9, the dianions of these elements do exist in many compounds because they are stabilized by chemical bonding.
**127.** 120, 170
**129.** Fr, [Rn] $7s^1$, >265, <376, >1.879, <29
  **a.** $Fr^+(aq)$, $OH^-(aq)$, $H_2(g)$
  **b.** $Fr_2O(s)$
  **c.** $FrCl(s)$
**131.** **a.** any group 6A element
  **b.** any group 5A element
  **c.** any group 1A element
**133.** **a.** true     **b.** true     **c.** false     **d.** true
**135.** Since Ca has valence electrons of $4s^2$, it has a relatively low ionization energy to lose two electrons. F has a highly exothermic electron affinity when gaining one electron but not a second electron because of its $2s^2 2p^5$ valence electrons. Therefore, calcium and fluoride combine in a 2:1 ratio.
**141.** **a.** First ionization energy increases from left to right because effective nuclear charge increases.
  **c.** The first ionization energy of P is about the same as S because the electrons in S are paired in the same orbital, so that even though effective nuclear charge is higher in S, the effect of the repulsion between the paired electrons makes the first ionization energies similar.
  **e.** $SiF_4$; This support student X's hypothesis, but not student Y's.

# Chapter 9

**35.** $1s^2 2s^2 2p^3$   ·N̈:

**37.** **a.** ·Äl·     **b.** $Na^+$     **c.** :C̈l:     **d.** $[:C̈l:]^-$
**39.** **a.** $Na^+[:F̈:]^-$     **b.** $Ca^{2+}[:Ö:]^{2-}$
  **c.** $Sr^{2+} 2[:B̈r:]^-$     **d.** $2 K^+[:Ö:]^{2-}$
**41.** **a.** SrSe     **b.** $BaCl_2$
  **c.** $Na_2S$     **d.** $Al_2O_3$

**43.** As the size of the alkaline earth metal ions increases, so does the distance between the metal cations and oxygen anions. Therefore, the magnitude of the lattice energy decreases accordingly because the potential energy decreases as the distance increases.
**45.** One factor of lattice energy is the product of the charges of the two ions. The product of the ion charges for CsF is $-1$, while that for BaO is $-4$. Because this product is four times greater, the lattice energy is also four times greater.
**47.** $-708$ kJ/mol
**49.** **a.** H:H, filled duets, 0 formal charge on both atoms
  **b.** :C̈l:C̈l:, filled octets, 0 formal charge on both atoms
  **c.** Ö=Ö, filled octets, 0 formal charge on both atoms
  **d.** :N≡N:, filled octets, 0 formal charge on both atoms

**51.** **a.** H—P̈—H with H below     **b.** :S̈—C̈l: with :C̈l: below

  **c.** H—Ï:     **d.** H—C—H with H above and H below

**53.** **a.** :F̈—S̈—F̈:     **b.** H—Si—H with H above and H below

  **c.** Ö=C—Ö: with H above C and H below     **d.** H—C—S̈—H with H above and H below C

**55.** **a.** pure covalent     **b.** polar covalent
  **c.** pure covalent     **d.** ionic bond
**57.** :C≡O:, 25%

**59.** **a.** :Ï—C—Ï: with :Ï: above and :Ï: below     **b.** :N≡N—Ö:

  **c.** H—Si—H with H above and H below     **d.** :C̈l—C—C̈l: with :O: above (double bond)

**61.** **a.** H—N̈=N̈—H     **b.** H—N̈—N̈—H with H below each N

  **c.** H—C≡C—H     **d.** H—C=C—H with H below each C

**63. a.** :Ö—Së=Ö: ⟷ :Ö=Së—Ö:
    −1 +1 0          0 +1 −1

**b.** [structure: carbonate resonance, C center with O groups, charges 0, −1]  ⟷  [structure]  ⟷

[structure]

**c.** [:C̈l—Ö:]⁻
      0   −1

**d.** [:Ö=N̈—Ö:]⁻ ⟷ [:Ö—N̈=Ö:]⁻
       0   0  −1        −1  0   0

**65.**   H          H
          |          |
      H—C=S̈     H—S=C̈    H₂CS is the better
          0  0        +2  −2
      structure.

**67.** :O≡C—Ö: does not provide a significant
        contribution to the resonance hybrid as it has a +1
        formal charge on a very electronegative atom (oxygen).

**69.**     H  Ö:              H  :Ö:⁻¹
            |  ‖               |  |
        H—C—C—Ö:⁻¹  ⟷  H—C—C=Ö:
            |                 |
            H                 H

**71.** N has a formal charge of +1; O has a formal charge
        of −1.

**73. a.**    :C̈l:
               |
              B
            /   \
         :C̈l    C̈l:

**b.** Ö=N̈—Ö: ⟷ :Ö—N̈=Ö:

**c.**     H
           |
           B
          / \
        H    H

**75. a.** [phosphate resonance structures, charge 3−] ⟷ [structure] ⟷

[structure] ⟷ [structure]

**b.** [:C≡N:]⁻
       −1   0

**c.** [sulfite resonance, SO₃²⁻] ⟷ [structure] ⟷

[structure]

**d.** [:Ö—C̈l—Ö:]⁻
        −1  +1  −1

**77. a.**        :F:
                   |
            :F̈    P—F̈:
              \   |
            :F̈    :F:
                   |
                  :F:

**b.** [:Ï—Ï—Ï:]⁻

**c.**     :F:
            |
        :F—S̈—F:
            |
           :F:

**d.**     :F:
            |
        :F—Ge—F:
            |
           :F:

**79.** H₃CCH₃, H₂CCH₂, HCCH
**81.** −128 kJ
**83.** −614 kJ

**85. a.**     :Ï:
                |
               B
              / \
            :Ï    Ï:

**b.** 2 K⁺[:S̈:]²⁻

**c.**     :O:
            ‖
        H—C—F̈:

**d.**     :B̈r:
            |
        :B̈r—P—B̈r:

**87. a.** Ba²⁺[carbonate structure]²⁻ ⟷

Ba²⁺[structure]²⁻ ⟷ Ba²⁺[structure]²⁻

**b.** Ca²⁺ 2[:Ö—H]⁻

**c.** K⁺[nitrate structure]⁻ ⟷ K⁺[structure]⁻ ⟷

K⁺[structure]⁻

**d.** Li⁺[:Ï—Ö:]⁻

**89. a.**

$$H-\overset{\overset{\displaystyle H}{|}}{C}-\overset{\overset{\displaystyle H}{|}}{C}-H$$
$$H-\underset{\underset{\displaystyle H}{|}}{C}-\underset{\underset{\displaystyle H}{|}}{C}-H$$

**b.**

(two resonance structures of cyclobutadiene)

**c.**

(cyclohexane structure)

**d.**

(two resonance structures of benzene)

**91.** $CH_2O_2$,

$$H-\overset{\overset{\displaystyle :O:}{\|}}{C}-\ddot{O}-H$$

**93.** The reaction is exothermic due to the energy released when the $Al_2O_3$ lattice forms.

**95.**

$$H-\ddot{O}-\overset{\overset{\displaystyle :O:^{0}}{\|}}{N}-\ddot{O}: \longleftrightarrow H-\ddot{O}-\overset{\overset{\displaystyle :\ddot{O}:^{-1}}{|}}{N}=\ddot{O} \longleftrightarrow$$
$$\quad 0 \quad 0 \ +1 \ -1 \qquad\qquad 0 \quad 0 \ +1 \quad 0$$

Most important

$$H-\ddot{O}=\overset{\overset{\displaystyle :\ddot{O}:^{-1}}{|}}{N}-\ddot{O}:$$
$$0 \ +1 \ +1 \ -1$$

**97.**

$$\left[\ddot{C}=N=\ddot{O}\right]^{-} \longleftrightarrow \left[:C\equiv\ddot{N}-\ddot{O}:\right]^{-}$$
$$\;\;-2 \ +1 \quad 0 \qquad\qquad -1 \ +1 \ -1$$

The fulminate ion is less stable because nitrogen is more electronegative than carbon and should therefore be terminal to accommodate the negative formal charge.

**99. a.** $H-\overset{\overset{\displaystyle }{}}{C}=\overset{\overset{\displaystyle }{}}{C}-H$ (with H H below the two carbons)

**b.**

$$H-\overset{\overset{\displaystyle H}{|}}{\underset{\underset{\displaystyle H}{|}}{C}}-\overset{\overset{\displaystyle }{}}{\underset{\underset{\displaystyle H}{|}}{\ddot{N}}}-H$$

**c.**

$$H-\overset{\overset{\displaystyle :O:}{\|}}{C}-H$$

**d.**

$$H-\overset{\overset{\displaystyle H}{|}}{\underset{\underset{\displaystyle H}{|}}{C}}-\overset{\overset{\displaystyle H}{|}}{\underset{\underset{\displaystyle H}{|}}{C}}-\ddot{O}-H$$

**e.**

$$H-\overset{\overset{\displaystyle :O:}{\|}}{C}-\ddot{O}-H$$

**101.**

Nonpolar — :S: H — Polar
$$H-\overset{\overset{\displaystyle \|}{}}{C}-\overset{\overset{\displaystyle |}{}}{\ddot{N}}-H$$
Nonpolar   Polar

**103. a.** $\left[\ddot{O}=\ddot{O}\cdot\right]^{-}$

**b.** $\left[:\ddot{O}:\right]^{-}$

**c.** $:\ddot{O}-H$

**d.**

$$H-\overset{\overset{\displaystyle H}{|}}{\underset{\underset{\displaystyle H}{|}}{C}}-\ddot{O}-\ddot{O}\cdot$$

**105.** $\Delta H_{rxn(H_2)} = -243 \text{ kJ/mol} = -121 \text{ kJ/g}$
$\Delta H_{rxn(CH_4)} = -802 \text{ kJ/mol} = -50.0 \text{ kJ/g}$
$CH_4$ yields more energy per mole, while $H_2$ yields more energy per gram.

**107. a.**

$$\ddot{O}=\overset{\overset{\displaystyle :O:}{\|}}{\underset{\underset{\displaystyle :O:}{\|}}{Cl}}-\ddot{O}-\overset{\overset{\displaystyle :O:}{\|}}{\underset{\underset{\displaystyle :O:}{\|}}{Cl}}=\ddot{O}$$

**b.**

$$H-\overset{\overset{\displaystyle :O:}{\|}}{\underset{\underset{\displaystyle :O:}{|}}{P}}-\ddot{O}-H$$
(with H below bottom O)

**c.**

$$H-\ddot{O}-\overset{\overset{\displaystyle :O:}{\|}}{\underset{\underset{\displaystyle :O:}{|}}{As}}-\ddot{O}-H$$
(with H below bottom O)

**109.** $Na^{+}F^{-}$, $Na^{+}O^{2-}$, $Mg^{2+}F^{-}$, $Mg^{2+}O^{2-}$, $Al^{3+}O^{2-}$

**111.** 333 kJ/mol

**113.** $H-C\equiv C-H$

**115.**

$$\ddot{\text{O}}=\ddot{\text{S}}=\ddot{\text{O}} + :\ddot{\text{O}}-\text{H} \longrightarrow \text{H}-\ddot{\text{O}}-\overset{\overset{\displaystyle ::\text{O}::}{\|}}{\underset{}{\text{S}}}-\ddot{\text{O}}:$$

$$\text{H}-\ddot{\text{O}}-\overset{\overset{\displaystyle ::\text{O}::}{\|}}{\underset{}{\text{S}}}-\ddot{\text{O}}: + \ddot{\text{O}}=\ddot{\text{O}} \longrightarrow$$

$$\ddot{\text{O}}-\overset{\overset{\displaystyle ::\text{O}::}{\|}}{\underset{}{\text{S}}}-\ddot{\text{O}}: + \text{H}-\ddot{\text{O}}-\ddot{\text{O}}\cdot$$

$$\ddot{\text{O}}-\overset{\overset{\displaystyle ::\text{O}::}{\|}}{\underset{}{\text{S}}}-\ddot{\text{O}}: + \text{H}-\ddot{\text{O}}-\text{H} \longrightarrow :\ddot{\text{O}}-\overset{\overset{\displaystyle \text{H} \;\; ::\text{O}::}{}}{\underset{\underset{\displaystyle \text{H}-\ddot{\text{O}}:}{|}}{\text{S}}}=\ddot{\text{O}}:$$

$\Delta H_{\text{rxn}} = -172 \text{ kJ}$

**117.** $r_{\text{HCl}} = 113 \text{ pm}$
$r_{\text{HF}} = 84 \text{ pm}$
These values are close to the accepted values.

**119.** 

$$:\text{P} \overset{\overset{\displaystyle \ddot{\text{P}}}{}}{\underset{\underset{\displaystyle \ddot{\text{P}}}{}}{\diamondsuit}} \text{P}:$$

**121.** 126 kJ/mol
**123.** The oxidation number of the S atoms bonded directly to hydrogen atoms is −1. The oxidation number of interior S atoms is 0.
**125.** 536 kJ
**127.** The compounds are energy rich because a great deal of energy is released when these compounds undergo a reaction that breaks weak bonds and forms strong ones.
**129.** The theory is successful because it allows us to predict and account for many chemical observations. The theory is limited because electrons cannot be treated as localized "dots."
**135. a.** The trend in lattice energy mirrors the trend in atomic radius shown in Figure b. The ions with smaller radii have higher (more negative) lattice energies.
   **c.** The lattice energy of $MnF_2$ is lower because the ionic radius of $Mn^{2+}$ is larger.
   **e.** $VF_2$ because the $F^-$ ion is smaller than the $Cl^-$ ion.

# Chapter 10

**31.** 4
**33. a.** 4 $e^-$ groups, 4 bonding groups, 0 lone pair
   **b.** 5 $e^-$ groups, 3 bonding groups, 2 lone pairs
   **c.** 6 $e^-$ groups, 5 bonding groups, 1 lone pair
**35. a.** $e^-$ geometry: tetrahedral molecular geometry: trigonal pyramidal idealized bond angle: 109.5°, deviation
   **b.** $e^-$ geometry: tetrahedral molecular geometry: bent idealized bond angle: 109.5°, deviation
   **c.** $e^-$ geometry: tetrahedral molecular geometry: tetrahedral idealized bond angle: 109.5°, deviation (due to large size of Cl compared to H)
   **d.** $e^-$ geometry: linear molecular geometry: linear idealized bond angle: 180°

**37.** $H_2O$ has a smaller bond angle due to lone pair–lone pair repulsions, the strongest electron group repulsion.

**39. a.** seesaw, 

$$\text{S} \overset{\overset{\displaystyle |}{\text{F}}}{\underset{\underset{\displaystyle |}{\text{F}}}{\diagup\!\!\!\diagdown}} \overset{\text{F}}{\underset{\blacktriangleright \text{F}}{}}$$

   **b.** T-shape, 

$$\overset{\text{F}}{\underset{\underset{\displaystyle \text{F}}{|}}{\text{Cl}-\text{F}}}$$

   **c.** linear, F—I—F
   **d.** square planar, 

$$\overset{\text{Br}}{\underset{\underset{\displaystyle \text{Br}}{|}}{\text{Br}-\text{I}-\text{Br}}}$$

**41. a.** linear, H—C≡C—H
   **b.** trigonal planar, 

$$\overset{\text{H}}{\underset{\text{H}}{}}\text{C}=\text{C}\overset{\text{H}}{\underset{\text{H}}{}}$$

   **c.** tetrahedral, 

$$\text{H}^{\text{IIII}}\overset{\overset{\displaystyle \text{H}}{}}{\text{C}}-\overset{}{\underset{\underset{\displaystyle \text{H}}{}}{\text{C}}}^{\text{IIII}}\text{H}$$

**43. a.** The lone pair will cause lone pair–bonding pair repulsions, pushing the three bonding pairs out of the same plane. The correct molecular geometry is trigonal pyramidal.
   **b.** The lone pair should take an equatorial position to minimize 90° bonding pair interactions. The correct molecular geometry is seesaw.
   **c.** The lone pairs should take positions on opposite sides of the central atom to reduce lone pair–lone pair interactions. The correct molecular geometry is square planar.

**45. a.** C: tetrahedral

$$\text{H}^{\text{IIII}}\overset{\overset{\displaystyle \text{O}^{\text{IIII}}\text{H}}{|}}{\text{C}}\overset{\text{H}}{\underset{\text{H}}{\blacktriangleleft}}$$

   O: bent

   **b.** C's: tetrahedral
      O: bent
   **c.** O's: bent

$$\overset{\text{H}}{\underset{}{}}\text{O}-\text{O}\overset{}{\underset{\text{H}}{}}$$

**47.** The vectors of the polar bonds in both $CO_2$ and $CCl_4$ oppose each other with equal magnitude and sum to 0.
**49.** $PF_3$, polar
$SBr_2$, nonpolar
$CHCl_3$, polar
$CS_2$, nonpolar

**51. a.** polar
   **b.** polar
   **c.** nonpolar
**53. a.** 0
   **b.** 3
   **c.** 1
**55.**

3s          3p

1s

1s

Expected bond angle = 90°

Valence bond theory is compatible with experimentally determined bond angle of 93.3° without hybrid orbitals.

1s

**57.**

2s     2p      Hybridization      $sp^3$

**59.** $sp^2$
**61. a.** $sp^3$

$\sigma$: Cl$(p)$ − C$(sp^3)$ [×4]

**b.** $sp^3$

$\sigma$: H$(s)$ − N$(sp^3)$ [×3]

**c.** $sp^3$

$\sigma$: F$(p)$ − O$(sp^3)$ [×2]

**d.** $sp$

$\pi$: O$(p)$ − C$(p)$ [×2]

$\sigma$: O$(p)$ − C$(sp)$ [×2]

**63. a.** $sp^2$

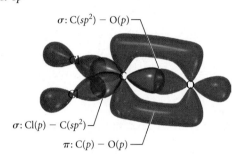

$\sigma$: C$(sp^2)$ − O$(p)$

$\sigma$: Cl$(p)$ − C$(sp^2)$

$\pi$: C$(p)$ − O$(p)$

**b.** $sp^3d^2$

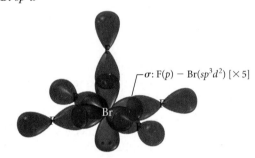

$\sigma$: F$(p)$ − Br$(sp^3d^2)$ [×5]

**c.** $sp^3d$

$\sigma$: F$(p)$ − Xe$(sp^3d)$ [×2]

**d.** $sp^3d$

$\sigma$: I$(p)$ − I$(sp^3d)$ [×2]

**65. a.** N's: $sp^2$

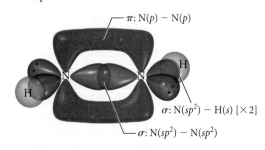

$\pi$: N$(p)$ − N$(p)$

$\sigma$: N$(sp^2)$ − H$(s)$ [×2]

$\sigma$: N$(sp^2)$ − N$(sp^2)$

*(continued on the next page)*

**b.** N's: $sp^3$

σ: H(s) − N($sp^3$) [×4]

σ: N($sp^3$) − N($sp^3$)

**c.** C: $sp^3$

N: $sp^3$

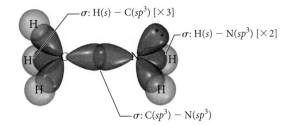

σ: H(s) − C($sp^3$) [×3]

σ: H(s) − N($sp^3$) [×2]

σ: C($sp^3$) − N($sp^3$)

**67.**

$sp^3$

$sp^3$

$sp^2$

**69.**

Constructive interference

**71.** $Be_2^+$    $\underline{1}$ $\sigma^*_{2s}$    $Be_2^-$

$\underline{1\!\!\downarrow}$ $\sigma_{2s}$

$\underline{1}$ $\sigma_{2p}$

$\underline{1\!\!\downarrow}$ $\sigma^*_{2s}$

$\underline{1\!\!\downarrow}$ $\sigma_{2s}$

bond order $Be_2^+$ = 1/2

bond order $Be_2^-$ = 1/2

Both will exist in gas phase.

**73.** Bonding

Antibonding

**75. a.**

| | $\sigma^*_{2p}$ |
| --- | --- |
| $\underline{\phantom{1}}$ $\underline{\phantom{1}}$ | $\pi^*_{2p}$ |
| $\underline{\phantom{1}}$ | $\sigma_{2p}$ |
| $\underline{\phantom{1}}$ $\underline{\phantom{1}}$ | $\pi_{2p}$ |
| $\underline{1\!\!\downarrow}$ | $\sigma^*_{2s}$ |
| $\underline{1\!\!\downarrow}$ | $\sigma_{2s}$ |

bond order = 0 diamagnetic

**b.**

| | $\sigma^*_{2p}$ |
| --- | --- |
| $\underline{\phantom{1}}$ $\underline{\phantom{1}}$ | $\pi^*_{2p}$ |
| $\underline{\phantom{1}}$ | $\sigma_{2p}$ |
| $\underline{1}$ $\underline{1}$ | $\pi_{2p}$ |
| $\underline{1\!\!\downarrow}$ | $\sigma^*_{2s}$ |
| $\underline{1\!\!\downarrow}$ | $\sigma_{2s}$ |

bond order = 1 paramagnetic

**c.**

| | $\sigma^*_{2p}$ |
| --- | --- |
| $\underline{\phantom{1}}$ $\underline{\phantom{1}}$ | $\pi^*_{2p}$ |
| $\underline{\phantom{1}}$ | $\sigma_{2p}$ |
| $\underline{1\!\!\downarrow}$ $\underline{1\!\!\downarrow}$ | $\pi_{2p}$ |
| $\underline{1\!\!\downarrow}$ | $\sigma^*_{2s}$ |
| $\underline{1\!\!\downarrow}$ | $\sigma_{2s}$ |

bond order = 2 diamagnetic

**d.**

| | $\sigma^*_{2p}$ |
| --- | --- |
| $\underline{\phantom{1}}$ $\underline{\phantom{1}}$ | $\pi^*_{2p}$ |
| $\underline{1}$ | $\sigma_{2p}$ |
| $\underline{1\!\!\downarrow}$ $\underline{1\!\!\downarrow}$ | $\pi_{2p}$ |
| $\underline{1\!\!\downarrow}$ | $\sigma^*_{2s}$ |
| $\underline{1\!\!\downarrow}$ | $\sigma_{2s}$ |

bond order = 2.5 paramagnetic

**77. a.** not stable    **b.** not stable

**c.** stable    **d.** not stable

**79.** $C_2^-$ has the highest bond order, the highest bond energy, and the shortest bond length.

**81.**

| | $\sigma^*_{2p}$ |
| --- | --- |
| $\underline{1\!\!\downarrow}$ $\underline{1\!\!\downarrow}$ | $\pi^*_{2p}$ |
| $\underline{1\!\!\downarrow}$ $\underline{1\!\!\downarrow}$ | $\pi_{2p}$ |
| $\underline{1\!\!\downarrow}$ | $\sigma_{2p}$ |
| $\underline{1\!\!\downarrow}$ | $\sigma^*_{2s}$ |
| $\underline{1\!\!\downarrow}$ | $\sigma_{2s}$ |

bond order = 3

**83. a.**

:Ö:
||
:F̈—C—F̈:

trigonal planar polar C: $sp^2$

**b.** :C̈l—S̈—S̈—C̈l:

bent polar S's: $sp^3$

**c.**

:F̈:
|
:F̈—S̈—F̈:
|
:F̈:

seesaw polar S: $sp^3d$

*(continued on the next page)*

**85. a.** $sp^3$, Bent; $sp^3$, Tetrahedral; $sp^3$, Tetrahedral; $sp^3$, Trigonal pyramidal; $sp^3$, Bent; $sp^2$, Trigonal planar

**b.** $sp^3$, Tetrahedral; $sp^3$, Tetrahedral; $sp^2$, Trigonal planar; $sp^3$, Trigonal pyramidal; $sp^2$, Trigonal planar; $sp^3$, Trigonal pyramidal; $sp^3$, Bent

**c.** $sp^3$, Bent; $sp^3$, Tetrahedral; $sp^3$, Bent; $sp^3$, Tetrahedral; $sp^2$, Trigonal planar; $sp^3$, Trigonal pyramidal

**87.** σ bonds: 25
π bonds: 4
lone pairs: on O's and N (without methyl group):
$sp^2$ orbitals
on N's (with methyl group): $sp^3$ orbitals

**89. a.** water soluble **b.** fat soluble
**c.** water soluble **d.** fat soluble

**91.**

| | | |
|---|---|---|
| — | | $\sigma_p^*$ |
| ⇅ | ⇅ | $\pi_p^*$ |
| ⇅ | ⇅ | $\pi_p$ |
| ⇅ | | $\sigma_p$ |
| ⇅ | | $\sigma_s^*$ |
| ⇅ | | $\sigma_s$ |

bond order = 1

**93.** BrF, unhybridized, linear

:B̈r—F̈:

$BrF_2^-$ has two bonds and three lone pairs on the central atom. The hybridization is $sp^3d$. The electron geometry is trigonal bipyramidal, with the three lone pairs equatorial. The molecular geometry is linear.

$$[:\ddot{F}—\dot{B}\dot{r}—\ddot{F}:]^-$$

$BrF_3$ has three bonds and two lone pairs on the central atom. The hybridization is $sp^3d$. The electron geometry

is trigonal bipyramidal, with the two lone pairs equatorial. The molecular geometry is T-shaped.

$$[:\ddot{F}—\dot{B}\dot{r}—\ddot{F}:]^-$$
:F̈:

$BrF_4^-$ has four bonds and two lone pairs on the central atom. The hybridization is $sp^3d^2$. The electron geometry is octahedral, with the two lone pairs on the same axis. The molecular geometry is square planar.

$BrF_5$ has five bonds and one lone pair on the central atom. The hybridization is $sp^3d^2$. The electron geometry is octahedral. The molecular geometry is square pyramidal.

**95.** The moments of the two Cl's cancel.

**97. a.** 10 **b.** 14 **c.** 2
**99.** According to valence bond theory, $CH_4$, $NH_3$, and $H_2O$ are all $sp^3$ hybridized. This hybridization results in a tetrahedral electron group configuration with a 109.5° bond angle. $NH_3$ and $H_2O$ deviate from this idealized bond angle because their lone electron pairs exist in their own $sp^3$ orbitals. The presence of lone pairs lowers the tendency for the central atom's orbitals to hybridize. As a result, as lone pairs are added, the bond angle moves further from the 109.5° hybrid angle and closer to the 90° unhybridized angle.
**101.** $NH_3$ is stable due to its bond order of 3.

| | | |
|---|---|---|
| — | — | Antibonding |
| | — | Antibonding |
| ⇅ | | Bonding |
| ⇅ | ⇅ | Bonding |
| ⇅ | | Bonding |

**103.** In $NO_2^+$, the central N has two electron groups, so the hybridization is $sp$ and the ONO angle is 180°. In $NO_2^-$ the central N has three electron groups, two bonds and one lone pair. The ideal hybridization is $sp^2$, but the ONO bond angle should close down a bit because of the lone pair. A bond angle around 115° is a good guess. In $NO_2$ there are three electron groups, but one group is a single electron. Again, the ideal hybridization would be $sp^2$, but since one unpaired electron

(continued on the next page)

must be much smaller than a lone pair or even a bonding pair, we predict that the ONO bond angle will spread and be greater than 120°. As a guess, the angle is probably significantly greater than 120°.

$$\left[\ddot{O}=N=\ddot{O}\right]^{+}$$

$$\left[:\ddot{O}=\ddot{N}-\ddot{O}:\right]^{-}$$

$$\ddot{O}=\ddot{N}-\ddot{O}:$$

**105.** In addition to the 2s and the three 2p orbitals, one more orbital is required to make five hybrid orbitals. The closest in energy is the 3s orbital. So the hybridization is $s^2p^3$. VSEPR predicts trigonal bipyramidal geometry for five identical substituents.

**107.**

H   :Ö
|    ‖
H—C—C—N̈—H
|    |
H    H

Lewis Structure

H   :Ö:⁻¹
|    |
H—C—C=N—H    +1
|    |
H    H

Resonance Structure

Terminal carbon is tetrahedral, central carbon is trigonal planar, nitrogen is trigonal pyramidal (but resonance structure is trigonal planar).

**109. a.** This is the best.
 **b.** This statement is similar to **a** but leaves out nonbonding lone-pair electron groups.
 **c.** Molecular geometries are not determined by overlapping orbitals but rather by the number and type of electron groups around each central atom.

**111.** Lewis theory defines a single bond, double bond, and triple bond as a sharing of two electrons, four electrons, and six electrons, respectively, between two atoms. Valence bond theory defines a single bond as a sigma overlap of two orbitals, a double bond as a single sigma bond combined with a pi bond, and a triple bond as a double bond with an additional pi bond. Molecular orbital theory defines a single bond, double bond, and triple bond as a bond order of 1, 2, or 3, respectively, between two atoms.

**118. a.** From top to bottom of column: 109.8°, 112.4°, 110.9°

# Chapter 11

**35. a.** dispersion
 **b.** dispersion, dipole–dipole, hydrogen bonding
 **c.** dispersion, dipole–dipole
 **d.** dispersion
**37. a.** dispersion, dipole–dipole
 **b.** dispersion, dipole–dipole, hydrogen bonding

 **c.** dispersion
 **d.** dispersion
**39. a, b, c, d,** Boiling point increases with increasing intermolecular forces. The molecules increase in their intermolecular forces as follows: **a,** dispersion forces; **b,** stronger dispersion forces (broader electron cloud); **c,** dispersion forces and dipole–dipole interactions; **d,** dispersion forces, dipole–dipole interactions, and hydrogen bonding.
**41. a.** $CH_3OH$, hydrogen bonding
 **b.** $CH_3CH_2OH$, hydrogen bonding
 **c.** $CH_3CH_3$, greater mass, broader electron cloud causes greater dispersion forces
**43. a.** $Br_2$, smaller mass results in weaker dispersion forces
 **b.** $H_2S$, lacks hydrogen bonding
 **c.** $PH_3$, lacks hydrogen bonding
**45. a.** not homogeneous
 **b.** homogeneous, dispersion, dipole–dipole, hydrogen bonding, ion–dipole
 **c.** homogeneous, dispersion
 **d.** homogeneous, dispersion, dipole–dipole, hydrogen bonding
**47.** Water. Surface tension increases with increasing intermolecular forces, and water can hydrogen-bond while acetone cannot.
**49.** compound A
**51.** When the tube is clean, water experiences adhesive forces with glass that are stronger than its cohesive forces, causing it to climb the surface of a glass tube. Water does not experience strong intermolecular forces with oil, so if the tube is coated in oil, the water's cohesive forces will be greater and it will not be attracted to the surface of the tube.
**53.** The water in the 12-cm dish will evaporate more quickly. The vapor pressure does not change, but the surface area does. The water in the dish evaporates more quickly because the greater surface area allows for more molecules to obtain enough energy at the surface and break free.
**55.** Water is more volatile than vegetable oil. When the water evaporates, the endothermic process results in cooling.
**57.** 0.405 L
**59.** 91 °C
**61.** $\Delta H_{vap} = 24.7$ kJ/mol, bp $= 239$ K
**63.** 41 torr
**65.** 27.5 kJ/mol
**67.** 22.0 kJ
**69.** 2.7 °C
**71.** 30.5 kJ
**73. a.** solid
 **b.** liquid
 **c.** gas
 **d.** supercritical fluid
 **e.** solid/liquid
 **f.** liquid/gas
 **g.** solid/liquid/gas

**75.** $N_2$ has a stable liquid phase at 1 atm.

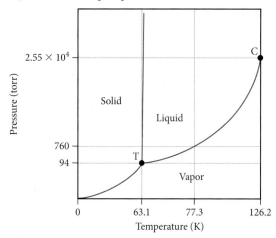

**77. a.** 0.027 mmHg     **b.** rhombic
**79.** Water has strong intermolecular forces. It is polar and experiences hydrogen bonding.
**81.** Water's exceptionally high specific heat capacity has a moderating effect on Earth's climate. Also, its high $\Delta H_{vap}$ causes water evaporation and condensation to have a strong effect on temperature.
**83.** The general trend is that melting point increases with increasing mass. This is because the electrons of the larger molecules are held more loosely and a stronger dipole moment can be induced more easily. HF is the exception to the rule. It has a relatively high melting point due to hydrogen bonding.
**85.** yes, 1.22 g
**87.** gas $\longrightarrow$ liquid $\longrightarrow$ solid

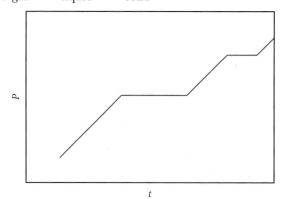

**89.** 26 °C
**91.**

**93.** $3.4 \times 10^3$ g $H_2O$

**95. a.** $CO_2(s) \longrightarrow CO_2(g)$ at 195 K
   **b.** $CO_2(s) \longrightarrow$ triple point at 216 K $\longrightarrow CO_2(g)$ just above 216 K
   **c.** $CO_2(s) \longrightarrow CO_2(l)$ at somewhat above 216 K $\longrightarrow CO_2(g)$ at around 250 K
   **d.** $CO_2(s) \longrightarrow CO_2(g) \longrightarrow$ supercritical fluid
**97.** Decreasing the pressure will decrease the temperature of liquid nitrogen. Because the nitrogen is boiling, its temperature must be constant at a given pressure. As the pressure decreases, the boiling point decreases, and therefore so does the temperature. If the pressure drops below the pressure of the triple point, the phase change will shift from vaporization to sublimation and the liquid nitrogen will become solid.
**99.** 70.7 L
**101.** 0.48 atm
**103.**

**105.** The water within a container with a larger surface area will evaporate more quickly because there is more surface area for the molecules to evaporate from. Vapor pressure is the pressure of the gas when it is in dynamic equilibrium with the liquid. The vapor pressure is dependent only on the substance and the temperature. The larger the surface area, the more quickly it will reach the dynamic state.
**107.** The triple point will be at a lower temperature since the fusion equilibrium line has a positive slope. This means that we will be increasing both temperature and pressure as we travel from the triple point to the normal melting point.
**109.** The liquid segment will have the least steep slope because it takes the most kJ/mol to raise the temperature of the phase.
**111.** There are substantial intermolecular attractions in the liquid but virtually none in the gas.
**117. a.** Molar mass correlates with boiling point for $H_2S$, $H_2Se$, and $Hg_2Te$, but not for $H_2O$.
   **c.** dispersion force
   **e.** Yes. Since polarizability correlates with the dispersion force and since polarizability increases as boiling point increases, the increasing dispersion force is responsible for the increasing boiling point.
   **g.**

# Chapter 12

**27.** 162 pm

**29. a.** 1    **b.** 2    **c.** 4

**31.** 68%

**33.** $l = 393$ pm, $d = 21.3$ g/cm$^3$

**35.** 134.5 pm

**37.** $6.0 \times 10^{23}$ atoms/mol

**39. a.** atomic
   **b.** molecular
   **c.** ionic
   **d.** atomic

**41.** LiCl($s$). The other three solids are held together by intermolecular forces, while LiCl is held together by stronger coulombic interactions between the cations and anions of the crystal lattice.

**43. a.** TiO$_2$($s$), ionic solid
   **b.** SiCl$_4$($s$), larger, stronger dispersion forces
   **c.** Xe($s$), larger, stronger dispersion forces
   **d.** CaO, ions have greater charge and therefore stronger coulombic forces

**45.** TiO$_2$

**47.** Cs:1(1) = 1
Cl:8(1/8) = 1
1:1
CsCl
Ba:8(1/8) + 6(1/2) = 4
Cl:8(1) = 8
4:8 = 1:2
BaCl$_2$

**49.** cesium chloride: none of these; barium(II) chloride: fluorite structure

**51.** face-centered cubic

**53. a.** nonoxide
   **b.** silicate
   **c.** nonoxide

**55.** boron oxide, B$_2$O$_3$

**57.** Ca: +2; Si: +2; O: −2

**59. a.** Zn($s$)

**61.** 0.807 mol orbitals

**63.** insulator

**65. a.** p-type
   **b.** n-type

**67.** yes, it has sufficient energy

**69.**

**71.**

**73.** H—C≡C—H

**75.**

**77.** CsCl has a higher melting point than AgI because of its higher coordination number. In CsCl, one anion bonds to eight cations (and vice versa), while in AgI, one anion bonds to only four cations.

**79. a.** 4$r$
   **b.** $c^2 = a^2 + b^2$   $c = 4r$, $a = l$, $b = l$
   $(4r)^2 = l^2 + l^2$
   $16r^2 = 2l^2$
   $8r^2 = l^2$
   $l = \sqrt[3]{8r^2}$
   $l = 2\sqrt{2}r$

**81.** 8 atoms/unit

**83.** 55.843 g/mol

**85.** 2.00 g/cm$^3$

**87.** body diagonal $= \sqrt{6}r$,
   radius $= (\sqrt{3} - \sqrt{2})r/\sqrt{2} = 0.2247r$

**89.** The higher-level electron transitions with their smaller energy gaps would give not off enough energy to create X-rays.

**91. a.**

**b.**

To obtain this structure, the monomer from part a would react in a head-to-head (or tail-to-tail) addition as opposed to the head-to-tail addition that leads to the structure shown in Table 12.3.

**93.** Because the structure is face-centered cubic, there are therefore four C$_{60}$ molecules per unit cell. Thus there must be $3 \times 4 = 12$ Rb atoms per unit cell, and all sites (tetrahedral and octahedral) are occupied.

**95.** The liquid must be cooled quickly in order to prevent the formation of an organized crystal structure, and instead achieve an amorphous product.

**97.** Both structures may be viewed as having essentially a face-centered cubic unit cell with half of the tetrahedral holes filled. Diamond, however, consists of only one type of atom (C) and is covalently bound, whereas zinc blende has S$^{2-}$ ions at the face-centered cubic sites and Zn$^{2+}$ ions in the tetrahedral holes, and is held together by ionic forces.

**99. d.** All of the above would likely lead to an increase in electrical conductivity.

**105. a.** Most of the metals that form hydrides have only one oxidation state.
   **c.** 4.6%
   **e.** heated

# Chapter 13

**29. a.** hexane, toluene, or $CCl_4$; dispersion forces
**b.** water, methanol; dispersion, dipole–dipole, hydrogen bonding
**c.** hexane, toluene, or $CCl_4$; dispersion forces
**d.** water, acetone, methanol, ethanol; dispersion, ion–dipole

**31.** $HOCH_2CH_2CH_2OH$

**33. a.** water; dispersion, dipole–dipole, hydrogen bonding
**b.** hexane; dispersion
**c.** water; dispersion, dipole–dipole
**d.** water; dispersion, dipole–dipole, hydrogen bonding

**35. a.** endothermic
**b.** The lattice energy is greater in magnitude than the heat of hydration.
**c.**

**d.** The solution forms because chemical systems tend toward greater entropy.

**37.** $-797$ kJ/mol

**39.** $\Delta H_{soln} = -6 \times 10^1$ kJ/mol, $-7$ kJ of energy evolved

**41.** unsaturated

**43.** About 31 g will precipitate.

**45.** Boiling water releases any $O_2$ dissolved in it. The solubility of gases decreases with increasing temperature.

**47.** As pressure increases, nitrogen will more easily dissolve in blood. To reverse this process, divers should ascend to lower pressures.

**49.** 1.1 g

**51.** 1.92 M, 2.0 $m$, 10.4%

**53.** 0.340 L

**55.** $1.6 \times 10^2$ g

**57.** $1.4 \times 10^4$ g

**59.** Add water to 7.31 mL of concentrated solution until a total volume of 1.15 L is acquired.

**61. a.** Add water to 3.73 g KCl to a volume of 100 mL.
**b.** Add 3.59 g KCl to 96.41 g $H_2O$.
**c.** Add 5.0 g KCl to 95 g $H_2O$.

**63. a.** 0.417 M   **b.** 0.444 $m$   **c.** 7.41% by mass
**d.** 0.00794   **e.** 0.794% by mole

**65.** 0.89 M

**67.** 15 $m$, 0.22

**69.** The level has decreased more in the beaker filled with pure water. The dissolved salt in the seawater decreases the vapor pressure and subsequently lowers the rate of vaporization.

**71.** 30.7 torr

**73. a.** $P_{hep} = 24.4$ torr, $P_{oct} = 5.09$ torr
**b.** 29.5 torr
**c.** 80.8% heptane by mass, 19.2% octane by mass
**d.** The vapor is richer in the more volatile component.

**75.** $P_{chl} = 51.9$ torr, $P_{ace} = 274$ torr, $P_{tot} = 326$ torr. The solution is not ideal. The chloroform–acetone interactions are stronger than the chloroform–chloroform and acetone–acetone interactions.

**77.** freezing point (fp) $= -1.27$ °C, bp $= 100.349$ °C

**79.** freezing point (fp) $= 1.0$ °C, boiling point (bp) $= 82.4$ °C

**81.** $1.8 \times 10^2$ g/mol

**83.** 26.1 atm

**85.** $6.36 \times 10^3$ g/mol

**87. a.** fp $= -0.558$ °C, bp $= 100.154$ °C
**b.** fp $= -1.98$ °C, bp $= 100.546$ °C
**c.** fp $= -2.5$ °C, bp $= 100.70$ °C

**89.** 157 g

**91. a.** $-0.632$ °C   **b.** 5.4 atm   **c.** 100.18 °C

**93.** 2.3

**95.** 3.4

**97.** 23.0 torr

**99.** Chloroform is polar and has stronger solute–solvent interactions than nonpolar carbon tetrachloride.

**101.** $\Delta H_{soln} = 51$ kJ/mol, $-8.7$ °C

**103.** $2.2 \times 10^{-3}$ M/atm

**105.** $1.3 \times 10^4$ L

**107.** 0.24 g

**109.** $-24$ °C

**111. a.** 1.1% by mass/$V$   **b.** 1.6% by mass/$V$
**c.** 5.3% by mass/$V$

**113.** 2.484

**115.** 0.229 atm

**117.** $\chi_{CHCl_3}$(original) $= 0.657$,
$P_{CHCl_3}$(condensed) $= 0.346$ atm

**119.** 1.74 M

**121.** $C_6H_{14}O_2$

**123.** 12 grams

**125.** $6.4 \times 10^{-3}$ L

**127.** 22.4% glucose by mass, 77.6% sucrose by mass

**129.** $P_{iso} = 0.131$ atm, $P_{pro} = 0.068$ atm. The major intermolecular attractions are between the OH groups. The OH group at the end of the chain in propyl alcohol is more accessible than the one in the middle of the chain in isopropyl alcohol. In addition, the molecular shape of propyl alcohol is a straight chain of carbon atoms, while that of isopropyl alcohol is a branched chain and is more like a ball. The contact area between two ball-like objects is smaller than that of two chain-like objects. The smaller contact area in isopropyl alcohol means the molecules don't attract each other as strongly as do those of propyl alcohol. As a result of both of these factors, the vapor pressure of isopropyl alcohol is higher.

**131.** 0.005 $m$

**133.** $Na_2CO_3$ 0.050 M, $NaHCO_3$ 0.075 M

**135.** The water should not be immediately cycled back into the river. As the water was warmed, dissolved oxygen would have been released, since the amount of a gas able to be dissolved into a liquid decreases as the temperature of the liquid increases. As such, the water returned to the river would lack dissolved oxygen if it was still hot. To preserve the dissolved oxygen necessary for the survival of fish and other aquatic life, the water must first be cooled.

**137. b.** NaCl
**144. a. i.** −2.39 °C          **ii.** −2.06 °C

**c.**

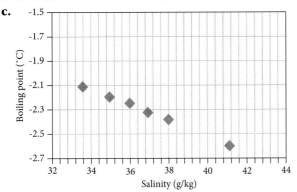

Figure c does not show the range of salinities actually present in seawater.

# Chapter 14

**25. a.** Rate $= -\dfrac{1}{2}\dfrac{\Delta[\text{HBr}]}{\Delta t} = \dfrac{\Delta[\text{H}_2]}{\Delta t} = \dfrac{\Delta[\text{Br}_2]}{\Delta t}$
   **b.** $1.8 \times 10^{-3}$ M/s          **c.** 0.040 mol $\text{Br}_2$

**27. a.** Rate $= -\dfrac{1}{2}\dfrac{\Delta[\text{A}]}{\Delta t} = -\dfrac{\Delta[\text{B}]}{\Delta t} = \dfrac{1}{3}\dfrac{\Delta[\text{C}]}{\Delta t}$

   **b.** $\dfrac{\Delta[\text{B}]}{\Delta t} = -0.0500$ M/s, $\dfrac{\Delta[\text{C}]}{\Delta t} = 0.150$ M/s

**29.**

| $\Delta[\text{Cl}_2]/\Delta t$ | $\Delta[\text{F}_2]/\Delta t$ | $\Delta[\text{ClF}_3]/\Delta t$ | Rate |
|---|---|---|---|
| −0.012 M/s | −0.036 M/s | 0.024 M/s | 0.012 M/s |

**31. a.** $0 \longrightarrow 10$ s : Rate $= 8.7 \times 10^{-3}$ M/s
   $40 \longrightarrow 50$ s : Rate $= 6.0 \times 10^{-3}$ M/s
   **b.** $1.4 \times 10^{-2}$ M/s
**33. a. i.** $1.0 \times 10^{-2}$ M/s          **ii.** $8.5 \times 10^{-3}$ M/s
   **iii.** 0.013 M/s

**b.**

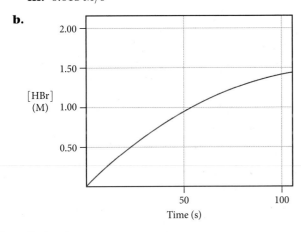

**35. a.** first order
   **b.**

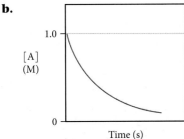

   **c.** Rate $= k[\text{A}]^1, k = 0.010$ s$^{-1}$

**37. a.** s$^{-1}$          **b.** M$^{-1}$ s$^{-1}$          **c.** M · s$^{-1}$
**39. a.** Rate $= k[\text{A}][\text{B}]^2$
   **b.** third order
   **c.** 2
   **d.** 4
   **e.** 1
   **f.** 8
**41.** second order, Rate $= 5.25$ M$^{-1}$ s$^{-1}$[A]$^2$
**43.** first order, Rate $= 0.065$ s$^{-1}$ [A]
**45.** Rate $= k[\text{NO}_2][\text{F}_2], k = 2.57$ M$^{-1}$ s$^{-1}$, second order
**47. a.** zero order
   **b.** first order
   **c.** second order
**49.** second order, $k = 2.25 \times 10^{-2}$ M$^{-1}$ s$^{-1}$,
   [AB] at 25 s $= 0.619$ M
**51.** first order, $k = 1.12 \times 10^{-2}$ s$^{-1}$, Rate $= 2.8 \times 10^{-3}$ M/s
**53. a.** $4.5 \times 10^{-3}$ s$^{-1}$
   **b.** Rate $= 4.5 \times 10^{-3}$ s$^{-1}$[A]
   **c.** $1.5 \times 10^2$ s
   **d.** [A] $= 0.0908$ M
**55. a.** $4.88 \times 10^3$ s
   **b.** $9.8 \times 10^3$ s
   **c.** $1.7 \times 10^3$ s
   **d.** 0.146 M at 200 s, 0.140 M at 500 s
**57.** $6.8 \times 10^8$ yr; $1.8 \times 10^{17}$ atoms
**59.**

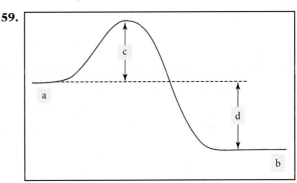

**61.** 17 s$^{-1}$
**63.** 61.90 kJ/mol
**65.** $E_a = 251$ kJ/mol, $A = 7.93 \times 10^{11}$ s$^{-1}$
**67.** $E_a = 23.0$ kJ/mol, $A = 8.05 \times 10^{10}$ M$^{-1}$s$^{-1}$
**69. a.** 122 kJ/mol
   **b.** 0.101 s$^{-1}$
**71.** 47.85 kJ/mol
**73. a**
**75.** The mechanism is valid.
**77. a.** $\text{Cl}_2(g) + \text{CHCl}_3(g) \longrightarrow \text{HCl}(g) + \text{CCl}_4(g)$
   **b.** Cl$(g)$, CCl$_3(g)$
   **c.** Rate $= k[\text{Cl}_2]^{1/2}[\text{CHCl}_3]$
**79.** Heterogeneous catalysts require a large surface area because catalysis can only happen at the surface. A greater surface area means greater opportunity for the substrate to react, which results in a faster reaction.
**81.** $10^{12}$
**83. a.** first order, $k = 0.0462$ hr$^{-1}$
   **b.** 15 hr
   **c.** $5.0 \times 10^1$ hr

**85.** 0.0531 M/s

**87.** rate = $4.5 \times 10^{-4}\,[CH_3CHO]^2$,
$$k = 4.5 \times 10^{-4}, 0.37\ \text{atm}$$

**89.** 219 torr

**91.** $1 \times 10^{-7}$ s

**93.** $1.6 \times 10^2$ seconds

**95. a.** 2

**b.**

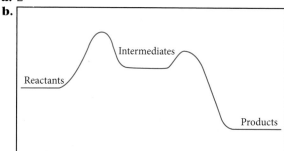

**c.** first step    **d.** exothermic

**97. a.** 5.41 s

**b.** 2.2 s for 25%, 5.4 s for 50%

**c.** 0.28 at 10 s, 0.077 at 20 s

**99. a.** $E_a = 89.5$ kJ/mol, $A = 4.22 \times 10^{11}\,s^{-1}$

**b.** $2.5 \times 10^{-5}\,M^{-1}\,s^{-1}$

**c.** $6.0 \times 10^{-4}$ M/s

**101. a.** No

**b.** No bond is broken, and the two radicals attract each other.

**c.** Formation of diatomic gases from atomic gases.

**103.** $1.35 \times 10^4$ years

**105. a.** Both are valid. For both, all steps sum to overall reaction, and the predicted rate law is consistent with experimental data.

**b.** Buildup of I($g$)

**107.** rate = $k_2[(k_1/k_{-1})[Br_2]]^{1/2}[H_2]$ The rate law is 3/2 order overall.

**109. a.** 0%

**b.** 25%

**c.** 33%

**111.** 174 kJ

**113. a.** second order

**b.** $CH_3NC + CH_3NC \underset{k_2}{\overset{k_1}{\rightleftharpoons}} CH_3NC^* + CH_3NC$    (fast)

$CH_3NC^* \xrightarrow{k_3} CH_3CN$    (slow)

Rate = $k_3[CH_3NC^*]$

$k_1[CH_3NC]^2 = k_2[CH_3NC^*][CH_3NC]$

$[CH_3NC^*] = \dfrac{k_1}{k_2}[CH_3NC]$

Rate = $k_3 \times \dfrac{k_1}{k_2}[CH_3NC]$

Rate = $k[CH_3NC]$

**115.** Rate = $k[A]^2$

Rate = $-\dfrac{d[A]}{dt}$

$\dfrac{d[A]}{dt} = -k[A]^2$

$2\dfrac{d[A]}{[A]^2} = k\,dt$

$$\int_{[A]_0}^{[A]} -\frac{1}{[A]^2}d[A] = \int_0^t k\,dt$$

$$\left[\frac{1}{[A]}\right]_{[A]_0}^{[A]} = k[t]_0^t$$

$$\frac{1}{[A]} - \frac{1}{[A]_0} = kt$$

$$\frac{1}{[A]} = kt + \frac{1}{[A]_0}$$

**117.** Rate = $k\,[CO][Cl_2]^{\frac{3}{2}}$

**119.** $[Cl_2] = 0.0084$ mol/L, $[NO] = 0.017$ mol/L

**121.** B is first order and A is second order.
B will be linear if you plot ln[B] versus time; A will be linear if you plot $1/[A]$ versus time.

**125. a.** First order; a plot of ln [% color] vs. time gives a straight line.

**c.** Yes. The the color is degraded more quickly at warmer temperatures.

**e.** 38.5 kJ/mol

# Chapter 15

**21. a.** $K = \dfrac{[SbCl_3][Cl_2]}{[SbCl_5]}$

**b.** $K = \dfrac{[NO]^2[Br_2]}{[BrNO]^2}$

**c.** $K = \dfrac{[CS_2][H_2]^4}{[CH_4][H_2S]^2}$

**d.** $K = \dfrac{[CO_2]^2}{[CO]^2[O_2]}$

**23.** The concentration of the reactants will be greater. No, this is not dependent on initial concentrations; it is dependent on the value of $K_c$.

**25. a.** figure v

**b.** The change in the decrease of reactants and increase of products would be faster.

**c.** No, catalysts affect kinetics, not equilibrium.

**27. a.** $4.42 \times 10^{-5}$, reactants favored

**b.** $1.50 \times 10^2$, products favored

**c.** $1.96 \times 10^{-9}$, reactants favored

**29.** $1.3 \times 10^{-29}$

**31. a.** $2.56 \times 10^{-23}$

**b.** $1.3 \times 10^{22}$

**c.** 81.9

**33. a.** $K_c = \dfrac{[HCO_3^-][OH^-]}{[CO_3^{2-}]}$    **b.** $K_c = [O_2]^3$

**c.** $K_c = \dfrac{[H_3O^+][F^-]}{[HF]}$    **d.** $K_c = \dfrac{[NH_4^+][OH^-]}{[NH_3]}$

**35.** 136

**37.**

| T (K) | [N₂] | [H₂] | [NH₃] | K_c |
|-------|------|------|-------|-----|
| 500 | 0.115 | 0.105 | 0.439 | $1.45 \times 10^3$ |
| 575 | 0.110 | 0.249 | 0.128 | 9.6 |
| 775 | 0.120 | 0.140 | $4.39 \times 10^{-3}$ | 0.0584 |

**39.** 234 torr
**41.** 18
**43.** $3.3 \times 10^2$
**45.** 764
**47.** More solid will form.
**49.** Additional solid will not dissolve.
**51. a.** [A] = 0.20 M, [B] = 0.80 M
**b.** [A] = 0.33 M, [B] = 0.67 M
**c.** [A] = 0.38 M, [B] = 1.2 M
**53.** [$N_2O_4$] = 0.0115 M, [$NO_2$] = 0.0770 M
**55.** 0.199 M
**57.** $1.9 \times 10^{-3}$ M
**59.** 7.84 torr
**61. a.** [A] = 0.38 M, [B] = 0.62 M, [C] = 0.62 M
**b.** [A] = 0.90 M, [B] = 0.095 M, [C] = 0.095 M
**c.** [A] = 1.0 M, [B] = $3.2 \times 10^{-3}$ M,
[C] = $3.2 \times 10^{-3}$ M
**63. a.** shift left
**b.** shift right
**c.** shift right
**65. a.** shift right
**b.** no effect
**c.** no effect
**d.** shift left
**67. a.** shift right
**b.** shift left
**c.** no effect
**69.** Increase temperature $\longrightarrow$ shift right, decrease temperature $\longrightarrow$ shift left. Increasing the temperature will increase the equilibrium constant.
**71. b, d**
**73. a.** $1.7 \times 10^2$
**b.** $\dfrac{[\text{Hb}-\text{CO}]}{[\text{Hb}-\text{O}_2]} = 0.85$ or 17/20

CO is highly toxic, as it blocks $O_2$ uptake by hemoglobin. CO at a level of 0.1% will replace nearly half of the $O_2$ in blood.
**75. a.** 1.68 atm
**b.** 1.41 atm
**77.** 0.406 g
**79. b, c, d**
**81.** 0.0144 atm
**83.** $3.1 \times 10^2$ g, 20% yield
**85.** 0.12 atm
**87.** 0.72 atm
**89.** 0.017 g
**91.** 0.226
**93. a.** 29.3
**b.** 169 torr
**95.** $P_{\text{NO}} = P_{\text{Cl}_2} = 429$ torr
**97.** $1.27 \times 10^{-2}$
**99.** $K_p = 5.1 \times 10^{-2}$
**101.** Yes, because the volume affects $Q$.
**103.** $a = 1, b = 2$
**105.** $K_c = 1$
**111. a.** For eqn 1, $K_p = 0.011$; For eqn 2, $K_p = 91$
**c.** yes, if the pressure was greater than 100 atm.

# Chapter 16

**33. a.** acid, $HNO_3(aq) \longrightarrow H^+(aq) + NO_3^-(aq)$
**b.** acid, $NH_4^+(aq) \rightleftharpoons H^+(aq) + NH_3(aq)$
**c.** base, $KOH(aq) \longrightarrow K^+(aq) + OH^-(aq)$
**d.** acid, $HC_2H_3O_2(aq) \rightleftharpoons H^+(aq) + C_2H_3O_2^-(aq)$
**35. a.** $\underset{\text{acid}}{H_2CO_3(aq)} + \underset{\text{base}}{H_2O(l)} \rightleftharpoons \underset{\text{conj. acid}}{H_3O^+(aq)} + \underset{\text{conj. base}}{HCO_3^-(aq)}$

**b.** $\underset{\text{base}}{NH_3(aq)} + \underset{\text{acid}}{H_2O(l)} \rightleftharpoons \underset{\text{conj. acid}}{NH_4^+(aq)} + \underset{\text{conj. base}}{OH^-(aq)}$

**c.** $\underset{\text{acid}}{HNO_3(aq)} + \underset{\text{base}}{H_2O(l)} \rightleftharpoons \underset{\text{conj. acid}}{H_3O^+(aq)} + \underset{\text{conj. base}}{NO_3^-(aq)}$

**d.** $\underset{\text{base}}{C_5H_5N(aq)} + \underset{\text{acid}}{H_2O(l)} \rightleftharpoons \underset{\text{conj. acid}}{C_5H_5NH^+(aq)} + \underset{\text{conj. base}}{OH^-(aq)}$
**37. a.** $Cl^-$ **b.** $HSO_3^-$ **c.** $CHO_2^-$ **d.** $F^-$
**39.** $H_2PO_4^-(aq) + H_2O(l) \rightleftharpoons HPO_4^{2-}(aq) + H_3O^+(aq)$
$H_2PO_4^-(aq) + H_2O(l) \rightleftharpoons H_3PO_4(aq) + OH^-(aq)$
**41. a.** strong
**b.** strong
**c.** strong
**d.** weak, $K_a = \dfrac{[H_3O^+][HSO_3^-]}{[H_2SO_3]}$
**43. a, b, c**
**45. a.** $F^-$
**b.** $NO_2^-$
**c.** $ClO^-$
**47. a.** $8.3 \times 10^{-7}$, basic
**b.** $1.2 \times 10^{-10}$, acidic
**c.** $2.9 \times 10^{-13}$, acidic
**49. a.** pH = 7.77, pOH = 6.23
**b.** pH = 7.00, pOH = 7.00
**c.** pH = 5.66, pOH = 8.34

**51.**

| [$H_3O^+$] | [$OH^-$] | pH | Acidic or Basic |
|---|---|---|---|
| $7.1 \times 10^{-4}$ | $1.4 \times 10^{-11}$ | 3.15 | Acidic |
| $3.7 \times 10^{-9}$ | $2.7 \times 10^{-6}$ | 8.43 | Basic |
| $7.9 \times 10^{-12}$ | $1.3 \times 10^{-3}$ | 11.1 | Basic |
| $6.3 \times 10^{-4}$ | $1.6 \times 10^{-11}$ | 3.20 | Acidic |

**53.** [$H_3O^+$] = $1.5 \times 10^{-7}$ M, pH = 6.81
**55.** pH = 1.36, 1.35, 1.34. A difference of 1 in the second significant digit in a concentration value produces a difference of 0.01 in pH. Therefore, the second significant digit in value of the concentration corresponds to the hundredths place in a pH value.
**57. a.** [$H_3O^+$] = 0.25 M, [$OH^-$] = $4.0 \times 10^{-14}$ M, pH = 0.60
**b.** [$H_3O^+$] = 0.015 M, [$OH^-$] = $6.7 \times 10^{-13}$ M, pH = 1.82
**c.** [$H_3O^+$] = 0.072 M, [$OH^-$] = $1.4 \times 10^{-13}$ M, pH = 1.14
**d.** [$H_3O^+$] = 0.105 M, [$OH^-$] = $9.5 \times 10^{-14}$ M, pH = 0.979
**59. a.** 1.8 g **b.** 0.57 g **c.** 0.045 g
**61.** 2.21
**63.** [$H_3O^+$] = $2.5 \times 10^{-3}$ M, pH = 2.59
**65. a.** 1.82 (approximation valid)
**b.** 2.18 (approximation breaks down)
**c.** 2.72 (approximation breaks down)

**67.** 2.75

**69.** $6.8 \times 10^{-6}$

**71.** 0.0063%

**73. a.** 0.42%   **b.** 0.60%
   **c.** 1.3%   **d.** 1.9%

**75.** $3.61 \times 10^{-5}$

**77. a.** pH = 1.88, percent ionization = 5.1%
   **b.** pH = 2.10, percent ionization = 7.9%
   **c.** pH = 2.26, percent ionization = 11%

**79. a.** 0.939   **b.** 1.07
   **c.** 2.19   **d.** 3.02

**81. a.** $[OH^-] = 0.15$ M, $[H_3O^+] = 6.7 \times 10^{-14}$ M,
   pH = 13.17, pOH = 0.83
   **b.** $[OH^+] = 0.003$ M, $[H_3O^+] = 3.3 \times 10^{-12}$ M,
   pH = 11.48, pOH = 2.52
   **c.** $[OH^-] = 9.6 \times 10^{-4}$ M, $[H_3O^+] = 1.0 \times 10^{-11}$ M,
   pH = 10.98, pOH = 3.02
   **d.** $[OH^-] = 8.7 \times 10^{-5}$ M, $[H_3O^+] = 1.1 \times 10^{-10}$ M,
   pH = 9.93, pOH = 4.07

**83.** 13.842

**85.** 0.104 L

**87. a.** $NH_3(aq) + H_2O(l) \rightleftharpoons NH_4^+(aq) + OH^-(aq)$,
$$K_b = \frac{[NH_4^+][OH^-]}{[NH_3]}$$
   **b.** $HCO_3^-(aq) + H_2O(l) \rightleftharpoons H_2CO_3(aq) + OH^-(aq)$,
$$K_b = \frac{[H_2CO_3][OH^-]}{[HCO_3^-]}$$
   **c.**
$CH_3NH_2(aq) + H_2O(l) \rightleftharpoons CH_3NH_3^+(aq) + OH^-(aq)$,
$$K_b = \frac{[CH_3NH_3^+][OH^-]}{[CH_3NH_2]}$$

**89.** $[OH^-] = 1.6 \times 10^{-3}$ M, pOH = 2.79, pH = 11.21

**91.** 7.48

**93.** $6.7 \times 10^{-7}$

**95. a.** neutral
   **b.** basic, $ClO^-(aq) + H_2O(l) \rightleftharpoons HClO(aq) + OH^-(aq)$
   **c.** basic, $CN^-(aq) + H_2O(l) \rightleftharpoons HCN(aq) + OH^-(aq)$
   **d.** neutral

**97.** $[OH^-] = 1.4 \times 10^{-6}$ M, pH = 8.16

**99. a.** acidic, $NH_4^+(aq) + H_2O(l) \rightleftharpoons NH_3(aq) + H_3O^+(aq)$
   **b.** neutral
   **c.** acidic, $Co(H_2O)_6^{3+}(aq) + H_2O(l) \rightleftharpoons$
   $Co(H_2O)_5(OH)^{2+}(aq) + H_3O^+(aq)$
   **d.** acidic, $CH_2NH_3^+(aq) + H_2O(l) \rightleftharpoons$
   $CH_2NH_2(aq) + H_3O^+(aq)$

**101. a.** acidic
   **b.** basic
   **c.** neutral
   **d.** acidic
   **e.** acidic

**103.** NaOH, NaHCO$_3$, NaCl, NH$_4$ClO$_2$, NH$_4$Cl

**105. a.** 5.13
   **b.** 8.87
   **c.** 7.0

**107.** $[K^+] = 0.15$ M, $[F^-] = 0.15$ M, $[HF] = 1.5 \times 10^{-6}$ M,
   $[OH^-] = 1.5 \times 10^{-6}$ M; $[H_3O^+] = 6.7 \times 10^{-9}$ M

**109.** $H_3PO_4(aq) + H_2O(l) \rightleftharpoons H_2PO_4^-(aq) + H_3O^+(aq)$,
$$K_{a_1} = \frac{[H_3O^+][H_2PO_4^-]}{[H_3PO_4]}$$
$H_2PO_4^-(aq) + H_2O(l) \rightleftharpoons HPO_4^{2-}(aq) + H_3O^+(aq)$,
$$K_{a_2} = \frac{[H_3O^+][HPO_4^{2-}]}{[H_2PO_4^-]}$$
$HPO_4^{2-}(aq) + H_2O(l) \rightleftharpoons PO_4^{3-}(aq) + H_3O^+(aq)$,
$$K_{a_3} = \frac{[H_3O^+][PO_4^{3-}]}{[HPO_4^{2-}]}$$

**111. a.** $[H_3O^+] = 0.048$ M, pH = 1.32
   **b.** $[H_3O^+] = 0.12$ M, pH = 0.92

**113.** $[H_2SO_3] = 0.418$ M
   $[HSO_3^-] = 0.082$ M
   $[SO_3^{2-}] = 6.4 \times 10^{-8}$ M
   $[H_3O^+] = 0.082$ M

**115. a.** $[H_3O^+] = 0.50$ M, pH = 0.30
   **b.** $[H_3O^+] = 0.11$ M, pH = 0.96 (x is small
   approximation breaks down)
   **c.** $[H_3O^+] = 0.059$ M, pH = 1.23

**117. a.** HCl, weaker bond
   **b.** HF, bond polarity
   **c.** H$_2$Se, weaker bond

**119. a.** H$_2$SO$_4$, more oxygen atoms bonded to S
   **b.** HClO$_2$, more oxygen atoms bonded to Cl
   **c.** HClO, Cl has higher electronegativity
   **d.** CCl$_3$COOH, Cl has higher electronegativity

**121.** $S^{2-}$, its conjugate acid (H$_2$S), is a weaker acid than H$_2$S.

**123. a.** Lewis acid
   **b.** Lewis acid
   **c.** Lewis base
   **d.** Lewis base

**125. a.** acid: $Fe^{3+}$, base: H$_2$O
   **b.** acid: $Zn^{2+}$, base: NH$_3$
   **c.** acid: BF$_3$, base: (CH$_3$)$_3$N

**127. a.** weak
   **b.** strong
   **c.** weak
   **d.** strong

**129.** If blood became acidic, the H$^+$ concentration would
   increase. According to Le Châtelier's principle, equilibrium would be shifted to the left and the concentration of oxygenated Hb would decrease.

**131.** All acid will be neutralized.

**133.** $[H_3O^+]$(Great Lakes) = $3 \times 10^{-5}$ M,
   $[H_3O^+]$(West Coast) = $4 \times 10^{-6}$ M. The rain over the
   Great Lakes is about eight times more concentrated.

**135.** 2.7

**137. a.** 2.000   **b.** 1.52   **c.** 12.95
   **d.** 11.12   **e.** 5.03

**139. a.** 1.260   **b.** 8.22   **c.** 0.824
   **d.** 8.57   **e.** 1.171

**141. a.** $CN^-(aq) + H^+(aq) \rightleftharpoons HCN(aq)$
   **b.** $NH_4^+(aq) + OH^-(aq) \rightleftharpoons NH_3(aq) + H_2O(l)$
   **c.** $CN^-(aq) + NH_4^+(aq) \rightleftharpoons HCN(aq) + NH_3(aq)$
   **d.** $HSO_4^-(aq) + C_2H_3O_2^-(aq) \rightleftharpoons$
   $SO_4^{2-}(aq) + HC_2H_3O_2(aq)$
   **e.** no reaction between the major species

**143.** 0.794

**145.** $K_a = 1.4 \times 10^{-4}$

**147.** 6.79

**149.** 2.14

**151.** $[A^-] = 4.5 \times 10^{-5}$ M
$[H^+] = 2.2 \times 10^{-4}$ M
$[HA_2^-] = 1.8 \times 10^{-4}$ M

**153.** 9.28

**155.** 50.1 g $NaHCO_3$

**157.** b

**159.** $CH_3COOH < CH_2ClCOOH < CHCl_2COOH < CCl_3COOH$

**166. a.** 

$$\left[ \begin{array}{c} \ddot{:}\ddot{O}: \ddot{:}O: \\ \parallel \phantom{-} \parallel \\ \ddot{:}O - S - S - \ddot{O}: \\ \phantom{--} \parallel \\ \phantom{----} :\ddot{O}: \end{array} \right]^{2-}$$ ; the left sulfur is tetrahedral

and the right sulfur is trigonal pyramidal.

**c.** 57 g sodium metabisulfate

**e.** 37 ppm

# Chapter 17

**27.** d

**29. a.** 3.62      **b.** 9.11

**31.** pure water: 2.1%, in $NaC_7H_5O_2$: 0.065%. The percent ionization in the sodium benzoate solution is much smaller because the presence of the benzoate ion shifts the equilibrium to the left.

**33. a.** 2.14      **b.** 8.32      **c.** 3.46

**35.** $HCl + NaC_2H_3O_2 \longrightarrow HC_2H_3O_2 + NaCl$
$NaOH + HC_2H_3O_2 \longrightarrow NaC_2H_3O_2 + H_2O$

**37. a.** 3.62      **b.** 9.11

**39. a.** 7.60      **b.** 11.18      **c.** 4.61

**41. a.** 3.86      **b.** 8.95

**43.** 3.5

**45.** 3.7 g

**47. a.** 4.74      **b.** 4.68      **c.** 4.81

**49. a.** initial 7.00 after 1.70
  **b.** initial 4.71 after 4.56
  **c.** initial 10.78 after 10.66

**51.** 1.2 g; 2.7 g

**53. a.** yes   **b.** no   **c.** yes
  **d.** no   **e.** no

**55. a.** 7.4
  **b.** 0.3 g
  **c.** 0.14 g

**57.** KClO/HClO = 0.79

**59. a.** does not exceed capacity
  **b.** does not exceed capacity
  **c.** does not exceed capacity
  **d.** does not exceed capacity

**61.  i. a.** pH = 8,
    **b.** pH = 7
   **ii. a.** weak acid,
    **b.** strong acid

**63. a.** 40.0 mL HI for both
  **b.** KOH: neutral, $CH_3NH_2$: acidic
  **c.** $CH_3NH_2$
  **d.** Titration of KOH with HI:

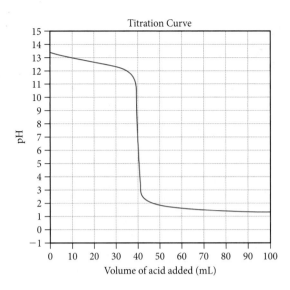

Titration Curve

Titration of $CH_3NH_2$ with HI:

Titration Curve

**65. a.** pH = 9, added base = 30 mL
  **b.** 0 mL
  **c.** 15 mL
  **d.** 30 mL
  **e.** 30 mL

**67. a.** 0.757      **b.** 30.6 mL      **c.** 1.038
  **d.** 7      **e.** 12.15

**69. a.** 13.06      **b.** 28.8 mL      **c.** 12.90
  **d.** 7      **e.** 2.07

**71. a.** 2.86      **b.** 16.8 mL      **c.** 4.37
  **d.** 4.74      **e.** 8.75      **f.** 12.17

**73. a.** 11.94      **b.** 29.2 mL      **c.** 11.33
  **d.** 10.64      **e.** 5.87      **f.** 1.90

**75.  i.** (a)
   **ii.** (b)

**77.** $pK_a = 3, 82$ g/mol

**79.** First equivalence: 22.7 mL
  Second equivalence: 45.4 mL

**81.** The indicator will appear red. The pH range is 4 to 6.

**83. a.** phenol red, $m$-nitrophenol
  **b.** alizarin, bromothymol blue, phenol red
  **c.** alizarin yellow R

**85. a.** $BaSO_4(s) \rightleftharpoons Ba^{2+}(aq) + SO_4^{2-}(aq)$,
$$K_{sp} = [Ba^{2+}][SO_4^{2-}]$$
   **b.** $PbBr_2(s) \rightleftharpoons Pb^{2+}(aq) + 2\,Br^-(aq)$,
$$K_{sp} = [Pb^{2+}][Br^-]^2$$
   **c.** $Ag_2CrO_4(s) \rightleftharpoons 2\,Ag^+(aq) + CrO_4^{2-}(aq)$,
$$K_{sp} = [Ag^+]^2[CrO_4^{2-}]$$

**87. a.** $7.31 \times 10^{-7}$ M   **b.** $3.72 \times 10^{-5}$ M
   **c.** $3.32 \times 10^{-4}$ M

**89. a.** $1.07 \times 10^{-21}$   **b.** $7.14 \times 10^{-7}$
   **c.** $7.44 \times 10^{-11}$

**91.** $AX_2$

**93.** $2.07 \times 10^{-5}$ g/100 mL

**95. a.** 0.0183 M   **b.** 0.00755 M   **c.** 0.00109 M

**97. a.** $5 \times 10^{14}$ M   **b.** $5 \times 10^8$ M   **c.** $5 \times 10^4$ M

**99. a.** more soluble, $CO_3^{2-}$ is basic
   **b.** more soluble, $S^{2-}$ is basic
   **c.** not, neutral
   **d.** not, neutral

**101.** precipitate will form, $CaF_2$

**103.** precipitate will form, $Mg(OH)_2$

**105. a.** 0.018 M
   **b.** $1.4 \times 10^{-7}$ M
   **c.** $1.1 \times 10^{-5}$ M

**107. a.** $BaSO_4$, $1.1 \times 10^{-8}$ M
   **b.** $3.0 \times 10^{-8}$ M

**109.** $8.7 \times 10^{-10}$ M

**111.** $5.6 \times 10^{16}$

**113.** 4.03

**115.** 3.57

**117.** HCl, 4.7 g

**119. a.** $NaOH(aq) + KHC_8H_4O_4(aq) \longrightarrow$
$$Na^+(aq) + K^+(aq) + C_8H_4O_4^{2-}(aq) + H_2O(l)$$
   **b.** 0.1046 M

**121.** 4.73

**123.** 176 g/mol; $1.0 \times 10^{-4}$

**125.** 14.2 L

**127.** $1.6 \times 10^{-7}$ M

**129.** $8.0 \times 10^{-8}$ M

**131.** 6.29

**133.** 0.172 M

**135.** The ratio by mass of dimethyl ammonium chloride to dimethyl amine needed is 3.6.

**137.** 0.18 M benzoic acid, 0.41 M sodium benzoate

**139.** 51.6 g

**141.** $1.8 \times 10^{-11}$ (based on this data)

**143. a.** $5.5 \times 10^{-25}$ M
   **b.** $5.5 \times 10^{-4}$ M

**145.** 1.38 L

**147.** 12.97

**149. a.** $pH < pK_a$   **b.** $pH > pK_a$
   **c.** $pH = pK_a$   **d.** $pH > pK_a$

**151. b**

**153. a.** no difference
   **b.** less soluble
   **c.** more soluble

**160. a.** $H_3C_6H_5O_7(aq) + OH^-(aq) \longrightarrow$
$$H_2O(l) + H_2C_6H_5O_7^-(aq)$$

**c.** $C_6H_5O_7^{3-}(aq) + H_3O^+(aq) \longrightarrow$
$$HC_6H_5O_7^{2-}(aq) + H_2O(l)$$
$HC_6H_5O_7^{2-}(aq) + OH^-(aq) \longrightarrow$
$$H_2O(l) + C_6H_5O_7^{3-}(aq)$$

**e.** The ratio should be $[K_3C_6H_5O_7]/[K_2HC_6H_5O_7] = 0.41$; $[K_3C_6H_5O_7] = 0.010$ M and $[K_2HC_6H_5O_7] = 0.024$ M

# Chapter 18

**27. a, c**

**29.** System B has the greatest entropy. There is only one energetically equivalent arrangement for System A. However, the particles of System B may exchange positions for a second energetically equivalent arrangement.

**31.** 29.2 J/K

**33.** −24.7 J/K

**35. a.** $\Delta S > 0$
   **b.** $\Delta S < 0$
   **c.** $\Delta S < 0$
   **d.** $\Delta S < 0$

**37. a.** $\Delta S_{sys} > 0$, $\Delta S_{surr} > 0$, spontaneous at all temperatures
   **b.** $\Delta S_{sys} < 0$, $\Delta S_{surr} < 0$, nonspontaneous at all temperatures
   **c.** $\Delta S_{sys} < 0$, $\Delta S_{surr} < 0$, nonspontaneous at all temperatures
   **d.** $\Delta S_{sys} > 0$, $\Delta S_{surr} > 0$, spontaneous at all temperatures

**39. a.** $1.29 \times 10^3$ J/K
   **b.** $5.00 \times 10^3$ J/K
   **c.** $-3.83 \times 10^2$ J/K
   **d.** $-1.48 \times 10^3$ J/K

**41. a.** −649 J/K, nonspontaneous
   **b.** 649 J/K, spontaneous
   **c.** 123 J/K, spontaneous
   **d.** −76 J/K, nonspontaneous

**43. a.** $1.93 \times 10^5$ J, nonspontaneous
   **b.** $-1.93 \times 10^5$ J, spontaneous
   **c.** $-3.7 \times 10^4$ J, spontaneous
   **d.** $4.7 \times 10^4$ J, nonspontaneous

**45.** $-2.247 \times 10^6$ J, spontaneous

**47.**

| $\Delta H$ | $\Delta S$ | $\Delta G$ | Low Temperature | High Temperature |
|---|---|---|---|---|
| − | + | − | Spontaneous | Spontaneous |
| − | − | Temperature dependent | Spontaneous | Nonspontaneous |
| + | + | Temperature dependent | Nonspontaneous | Spontaneous |
| + | − | + | Nonspontaneous | Nonspontaneous |

**49.** It increases.

**51. a.** $CO_2(g)$, greater molar mass and complexity
   **b.** $CH_3OH(g)$, gas phase
   **c.** $CO_2(g)$, greater molar mass and complexity
   **d.** $SiH_4(g)$, greater molar mass
   **e.** $CH_3CH_2CH_3(g)$, greater complexity
   **f.** $NaBr(aq)$, aqueous

**53. a.** He, Ne, $SO_2$, $NH_3$, $CH_3CH_2OH$. From He to Ne there is an increase in molar mass; beyond that, the molecules increase in complexity.

   **b.** $H_2O(s)$, $H_2O(l)$, $H_2O(g)$; increase in entropy in going from solid to liquid to gas phase.

   **c.** $CH_4$, $CF_4$, $CCl_4$; increasing entropy with increasing molar mass.

**55. a.** $-120.8$ J/K, decrease in moles of gas

   **b.** $133.9$ J/K, increase in moles of gas

   **c.** $-42.0$ J/K, small change because moles of gas stay constant

   **d.** $-390.8$ J/K, decrease in moles of gas

**57.** $-89.3$ J/K, decrease in moles of gas

**59.** $\Delta H^\circ_{rxn} = -1277$ kJ, $\Delta S^\circ_{rxn} = 313.6$ J/K, $\Delta G^\circ_{rxn} = -1.370 \times 10^3$ kJ; yes

**61. a.** $\Delta H^\circ_{rxn} = 57.2$ kJ, $\Delta S^\circ_{rxn} = 175.8$ J/K, $\Delta G^\circ_{rxn} = 4.8 \times 10^3$ J/mol; nonspontaneous, becomes spontaneous at high temperatures

   **b.** $\Delta H^\circ_{rxn} = 176.2$ kJ, $\Delta S^\circ_{rxn} = 285.1$ J/K, $\Delta G^\circ_{rxn} = 91.2$ kJ; nonspontaneous, becomes spontaneous at high temperatures

   **c.** $\Delta H^\circ_{rxn} = 98.8$ kJ, $\Delta S^\circ_{rxn} = 141.5$ J/K, $\Delta G^\circ_{rxn} = 56.6$ kJ; nonspontaneous, becomes spontaneous at high temperatures

   **d.** $\Delta H^\circ_{rxn} = -91.8$ kJ, $\Delta S^\circ_{rxn} = -198.1$ J/K, $\Delta G^\circ_{rxn} = -32.8$ kJ; spontaneous

**63. a.** 2.8 kJ   **b.** 91.2 kJ   **c.** 56.4 kJ   **d.** $-32.8$ kJ
   Values are comparable. The method using $\Delta H^\circ$ and $\Delta S^\circ$ can be used to determine how $\Delta G^\circ$ changes with temperature.

**65. a.** $-72.5$ kJ, spontaneous

   **b.** $-11.4$ kJ, spontaneous

   **c.** 9.1 kJ, nonspontaneous

**67.** $-29.4$ kJ

**69. a.** 19.3 kJ   **b. (i)** 2.9 kJ   **(ii)** $-2.9$ kJ

   **c.** The partial pressure of iodine is very low.

**71.** 11.9 kJ

**73. a.** $1.48 \times 10^{90}$   **b.** $2.09 \times 10^{-26}$

**75. a.** $-24.8$ kJ   **b.** 0   **c.** 9.4 kJ

**77. a.** $1.90 \times 10^{47}$   **b.** $1.51 \times 10^{-13}$

**79.** $\Delta H^\circ = 50.6$ kJ   $\Delta S^\circ = 226$ J·K

**81.** 4.8

**83. a.** +   **b.** −   **c.** −

**85. a.** $\Delta G^\circ = 175.2$ kJ, $K = 1.95 \times 10^{-31}$, nonspontaneous

   **b.** 133 kJ, yes

**87.** $Cl_2$: $\Delta H^\circ_{rxn} = -182.1$ kJ, $\Delta S^\circ_{rxn} = -134.4$ J/K, $\Delta G^\circ_{rxn} = -142.0$ kJ   $K = 7.94 \times 10^{24}$

   $Br_2$: $\Delta H^\circ_{rxn} = -121.6$ kJ, $\Delta S^\circ_{rxn} = -134.2$ J/K, $\Delta G^\circ_{rxn} = -81.6$ kJ   $K = 2.02 \times 10^{14}$

   $I_2$: $\Delta H^\circ_{rxn} = -48.3$ kJ, $\Delta S^\circ_{rxn} = -132.2$ J/K, $\Delta G^\circ_{rxn} = -8.9$ kJ   $K = 37$

   $Cl_2$ is the most spontaneous, $I_2$ is the least. Spontaneity is determined by the standard enthalpy of formation of the dihalogenated ethane. Higher temperatures make the reactions less spontaneous.

**89. a.** 107.8 kJ

   **b.** $5.0 \times 10^{-7}$ atm

   **c.** spontaneous at higher temperatures, $T = 923.4$ K

**91. a.** $2.22 \times 10^5$   **b.** 94.4 mol

**93. a.** $\Delta G^\circ = -689.6$ kJ, $\Delta G^\circ$ becomes less negative

   **b.** $\Delta G^\circ = -665.2$ kJ, $\Delta G^\circ$ becomes less negative

   **c.** $\Delta G^\circ = -632.4$ kJ, $\Delta G^\circ$ becomes less negative

   **d.** $\Delta G^\circ = -549.3$ kJ, $\Delta G^\circ$ becomes less negative

**95.** With one exception, the formation of any oxide of nitrogen at 298 K requires more moles of gas as reactants than are formed as products. For example, 1 mol of $N_2O$ requires 0.5 mol of $O_2$ and 1 mol of $N_2$, 1 mol of $N_2O_3$ requires 1 mol of $N_2$ and 1.5 mol of $O_2$, and so on. The exception is NO, where 1 mol of NO requires 0.5 mol of $O_2$ and 0.5 mol of $N_2$:

$$\frac{1}{2}N_2(g) + \frac{1}{2}O_2(g) \longrightarrow NO(g)$$

This reaction has a positive $\Delta S$ because what is essentially mixing of the N and O has taken place in the product.

**97.** 15.0 kJ

**99. a.** Positive, the process is spontaneous. It is slow unless a spark is applied.

   **b.** Positive, although the change in the system is not spontaneous; the overall change, which includes such processes as combustion or water flow to generate electricity, is spontaneous.

   **c.** Positive, the acorn-oak tree system is becoming more ordered, so the processes associated with growth are not spontaneous. But they are driven by spontaneous processes such as the generation of heat by the sun and the reactions that produce energy in the cell.

**101.** At 18.3 mmHg $\Delta G = 0$, at 760 mmHg $\Delta G^\circ = 55.4$ kJ

**103. a.** $3.24 \times 10^{-3}$

   **b.**
   $$NH_3 + ATP + H_2O \longrightarrow NH_3 - P_i + ADP$$
   $$NH_3 - P_i + C_5H_8O_4N^- \longrightarrow C_5H_9O_3N_2 + P_i + H_2O$$
   $$\overline{NH_3 + C_5H_8O_4N^- + ATP \longrightarrow C_5H_9O_3N_2 + ADP + P_i}$$
   $$\Delta G^\circ = -16.3 \text{ kJ}, K = 7.20 \times 10^2$$

**105. a.** $-95.3$ kJ/mol. Since the number of moles of reactants and products is the same, the decrease in volume affects the entropy of both equally, so there is no change in $\Delta G$.

   **b.** 102.8 kJ/mol. The entropy of the reactants (1.5 mol) is decreased more than the entropy of the product (1 mol). Since the product is relatively more favored at lower volume, $\Delta G$ is less positive.

   **c.** 204.2 kJ/mol. The entropy of the product (1 mol) is decreased more than the entropy of the reactant (0.5 mol). Since the product is relatively less favored, $\Delta G$ is more positive.

**107.** $\Delta H^\circ = -93$ kJ,   $\Delta S^\circ = -2.0 \times 10^2$ J/K

**109.** $\Delta S_{vap}$ diethyl ether $= 86.1$ J/mol K, $\Delta S_{vap}$ acetone $= 88.4$ J/mol K, $\Delta S_{vap}$ benzene $= 87.3$ J/mol K, $\Delta S_{vap}$ chloroform $= 88.0$ J/mol K. Because water and ethanol hydrogen bond, they are more ordered in

the liquid and we expect $\Delta S_{vap}$ to be more positive.
Ethanol $38600/351.0 = 38600/351.0 = 110\,J/mol\,K$,
$H_2O = 40700/373.2 = 109\,J/mol\,K$

**111. c**

**113. b**

**115. c**

**117.** $\Delta G^\circ_{rxn}$ is negative and $\Delta G_{rxn}$ is positive.

**122. a.** 121 kJ; endothermic

    **c.** 362 J/K

    **e.**

 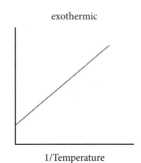

endothermic          exothermic

In K / 1/Temperature     1/Temperature

# Chapter 19

**37. a.** $3\,K(s) + Cr^{3+}(aq) \longrightarrow Cr(s) + 3\,K^+(aq)$

    **b.** $2\,Al(s) + 3\,Fe^{2+}(aq) \longrightarrow 2\,Al^{3+}(aq) + 3\,Fe(s)$

    **c.** $2\,BrO_3^-(aq) + 3\,N_2H_4(g) \longrightarrow$
$$2\,Br^-(aq) + 3\,N_2(g) + 6\,H_2O(l)$$

**39. a.** $PbO_2(s) + 2\,I^-(aq) + 4\,H^+(aq) \longrightarrow$
$$Pb^{2+}(aq) + I_2(s) + 2\,H_2O(l)$$

    **b.** $5\,SO_3^{2-}(aq) + 2\,MnO_4^-(aq) + 6\,H^+(aq) \longrightarrow$
$$5\,SO_4^{2-}(aq) + 2\,Mn^{2+}(aq) + 3\,H_2O(l)$$

    **c.** $S_2O_3^{2-}(aq) + 4\,Cl_2(g) + 5\,H_2O(l) \longrightarrow$
$$2\,SO_4^{2-}(aq) + 8\,Cl^-(aq) + 10\,H^+(aq)$$

**41. a.** $H_2O_2(aq) + 2\,ClO_2(aq) + 2\,OH^-(aq) \longrightarrow$
$$O_2(g) + 2\,ClO_2^-(aq) + 2\,H_2O(l)$$

    **b.** $Al(s) + MnO_4^-(aq) + 2\,H_2O(l) \longrightarrow$
$$Al(OH)_4^-(aq) + MnO_2(s)$$

    **c.** $Cl_2(g) + 2\,OH^-(aq) \longrightarrow$
$$Cl^-(aq) + ClO^-(aq) + H_2O(l)$$

**43. a.**

**b.**

**c.**

**45. a.** 0.93 V     **b.** 0.41 V     **c.** 1.99 V

**47. a, c, d**

    **b.** $Cr(s) + Fe^{3+}(aq) \longrightarrow Cr^{3+}(aq) + Fe(s)$, $E^\circ_{cell} = 0.69$ V

**49. a.** $Pb(s)\,|\,Pb^{2+}(aq)\,\|\,Ag^+(aq)\,|\,Ag(s)$

    **b.** $Pt(s), I_2(s)\,|\,I^-(aq)\,\|\,ClO_2^-(aq)\,|\,ClO_2(g)\,|\,Pt(s)$

    **c.** $Zn(s)\,|\,Zn^{2+}(aq)\,\|\,H_2O(l)\,|\,H^+(aq)\,|\,O_2(g)\,|\,Pt(s)$

**51.**

$$3\,Sn(s) + 2\,NO_3^-(aq) + 8\,H^+(aq) \longrightarrow$$
$$3\,Sn^{2+}(aq) + 2\,NO(g) + 4\,H_2O(l),\ E_{cell}^{\circ} = 1.10\,V$$

**53. b, c** occur spontaneously in the forward direction.

**55.** aluminum

**57. a.** yes, $2\,Al(s) + 6\,H^+(aq) \longrightarrow 2\,Al^{3+}(aq) + 3\,H_2(g)$
   **b.** no
   **c.** yes, $Pb(s) + 2\,H^+(aq) \longrightarrow Pb^{2+}(aq) + H_2(g)$

**59. a.** yes, $3\,Cu(s) + 2\,NO_3^-(aq) + 8\,H^+(aq) \longrightarrow$
$$3\,Cu^{2+}(aq) + 2\,NO(g) + 4\,H_2O(l)$$
   **b.** no

**61. a.** $-1.70\,V$, nonspontaneous
   **b.** $1.97\,V$, spontaneous
   **c.** $-1.51\,V$, nonspontaneous

**63. a**

**65. a.** $-432\,kJ$     **b.** $52\,kJ$     **c.** $-1.7 \times 10^2\,kJ$

**67. a.** $5.31 \times 10^{75}$    **b.** $7.7 \times 10^{-10}$    **c.** $6.3 \times 10^{29}$

**69.** $5.6 \times 10^5$

**71.** $\Delta G^{\circ} = -7.97\,kJ$, $E_{cell}^{\circ} = 0.041\,V$

**73. a.** $1.04\,V$     **b.** $0.97\,V$     **c.** $1.11\,V$

**75.** $1.87\,V$

**77. a.** $0.56\,V$     **b.** $0.52\,V$
   **c.** $[Ni^{2+}] = 0.003\,M$, $[Zn^{2+}] = 1.60\,M$

**79.**

$$\boxed{\begin{array}{c}Zn(s) \longrightarrow \\ Zn^{2+}(aq) + 2\,e^-\end{array}}\qquad\boxed{\begin{array}{c}Zn^{2+}(aq) + 2\,e^- \longrightarrow \\ Zn(s)\end{array}}$$

**81.** $\dfrac{[Sn^{2+}](ox)}{[Sn^{2+}](red)} = 4.2 \times 10^{-4}$

**83.** $0.3762$

**85.** $1.038\,V$

**87. a, c** would prevent the corrosion of iron

**89.**

minimum voltage $= 0.17\,V$

**91.** oxidation: $2\,Br^-(l) \longrightarrow Br_2(g) + 2\,e^-$
   reduction: $K^+(l) + e^- \longrightarrow K(l)$

**93.** oxidation: $2\,Br^-(l) \longrightarrow Br_2(g) + 2\,e^-$
   reduction: $K^+(l) + e^- \longrightarrow K(l)$

**95. a.** anode: $2\,Br^- \longrightarrow Br_2(l) + 2\,e^-$
       cathode: $2\,H_2O(l) + 2\,e^- \longrightarrow H_2(g) + 2\,OH^-(aq)$
   **b.** anode: $2\,I^-(aq) \longrightarrow I_2(s) + 2\,e^-$
         cathode: $Pb^{2+}(aq) + 2\,e^- \longrightarrow Pb(s)$
   **c.** anode: $2\,H_2O(l) \longrightarrow O_2(g) + 4\,H^+(aq) + 4\,e^-$
       cathode: $2\,H_2O(l) + 2\,e^- \longrightarrow H_2(g) + 2\,OH^-(aq)$

**97.**

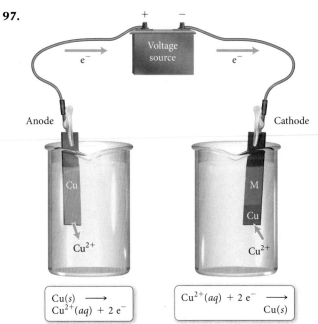

$$\boxed{\begin{array}{c}Cu(s) \longrightarrow \\ Cu^{2+}(aq) + 2\,e^-\end{array}}\qquad\boxed{\begin{array}{c}Cu^{2+}(aq) + 2\,e^- \longrightarrow \\ Cu(s)\end{array}}$$

**99.** $1.8 \times 10^2$ s

**101.** $1.2 \times 10^3$ A    $E_{cell} = 0.0725$ V, K = 282

**103.**

| [B²⁺] | [A²⁺] | Q | $E_{cell}$ | $\Delta G_{rxn}$ |
|---|---|---|---|---|
| 1.00 | 1.0 | 1.0 | 0.0725 V | −14.0 kJ |
| 1.00 | $1.00 \times 10^{-4}$ | $1.00 \times 10^{-4}$ | 0.191 | −36.8 kJ |
| $1.00 \times 10^{-4}$ | 1.0 | $1.00 \times 10^{4}$ | −0.0458 | 8.82 kJ |
| $3.52 \times 10^{-3}$ | 1.0 | 284 | 0 | 0 |

**105.** $2 MnO_4^-(aq) + 5 Zn(s) + 16 H^+(aq) \longrightarrow$
$$2 Mn^{2+}(aq) + 5 Zn^{2+}(aq) + 8 H_2O(l)$$
34.9 mL

**107.** The drawing should show that several Al atoms dissolve into solution as $Al^{3+}$ ions and that several $Cu^{2+}$ ions are deposited on the Al surface as solid Cu.

**109. a.** 68.3 mL

**b.** cannot be dissolved

**c.** cannot be dissolved

**111.** 0.25

**113.** There are no paired reactions that produce more than about 5 or 6 V.

**115. a.** 2.83 V    **b.** 2.71 V    **c.** 16 hr

**117.** 176 hr

**119.** 0.71 V

**121. a.** $\Delta G° = 461$ kJ, $K = 1.4 \times 10^{-81}$

**b.** $\Delta G° = 2.7 \times 10^2$ kJ, $K = 2.0 \times 10^{-48}$

**123.** $MCl_4$

**125.** 51.3%

**127.** pH = 0.85

**129.** 0.83 M

**131.** $4.1 \times 10^5$ L

**133.** 435 s

**135.** 8.39% U

**137.** The overall cell reaction for both cells is
$2 Cu^+(aq) \longrightarrow Cu^{2+}(aq) + Cu(s)$. The difference in $E°$ is because $n = 1$ for the first cell and $n = 2$ for the second cell. For both cells, $\Delta G° = -35.1$ kJ.

**139.** a

**141.** $\Delta G°_{rxn}$ is positive and $E°_{cell}$ is negative.

**147. a.**

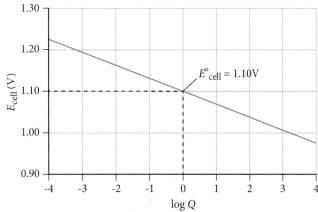

**c.** Any combination in which $[Zn^{2+}]/[Cu^{2+}] = 0.020$ will generate a cell potential = 1.15 V.

**e.** The $Cu^{2+}$ concentrations decreases and $Zn^{2+}$ concentration increases.

# Chapter 20

**31. a.** $^{234}_{92}U \longrightarrow ^4_2He + ^{230}_{90}Th$    **b.** $^{230}_{90}Th \longrightarrow ^4_2He + ^{226}_{88}Ra$

**c.** $^{214}_{82}Pb \longrightarrow ^0_{-1}e + ^{214}_{83}Bi$    **d.** $^{13}_7N \longrightarrow ^0_{+1}e + ^{13}_6C$

**e.** $^{51}_{24}Cr + ^0_{-1}e \longrightarrow ^{51}_{23}V$

**33.** $^{232}_{90}Th \longrightarrow ^4_2He + ^{228}_{88}Ra$    $^{228}_{88}Ra \longrightarrow ^0_{-1}e + ^{228}_{89}Ac$
$^{228}_{89}Ac \longrightarrow ^0_{-1}e + ^{228}_{90}Th$    $^{228}_{90}Th \longrightarrow ^4_2He + ^{224}_{88}Ra$

**35. a.** $^{221}_{87}Fr$    **b.** $^0_{-1}e$    **c.** $^0_{+1}e$    **d.** $^0_{-1}e$

**37. a.** stable, $N/Z$ ratio is close to 1, acceptable for low $Z$ atoms.

**b.** not stable, $N/Z$ ratio is much too high for low $Z$ atom.

**c.** not stable, $N/Z$ ratio is less than 1, much too low.

**d.** stable, $N/Z$ ratio is acceptable for this $Z$.

**39.** Sc, V, and Mn, each have odd numbers of protons. Atoms with an odd number of protons typically have less stable isotopes than those with an even number of protons.

**41. a.** beta decay    **b.** positron emission

**c.** positron emission    **d.** positron emission

**43. a.** Cs-125    **b.** Fe-62

**45.** $2.34 \times 10^9$ years

**47.** 0.57 g

**49.** 10.8 hr

**51.** $2.66 \times 10^3$ yr

**53.** $2.4 \times 10^4$ yr

**55.** $2.7 \times 10^9$ yr

**57.** $^{235}_{92}U + ^1_0n \longrightarrow ^{144}_{54}Xe + ^{90}_{38}Sr + 2 ^1_0n$

**59.** $^2_1H + ^2_1H \longrightarrow ^3_2He + ^1_0n$

**61.** $^{238}_{92}U + ^1_0n \longrightarrow ^{239}_{92}U$
$^{239}_{92}U \longrightarrow ^{239}_{93}Np + ^0_{-1}e$
$^{239}_{93}Np \longrightarrow ^{239}_{94}Pu + ^0_{-1}e$

**63.** $^{249}_{98}Cf + ^{12}_6C \longrightarrow ^{257}_{104}Rf + 4 ^1_0n$

**65.** $9.0 \times 10^{13}$ J

**67. a.** mass defect = 0.13701 amu binding energy = 7.976 MeV/nucleon

**b.** mass defect = 0.54369 amu binding energy = 8.732 MeV/nucleon

**c.** mass defect = 1.16754 amu binding energy = 8.431 MeV/nucleon

**69.** $7.228 \times 10^{10}$ J/g U-235

**71.** $7.84 \times 10^{10}$ J/g H-2

**73.** radiation: 25 J, fall: 370 J

**75.** 68 mi

**77. a.** $^1_1p + ^9_4Be \longrightarrow ^6_3Li + ^4_2He$    $1.03 \times 10^{11}$ J/mol

**b.** $^{209}_{83}Bi + ^{64}_{28}Ni \longrightarrow ^{272}_{111}Rg + ^1_0n$    $1.141 \times 10^{13}$ J/mol

**c.** $^{179}_{74}W + ^0_{-1}e \longrightarrow ^{179}_{73}Ta$    $7.59 \times 10^{10}$ J/mol

**79. a.** $^{114}_{44}Ru \longrightarrow ^0_{-1}e + ^{114}_{45}Rh$    **b.** $^{216}_{88}Ra \longrightarrow ^0_{+1}e + ^{216}_{87}Fr$

**c.** $^{58}_{30}Zn \longrightarrow ^0_{11}e + ^{58}_{29}Cu$    **d.** $^{31}_{10}Ne \longrightarrow ^0_{-1}e + ^{31}_{11}Na$

**81.** $2.9 \times 10^{21}$ beta emissions, 3700 Ci

**83.** $1.6 \times 10^{-5}$ L

**85.** $4.94 \times 10^7$ kJ/mol

**87.** 7.72 MeV

**89.** $^{14}N$

**91.** 0.15%

**93.** $1.24 \times 10^{21}$ atoms

**95.** $2.42 \times 10^{-12}$ m

**97.** $-0.7$ MeV, there is no coulombic barrier for collision with a neutron.

**99. a.** $1.164 \times 10^{10}$ kJ

    **b.** 0.1299 g

**101.** U-235 forms Pb-207 in seven $\alpha$-decays and four $\beta$-decays, and Th-232 forms Pb-208 in six $\alpha$-decays and four $\beta$-decays.

**103.** $3.0 \times 10^2$ K

**105.** $^{21}_{9}\text{F} \longrightarrow {}^{21}_{10}\text{Ne} + {}^{0}_{-1}\text{e}$

**107.** Nuclide A is more dangerous because the half-life is shorter (18.5 days) and so it decays faster.

**109.** Iodine is used by the thyroid gland to make hormones. Normally we ingest iodine in foods, especially iodized salt. The thyroid gland cannot tell the difference between stable and radioactive iodine and will absorb both. KI tablets work by blocking radioactive iodine from entering the thyroid. When a person takes KI, the stable iodine in the tablet gets absorbed by the thyroid. Because KI contains so much stable iodine, the thyroid gland becomes "full" and cannot absorb any more iodine—either stable or radioactive—for the next 24 hours.

**115. a.** 6.0 h

    **c.** Missing masses (top to bottom): 0.0301, 0.0190, 0.0151;  Missing emssions (top to bottom): 2390, 1510, 1200

# Chapter 21

**33. a.** alkane     **b.** alkene

    **c.** alkyne     **d.** alkene

**35.** $CH_3—CH_2—CH_2—CH_2—CH_2—CH_2—CH_3$

$$CH_3—\overset{\displaystyle CH_3}{\underset{\displaystyle |}{CH}}—CH_2—CH_2—CH_2—CH_3$$

$$CH_3—CH_2—\overset{\displaystyle CH_3}{\underset{\displaystyle |}{CH}}—CH_2—CH_2—CH_3$$

$$CH_3—\overset{CH_3}{\underset{|}{CH}}—\overset{CH_3}{\underset{|}{CH}}—CH_2—CH_3$$

$$CH_3—CH_2—\overset{\displaystyle CH_3}{\underset{\displaystyle \underset{\displaystyle CH_3}{|}}{\overset{|}{C}}}—CH_2—CH_3$$

$$H_3C—\overset{\displaystyle CH_3}{\underset{\displaystyle \underset{\displaystyle CH_3}{|}}{\overset{|}{C}}}—CH_2—CH_2—CH_3$$

$$H_3C—\overset{CH_3}{\underset{|}{CH}}—CH_2—\overset{CH_3}{\underset{|}{CH}}—CH_3$$

$$H_3C—\overset{\displaystyle CH_3}{\underset{\displaystyle \underset{\displaystyle CH_3\;\;CH_3}{|}}{\overset{|}{C}}}—CH—CH_3$$

$$H_3C—CH_2—\overset{\displaystyle CH_2}{\underset{\displaystyle \underset{\displaystyle CH_3}{|}}{CH}}—CH_2—CH_3$$

**37. a.** no   **b.** yes   **c.** yes   **d.** no

**39. a.** enantiomers   **b.** same   **c.** enantiomers

**41. a.** pentane

    **b.** 2-methylbutane

    **c.** 4-isopropyl-2-methylheptane

    **d.** 4-ethyl-2-methylhexane

**43. a.** $CH_3—CH_2—\overset{\displaystyle |}{\underset{\displaystyle CH_2—CH_3}{CH}}—CH_2—CH_2—CH_3$

    **b.** $CH_3—CH_2—\overset{\displaystyle CH_3}{\underset{\displaystyle \underset{\displaystyle CH_2—CH_3}{|}}{\overset{|}{C}}}—CH_2—CH_3$

    **c.** $CH_3—\overset{CH_3}{\underset{|}{CH}}—\overset{CH_3}{\underset{|}{CH}}—CH_3$

    **d.** $CH_3—\overset{\displaystyle CH_3}{\underset{\displaystyle \underset{\displaystyle CH_3}{|}}{\overset{|}{C}}}—CH_2—\overset{\displaystyle |}{\underset{\displaystyle CH_2—CH_3}{CH}}—CH_2—CH_2—\overset{\displaystyle CH_2—CH_3}{\underset{\displaystyle |}{CH}}—CH_2—CH_3$

**45. a.** $CH_3CH_2CH_3 + 5\,O_2 \longrightarrow 3\,CO_2 + 4\,H_2O$

    **b.** $CH_3CH_2CH{=}CH_2 + 6\,O_2 \longrightarrow 4\,CO_2 + 4\,H_2O$

    **c.** $2\,CH{\equiv}CH + 5\,O_2 \longrightarrow 4\,CO_2 + 2\,H_2O$

**47. a.** $CH_3CH_2Br$

    **b.** $CH_3CH_2CH_2Cl,\ CH_3CHClCH_3$

    **c.** $CHCl_2Br$

    **d.**

$$CH_3—\overset{\displaystyle H}{\underset{\displaystyle \underset{\displaystyle CH_3}{|}}{\overset{|}{C}}}—CH_2—Cl$$

$$CH_3—\overset{\displaystyle Cl}{\underset{\displaystyle \underset{\displaystyle CH_3}{|}}{\overset{|}{C}}}—CH_3$$

**49.** $CH_2{=}CH—CH_2—CH_2—CH_2—CH_3$

    $CH_3—CH{=}CH—CH_2—CH_2—CH_3$

    $CH_3—CH_2—CH{=}CH—CH_2—CH_3$

**51. a.** 1-butene

    **b.** 3,4-dimethyl-2-pentene

    **c.** 3-isopropyl-1-hexene

    **d.** 2,4-dimethyl-3-hexene

**53. a.** 2-butyne

    **b.** 4,4-dimethyl-2-hexyne

    **c.** 3-isopropyl-1-hexyne

    **d.** 3,6-dimethyl-4-nonyne

**55. a.** $CH_3-CH_2-CH-C\equiv C-CH_2-CH_2-CH_3$

**b.** $CH_3-CH_2-CH$
$\quad\quad\quad\quad\quad \| $
$\quad\quad\quad\quad\quad CH-CH_2-CH_2-CH_2-CH_2-CH_3$

**c.** $CH\equiv C-\overset{\overset{\displaystyle CH_3}{|}}{\underset{\underset{\displaystyle CH_3}{|}}{C}}-CH_2-CH_3$

**d.** $CH_3-CH=\overset{\overset{\displaystyle CH_3}{|}}{C}-CH_2-\overset{\overset{\displaystyle CH_3}{|}}{CH}-\underset{\underset{\displaystyle CH_2-CH_3}{|}}{CH}-CH_3$

**57. a.** $CH_3-\underset{\underset{\displaystyle Cl}{|}}{CH}-\underset{\underset{\displaystyle Cl}{|}}{CH}-CH_3$

**b.** $CH_3-\underset{\underset{\displaystyle CH_3}{|}}{CH}-CH_2-\underset{\underset{\displaystyle Br}{|}}{CH}-CH_3$ +
$\quad\quad\quad CH_3-\underset{\underset{\displaystyle CH_3}{|}}{CH}-\underset{\underset{\displaystyle Br}{|}}{CH}-CH_2-CH_3$

**c.** $CH_3-CH_2-\underset{\underset{\displaystyle Br}{|}}{CH}-\underset{\underset{\displaystyle Br}{|}}{CH}-CH_3$

**d.** $CH_3-\underset{\underset{\displaystyle CH_3}{|}}{CH}-CH_2-\overset{\overset{\displaystyle CH_3}{|}}{\underset{\underset{\displaystyle Cl}{|}}{C}}-CH_3$

**59. a.** $CH_2=CH-CH_3 + H_2 \longrightarrow CH_3-CH_2-CH_3$

**b.** $CH_3-\underset{\underset{\displaystyle CH_3}{|}}{CH}-CH=CH_2 + H_2 \longrightarrow$
$\quad\quad\quad\quad\quad\quad CH_3-\underset{\underset{\displaystyle CH_3}{|}}{CH}-CH_2-CH_3$

**c.** $CH_3-\underset{\underset{\displaystyle CH_3}{|}}{CH}-\underset{\underset{\displaystyle CH_3}{|}}{C}=CH_2 + H_2 \longrightarrow$
$\quad\quad\quad\quad\quad\quad CH_3-\underset{\underset{\displaystyle CH_3}{|}}{CH}-\underset{\underset{\displaystyle CH_3}{|}}{CH}-CH_3$

**61. a.** methylbenzene or toluene
**b.** bromobenzene
**c.** chlorobenzene
**63. a.** 3,5-dimethyl-7-phenylnonane
**b.** 2-phenyl-3-octene
**c.** 4,5-dimethyl-6-phenyl-2-octyne
**65. a.** 1,4-dibromobenzene or *p*-dibromobenzene
**b.** 1,3-diethylbenzene or *m*-diethylbenzene
**c.** 1-chloro-2-fluorobenzene or *o*-chlorofluorobenzene

**67. a.** $CH_3-CH-CH_3$ **b.** **c.**

**69. a.** + HBr **b.** $CH_3-CH-CH_3$ + HCl

**71. a.** 1-propanol **b.** 4-methyl-2-hexanol
**c.** 2,6-dimethyl-4-heptanol
**d.** 3-methyl-3-pentanol
**73. a.** $CH_3CH_2CH_2Br + H_2O$ **b.** $CH_3-\underset{\underset{\displaystyle CH_3}{|}}{C}=CH_2 + H_2O$

**c.** $CH_3CH_2ONa + \frac{1}{2}H_2$ **d.** $CH_3-\overset{\overset{\displaystyle CH_3}{|}}{\underset{\underset{\displaystyle CH_3}{|}}{C}}-CH_2-\overset{\overset{\displaystyle O}{\|}}{C}-OH$

**75. a.** butanone **b.** pentanal
**c.** 3,5,5-trimethylhexanal
**d.** 4-methyl-2-hexanone

**77.** $CH_3-CH_2-CH_2-\overset{\overset{\displaystyle OH}{|}}{\underset{\underset{\displaystyle H}{|}}{C}}-C\equiv N$

**79. a.** methylbutanoate **b.** propanoic acid
**c.** 5-methylhexanoic acid **d.** ethylpentanoate

**81. a.** $CH_3-CH_2-CH_2-CH_2-\overset{\overset{\displaystyle O}{\|}}{C}-O-CH_2-CH_3 + H_2O$

**b.** + $H_2O$

**83. a.** ethyl propyl ether **b.** ethyl pentyl ether
**c.** dipropyl ether **d.** butyl ethyl ether
**85. a.** diethylamine **b.** methylpropylamine
**c.** butylmethylpropylamine
**87. a.** acid–base, $(CH_3)_2NH_2^+(aq) + Cl^-(aq)$
**b.** condensation, $CH_3CH_2CONHCH_2CH_3(aq) + H_2O$
**c.** acid–base, $CH_3NH_3^+(aq) + HSO_4^-(aq)$
**89. a.** ester, methyl 3-methylbutanoate
**b.** ether, ethyl 2-methylbutyl ether
**c.** aromatic, 1-ethyl-3-methylbenzene or
*m*-ethylmethylbenzene
**d.** alkyne, 5-ethyl-4-methyl-2-heptyne
**e.** aldehyde, butanal
**f.** alcohol, 2-methyl-1-propanol
**91. a.** 5-isobutyl-3-methylnonane
**b.** 5-methyl-3-hexanone
**c.** 3-methyl-2-butanol
**d.** 4-ethyl-3,5-dimethyl-1-hexyne
**93. a.** isomers
**b.** isomers
**c.** same
**95.** 558 g

**97. a.** combustion
**b.** alkane substitution
**c.** alcohol elimination
**d.** aromatic substitution

**99. a.** $CH_3-CH_2-CH-CH=CH_2$
with $CH_3$ branch
Can exist as a stereoisomer

**b.** $CH_3-CH=C-CH_2-CH-CH_3$
with $CH_3$ branches
Can exist as a stereoisomer

**c.** $H_3C-CH=C-CH_2-CH_2-CH_3$
with $CH_2CH_2CH_3$ branch
Can exist as a stereoisomer

**101.   1.**   $H_3C-CH_2-CH_2-\overset{\displaystyle O}{\overset{\|}{CH}}$
Aldehyde

**2.**   $H_3C-\overset{\displaystyle O}{\overset{\|}{C}}-CH_2-CH_3$
Ketone

**3.**   $H_3C-CH=CH-O-CH_3$
Alkene, ether

**4.**   $H_2C=CH-O-CH_2-CH_3$
Alkene, ether

**5.**   $H_2C=CH-CH_2-O-CH_3$
Alkene, ether

**6.**   $H_3C-CH=CH-CH_2-OH$
Alkene, alcohol

**7.**   $H_3C-\overset{\displaystyle}{\underset{OH}{C}}=CH-CH_3$
Alkene, alcohol

**8.**   $H_3C-CH_2-CH=\overset{\displaystyle}{\underset{OH}{CH}}$
Alkene, alcohol

**9.**   $H_3C-CH_2-\overset{\displaystyle}{\underset{OH}{C}}=CH_2$
Alkene, alcohol

**10.**   $H_2C=CH-\overset{\displaystyle}{\underset{OH}{CH}}-CH_3$
Alkene, alcohol

**11.**   $H_2C=CH-CH_2-CH_2-OH$
Alkene, alcohol

**103.** In the acid form of the carboxylic acid, electron withdrawal by the $C=O$ enhances acidity. The conjugate base, the carboxylate anion, is stabilized by resonance, so the two O atoms are equivalent and bear the negative charge equally.

**105. a.**

**b.**

**c.**

**107. a.** $3:1$

**b.** 2° hydrogen atoms are more reactive. The reactivity of 2° hydrogens to 1° hydrogens is $11:3$.

**109.**

**111.** The first propagation step for F is very rapid and exothermic because of the strength of the $H-F$ bond that forms. For I the first propagation step is endothermic and slow because the $H-I$ bond that forms is relatively weak.

**113.**

**115.**

$H_3C-\overset{\displaystyle CH_3}{\underset{\displaystyle CH_3}{\overset{|}{\underset{|}{C}}}}-\overset{\displaystyle CH_3}{\underset{\displaystyle CH_3}{\overset{|}{\underset{|}{C}}}}-CH_3$

2,2,3,3-tetramethylbutane

**121. a.** Apple: methylbutanoate;  Pear: propyl ethanoate (or propyl acetate);  Pineapple: ethyl butanoate

**c.**

Apple

$$\text{CH}_3\text{CH}_2\text{CH}_2\text{CH}_2\text{OH} \xrightarrow[\text{H}_2\text{SO}_4]{\text{Na}_2\text{Cr}_2\text{O}_7}$$

$$\text{butanoic acid} + \text{CH}_3\text{OH} \xrightarrow{\text{H}_2\text{SO}_4} \text{methyl butanoate} + \text{H}_2\text{O}$$

Pear

$$\text{CH}_3\text{CH}_2\text{OH} \xrightarrow[\text{H}_2\text{SO}_4]{\text{Na}_2\text{Cr}_2\text{O}_7}$$

$$\text{acetic acid} + \text{propanol} \xrightarrow{\text{H}_2\text{SO}_4} \text{propyl acetate} + \text{H}_2\text{O}$$

Pineapple

$$\text{butanol} \xrightarrow[\text{H}_2\text{SO}_4]{\text{Na}_2\text{Cr}_2\text{O}_7}$$

$$\text{butanoic acid} + \text{ethanol} \xrightarrow{\text{H}_2\text{SO}_4} \text{ethyl butanoate} + \text{H}_2\text{O}$$

**e.** Butyl acetate

pentyl acetate

ethyl pentanoate

pentyl pentanoate

pentyl hexanoate

# Chapter 22

**31. c.** saturated fatty acid
**d.** steroid
**33. a.** saturated fatty acid
**b.** not a fatty acid
**c.** not a fatty acid
**d.** monounsaturated fatty acid

**35.**

$$\begin{array}{l} \text{H}_2\text{C}-\text{OH} \\ \text{HO}-\text{CH} \\ \text{H}_2\text{C}-\text{OH} \end{array} + 3\ \text{H}_3\text{C}(\text{CH}_2)_4-(\text{CH}=\text{CHCH}_2)_2(\text{CH}_2)_6-\text{C}-\text{OH} \longrightarrow$$

$$\text{H}_2\text{C}-\text{O}-\overset{\text{O}}{\overset{\|}{\text{C}}}-(\text{CH}_2)_6-(\text{CH}_2\text{CH}=\text{CH})_2-(\text{CH}_2)_4-\text{CH}_3$$
$$\text{H}-\text{C}-\text{O}-\overset{\text{O}}{\overset{\|}{\text{C}}}-(\text{CH}_2)_6-(\text{CH}_2\text{CH}=\text{CH})_2-(\text{CH}_2)_4-\text{CH}_3$$
$$\text{H}_2\text{C}-\text{O}-\overset{\text{O}}{\overset{\|}{\text{C}}}-(\text{CH}_2)_6-(\text{CH}_2\text{CH}=\text{CH})_2-(\text{CH}_2)_4-\text{CH}_3$$

Triglyceride is expected to be an oil.

**37. a.** monosaccharide
**c.** disaccharide
**39. a.** aldose, hexose          **b.** aldose, pentose
**c.** ketose, tetrose          **d.** aldose, tetrose
**41. a.** 5     **b.** 3     **c.** 1     **d.** 3

**43.**

$$\begin{array}{l} \text{CHO} \\ \text{H}-\text{C}-\text{OH} \\ \text{HO}-\text{C}-\text{H} \\ \text{H}-\text{C}-\text{OH} \\ \text{H}-\text{C}-\text{OH} \\ \text{H}_2\text{C}-\text{OH} \end{array}$$

**45.**

**47.**

Glucose          Fructose

**49. a.**
$$\text{H}_3\text{N}^+-\overset{\text{HO}-\text{CH}-\text{CH}_3}{\underset{}{\text{CH}}}-\overset{\text{O}}{\overset{\|}{\text{C}}}-\text{O}^-$$

**b.**
$$\text{H}_3\text{N}^+-\overset{\text{CH}_3}{\underset{}{\text{CH}}}-\overset{\text{O}}{\overset{\|}{\text{C}}}-\text{O}^-$$

**c.**
$$\text{H}_3\text{N}^+-\text{CH}-\overset{\text{O}}{\overset{\|}{\text{C}}}-\text{O}^- ,\ \text{CH}_2 ,\ \text{H}_3\text{C}-\text{CH}-\text{CH}_3$$

**d.**
$$\text{H}_3\text{N}^+-\text{CH}-\overset{\text{O}}{\overset{\|}{\text{C}}}-\text{O}^- ,\ (\text{CH}_2)_4-\text{NH}_2$$

**51.**

**53.** 6, Ser-Gly-Cys, Ser-Cys-Gly, Gly-Ser-Cys, Gly-Cys-Ser, Cys-Ser-Gly, Cys-Gly-Ser

**55.**

**57. a.**

**b.**

**c.**

**59.** tertiary

**61.** primary

**63. a.** A

   **b.** T

**65.**

**67.** A C A T G C G

**69.** 154 codons, 462 nucleotides

**71. a.** protein]

   **b.** carbohydrate

   **c.** lipid

**73.** A codon is composed of three nucleotides. A codon codes for a specific amino acid, while a gene codes for an entire protein.

**75.**

**77.** valine, leucine, isoleucine, phenylalanine

**79.** Gly-Arg-Ala-Leu-Phe-Gly-Asn-Lys-Trp-Glu-Cys

**81.**

**83.** As the temperature increases, the favorable entropy for uncoiling a chain becomes dominant. On cooling, the favorable enthalpy of forming hydrogen bonds between paired bases is dominant.

**85.** When the fake thymine nucleotide is added to the replicating DNA, the chain cannot continue to form because the $-N \equiv N^+ = NH$ group on the sugar prevents future phosphate linkages.

**87.** $V_{max} = 47.6, K_t = 1.68$

**89.** $H_3N^+CH_2COO^- + H^+ \rightleftharpoons$

$$(H_3N^+CH_2COOH\ [HA]/[A^-] = 2,$$
$$H_3N^+CH_2COO^- \rightleftharpoons H_2NCH_2COO^- + H^+ [HA]/[A^-]$$
$$= 0.4, pH = 6.0$$

**91.** A three-base codon codes for a single amino acid. If there are only three bases, there could be 27 different three-base codon arrangements. Therefore, you could theoretically code for the 20 different amino acids needed.

**98. a.**

**c.**

# Chapter 23

**15. a.** +4 **b.** +4 **c.** +4

**17.** $Ca_3Al_2(SiO_4)_3$

**19.** 4

**21.** tetrahedrons stand alone, orthosilicates

**23.** amphibole or double-chain structure; $Ca^{2+}, Mg^{2+}, Fe^{2+}, Al^{3+}$

**25.** 950 g

**27.** $NCl_3$ has a lone pair that $BCl_3$ lacks, giving it a trigonal pyramidal shape, as opposed to $BCl_3$'s trigonal planar shape.

**29. a.** 6 vertices, 8 faces
**b.** 12 vertices, 20 faces

**31.** *closo*-Boranes have the formula $B_nH_n^{2-}$ and form fully closed polyhedra, *nido*-boranes have the formula $B_nH_{n+4}$ and consist of a cage missing a corner, and *arachno*-boranes have the formula $B_nH_{n+6}$ and consist of a cage missing two or three corners.

**33.** Activated charcoal consists of fine particles, rather than a lump of charcoal, and subsequently has a much higher surface area.

**35.** Ionic carbides are composed of carbon, generally in the form of the carbide ion, $C_2^{2-}$, and low-electronegativity metals, such as the alkali and alkaline earth metals. Covalent carbides are composed of carbon and low-electronegativity nonmetals or metalloids, such as silicon.

**37. a.** solid $\longrightarrow$ gas
**b.** gas $\longrightarrow$ liquid $\longrightarrow$ solid
**c.** solid $\longrightarrow$ gas

**39. a.** $CO(g) + CuO(s) \longrightarrow CO_2(g) + Cu(s)$
**b.** $SiO_2(s) + 3\,C(s) \longrightarrow SiC(s) + 2\,CO(g)$
**c.** $S(s) + CO(g) \longrightarrow COS(g)$

**41. a.** +2 **b.** +4 **c.** +4/3

**43.** Fixing nitrogen refers to converting $N_2$ to a nitrogen-containing compound.

**45.** White phosphorus consists of $P_4$ molecules in a tetrahedral shape with the atoms at the corners of the tetrahedron. This allotrope is unstable because of the strain from the bond angles. Red phosphorus is much more stable because one bond of the tetrahedron is broken, allowing the phosphorus atoms to make chains with bond angles that are less strained.

**47.** saltpeter: 13.86% N by mass Chile saltpeter: 16.48% N by mass

**49.** $HN_3$ has a positive $\Delta G_f^\circ$, meaning that it spontaneously decomposes into $H_2$ and $N_2$ at room temperature. There are no temperatures at which $HN_3$ will be stable. $\Delta H_f$ is positive and $\Delta S_f$ is negative, so $\Delta G_f$ will always be negative.

**51. a.** $NH_4NO_3(aq) + \text{heat} \longrightarrow N_2O(g) + 2\,H_2O(l)$
**b.** $3\,NO_2(g) + H_2O(l) \longrightarrow 2\,HNO_3(l) + NO(g)$
**c.** $2\,PCl_3(l) + O_2(g) \longrightarrow 2\,POCl_3(l)$

**53.** $NO_3^-, NO_2^-, N_3^-, N_2H_5^+, NH_4^+$

**55.**

Trigonal pyramidal     Trigonal bipyramidal

**57.** $CO(NH_2)_2 + 2\,H_2O \longrightarrow (NH_4)_2CO_3$
14 g

**59.** $P_4O_6$ forms if there is only a limited amount of oxygen available, while $P_4O_{10}$ will form with greater amounts of oxygen.

**61.** The major source of oxygen is the fractionation of air by which air is cooled and liquefied and oxygen is separated from the other components.

**63. a.** superoxide **b.** oxide **c.** peroxide

**65.** Initially, liquid sulfur becomes less viscous when heated because the $S_8$ rings have greater thermal energy, which overcomes intermolecular forces. Above 150 °C the rings break and the broken rings entangle one another, causing greater viscosity.

**67. a.** $4 \times 10^{-22}$ g
**b.** $4.0 \times 10^{-19}$ g

**69.** $2\,FeS_2(s) \xrightarrow{\text{heat}} 2\,FeS(s) + S_2(g)$
510 L

**71. a.** +2, linear **b.** +6, octahedral
**c.** +6, square pyramidal

**73.** $Cl_2(g) + 2\,Br^-(aq) \longrightarrow 2\,Cl^-(aq) + Br_2(l)$
Oxidizing agent: $Cl_2$
Reducing agent: $Br^-$

**75.** No, there is not enough HF to dissolve all of the $SiO_2$. HF is the limiting reagent. 1.6 g $SiO_2$.

**77.** 8 kg from lignite, 40 kg from bituminous

**79.** Chlorine is much more electronegative than iodine, allowing it to withdraw an electron and ionize in solution much more easily.

**81. a.** $rate_{HCl}/rate_{Cl_2} = 1.395$
**b.** $rate_{HCl}/rate_{HF} = 0.7407$
**c.** $rate_{HCl}/rate_{HI} = 1.873$

**83.** $4\,Na_2O_2 + 3\,Fe \longrightarrow 4\,Na_2O + Fe_3O_4$

**85.** The bond length of the $O_2$ species increases as electrons are added because they are added to the $\pi^*$ antibonding orbital. $O_2^{2-}$ is diamagnetic.

**87.** 2.0 mol of C—C bonds, 715 kJ/mol, $6.9 \times 10^2$ kJ/mol. This value calculated from the bond energy is too low because it doesn't include van der Waals attractions between C atoms not directly bonded to each other.

**89.** $-50$ kJ/mol

**91. a.** $-13.6$ kJ/mol      **b.** $-11.0$ kJ/mol
    **c.** $-24.8$ kJ/mol
    $Fe_2O_3$ is the most exothermic because it has the highest oxidation state and is therefore able to oxidize the most CO per mol Fe.

**93. a.** $\ddot{O}{=}C{=}C{=}C{=}\ddot{O}$      **b.** $sp$
    **c.** $-92$ kJ/mol

**95. a.** $7.6 \times 10^{-22}$      **b.** $1.2 \times 10^{-8}$
    **c.** $[N_2H_4] = 0.009$ M, $[N_2H_5^+] = 0.0025$ M,
    $[N_2H_6^{2+}] = 7.0 \times 10^{-13}$ M

**97.** The acid is

**a.** HO—$\overset{\displaystyle\cdot\cdot}{N}$≡$\underset{\displaystyle\cdot\cdot}{N}$—OH

and the base is

**b.** $H_2N-\overset{\displaystyle O \atop \|}{\underset{\displaystyle O^-}{N^+}}$

The acid is weaker than nitrous acid because of electron donation by resonance in contributing structures such as

**c.** HO—$\overset{\ominus}{\underset{\displaystyle\cdot\cdot}{\ddot{N}}}$—$\underset{\displaystyle\cdot\cdot}{N}{=}\overset{\oplus}{\ddot{O}H}$

The base is weaker than ammonia because of electron withdrawal by the electronegative nitro group.

**99.** The triple bond in nitrogen is much stronger than the double bond in oxygen, so it is much harder to break. This makes it less likely that the bond in nitrogen will be broken.

**101.** Sodium dinitrogen phosphate ($NaH_2PO_4$) can act as a weak base or a weak acid. A buffer can be made by mixing it with either $Na_2HPO_4$ or with $Na_3PO_4$, depending on the desired pH of the buffer solution.

**103.** F is extremely small and so there is a huge driving force to fill the octet by adding an electron, giving a $-1$ oxidation state. Other halogens have access to $d$ orbitals, which allows for more hybridization and oxidation state options.

**105.** $SO_3$ cannot be a reducing agent because the oxidation state of S is $+6$, the highest possible oxidation state for S. Reducing agents need to be able to be oxidized. $SO_2$ can be a reducing agent or an oxidizing agent because the oxidation state of S is $+4$.

**112. a.** The solubility of $H_2S$ in water decreases as temperature increases.
    **c.** No
    **e.** Copper(II)
    **f.** $4.8 \times 10^3$ g CuS

# Chapter 24

**15.** Metals are typically opaque, are good conductors of heat and electricity, and are ductile and malleable, meaning they can be drawn into wires and flattened into sheets.

**17.** aluminum, iron, calcium, magnesium, sodium, potassium

**19.** Fe: hematite ($Fe_2O_3$), magnetite ($Fe_3O_4$)
Hg: cinnabar (HgS)
V: vanadite [$Pb_5(VO_4)Cl$], carnotite [$K_2(UO_2)_2(VO_4)_2 \cdot 3\,H_2O$]
Nb: columbite [$Fe(NbO_3)_2$]

**21.** $MgCO_3(s) + heat \longrightarrow MgO(s) + CO_2(g)$
$Mg(OH)_2(s) + heat \longrightarrow MgO(s) + H_2O(g)$

**23.** The flux is a material that will react with the gangue to form a substance with a low melting point. MgO is the flux.

**25.** Hydrometallurgy is used to separate metals from ores by selectively dissolving the metal in a solution, filtering out impurities, and then reducing the metal to its elemental form.

**27.** The Bayer process is a hydrometallurgical process by which $Al_2O_3$ is selectively dissolved, leaving other oxides as solids. The soluble form of aluminum is $Al(OH)_4^-$.

**29.** Sponge powdered iron contains many small holes in the iron particles due to the escaping of the oxygen when the iron is reduced. Water-atomized powdered iron has much smoother and denser particles as the powder is formed from molten iron.

**31. a.** 50% Cr, 50% V by moles; 50.5% Cr, 49.5% V by mass
    **b.** 25% Fe, 75% V by moles; 26.8% Fe, 73.2% V by mass
    **c.** 25% Cr, 25% Fe, 50% V by moles; 24.8% Cr, 26.6% Fe, 48.6% V by mass

**33.** Cr and Fe are very close to each other in mass, so their respective atomic radii are probably close enough to form an alloy. Also, they both form body-centered cubic structures.

**35.** A: solid, 20% Cr, 80% Fe
B: liquid, 50% Cr, 50% Fe

**37.** A: solid (20% Co and 80% Cu overall. Two phases; one is the Cu structure with 4% Co, and the other is the Co structure with 7% Cu. There will be more of the Cu structure).
B: solid (Co structure), 90% Co, 10% Cu

**39.** C would fill interstitial holes; Mn and Si would substitute for Fe.

**41. a.** $Mo_2N$      **b.** $CrH_2$

**43. a.** zinc      **b.** copper      **c.** manganese

**45.** $-19.4$ kJ/mol

**47.** When Cr is added to steel, it reacts with oxygen in steel to prevent it from rusting. A Cr steel alloy would be used in any situation where the steel might be easily oxidized, such as when it comes in contact with water.

**49.** rutile: 33.3% Ti by moles, 59.9% Ti by mass
ilmenite: 20.0% Ti by moles, 31.6% Ti by mass

**51.** Titanium must be arc-melted in an inert atmosphere because the high temperature and flow of electrons would cause the metal to oxidize in a normal atmosphere.

**53.** $TiO_2$ is the most important industrial product of titanium, and it is often used as a pigment in white paint.

**55.** The Bayer process is a hydrometallurgical process used to separate $Al_2O_3$ from other oxides. The $Al_2O_3$ is selectively dissolved by hot, concentrated NaOH. The other oxides are removed as solids, and the $Al_2O_3$ precipitates out of solution when the solution is neutralized.

**57.** cobalt and tungsten

**59.** 3.3 kg Fe, 2.0 kg Ti

**61.** Four atoms surround a tetrahedral hole, and six atoms surround an octahedral hole. The octahedral hole is larger because it is surrounded by a greater number of atoms.

**63.** Mn has one more electron orbital available for bonding than does chromium.

**65.** Ferromagnetic atoms, like paramagnetic ones, have unpaired electrons. However, in ferromagnetic atoms, these electrons align with their spin oriented in the same direction, resulting in a permanent magnetic field.

**67.** The nuclear charge of the last three is relatively high because of the lanthanide series in which the $4f$ subshell falls between them and the other six metals of the group.

**69. a.** 16.0 cm  **b.** 4.95 cm  **c.** 14%

**71.** 92%

**73.** $5.4 \times 10^7$

**75.** First, roast to form the oxide.
$4\ CoAsS(s) + 9\ O_2(g) \longrightarrow$
$\qquad 4\ CoO(s) + 4\ SO_2(g) + As_4O_6(s)$
Then reduce the oxide with coke.
$CoO(s) + C(s) \longrightarrow Co(s) + CO(g)$
The oxides of arsenic are relatively volatile and can be separated, but they are poisonous.

**77.** Au and Ag are found in elemental form because of their low reactivity. Na and Ca are group 1 and group 2 metals, respectively, and are highly reactive as they readily lose their valence electrons to obtain octets.

**84. a.** Because they are stronger and harder.
  **c.** Tin, 232 °C; Lead 328 °C
  **e.** Because the melting range is neither too sharp nor too broad. It has a pasty range that helps it to set, but then hardens quickly.

# Chapter 25

**17. a.** [Ar] $4s^23d^8$, [Ar] $3d^8$
  **b.** [Ar] $4s^23d^5$, [Ar] $3d^3$
  **c.** [Kr] $5s^24d^1$, [Kr] $5s^14d^1$

**d.** [Xe] $6s^24f^{14}5d^3$, [Xe] $4f^{14}5d^3$

**19. a.** +5  **b.** +7  **c.** +4

**21. a.** +3, 6  **b.** +2, 6
  **c.** +2, 4  **d.** +1, 2

**23. a.** hexaaquachromium(III)
  **b.** tetracyanocuprate(II)
  **c.** pentaaminebromoiron(III) sulfate
  **d.** aminetetraaquahydroxycobalt(III) chloride

**25. a.** $[Cr(NH_3)_6]^{3+}$
  **b.** $K_3[Fe(CN)_6]$
  **c.** $[Cu(en)(SCN)_2]$
  **d.** $[Pt(H_2O)_4][PtCl_6]$

**27. a.** $[Co(NH_3)_3(CN)_3]$, triaminetricyanocobalt(III)
  **b.** $[Cr(en)_3]^{3+}$, tris(ethylenediamine)chromium(III)

**29.**

**31.** $[Fe(H_2O)_5Cl]Cl \cdot H_2O$, pentaaquachloroiron(II) chloride monohydrate
$[Fe(H_2O)_4Cl_2] \cdot 2\ H_2O$, tetraaquadichloroiron(II) dihydrate

**33. b, c, e**

**35. a.** 3
  **b.** No geometric isomers.

**37. a.**

**b.**

**39.** *cis* isomer is optically active

**41. a.**
  **b.**
  **c.**
  **d.**

**43.** 163 kJ/mol

**45.** $[Co(CN)_6]^{3-} \longrightarrow$ 290 nm, colorless
$[Co(NH_3)_6]^{3+} \longrightarrow$ 440 nm, yellow
$[CoF_6]^{3-} \longrightarrow$ 770 nm, green

**47.** weak

**49. a.** 4
**b.** 3
**c.** 1

**51.** 3

**53.** porphyrin

**55.** Water is a weak field ligand that forms a high-spin complex with hemoglobin. Because deoxyhemoglobin is a weak field, it absorbs large-wavelength light and appears blue. Oxyhemoglobin is a low-spin complex and absorbs small-wavelength light, so $O_2$ must be a strong field ligand.

**57. a.** [Ar] $4s^1 3d^5$, [Ar] $3d^5$, [Ar] $3d^4$, [Ar] $3d^3$
**b.** [Ar] $4s^1 3d^{10}$, [Ar] $3d^{10}$, [Ar] $3d^9$

**59. a.** H—N̈—H
|
H

**b.** [S̈=C=N̈]⁻

**c.** Ö / H   H

**61.** $[MA_2B_2C_2]$ all cis; A trans and B and C cis; B trans and A and C cis; C trans and A and B cis; all trans.
$[MA_2B_3C]$ will have fac–mer isomers.
$[MAB_2C_3]$ will have fac–mer isomers.
$[MAB_3C_2]$ will have fac–mer isomers.
$[MA_3B_2C]$ will have fac–mer isomers.
$[MA_2BC_3]$ will have fac–mer isomers.
$[MA_3BC_2]$ will have fac–mer isomers.
$[MABC_2]$ will have AB cis–trans isomers.
$[MAB_4C]$ will have AC cis–trans isomers.
$[MA_4BC]$ will have BC cis–trans isomers.
$[MABC_4]$ will have AB cis–trans isomers.

**63.** , optical isomers

**65.** ⥮ ↑ ↑ , paramagnetic

**67.**

1. (Ru complex: NH₃, H₂O, Cl⁻, Cl⁻, NH₃, H₂O)
2. (Ru complex: NH₃, H₂O, NH₃, Cl⁻, Cl⁻, H₂O)
3. (Ru complex: NH₃, H₂O, NH₃, Cl⁻, H₂O, Cl⁻)
4. (Ru complex: Cl⁻, H₂O, NH₃, NH₃, H₂O, Cl⁻)
5. (Ru complex: Cl⁻, H₂O, NH₃, H₂O, Cl⁻, NH₃)

**Only structure 3. is chiral. This is its mirror image.**

(Ru complex: NH₃, H₂O, NH₃, H₂O, Cl⁻, Cl⁻)

**69.** (Pt complex: Cl, P(CH₃)₃, Cl, P(CH₃)₃)
*cis*-dichlorobis (trimethyl phosphine) platinum(II)

(Pt complex: Cl, P(CH₃)₃, (CH₃)₃P, Cl)
*trans*-dichlorobis (trimethyl phosphine) platinum(II)

**71.** _____ $d_{z^2}$
_____ _____ $d_{x^2-y^2}$ and $d_{xy}$
_____ _____ $d_{xz}$ and $d_{yz}$

**73. a.** $2 \times 10^{-8}$ M
**b.** $6.6 \times 10^{-3}$ M
**c.** NiS will dissolve more easily in the ammonia solution because the formation of the complex ion is favorable, removing $Ni^{2+}$ ions from the solution and allowing more NiS to dissolve.

**75.** Prepare a solution that contains both $[MCl_6]^{3-}$ and $[MBr_6]^{3-}$ and see if any complex ions that contain both Cl and Br form. If they do, it would demonstrate that these complexes are labile.

**77.** pH = 10.1

**79.** Au

**86. a.** red
**c.** smaller (because it absorbs at longer wavelengths)
**e.** The crystal fields splitting energies of the two complexes must be similar (because they have the same color).

# APPENDIX IV Answers to In-Chapter Practice Problems

## Chapter 1

**1.1. a.** The composition of the copper is not changing; thus, being hammered flat is a physical change that signifies a physical property.
  **b.** The dissolution and color change of the nickel indicate that it is undergoing a chemical change and exhibiting a chemical property.
  **c.** Sublimation is a physical change indicative of a physical property.
  **d.** When a match ignites, a chemical change begins as the match reacts with oxygen to form carbon dioxide and water. Flammability is a chemical property.
**1.2. a.** 29.8 °C
  **b.** 302.9 K
**1.3.** 21.4 g/cm$^3$. This matches the density of platinum.
**1.3.** *For More Practice* 4.50 g/cm$^3$ The metal is titanium.
**1.4.** The thermometer shown has markings every 1 °F; thus, the first digit of uncertainty is 0.1. The answer is 103.1 °F.
**1.5. a.** Each figure in this number is significant by rule 1: three significant figures.
  **b.** This is a defined quantity that has an unlimited number of significant figures.
  **c.** Both 1's are significant (rule 1), and the interior zero is significant as well (rule 2): three significant figures.
  **d.** Only the two 9's are significant, the leading zeroes are not (rule 3): two significant figures.
  **e.** There are five significant figures because the 1, 4, and 5 are nonzero (rule 1) and the trailing zeroes are after a decimal point so they are significant as well (rule 4).
  **f.** The number of significant figures is ambiguous because the trailing zeroes occur before an implied decimal point (rule 4). Assume two significant figures.
**1.6. a.** 0.381
  **b.** 121.0
  **c.** 1.174
  **d.** 8
**1.7.** 3.15 yd
**1.8.** 2.446 gal
**1.9.** $1.61 \times 10^6$ cm$^3$
**1.9.** *For More Practice* $3.23 \times 10^3$ kg
**1.10.** 1.03 kg
**1.10.** *For More Practice* $2.9 \times 10^{-2}$ cm$^3$
**1.11.** 0.855 cm
**1.12.** 2.70 g/cm$^3$

## Chapter 2

**2.1.** For the first sample:
$$\frac{\text{mass of oxygen}}{\text{mass of carbon}} = \frac{17.2 \text{ g O}}{12.9 \text{ g C}} = 1.33 \text{ or } 1.33:1$$

For the second sample:
$$\frac{\text{mass of oxygen}}{\text{mass of carbon}} = \frac{10.5 \text{ g O}}{7.88 \text{ g C}} = 1.33 \text{ or } 1.33:1$$

The ratios of oxygen to carbon are the same in the two samples of carbon monoxide, so these results are consistent with the law of definite proportions.
**2.2.** Mass of hydrogen to 1 g of oxygen in water/mass of hydrogen to 1 g of oxygen in hydrogen peroxide $= 0.125/0.0625 = 2.00$.
The ratio of the mass of hydrogen from one compound to the mass of hydrogen in the other is equal to 2. This is a simple whole number and therefore consistent with the law of multiple proportions.
**2.3. a.** $Z = 6, A = 13, {}^{13}_{6}\text{C}$
  **b.** 19 protons, 20 neutrons
**2.4. a.** N$^{3-}$
  **b.** Rb$^+$
**2.5.** 24.31 amu
**2.5.** *For More Practice* 70.92 amu
**2.6.** $4.65 \times 10^{-2}$ mol Ag
**2.7.** 0.563 mol Cu
**2.7.** *For More Practice* 22.6 g Ti
**2.8.** $1.3 \times 10^{22}$ C atoms
**2.8.** *For More Practice* 6.87 g W
**2.9.** $l = 1.72$ cm
**2.9.** *For More Practice* $2.90 \times 10^{24}$ Cu atoms

## Chapter 3

**3.1. a.** C$_5$H$_{12}$
  **b.** HgCl
  **c.** CH$_2$O
**3.2. a.** molecular element
  **b.** molecular compound
  **c.** atomic element
  **d.** ionic compound
  **e.** ionic compound
**3.3.** K$_2$S
**3.4.** AlN
**3.5.** silver nitride
**3.5.** *For More Practice* Rb$_2$S
**3.6.** iron(II) sulfide
**3.6.** *For More Practice* RuO$_2$
**3.7.** tin(II) chlorate

**3.7.** *For More Practice* $Co_3(PO_4)_2$
**3.8.** dinitrogen pentoxide
**3.8.** *For More Practice* $PBr_3$
**3.9.** hydrofluoric acid
**3.10.** nitrous acid
**3.10.** *For More Practice* $HClO_4$
**3.11.** sulfurous acid
**3.12.** 164.10 amu
**3.13.** $5.839 \times 10^{20}$ $C_{13}H_{18}O_2$ molecules
**3.13.** *For More Practice* 1.06 g $H_2O$
**3.14.** 53.29%
**3.14.** *For More Practice* 74.19% Na
**3.15.** 83.9 g $Fe_2O_3$
**3.15.** *For More Practice* 8.6 g Na
**3.16.** 4.0 g O
**3.16.** *For More Practice* 3.60 g C
**3.17.** $CH_2O$
**3.18.** $C_{13}H_{18}O_2$
**3.19.** $C_6H_6$
**3.19.** *For More Practice* $C_2H_8N_2$
**3.20.** $C_2H_5$
**3.21.** $C_2H_4O$
**3.22.** $SiO_2(s) + 3\ C(s) \longrightarrow SiC(s) + 2\ CO(g)$
**3.23.** $2\ C_2H_6(g) + 7\ O_2(g) \longrightarrow 4\ CO_2(g) + 6\ H_2O(g)$
**3.24.** $Pb(NO_3)_2(aq) + 2\ KCl(aq) \longrightarrow$
$$PbCl_2(s) + 2\ KNO_3(aq)$$

# Chapter 4

**4.1.** 4.08 g HCl
**4.2.** 22 kg $HNO_3$
**4.3.** $H_2$ is the limiting reagent, since it produces the least amount of $NH_3$. Therefore, 29.4 kg $NH_3$ is the theoretical yield.
**4.4.** CO is the limiting reagent, since it only produces 114 g Fe. Therefore, 114 g Fe is the theoretical yield: percentage yield = 63.4% yield.
**4.5.** 0.214 M $NaNO_3$
**4.5.** *For More Practice* 44.6 g KBr
**4.6.** 402 g $C_{12}H_{22}O_{11}$
**4.6.** *For More Practice* 221 mL of KCl solution
**4.7.** 667 mL
**4.7.** *For More Practice* 0.105 L
**4.8.** 51.4 mL $HNO_3$ solution
**4.8.** *For More Practice* 0.170 g $CO_2$
**4.9.** **a.** insoluble
     **b.** insoluble
     **c.** soluble
     **d.** soluble
**4.10.** $NH_4Cl(aq) + Fe(NO_3)_3(aq) \longrightarrow$ NO REACTION
**4.11.** $2\ NaOH(aq) + CuBr_2(aq) \longrightarrow$
$$Cu(OH)_2(s) + 2\ NaBr(aq)$$
**4.12.** $2\ H^+(aq) + 2\ I^-(aq) + Ba^{2+}(aq) + 2\ OH^-(aq) \longrightarrow$
$$2\ H_2O(l) + Ba^{2+}(aq) + 2\ I^-(aq)$$
$$H^+(aq) + OH^-(aq) \longrightarrow H_2O(l)$$

**4.13.** Molecular equation:
$$HBr(aq) + LiOH(aq) \longrightarrow H_2O(l) + LiBr(aq)$$
Net ionic equation:
$$H^+(aq) + OH^-(aq) \longrightarrow H_2O(l)$$
**4.14.** $HCHO_2(aq) + OH^-(aq) \longrightarrow H_2O(l) + CHO_2^-(aq)$
**4.15.** $9.03 \times 10^{-2}$ M $H_2SO_4$
**4.15.** *For More Practice* 24.5 mL NaOH solution
**4.16.** $2\ HBr(aq) + K_2SO_3(aq) \longrightarrow$
$$H_2O(l) + SO_2(g) + 2\ KBr(aq)$$
**4.16.** *For More Practice* $2\ H^+(aq) + S^{2-}(aq) \longrightarrow H_2S(g)$
**4.17.** **a.** $Cr = 0$
     **b.** $Cr^{3+} = +3$
     **c.** $Cl^- = -1, C = +4$
     **d.** $Br = -1, Sr = +2$
     **e.** $O = -2, S = +6$
     **f.** $O = -2, N = +5$
**4.18.** Sn is oxidized and N is reduced.
**4.18.** *For More Practice*  **b.** Reaction b is the only redox reaction. Al is oxidized and O is reduced.
**4.19.** **a.** This is a redox reaction in which Li is the reducing agent (it is oxidized) and $Cl_2$ is the oxidizing reagent (it is reduced).
     **b.** This is a redox reaction in which Al is the reducing agent and $Sn^{2+}$ is the oxidizing agent.
     **c.** This is not a redox reaction because no oxidation states change.
     **d.** This is a redox reaction in which C is the reducing agent and $O_2$ is the oxidizing agent.
**4.20.** $2\ C_2H_5SH(l) + 9\ O_2(g) \longrightarrow$
$$4\ CO_2(g) + 2\ SO_2(g) + 6\ H_2O(g)$$

# Chapter 5

**5.1.** 15.0 psi
**5.1.** *For More Practice* 80.6 kPa
**5.2.** 2.1 atm at a depth of approximately 11 m
**5.3.** 123 mL
**5.4.** 11.3 L
**5.5.** 1.63 atm, 23.9 psi
**5.6.** 16.1 L
**5.6.** *For More Practice* 976 mmHg
**5.7.** $d = 4.91$ g/L
**5.7.** *For More Practice* 44.0 g/mol
**5.8.** 70.7 g/mol
**5.9.** 0.0610 mol $H_2$
**5.10.** 4.2 atm
**5.11.** 12.0 mg $H_2$
**5.12.** 82.3 g $Ag_2O$
**5.12.** *For More Practice* 7.10 g $Ag_2O$
**5.13.** 6.53 L $O_2$
**5.14.** $u_{rms} = 238$ m/s
**5.15.** $\dfrac{\text{rate}_{H_2}}{\text{rate}_{Kr}} = 6.447$

# Chapter 6

**6.1.** $\Delta E = 71 \, J$

**6.2.** $C_s = 0.38 \dfrac{J}{g \cdot {}^\circ C}$

The specific heat capacity of gold is $0.128 \, J/g \cdot {}^\circ C$; therefore, the rock cannot be pure gold.

**6.2.** *For More Practice* $T_f = 42.1 \, {}^\circ C$

**6.3.** 37.8 grams Cu

**6.4.** $-122 \, J$

**6.4.** *For More Practice* $\Delta E = -998 \, J$

**6.5.** $\Delta E_{reaction} = -3.91 \times 10^3 \, kJ/mol \, C_6H_{14}$

**6.5.** *For More Practice* $C_{cal} = 4.55 \dfrac{kJ}{{}^\circ C}$

**6.6.**  **a.** endothermic, positive $\Delta H$
  **b.** endothermic, positive $\Delta H$
  **c.** exothermic, negative $\Delta H$

**6.7.** $-2.06 \times 10^3 \, kJ$

**6.7.** *For More Practice* 33 g $C_4H_{10}$
  99 g $CO_2$

**6.8.** $\Delta H_{rxn} = -68 \, kJ$

**6.9.** $N_2O(g) + NO_2(g) \longrightarrow 3 \, NO(g), \quad \Delta H_{rxn} = +157.6 \, kJ$

**6.9.** *For More Practice*
  $3 \, H_2(g) + O_3(g) \longrightarrow 3 \, H_2O(g), \quad \Delta H_{rxn} = -868.1 \, kJ$

**6.10.**  **a.** $Na(s) + \dfrac{1}{2}Cl_2(g) \longrightarrow$
$$NaCl(s), \Delta H_f^\circ = -411.2 \, kJ/mol$$
  **b.** $Pb(s) + N_2(g) + 3 \, O_2(g) \longrightarrow$
$$Pb(NO_3)_2(s), \Delta H_f^\circ = -451.9 \, kJ/mol$$

**6.11.** $\Delta H_{rxn}^\circ = -851.5 \, kJ$

**6.12.** $\Delta H_{rxn}^\circ = -1648.4 \, kJ$
  111 kJ emitted ($-111 \, kJ$)

**6.13.** $1.2 \times 10^2 \, kg \, CO_2$

# Chapter 7

**7.1.** $5.83 \times 10^{14} \, s^{-1}$

**7.2.** $2.64 \times 10^{20}$ photons

**7.2.** *For More Practice* 435 nm

**7.3.**  **a.** blue < green < red
  **b.** red < green < blue
  **c.** red < green < blue

**7.4.** $6.1 \times 10^6 \, m/s$

**7.5.** For the $5d$ orbitals:
  $n = 5$
  $l = 2$
  $m_l = -2, -1, 0, 1, 2$
  The 5 integer values for $m_l$ signify that there are five $5d$ orbitals.

**7.6.**  **a.** $l$ cannot equal 3 if $n = 3$. $l = 2$
  **b.** $m_l$ cannot equal $-2$ if $l = -1$. Possible values for $m_l = -1, 0,$ or 1
  **c.** $l$ cannot be 1 if $n = 1$. $l = 0$

**7.7.** 397 nm

**7.7.** *For More Practice* $n = 1$

# Chapter 8

**8.1.**  **a.** Cl $1s^2 2s^2 2p^6 3s^2 3p^5$ or [Ne] $3s^2 3p^5$
  **b.** Si $1s^2 2s^2 2p^6 3s^2 3p^2$ or [Ne] $3s^2 3p^2$
  **c.** Sr $1s^2 2s^2 2p^6 3s^2 3p^6 4s^2 3d^{10} 4p^6 5s^2$ or [Kr] $5s^2$
  **d.** O $1s^2 2s^2 2p^4$ or [He] $2s^2 2p^4$

**8.2.** There are no unpaired electrons.

**8.3.** $1s^2 2s^2 2p^6 3s^2 3p^3$ or [Ne] $3s^2 3p^3$. The 5 electrons in the $3s^2 3p^3$ orbitals are the valence electrons, while the 10 electrons in the $1s^2 2s^2 2p^6$ orbitals belong to the core.

**8.4.** [Xe] $6s^2 4f^{14} 5d^{10} 6p^3$

**8.4.** *For More Practice* [Kr] $5s^2 4d^{10} 5p^5$

**8.5.**  **a.** Sn
  **b.** cannot predict
  **c.** W
  **d.** Se

**8.5.** *For More Practice* Rb > Ca > Si > S > F

**8.6.**  **a.** [Ar] $4s^0 3d^7$. $Co^{2+}$ is paramagnetic.

  Co²⁺  [Ar]  ▢ | ⬆⬇ ⬆⬇ ⬆ ⬆ ⬆
                 4s      3d

  **b.** [He] $2s^2 2p^6$. $N^{3-}$ is diamagnetic.

  N³⁻  [He]  ⬆⬇ | ⬆⬇ ⬆⬇ ⬆⬇
             2s      2p

  **c.** [Ne] $3s^2 3p^6$. $Ca^{2+}$ is diamagnetic.

  Ca²⁺  [Ne]  ⬆⬇ | ⬆⬇ ⬆⬇ ⬆⬇
              3s      3p

**8.7.**  **a.** K  **b.** $F^-$  **c.** $Cl^-$

**8.8.**  **a.** I  **b.** Ca
  **c.** cannot predict  **d.** F

**8.8.** *For More Practice* F > S > Si > Ca > Rb

**8.9.**  **a.** Sn
  **b.** cannot predict based on simple trends (Po is larger)
  **c.** Bi
  **d.** B

**8.9.** *For More Practice* Cl < Si < Na < Rb

**8.10.**  **a.** $2 \, Al(s) + 3 \, Cl_2(g) \longrightarrow 2 \, AlCl_3(s)$
  **b.** $2 \, Li(s) + 2 \, H_2O(l) \longrightarrow$
$$2 \, Li^+(aq) + 2 \, OH^-(aq) + H_2(g)$$
  **c.** $H_2(g) + Br_2(l) \longrightarrow 2 \, HBr(g)$

# Chapter 9

**9.1.** $Mg_3N_2$

**9.2.** KI < LiBr < CaO

**9.2.** *For More Practice* $MgCl_2$

**9.3.**  **a.** pure covalent
  **b.** ionic
  **c.** polar covalent

**9.4.** $:C \equiv O:$

**9.5.**
$$\begin{array}{c} :\!O\!: \\ \parallel \\ H\!-\!C\!-\!H \end{array}$$

**9.6.** $\left[:\!\ddot{C}l\!:\!\ddot{O}\!:\right]^{-}$

**9.7.** $\left[:\!\ddot{O}\!=\!\ddot{N}\!-\!\ddot{O}\!:\right]^{-} \longleftrightarrow \left[:\!\ddot{O}\!-\!\ddot{N}\!=\!\ddot{O}\!:\right]^{-}$

**9.8.**

| Structure | A | | B | | C | | |
|---|---|---|---|---|---|---|---|
| | $:\!\ddot{N}\!=\!N\!=\!\ddot{O}\!:$ | | $:\!N\!\equiv\!N\!-\!\ddot{\ddot{O}}\!:$ | | $:\!\ddot{N}\!-\!N\!\equiv\!O\!:$ | | |
| number of valence e⁻ | 5 | 5 | 6 | 5 | 5 | 6 | 5 | 5 | 6 |
| number of nonbonding e⁻ | −4 | −0 | −4 | −2 | −0 | −6 | −6 | −0 | −2 |
| $\frac{1}{2}$(number of bonding e⁻) | −2 | −4 | −2 | −3 | −4 | −1 | −1 | −4 | −3 |
| **Formal charge** | **−1** | **−1** | **0** | **0** | **+1** | **−1** | **−2** | **+1** | **+1** |

Structure B contributes the most to the correct overall structure of $N_2O$.

**9.8.** *For More Practice* The nitrogen is +1, the singly bonded oxygen atoms are −1, and the double-bonded oxygen atom has no formal charge.

**9.9.**
$$H\!-\!\overset{|}{\underset{|}{C}}\!-\!\overset{+1}{N}\!\equiv\!N\!: \longleftrightarrow H\!-\!\overset{|}{C}\!=\!\overset{+1}{N}\!=\!\overset{-1}{\ddot{N}}\!:$$
with H atoms on carbon and −1 on C

**9.10.**
$$\begin{array}{c} :\!\ddot{F}\!: \\ | \\ :\!\ddot{F}\!-\!\ddot{X}e\!-\!\ddot{F}\!: \\ | \\ :\!\ddot{F}\!: \end{array}$$

**9.10.** *For More Practice*
$$\begin{array}{c} H \\ | \\ :\!O\!: \\ | \\ H\!-\!\ddot{O}\!-\!\overset{0}{P}\!-\!\ddot{O}\!-\!H \\ \parallel \\ :\!\ddot{O}\!:^{0} \end{array}$$

**9.11.** $CH_3OH(g) + \dfrac{3}{2}O_2(g) \longrightarrow CO_2(g) + 2\,H_2O(g)$

$\Delta H_{rxn} = -641\ kJ$

**9.11.** *For More Practice* $\Delta H_{rxn} = -8.0 \times 10^1\ kJ$

# Chapter 10

**10.1.** tetrahedral

$$\begin{array}{c} :\!\ddot{C}l\!: \\ | \\ :\!\ddot{C}l\!-\!C\!-\!\ddot{C}l\!: \\ | \\ :\!\ddot{C}l\!: \end{array}$$

**10.2.** bent
**10.3.** linear

**10.4.**

| Atom | Number of Electron Groups | Number of Lone Pairs | Molecular Geometry |
|---|---|---|---|
| Carbon (left) | 4 | 0 | Tetrahedral |
| Carbon (right) | 3 | 0 | Trigonal planar |
| Oxygen | 4 | 2 | Bent |

**10.5.** The molecule is nonpolar.

**10.6.** The xenon atom has six electron groups and therefore has an octahedral electron geometry. An octahedral electron geometry corresponds to $sp^3d^2$ hybridization (refer to Table 10.3).

**10.7.** Since there are only two electron groups around the central atom (C), the electron geometry is linear. According to Table 10.3, the corresponding hybridization on the carbon atom is $sp$.

**10.8.** Since there are only two electron groups about the central atom (C), the electron geometry is linear. The hybridization on C is $sp$ (refer to Table 10.3).

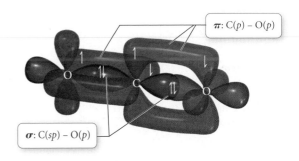

**10.8.** *For More Practice* There are five electron groups about the central atom (I); therefore, the electron geometry is trigonal bipyramidal and the corresponding hybridization of I is $sp^3d$ (refer to Table 10.3).

**10.9.** $H_2^+$ bond order $= +\dfrac{1}{2}$

Since the bond order is positive, the $H_2^+$ ion should be stable; however, the bond order of $H_2^+$ is lower than the bond order of $H_2$ (bond order $= 1$). Therefore, the bond in $H_2^+$ is weaker than in $H_2$.

**10.10.** The bond order of $N_2^+$ is 2.5, which is lower than that of the $N_2$ molecule (bond order $= 3$); therefore, the bond is weaker. The MO diagram shows that the $N_2^+$ ion has one unpaired electron and is therefore paramagnetic.

$\sigma_{2p}^*$

$\pi_{2p}^*$

$\sigma_{2p}$ $\boxed{\uparrow}$

$\pi_{2p}$ $\boxed{\uparrow\downarrow}\,\boxed{\uparrow\downarrow}$

$\sigma_{2s}^*$ $\boxed{\uparrow\downarrow}$

$\sigma_{2s}$ $\boxed{\uparrow\downarrow}$

**10.10.** *For More Practice* The bond order of $Ne_2$ is 0, which indicates that dineon does not exist.

**10.11.** The bond order of NO is $+2.5$. The MO diagram shows that the ion has one unpaired electron and is therefore paramagnetic.

# Chapter 11

**11.1. b, c**

**11.2.** HF has a higher boiling point than HCl because, unlike HCl, HF is able to form hydrogen bonds. The hydrogen bond is the strongest of the intermolecular forces and requires more energy to break.

**11.3.** $5.83 \times 10^3$ kJ

**11.3.** *For More Practice* 49 °C

**11.4.** 33.8 KJ/mol

**11.5.** $7.04 \times 10^3$ torr

# Chapter 12

**12.1.** 29.4°

**12.2.** 78.5%

**12.3.** $3.24 \times 10^{-23}\,cm^3$

**12.4.** $7.18\dfrac{g}{cm^3}$

**12.5.** Metallic

# Chapter 13

**13.1. a.** not soluble
   **b.** soluble

**c.** not soluble
**d.** not soluble

**13.2.** $2.7 \times 10^{-4}$ M

**13.3.** $42.5$ g $C_{12}H_{22}O_{11}$

**13.3.** *For More Practice* $3.3 \times 10^4$ L

**13.4. a.** $M = 0.415$ M
   **b.** $m = 0.443$ m
   **c.** % by mass $= 13.2\%$
   **d.** $\chi C_{12}H_{22}O_{11} = 0.00793$
   **e.** mole percent $= 0.793\%$

**13.5.** $0.600$ M

**13.5.** *For More Practice* $0.651$ m

**13.6.** $22.5$ torr

**13.6.** *For More Practice* $0.144$

**13.7. a.** $P_{benzene} = 26.6$ torr
      $P_{toluene} = 20.4$ torr
   **b.** $47.0$ torr
   **c.** 52.5% benzene; 47.5% toluene
      The vapor will be richer in the more volatile component, which in this case is benzene.

**13.8.** $T_f = -4.8\,°C$

**13.9.** $101.84\,°C$

**13.10.** $11.8$ atm

**13.11.** $-0.60\,°C$

**13.12.** $0.014$ mol NaCl

# Chapter 14

**14.1.** $\dfrac{\Delta[H_2O_2]}{\Delta t} = -4.40 \times 10^{-3}\,M/s$

$\dfrac{\Delta[I_3^-]}{\Delta t} = 4.40 \times 10^{-3}\,M/s$

**14.2. a.** Rate $= k[CHCl_3][Cl_2]^{1/2}$. (Fractional-order reactions are not common but are occasionally observed.)
   **b.** $3.5\,M^{-1/2}\cdot s^{-1}$

**14.3.** $5.78 \times 10^{-2}$ M

**14.4.** $0.0277$ M

**14.5.** $1.64 \times 10^{-3}$ M

**14.6.** $79.2$ s

**14.7.** $2.07 \times 10^{-5}\dfrac{L}{mol\cdot s}$

**14.8.** $6.13 \times 10^{-4}\dfrac{L}{mol\cdot s}$

**14.9.** $2\,A + B \longrightarrow A_2B$
   Rate $= k[A]^2$

# Chapter 15

**15.1.** $K = \dfrac{[CO_2]^3[H_2O]^4}{[C_3H_8][O_2]^5}$

**15.2.** $2.1 \times 10^{-13}$

**15.2.** *For More Practice* $1.4 \times 10^2$

**15.3.** $6.2 \times 10^2$

**15.4.** $K_c = \dfrac{[Cl_2]^2}{[HCl]^4[O_2]}$

**15.5.** 9.4

**15.6.** $1.1 \times 10^{-6}$

**15.7.** $Q_c = 0.0196$
Reaction proceeds to the left.

**15.8.** 0.033 M

**15.9.** $[N_2] = 4.45 \times 10^{-3}$ M
$[O_2] = 4.45 \times 10^{-3}$ M
$[NO] = 1.1 \times 10^{-3}$ M

**15.10.** $[N_2O_4] = 0.005$ M
$[NO_2] = 0.041$ M

**15.11.** $P_{I_2} = 0.0027$ atm
$P_{Cl_2} = 0.0027$ atm
$P_{ICl_2} = 0.246$ atm

**15.12.** $1.67 \times 10^{-7}$ M

**15.13.** $6.78 \times 10^{-6}$ M

**15.14.** Adding $Br_2$ increases the concentration of $Br_2$, causing a shift to the left (away from the $Br_2$). Adding BrNO increases the concentration of BrNO, causing a shift to the right.

**15.15.** Decreasing the volume causes the reaction to shift right. Increasing the volume causes the reaction to shift left.

**15.16.** If we increase the temperature, the reaction shifts to the left. If we decrease the temperature, the reaction shifts to the right.

# Chapter 16

**16.1.** **a.** $H_2O$ donates a proton to $C_5H_5N$, making it the acid. The conjugate base is therefore $OH^-$. Since $C_5H_5N$ accepts the proton, it is the base and becomes the conjugate acid $C_5H_5NH^+$.
**b.** Since $HNO_3$ donates a proton to $H_2O$, it is the acid, making $NO_3^-$ the conjugate base. Since $H_2O$ is the proton acceptor, it is the base and becomes the conjugate acid, $H_3O^+$.

**16.2.** **a.** Since $[H_3O^+] < [OH^-]$ the solution is basic.
**b.** $[H_3O^+] = 1.0 \times 10^{-7}$ M
Neutral solution.
**c.** $[H_3O^+] = 1.2 \times 10^{-5}$ M
Since $[H_3O^+] > [OH^-]$ the solution is acidic.

**16.3.** **a.** 8.02 (basic)
**b.** 11.85 (basic)

**16.4.** $4.3 \times 10^{-9}$ M

**16.5.** $9.4 \times 10^{-3}$ M

**16.6.** 3.28

**16.7.** 2.72

**16.8.** $1.8 \times 10^{-6}$

**16.9.** 0.85%

**16.10.** $2.0 \times 10^{-7}$ M

**16.11.** $[OH^-] = 0.020$ M  pH = 12.30

**16.12.** $[OH^-] = 1.2 \times 10^{-2}$ M  pH = 12.08

**16.13.** **a.** weak base        **b.** pH-neutral

**16.14.** 9.07

**16.15.** **a.** pH-neutral
**b.** weak acid
**c.** weak acid

**16.16.** **a.** basic
**b.** acidic
**c.** pH-neutral
**d.** acidic

**16.17.** 3.83

**16.18.** $[SO_4^{2-}] = 0.00386$ M
pH = 1.945

**16.19.** $5.6 \times 10^{-11}$ M

# Chapter 17

**17.1.** 4.44

**17.1.** *For More Practice* 3.44

**17.2.** 9.14

**17.3.** 4.87

**17.3.** *For More Practice* 4.65

**17.4.** 9.68

**17.4.** *For More Practice* 9.56

**17.5.** hypochlorous acid (HClO); 2.4 g NaClO

**17.6.** 1.74

**17.7.** 18.08

**17.8.** $2.30 \times 10^{-6}$ M

**17.9.** $5.3 \times 10^{-13}$

**17.10.** $1.21 \times 10^{-5}$ M

**17.11.** $FeCO_3$ will be more soluble in an acidic solution than $PbBr_2$ because the $CO_3^{2-}$ ion is a basic anion, whereas $Br^-$ is the conjugate base of a strong acid (HBr) and is therefore pH-neutral.

**17.12.** $Q > K_{sp}$; therefore, a precipitate forms.

**17.13.** $2.9 \times 10^{-6}$ M

**17.14.** **a.** AgCl precipitates first; $[NaCl] = 7.1 \times 10^{-9}$ M
**b.** $[Ag^+]$ is $1.5 \times 10^{-8}$ M when $PbCl_2$ begins to precipitate, and $[Pb^{2+}]$ is 0.085 M.

**17.15.** $9.6 \times 10^{-6}$ M

# Chapter 18

**18.1.** **a.** positive
**b.** negative
**c.** positive

**18.2.** 15.2 J/K

**18.3.** **a.** $-548$ J/K
**b.** $\Delta S_{sys}$ is negative.
**c.** $\Delta S_{univ}$ is negative, and the reaction is not spontaneous.

**18.3.** *For More Practice* 375 K

**18.4.** $\Delta G = -101.6 \times 10^3$ J
Therefore, the reaction is spontaneous. Since both $\Delta H$ and $\Delta S$ are negative, as the temperature increases $\Delta G$ will become more positive.

**18.5.** $-153.2$ J/K

**18.6.** $\Delta G^\circ_{rxn} = -36.3$ kJ
Since $\Delta G^\circ_{rxn}$ is negative, the reaction is spontaneous at this temperature.

**18.7.** $\Delta G^\circ_{rxn} = -42.1$ kJ
Since the value of $\Delta G^\circ_{rxn}$ at the lowered temperature is more negative (or less positive) (which is $-36.3$ kJ), the reaction is more spontaneous.

**18.8.** $\Delta G^\circ_{rxn} = -689.6$ kJ

Since $\Delta G^\circ_{rxn}$ is negative, the reaction is spontaneous at this temperature.

**18.8.** *For More Practice* $\Delta G^\circ_{rxn} = -689.7$ kJ (at 25°)

The value calculated for $\Delta G^\circ_{rxn}$ from the tabulated values ($-689.6$ kJ) is the same, to within 1 in the least significant digit, as the value calculated using the equation for $\Delta G^\circ_{rxn}$.

$\Delta G^\circ_{rxn} = -649.7$ kJ (at 500.0 K)

You could not calculate $\Delta G^\circ_{rxn}$ at 500.0 K using tabulated $\Delta G^\circ_f$ values because the tabulated values of free energy are calculated at a standard temperature of 298 K, much lower than 500 K.

**18.9.** $+107.1$ kJ

**18.10.** $\Delta G_{rxn} = -129$ kJ

The reaction is more spontaneous under these conditions than under standard conditions because $\Delta G_{rxn}$ is more negative than $\Delta G^\circ_{rxn}$.

**18.11.** $-10.9$ kJ

# Chapter 19

**19.1.** $2\,Cr(s) + 4\,H^+(aq) \longrightarrow 2\,Cr^{2+}(aq) + 2\,H_2(g)$

**19.2.** $Cu(s) + 4\,H^+(aq) + 2\,NO_3^-(aq) \longrightarrow$
$$Cu^{2+}(aq) + 2\,NO_2(g) + 2\,H_2O(l)$$

**19.3.** $3\,ClO^-(aq) + 2\,Cr(OH)_4^-(aq) + 2\,OH^-(aq) \longrightarrow$
$$3\,Cl^-(aq) + 2\,CrO_4^{2-}(aq) + 5\,H_2O(l)$$

**19.4.** $+0.60$ V

**19.5.** **a.** The reaction *will* be spontaneous as written.

**b.** The reaction *will not* be spontaneous as written.

**19.6.** $\Delta G^\circ = -3.63 \times 10^5$ J

Since $\Delta G^\circ$ is negative, the reaction is spontaneous.

**19.7.** $4.5 \times 10^3$

**19.8.** 1.08 V

**19.9.** *Anode:* $2\,H_2O(l) \longrightarrow O_2(g) + 4\,H^+(aq) + 4\,e^-$

*Cathode:* $2\,H_2O(l) + 2\,e^- \longrightarrow H_2(g) + 2\,OH^-(aq)$

**19.10.** $6.0 \times 10^1$ min

# Chapter 20

**20.1.** $^{216}_{84}Po \longrightarrow ^{212}_{82}Pb + ^4_2He$

**20.2.** **a.** $^{235}_{92}U \longrightarrow ^{231}_{90}Th + ^4_2He$

$^{231}_{90}Th \longrightarrow ^{231}_{91}Pa + ^0_{-1}e$

$^{231}_{91}Pa \longrightarrow ^{227}_{89}Ac + ^4_2He$

**b.** $^{22}_{11}Na \longrightarrow ^{22}_{10}Ne + ^0_{+1}e$

**c.** $^{76}_{36}Kr + ^0_{-1}e \longrightarrow ^{76}_{35}Br$

**20.2.** *For More Practice* Positron emission
$(^{40}_{19}K \longrightarrow ^{40}_{18}Ar + ^0_{+1}e)$ or electron capture
$(^{40}_{19}K + ^0_{+1}e \longrightarrow ^{40}_{18}Ar)$

**20.3.** **a.** positron emission

**b.** beta decay

**c.** positron emission

**20.4.** 10.7 yr

**20.5.** $t = 964$ yr

No, the C-14 content suggests that the scroll is from about A.D. 1000, not 500 B.C.

**20.6.** $1.0 \times 10^9$ yr

**20.7.** Mass defect $= 1.934$ amu
Nuclear binding energy $= 7.569$ MeV/nucleon

# Chapter 21

**21.1.**

**21.2.** 3-methylhexane

**21.3.** 3,5-dimethylheptane

**21.4.** 2,3,5-trimethylhexane

**21.5.** **a.** 4,4-dimethyl-2-pentyne

**b.** 3-ethyl-4,6-dimethyl-1-heptene

**21.6.** **a.** 2-methylbutane

**b.** 2-chloro-3-methylbutane

**21.7.** **a.** Alcohol reacting with an active metal.

$CH_3CH_2OH + Na \longrightarrow CH_3CH_2ONa + \frac{1}{2}H_2$

**b.** dehydration reaction

# Chapter 22

**22.1.** Fructose exhibits optical isomerism. It contains three chiral carbons.

**22.2.**

# Chapter 23

**23.1.** $KAlSi_3O_8$

**23.2.** $x = 2$

**23.3.** Orthosilicate (or neosilicate): each of the two Be ions has a charge of 2+ for a total of 4+, and the $SiO_4$ unit has a charge of 4−.

**23.4.** Inosilicate (or pyroxene): Ca and Mg each have a charge of 2+ for a total of 4+, and the $Si_2O_6$ unit has a charge of 4− (two $SiO_3{}^{2-}$ units).

**23.5.** $2\,H_2S(g) + 3\,O_2(g) \longrightarrow 2\,H_2O(g) + SO_2(g)$
S changes from the −2 to +4 oxidation state.

**23.6.** The oxidation state for Cl is +7 in $ClO_4{}^-$ and −1 in $Cl^-$.

**23.7.** $I_2(s) + 5\,F_2(g) \longrightarrow 2\,IF_5(g)$

**23.8.** The electron geometry is tetrahedral and the shape is bent for $ICl_2{}^+$.

**23.9.** The electron geometry is octahedral for $BrF_5$, and the molecular geometry is square pyramidal.

**23.10.** The oxidation number changes from −1 to 0 for the oxidation of the Cl in HCl to $Cl_2$ and from +5 to +4 for the reduction of the Cl in $NaClO_3$ to $ClO_2$. The oxidizing agent is $NaClO_3$, and the reducing agent is HCl.

# Chapter 24

**24.1.** At 50 mol % Ni and 1000 °C, this is a solid phase with half of the atoms each Ni and Cu.

**24.2.** At 50 mol % Ni and 1400 °C, this is a liquid phase with half of the atoms each Ni and Cu.

**24.3.** At 900 °C and 60 mol % Cr, this is a two-phase region with more Ni-rich face-centered cubic crystals than Cr-rich body-centered cubic crystals. The Ni-rich phase is about 42 mol % Cr and 58 mol % Ni. The Cr-rich phase is about 94 mol % Cr and 6 mol % Ni.

**24.4.** At 900 °C and 98 mol % Cr, this is a single-phase region with 100 mol % of the Cr-rich body-centered cubic crystals that contains 2% Ni.

# Chapter 25

**25.1.** $[Xe]6s^2 4f^{14} 5d^6$

**25.2.** $[Kr]5s^0 4d^3$ or $[Kr]4d^3$

**25.3.** pentaamminecarbonylmanganese(II) sulfate

**25.4.** sodium tetrachloroplatinate(II)

**25.5.** The complex ion $[Cr(H_2O)_3Cl_3]^+$ fits the general formula $MA_3B_3$, which results in fac and mer isomers.

Fac              Mer

**25.6.** The oxalate ligand is a small bidentate ligand, so it will have to occupy two adjacent (cis) positions of the octahedron. There are three ways to arrange the two $NH_3$ and two $Cl^-$ ligands in the four remaining positions. One has both $NH_3$ and both $Cl^-$ in cis positions (cis isomer). Another has the $NH_3$ ligands in a trans arrangement with both $Cl^-$ in cis positions (*trans*-ammine isomer). The third has both $NH_3$ ligands cis and the $Cl^-$ ligands trans (*trans*-chloro isomer).

Cis isomer

Trans (in $NH_3$)          Trans (in $Cl^-$)

**25.7.** Both the fac and mer isomers are superimposable (by rotating 180°) on their mirror images, so neither one is optically active.

**25.8.** 288 kJ/mol

**25.9.** five unpaired electrons

**25.10.** one unpaired electron

# GLOSSARY

**accuracy** A term that refers to how close a measured value is to the actual value. (1.7)

**acid** A molecular compound that is able to donate an $H^+$ ion (proton) when dissolved in water, thereby increasing the concentration of $H_3O^+$. (3.6)

**acid ionization constant ($K_a$)** The equilibrium constant for the ionization reaction of a weak acid; used to compare the relative strengths of weak acids. (16.4)

**acid–base reaction (neutralization reaction)** A reaction in which an acid reacts with a base and the two neutralize each other, producing water. (4.8)

**acid–base titration** A laboratory procedure in which a basic (or acidic) solution of unknown concentration is reacted with an acidic (or basic) solution of known concentration, in order to determine the concentration of the unknown. (17.4)

**acidic solution** A solution containing an acid that creates additional $H_3O^+$ ions, causing $[H_3O^+]$ to increase. (16.5)

**activated carbon** Very fine carbon particles with high surface area. (23.5)

**activated complex (transition state)** A high-energy intermediate state between reactant and product. (14.5)

**activation energy** An energy barrier in a chemical reaction that must be overcome for the reactants to be converted into products. (14.5)

**active site** The specific area of an enzyme at which catalysis occurs. (14.7)

**actual yield** The amount of product actually produced by a chemical reaction. (4.3)

**addition polymer** A polymer in which the monomers link together without the elimination of any atoms. (12.9)

**addition reaction** A type of organic reaction in which two substituents are added across a double bond. (21.10)

**alcohol** A member of the family of organic compounds that contain a hydroxyl functional group (—OH). (3.12, 21.9)

**aldehyde** A member of the family of organic compounds that contain a carbonyl functional group (C=O) bonded to two R groups, one of which is a hydrogen atom. (21.10)

**aldose** A sugar that is an aldehyde. (22.3)

**aliphatic hydrocarbon** A hydrocarbon—including alkanes, alkenes, and alkynes—that does not contain an aromatic ring. (21.3)

**alkali metals** Highly reactive metals in group 1A of the periodic table. (2.7)

**alkaline battery** A dry-cell battery that employs slightly different half-reactions in a basic medium. (19.7)

**alkaline earth metals** Fairly reactive metals in group 2A of the periodic table. (2.7)

**alkaloid** Organic bases found in plants; they are often poisonous. (16.2)

**alkane** A hydrocarbon containing only single bonds. (3.12, 21.3)

**alkene** A hydrocarbon containing one or more carbon–carbon double bonds. (3.12, 21.3)

**alkyne** A hydrocarbon containing one or more carbon–carbon triple bonds. (3.12, 21.3)

**alloy** A metallic material that contains more than one element. (24.4)

**alpha ($\alpha$) decay** The form of radioactive decay that occurs when an unstable nucleus emits a particle composed of two protons and two neutrons. (20.3)

**alpha ($\alpha$) particle** A particle released during alpha decay; equivalent to a helium-4 nucleus. (20.3)

**$\alpha$-helix** A pattern in the secondary structure of a protein that occurs when the amino acid chain is wrapped tightly in a coil with the side chains extending outward. (22.5)

**aluminosilicates** Members of a family of compounds in which aluminum atoms substitute for silicon atoms in some of the silicon lattice sites of the silica structure. (23.3)

**amine** An organic compound containing nitrogen and derived from ammonia by replacing one or more hydrogen atoms on ammonia with an alkyl group. (21.3)

**amino acid** An organic compound that contains a carbon atom, called the $\alpha$-carbon, bonded to four different groups: an amine group, an R group, a carboxylic acid group, and a hydrogen atom. (22.4)

**ammonia** $NH_3$, the strong smelling compound in which nitrogen displays its lowest oxidation state (−3). (23.6)

**amorphous** Describes a solid in which atoms or molecules do not have any long-range order. (1.3, 11.2)

**ampere (A)** The SI unit for electrical current; 1 A = 1 C/s. (19.3)

**amphoteric** Able to act as either an acid or a base. (16.3)

**amplitude** The vertical height of a crest (or depth of a trough) of a wave; a measure of wave intensity. (7.2)

**angular momentum quantum number ($l$)** An integer that determines the shape of an orbital. (7.5)

**anion** A negatively charged ion. (2.6)

**anode** The electrode in an electro-chemical cell at which oxidation occurs; electrons flow away from the anode. (19.3)

**antibonding orbital** A molecular orbital that is higher in energy than any of the atomic orbitals from which it was formed. (10.8)

**aqueous solution** A solution in which water acts as the solvent. (4.4, 13.2)

**arachno-boranes** Boranes with the formula $B_nH_{n+6}$, consisting of a cage of boron atoms that is missing two or three corners. (23.4)

**arc-melting**  A method in which a solid metal is melted with an arc from a high-voltage electric source in a controlled atmosphere to prevent oxidation. (24.5)

**aromatic hydrocarbon**  Any hydrocarbon containing a stabilized planar ring of carbon atoms, most commonly a benzene ring. (21.3)

**Arrhenius definitions (of acids and bases)**  According to these definitions, an acid is defined as a substance that produces $H^+$ ions in aqueous solution and a base is defined as a substance that produces $OH^-$ ions in aqueous solution. (4.8, 16.3)

**Arrhenius equation**  An equation that relates the rate constant of a reaction to the temperature, the activation energy, and the frequency factor; $k = Ae^{\frac{-E_a}{RT}}$. (14.5)

**Arrhenius plot**  A plot of the natural log of the rate constant ($\ln k$) versus the inverse of the temperature in kelvins ($1/T$) that yields a straight line with a slope of $-E_a/R$ and a $y$-intercept of $\ln A$. (14.5)

**atmosphere (atm)**  A unit of pressure based on the average pressure of air at sea level; 1 atm = 101,325 Pa. (5.2)

**atom**  A submicroscopic particle that constitutes the fundamental building block of ordinary matter; the smallest identifiable unit of an element. (1.1)

**atomic element**  Elements that exist in nature with single atoms as their basic units. (3.4)

**atomic mass (atomic weight)**  The average mass in amu of the atoms of a particular element based on the relative abundance of the various isotopes; numerically equivalent to the mass in grams of one mole of the element. (2.8)

**atomic mass unit (amu)**  A unit used to express the masses of atoms and subatomic particles, defined as $1/12$ the mass of a carbon atom containing 6 protons and 6 neutrons. (2.6)

**atomic number (Z)**  The number of protons in an atom; the atomic number defines the element. (2.6)

**atomic radius**  The average bonding radius of an atom determined from measurements on a large number of elements and compounds. (8.6)

**atomic solids**  Solids whose composite units are atoms; they include nonbonding atomic solids, metallic atomic solids, and network covalent solids. (12.4)

**atomic theory**  The theory that each element is composed of tiny indestructible particles called atoms, that all atoms of a given element have the same mass and other properties, and that atoms combine in simple, whole-number ratios to form compounds. (1.2, 2.3)

**aufbau principle**  The principle that indicates the pattern of orbital filling in an atom. (8.3)

**autoionization**  The process by which water acts as an acid and a base with itself. (16.5)

**Avogadro's law**  The law that states that the volume of a gas is directly proportional to its amount in moles ($V \propto n$). (5.3)

**Avogadro's number**  The number of $^{12}C$ atoms in exactly 12 g of $^{12}C$; equal to $6.0221421 \times 10^{23}$. (2.9)

**balanced chemical equation**  A symbolic representation of a chemical reaction; a balanced equation contains equal numbers of the atoms of each element on both sides of the equation. (3.11)

**ball-and-stick molecular model**  A representation of the arrangement of atoms in a molecule that shows how the atoms are bonded to each other and the overall shape of the molecule. (3.3)

**band gap**  An energy gap that exists between the valence band and conduction band of semiconductors and insulators. (12.8)

**band theory**  A model for bonding in atomic solids based on molecular orbital theory in which atomic orbitals combine and become delocalized over the entire crystal. (12.8)

**barometer**  An instrument used to measure atmospheric pressure. (5.2)

**base ionization constant ($K_b$)**  The equilibrium constant for the ionization reaction of a weak base; used to compare the relative strengths of weak bases. (16.7)

**basic solution**  A solution containing a base that creates additional $OH^-$ ions, causing the $[OH^-]$ to increase. (16.5)

**bent geometry**  A molecular geometry in which the angle between three bonded atoms is approximately 109° (but could be less). (10.3)

**beta ($\beta$) decay**  The form of radioactive decay that occurs when an unstable nucleus emits an electron. (20.3)

**beta ($\beta$) particle**  A medium-energy particle released during beta decay; equivalent to an electron. (20.3)

**$\beta$-pleated sheet**  A pattern in the secondary structure of a protein that occurs when the amino acid chain is extended and forms a zigzag pattern. (22.5)

**bidentate**  Describes ligands that donate two electron pairs to the central metal. (25.3)

**bimolecular**  An elementary step in a reaction that involves two particles, either the same species or different, that collide and go on to form products. (14.6)

**binary acid**  An acid composed of hydrogen and a nonmetal. (3.6)

**binary compound**  A compound that contains only two different elements. (3.5)

**biochemistry**  The study of the chemistry occurring in living organisms. (22.1)

**biological effectiveness factor (RBE)**  A correction factor multiplied by the dose of radiation exposure in rad to obtain the dose in rem. (20.11)

**black phosphorus**  An allotrope of phosphorus with a structure similar to that of graphite; the most thermodynamically stable form. (23.6)

**body-centered cubic**  A unit cell that consists of a cube with one atom at each corner and one atom at the center of the cube. (12.3)

**boiling point** The temperature at which the vapor pressure of a liquid equals the external pressure. (11.5)

**boiling point elevation** The effect of a solute that causes a solution to have a higher boiling point than the pure solvent. (13.6)

**bomb calorimeter** A piece of equipment designed to measure $\Delta E_{rxn}$ for combustion reactions at constant volume. (6.5)

**bond energy** The energy required to break 1 mol of the bond in the gas phase. (9.10)

**bond length** The average length of a bond between two particular atoms in a variety of compounds. (9.10)

**bond order** For a molecule, the number of electrons in bonding orbitals minus the number of electrons in nonbonding orbitals divided by two; a positive bond order implies that the molecule is stable. (10.8)

**bonding orbital** A molecular orbital that is lower in energy than any of the atomic orbitals from which it was formed. (10.8)

**bonding pair** A pair of electrons shared between two atoms. (9.5)

**boranes** Compounds composed of boron and hydrogen. (23.4)

**Born–Haber cycle** A hypothetical series of steps based on Hess's law that represents the formation of an ionic compound from its constituent elements. (9.4)

**borosilicate glass (Pyrex®)** A type of glass containing boric oxide ($B_2O_3$). Borosilicate glass expands less when heated and can withstand heating and cooling cycles without shattering. (12.7)

**Boyle's law** The law that states that volume of a gas is inversely proportional to its pressure $\left( V \propto \dfrac{1}{P} \right)$. (5.3)

**brass** A widely used alloy that contains copper and zinc. (24.5)

**Brønsted–Lowry definitions (of acids and bases)** According to these definitions, an acid is defined as a proton ($H^+$ ion) donor and a base is defined as a proton acceptor. (16.3)

**bronze** An alloy of copper and tin that has been used for thousands of years. (24.5)

**buffer** A solution containing significant amounts of both a weak acid and its conjugate base (or a weak base and its conjugate acid) that resists pH change by neutralizing added acid or added base. (17.2)

**buffer capacity** The amount of acid or base that can be added to a buffer without destroying its effectiveness. (17.3)

**calcination** Heating an ore in order to decompose it and drive off a volatile product. (24.3)

**calorie (cal)** A unit of energy defined as the amount of energy required to raise 1 g of water 1°C; equal to 4.184 J. (6.2)

**Calorie (Cal)** Shorthand notation for the kilocalorie (kcal), or 1000 calories; also called the nutritional calorie, the unit of energy used on nutritional labels. (6.2)

**calorimetry** The experimental procedure used to measure the heat evolved in a chemical reaction. (6.5)

**capillary action** The ability of a liquid to flow against gravity up a narrow tube due to adhesive and cohesive forces. (11.4)

**carbides** Binary compounds composed of carbon and a low-electronegativity element. (23.5)

**carbohydrate** A polyhydroxyl aldehyde or ketone. (22.3)

**carbon black** A fine powdered form of carbon. (23.5)

**carbonyl group** A functional group consisting of a carbon atom double-bonded to an oxygen atom ($C=O$). (21.10)

**carboxylic acid** An organic acid containing the functional group $—COOH$. (16.2, 21.11)

**catalyst** A substance that is not consumed in a chemical reaction that increases the rate of the reaction by providing an alternate mechanism in which the rate-determining step has a smaller activation energy. (14.7)

**cathode** The electrode in an electrochemical cell at which reduction occurs; electrons flow toward the cathode. (19.3)

**cathode ray** A stream of electrons produced when a high electrical voltage is applied between two electrodes within a partially evacuated tube. (2.4)

**cathode ray tube** An evacuated tube containing charged plates to accelerate and view electron beams. (2.4)

**cation** A positively charged ion. (2.6)

**cell potential (cell emf) ($E_{cell}$)** The potential difference between the cathode and the anode in an electrochemical cell. (19.3)

**cellulose** A polysaccharide that consists of glucose units bonded together by β-glycosidic linkages; the main structural component of plants, and the most abundant organic substance on Earth. (22.3)

**Celsius (°C) scale** The temperature scale most often used by scientists (and by most countries other than the United States), on which pure water freezes at 0 °C and boils at 100 °C (at 1 atm pressure). (1.6)

**ceramics** Inorganic metallic solids that are prepared from powders usually mixed with water, formed into the desired shape, and then heated. (12.7)

**chain reaction** A series of reactions in which previous reactions cause future ones; in a fission bomb, neutrons produced by the fission of one uranium nucleus induce fission in other uranium nuclei. (20.7)

**charcoal** A fuel similar to coal that is made by heating wood in the absence of air. (23.5)

**Charles's law** The law that states that the volume of a gas is directly proportional to its temperature ($V \propto T$). (5.3)

**chelate** A complex ion that contains either a bi- or polydentate ligand. (25.3)

**chelating agent** The coordinating ligand of a chelate. (25.3)

**chemical bond** The sharing or transfer of electrons to attain stable electron configurations for the bonding atoms. (9.3)

**chemical change** A change that alters the molecular composition of a substance; see also *chemical reaction*. (1.4)

**chemical energy** The energy associated with the relative positions of electrons and nuclei in atoms and molecules. (6.2)

**chemical equation** A symbolic representation of a chemical reaction; a balanced equation contains equal numbers of the atoms of each element on both sides of the equation. (3.11)

**chemical formula** A symbolic representation of a compound that indicates the elements present in the compound and the relative number of atoms of each. (3.3)

**chemical property** A property that a substance displays only by changing its composition via a chemical change. (1.4)

**chemical reaction** A process by which one or more substances convert to one or more different substances; see also *chemical change*. (3.11)

**chemical symbol** A one- or two-letter abbreviation for an element; listed directly below an element's atomic number on the periodic table. (2.6)

**chemistry** The science that seeks to understand the behavior of matter by studying the behavior of atoms and molecules. (1.1)

**chiral** Describes an environment or a molecule that is not superimposable on its mirror image; chiral molecules exhibit optical isomerism. (21.3)

**chromosome** The DNA-containing structures that occur in the nuclei of living cells. (22.6)

**cis–trans (geometric) isomerism** Cis-isomers have two functional groups (or R groups) on the same side of a bond and trans-isomers have them on opposite sides of a bond. (21.5)

**Claus process** An industrial process for obtaining sulfur through the oxidation of hydrogen sulfide. (23.8)

**Clausius–Clapeyron equation** An equation that displays the exponential relationship between vapor pressure and temperature;

$$\ln(P_{vap}) = \frac{-\Delta H_{vap}}{R}\left(\frac{1}{T}\right) + \ln\beta. \ (11.5)$$

**clay** A powdered form of a mineral (such as an aluminosilicate) mixed with water. (12.7)

***closo*-boranes** Boranes that have the formula $B_{12}H_{12}{}^{2-}$ and form the full icosohedral shape. (23.4)

**coal** A solid, black fuel with high carbon content that is the product of the decomposition of ancient plant material. (23.5)

**codon** A sequence of three bases in a nucleic acid that codes for one amino acid. (22.6)

**coffee-cup calorimeter** A piece of equipment designed to measure $\Delta H_{rxn}$ for reactions at constant pressure. (6.7)

**coke** A solid formed by heating coal in the absence of air that consists primarily of carbon and ash. (23.5)

**colligative property** A property that depends on the amount of a solute but not on the type. (13.6)

**collision frequency** A factor in collision theory that represents the number of collisions that occur per unit time. (14.5)

**collision model** A model of chemical reactions in which a reaction occurs after a sufficiently energetic collision between two reactant molecules. (14.5)

**colloidal dispersion (colloid)** A mixture in which a dispersed substance is finely dived but not truly dissolved in a dispersing medium. (13.8)

**combustion analysis** A method of obtaining empirical formulas for unknown compounds, especially those containing carbon and hydrogen, by burning a sample of the compound in pure oxygen and analyzing the products of the combustion reaction. (3.10)

**combustion reaction** A type of chemical reaction in which a substance combines with oxygen to form one or more oxygen-containing compounds; the reaction often causes the evolution of heat and light in the form of a flame. (3.11)

**common ion effect** The tendency for a common ion to decrease the solubility of an ionic compound or to decrease the ionization of a weak acid or weak base. (17.2)

**common name** A traditional name of a compound that gives little or no information about its chemical structure; for example, the common name of $NaHCO_3$ is "baking soda." (3.5)

**complementary** Capable of precise pairing; in particular, the bases of nucleic acids. (22.6)

**complementary properties** Properties that exclude one another, that is, the more you know about one, the less you know about the other; for example, the wave nature and particle nature of the electron are complementary. (7.4)

**complete ionic equation** An equation that lists individually all of the ions present as either reactants or products in a chemical reaction. (4.7)

**complex carbohydrate** Another term for a polysaccharide based on the fact that it is made up of many simple sugars. (22.3)

**complex ion** An ion that contains a central metal ion that is bound to one or more ligands. (17.8, 25.3)

**composition** The basic components that make up a substance. (1.3)

**compound** A substance composed of two or more elements in fixed, definite proportions. (1.3)

**concentrated solution** A solution that contains a large amount of solute relative to the amount of solvent. (4.4, 13.5)

**concrete** A mixture of Portland cement combined with sand and pebbles. (12.7)

**condensation** The phase transition from gas to liquid. (11.5)

**condensation polymer** A polymer formed by elimination of an atom or small group of atoms (usually water) between pairs of monomers during polymerization. (12.9)

**condensation reaction** A reaction in which two or more organic compounds are joined, often with the loss of water or some other small molecule. (21.11)

**conduction band**   In band theory, the band of energy levels that forms from antibonding molecular orbitals. (12.8)

**conjugate acid**   Any base to which a proton has been added. (16.3)

**conjugate acid–base pair**   Two substances related to each other by the transfer of a proton. (16.3)

**conjugate base**   Any acid from which a proton has been removed. (16.3)

**constructive interference**   The interaction of waves from two sources that align with overlapping crests, resulting in a wave of greater amplitude. (7.2)

**contact process**   An industrial method for the production of sulfuric acid. (23.8)

**conversion factor**   A factor used to convert between two different units; a conversion factor can be constructed from any two quantities known to be equivalent. (1.8)

**coordinate covalent bond**   The bond formed when a ligand donates electrons to an empty orbital of a metal in a complex ion. (25.3)

**coordination compound**   A neutral compound made when a complex ion combines with one or more counterions. (25.3)

**coordination isomers**   Isomers of complex ions that occur when a coordinated ligand exchanges places with the uncoordinated counter-ion. (25.4)

**coordination number (secondary valence)**   The number of molecules or ions directly bound to the metal atom in a complex ion. (25.3)

**coordination number (in a crystal structure)**   The number of atoms with which each atom in a crystal lattice is in direct contact. (12.3)

**core electrons**   Those electrons in a complete principal energy level and those in complete $d$ and $f$ sublevels. (8.4)

**corrosion**   The gradual, nearly always undesired, oxidation of metals that occurs when they are exposed to oxidizing agents in the environment. (19.9)

**Coulomb's law**   A scientific law that states the potential energy between two charged particles is proportional to the product of the charges divided by the distance that separates the charges. (8.3)

**covalent bond**   A chemical bond in which two atoms share electrons that interact with the nuclei of both atoms, lowering the potential energy of each through electrostatic interactions. (3.2, 9.2)

**covalent carbides**   Binary compounds composed of carbon combined with low-electronegativity nonmetals or metalloids. (23.5)

**covalent radius (bonding atomic radius)**   Defined in nonmetals as one-half the distance between two atoms bonded together, and in metals as one-half the distance between two adjacent atoms in a crystal of the metal. (8.6)

**critical mass**   The necessary amount of a radioactive isotope required to produce a self-sustaining fission reaction. (20.7)

**critical point**   The temperature and pressure above which a supercritical fluid exists. (11.8)

**critical pressure**   The pressure required to bring about a transition to a liquid at the critical temperature. (11.5)

**critical temperature**   The temperature above which a liquid cannot exist, regardless of pressure. (11.5)

**crystal field theory**   A bonding model for coordination compounds that focuses on the interactions between ligands and the central metal ion. (25.1)

**crystalline**   Describes a solid in which atoms, molecules, or ions are arranged in patterns with long-range, repeating order. (1.3, 11.2)

**crystalline lattice**   The regular arrangement of atoms in a crystalline solid. (12.3)

**cubic closest packing**   A closest-packed arrangement in which the third layer of atoms is offset from the first; the same structure as the face-centered cubic. (12.3)

**cyclotron**   A particle accelerator in which a charged particle is accelerated in an evacuated ring-shaped tube by an alternating voltage applied to each semi-circular half of the ring. (20.10)

**Dalton's law of partial pressures**   The law stating that the sum of the partial pressures of the components in a gas mixture must equal the total pressure. (5.6)

**de Broglie relation**   The observation that the wavelength of a particle is inversely proportional to its momentum; $\lambda = \dfrac{h}{mv}$. (7.4)

**decanting**   A method of separating immiscible liquids by pouring the top layer into another container. (1.3)

**degenerate**   A term describing two or more electron orbitals with the same value of $n$ that have the same energy. (8.3)

**density ($d$)**   The ratio of an object's mass to its volume. (1.6)

**deposition**   The phase transition from gas to solid. (11.6)

**derived unit**   A unit that is a combination of other base units. For example, the SI unit for speed is meters per second (m/s), a derived unit. (1.6)

**destructive interference**   The interaction of waves from two sources that are aligned so the crest of one overlaps the trough of the other, resulting in cancellation. (7.2)

**deterministic**   A characteristic of the classical laws of motion, which imply that present circumstances determine future events. (7.4)

**dextrorotatory**   Capable of rotating the plane of polarization of light clockwise. (21.3)

**diamagnetic**   The state of an atom or ion that contains only paired electrons and is, therefore, slightly repelled by an external magnetic field. (8.7)

**diamond**   An elemental form of carbon with a crystal structure that consists of carbon atoms connected to four other carbon atoms at the corners of a tetrahedron, creating a strong network covalent solid. (12.6)

**diffraction** The phenomena by which a wave emerging from an aperture spreads out to form a new wave front. (7.2)

**diffusion** The process by which a gas spreads through a space occupied by another gas. (5.9)

**dilute solution** A solution that contains a very small amount of solute relative to the amount of solvent. (4.4, 13.5)

**dimensional analysis** The use of units as a guide to solving problems. (1.8)

**dimer** The product that forms from the reaction of two monomers. (12.9)

**diode** A device that allows the flow of electrical current in only one direction. (12.8)

**dipeptide** Two amino acids linked together. (22.4)

**dipole moment ($\mu$)** A measure of the separation of positive and negative charge in a molecule. (9.6)

**dipole–dipole force** An intermolecular force exhibited by polar molecules that results from the uneven charge distribution. (11.3)

**diprotic acid** An acid that contains two ionizable protons. (4.8, 16.4)

**disaccharide** A carbohydrate composed of two monosaccharides. (22.3)

**dispersion force** An intermolecular force (also referred to as London force) exhibited by all atoms and molecules that results from fluctuations in the electron distribution. (11.3)

**distillation** The process by which mixtures of miscible liquids are separated by heating the mixture to boil off the more volatile liquid. The vaporized component is then recondensed and collected in a separate flask. (1.3)

**disubstituted benzene** A benzene in which two hydrogen atoms have been replaced by other atoms. (21.7)

**dose** The amount of energy absorbed by bodily tissues as a result of exposure to radiation. (20.11)

**double bond** The bond that forms when two electrons are shared between two atoms. (9.5)

**dry-cell battery** A battery that does not contain a large amount of liquid water, often using the oxidation of zinc and the reduction of $MnO_2$ to provide the electrical current. (19.7)

**duet** A Lewis symbol with two dots, signifying a filled outer electron shell for the elements H and He. (9.3)

**dynamic equilibrium** The point at which the rate of the reverse reaction or process equals the rate of the forward reaction or process. (11.5, 13.4, 15.2)

**effective nuclear charge ($Z_{eff}$)** The actual nuclear charge experienced by an electron, defined as the charge of the nucleus plus the charge of the shielding electrons. (8.3)

**effusion** The process by which a gas escapes from a container into a vacuum through a small hole. (5.9)

**electrical charge** A fundamental property of certain particles that causes them to experience a force in the presence of electric fields. (2.4)

**electrical current** The flow of electric charge. (19.3)

**electrochemical cell** A device in which a chemical reaction either produces or is carried out by an electrical current. (19.3)

**electrode** A conductive surface through which electrons can enter or leave a half-cell in an electrochemical cell. (19.3)

**electrolysis** The process in which electrical current drives an otherwise nonspontaneous redox reaction. (19.8)

**electrolyte** A substance that dissolves in water to form solutions that conduct electricity. (4.5)

**electrolytic cell** An electrochemical cell that uses electrical current to drive a nonspontaneous chemical reaction. (19.3)

**electromagnetic radiation** A form of energy embodied in oscillating electric and magnetic fields. (7.2)

**electromagnetic spectrum** The range of the wavelengths of all possible electromagnetic radiation. (7.2)

**electrometallurgy** The use of electrolysis to produce metals from their compounds. (24.3)

**electromotive force (emf)** The force that results in the motion of electrons due to a difference in potential. (19.3)

**electron** A negatively charged, low-mass particle found outside the nucleus of all atoms that occupies most of the atom's volume but contributes almost none of its mass. (2.4)

**electron affinity (EA)** The energy change associated with the gaining of an electron by an atom in its gaseous state. (8.8)

**electron capture** The form of radioactive decay that occurs when a nucleus assimilates an electron from an inner orbital. (20.3)

**electron configuration** A notation that shows the particular orbitals that are occupied by electrons in an atom. (8.3)

**electron geometry** The geometrical arrangement of electron groups in a molecule. (10.3)

**electron groups** A general term for lone pairs, single bonds, multiple bonds, or lone electrons in a molecule. (10.2)

**electronegativity** An atom's ability to attract electrons to itself in a covalent bond. (9.6)

**element** A substance that cannot be chemically broken down into simpler substances. (1.3)

**elementary step** An individual step in a reaction mechanism. (14.6)

**emission spectrum** The range of wavelengths emitted by a particular element; used to identify the element. (7.3)

**empirical formula** A chemical formula that shows the simplest whole number ratio of atoms in the compound. (3.3)

**empirical formula molar mass** The sum of the masses of all the atoms in an empirical formula. (3.10)

**enantiomers (optical isomers)** Two molecules that are nonsuperimposable mirror images of one another. (21.3, 25.4)

**endothermic reaction** A chemical reaction that absorbs heat from its surroundings; for an endothermic reaction, $\Delta H > 0$. (6.6)

**endpoint**   The point of pH change where an indicator changes color. (17.4)

**energy**   The capacity to do work. (1.5, 6.2)

**English system**   The system of units used in the United States and various other countries in which the inch is the unit of length, the pound is the unit of force, and the ounce is the unit of mass. (1.6)

**enthalpy (*H*)**   The sum of the internal energy of a system and the product of its pressure and volume; the energy associated with the breaking and forming of bonds in a chemical reaction. (6.6)

**enthalpy (heat) of reaction (Δ*H*ᵣₓₙ)**   The enthalpy change for a chemical reaction. (6.5)

**enthalpy of solution (Δ*H*ₛₒₗₙ)**   The enthalpy change associated with the formation of a solution. (13.3)

**entropy**   A thermodynamic function that is proportional to the number of energetically equivalent ways to arrange the components of a system to achieve a particular state; a measure of the energy randomization or energy dispersal in a system. (13.2, 18.3)

**enzyme**   A biochemical catalyst made of protein that increases the rates of biochemical reactions. (14.7, 22.4)

**equilibrium constant (*K*)**   The ratio, at equilibrium, of the concentrations of the products of a reaction raised to their stoichiometric coefficients to the concentrations of the reactants raised to their stoichiometric coefficients. (15.3)

**equivalence point**   The point in a titration at which the added solute completely reacts with the solute present in the solution; for acid–base titrations, the point at which the amount of acid is stoichiometrically equal to the amount of base in solution. (4.8, 17.4)

**ester**   A family of organic compounds with the general structure R—COO—R. (21.11)

**ester linkage**   The bonds that form between a carboxylic acid and an alcohol to form an ester, such as those in triglycerides. (22.2)

**ether**   A member of the family of organic compounds of the form R—O—R′. (21.12)

**exact numbers**   Numbers that have no uncertainty and thus do not limit the number of significant figures in any calculation. (1.7)

**exothermic reaction**   A chemical reaction that releases heat to its surroundings; for an exothermic reaction, Δ*H* < 0. (6.6)

**experiment**   A highly controlled procedure designed to generate observations that may support a hypothesis or prove it wrong. (1.2)

**exponential factor**   A number between zero and one that represents the fraction of molecules that have enough energy to make it over the activation barrier on a given approach. (14.5)

**exposure**   The number of radioactive decay events to which a person is exposed. (20.11)

**extensive property**   A property that depends on the amount of a given substance, such as mass. (1.6)

**extractive metallurgy**   The process by which an elemental metal is extracted from the compounds in which it is found. (24.3)

**face-centered cubic**   A crystal structure that has a unit cell that consists of a cube with one atom at each corner and one atom in the center of every face. (12.3)

**Fahrenheit (°F) scale**   The temperature scale that is most familiar in the United States, on which pure water freezes at 32 °F and boils at 212 °F at 1 atm of pressure. (1.6)

**family**   A group of organic compounds with the same functional group. (3.12)

**family (group)**   Columns within the main group elements in the periodic table that contain elements that exhibit similar chemical properties. (2.7)

**Faraday's constant (*F*)**   The charge in coulombs of 1 mol of electrons;
$$F = \frac{96,485\ \text{C}}{\text{mol e}^-}.\ (19.5)$$

**fatty acid**   A carboxylic acid with a long hydrocarbon tail. (22.2)

**ferromagnetic**   The state of an atom or ion that is very strongly attracted by an external magnetic field. (24.5)

**fibrous protein**   A protein with a relatively linear structure; fibrous proteins tend to be insoluble in aqueous solutions. (22.5)

**filtration**   A procedure used to separate a mixture composed of an insoluble solid and a liquid by pouring the mixture through filter paper or some other porous membrane or layer. (1.3)

**first law of thermodynamics**   The law stating that the total energy of the universe is constant. (6.3)

**flux**   In pyrometallurgy, material that reacts with the gangue to form a substance with a low melting point. (24.3)

**formal charge**   The charge that an atom in a Lewis structure would have if all the bonding electrons were shared equally between the bonded atoms. (9.8)

**formation constant (*K*f)**   The equilibrium constant associated with reactions for the formation of complex ions. (17.8)

**formula mass**   The average mass of a molecule of a compound in amu. (3.8)

**formula unit**   The smallest, electrically neutral collection of ions in an ionic compound. (3.4)

**Frasch process**   An industrial process for the recovery of sulfur that uses superheated water to liquefy sulfur deposits in Earth's crust and bring the molten sulfur to the surface. (23.8)

**free energy of formation (Δ*G*°f)**   The change in free energy when 1 mol of a compound forms from its constituent elements in their standard states. (18.8)

**free radical**   A molecule or ion with an odd number of electrons in its Lewis structure. (9.9)

**freezing**   The phase transition from liquid to solid. (11.6)

**freezing point depression**   The effect of a solute that causes a solution to have a lower melting point than the pure solvent. (13.6)

**frequency ($\nu$)**   For waves, the number of cycles (or complete wavelengths) that pass through a stationary point in one second. (7.2)

**frequency factor**   The number of times that reactants approach the activation energy per unit time. (14.5)

**fuel cell**   A voltaic cell that uses the oxidation of hydrogen and the reduction of oxygen, forming water, to provide electrical current. (19.7)

**fullerenes**   Carbon clusters, such as $C_{60}$, bonded in roughly spherical shapes containing from 36 to over 100 carbon atoms. (12.6)

**functional group**   A characteristic atom or group of atoms that imparts certain chemical properties to an organic compound. (3.12)

**gamma ($\gamma$) ray**   The form of electromagnetic radiation with the shortest wavelength and highest energy. (7.2, 20.3)

**gamma ($\gamma$) ray emission**   The form of radioactive decay that occurs when an unstable nucleus emits extremely high frequency electromagnetic radiation. (20.3)

**gangue**   The undesirable minerals that are separated from specific ores. (24.3)

**gas**   A state of matter in which atoms or molecules have a great deal of space between them and are free to move relative to one another; lacking a definite shape or volume, a gas conforms to those of its container. (1.3)

**gas-evolution reaction**   A reaction in which two aqueous solutions are mixed and a gas forms, resulting in bubbling. (4.8)

**Geiger-Müller counter**   A device used to detect radioactivity, which uses argon atoms that become ionized in the presence of energetic particles to produce an electrical signal. (20.5)

**gene**   A sequence of codons within a DNA molecule that codes for a single protein. (22.6)

**geometric (cis–trans) isomerism**   A form of stereoisomerism involving the orientation of functional groups in a molecule that contains bonds incapable of rotating. (21.5)

**geometric isomers**   For complex ions, isomers that result when the ligands bonded to the metal have a different spatial arrangement. (25.4)

**Gibbs free energy ($G$)**   A thermodynamic state function related to enthalpy and entropy by the equation $G = H - TS$; chemical systems tend toward lower Gibbs free energy, also called the *chemical potential*. (18.6)

**glass**   An amorphous form of silica. (12.7)

**globular protein**   A protein that folds into a roughly spherical shape so that its polar side chains are oriented outward and its nonpolar side chains toward the interior; globular proteins tend to be soluble in water. (22.5)

**glycogen**   A highly branched form of starch. (22.3)

**glycolipid**   A triglyceride composed of a fatty acid, a hydrocarbon chain, and a sugar molecule as the polar section. (22.2)

**glycosidic linkage**   A bond between carbohydrates that results from a dehydration reaction. (22.3)

**graphene**   A new material that consists of a single sheet of carbon atoms (similar to a sheet from graphite) one atom thick. (12.1)

**graphite**   An elemental form of carbon consisting of flat sheets of carbon atoms, bonded together as interconnected hexagonal rings held together by intermolecular forces, that can easily slide past each other. (12.6)

**Graham's law of effusion**   A scientific law that states the ratio of effusion rates of two gases is equal to the square root of the inverse of their molar masses. (5.9)

**ground state**   The lowest energy state in an atom, ion, or molecule. (8.3)

**Haber-Bosch process**   The industrial process for producing ammonia from nitrogen gas and hydrogen gas. (23.6)

**half-cell**   One half of an electrochemical cell where either oxidation or reduction occurs. (19.3)

**half-life ($t_{1/2}$)**   The time required for the concentration of a reactant or the amount of a radioactive isotope to fall to one-half of its initial value. (14.4)

**halogens**   Highly reactive nonmetals in group 7A of the periodic table. (2.7)

**heat ($q$)**   The flow of energy caused by a temperature difference. (6.2)

**heat capacity ($C$)**   The quantity of heat required to change a system's temperature by 1 °C. (6.4)

**heat of fusion ($\Delta H_{fus}$)**   The amount of heat required to melt 1 mole of a solid. (11.6)

**heat of hydration ($\Delta H_{hydration}$)**   The enthalpy change that occurs when 1 mol of gaseous solute ions is dissolved in water. (13.3)

**heat (or enthalpy) of sublimation ($\Delta H_{sub}$)**   The enthalpy change associated with the sublimation of a substance. (6.6)

**heat (or enthalpy) of vaporization ($\Delta H_{vap}$)**   The amount of heat required to vaporize one mole of a liquid to a gas. (11.5)

**Heisenberg's uncertainty principle**   The principle stating that due to the wave-particle duality, it is fundamentally impossible to precisely determine both the position and velocity of a particle at a given moment in time. (7.4)

**Henderson–Hasselbalch equation**   An equation used to easily calculate the pH of a buffer solution from the initial concentrations of the buffer components, assuming that the "*x is small*" approximation is valid; $pH = pK_a + \log \dfrac{[base]}{[acid]}$. (17.2)

**Henry's law**   An equation that expresses the relationship between solubility of a gas and pressure; $S_{gas} = k_H P_{gas}$. (13.4)

**Hess's law**   The law stating that if a chemical equation can be expressed as the sum of a series of steps, then $\Delta H_{rxn}$ for the overall equation is the sum of the heats of reactions for each step. (6.8)

**heterogeneous catalysis**   Catalysis in which the catalyst and the reactants exist in different phases. (14.7)

**heterogeneous mixture**   A mixture in which the composition varies from one region to another. (1.3)

**hexagonal closest packing**   A closest-packed arrangement in which the atoms of the third layer align exactly over those in the first layer. (12.3)

**hexose**   A six-carbon sugar. (22.3)

**high-spin complex**   A complex ion with weak field ligands that have the same number of unpaired electrons as the free metal ion. (25.5)

**homogeneous catalysis**   Catalysis in which the catalyst exists in the same phase as the reactants. (14.7)

**homogeneous mixture**   A mixture with the same composition throughout. (1.3)

**Hund's rule**   The principle stating that when electrons fill degenerate orbitals, they first fill them singly with parallel spins. (8.3)

**hybrid orbitals**   Orbitals formed from the combination of standard atomic orbitals that correspond more closely to the actual distribution of electrons in a chemically bonded atom. (10.7)

**hybridization**   A mathematical procedure in which standard atomic orbitals are combined to form new, hybrid orbitals. (10.7)

**hydrate**   An ionic compound that contains a specific number of water molecules associated with each formula unit. (3.5)

**hydrazine**   $N_2H_4$, a nitrogen and hydrogen compound in which nitrogen has a negative oxidation state ($-2$). (23.6)

**hydrocarbon**   An organic compound that contains only carbon and hydrogen. (3.12)

**hydrogen azide (HN$_3$)**   A nitrogen and hydrogen compound with a higher hydrogen-to-nitrogen ratio than ammonia or hydrazine. (23.6)

**hydrogen bond**   A strong dipole–dipole attractive force between a hydrogen bonded to O, N, or F and one of these electronegative atoms on a neighboring molecule. (11.3)

**hydrogen bonding**   The forming of a hydrogen bond. (11.3)

**hydrogenation**   The catalyzed addition of hydrogen to alkene double bonds to make single bonds. (14.7)

**hydrolysis**   The splitting of a chemical bond with water, resulting in the addition of H and OH to the products. (22.3)

**hydrometallurgy**   The use of an aqueous solution to extract metals from their ores. (24.3)

**hydronium ion**   $H_3O^+$; the ion formed from the association of a water molecule with an $H^+$ ion donated by an acid. (4.8, 16.3)

**hydroxyl group**   An —OH group. (21.9)

**hypothesis**   A tentative interpretation or explanation of an observation. A good hypothesis is *falsifiable*. (1.2)

**hypoxia**   A physiological condition caused by low levels of oxygen, marked by dizziness, headache, shortness of breath, and eventually unconsciousness or even death in severe cases. (5.6)

**ideal gas**   A theoretical gas that follows the ideal gas law under all conditions. (5.4)

**ideal gas constant**   The proportionality constant of the ideal gas law, $R$, equal to $8.314 \, \text{J/mol} \cdot \text{K}$ or $0.08206 \, \text{L} \cdot \text{atm/mol} \cdot \text{K}$. (5.4)

**ideal gas law**   The law that combines the relationships of Boyle's, Charles's, and Avogadro's laws into one comprehensive equation of state with the proportionality constant $R$ in the form $PV = nRT$. (5.4)

**ideal solution**   A solution that follows Raoult's law at all concentrations for both solute and solvent. (13.6)

**indeterminacy**   The principle asserting that present circumstances do not necessarily determine future evens in the quantum-mechanical realm. (7.4)

**indicator**   A dye whose color depends on the pH of the solution it is dissolved in; often used to detect the endpoint of a titration. (4.8, 17.4)

**infrared (IR) radiation**   Electromagnetic radiation emitted from warm objects, with wavelengths slightly larger than those of visible light. (7.2)

**insoluble**   Incapable of dissolving in water or being extremely difficult of solution. (4.5)

**integrated rate law**   A relationship between the concentrations of the reactants in a chemical reaction and time. (14.4)

**intensive property**   A property such as density that is independent of the amount of a given substance. (1.6)

**interference**   The superposition of two or more waves overlapping in space, resulting in either an increase in amplitude (constructive interference) or a decrease in amplitude (destructive interference). (7.2)

**interhalogen compounds**   A class of covalent compounds that contain two different halogens. (23.9)

**internal energy ($E$)**   The sum of the kinetic and potential energies of all of the particles that compose a system. (6.3)

**International System of Units (SI)**   The standard unit system used by scientists, based on the metric system. (1.6)

**interstitial alloy**   An alloy in which small, usually nonmetallic atoms fit between the metallic atoms of a crystal. (24.4)

**ion**   An atom or molecule with a net charge caused by the loss or gain of electrons. (2.6)

**ion product constant for water ($K_w$)**   The equilibrium constant for the autoionization of water. (16.5)

**ion–dipole force**   An intermolecular force between an ion and the oppositely charged end of a polar molecule. (11.3)

**ionic bond**   A chemical bond formed between two oppositely charged ions, generally a metallic cation and a nonmetallic anion, that are attracted to each other by electrostatic forces. (3.2, 9.2)

**ionic carbides** Binary compounds composed of carbon combined with low-electronegativity metals. (23.5)

**ionic compound** A compound composed of cations and anions bound together by electrostatic attraction. (3.2)

**ionic solids** Solids whose composite units are ions; they generally have high melting points. (12.4)

**ionization energy (IE)** The energy required to remove an electron from an atom or ion in its gaseous state. (8.7)

**ionizing power** The ability of radiation to ionize other molecules and atoms. (20.3)

**isotopes** Atoms of the same element with the same number of protons but different numbers of neutrons and consequently different masses. (2.6)

**joule (J)** The SI unit for energy: equal to $1 \text{ kg} \cdot \text{m}^2/\text{s}^2$. (6.2)

**kelvin (K)** The SI standard unit of temperature. (1.6)

**Kelvin scale** The temperature scale that assigns 0 K ($-273$ °C or $-459$ °F) to the coldest temperature possible, absolute zero, which is the temperature at which molecular motion virtually stops. (1.6)

**ketone** A member of the family of organic compounds that contain a carbonyl functional group ($C=O$) bonded to two R groups, neither of which is a hydrogen atom. (21.10)

**ketose** A sugar that is a ketone. (22.3)

**kilogram (kg)** The SI standard unit of mass defined as the mass of a block of metal kept at the International Bureau of Weights and Measures at Sèvres, France. (1.6)

**kilowatt-hour (kWh)** An energy unit used primarily to express large amounts of energy produced by the flow of electricity; equal to $3.60 \times 10^6$ J. (6.2)

**kinetic energy** The energy associated with motion of an object. (1.5, 6.2)

**kinetic molecular theory** A model of an ideal gas as a collection of point particles in constant motion

undergoing completely elastic collisions. (5.8)

**lanthanide contraction** The trend toward leveling off in size of the atoms in the third and fourth transition rows due to the ineffective shielding of the $f$ sublevel electrons. (25.2)

**lattice energy** The energy associated with forming a crystalline lattice from gaseous ions. (9.4)

**law** see *scientific law*. (1.2)

**law of conservation of energy** A law stating that energy can neither be created nor destroyed, only converted from one form to another. (1.5, 6.2)

**law of conservation of mass** A law stating that matter is neither created nor destroyed in a chemical reaction. (1.2, 2.3)

**law of definite proportions** A law stating that all samples of a given compound have the same proportions of their constituent elements. (2.3)

**law of mass action** A law regarding the relationship between the balanced chemical equation and the expression of the equilibrium constant. (15.3)

**law of multiple proportions** A law stating that when two elements (A and B) form two different compounds, the masses of element B that combine with one gram of element A can be expressed as a ratio of small whole numbers. (2.3)

**Le Châtelier's principle** The principle stating that when a chemical system at equilibrium is disturbed, the system shifts in a direction that minimizes the disturbance. (15.9)

**leaching** The process by which a metal is separated out of a mixture by selectively dissolving the metal into solution. (24.3)

**leaded glass** A type of glass (often called *crystal*) that contains PbO. This type of glass has a higher index of refraction (it bends light more than ordinary glass), which results in more brilliant-looking glassware. (12.7)

**lead–acid storage battery** A battery that uses the oxidation of lead and the reduction of lead(IV) oxide

in sulfuric acid to provide electrical current. (19.7)

**lever rule** The rule that states that in a two-phase region, whichever phase is closer to the composition of the alloy is the more abundant phase. (24.4)

**levorotatory** Capable of rotating the polarization of light counter-clockwise. (21.3)

**Lewis acid** An atom, ion, or molecule that is an electron pair acceptor. (16.11)

**Lewis base** An atom, ion, or molecule that is an electron pair donor. (16.11)

**Lewis electron-dot structures (Lewis structures)** A drawing that represents chemical bonds between atoms in molecules as shared or transferred electrons; the valence electrons of atoms are represented as dots in a Lewis structure. (9.1)

**Lewis model** A simple model of chemical bonding using diagrams that represent bonds between atoms as lines or pairs of dots. According to this model, atoms bond together to obtain stable octets (eight valence electrons). (9.1)

**Lewis symbol** The symbol of an element surrounded with dots representing the element's valence electrons. (9.3)

**ligand** A neutral molecule or an ion that acts as a Lewis base with the central metal ion in a complex ion. (17.8, 25.3)

**limiting reactant** The reactant that has the smallest stoichiometric amount in a reactant mixture and consequently limits the amount of product in a chemical reaction. (4.3)

**linear accelerator** A particle accelerator in which a charged particle is accelerated in an evacuated tube by a potential difference between the ends of the tube or by alternating charges in sections of the tube. (20.10)

**linear geometry** The molecular geometry of three atoms with a 180° bond angle due to the repulsion of two electron groups. (10.2)

**linkage isomers** Isomers of complex ions that occur when some ligands coordinate to the metal in different ways. (25.4)

**lipid**   A member of the class of biochemical compounds that are insoluble in water but soluble in nonpolar solvents, including fatty acids, triglycerides, and steroids. (22.2)

**lipid bilayer**   A double-layered structure made of phospholipids or glycolipids, in which the polar heads of the molecules interact with the environment and the nonpolar tails interact with each other; a component of many cellular membranes. (22.2)

**liquid**   A state of matter in which atoms or molecules pack about as closely as they do in solid matter but are free to move relative to each other; liquids have a fixed volume but not a fixed shape. (1.3)

**liter (L)**   A unit of volume equal to $1000 \ cm^3$ or 1.057 qt. (1.6)

**lithium ion battery**   A battery that produces electrical current in the form of motion of lithium ions from the anode to the cathode. (19.7)

**lone pair**   A pair of electrons associated with only one atom. (9.5)

**low-spin complex**   A complex ion with strong field ligands that have fewer unpaired electrons than the free metal ion. (25.5)

**magic numbers**   Certain numbers of nucleons ($N$ or $Z$ = 2, 8, 20, 28, 50, 82, and N = 126) that confer unique stability. (20.4)

**magnetic quantum number ($m_l$)**   An integer that specifies the orientation of an orbital. (7.5)

**main-group elements**   Those elements found in the $s$ or $p$ blocks of the periodic table, whose properties tend to be predictable based on their position in the table. (2.7, 23.2)

**manometer**   An instrument used to determine the pressure of a gaseous sample, consisting of a liquid-filled U-shaped tube with one end exposed to the ambient pressure and the other end connected to the sample. (5.2)

**mass**   A measure of the quantity of matter making up an object. (1.6)

**mass defect**   The difference in mass between the nucleus of an atom and the sum of the separated particles that make up that nucleus. (20.8)

**mass number (A)**   The sum of the number of protons and neutrons in an atom. (2.6)

**mass percent composition (mass percent)**   An element's percentage of the total mass of a compound containing the element. (3.9)

**mass spectrometry**   An experimental method of determining the precise mass and relative abundance of isotopes in a given sample using an instrument called a *mass spectrometer*. (2.8)

**matter**   Anything that occupies space and has mass. (1.3)

**mean free path**   The average distance that a molecule in a gas travels between collisions. (5.9)

**melting (fusion)**   The phase transition from solid to liquid. (11.6)

**melting point**   The temperature at which the molecules of a solid have enough thermal energy to overcome intermolecular forces and become a liquid. (11.6)

**metal**   A member of the class of elements that are generally good conductors of heat and electricity, malleable, ductile, lustrous, and tend to lose electrons during chemical changes. (2.7)

**metallic atomic solids**   Atomic solids held together by metallic bonds; they have variable melting points. (12.4)

**metallic bonding**   The type of bonding that occurs in metal crystals, in which metal atoms donate their electrons to an electron sea, delocalized over the entire crystal lattice. (9.2)

**metallic carbides**   Binary compounds composed of carbon combined with metals that have a metallic lattice with holes small enough to fit carbon atoms. (23.5)

**metalloid**   A member of the class of elements found on the boundary between the metals and nonmetals of the periodic table, with properties intermediate between those of both groups; also called *semimetals*. (2.7)

**metallurgy**   The part of chemistry that includes all the processes associated with mining, separating, and refining metals and the subsequent production of pure metals and mixtures of metals called alloys. (24.1)

**meter (m)**   The SI standard unit of length; equivalent to 39.37 inches. (1.6)

**metric system**   The system of measurements used in most countries in which the meter is the unit of length, the kilogram is the unit of mass, and the second is the unit of time. (1.6)

**microwaves**   Electromagnetic radiation with wavelengths slightly longer than those of infrared radiation; used for radar and in microwave ovens. (7.2)

**milliliter (mL)**   A unit of volume equal to $10^{-3}$ L or $1 \ cm^3$. (1.6)

**millimeter of mercury (mmHg)**   A common unit of pressure referring to the air pressure required to push a column of mercury to a height of 1 mm in a barometer; 760 mmHg = 1 atm. (5.2)

**mineral**   A homogeneous, naturally occurring, crystalline inorganic solid. (23.3, 24.2)

**miscibility**   The ability to mix without separating into two phases. (11.3)

**miscible**   The ability of two or more substances to be soluble in each other in all proportions. (13.2)

**mixture**   A substance composed of two or more different types of atoms or molecules that can be combined in variable proportions. (1.3)

**molality ($m$)**   A means of expressing solution concentration as the number of moles of solute per kilogram of solvent. (13.5)

**molar heat capacity**   The amount of heat required to raise the temperature of one mole of a substance by 1 °C. (6.4)

**molar mass**   The mass in grams of one mole of atoms of an element; numerically equivalent to the atomic mass of the element in amu. (2.9)

**molar solubility**   The solubility of a compound in units of moles per liter. (17.5)

**molar volume** The volume occupied by one mole of a gas; the molar volume of an ideal gas at STP is 22.4 L. (5.5)

**molarity (M)** A means of expressing solution concentration as the number of moles of solute per liter of solution. (4.4, 13.5)

**mole (mol)** A unit defined as the amount of material containing $6.0221421 \times 10^{23}$ (Avogadro's number) particles. (2.9)

**mole fraction ($\chi_A$)** The number of moles of a component in a mixture divided by the total number of moles in the mixture. (5.6)

**mole fraction ($\chi_{solute}$)** A means of expressing solution concentration as the number of moles of solute per moles of solution. (13.5)

**mole percent** A means of expressing solution concentration as the mole fraction multiplied by 100%. (13.5)

**molecular compound** Compounds composed of two or more covalently bonded nonmetals. (3.2)

**molecular element** Those elements that exist in nature with diatomic or polyatomic molecules as their basic unit. (3.4)

**molecular equation** An equation showing the complete neutral formula for each compound in a reaction. (4.7)

**molecular formula** A chemical formula that shows the actual number of atoms of each element in a molecule of a compound. (3.3)

**molecular geometry** The geometrical arrangement of atoms in a molecule. (10.3)

**molecular orbital theory** An advanced model of chemical bonding in which electrons reside in molecular orbitals delocalized over the entire molecule. In the simplest version, the molecular orbitals are simply linear combinations of atomic orbitals. (10.8)

**molecular solids** Solids whose composite units are molecules; they generally have low melting points. (12.4)

**molecularity** The number of reactant particles involved in an elementary step. (14.6)

**molecule** Two or more atoms joined chemically in a specific geometrical arrangement. (1.1)

**monodentate** Describes ligands that donate only one electron pair to the central metal. (25.3)

**monomer** A repeating unit in a polymer. (12.9)

**monoprotic acid** An acid that contains only one ionizable proton. (16.4)

**monosaccharide** The simplest carbohydrates, with three to eight carbon atoms and only one aldehyde or ketone group. (22.3)

**nanotubes** Long, tubular structures consisting of interconnected $C_6$ rings. (12.6)

**natural abundance** The relative percentage of a particular isotope in a naturally occurring sample with respect to other isotopes of the same element. (2.6)

**Nernst equation** The equation relating the cell potential of an electrochemical cell to the standard cell potential and the reaction quotient; $E_{cell} = E_{cell}^{\circ} - \dfrac{0.0592\ \text{V}}{n} \log Q$. (19.6)

**net ionic equation** An equation that shows only the species that actually change during the reaction. (4.7)

**network covalent atomic solids** Atomic solids held together by covalent bonds; they have high melting points. (12.4)

**neutral** The state of a solution where the concentrations of $H_3O^+$ and $OH^-$ are equal. (16.5)

**neutron** An electrically neutral subatomic particle found in the nucleus of an atom, with a mass almost equal to that of a proton. (2.5)

**nickel–cadmium (NiCad) battery** A battery that consists of an anode composed of solid cadmium and a cathode composed of NiO(OH)(s) in a KOH solution. (19.7)

**nickel–metal hydride (NiMH) battery** A battery that uses the same cathode reaction as the NiCad battery but a different anode reaction, the oxidation of hydrogens in a metal alloy. (19.7)

**nido-boranes** Boranes that have the formula $B_nH_{n+4}$ and consist of a cage of boron atoms missing one corner. (23.4)

**nitrogen narcosis** A physiological condition caused by an increased partial pressure of nitrogen, resulting in symptoms similar to those of intoxication. (5.6)

**noble gases** The group 8A elements, which are largely unreactive (inert) due to their stable filled p orbitals. (2.7)

**node** A point where the wave function ($\psi$), and therefore the probability density ($\psi^2$) and radial distribution function, all go through zero (7.6)

**nonbonding atomic solids** Atomic solids held together by dispersion forces; they have low melting points. (12.4)

**nonbonding electrons** Electrons in a Lewis structure that are not in a chemical bond; also called lone pair electrons. (9.5)

**nonbonding orbital** An orbital whose electrons remain localized on an atom. (10.8)

**nonelectrolyte** A compound that does not dissociate into ions when dissolved in water. (4.5)

**nonmetal** A member of the class of elements that tend to be poor conductors of heat and electricity and usually gain electrons during chemical reactions. (2.7)

**nonvolatile** Not easily vaporized. (11.5)

**normal boiling point** The temperature at which the vapor pressure of a liquid equals 1 atm. (11.5)

**n-type semiconductor** A semiconductor that employs negatively charged electrons in the conduction band as the charge carriers. (12.8)

**nuclear binding energy** The amount of energy that would be required to break apart the nucleus into its component nucleons. (20.8)

**nuclear equation** An equation that represents nuclear processes such as radioactivity. (20.3)

**nuclear fission** The splitting of the nucleus of an atom, resulting in a tremendous release of energy. (20.7)

**nuclear fusion** The combination of two light nuclei to form a heavier one. (20.9)

**nuclear theory** The theory that most of the atom's mass and all of its positive charge is contained in a small, dense nucleus. (2.5)

**nucleic acid** A biological polymer composed of nucleotide units that carries genetic information. (22.6)

**nucleons** The particles that compose the nucleus and that are protons and neutrons. (20.4)

**nucleotides** The individual units composing nucleic acids; each consists of a phosphate group, a sugar, and a nitrogenous base. (22.6)

**nucleus** The very small, dense core of the atom that contains most of the atom's mass and all of its positive charge; composed of protons and neutrons. (2.5)

**nuclide** A particular isotope of an atom. (20.3)

**octahedral geometry** The molecular geometry of seven atoms with 90° bond angles. (10.2)

**octahedral hole** A space that exists in the middle of six atoms on two adjacent close-packed sheets of atoms in a crystal lattice. (24.4)

**octet** The eight dots around atoms in a Lewis structure that signify a filled outer electron shell for $s$ and $p$ block elements. (9.3)

**octet rule** The tendency for most bonded atoms to possess or share eight electrons in their outer shell to obtain stable electron configurations and lower their potential energy. (9.3)

**optical isomers (enantiomers)** Two molecules that are nonsuperimposable mirror images of one another. (21.3, 25.4)

**orbital** A probability distribution map, based on the quantum-mechanical model of the atom, used to describe the likely position of an electron in an atom; also an allowed energy state for an electron. (7.5)

**orbital diagram** A diagram that gives information similar to an electron configuration but symbolizes an electron as an arrow in a box representing an orbital, with the arrow's direction denoting the electron's spin. (8.3)

**ore** A rock that contains a high concentration of a specific mineral. (24.2)

**organic chemistry** The study of carbon-based compounds. (21.1)

**organic compound** A compound composed of carbon and hydrogen and possibly a few other elements, including nitrogen, oxygen, and sulfur. (3.12)

**organic molecule** A molecule containing carbon combined with several other elements including hydrogen, nitrogen, oxygen, or sulfur. (21.1)

**orientation factor** A factor in collision theory (usually between 0 and 1) that represents the fraction of collisions having an orientation that can allow the reaction to occur. (14.5)

**orthosilicates** Silicates in which tetrahedral $SO_4{}^{4-}$ ions stand alone. (23.3)

**osmosis** The flow of solvent from a solution of lower solute concentration to one of higher solute concentration. (13.6)

**osmotic pressure** The pressure required to stop osmotic flow. (13.6)

**Ostwald process** An industrial process used for commercial preparation of nitric acid. (23.6)

**overall order** The sum of the orders of all reactants in a chemical reaction. (14.3)

**oxidation** The loss of one or more electrons; also the gaining of oxygen or the loss of hydrogen. (4.9)

**oxidation state (oxidation number)** A positive or negative whole number that represents the "charge" an atom in a compound would have if all shared electrons were assigned to the atom with a greater attraction for those electrons. (4.9)

**oxidation–reduction (redox) reaction** Reactions in which electrons are transferred from one reactant to another and the oxidation states of certain atoms are changed. (4.9)

**oxidizing agent** A substance that causes the oxidation of another substance; an oxidizing agent gains electrons and is reduced. (4.9)

**oxyacid** An acid composed of hydrogen and an oxyanion. (3.6)

**oxyanion** A polyatomic anion containing a nonmetal covalently bonded to one or more oxygen atoms. (3.5)

**oxygen toxicity** A physiological condition caused by an increased level of oxygen in the blood, resulting in muscle twitching, tunnel vision, and convulsions. (5.6)

**ozone** $O_3$, an allotrope of oxygen that is a toxic blue diamagnetic gas with a strong odor. (23.7)

**packing efficiency** The percentage of volume of a unit cell occupied by the atoms, assumed to be spherical. (12.3)

**paramagnetic** The state of an atom or ion that contains unpaired electrons and is, therefore, attracted by an external magnetic field. (8.7)

**partial pressure ($P_n$)** The pressure due to any individual component in a gas mixture. (5.6)

**parts by mass** A unit for expressing solution concentration as the mass of the solute divided by the mass of the solution multiplied by a multiplication factor. (13.5)

**parts by volume** A unit for expressing solution concentration as the volume of the solute divided by the volume of the solution multiplied by a multiplication factor. (13.5)

**parts per billion (ppb)** A unit for expressing solution concentration in parts by mass in which the multiplication factor is $10^9$. (13.5)

**parts per million (ppm)** A unit for expressing solution concentration in parts by mass in which the multiplication factor is $10^6$. (13.5)

**pascal (Pa)** The SI unit of pressure, defined as $1 \text{ N/m}^2$. (5.2)

**Pauli exclusion principle** The principle that no two electrons in an atom can have the same four quantum numbers. (8.3)

**penetrating power**    The ability of radiation to penetrate matter. (20.3)

**penetration**    The phenomenon of some higher-level atomic orbitals having significant amounts of probability within the space occupied by orbitals of lower energy level. For example, the 2s orbital penetrates into the 1s orbital. (8.3)

**peptide bond**    The bond that forms between the amine end of one amino acid and the carboxylic end of another. (22.4)

**percent by mass**    A unit for expressing solution concentration in parts by mass with a multiplication factor of 100%. (13.5)

**percent ionic character**    The ratio of a bond's actual dipole moment to the dipole moment it would have if the electron were transferred completely from one atom to the other, multiplied by 100%. (9.6)

**percent ionization**    The concentration of ionized acid in a solution divided by the initial concentration of acid multiplied by 100%. (16.6)

**percent yield**    The percentage of the theoretical yield of a chemical reaction that is actually produced; the ratio of the actual yield to the theoretical yield multiplied by 100%. (4.3)

**periodic law**    A law based on the observation that when the elements are arranged in order of increasing mass, certain sets of properties recur periodically. (2.7)

**periodic property**    A property of an element that is predictable based on an element's position in the periodic table. (8.1)

**permanent dipole**    A permanent separation of charge; a molecule with a permanent dipole always has a slightly negative charge at one end and a slightly positive charge at the other. (11.3)

**pH**    The negative log of the concentration of $H_3O^+$ in a solution; the pH scale is a compact way to specify the acidity of a solution. (16.5)

**phase**    With regard to waves and orbitals, the phase is the sign of the amplitude of the wave, which can be positive or negative. (7.6)

**phase diagram**    A map of the phase of a substance as a function of pressure and temperature. (11.8)

**phenyl group**    A benzene ring treated as a substituent. (21.7)

**phosphine**    $PH_3$; a colorless, poisonous gas that smells like decaying fish and has an oxidation state of $-3$ for phosphorus. (23.6)

**phospholipid**    Compound similar in structure to a triglyceride but in which one fatty acid is replaced by a phosphate group. (22.2)

**phosphorescence**    The long-lived emission of light that sometimes follows the absorption of light by certain atoms and molecules. (20.2)

**photoelectric effect**    The observation that many metals emit electrons when light falls upon them. (7.2)

**photon (quantum)**    The smallest possible packet of electromagnetic radiation with an energy equal to $h\nu$. (7.2)

**physical change**    A change that alters only the state or appearance of a substance but not its chemical composition. (1.4)

**physical property**    A property that a substance displays without changing its chemical composition. (1.4)

**pi ($\pi$) bond**    The bond that forms between two p orbitals that overlap side to side. (10.7)

**p–n junctions**    Tiny areas in electronic circuits that have p-type semiconductors on one side and n-type on the other. (12.8)

**polar covalent bond**    A covalent bond between two atoms with significantly different electronegativities, resulting in an uneven distribution of electron density. (9.6)

**polyatomic ion**    An ion composed of two or more atoms. (3.4)

**polydentate**    Describes ligands that donate more than one electron pair to the central metal. (25.3)

**polymer**    A long chain-like molecule composed of repeating units called monomers. (12.9)

**polymorphs**    The different crystal structures that can sometimes exist for the same compound. (12.4)

**polypeptide**    A chain of amino acids joined together by peptide bonds. (22.4)

**polyprotic acid**    An acid that contains more than one ionizable proton and releases them sequentially. (4.8, 16.9)

**polysaccharide**    A long, chainlike molecule composed of many monosaccharide units bonded together. (22.3)

**Portland cement**    A powdered mixture consisting mostly of limestone ($CaCO_3$) and silica ($SiO_2$), with smaller amounts of alumina ($Al_2O_3$), iron(III) oxide ($Fe_2O_3$), and gypsum ($CaSO_4 \cdot 2\,H_2O$). (12.7)

**positron**    The particle released in positron emission; equal in mass to an electron but opposite in charge. (20.3)

**positron emission**    The form of radioactive decay that occurs when an unstable nucleus emits a positron. (20.3)

**positron emission tomography (PET)**    A specialized imaging technique that employs positron-emitting nuclides, such as fluorine-18, as a radiotracer. (20.12)

**potential difference**    A measure of the difference in potential energy (usually in joules) per unit of charge (coulombs). (19.3)

**potential energy**    The energy associated with the position or composition of an object. (1.5, 6.2)

**powder metallurgy**    A process by which metallic components are made from powdered metal. (24.3)

**precipitate**    A solid, insoluble ionic compound that forms in, and separates from, a solution. (4.6)

**precipitation reaction**    A reaction in which a solid, insoluble product forms when two solutions are mixed. (4.6)

**precision**    A term that refers to how close a series of measurements are to one another or how reproducible they are. (1.7)

**prefix multipliers**    Multipliers that change the value of the unit by powers of ten. (1.6)

**pressure**  A measure of force exerted per unit area; in chemistry, most commonly the force exerted by gas molecules as they strike the surfaces around them. (5.1)

**pressure–volume work**  The work that occurs when a volume change takes place against an external pressure. (6.4)

**primary structure**  The sequence of amino acids in a protein chain. (22.5)

**primary valence**  The oxidation state on the central metal atom in a complex ion. (25.3)

**principal level (shell)**  The group of orbitals with the same value of $n$. (7.5)

**principal quantum number ($n$)**  An integer that specifies the overall size and energy of an orbital. The higher the quantum number $n$, the greater the average distance between the electron and the nucleus and the higher its energy. (7.5)

**probability density**  The probability (per unit volume) of finding the electron at a point in space as expressed by a three-dimensional plot of the wave function squared ($\psi^2$). (7.6)

**products**  The substances produced in a chemical reaction; they appear on the right-hand side of a chemical equation. (3.10)

**proton**  A positively charged subatomic particle found in the nucleus of an atom. (2.5)

**p-type semiconductor**  A semiconductor that employs positively charged "holes" in the valence band as the charge carriers. (12.8)

**pure substance**  A substance composed of only one type of atom or molecule. (1.3)

**pyrometallurgy**  A technique of extractive metallurgy in which heat is used to extract a metal from its mineral. (24.3)

**pyrosilicates**  Silicates in which two $SO_4^{4-}$ tetrahedral ions share a corner. (23.3)

**pyroxenes**  Silicates in which $SO_4^{4-}$ tetrahedral ions bond together to form chains. (23.3)

**qualitative analysis**  A systematic way to determine the ions present in an unknown solution. (17.7)

**quantitative analysis**  A systematic way to determine the amounts of substances in a solution or mixture. (17.7)

**quantum number**  One of four interrelated numbers that determine the shape and energy of orbitals, as specified by a solution of the Schrödinger equation. (7.5)

**quantum-mechanical model**  A model that explains the behavior of absolutely small particles such as electrons and photons. (7.1)

**quartz**  A silicate crystal that has a formula unit of $SiO_2$. (12.6, 23.3)

**quaternary structure**  The way that subunits fit together in a multimeric protein. (22.5)

**racemic mixture**  An equimolar mixture of two optical isomers that does not rotate the plane of polarization of light at all. (21.3)

**radial distribution function**  A mathematical function (corresponding to a specific orbital) that represents the total probability of finding an electron within a thin spherical shell at a distance $r$ from the nucleus. (7.6)

**radio waves**  The form of electromagnetic radiation with the longest wavelengths and smallest energy. (7.2)

**radioactive**  The state of those unstable atoms that emit subatomic particles or high-energy electromagnetic radiation. (20.1)

**radioactivity**  The emission of subatomic particles or high-energy electromagnetic radiation by the unstable nuclei of certain atoms. (2.5, 20.1)

**radiocarbon dating**  A form of radiometric dating based on the C-14 isotope. (20.6)

**radiometric dating**  A technique used to estimate the age of rocks, fossils, or artifacts that depends on the presence of radioactive isotopes and their predictable decay with time. (20.6)

**radiotracer**  A radioactive nuclide that has been attached to a compound or introduced into a mixture in order to track the movement of the compound or mixture within the body. (20.12)

**random coil**  A section of a protein's secondary structure that has less-regular patterns than $\alpha$-helixes or $\beta$-pleated sheets. (22.5)

**random error**  Error that has equal probability of being too high or too low. (1.7)

**Raoult's law**  An equation used to determine the vapor pressure of a solution; $P_{soln} = X_{solv}P^\circ_{solv}$. (13.6)

**rate constant ($k$)**  A constant of proportionality in the rate law. (14.3)

**rate law**  A relationship between the rate of a reaction and the concentration of the reactants. (14.3)

**rate-determining step**  The step in a reaction mechanism that occurs much more slowly than any of the other steps. (14.6)

**reactants**  The starting substances of a chemical reaction; they appear on the left-hand side of a chemical equation. (3.11)

**reaction intermediates**  Species that form in one step of a reaction mechanism and are consumed in another. (14.6)

**reaction mechanism**  A series of individual chemical steps by which an overall chemical reaction occurs. (14.6)

**reaction order ($n$)**  A value in the rate law that determines how the rate depends on the concentration of the reactants. (14.3)

**reaction quotient ($Q_c$)**  The ratio, at any point in the reaction, of the concentrations of the products of a reaction raised to their stoichiometric coefficients to the concentrations of the reactants raised to their stoichiometric coefficients. (15.7)

**recrystallization**  A technique used to purify solids in which the solid is put into hot solvent until the solution is saturated; when the solution cools, the purified solute comes out of solution. (13.4)

**red phosphorus**   An allotrope of phosphorus similar in structure to white phosphorus but in which one of the bonds between two phosphorus atoms in the tetrahedron is broken; red phosphorus is more stable than white. (23.6)

**reducing agent**   A substance that causes the reduction of another substance; a reducing agent loses electrons and is oxidized. (4.9)

**reduction**   The gaining of one or more electrons; also the gaining of hydrogen or the loss of oxygen. (4.9)

**refining**   A process in which the crude material is purified. (24.3)

**rem**   A unit of the dose of radiation exposure that stands for roentgen equivalent man, where a roentgen is defined as the amount of radiation that produces $2.58 \times 10^{-4}$ C of charge per kg of air. (20.11)

**resonance hybrid**   The actual structure of a molecule that is intermediate between two or more resonance structures. (9.8)

**resonance structures**   Two or more valid Lewis structures that are shown with double-headed arrows between them to indicate that the actual structure of the molecule is intermediate between them. (9.8)

**reversible**   As applied to a reaction, the ability to proceed in either the forward or the reverse direction. (15.2)

**reversible process**   A process that reverses direction upon an infinitesimally small change in some property (such as temperature or pressure) related to the reaction. (18.7)

**roasting**   Heating that causes a chemical reaction between a furnace atmosphere and a mineral in order to process ores. (24.3)

**salt**   An ionic compound formed in a neutralization reaction by the replacement of an $H^+$ ion from the acid with a cation from the base. (4.8)

**salt bridge**   An inverted, U-shaped tube containing a strong electrolyte (such as $KNO_3$) that connects two half-cells, allowing a flow of ions that neutralizes charge buildup. (19.3)

**saturated fat**   A triglyceride with no double bonds in the hydrocarbon chain; saturated fats tend to be solid at room temperature. (22.2)

**saturated hydrocarbon**   A hydrocarbon containing no double bonds in the carbon chain. (21.4)

**saturated solution**   A solution in which the dissolved solute is in dynamic equilibrium with any undissolved solute; any added solute will not dissolve in a saturated solution. (13.4)

**scientific law**   A brief statement or equation that summarizes past observations and predicts future ones. (1.2)

**scintillation counter**   A device that detects radioactivity using a material that emits ultraviolet or visible light in response to excitation by energetic particles. (20.5)

**second (s)**   The SI standard unit of time, defined as the duration of 9,192,631,770 periods of the radiation emitted from a certain transition in a cesium-133 atom. (1.6)

**second law of thermodynamics**   A law stating that for any spontaneous process, the entropy of the universe increases ($\Delta S_{univ} > 0$). (18.3)

**secondary structure**   The regular periodic or repeating patterns in the arrangement of protein chains. (22.5)

**secondary valence**   The number of molecules or ions directly bound to the metal atom in a complex ion; also called the *coordination number*. (25.3)

**seesaw geometry**   The molecular geometry of a molecule with trigonal bipyramidal electron geometry and one lone pair in an axial position. (10.3)

**selective precipitation**   A process involving the addition of a reagent to a solution that forms a precipitate with one of the dissolved ions but not the others. (17.6)

**semiconductor**   A material with intermediate electrical conductivity that can be changed and controlled. (2.7)

**semipermeable membrane**   A membrane that selectively allows some substances to pass through but not others. (13.6)

**shielding**   The effect on an electron of repulsion by electrons in lower-energy orbitals that screen it from the full effects of nuclear charge. (8.3)

**sigma ($\sigma$) bond**   The resulting bond that forms between a combination of any two $s$, $p$, or hybridized orbitals that overlap end to end. (10.7)

**significant figures (significant digits)**   In any reported measurement, the non-place-holding digits that indicate the precision of the measured quantity. (1.7)

**silica**   A silicate crystal that has a formula unit of $SiO_2$; also called *quartz*. (12.6, 23.3)

**silicates**   Covalent atomic solids that contain silicon, oxygen, and various metal atoms. (12.6, 23.3)

**simple cubic**   A unit cell that consists of a cube with one atom at each corner. (12.3)

**slag**   In pyrometallurgy, the waste liquid solution that is formed between the flux and gangue; usually a silicate material. (24.3)

**smelting**   A form of roasting in which the product is liquefied, which aids in separation. (24.3)

**soda-lime glass**   A type of glass that is about 70% $SiO_2$ with the balance being mostly $Na_2O$ and $CaO$. This type of glass is transparent to visible light (not ultraviolet) and has a high thermal expansion, but it is less expensive to make and form into desired shapes than vitreous silica. (12.7)

**solid**   A state of matter in which atoms or molecules are packed close to one another in fixed locations with definite volume. (1.3)

**solubility**   The amount of a substance that will dissolve in a given amount of solvent. (13.2)

**solubility product constant ($K_{sp}$)**   The equilibrium expression for a chemical equation representing the dissolution of a slightly to moderately soluble ionic compound. (17.5)

**soluble**  Able to dissolve to a significant extent, usually in water. (4.5)

**solute**  The minority component of a solution. (4.4, 13.1)

**solution**  A homogenous mixture of two substances. (4.4, 13.1)

**solvent**  The majority component of a solution. (4.4, 13.1)

**soot**  An amorphous form of carbon that forms during the incomplete combustion of hydrocarbons. (23.5)

**space-filling molecular model**  A representation of a molecule that shows how the atoms fill the spaces between them. (3.3)

**specific heat capacity ($C_s$)**  The amount of heat required to raise the temperature of 1 g of a substance by 1 °C. (6.4)

**spectator ion**  Ions in a complete ionic equation that do not participate in the reaction and therefore remain in solution. (4.7)

**spin quantum number ($m_s$)**  The fourth quantum number, which denotes the electron's spin as either $^1/_2$ (up arrow) or $-^1/_2$ (down arrow). (7.5)

**spontaneous process**  A process that occurs without ongoing outside intervention. (18.2)

**square planar geometry**  The molecular geometry of a molecule with octahedral electron geometry and two lone pairs. (10.3)

**square pyramidal geometry**  The molecular geometry of a molecule with octahedral electron geometry and one lone pair. (10.3)

**standard cell potential (standard emf) ($E°_{cell}$)**  The cell potential for a system in standard states (solute concentration of 1 M and gaseous reactant partial pressure of 1 atm). (19.3)

**standard electrode potential**  The potential of a half-cell in an electrochemical cell. (19.4)

**standard enthalpy change ($\Delta H°$)**  The change in enthalpy for a process when all reactants and products are in their standard states. (6.9)

**standard enthalpy (heat) of formation ($\Delta H°_f$)**  The change in enthalpy when 1 mol of a compound forms from its constituent elements in their standard states. (6.9)

**standard entropy change for a reaction ($\Delta S°_{rxn}$)**  The change in entropy for a process in which all reactants and products are in their standard states. (18.7)

**standard free energy change ($\Delta G°_{rxn}$)**  The change in free energy for a process when all reactants and products are in their standard states. (18.8)

**Standard Hydrogen Electrode (SHE)**  The half-cell consisting of an inert platinum electrode immersed in 1 M HCl with hydrogen gas at 1 atm bubbling through the solution; used as the standard of a cell potential of zero. (19.4)

**standard molar entropy ($S°$)**  A measure of the energy dispersed into 1 mol of a substance at a particular temperature. (18.7)

**standard state**  For a gas the standard state is the pure gas at a pressure of exactly 1 atm; for a liquid or solid the standard state is the pure substance in its most stable form at a pressure of 1 atm and the temperature of interest (often taken to be 25 °C); for a substance in solution the standard state is a concentration of exactly 1 M. (6.9)

**standard temperature and pressure (STP)**  The conditions of $T = 0$ °C (273 K) and $P = 1$ atm; used primarily in reference to a gas. (5.5)

**starch**  A polysaccharide that consists of glucose units bonded together by $\alpha$-glycosidic linkages; the main energy storage medium for plants. (22.3)

**state**  A classification of the form of matter as a solid, liquid, or gas. (1.3)

**state function**  A function whose value depends only on the state of the system, not on how the system got to that state. (6.3)

**stereoisomers**  Molecules in which the atoms are bonded in the same order but have a different spatial arrangement. (21.3, 25.4)

**steroid**  A lipid composed of four fused hydrocarbon rings. (22.2)

**stock solution**  A highly concentrated form of a solution used in laboratories to make less concentrated solutions via dilution. (4.4)

**stoichiometry**  The numerical relationships between amounts of reactants and products in a balanced chemical equation. (4.2)

**strong acid**  An acid that completely ionizes in solution. (4.5, 16.4)

**strong base**  A base that completely dissociates in solution. (16.7)

**strong electrolyte**  A substance that completely dissociates into ions when dissolved in water. (4.5)

**strong force**  Of the four fundamental forces of physics, the one that is the strongest but acts over the shortest distance; the strong force is responsible for holding the protons and neutrons together in the nucleus of an atom. (20.4)

**strong-field complex**  A complex ion in which the crystal field splitting is large. (25.5)

**structural formula**  A molecular formula that shows how the atoms in a molecule are connected or bonded to each other. (3.3, 21.3)

**structural isomers**  Molecules with the same molecular formula but different structures. (21.3, 25.4)

**sublevel (subshell)**  Those orbitals in the same principle level with the same value of $n$ and $l$. (7.5)

**sublimation**  The phase transition from solid to gas. (11.6)

**substance**  A specific instance of matter. (1.3)

**substitutional alloy**  An alloy in which one metal atom substitutes for another in the crystal structure. (24.4)

**substrate**  The reactant molecule of a biochemical reaction that binds to an enzyme at the active site. (14.7)

**supersaturated solution**  An unstable solution in which more than the equilibrium amount of solute is dissolved. (13.4)

**surface tension**  The energy required to increase the surface area of a liquid by a unit amount; responsible for the tendency of liquids to minimize their surface area, giving rise to a membrane-like surface. (11.4)

**surroundings**    In thermodynamics, everything in the universe that exists outside the system under investigation. (6.2)

**system**    In thermodynamics, the portion of the universe that is singled out for investigation. (6.2)

**systematic error**    Error that tends toward being consistently either too high or too low. (1.7)

**systematic name**    An official name for a compound, based on well-established rules, that can be determined by examining its chemical structure. (3.5)

**temperature**    A measure of the average kinetic energy of the atoms or molecules that compose a sample of matter. (1.6)

**termolecular**    An elementary step of a reaction in which three particles collide and go on to form products. (14.6)

**tertiary structure**    The large-scale bends and folds produced by interactions between the R groups of amino acids that are separated by large distances in the linear sequence of a protein chain. (22.5)

**tetrahedral geometry**    The molecular geometry of five atoms with $109.5°$ bond angles. (10.2)

**tetrahedral hole**    A space that exists directly above the center point of three closest-packed metal atoms in one plane and a fourth metal located directly above the center point in the adjacent plane in a crystal lattice. (24.4)

**theoretical yield**    The greatest possible amount of product that can be made in a chemical reaction based on the amount of limiting reactant. (4.3)

**theory**    A proposed explanation for observations and laws, based on well-established and tested hypotheses, that presents a model of the way nature works and predicts behavior beyond the observations and laws on which it was based. (1.2)

**thermal energy**    A type of kinetic energy associated with the temperature of an object, arising from the motion of individual atoms or molecules in the object; see also *heat*. (1.5, 6.2)

**thermal equilibrium**    The point at which there is no additional net transfer of heat between a system and its surroundings. (6.4)

**thermochemistry**    The study of the relationship between chemistry and energy. (6.1)

**thermodynamics**    The general study of energy and its interconversions. (6.3)

**thermoluminescent dosimeter**    A device used to measure the dose of radiation to which a person is exposed. (20.5)

**third law of thermodynamics**    The law stating that the entropy of a perfect crystal at absolute zero (0 K) is zero. (18.7)

**titration**    A laboratory procedure in which a substance in a solution of known concentration is reacted with another substance in a solution of unknown concentration in order to determine the unknown concentration; see also *acid–base titration*. (4.8)

**torr**    A unit of pressure named after Evangelista Torricelli and equivalent to 1/760 of an atmosphere (atm). (5.2)

**transition elements (transition metals)**    Those elements found in the *d* block of the periodic table whose properties tend to be less predictable based simply on their position in the table. (2.7)

**transmutation**    The transformation of one element into another as a result of nuclear reactions. (20.10)

**triglyceride**    Triesters composed of glycerol with three fatty acids attached. (21.2)

**trigonal bipyramidal geometry**    The molecular geometry of six atoms with $120°$ bond angles between the three equatorial electron groups and $90°$ bond angles between the two axial electron groups and the trigonal plane. (10.2)

**trigonal planar geometry**    The molecular geometry of four atoms with $120°$ bond angles in a plane. (10.2)

**trigonal pyramidal geometry**    The molecular geometry of a molecule with tetrahedral electron geometry and one lone pair. (10.3)

**triple bond**    The bond that forms when three electron pairs are shared between two atoms. (9.5)

**triple point**    The unique set of conditions at which all three phases of a substance are equally stable and in equilibrium. (11.8)

**triprotic acid**    An acid that contains three ionizable protons. (16.4)

**T-shaped geometry**    The molecular geometry of a molecule with trigonal bipyramidal electron geometry and two lone pairs in axial positions. (10.3)

**two-phase region**    The region between the two phases in a metal alloy phase diagram, where the amount of each phase depends upon the composition of the alloy. (24.4)

**Tyndall effect**    The scattering of light by a colloidal dispersion. (13.8)

**ultraviolet (UV) radiation**    Electromagnetic radiation with slightly smaller wavelengths than visible light. (7.2)

**unimolecular**    Describes a reaction that involves only one particle that goes on to form products. (14.6)

**unit cell**    The smallest divisible unit of a crystal that, when repeated in three dimensions, reproduces the entire crystal lattice. (12.3)

**units**    Standard quantities used to specify measurements. (1.6)

**unsaturated fat**    A triglyceride with one or more double bonds in the hydrocarbon chain; unsaturated fats tend to be liquid at room temperature. (22.2)

**unsaturated hydrocarbon**    A hydrocarbon that includes one or more double or triple bonds. (21.5)

**unsaturated solution**    A solution containing less than the equilibrium amount of solute; any added solute will dissolve until equilibrium is reached. (13.4)

**valence band**   In band theory, the band of energy levels that forms from bonding molecular orbitals. (12.8)

**valence bond theory**   An advanced model of chemical bonding in which electrons reside in quantum-mechanical orbitals localized on individual atoms that are a hybridized blend of standard atomic orbitals; chemical bonds result from an overlap of these orbitals. (10.6)

**valence electrons**   Those electrons that are important in chemical bonding. For main-group elements, the valence electrons are those in the outermost principal energy level. (8.4)

**valence shell electron pair repulsion (VSEPR) theory**   A theory that allows prediction of the shapes of molecules based on the idea that electrons—either as lone pairs or as bonding pairs—repel one another. (10.2)

**van der Waals equation**   The extrapolation of the ideal gas law that considers the effects of intermolecular forces and particle volume in a nonideal gas;
$$P + a\left(\frac{n}{V}\right)^2 \times (V - nb) = nRT. \text{ (5.10)}$$

**van der Waals radius (nonbonding atomic radius)**   Defined as one-half the distance between the centers of adjacent, nonbonding atoms in a crystal. (8.6)

**van't Hoff factor ($i$)**   The ratio of moles of particles in a solution to moles of formula units dissolved. (13.7)

**vapor pressure**   The partial pressure of a vapor in dynamic equilibrium with its liquid. (5.6, 11.5)

**vapor pressure lowering ($\Delta P$)**   The change in vapor pressure that occurs in pure substance upon addition of a solute. (13.6)

**vaporization**   The phase transition from liquid to gas. (11.5)

**viscosity**   A measure of the resistance of a liquid to flow. (11.4)

**visible light**   Those frequencies of electromagnetic radiation that can be detected by the human eye. (7.2)

**vitreous (or fused) silica**   A type of glass that is hard, resists high temperatures, has a low thermal expansion, and is transparent to both visible light and ultraviolet light. (12.7)

**volatile**   Tending to vaporize easily. (1.3, 11.5)

**volt**   A unit of of potential difference defined as 1 J/C. (19.3)

**voltaic (galvanic) cell**   An electrochemical cell that produces electrical current from a spontaneous chemical reaction. (19.3)

**volume ($V$)**   A measure of space. Any unit of length, when cubed (raised to the third power), becomes a unit of volume. (1.6)

**washing soda**   The hydrated crystal of sodium carbonate, $Na_2CO_3 \cdot 10H_2O$. (23.5)

**wave function ($\psi$)**   A mathematical function that describes the wavelike nature of the electron. (7.5)

**wavelength ($\lambda$)**   The distance between adjacent crests of a wave. (7.2)

**weak acid**   An acid that does not completely ionize in water. (4.5, 16.4)

**weak base**   A base that only partially ionizes in water. (16.7)

**weak electrolyte**   A substance that does not completely ionize in water and only weakly conducts electricity in solution. (4.5)

**weak-field complex**   A complex ion in which the crystal field splitting is small. (25.5)

**white phosphorus**   An unstable allotrope of phosphorus consisting of $P_4$ molecules in a tetrahedral shape, with the phosphorus atoms at the corners of the tetrahedron. (23.6)

**work ($w$)**   The result of a force acting through a distance. (1.5, 6.2)

**X-ray crystallography**   The process of using X-ray diffraction to determine the structure of a crystal. (12.2)

**X-ray diffraction**   A powerful laboratory technique that allows for the determination of the arrangement of atoms in a crystal and the measuring of the distance between them. (12.2)

**X-ray**   Electromagnetic radiation with wavelengths slightly longer than those of gamma rays; used to image bones and internal organs. (7.2)

# PHOTO AND TEXT CREDITS

**Chapter 1** Page 1: Eisntein, Albert. "Physics and Reality" (1936). Ideas and Opinions. Trans. Sonja Bargmann. New York: Bonanza, 1954. Print. Page 3: Lavoisier, Antoine. Elementary Treatise of Chemistry. Trans. Robert Kerr. Paris: Chez Cuchet, 1789. Reprinted Bruxelles: Cultures et Civilisations, 1965. Print. Page 3 (TL): Louella Folsom/Fotolia. Page 3 (TR): Oneo/Fotolia. Page 3 (BR): Tomas Abad/Alamy. Page 6 (ML): Oneo/Fotolia. Page 7 (BL): Al Messerschmidt/WireImage/Getty Images. Page 7 (BCL): Fotolia. Page 7 (BCR): YinYang/iStock/Getty Images. Page 7(BR): Denis Larkin/iStock/Getty Images. Page 9 (BC): Michael Dalton/Fundamental Photographs. Page 10 (ML): Siede Preis/Photodisc/Getty Images. Page 11 (TC): Charles D. Winters/Science Source. Page 11 (C): Renn Sminkey/Pearson Education. Page 11 (BC): Clark Brennan/Alamy. Page 13: NASA's Metric Confusion. CC-TECH. 03 September 1999., Available at available at http://articles.cnn.com/1999-09-30/tech/9909_30_mars.metric_1_mars-orbiter-climate-orbiter-spacecraft-team?_s=PM:TECH. Page 13 (BR): Image courtesy of NASA. Page 14 (C): Siede Preis/Photodisc/Getty Images. Page 14 (BL): Richard Megna/Fundamental Photographs. Page 16 (TL): Marty Honig/Photodisc/Getty Images. Page 16 (TCL): Kevron2001/Fotolia. Page 16 (TCR): Andraz Gregoric/iStock/Getty Images. Page 16 (TR): Byron W Moore/Shutterstock. Page 20 (TR): VR Photos/Shutterstock. Page 21 (TR): Richard Megna/Fundamental Photographs. Page 21 (MR): Richard Megna/Fundamental Photographs. Page 21 (CBR): Richard Megna/Fundamental Photographs. Page 21 (BR): Stacey Stambaugh/Pearson Education. Page 38 (TL): Richard Megna/Fundamental Photographs. Page 38 (TR): Richard Megna/Fundamental Photographs. Page 38 (MR): "Roman Sigaev/iStock/Getty Images". Page 39 (BL): Stacey Stambaugh/Pearson Education. Page 39 (BC): Stacey Stambaugh/Pearson Education. Page 39 (BR): Stacey Stambaugh/Pearson Education. Page 39 (TL): Stacey Stambaugh/Pearson Education. Page 39 (TC): Warren Rosenberg/Fundamental Photographs. Page 39 (TR): Warren Rosenberg/Fundamental Photographs. Page 41 (BL): NASA. Page 43: Elmas, Gülnur Mertoglu. "The content of heavy metals in the recycled paper for use in the packaging" Istanbul University,Forest Faculty,Forest Product Chemistry and Technology and Tanase, Ion Gh.. et al., "Estimation of the uncertainty of the measurement results of some trace levels elements in document paper samples using ICP-MS" The Royal Society of Chemistry. 08 January 2015. Print. Page 43: Elmas, Gülnur Mertoglu. "The content of heavy metals in the recycled paper for use in the packaging" Istanbul University,Forest Faculty,Forest Product Chemistry and Technology and Tanase, Ion Gh.. et al., "Estimation of the uncertainty of the measurement results of some trace levels elements in document paper samples using ICP-MS" The Royal Society of Chemistry. 08 January 2015. Print. All other Figures and Tables copyright Pearson Education, Inc. **Chapter 2** Page 40: Dalton, John. A New System of Chemical Philosophy. 3 Vol. Manchester, 1808, 1810, 1827. Print. Page 46: Oseen, C.W. "Nobel Prize in Physics 1926-Presentation Speech". Nobelprize.org. Nobel Media AB 2014. Print,. Page 46 (ML): Image originally created by IBM Corporation. Page 47: Democritus of Abdera. Diogenes Laertius IX, 44. Trans. R. D. Hicks (1925), Vol. 2, 453-5. Print. Page 48 (TL): Charles D. Winters/Science Source. Page 48 (TC): Charles D. Winters/Science Source. Page 48 (TR): Joseph Calev/Shutterstock. Page 50: Feynman, Richard. The Feynman Lectures on Physics; Volume 1. USA: Addison Wesley. ISBN 0-201-02115-3. Print. Page 51: Thomson, J.J. "On the Cathode Rays", Proceedings of the Cambridge Philosophical Society, 9, 243. 08 Februaray 1987. Print. Page 51 (BR): Richard Megna/Fundamental Photographs. Page 54: Glasstone, Samuel (1958). Sourcebook on Atomic Energy, 2nd edition, page 93. Princeton, NJ: Van Nostrand. Print. Page 55 (BR): Richard Megna/Fundamental Photographs. Page 56 (TL): Jeremy Woodhouse/Photodisc/Getty Images. Page 56

(BC): Steve Cole/Getty Images. Page 57 (BR): Library of Congress Prints and Photographs Division. Page 60 (MR): NASA. Page 62 (TL): Veronika Roosimaa/iStock/Getty Image. Page 62 (BL): Matthias Zepper. Page 62 (BLM): Charles D. Winters/Science Source. Page 62 (BML): Steffen Foerster/Shutterstock. Page 62 (BM): Clive Streeter/Dorling Kindersley Ltd. Page 62 (BTM): Harry Taylor/Dorling Kindersley Ltd. Page 62 (BTR): Gary Ombler/Dorling Kindersley Ltd. Page 62 (BRT): Tom Grundy/Shutterstock. Page 62 (BRM): Charles D. Winters/Science Source. Page 62 (BR): Perennou Nuridsany/Science Source. Page 63 (BR): Richard Megna/Fundamental Photographs. Page 63 (BR): Martyn F. Chillmaid/Science Source. Page 64 (TL): Charles D. Winters/Science Source. Page 64 (BL): Charles D. Winters/Science Source. Page 66: IUPAC Periodic Table of the Elements, courtesy of International Union of Pure & Applied Chemistry (IUPAC). Page 70 (TL): Stacey Stambaugh/Pearson Education. Page 70 (BL): Stacey Stambaugh/Pearson Education. Page 77 (T): Jeremy Woodhouse/Photodisc/Getty Images. Page 77 (B): Stacey Stambaugh/Pearson Education. Page 81 (TR): Image originally created by IBM Corporation. All other Figures and Tables copyright Pearson Education, Inc. **Chapter 3** Page 86: Crick, Francis. What Mad Pursuit: A Personal View of Scientific Discovery. New York: Basic Books Inc., 1988. Print. Page 88 (BC): Malerapaso/Getty Images. Page 89 (BL): Richard Megna/Fundamental Photographs. Page 89 (BC): Charles Falco/Science Source. Page 89 (BR): Charles D. Winters/Science Source. Page 94 (ML): Charles D. Winters/Science Source. Page 94 (BC): Charles Falco/Science Source. Page 96 (ML): Renn Sminkey/Pearson Education. Page 96 (MR): Renn Sminkey/Pearson Education. Page 96 (BC): Sidney Moulds/Science Source. Page 96 (BR): Basement Stock/Alamy. Page 101 (ML): Richard Megna/Fundamental Photographs. Page 101 (MR): Richard Megna/Fundamental Photographs. Page 103 (MR): Shutterstock. Page 103 (BR): Richard Megna/Fundamental Photographs. Page 104 (BL): Andre Jenny/Stock Connection Worldwide/Newscom. Page 104 (BR): Richard Megna/Fundamental Photographs. Page 109 (MR): NASA. Page 111 (BR): Richard Megna/Fundamental Photographs. Page 114 (TR): David R. Frazier/Getty Images. Page 123 (MR): Sinelev/Shutterstock. Page 124 (BL): Lightpoet/Fotolia. Page 129 (ML): Renn Sminkey/Pearson Education. Page 129 (MR): Renn Sminkey/Pearson Education. Page 129 (B): Shutterstock. Page 136: Tans, Dr. Pieter NOAA/ESRL (www.esrl.noaa.gov/gmd/ccgg/trends/) and Dr. Ralph Keeling, Scripps Institution of Oceanography (scrippsco2.ucsd.edu/). Print. All other Figures and Tables copyright Pearson Education, Inc. **Chapter 4** Page 138: Pauling, Linus (1901–1994). Print. Page 161 (MR): Richard Megna/Fundamental Photographs. Page 160 (BL): Richard Megna/Fundamental Photographs. Page 160 (BC): Richard Megna/Fundamental Photographs. Page 160 (BR): "Richard Megna/Fundamental Photographs". Page 162 (ML): Martyn F. Chillmaid/Science Source. Page 163 (ML): Kindersley, Dorling. Ltd. Print. Page 164 (ML): Richard Megna/Fundamental Photographs. Page 168 (TL): Chip Clark/Fundamental Photographs. Page 169 (B): Richard Megna/Fundamental Photographs. Page 172 (TL): Richard Megna/Fundamental Photographs. Page 172 (TC): Richard Megna/Fundamental Photographs. Page 172 (TR): Richard Megna/Fundamental Photographs. Page 174 (ML): Richard Megna/Fundamental Photographs. Page 175 (MR): Charles D. Winters/Science Source. Page 176 (ML): Richard Megna/Fundamental Photographs. Page 176 (MR): Richard Megna/Fundamental Photographs. Page 181 (TR): Neil Guegan/Image Source. Page 194: Reprinted courtesy of ML2R Consultancy. All other Figures and Tables copyright Pearson Education, Inc. **Chapter 5** Page 196: James Clerk Maxwell, The Scientific Papers of James Clerk Maxwell, Vol. II. Cambridge: Cambridge University Press. 1890. Print. Page 201 (TR): Patrick Watson/Pearson Education. Inc. Page 206 (TL): Carlos Caetano/Shutterstock.

Page 655 (MR): Image courtesy of NASA. Page 656 (TL): Image courtesy of NASA. Page 656 (TR): Image courtesy of NASA. All other Figures and Tables copyright Pearson Education, Inc. **Chapter 15** Page 674: Châtelier, Henry Le (1850–1936). Page 676 (TL): Kenneth Eward Illustration. Page 682 (BL): Vshivkova/Shutterstock. Page 709 (TL): Richard Megna/Fundamental Photographs. Page 709 (TR): Richard Megna/Fundamental Photographs. Page 720: Based on Figure 1 "CO2 Flow Measurement Key in CCS Schemes" Power Engineering International, Vol. 18, Issue 4. 2010. Page 720: "Springer-Verlag, Chemical Petrology, "The terrestrial planets" 1977, P. 113, Robert F. Mueller, Surendra K. Saxena, © Springer-Verlag New York Inc. 1986. With permission of Springer Science+Business Media". All other Figures and Tables copyright Pearson Education, Inc. **Chapter 16** Page 722: Huheey, James, E (1935–). Page 725 (BR): Ivaylo Ivanov/Shutterstock. Page 725 (BL): Igor Dutina/Getty Images. Page 725 (BC): Kovaleva_ka/Fotolia. Page 736 (BR): David M. Martin/Science Source. Page 765: Source: U.S. EIA Annual Energy Review 2015. Page 766: Source: National Atmospheric Deposition Program, National Trends Network. Page 766 (BL): Richard Megna/Fundamental Photographs. Page 766 (BR): Karol Kozlowski/Shutterstock. Page 766 (TL): National Atmospheric Deposition Program. Page 767: Source: U.S. EPA Air Trends. Page 775 (TR): Eco Images/Getty Images. Page 777: Source: Ben Rotter, www.brsquared.org. Reprinted with permission. Page 777: Data from Tim Donahue and Sabrina Lueck, "Practical White Wine Production: Theory and Practice," College Cellars of Walla Walla. Author research and data supplemented. All other Figures and Tables copyright Pearson Education, Inc. **Chapter 17** Page 778: Witt, Otto N. (1853–1915). Page 782 (CL): Richard Megna/Fundamental Photographs. Page 782 (C): Richard Megna/Fundamental Photographs. Page 782 (CR): Richard Megna/Fundamental Photographs. Page 794 (MR): Ronen/Fotolia. Page 795 (BL): Richard Megna/Fundamental Photographs. Page 795 (BC): Richard Megna/Fundamental Photographs. Page 795 (BR): Richard Megna/Fundamental Photographs. Page 811 (BL): Bernhard Edmaier/Science Source. Page 811 (BR): Sheila Terry/Science Source. Page 814 (ML): Santi Rodríguez/Fotolia. Page 815 (TR): Richard Megna/Fundamental Photographs. Page 815 (BL): Charles D. Winters/Science Source. Page 815 (BR): Charles D. Winters/Science Source. Page 820 (TL): Charles D. Winters/Science Source. Page 820 (BL): Jerry Mason/Science Source. Page 820 (BR): Jerry Mason/Science Source. Page 823 (TL): Richard Megna/Fundamental Photographs. Page 823 (BL): Richard Megna/Fundamental Photographs. Page 825 (TL): Richard Megna/Fundamental Photographs. Page 825 (R): Richard Megna/Fundamental Photographs. Page 825 (L): Richard Megna/Fundamental Photographs. Page 828 (ML): Richard Megna/Fundamental Photographs. Page 837: Sources: Data from Nour, V., Trandafir, I., Ionica, M. E. HPLC Organic Acid Analysis in Different Citrus Juices under Reversed Phase Conditions. Not. Bot. Hort. Agrobot. Cluj 2010, 38, 1, pp. 44–48; Weast, R. C., Ed. Handbook of Chemistry and Physics, The Chemical Rubber Company, Ohio, U.S. 1968; NCIB-National Center for Biotechnology Information. Compound Summary for Citric Acid,http://pubchem.ncbi.nlm.nih.gov/summary/summary.cgi?cid=311#x94 (accessed Oct 2, 2014); Penniston, K. L., Nakada, S. Y., Holmes, R. P., Assimos, D. G. Quantitative Assessment of Citric Acid in Lemon Juice, Lime Juice, and Commercially-Available Fruit Juice Products. J Endourol. Mar 2008; 22, 3, pp. 567–570.]. All other Figures and Tables copyright Pearson Education, Inc. **Chapter 18** Page 838: Clausius, Rudolf. Page 840 (TL): Eryrie/Alamy. Page 840 (BL): KtD/Fotolia. Page 841 (TC): Arthur S. Aubry/Getty Images. Page 841 (CL): Zurijeta/Shutterstock. Page 841 (CT): Getty Images. Page 841 (CTR): Martin Green/Fotolia. Page 841 (CB): A.G.Siemens. Page 841 (CBR): Scott T. Baxter/Getty Images. Page 841 (BR): Shubroto Chattopadhyay/Spirit/Corbis. Page 841 (BL): Dorling Kindersley, Ltd. Page 842 (BL): BlackJack3D/Getty Images. Page 842 (BR): Daniel Schwen. Page 843 (BL): Pulse/Corbis. Page 844 (T): Paradoxdes/Fotolia. Page 844 (M): Richard Megna/Fundamental Photographs. Page 845 (TR): Image provided courtesy of Johan Kocur. Page 849 (BL): Jan Will/Fotolia.

Page 870 (TL): Ryan McVay/Getty Images. All other Figures and Tables copyright Pearson Education, Inc. **Chapter 19** Page 888: Faraday, Michael (1791–1867). Page 890 (ML): Toyota Motor Corporate Services of N. America,Inc. Page 894 (TL): Richard Megna/Fundamental Photographs. Page 894 (TR): Richard Megna/Fundamental Photographs. Page 895 (BR): Alejandro Díaz Díez/AGE Fotostock. Page 905 (MR): Richard Megna/Fundamental Photographs. Page 915 (MR): Dorling Kindersley, Ltd. Page 916 (ML): Wojtek/Fotolia. Page 918 (MR): NielsDK/Newscom. Page 922 (ML): Charles D. Winters/Science Source. Page 925 (BR): StockPhotosArt/Shutterstock. Page 926 (TL): Donovan Reese/Getty Images. Page 926 (ML): Alan Pappe/Getty Images. Page 927 (TR): Bsani/Fotolia. All other Figures and Tables copyright Pearson Education, Inc. **Chapter 20** Page 938: Curie, Marie (1867–1934). Page 940 (BL): SPL/Science Source. Page 940 (ML): Martyn F. Chillmaid/Science Source. Page 941 (BR): Klaus Guldbrandsen/Science Source. Page 941 (TR): AFP/Getty Images. Page 946: Graph reprinted courtesy of Dr. Jeff C. Bryan. Page 948 (BL): Image courtesy of the United States Air Force. Page 949 (TR): Hank Morgan/Science Source. Page 952 (ML): Baz Ratner/Reuters. Page 953 (TR): Steve Smith/Shutterstock. Page 954 (MR): Antonio Calan ni/AP Images. Page 956 (BL): Pictorial Press Ltd/Alamy. Page 957 (C): Franklin D. Roosevelt Library. Page 958: Einstein–Szilárd letter. Written by Leó Szilárd and signed by Albert Einstein. Sent to Franklin D. Roosevelt. August 2, 1939. Available from The U.S. National Archives and Records Administration. Page 958 (ML): Stocktrek/Photodisc/Getty Images. Page 958 (MR): Elena Aliaga/Getty Images. Page 959 (MR): STR/AP Images. Page 963 (TR): General Atomics MS 15-124. Page 963 (MR): Fox Photos/Stringer/Hulton Archive/Getty Images. Page 964 (TL): David Parker/Science Source. Page 964 (BL): Fermilab Visual Media Services. Page 966-table: Source: Department of Health and Human Services, National Institutes of Health. Page 967 (MR): Custom Medical Stock Photo/Newscom. Page 968 (ML): SPL/Science Source. Page 968 (B): Centre Oscar Lambret/Science Source. Page 969 (TR): Cordelia Molloy/Science Source. Page 972 (C): Baz Ratner/Reuters. Page 976 (MR): Craig Ellenwood/Alamy. All other Figures and Tables copyright Pearson Education, Inc. **Chapter 21** Page 978: Wohler, Friedrich (1800–1882). Page 981 (MR): Photos.com/Getty Images. Page 992 (BL): PzAxe/Shutterstock. Page 993 (MR): John Casey/Fotolia. Page 1007 (BC): Lyudmila Suvorova/Fotolia. Page 1008 (TC): Andreas Von Einsiedel/Dorling Kindersley, Ltd. Page 1010 (TC): Tim UR/Fotolia. Page 1010 (BL): Yasonya/Shutterstock. Page 1016 (BC): Lurii Kachkovskyi/Shutterstock. All other Figures and Tables copyright Pearson Education, Inc. **Chapter 22** Page 1028: Crick, Francis H. C. (1916–2004). Page 1033 (ML): David Murray/Dorling Kindersley, Ltd. Page 1044 (TL): Science Source. Page 1049 (BR): Adrian T Sumner/Getty Images. Page 1050 (ML): Image courtesy of National Cancer Institute. Page 1055 (BC): Science Source. Page 1061: European Journal of Biochemistry, Vol. 114, p.653-661 (1981). All other Figures and Tables copyright Pearson Education, Inc. **Chapter 23** Page 1062: Fuller, Buckminster R. (1895–1983). Page 1068 (TL): SCIMAT/Science Source. Page 1068 (ML): Chip Clark/Fundamental Photographs. Page 1073 (MR): Klikk/Fotolia. Page 1074 (BL): ASM International. Page 1076 (TL): Veer Incorporated. Page 1083 (MR): National Oceanic and Atmospheric Administration (NOAA). Page 1086 (ML): Fundamental Photographs. Page 1086 (TC): Eisenger, Chris. Page 1086 (TR): Richard Megna/Fundamental Photographs. Page 1088 (BL): Martyn F. Chillmaid/Science Source. All other Figures and Tables copyright Pearson Education, Inc. **Chapter 24** Page 1100: William Henrry (1774–1836). Page 1102 (TL): SPL/Science Source. Page 1103 (TR): Bruno Barbier/Robert Harding Picture Library Ltd/Alamy. Page 1103 (TC) : Dorling Kindersley Ltd. Page 1107 (BR): Metal Powder Industries Federation. Page 1114 (ML): Richard Megna/Fundamental Photographs. Page 1115 (TR): Joseph P. Sinnot/Fundamental Photographs. Page 1117 (BR): Scol22/Fotolia. Page 1117 (TR): Demetrio Carrasco/Dorling Kindersley Ldt. Page 1117 (MC): Vaclav Janousek/Fotolia. Page 1117 (MR): Dirk Hoffmann/Fotolia. Page 1118 (ML): Heatbath Corporation. All other

# INDEX

## A

A. *See* Ampere; Mass number
A. *See* Frequency factor
A. O. Smith Company, 1107
Absolute concentration, 792, 794
Absolute temperature scale (Kelvin scale), 15–16, 206
Absolute zero, 205, 860
Absorption spectra, 310–311
Accuracy, 25–26
Acetaldehyde (ethanal), 126, 1007–1009
Acetate, 100
Acetic acid
   acid ionization constant for, 731
   buffering by, 781, 782
   as carboxylic acid, 126
   double bonds in, 124
   formula and uses of, 1009
   reaction between salicylic acid and, 1011
   strength of, 160, 169, 730
   structure of, 725
   uses of, 724
Acetone, 39, 126
   boiling point of, 504
   charge density plot for, 1009
   dipole–dipole force in, 492–493
   formation of, 1008
   heat of fusion of, 514
   heat of vaporization of, 504
   intermolecular forces in, 575
   as solvent, 576
   structure of, 1007
   volatility of, 503
Acetonitrile, 643–644
Acetylene (ethyne)
   from calcium carbide, 1074
   formula of, 125, 994
   hybrid orbitals of, 455–456
   Lewis structure of, 980
   representations of, 93
   structure and properties of, 993
   triple bonds in, 124
Acetylide ion, 1073
Acetylsalicylic acid, 479, 1011
Acid(s). *See also* Acid–base chemistry; Acid–base titration
   amino. *See* Amino acids
   binary, 103–104, 762
   carboxylic, 126, 725, 1009–1012
   dilution of, 154
   diprotic, 168, 729, 757, 760–761, 805–806
   fatty, 1030–1032
   inorganic nomenclature flowchart, 105–106
   metals dissolved in, 103, 905
   naming, 103–105
   nucleic. *See* Nucleic acids
   oxoacids, 104–105, 763
   polyprotic, 168, 757–761, 805–806
   rusting promoted by, 926

   strong, 160, 729, 737, 743–744, 795–799, 805
   weak. *See* Weak acid(s)
Acid anhydride, 1011
Acid–base chemistry, 168–173, 722–777. *See also* Buffers; pH
   acid–base properties of ions, 750–757
   acid ionization constant, 731–732
   acid rain, 765–767
   acid strength, 729–732, 762–763
   addition to buffer, 786–790
   Arrhenius model of, 168–169, 726–727
   autoionization of water, 732–734
   base solutions, 746–750
   Brønsted–Lowry model of, 727–729, 762, 764
   definitions of, 726–729
   heartburn and, 723–724
   Lewis model of, 764–765, 1071
   molecular structure, 762–763
   nature of, 724–726
   neutralization reactions, 1010–1011
   pOH and other p scales, 735–736
   polyprotic acids, 757–761
   salt solutions as acidic, basic, or neutral, 755–757
   writing equations, 170–171
Acid–base titration, 795–809
   endpoint of, 806, 807
   indicators, 171–173, 806–809
   of polyprotic acids, 805–806
   of strong acids with strong bases, 795–799
   titration/pH curve, 795–806
   of weak acids with strong bases, 800–805
   of weak bases with strong acids, 805
Acid dissociation constant. *See* Acid ionization constant(s) ($K_a$)
Acidic solutions, 733, 734, 736, 891–892
Acid ionization constant(s) ($K_a$), 731–732
   magnitude of, 732
   for polyprotic acids, 757–758
   for solutions with anions acting as weak bases, 752
   for weak acids, 737, 741
Acidosis, 780
Acid rain, 104, 143–144, 281, 765–767, 1088
Acid rebound, 750
Acid reflux, 723
Acquired immune deficiency syndrome (AIDS), 383–384
Actinides (inner transition elements), 347–348, 350–351
Activated carbon (activated charcoal), 1073
Activated complex (transition state), 643
Activation energy (activation barrier), 643–647, 653
Active metals, reaction of alcohols with, 1005
Active site, 384, 428, 656, 658
Actual yield, 146

Acute concern, acid rain as, 767
Acute radiation damage, 965
Addition polymers, 558
Addition reactions, 998–999, 1008–1009
Adducts, 764
Adenine, 498, 499, 1047, 1048
Adenosine triphosphate (ATP), 883
Adhesive forces, 502
Adipic acid, 559
*Adventure Thru Inner Space* (ride), 1–2
Aerosols, 209, 609
AES (auger electron spectroscopy), 332–333
AIDS (acquired immune deficiency syndrome), 383–384
Air
   composition of dry, 212, 214, 215
   fractionation of, 1083
   separation with, 1104
Air bags, automobile, 242
Air pollution, 281–282
Alanine, 1039, 1040
Alchemists, 963
Alcohol(s), 126, 577, 1003–1006
Alcoholic beverages, 1004
Aldehydes, 126, 1003, 1006–1009, 1035
Aldohexose, 1035
Aldose, 1035
Aldrin, 590
Aliphatic hydrocarbons, 982. *See also* Alkanes; Alkenes; Alkynes
Aliquot, 628
Alizarin, 809
Alizarin yellow R, 809
Alkali metals. *See* Group 1A metals
Alkaline batteries, 915
Alkaline earth metals. *See* Group 2A metals
Alkaloids, 726
Alkanes, 124, 982, 988–992
   *n*-alkanes, 988–989
   boiling points of, 491–492
   naming, 989–992
   reactions of, 997–998
   viscosity of, 501
Alka-Seltzer, 1076
Alkenes, 124, 656, 982, 992–997
Alkyl groups, 989–990
Alkynes, 124, 982, 992–999
Allotropes, 861, 1070, 1077, 1086
Alloys, 1102, 1108–1118
   copper, 1117, 1118
   interstitial, 1108, 1111–1113
   with limited solubility, 1110–1111
   nickel, 1118
   steel, 1114–1116
   substitutional, 1108–1110
   zinc, 1118
α-Carbon, 1039
Alpha (α) decay, 941–943, 945, 946
α-helix, 1043, 1044
Alpha (α) particles, 53, 54, 942
Altitude, 254, 508
Aluminosilicates, 553, 1066–1067

# Conversion Factors and Relationships

## Length
SI unit: meter (m)

$1\text{ m} = 1.0936\text{ yd}$

$1\text{ cm} = 0.39370\text{ in}$

$1\text{ in} = 2.54\text{ cm (exactly)}$

$1\text{ km} = 0.62137\text{ mi}$

$1\text{ mi} = 5280\text{ ft}$

$\quad\quad = 1.6093\text{ km}$

$1\text{ Å} = 10^{-10}\text{ m}$

## Temperature
SI unit: kelvin (K)

$0\text{ K} = -273.15\,°\text{C}$

$\quad\quad = -459.67\,°\text{F}$

$\text{K} = °\text{C} + 273.15$

$°\text{C} = \dfrac{(°\text{F} - 32)}{1.8}$

$°\text{F} = 1.8\,(°\text{C}) + 32$

## Energy (derived)
SI unit: joule (J)

$1\text{ J} = 1\text{ kg}\cdot\text{m}^2/\text{s}^2$

$\quad\quad = 0.23901\text{ cal}$

$\quad\quad = 1\text{ C}\cdot\text{V}$

$\quad\quad = 9.4781 \times 10^{-4}\text{ Btu}$

$1\text{ cal} = 4.184\text{ J}$

$1\text{ eV} = 1.6022 \times 10^{-19}\text{ J}$

## Pressure (derived)
SI unit: pascal (Pa)

$1\text{ Pa} = 1\text{ N}/\text{m}^2$

$\quad\quad = 1\text{ kg}/(\text{m}\cdot\text{s}^2)$

$1\text{ atm} = 101{,}325\text{ Pa}$

$\quad\quad = 760\text{ torr}$

$\quad\quad = 14.70\text{ lb}/\text{in}^2$

$1\text{ bar} = 10^5\text{ Pa}$

$1\text{ torr} = 1\text{ mmHg}$

## Volume (derived)
SI unit: cubic meter (m$^3$)

$1\text{ L} = 10^{-3}\text{ m}^3$

$\quad\quad = 1\text{ dm}^3$

$\quad\quad = 10^3\text{ cm}^3$

$\quad\quad = 1.0567\text{ qt}$

$1\text{ gal} = 4\text{ qt}$

$\quad\quad = 3.7854\text{ L}$

$1\text{ cm}^3 = 1\text{ mL}$

$1\text{ in}^3 = 16.39\text{ cm}^3$

$1\text{ qt} = 32\text{ fluid oz}$

## Mass
SI unit: kilogram (kg)

$1\text{ kg} = 2.2046\text{ lb}$

$1\text{ lb} = 453.59\text{ g}$

$\quad\quad = 16\text{ oz}$

$1\text{ amu} = 1.66053873 \times 10^{-27}\text{ kg}$

$1\text{ ton} = 2000\text{ lb}$

$\quad\quad = 907.185\text{ kg}$

$1\text{ metric ton} = 1000\text{ kg}$

$\quad\quad = 2204.6\text{ lb}$

## Geometric Relationships

$\pi = 3.14159\ldots$

Circumference of a circle $= 2\pi r$

Area of a circle $= \pi r^2$

Surface area of a sphere $= 4\pi r^2$

Volume of a sphere $= \dfrac{4}{3}\pi r^3$

Volume of a cylinder $= \pi r^2 h$

## Fundamental Constants

| | | |
|---|---|---|
| Atomic mass unit | 1 amu | $= 1.66053873 \times 10^{-27}\text{ kg}$ |
| | 1 g | $= 6.02214199 \times 10^{23}\text{ amu}$ |
| Avogadro's number | $N_A$ | $= 6.02214179 \times 10^{23}/\text{mol}$ |
| Bohr radius | $a_0$ | $= 5.29177211 \times 10^{-11}\text{ m}$ |
| Boltzmann's constant | $k$ | $= 1.38065052 \times 10^{-23}\text{ J/K}$ |
| Electron charge | $e$ | $= 1.60217653 \times 10^{-19}\text{ C}$ |
| Faraday's constant | $F$ | $= 9.64853383 \times 10^4\text{ C/mol}$ |
| Gas constant | $R$ | $= 0.08205821\ (\text{L}\cdot\text{atm}/(\text{mol}\cdot\text{K}))$ |
| | | $= 8.31447215\ \text{J}/(\text{mol}\cdot\text{K})$ |
| Mass of an electron | $m_e$ | $= 5.48579909 \times 10^{-4}\text{ amu}$ |
| | | $= 9.10938262 \times 10^{-31}\text{ kg}$ |
| Mass of a neutron | $m_n$ | $= 1.00866492\text{ amu}$ |
| | | $= 1.67492728 \times 10^{-27}\text{ kg}$ |
| Mass of a proton | $m_p$ | $= 1.00727647\text{ amu}$ |
| | | $= 1.67262171 \times 10^{-27}\text{ kg}$ |
| Planck's constant | $h$ | $= 6.62606931 \times 10^{-34}\text{ J}\cdot\text{s}$ |
| Speed of light in vacuum | $c$ | $= 2.99792458 \times 10^8\text{ m/s (exactly)}$ |

## SI Unit Prefixes

| a | f | p | n | $\mu$ | m | c | d | k | M | G | T | P | E |
|---|---|---|---|---|---|---|---|---|---|---|---|---|---|
| atto | femto | pico | nano | micro | milli | centi | deci | kilo | mega | giga | tera | peta | exa |
| $10^{-18}$ | $10^{-15}$ | $10^{-12}$ | $10^{-9}$ | $10^{-6}$ | $10^{-3}$ | $10^{-2}$ | $10^{-1}$ | $10^3$ | $10^6$ | $10^9$ | $10^{12}$ | $10^{15}$ | $10^{18}$ |